ALGEBRA'S COMMON GRAPHS

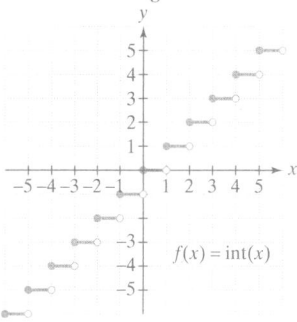

TRANSFORMATIONS

In each case, c represents a positive real number.

Function		Draw the graph of f and:
Vertical translations	$\begin{cases} y = f(x) + c \\ y = f(x) - c \end{cases}$	Shift f upward c units. Shift f downward c units.
Horizontal translations	$\begin{cases} y = f(x - c) \\ y = f(x + c) \end{cases}$	Shift f to the right c units. Shift f to the left c units.
Reflections	$\begin{cases} y = -f(x) \\ y = f(-x) \end{cases}$	Reflect f about the x-axis. Reflect f about the y-axis.
Stretching or Shrinking	$\begin{cases} y = cf(x); c > 1 \\ y = cf(x); 0 < x < 1 \end{cases}$	Stretch f, multiplying each of its y-values by c. Shrink f, multiplying each of its y-values by c.

DISTANCE AND MIDPOINT FORMULAS

1. The distance from (x_1, y_1) to (x_2, y_2) is
$$\sqrt{(x_2 - x_1)^2 + (y_2 - y_1)^2}.$$

2. The midpoint of the line segment with endpoints (x_1, y_1) and (x_2, y_2) is
$$\left(\frac{x_1 + x_2}{2}, \frac{y_1 + y_2}{2} \right).$$

QUADRATIC FORMULA

The solutions to $ax^2 + bx + c = 0$ with $a \neq 0$ are
$$x = \frac{-b \pm \sqrt{b^2 - 4ac}}{2a}.$$

FUNCTIONS

1. Linear Function: $f(x) = mx + b$
   Graph is a line with slope m and y-intercept b.

2. Quadratic Function: $f(x) = ax^2 + bx + c, a \neq 0$

Graph is a parabola with vertex at $x = -\dfrac{b}{2a}$.

Quadratic Function: $f(x) = a(x - h)^2 + k$
In this form, the parabola's vertex is (h, k).

3. nth-Degree Polynomial Function: $f(x) =$
$a_n x^n + a_{n-1} x^{n-1} + a_{n-2} x^{n-2} + \cdots + a_1 x + a_0, a_n \neq 0$
For n odd and $a_n > 0$, graph falls to the left and rises to the right.
For n odd and $a_n < 0$, graph rises to the left and falls to the right.
For n even and $a_n > 0$, graph rises to the left and to the right.
For n even and $a_n < 0$, graph falls to the left and to the right.

4. Rational Function: $f(x) = \dfrac{p(x)}{q(x)}$, $p(x)$ and $q(x)$ are polynomials, $q(x) \neq 0$

5. Exponential Function: $f(x) = b^x, b > 0, b \neq 1$
Graphs:

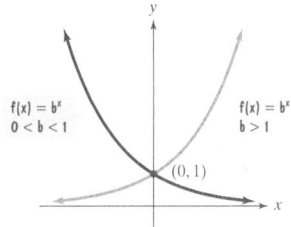

6. Logarithmic Function: $f(x) = \log_b x, b > 0, b \neq 1$
$y = \log_b x$ is equivalent to $x = b^y$.
Graph:

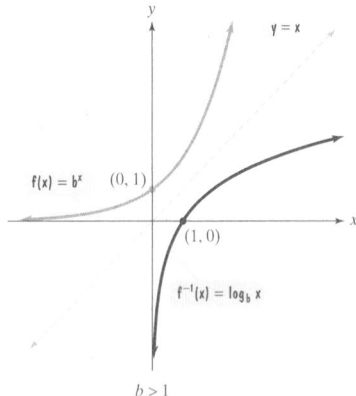

PROPERTIES OF LOGARITHMS

1. $\log_b(MN) = \log_b M + \log_b N$

2. $\log_b\left(\dfrac{M}{N}\right) = \log_b M - \log_b N$

3. $\log_b M^p = p \log_b M$

4. $\log_b M = \dfrac{\log_a M}{\log_a b} = \dfrac{\ln M}{\ln b} = \dfrac{\log M}{\log b}$

5. $\log_b b^x = x$; $\ln e^x = x$

6. $b^{\log_b x} = x$; $e^{\ln x} = x$

INVERSE OF A 2 × 2 MATRIX

If $A = \begin{bmatrix} a & b \\ c & d \end{bmatrix}$, then $A^{-1} = \dfrac{1}{ad - bc} \begin{bmatrix} d & -b \\ -c & a \end{bmatrix}$, where $ad - bc \neq 0$.

CRAMER'S RULE

If

$$a_{11}x_1 + a_{12}x_2 + a_{13}x_3 + \cdots + a_{1n}x_n = b_1$$
$$a_{21}x_1 + a_{22}x_2 + a_{23}x_3 + \cdots + a_{2n}x_n = b_2$$
$$a_{31}x_1 + a_{32}x_2 + a_{33}x_3 + \cdots + a_{3n}x_n = b_3$$
$$\vdots$$
$$a_{n1}x_1 + a_{n2}x_2 + a_{n3}x_3 + \cdots + a_{nn}x_n = b_n$$

then $x_i = \dfrac{D_i}{D}, D \neq 0$.

D: determinant of the system's coefficients

D_i: determinant in which coefficients of x_i are replaced by $b_1, b_2, b_3, \ldots, b_n$.

CONIC SECTIONS

Circle

Ellipse

Hyperbola

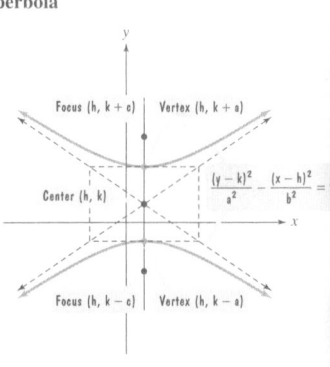

(continued on inside back cover)

SECOND EDITION

Algebra &
Trigonometry

Robert Blitzer
Miami-Dade Community College

PRENTICE HALL Upper Saddle River, NJ 07458

Library of Congress Cataloging-in-Publication Data

Blitzer, Robert.
    Algebra & trigonometry / Robert Blitzer. — 2nd ed.
      p. cm.
    Includes index.
    ISBN 0-13-101359-9
      1. Algebra.   2. Trigonometry I. Title: Algebra and trigonometry. II. Title.
    QA152.3.B63 2004
    512'.13—dc21   2002190814

Senior Acquisitions Editor: *Eric Frank*
Associate Editor : *Dawn Murrin*
Editor-in-Chief : *Sally Yagan*
Vice President/Director of Production and Manufacturing: *David W. Riccardi*
Executive Managing Editor: *Kathleen Schiaparelli*
Senior Managing Editor: *Linda Mihatov Behrens*
Production Management: *Elm Street Publishing Services, Inc.*
Production Assistant: *Nancy Bauer*
Assistant Managing Editor, Math Media Production: *John Matthews*
Media Production Editor: *Donna Crilly*
Manufacturing Buyer: *Michael Bell*
Manufacturing Manager: *Trudy Pisciotti*
Senior Marketing Manager: *Halee Dinsey*
Marketing Assistant: *Rachel Beckman*
Editorial Assistant/Supplements Editor: *Aja Shevelew*
Art Director: *Jon Boylan*
Interior/Cover Designer: *Maureen Eide*
Art Editor: *Thomas Benfatti*
Creative Director: *Carole Anson*
Director of Creative Services: *Paul Belfanti*
Director, Image Resource Center: *Melinda Reo*
Manager, Rights and Permissions: *Zina Arabia*
Interior Image Specialist: *Beth Boyd-Brenzel*
Cover Image Specialist: *Karen Sanatar*
Image Permission Coordinator/Photo Researcher: *Elaine Soares*
Cover Art: *Jalapeño pepper, photographed by John E. Kelly, copyright Getty Images*
Art Studio: *Artworks:*
 Managing Editor, AV Production & Management: *Patty Burns*
 Production Manager: *Ronda Whitson*
 Manager, Production Technologies: *Matt Haas*
 Project Coordinator: *Jessica Einsig*
 Illustrators: *Kathryn Anderson, Mark Landis*
 Quality Assurance: *Pamela Taylor, Timothy Nguyen*
Formatting Manager: *Jim Sullivan*
Assistant Manager of Formatting: *Allyson Graesser*
 Formatters: *Clara Bartunek, Beth Gschwind, Julita Nazario, Nancy Thompson, Judith Wilkens*

 © 2004, 2001 by Pearson Education, Inc.
Pearson Education, Inc.
Upper Saddle River, New Jersey 07458

Printed in the United States of America

10 9 8 7 6 5 4 3 2

ISBN 0-13-101359-9

Pearson Education LTD., *London*
Pearson Education Australia PTY, Limited, *Sydney*
Pearson Education Singapore, Pte. Ltd
Pearson Education North Asia Ltd, *Hong Kong*
Pearson Education Canada, Ltd., *Toronto*
Pearson Educación de Mexico, S.A. de C.V.
Pearson Education -- Japan, *Tokyo*
Pearson Education Malaysia, Pte. Ltd

Contents

Preface   vii
To the Student   xii
Applications Index   xx

Chapter P

Prerequisites: Fundamental Concepts of Algebra   1

P.1 Real Numbers and Algebraic Expressions 2
P.2 Exponents and Scientific Notation 13
P.3 Radicals and Rational Exponents 24
P.4 Polynomials 36
P.5 Factoring Polynomials 48
P.6 Rational Expressions 59
   Summary 71
   Review Exercises or Algebra Skills Diagnostic Test 71
   Chapter P Test 73

Chapter 1

Equations, Inequalities, and Mathematical Models   75

1.1 Graphs and Graphing Utilities 76
1.2 Linear Equations 84
1.3 Formulas and Applications 95
1.4 Complex Numbers 108
1.5 Quadratic Equations 114
1.6 Other Types of Equations 131
1.7 Linear Inequalities 144
1.8 Quadratic and Rational Inequalities 157
   Summary 168
   Review Exercises 170
   Chapter 1 Test 173

Chapter 2

Functions and Graphs   175

2.1 Lines and Slope 176
2.2 Distance and Midpoint Formulas; Circles 193
2.3 Basics of Functions 201
2.4 Graphs of Functions 214
2.5 Transformations of Functions 235
2.6 Combinations of Functions; Composite Functions 248
2.7 Inverse Functions 260
Summary 270
Review Exercises 272
Chapter 2 Test 276
Cumulative Review Exercises 278

Chapter 3

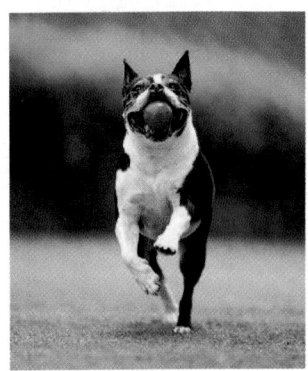

Polynomial and Rational Functions   279

3.1 Quadratic Functions 280
3.2 Polynomial Functions and Their Graphs 293
3.3 Dividing Polynomials; Remainder and Factor Theorems 304
3.4 Zeros of Polynomial Functions 315
3.5 More on Zeros of Polynomial Functions 325
3.6 Rational Functions and Their Graphs 335
3.7 Modeling Using Variation 353
Summary 366
Review Exercises 368
Chapter 3 Test 371
Cumulative Review Exercises 372

Chapter 4

Exponential and Logarithmic Functions   373

4.1 Exponential Functions 374
4.2 Logarithmic Functions 385
4.3 Properties of Logarithms 398
4.4 Exponential and Logarithmic Equations 407
4.5 Modeling with Exponential and Logarithmic Functions 418
Summary 432
Review Exercises 433
Chapter 4 Test 437
Cumulative Review Exercises 437

Chapter 5

Trigonometric Functions   438

5.1 Angles and Their Measure 439
5.2 Right Triangle Trigonometry 453
5.3 Trigonometric Functions of Any Angle 469
5.4 Trigonometric Functions of Real Numbers; Periodic Functions 479
5.5 Graphs of Sine and Cosine Functions 487
5.6 Graphs of Other Trigonometric Functions 509
5.7 Inverse Trigonometric Functions 521

5.8 Applications of Trigonometric Functions 537
Summary 548
Review Exercises 551
Chapter 5 Test 554
Cumulative Review Exercises 555

Chapter 6

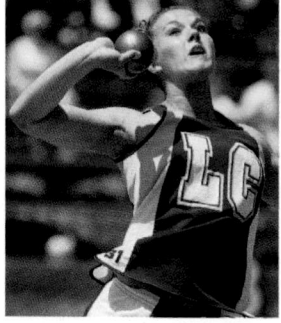

Analytic Trigonometry 556

6.1 Verifying Trigonometric Identities 557
6.2 Sum and Difference Formulas 567
6.3 Double-Angle and Half-Angle Formulas 577
6.4 Product-to-Sum and Sum-to-Product Formulas 588
6.5 Trigonometric Equations 596
Summary 606
Review Exercises 607
Chapter 6 Test 609
Cumulative Review Exercises 609

Chapter 7

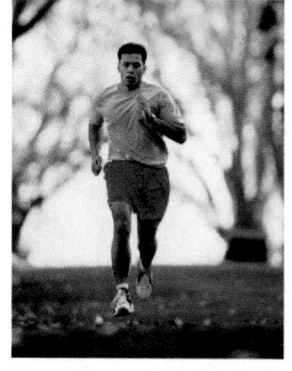

Additional Topics in Trigonometry 610

7.1 The Law of Sines 611
7.2 The Law of Cosines 623
7.3 Polar Coordinates 631
7.4 Graphs of Polar Equations 642
7.5 Complex Numbers in Polar Form; DeMoivre's Theorem 654
7.6 Vectors 667
7.7 The Dot Product 681
Summary 690
Review Exercises 693
Chapter 7 Test 696
Cumulative Review Exercises 696

Chapter 8

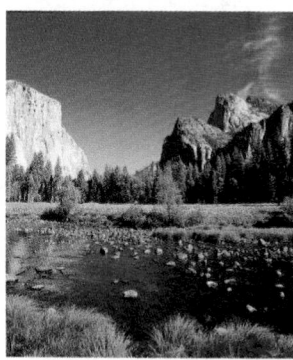

Systems of Equations and Inequalities 698

8.1 Systems of Linear Equations in Two Variables 699
8.2 Systems of Linear Equations in Three Variables 716
8.3 Partial Fractions 725
8.4 Systems of Nonlinear Equations in Two Variables 734
8.5 Systems of Inequalities 743
8.6 Linear Programming 753
Summary 762
Review Exercises 763
Chapter 8 Test 766
Cumulative Review Exercises 767

Chapter 9

Matrices and Determinants   768

9.1 Matrix Solutions to Linear Systems 769
9.2 Inconsistent and Dependent Systems and Their Applications 783
9.3 Matrix Operations and Their Applications 792
9.4 Multiplicative Inverses of Matrices and Matrix Equations 807
9.5 Determinants and Cramer's Rule 822
Summary 835
Review Exercises 836
Chapter 9 Test 839
Cumulative Review Exercises 839

Chapter 10

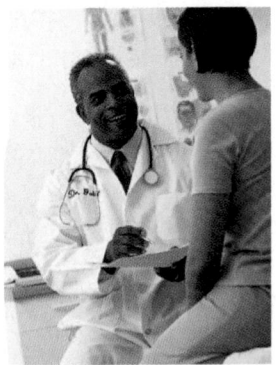

Conic Sections and Analytic Geometry   841

10.1 The Ellipse 842
10.2 The Hyperbola 855
10.3 The Parabola 871
10.4 Rotation of Axes 884
10.5 Parametric Equations 897
10.6 Conic Sections in Polar Coordinates 908
Summary 919
Review Exercises 921
Chapter 10 Test 923
Cumulative Review Exercises 924

Chapter 11

Sequences, Induction, and Probability   925

11.1 Sequences and Summation Notation 926
11.2 Arithmetic Sequences 937
11.3 Geometric Sequences 946
11.4 Mathematical Induction 960
11.5 The Binomial Theorem 969
11.6 Counting Principles, Permutations, and Combinations 977
11.7 Probability 988
Summary 1001
Review Exercises 1003
Chapter 11 Test 1006
Cumulative Review Exercises 1007

Appendix

Where Did That Come From? Selected Proofs   A1

Answers to Selected Exercises AA1
Subject Index I1
Photo Credits P1

Preface

've written **Algebra and Trigonometry, Second Edition** to help diverse students, with different backgrounds and future goals to succeed. The book has three fundamental goals:

1. To help students acquire a solid foundation in algebra and trigonometry, preparing them for other courses such as calculus, business calculus, and finite mathematics.
2. To show students how algebra and trigonometry can model and solve authentic real-world problems.
3. To enable students to develop problem-solving skills, while fostering critical thinking, within an interesting setting.

One major obstacle in the way of achieving these goals is the fact that very few students actually read their textbook. This has been a regular source of frustration for me and my colleagues in the classroom. Anectodal evidence gathered over years highlights two basic reasons that students do not take advantage of their textbook:

- "I'll never use this information."
- "I can't follow the explanations."

As a result, I've written every page of this book with the intent of eliminating these two objections. See the book's Walkthrough, beginning on page xiv for the ideas and tools I've used to do so.

A Brief Note on Technology

Technology, and specifically the use of a graphing utility, is covered thoroughly, although its coverage by an instructor is optional. If you require the use of a graphing utility in the course, you will find support for this approach, particularly in the wide selection of clearly designated technology exercises in each exercise set. If you wish to minimize or eliminate the discussion or use of a graphing utility, the book is written to enable you to do so. Regardless of the role technology plays in your course, the technology boxes with TI-83 screens that appear throughout the book should allow your students to understand what graphing utilities can do, enabling them to visualize, verify, or explore what they have already graphed or manipulated by hand. The book's technology coverage is intended to reinforce, but never replace, algebraic solutions.

What's New in the Second Edition

General Changes to the Second Edition

New Applications and Updated Real-World Data. Many new, innovative applications, supported by data that extend as far up to the present as possible, appear throughout the book.

Expanded Exercise Sets. There are new problems in many of the exercise sets. Some of these problems provide instructors with the *option* of creating assignments that take practice and application exercises from a basic level to a *more challenging level* than in the previous edition. In order to update applications and provide users with an ongoing selection of novel applications, many application problems from the previous edition were replaced with new exercises.

New Section Openers and Enrichment Essays. The Second Edition contains a variety of new section openers and enrichment essays, ranging from the five all-time celebrity winners on *Jeopardy!* (Section 2.3 opening scenario) to a comparison between the probability of dying and the probability of winning Florida's lottery (Section 11.7 essay).

Increased Study Tip Boxes. The book's study tip boxes offer suggestions for problem solving, point out common errors to avoid, and provide informal hints and suggestions. These invaluable hints, including suggestions for review in preparation for the section ahead, appear in greater abundance in the Second Edition.

Expanded Technology. An increase in the number of *optional* technology boxes in the Second Edition illustrates the many capabilities of graphing utilities that go beyond just graphing functions.

New Chapter Review Grids. The chapter summaries, presented as outlines in the previous edition, are now organized into two-column review charts. The left column summarizes the definitions and concepts for every section of the chapter. The right column refers students to examples (by example number and page number) that illustrate these key concepts.

Expanded Supplements Package. The Second Edition is supported by a wealth of supplements designed for added effectiveness and efficiency, many of these new to this edition. (New supplements include MathPak 5 tutorial software now with trigonometry content and a diagnostic component; PH GradeAssist—an automated homework/assessment creation, delivery, and grading system; Instructor Resource CD ROM – contains all supplements for instructors in one easy location; and more.) See page x for details, under "Supplements" or ask your Prentice Hall representative for information.

Specific Content and Organizational Changes to the Second Edition

Section P.5 (Factoring Polynomials) now contains a brief discussion on factoring algebraic expressions containing fractional and negative exponents. This skill is helpful to students going on to calculus.

The discussion of complex numbers was moved from Chapter P, the prerequisites chapter, to Chapter 1, Section 1.4. This change enables students to immediately apply their understanding of complex numbers to their work in solving quadratic equations (Section 1.5).

The discussion of graphs and graphing utilities was moved from Chapter P to Chapter 1, Section 1.1. This nicely sets the stage for using graphing to support the algebraic work on solving equations and inequalities developed in Chapter 1.

Section 2.1 (Lines and Slope) now contains a discussion on parallel and perpendicular lines. In the previous edition, this material appeared in a section that also discussed circles. Presenting parallel and perpendicular lines in the section on lines and slope results in a more complete and unified discussion.

Section 2.2 (Distance and Midpoint Formulas; Circles) is now primarily devoted to circles. Distance and midpoint formulas, presented in Chapter P in the previous edition, were moved to this section because students need to use the distance formula to develop the formula for a circle.

Section 2.3 (Basics of Functions) now contains the definition of the difference quotient with an illustrative example.

Section 2.4 (Graphs of Functions) contains a new presentation on relative maximum and relative minimum values of a function, a natural outgrowth of the section's material on increasing and decreasing functions. There is also a new discussion of a function's average rate of change, with a relationship to the difference quotient from the previous section. These new topics are extremely important to students going on to calculus. They provide all students with an increased understanding of functions' graphs and how those graphs are changing.

Section 2.5 (Transformations of Functions) is now devoted exclusively to transformations, a difficult topic for many students. Unlike the previous edition, the section does not contain material on combinations of functions.

Section 2.6 (Combinations of Functions; Composite Functions) now includes combinations of functions and composite functions in one section. The section now tells a more coherent story – how to create new functions from given functions. With the section's emphasis on composite functions, new discussions on determining domains for composite functions and writing functions as compositions have been added.

Section 2.7 (Inverse Functions) is now devoted exclusively to the topic of inverse functions. This should appeal to users who prefer to cover inverse functions in Chapter 4, after Section 4.1 (Exponential Functions) and before Section 4.2 (Logarithmic Functions).

Section 3.6 (Rational Functions and Their Graphs) now contains a general discussion on cost and average cost functions. This change makes it possible for students to model these functions from verbal conditions before exploring the behavior of their graphs.

Section 7.6 (Vectors) and **Section 7.7 (The Dot Product)** have been slightly reorganized. The discussion on writing a vector in terms of its magnitude and direction was moved from Section 7.7 to Section 7.6. This move allows Section 7.7 to be devoted, without interruption, to the dot product. It also allows for a more thorough presentation in Section 7.6 on finding resultant forces by expressing a vector in terms of its magnitude and direction.

Section 8.1 (Systems of Linear Equations in Two Variables) contains a new application involving cost functions, revenue functions, and break-even points. This topic is important to business majors and gives students further practice in developing functions that model verbal conditions.

Section 9.3 (Matrix Operations and Their Applications) now includes a brief discussion on solving matrix equations. This topic serves as a nice application of matrix addition, scalar multiplication, and their properties.

Section 10.3 (The Parabola) now uses the latus rectum as part of the graphing strategy.

Section 11.5 (The Binomial Theorem) now gives the formula for the $(r + 1)$st term, rather than the rth term, of the expansion of $(a + b)^n$. Many students find the formula for the $(r + 1)$st term easier to work with when finding a particular term in a binomial expansion.

Supplements

Student Supplements	Instructor Supplements
Student Solutions Manual Fully worked solutions to odd-numbered exercises. 0-13-101917-1	**Instructor's Solutions Manual** Fully worked solutions to all exercises in the text. 0-13-101996-1
CD Lecture Series More than 10 CD-ROMs contain 20 minutes of lectures and tutorials per textbook section. Objectives are reviewed and key examples from the textbook are worked out. These are available for students to purchase alone, or in a package with their book. 0-13-101920-1	**Instructor's Edition with Instructor's Resource CD** • Provides answers to all exercises in the back of the text. • Includes Instructor's Resource CD containing TestGen-EQ, Instructor's Solutions Manual, Additional Chapter Projects and Test Item File. ISM and TIF are passcode-protected. 0-13-101360-2
VHS Lecture Series Same content as CD Lecture Series in VHS format. Instructors can order the VHS videos and make them available to students in the library or media lab. NOTE: The VHS videos only feature the Right Triangle for trigonometry. 0-13-101997-X	**Test Item File** A printed test bank derived from TestGen-EQ. 0-13-101918-X
PH Tutor Center Provides students with help when they need it most—while they're doing their homework. PH math tutors (trained college instructors) provide help via a toll-free phone number, fax, and email. 0-13-064604-0	**TestGen-EQ** Test-generating software enabling instructors to create tests from the text section objectives. Many questions are algorithmically generated, allowing for unlimited versions of any test. Instructors may also edit problems or create their own. 0-13-101363-7

MathPak 5.0

- Features MathPro 5.0, an online, customizable tutorial and assessment software package, integrated with the text at the learning objective level. The easy-to-use gradebook enables instructors to track and evaluate student performance on tutorial work, quizzes and tests. An optional Diagnostics module allows students to identify weaknesses in prerequisite material, and to have a customized set of tutorials provided to them for additional practice on those identified weaknesses.

- Includes access to a website containing the Student Solutions Manual, Online Graphing Calculator Help, PowerPoint slides used in the lecture videos, and quizzes and tests allowing students to assess their skills and comprehension of the material.

Student Version: 0-13-140794-5
Instructor Version: 0-13-101921-X

PH Grade Assist

This online homework and assessment program enables instructors to create customized homework and tests by choosing problems from the text or algorithmic versions of those problems or creating their own problems. PHGA supports multiple question types including free response. The built-in parser is sophisticated, grading student responses while recognizing algebraic, numeric, and unit equivalents. The gradebook also allows instructors to easily track student performance.

Student Version: 0-13-140321-4
Instructor Version: 0-13-140322-2

Companion Website

Free website to all text users provides quizzes, chapter tests, PowerPoint slides available for download, and Online Graphing Calculator Help.

URL: www.prenhall.com/blitzer

Acknowledgments

I wish to express my appreciation to all of the reviewers of my precalculus series for their helpful criticisms and suggestions, frequently transmitted with wit, humor, and intelligence. In particular, I would like to thank the following for reviewing **College Algebra**, **Algebra and Trigonometry,** and **Precalculus.**

Reviewers for the Previous Editions

Kayoko Yates Barnhill, *Clark College*
Lloyd Best, *Pacific Union College*
Diana Colt, *University of Minnesota-Duluth*
Yvelyne Germain-McCarthy, *University of New Orleans*
Cynthia Glickman, *Community College of Southern Nevada*
Sudhir Kumar Goel, *Valdosta State University*
Donald Gordon, *Manatee Community College*
David L. Gross, *University of Connecticut*
Joel K. Haack, *University of Northern Iowa*
Mike Hall, *Univeristy of Mississippi*
Christopher N. Hay-Jahans, *University of South Dakota*
Celeste Hernandez, *Richland College*
Winfield A. Ihlow, *SUNY College at Oswego*
Nancy Raye Johnson, *Manatee Community College*
James Miller, *West Virginia University*
Debra A. Pharo, *Northwestern Michigan College*
Gloria Phoenix, *North Carolina Agricultural and Technical State University*
Juha Pohjanpelto, *Oregon State University*
Richard E. Van Lommel, *California State University-Sacramento*
Dan Van Peursem, *University of South Dakota*
David White, *The Victoria College*

Reviewers for the Current Edition

Timothy Beaver, *Isothermal Community College*
Bill Burgin, *Gaston College*
Jimmy Chang, *St. Petersburg College*
Donna Densmore, *Bossier Parish Community College*
Disa Enegren, *Rose State College*
Nancy Fisher, *University of Alabama*
Jeremy Haefner, *University of Colorado*
Joyce Hague, *University of Wisconsin at River Falls*
Mary Leesburg, *Manatee Community College*
Christine Heinecke Lehmann, *Purdue University North Central*
Alexander Levichev, *Boston University*
Zongzhu Lin, *Kansas State University*
Benjamin Marlin, *Northwestern Oklahoma State University*
Marilyn Massey, *Collin County Community College*
David Platt, *Front Range Community College*
Janice Rech, *University of Nebraska at Omaha*
Judith Salmon, *Fitchburg State College*
Cynthia Schultz, *Illinois Valley Community College*
Chris Stump, *Bethel College*
Pamela Trim, *Southwest Tennessee Community College*
Chris Turner, *Arkansas State University*
Philip Van Veldhuizen, *University of Nevada at Reno*
Tracy Wienckowski, *Univesity of Buffalo*

Additional acknowledgements are extended to Pat Foard, for the Herculean task of preparing the solutions manuals; Teri Lovelace and the team at LaurelTech for preparing the answer section and serving as accuracy checker; Jim Sullivan, Allyson Graesser, and the Prentice Hall formatting team, for the countless hours they spent paging the book; and Ingrid Mount of Elm Street Publishing Services, whose talents as supervisor of production kept every aspect of this complex project moving through its many stages.

I would like to thank my editor at Prentice Hall, Eric Frank, and Associate editor, Dawn Murrin, who guided and coordinated the book from manuscript through production. Thanks to the wonderful team of designers, including Jonathan Boylan and Maureen Eide, for the beautiful covers and interior design. Finally, thanks to Halee Dinsey and Rachel Beckman, for your innovative marketing efforts, and to the entire Prentice Hall sales force for your confidence and enthusiasm about the book.

To the Student

I've written this book so that you can learn about the power of algebra and trigonometry and how it relates directly to your life outside the classroom. All concepts are carefully explained, important definitions and procedures are set off in boxes, and worked-out examples that present solutions in a step-by-step manner appear in every section. Each example is followed by a similar matched problem, called a Check Point, for you to try so that you can actively participate in the learning process as you read the book. (Answers to all Check Points appear in the back of the book.) Study Tips offer hints and suggestions and often point out common errors to avoid. A great deal of attention has been given to applying algebra and trigonometry to your life to make your learning experience both interesting and relevant.

As you begin your studies, I would like to offer some specific suggestions for using this book and for being successful in this course:

1. **Attend all lectures.** No book is intended to be a substitute for valuable insights and interactions that occur in the classroom. In addition to arriving for lectures on time and being prepared, you will find it useful to read the section before it is covered in the lecture. This will give you a clear idea of the new material that will be discussed.

2. **Read the book.** Read each section with pen (or pencil) in hand. Move through the illustrative examples with great care. These worked-out examples provide a model for doing exercises in the exercise sets. As you proceed through the reading, do not give up if you do not understand every single word. Things will become clearer as you read on and see how various procedures are applied to specific worked-out examples.

3. **Work problems every day and check your answers.** The way to learn mathematics is by doing mathematics, which means working the Check Points and assigned exercises in the exercise sets. The more exercises you work, the better you will understand the material.

4. **Prepare for chapter exams.** After completing a chapter, study the summary, work the exercises in the Chapter Review, and work the exercises in the Chapter Test. Answers to all these exercises are given in the back of the book.

5. **Use the supplements available with this book.** A solutions manual containing worked-out solutions to the book's odd-numbered exercises, all review exercises, and all Check Points; a dynamic web page; and videotapes and CD-ROMs created for every section of the book are among the supplements created to help you tap into the power of mathematics. Ask your instructor or bookstore which supplements are available and where you can find them.

I wrote this book in beautiful and pristine Point Reyes National Seashore, north of San Francisco. It was my hope to convey the beauty of mathematics using nature as a source of inspiration and creativity. Enjoy the pages that follow as you empower yourself with the algebra and trigonometry needed to succeed in college, your career, and your life.

Regards,

Bob

Robert Blitzer

About the Author

Bob Blitzer is a native of Manhattan and received a Bachelor of Arts degree with dual majors in mathematics and psychology (minor: English literature) from the City College of New York. His unusual combination of academic interests led him toward a Master of Arts in mathematics from the University of Miami and a doctorate in behavioral sciences from Nova University. Bob is most energized by teaching mathematics and has taught a variety of mathematics courses at Miami-Dade Community College for nearly 30 years. He has received numerous teaching awards, including Innovator of the Year from the League for Innovations in the Community College, and was among

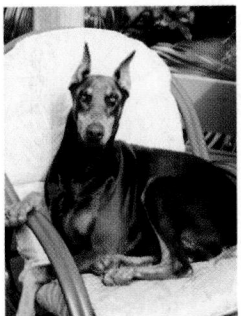

the first group of recipients at Miami-Dade Community College for an endowed chair based on excellence in the classroom. In addition to *Algebra and Trigonometry*, Bob has written *Introductory Algebra for College Students*, *Intermediate Algebra for College Students*, *Introductory and Intermediate Algebra for College Students*, *Algebra for College Students*, *Thinking Mathematically*, *College Algebra*, and *Precalculus*, all published by Prentice Hall.

Finally, Bob loves to spend time with his pal, Harley, pictured to the right. He's so cute (Harley, not Bob) that we couldn't resist including him.

Why Blitzer's Algebra & Trigonometry, 2nd Edition?

This text was written to address students' most commonly cited reasons for not using their texts:

- "I'll never use this information ("When will I use this?")"
- "I can't follow the explanations."

"When Will I Use This?"

This text integrates over 500 dynamic applications that connect mathematics to the entire spectrum of students' interests.

The interesting and diverse applications . . .

- represent a wide range of disciplines.

Page 214

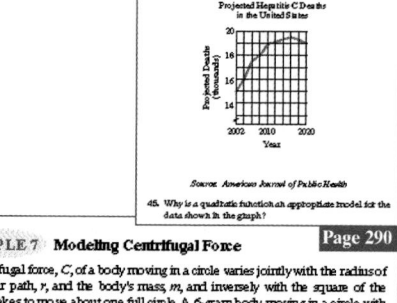

Page 314

Page 290

Page 362

- feature unique and interesting data that show students that mathematics can be applied in many settings.

Page 413

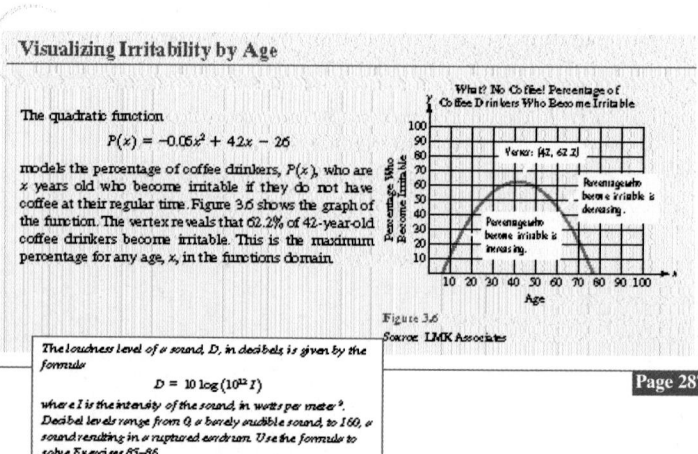

Page 287

Page 396

xiv

- are driven by real and sourced data, illustrating the power of algebra and trigonometry to model contemporary issues and problems.

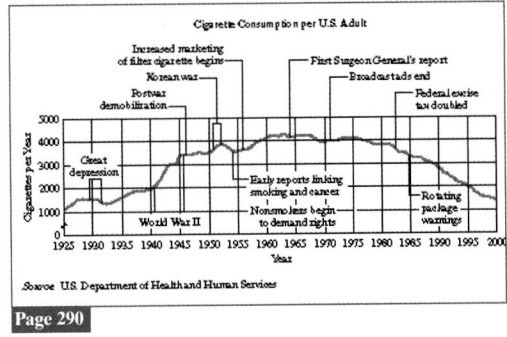

Page 290

Unique Chapter and Section Opening Vignettes

- Each chapter and section begins with a vignette highlighting an everyday scenario, posing a question about it, and exploring how the chapter section subject can be applied to answer the question.
- These are revisited in the course of the chapter or section in an example, discussion, or exercise.

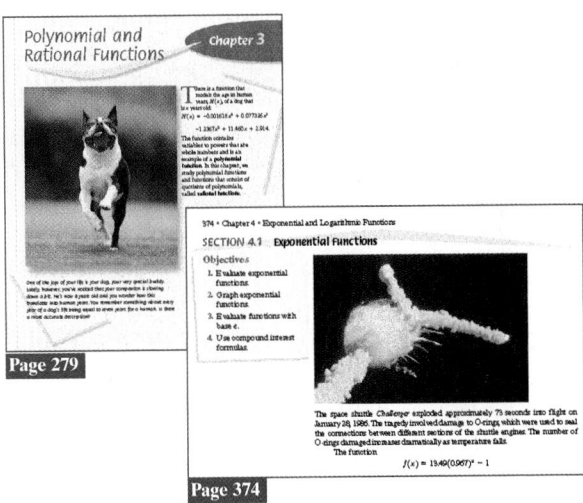

Page 279

Page 374

Enrichment Essay Literary (and other interesting asides)

- Enrichment Essays provide historical, interdisciplinary, and otherwise interesting connections to the math under study, showing students that math is an interesting and dynamic discipline.

Page 306

Approximately 32 years after 1979, in the year 2011, there will be 1.5 million U.S. workers in the environmental industry.

 Check Point 10   Use the formula in Example 10 to find by which year there will be two million, or 2000 thousand, U.S. workers in the environmental industry.

Page 415

WHEN WILL I USE THIS?

XV

"I Can't Follow the Explanations."

Clear & Friendly Writing Style

- Blitzer's language is clear, direct, and simple. He breaks down concepts in a conversational style, providing analogies and drawing connections to students' experiences whenever possible.

> Yogi Berra, catcher and renowned hitter for the New York Yankees (1946–1963), said it best: "Prediction is very hard, especially when it's about the future." At the start of the twenty-first century, we are plagued by questions about the environment. Will we run out of gas? How hot will it get? Will there be neighborhoods where the air is pristine? Can we make garbage disappear? Will there be any wilderness left? Which wild animals will become extinct? These concerns have led to the growth of the environmental industry in the United States.
>
> **EXAMPLE 10**   **The Growth of the Environmental Industry**

Page 414

Detailed Illustrations, Examples, and Check Points
Examples:

Page 305/306

- are abundant, because students learn by example.
- are thoroughly annotated to the right of the algebraic steps. These annotations are in a conversational style, providing the voice of an instructor in the book, explaining key steps and ideas as the problem is solved.
- offer students the opportunity to stop and test their understanding of the example by working a similar exercise immediately following that example, called a Check Point.
- The answers to the Check Points are provided in the answer section.

Exercise Sets that Precisely Parallel Examples (pages 244-247)

- An extensive collection of exercises is included at the end of each section.
- Exercises are organized by level within six category types: **Practice Exercises, Application Exercises, Writing in Mathematics, Technology Exercises, Critical Thinking Exercises,** and **Group Exercises.**
- The order of the practice exercises is exactly the same as the order of the section's illustrative examples. This parallel order enables students to refer to the titled examples and their detailed explanations to achieve success working the practice exercises.

Explanatory Voice Balloons

Voice balloons are used in a variety of ways to demystify mathematics. They:

- translate mathematical ideas into plain English.
- help clarify problem-solving procedures.
- present alternative ways of understanding concepts.
- connect complex problems to the basic concepts students have already learned.

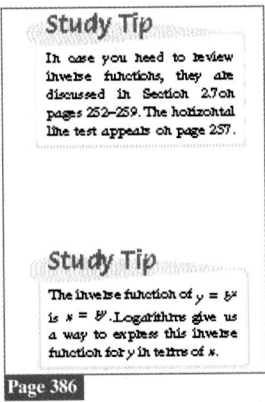

Page 100

Study Tips

Study Tip boxes:

- appear in abundance throughout the book.
- offer suggestions for problem solving.
- point out common mistakes.
- provide informal tips and suggestions.

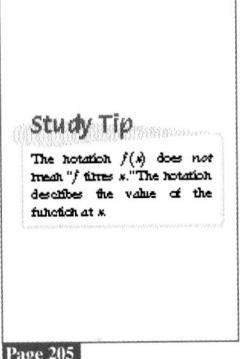

Page 386

Page 205

Clearly Stated Section Objectives

Learning objectives:

- are clearly stated at the beginning of each section.
- help students recognize and focus on the most important ideas.
- appear in the margin at their point of use.
- form the foundation for the algorithms in MathPro (tutorial software) and in TestGen-EQ (test generator software).

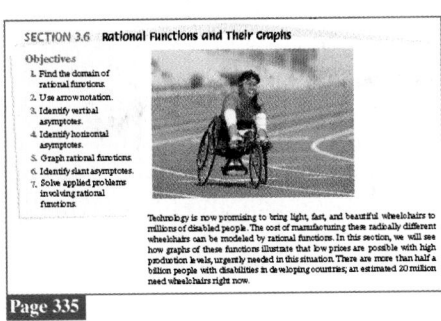

Page 335

Chapter Review Grids

- summarize definitions and concepts for every section of the chapter.
- refer students to the examples that illustrate these key concepts.

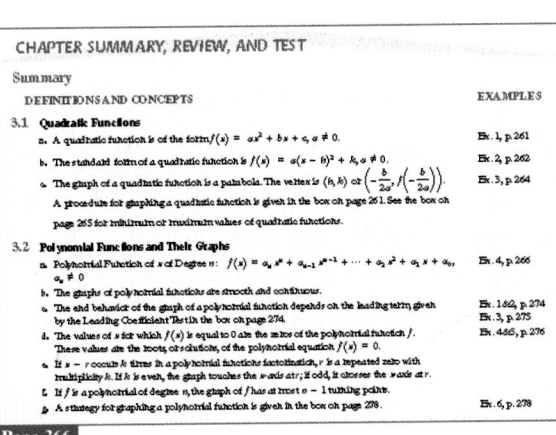

Page 366

xvii

TUTORIAL

Blitzer M@thP@k

An Integrated Learning Environment

Today's textbooks offer a wide variety of ancillary materials to students, from solutions manuals to tutorial software to text-specific Websites. Making the most of all of these resources can be difficult. Blitzer **M@thP@k** helps students get it together. **M@thP@k** seamlessly integrates the following key products into **an integrated learning environment:**

MathPro 5

MathPro 5 is online, customizable tutorial software integrated with the text at the Learning Objective level. MathPro 5's "watch" feature integrates lecture videos into the algorithmic tutorial environment. The easy-to-use course management system enables instructors to track and assess student performance on tutorial work, quizzes, and tests. A robust reports wizard provides a grade book, individual student reports, and class summaries. The customizable syllabus allows instructors to remove and reorganize chapters, sections, and objectives. MathPro 5's messaging system enhances communication between students and instructors. The combination of MathPro 5's richly integrated tutorial, testing, and robust course management tools provides an unparalleled tutorial experience for students, and new assessment and time-saving tools for instructors.

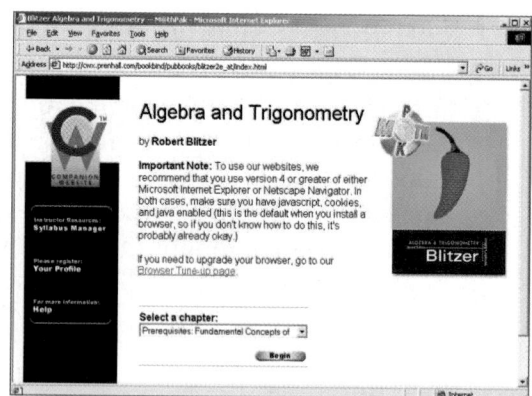

The Blitzer M@thP@k Website

This robust passcode-protected site features quizzes, homework starters, live animated examples, graphing calculator manuals, and much more. It offers the student many ways to test and reinforce their understanding of the course material.

Student Solutions Manual

The *Student Solutions Manual* offers thorough, accurate solutions that are consistent with the precise mathematics found in the text.

GREAT SUPPORT

Blitzer M@thP@k.
Helping Students Get it Together.

HOMEWORK

Problem Solving for Students.
Solving Problems for You.

Students need to practice solving problems—The more they practice, the better problem solvers they become. Professors want relief from the tedium of grading.

That's why we created **PH GradeAssist**. It's...

✓ online—available anytime, anywhere.
✓ text-specific—tied directly to your Prentice Hall Precalculus or Calculus text.
✓ algorithmic—contains unlimited questions and assignments for practice and assessment.
✓ customizable—completely unique to your course—edit questions and add yours.

How does PH GradeAssist work for the instructor?

• You create quizzes or homework assignments from question banks specific to your text. Choose the problems you prefer, edit them, or add your own.
• Your students go online and work the assignments that you have created.
• The problems let students work with real math, not just multiple choice.
• Many problems are algorithmically generated, so each student gets a slightly different problem with a different answer.
• PH GradeAssist scores these assignments for you, using a sophisticated math parser, which recognizes algebraic, numeric, and unit equivalents.
• Results can be easily accessed in a central gradebook.

For a demonstration, contact your local Prentice Hall representative or visit us online at www.prenhall.com/phga

xix

GREAT SUPPORT

Applications Index

Abominable Snowman footprints, 355
Acid rain, 417 (Exercises 69–70)
Actors selected for workshop, 1005 (Exercise 78)
Adult height percentage for girl, 396 (Exercises 81–82), 397 (Exercise 108)
Aerobic exercise and target zone, 750, 752 (Exercise 58)
Age of uncle and wife, 107 (Exercise 85)
Age trends in professional productivity, 323 (Exercises 43–44)
AIDS, 293, 294, 295, 431 (Exercises 51, 53), 724 (Exercise 39), 931
Airline departure lineups, 1005 (Exercise 81)
Airline fares, 276 (Exercise 3)
Airplane landing, 767 (Exercise 20)
Airplane speed/compass heading relative to ground, 681 (Exercise 90)
Airplane ticket cost, 364 (Exercise 33)
Alcohol consumption/heart disease, 192 (Exercise 96)
Alcohol safety program, 761 (Exercise 21)
Alligator, 172 (Exercises 73–74), 781 (Exercise 45)
Alzheimer's, 170 (Exercises 9–11)
America's Cup, 642
Amplitude and period of ball's motion, 576 (Exercise 67)
Angle measures, 715 (Exercise 67), 724 (Exercise 35), 1008 (Exercise 39)
Angle of depression, 469 (Exercise 87), 546–47 (Exercises 33–34), 621 (Exercise 44)
Angle of elevation, 463–64, 465, 467 (Exercise 64), 468 (Exercises 65–70), 469 (Exercise 87), 519–20 (Exercise 46–47), 538, 539, 546 (Exercises 29–32), 547 (Exercises 35–38), 548 (Exercise 61), 552 (Exercise 32), 553 (Exercises 101–102), 554 (Exercise 16), 555 (Exercise 20), 597, 606 (Exercises 85–86), 609 (Exercises 19, 64), 621 (Exercises 43, 44, 47), 622 (Exercises 50, 64)
Angular speed, 448–49, 452 (Exercise 81), 551 (Exercise 17)
Angular velocity, 452 (Exercise 95)
Annuities, 952–54
Arab-American faiths, 989
Area codes, 986 (Exercise 27)
Area
  of region under curve, 535–36 (Exercises 77–78)
  of shaded regions, 58 (Exercises 99–100)
  of shaded figures, 47 (Exercises 89–90), 48 (Exercise 104)
Astronomical units, 917 (Exercises 21–22)
Athletic enthusiasm, 47 (Exercise 86)
Automobiles
  accidents per day, 838 (Exercise 52)
  costs, 416 (Exercise 65), 417 (Exercise 73)
  depreciation, 213 (Exercise 92)
  headlights, 922 (Exercise 36), 923 (Exercise 10)
  inventory of, 837 (Exercise 32)
  mechanic's time, 155 (Exercise 110)
  in motion and brake application, 723 (Exercise 28)
  ordering, 985 (Exercise 21)
  race, 986 (Exercise 36), 987 (Exercise 49)
  rentals, 105 (Exercises 31–32), 144, 152–53
  value, 936 (Exercise 74)

Babies born out of wedlock, 105 (Exercises 33–34), 107 (Exercise 81)
Ball's distance from rest position, 605 (Exercises 83–84), 605
Ball's height, 164, 368 (Exercise 5), 696 (Exercise 14), 723 (Exercise 27), 1008 (Exercise 29)
Band lineups, 986 (Exercise 39)

Bank/credit union charges, 155 (Exercise 104)
Bariatric surgery, 975 (Exercise 51–52)
Baseball, 129 (Exercise 107), 906 (Exercises 59–60), 907 (Exercise 75)
Bearing
  and airplane flights, 554 (Exercise 106), 627, 629 (Exercise 19), 693 (Exercise 19)
  of boat/ship, 541, 547 (Exercises 39–40, 42, 44, 46), 618–19, 620 (Exercise 39), 622 (Exercise 51), 629 (Exercises 31, 37), 630 (Exercise 38), 696 (Exercise 20)
  from house, 547 (Exercise 43)
  and illegal radio station location, 620 (Exercise 40)
  and island navigation, 629 (Exercises 35–36)
  from one point to another, 554 (Exercise 18)
  and plane's speed/direction and angle of wind, 680 (Exercises 69–70)
  and trail system, 542
Berlin Airlift, 753, 761 (Exercise 20)
Biorhythms, 506 (Exercises 59–66), 507 (Exercise 81), 508 (Exercise 98)
Birthdays, 269 (Exercise 43), 1001 (Exercise 66)
Blood alcohol concentration and car accidents, 412–13, 416 (Exercises 53–54), 417 (Exercise 73)
Boat clearance and parabolic arch, 883 (Exercise 55)
Boat on tilted ramp, 685
Boat's velocity/speed/direction angle, 695 (Exercise 95)
Body-mass index, 365 (Exercise 43), 752 (Exercises 60–61)
Body temperature variations, 553 (Exercise 64)
Bookcase dimensions, 106 (Exercises 45–46)
Book club selection, 987 (Exercise 50)
Book shelving, 981
Books taken on vacation, 986 (Exercise 43), 1007 (Exercise 18)
Box dimensions, 323 (Exercise 47)
Box sizes, 130 (Exercises 113–114)
Box weight and magnitude, 679 (Exercises 61–62)
Brain activity, 520 (Exercise 58)
Break-even point, 555 (Exercise 17), 709–10, 712–13 (Exercises 47–50), 763 (Exercise 6), 766 (Exercise 13)
Breathing cycles, 487, 503
Bridge coupon book, 105 (Exercise 36)
Bridge tower cables, 883 (Exercise 53), 922 (Exercise 37)
Building height, 172 (Exercise 76)
Bus coupon book, 171 (Exercise 37)
Bus fares, 105 (Exercise 35)

Cabin dimensions, 693 (Exercise 17)
Cable lengths, 143 (Exercise 91)
Cable service plans, 924 (Exercise 20)
Cable television in households, 396 (Exercise 84)
Calorie-nutrient information, 764 (Exercise 10)
Calories burned per hour, 769
Calories in fast foods, 714 (Exercises 57–58)
Canoe weight, 356
Carbon dating, 429 (Exercises 17–18)
  and bones, 437 (Exercise 21)
  and cave paintings, 436 (Exercise 71)
  and Dead Sea Scrolls, 422–23
Cardboard, 742 (Exercise 54)
Cardiovascular disease and probability, 999 (Exercises 31–32)
Carousel animals, 448, 452 (Exercise 84), 641 (Exercises 73–74)
Carrot cake supply and demand, 742 (Exercise 58)

Casting of male roles, 986 (Exercise 38)
Causes of death in U.S., 1006 (Exercise 93)
CDs sold in U.S., 935 (Exercise 61)
CDs taken on vacation, 1005 (Exercise 79)
Celebrity Jeopardy all-time highest earners, 201, 202, 203
Celebrity thermometers, 192 (Exercise 100)
Centrifugal force, 362–63
Challenger explosion, 374, 375, 384 (Exercise 65)
Changing pattern of work in U.S., 1004 (Exercise 28)
Charitable giving, 277 (Exercise 23)
Checking account, 156 (Exercise 125), 173 (Exercise 109)
Checkout line, 987 (Exercise 70)
Chernobyl nuclear explosion, 383 (Exercise 46)
Chicken and red meat consumption, 715 (Exercises 77–78)
Child pulling sled, 679 (Exercise 54)
Child raising costs, 171 (Exercises 32)
Children driven to zoo, 986 (Exercise 45)
Child's drug dosage, 69 (Exercise 66)
Child's vocabulary and child's age, 93 (Exercise 72)
Cholesterol
  content, 714 (Exercises 59–60)
  intake, 752 (Exercise 56), 766 (Exercise 12)
  heart disease and, 93 (Exercise 71)
Cigarette consumption, 154 (Exercises 93–94), 290 (Exercise 41)
Cigarettes and lung cancer, 188
Cigarette smokers, 12 (Exercise 79)
Circular driveway, 200 (Exercise 58)
Club members for conference attendance, 987 (Exercise 53)
Club officers, 986 (Exercise 33), 1005 (Exercise 77)
Cocaine testing, 1000 (Exercise 63)
Coffee drinkers and irritability, 287
Coin toss, 155 (Exercise 99), 989, 996, 998 (Exercise 19–20)
Cold virus, 36, 37, 48 (Exercise 98), 302 (Exercise 52)
Collection of objects, 987 (Exercise 72)
College costs, 191 (Exercise 90), 334–35 (Exercise 69)
College enrollments and gender, 174 (Exercises 28–29)
Comedy club schedule, 986 (Exercise 29), 987 (Exercise 56)
Comet, 742 (Exercise 47)
Commercial lot cost, 694 (Exercise 21)
Commercial real estate lot cost, 630 (Exercises 43–44)
Committee choices, 986 (Exercise 42)
Committee formation, 984
Commutative property, 12 (Exercise 75)
Commuting speed, 69 (Exercise 69)
Compound interest, 380–81, 383 (Exercises 41–44), 384 (Exercises 56, 61), 385 (Exercise 68), 413–14, 416 (Exercises 57–64), 418 (Exercise 88), 434 (Exercises 9–10), 435 (Exercises 67–69), 609 (Exercise 15), 767 (Exercise 14), 936 (Exercises 65–66), 1005 (Exercise 54)
Computation problems and maximum scores, 761 (Exercise 22)
Computers
  assembly time, 371 (Exercise 74)
  depreciation, 191 (Exercise 80)
  infection rates, 213 (Exercise 80)
  prices/discounts, 259 (Exercise 69), 261
  sale, 57 (Exercise 95), 253
  in use, 723–24 (Exercise 30)
Concert attendance, 171 (Exercise 39)
Condominium sales, 714 (Exercise 62)
Cone's volume, 363
Constant phenomenons, 191 (Exercise 91)

Contraceptive prevalence and births per woman, 191 (Exercise 76)
Coronary heart disease, 430 (Exercises 27–30, 50)
Corporate board of directors, 986 (Exercise 34)
Corporate profits, 272 (Exercise 11)
Cost function, 370 (Exercise 65), 712–13 (Exercises 47–50), 763 (Exercise 6), 766 (Exercise 13)
Course schedule planning, 978
Current for resistance, 364 (Exercise 42)

Dart thrown, 1001 (Exercise 64)
Daylight hours, 486 (Exercises 13–14), 505, 506–7 (Exercises 67, 69), 520 (Exercise 48), 605 (Exercises 81–82)
Death rate/hours of sleep, 46 (Exercise 84)
Deaths in U.S. from car accidents and gunfire, 713 (Exercise 53)
Decay model and carbon–14, 429 (Exercises 13–14)
Decibel levels, 371 (Exercise 73), 396 (Exercises 85–86), 406 (Exercise 83), 437 (Exercise 18)
Deck of 52 cards and probability, 991, 993–94, 999 (Exercises 39–40), 1006 (Exercises 86–87), 1007 (Exercise 21)
Defective products, 156 (Exercise 128)
Degree-days, 945 (Exercise 72)
Degrees/radians and clock movement, 451 (Exercises 71–72)
Demand line, 191 (Exercise 81)
Department store, 259 (Exercise 68)
Desk manufacturing, 781 (Exercise 50)
Die roll, 990, 998 (Exercises 9–14), 998 (Exercises 23–24), 999 (Exercises 37–38), 1000 (Exercises 45–48), 1006 (Exercises 84–85)
Digital photography, 792, 802–3, 805 (Exercises 45–46), 806 (Exercise 62), 807 (Exercise 67), 837 (Exercises 30–31)
Dimensions of football field, 101–2
Direction angle, 681 (Exercise 92)
Disaster shipments, 761 (Exercise 19)
Distance
  across lake, 467 (Exercise 63), 552 (Exercise 31), 629 (Exercises 33–34)
  between oil platform and ends of beach, 620 (Exercise 41)
  between pitcher's mound and base, 630 (Exercises 41–42)
  between points on banks of river, 621 (Exercise 42)
  between points on Earth, 451–52 (Exercises 77–80, 103)
  between rotating beacon of light and given point, 519 (Exercise 45), 520 (Exercise 50)
  between tip of hour hand and ceiling, 506 (Exercise 68)
  between tips of clock hands, 630 (Exercise 54)
  and skydivers' falling time, 358
Diving board, 520 (Exercise 49)
Divorce rate, 368 (Exercise 6), 764 (Exercise 14)
Divorces by number of years of marriage, 83 (Exercises 45–48)
Dog's age in human years, 279, 304, 333 (Exercises 45–46)
Dominoes, 962, 968 (Exercise 32)
Downhill skiing, 189 (Exercise 69)
Dress price, 107 (Exercise 84)
Driver's age
  and fatal crashes, 128–29 (Exercises 99–100)
  and number of accidents, 232 (Exercises 91–93), 287
Driving rate, 359, 364 (Exercise 39), 371 (Exercise 72)

Drug concentration in blood, 323 (Exercise 46)
Drug convictions, 69 (Exercise 68), 70 (Exercise 82)
Drug testing, 36 (Exercise 117), 985 (Exercise 17), 987 (Exercise 51)
Duration of storm, 35 (Exercise 109)

Earnings
  and educational attainment, 713–14 (Exercises 55–56)
  and feeling of being underpaid, 46 (Exercise 83)
Earthquake intensity, 385, 392–93, 435 (Exercise 41)
Earthquake victims shipment, 754, 755
Earth's radius, 548 (Exercise 60)
Eccentricity and planet's orbit, 918 (Exercise 42)
Education fund, 959 (Exercise 70)
Election ballot, 986 (Exercise 41)
Electrical resistance, 365 (Exercise 48), 1008 (Exercise 30)
Electric bill, 370 (Exercise 70)
Elephant's weight, 417 (Exercise 86)
Elevator capacity, 155 (Exercises 107–108), 751 (Exercise 55)
Elk population, 302 (Exercise 51)
Elliptical ceiling in Statuary Hall, 854 (Exercise 59)
Elliptical orbits, 854 (Exercise 70)
Elliptical whispering room, 854 (Exercise 60)
Elliptipool, 854 (Exercise 68), 921 (Exercise 14)
Emotional fluctuations, 486 (Exercise 15)
Emotions graphs, 234 (Exercise 122)
Enclosure's dimensions, 739–41
Environmental industry growth, 414–15
Environmental pollution costs, 59
Equation of ellipse and circles, 855 (Exercise 73)
Equilibrium price, 715 (Exercise 76)
Evaporation, 35 (Exercise 106)
Events and probability, 999 (Exercises 33–36), 1000 (Exercises 57–62)
Examination grades, 155 (Exercise 109), 173 (Exercise 110), 278 (Exercise 20)
Exercise survey, 998 (Exercises 1–4)
Exercising and headaches, 272–73 (Exercise 12)
Explosion recorded by microphones, 866–67, 869 (Exercise 51)

Falling object, 35 (Exercise 104)
Family(ies)
  composition of and probability, 1006 (Exercise 94)
  incomes of, 171 (Exercise 31)
  sample of, 998 (Exercises 21–22)
  with nine girls in a row, 997
Farms in U.S., 288
Federal budget projections, 189–90 (Exercises 70–71)
Federal Express aircraft purchases, 761 (Exercise 23)
Fencing for enclosed area, 291 (Exercises 47–48), 292 (Exercise 70)
Ferris wheel, 200 (Exercise 59), 486 (Exercise 26)
Fetal weight, 172 (Exercise 72)
Field's area/fencing, 368 (Exercise 7)
Fingernail growth, 364 (Exercise 31)
Fire sighting, 547 (Exercise 41)
Fireworks' height, 290 (Exercise 43)
Five-sentence paragraphs, 986 (Exercise 31)
Flagpole lean, 622 (Exercise 52)
Flight time, 200 (Exercise 57)
Floods, 1006 (Exercise 95)
Floor dimensions, 172 (Exercise 75)
Floor dimensions/dimensions of square for pool, 742 (Exercise 53)
Florida's population growth, 948–49
Flu epidemic/outbreak, 384 (Exercise 57), 424, 429 (Exercise 25), 436 (Exercise 72), 959 (Exercise 84)
Flu inoculation, 69 (Exercise 65), 350 (Exercise 70)

Food group consumption, 944–45 (Exercises 51–54)
Food Stamp Program, 129 (Exercises 101–104), 279, 287
Football
  field dimensions, 171 (Exercise 36)
  height of, 290 (Exercise 44), 781 (Exercise 46)
  position, 922–23 (Exercise 59)
  speed of, 679 (Exercise 53)
Forces in equilibrium, 680 (Exercises 63–66)
Foreign-born Americans, 208, 209
Frame width, 130 (Exercise 117)
Free-falling object, 131 (Exercise 132)
Fuel efficiency, 214, 292 (Exercise 65)
Functions described by series of trigonometric functions, 594–95 (Exercise 49)

Galápagos Islands, 131, 134
Garbage production, 356–57
Garden dimensions, 278 (Exercise 19)
Garden length/width, 129 (Exercise 111)
Gasoline sold, 689 (Exercise 39)
Gas volume, 364 (Exercise 40)
Gay discharges from military, 722 (Exercise 25)
Gay service members, 83 (Exercise 56)
Gender ratios in U.S., 924 (Exercise 15)
Geometric figures, 725 (Exercise 44)
Glenn's orbit in Friendship 7, 917 (Exercises 23–24)
Global warming, 104 (Exercise 26), 306, 509, 520–21 (Exercise 73)
Goals encoded, 820 (Exercise 55)
Golden rectangle, 35 (Exercise 105)
Grading system, 806 (Exercise 51)
Gravitational force, 365 (Exercise 46)
Gravity model, 365 (Exercise 57)
Gray wolf population, 212 (Exercise 79)
Greenhouse effect, 435 (Exercise 65)
Guy wires lengths, 630 (Exercises 39–40)

Half-life of radioactive substances, 429 (Exercises 15–16), 840 (Exercise 13)
Health care spending, 23 (Exercise 95)
Health rating, 1000 (Exercises 51–52)
Heart rate, 104 (Exercises 21–22), 166 (Exercise 53)
Heat generation, 365 (Exercise 61)
Heating system costs, 174 (Exercise 31)
Heat loss rate, 555 (Exercise 19)
Height child will attain as adult, 75
Height of rock, 58 (Exercise 97)
Height of football, 82 (Exercise 39–44)
Hepatitis C, 290 (Exercises 45–46)
High school graduates, 58 (Exercise 96)
HIV virus, 369 (Exercise 13)
HMO enrollments, 383–84 (Exercises 53–54)
Home savings, 959 (Exercise 72)
Hotel room types, 714 (Exercise 61)
Hours in traffic, 104 (Exercise 29)
Households online (U.S.), 213 (Exercises 81–84)
House positioning, 869 (Exercise 53)
House sales prices, 428 (Exercises 9)
Hubble Telescope, 365 (Exercise 62)
Hurricanes
  and barometric pressure, 417 (Exercises 83–84)
  and probability, 1000 (Exercise 53)

Ice cream flavors, 987 (Exercises 57–58)
Ice cream store drinks, 985 (Exercise 23)
Identical twins, 715 (Exercise 83)
Illumination and given distance, 365 (Exercises 45, 60)
Incidence ratio, 350–51 (Exercises 71–74), 352 (Exercise 84)
Income, 106 (Exercises 49–50)
Infant mortality, 713 (Exercise 54)
Infant's dosage for medication, 609 (Exercise 18)
Inflation rate, 383 (Exercises 49–50)
Inn charges, 106 (Exercise 53)
Interfaith group, 996
Internet marketing consulting firm, 981

Internet use, 212 (Exercise 75)
Investments, 105 (Exercises 39–42), 171 (Exercise 35), 174 (Exercise 32), 278 (Exercise 17)
  choices, 381–82, 384 (Exercise 66), 437 (Exercise 17)
  IRA value, 959 (Exercise 69)
  maximized returns on, 762 (Exercise 33)
  rates, 724 (Exercises 33–34)
  risks, 752 (Exercise 57)

Jeans price/discount, 259 (Exercise 70)
Jewish American population (U.S.), 232 (Exercises 83–90)
Jobs
  applicants, 1007 (Exercise 17)
  multiple holdings of, 292 (Exercise 64)
  offers, 945 (Exercise 60), 946, 958 (Exercise 60), 958 (Exercise 66)

Kidney stone disintegration, 841, 850–51
Kinetic energy, 365 (Exercise 47)

Labor costs, 106 (Exercises 47–48)
Ladder against house, 129 (Exercise 108)
Land owned by federal government, 268–69 (Exercise 42)
Language extinction, 764 (Exercise 15)
Lawyers in U.S., 213 (Exercises 85–88)
Learning curve, 372 (Exercise 19)
Learning rate, 697 (Exercise 20)
Leisure activities, 154 (Exercises 85–92)
Length of arc on circle, 551 (Exercise 16), 554 (Exercise 3)
Lengths, 765 (Exercise 38)
Letter combinations, 986 (Exercise 40)
License plates, 980
Life expectancy
  for African-American females, 416 (Exercise 66)
  by year of birth, 95, 96
Life satisfaction, 181–82
Light's intensity, 372 (Exercise 20)
Linear programming in business, 762 (Exercise 36)
Linear speed, 449–50, 452 (Exercises 82–83, 94), 551 (Exercise 18)
Literacy, 176, 191 (Exercise 77)
Little League baseball, 623, 980
Loan payments, 70 (Exercise 83)
Long-distance carrier selection, 98–99
Longest-running television shows, 212 (Exercises 76–78)
Lost hikers, 93 (Exercise 74)
Lot's dimensions, 714 (Exercises 63–64)
Lotteries, 986 (Exercise 47), 992, 993, 998 (Exercises 25–26), 1001 (Exercise 65), 1006 (Exercise 90), 1007 (Exercise 20)
LPs sold in U.S., 935 (Exercise 62)
Lunch menus, 761 (Exercise 18), 985 (Exercise 24)

Mach number, 364 (Exercise 35), 586–87 (Exercises 70–71)
Magnitude
  and direction, 679 (Exercises 59–60), 695 (Exercise 94), 696 (Exercise 21)
  of force, 690 (Exercise 51–52)
Mailing costs, 175, 226, 233 (Exercise 96), 234 (Exercise 121)
Male actors for Rent, 1005 (Exercise 80)
Mandelbrot set, 666 (Exercises 77–78, 89)
Manufacturing and testing times, 791 (Exercise 32)
Manufacturing costs, 12 (Exercise 94)
Marching band, 715 (Exercise 84)
Marijuana use, 48 (Exercise 99)
Marital satisfaction for families with children, 302–3 (Exercises 54, 68)
Marriage ages for women (U.S.), 80–81, 269 (Exercise 44)
Marriage to someone other than the perfect mate, 781 (Exercises 47–48)
Married men employed in U.S., 437 (Exercise 19)
Mass
  of hydrogen atoms, 23 (Exercise 98)
  of oxygen molecules, 23 (Exercise 97)
Mathematics department, 1000 (Exercises 43–44)

Median heights, 246 (Exercise 57)
Medicare costs, 259 (Exercise 77)
Medicare spending, 166 (Exercises 54–56)
Melodies on touch-tone phone, 594 (Exercises 35–36, 42, 47, 48)
Men in U.S. labor force, 938
Message encoding/decoding, 807, 817–18
Migratory birds and power expenditure, 232–33 (Exercise 94)
Mile records, 269 (Exercise 46)
Miles per gallon, 291 (Exercise 57)
Million dollar earnings, 72 (Exercise 30)
Minimum wage, 419–21, 430 (Exercises 45–49)
Miscarriages by age, 431 (Exercise 52)
Moon weight of person, 364 (Exercise 34)
Morphing, 235, 242, 247 (Exercise 75)
Moth eggs, 315–16, 323 (Exercise 45)
Mountain bike, 350 (Exercise 67)
Mountain height, 172 (Exercise 91)
Moustache length, 355
Movies
  costs in making of, 104 (Exercise 27)
  rankings, 986 (Exercise 36)
  screens for in U.S., 190 (Exercise 75)
  spending on, 72 (Exercise 31)
Multiple-choice test, 979, 985 (Exercise 25), 986 (Exercise 26), 1007 (Exercise 23)
Multiplier effect, 956
Music majors, 107 (Exercise 83)
Music synthesizers, 537

National debt, 19, 21
Natural disaster costs, 105 (Exercise 30)
Newspaper readership, 171 (Exercise 30), 191 (Exercise 79)
New York City Marathon, 23 (Exercise 96)
New York State Lottery, 986 (Exercise 48)
Night club schedule, 986 (Exercise 30)
Nonviolent prisoners, 370 (Exercise 69)
Number of people living alone, 186–87
Nursing home costs, 937, 943
Nutritional content, 781 (Exercise 49), 791 (Exercise 31)

Objective function's maximum/minimum value, 765 (Exercise 56)
Object's distance, 554 (Exercises 109–110)
Object's falling time, 364 (Exercise 38)
Object's height above ground, 165
Officers for Internet company, 983
Oil price and supply, 361
Optimum heart rate, 12 (Exercise 81)

Pacific salmon, 104 (Exercises 23–24)
Parabolic path of home run, 897
Parallelogram's length of sides, 630 (Exercise 52)
Parking lot, 129 (Exercise 112)
Passion/commitment and functions of time, 233 (Exercise 97)
Password formation, 985 (Exercise 19)
Paycheck amounts, 106 (Exercise 56)
Pay raises, 945 (Exercises 55–56)
Payroll, 1005 (Exercise 55)
Peanuts cartoon, 35 (Exercise 107), 106 (Exercise 51)
Pen and pad costs, 1008 (Exercise 28)
Pen brands, 985 (Exercise 22)
Pen choices, 1005 (Exercise 75)
Pendulum swings, 958 (Exercises 67–68)
People with colds, 47 (Exercise 87)
Percentage of people with college degrees, 806 (Exercise 49)
Perigee and apogee of satellite's elliptical orbit, 855 (Exercise 72)
Personal income (U.S.), 23 (Exercise 94)
Pest-eradication, 960 (Exercise 90)
Pet population, 97–98
Phone calls, 353, 365 (Exercise 49)
Physician visits, 233 (Exercise 109)
Pier length, 621 (Exercise 46)
Pizza toppings, 979
Planes
  airspeed, 679 (Exercise 56)
  rate in still air and rate of wind, 715 (Exercise 66)

runway approaches, 679 (Exercise 55)
speed/wind speed, 680 (Exercise 67)
standbys, 986 (Exercise 44)
Planetary motion models, 916, 918 (Exercise 37)
Planetary years, 143 (Exercises 89–90)
Poker hands, 984–85, 999 (Exercise 29)
Police officers convicted, 114, 126
Police work and African-American community, 1006 (Exercises 91–92)
Pollution levels, 836 (Exercise 8)
Pop icons, 982
Population
 Africa, 421
 age 65 and over, 436 (Exercise 76)
 bird, 429 (Exercise 20), 429 (Exercise 26)
 California, 958 (Exercise 61)
 China and India, 104 (Exercise 28)
 doubling/tripling of, 429 (Exercises 21–24)
 elk, 437 (Exercise 22)
 Hispanic, 436 (Exercise 70), 1006 (Exercises 82–83)
 India, 104 (Exercise 28), 428 (Exercises 5–8)
 Iraq, 1005 (Exercise 52)
 Israel and Palestinian, 428–29 (Exercises 10–12)
 Los Angeles, CA, 435 (Exercise 64)
 Mexico, 383 (Exercise 45)
 modeling, 187–88
 percentage that is foreign-born (in U.S.), 723 (Exercise 26)
 Texas, 958 (Exercise 62)
 United States, 428 (Exercises 1–4)
 world, 436 (Exercises 73–75), 998 (Exercises 5–8)
Position formula application, 167 (Exercise 77), 173 (Exercise 115)
Potassium-40 dating, 429 (Exercise 19)
Preferred age in mate, 260, 268 (Exercise 41)
Prescription costs, 190 (Exercise 72)
Price before reduction, 106 (Exercise 53)
Prison inmates, 129 (Exercises 105–106), 333–34 (Exercises 48–49)
Probability of dying at any given age, 999 (Exercises 27–28)
Problems solved, 107 (Exercise 86)
Profits, 155 (Exercises 105–106), 371 (Exercise 4)
 on bookshelves and desks, 754, 755
 gas station, 803–4
 maximum daily, 782 (Exercise 62)
 on mountain bikes, 760 (Exercise 17)
 on skis, 767 (Exercise 16)
 on televisions, 760 (Exercise 15)
 on tents, 766 (Exercise 61)
 on paper, 766 (Exercise 60)
Projectile's height, 166 (Exercises 49–52)
Pyramid's volume, 371 (Exercise 75)

Racing cyclists, 24
Radio programming, 986 (Exercise 35)
Radio signals and ship's location, 869 (Exercise 52), 921 (Exercise 26)
Radio station, 986 (Exercise 28)
Raffle tickets, 985 (Exercises 18, 20), 987 (Exercises 52, 54)
Railroad curve, 452 (Exercise 102)
Railway crossing sign, 451 (Exercise 75)
Rain gutter depth, 130 (Exercise 115), 291 (Exercise 49)
Random selection from group, 996
Rational Zero Theorem, 324 (Exercise 66)
Razor blade sales, 724 (Exercise 32)
Recommended weights for given heights, 104 (Exercise 25), 106 (Exercise 55)
Record depth for breath-held diving, 93 (Exercise 73)
Recorded music, 782 (Exercise 56)
Record revolutions, 364 (Exercise 36)
Rectangles
 dimensions of, 765 (Exercise 36), 767 (Exercises 15, 19)

length/width of, 314 (Exercise 43), 742 (Exercises 49–50)
Redwood tree heights, 622 (Exercise 49)
Reflective properties of parabolic surfaces, 880, 881, 883 (Exercises 49–50)
Regions resulting from ranges of temperature/precipitation, 751 (Exercises 53–54)
Reindeer population, 303 (Exercise 67)
Responses in learning trials, 424–25
Resultant force, 676–78, 679 (Exercises 59–60), 681 (Exercise 91), 695 (Exercise 94), 696 (Exercise 21)
Retention of information, 384 (Exercise 55), 397 (Exercises 87, 101), 416–17 (Exercises 67–68), 435 (Exercise 42), 437 (Exercise 20)
Retirement funds, 407, 960 (Exercise 91)
Revenue and cost functions, 708
Revenue function, 712–13 (Exercises 47–50), 763 (Exercise 6), 766 (Exercise 13)
Reversibility of thought, 48
Richest people, 723 (Exercise 29)
Right triangles, solving, 537–38, 553 (Exercises 96–99), 554 (Exercise 15)
Road lengths, 143 (Exercise 92)
Rock concert, 752 (Exercise 58)
Rollerblade monthly sales, 361–62
Rolling motion, 905, 907 (Exercise 78)
Roulette wheel, 997
Rowing rate with/without current, 714 (Exercise 65)
Rug length and width, 742 (Exercise 52)
Runner's pulse, 417 (Exercise 85)
Running shoes analysis, 610
Running shoes, 350 (Exercise 68)
Running track, 200 (Exercise 60)

Sailing speed, 641 (Exercises 75–77, 86), 653 (Exercises 35–39)
Salaries
 average weekly, 171 (Exercise 33)
 baseball players, 938, 958 (Exercise 59), 960 (Exercise 92)
 hourly, 437 (Exercise 19)
 lifetime, 952, 958 (Exercise 65)
 and number of hours worked, 354–55
 options, 925, 945 (Exercise 60), 1008 (Exercise 26)
 total, 945 (Exercise 59), 1004 (Exercise 29), 1005 (Exercise 53), 1007 (Exercise 14)
Sales earnings, 171 (Exercise 38), 174 (Exercise 35), 278 (Exercise 18)
Sandwich costs, 291 (Exercise 50)
Satellite dish
 and dish depth, 883 (Exercise 56)
 and receiver placement, 883 (Exercise 51), 884 (Exercise 62), 922 (Exercise 38)
Savings, total, 958 (Exercises 58, 63–64)
Scattering experiments, 869 (Exercise 54)
Semielliptical arch, 921 (Exercise 12)
Senate members, 986 (Exercise 46)
Seniors' science test scores, 142 (Exercises 85–86)
Shared birthdays, 588
Sheet metal for tin can, 58 (Exercise 98)
Shipping costs, 275 (Exercise 58)
Ship tracking system, 742 (Exercise 48)
Shot-put ring, 621 (Exercise 45)
Shot put throw, 556
Simple harmonic motion, 542, 543–45, 547 (Exercises 47–50), 548 (Exercise 57), 554 (Exercises 19, 107–108), 696 (Exercise 15), 1008 (Exercise 31)
Simple interest, 99–100
Skid marks, 35 (Exercise 103)
Skull classification, 69 (Exercise 67)
Skydiving, 157, 370 (Exercise 71)
Sleep and death rate, 290 (Exercise 42), 721
Sleeping times, 173 (Exercise 107)

Snack food consumption, 764 (Exercise 8)
Snow fall, 83 (Exercises 61–64)
Soccer field dimensions, 105 (Exercise 44), 1008 (Exercise 27)
Social Security beneficiaries/number of workers, 190 (Exercise 73)
Sodium-iodine, 213 (Exercise 94)
Solid waste recovery/generation, 945 (Exercises 57–58), 946 (Exercise 74)
Spaceguard Survey, 870 (Exercise 61)
Speeding fines, 84, 86
Spending per uniformed member of military, 274 (Exercise 53)
Spinner, 999 (Exercises 41–42), 1006 (Exercises 88–89), 1007 (Exercise 24)
Sporting event, 396 (Exercise 83)
Spray can pressure, 360
Spread of rumor, 383 (Exercises 47–48)
Stadium seats, 945 (Exercise 62)
State lotteries, 1001 (Exercise 67)
Stereo headphones, 47 (Exercise 88)
Stock possibilities, 1005 (Exercise 76)
Stonehenge, 468 (Exercise 81)
Student loans, 246 (Exercise 58)
Students and random selection, 1007 (Exercise 22)
Student's earnings, 760 (Exercise 16)
Submarine pressure, 357
Super Bowl viewers, 155 (Exercise 120)
Supply and demand, 710–11, 713 (Exercises 51–52), 763–64 (Exercise 7)
Supply-side economics, 314 (Exercise 44), 315 (Exercise 55)
Survivors by age, 142 (Exercises 87–88)
Suspension bridge, 883 (Exercise 54)

Task mastery, 406 (Exercise 84), 435 (Exercise 43)
Taxes, 213 (Exercises 889–90)
 bills, 155 (Exercise 103)
 burden of, 23 (Exercise 93), 248–49
 filing, 35 (Exercise 108)
 rebate and multiplier effect, 959 (Exercise 75)
Tax Sheltered Annuity, 959 (Exercise 71)
Teachers' earnings, 12 (Exercise 80), 940–41
Tele-immersion, 768
Telephone
 calling plans, 155 (Exercise 102), 171 (Exercise 34)
 costs, 233 (Exercise 95)
 numbers, 979, 1007 (Exercise 19)
Television
 dimensions of, 127
 programming, 986 (Exercise 32)
 programs with greatest viewing percentage, 154–55 (Exercises 97–98)
 screen length/width, 742 (Exercise 51)
Temperature
 average, 191 (Exercise 78)
 of cake, 431 (Exercise 54)
 of coffee, 434 (Exercise 11)
 conversions, 154 (Exercise 95), 173 (Exercise 108)
 global mean, 154 (Exercise 96)
 increase/decrease/constant, 219–20
 monthly average, 155 (Exercise 100), 507 (Exercise 70), 508 (Exercise 93–94)
 windchill, 136
Tennis club options, 107 (Exercise 82)
Tennis court, 105 (Exercise 43)
Territorial area, 31, 36 (Exercise 118)
Textbook sales, 350 (Exercise 69)
Theater seats, 945 (Exercise 61), 1004 (Exercise 30)
Thefts in U.S., 302 (Exercise 53)
Throwing events, 586 (Exercise 69)
Thunder after seeing lightning, 354
Tickets sold, 724 (Exercise 31)
Tidal cycles, 486 (Exercise 16), 504–5, 507 (Exercises 71–72)
Tile border, 131 (Exercise 133)
Tip of minute hand of clock, 451 (Exercises 73–74)

Toll-free 800 numbers usage, 555 (Exercise 18)
Total cost, 213 (Exercise 91)
Total economic impact, 959 (Exercise 73)
Toxic chemicals, 370 (Exercise 66)
Traffic control, 783, 787–89, 790 (Exercises 25–29), 791 (Exercises 36, 38), 837 (Exercise 13)
Travel rate, 693 (Exercise 18)
Travel time, 213 (Exercise 93)
Triangles
 finding angle of, 539, 554 (Exercise 17)
 finding side of, 538, 552 (Exercises 27–29)
Triangular land, 693 (Exercise 20)
Trigonometric identities, 609 (Exercise 20)
Trucks
 delivery through tunnel, 842, 851–52, 854 (Exercises 57–58)
 passage through semielliptic archway, 921 (Exercise 13)
 rentals, 155 (Exercise 101), 924 (Exercise 19)
Tuning fork, 576 (Exercise 68)

Unemployment rate, 173 (Exercise 3)
Unhealthy air days, 174 (Exercise 30)
United Nations, 176

Vacation days, 202, 218–19
Vacation packages, 764 (Exercise 9)
Velocity of air flow, 605 (Exercises 79–80)
Veterinary costs, 258 (Exercises 59–62)
Video game sales, 353 (Exercise 97)
Video rentals, 689 (Exercise 40)
Viewing angle, 521, 535 (Exercises 73–76), 536 (Exercises 89, 99)
Violent crime, 47 (Exercise 85)
Vitamin content, 790 (Exercise 30)
Volume
 of box, 323 (Exercise 48)
 at given temperature, 365 (Exercise 44)
 of solid figure, 324 (Exercise 65)
Voter choices, 983
Voters by age group, 806 (Exercise 50)
Vulture's height, 274 (Exercise 57)

Wage gap, 303 (Exercise 55)
Wages and hours worked, 364 (Exercise 32)
Walking speed, 394–95, 397 (Exercise 94)
Wall's height, 622 (Exercise 48)
Wardrobe selection, 977–78
Warehouse plans, 105 (Exercises 37–38)
Weight
 average, 190 (Exercise 74)
 of man, 364 (Exercise 37)
 pine board and support of, 364 (Exercise 40)
Weightlifting, 373, 430 (Exercise 42), 431 (Exercise 55)
Welfare, 936 (Exercises 63–64)
Wheelchair manufacturing costs, 166–67 (Exercises 57–58), 335, 347–48
Wheel movement, 451 (Exercise 76)
Whispering gallery, 923 (Exercise 9)
Will distribution, 107 (Exercise 87)
Windchill temperature, 136
Wind pressure, 365 (Exercise 59)
Wing span of jet, 622 (Exercise 65)
Wire attached to pole, 174 (Exercise 34)
Wire lengths, 129 (Exercises 109–110), 130 (Exercise 116)
Words remembered over time, 351 (Exercise 75)
Work
 amount done by given force, 687–88, 689 (Exercises 41–48), 690 (Exercise 49–50, 60–61), 695 (Exercise 104–105), 696 (Exercise 22)
 challenges with, 936 (Exercise 75)
 hours, 96, 1008 (Exercise 25)
World population, 258 (Exercises 63–66), 379–80, 384 (Exercise 64), 418, 419, 427

Prerequisites: Fundamental Concepts of Algebra

This chapter reviews fundamental concepts of algebra that are prerequisites for the study of college algebra. Algebra, like all of mathematics, provides the tools to help you recognize, classify, and explore the hidden patterns of your world, revealing its underlying structure. Throughout the new millennium, literacy in algebra will be a prerequisite for functioning in a meaningful way personally, professionally, and as a citizen.

Listening to the radio on the way to work, you hear candidates in the upcoming election discussing the problem of the country's 5.6 trillion dollar deficit. It seems like this is a real problem, but then you realize that you don't really know what that number means. How can you look at this deficit in the proper perspective? If the national debt were evenly divided among all citizens of the country, how much would each citizen have to pay? Does the deficit seem like such a significant problem now?

SECTION P.1 Real Numbers and Algebraic Expressions

Objectives

1. Recognize subsets of the real numbers.
2. Use inequality symbols.
3. Evaluate absolute value.
4. Use absolute value to express distance.
5. Evaluate algebraic expressions.
6. Identify properties of the real numbers.
7. Simplify algebraic expressions.

The U.N. Building is designed with three golden rectangles.

The United Nations Building in New York was designed to represent its mission of promoting world harmony. Viewed from the front, the building looks like three rectangles stacked upon each other. In each rectangle, the ratio of the width to height is $\sqrt{5} + 1$ to 2, approximately 1.618 to 1. The ancient Greeks believed that such a rectangle, called a **golden rectangle,** was the most visually pleasing of all rectangles.

The ratio 1.618 to 1 is approximate because $\sqrt{5}$ is an irrational number, a special kind of real number. Irrational? Real? Let's make sense of all this by describing the kinds of numbers you will encounter in this course.

1 Recognize subsets of the real numbers.

The Set of Real Numbers

Before we describe the set of real numbers, let's be sure you are familiar with some basic ideas about sets. A **set** is a collection of objects whose contents can be clearly determined. The objects in a set are called the **elements** of the set. For example, the set of numbers used for counting can be represented by

$$\{1, 2, 3, 4, 5, \dots\}.$$

The braces, { }, indicate that we are representing a set. This form of representing a set uses commas to separate the elements of the set. The set of numbers used for counting is called the set of **natural numbers.** The three dots after the 5 indicate that there is no final element and that the listing goes on forever.

The sets that make up the real numbers are summarized in Table P.1. We refer to these sets as **subsets** of the real numbers, meaning that all elements in each subset are also elements in the set of real numbers.

Notice the use of the symbol $\approx$ in the examples of irrational numbers. The symbol means "is approximately equal to." Thus,

$$\sqrt{2} \approx 1.414214.$$

We can verify that this is only an approximation by multiplying 1.414214 by itself. The product is very close to, but not exactly, 2:

$$1.414214 \times 1.414214 = 2.0000012378.$$

Technology

A calculator with a square root key gives a decimal approximation for $\sqrt{2}$, not the exact value.

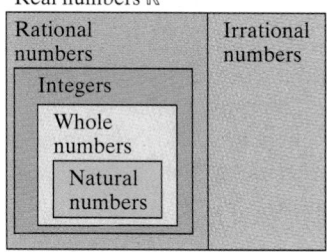

Real numbers ℝ

This diagram shows that every real number is rational or irrational.

Table P.1 Important Subsets of the Real Numbers

Name	Description	Examples
Natural numbers ℕ	$\{1, 2, 3, 4, 5, \ldots\}$ These numbers are used for counting.	$2, 3, 5, 17$
Whole numbers 𝕎	$\{0, 1, 2, 3, 4, 5, \ldots\}$ The set of whole numbers is formed by adding 0 to the set of natural numbers.	$0, 2, 3, 5, 17$
Integers ℤ	$\{\ldots, -5, -4, -3, -2, -1, 0, 1, 2, 3, 4, 5, \ldots\}$ The set of integers is formed by adding negatives of the natural numbers to the set of whole numbers.	$-17, -5, -3, -2, 0,$ $2, 3, 5, 17$
Rational numbers ℚ	The set of rational numbers is the set of all numbers which can be expressed in the form $\frac{a}{b}$, where a and b are integers and b is not equal to 0, written $b \neq 0$. Rational numbers can be expressed as terminating or repeating decimals.	$-17 = \frac{-17}{1}, -5 = \frac{-5}{1}, -3, -2,$ $0, 2, 3, 5, 17,$ $\frac{2}{5} = 0.4,$ $\frac{-2}{3} = -0.6666\cdots = -0.\overline{6}$
Irrational numbers 𝕀	This is the set of all numbers whose decimal representations are neither terminating nor repeating. Irrational numbers cannot be expressed as a quotient of integers.	$\sqrt{2} \approx 1.414214$ $-\sqrt{3} \approx -1.73205$ $\pi \approx 3.142$ $-\frac{\pi}{2} \approx -1.571$

Study Tip

Not all square roots are irrational numbers. For example, $\sqrt{25} = 5$ because $5 \times 5 = 25$. Thus, $\sqrt{25}$ is a natural number, a whole number, an integer, and a rational number $\left(\sqrt{25} = \frac{5}{1}\right)$.

The set of **real numbers** is formed by combining the rational numbers and the irrational numbers. Thus, every real number is either rational or irrational.

The Real Number Line

The **real number line** is a graph used to represent the set of real numbers. An arbitrary point, called the **origin,** is labeled 0; units to the right of the origin are **positive** and units to the left of the origin are **negative.** The real number line is shown in Figure P.1.

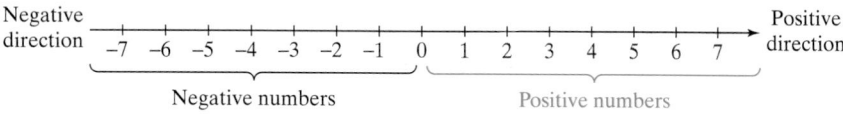

Figure P.1 The real number line

Real numbers are **graphed** on a number line by placing a dot at the correct location for each number. The integers are easiest to locate. In Figure P.2, we've graphed the integers −3, 0, and 4.

Figure P.2 Graphing −3, 0, and 4 on a number line

Every real number corresponds to a point on the number line and every point on the number line corresponds to a real number. We say there is a **one-to-one correspondence** between all the real numbers and all points on a real number line. If you draw a point on the real number line corresponding to a real number, you are **plotting** the real number. In Figure P.2, we are plotting the real numbers −3, 0, and 4.

2 Use inequality symbols.

Figure P.3

Study Tip

The symbols $<$ and $>$ always point to the lesser of the two real numbers when the inequality is true.

$2 < 5$ The symbol points to 2, the lesser number.

$5 > 2$ The symbol points to 2, the lesser number.

Ordering the Real Numbers

On the real number line, the real numbers increase from left to right. The lesser of two real numbers is the one farther to the left on a number line. The greater of two real numbers is the one farther to the right on a number line.

Look at the number line in Figure P.3. The integers 2 and 5 are plotted. Observe that 2 is to the left of 5 on the number line. This means that 2 is less than 5:

$2 < 5$:   2 is less than 5 because 2 is to the *left* of 5 on the number line.

In Figure P.3, we can also observe that 5 is to the right of 2 on the number line. This means that 5 is greater than 2:

$5 > 2$:   5 is greater than 2 because 5 is to the right of 2 on the number line.

The symbols $<$ and $>$ are called **inequality symbols.** They may be combined with an equal sign, as shown in the following table:

Symbols	Meaning	Example	Explanation
$a \leq b$	a is less than or equal to b.	$3 \leq 7$	Because $3 < 7$
		$7 \leq 7$	Because $7 = 7$
$b \geq a$	b is greater than or equal to a.	$7 \geq 3$	Because $7 > 3$
		$-5 \geq -5$	Because $-5 = -5$

3 Evaluate absolute value.

Figure P.4 Absolute value as the distance from 0

Absolute Value

The **absolute value** of a real number a, denoted by $|a|$, is the distance from 0 to a on the number line. This distance is always taken to be nonnegative. For example, the real number line in Figure P.4 shows that

$$|-3| = 3 \quad \text{and} \quad |5| = 5.$$

The absolute value of -3 is 3 because -3 is 3 units from 0 on the number line. The absolute value of 5 is 5 because 5 is 5 units from 0 on the number line. The absolute value of a positive real number or 0 is the number itself. The absolute value of a negative real number, such as -3, is the number without the negative sign.

We can define the absolute value of the real number x without referring to a number line. The algebraic definition of the absolute value of x is given as follows:

Definition of Absolute Value

$$|x| = \begin{cases} x & \text{if } x \geq 0 \\ -x & \text{if } x < 0 \end{cases}$$

If x is nonnegative (that is $x \geq 0$), the absolute value of x is the number itself. For example,

$$|5| = 5 \qquad |\pi| = \pi \qquad \left|\frac{1}{3}\right| = \frac{1}{3} \qquad |0| = 0.$$

Zero is the only number whose absolute value is 0.

If x is a negative number (that is, $x < 0$), the absolute value of x is the opposite of x. This makes the absolute value positive. For example,

$$|-3| = -(-3) = 3 \qquad |-\pi| = -(-\pi) = \pi \qquad \left|-\frac{1}{3}\right| = -\left(-\frac{1}{3}\right) = \frac{1}{3}.$$

This middle step is usually omitted.

EXAMPLE 1   Evaluating Absolute Value

Rewrite each expression without absolute value bars:

a. $\left|\sqrt{3} - 1\right|$ **b.** $\left|2 - \pi\right|$ **c.** $\dfrac{|x|}{x}$ if $x < 0$.

Solution

a. Because $\sqrt{3} \approx 1.7$, the expression inside the absolute value bars, $\sqrt{3} - 1$, is positive. The absolute value of a positive number is the number itself. Thus,
$$\left|\sqrt{3} - 1\right| = \sqrt{3} - 1.$$

b. Because $\pi \approx 3.14$, the number inside the absolute value bars, $2 - \pi$, is negative. The absolute value of x when $x < 0$ is $-x$. Thus,
$$\left|2 - \pi\right| = -(2 - \pi) = \pi - 2.$$

c. If $x < 0$, then $|x| = -x$. Thus,
$$\frac{|x|}{x} = \frac{-x}{x} = -1.$$

Study Tip

After working each Check Point, check your answer in the answer section before continuing your reading.

Discovery

Verify the triangle inequality if $a = 4$ and $b = 5$. Verify the triangle inequality if $a = 4$ and $b = -5$.

When does equality occur in the triangle inequality and when does inequality occur? Verify your observation with additional number pairs.

Check Point 1   Rewrite each expression without absolute value bars:

a. $\left|1 - \sqrt{2}\right|$ **b.** $\left|\pi - 3\right|$ **c.** $\dfrac{|x|}{x}$ if $x > 0$.

Listed below are several basic properties of absolute value. Each of these properties can be derived from the definition of absolute value.

Properties of Absolute Value

For all real numbers a and b,

1. $|a| \geq 0$ 2. $|-a| = |a|$ 3. $a \leq |a|$

4. $|ab| = |a||b|$ 5. $\left|\dfrac{a}{b}\right| = \dfrac{|a|}{|b|}$, $b \neq 0$

6. $|a + b| \leq |a| + |b|$ (called the triangle inequality)

4 Use absolute value to express distance.

Distance between Points on a Real Number Line

Absolute value is used to find the distance between two points on a real number line. If a and b are any real numbers, the **distance between a and b** is the absolute value of their difference. For example, the distance between 4 and 10 is 6. Using absolute value, we find this distance in one of two ways:

$$|10 - 4| = |6| = 6 \quad \text{or} \quad |4 - 10| = |-6| = 6.$$

The distance between 4 and 10 on the real number line is 6.

Notice that we obtain the same distance regardless of the order in which we subtract.

> **Distance between Two Points on the Real Number Line**
>
> If a and b are any two points on a real number line, then the distance between a and b is given by
>
> $$|a - b| \text{ or } |b - a|.$$

EXAMPLE 2 Distance between Two Points on a Number Line

Find the distance between −5 and 3 on the real number line.

Solution   Because the distance between a and b is given by $|a - b|$, the distance between −5 and 3 is

$$|-5 - 3| = |-8| = 8.$$

$$a = -5 \qquad b = 3$$

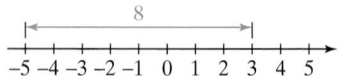

Figure P.5 The distance between −5 and 3 is 8.

Figure P.5 verifies that there are 8 units between −5 and 3 on the real number line. We obtain the same distance if we reverse the order of the subtraction:

$$|3 - (-5)| = |8| = 8.$$

> **Check Point 2**   Find the distance between −4 and 5 on the real number line.

Algebraic Expressions

Algebra uses letters, such as x and y, to represent real numbers. Such letters are called **variables.** For example, imagine that you are basking in the sun on the beach. We can let x represent the number of minutes that you can stay in the sun without burning with no sunscreen. With a number 6 sunscreen, exposure time without burning is six times as long, or 6 times x. This can be written $6 \cdot x$, but it is usually expressed as $6x$. Placing a number and a letter next to one another indicates multiplication.

Notice that $6x$ combines the number 6 and the variable x using the operation of multiplication. A combination of variables and numbers using the operations of addition, subtraction, multiplication, or division, as well as powers or roots (which are discussed later in this chapter), is called an **algebraic expression.** Here are some examples of algebraic expressions:

$$x + 6, \quad x - 6, \quad 6x, \quad \frac{x}{6}, \quad 3x + 5, \quad \sqrt{x} + 7.$$

5 Evaluate algebraic expressions.

Evaluating Algebraic Expressions

Evaluating an algebraic expression means to find the value of the expression for a given value of the variable. For example, we can evaluate $6x$ (from the sunscreen example) when $x = 15$. We substitute 15 for x. We obtain $6 \cdot 15$, or 90. This means if you can stay in the sun for 15 minutes without burning when you don't put on any lotion, then with a number 6 lotion, you can "cook" for 90 minutes without burning.

Many algebraic expressions involve more than one operation. Evaluating an algebraic expression without a calculator involves carefully applying the following order of operations agreement:

The Order of Operations Agreement

1. Perform operations within the innermost parentheses and work outward. If the algebraic expression involves a fraction, treat the numerator and the denominator as if they were each enclosed in parentheses.
2. Evaluate all exponential expressions.
3. Perform multiplications and divisions as they occur, working from left to right.
4. Perform additions and subtractions as they occur, working from left to right.

EXAMPLE 3   Evaluating an Algebraic Expression

The algebraic expression $2.35x + 179.5$ describes the population of the United States, in millions, x years after 1960. Evaluate the expression for $x = 40$. Describe what the answer means in practical terms.

Solution   We begin by substituting 40 for x. Because $x = 40$, we will be finding the U.S. population 40 years after 1960, in the year 2000.

$$2.35x + 179.5$$

Replace x with 40.

$$= 2.35(40) + 179.5$$
$$= 94 + 179.5 \qquad \text{Perform the multiplication: } 2.35(40) = 94.$$
$$= 273.5 \qquad \text{Perform the addition.}$$

According to the given algebraic expression, in 2000 the population of the United States was 273.5 million.

According to the U.S. Bureau of the Census, in 2000 the population of the United States was 281.4 million. Notice that the algebraic expression in Example 3 provides an approximate, but not an exact, description of the actual population.

Check Point 3   Evaluate the algebraic expression in Example 3 for $x = 30$. Describe what your answer means in practical terms.

⑥ Identify properties of the real numbers.

Properties of Real Numbers and Algebraic Expressions

When you use your calculator to add two real numbers, you can enter them in any order. The fact that two real numbers can be added in any order is called the **commutative property of addition.** You probably use this property, as well as other properties of real numbers listed in Table P.2 on the next page, without giving it much thought. The properties of the real numbers are especially useful when working with algebraic expressions. For each property listed in Table P.2, a, b, and c represent real numbers, variables, or algebraic expressions.

The Associative Property and the English Language

In the English language, phrases can take on different meanings depending on the way the words are associated with commas.

Here are two examples.

- *Woman, without her man, is nothing.*
  Woman, without her, man is nothing.

- *What's the latest dope?*
  What's the latest, dope?

Table P.2 Properties of the Real Numbers

Name	Meaning	Examples
Commutative Property of Addition	Two real numbers can be added in any order. $a + b = b + a$	• $13 + 7 = 7 + 13$ • $13x + 7 = 7 + 13x$
Commutative Property of Multiplication	Two real numbers can be multiplied in any order. $ab = ba$	• $\sqrt{2} \cdot \sqrt{5} = \sqrt{5} \cdot \sqrt{2}$ • $x \cdot 6 = 6x$
Associative Property of Addition	If three real numbers are added, it makes no difference which two are added first. $(a + b) + c = a + (b + c)$	• $3 + (8 + x) = (3 + 8) + x$ $\qquad\qquad\quad = 11 + x$
Associative Property of Multiplication	If three real numbers are multiplied, it makes no difference which two are multiplied first. $(a \cdot b) \cdot c = a \cdot (b \cdot c)$	• $-2(3x) = (-2 \cdot 3)x = -6x$
Distributive Property of Multiplication over Addition	Multiplication distributes over addition. $a \cdot (b + c) = a \cdot b + a \cdot c$	• $7(4 + \sqrt{3}) = 7 \cdot 4 + 7 \cdot \sqrt{3}$ $\qquad\qquad\quad = 28 + 7\sqrt{3}$ • $5(3x + 7) = 5 \cdot 3x + 5 \cdot 7$ $\qquad\qquad\quad = 15x + 35$
Identity Property of Addition	Zero can be deleted from a sum. $a + 0 = a$ $0 + a = a$	• $\sqrt{3} + 0 = \sqrt{3}$ • $0 + 6x = 6x$
Identity Property of Multiplication	One can be deleted from a product. $a \cdot 1 = a$ $1 \cdot a = a$	• $1 \cdot \pi = \pi$ • $13x \cdot 1 = 13x$
Inverse Property of Addition	The sum of a real number and its additive inverse gives 0, the additive identity. $a + (-a) = 0$ $(-a) + a = 0$	• $\sqrt{5} + (-\sqrt{5}) = 0$ • $-\pi + \pi = 0$ • $6x + (-6x) = 0$ • $(-4y) + 4y = 0$
Inverse Property of Multiplication	The product of a nonzero real number and its multiplicative inverse gives 1, the multiplicative identity. $a \cdot \dfrac{1}{a} = 1, \quad a \neq 0$ $\dfrac{1}{a} \cdot a = 1, \quad a \neq 0$	• $7 \cdot \dfrac{1}{7} = 1$ • $\left(\dfrac{1}{x - 3}\right)(x - 3) = 1, \ x \neq 3$

Commutative Words and Sentences

The commutative property states that a change in order produces no change in the answer. The words and sentences listed here suggest a characteristic of the commutative property; they read the same from left to right and from right to left!

dad	Draw, o coward!	Revolting is error. Resign it, lover.
repaper	Dennis sinned.	Naomi, did I moan?
never odd or even	Ma is a nun, as I am.	Al lets Della call Ed Stella.

The properties in Table P.2 apply to the operations of addition and multiplication. Subtraction and division are defined in terms of addition and multiplication.

Definitions of Subtraction and Division

Let a and b represent real numbers.

Subtraction: $a - b = a + (-b)$
We call $-b$ the **additive inverse** or **opposite** of b.

Division: $a \div b = a \cdot \frac{1}{b}$, where $b \neq 0$
We call $\frac{1}{b}$ the **multiplicative inverse** or **reciprocal** of b. The quotient of a and b, $a \div b$, can be written in the form $\frac{a}{b}$, where a is the **numerator** and b the **denominator** of the fraction.

Because subtraction is defined in terms of adding an inverse, the distributive property can be applied to subtraction:

$$a(b - c) = ab - ac$$

$$(b - c)a = ba - ca.$$

For example,

$$4(2x - 5) = 4 \cdot 2x - 4 \cdot 5 = 8x - 20.$$

7 Simplify algebraic expressions.

Simplifying Algebraic Expressions

The **terms** of an algebraic expression are those parts that are separated by addition. For example, consider the algebraic expression

$$7x - 9y - 3,$$

which can be expressed as

$$7x + (-9y) + (-3).$$

This expression contains three terms, namely $7x$, $-9y$, and -3.

The numerical part of a term is called its **numerical coefficient.** In the term $7x$, the 7 is the numerical coefficient. In the term $-9y$, the -9 is the numerical coefficient.

A term that consists of just a number is called a **constant term.** The constant term of $7x - 9y - 3$ is -3.

A term indicates a product. The expressions that are multiplied to form the term are called its **factors. Like terms** have the same variable factors with the same exponents on the variables. For example, $7x$ and $3x$ are like terms because they have the same variable factor, x. The distributive property (in reverse) can be used to add these terms:

$$7x + 3x = (7 + 3)x = 10x.$$

Study Tip

To add like terms, add their numerical coefficients. Use this result as the numerical coefficient of the terms' common variable(s).

An algebraic expression is **simplified** when parentheses have been removed and like terms have been combined.

EXAMPLE 4   Simplifying an Algebraic Expression

Simplify: $6(2x - 4y) + 10(4x + 3y)$.

Solution

$$6(2x - 4y) + 10(4x + 3y)$$

$= 6 \cdot 2x - 6 \cdot 4y + 10 \cdot 4x + 10 \cdot 3y$	Use the distributive property to remove the parentheses.
$= 12x - 24y + 40x + 30y$	Multiply.
$= (12x + 40x) + (30y - 24y)$	Group like terms.
$= 52x + 6y$	Combine like terms.

Check Point 4   Simplify: $7(4x - 3y) + 2(5x + y)$.

Properties of Negatives

The distributive property can be extended to cover more than two terms within parentheses. For example,

> This sign represents subtraction.

> This sign tells us that −3 is negative.

$$-3(4x - 2y + 6) = -3 \cdot 4x - (-3) \cdot 2y - 3 \cdot 6$$
$$= -12x - (-6y) - 18$$
$$= -12x + 6y - 18.$$

The voice balloons illustrate that negative signs can appear side by side. They can represent the operation of subtraction or the fact that a real number is negative. Here is a list of properties of negatives and how they are applied to algebraic expressions:

Properties of Negatives

Let a and b represent real numbers, variables, or algebraic expressions.

Property	Examples
1. $(-1)a = -a$	$(-1)4xy = -4xy$
2. $-(-a) = a$	$-(-6y) = 6y$
3. $(-a)b = -ab$	$(-7)4xy = -7 \cdot 4xy = -28xy$
4. $a(-b) = -ab$	$5x(-3y) = -5x \cdot 3y = -15xy$
5. $-(a + b) = -a - b$	$-(7x + 6y) = -7x - 6y$
6. $-(a - b) = -a + b$	$-(3x - 7y) = -3x + 7y$
$\quad\quad\quad = b - a$	$\quad\quad\quad = 7y - 3x$

Do you notice that properties 5 and 6 in the box are related? In general, expressions within parentheses that are preceded by a negative can be simplified by dropping the parentheses and changing the sign of every term inside the parentheses. For example,

$$-(3x - 2y + 5z - 6) = -3x + 2y - 5z + 6.$$

EXERCISE SET P.1

Practice Exercises

In Exercises 1–4, list all numbers from the given set that are **a.** *natural numbers,* **b.** *whole numbers,* **c.** *integers,* **d.** *rational numbers,* **e.** *irrational numbers.*

1. $\left\{-9, -\frac{4}{5}, 0, 0.25, \sqrt{3}, 9.2, \sqrt{100}\right\}$

2. $\left\{-7, -0.\overline{6}, 0, \sqrt{49}, \sqrt{50}\right\}$

3. $\left\{-11, -\frac{5}{6}, 0, 0.75, \sqrt{5}, \pi, \sqrt{64}\right\}$

4. $\left\{-5, -0.\overline{3}, 0, \sqrt{2}, \sqrt{4}\right\}$

5. Give an example of a whole number that is not a natural number.

6. Give an example of a rational number that is not an integer.

7. Give an example of a number that is an integer, a whole number, and a natural number.

8. Give an example of a number that is a rational number, an integer, and a real number.

Determine whether each statement in Exercises 9–14 is true or false.

9. $-13 \le -2$ **10.** $-6 > 2$

11. $4 \ge -7$ **12.** $-13 < -5$

13. $-\pi \ge -\pi$ **14.** $-3 > -13$

In Exercises 15–24, rewrite each expression without absolute value bars.

15. $|300|$ **16.** $|-203|$

17. $|12 - \pi|$ **18.** $|7 - \pi|$

19. $|\sqrt{2} - 5|$ **20.** $|\sqrt{5} - 13|$

21. $\dfrac{-3}{|-3|}$ **22.** $\dfrac{-7}{|-7|}$

23. $||-3| - |-7||$ **24.** $||-5| - |-13||$

In Exercises 25–30, evaluate each algebraic expression for $x = 2$ *and* $y = -5$.

25. $|x + y|$ **26.** $|x - y|$

27. $|x| + |y|$ **28.** $|x| - |y|$

29. $\dfrac{y}{|y|}$ **30.** $\dfrac{|x|}{x} + \dfrac{|y|}{y}$

In Exercises 31–38, express the distance between the given numbers using absolute value. Then find the distance by evaluating the absolute value expression.

31. 2 and 17 **32.** 4 and 15

33. -2 and 5 **34.** -6 and 8

35. -19 and -4 **36.** -26 and -3

37. -3.6 and -1.4 **38.** -5.4 and -1.2

In Exercises 39–48, evaluate each algebraic expression for the given value of the variable or variables.

39. $5x + 7$; $x = 4$ **40.** $9x + 6$; $x = 5$

41. $4(x + 3) - 11$; $x = -5$

42. $6(x + 5) - 13$; $x = -7$

43. $\dfrac{5}{9}(F - 32)$; $F = 77$ **44.** $\dfrac{5}{9}(F - 32)$; $F = 50$

45. $\dfrac{5(x + 2)}{2x - 14}$; $x = 10$ **46.** $\dfrac{7(x - 3)}{2x - 16}$; $x = 9$

47. $\dfrac{2x + 3y}{x + 1}$; $x = -2$ and $y = 4$

48. $\dfrac{2x + y}{xy - 2x}$; $x = -2$ and $y = 4$

In Exercises 49–58, state the name of the property illustrated.

49. $6 + (-4) = (-4) + 6$

50. $11 \cdot (7 + 4) = 11 \cdot 7 + 11 \cdot 4$

51. $6 + (2 + 7) = (6 + 2) + 7$

52. $6 \cdot (2 \cdot 3) = 6 \cdot (3 \cdot 2)$

53. $(2 + 3) + (4 + 5) = (4 + 5) + (2 + 3)$

54. $7 \cdot (11 \cdot 8) = (11 \cdot 8) \cdot 7$

55. $2(-8 + 6) = -16 + 12$

56. $-8(3 + 11) = -24 + (-88)$

57. $\dfrac{1}{(x + 3)}(x + 3) = 1$, $x \ne -3$

58. $(x + 4) + [-(x + 4)] = 0$

In Exercises 59–68, simplify each algebraic expression.

59. $5(3x + 4) - 4$ **60.** $2(5x + 4) - 3$

61. $5(3x - 2) + 12x$ **62.** $2(5x - 1) + 14x$

63. $7(3y - 5) + 2(4y + 3)$ **64.** $4(2y - 6) + 3(5y + 10)$

65. $5(3y - 2) - (7y + 2)$ **66.** $4(5y - 3) - (6y + 3)$
67. $7 - 4[3 - (4y - 5)]$ **68.** $6 - 5[8 - (2y - 4)]$

In Exercises 69–74, write each algebraic expression without parentheses.

69. $-(-14x)$ **70.** $-(-17y)$
71. $-(2x - 3y - 6)$ **72.** $-(5x - 13y - 1)$
73. $\frac{1}{3}(3x) + [(4y) + (-4y)]$ **74.** $\frac{1}{2}(2y) + [(-7x) + 7x]$

Application Exercises

75. Are first putting on your left shoe and then putting on your right shoe commutative?
76. Are first getting undressed and then taking a shower commutative?
77. Give an example of two things that you do that are not commutative.
78. Give an example of two things that you do that are commutative.
79. The algebraic expression $81 - 0.6x$ approximates the percentage of American adults who smoked cigarettes x years after 1900. Evaluate the expression for $x = 100$. Describe what the answer means in practical terms.
80. The algebraic expression $1527x + 31,290$ approximates average yearly earnings for elementary and secondary teachers in the United States x years after 1990. Evaluate the algebraic expression for $x = 10$. Describe what the answer means in practical terms.
81. The optimum heart rate is the rate that a person should achieve during exercise for the exercise to be most beneficial. The algebraic expression

$$0.6(220 - a)$$

describes a person's optimum heart rate, in beats per minute, where a represents the age of the person.
  a. Use the distributive property to rewrite the algebraic expression without parentheses.
  b. Use each form of the algebraic expression to determine the optimum heart rate for a 20-year-old runner.

Writing in Mathematics

Writing about mathematics will help you learn mathematics. For all writing exercises in this book, use complete sentences to respond to the question. Some writing exercises can be answered in a sentence; others require a paragraph or two. You can decide how much you need to write as long as your writing clearly and directly answers the question in the exercise. Standard references such as a dictionary and a thesaurus should be helpful.

82. How do the whole numbers differ from the natural numbers?

83. Can a real number be both rational and irrational? Explain your answer.
84. If you are given two real numbers, explain how to determine which one is the lesser.
85. How can $\frac{|x|}{x}$ be equal to 1 or -1?
86. What is an algebraic expression? Give an example with your explanation.
87. Why is $3(x + 7) - 4x$ not simplified? What must be done to simplify the expression?
88. You can transpose the letters in the word "conversation" to form the phrase "voices rant on." From "total abstainers" we can form "sit not at ale bars." What two algebraic properties do each of these transpositions (called anagrams) remind you of? Explain your answer.

Critical Thinking Exercises

89. Which one of the following statements is true?
  a. Every rational number is an integer.
  b. Some whole numbers are not integers.
  c. Some rational numbers are not positive.
  d. Irrational numbers cannot be negative.
90. Which of the following is true?
  a. The term x has no numerical coefficient.
  b. $5 + 3(x - 4) = 8(x - 4) = 8x - 32$
  c. $-x - x = -x + (-x) = 0$
  d. $x - 0.02(x + 200) = 0.98x - 4$

In Exercises 91–93, insert either $<$ or $>$ in the box between the numbers to make the statement true.

91. $\sqrt{2} \,\square\, 1.5$ **92.** $-\pi \,\square\, -3.5$

93. $-\dfrac{3.14}{2} \,\square\, -\dfrac{\pi}{2}$

94. A business that manufactures small alarm clocks has a weekly fixed cost of $5000. The average cost per clock for the business to manufacture x clocks is described by

$$\frac{0.5x + 5000}{x}.$$

  a. Find the average cost when $x = 100$, 1000, and 10,000.
  b. Like all other businesses, the alarm clock manufacturer must make a profit. To do this, each clock must be sold for at least 50¢ more than what it costs to manufacture. Due to competition from a larger company, the clocks can be sold for $1.50 each and no more. Our small manufacturer can only produce 2000 clocks weekly. Does this business have much of a future? Explain.

SECTION P.2   *Exponents and Scientific Notation*

Objectives

1. Understand and use integer exponents.
2. Use properties of exponents.
3. Simplify exponential expressions.
4. Use scientific notation.

1 Understand and use integer exponents.

Powers of Ten

$$10 = 10^1$$
$$100 = 10^2$$
$$1000 = 10^3$$
$$10,000 = 10^4$$
$$100,000 = 10^5$$
$$1,000,000 = 10^6 \quad \text{million}$$
$$10,000,000 = 10^7$$
$$100,000,000 = 10^8$$
$$1,000,000,000 = 10^9 \quad \text{billion}$$

Technology

You can use a calculator to evaluate exponential expressions. For example, to evaluate 5^3, press the following keys:

Many Scientific Calculators

5 $\boxed{y^x}$ 3 $\boxed{=}$

Many Graphing Calculators

5 $\boxed{\wedge}$ 3 $\boxed{\text{ENTER}}$.

Although calculators have special keys to evaluate powers of ten and squaring bases, you can always use one of the sequences shown here.

Although people do a great deal of talking, the total output since the beginning of gabble to the present day, including all baby talk, love songs, and congressional debates, only amounts to about 10 million billion words. This can be expressed as 16 factors of 10, or 10^{16} words.

Exponents such as 2, 3, 4, and so on are used to indicate repeated multiplication. For example,

$$10^2 = 10 \cdot 10 = 100,$$
$$10^3 = 10 \cdot 10 \cdot 10 = 1000, \quad 10^4 = 10 \cdot 10 \cdot 10 \cdot 10 = 10,000.$$

The 10 that is repeated when multiplying is called the **base.** The small numbers above and to the right of the base are called **exponents** or **powers.** The exponent tells the number of times the base is to be used when multiplying. In 10^3, the base is 10 and the exponent is 3.

Any number with an exponent of 1 is the number itself. Thus, $10^1 = 10$.

Multiplications that are expressed in exponential notation are read as follows:

10^1: "ten to the first power"

10^2: "ten to the second power" or "ten squared"

10^3: "ten to the third power" or "ten cubed"

10^4: "ten to the fourth power"

10^5: "ten to the fifth power"

Any real number can be used as the base. Thus,

$$7^2 = 7 \cdot 7 = 49 \text{ and } (-3)^4 = (-3)(-3)(-3)(-3) = 81.$$

The bases are 7 and -3, respectively. Do not confuse $(-3)^4$ and -3^4.

$$-3^4 = -3 \cdot 3 \cdot 3 \cdot 3 = -81$$

The negative is not taken to the power because it is not inside parentheses.

EXAMPLE 1   Evaluating an Exponential Expression

Evaluate: $(-2)^3 \cdot 3^2$.

Solution

$$(-2)^3 \cdot 3^2 = (-2)(-2)(-2) \cdot 3 \cdot 3 = -8 \cdot 9 = -72$$

This is $(-2)^3$, read "−2 cubed." This is 3^2, read "3 squared."

Check Point 1 Evaluate: $(-4)^3 \cdot 2^2$.

The formal algebraic definition of a natural number exponent summarizes our discussion:

Definition of a Natural Number Exponent

If b is a real number and n is a natural number,

Exponent

$$b^n = \underbrace{b \cdot b \cdot b \cdot \cdots \cdot b.}_{\text{b appears as a factor n times.}}$$

Base

b^n is read "the nth power of b" or "b to the nth power." Thus, the nth power of b is defined as the product of n factors of b.
    Furthermore, $b^1 = b$.

Negative Integers as Exponents

A nonzero base can be raised to a negative power using the following definition:

The Negative Exponent Rule

If b is any real number other than 0 and n is a natural number, then

$$b^{-n} = \frac{1}{b^n}.$$

EXAMPLE 2   Evaluating Expressions Containing Negative Exponents

Evaluate: **a.** 5^{-3} **b.** $\dfrac{1}{4^{-2}}$.

Solution

a. $5^{-3} = \dfrac{1}{5^3} = \dfrac{1}{5 \cdot 5 \cdot 5} = \dfrac{1}{125}$

b. $\dfrac{1}{4^{-2}} = \dfrac{1}{\dfrac{1}{4^2}} = 4^2 = 4 \cdot 4 = 16$

Study Tip

When a negative integer appears as an exponent, switch the position of the base (from numerator to denominator or from denominator to numerator) and make the exponent positive.

Check Point 2 Evaluate: **a.** 2^{-3} **b.** $\dfrac{1}{6^{-2}}$.

Zero as an Exponent

A nonzero base can be raised to the 0 power using the following definition:

The Zero Exponent Rule
If b is any real number other than 0,
$$b^0 = 1.$$

Here are three examples involving simplification using the zero exponent rule:

$$7^0 = 1 \quad (-5)^0 = 1 \quad -5^0 = -1.$$

Only 5 is raised to the zero power.

2 Use properties of exponents.

The Product Rule

Consider the multiplication of two exponential expressions, such as $2^4 \cdot 2^3$. We are multiplying 4 factors of 2 and 3 factors of 2. We have a total of 7 factors of 2. Thus,

$$2^4 \cdot 2^3 = 2^7.$$

We can quickly find the exponent on the product, 7, by adding 4 and 3, the original exponents. This suggests the following rule:

The Product Rule
$$b^m \cdot b^n = b^{m+n}$$

When multiplying exponential expressions with the same base, add the exponents. Use this sum as the exponent of the common base.

A Number with 369 Million Digits

The largest number that can be expressed with only three digits is

$$9^{(9^9)} \text{ or } 9^{387,420,489}.$$

This number begins with $428124773\ldots$, has 369 million digits, and would take around 70 years to read.

EXAMPLE 3 Using the Product Rule

Use the product rule to simplify each expression:

  a. $2^2 \cdot 2^3$ **b.** $4^2 \cdot 4^{-5}$ **c.** $x^{-3} \cdot x^7$.

Solution

  a. $2^2 \cdot 2^3 = 2^{2+3} = 2^5 = 32$ **b.** $4^2 \cdot 4^{-5} = 4^{2+(-5)} = 4^{-3} = \dfrac{1}{4^3} = \dfrac{1}{64}$

  c. $x^{-3} \cdot x^7 = x^{-3+7} = x^4$

Check Point 3 Use the product rule to simplify each expression:

  a. $3^3 \cdot 3^2$ **b.** $2^4 \cdot 2^{-7}$ **c.** $x^{-5} \cdot x^{11}$.

The Power Rule

The next property of exponents applies when an expression containing a power is itself raised to a power.

The Power Rule (Powers to Powers)

$$(b^m)^n = b^{mn}$$

When an exponential expression is raised to a power, multiply the exponents. Place the product of the exponents on the base and remove the parentheses.

EXAMPLE 4 Using the Power Rule

Use the power rule to simplify each expression:

  a. $(2^2)^3$ **b.** $(y^5)^6$ **c.** $(x^{-3})^4$.

Solution

  a. $(2^2)^3 = 2^{2 \cdot 3} = 2^6 = 64$ **b.** $(y^5)^6 = y^{5 \cdot 6} = y^{30}$

  c. $(x^{-3})^4 = x^{-3 \cdot 4} = x^{-12} = \dfrac{1}{x^{12}}$

Check Point 4 Use the power rule to simplify each expression:

  a. $(3^3)^2$ **b.** $(y^7)^4$ **c.** $(x^{-4})^2$.

The Quotient Rule

The next property of exponents applies when we are dividing exponential expressions with the same base.

The Quotient Rule

$$\frac{b^m}{b^n} = b^{m-n}, \ b \neq 0$$

When dividing exponential expressions with the same nonzero base, subtract the exponent in the denominator from the exponent in the numerator. Use this difference as the exponent of the common base.

EXAMPLE 5 Using the Quotient Rule

Use the quotient rule to simplify each expression:

a. $\dfrac{2^8}{2^4}$　　**b.** $\dfrac{x^3}{x^7}$　　**c.** $\dfrac{y^9}{y^{-5}}$.

Solution

a. $\dfrac{2^8}{2^4} = 2^{8-4} = 2^4 = 16$　　　　**b.** $\dfrac{x^3}{x^7} = x^{3-7} = x^{-4} = \dfrac{1}{x^4}$

c. $\dfrac{y^9}{y^{-5}} = y^{9-(-5)} = y^{9+5} = y^{14}$

Study Tip

$\dfrac{4^3}{4^5}$ and $\dfrac{4^5}{4^3}$ represent different numbers:

$\dfrac{4^3}{4^5} = 4^{3-5} = 4^{-2} = \dfrac{1}{4^2} = \dfrac{1}{16}$

$\dfrac{4^5}{4^3} = 4^{5-3} = 4^2 = 16.$

Check Point 5　Use the quotient rule to simplify each expression:

a. $\dfrac{3^6}{3^4}$　　**b.** $\dfrac{x^5}{x^{12}}$　　**c.** $\dfrac{y^2}{y^{-7}}$.

Products Raised to Powers

The next property of exponents applies when we are raising a product to a power.

Products to Powers

$$(ab)^n = a^n b^n$$

When a product is raised to a power, raise each factor to that power.

EXAMPLE 6 Raising a Product to a Power

Simplify: $(-2y)^4$.

Solution $(-2y)^4 = (-2)^4 y^4 = 16y^4$

Check Point 6　Simplify: $(-4x)^3$.

　　The rule for products of powers can be extended to cover three or more factors. For example,

$$(-2xy)^3 = (-2)^3 x^3 y^3 = -8x^3 y^3.$$

Quotients Raised to Powers

Our final exponential property applies when we are raising a quotient to a power.

Quotients to Powers

$$\left(\dfrac{a}{b}\right)^n = \dfrac{a^n}{b^n},\ b \neq 0$$

When a quotient is raised to a power, raise the numerator to that power and divide by the denominator to that power.

EXAMPLE 7 **Raising Quotients to Powers**

Simplify by raising each quotient to the given power:

a. $\left(\dfrac{2}{5}\right)^4$ **b.** $\left(-\dfrac{3}{x}\right)^3$.

Solution

a. $\left(\dfrac{2}{5}\right)^4 = \dfrac{2^4}{5^4} = \dfrac{16}{625}$ **b.** $\left(-\dfrac{3}{x}\right)^3 = \dfrac{(-3)^3}{x^3} = \dfrac{-27}{x^3}$

Check Point 7   Simplify: **a.** $\left(\dfrac{3}{4}\right)^3$ **b.** $\left(-\dfrac{2}{y}\right)^5$.

3 Simplify exponential expressions.

Simplifying Exponential Expressions

Properties of exponents are used to simplify exponential expressions. Here is a summary of the properties we have discussed.

Properties of Exponents			
1. $b^{-n} = \dfrac{1}{b^n}$	**2.** $b^0 = 1$	**3.** $b^m \cdot b^n = b^{m+n}$	**4.** $(b^m)^n = b^{mn}$
5. $\dfrac{b^m}{b^n} = b^{m-n}$	**6.** $(ab)^n = a^n b^n$		**7.** $\left(\dfrac{a}{b}\right)^n = \dfrac{a^n}{b^n}$

An exponential expression is **simplified** when

- No parentheses appear.
- No powers are raised to powers.
- Each base occurs only once.
- No negative exponents appear.

EXAMPLE 8   **Simplifying Exponential Expressions**

Simplify:

a. $(-3x^4y^5)^3$ **b.** $(-7xy^4)(-2x^5y^6)$ **c.** $\dfrac{-35x^2y^4}{5x^6y^{-8}}$ **d.** $\left(\dfrac{4x^2}{y}\right)^{-3}$.

Solution

a. $(-3x^4y^5)^3 = (-3)^3(x^4)^3(y^5)^3$ Raise each factor inside the parentheses to the third power.

$= (-3)^3 x^{4 \cdot 3} y^{5 \cdot 3}$ Multiply powers to powers.

$= -27x^{12}y^{15}$ $(-3)^3 = (-3)(-3)(-3) = -27$

b. $(-7xy^4)(-2x^5y^6) = (-7)(-2)xx^5y^4y^6$ Group factors with the same base.

$= 14x^{1+5}y^{4+6}$ When multiplying expressions with the same base, add the exponents.

$= 14x^6y^{10}$ Simplify.

c. $\dfrac{-35x^2y^4}{5x^6y^{-8}} = \left(\dfrac{-35}{5}\right)\left(\dfrac{x^2}{x^6}\right)\left(\dfrac{y^4}{y^{-8}}\right)$ Group factors with the same base.

$= -7x^{2-6}y^{4-(-8)}$ When dividing an expression with the same base, subtract the exponents.

$= -7x^{-4}y^{12}$ Simplify. Notice that $4 - (-8) = 4 + 8 = 12$.

$= \dfrac{-7y^{12}}{x^4}$ Move x^{-4}, the factor with the negative exponent, from the numerator to the denominator.

d. $\left(\dfrac{4x^2}{y}\right)^{-3} = \dfrac{4^{-3}(x^2)^{-3}}{y^{-3}}$ Raise each factor inside the parentheses to the -3 power.

$= \dfrac{4^{-3}x^{-6}}{y^{-3}}$ Multiply powers to powers.

$= \dfrac{y^3}{4^3x^6}$ Move factors with negative exponents from the numerator to the denominator (or vice versa) by changing the sign of the exponent.

$= \dfrac{y^3}{64x^6}$ $4^3 = 4 \cdot 4 \cdot 4 = 64$

Visualizing Powers of 3

The triangles contain 3, 3^2, 3^3, and 3^4 circles.

Check Point 8 Simplify:
a. $(2x^3y^6)^4$ **b.** $(-6x^2y^5)(3xy^3)$ **c.** $\dfrac{100x^{12}y^2}{20x^{16}y^{-4}}$ **d.** $\left(\dfrac{5x}{y^4}\right)^{-2}$.

Study Tip

Try to avoid the following common errors that can occur when simplifying exponential expressions.

Correct	Incorrect	Description of Error
$b^3 \cdot b^4 = b^7$	$b^3 \cdot b^4 = b^{12}$	The exponents should be added, not multiplied.
$3^2 \cdot 3^4 = 3^6$	$3^2 \cdot 3^4 = 9^6$	The common base should be retained, not multiplied.
$\dfrac{5^{16}}{5^4} = 5^{12}$	$\dfrac{5^{16}}{5^4} = 5^4$	The exponents should be subtracted, not divided.
$(4a)^3 = 64a^3$	$(4a)^3 = 4a^3$	Both factors should be cubed.
$b^{-n} = \dfrac{1}{b^n}$	$b^{-n} = -\dfrac{1}{b^n}$	Only the exponent should change sign.
$(a + b)^{-1} = \dfrac{1}{a + b}$	$(a + b)^{-1} = \dfrac{1}{a} + \dfrac{1}{b}$	The exponent applies to the entire expression $a + b$.

4 Use scientific notation.

Scientific Notation

The national debt of the United States is about $5.6 trillion. A stack of $1 bills equaling the national debt would rise to twice the distance from the Earth to the moon. Because a trillion is 10^{12}, the national debt can be expressed as

$$5.6 \times 10^{12}.$$

The number 5.6×10^{12} is written in a form called *scientific notation*. A number in **scientific notation** is expressed as a number greater than or equal to 1 and less than 10 multiplied by some power of 10. It is customary to use the multiplication symbol, $\times$, rather than a dot, to indicate multiplication in scientific notation. Here are two examples of numbers in scientific notation:

- Each day, 2.6×10^7 pounds of dust from the atmosphere settle on Earth.
- The diameter of a hydrogen atom is 1.016×10^{-8} centimeter.

We can use the exponent on the 10 to change a number in scientific notation to decimal notation. If the exponent is *positive*, move the decimal point in the number to the *right* the same number of places as the exponent. If the exponent is *negative*, move the decimal point in the number to the *left* the same number of places as the exponent.

EXAMPLE 9   Converting from Scientific to Decimal Notation

Write each number in decimal notation:

    a. 2.6×10^7 **b.** 1.016×10^{-8}.

Solution

a. We express 2.6×10^7 in decimal notation by moving the decimal point in 2.6 seven places to the right. We need to add six zeros.

$$2.6 \times 10^7 = 26,000,000$$

b. We express 1.016×10^{-8} in decimal notation by moving the decimal point in 1.016 eight places to the left. We need to add seven zeros to the right of the decimal point.

$$1.016 \times 10^{-8} = 0.00000001016$$

Check Point 9   Write each number in decimal notation:

    a. 7.4×10^9 **b.** 3.017×10^{-6}.

To convert from decimal notation to scientific notation, we reverse the procedure of Example 9.

- Move the decimal point in the given number to obtain a number greater than or equal to 1 and less than 10.

- The number of places the decimal point moves gives the exponent on 10; the exponent is positive if the given number is greater than 10 and negative if the given number is between 0 and 1.

EXAMPLE 10   Converting from Decimal Notation to Scientific Notation

Write each number in scientific notation:

    a. 4,600,000   **b.** 0.00023.

Solution

a. $4,600,000 = 4.6 \times 10^?$ *Decimal point moves 6 places.* $\longrightarrow 4.6 \times 10^6$

b. $0.00023 = 2.3 \times 10^{-?}$ *Decimal point moves 4 places.* $\longrightarrow 2.3 \times 10^{-4}$

Technology

You can use your calculator's EE (enter exponent) or EXP key to convert from decimal to scientific notation. Here is how it's done for 0.00023:

Many Scientific Calculators

Keystrokes	Display
.00023 EE =	2.3 − 04

Many Graphing Calculators

Use the mode setting for scientific notation.

Keystrokes	Display
.00023 ENTER	2.3E−4

Check Point 10 Write each number in scientific notation:

a. 7,410,000,000   **b.** 0.000000092.

Computations with Scientific Notation

The product and quotient rules for exponents can be used to multiply or divide numbers that are expressed in scientific notation. For example, here's how to find the product of 3.4×10^9 and 2×10^{-5}.

$$(3.4 \times 10^9)(2 \times 10^{-5}) = (3.4 \times 2) \times (10^9 \times 10^{-5})$$
$$= 6.8 \times 10^{9+(-5)}$$
$$= 6.8 \times 10^4 \quad \text{or} \quad 68,000$$

In our next example, we use the quotient of two numbers in scientific notation to help put a number into perspective. The number is our national debt. The United States began accumulating large deficits in the 1980s. To finance the deficit, the government had borrowed $5.6 trillion as of the end of 2000. The graph in Figure P.6 shows the national debt increasing over time.

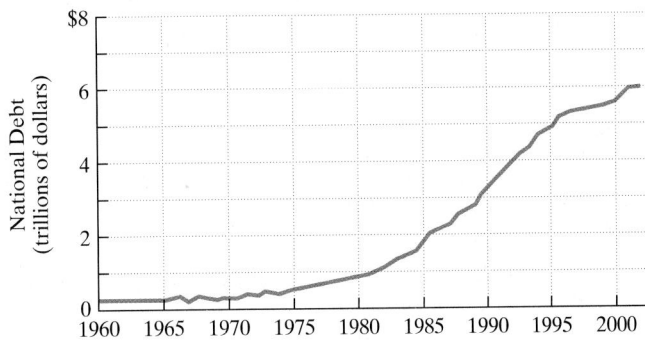

Figure P.6 The national debt

Source: Office of Management and Budget

EXAMPLE 11   The National Debt

As of the end of 2000, the national debt was $5.6 trillion, or 5.6×10^{12} dollars. At that time, the U.S. population was approximately 280,000,000 (280 million), or 2.8×10^8. If the national debt were evenly divided among every individual in the United States, how much would each citizen have to pay?

Solution   The amount each citizen must pay is the total debt, 5.6×10^{12} dollars, divided by the number of citizens, 2.8×10^8.

$$\frac{5.6 \times 10^{12}}{2.8 \times 10^8} = \left(\frac{5.6}{2.8}\right) \times \left(\frac{10^{12}}{10^8}\right)$$
$$= 2 \times 10^{12-8}$$
$$= 2 \times 10^4$$
$$= 20,000$$

Every U.S. citizen would have to pay about $20,000 to the federal government to pay off the national debt. A family of three would owe $60,000.

Check Point 11 In 2000, Americans spent 3.6×10^9 dollars on full-fat ice cream. At that time, the U.S. population was approximately 280 million, or 2.8×10^8. If ice cream spending is evenly divided, how much did each American spend?

EXERCISE SET P.2

Practice Exercises

Evaluate each exponential expression in Exercises 1–22.

1. $5^2 \cdot 2$

2. $6^2 \cdot 2$

3. $(-2)^6$

4. $(-2)^4$

5. -2^6

6. -2^4

7. $(-3)^0$

8. $(-9)^0$

9. -3^0

10. -9^0

11. 4^{-3}

12. 2^{-6}

13. $2^2 \cdot 2^3$

14. $3^3 \cdot 3^2$

15. $(2^2)^3$

16. $(3^3)^2$

17. $\dfrac{2^8}{2^4}$

18. $\dfrac{3^8}{3^4}$

19. $3^{-3} \cdot 3$

20. $2^{-3} \cdot 2$

21. $\dfrac{2^3}{2^7}$

22. $\dfrac{3^4}{3^7}$

Simplify each exponential expression in Exercises 23–64.

23. $x^{-2}y$

24. xy^{-3}

25. x^0y^5

26. x^7y^0

27. $x^3 \cdot x^7$

28. $x^{11} \cdot x^5$

29. $x^{-5} \cdot x^{10}$

30. $x^{-6} \cdot x^{12}$

31. $(x^3)^7$

32. $(x^{11})^5$

33. $(x^{-5})^3$

34. $(x^{-6})^4$

35. $\dfrac{x^{14}}{x^7}$

36. $\dfrac{x^{30}}{x^{10}}$

37. $\dfrac{x^{14}}{x^{-7}}$

38. $\dfrac{x^{30}}{x^{-10}}$

39. $(8x^3)^2$

40. $(6x^4)^2$

41. $\left(-\dfrac{4}{x}\right)^3$

42. $\left(-\dfrac{6}{y}\right)^3$

43. $(-3x^2y^5)^2$

44. $(-3x^4y^6)^3$

45. $(3x^4)(2x^7)$

46. $(11x^5)(9x^{12})$

47. $(-9x^3y)(-2x^6y^4)$

48. $(-5x^4y)(-6x^7y^{11})$

49. $\dfrac{8x^{20}}{2x^4}$

50. $\dfrac{20x^{24}}{10x^6}$

51. $\dfrac{25a^{13}b^4}{-5a^2b^3}$

52. $\dfrac{35a^{14}b^6}{-7a^7b^3}$

53. $\dfrac{14b^7}{7b^{14}}$

54. $\dfrac{20b^{10}}{10b^{20}}$

55. $(4x^3)^{-2}$

56. $(10x^2)^{-3}$

57. $\dfrac{24x^3y^5}{32x^7y^{-9}}$

58. $\dfrac{10x^4y^9}{30x^{12}y^{-3}}$

59. $\left(\dfrac{5x^3}{y}\right)^{-2}$

60. $\left(\dfrac{3x^4}{y}\right)^{-3}$

61. $\left(\dfrac{-15a^4b^2}{5a^{10}b^{-3}}\right)^3$

62. $\left(\dfrac{-30a^{14}b^8}{10a^{17}b^{-2}}\right)^3$

63. $\left(\dfrac{3a^{-5}b^2}{12a^3b^{-4}}\right)^0$

64. $\left(\dfrac{4a^{-5}b^3}{12a^3b^{-5}}\right)^0$

In Exercises 65–72, write each number in decimal notation.

65. 4.7×10^3

66. 9.12×10^5

67. 4×10^6

68. 7×10^6

69. 7.86×10^{-4}

70. 4.63×10^{-5}

71. 3.18×10^{-6}

72. 5.84×10^{-7}

In Exercises 73–80, write each number in scientific notation.

73. 3600

74. 2700

75. 220,000,000

76. 370,000,000,000

77. 0.027

78. 0.014

79. 0.000763

80. 0.000972

In Exercises 81–88, perform the indicated operation and express the answer in decimal notation.

81. $(2 \times 10^3)(3 \times 10^2)$

82. $(5 \times 10^2)(4 \times 10^4)$

83. $(4.1 \times 10^2)(3 \times 10^{-4})$

84. $(1.2 \times 10^3)(2 \times 10^{-5})$

85. $\dfrac{12 \times 10^6}{4 \times 10^2}$

86. $\dfrac{20 \times 10^{26}}{10 \times 10^{15}}$

87. $\dfrac{6.3 \times 10^3}{3 \times 10^5}$

88. $\dfrac{9.6 \times 10^2}{3 \times 10^{-3}}$

In Exercises 89–92, write each number in scientific notation and use scientific notation to perform the operation(s). Express the answer in scientific notation.

89. $\dfrac{480,000,000,000}{0.00012}$

90. $\dfrac{282,000,000,000}{0.00141}$

91. $\dfrac{0.00072 \times 0.003}{0.00024}$

92. $\dfrac{66,000 \times 0.001}{0.003 \times 0.002}$

Application Exercises

Use 10^{12} for one trillion and 2.8×10^8 for the U.S. population in 2000 to solve Exercises 93–95.

93. In 2000, the government collected approximately $1.9 trillion in taxes. What was the per capita tax burden, or the amount that each U.S. citizen paid in taxes? Round to the nearest hundred dollars.

94. In 2000, U.S. personal income was $8 trillion. What was the per capita income, or the income per U.S. citizen? Round to the nearest hundred dollars.

95. In the United States, we spend an average of $4000 per person each year on health care—the highest in the world. What do we spend each year on health care nationwide? Express the answer in scientific notation.

96. Approximately 2×10^4 people run in the New York City Marathon each year. Each runner runs a distance of 26 miles. Write the total distance covered by all the runners (assuming that each person completes the marathon) in scientific notation.

97. The mass of one oxygen molecule is 5.3×10^{-23} gram. Find the mass of 20,000 molecules of oxygen. Express the answer in scientific notation.

98. The mass of one hydrogen atom is 1.67×10^{-24} gram. Find the mass of 80,000 hydrogen atoms. Express the answer in scientific notation.

Writing in Mathematics

99. Describe what it means to raise a number to a power. In your description, include a discussion of the difference between -5^2 and $(-5)^2$.

100. Explain the product rule for exponents. Use $2^3 \cdot 2^5$ in your explanation.

101. Explain the power rule for exponents. Use $(3^2)^4$ in your explanation.

102. Explain the quotient rule for exponents. Use $\dfrac{5^8}{5^2}$ in your explanation.

103. Why is $(-3x^2)(2x^{-5})$ not simplified? What must be done to simplify the expression?

104. How do you know if a number is written in scientific notation?

105. Explain how to convert from scientific to decimal notation and give an example.

106. Explain how to convert from decimal to scientific notation and give an example.

Critical Thinking Exercises

107. Which one of the following is true?

    a. $4^{-2} < 4^{-3}$ **b.** $5^{-2} > 2^{-5}$

    c. $(-2)^4 = 2^{-4}$ **d.** $5^2 \cdot 5^{-2} > 2^5 \cdot 2^{-5}$

108. The mad Dr. Frankenstein has gathered enough bits and pieces (so to speak) for $2^{-1} + 2^{-2}$ of his creature-to-be. Write a fraction that represents the amount of his creature that must still be obtained.

109. If $b^A = MN$, $b^C = M$, and $b^D = N$, what is the relationship among A, C, and D?

Group Exercise

110. Putting Numbers into Perspective. A large number can be put into perspective by comparing it with another number. For example, we put the $5.6 trillion national debt into perspective by comparing it to the number of U.S. citizens. The total distance covered by all the runners in the New York City Marathon (Exercise 96) can be put into perspective by comparing this distance with, say, the distance from New York to San Francisco.

For this project, each group member should consult an almanac, a newspaper, or the World Wide Web to find a number greater than one million. Explain to other members of the group the context in which the large number is used. Express the number in scientific notation. Then put the number into perspective by comparing it with another number.

SECTION P.3 *Radicals and Rational Exponents*

Objectives

1. Evaluate square roots.
2. Use the product rule to simplify square roots.
3. Use the quotient rule to simplify square roots.
4. Add and subtract square roots.
5. Rationalize denominators.
6. Evaluate and perform operations with higher roots.
7. Understand and use rational exponents.

What is the maximum speed at which a racing cyclist can turn a corner without tipping over? The answer, in miles per hour, is given by the algebraic expression $4\sqrt{x}$, where x is the radius of the corner, in feet. Algebraic expressions containing roots describe phenomena as diverse as a wild animal's territorial area, evaporation on a lake's surface, and Albert Einstein's bizarre concept of how an astronaut moving close to the speed of light would barely age relative to friends watching from Earth. No description of your world can be complete without roots and radicals. In this section, we review the basics of radical expressions and the use of rational exponents to indicate radicals.

1 Evaluate square roots.

Square Roots

The **principal square root** of a nonnegative real number b, written $\sqrt{b}$, is that number whose square equals b. For example,

$$\sqrt{100} = 10 \text{ because } 10^2 = 100 \quad \text{and} \quad \sqrt{0} = 0 \text{ because } 0^2 = 0.$$

Observe that the principal square root of a positive number is positive and the principal square root of 0 is 0.

The symbol $\sqrt{}$ that we use to denote the principal square root is called a **radical sign**. The number under the radical sign is called the **radicand**. Together we refer to the radical sign and its radicand as a **radical.**

The following definition summarizes our discussion:

> **Definition of the Principal Square Root**
>
> If a is a nonnegative real number, the nonnegative number b such that $b^2 = a$, denoted by $b = \sqrt{a}$, is the **principal square root** of a.

In the real number system, negative numbers do not have square roots. For example, $\sqrt{-9}$ is not a real number because there is no real number whose square is -9.

If a number is nonnegative ($a \geq 0$), then $\left(\sqrt{a}\right)^2 = a$. For example,

$$\left(\sqrt{2}\right)^2 = 2, \quad \left(\sqrt{3}\right)^2 = 3, \quad \left(\sqrt{4}\right)^2 = 4, \quad \text{and} \left(\sqrt{5}\right)^2 = 5.$$

A number that is the square of a rational number is called a **perfect square.** For example,

64 is a perfect square because $64 = 8^2$.

$\frac{1}{9}$ is a perfect square because $\frac{1}{9} = \left(\frac{1}{3}\right)^2$.

The following rule can be used to find square roots of perfect squares:

> **Square Roots of Perfect Squares**
> $$\sqrt{a^2} = |a|$$

For example, $\sqrt{6^2} = 6$ and $\sqrt{(-6)^2} = |-6| = 6$.

② Use the product rule to simplify square roots.

The Product Rule for Square Roots

A square root is **simplified** when its radicand has no factors other than 1 that are perfect squares. For example, $\sqrt{500}$ is not simplified because it can be expressed as $\sqrt{100 \cdot 5}$ and $\sqrt{100}$ is a perfect square. The **product rule for square roots** can be used to simplify $\sqrt{500}$.

> **The Product Rule for Square Roots**
> If a and b represent nonnegative real numbers, then
> $$\sqrt{ab} = \sqrt{a}\,\sqrt{b} \text{ and } \sqrt{a}\,\sqrt{b} = \sqrt{ab}.$$
> The square root of a product is the product of the square roots.

Example 1 shows how the product rule is used to remove from the square root any perfect squares that occur as factors.

EXAMPLE 1 Using the Product Rule to Simplify Square Roots

Simplify: **a.** $\sqrt{500}$ **b.** $\sqrt{6x} \cdot \sqrt{3x}$.

Solution

a. $\sqrt{500} = \sqrt{100 \cdot 5}$ 100 is the largest perfect square factor of 500.
$= \sqrt{100}\,\sqrt{5}$ $\sqrt{ab} = \sqrt{a}\,\sqrt{b}$
$= 10\sqrt{5}$ $\sqrt{100} = 10$

b. We can simplify $\sqrt{6x} \cdot \sqrt{3x}$ using the power rule only if $6x$ and $3x$ represent nonnegative real numbers. Thus, $x \geq 0$.

$\sqrt{6x} \cdot \sqrt{3x} = \sqrt{6x \cdot 3x}$ $\sqrt{a}\,\sqrt{b} = \sqrt{ab}$
$= \sqrt{18x^2}$ Multiply.
$= \sqrt{9x^2 \cdot 2}$ 9 is the largest perfect square factor of 18.
$= \sqrt{9x^2}\,\sqrt{2}$ $\sqrt{ab} = \sqrt{a}\,\sqrt{b}$
$= \sqrt{9}\,\sqrt{x^2}\,\sqrt{2}$ Split $\sqrt{9x^2}$ into two square roots.
$= 3x\sqrt{2}$ $\sqrt{9} = 3$ (because $3^2 = 9$) and $\sqrt{x^2} = x$ because $x \geq 0$.

Check Point 1 Simplify:
a. $\sqrt{3^2}$ **b.** $\sqrt{5x} \cdot \sqrt{10x}$.

3 Use the quotient rule to simplify square roots.

The Quotient Rule for Square Roots

Another property for square roots involves division.

The Quotient Rule for Square Roots

If a and b represent nonnegative real numbers and $b \neq 0$, then

$$\frac{\sqrt{a}}{\sqrt{b}} = \sqrt{\frac{a}{b}} \quad \text{and} \quad \sqrt{\frac{a}{b}} = \frac{\sqrt{a}}{\sqrt{b}}.$$

The square root of a quotient is the quotient of the square roots.

EXAMPLE 2 Using the Quotient Rule to Simplify Square Roots

Simplify: **a.** $\sqrt{\dfrac{100}{9}}$ **b.** $\dfrac{\sqrt{48x^3}}{\sqrt{6x}}$.

Solution

a. $\sqrt{\dfrac{100}{9}} = \dfrac{\sqrt{100}}{\sqrt{9}} = \dfrac{10}{3}$

b. We can simplify the quotient of $\sqrt{48x^3}$ and $\sqrt{6x}$ using the quotient rule only if $48x^3$ and $6x$ represent nonnegative real numbers. Thus, $x \geq 0$.

$$\frac{\sqrt{48x^3}}{\sqrt{6x}} = \sqrt{\frac{48x^3}{6x}} = \sqrt{8x^2} = \sqrt{4x^2}\sqrt{2} = \sqrt{4}\sqrt{x^2}\sqrt{2} = 2x\sqrt{2}$$

$\sqrt{x^2} = x$ because $x \geq 0$.

Check Point 2 Simplify: **a.** $\sqrt{\dfrac{25}{16}}$ **b.** $\dfrac{\sqrt{150x^3}}{\sqrt{2x}}$.

4 Add and subtract square roots.

Adding and Subtracting Square Roots

Two or more square roots can be combined provided that they have the same radicand. Such radicals are called **like radicals**. For example,

$$7\sqrt{11} + 6\sqrt{11} = (7 + 6)\sqrt{11} = 13\sqrt{11}.$$

EXAMPLE 3 Adding and Subtracting Like Radicals

Add or subtract as indicated:
a. $7\sqrt{2} + 5\sqrt{2}$ **b.** $\sqrt{5x} - 7\sqrt{5x}$.

A Radical Idea: Time Is Relative

What does travel in space have to do with radicals? Imagine that in the future we will be able to travel at velocities approaching the speed of light (approximately 186,000 miles per second). According to Einstein's theory of relativity, time would pass more quickly on Earth than it would in the moving spaceship. The expression

$$R_f \sqrt{1 - \left(\frac{v}{c}\right)^2}$$

gives the aging rate of an astronaut relative to the aging rate of a friend on Earth, R_f. In the expression, v is the astronaut's speed and c is the speed of light. As the astronaut's speed approaches the speed of light, we can substitute c for v:

$$R_f \sqrt{1 - \left(\frac{v}{c}\right)^2} \quad \text{Let } v = c.$$

$$= R_f \sqrt{1 - 1^2}$$

$$= R_f \sqrt{0} = 0$$

Close to the speed of light, the astronaut's aging rate relative to a friend on Earth is nearly 0. What does this mean? As we age here on Earth, the space traveler would barely get older. The space traveler would return to a futuristic world in which friends and loved ones would be long dead.

Solution

a. $7\sqrt{2} + 5\sqrt{2} = (7 + 5)\sqrt{2}$ *Apply the distributive property.*

     $= 12\sqrt{2}$ *Simplify.*

b. $\sqrt{5x} - 7\sqrt{5x} = 1\sqrt{5x} - 7\sqrt{5x}$ *Write $\sqrt{5x}$ as $1\sqrt{5x}$.*

     $= (1 - 7)\sqrt{5x}$ *Apply the distributive property.*

     $= -6\sqrt{5x}$ *Simplify.*

> **Check Point 3** Add or subtract as indicated:
>
> **a.** $8\sqrt{13} + 9\sqrt{13}$ **b.** $\sqrt{17x} - 20\sqrt{17x}$.

In some cases, radicals can be combined once they have been simplified. For example, to add $\sqrt{2}$ and $\sqrt{8}$, we can write $\sqrt{8}$ as $\sqrt{4 \cdot 2}$ because 4 is a perfect square factor of 8.

$$\sqrt{2} + \sqrt{8} = \sqrt{2} + \sqrt{4 \cdot 2} = 1\sqrt{2} + 2\sqrt{2} = (1 + 2)\sqrt{2} = 3\sqrt{2}$$

EXAMPLE 4 Combining Radicals That First Require Simplification

Add or subtract as indicated:

a. $7\sqrt{3} + \sqrt{12}$ **b.** $4\sqrt{50x} - 6\sqrt{32x}$.

Solution

a. $7\sqrt{3} + \sqrt{12}$

     $= 7\sqrt{3} + \sqrt{4 \cdot 3}$ *Split 12 into two factors such that one is a perfect square.*

     $= 7\sqrt{3} + 2\sqrt{3}$ *$\sqrt{4 \cdot 3} = \sqrt{4}\sqrt{3} = 2\sqrt{3}$*

     $= (7 + 2)\sqrt{3}$ *Apply the distributive property. You will find that this step is usually done mentally.*

     $= 9\sqrt{3}$ *Simplify.*

b. $4\sqrt{50x} - 6\sqrt{32x}$

     $= 4\sqrt{25 \cdot 2x} - 6\sqrt{16 \cdot 2x}$ *25 is the largest perfect square factor of 50 and 16 is the largest perfect square factor of 32.*

     $= 4 \cdot 5\sqrt{2x} - 6 \cdot 4\sqrt{2x}$ *$\sqrt{25 \cdot 2} = \sqrt{25}\sqrt{2} = 5\sqrt{2}$ and $\sqrt{16 \cdot 2} = \sqrt{16}\sqrt{2} = 4\sqrt{2}$.*

     $= 20\sqrt{2x} - 24\sqrt{2x}$ *Multiply.*

     $= (20 - 24)\sqrt{2x}$ *Apply the distributive property.*

     $= -4\sqrt{2x}$ *Simplify.*

> **Check Point 4** Add or subtract as indicated:
>
> **a.** $5\sqrt{27} + \sqrt{12}$ **b.** $6\sqrt{18x} - 4\sqrt{8x}$.

5 Rationalize denominators.

Rationalizing Denominators

You can use a calculator to compare the approximate values for $\dfrac{1}{\sqrt{3}}$ and $\dfrac{\sqrt{3}}{3}$.

The two approximations are the same. This is not a coincidence:

$$\frac{1}{\sqrt{3}} = \frac{1}{\sqrt{3}} \cdot \boxed{\frac{\sqrt{3}}{\sqrt{3}}} = \frac{\sqrt{3}}{\sqrt{9}} = \frac{\sqrt{3}}{3}.$$

> Any number divided by itself is 1.
> Multiplication by 1 does not change the value of $\dfrac{1}{\sqrt{3}}$.

This process involves rewriting a radical expression as an equivalent expression in which the denominator no longer contains a radical. The process is called **rationalizing the denominator.** If the denominator contains the square root of a natural number that is not a perfect square, **multiply the numerator and denominator by the smallest number that produces the square root of a perfect square in the denominator.**

EXAMPLE 5   Rationalizing Denominators

Rationalize the denominator: **a.** $\dfrac{15}{\sqrt{6}}$ **b.** $\dfrac{12}{\sqrt{8}}$.

Solution

a. If we multiply numerator and denominator by $\sqrt{6}$, the denominator becomes $\sqrt{6} \cdot \sqrt{6} = \sqrt{36} = 6$. Therefore, we multiply by 1, choosing $\dfrac{\sqrt{6}}{\sqrt{6}}$ for 1.

$$\frac{15}{\sqrt{6}} = \frac{15}{\sqrt{6}} \cdot \frac{\sqrt{6}}{\sqrt{6}} = \frac{15\sqrt{6}}{\sqrt{36}} = \frac{15\sqrt{6}}{6} = \frac{5\sqrt{6}}{2}$$

> Multiply by 1.

> Simplify: $\dfrac{15}{6} = \dfrac{15 \div 3}{6 \div 3} = \dfrac{5}{2}$.

b. The *smallest* number that will produce a perfect square in the denominator of $\dfrac{12}{\sqrt{8}}$ is $\sqrt{2}$, because $\sqrt{8} \cdot \sqrt{2} = \sqrt{16} = 4$. We multiply by 1, choosing $\dfrac{\sqrt{2}}{\sqrt{2}}$ for 1.

$$\frac{12}{\sqrt{8}} = \frac{12}{\sqrt{8}} \cdot \frac{\sqrt{2}}{\sqrt{2}} = \frac{12\sqrt{2}}{\sqrt{16}} = \frac{12\sqrt{2}}{4} = 3\sqrt{2}$$

Check Point 5   Rationalize the denominator: **a.** $\dfrac{5}{\sqrt{3}}$ **b.** $\dfrac{6}{\sqrt{12}}$.

How can we rationalize a denominator if the denominator contains two terms? In general,

$$\left(\sqrt{a} + \sqrt{b}\right)\left(\sqrt{a} - \sqrt{b}\right) = \left(\sqrt{a}\right)^2 - \left(\sqrt{b}\right)^2 = a - b.$$

Notice that the product does not contain a radical. Here are some specific examples.

The Denominator Contains:	Multiply by:	The New Denominator Contains:
$7 + \sqrt{5}$	$7 - \sqrt{5}$	$7^2 - \left(\sqrt{5}\right)^2 = 49 - 5 = 44$
$\sqrt{3} - 6$	$\sqrt{3} + 6$	$\left(\sqrt{3}\right)^2 - 6^2 = 3 - 36 = -33$
$\sqrt{7} + \sqrt{3}$	$\sqrt{7} - \sqrt{3}$	$\left(\sqrt{7}\right)^2 - \left(\sqrt{3}\right)^2 = 7 - 3 = 4$

EXAMPLE 6 **Rationalizing a Denominator Containing Two Terms**

Rationalize the denominator: $\dfrac{7}{5 + \sqrt{3}}$.

Solution If we multiply the numerator and denominator by $5 - \sqrt{3}$, the denominator will not contain a radical. Therefore, we multiply by 1, choosing $\dfrac{5 - \sqrt{3}}{5 - \sqrt{3}}$ for 1.

$$\frac{7}{5 + \sqrt{3}} = \frac{7}{5 + \sqrt{3}} \cdot \frac{5 - \sqrt{3}}{5 - \sqrt{3}} = \frac{7\left(5 - \sqrt{3}\right)}{5^2 - \left(\sqrt{3}\right)^2} = \frac{7\left(5 - \sqrt{3}\right)}{25 - 3}$$

Multiply by 1.

$$= \frac{7\left(5 - \sqrt{3}\right)}{22} \quad \text{or} \quad \frac{35 - 7\sqrt{3}}{22}.$$

In either form of the answer, there is no radical in the denominator.

Check Point 6 Rationalize the denominator: $\dfrac{8}{4 + \sqrt{5}}$.

6 Evaluate and perform operations with higher roots.

Other Kinds of Roots

We define the **principal nth root** of a real number a, symbolized by $\sqrt[n]{a}$, as follows:

Definition of the Principal nth Root of a Real Number

$$\sqrt[n]{a} = b \text{ means that } b^n = a.$$

If n, the **index**, is even, then a is nonnegative ($a \geq 0$) and b is also nonnegative ($b \geq 0$). If n is odd, a and b can be any real numbers.

For example,

$$\sqrt[3]{64} = 4 \text{ because } 4^3 = 64 \quad \text{and} \quad \sqrt[5]{-32} = -2 \text{ because } (-2)^5 = -32.$$

The same vocabulary that we learned for square roots applies to nth roots. The symbol $\sqrt[n]{a}$ is called a **radical** and a is called the **radicand.**

A number that is the nth power of a rational number is called a **perfect nth power.** For example, 8 is a perfect third power, or perfect cube, because $8 = 2^3$. In general, one of the following rules can be used to find nth roots of perfect nth powers:

Finding nth Roots of Perfect nth Powers

If n is odd, $\sqrt[n]{a^n} = a$.
If n is even, $\sqrt[n]{a^n} = |a|$.

For example,

$$\sqrt[3]{(-2)^3} = -2 \quad \text{and} \quad \sqrt[4]{(-2)^4} = |-2| = 2.$$

Absolute value is not needed with odd roots, but is necessary with even roots.

The Product and Quotient Rules for Other Roots

The product and quotient rules apply to cube roots, fourth roots, and all higher roots.

The Product and Quotient Rules for nth Roots

For all real numbers, where the indicated roots represent real numbers,

$$\sqrt[n]{a} \cdot \sqrt[n]{b} = \sqrt[n]{ab} \quad \text{and} \quad \frac{\sqrt[n]{a}}{\sqrt[n]{b}} = \sqrt[n]{\frac{a}{b}}, \ b \neq 0.$$

EXAMPLE 7 Simplifying, Multiplying, and Dividing Higher Roots

Simplify: **a.** $\sqrt[3]{24}$ **b.** $\sqrt[4]{8} \cdot \sqrt[4]{4}$ **c.** $\sqrt[4]{\frac{81}{16}}$.

Solution

a. $\sqrt[3]{24} = \sqrt[3]{8 \cdot 3}$ — Find the largest perfect cube that is a factor of 24. $\sqrt[3]{8} = 2$, so 8 is a perfect cube and is the largest perfect cube factor of 24.

$$= \sqrt[3]{8} \cdot \sqrt[3]{3} \qquad \sqrt[n]{ab} = \sqrt[n]{a}\,\sqrt[n]{b}$$
$$= 2\sqrt[3]{3}$$

b. $\sqrt[4]{8} \cdot \sqrt[4]{4} = \sqrt[4]{8 \cdot 4} \qquad \sqrt[n]{a} \cdot \sqrt[n]{b} = \sqrt[n]{ab}$

$\quad = \sqrt[4]{32} \qquad$ Find the largest *perfect fourth power* that is a factor of 32.

$\quad = \sqrt[4]{16 \cdot 2} \qquad \sqrt[4]{16} = 2$, so 16 is a perfect fourth power and is the largest perfect fourth power that is a factor of 32.

$\quad = \sqrt[4]{16} \cdot \sqrt[4]{2} \qquad \sqrt[n]{ab} = \sqrt[n]{a} \cdot \sqrt[n]{b}$

$\quad = 2\sqrt[4]{2}$

c. $\sqrt[4]{\dfrac{81}{16}} = \dfrac{\sqrt[4]{81}}{\sqrt[4]{16}} \qquad \sqrt[n]{\dfrac{a}{b}} = \dfrac{\sqrt[n]{a}}{\sqrt[n]{b}}$

$\quad = \dfrac{3}{2} \qquad \sqrt[4]{81} = 3$ because $3^4 = 81$ and $\sqrt[4]{16} = 2$ because $2^4 = 16$.

Check Point 7 Simplify: **a.** $\sqrt[3]{40}$ **b.** $\sqrt[5]{8} \cdot \sqrt[5]{8}$ **c.** $\sqrt[3]{\dfrac{125}{27}}$.

We have seen that adding and subtracting square roots often involves simplifying terms. The same idea applies to adding and subtracting *n*th roots.

EXAMPLE 8 Combining Cube Roots

Subtract: $5\sqrt[3]{16} - 11\sqrt[3]{2}$.

Solution

$5\sqrt[3]{16} - 11\sqrt[3]{2}$

$= 5\sqrt[3]{8 \cdot 2} - 11\sqrt[3]{2} \qquad$ Because $16 = 8 \cdot 2$ and $\sqrt[3]{8} = 2$, 8 is the largest perfect cube that is a factor of 16.

$= 5 \cdot 2\sqrt[3]{2} - 11\sqrt[3]{2} \qquad \sqrt[3]{8 \cdot 2} = \sqrt[3]{8}\,\sqrt[3]{2} = 2\sqrt[3]{2}$

$= 10\sqrt[3]{2} - 11\sqrt[3]{2} \qquad$ Multiply.

$= (10 - 11)\sqrt[3]{2} \qquad$ Apply the distributive property.

$= -1\sqrt[3]{2}$ or $-\sqrt[3]{2} \qquad$ Simplify.

Check Point 8 Subtract: $3\sqrt[3]{81} - 4\sqrt[3]{3}$.

7 Understand and use rational exponents.

Rational Exponents

Animals in the wild have regions to which they confine their movement, called their territorial area. Territorial area, in square miles, is related to an animal's body weight. If an animal weighs W pounds, its territorial area is

$$W^{141/100}$$

square miles.

W to the *what* power?! How can we interpret the information given by this algebraic expression?

In the last part of this section, we turn our attention to rational exponents such as $\frac{141}{100}$ and their relationship to roots of real numbers.

Definition of Rational Exponents

If $\sqrt[n]{a}$ represents a real number and $n \geq 2$ is an integer, then

$$a^{1/n} = \sqrt[n]{a}.$$

Furthermore,

$$a^{-1/n} = \frac{1}{a^{1/n}} = \frac{1}{\sqrt[n]{a}}, a \neq 0.$$

EXAMPLE 9 Using the Definition of $a^{1/n}$

Simplify: **a.** $64^{1/2}$ **b.** $8^{1/3}$ **c.** $64^{-1/3}$.

Solution

a. $64^{1/2} = \sqrt{64} = 8$ **b.** $8^{1/3} = \sqrt[3]{8} = 2$

c. $64^{-1/3} = \dfrac{1}{64^{1/3}} = \dfrac{1}{\sqrt[3]{64}} = \dfrac{1}{4}$

Check Point 9 Simplify: **a.** $81^{1/2}$ **b.** $27^{1/3}$ **c.** $32^{-1/5}$.

Note that every rational exponent in Example 9 has a numerator of 1 or −1. We now define rational exponents with any integer in the numerator.

Definition of Rational Exponents

If $\sqrt[n]{a}$ represents a real number, $\dfrac{m}{n}$ is a rational number reduced to lowest terms, and $n \geq 2$ is an integer, then

$$a^{m/n} = \left(\sqrt[n]{a}\right)^m = \sqrt[n]{a^m}.$$

The exponent m/n consists of two parts: the denominator n is the root and the numerator m is the exponent. Furthermore,

$$a^{-m/n} = \frac{1}{a^{m/n}}.$$

EXAMPLE 10 Using the Definition of $a^{m/n}$

Simplify: **a.** $27^{2/3}$ **b.** $9^{3/2}$ **c.** $16^{-3/4}$.

Technology

Here are the calculator keystroke sequences for $27^{2/3}$:

Many Scientific Calculators

$27\ \boxed{y^x}\ \boxed{(}\ \boxed{2}\ \boxed{\div}\ \boxed{3}\ \boxed{)}\ \boxed{=}$

Many Graphing Calculators

$27\ \boxed{\wedge}\ \boxed{(}\ \boxed{2}\ \boxed{\div}\ \boxed{3}\ \boxed{)}\ \boxed{\text{ENTER}}$.

Solution

a. $27^{2/3} = \left(\sqrt[3]{27}\right)^2 = 3^2 = 9$

> The denominator of $\frac{2}{3}$ is the root and the numerator is the exponent.

b. $9^{3/2} = \left(\sqrt{9}\right)^3 = 3^3 = 27$

c. $16^{-3/4} = \dfrac{1}{16^{3/4}} = \dfrac{1}{\left(\sqrt[4]{16}\right)^3} = \dfrac{1}{2^3} = \dfrac{1}{8}$

Check Point 10   Simplify: **a.** $4^{3/2}$ **b.** $32^{-2/5}$.

Properties of exponents can be applied to expressions containing rational exponents.

EXAMPLE 11   **Simplifying Expressions with Rational Exponents**

Simplify using properties of exponents:

a. $(5x^{1/2})(7x^{3/4})$ **b.** $\dfrac{32x^{5/3}}{16x^{3/4}}$.

Solution

a. $(5x^{1/2})(7x^{3/4}) = 5 \cdot 7x^{1/2} \cdot x^{3/4}$ *Group factors with the same base.*

$\qquad\qquad\qquad\quad = 35x^{(1/2)+(3/4)}$ *When multiplying expressions with the same base, add the exponents.*

$\qquad\qquad\qquad\quad = 35x^{5/4}$ $\frac{1}{2} + \frac{3}{4} = \frac{2}{4} + \frac{3}{4} = \frac{5}{4}$

b. $\dfrac{32x^{5/3}}{16x^{3/4}} = \left(\dfrac{32}{16}\right)\left(\dfrac{x^{5/3}}{x^{3/4}}\right)$ *Group factors with the same base.*

$\qquad\qquad = 2x^{(5/3)-(3/4)}$ *When dividing expressions with the same base, subtract the exponents.*

$\qquad\qquad = 2x^{11/12}$ $\frac{5}{3} - \frac{3}{4} = \frac{20}{12} - \frac{9}{12} = \frac{11}{12}$

Check Point 11   Simplify: **a.** $(2x^{4/3})(5x^{8/3})$ **b.** $\dfrac{20x^4}{5x^{3/2}}$.

Rational exponents are sometimes useful for simplifying radicals by reducing their index.

EXAMPLE 12 Reducing the Index of a Radical

Simplify: $\sqrt[9]{x^3}$.

Solution $\sqrt[9]{x^3} = x^{3/9} = x^{1/3} = \sqrt[3]{x}$

Check Point 12 Simplify: $\sqrt[6]{x^3}$.

EXERCISE SET P.3

Practice Exercises

Evaluate each expression in Exercises 1–6 or indicate that the root is not a real number.

1. $\sqrt{36}$
2. $\sqrt{25}$
3. $\sqrt{-36}$
4. $\sqrt{-25}$
5. $\sqrt{(-13)^2}$
6. $\sqrt{(-17)^2}$

Use the product rule to simplify the expressions in Exercises 7–16. In Exercises 11–16, assume that variables represent nonnegative real numbers.

7. $\sqrt{50}$
8. $\sqrt{27}$
9. $\sqrt{45x^2}$
10. $\sqrt{125x^2}$
11. $\sqrt{2x} \cdot \sqrt{6x}$
12. $\sqrt{10x} \cdot \sqrt{8x}$
13. $\sqrt{x^3}$
14. $\sqrt{y^3}$
15. $\sqrt{2x^2} \cdot \sqrt{6x}$
16. $\sqrt{6x} \cdot \sqrt{3x^2}$

Use the quotient rule to simplify the expressions in Exercises 17–26. Assume that x > 0.

17. $\sqrt{\dfrac{1}{81}}$
18. $\sqrt{\dfrac{1}{49}}$
19. $\sqrt{\dfrac{49}{16}}$
20. $\sqrt{\dfrac{121}{9}}$
21. $\dfrac{\sqrt{48x^3}}{\sqrt{3x}}$
22. $\dfrac{\sqrt{72x^3}}{\sqrt{8x}}$
23. $\dfrac{\sqrt{150x^4}}{\sqrt{3x}}$
24. $\dfrac{\sqrt{24x^4}}{\sqrt{3x}}$
25. $\dfrac{\sqrt{200x^3}}{\sqrt{10x^{-1}}}$
26. $\dfrac{\sqrt{500x^3}}{\sqrt{10x^{-1}}}$

In Exercises 27–38, add or subtract terms whenever possible.

27. $7\sqrt{3} + 6\sqrt{3}$
28. $8\sqrt{5} + 11\sqrt{5}$
29. $6\sqrt{17x} - 8\sqrt{17x}$
30. $4\sqrt{13x} - 6\sqrt{13x}$
31. $\sqrt{8} + 3\sqrt{2}$
32. $\sqrt{20} + 6\sqrt{5}$
33. $\sqrt{50x} - \sqrt{8x}$
34. $\sqrt{63x} - \sqrt{28x}$
35. $3\sqrt{18} + 5\sqrt{50}$
36. $4\sqrt{12} - 2\sqrt{75}$
37. $3\sqrt{8} - \sqrt{32} + 3\sqrt{72} - \sqrt{75}$

38. $3\sqrt{54} - 2\sqrt{24} - \sqrt{96} + 4\sqrt{63}$

In Exercises 39–48, rationalize the denominator.

39. $\dfrac{1}{\sqrt{7}}$
40. $\dfrac{2}{\sqrt{10}}$
41. $\dfrac{\sqrt{2}}{\sqrt{5}}$
42. $\dfrac{\sqrt{7}}{\sqrt{3}}$
43. $\dfrac{13}{3 + \sqrt{11}}$
44. $\dfrac{3}{3 + \sqrt{7}}$
45. $\dfrac{7}{\sqrt{5} - 2}$
46. $\dfrac{5}{\sqrt{3} - 1}$
47. $\dfrac{6}{\sqrt{5} + \sqrt{3}}$
48. $\dfrac{11}{\sqrt{7} - \sqrt{3}}$

Evaluate each expression in Exercises 49–60, or indicate that the root is not a real number.

49. $\sqrt[3]{125}$
50. $\sqrt[3]{8}$
51. $\sqrt[3]{-8}$
52. $\sqrt[3]{-125}$
53. $\sqrt[4]{-16}$
54. $\sqrt[4]{-81}$
55. $\sqrt[4]{(-3)^4}$
56. $\sqrt[4]{(-2)^4}$
57. $\sqrt[5]{(-3)^5}$
58. $\sqrt[5]{(-2)^5}$
59. $\sqrt[5]{-\frac{1}{32}}$
60. $\sqrt[6]{\frac{1}{64}}$

Simplify the radical expressions in Exercises 61–68.

61. $\sqrt[3]{32}$
62. $\sqrt[3]{150}$
63. $\sqrt[3]{x^4}$
64. $\sqrt[3]{x^5}$
65. $\sqrt[3]{9} \cdot \sqrt[3]{6}$
66. $\sqrt[3]{12} \cdot \sqrt[3]{4}$
67. $\dfrac{\sqrt[5]{64x^6}}{\sqrt[5]{2x}}$
68. $\dfrac{\sqrt[4]{162x^5}}{\sqrt[4]{2x}}$

In Exercises 69–76, add or subtract terms whenever possible.

69. $4\sqrt[5]{2} + 3\sqrt[5]{2}$
70. $6\sqrt[5]{3} + 2\sqrt[5]{3}$
71. $5\sqrt[3]{16} + \sqrt[3]{54}$
72. $3\sqrt[3]{24} + \sqrt[3]{81}$
73. $\sqrt[3]{54xy^3} - y\sqrt[3]{128x}$
74. $\sqrt[3]{24xy^3} - y\sqrt[3]{81x}$
75. $\sqrt{2} + \sqrt[3]{8}$
76. $\sqrt{3} + \sqrt[3]{15}$

In Exercises 77–84, evaluate each expression without using a calculator.

77. $36^{1/2}$ **78.** $121^{1/2}$

79. $8^{1/3}$ **80.** $27^{1/3}$

81. $125^{2/3}$ **82.** $8^{2/3}$

83. $32^{-4/5}$ **84.** $16^{-5/2}$

In Exercises 85–94, simplify using properties of exponents.

85. $(7x^{1/3})(2x^{1/4})$ **86.** $(3x^{2/3})(4x^{3/4})$

87. $\dfrac{20x^{1/2}}{5x^{1/4}}$ **88.** $\dfrac{72x^{3/4}}{9x^{1/3}}$

89. $(x^{2/3})^3$ **90.** $(x^{4/5})^5$

91. $(25x^4y^6)^{1/2}$ **92.** $(125x^9y^6)^{1/3}$

93. $\dfrac{(3y^{1/4})^3}{y^{1/12}}$ **94.** $\dfrac{(2y^{1/5})^4}{y^{3/10}}$

In Exercises 95–102, simplify by reducing the index of the radical.

95. $\sqrt[4]{5^2}$ **96.** $\sqrt[4]{7^2}$

97. $\sqrt[3]{x^6}$ **98.** $\sqrt[4]{x^{12}}$

99. $\sqrt[6]{x^4}$ **100.** $\sqrt[9]{x^6}$

101. $\sqrt[9]{x^6y^3}$ **102.** $\sqrt[12]{x^4y^8}$

Application Exercises

103. The algebraic expression $2\sqrt{5L}$ is used to estimate the speed of a car prior to an accident, in miles per hour, based on the length of its skid marks, L, in feet. Find the speed of a car that left skid marks 40 feet long, and write the answer in simplified radical form.

104. The time, in seconds, that it takes an object to fall a distance d, in feet, is given by the algebraic expression $\sqrt{\dfrac{d}{16}}$. Find how long it will take a ball dropped from the top of a building 320 feet tall to hit the ground. Write the answer in simplified radical form.

105. The early Greeks believed that the most pleasing of all rectangles were golden rectangles whose ratio of width to height is

$$\frac{w}{h} = \frac{2}{\sqrt{5}-1}.$$

Rationalize the denominator for this ratio and then use a calculator to approximate the answer correct to the nearest hundredth.

106. The amount of evaporation, in inches per day, of a large body of water can be described by the algebraic expression

$$\frac{w}{20\sqrt{a}}$$

where

a = surface area of the water, in square miles

w = average wind speed of the air over the water, in miles per hour.

Determine the evaporation on a lake whose surface area is 9 square miles on a day when the wind speed over the water is 10 miles per hour.

107. In the Peanuts cartoon shown below, Woodstock appears to be working steps mentally. Fill in the missing steps that show how to go from $\dfrac{7\sqrt{2\cdot 2\cdot 3}}{6}$ to $\dfrac{7}{3}\sqrt{3}$.

PEANUTS reprinted by permission of United Feature Syndicate, Inc.

108. The algebraic expression $152a^{-1/5}$ describes the percentage of U.S. taxpayers who are a years old who file early. Evaluate the algebraic expression for $a = 32$. Describe what the answer means in practical terms.

109. The algebraic expression $0.07d^{3/2}$ describes the duration of a storm, in hours, whose diameter is d miles. Evaluate the algebraic expression for $d = 9$. Describe what the answer means in practical terms.

Writing in Mathematics

110. Explain how to simplify $\sqrt{10} \cdot \sqrt{5}$.

111. Explain how to add $\sqrt{3} + \sqrt{12}$.

112. Describe what it means to rationalize a denominator. Use both $\dfrac{1}{\sqrt{5}}$ and $\dfrac{1}{5+\sqrt{5}}$ in your explanation.

113. What difference is there in simplifying $\sqrt[3]{(-5)^3}$ and $\sqrt[4]{(-5)^4}$?

114. What does $a^{m/n}$ mean?

115. Describe the kinds of numbers that have rational fifth roots.

116. Why must a and b represent nonnegative numbers when we write $\sqrt{a} \cdot \sqrt{b} = \sqrt{ab}$? Is it necessary to use this restriction in the case of $\sqrt[3]{a} \cdot \sqrt[3]{b} = \sqrt[3]{ab}$? Explain.

 Technology Exercises

117. The algebraic expression
$$\frac{73t^{1/3} - 28t^{2/3}}{t}$$
describes the percentage of people in the United States applying for jobs t years after 1985 who tested positive for illegal drugs. Use a calculator to find the percentage who tested positive from 1986 through 2001. Round answers to the nearest hundredth of a percent. What trend do you observe for the percentage of potential employees testing positive for illegal drugs over time?

118. The territorial area of an animal in the wild is defined to be the area of the region to which the animal confines its movements. The algebraic expression $W^{1.41}$ describes the territorial area, in square miles, of an animal that weighs W pounds. Use a calculator to find the territorial area of animals weighing 25, 50, 150, 200, 250, and 300 pounds. What do the values indicate about the relationship between body weight and territorial area?

 Critical Thinking Exercises

119. Which one of the following is true?

    a. Neither $(-8)^{1/2}$ nor $(-8)^{1/3}$ represent real numbers.

    b. $\sqrt{x^2 + y^2} = x + y$

    c. $8^{-1/3} = -2$

    d. $2^{1/2} \cdot 2^{1/2} = 2$

In Exercises 120–121, fill in each box to make the statement true.

120. $(5 + \sqrt{\Box})(5 - \sqrt{\Box}) = 22$

121. $\sqrt[\Box]{x^\Box} = 5x^7$

122. Find exact value of $\sqrt{13 + \sqrt{2} + \dfrac{7}{3 + \sqrt{2}}}$ without the use of a calculator.

123. Place the correct symbol, $>$ or $<$, in the box between each of the given numbers. *Do not use a calculator.* Then check your result with a calculator.

    a. $3^{1/2} \Box 3^{1/3}$ **b.** $\sqrt{7} + \sqrt{18} \Box \sqrt{7 + 18}$

SECTION P.4 *Polynomials*

Objectives

1. Understand the vocabulary of polynomials.
2. Add and subtract polynomials.
3. Multiply polynomials.
4. Use FOIL in polynomial multiplication.
5. Use special products in polynomial multiplication.
6. Perform operations with polynomials in several variables.

This computer-simulated model of the common cold virus was developed by researchers at Purdue University. Their discovery of how the virus infects human cells could lead to more effective treatment for the illness.

Runny nose? Sneezing? You are probably familiar with the unpleasant onset of a cold. We "catch cold" when the cold virus enters our bodies, where it multiplies. Fortunately, at a certain point the virus begins to die. The algebraic expression $-0.75x^4 + 3x^3 + 5$ describes the billions of viral particles in our bodies after x days of invasion. The expression enables mathematicians to determine the day on which there is a maximum number of viral particles and, consequently, the day we feel sickest.

The algebraic expression $-0.75x^4 + 3x^3 + 5$ is an example of a polynomial. A **polynomial** is a single term or the sum of two or more terms containing variables with whole number exponents. This particular polynomial contains three terms. Equations containing polynomials are used in such diverse areas as science, business, medicine, psychology, and sociology. In this section, we review basic ideas about polynomials and their operations.

1 Understand the vocabulary of polynomials.

The Vocabulary of Polynomials

Consider the polynomial

$$7x^3 - 9x^2 + 13x - 6.$$

We can express this polynomial as

$$7x^3 + (-9x^2) + 13x + (-6).$$

The polynomial contains four terms. It is customary to write the terms in the order of descending powers of the variable. This is the **standard form** of a polynomial.

We begin this section by limiting our discussion to polynomials containing only one variable. Each term of a polynomial in x is of the form ax^n. The **degree** of ax^n is n. For example, the degree of the term $7x^3$ is 3.

Study Tip

We can express 0 in many ways, including $0x$, $0x^2$, and $0x^3$. It is impossible to assign a single exponent on the variable. This is why 0 has no defined degree.

The Degree of ax^n

If $a \neq 0$, the degree of ax^n is n. The degree of a nonzero constant is 0. The constant 0 has no defined degree.

Here is an example of a polynomial and the degree of each of its four terms:

$$6x^4 - 3x^3 + 2x - 5.$$

| degree 4 | degree 3 | degree 1 | degree of non-zero constant: 0 |

Notice that the exponent on x for the term $2x$ is understood to be 1: $2x^1$. For this reason, the degree of $2x$ is 1. You can think of -5 as $-5x^0$; thus, its degree is 0.

A polynomial which when simplified has exactly one term is called a **monomial.** A **binomial** is a simplified polynomial that has two terms, each with a different exponent. A **trinomial** is a simplified polynomial with three terms, each with a different exponent. Simplified polynomials with four or more terms have no special names.

The **degree of a polynomial** is the highest degree of all the terms of the polynomial. For example, $4x^2 + 3x$ is a binomial of degree 2 because the degree of the first term is 2, and the degree of the other term is less than 2. Also, $7x^5 - 2x^2 + 4$ is a trinomial of degree 5 because the degree of the first term is 5, and the degrees of the other terms are less than 5.

Up to now, we have used x to represent the variable in a polynomial. However, any letter can be used. For example,

- $7x^5 - 3x^3 + 8$ is a polynomial (in x) of degree 5.
- $6y^3 + 4y^2 - y + 3$ is a polynomial (in y) of degree 3.
- $z^7 + \sqrt{2}$ is a polynomial (in z) of degree 7.

Not every algebraic expression is a polynomial. Algebraic expressions whose variables do not contain whole number exponents such as

$$3x^{-2} + 7 \quad \text{and} \quad 5x^{3/2} + 9x^{1/2} + 2$$

are not polynomials. Furthermore, a quotient of polynomials such as

$$\frac{x^2 + 2x + 5}{x^3 - 7x^2 + 9x - 3}$$

is not a polynomial because the form of a polynomial involves only addition and subtraction of terms, not division.

We can tie together the threads of our discussion with the formal definition of a polynomial in one variable. In this definition, the coefficients of the terms are represented by a_n (read "a sub n"), a_{n-1} (read "a sub n minus 1"), a_{n-2}, and so on. The small letters to the lower right of each a are called **subscripts** and are *not exponents*. Subscripts are used to distinguish one constant from another when a large and undetermined number of such constants are needed.

Definition of a Polynomial in x

A **polynomial in x** is an algebraic expression of the form

$$a_n x^n + a_{n-1} x^{n-1} + a_{n-2} x^{n-2} + \cdots + a_1 x + a_0,$$

where $a_n, a_{n-1}, a_{n-2}, \ldots, a_1,$ and a_0 are real numbers, $a_n \neq 0$, and n is a nonnegative integer. The polynomial is of **degree n, a_n** is the **leading coefficient,** and a_0 is the **constant term.**

2 Add and subtract polynomials.

Adding and Subtracting Polynomials

Polynomials are added and subtracted by combining like terms. For example, we can combine the monomials $-9x^3$ and $13x^3$ using addition as follows:

$$-9x^3 + 13x^3 = (-9 + 13)x^3 = 4x^3.$$

EXAMPLE 1 Adding and Subtracting Polynomials

Perform the indicated operations and simplify:

a. $(-9x^3 + 7x^2 - 5x + 3) + (13x^3 + 2x^2 - 8x - 6)$
b. $(7x^3 - 8x^2 + 9x - 6) - (2x^3 - 6x^2 - 3x + 9).$

Solution

a. $(-9x^3 + 7x^2 - 5x + 3) + (13x^3 + 2x^2 - 8x - 6)$

$\quad = (-9x^3 + 13x^3) + (7x^2 + 2x^2)$ *Group like terms.*

$\quad\quad + (-5x - 8x) + (3 - 6)$

$\quad = 4x^3 + 9x^2 + (-13x) + (-3)$ *Combine like terms.*

$\quad = 4x^3 + 9x^2 - 13x - 3$ *Simplify.*

Study Tip

You can also arrange like terms in columns and combine vertically:

$$7x^3 - 8x^2 + 9x - 6$$
$$\underline{-2x^3 + 6x^2 + 3x - 9}$$
$$5x^3 - 2x^2 + 12x - 15$$

The like terms can be combined by adding their coefficients and keeping the same variable factor.

b. $(7x^3 - 8x^2 + 9x - 6) - (2x^3 - 6x^2 - 3x + 9)$

$= (7x^3 - 8x^2 + 9x - 6) + (-2x^3 + 6x^2 + 3x - 9)$ Rewrite subtraction as addition of the additive inverse. Be sure to change the sign of each term inside parentheses preceded by the negative sign.

$= (7x^3 - 2x^3) + (-8x^2 + 6x^2)$
$\quad + (9x + 3x) + (-6 - 9)$ Group like terms.

$= 5x^3 + (-2x^2) + 12x + (-15)$ Combine like terms.

$= 5x^3 - 2x^2 + 12x - 15$ Simplify.

Check Point 1   Perform the indicated operations and simplify:

a. $(-17x^3 + 4x^2 - 11x - 5) + (16x^3 - 3x^2 + 3x - 15)$

b. $(13x^3 - 9x^2 - 7x + 1) - (-7x^3 + 2x^2 - 5x + 9)$.

3   Multiply polynomials.

Multiplying Polynomials

The product of two monomials is obtained by using properties of exponents. For example,

$$(-8x^6)(5x^3) = -8 \cdot 5x^{6+3} = -40x^9.$$

Multiply coefficients and add exponents.

Furthermore, we can use the distributive property to multiply a monomial and a polynomial that is not a monomial. For example,

$$3x^4(2x^3 - 7x + 3) = 3x^4 \cdot 2x^3 - 3x^4 \cdot 7x + 3x^4 \cdot 3 = 6x^7 - 21x^5 + 9x^4.$$

monomial   trinomial

How do we multiply two polynomials if neither is a monomial? For example, consider

$$(2x + 3)(x^2 + 4x + 5).$$

binomial   trinomial

One way to perform this multiplication is to distribute $2x$ throughout the trinomial

$$2x(x^2 + 4x + 5)$$

and 3 throughout the trinomial

$$3(x^2 + 4x + 5).$$

Then combine the like terms that result.

Multiplying Polynomials when Neither is a Monomial

Multiply each term of one polynomial by each term of the other polynomial. Then combine like terms.

EXAMPLE 2 **Multiplying a Binomial and a Trinomial**

Multiply: $(2x + 3)(x^2 + 4x + 5)$.

Solution

$(2x + 3)(x^2 + 4x + 5)$

$= 2x(x^2 + 4x + 5) + 3(x^2 + 4x + 5)$ Multiply the trinomial by each term of the binomial.

$= 2x \cdot x^2 + 2x \cdot 4x + 2x \cdot 5 + 3 \cdot x^2 + 3 \cdot 4x + 3 \cdot 5$ Use the distributive property.

$= 2x^3 + 8x^2 + 10x + 3x^2 + 12x + 15$ Multiply the monomials: multiply coefficients and add exponents.

$= 2x^3 + 11x^2 + 22x + 15$ Combine like terms: $8x^2 + 3x^2 = 11x^2$ and $10x + 12x = 22x$.

Another method for solving Example 2 is to use a vertical format similar to that used for multiplying whole numbers.

$x^2 + 4x + 5$

$2x + 3$

$\overline{3x^2 + 12x + 15}$ $3(x^2 + 4x + 5)$

Write like terms in $2x^3 + 8x^2 + 10x$ $2x(x^2 + 4x + 5)$
the same column.

$\overline{2x^3 + 11x^2 + 22x + 15}$ Combine like terms.

Check Point 2 Multiply: $(5x - 2)(3x^2 - 5x + 4)$.

4 Use FOIL in polynomial multiplication.

The Product of Two Binomials: FOIL

Frequently we need to find the product of two binomials. We can use a method called FOIL, which is based on the distributive property, to do so. For example, we can find the product of the binomials $3x + 2$ and $4x + 5$ as follows:

$(3x + 2)(4x + 5) = 3x(4x + 5) + 2(4x + 5)$ First, distribute 3x over 4x + 5. Then distribute 2.

$= 3x(4x) + 3x(5) + 2(4x) + 2(5)$

$= 12x^2 + 15x + 8x + 10.$

Two binomials can be quickly multiplied by using the FOIL method, in which F represents the product of the **first** terms in each binomial, O represents the product of the **outside** terms, I represents the product of the two **inside** terms, and L represents the product of the **last,** or second, terms in each binomial.

first   last   F   O   I   L

$(3x + 2)(4x + 5) = 12x^2 + 15x + 8x + 10$

inside

outside

$= 12x^2 + 23x + 10$ Combine like terms.

In general, here is how to use the FOIL method to find the product of $ax + b$ and $cx + d$:

Using the FOIL Method to Multiply Binomials

$$(ax + b)(cx + d) = ax \cdot cx + ax \cdot d + b \cdot cx + b \cdot d$$

| Product of First terms | Product of Outside terms | Product of Inside terms | Product of Last terms |

EXAMPLE 3 Using the FOIL Method

Multiply: $(3x + 4)(5x - 3)$.

Solution

$$(3x + 4)(5x - 3) = 3x \cdot 5x + 3x(-3) + 4 \cdot 5x + 4(-3)$$
$$= 15x^2 - 9x + 20x - 12$$
$$= 15x^2 + 11x - 12 \quad \textit{Combine like terms.}$$

Check Point 3   Multiply: $(7x - 5)(4x - 3)$.

5 Use special products in polynomial multiplication.

Multiplying the Sum and Difference of Two Terms

We can use the FOIL method to multiply $A + B$ and $A - B$ as follows:

$$(A + B)(A - B) = A^2 - AB + AB - B^2 = A^2 - B^2.$$

Notice that the outside and inside products have a sum of 0 and the terms cancel. The FOIL multiplication provides us with a quick rule for multiplying the sum and difference of two terms, referred to as a special-product formula.

The Product of the Sum and Difference of Two Terms

$$(A + B)(A - B) = A^2 - B^2$$

| The product of the sum and the difference of the same two terms | is | the square of the first term minus the square of the second term. |

EXAMPLE 4 **Finding the Product of the Sum and Difference of Two Terms**

Find each product:

    a. $(4y + 3)(4y - 3)$ **b.** $(5a^4 + 6)(5a^4 - 6)$.

Solution   Use the special-product formula shown.

$$(A + B)(A - B) \quad = \quad A^2 \quad - \quad B^2$$

	First term squared	−	Second term squared	=	Product

    a. $(4y + 3)(4y - 3) \quad = \quad (4y)^2 \quad - \quad 3^2 \quad = 16y^2 - 9$

    b. $(5a^4 + 6)(5a^4 - 6) \quad = \quad (5a^4)^2 \quad - \quad 6^2 \quad = 25a^8 - 36$

Check Point 4   Find each product:

    a. $(7x + 8)(7x - 8)$ **b.** $(2y^3 - 5)(2y^3 + 5)$.

The Square of a Binomial

Let us find $(A + B)^2$, the square of a binomial sum. To do so, we begin with the FOIL method and look for a general rule.

	F	O	I	L

$$(A + B)^2 = (A + B)(A + B) = A \cdot A + A \cdot B + A \cdot B + B \cdot B$$
$$= A^2 + 2AB + B^2$$

This result implies the following rule, which is another example of a special-product formula:

Study Tip

Caution! The square of a sum is *not* the sum of the squares.

$(A + B)^2 \neq A^2 + B^2$

The middle term 2AB is missing

$(x + 3)^2 \neq x^2 + 9$

Incorrect

Show that $(x + 3)^2$ and $x^2 + 9$ are not equal by substituting 5 for x in each expression and simplifying.

The Square of a Binomial Sum

$$(A + B)^2 \quad = \quad A^2 \quad + \quad 2AB \quad + \quad B^2$$

The square of a binomial sum	is	first term squared	plus	2 times the product of the terms	plus	last term squared.

EXAMPLE 5 **Finding the Square of a Binomial Sum**

Square each binomial:

    a. $(x + 3)^2$ **b.** $(3x + 7)^2$.

Solution   Use the special-product formula shown.

$$(A + B)^2 = A^2 + 2AB + B^2$$

Square a Sum	(First Term)^2	$+$	2 · Product of the Terms	$+$	(Last Term)^2	$=$ Product
a. $(x + 3)^2 =$	x^2	$+$	$2 \cdot x \cdot 3$	$+$	3^2	$= x^2 + 6x + 9$
b. $(3x + 7)^2 =$	$(3x)^2$	$+$	$2(3x)(7)$	$+$	7^2	$= 9x^2 + 42x + 49$

Check Point 5   Square each binomial:

   a. $(x + 10)^2$ **b.** $(5x + 4)^2.$

Using the FOIL method on $(A - B)^2$, the square of a binomial difference, we obtain the following rule:

The Square of a Binomial Difference

$$(A - B)^2 = A^2 - 2AB + B^2$$

The square of a binomial difference   is   first term squared   minus   2 times the product of the terms   plus   last term squared.

EXAMPLE 6   Finding the Square of a Binomial Difference

Square each binomial:

   a. $(x - 4)^2$ **b.** $(5y - 6)^2.$

Solution   Use the special-product formula shown.

$$(A - B)^2 = A^2 - 2AB + B^2$$

Square a Difference	(First Term)^2	$-$	2 · Product of the Terms	$+$	(Last Term)^2	$=$ Product
a. $(x - 4)^2 =$	x^2	$-$	$2 \cdot x \cdot 4$	$+$	4^2	$= x^2 - 8x + 16$
b. $(5y - 6)^2 =$	$(5y)^2$	$-$	$2(5y)(6)$	$+$	6^2	$= 25y^2 - 60y + 36$

Check Point 6   Square each binomial:

   a. $(x - 9)^2$ **b.** $(7x - 3)^2.$

Special Products

There are several products that occur so frequently that it's convenient to memorize the form, or pattern, of these formulas.

Special Products

Let A and B represent real numbers, variables, or algebraic expressions.

Special Product	Example
Sum and Difference of Two Terms	
$(A + B)(A - B) = A^2 - B^2$	$(2x + 3)(2x - 3) = (2x)^2 - 3^2$
	$= 4x^2 - 9$
Squaring a Binomial	
$(A + B)^2 = A^2 + 2AB + B^2$	$(y + 5)^2 = y^2 + 2 \cdot y \cdot 5 + 5^2$
	$= y^2 + 10y + 25$
$(A - B)^2 = A^2 - 2AB + B^2$	$(3x - 4)^4$
	$= (3x)^2 - 2 \cdot 3x \cdot 4 + 4^2$
	$= 9x^2 - 24x + 16$
Cubing a Binomial	
$(A + B)^3 = A^3 + 3A^2B + 3AB^2 + B^3$	$(x + 4)^3$
	$= x^3 + 3x^2(4) + 3x(4)^2 + 4^3$
	$= x^3 + 12x^2 + 48x + 64$
$(A - B)^3 = A^3 - 3A^2B + 3AB^2 - B^3$	$(x - 2)^3$
	$= x^3 - 3x^2(2) + 3x(2)^2 - 2^3$
	$= x^3 - 6x^2 + 12x - 8$

Study Tip

Although it's convenient to memorize these forms, the FOIL method can be used on all five examples in the box. To cube $x + 4$, you can first square $x + 4$ using FOIL and then multiply this result by $x + 4$. In short, you do not necessarily have to utilize these special formulas. What is the advantage of knowing and using these forms?

6 Perform operations with polynomials in several variables.

Polynomials in Several Variables

The next time you visit the lumber yard and go rummaging through piles of wood, think *polynomials*, although polynomials a bit different from those we have encountered so far. The construction industry uses a polynomial in two variables to determine the number of board feet that can be manufactured from a tree with a diameter of x inches and a length of y feet. This polynomial is

$$\tfrac{1}{4}x^2y - 2xy + 4y.$$

In general, a **polynomial in two variables,** x and y, contains the sum of one or more monomials in the form ax^ny^m. The constant, a, is the **coefficient.** The exponents, n and m, represent whole numbers. The **degree** of the monomial ax^ny^m is $n + m$. We'll use the polynomial from the construction industry to illustrate these ideas.

The coefficients are $\tfrac{1}{4}$, -2, and 4.

$$\tfrac{1}{4}x^2y \qquad -2xy \qquad +4y$$

| Degree of monomial: $2 + 1 = 3$ | Degree of monomial: $1 + 1 = 2$ | Degree of monomial: $0 + 1 = 1$ |

The **degree of a polynomial in two variables** is the highest degree of all its terms. For the preceding polynomial, the degree is 3.

Polynomials containing two or more variables can be added, subtracted, and multiplied just like polynomials that contain only one variable.

EXAMPLE 7 Subtracting Polynomials in Two Variables

Subtract as indicated:

$$(5x^3 - 9x^2y + 3xy^2 - 4) - (3x^3 - 6x^2y - 2xy^2 + 3).$$

Solution

$(5x^3 - 9x^2y + 3xy^2 - 4) - (3x^3 - 6x^2y - 2xy^2 + 3)$

$= (5x^3 - 9x^2y + 3xy^2 - 4) + (-3x^3 + 6x^2y + 2xy^2 - 3)$

Change the sign of each term in the second polynomial and add the two polynomials.

$= (5x^3 - 3x^3) + (-9x^2y + 6x^2y) + (3xy^2 + 2xy^2) + (-4 - 3)$

Group like terms.

$= 2x^3 - 3x^2y + 5xy^2 - 7$ *Combine like terms by combining coefficients and keeping the same variable factors.*

Check Point 7 Subtract: $(x^3 - 4x^2y + 5xy^2 - y^3) - (x^3 - 6x^2y + y^3).$

EXAMPLE 8 Multiplying Polynomials in Two Variables

Multiply: **a.** $(x + 4y)(3x - 5y)$ **b.** $(5x + 3y)^2.$

Solution   We will perform the multiplication in part (a) using the FOIL method. We will multiply in part (b) using the formula for the square of a binomial sum, $(A + B)^2.$

a. $(x + 4y)(3x - 5y)$ *Multiply these binomials using the FOIL method.*

$\qquad$ F $\qquad$ O $\qquad$ I $\qquad$ L

$= (x)(3x) + (x)(-5y) + (4y)(3x) + (4y)(-5y)$
$= 3x^2 - 5xy + 12xy - 20y^2$
$= 3x^2 + 7xy - 20y^2$ *Combine like terms.*

$(A + B)^2 = A^2 + 2 \cdot A \cdot B + B^2$

b. $(5x + 3y)^2 = (5x)^2 + 2(5x)(3y) + (3y)^2$
$\qquad\qquad = 25x^2 + 30xy + 9y^2$

Check Point 8 Multiply:

a. $(7x - 6y)(3x - y)$ **b.** $(x^2 + 5y)^2.$

EXERCISE SET P.4

 Practice Exercises

In Exercises 1–4, is the algebraic expression a polynomial? If it is, write the polynomial in standard form.

1. $2x + 3x^2 - 5$ **2.** $2x + 3x^{-1} - 5$

3. $\dfrac{2x + 3}{x}$ **4.** $x^2 - x^3 + x^4 - 5$

In Exercises 5–8, find the degree of the polynomial.

5. $3x^2 - 5x + 4$ **6.** $-4x^3 + 7x^2 - 11$

7. $x^2 - 4x^3 + 9x - 12x^4 + 63$

8. $x^2 - 8x^3 + 15x^4 + 91$

In Exercises 9–14, perform the indicated operations. Write the resulting polynomial in standard form and indicate its degree.

9. $(-6x^3 + 5x^2 - 8x + 9) + (17x^3 + 2x^2 - 4x - 13)$

10. $(-7x^3 + 6x^2 - 11x + 13) + (19x^3 - 11x^2 + 7x - 17)$

11. $(17x^3 - 5x^2 + 4x - 3) - (5x^3 - 9x^2 - 8x + 11)$

12. $(18x^4 - 2x^3 - 7x + 8) - (9x^4 - 6x^3 - 5x + 7)$

13. $(5x^2 - 7x - 8) + (2x^2 - 3x + 7) - (x^2 - 4x - 3)$

14. $(8x^2 + 7x - 5) - (3x^2 - 4x) - (-6x^3 - 5x^2 + 3)$

In Exercises 15–58, find each product.

15. $(x + 1)(x^2 - x + 1)$ **16.** $(x + 5)(x^2 - 5x + 25)$

17. $(2x - 3)(x^2 - 3x + 5)$ **18.** $(2x - 1)(x^2 - 4x + 3)$

19. $(x + 7)(x + 3)$ **20.** $(x + 8)(x + 5)$

21. $(x - 5)(x + 3)$ **22.** $(x - 1)(x + 2)$

23. $(3x + 5)(2x + 1)$ **24.** $(7x + 4)(3x + 1)$

25. $(2x - 3)(5x + 3)$ **26.** $(2x - 5)(7x + 2)$

27. $(5x^2 - 4)(3x^2 - 7)$ **28.** $(7x^2 - 2)(3x^2 - 5)$

29. $(8x^3 + 3)(x^2 - 5)$ **30.** $(7x^3 + 5)(x^2 - 2)$

31. $(x + 3)(x - 3)$ **32.** $(x + 5)(x - 5)$

33. $(3x + 2)(3x - 2)$ **34.** $(2x + 5)(2x - 5)$

35. $(5 - 7x)(5 + 7x)$ **36.** $(4 - 3x)(4 + 3x)$

37. $(4x^2 + 5x)(4x^2 - 5x)$ **38.** $(3x^2 + 4x)(3x^2 - 4x)$

39. $(1 - y^5)(1 + y^5)$ **40.** $(2 - y^5)(2 + y^5)$

41. $(x + 2)^2$ **42.** $(x + 5)^2$

43. $(2x + 3)^2$ **44.** $(3x + 2)^2$

45. $(x - 3)^2$ **46.** $(x - 4)^2$

47. $(4x^2 - 1)^2$ **48.** $(5x^2 - 3)^2$

49. $(7 - 2x)^2$ **50.** $(9 - 5x)^2$

51. $(x + 1)^3$ **52.** $(x + 2)^3$

53. $(2x + 3)^3$ **54.** $(3x + 4)^3$

55. $(x - 3)^3$ **56.** $(x - 1)^3$

57. $(3x - 4)^3$ **58.** $(2x - 3)^3$

In Exercises 59–66, perform the indicated operations. Indicate the degree of the resulting polynomial.

59. $(5x^2y - 3xy) + (2x^2y - xy)$

60. $(-2x^2y + xy) + (4x^2y + 7xy)$

61. $(4x^2y + 8xy + 11) + (-2x^2y + 5xy + 2)$

62. $(7x^4y^2 - 5x^2y^2 + 3xy) + (-18x^4y^2 - 6x^2y^2 - xy)$

63. $(x^3 + 7xy - 5y^2) - (6x^3 - xy + 4y^2)$

64. $(x^4 - 7xy - 5y^3) - (6x^4 - 3xy + 4y^3)$

65. $(3x^4y^2 + 5x^3y - 3y) - (2x^4y^2 - 3x^3y - 4y + 6x)$

66. $(5x^4y^2 + 6x^3y - 7y) - (3x^4y^2 - 5x^3y - 6y + 8x)$

In Exercises 67–82, find each product.

67. $(x + 5y)(7x + 3y)$ **68.** $(x + 9y)(6x + 7y)$

69. $(x - 3y)(2x + 7y)$ **70.** $(3x - y)(2x + 5y)$

71. $(3xy - 1)(5xy + 2)$ **72.** $(7x^2y + 1)(2x^2y - 3)$

73. $(7x + 5y)^2$ **74.** $(9x + 7y)^2$

75. $(x^2y^2 - 3)^2$ **76.** $(x^2y^2 - 5)^2$

77. $(x - y)(x^2 + xy + y^2)$ **78.** $(x + y)(x^2 - xy + y^2)$

79. $(3x + 5y)(3x - 5y)$ **80.** $(7x + 3y)(7x - 3y)$

81. $(7xy^2 - 10y)(7xy^2 + 10y)$

82. $(3xy^2 - 4y)(3xy^2 + 4y)$

 Application Exercises

83. The polynomial $0.018x^2 - 0.757x + 9.047$ describes the amount, in thousands of dollars, that a person earning x thousand dollars a year feels underpaid. Evaluate the polynomial for $x = 40$. Describe what the answer means in practical terms.

84. The polynomial $104.5x^2 - 1501.5x + 6016$ describes the death rate per year, per 100,000 men, for men averaging x hours of sleep each night. Evaluate the polynomial for $x = 10$. Describe what the answer means in practical terms.

85. The polynomial $-1.45x^2 + 38.52x + 470.78$ describes the number of violent crimes in the United States, per 100,000 inhabitants, x years after 1975. Evaluate the polynomial for $x = 25$. Describe what the answer means in practical terms. How well does the polynomial describe the crime rate for the appropriate year shown in the bar graph?

Violent Crime in the United States

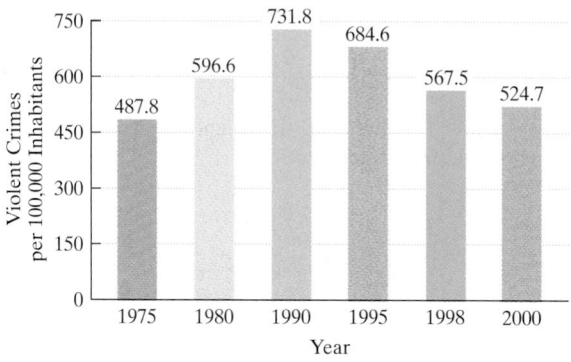

Source: F.B.I.

86. The polynomial $-0.02A^2 + 2A + 22$ is used by coaches to get athletes fired up so that they can perform well. The polynomial represents the performance level related to various levels of enthusiasm, from $A = 1$ (almost no enthusiasm) to $A = 100$ (maximum level of enthusiasm). Evaluate the polynomial for $A = 20$, $A = 50$, and $A = 80$. Describe what happens to performance as we get more and more fired up.

87. The number of people who catch a cold t weeks after January 1 is $5t - 3t^2 + t^3$. The number of people who recover t weeks after January 1 is $t - t^2 + \frac{1}{3}t^3$. Write a polynomial in standard form for the number of people who are still ill with a cold t weeks after January 1.

88. The weekly cost, in thousands of dollars, for producing x stereo headphones is $30x + 50$. The weekly revenue, in thousands of dollars, for selling x stereo headphones is $90x^2 - x$. Write a polynomial in standard form for the weekly profit, in thousands of dollars, for producing and selling x stereo headphones.

In Exercises 89–90, write a polynomial in standard form that represents the area of the shaded region of each figure.

89.

90.

 Writing in Mathematics

91. What is a polynomial in x?

92. Explain how to subtract polynomials.

93. Explain how to multiply two binomials using the FOIL method. Give an example with your explanation.

94. Explain how to find the product of the sum and difference of two terms. Give an example with your explanation.

95. Explain how to square a binomial difference. Give an example with your explanation.

96. Explain how to find the degree of a polynomial in two variables.

97. For Exercise 86, explain why performance levels do what they do as we get more and more fired up. If possible, describe an example of a time when you were too enthused and thus did poorly at something you were hoping to do well.

Technology Exercises

98. The common cold is caused by a rhinovirus. The polynomial

$$-0.75x^4 + 3x^3 + 5$$

describes the billions of viral particles in our bodies after x days of invasion. Use a calculator to find the number of viral particles after 0 days (the time of the cold's onset), 1 day, 2 days, 3 days, and 4 days. After how many days is the number of viral particles at a maximum and consequently the day we feel the sickest? By when should we feel completely better?

99. Using data from the National Institute on Drug Abuse, the polynomial

$$0.0032x^3 + 0.0235x^2 - 2.2477x + 61.1998$$

approximately describes the percentage of U.S. high school seniors in the class of x who had ever used marijuana, where x is the number of years after 1980. Use a calculator to find the percentage of high school seniors from the class of 1980 through the class of 2000 who had used marijuana. Round to the nearest tenth of a percent. Describe the trend in the data.

 Critical Thinking Exercises

In Exercises 100–103, perform the indicated operations.

100. $(x - y)^2 - (x + y)^2$

101. $[(7x + 5) + 4y][(7x + 5) - 4y]$

102. $[(3x + y) + 1]^2$

103. $(x + y)(x - y)(x^2 + y^2)$

104. Express the area of the plane figure shown as a polynomial in standard form.

SECTION P.5 *Factoring Polynomials*

Objectives

1. Factor out the greatest common factor of a polynomial.
2. Factor by grouping.
3. Factor trinomials.
4. Factor the difference of squares.
5. Factor perfect square trinomials.
6. Factor the sum and difference of cubes.
7. Use a general strategy for factoring polynomials.
8. Factor algebraic expressions containing fractional and negative exponents.

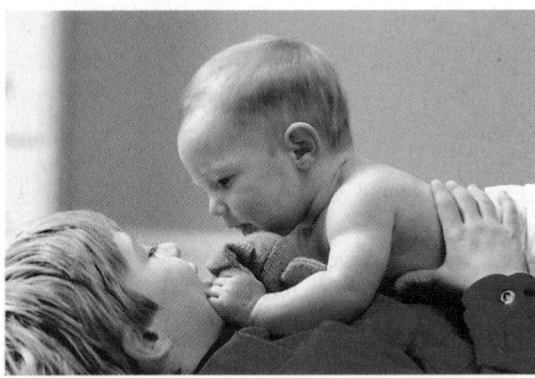

A two-year-old boy is asked, "Do you have a brother?" He answers, "Yes." "What is your brother's name?" "Tom." Asked if Tom has a brother, the two-year-old replies, "No." The child can go in the direction from self to brother, but he cannot reverse this direction and move from brother back to self.

As our intellects develop, we learn to reverse the direction of our thinking. Reversibility of thought is found throughout algebra. For example, we can multiply polynomials and show that

$$(2x + 1)(3x - 2) = 6x^2 - x - 2.$$

We can also reverse this process and express the resulting polynomial as

$$6x^2 - x - 2 = (2x + 1)(3x - 2).$$

Factoring is the process of writing a polynomial as the product of two or more polynomials. The factors of $6x^2 - x - 2$ are $2x + 1$ and $3x - 2$.

In this section, we will be **factoring over the set of integers,** meaning that the coefficients in the factors are integers. Polynomials that cannot be factored using integer coefficients are called **irreducible over the integers,** or **prime.**

The goal in factoring a polynomial is to use one or more factoring techniques until each of the polynomial's factors is prime or irreducible. In this situation, the polynomial is said to be **factored completely.**

We will now discuss basic techniques for factoring polynomials.

1 Factor out the greatest common factor of a polynomial.

Common Factors

In any factoring problem, the first step is to look for the *greatest common factor.* The **greatest common factor,** abbreviated GCF, is an expression of the highest degree that divides each term of the polynomial. The distributive property in the reverse direction

$$ab + ac = a(b + c)$$

can be used to factor out the greatest common factor.

EXAMPLE 1   Factoring out the Greatest Common Factor

Factor: **a.** $18x^3 + 27x^2$ **b.** $x^2(x + 3) + 5(x + 3)$.

Solution

a. We begin by determining the greatest common factor. 9 is the greatest integer that divides 18 and 27. Furthermore, x^2 is the greatest expression that divides x^3 and x^2. Thus, the greatest common factor of the two terms in the polynomial is $9x^2$.

$$18x^3 + 27x^2$$
$$= 9x^2(2x) + 9x^2(3)$$ Express each term as the product of the greatest common factor and its other factor.

$$= 9x^2(2x + 3)$$ Factor out the greatest common factor.

b. In this situation, the greatest common factor is the common binomial factor $(x + 3)$. We factor out this common factor as follows:

$$x^2(x + 3) + 5(x + 3) = (x + 3)(x^2 + 5).$$ Factor out the common binomial factor.

> **Study Tip**
>
> The variable part of the greatest common factor always contains the *smallest* power of a variable or algebraic expression that appears in all terms of the polynomial.

Check Point 1   Factor:

a. $10x^3 - 4x^2$ **b.** $2x(x - 7) + 3(x - 7)$.

2 Factor by grouping.

Factoring by Grouping

Some polynomials have only a greatest common factor of 1. However, by a suitable rearrangement of the terms, it still may be possible to factor. This process, called **factoring by grouping,** is illustrated in Example 2.

EXAMPLE 2   Factoring by Grouping

Factor: $x^3 + 4x^2 + 3x + 12$.

Solution   Group terms that have a common factor:

$$\boxed{x^3 + 4x^2} \quad + \quad \boxed{3x + 12}.$$

Common factor is x^2.　　Common factor is 3.

We now factor the given polynomial as follows.

$$x^3 + 4x^2 + 3x + 12$$
$$= (x^3 + 4x^2) + (3x + 12) \qquad \text{Group terms with common factors.}$$
$$= x^2(x + 4) + 3(x + 4) \qquad \text{Factor out the greatest common factor from the grouped terms. The remaining two terms have } x + 4 \text{ as a common binomial factor.}$$
$$= (x + 4)(x^2 + 3) \qquad \text{Factor } (x + 4) \text{ out of both terms.}$$

Thus, $x^3 + 4x^2 + 3x + 12 = (x + 4)(x^2 + 3)$. Check the factorization by multiplying the right side of the equation using the FOIL method. If the factorization is correct, you will obtain the original polynomial.

Discovery

In Example 2, group the terms as follows:
$$(x^3 + 3x) + (4x^2 + 12).$$
Factor out the greatest common factor from each group and complete the factoring process. Describe what happens. What can you conclude?

Check Point 2   Factor: $x^3 + 5x^2 - 2x - 10$.

3 Factor trinomials.

Factoring Trinomials

To factor a trinomial of the form $ax^2 + bx + c$, a little trial and error may be necessary.

A Strategy for Factoring $ax^2 + bx + c$

(Assume, for the moment, that there is no greatest common factor.)

1. Find two **First** terms whose product is ax^2:
$$(\Box x + \quad)(\Box x + \quad) = ax^2 + bx + c.$$

2. Find two **Last** terms whose product is c:
$$(x + \Box)(x + \Box) = ax^2 + bx + c.$$

3. By trial and error, perform steps 1 and 2 until the sum of the **Outside** product and **Inside** product is bx:
$$(\Box x + \Box)(\Box x + \Box) = ax^2 + bx + c.$$
$$\text{I}$$
$$\text{O}$$
(sum of O + I)

If no such combinations exist, the polynomial is prime.

EXAMPLE 3 Factoring Trinomials Whose Leading Coefficients Are 1

Factor: **a.** $x^2 + 6x + 8$ **b.** $x^2 + 3x - 18$.

Solution

a. The factors of the first term are x and x :

$$x^2 + 6x + 8 = (x \qquad)(x \qquad).$$

To find the second term of each factor, we must find two numbers whose product is 8 and whose sum is 6. From the table in the margin, we see that 4 and 2 are the required integers. Thus,

$$x^2 + 6x + 8 = (x + 4)(x + 2) \text{ or } (x + 2)(x + 4).$$

Factors of 8	8, 1	4, 2	−8, −1	−4, −2
Sum of Factors	9	6	−9	−6

This is the desired sum.

b. We begin with

$$x^2 + 3x - 18 = (x \qquad)(x \qquad).$$

To find the second term of each factor, we must find two numbers whose product is −18 and whose sum is 3. From the table in the margin, we see that 6 and −3 are the required integers. Thus,

$$x^2 + 3x - 18 = (x + 6)(x - 3)$$
$$\text{or} \quad (x - 3)(x + 6).$$

Factors of −18	18, −1	−18, 1	9, −2	−9, 2	6, −3	−6, 3
Sum of factors	17	−17	7	−7	3	−3

This is the desired sum.

Check Point 3 Factor:

a. $x^2 + 13x + 40$ **b.** $x^2 - 5x - 14$.

EXAMPLE 4 Factoring a Trinomial Whose Leading Coefficient Is Not 1

Factor: $8x^2 - 10x - 3$.

Solution

Step 1 Find two *First* terms whose product is $8x^2$.

$$8x^2 - 10x - 3 \stackrel{?}{=} (8x \qquad)(x \qquad)$$
$$8x^2 - 10x - 3 \stackrel{?}{=} (4x \qquad)(2x \qquad)$$

Step 2 Find two *Last* terms whose product is −3.   The possible factorizations are $1(-3)$ and $-1(3)$.

Step 3 Try various combinations of these factors.   The correct factorization of $8x^2 - 10x - 3$ is the one in which the sum of the *Outside* and *Inside* products is equal to $-10x$. Here is a list of the possible factorizations:

Possible Factorizations of $8x^2 - 10x - 3$	Sum of *Outside* and *Inside* Products (Should Equal $-10x$)
$(8x + 1)(x - 3)$	$-24x + x = -23x$
$(8x - 3)(x + 1)$	$8x - 3x = 5x$
$(8x - 1)(x + 3)$	$24x - x = 23x$
$(8x + 3)(x - 1)$	$-8x + 3x = -5x$
$(4x + 1)(2x - 3)$	$-12x + 2x = -10x$
$(4x - 3)(2x + 1)$	$4x - 6x = -2x$
$(4x - 1)(2x + 3)$	$12x - 2x = 10x$
$(4x + 3)(2x - 1)$	$-4x + 6x = 2x$

This is the required middle term.

Thus,
$$8x^2 - 10x - 3 = (4x + 1)(2x - 3) \quad \text{or} \quad (2x - 3)(4x + 1).$$

Show that this factorization is correct by multiplying the factors using the FOIL method. You should obtain the original trinomial.

Check Point 4 Factor: $6x^2 + 19x - 7$.

4 Factor the difference of squares.

Factoring the Difference of Two Squares

A method for factoring the difference of two squares is obtained by reversing the special product for the sum and difference of two terms.

The Difference of Two Squares

If A and B are real numbers, variables, or algebraic expressions, then
$$A^2 - B^2 = (A + B)(A - B).$$

In words: The difference of the squares of two terms factors as the product of a sum and a difference of those terms.

EXAMPLE 5 Factoring the Difference of Two Squares

Factor: **a.** $x^2 - 4$ **b.** $81x^2 - 49$.

Solution We must express each term as the square of some monomial. Then we use the formula for factoring $A^2 - B^2$.

a. $x^2 - 4 = x^2 - 2^2 = (x + 2)(x - 2)$

$$A^2 - B^2 = (A + B)(A - B)$$

b. $81x^2 - 49 = (9x)^2 - 7^2 = (9x + 7)(9x - 7)$

Check Point 5 Factor:

a. $x^2 - 81$ **b.** $36x^2 - 25$.

We have seen that a polynomial is factored completely when it is written as the product of prime polynomials. To be sure that you have factored completely, check to see whether the factors can be factored.

Study Tip

Factoring $x^4 - 81$ as
$$(x^2 + 9)(x^2 - 9)$$
is not a complete factorization. The second factor, $x^2 - 9$, is itself a difference of two squares and can be factored.

EXAMPLE 6 A Repeated Factorization

Factor completely: $x^4 - 81$.

Solution
$$x^4 - 81 = (x^2)^2 - 9^2 \qquad \text{Express as the difference of two squares.}$$
$$= (x^2 + 9)(x^2 - 9) \qquad \text{The factors are the sum and difference of the squared terms.}$$

$$= (x^2 + 9)(x^2 - 3^2) \quad \text{The factor } x^2 - 9 \text{ is the difference of two squares and can be factored.}$$

$$= (x^2 + 9)(x + 3)(x - 3) \quad \text{The factors of } x^2 - 9 \text{ are the sum and difference of the squared terms.}$$

Check Point 6   Factor completely: $81x^4 - 16$.

5 Factor perfect square trinomials.

Factoring Perfect Square Trinomials

Our next factoring technique is obtained by reversing the special products for squaring binomials. The trinomials that are factored using this technique are called **perfect square trinomials.**

> **Factoring Perfect Square Trinomials**
>
> Let A and B be real numbers, variables, or algebraic expressions.
>
> **1.** $A^2 + 2AB + B^2 = (A + B)^2$
>
> Same sign
>
> **2.** $A^2 - 2AB + B^2 = (A - B)^2$
>
> Same sign

The two items in the box show that perfect square trinomials come in two forms: one in which the middle term is positive and one in which the middle term is negative. Here's how to recognize a perfect square trinomial:

1. The first and last terms are squares of monomials or integers.
2. The middle term is twice the product of the expressions being squared in the first and last terms.

EXAMPLE 7   Factoring Perfect Square Trinomials

Factor: **a.** $x^2 + 6x + 9$ **b.** $25x^2 - 60x + 36$.

Solution

a. $x^2 + 6x + 9 = x^2 + 2 \cdot x \cdot 3 + 3^2 = (x + 3)^2$ The middle term has a positive sign.

$$A^2 + 2AB + B^2 = (A + B)^2$$

b. We suspect that $25x^2 - 60x + 36$ is a perfect square trinomial because $25x^2 = (5x)^2$ and $36 = 6^2$. The middle term can be expressed as twice the product of $5x$ and 6.

$$25x^2 - 60x + 36 = (5x)^2 - 2 \cdot 5x \cdot 6 + 6^2 = (5x - 6)^2$$

$$A^2 - 2AB + B^2 = (A - B)^2$$

Check Point 7 Factor:

      a. $x^2 + 14x + 49$ **b.** $16x^2 - 56x + 49$.

6 Factor the sum and difference of cubes.

Factoring the Sum and Difference of Two Cubes

We can use the following formulas to factor the sum or the difference of two cubes:

> **Factoring the Sum and Difference of Two Cubes**
>
> **1.** Factoring the Sum of Two Cubes
> $$A^3 + B^3 = (A + B)(A^2 - AB + B^2)$$
>
> **2.** Factoring the Difference of Two Cubes
> $$A^3 - B^3 = (A - B)(A^2 + AB + B^2)$$

EXAMPLE 8   Factoring Sums and Differences of Two Cubes

Factor: **a.** $x^3 + 8$ **b.** $64x^3 - 125$.

Solution

  a. $x^3 + 8 = x^3 + 2^3 = (x + 2)(x^2 - x \cdot 2 + 2^2) = (x + 2)(x^2 - 2x + 4)$

$$A^3 + B^3 = (A + B)(A^2 - AB + B^2)$$

  b. $64x^3 - 125 = (4x)^3 - 5^3 = (4x - 5)\left[(4x)^2 + (4x)(5) + 5^2\right]$

$$A^3 - B^3 = (A - B)(A^2 + AB + B^2)$$

$$= (4x - 5)(16x^2 + 20x + 25)$$

Check Point 8 Factor:

      a. $x^3 + 1$ **b.** $125x^3 - 8$.

7 Use a general strategy for factoring polynomials.

A Strategy for Factoring Polynomials

It is important to practice factoring a wide variety of polynomials so that you can quickly select the appropriate technique. The polynomial is factored completely when all its polynomial factors, except possibly for monomial factors, are prime. Because of the commutative property, the order of the factors does not matter.

A Strategy for Factoring a Polynomial

1. If there is a common factor, factor out the GCF.

2. Determine the number of terms in the polynomial and try factoring as follows:

    a. If there are two terms, can the binomial be factored by one of the following special forms?

        Difference of two squares: $A^2 - B^2 = (A + B)(A - B)$
        Sum of two cubes: $A^3 + B^3 = (A + B)(A^2 - AB + B^2)$
        Difference of two cubes: $A^3 - B^3 = (A - B)(A^2 + AB + B^2)$

    b. If there are three terms, is the trinomial a perfect square trinomial? If so, factor by one of the following special forms:

$$A^2 + 2AB + B^2 = (A + B)^2$$
$$A^2 - 2AB + B^2 = (A - B)^2.$$

    If the trinomial is not a perfect square trinomial, try factoring by trial and error.

    c. If there are four or more terms, try factoring by grouping.

3. Check to see if any factors with more than one term in the factored polynomial can be factored further. If so, factor completely.

EXAMPLE 9   Factoring a Polynomial

Factor: $2x^3 + 8x^2 + 8x$.

Solution

Step 1   **If there is a common factor, factor out the GCF.** Because $2x$ is common to all terms, we factor it out.

$$2x^3 + 8x^2 + 8x = 2x(x^2 + 4x + 4) \qquad \textit{Factor out the GCF.}$$

Step 2   **Determine the number of terms and factor accordingly.** The factor $x^2 + 4x + 4$ has three terms and is a perfect square trinomial. We factor using $A^2 + 2AB + B^2 = (A + B)^2$.

$$2x^3 + 8x^2 + 8x = 2x(x^2 + 4x + 4)$$

$$= 2x(x^2 + 2 \cdot x \cdot 2 + 2^2)$$

$$\overbrace{}^{A^2 \ + \ 2AB \ + \ B^2}$$

$$= 2x(x + 2)^2 \qquad\qquad A^2 + 2AB + B^2 = (A + B)^2$$

Step 3   **Check to see if factors can be factored further.** In this problem, they cannot. Thus,

$$2x^3 + 8x^2 + 8x = 2x(x + 2)^2.$$

Check Point 9   Factor: $3x^3 - 30x^2 + 75x$.

EXAMPLE 10   Factoring a Polynomial

Factor: $x^2 - 25a^2 + 8x + 16$.

Solution

Step 1   If there is a common factor, factor out the GCF. Other than 1 or –1, there is no common factor.

Step 2   Determine the number of terms and factor accordingly. There are four terms. We try factoring by grouping. Grouping into two groups of two terms does not result in a common binomial factor. Let's try grouping as a difference of squares.

$$x^2 - 25a^2 + 8x + 16$$
$$= (x^2 + 8x + 16) - 25a^2$$ Rearrange terms and group as a perfect square trinomial minus $25a^2$ to obtain a difference of squares.

$$= (x + 4)^2 - (5a)^2$$ Factor the perfect square trinomial.
$$= (x + 4 + 5a)(x + 4 - 5a)$$ Factor the difference of squares. The factors are the sum and difference of the expressions being squared.

Step 3   Check to see if factors can be factored further. In this case, they cannot, so we have factored completely.

Check Point 10   Factor: $x^2 - 36a^2 + 20x + 100$.

---

8 Factor algebraic expressions containing fractional and negative exponents.

Factoring Algebraic Expressions Containing Fractional and Negative Exponents

Although expressions containing fractional and negative exponents are not polynomials, they can be simplified using factoring techniques.

EXAMPLE 11   Factoring Involving Fractional and Negative Exponents

Factor and simplify: $x(x + 1)^{-3/4} + (x + 1)^{1/4}$.

Solution   The greatest common factor is $x + 1$ with the *smallest exponent* in the two terms. Thus, the greatest common factor is $(x + 1)^{-3/4}$.

$$x(x + 1)^{-3/4} + (x + 1)^{1/4}$$
$$= (x + 1)^{-3/4}x + (x + 1)^{-3/4}(x + 1)$$ Express each term as the product of the greatest common factor and its other factor.
$$= (x + 1)^{-3/4}[x + (x + 1)]$$ Factor out the greatest common factor.

$$= \frac{2x + 1}{(x + 1)^{3/4}}$$ $b^{-n} = \dfrac{1}{b^n}$

Check
Point
11
Factor and simplify: $x(x - 1)^{-1/2} + (x - 1)^{1/2}$.

EXERCISE SET P.5

Practice Exercises

In Exercises 1–10, factor out the greatest common factor.

1. $18x + 27$ **2.** $16x - 24$

3. $3x^2 + 6x$ **4.** $4x^2 - 8x$

5. $9x^4 - 18x^3 + 27x^2$ **6.** $6x^4 - 18x^3 + 12x^2$

7. $x(x + 5) + 3(x + 5)$ **8.** $x(2x + 1) + 4(2x + 1)$

9. $x^2(x - 3) + 12(x - 3)$ **10.** $x^2(2x + 5) + 17(2x + 5)$

In Exercises 11–16, factor by grouping.

11. $x^3 - 2x^2 + 5x - 10$ **12.** $x^3 - 3x^2 + 4x - 12$

13. $x^3 - x^2 + 2x - 2$ **14.** $x^3 + 6x^2 - 2x - 12$

15. $3x^3 - 2x^2 - 6x + 4$ **16.** $x^3 - x^2 - 5x + 5$

In Exercises 17–30, factor each trinomial, or state that the trinomial is prime.

17. $x^2 + 5x + 6$ **18.** $x^2 + 8x + 15$

19. $x^2 - 2x - 15$ **20.** $x^2 - 4x - 5$

21. $x^2 - 8x + 15$ **22.** $x^2 - 14x + 45$

23. $3x^2 - x - 2$ **24.** $2x^2 + 5x - 3$

25. $3x^2 - 25x - 28$ **26.** $3x^2 - 2x - 5$

27. $6x^2 - 11x + 4$ **28.** $6x^2 - 17x + 12$

29. $4x^2 + 16x + 15$ **30.** $8x^2 + 33x + 4$

In Exercises 31–40, factor the difference of two squares.

31. $x^2 - 100$ **32.** $x^2 - 144$

33. $36x^2 - 49$ **34.** $64x^2 - 81$

35. $9x^2 - 25y^2$ **36.** $36x^2 - 49y^2$

37. $x^4 - 16$ **38.** $x^4 - 1$

39. $16x^4 - 81$ **40.** $81x^4 - 1$

In Exercises 41–48, factor any perfect square trinomials, or state that the polynomial is prime.

41. $x^2 + 2x + 1$ **42.** $x^2 + 4x + 4$

43. $x^2 - 14x + 49$ **44.** $x^2 - 10x + 25$

45. $4x^2 + 4x + 1$ **46.** $25x^2 + 10x + 1$

47. $9x^2 - 6x + 1$ **48.** $64x^2 - 16x + 1$

In Exercises 49–56, factor using the formula for the sum or difference of two cubes.

49. $x^3 + 27$ **50.** $x^3 + 64$

51. $x^3 - 64$ **52.** $x^3 - 27$

53. $8x^3 - 1$ **54.** $27x^3 - 1$

55. $64x^3 + 27$ **56.** $8x^3 + 125$

In Exercises 57–84, factor completely, or state that the polynomial is prime.

57. $3x^3 - 3x$ **58.** $5x^3 - 45x$

59. $4x^2 - 4x - 24$ **60.** $6x^2 - 18x - 60$

61. $2x^4 - 162$ **62.** $7x^4 - 7$

63. $x^3 + 2x^2 - 9x - 18$ **64.** $x^3 + 3x^2 - 25x - 75$

65. $2x^2 - 2x - 112$ **66.** $6x^2 - 6x - 12$

67. $x^3 - 4x$ **68.** $9x^3 - 9x$

69. $x^2 + 64$ **70.** $x^2 + 36$

71. $x^3 + 2x^2 - 4x - 8$ **72.** $x^3 + 2x^2 - x - 2$

73. $y^5 - 81y$ **74.** $y^5 - 16y$

75. $20y^4 - 45y^2$ **76.** $48y^4 - 3y^2$

77. $x^2 - 12x + 36 - 49y^2$ **78.** $x^2 - 10x + 25 - 36y^2$

79. $9b^2x - 16y - 16x + 9b^2y$

80. $16a^2x - 25y - 25x + 16a^2y$

81. $x^2y - 16y + 32 - 2x^2$ **82.** $12x^2y - 27y - 4x^2 + 9$

83. $2x^3 - 8a^2x + 24x^2 + 72x$

84. $2x^3 - 98a^2x + 28x^2 + 98x$

In Exercises 85–94, factor and simplify each algebraic expression.

85. $x^{3/2} - x^{1/2}$ **86.** $x^{3/4} - x^{1/4}$

87. $4x^{-2/3} + 8x^{1/3}$ **88.** $12x^{-3/4} + 6x^{1/4}$

89. $(x + 3)^{1/2} - (x + 3)^{3/2}$

90. $(x^2 + 4)^{3/2} + (x^2 + 4)^{7/2}$

91. $(x + 5)^{-1/2} - (x + 5)^{-3/2}$

92. $(x^2 + 3)^{-2/3} + (x^2 + 3)^{-5/3}$

93. $(4x - 1)^{1/2} - \frac{1}{3}(4x - 1)^{3/2}$

94. $-8(4x + 3)^{-2} + 10(5x + 1)(4x + 3)^{-1}$

Application Exercises

95. Your computer store is having an incredible sale. The price on one model is reduced by 40%. Then the sale price is reduced by another 40%. If x is the computer's original price, the sale price can be represented by

$$(x - 0.4x) - 0.4(x - 0.4x).$$

a. Factor out $(x - 0.4x)$ from each term. Then simplify the resulting expression.

b. Use the simplified expression from part (a) to answer these questions: With a 40% reduction followed by a 40% reduction, is the computer selling at 20% of its original price? If not, at what percentage of the original price is it selling?

96. The polynomial $8x^2 + 20x + 2488$ describes the number, in thousands, of high school graduates in the United States x years after 1993.

a. According to this polynomial, how many students will graduate from U.S. high schools in 2003?

b. Factor the polynomial.

c. Use the factored form of the polynomial in part (b) to find the number of high school graduates in 2003. Do you get the same answer as you did in part (a)? If so, does this prove that your factorization is correct? Explain.

97. A rock is dropped from the top of a 256-foot cliff. The height, in feet, of the rock above the water after t seconds is described by the polynomial $256 - 16t^2$. Factor this expression completely.

98. The amount of sheet metal needed to manufacture a cylindrical tin can, that is, its surface area, S, is $S = 2\pi r^2 + 2\pi rh$. Express the surface area, S, in factored form.

In Exercises 99–100, find the formula for the area of the shaded region and express it in factored form.

99.

100.

Writing in Mathematics

101. Using an example, explain how to factor out the greatest common factor of a polynomial.

102. Suppose that a polynomial contains four terms. Explain how to use factoring by grouping to factor the polynomial.

103. Explain how to factor $3x^2 + 10x + 8$.

104. Explain how to factor the difference of two squares. Provide an example with your explanation.

105. What is a perfect square trinomial and how is it factored?

106. Explain how to factor $x^3 + 1$.

107. What does it mean to factor completely?

Critical Thinking Exercises

108. Which one of the following is true?

a. Because $x^2 + 1$ is irreducible over the integers, it follows that $x^3 + 1$ is also irreducible.

b. One correct factored form for $x^2 - 4x + 3$ is $x(x - 4) + 3$.

c. $x^3 - 64 = (x - 4)^3$

d. None of the above is true.

In Exercises 109–112, factor completely.

109. $x^{2n} + 6x^n + 8$ **110.** $-x^2 - 4x + 5$

111. $x^4 - y^4 - 2x^3y + 2xy^3$

112. $(x - 5)^{-1/2}(x + 5)^{-1/2} - (x + 5)^{1/2}(x - 5)^{-3/2}$

In Exercises 113–114, find all integers b so that the trinomial can be factored.

113. $x^2 + bx + 15$ **114.** $x^2 + 4x + b$

Group Exercise

115. Without looking at any factoring problems in the book, create five factoring problems. Make sure that some of your problems require at least two factoring techniques. Next, exchange problems with another person in your group. Work to factor your partner's problems. Evaluate the problems as you work: Are they too easy? Too difficult? Can the polynomials really be factored? Share your response with the person who wrote the problems. Finally, grade each other's work in factoring the polynomials. Each factoring problem is worth 20 points. You may award partial credit. If you take off points, explain why points are deducted and how you decided to take off a particular number of points for the error(s) that you found.

SECTION P.6 *Rational Expressions*

Objectives

1. Specify numbers that must be excluded from the domain of rational expressions.
2. Simplify rational expressions.
3. Multiply rational expressions.
4. Divide rational expressions.
5. Add and subtract rational expressions.
6. Simplify complex rational expressions.

How do we describe the costs of reducing environmental pollution? We often use algebraic expressions involving quotients of polynomials. For example, the algebraic expression

$$\frac{250x}{100 - x}$$

describes the cost, in millions of dollars, to remove x percent of the pollutants that are discharged into a river. Removing a modest percentage of pollutants, say 40%, is far less costly than removing a substantially greater percentage, such as 95%. We see this by evaluating the algebraic expression for $x = 40$ and $x = 95$.

Discovery

What happens if you try substituting 100 for x in

$$\frac{250x}{100 - x}?$$

What does this tell you about the cost of cleaning up all of the river's pollutants?

Evaluating $\dfrac{250x}{100 - x}$ for

$x = 40$:

Cost is $\dfrac{250(40)}{100 - 40} \approx 167.$

$x = 95$:

Cost is $\dfrac{250(95)}{100 - 95} = 4750.$

The cost increases from approximately $167 million to a possibly prohibitive $4750 million, or $4.75 billion. Costs spiral upward as the percentage of removed pollutants increases.

Many algebraic expressions that describe costs of environmental projects are examples of rational expressions. First we will define rational expressions. Then we will review how to perform operations with such expressions.

1 Specify numbers that must be excluded from the domain of rational expressions.

Rational Expressions

A **rational expression** is the quotient of two polynomials. Some examples are

$$\frac{x - 2}{4}, \quad \frac{4}{x - 2}, \quad \frac{x}{x^2 - 1}, \quad \text{and} \quad \frac{x^2 + 1}{x^2 + 2x - 3}.$$

The set of real numbers for which an algebraic expression is defined is the **domain** of the expression. Because rational expressions indicate division and division by zero is undefined, we must exclude numbers from a rational expression's domain that make the denominator zero.

EXAMPLE 1 Excluding Numbers from the Domain

Find all the numbers that must be excluded from the domain of each rational expression:

a. $\dfrac{4}{x-2}$ **b.** $\dfrac{x}{x^2-1}$.

Solution To determine the numbers that must be excluded from each domain, examine the denominators.

a. $\dfrac{4}{x-2}$ **b.** $\dfrac{x}{x^2-1} = \dfrac{x}{(x+1)(x-1)}$

| This denominator would equal zero if x = 2. | This factor would equal zero if x = −1. | This factor would equal zero if x = 1. |

For the rational expression in part (a), we must exclude 2 from the domain. For the rational expression in part (b), we must exclude both −1 and 1 from the domain. These excluded numbers are often written to the right of a rational expression.

$$\dfrac{4}{x-2}, x \neq 2 \qquad \dfrac{x}{x^2-1}, x \neq -1, x \neq 1$$

Check Point 1 Find all the numbers that must be excluded from the domain of each rational expression:

a. $\dfrac{7}{x+5}$ **b.** $\dfrac{x}{x^2-36}$.

2 Simplify rational expressions.

Simplifying Rational Expressions

A rational expression is **simplified** if its numerator and denominator have no common factors other than 1 or −1. The following procedure can be used to simplify rational expressions:

Simplifying Rational Expressions
1. Factor the numerator and denominator completely.
2. Divide both the numerator and denominator by the common factors.

EXAMPLE 2 Simplifying Rational Expressions

Simplify: **a.** $\dfrac{x^3+x^2}{x+1}$ **b.** $\dfrac{x^2+6x+5}{x^2-25}$.

Solution

a. $\dfrac{x^3 + x^2}{x + 1} = \dfrac{x^2(x + 1)}{x + 1}$ Factor the numerator. Because the denominator is $x + 1$, $x \neq -1$.

$= \dfrac{x^2 \cancel{(x + 1)}^{1}}{\cancel{x + 1}_{1}}$ Divide out the common factor, $x + 1$.

$= x^2, x \neq -1$ Denominators of 1 need not be written because $\frac{a}{1} = a$.

b. $\dfrac{x^2 + 6x + 5}{x^2 - 25} = \dfrac{(x + 5)(x + 1)}{(x + 5)(x - 5)}$ Factor the numerator and denominator. Because the denominator is $(x + 5)(x - 5)$, $x \neq -5$ and $x \neq 5$.

$= \dfrac{\cancel{(x + 5)}^{1}(x + 1)}{\cancel{(x + 5)}_{1}(x - 5)}$ Divide out the common factor, $x + 5$.

$= \dfrac{x + 1}{x - 5}, \quad x \neq -5, \quad x \neq 5$

Check Point 2 Simplify:

a. $\dfrac{x^3 + 3x^2}{x + 3}$ **b.** $\dfrac{x^2 - 1}{x^2 + 2x + 1}$.

3 Multiply rational expressions.

Multiplying Rational Expressions

The product of two rational expressions is the product of their numerators divided by the product of their denominators. Here is a step-by-step procedure for multiplying rational expressions:

Multiplying Rational Expressions

1. Factor all numerators and denominators completely.
2. Divide numerators and denominators by common factors.
3. Multiply the remaining factors in the numerator and multiply the remaining factors in the denominator.

EXAMPLE 3   Multiplying Rational Expressions

Multiply and simplify:

$$\dfrac{x - 7}{x - 1} \cdot \dfrac{x^2 - 1}{3x - 21}.$$

Solution

$$\frac{x-7}{x-1} \cdot \frac{x^2-1}{3x-21}$$

This is the given multiplication problem.

$$= \frac{x-7}{x-1} \cdot \frac{(x+1)(x-1)}{3(x-7)}$$

Factor all numerators and denominators. Because the denominator has factors of $x-1$ and $x-7$, $x \neq 1$ and $x \neq 7$.

$$= \frac{\overset{1}{\cancel{x-7}}}{\underset{1}{\cancel{x-1}}} \cdot \frac{(x+1)\overset{1}{\cancel{(x-1)}}}{3\underset{1}{\cancel{(x-7)}}}$$

Divide numerators and denominators by common factors.

$$= \frac{x+1}{3}, x \neq 1, x \neq 7$$

Multiply the remaining factors in the numerator and denominator.

These excluded numbers from the domain must also be excluded from the simplified expression's domain.

Check Point 3 Multiply and simplify:

$$\frac{x+3}{x^2-4} \cdot \frac{x^2-x-6}{x^2+6x+9}.$$

4 Divide rational expressions.

Dividing Rational Expressions

We find the quotient of two rational expressions by inverting the divisor and multiplying.

EXAMPLE 4 Dividing Rational Expressions

Divide and simplify:

$$\frac{x^2-2x-8}{x^2-9} \div \frac{x-4}{x+3}.$$

Solution

$$\frac{x^2-2x-8}{x^2-9} \div \frac{x-4}{x+3}$$

This is the given division problem.

$$= \frac{x^2-2x-8}{x^2-9} \cdot \frac{x+3}{x-4}$$

Invert the divisor and multiply.

$$= \frac{(x-4)(x+2)}{(x+3)(x-3)} \cdot \frac{x+3}{x-4}$$

Factor throughout. For nonzero denominators, $x \neq -3$, $x \neq 3$, and $x \neq 4$.

$$= \frac{\overset{1}{\cancel{(x-4)}}(x+2)}{\underset{1}{\cancel{(x+3)}}(x-3)} \cdot \frac{\overset{1}{\cancel{(x+3)}}}{\underset{1}{\cancel{(x-4)}}}$$

Divide numerators and denominators by common factors.

$$= \frac{x+2}{x-3}, x \neq -3, x \neq 3, x \neq 4$$

Multiply the remaining factors in the numerator and the denominator.

Check Point 4 Divide and simplify:

$$\frac{x^2-2x+1}{x^3+x} \div \frac{x^2+x-2}{3x^2+3}.$$

5 Add and subtract rational expressions.

Adding and Subtracting Rational Expressions with the Same Denominator

We add or subtract rational expressions with the same denominator by (1) adding or subtracting the numerators, (2) placing this result over the common denominator, and (3) simplifying, if possible.

EXAMPLE 5   Subtracting Rational Expressions with the Same Denominator

Subtract: $\dfrac{5x + 1}{x^2 - 9} - \dfrac{4x - 2}{x^2 - 9}$.

Solution

$$\frac{5x + 1}{x^2 - 9} - \frac{4x - 2}{x^2 - 9} = \frac{5x + 1 - (4x - 2)}{x^2 - 9}$$

Subtract numerators and include parentheses to indicate that both terms are subtracted. Place this difference over the common denominator.

$$= \frac{5x + 1 - 4x + 2}{x^2 - 9}$$

Remove parentheses and then change the sign of each term.

$$= \frac{x + 3}{x^2 - 9}$$

Combine like terms.

$$= \frac{\overset{1}{\cancel{x + 3}}}{\underset{1}{\cancel{(x + 3)}}(x - 3)}$$

Factor and simplify ($x \neq -3$ and $x \neq 3$).

$$= \frac{1}{x - 3}, \, x \neq -3, x \neq 3$$

> **Study Tip**
>
> Example 5 shows that when a numerator is being subtracted, we must subtract every term in that expression.

Check Point 5   Subtract: $\dfrac{x}{x + 1} - \dfrac{3x + 2}{x + 1}$.

Adding and Subtracting Rational Expressions with Different Denominators

Rational expressions that have no common factors in their denominators can be added or subtracted using one of the following properties:

$$\frac{a}{b} + \frac{c}{d} = \frac{ad + bc}{bd} \qquad \frac{a}{b} - \frac{c}{d} = \frac{ad - bc}{bd}, \, b \neq 0, d \neq 0.$$

The denominator, bd, is the product of the factors in the two denominators. Because we are looking at rational expressions that have no common factors in their denominators, the product bd gives the least common denominator.

EXAMPLE 6   Subtracting Rational Expressions Having No Common Factors in Their Denominators

Subtract: $\dfrac{x + 2}{2x - 3} - \dfrac{4}{x + 3}$.

Solution   We need to find the least common denominator. This is the product of the distinct factors in each denominator, namely $(2x - 3)(x + 3)$. We can therefore use the subtraction property given previously as follows:

$$\frac{a}{b} - \frac{c}{d} = \frac{ad - bc}{bd}$$

$$\frac{x + 2}{2x - 3} - \frac{4}{x + 3} = \frac{(x + 2)(x + 3) - (2x - 3)4}{(2x - 3)(x + 3)}$$

Observe that
$a = x + 2, b = 2x - 3,$
$c = 4,$ and $d = x + 3.$

$$= \frac{x^2 + 5x + 6 - (8x - 12)}{(2x - 3)(x + 3)}$$

Multiply.

$$= \frac{x^2 + 5x + 6 - 8x + 12}{(2x - 3)(x + 3)}$$

Remove parentheses and then change the sign of each term.

$$= \frac{x^2 - 3x + 18}{(2x - 3)(x + 3)}, x \neq \frac{3}{2}, x \neq -3$$

Combine like terms in the numerator.

Check Point 6 Add: $\dfrac{3}{x + 1} + \dfrac{5}{x - 1}$.

The **least common denominator,** or LCD, of several rational expressions is a polynomial consisting of the product of all prime factors in the denominators, with each factor raised to the greatest power of its occurrence in any denominator. When adding and subtracting rational expressions that have different denominators with one or more common factors in the denominators, it is efficient to find the least common denominator first.

Finding the Least Common Denominator

1. Factor each denominator completely.
2. List the factors of the first denominator.
3. Add to the list in step 2 any factors of the second denominator that do not appear in the list.
4. Form the product of each different factor from the list in step 3. This product is the least common denominator.

EXAMPLE 7   Finding the Least Common Denominator

Find the least common denominator of

$$\frac{7}{5x^2 + 15x} \quad \text{and} \quad \frac{9}{x^2 + 6x + 9}.$$

Solution

Step 1   Factor each denominator completely.

$$5x^2 + 15x = 5x(x + 3)$$
$$x^2 + 6x + 9 = (x + 3)^2$$

Step 2   List the factors of the first denominator.

$$5, x, (x + 3)$$

Step 3   Add any unlisted factors from the second denominator.   The second denominator is $(x + 3)^2$ or $(x + 3)(x + 3)$. One factor of $x + 3$ is already in our list, but the other factor is not. We add $x + 3$ to the list. We have

$$5, x, (x + 3), (x + 3).$$

Step 4   The least common denominator is the product of all factors in the final list.   Thus,

$$5x(x + 3)(x + 3), \quad \text{or} \quad 5x(x + 3)^2$$

is the least common denominator.

Check
Point
7   Find the least common denominator of

$$\frac{3}{x^2 - 6x + 9} \quad \text{and} \quad \frac{7}{x^2 - 9}.$$

Finding the least common denominator for two (or more) rational expressions is the first step needed to add or subtract the expressions.

> **Adding and Subtracting Rational Expressions That Have Different Denominators with Shared Factors**
>
> **1.** Find the least common denominator.
>
> **2.** Write all rational expressions in terms of the least common denominator. To do so, multiply both the numerator and the denominator of each rational expression by any factor(s) needed to convert the denominator into the least common denominator.
>
> **3.** Add or subtract the numerators, placing the resulting expression over the least common denominator.
>
> **4.** If necessary, simplify the resulting rational expression.

EXAMPLE 8   Adding Rational Expressions with Different Denominators

Add: $\dfrac{x + 3}{x^2 + x - 2} + \dfrac{2}{x^2 - 1}.$

Solution

Step 1   Find the least common denominator.   Start by factoring the denominators.

$$x^2 + x - 2 = (x + 2)(x - 1)$$
$$x^2 - 1 = (x + 1)(x - 1)$$

The factors of the first denominator are $x + 2$ and $x - 1$. The only factor from the second denominator that is not listed is $x + 1$. Thus, the least common denominator is

$$(x + 2)(x - 1)(x + 1).$$

Step 2 Write all rational expressions in terms of the least common denominator. We do so by multiplying both the numerator and the denominator by any factor(s) needed to convert the denominator into the least common denominator.

$$\frac{x + 3}{x^2 + x - 2} + \frac{2}{x^2 - 1}$$

$$= \frac{x + 3}{(x + 2)(x - 1)} + \frac{2}{(x + 1)(x - 1)}$$

The least common denominator is $(x + 2)(x - 1)(x + 1)$.

$$= \frac{(x + 3)(x + 1)}{(x + 2)(x - 1)(x + 1)} + \frac{2(x + 2)}{(x + 2)(x - 1)(x + 1)}$$

Multiply each numerator and denominator by the extra factor required to form $(x + 2)(x - 1)(x + 1)$, the least common denominator.

Step 3 Add numerators, putting this sum over the least common denominator.

$$= \frac{(x + 3)(x + 1) + 2(x + 2)}{(x + 2)(x - 1)(x + 1)}$$

$$= \frac{x^2 + 4x + 3 + 2x + 4}{(x + 2)(x - 1)(x + 1)}$$

Perform the multiplications in the numerator.

$$= \frac{x^2 + 6x + 7}{(x + 2)(x - 1)(x + 1)}, x \neq -2, x \neq 1, x \neq -1$$

Combine like terms in the numerator.

Step 4 If necessary, simplify. Because the numerator is prime, no further simplification is possible.

Check Point 8 Subtract: $\dfrac{x}{x^2 - 10x + 25} - \dfrac{x - 4}{2x - 10}$.

6 Simplify complex rational expressions.

Complex Rational Expressions

Complex rational expressions, also called **complex fractions,** have numerators or denominators containing one or more rational expressions. Here are two examples of such expressions:

$$\frac{1 + \dfrac{1}{x}}{1 - \dfrac{1}{x}}$$

Separate rational expressions occur in the numerator and denominator.

$$\frac{\dfrac{1}{x + h} - \dfrac{1}{x}}{h}.$$

Separate rational expressions occur in the numerator.

One method for simplifying a complex rational expression is to combine its numerator into a single expression and combine its denominator into a single expression. Then perform the division by inverting the denominator and multiplying.

EXAMPLE 9 Simplifying a Complex Rational Expression

Simplify: $\dfrac{1 + \dfrac{1}{x}}{1 - \dfrac{1}{x}}$.

Solution

$\dfrac{1 + \dfrac{1}{x}}{1 - \dfrac{1}{x}} = \dfrac{\dfrac{x}{x} + \dfrac{1}{x}}{\dfrac{x}{x} - \dfrac{1}{x}}, x \neq 0$

The terms in the numerator and in the denominator are each combined by performing the addition and subtraction. The least common denominator is x.

$= \dfrac{\dfrac{x + 1}{x}}{\dfrac{x - 1}{x}}$

Perform the addition in the numerator and the subtraction in the denominator.

$= \dfrac{x + 1}{x} \div \dfrac{x - 1}{x}$

Rewrite the main fraction bar as ÷.

$= \dfrac{x + 1}{x} \cdot \dfrac{x}{x - 1}$

Invert the divisor and multiply (x ≠ 0 and x ≠ 1).

$= \dfrac{x + 1}{\overset{}{\underset{1}{\cancel{x}}}} \cdot \dfrac{\overset{1}{\cancel{x}}}{x - 1}$

Divide a numerator and denominator by the common factor, x.

$= \dfrac{x + 1}{x - 1}, x \neq 0, x \neq 1$

Multiply the remaining factors in the numerator and in the denominator.

Check Point 9 Simplify: $\dfrac{\dfrac{1}{x} - \dfrac{3}{2}}{\dfrac{1}{x} + \dfrac{3}{4}}$.

A second method for simplifying a complex rational expression is to find the least common denominator of all the rational expressions in its numerator and denominator. Then multiply each term in its numerator and denominator by this least common denominator. Here we use this method to simplify the complex rational expression in Example 9.

$\dfrac{1 + \dfrac{1}{x}}{1 - \dfrac{1}{x}} = \dfrac{\left(1 + \dfrac{1}{x}\right)}{\left(1 - \dfrac{1}{x}\right)} \cdot \dfrac{x}{x}$

The least common denominator of all the rational expressions is x. Multiply the numerator and denominator by x. Because $\dfrac{x}{x} = 1$, we are not changing the complex fraction (x ≠ 0).

$= \dfrac{1 \cdot x + \dfrac{1}{x} \cdot x}{1 \cdot x - \dfrac{1}{x} \cdot x}$

Use the distributive property. Be sure to distribute x to every term.

$= \dfrac{x + 1}{x - 1}, x \neq 0, x \neq 1$

Multiply. The complex rational expression is now simplified.

EXERCISE SET P.6

 Practice Exercises

In Exercises 1–6, find all numbers that must be excluded from the domain of each rational expression.

1. $\dfrac{7}{x-3}$

2. $\dfrac{13}{x+9}$

3. $\dfrac{x+5}{x^2-25}$

4. $\dfrac{x+7}{x^2-49}$

5. $\dfrac{x-1}{x^2+11x+10}$

6. $\dfrac{x-3}{x^2+4x-45}$

In Exercises 7–14, simplify each rational expression. Find all numbers that must be excluded from the domain of the simplified rational expression.

7. $\dfrac{3x-9}{x^2-6x+9}$

8. $\dfrac{4x-8}{x^2-4x+4}$

9. $\dfrac{x^2-12x+36}{4x-24}$

10. $\dfrac{x^2-8x+16}{3x-12}$

11. $\dfrac{y^2+7y-18}{y^2-3y+2}$

12. $\dfrac{y^2-4y-5}{y^2+5y+4}$

13. $\dfrac{x^2+12x+36}{x^2-36}$

14. $\dfrac{x^2-14x+49}{x^2-49}$

In Exercises 15–32, multiply or divide as indicated.

15. $\dfrac{x-2}{3x+9}\cdot\dfrac{2x+6}{2x-4}$

16. $\dfrac{6x+9}{3x-15}\cdot\dfrac{x-5}{4x+6}$

17. $\dfrac{x^2-9}{x^2}\cdot\dfrac{x^2-3x}{x^2+x-12}$

18. $\dfrac{x^2-4}{x^2-4x+4}\cdot\dfrac{2x-4}{x+2}$

19. $\dfrac{x^2-5x+6}{x^2-2x-3}\cdot\dfrac{x^2-1}{x^2-4}$

20. $\dfrac{x^2+5x+6}{x^2+x-6}\cdot\dfrac{x^2-9}{x^2-x-6}$

21. $\dfrac{x^3-8}{x^2-4}\cdot\dfrac{x+2}{3x}$

22. $\dfrac{x^2+6x+9}{x^3+27}\cdot\dfrac{1}{x+3}$

23. $\dfrac{x+1}{3}\div\dfrac{3x+3}{7}$

24. $\dfrac{x+5}{7}\div\dfrac{4x+20}{9}$

25. $\dfrac{x^2-4}{x}\div\dfrac{x+2}{x-2}$

26. $\dfrac{x^2-4}{x-2}\div\dfrac{x+2}{4x-8}$

27. $\dfrac{4x^2+10}{x-3}\div\dfrac{6x^2+15}{x^2-9}$

28. $\dfrac{x^2+x}{x^2-4}\div\dfrac{x^2-1}{x^2+5x+6}$

29. $\dfrac{x^2-25}{2x-2}\div\dfrac{x^2+10x+25}{x^2+4x-5}$

30. $\dfrac{x^2-4}{x^2+3x-10}\div\dfrac{x^2+5x+6}{x^2+8x+15}$

31. $\dfrac{x^2+x-12}{x^2+x-30}\cdot\dfrac{x^2+5x+6}{x^2-2x-3}\div\dfrac{x+3}{x^2+7x+6}$

32. $\dfrac{x^3-25x}{4x^2}\cdot\dfrac{2x^2-2}{x^2-6x+5}\div\dfrac{x^2+5x}{7x+7}$

In Exercises 33–54, add or subtract as indicated.

33. $\dfrac{4x+1}{6x+5}+\dfrac{8x+9}{6x+5}$

34. $\dfrac{3x+2}{3x+4}+\dfrac{3x+6}{3x+4}$

35. $\dfrac{x^2-2x}{x^2+3x}+\dfrac{x^2+x}{x^2+3x}$

36. $\dfrac{x^2-4x}{x^2-x-6}+\dfrac{4x-4}{x^2-x-6}$

37. $\dfrac{4x-10}{x-2}-\dfrac{x-4}{x-2}$

38. $\dfrac{2x+3}{3x-6}-\dfrac{3-x}{3x-6}$

39. $\dfrac{x^2+3x}{x^2+x-12}-\dfrac{x^2-12}{x^2+x-12}$

40. $\dfrac{x^2-4x}{x^2-x-6}-\dfrac{x-6}{x^2-x-6}$

41. $\dfrac{3}{x+4}+\dfrac{6}{x+5}$

42. $\dfrac{8}{x-2}+\dfrac{2}{x-3}$

43. $\dfrac{3}{x+1}-\dfrac{3}{x}$

44. $\dfrac{4}{x}-\dfrac{3}{x+3}$

45. $\dfrac{2x}{x+2}+\dfrac{x+2}{x-2}$

46. $\dfrac{3x}{x-3}-\dfrac{x+4}{x+2}$

47. $\dfrac{x+5}{x-5}+\dfrac{x-5}{x+5}$

48. $\dfrac{x+3}{x-3}+\dfrac{x-3}{x+3}$

49. $\dfrac{4}{x^2+6x+9}+\dfrac{4}{x+3}$

50. $\dfrac{3}{5x+2}+\dfrac{5x}{25x^2-4}$

51. $\dfrac{3x}{x^2+3x-10}-\dfrac{2x}{x^2+x-6}$

52. $\dfrac{x}{x^2-2x-24}-\dfrac{x}{x^2-7x+6}$

53. $\dfrac{4x^2+x-6}{x^2+3x+2}-\dfrac{3x}{x+1}+\dfrac{5}{x+2}$

54. $\dfrac{6x^2+17x-40}{x^2+x-20}+\dfrac{3}{x-4}-\dfrac{5x}{x+5}$

In Exercise 55–64, simplify each complex rational expression.

55. $\dfrac{\dfrac{x}{3} - 1}{x - 3}$

56. $\dfrac{\dfrac{x}{4} - 1}{x - 4}$

57. $\dfrac{1 + \dfrac{1}{x}}{3 - \dfrac{1}{x}}$

58. $\dfrac{8 + \dfrac{1}{x}}{4 - \dfrac{1}{x}}$

59. $\dfrac{\dfrac{1}{x} + \dfrac{1}{y}}{x + y}$

60. $\dfrac{1 - \dfrac{1}{x}}{xy}$

61. $\dfrac{x - \dfrac{x}{x + 3}}{x + 2}$

62. $\dfrac{x - 3}{x - \dfrac{3}{x - 2}}$

63. $\dfrac{\dfrac{3}{x - 2} - \dfrac{4}{x + 2}}{\dfrac{7}{x^2 - 4}}$

64. $\dfrac{\dfrac{x}{x - 2} + 1}{\dfrac{3}{x^2 - 4} + 1}$

 Application Exercises

65. The rational expression

$$\frac{130x}{100 - x}$$

describes the cost, in millions of dollars, to inoculate x percent of the population against a particular strain of flu.

a. Evaluate the expression for $x = 40$, $x = 80$, and $x = 90$. Describe the meaning of each evaluation in terms of percentage inoculated and cost.

b. For what value of x is the expression undefined?

c. What happens to the cost as x approaches 100%? How can you interpret this observation?

66. Doctors use the rational expression

$$\frac{DA}{A + 12}$$

to determine the dosage of a drug prescribed for children. In this expression, A = child's age, and D = adult dosage. What is the difference in the child's dosage for a 7-year-old child and a 3-year-old child? Express the answer as a single rational expression in terms of D. Then describe what your answer means in terms of the variables in the rational expression.

67. Anthropologists and forensic scientists classify skulls using

$$\frac{L + 60W}{L} - \frac{L - 40W}{L}$$

where L is the skull's length and W is its width.

a. Express the classification as a single rational expression.

b. If the value of the rational expression in part (a) is less than 75, a skull is classified as long. A medium skull has a value between 75 and 80, and a round skull has a value over 80. Use your rational expression from part (a) to classify a skull that is 5 inches wide and 6 inches long.

68. The polynomial

$$6t^4 - 207t^3 + 2128t^2 - 6622t + 15,220$$

describes the annual number of drug convictions in the United States t years after 1984. The polynomial

$$28t^4 - 711t^3 + 5963t^2 - 1695t + 27,424$$

describes the annual number of drug arrests in the United States t years after 1984. Write a rational expression that describes the conviction rate for drug arrests in the United States t years after 1984.

69. The average speed on a round-trip commute having a one-way distance d is given by the complex rational expression

$$\frac{2d}{\dfrac{d}{r_1} + \dfrac{d}{r_2}}$$

in which r_1 and r_2 are the speeds on the outgoing and return trips, respectively. Simplify the expression. Then find the average speed for a person who drives from home to work at 30 miles per hour and returns on the same route averaging 20 miles per hour. Explain why the answer is not 25 miles per hour.

Writing in Mathematics

70. What is a rational expression?

71. Explain how to determine which numbers must be excluded from the domain of a rational expression.

72. Explain how to simplify a rational expression.

73. Explain how to multiply rational expressions.

74. Explain how to divide rational expressions.

75. Explain how to add or subtract rational expressions with the same denominators.

76. Explain how to add rational expressions having no common factors in their denominators. Use $\dfrac{3}{x+5} + \dfrac{7}{x+2}$ in your explanation.

77. Explain how to find the least common denominator for denominators of $x^2 - 100$ and $x^2 - 20x + 100$.

78. Describe two ways to simplify $\dfrac{\dfrac{3}{x} + \dfrac{2}{x^2}}{\dfrac{1}{x^2} + \dfrac{2}{x}}$.

Explain the error in Exercises 79–81. Then rewrite the right side of the equation to correct the error that now exists.

79. $\dfrac{1}{a} + \dfrac{1}{b} = \dfrac{1}{a+b}$ **80.** $\dfrac{1}{x} + 7 = \dfrac{1}{x+7}$

81. $\dfrac{a}{x} + \dfrac{a}{b} = \dfrac{a}{x+b}$

82. A politician claims that each year the conviction rate for drug arrests in the United States is increasing. Explain how to use the polynomials in Exercise 68 to verify this claim.

Technology Exercises

83. How much are your monthly payments on a loan? If P is the principal, or amount borrowed, i is the monthly interest rate, and n is the number of monthly payments, then the amount, A, of each monthly payment is

$$A = \frac{Pi}{1 - \dfrac{1}{(1+i)^n}}.$$

a. Simplify the complex rational expression for the amount of each payment.

b. You purchase a $20,000 automobile at 1% monthly interest to be paid over 48 months. How much do you pay each month? Use the simplified rational expression from part (a) and a calculator. Round to the nearest dollar.

Critical Thinking Exercises

84. Which one of the following is true?

a. $\dfrac{x^2 - 25}{x - 5} = x - 5$

b. $\dfrac{x}{y} \div \dfrac{y}{x} = 1$, if $x \neq 0$ and $y \neq 0$.

c. The least common denominator needed to find $\dfrac{1}{x} + \dfrac{1}{x+3}$ is $x + 3$.

d. The rational expression

$$\frac{x^2 - 16}{x - 4}$$

is not defined for $x = 4$. However, as x gets closer and closer to 4, the value of the expression approaches 8.

In Exercises 85–86, find the missing expression.

85. $\dfrac{3x}{x-5} + \dfrac{\boxed{}}{5-x} = \dfrac{7x+1}{x-5}$

86. $\dfrac{4}{x-2} - \boxed{} = \dfrac{2x+8}{(x-2)(x+1)}$

87. In one short sentence, five words or less, explain what

$$\frac{\dfrac{1}{x} + \dfrac{1}{x^2} + \dfrac{1}{x^3}}{\dfrac{1}{x^4} + \dfrac{1}{x^5} + \dfrac{1}{x^6}}$$

does to each number x.

CHAPTER SUMMARY, REVIEW, AND TEST

Summary: Basic Formulas

Definition of Absolute Value

$$|x| = \begin{cases} x & \text{if } x \geq 0 \\ -x & \text{if } x < 0 \end{cases}$$

Distance between Points a and b on a Number Line

$$|a - b| \quad \text{or} \quad |b - a|$$

Properties of Algebra

Commutative	$a + b = b + a, \quad ab = ba$
Associative	$(a + b) + c = a + (b + c)$
	$(ab)c = a(bc)$
Distributive	$a(b + c) = ab + ac$
Identity	$a + 0 = a \quad a \cdot 1 = a$
Inverse	$a + (-a) = 0 \quad a \cdot \dfrac{1}{a} = 1, a \neq 0$

Properties of Exponents

$$b^{-n} = \frac{1}{b^n}, \quad b^0 = 1, \quad b^m \cdot b^n = b^{m+n},$$

$$(b^m)^n = b^{mn}, \quad \frac{b^m}{b^n} = b^{m-n}, \quad (ab)^n = a^n b^n, \quad \left(\frac{a}{b}\right)^n = \frac{a^n}{b^n}$$

Product and Quotient Rules for nth Roots

$$\sqrt[n]{a} \cdot \sqrt[n]{b} = \sqrt[n]{ab} \qquad \frac{\sqrt[n]{a}}{\sqrt[n]{b}} = \sqrt[n]{\frac{a}{b}}$$

Rational Exponents

$$a^{1/n} = \sqrt[n]{a}, \quad a^{-1/n} = \frac{1}{a^{1/n}} = \frac{1}{\sqrt[n]{a}},$$

$$a^{m/n} = \left(\sqrt[n]{a}\right)^m = \sqrt[n]{a^m}, \quad a^{-m/n} = \frac{1}{a^{m/n}}$$

Special Products

$$(A + B)(A - B) = A^2 - B^2$$
$$(A + B)^2 = A^2 + 2AB + B^2$$
$$(A - B)^2 = A^2 - 2AB + B^2$$
$$(A + B)^3 = A^3 + 3A^2B + 3AB^2 + B^3$$
$$(A - B)^3 = A^3 - 3A^2B + 3AB^2 - B^3$$

Factoring Formulas

$$A^2 - B^2 = (A + B)(A - B)$$
$$A^2 + 2AB + B^2 = (A + B)^2$$
$$A^2 - 2AB + B^2 = (A - B)^2$$
$$A^3 + B^3 = (A + B)(A^2 - AB + B^2)$$
$$A^3 - B^3 = (A - B)(A^2 + AB + B^2)$$

Review Exercises

You can use these review exercises, like the review exercises at the end of each chapter, to test your understanding of the chapter's topics. However, you can also use these exercises as a prerequisite test to check your mastery of the fundamental algebra skills needed in this book.

P.1

1. Consider the set:

$$\left\{-17, -\tfrac{9}{13}, 0, 0.75, \sqrt{2}, \pi, \sqrt{81}\right\}.$$

List all numbers from the set that are **a.** natural numbers, **b.** whole numbers, **c.** integers, **d.** rational numbers, **e.** irrational numbers.

In Exercises 2–4, rewrite each expression without absolute value bars.

2. $|-103|$

3. $|\sqrt{2} - 1|$

4. $|3 - \sqrt{17}|$

5. Express the distance between the numbers -17 and 4 using absolute value. Then evaluate the absolute value.

In Exercises 6–7, evaluate each algebraic expression for the given value of the variable.

6. $\dfrac{5}{9}(F - 32)$; $F = 68$

7. $\dfrac{8(x + 5)}{3x + 8}$, $x = 2$

In Exercises 8–13, state the name of the property illustrated.

8. $3 + 17 = 17 + 3$

9. $(6 \cdot 3) \cdot 9 = 6 \cdot (3 \cdot 9)$

10. $\sqrt{3}(\sqrt{5} + \sqrt{3}) = \sqrt{15} + 3$

11. $(6 \cdot 9) \cdot 2 = 2 \cdot (6 \cdot 9)$

12. $\sqrt{3}(\sqrt{5} + \sqrt{3}) = (\sqrt{5} + \sqrt{3})\sqrt{3}$

13. $(3 \cdot 7) + (4 \cdot 7) = (4 \cdot 7) + (3 \cdot 7)$

In Exercises 14–15, simplify each algebraic expression.

14. $3(7x - 5y) - 2(4y - x + 1)$

15. $\frac{1}{5}(5x) + [(3y) + (-3y)] - (-x)$

P.2

Evaluate each exponential expression in Exercises 16–19.

16. $(-3)^3(-2)^2$

17. $2^{-4} + 4^{-1}$

18. $5^{-3} \cdot 5$

19. $\dfrac{3^3}{3^6}$

Simplify each exponential expression in Exercises 20–23.

20. $(-2x^4y^3)^3$

21. $(-5x^3y^2)(-2x^{-11}y^{-2})$

22. $(2x^3)^{-4}$

23. $\dfrac{7x^5y^6}{28x^{15}y^{-2}}$

In Exercises 24–25, write each number in decimal notation.

24. 3.74×10^4

25. 7.45×10^{-5}

In Exercises 26–27, write each number in scientific notation.

26. 3,590,000

27. 0.00725

In Exercises 28–29, perform the indicated operation and write the answer in decimal notation.

28. $(3 \times 10^3)(1.3 \times 10^2)$

29. $\dfrac{6.9 \times 10^3}{3 \times 10^5}$

30. If you earned $1 million per year ($10^6$), how long would it take to accumulate $1 billion ($10^9$)?

31. If the population of the United States is 2.8×10^8 and each person spends about $150 per year going to the movies (or renting movies), express the total annual spending on movies in scientific notation.

P.3

Use the product rule to simplify the expressions in Exercises 32–35. In Exercises 34–35, assume that variables represent nonnegative real numbers.

32. $\sqrt{300}$

33. $\sqrt{12x^2}$

34. $\sqrt{10x} \cdot \sqrt{2x}$

35. $\sqrt{r^3}$

Use the quotient rule to simplify the expressions in Exercises 36–37.

36. $\sqrt{\dfrac{121}{4}}$

37. $\dfrac{\sqrt{96x^3}}{\sqrt{2x}}$ (Assume that $x > 0$.)

In Exercises 38–40, add or subtract terms whenever possible.

38. $7\sqrt{5} + 13\sqrt{5}$

39. $2\sqrt{50} + 3\sqrt{8}$

40. $4\sqrt{72} - 2\sqrt{48}$

In Exercises 41–44, rationalize the denominator.

41. $\dfrac{30}{\sqrt{5}}$

42. $\dfrac{\sqrt{2}}{\sqrt{3}}$

43. $\dfrac{5}{6 + \sqrt{3}}$

44. $\dfrac{14}{\sqrt{7} - \sqrt{5}}$

Evaluate each expression in Exercises 45–48 or indicate that the root is not a real number.

45. $\sqrt[3]{125}$

46. $\sqrt[5]{-32}$

47. $\sqrt[4]{-125}$

48. $\sqrt[4]{(-5)^4}$

Simplify the radical expressions in Exercises 49–53.

49. $\sqrt[3]{81}$

50. $\sqrt[3]{y^5}$

51. $\sqrt[4]{8} \cdot \sqrt[4]{10}$

52. $4\sqrt[3]{16} + 5\sqrt[3]{2}$

53. $\dfrac{\sqrt[4]{32x^5}}{\sqrt[4]{16x}}$ (Assume that $x > 0$.)

In Exercises 54–59, evaluate each expression.

54. $16^{1/2}$

55. $25^{-1/2}$

56. $125^{1/3}$

57. $27^{-1/3}$

58. $64^{2/3}$

59. $27^{-4/3}$

In Exercises 60–62, simplify using properties of exponents.

60. $(5x^{2/3})(4x^{1/4})$

61. $\dfrac{15x^{3/4}}{5x^{1/2}}$

62. $(125x^6)^{2/3}$

63. Simplify by reducing the index of the radical: $\sqrt[6]{y^3}$.

P.4

In Exercises 64–65, perform the indicated operations. Write the resulting polynomial in standard form and indicate its degree.

64. $(-6x^3 + 7x^2 - 9x + 3) + (14x^3 + 3x^2 - 11x - 7)$

65. $(13x^4 - 8x^3 + 2x^2) - (5x^4 - 3x^3 + 2x^2 - 6)$

In Exercises 66–72, find each product.

66. $(3x - 2)(4x^2 + 3x - 5)$ **67.** $(3x - 5)(2x + 1)$

68. $(4x + 5)(4x - 5)$ **69.** $(2x + 5)^2$

70. $(3x - 4)^2$ **71.** $(2x + 1)^3$

72. $(5x - 2)^3$

In Exercises 73–74, perform the indicated operations. Indicate the degree of the resulting polynomial.

73. $(7x^2 - 8xy + y^2) + (-8x^2 - 9xy - 4y^2)$

74. $(13x^3y^2 - 5x^2y - 9x^2) - (-11x^3y^2 - 6x^2y + 3x^2 - 4)$

In Exercises 75–79, find each product.

75. $(x + 7y)(3x - 5y)$ **76.** $(3x - 5y)^2$

77. $(3x^2 + 2y)^2$ **78.** $(7x + 4y)(7x - 4y)$

79. $(a - b)(a^2 + ab + b^2)$

P.5

In Exercises 80–96, factor completely, or state that the polynomial is prime.

80. $15x^3 + 3x^2$ **81.** $x^2 - 11x + 28$

82. $15x^2 - x - 2$ **83.** $64 - x^2$

84. $x^2 + 16$ **85.** $3x^4 - 9x^3 - 30x^2$

86. $20x^7 - 36x^3$ **87.** $x^3 - 3x^2 - 9x + 27$

88. $16x^2 - 40x + 25$ **89.** $x^4 - 16$

90. $y^3 - 8$ **91.** $x^3 + 64$

92. $3x^4 - 12x^2$ **93.** $27x^3 - 125$

94. $x^5 - x$ **95.** $x^3 + 5x^2 - 2x - 10$

96. $x^2 + 18x + 81 - y^2$

In Exercises 97–99, factor and simplify each algebraic expression.

97. $16x^{-3/4} + 32x^{1/4}$

98. $(x^2 - 4)(x^2 + 3)^{1/2} - (x^2 - 4)^2(x^2 + 3)^{3/2}$

99. $12x^{-1/2} + 6x^{-3/2}$

P.6

In Exercises 100–102, simplify each rational expression. Also, list all numbers that must be excluded from the domain.

100. $\dfrac{x^3 + 2x^2}{x + 2}$ **101.** $\dfrac{x^2 + 3x - 18}{x^2 - 36}$

102. $\dfrac{x^2 + 2x}{x^2 + 4x + 4}$

In Exercises 103–105, multiply or divide as indicated.

103. $\dfrac{x^2 + 6x + 9}{x^2 - 4} \cdot \dfrac{x + 3}{x - 2}$ **104.** $\dfrac{6x + 2}{x^2 - 1} \div \dfrac{3x^2 + x}{x - 1}$

105. $\dfrac{x^2 - 5x - 24}{x^2 - x - 12} \div \dfrac{x^2 - 10x + 16}{x^2 + x - 6}$

In Exercises 106–109, add or subtract as indicated.

106. $\dfrac{2x - 7}{x^2 - 9} - \dfrac{x - 10}{x^2 - 9}$ **107.** $\dfrac{3x}{x + 2} + \dfrac{x}{x - 2}$

108. $\dfrac{x}{x^2 - 9} + \dfrac{x - 1}{x^2 - 5x + 6}$

109. $\dfrac{4x - 1}{2x^2 + 5x - 3} - \dfrac{x + 3}{6x^2 + x - 2}$

In Exercises 110–112, simplify each complex rational expression.

110. $\dfrac{\dfrac{1}{x} - \dfrac{1}{2}}{\dfrac{1}{3} - \dfrac{x}{6}}$ **111.** $\dfrac{3 + \dfrac{12}{x}}{1 - \dfrac{16}{x^2}}$ **112.** $\dfrac{3 - \dfrac{1}{x + 3}}{3 + \dfrac{1}{x + 3}}$

Chapter P Test

1. List all the rational numbers in this set:
$$\left\{-7, -\tfrac{4}{5}, 0, 0.25, \sqrt{3}, \sqrt{4}, \tfrac{22}{7}, \pi\right\}.$$

In Exercises 2–3, state the name of the property illustrated.

2. $3(2 + 5) = 3(5 + 2)$ **3.** $6(7 + 4) = 6 \cdot 7 + 6 \cdot 4$

4. Express in scientific notation: 0.00076.

Simplify each expression in Exercises 5–11.

5. $9(10x - 2y) - 5(x - 4y + 3)$

6. $\dfrac{30x^3y^4}{6x^9y^{-4}}$

7. $\sqrt{6r}\sqrt{3r}$ (Assume that $r \geq 0$.)

8. $4\sqrt{50} - 3\sqrt{18}$ **9.** $\dfrac{3}{5 + \sqrt{2}}$

10. $\sqrt[3]{16x^4}$ **11.** $\dfrac{x^2 + 2x - 3}{x^2 - 3x + 2}$

12. Evaluate: $27^{-5/3}$.

In Exercises 13–14, find each product.

13. $(2x - 5)(x^2 - 4x + 3)$ **14.** $(5x + 3y)^2$

In Exercises 15–20, factor completely, or state that the polynomial is prime.

15. $x^2 - 9x + 18$ **16.** $x^3 + 2x^2 + 3x + 6$

17. $25x^2 - 9$ **18.** $36x^2 - 84x + 49$

19. $y^3 - 125$ **20.** $x^2 + 10x + 25 - 9y^2$

21. Factor and simplify:

$$x(x + 3)^{-3/5} + (x + 3)^{2/5}.$$

In Exercises 22–25, perform the operations and simplify, if possible.

22. $\dfrac{2x + 8}{x - 3} \div \dfrac{x^2 + 5x + 4}{x^2 - 9}$ **23.** $\dfrac{x}{x + 3} + \dfrac{5}{x - 3}$

24. $\dfrac{2x + 3}{x^2 - 7x + 12} - \dfrac{2}{x - 3}$ **25.** $\dfrac{\dfrac{1}{x} - \dfrac{1}{3}}{\dfrac{1}{x}}$

Equations, Inequalities, and Mathematical Models

Formulas like those that describe the height a child will attain as an adult are frequently obtained from actual data. Formulas can be used to explain what is happening in the present and to make predictions about what might occur in the future. Knowing how to create and use formulas will help you recognize patterns, logic, and order in a world that can appear chaotic to the untrained eye. In many ways, algebra will provide you with a new way of looking at your world.

Sitting in the biology department office, you overhear two of the professors discussing the possible adult heights of their respective children. Looking at the blackboard that they've been writing on, you see that there are formulas that can estimate the height a child will attain as an adult. If the child is x years old and h inches tall, that child's adult height, H, in inches, is approximated by one of the following formulas:

Girls: $$H = \frac{h}{0.00028x^3 - 0.0071x^2 + 0.0926x + 0.3524}$$

Boys: $$H = \frac{h}{0.00011x^3 - 0.0032x^2 + 0.0604x + 0.3796}.$$

SECTION 1.1 *Graphs and Graphing Utilities*

Objectives

1. Plot points in the rectangular coordinate system.
2. Graph equations in the rectangular coordinate system.
3. Interpret information about a graphing utility's viewing rectangle.
4. Use a graph to determine intercepts.
5. Interpret information given by graphs.

The beginning of the seventeenth century was a time of innovative ideas and enormous intellectual progress in Europe. English theatergoers enjoyed a succession of exciting new plays by Shakespeare. William Harvey proposed the radical notion that the heart was a pump for blood rather than the center of emotion. Galileo, with his new-fangled invention called the telescope, supported the theory of Polish astronomer Copernicus that the sun, not the Earth, was the center of the solar system. Monteverdi was writing the world's first grand operas. French mathematicians Pascal and Fermat invented a new field of mathematics called probability theory.

Into this arena of intellectual electricity stepped French aristocrat René Descartes (1596–1650). Descartes, propelled by the creativity surrounding him, developed a new branch of mathematics that brought together algebra and geometry in a unified way—a way that visualized numbers as points on a graph, equations as geometric figures, and geometric figures as equations. This new branch of mathematics, called *analytic geometry*, established Descartes as one of the founders of modern thought and among the most original mathematicians and philosophers of any age. We begin this section by looking at Descartes's deceptively simple idea, called the **rectangular coordinate system** or (in his honor) the **Cartesian coordinate system.**

1 Plot points in the rectangular coordinate system.

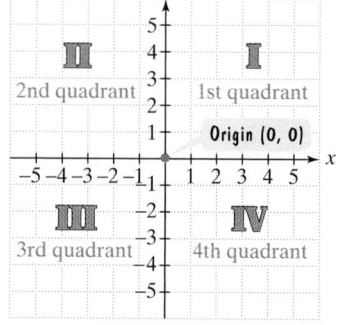

Figure 1.1 The rectangular coordinate system

Points and Ordered Pairs

Descartes used two number lines that intersect at right angles at their zero points, as shown in Figure 1.1. The horizontal number line is the **x-axis.** The vertical number line is the **y-axis.** The point of intersection of these axes is their zero points, called the **origin.** Positive numbers are shown to the right and above the origin. Negative numbers are shown to the left and below the origin. The axes divide the plane into four quarters, called **quadrants.** The points located on the axes are not in any quadrant.

Each point in the rectangular coordinate system corresponds to an **ordered pair** of real numbers, (x, y). Examples of such pairs are $(4, 2)$ and $(-5, -3)$. The first number in each pair, called the **x-coordinate,** denotes the distance and direction from the origin along the x-axis. The second number, called the **y-coordinate,** denotes vertical distance and direction along a line parallel to the y-axis or along the y-axis itself.

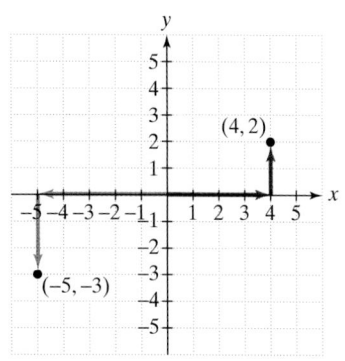

Figure 1.2 Plotting $(4, 2)$ and $(-5, -3)$

Figure 1.2 shows how we **plot,** or locate, the points corresponding to the ordered pairs $(4, 2)$ and $(-5, -3)$. We plot $(4, 2)$ by going 4 units from 0 to the right along the x-axis. Then we go 2 units up parallel to the y-axis. We plot $(-5, -3)$ by going 5 units from 0 to the left along the x-axis and 3 units down parallel to the y-axis. The phrase "the point corresponding to the ordered pair $(-5, -3)$" is often abbreviated as "the point $(-5, -3)$."

EXAMPLE 1 Plotting Points in the Rectangular Coordinate System

Plot the points: $A(-3, 5)$, $B(2, -4)$, $C(5, 0)$, $D(-5, -3)$, $E(0, 4)$, and $F(0, 0)$.

Solution See Figure 1.3. We move from the origin and plot the points in the following way:

$A(-3, 5)$:   3 units left, 5 units up

$B(2, -4)$:   2 units right, 4 units down

$C(5, 0)$:   5 units right, 0 units up or down

$D(-5, -3)$: 5 units left, 3 units down

$E(0, 4)$:   0 units right or left, 4 units up

$F(0, 0)$:   0 units right or left, 0 units up or down

Figure 1.3 Plotting points

The phrase *ordered pair* is used because **order is important.** For example, the points $(2, 5)$ and $(5, 2)$ are not the same. To plot $(2, 5)$, move 2 units right and 5 units up. To plot $(5, 2)$, move 5 units right and 2 units up. The points $(2, 5)$ and $(5, 2)$ are in different locations. **The order in which coordinates appear makes a difference in a point's location.**

Check Point 1 Plot the points:
$$A(-2, 4),\ B(4, -2),\ C(-3, 0),\ \text{and } D(0, -3).$$

Graphs of Equations

A relationship between two quantities can be expressed as an **equation in two variables,** such as

$$y = x^2 - 4.$$

2 Graph equations in the rectangular coordinate system.

A **solution** of this equation is an ordered pair of real numbers with the following property: When the x-coordinate is substituted for x and the y-coordinate is substituted for y in the equation, we obtain a true statement. For example, if we let $x = 3$, then $y = 3^2 - 4 = 9 - 4 = 5$. The ordered pair $(3, 5)$ is a solution of the equation $y = x^2 - 4$. We also say that $(3, 5)$ **satisfies** the equation.

We can generate as many ordered-pair solutions as desired of $y = x^2 - 4$ by substituting numbers for x and then finding the values for y. The **graph of the equation** is the set of all points whose coordinates satisfy the equation.

One method for graphing an equation such as $y = x^2 - 4$ is the **point-plotting method.** First, we find several ordered pairs that are solutions of the equation. Next, we plot these ordered pairs as points in the rectangular coordinate system. Finally, we connect the points with a smooth curve or line. This often gives us a picture of all ordered pairs that satisfy the equation.

EXAMPLE 2 **Graphing an Equation Using the Point-Plotting Method**

Graph $y = x^2 - 4$. Select integers for x, starting with -3 and ending with 3.

Solution For each value of x we find the corresponding value for y.

x	$y = x^2 - 4$	(x, y)
-3	$y = (-3)^2 - 4 = 9 - 4 = 5$	$(-3, 5)$
-2	$y = (-2)^2 - 4 = 4 - 4 = 0$	$(-2, 0)$
-1	$y = (-1)^2 - 4 = 1 - 4 = -3$	$(-1, -3)$
0	$y = 0^2 - 4 = 0 - 4 = -4$	$(0, -4)$
1	$y = 1^2 - 4 = 1 - 4 = -3$	$(1, -3)$
2	$y = 2^2 - 4 = 4 - 4 = 0$	$(2, 0)$
3	$y = 3^2 - 4 = 9 - 4 = 5$	$(3, 5)$

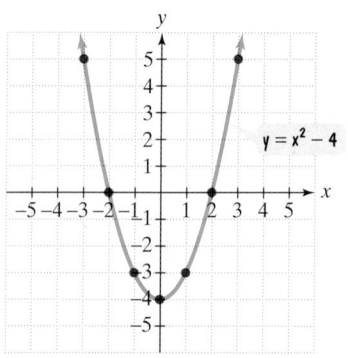

Figure 1.4 The graph of $y = x^2 - 4$

Now we plot the ordered pairs that are solutions of the equation and join the points with a smooth curve, as shown in Figure 1.4. The graph of $y = x^2 - 4$ is a curve where the part of the graph to the right of the y-axis is a reflection of the part to the left of it, and vice versa. The arrows on the left and the right of the curve indicate that it extends indefinitely in both directions.

Check Point 2 Graph $y = 2x - 4$. Select integers for x, starting with -1 and ending with 3.

Do you see a difference between the equations in Example 2 and Check Point 2? The equation in Example 2, $y = x^2 - 4$, involves a polynomial of degree 2. All such equations have graphs that are shaped like cups, such as the graph in Figure 1.4. These U-shaped "cups" can open upward, like the one in Figure 1.4, or downward. By contrast, the equation in Check Point 2, $y = 2x - 4$, involves a polynomial of degree 1. All such equations have graphs that are straight lines.

Study Tip

In Chapters 2 and 3, we will be studying graphs of equations in two variables in which

$$y = \text{a polynomial in } x.$$

Do not be concerned that we have not yet learned techniques, other than plotting points, for graphing such equations. As you solve some of the equations in this chapter, we will display graphs simply to enhance your visual understanding of your work. For now, think of graphs of first-degree polynomials as lines and graphs of second-degree polynomials as symmetric U-shaped cups.

3 Interpret information about a graphing utility's viewing rectangle.

Graphing Equations Using a Graphing Utility

Graphing calculators or graphing software packages for computers are referred to as **graphing utilities** or graphers. A graphing utility is a powerful tool that quickly generates the graph of an equation in two variables. Figure 1.5 shows two such graphs for the equations in Example 2 and Check Point 2.

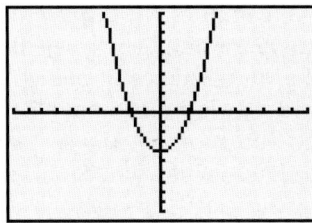

Figure 1.5(a)
The graph of $y = x^2 - 4$

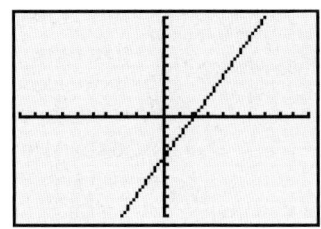

Figure 1.5(b)
The graph of $y = 2x - 4$

What differences do you notice between these graphs and graphs that we draw by hand? They do seem a bit "jittery." Arrows do not appear on the left and right ends of the graphs. Furthermore, numbers are not given along the axes. For both graphs in Figure 1.5, the x-axis extends from -10 to 10 and the y-axis also extends from -10 to 10. The distance represented by each consecutive tick mark is one unit. We say that the **viewing rectangle** is $[-10, 10, 1]$ by $[-10, 10, 1]$.

$$[-10, \qquad 10, \qquad 1] \quad \text{by} \quad [-10, \qquad 10, \qquad 1]$$

| The minimum x-value along the x-axis is −10. | The maximum x-value along the x-axis is 10. | Distance between consecutive tick marks on the x-axis is one unit. | The minimum y-value along the y-axis is −10. | The maximum y-value along the y-axis is 10. | Distance between consecutive tick marks on the y-axis is one unit. |

To graph an equation in x and y using a graphing utility, enter the equation and specify the size of the viewing rectangle. The size of the viewing rectangle sets minimum and maximum values for both the x- and y-axes. Enter these values, as well as the values between consecutive tick marks on the respective axes. The $[-10, 10, 1]$ by $[-10, 10, 1]$ viewing rectangle used in Figure 1.5 is called the **standard viewing rectangle.**

EXAMPLE 3   Understanding the Viewing Rectangle

What is the meaning of a $[-2, 3, 0.5]$ by $[-10, 20, 5]$ viewing rectangle?

Solution   We begin with $[-2, 3, 0.5]$, which describes the x-axis. The minimum x-value is -2 and the maximum x-value is 3. The distance between consecutive tick marks is 0.5.

Next, consider $[-10, 20, 5]$, which describes the y-axis. The minimum y-value is -10 and the maximum y-value is 20. The distance between consecutive tick marks is 5.

Figure 1.6 illustrates a $[-2, 3, 0.5]$ by $[-10, 20, 5]$ viewing rectangle. To make things clearer, we've placed numbers by each tick mark. These numbers do not appear on the axes when you use a graphing utility to graph an equation.

Figure 1.6 A $[-2, 3, 0.5]$ by $[-10, 20, 5]$ viewing rectangle

Check Point 3   What is the meaning of a $[-100, 100, 50]$ by $[-80, 80, 10]$ viewing rectangle? Create a figure like the one in Figure 1.6 that illustrates this viewing rectangle.

On most graphing utilities, the display screen is about two-thirds as high as it is wide. By using a square setting, you can make the distance of one unit along the x-axis the same as the distance of one unit along the y-axis. (This does not occur in the standard viewing rectangle.) Graphing utilities can also *zoom in* and *zoom out*. When you zoom in, you see a smaller portion of the graph, but you see it in greater detail. When you zoom out, you see a larger portion of the graph. Thus, zooming out may help you to develop a better understanding of the overall character of the graph. With practice, you will become more comfortable with graphing equations in two variables using your graphing utility. You will also develop a better sense of the size of the viewing rectangle that will reveal needed information about a particular graph.

4 Use a graph to determine intercepts.

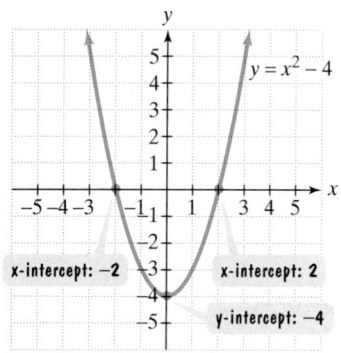

Figure 1.7 Intercepts of $y = x^2 - 4$

Intercepts

An **x-intercept** of a graph is the x-coordinate of a point where the graph intersects the x-axis. For example, look at the graph of $y = x^2 - 4$ in Figure 1.7. The graph crosses the x-axis at $(-2, 0)$ and $(2, 0)$. Thus, the x-intercepts are -2 and 2. **The y-coordinate corresponding to a graph's x-intercept is always zero.**

A **y-intercept** of a graph is the y-coordinate of a point where the graph intersects the y-axis. The graph of $y = x^2 - 4$ in Figure 1.7 shows that the graph crosses the y-axis at $(0, -4)$. Thus, the y-intercept is -4. **The x-coordinate corresponding to a graph's y-intercept is always zero.**

Figure 1.8 illustrates that a graph may have no intercepts or several intercepts.

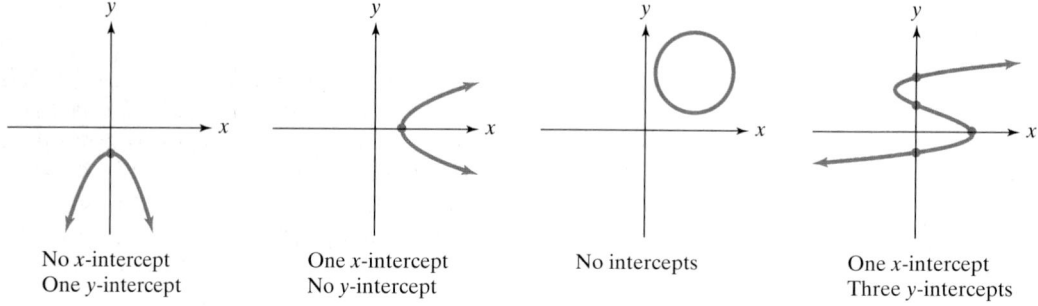

No x-intercept
One y-intercept

One x-intercept
No y-intercept

No intercepts

One x-intercept
Three y-intercepts

Figure 1.8

5 Interpret information given by graphs.

Interpreting Information Given by Graphs

Magazines and newspapers often display information using **line graphs** like the one in Figure 1.9. The graph shows the average age at which women in the United States married for the first time over a 110-year period. The years are listed on the horizontal axis and the ages are listed on the vertical axis.

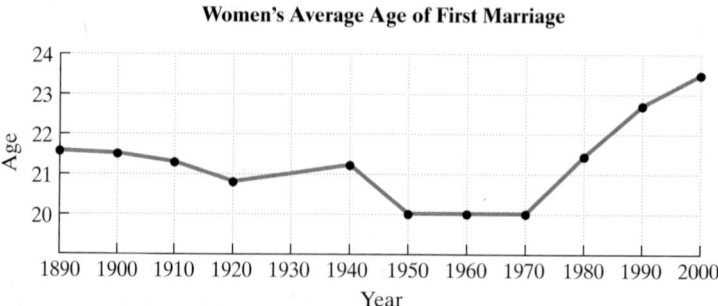

Women's Average Age of First Marriage

Figure 1.9 Average age at which U.S. women married for the first time

Source: U.S. Census Bureau

Like the graph in Figure 1.9, line graphs are often used to illustrate trends over time. Some measure of time, such as months or years, frequently appears on the horizontal axis. Amounts are generally listed on the vertical axis.

A line graph displays information in the first quadrant of a rectangular coordinate system. By identifying points on line graphs and their coordinates, you can interpret specific information given by the graph.

For example, Figure 1.10 shows how to find the average age at which women married for the first time in 1930. (Only the part of the graph that reveals what occurred through about 1940 is shown in the margin because we are interested in 1930.)

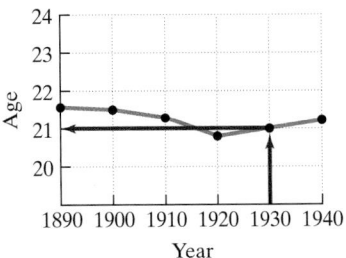

Figure 1.10 In 1930, women were 21 on average when they married for the first time.

Step 1 Locate 1930 on the horizontal axis.

Step 2 Locate the point above 1930.

Step 3 Read across to the corresponding age on the vertical axis.

The age is 21. The coordinates (1930, 21) tell us that in 1930, women in the United States married for the first time at an average age of 21.

EXAMPLE 4   Applying Estimation Techniques to a Line Graph

Use Figure 1.9 to estimate the maximum average age at which U.S. women married for the first time. When did this occur?

Figure 1.9 Shown again to show only 1960–2000

Solution The maximum average age at which U.S. women married for the first time can be found by locating the highest point on the graph. This point lies above 2000 on the horizontal axis. Read across to the corresponding age on the vertical axis. The age falls approximately midway between 23 and 24, at $23\frac{1}{2}$. The coordinates of the point are approximately $(2000, 23\frac{1}{2})$. Thus, according to the graph, the maximum average age at which U.S. women married for the first time is about $23\frac{1}{2}$. This occurred in 2000. Take another look at the complete line graph in Figure 1.9 at the bottom of page 80 that includes the years 1890 through 2000. Can you see that $23\frac{1}{2}$ is the oldest average age of first marriage over the 110-year period?

Check Point 4   Use the complete line graph in Figure 1.9 to estimate the maximum average age, for the period from 1900 through 1950, at which U.S. women married for the first time. When did this occur?

EXERCISE SET 1.1

Practice Exercises

In Exercises 1–12, plot the given point in a rectangular coordinate system.

1. $(1, 4)$ **2.** $(2, 5)$

3. $(-2, 3)$ **4.** $(-1, 4)$

5. $(-3, -5)$ **6.** $(-4, -2)$

7. $(4, -1)$ **8.** $(3, -2)$

9. $(-4, 0)$ **10.** $(0, -3)$

11. $(\frac{7}{2}, -\frac{3}{2})$ **12.** $(-\frac{5}{2}, \frac{3}{2})$

Graph each equation in Exercises 13–28. Let $x = -3, -2, -1,$ 0, 1, 2, and 3.

13. $y = x^2 - 2$ **14.** $y = x^2 + 2$

15. $y = x - 2$ **16.** $y = x + 2$

17. $y = 2x + 1$ **18.** $y = 2x - 4$

19. $y = -\frac{1}{2}x$ **20.** $y = -\frac{1}{2}x + 2$

21. $y = |x|$ **22.** $y = 2|x|$

23. $y = |x| + 1$ **24.** $y = |x| - 1$

25. $y = 4 - x^2$ **26.** $y = 9 - x^2$

27. $y = x^3$ **28.** $y = x^3 - 1$

In Exercises 29–32, match the viewing rectangle with the correct figure. Then label the tick marks in the figure to illustrate this viewing rectangle.

29. $[-5, 5, 1]$ by $[-5, 5, 1]$

30. $[-10, 10, 2]$ by $[-4, 4, 2]$

31. $[-20, 80, 10]$ by $[-30, 70, 10]$

32. $[-40, 40, 20]$ by $[-1000, 1000, 100]$

(a)

(b)

(c)

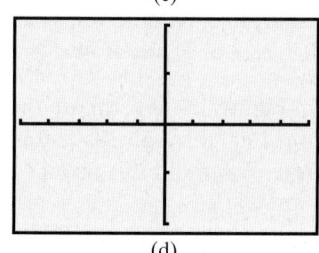

(d)

In Exercises 33–38, use the graph and **a.** determine the x-intercepts, if any; **b.** determine the y-intercepts, if any. For each graph, tick marks along the axes represent one unit each.

33.

34.

35.

36.

37.

38.

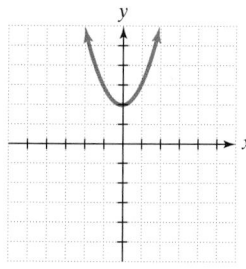

Application Exercises

A football is thrown by a quarterback to a receiver. The points in the figure show the height of the football, in feet, above the ground in terms of its distance, in yards, from the quarterback. Use this information to solve Exercises 39–44.

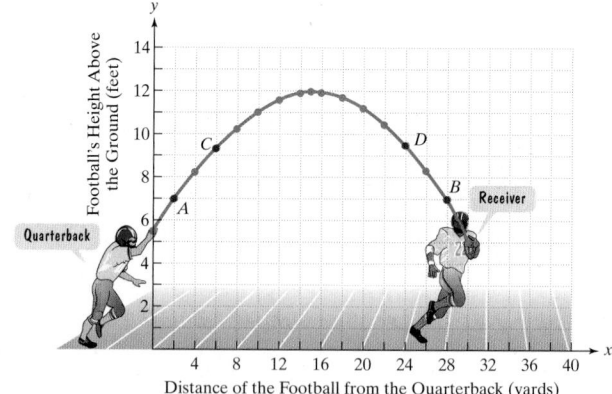

Distance of the Football from the Quarterback (yards)

39. Find the coordinates of point A. Then interpret the coordinates in terms of the information given.

40. Find the coordinates of point B. Then interpret the coordinates in terms of the information given.

41. Estimate the coordinates of point C.

42. Estimate the coordinates of point D.

43. What is the football's maximum height? What is its distance from the quarterback when it reaches its maximum height?

44. What is the football's height when it is caught by the receiver? What is the receiver's distance from the quarterback when he catches the football?

The graph shows the percent distribution of divorces in the United States by number of years of marriage. Use the graph to solve Exercises 45–48.

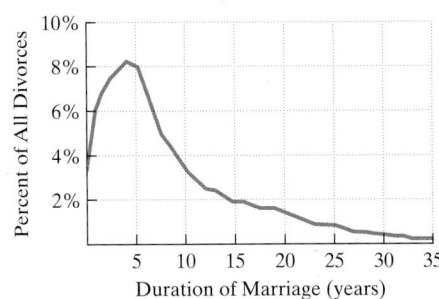

Percent Distribution of Divorces by Number of Years of Marriage

Source: Divorce Center

45. During which years of marriage is the chance of divorce increasing?

46. During which years of marriage is the chance of divorce decreasing?

47. During which year of marriage is the chance of divorce the highest? Estimate, to the nearest percent, the percentage of divorces that occur during this year.

48. During which year of marriage is the chance of divorce the lowest? Estimate, to the nearest percent, the percentage of divorces that occur during this year.

Writing in Mathematics

49. What is the rectangular coordinate system?

50. Explain how to plot a point in the rectangular coordinate system. Give an example with your explanation.

51. Explain why $(5, -2)$ and $(-2, 5)$ do not represent the same point.

52. Explain how to graph an equation in the rectangular coordinate system.

53. What does a $[-20, 2, 1]$ by $[-4, 5, 0.5]$ viewing rectangle mean?

54. Describe the trend shown in the graph for Exercises 45–48. What explanations can you offer for this trend?

Technology Exercises

55. Use a graphing utility to verify each of your hand-drawn graphs in Exercises 13–28. Experiment with the size of the viewing rectangle to make the graph displayed by the graphing utility resemble your hand-drawn graph as much as possible.

56. The stated intent of the 1994 "don't ask, don't tell" policy was to reduce the number of discharges of gay men and lesbians from the military. The equation

$$y = 45.48x^2 - 334.35x + 1237.9$$

describes the number of gay service members, y, discharged from the military for homosexuality x years after 1990. Graph the equation in a $[0, 10, 1]$ by $[0, 2200, 200]$ viewing rectangle. Then describe something about the relationship between x and y that is revealed by looking at the graph that is not obvious from the equation. What does the graph reveal about the success or lack of success of "don't ask, don't tell"?

A graph of an equation is a complete graph *if it shows all of the important features of the graph. Use a graphing utility to graph the equations in Exercises 57–59 in each of the given viewing rectangles. Then choose which viewing rectangle gives a complete graph.*

57. $y = x^2 + 10$
  a. $[-5, 5, 1]$ by $[-5, 5, 1]$
  b. $[-10, 10, 1]$ by $[-10, 10, 1]$
  c. $[-10, 10, 1]$ by $[-50, 50, 1]$

58. $y = 0.1x^4 - x^3 + 2x^2$
  a. $[-5, 5, 1]$ by $[-8, 2, 1]$
  b. $[-10, 10, 1]$ by $[-10, 10, 1]$
  c. $[-8, 16, 1]$ by $[-16, 8, 1]$

59. $y = x^3 - 30x + 20$
  a. $[-10, 10, 1]$ by $[-10, 10, 1]$
  b. $[-10, 10, 1]$ by $[-50, 50, 10]$
  c. $[-10, 10, 1]$ by $[-50, 100, 10]$

Critical Thinking Exercises

60. Which one of the following is true?
  a. If the coordinates of a point satisfy the inequality $xy > 0$, then (x, y) must be in quadrant I.
  b. The ordered pair $(2, 5)$ satisfies $3y - 2x = -4$.
  c. If a point is on the x-axis, it is neither up nor down, so $x = 0$.
  d. None of the above is true.

In Exercises 61–64, match the story with the correct figure. The figures are labeled (a), (b), (c), and (d).

61. As the blizzard got worse, the snow fell harder and harder.

62. The snow fell more and more softly.

63. It snowed hard, but then it stopped. After a short time, the snow started falling softly.

64. It snowed softly, and then it stopped. After a short time, the snow started falling hard.

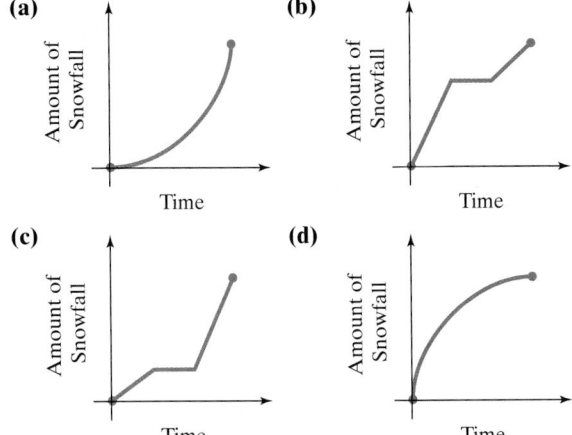

SECTION 1.2 *Linear Equations*

Objectives

1. Solve linear equations in one variable.
2. Solve equations with constants in denominators.
3. Solve equations with variables in denominators.
4. Recognize identities, conditional equations, and inconsistent equations.

Unfortunately, many of us have been fined for driving over the speed limit. The amount of the fine depends on how fast we are speeding. Suppose that a highway has a speed limit of 60 miles per hour. The amount that speeders are fined, F, is described by the statement of equality

$$F = 10x - 600$$

where x is the speed, in miles per hour. We can use this statement to determine the fine, F, for a speeder traveling at, say, 70 miles per hour. We substitute 70 for x in the given statement and then find the corresponding value for F.

$$F = 10(70) - 600 = 700 - 600 = 100$$

Thus, a person caught driving 70 miles per hour gets a $100 fine.

A friend, whom we shall call Leadfoot, borrows your car and returns a few hours later with a $400 speeding fine. Leadfoot is furious, protesting that the car was barely driven over the speed limit. Should you believe Leadfoot?

In order to decide if Leadfoot is telling the truth, use $F = 10x - 600$. Leadfoot was fined $400, so substitute 400 for F:

$$400 = 10x - 600.$$

In Example 1, we will find the value for x. This variable represents Leadfoot's speed, which resulted in the $400 fine.

An **equation** consists of two algebraic expressions joined by an equal sign. Thus, $400 = 10x - 600$ is an example of an equation. The equal sign divides the equation into two parts, the left side and the right side:

$$\boxed{400} \quad = \quad \boxed{10x - 600}.$$

Left side   Right side

The two sides of an equation can be reversed. So, we can also express this equation as

$$10x - 600 = 400.$$

Notice that the highest exponent on the variable is 1. Such an equation is called a *linear equation in one variable*. In this section, we will study how to solve linear equations.

1 Solve linear equations in one variable.

Solving Linear Equations in One Variable

We begin with a general definition of a linear equation in one variable.

> ### Definition of a Linear Equation
> A **linear equation in one variable** x is an equation that can be written in the form
> $$ax + b = 0$$
> where a and b are real numbers, and $a \neq 0$.

An example of a linear equation in one variable is $4x + 12 = 0$. **Solving an equation** in x involves determining all values of x that result in a true statement when substituted into the equation. Such values are **solutions,** or **roots,** of the equation. For example, substitute -3 into $4x + 12 = 0$. We obtain $4(-3) + 12 = 0$, or $-12 + 12 = 0$. This simplifies to the true statement $0 = 0$. Thus, -3 is a solution of the equation $4x + 12 = 0$. We also say that -3 **satisfies** the equation $4x + 12 = 0$, because when we substitute -3 for x, a true statement results. The set of all such solutions is called the equation's **solution set.** For example, the solution set of the equation $4x + 12 = 0$ is $\{-3\}$.

Equations that have the same solution set are called **equivalent equations.** For example, the equations $4x + 12 = 0$, $4x = -12$, and $x = -3$ are equivalent equations because the solution set for each is $\{-3\}$. To solve a linear equation in x, we transform the equation into an equivalent equation one or more times. Our final equivalent equation should be in the form $x = d$, where d is a real number. By inspection, we can see that the solution set of this equation is $\{d\}$.

To generate equivalent equations, we will use the following principles:

Study Tip

We can solve equations such as $3(x - 6) = 5x$ for a variable. However, we cannot solve for a variable in an algebraic expression such as $3(x - 6)$. We *simplify* algebraic expressions.

Correct

Simplify: $3(x - 6)$.
$$3(x - 6) = 3x - 18$$

Incorrect

~~Simplify: $3(x - 6)$.~~
~~$3(x - 6) = 0$~~
~~$3x - 18 = 0$~~
~~$3x - 18 + 18 = 0 + 18$~~
~~$3x = 18$~~
~~$\dfrac{3x}{3} = \dfrac{18}{3}$~~
~~$x = 6$~~

Generating Equivalent Equations

An equation can be transformed into an equivalent equation by one or more of the following operations:

Example

1. Simplify an expression by removing grouping symbols and combining like terms.

   • $3(x - 6) = 6x - x$
   $3x - 18 = 5x$

2. Add (or subtract) the same real number or variable expression on *both* sides of the equation.

   • $3x - 18 = 5x$ Subtract 3x from both sides of the equation.
   $3x - 18 - 3x = 5x - 3x$
   $-18 = 2x$

3. Multiply (or divide) on *both* sides of the equation by the same *nonzero* quantity.

   • $-18 = 2x$ Divide both sides of the equation by 2.
   $\dfrac{-18}{2} = \dfrac{2x}{2}$
   $-9 = x$

4. Interchange the two sides of the equation.

   • $-9 = x$
   $x = -9$

If you look closely at the equations in the box, you will notice that we have solved the equation $3(x - 6) = 6x - x$. The final equation, $x = -9$, with x isolated by itself on the left side, shows that $\{-9\}$ is the solution set. The idea in solving a linear equation is to get the variable by itself on one side of the equal sign and a number by itself on the other side.

EXAMPLE 1 Solving a Linear Equation (Is Leadfoot Telling the Truth?)

Solve the equation: $10x - 600 = 400$.

Study Tip

If your proposed solution is incorrect, you will get a false statement when you check your answer. For example, 65 is not a solution of $10x - 600 = 400$. Look what happens when we substitute 65 for x:

$$10x - 600 = 400$$
$$10(65) - 600 \stackrel{?}{=} 400$$
$$650 - 600 \stackrel{?}{=} 400$$
$$50 = 400 \quad \text{False.}$$

Solution   Remember that x represents Leadfoot's speed that resulted in the \$400 fine. Our goal is to get x by itself on the left side. We do this by adding 600 to both sides to get $10x$ by itself. Then we isolate x from $10x$ by dividing both sides of the equation by 10.

$10x - 600 = 400$	This is the given equation.
$10x - 600 + 600 = 400 + 600$	Add 600 to both sides.
$10x = 1000$	Combine like terms.
$\dfrac{10x}{10} = \dfrac{1000}{10}$	Divide both sides by 10.
$x = 100$	Simplify.

Can this possibly be correct? Was Leadfoot doing 100 miles per hour in the car he borrowed from you? To find out, check the proposed solution, 100, in the original equation. In other words, evaluate for $x = 100$.

Check 100:

$10x - 600 = 400$	This is the original equation.
$10(100) - 600 \stackrel{?}{=} 400$	Substitute 100 for x. The question mark indicates that we do not yet know if the two sides are equal.
$1000 - 600 \stackrel{?}{=} 400$	Multiply: $10(100) = 1000$.
$400 = 400$	Subtract: $1000 - 600 = 400$.

This statement is true.

The true statement $400 = 400$ indicates that 100 is the solution. This verifies that the solution set is $\{100\}$. Leadfoot was doing an outrageous 100 miles per hour, and lied by claiming that your car was barely driven over the speed limit.

Check Point 1   Solve and check: $5x - 8 = 72$.

We now present a step-by-step procedure for solving a linear equation in one variable. Not all of these steps are necessary to solve every equation.

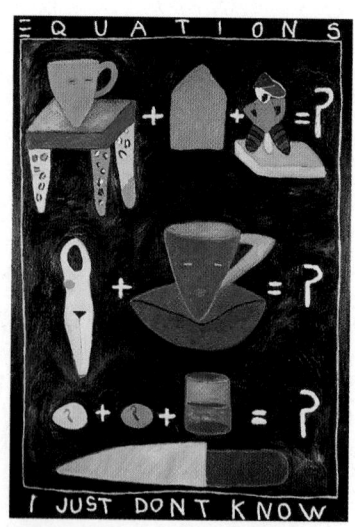

The compact, symbolic notation of algebra enables us to use a clear step-by-step method for solving equations, designed to avoid the confusion shown in the painting.

Solving a Linear Equation

1. Simplify the algebraic expression on each side.
2. Collect the variable terms on one side and the constant terms on the other side.
3. Isolate the variable and solve.
4. Check the proposed solution in the original equation.

EXAMPLE 2   Solving a Linear Equation

Solve the equation: $2(x - 3) - 17 = 13 - 3(x + 2)$.

Solution

Step 1   Simplify the algebraic expression on each side.

$$2(x - 3) - 17 = 13 - 3(x + 2) \qquad \text{This is given equation.}$$
$$2x - 6 - 17 = 13 - 3x - 6 \qquad \text{Use the distributive property.}$$
$$2x - 23 = -3x + 7 \qquad \text{Combine like terms.}$$

Discovery

Solve the equation in Example 2 by collecting terms with the variable on the right and numerical terms on the left. What do you observe?

Step 2   Collect variable terms on one side and constant terms on the other side.   We will collect variable terms on the left by adding $3x$ to both sides. We will collect the numbers on the right by adding 23 to both sides.

$$2x - 23 + 3x = -3x + 7 + 3x \qquad \text{Add 3x to both sides.}$$
$$5x - 23 = 7 \qquad \text{Simplify.}$$
$$5x - 23 + 23 = 7 + 23 \qquad \text{Add 23 to both sides.}$$
$$5x = 30 \qquad \text{Simplify.}$$

Step 3   Isolate the variable and solve.   We isolate x by dividing both sides by 5.

$$\frac{5x}{5} = \frac{30}{5} \qquad \text{Divide both sides by 5.}$$
$$x = 6 \qquad \text{Simplify.}$$

Step 4   Check the proposed solution in the original equation.   Substitute 6 for x in the original equation.

$$2(x - 3) - 17 = 13 - 3(x + 2) \qquad \text{This is the original equation.}$$
$$2(6 - 3) - 17 \overset{?}{=} 13 - 3(6 + 2) \qquad \text{Substitute 6 for x.}$$
$$2(3) - 17 \overset{?}{=} 13 - 3(8) \qquad \text{Simplify inside parentheses.}$$
$$6 - 17 \overset{?}{=} 13 - 24 \qquad \text{Multiply.}$$
$$-11 = -11 \qquad \text{Subtract.}$$

The true statement $-11 = -11$ verifies that the solution set is $\{6\}$.

Technology

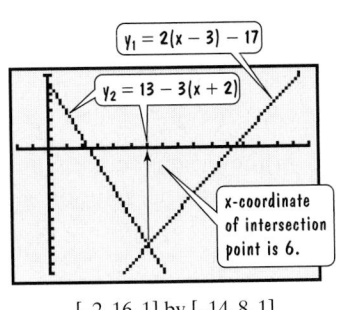

$y_1 = 2(x - 3) - 17$

$y_2 = 13 - 3(x + 2)$

x-coordinate of intersection point is 6.

$[-2, 16, 1]$ by $[-14, 8, 1]$

You can use a graphing utility to check the solution to a linear equation in one variable. **Graph the left side and graph the right side. The solution is the x-coordinate of the point where the graphs intersect.** For example, to verify that 6 is the solution of

$$2(x - 3) - 17 = 13 - 3(x + 2),$$

graph these two equations in the same viewing rectangle:

$$y_1 = 2(x - 3) - 17$$
$$\text{and} \quad y_2 = 13 - 3(x + 2).$$

Choose a large enough viewing rectangle so that you can see where the graphs intersect. The viewing rectangle on the left shows that the x-coordinate of the intersection point is 6, verifying that $\{6\}$ is the solution set for the equation in Example 2.

Check Point 2 Solve and check: $4(2x + 1) - 29 = 3(2x - 5)$.

2 Solve equations with constants in denominators.

Linear Equations with Fractions

Equations are easier to solve when they do not contain fractions. How do we solve equations involving fractions? We begin by multiplying both sides of the equation by the least common denominator of all fractions in the equation. The least common denominator is the smallest number that all the denominators will divide into. Multiplying every term on both sides of the equation by the least common denominator will eliminate the fractions in the equation. Example 3 shows how we "clear an equation of fractions."

EXAMPLE 3   Solving a Linear Equation Involving Fractions

Solve the equation: $\dfrac{3x}{2} = \dfrac{x}{5} - \dfrac{39}{5}$.

Solution   The denominators are 2, 5, and 5. The smallest number that is divisible by 2, 5, and 5 is 10. We begin by multiplying both sides of the equation by 10, the least common denominator.

$$\frac{3x}{2} = \frac{x}{5} - \frac{39}{5}$$ This is the given equation.

$$10 \cdot \frac{3x}{2} = 10\left(\frac{x}{5} - \frac{39}{5}\right)$$ Multiply both sides by 10.

$$10 \cdot \frac{3x}{2} = 10 \cdot \frac{x}{5} - 10 \cdot \frac{39}{5}$$ Use the distributive property and multiply each term by 10.

$$\overset{5}{\cancel{10}} \cdot \frac{3x}{\underset{1}{\cancel{2}}} = \overset{2}{\cancel{10}} \cdot \frac{x}{\underset{1}{\cancel{5}}} - \overset{2}{\cancel{10}} \cdot \frac{39}{\underset{1}{\cancel{5}}}$$ Divide out common factors in each multiplication.

$$15x = 2x - 78$$ Complete the multiplications. The fractions are now cleared.

At this point, we have an equation similar to those we previously solved. Collect the variable terms on one side and the constant terms on the other side.

$$15x - 2x = 2x - 2x - 78$$ Subtract 2x to get the variable terms on the left.

$$13x = -78$$ Simplify.

Isolate x by dividing both sides by 13.

$$\frac{13x}{13} = \frac{-78}{13}$$ Divide both sides by 13.

$$x = -6$$ Simplify.

Check the proposed solution. Substitute -6 for x in the original equation. You should obtain $-9 = -9$. This true statement verifies that the solution set is $\{-6\}$.

Check Point 3 Solve and check: $\dfrac{x}{4} = \dfrac{2x}{3} + \dfrac{5}{6}$.

3 Solve equations with variables in denominators.

Equations Involving Rational Expressions

In Example 3 we solved a linear equation with constants in denominators. Now, let's consider an equation such as

$$\frac{1}{x} = \frac{1}{5} + \frac{3}{2x}.$$

Can you see how this equation differs from the fractional equation that we solved earlier? The variable, x, appears in two of the denominators. The procedure for solving this equation still involves multiplying each side by the least common denominator. However, we must avoid any values of the variable that make a denominator zero. For example, examine the denominators in the equation

$$\frac{1}{x} = \frac{1}{5} + \frac{3}{2x}.$$

> This denominator would equal zero if x = 0.

> This denominator would equal zero if x = 0.

We see that x cannot equal zero. With this in mind, let's solve the equation.

EXAMPLE 4 **Solving an Equation Involving Rational Expressions**

Solve: $\dfrac{1}{x} = \dfrac{1}{5} + \dfrac{3}{2x}.$

Solution The denominators are x, 5, and $2x$. The least common denominator is $10x$. We begin by multiplying both sides of the equation by $10x$. We will also write the restriction that x cannot equal zero to the right of the equation.

$$\frac{1}{x} = \frac{1}{5} + \frac{3}{2x}, \quad x \neq 0 \qquad \text{This is the given equation.}$$

$$10x \cdot \frac{1}{x} = 10x\left(\frac{1}{5} + \frac{3}{2x}\right) \qquad \text{Multiply both sides by 10x.}$$

$$10x \cdot \frac{1}{x} = 10x \cdot \frac{1}{5} + 10x \cdot \frac{3}{2x} \qquad \text{Use the distributive property and multiply each term by 10x.}$$

$$10\cancel{x} \cdot \frac{1}{\cancel{x}} = \overset{2}{\cancel{10}}x \cdot \frac{1}{\cancel{5}} + \overset{5}{\cancel{10}}x \cdot \frac{3}{\underset{1}{\cancel{2}}\cancel{x}} \qquad \text{Divide out common factors in each multiplication.}$$

$$10 = 2x + 15 \qquad \text{Complete the multiplications.}$$

Observe that the resulting equation,

$$10 = 2x + 15,$$

is now cleared of fractions. With the variable term, $2x$, already on the right, we will collect constant terms on the left by subtracting 15 from both sides.

$$10 - 15 = 2x + 15 - 15 \qquad \text{Subtract 15 from both sides.}$$

$$-5 = 2x \qquad \text{Simplify.}$$

Finally, we isolate the variable, x, in $-5 = 2x$ by dividing both sides by 2.

$$\frac{-5}{2} = \frac{2x}{2} \qquad \text{Divide both sides by 2.}$$

$$-\frac{5}{2} = x \qquad \text{Simplify.}$$

We check our solution by substituting $-\frac{5}{2}$ into the original equation or by using a calculator. With a calculator, evaluate each side of the equation for $x = -\frac{5}{2}$, or for $x = -2.5$. Note that the original restriction that $x \neq 0$ is met. The solution set is $\{-\frac{5}{2}\}$.

Check Point 4 Solve: $\dfrac{5}{2x} = \dfrac{17}{18} - \dfrac{1}{3x}$.

EXAMPLE 5 Solving an Equation Involving Rational Expressions

Solve: $\dfrac{x}{x-3} = \dfrac{3}{x-3} + 9$.

Solution We must avoid any values of the variable x that make a denominator zero.

$$\frac{x}{x-3} = \frac{3}{x-3} + 9$$

These denominators are zero if $x - 3 = 0$, or equivalently, if $x = 3$.

We see that x cannot equal 3. With denominators of $x - 3$, $x - 3$, and 1, the least common denominator is $x - 3$. We multiply both sides of the equation by $x - 3$. We also write the restriction that x cannot equal 3 to the right of the equation.

$$\frac{x}{x-3} = \frac{3}{x-3} + 9, \quad x \neq 3 \qquad \text{This is the given equation.}$$

$$(x - 3) \cdot \frac{x}{x-3} = (x - 3)\left[\frac{3}{x-3} + 9\right] \qquad \begin{array}{l}\text{Multiply both sides by}\\ x - 3.\end{array}$$

$$(x - 3) \cdot \frac{x}{x-3} = (x - 3) \cdot \frac{3}{x-3} + (x - 3) \cdot 9 \qquad \begin{array}{l}\text{Use the distributive}\\ \text{property.}\end{array}$$

$$\cancel{(x-3)} \cdot \frac{x}{\cancel{x-3}} = \cancel{(x-3)} \cdot \frac{3}{\cancel{x-3}} + (x - 3) \cdot 9 \qquad \begin{array}{l}\text{Divide out common}\\ \text{factors in each}\\ \text{multiplication.}\end{array}$$

$$x = 3 + (x - 3) \cdot 9 \qquad \text{Simplify.}$$

The resulting equation, which can be expressed as

$$x = 3 + 9(x - 3),$$

is cleared of fractions. We now solve for x.

$$x = 3 + 9x - 27 \qquad \text{Use the distributive property.}$$

$$x = 9x - 24 \qquad \text{Combine numerical terms.}$$

$$x - 9x = 9x - 24 - 9x \qquad \text{Subtract 9x from both sides.}$$

$$-8x = -24 \qquad \text{Simplify.}$$

$$\frac{-8x}{-8} = \frac{-24}{-8} \qquad \text{Solve for x, dividing both sides by } -8.$$

$$x = 3 \qquad \text{Simplify.}$$

Study Tip

Reject any proposed solution that causes any denominator in an equation to equal 0.

The proposed solution, 3, is *not* a solution because of the restriction that $x \neq 3$. There is *no solution to this equation.* The solution set for this equation contains no elements and is called the empty set, written $\varnothing$.

Check Point 5 Solve: $\dfrac{x}{x - 2} = \dfrac{2}{x - 2} - \dfrac{2}{3}.$

4 Recognize identities, conditional equations, and inconsistent equations.

Types of Equations

We tend to place things in categories, allowing us to order and structure the world. For example, you can categorize yourself by your age group, your ethnicity, your academic major, or your gender. Equations can be placed into categories that depend on their solution sets.

An equation that is true for all real numbers for which both sides are defined is called an **identity.** An example of an identity is

$$x + 3 = x + 2 + 1.$$

Every number plus 3 is equal to that number plus 2 plus 1. Therefore, the solution set to this equation is the set of all real numbers. Another example of an identity is

$$\frac{2x}{x} = 2.$$

Because division by 0 is undefined, this equation is true for all real number values of x except 0. The solution set is the set of nonzero real numbers.

An equation that is not an identity, but that is true for at least one real number, is called a **conditional equation.** The equation $10x - 600 = 400$ is an example of a conditional equation. The equation is not an identity and is true only if x is 100.

An **inconsistent equation** is an equation that is not true for even one real number. An example of an inconsistent equation is

$$x = x + 7.$$

There is no number that is equal to itself plus 7. Some inconsistent equations are less obvious than this. Consider the equation in Example 5,

$$\frac{x}{x - 3} = \frac{3}{x - 3} + 9.$$

This equation is not true for any real number and has no solution. Thus, it is inconsistent.

EXAMPLE 6 Categorizing an Equation

Determine whether the equation

$$2(x + 1) = 2x + 3$$

is an identity, a conditional equation, or an inconsistent equation.

Technology

The graphs of $y_1 = 2(x + 1)$ and $y_2 = 2x + 3$ are parallel lines with no intersection point. This shows that the equation

$$2(x + 1) = 2x + 3$$

has no solution and is inconsistent.

$[-5, 2, 1]$ by $[-5, 5, 1]$

Solution Let's see what happens if we try solving $2(x + 1) = 2x + 3$. Applying the distributive property on the left side, we obtain

$$2x + 2 = 2x + 3.$$

Does something look strange? Can doubling a number and increasing the product by 2 give the same result as doubling the same number and increasing the product by 3? No. Let's continue solving the equation by subtracting $2x$ from both sides.

$$2x + 2 - 2x = 2x + 3 - 2x$$
$$2 = 3$$

The false statement $2 = 3$ verifies that the given equation is inconsistent.

Check Point 6 Determine whether the equation

$$2(x + 1) = 2x + 2$$

is an identity, a conditional equation, or an inconsistent equation.

EXERCISE SET 1.2

Practice Exercises

In Exercises 1–16, solve and check each linear equation.

1. $7x - 5 = 72$

2. $6x - 3 = 63$

3. $11x - (6x - 5) = 40$

4. $5x - (2x - 10) = 35$

5. $2x - 7 = 6 + x$

6. $3x + 5 = 2x + 13$

7. $7x + 4 = x + 16$

8. $13x + 14 = 12x - 5$

9. $3(x - 2) + 7 = 2(x + 5)$

10. $2(x - 1) + 3 = x - 3(x + 1)$

11. $3(x - 4) - 4(x - 3) = x + 3 - (x - 2)$

12. $2 - (7x + 5) = 13 - 3x$

13. $16 = 3(x - 1) - (x - 7)$

14. $5x - (2x + 2) = x + (3x - 5)$

15. $25 - [2 + 5y - 3(y + 2)] =$
$$-3(2y - 5) - [5(y - 1) - 3y + 3]$$

16. $45 - [4 - 2y - 4(y + 7)] =$
$$-4(1 + 3y) - [4 - 3(y + 2) - 2(2y - 5)]$$

Exercises 17–30 contain equations with constants in denominators. Solve each equation.

17. $\dfrac{x}{3} = \dfrac{x}{2} - 2$

18. $\dfrac{x}{5} = \dfrac{x}{6} + 1$

19. $20 - \dfrac{x}{3} = \dfrac{x}{2}$

20. $\dfrac{x}{5} - \dfrac{1}{2} = \dfrac{x}{6}$

21. $\dfrac{3x}{5} = \dfrac{2x}{3} + 1$

22. $\dfrac{x}{2} = \dfrac{3x}{4} + 5$

23. $\dfrac{3x}{5} - x = \dfrac{x}{10} - \dfrac{5}{2}$

24. $2x - \dfrac{2x}{7} = \dfrac{x}{2} + \dfrac{17}{2}$

25. $\dfrac{x + 3}{6} = \dfrac{3}{8} + \dfrac{x - 5}{4}$

26. $\dfrac{x + 1}{4} = \dfrac{1}{6} + \dfrac{2 - x}{3}$

27. $\dfrac{x}{4} = 2 + \dfrac{x - 3}{3}$

28. $5 + \dfrac{x - 2}{3} = \dfrac{x + 3}{8}$

29. $\dfrac{x + 1}{3} = 5 - \dfrac{x + 2}{7}$

30. $\dfrac{3x}{5} - \dfrac{x - 3}{2} = \dfrac{x + 2}{3}$

*Exercises 31–50 contain equations with variables in denominators. For each equation, **a.** Write the value or values of the variable that make a denominator zero. These are the restrictions on the variable. **b.** Keeping the restrictions in mind, solve the equation.*

31. $\dfrac{4}{x} = \dfrac{5}{2x} + 3$

32. $\dfrac{5}{x} = \dfrac{10}{3x} + 4$

33. $\dfrac{2}{x} + 3 = \dfrac{5}{2x} + \dfrac{13}{4}$

34. $\dfrac{7}{2x} - \dfrac{5}{3x} = \dfrac{22}{3}$

35. $\dfrac{2}{3x} + \dfrac{1}{4} = \dfrac{11}{6x} - \dfrac{1}{3}$

36. $\dfrac{5}{2x} - \dfrac{8}{9} = \dfrac{1}{18} - \dfrac{1}{3x}$

37. $\dfrac{x - 2}{2x} + 1 = \dfrac{x + 1}{x}$

38. $\dfrac{4}{x} = \dfrac{9}{5} - \dfrac{7x - 4}{5x}$

39. $\dfrac{1}{x - 1} + 5 = \dfrac{11}{x - 1}$

40. $\dfrac{3}{x + 4} - 7 = \dfrac{-4}{x + 4}$

41. $\dfrac{8x}{x + 1} = 4 - \dfrac{8}{x + 1}$

42. $\dfrac{2}{x - 2} = \dfrac{x}{x - 2} - 2$

43. $\dfrac{3}{2x - 2} + \dfrac{1}{2} = \dfrac{2}{x - 1}$

44. $\dfrac{3}{x + 3} = \dfrac{5}{2x + 6} + \dfrac{1}{x - 2}$

45. $\dfrac{3}{x + 2} + \dfrac{2}{x - 2} = \dfrac{8}{(x + 2)(x - 2)}$

46. $\dfrac{5}{x + 2} + \dfrac{3}{x - 2} = \dfrac{12}{(x + 2)(x - 2)}$

47. $\dfrac{2}{x + 1} - \dfrac{1}{x - 1} = \dfrac{2x}{x^2 - 1}$

48. $\dfrac{4}{x + 5} + \dfrac{2}{x - 5} = \dfrac{32}{x^2 - 25}$

49. $\dfrac{1}{x - 4} - \dfrac{5}{x + 2} = \dfrac{6}{x^2 - 2x - 8}$

50. $\dfrac{6}{x + 3} - \dfrac{5}{x - 2} = \dfrac{-20}{x^2 + x - 6}$

In Exercises 51–58, determine whether each equation is an identity, a conditional equation, or an inconsistent equation.

51. $4(x - 7) = 4x - 28$ **52.** $4(x - 7) = 4x + 28$

53. $2x + 3 = 2x - 3$ **54.** $\dfrac{7x}{x} = 7$

55. $4x + 5x = 8x$ **56.** $8x + 2x = 9x$

57. $\dfrac{2x}{x - 3} = \dfrac{6}{x - 3} + 4$ **58.** $\dfrac{3}{x - 3} = \dfrac{x}{x - 3} + 3$

The equations in Exercises 59–70 combine the types of equations we have discussed in this section. Solve each equation or state that it is true for all real numbers or no real numbers.

59. $\dfrac{x + 5}{2} - 4 = \dfrac{2x - 1}{3}$ **60.** $\dfrac{x + 2}{7} = 5 - \dfrac{x + 1}{3}$

61. $\dfrac{2}{x - 2} = 3 + \dfrac{x}{x - 2}$ **62.** $\dfrac{6}{x + 3} + 2 = \dfrac{-2x}{x + 3}$

63. $8x - (3x + 2) + 10 = 3x$

64. $2(x + 2) + 2x = 4(x + 1)$

65. $\dfrac{2}{x} + \dfrac{1}{2} = \dfrac{3}{4}$ **66.** $\dfrac{3}{x} - \dfrac{1}{6} = \dfrac{1}{3}$

67. $\dfrac{4}{x - 2} + \dfrac{3}{x + 5} = \dfrac{7}{(x + 5)(x - 2)}$

68. $\dfrac{1}{x - 1} = \dfrac{1}{(2x + 3)(x - 1)} + \dfrac{4}{2x + 3}$

69. $\dfrac{4x}{x + 3} - \dfrac{12}{x - 3} = \dfrac{4x^2 + 36}{x^2 - 9}$

70. $\dfrac{4}{x^2 + 3x - 10} - \dfrac{1}{x^2 + x - 6} = \dfrac{3}{x^2 - x - 12}$

Application Exercises

71. The equation $d = 5000c - 525{,}000$ describes the relationship between the annual number of deaths, d, in the United States from heart disease and the average cholesterol level, c, of blood. (Cholesterol level, c, is expressed in milligrams per deciliter of blood.)

a. In 2000, 725,000 Americans died from heart disease. Substitute 725,000 for d in the given equation and then solve for c to determine the average cholesterol level in 2000.

b. Suppose that the average cholesterol level for people in the United States could be reduced to 180. Substitute 180 for c in the given equation and then compute the value for d to determine the number of annual deaths from heart disease with this reduced cholesterol level. Compared to the number of deaths in 2000, how many lives would be saved by this cholesterol reduction?

72. There is a relationship between the vocabulary of a child and the child's age. The equation $60A - V = 900$ describes this relationship, where A is the age of the child, in months, and V is the number of words that the child uses. Suppose that a child uses 1500 words. Determine the child's age, in months.

73. The equation

$$p = 15 + \dfrac{15d}{33}$$

describes the pressure of sea water, p, in pounds per square foot, at a depth of d feet below the surface. The record depth for breath-held diving, by Francisco Ferreras (Cuba) off Grand-Bahama Island, on November 14, 1993, involved pressure of 201 pounds per square foot. To what depth did Ferreras descend on this ill-advised venture? (He was underwater for 2 minutes and 9 seconds!)

74. The equation $P = -0.5d + 100$ describes the percentage, P, of lost hikers found in search and rescue missions when members of the search team walk parallel to one another separated by a distance of d yards. If a search and rescue team finds 70% of lost hikers, find the parallel distance of separation between members of the search party.

Writing in Mathematics

75. What is a linear equation in one variable? Give an example of this type of equation.

76. What does it mean to solve an equation?

77. What is the solution set of an equation?

78. What are equivalent equations? Give an example.

79. What is the difference between solving an equation such as $2(x - 4) + 5x = 34$ and simplifying an algebraic expression such as $2(x - 4) + 5x$? If there is a difference, which topic should be taught first? Why?

80. Suppose that you solve $\dfrac{x}{5} - \dfrac{x}{2} = 1$ by multiplying both sides by 20, rather than the least common denominator of 5 and 2 (namely, 10). Describe what happens. If you get the correct solution, why do you think we clear the equation of fractions by multiplying by the *least* common denominator?

81. Suppose you are an algebra teacher grading the following solution on an examination:

$$-3(x - 6) = 2 - x$$
$$-3x - 18 = 2 - x$$
$$-2x - 18 = 2$$
$$-2x = -16$$
$$x = 8.$$

You should note that 8 checks, and the solution set is $\{8\}$. The student who worked the problem therefore wants full credit. Can you find any errors in the solution? If full credit is 10 points, how many points should you give the student? Justify your position.

82. Explain how to determine the restrictions on the variable for the equation

$$\frac{3}{x + 5} + \frac{4}{x - 2} = \frac{7}{(x + 5)(x - 2)}.$$

83. What is an identity? Give an example.

84. What is a conditional equation? Give an example.

85. What is an inconsistent equation? Give an example.

 Technology Exercises

For Exercises 86–89, use your graphing utility to graph each side of the equations in the same viewing rectangle. Based on the resulting graph, label each equation as conditional, inconsistent, or an identity. If the equation is conditional, use the x-coordinate of the intersection point to find the solution set. Verify this value by direct substitution into the equation.

86. $2(x - 6) + 3x = x + 6$

87. $9x + 3 - 3x = 2(3x + 1)$

88. $2\left(x + \frac{1}{2}\right) = 5x + 1 - 3x$

89. $\dfrac{2x - 1}{3} - \dfrac{x - 5}{6} = \dfrac{x - 3}{4}$

 Critical Thinking Exercises

90. Which one of the following is true?
 a. The equation $-7x = x$ has no solution.
 b. The equations $\dfrac{x}{x - 4} = \dfrac{4}{x - 4}$ and $x = 4$ are equivalent.
 c. The equations $3y - 1 = 11$ and $3y - 7 = 5$ are equivalent.
 d. If a and b are any real numbers, then $ax + b = 0$ always has one number in its solution set.

91. Solve for x: $ax + b = c$.

92. Write three equations that are equivalent to $x = 5$.

93. If x represents a number, write an English sentence about the number that results in an inconsistent equation.

94. Find b such that $\dfrac{7x + 4}{b} + 13 = x$ will have a solution set given by $\{-6\}$.

95. Find b such that $\dfrac{4x - b}{x - 5} = 3$ will have a solution set given by $\varnothing$.

Group Exercise

96. In your group, describe the best procedure for solving the following equation:

$$0.47x + \frac{19}{4} = -0.2 + \frac{2}{5}x.$$

Use this procedure to actually solve the equation. Then compare procedures with other groups working on this problem. Which group devised the most streamlined method?

SECTION 1.3 *Formulas and Applications*

Objectives

1. Solve problems using formulas.
2. Use linear equations to solve problems.
3. Solve for a variable in a formula.

Could you live to be 125? The number of Americans ages 100 or older could approach 850,000 by 2050. Some scientists predict that by 2100, our descendants could live to be 200 years of age. In this section, we will see how equations can be used to make these kinds of predictions as we turn to applications of linear equations.

1 Solve problems using formulas.

Formulas and Modeling Data

The graph in Figure 1.11 shows life expectancy in the United States by year of birth. For example, we can use the graph to find life expectancy for women born in 1980. Find the two bars for 1980 and then look at the bar on the right, representing females. The number printed on this bar is 77.4. Thus, the life expectancy for women born in 1980 is 77.4 years.

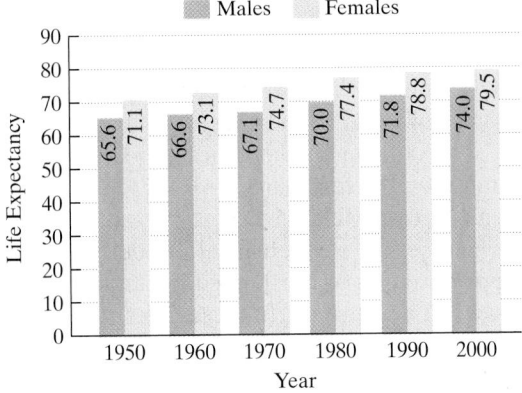

Life Expectancy by Year of Birth

Source: U.S. Bureau of the Census

Figure 1.11 Life expectancy by year of birth

The data for U.S. women in Figure 1.11 can be approximated by the equation

$$E = 0.177t + 71.35$$

where the variable E represents life expectancy for women born t years after 1950. This equation is an example of a *formula*. A **formula** is an equation that uses letters to express a relationship between two or more variables. The given formula expresses the relationship between the number of years born after 1950, t, and life expectancy for U.S. women, E.

EXAMPLE 1 Using a Formula

Use the formula

$$E = 0.177t + 71.35$$

to determine the year of birth for which U.S. women can expect to live 82 years.

Solution   We are given that the life expectancy for women is 82 years, so substitute 82 for E in the formula and solve for t.

$E = 0.177t + 71.35$	This is the given formula.
$82 = 0.177t + 71.35$	Replace E with 82 and solve for t.
$82 - 71.35 = 0.177t + 71.35 - 71.35$	Isolate the term containing t by subtracting 71.35 from both sides.
$10.65 = 0.177t$	Simplify.
$\dfrac{10.65}{0.177} = \dfrac{0.177t}{0.177}$	Divide both sides by 0.177.
$60 \approx t$	Simplify. Round to the nearest whole number.

The formula indicates that U.S. women born approximately 60 years after 1950, or in 2010, can expect to live 82 years.

The process of finding equations and formulas to describe real-world phenomena is called **mathematical modeling.** Such equations and formulas, together with the meaning assigned to the variables, are called **mathematical models.** One method of creating a mathematical model is to use available data and construct an equation that describes the behavior of the data. For example, consider the formula

$$E = 0.177t + 71.35$$

in which E is the life expectancy of the U.S. women born t years after 1950. This formula, or mathematical model, can be obtained from the data for women's life expectancy given in the bar graph in Figure 1.11 on the previous page. In Chapter 2, you will learn a modeling technique that will enable you to obtain the formula.

In creating mathematical models from data, we strive for both accuracy and simplicity. The formula $E = 0.177t + 71.35$ is relatively simple to use, but as we can see from Table 1.1, it is not an entirely accurate description of the data. Sometimes a mathematical model gives an estimate that is not a good approximation or is extended too far into the future, resulting in a prediction that does not make sense. In these cases, we say that **model breakdown** has occurred.

Table 1.1   Life Expectancy for U.S. Women

Birth Year	Actual Value	Value Predicted by $E = 0.177t + 71.35$
1950	71.1	71.35
1960	73.1	73.12
1970	74.7	74.89
1980	77.4	76.66
1990	78.8	78.43
2000	79.5	80.2

Check Point 1 The formula $W = 0.3x + 46.6$ models the average number of hours per week, W, that Americans worked x years after 1980. When will we average 55 hours of work per week?

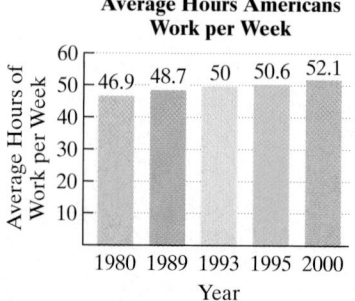

Average Hours Americans Work per Week

Source: U.S.A. Today

2 Use linear equations to solve problems.

Problem Solving with Linear Equations

Americans love their pets. The number of cats in the United States exceeds the number of dogs by 7.5 million. The number of cats and dogs combined is 114.7 million. So, how many dogs and cats are there in the United States?

Before answering the question, let's see if we can write a critical sentence that describes, or *models*, the problems conditions. The **verbal model** is

The number of dogs in the U.S.	plus	The number of cats in the U.S.	equals	114.7 million.
?	+	?	=	114.7 (million).

The question marks under the voice balloons indicate that we need algebraic expressions for these unknowns. Once we obtain these expressions, we will have an equation that models the verbal conditions. Because we are finding equations to describe real-world phenomena, we are engaged in mathematical modeling. The resulting equation, or mathematical model, is formed from a verbal model. Earlier, we mentioned that a mathematical model can be formed using actual data.

Here is a step-by-step strategy for solving problems using mathematical models that are created from verbal models:

Strategy for Problem Solving

Step 1 Read the problem carefully. Attempt to state the problem in your own words and state what the problem is looking for. Let x (or any variable) represent one of the quantities in the problem.

Step 2 If necessary, write expressions for any other unknown quantities in the problem in terms of x.

Step 3 Form a verbal model of the problems conditions and then write an equation in x that translates the verbal model.

Step 4 Solve the equation and answer the question in the problem.

Step 5 Check the proposed solution in the *original wording* of the problem, not in the equation obtained from the words.

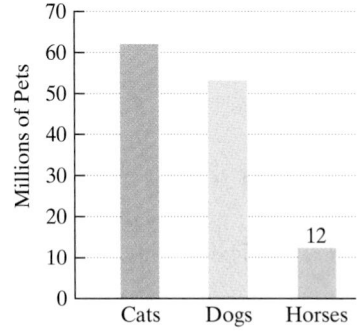

U.S. Pet Population

Source: American Veterinary Medical Association

Americans spend more than $21 billion a year on their pets. 31.4% of households have cats and 34.3% have dogs.

EXAMPLE 2 Pet Population

The number of cats in the United States exceeds the number of dogs by 7.5 million. The number of cats and dogs combined is 114.7 million. Determine the number of dogs and cats in the United States.

Solution

Step 1 Let x represent one of the quantities. We know something about the number of cats; the cat population exceeds the dog population by 7.5 million. This means that there are 7.5 million more cats than dogs. We will let

x = the number, in millions, of dogs in the United States.

Step 2 Represent other quantities in terms of x. The other unknown quantity is the number of cats. Because there are 7.5 million more cats than dogs, let

$x + 7.5$ = the number, in millions, of cats in the United States.

Step 3 Write an equation in x that describes the conditions. The number of cats and dogs combined is 114.7 million.

The number (in millions) of dogs in the U.S.	plus	the number (in millions) of cats in the U.S.	equals	114.7 million.
x	$+$	$x + 7.5$	$=$	114.7

Step 4 Solve the equation and answer the question.

$$x + x + 7.5 = 114.7 \qquad \text{This is the equation that models the verbal conditions.}$$

$$2x + 7.5 = 114.7 \qquad \text{Combine like terms on the left side.}$$

$$2x + 7.5 - 7.5 = 114.7 - 7.5 \qquad \text{Subtract 7.5 from both sides.}$$

$$2x = 107.2 \qquad \text{Simplify.}$$

$$\frac{2x}{2} = \frac{107.2}{2} \qquad \text{Divide both sides by 2.}$$

$$x = 53.6 \qquad \text{Simplify.}$$

Because x represents the number, in millions, of dogs, there are 53.6 million dogs in the United States. Because $x + 7.5$ represents the number, in millions, of cats, there are $53.6 + 7.5$, or 61.1 million cats in the United States.

Step 5 Check the proposed solution in the original wording of the problem. The problem states that the number of cats and dogs combined is 114.7 million. By adding 53.6 million, the dog population, and 61.1 million, the cat population, we do, indeed, obtain a sum of 114.7 million.

 Check Point 2 Two of the top-selling music albums of all time are *Jagged Little Pill* (Alanis Morissette) and *Saturday Night Fever* (Bee Gees). The Morissette album sold 5 million more copies than that of the Bee Gees. Combined, the two albums sold 27 million copies. Determine the number of sales for each of the albums.

EXAMPLE 3 Selecting a Long-Distance Carrier

You are choosing between two long-distance telephone plans. Plan A has a monthly fee of $20 with a charge of $0.05 per minute for all long-distance calls. Plan B has a monthly fee of $5 with a charge of $0.10 per minute for all long-distance calls. For how many minutes of long-distance calls will the costs for the two plans be the same?

Solution

Step 1 Let x represent one of the quantities. Let

$$x = \text{the number of minutes of long-distance calls for the two plans to cost the same.}$$

Step 2 Represent other quantities in terms of x. There are no other unknown quantities, so we can skip this step.

Step 3 Write an equation in x that describes the conditions. The monthly cost for plan A is the monthly fee, $20, plus the per minute charge, $0.05, times the number of minutes of long-distance calls, x. The monthly cost for plan B is the monthly fee, $5, plus the per-minute charge, $0.10, times the number of minutes of long-distance calls, x.

| The monthly cost for plan A | must equal | the monthly cost for plan B. |

$$20 + 0.05x = 5 + 0.10x$$

Step 4 Solve the equation and answer the question.

$20 + 0.05x = 5 + 0.10x$	This is the equation that models the verbal conditions.
$20 + 0.05x - 0.05x = 5 + 0.10x - 0.05x$	Subtract 0.05x from both sides.
$20 = 5 + 0.05x$	Simplify.
$20 - 5 = 5 + 0.05x - 5$	Subtract 5 from both sides.
$15 = 0.05x$	Simplify.
$\dfrac{15}{0.05} = \dfrac{0.05x}{0.05}$	Divide both sides by 0.05.
$300 = x$	Simplify.

Because x represents the number of minutes of long-distance calls for the two plans to cost the same, the costs will be the same with 300 minutes of long-distance calls.

Step 5 Check the proposed solution in the original wording of the problem.
The problem states that the costs for the two plans should be the same. Let's see if they are with 300 minutes of long-distance calls:

$$\text{Cost for plan A} = \$20 + \$0.05(300) = \$20 + \$15 = \$35$$

| Monthly fee | Per-minute charge |

$$\text{Cost for plan B} = \$5 + \$0.10(300) = \$5 + \$30 = \$35$$

With 300 minutes, or 5 hours, of long-distance chatting, both plans cost $35 for the month. Thus, the proposed solution, 300 minutes, satisfies the problems conditions.

Check Point 3 You are choosing between two long-distance telephone plans. Plan A has a monthly fee of $15 with a charge of $0.08 per minute for all long-distance calls. Plan B has a monthly fee of $3 with a charge of $0.12 per minute for all long-distance calls. For how many minutes of long-distance calls will the costs for the two plans be the same?

Our next example involves simple interest. The annual simple interest that an investment earns is given by the formula

$$I = Pr$$

where I is the simple interest, P is the principal, and r is the simple interest rate, expressed in decimal form. Suppose, for example, that you deposit $2000 ($P = 2000$) in a savings account that has a simple interest rate of 6% ($r = 0.06$). The annual simple interest is computed as follows:

$$I = Pr = (2000)(0.06) = 120.$$

The annual interest is $120.

EXAMPLE 4   Solving a Simple Interest Problem

You inherit $16,000 with the stipulation that for the first year the money must be placed in two investments paying 6% and 8% annual interest, respectively. How much should be invested at each rate if the total interest earned for the year is to be $1180?

Solution

Step 1   Let x represent one of the quantities.

Let x = the amount invested at 6%.

Step 2   Represent other quantities in terms of x.   The other quantity that we seek is the amount to be invested at 8%. Because the total amount to be invested is $16,000, and we already used up x,

$16,000 - x$ = the amount invested at 8%.

Step 3   Write an equation in x that describes the conditions.   The interest for the two investments combined must be $1180. Interest is Pr or rP for each investment.

Interest from 6% investment	plus	interest from 8% investment	is	$1180.
$0.06x$	$+$	$0.08(16{,}000 - x)$	$=$	1180
rate times principal		rate times principal		

Step 4   Solve the equation and answer the question.

$$0.06x + 0.08(16{,}000 - x) = 1180 \qquad \text{This is the equation that models the verbal conditions.}$$

$$0.06x + 1280 - 0.08x = 1180 \qquad \text{Use the distributive property.}$$

$$-0.02x + 1280 = 1180 \qquad \text{Combine like terms.}$$

$$-0.02x + 1280 - 1280 = 1180 - 1280 \qquad \text{Subtract 1280 from both sides.}$$

$$-0.02x = -100 \qquad \text{Simplify.}$$

$$\frac{-0.02x}{-0.02} = \frac{-100}{-0.02} \qquad \text{Divide both sides by } -0.02.$$

$$x = 5000 \qquad \text{Simplify.}$$

Because x represents the amount invested at 6%, $5000 should be invested at 6%. Because $16,000 - x$ represents the amount invested at 8%, $16,000 - $5000, or $11,000, should be invested at 8%.

Step 5   Check the proposed solution in the original wording of the problem. The problem states that the total interest should be $1180. The interest earned on $5000 at 6% is ($5000)(0.06), or $300. The interest earned on $11,000 at 8% is ($11,000)(0.08), or $880. The total interest is $300 + $880, or $1180, exactly as it should be.

Check Point 4   Suppose that you invest $25,000, part at 9% simple interest and the remainder at 12%. If the total yearly interest from these investments was $2550, find the amount invested at each rate.

Solving geometry problems usually requires a knowledge of basic geometric ideas and formulas. Formulas for area, perimeter, and volume are given in Table 1.2.

Table 1.2 Common Formulas for Area, Perimeter, and Volume

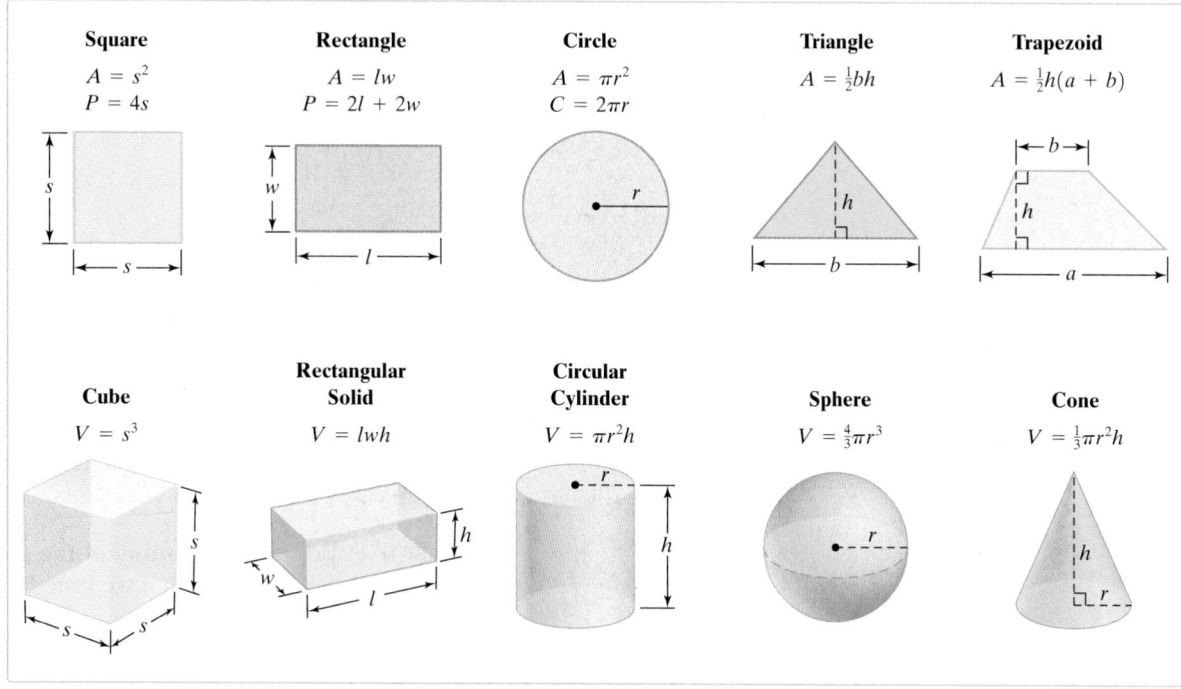

We will be using the formula for the perimeter of a rectangle, $P = 2l + 2w$, in our next example. A helpful verbal model for this formula is 2 times length plus 2 times width is a rectangles perimeter.

EXAMPLE 5 Finding the Dimensions of an American Football Field

The length of an American football field is 200 feet more than the width. If the perimeter of the field is 1040 feet, what are its dimensions?

Solution

Step 1   Let x represent one of the quantities.   We know something about the length; the length is 200 feet more than the width. We will let

$$x = \text{the width.}$$

Step 2   Represent other quantities in terms of x.   Because the length is 200 feet more than the width, let

$$x + 200 = \text{the length.}$$

Figure 1.12 illustrates an American football field and its dimensions.

Step 3   Write an equation in x that describes the conditions.   Because the perimeter of the field is 1040 feet,

Twice the length	plus	twice the width	is	the perimeter.
$2(x + 200)$	$+$	$2 \cdot x$	$=$	$1040.$

Figure 1.12 An American football field

Figure 1.12, repeated

Step 4   Solve the equation and answer the question.

$$2(x + 200) + 2x = 1040$$ This is the equation that models the verbal conditions.

$$2x + 400 + 2x = 1040$$ Apply the distributive property.

$$4x + 400 = 1040$$ Combine like terms: $2x + 2x = 4x$.

$$4x + 400 - 400 = 1040 - 400$$ Subtract 400 from both sides.

$$4x = 640$$ Simplify.

$$\frac{4x}{4} = \frac{640}{4}$$ Divide both sides by 4.

$$x = 160$$ Simplify.

Thus,

$$\text{width} = x = 160.$$

$$\text{length} = x + 200 = 160 + 200 = 360.$$

The dimensions of an American football field are 360 feet by 160 feet. (The 360-foot length is usually described as 120 yards.)

Step 5   Check the proposed solution in the original wording of the problem.
The perimeter of the football field using the dimensions that we found is

$$2(360\ \text{feet}) + 2(160\ \text{feet}) = 720\ \text{feet} + 320\ \text{feet} = 1040\ \text{feet}.$$

Because the problems wording tells us that the perimeter is 1040 feet, our dimensions are correct.

> **Check Point 5**   The length of a rectangular basketball court is 44 feet more than the width. If the perimeter of the basketball court is 288 feet, what are its dimensions?

3 Solve for a variable in a formula.

Solving for a Variable in a Formula

When working with formulas, such as the geometric formulas shown in Table 1.2 on the previous page, it is often necessary to solve for a specified variable. This is done by isolating the specified variable on one side of the equation. Begin by isolating all terms with the specified variable on one side of the equation and all terms without the specified variable on the other side. The next example shows how to do this.

EXAMPLE 6   Solving for a Variable in a Formula

Solve the formula $2l + 2w = P$ for w.

Solution   First, isolate $2w$ on the left by subtracting $2l$ from both sides. Then solve for w by dividing both sides by 2.

> We need to isolate w.

$$2l + 2w = P$$ This is the given formula.

$$2l - 2l + 2w = P - 2l \qquad \text{Isolate } 2w \text{ by subtracting } 2l \text{ from both sides.}$$

$$2w = P - 2l \qquad \text{Simplify.}$$

$$\frac{2w}{2} = \frac{P - 2l}{2} \qquad \text{Isolate } w \text{ by dividing both sides by 2.}$$

You can divide both P and $2l$ by 2, expressing the answer as $w = \frac{P}{2} - l$.

$$w = \frac{P - 2l}{2} \qquad \text{Simplify.}$$

Check Point 6   Solve $y = mx + b$ for m.

EXAMPLE 7   Solving for a Variable That Occurs Twice in a Formula

Study Tip

You cannot solve $A = P + Prt$ for P by subtracting Prt from both sides and writing

$$A - Prt = P.$$

When a formula is solved for a specified variable, that variable must be isolated on one side. The variable P occurs on both sides of

$$A - Prt = P.$$

Solve the formula $A = P + Prt$ for P.

Solution   Notice that all terms with P already occur on the right side of the equation. Factor P from the two terms on the right to isolate P.

$$A = P + Prt \qquad \text{This is the given formula.}$$

$$A = P(1 + rt) \qquad \text{Factor } P \text{ on the right side of the equation.}$$

$$\frac{A}{1 + rt} = \frac{P(1 + rt)}{1 + rt} \qquad \text{Divide both sides by } 1 + rt.$$

$$\frac{A}{1 + rt} = P \qquad \text{Simplify: } \frac{P(1 + rt)}{(1 + rt)} = \frac{P}{1} = P.$$

Check Point 7   Solve the formula $P = C + MC$ for C.

EXERCISE SET 1.3

Practice Exercises

In Exercises 1–14, let x represent the number. Write each English phrase as an algebraic expression.

1. The sum of a number and 9

2. A number increased by 13

3. A number subtracted from 20

4. 13 less than a number

5. 8 decreased by 5 times a number

6. 14 less than the product of 6 and a number

7. The quotient of 15 and a number

8. The quotient of a number and 15

9. The sum of twice a number and 20

10. Twice the sum of a number and 20

11. 30 subtracted from 7 times a number

12. The quotient of 12 and a number, decreased by 3 times the number

13. Four times the sum of a number and 12

14. Five times the difference of a number and 6

In Exercises 15–20, let x represent the number. Use the given conditions to write an equation. Solve the equation and find the number.

15. A number increased by 40 is equal to 450. Find the number.

16. The sum of a number and 29 is 54. Find the number.

17. Seven subtracted from five times a number is 123. Find the number.

18. Eight subtracted from six times a number is 184. Find the number.

19. Nine times a number is 30 more than three times that number. Find the number.

20. Five more than four times a number is that number increased by 35. Find the number.

Application Exercises

Medical researchers have found that the desirable heart rate, R, in beats per minute, for beneficial exercise is modeled by the formulas

$$R = 143 - 0.65A \quad \text{for women}$$

$$R = 165 - 0.75A \quad \text{for men}$$

where A is the person's age. Use these formulas to solve Exercises 21–22.

21. If the desirable heart rate for a woman is 117 beats per minute, how old is she? How is the solution shown on the accompanying line graph?

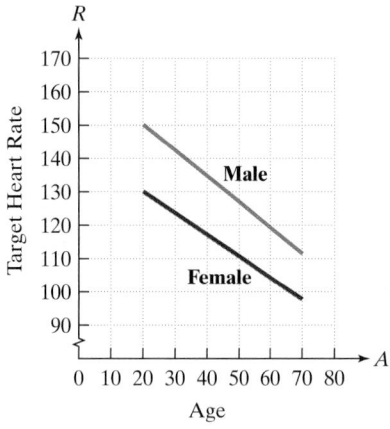

22. If the desirable heart rate for a man is 147 beats per minute, how old is he? How is the solution shown on the line graph?

Growth in human populations and economic activity threatens the continued existence of salmon in the Pacific Northwest. The bar graph shows the Pacific salmon population for various years. The data can be modeled by the formula

$$P = -0.22t + 9.6$$

in which P is the salmon population, in millions, t years after 1960. Use the formula to solve Exercises 23–24. Round to the nearest year.

Pacific Salmon Population

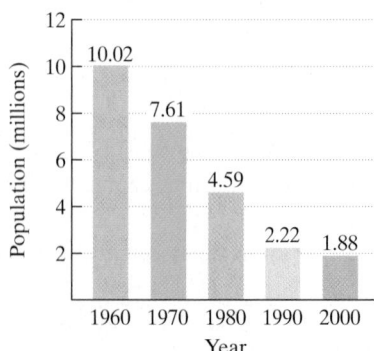

Source: U.S. Department of the Interior

23. When will the salmon population be reduced to 0.5 million?

24. When will there be no Pacific salmon?

25. The formula

$$\frac{W}{2} - 3H = 53$$

models the recommended weight W, in pounds, for a male, where H represents the man's height, in inches, over 5 feet. What is the recommended weight for a man who is 6 feet, 3 inches tall?

26. The International Panel on Climate Change is a U.N.-sponsored body made up of more than 1500 leading experts from 60 nations. According to their recent findings, increased levels of atmospheric carbon dioxide are affecting our climate. Global warming is under way and the effects could be catastrophic. The formula $C = 1.44t + 280$ models carbon dioxide concentration, C, in parts per million, t years after 1939. The preindustrial carbon dioxide concentration of 280 parts per million remained fairly constant until World War II, increasing after that due primarily to the burning of fossil fuels related to energy consumption. When will the concentration be double the preindustrial level? Round to the nearest year.

In Exercises 27–56, use the five-step strategy given in the box on page 97 to solve each problem.

27. Two of the most expensive movies ever made were *Titanic* and *Waterworld*. The cost to make *Titanic* exceeded the cost to make *Waterworld* by $40 million. The combined cost to make the two movies was $360 million. Find the cost of making each of these movies.

28. In 2001, the most populous countries in the world were China and India. In that year, China's population exceeded India's by 260 million. Combined, the two countries had a population of 2310 million. Determine the 2001 population for China and India.

29. Each year, Americans in 68 urban areas waste almost 7 billion gallons of fuel sitting in traffic. The bar graph shows the number of hours in traffic per year for the average motorist in ten cities. The average motorist in Los Angeles spends 32 hours less than twice that of the average motorist in Miami stuck in traffic each year. Together, the average motorist in Miami and the average motorist in Los Angeles spend 139 hours per year in traffic. How many hours are wasted in traffic by the average motorist in Los Angeles and Miami?

Hours in Traffic per Year for the Average Motorist

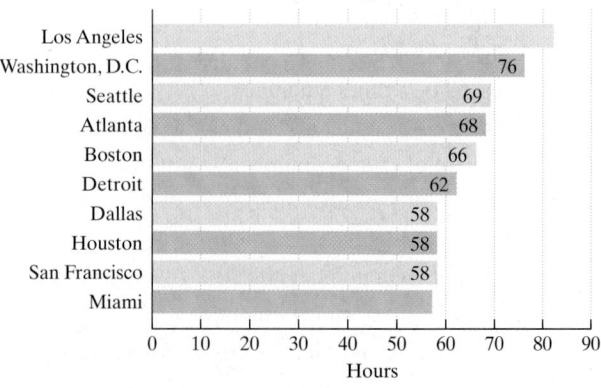

Source: Texas Transportation Institute

30. The graph shows the five costliest natural disasters in U.S. history. The cost of the Northridge, California, earthquake exceeded Hurricane Hugo by $5.5 billion and the cost of Hurricane Andrew exceeded twice that of Hugo by $6 billion. The combined cost of the three natural disasters was $39.5 billion. Determine the cost of each.

Costliest Natural Disasters in U.S. History

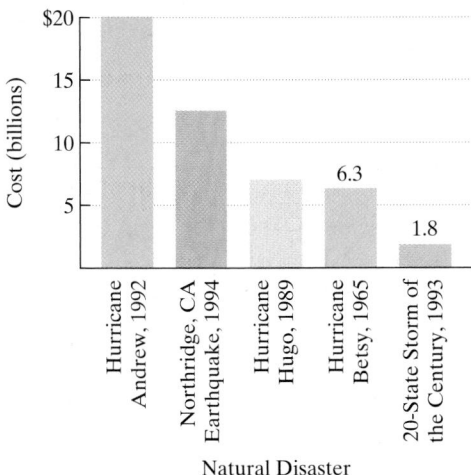

Source: Federal Emergency Management Agency

31. A car rental agency charges $200 per week plus $0.15 per mile to rent a car. How many miles can you travel in one week for $320?

32. A car rental agency charges $180 per week plus $0.25 per mile to rent a car. How many miles can you travel in one week for $395?

According to the National Center for Health Statistics, in 1990, 28% of babies in the United States were born to parents who were not married. Throughout the 1990s, this increased by approximately 0.6% per year. Use this information to solve Exercises 33–34.

33. If this trend continues, in which year will 37% of babies be born out of wedlock?

34. If this trend continues, in which year will 40% of babies be born out of wedlock?

35. The bus fare in a city is $1.25. People who use the bus have the option of purchasing a monthly coupon book for $21.00. With the coupon book, the fare is reduced to $0.50.
   a. Let x represent the number of times in a month the bus is used. Write algebraic expressions for the total monthly costs of using the bus x times both with and without the coupon book.
   b. Determine the number of times in a month the bus must be used so that the total monthly cost without the coupon book is the same as the total monthly cost with the coupon book.

36. A coupon book for a bridge costs $21 per month. The toll for the bridge is normally $2.50, but it is reduced to $1 for people who have purchased the coupon book.

   a. Let x represent the number of times in a month the bridge is used. Write algebraic expressions for the total monthly costs of using the bridge x times both with and without the coupon book.
   b. Determine the number of times in a month the bridge must be crossed so that the total monthly cost without the coupon book is the same as the total monthly cost with the coupon book.

37. You are choosing between two plans at a discount warehouse. Plan A offers an annual membership fee of $100 and you pay 80% of the manufacturer's recommended list price. Plan B offers an annual membership fee of $40 and you pay 90% of the manufacturer's recommended list price. How many dollars of merchandise would you have to purchase in a year to pay the same amount under both plans? What will be the cost for each plan?

38. You are choosing between two plans at a discount warehouse. Plan A offers an annual membership fee of $300 and you pay 70% of the manufacturer's recommended list price. Plan B offers an annual membership fee of $40 and you pay 90% of the manufacturer's recommended list price. How many dollars of merchandise would you have to purchase in a year to pay the same amount under both plans? What will be the cost for each plan?

39. Your grandmother needs your help. She has $50,000 to invest. Part of this money is to be invested in noninsured bonds paying 15% annual interest. The rest of this money is to be invested in a government-insured certificate of deposit paying 7% annual interest. She told you that she requires $6000 per year in extra income from both of these investments. How much money should be placed in each investment?

40. You inherit $18,750 with the stipulation that for the first year the money must be placed in two investments paying 10% and 12% annual interest, respectively. How much should be invested at each rate if the total interest earned for the year is to be $2117?

41. Things did not go quite as planned. You invested $8000, part of it in stock that paid 12% annual interest. However, the rest of the money suffered a 5% loss. If the total annual income from both investments was $620, how much was invested at each rate?

42. Things did not go quite as planned. You invested $12,000, part of it in stock that paid 14% annual interest. However, the rest of the money suffered a 6% loss. If the total annual income from both investments was $680, how much was invested at each rate?

43. The length of the rectangular tennis court at Wimbledon is 6 feet longer than twice the width. If the court's perimeter is 228 feet, what are the court's dimensions?

44. A rectangular soccer field is twice as long as it is wide. If the perimeter of the soccer field is 300 yards, what are its dimensions?

45. The height of the bookcase in the figure is 3 feet longer than the length of a shelf. If 18 feet of lumber is available for the entire unit, find the length and height of the unit.

shelf length

46. A bookcase is to be constructed as shown in the figure. The length is to be 3 times the height. If 60 feet of lumber is available for the entire unit, find the length and height of the bookcase.

length

47. An automobile repair shop charged a customer $448, listing $63 for parts and the remainder for labor. If the cost of labor is $35 per hour, how many hours of labor did it take to repair the car?

48. A repair bill on a yacht came to $1603, including $532 for parts and the remainder for labor. If the cost of labor is $63 per hour, how many hours of labor did it take to repair the yacht?

The graph shows median, or average, income by level of education. Exercises 49–50 use the information in the bar graph.

Income by Level of Education

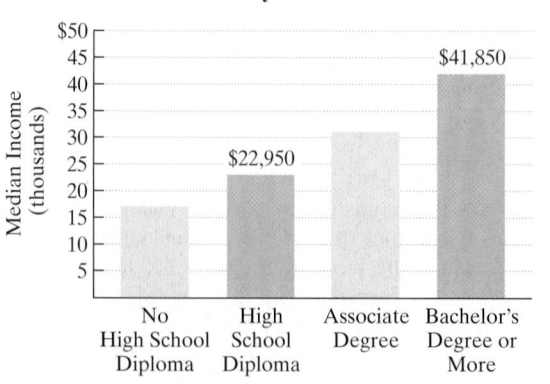

Source: U.S. Department of Commerce

49. The annual salary for people with a bachelor's degree or more is an increase of 35% of the annual salary for people with an associate degree. What is the average annual salary with an associate degree?

50. The annual salary for high school graduates is an increase of 35% of the annual salary for people without a high school diploma. What is the average annual salary without a high school diploma?

51. Answer the question in the following *Peanuts* cartoon strip. (*Note:* You may not use the answer given in the cartoon!)

PEANUTS reprinted by permission of United Features Syndicate, Inc.

52. After a graphing calculator's price is reduced by $\frac{1}{3}$ of its original price, you purchase it for $64. What was the graphing calculator's price before the reduction?

53. After a 12% price reduction, a car sold for $17,600. What was the car's price before the reduction?

54. Including 8% sales tax, an inn charges $162 per night. Find the inn's nightly cost before the tax is added.

55. An HMO pamphlet contains the following recommended weight for women: "Give yourself 100 pounds for the first 5 feet plus 5 pounds for every inch over 5 feet tall." Using this description, what height corresponds to a recommended weight of 135 pounds?

56. A job pays an annual salary of $33,150, which includes a holiday bonus of $750. If paychecks are issued twice a month, what is the gross amount for each paycheck?

In Exercises 57–76, solve each formula for the specified variable. Do you recognize the formula? If so, what does it describe?

57. $A = lw$ for w

58. $D = RT$ for R

59. $A = \frac{1}{2}bh$ for b

60. $V = \frac{1}{3}Bh$ for B

61. $I = Prt$ for P

62. $C = 2\pi r$ for r

63. $E = mc^2$ for m

64. $V = \pi r^2 h$ for h

65. $T = D + pm$ for p

66. $P = C + MC$ for M

67. $A = \frac{1}{2}h(a + b)$ for a

68. $A = \frac{1}{2}h(a + b)$ for b

69. $S = P + Prt$ for r

70. $S = P + Prt$ for t

71. $B = \dfrac{F}{S - V}$ for S

72. $S = \dfrac{C}{1 - r}$ for r

73. $IR + Ir = E$ for I

74. $A = 2lw + 2lh + 2wh$ for h

75. $\dfrac{1}{p} + \dfrac{1}{q} = \dfrac{1}{f}$ for f

76. $\dfrac{1}{R} = \dfrac{1}{R_1} + \dfrac{1}{R_2}$ for R_1

Writing in Mathematics

77. What is a formula?

78. We discussed formulas in this section after we considered procedures for solving linear equations. Doesn't working with a formula simply mean substituting given numbers into the formula and using the order of operations? Is it necessary to know how to solve equations to work with formulas? Explain.

79. In your own words, describe a step-by-step approach for solving algebraic word problems.

80. Did you have some difficulties solving some of the problems that were assigned in this exercise set? Discuss what you did if this happened to you. Did your course of action enhance your ability to solve algebraic word problems?

Technology Exercises

81. The formula $y = 28 + 0.6x$ models the percentage, y, of U.S. babies born out of wedlock x years after 1990. Graph the formula in a $[0, 20, 5]$ by $[0, 50, 10]$ viewing rectangle. Then use the ⌈TRACE⌉ or ⌈ZOOM⌉ feature to verify your answer in Exercise 33 or 34.

82. A tennis club offers two payment options. Members can pay a monthly fee of $30 plus $5 per hour for court rental time. The second option has no monthly fee, but court time costs $7.50 per hour.
  a. Write a mathematical model representing total monthly costs for each option for x hours of court rental time.
  b. Use a graphing utility to graph the two models in a $[0, 15, 1]$ by $[0, 120, 20]$ viewing rectangle.
  c. Use your utility's trace or intersection feature to determine where the two graphs intersect. Describe what the coordinates of this intersection point represent in practical terms.
  d. Verify part (c) using an algebraic approach by setting the two models equal to one another and determining how many hours one has to rent the court so that the two plans result in identical monthly costs.

Critical Thinking Exercises

83. At the north campus of a performing arts school, 10% of the students are music majors. At the south campus, 90% of the students are music majors. The campuses are merged into one east campus. If 42% of the 1000 students at the east campus are music majors, how many students did the north and south campuses have before the merger?

84. The price of a dress is reduced by 40%. When the dress still does not sell, it is reduced by 40% of the reduced price. If the price of the dress after both reductions is $72, what was the original price?

85. In a film, the actor Charles Coburn plays an elderly "uncle" character criticized for marrying a woman when he is 3 times her age. He wittily replies, "Ah, but in 20 years time I shall only be twice her age." How old is the "uncle" and the woman?

86. Suppose that we agree to pay you 8¢ for every problem in this chapter that you solve correctly and fine you 5¢ for every problem done incorrectly. If at the end of 26 problems we do not owe each other any money, how many problems did you solve correctly?

87. It was wartime when the Ricardos found out Mrs. Ricardo was pregnant. Ricky Ricardo was drafted and made out a will, deciding that $14,000 in a savings account was to be divided between his wife and his child-to-be. Rather strangely, and certainly with gender bias, Ricky stipulated that if the child were a boy, he would get twice the amount of the mother's portion. If it were a girl, the mother would get twice the amount the girl was to receive. We'll never know what Ricky was thinking of, for (as fate would have it) he did not return from war. Mrs. Ricardo gave birth to twins—a boy and a girl. How was the money divided?

88. Solve for C: $V = C - \dfrac{C - S}{L} N.$

Group Exercise

89. One of the best ways to learn how to *solve* a word problem in algebra is to *design* word problems of your own. Creating a word problem makes you very aware of precisely how much information is needed to solve the problem. You must also focus on the best way to present information to a reader and on how much information to give. As you write your problem, you gain skills that will help you solve problems created by others.

   The group should design five different word problems that can be solved using linear equations. All of the problems should be on different topics. For example, the group should not have more than one problem on simple interest. The group should turn in both the problems and their algebraic solutions.

SECTION 1.4 *Complex Numbers*

Objectives

1. Add and subtract complex numbers.
2. Multiply complex numbers.
3. Divide complex numbers.
4. Perform operations with square roots of negative numbers.

THE KID WHO LEARNED ABOUT MATH ON THE STREET

If you divide 6,973 by 0, you die.

Once, this guy tried to find the square root of -9, and his eyeballs turned black.

This girl my brother knows found out exactly what π equals, but she went nuts.

Who is this kid warning us about our eyeballs turning black if we attempt to find the square root of −9? Don't believe what you hear on the street. Although square roots of negative numbers are not real numbers, they do play a significant role in algebra. In this section, we move beyond the real numbers and discuss square roots with negative radicands.

The Imaginary Unit *i*

In the next section, we'll be studying equations whose solutions involve the square roots of negative numbers. Because the square of a real number is never negative, there is no real number x such that $x^2 = -1$. To provide a setting in which such equations have solutions, mathematicians invented an expanded system of numbers, the complex numbers. The *imaginary number i*, defined to be a solution of the equation $x^2 = -1$, is the basis of this new set.

> **The Imaginary Unit *i***
>
> The **imaginary unit** *i* is defined as
>
> $$i = \sqrt{-1}, \quad \text{where} \quad i^2 = -1.$$

Using the imaginary unit *i*, we can express the square root of any negative number as a real multiple of *i*. For example,

$$\sqrt{-25} = i\sqrt{25} = 5i.$$

We can check this result by squaring $5i$ and obtaining -25.

$$(5i)^2 = 5^2 i^2 = 25(-1) = -25$$

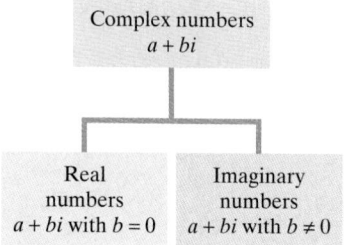

Figure 1.13 The complex number system

A new system of numbers, called *complex numbers*, is based on adding multiples of i, such as $5i$, to the real numbers.

Complex Numbers

The set of all numbers in the form

$$a + bi$$

with real numbers a and b, and i, the imaginary unit, is called the set of **complex numbers.** The real number a is called the **real part,** and the real number b is called the **imaginary part,** of the complex number $a + bi$. If $b \neq 0$, then the complex number is called an **imaginary number** (Figure 1.13). An imaginary number in the form bi is called a **pure imaginary number.**

Here are some examples of complex numbers. Each number can be written in the form $a + bi$.

$$-4 + 6i \qquad\qquad 2i = 0 + 2i \qquad\qquad 3 = 3 + 0i$$

| a, the real part, is −4. | b, the imaginary part, is 6. | a, the real part, is 0. | b, the imaginary part, is 2. | a, the real part, is 3. | b, the imaginary part, is 0. |

Can you see that b, the imaginary part, is not zero in the first two complex numbers? Because $b \neq 0$, these complex numbers are imaginary numbers. Furthermore, the imaginary number $2i$ is a pure imaginary number. By contrast, the imaginary part of the complex number on the right is zero. This complex number is not an imaginary number. The number 3, or $3 + 0i$, is a real number.

A complex number is said to be **simplified** if it is expressed in the **standard form** $a + bi$. If b is a radical, we usually write i before b. For example, we write $7 + i\sqrt{5}$ rather than $7 + \sqrt{5}i$, which could easily be confused with $7 + \sqrt{5i}$.

Expressed in standard form, two complex numbers are equal if and only if their real parts are equal and their imaginary parts are equal.

Equality of Complex Numbers

$a + bi = c + di$ if and only if $a = c$ and $b = d$.

1 Add and subtract complex numbers.

Operations with Complex Numbers

The form of a complex number $a + bi$ is like the binomial $a + bx$. Consequently, we can add, subtract, and multiply complex numbers using the same methods we used for binomials, remembering that $i^2 = -1$.

Adding and Subtracting Complex Numbers

1. $(a + bi) + (c + di) = (a + c) + (b + d)i$

In words, this says that you add complex numbers by adding their real parts, adding their imaginary parts, and expressing the sum as a complex number.

2. $(a + bi) - (c + di) = (a - c) + (b - d)i$

In words, this says that you subtract complex numbers by subtracting their real parts, subtracting their imaginary parts, and expressing the difference as a complex number.

EXAMPLE 1 Adding and Subtracting Complex Numbers

Perform the indicated operations, writing the result in standard form:

a. $(5 - 11i) + (7 + 4i)$ **b.** $(-5 + 7i) - (-11 - 6i)$.

Study Tip

The following examples, using the same integers as in Example 1, show how operations with complex numbers are just like operations with polynomials.

a. $(5 - 11x) + (7 + 4x)$
$= 12 - 7x$

b. $(-5 + 7x) - (-11 - 6x)$
$= -5 + 7x + 11 + 6x$
$= 6 + 13x$

Solution

a. $(5 - 11i) + (7 + 4i)$
$= 5 - 11i + 7 + 4i$ *Remove the parentheses.*
$= 5 + 7 - 11i + 4i$ *Group real and imaginary terms.*
$= (5 + 7) + (-11 + 4)i$ *Add real parts and add imaginary parts.*
$= 12 - 7i$ *Simplify.*

b. $(-5 + 7i) - (-11 - 6i)$
$= -5 + 7i + 11 + 6i$ *Remove the parentheses.*
$= -5 + 11 + 7i + 6i$ *Group real and imaginary terms.*
$= (-5 + 11) + (7 + 6)i$ *Add real parts and add imaginary parts.*
$= 6 + 13i$ *Simplify.*

> **Check Point 1**   Add or subtract as indicated:
>
> **a.** $(5 - 2i) + (3 + 3i)$ **b.** $(2 + 6i) - (12 - 4i)$.

2   Multiply complex numbers.

Multiplication of complex numbers is performed the same way as multiplication of polynomials, using the distributive property and the FOIL method. After completing the multiplication, we replace i^2 with -1. This idea is illustrated in the next example.

EXAMPLE 2 Multiplying Complex Numbers

Find the products:

a. $4i(3 - 5i)$ **b.** $(7 - 3i)(-2 - 5i)$.

Solution

a. $4i(3 - 5i) = 4i(3) - 4i(5i)$ *Distribute 4i throughout the parentheses.*
$= 12i - 20i^2$ *Multiply.*
$= 12i - 20(-1)$ *Replace i^2 with -1.*
$= 20 + 12i$ *Simplify to 12i + 20 and write in standard form.*

b. $(7 - 3i)(-2 - 5i)$

      F   O   I   L

$= -14 - 35i + 6i + 15i^2$ *Use the FOIL method.*
$= -14 - 35i + 6i + 15(-1)$ *$i^2 = -1$*
$= -14 - 15 - 35i + 6i$ *Group real and imaginary terms.*
$= -29 - 29i$ *Combine real and imaginary terms.*

> **Check Point 2**   Find the products:
>
> **a.** $7i(2 - 9i)$ **b.** $(5 + 4i)(6 - 7i)$.

3   Divide complex numbers.

Complex Conjugates and Division

It is possible to multiply complex numbers and obtain a real number. This occurs when we multiply $a + bi$ and $a - bi$.

$$(a + bi)(a - bi) = \overset{F}{a^2} - \overset{O}{abi} + \overset{I}{abi} - \overset{L}{b^2 i^2} \qquad \text{Use the FOIL method.}$$

$$= a^2 - b^2(-1) \qquad\qquad i^2 = -1$$

$$= a^2 + b^2 \qquad\qquad\qquad \text{Notice that this product eliminates } i.$$

For the complex number $a + bi$, we define its *complex conjugate* to be $a - bi$. The multiplication of complex conjugates results in a real number.

Conjugate of a Complex Number

The **complex conjugate** of the number $a + bi$ is $a - bi$, and the complex conjugate of $a - bi$ is $a + bi$. The multiplication of complex conjugates gives a real number.

$$(a + bi)(a - bi) = a^2 + b^2$$
$$(a - bi)(a + bi) = a^2 + b^2$$

Complex conjugates are used when dividing complex numbers. By multiplying the numerator and the denominator of the division by the complex conjugate of the denominator, you will obtain a real number in the denominator.

EXAMPLE 3   Using Complex Conjugates to Divide Complex Numbers

Divide and express the result in standard form: $\dfrac{7 + 4i}{2 - 5i}$.

Solution   The complex conjugate of the denominator, $2 - 5i$, is $2 + 5i$. Multiplication of both the numerator and the denominator by $2 + 5i$ will eliminate i from the denominator.

$$\frac{7 + 4i}{2 - 5i} = \frac{(7 + 4i)}{(2 - 5i)} \cdot \frac{(2 + 5i)}{(2 + 5i)} \qquad \begin{array}{l}\text{Multiply the numerator and the denominator by}\\ \text{the complex conjugate of the denominator.}\end{array}$$

$$= \frac{\overset{F}{14} + \overset{O}{35i} + \overset{I}{8i} + \overset{L}{20i^2}}{2^2 + 5^2} \qquad \begin{array}{l}\text{Use the FOIL method in the numerator and}\\ (a - bi)(a + bi) = a^2 + b^2 \text{ in the denominator.}\end{array}$$

$$= \frac{14 + 43i + 20(-1)}{29} \qquad \text{Combine imaginary terms and replace } i^2 \text{ with } -1.$$

$$= \frac{-6 + 43i}{29} \qquad \text{Combine real terms in the numerator.}$$

$$= -\frac{6}{29} + \frac{43}{29} i \qquad \text{Express the answer in standard form.}$$

Observe that the quotient is expressed in the standard form $a + bi$, with $a = -\frac{6}{29}$ and $b = \frac{43}{29}$.

Complex Numbers on a Postage Stamp

This stamp honors the work done by the German mathematician Carl Friedrich Gauss (1777–1855) with complex numbers. Gauss represented complex numbers as points in the plane.

Check Point 3 Divide and express the result in standard form: $\dfrac{5 + 4i}{4 - 2i}$.

4 Perform operations with square roots of negative numbers.

Roots of Negative Numbers

The square of $4i$ and the square of $-4i$ both result in -16.

$$(4i)^2 = 16i^2 = 16(-1) = -16 \qquad (-4i)^2 = 16i^2 = 16(-1) = -16$$

Consequently, in the complex number system -16 has two square roots, namely, $4i$ and $-4i$. We call $4i$ the **principal square root** of -16.

> **Principal Square Root of a Negative Number**
>
> For any positive number real number b, the **principal square root** of the negative number $-b$ is defined by
> $$\sqrt{-b} = i\sqrt{b}.$$

EXAMPLE 4   Operations Involving Square Roots of Negative Numbers

Perform the indicated operations and write the result in standard form:

a. $\sqrt{-18} - \sqrt{-8}$ **b.** $\left(-1 + \sqrt{-5}\right)^2$ **c.** $\dfrac{-25 + \sqrt{-50}}{15}$.

Solution   Begin by expressing all square roots of negative numbers in terms of i.

a. $\sqrt{-18} - \sqrt{-8} = i\sqrt{18} - i\sqrt{8} = i\sqrt{9 \cdot 2} - i\sqrt{4 \cdot 2}$

$$= 3i\sqrt{2} - 2i\sqrt{2} = i\sqrt{2}$$

$(A + B)^2 = A^2 + 2\,A\,B + B^2$

b. $\left(-1 + \sqrt{-5}\right)^2 = \left(-1 + i\sqrt{5}\right)^2 = (-1)^2 + 2(-1)(i\sqrt{5}) + (i\sqrt{5})^2$

$$= 1 - 2i\sqrt{5} + 5i^2$$
$$= 1 - 2i\sqrt{5} + 5(-1)$$
$$= -4 - 2i\sqrt{5}$$

c. $\dfrac{-25 + \sqrt{-50}}{15}$

$= \dfrac{-25 + i\sqrt{50}}{15}$ $\sqrt{-b} = i\sqrt{b}$

$= \dfrac{-25 + 5i\sqrt{2}}{15}$ $\sqrt{50} = \sqrt{25 \cdot 2} = 5\sqrt{2}$

$= \dfrac{-25}{15} + \dfrac{5i\sqrt{2}}{15}$ Write the complex number in standard form.

$= -\dfrac{5}{3} + i\dfrac{\sqrt{2}}{3}$ Simplify.

Study Tip

Do not apply the properties
$$\sqrt{b}\sqrt{c} = \sqrt{bc}$$
and
$$\frac{\sqrt{b}}{\sqrt{c}} = \sqrt{\frac{b}{c}}$$
to the pure imaginary numbers because these properties can only be used when b and c are positive.

Correct:
$$\sqrt{-25}\sqrt{-4} = i\sqrt{25}\,i\sqrt{4}$$
$$= (5i)(2i)$$
$$= 10i^2$$
$$= 10(-1)$$
$$= -10$$

Incorrect:
$$\sqrt{-25}\sqrt{-4} = \sqrt{(-25)(-4)}$$
$$= \sqrt{100}$$
$$= 10$$

One way to avoid confusion is to represent square roots of negative numbers in terms of i before performing any operations.

Check Point 4 Perform the indicated operations and write the result in standard form:

a. $\sqrt{-27} + \sqrt{-48}$ **b.** $\left(-2 + \sqrt{-3}\right)^2$ **c.** $\dfrac{-14 + \sqrt{-12}}{2}$.

EXERCISE SET 1.4

 Practice Exercises

In Exercises 1–8, add or subtract as indicated and write the result in standard form.

1. $(7 + 2i) + (1 - 4i)$ **2.** $(-2 + 6i) + (4 - i)$

3. $(3 + 2i) - (5 - 7i)$ **4.** $(-7 + 5i) - (-9 - 11i)$

5. $6 - (-5 + 4i) - (-13 - 11i)$

6. $7 - (-9 + 2i) - (-17 - 6i)$

7. $8i - (14 - 9i)$ **8.** $15i - (12 - 11i)$

In Exercises 9–20, find each product and write the result in standard form.

9. $-3i(7i - 5)$ **10.** $-8i(2i - 7)$

11. $(-5 + 4i)(3 + 7i)$ **12.** $(-4 - 8i)(3 + 9i)$

13. $(7 - 5i)(-2 - 3i)$ **14.** $(8 - 4i)(-3 + 9i)$

15. $(3 + 5i)(3 - 5i)$ **16.** $(2 + 7i)(2 - 7i)$

17. $(-5 + 3i)(-5 - 3i)$ **18.** $(-7 - 4i)(-7 + 4i)$

19. $(2 + 3i)^2$ **20.** $(5 - 2i)^2$

In Exercises 21–28, divide and express the result in standard form.

21. $\dfrac{2}{3 - i}$ **22.** $\dfrac{3}{4 + i}$

23. $\dfrac{2i}{1 + i}$ **24.** $\dfrac{5i}{2 - i}$

25. $\dfrac{8i}{4 - 3i}$ **26.** $\dfrac{-6i}{3 + 2i}$

27. $\dfrac{2 + 3i}{2 + i}$ **28.** $\dfrac{3 - 4i}{4 + 3i}$

In Exercises 29–44, perform the indicated operations and write the result in standard form.

29. $\sqrt{-64} - \sqrt{-25}$ **30.** $\sqrt{-81} - \sqrt{-144}$

31. $5\sqrt{-16} + 3\sqrt{-81}$ **32.** $5\sqrt{-8} + 3\sqrt{-18}$

33. $(-2 + \sqrt{-4})^2$ **34.** $(-5 - \sqrt{-9})^2$

35. $(-3 - \sqrt{-7})^2$ **36.** $(-2 + \sqrt{-11})^2$

37. $\dfrac{-8 + \sqrt{-32}}{24}$ **38.** $\dfrac{-12 + \sqrt{-28}}{32}$

39. $\dfrac{-6 - \sqrt{-12}}{48}$ **40.** $\dfrac{-15 - \sqrt{-18}}{33}$

41. $\sqrt{-8}(\sqrt{-3} - \sqrt{5})$ **42.** $\sqrt{-12}(\sqrt{-4} - \sqrt{2})$

43. $(3\sqrt{-5})(-4\sqrt{-12})$ **44.** $(3\sqrt{-7})(2\sqrt{-8})$

Writing in Mathematics

45. What is i?

46. Explain how to add complex numbers. Provide an example with your explanation.

47. Explain how to multiply complex numbers and give an example.

48. What is the complex conjugate of $2 + 3i$? What happens when you multiply this complex number by its complex conjugate?

49. Explain how to divide complex numbers. Provide an example with your explanation.

50. A stand-up comedian uses algebra in some jokes, including one about a telephone recording that announces "You have just reached an imaginary number. Please multiply by i and dial again." Explain the joke.

Explain the error in Exercises 51–52.

51. $\sqrt{-9} + \sqrt{-16} = \sqrt{-25} = i\sqrt{25} = 5i$

52. $(\sqrt{-9})^2 = \sqrt{-9} \cdot \sqrt{-9} = \sqrt{81} = 9$

Critical Thinking Exercises

53. Which one of the following is true?

  a. Some irrational numbers are not complex numbers.

  b. $(3 + 7i)(3 - 7i)$ is an imaginary number.

  c. $\dfrac{7 + 3i}{5 + 3i} = \dfrac{7}{5}$

  d. In the complex number system, $x^2 + y^2$ (the sum of two squares) can be factored as $(x + yi)(x - yi)$.

In Exercises 54–56, perform the indicated operations and write the result in standard form.

54. $(8 + 9i)(2 - i) - (1 - i)(1 + i)$

55. $\dfrac{4}{(2 + i)(3 - i)}$ **56.** $\dfrac{1 + i}{1 + 2i} + \dfrac{1 - i}{1 - 2i}$

57. Evaluate $x^2 - 2x + 2$ for $x = 1 + i$.

SECTION 1.5 *Quadratic Equations*

Objectives

1. Solve quadratic equations by factoring.
2. Solve quadratic equations by the square root method.
3. Solve quadratic equations by completing the square.
4. Solve quadratic equations using the quadratic formula.
5. Use the discriminant to determine the number and type of solutions.
6. Determine the most efficient method to use when solving a quadratic equation.
7. Solve problems modeled by quadratic equations.

Serpico, 1973, starring Al Pacino, is a movie about police corruption.

In 2000, a police scandal shocked Los Angeles. A police officer who had been convicted of stealing cocaine held as evidence described how members of his unit behaved in ways that resembled the gangs they were targeting, assaulting and framing innocent people.

Is police corruption on the rise? The graph in Figure 1.14 shows the number of convictions of police officers throughout the United States for seven years.

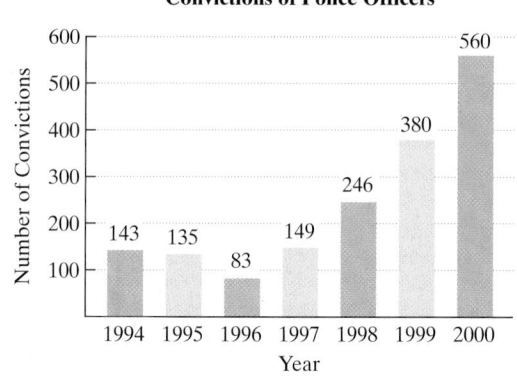

Convictions of Police Officers

Figure 1.14

Source: F.B.I.

The data can be modeled by the formula

$$N = 23.4x^2 - 259.1x + 815.8$$

where N is the number of police officers convicted of felonies x years after 1990. If present trends continue, in which year will 1000 police officers be convicted of felonies? To answer the question, it is necessary to substitute 1000 for N in the formula and solve for x, the number of years after 1990:

$$1000 = 23.4x^2 - 259.1x + 815.8.$$

Do you see how this equation differs from a linear equation? The exponent on x is 2. Solving such an equation involves finding the set of numbers that make the equation a true statement. In this section, we study a number of methods for solving equations in the form $ax^2 + bx + c = 0$. We also look at applications of these equations.

The General Form of a Quadratic Equation

We begin by defining a quadratic equation.

Definition of a Quadratic Equation

A **quadratic equation** in x is an equation that can be written in the **general form**

$$ax^2 + bx + c = 0$$

where a, b, and c are real numbers, with $a \neq 0$. A quadratic equation in x is also called a **second-degree polynomial equation** in x.

An example of a quadratic equation in general form is $x^2 - 7x + 10 = 0$. The coefficient of x^2 is $1\,(a = 1)$, the coefficient of x is $-7(b = -7)$, and the constant term is $10\,(c = 10)$.

1 Solve quadratic equations by factoring.

Solving Quadratic Equations by Factoring

We can factor the left side of the quadratic equation $x^2 - 7x + 10 = 0$. We obtain $(x - 5)(x - 2) = 0$. If a quadratic equation has zero on one side and a factored expression on the other side, it can be solved using the **zero-product principle.**

The Zero-Product Principle

If the product of two algebraic expressions is zero, then at least one of the factors is equal to zero.

If $AB = 0$, then $A = 0$ or $B = 0$.

For example, consider the equation $(x - 5)(x - 2) = 0$. According to the zero-product principle, this product can be zero only if at least one of the factors is zero. We set each individual factor equal to zero and solve each resulting equation for x.

$$(x - 5)(x - 2) = 0$$
$$x - 5 = 0 \quad \text{or} \quad x - 2 = 0$$
$$x = 5 \qquad\qquad x = 2$$

We can check each of these proposed solutions in the original quadratic equation, $x^2 - 7x + 10 = 0$.

Check 5:
$$5^2 - 7 \cdot 5 + 10 \stackrel{?}{=} 0$$
$$25 - 35 + 10 \stackrel{?}{=} 0$$
$$0 = 0 \checkmark$$

Check 2:
$$2^2 - 7 \cdot 2 + 10 \stackrel{?}{=} 0$$
$$4 - 14 + 10 \stackrel{?}{=} 0$$
$$0 = 0 \checkmark$$

The resulting true statements, indicated by the checks, show that the solutions are 5 and 2. The solution set is $\{5, 2\}$. Note that with a quadratic equation, we can have two solutions, compared to the conditional linear equation that had one.

Solving a Quadratic Equation by Factoring

1. If necessary, rewrite the equation in the form $ax^2 + bx + c = 0$, moving all terms to one side, thereby obtaining zero on the other side.
2. Factor.   (continues on the next page)

Solving a Quadratic Equation by Factoring (continued)

3. Apply the zero-product principle, setting each factor equal to zero.

4. Solve the equations in step 3.

5. Check the solutions in the original equation.

EXAMPLE 1 Solving Quadratic Equations by Factoring

Solve by factoring:

a. $4x^2 - 2x = 0$ **b.** $2x^2 + 7x = 4$.

Solution

a. We begin with $4x^2 - 2x = 0$.

Step 1 Move all terms to one side and obtain zero on the other side. All terms are already on the left and zero is on the other side, so we can skip this step.

Step 2 Factor. We factor out $2x$ from the two terms on the left side.

$$4x^2 - 2x = 0 \quad \text{This is the given equation.}$$

$$2x(2x - 1) = 0 \quad \text{Factor.}$$

Steps 3 and 4 Set each factor equal to zero and solve the resulting equations.

$$2x = 0 \qquad \text{or} \qquad 2x - 1 = 0$$

$$x = 0 \qquad\qquad\qquad 2x = 1$$

$$x = \tfrac{1}{2}$$

Step 5 Check the solutions in the original equation.

Check 0:	Check $\tfrac{1}{2}$:
$4x^2 - 2x = 0$	$4x^2 - 2x = 0$
$4 \cdot 0^2 - 2 \cdot 0 \overset{?}{=} 0$	$4\left(\tfrac{1}{2}\right)^2 - 2\left(\tfrac{1}{2}\right) \overset{?}{=} 0$
$0 - 0 \overset{?}{=} 0$	$4\left(\tfrac{1}{4}\right) - 2\left(\tfrac{1}{2}\right) \overset{?}{=} 0$
$0 = 0 \checkmark$	$1 - 1 \overset{?}{=} 0$
	$0 = 0 \checkmark$

The solution set is $\left\{0, \tfrac{1}{2}\right\}$.

b. Next, we solve $2x^2 + 7x = 4$.

Step 1 Move all terms to one side and obtain zero on the other side. Subtract 4 from both sides and write the equation in general form.

$$2x^2 + 7x = 4 \quad \text{This is the given equation.}$$

$$2x^2 + 7x - 4 = 4 - 4 \quad \text{Subtract 4 from both sides.}$$

$$2x^2 + 7x - 4 = 0 \quad \text{Simplify.}$$

Step 2 Factor.

$$2x^2 + 7x - 4 = 0$$

$$(2x - 1)(x + 4) = 0$$

Steps 3 and 4 **Set each factor equal to zero and solve each resulting equation.**

$$2x - 1 = 0 \qquad \text{or} \qquad x + 4 = 0$$
$$2x = 1 \qquad\qquad\qquad x = -4$$
$$x = \tfrac{1}{2}$$

Step 5 **Check the solutions in the original equation.**

Check $\tfrac{1}{2}$:

$$2x^2 + 7x = 4$$
$$2\left(\tfrac{1}{2}\right)^2 + 7\left(\tfrac{1}{2}\right) \overset{?}{=} 4$$
$$\tfrac{1}{2} + \tfrac{7}{2} \overset{?}{=} 4$$
$$4 = 4 \checkmark$$

Check -4:

$$2x^2 + 7x = 4$$
$$2(-4)^2 + 7(-4) \overset{?}{=} 4$$
$$32 + (-28) \overset{?}{=} 4$$
$$4 = 4 \checkmark$$

The solution set is $\left\{-4, \tfrac{1}{2}\right\}$.

Check Point 1 Solve by factoring:

a. $3x^2 - 9x = 0$ **b.** $2x^2 + x = 1$.

Technology

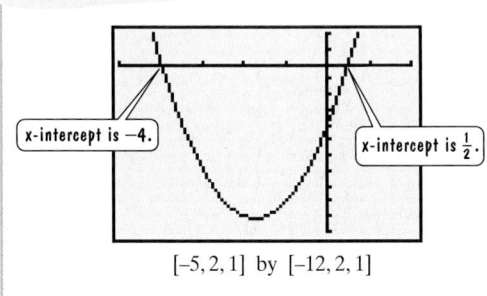

x-intercept is -4.

x-intercept is $\tfrac{1}{2}$.

$[-5, 2, 1]$ by $[-12, 2, 1]$

You can use a graphing utility to check the real solutions of a quadratic equation. **The solutions of $ax^2 + bx + c = 0$ correspond to the x-intercepts of the graph of $y = ax^2 + bx + c$.** For example, to check the solutions of $2x^2 + 7x = 4$, or $2x^2 + 7x - 4 = 0$, graph $y = 2x^2 + 7x - 4$. The cuplike U-shaped graph is shown on the left. Note that it is important to have all nonzero terms on one side of the quadratic equation before entering it into the graphing utility. The x-intercepts are -4 and $\tfrac{1}{2}$, and the graph of $y = 2x^2 + 7x - 4$ passes through $(-4, 0)$ and $\left(\tfrac{1}{2}, 0\right)$. This verifies that $\left\{-4, \tfrac{1}{2}\right\}$ is the solution set of $2x^2 + 7x - 4 = 0$.

2 Solve quadratic equations by the square root method.

Solving Quadratic Equations by the Square Root Method

Quadratic equations of the form $u^2 = d$, where $d > 0$ and u is an algebraic expression, can be solved by the **square root method.** First, isolate the squared expression u^2 on one side of the equation and the number d on the other side. Then take the square root of both sides. Remember, there are two numbers whose square is d. One number is positive and one is negative.

We can use factoring to verify that $u^2 = d$ has two solutions.

$$u^2 = d \qquad \text{\small This is the given equation.}$$
$$u^2 - d = 0 \qquad \text{\small Move all terms to one side and obtain zero on the other side.}$$
$$(u + \sqrt{d})(u - \sqrt{d}) = 0 \qquad \text{\small Factor.}$$
$$u + \sqrt{d} = 0 \qquad \text{or} \qquad u - \sqrt{d} = 0 \qquad \text{\small Set each factor equal to zero.}$$
$$u = -\sqrt{d} \qquad\qquad u = \sqrt{d} \qquad \text{\small Solve the resulting equations.}$$

Because the solutions differ only in sign, we can write them in abbreviated notation as $u = \pm\sqrt{d}$. We read this as "u equals positive or negative the square root of d" or "u equals plus or minus the square root of d."

Now that we have verified these solutions, we can solve $u^2 = d$ directly by taking square roots. This process is called **the square root method.**

The Square Root Method

If u is an algebraic expression and d is a positive real number, then $u^2 = d$ has exactly two solutions:

$$\text{If } u^2 = d, \text{ then } u = \sqrt{d} \text{ or } u = -\sqrt{d}.$$

Equivalently,

$$\text{If } u^2 = d, \text{ then } u = \pm\sqrt{d}.$$

EXAMPLE 2 **Solving Quadratic Equations by the Square Root Method**

Solve by the square root method:

a. $4x^2 = 20$ **b.** $(x - 2)^2 = 6.$

Solution

a. In order to apply the square root method, we need a squared expression by itself on one side of the equation.

$$4x^2 = 20$$

> We want x^2 by itself.

We can get x^2 by itself if we divide both sides by 4.

$$\frac{4x^2}{4} = \frac{20}{4}$$
$$x^2 = 5$$

Now, we can apply the square root method.

$$x = \pm\sqrt{5}$$

By checking both values in the original equation, we can confirm that the solution set is $\{-\sqrt{5}, \sqrt{5}\}$.

b. $(x - 2)^2 = 6$

> The squared expression is by itself.

With the squared expression by itself, we can apply the square root method.

$$x - 2 = \pm\sqrt{6}$$

We solve for x by adding 2 to both sides.

$$x = 2 \pm \sqrt{6}$$

By checking both values in the original equation, we can confirm that the solution set is $\{2 + \sqrt{6}, 2 - \sqrt{6}\}$ or $\{2 \pm \sqrt{6}\}$.

Check Point 2 Solve by the square root method:

a. $3x^2 = 21$ **b.** $(x + 5)^2 = 11$.

3 Solve quadratic equations by completing the square.

Completing the Square

How do we solve an equation in the form $ax^2 + bx + c = 0$ if the trinomial $ax^2 + bx + c$ cannot be factored? We cannot use the zero-product principle in such a case. However, we can convert the equation into an equivalent equation that can be solved using the square root method. This is accomplished by **completing the square.**

Completing the Square

If $x^2 + bx$ is a binomial, then by adding $\left(\frac{b}{2}\right)^2$, which is the square of half the coefficient of x, a perfect square trinomial will result. That is,

$$x^2 + bx + \left(\frac{b}{2}\right)^2 = \left(x + \frac{b}{2}\right)^2.$$

EXAMPLE 3 Completing the Square

What term should be added to the binomial $x^2 + 8x$ so that it becomes a perfect square trinomial? Then write and factor the trinomial.

Solution The term that should be added is the square of half the coefficient of x. The coefficient of x is 8. Thus, we will add $\left(\frac{8}{2}\right)^2 = 4^2$. A perfect square trinomial is the result.

$$x^2 + 8x + 4^2 = x^2 + 8x + 16 = (x + 4)^2$$

$(\text{half})^2$

Check Point 3 What term should be added to the binomial $x^2 - 14x$ so that it becomes a perfect square trinomial? Then write and factor the trinomial.

We can solve any quadratic equation by completing the square. If the coefficient of the x^2-term is one, we add the square of half the coefficient of x to both sides of the equation. **When you add a constant term to one side of the equation to complete the square, be certain to add the same constant to the other side of the equation.** These ideas are illustrated in Example 4.

EXAMPLE 4 Solving a Quadratic Equation by Completing the Square

Solve by completing the square: $x^2 - 6x + 2 = 0$.

Solution   We begin the procedure of solving $x^2 - 6x + 2 = 0$ by isolating the binomial, $x^2 - 6x$, so that we can complete the square. Thus, we subtract 2 from both sides of the equation.

$$x^2 - 6x + 2 = 0 \qquad \text{This is the given equation.}$$

$$x^2 - 6x + 2 - 2 = 0 - 2 \qquad \text{Subtract 2 from both sides.}$$

$$x^2 - 6x = -2 \qquad \text{Simplify.}$$

We need to add a constant to this binomial that will make it a perfect square trinomial.

What constant should we add? Add the square of half the coefficient of x.

$$x^2 - 6x = -2$$

-6 is the coefficient of x.

$$\left(\frac{-6}{2}\right)^2 = (-3)^2 = 9$$

Study Tip

When factoring perfect square trinomials, the constant in the factorization is always half the coefficient of x.

$$x^2 - 6x + 9 = (x - 3)^2$$

Half the coefficient of x, −6, is −3.

Thus, we need to add 9 to $x^2 - 6x$. In order to obtain an equivalent equation, we must add 9 to both sides.

$$x^2 - 6x = -2 \qquad \text{This is the quadratic equation with the binomial isolated.}$$

$$x^2 - 6x + 9 = -2 + 9 \qquad \text{Add 9 to both sides to complete the square.}$$

$$(x - 3)^2 = 7 \qquad \text{Factor the perfect square trinomial.}$$

In this step we have converted our equation into one that can be solved by the square root method.

$$x - 3 = \pm\sqrt{7} \qquad \text{Apply the square root method.}$$

$$x = 3 \pm \sqrt{7} \qquad \text{Add 3 to both sides.}$$

The solution set is $\{3 + \sqrt{7}, 3 - \sqrt{7}\}$ or $\{3 \pm \sqrt{7}\}$.

Check Point 4   Solve by completing the square: $x^2 - 2x - 2 = 0$.

If the coefficient of the x^2-term in a quadratic equation is not one, you must divide each side of the equation by this coefficient before completing the square. For example, to solve $3x^2 - 2x - 4 = 0$ by completing the square, first divide every term by 3:

$$\frac{3x^2}{3} - \frac{2x}{3} - \frac{4}{3} = \frac{0}{3}$$

$$x^2 - \frac{2}{3}x - \frac{4}{3} = 0.$$

Now that the coefficient of x^2 is one, we can solve by completing the square using the method of Example 4.

4 Solve quadratic equations using the quadratic formula.

Solving Quadratic Equations Using the Quadratic Formula

We can use the method of completing the square to derive a formula that can be used to solve all quadratic equations. The derivation given here also shows a particular quadratic equation, $3x^2 - 2x - 4 = 0$, to specifically illustrate each of the steps.

Deriving the Quadratic Formula

General Form of a Quadratic Equation	Comment	A Specific Example
$ax^2 + bx + c = 0, \quad a > 0$	This is the given equation.	$3x^2 - 2x - 4 = 0$
$x^2 + \dfrac{b}{a}x + \dfrac{c}{a} = 0$	Divide both sides by the coefficient of x^2.	$x^2 - \dfrac{2}{3}x - \dfrac{4}{3} = 0$
$x^2 + \dfrac{b}{a}x = -\dfrac{c}{a}$	Isolate the binomial by adding $-\dfrac{c}{a}$ on both sides.	$x^2 - \dfrac{2}{3}x = \dfrac{4}{3}$
$x^2 + \dfrac{b}{a}x + \left(\dfrac{b}{2a}\right)^2 = -\dfrac{c}{a} + \left(\dfrac{b}{2a}\right)^2$ $\underset{\text{(half)}^2}{\underline{\qquad\uparrow\qquad}}$	Complete the square. Add the square of half the coefficient of x to both sides.	$x^2 - \dfrac{2}{3}x + \left(-\dfrac{1}{3}\right)^2 = \dfrac{4}{3} + \left(-\dfrac{1}{3}\right)^2$ $\underset{\text{(half)}^2}{\underline{\qquad\uparrow\qquad}}$
$x^2 + \dfrac{b}{a}x + \dfrac{b^2}{4a^2} = -\dfrac{c}{a} + \dfrac{b^2}{4a^2}$		$x^2 - \dfrac{2}{3}x + \dfrac{1}{9} = \dfrac{4}{3} + \dfrac{1}{9}$
$\left(x + \dfrac{b}{2a}\right)^2 = -\dfrac{c}{a}\cdot\dfrac{4a}{4a} + \dfrac{b^2}{4a^2}$	Factor on the left side and obtain a common denominator on the right side.	$\left(x - \dfrac{1}{3}\right)^2 = \dfrac{4}{3}\cdot\dfrac{3}{3} + \dfrac{1}{9}$
$\left(x + \dfrac{b}{2a}\right)^2 = \dfrac{-4ac + b^2}{4a^2}$	Add fractions on the right side.	$\left(x - \dfrac{1}{3}\right)^2 = \dfrac{12 + 1}{9}$
$\left(x + \dfrac{b}{2a}\right)^2 = \dfrac{b^2 - 4ac}{4a^2}$		$\left(x - \dfrac{1}{3}\right)^2 = \dfrac{13}{9}$
$x + \dfrac{b}{2a} = \pm\sqrt{\dfrac{b^2 - 4ac}{4a^2}}$	Apply the square root method.	$x - \dfrac{1}{3} = \pm\sqrt{\dfrac{13}{9}}$
$x + \dfrac{b}{2a} = \pm\dfrac{\sqrt{b^2 - 4ac}}{2a}$	Take the square root of the quotient, simplifying the denominator.	$x - \dfrac{1}{3} = \pm\dfrac{\sqrt{13}}{3}$
$x = \dfrac{-b}{2a} \pm \dfrac{\sqrt{b^2 - 4ac}}{2a}$	Solve for x by subtracting $\dfrac{b}{2a}$ from both sides.	$x = \dfrac{1}{3} \pm \dfrac{\sqrt{13}}{3}$
$x = \dfrac{-b \pm \sqrt{b^2 - 4ac}}{2a}$	Combine fractions on the right.	$x = \dfrac{1 \pm \sqrt{13}}{3}$

The formula shown at the bottom of the left column is called the *quadratic formula*. A similar proof shows that the same formula can be used to solve quadratic equations if a, the coefficient of the x^2-term, is negative.

To Die at Twenty

Can the equations

$$7x^5 + 12x^3 - 9x + 4 = 0$$

and

$$8x^6 - 7x^5 + 4x^3 - 19 = 0$$

be solved using a formula similar to the quadratic formula? The first equation has five solutions and the second has six solutions, but they cannot be found using a formula. How do we know? In 1832, a 20-year-old Frenchman, Evariste Galois, wrote down a proof showing that there is no general formula to solve equations when the exponent on the variable is 5 or greater. Galois was jailed as a political activist several times while still a teenager. The day after his brilliant proof he fought a duel over a woman. The duel was a political setup. As he lay dying, Galois told his brother, Alfred, of the manuscript that contained his proof: "Mathematical manuscripts are in my room. On the table. Take care of my work. Make it known. Important. Don't cry, Alfred. I need all my courage—to die at twenty." (Our source is Leopold Infeld's biography of Galois, *Whom the Gods Love*. Some historians, however, dispute the story of Galois's ironic death the very day after his algebraic proof. Mathematical truths seem more reliable than historical ones!)

The Quadratic Formula

The solutions of a quadratic equation in standard form $ax^2 + bx + c = 0$, with $a \neq 0$, are given by the **quadratic formula**

$$x = \frac{-b \pm \sqrt{b^2 - 4ac}}{2a}.$$

x equals negative b, plus or minus the square root of b² – 4ac, all divided by 2a.

To use the quadratic formula, write the quadratic equation in general form if necessary. Then determine the numerical values for a (the coefficient of the squared term), b (the coefficient of the x-term), and c (the constant term). Substitute the values of a, b, and c in the quadratic formula and evaluate the expression. The $\pm$ sign indicates that there are two solutions of the equation.

EXAMPLE 5   Solving a Quadratic Equation Using the Quadratic Formula

Solve using the quadratic formula: $2x^2 - 6x + 1 = 0$.

Solution   The given equation is in general form. Begin by identifying the values for a, b, and c.

$$2x^2 - 6x + 1 = 0$$

$a = 2 \qquad b = -6 \qquad c = 1$

$$x = \frac{-b \pm \sqrt{b^2 - 4ac}}{2a}$$

Use the quadratic formula.

$$= \frac{-(-6) \pm \sqrt{(-6)^2 - 4(2)(1)}}{2 \cdot 2}$$

Substitute the values for a, b, and c: a = 2, b = −6, and c = 1.

$$= \frac{6 \pm \sqrt{36 - 8}}{4}$$

−(−6) = 6 and (−6)² = (−6)(−6) = 36.

$$= \frac{6 \pm \sqrt{28}}{2}$$

Complete the subtraction under the radical.

$$= \frac{6 \pm 2\sqrt{7}}{4}$$

$\sqrt{28} = \sqrt{4 \cdot 7} = \sqrt{4}\sqrt{7} = 2\sqrt{7}$

$$= \frac{2(3 \pm \sqrt{7})}{4}$$

Factor out 2 from the numerator.

$$= \frac{3 \pm \sqrt{7}}{2}$$

Divide the numerator and denominator by 2.

The solution set is $\left\{ \dfrac{3 + \sqrt{7}}{2}, \dfrac{3 - \sqrt{7}}{2} \right\}$ or $\left\{ \dfrac{3 \pm \sqrt{7}}{2} \right\}$.

Check Point 5   Solve using the quadratic formula:

$$2x^2 + 2x - 1 = 0.$$

We have seen that a graphing utility can be used to check the solutions of the quadratic equation $ax^2 + bx + c = 0$. The x-intercepts of the graph of

Figure 1.15 This graph has no *x*-intercepts.

$y = ax^2 + bx + c$ are the solutions. However, take a look at the graph of $y = 3x^2 - 2x + 4$, shown in Figure 1.15. Notice that the graph has no *x*-intercepts. Can you guess what this means about the solutions of the quadratic equation $3x^2 - 2x + 4 = 0$? If you're not sure, we'll answer this question in the next example.

EXAMPLE 6 Solving a Quadratic Equation Using the Quadratic Formula

Solve using the quadratic formula: $3x^2 - 2x + 4 = 0$.

Solution The given equation is in general form. Begin by identifying the values for *a*, *b*, and *c*.

$$3x^2 - 2x + 4 = 0$$

$$a = 3 \qquad b = -2 \qquad c = 4$$

$$x = \frac{-b \pm \sqrt{b^2 - 4ac}}{2a}$$ Use the quadratic formula.

$$= \frac{-(-2) \pm \sqrt{(-2)^2 - 4(3)(4)}}{2(3)}$$ Substitute the values for *a*, *b*, and *c*: $a = 3$, $b = -2$, and $c = 4$.

$$= \frac{2 \pm \sqrt{4 - 48}}{6}$$ $-(-2) = 2$ and $(-2)^2 = (-2)(-2) = 4$.

$$= \frac{2 \pm \sqrt{-44}}{6}$$ Subtract under the radical. Because the number under the radical sign is negative, the solutions will not be real numbers.

$$= \frac{2 \pm 2i\sqrt{11}}{6}$$ $\sqrt{-44} = \sqrt{4(11)(-1)}$
$= 2i\sqrt{11}$

$$= \frac{2(1 \pm i\sqrt{11})}{6}$$ Factor 2 from the numerator.

$$= \frac{1 \pm i\sqrt{11}}{3}$$ Divide numerator and denominator by 2.

$$= \frac{1}{3} \pm i\frac{\sqrt{11}}{3}$$ Write the complex numbers in standard form.

The solutions are complex conjugates, and the solution set is
$$\left\{ \frac{1}{3} + i\frac{\sqrt{11}}{3}, \frac{1}{3} - i\frac{\sqrt{11}}{3} \right\} \text{ or } \left\{ \frac{1}{3} \pm i\frac{\sqrt{11}}{3} \right\}.$$

If $ax^2 + bx + c = 0$ has complex imaginary solutions, the graph of $y = ax^2 + bx + c$ will not have *x*-intercepts. This is illustrated by the imaginary solutions of $3x^2 - 2x + 4 = 0$ in Example 6 and the graph in Figure 1.15.

Study Tip

Checking irrational and complex imaginary solutions can be time-consuming. The solutions given by the quadratic formula are always correct, unless you have made a careless error. Checking for computational errors or errors in simplification is sufficient.

Check Point 6   Solve using the quadratic formula:
$$x^2 - 2x + 2 = 0.$$

5 Use the discriminant to determine the number and type of solutions.

The Discriminant

The quantity $b^2 - 4ac$, which appears under the radical sign in the quadratic formula, is called the **discriminant.** In Example 5 the discriminant was 28, a positive number that is not a perfect square. The equation had two solutions that were irrational numbers. In Example 6, the discriminant was −44, a negative number. The equation had solutions that were complex imaginary numbers. These observations are generalized in Table 1.3.

Table 1.3   The Discriminant and the Kinds of Solutions to $ax^2 + bx + c = 0$

Discriminant $b^2 - 4ac$	Kinds of Solutions to $ax^2 + bx + c = 0$	Graph of $y = ax^2 + bx + c$
$b^2 - 4ac > 0$	**Two unequal real solutions;** if $a, b,$ and c are rational numbers and the discriminant is a perfect square, the solutions are rational. If the discriminant is not a perfect square, the solutions are irrational.	Two x-intercepts
$b^2 - 4ac = 0$	**One solution(a repeated solution) that is a real number;** If $a, b,$ and c are rational numbers, the repeated solution is also a rational number.	One x-intercepts
$b^2 - 4ac < 0$	**No real solution; two complex imaginary solutions;** The solutions are complex conjugates.	No x-intercepts

EXAMPLE 7   Using the Discriminant

Compute the discriminant of $4x^2 - 8x + 1 = 0$. What does the discriminant indicate about the number and type of solutions?

Solution   Begin by identifying the values for $a, b,$ and c.

$$4x^2 - 8x + 1 = 0$$

$$a = 4 \quad b = -8 \quad c = 1$$

Substitute and compute the discriminant:

$$b^2 - 4ac = (-8)^2 - 4 \cdot 4 \cdot 1 = 64 - 16 = 48.$$

The discriminant is 48. Because the discriminant is positive, the equation $4x^2 - 8x + 1 = 0$ has two unequal real solutions.

Check Point 7 Compute the discriminant of $3x^2 - 2x + 5 = 0$. What does the discriminant indicate about the number and type of solutions?

6 Determine the most efficient method to use when solving a quadratic equation.

Determining Which Method to Use

All quadratic equations can be solved by the quadratic formula. However, if an equation is in the form $u^2 = d$, such as $x^2 = 5$ or $(2x + 3)^2 = 8$, it is faster to use the square root method, taking the square root of both sides. If the equation is not in the form $u^2 = d$, write the quadratic equation in general form $(ax^2 + bx + c = 0)$. Try to solve the equation by the factoring method. If $ax^2 + bx + c$ cannot be factored, then solve the quadratic equation by the quadratic formula.

Because we used the method of completing the square to derive the quadratic formula, we no longer need it for solving quadratic equations. However, we will use completing the square later in the book to help graph certain kinds of equations.

Table 1.4 summarizes our observations about which technique to use when solving a quadratic equation.

Table 1.4 Determining the Most Efficient Technique to Use When Solving a Quadratic Equation

Description and Form of the Quadratic Equation	Most Efficient Solution Method	Example
$ax^2 + bx + c = 0$ and $ax^2 + bx + c$ can be factored easily.	Factor and use the zero-product principle.	$3x^2 + 5x - 2 = 0$ $(3x - 1)(x + 2) = 0$ $3x - 1 = 0$ or $x + 2 = 0$ $x = \dfrac{1}{3}$ $x = -2$
$ax^2 + c = 0$ The quadratic equation has no x-term. $(b = 0)$	Solve for x^2 and apply the square root method.	$4x^2 - 7 = 0$ $4x^2 = 7$ $x^2 = \dfrac{7}{4}$ $x = \pm\dfrac{\sqrt{7}}{2}$
$(ax + c)^2 = d$; $ax + c$ is a first-degree polynomial.	Use the square root method.	$(x + 4)^2 = 5$ $x + 4 = \pm\sqrt{5}$ $x = -4 \pm \sqrt{5}$
$ax^2 + bx + c = 0$ and $ax^2 + bx + c$ cannot be factored or the factoring is too difficult.	Use the quadratic formula: $x = \dfrac{-b \pm \sqrt{b^2 - 4ac}}{2a}$.	$x^2 - 2x - 6 = 0$ $a = 1$ $b = -2$ $c = -6$ $x = \dfrac{-(-2) \pm \sqrt{(-2)^2 - 4(1)(-6)}}{2(1)}$ $= \dfrac{2 \pm \sqrt{4 - 4(1)(-6)}}{2(1)}$ $= \dfrac{2 \pm \sqrt{28}}{2} = \dfrac{2 \pm \sqrt{4}\sqrt{7}}{2}$ $= \dfrac{2 \pm 2\sqrt{7}}{2} = \dfrac{2(1 \pm \sqrt{7})}{2}$ $= 1 \pm \sqrt{7}$

7 Solve problems modeled by quadratic equations.

Convictions of Police Officers

Figure 1.14, repeated

Source: F.B.I.

Applications

We opened this section with a graph (Figure 1.14, repeated in the margin) showing the number of convictions of police officers throughout the United States from 1994 through 2000. The data can be modeled by the formula

$$N = 23.4x^2 - 259.1x + 815.8$$

where N is the number of police officers convicted of felonies x years after 1990. Notice that this formula contains an expression in the form $ax^2 + bx + c$ on the right side. If a formula contains such an expression, we can write and solve a quadratic equation to answer questions about the variable x. Our next example shows how this is done.

EXAMPLE 8   Convictions of Police Officers

Use the formula $N = 23.4x^2 - 259.1x + 815.8$ to answer this question: In which year will 1000 police officers be convicted of felonies?

Solution   Because we are interested in 1000 convictions, we substitute 1000 for N in the given formula. Then we solve for x, the number of years after 1990.

$N = 23.4x^2 - 259.1x + 815.8$ This is the given formula.

$1000 = 23.4x^2 - 259.1x + 815.8$ Substitute 1000 for N.

$0 = 23.4x^2 - 259.1x - 184.2$ Subtract 1000 from both sides and write the quadratic equation in

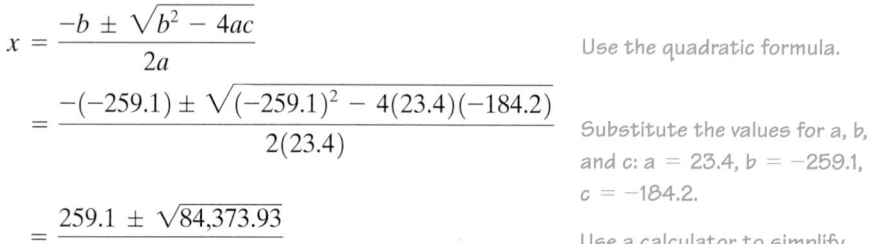

general form.

Because the trinomial on the right side of the equation is prime, we solve using the quadratic formula.

$$x = \frac{-b \pm \sqrt{b^2 - 4ac}}{2a}$$

Use the quadratic formula.

$$= \frac{-(-259.1) \pm \sqrt{(-259.1)^2 - 4(23.4)(-184.2)}}{2(23.4)}$$

Substitute the values for a, b, and c: a = 23.4, b = −259.1, c = −184.2.

$$= \frac{259.1 \pm \sqrt{84,373.93}}{46.8}$$

Use a calculator to simplify the radicand.

Thus,

$$x = \frac{259.1 + \sqrt{84,373.93}}{46.8} \quad \text{or} \quad x = \frac{259.1 - \sqrt{84,373.93}}{46.8}$$

$$x \approx 12 \qquad\qquad\qquad x \approx -1$$

Use a calculator and round to the nearest integer.

The model describes the number of convictions x years *after* 1990. Thus, we are interested only in the positive solution, 12. This means that approximately 12 years after 1990, in 2002, 1000 police officers will be convicted of felonies.

Technology

On most calculators, here is how to approximate

$$\frac{259.1 + \sqrt{84,373.93}}{46.8}:$$

Many Scientific Calculators

(259.1 + 84373.93 √

) ÷ 46.8 =

Many Graphing Calculators

(259.1 + √ 84373.93

) ÷ 46.8 ENTER.

Similar keystrokes can be used to approximate the other irrational solution in Example 8.

Check Point 8   Use the formula in Example 8 to answer this question: In which year after 1993 were 250 police officers convicted of felonies? How well does the formula model the actual number of convictions for that year shown in Figure 1.14?

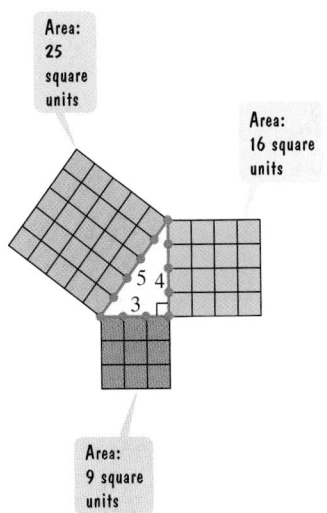

Area: 25 square units

Area: 16 square units

5 4
3

Area: 9 square units

Figure 1.16 The area of the large square equals the sum of the areas of the smaller squares.

In our next example, we will be using the *Pythagorean Theorem* to obtain a verbal model. The ancient Greek philosopher and mathematician Pythagoras (approximately 582–500 B.C.) founded a school whose motto was "All is number." Pythagoras is best remembered for his work with the **right triangle,** a triangle with one angle measuring 90°. The side opposite the 90° angle is called the **hypotenuse.** The other sides are called **legs.** Pythagoras found that if he constructed squares on each of the legs, as well as a larger square on the hypotenuse, the sum of the areas of the smaller squares is equal to the area of the larger square. This is illustrated in Figure 1.16.

This relationship is usually stated in terms of the lengths of the three sides of a right triangle and is called the **Pythagorean Theorem.**

The Pythagorean Theorem

The sum of the squares of the lengths of the legs of a right triangle equals the square of the length of the hypotenuse.

If the legs have lengths a and b, and the hypotenuse has length c, then

$$a^2 + b^2 = c^2.$$

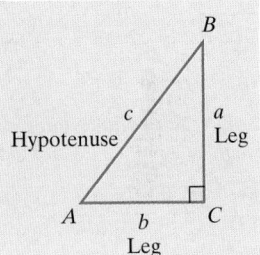

EXAMPLE 9   Using the Pythagorean Theorem

In a 25-inch television set, the length of the screen's diagonal is 25 inches. If the screen's height is 15 inches, what is its width?

Solution   Figure 1.17 shows a right triangle that is formed by the height, width, and diagonal. We can find w, the screen's width, using the Pythagorean Theorem.

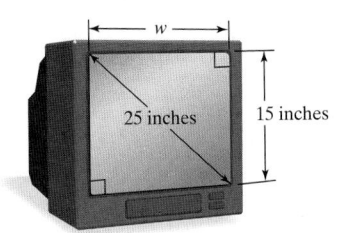

w

25 inches   15 inches

Figure 1.17 A right triangle is formed by the television's height, width, and diagonal.

(Leg)^2	plus	(Leg)^2	equals	(Hypotenuse)^2	
w^2	$+$	15^2	$=$	25^2	This is the equation resulting from the Pythagorean Theorem.

The equation $w^2 + 15^2 = 25^2$ can be solved most efficiently by the square root method.

$$w^2 + 15^2 = 25^2$$ This is the equation that models the verbal conditions.

$$w^2 + 225 = 625$$ Square 15 and 25.

$$w^2 + 225 - 225 = 625 - 225$$ Isolate w^2 by subtracting 225 from both sides.

$$w^2 = 400$$ Simplify.

$$w = \pm\sqrt{400}$$ Apply the square root method.

$$w = \pm 20$$ Simplify.

Because w represents the width of the television's screen, this dimension must be positive. We reject -20. Thus, the width of the television is 20 inches.

Check Point 9   What is the width in a 15-inch television set whose height is 9 inches?

EXERCISE SET 1.5

Practice Exercises

Solve each equation in Exercises 1–14 by factoring.

1. $x^2 - 3x - 10 = 0$ **2.** $x^2 - 13x + 36 = 0$

3. $x^2 = 8x - 15$ **4.** $x^2 = -11x - 10$

5. $6x^2 + 11x - 10 = 0$ **6.** $9x^2 + 9x + 2 = 0$

7. $3x^2 - 2x = 8$ **8.** $4x^2 - 13x = -3$

9. $3x^2 + 12x = 0$ **10.** $5x^2 - 20x = 0$

11. $2x(x - 3) = 5x^2 - 7x$ **12.** $16x(x - 2) = 8x - 25$

13. $7 - 7x = (3x + 2)(x - 1)$

14. $10x - 1 = (2x + 1)^2$

Solve each equation in Exercises 15–26 by the square root method.

15. $3x^2 = 27$ **16.** $5x^2 = 45$

17. $5x^2 + 1 = 51$ **18.** $3x^2 - 1 = 47$

19. $(x + 2)^2 = 25$ **20.** $(x - 3)^2 = 36$

21. $(3x + 2)^2 = 9$ **22.** $(4x - 1)^2 = 16$

23. $(5x - 1)^2 = 7$ **24.** $(8x - 3)^2 = 5$

25. $(3x - 4)^2 = 8$ **26.** $(2x + 8)^2 = 27$

In Exercises 27–38, determine the constant that should be added to the binomial so that it becomes a perfect square trinomial. Then write and factor the trinomial.

27. $x^2 + 12x$ **28.** $x^2 + 16x$

29. $x^2 - 10x$ **30.** $x^2 - 14x$

31. $x^2 + 3x$ **32.** $x^2 + 5x$

33. $x^2 - 7x$ **34.** $x^2 - 9x$

35. $x^2 - \dfrac{2}{3}x$ **36.** $x^2 + \dfrac{4}{5}x$

37. $x^2 - \dfrac{1}{3}x$ **38.** $x^2 - \dfrac{1}{4}x$

Solve each equation in Exercises 39–54 by completing the square.

39. $x^2 + 6x = 7$ **40.** $x^2 + 6x = -8$

41. $x^2 - 2x = 2$ **42.** $x^2 + 4x = 12$

43. $x^2 - 6x - 11 = 0$ **44.** $x^2 - 2x - 5 = 0$

45. $x^2 + 4x + 1 = 0$ **46.** $x^2 + 6x - 5 = 0$

47. $x^2 + 3x - 1 = 0$ **48.** $x^2 - 3x - 5 = 0$

49. $2x^2 - 7x + 3 = 0$ **50.** $2x^2 + 5x - 3 = 0$

51. $4x^2 - 4x - 1 = 0$ **52.** $2x^2 - 4x - 1 = 0$

53. $3x^2 - 2x - 2 = 0$ **54.** $3x^2 - 5x - 10 = 0$

Solve each equation in Exercises 55–64 using the quadratic formula.

55. $x^2 + 8x + 15 = 0$ **56.** $x^2 + 8x + 12 = 0$

57. $x^2 + 5x + 3 = 0$ **58.** $x^2 + 5x + 2 = 0$

59. $3x^2 - 3x - 4 = 0$ **60.** $5x^2 + x - 2 = 0$

61. $4x^2 = 2x + 7$ **62.** $3x^2 = 6x - 1$

63. $x^2 - 6x + 10 = 0$ **64.** $x^2 - 2x + 17 = 0$

Compute the discriminant of each equation in Exercises 65–72. What does the discriminant indicate about the number and type of solutions?

65. $x^2 - 4x - 5 = 0$ **66.** $4x^2 - 2x + 3 = 0$

67. $2x^2 - 11x + 3 = 0$ **68.** $2x^2 + 11x - 6 = 0$

69. $x^2 - 2x + 1 = 0$ **70.** $3x^2 = 2x - 1$

71. $x^2 - 3x - 7 = 0$ **72.** $3x^2 + 4x - 2 = 0$

Solve each equation in Exercises 73–98 by the method of your choice.

73. $2x^2 - x = 1$ **74.** $3x^2 - 4x = 4$

75. $5x^2 + 2 = 11x$ **76.** $5x^2 = 6 - 13x$

77. $3x^2 = 60$ **78.** $2x^2 = 250$

79. $x^2 - 2x = 1$ **80.** $2x^2 + 3x = 1$

81. $(2x + 3)(x + 4) = 1$ **82.** $(2x - 5)(x + 1) = 2$

83. $(3x - 4)^2 = 16$ **84.** $(2x + 7)^2 = 25$

85. $3x^2 - 12x + 12 = 0$ **86.** $9 - 6x + x^2 = 0$

87. $4x^2 - 16 = 0$ **88.** $3x^2 - 27 = 0$

89. $x^2 - 6x + 13 = 0$ **90.** $x^2 - 4x + 29 = 0$

91. $x^2 = 4x - 7$ **92.** $5x^2 = 2x - 3$

93. $2x^2 - 7x = 0$ **94.** $2x^2 + 5x = 3$

95. $\dfrac{1}{x} + \dfrac{1}{x + 2} = \dfrac{1}{3}$ **96.** $\dfrac{1}{x} + \dfrac{1}{x + 3} = \dfrac{1}{4}$

97. $\dfrac{2x}{x - 3} + \dfrac{6}{x + 3} = -\dfrac{28}{x^2 - 9}$

98. $\dfrac{3}{x - 3} + \dfrac{5}{x - 4} = \dfrac{x^2 - 20}{x^2 - 7x + 12}$

Application Exercises

    A driver's age has something to do with his or her chance of getting into a fatal car crash. The bar graph shows the number of fatal vehicle crashes per 100 million miles driven for drivers of various age groups. For example, 25-year-old drivers are involved in 4.1 fatal crashes per 100 million miles driven. Thus, when a group of 25-year-old Americans have driven a total of 100 million miles, approximately 4 have been in accidents in which someone died.

Age of U.S. Drivers and Fatal Crashes

Source: Insurance Institute for Highway Safety

The number of fatal vehicle crashes per 100 million miles, N, for drivers of age x can be modeled by the formula

$$N = 0.013x^2 - 1.19x + 28.24.$$

Use the formula to solve Exercises 99–100.

99. What age groups are expected to be involved in 10 fatal crashes per 100 million miles driven? How well does the formula model the trend in the actual data shown in the bar graph?

100. What age groups are expected to be involved in 3 fatal crashes per 100 million miles driven? How well does the formula model the trend in the actual data shown in the bar graph?

The Food Stamp Program is America's first line of defense against hunger for millions of families. Over half of all participants are children; one out of six is a low-income older adult. Exercises 101–104 involve the number of participants in the program from 1990 through 2000.
The formula

$$y = -\frac{1}{2}x^2 + 4x + 19$$

models the number of people, y, in millions, receiving food stamps x years after 1990. Use the formula to solve Exercises 101–102.

101. In which year did 27 million people receive food stamps?

102. In which years did 19 million people receive food stamps?

The graph of the formula in Exercises 101–102 is shown. Use the graph to solve Exercises 103–104.

Number of People Receiving Food Stamps

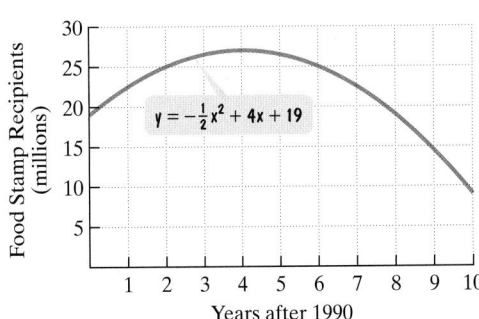

Source: New York Times

103. Identify your solution in Exercise 101 as a point on the graph. Describe what is significant about this point.

104. Identify your solution in Exercise 102 as one or more points on the graph. Then describe the trend shown by the graph.

The formula

$$N = 2x^2 + 22x + 320$$

models the number of inmates, N, in thousands, in U.S. state and federal prisons x years after 1980. The graph of the formula is shown in a [0, 20, 1] by [0, 1600, 100] viewing rectangle at the top of the next column. Use the formula to solve Exercises 105–106.

[0, 20, 1] by [0, 1600, 100]

105. In which year were there 740 thousand inmates in U.S. state and federal prisons? Identify the solution as a point on the graph shown.

106. In which year were 1100 thousand inmates in U.S. state and federal prisons? Identify the solution as a point on the graph shown?

107. A baseball diamond is actually a square with 90-foot sides. What is the distance from home plate to second base?

108. A 20-foot ladder is 15 feet from the house. How far up the house does the ladder reach?

109. An 8-foot tree is supported by two wires that extend from the top of the tree to a point on the ground located 15 feet from the base of the tree. Find the total length of the two support wires.

110. A vertical pole is supported by three wires. Each wire is 13 yards long and is anchored 5 yards from the base of the pole. How far up the pole will the wires be attached?

111. The length of a rectangular garden is 5 feet greater than the width. The area of the garden is 300 square feet. Find the length and the width.

112. A rectangular parking lot has a length that is 3 yards greater than the width. The area of the rectangular lot is 180 square yards. Find the length and the width.

113. A machine produces open boxes using square sheets of metal. The figure illustrates that the machine cuts equal-sized squares measuring 2 inches on a side from the corners and then shapes the metal into an open box by turning up the sides. If each box must have a volume of 200 cubic inches, find the length of the side of the open square-bottom box.

114. A machine produces open boxes using square sheets of metal. The machine cuts equal-sized squares measuring 3 inches on a side from the corners and then shapes the metal into an open box by turning up the sides. If each box must have a volume of 75 cubic inches, find the length of the side of the open square-bottom box.

115. A rain gutter is made from sheets of aluminum that are 20 inches wide. As shown in the figure, the edges are turned up to form right angles. Determine the depth of the gutter that will allow a cross-sectional area of 13 square inches. Show that there are two different solutions to the problem. Round to the nearest tenth of an inch.

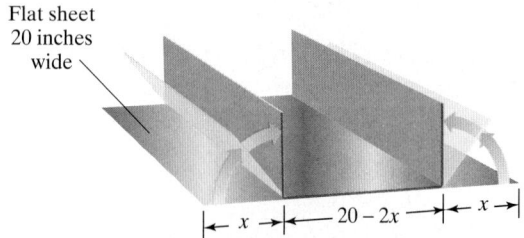

116. A piece of wire is 8 inches long. The wire is cut into two pieces and then each piece is bent into a square. Find the length of each piece if the sum of the areas of these squares is to be 2 square inches.

117. A painting measuring 10 inches by 16 inches is surrounded by a frame of uniform width. If the combined area of the painting and the frame is 280 square inches, determine the width of the frame.

Writing in Mathematics

118. What is a quadratic equation?

119. Explain how to solve $x^2 + 6x + 8 = 0$ using factoring and the zero-product principle.

120. Explain how to solve $x^2 + 6x + 8 = 0$ by completing the square.

121. Explain how to solve $x^2 + 6x + 8 = 0$ using the quadratic formula.

122. How is the quadratic formula derived?

123. What is the discriminant and what information does it provide about a quadratic equation?

124. If you are given a quadratic equation, how do you determine which method to use to solve it?

125. If $(x + 2)(x - 4) = 0$ indicates that $x + 2 = 0$ or $x - 4 = 0$, explain why $(x + 2)(x - 4) = 6$ does not mean $x + 2 = 6$ or $x - 4 = 6$. Could we solve the equation using $x + 2 = 3$ and $x - 4 = 2$ because $3 \cdot 2 = 6$?

126. Describe the trend shown by the data for the convictions of police officers in the graph in Figure 1.14 on page 114. Do you believe that this trend is likely to continue or might something occur that would make it impossible to extend the model into the future? Explain your answer.

Technology Exercises

127. If you have access to a calculator that solves quadratic equations, consult the owner's manual to determine how to use this feature. Then use your calculator to solve any five of the equations in Exercises 55–64.

128. Use a graphing utility and x-intercepts to verify any of the real solutions that you obtained for three of the quadratic equations in Exercises 55–64.

Critical Thinking Exercises

129. Which one of the following is true?

  a. The equation $(2x - 3)^2 = 25$ is equivalent to $2x - 3 = 5$.

  b. Every quadratic equation has two distinct numbers in its solution set.

  c. A quadratic equation whose coefficients are real numbers can never have a solution set containing one real number and one complex nonreal number.

  d. The equation $ax^2 + c = 0$ cannot be solved by the quadratic formula.

130. Solve the equation: $x^2 + 2\sqrt{3}x - 9 = 0$.

131. Write a quadratic equation in general form whose solution set is $\{-3, 5\}$.

132. A person throws a rock upward from the edge of an 80-foot cliff. The height, h, in feet, of the rock above the water at the bottom of the cliff after t seconds is described by the formula

$$h = -16t^2 + 64t + 80.$$

How long will it take for the rock to reach the water?

133. A rectangular swimming pool is 12 meters long and 8 meters wide. A tile border of uniform width is to be built around the pool using 120 square meters of tile. The tile is from a discontinued stock (so no additional materials are available), and all 120 square meters are to be used. How wide should the border be? Round to the nearest tenth of a meter. If zoning laws require at least a 2-meter-wide border around the pool, can this be done with the available tile?

 Group Exercise

134. Each group member should find an "intriguing" algebraic formula that contains an expression in the form $ax^2 + bx + c$ on one side. Consult college algebra books or liberal arts mathematics books to do so. Group members should select four of the formulas. For each formula selected, write and solve a problem similar to Exercises 99–102 in this exercise set.

SECTION 1.6   *Other Types of Equations*

Objectives

1. Solve polynomial equations by factoring.
2. Solve radical equations.
3. Solve equations with rational exponents.
4. Solve equations that are quadratic in form.
5. Solve equations involving absolute value.

Marine iguanas of the Galápagos Islands

The Galápagos Islands are a volcanic chain of islands lying 600 miles west of Ecuador. They are famed for their extraordinary wildlife, which includes a rare flightless cormorant, marine iguanas, and giant tortoises weighing more than 600 pounds. It was here that naturalist Charles Darwin began to formulate his theory of evolution. Darwin made an enormous collection of the islands' plant species. The formula

$$S = 28.5\sqrt[3]{x}$$

describes the number of plant species, S, on the various islands of the Galápagos chain in terms of the area, x, in square miles, of a particular island.

How can we find the area of a Galápagos island with 57 species of plants? Substitute 57 for S in the formula and solve for x:

$$57 = 28.5\sqrt[3]{x}.$$

The resulting equation contains a variable in the radicand and is called a *radical equation*. In this section, in addition to radical equations, we will show you how to solve certain kinds of polynomial equations, equations involving rational exponents, and equations involving absolute value.

1 Solve polynomial equations by factoring.

Polynomial Equations

The linear and quadratic equations that we studied in the first part of this chapter can be thought of as polynomial equations of degrees 1 and 2, respectively. By contrast, consider the following polynomial equations of degree greater than 2:

$$3x^4 = 27x^2 \qquad\qquad x^3 + x^2 = 4x + 4$$

> This equation is of degree 4 because 4 is the largest exponent.

> This equation is of degree 3 because 3 is the largest exponent.

We can solve these equations by moving all terms to one side, thereby obtaining zero on the other side. We then use factoring and the zero-product principle.

EXAMPLE 1 Solving a Polynomial Equation by Factoring

Solve by factoring: $3x^4 = 27x^2$.

Solution

Step 1 Move all terms to one side and obtain zero on the other side. Subtract $27x^2$ from both sides.

$$3x^4 = 27x^2 \qquad \text{This is the given equation.}$$
$$3x^4 - 27x^2 = 27x^2 - 27x^2 \qquad \text{Subtract } 27x^2 \text{ from both sides.}$$
$$3x^4 - 27x^2 = 0 \qquad \text{Simplify.}$$

Step 2 Factor. We can factor $3x^2$ from each term.

$$3x^4 - 27x^2 = 0$$
$$3x^2(x^2 - 9) = 0$$

Steps 3 and 4 Set each factor equal to zero and solve the resulting equations.

$$3x^2 = 0 \qquad \text{or} \qquad x^2 - 9 = 0$$
$$x^2 = 0 \qquad\qquad\qquad x^2 = 9$$
$$x = \pm\sqrt{0} \qquad\qquad\qquad x = \pm\sqrt{9}$$
$$x = 0 \qquad\qquad\qquad\qquad x = \pm 3$$

Step 5 Check the solutions in the original equation. Check the three solutions, $0, -3,$ and 3, by substituting them into the original equation. Can you verify that the solution set is $\{-3, 0, 3\}$?

Study Tip

In solving $3x^4 = 27x^2$, be careful not to divide both sides by x^2. If you do, you'll lose 0 as a solution. In general, do not divide both sides of an equation by a variable because that variable might take on the value 0 and you cannot divide by 0.

Check Point 1 Solve by factoring: $4x^4 = 12x^2$.

EXAMPLE 2   Solving a Polynomial Equation by Factoring

Solve by factoring: $x^3 + x^2 = 4x + 4$.

Solution

Step 1   Move all terms to one side and obtain zero on the other side.   Subtract $4x$ and subtract 4 from both sides.

$$x^3 + x^2 = 4x + 4 \qquad \text{This is the given equation.}$$

$$x^3 + x^2 - 4x - 4 = 4x + 4 - 4x - 4 \quad \text{Subtract 4x and 4 from both sides.}$$

$$x^3 + x^2 - 4x - 4 = 0 \qquad \text{Simplify.}$$

Step 2   Factor.   Because there are four terms, we use factoring by grouping. Group terms that have a common factor.

$$\boxed{x^3 + x^2} + \boxed{-4x - 4} = 0$$

Common factor is x^2.   Common factor is −4.

$$x^2(x + 1) - 4(x + 1) = 0 \quad \text{Factor } x^2 \text{ from the first two terms and } -4 \text{ from the last two terms.}$$

$$(x + 1)(x^2 - 4) = 0 \quad \text{Factor out the common binomial, } x + 1, \text{ from each term.}$$

Steps 3 and 4   Set each factor equal to zero and solve the resulting equations.

$$x + 1 = 0 \qquad \text{or} \qquad x^2 - 4 = 0$$
$$x = -1 \qquad\qquad\qquad x^2 = 4$$
$$x = \pm\sqrt{4} = \pm 2$$

Step 5   Check the solutions in the original equation.   Check the three solutions, $-1, -2,$ and 2, by substituting them into the original equation. Can you verify that the solution set is $\{-2, -1, 2\}$?

Technology

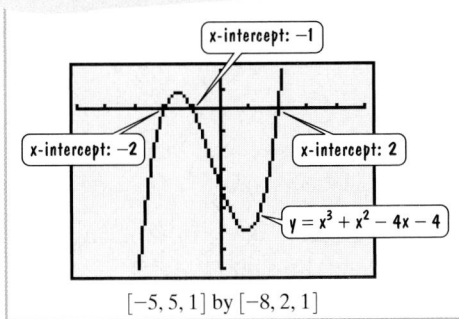

x-intercept: −1

x-intercept: −2

x-intercept: 2

$y = x^3 + x^2 - 4x - 4$

$[-5, 5, 1]$ by $[-8, 2, 1]$

You can use a graphing utility to check the solutions of $x^3 + x^2 - 4x - 4 = 0$. Graph $y = x^3 + x^2 - 4x - 4$, as shown on the left. The x-intercepts are $-2, -1,$ and 2, corresponding to the equation's solutions.

> **Check Point 2**   Solve by factoring: $2x^3 + 3x^2 = 8x + 12$.

2 Solve radical equations.

Equations Involving Radicals

A **radical equation** is an equation in which the variable occurs in a square root, cube root, or any higher root. An example of a radical equation is

$$28.5\sqrt[3]{x} = 57.$$

The variable occurs in a cube root.

The equation $28.5\sqrt[3]{x} = 57$ can be used to find the area, x, in square miles, of a Galápagos island with 57 species of plants. First, we isolate the radical by dividing both sides of the equation by 28.5.

$$\frac{28.5\sqrt[3]{x}}{28.5} = \frac{57}{28.5}$$

$$\sqrt[3]{x} = 2$$

Next, we eliminate the radical by raising each side of the equation to a power equal to the index of the radical. Because the index is 3, we cube both sides of the equation.

$$(\sqrt[3]{x})^3 = 2^3$$

$$x = 8$$

Thus, a Galápagos island with 57 species of plants has an area of 8 square miles.

The Galápagos equation shows that solving equations involving radicals involves raising both sides of the equation to a power equal to the radicals index. All solutions of the original equation are also solutions of the resulting equation. However, the resulting equation may have some extra solutions that do not satisfy the original equation. Because the resulting equation may not be equivalent to the original equation, we must check each proposed solution by substituting it into the original equation. Let's see exactly how this works.

Study Tip

Be sure to square *both sides* of an equation. Do *not* square each term.

Correct:

$$(\sqrt{26-11})^2 = (4-x)^2$$

Incorrect:

~~$(\sqrt{26-11})^2 = 4^2 - x^2$~~

Technology

The graph of

$$y = x + \sqrt{26-11x} - 4$$

is shown in a $[-10, 3, 1]$ by $[-4, 3, 1]$ viewing rectangle. The x-intercepts are -5 and 2, verifying $\{-5, 2\}$ as the solution set of

$$x + \sqrt{26-11x} = 4.$$

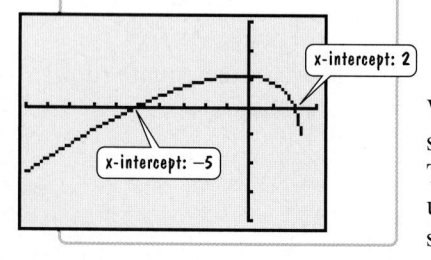

EXAMPLE 3   Solving an Equation Involving a Radical

Solve: $x + \sqrt{26 - 11x} = 4$.

Solution   To solve this equation, we isolate the radical expression $\sqrt{26-11x}$ on one side of the equation. By squaring both sides of the equation, we can then eliminate the square root.

$x + \sqrt{26 - 11x} = 4$	This is the given equation.
$x + \sqrt{26 - 11x} - x = 4 - x$	Isolate the radical by subtracting x from both sides.
$\sqrt{26 - 11x} = 4 - x$	Simplify.
$(\sqrt{26 - 11x})^2 = (4-x)^2$	Square both sides.
$26 - 11x = 16 - 8x + x^2$	Use $(A-B)^2 = A^2 - 2AB + B^2$ to square $4 - x$.

Next, we need to write this quadratic equation in general form. We can obtain zero on the left side by subtracting 26 and adding $11x$ on both sides.

$26 - 26 - 11x + 11x = 16 - 26 - 8x + 11x + x^2$	
$0 = x^2 + 3x - 10$	Simplify.
$0 = (x+5)(x-2)$	Factor.
$x + 5 = 0$ $\quad$ or $\quad$ $x - 2 = 0$	Set each factor equal to zero.
$x = -5$ $\qquad\qquad$ $x = 2$	Solve for x.

We have not completed the solution process. Although -5 and 2 satisfy the squared equation, there is no guarantee that they satisfy the original equation. Thus, we must check the proposed solutions. We can do this using a graphing utility (see the technology box in the margin) or by substituting both proposed solutions into the given equation.

CHECK −5:

$x + \sqrt{26 - 11x} = 4$

$-5 + \sqrt{26 - 11(-5)} \stackrel{?}{=} 4$

$-5 + \sqrt{81} \stackrel{?}{=} 4$

$-5 + 9 \stackrel{?}{=} 4$

$4 = 4 \checkmark$

CHECK 2:

$x + \sqrt{26 - 11x} = 4$

$2 + \sqrt{26 - 11 \cdot 2} \stackrel{?}{=} 4$

$2 + \sqrt{4} \stackrel{?}{=} 4$

$2 + 2 \stackrel{?}{=} 4$

$4 = 4 \checkmark$

The solution set is $\{-5, 2\}$.

Check Point 3 Solve and check: $\sqrt{6x + 7} - x = 2$.

When solving a radical equation, extra solutions may be introduced when you raise both sides of the equation to an even power. Such solutions, which are not solutions of the given equation, are called **extraneous solutions.**

The solution of radical equations with two or more square root expressions involves isolating a radical, squaring both sides, and then repeating this process. Let's consider an equation containing two square root expressions.

Study Tip

When solving equations by raising both sides to an even power, don't forget to check for extraneous solutions. Here is a simple example:

$x = 4$

$x^2 = 16$ Square both sides.

$x = \pm\sqrt{16}$ Use the square root method.

$x = \pm 4$.

However, -4 does not check in $x = 4$. Thus, -4 is an extraneous solution.

EXAMPLE 4   Solving an Equation Involving Two Radicals

Solve: $\sqrt{3x + 1} - \sqrt{x + 4} = 1$.

Solution

$\sqrt{3x + 1} - \sqrt{x + 4} = 1$ This is the given equation.

$\sqrt{3x + 1} = \sqrt{x + 4} + 1$ Isolate one of the radicals by adding $\sqrt{x + 4}$ to both sides.

$\left(\sqrt{3x + 1}\right)^2 = \left(\sqrt{x + 4} + 1\right)^2$ Square both sides.

Squaring the expression on the right side of the equation can be a bit tricky. We need to use the formula

$$(A + B)^2 = A^2 + 2AB + B^2.$$

Focusing on just the right side, here is how the squaring is done:

$$(A + B)^2 = A^2 + 2 \cdot A \cdot B + B^2$$

$$\left(\sqrt{x + 4} + 1\right)^2 = \left(\sqrt{x + 4}\right)^2 + 2 \cdot \sqrt{x + 4} \cdot 1 + 1^2.$$

This simplifies to $x + 4 + 2\sqrt{x + 4} + 1$. Thus, our equation $\left(\sqrt{3x + 1}\right)^2 = \left(\sqrt{x + 4} + 1\right)^2$ can be written as follows:

$3x + 1 = x + 4 + 2\sqrt{x + 4} + 1.$

$3x + 1 = x + 5 + 2\sqrt{x + 4}$ Combine numerical terms on the right.

$2x - 4 = 2\sqrt{x + 4}$ Isolate $2\sqrt{x + 4}$, the radical term, by subtracting $x + 5$ from both sides.

$(2x - 4)^2 = \left(2\sqrt{x + 4}\right)^2$ Square both sides.

Discovery

Divide each side of

$2x - 4 = 2\sqrt{x + 4}$

by 2 before squaring both sides. Solve the resulting equation. How does your solution compare to the one shown?

$$4x^2 - 16x + 16 = 4(x + 4)$$
Use $(A - B)^2 = A^2 - 2AB + B^2$ to square the left side. Use $(AB)^2 = A^2 B^2$ to square the right side.

$$4x^2 - 16x + 16 = 4x + 16$$
Use the distributive property.

$$4x^2 - 20x = 0$$
Write the quadratic equation in general form by subtracting $4x + 16$ from both sides.

$$4x(x - 5) = 0$$
Factor.

$$4x = 0 \quad \text{or} \quad x - 5 = 0$$
Set each factor equal to zero.

$$x = 0 \qquad\qquad x = 5$$
Solve for x.

Technology

The graph of
$$y = \sqrt{3x + 1} - \sqrt{x + 4} - 1$$
has only one x-intercept at 5. This verifies that the solution set of $\sqrt{3x + 1} - \sqrt{x + 4} = 1$ is $\{5\}$.

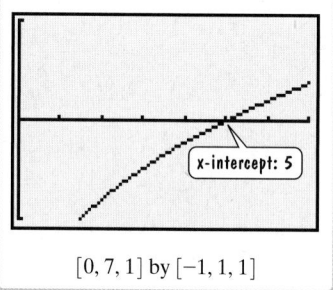

x-intercept: 5

$[0, 7, 1]$ by $[-1, 1, 1]$

Complete the solution process by checking both proposed solutions. We can do this using a graphing utility (see the technology box in the margin) or by substituting both proposed solutions in the given equation.

Check 0:
$$\sqrt{3x + 1} - \sqrt{x + 4} = 1$$
$$\sqrt{3 \cdot 0 + 1} - \sqrt{0 + 4} \overset{?}{=} 1$$
$$\sqrt{1} - \sqrt{4} \overset{?}{=} 1$$
$$1 - 2 \overset{?}{=} 1$$
$$-1 = 1 \quad \text{False}$$

Check 5:
$$\sqrt{3x + 1} - \sqrt{x + 4} = 1$$
$$\sqrt{3 \cdot 5 + 1} - \sqrt{5 + 4} \overset{?}{=} 1$$
$$\sqrt{16} - \sqrt{9} \overset{?}{=} 1$$
$$4 - 3 \overset{?}{=} 1$$
$$1 = 1 \checkmark$$

The false statement $-1 = 1$ indicates that 0 is not a solution. It is an extraneous solution brought about by squaring each side of the equation. The only solution is 5, and the solution set is $\{5\}$.

Check Point 4
Solve and check: $\sqrt{x + 5} - \sqrt{x - 3} = 2$.

Radicals and Windchill

The way that we perceive the temperature on a cold day depends on both air temperature and wind speed. The windchill temperature is what the air temperature would have to be with no wind to achieve the same chilling effect on the skin. The formula that describes windchill temperature, W, in terms of the velocity of the wind, v, in miles per hour, and the actual air temperature, t, in degrees Fahrenheit, is

$$W = 91.4 - \frac{(10.5 + 6.7\sqrt{v} - 0.45v)(457 - 5t)}{110}.$$

Use your calculator to describe how cold the air temperature feels (that is, the windchill temperature) when the temperature is 15° Fahrenheit and the wind is 5 miles per hour. Contrast this with a temperature of 40° Fahrenheit and a wind blowing at 50 miles per hour.

3 Solve equations with rational exponents.

Because $\sqrt[n]{b}$ can be expressed as $b^{1/n}$, radical equations can be written using rational exponents. For example, the Galápagos equation

$$28.5\sqrt[3]{x} = 57$$

can be written

$$28.5x^{1/3} = 57.$$

We solve this equation exactly as we did when it was expressed in radical form. First, isolate $x^{1/3}$.

$$\frac{28.5x^{1/3}}{28.5} = \frac{57}{28.5}$$

$$x^{1/3} = 2$$

Complete the solution process by raising both sides to the third power.

$$(x^{1/3})^3 = 2^3$$

$$x = 8$$

Solving Radical Equations of the Form $x^{m/n} = k$

Assume that m and n are positive integers, $\frac{m}{n}$ is in lowest terms, and k is a real number.

1. Isolate the expression with the rational exponent.
2. Raise both sides of the equation to the $\frac{n}{m}$ power.

If m is even:	If m is odd:
$x^{m/n} = k$	$x^{m/n} = k$
$(x^{m/n})^{n/m} = \pm k^{n/m}$	$(x^{m/n})^{n/m} = k^{n/m}$
$x = \pm k^{n/m}$	$x = k^{n/m}$

It is incorrect to insert the $\pm$ symbol when the numerator of the exponent is odd. An odd index has only one root.

3. Check all proposed solutions in the original equation to find out if they are actual solutions or extraneous solutions.

EXAMPLE 5 **Solving Equations Involving Rational Exponents**

Solve:

a. $3x^{3/4} - 6 = 0$ **b.** $x^{2/3} - \frac{3}{4} = -\frac{1}{2}$.

Solution

a. Our goal is to isolate $x^{3/4}$. Then we can raise both sides of the equation to the $\frac{4}{3}$ power because $\frac{4}{3}$ is the reciprocal of $\frac{3}{4}$.

$3x^{3/4} - 6 = 0$ This is the given equation; we will isolate $x^{3/4}$.

$3x^{3/4} = 6$ Add 6 to both sides.

$\dfrac{3x^{3/4}}{3} = \dfrac{6}{3}$ Divide both sides by 3.

$x^{3/4} = 2$ Simplify.

$(x^{3/4})^{4/3} = 2^{4/3}$ Raise both sides to the $\frac{4}{3}$ power. Because $\frac{m}{n} = \frac{3}{4}$ and m is odd, we do not use the $\pm$ symbol.

$x = 2^{4/3}$ Simplify the left side: $(x^{3/4})^{4/3} = x^{\frac{3 \cdot 4}{4 \cdot 3}} = x^{\frac{12}{12}} = x^1 = x.$

The proposed solution is $2^{4/3}$. Complete the solution process by checking this value in the given equation.

$$3x^{3/4} - 6 = 0 \qquad \text{This is the original equation.}$$

$$3(2^{4/3})^{3/4} - 6 \overset{?}{=} 0 \qquad \text{Substitute the proposed solution.}$$

$$3 \cdot 2 - 6 \overset{?}{=} 0 \qquad (2^{4/3})^{3/4} = 2^{\frac{4\cdot3}{3\cdot4}} = 2^{\frac{12}{12}} = 2^1 = 2.$$

$$0 = 0 \checkmark \qquad \text{The true statement shows that } 2^{4/3} \text{ is a solution.}$$

The solution is $2^{4/3} = \sqrt[3]{2^4} \approx 2.52$. The solution set is $\{2^{4/3}\}$.

b. To solve $x^{2/3} - \frac{3}{4} = -\frac{1}{2}$, our goal is to isolate $x^{2/3}$. Then we can raise both sides of the equation to the $\frac{3}{2}$ power because $\frac{3}{2}$ is the reciprocal of $\frac{2}{3}$.

$$x^{2/3} - \frac{3}{4} = -\frac{1}{2} \qquad \text{This is the given equation.}$$

$$x^{2/3} = \frac{1}{4} \qquad \text{Add } \frac{3}{4} \text{ to both sides.}$$

$$(x^{2/3})^{3/2} = \pm\left(\frac{1}{4}\right)^{3/2} \qquad \begin{array}{l}\text{Raise both sides to the } \frac{3}{2} \text{ power. Because } \frac{m}{n} = \frac{2}{3}\\ \text{and } m \text{ is even, the } \pm \text{ symbol is necessary.}\end{array}$$

$$x = \pm\frac{1}{8} \qquad \left(\frac{1}{4}\right)^{3/2} = \left(\sqrt{\frac{1}{4}}\right)^3 = \left(\frac{1}{2}\right)^3 = \frac{1}{8}$$

Take a moment to verify that the solution set is $\left\{-\frac{1}{8}, \frac{1}{8}\right\}$.

Check Point 5 Solve and check:

a. $5x^{3/2} - 25 = 0$ **b.** $x^{2/3} - 8 = -4$.

4 Solve equations that are quadratic in form.

Equations That Are Quadratic in Form

Some equations that are not quadratic can be written as quadratic equations using an appropriate substitution. Here are some examples:

Given Equation	Substitution	New Equation
$x^4 - 8x^2 - 9 = 0$ or $(x^2)^2 - 8x^2 - 9 = 0$	$t = x^2$	$t^2 - 8t - 9 = 0$
$5x^{2/3} + 11x^{1/3} + 2 = 0$ or $5(x^{1/3})^2 + 11x^{1/3} + 2 = 0$	$t = x^{1/3}$	$5t^2 + 11t + 2 = 0$

An equation that is **quadratic in form** is one that can be expressed as a quadratic equation using an appropriate substitution. Both of the preceding given equations are quadratic in form.

Equations that are quadratic in form contain an expression to a power, the same expression to that power squared, and a constant term. By letting t equal the expression to the power, a quadratic equation in t will result. Now it's easy. Solve this quadratic equation for t. Finally, use your substitution to find the values for the variable in the given equation. Example 6 shows how this is done.

EXAMPLE 6 Solving an Equation Quadratic in Form

Solve: $x^4 - 8x^2 - 9 = 0$.

Technology

The graph of
$$y = x^4 - 8x^2 - 9$$
has x-intercepts at -3 and 3. This verifies that the real solutions of
$$x^4 - 8x^2 - 9 = 0$$
are -3 and 3. The imaginary solutions, $-i$ and i, are not shown as intercepts.

| x-intercept: -3 | x-intercept: 3 |

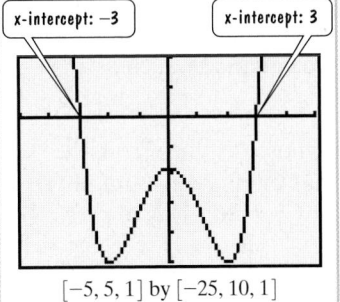

$[-5, 5, 1]$ by $[-25, 10, 1]$

Solution Notice that the equation contains an expression to a power, x^2, the same expression to that power squared, x^4 or $(x^2)^2$, and a constant term, -9. We let t equal the expression to the power. Thus,

$$\text{let } t = x^2.$$

Now we write the given equation as a quadratic equation in t and solve for t.

$$x^4 - 8x^2 - 9 = 0 \qquad \textit{This is the given equation.}$$
$$(x^2)^2 - 8x^2 - 9 = 0 \qquad \textit{The given equation contains } x^2 \textit{ and } x^2 \textit{ squared.}$$
$$t^2 - 8t - 9 = 0 \qquad \textit{Replace } x^2 \textit{ with } t.$$
$$(t - 9)(t + 1) = 0 \qquad \textit{Factor.}$$
$$t - 9 = 0 \quad \text{or} \quad t + 1 = 0 \qquad \textit{Apply the zero-product principle.}$$
$$t = 9 \qquad\qquad t = -1 \qquad \textit{Solve for } t.$$

We're not done! Why not? We were asked to solve for x and we have values for t. We use the original substitution, $t = x^2$, to solve for x. Replace t with x^2 in each equation shown, namely $t = 9$ and $t = -1$.

$$x^2 = 9 \qquad\qquad x^2 = -1$$
$$x = \pm\sqrt{9} \qquad\qquad x = \pm\sqrt{-1}$$
$$x = \pm 3 \qquad\qquad x = \pm i$$

The solution set is $\{-3, 3, -i, i\}$.

Check Point 6 Solve: $x^4 - 5x^2 + 6 = 0$.

EXAMPLE 7 Solving an Equation Quadratic in Form

Solve: $5x^{2/3} + 11x^{1/3} + 2 = 0$.

Solution Notice that the equation contains an expression to a power, $x^{1/3}$, the same expression to that power squared, $x^{2/3}$ or $(x^{1/3})^2$, and a constant term, 2. We let t equal the expression to the power. Thus,

$$\text{let } t = x^{1/3}.$$

Now we write the given equation as a quadratic equation in t and solve for t.

$$5x^{2/3} + 11x^{1/3} + 2 = 0 \qquad \textit{This is the given equation.}$$
$$5(x^{1/3})^2 + 11x^{1/3} + 2 = 0 \qquad \textit{The given equation contains } x^{1/3} \textit{ and } x^{1/3} \textit{ squared.}$$
$$5t^2 + 11t + 2 = 0 \qquad \textit{Replace } x^{1/3} \textit{ with } t.$$
$$(5t + 1)(t + 2) = 0 \qquad \textit{Factor.}$$
$$5t + 1 = 0 \quad \text{or} \quad t + 2 = 0 \qquad \textit{Set each factor equal to 0.}$$
$$5t = -1 \qquad\qquad t = -2 \quad \textit{Solve for } t.$$
$$t = -\frac{1}{5}$$

Use the original substitution, $t = x^{1/3}$, to solve for x. Replace t with $x^{1/3}$ in each of the preceding equations, namely $t = -\frac{1}{5}$ and $t = -2$.

$$x^{1/3} = -\frac{1}{5} \qquad\qquad x^{1/3} = -2 \qquad \text{Replace } t \text{ with } x^{1/3}.$$

$$\left(x^{1/3}\right)^3 = \left(-\frac{1}{5}\right)^3 \qquad \left(x^{1/3}\right)^3 = (-2)^3 \qquad \text{Solve for } x \text{ by cubing both sides of each equation.}$$

$$x = -\frac{1}{125} \qquad\qquad x = -8$$

Check these values to verify that the solution set is $\{-\frac{1}{125}, -8\}$.

Check Point 7 Solve: $3x^{2/3} - 11x^{1/3} - 4 = 0$.

5 Solve equations involving absolute value.

Equations Involving Absolute Value

We have seen that the absolute value of x, $|x|$, describes the distance of x from zero on a number line. Now consider **absolute value equations,** such as

$$|x| = 2.$$

$|-2| = 2 \qquad |2| = 2$

Figure 1.18 If $|x| = 2$, then $x = 2$ or $x = -2$.

This means that we must determine real numbers whose distance from the origin on the number line is 2. Figure 1.18 shows that there are two numbers such that $|x| = 2$, namely, 2 or -2. We write $x = 2$ or $x = -2$. This observation can be generalized as follows:

> **Rewriting an Absolute Value Equation without Absolute Value Bars**
> If c is a positive real number and X represents any algebraic expression, then $|X| = c$ is equivalent to $X = c$ or $X = -c$.

Technology

You can use a graphing utility to verify the solution of an absolute value equation. Consider, for example,

$$|2x - 3| = 11$$

Graph $y_1 = |2x - 3|$ and $y_2 = 11$. The graphs are shown in a $[-10, 10, 1]$ by $[-1, 15, 1]$ viewing rectangle. The x-coordinates of the intersection points are -4 and 7, verifying that $\{-4, 7\}$ is the solution set.

$y_2 = 11$

$y_1 = |2x - 3|$

$-4 \qquad 7$

$[-10, 10, 1]$ by $[-1, 15, 1]$

EXAMPLE 8 Solving an Equation Involving Absolute Value

Solve: $|2x - 3| = 11$.

Solution

$$|2x - 3| = 11 \qquad \text{This is the given equation.}$$

$2x - 3 = 11 \quad$ or $\quad 2x - 3 = -11 \qquad$ Rewrite the equation without absolute value bars.

$$2x = 14 \qquad\qquad 2x = -8 \qquad \text{Add 3 to both sides of each equation.}$$

$$x = 7 \qquad\qquad x = -4 \qquad \text{Divide both sides of each equation by 2.}$$

Check 7:	**Check -4:**	
$\|2x - 3\| = 11$	$\|2x - 3\| = 11$	This is the original equation.
$\|2(7) - 3\| \overset{?}{=} 11$	$\|2(-4) - 3\| \overset{?}{=} 11$	Substitute the proposed solutions.
$\|14 - 3\| \overset{?}{=} 11$	$\|-8 - 3\| \overset{?}{=} 11$	Perform operations inside the absolute value bars.
$\|11\| \overset{?}{=} 11$	$\|-11\| \overset{?}{=} 11$	
$11 = 11 \checkmark$	$11 = 11 \checkmark$	These true statements indicate that 7 and -4 are solutions.

The solution set is $\{-4, 7\}$.

Check Point 8
Solve: $|2x - 1| = 5$.

The absolute value of a number is never negative. Thus, if X is an algebraic expression and c is a negative number, then $|X| = c$ has no solution. For example, the equation $|3x - 6| = -2$ has no solution because $|3x - 6|$ cannot be negative. The solution set is $\varnothing$, the empty set.

The absolute value of 0 is 0. Thus, if X is an algebraic expression and $|X| = 0$, the solution is found by solving $X = 0$. For example, the solution of $|x - 2| = 0$ is obtained by solving $x - 2 = 0$. The solution is 2 and the solution set is $\{2\}$.

To solve some absolute value equations, it is necessary to first isolate the expression containing the absolute value symbols. For example, consider the equation

$$3|2x - 3| - 8 = 25.$$

We need to isolate $|2x - 3|$.

How can we isolate $|2x - 3|$? Add 8 to both sides of the equation and then divide both sides by 3.

$$3|2x - 3| - 8 = 25 \qquad \text{This is the given equation.}$$
$$3|2x - 3| = 33 \qquad \text{Add 8 to both sides.}$$
$$|2x - 3| = 11 \qquad \text{Divide both sides by 3.}$$

This results in the equation we solved in Example 8.

EXERCISE SET 1.6

Practice Exercises

Solve each polynomial equation in Exercises 1–10 by factoring and then using the zero-product principle.

1. $3x^4 - 48x^2 = 0$ **2.** $5x^4 - 20x^2 = 0$

3. $3x^3 + 2x^2 = 12x + 8$ **4.** $4x^3 - 12x^2 = 9x - 27$

5. $2x - 3 = 8x^3 - 12x^2$ **6.** $x + 1 = 9x^3 + 9x^2$

7. $4y^3 - 2 = y - 8y^2$ **8.** $9y^3 + 8 = 4y + 18y^2$

9. $2x^4 = 16x$ **10.** $3x^4 = 81x$

Solve each radical equation in Exercises 11–28. Check all proposed solutions.

11. $\sqrt{3x + 18} = x$ **12.** $\sqrt{20 - 8x} = x$

13. $\sqrt{x + 3} = x - 3$ **14.** $\sqrt{x + 10} = x - 2$

15. $\sqrt{2x + 13} = x + 7$ **16.** $\sqrt{6x + 1} = x - 1$

17. $x - \sqrt{2x + 5} = 5$ **18.** $x - \sqrt{x + 11} = 1$

19. $\sqrt{3x + 10} = x + 4$ **20.** $\sqrt{x} - 3 = x - 9$

21. $\sqrt{x + 8} - \sqrt{x - 4} = 2$

22. $\sqrt{x + 5} - \sqrt{x - 3} = 2$

23. $\sqrt{x - 5} - \sqrt{x - 8} = 3$

24. $\sqrt{2x - 3} - \sqrt{x - 2} = 1$

25. $\sqrt{2x + 3} + \sqrt{x - 2} = 2$

26. $\sqrt{x + 2} + \sqrt{3x + 7} = 1$

27. $\sqrt{3\sqrt{x + 1}} = \sqrt{3x - 5}$

28. $\sqrt{1 + 4\sqrt{x}} = 1 + \sqrt{x}$

Solve and check each equation with rational exponents in Exercises 29–38.

29. $x^{3/2} = 8$ **30.** $x^{3/2} = 27$

31. $(x - 4)^{3/2} = 27$ **32.** $(x + 5)^{3/2} = 8$

33. $6x^{5/2} - 12 = 0$ **34.** $8x^{5/3} - 24 = 0$

35. $(x - 4)^{2/3} = 16$ **36.** $(x + 5)^{2/3} = 4$

37. $(x^2 - x - 4)^{3/4} - 2 = 6$

38. $(x^2 - 3x + 3)^{3/2} - 1 = 0$

Solve each equation in Exercises 39–58 by making an appropriate substitution.

39. $x^4 - 5x^2 + 4 = 0$

40. $x^4 - 13x^2 + 36 = 0$

41. $9x^4 = 25x^2 - 16$

42. $4x^4 = 13x^2 - 9$

43. $x - 13\sqrt{x} + 40 = 0$

44. $2x - 7\sqrt{x} - 30 = 0$

45. $x^{-2} - x^{-1} - 20 = 0$

46. $x^{-2} - x^{-1} - 6 = 0$

47. $x^{2/3} - x^{1/3} - 6 = 0$

48. $2x^{2/3} + 7x^{1/3} - 15 = 0$

49. $x^{3/2} - 2x^{3/4} + 1 = 0$

50. $x^{2/5} + x^{1/5} - 6 = 0$

51. $2x - 3x^{1/2} + 1 = 0$

52. $x + 3x^{1/2} - 4 = 0$

53. $(x - 5)^2 - 4(x - 5) - 21 = 0$

54. $(x + 3)^2 + 7(x + 3) - 18 = 0$

55. $(x^2 - x)^2 - 14(x^2 - x) + 24 = 0$

56. $(x^2 - 2x)^2 - 11(x^2 - 2x) + 24 = 0$

57. $\left(y - \dfrac{8}{y}\right)^2 + 5\left(y - \dfrac{8}{y}\right) - 14 = 0$

58. $\left(y - \dfrac{10}{y}\right)^2 + 6\left(y - \dfrac{10}{y}\right) - 27 = 0$

In Exercises 59–74, solve each absolute value equation or indicate the equation has no solution.

59. $|x| = 8$

60. $|x| = 6$

61. $|x - 2| = 7$

62. $|x + 1| = 5$

63. $|2x - 1| = 5$

64. $|2x - 3| = 11$

65. $2|3x - 2| = 14$

66. $3|2x - 1| = 21$

67. $7|5x| + 2 = 16$

68. $7|3x| + 2 = 16$

69. $|x + 1| + 5 = 3$

70. $|x + 1| + 6 = 2$

71. $|2x - 1| + 3 = 3$

72. $|3x - 2| + 4 = 4$

Hint for Exercises 73–74: Absolute value expressions are equal when the expressions inside the absolute value bars are equal to or opposites of each other.

73. $|3x - 1| = |x + 5|$

74. $|2x - 7| = |x + 3|$

Solve each equation in Exercises 75–84 by the method of your choice.

75. $x + 2\sqrt{x} - 3 = 0$

76. $x^3 + 3x^2 - 4x - 12 = 0$

77. $(x + 4)^{3/2} = 8$

78. $(x^2 - 1)^2 - 2(x^2 - 1) = 3$

79. $\sqrt{4x + 15} - 2x = 0$

80. $x^{2/5} - 1 = 0$

81. $|x^2 + 2x - 36| = 12$

82. $\sqrt{3x + 1} - \sqrt{x - 1} = 2$

83. $x^3 - 2x^2 = x - 2$

84. $|x^2 + 6x + 1| = 8$

Application Exercises

First the good news: The graph shows that U.S. seniors' scores in standard testing in science have improved since 1982. Now the bad news: The highest possible score is 500, and in 1970, the average test score was 304.

U.S. Seniors' Test Scores in Science

Source: National Assessment of Educational Progress

The formula

$$S = 4\sqrt{x} + 280$$

models the average science test score, S, x years after 1982. Use the formula to solve Exercises 85–86.

85. When will the average science score return to the 1970 average of 304?

86. When will the average science test score be 300?

Out of a group of 50,000 births, the number of people, y, surviving to age x is modeled by the formula

$$y = 5000\sqrt{100 - x}.$$

The graph of the formula is shown. Use the formula to solve Exercises 87–88.

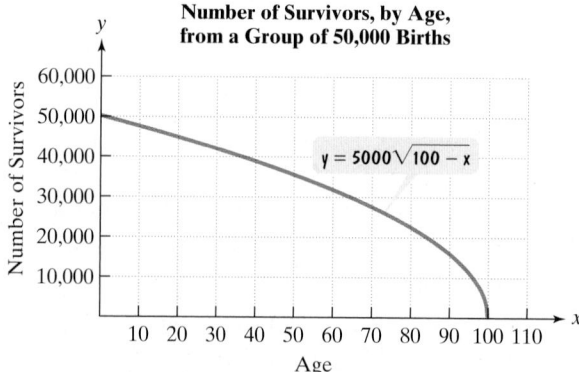

Number of Survivors, by Age, from a Group of 50,000 Births

$y = 5000\sqrt{100 - x}$

87. To what age will 40,000 people in the group survive? Identify the solution as a point on the graph of the formula.

88. To what age will 35,000 people in the group survive? Identify the solution as a point on the graph of the formula.

For each planet in our solar system, its year is the time it takes the planet to revolve once around the sun. The formula

$$E = 0.2x^{3/2}$$

models the number of Earth days in a planet's year, E, where x is the average distance of the planet from the sun, in millions of kilometers. Use the formula to solve Exercises 89–90.

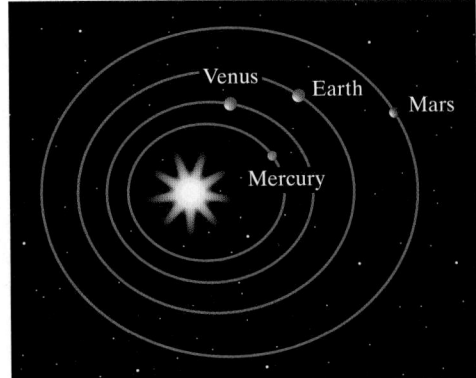

89. We, of course, have 365 Earth days in our year. What is the average distance of Earth from the sun? Use a calculator and round to the nearest million kilometers.

90. There are approximately 88 Earth days in the year of the planet Mercury. What is the average distance of Mercury from the sun? Use a calculator and round to the nearest million kilometers.

Use the Pythagorean Theorem to solve Exercises 91–92.

91. Two vertical poles of lengths 6 feet and 8 feet stand 10 feet apart (see the figure). A cable reaches from the top of one pole to some point on the ground between the poles and then to the top of the other pole. Where should this point be located to use 18 feet of cable?

92. Towns *A* and *B* are located 6 miles and 3 miles, respectively, from a major expressway. The point on the expressway closest to town *A* is 12 miles from the point on the expressway closest to town *B*. Two new roads are to be built from *A* to the expressway and then to *B*. (See the figure at the top of the next column.)

a. Find *x* if the length of the new roads is 15 miles.

b. Write a verbal description for the road crew telling them where to position the new roads based on your answer to part (a).

Writing in Mathematics

93. Without actually solving the equation, give a general description of how to solve $x^3 - 5x^2 - x + 5 = 0$.

94. In solving $\sqrt{3x + 4} - \sqrt{2x + 4} = 2$, why is it a good idea to isolate a radical term? What if we don't do this and simply square each side? Describe what happens.

95. What is an extraneous solution to a radical equation?

96. Explain how to recognize an equation that is quadratic in form. Provide two original examples with your explanation.

97. Describe two methods for solving this equation: $x - 5\sqrt{x} + 4 = 0$.

98. Explain how to solve an equation involving absolute value.

99. Explain why the procedure that you explained in Exercise 98 does not apply to the equation $|x - 2| = -3$. What is the solution set for this equation?

100. Describe the trend shown by the graph in Exercises 87–88. When is the rate of decrease most rapid? What does this mean about survival rate by age?

Technology Exercises

In Exercises 101–103, use a graphing utility and the graph's x-intercepts to solve each equation. Check by direct substitution. A viewing rectangle is given.

101. $x^3 + 3x^2 - x - 3 = 0$
$[-6, 6, 1]$ by $[-6, 6, 1]$

102. $-x^4 + 4x^3 - 4x^2 = 0$
$[-6, 6, 1]$ by $[-9, 2, 1]$

103. $\sqrt{2x + 13} - x - 5 = 0$
$[-5, 5, 1]$ by $[-5, 5, 1]$

104. Use a graphing utility to obtain the graph of the formula in Exercises 87–88. Then use the TRACE feature to trace along the curve until you reach the point that visually shows the solution to Exercise 87 or 88.

Critical Thinking Exercises

105. Which one of the following is true?

    a. Squaring both sides of $\sqrt{y + 4} + \sqrt{y - 1} = 5$ leads to $y + 4 + y - 1 = 25$, an equation with no radicals.

    b. The equation $(x^2 - 2x)^9 - 5(x^2 - 2x)^3 + 6 = 0$ is quadratic in form and should be solved by letting
$$t = (x^2 - 2x)^3.$$

    c. If a radical equation has two proposed solutions and one of these values is not a solution, the other value is also not a solution.

    d. None of these statements is true.

106. Solve: $\sqrt{6x - 2} = \sqrt{2x + 3} - \sqrt{4x - 1}$.

107. Solve *without* squaring both sides:
$$5 - \frac{2}{x} = \sqrt{5 - \frac{2}{x}}.$$

108. Solve for x: $\sqrt[3]{x\sqrt{x}} = 9$.

109. Solve for x: $x^{5/6} + x^{2/3} - 2x^{1/2} = 0$.

SECTION 1.7   *Linear Inequalities*

Objectives

1. Graph an inequality's solution set.

2. Use set-builder and interval notations.

3. Use properties of inequalities to solve inequalities.

4. Solve compound inequalities.

5. Solve inequalities involving absolute value.

Rent-a-Heap, a car rental company, charges \$125 per week plus \$0.20 per mile to rent one of their cars. Suppose you are limited by how much money you can spend for the week: You can spend at most \$335. If we let x represent the number of miles you drive the heap in a week, we can write an inequality that models the given conditions.

The weekly charge of \$125	plus	the charge of \$0.20 per mile for x miles	must be less than or equal to	\$335.
125	+	0.20x	≤	335

Using the commutative property of addition, we can express this inequality as $0.20x + 125 \le 335$. The form of this inequality is $ax + b \le c$, with $a = 0.20$, $b = 125$, and $c = 335$. Any inequality in this form is called a **linear inequality in one variable.** The greatest exponent on the variable in such an inequality is 1. The symbol between $ax + b$ and c can be $\le$ (is less than or equal to), $<$ (is less than), $\ge$ (is greater than or equal to), or $>$ (is greater than).

In this section, we will study how to solve linear inequalities such as $0.20x + 125 \leq 335$. **Solving an inequality** is the process of finding the set of numbers that make the inequality a true statement. These numbers are called the **solutions** of the inequality, and we say that they **satisfy** the inequality. The set of all solutions is called the **solution set** of the inequality. We begin by discussing how to graph and how to represent these solution sets.

1 Graph an inequality's solution set.

Graphs of Inequalities; Interval Notation

There are infinitely many solutions to the inequality $x > -4$, namely all real numbers that are greater than -4. Although we cannot list all the solutions, we can make a drawing on a number line that represents these solutions. Such a drawing is called the **graph of the inequality.**

Graphs of solutions to linear inequalities are shown on a number line by shading all points representing numbers that are solutions. Parentheses indicate endpoints that are not solutions. Square brackets indicate endpoints that are solutions.

EXAMPLE 1 Graphing Inequalities

Graph the solutions of:

    a. $x < 3$ **b.** $x \geq -1$ **c.** $-1 < x \leq 3$.

Study Tip

An inequality symbol points to the smaller number. Thus, another way to express $x < 3$ (x is less than 3) is $3 > x$ (3 is greater than x).

Solution

a. The solutions of $x < 3$ are all real numbers that are less than 3. They are graphed on a number line by shading all points to the left of 3. The parenthesis at 3 indicates that 3 is not a solution, but numbers such as 2.9999 and 2.6 are. The arrow shows that the graph extends indefinitely to the left.

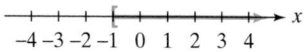

b. The solutions of $x \geq -1$ are all real numbers that are greater than or equal to -1. We shade all points to the right of -1 and the point for -1 itself. The bracket at -1 shows that -1 is a solution of the given inequality. The arrow shows that the graph extends indefinitely to the right.

c. The inequality $-1 < x \leq 3$ is read "-1 is less than x and x is less than or equal to 3," or "x is greater than -1 and less than or equal to 3." The solutions of $-1 < x \leq 3$ are all real numbers between -1 and 3, not including -1 but including 3. The parenthesis at -1 indicates that -1 is not a solution. By contrast, the bracket at 3 shows that 3 is a solution. Shading indicates the other solutions.

Check Point 1

Graph the solutions of:

a. $x \leq 2$ **b.** $x > -4$ **c.** $2 \leq x < 6$.

2 Use set-builder and interval notations.

Now that we know how to graph the solution set of an inequality such as $x > -4$, let's see how to represent the solution set. One method is with **set-builder notation.** Using this method, the solution set of $x > -4$ can be expressed as

$$\{x \mid x > -4\}.$$

The set of all x	such that

We read this as "the set of all real numbers x such that x is greater than -4."

Another method used to represent solution sets of inequalities is **interval notation.** Using this notation, the solution set of $x > -4$ is expressed as $(-4, \infty)$. The parenthesis at -4 indicates that -4 is not included in the interval. The infinity symbol, ∞, does not represent a real number. It indicates that the interval extends indefinitely to the right.

Table 1.5 lists nine possible types of intervals used to describe subsets of real numbers.

Table 1.5   Intervals on the Real Number Line

Let a and b be real numbers such that $a < b$.

Interval Notation	Set-Builder Notation	Graph
(a, b)	$\{x \mid a < x < b\}$	
$[a, b]$	$\{x \mid a \leq x \leq b\}$	
$[a, b)$	$\{x \mid a \leq x < b\}$	
$(a, b]$	$\{x \mid a < x \leq b\}$	
(a, ∞)	$\{x \mid x > a\}$	
$[a, \infty)$	$\{x \mid x \geq a\}$	
$(-\infty, b)$	$\{x \mid x < b\}$	
$(-\infty, b]$	$\{x \mid x \leq b\}$	
$(-\infty, \infty)$	$\mathbb{R}$ (set of all real numbers)	

EXAMPLE 2   Intervals and Inequalities

Express the intervals in terms of inequalities and graph:

a. $(-1, 4]$ **b.** $[2.5, 4]$ **c.** $(-4, \infty)$.

Solution

a. $(-1, 4] = \{x \mid -1 < x \leq 4\}$

b. $[2.5, 4] = \{x \mid 2.5 \leq x \leq 4\}$

c. $(-4, \infty) = \{x \mid x > -4\}$

Check Point 2   Express the intervals in terms of inequalities and graph:
   a. $[-2, 5)$ **b.** $[1, 3.5]$ **c.** $(-\infty, -1)$.

---

3 Use properties of inequalities to solve inequalities.

Solving Linear Inequalities

Back to our question: How many miles can you drive your Rent-a-Heap car if you can spend at most \$335 per week? We answer the question by solving

$$0.20x + 125 \le 335$$

for x. The solution procedure is nearly identical to that for solving

$$0.20x + 125 = 335.$$

Our goal is to get x by itself on the left side. We do this by subtracting 125 from both sides to isolate $0.20x$:

$$0.20x + 125 \le 335$$
$$0.20x + 125 - 125 \le 335 - 125$$
$$0.20x \le 210.$$

Finally, we isolate x from $0.20x$ by dividing both sides of the inequality by 0.20:

$$\frac{0.20x}{0.20} \le \frac{210}{0.20}$$
$$x \le 1050.$$

With at most \$335 per week to spend, you can travel at most 1050 miles.

We started with the inequality $0.20x + 125 \le 335$ and obtained the inequality $x \le 1050$ in the final step. Both of these inequalities have the same solution set, namely $\{x \mid x \le 1050\}$. Inequalities such as these, with the same solution set, are said to be **equivalent.**

We isolated x from $0.20x$ by dividing both sides of $0.20x \le 210$ by 0.20, a positive number. Let's see what happens if we divide both sides of an inequality by a negative number. Consider the inequality $10 < 14$. Divide 10 and 14 by -2:

$$\frac{10}{-2} = -5 \quad \text{and} \quad \frac{14}{-2} = -7.$$

Because -5 lies to the right of -7 on the number line, -5 is greater than -7:

$$-5 > -7.$$

Notice that the direction of the inequality symbol is reversed:

$$10 < 14$$
$$-5 > -7.$$

In general, **when we multiply or divide both sides of an inequality by a negative number, the direction of the inequality symbol is reversed.** When we reverse the direction of the inequality symbol, we say that we change the *sense* of the inequality.

We can isolate a variable in a linear inequality the same way we can isolate a variable in a linear equation. The following properties are used to create equivalent inequalities:

Study Tip

English phrases such as "at least" and "at most" can be modeled by inequalities.

English Sentence	Inequality
x is at least 5.	$x \ge 5$
x is at most 5.	$x \le 5$
x is between 5 and 7.	$5 < x < 7$
x is no more than 5.	$x \le 5$
x is no less than 5.	$x \ge 5$

Properties of Inequalities

Property	The Property in Words	Example
Addition and Subtraction Properties If $a < b$, then $a + c < b + c$. If $a < b$, then $a - c < b - c$.	If the same quantity is added to or subtracted from both sides of an inequality, the resulting inequality is equivalent to the original one.	$2x + 3 < 7$ Subtract 3: $2x + 3 - 3 < 7 - 3$. Simplify: $2x < 4$.
Positive Multiplication and Division Properties If $a < b$ and c is positive, then $ac < bc$. If $a < b$ and c is positive, then $\dfrac{a}{c} < \dfrac{b}{c}$.	If we multiply or divide both sides of an inequality by the same positive quantity, the resulting inequality is equivalent to the original one.	$2x < 4$ Divide by 2: $\dfrac{2x}{2} < \dfrac{4}{2}$. Simplify: $x < 2$.
Negative Multiplication and Division Properties If $a < b$ and c is negative, then $ac > bc$. If $a < b$ and c is negative, then $\dfrac{a}{c} > \dfrac{b}{c}$.	If we multiply or divide both sides of an inequality by the same negative quantity and reverse the direction of the inequality symbol, the result is an equivalent inequality.	$-4x < 20$ Divide by -4 and reverse the sense of the inequality: $\dfrac{-4x}{-4} > \dfrac{20}{-4}$. Simplify: $x > -5$.

EXAMPLE 3   Solving a Linear Inequality

Solve and graph the solution set on a number line:
$$3 - 2x < 11.$$

Solution

$3 - 2x < 11$	This is the given inequality.
$3 - 2x - 3 < 11 - 3$	Subtract 3 from both sides.
$-2x < 8$	Simplify.
$\dfrac{-2x}{-2} > \dfrac{8}{-2}$	Divide both sides by -2 and reverse the sense of the inequality.
$x > -4$	Simplify.

Discovery

As a partial check, select one number from the solution set for the inequality in Example 3. Substitute that number into the original inequality. Perform the resulting computations. You should obtain a true statement.

Is it possible to perform a partial check using a number that is not in the solution set? What should happen in this case? Try doing this.

The solution set consists of all real numbers that are greater than -4, expressed as $\{x \mid x > -4\}$ in set-builder notation. The interval notation for this solution set is $(-4, \infty)$. The graph of the solution set is shown as follows:

Check Point 3   Solve and graph the solution set on a number line:
$$2 - 3x \le 5.$$

EXAMPLE 4   Solving a Linear Inequality

Solve and graph the solution set: $7x + 15 \geq 13x + 51$.

Study Tip

You can solve

$$7x + 15 \geq 13x + 51$$

by isolating x on the right side. Subtract $7x$ from both sides.

$$7x + 15 - 7x$$
$$\geq 13x + 51 - 7x$$
$$15 \geq 6x + 51$$

Now subtract 51 from both sides.

$$15 - 51 \geq 6x + 51 - 51$$
$$-36 \geq 6x$$

Finally, divide both sides by 6.

$$\frac{-36}{6} \geq \frac{6x}{6}$$
$$-6 \geq x$$

This last inequality means the same thing as

$$x \leq -6.$$

Solution   We will collect variable terms on the left and constant terms on the right.

$7x + 15$	$\geq$	$13x + 51$	This is the given inequality.
$7x + 15 - 13x$	$\geq$	$13x + 51 - 13x$	Subtract 13x from both sides.
$-6x + 15$	$\geq$	51	Simplify.
$-6x + 15 - 15$	$\geq$	$51 - 15$	Subtract 15 from both sides.
$-6x$	$\geq$	36	Simplify.
$\dfrac{-6x}{-6}$	$\leq$	$\dfrac{36}{-6}$	Divide both sides by -6 and reverse the sense of the inequality.
x	$\leq$	-6	Simplify.

The solution set consists of all real numbers that are less than or equal to -6, expressed as $\{x \mid x \leq -6\}$. The interval notation for this solution set is $(-\infty, -6]$. The graph of the solution set is shown as follows:

$$\xleftarrow{\quad}\begin{array}{ccccccccccc} & & &] & & & & & & & \\ \hline -8 & -7 & -6 & -5 & -4 & -3 & -2 & -1 & 0 & 1 & 2 \end{array}\xrightarrow{\quad} x$$

Check Point 4   Solve and graph the solution set: $6 - 3x \leq 5x - 2$.

Technology

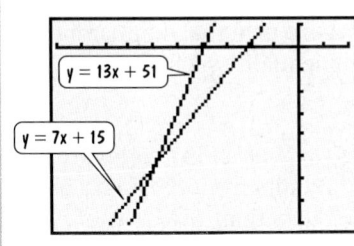

You can use a graphing utility to verify that $(-\infty, -6]$ is the solution set of

$$7x + 15 \qquad \geq \qquad 13x + 51.$$

For what values of x does the graph of $y = 7x + 15$	lie above or on	the graph of $y = 13x + 51$?

The graphs are shown on the left in a $[-10, 2, 1]$ by $[-40, 5, 5]$ viewing rectangle. Look closely at the graphs. Can you see that the graph of $y = 7x + 15$ lies above or on the graph of $y = 13x + 51$ when $x \leq -6$, or on the interval $(-\infty, -6]$?

4 Solve compound inequalities.

Solving Compound Inequalities

We now consider two inequalities such as

$$-3 < 2x + 1 \quad \text{and} \quad 2x + 1 \leq 3$$

expressed as a **compound inequality**

$$-3 < 2x + 1 \leq 3.$$

The word "and" does not appear when the inequality is written in the shorter form, although it is implied. The shorter form enables us to solve both inequalities at once. By performing the same operation on all three parts of the inequality, our goal is to **isolate x in the middle**.

Technology

To check Example 5, graph each part of

$$-3 < 2x + 1 \le 3.$$

The figure shows that the graph of $y_2 = 2x + 1$ lies above the graph of $y_1 = -3$ and on or below the graph of $y_3 = 3$ when x is in the interval $(-2, 1]$.

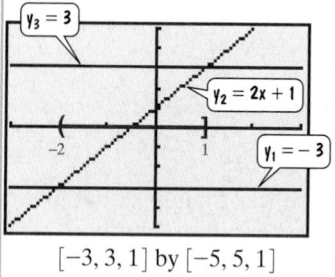

$[-3, 3, 1]$ by $[-5, 5, 1]$

EXAMPLE 5   Solving a Compound Inequality

Solve and graph the solution set:

$$-3 < 2x + 1 \le 3.$$

Solution   We would like to isolate x in the middle. We can do this by first subtracting 1 from all three parts of the compound inequality. Then we isolate x from $2x$ by dividing all three parts of the inequality by 2.

$-3 < 2x + 1 \le 3$	This is the given inequality.
$-3 - 1 < 2x + 1 - 1 \le 3 - 1$	Subtract 1 from all three parts.
$-4 < 2x \le 2$	Simplify.
$\dfrac{-4}{2} < \dfrac{2x}{2} \le \dfrac{2}{2}$	Divide each part by 2.
$-2 < x \le 1$	Simplify.

The solution set consists of all real numbers greater than -2 and less than or equal to 1, represented by $\{x \mid -2 < x \le 1\}$ in set-builder notation and $(-2, 1]$ in interval notation. The graph is shown as follows:

Check Point 5   Solve and graph the solution set: $1 \le 2x + 3 < 11$.

---

5   Solve inequalities involving absolute value.

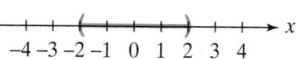

Figure 1.19 $|x| < 2$, so $-2 < x < 2$.

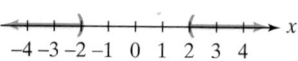

Figure 1.20 $|x| > 2$, so $x < -2$ or $x > 2$.

Study Tip

In the $|X| < c$ case, we have one compound inequality to solve. In the $|X| > c$ case, we have two separate inequalities to solve.

Solving Inequalities with Absolute Value

We know that $|x|$ describes the distance of x from zero on a real number line. We can use this geometric interpretation to solve an inequality such as

$$|x| < 2.$$

This means that the distance of x from 0 is *less than* 2, as shown in Figure 1.19. The interval shows values of x that lie less than 2 units from 0. Thus, x can lie between -2 and 2. That is, x is greater than -2 and less than 2. We write $(-2, 2)$ or $\{x \mid -2 < x < 2\}$.

Some absolute value inequalities use the "greater than" symbol. For example, $|x| > 2$ means that the distance of x from 0 is *greater than* 2, as shown in Figure 1.20. Thus, x can be less than -2 *or* greater than 2. We write $x < -2$ or $x > 2$.

These observations suggest the following principles for solving inequalities with absolute value:

Solving an Absolute Value Inequality

If X is an algebraic expression and c is a positive number,

1. The solutions of $|X| < c$ are the numbers that satisfy $-c < X < c$.

2. The solutions of $|X| > c$ are the numbers that satisfy $X < -c$ or $X > c$.

These rules are valid if $<$ is replaced by $\le$ and $>$ is replaced by $\ge$.

EXAMPLE 6   **Solving an Absolute Value Inequality with** $<$

Solve and graph the solution set: $|x - 4| < 3$.

Solution

$|X| < c$ means $-c < X < c$.

$|x - 4| < 3$ means $-3 < x - 4 < 3$.

We solve the compound inequality by adding 4 to all three parts.

$$-3 < x - 4 < 3$$
$$-3 + 4 < x - 4 + 4 < 3 + 4$$
$$1 < x < 7$$

The solution set is all real numbers greater than 1 and less than 7, denoted by $\{x \mid 1 < x < 7\}$ or $(1, 7)$. The graph of the solution set is shown as follows:

Check Point 6   Solve and graph the solution set: $|x - 2| < 5$.

EXAMPLE 7   **Solving an Absolute Value Inequality with** $\geq$

Solve and graph the solution set: $|2x + 3| \geq 5$.

Solution $|X| \geq c$ means $X \leq -c$ or $X \geq c$.

$|2x + 3| \geq 5$ means $2x + 3 \leq -5$ or $2x + 3 \geq 5$.

We solve each of these inequalities separately.

$2x + 3 \leq -5$	or	$2x + 3 \geq 5$	These are the inequalities without absolute value bars.
$2x + 3 - 3 \leq -5 - 3$	or	$2x + 3 - 3 \geq 5 - 3$	Subtract 3 from both sides.
$2x \leq -8$	or	$2x \geq 2$	Simplify.
$\dfrac{2x}{2} \leq \dfrac{-8}{2}$	or	$\dfrac{2x}{2} \geq \dfrac{2}{2}$	Divide both sides by 2.
$x \leq -4$	or	$x \geq 1$	Simplify.

The solution set is $\{x \mid x \leq -4 \text{ or } x \geq 1\}$, that is, all x in $(-\infty, -4]$ or $[1, \infty)$. The graph of the solution set is shown as follows:

Study Tip

The graph of the solution set for $|X| > c$ will be divided into two intervals. The graph of the solution set for $|X| < c$ will be a single interval.

Check Point 7 Solve and graph the solution set: $|2x - 5| \geq 3$.

Applications

Our next example shows how to use an inequality to select the better deal between two pricing options. We will use our five-step strategy for solving problems using mathematical models.

EXAMPLE 8 Creating and Comparing Mathematical Models

Acme Car rental agency charges $4 a day plus $0.15 a mile, whereas Interstate rental agency charges $20 a day and $0.05 a mile. Under what conditions is the daily cost of an Acme rental a better deal than an Interstate rental?

Solution

Step 1 Let x represent one of the quantities. We are looking for the number of miles driven in a day to make Acme the better deal. Thus,

$$\text{let } x = \text{ the number of miles driven in a day.}$$

Step 2 Represent other quantities in terms of x. We are not asked to find another quantity, so we can skip this step.

Step 3 Write an inequality in x that describes the conditions.

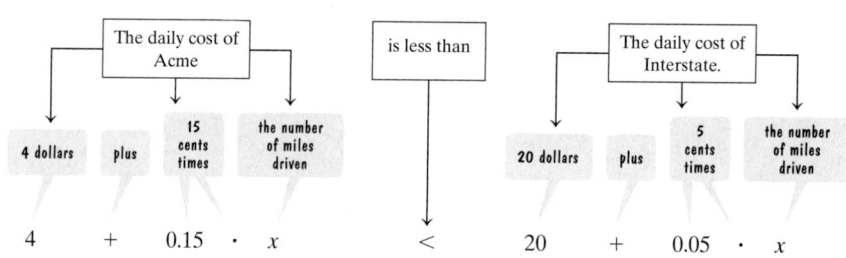

Step 4 Solve the inequality and answer the question.

$4 + 0.15x < 20 + 0.05x$	This is the inequality that models the verbal conditions.
$4 + 0.15x - 0.05x < 20 + 0.05x - 0.05x$	Subtract 0.05x from both sides.
$4 + 0.1x < 20$	Simplify.
$4 + 0.1x - 4 < 20 - 4$	Subtract 4 from both sides.
$0.1x < 16$	Simplify.
$\dfrac{0.1x}{0.1} < \dfrac{16}{0.1}$	Divide both sides by 0.1.
$x < 160$	Simplify.

Thus, driving fewer than 160 miles per day makes Acme the better deal.

Step 5 **Check the proposed solution in the original wording of the problem.**
One way to do this is to take a mileage less than 160 miles per day to see if Acme is the better deal. Suppose that 150 miles are driven in a day.

$$\text{Cost for Acme} = 4 + 0.15(150) = 26.50$$

$$\text{Cost for Interstate} = 20 + 0.05(150) = 27.50$$

Acme has a lower daily cost, making it the better deal.

Check Point 8 A car can be rented from Basic Rental for $260 per week with no extra charge for mileage. Continental charges $80 per week plus 25 cents for each mile driven to rent the same car. Under what conditions is the rental cost for Basic Rental a better deal than Continental's?

EXERCISE SET 1.7

Practice Exercises

In Exercises 1–12, graph the solutions of each inequality on a number line.

1. $x > 6$

2. $x > -2$

3. $x < -4$

4. $x < 0$

5. $x \geq -3$

6. $x \geq -5$

7. $x \leq 4$

8. $x \leq 7$

9. $-2 < x \leq 5$

10. $-3 \leq x < 7$

11. $-1 < x < 4$

12. $-7 \leq x \leq 0$

In Exercises 13–26, express each interval in terms of an inequality and graph the interval on a number line.

13. $(1, 6]$

14. $(-2, 4]$

15. $[-5, 2)$

16. $[-4, 3)$

17. $[-3, 1]$

18. $[-2, 5]$

19. $(2, \infty)$

20. $(3, \infty)$

21. $[-3, \infty)$

22. $[-5, \infty)$

23. $(-\infty, 3)$

24. $(-\infty, 2)$

25. $(-\infty, 5.5)$

26. $(-\infty, 3.5]$

Solve each linear inequality in Exercises 27–48 and graph the solution set on a number line. Express the solution set using interval notation.

27. $5x + 11 < 26$

28. $2x + 5 < 17$

29. $3x - 7 \geq 13$

30. $8x - 2 \geq 14$

31. $-9x \geq 36$

32. $-5x \leq 30$

33. $8x - 11 \leq 3x - 13$

34. $18x + 45 \leq 12x - 8$

35. $4(x + 1) + 2 \geq 3x + 6$

36. $8x + 3 > 3(2x + 1) + x + 5$

37. $2x - 11 < -3(x + 2)$

38. $-4(x + 2) > 3x + 20$

39. $1 - (x + 3) \geq 4 - 2x$

40. $5(3 - x) \leq 3x - 1$

41. $\dfrac{x}{4} - \dfrac{3}{5} \leq \dfrac{x}{2} + 1$

42. $\dfrac{3x}{10} + 1 \geq \dfrac{1}{5} - \dfrac{x}{10}$

43. $1 - \dfrac{x}{2} > 4$

44. $7 - \dfrac{4}{5}x < \dfrac{3}{5}$

45. $\dfrac{x - 4}{6} \geq \dfrac{x - 2}{9} + \dfrac{5}{18}$

46. $\dfrac{4x - 3}{6} + 2 \geq \dfrac{2x - 1}{12}$

47. $4(3x - 2) - 3x < 3(1 + 3x) - 7$

48. $3(x - 8) - 2(10 - x) > 5(x - 1)$

Solve each inequality in Exercises 49–56 and graph the solution set on a number line. Express the solution set using interval notation.

49. $6 < x + 3 < 8$

50. $7 < x + 5 < 11$

51. $-3 \leq x - 2 < 1$

52. $-6 < x - 4 \leq 1$

53. $-11 < 2x - 1 \leq -5$

54. $3 \leq 4x - 3 < 19$

55. $-3 \leq \dfrac{2}{3}x - 5 < -1$

56. $-6 \leq \dfrac{1}{2}x - 4 < -3$

Solve each inequality in Exercises 57–84 by first rewriting each one as an equivalent inequality without absolute value bars. Graph the solution set on a number line. Express the solution set using interval notation.

57. $|x| < 3$

58. $|x| < 5$

59. $|x - 1| \leq 2$

60. $|x + 3| \leq 4$

61. $|2x - 6| < 8$

62. $|3x + 5| < 17$

63. $|2(x - 1) + 4| \leq 8$

64. $|3(x - 1) + 2| \leq 20$

65. $\left|\dfrac{2y + 6}{3}\right| < 2$

66. $\left|\dfrac{3(x - 1)}{4}\right| < 6$

67. $|x| > 3$

68. $|x| > 5$

69. $|x - 1| \geq 2$

70. $|x + 3| \geq 4$

71. $|3x - 8| > 7$

72. $|5x - 2| > 13$

73. $\left| \dfrac{2x + 2}{4} \right| \geq 2$

74. $\left| \dfrac{3x - 3}{9} \right| \geq 1$

75. $\left| 3 - \dfrac{2}{3}x \right| > 5$

76. $\left| 3 - \dfrac{3}{4}x \right| > 9$

77. $3|x - 1| + 2 \geq 8$

78. $-2|4 - x| \geq -4$

79. $3 < |2x - 1|$

80. $5 \geq |4 - x|$

81. $12 < \left| -2x + \dfrac{6}{7} \right| + \dfrac{3}{7}$

82. $1 < \left| x - \dfrac{11}{3} \right| + \dfrac{7}{3}$

83. $4 + \left| 3 - \dfrac{x}{3} \right| \geq 9$

84. $\left| 2 - \dfrac{x}{2} \right| - 1 \leq 1$

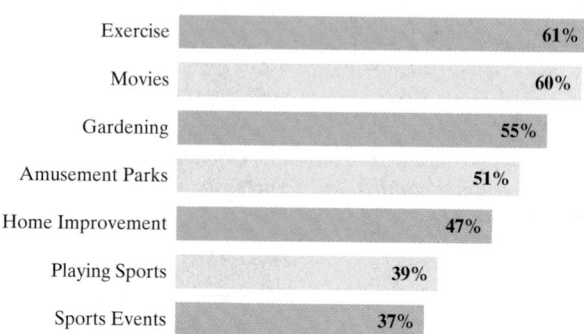

Application Exercises

The bar graph shows how we spend our leisure time. Let x represent the percentage of the population regularly participating in an activity. In Exercises 85–92, write the name or names of the activity described by the given inequality or interval.

Percentage of U.S. Population Participating in Each Activity on a Regular Basis

Exercise — 61%
Movies — 60%
Gardening — 55%
Amusement Parks — 51%
Home Improvement — 47%
Playing Sports — 39%
Sports Events — 37%

Source: U.S. Census Bureau

85. $x < 40\%$

86. $x < 50\%$

87. $[51\%, 61\%]$

88. $[47\%, 60\%]$

89. $(51\%, 61\%)$

90. $(47\%, 60\%)$

91. $(39\%, 55\%]$

92. $(37\%, 47\%]$

The line graph at the top of the next column shows the declining consumption of cigarettes in the United States. The data shown by the graph can be modeled by

$$N = 550 - 9x$$

where N is the number of cigarettes consumed, in billions, x

years after 1988. Use this formula to solve Exercises 93–94.

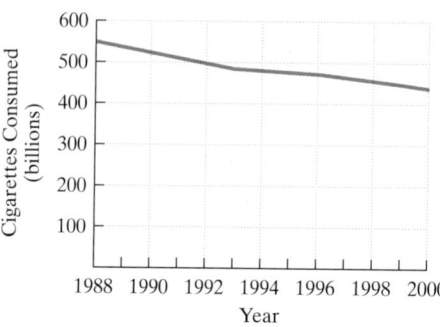

Consumption of Cigarettes in the U.S.

Source: Economic Research Service, USDA

93. How many years after 1988 will cigarette consumption be less than 370 billion cigarettes each year? Which years does this describe?

94. Describe how many years after 1988 cigarette consumption will be less than 325 billion cigarettes each year. Which years are included in your description?

95. The formula for converting Fahrenheit temperature, F, to Celsius temperature, C, is

$$C = \frac{5}{9}(F - 32).$$

If Celsius temperature ranges from 15° to 35°, inclusive, what is the range for the Fahrenheit temperature? Use interval notation to express this range.

96. The formula

$$T = 0.01x + 56.7$$

models the global mean temperature, T, in degrees Fahrenheit, of Earth x years after 1905. For which range of years was the global mean temperature at least 56.7°F and at most 57.2°F?

The three television programs viewed by the greatest percentage of U.S. households in the twentieth century are shown in the table. The data are from a random survey of 4000 TV households by Nielsen Media Research. In Exercises 97–98, let x represent the actual viewing percentage in the U.S. population.

TV Programs with the Greatest U.S. Audience Viewing Percentage of the Twentieth Century

Program	Viewing Percentage in Survey
1. "M*A*S*H" Feb. 28, 1983	60.2%
2. "Dallas" Nov. 21, 1980	53.3%
3. "Roots" Part 8 Jan. 30, 1977	51.1%

Source: Nielsen Media Research

97. The inequality $|x - 60.2| \le 1.6$ describes the actual viewing percentage for "M*A*S*H" in the U.S. population. Solve the inequality and interpret the solution. Explain why the survey's *margin of error* is ±1.6%.

98. The inequality $|x - 51.1| \le 1.6$ describes the actual viewing percentage for "Roots" Part 8 in the U.S. population. Solve the inequality and interpret the solution. Explain why the survey's *margin of error* is ±1.6%.

99. If a coin is tossed 100 times, we would expect approximately 50 of the outcomes to be heads. It can be demonstrated that a coin is unfair if h, the number of outcomes that result in heads, satisfies $\left|\dfrac{h - 50}{5}\right| \ge 1.645$. Describe the number of outcomes that determine an unfair coin that is tossed 100 times.

100. The inequality $|T - 57| \le 7$ describes the range of monthly average temperature, T, in degrees Fahrenheit, for San Francisco, California. The inequality $|T - 50| \le 22$ describes the range of monthly average temperature, T, in degrees Fahrenheit, for Albany, New York. Solve each inequality and interpret the solution. Then describe at least three differences between the monthly average temperatures for the two cities.

In Exercises 101–110, use the five-step strategy for solving word problems. Give a linear inequality that models the verbal conditions and then solve the problem.

101. A truck can be rented from Basic Rental for $50 a day plus $0.20 per mile. Continental charges $20 per day plus $0.50 per mile to rent the same truck. How many miles must be driven in a day to make the rental cost for Basic Rental a better deal then Continental's?

102. You are choosing between two long-distance telephone plans. Plan A has a monthly fee of $15 with a charge of $0.08 per minute for all long-distance calls. Plan B has a monthly fee of $3 with a charge of $0.12 per minute for all long-distance calls. How many minutes of long-distance calls in a month make plan A the better deal?

103. A city commission has proposed two tax bills. The first bill requires that a homeowner pay $1800 plus 3% of the assessed home value in taxes. The second bill requires taxes of $200 plus 8% of the assessed home value. What price range of home assessment would make the first bill a better deal?

104. A local bank charges $8 per month plus 5¢ per check. The credit union charges $2 per month plus 8¢ per check. How many check should be written each month to make the credit union a better deal?

105. A company manufactures and sells blank audiocassette tapes. The weekly fixed cost is $10,000 and it cost $0.40 to produce each tape. The selling price is $2.00 per tape. How many tapes must be produced and sold each week for the company to have a profit gain?

106. A company manufactures and sells personalized stationery. The weekly fixed cost is $3000 and it cost $3.00 to produce each package of stationery. The selling price is $5.50 per package. How many packages of stationery must be produced and sold each week for the company to have a profit gain?

107. An elevator at a construction site has a maximum capacity of 2800 pounds. If the elevator operator weighs 265 pounds and each cement bag weighs 65 pounds, how many bags of cement can be safely lifted on the elevator in one trip?

108. An elevator at a construction site has a maximum capacity of 3000 pounds. If the elevator operator weighs 245 pounds and each cement bag weighs 95 pounds, how many bags of cement can be safely lifted on the elevator in one trip?

109. On two examinations, you have grades of 86 and 88. There is an optional final examination, which counts as one grade. You decide to take the final in order to get a course grade of A, meaning a final average of at least 90.
   a. What must you get on the final to earn an A in the course?
   b. By taking the final, if you do poorly, you might risk the B that you have in the course based on the first two exam grades. If your final average is less than 80, you will lose your B in the course. Describe the grades on the final that will cause this happen.

110. Parts for an automobile repair cost $175. The mechanic charges $34 per hour. If you receive an estimate for at least $226 and at most $294 for fixing the car, what is the time interval that the mechanic will be working on the job?

Writing in Mathematics

111. When graphing the solutions of an inequality, what does a parenthesis signify? What does a bracket signify?

112. When solving an inequality, when is it necessary to change the sense of the inequality? Give an example.

113. Describe ways in which solving a linear inequality is similar to solving a linear equation.

114. Describe ways in which solving a linear inequality is different than solving a linear equation.

115. What is a compound inequality and how is it solved?

116. Describe how to solve an absolute value inequality involving the symbol $<$. Give an example.

117. Describe how to solve an absolute value inequality involving the symbol $>$. Give an example.

118. Explain why $|x| < -4$ has no solution.

119. Describe the solution set of $|x| > -4$.

120. The formula
$$V = 3.5x + 120$$
models Super Bowl viewers, V, in millions, x years after 1990. Use the formula to write a word problem that can be solved using a linear inequality. Then solve the problem.

Technology Exercises

In Exercises 121–122, solve each inequality using a graphing utility. Graph each side separately. Then determine the values of x for which the graph on the left side lies above the graph on the right side.

121. $-3(x - 6) > 2x - 2$ **122.** $-2(x + 4) > 6x + 16$

Use the same technique employed in Exercises 121–122 to solve each inequality in Exercises 123–124. In each case, what conclusion can you draw? What happens if you try solving the inequalities algebraically?

123. $12x - 10 > 2(x - 4) + 10x$

124. $2x + 3 > 3(2x - 4) - 4x$

125. A bank offers two checking account plans. Plan A has a base service charge of $4.00 per month plus 10¢ per check. Plan B charges a base service charge of $2.00 per month plus 15¢ per check.

   a. Write models for the total monthly costs for each plan if x checks are written.

   b. Use a graphing utility to graph the models in the same $[0, 50, 10]$ by $[0, 10, 1]$ viewing rectangle.

   c. Use the graphs (and the $\boxed{\text{TRACE}}$ or intersection feature) to determine for what number of checks per month plan A will be better than plan B.

   d. Verify the result of part (c) algebraically by solving an inequality.

Critical Thinking Exercises

126. Which one of the following is true?

   a. The first step in solving $|2x - 3| > -7$ is to rewrite the inequality as $2x - 3 > -7$ or $2x - 3 < 7$.

   b. The smallest real number in the solution set of $2x > 6$ is 4.

   c. All irrational numbers satisfy $|x - 4| > 0$.

   d. None of these statements is true.

127. What's wrong with this argument? Suppose x and y represent two real numbers, where $x > y$.

$2 > 1$	This is a true statement.
$2(y - x) > 1(y - x)$	Multiply both sides by $y - x$.
$2y - 2x > y - x$	Use the distributive property.
$y - 2x > -x$	Subtract y from both sides.
$y > x$	Add 2x to both sides.

The final inequality, $y > x$, is impossible because we were initially given $x > y$.

128. The graphs of $y = 6$, $y = 3(-x - 5) - 9$, and $y = 0$ are shown in the figure. The graphs were obtained using a graphing utility and a $[-12, 1, 1]$ by $[-2, 8, 1]$ viewing rectangle. Use the graphs to write the solution set for the compound inequality. Express the solution set using interval notation.

$$0 < 3(-x - 5) - 9 < 6.$$

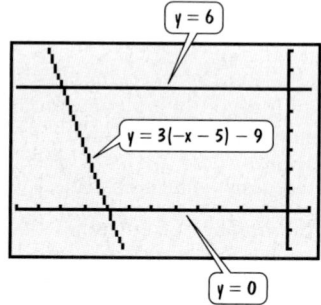

129. The percentage, p, of defective products manufactured by a company is given by $|p - 0.3\%| \leq 0.2\%$. If 100,000 products are manufactured and the company offers a $5 refund for each defective product, describe the company's cost for refunds.

Group Exercise

130. Each group member should research one situation that provides two different pricing options. These can involve areas such as public transportation options (with or without coupon books) or long-distance telephone plans or anything of interest. Be sure to bring in all the details for each option. At a second group meeting, select the two pricing situations that are most interesting and relevant. Using each situation, write a word problem about selecting the better of the two options. The word problem should be one that can be solved using a linear inequality. The group should turn in the two problems and their solutions.

SECTION 1.8 *Quadratic and Rational Inequalities*

Objectives

1. Solve quadratic inequalities.
2. Solve rational inequalities.
3. Solve problems modeled by nonlinear inequalities.

Not afraid of heights and cutting-edge excitement? How about sky diving? Behind your exhilarating experience is the world of algebra. After you jump from the airplane, your height above the ground at every instant of your fall can be described by a formula involving a variable that is squared. At some point, you'll need to open your parachute. How can you determine when you must do so? Let x represent the number of seconds you are falling. You can compute when to open the parachute by solving an inequality that takes on the form $ax^2 + bx + c < 0$. Such an inequality is called a **quadratic inequality.**

Definition of a Quadratic Inequality

A **quadratic inequality** is any inequality that can be put in one of the forms

$$ax^2 + bx + c < 0 \qquad ax^2 + bx + c > 0$$
$$ax^2 + bx + c \leq 0 \qquad ax^2 + bx + c \geq 0$$

where a, b, and c are real numbers and $a \neq 0$.

In this section we establish the basic techniques for solving quadratic inequalities. We will use these techniques to solve inequalities containing quotients, called **rational inequalities.** Finally, we will consider a formula that models the position of any free-falling object. As a sky diver, you could be that free-falling object!

1 Solve quadratic inequalities.

Solving Quadratic Inequalities

Graphs can help us to visualize the solutions of quadratic inequalities. The cuplike graph of $y = x^2 - 7x + 10$ is shown in Figure 1.21. The x-intercepts, 2 and 5, are **boundary points** between where the graph lies above the x-axis, shown in blue, and where the graph lies below the x-axis, shown in red. These boundary points play a critical role in solving quadratic inequalities.

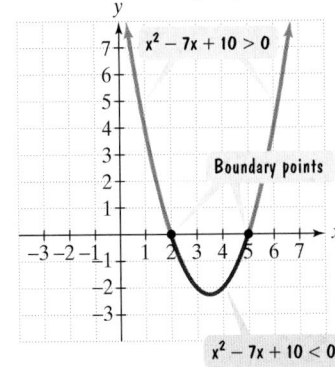

Figure 1.21

Procedure for Solving Quadratic Inequalities

1. Express the inequality in the general form
$$ax^2 + bx + c > 0 \quad \text{or} \quad ax^2 + bx + c < 0.$$

2. Solve the equation $ax^2 + bx + c = 0$. The real solutions are the **boundary points.**

3. Locate these boundary points on a number line, thereby dividing the number line into **test intervals.**

4. Choose one representative number within each test interval. If substituting that value into the original inequality produces a true statement, then all real numbers in the test interval belong to the solution set. If substituting that value into the original inequality produces a false statement, then no real numbers in the test interval belong to the solution set.

5. Write the solution set, selecting the interval(s) that produced a true statement. The graph of the solution set on a number line usually appears as

This procedure is valid if $<$ is replaced by $\leq$ and $>$ is replaced by $\geq$.

Study Tip

The five-step procedure for solving quadratic inequalities does not require graphing equations of the form $y = ax^2 + bx + c$. As we have done throughout the chapter, we'll show you these cuplike U-shaped graphs to enhance your visual understanding of solution sets.

EXAMPLE 1   Solving a Quadratic Inequality

Solve and graph the solution set on a real number line: $x^2 - 7x + 10 < 0$.

Solution

Step 1   Write the inequality in general form.   The inequality is given in this form, so this step has been done for us.

Step 2   Solve the related quadratic equation.   This equation is obtained by replacing the inequality sign by an equal sign. Thus, we will solve $x^2 - 7x + 10 = 0$.

$$x^2 - 7x + 10 = 0 \quad \text{This is the related quadratic equation.}$$

$$(x - 2)(x - 5) = 0 \quad \text{Factor.}$$

$$x - 2 = 0 \quad \text{or} \quad x - 5 = 0 \quad \text{Set each factor equal to 0.}$$

$$x = 2 \quad \text{or} \quad x = 5 \quad \text{Solve for x.}$$

The boundary points are 2 and 5.

Step 3   Locate the boundary points on a number line.   The number line with the boundary points is shown as follows:

The boundary points divide the number line into three test intervals, namely $(-\infty, 2), (2, 5)$, and $(5, \infty)$.

Step 4 **Take one representative number within each test interval and substitute that number into the original inequality.**

Test Interval	Representative Number	Substitute into $x^2 - 7x + 10 < 0$	Conclusion
$(-\infty, 2)$	0	$0^2 - 7 \cdot 0 + 10 \overset{?}{<} 0$ $10 < 0,$ False	$(-\infty, 2)$ does not belong to the solution set.
$(2, 5)$	3	$3^2 - 7 \cdot 3 + 10 \overset{?}{<} 0$ $9 - 21 + 10 \overset{?}{<} 0$ $-2 < 0,$ True	$(2, 5)$ belongs to the solution set.
$(5, \infty)$	6	$6^2 - 7 \cdot 6 + 10 \overset{?}{<} 0$ $36 - 42 + 10 \overset{?}{<} 0$ $4 < 0,$ False	$(5, \infty)$ does not belong to the solution set.

Step 5 **The solution set consists of the intervals that produce a true statement.** Our analysis shows that the solution set is the interval $(2, 5)$. The graph in Figure 1.22 confirms that $x^2 - 7x + 10 < 0$ (lies below the x-axis) in this interval. The graph of the solution set on a number line is shown as follows:

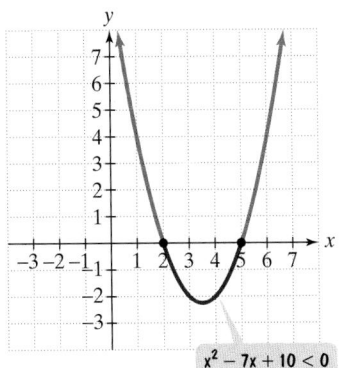

Figure 1.22 The graph lies below the x-axis between the boundary points 2 and 5, in the interval $(2, 5)$.

$x^2 - 7x + 10 < 0$

Check Point 1 Solve and graph the solution set on a real number line:
$$x^2 + 2x - 3 < 0.$$

EXAMPLE 2 Solving a Quadratic Inequality

Solve and graph the solution set: $2x^2 + x \geq 15$.

Solution

Step 1 **Write the inequality in general form.** We can write $2x^2 + x \geq 15$ in standard form by subtracting 15 from both sides. This will give us zero on the right.

$$2x^2 + x - 15 \geq 15 - 15$$
$$2x^2 + x - 15 \geq 0$$

Step 2 **Solve the related quadratic equation.** This equation is obtained by replacing the inequality sign by an equal sign. Thus, we will solve $2x^2 + x - 15 = 0$.

$$2x^2 + x - 15 = 0 \quad \text{\small This is the related quadratic equation.}$$
$$(2x - 5)(x + 3) = 0 \quad \text{\small Factor.}$$
$$2x - 5 = 0 \quad \text{or} \quad x + 3 = 0 \quad \text{\small Set each factor equal to 0.}$$
$$x = \tfrac{5}{2} \quad \text{or} \quad x = -3 \quad \text{\small Solve for x.}$$

The boundary points are -3 and $\tfrac{5}{2}$.

Step 3   Locate the boundary points on a number line.   The number line with the boundary points is shown as follows:

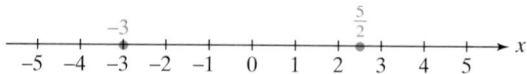

The boundary points divide the number line into three test intervals. Including the boundary points (because of the given greater than or *equal to* sign), the intervals are $(-\infty, -3]$, $[-3, \frac{5}{2}]$, and $[\frac{5}{2}, \infty)$.

Step 4   Take one representative number within each test interval and substitute that number into the original inequality.

Test Interval	Representative Number	Substitute into $2x^2 + x \geq 15$	Conclusion
$(-\infty, -3]$	-4	$2(-4)^2 + (-4) \overset{?}{\geq} 15$ $28 \geq 15$, True	$(-\infty, -3]$ belongs to the solution set.
$\left[-3, \frac{5}{2}\right]$	0	$2 \cdot 0^2 + 0 \overset{?}{\geq} 15$ $0 \geq 15$, False	$\left[-3, \frac{5}{2}\right]$ does not belong to the solution set.
$\left[\frac{5}{2}, \infty\right)$	3	$2 \cdot 3^2 + 3 \overset{?}{\geq} 15$ $21 \geq 15$, True	$\left[\frac{5}{2}, \infty\right)$ belongs to

Step 5   The solution set consists of the intervals that produce a true statement. Our analysis shows that the solution set is

$$(-\infty, -3] \text{ or } \left[\frac{5}{2}, \infty\right).$$

The graph of the solution set on a number line is shown as follows:

Check Point 2   Solve and graph the solution set: $x^2 - x \geq 20$.

Technology

The solution set for

$$2x^2 + x \geq 15$$

or, equivalently,

$$2x^2 + x - 15 \geq 0$$

can be verified with a graphing utility. The graph of $y = 2x^2 + x - 15$ was obtained using a $[-10, 10, 1]$ by $[-16, 6, 1]$ viewing rectangle. The graph lies above or on the *x*-axis, representing $\geq$, for all *x* in $(-\infty, -3]$ or $[\frac{5}{2}, \infty)$.

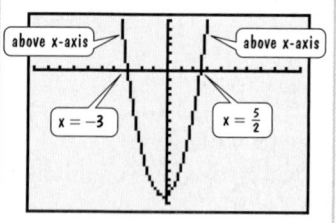

2 Solve rational inequalities.

Solving Rational Inequalities

Inequalities that involve quotients can be solved in the same manner as quadratic inequalities. For example, the inequalities

$$(x + 3)(x - 7) > 0 \quad \text{and} \quad \frac{x + 3}{x - 7} > 0$$

are similar in that both are positive under the same conditions. To be positive, each of these inequalities must have two positive linear expressions

$$x + 3 > 0 \quad \text{and} \quad x - 7 > 0$$

or two negative linear expressions

$$x + 3 < 0 \quad \text{and} \quad x - 7 < 0.$$

Consequently, we solve $\dfrac{x + 3}{x - 7} > 0$ using boundary points to divide the number line into test intervals. Then we select one representative number in each interval to determine whether that interval belongs to the solution set. Example 3 illustrates how this is done.

EXAMPLE 3   **Using Test Numbers to Solve a Rational Inequality**

Solve and graph the solution set: $\dfrac{x + 3}{x - 7} > 0.$

Study Tip

Many students want to solve

$$\frac{x + 3}{x - 7} > 0$$

by first multiplying both sides by $x - 7$ to clear fractions. This is incorrect. The problem is that $x - 7$ contains a variable and can be positive or negative, depending on the value of x. Thus, we do not know whether or not to reverse the sense of the inequality.

Solution   We begin by finding values of x that make the numerator and denominator 0.

$$x + 3 = 0 \qquad x - 7 = 0 \qquad \text{Set the numerator and denominator equal to 0.}$$

$$x = -3 \qquad\quad x = 7 \qquad \text{Solve.}$$

The boundary points are -3 and 7. We locate these numbers on a number line as follows:

These boundary points divide the number line into three test intervals, namely $(-\infty, -3)$, $(-3, 7)$, and $(7, \infty)$. Now, we take one representative number from each test interval and substitute that number into the original inequality.

Test Interval	Representative Number	Substitute into $\dfrac{x + 3}{x - 7} > 0$	Conclusion
$(-\infty, -3)$	-4	$\dfrac{-4 + 3}{-4 - 7} \overset{?}{>} 0$ $\dfrac{-1}{-11} \overset{?}{>} 0$ $\dfrac{1}{11} > 0, \text{True}$	$(-\infty, -3)$ belongs to the solution set.
$(-3, 7)$	0	$\dfrac{0 + 3}{0 - 7} \overset{?}{>} 0$ $-\dfrac{3}{7} > 0, \text{False}$	$(-3, 7)$ does not belong to the solution set.
$(7, \infty)$	8	$\dfrac{8 + 3}{8 - 7} \overset{?}{>} 0$ $11 > 0, \text{True}$	$(7, \infty)$ belongs to the solution set.

Our analysis shows that the solution set is

$$(-\infty, -3) \quad \text{or} \quad (7, \infty).$$

The graph of the solution set on a number line is shown as follows:

Check Point 3   Solve and graph the solution set: $\dfrac{x-5}{x+2} > 0$.

The first step in solving a rational inequality is to bring all terms to one side, obtaining zero on the other side. Then express the nonzero side as a single quotient. At this point, we follow the same procedure as in Example 3, finding values of the variable that make the numerator and denominator 0. These values serve as boundary points that separate the number line into intervals.

<table>
<tr><td>

Study Tip

Do not begin solving

$$\frac{x+1}{x+3} \le 2$$

by multiplying both sides by $x + 3$. We do not know if $x + 3$ is positive or negative. Thus, we do not know whether or not to reverse the sense of the inequality.

</td><td>

EXAMPLE 4   Solving a Rational Inequality

Solve and graph the solution set: $\dfrac{x+1}{x+3} \le 2$.

Solution

Step 1   Express the inequality so that one side is zero and the other side is a single quotient.   We subtract 2 from both sides to obtain zero on the right.

$$\frac{x+1}{x+3} \le 2 \qquad \text{This is the given inequality.}$$

$$\frac{x+1}{x+3} - 2 \le 0 \qquad \text{Subtract 2 from both sides, obtaining 0 on the right.}$$

$$\frac{x+1}{x+3} - \frac{2(x+3)}{x+3} \le 0 \qquad \text{The least common denominator is } x + 3. \text{ Express 2 in terms of this denominator.}$$

$$\frac{x+1-2(x+3)}{x+3} \le 0 \qquad \text{Subtract rational expressions.}$$

$$\frac{x+1-2x-6}{x+3} \le 0 \qquad \text{Apply the distributive property.}$$

$$\frac{-x-5}{x+3} \le 0 \qquad \text{Simplify.}$$

Step 2   Find boundary points by setting the numerator and the denominator equal to zero.

$$-x - 5 = 0 \qquad x + 3 = 0 \qquad \text{Set the numerator and denominator equal to 0. These are the values that make the previous quotient zero or undefined.}$$

$$x = -5 \qquad x = -3 \qquad \text{Solve for } x.$$

</td></tr>
</table>

Study Tip

Any values obtained by setting a denominator equal to zero should never be included in the solution set of a rational inequality. Division by zero is undefined.

The boundary points are -5 and -3. Because equality is included in the given less-than-or-equal-to symbol, we include the value of x that causes the quotient $\dfrac{-x-5}{x+3}$ to be zero. Thus, -5 is included in the solution set. By contrast, we do not include -3 in the solution set because -3 makes the denominator zero.

Step 3 Locate boundary points on a number line. The number line, with the boundary points, is shown as follows:

The open dot at −3 indicates −3 is not
to be included in the solution set.
We can't divide by zero.

The boundary points divide the number line into three test intervals, namely $(-\infty, -5]$, $[-5, -3)$, and $(-3, \infty)$.

Step 4 Take one representative number within each test interval and substitute that number into the original inequality.

Test Interval	Representative Number	Substitute into $\dfrac{x + 1}{x + 3} \leq 2$	Conclusion
$(-\infty, -5]$	−6	$\dfrac{-6 + 1}{-6 + 3} \overset{?}{\leq} 2$ $\dfrac{5}{3} \leq 2$, True	$(-\infty, -5]$ belongs to the solution set.
$[-5, -3)$	−4	$\dfrac{-4 + 1}{-4 + 3} \overset{?}{\leq} 2$ $3 \leq 2$, False	$[-5, -3)$ does not belong to the solution set.
$(-3, \infty)$	0	$\dfrac{0 + 1}{0 + 3} \overset{?}{\leq} 2$ $\dfrac{1}{3} \leq 2$, True	$(-3, \infty)$ belongs to the solution set.

Step 5 The solution set consists of the intervals that produce a true statement. Our analysis shows that the solution set is

$$(-\infty, -5] \quad \text{or} \quad (-3, \infty).$$

The graph of the solution set on a number line is shown as follows:

Discovery

Because $(x + 3)^2$ is positive, it is possible so solve

$$\frac{x + 1}{x + 3} \leq 2$$

by first multiplying both sides by $(x + 3)^2$ (where $x \neq -3$). This will not reverse the sense of the inequality and will clear the fraction. Try using this solution method and compare it to the solution on pages 162–163.

Check Point 4 Solve and graph the solution set: $\dfrac{2x}{x + 1} \leq 1.$

3 Solve problems modeled by nonlinear inequalities.

Applications

We are surrounded by evidence that the world is profoundly mathematical. For example, did you know that every time you throw an object vertically upward, its changing height above the ground can be described by a mathematical formula? The same formula can be used to describe objects that are falling, such as the sky divers shown in the opening to this section.

The Position Formula for a Free-Falling Object Near Earth's Surface

An object that is falling or vertically projected into the air has its height above the ground, s, in feet, given by

$$s = -16t^2 + v_0t + s_0$$

where v_0 is the original velocity (initial velocity) of the object, in feet per second, t is the time that the object is in motion, in seconds, and s_0 is the original height (initial height) of the object, in feet.

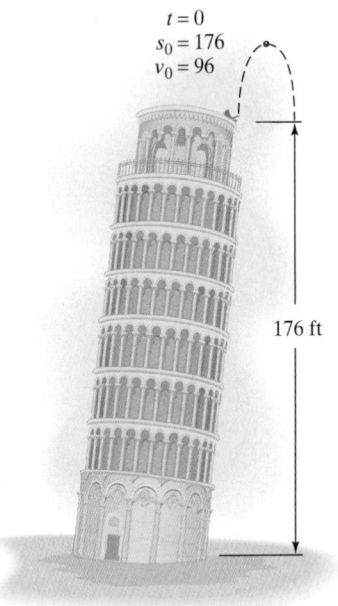

$t = 0$
$s_0 = 176$
$v_0 = 96$

176 ft

Figure 1.23 Throwing a ball from 176 feet with a velocity of 96 feet per second

In Example 5, we solve a quadratic inequality in a problem about the position of a free-falling object.

EXAMPLE 5 Using the Position Model

A ball is thrown vertically upward from the top of the Leaning Tower of Pisa (176 feet high) with an initial velocity of 96 feet per second (Figure 1.23). During which time period will the ball's height exceed that of the tower?

Solution

$s = -16t^2 + v_0t + s_0$ This is the position formula for a free-falling object.

$s = -16t^2 + 96t + 176$ Because v_0 (initial velocity) = 96 and s_0 (initial position) = 176, substitute these values into the formula.

When will the ball's height   exceed that   of the tower?

$-16t^2 + 96t + 176$ >   176

$-16t^2 + 96t + 176 > 176$ This is the inequality implied by the problem's question. We must find t.

$-16t^2 + 96t > 0$ Subtract 176 from both sides.

$-16t^2 + 96t = 0$ Solve the related quadratic equation.

$-16t(t - 6) = 0$ Factor.

$-16t = 0$ or $t - 6 = 0$ Set each factor equal to 0.

$t = 0$ $t = 6$ Solve for t. The boundary points are 0 and 6.

Locate these values on a number line, with $t \geq 0$.

The intervals are $(-\infty, 0)$, $(0, 6)$ and $(6, \infty)$. For our purposes, the mathematical model is useful only from $t = 0$ until the ball hits the ground. (By setting $-16t^2 + 96t + 176$ equal to zero, we find $t \approx 7.47$; the ball hits the ground after approximately 7.47 seconds.) Thus, we use $(0, 6)$ and $(6, 7.47)$ for our test intervals.

Technology

The graphs of
$$y_1 = -16x^2 + 96x + 176$$
and
$$y_2 = 176$$
are shown in a
$$[0, 8, 1] \text{ by } [0, 320, 32]$$

seconds in motion | height, in feet

viewing rectangle. The graphs show that the ball's height exceeds that of the tower between 0 and 6 seconds.

Height of Ball:
$y_1 = -16x^2 + 96x + 176$

320

Height (feet)

Height of Tower:
$y_2 = 176$

Ball hits ground after 7.47 seconds.

0   6
Seconds in Motion
$[0, 8, 1]$ by $[0, 320, 32]$

Test Interval	Representative Number	Substitute into $-16t^2 + 96t + 176 > 176$	Conclusion
$(0, 6)$	1	$-16 \cdot 1^2 + 96 \cdot 1 + 176 \overset{?}{>} 176$ $256 > 176$, True	$(0, 6)$ belongs to the solution set.
$(6, 7.47)$	7	$-16 \cdot 7^2 + 96 \cdot 7 + 176 \overset{?}{>} 176$ $64 > 176$, False	$(6, 7.47)$ does not belong to the solution set.

The ball's height exceeds that of the tower between 0 and 6 seconds, excluding $t = 0$ and $t = 6$.

Check Point 5 An object is propelled straight up from ground level with an initial velocity of 80 feet per second. Its height at time t is described by

$$s = -16t^2 + 80t$$

where the height, s, is measured in feet and the time, t, is measured in seconds. In which time interval will the object be more than 64 feet above the ground?

EXERCISE SET 1.8

Practice Exercises

Solve each quadratic inequality in Exercises 1–28, and graph the solution set on a real number line. Express each solution set in interval notation.

1. $(x - 4)(x + 2) > 0$ **2.** $(x + 3)(x - 5) > 0$

3. $(x - 7)(x + 3) \le 0$ **4.** $(x + 1)(x - 7) \le 0$

5. $x^2 - 5x + 4 > 0$ **6.** $x^2 - 4x + 3 < 0$

7. $x^2 + 5x + 4 > 0$ **8.** $x^2 + x - 6 > 0$

9. $x^2 - 6x + 9 < 0$ **10.** $x^2 - 2x + 1 > 0$

11. $x^2 - 6x + 8 \le 0$ **12.** $x^2 - 2x - 3 \ge 0$

13. $3x^2 + 10x - 8 \le 0$ **14.** $9x^2 + 3x - 2 \ge 0$

15. $2x^2 + x < 15$ **16.** $6x^2 + x > 1$

17. $4x^2 + 7x < -3$ **18.** $3x^2 + 16x < -5$

19. $5x \le 2 - 3x^2$ **20.** $4x^2 + 1 \ge 4x$

21. $x^2 - 4x \ge 0$ **22.** $x^2 + 2x < 0$

23. $2x^2 + 3x > 0$ **24.** $3x^2 - 5x \le 0$

25. $-x^2 + x \ge 0$ **26.** $-x^2 + 2x \ge 0$

27. $|x^2 + 2x - 36| > 12$ **28.** $|x^2 + 6x + 1| > 8$

Solve each rational inequality in Exercises 29–48, and graph the solution set on a real number line. Express each solution set in interval notation.

29. $\dfrac{x - 4}{x + 3} > 0$ **30.** $\dfrac{x + 5}{x - 2} > 0$

31. $\dfrac{x + 3}{x + 4} < 0$ **32.** $\dfrac{x + 5}{x + 2} < 0$

33. $\dfrac{-x + 2}{x - 4} \geq 0$

34. $\dfrac{-x - 3}{x + 2} \leq 0$

35. $\dfrac{4 - 2x}{3x + 4} \leq 0$

36. $\dfrac{3x + 5}{6 - 2x} \geq 0$

37. $\dfrac{x}{x - 3} > 0$

38. $\dfrac{x + 4}{x} > 0$

39. $\dfrac{x + 1}{x + 3} < 2$

40. $\dfrac{x}{x - 1} > 2$

41. $\dfrac{x + 4}{2x - 1} \leq 3$

42. $\dfrac{1}{x - 3} < 1$

43. $\dfrac{x - 2}{x + 2} \leq 2$

44. $\dfrac{x}{x + 2} \geq 2$

45. $\dfrac{3}{x + 3} > \dfrac{3}{x - 2}$

46. $\dfrac{1}{x + 1} > \dfrac{2}{x - 1}$

47. $\dfrac{x^2 - x - 2}{x^2 - 4x + 3} > 0$

48. $\dfrac{x^2 - 3x + 2}{x^2 - 2x - 3} > 0$

 Application Exercises

Use the position formula

$$s = -16t^2 + v_0 t + s_0$$

(v_0 = initial velocity, s_0 = initial position, t = time)

to answer Exercises 49–52. If necessary, round answers to the nearest hundredth of a second.

49. A projectile is fired straight upward from ground level with an initial velocity of 80 feet per second. During which interval of time will the projectile's height exceed 96 feet?

50. A projectile is fired straight upward from ground level with an initial velocity of 128 feet per second. During which interval of time will the projectile's height exceed 128 feet?

51. A ball is thrown vertically upward with a velocity of 64 feet per second from the top edge of a building 80 feet high. For how long is the ball higher than 96 feet?

52. A diver leaps into the air at 20 feet per second from a diving board that is 10 feet above the water. For how many seconds is the diver at least 12 feet above the water?

53. The formula

$$H = \frac{15}{8} x^2 - 30x + 200$$

models heart rate, H, in beats per minute, x minutes after a strenuous workout.

a. What is the heart rate immediately following the workout?

b. According to the model, during which intervals of time after the strenuous workout does the heart rate exceed 110 beats per minute? For which of these intervals has model breakdown occurred? Which interval provides a more realistic answer? How did you determine this?

The bar graph at the top of the next column shows the cost of Medicare, in billions of dollars, projected through 2005. The data can be modeled by

a linear model, $y = 27x + 163$;

a quadratic model, $y = 1.2x^2 + 15.2x + 181.4$.

In each formula, x represent the number of years after 1995 and y represents Medicare spending, in billions of dollars. Use these formulas to solve Exercises 54–56.

Medicare Spending

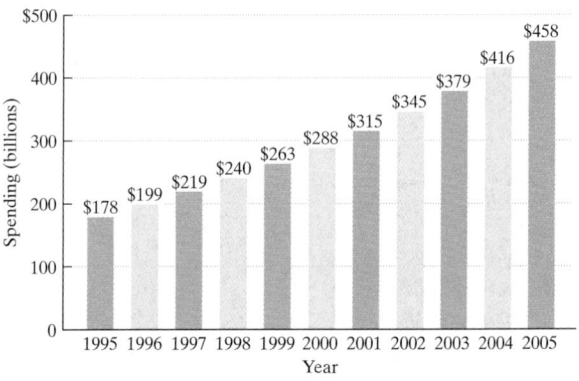

Source: Congressional Budget Office

54. The graph indicates that Medicare spending will reach $458 billion in 2005. Find the amount predicted by each of the formulas for that year. How well do the formulas model the value in the graph? Which formula serves as a better model for that year?

55. For which years does the quadratic model indicate that Medicare spending will exceed $536.6 billion?

56. For which years does the quadratic model indicate that Medicare spending will exceed $629.4 billion?

A company manufactures wheelchairs. The average cost, y, of producing x wheelchairs per month is given by

$$y = \frac{500,000 + 400x}{x}.$$

The graph of the formula is shown. Use the formula to solve Exercises 57–58.

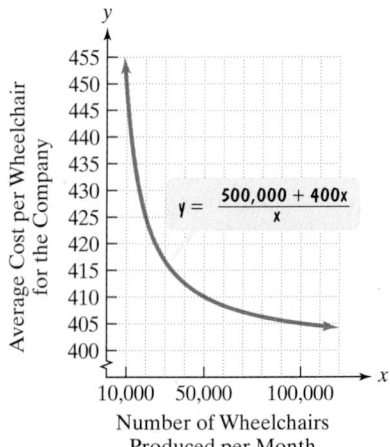
$$y = \frac{500,000 + 400x}{x}$$

57. Describe the company's production level so that the average cost of producing each wheelchair does not exceed $425. Use a rational inequality to solve the problem. Then explain how your solution is shown on the graph.

58. Describe the company's production level so that the average cost of producing each wheelchair does not exceed $410. Use a rational inequality to solve the problem. Then explain how your solution is shown on the graph.

Writing in Mathematics

59. What is a quadratic inequality?

60. What is a rational inequality?

61. Describe similarities and differences between the solutions of

$$(x - 2)(x + 5) \geq 0 \quad \text{and} \quad \frac{x - 2}{x + 5} \geq 0.$$

Technology Exercises

Solve each inequality in Exercises 62–65 using a graphing utility.

62. $x^2 + 3x - 10 > 0$

63. $2x^2 + 5x - 3 \leq 0$

64. $x^3 + x^2 - 4x - 4 > 0$

65. $\dfrac{x - 4}{x - 1} \leq 0$

Critical Thinking Exercises

66. Which one of the following is true?

a. The solution set of $x^2 > 25$ is $(5, \infty)$.

b. The inequality $\dfrac{x - 2}{x + 3} < 2$ can be solved by multiplying both sides by $x + 3$, resulting in the equivalent inequality $x - 2 < 2(x + 3)$.

c. $(x + 3)(x - 1) \geq 0$ and $\dfrac{x + 3}{x - 1} \geq 0$ have the same solution set.

d. None of these statements is true.

67. Write a quadratic inequality whose solution set is $[-3, 5]$.

68. Write a rational inequality whose solution set is $(-\infty, -4)$ or $[3, \infty)$.

In Exercises 69–72, use inspection to describe each inequality's solution set. Do not solve any of the inequalities.

69. $(x - 2)^2 > 0$

70. $(x - 2)^2 \leq 0$

71. $(x - 2)^2 < -1$

72. $\dfrac{1}{(x - 2)^2} > 0$

In Exercises 73–74, use the method for solving quadratic inequalities to solve each higher-order polynomial inequality.

73. $x^3 + x^2 - 4x - 4 > 0$

74. $x^3 + 2x^2 - x - 2 \geq 0$

75. The graphing utility screen shows the graph of $y = 4x^2 - 8x + 7$.

$y = 4x^2 - 8x + 7$

a. Use the graph to describe the solution set of $4x^2 - 8x + 7 > 0$.

b. Use the graph to describe the solution set of $4x^2 - 8x + 7 < 0$.

c. Use an algebraic approach to verify each of your descriptions in parts (a) and (b).

76. The graphing utility screen shows the graph of $y = \sqrt{27 - 3x^2}$. Write and solve a quadratic inequality that explains why the graph only appears for $-3 \leq x \leq 3$.

$y = \sqrt{27 - 3x^2}$

Group Exercise

77. This exercise is intended as a group learning experience and is appropriate for groups of three to five people. Before working on the various parts of the problem, reread the description of the position formula on page 164.

a. Drop a ball from a height of 3 feet, 6 feet, and 12 feet. Record the number of seconds it takes for the ball to hit the ground.

b. For each of the three initial positions, use the position formula to determine the time required for the ball to hit the ground.

c. What factors might result in differences between the times that you recorded and the times indicated by the formula?

d. What appears to be happening to the time required for a free-falling object to hit the ground as its initial height is doubled? Verify this observation algebraically and with a graphing utility.

e. Repeat part (a) using a sheet of paper rather than a ball. What differences do you observe? What factor seems to be ignored in the position formula?

f. What is meant by the acceleration of gravity and how does this number appear in the position formula for a free-falling object?

CHAPTER SUMMARY, REVIEW, AND TEST

Summary

DEFINITIONS AND CONCEPTS	EXAMPLES

1.1 Graphs and Graphing Utilities

a. The rectangular coordinate system consists of a horizontal number line, the x-axis, and a vertical number line, the y-axis, intersecting at their zero points, the origin. Each point in the system corresponds to an ordered pair of real numbers (x,y). The first number in the pair is the x-coordinate; the second number is the y-coordinate. See Figure 1.1 on page 76.

Ex. 1, p. 77

b. An ordered pair is a solution of an equation in two variables if replacing the variables by the corresponding coordinates results in a true statement. The ordered pair is said to satisfy the equation. The graph of the equation is the set of all points whose coordinates satisfy the equation. One method for graphing an equation is to plot ordered-pair solutions and connect them with a smooth curve or line.

Ex. 2, p. 78

c. An x-intercept of a graph is the x-coordinate of a point where the graph intersects the x-axis. The y-coordinate corresponding to a graph's x-intercept is always zero.

d. A y-intercept of a graph is the y-coordinate of a point where the graph intersects the y-axis. The x-coordinate corresponding to a graph's y-intercept is always zero.

1.2 Linear Equations

a. A linear equation in one variable x can be written in the form $ax + b = 0, a \neq 0$.

b. The procedure for solving a linear equation is given in the box on page 86.

Ex. 1, p. 86;
Ex. 2, p. 87

c. If an equation contains fractions, begin by multiplying both sides by the least common denominator, thereby clearing fractions.

Ex. 3, p. 88

d. If an equation contains rational expressions with variable denominators, avoid in the solution set any values of the variable that make a denominator zero.

Ex. 4, p. 89;
Ex. 5, p. 90

e. An identity is an equation that is true for all real numbers for which both sides are defined. A conditional equation is not an identity and is true for at least one real number. An inconsistent equation is an equation that is not true for even one real number.

Ex. 6, p. 91

1.3 Formulas and Applications

a. A formula is an equation that uses letters to express a relationship between two or more variables.

Ex. 1, p. 96

b. Mathematical modeling is the process of finding equations and formulas to describe real-world phenomena. Such equations and formulas, together with the meaning assigned to the variables, are called mathematical models. Mathematical models can be formed from verbal models or from actual data.

c. A five-step procedure for solving problems using mathematical models is given in the box on page 97.

Ex. 2, p. 97;
Ex. 3, p. 98;
Ex. 4, p. 100;
Ex. 5, p. 101

1.4 Complex Numbers

a. The imaginary unit i is defined as

$$i = \sqrt{-1}, \text{ where } i^2 = -1.$$

The set of numbers in the form $a + bi$ is called the set of complex numbers; a is the real part and b is the imaginary part. If $b = 0$, the complex number is a real number. If $b \neq 0$, the complex number is an imaginary number. Complex numbers in the form bi are called pure imaginary numbers.

b. Rules for adding and subtracting complex numbers are given in the box on page 109.

Ex. 1, p. 110

c. To multiply complex numbers, multiply as if they are polynomials. After completing the multiplication, replace i^2 with -1.

Ex. 2, p. 110

DEFINITIONS AND CONCEPTS	EXAMPLES

d. The complex conjugate of $a + bi$ is $a - bi$ and vice versa. The multiplication of complex conjugates gives a real number:

$$(a + bi)(a - bi) = a^2 + b^2.$$

e. To divide complex numbers, multiply the numerator and the denominator by the complex conjugate of the denominator.

Ex. 3, p. 111

f. When performing operations with square roots of negative numbers, begin by expressing all square roots in terms of i. The principal square root of $-b$ is defined by

Ex. 4, p. 112

$$\sqrt{-b} = i\sqrt{b}.$$

1.5 Quadratic Equations

a. A quadratic equation in x can be written in the general form $ax^2 + bx + c = 0, a \neq 0$.

b. The procedure for solving a quadratic equation by factoring and the zero-product principle is given in the box on pages 115–116.

Ex. 1, p. 116

c. The procedure for solving a quadratic equation by the square root method is given in the box on page 118.

Ex. 2, p. 118

d. All quadratic equations can be solved by completing the square. Isolate the binomial with the two variable terms on one side of the equation. If the coefficient of the x^2-term is not one, divide each side of the equation by this coefficient. Then add the square of half the coefficient of x to both sides.

Ex. 4, p. 119

e. All quadratic equations can be solved by the quadratic formula

Ex. 5, p. 122;
Ex. 6, p. 123

$$x = \frac{-b \pm \sqrt{b^2 - 4ac}}{2a}.$$

The formula is derived by completing the square of the equation $ax^2 + bx + c = 0$.

f. The discriminant, $b^2 - 4ac$, indicates the number and type of solutions to the quadratic equation $ax^2 + bx + c = 0$, shown in Table 1.3 on page 124.

Ex. 7, p. 124

g. Table 1.4 on page 125 shows the most efficient technique to use when solving a quadratic equation.

Ex. 8, p. 126;
Ex. 9, p. 127

1.6 Other Types of Equations

a. Some polynomial equations of degree 3 or greater can be solved by moving all terms to one side, obtaining zero on the other side, factoring, and using the zero-product principle. Factoring by grouping is often used.

Ex. 1, p. 132;
Ex. 2, p. 133

b. A radical equation is an equation in which the variable occurs in a square root, cube root, and so on. A radical equation can be solved by isolating the radical and raising both sides of the equation to a power equal to the radicals index. When raising both sides to an even power, check all proposed solutions in the original equation. Eliminate extraneous solutions from the solution set.

Ex. 3, p. 134;
Ex. 4, p. 135

c. A radical equation with rational exponents can be solved by isolating the expression with the rational exponent and raising both sides of the equation to a power that is the reciprocal of the rational exponent. See the details in the box on page 137.

Ex. 5, p. 137

d. An equation is quadratic in form if it can be written in the form $at^2 + bt + c = 0$, where t is an algebraic expression and $a \neq 0$. Solve for t and use the substitution that resulted in this equation to find the values for the variable in the given equation.

Ex. 6, p. 139;
Ex. 7, p. 139

e. Absolute value equations in the form $|X| = c, c > 0$, can be solved by rewriting the equation without absolute value bars: $X = c$ or $X = -c$.

Ex. 8, p. 140

DEFINITIONS AND CONCEPTS   **EXAMPLES**

1.7 Linear Inequalities

a. A linear inequality in one variable x can be expressed as   Ex. 1, p. 144
$ax + b \leq c$, $ax + b < c$, $ax + b \geq c$, or $ax + b > c$, $a \neq 0$.

b. Graphs of solutions to inequalities are shown on a number line by shading all points representing   Ex. 2, p. 146
numbers that are solutions. Parentheses exclude endpoints and square brackets include endpoints.

c. Solution sets of inequalities can be expressed in set-builder or interval notation. Table 1.5 on   Ex. 3, p. 148;
page 146 compares the notations.   Ex. 4, p. 149

d. A linear inequality is solved using a procedure similar to solving a linear equation. However,   Ex. 5, p. 150
when multiplying or dividing by a negative number, reverse the sense of the inequality.

e. A compound inequality with three parts can be solved by isolating x in the middle.

f. Inequalities involving absolute value can be solved by rewriting the inequalities without absolute   Ex. 6, p. 151;
value bars. The ways to do this are shown in the box on page 150.   Ex. 7, p. 151

1.8 Quadratic and Rational Inequalities

a. A quadratic inequality can be expressed as
$ax^2 + bx + c < 0$, $ax^2 + bx + c > 0$,
$ax^2 + bx + c \leq 0$, or $ax^2 + bx + c \geq 0$, $a \neq 0$.   Ex. 1, p. 158;

b. A procedure for solving quadratic inequalities is given in the box on page 158.   Ex. 2, p. 159

c. Inequalities involving quotients are called rational inequalities. The procedure for solving such   Ex. 3, p. 161;
inequalities begins with expressing them so that one side is zero and the other side is a single   Ex. 4, p. 162
quotient. Find boundary points by setting the numerator and denominator equal to zero. Then
follow a procedure similar to that for solving quadratic inequalities.

Review Exercises

1.1

Graph each equation in Exercises 1–4.
Let $x = -3, -2, -1, 0, 1, 2$, and 3.

1. $y = 2x - 2$ **2.** $y = x^2 - 3$
3. $y = x$ **4.** $y = |x| - 2$

5. What does a $[-20, 40, 10]$ by $[-5, 5, 1]$ viewing rectangle
mean? Draw axes with tick marks and label the tick
marks to illustrate this viewing rectangle.

In Exercises 6–8, use the graph and determine the
x-intercepts, if any, and the y-intercepts, if any. For each
graph, tick marks along the axes represent one unit each.

6.   **7.**

8.

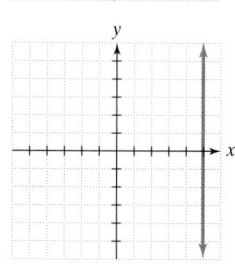

The caseload of Alzheimer's disease in the United States is
expected to explode as baby boomers head into their later
years. The graph shows the percentage of Americans with the
disease, by age. Use the graph to solve Exercises 9–11.

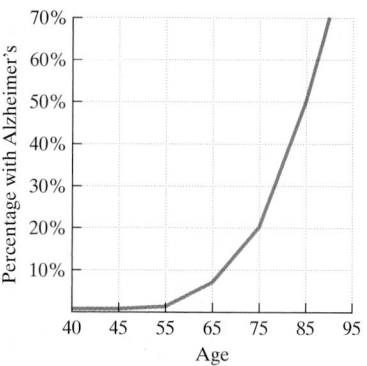

Alzheimer's Prevalence in the U.S., by Age

Source: Centers for Disease Control

9. What percentage of Americans who are 75 have Alzheimer's
disease?

10. What age represents 50% prevalence of Alzheimer's disease?

11. Describe the trend shown by the graph.

1.2

In Exercises 12–17, solve and check each linear equation.

12. $2x - 5 = 7$ **13.** $5x + 20 = 3x$
14. $7(x - 4) = x + 2$ **15.** $1 - 2(6 - x) = 3x + 2$

16. $2(x - 4) + 3(x + 5) = 2x - 2$

17. $2x - 4(5x + 1) = 3x + 17$

Exercises 18–22 contain equations with constants in denominators. Solve each equation and check by the method of your choice.

18. $\dfrac{2x}{3} = \dfrac{x}{6} + 1$

19. $\dfrac{x}{2} - \dfrac{1}{10} = \dfrac{x}{5} + \dfrac{1}{2}$

20. $\dfrac{2x}{3} = 6 - \dfrac{x}{4}$

21. $\dfrac{x}{4} = 2 + \dfrac{x - 3}{3}$

22. $\dfrac{3x + 1}{3} - \dfrac{13}{2} = \dfrac{1 - x}{4}$

Exercises 23–26 contain equations with variables in denominators. **a.** *List the value or values representing restriction(s) on the variable.* **b.** *Solve the equation.*

23. $\dfrac{9}{4} - \dfrac{1}{2x} = \dfrac{4}{x}$

24. $\dfrac{7}{x - 5} + 2 = \dfrac{x + 2}{x - 5}$

25. $\dfrac{1}{x - 1} - \dfrac{1}{x + 1} = \dfrac{2}{x^2 - 1}$

26. $\dfrac{4}{x + 2} + \dfrac{2}{x - 4} = \dfrac{30}{x^2 - 2x - 8}$

In Exercises 27–29, determine whether each equation is an identity, a conditional equation, or an inconsistent equation.

27. $\dfrac{1}{x + 5} = 0$

28. $7x + 13 = 4x - 10 + 3x + 23$

29. $7x + 13 = 3x - 10 + 2x + 23$

1.3

30. The percentage, P, of U.S. adults who read the daily newspaper can be modeled by the formula

$$P = -0.7x + 80$$

where x is the number of years after 1965. In which year will 52% of U.S. adults read the daily newspaper?

31. Suppose you were to list in order, from least to most, the family income for every U.S. family. The median income is the income in the middle of this list of ranked data. This income can be modeled by the formula

$$I = 1321.7(x - 1980) + 21{,}153.$$

In this formula, I represents median family income in the United States and x is the actual year, beginning in 1980. When was the median income $47,587?

In Exercises 32–39, use the five-step strategy given in the box on page 97 to solve each problem.

32. The cost of raising a child through the age of 17 varies by income group. The cost in middle-income families exceeds that of low-income families by $63 thousand, and the cost of high-income families is $3 thousand less than twice that of low-income families. Three children, one in a low-income family, one in a middle-income family, and one in a high-income family, will cost a total of $756 thousand to raise through the age of 17. Find the cost of raising a child in

each of the three income groups. (*Source: The World Almanac;* low annual income is less than $36,800, middle is $36,800–$61,900, and high exceeds $61,900.)

33. In 2000, the average weekly salary for workers in the United States was $567. If this amount is increasing by $15 yearly, in how many years after 2000 will the average salary reach $702. In which year will that be?

34. You are choosing between two long-distance telephone plans. One plan has a monthly fee of $15 with a charge of $0.05 per minute. The other plan has a monthly fee of $5 with a charge of $0.07 per minute. For how many minutes of long-distance calls will the costs for the two plans be the same?

35. You inherit $10,000 with the stipulation that for the first year the money must be placed in two investments paying 8% and 12% annual interest, respectively. How much should be invested at each rate if the total interest earned for the year is to be $950?

36. The length of a rectangular football field is 14 meters more than twice the width. If the perimeter is 346 meters, find the field's dimensions.

37. The bus fare in a city is $1.50. People who use the bus have the option of purchasing a monthly coupon book for $25.00. With the coupon book, the fare is reduced to $0.25. Determine the number of times in a month the bus must be used so that the total monthly cost without the coupon book is the same as the total monthly cost with the coupon book.

38. A salesperson earns $300 per week plus 5% commission of sales. How much must be sold to earn $800 in a week?

39. A study entitled *Performing Arts—The Economic Dilemma* documents the relationship between the number of concerts given by a major orchestra and the attendance per concert. For each additional concert given per year, attendance per concert drops by approximately eight people. If 50 concerts are given, attendance per concert is 2987 people. How many concerts should be given to ensure an audience of 2627 people at each concert?

In Exercises 40–42, solve each formula for the specified variable.

40. $V = \dfrac{1}{3} Bh$ for h

41. $F = f(1 - M)$ for M

42. $T = gr + gvt$ for g

1.4

In Exercises 43–52, perform the indicated operations and write the result in standard form.

43. $(8 - 3i) - (17 - 7i)$

44. $4i(3i - 2)$

45. $(7 - 5i)(2 + 3i)$

46. $(3 - 4i)^2$

47. $(7 + 8i)(7 - 8i)$

48. $\dfrac{6}{5 + i}$

49. $\dfrac{3 + 4i}{4 - 2i}$

50. $\sqrt{-32} - \sqrt{-18}$

51. $(-2 + \sqrt{-100})^2$

52. $\dfrac{4 + \sqrt{-8}}{2}$

1.5

Solve each equation in Exercises 53–54 by factoring.

53. $2x^2 + 15x = 8$ **54.** $5x^2 + 20x = 0$

Solve each equation in Exercises 55–56 by the square root method.

55. $2x^2 - 3 = 125$ **56.** $(3x - 4)^2 = 18$

In Exercises 57–58, determine the constant that should be added to the binomial so that it becomes a perfect square trinomial. Then write and factor the trinomial.

57. $x^2 + 20x$ **58.** $x^2 - 3x$

Solve each equation in Exercises 59–60 by completing the square.

59. $x^2 - 12x + 27 = 0$ **60.** $3x^2 - 12x + 11 = 0$

Solve each equation in Exercises 61–63 using the quadratic formula.

61. $x^2 = 2x + 4$ **62.** $x^2 - 2x + 19 = 0$
63. $2x^2 = 3 - 4x$

Compute the discriminant of each equation in Exercises 64–65. What does the discriminant indicate about the number and type of solutions?

64. $x^2 - 4x + 13 = 0$ **65.** $9x^2 = 2 - 3x$

Solve each equation in Exercises 66–71 by the method of your choice.

66. $2x^2 - 11x + 5 = 0$ **67.** $(3x + 5)(x - 3) = 5$
68. $3x^2 - 7x + 1 = 0$ **69.** $x^2 - 9 = 0$
70. $(x - 3)^2 - 25 = 0$ **71.** $3x^2 - x + 2 = 0$

72. The weight of a human fetus is modeled by the formula $W = 3t^2$, where W is the weight, in grams, and t is the time, in weeks, $0 \le t \le 39$. After how many weeks does the fetus weigh 1200 grams?

73. The alligator, an endangered species, is the subject of a protection program. The formula
$$P = -10x^2 + 475x + 3500$$
models the alligator population, P, after x years of the protection program, where $0 \le x \le 12$. After how many years is the population up to 7250?

74. The graph of the alligator population described in Exercise 73 is shown over time. Identify your solution in Exercise 73 as a point on the graph.

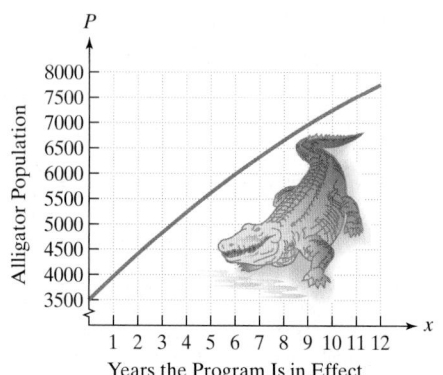

Years the Program Is in Effect

75. An architect is allowed 15 square yards of floor space to add a small bedroom to a house. Because of the room's design in relationship to the existing structure, the width of the rectangular floor must be 7 yards less than two times the length. Find the length and width of the rectangular floor that the architect is permitted.

76. A building casts a shadow that is double the length of its height. If the distance from the end of the shadow to the top of the building is 300 meters, how high is the building? Round to the nearest meter.

1.6

Solve each polynomial equation in Exercises 77–78.

77. $2x^4 = 50x^2$ **78.** $2x^3 - x^2 - 18x + 9 = 0$

Solve each radical equation in Exercises 79–80.

79. $\sqrt{2x - 3} + x = 3$ **80.** $\sqrt{x - 4} + \sqrt{x + 1} = 5$

Solve the equations with rational exponents in Exercises 81–82.

81. $3x^{3/4} - 24 = 0$ **82.** $(x - 7)^{2/3} = 25$

Solve each equation in Exercises 83–84 by making an appropriate substitution.

83. $x^4 - 5x^2 + 4 = 0$ **84.** $x^{1/2} + 3x^{1/4} - 10 = 0$

Solve the equations containing absolute value in Exercises 85–86.

85. $|2x + 1| = 7$ **86.** $2|x - 3| - 6 = 10$

Solve each equation in Exercises 87–90 by the method of your choice.

87. $3x^{4/3} - 5x^{2/3} + 2 = 0$ **88.** $2\sqrt{x - 1} = x$
89. $|2x - 5| - 3 = 0$ **90.** $x^3 + 2x^2 = 9x + 18$

91. The distance to the horizon that you can see, D, in miles, from the top of a mountain H feet high is modeled by the formula $D = \sqrt{2H}$. You've hiked to the top of a mountain with views extending 50 miles to the horizon. How high is the mountain?

1.7

In Exercises 92–94, graph the solutions of each inequality on a number line.

92. $x > 5$ **93.** $x \le 1$ **94.** $-3 \le x < 0$

In Exercises 95–97, express each interval in terms of an inequality, and graph the interval on a number line.

95. $(-2, 3]$ **96.** $[-1.5, 2]$ **97.** $(-1, \infty)$

Solve each linear inequality in Exercises 98–103 and graph the solution set on a number line. Express each solution set in interval notation.

98. $-6x + 3 \le 15$ **99.** $6x - 9 \ge -4x - 3$
100. $\dfrac{x}{3} - \dfrac{3}{4} - 1 > \dfrac{x}{2}$ **101.** $6x + 5 > -2(x - 3) - 25$
102. $3(2x - 1) - 2(x - 4) \ge 7 + 2(3 + 4x)$
103. $7 < 2x + 3 \le 9$

Solve each inequality in Exercises 104–106 by first rewriting each one as an equivalent inequality without absolute value bars. Graph the solution set on a number line. Express each solution set in interval notation.

104. $|2x + 3| \leq 15$

105. $\left|\dfrac{2x + 6}{3}\right| > 2$

106. $|2x + 5| - 7 \geq -6$

107. Approximately 90% of the population sleeps h hours daily, where h is modeled by the inequality $|h - 6.5| \leq 1$. Write a sentence describing the range for the number of hours that most people sleep. Do *not* use the phrase "absolute value" in your description.

108. The formula for converting Fahrenheit temperature, F, to Celsius temperature, C, is $C = \frac{5}{9}(F - 32)$. If Celsius temperature ranges from $10°$ to $25°$, inclusive, what is the range for the Fahrenheit temperature?

109. A person can choose between two charges on a checking account. The first method involves a fixed cost of $11 per month plus 6¢ for each check written. The second method involves a fixed cost of $4 per month plus 20¢ for each check written. How many checks should be written to make the first method a better deal?

110. A student has grades on three examinations of 75, 80, and 72. What must the student earn on a fourth examination in order to have an average of at least 80?

1.8

Solve each quadratic inequality in Exercises 111–112, and graph the solution set on a real number line. Express each solution set in interval notation.

111. $2x^2 + 7x \leq 4$

112. $2x^2 > 6x - 3$

Solve each rational inequality in Exercises 113–114, and graph the solution set on a real number line. Express each solution set in interval notation.

113. $\dfrac{x - 6}{x + 2} > 0$

114. $\dfrac{x + 3}{x - 4} \leq 5$

115. Use the position formula

$$s = -16t^2 + v_0 t + s_0$$

initial velocity   initial height

to solve this problem. A projectile is fired vertically upward from ground level with an initial velocity of 48 feet per second. During which time period will the projectile's height exceed 32 feet?

Chapter 1 Test

1. Graph $y = x^2 - 4$ by letting x equal integers from -3 through 3.

2. The graph of $y = -\frac{3}{2}x + 3$ is shown in a $[-6, 6, 1]$ by $[-6, 6, 1]$ viewing rectangle. Determine the x-intercepts, if any, and the y-intercepts, if any.

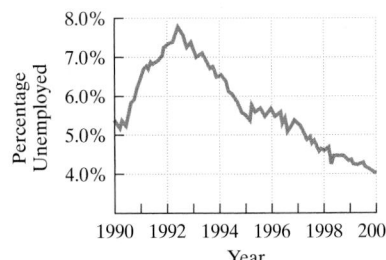

$y = -\frac{3}{2}x + 3$

3. The graph shows the unemployment rate in the United States from 1990 through 2000. For the period shown, during which year did the unemployment rate reach a maximum? Estimate the percentage of the work force unemployed, to the nearest tenth of a percent, at that time.

U.S. Unemployment Rate

Percentage Unemployed

8.0%
7.0%
6.0%
5.0%
4.0%

1990 1992 1994 1996 1998 2000
Year

Source: Bureau of Labor Statistics

Find the solution set for each equation in Exercises 4–16.

4. $7(x - 2) = 4(x + 1) - 21$

5. $\dfrac{2x - 3}{4} = \dfrac{x - 4}{2} - \dfrac{x + 1}{4}$

6. $\dfrac{2}{x - 3} - \dfrac{4}{x + 3} = \dfrac{8}{x^2 - 9}$

7. $2x^2 - 3x - 2 = 0$

8. $(3x - 1)^2 = 75$

9. $x(x - 2) = 4$

10. $4x^2 = 8x - 5$

11. $x^3 - 4x^2 - x + 4 = 0$

12. $\sqrt{x - 3} + 5 = x$

13. $\sqrt{x + 4} + \sqrt{x - 1} = 5$

14. $5x^{3/2} - 10 = 0$

15. $x^{2/3} - 9x^{1/3} + 8 = 0$

16. $\left|\dfrac{2}{3}x - 6\right| = 2$

Solve each inequality in Exercises 17–22. Express the answer in interval notation and graph the solution set on a number line.

17. $3(x + 4) \geq 5x - 12$

18. $\dfrac{x}{6} + \dfrac{1}{8} \leq \dfrac{x}{2} - \dfrac{3}{4}$

19. $-3 \leq \dfrac{2x + 5}{3} < 6$

20. $|3x + 2| \geq 3$

21. $x^2 < x + 12$

22. $\dfrac{2x + 1}{x - 3} > 3$

In Exercises 23–25, perform the indicated operations and write the result in standard form.

23. $(6 - 7i)(2 + 5i)$

24. $\dfrac{5}{2 - i}$

25. $2\sqrt{-49} + 3\sqrt{-64}$

In Exercises 26–27, solve each formula for the specified variable.

26. $V = \dfrac{1}{3} lwh$ for h

27. $y - y_1 = m(x - x_1)$ for x

The male minority? The graphs show enrollment in U.S. colleges, with projections from 2000 to 2009. The trend indicated by the graphs is among the hottest topics of debate among college-admission officers. Some private liberal arts colleges have quietly begun special efforts to recruit men— including admissions preferences for them.

Enrollment in U.S. Colleges

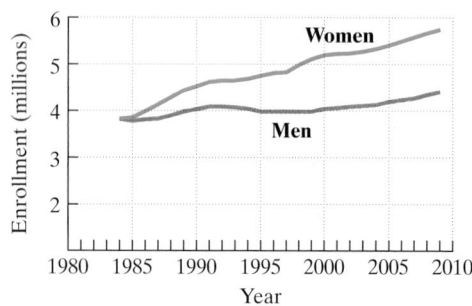

Source: U.S. Department of Education

Exercises 28–29 are based on the data shown by the graphs.

28. The data for the men can be modeled by the formula

$$N = 0.01x + 3.9$$

where N represents enrollment, in millions, x years after 1984. According to the formula, when will the projected enrollment for men be 4.1 million? How well does the formula describe enrollment for that year shown by the line graph?

29. The data for the women can be modeled from the following verbal description:

In 1984, 4.1 million women were enrolled. Female enrollment has increased by 0.07 million per year since then.

According to the verbal model, when will the projected enrollment for women be 5.71 million? How well does the verbal model describe enrollment for that year shown by the line graph?

30. On average, the number of unhealthy air days per year in Los Angeles exceeds three times that of New York City by 48 days. If Los Angeles and New York City combined have 268 unhealthy air days per year, determine the number of unhealthy days for the two cities. (*Source:* U.S. Environmental Protection Agency)

31. The costs for two different kinds of heating systems for a three-bedroom home are given in the following table. After how many years will total costs for solar heating and electric heating be the same? What will be the cost at that time?

System	Cost to Install	Operating Cost/Year
Solar	$29,700	$150
Electric	$5000	$1100

32. You placed $10,000 in two investments paying 8% and 10% annual interest, respectively. At the end of the year, the total interest from these investments was $940. How much was invested at each rate?

33. The length of a rectangular carpet is 4 feet greater than twice its width. If the area is 48 square feet, find the carpet's length and width.

34. A vertical pole is to be supported by a wire that is 26 feet long and anchored 24 feet from the base of the pole. How far up the pole should the wire be attached?

35. You take a summer job selling medical supplies. You are paid $600 per month plus 4% of the sales price of all the supplies you sell. If you want to earn more than $2500 per month, what value of medical supplies must you sell?

Functions and Graphs

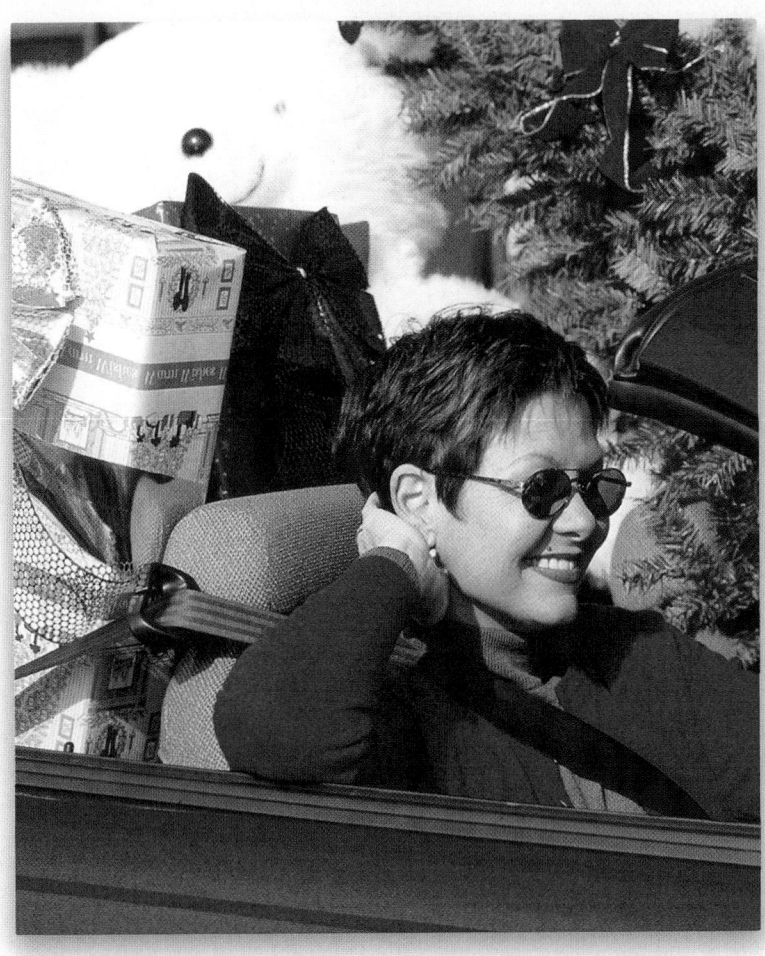

The cost of mailing a package depends on its weight. The probability that you and another person in a room share the same birthday depends on the number of people in the room. In both these situations, the relationship between variables can be described by a *function*. Understanding this concept will give you a new perspective on many ordinary situations.

'Tis the season and you've waited until the last minute to mail your holiday gifts. Your only option is overnight express mail. You realize that the cost of mailing a gift depends on its weight, but the mailing costs seem somewhat odd. Your packages that weigh 1.1 pounds, 1.5 pounds, and 2 pounds cost $15.75 each to send overnight. Packages that weigh 2.01 pounds and 3 pounds cost you $18.50 each. Finally, your heaviest gift is barely over 3 pounds and its mailing cost is $21.25. What sort of system is this in which costs increase by $2.75, stepping from $15.75 to $18.50 and from $18.50 to $21.25?

SECTION 2.1 *Lines and Slope*

Objectives

1. Compute a line's slope.
2. Write the point-slope equation of a line.
3. Write and graph the slope-intercept equation of a line.
4. Recognize equations of horizontal and vertical lines.
5. Recognize and use the general form of a line's equation.
6. Find slopes and equations of parallel and perpendicular lines.
7. Model data with linear equations.

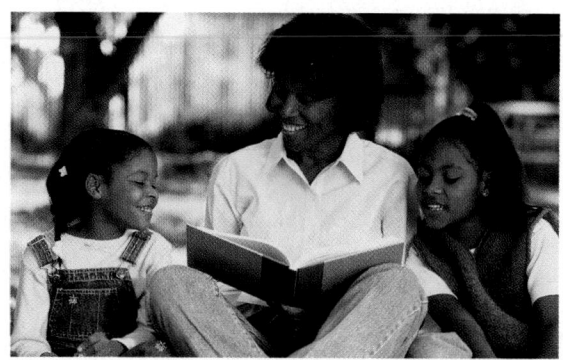

Is there a relationship between literacy and child mortality? As the percentage of adult females who are literate increases, does the mortality of children under five decrease? Figure 2.1, based on data from the United Nations, indicates that this is, indeed, the case. Each point in the figure represents one country.

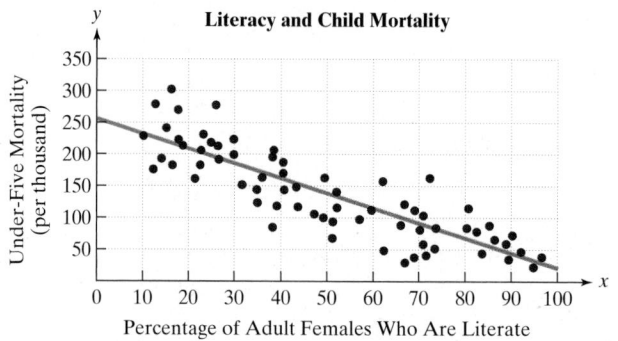

Figure 2.1
Source: United Nations

Data presented in a visual form as a set of points is called a **scatter plot.** Also shown in Figure 2.1 is a line that passes through or near the points. A line that best fits the data points in a scatter plot is called a **regression line.** By writing the equation of this line, we can obtain a model of the data and make predictions about child mortality based on the percentage of adult females in a country who are literate.

Data often fall on or near a line. In this section we will use equations to model such data and make predictions. We begin with a discussion of a line's steepness.

1 Calculate a line's slope.

The Slope of a Line

Mathematicians have developed a useful measure of the steepness of a line, called the *slope* of the line. Slope compares the vertical change (the **rise**) to the horizontal change (the **run**) when moving from one fixed point to another along the line. To calculate the slope of a line, we use a ratio that compares the change in y (the rise) to the corresponding change in x (the run).

Definition of Slope

The **slope** of the line through the distinct points (x_1, y_1) and (x_2, y_2) is

$$\frac{\text{Change in } y}{\text{Change in } x} = \frac{\text{Rise}}{\text{Run}}$$

$$= \frac{y_2 - y_1}{x_2 - x_1}$$

where $x_2 - x_1 \neq 0$.

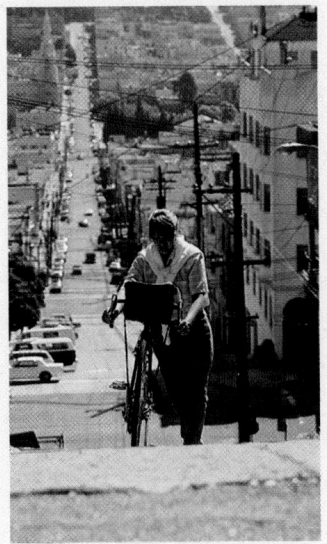

Slope and the Streets of San Francisco

San Francisco's Filbert Street has a slope of 0.613, meaning that for every horizontal distance of 100 feet, the street ascends 61.3 feet vertically. With its 31.5° angle of inclination, the street is too steep to pave and is only accessible by wooden stairs.

It is common notation to let the letter m represent the slope of a line. The letter m is used because it is the first letter of the French verb *monter*, meaning to rise, or to ascend.

EXAMPLE 1 Using the Definition of Slope

Find the slope of the line passing through each pair of points:

    a. $(-3, -1)$ and $(-2, 4)$ **b.** $(-3, 4)$ and $(2, -2)$.

Solution

    a. Let $(x_1, y_1) = (-3, -1)$ and $(x_2, y_2) = (-2, 4)$. We obtain a slope of

$$m = \frac{\text{Change in } y}{\text{Change in } x} = \frac{y_2 - y_1}{x_2 - x_1} = \frac{4 - (-1)}{-2 - (-3)} = \frac{5}{1} = 5.$$

The situation is illustrated in Figure 2.2(a). The slope of the line is 5, indicating that there is a vertical change, a rise, of 5 units for each horizontal change, a run, of 1 unit. The slope is positive, and the line rises from left to right.

Study Tip

When computing slope, it makes no difference which point you call (x_1, y_1) and which point you call (x_2, y_2). If we let $(x_1, y_1) = (-2, 4)$ and $(x_2, y_2) = (-3, -1)$, the slope is still 5:

$$m = \frac{y_2 - y_1}{x_2 - x_1} = \frac{-1 - 4}{-3 - (-2)} = \frac{-5}{-1} = 5.$$

However, you should not subtract in one order in the numerator $(y_2 - y_1)$ and then in a different order in the denominator $(x_1 - x_2)$. The slope is *not*

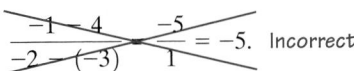

                                    $= -5.$ Incorrect

    b. We can let $(x_1, y_1) = (-3, 4)$ and $(x_2, y_2) = (2, -2)$. The slope of the line shown in Figure 2.2(b) is computed as follows:

$$m = \frac{-2 - 4}{2 - (-3)} = \frac{-6}{5} = -\frac{6}{5}.$$

The slope of the line is $-\frac{6}{5}$. For every vertical change of -6 units (6 units down), there is a corresponding horizontal change of 5 units. The slope is negative and the line falls from left to right.

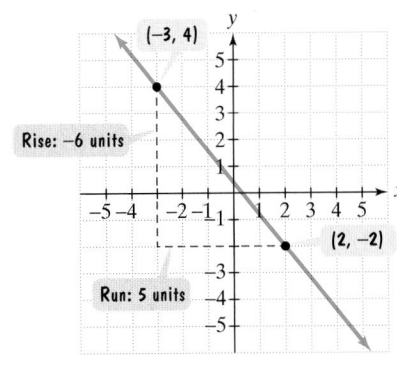

Figure 2.2 Visualizing slope   **(a)**   **(b)**

Check Point 1 Find the slope of the line passing through each pair of points:

a. $(-3, 4)$ and $(-4, -2)$ **b.** $(4, -2)$ and $(-1, 5)$.

Example 1 illustrates that a line with a positive slope is rising from left to right and a line with a negative slope is falling from left to right. By contrast, a horizontal line neither rises nor falls and has a slope of zero. A vertical line has no horizontal change, so $x_2 - x_1 = 0$ in the formula for slope. Because we cannot divide by zero, the slope of a vertical line is undefined. This discussion is summarized in Table 2.1.

Table 2.1   Possibilities for a Line's Slope

Positive Slope	Negative Slope	Zero Slope	Undefined Slope
$m > 0$	$m < 0$	$m = 0$	m is undefined.
Line rises from left to right.	Line falls from left to right.	Line is horizontal.	Line is vertical.

2 Write the point-slope equation of a line.

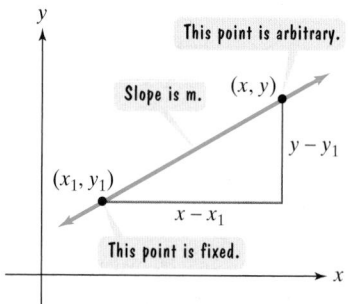

Figure 2.3 A line passing through (x_1, y_1) with slope m

The Point-Slope Form of the Equation of a Line

We can use the slope of a line to obtain various forms of the line's equation. For example, consider a nonvertical line with slope m that contains the point (x_1, y_1). Now, let (x, y) represent any other point on the line, shown in Figure 2.3. Keep in mind that the point (x, y) is arbitrary and is not in one fixed position. By contrast, the point (x_1, y_1) is fixed. Regardless of where the point (x, y) is located, the shape of the triangle in Figure 2.3 remains the same. Thus, the ratio for slope stays a constant m. This means that for all points along the line,

$$m = \frac{y - y_1}{x - x_1}, \quad x \neq x_1.$$

We can clear the fraction by multiplying both sides by $x - x_1$.

$$m(x - x_1) = \frac{y - y_1}{x - x_1} \cdot x - x_1$$

$$m(x - x_1) = y - y_1 \qquad \text{Simplify.}$$

Now, if we reverse the two sides, we obtain the *point-slope form* of the equation of a line.

Point-Slope Form of the Equation of a Line

The **point-slope equation** of a nonvertical line with slope m that passes through the point (x_1, y_1) is

$$y - y_1 = m(x - x_1).$$

For example, an equation of the line passing through $(1, 5)$ with slope 2 ($m = 2$) is

$$y - 5 = 2(x - 1).$$

After we obtain the point-slope form of a line, it is customary to express the equation with y isolated on one side of the equal sign. Example 2 illustrates how this is done.

EXAMPLE 2 Writing the Point-Slope Equation of a Line

Write the point-slope form of the equation of the line passing through $(-1, 3)$ with slope 4. Then solve the equation for y.

Solution   We use the point-slope equation of a line with $m = 4$, $x_1 = -1$, and $y_1 = 3$.

$$y - y_1 = m(x - x_1) \qquad \text{This is the point-slope form of the equation.}$$

$$y - 3 = 4[x - (-1)] \qquad \text{Substitute the given values.}$$

$$y - 3 = 4(x + 1) \qquad \begin{array}{l}\text{We now have the point-slope form of the}\\\text{equation for the given line.}\end{array}$$

We can solve this equation for y by applying the distributive property on the right side.

$$y - 3 = 4x + 4$$

Finally, we add 3 to both sides.

$$y = 4x + 7$$

Check Point 2   Write the point-slope form of the equation of the line passing through $(2, -5)$ with slope 6. Then solve the equation for y.

EXAMPLE 3 Writing the Point-Slope Equation of a Line

Write the point-slope form of the equation of the line passing through the points $(4, -3)$ and $(-2, 6)$. (See Figure 2.4.) Then solve the equation for y.

Solution   To use the point-slope form, we need to find the slope. The slope is the change in the y-coordinates divided by the corresponding change in the x-coordinates.

$$m = \frac{6 - (-3)}{-2 - 4} = \frac{9}{-6} = -\frac{3}{2} \qquad \begin{array}{l}\text{This is the definition of slope using } (4, -3)\\\text{and } (-2, 6).\end{array}$$

We can take either point on the line to be (x_1, y_1). Let's use $(x_1, y_1) = (4, -3)$. Now, we are ready to write the point-slope equation.

$$y - y_1 = m(x - x_1) \qquad \text{This is the point-slope form of the equation.}$$

$$y - (-3) = -\tfrac{3}{2}(x - 4) \qquad \text{Substitute: } (x_1, y_1) = (4, -3) \text{ and } m = -\tfrac{3}{2}.$$

$$y + 3 = -\tfrac{3}{2}(x - 4) \qquad \text{Simplify.}$$

We now have the point-slope form of the equation of the line shown in Figure 2.4. Now, we solve this equation for y.

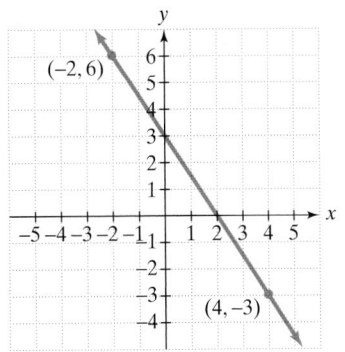

Figure 2.4 Write the point-slope equation of this line.

Discovery

You can use either point for (x_1, y_1) when you write a line's point-slope equation. Rework Example 3 using $(-2, 6)$ for (x_1, y_1). Once you solve for y, you should still obtain

$$y = -\tfrac{3}{2}x + 3.$$

$$y + 3 = -\tfrac{3}{2}(x - 4) \qquad \text{This is the point-slope form of the equation.}$$
$$y + 3 = -\tfrac{3}{2}x + 6 \qquad \text{Use the distributive property.}$$
$$y = -\tfrac{3}{2}x + 3 \qquad \text{Subtract 3 from both sides.}$$

Check Point 3 Write the point-slope form of the equation of the line passing through the points $(-2, -1)$ and $(-1, -6)$. Then solve the equation for y.

3 Write and graph the slope-intercept equation of a line.

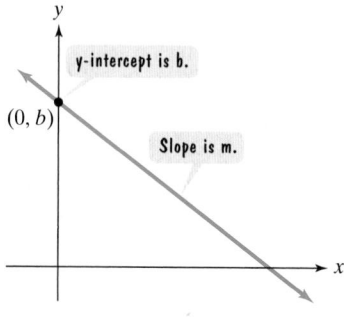

Figure 2.5 A line with slope m and y-intercept b

The Slope-Intercept Form of the Equation of a Line

Let's write the point-slope form of the equation of a nonvertical line with slope m and y-intercept b. The line is shown in Figure 2.5. Because the y-intercept is b, the line passes through $(0, b)$. We use the point-slope form with $x_1 = 0$ and $y_1 = b$.

$$y - y_1 = m(x - x_1)$$

Let $y_1 = b.$ Let $x_1 = 0.$

We obtain

$$y - b = m(x - 0).$$

Simplifying on the right side gives us

$$y - b = mx.$$

Finally, we solve for y by adding b to both sides.

$$y = mx + b$$

Thus, if a line's equation is written with y isolated on one side, the x-coefficient is the line's slope and the constant term is the y-intercept. This form of a line's equation is called the *slope-intercept form* of a line.

Slope-Intercept Form of the Equation of a Line

The **slope-intercept equation** of a nonvertical line with slope m and y-intercept b is

$$y = mx + b.$$

EXAMPLE 4 Graphing by Using the Slope and y-Intercept

Graph the line whose equation is $y = \tfrac{2}{3}x + 2$.

Solution The equation of the line is in the form $y = mx + b$. We can find the slope, m, by identifying the coefficient of x. We can find the y-intercept, b, by identifying the constant term.

$$y = \frac{2}{3}x + 2$$

The slope is $\frac{2}{3}$.   The y-intercept is 2.

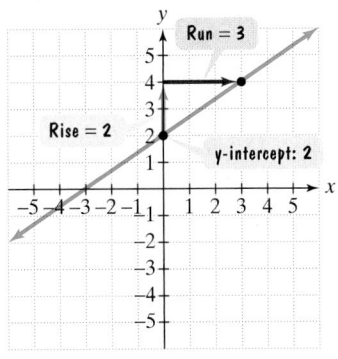

Figure 2.6 The graph of $y = \frac{2}{3}x + 2$

We need two points in order to graph the line. We can use the y-intercept, 2, to obtain the first point $(0, 2)$. Plot this point on the y-axis, shown in Figure 2.6.

We know the slope and one point on the line. We can use the slope, $\frac{2}{3}$, to determine a second point on the line. By definition,

$$m = \frac{2}{3} = \frac{\text{Rise}}{\text{Run}}.$$

We plot the second point on the line by starting at $(0, 2)$, the first point. Based on the slope, we move 2 units *up* (the rise) and 3 units to the *right* (the run). This puts us at a second point on the line, $(3, 4)$, shown in Figure 2.6.

We use a straightedge to draw a line through the two points. The graph of $y = \frac{2}{3}x + 2$ is shown in Figure 2.6.

Graphing $y = mx + b$ by Using the Slope and y-Intercept

1. Plot the y-intercept on the y-axis. This is the point $(0, b)$.
2. Obtain a second point using the slope, m. Write m as a fraction, and use rise over run, starting at the point containing the y-intercept, to plot this point.
3. Use a straightedge to draw a line through the two points. Draw arrowheads at the ends of the line to show that the line continues indefinitely in both directions.

Check Point 4 Graph the line whose equation is $y = \frac{3}{5}x + 1$.

4 Recognize equations of horizontal and vertical lines.

Equations of Horizontal and Vertical Lines

Some things change very little. For example, Figure 2.7 shows that the percentage of people in the United States satisfied with their lives remains relatively constant for all age groups. Shown in the figure is a horizontal line that passes near most tops of the six bars.

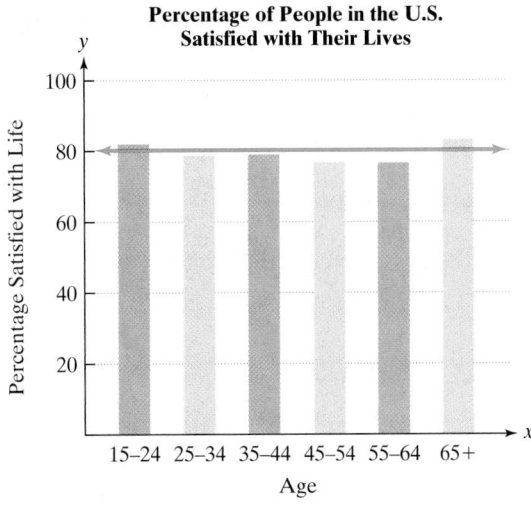

Figure 2.7
Source: Culture Shift in Advanced Industrial Society, Princeton University Press

We can use $y = mx + b$, the slope-intercept form of a line's equation, to write the equation of the horizontal line in Figure 2.7. We need the line's slope, m, and its y-intercept, b. Because the line is horizontal, $m = 0$. The line intersects the y-axis at $(0, 80)$, so its y-intercept is 80: $b = 80$.

Thus, an equation in the form $y = mx + b$ that models the percentage, y, of people at age x satisfied with their lives is

$$y = 0x + 80, \quad \text{or} \quad y = 80.$$

The percentage of people satisfied with their lives remains relatively constant in the United States for all age groups, at approximately 80%.

In general, if a line is horizontal, its slope is zero: $m = 0$. Thus, the equation $y = mx + b$ becomes $y = b$, where b is the y-intercept. All horizontal lines have equations of the form $y = b$.

EXAMPLE 5   Graphing a Horizontal Line

Graph $y = -4$ in the rectangular coordinate system.

Solution All points on the graph of $y = -4$ have a value of y that is always -4. No matter what the x-coordinate is, the y-coordinate for every point on the line is -4. Let us select three of the possible values for x: $-2, 0,$ and 3. So, three of the points on the graph $y = -4$ are $(-2, -4)$, $(0, -4)$, and $(3, -4)$. Plot each of these points. Drawing a line that passes through the three points gives the horizontal line shown in Figure 2.8.

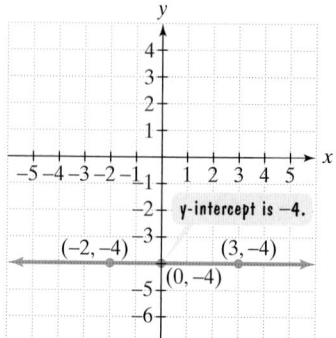

Figure 2.8 The graph of $y = -4$

Check Point 5 Graph $y = 3$ in the rectangular coordinate system.

Next, let's see what we can discover about the graph of an equation of the form $x = a$ by looking at an example.

EXAMPLE 6   Graphing a Vertical Line

Graph $x = 5$ in the rectangular coordinate system.

Solution All points on the graph of $x = 5$ have a value of x that is always 5. No matter what the y-coordinate is, the corresponding x-coordinate for every point on the line is 5. Let us select three of the possible values of y: $-2, 0,$ and 3. So, three of the points on the graph of $x = 5$ are $(5, -2)$, $(5, 0)$, and $(5, 3)$. Plot each of these points. Drawing a line that passes through the three points gives the vertical line shown in Figure 2.9.

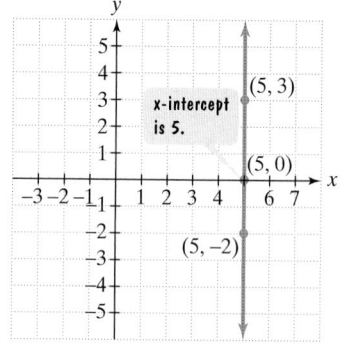

Figure 2.9 The graph $x = 5$

Horizontal and Vertical Lines

The graph of $y = b$ is a horizontal line. The y-intercept is b.

The graph of $x = a$ is a vertical line. The x-intercept is a.

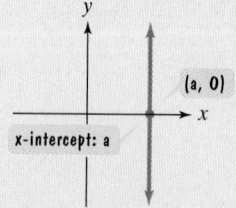

Check Point 6 Graph $x = -1$ in the rectangular coordinate system.

5 Recognize and use the general form of a line's equation.

The General Form of the Equation of a Line

The vertical line whose equation is $x = 5$ cannot be written in slope-intercept form, $y = mx + b$, because its slope is undefined. However, every line has an equation that can be expressed in the form $Ax + By + C = 0$. For example, $x = 5$ can be expressed as $1x + 0y - 5 = 0$, or $x - 5 = 0$. The equation $Ax + By + C = 0$ is called the *general form* of the equation of a line.

General Form of the Equation of a Line

Every line has an equation that can be written in the **general form**

$$Ax + By + C = 0$$

where A, B, and C are real numbers, and A and B are not both zero.

If the equation of a line is given in general form, it is possible to find the slope, m, and the y-intercept, b, for the line. We solve the equation for y, transforming it into the slope-intercept form $y = mx + b$. In this form, the coefficient of x is the slope of the line, and the constant term is its y-intercept.

EXAMPLE 7   Finding the Slope and the y-Intercept

Find the slope and the y-intercept of the line whose equation is $2x - 3y + 6 = 0$.

Solution   The equation is given in general form. We begin by rewriting it in the form $y = mx + b$. We need to solve for y.

$$2x - 3y + 6 = 0 \qquad \text{This is the given equation.}$$
$$2x + 6 = 3y \qquad \text{To isolate the y-term, add 3y to both sides.}$$
$$3y = 2x + 6 \qquad \text{Reverse the two sides. (This step is optional.)}$$
$$y = \frac{2}{3}x + 2 \qquad \text{Divide both sides by 3.}$$

The coefficient of x, $\frac{2}{3}$, is the slope and the constant term, 2, is the y-intercept. This is the form of the equation that we graphed in Figure 2.6 on page 181.

Check Point 7   Find the slope and the y-intercept of the line whose equation is $3x + 6y - 12 = 0$. Then use the y-intercept and the slope to graph the equation.

We've covered a lot of territory. Let's take a moment to summarize the various forms for equations of lines.

Equations of Lines

1. Point-slope form:	$y - y_1 = m(x - x_1)$
2. Slope-intercept form:	$y = mx + b$
3. Horizontal line:	$y = b$
4. Vertical line:	$x = a$
5. General form:	$Ax + By + C = 0$

6 Find slopes and equations of parallel and perpendicular lines.

Parallel and Perpendicular Lines

Two nonintersecting lines that lie in the same plane are **parallel.** If two lines do not intersect, the ratio of the vertical change to the horizontal change is the same for each line. Because two parallel lines have the same "steepness," they must have the same slope.

> ### Slope and Parallel Lines
>
> **1.** If two nonvertical lines are parallel, then they have the same slope.
> **2.** If two distinct nonvertical lines have the same slope, then they are parallel.
> **3.** Two distinct vertical lines, both with undefined slopes, are parallel.

EXAMPLE 8 Writing Equations of a Line Parallel to a Given Line

Write an equation of the line passing through $(-3, 2)$ and parallel to the line whose equation is $y = 2x + 1$. Express the equation in point-slope form and slope-intercept form.

Solution   The situation is illustrated in Figure 2.10. We are looking for the equation of the line shown on the left. How do we obtain this equation? Notice that the line passes through the point $(-3, 2)$. Using the point-slope form of the line's equation, we have $x_1 = -3$ and $y_1 = 2$.

$$y - y_1 = m(x - x_1)$$

$$y_1 = 2 \qquad x_1 = -3$$

The equation of this line is given: $y = 2x + 1$.

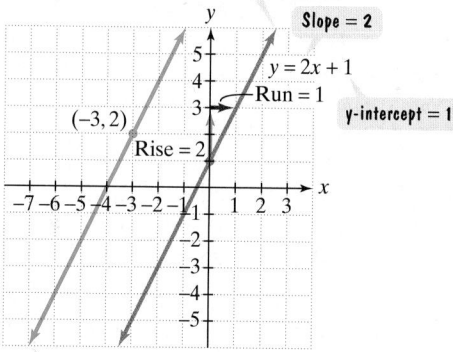

We must write the equation of this line.

Figure 2.10 Writing equations of a line parallel to a given line

Now, the only thing missing from the equation is m, the slope of the line on the left. Do we know anything about the slope of either line in Figure 2.10? The answer is yes; we know the slope of the line on the right, whose equation is given.

$$y = 2x + 1$$

The slope of the line on the right in Figure 2.10 is 2.

Parallel lines have the same slope. Because the slope of the line with the given equation is $2, m = 2$ for the line whose equation we must write.

$$y - y_1 = m(x - x_1)$$

$$y_1 = 2 \qquad m = 2 \qquad x_1 = -3$$

The point-slope form of the line's equation is

$$y - 2 = 2[x - (-3)] \text{ or}$$

$$y - 2 = 2(x + 3).$$

Solving for y, we obtain the slope-intercept form of the equation.

$$y - 2 = 2x + 6 \qquad \text{Apply the distributive property.}$$

$$y = 2x + 8 \qquad \text{Add 2 to both sides. This is the slope-intercept form,}$$
$$\qquad\qquad\qquad\qquad y = mx + b, \text{ of the equation.}$$

Check Point 8   Write an equation of the line passing through $(-2, 5)$ and parallel to the line whose equation is $y = 3x + 1$. Express the equation in point-slope form and slope-intercept form.

Two lines that intersect at a right angle (90°) are said to be **perpendicular,** shown in Figure 2.11. There is a relationship between the slopes of perpendicular lines.

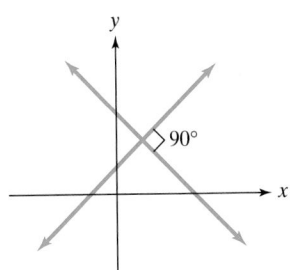

Figure 2.11 Perpendicular lines

Slope and Perpendicular Lines

1. If two nonvertical lines are perpendicular, then the product of their slopes is -1.
2. If the product of the slopes of two lines is -1, then the lines are perpendicular.
3. A horizontal line having zero slope is perpendicular to a vertical line having undefined slope.

An equivalent way of stating this relationship is to say that one line is perpendicular to another line if its slope is the *negative reciprocal* of the slope of the other. For example, if a line has slope 5, any line having slope $-\frac{1}{5}$ is perpendicular to it. Similarly, if a line has slope $-\frac{3}{4}$, any line having slope $\frac{4}{3}$ is perpendicular to it.

EXAMPLE 9   Finding the Slope of a Line Perpendicular to a Given Line

Find the slope of any line that is perpendicular to the line whose equation is $x + 4y - 8 = 0$.

Solution We begin by writing the equation of the given line, $x + 4y - 8 = 0$, in slope-intercept form. Solve for y.

$$x + 4y - 8 = 0 \qquad \text{This is the given equation.}$$
$$4y = -x + 8 \qquad \text{To isolate the y-term, subtract x and add 8 on both sides.}$$
$$y = -\tfrac{1}{4}x + 2 \qquad \text{Divide both sides by 4.}$$

> Slope is $-\dfrac{1}{4}$.

The given line has slope $-\frac{1}{4}$. Any line perpendicular to this line has a slope that is the negative reciprocal of $-\frac{1}{4}$. Thus, the slope of any perpendicular line is 4.

> **Check Point 9** Find the slope of any line that is perpendicular to the line whose equation is $x + 3y - 12 = 0$.

7 Model data with linear equations.

Applications

Slope is defined as the ratio of a change in y to a corresponding change in x. Our next example shows how slope can be interpreted as a **rate of change** in an applied situation.

EXAMPLE 10 Slope as a Rate of Change

A best guess at the look of our nation in the next decade indicates that the number of men and women living alone will increase each year. Figure 2.12 shows line graphs for the number of U.S. men and women living alone, projected through 2010. Find the slope of the line segment for the women. Describe what the slope represents.

Solution We let x represent a year and y the number of women living alone in that year. The two points shown on the line segment for women have the following coordinates:

$$(1995, 14) \qquad \text{and} \qquad (2010, 17).$$

> In 1995, 14 million U.S. women lived alone.

> In 2010, 17 million U.S. women are projected to live alone.

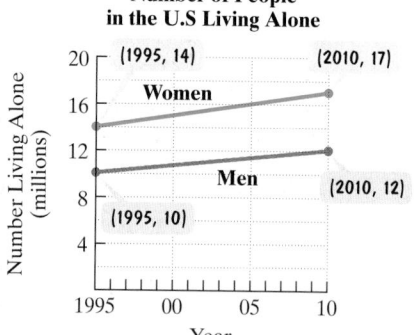

Figure 2.12 *Source:* Forrester Research

Now we compute the slope:

$$m = \frac{\text{Change in } y}{\text{Change in } x} = \frac{17 - 14}{2010 - 1995}$$

> The unit in the numerator is million people.

> The unit in the denominator is year.

$$= \frac{3}{15} = \frac{1}{5} = \frac{0.2 \text{ million people}}{\text{year}}.$$

The slope indicates that the number of U.S. women living alone is projected to increase by 0.2 million each year. The rate of change is 0.2 million women per year.

> **Check Point 10** Use the graph in Example 10 to find the slope of the line segment for the men. Express the slope correct to two decimal places and describe what it represents.

If an equation in slope-intercept form models relationships between variables, then the slope and y-intercept have physical interpretations. For the equation $y = mx + b$, the y-intercept, b, tells us what is happening to y when x is 0. If x represents time, the y-intercept describes the value of y at the beginning, or when time equals 0. The slope represents the rate of change in y per unit change in x.

Using these ideas, we can develop a model for the data for women living alone, shown in Figure 2.12 on the previous page. We let x = the number of years after 1995. At the beginning of our data, or 0 years after 1995, 14 million women lived alone. Thus, $b = 14$. In Example 10, we found that $m = 0.2$ (rate of change is 0.2 million women per year). An equation of the form $y = mx + b$ that models the data is

$$y = 0.2x + 14,$$

where y is the number, in millions, of U.S. women living alone x years after 1995.

Linear equations are useful for modeling data in scatter plots that fall on or near a line. For example, Table 2.2 gives the population of the United States, in millions, in the indicated year. The data are displayed in a scatter plot as a set of six points in Figure 2.13.

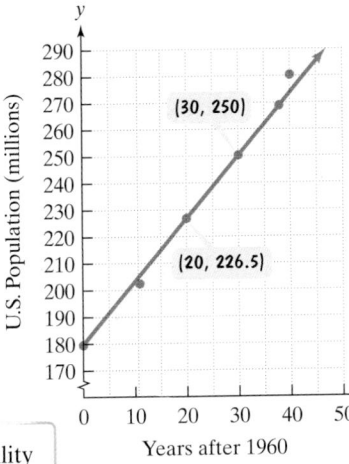

Figure 2.13

Table 2.2

Year	x (Years after 1960)	y (U.S. Population) (in millions)
1960	0	179.3
1970	10	203.3
1980	20	226.5
1990	30	250.0
1998	38	268.9
2000	40	281.4

Technology

You can use a graphing utility to obtain a model for a scatter plot in which the data points fall on or near a straight line. The line that best fits the data is called the **regression line.** After entering the data in Table 2.2, a graphing utility displays a scatter plot of the data and the regression line.

$[-5, 45, 5]$ by $[170, 285, 10]$

Also displayed is the regression line's equation.

```
LinReg
y=ax+b
a=2.45748031496
b=178.377952756
```

Also shown in Figure 2.13 is a line that passes through or near the six points. By writing the equation of this line, we can obtain a model of the data and make predictions about the population of the United States in the future.

EXAMPLE 11   Modeling U.S. Population

Write the slope-intercept equation of the line shown in Figure 2.13. Use the equation to predict U.S. population in 2010.

Solution   The line in Figure 2.13 passes through (20, 226.5) and (30, 250). We start by finding the slope.

$$m = \frac{\text{Change in } y}{\text{Change in } x} = \frac{250 - 226.5}{30 - 20} = \frac{23.5}{10} = 2.35$$

The slope indicates that the rate of change in the U.S. population is 2.35 million people per year. Now we write the line's slope-intercept equation.

$$y - y_1 = m(x - x_1)$$ Begin with the point-slope form.

$$y - 250 = 2.35(x - 30)$$ Either ordered pair can be (x_1, y_1). Let $(x_1, y_1) = (30, 250)$. From above, $m = 2.35$.

$$y - 250 = 2.35x - 70.5$$ Apply the distributive property on the right.

$$y = 2.35x + 179.5$$ Add 250 to both sides and solve for y.

A linear equation that models U.S. population, y, in millions, x years after 1960 is
$$y = 2.35x + 179.5.$$

Now, let's use this equation to predict U.S. population in 2010. Because 2010 is 50 years after 1960, substitute 50 for x and compute y.
$$y = 2.35(50) + 179.5 = 297$$

Our equation predicts that the population of the United States in the year 2010 will be 297 million. (The projected figure from the U.S. Census Bureau is 297.716 million.)

> **Check Point 11** Use the data points $(10, 203.3)$ and $(20, 226.5)$ from Table 2.2 to write an equation that models U.S. population x years after 1960. Use the equation to predict U.S. population in 2020.

Cigarettes and Lung Cancer

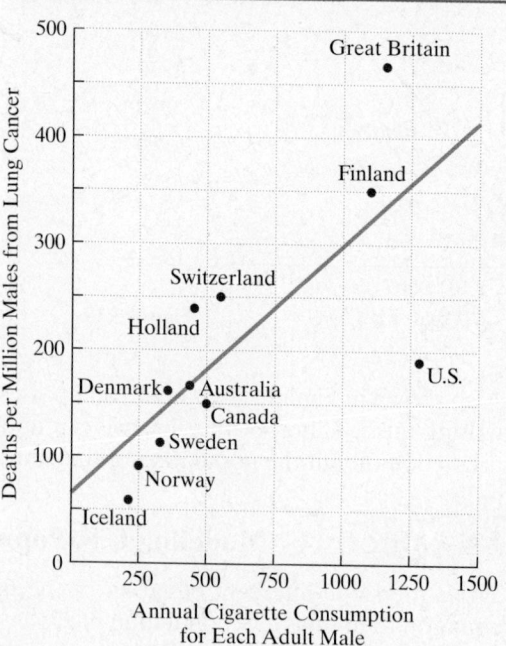

This scatter plot shows a relationship between cigarette consumption among males and deaths due to lung cancer per million males. The data are from 11 countries and date back to a 1964 report by the U.S. Surgeon General. The scatter plot can be modeled by a line whose slope indicates an increasing death rate from lung cancer with increased cigarette consumption. At that time, the tobacco industry argued that in spite of this regression line, tobacco use is not the cause of cancer. Recent data do, indeed, show a causal effect between tobacco use and numerous diseases.

Source: Smoking and Health, Washington, D.C., 1964

EXERCISE SET 2.1

Practice Exercises

In Exercises 1–10, find the slope of the line passing through each pair of points or state that the slope is undefined. Then indicate whether the line through the points rises, falls, is horizontal, or is vertical.

1. $(4, 7)$ and $(8, 10)$
2. $(2, 1)$ and $(3, 4)$
3. $(-2, 1)$ and $(2, 2)$
4. $(-1, 3)$ and $(2, 4)$
5. $(4, -2)$ and $(3, -2)$
6. $(4, -1)$ and $(3, -1)$
7. $(-2, 4)$ and $(-1, -1)$
8. $(6, -4)$ and $(4, -2)$
9. $(5, 3)$ and $(5, -2)$
10. $(3, -4)$ and $(3, 5)$

In Exercises 11–38, use the given conditions to write an equation for each line in point-slope form and slope-intercept form.

11. Slope $= 2$, passing through $(3, 5)$
12. Slope $= 4$, passing through $(1, 3)$
13. Slope $= 6$, passing through $(-2, 5)$
14. Slope $= 8$, passing through $(4, -1)$
15. Slope $= -3$, passing through $(-2, -3)$

16. Slope $= -5$, passing through $(-4, -2)$

17. Slope $= -4$, passing through $(-4, 0)$

18. Slope $= -2$, passing through $(0, -3)$

19. Slope $= -1$, passing through $\left(-\frac{1}{2}, -2\right)$

20. Slope $= -1$, passing through $\left(-4, -\frac{1}{4}\right)$

21. Slope $= \frac{1}{2}$, passing through the origin

22. Slope $= \frac{1}{3}$, passing through the origin

23. Slope $= -\frac{2}{3}$, passing through $(6, -2)$

24. Slope $= -\frac{3}{5}$, passing through $(10, -4)$

25. Passing through $(1, 2)$ and $(5, 10)$

26. Passing through $(3, 5)$ and $(8, 15)$

27. Passing through $(-3, 0)$ and $(0, 3)$

28. Passing through $(-2, 0)$ and $(0, 2)$

29. Passing through $(-3, -1)$ and $(2, 4)$

30. Passing through $(-2, -4)$ and $(1, -1)$

31. Passing through $(-3, -2)$ and $(3, 6)$

32. Passing through $(-3, 6)$ and $(3, -2)$

33. Passing through $(-3, -1)$ and $(4, -1)$

34. Passing through $(-2, -5)$ and $(6, -5)$

35. Passing through $(2, 4)$ with x-intercept $= -2$

36. Passing through $(1, -3)$ with x-intercept $= -1$

37. x-intercept $= -\frac{1}{2}$ and y-intercept $= 4$

38. x-intercept $= 4$ and y-intercept $= -2$

In Exercises 39–46, give the slope and y-intercept of each line whose equation is given. Then graph the line.

39. $y = 2x + 1$

40. $y = 3x + 2$

41. $y = -2x + 1$

42. $y = -3x + 2$

43. $y = \frac{3}{4}x - 2$

44. $y = \frac{3}{4}x - 3$

45. $y = -\frac{3}{5}x + 7$

46. $y = -\frac{2}{5}x + 6$

In Exercises 47–52, graph each equation in the rectangular coordinate system.

47. $y = -2$

48. $y = 4$

49. $x = -3$

50. $x = 5$

51. $y = 0$

52. $x = 0$

In Exercises 53–60,

    a. *Rewrite the given equation in slope-intercept form.*

    b. *Give the slope and y-intercept.*

    c. *Graph the equation.*

53. $3x + y - 5 = 0$

54. $4x + y - 6 = 0$

55. $2x + 3y - 18 = 0$

56. $4x + 6y + 12 = 0$

57. $8x - 4y - 12 = 0$

58. $6x - 5y - 20 = 0$

59. $3x - 9 = 0$

60. $4y + 28 = 0$

In Exercises 61–68, use the given conditions to write an equation for each line in point-slope form and slope-intercept form.

61. Passing through $(-8, -10)$ and parallel to the line whose equation is $y = -4x + 3$

62. Passing through $(-2, -7)$ and parallel to the line whose equation is $y = -5x + 4$

63. Passing through $(2, -3)$ and perpendicular to the line whose equation is $y = \frac{1}{5}x + 6$

64. Passing through $(-4, 2)$ and perpendicular to the line whose equation is $y = \frac{1}{3}x + 7$

65. Passing through $(-2, 2)$ and parallel to the line whose equation is $2x - 3y - 7 = 0$

66. Passing through $(-1, 3)$ and parallel to the line whose equation is $3x - 2y - 5 = 0$

67. Passing through $(4, -7)$ and perpendicular to the line whose equation is $x - 2y - 3 = 0$

68. Passing through $(5, -9)$ and perpendicular to the line whose equation is $x + 7y - 12 = 0$

Application Exercises

69. The scatter plot shows that from 1985 to 2001, the number of Americans participating in downhill skiing remained relatively constant. Write an equation that models the number of participants in downhill skiing, y, in millions, for this period.

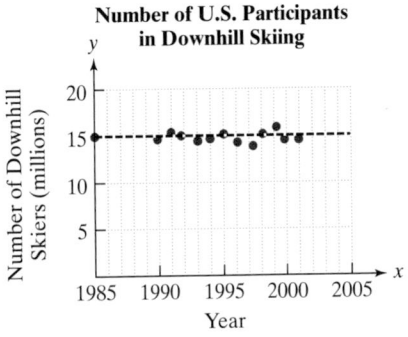

Number of U.S. Participants in Downhill Skiing

Source: National Ski Areas Association

If talk about a federal budget surplus sounded too good to be true, that's because it probably was. The Congressional Budget Office's estimates for 2010 range from a $1.2 trillion budget surplus to a $286 billion deficit. Use the information provided by the Congressional Budget Office graphs to solve Exercises 70–71.

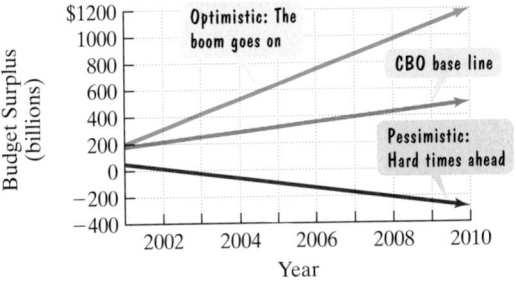

Federal Budget Projections

Source: Congressional Budget Office

70. Turn back a page and look at the line that indicates hard times ahead. Find the slope of this line using (2001, 50) and (2010, −286). Use a calculator and round to the nearest whole number. Describe what the slope represents.

71. Turn back a page and look at the line that indicates the boom goes on. Find the slope of this line using (2001, 200) and (2010, 1200). Use a calculator and round to the nearest whole number. Describe what the slope represents.

72. Horrified at the cost the last time you needed a prescription drug? The graph shows that the cost of the average retail prescription has been rising steadily since 1991.

Average Cost of a Retail Prescription

Source: *Newsweek*

a. According to the graph, what is the *y*-intercept? Describe what this represents in this situation.
b. Use the coordinates of the two points shown to compute the slope. What does this mean about the cost of the average retail prescription?
c. Write a linear equation in slope-intercept form that models the cost of the average retail prescription, *y*, *x* years after 1991.
d. Use your model from part (c) to predict the cost of the average retail prescription in 2010.

73. For 61 years, Social Security has been a huge success. It is the primary source of income for 66% of Americans over 65 and the only thing that keeps 42% of the elderly from poverty. However, the number of workers per Social Security beneficiary has been declining steadily since 1950.

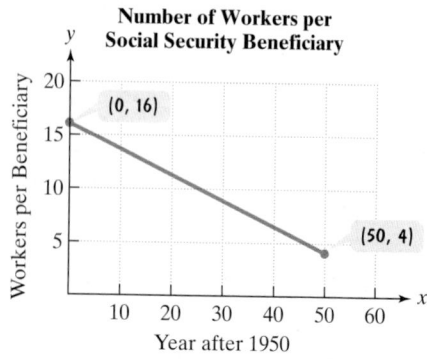

Number of Workers per Social Security Beneficiary

Source: Social Security Administration

a. According to the graph, what is the *y*-intercept? Describe what this represents in this situation.
b. Use the coordinates of the two points shown to compute the slope. What does this mean about the number of workers per beneficiary?
c. Write a linear equation in slope-intercept form that models the number of workers per beneficiary, *y*, *x* years after 1950.
d. Use your model from part (c) to predict number of workers per beneficiary in 2010. For every 8 workers, how many beneficiaries will there be?

74. We seem to be fed up with being lectured at about our waistlines. The points in the graph show the average weight of American adults from 1990 through 2000. Also shown is a line that passes through or near the points.

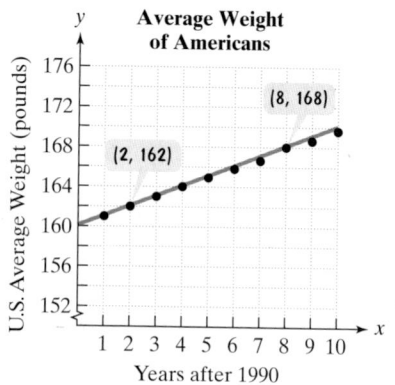

Average Weight of Americans

Source: Diabetes Care

a. Use the two points whose coordinates are shown by the voice balloons to find the point-slope equation of the line that models average weight of Americans, *y*, in pounds, *x* years after 1990.
b. Write the equation in part (a) in slope-intercept form.
c. Use the slope-intercept equation to predict the average weight of Americans in 2008.

75. Films may not be getting any better, but in this era of moviegoing, the number of screens available for new films and the classics has exploded. The points in the graph show the number of screens in the United States from 1995 through 2000. Also shown is a line that passes through or near the points.

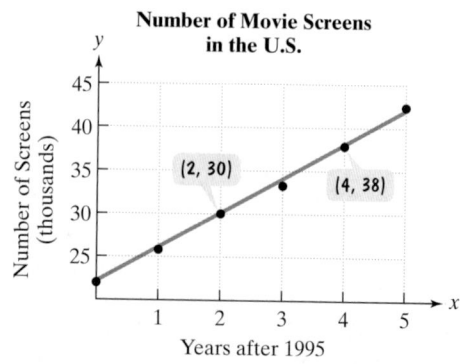

Number of Movie Screens in the U.S.

Source: Motion Picture Association of America

a. Use the two points whose coordinates are shown by the voice balloons to find the point-slope equation of the line that models the number of screens, y, in thousands, x years after 1995.

b. Write the equation in part (a) in slope-intercept form.

c. Use the slope-intercept equation to predict the number of screens, in thousands, in 2008.

76. The scatter plot shows the relationship between the percentage of married women of child-bearing age using contraceptives and the births per woman in selected countries. Also shown is the regression line. Use two points on this line to write both its point-slope and slope-intercept equations. Then find the number of births per woman if 90% of married women of child-bearing age use contraceptives.

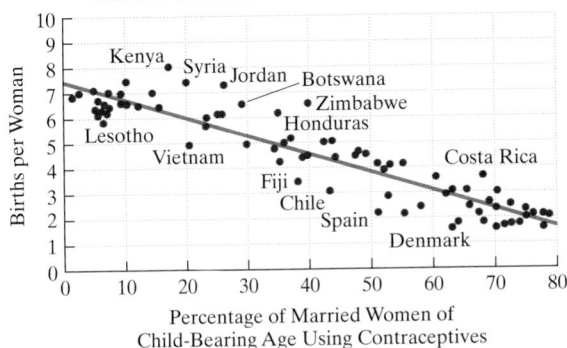

Contraceptive Prevalence and Births per Woman, Selected Countries

Source: Population Reference Bureau

77. Shown, again, is the scatter plot that indicates a relationship between the percentage of adult females in a country who are literate and the mortality of children under five. Also shown is a line that passes through or near the points. Find a linear equation that models the data by finding the slope-intercept equation of the line. Use the model to make a prediction about child mortality based on the percentage of adult females in a country who are literate.

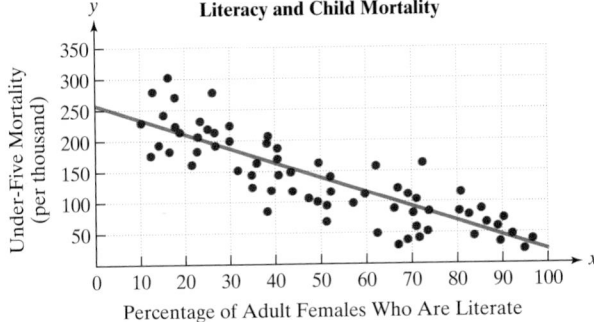

Literacy and Child Mortality

Source: United Nations

In Exercises 78–80, find a linear equation in slope-intercept form that models the given description. Describe what each variable in your model represents. Then use the model to make a prediction.

78. In 1995, the average temperature of Earth was 57.7°F and has increased at a rate of 0.01°F per year since then.

79. In 1995, 60% of U.S. adults read a newspaper and this percentage has decreased at a rate of 0.7% per year since then.

80. A computer that was purchased for $4000 is depreciating at a rate of $950 per year.

81. A business discovers a linear relationship between the number of shirts it can sell and the price per shirt. In particular, 20,000 shirts can be sold at $19 each, and 2000 of the same shirts can be sold at $55 each. Write the slope-intercept equation of the *demand line* that models the number of shirts that can be sold, y, at a price of x dollars. Then determine the number of shirts that can be sold at $50 each.

Writing in Mathematics

82. What is the slope of a line and how is it found?

83. Describe how to write the equation of a line if two points along the line are known.

84. Explain how to derive the slope-intercept form of a line's equation, $y = mx + b$, from the point-slope form

$$y - y_1 = m(x - x_1).$$

85. Explain how to graph the equation $x = 2$. Can this equation be expressed in slope-intercept form? Explain.

86. Explain how to use the general form of a line's equation to find the line's slope and y-intercept.

87. If two lines are parallel, describe the relationship between their slopes.

88. If two lines are perpendicular, describe the relationship between their slopes.

88. If you know a point on a line and you know the equation of a line perpendicular to this line, explain how to write the line's equation.

90. A formula in the form $y = mx + b$ models the cost, y, of a four-year college x years after 2003. Would you expect m to be positive, negative, or zero? Explain your answer.

91. We saw that the percentage of people satisfied with their lives remains relatively constant for all age groups. Exercise 69 showed that the number of skiers in the United States has remained relatively constant over time. Give another example of a real-world phenomenon that has remained relatively constant. Try writing an equation that models this phenomenon.

Technology Exercises

Use a graphing utility to graph each equation in Exercises 92–95. Then use the TRACE *feature to trace along the line and find the coordinates of two points. Use these points to compute the line's slope. Check your result by using the coefficient of x in the line's equation.*

92. $y = 2x + 4$

93. $y = -3x + 6$

94. $y = -\frac{1}{2}x - 5$

95. $y = \frac{3}{4}x - 2$

96. Is there a relationship between alcohol from moderate wine consumption and heart disease death rate? The table gives data from 19 developed countries.

France

Country	A	B	C	D	E	F	G
Liters of alcohol from drinking wine, per person, per year (x)	2.5	3.9	2.9	2.4	2.9	0.8	9.1
Deaths from heart disease, per 100,000 people per year (y)	211	167	131	191	220	297	71

U.S.

Country	H	I	J	K	L	M	N	O	P	Q	R	S
(x)	0.8	0.7	7.9	1.8	1.9	0.8	6.5	1.6	5.8	1.3	1.2	2.7
(y)	211	300	107	167	266	227	86	207	115	285	199	172

Source: New York Times, December 28, 1994

a. Use the statistical menu of your graphing utility to enter the 19 ordered pairs of data items shown in the table.

b. Use the DRAW menu and the scatter plot capability to draw a scatter plot of the data.

c. Select the linear regression option. Use your utility to obtain values for a and b for the equation of the regression line, $y = ax + b$. You may also be given a **correlation coefficient,** r. Values of r close to 1 indicate that the points can be described by a linear relationship and the regression line has a positive slope. Values of r close to -1 indicate that the points can be described by a linear relationship and the regression line has a negative slope. Values of r close to 0 indicate no linear relationship between the variables. In this case, a linear model does not accurately describe the data.

d. Use the appropriate sequence (consult your manual) to graph the regression equation on top of the points in the scatter plot.

Critical Thinking Exercises

97. Which one of the following is true?
a. A linear equation with nonnegative slope has a graph that rises from left to right.
b. The equations $y = 4x$ and $y = -4x$ have graphs that are perpendicular lines.
c. The line whose equation is $5x + 6y - 30 = 0$ passes through the point $(6, 0)$ and has slope $-\frac{5}{6}$.
d. The graph of $y = 7$ in the rectangular coordinate system is the single point $(7, 0)$.

98. Prove that the equation of a line passing through $(a, 0)$ and $(0, b)$ $(a \neq 0, b \neq 0)$ can be written in the form $\frac{x}{a} + \frac{y}{b} = 1$. Why is this called the *intercept form* of a line?

99. Use the figure shown to make the following lists.
a. List the slopes m_1, m_2, m_3, and m_4 in order of decreasing size.
b. List the y-intercepts b_1, b_2, b_3, and b_4 in order of decreasing size.

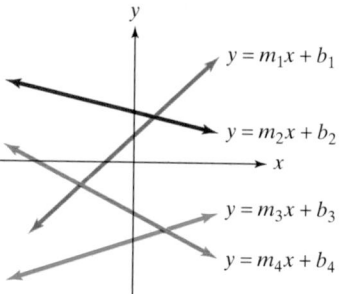

100. Excited about the success of celebrity stamps, post office officials were rumored to have put forth a plan to institute two new types of thermometers. On these new scales, $°E$ represents degrees Elvis and $°M$ represents degrees Madonna. If it is known that $40°E = 25°M$, $280°E = 125°M$, and degrees Elvis is linearly related to degrees Madonna, write an equation expressing E in terms of M.

Group Exercise

101. Group members should consult an almanac, newspaper, magazine, or the Internet to find data that lie approximately on or near a straight line. Working by hand or using a graphing utility, construct a scatter plot for the data. If working by hand, draw a line that approximately fits the data and then write its equation. If using a graphing utility, obtain the equation of the regression line. Then use the equation of the line to make a prediction about what might happen in the future. Are there circumstances that might affect the accuracy of this prediction? List some of these circumstances.

SECTION 2.2   *Distance and Midpoint Formulas; Circles*

Objectives

1. Find the distance between two points.
2. Find the midpoint of a line segment.
3. Write the standard form of a circle's equation.
4. Give the center and radius of a circle whose equation is in standard form.
5. Convert the general form of a circle's equation to standard form.

It's a good idea to know your way around a circle. Clocks, angles, maps, and compasses are based on circles. Circles occur everywhere in nature: in ripples on water, patterns on a butterfly's wings, and cross sections of trees. Some consider the circle to be the most pleasing of all shapes.

The rectangular coordinate system gives us a unique way of knowing a circle. It enables us to translate a circle's geometric definition into an algebraic equation. To do this, we must first develop a formula for the distance between any two points in rectangular coordinates.

1 Find the distance between two points.

The Distance Formula

Using the Pythagorean Theorem, we can find the distance between the two points $P_1(x_1, y_1)$ and $P_2(x_2, y_2)$ in the rectangular coordinate system. The two points are illustrated in Figure 2.14.

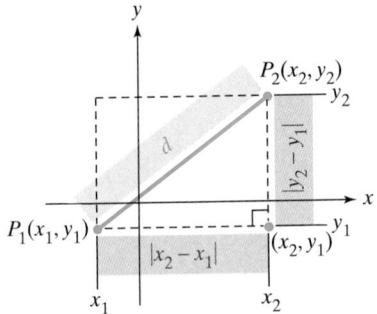

Figure 2.14

The distance that we need to find is represented by d and shown in blue. Notice that the distance between two points on the dashed horizontal line is the absolute value of the difference between the x-coordinates of the two points. This distance, $|x_2 - x_1|$, is shown in pink. Similarly, the distance between two points on the dashed vertical line is the absolute value of the difference between the y-coordinates of the two points. This distance, $|y_2 - y_1|$, is also shown in pink.

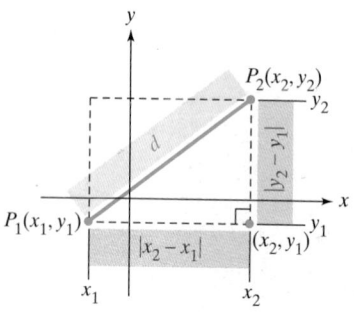

Figure 2.14, repeated

Because the dashed lines are horizontal and vertical, a right triangle is formed. Thus, we can use the Pythagorean Theorem to find distance d. By the Pythagorean Theorem,

$$d^2 = |x_2 - x_1|^2 + |y_2 - y_1|^2$$

$$d = \sqrt{|x_2 - x_1|^2 + |y_2 - y_1|^2}$$

$$d = \sqrt{(x_2 - x_1)^2 + (y_2 - y_1)^2}.$$

This result is called the **distance formula.**

The Distance Formula

The distance, d, between the points (x_1, y_1) and (x_2, y_2) in the rectangular coordinate system is

$$d = \sqrt{(x_2 - x_1)^2 + (y_2 - y_1)^2}.$$

When using the distance formula, it does not matter which point you call (x_1, y_1) and which you call (x_2, y_2).

EXAMPLE 1 Using the Distance Formula

Find the distance between $(-1, -3)$ and $(2, 3)$.

Solution   Letting $(x_1, y_1) = (-1, -3)$ and $(x_2, y_2) = (2, 3)$, we obtain

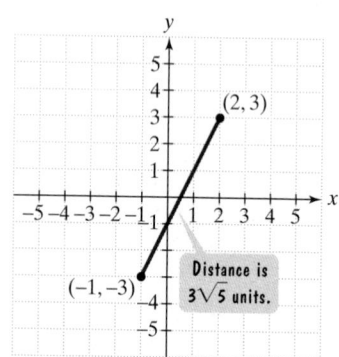

Figure 2.15 Finding the distance between two points

$$d = \sqrt{(x_2 - x_1)^2 + (y_2 - y_1)^2} \qquad \text{Use the distance formula.}$$

$$= \sqrt{[2 - (-1)]^2 + [3 - (-3)]^2} \qquad \text{Substitute the given values.}$$

$$= \sqrt{(2 + 1)^2 + (3 + 3)^2} \qquad \text{Apply the definition of subtraction within the grouping symbols.}$$

$$= \sqrt{3^2 + 6^2} \qquad \text{Perform the resulting additions.}$$

$$= \sqrt{9 + 36} \qquad \text{Square 3 and 6.}$$

$$= \sqrt{45} \qquad \text{Add.}$$

$$= 3\sqrt{5} \approx 6.71. \qquad \sqrt{45} = \sqrt{9 \cdot 5} = \sqrt{9}\sqrt{5} = 3\sqrt{5}$$

The distance between the given points is $3\sqrt{5}$ units, or approximately 6.71 units. The situation is illustrated in Figure 2.15.

Check Point 1   Find the distance between $(2, -2)$ and $(5, 2)$.

2 Find the midpoint of a line segment.

The Midpoint Formula

The distance formula can be used to derive a formula for finding the midpoint of a line segment between two given points. The formula is given as follows:

The Midpoint Formula

Consider a line segment whose endpoints are (x_1, y_1) and (x_2, y_2). The coordinates of the segment's midpoint are

$$\left(\frac{x_1 + x_2}{2}, \frac{y_1 + y_2}{2} \right).$$

To find the midpoint, take the average of the two x-coordinates and the average of the two y-coordinates.

EXAMPLE 2 Using the Midpoint Formula

Find the midpoint of the line segment with endpoints $(1, -6)$ and $(-8, -4)$.

Solution To find the coordinates of the midpoint, we average the coordinates of the endpoints.

$$\text{Midpoint} = \left(\frac{1 + (-8)}{2}, \frac{-6 + (-4)}{2} \right) = \left(\frac{-7}{2}, \frac{-10}{2} \right) = \left(-\frac{7}{2}, -5 \right)$$

Average the x-coordinates.   Average the y-coordinates.

Figure 2.16 Finding a line segment's midpoint

Figure 2.16 illustrates that the point $\left(-\frac{7}{2}, -5\right)$ is midway between the points $(1, -6)$ and $(-8, -4)$.

Check Point 2 Find the midpoint of the line segment with endpoints $(1, 2)$ and $(7, -3)$.

Circles

Our goal is to translate a circle's geometric definition into an equation. We begin with this geometric definition.

Definition of a Circle

A **circle** is the set of all points in a plane that are equidistant from a fixed point, called the **center.** The fixed distance from the circle's center to any point on the circle is called the **radius.**

Figure 2.17 is our starting point for obtaining a circle's equation. We've placed the circle into a rectangular coordinate system. The circle's center is (h, k) and its radius is r. We let (x, y) represent the coordinates of any point on the circle.

What does the geometric definition of a circle tell us about point (x, y) in Figure 2.17? The point is on the circle if and only if its distance from the center is r. We can use the distance formula to express this idea algebraically:

The distance between (x, y) and (h, k) is always r.

$$\sqrt{(x - h)^2 + (y - k)^2} \quad = \quad r$$

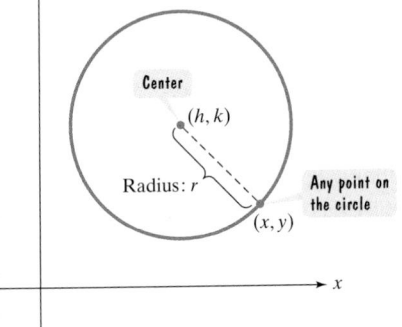

Figure 2.17 A circle centered at (h, k) with radius r

Squaring both sides of $\sqrt{(x-h)^2 + (y-k)^2} = r$ yields the *standard form of the equation of a circle.*

The Standard Form of the Equation of a Circle

The **standard form of the equation of a circle** with center (h, k) and radius r is
$$(x - h)^2 + (y - k)^2 = r^2.$$

---

3 Write the standard form of a circle's equation.

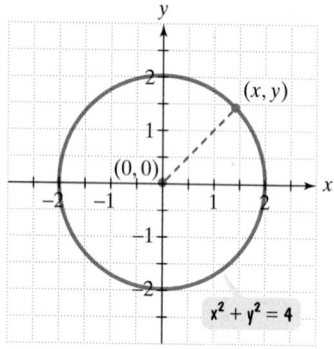

Figure 2.18 The graph of $x^2 + y^2 = 4$

EXAMPLE 3   Finding the Standard Form of a Circle's Equation

Write the standard form of the equation of the circle with center $(0, 0)$ and radius 2. Graph the circle.

Solution The center is $(0, 0)$. Because the center is represented as (h, k) in the standard form of the equation, $h = 0$ and $k = 0$. The radius is 2, so we will let $r = 2$ in the equation.

$(x - h)^2 + (y - k)^2 = r^2$ This is the standard form of a circle's equation.

$(x - 0)^2 + (y - 0)^2 = 2^2$ Substitute 0 for h, 0 for k, and 2 for r.

$x^2 + y^2 = 4$ Simplify.

The standard form of the equation of the circle is $x^2 + y^2 = 4$. Figure 2.18 shows the graph.

Check Point 3   Write the standard form of the equation of the circle with center $(0, 0)$ and radius 4.

Technology

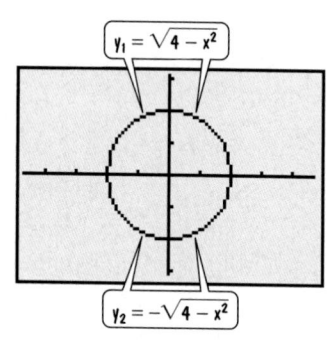

To graph a circle with a graphing utility, first solve the equation for y.

$$x^2 + y^2 = 4$$

$$y^2 = 4 - x^2$$

$$y = \pm\sqrt{4 - x^2}$$

Graph the two equations

$$y_1 = \sqrt{4 - x^2} \quad \text{and} \quad y_2 = -\sqrt{4 - x^2}$$

in the same viewing rectangle. The graph of $y_1 = \sqrt{4 - x^2}$ is the top semicircle because y is always positive. The graph of $y_2 = -\sqrt{4 - x^2}$ is the bottom semicircle because y is always negative. Use a [ZOOM SQUARE] setting so that the circle looks like a circle. (Many graphing utilities have problems connecting the two semicircles because the segments directly across horizontally from the center become nearly vertical.)

Example 3 and Check Point 3 involved circles centered at the origin. The standard form of the equation of all such circles is $x^2 + y^2 = r^2$, where r is the circle's radius. Now, let's consider a circle whose center is not at the origin.

EXAMPLE 4   Finding the Standard Form of a Circle's Equation

Write the standard form of the equation of the circle with center $(-2, 3)$ and radius 4.

Solution   The center is $(-2, 3)$. Because the center is represented as (h, k) in the standard form of the equation, $h = -2$ and $k = 3$. The radius is 4, so we will let $r = 4$ in the equation.

$$(x - h)^2 + (y - k)^2 = r^2 \qquad \text{This is the standard form of a circle's equation.}$$
$$[x - (-2)]^2 + (y - 3)^2 = 4^2 \qquad \text{Substitute } -2 \text{ for } h, 3 \text{ for } k, \text{ and } 4 \text{ for } r.$$
$$(x + 2)^2 + (y - 3)^2 = 16 \qquad \text{Simplify.}$$

The standard form of the equation of the circle is $(x + 2)^2 + (y - 3)^2 = 16$.

> **Check Point 4**   Write the standard form of the equation of the circle with center $(5, -6)$ and radius 10.

4 Give the center and radius of a circle whose equation is in standard form.

EXAMPLE 5   Using the Standard Form of a Circle's Equation to Graph the Circle

Find the center and radius of the circle whose equation is

$$(x - 2)^2 + (y + 4)^2 = 9$$

and graph the equation.

Solution   In order to graph the circle, we need to know its center, (h, k), and its radius, r. We can find the values for h, k, and r by comparing the given equation to the standard form of the equation of a circle.

$$(x - 2)^2 + (y + 4)^2 = 9$$
$$(x - 2)^2 + [y - (-4)]^2 = 3^2$$

This is $(x - h)^2$, with $h = 2$.	This is $(y - k)^2$, with $k = -4$.	This is r^2, with $r = 3$.

We see that $h = 2$, $k = -4$, and $r = 3$. Thus, the circle has center $(h, k) = (2, -4)$ and a radius of 3 units. To graph this circle, first plot the center $(2, -4)$. Because the radius is 3, you can locate at least four points on the circle by going out three units to the right, to the left, up, and down from the center.

The points three units to the right and to the left of $(2, -4)$ are $(5, -4)$ and $(-1, -4)$, respectively. The points three units up and down from $(2, -4)$ are $(2, -1)$ and $(2, -7)$, respectively.

Using these points, we obtain the graph in Figure 2.19.

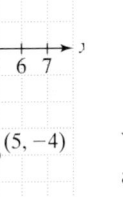

Figure 2.19 The graph of $(x - 2)^2 + (y + 4)^2 = 9$

> **Check Point 5**   Find the center and radius of the circle whose equation is
> $$(x + 3)^2 + (y - 1)^2 = 4$$
> and graph the equation.

If we square $x - 2$ and $y + 4$ in the standard form of the equation from Example 5, we obtain another form for the circle's equation.

$$(x - 2)^2 + (y + 4)^2 = 9 \qquad \text{This is the standard form of the equation from Example 5.}$$

$$x^2 - 4x + 4 + y^2 + 8y + 16 = 9 \qquad \text{Square } x - 2 \text{ and } y + 4.$$

$$x^2 + y^2 - 4x + 8y + 20 = 9 \qquad \text{Combine numerical terms and rearrange terms.}$$

$$x^2 + y^2 - 4x + 8y + 11 = 0 \qquad \text{Subtract 9 from both sides.}$$

This result suggests that an equation in the form $x^2 + y^2 + Dx + Ey + F = 0$ can represent a circle. This is called the *general form of the equation of a circle*.

> **The General Form of the Equation of a Circle**
> The **general form of the equation of a circle** is
> $$x^2 + y^2 + Dx + Ey + F = 0.$$

5 Convert the general form of a circle's equation to standard form.

We can convert the general form of the equation of a circle to the standard form $(x - h)^2 + (y - k)^2 = r^2$. We do so by completing the square on x and y. Let's see how this is done.

EXAMPLE 6 Converting the General Form of a Circle's Equation to Standard Form and Graphing the Circle

Study Tip

To review completing the square, see Section 1.5, pages 119–120.

Write in standard form and graph: $x^2 + y^2 + 4x - 6y - 23 = 0$.

Solution Because we plan to complete the square on both x and y, let's rearrange terms so that x-terms are arranged in descending order, y-terms are arranged in descending order, and the constant term appears on the right.

$$x^2 + y^2 + 4x - 6y - 23 = 0 \qquad \text{This is the given equation.}$$

$$(x^2 + 4x \quad) + (y^2 - 6y \quad) = 23 \qquad \text{Rewrite in anticipation of completing the square.}$$

$$(x^2 + 4x + 4) + (y^2 - 6y + 9) = 23 + 4 + 9 \qquad \text{Complete the square on x:}$$
$$\tfrac{1}{2} \cdot 4 = 2 \text{ and } 2^2 = 4, \text{ so add 4 to both sides. Complete the square on y: } \tfrac{1}{2}(-6) = -3 \text{ and } (-3)^2 = 9, \text{ so add 9 to both sides.}$$

> Remember that numbers added on the left side must also be added on the right side.

$$(x + 2)^2 + (y - 3)^2 = 36 \qquad \text{Factor on the left and add on the right.}$$

This last equation is in standard form. We can identify the circle's center and radius by comparing this equation to the standard form of the equation of a circle, $(x - h)^2 + (y - k)^2 = r^2$.

$$(x + 2)^2 + (y - 3)^2 = 36$$
$$[x - (-2)]^2 + (y - 3)^2 = 6^2$$

| This is $(x - h)^2$, with h = −2. | This is $(y - k)^2$, with k = 3. | This is r^2, with r = 6. |

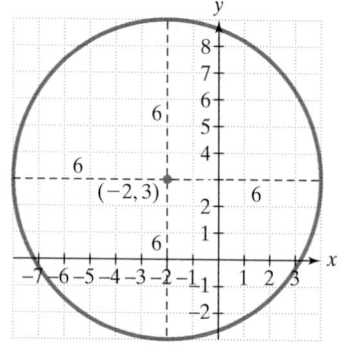

Figure 2.20 The graph of $(x + 2)^2 + (y - 3)^2 = 36$

We use the center, $(h, k) = (-2, 3)$, and the radius, $r = 6$, to graph the circle. The graph is shown in Figure 2.20.

Technology

To graph $x^2 + y^2 + 4x - 6y - 23 = 0$, rewrite the equation as a quadratic equation in y.

$$y^2 - 6y + (x^2 + 4x - 23) = 0$$

Now solve for y using the quadratic formula, with $a = 1$, $b = -6$, and $c = x^2 + 4x - 23$.

$$y = \frac{-b \pm \sqrt{b^2 - 4ac}}{2a} = \frac{-(-6) \pm \sqrt{(-6)^2 - 4 \cdot 1(x^2 + 4x - 23)}}{2 \cdot 1} = \frac{6 \pm \sqrt{36 - 4(x^2 + 4x - 23)}}{2}$$

Because we will enter these equations, there is no need to simplify. Enter

$$y_1 = \frac{6 + \sqrt{36 - 4(x^2 + 4x - 23)}}{2}$$

and

$$y_2 = \frac{6 - \sqrt{36 - 4(x^2 + 4x - 23)}}{2}.$$

Use a ZOOM SQUARE setting. The graph is shown on the right.

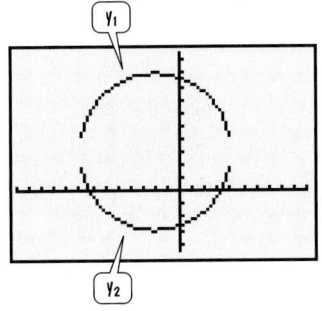

Check Point 6 Write in standard form and graph:
$$x^2 + y^2 + 4x - 4y - 1 = 0.$$

EXERCISE SET 2.2

Practice Exercises

In Exercises 1–18, find the distance between each pair of points. If necessary, round answers to two decimals places.

1. $(2, 3)$ and $(14, 8)$
2. $(5, 1)$ and $(8, 5)$
3. $(4, 1)$ and $(6, 3)$
4. $(2, 3)$ and $(3, 5)$
5. $(0, 0)$ and $(-3, 4)$
6. $(0, 0)$ and $(3, -4)$
7. $(-2, -6)$ and $(3, -4)$
8. $(-4, -1)$ and $(2, -3)$
9. $(0, -3)$ and $(4, 1)$
10. $(0, -2)$ and $(4, 3)$
11. $(3.5, 8.2)$ and $(-0.5, 6.2)$
12. $(2.6, 1.3)$ and $(1.6, -5.7)$
13. $(0, -\sqrt{3})$ and $(\sqrt{5}, 0)$
14. $(0, -\sqrt{2})$ and $(\sqrt{7}, 0)$
15. $(3\sqrt{3}, \sqrt{5})$ and $(-\sqrt{3}, 4\sqrt{5})$
16. $(2\sqrt{3}, \sqrt{6})$ and $(-\sqrt{3}, 5\sqrt{6})$
17. $\left(\frac{7}{3}, \frac{1}{5}\right)$ and $\left(\frac{1}{3}, \frac{6}{5}\right)$
18. $\left(-\frac{1}{4}, -\frac{1}{7}\right)$ and $\left(\frac{3}{4}, \frac{6}{7}\right)$

In Exercises 19–30, find the midpoint of each line segment with the given endpoints.

19. $(6, 8)$ and $(2, 4)$
20. $(10, 4)$ and $(2, 6)$
21. $(-2, -8)$ and $(-6, -2)$
22. $(-4, -7)$ and $(-1, -3)$
23. $(-3, -4)$ and $(6, -8)$
24. $(-2, -1)$ and $(-8, 6)$
25. $\left(-\frac{7}{2}, \frac{3}{2}\right)$ and $\left(-\frac{5}{2}, -\frac{11}{2}\right)$
26. $\left(-\frac{2}{5}, \frac{7}{15}\right)$ and $\left(-\frac{2}{5}, -\frac{4}{15}\right)$
27. $(8, 3\sqrt{5})$ and $(-6, 7\sqrt{5})$
28. $(7\sqrt{3}, -6)$ and $(3\sqrt{3}, -2)$
29. $(\sqrt{18}, -4)$ and $(\sqrt{2}, 4)$
30. $(\sqrt{50}, -6)$ and $(\sqrt{2}, 6)$

In Exercises 31–40, write the standard form of the equation of the circle with the given center and radius.

31. Center $(0, 0)$, $r = 7$
32. Center $(0, 0)$, $r = 8$
33. Center $(3, 2)$, $r = 5$
34. Center $(2, -1)$, $r = 4$
35. Center $(-1, 4)$, $r = 2$
36. Center $(-3, 5)$, $r = 3$
37. Center $(-3, -1)$, $r = \sqrt{3}$

38. Center $(-5, -3), r = \sqrt{5}$

39. Center $(-4, 0), r = 10$

40. Center $(-2, 0), r = 6$

In Exercises 41–48, give the center and radius of the circle described by the equation and graph each equation.

41. $x^2 + y^2 = 16$

42. $x^2 + y^2 = 49$

43. $(x - 3)^2 + (y - 1)^2 = 36$

44. $(x - 2)^2 + (y - 3)^2 = 16$

45. $(x + 3)^2 + (y - 2)^2 = 4$

46. $(x + 1)^2 + (y - 4)^2 = 25$

47. $(x + 2)^2 + (y + 2)^2 = 4$

48. $(x + 4)^2 + (y + 5)^2 = 36$

In Exercises 49–56, complete the square and write the equation in standard form. Then give the center and radius of each circle and graph the equation.

49. $x^2 + y^2 + 6x + 2y + 6 = 0$

50. $x^2 + y^2 + 8x + 4y + 16 = 0$

51. $x^2 + y^2 - 10x - 6y - 30 = 0$

52. $x^2 + y^2 - 4x - 12y - 9 = 0$

53. $x^2 + y^2 + 8x - 2y - 8 = 0$

54. $x^2 + y^2 + 12x - 6y - 4 = 0$

55. $x^2 - 2x + y^2 - 15 = 0$

56. $x^2 + y^2 - 6y - 7 = 0$

Application Exercises

57. A rectangular coordinate system with coordinates in miles is placed on the map in the figure shown. Bangkok has coordinates $(-115, 170)$ and Phnom Penh has coordinates $(65, 70)$. How long will it take a plane averaging 400 miles per hour to fly directly from one city to the other? Round to the nearest tenth of an hour. Approximately how many minutes is the flight?

58. We refer to the driveway in the figure shown as being *circular*, meaning that it is bounded by two circles. The figure indicates that the radius of the larger circle is 52 feet and the radius of the smaller circle is 38 feet.

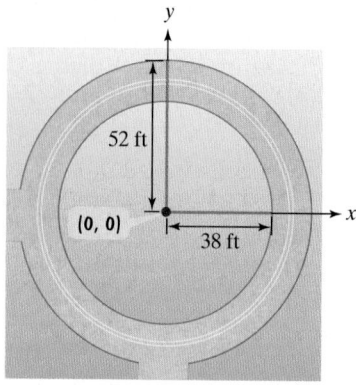

a. Use the coordinate system shown to write the equation of the smaller circle.

b. Use the coordinate system shown to write the equation of the larger circle.

59. The ferris wheel in the figure has a radius of 68 feet. The clearance between the wheel and the ground is 14 feet. The rectangular coordinate system shown has its origin on the ground directly below the center of the wheel. Use the coordinate system to write the equation of the circular wheel.

60. The circle formed by the middle lane of a circular running track can be described algebraically by $x^2 + y^2 = 4$, where all measurements are in miles. If you run around the track's middle lane twice, approximately how many miles have you covered?

Writing in Mathematics

61. In your own words, describe how to find the distance between two points in the rectangular coordinate system.

62. In your own words, describe how to find the midpoint of a line segment if its endpoints are known.

63. What is a circle? Without using variables, describe how the definition of a circle can be used to obtain a form of its equation.

64. Give an example of a circle's equation in standard form. Describe how to find the center and radius for this circle.

65. How is the standard form of a circle's equation obtained from its general form?

66. Does $(x - 3)^2 + (y - 5)^2 = 0$ represent the equation of a circle? If not, describe the graph of this equation.

67. Does $(x - 3)^2 + (y - 5)^2 = -25$ represent the equation of a circle? What sort of set is the graph of this equation?

 Technology Exercises

In Exercises 68–70, use a graphing utility to graph each circle whose equation is given.

68. $x^2 + y^2 = 25$

69. $(y + 1)^2 = 36 - (x - 3)^2$

70. $x^2 + 10x + y^2 - 4y - 20 = 0$

 Critical Thinking Exercises

71. Which one of the following is true?

 a. The equation of the circle whose center is at the origin with radius 16 is $x^2 + y^2 = 16$.

 b. The graph of $(x - 3)^2 + (y + 5)^2 = 36$ is a circle with radius 6 centered at $(-3, 5)$.

 c. The graph of $(x - 4) + (y + 6) = 25$ is a circle with radius 5 centered at $(4, -6)$.

 d. None of the above is true.

72. Show that the points $A(1, 1 + d)$, $B(3, 3 + d)$, and $C(6, 6 + d)$ are collinear (lie along a straight line) by showing that the distance from A to B plus the distance from B to C equals the distance from A to C.

73. Prove the midpoint formula by using the following procedure.

 a. Show that the distance between (x_1, y_1) and $\left(\dfrac{x_1 + x_2}{2}, \dfrac{y_1 + y_2}{2}\right)$ is equal to the distance between (x_2, y_2) and $\left(\dfrac{x_1 + x_2}{2}, \dfrac{y_1 + y_2}{2}\right)$.

 b. Use the procedure from Exercise 72 and the distances from part (a) to show that the points (x_1, y_1), $\left(\dfrac{x_1 + x_2}{2}, \dfrac{y_1 + y_2}{2}\right)$, and (x_2, y_2) are collinear.

In Exercises 74–75, write the standard form and the general form of the equation of each circle.

74. Center at $(3, -5)$ and passing through the point $(-2, 1)$

75. Passing through $(-7, 2)$ and $(1, 2)$; these points are endpoints of the diameter, the line that passes through the circle's center.

76. Find the area of the donut-shaped region bounded by the graphs of $(x - 2)^2 + (y + 3)^2 = 25$ and $(x - 2)^2 + (y + 3)^2 = 36$.

77. A **tangent line** to a circle is a line that intersects the circle at exactly one point. The tangent line is perpendicular to the radius of the circle at this point of contact. Write the point-slope equation of a line tangent to the circle whose equation is $x^2 + y^2 = 25$ at the point $(3, -4)$.

SECTION 2.3   *Basics of Functions*

Objectives

1. Find the domain and range of a relation.

2. Determine whether a relation is a function.

3. Determine whether an equation represents a function.

4. Evaluate a function.

5. Find and simplify a function's difference quotient.

6. Understand and use piecewise functions.

7. Find the domain of a function.

Jerry Orbach	$34,000
Charles Shaugnessy	$31,800
Andy Richter	$29,400
Norman Schwarzkopf	$28,000
Jon Stewart	$28,000

The answer: See the above list. The question: Who are *Celebrity Jeopardy's* five all-time highest earners? The list indicates a correspondence between the five all-time highest earners and their winnings. We can write this correspondence using a set of ordered pairs:

{(Orbach, $34,000), (Shaugnessy, $31,800), (Richter, $29,400),
(Schwarzkopf, $28,000), (Stewart, $28,000)}.

1 Find the domain and range of a relation.

The mathematical term for a set of ordered pairs is a *relation*.

> **Definition of a Relation**
>
> A **relation** is any set of ordered pairs. The set of all first components of the ordered pairs is called the **domain** of the relation, and the set of all second components is called the **range** of the relation.

EXAMPLE 1   Finding the Domain and Range of a Relation

Find the domain and range of the relation:

$$\{(\text{Orbach, \$34,000}), (\text{Shaugnessy, \$31,800}), (\text{Richter, \$29,400}),$$
$$(\text{Schwarzkopf, \$28,000}), (\text{Stewart, \$28,000})\}.$$

Solution   The domain is the set of all first components. Thus, the domain is

{Orbach, Shaugnessy, Richter, Schwarzkopf, Stewart}.

The range is the set of all second components. Thus, the range is

$$\{\$34,000, \$31,800, \$29,400, \$28,000\}.$$

Check Point 1   Find the domain and the range of the relation:

$$\{(5, 12.8), (10, 16.2), (15, 18.9), (20, 20.7), (25, 21.8)\}.$$

As you worked Check Point 1, did you wonder if there was a rule that assigned the "inputs" in the domain to the "outputs" in the range? For example, for the ordered pair (15, 18.9), how does the output 18.9 depend on the input 15? Think paid vacation days! The first number in each ordered pair is the number of years a full-time employee has been employed by a medium to large U.S. company. The second number is the average number of paid vacation days each year. Consider, for example, the ordered pair (15, 18.9).

(15, 18.9)

After 15 years   Workers average 18.9 paid vacation days each year

The relation in the vacation-days example can be pictured as follows:

```
     5 ───────→ 12.8
    10 ───────→ 16.2
    15 ───────→ 18.9
    20 ───────→ 20.7
    25 ───────→ 21.8

  Domain        Range
```

A scatter plot, like the one shown in Figure 2.21, is another way to represent the relation.

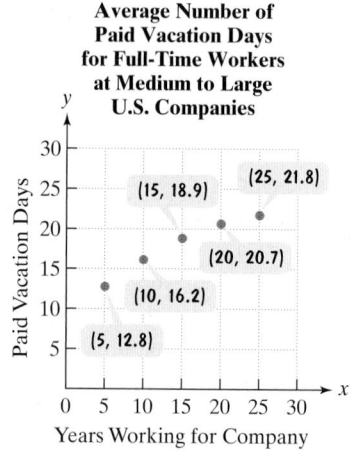

Average Number of Paid Vacation Days for Full-Time Workers at Medium to Large U.S. Companies

Figure 2.21 The graph of a relation showing a correspondence between years with a company and paid vacation days

Source: Bureau of Labor Statistics

2 Determine whether a relation is a function.

Jerry Orbach	$34,000
Charles Shaugnessy	$31,800
Andy Richter	$29,400
Norman Schwarzkopf	$28,000
Jon Stewart	$28,000

Functions

Shown, again, in the margin are *Celebrity Jeopardy's* five all-time highest winners and their winnings. We've used this information to define two relations. Figure 2.22(a) shows a correspondence between winners and their winnings. Figure 2.22(b) shows a correspondence between winnings and winners.

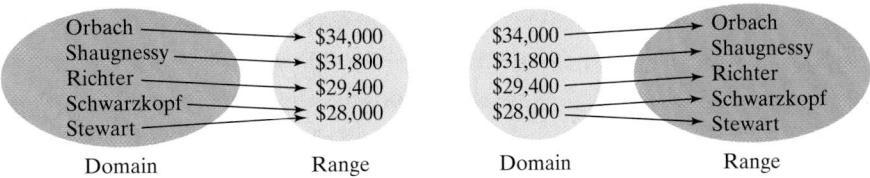

Figure 2.22(a)
Winners correspond to winnings

Figure 2.22(b)
Winnings correspond to winners

A relation in which each member of the domain corresponds to exactly one member of the range is a **function**. Can you see that the relation in Figure 2.22(a) is a function? Each winner in the domain corresponds to exactly one winning amount in the range. If we know the winner, we can be sure of the amount won. Notice that more than one element in the domain can correspond to the same element in the range. (Schwarzkopf and Stewart both won $28,000.)

Is the relation in Figure 2.22(b) a function? Does each member of the domain correspond to precisely one member of the range? This relation is not a function because there is a member of the domain that corresponds to two members of the range:

($28,000, Schwarzkopf)   ($28,000, Stewart).

The member of the domain, $28,000, corresponds to both Schwarzkopf and Stewart in the range. If we know the amount won, $28,000, we cannot be sure of the winner. Because **a function is a relation in which no two ordered pairs have the same first component and different second components,** the ordered pairs ($28,000, Schwarzkopf) and ($28,000, Stewart) are not ordered pairs of a function.

Same first component

($28,000, Schwarzkopf)   ($28,000, Stewart)

Different second components

Definition of a Function

A **function** is a correspondence from a first set, called the **domain,** to a second set, called the **range,** such that each element in the domain corresponds to *exactly one* element in the range.

Example 2 illustrates that not every correspondence between sets is a function.

EXAMPLE 2   Determining Whether a Relation is a Function

Determine whether each relation is a function:

a. $\{(1, 6), (2, 6), (3, 8), (4, 9)\}$ **b.** $\{(6, 1), (6, 2), (8, 3), (9, 4)\}$.

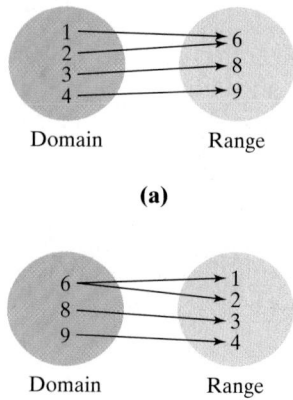

Domain   Range

(a)

Domain   Range

(b)

Figure 2.23

Study Tip

The word "range" can mean many things, from a chain of mountains to a cooking stove. For functions, it means the set of all function values. For graphing utilities, it means the setting used for the viewing rectangle. Try not to confuse these meanings.

3 Determine whether an equation represents a function.

Solution We will make a figure for each relation that shows the domain and the range.

a. We begin with the relation $\{(1, 6), (2, 6), (3, 8), (4, 9)\}$. Figure 2.23(a) shows that every element in the domain corresponds to exactly one element in the range. The element 1 in the domain corresponds to the element 6 in the range. Furthermore, 2 corresponds to 6, 3 corresponds to 8, and 4 corresponds to 9. No two ordered pairs in the given relation have the same first component and different second components. Thus, the relation is a function.

b. We now consider the relation $\{(6, 1), (6, 2), (8, 3), (9, 4)\}$. Figure 2.23(b) shows that 6 corresponds to both 1 and 2. If any element in the domain corresponds to more than one element in the range, the relation is not a function. This relation is not a function; two ordered pairs have the same first component and different second components.

Same first components

$(6, 1)$ $(6, 2)$

Different second components

Look at Figure 2.23 again. The fact that 1 and 2 in the domain have the same image, 6, in the range does not violate the definition of a function. **A function can have two different first components with the same second component.** By contrast, a relation is not a function when two different ordered pairs have the same first component and different second components. Thus, the relation in Example 2(b) is not a function.

Check Point 2 Determine whether each relation is a function:
a. $\{(1, 2), (3, 4), (5, 6), (5, 8)\}$
b. $\{(1, 2), (3, 4), (6, 5), (8, 5)\}$.

Functions as Equations

Functions are usually given in terms of equations rather than as sets of ordered pairs. For example, here is an equation that models paid vacation days each year as a function of years working for a company:

$$y = -0.016x^2 + 0.93x + 8.5.$$

The variable x represents years working for a company. The variable y represents the average number of vacation days each year. The variable y is a function of the variable x. For each value of x, there is one and only one value of y. The variable x is called the **independent variable** because it can be assigned any value from the domain. Thus, x can be assigned any positive integer representing the number of years working for a company. The variable y is called the **dependent variable** because its value depends on x. Paid vacation days depend on years working for a company. The value of the dependent variable, y, is calculated after selecting a value for the independent variable, x.

We have seen that not every set of ordered pairs defines a function. Similarly, not all equations with the variables x and y define a function. If an equation is solved for y and more than one value of y can be obtained for a given x, then the equation does not define y as a function of x.

EXAMPLE 3   **Determining Whether an Equation Represents a Function**

Determine whether each equation defines y as a function of x:

a. $x^2 + y = 4$ **b.** $x^2 + y^2 = 4$.

Solution   Solve each equation for y in terms of x. If two or more values of y can be obtained for a given x, the equation is not a function.

a.

$x^2 + y = 4$	This is the given equation.
$x^2 + y - x^2 = 4 - x^2$	Solve for y by subtracting x² from both sides.
$y = 4 - x^2$	Simplify.

From this last equation we can see that for each value of x, there is one and only one value of y. For example, if $x = 1$, then $y = 4 - 1^2 = 3$. The equation defines y as a function of x.

b.

$x^2 + y^2 = 4$	This given equation describes a circle.
$x^2 + y^2 - x^2 = 4 - x^2$	Isolate y² by subtracting x² from both sides.
$y^2 = 4 - x^2$	Simplify.
$y = \pm\sqrt{4 - x^2}$	Apply the square root method.

The $\pm$ in this last equation shows that for certain values of x (all values between -2 and 2), there are two values of y. For example, if $x = 1$, then $y = \pm\sqrt{4 - 1^2} = \pm\sqrt{3}$. For this reason, the equation does not define y as a function of x.

Check Point 3   Solve each equation for y and then determine whether the equation defines y as a function of x:
a. $2x + y = 6$ **b.** $x^2 + y^2 = 1$.

4   Evaluate a function.

Function Notation

When an equation represents a function, the function is often named by a letter such as $f, g, h, F, G,$ or H. Any letter can be used to name a function. Suppose that f names a function. Think of the domain as the set of the function's inputs and the range as the set of the function's outputs. As shown in Figure 2.24, the input is represented by x and the output by $f(x)$. The special notation $f(x)$, read "f of x" or "f at x," represents the **value of the function at the number x.**

Study Tip

The notation $f(x)$ does *not* mean "f times x." The notation describes the value of the function at x.

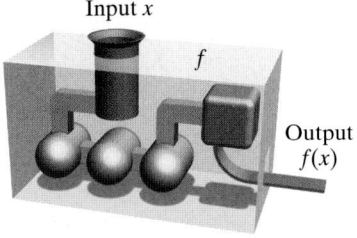

Figure 2.24 A function as a machine with inputs and outputs

Let's make this clearer by considering a specific example. We know that the equation

$$y = -0.016x^2 + 0.93x + 8.5$$

defines y as a function of x. We'll name the function f. Now, we can apply our new function notation.

Input	Output	Equation	We read this equation as "f of x equals $-0.016x^2 + 0.93x + 8.5$."
x	$f(x)$	$f(x) = -0.016x^2 + 0.93x + 8.5$	

Technology

Graphing utilities can be used to evaluate functions. The screens below show the evaluation of

$$f(x) = -0.016x^2 + 0.93x + 8.5$$

at 10 on a T1-83 graphing calculator. The function f is named Y_1.

```
Plot1  Plot2  Plot3
\Y1 -.016X²+0.93
X+8.5
\Y2=
\Y3=
\Y4=
\Y5=
\Y6=
```

```
Y1(10)
            16.2
```

Suppose we are interested in finding $f(10)$, the function's output when the input is 10. To find the value of the function at 10, we substitute 10 for x. We are **evaluating the function** at 10.

$f(x) = -0.016x^2 + 0.93x + 8.5$ This is the given function.

$f(10) = -0.016(10)^2 + 0.93(10) + 8.5$ Replace each occurrence of x with 10.

$= -0.016(100) + 0.93(10) + 8.5$ Evaluate the exponential expression: $10^2 = 100$.

$= -1.6 + 9.3 + 8.5$ Perform the multiplications.

$= 16.2$ Add from left to right.

The statement $f(10) = 16.2$, read "f of 10 equals 16.2," tells us that the value of the function at 10 is 16.2. When the function's input is 10, its output is 16.2 (After 10 years, workers average 16.2 vacation days each year.) To find other function values, such as $f(15)$, $f(20)$, or $f(23)$, substitute the specified input values for x into the function's equation.

If a function is named f and x represents the independent variable, the notation $f(x)$ corresponds to the y-value for a given x. Thus,

$$f(x) = -0.016x^2 + 0.93x + 8.5 \quad \text{and} \quad y = -0.016x^2 + 0.93x + 8.5$$

define the same function. This function may be written as

$$y = f(x) = -0.016x^2 + 0.93x + 8.5.$$

EXAMPLE 4   Evaluating a Function

If $f(x) = x^2 + 3x + 5$, evaluate:

   a. $f(2)$ **b.** $f(x + 3)$ **c.** $f(-x)$.

Solution   We substitute 2, $x + 3$, and $-x$ for x in the definition of f. When replacing x with a variable or an algebraic expression, you might find it helpful to think of the function's equation as

$$f(\,x\,) = \,x^2 + 3\,x\, + 5.$$

   a. We find $f(2)$ by substituting 2 for x in the equation.

$$f(\,2\,) = \,2^2 + 3 \cdot 2\, + 5 = 4 + 6 + 5 = 15$$

   Thus, $f(2) = 15$.

   b. We find $f(x + 3)$ by substituting $x + 3$ for x in the equation.

$$f(\,x + 3\,) = \,(x + 3)^2 + 3\,(x + 3)\, + 5$$

Equivalently,

$$f(x + 3) = (x + 3)^2 + 3(x + 3) + 5$$

$$= x^2 + 6x + 9 + 3x + 9 + 5 \quad \text{Square } x + 3 \text{ using}$$

$(A + B)^2 = A^2 + 2AB + B^2.$
Distribute 3 throughout the parentheses.

$$= x^2 + 9x + 23. \qquad \text{Combine like terms.}$$

Discovery

Using $f(x) = x^2 + 3x + 5$ and the answers in parts (b) and (c):

1. Is $f(x + 3)$ equal to $f(x) + f(3)$?
2. Is $f(-x)$ equal to $-f(x)$?

c. We find $f(-x)$ by substituting $-x$ for x in the equation.

$$f(-x) = (-x)^2 + 3(-x) + 5$$

Equivalently,

$$f(-x) = (-x)^2 + 3(-x) + 5$$

$$= x^2 - 3x + 5.$$

Check Point 4   If $f(x) = x^2 - 2x + 7$, evaluate:

a. $f(-5)$ **b.** $f(x + 4)$ **c.** $f(-x)$.

5 Find and simplify a function's difference quotient.

Functions and Difference Quotients

We have seen how slope can be interpreted as a rate of change. In the next section, we will be studying the average rate of change of a function. A ratio, called the *difference quotient*, plays an important role in understanding the rate at which functions change.

Definition of a Difference Quotient

The expression

$$\frac{f(x + h) - f(x)}{h}$$

for $h \neq 0$ is called the **difference quotient**.

EXAMPLE 5   Evaluating and Simplifying a Difference Quotient

If $f(x) = x^2 + 3x + 5$, find and simplify:

a. $f(x + h)$ **b.** $\dfrac{f(x + h) - f(x)}{h}, h \neq 0.$

Solution

a. We find $f(x + h)$ by replacing x with $x + h$ each time that x appears in the equation.

$$f(x) \quad = \quad x^2 \quad + \quad 3x \quad + \quad 5$$

Replace x with $x + h$.   Replace x with $x + h$.   Replace x with $x + h$.   Copy the 5. There is no x in this term.

$$f(x + h) \quad = \quad (x + h)^2 \quad + \quad 3(x + h) \quad + \quad 5$$

$$= x^2 + 2xh + h^2 + 3x + 3h \quad + \quad 5$$

b. Using our result from part (a), we obtain the following:

> This is f(x + h) from part (a).

> This is f(x) from the given equation.

$$\frac{f(x+h) - f(x)}{h} = \frac{x^2 + 2xh + h^2 + 3x + 3h + 5 - (x^2 + 3x + 5)}{h}$$

$$= \frac{x^2 + 2xh + h^2 + 3x + 3h + 5 - x^2 - 3x - 5}{h}$$

Remove parentheses and change the sign of each term in the parentheses.

$$= \frac{(x^2 - x^2) + (3x - 3x) + (5 - 5) + 2xh + h^2 + 3h}{h}$$

Group like terms.

$$= \frac{2xh + h^2 + 3h}{h}$$

Simplify.

$$= \frac{h(2x + h + 3)}{h}$$

Factor h from the numerator.

$$= 2x + h + 3, h \neq 0.$$

Cancel identical factors of h in the numerator and denominator.

Check Point 5 If $f(x) = x^2 - 7x + 3$, find and simplify:

a. $f(x + h)$ **b.** $\dfrac{f(x+h) - f(x)}{h}, h \neq 0.$

⑥ Understand and use piecewise functions.

Piecewise Functions

The early part of the twentieth century was the golden age of immigration in America. More than 13 million people migrated to the United States between 1900 and 1914. By 1910, foreign-born residents accounted for 15% of the total U.S. population. The graph in Figure 2.25 shows the percentage of Americans who were foreign born throughout the twentieth century.

We can model the data from 1910 through 2000 with two equations, one from 1910 through 1970, years in which the percentage was decreasing, and one from 1970 through 2000, years in which the percentage was increasing. These two trends can be approximated by the function

$$P(t) = \begin{cases} -\dfrac{11}{60}t + 15 & \text{if } 0 \leq t < 60 \\[2mm] \dfrac{1}{5}t - 8 & \text{if } 60 \leq t \leq 90 \end{cases}$$

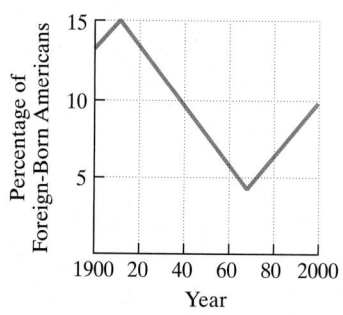

Percentage of Americans Who Were Foreign Born in the Twentieth Century

Percentage of Foreign-Born Americans

Figure 2.25 *Source*: U.S. Census Bureau

in which t represents the number of years after 1910 and $P(t)$ is the percentage of foreign-born Americans. A function that is defined by two (or more) equations over a specified domain is called a **piecewise function.**

EXAMPLE 6   Evaluating a Piecewise Function

Use the function $P(t)$, described previously, to find and interpret:

   a. $P(30)$ **b.** $P(80)$.

Solution

   a. To find $P(30)$, we let $t = 30$. Because 30 is less than 60, we use the first line of the piecewise function.

$$P(t) = -\tfrac{11}{60}t + 15 \qquad \text{This is the function's equation for } 0 \le t < 60.$$

$$P(30) = -\tfrac{11}{60} \cdot 30 + 15 \qquad \text{Replace } t \text{ with } 60.$$

$$= 9.5$$

   This means that 30 years after 1910, in 1940, 9.5% of Americans were foreign born.

   b. To find $P(80)$, we let $t = 80$. Because 80 is between 60 and 90, we use the second line of the piecewise function.

$$P(t) = \frac{1}{5}t - 8 \qquad \text{This is the function's equation for } 60 \le t \le 90.$$

$$P(80) = \tfrac{1}{5} \cdot 80 - 8 \qquad \text{Replace } t \text{ with } 80.$$

$$= 8$$

   This means that 80 years after 1910, in 1990, 8% of Americans were foreign born.

Check Point 6   If $f(x) = \begin{cases} x^2 + 3 & \text{if } x < 0 \\ 5x + 3 & \text{if } x \ge 0 \end{cases}$, find:

   a. $f(-5)$ **b.** $f(6)$.

7 Find the domain of a function.

The Domain of a Function

Let's reconsider the function that models the percentage of foreign-born Americans t years after 1910, up through and including 2000. The domain of this function is

$$\{0, \quad 1, \quad 2, \quad 3, \quad \dots, \quad 90\}.$$

| 0 years after 1910 is 1910. | 3 years after 1910 is 1913. | 90 years after 1910 brings the domain up to the year 2000. |

Functions that model data often have their domains explicitly given along with the function's equation. However, for most functions, only an equation is given, and the domain is not specified. In cases like this, the domain of f is the largest set of real numbers for which the value of $f(x)$ is a real number. For example, consider the function

$$f(x) = \frac{1}{x - 3}.$$

Because division by 0 is undefined (and not a real number), the denominator $x - 3$ cannot be 0. Thus, x cannot equal 3. The domain of the function consists of all real numbers other than 3, represented by $\{x \mid x \ne 3\}$. We say that f is not defined at 3, or $f(3)$ does not exist.

Just as the domain of a function must exclude real numbers that cause division by zero, it must also exclude real numbers that result in an even root of a negative number. For example, consider the function

$$g(x) = \sqrt{x}.$$

The equation tells us to take the square root of x. Because only nonnegative numbers have real square roots, the expression under the radical sign, x, must be greater than or equal to 0. The domain of g is $\{x \mid x \geq 0\}$, or the interval $[0, \infty)$.

Finding a Function's Domain

If a function f does not model data or verbal conditions, its domain is the largest set of real numbers for which the value of $f(x)$ is a real number. Exclude from a function's domain real numbers that cause division by zero and real numbers that result in an even root of a negative number.

EXAMPLE 7   Finding the Domain of a Function

Find the domain of each function:

a. $f(x) = x^2 - 7x$ **b.** $g(x) = \dfrac{6x}{x^2 - 9}$ **c.** $h(x) = \sqrt{3x + 12}$.

Solution

a. The function $f(x) = x^2 - 7x$ contains neither division nor an even root. The domain of f is the set of all real numbers.

b. The function $g(x) = \dfrac{6x}{x^2 - 9}$ contains division. Because division by 0 is undefined, we must exclude from the domain values of x that cause $x^2 - 9$ to be 0. Thus, x cannot equal -3 or 3. The domain of g is $\{x \mid x \neq -3, x \neq 3\}$.

c. The function $h(x) = \sqrt{3x + 12}$ contains an even root. Because only nonnegative numbers have real square roots, the quantity under the radical sign, $3x + 12$, must be greater than or equal to 0.

$$3x + 12 \geq 0$$

$$3x \geq -12$$

$$x \geq -4$$

The domain of h is $\{x \mid x \geq -4\}$, or the interval $[-4, \infty)$.

Technology

You can graph a function and often get hints about its domain. For example, $h(x) = \sqrt{3x + 12}$, or $y = \sqrt{3x + 12}$, appears only for $x \geq -4$, verifying $[-4, \infty)$ as the domain.

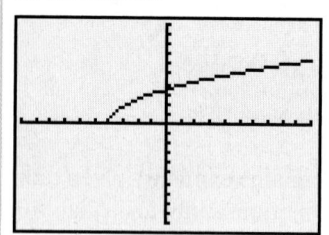

Check Point 7   Find the domain of each function:

a. $f(x) = x^2 + 3x - 17$ **b.** $g(x) = \dfrac{5x}{x^2 - 49}$

c. $h(x) = \sqrt{9x - 27}$.

EXERCISE SET 2.3

Practice Exercises

In Exercises 1–8, determine whether each relation is a function. Give the domain and range for each relation.

1. $\{(1, 2), (3, 4), (5, 5)\}$　　**2.** $\{(4, 5), (6, 7), (8, 8)\}$

3. $\{(3, 4), (3, 5), (4, 4), (4, 5)\}$

4. $\{(5, 6), (5, 7) (6, 6), (6, 7)\}$

5. $\{(-3, -3), (-2, -2), (-1, -1), (0, 0)\}$

6. $\{(-7, -7), (-5, -5), (-3, -3), (0, 0)\}$

7. $\{(1, 4), (1, 5), (1, 6)\}$　　**8.** $\{(4, 1), (5, 1), (6, 1)\}$

In Exercises 9–20, determine whether each equation defines y as a function of x.

9. $x + y = 16$ 　　　　**10.** $x + y = 25$

11. $x^2 + y = 16$ 　　　**12.** $x^2 + y = 25$

13. $x^2 + y^2 = 16$ 　　　**14.** $x^2 + y^2 = 25$

15. $x = y^2$ 　　　　　**16.** $4x = y^2$

17. $y = \sqrt{x + 4}$ 　　　**18.** $y = -\sqrt{x + 4}$

19. $x + y^3 = 8$ 　　　**20.** $x + y^3 = 27$

In Exercises 21–32, evaluate each function at the given values of the independent variable and simplify.

21. $f(x) = 4x + 5$
　　a. $f(6)$ 　　**b.** $f(x + 1)$ 　　**c.** $f(-x)$

22. $f(x) = 3x + 7$
　　a. $f(4)$ 　　**b.** $f(x + 1)$ 　　**c.** $f(-x)$

23. $g(x) = x^2 + 2x + 3$
　　a. $g(-1)$ 　　**b.** $g(x + 5)$ 　　**c.** $g(-x)$

24. $g(x) = x^2 - 10x - 3$
　　a. $g(-1)$ 　　**b.** $g(x + 2)$ 　　**c.** $g(-x)$

25. $h(x) = x^4 - x^2 + 1$
　　a. $h(2)$ 　　**b.** $h(-1)$
　　c. $h(-x)$ 　　**d.** $h(3a)$

26. $h(x) = x^3 - x + 1$
　　a. $h(3)$ 　　**b.** $h(-2)$
　　c. $h(-x)$ 　　**d.** $h(3a)$

27. $f(r) = \sqrt{r + 6} + 3$
　　a. $f(-6)$ 　　**b.** $f(10)$ 　　**c.** $f(x - 6)$

28. $f(r) = \sqrt{25 - r} - 6$
　　a. $f(16)$ 　　**b.** $f(-24)$ 　　**c.** $f(25 - 2x)$

29. $f(x) = \dfrac{4x^2 - 1}{x^2}$
　　a. $f(2)$ 　　**b.** $f(-2)$ 　　**c.** $f(-x)$

30. $f(x) = \dfrac{4x^3 + 1}{x^3}$
　　a. $f(2)$ 　　**b.** $f(-2)$ 　　**c.** $f(-x)$

31. $f(x) = \dfrac{x}{|x|}$
　　a. $f(6)$ 　　**b.** $f(-6)$ 　　**c.** $f(r^2)$

32. $f(x) = \dfrac{|x + 3|}{x + 3}$
　　a. $f(5)$ 　　**b.** $f(-5)$ 　　**c.** $f(-9 - x)$

In Exercises 33–44, find and simplify the difference quotient

$$\frac{f(x + h) - f(x)}{h}, \quad h \neq 0$$

for the given function.

33. $f(x) = 4x$ 　　　　**34.** $f(x) = 7x$

35. $f(x) = 3x + 7$ 　　**36.** $f(x) = 6x + 1$

37. $f(x) = x^2$ 　　　　**38.** $f(x) = 2x^2$

39. $f(x) = x^2 - 4x + 3$ 　**40.** $f(x) = x^2 - 5x + 8$

41. $f(x) = 6$ 　　　　**42.** $f(x) = 7$

43. $f(x) = \dfrac{1}{x}$ 　　　**44.** $f(x) = \dfrac{1}{2x}$

In Exercises 45–50, evaluate each piecewise function at the given values of the independent variable.

45. $f(x) = \begin{cases} 3x + 5 & \text{if } x < 0 \\ 4x + 7 & \text{if } x \geq 0 \end{cases}$
　　a. $f(-2)$ 　　**b.** $f(0)$ 　　**c.** $f(3)$

46. $f(x) = \begin{cases} 6x - 1 & \text{if } x < 0 \\ 7x + 3 & \text{if } x \geq 0 \end{cases}$
　　a. $f(-3)$ 　　**b.** $f(0)$ 　　**c.** $f(4)$

47. $g(x) = \begin{cases} x + 3 & \text{if } x \geq -3 \\ -(x + 3) & \text{if } x < -3 \end{cases}$
　　a. $g(0)$ 　　**b.** $g(-6)$ 　　**c.** $g(-3)$

48. $g(x) = \begin{cases} x + 5 & \text{if } x \geq -5 \\ -(x + 5) & \text{if } x < -5 \end{cases}$
　　a. $g(0)$ 　　**b.** $g(-6)$ 　　**c.** $g(-5)$

49. $h(x) = \begin{cases} \dfrac{x^2 - 9}{x - 3} & \text{if } x \neq 3 \\ 6 & \text{if } x = 3 \end{cases}$
　　a. $h(5)$ 　　**b.** $h(0)$ 　　**c.** $h(3)$

50. $h(x) = \begin{cases} \dfrac{x^2 - 25}{x - 5} & \text{if } x \neq 5 \\ 10 & \text{if } x = 5 \end{cases}$
　　a. $h(7)$ 　　**b.** $h(0)$ 　　**c.** $h(5)$

In Exercises 51–74, find the domain of each function.

51. $f(x) = 4x^2 - 3x + 1$

52. $f(x) = 8x^2 - 5x + 2$

53. $g(x) = \dfrac{3}{x - 4}$

54. $g(x) = \dfrac{2}{x + 5}$

55. $h(x) = \dfrac{7x}{x^2 - 16}$

56. $h(x) = \dfrac{12x}{x^2 - 36}$

57. $f(x) = \dfrac{2}{(x + 3)(x - 7)}$

58. $f(x) = \dfrac{15}{(x + 8)(x - 3)}$

59. $H(r) = \dfrac{4}{r^2 + 11r + 24}$

60. $H(r) = \dfrac{5}{6r^2 + r - 2}$

61. $f(t) = \dfrac{3}{t^2 + 4}$

62. $f(t) = \dfrac{5}{t^2 + 9}$

63. $f(x) = \sqrt{x - 3}$

64. $f(x) = \sqrt{x + 2}$

65. $f(x) = \dfrac{1}{\sqrt{x - 3}}$

66. $f(x) = \dfrac{1}{\sqrt{x + 2}}$

67. $g(x) = \sqrt{5x + 35}$

68. $g(x) = \sqrt{7x - 70}$

69. $f(x) = \sqrt{24 - 2x}$

70. $f(x) = \sqrt{84 - 6x}$

71. $f(x) = \sqrt{x^2 - 5x - 14}$

72. $f(x) = \sqrt{x^2 - 5x - 24}$

73. $f(x) = \dfrac{\sqrt{x - 2}}{x - 5}$

74. $f(x) = \dfrac{\sqrt{x - 3}}{x - 6}$

 Application Exercises

75. The bar graph shows the percentage of people in the United States using the Internet by education level. Write five ordered pairs for

(education level, percentage using Internet)

as a relation. Find the domain and the range of the relation. Is this relation a function? Explain your answer.

Internet Use in the U.S. by Education Level

Education Level
(1 = less than high school, 2 = high school graduate, 3 = some college, 4 = college graduate, 5 = advanced degree)

Source: U.C.L.A. Center for Communication Policy

The table shows the ten longest-running television shows of the twentieth century. Use the information in the table to solve Exercises 76–78.

Ten Longest-Running National Network TV Series of the Twentieth Century

Program	Number of Seasons the Show Ran
"Walt Disney"	33
"60 Minutes"	33
"The Ed Sullivan Show"	24
"Gunsmoke"	20
"The Red Skelton Show"	20
"Meet the Press"	18
"What's My Line?"	18
"I've Got a Secret"	17
"Lassie"	17
"The Lawrence Welk Show"	17

Source: Nielsen Media Research

76. Consider the relation for which the domain represents the ten longest-running series and the range represents the number of seasons the series ran. Is this relation a function? Explain your answer.

77. Consider the relation for which the domain represents the number of seasons the ten longest-running series ran and the range represents the ten longest-running series. Is this relation a function? Explain your answer.

78. Use your answers from Exercises 76 and 77 to answer the following question: If the components in a function's ordered pairs are reversed, must the resulting relation also be a function?

79. The function

$$P(x) = 0.72x^2 + 9.4x + 783$$

models the gray wolf population in the United States, $P(x)$, x years after 1960. Find and interpret $P(30)$. How well does the function model the actual value shown in the bar graph?

Gray Wolf Population (to the Nearest Hundred)

Year

Source: U.S. Department of the Interior

80. As the use of the Internet increases, so has the number of computer infections from viruses. The function

$$N(x) = 0.2x^2 - 1.2x + 2$$

models the number of infections per month for every 1000 computers, $N(x)$, x years after 1990. Find and interpret $N(10)$. How well does the function model the actual value shown in the bar graph?

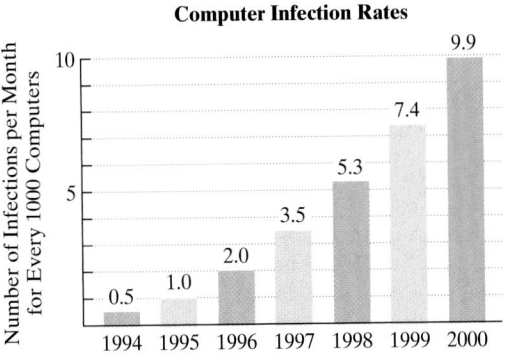

Computer Infection Rates

Source: Jupiter Communications

The function

$$P(t) = 6.85\sqrt{t} + 19$$

models the percentage of U.S. households online, $P(t)$, t years after 1997. In Exercises 81–84, use this function to find and interpret the given expression. If necessary, use a calculator and round to the nearest whole percent. How well does the function model the actual data shown in the bar graph?

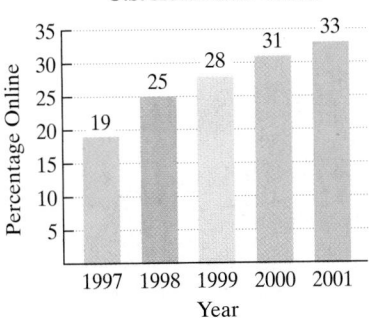

Percentage of U.S. Households Online

Source: Forrester Research

81. $P(0)$ **82.** $P(4)$

83. $P(3) - P(1)$ **84.** $P(2) - P(1)$

The number of lawyers in the United States can be modeled by the function

$$f(x) = \begin{cases} 6.5x + 200 & \text{if } 0 \le x < 23 \\ 26.2x - 252 & \text{if } x \ge 23 \end{cases}$$

where x represents the number of years after 1951 and $f(x)$ represents the number of lawyers, in thousands. In Exercises 85–88, use this function to find and interpret each of the following.

85. $f(0)$ **86.** $f(10)$
87. $f(50)$ **88.** $f(60)$

During a particular year, the taxes owed, $T(x)$, in dollars, filing separately with an adjusted gross income of x dollars is given by the piecewise function

$$T(x) = \begin{cases} 0.15x & \text{if } 0 \le x < 17{,}900 \\ 0.28(x - 17{,}900) + 2685 & \text{if } 17{,}900 \le x < 43{,}250 \\ 0.31(x - 43{,}250) + 9783 & \text{if } x \ge 43{,}250 \end{cases}$$

In Exercises 89–90, use this function to find and interpret each of the following.

89. $T(40{,}000)$ **90.** $T(70{,}000)$

In Exercises 91–94, you will be developing functions that model given conditions.

91. A company that manufactures bicycles has a fixed cost of $100,000. It costs $100 to produce each bicycle. The total cost for the company is the sum of its fixed cost and variable costs. Write the total cost, C, as a function of the number of bicycles produced. Then find and interpret $C(90)$.

92. A car was purchased for $22,500. The value of the car decreases by $3200 per year for the first six years. Write a function that describes the value of the car, V, after x years, where $0 \le x \le 7$. Then find and interpret $V(3)$.

93. You commute to work a distance of 40 miles and return on the same route at the end of the day. Your average rate on the return trip is 30 miles per hour faster than your average rate on the outgoing trip. Write the total time, T, in hours, devoted to your outgoing and return trips as a function of your rate on the outgoing trip. Then find and interpret $T(30)$. Hint:

$$\text{Time traveled} = \frac{\text{Distance traveled}}{\text{Rate of travel}}.$$

94. A chemist working on a flu vaccine needs to mix a 10% sodium-iodine solution with a 60% sodium-iodine solution to obtain a 50-milliliter mixture. Write the amount of sodium iodine in the mixture, S, in milliliters, as a function of the number of milliliters of the 10% solution used. Then find and interpret $S(30)$.

Writing in Mathematics

95. If a relation is represented by a set of ordered pairs, explain how to determine whether the relation is a function.

96. How do you determine if an equation in x and y defines y as a function of x?

97. A student in introductory algebra hears that functions are studied in subsequent algebra courses. The student asks you what a function is. Provide the student with a clear, relatively concise response.

98. Describe one advantage of using $f(x)$ rather than y in a function's equation.

99. Explain how to find the difference quotient,

$$\frac{f(x + h) - f(x)}{h}, \text{ if a function's equation is given.}$$

100. What is a piecewise function?

101. How is the domain of a function determined?

102. For people filing a single return, federal income tax is a function of adjusted gross income because for each value of adjusted gross income there is a specific tax to be paid. On the other hand, the price of a house is not a function of the lot size on which the house sits because houses on same-sized lots can sell for many different prices.

    a. Describe an everyday situation between variables that is a function.

    b. Describe an everyday situation between variables that is not a function.

Technology Exercises

Use a graphing utility to find the domain of each function in Exercises 103–105. Then verify your observation algebraically.

103. $f(x) = \sqrt{x - 1}$ **104.** $g(x) = \sqrt{2x + 6}$

105. $h(x) = \sqrt{15 - 3x}$

Critical Thinking Exercises

106. Write a function defined by an equation in x whose domain is $\{x \mid x \neq -4, x \neq 11\}$.

107. Write a function defined by an equation in x whose domain is $[-6, \infty)$.

108. Give an example of an equation that does not define y as a function of x but that does define x as a function of y.

109. If $f(x) = ax^2 + bx + c$ and $r_1 = \dfrac{-b + \sqrt{b^2 - 4ac}}{2a}$, find $f(r_1)$ without doing any algebra and explain how you arrived at your result.

Group Exercise

110. Almanacs, newspapers, magazines, and the Internet contain bar graphs and line graphs that describe how things are changing over time. For example, the graphs in Exercises 79–82 show how various phenomena are changing over time. Find a bar or line graph showing yearly changes that you find intriguing. Describe to the group what interests you about this data. The group should select their two favorite graphs. For each graph selected:

    a. Rewrite the data so that they are presented as a relation in the form of a set of ordered pairs.

    b. Determine whether the relation in part (a) is a function. Explain why the relation is a function, or why it is not.

SECTION 2.4   *Graphs of Functions*

Objectives

1. Graph functions by plotting points.
2. Obtain information about a function from its graph.
3. Use the vertical line test to identify functions.
4. Identify intervals on which a function increases, decreases, or is constant.
5. Use graphs to locate relative maxima or minima.
6. Find a function's average rate of change.
7. Identify even or odd functions and recognize their symmetries.
8. Graph step functions.

Have you ever seen a gas-guzzling car from the 1950s, with its huge fins and overstated design? The worst year for automobile fuel efficiency was 1958, when cars averaged a dismal 12.4 miles per gallon. The function

$$f(x) = 0.0075x^2 - 0.2672x + 14.8$$

models the average number of miles per gallon for U.S. automobiles, $f(x)$, x years after 1940. If we could see the graph of the function's equation, we would get a much better idea of the relationship between time and fuel efficiency. In this section, we will learn how to use the graph of a function to obtain useful information about the function.

Graphs of Functions

A graph enables us to visualize a function's behavior. The graph shows the relationship between the function's two variables more clearly than the function's equation does. The **graph of a function** is the graph of its ordered pairs. For example, the graph of $f(x) = \sqrt{x}$ is the set of points (x, y) in the rectangular coordinate system satisfying the equation $y = \sqrt{x}$. Thus, one way to graph a function is by plotting several of its ordered pairs and drawing a line or smooth curve through them. With the function's graph, we can picture its domain on the x-axis and its range on the y-axis. Our first example illustrates how this is done.

1 Graph functions by plotting points.

EXAMPLE 1   Graphing a Function by Plotting Points

Graph $f(x) = x^2 + 1$. To do so, use integer values of x from the set $\{-3, -2, -1, 0, 1, 2, 3\}$ to obtain seven ordered pairs. Plot each ordered pair and draw a smooth curve through the points. Use the graph to specify the function's domain and range.

Solution   The graph of $f(x) = x^2 + 1$ is, by definition, the graph of $y = x^2 + 1$. We begin by setting up a partial table of coordinates.

x	$f(x) = x^2 + 1$	(x, y) or $(x, f(x))$
-3	$f(-3) = (-3)^2 + 1 = 10$	$(-3, 10)$
-2	$f(-2) = (-2)^2 + 1 = 5$	$(-2, 5)$
-1	$f(-1) = (-1)^2 + 1 = 2$	$(-1, 2)$
0	$f(0) = 0^2 + 1 = 1$	$(0, 1)$
1	$f(1) = 1^2 + 1 = 2$	$(1, 2)$
2	$f(2) = 2^2 + 1 = 5$	$(2, 5)$
3	$f(3) = 3^2 + 1 = 10$	$(3, 10)$

Now, we plot the seven points and draw a smooth curve through them, as shown in Figure 2.26. The graph of f has a cuplike shape. The points on the graph of f have x-coordinates that extend indefinitely to the left and to the right. Thus, the domain consists of all real numbers, represented by $(-\infty, \infty)$. By contrast, the points on the graph have y-coordinates that start at 1 and extend indefinitely upward. Thus, the range consists of all real numbers greater than or equal to 1, represented by $[1, \infty)$.

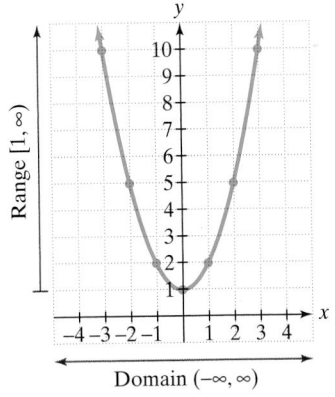

Figure 2.26 The graph of $f(x) = x^2 + 1$

Check Point 1   Graph $f(x) = x^2 - 2$, using integers from -3 to 3 for x in the partial table of coordinates. Use the graph to specify the function's domain and range.

Technology

Does your graphing utility have a TABLE feature? If so, you can use it to create tables of coordinates for a function. You will need to enter the equation of the function and specify the starting value for x, TblStart, and the increment between successive x-values, ΔTbl. For the table of coordinates in Example 1, we start the table at $x = -3$ and increment by 1. Using the up- or down-arrow keys, you can scroll through the table and determine as many ordered pairs of the graph as desired.

2 Obtain information about a function from its graph.

Obtaining Information from Graphs

You can obtain information about a function from its graph. At the right or left of a graph, you will find closed dots, open dots, or arrows.

- A closed dot indicates that the graph does not extend beyond this point and the point belongs to the graph.

- An open dot indicates that the graph does not extend beyond this point and the point does not belong to the graph.

- An arrow indicates that the graph extends indefinitely in the direction in which the arrow points.

EXAMPLE 2   Obtaining Information from a Function's Graph

Use the graph of the function f, shown in Figure 2.27, to answer the following questions:

a. What are the function values $f(-1)$ and $f(1)$?

b. What is the domain of f?

c. What is the range of f?

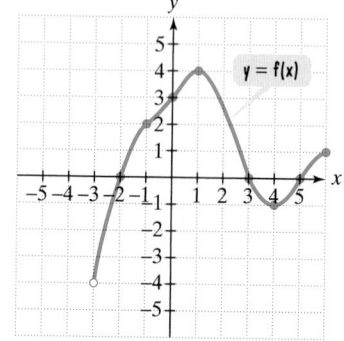

Figure 2.27

Solution

a. Because $(-1, 2)$ is a point on the graph of f, the y-coordinate, 2, is the value of the function at the x-coordinate, -1. Thus, $f(-1) = 2$. Similarly, because $(1, 4)$ is also a point on the graph of f, this indicates that $f(1) = 4$.

b. The open dot on the left shows that $x = -3$ is not in the domain of f. By contrast, the closed dot on the right shows that $x = 6$ is in the domain of f. We determine the domain of f by noticing that the points on the graph of f have x-coordinates between -3, excluding -3, and 6, including 6. For each number x between -3 and 6, there is a point $(x, f(x))$ on the graph. Thus, the domain of f is $\{x \mid -3 < x \le 6\}$, or the interval $(-3, 6]$.

c. The points on the graph all have y-coordinates between -4, not including -4, and 4, including 4. The graph does not extend below $y = -4$ or above $y = 4$. Thus, the range of f is $\{y \mid -4 < y \le 4\}$, or the interval $(-4, 4]$.

Check Point 2 Use the graph of function f, shown below, to find $f(4)$, the domain, and the range.

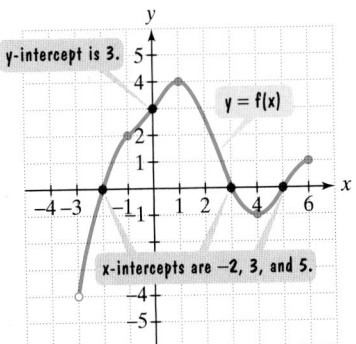

Figure 2.28 Identifying intercepts

3 Use the vertical line test to identify functions.

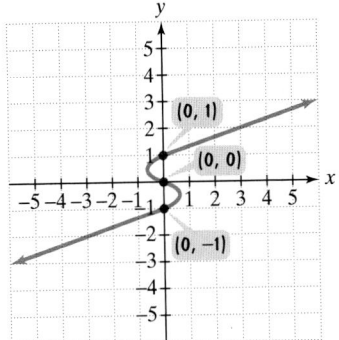

Figure 2.29 y is not a function of x because 0 is paired with three values of y, namely, 1, 0, and −1.

Figure 2.28 illustrates how we can identify a graph's intercepts. To find the x-intercepts, look for the points at which the graph crosses the x-axis. There are three such points: $(-2, 0)$, $(3, 0)$, and $(5, 0)$. Thus, the x-intercepts are $-2, 3$, and 5. We express this in function notation by writing $f(-2) = 0$, $f(3) = 0$, and $f(5) = 0$. We say that $-2, 3$, and 5 are the *zeros of the function*. The **zeros of a function**, f, are the x-values for which $f(x) = 0$.

To find the y-intercept, look for the point at which the graph crosses the y-axis. This occurs at $(0, 3)$. Thus, the y-intercept is 3. We express this in function notation by writing $f(0) = 3$.

By the definition of a function, for each value of x we can have at most one value for y. What does this mean in terms of intercepts? **A function can have more than one x-intercept but at most one y-intercept.**

The Vertical Line Test

Not every graph in the rectangular coordinate system is the graph of a function. The definition of a function specifies that no value of x can be paired with two or more different values of y. Consequently, if a graph contains two or more different points with the same first coordinate, the graph cannot represent a function. This is illustrated in Figure 2.29. Observe that points sharing a common first coordinate are vertically above or below each other.

This observation is the basis of a useful test for determining whether a graph defines y as a function of x. The test is called the **vertical line test.**

The Vertical Line Test for Functions

If any vertical line intersects a graph in more than one point, the graph does not define y as a function of x.

EXAMPLE 3   Using the Vertical Line Test

Use the vertical line test to identify graphs in which y is a function of x.

a.

b.

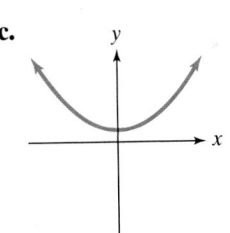
c.

d.

Solution y is a function of x for the graphs in **b** and **c.**

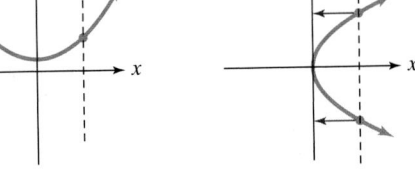

a.

y is **not a function of** x. Two values of y correspond to an x-value.

b.

y is **a function of** x.

c.

y is **a function of** x.

d.

y is **not a function of** x. Two values of y correspond to an x-value.

Check Point 3 Use the vertical line test to identify graphs in which y is a function of x.

a.

b.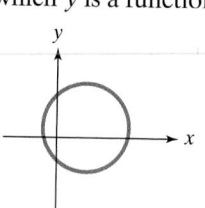

c.

Average Number of Paid Vacation Days for Full-Time Workers at Medium to Large U.S. Companies

$f(x) = -0.016x^2 + 0.93x + 8.5$

Paid Vacation Days / Years Working for Company

Figure 2.30 *Source*: Bureau of Labor Statistics

EXAMPLE 4 Analyzing the Graph of a Function

The function

$$f(x) = -0.016x^2 + 0.93x + 8.5$$

models the average number of paid vacation days each year, $f(x)$, for full-time workers at medium to large U.S. companies after x years. The graph of f is shown in Figure 2.30.

a. Explain why f represents the graph of a function.
b. Use the graph to find a reasonable estimate of $f(5)$.
c. For what value of x is $f(x) = 20$?
d. Describe the general trend shown by the graph.

Solution

a. No vertical line intersects the graph of f more than once. By the vertical line test, f represents the graph of a function.

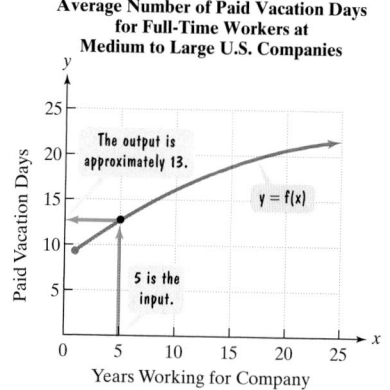

Average Number of Paid Vacation Days for Full-Time Workers at Medium to Large U.S. Companies

b. To find $f(5)$, or f of 5, we locate 5 on the x-axis. The figure shows the point on the graph of f for which 5 is the first coordinate. From this point, we look to the y-axis to find the corresponding y-coordinate. A reasonable estimate of the y-coordinate is 13. Thus, $f(5) \approx 13$. After 5 years, a worker can expect approximately 13 paid vacation days.

Average Number of Paid Vacation Days for Full-Time Workers at Medium to Large U.S. Companies

c. To find the value of x for which $f(x) = 20$, we locate 20 on the y-axis. The figure shows that there is one point on the graph of f for which 20 is the second coordinate. From this point, we look to the x-axis to find the corresponding x-coordinate. A reasonable estimate of the x-coordinate is 18. Thus, $f(x) = 20$ for $x \approx 18$. A worker with 20 paid vacation days has been with the company approximately 18 years.

d. The graph of f is rising from left to right. This shows that paid vacation days increase as time with the company increases. However, the rate of increase is slowing down as the graph moves to the right. This means that the increase in paid vacation days takes place more slowly the longer an employee is with the company.

Check Point 4

a. Use the graph of f in Figure 2.30 to find a reasonable estimate of $f(10)$.

b. For what value of x is $f(x) = 15$? Round to the nearest whole number.

Increasing and Decreasing Functions

Identify intervals on which a function increases, decreases, or is constant.

Too late for that flu shot now! It's only 8 A.M. and you're feeling lousy. Your temperature is 101°F. Fascinated by the way that algebra models the world (your author is projecting a bit here), you decide to construct graphs showing your body temperature as a function of the time of day. You decide to let x represent the number of hours after 8 A.M. and $f(x)$ your temperature at time x.

At 8 A.M. your temperature is 101°F and you are not feeling well. However, your temperature starts to decrease. It reaches normal (98.6°F) by 11 A.M. Feeling energized, you construct the graph shown on the right, indicating decreasing temperature for $\{x \mid 0 < x < 3\}$, or on the interval $(0, 3)$.

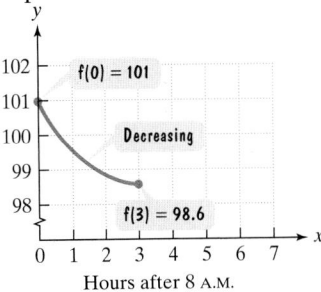

Temperature decreases on $(0, 3)$, reaching 98.6° by 11 A.M.

Did creating that first graph drain you of your energy? Your temperature starts to rise after 11 A.M. By 1 P.M., 5 hours after 8 A.M., your temperature reaches 100°F. However, you keep plotting points on your graph. At right, we can see that your temperature increases for $\{x \mid 3 < x < 5\}$, or on the interval $(3, 5)$.

The graph of f is decreasing to the left of $x = 3$ and increasing to the right of $x = 3$. Thus, your temperature 3 hours after 8 A.M. was at a relative minimum. Your relative minimum temperature was 98.6°.

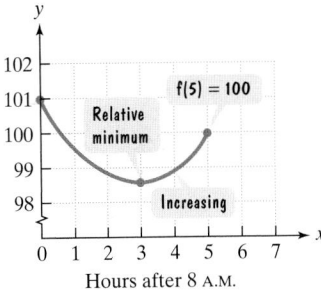

Temperature increases on $(3, 5)$.

By 3 P.M., your temperature is no worse than it was at 1 P.M.: It is still 100°F. (Of course, it's no better, either.) Your temperature remained the same, or constant, for $\{x \mid 5 < x < 7\}$, or on the interval $(5, 7)$.

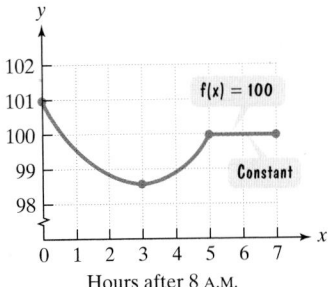

Temperature remains constant at 100° on $(5, 7)$.

The time-temperature flu scenario illustrates that a function f is increasing when its graph rises, decreasing when its graph falls, and remains constant when it neither rises nor falls. Let's now provide a more algebraic description for these intuitive concepts.

Increasing, Decreasing, and Constant Functions

1. A function is **increasing** on an open interval, I, if for any x_1 and x_2 in the interval, where $x_1 < x_2$, then $f(x_1) < f(x_2)$.
2. A function is **decreasing** on an open interval, I, if for any x_1 and x_2 in the interval, where $x_1 < x_2$, then $f(x_1) > f(x_2)$.
3. A function is **constant** on an open interval, I, if for any x_1 and x_2 in the interval, where $x_1 < x_2$, then $f(x_1) = f(x_2)$.

Increasing   **Decreasing**   **Constant**

(1) For $x_1 < x_2$ in I, $f(x_1) < f(x_2)$; f is increasing on I.

(2) For $x_1 < x_2$ in I, $f(x_1) > f(x_2)$; f is decreasing on I.

(3) For $x_1 < x_2$ in I, $f(x_1) = f(x_2)$; f is constant on I.

EXAMPLE 5 **Intervals on Which a Function Increases, Decreases, or Is Constant**

Give the intervals on which each function whose graph is shown is increasing, decreasing, or constant.

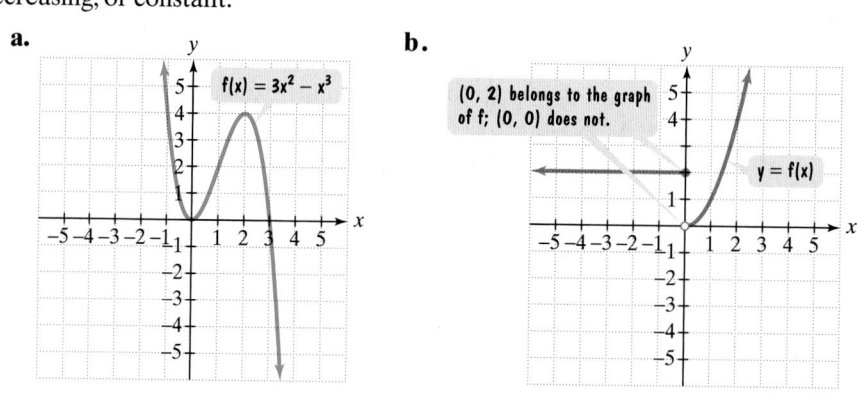

a. $f(x) = 3x^2 - x^3$

b. $(0, 2)$ belongs to the graph of f; $(0, 0)$ does not. $y = f(x)$

Solution

a. The function is decreasing on the interval $(-\infty, 0)$, increasing on the interval $(0, 2)$, and decreasing on the interval $(2, \infty)$.

b. Although the function's equations are not given, the graph indicates that the function is defined in two pieces. The part of the graph to the left of the y-axis shows that the function is constant on the interval $(-\infty, 0)$. The part to the right of the y-axis shows that the function is increasing on the interval $(0, \infty)$.

Check Point 5 Give the intervals on which the function whose graph is shown is increasing, decreasing, or constant.

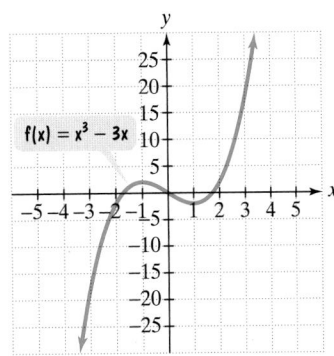

5 Use graphs to locate relative maxima or minima.

Relative Maxima and Relative Minima

The points at which a function changes its increasing or decreasing behavior can be used to find the *relative maximum* or *relative minimum* values of the function. For example, consider the function with which we opened this section:

$$f(x) = 0.0075x^2 - 0.2672x + 14.8.$$

Recall that the function models the average number of miles per gallon of U.S. automobiles, $f(x)$, x years after 1940. The graph of this function is shown as a continuous curve in Figure 2.31. (It can also be shown as a series of points, each point representing a year and miles per gallon for that year.)

The graph of f is decreasing to the left of $x = 18$ and increasing to the right of $x = 18$. Thus, 18 years after 1940, in 1958, fuel efficiency was at a minimum. We say that the relative minimum fuel efficiency is $f(18)$, or approximately 12.4 miles per gallon. Mathematicians use the word "relative" to suggest that relative to an open interval about 18, the value $f(18)$ is smallest.

Figure 2.31 Fuel efficiency of U.S. automobiles over time

Study Tip

The word *local* is sometimes used instead of *relative* when describing maxima or minima. If f has a relative, or local, maximum at a, $f(a)$ is greater than the values of f near a. If f has a relative, or local, minimum at b, $f(b)$ is less than the values of f near b.

Definitions of Relative Maximum and Relative Minimum

1. A function value $f(a)$ is a **relative maximum** of f if there exists an open interval about a such that $f(a) > f(x)$ for all x in the open interval.
2. A function value $f(b)$ is a **relative minimum** of f if there exists an open interval about b such that $f(b) < f(x)$ for all x in the open interval.

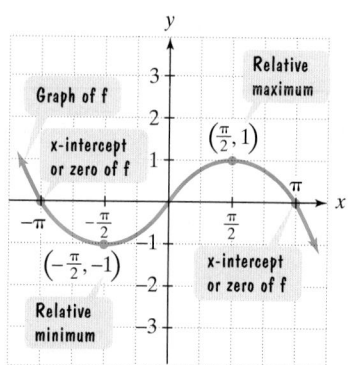

Figure 2.32 Using a graph to locate where f has a relative maximum or minimum

If the graph of a function is given, we can often visually locate the number(s) at which the function has a relative maximum or a relative minimum. For example, the graph of f in Figure 2.32 shows that:

• f has a relative maximum at $\dfrac{\pi}{2}$.

  The relative maximum is $f\left(\dfrac{\pi}{2}\right) = 1$.

• f has a relative minimum at $-\dfrac{\pi}{2}$.

  The relative minimum is $f\left(-\dfrac{\pi}{2}\right) = -1$.

Notice that f does not have a relative maximum or minimum at $-\pi$ and π, the x-intercepts, or zeros, of the function.

6 Find a function's average rate of change.

The Average Rate of Change of a Function

We have seen that the slope of a line can be interpreted as its rate of change. If the graph of a function is not a straight line, we speak of an **average rate of change** between any two points on its graph. To find the average rate of change, calculate the slope of the line containing the two points. This line is called a **secant line.**

The Average Rate of Change of a Function

Let $(x_1, f(x_1))$ and $(x_2, f(x_2))$ be distinct points on the graph of a function f. (See Figure 2.33.) The **average rate of change of f** from x_1 to x_2 is

$$\frac{f(x_2) - f(x_1)}{x_2 - x_1}.$$

Figure 2.33

EXAMPLE 6   Finding the Average Rate of Change

Find the average rate of change of $f(x) = x^2$ from:

a. $x_1 = 0$ to $x_2 = 1$ **b.** $x_1 = 1$ to $x_2 = 2$ **c.** $x_1 = -2$ to $x_2 = 0$.

Solution

a. The average rate of change of $f(x) = x^2$ from $x_1 = 0$ to $x_2 = 1$ is

$$\frac{f(x_2) - f(x_1)}{x_2 - x_1} = \frac{f(1) - f(0)}{1 - 0} = \frac{1^2 - 0^2}{1} = 1.$$

Figure 2.34(a) shows the secant line of $f(x) = x^2$ from $x_1 = 0$ to $x_2 = 1$. The average rate of change is positive, and the function is increasing on the interval $(0, 1)$.

b. The average rate of change of $f(x) = x^2$ from $x_1 = 1$ to $x_2 = 2$ is

$$\frac{f(x_2) - f(x_1)}{x_2 - x_1} = \frac{f(2) - f(1)}{2 - 1} = \frac{2^2 - 1^2}{1} = 3.$$

Figure 2.34(b) shows the secant line of $f(x) = x^2$ from $x_1 = 1$ to $x_2 = 2$. The average rate of change is positive, and the function is increasing on the interval $(1, 2)$. Can you see that the graph rises more steeply on the interval $(1, 2)$ than on $(0, 1)$? This is because the average rate of change from $x_1 = 1$ to $x_2 = 2$ is greater than the average rate of change from $x_1 = 0$ to $x_2 = 1$.

c. The average rate of change of $f(x) = x^2$ from $x_1 = -2$ to $x_2 = 0$ is

$$\frac{f(x_2) - f(x_1)}{x_2 - x_1} = \frac{f(0) - f(-2)}{0 - (-2)} = \frac{0^2 - (-2)^2}{2} = \frac{-4}{2} = -2.$$

Figure 2.34(c) shows the secant line of $f(x) = x^2$ from $x_1 = -2$ to $x_2 = 0$. The average rate of change is negative, and the function is decreasing on the interval $(-2, 0)$.

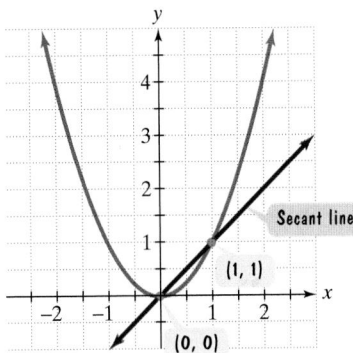

Figure 2.34(a) The secant line of $f(x) = x^2$ from $x_1 = 0$ to $x_2 = 1$

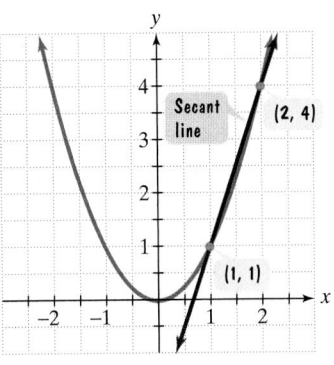

Figure 2.34(b) The secant line of $f(x) = x^2$ from $x_1 = 1$ to $x_2 = 2$

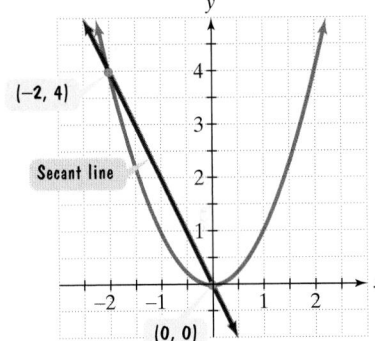

Figure 2.34(c) The secant line of $f(x) = x^2$ from $x_1 = -2$ to $x_2 = 0$

Check Point 6

Find the average rate of change of $f(x) = x^3$ from

a. $x_1 = 0$ to $x_2 = 1$

b. $x_1 = 1$ to $x_2 = 2$

c. $x_1 = -2$ to $x_2 = 0$.

Suppose we are interested in the average rate of change of f from $x_1 = x$ to $x_2 = x + h$. In this case, the average rate of change is

$$\frac{f(x_2) - f(x_1)}{x_2 - x_1} = \frac{f(x + h) - f(x)}{x + h - x} = \frac{f(x + h) - f(x)}{h}.$$

Do you recognize the last expression? It is the difference quotient that you used in the previous section to practice evaluating functions. Thus, the difference quotient gives the average rate of change of a function from x to $x + h$. In the difference quotient, h is thought of as a number very close to 0. In this way, the average rate of change can be found for a very short interval.

7 Identify even or odd functions and recognize their symmetries.

Even and Odd Functions and Symmetry

Is beauty in the eye of the beholder? Or are there certain objects (or people) that are so well balanced and proportioned that they are universally pleasing to the eye? What constitutes an attractive human face? In Figure 2.35, we've drawn lines between paired features and marked the midpoints. Notice how the features line up almost perfectly. Each half of the face is a mirror image of the other half through the white vertical line.

Did you know that graphs of some equations exhibit exactly the kind of symmetry shown by the attractive face in Figure 2.35? The word *symmetry* comes from the Greek *symmetria*, meaning "the same measure." We can identify graphs with symmetry by looking at a function's equation and determining if the function is *even* or *odd*.

Figure 2.35 To most people, an attractive face is one in which each half is an almost perfect mirror image of the other half

---

Definition of Even and Odd Functions

The function f is an **even function** if

$$f(-x) = f(x) \quad \text{for all } x \text{ in the domain of } f.$$

The right side of the equation of an even function does not change if x is replaced with $-x$.

The function f is an **odd function** if

$$f(-x) = -f(x) \quad \text{for all } x \text{ in the domain of } f.$$

Every term in the right side of the equation of an odd function changes sign if x is replaced with $-x$.

EXAMPLE 7   Identifying Even or Odd Functions

Identify each of the following functions as even, odd, or neither:

a. $f(x) = x^3$ **b.** $g(x) = x^4 - 2x^2$ **c.** $h(x) = x^2 + 2x + 1$.

Solution   In each case, replace x with $-x$ and simplify. If the right side of the equation stays the same, the function is even. If every term on the right changes sign, the function is odd.

a. We use the given function's equation, $f(x) = x^3$, to find $f(-x)$.

Use $f(x) = x^3$.

Replace x with −x.   Replace x with −x.

$$f(-x) = (-x)^3 = (-x)(-x)(-x) = -x^3$$

There is only one term in the equation $f(x) = x^3$, and the term changed signs when we replaced x with $-x$. Because $f(-x) = -f(x)$, f is an odd function.

b. We use the given function's equation, $g(x) = x^4 - 2x^2$, to find $g(-x)$.

Use $g(x) = x^4 - 2x^2$.

> Replace x with −x.

$$g(-x) = (-x)^4 - 2(-x)^2 = (-x)(-x)(-x)(-x) - 2(-x)(-x)$$
$$= x^4 - 2x^2$$

The right side of the equation of the given function, $g(x) = x^4 - 2x^2$, did not change when we replaced x with $-x$. Because $g(-x) = g(x)$, g is an even function.

c. We use the given function's equation, $h(x) = x^2 + 2x + 1$, to find $h(-x)$.

Use $h(x) = x^2 + 2x + 1$.

> Replace x with −x.

$$h(-x) = (-x)^2 + 2(-x) + 1 = x^2 - 2x + 1$$

The right side of the equation of the given function, $h(x) = x^2 + 2x + 1$, changed when we replaced x with $-x$. Thus, $h(-x) \neq h(x)$, so h is not an even function. The sign of *each* of the three terms in the equation for $h(x)$ did not change when we replaced x with $-x$. Only the second term changed signs. Thus, $h(-x) \neq -h(x)$, so h is not an odd function. We conclude that h is neither an even nor an odd function.

> **Check Point 7** Determine whether each of the following functions is even, odd, or neither:
>
> **a.** $f(x) = x^2 + 6$ **b.** $g(x) = 7x^3 - x$ **c.** $h(x) = x^5 + 1$.

Now, let's see what even and odd functions tell us about a function's graph. Begin with the even function $f(x) = x^2 - 4$, shown in Figure 2.36. The function is even because

$$f(-x) = (-x)^2 - 4 = x^2 - 4 = f(x).$$

Examine the pairs of points shown, such as $(3, 5)$ and $(-3, 5)$. Notice that we obtain the same y-coordinate whenever we evaluate the function at a value of x and the value of its opposite, $-x$. Like the attractive face, each half of the graph is a mirror image of the other half through the y-axis. If we were to fold the paper along the y-axis, the two halves of the graph would coincide. This causes the graph to be *symmetric with respect to the y-axis*. A graph is **symmetric with respect to the y-axis** if, for every point (x, y) on the graph, the point $(-x, y)$ is also on the graph. All even functions have graphs with this kind of symmetry.

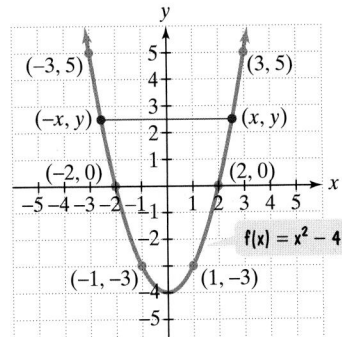

Figure 2.36 y-axis symmetry with $f(-x) = f(x)$

Even Functions and y-Axis Symmetry

The graph of an even function in which $f(-x) = f(x)$ is symmetric with respect to the y-axis.

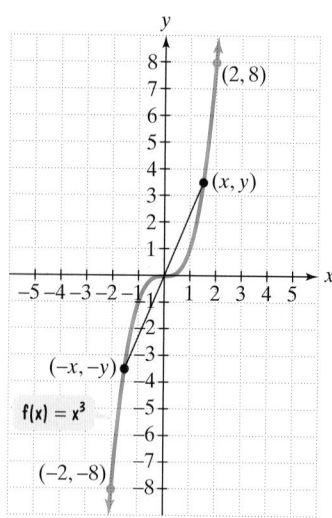

Figure 2.37 Origin symmetry with $f(-x) = -f(x)$

Now, consider the graph of the function $f(x) = x^3$. In Example 5, we saw that $f(-x) = -f(x)$, so this is an odd function. Although the graph in Figure 2.37 is not symmetric with respect to the y-axis, it is symmetric in another way. Look at the pairs of points, such as $(2, 8)$ and $(-2, -8)$. For each point (x, y) on the graph, the point $(-x, -y)$ is also on the graph. The points $(2, 8)$ and $(-2, -8)$ are reflections of one another in the origin. This means that:

- the points are the same distance from the origin, and
- the points lie on a line through the origin.

A graph is **symmetric with respect to the origin** if, for every point (x, y) on the graph, the point $(-x, -y)$ is also on the graph. Observe that the first- and third-quadrant portions of $f(x) = x^3$ are reflections of one another with respect to the origin. Notice that $f(x)$ and $f(-x)$ have opposite signs, so that $f(-x) = -f(x)$. All odd functions have graphs with origin symmetry.

Odd Functions and Origin Symmetry

The graph of an odd function in which $f(-x) = -f(x)$ is symmetric with respect to the origin.

8 Graph step functions.

Step Functions

Have you ever mailed a letter that seemed heavier than usual? Perhaps you worried that the letter would not have enough postage. Costs for mailing a letter weighing up to 5 ounces are given in Table 2.3. If your letter weighs an ounce or less, the cost is $0.37. If your letter weighs 1.05 ounces, 1.50 ounces, 1.90 ounces, or 2.00 ounces, the cost "steps" to $0.60. The cost does not take on any value between $0.37 and $0.60. If your letter weighs 2.05 ounces, 2.50 ounces, 2.90 ounces, or 3 ounces, the cost "steps" to $0.83. Cost increases are $0.23 per step.

Now, let's see what the graph of the function that models this situation looks like. Let

Table 2.3 Cost of First-Class Mail (Effective June 30, 2002)

Weight Not Over	Cost
1 ounce	$0.37
2 ounces	0.60
3 ounces	0.83
4 ounces	1.06
5 ounces	1.29

Source: U.S. Postal Service

$$x = \text{the weight of the letter, in ounces, and}$$

$$y = f(x) = \text{the cost of mailing a letter weighing } x \text{ ounces.}$$

The graph is shown in Figure 2.38. Notice how it consists of a series of steps that jump vertically 0.23 unit at each integer. The graph is constant between each pair of consecutive integers.

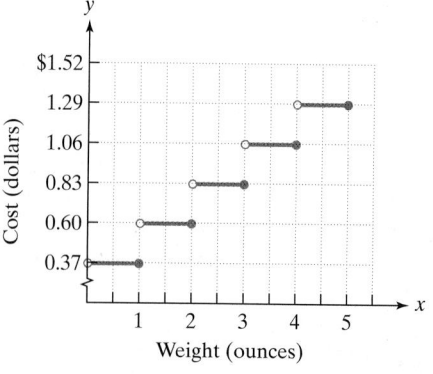

Figure 2.38

Mathematicians have defined functions that describe situations where function values graphically form discontinuous steps. One such function is called the **greatest integer function,** symbolized by $\text{int}(x)$ or $[x]$. And what is $\text{int}(x)$?

$$\text{int}(x) = \text{the greatest integer that is less than or equal to } x.$$

For example,

$$\text{int}(1) = 1, \quad \text{int}(1.3) = 1, \quad \text{int}(1.5) = 1, \quad \text{int}(1.9) = 1.$$

1 is the greatest integer that is less than or equal to 1, 1.3, 1.5, and 1.9.

Here are some additional examples:

$$\text{int}(2) = 2, \quad \text{int}(2.3) = 2, \quad \text{int}(2.5) = 2, \quad \text{int}(2.9) = 2.$$

2 is the greatest integer that is less than or equal to 2, 2.3, 2.5, and 2.9.

Notice how we jumped from 1 to 2 in the function values for $\text{int}(x)$. In particular,

$$\text{If } 1 \le x < 2, \quad \text{then} \quad \text{int}(x) = 1.$$
$$\text{If } 2 \le x < 3, \quad \text{then} \quad \text{int}(x) = 2.$$

The graph of $f(x) = \text{int}(x)$ is shown in Figure 2.39. The graph of the greatest integer function jumps vertically one unit at each integer. However, the graph is constant between each pair of consecutive integers. The rightmost horizontal step shown in the graph illustrates that

$$\text{If } 5 \le x < 6, \quad \text{then} \quad \text{int}(x) = 5.$$

In general,

$$\text{If } n \le x < n + 1, \text{ where } n \text{ is an integer,} \quad \text{then} \quad \text{int}(x) = n.$$

By contrast to the graph for the cost of first-class mail, the graph of the greatest integer function includes the point on the left of each horizontal step, but does not include the point on the right. The domain of $f(x) = \text{int}(x)$ is the set of all real numbers, $(-\infty, \infty)$. The range is the set of all integers.

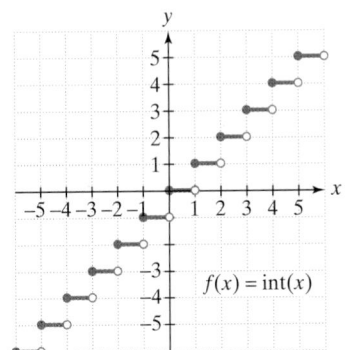

Figure 2.39 The graph of the greatest integer function

Technology

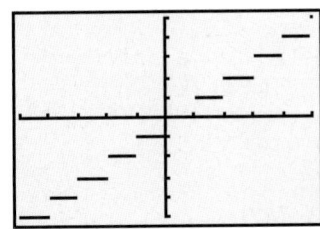

The graph of $f(x) = \text{int}(x)$, shown on the left, was obtained with a graphing utility. By graphing in "dot" mode, we can see the discontinuities at the integers. By looking at the graph, it is impossible to tell that, for each step, the point on the left is included and the point on the right is not. We must trace along the graph to obtain such information.

EXERCISE SET 2.4

Practice Exercises

Graph the function in Exercises 1–14. Use the integer values of x given to the right of the function to obtain ordered pairs. Use the graph to specify the function's domain and range.

1. $f(x) = x^2 + 2$ $x = -3, -2, -1, 0, 1, 2, 3$

2. $f(x) = x^2 - 1$ $x = -3, -2, -1, 0, 1, 2, 3$

3. $g(x) = \sqrt{x} - 1$ $x = 0, 1, 4, 9$

4. $g(x) = \sqrt{x} + 2$ $x = 0, 1, 4, 9$

5. $h(x) = \sqrt{x - 1}$ $x = 1, 2, 5, 10$

6. $h(x) = \sqrt{x + 2}$ $x = -2, -1, 2, 7$

7. $f(x) = |x| - 1$ $x = -3, -2, -1, 0, 1, 2, 3$

8. $f(x) = |x| + 1$ $x = -3, -2, -1, 0, 1, 2, 3$

9. $g(x) = |x - 1|$ $x = -3, -2, -1, 0, 1, 2, 3$

10. $g(x) = |x + 1|$ $x = -3, -2, -1, 0, 1, 2, 3$

11. $f(x) = 5$ $x = -3, -2, -1, 0, 1, 2, 3$

12. $f(x) = 3$ $x = -3, -2, -1, 0, 1, 2, 3$

13. $f(x) = x^3 - 2$ $x = -2, -1, 0, 1, 2$

14. $f(x) = x^3 + 2$ $x = -2, -1, 0, 1, 2$

In Exercises 15–30, use the graph to determine **a.** *the function's domain;* **b.** *the function's range;* **c.** *the x-intercepts, if any;* **d.** *the y-intercept, if any; and* **e.** *the function values indicated below some of the graphs.*

15.

16.

17.

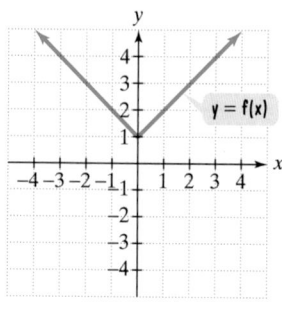

$f(-1) = ?$ $f(3) = ?$

18.

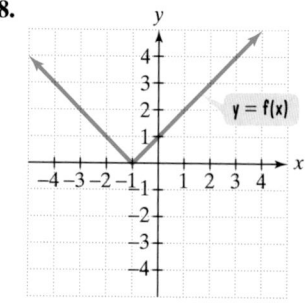

$f(-4) = ?$ $f(3) = ?$

19.

$f(3) = ?$

20.

$f(-5) = ?$

21.

$$f(4) = ?$$

22.

$$f(3) = ?$$

23.

$$f(-1) = ?$$

24.

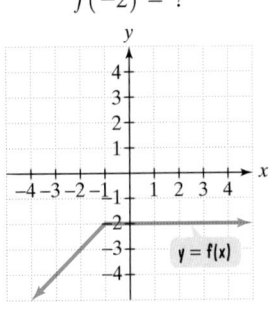

$$f(-2) = ?$$

25.

$$f(-4) = ? \quad f(4) = ?$$

26.

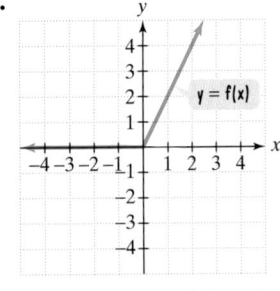

$$f(-2) = ? \quad f(2) = ?$$

27.

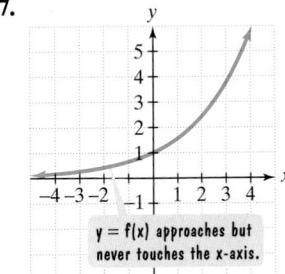

$y = f(x)$ approaches but never touches the x-axis.

28.

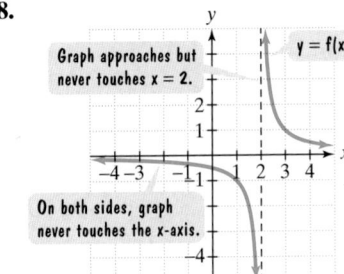

Graph approaches but never touches x = 2.

$y = f(x)$

On both sides, graph never touches the x-axis.

29.

30.

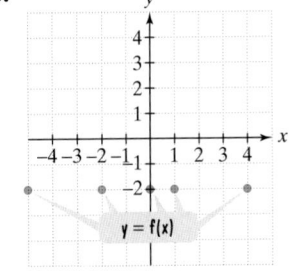

In Exercises 31–38, use the vertical line test to identify graphs in which y is a function of x.

31.

32.

33.

34.

35.

36.

37.

38.

47.

48.

49.

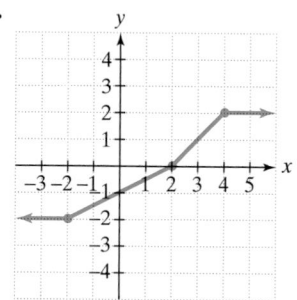

In Exercises 39–50, use the graph to determine:

   a. *intervals on which the function is increasing, if any.*

   b. *intervals on which the function is decreasing, if any.*

   c. *intervals on which the function is constant, if any.*

39. Use the graph in Exercise 15.

40. Use the graph in Exercise 16.

41. Use the graph in Exercise 21.

42. Use the graph in Exercise 22.

43. Use the graph in Exercise 23.

44. Use the graph in Exercise 24.

45. Use the graph in Exercise 25.

46. Use the graph in Exercise 26.

50.

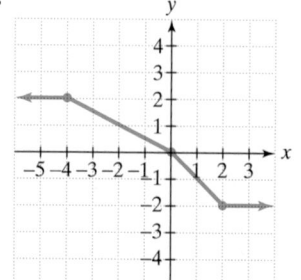

In Exercises 51–54, the graph of a function f is given. Use the graph to find:

   a. The numbers, if any, at which f has a relative maximum. What are these relative maxima?

   b. The numbers, if any, at which f has a relative minimum. What are these relative minima?

51.

52.

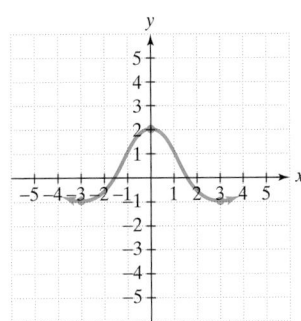

53.

$f(x) = 2x^3 + 3x^2 - 12x + 1$

$[-4, 4, 1]$ by $[-15, 25, 5]$

54.

$f(x) = 2x^3 - 15x^2 + 24x + 19$

$[-2, 6, 1]$ by $[-15, 35, 5]$

In Exercises 55–60, find the average rate of change of the function from x_1 to x_2.

55. $f(x) = 3x$ from $x_1 = 0$ to $x_2 = 5$

56. $f(x) = 6x$ from $x_1 = 0$ to $x_2 = 4$

57. $f(x) = x^2 + 2x$ from $x_1 = 3$ to $x_2 = 5$

58. $f(x) = x^2 - 2x$ from $x_2 = 3$ to $x_2 = 6$

59. $f(x) = \sqrt{x}$ from $x_1 = 4$ to $x_2 = 9$

60. $f(x) = \sqrt{x}$ from $x_1 = 9$ to $x_2 = 16$

In Exercises 61–72, determine whether each function is even, odd, or neither.

61. $f(x) = x^3 + x$ **62.** $f(x) = x^3 - x$

63. $g(x) = x^2 + x$ **64.** $g(x) = x^2 - x$

65. $h(x) = x^2 - x^4$ **66.** $h(x) = 2x^2 + x^4$

67. $f(x) = x^2 - x^4 + 1$ **68.** $f(x) = 2x^2 + x^4 + 1$

69. $f(x) = \frac{1}{5}x^6 - 3x^2$ **70.** $f(x) = 2x^3 - 6x^5$

71. $f(x) = x\sqrt{1 - x^2}$ **72.** $f(x) = x^2\sqrt{1 - x^2}$

In Exercises 73–76, use possible symmetry to determine whether each graph is the graph of an even function, an odd function, or a function that is neither even nor odd.

73.

74.

75.

76.

In Exercises 77–82, if $f(x) = int(x)$, find each function value.

77. $f(1.06)$

78. $f(2.99)$

79. $f(\frac{1}{3})$

80. $f(-1.5)$

81. $f(-2.3)$

82. $f(-99.001)$

Application Exercises

The figure shows the percentage of the U.S. population made up of Jewish Americans, $f(x)$, as a function of time, x, where x is the number of years after 1900. Use the graph to solve Exercises 83–90.

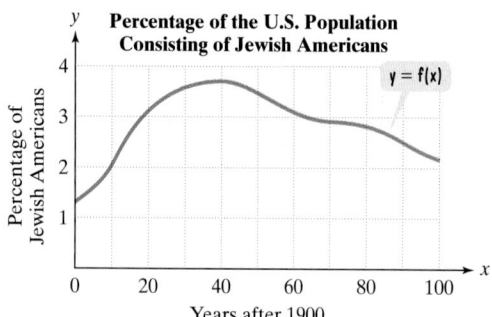

Source: American Jewish Yearbook

83. Use the graph to find a reasonable estimate of $f(60)$. What does this mean in terms of the variables in this situation?

84. Use the graph to find a reasonable estimate of $f(100)$. What does this mean in terms of the variables in this situation?

85. For what value or values of x is $f(x) = 3$? Round to the nearest year. What does this mean in terms of the variables in this situation?

86. For what value or values of x is $f(x) = 2.5$? Round to the nearest year. What does this mean in terms of the variables in this situation?

87. In which year did the percentage of Jewish Americans in the U.S. population reach a maximum? What is a reasonable estimate of the percentage for that year?

88. In which year was the percentage of Jewish Americans in the U.S. population at a minimum? What is a reasonable estimate of the percentage for that year?

89. Explain why f represents the graph of a function.

90. Describe the general trend shown by the graph.

The function

$$f(x) = 0.4x^2 - 36x + 1000$$

models the number of accidents, $f(x)$, per 50 million miles driven as a function of the driver's age, x, in years, where x includes drivers from ages 16 through 74. The graph of f is shown. Use the graph of f, and possibly the equation, to solve Exercises 91–93.

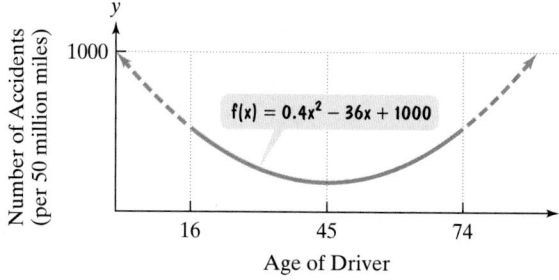

91. State the intervals on which the function is increasing and decreasing and describe what this means in terms of the variables modeled by the function.

92. For what value of x does the graph reach its lowest point? What is the minimum value of y? Describe the practical significance of this minimum value.

93. Use the graph to identify two different ages for which drivers have the same number of accidents. Use the equation for f to find the number of accidents for drivers at each of these ages.

94. Based on a study by Vance Tucker (*Scientific American*, May 1969), the power expenditure of migratory birds in flight is a function of their flying speed, x, in miles per hour, modeled by $f(x) = 0.67x^2 - 27.74x + 387$. Power expenditure, $f(x)$, is measured in calories, and migratory birds generally fly between 12 and 30 miles per hour. The graph of f is shown in the figure on the next page, with a domain of $[12, 30]$.

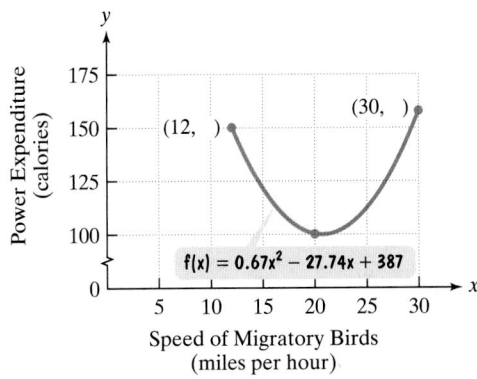

Speed of Migratory Birds
(miles per hour)

a. State the intervals on which the function is increasing and decreasing and describe what this means in terms of the variables modeled by the function.

b. For what approximate value of x does the graph reach its lowest point? What is the minimum value of y? Describe the practical significance of this minimum value.

95. The cost of a telephone call between two cities is $0.10 for the first minute and $0.05 for each additional minute or portion of a minute. Draw a graph of the cost, C, in dollars, of the phone call as a function of time, t, in minutes, on the interval $(0, 5]$.

96. A cargo service charges a flat fee of $4 plus $1 for each pound or fraction of a pound to mail a package. Let $C(x)$ represent the cost to mail a package that weighs x pounds. Graph the cost function on the interval $(0, 5]$.

97. Researchers at Yale University have suggested that levels of passion and commitment in human relations are functions of time. Based on the shapes of the graphs shown, which do you think depicts passion and which represents commitment? Explain how you arrived at your answer.

Writing in Mathematics

98. Discuss one disadvantage to using point plotting as a method for graphing functions.

99. Explain how to use a function's graph to find the function's domain and range.

100. Explain how the vertical line test is used to determine whether a graph is a function.

101. What does it mean if function f is increasing on an interval?

102. Suppose that a function f is increasing on (a, b) and decreasing on (b, c). Describe what occurs at $x = b$. What does the function value $f(b)$ represent?

103. What is a secant line?

104. What is the average rate of change of a function?

105. If you are given a function's equation, how do you determine if the function is even, odd, or neither?

106. If you are given a function's graph, how do you determine if the function is even, odd, or neither?

107. What is a step function? Give an example of an everyday situation that can be modeled using such a function. Do not use the cost-of-mail example.

108. Explain how to find int(-3.000004).

Technology Exercises

109. The function

$$f(x) = -0.00002x^3 + 0.008x^2 - 0.3x + 6.95$$

models the number of annual physician visits, $f(x)$, by a person of age x. Graph the function in a $[0, 100, 5]$ by $[0, 40, 2]$ viewing rectangle. What does the shape of the graph indicate about the relationship between one's age and the number of annual physician visits? Use the $\boxed{\text{TRACE}}$ or minimum function capability to find the coordinates of the minimum point on the graph of the function. What does this mean?

In Exercises 110–115, use a graphing utility to graph each function. Use a $[-5, 5, 1]$ by $[-5, 5, 1]$ viewing rectangle. Then find the intervals on which the function is increasing, decreasing, or constant.

110. $f(x) = x^3 - 6x^2 + 9x + 1$ **111.** $g(x) = |4 - x^2|$

112. $h(x) = |x - 2| + |x + 2|$ **113.** $f(x) = x^{1/3}(x - 4)$

114. $g(x) = x^{2/3}$ **115.** $h(x) = 2 - x^{2/5}$

116. a. Graph the functions $f(x) = x^n$ for $n = 2, 4$, and 6 in a $[-2, 2, 1]$ by $[-1, 3, 1]$ viewing rectangle.

b. Graph the functions $f(x) = x^n$ for $n = 1, 3$, and 5 in a $[-2, 2, 1]$ by $[-2, 2, 1]$ viewing rectangle.

c. If n is even, where is the graph of $f(x) = x^n$ increasing and where is it decreasing?

d. If n is odd, what can you conclude about the graph of $f(x) = x^n$ in terms of increasing or decreasing behavior?

e. Graph all six functions in a $[-1, 3, 1]$ by $[-1, 3, 1]$ viewing rectangle. What do you observe about the graphs in terms of how flat or how steep they are?

Critical Thinking Exercises

117. Which one of the following is true based on the graph of f in the figure?

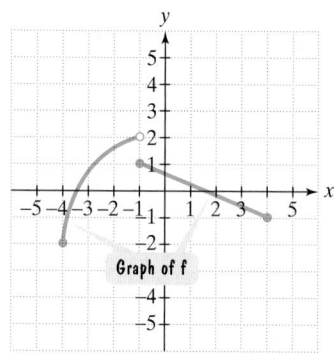

Graph of f

a. The domain of f is $[-4, 1)$ or $(1, 4]$.
b. The range of f is $[-2, 2]$.
c. $f(-1) - f(4) = 2$
d. $f(0) = 2.1$

118. Sketch the graph of f using the following properties. (More than one correct graph is possible.) f is a piecewise function that is decreasing on $(-\infty, 2), f(2) = 0, f$ is increasing on $(2, \infty)$, and the range of f is $[0, \infty)$.

119. Define a piecewise function on the intervals $(-\infty, 2]$, $(2, 5)$, and $[5, \infty)$ that does not "jump" at 2 or 5 such that one piece is a constant function, another piece is an increasing function, and the third piece is a decreasing function.

120. Suppose that $h(x) = \dfrac{f(x)}{g(x)}$. The function f can be even, odd, or neither. The same is true for the function g.

a. Under what conditions is h definitely an even function?

b. Under what conditions is h definitely an odd function?

121. Take another look at the cost of first-class mail and its graph (Table 2.3 and Figure 2.38 on page 226. Change the description of the heading in the left column of Table 2.3 so that the graph includes the point on the left of each horizontal step, but does not include the point on the right.

Group Exercise

122. In Exercise 97, passion and commitment are graphed over time. For this activity, you will be creating a graph of a particular experience that involved your feelings of love, anger, sadness, or any other emotion you choose. The horizontal axis should be labeled time and the vertical axis the emotion you are graphing. You will not be using your algebra skills to create your graph; however, you should try to make the graph as precise as possible. You may use negative numbers on the vertical axis, if appropriate. After each group member has created a graph, pool together all of the graphs and study them to see if there are any similarities in the graphs for a particular emotion or for all emotions.

SECTION 2.5 *Transformations of Functions*

Objectives

1. Recognize graphs of common functions.
2. Use vertical shifts to graph functions.
3. Use horizontal shifts to graph functions.
4. Use reflections to graph functions.
5. Use vertical stretching and shrinking to graph functions.
6. Graph functions involving a sequence of transformations.

Have you seen *Terminator 2*, *The Mask*, or *The Matrix*? These were among the first films to use spectacular effects in which a character or object having one shape was transformed in a fluid fashion into a quite different shape. The name for such a transformation is **morphing.** The effect allows a real actor to be seamlessly transformed into a computer-generated animation. The animation can be made to perform impossible feats before it is morphed back to the conventionally filmed image.

Like transformed movie images, the graph of one function can be turned into the graph of a different function. To do this, we need to rely on a function's equation. Knowing that a graph is a transformation of a familiar graph makes graphing easier.

1 Recognize graphs of common functions.

Graphs of Common Functions

Table 2.4 below and on page 236 gives names to six frequently encountered functions in algebra. The table shows each function's graph and lists characteristics of the function. Study the shape of each graph and take a few minutes to verify the function's characteristics from its graph. Knowing these graphs is essential for analyzing their transformations into more complicated graphs.

Table 2.4 Algebra's Common Graphs

Constant Function	Identity Function	Standard Quadratic Function
		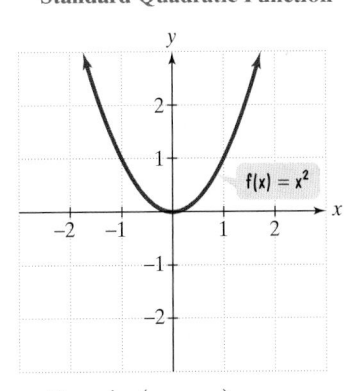
• Domain: $(-\infty, \infty)$	• Domain: $(-\infty, \infty)$	• Domain: $(-\infty, \infty)$
• Range: the single number c	• Range: $(-\infty, \infty)$	• Range: $[0, \infty)$
• Constant on $(-\infty, \infty)$	• Increasing on $(-\infty, \infty)$	• Decreasing on $(-\infty, 0)$ and increasing on $(0, \infty)$
• Even function	• Odd function	• Even function

Table 2.4   Algebra's Common Graphs *(continued)*

Standard Cubic Function	Square Root Function	Absolute Value Function
		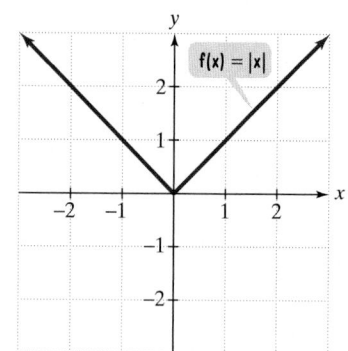
• Domain: $(-\infty, \infty)$	• Domain: $[0, \infty)$	• Domain: $(-\infty, \infty)$
• Range: $(-\infty, \infty)$	• Range: $[0, \infty)$	• Range: $[0, \infty)$
• Increasing on $(-\infty, \infty)$	• Increasing on $(0, \infty)$	• Decreasing on $(-\infty, 0)$ and increasing on $(0, \infty)$
• Odd function	• Neither even nor odd	• Even function

Discovery

The study of how changing a function's equation can affect its graph can be explored with a graphing utility. Use your graphing utility to verify the hand-drawn graphs as you read this section.

2 Use vertical shifts to graph functions.

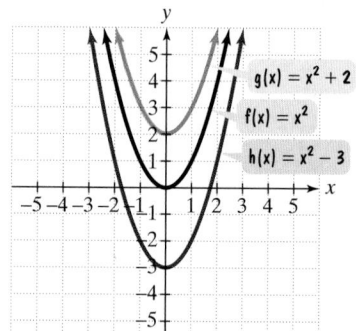

Figure 2.40 Vertical shifts

The graph of $f(x) = x^2$ can be gradually morphed into the graph of $g(x) = x^2 + 2$ by using animation to graph $f(x) = x^2 + c$ for $0 \leq c \leq 2$. By selecting many values for c, we can create an animated sequence in which change appears to occur continuously.

Vertical Shifts

Let's begin by looking at three graphs whose shapes are the same. Figure 2.40 shows the graphs. The black graph in the middle is the standard quadratic function, $f(x) = x^2$. Now, look at the blue graph on the top. The equation of this graph, $g(x) = x^2 + 2$, adds 2 to the right side of $f(x) = x^2$. What effect does this have on the graph of f? It shifts the graph vertically up by 2 units.

$$g(x) = x^2 + 2 = f(x) + 2$$

The graph of g   *shifts the graph of f up 2 units.*

Finally, look at the red graph on the bottom of Figure 2.40. The equation of this graph, $h(x) = x^2 - 3$, subtracts 3 from the right side of $f(x) = x^2$. What effect does this have on the graph of f? It shifts the graph vertically down by 3 units.

$$h(x) = x^2 - 3 = f(x) - 3$$

The graph of h   *shifts the graph of f down 3 units.*

In general, if c is positive, $y = f(x) + c$ shifts the graph of f upward c units and $y = f(x) - c$ shifts the graph of f downward c units. These are called **vertical shifts** of the graph of f.

Vertical Shifts

Let f be a function and c a positive real number.

- The graph of $y = f(x) + c$ is the graph of $y = f(x)$ shifted c units vertically upward.
- The graph of $y = f(x) - c$ is the graph of $y = f(x)$ shifted c units vertically downward.

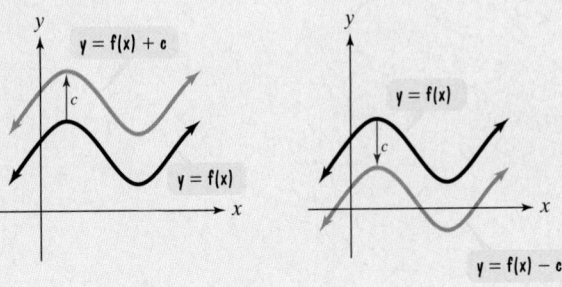

EXAMPLE 1   Vertical Shift Down

Use the graph of $f(x) = |x|$ to obtain the graph of $g(x) = |x| - 4$.

Solution   The graph of $g(x) = |x| - 4$ has the same shape as the graph of $f(x) = |x|$. However, it is shifted down vertically 4 units. We have constructed a table showing some of the coordinates for f and g. The graphs of f and g are shown in Figure 2.41.

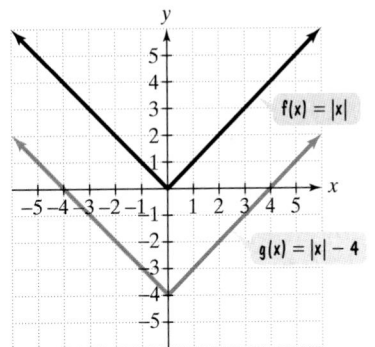

x	$y = f(x) = |x|$	$(x, f(x))$	$y = g(x)$ $= |x| - 4 = f(x) - 4$	$(x, g(x))$
-2	$|-2| = 2$	$(-2, 2)$	$|-2| - 4 = -2$	$(-2, -2)$
-1	$|-1| = 1$	$(-1, 1)$	$|-1| - 4 = -3$	$(-1, -3)$
0	$|0| = 0$	$(0, 0)$	$|0| - 4 = -4$	$(0, -4)$
1	$|1| = 1$	$(1, 1)$	$|1| - 4 = -3$	$(1, -3)$
2	$|2| = 2$	$(2, 2)$	$|2| - 4 = -2$	$(2, -2)$

Figure 2.41

Check Point 1   Use the graph of $f(x) = |x|$ to obtain the graph of $g(x) = |x| + 3$.

3 Use horizontal shifts to graph functions.

Horizontal Shifts

We return to the graph of $f(x) = x^2$, the standard quadratic function. In Figure 2.42 on the next page, the graph of function f is in the middle of the three graphs. Turn the page and verify this observation.

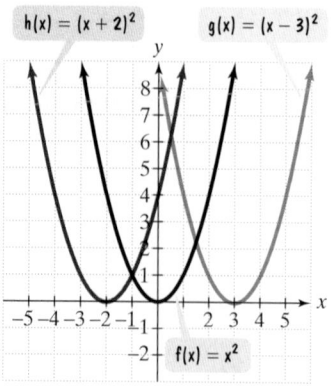

Figure 2.42 Horizontal shifts

By contrast to the vertical shift situation, this time there are graphs to the left and to the right of the graph of f. Look at the blue graph on the right. The equation of this graph, $g(x) = (x - 3)^2$, subtracts 3 from each value of x in the domain of $f(x) = x^2$. What effect does this have on the graph of f? It shifts the graph horizontally to the right by 3 units.

$$g(x) = (x - 3)^2 = f(x - 3)$$

The graph of g   shifts the graph of f **3** units to the right.

Now, look at the red graph on the left in Figure 2.42. The equation of this graph, $h(x) = (x + 2)^2$, adds 2 to each value of x in the domain of $f(x) = x^2$. What effect does this have on the graph of f? It shifts the graph horizontally to the left by 2 units.

$$h(x) = (x + 2)^2 = f(x + 2)$$

The graph of h   shifts the graph of f **2** units to the left.

In general, if c is positive, $y = f(x + c)$ shifts the graph of f to the left c units and $y = f(x - c)$ shifts the graph of f to the right c units. These are called **horizontal shifts** of the graph of f.

Study Tip

We know that positive numbers are to the right of zero on a number line and negative numbers are to the left of zero. This positive-negative orientation does not apply to horizontal shifts. A *positive* number causes a shift to the *left* and a *negative* number causes a shift to the *right*.

Horizontal Shifts

Let f be a function and c a positive real number.

- The graph of $y = f(x + c)$ is the graph of $y = f(x)$ shifted to the left c units.
- The graph of $y = f(x - c)$ is the graph of $y = f(x)$ shifted to the right c units.

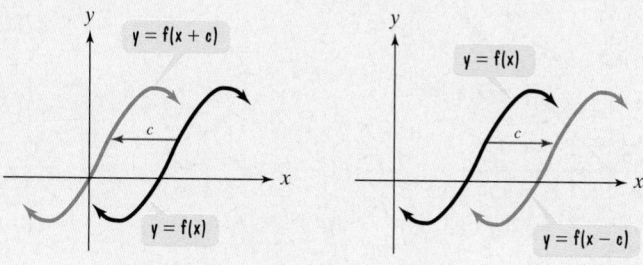

EXAMPLE 2   Horizontal Shift to the Left

Use the graph of $f(x) = \sqrt{x}$ to obtain the graph of $g(x) = \sqrt{x + 5}$.

Solution   Compare the equations for $f(x) = \sqrt{x}$ and $g(x) = \sqrt{x + 5}$. The equation for g adds 5 to each value of x in the domain of f.

$$y = g(x) = \sqrt{x + 5} = f(x + 5)$$

The graph of g   shifts the graph of f **5** units to the left.

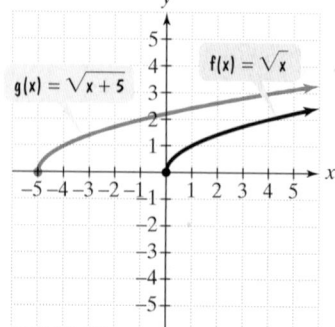

Figure 2.43 Shifting $f(x) = \sqrt{x}$ five units left

The graph of $g(x) = \sqrt{x + 5}$ has the same shape as the graph of $f(x) = \sqrt{x}$. However, it is shifted horizontally to the left 5 units. We have created tables on the next page showing some of the coordinates for f and g. As shown in Figure 2.43, every point in the graph of g is exactly 5 units to the left of a corresponding point on the graph of f.

x	$y = f(x) = \sqrt{x}$	$(x, f(x))$
0	$\sqrt{0} = 0$	$(0, 0)$
1	$\sqrt{1} = 1$	$(1, 1)$
4	$\sqrt{4} = 2$	$(4, 2)$

x	$y = g(x) = \sqrt{x + 5}$	$(x, g(x))$
-5	$\sqrt{-5 + 5} = \sqrt{0} = 0$	$(-5, 0)$
-4	$\sqrt{-4 + 5} = \sqrt{1} = 1$	$(-4, 1)$
-1	$\sqrt{-1 + 5} = \sqrt{4} = 2$	$(-1, 2)$

Check Point 2 Use the graph of $f(x) = \sqrt{x}$ to obtain the graph of $g(x) = \sqrt{x - 4}$.

Some functions can be graphed by combining horizontal and vertical shifts. These functions will be variations of a function whose equation you know how to graph, such as the standard quadratic function, the standard cubic function, the square root function, or the absolute value function.

In our next example, we will use the graph of the standard quadratic function, $f(x) = x^2$, to obtain the graph of $h(x) = (x + 1)^2 - 3$. We will graph three functions:

$$f(x) = x^2 \qquad g(x) = (x + 1)^2 \qquad h(x) = (x + 1)^2 - 3.$$

Start by graphing the standard quadratic function.

Shift the graph of f horizontally one unit to the left.

Shift the graph of g vertically down 3 units.

Discovery

Work Example 3 by first shifting the graph of $f(x) = x^2$ three units down, graphing $g(x) = x^2 - 3$. Now, shift this graph one unit left to graph $h(x) = (x + 1)^2 - 3$. Did you obtain the graph in Figure 2.44(c)? What can you conclude?

EXAMPLE 3 Combining Horizontal and Vertical Shifts

Use the graph of $f(x) = x^2$ to obtain the graph of $h(x) = (x + 1)^2 - 3$.

Solution

Step 1   Graph $f(x) = x^2$. The graph of the standard quadratic function is shown in Figure 2.44(a). We've identified three points on the graph.

Step 2   Graph $g(x) = (x + 1)^2$. Because we add 1 to each value of x in the domain of the standard quadratic function, $f(x) = x^2$, we shift the graph of f horizontally one unit to the left. This is shown in Figure 2.44(b). Notice that every point in the graph in Figure 2.44(b) has an x-coordinate that is one less than the x-coordinate for the corresponding point in the graph in Figure 2.44(a).

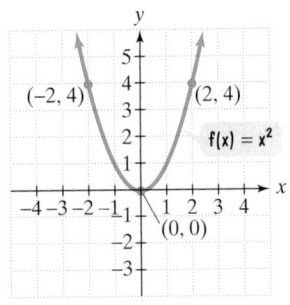

(a) The graph of $f(x) = x^2$

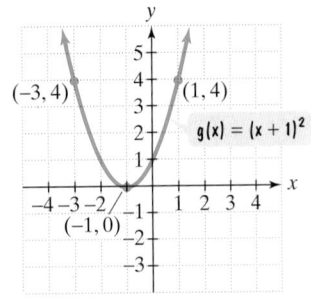

(b) The graph of $g(x) = (x + 1)^2$

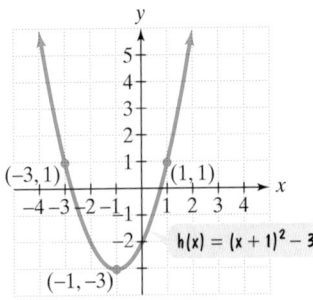

(c) The graph of $h(x) = (x + 1)^2 - 3$

Figure 2.44

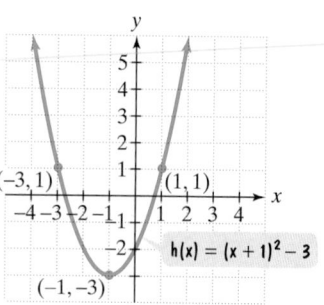

(a) The graph of $f(x) = x^2$ **(b)** The graph of $g(x) = (x + 1)^2$ **(c)** The graph of $h(x) = (x + 1)^2 - 3$

Figure 2.44, repeated

Step 3 Graph $h(x) = (x + 1)^2 - 3$. Because we subtract 3, we shift the graph in Figure 2.44(b) vertically down 3 units. The graph is shown in Figure 2.44(c). Notice that every point in the graph in Figure 2.44(c) has a y-coordinate that is three less than the y-coordinate of the corresponding point in the graph in Figure 2.44(b).

> **Check Point 3** Use the graph of $f(x) = \sqrt{x}$ to obtain the graph of
> $h(x) = \sqrt{x - 1} - 2$.

4 Use reflections to graph functions.

Reflections of Graphs

This photograph shows a reflection of an old bridge in a Maryland river. This perfect reflection occurs because the surface of the water is absolutely still. A mild breeze rippling the water's surface would distort the reflection.

Is it possible for graphs to have mirror-like qualities? Yes. Figure 2.45 shows the graphs of $f(x) = x^2$ and $g(x) = -x^2$. The graph of g is a **reflection about the x-axis** of the graph of f. In general, the graph of $y = -f(x)$ reflects the graph of f about the x-axis. Thus, the graph of g is a reflection of the graph of f about the x-axis because

$$g(x) = -x^2 = -f(x).$$

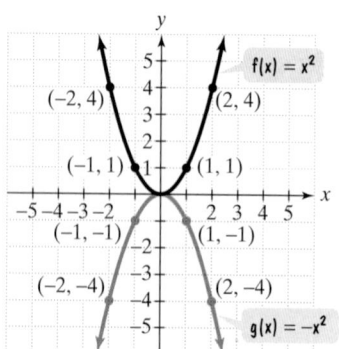

Figure 2.45 Reflections about the x-axis

Reflection about the x-Axis

The graph of $y = -f(x)$ is the graph of $y = f(x)$ reflected about the x-axis.

EXAMPLE 4   Reflection about the *x*-Axis

Use the graph of $f(x) = \sqrt{x}$ to obtain the graph of $g(x) = -\sqrt{x}$.

Solution   Compare the equations for $f(x) = \sqrt{x}$ and $g(x) = -\sqrt{x}$. The graph of *g* is a reflection about the *x*-axis of the graph of *f* because

$$g(x) = -\sqrt{x} = -f(x).$$

We have created a table showing some of the coordinates for *f* and *g*. The graphs of *f* and *g* are shown in Figure 2.46.

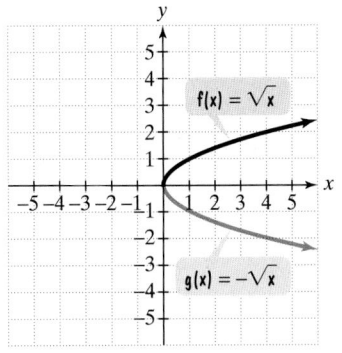

Figure 2.46 Reflecting $f(x) = \sqrt{x}$ about the *x*-axis

x	$f(x) = \sqrt{x}$	$(x, f(x))$	$g(x) = -\sqrt{x}$	$(x, g(x))$
0	$\sqrt{0} = 0$	$(0, 0)$	$-\sqrt{0} = 0$	$(0, 0)$
1	$\sqrt{1} = 1$	$(1, 1)$	$-\sqrt{1} = -1$	$(1, -1)$
4	$\sqrt{4} = 2$	$(4, 2)$	$-\sqrt{4} = -2$	$(4, -2)$

> **Check Point 4**   Use the graph of $f(x) = |x|$ to obtain the graph of $g(x) = -|x|$.

It is also possible to reflect graphs about the *y*-axis.

Reflection about the *y*-Axis

The graph of $y = f(-x)$ is the graph of $y = f(x)$ reflected about the *y*-axis.

EXAMPLE 5   Reflection about the *y*-Axis

Use the graph of $f(x) = \sqrt{x}$ to obtain the graph of $h(x) = \sqrt{-x}$.

Solution   Compare the equations for $f(x) = \sqrt{x}$ and $h(x) = \sqrt{-x}$. The graph of *h* is a reflection about the *y*-axis of the graph of *f* because

$$h(x) = \sqrt{-x} = f(-x).$$

We have created tables showing some of the coordinates for *f* and *h*. The graphs of *f* and *h* are shown in Figure 2.47.

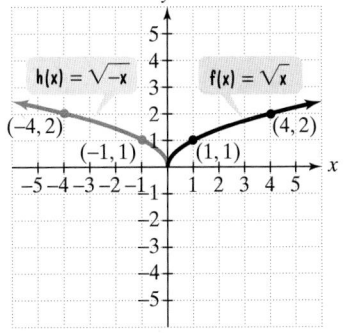

Figure 2.47 Reflecting $f(x) = \sqrt{x}$ about the *y*-axis

x	$f(x) = \sqrt{x}$	$(x, f(x))$
0	$\sqrt{0} = 0$	$(0, 0)$
1	$\sqrt{1} = 1$	$(1, 1)$
4	$\sqrt{4} = 2$	$(4, 2)$

x	$h(x) = \sqrt{-x}$	$(x, h(x))$
0	$\sqrt{-0} = \sqrt{0} = 0$	$(0, 0)$
-1	$\sqrt{-(-1)} = \sqrt{1} = 1$	$(-1, 1)$
-4	$\sqrt{-(-4)} = \sqrt{4} = 2$	$(-4, 2)$

Figure 2.48

> **Check Point 5**   Use the graph of $f(x) = \sqrt{x-1}$ in Figure 2.48 to obtain the graph of $h(x) = \sqrt{-x-1}$.

5 Use vertical stretching and shrinking to graph functions.

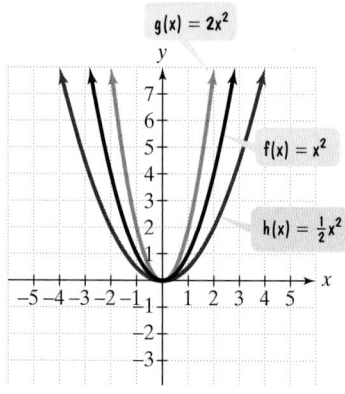

Figure 2.49 Stretching and shrinking $f(x) = x^2$

Vertical Stretching and Shrinking

Morphing does much more than move an image horizontally, vertically, or about an axis. An object having one shape is transformed into a different shape. Horizontal shifts, vertical shifts, and reflections do not change the basic shape of a graph. How can we shrink and stretch graphs, thereby altering their basic shapes?

Look at the three graphs in Figure 2.49. The black graph in the middle is the graph of the standard quadratic function, $f(x) = x^2$. Now, look at the blue graph on the top. The equation of this graph is $g(x) = 2x^2$. Thus, for each x, the y-coordinate of g is 2 times as large as the corresponding y-coordinate on the graph of f. The result is a narrower graph. We say that the graph of g is obtained by vertically *stretching* the graph of f. Now, look at the red graph on the bottom. The equation of this graph is $h(x) = \frac{1}{2}x^2$, or $h(x) = \frac{1}{2}f(x)$. Thus, for each x, the y-coordinate of h is one-half as large as the corresponding y-coordinate on the graph of f. The result is a wider graph. We say that the graph of h is obtained by vertically *shrinking* the graph of f.

These observations can be summarized as follows:

> ### Stretching and Shrinking Graphs
>
> Let f be a function and c a positive real number.
> - If $c > 1$, the graph of $y = cf(x)$ is the graph of $y = f(x)$ vertically stretched by multiplying each of its y-coordinates by c.
> - If $0 < c < 1$, the graph of $y = cf(x)$ is the graph of $y = f(x)$ vertically shrunk by multiplying each of its y-coordinates by c.

EXAMPLE 6 Vertically Stretching a Graph

Use the graph of $f(x) = |x|$ to obtain the graph of $g(x) = 2|x|$.

Solution The graph of $g(x) = 2|x|$ is obtained by vertically stretching the graph of $f(x) = |x|$. We have constructed a table showing some of the coordinates for f and g. Observe that the y-coordinate on the graph of g is twice as large as the corresponding y-coordinate on the graph of f. The graphs of f and g are shown in Figure 2.50.

x	$f(x) = |x|$	$(x, f(x))$	$g(x) = 2|x| = 2f(x)$	$(x, g(x))$
-2	$|-2| = 2$	$(-2, 2)$	$2|-2| = 4$	$(-2, 4)$
-1	$|-1| = 1$	$(-1, 1)$	$2|-1| = 2$	$(-1, 2)$
0	$|0| = 0$	$(0, 0)$	$2|0| = 0$	$(0, 0)$
1	$|1| = 1$	$(1, 1)$	$2|1| = 2$	$(1, 2)$
2	$|2| = 2$	$(2, 2)$	$2|2| = 4$	$(2, 4)$

Figure 2.50 Stretching $f(x) = |x|$

Check Point 6 Use the graph of $f(x) = |x|$ to obtain the graph of $g(x) = 3|x|$.

EXAMPLE 7 Vertically Shrinking a Graph

Use the graph of $f(x) = |x|$ to obtain the graph of $h(x) = \frac{1}{2}|x|$.

Solution The graph of $h(x) = \frac{1}{2}|x|$ is obtained by vertically shrinking the graph of $f(x) = |x|$. We have constructed a table showing some of the coordinates for f and h. Observe that the y-coordinate on the graph of h is one-half the corresponding y-coordinate on the graph of f. The graphs of f and h are shown in Figure 2.51.

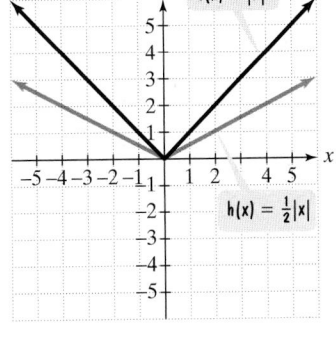

Figure 2.51 Shrinking $f(x) = |x|$

x	$f(x) = |x|$	$(x, f(x))$	$h(x) = \frac{1}{2}|x| = \frac{1}{2}f(x)$	$(x, h(x))$
-2	$|-2| = 2$	$(-2, 2)$	$\frac{1}{2}|-2| = 1$	$(-2, 1)$
-1	$|-1| = 1$	$(-1, 1)$	$\frac{1}{2}|-1| = \frac{1}{2}$	$(-1, \frac{1}{2})$
0	$|0| = 0$	$(0, 0)$	$\frac{1}{2}|0| = 0$	$(0, 0)$
1	$|1| = 1$	$(1, 1)$	$\frac{1}{2}|1| = \frac{1}{2}$	$(1, \frac{1}{2})$
2	$|2| = 2$	$(2, 2)$	$\frac{1}{2}|2| = 1$	$(2, 1)$

> **Check Point 7** Use the graph of $f(x) = |x|$ to obtain the graph of $h(x) = \frac{1}{4}|x|$.

6 Graph functions involving a sequence of transformations.

Sequences of Transformations

Table 2.5 summarizes the procedures for transforming the graph of $y = f(x)$.

Table 2.5 Summary of Transformations
In each case, c represents a positive real number.

To Graph:	Draw the Graph of f and:	Changes in the Equation of $y = f(x)$
Vertical shifts $y = f(x) + c$ $y = f(x) - c$	Raise the graph of f by c units. Lower the graph of f by c units.	c is added to $f(x)$. c is subtracted from $f(x)$.
Horizontal shifts $y = f(x + c)$ $y = f(x - c)$	Shift the graph of f to the left c units. Shift the graph of f to the right c units.	x is replaced with $x + c$. x is replaced with $x - c$.
Reflection about the x-axis $y = -f(x)$	Reflect the graph of f about the x-axis.	$f(x)$ is multiplied by -1.
Reflection about the y-axis $y = f(-x)$	Reflect the graph of f about the y-axis.	x is replaced with $-x$.
Vertical stretching or shrinking $y = cf(x), c > 1$ $y = cf(x), 0 < c < 1$	Multiply each y-coordinate of $y = f(x)$ by c, vertically stretching the graph of f. Multiply each y-coordinate of $y = f(x)$ by c, vertically shrinking the graph of f.	$f(x)$ is multiplied by $c, c > 1$. $f(x)$ is multiplied by $c, 0 < c < 1$.

A function involving more than one transformation can be graphed by performing transformations in the following order:

1. Horizontal shifting
2. Vertical stretching or shrinking
3. Reflecting
4. Vertical shifting

EXAMPLE 8   Graphing Using a Sequence of Transformations

Use the graph of $f(x) = \sqrt{x}$ to graph $g(x) = \sqrt{1 - x} + 3$.

Solution   The following sequence of steps is illustrated in Figure 2.52. We begin with the graph of $f(x) = \sqrt{x}$.

Step 1   Horizontal Shifting   Graph $y = \sqrt{x + 1}$. Because x is replaced with $x + 1$, the graph of $f(x) = \sqrt{x}$ is shifted 1 unit to the left.

Step 2   Vertical Stretching or Shrinking   Because the equation $y = \sqrt{x + 1}$ is not multiplied by a constant in $g(x) = \sqrt{1 - x} + 3$, no stretching or shrinking is involved.

Step 3   Reflecting   We are interested in graphing $y = \sqrt{1 - x} + 3$, or $y = \sqrt{-x + 1} + 3$. We have now graphed $y = \sqrt{x + 1}$. We can graph $y = \sqrt{-x + 1}$ by noting that x is replaced with $-x$. Thus, we graph $y = \sqrt{-x + 1}$ by reflecting the graph of $y = \sqrt{x + 1}$ about the y-axis.

Step 4   Vertical Shifting   We can use the graph of $y = \sqrt{1 - x}$ to get the graph of $g(x) = \sqrt{1 - x} + 3$. Because 3 is added, shift the graph of $y = \sqrt{1 - x}$ up 3 units.

   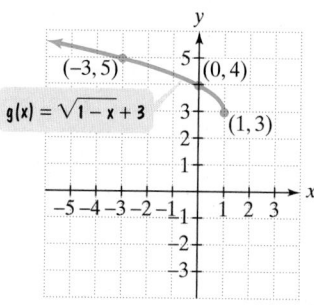

Figure 2.52 Using $f(x) = \sqrt{x}$ to graph $g(x) = \sqrt{1 - x} + 3$

Check Point 8   Use the graph of $f(x) = x^2$ to graph $g(x) = -(x - 2)^2 + 3$.

EXERCISE SET 2.5

Practice Exercises

In Exercises 1–10, begin by graphing the standard quadratic function, $f(x) = x^2$. Then use transformations of this graph to graph the given function.

1. $g(x) = x^2 - 2$

2. $g(x) = x^2 - 1$

3. $g(x) = (x - 2)^2$

4. $g(x) = (x - 1)^2$

5. $h(x) = -(x - 2)^2$

6. $h(x) = -(x - 1)^2$

7. $h(x) = (x - 2)^2 + 1$

8. $h(x) = (x - 1)^2 + 2$

9. $g(x) = 2(x - 2)^2$

10. $g(x) = \frac{1}{2}(x - 1)^2$

In Exercises 11–22, begin by graphing the square root function, $f(x) = \sqrt{x}$. Then use transformations of this graph to graph the given function.

11. $g(x) = \sqrt{x} + 2$ **12.** $g(x) = \sqrt{x} + 1$

13. $g(x) = \sqrt{x + 2}$ **14.** $g(x) = \sqrt{x + 1}$

15. $h(x) = -\sqrt{x + 2}$ **16.** $h(x) = -\sqrt{x + 1}$

17. $h(x) = \sqrt{-x + 2}$ **18.** $h(x) = \sqrt{-x + 1}$

19. $g(x) = \frac{1}{2}\sqrt{x + 2}$ **20.** $g(x) = 2\sqrt{x + 1}$

21. $h(x) = \sqrt{x + 2} - 2$ **22.** $h(x) = \sqrt{x + 1} - 1$

In Exercises 23–34, begin by graphing the absolute value function, $f(x) = |x|$. Then use transformations of this graph to graph the given function.

23. $g(x) = |x| + 4$ **24.** $g(x) = |x| + 3$

25. $g(x) = |x + 4|$ **26.** $g(x) = |x + 3|$

27. $h(x) = |x + 4| - 2$ **28.** $h(x) = |x + 3| - 2$

29. $h(x) = -|x + 4|$ **30.** $h(x) = -|x + 3|$

31. $g(x) = -|x + 4| + 1$ **32.** $g(x) = -|x + 4| + 2$

33. $h(x) = 2|x + 4|$ **34.** $h(x) = 2|x + 3|$

In Exercises 35–44, begin by graphing the standard cubic function, $f(x) = x^3$. Then use transformations of this graph to graph the given function.

35. $g(x) = x^3 - 3$ **36.** $g(x) = x^3 - 2$

37. $g(x) = (x - 3)^3$ **38.** $g(x) = (x - 2)^3$

39. $h(x) = -x^3$ **40.** $h(x) = -(x - 2)^3$

41. $h(x) = \frac{1}{2}x^3$ **42.** $h(x) = \frac{1}{4}x^3$

43. $r(x) = (x - 3)^3 + 2$ **44.** $r(x) = (x - 2)^3 + 1$

In Exercises 45–52, use the graph of the function f to sketch the graph of the given function g.

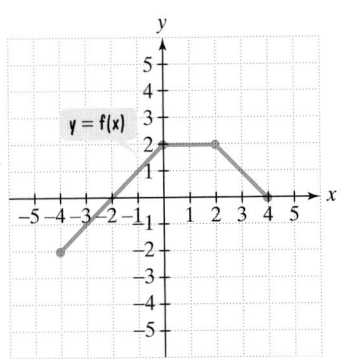

45. $g(x) = f(x) + 1$ **46.** $g(x) = f(x) + 2$

47. $g(x) = f(x + 1)$ **48.** $g(x) = f(x + 2)$

49. $g(x) = -f(x)$ **50.** $g(x) = \frac{1}{2}f(x)$

51. $g(x) = \frac{1}{2}f(x + 1)$ **52.** $g(x) = -f(x + 2)$

In Exercises 53–56, write a possible equation for the function whose graph is shown. Each graph shows a transformation of a common function.

53.

[-2, 8, 1] by [-1, 4, 1]

54.

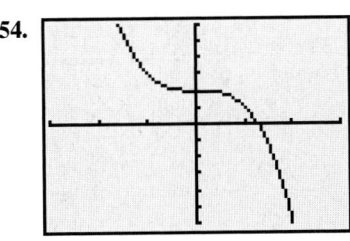

[-3, 3, 1] by [-6, 6, 1]

55.

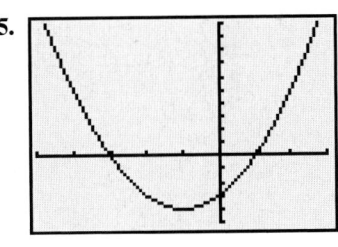

[-5, 3, 1] by [-5, 10, 1]

56.

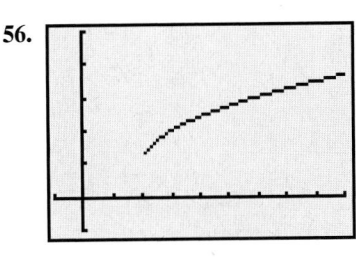

[-1, 9, 1] by [-1, 5, 1]

Application Exercises

57. The function $f(x) = 2.9\sqrt{x} + 20.1$ models the median height, $f(x)$, in inches, of boys who are x months of age. The graph of f is shown.

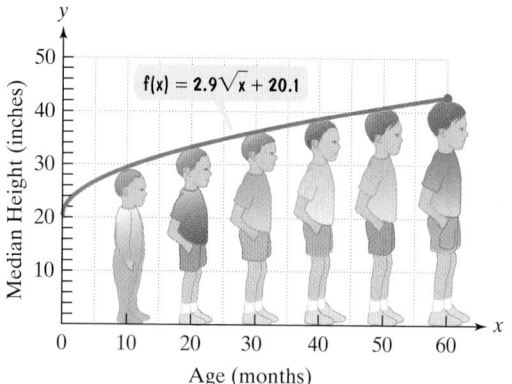

$f(x) = 2.9\sqrt{x} + 20.1$

a. Describe how the graph can be obtained using transformations of the square root function $f(x) = \sqrt{x}$.
b. According to the model, what is the median height of boys who are 48 months, or four years, old? Use a calculator and round to the nearest tenth of an inch. The actual median height for boys at 48 months is 40.8 inches. How well does the model describe the actual height?
c. Use the model to find the average rate of change, in inches per month, between birth and 10 months. Round to the nearest tenth.
d. Use the model to find the average rate of change, in inches per month, between 50 and 60 months. Round to the nearest tenth. How does this compare with your answer in part (c)? How is this difference shown by the graph?

58. The graph shows the amount of money, in billions of dollars, of new student loans from 1993 through 2000.

Amount of New Student Loans

Source: U.S. Department of Education

The data shown can be modeled by the function $f(x) = 6.75\sqrt{x} + 12$, where $f(x)$ is the amount, in billions of dollars, of new student loans x years after 1993.

a. Describe how the graph of f can be obtained using transformations of the square root function $f(x) = \sqrt{x}$. Then sketch the graph of f over the interval $0 \le x \le 9$. If applicable, use a graphing utility to verify your hand-drawn graph.
b. According to the model, how much was loaned in 2000? Round to the nearest tenth of a billion. How well does the model describe the actual data?
c. Use the model to find the average rate of change, in billions of dollars per year, between 1993 and 1995. Round to the nearest tenth.
d. Use the model to find the average rate of change, in billions of dollars per year, between 1998 and 2000. Round to the nearest tenth. How does this compare with you answer in part (c)? How is this difference shown by your graph?
e. Rewrite the function so that it represents the amount, $f(x)$, in billions of dollars, of new student loans x years after 1995.

Writing in Mathematics

59. What must be done to a function's equation so that its graph is shifted vertically upward?
60. What must be done to a function's equation so that its graph is shifted horizontally to the right?
61. What must be done to a function's equation so that its graph is reflected about the x-axis?
62. What must be done to a function's equation so that its graph is reflected about the y-axis?
63. What must be done to a function's equation so that its graph is stretched?

Technology Exercises

64. a. Use a graphing utility to graph $f(x) = x^2 + 1$.
b. Graph $f(x) = x^2 + 1$, $g(x) = f(2x)$, $h(x) = f(3x)$, and $k(x) = f(4x)$ in the same viewing rectangle.
c. Describe the relationship among the graphs of f, g, h, and k, with emphasis on different values of x for points on all four graphs that give the same y-coordinate.
d. Generalize by describing the relationship between the graph of f and the graph of g, where $g(x) = f(cx)$ for $c > 1$.
e. Try out your generalization by sketching the graphs of $f(cx)$ for $c = 1$, $c = 2$, $c = 3$, and $c = 4$ for a function of your choice.

65. a. Use a graphing utility to graph $f(x) = x^2 + 1$.
b. Graph $f(x) = x^2 + 1$, and $g(x) = f(\frac{1}{2}x)$, and $h(x) = f(\frac{1}{4}x)$ in the same viewing rectangle.
c. Describe the relationship among the graphs of f, g, and h, with emphasis on different values of x for points on all three graphs that give the same y-coordinate.
d. Generalize by describing the relationship between the graph of f and the graph of g, where $g(x) = f(cx)$ for $0 < c < 1$.
e. Try out your generalization by sketching the graphs of $f(cx)$ for $c = 1$, and $c = \frac{1}{2}$, and $c = \frac{1}{4}$ for a function of your choice.

Critical Thinking Exercises

66. Which one of the following is true?

   a. If $f(x) = |x|$ and $g(x) = |x + 3| + 3$, then the graph of g is a translation of three units to the right and three units upward of the graph of f.

   b. If $f(x) = -\sqrt{x}$ and $g(x) = \sqrt{-x}$, then f and g have identical graphs.

   c. If $f(x) = x^2$ and $g(x) = 5(x^2 - 2)$, then the graph of g can be obtained from the graph of f by stretching f five units followed by a downward shift of two units.

   d. If $f(x) = x^3$ and $g(x) = -(x - 3)^3 - 4$, then the graph of g can be obtained from the graph of f by moving f three units to the right, reflecting in the x-axis, and then moving the resulting graph down four units.

In Exercises 67–70, functions f and g are graphed in the same rectangular coordinate system. If g is obtained from f through a sequence of transformations, find an equation for g.

67.

68.

69.

70.

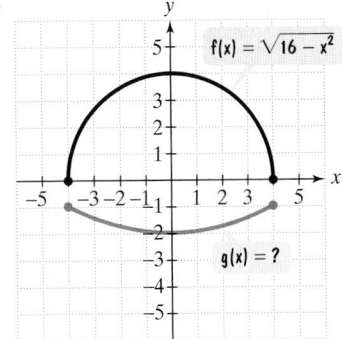

For Exercises 71–74, assume that (a, b) is a point on the graph of f. What is the corresponding point on the graph of each of the following functions?

71. $y = f(-x)$ **72.** $y = 2f(x)$

73. $y = f(x - 3)$ **74.** $y = f(x) - 3$

Group Exercise

75. This activity is a group research project on morphing and should result in a presentation made by group members to the entire class. Be sure to include morphing images that will intrigue class members. You should have no problem finding an array of fascinating images online. Also include a discussion of films using spectacular morphing effects. Rent videos of these films and show appropriate excerpts.

SECTION 2.6 *Combinations of Functions; Composite Functions*

Objectives

1. Combine functions arithmetically, specifying domains.
2. Form composite functions.
3. Determine domains for composite functions.
4. Write functions as compositions.

They say a fool and his money are soon parted and the rest of us just wait to be taxed. It's hard to believe that the United States was a low-tax country in the early part of the twentieth century. Figure 2.53 shows how the tax burden has grown since then. We can use the information shown to illustrate how two functions can be combined to form a new function. In this section, you will learn how to combine functions to obtain new functions.

U.S. Per Capita Tax Burden in 2000 Dollars

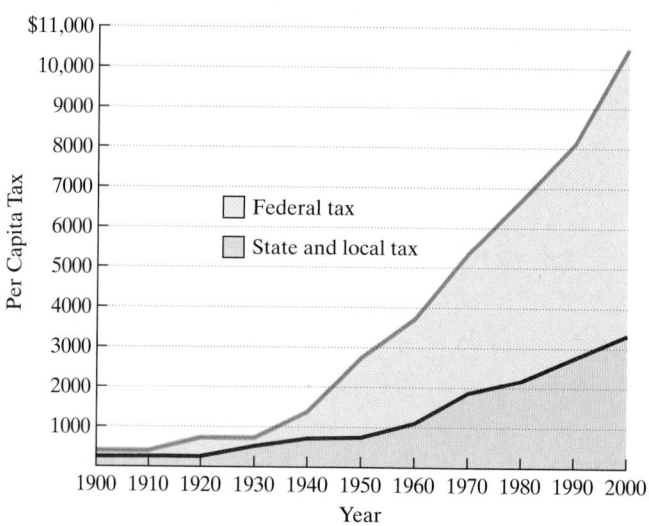

Figure 2.53 *Source:* Tax Foundation

1 Combine functions arithmetically, specifying domains.

Combinations of Functions

To begin our discussion, take a look at the information shown for the year 2000. The total per capita tax burden is approximately $10,500. The per capita state and local tax is approximately $3400. The per capita federal tax is the difference between these amounts.

$$\text{Per capita federal tax} = \$10,500 - \$3400 = \$7100$$

We can think of this subtraction as the subtraction of function values. We do this by introducing the following functions:

Let $T(x)$ = total per capita tax in year x.

Let $S(x)$ = per capita state and local tax in year x.

Using Figure 2.53, we see that

$$T(2000) = \$10{,}500 \quad \text{and} \quad S(2000) = \$3400.$$

We can subtract these function values by introducing a new function, $T - S$, defined by the subtraction of $T(x)$ and $S(x)$. Thus,

$$(T - S)(x) = T(x) - S(x) = \quad \begin{array}{l} \text{total per capita tax in year x} \\ \text{minus state and local per} \\ \text{capita tax in year x.} \end{array}$$

For example,

$$(T - S)(2000) = T(2000) - S(2000) = \$10{,}500 - \$3400 = \$7100.$$

> In 2000, the difference between total tax and state and local tax was $7100. This is the per capita federal tax.

Figure 2.53 illustrates that information involving differences of functions often appears in graphs seen in newspapers and magazines. Like numbers and algebraic expressions, two functions can be added, subtracted multiplied, or divided as long as there are numbers common to the domains of both functions. The common domain for functions T and S in Figure 2.53 is

$$\{1900, 1901, 1902, 1903, \ldots, 2000\}.$$

Because functions are usually given as equations, we perform operations by carrying out these operations with the algebraic expressions that appear on the right side of the equations. For example, we can combine the following two functions using addition:

$$f(x) = 2x + 1 \quad \text{and} \quad g(x) = x^2 - 4.$$

To do so, we add the terms to the right of the equal sign for *f(x)* to the terms to the right of the equal sign for *g(x)*. Here is how it's done:

$$\begin{aligned} (f + g)(x) &= f(x) + g(x) \\ &= (2x + 1) + (x^2 - 4) \quad &&\text{Add terms for f(x) and g(x).} \\ &= 2x - 3 + x^2 \quad &&\text{Combine like terms.} \\ &= x^2 + 2x - 3 \quad &&\text{Arrange terms in descending powers of x.} \end{aligned}$$

The name of this new function is $f + g$. Thus, the sum $f + g$ is the function defined by $(f + g)(x) = x^2 + 2x - 3$. The domain of $f + g$ consists of the numbers x that are in the domain of f and in the domain of g. Because neither f nor g contains division or even roots, the domain of each function is the set of all real numbers. Thus, the domain of $f + g$ is also the set of all real numbers.

EXAMPLE 1 Finding the Sum of Two Functions

Let $f(x) = x^2 - 3$ and $g(x) = 4x + 5$. Find:

a. $(f + g)(x)$ **b.** $(f + g)(3)$.

Solution

a. $(f + g)(x) = f(x) + g(x) = (x^2 - 3) + (4x + 5) = x^2 + 4x + 2$. Thus, $(f + g)(x) = x^2 + 4x + 2$.

b. We find $(f + g)(3)$ by substituting 3 for x in the equation for $f + g$.

$(f + g)(x) = x^2 + 4x + 2$ *This is the equation for f + g.*

Substitute 3 for x.

$(f + g)(3) = 3^2 + 4 \cdot 3 + 2 = 9 + 12 + 2 = 23$

Check Point 1 Let $f(x) = 3x^2 + 4x - 1$ and $g(x) = 2x + 7$. Find:

a. $(f + g)(x)$ **b.** $(f + g)(4)$.

Here is a general definition for function addition:

The Sum of Functions

Let f and g be two functions. The **sum $f + g$** is the function defined by

$$(f + g)(x) = f(x) + g(x).$$

The domain of $f + g$ is the set of all real numbers that are common to the domain of f and the domain of g.

EXAMPLE 2 Adding Functions and Determining the Domain

Let $f(x) = \sqrt{x + 3}$ and $g(x) = \sqrt{x - 2}$. Find:

a. $(f + g)(x)$ **b.** the domain of $f + g$.

Solution

a. $(f + g)(x) = f(x) + g(x) = \sqrt{x + 3} + \sqrt{x - 2}$

b. The domain of $f + g$ is the set of all real numbers that are common to the domain of f and the domain of g. Thus, we must find the domains of f and g. We will do so for f first.

Note that $f(x) = \sqrt{x + 3}$ is a function involving the square root of $x + 3$. Because the square root of a negative quantity is not a real number, the value of $x + 3$ must be nonnegative. Thus, the domain of f is all x such that $x + 3 \geq 0$. Equivalently, the domain is $\{x \mid x \geq -3\}$, or $[-3, \infty)$.

Likewise, $g(x) = \sqrt{x - 2}$ is also a square root function. Because the square root of a negative quantity is not a real number, the value of $x - 2$ must be nonnegative. Thus, the domain of g is all x such that $x - 2 \geq 0$. Equivalently, the domain is $\{x \mid x \geq 2\}$, or $[2, \infty)$.

Now, we can use a number line to determine the domain of $f + g$. Figure 2.54 shows the domain of f in blue and the domain of g in red. Can you see that all real numbers greater than or equal to 2 are common to both domains? This is shown in purple on the number line. Thus, the domain of $f + g$ is $[2, \infty)$.

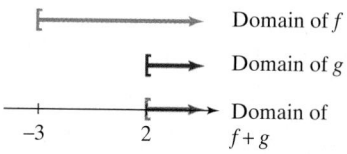

Domain of f

Domain of g

Domain of $f + g$

Figure 2.54 Finding the domain of the sum $f + g$

Technology

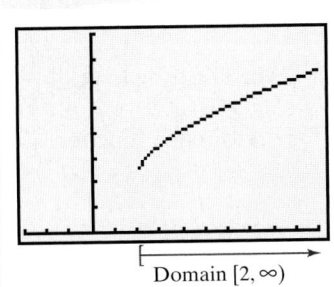

Domain $[2, \infty)$

The graph on the left is the graph of
$$y = \sqrt{x + 3} + \sqrt{x - 2}$$
in a $[-3, 10, 1]$ by $[0, 8, 1]$ viewing rectangle. The graph reveals what we discovered algebraically in Example 2(b). The domain of this function is $[2, \infty)$.

Check Point 2 Let $f(x) = \sqrt{x - 3}$ and $g(x) = \sqrt{x + 1}$. Find:
  a. $(f + g)(x)$ **b.** the domain of $f + g$.

We can also combine functions using subtraction, multiplication, and division by performing operations with the algebraic expressions that appear on the right side of the equations. For example, the functions $f(x) = x + 3$ and $g(x) = x - 1$ can be combined to form the difference, product, and quotient of f and g. Here's how it's done.

Difference: f − g
$$(f - g)(x) = f(x) - g(x)$$
$$= (x + 3) - (x - 1) = x + 3 - x + 1 = 4$$

Product: fg
$$(fg)(x) = f(x) \cdot g(x)$$
$$= (x + 3)(x - 1) = x^2 + 2x - 3$$

Quotient: $\frac{f}{g}$
$$\left(\frac{f}{g}\right)(x) = \frac{f(x)}{g(x)} = \frac{x + 3}{x - 1}, \quad x \neq 1$$

Just like the domain for $f + g$, the domain for each of these functions consists of all real numbers that are common to the domains of f and g. In the case of the quotient function $\dfrac{f(x)}{g(x)}$, we must remember not to divide by 0, so we add the further restriction that $g(x) \neq 0$.

The following definitions summarize our discussion:

Definitions: Sum, Difference, Product, and Quotient of Functions

Let f and g be two functions. The **sum** $f + g$, the **difference** $f - g$, the **product** fg, and the **quotient** $\frac{f}{g}$ are functions whose domains are the set of all real numbers common to the domains of f and g, defined as follows:

1. Sum: $(f + g)(x) = f(x) + g(x)$

2. Difference: $(f - g)(x) = f(x) - g(x)$

3. Product: $(fg)(x) = f(x) \cdot g(x)$

4. Quotient: $\left(\dfrac{f}{g}\right)(x) = \dfrac{f(x)}{g(x)}$, provided $g(x) \neq 0$

EXAMPLE 3   Combining Functions

If $f(x) = 2x - 1$ and $g(x) = x^2 + x - 2$, find:

a. $(f - g)(x)$ **b.** $(fg)(x)$ **c.** $\left(\dfrac{f}{g}\right)(x)$.

Determine the domain for each function.

Solution

a. $(f - g)(x) = f(x) - g(x)$ This is the definition of the difference $f - g$.

$\qquad = (2x - 1) - (x^2 + x - 2)$ Subtract $g(x)$ from $f(x)$.

$\qquad = 2x - 1 - x^2 - x + 2$ Perform the subtraction.

$\qquad = -x^2 + x + 1$ Combine like terms and arrange terms in descending powers of x.

b. $(fg)(x) = f(x) \cdot g(x)$ This is the definition of the product fg.

$\qquad = (2x - 1)(x^2 + x - 2)$ Multiply $f(x)$ and $g(x)$.

$\qquad = 2x(x^2 + x - 2) - 1(x^2 + x - 2)$ Multiply each term in the second factor by 2x and −1, respectively.

$\qquad = 2x^3 + 2x^2 - 4x - x^2 - x + 2$ Use the distributive property.

$\qquad = 2x^3 + (2x^2 - x^2) + (-4x - x) + 2$ Rearrange terms so that like terms are adjacent.

$\qquad = 2x^3 + x^2 - 5x + 2$ Combine like terms.

c. $\left(\dfrac{f}{g}\right)(x) = \dfrac{f(x)}{g(x)}$ This is the definition of the quotient $\dfrac{f}{g}$.

$\qquad = \dfrac{2x - 1}{x^2 + x - 2}$ Divide the algebraic expressions for $f(x)$ and $g(x)$.

Study Tip

If the function $\frac{f}{g}$ can be simplified, determine the domain *before* simplifying.

EXAMPLE:

$f(x) = x^2 - 4$ and

$g(x) = x - 2$

$\left(\dfrac{f}{g}\right)(x) = \dfrac{x^2 - 4}{x - 2}$

Domain of $\left(\dfrac{f}{g}\right)$ is $\{x | x \neq 2\}$.

$= \dfrac{\overset{1}{(x + 2)\cancel{(x - 2)}}}{\underset{1}{\cancel{(x - 2)}}} = x + 2$

Because the equations for f and g do not involve division or contain even roots, the domain of both f and g is the set of all real numbers. Thus, the domain of $f - g$ and fg is the set of all real numbers. However, for $\frac{f}{g}$, the denominator cannot equal zero. We can factor the denominator as follows:

$$\left(\dfrac{f}{g}\right)(x) = \dfrac{2x - 1}{x^2 + x - 2} = \dfrac{2x - 1}{(x + 2)(x - 1)}.$$

Because $x + 2 \neq 0$, $x \neq -2$.   Because $x - 1 \neq 0$, $x \neq 1$.

We see that the domain for $\frac{f}{g}$ is the set of all real numbers except -2 and 1: $\{x | x \neq -2, x \neq 1\}$.

Check Point 3   If $f(x) = x - 5$ and $g(x) = x^2 - 1$, find:

a. $(f - g)(x)$ **b.** $(fg)(x)$ **c.** $\left(\dfrac{f}{g}\right)(x)$.

Determine the domain for each function.

② Form composite functions.

Composite Functions

There is another way of combining two functions. To help understand this new combination, suppose that your computer store is having a sale. The models that are on sale cost either $300 less than the regular price or 85% of the regular price. If x represents the computer's regular price, both discounts can be described with the following functions:

$$f(x) = x - 300 \qquad\qquad g(x) = 0.85x.$$

> The computer is on sale for $300 less than its regular price.

> The computer is on sale for 85% of its regular price.

At the store, you bargain with the salesperson. Eventually, she makes an offer you can't refuse: The sale price is 85% of the regular price followed by a $300 reduction:

$$0.85x - 300.$$

> 85% of the regular price

> followed by a $300 reduction

In terms of functions f and g, this offer can be obtained by taking the output of $g(x) = 0.85x$, namely $0.85x$, and using it as the input of f:

$$f(x) = x - 300$$

> Replace x with 0.85x, the output of g(x) = 0.85x.

$$f(0.85x) = 0.85x - 300.$$

Because $0.85x$ is $g(x)$, we can write this last equation as

$$f(g(x)) = 0.85x - 300.$$

We read this equation as "f of g of x is equal to $0.85x - 300$." We call $f(g(x))$ the *composition of the function f with g*, or a *composite function*. This composite function is written $f \circ g$. Thus,

$$(f \circ g)(x) = f(g(x)) = 0.85x - 300.$$

Like all functions, we can evaluate $f \circ g$ for a specified value of x in the function's domain. For example, here's how to find the value of this function at 1400:

$$(f \circ g)(x) = 0.85x - 300 \quad \text{This composite function describes the offer you cannot refuse.}$$

> Replace x with 1400.

$$(f \circ g)(1400) = 0.85(1400) - 300 = 1190 - 300 = 890.$$

This means that a computer that regularly sells for $1400 is on sale for $890 subject to both discounts.

Before you run out to buy a new computer, let's generalize our discussion of the computer's double discount and define the composition of any two functions.

The Composition of Functions

The **composition of the function f with g** is denoted by $f \circ g$ and is defined by the equation

$$(f \circ g)(x) = f(g(x)).$$

The domain of the **composite function $f \circ g$** is the set of all x such that

1. x is in the domain of g and
2. $g(x)$ is in the domain of f.

The composition of f with g, $f \circ g$, is pictured as a machine with inputs and outputs in Figure 2.55. The diagram indicates that the output of g, or $g(x)$, becomes the input for "machine" f. If $g(x)$ is not in the domain of f, it cannot be input into machine f, and so $g(x)$ must be discarded.

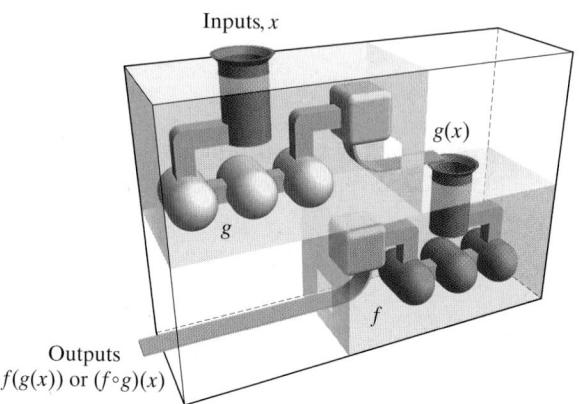

Figure 2.55 Inputting one function into a second function

EXAMPLE 4   Forming Composite Functions

Given $f(x) = 3x - 4$ and $g(x) = x^2 + 6$, find:

a. $(f \circ g)(x)$ **b.** $(g \circ f)(x)$.

Solution

a. We begin with $(f \circ g)(x)$, the composition of f with g. Because $(f \circ g)(x)$ means $f(g(x))$, we must replace each occurrence of x in the equation for f with $g(x)$.

$$f(x) = 3x - 4 \qquad \text{This is the given equation for } f.$$

Replace x with $g(x)$.

$$(f \circ g)(x) = f(g(x)) = 3g(x) - 4$$
$$= 3(x^2 + 6) - 4 \quad \text{Because } g(x) = x^2 + 6, \text{ replace } g(x) \text{ with } x^2 + 6.$$
$$= 3x^2 + 18 - 4 \quad \text{Use the distributive property.}$$
$$= 3x^2 + 14 \qquad \text{Simplify.}$$

Thus, $(f \circ g)(x) = 3x^2 + 14$.

b. Next, we find $(g \circ f)(x)$, the composition of g with f. Because $(g \circ f)(x)$ means $g(f(x))$, we must replace each occurrence of x in the equation for g with $f(x)$.

$$g(x) = x^2 + 6 \qquad \textit{This is the given equation for g.}$$

Replace x with f(x).

$$(g \circ f)(x) = g(f(x)) = (f(x))^2 + 6$$

$$= (3x - 4)^2 + 6 \qquad \textit{Because f(x) = 3x − 4, replace f(x) with 3x − 4.}$$

$$= 9x^2 - 24x + 16 + 6 \qquad \textit{Use (A − B)}^2 = A^2 − 2AB + B^2 \textit{ to square 3x − 4.}$$

$$= 9x^2 - 24x + 22 \qquad \textit{Simplify.}$$

Thus, $(g \circ f)(x) = 9x^2 - 24x + 22$. Notice that $(f \circ g)(x)$ is not the same function as $(g \circ f)(x)$.

Check Point 4 Given $f(x) = 5x + 6$ and $g(x) = x^2 - 1$, find:

a. $(f \circ g)(x)$ **b.** $(g \circ f)(x)$.

3 Determine domains for composite functions.

We need to be careful in determining the domain for the composite function

$$(f \circ g)(x) = f(g(x)).$$

The following values must be excluded from the input x:

- If x is not in the domain of g, it must not be in the domain of $f \circ g$.
- Any x for which $g(x)$ is not in the domain of f must not be in the domain of $f \circ g$.

EXAMPLE 5 **Forming a Composite Function and Finding Its Domain**

Given $f(x) = \dfrac{2}{x-1}$ and $g(x) = \dfrac{3}{x}$, find:

a. $(f \circ g)(x)$ **b.** the domain of $f \circ g$.

Solution

a. Because $(f \circ g)(x)$ means $f(g(x))$, we must replace x in $f(x) = \dfrac{2}{x-1}$ with $g(x)$.

Study Tip

The procedure for simplifying complex fractions can be found in Section P.6, pages 66–67.

$$(f \circ g)(x) = f(g(x)) = \frac{2}{g(x) - 1} = \frac{2}{\frac{3}{x} - 1} = \frac{2}{\frac{3}{x} - 1} \cdot \frac{x}{x} = \frac{2x}{3 - x}$$

$g(x) = \dfrac{3}{x}$

Simplify the complex fraction by multiplying by $\frac{x}{x}$, or 1.

Thus, $(f \circ g)(x) = \dfrac{2x}{3 - x}$.

b. We determine the domain of $(f \circ g)(x)$ in two steps.

Rules for Excluding Numbers from the Domain of $(f \circ g)(x) = f(g(x))$	Applying the Rules to $f(x) = \dfrac{2}{x-1}$ and $g(x) = \dfrac{3}{x}$
If x is not in the domain of g, it must not be in the domain of $f \circ g$.	The domain of g is $\{x \mid x \neq 0\}$. Thus, 0 must be excluded from the domain of $f \circ g$.
Any x for which $g(x)$ is not in the domain of f must not be in the domain of $f \circ g$.	The domain of f is $\{x \mid x \neq 1\}$. This means we must exclude from the domain of $f \circ g$ any x for which $g(x) = 1$. $\dfrac{3}{x} = 1$ Set g(x) equal to 1. $3 = x$ Multiply both sides by x. 3 must be excluded from the domain of $f \circ g$.

The domain of $f \circ g$ is $\{x \mid x \neq 0 \text{ and } x \neq 3\}$.

Check Point 5 Given $f(x) = \dfrac{4}{x+2}$ and $g(x) = \dfrac{1}{x}$, find:

  a. $(f \circ g)(x)$ **b.** the domain of $f \circ g$.

4 Write functions as compositions

Decomposing Functions

When you form a composite function, you "compose" two functions to form a new function. It is also possible to reverse this process. That is, you can "decompose" a given function and express it as a composition of two functions. Although there is more than one way to do this, there is often a "natural" selection that comes to mind first. For example, consider the function h defined by

$$h(x) = (3x^2 - 4x + 1)^5.$$

The function h takes $3x^2 - 4x + 1$ and raises it to the power 5. A natural way to write h as a composition of two functions is to raise the function $g(x) = 3x^2 - 4x + 1$ to the power 5. Thus, if we let

$$f(x) = x^5 \text{ and } g(x) = 3x^2 - 4x + 1, \text{ then}$$
$$(f \circ g)(x) = f(g(x)) = f(3x^2 - 4x + 1) = (3x^2 - 4x + 1)^5.$$

EXAMPLE 6 Writing a Function as a Composition

Express as a composition of two functions:

$$h(x) = \sqrt[3]{x^2 + 1}.$$

Solution The function h takes $x^2 + 1$ and takes its cube root. A natural way to write h as a composition of two functions is to take the cube root of the function $g(x) = x^2 + 1$. Thus, we let

$$f(x) = \sqrt[3]{x} \text{ and } g(x) = x^2 + 1.$$

Study Tip

Suppose the form of function h is

$h(x) = (\text{algebraic expression})^{\text{power}}$.

Function h can be expressed as a composition, $f \circ g$, using

$f(x) = x^{\text{power}}$

$g(x) = \text{algebraic expression}$.

We can check this composition by finding $(f \circ g)(x)$. This should give the original function, namely $h(x) = \sqrt[3]{x^2 + 1}$.

$$(f \circ g)(x) = f(g(x)) = f(x^2 + 1) = \sqrt[3]{x^2 + 1} = h(x)$$

Check Point 6 Express as a composition of two functions:
$$h(x) = \sqrt{x^2 + 5}.$$

EXERCISE SET 2.6

Practice Exercises

1. If $f(x) = 2x^2 - 5$ and $g(x) = 3x + 7$, find:
  a. $(f + g)(x)$ **b.** $(f + g)(4)$.

2. If $f(x) = 3x^2 - 2x + 1$ and $g(x) = 4x - 1$, find:
  a. $(f + g)(x)$ **b.** $(f + g)(5)$.

3. If $f(x) = \sqrt{x - 6}$ and $g(x) = \sqrt{x + 2}$, find:
  a. $(f + g)(x)$ **b.** the domain of $f + g$.

4. If $f(x) = \sqrt{x - 8}$ and $g(x) = \sqrt{x + 5}$, find:
  a. $(f + g)(x)$ **b.** the domain of $f + g$.

In Exercises 5–16, find $f + g, f - g, fg,$ and $\frac{f}{g}$. Determine the domain for each function.

5. $f(x) = 2x + 3, \quad g(x) = x - 1$

6. $f(x) = 3x - 4, \quad g(x) = x + 2$

7. $f(x) = x - 5, \quad g(x) = 3x^2$

8. $f(x) = x - 6, \quad g(x) = 5x^2$

9. $f(x) = 2x^2 - x - 3, \quad g(x) = x + 1$

10. $f(x) = 6x^2 - x - 1, \quad g(x) = x - 1$

11. $f(x) = \sqrt{x}, \quad g(x) = x - 4$

12. $f(x) = \dfrac{1}{x}, \quad g(x) = x - 5$

13. $f(x) = 2 + \dfrac{1}{x}, \quad g(x) = \dfrac{1}{x}$

14. $f(x) = 6 - \dfrac{1}{x}, \quad g(x) = \dfrac{1}{x}$

15. $f(x) = \sqrt{x + 4}, \quad g(x) = \sqrt{x - 1}$

16. $f(x) = \sqrt{x + 6}, \quad g(x) = \sqrt{x - 3}$

In Exercises 17–28, find:
  a. $(f \circ g)(x)$
  b. $(g \circ f)(x)$
  c. $(f \circ g)(2)$.

17. $f(x) = 2x, \quad g(x) = x + 7$

18. $f(x) = 3x, \quad g(x) = x - 5$

19. $f(x) = x + 4, \quad g(x) = 2x + 1$

20. $f(x) = 5x + 2, \quad g(x) = 3x - 4$

21. $f(x) = 4x - 3, \quad g(x) = 5x^2 - 2$

22. $f(x) = 7x + 1, \quad g(x) = 2x^2 - 9$

23. $f(x) = x^2 + 2, \quad g(x) = x^2 - 2$

24. $f(x) = x^2 + 1, \quad g(x) = x^2 - 3$

25. $f(x) = \sqrt{x}, \quad g(x) = x - 1$

26. $f(x) = \sqrt{x}, \quad g(x) = x + 2$

27. $f(x) = 2x - 3, \quad g(x) = \dfrac{x + 3}{2}$

28. $f(x) = 6x - 3, \quad g(x) = \dfrac{x + 3}{6}$

In Exercises 29–38, find:
  a. $(f \circ g)(x)$ **b.** the domain of $f \circ g$.

29. $f(x) = \dfrac{2}{x + 3}, \quad g(x) = \dfrac{1}{x}$

30. $f(x) = \dfrac{5}{x + 4}, \quad g(x) = \dfrac{1}{x}$

31. $f(x) = \dfrac{x}{x + 1}, \quad g(x) = \dfrac{4}{x}$

32. $f(x) = \dfrac{x}{x + 5}, \quad g(x) = \dfrac{6}{x}$

33. $f(x) = \sqrt{x}, \quad g(x) = x + 3$

34. $f(x) = \sqrt{x}, \quad g(x) = x - 3$

35. $f(x) = x^2 + 4, \quad g(x) = \sqrt{1 - x}$

36. $f(x) = x^2 + 1, \quad g(x) = \sqrt{2 - x}$

37. $f(x) = 4 - x^2, \quad g(x) = \sqrt{x^2 - 4}$

38. $f(x) = 9 - x^2, \quad g(x) = \sqrt{x^2 - 9}$

In Exercises 39–46, express the given function h as a composition of two functions f and g so that $h(x) = (f \circ g)(x)$.

39. $h(x) = (3x - 1)^4$

40. $h(x) = (2x - 5)^3$

41. $h(x) = \sqrt[3]{x^2 - 9}$

42. $h(x) = \sqrt{5x^2 + 3}$

43. $h(x) = |2x - 5|$

44. $h(x) = |3x - 4|$

45. $h(x) = \dfrac{1}{2x - 3}$

46. $h(x) = \dfrac{1}{4x + 5}$

In Exercises 47–58, use the graphs of f and g to evaluate each function.

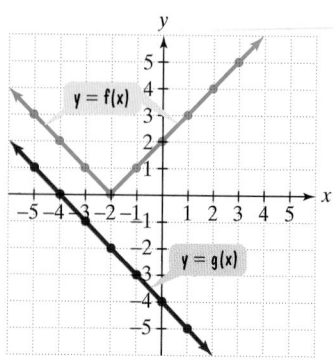

47. $(f + g)(-3)$

48. $(f + g)(-4)$

49. $(f - g)(2)$

50. $(g - f)(2)$

51. $\left(\dfrac{f}{g}\right)(-6)$

52. $\left(\dfrac{f}{g}\right)(-5)$

53. $(fg)(-4)$

54. $(fg)(-2)$

55. $(f \circ g)(2)$

56. $(f \circ g)(1)$

57. $(g \circ f)(0)$

58. $(g \circ f)(-1)$

Application Exercises

It seems that Phideau's medical bills are costing us an arm and a paw. The graph shows veterinary costs, in billions of dollars, for dogs and cats in five selected years. Let

$D(x)$ = *veterinary costs, in billions of dollars, for dogs in year x*

$C(x)$ = *veterinary costs, in billions of dollars, for cats in year x.*

Use the graph to solve Exercises 59–62.

Veterinary Costs in the U.S.

Dogs Cats

Cost (billions)

$15

12

9

6

3

1983 1987 1991 1996 2000

Year

Source: American Veterinary Medical Association

59. Find an estimate of $(D + C)(2000)$. What does this mean in terms of the variables in this situation?

60. Find an estimate of $(D - C)(2000)$. What does this mean in terms of the variables in this situation?

61. Using the information shown in the graph, what is the domain of $D + C$?

62. Using the information shown in the graph, what is the domain of $D - C$?

Consider the following functions:

$f(x)$ = *population of the world's more developed regions in year x*

$g(x)$ = *population of the world's less developed regions in year x*

$h(x)$ = *total world population in year x.*

Use these functions and the graph shown to answer Exercises 63–66.

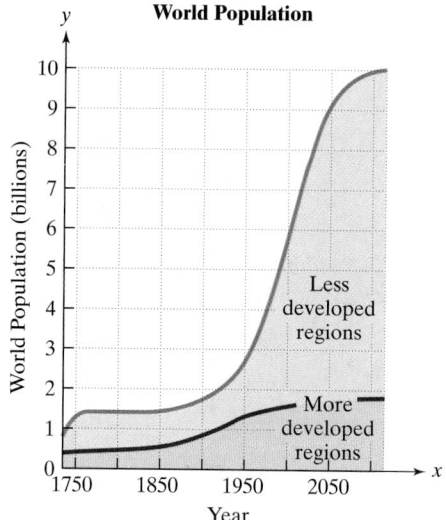

World Population

Less developed regions

More developed regions

Source: Population Reference Bureau

63. What does the function $f + g$ represent?

64. What does the function $h - g$ represent?

65. Use the graph to estimate $(f + g)(2000)$.

66. Use the graph to estimate $(h - g)(2000)$.

67. A company that sells radios has a yearly fixed cost of $600,000. It costs the company $45 to produce each radio. Each radio will sell for $65. The company's costs and revenue are modeled by the following functions:

$C(x) = 600,000 + 45x$ This function models the company's costs.

$R(x) = 65x.$ This function models the company's revenue.

Find and interpret $(R - C)(20,000)$, $(R - C)(30,000)$ and $(R - C)(40,000)$.

68. A department store has two locations in a city. From 1998 through 2002, the profits for each of the store's two branches are modeled by the functions $f(x) = -0.44x + 13.62$ and $g(x) = 0.51x + 11.14$. In each model, x represents the number of years after 1998 and f and g represent the profit, in millions of dollars.
  a. What is the slope of f? Describe what this means.
  b. What is the slope of g? Describe what this means.
  c. Find $f + g$. What is the slope of this function? What does this mean?

69. The regular price of a computer is x dollars. Let $f(x) = x - 400$ and $g(x) = 0.75x$.
  a. Describe what the functions f and g model in terms of the price of the computer.
  b. Find $(f \circ g)(x)$ and describe what this models in terms of the price of the computer.
  c. Repeat part (b) for $(g \circ f)(x)$.
  d. Which composite function models the greater discount on the computer, $f \circ g$ or $g \circ f$? Explain.

70. The regular price of a pair of jeans is x dollars. Let $f(x) = x - 5$ and $g(x) = 0.6x$.
  a. Describe what functions f and g model in terms of the price of the jeans.
  b. Find $(f \circ g)(x)$ and describe what this models in terms of the price of the jeans.
  c. Repeat part (b) for $(g \circ f)(x)$.
  d. Which composite function models the greater discount on the jeans, $f \circ g$ or $g \circ f$? Explain.

Writing in Mathematics

71. If equations for functions f and g are given, explain how to find $f + g$.

72. If the equations of two functions are given, explain how to obtain the quotient function and its domain.

73. If equations for functions f and g are given, describe two ways to find $(f - g)(3)$.

74. Explain how to use the graphs in Figure 2.53 on page 248 to estimate the per capita federal tax for any one of the years shown on the horizontal axis.

75. Describe a procedure for finding $(f \circ g)(x)$. What is the name of this function?

76. Describe the values of x that must be excluded from the domain of $(f \circ g)(x)$.

Technology Exercises

77. The function $f(t) = -0.14t^2 + 0.51t + 31.6$ models the U.S. population ages 65 and older, $f(t)$, in millions, t years after 1990. The function $g(t) = 0.54t^2 + 12.64t + 107.1$ models the total yearly cost of Medicare, $g(t)$, in billions of dollars, t years after 1990. Graph the function $\dfrac{g}{f}$ in a $[0, 15, 1]$ by $[0, 60, 1]$ viewing rectangle. What does the shape of the graph indicate about the per capita costs of Medicare for the U.S. population ages 65 and over with increasing time?

78. Graph $y_1 = x^2 - 2x, y_2 = x$, and $y_3 = y_1 \div y_2$ in the same $[-10, 10, 1]$ by $[-10, 10, 1]$ viewing rectangle. Then use the $\boxed{\text{TRACE}}$ feature to trace along y_3. What happens at $x = 0$? Explain why this occurs.

79. Graph $y_1 = x^2 - 4$, $y_2 = \sqrt{4 - x^2}$, and $y_3 = y_2^2 - 4$ in the same $[-5, 5, 1]$ by $[-5, 5, 1]$ viewing rectangle. If y_1 represents f and y_2 represents g, use the graph of y_3 to find the domain of $f \circ g$. Then verify your observation algebraically.

Critical Thinking Exercises

80. Which one of the following is true?
  a. If $f(x) = x^2 - 4$ and $g(x) = \sqrt{x^2 - 4}$, then $(f \circ g)(x) = -x^2$ and $(f \circ g)(5) = -25$.
  b. There can never be two functions f and g, where $f \neq g$, for which $(f \circ g)(x) = (g \circ f)(x)$.
  c. If $f(7) = 5$ and $g(4) = 7$ then $(f \circ g)(4) = 35$.
  d. If $f(x) = \sqrt{x}$ and $g(x) = 2x - 1$, then $(f \circ g)(5) = g(2)$.

81. Prove that if f and g are even functions, then fg is also an even function.

82. Define two functions f and g so that $f \circ g = g \circ f$.

83. Use the graphs given in Exercises 63–66 to create a graph that shows the population, in billions, of less developed regions from 1950 through 2050.

Group Exercise

84. Consult an almanac, newspaper, magazine, or the Internet to find data displayed in a graph in the style of Figure 2.53 on page 248. Using the two graphs that group members find most interesting, introduce two functions that are related to the graphs. Then write and solve a problem involving function subtraction for each selected graph. If you are not sure where to begin, reread page 248–249 or look at Exercises 63–66 in this exercise set.

SECTION 2.7 Inverse Functions

Objectives

1. Verify inverse functions.
2. Find the inverse of a function.
3. Use the horizontal line test to determine if a function has an inverse function.
4. Use the graph of a one-to-one function to graph its inverse function.

In most societies, women say they prefer to marry men who are older than themselves, whereas men say they prefer women who are younger. Evolutionary psychologists attribute these preferences to female concern with a partner's material resources and male concern with a partner's fertility (*Source:* David M. Buss, *Psychological Inquiry*, 6, 1–30). When the man is considerably older than the woman, people rarely comment. However, when the woman is older, as in the relationship between actors Susan Sarandon and Tim Robbins, people take notice.

Figure 2.56 shows the preferred age in a mate in five selected countries. We can focus on the data for the women and define a function.

Preferred Age in a Mate

Figure 2.56

Source: Carole Wade and Carol Tavris, *Psychology Sixth Edition*, Prentice Hall, 2000

Let the domain of the function be the set of the five countries shown in the graph. Let the range be the set of the average number of years women in each of the respective countries prefer men who are older than themselves. The function can be written as follows:

$$f:\{(\text{Zambia}, 4.2), (\text{Colombia}, 4.5), (\text{Poland}, 3.3), (\text{Italy}, 3.3), (\text{U.S.}, 2.5)\}.$$

Now let's "undo" f by interchanging the first and second components in each of its ordered pairs. Switching the inputs and outputs of f, we obtain the following relation:

Same first component

Undoing f:$\{(4.2, \text{Zambia}), (4.5, \text{Colombia}), (3.3, \text{Poland}), (3.3, \text{Italy}), (2.5, \text{U.S.})\}$.

Different second components

Can you see that this relation is not a function? Two of its ordered pairs have the same first component and different second components. This violates the definition of a function.

If a function f is a set of ordered pairs, (x, y), then the changes produced by f can be "undone" by reversing the components of all the ordered pairs. The resulting relation, (y, x), may or may not be a function. In this section, we will develop these ideas by studying functions whose compositions have a special "undoing" relationship.

Inverse Functions

Here are two functions that describe situations related to the price of a computer, x:

$$f(x) = x - 300 \qquad g(x) = x + 300.$$

Function f subtracts \$300 from the computer's price and function g adds \$300 to the computer's price. Let's see what $f(g(x))$ does. Put $g(x)$ into f:

$$f(x) = x - 300 \qquad \text{This is the given equation for f.}$$

Replace x with g(x).

$$f(g(x)) = g(x) - 300$$
$$= x + 300 - 300 \qquad \text{Because g(x) = x + 300, replace g(x) with}$$
$$\text{x + 300.}$$

$$= x. \qquad \text{This is the computer's original price.}$$

Using $f(g(x))$, the computer's price, x, went through two changes: the first, an increase; the second, a decrease:

$$x + 300 - 300.$$

The final price of the computer, x, is identical to its starting price, x.

In general, if the changes made to x by function g are undone by the changes made by function f, then

$$f(g(x)) = x.$$

Assume, also, that this "undoing" takes place in the other direction:

$$g(f(x)) = x.$$

Under these conditions, we say that each function is the *inverse function* of the other. The fact that g is the inverse of f is expressed by renaming g as f^{-1}, read "f-inverse." For example, the inverse functions

$$f(x) = x - 300 \qquad g(x) = x + 300$$

are usually named as follows:

$$f(x) = x - 300 \qquad f^{-1}(x) = x + 300.$$

With these ideas in mind, we present the formal definition of the inverse of a function:

Definition of the Inverse of a Function

Let f and g be two functions such that

$$f(g(x)) = x \qquad \text{for every } x \text{ in the domain of } g$$

and

$$g(f(x)) = x \qquad \text{for every } x \text{ in the domain of } f.$$

The function g is the **inverse of the function** f, and is denoted by f^{-1} (read "f-inverse"). Thus, $f(f^{-1}(x)) = x$ and $f^{-1}(f(x)) = x$. The domain of f is equal to the range of f^{-1}, and vice versa.

1 Verify inverse functions.

EXAMPLE 1 Verifying Inverse Functions

Show that each function is an inverse of the other:

$$f(x) = 5x \qquad \text{and} \qquad g(x) = \frac{x}{5}.$$

Solution To show that f and g are inverses of each other, we must show that $f(g(x)) = x$ and $g(f(x)) = x$. We begin with $f(g(x))$.

$$f(x) = 5x \qquad \text{This is the given equation for f.}$$

Replace x with g(x).

$$f(g(x)) = 5g(x) = 5\left(\frac{x}{5}\right) = x$$

Next, we find $g(f(x))$.

$$g(x) = \frac{x}{5} \qquad \text{This is the given equation for g.}$$

Replace x with f(x).

$$g(f(x)) = \frac{f(x)}{5} = \frac{5x}{5} = x$$

Because g is the inverse of f (and vice versa), we can use inverse notation and write

$$f(x) = 5x \qquad \text{and} \qquad f^{-1}(x) = \frac{x}{5}.$$

Notice how f^{-1} undoes the change produced by f: f changes x by multiplying by 5 and f^{-1} undoes this by dividing by 5.

Check Point 1 Show that each function is an inverse of the other:

$$f(x) = 7x \qquad \text{and} \qquad g(x) = \frac{x}{7}.$$

EXAMPLE 2 Verifying Inverse Functions

Show that each function is an inverse of the other:

$$f(x) = 3x + 2 \quad \text{and} \quad g(x) = \frac{x - 2}{3}.$$

Solution To show that f and g are inverses of each other, we must show that $f(g(x)) = x$ and $g(f(x)) = x$. We begin with $f(g(x))$.

$$f(x) = 3x + 2 \qquad \textit{This is the equation for f.}$$

Replace x with g(x).

$$f(g(x)) = 3g(x) + 2 = 3\left(\frac{x - 2}{3}\right) + 2 = x - 2 + 2 = x$$

Next, we find $g(f(x))$.

$$g(x) = \frac{x - 2}{3} \qquad \textit{This is the equation for g.}$$

Replace x with f (x).

$$g(f(x)) = \frac{f(x) - 2}{3} = \frac{(3x + 2) - 2}{3} = \frac{3x}{3} = x$$

Because g is the inverse of f (and vice versa), we can use inverse notation and write

$$f(x) = 3x + 2 \quad \text{and} \quad f^{-1}(x) = \frac{x - 2}{3}.$$

Notice how f^{-1} undoes the changes produced by f: f changes x by *multiplying* by 3 and *adding* 2, and f^{-1} undoes this by *subtracting* 2 and *dividing* by 3. This "undoing" process is illustrated in Figure 2.57.

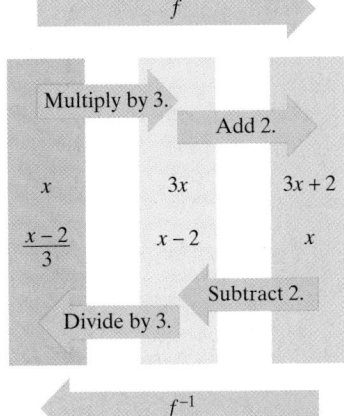

Figure 2.57 f^{-1} undoes the changes produced by f.

Multiply by 3.
Add 2.
x $3x$ $3x + 2$
$\dfrac{x - 2}{3}$ $x - 2$ x
Subtract 2.
Divide by 3.
f
f^{-1}

Check Point 2 Show that each function is an inverse of the other:

$$f(x) = 4x - 7 \quad \text{and} \quad g(x) = \frac{x + 7}{4}.$$

2 Find the inverse of a function.

Finding the Inverse of a Function

The definition of the inverse of a function tells us that the domain of f is equal to the range of f^{-1}, and vice versa. This means that if the function f is the set of ordered pairs (x, y), then the inverse of f is the set of ordered pairs (y, x). If a function is defined by an equation, we can obtain the equation for f^{-1}, the inverse of f, by interchanging the role of x and y in the equation for the function f.

Finding the Inverse of a Function

The equation for the inverse of a function f can be found as follows:

1. Replace $f(x)$ with y in the equation for $f(x)$.
2. Interchange x and y.
3. Solve for y. If this equation does not define y as a function of x, the function f does not have an inverse function and this procedure ends. If this equation does define y as a function of x, the function f has an inverse function.
4. If f has an inverse function, replace y in step 3 by $f^{-1}(x)$. We can verify our result by showing that $f(f^{-1}(x)) = x$ and $f^{-1}(f(x)) = x$.

Study Tip

The procedure for finding a function's inverse uses a *switch-and-solve* strategy. Switch x and y, then solve for y.

EXAMPLE 3   Finding the Inverse of a Function

Find the inverse of $f(x) = 7x - 5$.

Solution

Step 1   Replace $f(x)$ with y:

$$y = 7x - 5$$

Step 2   Interchange x and y:

$$x = 7y - 5 \quad \text{This is the inverse function.}$$

Step 3   Solve for y:

$$x + 5 = 7y \qquad \text{Add 5 to both sides.}$$

$$\frac{x + 5}{7} = y \qquad \text{Divide both sides by 7.}$$

Discovery

In Example 3, we found that if $f(x) = 7x - 5$, then

$$f^{-1}(x) = \frac{x + 5}{7}.$$

Verify this result by showing that

$$f(f^{-1}(x)) = x$$

and

$$f^{-1}(f(x)) = x.$$

Step 4   Replace y with $f^{-1}(x)$:

$$f^{-1}(x) = \frac{x + 5}{7} \qquad \text{The equation is written with } f^{-1} \text{ on the left.}$$

Thus, the inverse of $f(x) = 7x - 5$ is $f^{-1}(x) = \dfrac{x + 5}{7}$.

The inverse function, f^{-1}, undoes the changes produced by f. f changes x by multiplying by 7 and subtracting 5. f^{-1} undoes this by adding 5 and dividing by 7.

Check Point 3   Find the inverse of $f(x) = 2x + 7$.

EXAMPLE 4   Finding the Equation of the Inverse

Find the inverse of $f(x) = x^3 + 1$.

Solution

Step 1   Replace $f(x)$ with y: $y = x^3 + 1$

Step 2   Interchange x and y: $x = y^3 + 1$

Step 3   Solve for y:
$$x - 1 = y^3$$
$$\sqrt[3]{x - 1} = \sqrt[3]{y^3}$$
$$\sqrt[3]{x - 1} = y$$

Step 4   Replace y with $f^{-1}(x)$: $f^{-1}(x) = \sqrt[3]{x - 1}$.

Thus, the inverse of $f(x) = x^3 + 1$ is $f^{-1}(x) = \sqrt[3]{x - 1}$.

Check Point 4   Find the inverse of $f(x) = 4x^3 - 1$.

3 Use the horizontal line test to determine if a function has an inverse function.

The Horizontal Line Test and One-to-One Functions

Let's see what happens if we try to find the inverse of the standard quadratic function, $f(x) = x^2$.

Step 1 **Replace $f(x)$ with y:** $y = x^2$.

Step 2 **Interchange x and y:** $x = y^2$.

Step 3 **Solve for y:**   We apply the square root method to solve $y^2 = x$ for y.

We obtain

$$y = \pm\sqrt{x}.$$

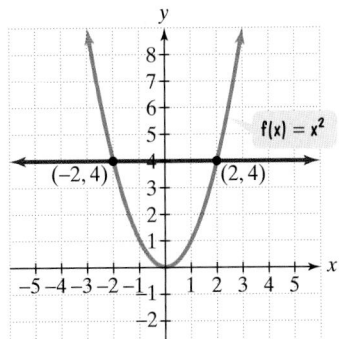

The $\pm$ in this last equation shows that for certain values of x (all positive real numbers), there are two values of y. Because this equation does not represent y as a function of x, the standard quadratic function does not have an inverse function.

    Can we look at the graph of a function and tell if it represents a function with an inverse? Yes. The graph of the standard quadratic function is shown in Figure 2.58. Four units above the x-axis, a horizontal line is drawn. This line intersects the graph at two of its points, $(-2, 4)$ and $(2, 4)$. Because inverse functions have ordered pairs with the coordinates reversed, let's see what happens if we reverse these coordinates. We obtain $(4, -2)$ and $(4, 2)$. A function provides exactly one output for each input. However, the input 4 is associated with two outputs, -2 and 2. The points $(4, -2)$ and $(4, 2)$ do not define a function.

    If any horizontal line, such as the one in Figure 2.58, intersects a graph at two or more points, these points will not define a function when their coordinates are reversed. This suggests the **horizontal line test** for inverse functions:

Figure 2.58 The horizontal line intersects the graph twice.

Discovery

How might you restrict the domain of $f(x) = x^2$, graphed in Figure 2.58, so that the remaining portion of the graph passes the horizontal line test?

The Horizontal Line Test For Inverse Functions

A function f has an inverse that is a function, f^{-1}, if there is no horizontal line that intersects the graph of the function f at more than one point.

EXAMPLE 5   Applying the Horizontal Line Test

Which of the following graphs represent functions that have inverse functions?

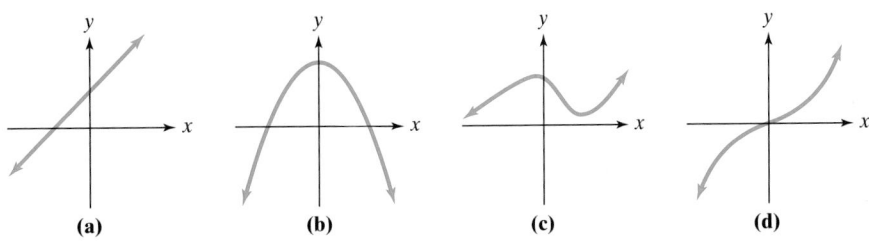

(a)   **(b)**   **(c)**   **(d)**

Solution   Can you see that horizontal lines can be drawn in parts (b) and (c) that intersect the graphs more than once? This is illustrated in the figure at the top of the next page. These graphs do not pass the horizontal line test. The graphs in parts (b) and (c) are not the graphs of functions with inverse functions. By contrast, no horizontal line can be drawn in parts (a) and (d) that intersect the graphs more than once. These graphs pass the horizontal line test. Thus, the graphs in parts (a) and (d) represent functions that have inverse functions.

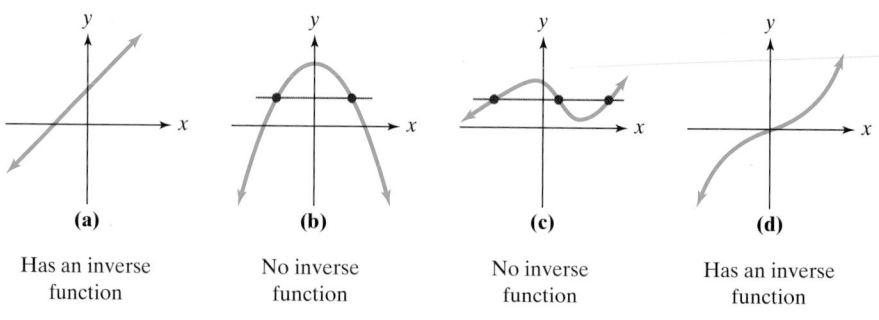

(a)	(b)	(c)	(d)
Has an inverse function	No inverse function	No inverse function	Has an inverse function

Check Point 5   Which of the following graphs represent functions that have inverse functions?

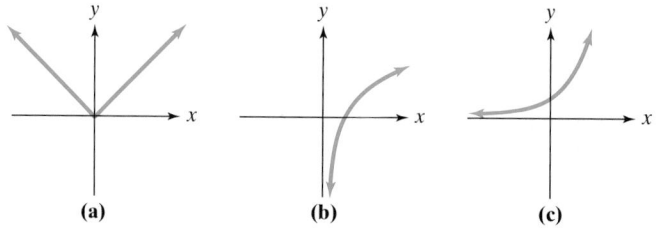

(a)   (b)   (c)

A function passes the horizontal line test when no two different ordered pairs have the same second component. This means that if $x_1 \neq x_2$, then $f(x_1) \neq f(x_2)$. Such a function is called a **one-to-one function.** Thus, a one-to-one function is a function in which no two different ordered pairs have the same second component. Only one-to-one functions have inverse functions. Any function that passes the horizontal line test is a one-to-one function. Any one-to-one function has a graph that passes the horizontal line test.

4   Use the graph of a one-to-one function to graph its inverse function.

Graphs of f and f^{-1}

There is a relationship between the graph of a one-to-one function, f, and its inverse, f^{-1}. Because inverse functions have ordered pairs with the coordinates reversed, if the point (a, b) is on the graph of f, then the point (b, a) is on the graph of f^{-1}. The points (a, b) and (b, a) are symmetric with respect to the line $y = x$. Thus, **the graph of f^{-1} is a reflection of the graph of f about the line $y = x$.** This is illustrated in Figure 2.59.

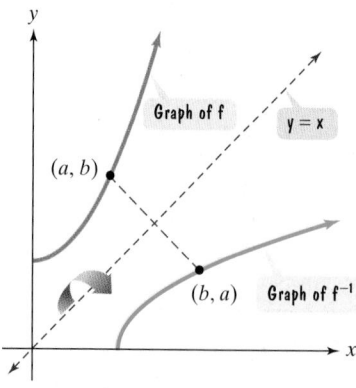

Figure 2.59 The graph of f^{-1} is a reflection of the graph of f about $y = x$.

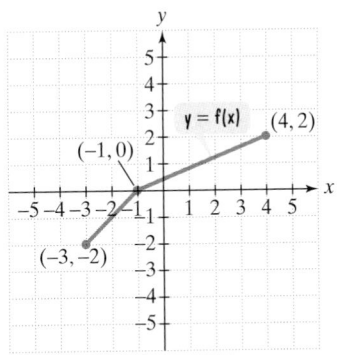

Figure 2.60

EXAMPLE 6 Graphing the Inverse Function

Use the graph of f in Figure 2.60 to draw the graph of its inverse function.

Solution We begin by noting that no horizontal line intersects the graph of f at more than one point, so f does have an inverse function. Because the points $(-3, -2), (-1, 0)$, and $(4, 2)$ are on the graph of f, the graph of the inverse function, f^{-1}, has points with these ordered pairs reversed. Thus, $(-2, -3), (0, -1)$, and $(2, 4)$ are on the graph of f^{-1}. We can use these points to graph f^{-1}. The graph of f^{-1} is shown in Figure 2.61. Note that the graph of f^{-1} is the reflection of the graph of f about the line $y = x$.

Check Point 6 Use the graph of f in the figure below to draw the graph of its inverse function.

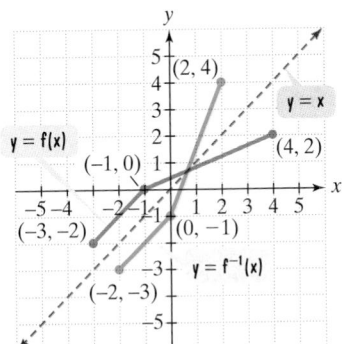

Figure 2.61 The graphs of f and f^{-1}

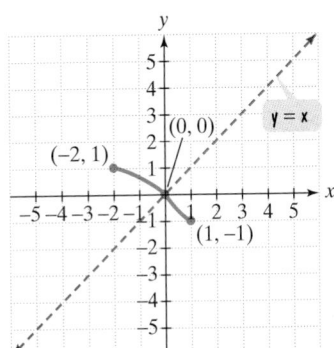

EXERCISE SET 2.7

Practice Exercises

In Exercises 1–10, find $f(g(x))$ and $g(f(x))$ and determine whether each pair of functions f and g are inverses of each other.

1. $f(x) = 4x$ and $g(x) = \dfrac{x}{4}$

2. $f(x) = 6x$ and $g(x) = \dfrac{x}{6}$

3. $f(x) = 3x + 8$ and $g(x) = \dfrac{x - 8}{3}$

4. $f(x) = 4x + 9$ and $g(x) = \dfrac{x - 9}{4}$

5. $f(x) = 5x - 9$ and $g(x) = \dfrac{x + 5}{9}$

6. $f(x) = 3x - 7$ and $g(x) = \dfrac{x + 3}{7}$

7. $f(x) = \dfrac{3}{x - 4}$ and $g(x) = \dfrac{3}{x} + 4$

8. $f(x) = \dfrac{2}{x - 5}$ and $g(x) = \dfrac{2}{x} + 5$

9. $f(x) = -x$ and $g(x) = -x$

10. $f(x) = \sqrt[3]{x - 4}$ and $g(x) = x^3 + 4$

The functions in Exercises 11–30 are all one-to-one. For each function:
- **a.** *Find an equation for $f^{-1}(x)$, the inverse function.*
- **b.** *Verify that your equation is correct by showing that $f(f^{-1}(x)) = x$ and $f^{-1}(f(x)) = x$.*

11. $f(x) = x + 3$

12. $f(x) = x + 5$

13. $f(x) = 2x$

14. $f(x) = 4x$

15. $f(x) = 2x + 3$

16. $f(x) = 3x - 1$

17. $f(x) = x^3 + 2$

18. $f(x) = x^3 - 1$

19. $f(x) = (x + 2)^3$

20. $f(x) = (x - 1)^3$

21. $f(x) = \dfrac{1}{x}$

22. $f(x) = \dfrac{2}{x}$

23. $f(x) = \sqrt{x}$

24. $f(x) = \sqrt[3]{x}$

25. $f(x) = x^2 + 1$, for $x \geq 0$

26. $f(x) = x^2 - 1$, for $x \geq 0$

27. $f(x) = \dfrac{2x + 1}{x - 3}$

28. $f(x) = \dfrac{2x - 3}{x + 1}$

29. $f(x) = \sqrt[3]{x - 4} + 3$

30. $f(x) = x^{3/5}$

Which graphs in Exercises 31–36 represent functions that have inverse functions?

31.

32.

33.

34.

35.

36.

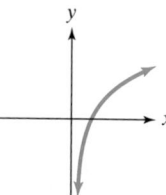

In Exercises 37–40, use the graph of f to draw the graph of its inverse function.

37.

38.

39.

40.

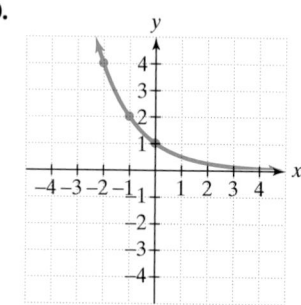

Application Exercises

41. Refer of Figure 2.56 on page 260. Recall that the bar graphs in the figure show the preferred age in a mate in five selected countries.

　a. Consider a function, f, whose domain is the set of the five countries shown in the graph. Let the range be the set of the average number of years men in each of the respective countries prefer women who are younger than themselves. (You will need to use the graph to estimate these values. Assume that the bars for Poland and Italy have the same length. Round to the nearest tenth of a year.) Write function f as a set of ordered pairs.

　b. Write the relation that is the inverse of f as a set of ordered pairs. Is this relation a function? Explain your answer.

42. The bar graph shows the percentage of land owned by the federal government in western states in which the government owns at least half of the land.

Percentage of Land Owned by the Federal Government

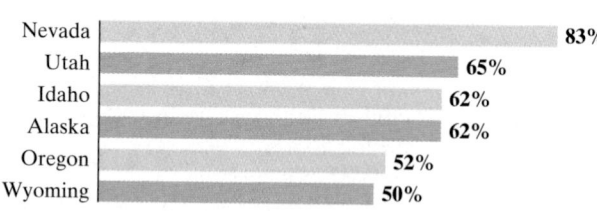

Source: Bureau of Land Management

a. Consider a function, f, whose domain is the set of six states shown. Let the range be the percentage of land owned by the federal government in each of the respective states. Write the function f as a set of ordered pairs.

b. Write the relation that is the inverse of f as a set of ordered pairs. Is this relation a function? Explain your answer.

43. The graph represents the probability of two people in the same room sharing a birthday as a function of the number of people in the room. Call the function f.

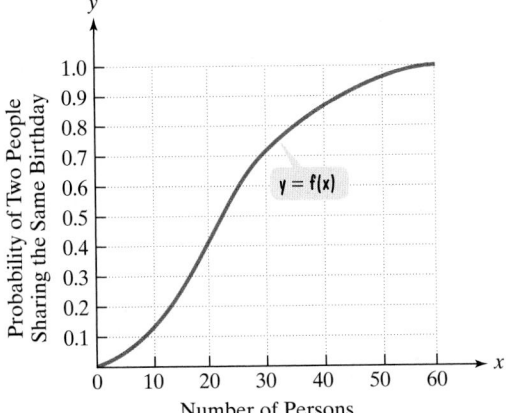

a. Explain why f has an inverse that is a function.

b. Describe in practical terms the meaning of $f^{-1}(0.25)$, $f^{-1}(0.5)$, and $f^{-1}(0.7)$.

44. The graph shows the average age at which women in the United States marry for the first time over a 110-year period.

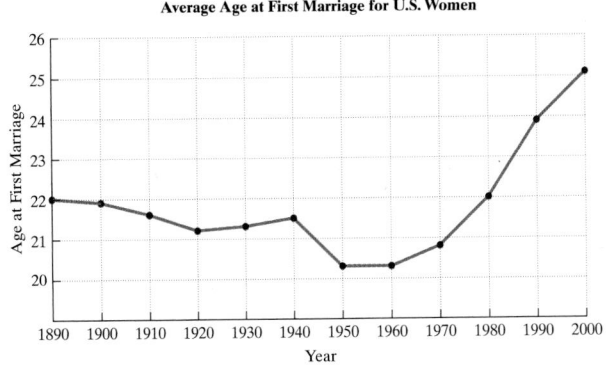

Average Age at First Marriage for U.S. Women

Source: U.S. Census Bureau

a. Does this graph have an inverse that is a function? What does this mean about the average age at which U.S. women marry during the period shown?

b. Identify two or more years in which U.S. women married for the first time at the same average age. What is a reasonable estimate of this average age?

45. The formula

$$y = f(x) = \frac{9}{5}x + 32$$

is used to convert from x degrees Celsius to y degrees Fahrenheit. The formula

$$y = g(x) = \frac{5}{9}(x - 32)$$

is used to convert from x degrees Fahrenheit to y degrees Celsius. Show that f and g are inverse functions.

46. One yardstick for measuring how steadily—if slowly— athletic performance improved is the mile run. In 1923, the record for the mile was a comparatively sleepy 4 minutes, 10.4 seconds. In 1954, Roger Bannister of Britain cracked the 4-minute mark, coming in at 3 minutes, 59.4 seconds. In the half-century since, about 0.3 second per year has been shaved off Bannister's record.

Mile Records			
1886	4:12.3	1958	3:54.5
1923	4:10.4	1966	3:51.3
1933	4:07.6	1979	3:48.9
1945	4:01.3	1985	3:46.3
1954	3:59.4	1999	3:43.1

Source: U.S.A. Track and Field

a. Consider the following information:
- In 1954, the record was 3 minutes, 59.4 seconds, or 239.4 seconds.
- The record has decreased by 0.3 second per year since then.

Use this information to write a function, f, that models the mile record, $f(x)$, in seconds, x years after 1954.

b. Find the inverse of the mile-record function. Describe what each variable in the inverse function represents.

c. According to the inverse model, how many years after 1954 will someone run a 3-minute, or 180-second, mile? In which year will this occur?

Writing in Mathematics

47. Explain how to determine if two functions are inverses of each other.

48. Describe how to find the inverse of a one-to-one function.

49. What is the horizontal line test and what does it indicate?

50. Describe how to use the graph of a one-to-one function to draw the graph of its inverse function.

51. How can a graphing utility be used to visually determine if two functions are inverses of each other?

Technology Exercises

In Exercises 52–60, use a graphing utility to graph the function. Use the graph to determine whether the function has an inverse that is a function (that is, whether the function is one-to-one).

52. $f(x) = x^2 - 1$

53. $f(x) = \sqrt[3]{2 - x}$

54. $f(x) = \dfrac{x^3}{2}$

55. $f(x) = \dfrac{x^4}{4}$

56. $f(x) = \text{int}(x - 2)$

57. $f(x) = |x - 2|$

58. $f(x) = (x - 1)^3$

59. $f(x) = -\sqrt{16 - x^2}$

60. $f(x) = x^3 + x + 1$

In Exercises 61–63, use a graphing utility to graph f and g in the same viewing rectangle. In addition, graph the line y = x and visually determine if f and g are inverses.

61. $f(x) = 4x + 4, \ g(x) = 0.25x - 1$

62. $f(x) = \dfrac{1}{x} + 2, \ g(x) = \dfrac{1}{x - 2}$

63. $f(x) = \sqrt[3]{x} - 2, \ g(x) = (x + 2)^3$

Critical Thinking Exercises

64. Which one of the following is true?

    a. The inverse of $\{(1, 4), (2, 7)\}$ is $\{(2, 7), (1, 4)\}$.

    b. The function $f(x) = 5$ is one-to-one.

    c. If $f(x) = 3x$, then $f^{-1}(x) = \dfrac{1}{3x}$.

    d. The domain of f is the same as the range of f^{-1}.

65. If $f(x) = 3x$ and $g(x) = x + 5$, find $(f \circ g)^{-1}(x)$ and $(g^{-1} \circ f^{-1})(x)$.

66. Show that

$$f(x) = \frac{3x - 2}{5x - 3}$$

is its own inverse.

67. *Freedom 7* was the spacecraft that carried the first American into space in 1961. Total flight time was 15 minutes, and the spacecraft reached a maximum height of 116 miles. Consider a function, s, that expresses *Freedom 7*'s height, $s(t)$, in miles, after t minutes. Is s a one-to-one function? Explain your answer.

68. If $f(2) = 6$, find x satisfying $8 + f^{-1}(x - 1) = 10$.

Group Exercise

69. In Tom Stoppard's play *Arcadia*, the characters dream and talk about mathematics, including ideas involving graphing, composite functions, symmetry, and lack of symmetry in things that are tangled, mysterious, and unpredictable. Group members should read the play. Present a report on the ideas discussed by the characters that are related to concepts that we studied in this chapter. Bring in a copy of the play and read appropriate excerpts.

CHAPTER SUMMARY, REVIEW, AND TEST

Summary

DEFINITIONS AND CONCEPTS	EXAMPLES
2.1 Lines and Slope	
a. The slope, m, of the line through (x_1, y_1) and (x_2, y_2) is $m = \dfrac{y_2 - y_1}{x_2 - x_1}$.	Ex. 1, p.177
b. Equations of lines include point-slope form, $y - y_1 = m(x - x_1)$, slope-intercept form, $y = mx + b$, and general form, $Ax + By + C = 0$. The equation of a horizontal line is $y = b$; a vertical line is $x = a$.	Ex. 2 & 3, p.179; Ex. 5 & 6, p.182
c. Parallel lines have equal slopes. Perpendicular lines have slopes that are negative reciprocals.	Ex. 8 & 9, p.184–185
2.2 Distance and Midpoint Formulas; Circles	
a. The distance, d, between the points (x_1, y_1) and (x_2, y_2) is given by $d = \sqrt{(x_2 - x_1)^2 + (y_2 - y_1)^2}$.	Ex. 1, p.194
b. The midpoint of the line segment whose endpoints are (x_1, y_1) and (x_2, y_2) is the point with coordinates $\left(\dfrac{x_1 + x_2}{2}, \dfrac{y_1 + y_2}{2} \right)$.	Ex. 2, p.195

c. The standard form of the equation of a circle with center (h, k) and radius r is
$(x - h)^2 + (y - k)^2 = r^2$.

Ex. 3, p. 196
Ex. 4 & 5, p. 197

d. The general form of the equation of a circle is $x^2 + y^2 + Dx + Ey + F = 0$.

e. To convert from the general form to the standard form of a circle's equation, complete the square on x and y.

Ex. 6, p. 198

2.3 Basics of Functions

a. A relation is any set of ordered pairs. The set of first components is the domain and the set of second components is the range.

Ex. 1, p. 202

b. A function is a correspondence from a first set, called the domain, to a second set, called the range, such that each element in the domain corresponds to exactly one element in the range. If any element in a relation's domain corresponds to more than one element in the range, the relation is not a function.

Ex. 2, p. 203

c. Functions are usually given in terms of equations involving x and y, in which x is the independent variable and y is the dependent variable. If an equation is solved for y and more than one value of y can be obtained for a given x, then the equation does not define y as a function of x. If an equation defines a function, the value of the function at x, $f(x)$, often replaces y.

Ex. 3, p. 205

d. The difference quotient is
$$\frac{f(x + h) - f(x)}{h}, h \neq 0.$$

Ex. 5, p. 207

e. If a function f does not model data or verbal conditions, its domain is the largest set of real numbers for which the value of $f(x)$ is a real number. Exclude from the function's domain real numbers that cause division by zero and real numbers that result in an even root of a negative number.

Ex. 7, p. 210

2.4 Graphs of Functions

a. The graph of a function is the graph of its ordered pairs.

Ex. 1, p. 215

b. The vertical line test for functions: If any vertical line intersects a graph in more than one point, the graph does not define y as a function of x.

Ex. 3, p. 217

c. A function is increasing on intervals where its graph rises, decreasing on intervals where it falls, and constant on intervals where it neither rises nor falls. Precise definitions are given in the box on page 220.

Ex. 5, p. 220

d. If the graph of a function is given, we can often visually locate the number(s) at which the function has a relative maximum or relative minimum. Precise definitions are given in the box on page 221.

Fig 2.32, p. 222

e. The average rate of change of f from x_1 to x_2 is
$$\frac{f(x_2) - f(x_1)}{x_2 - x_1}.$$

Ex. 6, p. 222

f. The graph of an even function in which $f(-x) = f(x)$ is symmetric with respect to the y-axis. The graph of an odd function in which $f(-x) = -f(x)$ is symmetric with respect to the origin.

Ex. 7, p. 224

g. The graph of $f(x) = \text{int}(x)$, where $\text{int}(x)$ is the greatest integer that is less than or equal to x, has function values that form discontinuous steps, shown in Figure 2.39 on page 227. If $n \leq x < n + 1$, where n is an integer, then $\text{int}(x) = n$.

2.5 Transformations of Functions

a. Table 2.4 on pages 235–236 shows the graphs of the constant function, $f(x) = c$, the identity function, $f(x) = x$, the standard quadratic function, $f(x) = x^2$, the standard cubic function, $f(x) = x^3$, the square root function, $f(x) = \sqrt{x}$, and the absolute value function, $f(x) = |x|$. The table also lists characteristics of each function.

b. Table 2.5 on page 243 summarizes how to graph a function using vertical shifts, $y = f(x) \pm c$, horizontal shifts, $y = f(x \pm c)$, reflections about the x-axis, $y = -f(x)$, reflections about the y-axis, $y = f(-x)$, vertical stretching, $y = cf(x)$, $c > 1$, and vertical shrinking, $y = cf(x)$, $0 < c < 1$.

Ex. 1 & 2, p. 237–238;
Ex. 3, p. 239;
Ex. 4–7, p. 241–243
Ex. 8, p. 244

c. A function involving more than one transformation can be graphed in the following order:
(1) horizontal shifting; (2) vertical stretching or shrinking; (3) reflecting; (4) vertical shifting.

a. Draw a line that fits the data so that the spread of the data points around the line is as small as possible.

b. Use the coordinates of two points along your line to write its point-slope and slope-intercept equations.

c. Use the equation in part (b) to predict the number of headaches per month for a person exercising 130 minutes per week.

In Exercises 13–14, use the given conditions to write an equation for each line in point-slope form and slope-intercept form.

13. Passing through $(4, -7)$ and parallel to the line whose equation is $3x + y - 9 = 0$

14. Passing through $(-3, 6)$ and perpendicular to the line whose equation is $y = \frac{1}{3}x + 4$

2.2

In Exercises 15–16, find the distance between each pair of points. If necessary, round answers to two decimal places.

15. $(-2, -3)$ and $(3, 9)$ **16.** $(-4, 3)$ and $(-2, 5)$

In Exercises 17–18, find the midpoint of each line segment with the given endpoints.

17. $(2, 6)$ and $(-12, 4)$ **18.** $(4, -6)$ and $(-15, 2)$

In Exercises 19–20, write the standard form of the equation of the circle with the given center and radius.

19. Center $(0, 0)$, $r = 3$ **20.** Center $(-2, 4)$, $r = 6$

In Exercises 21–23, give the center and radius of each circle and graph its equation.

21. $x^2 + y^2 = 1$ **22.** $(x + 2)^2 + (y - 3)^2 = 9$

23. $x^2 + y^2 - 4x + 2y - 4 = 0$

2.3

In Exercises 24–26, determine whether each relation is a function. Give the domain and range for each relation.

24. $\{(2, 7), (3, 7), (5, 7)\}$ **25.** $\{(1, 10)\ (2, 500),\ (13, \pi)\}$

26. $\{(12, 13), (14, 15), (12, 19)\}$

In Exercises 27–29, determine whether each equation defines y as a function of x.

27. $2x + y = 8$ **28.** $3x^2 + y = 14$

29. $2x + y^2 = 6$

In Exercises 30–33, evaluate each function at the given values of the independent variable and simplify.

30. $f(x) = 5 - 7x$
  a. $f(4)$ **b.** $f(x + 3)$ **c.** $f(-x)$

31. $g(x) = 3x^2 - 5x + 2$
  a. $g(0)$ **b.** $g(-2)$
  c. $g(x - 1)$ **d.** $g(-x)$

32. $g(x) = \begin{cases} \sqrt{x - 4} & \text{if } x \geq 4 \\ 4 - x & \text{if } x < 4 \end{cases}$
  a. $g(13)$ **b.** $g(0)$ **c.** $g(-3)$

33. $f(x) = \begin{cases} \dfrac{x^2 - 1}{x - 1} & \text{if } x \neq 1 \\ 12 & \text{if } x = 1 \end{cases}$
  a. $f(-2)$ **b.** $f(1)$ **c.** $f(2)$

In Exercises 34–35, find and simplify the difference quotient

$$\frac{f(x + h) - f(x)}{h}, \quad h \neq 0$$

for the given function.

34. $f(x) = 8x - 11$ **35.** $f(x) = x^2 - 13x + 5$

In Exercises 36–40, find the domain of each function.

36. $f(x) = x^2 + 6x - 3$ **37.** $g(x) = \dfrac{4}{x - 7}$

38. $h(x) = \sqrt{8 - 2x}$ **39.** $f(x) = \dfrac{x}{x^2 - 1}$

40. $g(x) = \dfrac{\sqrt{x - 2}}{x - 5}$

2.4

Graph the functions in Exercises 41–42. Use the integer values of x given to the right of the function to obtain the ordered pairs. Use the graph to specify the function's domain and range.

41. $f(x) = x^2 - 4x + 4$ $x = -1, 0, 1, 2, 3, 4$

42. $f(x) = |2 - x|$ $x = -1, 0, 1, 2, 3, 4$

*In Exercises 43–45, use the graph to determine **a.** the function's domain; **b.** the function's range; **c.** the x-intercepts, if any; **d.** the y-intercept, if any; **e.** intervals on which the function is increasing, decreasing, or constant; and **f.** the function values indicated below the graphs.*

43.

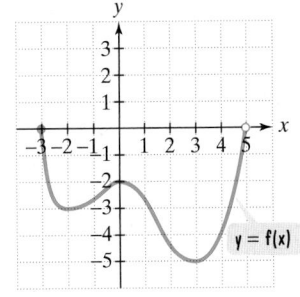

$f(-2) = ?\ f(3) = ?$

44.

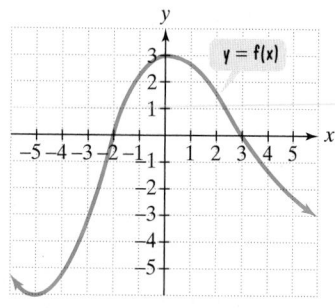

$f(-2) = ?$ $f(6) = ?$

45.

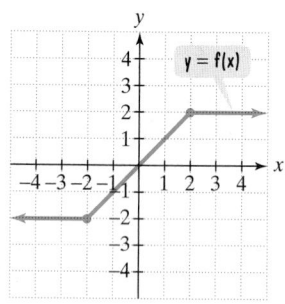

$f(-9) = ?$ $f(14) = ?$

In Exercises 46–47, find:

a. *The numbers, if any, at which f has a relative maximum. What are these relative maxima?*
b. *The numbers, if any, at which f has a relative minimum. What are these relative minima?*

46. Use the graph in Exercise 43.

47. Use the graph in Exercise 44.

In Exercises 48–51, use the vertical line test to identify graphs in which y is a function of x.

48.

49.

50.

51.

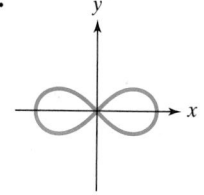

52. Find the average rate of change of $f(x) = x^2 - 4x$ from $x_1 = 5$ to $x_2 = 9$.

53. The graph shows annual spending per uniformed member of the U.S. military in inflation-adjusted dollars. Find the average rate of change of spending per year from 1955 through 2000. Round to the nearest dollar per year.

Source: Center for Strategic and Budgetary Assessments

In Exercises 54–56, determine whether each function is even, odd, or neither. State each function's symmetry. If you are using a graphing utility, graph the function and verify its possible symmetry.

54. $f(x) = x^3 - 5x$ **55.** $f(x) = x^4 - 2x^2 + 1$
56. $f(x) = 2x\sqrt{1 - x^2}$

57. The graph shows the height, in meters, of a vulture in terms of its time, in seconds, in flight.

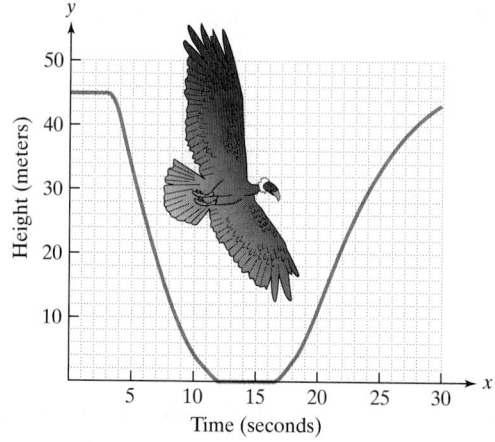

a. Is the vulture's height a function of time? Use the graph to explain why or why not.
b. On which interval is the function decreasing? Describe what this means in practical terms.
c. On which intervals is the function constant? What does this mean for each of these intervals?
d. On which interval is the function increasing? What does this mean?

58. A cargo service charges a flat fee of $5 plus $1.50 for each pound or fraction of a pound. Graph shipping cost, $C(x)$, in dollars, as a function of weight, x, in pounds, for $0 < x \leq 5$.

2.5

In Exercises 59–61, begin by graphing the standard quadratic function, $f(x) = x^2$. Then use transformations of this graph to graph the given function.

59. $g(x) = x^2 + 2$ **60.** $h(x) = (x + 2)^2$
61. $r(x) = -(x + 1)^2$

In Exercises 62–64, begin by graphing the square root function, $f(x) = \sqrt{x}$. Then use transformations of this graph to graph the given function.

62. $g(x) = \sqrt{x + 3}$ **63.** $h(x) = \sqrt{3 - x}$
64. $r(x) = 2\sqrt{x + 2}$

In Exercises 65–67, begin by graphing the absolute value function, $f(x) = |x|$. Then use transformations of this graph to graph the given function.

65. $g(x) = |x + 2| - 3$ **66.** $h(x) = -|x - 1| + 1$
67. $r(x) = \frac{1}{2}|x + 2|$

In Exercises 68–70, begin by graphing the standard cubic function, $f(x) = x^3$. Then use transformations of this graph to graph the given function.

68. $g(x) = \frac{1}{2}(x - 1)^3$ **69.** $h(x) = -(x + 1)^3$
70. $r(x) = \frac{1}{4}x^3 - 1$

In Exercises 71–73, use the graph of the function f to sketch the graph of the given function g.

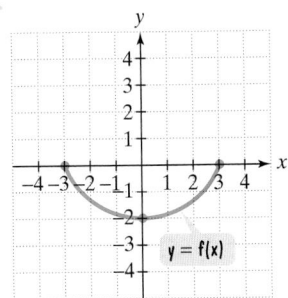

$y = f(x)$

71. $g(x) = f(x + 2) + 3$ **72.** $g(x) = \frac{1}{2}f(x - 1)$
73. $g(x) = -2 + 2f(x + 2)$

2.6

In Exercises 74–76, find $f + g, f - g, fg,$ and $\frac{f}{g}$. Determine the domain for each function.

74. $f(x) = 3x - 1, \quad g(x) = x - 5$
75. $f(x) = x^2 + x + 1, \quad g(x) = x^2 - 1$
76. $f(x) = \sqrt{x + 7}, \quad g(x) = \sqrt{x - 2}$

In Exercises 77–78, find **a.** $(f \circ g)(x)$; **b.** $(g \circ f)(x)$;
c. $(f \circ g)(3)$.

77. $f(x) = x^2 + 3, \quad g(x) = 4x - 1$
78. $f(x) = \sqrt{x}, \quad g(x) = x + 1$

In Exercises 79–80, find **a.** $(f \circ g)(x)$; **b.** *the domain of* $(f \circ g)$.

79. $f(x) = \dfrac{x + 1}{x - 2}, \quad g(x) = \dfrac{1}{x}$
80. $f(x) = \sqrt{x - 1}, \quad g(x) = x + 3$

In Exercises 81–82, express the given function h as a composition of two functions f and g so that $h(x) = (f \circ g)(x)$.

81. $h(x) = (x^2 + 2x - 1)^4$ **82.** $h(x) = \sqrt[3]{7x + 4}$

2.7

In Exercises 83–84, find $f(g(x))$ and $g(f(x))$ and determine whether each pair of functions f and g are inverses of each other.

83. $f(x) = \dfrac{3}{5}x + \dfrac{1}{2}$ and $g(x) = \dfrac{5}{3}x - 2$

84. $f(x) = 2 - 5x$ and $g(x) = \dfrac{2 - x}{5}$

The functions in Exercises 85–87 are all one-to-one. For each function:
   a. *Find an equation for $f^{-1}(x)$, the inverse function.*
   b. *Verify that your equation is correct by showing that $f(f^{-1}(x)) = x$ and $f^{-1}(f(x)) = x$.*

85. $f(x) = 4x - 3$ **86.** $f(x) = \sqrt{x + 2}$
87. $f(x) = 8x^3 + 1$

Which graphs in Exercises 88–91 represent functions that have inverse functions?

88.

89.

90.

91.

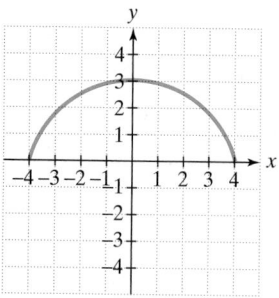

92. Use the graph of f in the figure shown to draw the graph of its inverse function.

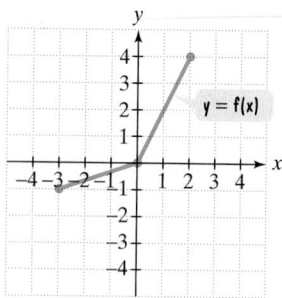

Chapter 2 Test

In Exercises 1–2, use the given conditions to write an equation for each line in point-slope form and slope-intercept form.

1. Passing through $(2, 1)$ and $(-1, -8)$

2. Passing through $(-4, 6)$ and perpendicular to the line whose equation is $y = -\frac{1}{4}x + 5$

3. Strong demand plus higher fuel and labor costs are driving up the price of flying. The graph shows the national averages for one-way fares. Also shown is a line that models that data.

National Averages for One-Way Airline Fares: Business Travel

Source: American Express

   a. Use the two points whose coordinates are shown by the voice balloons to write the slope-intercept equation of the line that models the average one-way fare, y, in dollars, x years after 1995.

   b. According to the model, what will the national average for one-way fares be in 2008?

4. Give the center and radius of the circle whose equation is $x^2 + y^2 + 4x - 6y - 3 = 0$ and graph the equation.

5. List by letter all relations that are not functions.

   a. $\{(7, 5), (8, 5), (9, 5)\}$
   b. $\{(5, 7), (5, 8), (5, 9)\}$
   c.

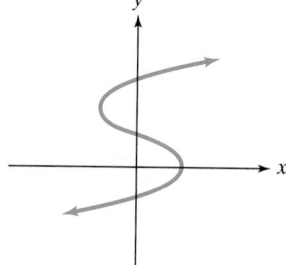

   d. $x^2 + y^2 = 100$
   e.

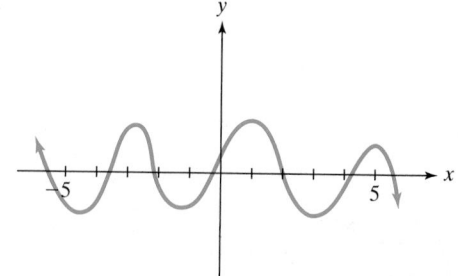

6. If $f(x) = x^2 - 2x + 5$, find $f(x - 1)$ and simplify.

7. If $g(x) = \begin{cases} \sqrt{x-3} & \text{if } x \geq 3 \\ 3-x & \text{if } x < 3 \end{cases}$, find $g(-1)$ and $g(7)$.

8. If $f(x) = \sqrt{12-3x}$, find the domain of f.

9. If $f(x) = x^2 + 11x - 7$, find and simplify the difference quotient $\dfrac{f(x+h) - f(x)}{h}$.

10. Use the graph of function f to answer the following questions.

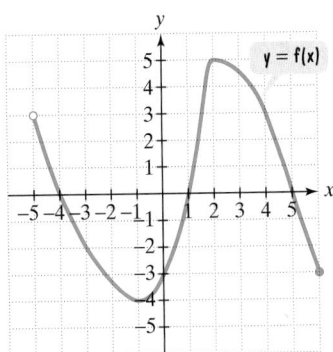

a. What is $f(4) - f(-3)$?
b. What is the domain of f?
c. What is the range of f?
d. On which interval or intervals is f increasing?
e. On which interval or intervals is f decreasing?
f. For what number does f have a relative maximum? What is the relative maximum?
g. For what number does f have a relative minimum? What is the relative minimum?
h. What are the x-intercepts?
i. What is the y-intercept?

11. Find the average rate of change of $f(x) = 3x^2 - 5$ from $x_1 = 6$ to $x_2 = 10$.

12. Determine whether $f(x) = x^4 - x^2$ is even, odd, or neither. Use your answer to explain why the graph in the figure shown cannot be the graph of f.

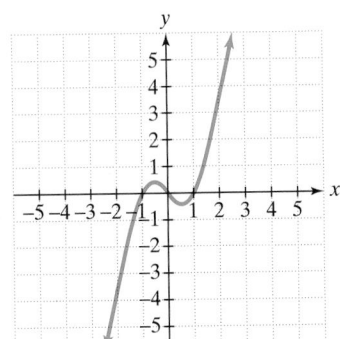

13. The figure at the top of the next column shows how the graph of $h(x) = -2(x-3)^2$ is obtained from the graph of $f(x) = x^2$. Describe this process, using the graph of g in your description.

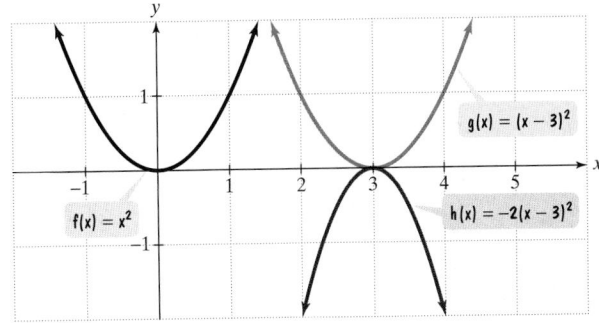

14. Begin by graphing the absolute value function, $f(x) = |x|$. Then use transformations of this graph to graph $g(x) = \frac{1}{2}|x+1| + 3$.

If $f(x) = x^2 + 3x - 4$ and $g(x) = 5x - 2$, find each function or function value in Exercises 15–19.

15. $(f - g)(x)$

16. $\left(\dfrac{f}{g}\right)(x)$ and its domain

17. $(f \circ g)(x)$

18. $(g \circ f)(x)$

19. $f(g(2))$

20. If $f(x) = \dfrac{7}{x-4}$ and $g(x) = \dfrac{2}{x}$, find $(f \circ g)(x)$ and the domain of $f \circ g$.

21. Express $h(x) = (2x + 13)^7$ as a composition of two functions f and g so that $h(x) = (f \circ g)(x)$.

22. If $f(x) = \sqrt{x-2}$, find the equation for $f^{-1}(x)$. Then verify that your equation is correct by showing that $f(f^{-1}(x)) = x$ and $f^{-1}(f(x)) = x$.

23. A function f models the amount given to charity as a function of income. The graph of f is shown in the figure.

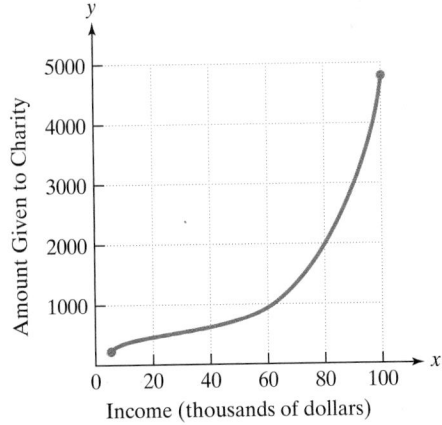

a. Explain why f has an inverse that is a function.
b. Find $f(80)$.
c. Describe in practical terms the meaning of $f^{-1}(2000)$.

24. Use a graphing utility to graph $f(x) = \dfrac{x^3}{3} + x^2 - 15x + 3$

in a $[-10, 10, 1]$ by $[-30, 70, 10]$ viewing rectangle. Use the graph to answer the following questions.

a. Is f one-to-one? Explain.

b. Is f even, odd, or neither? Explain.

c. What is the range of f?

d. On which interval or intervals is f increasing?

e. On which interval or intervals is f decreasing?

f. For what number does f have a relative maximum? What is the relative maximum?

g. For what number does f have a relative minimum? What is the relative minimum?

Cumulative Review Exercises (Chapters P–2)

Simplify each expression in Exercises 1 and 2.

1. $\dfrac{4x^2 y}{2x^5 y^{-3}}$

2. $\dfrac{5}{4\sqrt{2}}$

3. Factor: $x^3 - 4x^2 + 2x - 8$.

In Exercises 4 and 5, perform the operations and simplify.

4. $\dfrac{x - 3}{x + 4} + \dfrac{x}{x - 2}$

5. $\dfrac{4 + \dfrac{2}{x}}{4 - \dfrac{2}{x}}$

Solve each equation in Exercises 6–9.

6. $(x + 3)(x - 4) = 8$

7. $3(4x - 1) = 4 - 6(x - 3)$

8. $\sqrt{x} + 2 = x$

9. $x^{2/3} - x^{1/3} - 6 = 0$

Solve each inequality in Exercises 10 and 11. Express the answer in interval notation.

10. $\dfrac{x}{2} - 3 \le \dfrac{x}{4} + 2$

11. $\dfrac{x + 3}{x - 2} \le 2$

12. Write the point-slope form and the slope-intercept form of the line passing through $(-2, 5)$ and perpendicular to the line whose equation is $y = -\frac{1}{4} x + \frac{1}{3}$.

13. Graph $f(x) = \sqrt{x}$ and then use transformations of this graph to graph $g(x) = \sqrt{x - 3} + 4$ in the same rectangular coordinate system.

14. If $f(x) = 2 + \sqrt{x - 3}$, find the equation for $f^{-1}(x)$.

15. If $f(x) = 3 - x^2$, find $\dfrac{f(x + h) - f(x)}{h}$ and simplify.

16. Solve for c: $A = \dfrac{cd}{c + d}$.

17. You invested $6000 in two accounts paying 7% and 9% annual interest, respectively. At the end of the year, the total interest from these investments was $510. How much was invested at each rate?

18. For a summer sales job, you are choosing between two pay arrangements: a weekly salary of $200 plus 5% commission on sales, or a straight 15% commission. For how many dollars of sales will the earnings be the same regardless of the pay arrangement?

19. The length of a rectangular garden is 2 feet more than twice its width. If 22 feet of fencing is needed to enclose the garden, what are its dimensions?

20. On the first five tests you have scores of 61, 95, 71, 83, and 80. The last test, a final exam, counts as two grades. What score do you need on the final in order to have an average score of 80?

Polynomial and Rational Functions

There is a function that models the age in human years, $H(x)$, of a dog that is x years old:

$$H(x) = -0.001618x^4 + 0.077326x^3$$
$$-1.2367x^2 + 11.460x + 2.914.$$

The function contains variables to powers that are whole numbers and is an example of a **polynomial function.** In this chapter, we study polynomial functions and functions that consist of quotients of polynomials, called **rational functions.**

One of the joys of your life is your dog, your very special buddy. Lately, however, you've noticed that your companion is slowing down a bit. He's now 8 years old and you wonder how this translates into human years. You remember something about every year of a dog's life being equal to seven years for a human. Is there a more accurate description?

SECTION 3.1 *Quadratic Functions*

Objectives

1. Recognize characteristics of parabolas.
2. Graph parabolas.
3. Solve problems involving minimizing or maximizing quadratic functions.

The Food Stamp Program is the first line of defense against hunger for millions of American families. The program provides benefits for eligible participants to purchase approved food items at approved food stores. Over half of all participants are children; one out of six is a low-income older adult. The function

$$f(x) = -0.5x^2 + 4x + 19$$

models the number of people, $f(x)$, in millions, receiving food stamps x years after 1990. For example, to find the number of food stamp recipients in 2000, substitute 10 for x because 2000 is 10 years after 1990:

$$f(10) = -0.5(10)^2 + 4(10) + 19 = 9.$$

Thus, in 2000, there were 9 million food stamp recipients.

The function $f(x) = -0.5x^2 + 4x + 19$ is an example of a *quadratic function*. A **quadratic function** is any function of the form

$$f(x) = ax^2 + bx + c$$

where a, b, and c are real numbers and $a \neq 0$. A quadratic function is a polynomial function whose highest power is 2. In this section, we will study quadratic functions and their graphs.

1 Recognize characteristics of parabolas.

Graphs of Quadratic Functions

The graph of any quadratic function is called a **parabola.** Parabolas are shaped like cups, as shown in Figure 3.1. If the coefficient of x^2 (the value of a in $ax^2 + bx + c$) is positive, the parabola opens upward. If the coefficient of x^2 is negative, the graph opens downward. The **vertex** (or turning point) of the parabola is the minimum point on the graph when it opens upward, and the maximum point on the graph when it opens downward.

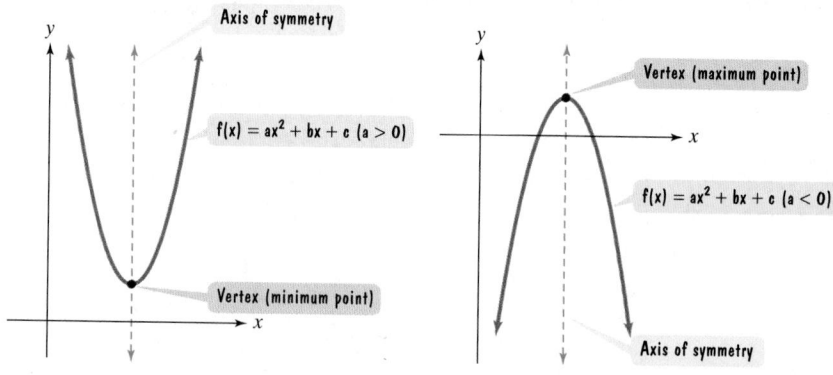

Figure 3.1 Characteristics of parabolas

$a > 0$: Parabola opens upward. $a < 0$: Parabola opens downward.

Look at the unusual image of the word "mirror" shown below. The artist, Scott Kim, has created the image so that the two halves of the whole are mirror images of each other. A parabola shares this kind of symmetry, in which a line through the vertex divides the figure in half. Parabolas are symmetric with respect to this line, called the **axis of symmetry.** The movements of gymnasts, divers, and swimmers can approximate this symmetry. If a parabola is folded along its axis of symmetry, the two halves match exactly.

2 Graph parabolas.

Graphing Quadratic Functions in Standard Form

In Section 2.5, we applied a series of transformations to the graph of $f(x) = x^2$. The graph of this function is a parabola. The vertex for this parabola is $(0, 0)$. In Figure 3.2(a), the graph of $f(x) = ax^2$ for $a > 0$ is shown in black; it opens *upward.* In Figure 3.2(b), the graph of $f(x) = ax^2$ for $a < 0$ is shown in black; it opens *downward.*

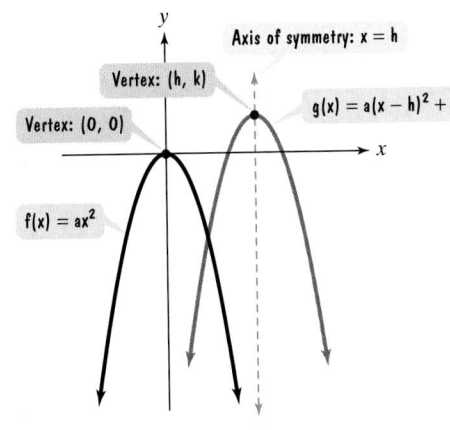

Figure 3.2 Transformations of $f(x) = ax^2$

(a) $a > 0$: Parabola opens upward.

(b) $a < 0$: Parabola opens downward.

Figure 3.2 also shows the graphs of $g(x) = a(x - h)^2 + k$ in blue. Compare these graphs to those of $f(x) = ax^2$. Observe that h determines the horizontal shift and k determines the vertical shift of the graph of $f(x) = ax^2$:

$$g(x) = a(x - h)^2 + k.$$

> If $h > 0$, it shifts the graph of $f(x) = ax^2$ h units to the right.

> If $k > 0$, it shifts the graph of $y = a(x-h)^2$ k units up.

Consequently, the vertex $(0, 0)$ on the black graph of $f(x) = ax^2$ moves to the point (h, k) on the blue graph of $g(x) = a(x - h)^2 + k$. The axis of symmetry is the vertical line whose equation is $x = h$.

The form of the expression for g is convenient because it immediately identifies the vertex of the parabola as (h, k). This is the **standard form** of a quadratic function.

The Standard Form of a Quadratic Function

The quadratic function

$$f(x) = a(x - h)^2 + k, \qquad a \neq 0$$

is in **standard form**. The graph of f is a parabola whose vertex is the point (h, k). The parabola is symmetric with respect to the line $x = h$. If $a > 0$, the parabola opens upward; if $a < 0$, the parabola opens downward.

The sign of a in $f(x) = a(x - h)^2 + k$ determines whether the parabola opens upward or downward. Furthermore, if $|a|$ is small, the parabola opens more widely than if $|a|$ is large. Here is a general procedure for graphing parabolas whose equations are in standard form:

Graphing Quadratic Functions with Equations in Standard Form

To graph $f(x) = a(x - h)^2 + k$,

1. Determine whether the parabola opens upward or downward. If $a > 0$, it opens upward. If $a < 0$, it opens downward.
2. Determine the vertex of the parabola. The vertex is (h, k).
3. Find any x-intercepts by replacing $f(x)$ with 0. Solve the resulting quadratic equation for x.
4. Find the y-intercept by replacing x with 0.
5. Plot the intercepts and vertex. Connect these points with a smooth curve that is shaped like a cup. Draw a dashed vertical line for the axis of symmetry.

EXAMPLE 1   Graphing a Quadratic Function in Standard Form

Graph the quadratic function $f(x) = -2(x - 3)^2 + 8$.

Solution   We can graph this function by following the steps in the preceding box. We begin by identifying values for a, h, and k.

Standard form $f(x) = a(x - h)^2 + k$

$$a = -2 \qquad h = 3 \qquad k = 8$$

Given equation $f(x) = -2(x - 3)^2 + 8$

Step 1   Determine how the parabola opens.   Note that a, the coefficient of x^2, is -2. Thus, $a < 0$; this negative value tells us that the parabola opens downward.

Step 2   Find the vertex.   The vertex of the parabola is (h, k). Because $h = 3$ and $k = 8$, the parabola's vertex is $(3, 8)$.

Step 3 **Find the x-intercepts.** Replace $f(x)$ with 0 in $f(x) = -2(x - 3)^2 + 8$.

$$0 = -2(x - 3)^2 + 8 \qquad \text{Find x-intercepts, setting f(x) equal to 0.}$$

$$2(x - 3)^2 = 8 \qquad \text{Solve for x. Add } 2(x - 3)^2 \text{ to both sides of the equation.}$$

$$(x - 3)^2 = 4 \qquad \text{Divide both sides by 2.}$$

$$(x - 3) = \pm\sqrt{4} \qquad \text{Apply the square root method. If } (x - c)^2 = d, \text{ then } x - c = \pm\sqrt{d}.$$

$$x - 3 = -2 \quad \text{or} \quad x - 3 = 2 \qquad \text{Express as two separate equations.}$$

$$x = 1 \quad \text{or} \quad x = 5 \qquad \text{Add 3 to both sides in each equation.}$$

The x-intercepts are 1 and 5. The parabola passes through $(1, 0)$ and $(5, 0)$.

Step 4 **Find the y-intercept.** Replace x with 0 in $f(x) = -2(x - 3)^2 + 8$.

$$f(0) = -2(0 - 3)^2 + 8 = -2(-3)^2 + 8 = -2(9) + 8 = -10$$

The y-intercept is -10. The parabola passes through $(0, -10)$.

Step 5 **Graph the parabola.** With a vertex at $(3, 8)$, x-intercepts at 1 and 5, and a y-intercept at -10, the graph of f is shown in Figure 3.3. The axis of symmetry is the vertical line whose equation is $x = 3$.

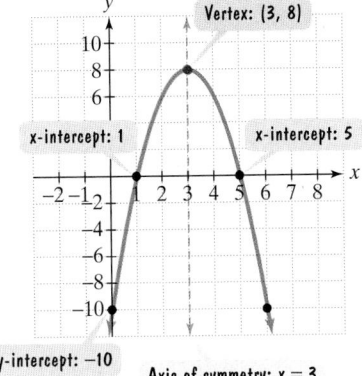

Figure 3.3 The graph of $f(x) = -2(x - 3)^2 + 8$

Check Point 1 Graph the quadratic function $f(x) = -(x - 1)^2 + 4$.

EXAMPLE 2 Graphing a Quadratic Function in Standard Form

Graph the quadratic function $f(x) = (x + 3)^2 + 1$.

Solution We begin by finding values for a, h, and k.

Standard form		$f(x) = a(x - h)^2 + k$
Given equation		$f(x) = (x + 3)^2 + 1$
	or	$f(x) = 1(x - (-3))^2 + 1$

$a = 1 \qquad h = -3 \qquad k = 1$

Step 1 **Determine how the parabola opens.** Note that a, the coefficient of x^2, is 1. Thus, $a > 0$; this positive value tells us that the parabola opens upward.

Step 2 **Find the vertex.** The vertex of the parabola is (h, k). Because $h = -3$ and $k = 1$, the parabola's vertex is $(-3, 1)$.

Step 3 Find the *x*-intercepts. Replace $f(x)$ with 0 in $f(x) = (x + 3)^2 + 1$. Because the vertex is $(-3, 1)$, which lies above the *x*-axis, and the parabola opens upward, it appears that this parabola has no *x*-intercepts. We can verify this observation algebraically.

$$0 = (x + 3)^2 + 1$$ Find possible x-intercepts, setting f(x) equal to 0.

$$-1 = (x + 3)^2$$ Solve for x. Subtract 1 from both sides.

$$x + 3 = \pm\sqrt{-1}$$ Apply the square root method.

$$x + 3 = \pm i$$ Recall that $\sqrt{-1} = i$, an imaginary number.

$$x = -3 \pm i$$ Subtract 3 from both sides.

Because this equation has no real solutions, the parabola has no *x*-intercepts.

Step 4 Find the *y*-intercept. Replace *x* with 0 in $f(x) = (x + 3)^2 + 1$.

$$f(0) = (0 + 3)^2 + 1 = 3^2 + 1 = 9 + 1 = 10$$

The *y*-intercept is 10. The parabola passes through $(0, 10)$.

Step 5 Graph the parabola. With a vertex at $(-3, 1)$, no *x*-intercepts, and a *y*-intercept at 10, the graph of *f* is shown in Figure 3.4. The axis of symmetry is the vertical line whose equation is $x = -3$.

Figure 3.4 The graph of $g(x) = (x + 3)^2 + 1$

Check Point 2 Graph the quadratic function $f(x) = (x - 2)^2 + 1$.

Graphing Quadratic Functions in the Form $f(x) = ax^2 + bx + c$

Quadratic functions are frequently expressed in the form $f(x) = ax^2 + bx + c$. How can we identify the vertex of a parabola whose equation is in this form? By completing the square, we can find a way to describe the vertex in terms of *a* and *b*.

$$f(x) = ax^2 + bx + c$$

$$= a\left(x^2 + \frac{b}{a}x\right) + c$$ Factor out a from $ax^2 + bx$.

$$= a\left(x^2 + \frac{b}{a}x + \frac{b^2}{4a^2}\right) + c - a\left(\frac{b^2}{4a^2}\right)$$

Complete the square by adding the square of half the coefficient of x.

By completing the square, we added $a \cdot \frac{b^2}{4a^2}$. To avoid changing the function's equation, we must subtract this term.

$$= a\left(x + \frac{b}{2a}\right)^2 + c - \frac{b^2}{4a}$$ Write the trinomial as the square of a binomial and simplify the constant term.

Compare this form of the equation with a quadratic function's standard form.

Standard form $\qquad f(x) = a(x - h)^2 + k$

$$h = -\frac{b}{2a} \qquad k = c - \frac{b^2}{4a}$$

Equation under discussion $\quad f(x) = a\left(x - \left(-\frac{b}{2a}\right)\right)^2 + c - \frac{b^2}{4a}$

The important part of this observation is that h, the x-coordinate of the vertex, is $-\dfrac{b}{2a}$. The y-coordinate can be found by evaluating the function at $-\dfrac{b}{2a}$.

The Vertex of a Parabola Whose Equation Is $f(x) = ax^2 + bx + c$
Consider the parabola defined by the quadratic function
$f(x) = ax^2 + bx + c$. The parabola's vertex is $\left(-\dfrac{b}{2a}, f\left(-\dfrac{b}{2a}\right)\right)$.

We can apply our five-step procedure and graph parabolas in the form $f(x) = ax^2 + bx + c$. The only step that is different is how we determine the vertex.

EXAMPLE 3 **Graphing a Quadratic Function in the Form $f(x) = ax^2 + bx + c$**

Graph the quadratic function $f(x) = -x^2 + 4x - 1$.
Solution

Step 1 Determine how the parabola opens. Note that a, the coefficient of x^2, is -1. Thus, $a < 0$; this negative value tells us that the parabola opens downward.

Step 2 Find the vertex. We know that the x-coordinate of the vertex is $x = -\dfrac{b}{2a}$. We identify $a, b,$ and c in $f(x) = ax^2 + bx + c$.

$$f(x) = -x^2 + 4x - 1$$

$$a = -1 \qquad b = 4 \qquad c = -1$$

Substitute the values of a and b into the equation for the x-coordinate:

$$x = -\frac{b}{2a} = -\frac{4}{2(-1)} = \frac{-4}{-2} = 2.$$

The x-coordinate of the vertex is 2. We substitute 2 for x in $f(x) = -x^2 + 4x - 1$, the equation of the function, to find the y-coordinate:

$$f(2) = -2^2 + 4 \cdot 2 - 1 = -4 + 8 - 1 = 3.$$

The vertex is $(2, 3)$.

Step 3   Find the x-intercepts.   Replace $f(x)$ with 0 in $f(x) = -x^2 + 4x - 1$. We obtain $0 = -x^2 + 4x - 1$ or $-x^2 + 4x - 1 = 0$. This equation cannot be solved by factoring. We will use the quadratic formula to solve it.

$$a = -1, \qquad b = 4, \qquad c = -1$$

$$x = \frac{-b \pm \sqrt{b^2 - 4ac}}{2a} = \frac{-4 \pm \sqrt{4^2 - 4(-1)(-1)}}{2(-1)} = \frac{-4 \pm \sqrt{16 - 4}}{-2}$$

$$x = \frac{-4 - \sqrt{12}}{-2} \approx 3.7 \quad \text{or} \quad x = \frac{-4 + \sqrt{12}}{-2} \approx 0.3$$

The x-intercepts are approximately 0.3 and 3.7. The parabola passes through the corresponding points, which we approximate as $(0.3, 0)$ and $(3.7, 0)$.

Step 4   Find the y-intercept.   Replace x with 0 in $f(x) = -x^2 + 4x - 1$.

$$f(0) = -0^2 + 4 \cdot 0 - 1 = -1$$

The y-intercept is -1. The parabola passes through $(0, -1)$.

Step 5   Graph the parabola.   With a vertex at $(2, 3)$, x-intercepts at approximately 0.3 and 3.7, and a y-intercept at -1, the graph of f is shown in Figure 3.5. The axis of symmetry is the vertical line whose equation is $x = 2$.

Figure 3.5 The graph of $f(x) = -x^2 + 4x - 1$

Check Point 3   Graph the quadratic function $f(x) = x^2 - 2x - 3$.

3   Solve problems involving minimizing or maximizing quadratic functions.

Applications of Quadratic Functions

When did the maximum number of Americans participate in the food stamp program? What is the age of a driver having the least number of car accidents? How do people launching fireworks know when they should explode to be viewed at the greatest possible height? The answers to these questions involve finding the maximum or minimum value of quadratic functions.

Consider the quadratic function $f(x) = ax^2 + bx + c$. If $a > 0$, the parabola opens upward and the vertex is its lowest point. If $a < 0$, the parabola opens downward and the vertex is its highest point. The x-coordinate of the vertex is $-\dfrac{b}{2a}$. Thus, we can find the minimum or maximum value of f by evaluating the quadratic function at $x = -\dfrac{b}{2a}$.

Minimum and Maximum: Quadratic Functions

Consider $f(x) = ax^2 + bx + c$.

1. If $a > 0$, then f has a minimum that occurs at $x = -\dfrac{b}{2a}$.
   This minimum value is $f\left(-\dfrac{b}{2a}\right)$.

2. If $a < 0$, then f has a maximum that occurs at $x = -\dfrac{b}{2a}$.
   This maximum value is $f\left(-\dfrac{b}{2a}\right)$.

EXAMPLE 4   An Application: The Food Stamp Program

The function

$$f(x) = -0.5x^2 + 4x + 19$$

models the number of people, $f(x)$, in millions, receiving food stamps x years after 1990. (*Source: New York Times*) In which year was this number at a maximum? How many food stamp recipients were there for that year?

Solution   The quadratic function is in the form $f(x) = ax^2 + bx + c$ with $a = -0.5$ and $b = 4$. Because $a < 0$, the function has a maximum value that occurs at $x = -\dfrac{b}{2a}$.

$$x = -\frac{b}{2a} = -\frac{4}{2(-0.5)} = \frac{-4}{-1} = 4$$

This means that the number of people receiving food stamps was at a maximum 4 years after 1990, in 1994. The number of recipients, in millions, for that year was

$$f(4) = -0.5(4)^2 + 4(4) + 19 = -8 + 16 + 19 = 27.$$

In 1994, the number of people receiving food stamps reached a maximum of 27 million.

Technology

The graph of the function modeling the millions of food stamp recipients

$$f(x) = -0.5x^2 + 4x + 19$$

is shown in a $[0, 10, 1]$ by $[0, 35, 5]$ viewing rectangle. The maximum function feature verifies that 4 years after 1990, the number of recipients reached a maximum of 27 million. Notice that x gives the location of the maximum and y gives the maximum value.

Check Point 4   The function $f(x) = 0.4x^2 - 36x + 1000$ models the number of accidents, $f(x)$, per 50 million miles driven, in terms of a driver's age, x, in years, where $16 \le x \le 74$. What is the age of a driver having the least number of car accidents? What is the minimum number of car accidents per 50 million miles driven?

Visualizing Irritability by Age

The quadratic function

$$P(x) = -0.05x^2 + 4.2x - 26$$

models the percentage of coffee drinkers, $P(x)$, who are x years old who become irritable if they do not have coffee at their regular time. Figure 3.6 shows the graph of the function. The vertex reveals that 62.2% of 42-year-old coffee drinkers become irritable. This is the maximum percentage for any age, x, in the function's domain.

Figure 3.6

Source: LMK Associates

Technology

We've come a long way from the small nation of "embattled farmers" who launched the American Revolution. In the early days of our Republic, 95% of the population was involved in farming. The graph in Figure 3.7 shows the number of farms in the United States from 1850 through 2010 (projected). Because the graph is shaped like a cup, with an increasing number of farms from 1850 to 1910 and a decreasing number of farms from 1910 to 2010, a quadratic function is an appropriate model for the data. You can use the statistical menu of a graphing utility to enter the data in Figure 3.7. We entered the data using (number of decades after 1850, millions of U.S. farms). The data are shown to the right of Figure 3.7.

Number of U. S. Farms, 1850–2010

Figure 3.7

Source: U.S. Bureau of the Census

Data:
(0, 2.3), (2, 3.3), (4, 5.1),
(6, 6.7), (8, 6.4), (10, 5.8),
(12, 3.6), (14, 2.9), (16, 2.3)

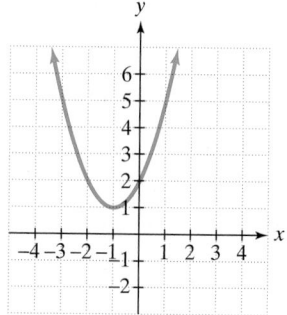

Upon entering the QUADratic REGression program, we obtain the results shown in the screen. Thus, the quadratic function of best fit is

$$f(x) = -0.064x^2 + 0.99x + 2.2$$

where x represents the number of decades after 1850 and $f(x)$ represents the number of U.S. farms, in millions.

EXERCISE SET 3.1

Practice Exercises

In Exercises 1–4, the graph of a quadratic function is given. Write the function's equation, selecting from the following options.

$f(x) = (x + 1)^2 - 1$ $g(x) = (x + 1)^2 + 1$
$h(x) = (x - 1)^2 + 1$ $j(x) = (x - 1)^2 - 1$

1.

2.

3.

4.

8.

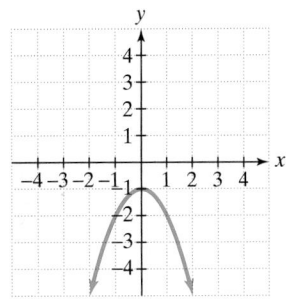

In Exercises 5–8, the graph of a quadratic function is given. Write the function's equation, selecting from the following options.

$$f(x) = x^2 + 2x + 1 \qquad g(x) = x^2 - 2x + 1$$
$$h(x) = x^2 - 1 \qquad\qquad j(x) = -x^2 - 1$$

In Exercises 9–16, find the coordinates of the vertex for the parabola defined by the given quadratic function.

9. $f(x) = 2(x - 3)^2 + 1$ **10.** $f(x) = -3(x - 2)^2 + 12$

11. $f(x) = -2(x + 1)^2 + 5$ **12.** $f(x) = -2(x + 4)^2 - 8$

13. $f(x) = 2x^2 - 8x + 3$ **14.** $f(x) = 3x^2 - 12x + 1$

15. $f(x) = -x^2 - 2x + 8$ **16.** $f(x) = -2x^2 + 8x - 1$

5.

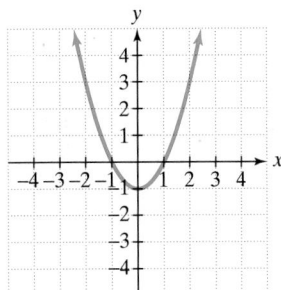

In Exercises 17–34, use the vertex and intercepts to sketch the graph of each quadratic function. Give the equation of the parabola's axis of symmetry. Use the graph to determine the function's domain and range.

17. $f(x) = (x - 4)^2 - 1$ **18.** $f(x) = (x - 1)^2 - 2$

19. $f(x) = (x - 1)^2 + 2$ **20.** $f(x) = (x - 3)^2 + 2$

21. $y - 1 = (x - 3)^2$ **22.** $y - 3 = (x - 1)^2$

6.

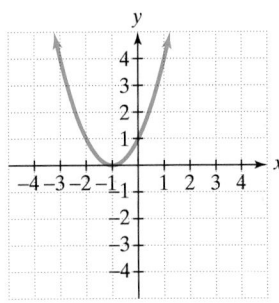

23. $f(x) = 2(x + 2)^2 - 1$ **24.** $f(x) = \frac{5}{4} - \left(x - \frac{1}{2}\right)^2$

25. $f(x) = 4 - (x - 1)^2$ **26.** $f(x) = 1 - (x - 3)^2$

27. $f(x) = x^2 - 2x - 3$ **28.** $f(x) = x^2 - 2x - 15$

29. $f(x) = x^2 + 3x - 10$ **30.** $f(x) = 2x^2 - 7x - 4$

31. $f(x) = 2x - x^2 + 3$ **32.** $f(x) = 5 - 4x - x^2$

33. $f(x) = 2x - x^2 - 2$ **34.** $f(x) = 6 - 4x + x^2$

7.

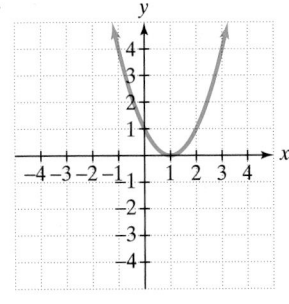

In Exercises 35–40, determine, without graphing, whether the given quadratic function has a minimum value or a maximum value. Then find the coordinates of the minimum or the maximum point.

35. $f(x) = 3x^2 - 12x - 1$ **36.** $f(x) = 2x^2 - 8x - 3$

37. $f(x) = -4x^2 + 8x - 3$ **38.** $f(x) = -2x^2 - 12x + 3$

39. $f(x) = 5x^2 - 5x$ **40.** $f(x) = 6x^2 - 6x$

Application Exercises

Cigarette Consumption per U.S. Adult

Source: U.S. Department of Health and Human Services

41. Please see the graph above. The function

$$f(x) = -3.1x^2 + 51.4x + 4024.5$$

models the average annual per capita consumption of cigarettes, $f(x)$, by Americans 18 and older x years after 1960. According to this model, in which year did cigarette consumption per capita reach a maximum? What was the consumption for that year? Does this accurately model what actually occurred as shown by the graph above?

42. The function

$$f(x) = 104.5x^2 - 1501.5x + 6016$$

models the death rate per year per 100,000 males, $f(x)$, for U.S. men who average x hours of sleep each night. How many hours of sleep, to the nearest tenth of an hour, corresponds to the minimum death rate? What is this minimum death rate, to the nearest whole number?

43. Fireworks are launched into the air. The quadratic function

$$s(t) = -16t^2 + 200t + 4$$

models the fireworks' height, $s(t)$, in feet, t seconds after they are launched. When should the fireworks explode so that they go off at the greatest height? What is that height?

44. A football is thrown by a quarterback to a receiver 40 yards away. The quadratic function

$$s(t) = -0.025t^2 + t + 5$$

models the football's height above the ground, $s(t)$, in feet, when it is t yards from the quarterback. How many yards from the quarterback does the football reach its greatest height? What is that height?

In the United States, HCV, or hepatitis C virus, is four times as widespread as HIV. Few of the nation's three to four million carriers have any idea they are infected. The graph shows the projected mortality, in thousands, from the virus. Use the information in the graph to solve Exercises 45–46.

Projected Hepatitis C Deaths in the United States

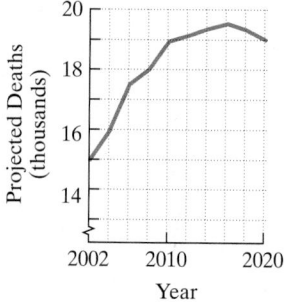

Source: *American Journal of Public Health*

45. Why is a quadratic function an appropriate model for the data shown in the graph?

46. Suppose that a quadratic function is used to model the data shown with ordered pairs representing (number of years after 2002, thousands of hepatitis C deaths). Determine, without obtaining an actual quadratic function that models the data, the approximate coordinates of the vertex for the function's graph. Describe what this means in practical terms. Use the word "maximum" in your description.

47. You have 120 feet of fencing to enclose a rectangular plot that borders on a river. If you do not fence the side along the river, find the length and width of the plot that will maximize the area. What is the largest area that can be enclosed?

48. The figure shown indicates that you have 100 yards of fencing to enclose a rectangular area. Find the dimensions of the rectangle that maximize the enclosed area. What is the maximum area?

49. A rain gutter is made from sheets of aluminum that are 20 inches wide. As shown in the figure, the edges are turned up to form right angles. Determine the depth of the gutter that will maximize its cross-sectional area and allow the greatest amount of water to flow.

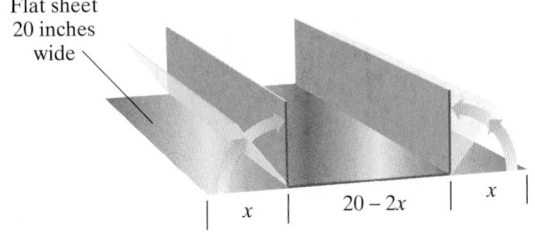

50. Hunky Beef, a local sandwich store, has a fixed weekly cost of $525.00, and variable costs for making a roast beef sandwich are $0.55.

a. Let x represent the number of roast beef sandwiches made and sold each week. Write the weekly cost function, C, for Hunky Beef. (*Hint:* The cost function is the sum of fixed and variable costs.)

b. The function $R(x) = -0.001x^2 + 3x$ describes the money, in dollars, that Hunky Beef takes in each week from the sale of x roast beef sandwiches. Use this revenue function and the cost function from part (a) to write the stores weekly profit function, P. (*Hint:* The profit function is the difference between revenue and cost functions.)

c. Use the store's profit function to determine the number of roast beef sandwiches it should make and sell each week to maximize profit. What is the maximum weekly profit?

Writing in Mathematics

51. What is a quadratic function?

52. What is a parabola? Describe its shape.

53. Explain how to decide whether a parabola opens upward or downward.

54. Describe how to find a parabola's vertex if its equation is expressed in standard form. Give an example.

55. Describe how to find a parabola's vertex if its equation is in the form $f(x) = ax^2 + bx + c$. Use $f(x) = x^2 - 6x + 8$ as an example.

56. A parabola that opens upward has its vertex at $(1, 2)$. Describe as much as you can about the parabola based on this information. Include in your discussion the number of x-intercepts (if any) for the parabola.

57. The quadratic function

$$f(x) = -0.018x^2 + 1.93x - 25.34$$

describes the miles per gallon, $f(x)$, of a Ford Taurus driven at x miles per hour. Suppose that you own a Ford Taurus. Describe how you can use this function to save money.

Technology Exercises

58. Use a graphing utility to verify any five of your hand-drawn graphs in Exercises 17–34.

59. a. Use a graphing utility to graph $y = 2x^2 - 82x + 720$ in a standard viewing rectangle. What do you observe?

b. Find the coordinates of the vertex for the given quadratic function.

c. The answer to part (b) is $(20.5, -120.5)$. Because the leading coefficient of the given function (2) is positive, the vertex is a minimum point on the graph. Use this fact to help find a viewing rectangle that will give a relatively complete picture of the parabola. With an axis of symmetry at $x = 20.5$, the setting for x should extend past this, so try Xmin = 0 and Xmax = 30. The setting for y should include (and probably go below) the y-coordinate of the graph's minimum point, so try Ymin = -130. Experiment with Ymax until your utility shows the parabola's major features.

d. In general, explain how knowing the coordinates of a parabola's vertex can help determine a reasonable viewing rectangle on a graphing utility for obtaining a complete picture of the parabola.

In Exercises 60–63, find the vertex for each parabola. Then determine a reasonable viewing rectangle on your graphing utility and use it to graph the quadratic function.

60. $y = -0.25x^2 + 40x$

61. $y = -4x^2 + 20x + 160$

62. $y = 5x^2 + 40x + 600$

63. $y = 0.01x^2 + 0.6x + 100$

64. The function $y = 0.011x^2 - 0.097x + 4.1$ models the number of people in the United States, y, in millions, holding more than one job x years after 1970. Use a graphing utility to graph the function in a $[0, 20, 1]$ by $[3, 6, 1]$ viewing rectangle. TRACE along the curve or use your utility's minimum value feature to approximate the coordinates of the parabola's vertex. Describe what this represents in practical terms.

65. The following data show fuel efficiency, in miles per gallon, for all U.S. automobiles in the indicated year.

x (Years after 1940)	y (Average Number of Miles per Gallon for U.S. Automobiles)
1940: 0	14.8
1950: 10	13.9
1960: 20	13.4
1970: 30	13.5
1980: 40	15.9
1990: 50	20.2
1998: 58	21.8

Source: U.S. Department of Transportation.

a. Use a graphing utility to draw a scatter plot of the data. Explain why a quadratic function is appropriate for modeling these data.

b. Use the quadratic regression feature to find the quadratic function that best fits the data.

c. Use the model in part (b) to determine the worst year for automobile fuel efficiency. What was the average number of miles per gallon for that year?

d. Use a graphing utility to draw a scatter plot of the data and graph the quadratic function of best fit on the scatter plot.

 Critical Thinking Exercises

66. Which one of the following is true?

a. No quadratic functions have a range of $(-\infty, \infty)$.

b. The vertex of the parabola described by $f(x) = 2(x - 5)^2 - 1$ is $(5, 1)$.

c. The graph of $f(x) = -2(x + 4)^2 - 8$ has one y-intercept and two x-intercepts.

d. The maximum value of y for the quadratic function $f(x) = -x^2 + x + 1$ is 1.

67. What explanations can you offer for your answer to Exercise 41? Use a graphing utility to graph f. Do you agree with the long-term predictions made by the graph? Explain.

In Exercises 68–69, find the axis of symmetry for each parabola whose equation is given. Use the axis of symmetry to find a second point on the parabola whose y-coordinate is the same as the given point.

68. $f(x) = 3(x + 2)^2 - 5;$ $(-1, -2)$

69. $f(x) = (x - 3)^2 + 2;$ $(6, 11)$

70. A rancher has 1000 feet of fencing to construct six corrals, as shown in the figure. Find the dimensions that maximize the enclosed area. What is the maximum area?

 Group Exercise

71. Each group member should consult an almanac, newspaper, magazine, or the Internet to find data that can be modeled by a quadratic function. Group members should select the two sets of data that are most interesting and relevant. For each data set selected:

a. Use the quadratic regression feature of a graphing utility to find the quadratic function that best fits the data.

b. Use the equation of the quadratic function to make a prediction from the data. What circumstances might affect the accuracy of your prediction?

c. Use the equation of the quadratic function to write and solve a problem involving maximizing or minimizing the function.

SECTION 3.2 *Polynomial Functions and Their Graphs*

Objectives

1. Recognize characteristics of graphs of polynomial functions.
2. Determine end behavior.
3. Use factoring to find zeros of polynomial functions.
4. Identify the multiplicity of a zero.
5. Understand the relationship between degree and turning points.
6. Graph polynomial functions.

Magnified 6000 times, this color-scanned image shows a T-lymphocyte blood cell (green) infected with the HIV virus (red). Depletion of the number of T-cells causes destruction of the immune system.

In 1980, U.S. doctors diagnosed 41 cases of a rare form of cancer, Kaposi's sarcoma, that involved skin lesions, pneumonia, and severe immunological deficiencies. All cases involved gay men ranging in age from 26 to 51. By the end of 2000, approximately 775,000 Americans, straight and gay, male and female, old and young, were infected with the HIV virus.

Modeling AIDS-related data and making predictions about the epidemic's havoc is serious business. Changing circumstances and unforeseen events have resulted in models that are not particularly useful over long periods of time. For example, the function

$$f(x) = -143x^3 + 1810x^2 - 187x + 2331$$

models the number of AIDS cases diagnosed in the United States x years after 1983. The model was obtained using cases diagnosed from 1983 through 1991. Figure 3.8 shows the graph of f from 1983 through 1991 in a $[0, 8, 1]$ by $[0, 50,000, 5000]$ viewing rectangle. The function used to describe the number of new AIDS cases in the United States over a limited period of time is an example of a *polynomial function*.

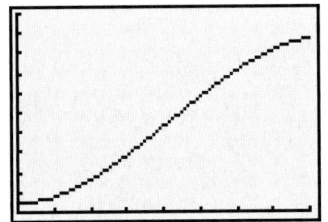

Figure 3.8 The graph of a function modeling the number of new AIDS cases in the U.S. from 1983 through 1991

Definition of a Polynomial Function

Let n be a nonnegative integer and let $a_n, a_{n-1}, \ldots, a_2, a_1, a_0$, be real numbers, with $a_n \neq 0$. The function defined by

$$f(x) = a_n x^n + a_{n-1} x^{n-1} + \cdots + a_2 x^2 + a_1 x + a_0$$

is called a **polynomial function of x of degree n.** The number a_n, the coefficient of the variable to the highest power, is called the **leading coefficient.**

A constant function $f(x) = c$, where $c \neq 0$, is a polynomial function of degree 0. A linear function $f(x) = mx + b$, where $m \neq 0$, is a polynomial function of degree 1. A quadratic function $f(x) = ax^2 + bx + c$, where $a \neq 0$, is a polynomial function of degree 2. In this section, we focus on polynomial functions of degree 3 or higher.

1 Recognize characteristics of graphs of polynomial functions.

Smooth, Continuous Graphs

Polynomial functions of degree 2 or less have graphs that are either parabolas or lines. We can graph such functions by plotting points. We can also graph polynomial functions of degree 3 or higher by plotting points. However, the process is rather tedious: Many points must be plotted. It may be easier to use a graphing utility for such functions. Regardless of the graphing method you use, you will find an ability to recognize the basic features of polynomial functions helpful. For example, they may help you choose an appropriate viewing rectangle for a graphing utility.

Two important features of the graphs of polynomial functions are that they are *smooth* and *continuous*. By **smooth,** we mean that the graph contains only rounded curves with no sharp corners. By **continuous,** we mean that the graph has no breaks and can be drawn without lifting your pencil from the rectangular coordinate system. These ideas are illustrated in Figure 3.9.

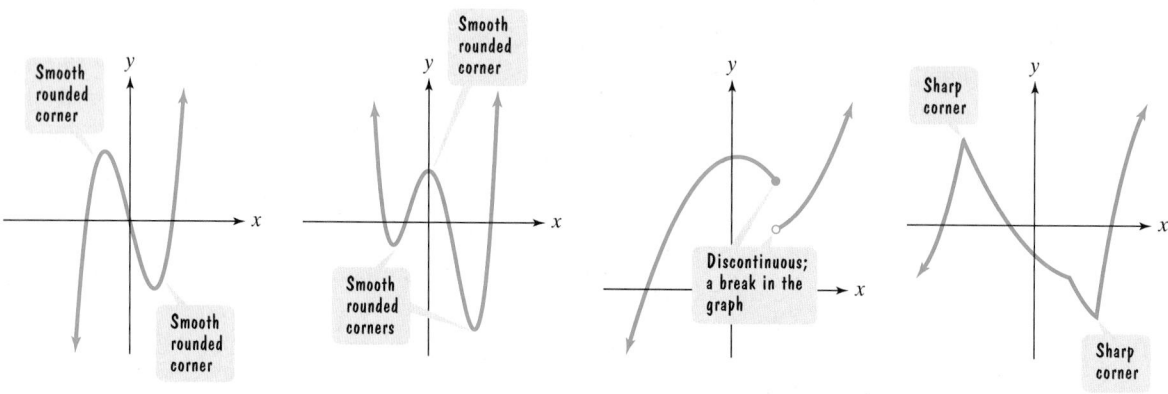

Figure 3.9 Recognizing graphs of polynomial functions

2 Determine end behavior.

Figure 3.10 By extending the viewing rectangle, y is eventually negative and the function no longer models the number of AIDS cases.

End Behavior of Polynomial Functions

Figure 3.10 shows the graph of the function

$$f(x) = -143x^3 + 1810x^2 - 187x + 2331,$$

which models U.S. AIDS cases from 1983 through 1991. Look what happens to the graph when we extend the year up through 1998 with a $[0, 15, 1]$ by $[-5000, 50,000, 5000]$ viewing rectangle. By year 13 (1996), the values of y are negative and the function no longer models AIDS cases. We've added an arrow to the graph at the far right to emphasize that it continues to decrease without bound. It is this far-right *end behavior* of the graph that makes it inappropriate for modeling AIDS cases into the future.

The behavior of a graph of a function to the far left or the far right is called its **end behavior.** Although the graph of a polynomial function may have intervals where it increases or decreases, the graph will eventually rise or fall without bound as it moves far to the left or far to the right.

How can you determine whether the graph of a polynomial function goes up or down at each end? The end behavior of a polynomial function

$$f(x) = a_n x^n + a_{n-1} x^{n-1} + \cdots + a_1 x + a_0$$

depends upon the leading term $a_n x^n$. In particular, the sign of the leading coefficient, a_n, and the degree, n, of the polynomial function reveal its end behavior. In terms of end behavior, only the term of highest degree counts, summarized by the **Leading Coefficient Test.**

The Leading Coefficient Test

As x increases or decreases without bound, the graph of the polynomial function

$$f(x) = a_n x^n + a_{n-1} x^{n-1} + a_{n-2} x^{n-2} + \cdots + a_1 x + a_0 \quad (a_n \neq 0)$$

eventually rises or falls. In particular,

1. For n odd:

2. For n even:

If the leading coefficient is positive, the graph falls to the left and rises to the right.

$a_n > 0$

If the leading coefficient is negative, the graph rises to the left and falls to the right.

$a_n < 0$

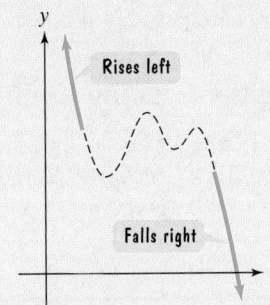

If the leading coefficient is positive, the graph rises to the left and to the right.

$a_n > 0$

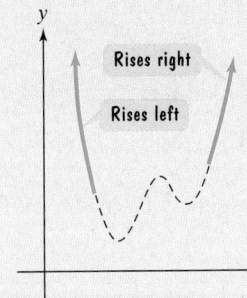

If the leading coefficient is negative, the graph falls to the left and to the right.

$a_n < 0$

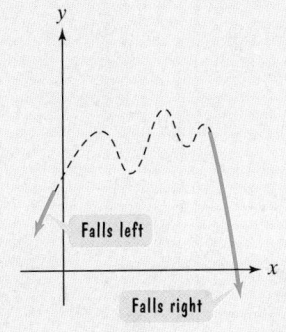

Study Tip

Odd-degree polynomial functions have graphs with opposite behavior at each end. Even-degree polynomial functions have graphs with the same behavior at each end.

EXAMPLE 1   Using the Leading Coefficient Test

Use the Leading Coefficient Test to determine the end behavior of the graph of

$$f(x) = x^3 + 3x^2 - x - 3.$$

Solution   Because the degree is odd ($n = 3$) and the leading coefficient, 1, is positive, the graph falls to the left and rises to the right, as shown in Figure 3.11.

Check Point 1   Use the Leading Coefficient Test to determine the end behavior of the graph of $f(x) = x^4 - 4x^2$.

Figure 3.11 The graph of $f(x) = x^3 + 3x^2 - x - 3$

EXAMPLE 2   Using the Leading Coefficient Test

Use end behavior to explain why

$$f(x) = -143x^3 + 1810x^2 - 187x + 2331$$

is only an appropriate model for AIDS cases for a limited time period.

Solution   Because the degree is odd ($n = 3$) and the leading coefficient, -143, is negative, the graph rises to the left and falls to the right. The fact that it falls to the right indicates at some point the number of AIDS cases will be negative, an impossibility. If a function has a graph that decreases without bound over time, it will not be capable of modeling nonnegative phenomena over long time periods.

> **Check Point 2**
>
> The polynomial function
>
> $$f(x) = -0.27x^3 + 9.2x^2 - 102.9x + 400$$
>
> models the ratio of students to computers, $f(x)$, in U.S. public schools x years after 1980. Use end behavior to determine whether this function could be an appropriate model for computers in the classroom well into the twenty-first century. Explain your answer.

If you use a graphing utility to graph a polynomial function, it is important to select a viewing rectangle that accurately reveals the graph's end behavior. If the viewing rectangle is too small, it may not accurately show the end behavior.

EXAMPLE 3 Using the Leading Coefficient Test

The graph of $f(x) = -x^4 + 8x^3 + 4x^2 + 2$ was obtained with a graphing utility using a $[-8, 8, 1]$ by $[-10, 10, 1]$ viewing rectangle. The graph's is shown in Figure 3.12(a). Does the graph show the end behavior of the function?

Figure 3.12 (a) $[-8, 8, 1]$ by $[-10, 10, 1]$ (b) $[-10, 10, 1]$ by $[-1000, 750, 250]$

Solution Note that the degree is even $(n = 4)$ and the leading coefficient, -1, is negative. Even-degree polynomial functions have graphs with the same behavior at each end. The Leading Coefficient Test indicates that the graph should fall to the left and fall to the right. The graph in Figure 3.12(a) is falling to the left, but it is not falling to the right. Therefore, the graph is not complete enough to show end behavior. A more complete graph of the function is shown in a larger viewing rectangle in Figure 3.12(b).

Figure 3.13

> **Check Point 3**
>
> The graph of $f(x) = x^3 + 13x^2 + 10x - 4$ is shown in a standard viewing rectangle in Figure 3.13. Use the Leading Coefficient Test to determine whether the graph shows the end behavior of the function. Explain your answer.

3 Use factoring to find zeros of polynomial functions.

Zeros of Polynomial Functions

If f is a polynomial function, then the values of x for which $f(x)$ is equal to 0 are called the **zeros** of f. These values of x are the **roots,** or **solutions,** of the polynomial equation $f(x) = 0$. Each real root of the polynomial equation appears as an x-intercept of the graph of the polynomial function.

EXAMPLE 4   Finding Zeros of a Polynomial Function

Find all zeros of $f(x) = x^3 + 3x^2 - x - 3$.

Solution   By definition, the zeros are the values of x for which $f(x)$ is equal to 0. Thus, we set $f(x)$ equal to 0:

$$f(x) = x^3 + 3x^2 - x - 3 = 0.$$

We solve the polynomial equation $x^3 + 3x^2 - x - 3 = 0$ for x as follows:

$x^3 + 3x^2 - x - 3 = 0$	This is the equation needed to find the function's zeros.
$x^2(x + 3) - 1(x + 3) = 0$	Factor x^2 from the first two terms and -1 from the last two terms.
$(x + 3)(x^2 - 1) = 0$	A common factor of $x + 3$ is factored from the expression.
$x + 3 = 0$ or $x^2 - 1 = 0$	Set each factor equal to 0.
$x = -3$ $x^2 = 1$	Solve for x.
$x = \pm 1$	Remember that if $x^2 = d$, then $x = \pm\sqrt{d}$.

The zeros of f are $-3, -1$, and 1. The graph of f in Figure 3.14 shows that each zero is an x-intercept.

Technology

The graph of

$$f(x) = x^3 + 3x^2 - x - 3$$

is shown in a $[-6, 6, 1]$ by $[-6, 6, 1]$ viewing rectangle. The x-intercepts indicate that f has three real zeros,

$$-3, -1, \text{ and } 1.$$

The table provides numerical support for the three real zeros shown by the graph.

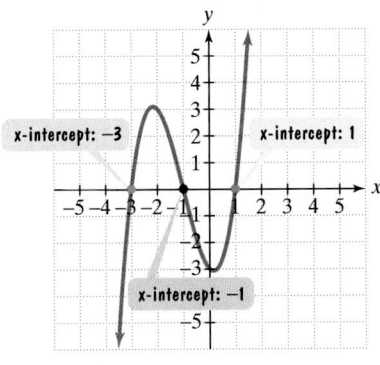

Figure 3.14 The real zeros of $f(x) = x^3 + 3x^2 - x - 3$, namely $-3, -1$, and 1, are the x-intercepts for the graph of f.

> **Check Point 4**   Find all zeros of $f(x) = x^3 + 2x^2 - 4x - 8$.

EXAMPLE 5   Finding Zeros of a Polynomial Function

Find all zeros of $f(x) = -x^4 + 4x^3 - 4x^2$.

Solution   We find the zeros of f by setting $f(x)$ equal to 0 and solving the resulting equation.

$-x^4 + 4x^3 - 4x^2 = 0$	We now have a polynomial equation.
$x^4 - 4x^3 + 4x^2 = 0$	Multiply both sides by -1. This step is optional.
$x^2(x^2 - 4x + 4) = 0$	Factor out x^2.
$x^2(x - 2)^2 = 0$	Factor completely.
$x^2 = 0$ or $(x - 2)^2 = 0$	Set each factor equal to 0.
$x = 0$ $x = 2$	Solve for x.

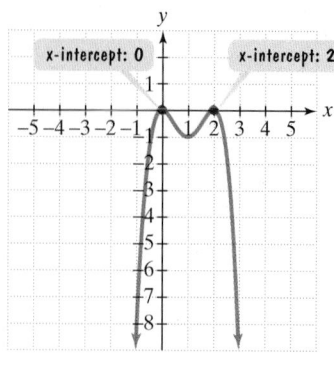

Figure 3.15 The zeros of $f(x) = -x^4 + 4x^3 - 4x^2$, namely 0 and 2, are the x-intercepts for the graph of f.

The zeros of $f(x) = -x^4 + 4x^3 - 4x^2$ are 0 and 2. The graph of f, shown in Figure 3.15, has x-intercepts at 0 and 2.

> **Check Point 5** Find all zeros of $f(x) = x^4 - 4x^2$.

We can use the results of factoring to express a polynomial as a product of factors. For instance, in Example 5, we can use our factoring to express the function's equation as follows:

$$f(x) = -x^4 + 4x^3 - 4x^2 = -(x^4 - 4x^3 + 4x^2) = -x^2(x - 2)^2.$$

The factor x occurs twice: $x^2 = x \cdot x$.

The factor $(x - 2)$ occurs twice: $(x - 2)^2 = (x - 2)(x - 2)$.

4 Identify the multiplicity of a zero.

Notice that each factor occurs twice. In factoring the equation for the polynomial function f, if the same factor $x - r$ occurs k times, but not $k + 1$ times, we call r a **repeated zero with multiplicity k.** For the polynomial function

$$f(x) = -x^2(x - 2)^2,$$

0 and 2 are both repeated zeros with multiplicity 2. For the polynomial function

$$f(x) = 4(x - 5)(x + 2)^3 \left(x - \tfrac{1}{4}\right)^4,$$

5 is a zero with multiplicity 1, -2 is a repeated zero with multiplicity 3, and $\tfrac{1}{4}$ is a repeated zero with multiplicity 4.

The multiplicity of a zero tells us if the graph of a polynomial function touches the x-axis at the zero and turns around, or crosses the x-axis at the zero. For example, look again at the graph of $f(x) = -x^4 + 4x^3 - 4x^2$ in Figure 3.15. Each zero, 0 and 2, is a repeated zero with multiplicity 2. The graph of f touches, but does not cross, the x-axis at each of these zeros of even multiplicity. By contrast, a graph crosses the x-axis at zeros of odd multiplicity.

> ### Multiplicity and x-Intercepts
> If r is a zero of even multiplicity, then the graph **touches** the x-axis and turns around at r. If r is a zero of odd multiplicity, then the graph **crosses** the x-axis at r. Regardless of whether a zero is even or odd, graphs tend to flatten out at zeros with multiplicity greater than one.

Turning Points of Polynomial Functions

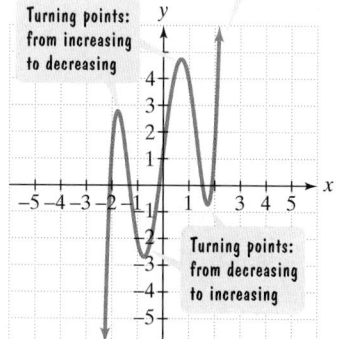

Figure 3.16 Graph with four turning points

5 Understand the relationship between degree and turning points.

The graph of $f(x) = x^5 - 6x^3 + 8x + 1$ is shown in Figure 3.16. The graph has four smooth **turning points.** At each turning point, the graph changes direction from increasing to decreasing or vice versa. The function value at a turning point is either a relative maximum of f or a relative minimum of f. The given equation has 5 as its greatest exponent and is therefore a polynomial function of degree 5. Notice that the graph has four turning points. In general, **if f is a polynomial of degree n, then the graph of f has at most $n - 1$ turning points.**

6 Graph polynomial functions.

A Strategy for Graphing Polynomial Functions

Here's a general strategy for graphing a polynomial function. A graphing utility is a valuable complement to this strategy. Some of the steps listed in the following box will help you to select a viewing rectangle that shows the important parts of the graph.

Graphing a Polynomial Function

$$f(x) = a_n x^n + a_{n-1} x^{n-1} + a_{n-2} x^{n-2} + \cdots + a_1 x + a_0, \quad a_n \neq 0$$

1. Use the Leading Coefficient Test to determine the graph's end behavior.
2. Find x-intercepts by setting $f(x) = 0$ and solving the resulting polynomial equation. If there is an x-intercept at r as a result of $(x - r)^k$ in the complete factorization of $f(x)$, then
   a. If k is even, the graph touches the x-axis at r and turns around.
   b. If k is odd, the graph crosses the x-axis at r.
   c. If $k > 1$, the graph flattens out at $(r, 0)$.
3. Find the y-intercept by computing $f(0)$.
4. Use symmetry, if applicable, to help draw the graph:
   a. y-axis symmetry: $f(-x) = f(x)$
   b. Origin symmetry: $f(-x) = -f(x)$
5. Use the fact that the maximum number of turning points of the graph is $n - 1$ to check whether it is drawn correctly.

EXAMPLE 6 Graphing a Polynomial Function

Graph: $f(x) = x^4 - 2x^2 + 1$.

Solution

Step 1 Determine end behavior. Because the degree is even ($n = 4$) and the leading coefficient, 1, is positive, the graph rises to the left and rises to the right.

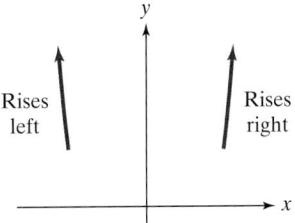

Step 2 Find x-intercepts (zeros of the function) by setting $f(x) = 0$.

$$x^4 - 2x^2 + 1 = 0$$

$(x^2 - 1)(x^2 - 1) = 0$ Factor.

$(x + 1)(x - 1)(x + 1)(x - 1) = 0$ Factor completely.

$(x + 1)^2(x - 1)^2 = 0$ Express the factoring in a more compact notation.

$(x + 1)^2 = 0$ or $(x - 1)^2 = 0$ Set each factor equal to 0.

$x = -1$ $x = 1$ Solve for x.

We see that -1 and 1 are both repeated zeros with multiplicity 2. Because of the even multiplicity, the graph touches the x-axis at -1 and 1 and turns around. Furthermore, the graph tends to flatten out at these zeros with multiplicity greater than one.

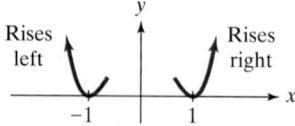

Step 3   Find the y-intercept by computing $f(0)$. We use $f(x) = x^4 - 2x^2 + 1$ and compute $f(0)$.

$$f(0) = 0^4 - 2 \cdot 0^2 + 1 = 1$$

There is a y-intercept at 1, so the graph passes through $(0, 1)$:

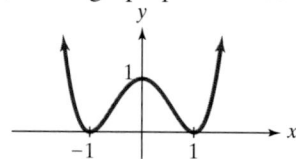

Step 4   Use possible symmetry to help draw the graph.   Our partial graph suggests y-axis symmetry. Let's verify this by finding $f(-x)$.

$$f(x) \quad = x^4 \quad - \quad 2x^2 \quad + \quad 1$$

Replace x with $-x$.

$$f(-x) = (-x)^4 - 2(-x)^2 + 1 = x^4 - 2x^2 + 1$$

Because $f(-x) = f(x)$, the graph of f is symmetric with respect to the y-axis. Figure 3.17 shows the graph of $f(x) = x^4 - 2x^2 + 1$.

Step 5   Use the fact that the maximum number of turning points of the graph is $n - 1$ to check whether it is drawn correctly.   Because $n = 4$, the maximum number of turning points is $4 - 1$, or 3. Because the graph in Figure 3.17 has three turning points, we have not violated the maximum number possible.

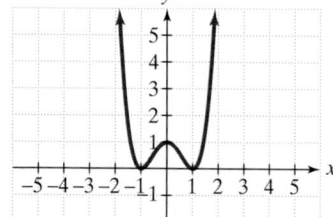

Figure 3.17 The graph of $f(x) = x^4 - 2x^2 + 1$

Check Point 6   Use the five-step strategy to graph $f(x) = x^3 - 3x^2$.

EXERCISE SET 3.2

Practice Exercises

In Exercises 1–10, determine which functions are polynomial functions. For those that are, identify the degree.

1. $f(x) = 5x^2 + 6x^3$

2. $f(x) = 7x^2 + 9x^4$

3. $g(x) = 7x^5 - \pi x^3 + \frac{1}{5}x$

4. $g(x) = 6x^7 + \pi x^5 + \frac{2}{3}x$

5. $h(x) = 7x^3 + 2x^2 + \frac{1}{x}$

6. $h(x) = 8x^3 - x^2 + \frac{2}{x}$

7. $f(x) = x^{1/2} - 3x^2 + 5$

8. $f(x) = x^{1/3} - 4x^2 + 7$

9. $f(x) = \dfrac{x^2 + 7}{x^3}$

10. $f(x) = \dfrac{x^2 + 7}{3}$

In Exercises 11–14, identify which graphs are not those of polynomial functions.

11.

12.

13.

14.

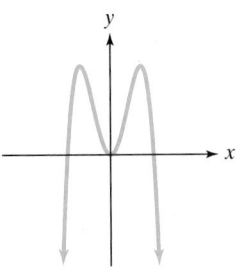

In Exercises 15–20, use the Leading Coefficient Test to determine the end behavior of the graph of the given polynomial function. Then use this end behavior to match the polynomial function with its graph. [The graphs are labeled (a) through (f).]

15. $f(x) = -x^4 + x^2$ **16.** $f(x) = x^3 - 4x^2$

17. $f(x) = (x - 3)^2$

18. $f(x) = -x^3 - x^2 + 5x - 3$

19. $f(x) = x - 3$ **20.** $f(x) = (x + 1)^2(x - 1)^2$

(a)

(b)

(c)

(d)

(e)

(f)

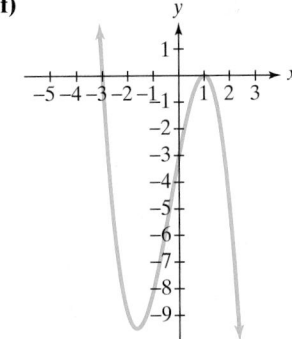

In Exercises 21–26, use the Leading Coefficient Test to determine the end behavior of the graph of the polynomial function.

21. $f(x) = 5x^3 + 7x^2 - x + 9$

22. $f(x) = 11x^3 - 6x^2 + x + 3$

23. $f(x) = 5x^4 + 7x^2 - x + 9$

24. $f(x) = 11x^4 - 6x^2 + x + 3$

25. $f(x) = -5x^4 + 7x^2 - x + 9$

26. $f(x) = -11x^4 - 6x^2 + x + 3$

In Exercises 27–34, find the zeros for each polynomial function and give the multiplicity for each zero. State whether the graph crosses the x-axis, or touches the x-axis and turns around, at each zero.

27. $f(x) = 2(x - 5)(x + 4)^2$

28. $f(x) = 3(x + 5)(x + 2)^2$

29. $f(x) = 4(x - 3)(x + 6)^3$

30. $f(x) = -3(x + \frac{1}{2})(x - 4)^3$

31. $f(x) = x^3 - 2x^2 + x$

32. $f(x) = x^3 + 4x^2 + 4x$

33. $f(x) = x^3 + 7x^2 - 4x - 28$

34. $f(x) = x^3 + 5x^2 - 9x - 45$

In Exercises 35–50,

    a. *Use the Leading Coefficient Test to determine the graph's end behavior.*

    b. *Find x-intercepts by setting $f(x) = 0$ and solving the resulting polynomial equation. State whether the graph crosses the x-axis, or touches the x-axis and turns around, at each intercept.*

    c. *Find the y-intercept by setting x equal to 0 and computing $f(0)$.*

    d. *Determine whether the graph has y-axis symmetry, origin symmetry, or neither.*

    e. *If necessary, find a few additional points and graph the function. Use the fact that the maximum number of turning points of the graph is $n - 1$ to check whether it is drawn correctly.*

35. $f(x) = x^3 + 2x^2 - x - 2$

36. $f(x) = x^3 + x^2 - 4x - 4$

37. $f(x) = x^4 - 9x^2$ **38.** $f(x) = x^4 - x^2$

39. $f(x) = -x^4 + 16x^2$ **40.** $f(x) = -x^4 + 4x^2$

41. $f(x) = x^4 - 2x^3 + x^2$ **42.** $f(x) = x^4 - 6x^3 + 9x^2$

43. $f(x) = -2x^4 + 4x^3$ **44.** $f(x) = -2x^4 + 2x^3$

45. $f(x) = 6x^3 - 9x - x^5$ **46.** $f(x) = 6x - x^3 - x^5$

47. $f(x) = 3x^2 - x^3$ **48.** $f(x) = \frac{1}{2} - \frac{1}{2}x^4$

49. $f(x) = -3(x - 1)^2(x^2 - 4)$

50. $f(x) = -2(x - 4)^2(x^2 - 25)$

Application Exercises

51. A herd of 100 elk is introduced to a small island. The number of elk, $N(t)$, after t years is described by the polynomial function $N(t) = -t^4 + 21t^2 + 100$.

    a. Use the Leading Coefficient Test to determine the graph's end behavior to the right. What does this mean about what will eventually happen to the elk population?

    b. Graph the function.

    c. Graph only the portion of the function that serves as a realistic model for the elk population over time. When does the population become extinct?

52. The common cold is caused by a rhinovirus. After x days of invasion by the viral particles, the number of particles in our bodies, $f(x)$, in billions, can be modeled by the polynomial function

$$f(x) = -0.75x^4 + 3x^3 + 5.$$

Use the Leading Coefficient Test to determine the graph's end behavior to the right. What does this mean about the number of viral particles in our bodies over time?

53. The polynomial function

$$f(x) = -0.87x^3 + 0.35x^2 + 81.62x + 7684.94$$

models the number of thefts, $f(x)$, in thousands, in the United States x years after 1987. Will this function be useful in modeling the number of thefts over an extended period of time? Explain your answer.

54. The graphs show the percentage of husbands and wives with one or more children who said their marriage was going well "all the time" at various stages in their relationships.

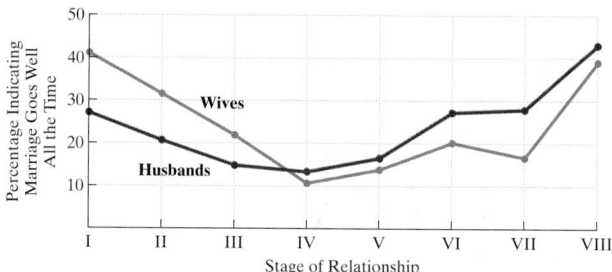

Marital Satisfaction for Families with Children

Source: Rollins, B., & Feldman, H. (1970), Marital satisfaction over the family life cycle. *Journal of Marriage and the Family,* 32, 20–28.

    Stage   I: Beginning families

    Stage   II: Child-bearing families

    Stage   III: Families with preschool children

    Stage   IV: Families with school-age children

Stage   V: Families with teenagers
Stage   VI: Families with adult children leaving home
Stage VII: Families in the middle years
Stage VIII: Aging families

a. Between which stages was marital satisfaction for wives decreasing?

b. Between which stages was marital satisfaction for wives increasing?

c. How many turning points (from decreasing to increasing or from increasing to decreasing) are shown in the graph for wives?

d. Suppose that a polynomial function is used to model the data shown in the graph for wives using

(stage in the relationship, percentage indicating that the marriage was going well all the time).

Use the number of turning points to determine the degree of the polynomial function of best fit.

e. For the model in part (d), should the leading coefficient of the polynomial function be positive or negative? Explain your answer.

55. The bar graph below shows women's earnings as a percentage of men's from 1970 through 2000. Suppose that a polynomial function is used to model the data shown using

(number of years after 1970, women's earnings as a percentage of men's).

Determine the degree of the polynomial function of best fit. Should the leading coefficient be positive or negative? Explain your answer.

The Wage Gap in the United States

Source: U.S. Women's Bureau

Writing in Mathematics

56. What is a polynomial function?

57. What do we mean when we describe the graph of a polynomial function as smooth and continuous?

58. What is meant by the end behavior of a polynomial function?

59. Explain how to use the Leading Coefficient Test to determine the end behavior of a polynomial function.

60. Why is a third-degree polynomial function with a negative leading coefficient not appropriate for modeling non-negative real-world phenomena over a long period of time?

61. What are the zeros of a polynomial function and how are they found?

62. Explain the relationship between the multiplicity of a zero and whether or not the graph crosses or touches the x-axis at that zero.

63. Explain the relationship between the degree of a polynomial function and the number of turning points on its graph.

64. Can the graph of a polynomial function have no x-intercepts? Explain.

65. Can the graph of a polynomial function have no y-intercept? Explain.

66. Describe a strategy for graphing a polynomial function. In your description, mention intercepts, the polynomial's degree, and turning points.

67. In a favorable habitat and without natural predators, a population of reindeer is introduced to an island preserve. The reindeer population, $f(t)$, t years after their introduction is modeled by the polynomial function $f(t) = -0.125t^5 + 3.125t^4 + 4000$. Discuss the growth and decline of the reindeer population. Describe the factors that might contribute to this population model.

68. The graphs shown in Exercise 54 indicate that marital satisfaction tends to be greatest at the beginning and at the end of the stages in the relationship, with a decline occurring in the middle. What explanations can you offer for this trend?

 Technology Exercises

69. Use a graphing utility to verify any five of the graphs that you drew by hand in Exercises 35–50.

Write a polynomial function that imitates the end behavior of each graph in Exercises 70–73. The dashed portions of the graphs indicate that you should focus only on imitating the left and right behavior of the graph and can be flexible about what occurs between the left and right ends. Then use your graphing utility to graph the polynomial function and verify that you imitated the end behavior shown in the given graph.

70.   **71.**

72. 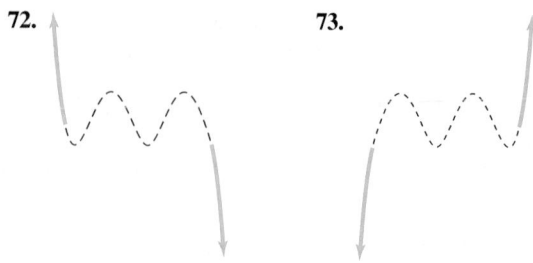 **73.**

In Exercises 74–77, use a graphing utility with a viewing rectangle large enough to show end behavior to graph each polynomial function.

74. $f(x) = x^3 + 13x^2 + 10x - 4$

75. $f(x) = -2x^3 + 6x^2 + 3x - 1$

76. $f(x) = -x^4 + 8x^3 + 4x^2 + 2$

77. $f(x) = -x^5 + 5x^4 - 6x^3 + 2x + 20$

In Exercises 78–79, use a graphing utility to graph f and g in the same viewing rectangle. Then use the $\boxed{\text{ZOOM OUT}}$ feature to show that f and g have identical end behavior.

78. $f(x) = x^3 - 6x + 1, \quad g(x) = x^3$

79. $f(x) = -x^4 + 2x^3 - 6x, \quad g(x) = -x^4$

 Critical Thinking Exercises

80. Which one of the following is true?

a. If $f(x) = -x^3 + 4x$, then the graph of f falls to the left and to the right.

b. A mathematical model that is a polynomial of degree n whose leading term is $a_n x^n$, n odd and $a_n < 0$, is ideally suited to describe nonnegative phenomena over unlimited periods of time.

c. There is more than one third-degree polynomial function with the same three x-intercepts.

d. The graph of a function with origin symmetry can rise to the left and to the right.

Use the descriptions in Exercises 81–82 to write an equation of a polynomial function with the given characteristics. Use a graphing utility to graph your function to see if you are correct. If not, modify the function's equation and repeat this process.

81. Crosses the x-axis at -4, 0, and 3; lies above the x-axis between -4 and 0; lies below the x-axis between 0 and 3

82. Touches the x-axis at 0 and crosses the x-axis at 2; lies below the x-axis between 0 and 2

 Group Exercise

83. This exercise is based on the group's work in Exercise 71 of Exercise Set 3.1. For the two data sets that the group selected:

a. Use the polynomial regression feature of a graphing utility to find the third-degree polynomial function that best fits the data.

b. Use this function to repeat the predictions that you made with the quadratic function. How do these predictions compare with those that you obtained previously?

c. For each data set, describe whether the quadratic function or the third-degree function is a better fit. Use a graphing utility, a scatter plot of the data, and the function of best fit drawn on the scatter plot to help determine which function is the better fit.

SECTION 3.3   Dividing Polynomials; Remainder and Factor Theorems

Objectives

1. Use long division to divide polynomials.

2. Use synthetic division to divide polynomials.

3. Evaluate a polynomial using the Remainder Theorem.

4. Use the Factor Theorem to solve a polynomial equation.

For those of you who are dog lovers, you might still be thinking of the polynomial function that models the age in human years, $H(x)$, of a dog that is x years old, namely

$$H(x) = -0.001618x^4 + 0.077326x^3 - 1.2367x^2 + 11.460x + 2.914.$$

Suppose that you are in your twenties, say 25. What is Fido's equivalent age? To answer this question, we must substitute 25 for $H(x)$ and solve the resulting polynomial equation for x:

$$25 = -0.001618x^4 + 0.077326x^3 - 1.2367x^2 + 11.460x + 2.914.$$

How can we solve such an equation? You might begin by subtracting 25 from both sides to obtain zero on one side. But then what? The factoring that we used in the previous section will not work in this situation.

In Sections 3.4 and 3.5, we will present techniques for solving certain kinds of polynomial equations. These techniques will further enhance your ability to manipulate algebraically the polynomial functions that model your world. Because these techniques are based on understanding polynomial division, in this section we look at two methods for dividing polynomials.

1 Use long division to divide polynomials.

Long Division of Polynomials and the Division Algorithm

We begin by looking at division by a polynomial containing more than one term, such as

$$x + 3 \overline{)\, x^2 + 10x + 21.}$$

Divisor has two terms.   Dividend has three terms.

When a divisor has more than one term, the four steps used to divide whole numbers—**divide**, **multiply**, **subtract**, **bring down the next term**—form the repetitive procedure for polynomial long division.

EXAMPLE 1   Long Division of Polynomials

Divide $x^2 + 10x + 21$ by $x + 3$.

Solution The following steps illustrate how polynomial division is very similar to numerical division.

$$x + 3 \overline{)\, x^2 + 10x + 21}$$

Arrange the terms of the dividend ($x^2 + 10x + 21$) and the divisor ($x + 3$) in descending powers of x.

$$\begin{array}{r} x \\ x + 3 \overline{)\, x^2 + 10x + 21} \end{array}$$

Divide x^2 (the first term in the dividend) by x (the first term in the divisor): $\dfrac{x^2}{x} = x$. Align like terms.

times

$$\begin{array}{r} x \\ x + 3 \overline{)\, x^2 + 10x + 21} \\ x^2 + 3x \end{array}$$

equals

Multiply each term in the divisor x + 3 by x, aligning terms of the product under like terms in the dividend.

$$\begin{array}{r} x \\ x + 3 \overline{)\, x^2 + 10x + 21} \\ \ominus x^2 + \ominus 3x \\ \hline 7x \end{array}$$

Change signs of the polynomial being subtracted.

Subtract $x^2 + 3x$ from $x^2 + 10x$ by changing the sign of each term in the lower expression and adding.

$$\begin{array}{r} x \\ x + 3 \overline{)\, x^2 + 10x + 21} \\ x^2 + 3x \downarrow \\ \hline 7x + 21 \end{array}$$

Bring down 21 from the original dividend and add algebraically to form a new dividend.

$$
\begin{array}{r}
x + 7 \\
x + 3 \overline{\smash{)}\, x^2 + 10x + 21} \\
\underline{x^2 + 3x} \\
7x + 21
\end{array}
$$

Find the second term of the quotient. **Divide** the first term of $7x + 21$ by x, the first term of the divisor: $\dfrac{7x}{x} = 7.$

$$
\begin{array}{r}
x + 7 \\
x + 3 \overline{\smash{)}\, x^2 + 10x + 21} \\
\underline{x^2 + 3x} \\
7x + 21 \\
\underline{7x + 21} \\
0
\end{array}
$$

times

equals

Multiply the divisor $x + 3$ by 7, aligning under like terms in the new dividend. Then **subtract** to obtain the remainder of 0.

Remainder

The quotient is $x + 7$. Because the remainder is 0, we can conclude that $x + 3$ is a factor of $x^2 + 10x + 21$ and

$$
\frac{x^2 + 10x + 21}{x + 3} = x + 7.
$$

Check Point 1   Divide $x^2 + 14x + 45$ by $x + 9$.

Before considering additional examples, let's summarize the general procedure for dividing one polynomial by another.

Long Division of Polynomials

1. **Arrange the terms** of both the dividend and the divisor in descending powers of the variable.

2. **Divide** the first term in the dividend by the first term in the divisor. The result is the first term of the quotient.

3. **Multiply** every term in the divisor by the first term in the quotient. Write the resulting product beneath the dividend with like terms lined up.

4. **Subtract** the product from the dividend.

5. **Bring down** the next term in the original dividend and write it next to the remainder to form a new dividend.

6. Use this new expression as the dividend and repeat this process until the remainder can no longer be divided. This will occur when the degree of the remainder (the highest exponent on a variable in the remainder) is less than the degree of the divisor.

In our next long division, we will obtain a nonzero remainder.

EXAMPLE 2   Long Division of Polynomials

Divide $4 - 5x - x^2 + 6x^3$ by $3x - 2$.

Solution   We begin by writing the divisor and dividend in descending powers of x.

Is It Hot in Here, or Is It Just Me?

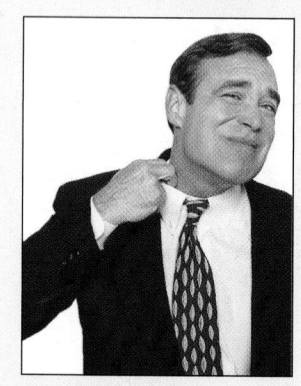

In the 1980s, a rising trend in global surface temperature was observed, and the term "global warming" was coined. Scientists are more convinced than ever that burning coal, oil, and gas results in a buildup of gases and particles that trap heat and raise the planet's temperature. The average increase in global surface temperature, $T(x)$, in degrees Centigrade, x years after 1980 can be modeled by the polynomial function

$$
T(x) = \frac{21}{5,000,000}\, x^3
$$
$$
- \frac{127}{1,000,000}\, x^2 + \frac{1293}{50,000}\, x.
$$

Use your graphing utility to graph the function in a $[0, 60, 3]$ by $[0, 2, 0.1]$ viewing rectangle. (Place parentheses around each fractional coefficient when you enter the equation.) What do you observe about global warming through the year 2040?

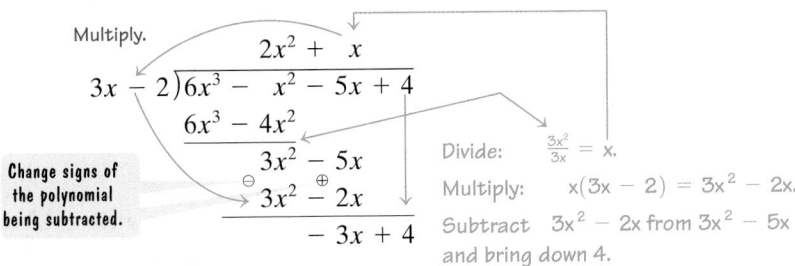

Now we divide $3x^2$ by $3x$ to obtain x, multiply x and the divisor, and subtract.

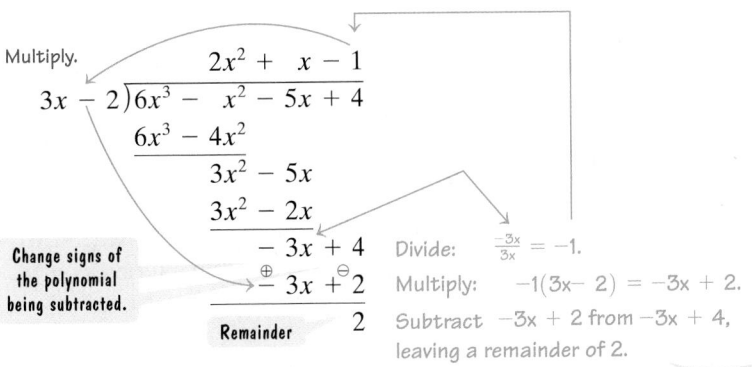

Now we divide $-3x$ by $3x$ to obtain -1, multiply -1 and the divisor, and subtract.

In Example 2, the quotient is $2x^2 + x - 1$ and the remainder is 2. When there is a nonzero remainder, as in this example, list the quotient, plus the remainder above the divisor. Thus,

Dividend   Quotient   Remainder

$$\frac{6x^3 - x^2 - 5x + 4}{3x - 2} = 2x^2 + x - 1 + \frac{2}{3x - 2}.$$

Divisor   Divisor

Multiplying both sides of this equation by $3x - 2$ results in the following equation:

$$6x^3 - x^2 - 5x + 4 = (3x - 2)(2x^2 + x - 1) + 2.$$

Dividend   Divisor   Quotient   Remainder

Polynomial long division is checked by multiplying the divisor with the quotient and then adding the remainder. This should give the dividend. The process illustrates the **Division Algorithm.**

The Division Algorithm

If $f(x)$ and $d(x)$ are polynomials, with $d(x) \neq 0$, and the degree of $d(x)$ is less than or equal to the degree of $f(x)$, then there exist unique polynomials $q(x)$ and $r(x)$ such that

$$f(x) \quad = \quad d(x) \quad \cdot \quad q(x) \quad + \quad r(x).$$

| Dividend | Divisor | Quotient | Remainder |

The remainder, $r(x)$, equals 0 or it is of degree less than the degree of $d(x)$. If $r(x) = 0$, we say that $d(x)$ **divides evenly** into $f(x)$ and that $d(x)$ and $q(x)$ are **factors** of $f(x)$.

Check Point 2   Divide $7 - 11x - 3x^2 + 2x^3$ by $x - 3$. Use the remainder to express your result in fractional form.

If a power of x is missing in either a dividend or a divisor, add that power of x with a coefficient of 0 and then divide. In this way, like terms will be aligned as you carry out the long division.

EXAMPLE 3   Long Division of Polynomials

Divide $6x^4 + 5x^3 + 3x - 5$ by $3x^2 - 2x$.

Solution   We write the dividend, $6x^4 + 5x^3 + 3x - 5$, as $6x^4 + 5x^3 + 0x^2 + 3x - 5$ to keep all like terms aligned.

The division process is finished because the degree of $7x - 5$, which is 1, is less than the degree of the divisor $3x^2 - 2x$, which is 2. The answer is

$$\frac{6x^4 + 5x^3 + 3x - 5}{3x^2 - 2x} = 2x^2 + 3x + 2 + \frac{7x - 5}{3x^2 - 2x}.$$

Check Point 3   Divide $2x^4 + 3x^3 - 7x - 10$ by $x^2 - 2x$.

2 Use synthetic division to divide polynomials.

Dividing Polynomials Using Synthetic Division

We can use **synthetic division** to divide polynomials if the divisor is of the form $x - c$. This method provides a quotient more quickly than long division. Let's compare the two methods showing $x^3 + 4x^2 - 5x + 5$ divided by $x - 3$.

Long Division

$$
\begin{array}{r}
x^2 + 7x + 16 \\
x - 3 \overline{)\,x^3 + 4x^2 - 5x + 5} \\
\ominus x^3 \overset{\oplus}{-} 3x^2 \\
\hline
7x^2 - 5x \\
\ominus 7x^2 \overset{\oplus}{-} 21x \\
\hline
16x + 5 \\
\ominus 16x \overset{\oplus}{-} 48 \\
\hline
53
\end{array}
$$

Divisor
$x - c$;
$c = 3$

Quotient

Dividend

Remainder

Synthetic Division

$$
\begin{array}{r|rrrr}
3 & 1 & 4 & -5 & 5 \\
  &   & 3 & 21 & 48 \\
\hline
  & 1 & 7 & 16 & 53
\end{array}
$$

Notice the relationship between the polynomials in the long division process and the numbers that appear in synthetic division.

The divisor is $x - 3$. This is 3, or c, in $x - c$.

These are the coefficients of the dividend $x^3 + 4x^2 - 5x + 5$.

$$
\begin{array}{r|rrrr}
3 & 1 & 4 & -5 & 5 \\
  &   & 3 & 21 & 48 \\
\hline
  & 1 & 7 & 16 & 53
\end{array}
$$

These are the coefficients of the quotient $x^2 + 7x + 16$.

This is the remainder.

Now let's look at the steps involved in synthetic division.

Synthetic Division

To divide a polynomial by $x - c$,

Example

$x - 3\overline{)\,x^3 + 4x^2 - 5x + 5}$

1. Arrange polynomials in descending powers, with a 0 coefficient for any missing term.

2. Write c for the divisor, $x - c$. To the right, write the coefficients of the dividend.

$$
\begin{array}{r|rrrr}
3 & 1 & 4 & -5 & 5
\end{array}
$$

3. Write the leading coefficient of the dividend on the bottom row.

$$
\begin{array}{r|rrrr}
3 & 1 & 4 & -5 & 5 \\
\hline
  & 1
\end{array}
$$
Bring down 1.

4. Multiply c (in this case, 3) times the value just written on the bottom row. Write the product in the next column in the second row.

$$
\begin{array}{r|rrrr}
3 & 1 & 4 & -5 & 5 \\
  &   & 3 \\
\hline
  & 1
\end{array}
$$
Multiply by 3: $3 \cdot 1 = 3$.

5. Add the values in this new column, writing the sum in the bottom row.

$$
\begin{array}{r|rrrr}
3 & 1 & 4 & -5 & 5 \\
  &   & 3 \\
\hline
  & 1 & 7
\end{array}
$$
Add.

6. Repeat this series of multiplications and additions until all columns are filled in.

$$\begin{array}{r|rrrr} 3 & 1 & 4 & -5 & 5 \\ & & 3 & 21 & \\ \hline & 1 & 7 & 16 & \end{array} \text{ Add.}$$

Multiply by 3: $3 \cdot 7 = 21$.

$$\begin{array}{r|rrrr} 3 & 1 & 4 & -5 & 5 \\ & & 3 & 21 & 48 \\ \hline & 1 & 7 & 16 & 53 \end{array} \text{ Add.}$$

Multiply by 3: $3 \cdot 16 = 48$.

7. Use the numbers in the last row to write the quotient, plus the remainder above the divisor. **The degree of the first term of the quotient is one less than the degree of the first term of the dividend.** The final value in this row is the remainder.

Written from the last row of the synthetic division

$$x - 3 \overline{)x^3 + 4x^2 - 5x + 5} \qquad 1x^2 + 7x + 16 + \frac{53}{x - 3}$$

EXAMPLE 4 Using Synthetic Division

Use synthetic division to divide $5x^3 + 6x + 8$ by $x + 2$.

Solution The divisor must be in the form $x - c$. Thus, we write $x + 2$ as $x - (-2)$. This means that $c = -2$. Writing a 0 coefficient for the missing x^2-term in the dividend, we can express the division as follows:

$$x - (-2) \overline{)5x^3 + 0x^2 + 6x + 8}.$$

Now we are ready to set up the problem so that we can use synthetic division.

Use the coefficients of the dividend $5x^3 + 0x^2 + 6x + 8$ in descending powers of x.

This is c in x−(−2).

$$\begin{array}{r|rrrr} -2 & 5 & 0 & 6 & 8 \end{array}$$

We begin the synthetic division process by bringing down 5. This is followed by a series of multiplications and additions.

1. Bring down 5.
$$\begin{array}{r|rrrr} -2 & 5 & 0 & 6 & 8 \\ & \downarrow & & & \\ \hline & 5 & & & \end{array}$$

2. Multiply: $-2(5) = -10.$
$$\begin{array}{r|rrrr} -2 & 5 & 0 & 6 & 8 \\ & & -10 & & \\ \hline & 5 & & & \end{array}$$
Multiply by -2.

3. Add: $0 + (-10) = -10.$
$$\begin{array}{r|rrrr} -2 & 5 & 0 & 6 & 8 \\ & & -10 & & \text{Add.} \\ \hline & 5 & -10 & & \end{array}$$

4. Multiply: $-2(-10) = 20.$
$$\begin{array}{r|rrrr} -2 & 5 & 0 & 6 & 8 \\ & & -10 & 20 & \\ \hline & 5 & -10 & & \end{array}$$
Multiply by -2.

5. Add: $6 + 20 = 26.$
$$\begin{array}{r|rrrr} -2 & 5 & 0 & 6 & 8 \\ & & -10 & 20 & \text{Add.} \\ \hline & 5 & -10 & 26 & \end{array}$$

6. Multiply: $-2(26) = -52.$
$$\begin{array}{r|rrrr} -2 & 5 & 0 & 6 & 8 \\ & & -10 & 20 & -52 \\ \hline & 5 & -10 & 26 & \end{array}$$
Multiply by -2.

7. Add: $8 + (-52) = -44.$
$$\begin{array}{r|rrrr} -2 & 5 & 0 & 6 & 8 \\ & & -10 & 20 & -52 \\ \hline & 5 & -10 & 26 & -44 \end{array} \text{ Add.}$$

The numbers in the last row represent the coefficients of the quotient and the remainder. The degree of the first term of the quotient is one less than that of the dividend. Because the degree of the dividend, $5x^3 + 6x + 8$, is 3, the degree of the quotient is 2. This means that the 5 in the last row represents $5x^2$.

$$\begin{array}{r|rrrr} -2 & 5 & 0 & 6 & 8 \\ & & -10 & 20 & -52 \\ \hline & 5 & -10 & 26 & -44 \end{array}$$

The quotient is $5x^2 - 10x + 26$. The remainder is -44.

Thus,

$$x + 2 \overline{)5x^3 + 6x + 8} = 5x^2 - 10x + 26 - \frac{44}{x + 2}$$

Check Point 4 Use synthetic division to divide

$$x^3 - 7x - 6 \text{ by } x + 2.$$

3 Evaluate a polynomial using the Remainder Theorem.

The Remainder Theorem

Let's consider the Division Algorithm when the dividend, $f(x)$, is divided by $x - c$. In this case, the remainder must be a constant because its degree is less than one, the degree of $x - c$.

$$f(x) = (x - c)q(x) + r$$

Dividend   Divisor   Quotient   The remainder, r, is a constant when dividing by x - c.

Now let's evaluate f at c.

$$f(c) = (c - c)q(c) + r \quad \text{Find } f(c) \text{ by letting } x = c \text{ in } f(x) = (x-c)q(x) + r.$$
$$f(c) = 0 \cdot q(c) + r \quad c - c = 0$$
$$f(c) = r \quad 0 \cdot q(c) = 0 \text{ and } 0 + r = r.$$

What does this last equation mean? If a polynomial is divided by $x - c$, the remainder is the value of the polynomial at c. This result is called the **Remainder Theorem.**

The Remainder Theorem
If the polynomial $f(x)$ is divided by $x - c$, then the remainder is $f(c)$.

Example 5 shows how we can use the Remainder Theorem to evaluate a polynomial function at 2. Rather than substituting 2 for x, we divide the function by $x - 2$. The remainder is $f(2)$.

EXAMPLE 5 **Using the Remainder Theorem to Evaluate a Polynomial Function**

Given $f(x) = x^3 - 4x^2 + 5x + 3$, use the Remainder Theorem to find $f(2)$.

Solution By the Remainder Theorem, if $f(x)$ is divided by $x - 2$, then the remainder is $f(2)$. We'll use synthetic division to divide.

$$
\begin{array}{r|rrrr}
2 & 1 & -4 & 5 & 3 \\
  &   & 2 & -4 & 2 \\
\hline
  & 1 & -2 & 1 & 5 \quad \text{Remainder}
\end{array}
$$

The remainder, 5, is the value of $f(2)$. Thus, $f(2) = 5$. We can verify that this is correct by evaluating $f(2)$ directly. Using $f(x) = x^3 - 4x^2 + 5x + 3$, we obtain

$$f(2) = 2^3 - 4 \cdot 2^2 + 5 \cdot 2 + 3 = 8 - 16 + 10 + 3 = 5.$$

Check Point 5 Given $f(x) = 3x^3 + 4x^2 - 5x + 3$, use the Remainder Theorem to find $f(-4)$.

4 Use the Factor Theorem to solve a polynomial equation.

The Factor Theorem

Let's look again at the Division Algorithm when the divisor is of the form $x - c$.

$$f(x) = (x - c)q(x) + r$$

| Dividend | Divisor | Quotient | Constant remainder |

By the Remainder Theorem, the remainder r is $f(c)$, so we can substitute $f(c)$ for r:

$$f(x) = (x - c)q(x) + f(c).$$

Notice that if $f(c) = 0$, then

$$f(x) = (x - c)q(x)$$

so that $x - c$ is a factor of $f(x)$. This means that for the polynomial function $f(x)$, if $f(c) = 0$, then $x - c$ is a factor of $f(x)$.

Let's reverse directions and see what happens if $x - c$ is a factor of $f(x)$. This means that

$$f(x) = (x - c)q(x).$$

If we replace x with c, we obtain

$$f(c) = (c - c)q(c) = 0.$$

Thus, if $x - c$ is a factor of $f(x)$, then $f(c) = 0$.

We have proved a result known as the **Factor Theorem.**

The Factor Theorem

Let $f(x)$ be a polynomial.

a. If $f(c) = 0$, then $x - c$ is a factor of $f(x)$.

b. If $x - c$ is a factor of $f(x)$, then $f(c) = 0$.

The example that follows shows how the Factor Theorem can be used to solve a polynomial equation.

EXAMPLE 6   Using the Factor Theorem

Solve the equation $2x^3 - 3x^2 - 11x + 6 = 0$ given that 3 is a zero of $f(x) = 2x^3 - 3x^2 - 11x + 6$.

Solution   We are given that $f(3) = 0$. The Factor Theorem tells us that $x - 3$ is a factor of $f(x)$. We'll use synthetic division to divide $f(x)$ by $x - 3$.

$$\begin{array}{r|rrrr} 3 & 2 & -3 & -11 & 6 \\ & & 6 & 9 & -6 \\ \hline & 2 & 3 & -2 & 0 \end{array}$$

$$\begin{array}{r} 2x^2 + 3x - 2 \\ x - 3 \overline{)2x^3 - 3x^2 - 11x + 6} \end{array}$$

The remainder, 0, verifies that x − 3 is a factor of $2x^3 - 3x^2 - 11x + 6$.

Equivalently,

$$2x^3 - 3x^2 - 11x + 6 = (x - 3)(2x^2 + 3x - 2).$$

Now we can solve the polynomial equation.

$2x^3 - 3x^2 - 11x + 6 = 0$ This is the given equation.

$(x - 3)(2x^2 + 3x - 2) = 0$ Factor using the result from the synthetic division.

$(x - 3)(2x - 1)(x + 2) = 0$ Factor the trinomial.

$x - 3 = 0$ or $2x - 1 = 0$ or $x + 2 = 0$ Set each factor equal to 0.

$x = 3$ $x = \frac{1}{2}$ $x = -2$ Solve for x.

The solution set is $\left\{-2, \frac{1}{2}, 3\right\}$.

Based on the Factor Theorem, the following statements are useful in solving polynomial equations:

1. If $f(x)$ is divided by $x - c$ and the remainder is zero, then c is a zero of f and c is a root of the polynomial equation $f(x) = 0$.

2. If $f(x)$ is divided by $x - c$ and the remainder is zero, then $x - c$ is a factor of $f(x)$.

Check Point 6   Solve the equation $15x^3 + 14x^2 - 3x - 2 = 0$ given that -1 is a zero of $f(x) = 15x^3 + 14x^2 - 3x - 2$.

EXERCISE SET 3.3

Practice Exercises

In Exercises 1–16, divide using long division. State the quotient, q(x), and the remainder, r(x).

1. $(x^2 + 8x + 15) \div (x + 5)$

2. $(x^2 + 3x - 10) \div (x - 2)$

3. $(x^3 + 5x^2 + 7x + 2) \div (x + 2)$

4. $(x^3 - 2x^2 - 5x + 6) \div (x - 3)$

5. $(6x^3 + 7x^2 + 12x - 5) \div (3x - 1)$

6. $(6x^3 + 17x^2 + 27x + 20) \div (3x + 4)$

7. $(12x^2 + x - 4) \div (3x - 2)$

8. $(4x^2 - 8x + 6) \div (2x - 1)$

9. $\dfrac{2x^3 + 7x^2 + 9x - 20}{x + 3}$

10. $\dfrac{3x^2 - 2x + 5}{x - 3}$

11. $\dfrac{4x^4 - 4x^2 + 6x}{x - 4}$

12. $\dfrac{x^4 - 81}{x - 3}$

13. $\dfrac{6x^3 + 13x^2 - 11x - 15}{3x^2 - x - 3}$

14. $\dfrac{x^4 + 2x^3 - 4x^2 - 5x - 6}{x^2 + x - 2}$

15. $\dfrac{18x^4 + 9x^3 + 3x^2}{3x^2 + 1}$　　**16.** $\dfrac{2x^5 - 8x^4 + 2x^3 + x^2}{2x^3 + 1}$

In Exercises 17–32, divide using synthetic division.

17. $(2x^2 + x - 10) \div (x - 2)$

18. $(x^2 + x - 2) \div (x - 1)$

19. $(3x^2 + 7x - 20) \div (x + 5)$

20. $(5x^2 - 12x - 8) \div (x + 3)$

21. $(4x^3 - 3x^2 + 3x - 1) \div (x - 1)$

22. $(5x^3 - 6x^2 + 3x + 11) \div (x - 2)$

23. $(6x^5 - 2x^3 + 4x^2 - 3x + 1) \div (x - 2)$

24. $(x^5 + 4x^4 - 3x^2 + 2x + 3) \div (x - 3)$

25. $(x^2 - 5x - 5x^3 + x^4) \div (5 + x)$

26. $(x^2 - 6x - 6x^3 + x^4) \div (6 + x)$

27. $\dfrac{x^5 + x^3 - 2}{x - 1}$　　**28.** $\dfrac{x^7 + x^5 - 10x^3 + 12}{x + 2}$

29. $\dfrac{x^4 - 256}{x - 4}$　　**30.** $\dfrac{x^7 - 128}{x - 2}$

31. $\dfrac{2x^5 - 3x^4 + x^3 - x^2 + 2x - 1}{x + 2}$

32. $\dfrac{x^5 - 2x^4 - x^3 + 3x^2 - x + 1}{x - 2}$

33. Given $f(x) = 2x^3 - 11x^2 + 7x - 5$, use the Remainder Theorem to find $f(4)$.

34. Given $f(x) = x^3 - 7x^2 + 5x - 6$, use the Remainder Theorem to find $f(3)$.

35. Given $f(x) = 7x^4 - 3x^3 + 6x + 9$, use the Remainder Theorem to find $f(-5)$.

36. Given $f(x) = 3x^4 + 6x^3 - 2x + 4$, use the Remainder Theorem to find $f(-4)$.

37. Use synthetic division to divide $f(x) = x^3 - 4x^2 + x + 6$ by $x + 1$. Use the result to find all zeros of f.

38. Use synthetic division to divide $f(x) = x^3 - 2x^2 - x + 2$ by $x + 1$. Use the result to find all zeros of f.

39. Solve the equation $2x^3 - 5x^2 + x + 2 = 0$ given that 2 is a zero of $f(x) = 2x^3 - 5x^2 + x + 2$.

40. Solve the equation $2x^3 - 3x^2 - 11x + 6 = 0$ given that -2 is a zero of $f(x) = 2x^3 - 3x^2 - 11x + 6$.

41. Solve the equation $12x^3 + 16x^2 - 5x - 3 = 0$ given that $-\frac{3}{2}$ is a root.

42. Solve the equation $3x^3 + 7x^2 - 22x - 8 = 0$ given that $-\frac{1}{3}$ is a root.

Application Exercises

43. A rectangle with length $2x + 5$ inches has an area of $2x^4 + 15x^3 + 7x^2 - 135x - 225$ square inches. Write a polynomial that represents its width.

44. If you travel a distance of $x^3 + 3x^2 + 5x + 3$ miles at a rate of $x + 1$ miles per hour, write a polynomial that represents the number of hours you traveled.

During the 1980s, the controversial economist Arthur Laffer promoted the idea that tax increases lead to a reduction in government revenue. Called supply-side economics, the theory uses function such as

$$f(x) = \dfrac{80x - 8000}{x - 110}, \ 30 \le x \le 100.$$

This function models the government tax revenue, $f(x)$, in tens of billions of dollars, in terms of the tax rate, x. The graph of the function is shown. It illustrates tax revenue decreasing quite dramatically as the tax rate increases. At a tax rate of (gasp) 100%, the government takes all our money and no one has an incentive to work. With no income earned, zero dollars in tax revenue is generated.

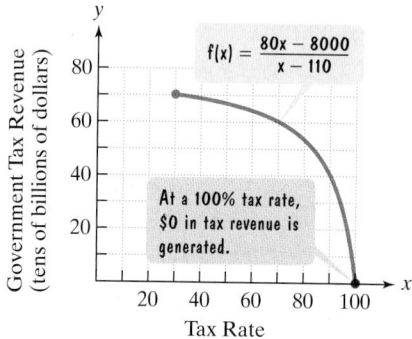

Use function f and its graph to solve Exercises 45–46.

45. a. Find and interpret $f(30)$. Identify the solution as a point on the graph of the function.

　b. Rewrite the function by using long division to perform
$$(80x - 8000) \div (x - 110).$$
Then use this new form of the function to find $f(30)$. Do you obtain the same answer as you did in part (a)?

　c. Is f a polynomial function? Explain your answer.

46. a. Find and interpret $f(40)$. Identify the solution as a point on the graph of the function.

　b. Rewrite the function by using long division to perform
$$(80x - 8000) \div (x - 110).$$
Then use this new form of the function to find $f(40)$. Do you obtain the same answer as you did in part (a)?

　c. Is f a polynomial function? Explain your answer.

Writing in Mathematics

47. Explain how to perform long division of polynomials. Use $2x^3 - 3x^2 - 11x + 7$ divided by $x - 3$ in your explanation.

48. In your own words, state the Division Algorithm.

49. How can the Division Algorithm be used to check the quotient and remainder in a long division problem?

50. Explain how to perform synthetic division. Use the division problem in Exercise 47 to support your explanation.

51. State the Remainder Theorem.

52. Explain how the Remainder Theorem can be used to find $f(-6)$ if $f(x) = x^4 + 7x^3 + 8x^2 + 11x + 5$. What advantage is there to using the Remainder Theorem in this situation rather than evaluating $f(-6)$ directly?

53. How can the Factor Theorem be used to determine if $x - 1$ is a factor of $x^3 - 2x^2 - 11x + 12$?

54. If you know that -2 is a zero of
$$f(x) = x^3 + 7x^2 + 4x - 12,$$
explain how to solve the equation
$$x^3 + 7x^2 + 4x - 12 = 0.$$

55. The idea of supply-side economics (see Exercises 45–46) is that an increase in the tax rate may actually reduce government revenue. What explanation can you offer for this theory?

 Technology Exercises

In Exercises 56–59, use a graphing utility to determine if the division has been performed correctly Graph the function on each side of the equation in the same viewing rectangle. If the graphs do not coincide, correct the expression on the right side by using polynomial long division. Then verify your correction using the graphing utility.

56. $(6x^2 + 16x + 8) \div (3x + 2) = 2x + 4, x \neq -\frac{2}{3}$

57. $\dfrac{x^4 + 6x^3 + 6x^2 - 10x - 3}{x^2 + 2x - 3} = x^2 + 4x + 1,$
$x \neq -3, x \neq 1$

58. $\dfrac{2x^3 - 3x^2 - 3x + 4}{x - 1} = 2x^2 - x + 4, x \neq 1$

59. $\dfrac{3x^4 + 4x^3 - 32x^2 - 5x - 20}{x + 4} = 3x^3 + 8x^2 - 5,$
$x \neq -4$

 Critical Thinking Exercises

60. Which one of the following is true?
 a. If a trinomial in x of degree 6 is divided by a trinomial in x of degree 3, the degree of the quotient is 2.
 b. Synthetic division could not be used to find the quotient of $10x^3 - 6x^2 + 4x - 1$ and $x - \frac{1}{2}$.
 c. Any problem that can be done by synthetic division can also be done by the method for long division of polynomials.
 d. If a polynomial long-division problem results in a remainder that is a whole number, then the divisor is a factor of the dividend.

61. Find k so that $4x + 3$ is a factor of
$$20x^3 + 23x^2 - 10x + k.$$

62. When $2x^2 - 7x + 9$ is divided by a polynomial, the quotient is $2x - 3$ and the remainder is 3. Find the polynomial.

63. Find the quotient of $x^{3n} + 1$ and $x^n + 1$.

64. Synthetic division is a process for dividing a polynomial by $x - c$. The coefficient of x is 1. How might synthetic division be used if you are dividing by $2x - 4$?

SECTION 3.4 *Zeros of Polynomial Functions*

Objectives

1. Use the Rational Zero Theorem to find possible rational zeros.

2. Find zeros of a polynomial function.

3. Solve polynomial equations.

4. Use Descartes's Rule of Signs.

A moth has moved into your closet. She appeared in your bedroom at night, but somehow her relatively stout body escaped your clutches. Within a few weeks, swarms of moths in your tattered wardrobe suggest that Mama Moth was in the family way. There must be at least 200 critters nesting in every crevice of your clothing.

Two hundred plus moth-tykes from one female moth; is this possible? Indeed it is. The number of eggs, N, in a female moth is a function of her abdominal width, W, in millimeters, modeled by

$$N(W) = 14W^3 - 17W^2 - 16W + 34$$

for $1.5 \leq W \leq 3.5$. Because there are 200 moths feasting on your favorite sweaters, Mama's abdominal width can be estimated by finding the roots of the polynomial equation

$$14W^3 - 17W^2 - 16W + 34 = 200.$$

With mathematics present even in your quickly disappearing attire, we move from rags to polynomial equations. The process of solving such equations begins with listing possibilities for Mama Moth's abdominal width. To do this, we turn to a theorem that plays an important role in finding zeros of polynomial functions.

1 Use the Rational Zero Theorem to find possible rational zeros.

The Rational Zero Theorem

The Rational Zero Theorem gives a list of all possible rational zeros of a polynomial function. Equivalently, the theorem gives all possible rational roots of a polynomial equation. Not every number in the list will be a zero of the function, but every rational zero of the polynomial function will appear somewhere in the list.

The Rational Zero Theorem

If $f(x) = a_n x^n + a_{n-1} x^{n-1} + \cdots + a_1 x + a_0$ has *integer* coefficients and $\dfrac{p}{q}$ (where $\dfrac{p}{q}$ is reduced) is a rational zero, then p is a factor of the constant term, a_0, and q is a factor of the leading coefficient, a_n.

You can explore the "why" behind the Rational Zero Theorem in Exercise 66 of Exercise Set 3.4. For now, let's see if we can figure out what the theorem tells us about possible rational zeros. In order to use the theorem, list all the integers that are factors of the constant term, a_0. Then list all the integers that are factors of the leading coefficient, a_n. Finally list all possible rational zeros:

$$\text{Possible rational zeros} = \frac{\text{Factors of the constant term}}{\text{Factors of the leading coefficient}}.$$

EXAMPLE 1   Using the Rational Zero Theorem

List all possible rational zeros of $f(x) = -x^4 + 3x^2 + 4$.

Solution   The constant term is 4. We list all of its factors: $\pm 1, \pm 2, \pm 4$. The leading coefficient is -1. Its factors are ± 1.

Factors of the constant term:	$\pm 1, \quad \pm 2, \quad \pm 4$
Factors of the leading coefficient:	± 1

Because

$$\text{Possible rational zeros} = \frac{\text{Factors of the constant term}}{\text{Factors of the leading coefficient}},$$

we must take each number in the first row, $\pm 1, \pm 2, \pm 4$, and divide by each number in the second row, ± 1.

Study Tip

Always keep in mind the relationship among zeros, roots, and x-intercepts. The zeros of function f are the roots, or solutions, of the equation $f(x) = 0$. Furthermore, the real zeros, or real roots, are the x-intercepts of the graph of f.

$$\text{Possible rational zeros} = \frac{\text{Factors of } 4}{\text{Factors of } -1} = \frac{\pm 1, \pm 2, \pm 4}{\pm 1} = \pm 1, \quad \pm 2, \quad \pm 4$$

Divide ± 1 by ± 1. Divide ± 2 by ± 1. Divide ± 4 by ± 1.

There are six possible rational zeros. The graph of $f(x) = -x^4 + 3x^2 + 4$ is shown in Figure 3.18. The x-intercepts are -2 and 2. Thus, -2 and 2 are the actual rational zeros.

−2 is a rational zero. 2 is a rational zero.

Figure 3.18 The graph of $f(x) = -x^4 + 3x^2 + 4$ shows that -2 and 2 are rational zeros.

Check Point 1 List all possible rational zeros of
$$f(x) = x^3 + 2x^2 - 5x - 6.$$

EXAMPLE 2 Using the Rational Zero Theorem

List all possible rational zeros of $f(x) = 15x^3 + 14x^2 - 3x - 2$.

Solution The constant term is -2 and the leading coefficient is 15.

$$\text{Possible rational zeros} = \frac{\text{Factors of the constant term, } -2}{\text{Factors of the leading coefficient, } 15}$$

$$= \frac{\pm 1, \pm 2}{\pm 1, \pm 3, \pm 5, \pm 15}$$

$$= \pm 1, \quad \pm 2, \quad \pm \tfrac{1}{3}, \quad \pm \tfrac{2}{3}, \quad \pm \tfrac{1}{5}, \quad \pm \tfrac{2}{5}, \quad \pm \tfrac{1}{15}, \quad \pm \tfrac{2}{15}$$

Divide ± 1 and ± 2 by ± 1. Divide ± 1 and ± 2 by ± 3. Divide ± 1 and ± 2 by ± 5. Divide ± 1 and ± 2 by ± 15.

There are 16 possible rational zeros. The actual solution set of
$$15x^3 + 14x^2 - 3x - 2 = 0$$
is $\left\{-1, -\tfrac{1}{3}, \tfrac{2}{5}\right\}$, which contains 3 of the 16 possible zeros.

Check Point 2 List all possible rational zeros of
$$f(x) = 4x^5 + 12x^4 - x - 3.$$

2 Find zeros of a polynomial function.

How do we determine which (if any) of the possible rational zeros are rational zeros of the polynomial function? To find the first rational zero, we can use a trial-and-error process involving synthetic division: If $f(x)$ is divided by $x - c$ and the remainder is zero, then c is a zero of f. After we identify the first rational zero, we use the result of the synthetic division to factor the original polynomial. Then we set each factor equal to zero to identify any additional rational zeros.

EXAMPLE 3   Finding Zeros of a Polynomial Function

Find all rational zeros of $f(x) = x^3 + 2x^2 - 5x - 6$.

Solution   We begin by listing all possible rational zeros.

Possible rational zeros

$$= \frac{\text{Factors of the constant term, } -6}{\text{Factors of the leading coefficient, } 1} = \frac{\pm 1, \pm 2, \pm 3, \pm 6}{\pm 1} = \pm 1, \pm 2, \pm 3, \pm 6$$

> Divide the eight numbers in the numerator by ±1.

Now we will use synthetic division to see if we can find a rational zero among the possible rational zeros $\pm 1, \pm 2, \pm 3, \pm 6$. Keep in mind that if $f(x)$ is divided by $x - c$ and the remainder is zero, then c is a zero of f. Let's start by testing 1. If 1 is not a rational zero, then we will test other possible rational zeros.

Test 1

Possible rational zero

Coefficients of $f(x) = x^3 + 2x^2 - 5x - 6$

```
1 | 1   2   -5   -6
  |     1    3   -2
  ------------------
    1   3   -2   -8
```

> The nonzero remainder shows that 1 is not a zero.

Test 2

Possible rational zero

Coefficients of $f(x) = x^3 + 2x^2 - 5x - 6$

```
2 | 1   2   -5   -6
  |     2    8    6
  ------------------
    1   4    3    0
```

> The zero remainder shows that 2 is a zero.

The zero remainder tells us that 2 is a zero of the polynomial function $f(x) = x^3 + 2x^2 - 5x - 6$. Equivalently, 2 is a solution, or root, of the polynomial equation $x^3 + 2x^2 - 5x - 6 = 0$. Thus, $x - 2$ is a factor of the polynomial.

$x^3 + 2x^2 - 5x - 6 = 0$ Finding the zeros of $f(x) = x^3 + 2x^2 - 5x - 6$ is the same as finding the roots of this equation.

$(x - 2)(x^2 + 4x + 3) = 0$ Factor using the result from the synthetic division.

$(x - 2)(x + 3)(x + 1) = 0$ Factor completely.

$x - 2 = 0$ or $x + 3 = 0$ or $x + 1 = 0$ Set each factor equal to zero.

$x = 2$ $x = -3$ $x = -1$ Solve for x.

The solution set is $\{-3, -1, 2\}$. The rational zeros of f are $-3, -1$, and 2.

Check Point 3   Find all rational zeros of

$$f(x) = x^3 + 8x^2 + 11x - 20.$$

Our work in Example 3 involved solving a third-degree equation. We found one factor by synthetic division and then factored the remaining quadratic factor. If the degree of a polynomial function or equation is 4 or higher, it is often necessary to find more than one linear factor by synthetic division.

One way to speed up the process of finding the first zero is to graph the function. Any x-intercept is a zero.

3 Solve polynomial equations.

EXAMPLE 4   Solving a Polynomial Equation

Solve: $x^4 - 6x^2 - 8x + 24 = 0$.

Solution   Recall that we refer to the *zeros* of a polynomial function and the *roots* of a polynomial equation. Because we are given an equation, we will use the word "roots," rather than "zeros," in the solution process. We begin by listing all possible rational roots.

Possible rational roots

$$= \frac{\text{Factors of the constant term, 24}}{\text{Factors of the leading coefficient, 1}}$$

$$= \frac{\pm 1, \pm 2, \pm 3, \pm 4, \pm 6, \pm 8, \pm 12, \pm 24}{\pm 1} = \pm 1, \pm 2, \pm 3, \pm 4, \pm 6, \pm 8, \pm 12, \pm 24$$

The graph of $f(x) = x^4 - 6x^2 - 8x + 24$ is shown in Figure 3.19. Because the x-intercept is 2, we will test 2 by synthetic division and show that it is a root of the given equation.

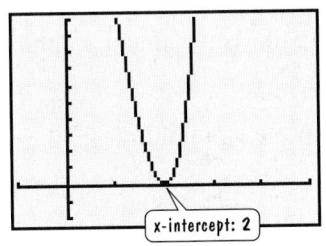

Figure 3.19 The graph of $f(x) = x^4 - 6x^2 - 8x + 24$ in a $[-1, 5, 1]$ by $[-2, 10, 1]$ viewing rectangle

$$
\begin{array}{r|rrrrr}
2 & 1 & 0 & -6 & -8 & 24 \\
  &   & 2 & 4 & -4 & -24 \\
\hline
  & 1 & 2 & -2 & -12 & 0 \\
\end{array}
$$

Careful!
$x^4 - 6x^2 - 8x + 24 = x^4 + 0x^3 - 6x^2 - 8x + 24$

The zero remainder indicates that 2 is a root of $x^4 - 6x^2 - 8x + 24 = 0$.

Now we can rewrite the given equation in factored form.

$$x^4 - 6x^2 - 8x + 24 = 0 \quad \text{This is the given equation.}$$

$$(x - 2)(x^3 + 2x^2 - 2x - 12) = 0 \quad \text{This is the result obtained from the synthetic division.}$$

$$x - 2 = 0 \quad \text{or} \quad x^3 + 2x^2 - 2x - 12 = 0 \quad \text{Set each factor equal to 0.}$$

We can use the same approach to look for rational roots of the polynomial equation $x^3 + 2x^2 - 2x - 12 = 0$, listing all possible rational roots. However, take a second look at the graph in Figure 3.19. Because the graph turns around at 2, this means that 2 is a root of even multiplicity. Thus, 2 must also be a root of $x^3 + 2x^2 - 2x - 12 = 0$, confirmed by the following synthetic division.

$$
\begin{array}{r|rrrr}
2 & 1 & 2 & -2 & -12 \\
  &   & 2 & 8 & 12 \\
\hline
  & 1 & 4 & 6 & 0 \\
\end{array}
$$

These are the coefficients of $x^3 + 2x^2 - 2x - 12 = 0$.

The zero remainder indicates that 2 is a root of $x^3 + 2x^2 - 2x - 12 = 0$.

Now we can solve the original equation as follows:

$$x^4 - 6x^2 - 8x + 24 = 0 \quad \text{This is the given equation.}$$

$$(x - 2)(x^3 + 2x^2 - 2x - 12) = 0 \quad \text{This factorization is obtained from the first synthetic division.}$$

$$(x - 2)(x - 2)(x^2 + 4x + 6) = 0 \quad \text{This factorization is obtained from the second synthetic division.}$$

$$x - 2 = 0 \quad \text{or} \quad x - 2 = 0 \quad \text{or} \quad x^2 + 4x + 6 = 0 \quad \text{Set each factor equal to 0.}$$

$$x = 2 \qquad\qquad x = 2 \qquad\quad x^2 + 4x + 6 = 0 \quad \text{Solve.}$$

We can use the quadratic formula to solve $x^2 + 4x + 6 = 0$.

$$x = \frac{-b \pm \sqrt{b^2 - 4ac}}{2a}$$

We use the quadratic formula because $x^2 + 4x + 6$ cannot be factored.

$$= \frac{-4 \pm \sqrt{4^2 - 4(1)(6)}}{2(1)}$$

Let $a = 1$, $b = 4$, and $c = 6$.

$$= \frac{-4 \pm \sqrt{-8}}{2}$$

Multiply and subtract under the radical.

$$= \frac{-4 \pm 2i\sqrt{2}}{2}$$

$\sqrt{-8} = \sqrt{4(2)(-1)} = 2i\sqrt{2}$

$$= -2 \pm i\sqrt{2}$$

Simplify.

Figure 3.19, repeated

The solution set of the original equation, $x^4 - 6x^2 - 8x + 24 = 0$, is $\{2, -2 - i\sqrt{2}, -2 + i\sqrt{2}\}$. The graph in Figure 3.19 illustrates that a graphing utility does not reveal the two imaginary roots.

In Example 4, 2 is a repeated root of the equation with multiplicity 2. The example illustrates two general properties:

Properties of Polynomial Equations

1. If a polynomial equation is of degree n, then counting multiple roots separately, the equation has n roots.

2. If $a + bi$ is a root of a polynomial equation $(b \neq 0)$, then the complex imaginary number $a - bi$ is also a root. Complex imaginary roots, if they exist, occur in conjugate pairs.

These ideas will be developed in more detail in the next section.

Check Point 4 Solve: $x^4 - 6x^3 + 22x^2 - 30x + 13 = 0$.

4 Use Descartes's Rule of Signs.

Descartes's Rule of Signs

Because an nth-degree polynomial equation might have roots that are imaginary numbers, we should note that such an equation can have *at most* n real roots. **Descartes's Rule of Signs** provides even more specific information about the number of real zeros that a polynomial can have. The rule is based on considering *variations in sign* between consecutive coefficients. For example, the function

$$f(x) = 3x^7 - 2x^5 - x^4 + 7x^2 + x - 3$$

has three sign changes.

An equation can have as many true [positive] roots as it contains changes of sign, from plus to minus or from minus to plus. ... René Descartes (1596–1650) in *La Géométrie* (1637)

Descartes's Rule of Signs

Let $f(x) = a_n x^n + a_{n-1} x^{n-1} + \cdots + a_2 x^2 + a_1 x + a_0$ be a polynomial with real coefficients.

1. The number of *positive real zeros* of f is either equal to the number of sign changes of $f(x)$ or is less than that number by an even integer. If there is only one variation in sign, there is exactly one positive real zero.

2. The number of *negative real zeros* of f is either equal to the number of sign changes of $f(-x)$ or is less than that number by an even integer. If $f(-x)$ has only one variation in sign, then f has exactly one negative real zero.

EXAMPLE 5   Using Descartes's Rule of Signs

Determine the possible number of positive and negative real zeros of $f(x) = x^3 + 2x^2 + 5x + 4$.

Solution

1. To find possibilities for positive real zeros, count the number of sign changes in the equation for $f(x)$. Because all the terms are positive, there are no variations in sign. Thus, there are no positive real zeros.

2. To find possibilities for negative real zeros, count the number of sign changes in the equation for $f(-x)$. We obtain this equation by replacing x with $-x$ in the given function.

$$f(x) = x^3 + 2x^2 + 5x + 4 \quad \text{This is the given polynomial function.}$$

Replace x with $-x$.

$$f(-x) = (-x)^3 + 2(-x)^2 + 5(-x) + 4$$
$$= -x^3 + 2x^2 - 5x + 4$$

Now count the sign changes.

$$f(-x) = -x^3 + 2x^2 - 5x + 4$$
$$\qquad\qquad\; 1 \quad\; 2 \quad\; 3$$

There are three variations in sign. The number of negative real zeros of f is either equal to the number of sign changes, 3, or is less than this number by an even integer. This means that there are either 3 negative real zeros or $3 - 2 = 1$ negative real zero.

What do the results of Example 5 mean in terms of solving
$$x^3 + 2x^2 + 5x + 4 = 0?$$

Without using Descartes's Rule of Signs, we list possible rational roots as follows:

Possible rational roots

$$= \frac{\text{Factors of the constant term, 4}}{\text{Factors of the leading coefficient, 1}} = \frac{\pm 1, \pm 2, \pm 4}{\pm 1} = \pm 1, \quad \pm 2, \quad \pm 4.$$

However, Descartes's Rule of Signs informed us that $f(x) = x^3 + 2x^2 + 5x + 4$ has no positive real zeros. Thus, the polynomial equation $x^3 + 2x^2 + 5x + 4 = 0$ has no positive real roots. This means that we can eliminate the positive numbers from our list of possible rational roots. Possible rational roots include only $-1, -2$, and -4. We can use synthetic division to test the three possible rational roots. Our test on two of the three rational roots is shown on the next page.

$$\begin{array}{r|rrrr} -1 & 1 & 2 & 5 & 4 \\ & & -1 & -1 & -4 \\ \hline & 1 & 1 & 4 & 0 \end{array}$$

The zero remainder shows that −1 is a root.

$$\begin{array}{r|rrrr} -2 & 1 & 2 & 5 & 4 \\ & & -2 & 0 & -10 \\ \hline & 1 & 0 & 5 & -6 \end{array}$$

The nonzero remainder shows that −2 is not a root.

By solving the equation $x^3 + 2x^2 + 5x + 4 = 0$, you will find that this equation of degree 3 has three roots. One root is −1, and the other two roots are imaginary numbers in a conjugate pair. Verify this by completing the solution process.

Check Point 5 Determine the possible number of positive and negative real zeros of $f(x) = x^4 - 14x^3 + 71x^2 - 154x + 120$.

EXERCISE SET 3.4

Practice Exercises

In Exercises 1–8, use the Rational Zero Theorem to list all possible rational zeros for each given function.

1. $f(x) = x^3 + x^2 - 4x - 4$

2. $f(x) = x^3 + 3x^2 - 6x - 8$

3. $f(x) = 3x^4 - 11x^3 - x^2 + 19x + 6$

4. $f(x) = 2x^4 + 3x^3 - 11x^2 - 9x + 15$

5. $f(x) = 4x^4 - x^3 + 5x^2 - 2x - 6$

6. $f(x) = 3x^4 - 11x^3 - 3x^2 - 6x + 8$

7. $f(x) = x^5 - x^4 - 7x^3 + 7x^2 - 12x - 12$

8. $f(x) = 4x^5 - 8x^4 - x + 2$

In Exercises 9–14,
 a. *List all possible rational zeros.*
 b. *Use synthetic division to test the possible rational zeros and find an actual zero.*
 c. *Use the zero from part (b) to find all the zeros of the polynomial function.*

9. $f(x) = x^3 + x^2 - 4x - 4$

10. $f(x) = x^3 - 2x^2 - 11x + 12$

11. $f(x) = 2x^3 - 3x^2 - 11x + 6$

12. $f(x) = 2x^3 - 5x^2 + x + 2$

13. $f(x) = 3x^3 + 7x^2 - 22x - 8$

14. $f(x) = 3x^3 + 8x^2 - 15x + 4$

In Exercises 15–22,
 a. *List all possible rational roots.*
 b. *Use synthetic division to test the possible rational roots and find an actual root.*
 c. *Use the root from part (b) and solve the equation.*

15. $x^3 - 2x^2 - 11x + 12 = 0$

16. $x^3 - 2x^2 - 7x - 4 = 0$

17. $x^3 - 10x - 12 = 0$

18. $x^3 - 5x^2 + 17x - 13 = 0$

19. $6x^3 + 25x^2 - 24x + 5 = 0$

20. $2x^3 - 5x^2 - 6x + 4 = 0$

21. $x^4 - 2x^3 - 5x^2 + 8x + 4 = 0$

22. $x^4 - 2x^2 - 16x - 15 = 0$

In Exercises 23–28, use Descartes's Rule of Signs to determine the possible number of positive and negative real zeros for each given function.

23. $f(x) = x^3 + 2x^2 + 5x + 4$

24. $f(x) = x^3 + 7x^2 + x + 7$

25. $f(x) = 5x^3 - 3x^2 + 3x - 1$

26. $f(x) = -2x^3 + x^2 - x + 7$

27. $f(x) = 2x^4 - 5x^3 - x^2 - 6x + 4$

28. $f(x) = 4x^4 - x^3 + 5x^2 - 2x - 6$

In Exercises 29–42, find all zeros of the polynomial function or solve the given polynomial equation. Use the Rational Zero Theorem, Descartes's Rule of Signs, and possibly the graph of the polynomial function shown by a graphing utility as an aid in obtaining the first zero or the first root.

29. $f(x) = x^3 - 4x^2 - 7x + 10$

30. $f(x) = x^3 + 12x^2 + 21x + 10$

31. $2x^3 - x^2 - 9x - 4 = 0$

32. $3x^3 - 8x^2 - 8x + 8 = 0$

33. $f(x) = x^4 - 2x^3 + x^2 + 12x + 8$

34. $f(x) = x^4 - 4x^3 - x^2 + 14x + 10$

35. $x^4 - 3x^3 - 20x^2 - 24x - 8 = 0$

36. $x^4 - x^3 + 2x^2 - 4x - 8 = 0$

37. $f(x) = 3x^4 - 11x^3 - x^2 + 19x + 6$

38. $f(x) = 2x^4 + 3x^3 - 11x^2 - 9x + 15$

39. $4x^4 - x^3 + 5x^2 - 2x - 6 = 0$

40. $3x^4 - 11x^3 - 3x^2 - 6x + 8 = 0$

41. $2x^5 + 7x^4 - 18x^2 - 8x + 8 = 0$

42. $4x^5 + 12x^4 - 41x^3 - 99x^2 + 10x + 24 = 0$

Application Exercises

The graphs are based on a study of the percentage of professional works completed in each age decade of life by 738 people who lived to be at least 79. Use the graphs to solve Exercises 43–44.

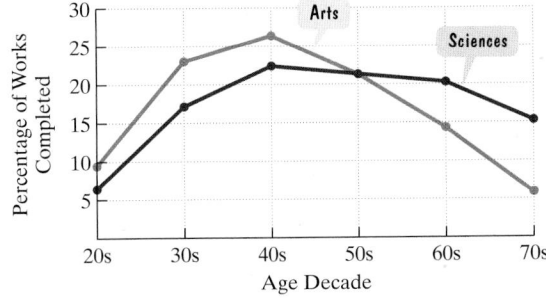

Age Trends in Professional Productivity

Source: Dennis, W. (1966), Creative productivity between the ages of 20 and 80 years. *Journal of Gerontology*, 21, 1–8.

43. Suppose that a polynomial function f is used to model the data shown in the graph for the arts using

(age decade, percentage of works completed).

a. Use the graph to solve the polynomial equation $f(x) = 27$. Describe what this means in terms of an age decade and productivity.

b. Describe the degree and the leading coefficient of a function f that can be used to model the data in the graph.

44. Suppose that a polynomial function g is used to model the data shown in the graph for the sciences using

(age decade, percentage of works completed).

a. Use the graph to solve the polynomial equation $g(x) = 20$. Find only the meaningful value of x and then describe what this means in terms of an age decade and productivity.

b. Describe the degree and the leading coefficient of a function g that can be used to model the data in the graph.

45. The number of eggs, N, in a female moth is a function of her abdominal width, W, in millimeters, modeled by $N = 14W^3 - 17W^2 - 16W + 34$, for $1.5 \le W \le 3.5$. What is the abdominal width when there are 211 eggs?

46. The concentration of a drug, $f(x)$, in parts per million, in a patient's blood x hours after the drug is administered is given by the function

$$f(x) = -x^4 + 12x^3 - 58x^2 + 132x.$$

How many hours after the drug is administered will it be eliminated from the bloodstream?

47. The width of a rectangular box is twice the height and the length is 7 inches more than the height. If the volume is 72 cubic inches, find the dimensions of the box.

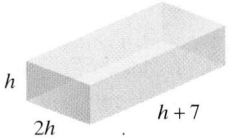

48. A box with an open top is formed by cutting squares out of the corners of a rectangular piece of cardboard 10 inches by 8 inches and then folding up the sides. If x represents the length of the side of the square cut from each corner of the rectangle, what size square must be cut if the volume of the box is to be 48 cubic inches?

Writing in Mathematics

49. Describe how to find the possible rational zeros of a polynomial function.

50. Describe how to use Descartes's Rule of Signs to determine the possible number of positive real zeros of a polynomial function.

51. Describe how to use Descartes's Rule of Signs to determine the possible number of negative roots of a polynomial equation.

52. Why must every polynomial equation of degree 3 have at least one real root?

53. Explain why the equation $x^4 + 6x^2 + 2 = 0$ has no rational roots.

54. Suppose $\frac{3}{4}$ is a root of a polynomial equation. What does this tell us about the leading coefficient and the constant term in the equation?

55. Use the graphs for Exercises 43–44 to describe one similarity and one difference between age trends in professional productivity in the arts and the sciences.

Technology Exercises

The equations in Exercises 56–59 have real roots that are rational. Use the Rational Zero Theorem to list all possible rational roots. Then graph the polynomial function in the given viewing rectangle to determine which possible rational roots are actual roots of the equation.

56. $2x^3 - 15x^2 + 22x + 15 = 0$; $[-1, 6, 1]$ by $[-50, 50, 1]$

57. $6x^3 - 19x^2 + 16x - 4 = 0$; $[0, 2, 1]$ by $[-3, 2, 1]$

58. $2x^4 + 7x^3 - 4x^2 - 27x - 18 = 0$; $[-4, 3, 1]$ by $[-45, 45, 1]$

59. $4x^4 + 4x^3 + 7x^2 - x - 2 = 0$; $[-2, 2, 1]$ by $[-5, 5, 1]$

60. Use Descartes's Rule of Signs to determine the possible number of positive and negative real zeros of $f(x) = 3x^4 + 5x^2 + 2$. What does this mean in terms of the graph of f? Verify your result by using a graphing utility to graph f.

61. Use Descartes's Rule of Signs to determine the possible number of positive and negative real zeros of $f(x) = x^5 - x^4 + x^3 - x^2 + x - 8$. Verify your result by using a graphing utility to graph f.

62. Determine a number of polynomial functions of odd degree and graph each function. Is it possible for the graph to have no real zeros? Explain. Try doing the same thing for polynomial functions of even degree. Now is it possible to have no real zeros?

Critical Thinking Exercises

63. Which one of the following is true?

   a. The equation $x^3 + 5x^2 + 6x + 1 = 0$ has one positive real root.

   b. Descartes's Rule of Signs gives the exact number of

positive and negative real roots for a polynomial equation.

   c. Every polynomial equation of degree 3 has at least one rational root.

   d. None of the above is true.

64. Give an example of a polynomial equation that has no real roots. Describe how you obtained the equation.

65. If the volume of the solid shown in the figure is 208 cubic inches, find the value of x.

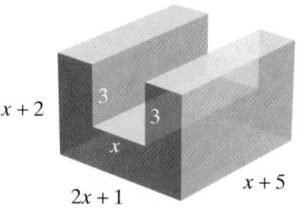

66. In this exercise, we lead you through the steps involved in the proof of the Rational Zero Theorem. Consider the polynomial equation

$$a_n x^n + a_{n-1} x^{n-1} + a_{n-2} x^{n-2} + \cdots + a_1 x + a_0 = 0$$

where $\dfrac{p}{q}$ is a rational root reduced to lowest terms.

   a. Substitute $\dfrac{p}{q}$ for x in the equation and show that the equation can be written as

$$a_n p^n + a_{n-1} p^{n-1} q$$
$$+ a_{n-2} p^{n-2} q^2 + \cdots + a_1 p q^{n-1} = -a_0 q^n.$$

   b. Why is p a factor of the left side of the equation?

   c. Because p divides the left side, it must also divide the right side. However, because $\dfrac{p}{q}$ is reduced to lowest terms, p cannot divide q. Thus, p and q have no common factors other than -1 and 1. Because p does divide the right side and it is not a factor of q^n, what can you conclude?

   d. Rewrite the equation from part (a) with all terms containing q on the left and the term that does not have a factor of q on the right. Use an argument that parallels parts (b) and (c) to conclude that q is a factor of a_n.

SECTION 3.5 *More On Zeros of Polynomial Functions*

Objectives

1. Find bounds for the roots of a polynomial equation.
2. Approximate real zeros.
3. Use conjugate roots to solve a polynomial equation.
4. Use the Linear Factorization Theorem to factor a polynomial.
5. Find polynomials with given zeros.

You stole my formula!

Tartaglia's Secret Formula for One Solution of $x^3 + mx = n$

$$x = \sqrt[3]{\sqrt{\left(\frac{n}{2}\right)^2 + \left(\frac{m}{3}\right)^3} + \frac{n}{2}}$$
$$-\sqrt[3]{\sqrt{\left(\frac{n}{2}\right)^2 + \left(\frac{m}{3}\right)^3} - \frac{n}{2}}$$

Popularizers of mathematics are sharing bizarre stories that are giving math a secure place in popular culture. One episode, able to compete with the wildest fare served up by television talk shows and the tabloids, involves three Italian mathematicians and, of all things, zeros of polynomial functions.

Tartaglia (1499–1557), poor and starving, has found a formula that gives a root for a third-degree polynomial equation. Cardano (1501–1576) begs Tartaglia to reveal the secret formula, wheedling it from him with the promise he will find the impoverished Tartaglia a patron. Then Cardano publishes his famous work *Ars Magna*, in which he presents Tartaglia's formula as his own. Cardano uses his most talented student, Ferrari (1522–1565), who derived a formula for a root of a fourth-degree polynomial equation, to falsely accuse Tartaglia of plagiarism. The dispute becomes violent and Tartaglia is fortunate to escape alive.

The noise from this "You Stole My Formula" episode is quieted by the work of French mathematician Evariste Galois (1811–1832). Galois proved that there is no general formula for finding roots of polynomial equations of degree 5 or higher. There are, of course, methods for finding roots. In this section, we continue our study of methods for finding zeros of polynomial functions.

1 Find bounds for the roots of a polynomial equation.

Upper and Lower Bounds for Roots

The **Upper and Lower Bound Theorem** helps us rule out many of a polynomial equations possible rational roots. Figure 3.20 illustrates that a is a **lower bound** and b is an **upper bound** for the roots of $f(x) = 0$ because every real root c of the equation satisfies $a \le c \le b$.

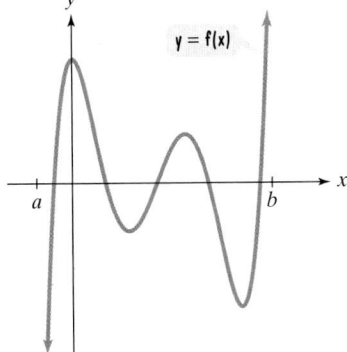

Figure 3.20 b is an upper bound and a is a lower bound for the real roots of $f(x) = 0$.

The Upper and Lower Bound Theorem

Let $f(x)$ be a polynomial with real coefficients and a positive leading coefficient, and let a and b be nonzero real numbers.

1. Divide $f(x)$ by $x - b$ (where $b > 0$) using synthetic division. If the last row containing the quotient and remainder has no negative numbers, then b is an **upper bound** for the real roots of $f(x) = 0$.
2. Divide $f(x)$ by $x - a$ (where $a < 0$) using synthetic division. If the last row containing the quotient and remainder has numbers that alternate in sign (zero entries count as positive or negative), then a is a **lower bound** for the real roots of $f(x) = 0$.

EXAMPLE 1   Finding Bounds for the Roots

Show that all the real roots of the equation $8x^3 + 10x^2 - 39x + 9 = 0$ lie between -3 and 2.

Solution   We begin by showing that 2 is an upper bound for the real roots. Divide the polynomial by $x - 2$. If all the numbers in the bottom row of the synthetic division are nonnegative, then 2 is an upper bound.

$$
\begin{array}{r|rrrr}
2 & 8 & 10 & -39 & 9 \\
  &   & 16 & 52 & 26 \\
\hline
  & 8 & 26 & 13 & 35
\end{array}
$$

All numbers in this row are nonnegative.

The nonnegative entries in the last row verify that 2 is an upper bound. Next, we show that -3 is a lower bound for the real roots. Divide the polynomial by $x - (-3)$, or $x + 3$. If the numbers in the bottom row of the synthetic division alternate in sign, then -3 is a lower bound. Remember that the number zero can be considered positive or negative.

$$
\begin{array}{r|rrrr}
-3 & 8 & 10 & -39 & 9 \\
   &   & -24 & 42 & -9 \\
\hline
   & 8 & -14 & 3   & 0
\end{array}
$$

Counting 0 as negative, the signs alternate: $+, -, +, -$.

By the Upper and Lower Bound Theorem, the alternating signs in the last row indicate that -3 is a lower bound for the roots. (The zero remainder indicates that -3 is also a root.)

Check Point 1   Show that all the real roots of the equation $2x^3 + 11x^2 - 7x - 6 = 0$ lie between -7 and 2.

How might the Upper and Lower Bound Theorem be helpful in solving a polynomial equation? Consider the equation

$$x^4 + 3x^3 - 27x^2 + 3x - 28 = 0.$$

With a leading coefficient of 1 and a constant term of -28, the possible rational roots are

$$\pm 1, \quad \pm 2, \quad \pm 4, \quad \pm 7, \quad \pm 14, \quad \pm 28.$$

We begin testing for an actual root using synthetic division. The following divisions indicate that 1 and 2 are not roots because of the nonzero remainders. However, something interesting happens when testing 4.

$$
\begin{array}{r|rrrrr}
1 & 1 & 3 & -27 & 3 & -28 \\
  &   & 1 & 4   & -23 & -20 \\
\hline
  & 1 & 4 & -23 & -20 & -48
\end{array}
\qquad
\begin{array}{r|rrrrr}
2 & 1 & 3 & -27 & 3 & -28 \\
  &   & 2 & 10 & -34 & -62 \\
\hline
  & 1 & 5 & -17 & -31 & -90
\end{array}
$$

$$
\begin{array}{r|rrrrr}
4 & 1 & 3 & -27 & 3 & -28 \\
  &   & 4 & 28 & 4 & 28 \\
\hline
  & 1 & 7 & 1   & 7 & 0
\end{array}
$$

Nonnegative numbers

4 is a root of the equation because the remainder is 0.

4 is an upper bound for the roots of the equation.

Notice that 4 is both a root and an upper bound for the roots. Should you take the time to use synthetic division and test 7, 14, and 28? There is no need to do this because all three numbers exceed 4, the upper bound for the roots. Thus, 7, 14, and 28 cannot be roots of the equation.

Technology

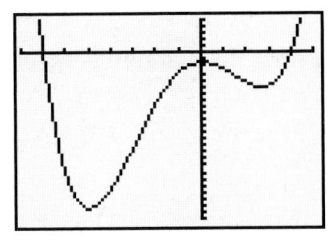

Figure 3.21

The Upper and Lower Bound Theorem and your knowledge of polynomial functions can help you to find a reasonable range setting when using your graphing utility. Consider

$$f(x) = x^4 + 3x^3 - 27x^2 + 3x - 28.$$

Based on our discussion, 4 is a zero and an upper bound for the zeros. We can also use synthetic division to show that -7 is a zero and a lower bound for the zeros. We can use these lower and upper bounds to determine Xmin and Xmax. We'll go one unit to the left and to the right of these bounds and use $[-8, 5, 1]$. Now, how do we determine Ymin and Ymax? Let's see what kinds of values of y we obtain when we evaluate the function between -8 and 5. Using synthetic division, direct substitution, or the table feature of some graphing utilities, we have $f(-6) = -370$, $f(-5) = -468$, $f(0) = -28$, and $f(3) = -100$. These evaluations suggest that we can use -500 for Ymin and 100 for Ymax. The graph of $f(x) = x^4 + 3x^3 - 27x^2 + 3x - 28$ is shown in a $[-8, 5, 1]$ by $[-500, 100, 20]$ viewing rectangle in Figure 3.21. Because the degree is even ($n = 4$) and the leading coefficient, 1, is positive, the graph should rise to the left and to the right. This is precisely what occurs in Figure 3.21. Our work in obtaining this complete graph is an excellent illustration of the fact that technology complements human knowledge and is not intended to replace it.

2 Approximate real zeros.

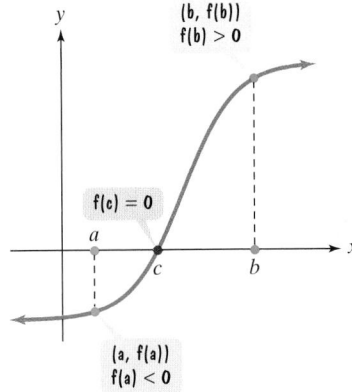

Figure 3.22 The graph must cross the x-axis at some value between a and b.

The Intermediate Value Theorem

We can find decimal approximations for real zeros of polynomial functions using a graphing utility. The **Intermediate Value Theorem** tells us of the existence of real zeros and how to approximate them. The idea behind the theorem is illustrated in Figure 3.22. The figure shows that if $(a, f(a))$ lies below the x-axis and $(b, f(b))$ lies above the x-axis, the smooth, continuous graph of a polynomial function f must cross the x-axis at some value c between a and b. This value is a real zero for the function.

These observations are summarized in the **Intermediate Value Theorem.**

The Intermediate Value Theorem for Polynomials

Let f be a polynomial function with real coefficients. If $f(a)$ and $f(b)$ have opposite signs, then there is at least one value of c between a and b for which $f(c) = 0$. Equivalently, the equation $f(x) = 0$ has at least one real root between a and b.

EXAMPLE 2 Approximating a Real Zero

a. Show that the polynomial function $f(x) = x^3 - 2x - 5$ has a real zero between 2 and 3.

b. Use the Intermediate Value Theorem to find an approximation for this real zero to the nearest tenth.

Solution

a. Let us evaluate $f(x)$ at 2 and 3. If $f(2)$ and $f(3)$ have opposite signs, then there is a real zero between 2 and 3. Using $f(x) = x^3 - 2x - 5$, we obtain

$$f(2) = 2^3 - 2 \cdot 2 - 5 = 8 - 4 - 5 = -1$$

> f(2) is negative.

and

$$f(3) = 3^3 - 2 \cdot 3 - 5 = 27 - 6 - 5 = 16.$$

> f(3) is positive.

This sign change shows that the polynomial function has a real zero between 2 and 3.

b. A numerical approach is to evaluate f at successive tenths between 2 and 3, looking for a sign change. This sign change will place the real zero between a pair of successive tenths.

x	$f(x) = x^3 - 2x - 5$	
2	$f(2) = 2^3 - 2(2) - 5$	$= -1$
2.1	$f(2.1) = (2.1)^3 - 2(2.1) - 5 = 0.061$	

> Sign change

The sign change indicates that f has a real zero between 2 and 2.1. We now follow a similar procedure to locate the real zero between successive hundredths. We divide the interval $[2, 2.1]$ into ten equal subintervals. Then we evaluate f at each endpoint and look for a sign change.

x	$f(x) = x^3 - 2x - 5$	x	$f(x) = x^3 - 2x - 5$
2.00	$f(2.00) = -1$	2.06	$f(2.06) = -0.378184$
2.01	$f(2.01) = -0.899399$	2.07	$f(2.07) = -0.270257$
2.02	$f(2.02) = -0.797592$	2.08	$f(2.08) = -0.161088$
2.03	$f(2.03) = -0.694573$	2.09	$f(2.09) = -0.050671$
2.04	$f(2.04) = -0.590336$	2.1	$f(2.1) = 0.061$
2.05	$f(2.05) = -0.484875$		

> Sign change

The sign change indicates that f has a real zero between 2.09 and 2.1. Correct to the nearest tenth, the zero is 2.1.

Technology

Most graphing utilities have a zero or root feature that gives zeros to a great degree of accuracy. Using the graph shown below, the real zero of

$$f(x) = x^3 - 2x - 5,$$

correct to the nearest thousandth, is 2.095.

$y = x^3 - 2x - 5$

Zero
X=2.0945515 Y=0

$[-3, 3, 1]$ by $[-10, 10, 1]$

Check Point 2 Show that the polynomial function $f(x) = 3x^3 - 10x + 9$ has a real zero between -3 and -2.

3 Use conjugate roots to solve a polynomial equation.

The Fundamental Theorem of Algebra

We have seen that if a polynomial equation is of degree n, then counting multiple roots separately, the equation has n roots. Some of these roots may be imaginary numbers—that is, nonreal complex numbers—that occur in conjugate pairs, such as $2 + i$ and $2 - i$.

EXAMPLE 3 **Using Conjugate Roots to Solve a Polynomial Equation**

Solve $x^4 - 4x^3 + 3x^2 + 8x - 10 = 0$ given that $2 + i$ is a root.

Solution The degree of the given equation is 4. This means that there are four roots. One of the roots is $2 + i$. Because imaginary roots come in conjugate pairs, we know that $2 - i$ is a second root. By the Factor Theorem, both

$$[x - (2 + i)] \quad \text{and} \quad [x - (2 - i)]$$

are factors of the given polynomial. We multiply these known factors as follows:

$$[x - (2 + i)][x - (2 - i)]$$

F　　O　　I　　L

$$= x^2 - x(2 - i) - x(2 + i) + (2 + i)(2 - i) \quad \text{Multiply using the FOIL method.}$$
$$= x^2 - 2x + ix - 2x - ix + (4 - i^2) \quad \text{Continue multiplying.}$$
$$= x^2 - 2x + ix - 2x - ix + [4 - (-1)] \quad \text{Simplify using } i^2 = 1.$$
$$= x^2 - 4x + 5 \quad \text{Combine like terms.}$$

At this point we have only two of the four possible roots, $2 + i$ and $2 - i$. We can find the other two roots by factoring the given equation. The FOIL multiplication shows that $x^2 - 4x + 5$ is one of the factors. We can find the other factor(s) by dividing $x^2 - 4x + 5$ into the polynomial on the left side of the given equation.

$$\begin{array}{r} x^2 \qquad\qquad - 2 \\ x^2 - 4x + 5 \overline{\smash{)}\, x^4 - 4x^3 + 3x^2 + 8x - 10} \\ \underline{\ominus x^4 - \oplus 4x^3 + \ominus 5x^2} \\ -2x^2 + 8x - 10 \\ \underline{\oplus -2x^2 \ominus 8x \oplus 10} \\ 0 \end{array}$$

The zero remainder confirms that $x^2 - 4x + 5$ is a factor.

We can now solve the given equation.

$$x^4 - 4x^3 + 3x^2 + 8x - 10 = 0 \quad \text{This is the original equation.}$$
$$(x^2 - 4x + 5)(x^2 - 2) = 0 \quad \text{Factor using the result of the polynomial long division.}$$
$$x^2 - 4x + 5 = 0 \quad \text{or} \quad x^2 - 2 = 0 \quad \text{Set each factor equal to 0.}$$
$$x = 2 \pm i \qquad x = \pm\sqrt{2} \quad \text{Solve for x. We know the roots of the first equation, } x^2 - 4x + 5 = 0, \text{ by our previous analysis.}$$

The solution set is $\{2 \pm i, \pm\sqrt{2}\}$.

Check Point 3 Solve $x^4 - 8x^3 + 64x - 105 = 0$ given that $2 - i$ is a root.

The fact that a polynomial equation of degree n has n roots is a consequence of a theorem proved in 1799 by a 22-year-old student named Carl Friedrich Gauss in his doctoral dissertation. His result is called the **Fundamental Theorem of Algebra.**

Technology

The graph of
$$f(x) = x^4 - 4x^3 + 3x^2 + 8x - 10$$
is shown in a $[-4, 4, 1]$ by $[-15, 5, 1]$ viewing rectangle. The real roots of $f(x) = 0$, the equation in Example 3, are $-\sqrt{2}$ and $\sqrt{2}$. These appear as x-intercepts at approximately -1.4 and 1.4.

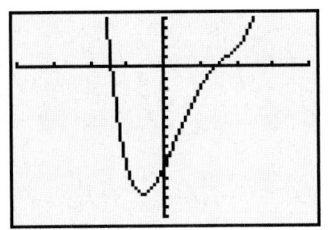

The Fundamental Theorem of Algebra

If $f(x)$ is a polynomial of degree n, where $n \geq 1$, then the equation $f(x) = 0$ has at least one complex root.

Suppose, for example, that $f(x) = 0$ represents a polynomial equation of degree n. By the Fundamental Theorem of Algebra, we know that this equation has at least one complex root; we'll call it c_1. By the Factor Theorem, we know that $x - c_1$ is a factor of $f(x)$. Therefore, we obtain

$$(x - c_1)q_1(x) = 0 \quad \text{The degree of the polynomial } q_1(x) \text{ is } n - 1.$$
$$x - c_1 = 0 \quad \text{or} \quad q_1(x) = 0. \quad \text{Set each factor equal to 0.}$$

If the degree of $q_1(x)$ is at least 1, by the Fundamental Theorem of Algebra the equation $q_1(x) = 0$ has at least one complex root. We'll call it c_2. The Factor Theorem gives us

$$q_1(x) = 0 \quad \text{The degree of } q_1(x) \text{ is } n - 1.$$
$$(x - c_2)q_2(x) = 0 \quad \text{The degree of } q_2(x) \text{ is } n - 2.$$
$$x - c_2 = 0 \quad \text{or} \quad q_2(x) = 0. \quad \text{Set each factor equal to 0.}$$

Let's see what we have up to this point, and then continue the process.

$$f(x) = 0 \quad \text{This is the original polynomial equation of degree } n.$$
$$(x - c_1)q_1(x) = 0 \quad \text{This is the result from our first application of the Fundamental Theorem.}$$
$$(x - c_1)(x - c_2)q_2(x) = 0 \quad \text{This is the result from our second application of the Fundamental Theorem.}$$

By continuing this process, we will obtain the product of n linear factors. Setting each of these linear factors equal to zero results in n complex roots. Thus, if $f(x)$ is a polynomial of degree n, where $n \geq 1$, then $f(x) = 0$ has exactly n roots, where roots are counted according to their multiplicity.

4 Use the Linear Factorization Theorem to factor a polynomial.

The Linear Factorization Theorem

In Example 3, we found that $x^4 - 4x^3 + 3x^2 + 8x - 10 = 0$ has $\{2 \pm i, \pm\sqrt{2}\}$ as a solution set. The polynomial can be factored over the complex nonreal numbers as follows:

$$f(x) = x^4 - 4x^3 + 3x^2 + 8x - 10 \quad \text{These are the four zeros.}$$
$$= [x - (2 + i)][x - (2 - i)](x + \sqrt{2})(x - \sqrt{2}).$$

These are four linear factors.

This fourth-degree polynomial has four linear factors. Just as an nth-degree polynomial equation has n roots, an nth-degree polynomial has n linear factors. This is formally stated as the **Linear Factorization Theorem.**

The Linear Factorization Theorem

If $f(x) = a_n x^n + a_{n-1} x^{n-1} + \cdots + a_1 x + a_0$, where $n \geq 1$ and $a_n \neq 0$, then

$$f(x) = a_n(x - c_1)(x - c_2)\cdots(x - c_n),$$

where $c_1, c_2, \ldots, c_n$ are complex numbers (possibly real and not necessarily distinct). In words: An nth-degree polynomial can be expressed as the product of a nonzero constant and n linear factors.

The Linear Factorization Theorem involves factors somewhat different than those you are used to seeing. For example, the polynomial $x^2 - 3$ is irreducible over the rational numbers. However, it can be factored over the real numbers as follows:

$$x^2 - 3 = (x + \sqrt{3})(x - \sqrt{3}).$$ Use $a^2 - b^2 = (a + b)(a - b)$ with $a = x$ and $b = \sqrt{3}$.

The polynomial $x^2 + 1$ is irreducible over the real numbers, but reducible over the complex imaginary numbers:

$$x^2 + 1 = (x + i)(x - i).$$

Study Tip

The sum of squares, irreducible over the real numbers, can be factored over the imaginary numbers as

$$a^2 + b^2 = (a + bi)(a - bi).$$

EXAMPLE 4 Factoring a Polynomial

Factor $x^4 - 3x^2 - 28$:

a. As the product of factors that are irreducible over the rational numbers.
b. As the product of factors that are irreducible over the real numbers.
c. In completely factored form involving complex imaginary numbers.

Solution

a. $x^4 - 3x^2 - 28 = (x^2 - 7)(x^2 + 4)$ Both quadratic factors are irreducible over the rational numbers.

b. $= (x + \sqrt{7})(x - \sqrt{7})(x^2 + 4)$ The third factor is still irreducible over the real numbers.

c. $= (x + \sqrt{7})(x - \sqrt{7})(x + 2i)(x - 2i)$ This is the completely factored form using complex imaginary numbers.

Check Point 4   Factor $x^4 - 4x^2 - 5$ as the product of factors that are irreducible over **a.** the rational numbers; **b.** the real numbers; **c.** the complex imaginary numbers.

5 Find polynomials with given zeros.

Reversing Things: Finding Polynomials when the Zeros Are Given

Many of our problems involving polynomial functions and polynomial equations dealt with the process of finding zeros and roots. The Linear Factorization Theorem enables us to reverse this process, finding a polynomial function when the zeros are given.

EXAMPLE 5 Finding a Polynomial Function with Given Zeros

Find a fourth-degree polynomial function $f(x)$ with real coefficients that has $-2, 2$, and i as zeros and such that $f(3) = -150$.

Solution Because i is a zero and the polynomial has real coefficients, the conjugate, $-i$, must also be a zero. We can now use the Linear Factorization Theorem.

Technology

The graph of $f(x) = -3x^4 + 9x^2 + 12$, shown in a $[-3, 3, 1]$ by $[-200, 20, 20]$ viewing rectangle, verifies that -2 and 2 are real zeros. By tracing along the curve, we can check that $f(3) = -150$.

-2 is a zero.

2 is a zero.

$$f(x) = a_n(x - c_1)(x - c_2)(x - c_3)(x - c_4)$$ This is the linear factorization for a fourth-degree polynomial.

$$= a_n(x + 2)(x - 2)(x - i)(x + i)$$ Use the given zeros: $c_1 = -2$, $c_2 = 2$, $c_3 = i$, and, from above, $c_4 = -i$.

$$= a_n(x^2 - 4)(x^2 + 1)$$ Multiply.

$$f(x) = a_n(x^4 - 3x^2 - 4)$$ Complete the multiplication.

$$f(3) = a_n(3^4 - 3 \cdot 3^2 - 4) = -150$$ To find a_n, use the fact that $f(3) = -150$.

$$a_n(81 - 27 - 4) = -150$$ Solve for a_n.

$$50a_n = -150$$ Simplify: $81 - 27 - 4 = 50$.

$$a_n = -3$$ Divide both sides by 50.

Substituting -3 for a_n in the formula for $f(x)$, we obtain
$$f(x) = -3(x^4 - 3x^2 - 4).$$
Equivalently,
$$f(x) = -3x^4 + 9x^2 + 12.$$

Check Point 5   Find a third-degree polynomial function $f(x)$ with real coefficients that has -3 and i as zeros and such that $f(1) = 8$.

EXERCISE SET 3.5

Practice Exercises

Use the Upper and Lower Bound Theorem to solve Exercises 1–4.

1. Show that all the real roots of the equation $x^4 - 5x^3 + 11x^2 + 33x - 18 = 0$ lie between -4 and 7.

2. Show that all the real roots of the equation $x^4 + 11x^3 - 12x^2 + 6 = 0$ lie between -13 and 1.

3. Show that all the real roots of the equation $2x^3 + 5x^2 - 8x - 7 = 0$ lie between -4 and 2.

4. Show that all the real roots of the equation $2x^5 - 13x^3 + 2x - 5 = 0$ lie between -3 and 3.

5. Consider the equation $x^4 + 3x^3 + 2x^2 - 5x + 12 = 0$.
   a. List all possible rational roots.
   b. Determine whether 1 is a root using synthetic division. What two conclusions can you draw?
   c. Based on part (b), what possible rational roots can you eliminate?
   d. Determine whether -3 is a root using synthetic division. What two conclusions can you draw?
   e. Based on part (d), what possible rational roots can you eliminate?

6. Consider the equation
   $$2x^5 + 5x^4 - 8x^3 - 14x^2 + 6x + 9 = 0.$$
   a. List all possible rational roots.
   b. Determine whether $\frac{3}{2}$ is a root using synthetic division. What two conclusions can you draw?
   c. Based on part (b), what possible rational roots can you eliminate?

   d. Determine whether -3 is a root using synthetic division. What two conclusions can you draw?
   e. Based on part (d), what possible rational roots can you eliminate?

In Exercises 7–14, show that each polynomial has a real zero between the given integers. Then use the Intermediate Value Theorem to find an approximation for this zero to the nearest tenth. If applicable, use a graphing utility's zero feature to verify your answer.

7. $f(x) = x^3 - x - 1$; between 1 and 2
8. $f(x) = x^3 - 4x^2 + 2$; between 0 and 1
9. $f(x) = 2x^4 - 4x^2 + 1$; between -1 and 0
10. $f(x) = x^4 + 6x^3 - 18x^2$; between 2 and 3
11. $f(x) = x^3 + x^2 - 2x + 1$; between -3 and -2
12. $f(x) = x^5 - x^3 - 1$; between 1 and 2
13. $f(x) = 3x^3 - 10x + 9$; between -3 and -2
14. $f(x) = 3x^3 - 8x^2 + x + 2$; between 2 and 3

In Exercises 15–22, use the given root to find the solution set of the polynomial equation.

15. $x^3 - 2x^2 + 4x - 8 = 0$; $-2i$
16. $x^4 + 13x^2 + 36 = 0$; $3i$
17. $3x^3 - 7x^2 + 8x - 2 = 0$; $1 + i$
18. $x^3 - 7x^2 + 16x - 10 = 0$; $3 + i$
19. $x^4 - 6x^2 + 25 = 0$; $2 - i$
20. $x^4 - x^3 - 9x^2 + 29x - 60 = 0$; $1 + 2i$

21. $x^4 - 8x^3 + 64x - 105 = 0$; $2 - i$

22. $4x^4 - 28x^3 + 129x^2 - 130x + 125 = 0$; $3 - 4i$

In Exercises 23–28, factor each polynomial:

    a. *as the product of factors that are irreducible over the rational numbers.*

    b. *as the product of factors that are irreducible over the real numbers.*

    c. *in completely factored form involving complex nonreal, or imaginary, numbers.*

23. $x^4 - x^2 - 20$ **24.** $x^4 + 6x^2 - 27$

25. $x^4 + x^2 - 6$ **26.** $x^4 - 9x^2 - 22$

27. $x^4 - 2x^3 + x^2 - 8x - 12$
    (*Hint:* One factor is $x^2 + 4$.)

28. $x^4 - 4x^3 + 14x^2 - 36x + 45$
    (*Hint:* One factor is $x^2 + 9$.)

In Exercises 29–36, find an nth-degree polynomial function with real coefficients satisfying the given conditions. If you are using a graphing utility, use it to graph the function and verify the real zeros and the given function value.

29. $n = 3$; 1 and $5i$ are zeros; $f(-1) = -104$

30. $n = 3$; 4 and $2i$ are zeros; $f(-1) = -50$

31. $n = 3$; -5 and $4 + 3i$ are zeros; $f(2) = 91$

32. $n = 3$; 6 and $-5 + 2i$ are zeros; $f(2) = -636$

33. $n = 4$; i and $3i$ are zeros; $f(-1) = 20$

34. $n = 4$; $-2, -\frac{1}{2}$, and i are zeros; $f(1) = 18$

35. $n = 4$; $-2, 5$, and $3 + 2i$ are zeros; $f(1) = -96$

36. $n = 4$; $-4, \frac{1}{3}$, and $2 + 3i$ are zeros; $f(1) = 100$

In Exercises 37–44, find all the zeros of the function and write the polynomial as a product of linear factors.

37. $f(x) = x^3 - x^2 + 25x - 25$

38. $f(x) = x^3 - 10x^2 + 33x - 34$

39. $f(x) = x^3 - 8x^2 + 25x - 26$

40. $f(x) = x^3 - 8x^2 + 17x - 4$

41. $f(x) = x^4 + 37x^2 + 36$

42. $f(x) = x^4 + 8x^3 + 9x^2 - 10x + 100$

43. $f(x) = 16x^4 + 36x^3 + 16x^2 + x - 30$

44. $f(x) = 2x^4 - x^3 + 7x^2 - 4x - 4$

Application Exercises

We have seen that the polynomial function
    $H(x) = -0.001618x^4 + 0.077326x^3 - 1.2367x^2 + 11.460x + 2.914$ *models the age in human years, $H(x)$, of a dog that is x years old, where $x \geq 1$. Although the coefficients make it difficult to solve equations algebraically using this function, a graph of the function makes approximate solutions possible.*

Use the graph shown to solve Exercises 45–46. Round all answers to the nearest year.

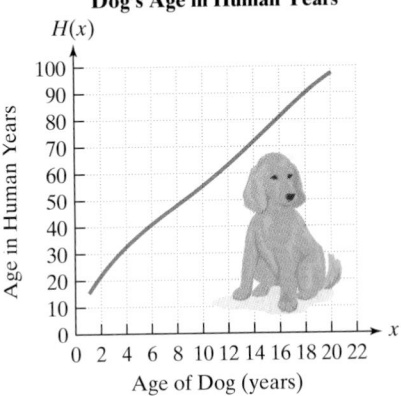

Dog's Age in Human Years

Source: U.C. Davis

45. If you are 25, what is the equivalent age for dogs?

46. If you are 90, what is the equivalent age for dogs?

47. Set up an equation to answer the question in either Exercise 45 or 46. Bring all terms to one side and obtain zero on the other side. What are some of the difficulties involved in solving this equation? Explain how the Intermediate Value Theorem can be used to verify the approximate solution that you obtained from the graph.

The United States has more people in prison, as well as more people in prison per capita, than any other western industrialized nation. The bar graph shows the number of inmates in U.S. state and federal prisons in seven selected years from 1985 through 2000.

Source: U.S. Justice Department

The data in the graph can be modeled by

    a linear function, $f(x) = 61.3x + 495$

    a quadratic function,
        $g(x) = -0.131x^2 + 63.27x + 491.6$

    a third-degree polynomial function,
        $h(x) = -0.219x^3 + 4.885x^2 + 35.14x + 503.14$.

For each of these functions, x represents the number of years after 1985 and the function value represents the number of inmates, in thousands. Use this information to solve Exercises 48–49.

48. The graph indicates that in 2000, there were 1382 thousand inmates. Substitute 1382 for $f(x)$ and $g(x)$ in the linear and quadratic models. Then solve each resulting equation to find how many years after 1985, to the nearest tenth of a year, inmate population was 1382 thousand. How well do the linear and quadratic functions serve as a model for 2000?

49. The graph indicates that in 2000, there were 1382 thousand inmates. Substitute 1382 for $h(x)$ in the third-degree model. Set the resulting equation equal to 0 and show that it has a real root between 14 and 15. Then use the Intermediate Value Theorem or a graphing utility's zero feature to find an approximation, to the nearest tenth, for this root. How well does the third-degree polynomial function serve as a model for 2000?

Writing in Mathematics

50. When testing a number using synthetic division, how do you know if it is an upper bound for the real roots?

51. When testing a number using synthetic division, how do you know if it is a lower bound for the real roots?

52. How do you show that a polynomial function has a real zero between two given numbers?

53. How does the linear factorization of $f(x)$, that is,
$$f(x) = a_n(x - c_1)(x - c_2)\cdots(x - c_n),$$
show that a polynomial equation of degree n has n roots?

Technology Exercises

54. Show that -1 is a lower bound of $f(x) = x^3 - 53x^2 + 103x - 51$. Show that 60 is an upper bound. Use this information and a graphing utility to draw a relatively complete graph of f.

In Exercises 55–56, use a graphing utility to determine upper and lower bounds for the zeros of f. Does synthetic division verify your observations?

55. $f(x) = 2x^3 + x^2 - 14x - 7$

56. $f(x) = 2x^4 - 7x^3 - 5x^2 + 28x - 12$

57. The function $f(x) = -0.00002x^3 + 0.008x^2 - 0.3x + 6.95$ models the number of annual physician visits, $f(x)$, by a person of age x.

  a. Graph the function for meaningful values of x and discuss what the graph reveals in terms of the variables described by the model.

  b. Use the zero or root feature of your graphing utility to find the age, to the nearest year, for the group that averages 13.43 annual physician visits.

  c. Verify part (b) using the graph of f.

Use a graphing utility to obtain a complete graph for each polynomial function in Exercises 58–61. Then determine the number of real zeros and the number of nonreal complex zeros for each function.

58. $f(x) = x^3 - 6x - 9$

59. $f(x) = 3x^5 - 2x^4 + 6x^3 - 4x^2 - 24x + 16$

60. $f(x) = 3x^4 + 4x^3 - 7x^2 - 2x - 3$

61. $f(x) = x^6 - 64$

Critical Thinking Exercises

In Exercises 62–65, what is the smallest degree that each polynomial could have?

62.

63.

64.

65.

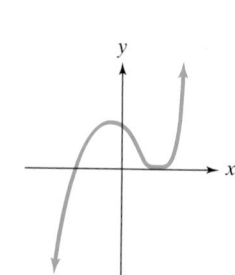

66. Explain why nonreal complex zeros are gained or lost in pairs in terms of graphs of polynomial functions.

67. Explain why a polynomial function of degree 20 cannot cross the x-axis exactly once.

68. Give an example of a function that is not subject to the Intermediate Value Theorem.

Group Exercise

69. The graph at the top of the next page shows costs for private and public four-year colleges projected through the year 2017. According to these projections, your daughter's college education at a private four-year school could cost about $250,000. This activity involves forming and using models from these data. Group members should begin by deciding whether to work with data for private or public colleges.

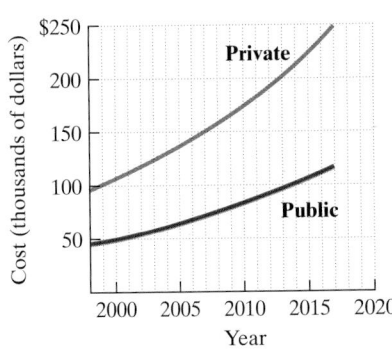

Cost of a Four-Year College

Source: U.S. Department of Education

a. Let $x = 0$ correspond to 1998, $x = 1$ to 1999, $x = 2$ to 2000, and so on up through $x = 19$ for 2017. Group members should use the chosen line graph to obtain a reasonable estimate for the cost of an education, y, in thousands of dollars, for each x.

b. Use the regression feature of a graphing utility to model the data for the cost of a four-year college x years after 1998 using a linear function, a quadratic function, and a third-degree polynomial function.

c. Use these functions to write and solve problems similar to Exercises 48 and 49.

d. Use these functions to make predictions well into the future. Which function, if any, seems to be most reasonable in its predicted cost? Of course, these are only predictions, subject to unforeseeable events. What events might render each of these models relatively useless over long periods of time?

SECTION 3.6 *Rational Functions and Their Graphs*

Objectives

1. Find the domain of rational functions.
2. Use arrow notation.
3. Identify vertical asymptotes.
4. Identify horizontal asymptotes.
5. Graph rational functions.
6. Identify slant asymptotes.
7. Solve applied problems involving rational functions.

Technology is now promising to bring light, fast, and beautiful wheelchairs to millions of disabled people. The cost of manufacturing these radically different wheelchairs can be modeled by rational functions. In this section, we will see how graphs of these functions illustrate that low prices are possible with high production levels, urgently needed in this situation. There are more than half a billion people with disabilities in developing countries; an estimated 20 million need wheelchairs right now.

① Find the domain of rational functions.

Rational Functions

Rational functions are quotients of polynomial functions. This means that rational functions can be expressed as

$$f(x) = \frac{p(x)}{q(x)}$$

where p and q are polynomial functions and $q(x) \neq 0$. The **domain** of a rational function is the set of all real numbers except the x-values that make the denominator zero. For example, the domain of the rational function

$$f(x) = \frac{x^2 + 7x + 9}{x(x - 2)(x + 5)}$$

This is $p(x)$.

This is $q(x)$.

is the set of all real numbers except $0, 2,$ and -5.

EXAMPLE 1 Finding the Domain of a Rational Function

Find the domain of each rational function:

a. $f(x) = \dfrac{x^2 - 9}{x - 3}$ **b.** $g(x) = \dfrac{x}{x^2 - 9}$ **c.** $h(x) = \dfrac{x + 3}{x^2 + 9}.$

Solution Rational functions contain division. Because division by 0 is undefined, we must exclude from the domain of each function values of x that cause the polynomial function in the denominator to be 0.

a. The denominator of $f(x) = \dfrac{x^2 - 9}{x - 3}$ is 0 if $x = 3$. Thus, x cannot equal 3. The domain of f consists of all real numbers except 3. We can express the domain in set-builder or interval notation:

$$\text{Domain of } f = \{x \mid x \neq 3\}$$
$$\text{Domain of } f = (-\infty, 3) \text{ or } (3, \infty).$$

b. The denominator of $g(x) = \dfrac{x}{x^2 - 9}$ is 0 if $x = -3$ or $x = 3$. Thus, the domain of g consists of all real numbers except -3 and 3. We can express the domain in set-builder or interval notation:

$$\text{Domain of } g = \{x \mid x \neq -3, x \neq 3\}$$
$$\text{Domain of } g = (-\infty, -3) \text{ or } (-3, 3) \text{ or } (3, \infty).$$

c. No real numbers cause the denominator of $h(x) = \dfrac{x + 3}{x^2 + 9}$ to equal 0. The domain of h consists of all real numbers.

$$\text{Domain of } h = (-\infty, \infty)$$

Study Tip

Because the domain of a rational function is the set of all real numbers except those for which the denominator is 0, you can identify such numbers by setting the denominator equal to 0 and solving for x. Exclude the resulting real values of x from the domain.

Check Point 1   Find the domain of each rational function:

a. $f(x) = \dfrac{x^2 - 25}{x - 5}$ **b.** $g(x) = \dfrac{x}{x^2 - 25}$ **c.** $h(x) = \dfrac{x + 5}{x^2 + 25}.$

(Ask your professor if a particular notation is preferred.)

2 Use arrow notation.

The most basic rational function is the **reciprocal function,** defined by $f(x) = \dfrac{1}{x}$. The denominator of the reciprocal function is zero when $x = 0$, so the domain of f is the set of all real numbers except 0.

Let's look at the behavior of f near the excluded value 0. We start by evaluating $f(x)$ to the left of 0.

x approaches 0 from the left.

x	-1	-0.5	-0.1	-0.01	-0.001
$f(x) = \dfrac{1}{x}$	-1	-2	-10	-100	-1000

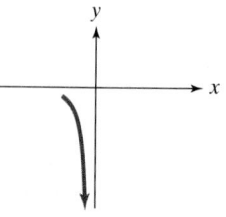

Mathematically, we say that "x approaches 0 from the left." From the table and the accompanying graph, on the bottom of the previous page, it appears that as x approaches 0 from the left, the function values, $f(x)$, decrease without bound. We say that "$f(x)$ approaches negative infinity." We use a special arrow notation to describe this situation symbolically:

$$\text{As } x \to 0^-, \ f(x) \to -\infty.$$

As x approaches 0 from the left, f(x) approaches negative infinity (that is, the graph falls).

Observe that the minus ($-$) superscript on the 0 $(x \to 0^-)$ is read "from the left."

Next, we evaluate $f(x)$ to the right of 0.

x approaches 0 from the right.

x	0.001	0.01	0.1	0.5	1
$f(x) = \dfrac{1}{x}$	1000	100	10	2	1

Mathematically, we say that "x approaches 0 from the right." From the table and the accompanying graph, it appears that as x approaches 0 from the right, the function values, $f(x)$, increase without bound. We say that "$f(x)$ approaches infinity." We again use a special arrow notation to describe this situation symbolically:

$$\text{As } x \to 0^+, \ f(x) \to \infty.$$

As x approaches 0 from the right, f(x) approaches infinity (that is, the graph rises).

Observe that the plus ($+$) superscript on the 0 $(x \to 0^+)$ is read "from the right."

Now let's see what happens to the function values, $f(x)$, as x gets farther away from the origin. The following tables suggest what happens to $f(x)$ as x increases or decreases without bound.

x increases without bound:

x	1	10	100	1000
$f(x) = \dfrac{1}{x}$	1	0.1	0.01	0.001

x decreases without bound:

x	-1	-10	-100	-1000
$f(x) = \dfrac{1}{x}$	-1	-0.1	-0.01	-0.001

Figure 3.23 illustrates the end behavior of $f(x) = \dfrac{1}{x}$ as x increases or decreases without bound. The function values, $f(x)$, are getting progressively closer to 0. This means that as x increases or decreases without bound, the graph of f is approaching the horizontal line $y = 0$ (that is, the x-axis). We use the arrow notation to describe this situation:

$$\text{As } x \to \infty, \ f(x) \to 0 \quad \text{and} \quad \text{as } x \to -\infty, \ f(x) \to 0.$$

As x approaches infinity (that is, increases without bound), f(x) approaches 0.

As x approaches negative infinity (that is, decreases without bound), f(x) approaches 0.

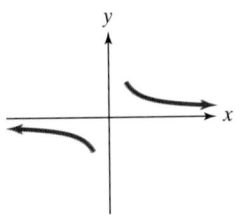

Figure 3.23 $f(x)$ approaches 0 as x increases or decreases without bound

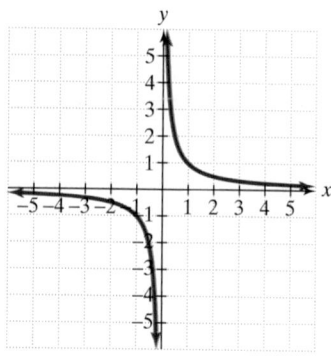

Figure 3.24 The graph of the reciprocal function $f(x) = \dfrac{1}{x}$

Thus, as x approaches infinity $(x \to \infty)$ or as x approaches negative infinity $(x \to -\infty)$, the function values are approaching zero: $f(x) \to 0$.

The graph of the reciprocal function $f(x) = \dfrac{1}{x}$ is shown in Figure 3.24. Unlike the graph of a polynomial function, the graph of the reciprocal function has a break in it and is composed of two distinct branches.

The arrow notation used throughout our discussion of the reciprocal function is summarized in the following box:

Arrow Notation

Symbol	Meaning
$x \to a^{+}$	x approaches a from the right.
$x \to a^{-}$	x approaches a from the left.
$x \to \infty$	x approaches infinity; that is, x increases without bound.
$x \to -\infty$	x approaches negative infinity; that is, x decreases without bound.

3 Identify vertical asymptotes.

Vertical Asymptotes of Rational Functions

Look again at the graph of $f(x) = \dfrac{1}{x}$. The curve approaches, but does not touch, the y-axis. The y-axis, or $x = 0$, is said to be a *vertical asymptote* of the graph. A rational function may have no vertical asymptotes, one vertical asymptote, or several vertical asymptotes. The graph of a rational function never intersects a vertical asymptote. We will use dashed lines to show asymptotes.

Definition of a Vertical Asymptote

The line $x = a$ is a **vertical asymptote** of the graph of a function f if $f(x)$ increases or decreases without bound as x approaches a.

As $x \to a^{+}$, $f(x) \to \infty$.   As $x \to a^{-}$, $f(x) \to \infty$.   As $x \to a^{+}$, $f(x) \to -\infty$.   As $x \to a^{-}$, $f(x) \to -\infty$.

Thus, as x approaches a from either the left or the right, $f(x) \to \infty$ or $f(x) \to -\infty$.

If the graph of a rational function has vertical asymptotes, they can be located using the following theorem:

Locating Vertical Asymptotes

If $f(x) = \dfrac{p(x)}{q(x)}$ is a rational function in which $p(x)$ and $q(x)$ have no common factors and a is a zero of $q(x)$, the denominator, then $x = a$ is a vertical asymptote of the graph of f.

EXAMPLE 2   **Finding the Vertical Asymptotes of a Rational Function**

Find the vertical asymptotes, if any, of the graph of each rational function:

a. $f(x) = \dfrac{x}{x^2 - 9}$ **b.** $g(x) = \dfrac{x + 3}{x^2 - 9}$ **c.** $h(x) = \dfrac{x + 3}{x^2 + 9}$.

Solution   Factoring is usually helpful in identifying zeros of denominators.

a.
$$f(x) = \frac{x}{x^2 - 9} = \frac{x}{(x + 3)(x - 3)}$$

> This factor is 0 if $x = -3$.

> This factor is 0 if $x = 3$.

There are no common factors in the numerator and the denominator. The zeros of the denominator are -3 and 3. Thus, the lines $x = -3$ and $x = 3$ are the vertical asymptotes for the graph of f.

b. We will use factoring to see if there are common factors.

$$g(x) = \frac{x + 3}{x^2 - 9} = \frac{(x + 3)}{(x + 3)(x - 3)} = \frac{1}{x - 3}$$

> There is a common factor, $x + 3$, so simplify.

> This denominator is 0 if $x = 3$.

The only zero of the denominator of $g(x)$ in simplified form is 3. Thus, the line $x = 3$ is the only vertical asymptote of the graph of g.

c. We cannot factor the denominator of $h(x)$ over the real numbers.

$$h(x) = \frac{x + 3}{x^2 + 9}$$

> No real numbers make this denominator 0.

The denominator has no real zeros. Thus, the graph of h has no vertical asymptotes.

Check Point 2   Find the vertical asymptotes, if any, of the graph of each rational function:

a. $f(x) = \dfrac{x}{x^2 - 1}$ **b.** $g(x) = \dfrac{x - 1}{x^2 - 1}$ **c.** $h(x) = \dfrac{x - 1}{x^2 + 1}$.

A value where the denominator of a function is zero does not necessarily result in a vertical asymptote. There is a hole corresponding to $x = a$, and not a vertical asymptote, in the graph of a function under the following conditions: The value a causes the denominator to be zero, but there is a reduced form of the functions equation in which a does not cause the denominator to be zero. Consider, for example, the function

$$f(x) = \frac{x^2 - 9}{x - 3}.$$

Because the denominator is zero when $x = 3$, the functions domain is all real numbers except 3. However, there is a reduced form of the equation in which 3 does not cause the denominator to be zero:

$$f(x) = \frac{x^2 - 9}{x - 3} = \frac{(x + 3)(x - 3)}{x - 3} = x + 3, \quad x \neq 3$$

Denominator is zero at x = 3.

In this reduced form, 3 does not result in a zero denominator.

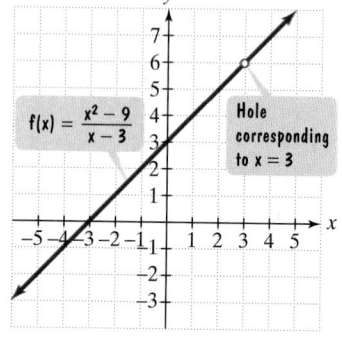

Figure 3.25 A graph with a hole corresponding to the denominator's zero

Figure 3.25 shows that the graph has a hole corresponding to $x = 3$. Graphing utilities do not show this feature of the graph.

Horizontal Asymptotes of Rational Functions

4 Identify horizontal asymptotes.

Figure 3.24 shows the graph of the reciprocal function $f(x) = \frac{1}{x}$. As $x \to \infty$ and as $x \to -\infty$, the function values are approaching 0: $f(x) \to 0$. The line $y = 0$ (that is, the x-axis) is a *horizontal asymptote* of the graph. Many, but not all, rational functions have horizontal asymptotes.

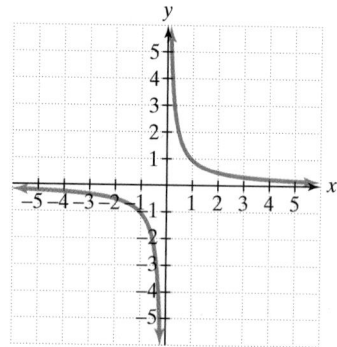

Figure 3.24 The graph of the reciprocal function $f(x) = \frac{1}{x}$, repeated

Definition of a Horizontal Asymptote

The line $y = b$ is a **horizontal asymptote** of the graph of a function f if $f(x)$ approaches b as x increases or decreases without bound.

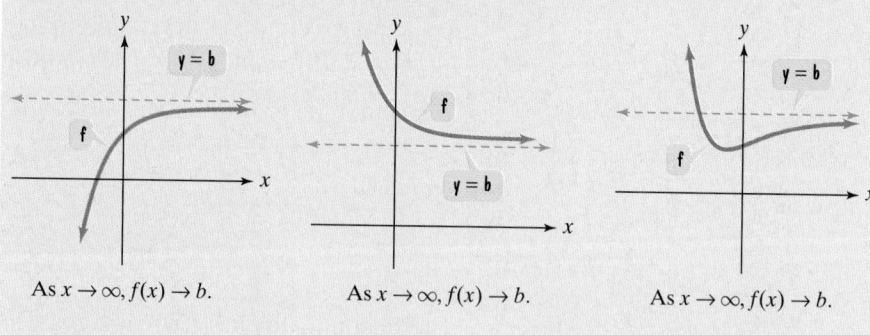

As $x \to \infty$, $f(x) \to b$.   As $x \to \infty$, $f(x) \to b$.   As $x \to \infty$, $f(x) \to b$.

Recall that a rational function may have several vertical asymptotes. By contrast, it can have at most one horizontal asymptote. Although a graph can never intersect a vertical asymptote, it may cross its horizontal asymptote.

If the graph of a rational function has a horizontal asymptote, it can be located using the following theorem:

Locating Horizontal Asymptotes

Let f be the rational function given by

$$f(x) = \frac{a_n x^n + a_{n-1} x^{n-1} + \cdots + a_1 x + a_0}{b_m x^m + b_{m-1} x^{m-1} + \cdots + b_1 x + b_0}, \quad a_n \neq 0, b_m \neq 0.$$

The degree of the numerator is n. The degree of the denominator is m.

1. If $n < m$, the x-axis, or $y = 0$, is the horizontal asymptote of the graph of f.

2. If $n = m$, the line $y = \dfrac{a_n}{b_m}$ is the horizontal asymptote of the graph of f.

3. If $n > m$, the graph of f has no horizontal asymptote.

EXAMPLE 3   **Finding the Horizontal Asymptote of a Rational Function**

Find the horizontal asymptote, if any, of the graph of each rational function:

a. $f(x) = \dfrac{4x}{2x^2 + 1}$ **b.** $g(x) = \dfrac{4x^2}{2x^2 + 1}$ **c.** $h(x) = \dfrac{4x^3}{2x^2 + 1}.$

Solution

a. $f(x) = \dfrac{4x}{2x^2 + 1}$

The degree of the numerator, 1, is less than the degree of the denominator, 2. Thus, the graph of f has the x-axis as a horizontal asymptote [see Figure 3.26(a)]. The equation of the horizontal asymptote is $y = 0$.

b. $g(x) = \dfrac{4x^2}{2x^2 + 1}$

The degree of the numerator, 2, is equal to the degree of the denominator, 2. The leading coefficients of the numerator and denominator, 4 and 2, are used to obtain the equation of the horizontal asymptote. The equation of the horizontal asymptote is $y = \frac{4}{2}$ or $y = 2$ [see Figure 3.26(b)].

c. $h(x) = \dfrac{4x^3}{2x^2 + 1}$

The degree of the numerator, 3, is greater than the degree of the denominator, 2. Thus, the graph of h has no horizontal asymptote [see Figure 3.26(c)].

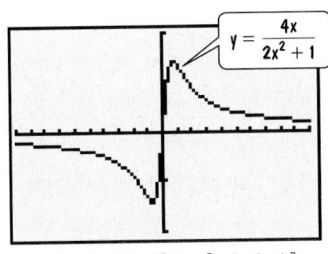

$[-10, 10, 1]$ by $[-2, 2, 1]$
(a) The horizontal asymptote of the graph is $y = 0$.

$[-10, 10, 1]$ by $[-1, 3, 1]$
(b) The horizontal asymptote of the graph is $y = 2$.

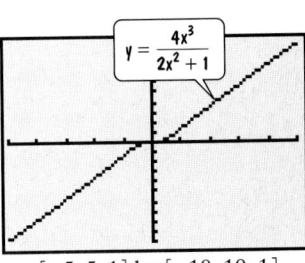

$[-5, 5, 1]$ by $[-10, 10, 1]$
(c) The graph has no horizontal asymptote.

Figure 3.26

Check Point 3 Find the horizontal asymptote, if any, of the graph of each rational function:

a. $f(x) = \dfrac{9x^2}{3x^2 + 1}$ **b.** $g(x) = \dfrac{9x}{3x^2 + 1}$ **c.** $h(x) = \dfrac{9x^3}{3x^2 + 1}$.

5 Graph rational functions.

Graphing Rational Functions

Here are some suggestions for graphing rational functions:

Strategy for Graphing a Rational Function

Suppose that

$$f(x) = \frac{p(x)}{q(x)},$$

where p and q are polynomial functions with no common factors.

1. Determine whether the graph of f has symmetry.

$$f(-x) = f(x): \text{ } y\text{-axis symmetry}$$
$$f(-x) = -f(x): \text{ origin symmetry}$$

2. Find the y-intercept (if there is one) by evaluating $f(0)$.
3. Find the x-intercepts (if there are any) by solving the equation $p(x) = 0$.
4. Find any vertical asymptote(s) by solving the equation $q(x) = 0$.
5. Find the horizontal asymptote (if there is one) using the rule for determining the horizontal asymptote of a rational function.
6. Plot at least one point between and beyond each x-intercept and vertical asymptote.
7. Use the information obtained previously to graph the function between and beyond the vertical asymptotes.

EXAMPLE 4   Graphing a Rational Function

Graph: $f(x) = \dfrac{2x}{x - 1}$.

Solution

Step 1   Determine symmetry.

$$f(-x) = \frac{2(-x)}{-x - 1} = \frac{-2x}{-x - 1} = \frac{2x}{x + 1}$$

Because $f(-x)$ does not equal $f(x)$ or $-f(x)$, the graph has neither y-axis nor origin symmetry.

Step 2   Find the y-intercept.   Evaluate $f(0)$.

$$f(0) = \frac{2 \cdot 0}{0 - 1} = \frac{0}{-1} = 0$$

The y-intercept is 0, and so the graph passes through the origin.

The function to be graphed,

$$f(x) = \frac{2x}{x - 1}, \text{repeated}$$

Step 3 Find x-intercept(s). This is done by solving $p(x) = 0$.

$$2x = 0 \quad \text{Set the numerator equal to 0.}$$
$$x = 0$$

There is only one x-intercept. This verifies that the graph passes through the origin.

Step 4 Find the vertical asymptote(s). Solve $q(x) = 0$, thereby finding zeros of the denominator.

$$x - 1 = 0 \quad \text{Set the denominator equal to 0.}$$
$$x = 1$$

The equation of the vertical asymptote is $x = 1$.

Step 5 Find the horizontal asymptote. Because the numerator and denominator have the same degree, the leading coefficients of the numerator and denominator, 2 and 1, are used to obtain the equation of the horizontal asymptote.

$$y = \frac{2}{1} = 2.$$

The equation of the horizontal asymptote is $y = 2$.

Step 6 Plot points between and beyond each x-intercept and vertical asymptote. With an x-intercept at 0 and a vertical asymptote at $x = 1$, we evaluate the function at $-2, -1, \frac{1}{2}, 2,$ and 4.

x	-2	-1	$\dfrac{1}{2}$	2	4
$f(x) = \dfrac{2x}{x - 1}$	$\dfrac{4}{3}$	1	-2	4	$\dfrac{8}{3}$

Figure 3.27 shows these points, the y-intercept, the x-intercept, and the asymptotes.

Step 7 Graph the function. The graph of $f(x) = \dfrac{2x}{x - 1}$ is shown in Figure 3.28.

Technology

The graph of $y = \dfrac{2x}{x - 1}$, obtained using the $\boxed{\text{DOT}}$ mode in a $[-6, 6, 1]$ by $[-6, 6\ 1]$ viewing rectangle, verifies that our hand-drawn graph is correct.

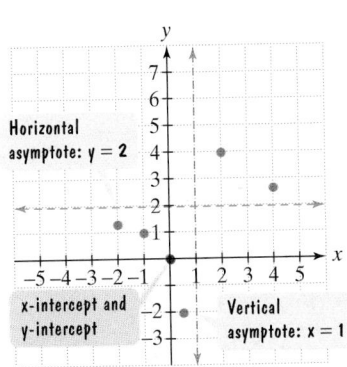

Figure 3.27 Preparing to graph the rational function $f(x) = \dfrac{2x}{x - 1}$

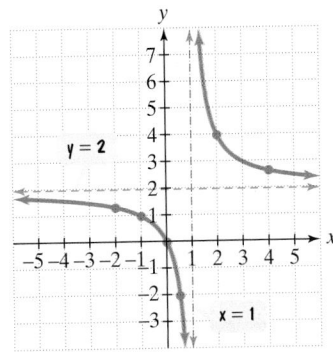

Figure 3.28 The graph of $f(x) = \dfrac{2x}{x - 1}$

Check Point 4 Graph: $f(x) = \dfrac{3x}{x - 2}.$

EXAMPLE 5   Graphing a Rational Function

Graph: $f(x) = \dfrac{3x^2}{x^2 - 4}$.

Solution

Step 1   Determine symmetry: $f(-x) = \dfrac{3(-x)^2}{(-x)^2 - 4} = \dfrac{3x^2}{x^2 - 4} = f(x)$: The graph of f is symmetric with respect to the y-axis.

Step 2   Find the y-intercept: $f(0) = \dfrac{3 \cdot 0^2}{0^2 - 4} = \dfrac{0}{-4} = 0$: The y-intercept is 0.

Step 3   Find the x-intercept: $3x^2 = 0$, so $x = 0$: The x-intercept is 0.

Step 4   Find the vertical asymptotes: Set $q(x) = 0$.

$$x^2 - 4 = 0 \qquad \text{Set the denominator equal to 0.}$$
$$x^2 = 4$$
$$x = \pm 2$$

The vertical asymptotes are $x = -2$ and $x = 2$.

Step 5   Find the horizontal asymptote: The horizontal asymptote is $y = \frac{3}{1} = 3$.

Step 6   Plot points between and beyond the x-intercept and the vertical asymptotes. With an x-intercept at 0 and vertical asymptotes at $x = -2$ and $x = 2$, we evaluate the function at $-3, -1, 1, 3,$ and 4.

Study Tip

Because the graph has y-axis symmetry, it is not necessary to evaluate the even function at −3 and again at 3.
$$f(-3) = f(3) = \tfrac{27}{5}$$
This also applies to evaluation at −1 and 1.

x		−3	−1	1	3	4
$f(x) = \dfrac{3x^2}{x^2 - 4}$		$\dfrac{27}{5}$	−1	−1	$\dfrac{27}{5}$	4

Figure 3.29 shows these points, the y-intercept, the x-intercept, and the asymptotes.

Step 7   Graph the function. The graph of $f(x) = \dfrac{3x^2}{x^2 - 4}$ is shown in Figure 3.30. The y-axis symmetry is now obvious.

Technology

The graph of $y = \dfrac{3x^2}{x^2 - 4}$, generated by a graphing utility, verifies that our hand-drawn graph is correct.

$[-6, 6, 1]$ by $[-6, 6, 1]$

Figure 3.29 Preparing to graph $f(x) = \dfrac{3x^2}{x^2 - 4}$

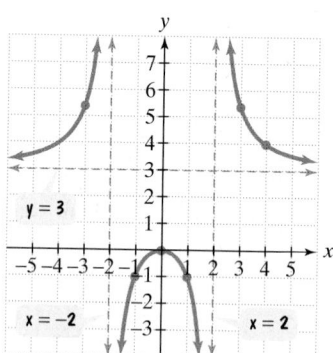

Figure 3.30 The graph of $f(x) = \dfrac{3x^2}{x^2 - 4}$

Check Point 5 Graph: $f(x) = \dfrac{2x^2}{x^2 - 9}$.

Example 6 illustrates that not every rational function has vertical and horizontal asymptotes.

EXAMPLE 6   Graphing a Rational Function

Graph: $f(x) = \dfrac{x^4}{x^2 + 1}$.

Solution

Step 1   Determine symmetry: $f(-x) = \dfrac{(-x)^4}{(-x)^2 + 1} = \dfrac{x^4}{x^2 + 1} = f(x)$:

The graph of f is symmetric with respect to the y-axis.

Step 2   Find the y-intercept: $f(0) = \dfrac{0^4}{0^2 + 1} = \dfrac{0}{1} = 0$: The y-intercept is 0.

Step 3   Find the x-intercept: $x^4 = 0$, so $x = 0$: The x-intercept is 0.

Step 4   Find the vertical asymptote: Set $q(x) = 0$.

$$x^2 + 1 = 0 \qquad \textit{Set the denominator equal to 0.}$$

$$x^2 = -1$$

Although this equation has imaginary roots ($x = \pm i$), there are no real roots. Thus, there is no vertical asymptote.

Step 5   Find the horizontal asymptote: Because the degree of the numerator, 4, is greater than the degree of the denominator, 2, there is no horizontal asymptote.

Step 6   Plot points between and beyond the x-intercept and the vertical asymptotes. With an x-intercept at 0 and no vertical asymptotes, let's look at function values at $-2, -1, 1,$ and 2. You can evaluate the function at 1 and 2. Use y-axis symmetry to obtain function values at -1 and -2:

$$f(-1) = f(1) \text{ and } f(-2) = f(2).$$

x	-2	-1	1	2
$f(x) = \dfrac{x^4}{x^2 + 1}$	$\dfrac{16}{5}$	$\dfrac{1}{2}$	$\dfrac{1}{2}$	$\dfrac{16}{5}$

Step 7   Graph the function. Figure 3.31 shows the graph of f using the points obtained from the table and y-axis symmetry. Notice that as x approaches infinity or negative infinity ($x \to \infty$ or $x \to -\infty$), the function values, $f(x)$, are getting larger without bound $\left[f(x) \to \infty\right]$.

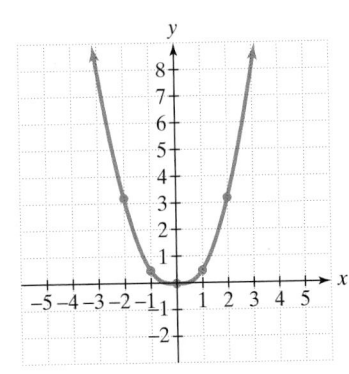

Figure 3.31 The graph of $f(x) = \dfrac{x^4}{x^2 + 1}$

Check Point 6 Graph: $f(x) = \dfrac{x^4}{x^2 + 2}$.

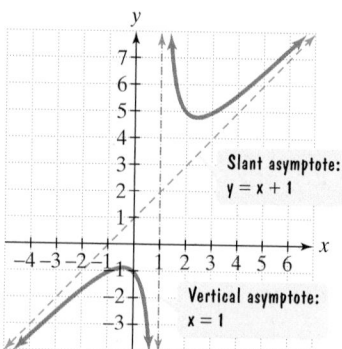

6 Identify slant asymptotes.

Slant Asymptotes

Examine the graph of

$$f(x) = \frac{x^2 + 1}{x - 1}$$

shown in Figure 3.32. Note that the degree of the numerator, 2, is greater than the degree of the denominator, 1. Thus, the graph of this function has no horizontal asymptote. However, the graph has a **slant asymptote, $y = x + 1$.**

    The graph of a rational function has a slant asymptote if the degree of the numerator is one more than the degree of the denominator. The equation of the slant asymptote can be found by division. For example, to find the slant asymptote for the graph of $f(x) = \dfrac{x^2 + 1}{x - 1}$, divide $x - 1$ into $x^2 + 1$:

Figure 3.32 The graph of $f(x) = \dfrac{x^2 + 1}{x - 1}$ with a slant asymptote

$$\underline{1}\!\rfloor \quad \begin{array}{ccc} 1 & 0 & 1 \\ & 1 & 1 \\ \hline 1 & 1 & 2 \end{array}$$

$$\begin{array}{r} 1x + 1 + \dfrac{2}{x-1} \\ x - 1 \overline{\smash{\big)}\, x^2 + 0x + 1} \end{array}$$

Remainder

Observe that

$$f(x) = \frac{x^2 + 1}{x - 1} = \underbrace{x + 1}_{\text{Slant asymptote:}\ y = x+1} + \frac{2}{x - 1}$$

If $|x| \to \infty$, the value of $\dfrac{2}{x - 1}$ is approximately 0. Thus, when $|x|$ is large, the function is very close to $y = x + 1 + 0$. This means that as $x \to \infty$ or as $x \to -\infty$, the graph of f gets closer and closer to the line whose equation is $y = x + 1$. The line $y = x + 1$ is a slant asymptote of the graph.

    In general, if $f(x) = \dfrac{p(x)}{q(x)}$, p and q have no common factors, and the degree of p is one greater than the degree of q, find the slant asymptote by dividing $q(x)$ into $p(x)$. The division will take the form

$$\frac{p(x)}{q(x)} = \underbrace{mx + b}_{\substack{\text{Slant asymptote:}\\ y = mx + b}} + \frac{\text{remainder}}{q(x)}.$$

The equation of the slant asymptote is obtained by dropping the term with the remainder. Thus, the equation of the slant asymptote is $y = mx + b$.

EXAMPLE 7 Finding the Slant Asymptote of a Rational Function

Find the slant asymptote of $f(x) = \dfrac{x^2 - 4x - 5}{x - 3}$.

Solution Because the degree of the numerator, 2, is exactly one more than the degree of the denominator, 1, and $x - 3$ is not a factor of $x^2 - 4x - 5$, the graph of f has a slant asymptote. To find the equation of the slant asymptote, divide $x - 3$ into $x^2 - 4x - 5$:

$$\underline{3|}\ \ 1 \quad -4 \quad -5$$
$$\qquad\quad 3 \quad -3$$
$$\overline{\quad 1 \quad -1 \quad -8}$$

$$1x - 1 - \frac{8}{x - 3}$$
$$x - 3\overline{)x^2 - 4x - 5}$$

Drop the remainder term and you'll have the equation of the slant asymptote.

Remainder

The equation of the slant asymptote is $y = x - 1$. Using our strategy for graphing rational functions, the graph of $f(x) = \dfrac{x^2 - 4x - 5}{x - 3}$ is shown in Figure 3.33.

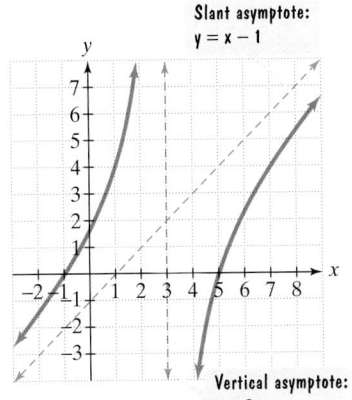

Slant asymptote: y = x − 1
Vertical asymptote: x = 3

Figure 3.33 The graph of $f(x) = \dfrac{x^2 - 4x - 5}{x - 3}$

Check Point 7 Find the slant asymptote of $f(x) = \dfrac{2x^2 - 5x + 7}{x - 2}$.

7 Solve applied problems involving rational functions.

Applications

There are numerous examples of asymptotic behavior in functions that describe real-world phenomena. Let's consider an example from the business world. The **cost function,** C, for a business is the sum of its fixed and variable costs:

$$C(x) = (\text{fixed cost}) + cx.$$

Cost per unit times the number of units produced, x

The **average cost** per unit for a company to produce x units is the sum of its fixed and variable costs divided by the number of units produced. The **average cost function** is a rational function that is denoted by $\bar{C}$. Thus,

$$\bar{C}(x) = \frac{(\text{fixed cost}) + cx}{x}.$$

Cost of producing x units: fixed plus variable costs

Number of units produced

EXAMPLE 8 Average Cost of Producing a Wheelchair

A company is planning to manufacture wheelchairs that are light, fast, and beautiful. Fixed monthly cost will be \$500,000, and it will cost \$400 to produce each radically innovative chair.

a. Write the cost function, C, of producing x wheelchairs.

b. Write the average cost function, $\bar{C}$, of producing x wheelchairs.

c. Find and interpret $\bar{C}(1000)$, $\bar{C}(10,000)$, and $\bar{C}(100,000)$.

d. What is the horizontal asymptote for the average cost function, $\bar{C}$? Describe what this represents for the company.

Solution

a. The cost function of producing x wheelchairs, C, is the sum of the fixed cost and the variable cost.

| Fixed cost is $500,000. | Variable cost: $400 for each wheelchair produced |

$$C(x) = 500,000 + 400x$$

b. The average cost function of producing x wheelchairs, $\bar{C}$, is the sum of fixed and variable costs divided by the number of wheelchairs produced.

$$\bar{C}(x) = \frac{500,000 + 400x}{x} \quad \text{or} \quad \bar{C}(x) = \frac{400x + 500,000}{x}$$

c. We evaluate $\bar{C}$ at 1000, 10,000, and 100,000, interpreting the results.

$$\bar{C}(1000) = \frac{400(1000) + 500,000}{1000} = 900$$

The average cost per wheelchair of producing 1000 wheelchairs per month is $900.

$$\bar{C}(10,000) = \frac{400(10,000) + 500,000}{10,000} = 450$$

The average cost per wheelchair of producing 10,000 wheelchairs per month is $450.

$$\bar{C}(100,000) = \frac{400(100,000) + 500,000}{100,000} = 405$$

The average cost per wheelchair of producing 100,000 wheelchairs per month is $405. Notice that with higher production levels, the cost of producing each wheelchair decreases.

d. We developed the average cost function

$$\bar{C}(x) = \frac{400x + 500,000}{x}$$

in which the degree of the numerator, 1, is equal to the degree of the denominator, 1. The leading coefficients of the numerator and denominator, 400 and 1, are used to obtain the equation of the horizontal asymptote. The equation of the horizontal asymptote is

$$y = \frac{400}{1} \quad \text{or} \quad y = 400.$$

The horizontal asymptote is shown in Figure 3.34. This means that the more wheelchairs produced per month, the closer the average cost per wheelchair for the company comes to $400. The least possible cost per wheelchair is approaching $400. Competitively low prices take place with high production levels, posing a major problem for small businesses.

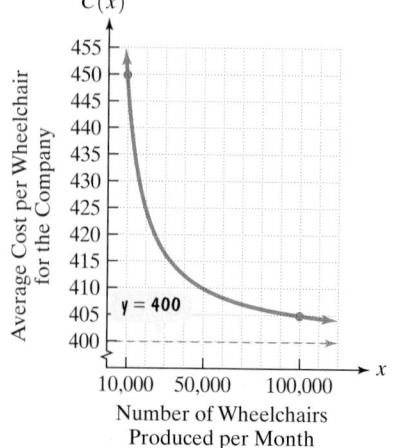

Figure 3.34 As production level increases, the average cost per wheelchair approaches $400.

> **Check Point 8**
>
> The time: the not-too-distant future. A new company is hoping to replace traditional computers and two-dimensional monitors with its virtual reality system. The fixed monthly cost will be $600,000, and it will cost $500 to produce each system.
>
> **a.** Write the cost function, C, of producing x virtual reality systems.
> **b.** Write the average cost function, $\bar{C}$, of producing x virtual reality systems.
> **c.** Find and interpret $\bar{C}(1000)$, $\bar{C}(10,000)$, and $\bar{C}(100,000)$.
> **d.** What is the horizontal asymptote for the average cost function, $\bar{C}$? Describe what this represents for the company.

EXERCISE SET 3.6

Practice Exercises

In Exercises 1–8, find the domain of each rational function.

1. $f(x) = \dfrac{5x}{x - 4}$

2. $f(x) = \dfrac{7x}{x - 8}$

3. $g(x) = \dfrac{3x^2}{(x - 5)(x + 4)}$

4. $g(x) = \dfrac{2x^2}{(x - 2)(x + 6)}$

5. $h(x) = \dfrac{x + 7}{x^2 - 49}$

6. $h(x) = \dfrac{x + 8}{x^2 - 64}$

7. $f(x) = \dfrac{x + 7}{x^2 + 49}$

8. $f(x) = \dfrac{x + 8}{x^2 + 64}$

Use the graph of the rational function in the figure shown to complete each statement in Exercises 9–14.

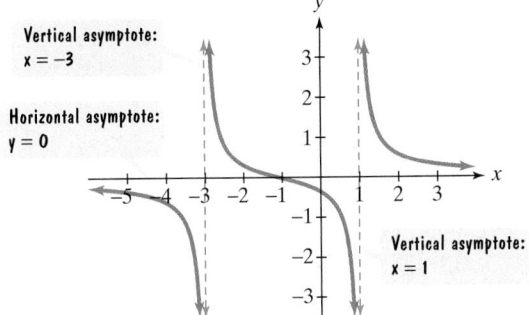

9. As $x \to -3^-$, $f(x) \to$ _____.

10. As $x \to -3^+$, $f(x) \to$ _____.

11. As $x \to 1^-$, $f(x) \to$ _____.

12. As $x \to 1^+$, $f(x) \to$ _____.

13. As $x \to -\infty$, $f(x) \to$ _____.

14. As $x \to \infty$, $f(x) \to$ _____.

Use the graph of the rational function in the figure shown to complete each statement in Exercises 15–20.

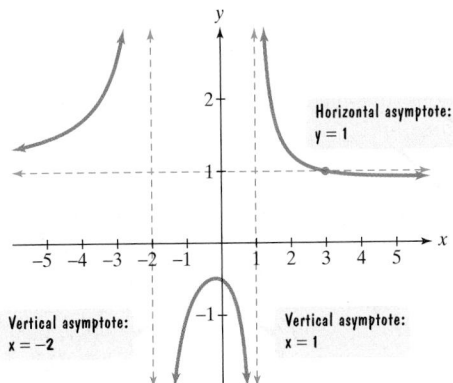

15. As $x \to 1^+$, $f(x) \to$ _____.

16. As $x \to 1^-$, $f(x) \to$ _____.

17. As $x \to -2^+$, $f(x) \to$ _____.

18. As $x \to -2^-$, $f(x) \to$ _____.

19. As $x \to \infty$, $f(x) \to$ _____.

20. As $x \to -\infty$, $f(x) \to$ _____.

In Exercises 21–28, find the vertical asymptotes, if any, of the graph of each rational function.

21. $f(x) = \dfrac{x}{x + 4}$

22. $f(x) = \dfrac{x}{x - 3}$

23. $g(x) = \dfrac{x + 3}{x(x + 4)}$

24. $g(x) = \dfrac{x + 3}{x(x - 3)}$

25. $h(x) = \dfrac{x}{x(x + 4)}$

26. $h(x) = \dfrac{x}{x(x - 3)}$

27. $r(x) = \dfrac{x}{x^2 + 4}$

28. $r(x) = \dfrac{x}{x^2 + 3}$

In Exercises 29–36, find the horizontal asymptote, if any, of the graph of each rational function.

29. $f(x) = \dfrac{12x}{3x^2 + 1}$

30. $f(x) = \dfrac{15x}{3x^2 + 1}$

31. $g(x) = \dfrac{12x^2}{3x^2 + 1}$

32. $g(x) = \dfrac{15x^2}{3x^2 + 1}$

33. $h(x) = \dfrac{12x^3}{3x^2 + 1}$

34. $h(x) = \dfrac{15x^3}{3x^2 + 1}$

35. $f(x) = \dfrac{-2x + 1}{3x + 5}$

36. $f(x) = \dfrac{-3x + 7}{5x - 2}$

In Exercises 37–58, follow the seven steps on page 342 to graph each rational function.

37. $f(x) = \dfrac{4x}{x - 2}$

38. $f(x) = \dfrac{3x}{x - 1}$

39. $f(x) = \dfrac{2x}{x^2 - 4}$

40. $f(x) = \dfrac{4x}{x^2 - 1}$

41. $f(x) = \dfrac{2x^2}{x^2 - 1}$

42. $f(x) = \dfrac{4x^2}{x^2 - 9}$

43. $f(x) = \dfrac{-x}{x + 1}$

44. $f(x) = \dfrac{-3x}{x + 2}$

45. $f(x) = -\dfrac{1}{x^2 - 4}$

46. $f(x) = -\dfrac{2}{x^2 - 1}$

47. $f(x) = \dfrac{2}{x^2 + x - 2}$

48. $f(x) = \dfrac{-2}{x^2 - x - 2}$

49. $f(x) = \dfrac{2x^2}{x^2 + 4}$

50. $f(x) = \dfrac{4x^2}{x^2 + 1}$

51. $f(x) = \dfrac{x + 2}{x^2 + x - 6}$

52. $f(x) = \dfrac{x - 4}{x^2 - x - 6}$

53. $f(x) = \dfrac{x^4}{x^2 + 2}$

54. $f(x) = \dfrac{2x^4}{x^2 + 1}$

55. $f(x) = \dfrac{x^2 + x - 12}{x^2 - 4}$ **56.** $f(x) = \dfrac{x^2}{x^2 + x - 6}$

57. $f(x) = \dfrac{3x^2 + x - 4}{2x^2 - 5x}$ **58.** $f(x) = \dfrac{x^2 - 4x + 3}{(x + 1)^2}$

In Exercises 59–66, **a.** *Find the slant asymptote of the graph of each rational function and* **b.** *Follow the seven-step strategy and use the slant asymptote to graph each rational function.*

59. $f(x) = \dfrac{x^2 - 1}{x}$ **60.** $f(x) = \dfrac{x^2 - 4}{x}$

61. $f(x) = \dfrac{x^2 + 1}{x}$ **62.** $f(x) = \dfrac{x^2 + 4}{x}$

63. $f(x) = \dfrac{x^2 + x - 6}{x - 3}$ **64.** $f(x) = \dfrac{x^2 - x + 1}{x - 1}$

65. $f(x) = \dfrac{x^3 + 1}{x^2 + 2x}$ **66.** $f(x) = \dfrac{x^3 - 1}{x^2 - 9}$

Application Exercises

67. A company is planning to manufacture mountain bikes. Fixed monthly cost will be $100,000 and it will cost $100 to produce each bicycle.

   a. Write the cost function, C, of producing x mountain bikes.

   b. Write the average cost function, $\bar{C}$, of producing x mountain bikes.

   c. Find and interpret $\bar{C}(500)$, $\bar{C}(1000)$, $\bar{C}(2000)$, and $\bar{C}(4000)$.

   d. What is the horizontal asymptote for the function, $\bar{C}$? Describe what this means in practical terms.

68. A company that manufactures running shoes has a fixed monthly cost of $300,000. It costs $30 to produce each pair of shoes.

   a. Write the cost function, C, of producing x pairs of shoes.

   b. Write the average cost function, $\bar{C}$, of producing x pairs of shoes.

   c. Find and interpret $\bar{C}(1000)$, $\bar{C}(10,000)$, and $\bar{C}(100,000)$.

   d. What is the horizontal asymptote for the average cost function, $\bar{C}$? Describe what this represents for the company.

69. Textbook sales at college stores have increased during the past two decades. The function

$$B(x) = 190.9x + 2413.99$$

models textbook sales, $B(x)$, in millions of dollars, x years after 1985. College enrollment has also increased. The function

$$E(x) = 0.234x + 12.54$$

models total college enrollment, $E(x)$, in millions, x years after 1985.

   a. Write a rational function that models the average amount of money spent on textbooks per college student, $M(x)$, in dollars per student, x years after 1985. The graph of M is shown in the figure.

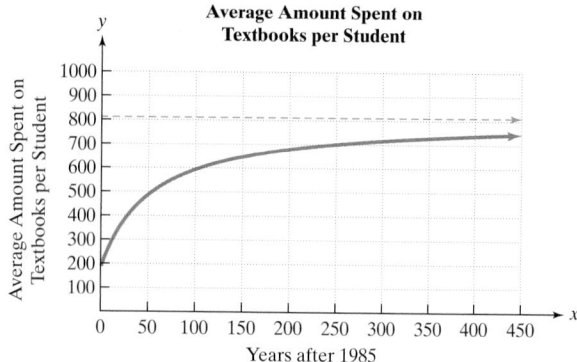

Average Amount Spent on Textbooks per Student

   b. Predict the average amount of money that will be spent on textbooks per college student in 2004. How is this shown on the graph of M?

   c. What is the horizontal asymptote for the function that models the average amount of money spent on textbooks per college student? Describe what this represents in practical terms.

70. The rational function

$$C(x) = \frac{130x}{100 - x}, \quad 0 \le x < 100,$$

describes the cost, $C(x)$, in millions of dollars, to inoculate $x\%$ of the population against a particular strain of flu.

   a. Find and interpret $C(20), C(40), C(60), C(80)$, and $C(90)$.

   b. What is the equation of the vertical asymptote? What does this mean in terms of the variables in the function?

   c. Graph the function.

Among all deaths from a particular disease, the percentage that are smoking related (21–39 cigarettes per day) is a function of the disease's **incidence ratio.** *The incidence ratio describes the number of times more likely smokers are than nonsmokers to die from the disease. The following table shows the incidence ratios for heart disease and lung cancer for two age groups.*

Incidence Ratios

	Heart Disease	Lung Cancer
Ages 55–64	1.9	10
Ages 65–74	1.7	9

Source: Alexander M. Walker, *Observations and Inference*, page 20.

For example, the incidence ratio of 9 in the table means that smokers between the ages of 65 and 74 are 9 times more likely than nonsmokers in the same group to die from lung cancer. The rational function

$$P(x) = \frac{100(x - 1)}{x}$$

models the percentage of smoking-related deaths among all deaths from a disease, $P(x)$, in terms of the disease's incidence ratio, x. The graph of the rational function is shown. Use this function to solve Exercises 71–74.

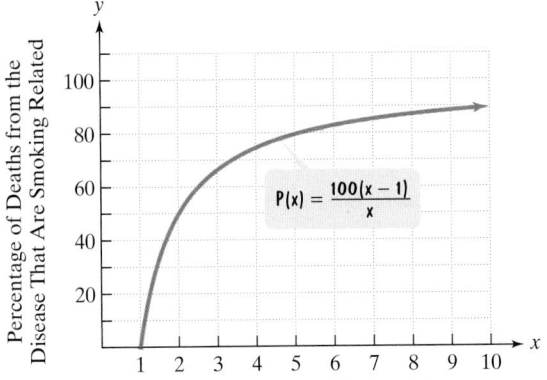

The Disease's Incidence Ratio:
The number of times more likely smokers are than nonsmokers to die from the disease

71. Find $P(10)$. Describe what this means in terms of the incidence ratio, 10, given in the table. Identify your solution as a point on the graph.

72. Find $P(9)$. Round to the nearest percent. Describe what this means in terms of the incidence ratio, 9, given in the table. Identify your solution as a point on the graph.

73. What is the horizontal asymptote of the graph? Describe what this means about the percentage of deaths caused by smoking with increasing incidence ratios.

74. According to the model and its graph, is there a disease for which all deaths are caused by smoking? Explain your answer.

75. Rational functions are often used to model how much we remember over time. In an experiment on memory, students in a language class are asked to memorize 40 vocabulary words in Latin, a language with which the students are not familiar. After studying the words for one day, the class is tested each day thereafter to see how many words they remember. The class average is taken and the results are graphed below.

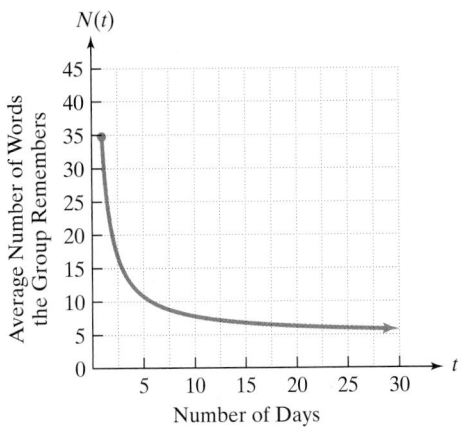

a. Use the graph to find a reasonable estimate of the number of Latin words remembered after 1 day, 5 days, and 15 days.

b. The function that models the number of Latin words remembered by the students after t days is given by

$$N(t) = \frac{5t + 30}{t}, \quad \text{where } t \geq 1.$$

Find $N(1)$, $N(5)$, and $N(15)$, comparing these values with your estimates from part (a).

c. What does the graph indicate about the number of Latin words remembered by the group over time?

d. Use the function in part (b) to find the horizontal asymptote for the graph. Describe what this horizontal asymptote means in terms of the variables modeled in this situation.

Writing in Mathematics

76. What is a rational function?

77. Use everyday language to describe the graph of a rational function f such that as $x \to -\infty$, $f(x) \to 3$.

78. Use everyday language to describe the behavior of a graph near its vertical asymptote if $f(x) \to \infty$ as $x \to -2^-$ and $f(x) \to -\infty$ as $x \to -2^+$.

79. If you are given the equation of a rational function, explain how to find the vertical asymptotes, if any, of the function's graph.

80. If you are given the equation of a rational function, explain how to find the horizontal asymptote, if any, of the function's graph.

81. Describe how to graph a rational function.

82. If you are given the equation of a rational function, how can you tell if the graph has a slant asymptote? If it does, how do you find its equation?

83. Is every rational function a polynomial function? Why or why not? Does a true statement result if the two adjectives *rational* and *polynomial* are reversed? Explain.

84. Although your friend has a family history of heart disease, he smokes, on average, 25 cigarettes per day. He sees the table showing incidence ratios for heart disease (see Exercises 71–74) and feels comfortable that they are less than 2, compared to 9 and 10 for lung cancer. He claims that all family deaths have been from heart disease, and decides not to give up smoking. Use the given function and its graph to describe some additional information not given in the table that might influence his decision.

Technology Exercises

85. Use a graphing utility to verify any five of your hand-drawn graphs in Exercises 37–66.

86. Use a graphing utility to verify your hand-drawn graph in Exercise 70.

87. Use a graphing utility to graph $y = \dfrac{1}{x}$, $y = \dfrac{1}{x^3}$, and $\dfrac{1}{x^5}$ in the same viewing rectangle. For odd values of n, how does changing n affect the graph of $y = \dfrac{1}{x^n}$?

88. Use a graphing utility to graph $y = \dfrac{1}{x^2}$, $y = \dfrac{1}{x^4}$, and $y = \dfrac{1}{x^6}$ in the same viewing rectangle. For even values of n, how does changing n affect the graph of $y = \dfrac{1}{x^n}$?

89. Use a graphing utility to graph

$$f(x) = \frac{x^2 - 4x + 3}{x - 2} \quad \text{and} \quad g(x) = \frac{x^2 - 5x + 6}{x - 2}.$$

What differences do you observe between the graph of f and g? How do you account for these differences?

90. The rational function

$$f(x) = \frac{27{,}725(x - 14)}{x^2 + 9} - 5x$$

models the number of arrests, $f(x)$, per 100,000 drivers, for driving under the influence of alcohol, as a function of a driver's age, x.

a. Graph the function in a $[0, 70, 5]$ by $[0, 400, 20]$ viewing rectangle.

b. Describe the trend shown by the graph.

c. Use the ZOOM and TRACE features or the maximum function feature of your graphing utility to find the age that corresponds to the greatest number of arrests. How many arrests, per 100,000 drivers, are there for this age group?

Critical Thinking Exercises

91. Which one of the following is true?

a. The graph of a rational function cannot have both a vertical and a horizontal asymptote.

b. It is not possible to have a rational function whose graph has no y-intercept.

c. The graph of a rational function can have three horizontal asymptotes.

d. The graph of a rational function can never cross a vertical asymptote.

92. Which one of the following is true?

a. The function $f(x) = \dfrac{1}{\sqrt{x - 3}}$ is a rational function.

b. The x-axis is a horizontal asymptote for the graph of $f(x) = \dfrac{4x - 1}{x + 3}$.

c. The number of televisions that a company can produce per week after t weeks of production is given by

$$N(t) = \frac{3000t^2 + 30{,}000t}{t^2 + 10t + 25}.$$

Using this model, the company will eventually be able to produce 30,000 televisions in a single week.

d. None of the given statements is true.

In Exercises 93–96, write the equation of a rational function $f(x) = \dfrac{p(x)}{q(x)}$ *having the indicated properties, in which the degrees of p and q are as small as possible. More than one correct function may be possible. Graph your function using a graphing utility to verify that it has the required properties.*

93. f has a vertical asymptote given by $x = 3$, a horizontal asymptote $y = 0$, y-intercept at -1, and no x-intercept.

94. f has vertical asymptotes given by $x = -2$ and $x = 2$, a horizontal asymptote $y = 2$, y-intercept at $\frac{9}{2}$, x-intercepts at -3 and 3, and y-axis symmetry.

95. f has a vertical asymptote given by $x = 1$, a slant asymptote whose equation is $y = x$, y-intercept at 2, and x-intercepts at -1 and 2.

96. f has no vertical, horizontal, or slant asymptotes, and no x-intercepts.

Group Exercise

97. Group members form the sales team for a company that makes computer video games. It has been determined that the rational function

$$f(x) = \frac{200x}{x^2 + 100}$$

models the monthly sales, $f(x)$, in thousands of games, of a new video game as a function of the number of months, x, after the game is introduced. The figure shows the graph of the function. What are the team's recommendations to the company in terms of how long the video game should be on the market before another new video game is introduced? What other factors might members want to take into account in terms of the recommendations? What

will eventually happen to sales, and how is this indicated by the graph? What does this have to do with a horizontal asymptote? What could the company do to change the behavior of this function and continue generating sales? Would this be cost effective?

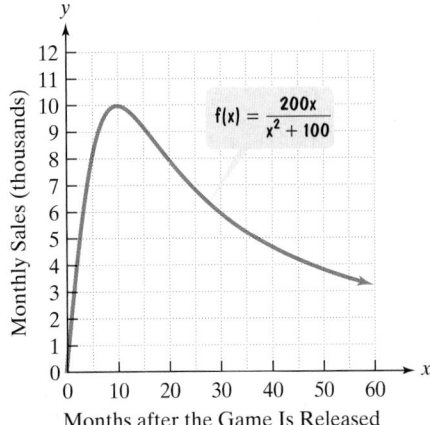

Monthly Sales of a New Video Game

SECTION 3.7 *Modeling Using Variation*

Objectives

1. Solve direct variation problems.
2. Solve inverse variation problems.
3. Solve combined variation problems.
4. Solve problems involving joint variation.

Have you ever wondered how telecommunication companies estimate the number of phone calls expected per day between two cities? The formula

$$N = \frac{400P_1 P_2}{d^2}$$

shows that the daily number of phone calls, N, increases as the populations of the cities, P_1 and P_2, in thousands, increase and decreases as the distance, d, between the cities increases.

Certain formulas occur so frequently in applied situations that they are given special names. Variation formulas show how one quantity changes in relation to other quantities. Quantities can vary *directly*, *inversely*, or *jointly*. In this section, we look at situations that can be modeled by each of these kinds of variation. And think of this: The next time you get one of those "all-circuits-are-busy" messages, you will be able to use a variation formula to estimate how many other callers you're competing with for those precious 8-cent minutes.

1 Solve direct variation problems.

Direct Variation

Because light travels faster than sound, during a thunderstorm we see lightning before we hear thunder. The formula

$$d = 1080t$$

describes the distance, in feet, of the storm's center if it takes t seconds to hear thunder after seeing lightning. Thus,

If $t = 1$,　$d = 1080 \cdot 1 = 1080$.　If it takes 1 second to hear thunder, the storm's center is 1080 feet away.

If $t = 2$,　$d = 1080 \cdot 2 = 2160$.　If it takes 2 seconds to hear thunder, the storm's center is 2160 feet away.

If $t = 3$,　$d = 1080 \cdot 3 = 3240$.　If it takes 3 seconds to hear thunder, the storm's center is 3240 feet away.

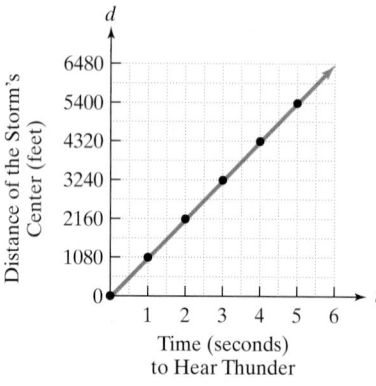

The graph of $d = 1080t$. Distance to a storm's center varies directly as the time it takes to hear thunder.

As the formula $d = 1080t$ illustrates, the distance to the storm's center is a constant multiple of how long it takes to hear the thunder. When the time is doubled, the storm's distance is doubled; when the time is tripled, the storm's distance is tripled; and so on. Because of this, the distance is said to *vary directly* as the time. The *equation of variation* is

$$d = 1080t.$$

Generalizing, we obtain the following statement:

Direct Variation

If a situation is described by an equation in the form

$$y = kx$$

where k is a nonzero constant, we say that **y varies directly as x or y is directly proportional to x.** The number k is called the **constant of variation** or the **constant of proportionality.**

EXAMPLE 1　Writing a Direct Variation Equation

A person's salary, S, varies directly as the number of hours worked, h.

a. Write an equation that expresses this relationship.
b. Margarita earns \$18 per hour. Substitute 18 for k, the constant of variation, in the equation in part (a) and write the equation for Margarita's salary.

Solution

a. We know that y varies directly as x is expressed as

$$y = kx.$$

By changing letters, we can write an equation that describes the following English statement: Salary, S, varies directly as the number of hours worked, h.

$$S = kh$$　We can also read this as "salary is directly proportional to hours worked."

b. Substituting 18 for k in $S = kh$, the direct variation equation, gives

$$S = 18h.$$

This equation describes Margarita's salary in terms of the number of hours she works. For example, if she works 10 hours, we can substitute 10 for h and determine her salary:

$$S = 18(10) = 180.$$

Her salary for working 10 hours is $180. Notice that, as the number of hours worked increases, the salary increases.

Check Point 1

A person's hair length, L, in inches, varies directly as the number of years it has been growing, N.

a. Write an equation that expresses this relationship.
b. The longest moustache on record was grown by Kalyan Sain of India. His moustache grew 4 inches each year. Substitute 4 for k, the constant of variation, in the equation in part (a) and write the equation for the length of Sain's moustache.
c. Sain grew his moustache for 17 years. Substitute 17 for N in the equation from part (b) and find its length.

In Example 1 and CheckPoint 1, the constants of variation, or proportionality, were given. If the constant of variation is not given, we can find it by substituting given values in the variation formula and solving for k. Example 2 shows how this is done.

EXAMPLE 2 Finding k, the Constant of Variation

Height, H, varies directly as foot length, F.

a. Write an equation that expresses this relationship.
b. Photographs of large footprints were published in 1951. Some speculated that these footprints were made by the Abominable Snowman. Each footprint was 23 inches long. The Abominable Snowman's height was determined to be 154.1 inches. (This is 12 feet, 10.1 inches, so it might not be a pleasant experience to run into this critter on a mellow hike through the woods!) Use $H = 154.1$ and $F = 23$ to find the constant of variation.

Solution

a. We know that y varies directly as x is expressed as

$$y = kx.$$

By changing letters, we can write an equation that describes the following English statement: Height, H, varies directly as foot length, F.

$$H = kF$$

Equivalently, height is directly proportional to foot length.

b. The Abominable Snowman's height is 154.1 inches, and foot length is 23 inches. Substitute 154.1 for H and 23 for F in the direct variation equation.

$$H = kF$$
$$154.1 = k \cdot 23$$

Solve for k, the constant of variation, by dividing both sides of the equation by 23:

$$\frac{154.1}{23} = \frac{k \cdot 23}{23}$$

$$6.7 = k.$$

Remember that 6.7 is also called the constant of proportionality.

Thus, the constant of variation is 6.7.

In Example 2, now that we know the constant of variation ($k = 6.7$), we can rewrite $H = kF$ using this constant. The equation of variation is

$$H = 6.7F.$$

We can use this equation to find other values. For example, if your foot length is 10 inches, your height is

$$H = 6.7(10) = 67,$$

or approximately 67 inches.

Check Point 2 The weight, W, of an aluminum canoe varies directly as its length, L.
  a. Write an equation that expresses this relationship.
  b. A 6-foot canoe weighs 75 pounds. Substitute 75 for W and 6 for L in the equation from part (a) and find k, the constant of variation.
  c. Substitute the value of k into your equation in part (a) and write the equation that describes the weight of this type of canoe in terms of its length.
  d. Use the equation from part (c) to find the weight of a 16-foot canoe of this type.

Our work up to this point provides a step-by-step procedure for solving variation problems. This procedure applies to direct variation problems as well as to the other kinds of variation problems that we will discuss.

Solving Variation Problems

1. Write an equation that describes the given English statement.
2. Substitute the given pair of values into the equation in step 1 and find the value of k.
3. Substitute the value of k into the equation in step 1.
4. Use the equation from step 3 to answer the problem's question.

EXAMPLE 3   Solving a Direct Variation Problem

The amount of garbage, G, varies directly as the population, P. Allegheny County, Pennsylvania, has a population of 1.3 million and creates 26 million pounds of garbage each week. Find the weekly garbage produced by New York City with a population of 7.3 million.

Solution

Step 1 Write an equation.   We know that y varies directly as x is expressed as

$$y = kx.$$

By changing letters, we can write an equation that describes the following English statement: Garbage production, G, varies directly as the population, P.

$$G = kP$$

> Equivalently, garbage production is directly proportional to the population.

Step 2 Use the given values to find k.   Allegheny County has a population of 1.3 million and creates 26 million pounds of garbage weekly. Substitute 26 for G and 1.3 for P in the direct variation equation. Then solve for k.

$G = kP$	This is the direct variation equation.
$26 = k \cdot 1.3$	$G = 26$ and $P = 1.3$.
$\dfrac{26}{1.3} = \dfrac{k \cdot 1.3}{1.3}$	Divide both sides by 1.3.
$20 = k$	Simplify.

Step 3 Substitute the value of k into the equation.

$G = kP$	Use the direct variation equation from step 1.
$G = 20P$	Replace k, the constant of variation, with 20.

Step 4 Answer the problem's question.   New York City has a population of 7.3 million. To find its weekly garbage production, substitute 7.3 for P in $G = 20P$ and solve for G.

$G = 20P$	Use the equation from step 3.
$G = 20(7.3)$	Substitute 7.3 for P.
$G = 146$	

The weekly garbage produced by New York City weighs approximately 146 million pounds.

Check Point 3   The pressure, P, of water on an object below the surface varies directly as its distance, D, below the surface. If a submarine experiences a pressure of 25 pounds per square inch 60 feet below the surface, how much pressure will it experience 330 feet below the surface?

The direct variation equation $y = kx$, or $f(x) = kx$, is a linear function. If $k > 0$, then the slope of the line is positive. Consequently, as x increases, y also increases.

A direct variation situation can involve variables to higher powers. For example, y can vary directly as x^2 $(y = kx^2)$ or as x^3 $(y = kx^3)$.

Direct Variation with Powers

y* varies directly as the *n*th power of *x if there exists some nonzero constant *k* such that

$$y = kx^n.$$

We also say that ***y* is directly proportional to the *n*th power of *x*.**

Direct variation with powers is modeled by polynomial functions. In our next example, the graph of the variation equation is the familiar parabola.

EXAMPLE 4 Solving a Direct Variation Problem

The distance, *s*, that a body falls from rest varies directly as the square of the time, *t*, of the fall. If skydivers fall 64 feet in 2 seconds, how far will they fall in 4.5 seconds?

Solution

Step 1 Write an equation. We know that *y* varies directly as the square of *x* is expressed as

$$y = kx^2.$$

By changing letters, we can write an equation that describes the following English statement: Distance, *s*, varies directly as the square of time, *t*, of the fall.

$$s = kt^2$$

Equivalently, distance is directly proportional to the square of time.

Step 2 Use the given values to find *k*. Skydivers fall 64 feet in 2 seconds. Substitute 64 for *s* and 2 for *t* in the direct variation equation. Then solve for *k*.

$$s = kt^2 \qquad \text{This is the direct variation equation.}$$
$$64 = k \cdot 2^2 \qquad s = 64 \text{ and } t = 2.$$
$$64 = 4k \qquad \text{Simplify.}$$
$$\frac{64}{4} = \frac{4k}{4} \qquad \text{Divide both sides by 4.}$$
$$16 = k \qquad \text{Simplify.}$$

Step 3 Substitute the value of *k* into the equation.

$$s = kt^2 \qquad \text{Use the direct variation equation from step 1.}$$
$$s = 16t^2 \qquad \text{Replace } k, \text{ the constant of variation, with 16.}$$

Step 4 Answer the problem's question. How far will the skydivers fall in 4.5 seconds? Substitute 4.5 for *t* in $s = 16t^2$ and solve for *s*.

$$s = 16(4.5)^2 = 16(20.25) = 324$$

Thus, in 4.5 seconds, skydivers will fall 324 feet.

We can express the variation equation from Example 4 in function notation, writing

$$s(t) = 16t^2.$$

The distance that a body falls from rest is a function of the time, *t*, of the fall. The parabola that is the graph of this quadratic function is shown in Figure 3.35. The graph increases rapidly from left to right, showing the effects of the acceleration of gravity.

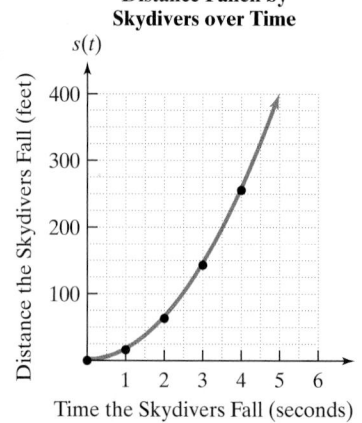

Distance Fallen by Skydivers over Time

Distance the Skydivers Fall (feet)

Time the Skydivers Fall (seconds)

Figure 3.35 The graph of $s(t) = 16t^2$

> **Check Point 4** The distance required to stop a car varies directly as the square of its speed. If 200 feet are required to stop a car traveling 60 miles per hour, how many feet are required to stop a car traveling 100 miles per hour?

2 Solve inverse variation problems.

Inverse Variation

The distance from Atlanta, Georgia, to Orlando, Florida, is 450 miles. The time that it takes to drive from Atlanta to Orlando depends on the rate at which one drives and is given by

$$\text{Time} = \frac{450}{\text{Rate}}.$$

For example, if you average 45 miles per hour, the time for the drive is

$$\text{Time} = \frac{450}{45} = 10,$$

or 10 hours. If you ignore speed limits and average 75 miles per hour, the time for the drive is

$$\text{Time} = \frac{450}{75} = 6,$$

or 6 hours. As your rate (or speed) increases, the time for the trip decreases and vice versa. This is illustrated in Figure 3.36.

We can express the time for the Atlanta–Orlando trip using t for time and r for rate:

$$t = \frac{450}{r}.$$

Figure 3.36

This equation is an example of an *inverse variation* equation. Time, t, *varies inversely* as rate, r. When two quantities vary inversely, and the constant of variation is positive, such as 450, one quantity increases as the other decreases, and vice versa.

Generalizing, we obtain the following statement:

Inverse Variation

If a situation is described by an equation in the form

$$y = \frac{k}{x}$$

where k is a nonzero constant, we say that y **varies inversely as x** or y is **inversely proportional to x.** The number k is called the **constant of variation.**

Notice that the inverse variation equation

$$y = \frac{k}{x}, \quad \text{or} \quad f(x) = \frac{k}{x},$$

is a rational function. For $k > 0$ and $x > 0$, the graph of the function takes on the shape shown in Figure 3.37.

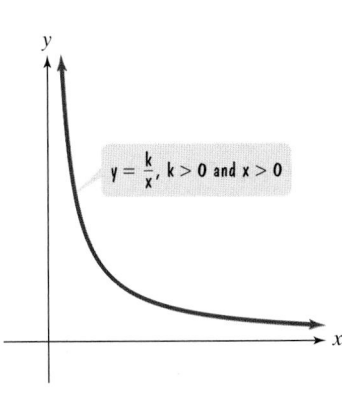

Figure 3.37 The graph of the inverse variation equation

We use the same procedure to solve inverse variation problems as we did to solve direct variation problems. Example 5 illustrates this procedure.

EXAMPLE 5   Solving an Inverse Variation Problem

When you use a spray can and press the valve at the top, you decrease the pressure of the gas in the can. This decrease of pressure causes the volume of the gas in the can to increase. Because the gas needs more room than is provided in the can, it expands in spray form through the small hole near the valve. In general, if the temperature is constant, the pressure, P, of a gas in a container varies inversely as the volume, V, of the container. The pressure of a gas sample in a container whose volume is 8 cubic inches is 12 pounds per square inch. If the sample expands to a volume of 22 cubic inches, what is the new pressure of the gas?

2P

P

2V

V

Doubling the pressure halves the volume.

Solution

Step 1   Write an equation. We know that y varies inversely as x is expressed as

$$y = \frac{k}{x}.$$

By changing letters, we can write an equation that describes the following English statement: The pressure, P, of a gas in a container varies inversely as the volume, V.

$$P = \frac{k}{V}$$

Equivalently, pressure is inversely proportional to volume.

Step 2   Use the given values to find k. The pressure of a gas sample in a container whose volume is 8 cubic inches is 12 pounds per square inch. Substitute 12 for P and 8 for V in the inverse variation equation. Then solve for k.

$$P = \frac{k}{V} \qquad \text{This is the inverse variation equation.}$$

$$12 = \frac{k}{8} \qquad P = 12 \text{ and } V = 8.$$

$$12 \cdot 8 = \frac{k}{8} \cdot 8 \qquad \text{Multiply both sides by 8.}$$

$$96 = k \qquad \text{Simplify.}$$

Step 3   Substitute the value of k into the equation.

$$P = \frac{k}{V} \qquad \text{Use the inverse variation equation from step 1.}$$

$$P = \frac{96}{V} \qquad \text{Replace } k, \text{ the constant of variation, with 96.}$$

Step 4   Answer the problem's question. We need to find the pressure when the volume expands to 22 cubic inches. Substitute 22 for V and solve for P.

$$P = \frac{96}{V} = \frac{96}{22} = 4\frac{4}{11}$$

Crude-Oil Price per Barrel

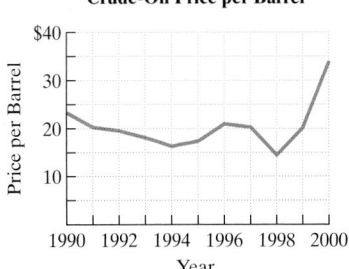

Year

Source: U.S. Department of Commerce

3 Solve combined variation problems

When the volume is 22 cubic inches, the pressure of the gas is $4\frac{4}{11}$ pounds per square inch.

> **Check Point 5** The price, P, of oil varies inversely as the supply, S. An OPEC nation sells oil for $19.50 per barrel when its daily production level is 4 million barrels. At what price will it sell oil if the daily production level is decreased to 3 million barrels?

Combined Variation

In a **combined variation** situation, direct and inverse variation occur at the same time. For example, as the advertising budget, A, of a company increases, its monthly sales, S, also increase. Monthly sales vary directly as the advertising budget:

$$S = kA.$$

By contrast, as the price of the company's product, P, increases, its monthly sales, S, decrease. Monthly sales vary inversely as the price of the product:

$$S = \frac{k}{P}.$$

We can combine these two variation equations into one combined equation:

$$S = \frac{kA}{P}.$$

The following example illustrates the application of combined variation.

EXAMPLE 6 Solving a Combined Variation Problem

The owners of Rollerblades Now determine that the monthly sales, S, of its skates vary directly as its advertising budget, A, and inversely as the price of the skates, P. When $60,000 is spent on advertising and the price of the skates is $40, the monthly sales are 12,000 pairs of rollerblades.

a. Write an equation of variation that describes this situation.

b. Determine monthly sales if the amount of the advertising budget is increased to $70,000.

Solution

a. Write an equation.

$$S = \frac{kA}{P}$$ Translate "sales vary directly as the advertising budget and inversely as the skates' price."

Use the given values to find k.

$$12,000 = \frac{k(60,000)}{40}$$ When $60,000 is spent on advertising ($A = 60,000$) and the price is $40 ($P = 40$), monthly sales are 12,000 units ($S = 12,000$).

$$12,000 = k \cdot 1500$$ Divide 60,000 by 40.

$$\frac{12,000}{1500} = \frac{k \cdot 1500}{1500}$$ Divide both sides of the equation by 1500.

$$8 = k$$ Simplify.

Using $k = 8$ and $S = \dfrac{kA}{P}$, the equation of variation that describes monthly sales is

$$S = \dfrac{8A}{P}.$$

b. The advertising budget is increased to $70,000, so $A = 70,000$. The skates' price is still $40, so $P = 40$.

$$S = \dfrac{8A}{P}$$ *This is the combined variation equation from part (a).*

$$S = \dfrac{8(70,000)}{40}$$ *Substitute 70,000 for A and 40 for P.*

$$S = 14,000$$ *Simplify.*

With a $70,000 advertising budget and $40 price, the company can expect to sell 14,000 pairs of rollerblades in a month (up from 12,000).

> **Check Point 6**
>
> The number of minutes needed to solve an exercise set of variation problems varies directly as the number of problems and inversely as the number of people working to solve the problems. It takes 4 people 32 minutes to solve 16 problems. How many minutes will it take 8 people to solve 24 problems?

4 Solve problems involving joint variation.

Joint Variation

Joint variation is a variation in which a variable varies directly as the product of two or more other variables. Thus, the equation $y = kxz$ is read "y varies jointly as x and z."

Joint variation plays a critical role in Isaac Newton's formula for gravitation:

$$F = G\dfrac{m_1 m_2}{d^2}.$$

The formula states that the force of gravitation, F, between two bodies varies jointly as the product of their masses, m_1 and m_2, and inversely as the square of the distance between them, d. (G is the gravitational constant.) The formula indicates that gravitational force exists between any two objects in the universe, increasing as the distance between the bodies decreases. One practical result is that the pull of the moon on the oceans is greater on the side of the Earth closer to the moon. This gravitational imbalance is what produces tides.

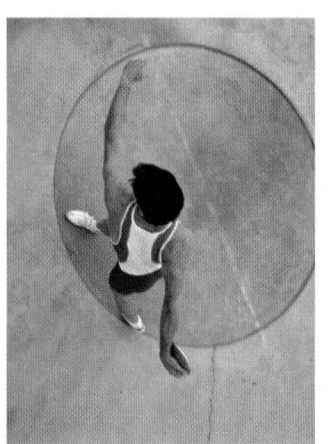

EXAMPLE 7   Modeling Centrifugal Force

The centrifugal force, C, of a body moving in a circle varies jointly with the radius of the circular path, r, and the body's mass, m, and inversely with the square of the time, t, it takes to move about one full circle. A 6-gram body moving in a circle with radius 100 centimeters at a rate of 1 revolution in 2 seconds has a centrifugal force of 6000 dynes. Find the centrifugal force of an 18-gram body moving in a circle with radius 100 centimeters at a rate of 1 revolution in 3 seconds.

Solution

$$C = \dfrac{krm}{t^2}$$ *Translate "Centrifugal force, C, varies jointly with radius, r, and mass, m, and inversely with the square of time, t."*

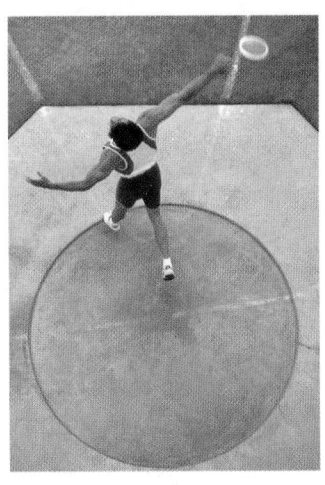

$$6000 = \frac{k(100)(6)}{2^2}$$ If $r = 100$, $m = 6$, and $t = 2$, then $C = 6000$.

$$40 = k$$ Solve for k.

$$C = \frac{40rm}{t^2}$$ Substitute 40 for k in the model for centrifugal force.

$$= \frac{40(100)(18)}{3^2}$$ Find C when $r = 100$, $m = 18$, and $t = 3$.

$$= 8000$$

The centrifugal force is 8000 dynes.

Check Point 7   The volume of a cone, V, varies jointly as its height, h, and the square of its radius, r. A cone with a radius measuring 6 feet and a height measuring 10 feet has a volume of 120π cubic feet. Find the volume of a cone having a radius of 12 feet and a height of 2 feet.

EXERCISE SET 3.7

Practice Exercises

In Exercises 1–12, write an equation that expresses each relationship. Use k as the constant of variation.

1. g varies directly as h.

2. v varies directly as r.

3. a is directly proportional to the square of b.

4. s is directly proportional to the cube of v.

5. r varies inversely as t.

6. w varies inversely as l.

7. a is inversely proportional to the cube of b.

8. y is inversely proportional to the square root of x.

9. r varies directly as s and inversely as v.

10. a varies directly as d and inversely as g.

11. s varies jointly as g and the square of t.

12. V varies jointly as h and the square of r.

In Exercises 13–22, determine the constant of variation for each stated condition.

13. y varies directly as x, and $y = 75$ when $x = 3$.

14. y varies directly as x, and $y = 55$ when $x = 11$.

15. y varies directly as x^2, and $y = 45$ when $x = 3$.

16. y varies directly as x^2, and $y = 72$ when $x = 6$.

17. W varies inversely as r, and $W = 500$ when $r = 10$.

18. T varies inversely as n, and $T = 7$ when $n = 12$.

19. A varies directly as B and inversely as C, and $A = 9$ when $B = 12$ and $C = 4$.

20. D varies directly as E and inversely as F, and $D = 6$ when $E = 12$ and $F = 10$.

21. a varies jointly as b and c, and $a = 72$ when $b = 18$ and $c = 2$.

22. z varies jointly as w and y, and $z = 38$ when $w = 38$ and $y = 2$.

Use the four-step procedure for solving variation problems given on page 356 to solve Exercises 23–30.

23. y varies directly as x. $y = 35$ when $x = 5$. Find y when $x = 12$.

24. y varies directly as x. $y = 55$ when $x = 5$. Find y when $x = 13$.

25. y varies inversely as x. $y = 10$ when $x = 5$. Find y when $x = 2$.

26. y varies inversely as x. $y = 5$ when $x = 3$. Find y when $x = 9$.

27. y is directly proportional to x and inversely proportional to the square of z. $y = 20$ when $x = 50$ and $z = 5$. Find y when $x = 3$ and $z = 6$.

28. a is directly proportional to b and inversely proportional to the square of c. $a = 7$ when $b = 9$ and $c = 6$. Find a when $b = 4$ and $c = 8$.

29. y varies jointly as x and z. $y = 25$ when $x = 2$ and $z = 5$. Find y when $x = 8$ and $z = 12$.

30. C varies jointly as A and T. $C = 175$ when $A = 2100$ and $T = 4$. Find C when $A = 2400$ and $T = 6$.

Application Exercises

31. A person's fingernail growth, G, in inches, varies directly as the number of weeks it has been growing, W.

   a. Write an equation that expresses this relationship.

   b. Fingernails grow at a rate of about 0.02 inch per week. Substitute 0.02 for k, the constant of variation, in the equation in part (a) and write the equation for fingernail growth.

   c. Substitute 52 for W to determine your fingernail length at the end of one year if for some bizarre reason you decided not to cut them and they did not break.

32. A person's wages, W, vary directly as the number of hours worked, h.

   a. Write an equation that expresses this relationship.

   b. For a 40-hour work week, Gloria earned $1400. Substitute 1400 for W and 40 for h in the equation from part (a) and find k, the constant of variation.

   c. Substitute the value of k into your equation in part (a) and write the equation that describes Gloria's wages in terms of the number of hours she works.

   d. Use the equation from part (c) to find Gloria's wages for 25 hours of work.

Use the four-step procedure for solving variation problems given on page 356 to solve Exercises 33–49.

33. The cost, C, of an airplane ticket varies directly as the number of miles, M, in the trip. A 3000-mile trip costs $400. What is the cost of a 450-mile trip?

34. An object's weight on the moon, M, varies directly as its weight on Earth, E. A person who weighs 55 kilograms on Earth weights 8.8 kilograms on the moon. What is the moon weight of a person who weighs 90 kilograms on Earth?

35. The Mach number is a measurement of speed named after the man who suggested it, Ernst Mach (1838–1916). The speed of an aircraft is directly proportional to its Mach number. Shown here are two aircraft. Use the figures for the Concorde to determine the Blackbird's speed.

Concorde
Mach 2.03
Speed = 1502.2 miles per hour

SR-71 Blackbird
Mach 3.3
Speed = ?

36. Do you still own records, or are you strictly a CD person? Record owners claim that the quality of sound on good vinyl surpasses that of a CD, although this is up for debate. This, however, is not debatable: The number of revolutions a record makes as it is being played is directly proportional to the time that it is on the turntable. A record that lasted 3 minutes made 135 revolutions. If a record takes 2.4 minutes to play, how many revolutions does it make?

37. If all men had identical body types, their weight would vary directly as the cube of their height. Shown is Robert Wadlow, who reached a record height of 8 feet 11 inches (107 inches) before his death at age 22. If a man who is 5 feet 10 inches tall (70 inches) with the same body type as Mr. Wadlow weighs 170 pounds, what was Robert Wadlow's weight shortly before his death?

38. The distance that an object falls varies directly as the square of the time it has been falling. An object falls 144 feet in 3 seconds. Find how far it will fall in 7 seconds.

39. The time that it takes you to get to campus varies inversely as your driving rate. Averaging 20 miles per hour in terrible traffic, it takes you 1.5 hours to get to campus. How long would the trip take averaging 60 miles per hour?

40. The weight that can be supported by a 2-inch by 4-inch piece of pine (called a 2-by-4) varies inversely as its length. A 10-foot 2-by-4 can support 500 pounds. What weight can be supported by a 5-foot 2-by-4?

41. The volume of a gas in a container at a constant temperature is inversely proportional to the pressure. If the volume is 32 cubic centimeters at a pressure of 8 pounds, find the pressure when the volume is 40 cubic centimeters.

42. The current in a circuit is inversely proportional to the resistance. The current is 20 amperes when the resistance is 5 ohms. Find the current for a resistance of 16 ohms.

43. A person's body-mass index is used to assess levels of fatness, with an index from 20 to 26 considered in the desirable range. The index varies directly as one's weight, in pounds, and inversely as one's height, in inches. A person who weighs 150 pounds and is 70 inches tall has an index of 21. What is the body-mass index of a person who weighs 240 pounds and is 74 inches tall? Because the index is rounded to the nearest whole number, do so and then determine if this person's level of fatness is in the desirable range.

44. The volume of a gas varies directly as its temperature and inversely as its pressure. At a temperature of 100 Kelvin and a pressure of 15 kilograms per square meter, the gas occupies a volume of 20 cubic meters. Find the volume at a temperature of 150 Kelvin and a pressure of 30 kilograms per square meter.

45. The intensity of illumination on a surface varies inversely as the square of the distance of the light source from the surface. The illumination from a source is 25 foot-candles at a distance of 4 feet. What is the illumination when the distance is 6 feet?

46. The gravitational force with which Earth attracts an object varies inversely with the square of the distance from the center of Earth. A gravitational force of 0.4 pound acts on an object 8000 miles from Earth's center. Find the force of attraction on an object 6000 miles from the center of Earth.

47. Kinetic energy varies jointly as the mass and the square of the velocity. A mass of 8 grams and velocity of 3 centimeters per second has a kinetic energy of 36 ergs. Find the kinetic energy for a mass of 4 grams and velocity of 6 centimeters per second.

48. The electrical resistance of a wire varies directly as its length and inversely as the square of its diameter. A wire of 720 feet with $\frac{1}{4}$-inch diameter has a resistance of $1\frac{1}{2}$ ohms. Find the resistance for 960 feet of the same kind of wire if its diameter is doubled.

49. The average number of phone calls between two cities in a day varies jointly as the product of their populations and inversely as the square of the distance between them. The population of Minneapolis is 2538 thousand and the population of Cincinnati is 1818 thousand. Separated by 108 miles, the average number of telephone calls per day between the two cities is 158,233. Find the average number of telephone calls per day between Orlando, Florida (population 1225 thousand) and Seattle, Washington (population 2970 thousand), two cities that are 3403 miles apart.

Writing in Mathematics

50. What does it mean if two quantities vary directly?

51. In your own words, explain how to solve a variation problem.

52. What does it mean if two quantities vary inversely?

53. Explain what is meant by combined variation. Give an example with your explanation.

54. Explain what is meant by joint variation. Give an example with your explanation.

In Exercises 55–56, describe in words the variation shown by the given equation.

55. $z = \dfrac{k\sqrt{x}}{y^2}$ **56.** $z = kx^2\sqrt{y}$

57. We have seen that the daily number of phone calls between two cities varies jointly as their populations and inversely as the square of the distance between them. This model, used by telecommunication companies to estimate the line capacities needed between various cities, is called the *gravity model*. Compare the model to Newton's formula for gravitation on page 362 and describe why the name *gravity model* is appropriate.

Technology Exercise

58. Use a graphing utility to graph any three of the variation equations in Exercises 33–42. Then TRACE along each curve and identify the point that corresponds to the problem's solution.

Critical Thinking Exercises

59. In a hurricane, the wind pressure varies directly as the square of the wind velocity. If wind pressure is a measure of a hurricane's destructive capacity, what happens to this destructive power when the wind speed doubles?

60. The illumination from a light source varies inversely as the square of the distance from the light source. If you raise a lamp from 15 inches to 30 inches over your desk, what happens to the illumination?

61. The heat generated by a stove element varies directly as the square of the voltage and inversely as the resistance. If the voltage remains constant, what needs to be done to triple the amount of heat generated?

62. Galileo's telescope brought about revolutionary changes in astronomy. A comparable leap in our ability to observe the universe took place as a result of the Hubble Space Telescope. The space telescope can see stars and galaxies whose brightness is $\frac{1}{50}$ of the faintest objects now observable using ground-based telescopes. Use the fact that the brightness of a point source, such as a star, varies inversely as the square of its distance from an observer to show that the space telescope can see about seven times farther than a ground-based telescope.

Group Exercise

63. Begin by deciding on a product that interests the group because you are now in charge of advertising this product. Members were told that the demand for the product varies directly as the amount spent on advertising and inversely as the price of the product. However, as more money is spent on advertising, the price of your product rises. Under what conditions would members recommend an increased expense in advertising? Once you've determined what your product is, write formulas for the given conditions and experiment with hypothetical numbers. What other factors might you take into consideration in terms of your recommendation? How do these factor affect the demand for your product?

CHAPTER SUMMARY, REVIEW, AND TEST

Summary

DEFINITIONS AND CONCEPTS	EXAMPLES

3.1 Quadratic Functions

a. A quadratic function is of the form $f(x) = ax^2 + bx + c, a \neq 0$.

b. The standard form of a quadratic function is $f(x) = a(x - h)^2 + k, a \neq 0$.

c. The graph of a quadratic function is a parabola. The vertex is (h, k) or $\left(-\dfrac{b}{2a}, f\left(-\dfrac{b}{2a}\right)\right)$.

   A procedure for graphing a quadratic function is given in the box on page 282.

Ex. 1, p. 282;
Ex. 2, p. 283;
Ex. 3, p. 285

d. See the box on page 286 for minimum or maximum values of quadratic functions.

Ex. 4, p. 287

3.2 Polynomial Functions and Their Graphs

a. Polynomial Function of x of Degree n: $f(x) = a_n x^n + a_{n-1} x^{n-1} + \cdots + a_2 x^2 + a_1 x + a_0$, $a_n \neq 0$

b. The graphs of polynomial functions are smooth and continuous.

c. The end behavior of the graph of a polynomial function depends on the leading term, given by the Leading Coefficient Test in the box on page 295.

Ex. 1&2, p. 295;
Ex. 3, p. 296

d. The values of x for which $f(x)$ is equal to 0 are the zeros of the polynomial function f. These values are the roots, or solutions, of the polynomial equation $f(x) = 0$.

Ex. 4&5, p. 297

e. If $x - r$ occurs k times in a polynomial function's factorization, r is a repeated zero with multiplicity k. If k is even, the graph touches the x-axis at r; if odd, it crosses the x-axis at r.

f. If f is a polynomial of degree n, the graph of f has at most $n - 1$ turning points.

g. A strategy for graphing a polynomial function is given in the box on page 299.

Ex. 6, p. 299

3.3 Dividing Polynomials; Remainder and Factor Theorems

a. Long division of polynomials is performed by dividing, multiplying, subtracting, bringing down the next term, and repeating this process until the degree of the remainder is less than the degree of the divisor. The details are given in the box on page 306.

Ex. 1, p. 305;
Ex. 2, p. 306;
Ex. 3, p. 308

b. The Division Algorithm: $f(x) = d(x)q(x) + r(x)$. The dividend is the product of the divisor and the quotient plus the remainder.

c. Synthetic division is used to divide a polynomial by $x - c$. The details are given in the box on pages 309–310.

Ex. 4, p. 310

d. The Remainder Theorem: If polynomial $f(x)$ is divided by $x - c$, then the remainder is $f(c)$.

Ex. 5, p. 312

e. The Factor Theorem: If $x - c$ is a factor of a polynomial function $f(x)$, c is a zero of f and a root of $f(x) = 0$. If c is a zero of f or a root of $f(x) = 0$, then $x - c$ is a factor of $f(x)$.

Ex. 6, p. 313

DEFINITIONS AND CONCEPTS	EXAMPLES

3.4 Zeros of Polynomial Functions

a. The Rational Zero Theorem states that possible rational zeros of a polynomial function $=$ $\dfrac{\text{Factors of the constant term}}{\text{Factors of the leading coefficient}}$. The theorem is stated in the box on page 316.

Ex. 1, p. 316;
Ex. 2, p. 317;
Ex. 3, p. 318;
Ex. 4, p. 319

b. Descartes's Rule of Signs: The number of positive real zeros of f equals the number of sign changes of $f(x)$ or is less than that number by an even integer. The number of negative real zeros of f applies a similar statement to $f(-x)$.

Ex. 5, p. 321

3.5 More on Zeros of Polynomial Functions

a. The Upper and Lower Bound Theorem: The number $b > 0$ is an upper bound for the real roots of $f(x) = 0$ if synthetic division of $f(x)$ by $x - b$ results in no negative numbers. The $a < 0$ is a lower bound if synthetic division by $x - a$ results in numbers that alternate in sign, number counting zero entries as positive or negative.

Ex. 1, p. 326

b. The Intermediate Value Theorem: If $f(a)$ and $f(b)$ have opposite signs, there is at least one value of c between a and b for which $f(c) = 0$.

Ex. 2, p. 327

c. Number of roots: If $f(x)$ is a polynomial of degree $n \geq 1$, then, counting multiple roots separately, the equation $f(x) = 0$ has n roots.

d. If $a + bi$ is a root of $f(x) = 0$, then $a - bi$ is also a root.

Ex. 3, p. 329

e. The Linear Factorization Theorem: An nth-degree polynomial can be expressed as the product of n linear factors. Thus, $f(x) = a_n(x - c_1)(x - c_2)\cdots(x - c_n)$.

Ex. 4, p. 331;
Ex. 5, p. 331

3.6 Rational Functions and Their Graphs

a. Rational function: $f(x) = \dfrac{p(x)}{q(x)}$; p and q are polynomial functions and $q(x) \neq 0$. The domain of f is the set of all real numbers excluding values of x that make $q(x)$ zero.

Ex. 1, p. 336

b. Arrow notation is summarized in the box on page 338.

c. The line $x = a$ is a vertical asymptote of the graph of f if $f(x)$ increases or decreases without bound as x approaches a. Vertical asymptotes are identified using the location theorem in the box on page 339.

Ex. 2, p. 339

d. The line $y = b$ is a horizontal asymptote of the graph of f if $f(x)$ approaches b as x increases or decreases without bound. Horizontal asymptotes are identified using the location theorem in the box on page 341.

Ex. 3, p. 341

e. A strategy for graphing rational functions is given in the box on page 342.

Ex. 4, p. 342;
Ex. 5, p. 344;
Ex. 6, p. 345

f. The graph of a rational function has a slant asymptote when the degree of the numerator is one more than the degree of the denominator. The equation of the slant asymptote is found using division and dropping the remainder term.

Ex. 7, p. 347

3.7 Modeling Using Variation

a. English Statement	Equation	
y varies directly as x. y is directly proportional to x.	$y = kx$	Ex. 3, p. 356
y varies directly as x^n. y is directly proportional to x^n.	$y = kx^n$	Ex. 4, p. 358

DEFINITIONS AND CONCEPTS		**EXAMPLES**

y varies inversely as x.
y is inversely proportional to x.

$$y = \frac{k}{x}$$

Ex. 5, p. 360
Ex. 6, p. 361

y varies inversely as x^n.
y is inversely proportional to x^n.

$$y = \frac{k}{x^n}$$

y varies jointly as x and z. $y = kxz$ Ex. 7, p. 362

b. A procedure for solving variation problems is given in the box on page 356.

Review Exercises

3.1

In Exercises 1–4, use the vertex and intercepts to sketch the graph of each quadratic function. Give the equation for the parabola's axis of symmetry.

1. $f(x) = -2(x - 1)^2 + 3$ **2.** $f(x) = (x + 4)^2 - 2$

3. $f(x) = -x^2 + 2x + 3$ **4.** $f(x) = 2x^2 - 4x - 6$

5. A person standing close to the edge on the top of an 80-foot building throws a ball vertically upward with an initial velocity of 64 feet per second. The function $s(t) = -16t^2 + 64t + 80$ describes the ball's height above the ground, $s(t)$, in feet, t seconds after it is thrown. After how many seconds does the ball reach its maximum height? What is the maximum height?

6. Suppose that a quadratic function is used to model the data shown in the graph using

(number of years after 1960, divorce rate per 1000 population).

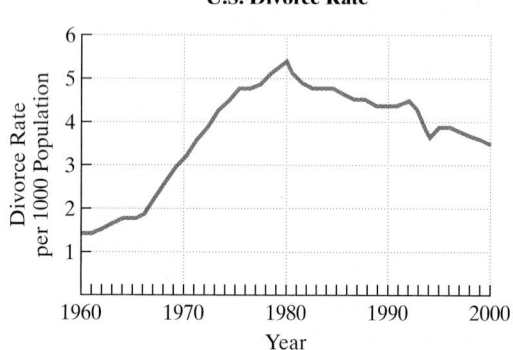

U.S. Divorce Rate

Source: U.S. Department of Health and Human Services

Determine, without obtaining an actual quadratic function that models the data, the approximate coordinates of the vertex for the function's graph. Describe what this means in practical terms. Use the word "maximum" in your description.

7. A field bordering a straight stream is to be enclosed. The side bordering the stream is not to be fenced. If 1000 yards of fencing material is to be used, what are the dimensions of the largest rectangular field that can be fenced? What is the maximum area?

3.2

In Exercises 8–11, use the Leading Coefficient Test to determine the end behavior of the graph of the given polynomial function. Then use this end behavior to match the polynomial function with its graph. [The graphs are labeled (a) through (d).]

8. $f(x) = -x^3 + 12x^2 - x$

9. $f(x) = x^6 - 6x^4 + 9x^2$

10. $f(x) = x^5 - 5x^3 + 4x$

11. $f(x) = -x^4 + 1$

a.

b.

c.   **d.**

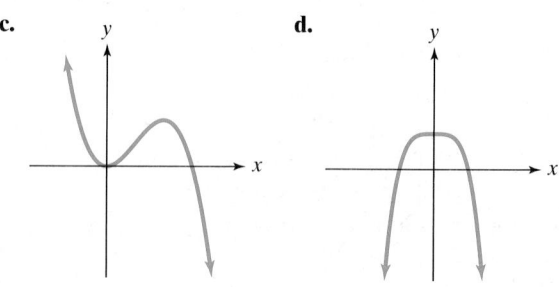

12. The function $f(x) = -0.0013x^3 + 0.78x^2 - 1.43x + 18.1$ models the percentage of U.S. families below the poverty level, $f(x)$, x years after 1960. Use end behavior to explain why the model is valid only for a limited period of time.

13. Despite a combination of drugs used to inhibit the growth of the HIV virus, a patient dies as a result of the virus overwhelming his body. Could the function $N(t) = -\frac{3}{4}t^4 + 3t^3 + 5$ model the number of viral particles, $N(t)$, in billions, in this patient's body over time? Use the graph's end behavior to the right to answer the question. Explain your answer.

In Exercises 14–15, find the zeros for each polynomial function and give the multiplicity of each zero. State whether the graph crosses or touches the x-axis at each zero.

14. $f(x) = -2(x-1)(x+2)^2(x+5)^3$

15. $f(x) = x^3 - 5x^2 - 25x + 125$

In Exercises 16–21,
   a. *Use the Leading Coefficient Test to determine the graph's end behavior.*
   b. *Determine whether the graph has y-axis symmetry, origin symmetry, or neither.*
   c. *Graph the function.*

16. $f(x) = x^3 - x^2 - 9x + 9$ **17.** $f(x) = 4x - x^3$

18. $f(x) = 2x^3 + 3x^2 - 8x - 12$ **19.** $f(x) = -x^4 + 25x^2$

20. $f(x) = -x^4 + 6x^3 - 9x^2$ **21.** $f(x) = 3x^4 - 15x^3$

3.3

In Exercises 22–24, divide using long division.

22. $(4x^3 - 3x^2 - 2x + 1) \div (x + 1)$

23. $(10x^3 - 26x^2 + 17x - 13) \div (5x - 3)$

24. $(4x^4 + 6x^3 + 3x - 1) \div (2x^2 + 1)$

In Exercises 25–26, divide using synthetic division.

25. $(3x^4 + 11x^3 - 20x^2 + 7x + 35) \div (x + 5)$

26. $(3x^4 - 2x^2 - 10x) \div (x - 2)$

27. Given $f(x) = 2x^3 - 7x^2 + 9x - 3$, use the Remainder Theorem to find $f(-13)$.

28. Use synthetic division to divide $f(x) = 2x^3 + x^2 - 13x + 6$ by $x - 2$. Use the result to find all zeros of f.

29. Solve the equation $x^3 - 17x + 4 = 0$ given that 4 is a root.

3.4

In Exercises 30–31, use the Rational Zero Theorem to list all possible rational zeros for each given function.

30. $f(x) = x^4 - 6x^3 + 14x^2 - 14x + 5$

31. $f(x) = 3x^5 - 2x^4 - 15x^3 + 10x^2 + 12x - 8$

In Exercises 32–33, use Descartes's Rule of Signs to determine the possible number of positive and negative real zeros for each given function.

32. $f(x) = 3x^4 - 2x^3 - 8x + 5$

33. $f(x) = 2x^5 - 3x^3 - 5x^2 + 3x - 1$

34. Use Descartes's Rule of Signs to explain why $2x^4 + 6x^2 + 8 = 0$ has no real roots.

For Exercises 35–40,
   a. *List all possible rational roots or rational zeros.*
   b. *Use Descartes's Rule of Signs to determine the possible number of positive and negative real roots or real zeros.*
   c. *Use synthetic division to test the possible rational roots or zeros and find an actual root or zero.*
   d. *Use the root or zero from part (c) to find all the zeros or roots.*

35. $f(x) = x^3 + 3x^2 - 4$

36. $f(x) = 6x^3 + x^2 - 4x + 1$

37. $8x^3 - 36x^2 + 46x - 15 = 0$

38. $x^4 - x^3 - 7x^2 + x + 6 = 0$

39. $4x^4 + 7x^2 - 2 = 0$

40. $f(x) = 2x^4 + x^3 - 9x^2 - 4x + 4$

3.5

41. Show that all real roots of the equation
$$2x^4 - 7x^3 - 5x^2 + 28x - 12 = 0$$
lie between -2 and 6. Use this result to list all possible rational roots.

42. Consider the equation $2x^4 - x^3 - 5x^2 + 10x + 12 = 0$.
   a. List all possible rational roots.
   b. Determine whether 2 is a root using synthetic division. In terms of bounds, what can you conclude?
   c. Determine whether -2 is a root using synthetic division. In terms of bounds, what can you conclude?
   d. Use the results of parts (b) and (c) to discard some of the possible rational roots from part (a). Now what are the possible rational roots?

In Exercises 43–44, show that the polynomial has a zero between the given integers. Then use the Intermediate Value Theorem to find an approximation for this zero to the nearest tenth. If applicable, use a graphing utility's zero feature to verify your answer.

43. $f(x) = x^3 - 2x - 1$; between 1 and 2

44. $f(x) = 3x^3 + 2x^2 - 8x + 7$; between -3 and -2

In Exercises 45–47, use the given root to find the solution set of the polynomial equation.

45. $4x^3 - 47x^2 + 232x + 61 = 0$; $6 + 5i$

46. $x^4 - 4x^3 + 16x^2 - 24x + 20 = 0$; $1 - 3i$

47. $2x^4 - 17x^3 + 137x^2 - 57x - 65 = 0$; $4 + 7i$

In Exercises 48–50, find an nth-degree polynomial function with real coefficients satisfying the given conditions. If you are using a graphing utility, graph the function and verify the real zeros and the given function value.

48. $n = 3$; 2 and $2 - 3i$ are zeros; $f(1) = -10$

49. $n = 4$; i is a zero; -3 is a zero of multiplicity 2; $f(-1) = 16$

50. $n = 4$; $-2, 3$, and $1 + 3i$ are zeros; $f(2) = -40$

In Exercises 51–52, find all the zeros of each polynomial function and write the polynomial as a product of linear factors.

51. $f(x) = 2x^4 + 3x^3 + 3x - 2$

52. $g(x) = x^4 - 6x^3 + x^2 + 24x + 16$

In Exercises 53–56, graphs of fifth-degree polynomial functions are shown. In each case, specify the number of real zeros and the number of imaginary zeros. Indicate whether there are any real zeros with multiplicity other than 1.

53.   **54.**

55.   **56.**

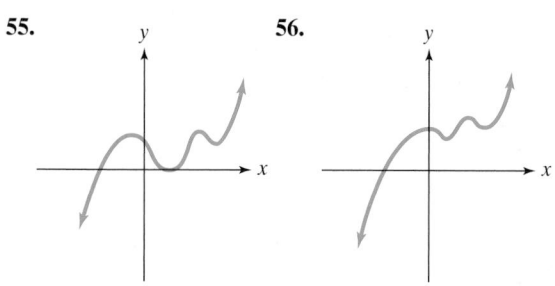

3.6

In Exercises 57–64, find the vertical asymptotes, if any, the horizontal asymptote, if there is one, and the slant asymptote, if there is one, of the graph of each rational function. Then graph the rational function.

57. $f(x) = \dfrac{2x}{x^2 - 9}$

58. $g(x) = \dfrac{2x - 4}{x + 3}$

59. $h(x) = \dfrac{x^2 - 3x - 4}{x^2 - x - 6}$

60. $r(x) = \dfrac{x^2 + 4x + 3}{(x + 2)^2}$

61. $y = \dfrac{x^2}{x + 1}$

62. $y = \dfrac{x^2 + 2x - 3}{x - 3}$

63. $f(x) = \dfrac{-2x^3}{x^2 + 1}$

64. $g(x) = \dfrac{4x^2 - 16x + 16}{2x - 3}$

65. A company is planning to manufacture affordable graphing calculators. Fixed monthly cost will be $50,000, and it will cost $25 to produce each calculator.

   a. Write the cost function, C, of producing x graphing calculators.

   b. Write the average cost function, $\bar{C}$, of producing x graphing calculators.

   c. Find and interpret $\bar{C}(50)$, $\bar{C}(100)$, $\bar{C}(1000)$, and $\bar{C}(100,000)$.

   d. What is the horizontal asymptote for this function, and what does it represent?

66. In Palo Alto, California, a government agency ordered computer-related companies to contribute to a monetary pool to clean up underground water supplies. (The companies had stored toxic chemicals in leaking underground containers.) The rational function

$$C(x) = \frac{200x}{100 - x}$$

models the cost, $C(x)$, in tens of thousands of dollars, for removing x percent of the contaminant's.

   a. Find and interpret $C(90) - C(50)$.

   b. What is the equation for the vertical asymptote? What does this mean in terms of the variables given by the function?

Exercises 67–68 involve rational functions that model the given situations. In each case, find the horizontal asymptote as $x \to \infty$ and then describe what this means in practical terms.

67. $f(x) = \dfrac{150x + 120}{0.05x + 1}$; the number of bass, $f(x)$, after x months in a lake that was stocked with 120 bass

68. $P(x) = \dfrac{72,900}{100x^2 + 729}$; the percentage, $P(x)$, of people in the United States with x years of education who are unemployed

69. The function $p(x) = 1.96x + 3.14$ models the number of nonviolent prisoners, $p(x)$, in thousands, in New York State prisons x years after 1980. The function $q(x) = 3.04x + 21.79$ models the total number of prisoners, $q(x)$, in thousands, in New York State prisons x years after 1980.

   a. Write a function that models the fraction of nonviolent prisoners in New York prisons x years after 1980.

   b. What is the equation of the horizontal asymptote associated with the function in part (a)? Describe what this means about the percentage, to the nearest tenth of a percent, of nonviolent prisoners in New York prisons over time.

   c. Use your equation in part (b) to explain why, in 1998, New York implemented a strategy where more nonviolent offenders are granted parole and more violent offenders are denied parole.

3.7

Solve the variation problems in Exercises 70–75.

70. An electric bill varies directly as the amount of electricity used. The bill for 1400 kilowatts of electricity is $98. What is the bill for 2200 kilowatts of electricity?

71. The distance that a body falls from rest is directly proportional to the square of the time of the fall. If skydivers fall 144 feet in 3 seconds, how far will they fall in 10 seconds?

72. The time it takes to drive a certain distance is inversely proportional to the rate of travel. If it takes 4 hours at 50 miles per hour to drive the distance, how long will it take at 40 miles per hour?

73. The loudness of a stereo speaker, measured in decibels, varies inversely as the square of your distance from the speaker. When you are 8 feet from the speaker, the loudness is 28 decibels. What is the loudness when you are 4 feet from the speaker?

74. The time required to assemble computers varies directly as the number of computers assembled and inversely as the number of workers. If 30 computers can be assembled by 6 workers in 10 hours, how long would it take 5 workers to assemble 40 computers?

75. The volume of a pyramid varies jointly as its height and the area of its base. A pyramid with a height of 15 feet and a base with an area of 35 square feet has a volume of 175 cubic feet. Find the volume of a pyramid with a height of 20 feet and a base with an area of 120 square feet.

Chapter 3 Test

In Exercises 1–2, use the vertex and intercepts to sketch the graph of each quadratic function. Give the equation for the parabola's axis of symmetry.

1. $f(x) = (x + 1)^2 + 4$ **2.** $f(x) = x^2 - 2x - 3$

3. Determine, without graphing, whether the quadratic function $f(x) = -2x^2 + 12x - 16$ has a minimum value or a maximum value. Then find the coordinates of the minimum or the maximum point.

4. The function $f(x) = -x^2 + 46x - 360$ models the daily profit, $f(x)$, in hundreds of dollars, for a company that manufactures x computers daily. How many computers should be manufactured each day to maximize profit? What is the maximum daily profit?

5. Consider the function $f(x) = x^3 - 5x^2 - 4x + 20$.

   a. Use factoring to find all zeros of f.

   b. Use the Leading Coefficient Test and the zeros of f to graph the function.

6. Use end behavior to explain why the graph cannot be the graph of $f(x) = x^5 - x$. Then use intercepts to explain why the graph cannot represent $f(x) = x^5 - x$.

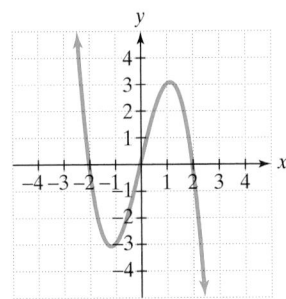

7. The graph of $f(x) = 6x^3 - 19x^2 + 16x - 4$ is shown in the figure at the top of the next column.

   a. Based on the graph of f, find the root of the equation $6x^3 - 19x^2 + 16x - 4 = 0$ that is an integer.

   b. Use synthetic division to find the other two roots of $6x^3 - 19x^2 + 16x - 4 = 0$.

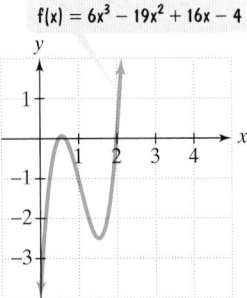

$f(x) = 6x^3 - 19x^2 + 16x - 4$

8. Use the Rational Zero Theorem to list all possible rational zeros of $f(x) = 2x^3 + 11x^2 - 7x - 6$.

9. Use Descartes's Rule of Signs to determine the possible number of positive and negative real zeros of $f(x) = 3x^5 - 2x^4 - 2x^2 + x - 1$.

10. Solve: $x^3 + 6x^2 - x - 30 = 0$.

11. Consider the function whose equation is given by $f(x) = 2x^4 - x^3 - 13x^2 + 5x + 15$.

   a. List all possible rational zeros.

   b. Use the graph of f in the figure shown and synthetic division to find all zeros of the function.

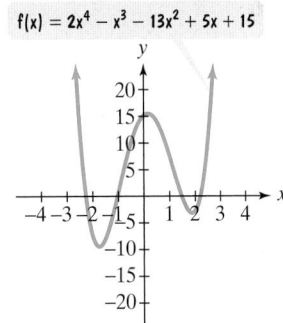

$f(x) = 2x^4 - x^3 - 13x^2 + 5x + 15$

12. Use the graph of $f(x) = 3x^4 + 4x^3 - 7x^2 - 2x - 3$ in the figure shown on the next page to find the smallest positive integer that is an upper bound and the largest negative integer that is a lower bound for the real roots of

$$3x^4 + 4x^3 - 7x^2 - 2x - 3 = 0.$$

Then use synthetic division to show that all the real roots of the equation lie between these integers.

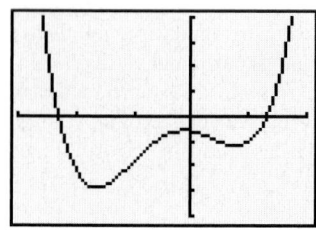

[−3, 2, 1] by [−4, 4, 1]

13. Solve $x^4 - 7x^3 + 18x^2 - 22x + 12 = 0$ given that $1 - i$ is a root.

14. Use the graph of $f(x) = x^3 + 3x^2 - 4$ in the figure shown to factor $x^3 + 3x^2 - 4$.

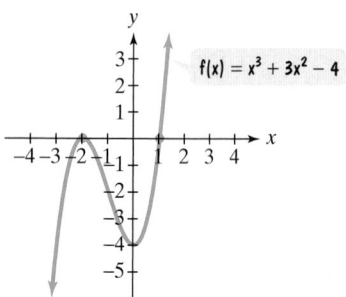

In Exercises 15–18, find the domain of each rational function and graph the function.

15. $f(x) = \dfrac{x}{x^2 - 16}$

16. $f(x) = \dfrac{x^2 - 9}{x - 2}$

17. $f(x) = \dfrac{x + 1}{x^2 + 2x - 3}$

18. $f(x) = \dfrac{4x^2}{x^2 + 3}$

19. Rational functions can be used to model learning. Many of these functions model the proportion of correct responses as a function of the number of trials of a particular task. One such model, called a learning curve, is

$$f(x) = \frac{0.9x - 0.4}{0.9x + 0.1},$$

where $f(x)$ is the proportion of correct responses after x trials. If $f(x) = 0$, there are no correct responses. If $f(x) = 1$, all responses are correct. The graph of the rational function is shown.

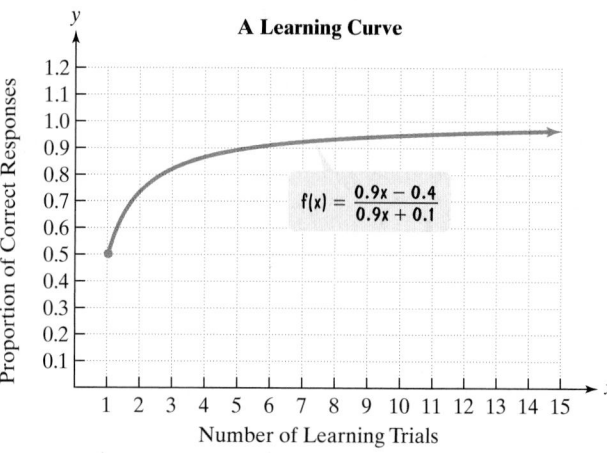

a. According to the graph, what proportion of responses are correct after 5 learning trials?

b. According to the graph, how many learning trials are necessary for 0.95 of the responses to be correct?

c. Use the function's equation to write the equation of the horizontal asymptote. What does this mean in terms of the variables modeled by the learning curve?

20. The intensity of light received at a source varies inversely as the square of the distance from the source. A particular light has an intensity of 20 foot-candles at 15 feet. What is the light's intensity at 10 feet?

Cumulative Review Exercises (Chapters P–3)

Simplify each expression in Exercises 1–3.

1. $\dfrac{1}{2 - \sqrt{3}}$

2. $3\left(x^2 - 3x + 1\right) - 2\left(3x^2 + x - 4\right)$

3. $3\sqrt{8} + 5\sqrt{50} - 4\sqrt{32}$

4. Factor completely: $x^7 - x^5$.

Solve each equation in Exercises 5–8.

5. $|2x - 1| = 3$

6. $3x^2 - 5x + 1 = 0$

7. $9 + \dfrac{3}{x} = \dfrac{2}{x^2}$

8. $x^3 + 2x^2 - 5x - 6 = 0$

Solve each inequality in Exercises 9–10. Express the answer in interval notation.

9. $|2x - 5| > 3$

10. $3x^2 > 2x + 5$

11. Give the center and radius. Then graph the equation:
$x^2 + y^2 - 2x + 4y - 4 = 0$.

12. Solve for t: $V = C(1 - t)$.

13. If $f(x) = \sqrt{45 - 9x}$, find the domain of f.

If $f(x) = x^2 + 2x - 5$ and $g(x) = 4x - 1$, find each function or function value in Exercises 14–16.

14. $(f - g)(x)$

15. $(f \circ g)(x)$

16. $g(f(-3))$

17. Consider the function $f(x) = x^3 - 4x^2 - x + 4$.

    a. Use factoring to find all zeros of f.

    b. Use the Leading Coefficient Test and the zeros of f to graph the function.

Graph each function in Exercises 18–20.

18. $f(x) = x^2 + 2x - 8$

19. $f(x) = x^2(x - 3)$

20. $f(x) = \dfrac{x - 1}{x - 2}$

Exponential and Logarithmic Functions

Chapter 4

What went wrong on the space shuttle *Challenger*? Will population growth lead to a future without comfort or individual choice? Can I put aside a small amount of money and have millions for early retirement? Why did I feel I was walking too slowly on my visit to New York City? Why are people in California at far greater risk from drunk drivers than from earthquakes? What is the difference between earthquakes measuring 6 and 7 on the Richter scale? And what can I hope to accomplish in weightlifting?

The functions that you will be learning about in this chapter will provide you with the mathematics for answering these questions. You will see how these remarkable functions enable us to predict the future and rediscover the past.

You've recently taken up weightlifting, recording the maximum number of pounds you can lift at the end of each week. At first your weight limit increases rapidly, but now you notice that this growth is beginning to level off. You wonder about a function that would serve as a mathematical model to predict the number of pounds you can lift as you continue the sport.

SECTION 4.1 *Exponential Functions*

Objectives

1. Evaluate exponential functions.
2. Graph exponential functions.
3. Evaluate functions with base e.
4. Use compound interest formulas.

The space shuttle *Challenger* exploded approximately 73 seconds into flight on January 28, 1986. The tragedy involved damage to O-rings, which were used to seal the connections between different sections of the shuttle engines. The number of O-rings damaged increases dramatically as temperature falls.

The function

$$f(x) = 13.49(0.967)^x - 1$$

models the number of O-rings expected to fail when the temperature is $x°$F. Can you see how this function is different from polynomial functions? The variable x is in the exponent. Functions whose equations contain a variable in the exponent are called **exponential functions.** Many real-life situations, including population growth, growth of epidemics, radioactive decay, and other changes that involve rapid increase or decrease, can be described using exponential functions.

Definition of the Exponential Function

The **exponential function f with base b** is defined by

$$f(x) = b^x \quad \text{or} \quad y = b^x$$

where b is a positive constant other than $1 \, (b > 0$ and $b \neq 1)$ and x is any real number.

Here are some examples of exponential functions:

$$f(x) = 2^x \qquad g(x) = 10^x \qquad h(x) = 3^{x+1}.$$

Base is 2.　　　　Base is 10.　　　　Base is 3.

Each of these functions has a constant base and a variable exponent. By contrast, the following functions are not exponential:

$$F(x) = x^2 \qquad G(x) = 1^x \qquad H(x) = x^x.$$

Variable is the base and not the exponent.　　The base of an exponential function must be a positive constant other than 1.　　Variable is both the base and the exponent.

Why is $G(x) = 1^x$ not classified as an exponential function? The number 1 raised to any power is 1. Thus, the function G can be written as $G(x) = 1$, which is a constant function.

1 Evaluate exponential functions.

You will need a calculator to evaluate exponential expressions. Most scientific calculators have a $\boxed{y^x}$ key. Graphing calculators have a $\boxed{\wedge}$ key. To evaluate expressions of the form b^x, enter the base b, press $\boxed{y^x}$ or $\boxed{\wedge}$, enter the exponent x, and finally press $\boxed{=}$ or $\boxed{\text{ENTER}}$.

EXAMPLE 1 Evaluating an Exponential Function

The exponential function $f(x) = 13.49(0.967)^x - 1$ describes the number of O-rings expected to fail, $f(x)$, when the temperature is $x°$F. On the morning the *Challenger* was launched, the temperature was $31°$F, colder than any previous experience. Find the number of O-rings expected to fail at this temperature.

Solution   Because the temperature was $31°$F, substitute 31 for x and evaluate the function.

$$f(x) = 13.49(0.967)^x - 1 \quad \text{This is the given function.}$$
$$f(31) = 13.49(0.967)^{31} - 1 \quad \text{Substitute 31 for x.}$$

Use a scientific or graphing calculator to evaluate $f(31)$. Press the following keys on your calculator to do this:

Scientific calculator: $13.49 \boxed{\times} .967 \boxed{y^x} 31 \boxed{-} 1 \boxed{=}$

Graphing calculator: $13.49 \boxed{\times} .967 \boxed{\wedge} 31 \boxed{-} 1 \boxed{\text{ENTER}}$.

The display should be approximately 3.7668627.

$$f(31) = 13.49(0.967)^{31} - 1 \approx 3.8 \approx 4$$

Thus, four O-rings are expected to fail at a temperature of $31°$F.

Check Point 1   Use the function in Example 1 to find the number of O-rings expected to fail at a temperature of $60°$F. Round to the nearest whole number.

2 Graph exponential functions.

Graphing Exponential Functions

We are familiar with expressions involving b^x, where x is a rational number. For example,

$$b^{1.7} = b^{17/10} = \sqrt[10]{b^{17}} \quad \text{and} \quad b^{1.73} = b^{173/100} = \sqrt[100]{b^{173}}.$$

However, note that the definition of $f(x) = b^x$ includes all real numbers for the domain x. You may wonder what b^x means when x is an irrational number, such as $b^{\sqrt{3}}$ or b^{π}. Using closer and closer approximations for $\sqrt{3}$ ($\sqrt{3} \approx 1.73205$), we can think of $b^{\sqrt{3}}$ as the value that has the successively closer approximations

$$b^{1.7}, b^{1.73}, b^{1.732}, b^{1.73205}, \ldots .$$

In this way, we can graph the exponential function with no holes, or points of discontinuity, at the irrational domain values.

EXAMPLE 2   Graphing an Exponential Function

Graph: $f(x) = 2^x$.

Solution   We begin by setting up a table of coordinates.

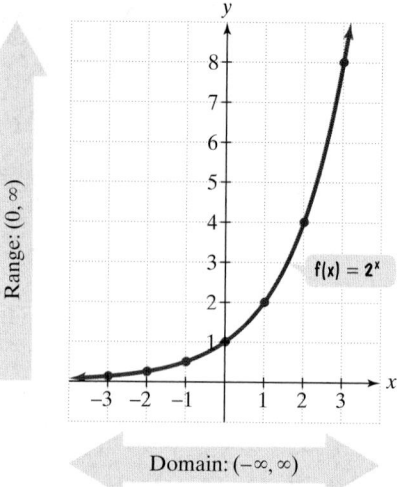

x	$f(x) = 2^x$
-3	$f(-3) = 2^{-3} = \frac{1}{8}$
-2	$f(-2) = 2^{-2} = \frac{1}{4}$
-1	$f(-1) = 2^{-1} = \frac{1}{2}$
0	$f(0) = 2^0 = 1$
1	$f(1) = 2^1 = 2$
2	$f(2) = 2^2 = 4$
3	$f(3) = 2^3 = 8$

Figure 4.1 The graph of $f(x) = 2^x$

We plot these points, connecting them with a continuous curve. Figure 4.1 shows the graph of $f(x) = 2^x$. Observe that the graph approaches, but never touches, the negative portion of the x-axis. Thus, the x-axis is a horizontal asymptote. The range is the set of all positive real numbers. Although we used integers for x in our table of coordinates, you can use a calculator to find additional points. For example, $f(0.3) = 2^{0.3} \approx 1.231$, $f(0.95) = 2^{0.95} \approx 1.932$. The points $(0.3, 1.231)$ and $(0.95, 1.932)$ approximately fit the graph.

Check Point 2   Graph: $f(x) = 3^x$.

Four exponential functions have been graphed in Figure 4.2. Compare the black and green graphs, where $b > 1$, to those in blue and red, where $b < 1$. When $b > 1$, the value of y increases as the value of x increases. When $b < 1$, the value of y decreases as the value of x increases. Notice that all four graphs pass through $(0, 1)$.

Study Tip

The graph of $y = \left(\frac{1}{2}\right)^x$, meaning $y = 2^{-x}$, is the graph of $y = 2^x$ reflected about the y-axis.

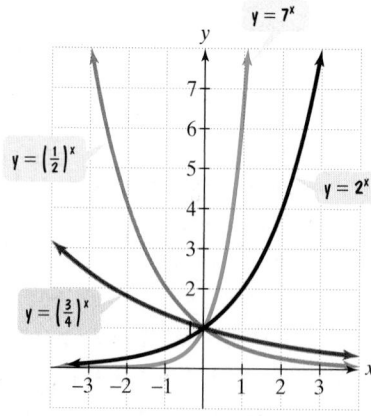

Figure 4.2 Graphs of four exponential functions

The graphs on the previous page illustrate the following general characteristics of exponential functions:

Characteristics of Exponential Functions of the Form $f(x) = b^x$

1. The domain of $f(x) = b^x$ consists of all real numbers. The range of $f(x) = b^x$ consists of all positive real numbers.
2. The graphs of all exponential functions of the form $f(x) = b^x$ pass through the point $(0, 1)$ because $f(0) = b^0 = 1 (b \neq 0)$. The y-intercept is 1.
3. If $b > 1$, $f(x) = b^x$ has a graph that goes up to the right and is an increasing function. The greater the value of b, the steeper the increase.
4. If $0 < b < 1$, $f(x) = b^x$ has a graph that goes down to the right and is a decreasing function. The smaller the value of b, the steeper the decrease.
5. $f(x) = b^x$ is one-to-one and has an inverse that is a function.
6. The graph of $f(x) = b^x$ approaches, but does not cross, the x-axis. The x-axis is a horizontal asymptote.

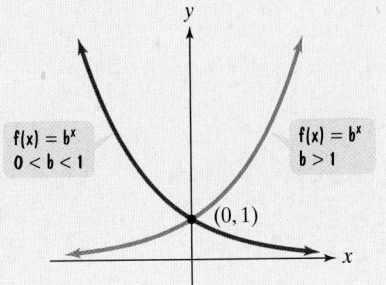

Transformations of Exponential Functions The graphs of exponential functions can be translated vertically or horizontally, reflected, stretched, or shrunk. We use the ideas of Section 2.5 to do so, as summarized in Table 4.1.

Table 4.1 Transformations Involving Exponential Functions
In each case, c represents a positive real number.

Transformation	Equation	Description
Vertical translation	$g(x) = b^x + c$	• Shifts the graph of $f(x) = b^x$ upward c units.
	$g(x) = b^x - c$	• Shifts the graph of $f(x) = b^x$ downward c units.
Horizontal translation	$g(x) = b^{x+c}$	• Shifts the graph of $f(x) = b^x$ to the left c units.
	$g(x) = b^{x-c}$	• Shifts the graph of $f(x) = b^x$ to the right c units.
Reflecting	$g(x) = -b^x$	• Reflects the graph of $f(x) = b^x$ about the x-axis.
	$g(x) = b^{-x}$	• Reflects the graph of $f(x) = b^x$ about the y-axis.
Vertical stretching or shrinking	$g(x) = cb^x$	• Stretches the graph of $f(x) = b^x$ if $c > 1$. • Shrinks the graph of $f(x) = b^x$ if $0 < c < 1$.

Using the information in Table 4.1 and a table of coordinates, you will obtain relatively accurate graphs that can be verified using a graphing utility.

EXAMPLE 3   Transformations Involving Exponential Functions

Use the graph of $f(x) = 3^x$ to obtain the graph of $g(x) = 3^{x+1}$.

Solution Examine Table 4.1. Note that the function $g(x) = 3^{x+1}$ has the general form $g(x) = b^{x+c}$, where $c = 1$. Thus, we graph $g(x) = 3^{x+1}$ by shifting the graph of $f(x) = 3^x$ *one* unit to the *left*. We construct a table showing some of the coordinates for f and g, selecting integers from -2 to 2 for x. The graphs of f and g are shown in Figure 4.3.

x	$f(x) = 3^x$	$g(x) = 3^{x+1}$
-2	$f(-2) = 3^{-2} = \frac{1}{9}$	$g(-2) = 3^{-2+1} = 3^{-1} = \frac{1}{3}$
-1	$f(-1) = 3^{-1} = \frac{1}{3}$	$g(-1) = 3^{-1+1} = 3^0 = 1$
0	$f(0) = 3^0 = 1$	$g(0) = 3^{0+1} = 3^1 = 3$
1	$f(1) = 3^1 = 3$	$g(1) = 3^{1+1} = 3^2 = 9$
2	$f(2) = 3^2 = 9$	$g(2) = 3^{2+1} = 3^3 = 27$

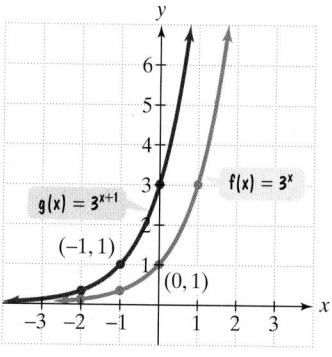

Figure 4.3 The graph of $g(x) = 3^{x+1}$ is the graph of $f(x) = 3^x$ shifted one unit to the left.

Check Point 3   Use the graph of $f(x) = 3^x$ to obtain the graph of $g(x) = 3^{x-1}$.

If an exponential function is translated upward or downward, the horizontal asymptote is shifted by the amount of the vertical shift.

EXAMPLE 4   Transformations Involving Exponential Functions

Use the graph of $f(x) = 2^x$ to obtain the graph of $g(x) = 2^x - 3$.

Solution The function $g(x) = 2^x - 3$ has the general form $g(x) = b^x - c$, where $c = 3$. Thus, we graph $g(x) = 2^x - 3$ by shifting the graph of $f(x) = 2^x$ *down three* units. We construct a table showing some of the coordinates for f and g, selecting integers from -2 to 2 for x.

The graphs of f and g are shown in Figure 4.4. Notice that the horizontal asymptote for f, the x-axis, is shifted down three units for the horizontal asymptote for g. As a result, $y = -3$ is the horizontal asymptote for g.

x	$f(x) = 2^x$	$g(x) = 2^x - 3$
-2	$f(-2) = 2^{-2} = \frac{1}{4}$	$g(-2) = 2^{-2} - 3 = \frac{1}{4} - 3 = -2\frac{3}{4}$
-1	$f(-1) = 2^{-1} = \frac{1}{2}$	$g(-1) = 2^{-1} - 3 = \frac{1}{2} - 3 = -2\frac{1}{2}$
0	$f(0) = 2^0 = 1$	$g(0) = 2^0 - 3 = 1 - 3 = -2$
1	$f(1) = 2^1 = 2$	$g(1) = 2^1 - 3 = 2 - 3 = -1$
2	$f(2) = 2^2 = 4$	$g(2) = 2^2 - 3 = 4 - 3 = 1$

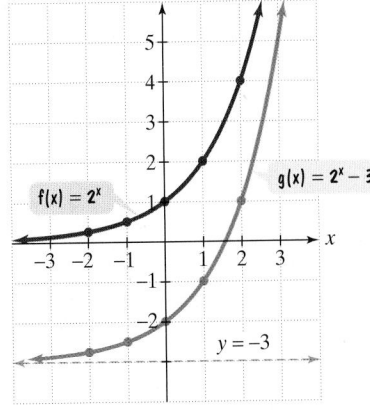

Figure 4.4 The graph of $g(x) = 2^x - 3$ is the graph of $f(x) = 2^x$ shifted down three units.

Check Point 4 Use the graph of $f(x) = 2^x$ to obtain the graph of $g(x) = 2^x + 1$.

3 Evaluate functions with base e.

The Natural Base e

An irrational number, symbolized by the letter e, appears as the base in many applied exponential functions. This irrational number is approximately equal to 2.72. More accurately,

$$e \approx 2.71828\ldots.$$

The number e is called the **natural base.** The function $f(x) = e^x$ is called the **natural exponential function.**

Use a scientific or graphing calculator with an $\boxed{e^x}$ key to evaluate e to various powers. For example, to find e^2, press the following keys on most calculators:

Scientific calculator: 2 $\boxed{e^x}$

Graphing calculator: $\boxed{e^x}$ 2 $\boxed{\text{ENTER}}$.

The display should be approximately 7.389.

$$e^2 \approx 7.389$$

The number e lies between 2 and 3. Because $2^2 = 4$ and $3^2 = 9$, it makes sense that e^2, approximately 7.389, lies between 4 and 9.

Because $2 < e < 3$, the graph of $y = e^x$ lies between the graphs of $y = 2^x$ and $y = 3^x$, shown in Figure 4.5.

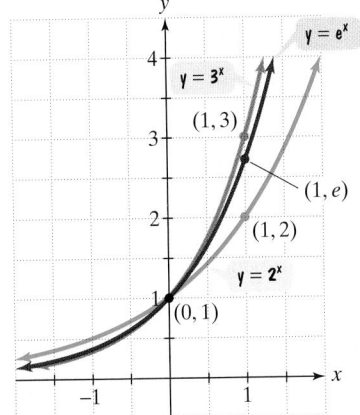

Figure 4.5 Graphs of three exponential functions

EXAMPLE 5   World Population

In a report entitled *Resources and Man*, the U.S. National Academy of Sciences concluded that a world population of 10 billion "is close to (if not above) the maximum that an intensely managed world might hope to support with some degree of comfort and individual choice." At the time the report was issued in 1969, world population was approximately 3.6 billion, with a growth rate of 2% per year. The function

$$f(x) = 3.6e^{0.02x}$$

describes world population, $f(x)$, in billions, x years after 1969. Use the function to find world population in the year 2020. Is there cause for alarm?

World Population, in Billions

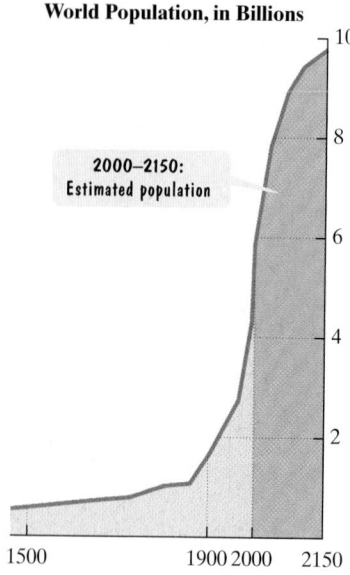

2000–2150: Estimated population

Source: U.N. Population Division

Solution Because 2020 is 51 years after 1969, we substitute 51 for x in $f(x) = 3.6e^{0.02x}$:

$$f(51) = 3.6e^{0.02(51)}.$$

Perform this computation on your calculator.

Scientific calculator: $\quad$ 3.6 $\boxed{\times}$ $\boxed{(}$ .02 $\boxed{\times}$ 51 $\boxed{)}$ $\boxed{e^x}$ $\boxed{=}$

Graphing calculator: $\quad$ 3.6 $\boxed{\times}$ $\boxed{e^x}$ $\boxed{(}$ .02 $\boxed{\times}$ 51 $\boxed{)}$ $\boxed{\text{ENTER}}$

The display should be approximately 9.9835012. Thus,

$$f(51) = 3.6e^{0.02(51)} \approx 9.98.$$

This indicates that world population in the year 2020 will be approximately 9.98 billion. Because this number is quite close to 10 billion, the given function suggests that there may be cause for alarm.

World population in 2000 was approximately 6 billion, but the growth rate was no longer 2%. It had slowed down to 1.3%. Using this current growth rate, exponential functions now predict a world population of 7.8 billion in the year 2020. Experts think the population may stabilize at 10 billion after 2200 if the deceleration in growth rate continues.

> **Check Point 5** The function $f(x) = 6e^{0.013x}$ describes world population, $f(x)$, in billions, x years after 2000 subject to a growth rate of 1.3% annually. Use the function to find world population in 2050.

4 Use compound interest formulas.

Compound Interest

We all want a wonderful life with fulfilling work, good health, and loving relationships. And let's be honest: Financial security wouldn't hurt! Achieving this goal depends on understanding how money in savings accounts grows in remarkable ways as a result of *compound interest*. **Compound interest** is interest computed on your original investment as well as on any accumulated interest.

Suppose a sum of money, called the **principal, P,** is invested at an annual percentage rate r, in decimal form, compounded once per year. Because the interest is added to the principal at year's end, the accumulated value, A, is

$$A = P + Pr = P(1 + r).$$

The accumulated amount of money follows this pattern of multiplying the previous principal by $(1 + r)$ for each successive year, as indicated in Table 4.2.

Table 4.2

Time in Years	Accumulated Value after Each Compounding
0	$A = P$
1	$A = P(1 + r)$
2	$A = P(1 + r)(1 + r) = P(1 + r)^2$
3	$A = P(1 + r)^2(1 + r) = P(1 + r)^3$
4	$A = P(1 + r)^3(1 + r) = P(1 + r)^4$
$\vdots$	$\vdots$
t	$A = P(1 + r)^t$

This formula gives the balance, A, that a principal, P, is worth after t years at interest rate r, compounded once a year.

n	$\left(1 + \dfrac{1}{n}\right)^n$
1	2
2	2.25
5	2.48832
10	2.59374246
100	2.704813829
1000	2.716923932
10,000	2.718145927
100,000	2.718268237
1,000,000	2.718280469
1,000,000,000	2.718281827

As n takes on increasingly large values, the expression $\left(1 + \dfrac{1}{n}\right)^n$ approaches e.

Most savings institutions have plans in which interest is paid more than once a year. If compound interest is paid twice a year, the compounding period is six months. We say that the interest is **compounded semiannually.** When compound interest is paid four times a year, the compounding period is three months and the interest is said to be **compounded quarterly.** Some plans allow for monthly compounding or daily compounding.

In general, when compound interest is paid n times a year, we say that there are n **compounding periods per year.** The formula $A = P(1 + r)^t$ can be adjusted to take into account the number of compounding periods in a year. If there are n compounding periods per year, the formula becomes

$$A = P\left(1 + \frac{r}{n}\right)^{nt}.$$

Some banks use **continuous compounding,** where the number of compounding periods increases infinitely (compounding interest every trillionth of a second, every quadrillionth of a second, etc.). As n, the number of compounding periods in a year, increases without bound, the expression $\left(1 + \dfrac{1}{n}\right)^n$ approaches e. As a result, the formula for continuous compounding is $A = Pe^{rt}$. Although continuous compounding sounds terrific, it yields only a fraction of a percent more interest over a year than daily compounding.

Formulas for Compound Interest

After t years, the balance, A, in an account with principal P and annual interest rate r (in decimal form) is given by the following formulas:

1. For n compoundings per year: $A = P\left(1 + \dfrac{r}{n}\right)^{nt}$

2. For continuous compounding: $A = Pe^{rt}$.

EXAMPLE 6   Choosing between Investments

You want to invest $8000 for 6 years, and you have a choice between two accounts. The first pays 7% per year, compounded monthly. The second pays 6.85% per year, compounded continuously. Which is the better investment?

Solution   The better investment is the one with the greater balance in the account after 6 years. Let's begin with the account with monthly compounding. We use the compound interest model with $P = 8000$, $r = 7\% = 0.07$, $n = 12$ (monthly compounding means 12 compoundings per year), and $t = 6$.

$$A = P\left(1 + \frac{r}{n}\right)^{nt} = 8000\left(1 + \frac{0.07}{12}\right)^{12 \cdot 6} \approx 12{,}160.84$$

The balance in this account after 6 years is $12,160.84. For the second investment option, we use the model for continuous compounding with $P = 8000$, $r = 6.85\% = 0.0685$, and $t = 6$.

$$A = Pe^{rt} = 8000e^{0.0685(6)} \approx 12{,}066.60$$

Technology

The graphs illustrate that as x increases, $\left(1 + \dfrac{1}{x}\right)^x$ approaches e.

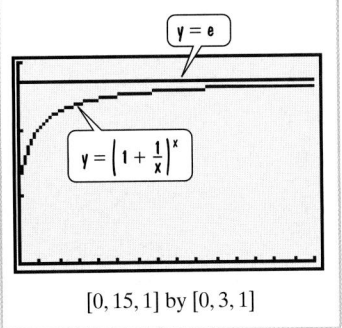

$[0, 15, 1]$ by $[0, 3, 1]$

The balance in this account after 6 years is $12,066.60, slightly less than the previous amount. Thus, the better investment is the 7% monthly compounding option.

Check Point 6 A sum of $10,000 is invested at an annual rate of 8%. Find the balance in the account after 5 years subject to **a.** quarterly compounding and **b.** continuous compounding.

EXERCISE SET 4.1

 Practice Exercises

In Exercises 1–10, approximate each number using a calculator. Round your answer to three decimal places.

1. $2^{3.4}$ **2.** $3^{2.4}$ **3.** $3^{\sqrt{5}}$ **4.** $5^{\sqrt{3}}$ **5.** $4^{-1.5}$

6. $6^{-1.2}$ **7.** $e^{2.3}$ **8.** $e^{3.4}$ **9.** $e^{-0.95}$ **10.** $e^{-0.75}$

In Exercises 11–18, graph each function by making a table of coordinates. If applicable, use a graphing utility to confirm your hand-drawn graph.

11. $f(x) = 4^x$

12. $f(x) = 5^x$

13. $g(x) = \left(\frac{3}{2}\right)^x$

14. $g(x) = \left(\frac{4}{3}\right)^x$

15. $h(x) = \left(\frac{1}{2}\right)^x$

16. $h(x) = \left(\frac{1}{3}\right)^x$

17. $f(x) = (0.6)^x$

18. $f(x) = (0.8)^x$

In Exercises 19–24, the graph of an exponential function is given. Select the function for each graph from the following options:

$$f(x) = 3^x, g(x) = 3^{x-1}, h(x) = 3^x - 1,$$
$$F(x) = -3^x, G(x) = 3^{-x}, H(x) = -3^{-x}.$$

19.

20.

21.

22.

23.

24.

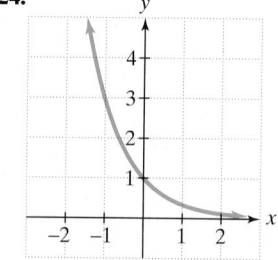

In Exercises 25–34, begin by graphing $f(x) = 2^x$. Then use transformations of this graph and a table of coordinates to graph the given function. If applicable, use a graphing utility to confirm your hand-drawn graphs.

25. $g(x) = 2^{x+1}$ **26.** $g(x) = 2^{x+2}$

27. $g(x) = 2^x - 1$ **28.** $g(x) = 2^x + 2$

29. $h(x) = 2^{x+1} - 1$ **30.** $h(x) = 2^{x+2} - 1$

31. $g(x) = -2^x$ **32.** $g(x) = 2^{-x}$

33. $g(x) = 2 \cdot 2^x$ **34.** $g(x) = \frac{1}{2} \cdot 2^x$

In Exercises 35–40, graph functions f and g in the same rectangular coordinate system. If applicable, use a graphing utility to confirm your hand-drawn graphs.

35. $f(x) = 3^x$ and $g(x) = 3^{-x}$

36. $f(x) = 3^x$ and $g(x) = -3^x$

37. $f(x) = 3^x$ and $g(x) = \frac{1}{3} \cdot 3^x$

38. $f(x) = 3^x$ and $g(x) = 3 \cdot 3^x$

39. $f(x) = (\frac{1}{2})^x$ and $g(x) = (\frac{1}{2})^{x-1} + 1$

40. $f(x) = (\frac{1}{2})^x$ and $g(x) = (\frac{1}{2})^{x-1} + 2$

Use the compound interest formulas $A = P\left(1 + \frac{r}{n}\right)^{nt}$ and $A = Pe^{rt}$ to solve Exercises 41–44. Round answers to the nearest cent.

41. Find the accumulated value of an investment of $10,000 for 5 years at an interest rate of 5.5% if the money is **a.** compounded semiannually; **b.** compounded quarterly; **c.** compounded monthly; **d.** compounded continuously.

42. Find the accumulated value of an investment of $5000 for 10 years at an interest rate of 6.5% if the money is **a.** compounded semiannually; **b.** compounded quarterly; **c.** compounded monthly; **d.** compounded continuously.

43. Suppose that you have $12,000 to invest. Which investment yields the greatest return over 3 years: 7% compounded monthly or 6.85% compounded continuously?

44. Suppose that you have $6000 to invest. Which investment yields the greatest return over 4 years: 8.25% compounded quarterly or 8.3% compounded semiannually?

Application Exercises

Use a calculator with a $\boxed{y^x}$ key or a $\boxed{\wedge}$ key to solve Exercises 45–52.

45. The exponential function $f(x) = 67.38(1.026)^x$ describes the population of Mexico, $f(x)$, in millions, x years after 1980.
   a. Substitute 0 for x and, without using a calculator, find Mexico's population in 1980.
   b. Substitute 27 for x and use your calculator to find Mexico's population in the year 2007 as predicted by this function.

   c. Find Mexico's population in the year 2034 as predicted by this function.
   d. Find Mexico's population in the year 2061 as predicted by this function.
   e. What appears to be happening to Mexico's population every 27 years?

46. The 1986 explosion at the Chernobyl nuclear power plant in the former Soviet Union sent about 1000 kilograms of radioactive cesium-137 into the atmosphere. The function $f(x) = 1000(0.5)^{x/30}$ describes the amount, $f(x)$, in kilograms, of cesium-137 remaining in Chernobyl x years after 1986. If even 100 kilograms of cesium-137 remain in Chernobyl's atmosphere, the area is considered unsafe for human habitation. Find $f(80)$ and determine if Chernobyl will be safe for human habitation by 2066.

It is 8:00 P.M. and West Side Story *is scheduled to begin. When the curtain does not go up, a rumor begins to spread through the 400-member audience: The lead roles of Tony and Maria might be understudied by Anthony Hopkins and Jodie Foster. The function*

$$f(x) = \frac{400}{1 + 399(0.67)^x}$$

models the number of people in the audience, f(x), who have heard the rumor x minutes after 8:00. Use this function to solve Exercises 47–48.

47. Evaluate $f(10)$ and describe what this means in practical terms.

48. Evaluate $f(20)$ and describe what this means in practical terms.

The formula $S = C(1 + r)^t$ models inflation, where C = the value today, r = the annual inflation rate, and S = the inflated value t years from now. Use this formula to solve Exercises 49–50.

49. If the inflation rate is 6%, how much will a house now worth $65,000 be worth in 10 years?

50. If the inflation rate is 3%, how much will a house now worth $110,000 be worth in 5 years?

51. A decimal approximation for $\sqrt{3}$ is 1.7320508. Use a calculator to find $2^{1.7}$, $2^{1.73}$, $2^{1.732}$, $2^{1.73205}$, and $2^{1.7320508}$. Now find $2^{\sqrt{3}}$. What do you observe?

52. A decimal approximation for π is 3.141593. Use a calculator to find 2^3, $2^{3.1}$, $2^{3.14}$, $2^{3.141}$, $2^{3.1415}$, $2^{3.14159}$, and $2^{3.141593}$. Now find 2^{π}. What do you observe?

The graph on the next page shows the number of Americans enrolled in HMOs, in millions, from 1992 through 2000. The data can be modeled by the exponential function

$$f(x) = 36.1e^{0.113x},$$

which describes enrollment in HMOs, f(x), in millions, x years after 1992. Use this function to solve Exercises 53–54.

Enrollment in HMOs

Source: Department of Health and Human Services

53. According to the model, how many Americans will be enrolled in HMOs in the year 2006? Round to the nearest tenth of a million.

54. According to the model, how many Americans will be enrolled in HMOs in the year 2008? Round to the nearest tenth of a million.

55. In college, we study large volumes of information—information that, unfortunately, we do not often retain for very long. The function

$$f(x) = 80e^{-0.5x} + 20$$

describes the percentage of information, $f(x)$, that a particular person remembers x weeks after learning the information.

a. Substitute 0 for x and, without using a calculator, find the percentage of information remembered at the moment it is first learned.

b. Substitute 1 for x and find the percentage of information that is remembered after 1 week.

c. Find the percentage of information that is remembered after 4 weeks.

d. Find the percentage of information that is remembered after one year (52 weeks).

56. In 1626, Peter Minuit convinced the Wappinger Indians to sell him Manhattan Island for $24. If the Native Americans had put the $24 into a bank account paying 5% interest, how much would the investment be worth in the year 2000 if interest were compounded
a. monthly?   **b.** continuously?

57. The function

$$N(t) = \frac{30,000}{1 + 20e^{-1.5t}}$$

describes the number of people, $N(t)$, who become ill with influenza t weeks after its initial outbreak in a town with 30,000 inhabitants. The horizontal asymptote in the graph at the top of the next column indicates that there is a limit to the epidemic's growth.

a. How many people became ill with the flu when the epidemic began? (When the epidemic began, $t = 0$.)

b. How many people were ill by the end of the third week?

c. Why can't the spread of an epidemic simply grow indefinitely? What does the horizontal asymptote shown in the graph indicate about the limiting size of the population that becomes ill?

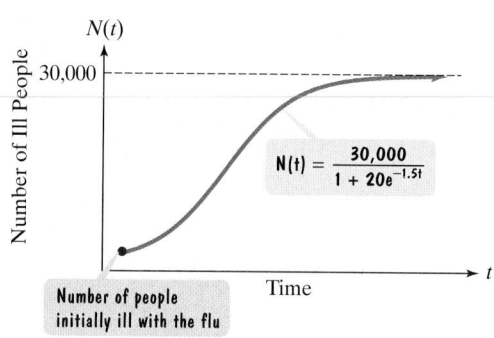

$$N(t) = \frac{30,000}{1 + 20e^{-1.5t}}$$

Number of people initially ill with the flu

 Writing in Mathematics

58. What is an exponential function?

59. What is the natural exponential function?

60. Use a calculator to evaluate $\left(1 + \dfrac{1}{x}\right)^x$ for $x = 10, 100, 1000, 10,000, 100,000,$ and $1,000,000$. Describe what happens to the expression as x increases.

61. Write an example similar to Example 6 on page 358 in which continuous compounding at a slightly lower yearly interest rate is a better investment than compounding n times per year.

62. Describe how you could use the graph of $f(x) = 2^x$ to obtain a decimal approximation for $\sqrt{2}$.

63. The exponential function $y = 2^x$ is one-to-one and has an inverse function. Try finding the inverse function by exchanging x and y and solving for y. Describe the difficulty that you encounter in this process. What is needed to overcome this problem?

64. In 2000, world population was approximately 6 billion with an annual growth rate of 1.3%. Discuss two factors that would cause this growth rate to slow down over the next ten years.

Technology Exercises

65. Graph $y = 13.49(0.967)^x - 1$, the function for the number of O-rings expected to fail at $x°F$, in a $[0, 90, 10]$ by $[0, 20, 5]$ viewing rectangle. If NASA engineers had used this function and its graph, is it likely they would have allowed the *Challenger* to be launched when the temperature was $31°F$? Explain.

66. You have $10,000 to invest. One bank pays 5% interest compounded quarterly and the other pays 4.5% interest compounded monthly.

a. Use the formula for compound interest to write a function for the balance in each account at any time t.

b. Use a graphing utility to graph both functions in an appropriate viewing rectangle. Based on the graphs, which bank offers the better return on your money?

67. a. Graph $y = e^x$ and $y = 1 + x + \dfrac{x^2}{2}$ in the same viewing rectangle.

b. Graph $y = e^x$ and $y = 1 + x + \dfrac{x^2}{2} + \dfrac{x^3}{6}$ in the same viewing rectangle.

c. Graph $y = e^x$ and $y = 1 + x + \dfrac{x^2}{2} + \dfrac{x^3}{6} + \dfrac{x^4}{24}$ in the same viewing rectangle.

d. Describe what you observe in parts (a)–(c). Try generalizing this observation.

Critical Thinking Exercises

68. Which one of the following is true?

a. As the number of compounding periods increases on a fixed investment, the amount of money in the account over a fixed interval of time will increase without bound.

b. The functions $f(x) = 3^{-x}$ and $g(x) = -3^x$ have the same graph.

c. $e = 2.718$

d. The functions $f(x) = \left(\frac{1}{3}\right)^x$ and $g(x) = 3^{-x}$ have the same graph.

69. The graphs labeled (a)–(d) in the figure represent $y = 3^x$, $y = 5^x$, $y = \left(\frac{1}{3}\right)^x$, and $y = \left(\frac{1}{5}\right)^x$, but not necessarily in that order. Which is which? Describe the process that enables you to make this decision.

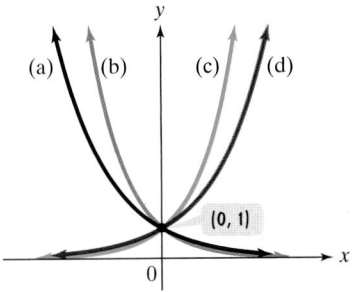

70. Graph $f(x) = 2^x$ and its inverse function in the same rectangular coordinate system.

71. The *hyperbolic cosine* and *hyperbolic sine* functions are defined by

$$\cosh x = \frac{e^x + e^{-x}}{2} \quad \text{and} \quad \sinh x = \frac{e^x - e^{-x}}{2}.$$

Prove that $(\cosh x)^2 - (\sinh x)^2 = 1$.

SECTION 4.2 *Logarithmic Functions*

Objectives

1. Change from logarithmic to exponential form.
2. Change from exponential to logarithmic form.
3. Evaluate logarithms.
4. Use basic logarithmic properties.
5. Graph logarithmic functions.
6. Find the domain of a logarithmic function.
7. Use common logarithms.
8. Use natural logarithms.

The earthquake that ripped through northern California on October 17, 1989, measured 7.1 on the Richter scale, killed more than 60 people, and injured more than 2400. Shown here is San Francisco's Marina district, where shock waves tossed houses off their foundations and into the street.

A higher measure on the Richter scale is more devastating than it seems because for each increase in one unit on the scale, there is a tenfold increase in the intensity of an earthquake. In this section, our focus is on the inverse of the exponential function, called the logarithmic function. The logarithmic function will help you to understand diverse phenomena, including earthquake intensity, human memory, and the pace of life in large cities.

Study Tip

In case you need to review inverse functions, they are discussed in Section 2.7 on pages 260–267. The horizontal line test appears on page 265.

The Definition of Logarithmic Functions

No horizontal line can be drawn that intersects the graph of an exponential function at more than one point. This means that the exponential function is one-to-one and has an inverse. The inverse function of the exponential function with base b is called the *logarithmic function with base b*.

> ### Definition of the Logarithmic Function
> For $x > 0$ and $b > 0, b \neq 1$,
> $$y = \log_b x \text{ is equivalent to } b^y = x.$$
> The function $f(x) = \log_b x$ is the **logarithmic function with base b**.

Study Tip

The inverse function of $y = b^x$ is $x = b^y$. Logarithms give us a way to express this inverse function for y in terms of x.

The equations

$$y = \log_b x \quad \text{and} \quad b^y = x$$

are different ways of expressing the same thing. The first equation is in **logarithmic form** and the second equivalent equation is in **exponential form.**

Notice that a **logarithm, y, is an exponent.** You should learn the location of the base and exponent in each form.

> ### Location of Base and Exponent in Exponential and Logarithmic Forms
>
> Exponent
>
> Logarithmic Form: $y = \log_b x$ Exponential Form: $b^y = x$
>
> Base
>
> Exponent
>
> Base

1 Change from logarithmic to exponential form.

EXAMPLE 1   Changing from Logarithmic to Exponential Form

Write each equation in its equivalent exponential form:

a. $2 = \log_5 x$ **b.** $3 = \log_b 64$ **c.** $\log_3 7 = y$.

Solution   We use the fact that $y = \log_b x$ means $b^y = x$.

a. $2 = \log_5 x$ means $5^2 = x$.   **b.** $3 = \log_b 64$ means $b^3 = 64$.

Logarithms are exponents.   Logarithms are exponents.

c. $\log_3 7 = y$ or $y = \log_3 7$ means $3^y = 7$.

Check Point 1   Write each equation in its equivalent exponential form:

a. $3 = \log_7 x$ **b.** $2 = \log_b 25$ **c.** $\log_4 26 = y$.

2 Change from exponential to logarithmic form.

EXAMPLE 2   Changing from Exponential to Logarithmic Form

Write each equation in its equivalent logarithmic form:

a. $12^2 = x$ **b.** $b^3 = 8$ **c.** $e^y = 9$.

Solution We use the fact that $b^y = x$ means $y = \log_b x$.

a. $12^2 = x$ means $2 = \log_{12} x$. **b.** $b^3 = 8$ means $3 = \log_b 8$.

> Exponents are logarithms.

> Exponents are logarithms.

c. $e^y = 9$ means $y = \log_e 9$.

Check Point 2 Write each equation in its equivalent logarithmic form:

a. $2^5 = x$ **b.** $b^3 = 27$ **c.** $e^y = 33$.

3 Evaluate logarithms.

Remembering that logarithms are exponents makes it possible to evaluate some logarithms by inspection. The logarithm of x with base b, $\log_b x$, is the exponent to which b must be raised to get x. For example, suppose we want to evaluate $\log_2 32$. We ask, 2 to what power gives 32? Because $2^5 = 32, \log_2 32 = 5$.

EXAMPLE 3   Evaluating Logarithms

Evaluate:

a. $\log_2 16$ **b.** $\log_3 9$ **c.** $\log_{25} 5$.

Solution

Logarithmic Expression	Question Needed for Evaluation	Logarithmic Expression Evaluated
a. $\log_2 16$	2 to what power gives 16?	$\log_2 16 = 4$ because $2^4 = 16$.
b. $\log_3 9$	3 to what power gives 9?	$\log_3 9 = 2$ because $3^2 = 9$.
c. $\log_{25} 5$	25 to what power gives 5?	$\log_{25} 5 = \frac{1}{2}$ because $25^{1/2} = \sqrt{25} = 5$.

Check Point 3 Evaluate:

a. $\log_{10} 100$ **b.** $\log_3 3$ **c.** $\log_{36} 6$.

4 Use basic logarithmic properties.

Basic Logarithmic Properties

Because logarithms are exponents, they have properties that can be verified using properties of exponents.

Basic Logarithmic Properties Involving One

1. $\log_b b = 1$ because 1 is the exponent to which b must be raised to obtain b. $\left(b^1 = b\right)$

2. $\log_b 1 = 0$ because 0 is the exponent to which b must be raised to obtain 1. $\left(b^0 = 1\right)$

EXAMPLE 4   Using Properties of Logarithms

Evaluate:

a. $\log_7 7$ **b.** $\log_5 1$.

Solution
a. Because $\log_b b = 1$, we conclude $\log_7 7 = 1$.
b. Because $\log_b 1 = 0$, we conclude $\log_5 1 = 0$.

> **Check Point 4**   Evaluate:
>
> **a.** $\log_9 9$ **b.** $\log_8 1$.

The inverse of the exponential function is the logarithmic function. Thus, if $f(x) = b^x$, then $f^{-1}(x) = \log_b x$. In Chapter 2, we saw how inverse functions "undo" one another. In particular,

$$f(f^{-1}(x)) = x \text{ and } f^{-1}(f(x)) = x.$$

Applying these relationships to exponential and logarithmic functions, we obtain the following **inverse properties of logarithms:**

Inverse Properties of Logarithms

For $b > 0$ and $b \neq 1$,

$$\log_b b^x = x \qquad \text{The logarithm with base } b \text{ of } b \text{ raised to a power equals that power.}$$

$$b^{\log_b x} = x \qquad b \text{ raised to the logarithm with base } b \text{ of a number equals that number.}$$

EXAMPLE 5   Using Inverse Properties of Logarithms

Evaluate:

a. $\log_4 4^5$ **b.** $6^{\log_6 9}$.

Solution
a. Because $\log_b b^x = x$, we conclude $\log_4 4^5 = 5$.
b. Because $b^{\log_b x} = x$, we conclude $6^{\log_6 9} = 9$.

> **Check Point 5**   Evaluate:
>
> **a.** $\log_7 7^8$ **b.** $3^{\log_3 17}$.

⑤ Graph logarithmic functions.

Graphs of Logarithmic Functions

How do we graph logarithmic functions? We use the fact that the logarithmic function is the inverse of the exponential function. This means that the logarithmic function reverses the coordinates of the exponential function. It also means that the graph of the logarithmic function is a reflection of the graph of the exponential function about the line $y = x$.

EXAMPLE 6 **Graphs of Exponential and Logarithmic Functions**

Graph $f(x) = 2^x$ and $g(x) = \log_2 x$ in the same rectangular coordinate system.

Solution We first set up a table of coordinates for $f(x) = 2^x$. Reversing, these coordinates gives the coordinates for the inverse function $g(x) = \log_2 x$.

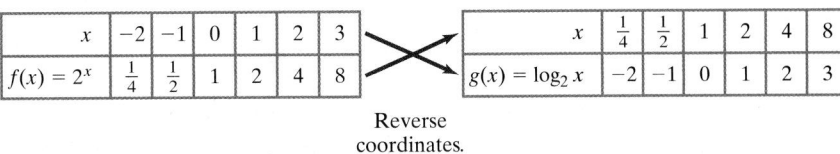

x	-2	-1	0	1	2	3
$f(x) = 2^x$	$\frac{1}{4}$	$\frac{1}{2}$	1	2	4	8

x	$\frac{1}{4}$	$\frac{1}{2}$	1	2	4	8
$g(x) = \log_2 x$	-2	-1	0	1	2	3

Reverse
coordinates.

We now plot the ordered pairs in both tables, connecting them with smooth curves. Figure 4.6 shows the graphs of $f(x) = 2^x$ and its inverse function $g(x) = \log_2 x$. The graph of the inverse can also be drawn by reflecting the graph of $f(x) = 2^x$ about the line $y = x$.

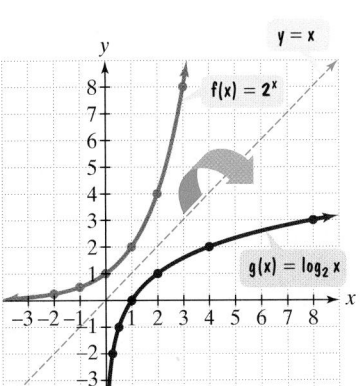

Figure 4.6 The graphs of $f(x) = 2^x$ and its inverse function

Check Point 6 Graph $f(x) = 3^x$ and $g(x) = \log_3 x$ in the same rectangular coordinate system.

Figure 4.7 illustrates the relationship between the graph of the exponential function, shown in blue, and its inverse, the logarithmic function, shown in red, for bases greater than 1 and for bases between 0 and 1.

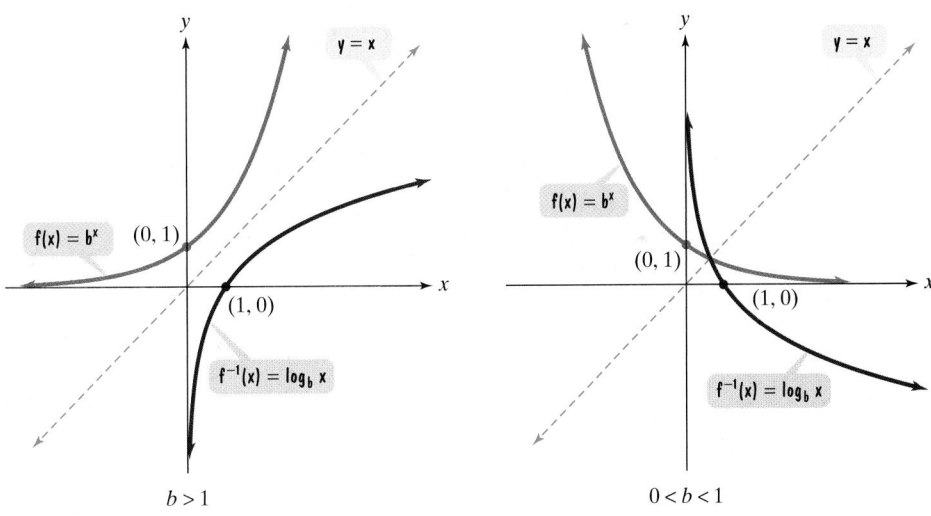

$b > 1$ $0 < b < 1$

Figure 4.7 Graphs of exponential and logarithmic functions

Discovery

Verify each of the four characteristics in the box for the red graphs in Figure 4.7.

Characteristics of the Graphs of Logarithmic Functions of the Form $f(x) = \log_b x$

- The x-intercept is 1. There is no y-intercept.
- The y-axis is a vertical asymptote.
- If $b > 1$, the function is increasing. If $0 < b < 1$, the function is decreasing.
- The graph is smooth and continuous. It has no sharp corners or gaps.

The graphs of logarithmic functions can be translated vertically or horizontally, reflected, stretched, or shrunk. We use the ideas of Section 2.5 to do so, as summarized in Table 4.3.

Table 4.3 Transformations Involving Logarithmic Functions
In each case, c represents a positive real number.

Transformation	Equation	Description
Vertical translation	$g(x) = \log_b x + c$	• Shifts the graph of $f(x) = \log_b x$ upward c units.
	$g(x) = \log_b x - c$	• Shifts the graph of $f(x) = \log_b x$ downward c units.
Horizontal translation	$g(x) = \log_b (x + c)$	• Shifts the graph of $f(x) = \log_b x$ to the left c units. Vertical asymptote: $x = -c$.
	$g(x) = \log_b (x - c)$	• Shifts the graph of $f(x) = \log_b x$ to the right c units. Vertical asymptote: $x = c$.
Reflecting	$g(x) = -\log_b x$	• Reflects the graph of $f(x) = \log_b x$ about the x-axis.
	$g(x) = \log_b (-x)$	• Reflects the graph of $f(x) = \log_b x$ about the y-axis.
Vertical stretching or shrinking	$g(x) = c \log_b x$	• Stretches the graph of $f(x) = \log_b x$ if $c > 1$. • Shrinks the graph of $f(x) = \log_b x$ if $0 < c < 1$.

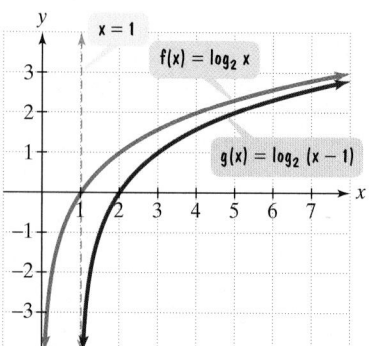

Figure 4.8 Shifting $f(x) = \log_2 x$ one unit to the right

For example, Figure 4.8 illustrates that the graph of $g(x) = \log_2 (x - 1)$ is the graph of $f(x) = \log_2 x$ shifted one unit to the right. If a logarithmic function is translated to the left or to the right, both the x-intercept and the vertical asymptote are shifted by the amount of the horizontal shift. In Figure 4.8, the x-intercept of f is 1. Because g is shifted one unit to the right, its x-intercept is 2. Also observe that the vertical asymptote for f, the y-axis, is shifted one unit to the right for the vertical asymptote for g. Thus, $x = 1$ is the vertical asymptote for g.

Here are some other examples of transformations of graphs of logarithmic functions:

- The graph of $g(x) = 3 + \log_4 x$ is the graph of $f(x) = \log_4 x$ shifted up three units, shown in Figure 4.9.
- The graph of $h(x) = -\log_2 x$ is the graph of $f(x) = \log_2 x$ reflected about the x-axis, shown in Figure 4.10.
- The graph of $r(x) = \log_2 (-x)$ is the graph of $f(x) = \log_2 x$ reflected about the y-axis, shown in Figure 4.11.

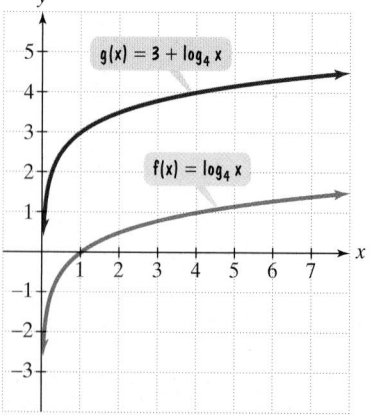

Figure 4.9 Shifting vertically up three units

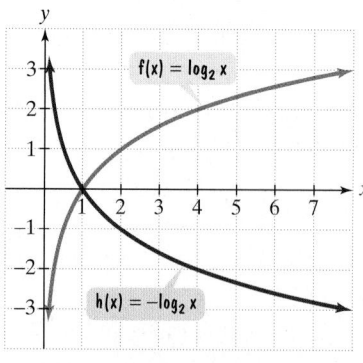

Figure 4.10 Reflection about the x-axis

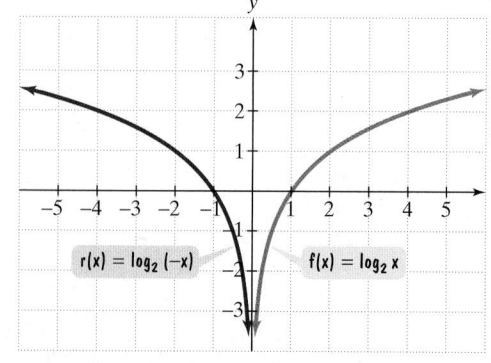

Figure 4.11 Reflection about the y-axis

6 Find the domain of a logarithmic function.

The Domain of a Logarithmic Function

In Section 4.1, we learned that the domain of an exponential function of the form $f(x) = b^x$ includes all real numbers and its range is the set of positive real numbers. Because the logarithmic function reverses the domain and the range of the exponential function, the **domain of a logarithmic function of the form $f(x) = \log_b x$ is the set of all positive real numbers.** Thus, $\log_2 8$ is defined because the value of x in the logarithmic expression, 8, is greater than zero and therefore is included in the domain of the logarithmic function $f(x) = \log_2 x$. However, $\log_2 0$ and $\log_2 (-8)$ are not defined because 0 and -8 are not positive real numbers and therefore are excluded from the domain of the logarithmic function $f(x) = \log_2 x$. In general, the domain of $f(x) = \log_b (x + c)$ consists of all x for which $x + c > 0$.

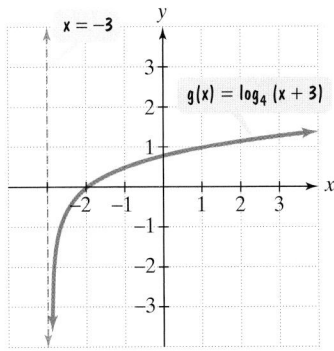

Figure 4.12 The domain of $g(x) = \log_4 (x + 3)$ is $(-3, \infty)$.

EXAMPLE 7 Finding the Domain of a Logarithmic Function

Find the domain of $g(x) = \log_4 (x + 3)$.

Solution The domain of g consists of all x for which $x + 3 > 0$. Solving this inequality for x, we obtain $x > -3$. Thus, the domain of g is $(-3, \infty)$. This is illustrated in Figure 4.12. The vertical asymptote is $x = -3$, and all points on the graph of g have x-coordinates that are greater than -3.

Check Point 7 Find the domain of $h(x) = \log_4 (x - 5)$.

7 Use common logarithms.

Common Logarithms

The logarithmic function with base 10 is called the **common logarithmic function.** The function $f(x) = \log_{10} x$ is usually expressed as $f(x) = \log x$. A calculator with a $\boxed{\text{LOG}}$ key can be used to evaluate common logarithms. Here are some examples:

Logarithm	Most Scientific Calculator Keystrokes	Most Graphing Calculator Keystrokes	Display (or Approximate Display)
$\log 1000$	1000 $\boxed{\text{LOG}}$	$\boxed{\text{LOG}}$ 1000 $\boxed{\text{ENTER}}$	3
$\log \dfrac{5}{2}$	$\boxed{(}\, 5 \boxed{\div} 2 \boxed{)}\, \boxed{\text{LOG}}$	$\boxed{\text{LOG}}\, \boxed{(}\, 5 \boxed{\div} 2 \boxed{)}\, \boxed{\text{ENTER}}$	0.39794
$\dfrac{\log 5}{\log 2}$	5 $\boxed{\text{LOG}}\, \boxed{\div}$ 2 $\boxed{\text{LOG}}\, \boxed{=}$	$\boxed{\text{LOG}}$ 5 $\boxed{\div}\, \boxed{\text{LOG}}$ 2 $\boxed{\text{ENTER}}$	2.32193
$\log(-3)$	3 $\boxed{+/-}\, \boxed{\text{LOG}}$	$\boxed{\text{LOG}}\, \boxed{(-)}$ 3 $\boxed{\text{ENTER}}$	$\boxed{\text{ERROR}}$

The error message given by many calculators for $\log (-3)$ is a reminder that the domain of every logarithmic function, including the common logarithmic function, is the set of positive real numbers.

Many real-life phenomena start with rapid growth, and then the growth begins to level off. This type of behavior can be modeled by logarithmic functions.

EXAMPLE 8 Modeling Height of Children

The percentage of adult height attained by a boy who is x years old can be modeled by

$$f(x) = 29 + 48.8 \log (x + 1)$$

where x represents the boy's age and $f(x)$ represents the percentage of his adult height. Approximately what percent of his adult height is a boy at age eight?

Solution   We substitute the boy's age, 8, for x and evaluate the function.

$$f(x) = 29 + 48.8 \log (x + 1) \quad \text{This is the given function.}$$
$$f(8) = 29 + 48.8 \log (8 + 1) \quad \text{Substitute 8 for x.}$$
$$= 29 + 48.8 \log 9 \quad \text{Graphing calculator keystrokes:}$$
$$\approx 76 \quad \boxed{29} \; \boxed{+} \; 48.8 \; \boxed{\text{LOG}} \; \boxed{9} \; \boxed{\text{ENTER}}$$

Thus, an 8-year-old boy is approximately 76% of his adult height.

Check Point 8   Use the function in Example 8 to answer this question: Approximately what percent of his adult height is a boy at age 10?

The basic properties of logarithms that were listed earlier in this section can be applied to common logarithms.

Properties of Common Logarithms

General Properties	Common Logarithm Properties
1. $\log_b 1 = 0$	**1.** $\log 1 = 0$
2. $\log_b b = 1$	**2.** $\log 10 = 1$
3. $\log_b b^x = x$	**3.** $\log 10^x = x$
4. $b^{\log_b x} = x$	**4.** $10^{\log x} = x$

The property $\log 10^x = x$ can be used to evaluate common logarithms involving powers of 10. For example,

$$\log 100 = \log 10^2 = 2, \quad \log 1000 = \log 10^3 = 3, \quad \text{and} \quad \log 10^{7.1} = 7.1.$$

EXAMPLE 9 Earthquake Intensity

The magnitude, R, on the Richter scale of an earthquake of intensity I is given by

$$R = \log \frac{I}{I_0}$$

where I_0 is the intensity of a barely felt zero-level earthquake. The earthquake that destroyed San Francisco in 1906 was $10^{8.3}$ times as intense as a zero-level earthquake. What was its magnitude on the Richter scale?

Solution Because the earthquake was $10^{8.3}$ times as intense as a zero-level earthquake, the intensity, I, is $10^{8.3}I_0$.

$$R = \log \frac{I}{I_0}$$ This is the formula for magnitude on the Richter scale.

$$R = \log \frac{10^{8.3}I_0}{I_0}$$ Substitute $10^{8.3}I_0$ for I.

$$= \log 10^{8.3}$$ Simplify.

$$= 8.3$$ Use the property $\log 10^x = x$.

San Francisco's 1906 earthquake registered 8.3 on the Richter scale.

Check Point 9 Use the formula in Example 9 to solve this problem. If an earthquake is 10,000 times as intense as a zero-level quake ($I = 10,000I_0$), what is its magnitude on the Richter scale?

8 Use natural logarithms.

Natural Logarithms

The logarithmic function with base e is called the **natural logarithmic function.** The function $f(x) = \log_e x$ is usually expressed as $f(x) = \ln x$, read "el en of x." A calculator with an $\boxed{\text{LN}}$ key can be used to evaluate natural logarithms.

Like the domain of all logarithmic functions, the domain of the natural logarithmic function is the set of all positive real numbers. Thus, the domain of $f(x) = \ln (x + c)$ consists of all x for which $x + c > 0$.

EXAMPLE 10 **Finding Domains of Natural Logarithmic Functions**

Find the domain of each function:

a. $f(x) = \ln (3 - x)$ **b.** $g(x) = \ln (x - 3)^2.$

Solution

a. The domain of f consists of all x for which $3 - x > 0$. Solving this inequality for x, we obtain $x < 3$. Thus, the domain of f is $\{x | x < 3\}$, or $(-\infty, 3)$. This is verified by the graph in Figure 4.13.

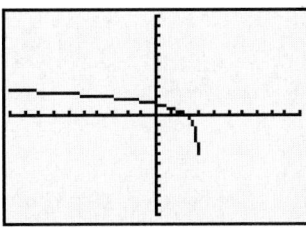

Figure 4.13 The domain of $f(x) = \ln (3 - x)$ is $(-\infty, 3)$.

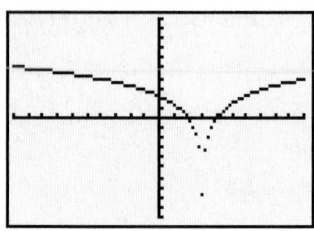

Figure 4.14 3 is excluded from the domain of $g(x) = \ln{(x - 3)^2}$.

b. The domain of g consists of all x for which $(x - 3)^2 > 0$. It follows that the domain of g is the set of all real numbers except 3. Thus, the domain of g is $\{x | x \neq 3\}$, or, in interval notation, $(-\infty, 3)$ or $(3, \infty)$. This is shown by the graph in Figure 4.14. To make it more obvious that 3 is excluded from the domain, we changed the $\boxed{\text{MODE}}$ to Dot.

Check Point 10   Find the domain of each function:

a. $f(x) = \ln{(4 - x)}$ **b.** $g(x) = \ln{x^2}$.

The basic properties of logarithms that were listed earlier in this section can be applied to natural logarithms.

Properties of Natural Logarithms

General Properties	**Natural Logarithm Properties**
1. $\log_b 1 = 0$	**1.** $\ln 1 = 0$
2. $\log_b b = 1$	**2.** $\ln e = 1$
3. $\log_b b^x = x$	**3.** $\ln e^x = x$
4. $b^{\log_b x} = x$	**4.** $e^{\ln x} = x$

The property $\ln e^x = x$ can be used to evaluate natural logarithms involving powers of e. For example,

$$\ln e^2 = 2, \quad \ln e^3 = 3, \quad \ln e^{7.1} = 7.1, \quad \text{and} \quad \ln \frac{1}{e} = \ln e^{-1} = -1.$$

EXAMPLE 11   Using Inverse Properties

Use inverse properties to simplify:

a. $\ln e^{7x}$ **b.** $e^{\ln 4x^2}$.

Solution

a. Because $\ln e^x = x$, we conclude that $\ln e^{7x} = 7x$.

b. Because $e^{\ln x} = x$, we conclude $e^{\ln 4x^2} = 4x^2$.

Check Point 11   Use inverse properties to simplify:

a. $\ln e^{25x}$ **b.** $e^{\ln \sqrt{x}}$.

EXAMPLE 12   Walking Speed and City Population

As the population of a city increases, the pace of life also increases. The formula

$$W = 0.35 \ln P + 2.74$$

models average walking speed, W, in feet per second, for a resident of a city whose population is P thousand. Find the average walking speed for people living in New York City with a population of 7323 thousand.

Solution We use the formula and substitute 7323 for P, the population in thousands.

$$W = 0.35 \ln P + 2.74 \qquad \text{This is the given formula.}$$
$$W = 0.35 \ln 7323 + 2.74 \qquad \text{Substitute 7323 for } P.$$
$$\approx 5.9 \qquad \text{Graphing calculator keystrokes:}$$
$$\boxed{0.35 \;\fbox{LN}\; 7323 \;\fbox{+}\; 2.74 \;\fbox{ENTER}}.$$

The average walking speed in New York City is approximately 5.9 feet per second.

Check Point 12 Use the formula $W = 0.35 \ln P + 2.74$ to find the average walking speed in Jackson, Mississippi, with a population of 197 thousand.

EXERCISE SET 4.2

Practice Exercises

In Exercises 1–8, write each equation in its equivalent exponential form.

1. $4 = \log_2 16$
2. $6 = \log_2 64$
3. $2 = \log_3 x$
4. $2 = \log_9 x$
5. $5 = \log_b 32$
6. $3 = \log_b 27$
7. $\log_6 216 = y$
8. $\log_5 125 = y$

In Exercises 9–20, write each equation in its equivalent logarithmic form.

9. $2^3 = 8$
10. $5^4 = 625$
11. $2^{-4} = \frac{1}{16}$
12. $5^{-3} = \frac{1}{125}$
13. $\sqrt[3]{8} = 2$
14. $\sqrt[3]{64} = 4$
15. $13^2 = x$
16. $15^2 = x$
17. $b^3 = 1000$
18. $b^3 = 343$
19. $7^y = 200$
20. $8^y = 300$

In Exercises 21–38, evaluate each expression without using a calculator.

21. $\log_4 16$
22. $\log_7 49$
23. $\log_2 64$
24. $\log_3 27$
25. $\log_7 \sqrt{7}$
26. $\log_6 \sqrt{6}$
27. $\log_2 \frac{1}{8}$
28. $\log_3 \frac{1}{9}$
29. $\log_{64} 8$
30. $\log_{81} 9$
31. $\log_5 5$
32. $\log_{11} 11$
33. $\log_4 1$
34. $\log_6 1$
35. $\log_5 5^7$
36. $\log_4 4^6$
37. $8^{\log_8 19}$
38. $7^{\log_7 23}$

39. Graph $f(x) = 4^x$ and $g(x) = \log_4 x$ in the same rectangular coordinate system.

40. Graph $f(x) = 5^x$ and $g(x) = \log_5 x$ in the same rectangular coordinate system.

41. Graph $f(x) = \left(\frac{1}{2}\right)^x$ and $g(x) = \log_{1/2} x$ in the same rectangular coordinate system.

42. Graph $f(x) = \left(\frac{1}{4}\right)^x$ and $g(x) = \log_{1/4} x$ in the same rectangular coordinate system.

In Exercises 43–48, the graph of a logarithmic function is given. Select the function for each graph from the following options:

$$f(x) = \log_3 x, \, g(x) = \log_3 (x - 1), \, h(x) = \log_3 x - 1,$$
$$F(x) = -\log_3 x, \, G(x) = \log_3 (-x), \, H(x) = 1 - \log_3 x.$$

43.

44.

45.

46.

47.

48.

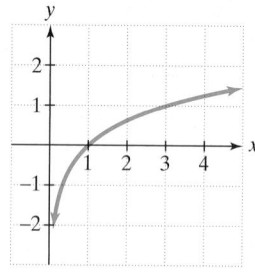

In Exercises 49–54, begin by graphing $f(x) = \log_2 x$. Then use transformations of this graph to graph the given function. What is the graph's x-intercept? What is the vertical asymptote?

49. $g(x) = \log_2 (x + 1)$ **50.** $g(x) = \log_2 (x + 2)$

51. $h(x) = 1 + \log_2 x$ **52.** $h(x) = 2 + \log_2 x$

53. $g(x) = \frac{1}{2} \log_2 x$ **54.** $g(x) = -2 \log_2 x$

In Exercises 55–60, find the domain of each logarithmic function.

55. $f(x) = \log_5 (x + 4)$ **56.** $f(x) = \log_5 (x + 6)$

57. $f(x) = \log (2 - x)$ **58.** $f(x) = \log (7 - x)$

59. $f(x) = \ln (x - 2)^2$ **60.** $f(x) = \ln (x - 7)^2$

In Exercises 61–74, evaluate each expression without using a calculator.

61. $\log 100$ **62.** $\log 1000$ **63.** $\log 10^7$

64. $\log 10^8$ **65.** $10^{\log 33}$ **66.** $10^{\log 53}$

67. $\ln 1$ **68.** $\ln e$ **69.** $\ln e^6$

70. $\ln e^7$ **71.** $\ln \dfrac{1}{e^6}$ **72.** $\ln \dfrac{1}{e^7}$

73. $e^{\ln 125}$ **74.** $e^{\ln 300}$

In Exercises 75–80, use inverse properties of logarithms to simplify each expression.

75. $\ln e^{9x}$ **76.** $\ln e^{13x}$ **77.** $e^{\ln 5x^2}$

78. $e^{\ln 7x^2}$ **79.** $10^{\log \sqrt{x}}$ **80.** $10^{\log \sqrt[3]{x}}$

Application Exercises

T he percentage of adult height attained by a girl who is x years old can be modeled by

$$f(x) = 62 + 35 \log (x - 4)$$

where x represents the girl's age (from 5 to 15) and $f(x)$ represents the percentage of her adult height. Use the function to solve Exercises 81–82.

81. Approximately what percent of her adult height is a girl at age 13?

82. Approximately what percent of her adult height is a girl at age ten?

83. The annual amount that we spend to attend sporting events can be modeled by

$$f(x) = 2.05 + 1.3 \ln x$$

where x represents the number of years after 1984 and $f(x)$ represents the total annual expenditures for admission to spectator sports, in billions of dollars. In 2000, approximately how much was spent on admission to spectator sports?

84. The percentage of U.S. households with cable television can be modeled by

$$f(x) = 18.32 + 15.94 \ln x$$

where x represents the number of years after 1979 and $f(x)$ represents the percentage of U.S. households with cable television. What percentage of U.S. households had cable television in 1990?

The loudness level of a sound, D, in decibels, is given by the formula

$$D = 10 \log (10^{12} I)$$

where I is the intensity of the sound, in watts per meter². Decibel levels range from 0, a barely audible sound, to 160, a sound resulting in a ruptured eardrum. Use the formula to solve Exercises 85–86.

85. The sound of a blue whale can be heard 500 miles away, reaching an intensity of 6.3×10^6 watts per meter². Determine the decibel level of this sound. At close range, can the sound of a blue whale rupture the human eardrum?

86. What is the decibel level of a normal conversation, 3.2×10^{-6} watt per meter²?

87. Students in a psychology class took a final examination. As part of an experiment to see how much of the course content they remembered over time, they took equivalent forms of the exam in monthly intervals thereafter. The average score for the group, $f(t)$, after t months was modeled by the function

$$f(t) = 88 - 15 \ln (t + 1), \quad 0 \leq t \leq 12.$$

a. What was the average score on the original exam?

b. What was the average score after 2 months? 4 months? 6 months? 8 months? 10 months? one year?

c. Sketch the graph of f (either by hand or with a graphing utility). Describe what the graph indicates in terms of the material retained by the students.

Writing in Mathematics

88. Describe the relationship between an equation in logarithmic form and an equivalent equation in exponential form.

89. What question can be asked to help evaluate $\log_3 81$?

90. Explain why the logarithm of 1 with base b is 0.

91. Describe the following property using words: $\log_b b^x = x$.

92. Explain how to use the graph of $f(x) = 2^x$ to obtain the graph of $g(x) = \log_2 x$.

93. Explain how to find the domain of a logarithmic function.

94. New York City is one of the world's great walking cities. Use the formula in Example 12 on page 394 to describe what frequently happens to tourists exploring the city by foot.

95. Logarithmic models are well suited to phenomena in which growth is initially rapid but then begins to level off. Describe something that is changing over time that can be modeled using a logarithmic function.

96. Suppose that a girl is 4′ 6″ at age 10. Explain how to use the function in Exercises 81–82 to determine how tall she can expect to be as an adult.

Technology Exercises

In Exercises 97–100, graph f and g in the same viewing rectangle. Then describe the relationship of the graph of g to the graph of f.

97. $f(x) = \ln x, g(x) = \ln (x + 3)$

98. $f(x) = \ln x, g(x) = \ln x + 3$

99. $f(x) = \log x, g(x) = -\log x$

100. $f(x) = \log x, g(x) = \log (x - 2) + 1$

101. Students in a mathematics class took a final examination. They took equivalent forms of the exam in monthly intervals thereafter. The average score, $f(t)$, for the group after t months was modeled by the human memory function $f(t) = 75 - 10 \log (t + 1)$, where $0 \leq t \leq 12$.

Use a graphing utility to graph the function. Then determine how many months will elapse before the average score falls below 65.

102. Graph f and g in the same viewing rectangle.

a. $f(x) = \ln (3x), g(x) = \ln 3 + \ln x$

b. $f(x) = \log (5x^2), g(x) = \log 5 + \log x^2$

c. $f(x) = \ln (2x^3), g(x) = \ln 2 + \ln x^3$

d. Describe what you observe in parts (a)–(c). Generalize this observation by writing an equivalent expression for $\log_b (MN)$, where $M > 0$ and $N > 0$.

e. Complete this statement: The logarithm of a product is equal to _____.

103. Graph each of the following functions in the same viewing rectangle and then place the functions in order from the one that increases most slowly to the one that increases most rapidly.

$$y = x, y = \sqrt{x}, y = e^x, y = \ln x, y = x^x, y = x^2$$

Critical Thinking Exercises

104. Which one of the following is true?

a. $\dfrac{\log_2 8}{\log_2 4} = \dfrac{8}{4}$

b. $\log (-100) = -2$

c. The domain of $f(x) = \log_2 x$ is $(-\infty, \infty)$.

d. $\log_b x$ is the exponent to which b must be raised to obtain x.

105. Without using a calculator, find the exact value of

$$\frac{\log_3 81 - \log_\pi 1}{\log_{2\sqrt{2}} 8 - \log 0.001}.$$

106. Solve for x: $\log_4[\log_3(\log_2 x)] = 0$.

107. Without using a calculator, determine which is the greater number: $\log_4 60$ or $\log_3 40$.

Group Exercise

108. This group exercise involves exploring the way we grow. Group members should create a graph for the function that models the percentage of adult height attained by a boy who is x years old, $f(x) = 29 + 48.8 \log (x + 1)$. Let $x = 1, 2, 3, \ldots, 12$, find function values, and connect the resulting points with a smooth curve. Then create a function that models the percentage of adult height attained by a girl who is x years old, $g(x) = 62 + 35 \log (x - 4)$. Let $x = 5, 6, 7, \ldots, 15$, find function values, and connect the resulting points with a smooth curve. Group members should then discuss similarities and differences in the growth patterns for boys and girls based on the graphs.

SECTION 4.3 *Properties of Logarithms*

Objectives

1. Use the product rule.
2. Use the quotient rule.
3. Use the power rule.
4. Expand logarithmic expressions.
5. Condense logarithmic expressions.
6. Use the change-of-base property.

We all learn new things in different ways. In this section, we consider important properties of logarithms. What would be the most effective way for you to learn about these properties? Would it be helpful to use your graphing utility and discover one of these properties for yourself? To do so, work Exercise 102 in Exercise Set 4.2 before continuing. Would the properties become more meaningful if you could see exactly where they come from? If so, you will find details of the proofs of many of these properties in the appendix. The remainder of our work in this chapter will be based on the properties of logarithms that you learn in this section.

1   Use the product rule.

The Product Rule

Properties of exponents correspond to properties of logarithms. For example, when we multiply with the same base, we add exponents:

$$b^m \cdot b^n = b^{m+n}.$$

This property of exponents, coupled with an awareness that a logarithm is an exponent, suggests the following property, called the **product rule:**

Discovery

We know that log 100,000 = 5. Show that you get the same result by writing 100,000 as $1000 \cdot 100$ and then using the product rule. Then verify the product rule by using other numbers whose logarithms are easy to find.

> **The Product Rule**
> Let b, M, and N be positive real numbers with $b \neq 1$.
> $$\log_b (MN) = \log_b M + \log_b N$$
> The logarithm of a product is the sum of the logarithms.

When we use the product rule to write a single logarithm as the sum of two logarithms, we say that we are **expanding a logarithmic expression.** For example, we can use the product rule to expand $\ln(4x)$:

$$\ln (7x) = \ln 7 + \ln x.$$

The logarithm of a product   is   the sum of the logarithms.

EXAMPLE 1 Using the Product Rule

Use the product rule to expand each logarithmic expression:

a. $\log_4 (7 \cdot 5)$ **b.** $\log (10x)$.

Solution

a. $\log_4 (7 \cdot 5) = \log_4 7 + \log_4 5$ *The logarithm of a product is the sum of the logarithms.*

b. $\log (10x) = \log 10 + \log x$ *The logarithm of a product is the sum of the logarithms. These are common logarithms with base 10 understood.*

$\qquad\qquad = 1 + \log x$ *Because $\log_b b = 1$, then $\log 10 = 1$.*

Check Point 1 Use the product rule to expand each logarithmic expression:

a. $\log_6 (7 \cdot 11)$ **b.** $\log (100x)$.

2 Use the quotient rule.

The Quotient Rule

When we divide with the same base, we subtract exponents:

$$\frac{b^m}{b^n} = b^{m-n}.$$

This property suggests the following property of logarithms, called the **quotient rule:**

Discovery

We know that $\log_2 16 = 4$. Show that you get the same result by writing 16 as $\frac{32}{2}$ and then using the quotient rule. Then verify the quotient rule using other numbers whose logarithms are easy to find.

The Quotient Rule

Let b, M, and N be positive real numbers with $b \neq 1$.

$$\log_b \left(\frac{M}{N}\right) = \log_b M - \log_b N$$

The logarithm of a quotient is the difference of the logarithms.

When we use the quotient rule to write a single logarithm as the difference of two logarithms, we say that we are **expanding a logarithmic expression.** For example, we can use the quotient rule to expand $\log \frac{x}{2}$:

$$\log \frac{x}{2} = \log x - \log 2.$$

The logarithm of a quotient | is | the difference of the logarithms.

EXAMPLE 2 Using the Quotient Rule

Use the quotient rule to expand each logarithmic expression:

a. $\log_7 \left(\frac{19}{x}\right)$ **b.** $\ln \left(\frac{e^3}{7}\right)$.

Solution

a. $\log_7\left(\dfrac{19}{x}\right) = \log_7 19 - \log_7 x$ *The logarithm of a quotient is the difference of the logarithms.*

b. $\ln\left(\dfrac{e^3}{7}\right) = \ln e^3 - \ln 7$ *The logarithm of a quotient is the difference of the logarithms. These are natural logarithms with base e understood.*

$\qquad\qquad = 3 - \ln 7$ *Because $\ln e^x = x$, then $\ln e^3 = 3$.*

> **Check Point 2** Use the quotient rule to expand each logarithmic expression:
> **a.** $\log_8\left(\dfrac{23}{x}\right)$ **b.** $\ln\left(\dfrac{e^5}{11}\right)$.

3 Use the power rule.

The Power Rule

When an exponential expression is raised to a power, we multiply exponents:
$$(b^m)^n = b^{mn}.$$

This property suggests the following property of logarithms, called the **power rule:**

> **The Power Rule**
>
> Let b and M be positive real numbers with $b \neq 1$, and let p be any real number.
>
> $$\log_b M^p = p \log_b M$$
>
> The logarithm of a number with an exponent is the product of the exponent and the logarithm of that number.

When we use the power rule to "pull the exponent to the front," we say that we are **expanding a logarithmic expression.** For example, we can use the power rule to expand $\ln x^2$:

$$\ln x^2 = 2 \ln x.$$

The logarithm of a number with an exponent   is   the product of the exponent and the logarithm of that number.

Figure 4.15 shows the graphs of $y = \ln x^2$ and $y = 2 \ln x$. Are $\ln x^2$ and $2 \ln x$ the same? The graphs illustrate that $y = \ln x^2$ and $y = 2 \ln x$ have different domains. The graphs are only the same if $x > 0$. Thus, we should write

$$\ln x^2 = 2 \ln x \text{ for } x > 0.$$

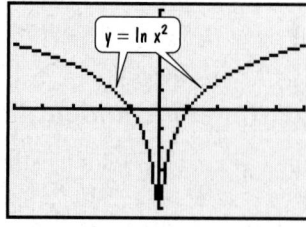

Domain: $(-\infty, 0)$ or $(0, \infty)$

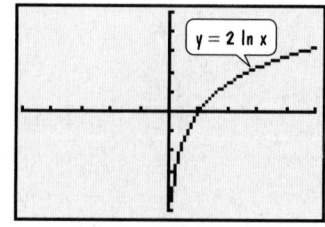

Domain: $(0, \infty)$

Figure 4.15 $\ln x^2$ and $2 \ln x$ have different domains.

Study Tip

The graphs show

$$y_1 = \ln(x + 3)$$

and

$$y_2 = \ln x + \ln 3.$$

The graphs are not the same. The graph of y_1 is the graph of the natural logarithmic function shifted 3 units to the left. By contrast, the graph of y_2 is the graph of the natural logarithmic function shifted upward by $\ln 3$, or about 1.1 units. Thus we see that

$$\ln(x + 3) \neq \ln x + \ln 3.$$

In general,

$$\log_b(M + N) \neq \log_b M + \log_b N.$$

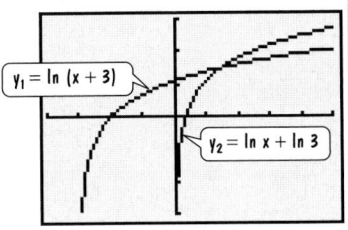

Try to avoid the following errors:

INCORRECT

$$\log_b(M + N) = \log_b M + \log_b N$$

$$\log_b(M - N) = \log_b M - \log_b N$$

$$\log_b(M \cdot N) = \log_b M \cdot \log_b N$$

$$\log_b\left(\frac{M}{N}\right) = \frac{\log_b M}{\log_b N}$$

$$\frac{\log_b M}{\log_b N} = \log_b M - \log_b N$$

4 Expand logarithmic expressions.

When expanding a logarithmic expression, you might want to determine whether the rewriting has changed the domain of the expression. For the rest of this section, assume that all variable and variable expressions represent positive numbers.

EXAMPLE 3   Using the Power Rule

Use the power rule to expand each logarithmic expression:

a. $\log_5 7^4$ **b.** $\ln \sqrt{x}$.

Solution

a. $\log_5 7^4 = 4 \log_5 7$ *The logarithm of a number with an exponent is the exponent times the logarithm of the number.*

b. $\ln \sqrt{x} = \ln x^{1/2}$ *Rewrite the radical using a rational exponent.*

$\qquad\quad\; = \frac{1}{2} \ln x$ *Use the power rule to bring the exponent to the front.*

Check Point 3   Use the power rule to expand each logarithmic expression:

a. $\log_6 3^9$ **b.** $\ln \sqrt[3]{x}$.

Expanding Logarithmic Expressions

It is sometimes necessary to use more than one property of logarithms when you expand a logarithmic expression. Properties for expanding logarithmic expressions are as follows:

Properties for Expanding Logarithmic Expressions

For $M > 0$ and $N > 0$:

1. $\log_b(MN) = \log_b M + \log_b N$ *Product rule*

2. $\log_b\left(\dfrac{M}{N}\right) = \log_b M - \log_b N$ *Quotient rule*

3. $\log_b M^p = p \log_b M$ *Power rule*

EXAMPLE 4   Expanding Logarithmic Expressions

Use logarithmic properties to expand each expression as much as possible:

a. $\log_b(x^2\sqrt{y})$ **b.** $\log_6\left(\dfrac{\sqrt[3]{x}}{36y^4}\right)$.

Solution   We will have to use two or more of the properties for expanding logarithms in each part of this example.

a. $\log_b(x^2\sqrt{y}) = \log_b(x^2 y^{1/2})$ *Use exponential notation.*

$\qquad\qquad\quad\; = \log_b x^2 + \log_b y^{1/2}$ *Use the product rule.*

$\qquad\qquad\quad\; = 2 \log_b x + \dfrac{1}{2} \log_b y$ *Use the power rule.*

b. $\log_6\left(\dfrac{\sqrt[3]{x}}{36y^4}\right) = \log_6\left(\dfrac{x^{1/3}}{36y^4}\right)$ *Use exponential notation.*

$$= \log_6 x^{1/3} - \log_6(36y^4) \qquad \text{\textit{Use the quotient rule.}}$$

$$= \log_6 x^{1/3} - \left(\log_6 36 + \log_6 y^4\right) \qquad \text{\textit{Use the product rule on}}$$
$\log_6(36y^4)$.

$$= \frac{1}{3}\log_6 x - \left(\log_6 36 + 4\log_6 y\right) \qquad \text{\textit{Use the power rule.}}$$

$$= \frac{1}{3}\log_6 x - \log_6 36 - 4\log_6 y \qquad \text{\textit{Apply the distributive property.}}$$

$$= \frac{1}{3}\log_6 x - 2 - 4\log_6 y \qquad \text{\textit{$\log_6 36 = 2$ because 2 is the}}$$
power to which we must raise
6 to get 36. ($6^2 = 36$)

Check Point 4   Use logarithmic properties to expand each expression as much as possible:

    a. $\log_b\left(x^4\sqrt[3]{y}\right)$ **b.** $\log_5 \dfrac{\sqrt{x}}{25y^3}.$

5   Condense logarithmic expressions.

Condensing Logarithmic Expressions

To **condense a logarithmic expression,** we write the sum or difference of two or more logarithmic expressions as a single logarithmic expression. We use the properties of logarithms to do so.

Study Tip

These properties are the same as those in the box on page 401. The only difference is that we've reversed the sides in each property from the previous box.

Properties for Condensing Logarithmic Expressions

For $M > 0$ and $N > 0$:

1. $\log_b M + \log_b N = \log_b(MN)$ *Product rule*

2. $\log_b M - \log_b N = \log_b\left(\dfrac{M}{N}\right)$ *Quotient rule*

3. $p\log_b M = \log_b M^p$ *Power rule*

EXAMPLE 5   **Condensing Logarithmic Expressions**

Write as a single logarithm:

    a. $\log_4 2 + \log_4 32$ **b.** $\log(4x - 3) - \log x.$

Solution

    a. $\log_4 2 + \log_4 32 = \log_4(2 \cdot 32)$ *Use the product rule.*

$$= \log_4 64 \qquad \text{\textit{We now have a single logarithm.}}$$
However, we can simplify.

$$= 3 \qquad \text{\textit{$\log_4 64 = 3$ because $4^3 = 64$.}}$$

    b. $\log(4x - 3) - \log x = \log\left(\dfrac{4x - 3}{x}\right)$ *Use the quotient rule.*

Check Point 5 Write as a single logarithm:

    a. $\log 25 + \log 4$ **b.** $\log (7x + 6) - \log x$.

Coefficients of logarithms must be 1 before you can condense them using the product and quotient rules. For example, to condense

$$2 \ln x + \ln (x + 1),$$

the coefficient of the first term must be 1. We use the power rule to rewrite the coefficient as an exponent:

> 1. Use the power rule to make the number in front an exponent.

$$2 \ln x + \ln (x + 1) = \ln x^2 + \ln (x + 1) = \ln [x^2(x + 1)].$$

> 2. Use the product rule. The sum of logarithms with coefficients 1 is the logarithm of the product.

EXAMPLE 6 Condensing Logarithmic Expressions

Write as a single logarithm:

    a. $\frac{1}{2} \log x + 4 \log (x - 1)$ **b.** $3 \ln (x + 7) - \ln x$

    c. $4 \log_b x - 2 \log_b 6 + \frac{1}{2} \log_b y$.

Solution

a. $\frac{1}{2} \log x + 4 \log (x - 1)$

$= \log x^{1/2} + \log (x - 1)^4$ Use the power rule so that all coefficients are 1.

$= \log [x^{1/2} (x - 1)^4]$ Use the product rule. The condensation can be expressed as $\log \left[\sqrt{x} \, (x - 1)^4 \right]$.

b. $3 \ln (x + 7) - \ln x$

$= \ln (x + 7)^3 - \ln x$ Use the power rule so that all coefficients are 1.

$= \ln \left[\dfrac{(x + 7)^3}{x} \right]$ Use the quotient rule.

c. $4 \log_b x - 2 \log_b 6 + \frac{1}{2} \log_b y$

$= \log_b x^4 - \log_b 6^2 + \log_b y^{1/2}$ Use the power rule so that all coefficients are 1.

$= (\log_b x^4 - \log_b 36) + \log_b y^{1/2}$ This optional step emphasizes the order of operations.

$= \log_b \left(\dfrac{x^4}{36} \right) + \log_b y^{1/2}$ Use the quotient rule.

$= \log_b \left(\dfrac{x^4}{36} \cdot y^{1/2} \right)$ or $\log \left(\dfrac{x^4 \sqrt{y}}{36} \right)$ Use the product rule.

Check Point 6 Write as a single logarithm:

    a. $2 \ln x + \frac{1}{3} \ln (x + 5)$ **b.** $2 \log (x - 3) - \log x$

    c. $\frac{1}{4} \log_b x - 2 \log_b 5 + 10 \log_b y$.

6 Use the change-of-base property.

The Change-of-Base Property

We have seen that calculators give the values of both common logarithms (base 10) and natural logarithms (base e). To find a logarithm with any other base, we can use the following change-of-base property:

> **The Change-of-Base Property**
>
> For any logarithmic bases a and b, and any positive number M,
> $$\log_b M = \frac{\log_a M}{\log_a b}.$$
>
> The logarithm of M with base b is equal to the logarithm of M with any new base divided by the logarithm of b with that new base.

In the change-of-base property, base b is the base of the original logarithm. Base a is a new base that we introduce. Thus, the change-of-base property allows us to change from base b to *any* new base a, as long as the newly introduced base is a positive number not equal to 1.

The change-of-base property is used to write a logarithm in terms of quantities that can be evaluated with a calculator. Because calculators contain keys for common (base 10) and natural (base e) logarithms, we will frequently introduce base 10 or base e.

Change-of-Base Property	Introducing Common Logarithms	Introducing Natural Logarithms
$\log_b M = \dfrac{\log_a M}{\log_a b}$	$\log_b M = \dfrac{\log_{10} M}{\log_{10} b}$	$\log_b M = \dfrac{\log_e M}{\log_e b}$
a is the new introduced base.	10 is the new introduced base.	e is the new introduced base.

Using the notations for common logarithms and natural logarithms, we have the following results:

> **The Change-of-Base Property: Introducing Common and Natural Logarithms**
>
> Introducing Common Logarithms   Introducing Natural Logarithms
> $$\log_b M = \frac{\log M}{\log b} \qquad\qquad \log_b M = \frac{\ln M}{\ln b}$$

EXAMPLE 7 Changing Base to Common Logarithms

Use common logarithms to evaluate $\log_5 140$.

Solution Because $\log_b M = \dfrac{\log M}{\log b}$,

$$\log_5 140 = \frac{\log 140}{\log 5}$$

$$\approx 3.07. \qquad \text{Use a calculator: } 140 \boxed{\text{LOG}} \ \boxed{\div} \ 5 \boxed{\text{LOG}}$$
$$\boxed{=} \text{ or } \boxed{\text{LOG}} \ 140 \ \boxed{\div} \ \boxed{\text{LOG}} \ 5 \ \boxed{\text{ENTER}}.$$

This means that $\log_5 140 \approx 3.07$.

Discovery

Find a reasonable estimate of $\log_5 140$ to the nearest whole number. 5 to what power is 140? Compare your estimate to the value obtained in Example 7.

Check Point 7 Use common logarithms to evaluate $\log_7 2506$.

EXAMPLE 8 Changing Base to Natural Logarithms

Use natural logarithms to evaluate $\log_5 140$.

Solution Because $\log_b M = \dfrac{\ln M}{\ln b}$,

$$\log_5 140 = \frac{\ln 140}{\ln 5}$$

$$\approx 3.07.$$

Use a calculator: 140 [LN] [÷] 5 [LN]
[=] or [LN] 140 [÷] [LN] 5 [ENTER].

We have again shown that $\log_5 140 \approx 3.07$.

Check Point 8 Use natural logarithms to evaluate $\log_7 2506$.

We can use the change-of-base property to graph logarithmic functions with bases other than 10 or e on a graphing utility. For example, Figure 4.16 shows the graphs of

$$y = \log_2 x \quad \text{and} \quad y = \log_{20} x$$

in a [0, 10, 1] by [−3, 3, 1] viewing rectangle. Because $\log_2 x = \dfrac{\ln x}{\ln 2}$ and $\log_{20} x = \dfrac{\ln x}{\ln 20}$, the functions can be entered as

$$y_1 = \boxed{\text{LN}} \; x \; \boxed{÷} \; \boxed{\text{LN}} \; 2$$

$$\text{and} \quad y_2 = \boxed{\text{LN}} \; x \; \boxed{÷} \; \boxed{\text{LN}} \; 20.$$

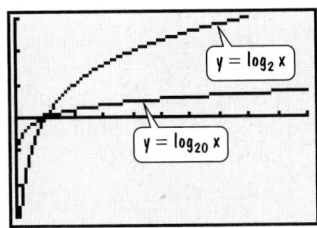

Figure 4.16 Using the change-of-base property to graph logarithmic functions

EXERCISE SET 4.3

Practice Exercises

In Exercises 1–40, use properties of logarithms to expand each logarithmic expression as much as possible. Where possible, evaluate logarithmic expressions without using a calculator.

1. $\log_5 (7 \cdot 3)$

2. $\log_8 (13 \cdot 7)$

3. $\log_7 (7x)$

4. $\log_9 (9x)$

5. $\log (1000x)$

6. $\log (10{,}000x)$

7. $\log_7 \left(\dfrac{7}{x} \right)$

8. $\log_9 \left(\dfrac{9}{x} \right)$

9. $\log \left(\dfrac{x}{100} \right)$

10. $\log \left(\dfrac{x}{1000} \right)$

11. $\log_4 \left(\dfrac{64}{y} \right)$

12. $\log_5 \left(\dfrac{125}{y} \right)$

13. $\ln \left(\dfrac{e^2}{5} \right)$

14. $\ln \left(\dfrac{e^4}{8} \right)$

15. $\log_b x^3$

16. $\log_b x^7$

17. $\log N^{-6}$

18. $\log M^{-8}$

19. $\ln \sqrt[5]{x}$

20. $\ln \sqrt[7]{x}$

21. $\log_b (x^2 y)$

22. $\log_b (xy^3)$

23. $\log_4 \left(\dfrac{\sqrt{x}}{64} \right)$

24. $\log_5 \left(\dfrac{\sqrt{x}}{25} \right)$

25. $\log_6 \left(\dfrac{36}{\sqrt{x+1}} \right)$

26. $\log_8 \left(\dfrac{64}{\sqrt{x+1}} \right)$

27. $\log_b \left(\dfrac{x^2 y}{z^2} \right)$

28. $\log_b \left(\dfrac{x^3 y}{z^2} \right)$

29. $\log \sqrt{100x}$

30. $\ln \sqrt{ex}$

31. $\log \sqrt[3]{\dfrac{x}{y}}$

32. $\log \sqrt[5]{\dfrac{x}{y}}$

33. $\log_b \left(\dfrac{\sqrt{x}y^3}{z^3} \right)$

34. $\log_b \left(\dfrac{\sqrt[3]{x}y^4}{z^5} \right)$

35. $\log_5 \sqrt[3]{\dfrac{x^2 y}{25}}$

36. $\log_2 \sqrt[5]{\dfrac{xy^4}{16}}$

37. $\ln \left[\dfrac{x^3 \sqrt{x^2+1}}{(x+1)^4} \right]$

38. $\ln \left[\dfrac{x^4 \sqrt{x^2+3}}{(x+3)^5} \right]$

39. $\log \left[\dfrac{10x^2 \sqrt[3]{1-x}}{7(x+1)^2} \right]$

40. $\log \left[\dfrac{100x^3 \sqrt[3]{5-x}}{3(x+7)^2} \right]$

In Exercises 41–70, use properties of logarithms to condense each logarithmic expression. Write the expression as a single logarithm whose coefficient is 1. Where possible, evaluate logarithmic expressions.

41. $\log 5 + \log 2$

42. $\log 250 + \log 4$

43. $\ln x + \ln 7$

44. $\ln x + \ln 3$

45. $\log_2 96 - \log_2 3$

46. $\log_3 405 - \log_3 5$

47. $\log (2x + 5) - \log x$

48. $\log (3x + 7) - \log x$

49. $\log x + 3 \log y$

50. $\log x + 7 \log y$

51. $\frac{1}{2} \ln x + \ln y$

52. $\frac{1}{3} \ln x + \ln y$

53. $2 \log_b x + 3 \log_b y$

54. $5 \log_b x + 6 \log_b y$

55. $5 \ln x - 2 \ln y$

56. $7 \ln x - 3 \ln y$

57. $3 \ln x - \frac{1}{3} \ln y$

58. $2 \ln x - \frac{1}{2} \ln y$

59. $4 \ln (x + 6) - 3 \ln x$

60. $8 \ln (x + 9) - 4 \ln x$

61. $3 \ln x + 5 \ln y - 6 \ln z$

62. $4 \ln x + 7 \ln y - 3 \ln z$

63. $\frac{1}{2} (\log x + \log y)$

64. $\frac{1}{3} (\log_4 x - \log_4 y)$

65. $\frac{1}{2} (\log_5 x + \log_5 y) - 2 \log_5 (x + 1)$

66. $\frac{1}{3} (\log_4 x - \log_4 y) + 2 \log_4 (x + 1)$

67. $\frac{1}{3} [2 \ln(x+5) - \ln x - \ln(x^2-4)]$

68. $\frac{1}{3} [5 \ln(x+6) - \ln x - \ln(x^2-25)]$

69. $\log x + \log 7 + \log(x^2 - 1) - \log(x + 1)$

70. $\log x + \log 15 + \log(x^2 - 4) - \log(x + 2)$

In Exercises 71–78, use common logarithms or natural logarithms and a calculator to evaluate to four decimal places.

71. $\log_5 13$

72. $\log_6 17$

73. $\log_{14} 87.5$

74. $\log_{16} 57.2$

75. $\log_{0.1} 17$

76. $\log_{0.3} 19$

77. $\log_\pi 63$

78. $\log_\pi 400$

In Exercises 79–82, use a graphing utility and the change-of-base property to graph each function.

79. $y = \log_3 x$

80. $y = \log_{15} x$

81. $y = \log_2 (x + 2)$

82. $y = \log_3 (x - 2)$

Application Exercises

83. The loudness level of a sound can be expressed by comparing the sound's intensity to the intensity of a sound barely audible to the human ear. The formula

$$D = 10(\log I - \log I_0)$$

describes the loudness level of a sound, D, in decibels, where I is the intensity of the sound, in watts per meter^2, and I_0 is the intensity of a sound barely audible to the human ear.

a. Express the formula so that the expression in parentheses is written as a single logarithm.

b. Use the form of the formula from part (a) to answer this question: If a sound has an intensity 100 times the intensity of a softer sound, how much larger on the decibel scale is the loudness level of the more intense sound?

84. The formula

$$t = \frac{1}{c} \left[\ln A - \ln(A - N) \right]$$

describes the time, t, in weeks, that it takes to achieve mastery of a portion of a task, where A is the maximum learning possible, N is the portion of the learning that is to be achieved, and c is a constant used to measure an individual's learning style.

a. Express the formula so that the expression in brackets is written as a single logarithm.

b. The formula is also used to determine how long it will take chimpanzees and apes to master a task. For example, a typical chimpanzee learning sign language can master a maximum of 65 signs. Use the form of the formula from part (a) to answer this question: How many weeks will it take a chimpanzee to master 30 signs if c for that chimp is 0.03?

Writing in Mathematics

85. Describe the product rule for logarithms and give an example.

86. Describe the quotient rule for logarithms and give an example.

87. Describe the power rule for logarithms and give an example.

88. Without showing the details, explain how to condense $\ln x - 2 \ln (x + 1)$.

89. Describe the change-of-base property and give an example.

90. Explain how to use your calculator to find $\log_{14} 283$.

91. You overhear a student talking about a property of logarithms in which division becomes subtraction. Explain what the student means by this.

92. Find $\ln 2$ using a calculator. Then calculate each of the following: $1 - \frac{1}{2}$; $1 - \frac{1}{2} + \frac{1}{3}$; $1 - \frac{1}{2} + \frac{1}{3} - \frac{1}{4}$; $1 - \frac{1}{2} + \frac{1}{3} - \frac{1}{4} + \frac{1}{5}$; .... Describe what you observe.

 Technology Exercises

93. a. Use a graphing utility (and the change-of-base property) to graph $y = \log_3 x$.

b. Graph $y = 2 + \log_3 x$, $y = \log_3(x + 2)$, and $y = -\log_3 x$ in the same viewing rectangle as $y = \log_3 x$. Then describe the change or changes that need to be made to the graph of $y = \log_3 x$ to obtain each of these three graphs.

94. Graph $y = \log x$, $y = \log(10x)$, and $y = \log(0.1x)$ in the same viewing rectangle. Describe the relationship among the three graphs. What logarithmic property accounts for this relationship?

95. Use a graphing utility and the change-of-base property to graph $y = \log_3 x$, $y = \log_{25} x$, and $y = \log_{100} x$ in the same viewing rectangle.

a. Which graph is on the top in the interval $(0, 1)$? Which is on the bottom?

b. Which graph is on the top in the interval $(1, \infty)$? Which is on the bottom?

c. Generalize by writing a statement about which graph is on top, which is on the bottom, and in which intervals, using $y = \log_b x$ where $b > 1$.

Disprove each statement in Exercises 96–100 by

a. *letting y equal a positive constant of your choice.*

b. *using a graphing utility to graph the function on each side of the equal sign. The two functions should have different graphs, showing that the equation is not true in general.*

96. $\log(x + y) = \log x + \log y$ **97.** $\log \dfrac{x}{y} = \dfrac{\log x}{\log y}$

98. $\ln(x - y) = \ln x - \ln y$ **99.** $\ln(xy) = (\ln x)(\ln y)$

100. $\dfrac{\ln x}{\ln y} = \ln x - \ln y$

 Critical Thinking Exercises

101. Which one of the following is true?

a. $\dfrac{\log_7 49}{\log_7 7} = \log_7 49 - \log_7 7$

b. $\log_b(x^3 + y^3) = 3 \log_b x + 3 \log_b y$

c. $\log_b(xy)^5 = (\log_b x + \log_b y)^5$

d. $\ln \sqrt{2} = \dfrac{\ln 2}{2}$

102. Use the change-of-base property to prove that
$$\log e = \frac{1}{\ln 10}.$$

103. If $\log 3 = A$ and $\log 7 = B$, find $\log_7 9$ in terms of A and B.

104. Write as a single term that does not contain a logarithm:
$$e^{\ln 8x^5 - \ln 2x^2}.$$

105. If $f(x) = \log_b x$, show that
$$\frac{f(x + h) - f(x)}{h} = \log_b\left(1 + \frac{h}{x}\right)^{1/h}, h \ne 0.$$

SECTION 4.4 Exponential and Logarithmic Equations

Objectives

1. Solve exponential equations.

2. Solve logarithmic equations.

3. Solve applied problems involving exponential and logarithmic equations.

Is an early retirement awaiting you?

You inherited $30,000. You'd like to put aside $25,000 and eventually have over half a million dollars for early retirement. Is this possible? In this section, you will see how techniques for solving equations with variable exponents provide an answer to the question.

1 Solve exponential equations.

Exponential Equations

An **exponential equation** is an equation containing a variable in an exponent. Examples of exponential equations include

$$3^x = 81, \quad 4^x = 15, \quad \text{and} \quad 40e^{0.6x} = 240.$$

Each side of the first equation can be expressed with the same base. Can you see that we can rewrite

$$3^x = 81 \quad \text{as} \quad 3^x = 3^4?$$

All exponential functions are one-to-one—that is, if b is a positive number other than 1 and $b^M = b^N$, then $M = N$. Because we have expressed $3^x = 81$ as $3^x = 3^4$, we conclude that $x = 4$. The equation's solution set is {4}.

Most exponential equations cannot be rewritten so that each side has the same base. Logarithms are extremely useful in solving such equations. The solution begins with isolating the exponential expression and taking the natural logarithm on both sides. Why can we do this? All logarithmic relations are functions. Thus, if M and N are positive real numbers and $M = N$, then $\log_b M = \log_b N$.

Using Natural Logarithms to Solve Exponential Equations

1. Isolate the exponential expression.
2. Take the natural logarithm on both sides of the equation.
3. Simplify using one of the following properties:

$$\ln b^x = x \ln b \quad \text{or} \quad \ln e^x = x.$$

4. Solve for the variable.

EXAMPLE 1   Solving an Exponential Equation

Solve: $4^x = 15$.

Solution   Because the exponential expression, 4^x, is already isolated on the left, we begin by taking the natural logarithm on both sides of the equation.

$4^x = 15$ This is the given equation.

$\ln 4^x = \ln 15$ Take the natural logarithm on both sides.

$x \ln 4 = \ln 15$ Use the power rule and bring the variable exponent to the front: $\ln b^x = x \ln b$.

$x = \dfrac{\ln 15}{\ln 4}$ Solve for x by dividing both sides by ln 4.

Discovery

The base that is used when taking the logarithm on both sides of an equation can be any base at all. Solve $4^x = 15$ by taking the common logarithm on both sides. Solve again, this time taking the logarithm with base 4 on both sides. Use the change-of-base property to show that the solutions are the same as the one obtained in Example 1.

We now have an exact value for x. We use the exact value for x in the equation's solution set. Thus, the equation's solution is $\dfrac{\ln 15}{\ln 4}$ and the solution set is $\left\{ \dfrac{\ln 15}{\ln 4} \right\}$.

We can obtain a decimal approximation by using a calculator: $x \approx 1.95$. Because $4^2 = 16$, it seems reasonable that the solution to $4^x = 15$ is approximately 1.95

Check Point 1   Solve: $5^x = 134$. Find the solution set and then use a calculator to obtain a decimal approximation to two decimal places for the solution.

EXAMPLE 2   Solving an Exponential Equation

Solve: $40e^{0.6x} = 240$.

Solution   We begin by dividing both sides by 40 to isolate the exponential expression, $e^{0.6x}$. Then we take the natural logarithm on both sides of the equation.

$40e^{0.6x} = 240$ This is the given equation.

$e^{0.6x} = 6$ Isolate the exponential factor by dividing both sides by 40.

$\ln e^{0.6x} = \ln 6$ Take the natural logarithm on both sides.

$0.6x = \ln 6$ Use the inverse property $\ln e^x = x$ on the left.

$x = \dfrac{\ln 6}{0.6} \approx 2.99$ Divide both sides by 0.6.

Thus, the solution of the equation is $\dfrac{\ln 6}{0.6} \approx 2.99$. Try checking this approximate solution in the original equation to verify that $\left\{ \dfrac{\ln 6}{0.6} \right\}$ is the solution set.

Check Point 2   Solve: $7e^{2x} = 63$. Find the solution set and then use a calculator to obtain a decimal approximation to two decimal places for the solution.

EXAMPLE 3   Solving an Exponential Equation

Solve: $5^{4x-7} - 3 = 10$

Solution   We begin by adding 3 to both sides to isolate the exponential expression, 5^{4x-7}. Then we take the natural logarithm on both sides of the equation.

$5^{4x-7} - 3 = 10$ This is the given equation.

$5^{4x-7} = 13$ Add 3 to both sides.

$\ln 5^{4x-7} = \ln 13$ Take the natural logarithm on both sides.

$(4x - 7)\ln 5 = \ln 13$ Use the power rule to bring the exponent to the front: $\ln M^p = p\ln M$.

$4x\ln 5 - 7\ln 5 = \ln 13$ Use the distributive property and distribute $\ln 5$ to both terms in parentheses.

$4x\ln 5 = \ln 13 + 7\ln 5$ Isolate the variable term by adding $7\ln 5$ to both sides.

$x = \dfrac{\ln 13 + 7\ln 5}{4\ln 5}$ Isolate x by dividing both sides by $4\ln 5$.

The solution set is $\left\{ \dfrac{\ln 13 + 7\ln 5}{4\ln 5} \right\}$. The solution is approximately 2.15.

Check Point 3
Solve: $6^{3x-4} - 7 = 2081$. Find the solution set and then use a calculator to obtain a decimal approximation to two decimal places for the solution.

EXAMPLE 4   Solving an Exponential Equation

Solve: $e^{2x} - 4e^x + 3 = 0$.

Solution   The given equation is quadratic in form. If $t = e^x$, the equation can be expressed as $t^2 - 4t + 3 = 0$. Because this equation can be solved by factoring, we factor to isolate the exponential term.

Technology

Shown below is the graph of $y = e^{2x} - 4e^x + 3$. There are two x-intercepts, one at 0 and one at approximately 1.10. These intercepts verify our algebraic solution.

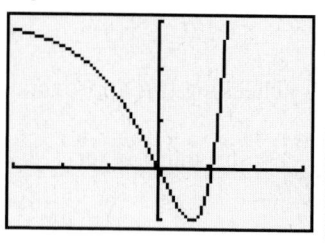

$e^{2x} - 4e^x + 3 = 0$	This is the given equation.
$(e^x - 3)(e^x - 1) = 0$	Factor on the left. Notice that if $t = e^x$, $t^2 - 4t + 3 = (t - 3)(t - 1)$.
$e^x - 3 = 0$ or $e^x - 1 = 0$	Set each factor equal to 0.
$e^x = 3$ $\quad$ $e^x = 1$	Solve for e^x.
$\ln e^x = \ln 3$ $\quad$ $x = 0$	Take the natural logarithm on both sides of the first equation. The equation on the right can be solved by inspection.
$x = \ln 3$	$\ln e^x = x$

The solution set is $\{0, \ln 3\}$. The solutions are 0 and approximately 1.10.

Check Point 4
Solve: $e^{2x} - 8e^x + 7 = 0$. Find the solution set and then use a calculator to obtain a decimal approximation to two decimal places, if necessary, for the solutions.

2 Solve logarithmic equations.

Logarithmic Equations

A **logarithmic equation** is an equation containing a variable in a logarithmic expression. Examples of logarithmic equations include

$$\log_4 (x + 3) = 2 \quad \text{and} \quad \ln (2x) = 3.$$

If a logarithmic equation is in the form $\log_b x = c$, we can solve the equation by rewriting it in its equivalent exponential form $b^c = x$. Example 5 illustrates how this is done.

EXAMPLE 5   Solving a Logarithmic Equation

Solve: $\log_4 (x + 3) = 2$.

Solution   We first rewrite the equation as an equivalent equation in exponential form using the fact that $\log_b x = c$ means $b^c = x$.

$$\log_4 (x + 3) = 2 \quad \text{means} \quad 4^2 = x + 3$$

Logarithms are exponents.

Technology

The graphs of
$y_1 = \log_4(x + 3)$ and $y_2 = 2$
have an intersection point
whose x-coordinate is 13. This
verifies that {13} is the solution
set for $\log_4(x + 3) = 2$.

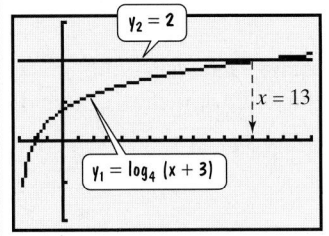

$[-3, 17, 1]$ by $[-2, 3, 1]$

Note:
Because

$$\log_b x = \frac{\ln x}{\ln b}$$

(change-of-base property),
we entered y_1 using

$$y_1 = \frac{\ln(x + 3)}{\ln 4}.$$

Now we solve the equivalent equation for x.

$$4^2 = x + 3 \qquad \text{This is the equation equivalent to } \log_4(x + 3) = 2.$$
$$16 = x + 3 \qquad \text{Square 4.}$$
$$13 = x \qquad \text{Subtract 3 from both sides.}$$

Check 13:

$$\log_4(x + 3) = 2 \qquad \text{This is the given logarithmic equation.}$$
$$\log_4(13 + 3) \stackrel{?}{=} 2 \qquad \text{Substitute 13 for x.}$$
$$\log_4 16 \stackrel{?}{=} 2$$
$$2 = 2 \checkmark \qquad \log_4 16 = 2 \text{ because } 4^2 = 16.$$

This true statement indicates that the solution set is {13}.

Check Point 5 Solve: $\log_2(x - 4) = 3$.

Logarithmic expressions are defined only for logarithms of positive real numbers. Always check proposed solutions of a logarithmic equation in the original equation. Exclude from the solution set any proposed solution that produces the logarithm of a negative number or the logarithm of 0.

To rewrite the logarithmic equation $\log_b x = c$ in the equivalent exponential form $b^c = x$, we need a single logarithm whose coefficient is one. It is sometimes necessary to use properties of logarithms to condense logarithms into a single logarithm. In the next example, we use the product rule for logarithms to obtain a single logarithmic expression on the left side.

EXAMPLE 6   Using the Product Rule to Solve a Logarithmic Equation

Solve: $\log_2 x + \log_2(x - 7) = 3$.

Solution

$$\log_2 x + \log_2(x - 7) = 3 \qquad \text{This is the given equation.}$$
$$\log_2[x(x - 7)] = 3 \qquad \begin{array}{l}\text{Use the product rule to obtain a single} \\ \text{logarithm: } \log_b M + \log_b N = \log_b(MN).\end{array}$$
$$2^3 = x(x - 7) \qquad \log_b x = c \text{ means } b^c = x.$$
$$8 = x^2 - 7x \qquad \begin{array}{l}\text{Apply the distributive property} \\ \text{on the right and evaluate } 2^3 \text{ on the left.}\end{array}$$
$$0 = x^2 - 7x - 8 \qquad \text{Set the equation equal to 0.}$$
$$0 = (x - 8)(x + 1) \qquad \text{Factor.}$$
$$x - 8 = 0 \quad \text{or} \quad x + 1 = 0 \qquad \text{Set each factor equal to 0.}$$
$$x = 8 \qquad\qquad x = -1 \qquad \text{Solve for x.}$$

Check 8:

$$\log_2 x + \log_2(x - 7) = 3$$
$$\log_2 8 + \log_2(8 - 7) \stackrel{?}{=} 3$$
$$\log_2 8 + \log_2 1 \stackrel{?}{=} 3$$
$$3 + 0 \stackrel{?}{=} 3$$
$$3 = 3 \checkmark$$

Check −1:

$$\log_2 x + \log_2(x - 7) = 3$$
$$\log_2(-1) + \log_2(-1 - 7) \stackrel{?}{=} 3$$

The number −1 does not check. Negative numbers do not have logarithms.

The solution set is {8}.

Check Point 6 Solve: $\log x + \log (x - 3) = 1$.

Equations involving natural logarithms can be solved using the inverse property $e^{\ln x} = x$. For example, to solve

$$\ln x = 5$$

we write both sides of the equation as exponents on base e:

$$e^{\ln x} = e^5.$$

This is called **exponentiating both sides** of the equation. Using the inverse property $e^{\ln x} = x$, we simplify the left side of the equation and obtain the solution:

$$x = e^5.$$

EXAMPLE 7 Solving an Equation with a Natural Logarithm

Solve: $3 \ln (2x) = 12$.

Solution

$3 \ln(2x) = 12$	This is the given equation.
$\ln (2x) = 4$	Divide both sides by 3.
$e^{\ln (2x)} = e^4$	Exponentiate both sides.
$2x = e^4$	Use the inverse property to simplify the left side: $e^{\ln x} = x$.
$x = \dfrac{e^4}{2} \approx 27.30$	Divide both sides by 2.

Check $\dfrac{e^4}{2}$:

$$3 \ln (2x) = 12 \quad \text{This is the given logarithmic equation.}$$
$$3 \ln \left[2\left(\frac{e^4}{2}\right)\right] \stackrel{?}{=} 12 \quad \text{Substitute } \frac{e^4}{2} \text{ for x.}$$
$$3 \ln e^4 \stackrel{?}{=} 12 \quad \text{Simplify: } \frac{2}{1} \cdot \frac{e^4}{2} = e^4.$$
$$3 \cdot 4 \stackrel{?}{=} 12 \quad \text{Because } \ln e^x = x, \text{ we conclude } \ln e^4 = 4.$$
$$12 = 12 \checkmark$$

This true statement indicates that the solution set is $\left\{\dfrac{e^4}{2}\right\}$.

Check Point 7 Solve: $4 \ln 3x = 8$.

3 Solve applied problems involving exponential and logarithmic equations.

Applications

Our first applied example provides a mathematical perspective on the old slogan "Alcohol and driving don't mix." In California, where 38% of fatal traffic crashes involve drinking drivers, it is illegal to drive with a blood alcohol concentration of 0.08 or higher. At these levels, drivers may be arrested and charged with driving under the influence.

EXAMPLE 8 Alcohol and Risk of a Car Accident

Medical research indicates that the risk of having a car accident increases exponentially as the concentration of alcohol in the blood increases. The risk is modeled by

$$R = 6e^{12.77x}$$

where x is the blood alcohol concentration and R, given as a percent, is the risk of having a car accident. What blood alcohol concentration corresponds to a 17% risk of a car accident?

Solution For a risk of 17%, we let $R = 17$ in the equation and solve for x, the blood alcohol concentration.

$R = 6e^{12.77x}$	This is the given equation.
$6e^{12.77x} = 17$	Substitute 17 for R and (optional) reverse the two sides of the equation.
$e^{12.77x} = \dfrac{17}{6}$	Isolate the exponential factor by dividing both sides by 6.
$\ln e^{12.77x} = \ln\left(\dfrac{17}{6}\right)$	Take the natural logarithm on both sides.
$12.77x = \ln\left(\dfrac{17}{6}\right)$	Use the inverse property $\ln e^x = x$ on the left.
$x = \dfrac{\ln\left(\dfrac{17}{6}\right)}{12.77} \approx 0.08$	Divide both sides by 12.77.

For a blood alcohol concentration of 0.08, the risk of a car accident is 17%. In many states, it is illegal to drive at this blood alcohol concentration.

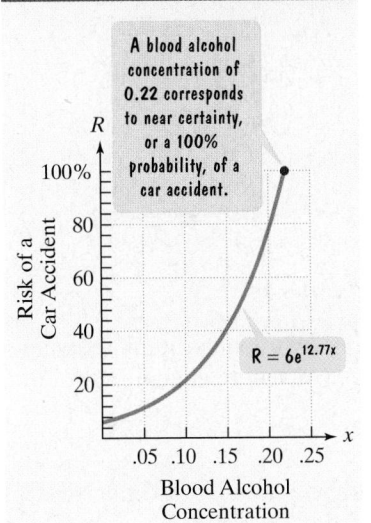

Visualizing the Relationship between Blood Alcohol Concentration and the Risk of a Car Accident

A blood alcohol concentration of 0.22 corresponds to near certainty, or a 100% probability, of a car accident.

$R = 6e^{12.77x}$

Risk of a Car Accident

Blood Alcohol Concentration

Check Point 8 Use the formula in Example 8 to answer this question: What blood alcohol concentration corresponds to a 7% risk of a car accident? (In many states, drivers under the age of 21 can lose their license for driving at this level.)

Suppose that you inherit $30,000. Is it possible to invest $25,000 and have over half a million dollars for early retirement? Our next example illustrates the power of compound interest.

EXAMPLE 9 Revisiting the Formula for Compound Interest

The formula

$$A = P\left(1 + \frac{r}{n}\right)^{nt}$$

describes the accumulated value, A, of a sum of money, P, the principal, after t years at annual percentage rate r (in decimal form) compounded n times a year. How long will it take $25,000 to grow to $500,000 at 9% annual interest compounded monthly?

Solution

$$A = P\left(1 + \frac{r}{n}\right)^{nt}$$ *This is the given formula.*

$$500{,}000 = 25{,}000\left(1 + \frac{0.09}{12}\right)^{12t}$$ *A (the desired accumulated value) = $500,000,*
P (the principal) = $25,000,
r (the interest rate) = 9% = 0.09, and n = 12
(monthly compounding).

Our goal is to solve the equation for t. Let's reverse the two sides of the equation and then simplify within parentheses.

$$25{,}000\left(1 + \frac{0.09}{12}\right)^{12t} = 500{,}000$$ *Reverse the two sides of the previous equation.*

$$25{,}000(1 + 0.0075)^{12t} = 500{,}000$$ *Divide within parentheses: $\frac{0.09}{12} = 0.0075$.*

$$25{,}000(1.0075)^{12t} = 500{,}000$$ *Add within parentheses.*

$$(1.0075)^{12t} = 20$$ *Divide both sides by 25,000.*

$$\ln(1.0075)^{12t} = \ln 20$$ *Take the natural logarithm on both sides.*

$$12t \ln(1.0075) = \ln 20$$ *Use the power rule to bring the exponent to the front: $\ln M^p = p \ln M$.*

$$t = \frac{\ln 20}{12 \ln 1.0075}$$ *Solve for t, dividing both sides by 12 ln 1.0075.*

$$\approx 33.4$$ *Use a calculator.*

After approximately 33.4 years, the $25,000 will grow to an accumulated value of $500,000. If you set aside the money at age 20, you can begin enjoying a life of leisure at about age 53.

> **Check Point 9**   How long, to the nearest tenth of a year, will it take $1000 to grow to $3600 at 8% annual interest compounded quarterly?

Playing Doubles: Interest Rates and Doubling Time

One way to calculate what your savings will be worth at some point in the future is to consider doubling time. Shown below is how long it takes for your money to double at different annual interest rates subject to continuous compounding.

Annual Interest Rate	Years to Double
5%	13.9 years
7%	9.9 years
9%	7.7 years
11%	6.3 years

Of course, the first problem is collecting some money to invest. The second problem is finding a reasonably safe investment with a return of 9% or more.

Yogi Berra, catcher and renowned hitter for the New York Yankees (1946–1963), said it best: "Prediction is very hard, especially when it's about the future." At the start of the twenty-first century, we are plagued by questions about the environment. Will we run out of gas? How hot will it get? Will there be neighborhoods where the air is pristine? Can we make garbage disappear? Will there be any wilderness left? Which wild animals will become extinct? These concerns have led to the growth of the environmental industry in the United States.

EXAMPLE 10   The Growth of the Environmental Industry

The formula

$$N = 461.87 + 299.4 \ln x$$

models the thousands of workers, N, in the environmental industry in the United States x years after 1979. By which year will there be 1,500,000, or 1500 thousand, U.S. workers in the environmental industry?

Solution   We substitute 1500 for N and solve for x, the number of years after 1979.

$$N = 461.87 + 299.4 \ln x$$ This is the given formula.

$$461.87 + 299.4 \ln x = 1500$$ Substitute 1500 for N and reverse the two sides of the equation.

Our goal is to isolate $\ln x$. We can then find x by exponentiating both sides of the equation, using the inverse property $e^{\ln x} = x$.

$$299.4 \ln x = 1038.13$$ Subtract 461.87 from both sides.

$$\ln x = \frac{1038.13}{299.4}$$ Divide both sides by 299.4.

$$e^{\ln x} = e^{1038.13/299.4}$$ Exponentiate both sides.

$$x = e^{1038.13/299.4}$$ $e^{\ln x} = x$

$$\approx 32$$ Use a calculator.

Approximately 32 years after 1979, in the year 2011, there will be 1.5 million U.S. workers in the environmental industry.

Check Point 10   Use the formula in Example 10 to find by which year there will be two million, or 2000 thousand, U.S. workers in the environmental industry.

EXERCISE SET 4.4

 Practice Exercises

Solve each exponential equation in Exercises 1–26. Express the solution set in terms of natural logarithms. Then use a calculator to obtain a decimal approximation, correct to two decimal places, for the solution.

1. $10^x = 3.91$ **2.** $10^x = 8.07$

3. $e^x = 5.7$ **4.** $e^x = 0.83$

5. $5^x = 17$ **6.** $19^x = 143$

7. $5e^x = 23$ **8.** $9e^x = 107$

9. $3e^{5x} = 1977$ **10.** $4e^{7x} = 10{,}273$

11. $e^{1-5x} = 793$ **12.** $e^{1-8x} = 7957$

13. $e^{5x-3} - 2 = 10{,}476$ **14.** $e^{4x-5} - 7 = 11{,}243$

15. $7^{x+2} = 410$ **16.** $5^{x-3} = 137$

17. $7^{0.3x} = 813$ **18.** $3^{x/7} = 0.2$

19. $5^{2x+3} = 3^{x-1}$ **20.** $7^{2x+1} = 3^{x+2}$

21. $e^{2x} - 3e^x + 2 = 0$ **22.** $e^{2x} - 2e^x - 3 = 0$

23. $e^{4x} + 5e^{2x} - 24 = 0$ **24.** $e^{4x} - 3e^{2x} - 18 = 0$

25. $3^{2x} + 3^x - 2 = 0$ **26.** $2^{2x} + 2^x - 12 = 0$

Solve each logarithmic equation in Exercises 27–44. Be sure to reject any value of x that produces the logarithm of a negative number or the logarithm of 0.

27. $\log_3 x = 4$ **28.** $\log_5 x = 3$

29. $\log_4(x + 5) = 3$ **30.** $\log_5(x - 7) = 2$

31. $\log_3(x - 4) = -3$ **32.** $\log_7(x + 2) = -2$

33. $\log_4(3x + 2) = 3$ **34.** $\log_2(4x + 1) = 5$

35. $\log_5 x + \log_5(4x - 1) = 1$

36. $\log_6(x + 5) + \log_6 x = 2$

37. $\log_3(x - 5) + \log_3(x + 3) = 2$

38. $\log_2(x - 1) + \log_2(x + 1) = 3$

39. $\log_2(x + 2) - \log_2(x - 5) = 3$

40. $\log_4(x + 2) - \log_4(x - 1) = 1$

41. $2 \log_3(x + 4) = \log_3 9 + 2$

42. $3 \log_2(x - 1) = 5 - \log_2 4$

43. $\log_2(x - 6) + \log_2(x - 4) - \log_2 x = 2$

44. $\log_2(x - 3) + \log_2 x - \log_2(x + 2) = 2$

Exercises 45–52 involve equations with natural logarithms. Solve each equation by isolating the natural logarithm and exponentiating both sides. Express the answer in terms of e. Then use a calculator to obtain a decimal approximation, correct to two decimal places, for the solution.

45. $\ln x = 2$ **46.** $\ln x = 3$

47. $5 \ln(2x) = 20$ **48.** $6 \ln(2x) = 30$

49. $6 + 2 \ln x = 5$ **50.** $7 + 3 \ln x = 6$

51. $\ln \sqrt{x + 3} = 1$ **52.** $\ln \sqrt{x + 4} = 1$

Application Exercises

Use the formula $R = 6e^{12.77x}$, where x is the blood alcohol concentration and R, given as a percent, is the risk of having a car accident, to solve Exercises 53–54.

53. What blood alcohol concentration corresponds to a 25% risk of a car accident?

54. What blood alcohol concentration corresponds to a 50% risk of a car accident?

55. The formula $A = 18.9e^{0.0055t}$ models the population of New York State, A, in millions, t years after 2000.
  a. What was the population of New York in 2000?
  b. When will the population of New York reach 19.6 million?

56. The formula $A = 15.9e^{0.0235t}$ models the population of Florida, A, in millions, t years after 2000.
  a. What was the population of Florida in 2000?
  b. When will the population of Florida reach 17.5 million?

In Exercises 57–60, complete the table for a savings account subjected to n compoundings yearly $\left[A = P\left(1 + \dfrac{r}{n}\right)^{nt} \right]$.
Round answers to one decimal place.

	Amount Invested	Number of Compounding Periods	Annual Interest Rate	Accumulated Amount	Time t in Years
57.	$12,500	4	5.75%	$20,000	
58.	$7250	12	6.5%	$15,000	
59.	$1000	360		$1400	2
60.	$5000	360		$9000	4

In Exercises 61–64, complete the table for a savings account subjected to continuous compounding $(A = Pe^{rt})$.
Round answers to one decimal place.

	Amount Invested	Annual Interest Rate	Accumulated Amount	Time t in Years
61.	$8000	8%	Double the amount invested	
62.	$8000		$12,000	2
63.	$2350		Triple the amount invested	7
64.	$17,425	4.25%	$25,000	

65. The function $f(x) = 15,557 + 5259 \ln x$ models the average cost of a new car, $f(x)$, in dollars, x years after 1989. When was the average cost of a new car $25,000?

66. The function $f(x) = 68.41 + 1.75 \ln x$ models the life expectancy, $f(x)$, in years, for African-American females born x years after 1969. In which birth year was life expectancy 73.7 years? Round to the nearest year.

The function $P(x) = 95 - 30 \log_2 x$ models the percentage, $P(x)$, of students who could recall the important features of a classroom lecture as a function of time, where x represents the number of days that have elapsed since the lecture was given. The figure shows the graph of the function. Use this information to solve Exercises 67–68. Round answers to one decimal place.

67. After how many days do only half the students recall the important features of the classroom lecture? (Let $P(x) = 50$ and solve for x.) Locate the point on the graph that conveys this information.

68. After how many days have all students forgotten the important features of the classroom lecture? (Let $P(x) = 0$ and solve for x.) Locate the point on the graph on the previous page that conveys this information.

The pH of a solution ranges from 0 to 14. An acid solution has a pH less than 7. Pure water is neutral and has a pH of 7. Normal, unpolluted rain has a pH of about 5.6. The pH of a solution is given by

$$pH = -\log x$$

where x represents the concentration of the hydrogen ions in the solution, in moles per liter. Use the formula to solve Exercises 69–70.

69. An environmental concern involves the destructive effects of acid rain. The most acidic rainfall ever had a pH of 2.4. What was the hydrogen ion concentration? Express the answer as a power of 10, and then round to the nearest thousandth.

70. The figure shows very acidic rain in the northeast United States. What is the hydrogen ion concentration of rainfall with a pH of 4.2? Express the answer as a power of 10, and then round to the nearest hundred-thousandth.

Acid Rain over Canada and the United States

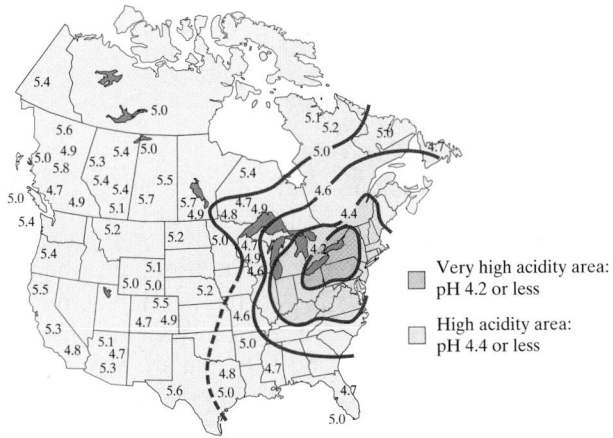

Very high acidity area: pH 4.2 or less

High acidity area: pH 4.4 or less

Source: National Atmospheric Program

 Writing in Mathematics

71. Explain how to solve an exponential equation. Use $3^x = 140$ in your explanation.

72. Explain how to solve a logarithmic equation. Use $\log_3(x - 1) = 4$ in your explanation.

73. In many states, a 17% risk of a car accident with a blood alcohol concentration of 0.08 is the lowest level for charging a motorist with driving under the influence. Do you agree with the 17% risk as a cutoff percentage, or do you feel that the percentage should be lower or higher?

Explain your answer. What blood alcohol concentration corresponds to what you believe is an appropriate percentage?

74. Have you purchased a new or used car recently? If so, describe if the function in Exercise 65 accurately models what you paid for your car. If there is a big difference between the figure given by the formula and the amount that you paid, how can you explain this difference?

Technology Exercises

In Exercises 75–82, use your graphing utility to graph each side of the equation in the same viewing rectangle. Then use the x-coordinate of the intersection point to find the equation's solution set. Verify this value by direct substitution into the equation.

75. $2^{x+1} = 8$ **76.** $3^{x+1} = 9$
77. $\log_3(4x - 7) = 2$ **78.** $\log_3(3x - 2) = 2$
79. $\log(x + 3) + \log x = 1$ **80.** $\log(x - 15) + \log x = 2$
81. $3^x = 2x + 3$ **82.** $5^x = 3x + 4$

Hurricanes are one of nature's most destructive forces. These low-pressure areas often have diameters of over 500 miles. The function $f(x) = 0.48 \ln(x + 1) + 27$ models the barometric air pressure, $f(x)$, in inches of mercury, at a distance of x miles from the eye of a hurricane. Use this function to solve Exercises 83–84.

83. Graph the function in a $[0, 500, 50]$ by $[27, 30, 1]$ viewing rectangle. What does the shape of the graph indicate about barometric air pressure as the distance from the eye increases?

84. Use an equation to answer this question: How far from the eye of a hurricane is the barometric air pressure 29 inches of mercury? Use the TRACE and ZOOM features or the intersect command of your graphing utility to verify your answer.

85. The function $P(t) = 145e^{-0.092t}$ models a runner's pulse, $P(t)$, in beats per minute, t minutes after a race, where $0 \le t \le 15$. Graph the function using a graphing utility. TRACE along the graph and determine after how many minutes the runner's pulse will be 70 beats per minute. Round to the nearest tenth of a minute. Verify your observation algebraically.

86. The function $W(t) = 2600(1 - 0.51e^{-0.075t})^3$ models the weight, $W(t)$, in kilograms, of a female African elephant at age t years. (1 kilogram $\approx$ 2.2 pounds) Use a graphing utility to graph the function. Then TRACE along the curve to estimate the age of an adult female elephant weighing 1800 kilograms.

Critical Thinking Exercises

87. Which one of the following is true?
  a. If $\log(x + 3) = 2$, then $e^2 = x + 3$.
  b. If $\log(7x + 3) - \log(2x + 5) = 4$, then in exponential form $10^4 = (7x + 3) - (2x + 5)$.

c. If $x = \dfrac{1}{k} \ln y$, then $y = e^{kx}$.

d. Examples of exponential equations include $10^x = 5.71$, $e^x = 0.72$, and $x^{10} = 5.71$.

88. If \$4000 is deposited into an account paying 3% interest compounded annually and at the same time \$2000 is deposited into an account paying 5% interest compounded annually, after how long will the two accounts have the same balance?

Solve each equation in Exercises 89–91. Check each proposed solution by direct substitution or with a graphing utility.

89. $(\ln x)^2 = \ln x^2$

90. $(\log x)(2 \log x + 1) = 6$

91. $\ln(\ln x) = 0$

Group Exercise

92. Research applications of logarithmic functions as mathematical models and plan a seminar based on your group's research. Each group member should research one of the following areas or any other area of interest: pH (acidity of solutions), intensity of sound (decibels), brightness of stars, consumption of natural resources, human memory, progress over time in a sport, profit over time. For the area that you select, explain how logarithmic functions are used and provide examples.

SECTION 4.5 *Modeling with Exponential and Logarithmic Functions*

Objectives

1. Model exponential growth and decay.

2. Use logistic growth models.

3. Model data with exponential and logarithmic functions.

4. Express an exponential model in base e.

The most casual cruise on the Internet shows how people disagree when it comes to making predictions about the effects of the world's growing population. Some argue that there is a recent slowdown in the growth rate, economies remain robust, and famines in Biafra and Ethiopia are aberrations rather than signs of the future. Others say that the 6 billion people on Earth is twice as many as can be supported in middle-class comfort, and the world is running out of arable land and fresh water. Debates about entities that are growing exponentially can be approached mathematically: We can create functions that model data and use these functions to make predictions. In this section we will show you how this is done.

1 Model exponential growth and decay.

Exponential Growth and Decay

One of algebra's many applications is to predict the behavior of variables. This can be done with *exponential growth* and *decay models*. With exponential growth or decay, quantities grow or decay at a rate directly proportional to their size. Populations that are growing exponentially grow extremely rapidly as they get larger because there are more adults to have offspring. For example, the **growth rate** for world population is 1.3%, or 0.013. This means that each year world population is 1.3% more than what it was in the previous year. In 2001, world population was approximately 6.2 billion. Thus, we compute the world population in 2002 as follows:

6.2 billion + 1.3% of 6.2 billion = 6.2 + (0.013)(6.2) = 6.2806.

This computation suggests that 6.2806 billion people will populate the world in 2002. The 0.0806 billion represents an increase of 80.6 million people from 2001 to 2002, the equivalent of the population of Germany. Using 1.3% as the annual growth rate, world population for 2003 is found in a similar manner:

6.2806 + 1.3% of 6.2806 = 6.2806 + (0.013)(6.2806) ≈ 6.3622.

This computation suggests that approximately 6.3622 billion people will populate the world in 2003.

The explosive growth of world population may remind you of the growth of money in an account subject to compound interest. Just as the growth rate for world population is multiplied by the population plus any increase in the population, a compound interest rate is multiplied by your original investment plus any accumulated interest. The balance in an account subject to continuous compounding and world population are special cases of an *exponential growth model*.

Study Tip

You have seen the formula for exponential growth before, but with different letters. It is the formula for compound interest with continous compounding.

$$A = Pe^{rt}$$

| Amount at time t | Principal is the original amount. | Interest rate is the growth rate. |

$$A = A_0 e^{kt}$$

Exponential Growth and Decay Models

The mathematical model for **exponential growth** or **decay** is given by

$$f(t) = A_0 e^{kt} \quad \text{or} \quad A = A_0 e^{kt}.$$

- **If $k > 0$, the function models the amount, or size, of a *growing* entity.** A_0 is the original amount, or size, of the growing entity at time $t = 0$, A is the amount at time t, and k is a constant representing the growth rate.
- **If $k < 0$, the function models the amount, or size, of a *decaying* entity.** A_0 is the original amount, or size, of the decaying entity at time $t = 0$, A is the amount at time t, and k is a constant representing the decay rate.

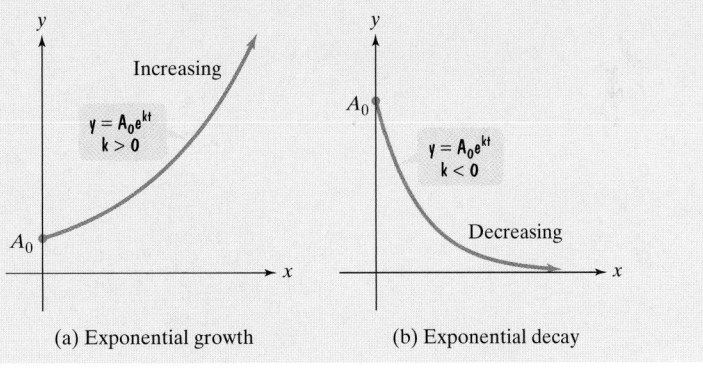

(a) Exponential growth   (b) Exponential decay

Sometimes we need to use given data to determine k, the rate of growth or decay. After we compute the value of k, we can use the formula $A = A_0 e^{kt}$ to make predictions. This idea is illustrated in our first two examples.

EXAMPLE 1   Modeling the Growth of the Minimum Wage

The graph in Figure 4.17 shows the growth of the minimum wage from 1970 through 2000. In 1970, the minimum wage was $1.60 per hour. By 2000, it had grown to $5.15 per hour.

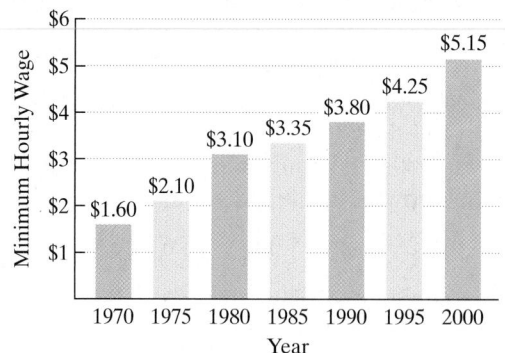

Figure 4.17 *Source:* U.S. Employment Standards Administration

a. Find the exponential growth function that models the data for 1970 through 2000.

b. By which year will the minimum wage reach $7.50 per hour?

Solution

a. We use the exponential growth model

$$A = A_0 e^{kt}$$

in which t is the number of years after 1970. This means that 1970 corresponds to $t = 0$. At that time the minimum wage was $1.60, so we substitute 1.6 for A_0 in the growth model:

$$A = 1.6e^{kt}.$$

We are given that $5.15 is the minimum wage in 2000. Because 2000 is 30 years after 1970, when $t = 30$ the value of A is 5.15. Substituting these numbers into the growth model will enable us to find k, the growth rate. We know that $k > 0$ because the problem involves growth.

$A = 1.6e^{kt}$	Use the growth model, $A = A_0 e^{kt}$, with $A_0 = 1.6$.
$5.15 = 1.6e^{k \cdot 30}$	When $t = 30$, $A = 5.15$. Substitute these numbers into the model.
$e^{30k} = \dfrac{5.15}{1.6}$	Isolate the exponential factor by dividing both sides by 1.6. We also reversed the sides.
$\ln e^{30k} = \ln \dfrac{5.15}{1.6}$	Take the natural logarithm on both sides.
$30k = \ln \dfrac{5.15}{1.6}$	Simplify the left side using $\ln e^x = x$.
$k = \dfrac{\ln \dfrac{5.15}{1.6}}{30} \approx 0.039$	Divide both sides by 30 and solve for k.

We substitute 0.039 for k in the growth model to obtain the exponential growth function for the minimum wage. It is

$$A = 1.6e^{0.039t}$$

where t is measured in years after 1970.

b. To find the year in which the minimum wage will reach $7.50 per hour, we substitute 7.5 for A in the model from part (a) and solve for t.

$$A = 1.6e^{0.039t} \qquad \text{This is the model from part (a).}$$

$$7.5 = 1.6e^{0.039t} \qquad \text{Substitute 7.5 for A.}$$

$$e^{0.039t} = \frac{7.5}{1.6} \qquad \text{Divide both sides by 1.6. We also reversed the sides.}$$

$$\ln e^{0.039t} = \ln \frac{7.5}{1.6} \qquad \text{Take the natural logarithm on both sides.}$$

$$0.039t = \ln \frac{7.5}{1.6} \qquad \text{Simplify on the left using } \ln e^x = x.$$

$$t = \frac{\ln \dfrac{7.5}{1.6}}{0.039} \approx 40 \qquad \text{Solve for t by dividing both sides by 0.039.}$$

Because 40 is the number of years after 1970, the model indicates that the minimum wage will reach $7.50 by 1970 + 40, or in the year 2010.

> **Check Point 1**
> In 1990, the population of Africa was 643 million and by 2000 it had grown to 813 million.
> **a.** Use the exponential growth model $A = A_0 e^{kt}$, in which t is the number of years after 1990, to find the exponential growth function that models the data.
> **b.** By which year will Africa's population reach 2000 million, or two billion?

Lying with Statistics

Benjamin Disraeli, Queen Victoria's prime minister, stated that there are "lies, damned lies, and statistics." The problem is not that data lie, but rather that liars use data. For example, the data in Example 1 create the impression that wages are on the rise and workers are better off each year. The graph in Figure 4.18 is more effective in creating an accurate picture. Why? It is adjusted for inflation and measured in constant 1996 dollars. Something else to think about: In predicting a minimum wage of $7.50 by 2010, are we using the best possible model for the data? We return to this issue in the exercise set.

Figure 4.18
Source: U.S. Employment Standards Administration

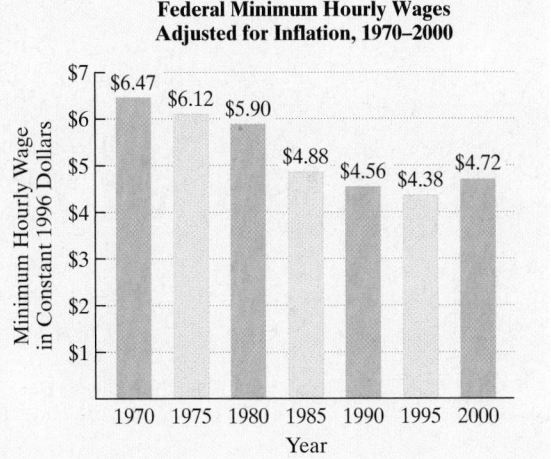

Carbon Dating and Artistic Development

The artistic community was electrified by the discovery in 1995 of spectacular cave paintings in a limestone cavern in France. Carbon dating of the charcoal from the site showed that the images, created by artists of remarkable talent, were 30,000 years old, making them the oldest cave paintings ever found. The artists seemed to have used the cavern's natural contours to heighten a sense of perspective. The quality of the painting suggests that the art of early humans did not mature steadily from primitive to sophisticated in any simple linear fashion.

Our next example involves exponential decay and its use in determining the age of fossils and artifacts. The method is based on considering the percentage of carbon-14 remaining in the fossil or artifact. Carbon-14 decays exponentially with a *half-life* of approximately 5715 years. The **half-life** of a substance is the time required for half of a given sample to disintegrate. Thus, after 5715 years a given amount of carbon-14 will have decayed to half the original amount. Carbon dating is useful for artifacts or fossils up to 80,000 years old. Older objects do not have enough carbon-14 left to date age accurately.

EXAMPLE 2 Carbon-14 Dating: The Dead Sea Scrolls

a. Use the fact that after 5715 years a given amount of carbon-14 will have decayed to half the original amount to find the exponential decay model for carbon-14.

b. In 1947, earthenware jars containing what are known as the Dead Sea Scrolls were found by an Arab Bedouin herdsman. Analysis indicated that the scroll wrappings contained 76% of their original carbon-14. Estimate the age of the Dead Sea Scrolls.

Solution   We begin with the exponential decay model $A = A_0 e^{kt}$. We know that $k < 0$ because the problem involves the decay of carbon-14. After 5715 years ($t = 5715$), the amount of carbon-14 present, A, is half the original amount A_0. Thus, we can substitute $\dfrac{A_0}{2}$ for A in the exponential decay model. This will enable us to find k, the decay rate.

a.

$$A = A_0 e^{kt}$$

 Begin with the exponential decay model .

$$\frac{A_0}{2} = A_0 e^{k \cdot 5715}$$

 After 5715 years ($t = 5715$), $A = \dfrac{A_0}{2}$ (because the amount present, A, is half the original amount, A_0).

$$\frac{1}{2} = e^{5715k}$$

 Divide both sides of the equation by A_0.

$$\ln \frac{1}{2} = \ln e^{5715k}$$

 Take the natural logarithm on both sides.

$$\ln \frac{1}{2} = 5715k$$

 Simplify the right side using $\ln e^x = x$.

$$k = \frac{\ln \dfrac{1}{2}}{5715} \approx -0.000121$$

 Divide both sides by 5715 and solve for k.

Substituting for k in the decay model, $A = A_0 e^{kt}$, the model for carbon-14 is

$$A = A_0 e^{-0.000121t}.$$

b.

$$A = A_0 e^{-0.000121t}$$

 This is the decay model for carbon-14.

$$0.76A_0 = A_0 e^{-0.000121t}$$

 A, the amount present, is 76% of the original amount, so $A = 0.76A_0$.

$$0.76 = e^{-0.000121t}$$

 Divide both sides of the equation by A_0.

$$\ln 0.76 = \ln e^{-0.000121t}$$

 Take the natural logarithm on both sides.

$$\ln 0.76 = \ln e^{-0.000121t}$$ We've repeated this equation from the bottom of the previous page.

$$\ln 0.76 = -0.000121t$$ Simplify the right side using $\ln e^x = x$.

$$t = \frac{\ln 0.76}{-0.000121} \approx 2268$$ Divide both sides by -0.000121 and solve for t.

The Dead Sea Scrolls are approximately 2268 years old plus the number of years between 1947 and the current year.

Check Point 2 Strontium-90 is a waste product from nuclear reactors. As a consequence of fallout from atmospheric nuclear tests, we all have a measurable amount of strontium-90 in our bones.

  a. Use the fact that after 28 years a given amount of strontium-90 will have decayed to half the original amount to find the exponential decay model for strontium-90.

  b. Suppose that a nuclear accident occurs and releases 60 grams of strontium-90 into the atmosphere. How long will it take for strontium-90 to decay to a level of 10 grams?

②   Use logistic growth models.

Logistic Growth Models

From population growth to the spread of an epidemic, nothing on Earth can grow exponentially indefinitely. Growth is always limited. This is shown in Figure 4.19 by the horizontal asymptote. The **logistic growth model** is an exponential function used to model situations in which growth is limited.

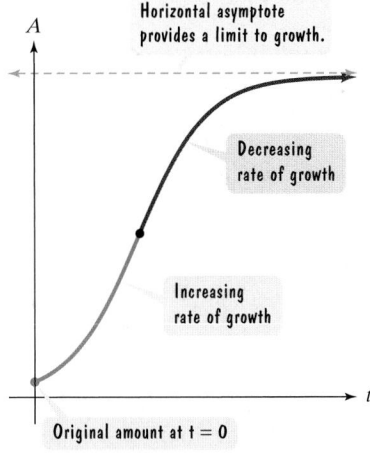

Horizontal asymptote provides a limit to growth.

Decreasing rate of growth

Increasing rate of growth

Original amount at t = 0

Figure 4.19 The logistic growth curve has a horizontal asymptote that limits the growth of A over time.

Logistic Growth Model

The mathematical model for limited logistic growth is given by

$$f(t) = \frac{c}{1 + ae^{-bt}} \quad \text{or} \quad A = \frac{c}{1 + ae^{-bt}}$$

where a, b, and c are constants, with $c > 0$ and $b > 0$.

As time increases $(t \to \infty)$, the expression ae^{-bt} in the model approaches 0, and A gets closer and closer to c. This means that $y = c$ is a horizontal asymptote for the graph of the function. Thus, the value of A can never exceed c and c represents the limiting size that A can attain.

EXAMPLE 3   Modeling the Spread of the Flu

The function

$$f(t) = \frac{30,000}{1 + 20e^{-1.5t}}$$

describes the number of people, $f(t)$, who have become ill with influenza t weeks after its initial outbreak in a town with 30,000 inhabitants.

a. How many people became ill with the flu when the epidemic began?

b. How many people were ill by the end of the fourth week?

c. What is the limiting size of $f(t)$, the population that becomes ill?

Solution

a. The time at the beginning of the flu epidemic is $t = 0$. Thus, we can find the number of people who were ill at the beginning of the epidemic by substituting 0 for t.

$$f(t) = \frac{30,000}{1 + 20e^{-1.5t}} \qquad \text{This is the given logistic growth function.}$$

$$f(0) = \frac{30,000}{1 + 20e^{-1.5(0)}} \qquad \text{When the epidemic began, } t = 0.$$

$$= \frac{30,000}{1 + 20} \qquad e^{-1.5(0)} = e^{0} = 1$$

$$\approx 1429$$

Approximately 1429 people were ill when the epidemic began.

b. We find the number of people who were ill at the end of the fourth week by substituting 4 for t in the logistic growth function.

$$f(t) = \frac{30,000}{1 + 20e^{-1.5t}} \qquad \text{Use the given logistic growth function.}$$

$$f(4) = \frac{30,000}{1 + 20e^{-1.5(4)}} \qquad \text{To find the number of people ill by the end of week four, let } t = 4.$$

$$= 28,583 \qquad \text{Use a calculator.}$$

Approximately 28,583 people were ill by the end of the fourth week. Compared with the number of people who were ill initially, 1429, this illustrates the virulence of the epidemic.

c. Recall that in the logistic growth model, $f(t) = \dfrac{c}{1 + ae^{-bt}}$, the constant c represents the limiting size that $f(t)$ can attain. Thus, the number in the numerator, 30,000, is the limiting size of the population that becomes ill.

Technology

The graph of the logistic growth function for the flu epidemic

$$y = \frac{30,000}{1 + 20e^{-1.5x}}$$

can be obtained using a graphing utility. We started x at 0 and ended at 10. This takes us to week 10. (In Example 3, we found that by week 4 approximately 28,583 people were ill.) We also know that 30,000 is the limiting size, so we took values of y up to 30,000. Using a $[0, 10, 1]$ by $[0, 30,000, 3000]$ viewing rectangle, the graph of the logistic growth function is shown below.

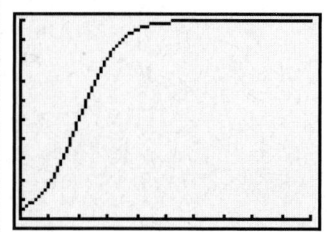

Check Point 3   In a learning theory project, psychologists discovered that

$$f(t) = \frac{0.8}{1 + e^{-0.2t}}$$

is a model for describing the proportion of correct responses, $f(t)$, after t learning trials.

Check Point 3 continued

a. Find the proportion of correct responses prior to learning trials taking place.

b. Find the proportion of correct responses after 10 learning trials.

c. What is the limiting size of $f(t)$, the proportion of correct responses, as continued learning trials take place?

3 Model data with exponential and logarithmic functions.

The Art of Modeling

Throughout this chapter, we have been working with models that were given. However, we can create functions that model data by observing patterns in scatter plots. Figure 4.20 shows scatter plots for data that are exponential or logarithmic.

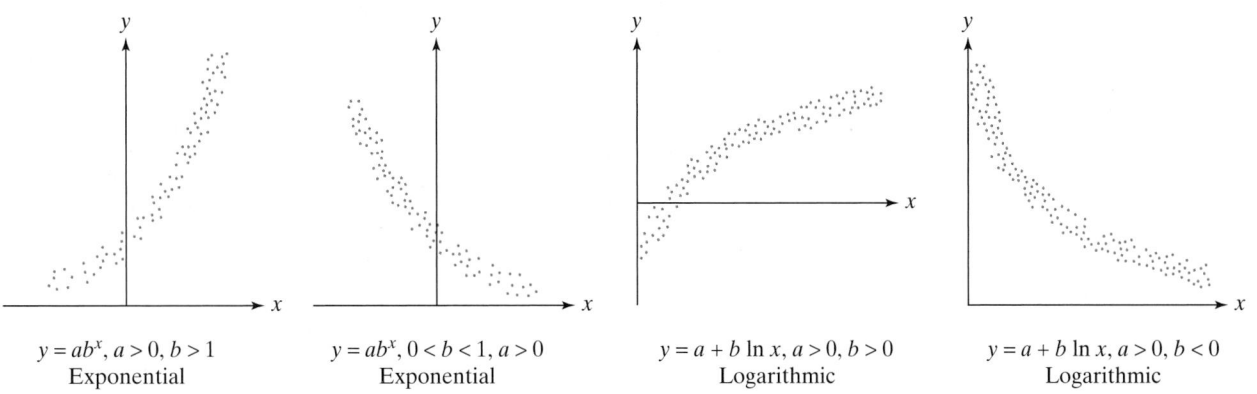

$y = ab^x, a > 0, b > 1$
Exponential

$y = ab^x, 0 < b < 1, a > 0$
Exponential

$y = a + b \ln x, a > 0, b > 0$
Logarithmic

$y = a + b \ln x, a > 0, b < 0$
Logarithmic

Figure 4.20 Scatter plots for exponential or logarithmic models

Graphing utilities can be used to find the equation of a function that is derived from data. For example, earlier in the chapter we encountered a function that modeled the size of a city and the average walking speed, in feet per second, of pedestrians. The function was derived from the data in Table 4.4. The scatter plot is shown in Figure 4.21.

Table 4.4

x, Population (thousands)	y, Walking Speed (feet per second)
5.5	3.3
14	3.7
71	4.3
138	4.4
342	4.8

Source: Mark and Helen Bornstein, "The Pace of Life"

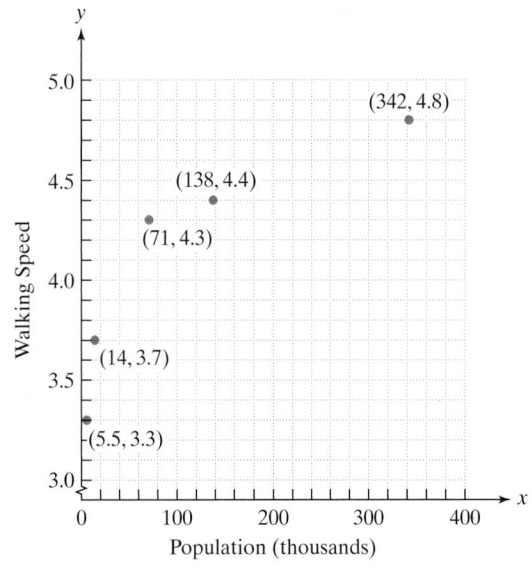

Figure 4.21 Scatter plot for data in Table 4.4

```
LnReg
 y=a+blnx
 a=2.734605225
 b=.3524772878
 r=.9958996083
```

Figure 4.22 A logarithmic model for the data in Table 4.4

Because the data in this scatter plot increase rapidly at first and then begin to level off a bit, the shape suggests that a logarithmic model might be a good choice. A graphing utility fits the data in Table 4.4 to a logarithmic model of the form $y = a + b \ln x$ by using the Natural Logarithmic REGression (LnReg) option (see Figure 4.22). From the figure, we see that the logarithmic model of the data, with numbers rounded to three decimal places, is

$$y = 2.735 + 0.352 \ln x.$$

The number r that appears in Figure 4.22 is called the **correlation coefficient** and is a measure of how well the model fits the data. The value of r is such that $-1 \leq r \leq 1$. A positive r means that as the x-values increase, so do the y-values. A negative r means that as the x-values increase, the y-values decrease. **The closer that r is to -1 or 1, the better the model fits the data.** Because r is approximately 0.996, the model

$$y = 2.735 + 0.352 \ln x$$

fits the data very well.

Now let's look at data whose scatter plot suggests an exponential model. The data in Table 4.5 indicate world population for six years. The scatter plot is shown in Figure 4.23.

Table 4.5

x, Year	y, World Population (billions)
1950	2.6
1960	3.1
1970	3.7
1980	4.5
1989	5.3
2001	6.2

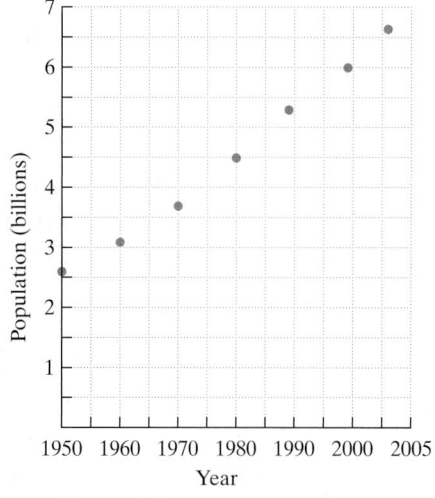

Figure 4.23 A scatter plot for data in Table 4.5

Because the data in this scatter plot have a rapidly increasing pattern, the shape suggests that an exponential model might be a good choice. (You might also want to try a linear model.) If you select the exponential option, you will use a graphing utility's Exponential REGression option. With this feature, a graphing utility fits the data to an exponential model of the form $y = ab^x$.

When computing an exponential model of the form $y = ab^x$, many graphing utilities rewrite the equation using logarithms. Because the domain of the logarithmic function is the set of positive numbers, **zero must not be a value for x** when using such utilities. What does this mean in terms of our data for world population that starts in the year 1950? We must start values of x after 0. Thus, we'll assign x to represent the number of years after 1949.

This gives us the data shown in Table 4.6. Using the Exponential REGression option, we obtain the equation in Figure 4.24.

Table 4.6

x, Numbers of Years after 1949		y, World Population (billions)
1	(1950)	2.6
11	(1960)	3.1
21	(1970)	3.7
31	(1980)	4.5
40	(1989)	5.3
52	(2001)	6.2

```
ExpReg
y=a*b^x
a=2.5731257405
b=1.01759434422
r=.998331654439
```

Figure 4.24 An exponential model for the data in Table 4.6

From Figure 4.24, we see that the exponential model of the data for world population, y, in billions, x years after 1949, with numbers rounded to three decimal places, is

$$y = 2.573(1.018)^x.$$

The correlation coefficient, r, is close to 1, indicating that the model fits the data very well.

Because $b = e^{\ln b}$, we can rewrite any model in the form $y = ab^x$ in terms of base e.

4 Express an exponential model in base e.

Expressing an Exponential Model in Base e

$$y = ab^x \quad \text{is equivalent to} \quad y = ae^{(\ln b) \cdot x}.$$

EXAMPLE 4 Rewriting an Exponential Model in Base e

Rewrite $y = 2.573(1.018)^x$ in terms of base e.

Solution

$y = ab^x$ is equivalent to $y = ae^{(\ln b) \cdot x}$.

$y = 2.573(1.018)^x$ is equivalent to $y = 2.573e^{(\ln 1.018) \cdot x}$.

Using $\ln 1.018 \approx 0.018$, the exponential growth model for world population, y, in billions, x years after 1949 is

$$y = 2.573e^{0.018x}.$$

In Example 4, we can replace y with A and x with t so that the model has the same letters as those in the exponential growth model $A = A_0e^{kt}$.

$A = A_0 e^{kt}$ — This is the exponential growth model.

$A = 2.573e^{0.018t}$ — This is the model for world population.

The value of k, 0.018, indicates a growth rate of 1.8%. Although this is an excellent model for the data, we must be careful about making projections about world population using this growth function. Why? World population growth rate is now 1.3%, not 1.8%, so our model will overestimate future populations.

Check Point 4 Rewrite $y = 4(7.8)^x$ in terms of base e. Express the answer in terms of a natural logarithm, and then round to three decimal places.

When using a graphing utility to model data, begin with a scatter plot, drawn either by hand or with the graphing utility, to obtain a general picture for the shape of the data. It might be difficult to determine which model best fits the data—linear, logarithmic, exponential, quadratic, or something else. If necessary, use your graphing utility to fit several models to the data. The best model is the one that yields the value r, the correlation coefficient, closest to 1 or -1. Finding a proper fit for data can be almost as much art as it is mathematics. In this era of technology, the process of creating models that best fit data is one that involves more decision making than computation.

EXERCISE SET 4.5

 ## Practice and Application Exercises

The exponential growth model $A = 203e^{0.011t}$ describes the population of the United States, A, in millions, t years after 1970. Use this model to solve Exercises 1–4.

1. What was the population of the United States in 1970?
2. By what percentage is the population of the United States increasing each year?
3. When will the U.S. population be 300 million?
4. When will the U.S. population be 350 million?

India is currently one of the world's fastest-growing countries. By 2040, the population of India will be larger than the population of China; by 2050, nearly one-third of the world's population will live in these two countries alone. The exponential growth model $A = 574e^{0.026t}$ describes the population of India, A, in millions, t years after 1974. Use this model to solve Exercises 5–8.

5. By what percentage is the population of India increasing each year?
6. What was the population of India in 1974?
7. When will India's population be 1624 million?
8. When will India's population be 2732 million?
9. Low interest rates, easy credit, and strong demand from new immigrants have driven up the average sales price of new one-family houses in the United States. In 1995, the average sales price was $158,700 and by 2000 it had increased to $207,200.
   a. Use the exponential growth model $A = A_0e^{kt}$, in which t is the number of years after 1995, to find the exponential growth function that models the data.
   b. According to your model, by which year will the average sales price of a new one-family house reach $300,000?

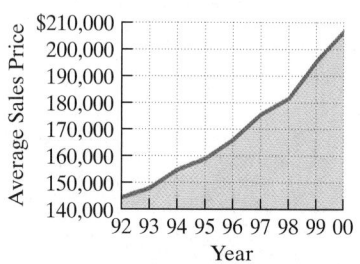

Average Sales Prices of New One-Family Houses

Source: U.S. Census Bureau

About the size of New Jersey, Israel has seen its population soar to more than 6 million since it was established. With the help of U.S. aid, the country now has a diversified economy rivaling those of other developed Western nations. By contrast, the Palestinians, living under Israeli occupation and a corrupt regime, endure bleak conditions. The graphs show that by 2050, Palestinians in the West Bank, Gaza Strip, and East Jerusalem will outnumber Israelis. Exercises 10–12 involve the projected growth of these two populations.

Population of Israel

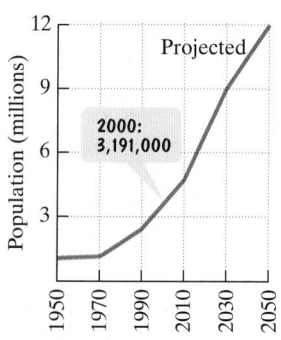

Palestinian Population in West Bank, Gaza, and East Jerusalem

Source: Newsweek

10. In 2000, the population of the Palestinians in the West Bank, Gaza Strip, and East Jerusalem was approximately 3.2 million and by 2050 it is projected to grow to 12 million. Use the exponential growth model $A = A_0 e^{kt}$, in which t is the number of years after 2000, to find the exponential growth function that models the data.

11. In 2000, the population of Israel was approximately 6.04 million and by 2050 it is projected to grow to 10 million. Use the exponential growth model $A = A_0 e^{kt}$, in which t is the number of years after 2000, to find an exponential growth function that models the data.

12. Use the growth models in Exercises 10 and 11 to determine the year in which the two populations will be the same.

An artifact originally had 16 grams of carbon-14 present. The decay model $A = 16 e^{-0.000121t}$ describes the amount of carbon-14 present, A, in grams, after t years. Use this model to solve Exercises 13–14.

13. How many grams of carbon-14 will be present after 5715 years?

14. How many grams of carbon-14 will be present after 11,430 years?

15. The half-life of the radioactive element krypton-91 is 10 seconds. If 16 grams of krypton-91 are initially present, how many grams are present after 10 seconds? 20 seconds? 30 seconds? 40 seconds? 50 seconds?

16. The half-life of the radioactive element plutonium-239 is 25,000 years. If 16 grams of plutonium-239 are initially present how many grams are present after 25,000 years? 50,000 years? 75,000 years? 100,000 years? 125,000 years?

Use the exponential decay model for carbon-14, $A = A_0 e^{-0.000121t}$, to solve Exercises 17–18.

17. Prehistoric cave paintings were discovered in a cave in France. The paint contained 15% of the original carbon-14. Estimate the age of the paintings.

18. Skeletons were found at a construction site in San Francisco in 1989. The skeletons contained 88% of the expected amount of carbon-14 found in a living person. In 1989, how old were the skeletons?

19. The August 1978 issue of *National Geographic* described the 1964 find of dinosaur bones of a newly discovered dinosaur weighing 170 pounds, measuring 9 feet, with a 6-inch claw on one toe of each hind foot. The age of the dinosaur was estimated using potassium-40 dating of rocks surrounding the bones.

　a. Potassium-40 decays exponentially with a half-life of approximately 1.31 billion years. Use the fact that after 1.31 billion years a given amount of potassium-40 will have decayed to half the original amount to show that the decay model for potassium-40 is given by $A = A_0 e^{-0.52912t}$, where t is in billions of years.

　b. Analysis of the rocks surrounding the dinosaur bones indicated that 94.5% of the original amount of potassium-40 was still present. Let $A = 0.945 A_0$ in the model in part (a) and estimate the age of the bones of the dinosaur.

20. A bird species in danger of extinction has a population that is decreasing exponentially $(A = A_0 e^{kt})$. Five years ago the population was at 1400 and today only 1000 of the birds are alive. Once the population drops below 100, the situation will be irreversible. When will this happen?

21. Use the exponential growth model, $A = A_0 e^{kt}$, to show that the time it takes a population to double (to grow from A_0 to $2A_0$) is given by $t = \dfrac{\ln 2}{k}$.

22. Use the exponential growth model, $A = A_0 e^{kt}$, to show that the time it takes a population to triple (to grow from A_0 to $3A_0$) is given by $t = \dfrac{\ln 3}{k}$.

Use the formula $t = \dfrac{\ln 2}{k}$ that gives the time for a population with a growth rate k to double to solve Exercises 23–24. Express each answer to the nearest whole year.

23. China is growing at a rate of 1.1% per year. How long will it take China to double its population?

24. Japan is growing at a rate of 0.3% per year. How long will it take Japan to double its population?

25. The logistic growth function

$$f(t) = \frac{100{,}000}{1 + 5000 e^{-t}}$$

describes the number of people, $f(t)$, who have become ill with influenza t weeks after its initial outbreak in a particular community.

　a. How many people became ill with the flu when the epidemic began?

　b. How many people were ill by the end of the fourth week?

　c. What is the limiting size of the population that becomes ill?

26. The logistic growth function

$$f(t) = \frac{500}{1 + 83.3 e^{-0.162t}}$$

describes the population, $f(t)$, of an endangered species of birds t years after they are introduced to a nonthreatening habitat.

　a. How many birds were initially introduced to the habitat?

　b. How many birds are expected in the habitat after 10 years?

　c. What is the limiting size of the bird population that the habitat will sustain?

The logistic growth function

$$P(x) = \frac{90}{1 + 271e^{-0.122x}}$$

models the percentage, $P(x)$, of Americans who are x years old with some coronary heart disease. Use the function to solve Exercises 27–30.

27. What percentage of 20-year-olds have some coronary heart disease?

28. What percentage of 80-year-olds have some coronary heart disease?

29. At what age is the percentage of some coronary heart disease 50%?

30. At what age is the percentage of some coronary heart disease 70%?

In Exercises 31–34, rewrite the equation in terms of base e. Express the answer in terms of a natural logarithm, and then round to three decimal places.

31. $y = 100(4.6)^x$
32. $y = 1000(7.3)^x$
33. $y = 2.5(0.7)^x$
34. $y = 4.5(0.6)^x$

 Writing in Mathematics

35. Nigeria has a growth rate of 0.031 or 3.1%. Describe what this means.

36 How can you tell if an exponential model describes exponential growth or exponential decay?

37. Suppose that a population that is growing exponentially increases from 800,000 people in 2003 to 1,000,000 people in 2006. Without showing the details, describe how to obtain the exponential growth function that models the data.

38. What is the half-life of a substance?

39. Describe a difference between exponential growth and logistic growth.

40. Describe the shape of a scatter plot that suggests modeling the data with an exponential function.

41. Based on the graphs in Exercises 10–12, for which population is an exponential growth function a better model? Explain your answer.

42. You take up weightlifting and record the maximum number of pounds you can lift at the end of each week. You start off with rapid growth in terms of the weight you can lift from week to week, but then the growth begins to level off. Describe how to obtain a function that models the number of pounds you can lift at the end of each week. How can you use this function to predict what might happen if you continue the sport?

43. Would you prefer that your salary be modeled exponentially or logarithmically? Explain your answer.

44. One problem with all exponential growth models is that nothing can grow exponentially forever. Describe factors that might limit the size of a population.

Technology Exercises

In Example 1 on page 420–421, we used two data points and an exponential function to model federal minimum wages that were not adjusted for inflation from 1970 through 2000. The data are shown again in the table. Use all seven data points to solve Exercises 45–49.

x, Number of Years after 1969	y, Federal Minimum Wage
1	1.60
6	2.10
11	3.10
16	3.35
21	3.80
26	4.25
31	5.15

45. Use your graphing utility's Exponential REGression option to obtain a model of the form $y = ab^x$ that fits the data. How well does the correlation coefficient, r, indicate that the model fits the data?

46. Use your graphing utility's Logarithmic REGression option to obtain a model of the form $y = a + b \ln x$ that fits the data. How well does the correlation coefficient, r, indicate that the model fits the data?

47. Use your graphing utility's Linear REGression option to obtain a model of the form $y = ax + b$ that fits the data. How well does the correlation coefficient, r, indicate that the model fits the data?

48. Use your graphing utility's Power REGression option to obtain a model of the form $y = ax^b$ that fits the data. How well does the correlation coefficient, r, indicate that the model fits the data?

49. Use the value of r in Exercises 45–48 to select the model of best fit. Use this model to predict by which year the minimum wage will reach $7.50. How does this answer compare to the year we found in Example 1, namely 2010? If you obtained a different year, how do you account for this difference?

50. In Exercises 27–30, you worked with the logistic growth function

$$P(x) = \frac{90}{1 + 271e^{-0.122x}}$$

which models the percentage, $P(x)$, of Americans who are x years old with some coronary heart disease. Use your graphing utility to graph the function in a $[0, 100, 10]$ by $[0, 100, 10]$ viewing rectangle. Describe as specifically as possible what the logistic curve indicates about aging and the percentage of Americans with coronary heart disease.

In Exercises 51–52, use a graphing utility to find the model that best fits the given data. Then use the model to make a reasonable prediction for a value that exceeds those shown on the graph's horizontal axis.

51.

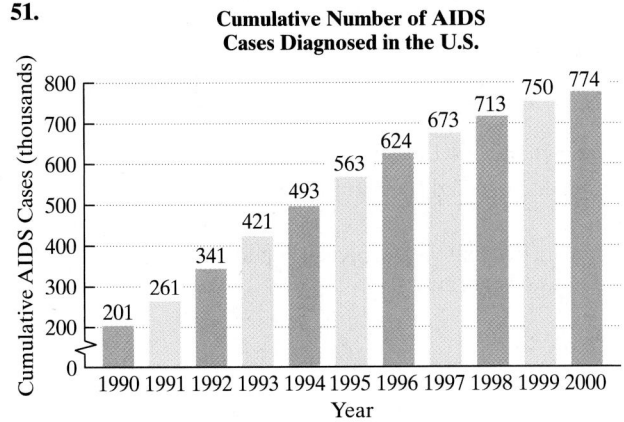

Cumulative Number of AIDS Cases Diagnosed in the U.S.

Source: Centers for Disease Control

52.

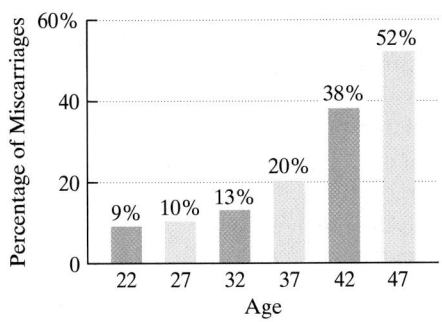

Percentage of Miscarriages, by Age

Source: *Time*

Critical Thinking Exercises

53. The World Health Organization makes predictions about the number of AIDS cases based on a compromise between a linear model and an exponential growth model. Explain why the World Health Organization does this.

54. Over a period of time, a hot object cools to the temperature of the surrounding air. This is described mathematically by

$$T = C + (T_0 - C)e^{-kt},$$

where t is the time it takes for an object to cool from temperature T_0 to temperature T, C is the surrounding air temperature, and k is a positive constant that is associated with the cooling object. A cake removed from the oven has a temperature of 210°F and is left to cool in a room that has a temperature of 70°F. After 30 minutes, the temperature of the cake is 140°F. What is the temperature of the cake after 40 minutes?

Group Exercise

55. This activity is intended for three or four people who would like to take up weightlifting. Each person in the group should record the maximum number of pounds that he or she can lift at the end of each week for the first 10 consecutive weeks. Use the Logarithmic REGression option of a graphing utility to obtain a model showing the amount of weight that group members can lift from week 1 through week 10. Graph each of the models in the same viewing rectangle to observe similarities and differences among weight-growth patterns of each member. Use the functions to predict the amount of weight that group members will be able to lift in the future. If the group continues to work out together, check the accuracy of these predictions.

56. Each group member should consult an almanac, newspaper, magazine, or the Internet to find data that can be modeled by exponential or logarithmic functions. Group members should select the two sets of data that are most interesting and relevant. For each data set selected, find a model that best fits the data. Each group member should make one prediction based on the model and then discuss a consequence of this prediction. What factors might change the accuracy of each prediction?

CHAPTER SUMMARY, REVIEW, AND TEST

Summary

DEFINITIONS AND CONCEPTS **EXAMPLES**

4.1 Exponential Functions

a. The exponential function with base b is defined by $f(x) = b^x$, where $b > 0$ and $b \neq 1$. Ex. 1, p. 375

b. Characteristics of exponential functions and graphs for $0 < b < 1$ and $b > 1$ are shown in the box on page 377. Ex. 2, p. 376

c. Transformations involving exponential functions are summarized in Table 4.1 on page 377. Exs. 3 & 4, p. 378

d. The natural exponential function $f(x) = e^x$. The irrational number e is called the natural base, where $e \approx 2.7183$. Ex. 5, p. 379

e. Formulas for compound interest: After t years, the balance, A, in an account with principal P and annual interest rate r (in decimal form) is given by one of the following formulas: Ex. 6, p. 381

    1. For n compoundings per year: $A = P\left(1 + \dfrac{r}{n}\right)^{nt}$

    2. For continuous compounding: $A = Pe^{rt}$

4.2 Logarithmic Functions

a. Definition of the logarithmic function: For $x > 0$ and $b > 0$, $b \neq 1$, $y = \log_b x$ is equivalent to $b^y = x$. The function $f(x) = \log_b x$ is the logarithmic function with base b. This function is the inverse function of the exponential function with base b. Ex. 1, p. 386; Ex. 2, p. 386; Ex. 3, p. 387

b. Graphs of logarithmic functions for $b > 1$ and $0 < b < 1$ are shown in Figure 4.7 on page 389. Characteristics of the graphs are summarized in the box that follows the figure. Ex. 6, p. 389

c. Transformations involving logarithmic functions are summarized in Table 4.3 on page 390.

d. The domain of a logarithmic function of the form $f(x) = \log_b x$ is the set of all positive real numbers. The domain of $f(x) = \log_b(x + c)$ consists of all x for which $x + c > 0$. Ex. 7, p. 391; Ex. 10, p. 391

e. Common and natural logarithms: $f(x) = \log x$ means $f(x) = \log_{10} x$ and is the common logarithmic function. $f(x) = \ln x$ means $f(x) = \log_e x$ and is the natural logarithmic function. Ex. 8, p. 392; Ex. 9, p. 392

f. Basic Logarithmic Properties

Base b $(b > 0, b \neq 1)$	Base 10 (Common Logarithms)	Base e (Natural Logarithms)	
$\log_b 1 = 0$	$\log 1 = 0$	$\ln 1 = 0$	Ex. 4, p. 388
$\log_b b = 1$	$\log 10 = 1$	$\ln e = 1$	Ex. 5, p. 388
$\log_b b^x = x$	$\log 10^x = 1$	$\ln e^x = x$	Ex. 11, p. 394
$b^{\log_b x} = x$	$10^{\log x} = x$	$e^{\ln x} = x$	Ex. 12, p. 394

4.3 Properties of Logarithms

a. *The Product Rule:* $\log_b(MN) = \log_b M + \log_b N$ Ex. 1, p. 399

b. *The Quotient Rule:* $\log_b\left(\dfrac{M}{N}\right) = \log_b M - \log_b N$ Ex. 2, p. 399

c. *The Power Rule:* $\log_b M^p = p \log_b M$ Ex. 3, p. 401

d. *The Change-of-Base Property:*

The General Property	Introducing Common Logarithms	Introducing Natural Logarithms	
$\log_b M = \dfrac{\log_a M}{\log_a b}$	$\log_b M = \dfrac{\log M}{\log b}$	$\log_b M = \dfrac{\ln M}{\ln b}$	Ex. 7, p. 404 Ex. 8, p. 405

DEFINITIONS AND CONCEPTS	EXAMPLES

4.4 Exponential and Logarithmic Equations

a. An exponential equation is an equation containing a variable in an exponent. The solution procedure involves isolating the exponential expression and taking the natural logarithm on both sides. The box on page 408 provides the details.

Ex.1, p. 408;
Ex.2, p. 409;
Ex.3, p. 409;
Ex.4, p. 410

b. A logarithmic equation is an equation containing a variable in a logarithmic expression. Logarithmic equations in the form $\log_b x = c$ can be solved by rewriting as $b^c = x$.

Ex.5, p. 410

c. When checking logarithmic equations, reject proposed solutions that produce the logarithm of a negative number or the logarithm of 0 in the original equation.

Ex.6, p. 411

d. Equations involving natural logarithms are solved by isolating the natural logarithm with coefficient 1 on one side and exponentiating both sides. Simplify using $e^{\ln x} = x$.

Ex.7, p. 412

4.5 Modeling with Exponential and Logarithmic Functions

a. Exponential growth and decay models are given by $A = A_0 e^{kt}$ in which t represents time, A_0 is the amount present at $t = 0$, and A is the amount present at time t. If $k > 0$, the model describes growth and k is the growth rate. If $k < 0$, the model describes decay and k is the decay rate.

Ex.1, p. 420;
Ex.2, p. 422

b. The logistic growth model, given by $A = \dfrac{c}{1 + ae^{-bt}}$, describes situations in which growth is limited. $y = c$ is a horizontal asymptote for the graph, and growth, A, can never exceed c.

Ex.3, p. 424

c. Scatter plots for exponential and logarithmic models are shown in Figure 4.20 on page 425. When using a graphing utility to model data, the closer that the correlation coefficient, r, is to -1 or 1, the better the model fits the data.

d. Expressing an Exponential Model in Base e: $y = ab^x$ is equivalent to $y = ae^{(\ln b) \cdot x}$.

Ex.4, p. 427

Review Exercises

4.1

In Exercises 1–4, the graph of an exponential function is given. Select the function for each graph from the following options:

$$f(x) = 4^x, g(x) = 4^{-x},$$
$$h(x) = -4^{-x}, r(x) = -4^{-x} + 3.$$

1.

2.

3.

4.

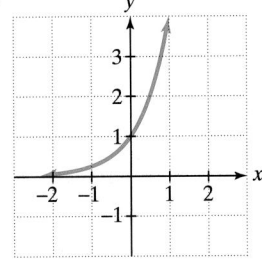

In Exercises 5–8, sketch by hand the graphs of the two functions in the same rectangular coordinate system. Use a table of coordinates to sketch the first function and transformations of this function with a table of coordinates to graph the second function.

5. $f(x) = 2^x$ and $g(x) = 2^{x-1}$

6. $f(x) = 3^x$ and $g(x) = 3^x - 1$

7. $f(x) = 3^x$ and $g(x) = -3^x$

8. $f(x) = \left(\frac{1}{2}\right)^x$ and $g(x) = \left(\frac{1}{2}\right)^{-x}$

Use the compound interest formulas to solve Exercises 9–10.

9. Suppose that you have \$5000 to invest. Which investment yields the greater return over 5 years: 5.5% compounded semiannually or 5.25% compounded monthly?

10. Suppose that you have \$14,000 to invest. Which investment yields the greater return over 10 years: 7% compounded monthly or 6.85% compounded continuously?

11. A cup of coffee is taken out of a microwave oven and placed in a room. The temperature, T, in degrees Fahrenheit, of the coffee after t minutes is modeled by the function $T = 70 + 130e^{-0.04855t}$. The graph of the function is shown in the figure.

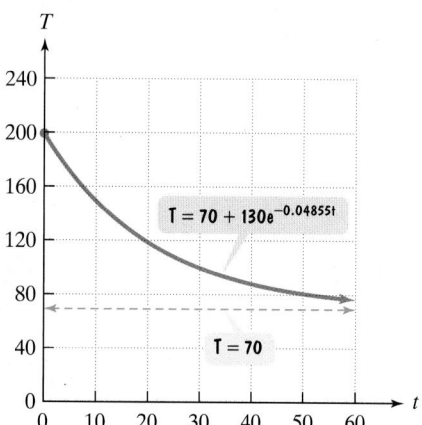

Use the graph to answer each of the following questions.

a. What was the temperature of the coffee when it was first taken out of the microwave?

b. What is a reasonable estimate of the temperature of the coffee after 20 minutes? Use your calculator to verify this estimate.

c. What is the limit of the temperature to which the coffee will cool? What does this tell you about the temperature of the room?

4.2

In Exercises 12–14, write each equation in its equivalent exponential form.

12. $\frac{1}{2} = \log_{49} 7$ **13.** $3 = \log_4 x$ **14.** $\log_3 81 = y$

In Exercises 15–17, write each equation in its equivalent logarithmic form.

15. $6^3 = 216$ **16.** $b^4 = 625$ **17.** $13^y = 874$

In Exercises 18–25, evaluate each expression without using a calculator. If evaluation is not possible, state the reason.

18. $\log_4 64$ **19.** $\log_5 \frac{1}{25}$ **20.** $\log_3 (-9)$

21. $\log_{16} 4$ **22.** $\log_{17} 17$ **23.** $\log_3 3^8$

24. $\ln e^5$ **25.** $\log_3 (\log_8 8)$

26. Graph $f(x) = 2^x$ and $g(x) = \log_2 x$ in the same rectangular coordinate system.

27. Graph $f(x) = \left(\frac{1}{3}\right)^x$ and $g(x) = \log_{1/3} x$ in the same rectangular coordinate system.

In Exercises 28–31, the graph of a logarithmic function is given. Select the function for each graph from the following options:

$$f(x) = \log x, \, g(x) = \log(-x),$$
$$h(x) = \log(2 - x), \, r(x) = 1 + \log(2 - x).$$

28.

29.

30.

31.

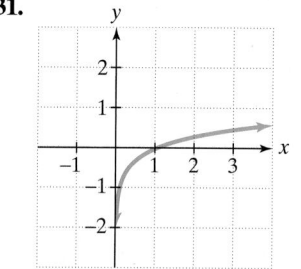

In Exercises 32–34, begin by graphing $f(x) = \log_2 x$. Then use transformations of this graph to graph the given function. What is the graph's x-intercept? What is the vertical asymptote?

32. $g(x) = \log_2(x - 2)$ **33.** $h(x) = -1 + \log_2 x$

34. $r(x) = \log_2(-x)$

In Exercises 35–37, find the domain of each logarithmic function.

35. $f(x) = \log_8(x + 5)$ **36.** $f(x) = \log(3 - x)$
37. $f(x) = \ln(x - 1)^2$

In Exercises 38–40, use inverse properties of logarithms to simplify each expression.

38. $\ln e^{6x}$ **39.** $e^{\ln \sqrt{x}}$ **40.** $10^{\log 4x^2}$

41. On the Richter scale, the magnitude, R, of an earthquake of intensity I is given by $R = \log \dfrac{I}{I_0}$, where I_0 is the intensity of a barely felt zero-level earthquake. If the intensity of an earthquake is $1000I_0$, what is its magnitude on the Richter scale?

42. Students in a psychology class took a final examination. As part of an experiment to see how much of the course content they remembered over time, they took equivalent forms of the exam in monthly intervals thereafter. The average score, $f(t)$, for the group after t months is modeled by the function $f(t) = 76 - 18 \log(t + 1)$, where $0 \le t \le 12$.

   a. What was the average score when the exam was first given?

   b. What was the average score after 2 months? 4 months? 6 months? 8 months? one year?

   c. Use the results from parts (a) and (b) to graph f. Describe what the shape of the graph indicates in terms of the material retained by the students.

43. The formula

$$t = \frac{1}{c} \ln\left(\frac{A}{A - N}\right)$$

describes the time, t, in weeks, that it takes to achieve mastery of a portion of a task. In the formula, A represents maximum learning possible, N is the portion of the learning that is to be achieved, and c is a constant used to measure an individual's learning style. A 50-year-old man decides to start running as a way to maintain good health. He feels that the maximum rate he could ever hope to achieve is 12 miles per hour. How many weeks will it take before the man can run 5 miles per hour if $c = 0.06$ for this person?

4.3

In Exercises 44–47, use properties of logarithms to expand each logarithmic expression as much as possible. Where possible, evaluate logarithmic expressions without using a calculator.

44. $\log_6(36x^3)$ **45.** $\log_4\left(\dfrac{\sqrt{x}}{64}\right)$

46. $\log_2\left(\dfrac{xy^2}{64}\right)$ **47.** $\ln\sqrt[3]{\dfrac{x}{e}}$

In Exercises 48–51, use properties of logarithms to condense each logarithmic expression. Write the expression as a single logarithm whose coefficient is 1.

48. $\log_b 7 + \log_b 3$ **49.** $\log 3 - 3\log x$

50. $3 \ln x + 4 \ln y$ **51.** $\frac{1}{2} \ln x - \ln y$

In Exercises 52–53, use common logarithms or natural logarithms and a calculator to evaluate to four decimal places.

52. $\log_6 72{,}348$ **53.** $\log_4 0.863$

4.4

Solve each exponential equation in Exercises 54–58. Express the answer in terms of natural logarithms. Then use a calculator to obtain a decimal approximation, correct to two decimal places, for the solution.

54. $8^x = 12{,}143$ **55.** $9e^{5x} = 1269$

56. $e^{12-5x} - 7 = 123$ **57.** $5^{4x+2} = 37{,}500$

58. $e^{2x} - e^x - 6 = 0$

Solve each logarithmic equation in Exercises 59–63.

59. $\log_4(3x - 5) = 3$

60. $\log_2(x + 3) + \log_2(x - 3) = 4$

61. $\log_3(x - 1) - \log_3(x + 2) = 2$

62. $\ln x = -1$ **63.** $3 + 4 \ln(2x) = 15$

64. The formula $A = 10.1e^{0.005t}$ models the population of Los Angeles, California, A, in millions, t years after 1992. If the growth rate continues into the future, when will the population reach 13 million?

65. The amount of carbon dioxide in the atmosphere, measured in parts per million, has been increasing as a result of the burning of oil and coal. The buildup of gases and particles traps heat and raises the planet's temperature, a phenomenon called the *greenhouse effect*. Carbon dioxide accounts for about half of the warming. The function $f(t) = 364(1.005)^t$ projects carbon dioxide concentration, $f(t)$, in parts per million, t years after 2000. Using the projections given by the function, when will the carbon dioxide concentration be double the preindustrial level of 280 parts per million?

66. The formula $\overline{C}(x) = 15{,}557 + 5259 \ln x$ models the average cost of a new car, $\overline{C}(x)$, x years after 1989. When will the average cost of a new car reach $30,000?

67. Use the formula for compound interest with n compoundings each year to solve this problem. How long, to the nearest tenth of a year, will it take $12,500 to grow to $20,000 at 6.5% annual interest compounded quarterly?

Use the formula for continuous compounding to solve Exercises 68–69.

68. How long, to the nearest tenth of a year, will it take $50,000 to triple in value at 7.5% annual interest compounded continuously?

69. What interest rate is required for an investment subject to continuous compounding to triple in 5 years?

4.5

70. According to the U.S. Bureau of the Census, in 1990 there were 22.4 million residents of Hispanic origin living in the United States. By 2000, the number had increased to 35.3 million. The exponential growth function $A = 22.4e^{kt}$ describes the U.S. Hispanic population, A, in millions, t years after 1990.

a. Find k, correct to three decimal places.

b. Use the resulting model to project the Hispanic resident population in 2010.

c. In which year will the Hispanic resident population reach 60 million?

71. Use the exponential decay model for carbon-14, $A = A_0 e^{-0.000121t}$, to solve this exercise. Prehistoric cave paintings were discovered in the Lascaux cave in France. The paint contained 15% of the original carbon-14. Estimate the age of the paintings at the time of the discovery.

72. The function

$$f(t) = \frac{500,000}{1 + 2499e^{-0.92t}}$$

models the number of people, $f(t)$, in a city who have become ill with influenza t weeks after its initial outbreak.

a. How many people became ill with the flu when the epidemic began?

b. How many people were ill by the end of the sixth week?

c. What is the limiting size of $f(t)$, the population that becomes ill?

In Exercises 73–74, rewrite the equation in terms of base e. Express the answer in terms of a natural logarithm, and then round to three decimal places.

73. $y = 73(2.6)^x$

74. $y = 6.5(0.43)^x$

75. The figure shows world population projections through the year 2150. The data are from the United Nations Family Planning Program and are based on optimistic or pessimistic expectations for successful control of human population growth. Suppose that you are interested in modeling these data using exponential, logarithmic, linear, and quadratic functions. Which function would you use to model each of the projections? Explain your choices. For the choice corresponding to a quadratic model, would your formula involve one with a positive or negative leading coefficient? Explain.

76. The figure shows the number of people in the United States age 65 and over, with projected figures for the year 2010 and beyond.

U. S. Population Age 65 and Over

Source: U.S. Bureau of the Census

Let x represent the number of years after 1899 and let y represent the U.S. population, in millions, age 65 and over. Use your graphing utility to find the model that best fits the data in the bar graph. Then use the model to find the projected U.S. population age 65 and over in 2050.

Chapter 4 Test

1. Graph $f(x) = 2^x$ and $g(x) = 2^{x+1}$ in the same rectangular coordinate system.

2. Graph $f(x) = \log_2 x$ and $g(x) = \log_2(x - 1)$ in the same rectangular coordinate system.

3. Write in exponential form: $\log_5 125 = 3$.

4. Write in logarithmic form: $\sqrt{36} = 6$.

5. Find the domain of $f(x) = \ln(3 - x)$.

In Exercises 6–7, use properties of logarithms to expand each logarithmic expression as much as possible. Where possible, evaluate logarithmic expressions without using a calculator.

6. $\log_4(64x^5)$

7. $\log_3\left(\dfrac{\sqrt[3]{x}}{81}\right)$

In Exercises 8–9, write each expression as a single logarithm.

8. $6 \log x + 2 \log y$

9. $\ln 7 - 3 \ln x$

10. Use a calculator to evaluate $\log_{15} 71$ to four decimal places.

In Exercises 11–16, solve each equation.

11. $5^x = 1.4$

12. $400e^{0.005x} = 1600$

13. $e^{2x} - 6e^x + 5 = 0$

14. $\log_6(4x - 1) = 3$

15. $\log x + \log(x + 15) = 2$

16. $2 \ln(3x) = 8$

17. Suppose you have $3000 to invest. Which investment yields the greater return over 10 years: 6.5% compounded semiannually or 6% compounded continuously? How much more (to the nearest dollar) is yielded by the better investment?

18. On the decibel scale, the loudness of a sound, D, in decibels, is given by $D = 10 \log \dfrac{I}{I_0}$, where I is the intensity of the sound, in watts per meter^2, and I_0 is the intensity of a sound barely audible to the human ear. If the intensity of a sound is $10^{12} I_0$, what is its loudness in decibels? (Such a sound is potentially damaging to the ear.)

19. The function
$$P(t) = 89.18e^{-0.004t}$$
models the percentage, $P(t)$, of married men in the United States who were employed t years after 1959.

   a. What percentage of married men were employed in 1959?

   b. Is the percentage of married men who are employed increasing or decreasing? Explain.

   c. In what year were 77% of U.S. married men employed?

20. The 1990 population of Europe was 509 million; in 2000, it was 729 million. Write the exponential growth function that describes the population of Europe, in millions, t years after 1990.

21. Use the exponential decay model for carbon-14, $A = A_0 e^{-0.000121t}$, to solve this exercise. Bones of a prehistoric man were discovered and contained 5% of the original amount of carbon-14. How long ago did the man die?

22. The logistic growth function
$$f(t) = \frac{140}{1 + 9e^{-0.165t}}$$
describes the population, $f(t)$, of an endangered species of elk t years after they were introduced to a nonthreatening habitat.

   a. How many elk were initially introduced to the habitat?

   b. How many elk are expected in the habitat after 10 years?

   c. What is the limiting size of the elk population that the habitat will sustain?

Cumulative Review Exercises (Chapters 1–4)

Solve each equation in Exercises 1–5.

1. $|3x - 4| = 2$

2. $\sqrt{2x - 5} - \sqrt{x - 3} = 1$

3. $x^4 + x^3 - 3x^2 - x + 2 = 0$

4. $e^{5x} - 32 = 96$

5. $\log_2(x + 5) + \log_2(x - 1) = 4$

Solve each inequality in Exercises 6–7. Express the answer in interval notation.

6. $14 - 5x \geq -6$

7. $|2x - 4| \leq 2$

8. Write the point-slope form and the slope-intercept form of the line passing through $(1, 3)$ and $(3, -3)$.

9. If $f(x) = x^2$ and $g(x) = x + 2$, find $(f \circ g)(x)$ and $(g \circ f)(x)$.

10. If $f(x) = 2x - 7$, find $f^{-1}(x)$.

11. Divide $x^3 + 5x^2 + 3x - 10$ by $x + 2$.

12. Use the Rational Zero Theorem to list all possible rational zeros for $f(x) = 4x^3 - 7x - 3$.

13. The value of y varies directly as the square of x. If $x = 3$ when $y = 12$, find y when $x = 15$.

14. Solve $x^3 - 4x^2 + 6x - 4 = 0$ given that $1 + i$ is a root.

In Exercises 15–18, graph each equation.

15. $(x - 3)^2 + (y + 2)^2 = 4$

16. $f(x) = (x - 2)^2 - 1$

17. $f(x) = \dfrac{x^2 - 1}{x^2 - 4}$

18. $f(x) = (x - 2)^2(x + 1)$

19. You are paid time-and-a-half for each hour worked over 40 hours a week. Last week you worked 50 hours and earned $660. What is your normal hourly salary?

20. The function $F(t) = 1 - k \ln(t + 1)$ models the fraction of people, $F(t)$, who remember all the words in a list of nonsense words t hours after memorizing the list. After 3 hours, only half the people could remember all the words. Determine the value of k and then predict the fraction of people in the group who will remember all the words after 6 hours. Round to three decimal places and then express the fraction with a denominator of 1000.

Trigonometric Functions

Have you had days where your physical, intellectual, and emotional potentials were all at their peak? Then there are those other days when we feel we should not even bother getting out of bed. Do our potentials run in oscillating cycles like the tides? Can they be described mathematically? In this chapter you will encounter functions that enable us to model phenomena that occur in cycles.

What a day! It started when you added two miles to your morning run. You've experienced a feeling of peak physical well-being ever since. College was wonderful: You actually enjoyed two difficult lectures and breezed through a math test that had you worried. Now you're having dinner with an old group of friends. You experience the warmth from bonds of friendship filling the room.

SECTION 5.1 *Angles and Their Measure*

Objectives

1. Recognize and use the vocabulary of angles.
2. Use degree measure.
3. Draw angles in standard position.
4. Find coterminal angles.
5. Find complements and supplements.
6. Use radian measure.
7. Convert between degrees and radians.
8. Find the length of a circular arc.
9. Use linear and angular speed to describe motion on a circular path.

The San Francisco Museum of Modern Art was constructed in 1995 to illustrate how art and architecture can enrich one another. The exterior involves geometric shapes, symmetry, and unusual facades. Although there are no windows, natural light streams in through a truncated cylindrical skylight that crowns the building. The architect worked with a scale model of the museum at the site and observed how light hit it during different times of the day. These observations were used to cut the cylindrical skylight at an angle that maximizes sunlight entering the interior.

Angles play a critical role in creating modern architecture. They are also fundamental in trigonometry. In this section, we begin our study of trigonometry by looking at angles and methods for measuring them.

1 Recognize and use the vocabulary of angles.

Figure 5.1 Clock with hands forming an angle

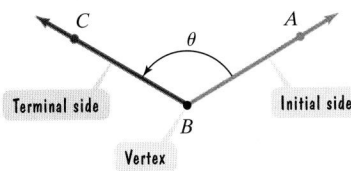

Figure 5.2 An angle; two rays with a common endpoint

Angles

The hour hand of a clock suggests a **ray,** a part of a line that has only one endpoint and extends forever in the opposite direction. An **angle** is formed by two rays that have a common endpoint. One ray is called the **initial side** and the other the **terminal side.**

A rotating ray is often a useful way to think about angles. The ray in Figure 5.1 rotates from 12 to 2. The ray pointing to 12 is the **initial side** and the ray pointing to 2 is the **terminal side.** The common endpoint of an angle's initial side and terminal side is the **vertex** of the angle.

Figure 5.2 shows an angle. The arrow near the vertex shows the direction and the amount of rotation from the initial side to the terminal side. Several methods can be used to name an angle. Lowercase Greek letters, such as α (alpha), β (beta), γ (gamma), and θ (theta), are often used.

An angle is in **standard position** if

- its vertex is at the origin of a rectangular coordinate system

and

- its initial side lies along the positive x-axis.

The angles in Figure 5.3 are both in standard position.

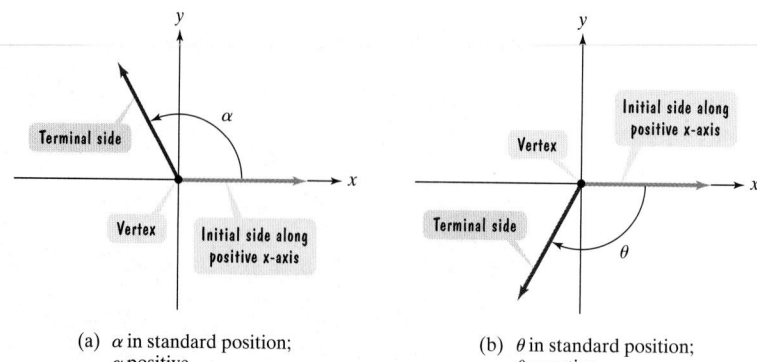

Figure 5.3 Two angles in standard position

(a) α in standard position; α positive

(b) θ in standard position; θ negative

When we see an initial side and a terminal side in place, there are two kinds of rotation that could have generated it. The arrow in Figure 5.3(a) indicates that the rotation from the initial side to the terminal side is in the counterclockwise direction. **Positive angles** are generated by counterclockwise rotation. Thus, angle α is positive. By contrast, the arrow in Figure 5.3(b) shows that the rotation from the initial side to the terminal side is in the clockwise direction. **Negative angles** are generated by clockwise rotation. Thus, angle θ is negative.

When an angle is in standard position, its terminal side can lie in a quadrant. We say that the angle **lies in that quadrant.** For example, in Figure 5.3(a), the terminal side of angle α lies in quadrant II. Thus, angle α lies in quadrant II. By contrast, in Figure 5.3(b), the terminal side of angle θ lies in quadrant III. Thus, angle θ lies in quadrant III.

Must all angles in standard position lie in a quadrant? The answer is no. The terminal side can lie on the x-axis or the y-axis. For example, angle β in Figure 5.4 has a terminal side that lies on the negative y-axis. An angle is called a **quadrantal angle** if its terminal side lies on the x-axis or the y-axis. Angle β in Figure 5.4 is an example of a quadrantal angle.

Figure 5.4 β is a quadrantal angle.

2 Use degree measure.

Measuring Angles Using Degrees

Angles are measured by determining the amount of rotation from the initial side to the terminal side. One way to measure angles is in **degrees,** symbolized by a small, raised circle °. Think of the hour hand of a clock. From 12 noon to 12 midnight, the hour hand moves around in a complete circle. By definition, the ray has rotated through 360 degrees, or 360°. Using 360° as the amount of rotation of a ray back onto itself, a degree, 1°, is $\frac{1}{360}$ of a complete rotation.

Figure 5.5 shows angles classified by their degree measurement. An **acute angle** measures less than 90° [see Figure 5.5(a)]. A **right angle,** one quarter of a complete rotation, measures 90° [Figure 5.5(b)]. Examine the right angle—do you see a small square at the vertex? This symbol is used to indicate a right angle. An **obtuse angle** measures more than 90°, but less than 180° [Figure 5.5(c)]. Finally, a **straight angle,** one-half a complete rotation, measures 180° [Figure 5.5(d)].

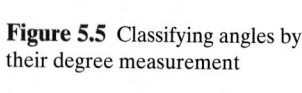

A complete 360° rotation

Figure 5.5 Classifying angles by their degree measurement

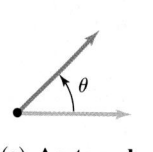

(a) **Acute angle** $(0° < \theta < 90°)$

(b) **Right angle** ($\frac{1}{4}$ rotation)

(c) **Obtuse angle** $(90° < \theta < 180°)$

(d) **Straight angle** ($\frac{1}{2}$ rotation)

3 Draw angles in standard position.

We will be using notation such as $\theta = 60°$ to refer to an angle θ whose measure is 60°. We also refer to *an angle of 60°* or a *60° angle*, rather than using the more precise (but cumbersome) phrase *an angle whose measure is 60°*.

Technology

Fractional parts of degrees are measured in minutes and seconds. One minute, written 1′, is $\frac{1}{60}$ degree: $1' = \frac{1}{60}°$.

One second, written 1″, is $\frac{1}{3600}$ degree: $1'' = \frac{1}{3600}°$.

For example,

$31° \, 47' \, 12''$

$= \left(31 + \dfrac{47}{60} + \dfrac{12}{3600}\right)°$

$\approx 31.787°.$

Many calculators have keys for changing an angle from degree, minute, second notation (D°M′S″) to a decimal form and vice versa.

EXAMPLE 1 Drawing Angles in Standard Position

Draw each angle in standard position:

a. a 45° angle   **b.** a 225° angle   **c.** a −135° angle   **d.** a 405° angle.

Solution   Because we are drawing angles in standard position, each vertex is at the origin and each initial side lies along the positive x-axis.

a. A 45° angle is half of a right angle. The angle lies in quadrant I and is shown in Figure 5.6(a).

b. A 225° angle is a positive angle. It has a counterclockwise rotation of 180° followed by a counterclockwise rotation of 45°. The angle lies in quadrant III and is shown in Figure 5.6(b).

c. A −135° angle is negative angle. It has a clockwise rotation of 90° followed by a clockwise rotation of 45°. The angle lies in quadrant III and is shown in Figure 5.6(c).

d. A 405° angle is a positive angle. It has a counterclockwise rotation of 360°, one complete rotation, followed by a counterclockwise rotation of 45°. The angle lies in quadrant I and is shown in Figure 5.6(d).

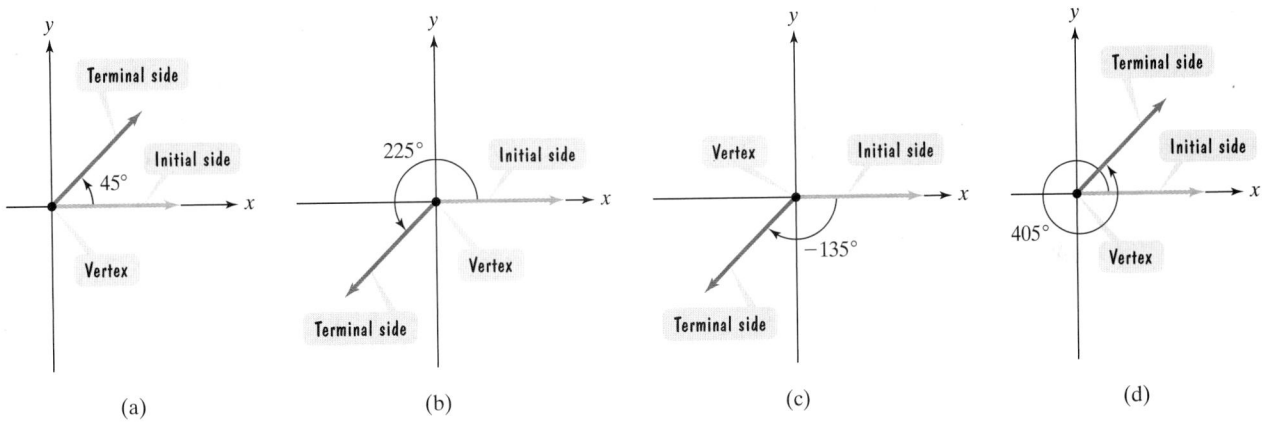

(a)   (b)   (c)   (d)

Figure 5.6 Four angles in standard position

Check Point 1   Draw each angle in standard position:

a. a 30° angle   **b.** a 210° angle

c. a −120° angle   **d.** a 390° angle.

4 Find coterminal angles.

Look at Figure 5.6 again. The 45° and 405° angles in parts (a) and (d) have the same initial and terminal sides. Similarly, the 225° and −135° angles in parts (b) and (c) have the same initial and terminal sides. Two angles with the same initial and terminal sides are called **coterminal angles.**

Every angle has infinitely many coterminal angles. Why? Think of an angle in standard position. One or more complete rotations of 360°, clockwise or counterclockwise, result in angles with the same initial and terminal sides as the original angle.

Coterminal Angles

An angle of $x°$ is coterminal with angles of

$$x° + k \cdot 360°$$

where k is an integer.

Two coterminal angles for an angle of $x°$ can be found by adding 360° to $x°$ and subtracting 360° from $x°$.

Counterclockwise Clocks

The counterclockwise rotation associated with positive angles was used in England to manufacture counterclockwise clocks. They ran backward but told the time perfectly correctly.

EXAMPLE 2 Finding Coterminal Angles

Assume the following angles are in standard position. Find a positive angle less than 360° that is coterminal with:

a. a 420° angle   **b.** a −120° angle.

Solution   We obtain the coterminal angle by adding or subtracting 360°. The requirement to obtain a positive angle less than 360° determines whether we should add or subtract.

a. For a 420° angle, subtract 360° to find a positive coterminal angle.

$$420° - 360° = 60°$$

A 60° angle is coterminal with a 420° angle. Figure 5.7(a) illustrates that these angles have the same initial and terminal sides.

b. For a −120° angle, add 360° to find a positive coterminal angle.

$$-120° + 360° = 240°$$

A 240° angle is coterminal with a −120° angle. Figure 5.7(b) illustrates that these angles have the same initial and terminal sides.

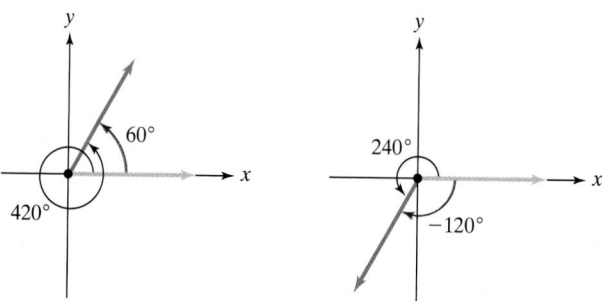

(a) Angles of 420° and 60° are coterminal.

(b) Angles of −120° and 240° are coterminal.

Figure 5.7 Pairs of coterminal angles

Check Point 2 Find a positive angle less than 360° that is coterminal with:
a. a 400° angle **b.** a −135° angle.

5 Find complements and supplements.

Two positive angles are **complements** if their sum is 90°. For example, angles of 70° and 20° are complements because 70° + 20° = 90°.

Two positive angles are **supplements** if their sum is 180°. For example, angles of 130° and 50° are supplements because 130° + 50° = 180°.

Finding Complements and Supplements

- For an $x°$ angle, the complement is a $90° - x°$ angle. Thus, the complement's measure is found by subtracting the angle's measure from 90°.
- For an $x°$ angle, the supplement is a $180° - x°$ angle. Thus, the supplement's measure is found by subtracting the angle's measure from 180°.

Because we use only positive angles for complements and supplements, some angles do not have complements and supplements.

EXAMPLE 3 Complements and Supplements

If possible, find the complement and the supplement of the given angle:
  a. $\theta = 62°$ **b.** $\alpha = 123°$.

Solution We find the complement by subtracting the angle's measure from 90°. We find the supplement by subtracting the angle's measure from 180°.

a. We begin with $\theta = 62°$.

$$\text{complement} = 90° - 62° = 28°$$
$$\text{supplement} = 180° - 62° = 118°$$

For a 62° angle, the complement is a 28° angle and the supplement is a 118° angle.

b. Now we turn to $\alpha = 123°$. For the angle's complement, we consider subtracting 123° from 90°. The difference is negative. Because we use only positive angles for complements, a 123° angle has no complement. It does, however, have a supplement.

$$\text{supplement} = 180° - 123° = 57°$$

The supplement of a 123° angle is a 57° angle.

Check Point 3 If possible, find the complement and the supplement of the given angle:
  a. $\theta = 78°$ **b.** $\alpha = 150°$.

6 Use radian measure.

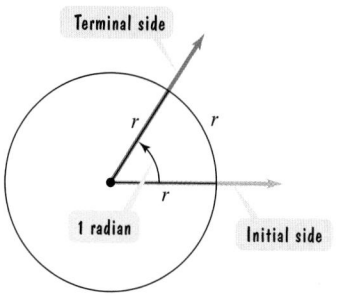

Figure 5.8 For a 1-radian angle, the intercepted arc and the radius are equal.

Measuring Angles Using Radians

Another way to measure angles is in *radians*. Let's first define an angle measuring **1 radian.** We use a circle of radius r. In Figure 5.8, we've constructed an angle whose vertex is at the center of the circle. Such an angle is called a **central angle.** Notice that this central angle intercepts an arc along the circle measuring r units. The radius of the circle is also r units. The measure of such an angle is 1 radian.

Definition of a Radian

One radian is the measure of the central angle of a circle that intercepts an arc equal in length to the radius of the circle.

The **radian measure** of any central angle is the length of the intercepted arc divided by the circle's radius. In Figure 5.9(a), the length of the arc intercepted by angle β is double the radius, r. We find the measure of angle β in radians by dividing the length of the intercepted arc by the radius.

$$\beta = \frac{\text{length of the intercepted arc}}{\text{radius}} = \frac{2r}{r} = 2$$

Thus, angle β measures 2 radians.

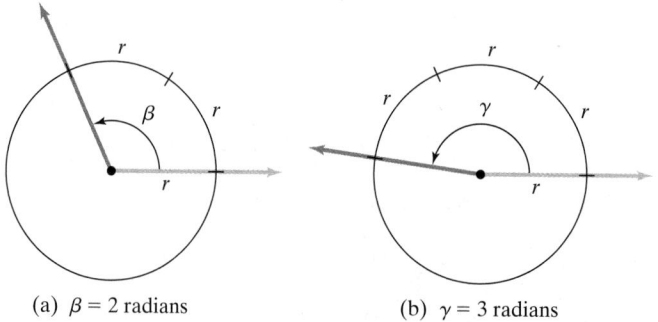

(a) $\beta = 2$ radians

(b) $\gamma = 3$ radians

Figure 5.9 Two central angles measured in radians

In Figure 5.9(b), the length of the intercepted arc is triple the radius, r. Let us find the measure of angle γ:

$$\gamma = \frac{\text{length of the intercepted arc}}{\text{radius}} = \frac{3r}{r} = 3.$$

Thus, angle γ measures 3 radians.

Radian Measure

Consider an arc of length s on a circle of radius r. The measure of the central angle, θ, that intercepts the arc is

$$\theta = \frac{s}{r} \text{ radians.}$$

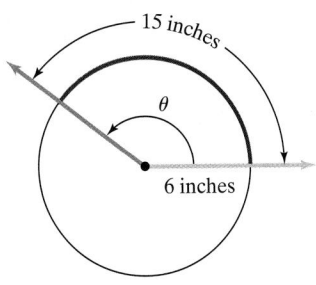

Figure 5.10

EXAMPLE 4 Computing Radian Measure

A central angle, θ, in a circle of radius 6 inches intercepts an arc of length 15 inches. What is the radian measure of θ?

Solution Angle θ is shown in Figure 5.10. The radian measure of a central angle is the length of the intercepted arc, s, divided by the circle's radius, r. The length of the intercepted arc is 15 inches: $s = 15$ inches. The circle's radius is 6 inches: $r = 6$ inches. Now we use the formula for radian measure to find the radian measure of θ.

$$\theta = \frac{s}{r} = \frac{15 \text{ inches}}{6 \text{ inches}} = 2.5$$

Thus, the radian measure of θ is 2.5.

Study Tip

Before applying the formula for radian measure, be sure that the same unit of length is used for the intercepted arc, s, and the radius, r.

In Example 4, notice that the units (inches) cancel when we use the formula for radian measure. We are left with a number with no units. Thus, if an angle θ has a measure of 2.5 radians, we can write $\theta = 2.5$ radians or $\theta = 2.5$. We will often include the word *radians* simply for emphasis. There should be no confusion as to whether radian or degree measure is being used. Why is this so? If θ has a degree measure of, say, 2.5°, we must include the degree symbol and write $\theta = 2.5°$, and *not* $\theta = 2.5$.

> **Check Point 4** A central angle, θ, in a circle of radius 12 feet intercepts an arc of length 42 feet. What is the radian measure of θ?

7 Convert between degrees and radians.

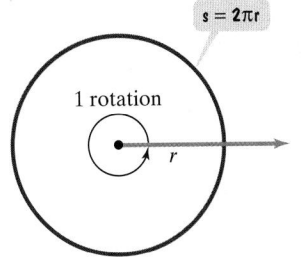

Figure 5.11 A complete rotation

Relationship between Degrees and Radians

How can we obtain a relationship between degrees and radians? We compare the number of degrees and the number of radians in one complete rotation, shown in Figure 5.11. We know that 360° is the amount of rotation of a ray back onto itself. The length of the intercepted arc is equal to the circumference of the circle. Thus, the radian measure of this central angle is the circumference of the circle divided by the circle's radius, r. The circumference of a circle of radius r is $2\pi r$. We use the formula for radian measure to find the radian measure of the 360° angle.

$$\theta = \frac{s}{r} = \frac{\text{the circle's circumference}}{r} = \frac{2\pi r}{r} = 2\pi$$

Because one complete rotation measures 360° and 2π radians,

$$360° = 2\pi \text{ radians.}$$

Dividing both sides by 2, we have

$$180° = \pi \text{ radians.}$$

Dividing this last equation by 180° or π gives the conversion rules that appear on the next page.

Study Tip

The unit you are converting *to* appears in the *numerator* of the conversion factor.

Conversion between Degrees and Radians

Using the basic relationship π radians $= 180°$,

1. To convert degrees to radians, multiply degrees by $\dfrac{\pi \text{ radians}}{180°}$.

2. To convert radians to degrees, multiply radians by $\dfrac{180°}{\pi \text{ radians}}$.

Angles that are fractions of a complete rotation are usually expressed in radian measure as fractional multiples of π, rather than as decimal approximations. For example, we write $\theta = \dfrac{\pi}{2}$ rather than using the decimal approximation $\theta \approx 1.57$.

EXAMPLE 5 Converting from Degrees to Radians

Convert each angle in degrees to radians:

    a. $30°$ **b.** $90°$ **c.** $-135°$.

Solution To convert degrees to radians, multiply by $\dfrac{\pi \text{ radians}}{180°}$. Observe how the degree units cancel.

    a. $30° = 30° \cdot \dfrac{\pi \text{ radians}}{180°} = \dfrac{30\pi}{180} \text{ radians} = \dfrac{\pi}{6} \text{ radians}$

    b. $90° = 90° \cdot \dfrac{\pi \text{ radians}}{180°} = \dfrac{90\pi}{180} \text{ radians} = \dfrac{\pi}{2} \text{ radians}$

    c. $-135° = -135° \cdot \dfrac{\pi \text{ radians}}{180°} = -\dfrac{135\pi}{180} \text{ radians} = -\dfrac{3\pi}{4} \text{ radians}$

> Divide the numerator and denominator by 45.

Check Point 5 Convert each angle in degrees to radians:

    a. $60°$ **b.** $270°$ **c.** $-300°$.

EXAMPLE 6 Converting from Radians to Degrees

Convert each angle in radians to degrees:

    a. $\dfrac{\pi}{3}$ radians   **b.** $-\dfrac{5\pi}{3}$ radians   **c.** 1 radian.

Solution To convert radians to degrees, multiply by $\dfrac{180°}{\pi \text{ radians}}$. Observe how the radian units cancel.

Study Tip

In Example 6(c), we see that 1 radian is approximately 57°. Keep in mind that a radian is much larger than a degree.

a. $\dfrac{\pi}{3}$ radians $= \dfrac{\pi \text{ radians}}{3} \cdot \dfrac{180°}{\pi \text{ radians}} = \dfrac{180°}{3} = 60°$

b. $-\dfrac{5\pi}{3}$ radians $= -\dfrac{5\pi \text{ radians}}{3} \cdot \dfrac{180°}{\pi \text{ radians}} = -\dfrac{5 \cdot 180°}{3} = -300°$

c. 1 radian $= 1 \text{ radian} \cdot \dfrac{180°}{\pi \text{ radians}} = \dfrac{180°}{\pi} \approx 57.3°$

Check Point 6

Convert each angle in radians to degrees:

a. $\dfrac{\pi}{4}$ radians   **b.** $-\dfrac{4\pi}{3}$ radians   **c.** 6 radians.

Figure 5.12 illustrates the degree and radian measures of angles that you will commonly see in trigonometry. Each angle is in standard position, so that the initial side lies along the positive *x*-axis. We will be using both degree and radian measure for these angles.

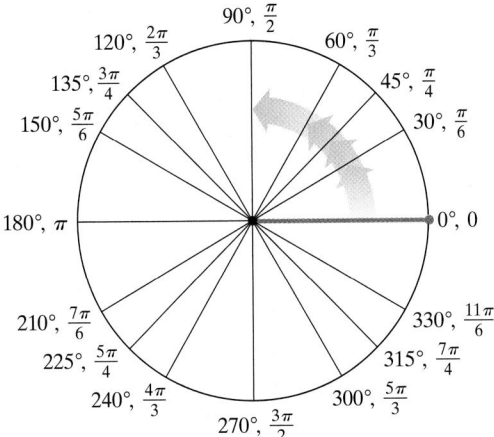

Figure 5.12 Degree and radian measures of selected angles

8   Find the length of a circular arc.

The Length of a Circular Arc

We can use the radian measure formula, $\theta = \dfrac{s}{r}$, to find the length of the arc of a circle. How do we do this? Remember that *s* represents the length of the arc intercepted by the central angle θ. Thus, by solving the formula for *s*, we have an equation for arc length.

The Length of a Circular Arc

Let *r* be the radius of a circle and θ the nonnegative radian measure of a central angle of the circle. The length of the arc intercepted by the central angle is

$$s = r\theta.$$

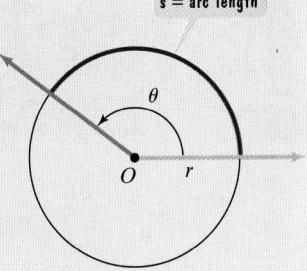

EXAMPLE 7 Finding the Length of a Circular Arc

A circle has a radius of 10 inches. Find the length of the arc intercepted by a central angle of 120°.

Study Tip

The unit used to describe the length of a circular arc is the same unit that is given in the circle's radius.

Solution The formula $s = r\theta$ can be used only when θ is expressed in radians.

Thus, we begin by converting 120° to radians. Multiply by $\dfrac{\pi \text{ radians}}{180°}$.

$$120° = 120° \cdot \frac{\pi \text{ radians}}{180°} = \frac{120\pi}{180} \text{ radians} = \frac{2\pi}{3} \text{ radians}$$

Now we can use the formula $s = r\theta$ to find the length of the arc. The circle's radius is 10 inches: $r = 10$ inches. The measure of the central angle, in radians, is $\dfrac{2\pi}{3}$: $\theta = \dfrac{2\pi}{3}$. The length of the arc intercepted by this central angle is

$$s = r\theta = (10 \text{ inches})\left(\frac{2\pi}{3}\right) = \frac{20\pi}{3} \text{ inches} \approx 20.94 \text{ inches.}$$

Check Point 7 A circle has a radius of 6 inches. Find the length of the arc intercepted by a central angle of 45°. Express arc length in terms of π. Then round your answer to two decimal places.

9 Use linear and angular speed to describe motion on a circular path.

Linear and Angular Speed

A carousel contains four circular rows of animals. As the carousel revolves, the animals in the outer row travel a greater distance per unit of time than those in the inner rows. These animals have a greater *linear speed* than those in the inner rows. By contrast, all animals, regardless of the row, complete the same number of revolutions per unit of time. All animals in the four circular rows travel at the same *angular speed*.

Using v for linear speed and ω (omega) for angular speed, we define these two kinds of speeds along a circular path as follows:

Definitions of Linear and Angular Speed

If a point is in motion on a circle of radius r through an angle of θ radians in time t, then its **linear speed** is

$$v = \frac{s}{t}$$

where s is the arc length given by $s = r\theta$, and its **angular speed** is

$$\omega = \frac{\theta}{t}.$$

The hard drive in a computer rotates at 3600 revolutions per minute. This angular speed, expressed in revolutions per minute, can also be expressed in revolutions per second, radians per minute, and radians per second. Using 2π radians $= 1$ revolution, we express the angular speed of a hard drive in radians per minute as follows:

3600 revolutions per minute

$$= \frac{3600 \text{ revolutions}}{1 \text{ minute}} \cdot \frac{2\pi \text{ radians}}{1 \text{ revolution}} = \frac{7200\pi \text{ radians}}{1 \text{ minute}}$$

$= 7200\pi$ radians per minute.

We can establish a relationship between the two kinds of speed by dividing both sides of the arc length formula, $s = r\theta$, by t:

$$\frac{s}{t} = \frac{r\theta}{t} = r\frac{\theta}{t}.$$

This expression defines linear speed.

This expression defines angular speed.

Thus, linear speed is the product of the radius and the angular speed.

Linear Speed in Terms of Angular Speed

The linear speed, v, of a point a distance r from the center of rotation is given by

$$v = r\omega$$

where ω is the angular speed in radians per unit of time.

EXAMPLE 8   **Finding Linear Speed**

A wind machine used to generate electricity has blades that are 10 feet in length (see Figure 5.13). The propeller is rotating at four revolutions per second. Find the linear speed, in feet per second, of the tips of the blades.

10 feet

Figure 5.13

Solution   We are given ω, the angular speed.

$$\omega = 4 \text{ revolutions per second}$$

We use the formula $v = r\omega$ to find v, the linear speed. Before applying the formula, we must express ω in radians per second.

$$\omega = \frac{4 \text{ revolutions}}{1 \text{ second}} \cdot \frac{2\pi \text{ radians}}{1 \text{ revolution}} = \frac{8\pi \text{ radians}}{1 \text{ second}} \quad \text{or} \quad \frac{8\pi}{1 \text{ second}}$$

The angular speed of the propeller is 8π radians per second. The linear speed is

$$v = r\omega = 10 \text{ feet} \cdot \frac{8\pi}{1 \text{ second}} = \frac{80\pi \text{ feet}}{\text{second}}.$$

The linear speed of the tips of the blades is 80π feet per second, which is approximately 251 feet per second.

Check Point 8   A 45-rpm record has an angular speed of 45 revolutions per minute. Find the linear speed, in inches per minute, at the point where the needle is 1.5 inches from the record's center.

Figure 5.13, repeated

EXERCISE SET 5.1

Practice Exercises

In Exercises 1–6, each angle is in standard position. Determine the quadrant in which the angle lies.

1. 145°   **2.** 285°

3. −100°   **4.** −110°

5. 362°   **6.** 364°

In Exercises 7–10, classify the angle as acute, right, straight, or obtuse.

7.

8.

9.

10.

In Exercises 11–18, draw each angle in standard position.

11. 135°   **12.** 120°

13. −150°   **14.** −240°

15. 420°   **16.** 450°

17. −90°   **18.** −270°

In Exercises 19–24, find a positive angle less than 360° that is coterminal with the given angle.

19. 395°   **20.** 415°

21. −150°   **22.** −160°

23. −45°   **24.** −40°

In Exercises 25–30, if possible, find the complement and the supplement of the given angle.

25. 52°   **26.** 85°

27. 37.4°   **28.** 47.6°

29. 111°   **30.** 95°

In Exercises 31–36, find the radian measure of the central angle of a circle of radius r that intercepts an arc of length s.

Radius, r	Arc length, s
31. 10 inches	40 inches
32. 5 feet	30 feet
33. 6 yards	8 yards
34. 8 yards	18 yards
35. 1 meter	400 centimeters
36. 1 meter	600 centimeters

In Exercises 37–44, convert each angle in degrees to radians. Express your answer as a multiple of π.

37. $45°$ **38.** $18°$

39. $135°$ **40.** $150°$

41. $300°$ **42.** $330°$

43. $-225°$ **44.** $-270°$

In Exercises 45–52, convert each angle in radians to degrees.

45. $\dfrac{\pi}{2}$ **46.** $\dfrac{\pi}{9}$

47. $\dfrac{2\pi}{3}$ **48.** $\dfrac{3\pi}{4}$

49. $\dfrac{7\pi}{6}$ **50.** $\dfrac{11\pi}{6}$

51. -3π **52.** -4π

In Exercises 53–58, convert each angle in degrees to radians. Round to two decimal places.

53. $18°$ **54.** $76°$

55. $-40°$ **56.** $-50°$

57. $200°$ **58.** $250°$

In Exercises 59–64, convert each angle in radians to degrees. Round to two decimal places.

59. 2 radians   **60.** 3 radians

61. $\dfrac{\pi}{13}$ radians   **62.** $\dfrac{\pi}{17}$ radians

63. -4.8 radians   **64.** -5.2 radians

In Exercises 65–68, find the length of the arc on a circle of radius r intercepted by a central angle θ. Express arc length in terms of π. Then round your answer to two decimal places.

Radius, r	Central angle, θ
65. 12 inches	$\theta = 45°$
66. 16 inches	$\theta = 60°$
67. 8 feet	$\theta = 225°$
68. 9 yards	$\theta = 315°$

In Exercises 69–70, express each angular speed in radians per second.

69. 6 revolutions per second

70. 20 revolutions per second

 Application Exercises

71. The minute hand of a clock moves from 12 to 2 o'clock, or $\frac{1}{6}$ of a complete revolution. Through how many degrees does it move? Through how many radians does it move?

72. The minute hand of a clock moves from 12 to 4 o'clock, or $\frac{1}{3}$ of a complete revolution. Through how many degrees does it move? Through how many radians does it move?

73. The minute hand of a clock is 8 inches long and moves from 12 to 2 o'clock. How far does the tip of the minute hand move? Express your answer in terms of π and then round to two decimal places.

74. The minute hand of a clock is 6 inches long and moves from 12 to 4 o'clock. How far does the tip of the minute hand move? Express your answer in terms of π and then round to two decimal places.

75. The figure shows a highway sign that warns of a railway crossing. The lines that form the cross pass through the circle's center and intersect at right angles. If the radius of the circle is 24 inches, find the length of each of the four arcs formed by the cross. Express your answer in terms of π and then round to two decimal places.

76. The radius of a wheel is 80 centimeters. If the wheel rotates through an angle of $60°$, how many centimeters does it move? Express your answer in terms of π and then round to two decimal places.

How do we measure the distance between two points, A and B, on Earth? We measure along a circle with a center, C, at the center of Earth. The radius of the circle is equal to the distance from C to the surface. Use the fact that Earth is a sphere of radius equal to approximately 4000 miles to solve Exercises 77–80.

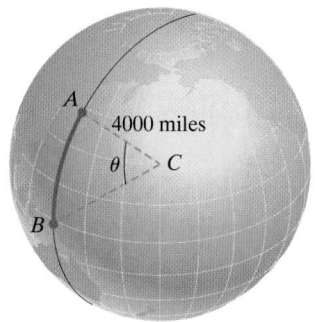

77. If two points, *A* and *B*, are 8000 miles apart, express angle θ in radians and in degrees.

78. If two points, *A* and *B*, are 10,000 miles apart, express angle θ in radians and in degrees.

79. If $\theta = 30°$, find the distance between *A* and *B* to the nearest mile.

80. If $\theta = 10°$, find the distance between *A* and *B* to the nearest mile.

81. The angular speed of a point on Earth is $\frac{\pi}{12}$ radians per hour. The Equator lies on a circle of radius approximately 4000 miles. Find the linear velocity, in miles per hour, of a point on the Equator.

82. A ferris wheel has a radius of 25 feet. The wheel is rotating at three revolutions per minute. Find the linear speed, in feet per minute, of this ferris wheel.

83. A water wheel has a radius of 12 feet. The wheel is rotating at 20 revolutions per minute. Find the linear speed, in feet per minute, of the water.

84. On a carousel, the outer row of animals is 20 feet from the center. The inner row of animals is 10 feet from the center. The carousel is rotating at 2.5 revolutions per minute. What is the difference, in feet per minute, in the linear speeds of the animals in the outer and inner rows? Round to the nearest foot per minute.

Writing in Mathematics

85. What is an angle?

86. What determines the size of an angle?

87. Describe an angle in standard position.

88. Explain the difference between positive and negative angles. What are coterminal angles?

89. Explain what is meant by one radian.

90. Explain how to find the radian measure of a central angle.

91. Describe how to convert an angle in degrees to radians.

92. Explain how to convert an angle in radians to degrees.

93. Explain how to find the length of a circular arc.

94. If a carousel is rotating at 2.5 revolutions per minute, explain how to find the linear speed of a child seated on one of the animals.

95. The angular velocity of a point on Earth is $\frac{\pi}{12}$ radians per hour. Describe what happens every 24 hours.

96. Have you ever noticed that we use the vocabulary of angles in everyday speech? Here is an example:

> My opinion about art museums took a 180° turn after visiting the San Francisco Museum of Modern Art.

Explain what this means. Then give another example of the vocabulary of angles in everyday use.

Technology Exercises

In Exercises 97–100, use the keys on your calculator or graphing utility for converting an angle in degrees, minutes, and seconds ($D°M'S''$) into decimal form, and vice versa.

In Exercises 97–98, convert each angle to a decimal in degrees. Round your answer to two decimal places.

97. 30°15′10″   **98.** 65°45′20″

In Exercises 99–100, convert each angle to $D°M'S''$ form. Round your answer to the nearest second.

99. 30.42°   **100.** 50.42°

Critical Thinking Exercises

101. If $\theta = \frac{3}{2}$, is this angle larger or smaller than a right angle?

102. A railroad curve is laid out on a circle. What radius should be used if the track is to change direction by 20° in a distance of 100 miles? Round your answer to the nearest mile.

103. Assuming Earth to be a sphere of radius 4000 miles, how many miles north of the Equator is Miami, Florida, if it is 26° north from the Equator? Round your answer to the nearest mile.

SECTION 5.2 *Right Triangle Trigonometry*

Objectives

1. Use right triangles to evaluate trigonometric functions.
2. Find function values for $30°\left(\dfrac{\pi}{6}\right)$, $45°\left(\dfrac{\pi}{4}\right)$, and $60°\left(\dfrac{\pi}{3}\right)$.
3. Recognize and use fundamental identities.
4. Use equal cofunctions of complements.
5. Evaluate trigonometric functions with a calculator.
6. Use right triangle trigonometry to solve applied problems.

In the last century, Ang Rita Sherpa climbed Mount Everest ten times, all without the use of bottled oxygen.

Mountain climbers have forever been fascinated by reaching the top of Mount Everest, sometimes with tragic results. The mountain, on Asia's Tibet-Nepal border, is Earth's highest, peaking at an incredible 29,035 feet. The heights of mountains can be found using **trigonometry.** The word *trigonometry* means *measurement of triangles.* Trigonometry is used in navigation, building, and engineering. For centuries, Muslims used trigonometry and the stars to navigate across the Arabian desert to Mecca, the birthplace of the prophet Muhammad, the founder of Islam. The ancient Greeks used trigonometry to record the locations of thousands of stars and worked out the motion of the Moon relative to Earth. Today, trigonometry is used to study the structure of DNA, the master molecule that determines how we grow from a single cell to a complex, fully developed adult.

1 Use right triangles to evaluate trigonometric functions.

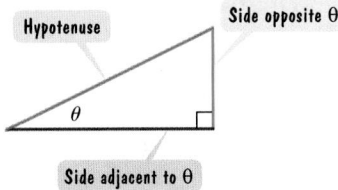

Figure 5.14 Naming a right triangle's sides from the point of view of an acute angle θ

The Six Trigonometric Functions

We begin the study of trigonometry by defining six functions, the six *trigonometric functions.* The inputs for these functions are measures of acute angles in right triangles. The outputs are the ratios of the lengths of the sides of right triangles.

Figure 5.14 shows a right triangle with one of its acute angles labeled θ. The side opposite the right angle is known as the **hypotenuse.** The other sides of the triangle are described by their position relative to the acute angle θ. One side is opposite θ and one is adjacent to θ.

The trigonometric functions have names that are words, rather than single letters such as f, g, and h. For example, the **sine of** θ is the length of the side opposite θ divided by the length of the hypotenuse:

$$\sin\theta = \frac{\text{length of side opposite }\theta}{\text{length of hypotenuse}}.$$

Input is the measure of an acute angle.

Output is the ratio of the lengths of the sides.

The ratio of lengths depends on angle θ and thus is a function of θ. The expression $\sin\theta$ really means $\sin(\theta)$, where sine is the name of the function and θ, the measure of an acute angle, is an input.

Here are the names of the six trigonometric functions, along with their abbreviations:

Name	Abbreviation	Name	Abbreviation
sine	sin	cosecant	csc
cosine	cos	secant	sec
tangent	tan	cotangent	cot

Now, let θ be an acute angle in a right triangle, shown in Figure 5.15. The length of the side opposite θ is a, the length of the side adjacent to θ is b, and the length of the hypotenuse is c.

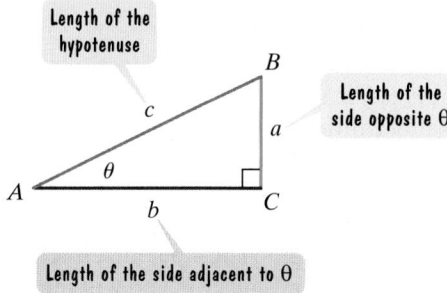

Figure 5.15

Right Triangle Definitions of Trigonometric Functions

See Figure 5.15. The six **trigonometric functions of the acute angle θ** are defined as follows:

$$\sin\theta = \frac{\text{length of side opposite angle } \theta}{\text{length of hypotenuse}} = \frac{a}{c}$$

$$\csc\theta = \frac{\text{length of hypotenuse}}{\text{length of side opposite angle } \theta} = \frac{c}{a}$$

$$\cos\theta = \frac{\text{length of side adjacent to angle } \theta}{\text{length of hypotenuse}} = \frac{b}{c}$$

$$\sec\theta = \frac{\text{length of hypotenuse}}{\text{length of side adjacent to angle } \theta} = \frac{c}{b}$$

$$\tan\theta = \frac{\text{length of side opposite angle } \theta}{\text{length of side adjacent to angle } \theta} = \frac{a}{b}$$

$$\cot\theta = \frac{\text{length of side adjacent to angle } \theta}{\text{length of side opposite angle } \theta} = \frac{b}{a}$$

Each of the trigonometric functions of the acute angle θ is positive. Observe that the functions in the second column in the box are the reciprocals of the corresponding functions in the first column.

Figure 5.16 on the next page shows four right triangles of varying sizes. In each of the triangles, θ is the same acute angle, measuring approximately 56.3°. All four of these similar triangles have the same shape and the lengths of corresponding sides are in the same ratio. In each triangle, the tangent function has the same value: $\tan\theta = \frac{3}{2}$.

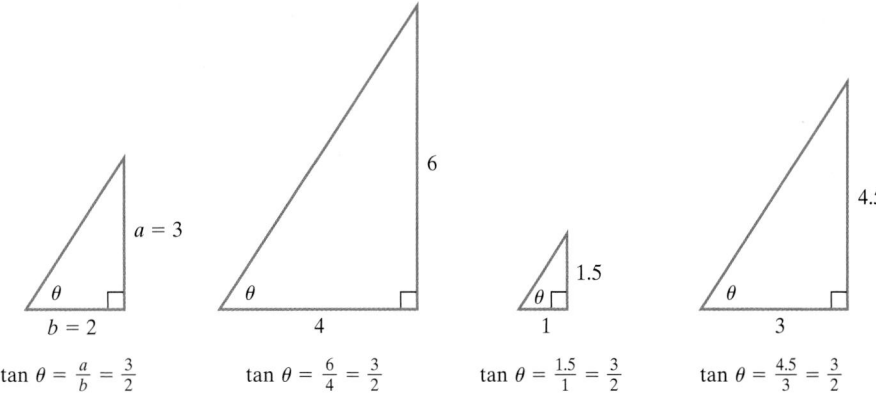

$$\tan \theta = \frac{a}{b} = \frac{3}{2} \qquad \tan \theta = \frac{6}{4} = \frac{3}{2} \qquad \tan \theta = \frac{1.5}{1} = \frac{3}{2} \qquad \tan \theta = \frac{4.5}{3} = \frac{3}{2}$$

Figure 5.16 A particular acute angle always gives the same ratio of opposite to adjacent sides.

In general, **the trigonometric function values of θ depend only on the size of angle θ, and not on the size of the triangle.**

Figure 5.17

EXAMPLE 1 Evaluating Trigonometric Functions

Find the value of each of the six trigonometric functions of θ in Figure 5.17.

Solution We need to find the values of the six trigonometric functions of θ. However, we must know the lengths of all three sides of the triangle ($a, b,$ and c) to evaluate all six functions. The values of a and b are given. We can use the Pythagorean Theorem, $c^2 = a^2 + b^2$, to find c.

$$a = 5 \qquad b = 12$$

$$c^2 = a^2 + b^2 = 5^2 + 12^2 = 25 + 144 = 169$$

$$c = \sqrt{169} = 13$$

Now that we know the lengths of the three sides of the triangle, we apply the definitions of the six trigonometric functions of θ. Referring to these lengths as opposite, adjacent, and hypotenuse, we have

$$\sin \theta = \frac{\text{opposite}}{\text{hypotenuse}} = \frac{5}{13} \qquad \csc \theta = \frac{\text{hypotenuse}}{\text{opposite}} = \frac{13}{5}$$

$$\cos \theta = \frac{\text{adjacent}}{\text{hypotenuse}} = \frac{12}{13} \qquad \sec \theta = \frac{\text{hypotenuse}}{\text{adjacent}} = \frac{13}{12}$$

$$\tan \theta = \frac{\text{opposite}}{\text{adjacent}} = \frac{5}{12} \qquad \cot \theta = \frac{\text{adjacent}}{\text{opposite}} = \frac{12}{5}.$$

Study Tip

The functions in the second column are reciprocals of those in the first column. You can obtain their values by exchanging the numerator and denominator of the corresponding ratios in the first column.

Check Point 1 Find the value of each of the six trigonometric functions of θ in the figure.

2 Find function values for $30° \left(\dfrac{\pi}{6} \right)$, $45° \left(\dfrac{\pi}{4} \right)$, and $60° \left(\dfrac{\pi}{3} \right)$.

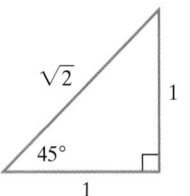

Figure 5.18 An isosceles right triangle

Function Values for Some Special Angles

A 45°, or $\dfrac{\pi}{4}$ radian, angle occurs frequently in trigonometry. How do we find the values of the trigonometric functions of 45°? We construct a right triangle with a 45° angle, shown in Figure 5.18. The triangle actually has two 45° angles. Thus, the triangle is isosceles—that is, it has two sides of the same length. Assume that each leg of the triangle has a length equal to 1. We can find the length of the hypotenuse using the Pythagorean Theorem.

$$(\text{length of hypotenuse})^2 = 1^2 + 1^2 = 2$$

$$\text{length of hypotenuse} = \sqrt{2}$$

With Figure 5.18, we can determine the trigonometric function values for 45°.

EXAMPLE 2   Evaluating Trigonometric Functions of 45°

Use Figure 5.18 to find $\sin 45°$, $\cos 45°$, and $\tan 45°$.

Solution   We apply the definitions of these three trigonometric functions.

$$\sin 45° = \frac{\text{length of side opposite } 45°}{\text{length of hypotenuse}} = \frac{1}{\sqrt{2}}$$

$$\cos 45° = \frac{\text{length of side adjacent to } 45°}{\text{length of hypotenuse}} = \frac{1}{\sqrt{2}}$$

$$\tan 45° = \frac{\text{length of side opposite } 45°}{\text{length of side adjacent to } 45°} = \frac{1}{1} = 1$$

Check Point 2   Use Figure 5.18 to find $\csc 45°$, $\sec 45°$, and $\cot 45°$.

When you worked Check Point 2, did you actually use Figure 5.18 or did you use reciprocals to find the values?

$$\csc 45° = \sqrt{2} \qquad \sec 45° = \sqrt{2} \qquad \cot 45° = 1$$

Take the reciprocal of $\sin 45° = \dfrac{1}{\sqrt{2}}$.

Take the reciprocal of $\cos 45° = \dfrac{1}{\sqrt{2}}$.

Take the reciprocal of $\tan 45° = \dfrac{1}{1}$.

We found that $\sin 45° = \dfrac{1}{\sqrt{2}}$ and $\cos 45° = \dfrac{1}{\sqrt{2}}$. This value is often expressed by rationalizing the denominator:

$$\frac{1}{\sqrt{2}} = \frac{1}{\sqrt{2}} \cdot \frac{\sqrt{2}}{\sqrt{2}} = \frac{\sqrt{2}}{2}.$$

We are multiplying by 1 and not changing the value of $\dfrac{1}{\sqrt{2}}$.

Thus, $\sin 45° = \dfrac{\sqrt{2}}{2}$ and $\cos 45° = \dfrac{\sqrt{2}}{2}$.

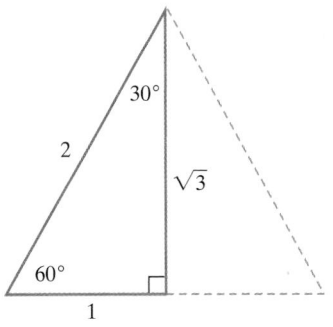

Figure 5.19 30°–60°–90° triangle

Two other angles that occur frequently in trigonometry are 30°, or $\dfrac{\pi}{6}$ radian, and 60°, or $\dfrac{\pi}{3}$ radian, angles. We can find the values of the trigonometric functions of 30° and 60° by using a right triangle. To form this right triangle, draw an equilateral triangle—that is a triangle with all sides the same length. Assume that each side has a length equal to 2. Now take half of the equilateral triangle. We obtain the right triangle in Figure 5.19. This right triangle has a hypotenuse of length 2 and a leg of length 1. The other leg has length a, which can be found using the Pythagorean Theorem.

$$a^2 + 1^2 = 2^2$$
$$a^2 + 1 = 4$$
$$a^2 = 3$$
$$a = \sqrt{3}$$

With the right triangle in Figure 5.19, we can determine the trigonometric functions for 30° and 60°.

EXAMPLE 3   Evaluating Trigonometric Functions of 30° and 60°

Use Figure 5.19 to find $\sin 60°$, $\cos 60°$, $\sin 30°$, and $\cos 30°$.

Solution   We begin with 60°. Use the angle on the lower left in Figure 5.19.

$$\sin 60° = \frac{\text{length of side opposite } 60°}{\text{length of hypotenuse}} = \frac{\sqrt{3}}{2}$$

$$\cos 60° = \frac{\text{length of side adjacent to } 60°}{\text{length of hypotenuse}} = \frac{1}{2}$$

To find $\sin 30°$ and $\cos 30°$, use the angle on the upper right in Figure 5.19.

$$\sin 30° = \frac{\text{length of side opposite } 30°}{\text{length of hypotenuse}} = \frac{1}{2}$$

$$\cos 30° = \frac{\text{length of side adjacent to } 30°}{\text{length of hypotenuse}} = \frac{\sqrt{3}}{2}$$

Check Point 3   Use Figure 5.19 to find $\tan 60°$ and $\tan 30°$. If a radical appears in a denominator, rationalize the denominator.

Because we will often use the function values of 30°, 45°, and 60°, you should learn to construct the right triangles shown in Figures 5.18 and 5.19. With sufficient practice, you will memorize the values in the box on the next page.

Sines, Cosines, and Tangents of Special Angles

$$\sin 30° = \sin \frac{\pi}{6} = \frac{1}{2} \qquad \cos 30° = \cos \frac{\pi}{6} = \frac{\sqrt{3}}{2} \qquad \tan 30° = \tan \frac{\pi}{6} = \frac{\sqrt{3}}{3}$$

$$\sin 45° = \sin \frac{\pi}{4} = \frac{\sqrt{2}}{2} \qquad \cos 45° = \cos \frac{\pi}{4} = \frac{\sqrt{2}}{2} \qquad \tan 45° = \tan \frac{\pi}{4} = 1$$

$$\sin 60° = \sin \frac{\pi}{3} = \frac{\sqrt{3}}{2} \qquad \cos 60° = \cos \frac{\pi}{3} = \frac{1}{2} \qquad \tan 60° = \tan \frac{\pi}{3} = \sqrt{3}$$

3 Recognize and use fundamental identities.

Fundamental Identities

Many relationships exist among the six trigonometric functions. These relationships are described using **trigonometric identities.** For example, $\csc \theta$ is defined as the reciprocal of $\sin \theta$. This relationship can be expressed by the identity

$$\csc \theta = \frac{1}{\sin \theta}.$$

This identity is one of six **reciprocal identities.**

Reciprocal Identities

$$\sin \theta = \frac{1}{\csc \theta} \qquad \cos \theta = \frac{1}{\sec \theta} \qquad \tan \theta = \frac{1}{\cot \theta}$$

$$\csc \theta = \frac{1}{\sin \theta} \qquad \sec \theta = \frac{1}{\cos \theta} \qquad \cot \theta = \frac{1}{\tan \theta}$$

Two other relationships that follow from the definitions of the trigonometric functions are called the **quotient identities.**

Quotient Identities

$$\tan \theta = \frac{\sin \theta}{\cos \theta} \qquad \cot \theta = \frac{\cos \theta}{\sin \theta}$$

If $\sin \theta$ and $\cos \theta$ are known, a quotient identity and three reciprocal identities make it possible to find the value of each of the four remaining trigonometric functions.

EXAMPLE 4   Using Quotient and Reciprocal Identities

Given $\sin \theta = \dfrac{1}{2}$ and $\cos \theta = \dfrac{\sqrt{3}}{2}$, find the value of each of the four remaining trigonometric functions.

Solution   We can find $\tan \theta$ by using the quotient identity that describes $\tan \theta$ as the quotient of $\sin \theta$ and $\cos \theta$.

$$\tan \theta = \frac{\sin \theta}{\cos \theta} = \frac{\dfrac{1}{2}}{\dfrac{\sqrt{3}}{2}} = \frac{1}{2} \cdot \frac{2}{\sqrt{3}} = \frac{1}{\sqrt{3}} = \frac{1}{\sqrt{3}} \cdot \frac{\sqrt{3}}{\sqrt{3}} = \frac{\sqrt{3}}{3}$$

Rationalize the denominator.

We use the reciprocal identities to find the value of each of the remaining three functions.

$$\csc \theta = \frac{1}{\sin \theta} = \frac{1}{\dfrac{1}{2}} = 2$$

$$\sec \theta = \frac{1}{\cos \theta} = \frac{1}{\dfrac{\sqrt{3}}{2}} = \frac{2}{\sqrt{3}} = \frac{2}{\sqrt{3}} \cdot \frac{\sqrt{3}}{\sqrt{3}} = \frac{2\sqrt{3}}{3}$$

Rationalize the denominator.

$$\cot \theta = \frac{1}{\tan \theta} = \frac{1}{\dfrac{1}{\sqrt{3}}} = \sqrt{3}$$

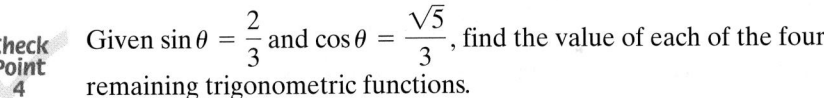 We found $\tan \theta = \dfrac{1}{\sqrt{3}}$. We could use $\tan \theta = \dfrac{\sqrt{3}}{3}$, but then we would have to rationalize the denominator.

Check Point 4   Given $\sin \theta = \dfrac{2}{3}$ and $\cos \theta = \dfrac{\sqrt{5}}{3}$, find the value of each of the four remaining trigonometric functions.

Other relationships among trigonometric functions follow from the Pythagorean Theorem. Using Figure 5.20, the Pythagorean Theorem states that

$$a^2 + b^2 = c^2.$$

To obtain ratios that correspond to trigonometric functions, divide both sides of this equation by c^2.

$$\frac{a^2}{c^2} + \frac{b^2}{c^2} = 1 \quad \text{or} \quad \left(\frac{a}{c}\right)^2 + \left(\frac{b}{c}\right)^2 = 1$$

In Figure 5.20, $\sin \theta = \dfrac{a}{c}$, so this is $(\sin \theta)^2$.   In Figure 5.20, $\cos \theta = \dfrac{b}{c}$, so this is $(\cos \theta)^2$.

Based on the observations in the voice balloons, we see that

$$(\sin \theta)^2 + (\cos \theta)^2 = 1.$$

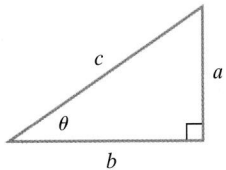

Figure 5.20

We will use the notation $\sin^2\theta$ for $(\sin\theta)^2$ and $\cos^2\theta$ for $(\cos\theta)^2$. With this notation, we can write the identity as

$$\sin^2\theta + \cos^2\theta = 1.$$

Two additional identities can be obtained from $a^2 + b^2 = c^2$ by dividing both sides by b^2 and a^2, respectively. The three identities are called the **Pythagorean identities.**

Pythagorean Identities

$$\sin^2\theta + \cos^2\theta = 1 \qquad 1 + \tan^2\theta = \sec^2\theta \qquad 1 + \cot^2\theta = \csc^2\theta$$

EXAMPLE 5 Using a Pythagorean Identity

Given that $\sin\theta = \frac{3}{5}$ and θ is an acute angle, find the value of $\cos\theta$ using a trigonometric identity.

Solution We can find the value of $\cos\theta$ by using the Pythagorean identity

$$\sin^2\theta + \cos^2\theta = 1.$$

$$\left(\frac{3}{5}\right)^2 + \cos^2\theta = 1 \qquad \text{We are given that } \sin\theta = \frac{3}{5}.$$

$$\frac{9}{25} + \cos^2\theta = 1 \qquad \text{Square } \frac{3}{5}: \left(\frac{3}{5}\right)^2 = \frac{3^2}{5^2} = \frac{9}{25}.$$

$$\cos^2\theta = 1 - \frac{9}{25} \qquad \text{Subtract } \frac{9}{25} \text{ from both sides.}$$

$$\cos^2\theta = \frac{16}{25} \qquad \text{Simplify: } 1 - \frac{9}{25} = \frac{25}{25} - \frac{9}{25} = \frac{16}{25}.$$

$$\cos\theta = \sqrt{\frac{16}{25}} = \frac{4}{5} \qquad \text{Because } \theta \text{ is an acute angle, } \cos\theta \text{ is positive.}$$

Thus, $\cos\theta = \frac{4}{5}$.

Check Point 5 Given that $\sin\theta = \frac{1}{2}$ and θ is an acute angle, find the value of $\cos\theta$ using a trigonometric identity.

4 Use equal cofunctions of complements.

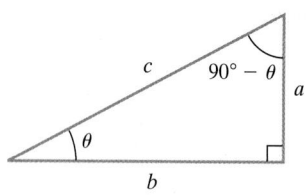

Figure 5.21

Trigonometric Functions and Complements

Another relationship among trigonometric functions is based on angles that are complements. Refer to Figure 5.21. Because the sum of the angles of any triangle is $180°$, in a right triangle the sum of the acute angles is $90°$. Thus, the acute angles are complements. If the degree measure of one acute angle is θ, then the degree measure of the other acute angle is $(90° - \theta)$. This angle is shown on the upper right in Figure 5.21.

Let's use Figure 5.21 to compare $\sin\theta$ and $\cos(90° - \theta)$.

$$\sin\theta = \frac{\text{length of side opposite } \theta}{\text{length of hypotenuse}} = \frac{a}{c}$$

$$\cos(90° - \theta) = \frac{\text{length of side adjacent to } (90° - \theta)}{\text{length of hypotenuse}} = \frac{a}{c}$$

Thus, $\sin\theta = \cos(90° - \theta)$. If two angles are complements, the sine of one equals the cosine of the other. Because of this relationship, the sine and cosine are called *cofunctions* of each other. The name *cosine* is a shortened form of the phrase *complement's sine*.

Any pair of trigonometric functions f and g for which

$$f(\theta) = g(90° - \theta) \quad \text{and} \quad g(\theta) = f(90° - \theta)$$

are called **cofunctions**. Using Figure 5.21, we can show that the tangent and cotangent are also cofunctions of each other. So are the secant and cosecant.

Cofunction Identities

The value of a trigonometric function of θ is equal to the cofunction of the complement of θ.

$$\sin\theta = \cos(90° - \theta) \qquad \cos\theta = \sin(90° - \theta)$$
$$\tan\theta = \cot(90° - \theta) \qquad \cot\theta = \tan(90° - \theta)$$
$$\sec\theta = \csc(90° - \theta) \qquad \csc\theta = \sec(90° - \theta)$$

If θ is in radians, replace $90°$ with $\dfrac{\pi}{2}$.

EXAMPLE 6

Find a cofunction with the same value as the given expression:

a. $\sin 72°$ **b.** $\csc \dfrac{\pi}{3}$.

Solution Because the value of a trigonometric function of θ is equal to the cofunction of the complement of θ, we need to find the complement of each angle. We do this by subtracting the angle's measure from $90°$ or its radian equivalent, $\dfrac{\pi}{2}$.

a. $\sin 72° = \cos(90° - 72°) = \cos 18°$

> We have a function and its cofunction.

b. $\csc \dfrac{\pi}{3} = \sec\left(\dfrac{\pi}{2} - \dfrac{\pi}{3}\right) = \sec\left(\dfrac{3\pi}{6} - \dfrac{2\pi}{6}\right) = \sec\dfrac{\pi}{6}$

> We have a cofunction and its function.

> Perform the subtraction using the least common denominator, 6.

Check Point 6 Find a cofunction with the same value as the given expression:

a. $\sin 46°$ **b.** $\cot \dfrac{\pi}{12}$.

---

5 Evaluate trigonometric functions with a calculator.

Using a Calculator to Evaluate Trigonometric Functions

The values of the trigonometric functions obtained with the special triangles are exact values. For most angles other than $30°$, $45°$, and $60°$, we approximate the value of each of the trigonometric functions using a calculator. The first step is to set the calculator to the correct *mode*, degrees or radians, depending on how the acute angle is measured.

Most calculators have keys marked $\boxed{\text{SIN}}$, $\boxed{\text{COS}}$, and $\boxed{\text{TAN}}$. For example, to find the value of sin 30°, set the calculator to the degree mode and enter 30 $\boxed{\text{SIN}}$ on most scientific calculators and $\boxed{\text{SIN}}$ 30 $\boxed{\text{ENTER}}$ on most graphing calculators. Consult the manual for your calculator.

To evaluate the cosecant, secant, and cotangent functions, use the key for the respective reciprocal function, $\boxed{\text{SIN}}$, $\boxed{\text{COS}}$, or $\boxed{\text{TAN}}$, and then use the reciprocal key. The reciprocal key is $\boxed{1/x}$ on many scientific calculators and $\boxed{x^{-1}}$ on many graphing calculators. For example, we can evaluate $\sec\dfrac{\pi}{12}$ using the following reciprocal relationship:

$$\sec\frac{\pi}{12} = \frac{1}{\cos\dfrac{\pi}{12}}.$$

Using the radian mode, enter one of the following keystroke sequences:

Many Scientific Calculators

$$\boxed{\pi}\ \boxed{\div}\ \boxed{12}\ \boxed{=}\ \boxed{\text{COS}}\ \boxed{1/x}$$

Many Graphing Calculators

$$\boxed{(}\ \boxed{\text{COS}}\ \boxed{(}\ \boxed{\pi}\ \boxed{\div}\ \boxed{12}\ \boxed{)}\ \boxed{)}\ \boxed{x^{-1}}\ \boxed{\text{ENTER}}.$$

Rounding the display to four decimal places, we obtain $\sec\dfrac{\pi}{12} = 1.0353$.

EXAMPLE 7 Evaluating Trigonometric Functions with a Calculator

Use a calculator to find the value to four decimal places:

a. cos 48.2°   **b.** cot 1.2.

Solution

Scientific Calculator Solution

Function	Mode	Keystrokes	Display, rounded to four decimal places
a. cos 48.2°	Degree	48.2 $\boxed{\text{COS}}$	0.6665
b. cot 1.2	Radian	1.2 $\boxed{\text{TAN}}$ $\boxed{1/x}$	0.3888

Graphing Calculator Solution

Function	Mode	Keystrokes	Display, rounded to four decimal places
a. cos 48.2°	Degree	$\boxed{\text{COS}}$ 48.2 $\boxed{\text{ENTER}}$	0.6665
b. cot 1.2	Radian	$\boxed{(}$ $\boxed{\text{TAN}}$ 1.2 $\boxed{)}$ $\boxed{x^{-1}}$ $\boxed{\text{ENTER}}$	0.3888

Check Point 7
Use a calculator to find the value to four decimal places:

a. $\sin 72.8°$ **b.** $\csc 1.5$.

6 Use right triangle trigonometry to solve applied problems.

Applications

Many applications of right triangle trigonometry involve the angle made with an imaginary horizontal line. As shown in Figure 5.22, an angle formed by a horizontal line and the line of sight to an object that is above the horizontal line is called the **angle of elevation.** The angle formed by a horizontal line and the line of sight to an object that is below the horizontal line is called the **angle of depression.** Transits and sextants are instruments used to measure such angles.

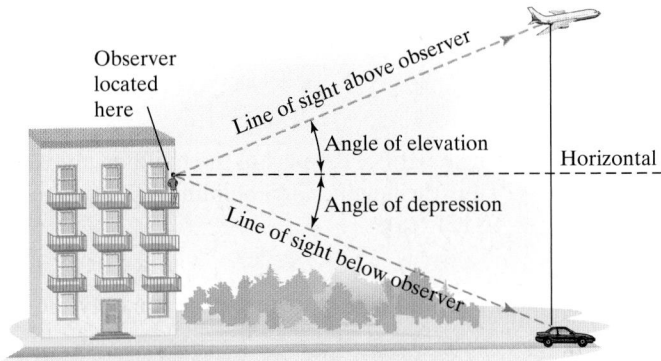

Figure 5.22

EXAMPLE 8   Problem Solving Using an Angle of Elevation

Sighting the top of a building, a surveyor measured the angle of elevation to be $22°$. The transit is 5 feet above the ground and 300 feet from the building. Find the building's height.

Solution   The situation is illustrated in Figure 5.23. Let a be the height of the portion of the building that lies above the transit. The height of the building is the transit's height, 5 feet, plus a. Thus, we need to identify a trigonometric function that will make it possible to find a. In terms of the $22°$ angle, we are looking for the side opposite the angle. The transit is 300 feet from the building, so the side adjacent to the $22°$ angle is 300 feet. Because we have a known angle, an unknown opposite side, and a known adjacent side, we select the tangent function.

Figure 5.23

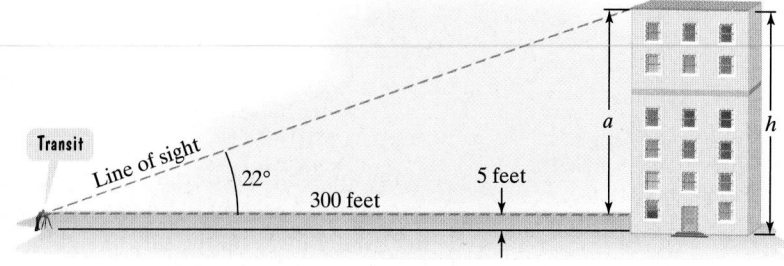

Figure 5.23, repeated

$$\tan 22° = \frac{a}{300}$$

Length of side opposite the 22° angle

Length of side adjacent to the 22° angle

$$a = 300 \tan 22°$$ Multiply both sides of the equation by 300.

$$a \approx 300(0.4040) \approx 121$$ Find $\tan 22°$ with a calculator in the degree mode.

The height of the part of the building above the transit is approximately 121 feet. Thus, the height of the building is determined by adding the transit's height, 5 feet, to 121 feet.

$$h \approx 5 + 121 = 126$$

The building's height is approximately 126 feet.

Check Point 8 The irregular blue shape in Figure 5.24 represents a lake. The distance across the lake, a, is unknown. To find this distance, a surveyor took the measurements shown in the figure. What is the distance across the lake?

Figure 5.24

If two sides of a right triangle are known, an appropriate trigonometric function can be used to find an acute angle θ in the triangle. You will also need to use the *inverse key* on a calculator. This key uses a function value to display the acute angle θ. For example, suppose that $\sin \theta = 0.866$. We can find θ in the degree mode by using the secondary *inverse sine* key, usually labeled $\boxed{\text{SIN}^{-1}}$.

Study Tip

$\boxed{\text{SIN}^{-1}}$ is not a button you will actually press. It is the secondary function for the button labeled $\boxed{\text{SIN}}$.

Many Scientific Calculators:

.866 $\boxed{\text{2nd}}$ $\boxed{\text{SIN}^{-1}}$

Many Graphing Calculators:

$\boxed{\text{2nd}}$ $\boxed{\text{SIN}^{-1}}$ .866 $\boxed{\text{ENTER}}$

The display should show approximately 59.99, which can be rounded to 60. Thus, if $\sin \theta = 0.866$, then $\theta \approx 60°$.

EXAMPLE 9   Determining the Angle of Elevation

A building that is 21 meters tall casts a shadow 25 meters long. Find the angle of elevation of the sun to the nearest degree.

Solution   The situation is illustrated in Figure 5.25. We are asked to find θ. We begin with the tangent function.

$$\tan\theta = \frac{\text{side opposite } \theta}{\text{side adjacent to } \theta} = \frac{21}{25}$$

We use a calculator in the degree mode to find θ.

Many Scientific Calculators:

$\boxed{(}\;\boxed{21}\;\boxed{\div}\;\boxed{25}\;\boxed{)}\quad\boxed{\text{2nd}}\;\boxed{\text{TAN}^{\text{-}1}}$

Many Graphing Calculators:

$\boxed{\text{2nd}}\;\boxed{\text{TAN}^{\text{-}1}}\;\boxed{(}\;\boxed{21}\;\boxed{\div}\;\boxed{25}\;\boxed{)}\;\boxed{\text{ENTER}}$

The display should show approximately 40. Thus, the angle of elevation of the sun is approximately 40°.

21 m

Angle of elevation

θ

25 m

Figure 5.25

Check Point 9   A flagpole that is 14 meters tall casts a shadow 10 meters long. Find the angle of elevation of the sun to the nearest degree.

The Mountain Man

In the 1930s, a *National Geographic* team headed by Brad Washburn used trigonometry to create a map of the 5000-square-mile region of the Yukon, near the Canadian border. The team started with aerial photography. By drawing a network of angles on the photographs, the approximate locations of the major mountains and their rough heights were determined. The expedition then spent three months on foot to find the exact heights. Team members established two base points a known distance apart, one directly under the mountain's peak. By measuring the angle of elevation from one of the base points to the peak, the tangent function was used to determine the peak's height. The Yukon expedition was a major advance in the way maps are made.

EXERCISE SET 5.2

Practice Exercises

In Exercises 1–8, use the Pythagorean Theorem to find the length of the missing side of each right triangle. Then find the value of each of the six trigonometric functions of θ.

1.

2.

3.

4.

5.

6.

7.

8.
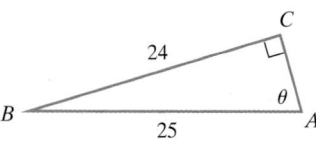

In Exercises 9–16, use the given triangles to evaluate each expression. If necessary, express the value without a square root in the denominator by rationalizing the denominator.

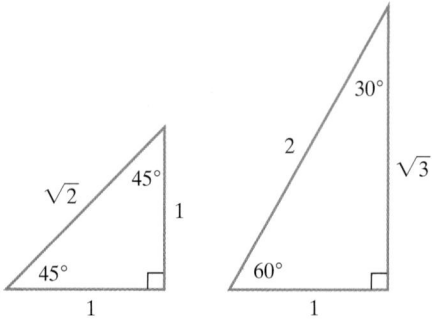

9. $\cos 30°$

10. $\tan 30°$

11. $\sec 45°$

12. $\csc 45°$

13. $\tan \dfrac{\pi}{3}$

14. $\cot \dfrac{\pi}{3}$

15. $\sin \dfrac{\pi}{4} - \cos \dfrac{\pi}{4}$

16. $\tan \dfrac{\pi}{4} + \csc \dfrac{\pi}{6}$

In Exercises 17–20, θ is an acute angle and $\sin\theta$ and $\cos\theta$ are given. Use identities to find $\tan\theta$, $\csc\theta$, $\sec\theta$, and $\cot\theta$. Where necessary, rationalize denominators.

17. $\sin\theta = \dfrac{8}{17}$, $\quad \cos\theta = \dfrac{15}{17}$

18. $\sin\theta = \dfrac{3}{5}$, $\quad \cos\theta = \dfrac{4}{5}$

19. $\sin\theta = \dfrac{1}{3}$, $\quad \cos\theta = \dfrac{2\sqrt{2}}{3}$

20. $\sin\theta = \dfrac{2}{3}$, $\quad \cos\theta = \dfrac{\sqrt{5}}{3}$

In Exercises 21–24, θ is an acute angle and $\sin\theta$ is given. Use the Pythagorean identity $\sin^2\theta + \cos^2\theta = 1$ to find $\cos\theta$.

21. $\sin\theta = \dfrac{6}{7}$

22. $\sin\theta = \dfrac{7}{8}$

23. $\sin\theta = \dfrac{\sqrt{39}}{8}$

24. $\sin\theta = \dfrac{\sqrt{21}}{5}$

In Exercises 25–30, use an identity to find the value of each expression. Do not use a calculator.

25. $\sin 37° \csc 37°$

26. $\cos 53° \sec 53°$

27. $\sin^2 \dfrac{\pi}{9} + \cos^2 \dfrac{\pi}{9}$

28. $\sin^2 \dfrac{\pi}{10} + \cos^2 \dfrac{\pi}{10}$

29. $\sec^2 23° - \tan^2 23°$

30. $\csc^2 63° - \cot^2 63°$

In Exercises 31–38, find a cofunction with the same value as the given expression.

31. $\sin 7°$

32. $\sin 19°$

33. $\csc 25°$

34. $\csc 35°$

35. $\tan \dfrac{\pi}{9}$

36. $\tan \dfrac{\pi}{7}$

37. $\cos \dfrac{2\pi}{5}$

38. $\cos \dfrac{3\pi}{8}$

In Exercises 39–48, use a calculator to find the value of the trigonometric function to four decimal places.

39. $\sin 38°$

40. $\cos 21°$

41. $\tan 32.7°$

42. $\tan 52.6°$

43. $\csc 17°$

44. $\sec 55°$

45. $\cos \dfrac{\pi}{10}$

46. $\sin \dfrac{3\pi}{10}$

47. $\cot \dfrac{\pi}{12}$

48. $\cot \dfrac{\pi}{18}$

In Exercises 49–54, find the measure of the side of the right triangle whose length is designated by a lowercase letter. Round answers to the nearest whole number.

49.

50.

51.

52.

53.

54.
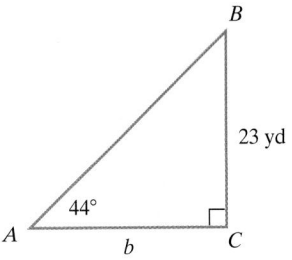

In Exercises 55–58, use a calculator to find the value of the acute angle θ to the nearest degree.

55. $\sin \theta = 0.2974$

56. $\cos \theta = 0.8771$

57. $\tan \theta = 4.6252$

58. $\tan \theta = 26.0307$

In Exercises 59–62, use a calculator to find the value of the acute angle θ in radians, rounded to three decimal places.

59. $\cos \theta = 0.4112$

60. $\sin \theta = 0.9499$

61. $\tan \theta = 0.4169$

62. $\tan \theta = 0.5117$

 Application Exercises

63. To find the distance across a lake, a surveyor took the measurements in the figure shown. Use these measurements to determine how far it is across the lake. Round to the nearest yard.

64. At a certain time of day, the angle of elevation of the sun is 40°. To the nearest foot, find the height of a tree whose shadow is 35 feet long.

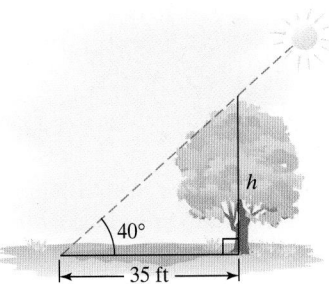

65. A tower that is 125 feet tall casts a shadow 172 feet long. Find the angle of elevation of the sun to the nearest degree.

66. The Washington Monument is 555 feet high. If you stand one quarter of a mile, or 1320 feet, from the base of the monument and look to the top, find the angle of elevation to the nearest degree.

67. A plane rises from take-off and flies at an angle of 10° with the horizontal runway. When it has gained 500 feet, find the distance, to the nearest foot, the plane has flown.

68. A road is inclined at an angle of 5°. After driving 5000 feet along this road, find the driver's increase in altitude. Round to the nearest foot.

69. A telephone pole, shown at the top of the next column, is 60 feet tall. A guy wire 75 feet long is attached from the ground to the top of the pole. Find the angle between the wire and the pole to the nearest degree.

70. A telephone pole is 55 feet tall. A guy wire 80 feet long is attached from the ground to the top of the pole. Find the angle between the wire and the pole to the nearest degree.

Writing in Mathematics

71. If you are given the lengths of the sides of a right triangle, describe how to find the sine of either acute angle.

72. Describe one similarity and one difference between the definitions of $\sin \theta$ and $\cos \theta$, where θ is an acute angle of a right triangle.

73. Describe the triangle used to find the trigonometric functions of 45°.

74. Describe the triangle used to find the trigonometric functions of 30° and 60°.

75. What is a trigonometric identity?

76. Use words (not an equation) to describe one of the reciprocal identities.

77. Use words (not an equation) to describe one of the quotient identities.

78. Use words (not an equation) to describe one of the Pythagorean identities.

79. Describe a relationship among trigonometric functions that is based on angles that are complements.

80. Describe what is meant by an angle of elevation and an angle of depression.

81. Stonehenge, the famous "stone circle" in England, was built between 2750 B.C. and 1300 B.C. using solid stone blocks weighing over 99,000 pounds each. It required 550 people to pull a single stone up a ramp inclined at a 9° angle. Describe how right triangle trigonometry can be used to determine the distance the 550 workers had to drag a stone in order to raise it to a height of 30 feet.

Technology Exercises

82. Use a calculator in the radian mode to fill in the values in the following table. Then draw a conclusion about $\dfrac{\sin \theta}{\theta}$ as θ approaches 0.

θ	0.4	0.3	0.2	0.1	0.01	0.001	0.0001	0.00001
$\sin \theta$								
$\dfrac{\sin \theta}{\theta}$								

83. Use a calculator in the radian mode to fill in the values in the following table. Then draw a conclusion about $\dfrac{\cos \theta - 1}{\theta}$ as θ approaches 0.

θ	0.4	0.3	0.2	0.1	0.01	0.001	0.0001	0.00001
$\cos \theta$								
$\dfrac{\cos \theta - 1}{\theta}$								

 Critical Thinking Exercises

84. Which one of the following is true?

a. $\dfrac{\tan 45°}{\tan 15°} = \tan 3°$

b. $\tan^2 15° - \sec^2 15° = -1$

c. $\sin 45° + \cos 45° = 1$

d. $\tan^2 5° = \tan 25°$

85. Explain why the sine or cosine of an acute angle cannot be greater than or equal to 1.

86. Describe what happens to the tangent of an acute angle as the angle gets close to 90°. What happens at 90°?

87. From the top of a 250-foot lighthouse, a plane is sighted overhead and a ship is observed directly below the plane. The angle of elevation of the plane is 22° and the angle of depression of the ship is 35°. Find **a.** the distance of the ship from the lighthouse; **b.** the plane's height above the water. Round to the nearest foot.

SECTION 5.3 Trigonometric Functions of Any Angle

Objectives

1. Use the definitions of trigonometric functions of any angle.
2. Use the signs of the trigonometric functions.
3. Find reference angles.
4. Use reference angles to evaluate trigonometric functions.

There is something comforting in the repetition of some of nature's patterns. The ocean level at a beach varies between high and low tide approximately every 12 hours. The number of hours of daylight oscillates from a maximum on the summer solstice, June 21, to a minimum on the winter solstice, December 21. Then it increases to the same maximum the following June 21. Some believe that cycles, called biorhythms, represent physical, emotional, and intellectual aspects of our lives. Throughout the remainder of this chapter, we will see how the trigonometric functions are used to model phenomena that occur again and again. To do this, we need to move beyond right triangles.

1 Use the definitions of trigonometric functions of any angle.

Trigonometric Functions of Any Angle

In the last section, we evaluated trigonometric functions of acute angles, such as that shown in Figure 5.26(a). Note that this angle is in standard position. The point $P = (x, y)$ is a point r units from the origin on the terminal side of θ. A right triangle is formed by drawing a perpendicular from $P = (x, y)$ to the x-axis. Note that y is the length of the side opposite θ and x is the length of the side adjacent to θ.

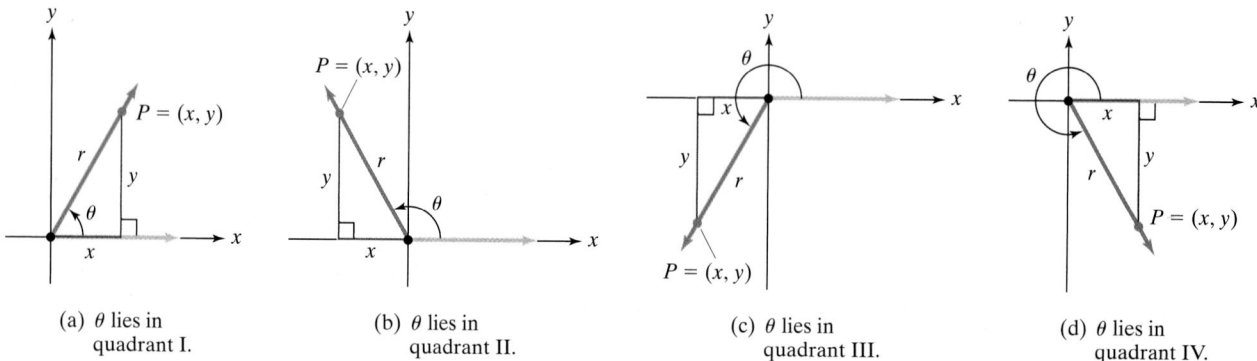

(a) θ lies in quadrant I.   (b) θ lies in quadrant II.   (c) θ lies in quadrant III.   (d) θ lies in quadrant IV.

Figure 5.26

Figures 5.26(b), (c), and (d) show angles in standard position, but they are not acute. We can extend our definitions of the six trigonometric functions to include such angles, as well as quadrantal angles. (Recall that a quadrantal angle has its terminal side on the x-axis or y-axis; such angles are *not* shown in Figure 5.26.) The point $P = (x, y)$ may be any point on the terminal side of the angle θ other than the origin, $(0, 0)$.

Definitions of Trigonometric Functions of Any Angle

Let θ be any angle in standard position, and let $P = (x, y)$ be a point on the terminal side of θ. If $r = \sqrt{x^2 + y^2}$ is the distance from $(0, 0)$ to (x, y), as shown in Figure 5.26, the **six trigonometric functions of θ** are defined by the following ratios:

$$\sin\theta = \frac{y}{r} \qquad \cos\theta = \frac{x}{r} \qquad \tan\theta = \frac{y}{x}, x \neq 0$$

$$\csc\theta = \frac{r}{y}, y \neq 0 \qquad \sec\theta = \frac{r}{x}, x \neq 0 \qquad \cot\theta = \frac{x}{y}, y \neq 0.$$

Because the point $P = (x, y)$ is any point on the terminal side of θ other than the origin, $(0, 0)$, $r = \sqrt{x^2 + y^2}$ cannot be zero. Examine the six trigonometric functions defined previously. Note that the denominator of the sine and cosine functions is r. Because $r \neq 0$, the sine and cosine functions are defined for any real value of the angle θ. This is not true for the other four trigonometric functions. Note that the denominator of the tangent and secant functions is x. These functions are not defined if $x = 0$. If the point $P = (x, y)$ is on the y-axis, then $x = 0$. Thus, the tangent and secant functions are undefined for all quadrantal angles with terminal sides on the positive or negative y-axis. Likewise, if $P = (x, y)$ is on the x-axis, then $y = 0$, and the cotangent and cosecant functions are undefined. The cotangent and cosecant functions are undefined for all quadrantal angles with terminal sides on the positive or negative x-axis.

EXAMPLE 1 Evaluating Trigonometric Functions

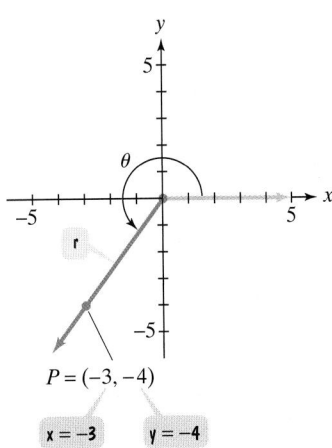

Figure 5.27

Let $P = (-3, -4)$ be a point on the terminal side of θ. Find each of the six trigonometric functions of θ.

Solution The situation is shown in Figure 5.27. We need values for $x, y,$ and r to evaluate all six trigonometric functions. We are given the values of x and y. Because $P = (-3, -4)$ is a point on the terminal side of θ, $x = -3$ and $y = -4$. Furthermore,

$$r = \sqrt{x^2 + y^2} = \sqrt{(-3)^2 + (-4)^2} = \sqrt{9 + 16} = \sqrt{25} = 5.$$

Now that we know $x, y,$ and r, we can find the six trigonometric functions of θ.

$$\sin\theta = \frac{y}{r} = \frac{-4}{5} = -\frac{4}{5}, \quad \cos\theta = \frac{x}{r} = \frac{-3}{5} = -\frac{3}{5}, \quad \tan\theta = \frac{y}{x} = \frac{-4}{-3} = \frac{4}{3}$$

$$\csc\theta = \frac{r}{y} = \frac{5}{-4} = -\frac{5}{4}, \quad \sec\theta = \frac{r}{x} = \frac{5}{-3} = -\frac{5}{3}, \quad \cot\theta = \frac{x}{y} = \frac{-3}{-4} = \frac{3}{4}$$

These ratios are the reciprocals of those shown directly above.

Check Point 1 Let $P = (4, -3)$ be a point on the terminal side of θ. Find each of the six trigonometric functions of θ.

How do we find the values of the trigonometric functions for a quadrantal angle? First, draw the angle in standard position. Second, choose a point P on the angle's terminal side. The trigonometric function values of θ depend only on the size of θ and not on the distance of point P from the origin. Thus, we choose a point that is 1 unit from the origin. Finally, apply the definition of the appropriate trigonometric function.

EXAMPLE 2 Trigonometric Functions of Quadrantal Angles

Evaluate, if possible, the sine function and the tangent function at the following four quadrantal angles:

a. $\theta = 0° = 0$ **b.** $\theta = 90° = \dfrac{\pi}{2}$ **c.** $\theta = 180° = \pi$ **d.** $\theta = 270° = \dfrac{3\pi}{2}$.

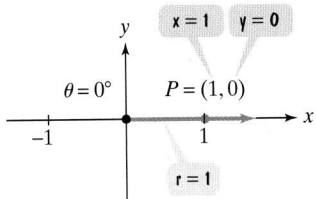

Solution

a. If $\theta = 0° = 0$ radians, then the terminal side of the angle is on the positive x-axis. Let us select the point $P = (1, 0)$ with $x = 1$ and $y = 0$. This point is 1 unit from the origin, so $r = 1$. Now that we know $x, y,$ and r, we can apply the definitions of the sine and tangent functions. (The figure on the left is repeated at the top of the next page.)

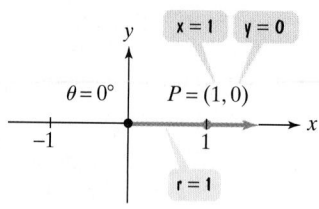

$$\sin 0° = \sin 0 = \frac{y}{r} = \frac{0}{1} = 0$$

$$\tan 0° = \tan 0 = \frac{y}{x} = \frac{0}{1} = 0$$

b. If $\theta = 90° = \dfrac{\pi}{2}$ radians, then the terminal side of the angle is on the positive y-axis. Let us select the point $P = (0, 1)$ with $x = 0$ and $y = 1$. This point is 1 unit from the origin, so $r = 1$. Now that we know x, y, and r, we can apply the definitions of the sine and tangent functions.

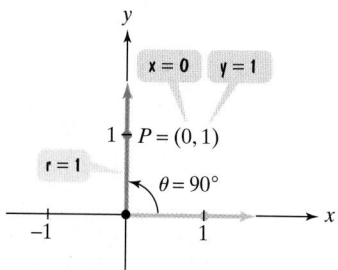

$$\sin 90° = \sin \frac{\pi}{2} = \frac{y}{r} = \frac{1}{1} = 1$$

$$\tan 90° = \tan \frac{\pi}{2} = \frac{y}{x} = \frac{1}{0}$$

Because division by 0 is undefined, $\tan 90°$ is undefined.

c. If $\theta = 180° = \pi$ radians, then the terminal side of the angle is on the negative x-axis. Let us select the point $P = (-1, 0)$ with $x = -1$ and $y = 0$. This point is 1 unit from the origin, so $r = 1$. Now that we know x, y, and r, we can apply the definitions of the sine and tangent functions.

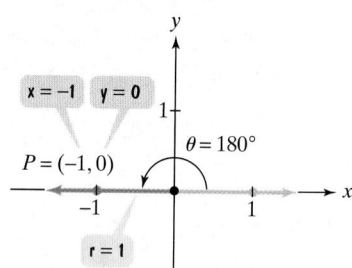

$$\sin 180° = \sin \pi = \frac{y}{r} = \frac{0}{1} = 0$$

$$\tan 180° = \tan \pi = \frac{y}{x} = \frac{0}{-1} = 0$$

d. If $\theta = 270° = \frac{3\pi}{2}$ radians, then the terminal side of the angle is on the negative y-axis. Let us select the point $P = (0, -1)$ with $x = 0$ and $y = -1$. This point is 1 unit from the origin, so $r = 1$. Now that we know x, y, and r, we can apply the definitions of the sine and tangent functions.

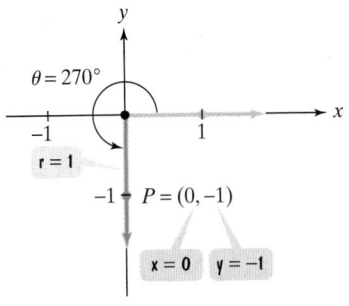

$$\sin 270° = \sin \frac{3\pi}{2} = \frac{y}{r} = \frac{-1}{1} = -1$$

$$\tan 270° = \tan \frac{3\pi}{2} = \frac{y}{x} = \frac{-1}{0}$$

Because division by 0 is undefined, tan 270° is undefined.

Check Point 2 Evaluate, if possible, the cosine function and the cosecant function at the following four quadrantal angles:

a. $\theta = 0° = 0$ **b.** $\theta = 90° = \frac{\pi}{2}$

c. $\theta = 180° = \pi$ **d.** $\theta = 270° = \frac{3\pi}{2}$.

2 Use the signs of the trigonometric functions.

The Signs of the Trigonometric Functions

In Example 2, we evaluated trigonometric functions of quadrantal angles. However, we will now return to the trigonometric functions of nonquadrantal angles. **If θ is not a quadrantal angle, the sign of a trigonometric function depends on the quadrant in which θ lies.** In all four quadrants, r is positive. However, x and y can be positive or negative. For example, if θ lies in quadrant II, x is negative and y is positive. Thus, the only positive ratios in this quadrant are $\frac{y}{r}$ and its reciprocal, $\frac{r}{y}$. These ratios are the function values for the sine and cosecant, respectively. In short, if θ lies in quadrant II, $\sin\theta$ and $\csc\theta$ are positive. The other four trigonometric functions are negative.

Figure 5.28 summarizes the signs of the trigonometric functions. If θ lies in quadrant I, all six functions are positive. If θ lies in quadrant II, only $\sin\theta$ and $\csc\theta$ are positive. If θ lies in quadrant III, only $\tan\theta$ and $\cot\theta$ are positive. Finally, if θ lies in quadrant IV, only $\cos\theta$ and $\sec\theta$ are positive. Observe that the positive functions in each quadrant occur in reciprocal pairs.

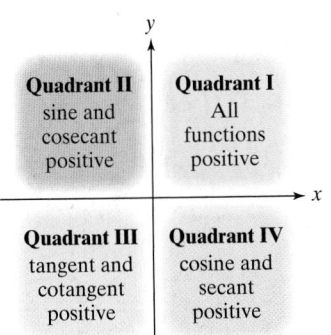

Figure 5.28 The signs of the trigonometric functions

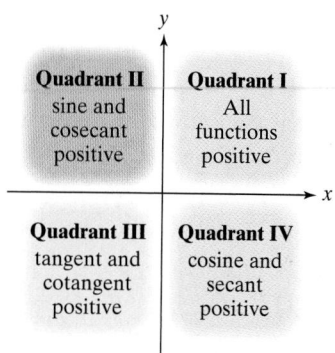

Figure 5.28 The signs of the trigonometric functions, repeated

EXAMPLE 3 Finding the Quadrant in Which an Angle Lies

If $\tan\theta < 0$ and $\cos\theta > 0$, name the quadrant in which angle θ lies.

Solution Because $\tan\theta < 0$, θ cannot lie in quadrant I; all the functions are positive in quadrant I. Furthermore, θ cannot lie in quadrant III; $\tan\theta$ is positive in quadrant III. Thus, with $\tan\theta < 0$, θ lies in quadrant II or quadrant IV. We are also given that $\cos\theta > 0$. Because quadrant IV is the only quadrant in which the cosine is positive and the tangent is negative, we conclude that θ lies in quadrant IV.

> **Check Point 3** If $\sin\theta < 0$ and $\cos\theta < 0$, name the quadrant in which angle θ lies.

EXAMPLE 4 Evaluating Trigonometric Functions

Given $\tan\theta = -\frac{2}{3}$ and $\cos\theta > 0$, find $\cos\theta$ and $\csc\theta$.

Solution Because the tangent is negative and the cosine is positive, θ lies in quadrant IV. This will help us to determine whether the negative sign in $\tan\theta = -\frac{2}{3}$ should be associated with the numerator or the denominator. Keep in mind that in quadrant IV, x is positive and y is negative. Thus,

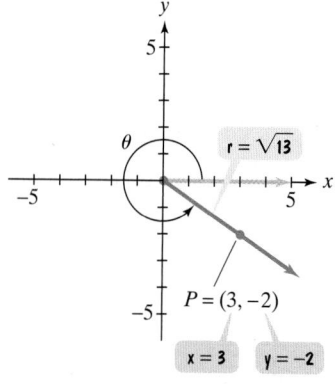

Figure 5.29 $\tan\theta = -\frac{2}{3}$ and $\cos\theta > 0$

> In quadrant IV, y is negative.

$$\tan\theta = -\frac{2}{3} = \frac{y}{x} = \frac{-2}{3}.$$

(See Figure 5.29.) Thus, $x = 3$ and $y = -2$. Furthermore,

$$r = \sqrt{x^2 + y^2} = \sqrt{3^2 + (-2)^2} = \sqrt{9 + 4} = \sqrt{13}.$$

Now that we know $x, y,$ and r, we can find $\cos\theta$ and $\csc\theta$.

$$\cos\theta = \frac{x}{r} = \frac{3}{\sqrt{13}} = \frac{3}{\sqrt{13}} \cdot \frac{\sqrt{13}}{\sqrt{13}} = \frac{3\sqrt{13}}{13} \qquad \csc\theta = \frac{r}{y} = \frac{\sqrt{13}}{-2} = -\frac{\sqrt{13}}{2}$$

> **Check Point 4** Given $\tan\theta = -\frac{1}{3}$ and $\cos\theta < 0$, find $\sin\theta$ and $\sec\theta$.

3 Find reference angles.

Reference Angles

We will often evaluate trigonometric functions of positive angles greater than 90° and all negative angles by making use of a positive acute angle. This positive acute angle is called a *reference angle*.

Definition of a Reference Angle

Let θ be a nonacute angle in standard position that lies in a quadrant. Its **reference angle** is the positive acute angle θ' formed by the terminal side of θ and the x-axis.

Figure 5.30 shows the reference angle for θ lying in quadrants II, III, and IV. Notice that the formula used to find θ', the reference angle, varies according to the quadrant in which θ lies. You may find it easier to find the reference angle for a given angle by making a figure that shows the angle in standard position. The acute angle formed by the terminal side of this angle and the x-axis is the reference angle.

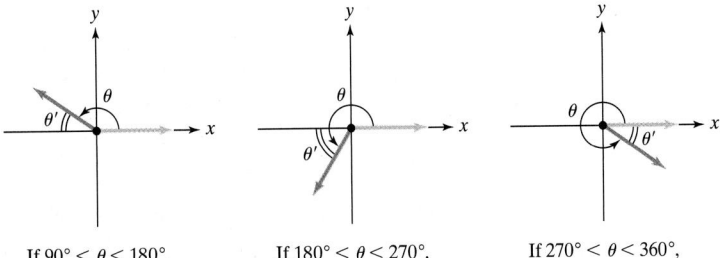

Figure 5.30 Reference angles, θ', for positive angles, θ, in quadrants II, III, and IV

If $90° < \theta < 180°$, then $\theta' = 180° - \theta$.

If $180° < \theta < 270°$, then $\theta' = \theta - 180°$.

If $270° < \theta < 360°$, then $\theta' = 360° - \theta$.

EXAMPLE 5   Finding Reference Angles

Find the reference angle, θ', for each of the following angles:

a. $\theta = 345°$ **b.** $\theta = \dfrac{5\pi}{6}$ **c.** $\theta = -135°$ **d.** $\theta = 2.5$.

Solution

a. A $345°$ angle in standard position is shown in Figure 5.31. Because $345°$ lies in quadrant IV, the reference angle is

$$\theta' = 360° - 345° = 15°.$$

Figure 5.31

b. Because $\dfrac{5\pi}{6}$ lies between $\dfrac{\pi}{2} = \dfrac{3\pi}{6}$ and $\pi = \dfrac{6\pi}{6}$, $\theta = \dfrac{5\pi}{6}$ lies in quadrant II. The angle is shown in Figure 5.32. The reference angle is

$$\theta' = \pi - \frac{5\pi}{6} = \frac{6\pi}{6} - \frac{5\pi}{6} = \frac{\pi}{6}.$$

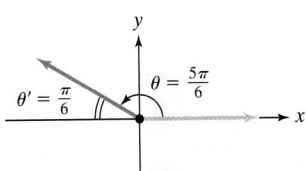

Figure 5.32

c. A $-135°$ angle in standard position is shown in Figure 5.33. The figure indicates that the positive acute angle formed by the terminal side of θ and the x-axis is $45°$. The reference angle is

$$\theta' = 45°.$$

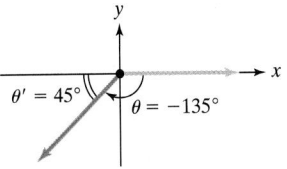

Figure 5.33

d. The angle $\theta = 2.5$ lies between $\dfrac{\pi}{2} \approx 1.57$ and $\pi \approx 3.14$. This means that $\theta = 2.5$ is in quadrant II, shown in Figure 5.34. The reference angle is

$$\theta' = \pi - 2.5 \approx 0.64.$$

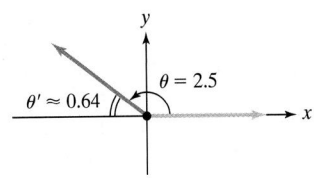

Figure 5.34

Check Point 5
Find the reference angle, θ', for each of the following angles:

a. $\theta = 210°$ **b.** $\theta = \dfrac{7\pi}{4}$ **c.** $\theta = -240°$ **d.** $\theta = 3.6$

The way that reference angles are defined makes them useful in evaluating trigonometric functions.

4 Use reference angles to evaluate trigonometric functions.

Using Reference Angles to Evaluate Trigonometric Functions

The values of the trigonometric functions of a given angle, θ, are the same as the values of the trigonometric functions of the reference angle, θ', except possibly for the sign. A function value of the acute reference angle, θ', is always positive. However, the same function value for θ may be positive or negative.

For example, we can use a reference angle, θ', to obtain an exact value for $\tan 120°$. The reference angle for $\theta = 120°$ is $\theta' = 180° - 120° = 60°$. We know the exact value of the tangent function of the reference angle: $\tan 60° = \sqrt{3}$. We also know that the value of a trigonometric function of a given angle, θ, is the same as that of its reference angle, θ', except possibly for the sign. Thus, we can conclude that $\tan 120°$ equals $-\sqrt{3}$ or $\sqrt{3}$.

What sign should we attach to $\sqrt{3}$? A $120°$ angle lies in quadrant II, where only the sine and cosecant are positive. Thus, the tangent function is negative for a $120°$ angle. Therefore,

> Prefix by a negative sign to show tangent is negative in quadrant II.

$$\tan 120° = -\tan 60° = -\sqrt{3}.$$

> The reference angle for $120°$ is $60°$.

In the previous section, we used two right triangles to find exact trigonometric values of $30°$, $45°$, and $60°$. Using a procedure similar to finding $\tan 120°$, we can now find the function values of all angles for which $30°$, $45°$, or $60°$ are reference angles.

A Procedure for Using Reference Angles to Evaluate Trigonometric Functions

The value of a trigonometric function of any angle θ is found as follows:

1. Find the associated reference angle, θ', and the function value for θ'.

2. Use the quadrant in which θ lies to prefix the appropriate sign to the function value in step 1.

Discovery

Draw the two right triangles involving $30°$, $45°$, and $60°$. Indicate the length of each side. Use these lengths to verify the function values for the reference angles in the solution to Example 6.

EXAMPLE 6   Using Reference Angles to Evaluate Trigonometric Functions

Use reference angles to find the exact value of each of the following trigonometric functions:

a. $\sin 135°$ **b.** $\cos \dfrac{4\pi}{3}$ **c.** $\cot\left(-\dfrac{\pi}{3}\right)$.

Solution

a. We use our two-step procedure to find $\sin 135°$.

Step 1   Find the reference angle, θ', and $\sin \theta'$.   Figure 5.35 shows 135° lies in quadrant II. The reference angle is

$$\theta' = 180° - 135° = 45°.$$

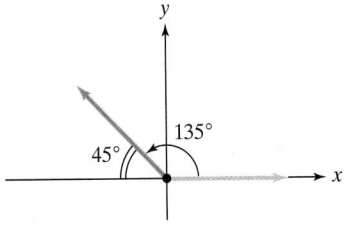

Figure 5.35 Reference angle for 135°

The function value for the reference angle is $\sin 45° = \dfrac{\sqrt{2}}{2}$.

Step 2   Use the quadrant in which θ lies to prefix the appropriate sign to the function value in step 1.   The angle $\theta = 135°$ lies in quadrant II. Because the sine is positive in quadrant II, we put a $+$ sign before the function value of the reference angle. Thus,

The sine is positive in quadrant II.

$$\sin 135° = +\sin 45° = \frac{\sqrt{2}}{2}.$$

The reference angle for 135° is 45°.

b. We use our two-step procedure to find $\cos \dfrac{4\pi}{3}$.

Step 1   Find the reference angle, θ', and $\cos \theta'$.   Figure 5.36 shows that $\theta = \dfrac{4\pi}{3}$ lies in quadrant III. The reference angle is

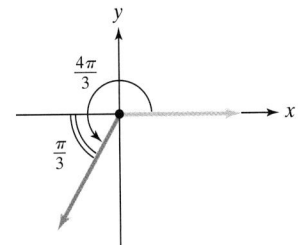

Figure 5.36 Reference angle for $\dfrac{4\pi}{3}$

$$\theta' = \frac{4\pi}{3} - \pi = \frac{4\pi}{3} - \frac{3\pi}{3} = \frac{\pi}{3}.$$

The function value for the reference angle is

$$\cos \frac{\pi}{3} = \frac{1}{2}.$$

Step 2   Use the quadrant in which θ lies to prefix the appropriate sign to the function value in step 1.   The angle $\theta = \dfrac{4\pi}{3}$ lies in quadrant III. Because only the tangent and cotangent are positive in quadrant III, the cosine is negative in this quadrant. We put a $-$ sign before the function value of the reference angle. Thus,

The cosine is negative in quadrant III.

$$\cos \frac{4\pi}{3} = -\cos \frac{\pi}{3} = -\frac{1}{2}.$$

The reference angle for $\dfrac{4\pi}{3}$ is $\dfrac{\pi}{3}$.

c. We use our two-step procedure to find $\cot\left(-\dfrac{\pi}{3}\right)$.

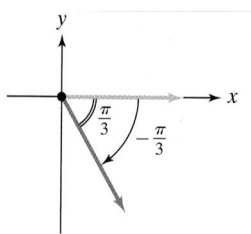

Figure 5.37 Reference angle for $-\dfrac{\pi}{3}$

Step 1 Find the reference angle, θ', and $\cot\theta'$. Figure 5.37 shows that $\theta = -\dfrac{\pi}{3}$ lies in quadrant IV. The reference angle is $\theta' = \dfrac{\pi}{3}$. The function value for the reference angle is $\cot\dfrac{\pi}{3} = \dfrac{\sqrt{3}}{3}$.

Step 2 Use the quadrant in which θ lies to prefix the appropriate sign to the function value in step 1. The angle $\theta = -\dfrac{\pi}{3}$ lies in quadrant IV. Because only the cosine and secant are positive in quadrant IV, the cotangent is negative in this quadrant. We put a $-$ sign before the function value of the reference angle. Thus,

The cotangent is negative in quadrant IV.

$$\cot\left(-\frac{\pi}{3}\right) = -\cot\frac{\pi}{3} = -\frac{\sqrt{3}}{3}.$$

The reference angle for $-\dfrac{\pi}{3}$ is $\dfrac{\pi}{3}$.

Check Point 6 Use reference angles to find the exact value of the following trigonometric functions:

a. $\sin 300°$ **b.** $\tan\dfrac{5\pi}{4}$ **c.** $\sec\left(-\dfrac{\pi}{6}\right)$.

EXERCISE SET 5.3

Practice Exercises

In Exercises 1–8, a point on the terminal side of angle θ is given. Find the exact value of each of the six trigonometric functions of θ.

1. $(-4, 3)$ **2.** $(-12, 5)$
3. $(2, 3)$ **4.** $(3, 7)$
5. $(3, -3)$ **6.** $(5, -5)$
7. $(-2, -5)$ **8.** $(-1, -3)$

In Exercises 9–16, evaluate the trigonometric function at the quadrantal angle, or state that the expression is undefined.

9. $\cos\pi$ **10.** $\tan\pi$
11. $\sec\pi$ **12.** $\csc\pi$
13. $\tan\dfrac{3\pi}{2}$ **14.** $\cos\dfrac{3\pi}{2}$
15. $\cot\dfrac{\pi}{2}$ **16.** $\tan\dfrac{\pi}{2}$

In Exercises 17–22, let θ be an angle in standard position. Name the quadrant in which θ lies.

17. $\sin\theta > 0, \quad \cos\theta > 0$
18. $\sin\theta < 0, \quad \cos\theta > 0$
19. $\sin\theta < 0, \quad \cos\theta < 0$
20. $\tan\theta < 0, \quad \sin\theta < 0$
21. $\tan\theta < 0, \quad \cos\theta < 0$
22. $\cot\theta > 0, \quad \sec\theta < 0$

In Exercises 23–34, find the exact value of each of the remaining trigonometric functions of θ.

23. $\cos\theta = -\dfrac{3}{5}, \quad \theta$ in quadrant III
24. $\sin\theta = -\dfrac{12}{13}, \quad \theta$ in quadrant III
25. $\sin\theta = \dfrac{5}{13}, \quad \theta$ in quadrant II
26. $\cos\theta = \dfrac{4}{5}, \quad \theta$ in quadrant IV
27. $\cos\theta = \dfrac{8}{17}, \quad 270° < \theta < 360°$
28. $\cos\theta = \dfrac{1}{3}, \quad 270° < \theta < 360°$
29. $\tan\theta = -\dfrac{2}{3}, \quad \sin\theta > 0$
30. $\tan\theta = -\dfrac{1}{3}, \quad \sin\theta > 0$
31. $\tan\theta = \dfrac{4}{3}, \quad \cos\theta < 0$
32. $\tan\theta = \dfrac{5}{12}, \quad \cos\theta < 0$
33. $\sec\theta = -3, \quad \tan\theta > 0$
34. $\csc\theta = -4, \quad \tan\theta > 0$

In Exercises 35–50, find the reference angle for each angle.

35. $160°$

36. $170°$

37. $205°$

38. $210°$

39. $355°$

40. $351°$

41. $\dfrac{7\pi}{4}$

42. $\dfrac{5\pi}{4}$

43. $\dfrac{5\pi}{6}$

44. $\dfrac{5\pi}{7}$

45. $-150°$

46. $-250°$

47. $-335°$

48. $-359°$

49. 4.7

50. 5.5

In Exercises 51–66, use reference angles to find the exact value of each expression. Do not use a calculator.

51. $\cos 225°$

52. $\sin 300°$

53. $\tan 210°$

54. $\sec 240°$

55. $\tan 420°$

56. $\tan 405°$

57. $\sin \dfrac{2\pi}{3}$

58. $\cos \dfrac{3\pi}{4}$

59. $\csc \dfrac{7\pi}{6}$

60. $\cot \dfrac{7\pi}{4}$

61. $\tan \dfrac{9\pi}{4}$

62. $\tan \dfrac{9\pi}{2}$

63. $\sin(-240°)$

64. $\sin(-225°)$

65. $\tan\left(-\dfrac{\pi}{4}\right)$

66. $\tan\left(-\dfrac{\pi}{6}\right)$

Writing in Mathematics

67. If you are given a point on the terminal side of angle θ, explain how to find $\sin\theta$.

68. Explain why $\tan 90°$ is undefined.

69. If $\cos\theta > 0$ and $\tan\theta < 0$, explain how to find the quadrant in which θ lies.

70. What is a reference angle? Give an example with your description.

71. Explain how reference angles are used to evaluate trigonometric functions. Give an example with your description.

SECTION 5.4 Trigonometric Functions of Real Numbers; Periodic Functions

Objectives

1. Use a unit circle to define trigonometric functions of real numbers.
2. Recognize the domain and range of sine and cosine functions.
3. Use even and odd trigonometric functions.
4. Use periodic properties.

Cycles govern many aspects of life—heartbeats, sleep patterns, seasons, and tides all follow regular, predictable cycles. In this section, we will see why trigonometric functions are used to model phenomena that occur in cycles. To do this, we need to move beyond angles and consider trigonometric functions of real numbers.

1 Use a unit circle to define trigonometric functions of real numbers.

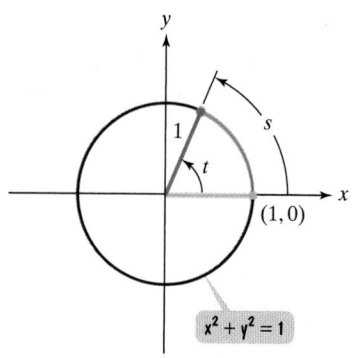

Figure 5.38 Unit circle with a central angle measuring t radians

Trigonometric Functions of Real Numbers

Thus far, we have considered trigonometric functions of angles measured in degrees or radians. To define trigonometric functions of real numbers, rather than angles, we use a unit circle. A **unit circle** is a circle of radius 1, with its center at the origin of a rectangular coordinate system. The equation of this unit circle is $x^2 + y^2 = 1$. Figure 5.38 shows a unit circle in which the central angle measures t radians. We can use the formula for the length of a circular arc, $s = r\theta$, to find the length of the intercepted arc.

$$s = r\theta = 1 \cdot t = t$$

The radius of a unit circle is 1.　　The radian measure of the central angle is t.

Thus, the length of the intercepted arc is t. This is also the radian measure of the central angle. Thus, **in a unit circle, the radian measure of the angle is equal to the measure of the intercepted arc.** Both are given by the same *real number t*.

In Figure 5.39, the radian measure of the angle and the length of the intercepted arc are both shown by t. Let $P = (x, y)$ denote the point on the unit circle that has arc length t from $(1, 0)$. Figure 5.39(a) shows that if t is positive, point P is reached by moving counterclockwise along the unit circle from $(1, 0)$. Figure 5.39(b) shows that if t is negative, point P is reached by moving clockwise along the unit circle from $(1, 0)$. For each real number t, there corresponds a point $P = (x, y)$ on the unit circle.

Figure 5.39

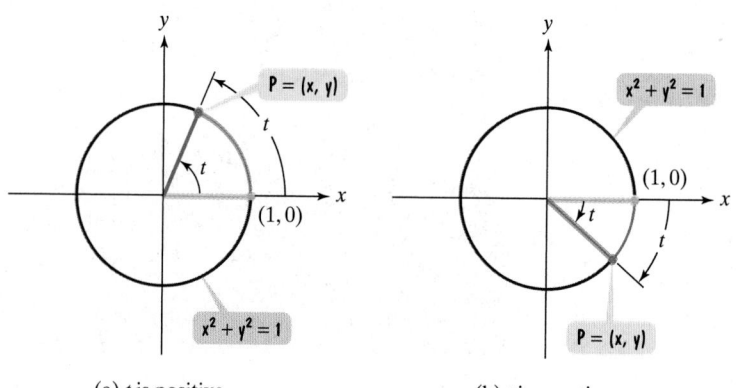

(a) t is positive.　　　　(b) t is negative.

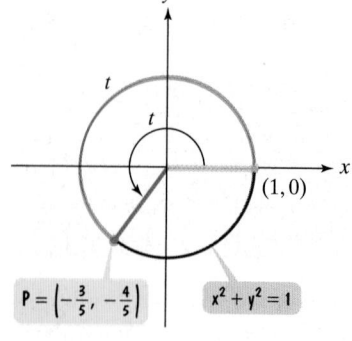

Figure 5.40

Using Figure 5.39, we define the cosine function at t as the x-coordinate of P and the sine function at t as the y-coordinate of P. Thus,

$$x = \cos t \quad \text{and} \quad y = \sin t.$$

For example, a point $P = (x, y)$ on the unit circle corresponding to a real number t is shown in Figure 5.40 for $\pi < t < \dfrac{3\pi}{2}$. We see that the coordinates of $P = (x, y)$ are $x = -\frac{3}{5}$ and $y = -\frac{4}{5}$. Because the cosine function is the x-coordinate of P and the sine function is the y-coordinate of P, the values of these trigonometric functions at the real number t are

$$\cos t = -\frac{3}{5} \quad \text{and} \quad \sin t = -\frac{4}{5}.$$

Definitions of the Trigonometric Functions in Terms of a Unit Circle

If t is a real number and $P = (x, y)$ is a point on the unit circle that corresponds to t, then

$$\sin t = y \qquad \cos t = x \qquad \tan t = \frac{y}{x}, x \neq 0$$

$$\csc t = \frac{1}{y}, y \neq 0 \qquad \sec t = \frac{1}{x}, x \neq 0 \qquad \cot t = \frac{x}{y}, y \neq 0.$$

Because this definition expresses function values in terms of coordinates of a point on a unit circle, the trigonometric functions are sometimes called the **circular functions.**

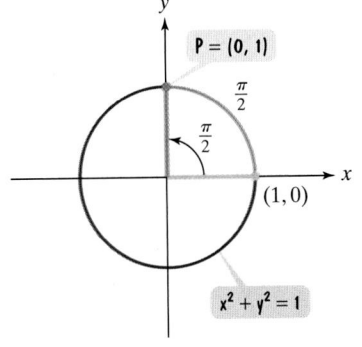

Figure 5.41

EXAMPLE 1 Finding Values of the Trigonometric Functions

Use Figure 5.41 to find the values of the trigonometric functions at $t = \dfrac{\pi}{2}$.

Solution The point P on the unit circle that corresponds to $t = \dfrac{\pi}{2}$ has coordinates $(0, 1)$. We use $x = 0$ and $y = 1$ to find the values of the trigonometric functions.

$$\sin \frac{\pi}{2} = y = 1 \qquad\qquad \cos \frac{\pi}{2} = x = 0$$

$$\csc \frac{\pi}{2} = \frac{1}{y} = \frac{1}{1} = 1 \qquad\qquad \cot \frac{\pi}{2} = \frac{x}{y} = \frac{0}{1} = 0$$

By definition, $\tan t = \dfrac{y}{x}$ and $\sec t = \dfrac{1}{x}$. Because $x = 0$, $\tan \dfrac{\pi}{2}$ and $\sec \dfrac{\pi}{2}$, are undefined.

Check Point 1 Use the figure on the right to find the values of the trigonometric functions at $t = \pi$.

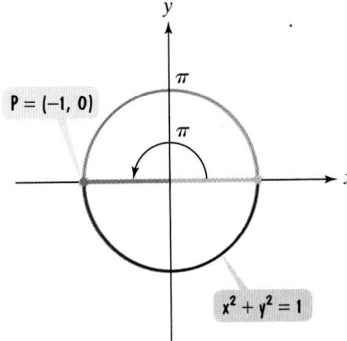

Domain and Range of Sine and Cosine Functions

2 Recognize the domain and range of sine and cosine functions.

The value of a trigonometric function at the real number t is its value at an angle of t radians. However, using real number domains, we can observe properties of trigonometric functions that are not as apparent using the angle approach. For example, the domain and range of each trigonometric function can be found from the unit circle definition. At this point, let's look only at the sine and cosine functions,

$$\sin t = y \quad \text{and} \quad \cos t = x.$$

Because t can be the radian measure of any angle or, equivalently, the measure of any intercepted arc, the domain of the sine function and the cosine function is the set of all real numbers. Because the radius of the unit circle is 1, we have

$$-1 \le x \le 1 \quad \text{and} \quad -1 \le y \le 1.$$

Therefore, with $x = \cos t$ and $y = \sin t$, we obtain

$$-1 \le \cos t \le 1 \quad \text{and} \quad -1 \le \sin t \le 1.$$

The range of the cosine and sine functions is $[-1, 1]$.

The Domain and Range of the Sine and Cosine Functions

The domain of the sine function and the cosine function is the set of all real numbers. The range of these functions is the set of all real numbers from -1 to 1, inclusive.

3 Use even and odd trigonometric functions.

Even and Odd Trigonometric Functions

In Chapter 2, we saw that a function is even if $f(-t) = f(t)$ and odd if $f(-t) = -f(t)$. We can use Figure 5.42 to show that the cosine function is an even function and the sine function is an odd function. By definition, the coordinates of the points P and Q in Figure 5.42 are as follows:

$$P: \ (\cos t, \sin t)$$

$$Q: \ (\cos(-t), \sin(-t)).$$

In Figure 5.42, the x-coordinates of P and Q are the same. Thus,

$$\cos(-t) = \cos t.$$

This shows that the cosine function is an even function. By contrast, the y-coordinates of P and Q are negatives of each other. Thus,

$$\sin(-t) = -\sin t.$$

This shows that the sine function is an odd function.

This argument is valid regardless of the length of t. Thus, the arc may terminate in any of the four quadrants or on any axis. Using the unit circle definition of the trigonometric functions, we obtain the following results:

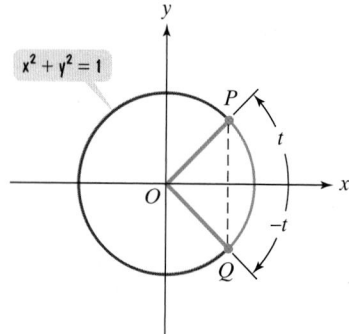

Figure 5.42

Even and Odd Trigonometric Functions

The cosine and secant functions are *even*.

$$\cos(-t) = \cos t \qquad\qquad \sec(-t) = \sec t$$

The sine, cosecant, tangent, and cotangent functions are *odd*.

$$\sin(-t) = -\sin t \qquad\qquad \csc(-t) = -\csc t$$

$$\tan(-t) = -\tan t \qquad\qquad \cot(-t) = -\cot t$$

EXAMPLE 2   Using Even and Odd Functions to Find Exact Values

Find the exact value of:

a. $\cos(-45°)$ **b.** $\tan\left(-\dfrac{\pi}{3}\right)$.

Solution

a. $\cos(-45°) = \cos 45° = \dfrac{\sqrt{2}}{2}$ **b.** $\tan\left(-\dfrac{\pi}{3}\right) = -\tan\dfrac{\pi}{3} = -\sqrt{3}$

> **Check Point 2**   Find the exact value of:
>
> **a.** $\cos(-60°)$ **b.** $\tan\left(-\dfrac{\pi}{6}\right)$.

4 Use periodic properties.

Periodic Functions

Certain patterns in nature repeat again and again. For example, the ocean level at a beach varies from low tide to high tide and then back to low tide approximately every 12 hours. If low tide occurs at noon, then high tide will be around 6 P.M. and low tide will occur again around midnight, and so on infinitely. If $f(t)$ represents the ocean level at the beach at any time t, then the level is the same 12 hours later. Thus,

$$f(t + 12) = f(t).$$

The word *periodic* means that this tidal behavior repeats infinitely. The *period*, 12 hours, is the time it takes to complete one full cycle.

> **Definition of a Periodic Function**
>
> A function f is **periodic** if there exists a positive number p such that
>
> $$f(t + p) = f(t)$$
>
> for all t in the domain of f. The smallest number p for which f is periodic is called the **period** of f.

The trigonometric functions are used to model periodic phenomena. Why? If we begin at any point P on the unit circle and travel a distance of 2π units along the perimeter, we will return to the same point P. Because the trigonometric functions are defined in terms of the coordinates of that point P, we obtain the following results:

Periodic Properties of the Sine and Cosine Functions

$$\sin(t + 2\pi) = \sin t \quad \text{and} \quad \cos(t + 2\pi) = \cos t$$

The sine and cosine functions are periodic functions and have period 2π.

EXAMPLE 3   Using Periodic Properties to Find Exact Values

Find the exact value of: **a.** $\tan 420°$ **b.** $\sin \dfrac{9\pi}{4}$.

Solution

a. $\tan 420° = \tan(60° + 360°) = \tan 60° = \sqrt{3}$

b. $\sin \dfrac{9\pi}{4} = \sin\left(\dfrac{\pi}{4} + 2\pi\right) = \sin \dfrac{\pi}{4} = \dfrac{\sqrt{2}}{2}$

Check Point 3 Find the exact value of:

a. $\cos 405°$ **b.** $\tan \dfrac{7\pi}{3}$.

Like the sine and cosine functions, the secant and cosecant functions have period 2π. However, the tangent and cotangent functions have a smaller period. If we begin at any point $P(x, y)$ on the unit circle and travel a distance of π units along the perimeter, we arrive at the point $(-x, -y)$. The tangent function, defined in terms of the coordinates of a point, is the same at (x, y) and $(-x, -y)$.

Tangent function at (x, y) $\dfrac{y}{x} = \dfrac{-y}{-x}$ Tangent function π radians later

We see that $\tan(t + \pi) = \tan t$. The same observations apply to the cotangent function.

Periodic Properties of the Tangent and Cotangent Functions

$$\tan(t + \pi) = \tan t \quad \text{and} \quad \cot(t + \pi) = \cot t$$

The tangent and cotangent functions are periodic functions and have period π.

Why do the trigonometric functions model phenomena that repeat *indefinitely*? By starting at point P on the unit circle and traveling a distance of 2π units, 4π units, 6π units, and so on, we return to the starting point P. Because the trigonometric functions are defined in terms of the coordinates of that point P, if we add (or subtract) multiples of 2π, the trigonometric values do not change. Furthermore, the trigonometric values for the tangent and cotangent functions do not change if we add (or subtract) multiples of π.

Repetitive Behavior of the Sine, Cosine, and Tangent Functions

For any integer n and real number t,

$$\sin(t + 2\pi n) = \sin t, \quad \cos(t + 2\pi n) = \cos t, \quad \text{and} \quad \tan(t + \pi n) = \tan t.$$

EXERCISE SET 5.4

Practice Exercises

In Exercises 1–4, a point $P(x, y)$ is shown on the unit circle corresponding to a real number t. Find the values of the trigonometric functions at t.

1.

2.

3.

4.

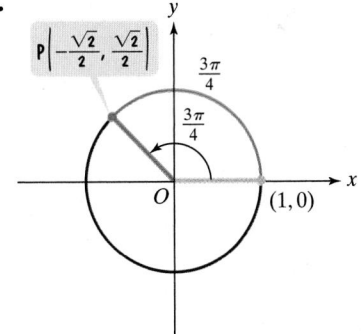

In Exercises 5–8, use even and odd properties of the trigonometric functions to find the exact value of each expression.

5. $\sin(-45°)$

6. $\tan(-45°)$

7. $\sec\left(-\dfrac{\pi}{3}\right)$

8. $\sec\left(-\dfrac{\pi}{6}\right)$

In Exercises 9–12, use periodic properties to find the exact value of each expression.

9. $\cos 585°$

10. $\cos 570°$

11. $\cot \dfrac{7\pi}{3}$

12. $\cot \dfrac{9\pi}{4}$

Application Exercises

13. The number of hours of daylight, H, on day t of any given year (on January 1, $t = 1$) in Fairbanks, Alaska, can be modeled by the function

$$H(t) = 12 + 8.3 \sin\left[\frac{2\pi}{365}(t - 80)\right].$$

  a. March 21, the 80th day of the year, is the spring equinox. Find the number of hours of daylight in Fairbanks on this day.

  b. June 21, the 172nd day of the year, is the summer solstice, the day with the maximum number of hours of daylight. To the nearest tenth of an hour, find the number of hours of daylight in Fairbanks on this day.

  c. December 21, the 355th day of the year, is the winter solstice, the day with the minimum number of hours of daylight. Find, to the nearest tenth of an hour, the number of hours of daylight in Fairbanks on this day.

14. The number of hours of daylight, H, on day t of any given year (on January 1, $t = 1$) in San Diego, California, can be modeled by the function

$$H(t) = 12 + 2.4 \sin\left[\frac{2\pi}{365}(t - 80)\right].$$

  a. March 21, the 80th day of the year, is the spring equinox. Find the number of hours of daylight in San Diego on this day.

  b. June 21, the 172nd day of the year, is the summer solstice, the day with the maximum number of hours of daylight. Find, to the nearest tenth of an hour, the number of hours of daylight in San Diego on this day.

  c. December 21, the 355th day of the year, is the winter solstice, the day with the minimum number of hours of daylight. To the nearest tenth of an hour, find the number of hours of daylight in San Diego on this day.

15. People who believe in biorhythms claim that there are three cycles that rule our behavior—the physical, emotional, and mental. Each is a sine function of a certain period. The function for our emotional fluctuations is

$$E = \sin \frac{\pi}{14} t$$

where t is measured in days starting at birth. Emotional fluctuations, E, are measured from -1 to 1, inclusive, with 1 representing peak emotional well-being, -1 representing the low for emotional well-being, and 0 representing feeling neither emotionally high nor low.

  a. Find E corresponding to $t = 7$, 14, 21, 28, and 35. Describe what you observe.

  b. What is the period of the emotional cycle?

16. The height of the water, H, in feet, at a boat dock t hours after 6 A.M. is given by

$$H = 10 + 4 \sin \frac{\pi}{6} t.$$

  a. Find the height of the water at the dock at 6 A.M., 9 A.M., noon, 6 P.M., midnight, and 3 A.M.

  b. When is low tide and when is high tide?

  c. What is the period of this function and what does this mean about the tides?

Writing in Mathematics

17. Why are the trigonometric functions sometimes called circular functions?

18. What is the range of the sine function? Use the unit circle to explain where this range comes from.

19. What do we mean by even trigonometric functions? Which of the six functions fall into this category?

20. What is a periodic function? Why are the sine and cosine functions periodic?

21. Explain how you can use the function for emotional fluctuations in Exercise 15 to determine good days for having dinner with your moody boss.

22. Describe a phenomenon that repeats infinitely. What is its period?

Critical Thinking Exercises

23. Find the exact value of $\cos 0° + \cos 1° + \cos 2° + \cos 3° + \cdots + \cos 179° + \cos 180°$.

24. If $f(x) = \sin x$ and $f(a) = \frac{1}{4}$, find the value of

$$f(a) + f(a + 2\pi) + f(a + 4\pi) + f(a + 6\pi).$$

25. If $f(x) = \sin x$ and $f(a) = \frac{1}{4}$, find the value of $f(a) + 2f(-a)$.

26. The seats of a ferris wheel are 40 feet from the wheel's center. When you get on the ride, your seat is 5 feet above the ground. How far above the ground are you after rotating through an angle of 765°? Round to the nearest foot.

SECTION 5.5 *Graphs of Sine and Cosine Functions*

Objectives

1. Understand the graph of $y = \sin x$.
2. Graph variations of $y = \sin x$.
3. Understand the graph of $y = \cos x$.
4. Graph variations of $y = \cos x$.
5. Use vertical shifts of sine and cosine curves.
6. Model periodic behavior.

Take a deep breath and relax. Many relaxation exercises involve slowing down our breathing. Some people suggest that the way we breathe affects every part of our lives. Did you know that graphs of trigonometric functions can be used to analyze the breathing cycle, which is our closest link to both life and death?

In this section, we use graphs of sine and cosine functions to visualize their properties. We use the traditional symbol x, rather than θ or t, to represent the independent variable. We use the symbol y for the dependent variable, or the function's value at x. Thus, we will be graphing $y = \sin x$ and $y = \cos x$ in rectangular coordinates. In all graphs of trigonometric functions, the independent variable, x, is measured in radians.

1 Understand the graph of $y = \sin x$.

The Graph of $y = \sin x$

The trigonometric functions can be graphed in a rectangular coordinate system by plotting points whose coordinates satisfy the function. Thus, we graph $y = \sin x$ by listing some points on the graph. Because the period of the sine function is 2π, we will graph the function on the interval $[0, 2\pi]$. The rest of the graph is made up of repetitions of this portion.

Table 5.1 lists some values of (x, y) on the graph of $y = \sin x, 0 \le x \le 2\pi$.

Table 5.1   Values of (x, y) on $y = \sin x$

x	0	$\dfrac{\pi}{6}$	$\dfrac{\pi}{3}$	$\dfrac{\pi}{2}$	$\dfrac{2\pi}{3}$	$\dfrac{5\pi}{6}$	π	$\dfrac{7\pi}{6}$	$\dfrac{4\pi}{3}$	$\dfrac{3\pi}{2}$	$\dfrac{5\pi}{3}$	$\dfrac{11\pi}{6}$	2π
$y = \sin x$	0	$\dfrac{1}{2}$	$\dfrac{\sqrt{3}}{2}$	1	$\dfrac{\sqrt{3}}{2}$	$\dfrac{1}{2}$	0	$-\dfrac{1}{2}$	$-\dfrac{\sqrt{3}}{2}$	-1	$-\dfrac{\sqrt{3}}{2}$	$-\dfrac{1}{2}$	0

As x increases from 0 to $\frac{\pi}{2}$, y increases from 0 to 1.

As x increases from $\frac{\pi}{2}$ to π, y decreases from 1 to 0.

As x increases from π to $\frac{3\pi}{2}$, y decreases from 0 to −1.

As x increases from $\frac{3\pi}{2}$ to 2π, y increases from −1 to 0.

In plotting the points obtained in Table 5.1, we will use the approximation $\dfrac{\sqrt{3}}{2} \approx 0.87$. Rather than approximating π, we will mark off units on the x-axis in terms of π. If we connect these points with a smooth curve, we obtain the graph shown in Figure 5.43 on the next page. The figure shows one period of the graph of $y = \sin x$.

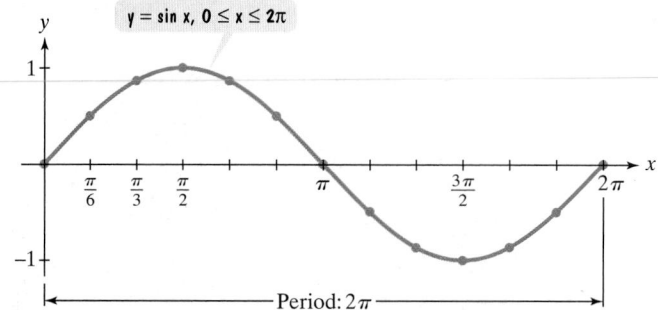

Figure 5.43 One period of the graph of $y = \sin x$

We can obtain a more complete graph of $y = \sin x$ by continuing the portion shown in Figure 5.43 to the left and to the right. The graph of the sine function, called a **sine curve,** is shown in Figure 5.44. Any part of the graph that corresponds to one period (2π) is one cycle of the graph of $y = \sin x$.

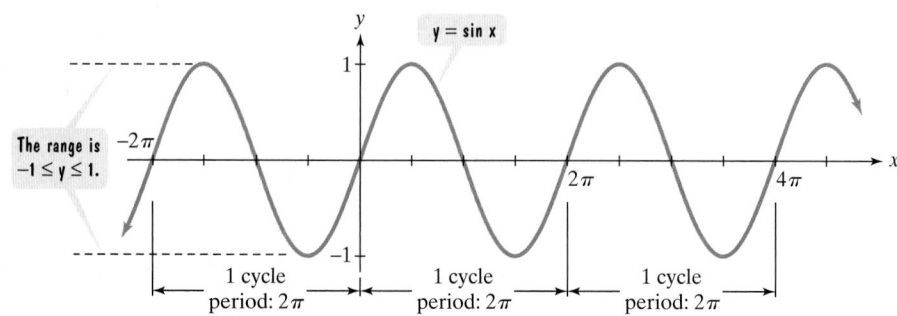

Figure 5.44 The graph of $y = \sin x$

The graph of $y = \sin x$ allows us to visualize some of the properties of the sine function.

- The domain is the set of all real numbers. The graph extends indefinitely to the left and to the right with no gaps or holes.
- The range consists of all numbers between -1 and 1, inclusive. The graph never rises above 1 or falls below -1.
- The period is 2π. The graph's pattern repeats in every interval of length 2π.
- The function is an odd function: $\sin(-x) = -\sin x$. This can be seen by observing that the graph is symmetric with respect to the origin.

2 Graph variations of $y = \sin x$.

Graphing Variations of $y = \sin x$

To graph variations of $y = \sin x$ by hand, it is helpful to find x-intercepts, maximum points, and minimum points. One complete cycle of the sine curve includes three x-intercepts, one maximum point, and one minimum point. The graph of $y = \sin x$ has x-intercepts at the beginning, middle, and end of its full period, shown in Figure 5.45. The curve reaches its maximum point $\frac{1}{4}$ of the way through the period. It reaches its minimum point $\frac{3}{4}$ of the way through the period. Thus, key points in graphing sine functions are

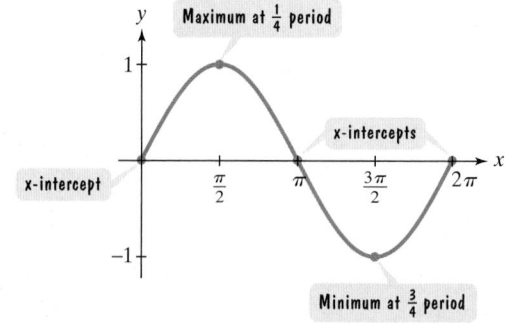

Figure 5.45 Key points in graphing the sine function

obtained by dividing the period into four equal parts. The x-coordinates of the five key points are as follows:

$$x_1 = \text{value of } x \text{ where the cycle begins}$$

$$x_2 = x_1 + \frac{\text{period}}{4}$$

$$x_3 = x_2 + \frac{\text{period}}{4}$$

Add "quarter-periods" to find successive value of x.

$$x_4 = x_3 + \frac{\text{period}}{4}$$

$$x_5 = x_4 + \frac{\text{period}}{4}.$$

The y-coordinates of the five key points are obtained by evaluating the given function at each of these values of x.

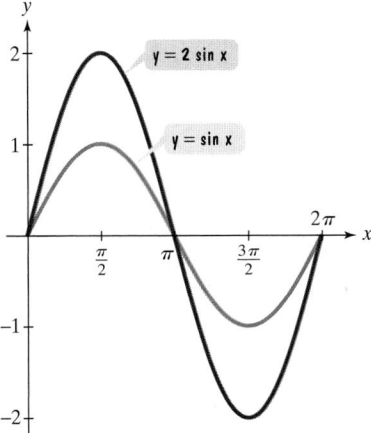

The graph of $y = \sin x$ forms the basis for graphing functions of the form

$$y = A \sin x.$$

For example, consider $y = 2 \sin x$, in which $A = 2$. We can obtain the graph of $y = 2 \sin x$ from that of $y = \sin x$ if we multiply each y-coordinate on the graph of $y = \sin x$ by 2. Figure 5.46 shows the graphs. The basic sine curve is *stretched* and ranges between -2 and 2, rather than between -1 and 1. However, both $y = \sin x$ and $y = 2 \sin x$ have a period of 2π.

In general, the graph of $y = A \sin x$ ranges between $-|A|$ and $|A|$. Thus, the range of the function is $-|A| \le y \le |A|$. If $|A| > 1$, the basic sine curve is *stretched*, as in Figure 5.46. If $|A| < 1$, the basic sine curve is *shrunk*. We call $|A|$ the **amplitude** of $y = A \sin x$. The maximum value of y on the graph of $y = A \sin x$ is $|A|$, the amplitude.

Figure 5.46 Comparing the graphs of $y = \sin x$ and $y = 2 \sin x$

Graphing Variations of $y = \sin x$

1. Identify the amplitude and the period.
2. Find the values of x for the five key points—the three x-intercepts, the maximum point, and the minimum point. Start with the value of x where the cycle begins and add quarter-periods—that is, $\frac{\text{period}}{4}$ —to find successive values of x.
3. Find the values of y for the five key points by evaluating the function at each value of x from step 2.
4. Connect the five key points with a smooth curve and graph one complete cycle of the given function.
5. Extend the graph in step 4 to the left or right as desired.

EXAMPLE 1 Graphing a Variation of $y = \sin x$

Determine the amplitude of $y = \frac{1}{2} \sin x$. Then graph $y = \sin x$ and $y = \frac{1}{2} \sin x$ for $0 \le x \le 2\pi$.

Solution

Step 1 Identify the amplitude and the period. The equation $y = \frac{1}{2} \sin x$ is of the form $y = A \sin x$ with $A = \frac{1}{2}$. Thus, the amplitude is $|A| = \frac{1}{2}$. This means that the maximum value of y is $\frac{1}{2}$ and the minimum value of y is $-\frac{1}{2}$. The period for both $y = \frac{1}{2} \sin x$ and $y = \sin x$ is 2π.

Step 2 Find the values of x for the five key points. We need to find the three x-intercepts, the maximum point, and the minimum point on the interval $[0, 2\pi]$. To do so, we begin by dividing the period, 2π, by 4.

$$\frac{\text{period}}{4} = \frac{2\pi}{4} = \frac{\pi}{2}$$

We start with the value of x where the cycle begins: $x = 0$. Now we add quarter-periods, $\frac{\pi}{2}$, to generate x-values for each of the key points. The five x-values are

$$x = 0, \quad x = 0 + \frac{\pi}{2} = \frac{\pi}{2}, \quad x = \frac{\pi}{2} + \frac{\pi}{2} = \pi,$$

$$x = \pi + \frac{\pi}{2} = \frac{3\pi}{2}, \quad x = \frac{3\pi}{2} + \frac{\pi}{2} = 2\pi.$$

Step 3 Find the values of y for the five key points. We evaluate the function at each value of x from step 2.

Value of x	Value of y: $y = \frac{1}{2}\sin x$	Coordinates of key point	
0	$y = \frac{1}{2}\sin 0 = \frac{1}{2} \cdot 0 = 0$	$(0, 0)$	
$\frac{\pi}{2}$	$y = \frac{1}{2}\sin \frac{\pi}{2} = \frac{1}{2} \cdot 1 = \frac{1}{2}$	$\left(\frac{\pi}{2}, \frac{1}{2}\right)$	maximum point
π	$y = \frac{1}{2}\sin \pi = \frac{1}{2} \cdot 0 = 0$	$(\pi, 0)$	
$\frac{3\pi}{2}$	$y = \frac{1}{2}\sin \frac{3\pi}{2} = \frac{1}{2}(-1) = -\frac{1}{2}$	$\left(\frac{3\pi}{2}, -\frac{1}{2}\right)$	minimum point
2π	$y = \frac{1}{2}\sin 2\pi = \frac{1}{2} \cdot 0 = 0$	$(2\pi, 0)$	

There are x-intercepts at 0, π, and 2π. The maximum and minimum points are indicated by the voice balloons.

Step 4 Connect the five key points with a smooth curve and graph one complete cycle of the given function. The five key points for $y = \frac{1}{2}\sin x$ are shown in Figure 5.47. By connecting the points with a smooth curve, the figure shows one complete cycle of $y = \frac{1}{2}\sin x$. Also shown is the graph of $y = \sin x$. The graph of $y = \frac{1}{2}\sin x$ shrinks the graph of $y = \sin x$.

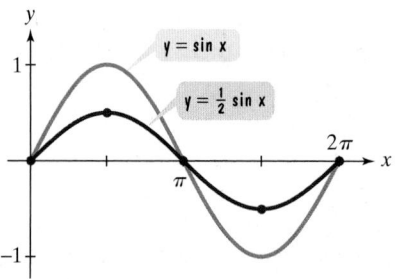

Figure 5.47 The graphs of $y = \sin x$ and $y = \frac{1}{2}\sin x$, $0 \le x \le 2\pi$

> **Check Point 1** Determine the amplitude of $y = 3 \sin x$. Then graph $y = \sin x$ and $y = 3 \sin x$ for $0 \le x \le 2\pi$.

EXAMPLE 2 Graphing a Variation of $y = \sin x$

Determine the amplitude of $y = -2 \sin x$. Then graph $y = \sin x$ and $y = -2 \sin x$ for $-\pi \le x \le 3\pi$.

Solution

Step 1 Identify the amplitude and the period. The equation $y = -2 \sin x$ is of the form $y = A \sin x$ with $A = -2$. Thus, the amplitude is $|A| = |-2| = 2$. This means that the maximum value of y is 2 and the minimum value of y is -2. Both $y = \sin x$ and $y = -2 \sin x$ have a period of 2π.

Step 2 Find the x-values for the five key points. Begin by dividing the period, 2π, by 4.

$$\frac{\text{period}}{4} = \frac{2\pi}{4} = \frac{\pi}{2}$$

Start with the value of x where the cycle begins: $x = 0$. Adding quarter-periods, $\frac{\pi}{2}$, the five x-values for the key points are

$$x = 0, \quad x = 0 + \frac{\pi}{2} = \frac{\pi}{2}, \quad x = \frac{\pi}{2} + \frac{\pi}{2} = \pi,$$

$$x = \pi + \frac{\pi}{2} = \frac{3\pi}{2}, \quad x = \frac{3\pi}{2} + \frac{\pi}{2} = 2\pi.$$

Step 3 Find the values of y for the five key points. We evaluate the function at each value of x from step 2.

Value of x	Value of y: $y = -2 \sin x$	Coordinates of key point	
0	$y = -2 \sin 0 = -2 \cdot 0 = 0$	$(0, 0)$	
$\dfrac{\pi}{2}$	$y = -2 \sin \dfrac{\pi}{2} = -2 \cdot 1 = -2$	$\left(\dfrac{\pi}{2}, -2\right)$	minimum point
π	$y = -2 \sin \pi = -2 \cdot 0 = 0$	$(\pi, 0)$	
$\dfrac{3\pi}{2}$	$y = -2 \sin \dfrac{3\pi}{2} = -2(-1) = 2$	$\left(\dfrac{3\pi}{2}, 2\right)$	maximum point
2π	$y = -2 \sin 2\pi = -2 \cdot 0 = 0$	$(2\pi, 0)$	

There are x-intercepts at 0, π, and 2π. The minimum and maximum points are indicated by the voice balloons.

Step 4 Connect the five key points with a smooth curve and graph one complete cycle of the given function. The five key points for $y = -2 \sin x$ are shown in Figure 5.48. By connecting the points with a smooth curve, the dark red portion shows one complete cycle of $y = -2 \sin x$. Also shown in dark blue is one complete cycle of the graph of $y = \sin x$.

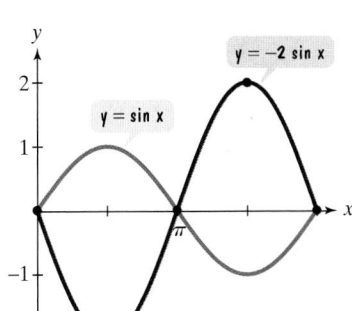

Figure 5.48 The graphs of $y = \sin x$ and $y = -2 \sin x$, $0 \le x \le 2\pi$

Step 5   Extend the graph in step 4 to the left or right as desired.   The dark red and dark blue portions of the graphs in Figure 5.48 are from 0 to 2π. In order to graph for $-\pi \leq x \leq 3\pi$, continue the pattern of each graph to the left and to the right. These extensions are shown by the lighter colors in Figure 5.48.

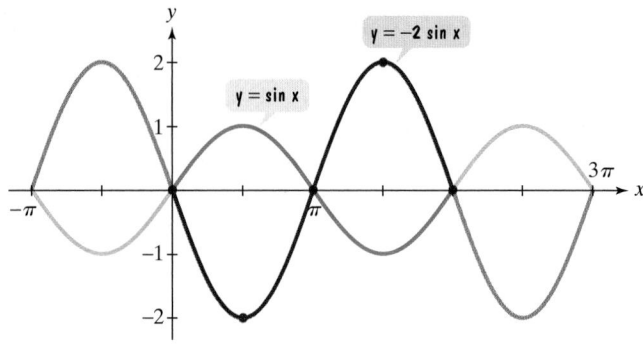

Figure 5.48, extended
The graphs of $y = \sin x$ and
$y = -2 \sin x, -\pi \leq x \leq 3\pi$

Check Point 2   Determine the amplitude of $y = -\frac{1}{2} \sin x$. Then graph $y = \sin x$ and $y = -\frac{1}{2} \sin x$ for $-\pi \leq x \leq 3\pi$.

Now let us examine the graphs of functions of the form $y = A \sin Bx$, where B is the coefficient of x. How do such graphs compare to those of functions of the form $y = A \sin x$? We know that $y = A \sin x$ completes one cycle from $x = 0$ to $x = 2\pi$. Thus, $y = A \sin Bx$ completes one cycle from $Bx = 0$ to $Bx = 2\pi$. Solve each of these equations for x.

$$Bx = 0 \qquad\qquad Bx = 2\pi$$

$$x = 0 \qquad\qquad x = \frac{2\pi}{B} \qquad \text{Divide both sides of each equation by B.}$$

This means that $y = A \sin Bx$ completes one cycle from 0 to $\frac{2\pi}{B}$. The period is $\frac{2\pi}{B}$.

Amplitudes and Periods

The graph of $y = A \sin Bx$ has

amplitude $= |A|$

period $= \dfrac{2\pi}{B}$.

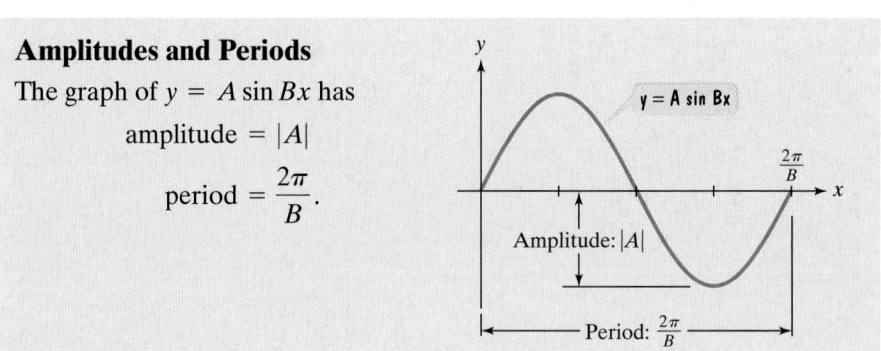

EXAMPLE 3   Graphing a Function of the Form $y = A \sin Bx$

Determine the amplitude and period of $y = 3 \sin 2x$. Then graph the function for $0 \leq x \leq 2\pi$.

Solution

Step 1   Identify the amplitude and the period.   The equation $y = 3 \sin 2x$ is of the form $y = A \sin Bx$ with $A = 3$ and $B = 2$.

$$\text{amplitude:} \quad |A| = |3| = 3$$

$$\text{period:} \quad \frac{2\pi}{B} = \frac{2\pi}{2} = \pi$$

The amplitude, 3, tells us that the maximum value of y is 3 and the minimum value of y is -3.

Step 2   Find the x-values for the five key points.   Begin by dividing the period, π, by 4.

$$\frac{\text{period}}{4} = \frac{\pi}{4}$$

Start with the value of x where the cycle begins: $x = 0$. Adding quarter-periods, $\frac{\pi}{4}$, the five x-values for the key points are

$$x = 0, \quad x = 0 + \frac{\pi}{4} = \frac{\pi}{4}, \quad x = \frac{\pi}{4} + \frac{\pi}{4} = \frac{\pi}{2},$$

$$x = \frac{\pi}{2} + \frac{\pi}{4} = \frac{3\pi}{4}, \quad x = \frac{3\pi}{4} + \frac{\pi}{4} = \pi.$$

Step 3   Find the values of y for the five key points.   We evaluate the function at each value of x from step 2.

Value of x	Value of y: $y = 3 \sin 2x$	Coordinates of key point	
0	$y = 3 \sin 2 \cdot 0$ $= 3 \sin 0 = 3 \cdot 0 = 0$	$(0, 0)$	
$\dfrac{\pi}{4}$	$y = 3 \sin 2 \cdot \dfrac{\pi}{4}$ $= 3 \sin \dfrac{\pi}{2} = 3 \cdot 1 = 3$	$\left(\dfrac{\pi}{4}, 3\right)$	maximum point
$\dfrac{\pi}{2}$	$y = 3 \sin 2 \cdot \dfrac{\pi}{2}$ $= 3 \sin \pi = 3 \cdot 0 = 0$	$\left(\dfrac{\pi}{2}, 0\right)$	
$\dfrac{3\pi}{4}$	$y = 3 \sin 2 \cdot \dfrac{3\pi}{4}$ $= 3 \sin \dfrac{3\pi}{2} = 3(-1) = -3$	$\left(\dfrac{3\pi}{4}, -3\right)$	minimum point
π	$y = 3 \sin 2 \cdot \pi$ $= 3 \sin 2\pi = 3 \cdot 0 = 0$	$(\pi, 0)$	

In the interval $[0, \pi]$, there are x-intercepts at $0, \dfrac{\pi}{2}$, and π. The maximum and minimum points are indicated by the voice balloons.

Step 4   Connect the five key points with a smooth curve and graph one complete cycle of the given function.   The five key points for $y = 3 \sin 2x$ are shown in Figure 5.49. By connecting the points with a smooth curve, the blue portion shows one complete cycle of $y = 3 \sin 2x$ from 0 to π.

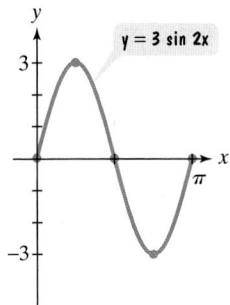

Figure 5.49 The graph of $y = 3 \sin 2x, 0 \le x \le \pi$

Technology

The graph of $y = 3 \sin 2x$ in a $\left[0, 2\pi, \dfrac{\pi}{2}\right]$ by $[-4, 4, 1]$ viewing rectangle verifies our hand-drawn graph in Figure 5.49.

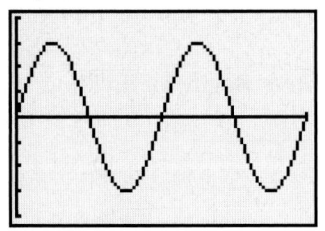

Step 5 Extend the graph in step 4 to the left or right as desired. The blue portion of the graph in Figure 5.49 is from 0 to π. In order to graph for $0 \le x \le 2\pi$, we continue this portion and extend the graph another full period to the right. This extension is shown in black in Figure 5.49.

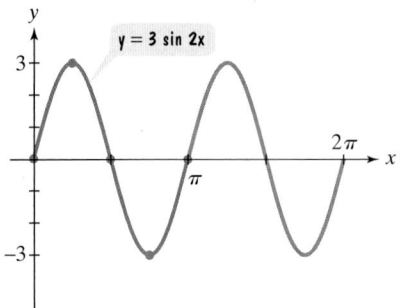

Figure 5.49, extended

> **Check Point 3** Determine the amplitude and period of $y = 2 \sin \frac{1}{2} x$. Then graph the function for $0 \le x \le 8\pi$.

Now let us examine the graphs of functions of the form $y = A \sin(Bx - C)$. How do such graphs compare to those of functions of the form $y = A \sin Bx$?

In both cases, the amplitude is $|A|$ and the period is $\dfrac{2\pi}{B}$. One complete cycle occurs if $Bx - C$ increases from 0 to 2π. This means that we can find an interval containing one cycle by solving the equations

$$Bx - C = 0 \quad \text{and} \quad Bx - C = 2\pi.$$

$$Bx = C \qquad\qquad Bx = C + 2\pi \qquad \text{Add } C \text{ to both sides in each equation.}$$

$$x = \frac{C}{B} \qquad\qquad x = \frac{C}{B} + \frac{2\pi}{B}. \qquad \text{Divide both sides by } B \text{ in each equation.}$$

This is the x-coordinate on the left where the cycle begins.

This is the x-coordinate on the right where the cycle ends. $\dfrac{2\pi}{B}$ is the period.

The voice balloon on the left indicates that $y = A \sin(Bx - C)$ shifts the graph of $y = A \sin Bx$ horizontally by $\dfrac{C}{B}$. Thus, the number $\dfrac{C}{B}$ is the **phase shift** associated with the graph.

The Graph of $y = A \sin(Bx - C)$

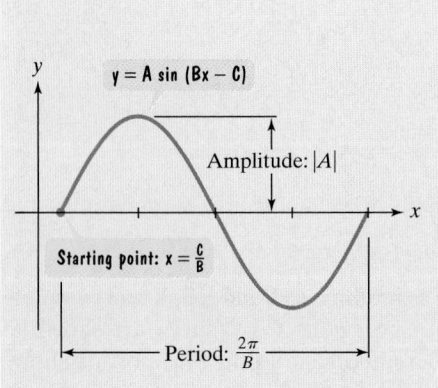

The graph of $y = A \sin(Bx - C)$ is obtained by horizontally shifting the graph of $y = A \sin Bx$ so that the starting point of the cycle is shifted from $x = 0$ to $x = \dfrac{C}{B}$. If $\dfrac{C}{B} > 0$, the shift is to the right. If $\dfrac{C}{B} < 0$, the shift is to the left. The number $\dfrac{C}{B}$ is called the **phase shift.**

$$\text{amplitude} = |A|$$

$$\text{period} = \frac{2\pi}{B}$$

EXAMPLE 4 **Graphing a Function of the Form $y = A \sin(Bx - C)$**

Determine the amplitude, period, and phase shift of $y = 4 \sin\left(2x - \dfrac{2\pi}{3}\right)$. Then graph one period of the function.

Solution

Step 1 **Identify the amplitude, the period, and the phase shift.** We must first identify values for A, B, and C.

> This equation is of the form
> $y = A \sin(Bx - C)$.

$$y = 4 \sin\left(2x - \dfrac{2\pi}{3}\right)$$

Using the voice balloon, we see that $A = 4$, $B = 2$, and $C = \dfrac{2\pi}{3}$.

amplitude: $\quad |A| = |4| = 4 \qquad$ The maximum y is 4 and the minimum is -4.

period: $\quad \dfrac{2\pi}{B} = \dfrac{2\pi}{2} = \pi \qquad$ Each cycle is completed in π radians.

phase shift: $\quad \dfrac{C}{B} = \dfrac{\frac{2\pi}{3}}{2} = \dfrac{2\pi}{3} \cdot \dfrac{1}{2} = \dfrac{\pi}{3} \qquad$ A cycle starts at $x = \frac{\pi}{3}$.

Step 2 **Find the x-values for the five key points.** Begin by dividing the period, π, by 4.

$$\dfrac{\text{period}}{4} = \dfrac{\pi}{4}$$

Study Tip

You can speed up the additions on the right by first writing the starting point, $\frac{\pi}{3}$, and the quarter-period, $\frac{\pi}{4}$, with a common denominator, 12.

starting point
$$= \dfrac{\pi}{3} = \dfrac{4\pi}{12}$$

quarter-period
$$= \dfrac{\pi}{4} = \dfrac{3\pi}{12}$$

Start with the value of x where the cycle begins: $x = \dfrac{\pi}{3}$. Adding quarter-periods, $\dfrac{\pi}{4}$, the five x-values for the key points are

$$x = \dfrac{\pi}{3}, \quad x = \dfrac{\pi}{3} + \dfrac{\pi}{4} = \dfrac{4\pi}{12} + \dfrac{3\pi}{12} = \dfrac{7\pi}{12},$$

$$x = \dfrac{7\pi}{12} + \dfrac{\pi}{4} = \dfrac{7\pi}{12} + \dfrac{3\pi}{12} = \dfrac{10\pi}{12} = \dfrac{5\pi}{6},$$

$$x = \dfrac{5\pi}{6} + \dfrac{\pi}{4} = \dfrac{10\pi}{12} + \dfrac{3\pi}{12} = \dfrac{13\pi}{12},$$

$$x = \dfrac{13\pi}{12} + \dfrac{\pi}{4} = \dfrac{13\pi}{12} + \dfrac{3\pi}{12} = \dfrac{16\pi}{12} = \dfrac{4\pi}{3}.$$

Step 3 **Find the values of y for the five key points.** We evaluate the function at each value of x from step 2.

Value of x	Value of y: $y = 4 \sin\left(2x - \dfrac{2\pi}{3}\right)$	Coordinates of key point	
$\dfrac{\pi}{3}$	$y = 4 \sin\left(2 \cdot \dfrac{\pi}{3} - \dfrac{2\pi}{3}\right)$ $= 4\sin 0 = 4 \cdot 0 = 0$	$\left(\dfrac{\pi}{3}, 0\right)$	
$\dfrac{7\pi}{12}$	$y = 4 \sin\left(2 \cdot \dfrac{7\pi}{12} - \dfrac{2\pi}{3}\right)$ $= 4 \sin\left(\dfrac{7\pi}{6} - \dfrac{2\pi}{3}\right)$ $= 4 \sin \dfrac{3\pi}{6} = 4 \sin \dfrac{\pi}{2} = 4 \cdot 1 = 4$	$\left(\dfrac{7\pi}{12}, 4\right)$	maximum point
$\dfrac{5\pi}{6}$	$y = 4 \sin\left(2 \cdot \dfrac{5\pi}{6} - \dfrac{2\pi}{3}\right)$ $= 4 \sin\left(\dfrac{5\pi}{3} - \dfrac{2\pi}{3}\right)$ $= 4 \sin \dfrac{3\pi}{3} = 4 \sin \pi = 4 \cdot 0 = 0$	$\left(\dfrac{5\pi}{6}, 0\right)$	
$\dfrac{13\pi}{12}$	$y = 4 \sin\left(2 \cdot \dfrac{13\pi}{12} - \dfrac{2\pi}{3}\right)$ $= 4 \sin\left(\dfrac{13\pi}{6} - \dfrac{4\pi}{6}\right)$ $= 4 \sin \dfrac{9\pi}{6} = 4 \sin \dfrac{3\pi}{2} = 4(-1) = -4$	$\left(\dfrac{13\pi}{12}, -4\right)$	minimum point
$\dfrac{4\pi}{3}$	$y = 4 \sin\left(2 \cdot \dfrac{4\pi}{3} - \dfrac{2\pi}{3}\right)$ $= 4 \sin \dfrac{6\pi}{3} = 4 \sin 2\pi = 4 \cdot 0 = 0$	$\left(\dfrac{4\pi}{3}, 0\right)$	

In the interval $\left[\dfrac{\pi}{3}, \dfrac{4\pi}{3}\right]$, there are x-intercepts at $\dfrac{\pi}{3}, \dfrac{5\pi}{6}$, and $\dfrac{4\pi}{3}$. The maximum and minimum points are indicated by the voice balloons.

Step 4 Connect the five key points with a smooth curve and graph one complete cycle of the given function. The key points and the graph of $y = 4 \sin\left(2x - \dfrac{2\pi}{3}\right)$ are shown in Figure 5.50.

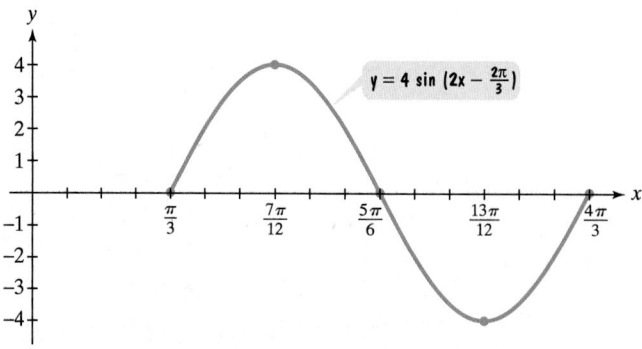

Figure 5.50

Check Point 4   Determine the amplitude, period, and phase shift of

$$y = 3 \sin\left(2x - \frac{\pi}{3}\right).$$ Then graph one period of the function.

3 Understand the graph of $y = \cos x$.

The Graph of $y = \cos x$

We graph $y = \cos x$ by listing some points on the graph. Because the period of the cosine function is 2π, we will concentrate on the graph of the basic cosine curve on the interval $[0, 2\pi]$. The rest of the graph is made up of repetitions of this portion. Table 5.2 lists some values of (x, y) on the graph of $y = \cos x$.

Table 5.2   Values of (x, y) on $y = \cos x$

x	0	$\dfrac{\pi}{6}$	$\dfrac{\pi}{3}$	$\dfrac{\pi}{2}$	$\dfrac{2\pi}{3}$	$\dfrac{5\pi}{6}$	π	$\dfrac{7\pi}{6}$	$\dfrac{4\pi}{3}$	$\dfrac{3\pi}{2}$	$\dfrac{5\pi}{3}$	$\dfrac{11\pi}{6}$	2π
$y = \cos x$	1	$\dfrac{\sqrt{3}}{2}$	$\dfrac{1}{2}$	0	$-\dfrac{1}{2}$	$-\dfrac{\sqrt{3}}{2}$	-1	$-\dfrac{\sqrt{3}}{2}$	$-\dfrac{1}{2}$	0	$\dfrac{1}{2}$	$\dfrac{\sqrt{3}}{2}$	1

As x increases from 0 to $\frac{\pi}{2}$, y decreases from 1 to 0.

As x increases from $\frac{\pi}{2}$ to π, y decreases from 0 to -1.

As x increases from π to $\frac{3\pi}{2}$, y increases from -1 to 0.

As x increases from $\frac{3\pi}{2}$ to 2π, y increases from 0 to 1.

Plotting the points in Table 5.2 and connecting them with a smooth curve, we obtain the graph shown in Figure 5.51. The portion of the graph in dark blue shows one complete period. We can obtain a more complete graph of $y = \cos x$ by extending this dark blue portion to the left and to the right.

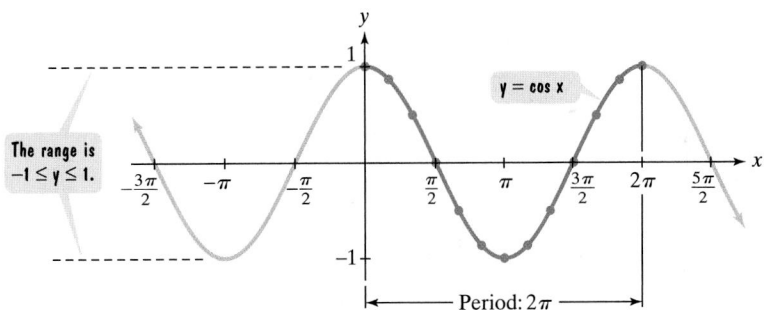

The range is $-1 \le y \le 1$.

Period: 2π

Figure 5.51 The graph of $y = \cos x$

The graph of $y = \cos x$ allows us to visualize some of the properties of the cosine function.

* The domain is the set of all real numbers. The graph extends indefinitely to the left and to the right with no gaps or holes.
* The range consists of all numbers between -1 and 1, inclusive. The graph never rises above 1 or falls below -1.
* The period is 2π. The graph's pattern repeats in every interval of length 2π.
* The function is an even function: $\cos(-x) = \cos x$. This can be seen by observing that the graph is symmetric with respect to the y-axis.

Turn back the page and take a second look at Figure 5.51. Can you see that the graph of $y = \cos x$ is the graph of $y = \sin x$ with a phase shift of $-\frac{\pi}{2}$ radians? If you trace along the curve from $x = -\frac{\pi}{2}$ to $x = \frac{3\pi}{2}$, you are tracing one complete cycle of the sine curve. This can be expressed as an identity:

$$\cos x = \sin\left(x + \frac{\pi}{2}\right).$$

Because of this similarity, the graphs of sine functions and cosine functions are called **sinusoidal graphs.**

4 Graph variations of $y = \cos x$.

Graphing Variations of $y = \cos x$

We use the same steps to graph variations of $y = \cos x$ as we did for graphing variations of $y = \sin x$. We will continue finding key points by dividing the period into four equal parts. Amplitudes, periods, and phase shifts play an important role when graphing by hand.

The Graph of $y = A \cos Bx$

The graph of $y = A \cos Bx$ has

$$\text{amplitude} = |A|$$

$$\text{period} = \frac{2\pi}{B}.$$

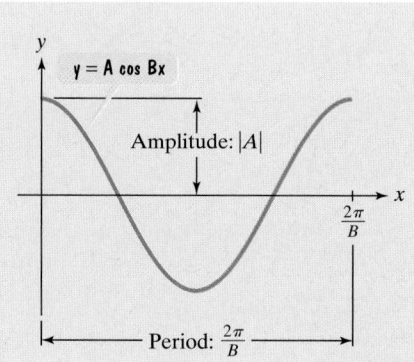

EXAMPLE 5 Graphing a Function of the Form $y = A \cos Bx$

Determine the amplitude and period of $y = -3 \cos \frac{\pi}{2} x$. Then graph the function for $-4 \le x \le 4$.

Solution

Step 1 Identify the amplitude and the period. The equation $y = -3 \cos \frac{\pi}{2} x$ is of the form $y = A \cos Bx$ with $A = -3$ and $B = \frac{\pi}{2}$.

amplitude: $|A| = |-3| = 3$ The maximum y is 3 and the minimum is −3.

period: $\dfrac{2\pi}{B} = \dfrac{2\pi}{\dfrac{\pi}{2}} = 2\pi \cdot \dfrac{2}{\pi} = 4$ Each cycle is completed in 4 radians.

Step 2 Find the x-values for the five key points. Begin by dividing the period, 4, by 4.

$$\frac{\text{period}}{4} = \frac{4}{4} = 1$$

Start with the value of x where the cycle begins: $x = 0$. Adding quarter-periods, 1, the five x-values for the key points are

$$x = 0, \quad x = 0 + 1 = 1, \quad x = 1 + 1 = 2, \quad x = 2 + 1 = 3, \quad x = 3 + 1 = 4$$

Step 3 Find the values of y for the five key points. We evaluate the function at each value of x from step 2.

Value of x	Value of y: $y = -3\cos\dfrac{\pi}{2}x$	Coordinates of key point	
0	$y = -3\cos\dfrac{\pi}{2}\cdot 0$ $= -3\cos 0 = -3\cdot 1 = -3$	$(0, -3)$	minimum point
1	$y = -3\cos\dfrac{\pi}{2}\cdot 1$ $= -3\cos\dfrac{\pi}{2} = -3\cdot 0 = 0$	$(1, 0)$	
2	$y = -3\cos\dfrac{\pi}{2}\cdot 2$ $= -3\cos\pi = -3(-1) = 3$	$(2, 3)$	maximum point
3	$y = -3\cos\dfrac{\pi}{2}\cdot 3$ $= -3\cos\dfrac{3\pi}{2} = -3(0) = 0$	$(3, 0)$	
4	$y = -3\cos\dfrac{\pi}{2}\cdot 4$ $= -3\cos 2\pi = -3(1) = -3$	$(4, -3)$	minimum point

In the interval $[0, 4]$, there are x-intercepts at 1 and 3. The minimum and maximum points are indicated by the voice balloons.

Step 4 Connect the five key points with a smooth curve and graph one complete cycle of the given function. The five key points for $y = -3\cos\dfrac{\pi}{2}x$ are shown in Figure 5.52. By connecting the points with a smooth curve, the blue portion shows one complete cycle of $y = -3\cos\dfrac{\pi}{2}x$ from 0 to 4.

Technology

The graph of $y = -3\cos\dfrac{\pi}{2}x$ in a $[-4, 4, 1]$ by $[-4, 4, 1]$ viewing rectangle verifies our hand-drawn graph in Figure 5.52.

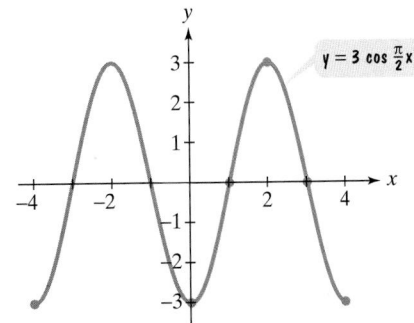

$y = 3\cos\frac{\pi}{2}x$

Figure 5.52

Step 5 Extend the graph in step 4 to the left or right as desired. The blue portion of the graph in Figure 5.52 is for x from 0 to 4. In order to graph for $-4 \le x \le 4$, we continue this portion and extend the graph another full period to the left. This extension is shown in black in Figure 5.52.

Check Point 5   Determine the amplitude and period of $y = -4 \cos \pi x$. Then graph the function for $-2 \le x \le 2$.

Finally, let us examine the graphs of functions of the form $y = A \cos(Bx - C)$.

Graphs of these functions shift the graph of $y = A \cos Bx$ horizontally by $\dfrac{C}{B}$.

The Graph of $y = A \cos(Bx - C)$

The graph of $y = A \cos(Bx - C)$ is obtained by horizontally shifting the graph of $y = A \cos Bx$ so that the starting point of the cycle is shifted from $x = 0$ to $x = \dfrac{C}{B}$. If $\dfrac{C}{B} > 0$, the shift is to the right. If $\dfrac{C}{B} < 0$, the shift is to the left. The number $\dfrac{C}{B}$ is called the **phase shift.**

$$\text{amplitude} = |A|$$

$$\text{period} = \frac{2\pi}{B}$$

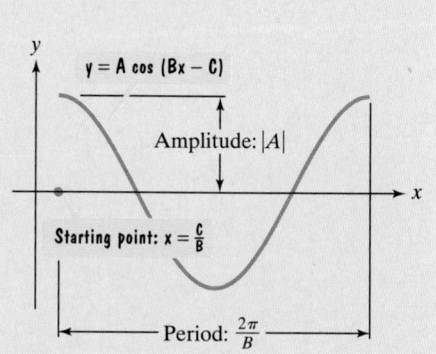

EXAMPLE 6   Graphing a Function of the Form $y = A \cos(Bx - C)$

Determine the amplitude, period, and phase shift of $y = \frac{1}{2}\cos(4x + \pi)$. Then graph one period of the function.

Solution

Step 1   Identify the amplitude, the period, and the phase shift. We must first identify values for A, B, and C. To do this, we need to express the equation in the form $y = A \cos(Bx - C)$. Thus, we write $y = \frac{1}{2}\cos(4x + \pi)$ as $y = \frac{1}{2}\cos[4x - (-\pi)]$. Now we can identify values for A, B, and C.

> This equation is of the form $y = A \cos(Bx - C)$.

$$y = \frac{1}{2}\cos[4x - (-\pi)]$$

Using the voice balloon, we see that $A = \frac{1}{2}$, $B = 4$, and $C = -\pi$.

amplitude: $|A| = \left|\dfrac{1}{2}\right| = \dfrac{1}{2}$ The maximum y is $\frac{1}{2}$ and the minimum is $-\frac{1}{2}$.

period: $\dfrac{2\pi}{B} = \dfrac{2\pi}{4} = \dfrac{\pi}{2}$ Each cycle is completed in $\frac{\pi}{2}$ radians.

phase shift: $\dfrac{C}{B} = -\dfrac{\pi}{4}$ A cycle starts at $x = -\frac{\pi}{4}$.

Step 2 Find the x-values for the five key points. Begin by dividing the period, $\dfrac{\pi}{2}$, by 4.

$$\frac{\text{period}}{4} = \frac{\dfrac{\pi}{2}}{4} = \frac{\pi}{8}$$

Start with the value of x where the cycle begins: $x = -\dfrac{\pi}{4}$. Adding quarter-periods, $\dfrac{\pi}{8}$, the five x-values for the key points are

$$x = -\frac{\pi}{4}, \quad x = -\frac{\pi}{4} + \frac{\pi}{8} = -\frac{2\pi}{8} + \frac{\pi}{8} = -\frac{\pi}{8}, \quad x = -\frac{\pi}{8} + \frac{\pi}{8} = 0,$$

$$x = 0 + \frac{\pi}{8} = \frac{\pi}{8}, \quad x = \frac{\pi}{8} + \frac{\pi}{8} = \frac{2\pi}{8} = \frac{\pi}{4}.$$

Step 3 Find the values of y for the five key points. Take a few minutes and use your calculator to evaluate the function at each value of x from step 2. Show that the key points are

$$\left(-\frac{\pi}{4}, \frac{1}{2}\right), \quad \left(-\frac{\pi}{8}, 0\right), \quad \left(0, -\frac{1}{2}\right), \quad \left(\frac{\pi}{8}, 0\right), \quad \text{and} \quad \left(\frac{\pi}{4}, \frac{1}{2}\right).$$

| maximum point | x-intercept at $-\frac{\pi}{8}$ | minimum point | x-intercept at $\frac{\pi}{8}$ | maximum point |

The graph of

$$y = \frac{1}{2}\cos(4x + \pi)$$

in a $\left[-\dfrac{\pi}{4}, \dfrac{\pi}{4}, \dfrac{\pi}{8}\right]$ by $[-1, 1, 1]$ viewing rectangle verifies our hand-drawn graph in Figure 5.53.

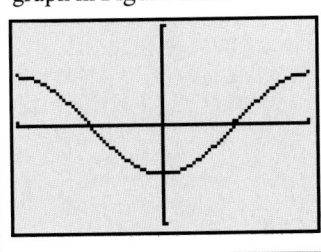

Step 4 Connect the five key points with a smooth curve and graph one complete cycle of the given function. The key points and the graph of $y = \frac{1}{2}\cos(4x + \pi)$ are shown in Figure 5.53.

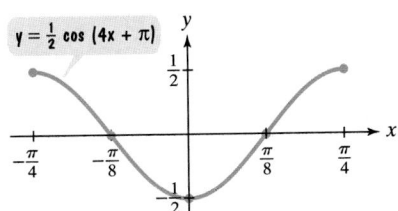

$y = \frac{1}{2} \cos(4x + \pi)$

Figure 5.53

Check Point 6 Determine the amplitude, period, and phase shift of $y = \frac{3}{2}\cos(2x + \pi)$. Then graph one period of the function.

5 Use vertical shifts of sine and cosine curves.

Vertical Shifts of Sinusoidal Graphs

We now look at sinusoidal graphs of

$$y = A\sin(Bx - C) + D \quad \text{and} \quad y = A\cos(Bx - C) + D.$$

The constant D causes vertical shifts in the graphs of $y = A\sin(Bx - C)$ and $y = A\cos(Bx - C)$. If D is positive, the shift is D units upward. If D is negative, the shift is D units downward. These vertical shifts result in sinusoidal graphs oscillating about the horizontal line $y = D$ rather than about the x-axis. Thus, the maximum y is $D + |A|$ and the minimum y is $D - |A|$.

EXAMPLE 7   A Vertical Shift

Graph one period of the function $y = \frac{1}{2}\cos x - 1$.

Solution   The graph of $y = \frac{1}{2}\cos x - 1$ is the graph of $y = \frac{1}{2}\cos x$ shifted one unit downward. The period of $y = \frac{1}{2}\cos x$ is 2π, which is also the period for the vertically shifted graph. The key points on the interval $[0, 2\pi]$ for $y = \frac{1}{2}\cos x - 1$ are found by first determining their x-coordinates. The quarter-period is $\dfrac{2\pi}{4}$, or $\dfrac{\pi}{2}$.

The cycle begins at $x = 0$. As always, we add quarter-periods to generate x-values for each of the key points. The five x-values are

$$x = 0, \quad x = 0 + \frac{\pi}{2} = \frac{\pi}{2}, \quad x = \frac{\pi}{2} + \frac{\pi}{2} = \pi,$$

$$x = \pi + \frac{\pi}{2} = \frac{3\pi}{2}, \quad x = \frac{3\pi}{2} + \frac{\pi}{2} = 2\pi.$$

The values of y for the five key points and their coordinates are determined as follows.

Value of x	Value of y: $y = \dfrac{1}{2}\cos x - 1$	Coordinates of key point
0	$y = \dfrac{1}{2}\cos 0 - 1$ $= \dfrac{1}{2} \cdot 1 - 1 = -\dfrac{1}{2}$	$\left(0, -\dfrac{1}{2}\right)$
$\dfrac{\pi}{2}$	$y = \dfrac{1}{2}\cos\dfrac{\pi}{2} - 1$ $= \dfrac{1}{2} \cdot 0 - 1 = -1$	$\left(\dfrac{\pi}{2}, -1\right)$
π	$y = \dfrac{1}{2}\cos \pi - 1$ $= \dfrac{1}{2}(-1) - 1 = -\dfrac{3}{2}$	$\left(\pi, -\dfrac{3}{2}\right)$
$\dfrac{3\pi}{2}$	$y = \dfrac{1}{2}\cos\dfrac{3\pi}{2} - 1$ $= \dfrac{1}{2} \cdot 0 - 1 = -1$	$\left(\dfrac{3\pi}{2}, -1\right)$
2π	$y = \dfrac{1}{2}\cos 2\pi - 1$ $= \dfrac{1}{2} \cdot 1 - 1 = -\dfrac{1}{2}$	$\left(2\pi, -\dfrac{1}{2}\right)$

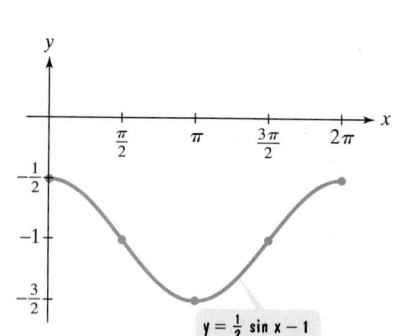

$y = \frac{1}{2} \sin x - 1$

Figure 5.54

The five key points for $y = \frac{1}{2}\cos x - 1$ are shown in Figure 5.54. By connecting the points with a smooth curve, we obtain one period of the graph.

Check Point 7   Graph one period of the function $y = 2\cos x + 1$.

6 Model periodic behavior.

Modeling Periodic Behavior

Our breathing consists of alternating periods of inhaling and exhaling. Each complete pumping cycle of the human heart can be described using a sine function. Our brain waves during deep sleep are sinusoidal. Viewed in this way, trigonometry becomes an intimate experience.

Some graphing utilities have a SINe REGression feature. This feature gives the sine function in the form $y = A \sin(Bx + C) + D$ of best fit for wavelike data. At least four data points must be used. However, it is not always necessary to use technology. In our next example, we use our understanding of sinusoidal graphs to model the process of breathing.

EXAMPLE 8 A Trigonometric Breath of Life

The graph in Figure 5.55 shows one complete normal breathing cycle. The cycle consists of inhaling and exhaling. It takes place every 5 seconds. Velocity of air flow is positive when we inhale and negative when we exhale. It is measured in liters per second. If y represents velocity of air flow after x seconds, find a function of the form $y = A \sin Bx$ that models air flow in a normal breathing cycle.

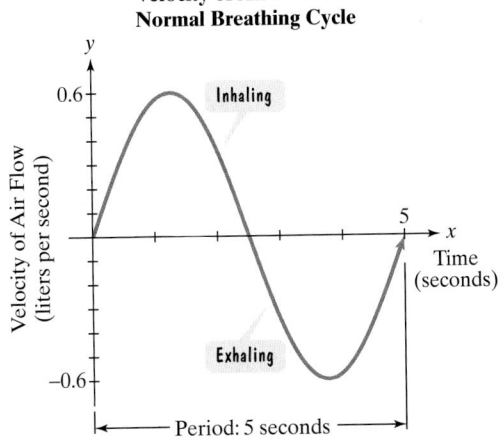

Figure 5.55

Solution We need to determine values for A and B in the equation $y = A \sin Bx$. A, the amplitude, is the maximum value of y. Figure 5.55 shows that this maximum value is 0.6. Thus, $A = 0.6$.

The value of B in $y = A \sin Bx$ can be found using the formula for the period: $\text{period} = \dfrac{2\pi}{B}$. The period of our breathing cycle is 5 seconds. Thus,

$$5 = \frac{2\pi}{B} \qquad \textit{Our goal is to solve this equation for B.}$$

$$5B = 2\pi \qquad \textit{Multiply both sides of the equation by B.}$$

$$B = \frac{2\pi}{5}. \qquad \textit{Divide both sides of the equation by 5.}$$

We see that $A = 0.6$ and $B = \dfrac{2\pi}{5}$. Substitute these values into $y = A \sin Bx$. The breathing cycle is modeled by

$$y = 0.6 \sin \frac{2\pi}{5} x.$$

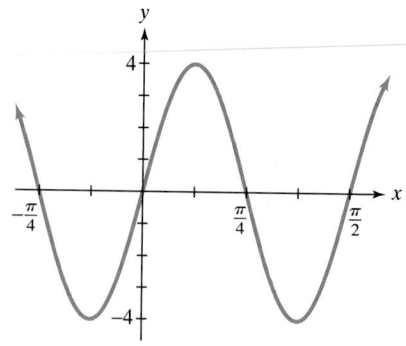

Check Point 8 Find an equation of the form $y = A \sin Bx$ that produces the graph shown in the figure on the right.

EXAMPLE 9 Modeling a Tidal Cycle

Figure 5.56 shows that the depth of water at a boat dock varies with the tides. The depth is 5 feet at low tide and 13 feet at high tide. On a certain day, low tide occurs at 4 A.M. and high tide at 10 A.M. If y represents the depth of the water, in feet, x hours after midnight, use a sine function of the form $y = A \sin(Bx - C) + D$ to model the water's depth.

Figure 5.56

Solution We need to determine values for $A, B, C,$ and D in the equation $y = A \sin(Bx - C) + D$. We can find these values using Figure 5.56. We begin with D.

To find D, we use the vertical shift. Because the water's depth ranges from a minimum of 5 feet to a maximum of 13 feet, the curve oscillates about the middle value, 9 feet. Thus, $D = 9$, which is the vertical shift.

At maximum depth, the water is 4 feet above 9 feet. Thus, A, the amplitude, is 4: $A = 4$.

To find B, we use the period. The blue portion of the graph shows that one complete tidal cycle occurs in $19 - 7$, or 12 hours. The period is 12. Thus,

$$12 = \frac{2\pi}{B} \qquad \text{\textit{Our goal is to solve this equation for B.}}$$

$$12B = 2\pi \qquad \text{\textit{Multiply both sides by B.}}$$

$$B = \frac{2\pi}{12} = \frac{\pi}{6}. \qquad \text{\textit{Divide both sides by 12.}}$$

To find C, we use the phase shift. The blue portion of the graph shows that the starting point of the cycle is shifted from 0 to 7. The phase shift, $\dfrac{C}{B}$, is 7.

$$7 = \frac{C}{B} \qquad \text{\textit{The phase shift of } } y = A\sin(Bx - C) \text{ \textit{is} } \frac{C}{B}.$$

$$7 = \frac{C}{\dfrac{\pi}{6}} \qquad \text{\textit{From above, we have } } B = \frac{\pi}{6}.$$

$$\frac{7\pi}{6} = C \qquad \text{\textit{Multiply both sides of the equation by } } \frac{\pi}{6}.$$

Technology

We can use a graphing utility to verify that the model in Example 9

$$y = 4 \sin\left(\frac{\pi}{6}x - \frac{7\pi}{6}\right) + 9$$

is correct. The graph of the function is shown in a $[0, 28, 4]$ by $[0, 15, 5]$ viewing rectangle.

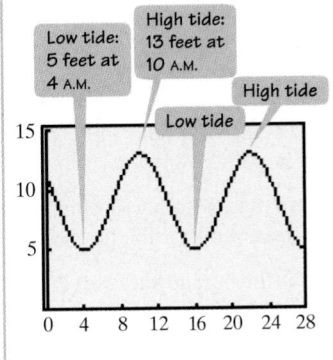

We see that $A = 4$, $B = \dfrac{\pi}{6}$, $C = \dfrac{7\pi}{6}$, and $D = 9$. Substitute these values into $y = A \sin(Bx - C) + D$. The water's depth, in feet, x hours after midnight is modeled by

$$y = 4 \sin\left(\frac{\pi}{6}x - \frac{7\pi}{6}\right) + 9.$$

Check Point 9

The figure shows the number of hours of daylight for a region that is 30° north of the equator. Hours of daylight are at a minimum of 10 hours in December. Hours of daylight are at a maximum of 14 hours in June. Let x represent the month of the year, with 1 for January, 2 for February, 3 for March, and 12 for December. If y represents the number of hours of daylight in month x, use a sine function of the form $y = A \sin(Bx - C) + D$ to model the hours of daylight.

EXERCISE SET 5.5

Practice Exercises

In Exercises 1–6, determine the amplitude of each function. Then graph the function and $y = \sin x$ in the same rectangular coordinate system for $0 \le x \le 2\pi$.

1. $y = 4 \sin x$

2. $y = 5 \sin x$

3. $y = \frac{1}{3} \sin x$

4. $y = \frac{1}{4} \sin x$

5. $y = -3 \sin x$

6. $y = -4 \sin x$

In Exercises 7–16, determine the amplitude and period of each function. Then graph one period of the function.

7. $y = \sin 2x$

8. $y = \sin 4x$

9. $y = 3 \sin \frac{1}{2} x$

10. $y = 2 \sin \frac{1}{4} x$

11. $y = 4 \sin \pi x$

12. $y = 3 \sin 2\pi x$

13. $y = -3 \sin 2\pi x$

14. $y = -2 \sin \pi x$

15. $y = -\sin \frac{2}{3} x$

16. $y = -\sin \frac{4}{3} x$

In Exercises 17–30, determine the amplitude, period, and phase shift of each function. Then graph one period of the function.

17. $y = \sin(x - \pi)$

18. $y = \sin\left(x - \frac{\pi}{2}\right)$

19. $y = \sin(2x - \pi)$

20. $y = \sin\left(2x - \frac{\pi}{2}\right)$

21. $y = 3 \sin(2x - \pi)$

22. $y = 3 \sin\left(2x - \dfrac{\pi}{2}\right)$

23. $y = \frac{1}{2} \sin\left(x + \dfrac{\pi}{2}\right)$

24. $y = \frac{1}{2} \sin(x + \pi)$

25. $y = -2 \sin\left(2x + \dfrac{\pi}{2}\right)$

26. $y = -3 \sin\left(2x + \dfrac{\pi}{2}\right)$

27. $y = 3 \sin(\pi x + 2)$

28. $y = 3 \sin(2\pi x + 4)$

29. $y = -2 \sin(2\pi x + 4\pi)$

30. $y = -3 \sin(2\pi x + 4\pi)$

In Exercises 31–34, determine the amplitude of each function. Then graph the function and $y = \cos x$ in the same rectangular coordinate system for $0 \le x \le 2\pi$.

31. $y = 2 \cos x$

32. $y = 3 \cos x$

33. $y = -2 \cos x$

34. $y = -3 \cos x$

In Exercises 35–42, determine the amplitude and period of each function. Then graph one period of the function.

35. $y = \cos 2x$

36. $y = \cos 4x$

37. $y = 4 \cos 2\pi x$

38. $y = 5 \cos 2\pi x$

39. $y = -4 \cos \frac{1}{2} x$

40. $y = -3 \cos \frac{1}{3} x$

41. $y = -\frac{1}{2} \cos \dfrac{\pi}{3} x$

42. $y = -\frac{1}{2} \cos \dfrac{\pi}{4} x$

In Exercises 43–50, determine the amplitude, period, and phase shift of each function. Then graph one period of the function.

43. $y = 3 \cos(2x - \pi)$

44. $y = 4 \cos(2x - \pi)$

45. $y = \frac{1}{2} \cos\left(3x + \dfrac{\pi}{2}\right)$

46. $y = \frac{1}{2} \cos(2x + \pi)$

47. $y = -3 \cos\left(2x - \dfrac{\pi}{2}\right)$

48. $y = -4 \cos\left(2x - \dfrac{\pi}{2}\right)$

49. $y = 2 \cos(2\pi x + 8\pi)$

50. $y = 3 \cos(2\pi x + 4\pi)$

In Exercises 51–58, use a vertical shift to graph one period of the function.

51. $y = \sin x + 2$

52. $y = \sin x - 2$

53. $y = \cos x - 3$

54. $y = \cos x + 3$

55. $y = 2 \sin \frac{1}{2} x + 1$

56. $y = 2 \cos \frac{1}{2} x + 1$

57. $y = -3 \cos 2\pi x + 2$

58. $y = -3 \sin 2\pi x + 2$

Application Exercises

In the theory of biorhythms, sine functions are used to measure a person's potential. You can obtain your biorhythm chart online by simply entering your date of birth, the date you want your biorhythm chart to begin, and the number of months you wish to be included in the plot. At the top of the next column is your author's chart, beginning January 25, 2003, when he was 21,093 days old. We all have cycles with the same amplitudes and periods as those shown here. Each of our three basic cycles begins at birth. Use the

biorhythm chart shown to solve Exercises 59–66. The longer tick marks correspond to the dates shown.

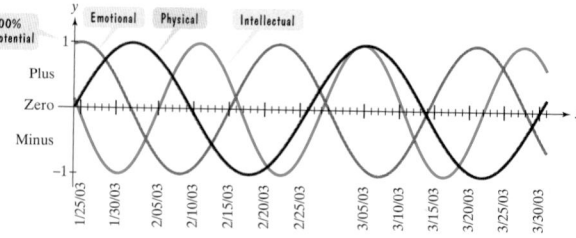

59. What is the period of the physical cycle?

60. What is the period of the emotional cycle?

61. What is the period of the intellectual cycle?

62. For the period shown, what is the worst day in February for your author to run in a marathon?

63. For the period shown, what is the best day in March for your author to meet an online friend for the first time?

64. For the period shown, what is the best day in February for your author to begin writing this trigonometry chapter?

65. If you extend these sinusoidal graphs to the end of the year, is there a day when your author should not even bother getting out of bed?

66. If you extend these sinusoidal graphs to the end of the year, are there any days where your author is at near-peak physical, emotional, and intellectual potential?

67. Rounded to the nearest hour, Los Angeles averages 14 hours of daylight in June, 10 hours in December, and 12 hours in March and September. Let x represent the number of months after June and y represent the number of hours of daylight in month x. Make a graph that displays the information from June of one year to June of the following year.

68. A clock with an hour hand that is 15 inches long is hanging on a wall. At noon, the distance between the tip of the hour hand and the ceiling is 23 inches. At 3 P.M., the distance is 38 inches; at 6 P.M., 53 inches; at 9 P.M., 38 inches; and at midnight the distance is again 23 inches. If y represents the distance between the tip of the hour hand and the ceiling x hours after noon, make a graph that displays the information for $0 \le x \le 24$.

69. The number of hours of daylight in Boston is given by

$$y = 3 \sin \frac{2\pi}{365} (x - 79) + 12$$

where x is the number of days after January 1.

a. What is the amplitude of this function?

b. What is the period of this function?

c. How many hours of daylight are there on the longest day of the year?

d. How many hours of daylight are there on the shortest day of the year?

e. Graph the function for one period, starting on January 1.

70. The average monthly temperature, y, in degrees Fahrenheit, for Juneau, Alaska, can be modeled by

$$y = 16 \sin\left(\frac{\pi}{6} x - \frac{2\pi}{3}\right) + 40,$$ where x is the month of

the year (January = 1, February = 2, ... December = 12). Graph the function for $1 \le x \le 12$. What is the highest average monthly temperature? In which month does this occur?

71. The figure shows the depth of water at the end of a boat dock. The depth is 6 feet at low tide and 12 feet at high tide. On a certain day, low tide occurs at 6 A.M. and high tide at noon. If y represents the depth of the water x hours after midnight, use a cosine function of the form $y = A \cos Bx + D$ to model the water's depth.

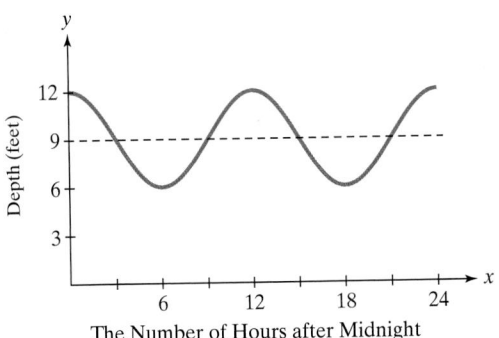

The Number of Hours after Midnight

72. The figure shows the depth of water at the end of a boat dock. The depth is 5 feet at high tide and 3 feet at low tide. On a certain day, high tide occurs at noon and low tide at 6 P.M. If y represents the depth of the water x hours after noon, use a cosine function of the form $y = A \cos Bx + D$ to model the water's depth.

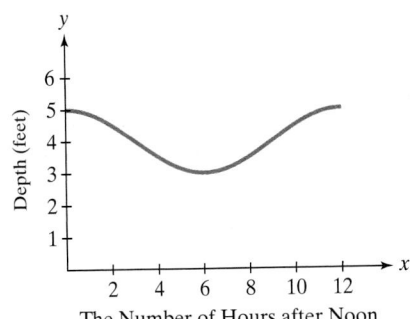

The Number of Hours after Noon

Writing in Mathematics

73. Without drawing a graph, describe the behavior of the basic sine curve.

74. What is the amplitude of the sine function? What does this tell you about the graph?

75. If you are given the equation of a sine function, how do you determine the period?

76. What does a phase shift indicate about the graph of a sine function? How do you determine the phase shift from the function's equation?

77. Describe a general procedure for obtaining the graph of $y = A \sin(Bx - C)$.

78. Without drawing a graph, describe the behavior of the basic cosine curve.

79. Describe a relationship between the graphs of $y = \sin x$ and $y = \cos x$.

80. Describe the relationship between the graphs of $y = A \cos(Bx - C)$ and $y = A \cos(Bx - C) + D$.

81. Biorhythm cycles provide interesting applications of sinusoidal graphs. But do you believe in the validity of biorhythms? Write a few sentences explaining why or why not.

Technology Exercises

82. Use a graphing utility to verify any five of the sine curves that you drew by hand in Exercises 7–30. The amplitude, period, and phase shift should help you to determine appropriate range settings.

83. Use a graphing utility to verify any five of the cosine curves that you drew by hand in Exercises 31–50.

84. Use a graphing utility to verify any two of the sinusoidal curves with vertical shifts that you drew in Exercises 51–58.

In Exercises 85–88, use a graphing utility to graph two periods of the function.

85. $y = 3 \sin(2x + \pi)$

86. $y = -2 \cos\left(2\pi x - \frac{\pi}{2}\right)$

87. $y = 0.2 \sin\left(\frac{\pi}{10} x + \pi\right)$

88. $y = 3 \sin(2x - \pi) + 5$

89. Use a graphing utility to graph $y = \sin x$ and $y = x - \frac{x^3}{6} + \frac{x^5}{120}$ in a $\left[-\pi, \pi, \frac{\pi}{2}\right]$ by $[-2, 2, 1]$ viewing rectangle. How do the graphs compare?

90. Use a graphing utility to graph $y = \cos x$ and $y = 1 - \dfrac{x^2}{2} + \dfrac{x^4}{24}$ in a $\left[-\pi, \pi, \dfrac{\pi}{2}\right]$ by $[-2, 2, 1]$ viewing rectangle. How do the graphs compare?

91. Use a graphing utility to graph

$$y = \sin x + \frac{\sin 2x}{2} + \frac{\sin 3x}{3} + \frac{\sin 4x}{4}$$

in a $\left[-2\pi, 2\pi, \dfrac{\pi}{2}\right]$ by $[-2, 2, 1]$ viewing rectangle. How do these waves compare to the smooth rolling waves of the basic sine curve?

92. Use a graphing utility to graph

$$y = \sin x - \frac{\sin 3x}{9} + \frac{\sin 5x}{25}$$

in a $\left[-2\pi, 2\pi, \dfrac{\pi}{2}\right]$ by $[-2, 2, 1]$ viewing rectangle. How do these waves compare to the smooth rolling waves of the basic sine curve?

93. The data show the average monthly temperatures for Washington, D.C.

  a. Use your graphing utility to draw a scatter plot of the data from $x = 1$ through $x = 12$.

  b. Use the SINe REGression feature to find the sinusoidal function of the form $y = A \sin (Bx + C) + D$ that best fits the data.

  c. Use your graphing utility to draw the sinusoidal function of best fit on the scatter plot.

x Month		Average Monthly Temperature, °F
1	(January)	34.6
2	(February)	37.5
3	(March)	47.2
4	(April)	56.5
5	(May)	66.4
6	(June)	75.6
7	(July)	80.0
8	(August)	78.5
9	(September)	71.3
10	(October)	59.7
11	(November)	49.8
12	(December)	39.4

Source: U.S. National Oceanic and Atmospheric Administration

94. Repeat Exercise 93 for data of your choice. The data can involve the average monthly temperatures for the region where you live or any data whose scatter plot takes the form of a sinusoidal function.

 Critical Thinking Exercises

Graph each function in Exercises 95–96 by hand.

95. $y = \sin x + \cos x$ for $0 \le x \le 2\pi$

96. $y = x + \cos x$ for $0 \le x \le \dfrac{5\pi}{2}$

97. Use the cosine function to find an equation of the graph in the figure shown.

 Group Exercise

98. This exercise is intended to provide some fun with biorhythms, regardless of whether you believe they have any validity. We will use each member's chart to determine biorhythmic compatibility. Before meeting, each group member should go online and obtain his or her biorhythm chart. The date of the group meeting is the date on which your chart should begin. Include 12 months in the plot. At the meeting, compare differences and similarities among the intellectual sinusoidal curves. Using these comparisons, each person should find the one other person with whom he or she would be most intellectually compatible.

SECTION 5.6 *Graphs of Other Trigonometric Functions*

Objectives

1. Understand the graph of $y = \tan x$.
2. Graph variations of $y = \tan x$.
3. Understand the graph of $y = \cot x$.
4. Graph variations of $y = \cot x$.
5. Understand the graphs of $y = \csc x$ and $y = \sec x$.
6. Graph variations of $y = \csc x$ and $y = \sec x$.

Recent advances in our understanding of climate have changed global warming from a subject for a disaster movie (the Statue of Liberty up to its chin in water) to a serious but manageable scientific and policy issue. Global warming is related to the burning of fossil fuels, which adds carbon dioxide to the atmosphere. In the new millennium, we will see whether our use of fossil fuels will add enough carbon dioxide to the atmosphere to change it (and our climate) in significant ways. In this section's exercise set, you will see how trigonometric graphs reveal interesting patterns in carbon dioxide concentration from 1990 through 2005. In the section itself, trigonometric graphs will reveal patterns involving the tangent, cotangent, secant, and cosecant functions.

1 Understand the graph of $y = \tan x$.

The Graph of $y = \tan x$

The properties of the tangent function discussed in Section 5.4 will help us determine its graph. Because the tangent function has properties that are different from sinusoidal functions, its graph differs significantly from those of sine and cosine. Properties of the tangent function include the following:

- The period is π. It is only necessary to graph $y = \tan x$ over an interval of length π. The remainder of the graph consists of repetitions of that graph at intervals of π.
- The tangent function is an odd function: $\tan(-x) = -\tan x$. The graph is symmetric with respect to the origin.
- The tangent function is undefined at $\frac{\pi}{2}$. The graph of $y = \tan x$ has a vertical asymptote at $x = \frac{\pi}{2}$.

We obtain the graph of $y = \tan x$ using some points on the graph and origin symmetry. Table 5.3 lists some values of (x, y) on the graph of $y = \tan x$ on the interval $\left[0, \frac{\pi}{2}\right)$.

Table 5.3   Values of (x, y) on $y = \tan x$

x	0	$\frac{\pi}{6}$	$\frac{\pi}{4}$	$\frac{\pi}{3}$	$\frac{5\pi}{12}$ $(75°)$	$\frac{17\pi}{36}$ $(85°)$	$\frac{89\pi}{180}$ $(89°)$	1.57	$\frac{\pi}{2}$
$y = \tan x$	0	$\frac{\sqrt{3}}{3} \approx 0.6$	1	$\sqrt{3} \approx 1.7$	3.7	11.4	57.3	1255.8	undefined

As x increases from 0 to $\frac{\pi}{2}$, y increases slowly at first, then more and more rapidly.

The graph in Figure 5.57(a) is based on our observation that as x increases from 0 to $\frac{\pi}{2}$, y increases slowly at first, then more and more rapidly. Notice that y increases without bound as x approaches $\frac{\pi}{2}$. As the figure shows, the graph of $y = \tan x$ has a vertical asymptote at $x = \frac{\pi}{2}$.

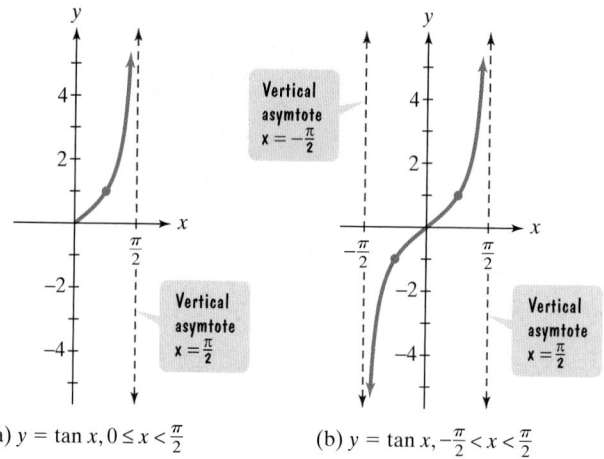

Figure 5.57 Graphing the tangent function

(a) $y = \tan x, 0 \le x < \frac{\pi}{2}$ (b) $y = \tan x, -\frac{\pi}{2} < x < \frac{\pi}{2}$

The graph of $y = \tan x$ can be completed on the interval $\left(-\frac{\pi}{2}, \frac{\pi}{2}\right)$ by using origin symmetry. Figure 5.57(b) shows the result of reflecting the graph in Figure 5.57(a) about the origin. The graph of $y = \tan x$ has another vertical asymptote at $x = -\frac{\pi}{2}$. Notice that y decreases without bound as x approaches $-\frac{\pi}{2}$.

Because the period of the tangent function is π radians, the graph in Figure 5.57(b) shows one complete period of $y = \tan x$. We obtain the complete graph of $y = \tan x$ by repeating the graph in Figure 5.57(b) to the left and right over intervals of π. The resulting graph and its main characteristics are shown in the following box:

The Tangent Curve: The Graph of $y = \tan x$ and Its Characteristics

Characteristics

- **Period:** π
- **Domain:** All real numbers except odd multiples of $\frac{\pi}{2}$
- **Range:** All real numbers
- **Vertical asymptotes** at odd multiples of $\frac{\pi}{2}$
- **An x-intercept** occurs midway between each pair of consecutive asymptotes.
- **Odd function** with origin symmetry
- Points on the graph $\frac{1}{4}$ and $\frac{3}{4}$ of the way between consecutive asymptotes have y-coordinates of -1 and 1.

2 Graph variations of $y = \tan x$.

Graphing Variations of $y = \tan x$

We use the characteristics of the tangent curve to graph tangent functions of the form $y = A \tan(Bx - C)$.

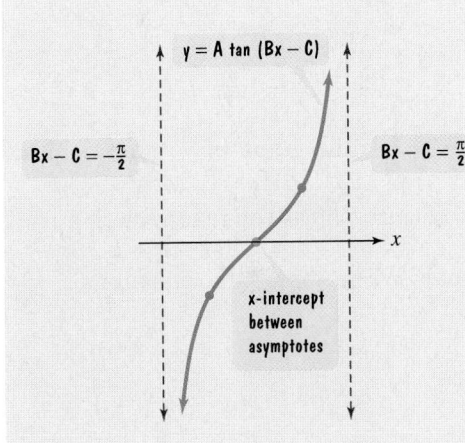

Graphing $y = A \tan(Bx - C)$

1. Find two consecutive asymptotes by setting the variable expression in the tangent equal to $-\dfrac{\pi}{2}$ and $\dfrac{\pi}{2}$ and solving

$$Bx - C = -\frac{\pi}{2} \quad \text{and} \quad Bx - C = \frac{\pi}{2}.$$

2. Identify an x-intercept, midway between the consecutive asymptotes.

3. Find the points on the graph $\dfrac{1}{4}$ and $\dfrac{3}{4}$ of the way between the consecutive asymptotes. These points have y-coordinates of $-A$ and A.

4. Use steps 1–3 to graph one full period of the function. Add additional cycles to the left or right as needed.

EXAMPLE 1 Graphing a Tangent Function

Graph $y = 2 \tan \dfrac{x}{2}$ for $-\pi < x < 3\pi$.

Solution Refer to Figure 5.58 as you read each step.

Step 1 Find two consecutive asymptotes. We must solve the equations

$$\frac{x}{2} = -\frac{\pi}{2} \quad \text{and} \quad \frac{x}{2} = \frac{\pi}{2}. \quad \text{\small Set the variable expression in the tangent equal to } -\frac{\pi}{2} \text{ and } \frac{\pi}{2}.$$

$$x = -\pi \qquad\qquad x = \pi \quad \text{\small Multiply both sides of each equation by 2.}$$

Thus, two consecutive asymptotes occur at $x = -\pi$ and $x = \pi$.

Step 2 Identify an x-intercept, midway between the consecutive asymptotes. Midway between $x = -\pi$ and $x = \pi$ is $x = 0$. An x-intercept is 0 and the graph passes through $(0, 0)$.

Step 3 Find points on the graph $\dfrac{1}{4}$ and $\dfrac{3}{4}$ of the way between the consecutive asymptotes. These points have y-coordinates of $-A$ and A. Because A, the coefficient of the tangent in $y = 2 \tan \dfrac{x}{2}$ is 2, these points have y-coordinates of -2 and 2. The graph passes through $\left(-\dfrac{\pi}{2}, -2\right)$ and $\left(\dfrac{\pi}{2}, 2\right)$.

Step 4 Use steps 1–3 to graph one full period of the function. We use the two consecutive asymptotes, $x = -\pi$ and $x = \pi$, an x-intercept of 0, and points midway between the x-intercept and asymptotes with y-coordinates of -2 and 2. We graph one period of $y = 2 \tan \dfrac{\pi}{2}$ from $-\pi$ to π. In order to graph for $-\pi < x < 3\pi$, we continue the pattern and extend the graph another full period to the right. The graph is shown in Figure 5.58.

Figure 5.58 The graph is shown for two full periods.

Check Point 1 Graph $y = 3 \tan 2x$ for $-\dfrac{\pi}{4} < x < \dfrac{3\pi}{4}$.

EXAMPLE 2 Graphing a Tangent Function

Graph two full periods of $y = \tan\left(x + \dfrac{\pi}{4}\right)$.

Solution The graph of $y = \tan\left(x + \dfrac{\pi}{4}\right)$ is the graph of $y = \tan x$ shifted horizontally to the left $\dfrac{\pi}{4}$ units. Refer to Figure 5.59 as you read each step.

Step 1 Find two consecutive asymptotes. We must solve the equations

$$x + \frac{\pi}{4} = -\frac{\pi}{2} \quad \text{and} \quad x + \frac{\pi}{4} = \frac{\pi}{2}.$$
Set the variable expression in the tangent equal to $-\dfrac{\pi}{2}$ and $\dfrac{\pi}{2}$.

$$x = -\frac{\pi}{4} - \frac{\pi}{2} \qquad\qquad x = -\frac{\pi}{4} + \frac{\pi}{2}$$
Subtract $\dfrac{\pi}{4}$ from both sides in each equation.

$$x = -\frac{3\pi}{4} \qquad\qquad x = \frac{\pi}{4}$$
Simplify.

Thus, two consecutive asymptotes occur at $x = -\dfrac{3\pi}{4}$ and $x = \dfrac{\pi}{4}$.

Step 2 Identify an x-intercept, midway between the consecutive asymptotes.

$$x\text{-intercept} = \frac{-\dfrac{3\pi}{4} + \dfrac{\pi}{4}}{2} = \frac{-\dfrac{2\pi}{4}}{2} = -\frac{2\pi}{8} = -\frac{\pi}{4}$$

An x-intercept is $-\dfrac{\pi}{4}$ and the graph passes through $\left(-\dfrac{\pi}{4}, 0\right)$.

Step 3 Find points on the graph $\dfrac{1}{4}$ and $\dfrac{3}{4}$ of the way between the consecutive asymptotes. These points have y-coordinates of $-A$ and A. Because A, the coefficient of the tangent in $y = \tan\left(x + \dfrac{\pi}{4}\right)$ is 1, these points have y-coordinates of -1 and 1. They are shown as blue dots in Figure 5.59.

Step 4 Use steps 1–3 to graph one full period of the function. We use the two consecutive asymptotes, $x = -\dfrac{3\pi}{4}$ and $x = \dfrac{\pi}{4}$, to graph one full period of $y = \tan\left(x + \dfrac{\pi}{4}\right)$ from $-\dfrac{3\pi}{4}$ to $\dfrac{\pi}{4}$. We graph two full periods by continuing the pattern and extending the graph another full period to the right. The graph is shown in Figure 5.59.

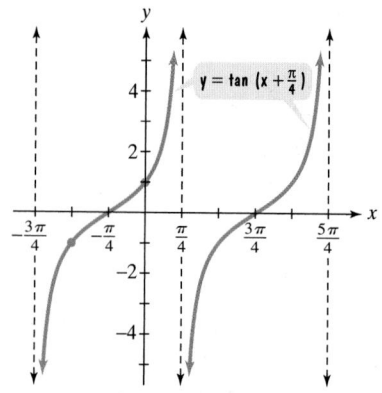

Figure 5.59 The graph is shown for two full periods.

Check Point 2 Graph two full periods of $y = \tan\left(x - \dfrac{\pi}{2}\right)$.

3 Understand the graph of $y = \cot x$.

The Graph of $y = \cot x$

Like the tangent function, the cotangent function, $y = \cot x$, has a period of π. The graph and its main characteristics are shown in the following box:

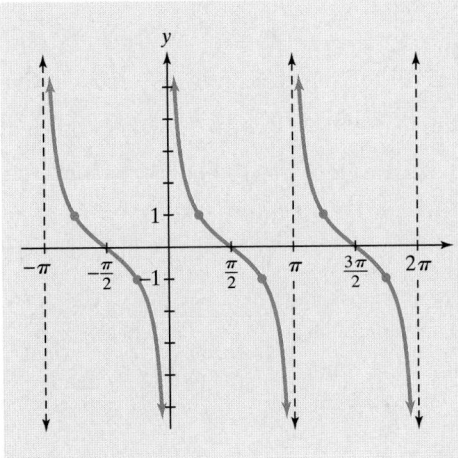

The Cotangent Curve: The Graph of $y = \cot x$ and Its Characteristics

Characteristics

- **Period:** π
- **Domain:** All real numbers except integral multiples of π
- **Range:** All real numbers
- **Vertical asymptotes** at integral multiples of π
- **An x-intercept** occurs midway between each pair of consecutive asymptotes.
- **Odd function** with origin symmetry
- Points on the graph $\frac{1}{4}$ and $\frac{3}{4}$ of the way between consecutive asymptotes have y-coordinates of 1 and -1.

4 Graph variations of $y = \cot x$.

Graphing Variations of $y = \cot x$

We use the characteristics of the cotangent curve to graph cotangent functions of the form $y = A \cot(Bx - C)$.

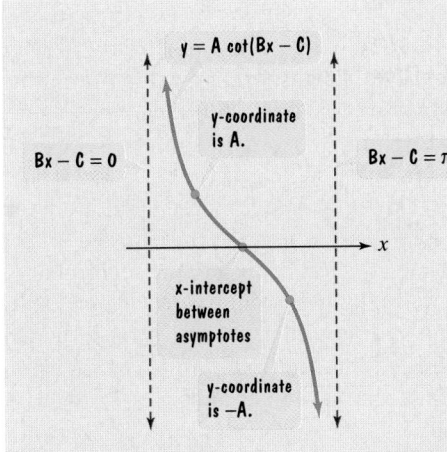

Graphing $y = A \cot(Bx - C)$

1. Find two consecutive asymptotes by setting the variable expression in the cotangent equal to 0 and π and solving

$$Bx - C = 0 \text{ and } Bx - C = \pi.$$

2. Identify an x-intercept, midway between the consecutive asymptotes.

3. Find the points on the graph $\frac{1}{4}$ and $\frac{3}{4}$ of the way between the consecutive asymptotes. These points have y-coordinates of A and $-A$.

4. Use steps 1–3 to graph one full period of the function. Add additional cycles to the left or right as needed.

EXAMPLE 3   Graphing a Cotangent Function

Graph $y = 3 \cot 2x$.

Solution   Refer to Figure 5.60, shown on the next page, as you read each step.

Step 1   Find two consecutive asymptotes. We must solve the equations

$2x = 0$ and $2x = \pi$.   *Set the variable expression in the cotangent equal to 0 and π.*

$x = 0$ $x = \dfrac{\pi}{2}$ *Divide both sides of each equation by 2.*

Two consecutive asymptotes occur at $x = 0$ and $x = \dfrac{\pi}{2}$.

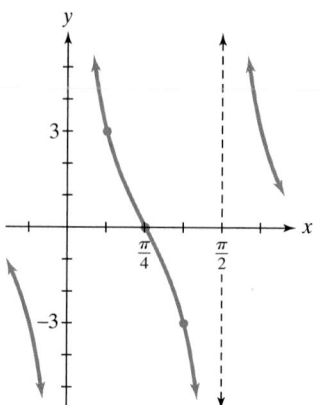

Figure 5.60 The graph of $y = 3 \cot 2x$

Step 2   Identify an x-intercept, midway between the consecutive asymptotes. Midway between $x = 0$ and $x = \dfrac{\pi}{2}$ is $x = \dfrac{\pi}{4}$. An x-intercept is $\dfrac{\pi}{4}$ and the graph passes through $\left(\dfrac{\pi}{4}, 0\right)$.

Step 3   Find points on the graph $\dfrac{1}{4}$ and $\dfrac{3}{4}$ of the way between consecutive asymptotes. These points have y-coordinates of A and $-A$. Because A, the coefficient of the cotangent in $y = 3 \cot 2x$ is 3, these points have y-coordinates of 3 and -3.

Step 4   Use steps 1–3 to graph one full period of the function. We use the two consecutive asymptotes, $x = 0$ and $x = \dfrac{\pi}{2}$, to graph one full period of $y = 3 \cot 2x$. This curve is repeated to the left and right, as shown in Figure 5.60.

Check Point 3   Graph $y = \dfrac{1}{2} \cot \dfrac{\pi}{2} x$.

5   Understand the graphs of $y = \csc x$ and $y = \sec x$.

The Graphs of $y = \csc x$ and $y = \sec x$

We obtain the graphs of the cosecant and secant curves by using the reciprocal identities

$$\csc x = \frac{1}{\sin x} \quad \text{and} \quad \sec x = \frac{1}{\cos x}.$$

The identity on the left tells us that the value of the cosecant function $y = \csc x$ at a given value of x equals the reciprocal of the corresponding value of the sine function, provided that the value of the sine function is not 0. If the value of $\sin x$ is 0, then at each of these values of x, the cosecant function is not defined. A vertical asymptote is associated with each of these values on the graph of $y = \csc x$.

We obtain the graph of $y = \csc x$ by taking reciprocals of the y-values in the graph of $y = \sin x$. Vertical asymptotes of $y = \csc x$ occur at the x-intercepts of $y = \sin x$. Likewise, we obtain the graph of $y = \sec x$ by taking the reciprocal of $y = \cos x$. Vertical asymptotes of $y = \sec x$ occur at the x-intercepts of $y = \cos x$. The graphs of $y = \csc x$ and $y = \sec x$ and their key characteristics are shown in the following boxes. We have used dashed red lines to first graph $y = \sin x$ and $y = \cos x$, drawing vertical asymptotes through the x-intercepts.

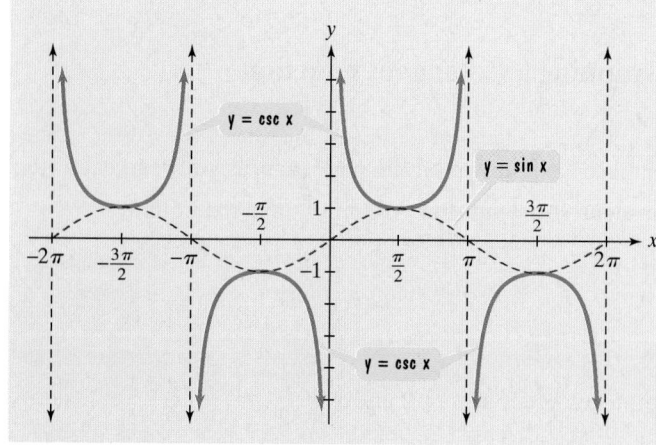

The Cosecant Curve: The Graph of $y = \csc x$ and Its Characteristics

Characteristics

- **Period:** 2π
- **Domain:** All real numbers except integral multiples of π
- **Range:** All real numbers y such that $y \le -1$ or $y \ge 1$
- **Vertical asymptotes** at integral multiples of π
- **Odd function,** $\csc(-x) = -\csc x$, with origin symmetry

The Secant Curve: The Graph of $y = \sec x$ and Its Characteristics

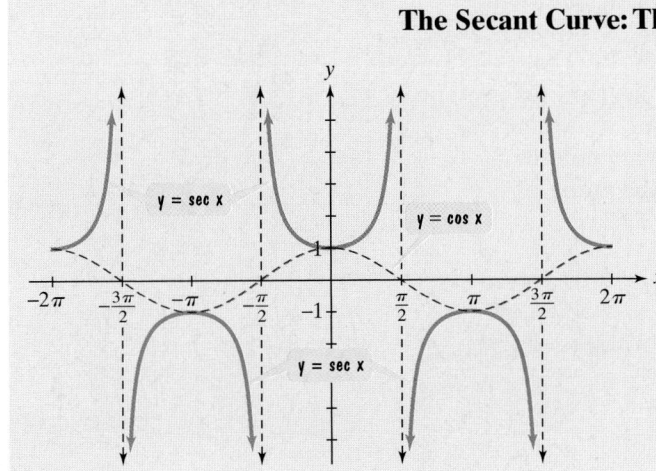

Characteristics

- **Period:** 2π
- **Domain:** All real numbers except odd multiples of $\dfrac{\pi}{2}$
- **Range:** All real numbers y such that $y \leq -1$ or $y \geq 1$
- **Vertical asymptotes** at odd multiples of $\dfrac{\pi}{2}$
- **Even function,** $\sec(-x) = \sec x$, with y-axis symmetry

6 Graph variations of $y = \csc x$ and $y = \sec x$.

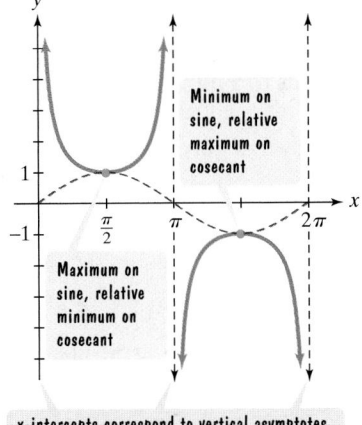

Minimum on sine, relative maximum on cosecant

Maximum on sine, relative minimum on cosecant

x-intercepts correspond to vertical asymptotes.

Figure 5.61

Graphing Variations of $y = \csc x$ and $y = \sec x$

We use graphs of reciprocal functions to obtain graphs of cosecant and secant functions. To graph a cosecant or secant curve, begin by graphing the reciprocal function. For example, to graph $y = 2\csc 2x$, we use the graph of $y = 2\sin 2x$. Likewise, to graph $y = -3\sec\dfrac{x}{2}$, we use the graph of $y = -3\cos\dfrac{x}{2}$.

Figure 5.61 illustrates how we use a sine curve to obtain a cosecant curve. Notice that

- x-intercepts on the red sine curve correspond to vertical asymptotes of the blue cosecant curve.
- A maximum point on the red sine curve corresponds to a minimum point on a continuous portion of the blue cosecant curve.
- A minimum point on the red sine curve corresponds to a maximum point on a continuous portion of the blue cosecant curve.

EXAMPLE 4   Using a Sine Curve to Obtain a Cosecant Curve

Use the graph of $y = 2\sin 2x$ in Figure 5.62 to obtain the graph of $y = 2\csc 2x$.

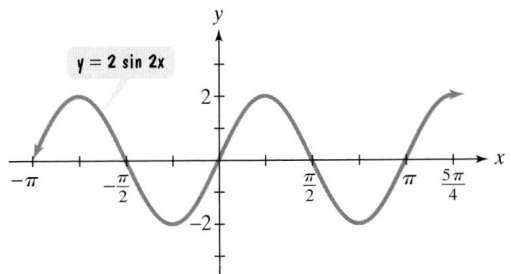

Figure 5.62

Solution   On the next page, we begin our work by showing the given graph, the graph of $y = 2\sin 2x$, using dashed red lines.

The x-intercepts of $y = 2\sin 2x$ correspond to the vertical asymptotes of $y = 2\csc 2x$. Thus, we draw vertical asymptotes through the x-intercepts, shown in Figure 5.63. Using the asymptotes as guides, we sketch the graph of $y = 2\csc 2x$ in Figure 5.63.

Figure 5.63 Using a sine curve to graph $y = 2\csc 2x$

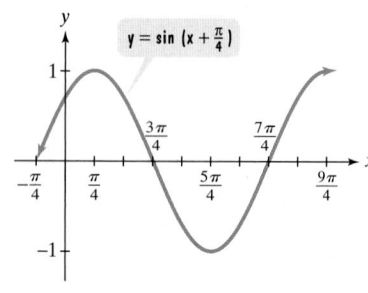

Check Point 4 Use the graph of $y = \sin\left(x + \dfrac{\pi}{4}\right)$, shown on the left, to obtain the graph of $y = \csc\left(x + \dfrac{\pi}{4}\right)$.

We use a cosine curve to obtain a secant curve in exactly the same way we used a sine curve to obtain a cosecant curve. Thus,

- x-intercepts on the cosine curve correspond to vertical asymptotes on the secant curve.
- A maximum point on the cosine curve corresponds to a minimum point on a continuous portion of the secant curve.
- A minimum point on the cosine curve corresponds to a maximum point on a continuous portion of the secant curve.

EXAMPLE 5 Graphing a Secant Function

Graph $y = -3\sec\dfrac{x}{2}$ for $-\pi < x < 5\pi$.

Solution We begin by graphing the reciprocal cosine function, $y = -3\cos\dfrac{x}{2}$. This equation is of the form $y = A\cos Bx$ with $A = -3$ and $B = \frac{1}{2}$.

amplitude: $\quad |A| = |-3| = 3$ *The maximum y is 3 and the minimum is −3.*

period: $\quad \dfrac{2\pi}{B} = \dfrac{2\pi}{\frac{1}{2}} = 4\pi$ *Each cycle is completed in 4π radians.*

We use quarter-periods, $\dfrac{4\pi}{4}$, or π, to find the x-values for the five key points. Starting with $x = 0$, the x-values are $0, \pi, 2\pi, 3\pi$, and 4π. Evaluating the function at each of these values of x, the key points are

$$(0, -3), (\pi, 0), (2\pi, 3), (3\pi, 0), \text{ and } (4\pi, -3).$$

We use these key points to graph $y = -3\cos\dfrac{x}{2}$ from 0 to 4π. In order to graph for $-\pi \le x \le 5\pi$, extend the graph π units to the left and π units to the right. The graph is shown using a dashed red line in Figure 5.64. Now use this dashed red graph to obtain the graph of the reciprocal function. Draw vertical asymptotes through the x-intercepts. Using these asymptotes as guides, the graph of $y = -3\sec\dfrac{x}{2}$ is shown in blue in Figure 5.64.

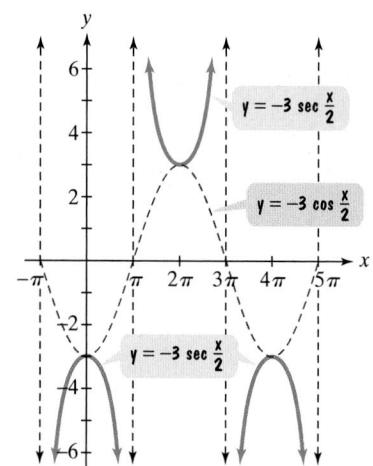

Figure 5.64 Using a cosine curve to graph $y = -3\sec\dfrac{x}{2}$

Check
Point
5

Graph $y = 2 \sec 2x$ for $-\dfrac{3\pi}{4} < x < \dfrac{3\pi}{4}$.

The Six Curves of Trigonometry

Table 5.4 summarizes the graphs of the six trigonometric functions. Below each of the graphs is a description of the domain, range, and period of the function.

Table 5.4   Graphs of the Six Trigonometric Functions

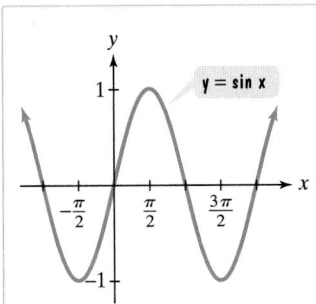

Domain: all real numbers

Range: $[-1, 1]$
Period: 2π

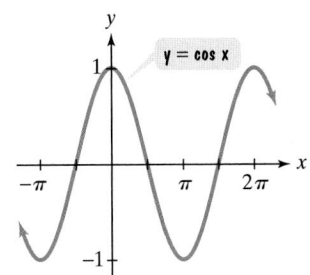

Domain: all real numbers

Range: $[-1, 1]$
Period: 2π

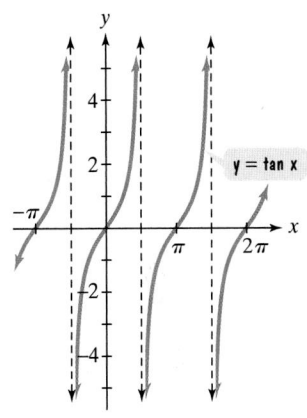

Domain: all real numbers
  except odd multiples of $\dfrac{\pi}{2}$
Range: all real numbers
Period: π

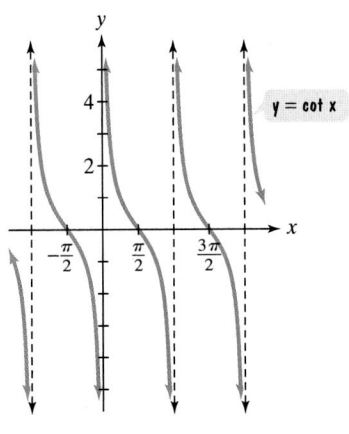

Domain: all real numbers
  except integral multiples of π

Range: all real numbers
Period: π

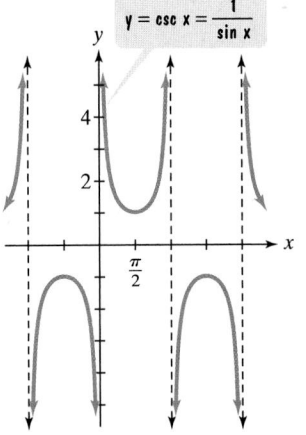

Domain: all real numbers
  except integral multiples of π

Range: $(-\infty, -1]$ or $[1, \infty)$
Period: 2π

Domain: all real numbers
  except odd multiples of $\dfrac{\pi}{2}$

Range: $(-\infty, -1]$ or $[1, \infty)$
Period: 2π

EXERCISE SET 5.6

Practice Exercises

In Exercises 1–4, the graph of a tangent function is given.
Select the equation for each graph from the following options:

$$y = \tan\left(x + \frac{\pi}{2}\right), \quad y = \tan(x + \pi), \quad y = -\tan x, \quad y = -\tan\left(x - \frac{\pi}{2}\right).$$

1.

2.

3.

4.
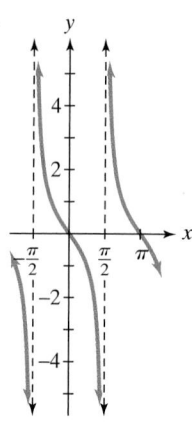

In Exercises 5–12, graph two periods of the given tangent function.

5. $y = 3 \tan \dfrac{x}{4}$

6. $y = 2 \tan \dfrac{x}{4}$

7. $y = \dfrac{1}{2} \tan 2x$

8. $y = 3 \tan 2x$

9. $y = -2 \tan \dfrac{1}{2} x$

10. $y = -3 \tan \dfrac{1}{2} x$

11. $y = \tan(x - \pi)$

12. $y = \tan\left(x + \dfrac{\pi}{2}\right)$

In Exercises 13–16, the graph of a cotangent function is given. Select the equation for each graph from the following options:

$$y = \cot\left(x + \frac{\pi}{2}\right), \quad y = \cot(x + \pi), \quad y = -\cot x, \quad y = -\cot\left(x - \frac{\pi}{2}\right).$$

13.

14.

15.

16.
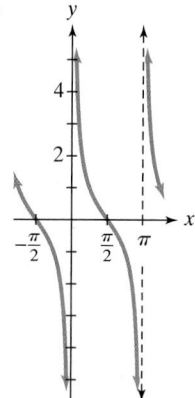

In Exercises 17–24, graph two periods of the given cotangent function.

17. $y = 2 \cot x$

18. $y = \dfrac{1}{2} \cot x$

19. $y = \dfrac{1}{2} \cot 2x$

20. $y = 2 \cot 2x$

21. $y = -3 \cot \dfrac{\pi}{2} x$

22. $y = -2 \cot \dfrac{\pi}{4} x$

23. $y = 3 \cot \left(x + \dfrac{\pi}{2} \right)$

24. $y = 3 \cot \left(x + \dfrac{\pi}{4} \right)$

In Exercises 25–28, use each graph to obtain the graph of the reciprocal function. Give the equation of the function for the graph that you obtain.

25.

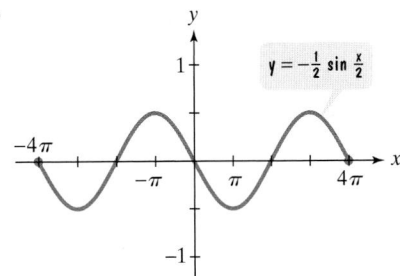

$y = -\tfrac{1}{2} \sin \tfrac{x}{2}$

26.

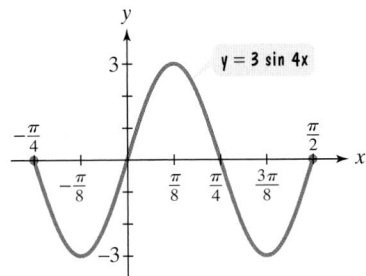

$y = 3 \sin 4x$

27.

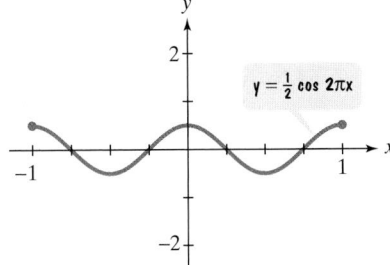

$y = \tfrac{1}{2} \cos 2\pi x$

28.

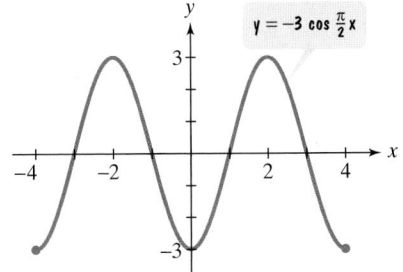

$y = -3 \cos \tfrac{\pi}{2} x$

In Exercises 29–44, graph two periods of the given cosecant or secant function.

29. $y = 3 \csc x$

30. $y = 2 \csc x$

31. $y = \dfrac{1}{2} \csc \dfrac{x}{2}$

32. $y = \dfrac{3}{2} \csc \dfrac{x}{4}$

33. $y = 2 \sec x$

34. $y = 3 \sec x$

35. $y = \sec \dfrac{x}{3}$

36. $y = \sec \dfrac{x}{2}$

37. $y = -2 \csc \pi x$

38. $y = -\dfrac{1}{2} \csc \pi x$

39. $y = -\dfrac{1}{2} \sec \pi x$

40. $y = -\dfrac{3}{2} \sec \pi x$

41. $y = \csc(x - \pi)$

42. $y = \csc \left(x - \dfrac{\pi}{2} \right)$

43. $y = 2 \sec(x + \pi)$

44. $y = 2 \sec \left(x + \dfrac{\pi}{2} \right)$

 Application Exercises

45. An ambulance with a rotating beacon of light is parked 12 feet from a building. The function

$$d = 12 \tan 2\pi t$$

describes the distance, d, in feet, of the rotating beacon from point C after t seconds.

a. Graph the function on the interval $[0, 2]$.

b. For what values of t in $[0, 2]$ is the function undefined? What does this mean in terms of the rotating beacon in the figure shown?

46. The angle of elevation from the top of a house to a jet flying 2 miles above the house is x radians. If d represents the horizontal distance, in miles, of the jet from the house, express d in terms of a trigonometric function of x. Then graph the function for $0 < x < \pi$.

47. Your best friend is marching with a band and has asked you to film her. The figure below shows that you have set yourself up 10 feet from the street where your friend will be passing from left to right. If d represents your distance, in feet, from your friend and x is the radian measure of the angle shown, express d in terms of a trigonometric function of x. Then graph the function for $-\dfrac{\pi}{2} < x < \dfrac{\pi}{2}$. Negative angles indicate that your marching buddy is on your left.

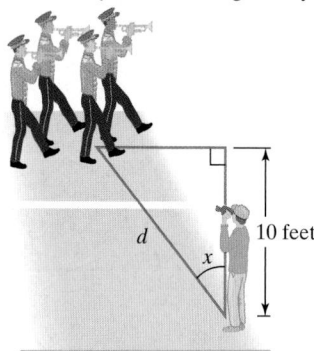

In Exercises 48–50, sketch a reasonable graph that models the given situation.

48. The number of hours of daylight per day in your hometown over a two-year period

49. The motion of a diving board vibrating 10 inches in each direction per second just after someone has dived off

50. The distance of a rotating beacon of light from a point on a wall (See the figure for Exercise 45.)

 Writing in Mathematics

51. Without drawing a graph, describe the behavior of the basic tangent curve.

52. If you are given the equation of a tangent function, how do you find consecutive asymptotes?

53. If you are given the equation of a tangent function, how do you identify an x-intercept?

54. Without drawing a graph, describe the behavior of the basic cotangent curve.

55. If you are given the equation of a cotangent function, how do you find consecutive asymptotes?

56. Explain how to determine the range of $y = \csc x$ from the graph. What is the range?

57. Explain how to use a sine curve to obtain a cosecant curve. Why can the same procedure be used to obtain a secant curve from a cosine curve?

58. Scientists record brain activity by attaching electrodes to the scalp and then connecting these electrodes to a machine. The record of brain activity recorded with this machine is shown in the three graphs at the top of the next column. Which trigonometric functions would be most appropriate for describing the oscillations in brain activity? Describe similarities and differences among these functions when modeling brain activity when awake, during dreaming sleep, and during nondreaming sleep.

Human Brain Activity

Technology Exercises

In working Exercises 59–62, describe what happens at the asymptotes on the graphing utility. Compare the graphs in the connected and dot modes.

59. Use a graphing utility to verify any two of the tangent curves that you drew by hand in Exercises 5–12.

60. Use a graphing utility to verify any two of the cotangent curves that you drew by hand in Exercises 17–24.

61. Use a graphing utility to verify any two of the cosecant curves that you drew by hand in Exercises 29–44.

62. Use a graphing utility to verify any two of the secant curves that you drew by hand in Exercises 29–44.

In Exercises 63–68, use a graphing utility to graph each function. Use a range setting so that the graph is shown for at least two periods.

63. $y = \tan \dfrac{x}{4}$

64. $y = \tan 4x$

65. $y = \cot 2x$

66. $y = \cot \dfrac{x}{2}$

67. $y = \dfrac{1}{2} \tan \pi x$

68. $y = \dfrac{1}{2} \tan(\pi x + 1)$

In Exercises 69–72, use a graphing utility to graph each pair of functions in the same viewing rectangle. Use a range setting so that the graphs are shown for at least two periods.

69. $y = 0.8 \sin \dfrac{x}{2}$ and $y = 0.8 \csc \dfrac{x}{2}$

70. $y = -2.5 \sin \dfrac{\pi}{3} x$ and $y = -2.5 \csc \dfrac{\pi}{3} x$

71. $y = 4 \cos\left(2x - \dfrac{\pi}{6}\right)$ and $y = 4 \sec\left(2x - \dfrac{\pi}{6}\right)$

72. $y = -3.5 \cos\left(\pi x - \dfrac{\pi}{6}\right)$ and $y = -3.5 \sec\left(\pi x - \dfrac{\pi}{6}\right)$

73. Carbon dioxide particles in our atmosphere trap heat and raise the planet's temperature. The resultant gradually increasing temperature is called the greenhouse effect. Carbon dioxide accounts for about half of global warming. The function

$$y = 2.5 \sin 2\pi x + 0.0216x^2 + 0.654x + 316$$

models carbon dioxide concentration, y, in parts per million, where $x = 0$ represents January 1960; $x = \frac{1}{12}$, February 1960; $x = \frac{2}{12}$, March 1960; $\cdots$, $x = 1$, January 1961; $x = \frac{13}{12}$, February 1961; and so on. Use a graphing utility to graph the function in a $[30, 45, 5]$ by $[310, 420, 5]$

viewing rectangle. Describe what the graph reveals about carbon dioxide concentration from 1990 through 2005.

74. Graph $y = \sin \dfrac{1}{x}$ in a $[-0.2, 0.2, 0.01]$ by $[-1.2, 1.2, 0.01]$

viewing rectangle. What is happening as x approaches 0 from the left or the right? Explain this behavior.

 Critical Thinking Exercises

In Exercises 75–76, write an equation for each blue graph.

75.

76.

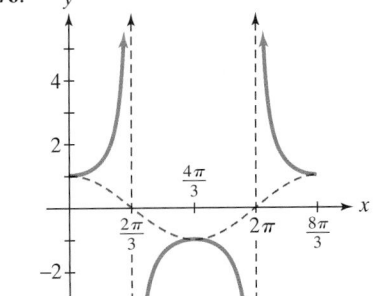

77. For $x > 0$, what effect does 2^{-x} in $y = 2^{-x} \sin x$ have on the graph of $y = \sin x$? What kind of behavior can be modeled by a function such as $y = 2^{-x} \sin x$?

SECTION 5.7   *Inverse Trigonometric Functions*

Objectives

1. Understand and use the inverse sine function.
2. Understand and use the inverse cosine function.
3. Understand and use the inverse tangent function.
4. Use a calculator to evaluate inverse trigonometric functions.
5. Find exact values of composite functions with inverse trigonometric functions.

You watched *The Matrix* on video and were impressed by the elaborate computer-generated effects. The movie is being shown again at a local theater, where you can experience its stunning visual force on a large screen. Where in the theater should you sit to maximize the film's visual impact? In this section's exercise set, you will see how an inverse trigonometric function can enhance your movie-going experiences.

Study Tip

Here are some helpful things to remember from our discussion of inverse functions in Section 2.7.

- If no horizontal line intersects the graph of a function more than once, the function is one-to-one and has an inverse function.
- If the point (a, b) is on the graph of f, then the point (b, a) is on the graph of the inverse function, denoted f^{-1}. The graph of f^{-1} is a reflection of the graph of f about the line $y = x$.

1 Understand and use the inverse sine function.

The Inverse Sine Function

Figure 5.65 shows the graph of $y = \sin x$. Can you see that every horizontal line that can be drawn between -1 and 1 intersects the graph infinitely many times? Thus, the sine function is not one-to-one and has no inverse function.

In Figure 5.66, we have taken a portion of the sine curve, restricting the domain of the sine function to $-\dfrac{\pi}{2} \leq x \leq \dfrac{\pi}{2}$. With this restricted domain, every horizontal line that can be drawn between -1 and 1 intersects the graph exactly once. Thus, the restricted function passes the horizontal line test and is one-to-one.

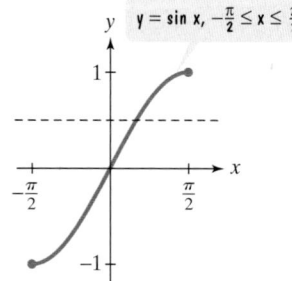

Figure 5.65 The horizontal line test shows that the sine function is not one-to-one and has no inverse function.

Figure 5.66 The restricted sine function passes the horizontal line test. It is one-to-one and has an inverse function.

On the restricted domain $-\dfrac{\pi}{2} \leq x \leq \dfrac{\pi}{2}$, $y = \sin x$ has an inverse function.

The inverse of the restricted sine function is called the **inverse sine function.** Two notations are commonly used to denote the inverse sine function:

$$y = \sin^{-1} x \quad \text{or} \quad y = \arcsin x.$$

In this book, we will use $y = \sin^{-1} x$. This notation has the same symbol as the inverse function notation $f^{-1}(x)$.

The Inverse Sine Function

The **inverse sine function,** denoted by $\sin^{-1}$, is the inverse of the restricted sine function $y = \sin x, -\dfrac{\pi}{2} \leq x \leq \dfrac{\pi}{2}$. Thus,

$$y = \sin^{-1} x \quad \text{means} \quad \sin y = x$$

where $-\dfrac{\pi}{2} \leq y \leq \dfrac{\pi}{2}$ and $-1 \leq x \leq 1$. We read $y = \sin^{-1} x$ as "y equals the inverse sine at x."

Figure 5.67 The restricted sine function

One way to graph $y = \sin^{-1} x$ is to take points on the graph of the restricted sine function and reverse the order of the coordinates. For example, Figure 5.67 shows that $\left(-\dfrac{\pi}{2}, -1\right)$, $(0, 0)$, and $\left(\dfrac{\pi}{2}, 1\right)$ are on the graph of the restricted sine function. Reversing the order of the coordinates gives $\left(-1, -\dfrac{\pi}{2}\right)$, $(0, 0)$, and $\left(1, \dfrac{\pi}{2}\right)$. We now use these three points to sketch the inverse sine function. The graph of $y = \sin^{-1} x$ is shown in Figure 5.68.

Another way to obtain the graph of $y = \sin^{-1} x$ is to reflect the graph of the restricted sine function about the line $y = x$, shown in Figure 5.69. The red graph is the restricted sine function and the blue graph is the graph of $y = \sin^{-1} x$.

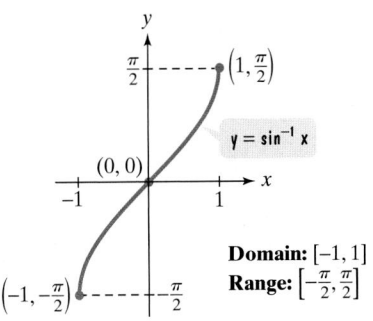

Figure 5.68 The graph of the inverse sine function

Figure 5.69 Using a reflection to obtain the graph of the inverse sine function

Exact values of $\sin^{-1} x$ can be found by thinking of **$\sin^{-1} x$ as the angle in the interval $\left[-\dfrac{\pi}{2}, \dfrac{\pi}{2}\right]$ whose sine is x.** For example, we can use the two points on the blue graph of the inverse sine function in Figure 5.69 and write

$$\sin^{-1}(-1) = -\frac{\pi}{2} \quad \text{and} \quad \sin^{-1} 1 = \frac{\pi}{2}.$$

The angle whose sine is -1 is $-\dfrac{\pi}{2}$.

The angle whose sine is 1 is $\dfrac{\pi}{2}$.

Because we are thinking of $\sin^{-1} x$ in terms of an angle, we will represent such an angle by θ.

Finding Exact Values of $\sin^{-1} x$.

1. Let $\theta = \sin^{-1} x$.

2. Rewrite $\theta = \sin^{-1} x$ as $\sin \theta = x$.

3. Use the exact values in Table 5.5, shown on the next page, to find the value of θ in $\left[-\dfrac{\pi}{2}, \dfrac{\pi}{2}\right]$ that satisfies $\sin \theta = x$.

Table 5.5 Exact Values for
$$\sin\theta, -\frac{\pi}{2} \le \theta \le \frac{\pi}{2}$$

θ	$\sin\theta$
$-\frac{\pi}{2}$	-1
$-\frac{\pi}{3}$	$-\frac{\sqrt{3}}{2}$
$-\frac{\pi}{4}$	$-\frac{\sqrt{2}}{2}$
$-\frac{\pi}{6}$	$-\frac{1}{2}$
0	0
$\frac{\pi}{6}$	$\frac{1}{2}$
$\frac{\pi}{4}$	$\frac{\sqrt{2}}{2}$
$\frac{\pi}{3}$	$\frac{\sqrt{3}}{2}$
$\frac{\pi}{2}$	1

EXAMPLE 1 Finding the Exact Value of an Inverse Sine Function

Find the exact value of $\sin^{-1}\frac{\sqrt{2}}{2}$.

Solution

Step 1 Let $\theta = \sin^{-1}x$. Thus,

$$\theta = \sin^{-1}\frac{\sqrt{2}}{2}.$$

We must find the angle θ, $-\frac{\pi}{2} \le \theta \le \frac{\pi}{2}$, whose sine equals $\frac{\sqrt{2}}{2}$.

Step 2 Rewrite $\theta = \sin^{-1}x$ as $\sin\theta = x$. Using the definition of the inverse sine function, we rewrite $\theta = \sin^{-1}\frac{\sqrt{2}}{2}$ as

$$\sin\theta = \frac{\sqrt{2}}{2}.$$

Step 3 Use the exact values in Table 5.5 to find the value of θ in $\left[-\frac{\pi}{2}, \frac{\pi}{2}\right]$ that satisfies $\sin\theta = x$. Table 5.5 shows that the only angle in the interval $\left[-\frac{\pi}{2}, \frac{\pi}{2}\right]$ that satisfies $\sin\theta = \frac{\sqrt{2}}{2}$ is $\frac{\pi}{4}$. Thus, $\theta = \frac{\pi}{4}$. Because θ, in step 1, represents $\sin^{-1}\frac{\sqrt{2}}{2}$, we conclude that

$$\sin^{-1}\frac{\sqrt{2}}{2} = \frac{\pi}{4}. \quad \text{The angle in } \left[-\frac{\pi}{2}, \frac{\pi}{2}\right] \text{ whose sine is } \frac{\sqrt{2}}{2} \text{ is } \frac{\pi}{4}.$$

Check Point 1 Find the exact value of $\sin^{-1}\frac{\sqrt{3}}{2}$.

EXAMPLE 2 Finding the Exact Value of an Inverse Sine Function

Find the exact value of $\sin^{-1}\left(-\frac{1}{2}\right)$.

Solution

Step 1 Let $\theta = \sin^{-1}x$. Thus,

$$\theta = \sin^{-1}\left(-\frac{1}{2}\right).$$

We must find the angle θ, $-\frac{\pi}{2} \le \theta \le \frac{\pi}{2}$, whose sine equals $-\frac{1}{2}$.

Step 2   **Rewrite $\theta = \sin^{-1}x$ as $\sin\theta = x$.** We rewrite $\theta = \sin^{-1}\left(-\dfrac{1}{2}\right)$ and obtain

$$\sin\theta = -\frac{1}{2}.$$

Step 3   **Use the exact values in Table 5.5 to find the value of θ in $\left[-\dfrac{\pi}{2}, \dfrac{\pi}{2}\right]$ that satisfies $\sin\theta = x$.** Table 5.5, on page 524, shows that the only angle in the interval $\left[-\dfrac{\pi}{2}, \dfrac{\pi}{2}\right]$ that satisfies $\sin\theta = -\dfrac{1}{2}$ is $-\dfrac{\pi}{6}$. Thus,

$$\sin^{-1}\left(-\frac{1}{2}\right) = -\frac{\pi}{6}.$$

Check Point 2   Find the exact value of $\sin^{-1}\left(-\dfrac{\sqrt{2}}{2}\right)$.

Some inverse sine expressions cannot be evaluated. Because the domain of the inverse sine function is $[-1, 1]$, it is only possible to evaluate $\sin^{-1}x$ for values of x in this domain. Thus, $\sin^{-1}3$ cannot be evaluated. There is no angle whose sine is 3.

2 Understand and use the inverse cosine function.

The Inverse Cosine Function

Figure 5.70 shows how we restrict the domain of the cosine function so that it becomes one-to-one and has an inverse function. Restrict the domain to the interval $[0, \pi]$, shown by the dark blue graph. Over this interval, the restricted cosine function passes the horizontal line test and has an inverse function.

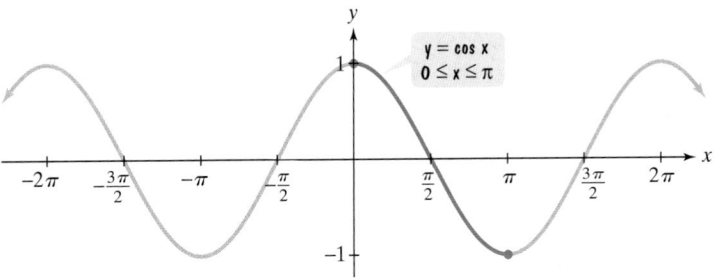

Figure 5.70 $y = \cos x$ is one-to-one on the interval $[0, \pi]$.

The Inverse Cosine Function

The **inverse cosine function,** denoted by $\cos^{-1}$, is the inverse of the restricted cosine function $y = \cos x, 0 \le x \le \pi$. Thus,

$$y = \cos^{-1}x \quad \text{means} \quad \cos y = x$$

where $0 \le y \le \pi$ and $-1 \le x \le 1$.

One way to graph $y = \cos^{-1} x$ is to take points on the graph of the restricted cosine function and reverse the order of the coordinates. For example, Figure 5.71 shows that $(0, 1)$, $\left(\dfrac{\pi}{2}, 0\right)$, and $(\pi, -1)$ are on the graph of the restricted cosine function. Reversing the order of the coordinates gives $(1, 0)$, $\left(0, \dfrac{\pi}{2}\right)$, and $(-1, \pi)$. We now use these three points to sketch the inverse cosine function. The graph of $y = \cos^{-1} x$ is shown in Figure 5.72. You can also obtain this graph by reflecting the graph of the restricted cosine function about the line $y = x$.

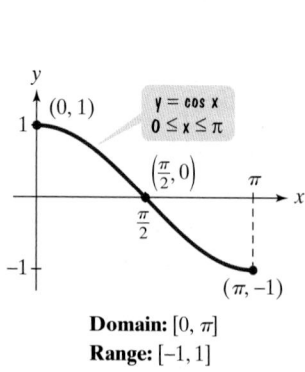

Domain: $[0, \pi]$
Range: $[-1, 1]$

Figure 5.71 The restricted cosine function

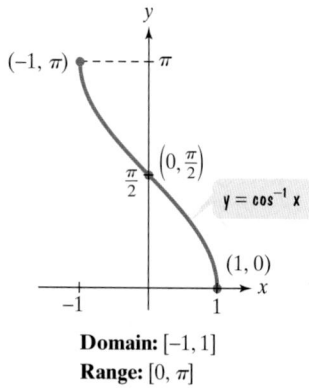

Domain: $[-1, 1]$
Range: $[0, \pi]$

Figure 5.72 The graph of the inverse cosine function

Table 5.6   Exact Values for $\cos\theta, 0 \le \theta \le \pi$

θ	$\cos\theta$
0	1
$\dfrac{\pi}{6}$	$\dfrac{\sqrt{3}}{2}$
$\dfrac{\pi}{4}$	$\dfrac{\sqrt{2}}{2}$
$\dfrac{\pi}{3}$	$\dfrac{1}{2}$
$\dfrac{\pi}{2}$	0
$\dfrac{2\pi}{3}$	$-\dfrac{1}{2}$
$\dfrac{3\pi}{4}$	$-\dfrac{\sqrt{2}}{2}$
$\dfrac{5\pi}{6}$	$-\dfrac{\sqrt{3}}{2}$
π	-1

Exact values of $\cos^{-1} x$ can be found by thinking of $\cos^{-1} x$ as **the angle in the interval $[0, \pi]$ whose cosine is x.** This time we will use Table 5.6, which shows exact values for $\cos\theta$ for θ in the interval $[0, \pi]$.

EXAMPLE 3   Finding the Exact Value of an Inverse Cosine Function

Find the exact value of $\cos^{-1}\left(-\dfrac{\sqrt{3}}{2}\right)$.

Solution

Step 1   Let $\theta = \cos^{-1} x$. Thus,

$$\theta = \cos^{-1}\left(-\dfrac{\sqrt{3}}{2}\right).$$

We must find the angle θ, $0 \le \theta \le \pi$, whose cosine equals $-\dfrac{\sqrt{3}}{2}$.

Step 2   Rewrite $\theta = \cos^{-1} x$ as $\cos\theta = x$. We obtain

$$\cos\theta = -\dfrac{\sqrt{3}}{2}.$$

Step 3   **Use the exact values in Table 5.6 to find the value of θ in $[0, \pi]$ that satisfies $\cos\theta = x$.** The table shows that the only angle in the interval $[0, \pi]$ that satisfies $\cos\theta = -\dfrac{\sqrt{3}}{2}$ is $\dfrac{5\pi}{6}$. Thus, $\theta = \dfrac{5\pi}{6}$ and

$$\cos^{-1}\left(-\frac{\sqrt{3}}{2}\right) = \frac{5\pi}{6}.$$ The angle in $[0, \pi]$ whose cosine is $-\dfrac{\sqrt{3}}{2}$ is $\dfrac{5\pi}{6}$.

> **Check Point 3**   Find the exact value of $\cos^{-1}\left(-\dfrac{1}{2}\right)$.

3   Understand and use the inverse tangent function.

The Inverse Tangent Function

Figure 5.73 shows how we restrict the domain of the tangent function so that it becomes one-to-one and has an inverse function. Restrict the domain to the interval $\left(-\dfrac{\pi}{2}, \dfrac{\pi}{2}\right)$, shown by the solid blue graph. Over this interval, the restricted tangent function passes the horizontal line test and has an inverse function.

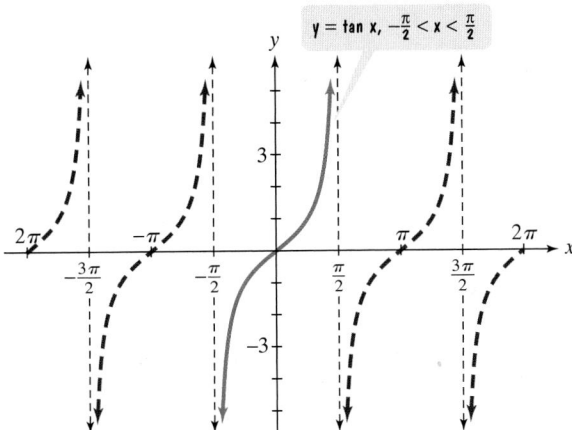

Figure 5.73 $y = \tan x$ is one-to-one on the interval $\left(-\dfrac{\pi}{2}, \dfrac{\pi}{2}\right)$.

> ## The Inverse Tangent Function
> The **inverse tangent function**, denoted by $\tan^{-1}$, is the inverse of the restricted tangent function $y = \tan x, -\dfrac{\pi}{2} < x < \dfrac{\pi}{2}$. Thus,
>
> $$y = \tan^{-1}x \quad \text{means} \quad \tan y = x$$
>
> where $-\dfrac{\pi}{2} < y < \dfrac{\pi}{2}$ and $-\infty < x < \infty$.

We graph $y = \tan^{-1} x$ by taking points on the graph of the restricted function and reversing the order of the coordinates. Figure 5.74 shows that $\left(-\dfrac{\pi}{4}, -1 \right)$, $(0, 0)$, and $\left(\dfrac{\pi}{4}, 1 \right)$ are on the graph of the restricted tangent function. Reversing the order gives $\left(-1, -\dfrac{\pi}{4} \right)$, $(0, 0)$, and $\left(1, \dfrac{\pi}{4} \right)$. We now use these three points to graph the inverse tangent function. The graph of $y = \tan^{-1} x$ is shown in Figure 5.75. Notice that the vertical asymptotes become horizontal asymptotes for the graph of the inverse function.

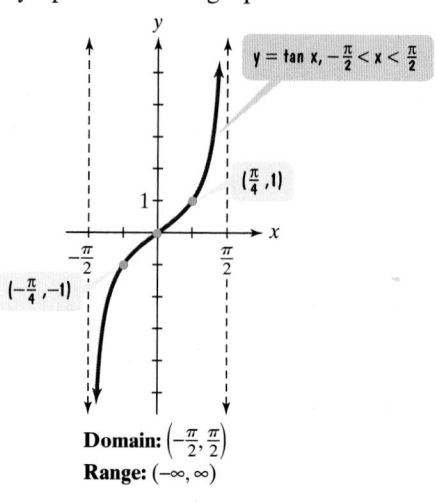

Figure 5.74 The restricted tangent function

Figure 5.75 The graph of the inverse tangent function

Exact values of $\tan^{-1} x$ can be found by thinking of $\mathbf{\tan^{-1} x}$ as **the angle in the interval** $\left(-\dfrac{\pi}{2}, \dfrac{\pi}{2} \right)$ **whose tangent is** x. We use Table 5.7, which shows exact values for $\tan \theta$ for θ in the interval $\left(-\dfrac{\pi}{2}, \dfrac{\pi}{2} \right)$.

Table 5.7 Exact Values for $\tan\theta, -\dfrac{\pi}{2} < \theta < \dfrac{\pi}{2}$

θ	$\tan \theta$
$-\dfrac{\pi}{2}$	Undefined
$-\dfrac{\pi}{3}$	$-\sqrt{3}$
$-\dfrac{\pi}{4}$	-1
$-\dfrac{\pi}{6}$	$-\dfrac{\sqrt{3}}{3}$
0	0
$\dfrac{\pi}{6}$	$\dfrac{\sqrt{3}}{3}$
$\dfrac{\pi}{4}$	1
$\dfrac{\pi}{3}$	$\sqrt{3}$
$\dfrac{\pi}{2}$	Undefined

EXAMPLE 4 Finding the Exact Value of an Inverse Tangent Function

Find the exact value of $\tan^{-1} \sqrt{3}$.

Solution

Step 1 Let $\theta = \tan^{-1} x$. Thus,
$$\theta = \tan^{-1} \sqrt{3}.$$

We must find the angle θ, $-\dfrac{\pi}{2} < \theta < \dfrac{\pi}{2}$, whose tangent equals $\sqrt{3}$.

Step 2 Rewrite $\theta = \tan^{-1} x$ as $\tan \theta = x$. We obtain $\tan \theta = \sqrt{3}$.

Step 3 Use the exact values in Table 5.7 to find the value of θ in $\left(-\dfrac{\pi}{2}, \dfrac{\pi}{2} \right)$ **that satisfies $\tan \theta = x$.** The table shows that the only angle in the interval $\left(-\dfrac{\pi}{2}, \dfrac{\pi}{2} \right)$ that satisfies $\tan \theta = \sqrt{3}$ is $\dfrac{\pi}{3}$. Thus, $\theta = \dfrac{\pi}{3}$ and

$$\tan^{-1} \sqrt{3} = \dfrac{\pi}{3}.$$ The angle in $\left(-\dfrac{\pi}{2}, \dfrac{\pi}{2} \right)$ whose tangent is $\sqrt{3}$ is $\dfrac{\pi}{3}$.

Check Point 4   Find the exact value of $\tan^{-1}(-1)$.

Table 5.8 summarizes the graphs of the three basic inverse trigonometric functions. Below each of the graphs is a description of the function's domain and range.

Table 5.8   Graphs of the Three Basic Inverse Trigonometric Functions

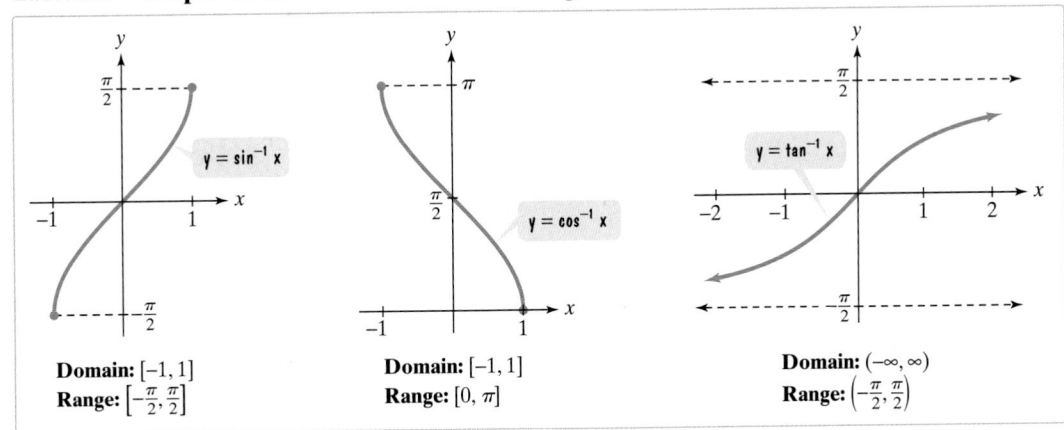

| Domain: $[-1, 1]$ | Domain: $[-1, 1]$ | Domain: $(-\infty, \infty)$ |
| Range: $\left[-\frac{\pi}{2}, \frac{\pi}{2}\right]$ | Range: $[0, \pi]$ | Range: $\left(-\frac{\pi}{2}, \frac{\pi}{2}\right)$ |

4 Use a calculator to evaluate inverse trigonometric functions.

Using a Calculator to Evaluate Inverse Trigonometric Functions

Calculators give approximate values of inverse trigonometric functions. Use the secondary keys marked $\boxed{\text{SIN}^{-1}}$, $\boxed{\text{COS}^{-1}}$, and $\boxed{\text{TAN}^{-1}}$. These keys are not buttons that you actually press. They are the secondary functions for the buttons labeled $\boxed{\text{SIN}}$, $\boxed{\text{COS}}$, and $\boxed{\text{TAN}}$, respectively. Consult your manual for the location of this feature.

EXAMPLE 5   Calculators and Inverse Trigonometric Functions

Use a calculator to find the value to four decimal places of:

a. $\sin^{-1}\dfrac{1}{4}$ **b.** $\tan^{-1}(-9.65)$.

Solution

Scientific Calculator Solution

Function	Mode	Keystrokes	Display, rounded to four places
a. $\sin^{-1}\dfrac{1}{4}$	Radian	$1\boxed{\div}4\boxed{=}\boxed{\text{SIN}^{-1}}$	0.2527
b. $\tan^{-1}(-9.65)$	Radian	$9.65\boxed{{}^+\!/_-}\boxed{\text{TAN}^{-1}}$	-1.4675

Graphing Calculator Solution

Function	Mode	Keystrokes	Display, rounded to four places
a. $\sin^{-1}\dfrac{1}{4}$	Radian	$\boxed{\text{SIN}^{-1}}\boxed{(}\boxed{1}\boxed{\div}\boxed{4}\boxed{)}\boxed{\text{ENTER}}$	0.2527
b. $\tan^{-1}(-9.65)$	Radian	$\boxed{\text{TAN}^{-1}}\boxed{(-)}9.65\boxed{\text{ENTER}}$	-1.4675

Check
Point
5 Use a calculator to find the value to four decimal places of:

a. $\cos^{-1}\dfrac{1}{3}$ **b.** $\tan^{-1}(-35.85)$.

What happens if you attempt to evaluate an inverse trigomometric function at a value that is not in its domain? In real number mode, most calculators will display an error message. For example, an error message can result if you attempt to approximate $\cos^{-1}3$. There is no angle whose cosine is 3. The domain of the inverse cosine function is $[-1, 1]$, and 3 does not belong to this domain.

5 Find exact values of composite functions with inverse trigonometric functions.

Composition of Functions Involving Inverse Trigonometric Functions

In our discussion of functions and their inverses in Section 2.7, we saw that
$$f(f^{-1}(x)) = x \quad \text{and} \quad f^{-1}(f(x)) = x.$$

x must be in the domain of f⁻¹. *x must be in the domain of f.*

We apply these properties to the sine, cosine, tangent, and their inverse functions to obtain the following properties:

Inverse Properties

The Sine Function and Its Inverse

$\sin(\sin^{-1} x) = x$ for every x in the interval $[-1, 1]$

$\sin^{-1}(\sin x) = x$ for every x in the interval $\left[-\dfrac{\pi}{2}, \dfrac{\pi}{2}\right]$

The Cosine Function and Its Inverse

$\cos(\cos^{-1} x) = x$ for every x in the interval $[-1, 1]$

$\cos^{-1}(\cos x) = x$ for every x in the interval $[0, \pi]$

The Tangent Function and Its Inverse

$\tan(\tan^{-1} x) = x$ for every real number x

$\tan^{-1}(\tan x) = x$ for every x in the interval $\left(-\dfrac{\pi}{2}, \dfrac{\pi}{2}\right)$

The restrictions on x in the inverse properties are a bit tricky. For example,
$$\sin^{-1}\left(\sin\dfrac{\pi}{4}\right) = \dfrac{\pi}{4}$$

$\sin^{-1}(\sin x) = x$ for x in $\left[-\dfrac{\pi}{2}, \dfrac{\pi}{2}\right]$. Observe that $\dfrac{\pi}{4}$ is in this interval.

Can we use $\sin^{-1}(\sin x) = x$ to find the exact value of $\sin^{-1}\left(\sin\dfrac{5\pi}{4}\right)$? Is $\dfrac{5\pi}{4}$ in the interval $\left[-\dfrac{\pi}{2}, \dfrac{\pi}{2}\right]$? No. Thus, to evaluate $\sin^{-1}\left(\sin\dfrac{5\pi}{4}\right)$, we must first find $\sin\dfrac{5\pi}{4}$.

$\dfrac{5\pi}{4}$ is in quadrant III, where the sine is negative.

$$\sin\frac{5\pi}{4} = -\sin\frac{\pi}{4} = -\frac{\sqrt{2}}{2}$$

The reference angle for $\dfrac{5\pi}{4}$ is $\dfrac{\pi}{4}$.

We evaluate $\sin^{-1}\left(\sin\dfrac{5\pi}{4}\right)$ as follows:

$$\sin^{-1}\left(\sin\frac{5\pi}{4}\right) = \sin^{-1}\left(-\frac{\sqrt{2}}{2}\right) = -\frac{\pi}{4} \quad \text{If necessary, see Table 5.5 on page 524.}$$

To determine how to evaluate the composition of functions involving inverse trigonometric functions, first examine the value of x. You can use the inverse properties in the box on the previous page only if x is in the specified interval.

EXAMPLE 6　Evaluating Compositions of Functions and Their Inverses

Find the exact value, if possible, of:

a. $\cos(\cos^{-1}0.6)$　　**b.** $\sin^{-1}\left(\sin\dfrac{3\pi}{2}\right)$　　**c.** $\cos(\cos^{-1}2\pi)$.

Solution

a. The inverse property $\cos(\cos^{-1}x) = x$ applies for every x in $[-1, 1]$. To evaluate $\cos(\cos^{-1}0.6)$, observe that $x = 0.6$. This value of x lies in $[-1, 1]$, which is the domain of the inverse cosine function. This means that we can use the inverse property $\cos(\cos^{-1}x) = x$. Thus,

$$\cos(\cos^{-1}0.6) = 0.6.$$

b. The inverse property $\sin^{-1}(\sin x) = x$ applies for every x in $\left[-\dfrac{\pi}{2}, \dfrac{\pi}{2}\right]$. To evaluate $\sin^{-1}\left(\sin\dfrac{3\pi}{2}\right)$, observe that $x = \dfrac{3\pi}{2}$. This value of x does not lie in $\left[-\dfrac{\pi}{2}, \dfrac{\pi}{2}\right]$. To evaluate this expression, we first find $\sin\dfrac{3\pi}{2}$.

$$\sin^{-1}\left(\sin\frac{3\pi}{2}\right) = \sin^{-1}(-1) = -\frac{\pi}{2} \quad \text{The angle in } \left[-\frac{\pi}{2}, \frac{\pi}{2}\right] \text{ whose sine is } -1 \text{ is } -\frac{\pi}{2}.$$

c. The inverse property $\cos(\cos^{-1}x) = x$ applies for every x in $[-1, 1]$. To attempt to evaluate $\cos(\cos^{-1}2\pi)$, observe that $x = 2\pi$. This value of x does not lie in $[-1, 1]$, which is the domain of the inverse cosine function. Thus, the expression $\cos(\cos^{-1}2\pi)$ is not defined because $\cos^{-1}2\pi$ is not defined.

Check Point 6　Find the exact value, if possible, of:

a. $\cos(\cos^{-1}0.7)$　　**b.** $\sin^{-1}(\sin\pi)$　　**c.** $\cos(\cos^{-1}\pi)$.

We can use points on terminal sides of angles in standard position to find exact values of expressions involving the composition of a function and a different inverse function. Here are two examples:

$$\cos\left(\tan^{-1}\frac{5}{12}\right) \qquad \cot\left[\sin^{-1}\left(-\frac{1}{3}\right)\right].$$

Inner part involves the angle in $\left(-\frac{\pi}{2},\frac{\pi}{2}\right)$ whose tangent is $\frac{5}{12}$.

Inner part involves the angle in $\left[-\frac{\pi}{2},\frac{\pi}{2}\right]$ whose sine is $-\frac{1}{3}$.

The inner part of each expression involves an angle. To evaluate such expressions, we represent such angles by θ. Then we use a sketch that illustrates our representation. Examples 7 and 8 show how to carry out such evaluations.

EXAMPLE 7 Evaluating a Composite Trigonometric Expression

Find the exact value of $\cos\left(\tan^{-1}\frac{5}{12}\right)$.

Solution We let θ represent the angle in $\left(-\frac{\pi}{2},\frac{\pi}{2}\right)$ whose tangent is $\frac{5}{12}$. Thus,

$$\theta = \tan^{-1}\frac{5}{12}.$$

Using the definition of the inverse tangent function, we can rewrite this as

$$\tan\theta = \frac{5}{12}.$$

Because $\tan\theta$ is positive, θ must be an angle in $\left(0,\frac{\pi}{2}\right)$. Thus, θ is a first-quadrant angle. Figure 5.76 shows a right triangle in quadrant I with

$$\tan\theta = \frac{5}{12}.$$
Side opposite θ, or y
Side adjacent to θ, or x

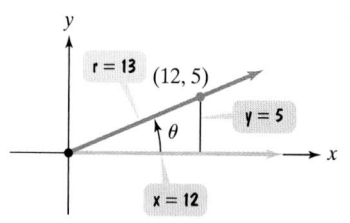

Figure 5.76 Representing $\tan\theta = \frac{5}{12}$

The hypotenuse of the triangle, r, or the distance from the origin to $(12, 5)$, is found using $r = \sqrt{x^2 + y^2}$.

$$r = \sqrt{x^2 + y^2} = \sqrt{12^2 + 5^2} = \sqrt{144 + 25} = \sqrt{169} = 13$$

We use the values for x and r to find the exact value of $\cos\left(\tan^{-1}\frac{5}{12}\right)$.

$$\cos\left(\tan^{-1}\frac{5}{12}\right) = \cos\theta = \frac{\text{side adjacent to }\theta,\text{ or }x}{\text{hypotenuse, or }r} = \frac{12}{13}$$

Check Point 7 Find the exact value of $\sin\left(\tan^{-1}\frac{3}{4}\right)$.

EXAMPLE 8 Evaluating a Composite Trigonometric Expression

Find the exact value of $\cot\left[\sin^{-1}\left(-\frac{1}{3}\right)\right]$.

Solution We let θ represent the angle in $\left[-\frac{\pi}{2},\frac{\pi}{2}\right]$ whose sine is $-\frac{1}{3}$. Thus,

$$\theta = \sin^{-1}\left(-\frac{1}{3}\right) \quad \text{and} \quad \sin\theta = -\frac{1}{3}.$$

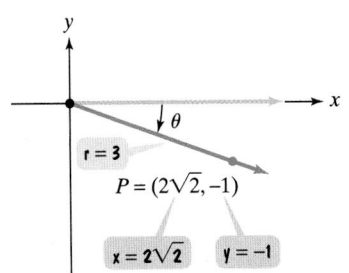

Figure 5.77 Representing $\sin\theta = -\frac{1}{3}$

Because $\sin\theta$ is negative in $\sin\theta = -\frac{1}{3}$, θ must be an angle in $\left[-\frac{\pi}{2}, 0\right)$. Thus, θ is a negative angle that lies in quadrant IV. Figure 5.77 shows angle θ in quadrant IV with

In quadrant IV, y is negative.

$$\sin\theta = -\frac{1}{3} = \frac{y}{r} = \frac{-1}{3}.$$

Thus, $y = -1$ and $r = 3$. The value of x can be found using $r = \sqrt{x^2 + y^2}$ or $x^2 + y^2 = r^2$.

$$x^2 + (-1)^2 = 3^2$$
$$x^2 + 1 = 9$$
$$x^2 = 8$$
$$x = \sqrt{8} = \sqrt{4 \cdot 2} = 2\sqrt{2} \qquad \text{Remember that x is positive in quadrant IV.}$$

We use values for x and y to find the exact value of $\cot\left[\sin^{-1}\left(-\frac{1}{3}\right)\right]$.

$$\cot\left[\sin^{-1}\left(-\frac{1}{3}\right)\right] = \cot\theta = \frac{x}{y} = \frac{2\sqrt{2}}{-1} = -2\sqrt{2}$$

Check Point 8 Find the exact value of $\cos\left[\sin^{-1}\left(-\frac{1}{2}\right)\right]$.

Some composite functions with inverse trigonometric functions can be simplified to algebraic expressions. To simplify such an expression, we represent the inverse trigonometric function in the expression by θ. Then we use a right triangle.

EXAMPLE 9 Simplifying an Expression Involving $\sin^{-1}x$

If $0 < x \leq 1$, write $\cos(\sin^{-1}x)$ as an algebraic expression in x.

Solution We let θ represent the angle in $\left[-\frac{\pi}{2}, \frac{\pi}{2}\right]$ whose sine is x. Thus,

$$\theta = \sin^{-1}x, \quad \text{and} \quad \sin\theta = x.$$

Because $0 < x \leq 1$, $\sin\theta$ is positive. Thus, θ is a first-quadrant angle and can be represented as an acute angle of a right triangle. Figure 5.78 shows a right triangle with

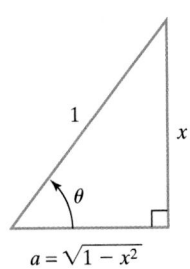

Figure 5.78 Representing $\sin\theta = x$

$$\sin\theta = x = \frac{x}{1}. \qquad \text{Side opposite } \theta$$
$$\text{Hypotenuse}$$

The third side, a in Figure 5.78, can be found using the Pythagorean Theorem.

$$a^2 + x^2 = 1^2 \qquad \text{Apply the Pythagorean Theorem to the right triangle in Figure 5.78.}$$

$$a^2 = 1 - x^2 \qquad \text{Subtract } x^2 \text{ from both sides.}$$

$$a = \sqrt{1 - x^2} \qquad \text{Solve for } a.$$

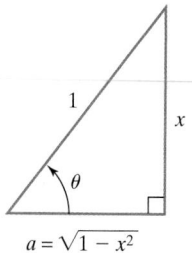

$a = \sqrt{1 - x^2}$

Figure 5.78, repeated

We use the right triangle in Figure 5.78 to write $\cos(\sin^{-1} x)$ as an algebraic expression.

$$\cos(\sin^{-1} x) = \cos\theta = \frac{\text{side adjacent to } \theta}{\text{hypotenuse}} = \frac{\sqrt{1 - x^2}}{1} = \sqrt{1 - x^2}$$

Check Point 9 If $x > 0$, write $\sec(\tan^{-1} x)$ as an algebraic expression in x.

The inverse secant function, $y = \sec^{-1} x$, is used in calculus. However, inverse cotangent and inverse cosecant functions are rarely used. Two of these remaining inverse trigonometric functions are briefly developed in the exercise set that follows.

EXERCISE SET 5.7

Practice Exercises

In Exercises 1–18, find the exact value of each expression.

1. $\sin^{-1}\dfrac{1}{2}$ **2.** $\sin^{-1} 0$ **3.** $\sin^{-1}\dfrac{\sqrt{2}}{2}$

4. $\sin^{-1}\dfrac{\sqrt{3}}{2}$ **5.** $\sin^{-1}\left(-\dfrac{1}{2}\right)$ **6.** $\sin^{-1}\left(-\dfrac{\sqrt{3}}{2}\right)$

7. $\cos^{-1}\dfrac{\sqrt{3}}{2}$ **8.** $\cos^{-1}\dfrac{\sqrt{2}}{2}$ **9.** $\cos^{-1}\left(-\dfrac{\sqrt{2}}{2}\right)$

10. $\cos^{-1}\left(-\dfrac{\sqrt{3}}{2}\right)$ **11.** $\cos^{-1} 0$ **12.** $\cos^{-1} 1$

13. $\tan^{-1}\dfrac{\sqrt{3}}{3}$ **14.** $\tan^{-1} 1$ **15.** $\tan^{-1} 0$

16. $\tan^{-1}(-1)$ **17.** $\tan^{-1}(-\sqrt{3})$ **18.** $\tan^{-1}\left(-\dfrac{\sqrt{3}}{3}\right)$

In Exercises 19–30, use a calculator to find the value of each expression rounded to two decimal places.

19. $\sin^{-1} 0.3$ **20.** $\sin^{-1} 0.47$

21. $\sin^{-1}(-0.32)$ **22.** $\sin^{-1}(-0.625)$

23. $\cos^{-1}\dfrac{3}{8}$ **24.** $\cos^{-1}\dfrac{4}{9}$

25. $\cos^{-1}\dfrac{\sqrt{5}}{7}$ **26.** $\cos^{-1}\dfrac{\sqrt{7}}{10}$

27. $\tan^{-1}(-20)$ **28.** $\tan^{-1}(-30)$

29. $\tan^{-1}(-\sqrt{473})$ **30.** $\tan^{-1}(-\sqrt{5061})$

In Exercises 31–46, find the exact value of each expression, if possible. Do not use a calculator.

31. $\sin(\sin^{-1} 0.9)$ **32.** $\cos(\cos^{-1} 0.57)$

33. $\sin^{-1}\left(\sin\dfrac{\pi}{3}\right)$ **34.** $\cos^{-1}\left(\cos\dfrac{2\pi}{3}\right)$

35. $\sin^{-1}\left(\sin\dfrac{5\pi}{6}\right)$ **36.** $\cos^{-1}\left(\cos\dfrac{4\pi}{3}\right)$

37. $\tan(\tan^{-1} 125)$ **38.** $\tan(\tan^{-1} 380)$

39. $\tan^{-1}\left[\tan\left(-\dfrac{\pi}{6}\right)\right]$ **40.** $\tan^{-1}\left[\tan\left(-\dfrac{\pi}{3}\right)\right]$

41. $\tan^{-1}\left(\tan\dfrac{2\pi}{3}\right)$ **42.** $\tan^{-1}\left(\tan\dfrac{3\pi}{4}\right)$

43. $\sin^{-1}(\sin\pi)$ **44.** $\cos^{-1}(\cos 2\pi)$

45. $\sin(\sin^{-1}\pi)$ **46.** $\cos(\cos^{-1} 3\pi)$

In Exercises 47–60, use a sketch to find the exact value of each expression.

47. $\cos\left(\sin^{-1}\frac{4}{5}\right)$ **48.** $\sin\left(\tan^{-1}\frac{7}{24}\right)$

49. $\tan\left(\cos^{-1}\frac{5}{13}\right)$ **50.** $\cot\left(\sin^{-1}\frac{5}{13}\right)$

51. $\tan\left[\sin^{-1}\left(-\frac{3}{5}\right)\right]$ **52.** $\cos\left[\sin^{-1}\left(-\frac{4}{5}\right)\right]$

53. $\sin\left(\cos^{-1}\dfrac{\sqrt{2}}{2}\right)$ **54.** $\cos\left(\sin^{-1}\frac{1}{2}\right)$

55. $\sec\left[\sin^{-1}\left(-\frac{1}{4}\right)\right]$ **56.** $\sec\left[\sin^{-1}\left(-\frac{1}{2}\right)\right]$

57. $\tan\left[\cos^{-1}\left(-\frac{1}{3}\right)\right]$ **58.** $\tan\left[\cos^{-1}\left(-\frac{1}{4}\right)\right]$

59. $\csc\left[\cos^{-1}\left(-\dfrac{\sqrt{3}}{2}\right)\right]$ **60.** $\sec\left[\sin^{-1}\left(-\dfrac{\sqrt{2}}{2}\right)\right]$

In Exercises 61–70, use a right triangle to write each expression as an algebraic expression. Assume that x is positive and in the domain of the given inverse trigonometric function.

61. $\tan(\cos^{-1} x)$ **62.** $\sin(\tan^{-1} x)$

63. $\cos(\sin^{-1} 2x)$ **64.** $\sin(\cos^{-1} 2x)$

65. $\cos\left(\sin^{-1}\dfrac{1}{x}\right)$

66. $\sec\left(\cos^{-1}\dfrac{1}{x}\right)$

67. $\cot\left(\tan^{-1}\dfrac{x}{\sqrt{3}}\right)$

68. $\cot\left(\tan^{-1}\dfrac{x}{\sqrt{2}}\right)$

69. $\sec\left(\sin^{-1}\dfrac{x}{\sqrt{x^2+4}}\right)$

70. $\cot\left(\sin^{-1}\dfrac{\sqrt{x^2-9}}{x}\right)$

71. a. Graph the restricte-d secant function, $y = \sec x$, by restricting x to the intervals $\left[0, \dfrac{\pi}{2}\right)$ and $\left(\dfrac{\pi}{2}, \pi\right]$.

b. Use the horizontal line test to explain why the restricted secant function has an inverse function.

c. Use the graph of the restricted secant function to graph $y = \sec^{-1} x$.

72. a. Graph the restricted cotangent function, $y = \cot x$, by restricting x to the interval $(0, \pi)$.

b. Use the horizontal line test to explain why the restricted cotangent function has an inverse function.

c. Use the graph of the restricted cotangent function to graph $y = \cot^{-1} x$.

 Application Exercises

73. Your neighborhood movie theater has a 25-foot-high screen located 8 feet above your eye level. If you sit too close to the screen, your viewing angle is too small, resulting in a distorted picture. By contrast, if you sit too far back, the image is quite small, diminishing the movie's visual impact. If you sit x feet back from the screen, your viewing angle, θ, is given by

$$\theta = \tan^{-1}\frac{33}{x} - \tan^{-1}\frac{8}{x}.$$

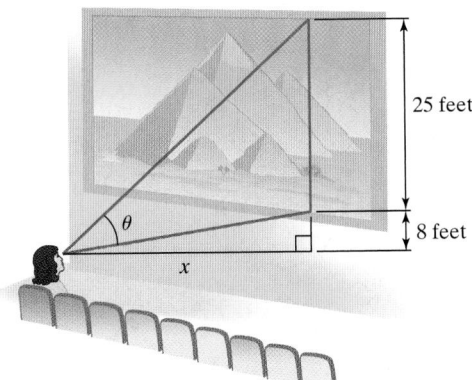

25 feet

8 feet

Find the viewing angle, in radians, at distances of 5 feet, 10 feet, 15 feet, 20 feet, and 25 feet.

74. The function $\theta = \tan^{-1}\dfrac{33}{x} - \tan^{-1}\dfrac{8}{x}$, described in Exercise 73, is graphed below in a $[0, 50, 10]$ by $[0, 1, 0.1]$ viewing rectangle. Use the graph to describe what happens to your viewing angle as you move farther back from the screen. How far back from the screen, to the nearest foot, should you sit to maximize your viewing angle? Verify this observation by finding the viewing angle one foot closer to the screen and one foot farther from the screen for this ideal viewing distance.

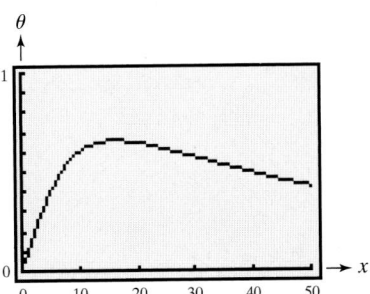

The formula

$$\theta = 2\tan^{-1}\frac{21.634}{x}$$

gives the viewing angle, θ, in radians, for a camera whose lens is x millimeters wide. Use this formula to solve Exercises 75–76.

75. Find the viewing angle, in radians and in degrees (to the nearest tenth of a degree), of a 28-millimeter lens.

76. Find the viewing angle, in radians and degrees (to the nearest tenth of a degree), of a 300-millimeter telephoto lens.

For years, mathematicians were challenged by the following problem: What is the area of a region under a curve between two values of x? The problem was solved in the seventeenth century with the development of integral calculus. Using calculus, the area of the region under $y = \dfrac{1}{x^2+1}$, above the x-axis, and between $x = a$ and $x = b$ is $\tan^{-1}b - \tan^{-1}a$. Use this result, shown in the figure, on the next page, to find

the area of the region under $y = \dfrac{1}{x^2 + 1}$, *above the x-axis,*
and between the values of a and b given in Exercises 77–78.

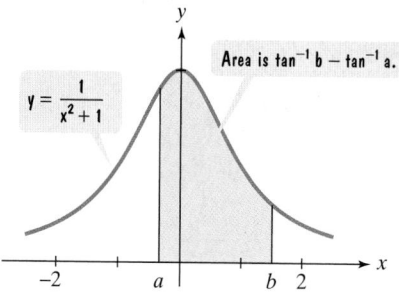

77. $a = 0$ and $b = 2$

78. $a = -2$ and $b = 1$

Writing in Mathematics

79. Explain why, without restrictions, no trigonometric function has an inverse function.

80. Describe the restriction on the sine function so that it has an inverse function.

81. How can the graph of $y = \sin^{-1} x$ be obtained from the graph of the restricted sine function?

82. Without drawing a graph, describe the behavior of the graph of $y = \sin^{-1} x$. Mention the function's domain and range in your description.

83. Describe the restriction on the cosine function so that it has an inverse function.

84. Without drawing a graph, describe the behavior of the graph of $y = \cos^{-1} x$. Mention the function's domain and range in your description.

85. Describe the restriction on the tangent function so that it has an inverse function.

86. Without drawing a graph, describe the behavior of the graph of $y = \tan^{-1} x$. Mention the function's domain and range in your description.

87. If $\sin^{-1}\left(\sin\dfrac{\pi}{3}\right) = \dfrac{\pi}{3}$, is $\sin^{-1}\left(\sin\dfrac{5\pi}{6}\right) = \dfrac{5\pi}{6}$? Explain your answer.

88. Explain how a right triangle can be used to find the exact value of $\sec\left(\sin^{-1}\frac{4}{5}\right)$.

89. Find the height of the screen and the number of feet that it is located above eye level in your favorite movie theater. Modify the formula given in Exercise 73 so that it applies to your theater. Then describe where in the theater you should sit so that a movie creates the greatest visual impact.

Technology Exercises

In Exercises 90–93, graph each pair of functions in the same viewing rectangle. Use your knowledge of the domain and range for the inverse trigonometric functions to select an appropriate viewing rectangle. What does the graph of the second equation in each exercise do to the graph of the first equation?

90. $y = \sin^{-1} x$ and $y = \sin^{-1} x + 2$

91. $y = \cos^{-1} x$ and $y = \cos^{-1} (x - 1)$

92. $y = \tan^{-1} x$ and $y = -2\tan^{-1} x$

93. $y = \sin^{-1} x$ and $y = \sin^{-1} (x + 2) + 1$

94. Graph $y = \tan^{-1} x$ and its two horizontal asymptotes in a $[-3, 3, 1]$ by $\left[-\pi, \pi, \dfrac{\pi}{2}\right]$ viewing rectangle. Then change the range setting to $[-50, 50, 5]$ by $\left[-\pi, \pi, \dfrac{\pi}{2}\right]$. What do you observe?

95. Graph $y = \sin^{-1} x + \cos^{-1} x$ in a $[-2, 2, 1]$ by $[0, 3, 1]$ viewing rectangle. What appears to be true about the sum of the inverse sine and inverse cosine for values between -1 and 1 inclusive?

Critical Thinking Exercises

96. Solve $y = 2\sin^{-1} (x - 5)$ for x in terms of y.

97. Solve for x: $2\sin^{-1} x = \dfrac{\pi}{4}$.

98. Prove that if $x > 0$, $\tan^{-1} x + \tan^{-1}\dfrac{1}{x} = \dfrac{\pi}{2}$.

99. Derive the formula for θ, your viewing angle at the movie theater, in Exercise 73. *Hint:* Use the figure shown and represent the acute angle on the left in the smaller right triangle by α. Find expressions for $\tan \alpha$ and $\tan (\alpha + \theta)$.

SECTION 5.8   *Applications of Trigonometric Functions*

Objectives

1. Solve a right triangle.
2. Solve problems involving bearings.
3. Model simple harmonic motion.

In the late 1960s, popular musicians were searching for new sounds. Film composers were looking for ways to create unique sounds as well. From these efforts, synthesizers that electronically reproduce musical sounds were born. From providing the backbone of today's most popular music to providing the strange sounds for the most experimental music, synthesizers are at the forefront of music technology.

If we did not understand the periodic nature of sinusoidal functions, the synthesizers used in almost all forms of music would not exist. In this section, we look at applications of trigonometric functions in right triangles and in modeling periodic phenomena such as sound.

Solve a right triangle.

Solving Right Triangles

Solving a right triangle means finding the missing lengths of its sides and the measurements of its angles. We will label right triangles so that side a is opposite angle A, side b is opposite angle B, and side c, the hypotenuse, is opposite right angle C. Figure 5.79 illustrates this labeling.

When solving a right triangle, we will use the sine, cosine, and tangent functions, rather than their reciprocals. Example 1 shows how to solve a right triangle when we know the length of a side and the measure of an acute angle.

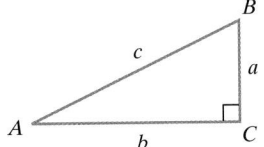

Figure 5.79 Labeling right triangles

EXAMPLE 1   Solving a Right Triangle

Solve the right triangle shown in Figure 5.80.

Solution   We begin by finding the measure of angle B. We do not need a trigonometric function to do so. Because $C = 90°$ and the sum of a triangle's angles is $180°$, we see that $A + B = 90°$. Thus,

$$B = 90° - A = 90° - 34.5° = 55.5°.$$

Now we need to find a. Because we have a known angle, an unknown opposite side, and a known adjacent side, we use the tangent function.

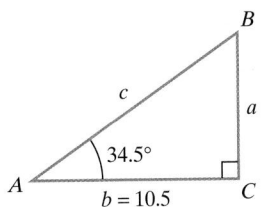

Figure 5.80 Find B, a, and c.

$$\tan 34.5° = \frac{a}{10.5}$$

> Side opposite the 34.5° angle

> Side adjacent to the 34.5° angle

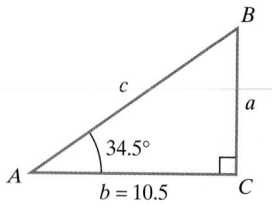

Figure 5.80, repeated

Now we multiply both sides of this equation by 10.5 and solve for a.

$$a = 10.5 \tan 34.5° \approx 7.22$$

Finally, we need to find c. Because we have a known angle, a known adjacent side, and an unknown hypotenuse, we use the cosine function.

$$\cos 34.5° = \frac{10.5}{c}$$

Side adjacent to the 34.5° angle

hypotenuse

Discovery

There is often more than one correct way to solve a right triangle. In Example 1, find a using angle $B = 55.5°$. Find c using the Pythagorean Theorem.

Now we multiply both sides of this equation by c and then solve for c.

$$c \cos 34.5° = 10.5 \qquad \text{Multiply both sides by } c.$$

$$c = \frac{10.5}{\cos 34.5°} \approx 12.74 \qquad \text{Divide both sides by } \cos 34.5° \text{ and solve for } c.$$

In summary, $B = 55.5°$, $a \approx 7.22$, and $c \approx 12.74$.

Check Point 1 In Figure 5.79, let $A = 62.7°$ and $a = 8.4$. Solve the right triangle, rounding lengths to two decimal places.

Trigonometry was first developed to measure heights and distances that were inconvenient or impossible to measure. In solving application problems, begin by making a sketch involving a right triangle that illustrates the problem's conditions. Then put your knowledge of solving right triangles to work and find the required distance or height.

EXAMPLE 2 Finding the Side of a Triangle

From a point on level ground 125 feet from the base of a tower, the angle of elevation is 57.2°. Approximate the height of the tower to the nearest foot.

Solution A sketch is shown in Figure 5.81, where a represents the height of the tower. In the right triangle, we have a known angle, an unknown opposite side, and a known adjacent side. Therefore, we use the tangent function.

$$\tan 57.2° = \frac{a}{125}$$

Side opposite the 57.2° angle

Side adjacent to the 57.2° angle

Now we multiply both sides of this equation by 125 and solve for a.

$$a = 125 \tan 57.2° \approx 194$$

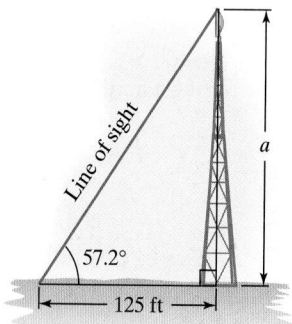

Figure 5.81 Determining height without using direct measurement

The tower is approximately 194 feet high.

Check Point 2 From a point on level ground 80 feet from the base of the Eiffel Tower, the angle of elevation is 85.4°. Approximate the height of the Eiffel Tower to the nearest foot.

Example 3 illustrates how to find the measure of an acute angle of a right triangle if the lengths of two sides are known.

EXAMPLE 3   Finding the Angle of a Triangle

A kite flies at a height of 30 feet when 65 feet of string is out. If the string is in a straight line, find the angle that it makes with the ground. Round to the nearest tenth of a degree.

Solution   A sketch is shown in Figure 5.82, where A represents the angle the string makes with the ground. In the right triangle, we have an unknown angle, a known opposite side, and a known hypotenuse. Therefore, we use the sine function.

$$\sin A = \frac{30}{65} \quad \begin{array}{l}\text{Side opposite } A\\ \text{hypotenuse}\end{array}$$

$$A = \sin^{-1}\frac{30}{65} \approx 27.5°$$

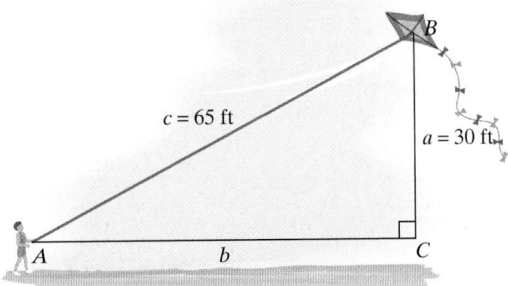

Figure 5.82 Flying a kite

The string makes an angle of approximately $27.5°$ with the ground.

Check Point 3   A guy wire is 13.8 yards long and is attached from the ground to a pole 6.7 yards above the ground. Find the angle, to the nearest tenth of a degree, that the wire makes with the ground.

EXAMPLE 4   Using Two Right Triangles to Solve a Problem

You are taking your first hot-air balloon ride. Your friend is standing on level ground, 100 feet away from your point of launch, making a video of the terrified look on your rapidly ascending face. How rapidly? At one instant, the angle of elevation from the video camera to your face is $31.7°$. One minute later, the angle of elevation is $76.2°$. How far did you travel during that minute?

Solution   A sketch that illustrates the problem is shown in Figure 5.83. We need to determine $b - a$, the distance traveled during the one-minute period. We find a using the small right triangle. Because we have a known angle, an unknown opposite side, and a known adjacent side, we use the tangent function.

$$\tan 31.7° = \frac{a}{100} \quad \begin{array}{l}\text{Side opposite the } 31.7° \text{ angle}\\ \text{Side adjacent to the } 31.7° \text{ angle}\end{array}$$

$$a = 100 \tan 31.7° \approx 61.8$$

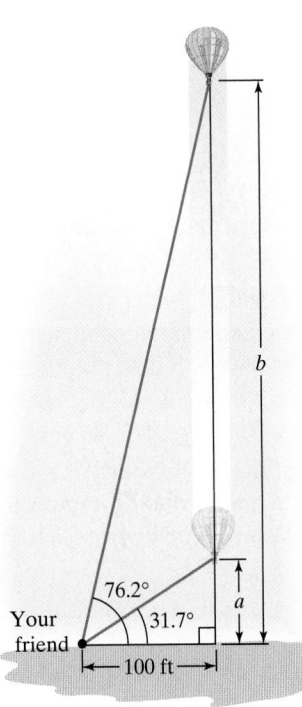

Figure 5.83 Ascending in a hot-air balloon

We find b using the tangent function in the large right triangle. Turn back the page and verify this observation in Figure 5.83.

$$\tan 76.2° = \frac{b}{100}$$
Side opposite the 76.2° angle
Side adjacent to the 76.2° angle

$$b = 100 \tan 76.2° \approx 407.1$$

The balloon traveled $407.1 - 61.8$, or approximately 345.3 feet, during the minute.

> **Check Point 4**
> You are standing on level ground 800 feet from Mt. Rushmore, looking at the sculpture of Abraham Lincoln's face. The angle of elevation to the bottom of the sculpture is 32° and the angle of elevation to the top is 35°. Find the height of the sculpture of Lincoln's face to the nearest tenth of a foot.

2 Solve problems involving bearings.

Trigonometry and Bearings

In navigation and surveying problems, the term *bearing* is used to specify the location of one point relative to another. The **bearing** from point O to point P is the acute angle between ray OP and a north-south line. Figure 5.84 illustrates some examples of bearings. The north-south line and the east-west line intersect at right angles.

Study Tip

The bearing from O to P can also be described using the phrase "the bearing of P from O."

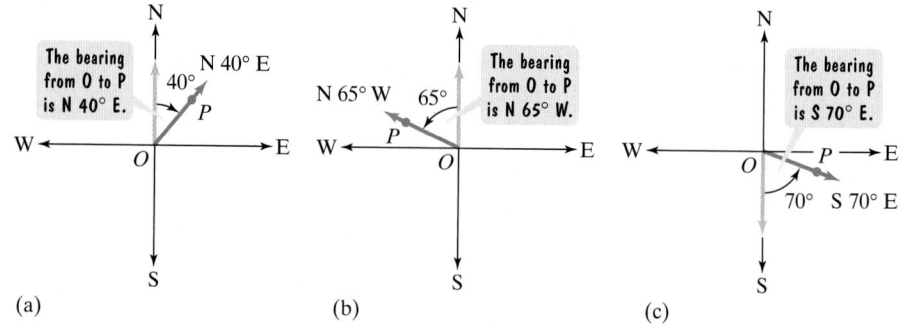

Figure 5.84 An illustration of three bearings

(a) (b) (c)

Each bearing has three parts: a letter (N or S), the measure of an acute angle, and a letter (E or W). Here's how we write a bearing:

- If the acute angle is measured from the *north side* of the north-south line, then we write N first. [See Figure 5.84(a).] If the acute angle is measured from the *south side* of the north-south line, then we write S first. [See Figure 5.84(c).]
- Second, we write the measure of the acute angle.
- If the acute angle is measured on the *east side* of the north-south line, then we write E last. [See Figure 5.84(a)]. If the acute angle is measured on the *west side* of the north-south line, then we write W last. [See Figure 5.84(b).]

EXAMPLE 5 Understanding Bearings

Use Figure 5.85 on the next page to find:

a. the bearing from O to B.

b. the bearing from O to A.

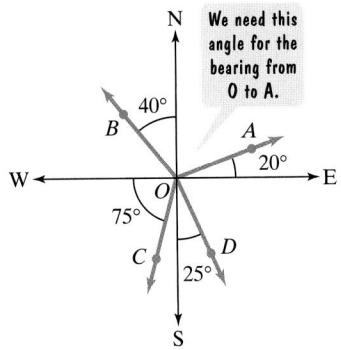

Figure 5.85 Finding bearings

Solution

a. To find the bearing from O to B, we need the acute angle between the ray OB and the north-south line through O. The measurement of this angle is given to be 40°. Figure 5.85 shows that the angle is measured from the north side of the north-south line and lies west of the north-south line. Thus, the bearing from O to B is N 40° W.

b. To find the bearing from O to A, we need the acute angle between the ray OA and the north-south line through O. This angle is specified by the voice balloon in Figure 5.85. Because of the given 20° angle, this angle measures $90° - 20°$, or 70°. This angle is measured from the north side of the north-south line. This angle is also east of the north-south line. Thus, the bearing from O to A is N 70° E.

Check Point 5 Use Figure 5.85 to find:

a. the bearing from O to D.

b. the bearing from O to C.

EXAMPLE 6 Finding the Bearing of a Boat

A boat leaves the entrance to a harbor and travels 25 miles on a bearing of N 42° E. Figure 5.86 shows that the captain then turns the boat 90°clockwise and travels 18 miles on a bearing of S 48° E. At that time:

a. How far is the boat from the harbor entrance?

b. What is the bearing of the boat from the harbor entrance?

Solution

a. The boat's distance from the harbor entrance is represented by c in Figure 5.86. Because we know the length of two sides of the right triangle, we find c using the Pythagorean Theorem. We have

$$c^2 = a^2 + b^2 = 25^2 + 18^2 = 949$$
$$c = \sqrt{949} \approx 30.8.$$

The boat is approximately 30.8 miles from the harbor entrance.

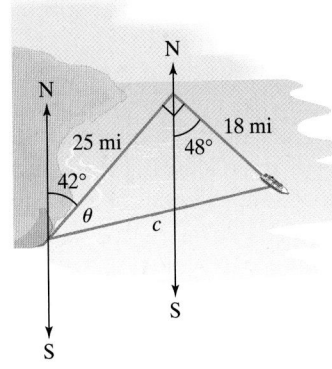

Figure 5.86 Finding a boat's bearing from the harbor entrance

Study Tip

When making a diagram showing bearings, draw a north-south line through each point at which a change in course occurs. The north side of the line lies above each point. The south side of the line lies below each point.

b. The bearing of the boat from the harbor entrance means the bearing from the entrance to the boat. Look at the north-south line passing through the harbor entrance on the left in Figure 5.86. The acute angle from this line to the ray on which the boat lies is $42° + \theta$. Because we are measuring the angle from the north side of the line and the boat is east of the harbor, its bearing from the harbor entrance is N$(42° + \theta)$E. To find θ, we use the right triangle shown in Figure 5.86 and the tangent function.

$$\tan\theta = \frac{\text{side opposite } \theta}{\text{side adjacent to } \theta} = \frac{18}{25}$$

$$\theta = \tan^{-1}\frac{18}{25}$$

We can use a calculator in degree mode to find the value of θ: $\theta \approx 35.8°$. Thus, $42° + \theta = 42° + 35.8° = 77.8°$. The bearing of the boat from the harbor entrance is N 77.8° E.

You leave the entrance to a system of hiking trails and hike 2.3 miles on a bearing of S 31° W. Then the trail turns 90° clockwise and you hike 3.5 miles on a bearing of N 59° W. At that time:

a. How far are you from the entrance to the trail system?

b. What is your bearing from the entrance to the trail system?

3 Model simple harmonic motion.

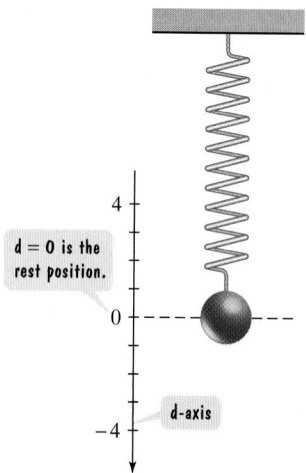

Figure 5.87 Using a d-axis to describe a ball's distance from its rest position

Simple Harmonic Motion

Because of their periodic nature, trigonometric functions are used to model phenomena that occur again and again. This includes vibratory or oscillatory motion, such as the motion of a vibrating guitar string, the swinging of a pendulum, or the bobbing of an object attached to a spring. Trigonometric functions are also used to describe radio waves from your favorite FM station, television waves from your not-to-be-missed weekly sitcom, and sound waves from your most-prized CDs.

To see how trigonometric functions are used to model vibratory motion, consider this: A ball is attached to a spring hung from the ceiling. You pull the ball down 4 inches and then release it. If we neglect the effects of friction and air resistance, the ball will continue bobbing up and down on the end of the spring. These up-and-down oscillations are called **simple harmonic motion.**

To better understand this motion, we use a d-axis, where d represents distance. This axis is shown in Figure 5.87. On this axis, the position of the ball before you pull it down is $d = 0$. This rest position is called the **equilibrium position.** Now you pull the ball down 4 inches to $d = -4$ and release it. Figure 5.88 shows a sequence of "photographs" taken at one-second time intervals illustrating the distance of the ball from its rest position, d.

The curve in Figure 5.88 shows how the ball's distance from its rest position changes over time. The curve is sinusoidal and the motion can be described using a cosine or a sine function.

Figure 5.88 A sequence of "photographs" showing the bobbing ball's distance from the rest position, taken at one-second intervals

Simple Harmonic Motion

An object that moves on a coordinate axis is in **simple harmonic motion** if its distance from the origin, d, at time t is given by either

$$d = a \cos \omega t \ \text{ or } \ d = a \sin \omega t.$$

The motion has **amplitude** $|a|$, the maximum displacement of the object from its rest position. The **period** of the motion is $\dfrac{2\pi}{\omega}$, where $\omega > 0$. The period gives the time it takes for the motion to go through one complete cycle.

In describing simple harmonic motion, the equation with the cosine function is used if the object is at its greatest distance from rest position, the origin, at $t = 0$. By contrast, the equation with the sine function is used if the object is at its rest position, the origin, at $t = 0$.

Diminishing Motion with Increasing Time

Due to friction and other resistive forces, the motion of an oscillating object decreases over time. The function

$$d = 3e^{-0.1t} \cos 2t$$

models this type of motion. The graph of the function is shown in a $t = [0, 10, 1]$ by $d = [-3, 3, 1]$ viewing rectangle. Notice how the amplitude is decreasing with time as the moving object loses energy.

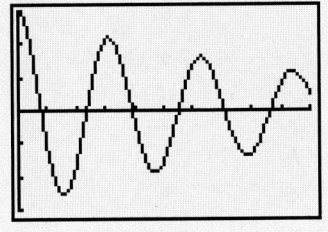

EXAMPLE 7 Finding an Equation for an Object in Simple Harmonic Motion

A ball on a spring is pulled 4 inches below its rest position and then released. The period of the motion is 6 seconds. Write the equation for the ball's simple harmonic motion.

Solution We need to write an equation that describes d, the distance of the ball from its rest position, after t seconds. (The motion is illustrated by the "photo" sequence in Figure 5.88 on page 542.) When the object is released ($t = 0$), the ball's distance from its rest position is 4 inches down. Because it is *down* 4 inches, d is negative: When $t = 0$, $d = -4$. Notice the greatest distance from rest position occurs at $t = 0$. Thus, we will use the equation with the cosine function,

$$d = a \cos \omega t,$$

to model the ball's simple harmonic motion.

Now we determine values for a and ω. Recall that $|a|$ is the maximum displacement. Because the ball initially moves down, $a = -4$.

The value of ω in $d = a \cos \omega t$ can be found using the formula for the period.

$$\text{period} = \frac{2\pi}{\omega} = 6 \qquad \text{We are given that the period of the motion is 6 seconds.}$$

$$2\pi = 6\omega \qquad \text{Multiply both sides by } \omega.$$

$$\omega = \frac{2\pi}{6} = \frac{\pi}{3} \qquad \text{Divide both sides by 6 and solve for } \omega.$$

We see that $a = -4$ and $\omega = \dfrac{\pi}{3}$. Substitute these values into $d = a \cos \omega t$. The equation for the ball's simple harmonic motion is

$$d = -4 \cos \frac{\pi}{3} t.$$

Modeling Music

Check Point 7 A ball on a spring is pulled 6 inches below its rest position and then released. The period for the motion is 4 seconds. Write the equation for the ball's simple harmonic motion.

The period of the harmonic motion in Example 7 was 6 seconds. It takes 6 seconds for the moving object to complete one cycle. Thus, $\frac{1}{6}$ of a cycle is completed every second. We call $\frac{1}{6}$ the *frequency* of the moving object. **Frequency** describes the number of complete cycles per unit time and is the reciprocal of the period.

Frequency of an Object in Simple Harmonic Motion

An object in simple harmonic motion given by

$$d = a \cos \omega t \ \text{ or } \ d = a \sin \omega t$$

has **frequency** f given by

$$f = \frac{\omega}{2\pi}, \omega > 0.$$

Equivalently,

$$f = \frac{1}{\text{period}}.$$

EXAMPLE 8   Analyzing Simple Harmonic Motion

Figure 5.89 shows a mass on a smooth table attached to a spring. The mass moves in simple harmonic motion described by

$$d = 10 \cos \frac{\pi}{6} t$$

with t measured in seconds and d in centimeters. Find (**a.**) the maximum displacement, (**b.**) the frequency, and (**c.**) the time required for one cycle.

Figure 5.89 A mass attached to a spring, moving in simple harmonic motion

Solution   We begin by identifying values for a and ω.

$$d = 10 \cos \frac{\pi}{6} t$$

The form of this equation is
$d = a \cos \omega t$
with $a = 10$ and $\omega = \frac{\pi}{6}$.

a. The maximum displacement from the rest position is the amplitude. Because $a = 10$, the maximum displacement is 10 centimeters.

b. The frequency, f, is

$$f = \frac{\omega}{2\pi} = \frac{\frac{\pi}{6}}{2\pi} = \frac{\pi}{6} \cdot \frac{1}{2\pi} = \frac{1}{12}.$$

The frequency is $\frac{1}{12}$ cycle (or oscillation) per second.

c. The time required for one cycle is the period.

$$\text{period} = \frac{2\pi}{\omega} = \frac{2\pi}{\frac{\pi}{6}} = 2\pi \cdot \frac{6}{\pi} = 12$$

The time required for one cycle is 12 seconds.

Check Point 8

An object moves in simple harmonic motion described by $d = 12 \cos \frac{\pi}{4} t$, where t is measured in seconds and d in centimeters. Find (**a.**) the maximum displacement, (**b.**) the frequency, and (**c.**) the time required for one cycle.

Resisting Damage of Simple Harmonic Motion

Simple harmonic motion from an earthquake caused this highway in Oakland, California, to collapse. By studying the harmonic motion of the soil under the highway, engineers learn to build structures that can resist damage.

EXERCISE SET 5.8

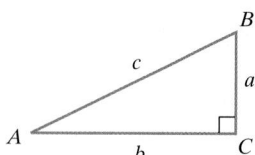

Practice Exercises

In Exercises 1–12, solve the right triangle shown in the figure. Round lengths to two decimal places and express angles to the nearest tenth of a degree.

A triangle with vertices A, B, C. Side c connects A and B, side a connects B and C (vertical), side b connects A and C. The right angle is at C.

1. $A = 23.5°, b = 10$ **2.** $A = 41.5°, b = 20$

3. $A = 52.6°, c = 54$ **4.** $A = 54.8°, c = 80$

5. $B = 16.8°, b = 30.5$ **6.** $B = 23.8°, b = 40.5$

7. $a = 30.4, c = 50.2$ **8.** $a = 11.2, c = 65.8$

9. $a = 10.8, b = 24.7$ **10.** $a = 15.3, b = 17.6$

11. $b = 2, c = 7$ **12.** $b = 4, c = 9$

Use the figure shown to solve Exercises 13–16.

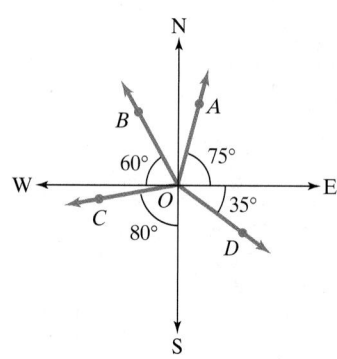

13. Find the bearing from O to A.

14. Find the bearing from O to B.

15. Find the bearing from O to C.

16. Find the bearing from O to D.

In Exercises 17–20, an object is attached to a coiled spring. The object is pulled down (negative direction from the rest position) and then released. Write an equation for the distance of the object from its rest position after t seconds.

Distance from rest position at $t = 0$	Amplitude	Period
17. 6 centimeters	6 centimeters	4 seconds
18. 8 inches	8 inches	2 seconds
19. 0	3 inches	1.5 seconds
20. 0	5 centimeters	2.5 seconds

In Exercises 21–28, an object moves in simple harmonic motion described by the given equation, where t is measured in seconds and d in inches. In each exercise, find:

a. *the maximum displacement*

b. *the frequency*

c. *the time required for one cycle.*

21. $d = 5 \cos \dfrac{\pi}{2} t$ **22.** $d = 10 \cos 2\pi t$

23. $d = -6 \cos 2\pi t$ **24.** $d = -8 \cos \dfrac{\pi}{2} t$

25. $d = \frac{1}{2} \sin 2t$ **26.** $d = \frac{1}{3} \sin 2t$

27. $d = -5 \sin \dfrac{2\pi}{3} t$ **28.** $d = -4 \sin \dfrac{3\pi}{2} t$

Application Exercises

29. The tallest television transmitting tower in the world is in North Dakota. From a point on level ground 5280 feet (one mile) from the base of the tower, the angle of elevation is 21.3°. Approximate the height of the tower to the nearest foot.

30. From a point on level ground 30 yards from the base of a building, the angle of elevation is 38.7°. Approximate the height of the building to the nearest foot.

31. The Statue of Liberty is approximately 305 feet tall. If the angle of elevation of a ship to the top of the statue is 23.7°, how far, to the nearest foot, is the ship from the statue's base?

32. A 200-foot cliff drops vertically into the ocean. If the angle of elevation of a ship to the top of the cliff is 22.3°, how far off shore, to the nearest foot, is the ship?

33. A helicopter hovers 1000 feet above a small island. The figure shows that the angle of depression from the helicopter to point P is 36°. How far off the coast, to the nearest foot, is the island?

34. A police helicopter is flying at 800 feet. A stolen car is sighted at an angle of depression of 72°. Find the distance of the stolen car, to the nearest foot, from a point directly below the helicopter.

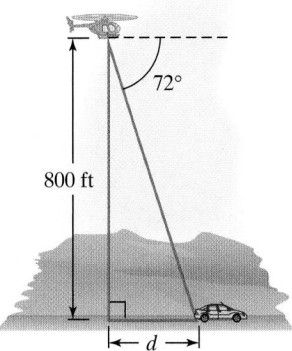

35. A wheelchair ramp is to be built beside the steps to the campus library. Find the angle of elevation of the 23-foot ramp, to the nearest tenth of a degree, if its final height is 6 feet.

36. A building that is 250 feet high casts a shadow 40 feet long. Find the angle of elevation, to the nearest tenth of a degree, of the sun at this time.

37. A hot-air balloon is rising vertically. The angle of elevation from a point on level ground 125 feet from the balloon to a point directly under the passenger compartment changes from 19.2° to 31.7°. How far, to the nearest tenth of a foot, does the balloon rise during this period?

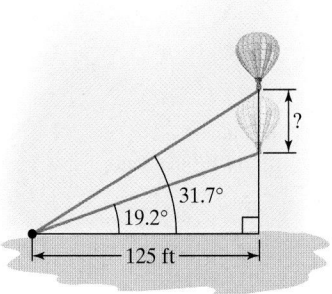

38. A flagpole is situated on top of a building. The angle of elevation from a point on level ground 330 feet from the building to the top of the flagpole is 63°. The angle of elevation from the same point to the bottom of the flagpole is 53°. Find the height of the flagpole to the nearest tenth of a foot.

39. A boat leaves the entrance to a harbor and travels 150 miles on a bearing of N 53° E. How many miles north and how many miles east from the harbor has the boat traveled?

40. A boat leaves the entrance to a harbor and travels 40 miles on a bearing of S 64° E. How many miles south and how many miles east from the harbor has the boat traveled?

41. A forest ranger sights a fire directly to the south. A second ranger, 7 miles east of the first ranger, also sights the fire. The bearing from the second ranger to the fire is S 28° W. How far, to the nearest tenth of a mile, is the first ranger from the fire?

42. A ship sights a lighthouse directly to the south. A second ship, 9 miles east of the first ship, also sights the lighthouse. The bearing from the second ship to the lighthouse is S 34° W. How far, to the nearest tenth of a mile, is the first ship from the lighthouse?

43. You leave your house and run 2 miles due west followed by 1.5 miles due north. At that time, what is your bearing from your house?

44. A ship is 9 miles east and 6 miles south of a harbor. What bearing should be taken to sail directly to the harbor?

45. A jet leaves a runway whose bearing is N 35° E from the control tower. After flying 5 miles, the jet turns 90° and flies on a bearing of S 55° E for 7 miles. At that time, what is the bearing of the jet from the control tower?

46. A ship leaves port with a bearing of S 40° W. After traveling 7 miles, the ship turns 90° and travels on a bearing of N 50° W for 11 miles. At that time, what is the bearing of the ship from port?

47. An object in simple harmonic motion has a frequency of $\frac{1}{2}$ oscillation per minute and an amplitude of 6 feet. Write an equation in the form $d = a \sin \omega t$ for the object's simple harmonic motion.

48. An object in simple harmonic motion has a frequency of $\frac{1}{4}$ oscillation per minute and an amplitude of 8 feet. Write an equation in the form $d = a \sin \omega t$ for the object's simple harmonic motion.

49. A piano tuner uses a tuning fork. If middle C has a frequency of 264 vibrations per second, write an equation in the form $d = \sin \omega t$ for the simple harmonic motion.

50. A radio station, 98.1 on the FM dial, has radio waves with a frequency of 98.1 million cycles per second. Write an equation in the form $d = \sin \omega t$ for the simple harmonic motion of the radio waves.

Writing in Mathematics

51. What does it mean to solve a right triangle?

52. Explain how to find one of the acute angles of a right triangle if two sides are known.

53. Describe a situation in which a right triangle and a trigonometric function are used to measure a height or distance that would otherwise be inconvenient or impossible to measure.

54. What is meant by the bearing from point O to point P? Give an example with your description.

55. What is simple harmonic motion? Give an example with your description.

56. Explain the period and the frequency of simple harmonic motion. How are they related?

57. Explain how the photograph of the damaged highway on page 545 illustrates simple harmonic motion.

Technology Exercises

The functions in Exercises 58–59 model motion in which the amplitude decreases with time due to friction or other resistive forces. Graph each function in the given viewing rectangle. How many complete oscillations occur on the time interval $0 \le x \le 10$?

58. $y = 4e^{-0.1} \cos 2x$ $[0, 10, 1]$ by $[-4, 4, 1]$

59. $y = -6e^{-0.09x} \cos 2\pi x$ $[0, 10, 1]$ by $[-6, 6, 1]$

Critical Thinking Exercises

60. The figure at the top of the next column shows a satellite circling 112 miles above Earth. When the satellite is directly above point B, angle A measures $76.6°$. Find Earth's radius to the nearest mile.

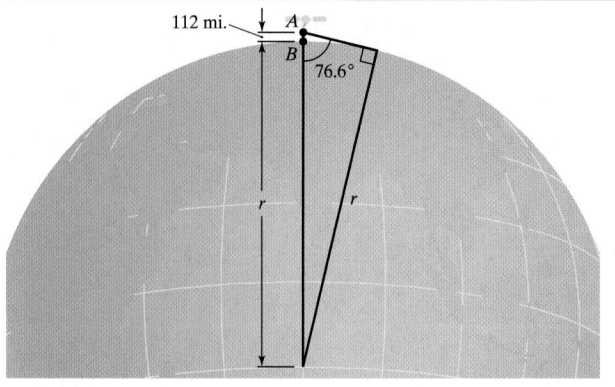

61. The figure shows that the angle of elevation to the top of the building changes from $20°$ to $40°$ as an observer advances 75 feet toward the building. Find the height of the building to the nearest foot.

Group Exercise

62. Music and mathematics have been linked over the centuries. Group members should research and present a seminar to the class on music and mathematics. Be sure to include the role of trigonometric functions in the music-mathematics link.

Chapter Summary, Review, and Test

Summary

DEFINITIONS AND CONCEPTS	EXAMPLES

5.1 Angles and Their Measure

a. An angle consists of two rays with a common endpoint, the vertex.

b. An angle is in standard position if its vertex is at the origin and its initial side lies along the positive x-axis. Figure 5.3 on page 440 shows positive and negative angles in standard position.   Ex. 1, p. 441

c. A quadrantal angle is one with its terminal side on the x-axis or the y-axis.

d. Angles can be measured in degrees. $1°$ is $\frac{1}{360}$ of a complete rotation.

e. Acute angles measure less than $90°$, right angles $90°$, obtuse angles more than $90°$ but less than $180°$, and straight angles $180°$.

DEFINITIONS AND CONCEPTS	EXAMPLES

f. Two angles with the same initial and terminal sides are called coterminal angles.   Ex. 2, p. 442

g. Two angles are complements if their sum is 90° and supplements if their sum is 180°. Only positive angles are used.   Ex. 3, p. 443

h. Angles can be measured in radians. One radian is the measure of the central angle when the intercepted arc and radius have the same length. In general, the radian measure of a central angle is the length of the intercepted arc divided by the circle's radius: $\theta = \dfrac{s}{r}$.   Ex. 4, p. 445

i. To convert from degrees to radians, multiply degrees by $\dfrac{\pi \text{ radians}}{180°}$. To convert from radians to degrees, multiply radians by $\dfrac{180°}{\pi \text{ radians}}$.   Ex. 5, p. 446; Ex. 6, p. 446

j. The arc length formula, $s = r\theta$, is described in the box on page 447.   Ex. 7, p. 448

k. The definitions of linear speed, $v = \dfrac{s}{t}$, and angular speed, $\omega = \dfrac{\theta}{t}$, are given in the box on page 448.

l. Linear speed is expressed in terms of angular speed by $v = r\omega$, where v is the linear speed of a point a distance r from the center of rotation and ω is the angular speed in radians per unit of time.   Ex. 8, p. 449

5.2 Right Triangle Trigonometry

a. The right triangle definitions of the six trigonometric functions are given in the box on page 454.   Ex. 1, p. 455

b. Function values for 30°, 45°, and 60° can be obtained using these special triangles.   Ex. 2, p. 456; Ex. 3, p. 457

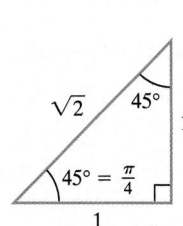

c. Fundamental Identities

1. Reciprocal Identities

$\sin\theta = \dfrac{1}{\csc\theta} \quad \cos\theta = \dfrac{1}{\sec\theta} \quad \tan\theta = \dfrac{1}{\cot\theta}$

$\csc\theta = \dfrac{1}{\sin\theta} \quad \sec\theta = \dfrac{1}{\cos\theta} \quad \cot\theta = \dfrac{1}{\tan\theta}$

Ex. 4, p. 459; Ex. 5, p. 460

2. Quotient Identities

$\tan\theta = \dfrac{\sin\theta}{\cos\theta} \quad \cot\theta = \dfrac{\cos\theta}{\sin\theta}$

3. Pythagorean Identities

$$\sin^2\theta + \cos^2\theta = 1$$
$$1 + \tan^2\theta = \sec^2\theta$$
$$1 + \cot^2\theta = \csc^2\theta$$

d. The value of a trigonometric function of θ is equal to the cofunction of the complement of θ. Cofunction identities are listed in the box on page 461.   Ex. 6, p. 461

DEFINITIONS AND CONCEPTS	EXAMPLES

5.3 Trigonometric Functions of Any Angle

a. Definitions of the trigonometric functions of any angle are given in the box on page 470.

Ex. 1, p. 471;
Ex. 2, p. 471

b. Signs of the trigonometric functions: All functions are positive in quadrant I. If θ lies in quadrant II, $\sin\theta$ and $\csc\theta$ are positive. If θ lies in quadrant III, $\tan\theta$ and $\cot\theta$ are positive. If θ lies in quadrant IV, $\cos\theta$ and $\sec\theta$ are positive.

Ex. 3, p. 474;
Ex. 4, p. 474

c. If θ is a nonacute angle in standard position that lies in a quadrant, its reference angle is the positive acute angle θ' formed by the terminal side of θ and the x-axis. The reference angle for a given angle can be found by making a sketch that shows the angle in standard position. Figure 5.30 on page 475 shows reference angles for θ in quadrants II, III, and IV.

Ex. 5, p. 475

d. The values of the trigonometric functions of a given angle are the same as the values of the functions of the reference angle, except possibly for the sign. A procedure for using reference angles to evaluate trigonometric functions is given in the box on page 476.

Ex. 6, p. 476

5.4 Trigonometric Functions of Real Numbers; Periodic Functions

a. Definitions of the trigonometric functions in terms of a unit circle are given in the box on page 481.

Ex. 1, p. 481

b. The cosine and secant functions are even:

Ex. 2, p. 483

$$\cos(-t) = \cos t, \quad \sec(-t) = \sec t.$$

The other trigonometric functions are odd:

$$\sin(-t) = -\sin t, \quad \csc(-t) = -\csc t,$$

$$\tan(-t) = -\tan t, \quad \cot(-t) = -\cot t.$$

c. If $f(t + p) = f(t)$, function f is periodic. The smallest p for which f is periodic is the period of f. The tangent and cotangent functions have period π. The other four trigonometric functions have period 2π.

Ex. 3, p. 484

5.5 and 5.6 Graphs of the Trigonometric Functions

a. Graphs of the six trigonometric functions, with a description of the domain, range, and period of each function, are given in Table 5.4 on page 517.

b. The graph of $y = A\sin(Bx - C)$ can be obtained using amplitude $= |A|$, period $= \dfrac{2\pi}{B}$ and phase shift $= \dfrac{C}{B}$. See the illustration in the box on page 494.

Ex. 1, p. 489;
Ex. 2, p. 491;
Ex. 3, p. 492;
Ex. 4, p. 495

c. The graph of $y = A\cos(Bx - C)$ can be obtained using amplitude $= |A|$, period $= \dfrac{2\pi}{B}$, and phase shift $= \dfrac{C}{B}$. See the illustration in the box on page 500.

Ex. 5, p. 498;
Ex. 6, p. 500

d. The constant D in $y = A\sin(Bx - C) + D$ and $y = A\cos(Bx - C) + D$ causes vertical shifts in the graphs in the preceding items (b) and (c). If $D > 0$, the shift is D units upward and if $D < 0$, the shift is D units downward. Oscillation is about $y = D$.

Ex. 7, p. 502

e. The graph of $y = A\tan(Bx - C)$ is obtained using the procedure in the box on page 511. Consecutive asymptotes $\left(\text{solve } Bx - C = -\dfrac{\pi}{2} \text{ and } Bx - C = \dfrac{\pi}{2}\right)$ and an x-intercept midway between them play a key role in the graphing process.

Ex. 1, p. 511;
Ex. 2, p. 512

f. The graph of $y = A\cot(Bx - C)$ is obtained using the procedure in the box on page 513. Consecutive asymptotes (solve $Bx - C = 0$ and $Bx - C = \pi$) and an x-intercept midway between them play a key role in the graphing process.

Ex. 3, p. 513

g. To graph a cosecant curve, begin by graphing the reciprocal sine curve. Draw vertical asymptotes through x-intercepts, using asymptotes as guides to sketch the graph. To graph a secant curve, first graph the reciprocal cosine curve and use the same procedure.

Ex. 4, p. 515;
Ex. 5, p. 516

DEFINITIONS AND CONCEPTS	EXAMPLES

5.7   Inverse Trigonometric Functions

a. On the restricted domain $-\dfrac{\pi}{2} \le x \le \dfrac{\pi}{2}$, $y = \sin x$ has an inverse function, defined in the box on page 522. Think of $\sin^{-1} x$ as the angle in $\left[-\dfrac{\pi}{2}, \dfrac{\pi}{2}\right]$ whose sine is x.

Ex. 1, p. 524;
Ex. 2, p. 524

b. On the restricted domain $0 \le x \le \pi$, $y = \cos x$ has an inverse function, defined in the box on page 525. Think of $\cos^{-1} x$ as the angle in $[0, \pi]$ whose cosine is x.

Ex. 3, p. 526

c. On the restricted domain $-\dfrac{\pi}{2} < x < \dfrac{\pi}{2}$, $y = \tan x$ has an inverse function, defined in the box on page 527. Think of $\tan^{-1} x$ as the angle in $\left(-\dfrac{\pi}{2}, \dfrac{\pi}{2}\right)$ whose tangent is x.

Ex. 4, p. 528

d. Graphs of the three basic inverse trigonometric functions, with a description of the domain and range of each function, are given in Table 5.8 on page 529.

e. Inverse properties are given in the box on page 530. Points on terminal sides of angles in standard position and right triangles are used to find exact values of the composition of a function and a different inverse function.

Ex. 6, p. 531;
Ex. 7, p. 532;
Ex. 8, p. 532;
Ex. 9, p. 533

5.8   Applications of Trigonometric Functions

a. Solving a right triangle means finding the missing lengths of its sides and the measurements of its angles. The Pythagorean Theorem, two acute angles whose sum is 90°, and appropriate trigonometric functions are used in this process.

Ex. 1, p. 537;
Ex. 2, p. 538;
Ex. 3, p. 539;
Ex. 4, p. 539

b. The bearing from point O to point P is the acute angle between ray OP and a north-south line.

Ex. 5, p. 540;
Ex. 6, p. 541

c. Simple harmonic motion, described in the box on page 543, is modeled by $d = a \cos \omega t$ or $d = a \sin \omega t$, with amplitude $= |a|$, period $= \dfrac{2\pi}{\omega}$ and frequency $= \dfrac{\omega}{2\pi} = \dfrac{1}{\text{period}}$.

Ex. 7, p. 543;
Ex. 8, p. 544

Review Exercises

5.1

In Exercises 1–4, draw each angle in standard position.

1. 190°

2. −135°

3. $\dfrac{5\pi}{6}$

4. $-\dfrac{2\pi}{3}$

In Exercises 5–6, find a positive angle less than 360° that is coterminal with the given angle.

5. 400°

6. −85°

In Exercises 7–8, if possible, find the complement and the supplement of the given angle.

7. 73°

8. $\dfrac{2\pi}{3}$

9. Find the radian measure of the central angle of a circle of radius 6 centimeters that intercepts an arc of length 27 centimeters.

In Exercises 10–12, convert each angle in degrees to radians. Express your answer as a multiple of π.

10. 15°

11. 120°

12. 315°

In Exercises 13–15, convert each angle in radians to degrees.

13. $\dfrac{5\pi}{3}$

14. $\dfrac{7\pi}{5}$

15. $-\dfrac{5\pi}{6}$

16. Find the length of the arc on a circle of radius 10 feet intercepted by a 135° central angle. Express arc length in terms of π. Then round your answer to two decimal places.

17. The angular speed of a propeller on a wind generator is 10.3 revolutions per minute. Express this angular speed in radians per minute.

18. The propeller of an airplane has a radius of 3 feet. The propeller is rotating at 2250 revolutions per minute. Find the linear speed, in feet per minute, of the tip of the propeller.

5.2

19. Use the triangle to find each of the six trigonometric functions of θ.

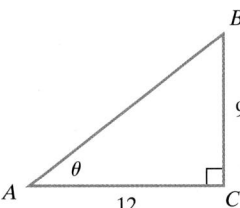

In Exercises 20–23, find the exact value of each expression. Do not use a calculator.

20. $\tan 60°$

21. $\cos \dfrac{\pi}{4}$

22. $\sec \dfrac{\pi}{6}$

23. $\sin^2 \dfrac{\pi}{5} + \cos^2 \dfrac{\pi}{5}$

24. If θ is an acute angle and $\sin \theta = \dfrac{2}{\sqrt{7}}$, use the identity $\sin^2 \theta + \cos^2 \theta = 1$ to find $\cos \theta$.

In Exercises 25–26, find a cofunction with the same value as the given expression.

25. $\sin 70°$

26. $\cos \dfrac{\pi}{2}$

In Exercises 27–29, find the measure of the side of the right triangle whose length is designated by a lowercase letter. Round answers to the nearest whole number.

27.

28.

29.

30. A hiker climbs for a half mile up a slope whose inclination is $17°$. How many feet of altitude, to the nearest foot, does the hiker gain?

31. To find the distance across a lake, a surveyor took the measurements in the figure shown. What is the distance across the lake? Round to the nearest meter.

32. When a six-foot pole casts a four-foot shadow, what is the angle of elevation of the sun? Round to the nearest whole degree.

5.3 and 5.4

In Exercises 33–34, a point on the terminal side of angle θ is given. Find the exact value of each of the six trigonometric functions of θ, or state that the function is undefined.

33. $(-1, -5)$

34. $(0, -1)$

In Exercises 35–36, let θ be an angle in standard position. Name the quadrant in which θ lies.

35. $\tan \theta > 0$ and $\sec \theta > 0$

36. $\tan \theta > 0$ and $\cos \theta < 0$

In Exercises 37–38, find the exact value of each of the remaining trigonometric functions of θ.

37. $\cos \theta = \frac{2}{5}, \sin \theta < 0$

38. $\tan \theta = -\frac{1}{3}, \sin \theta > 0$

In Exercises 39–41, find the reference angle for each angle.

39. $265°$

40. $\dfrac{5\pi}{8}$

41. $-410°$

In Exercises 42–50, find the exact value of each expression. Do not use a calculator.

42. $\sin 240°$

43. $\tan 120°$

44. $\sec \dfrac{7\pi}{4}$

45. $\cos \dfrac{11\pi}{6}$

46. $\cot (-210°)$

47. $\csc \left(-\dfrac{2\pi}{3}\right)$

48. $\sin \left(-\dfrac{\pi}{3}\right)$

49. $\sin 495°$

50. $\tan \dfrac{13\pi}{4}$

5.5

In Exercises 51–56, determine the amplitude and period of each function. Then graph one period of the function.

51. $y = 3 \sin 4x$

52. $y = -2 \cos 2x$

53. $y = 2 \cos \frac{1}{2} x$

54. $y = \dfrac{1}{2} \sin \dfrac{\pi}{3} x$

55. $y = -\sin \pi x$

56. $y = 3 \cos \dfrac{x}{3}$

In Exercises 57–61, determine the amplitude, period, and phase shift of each function. Then graph one period of the function.

57. $y = 2 \sin (x - \pi)$

58. $y = -3 \cos (x + \pi)$

59. $y = \frac{3}{2} \cos \left(2x + \frac{\pi}{4} \right)$

60. $y = \frac{5}{2} \sin \left(2x + \frac{\pi}{2} \right)$

61. $y = -3 \sin \left(\frac{\pi}{3} x - 3\pi \right)$

In Exercises 62–63, use a vertical shift to graph one period of the function.

62. $y = \sin 2x + 1$

63. $y = 2 \cos \frac{1}{3} x - 2$

64. The equation

$$y = 98.6 + 0.3 \sin \left(\frac{\pi}{12} x - \frac{11\pi}{12} \right)$$

models variation in body temperature, y, in °F, x hours after midnight.

   a. What is body temperature at midnight?

   b. What is the period of the body temperature cycle?

   c. When is body temperature highest? What is the body temperature at this time?

   d. When is body temperature lowest? What is the body temperature at this time?

   e. Graph one period of the body temperature function.

5.6

In Exercises 65–71, graph two full periods of the given tangent or cotangent function.

65. $y = 4 \tan 2x$

66. $y = -2 \tan \frac{\pi}{4} x$

67. $y = \tan (x + \pi)$

68. $y = -\tan \left(x - \frac{\pi}{4} \right)$

69. $y = 2 \cot 3x$

70. $y = -\frac{1}{2} \cot \frac{\pi}{2} x$

71. $y = 2 \cot \left(x + \frac{\pi}{2} \right)$

In Exercises 72–75, graph two full periods of the given cosecant or secant function.

72. $y = 3 \sec 2\pi x$

73. $y = -2 \csc \pi x$

74. $y = 3 \sec (x + \pi)$

75. $y = \frac{1}{2} \csc (x - \pi)$

5.7

In Exercises 76–93, find the exact value of each expression. Do not use a calculator.

76. $\sin^{-1} 1$

77. $\cos^{-1} 1$

78. $\tan^{-1} 1$

79. $\sin^{-1} \left(-\frac{\sqrt{3}}{2} \right)$

80. $\cos^{-1} \left(-\frac{1}{2} \right)$

81. $\tan^{-1} \left(-\frac{\sqrt{3}}{3} \right)$

82. $\cos \left(\sin^{-1} \frac{\sqrt{2}}{2} \right)$

83. $\sin (\cos^{-1} 0)$

84. $\tan \left[\sin^{-1} \left(-\frac{1}{2} \right) \right]$

85. $\tan \left[\cos^{-1} \left(-\frac{\sqrt{3}}{2} \right) \right]$

86. $\csc \left(\tan^{-1} \frac{\sqrt{3}}{3} \right)$

87. $\cos \left(\tan^{-1} \frac{3}{4} \right)$

88. $\sin \left(\cos^{-1} \frac{3}{5} \right)$

89. $\tan \left[\sin^{-1} \left(-\frac{3}{5} \right) \right]$

90. $\tan \left[\cos^{-1} \left(-\frac{4}{5} \right) \right]$

91. $\sin^{-1} \left(\sin \frac{\pi}{3} \right)$

92. $\sin^{-1} \left(\sin \frac{2\pi}{3} \right)$

93. $\sin^{-1} \left(\cos \frac{2\pi}{3} \right)$

In Exercises 94–95, use a right triangle to write each expression as an algebraic expression. Assume that x is positive and in the domain of the given inverse trigonometric function.

94. $\cos \left(\tan^{-1} \frac{x}{2} \right)$

95. $\sec \left(\sin^{-1} \frac{1}{x} \right)$

5.8

In Exercises 96–99, solve the right triangle shown in the figure. Round lengths to two decimal places and express angles to the nearest tenth of a degree.

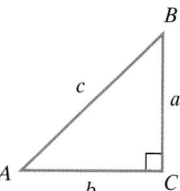

96. $A = 22.3°, c = 10$

97. $B = 37.4°, b = 6$

98. $a = 2, c = 7$

99. $a = 1.4, b = 3.6$

100. From a point on level ground 80 feet from the base of a building, the angle of elevation is 25.6°. Approximate the height of the building to the nearest foot.

101. Two buildings with flat roofs are 60 yards apart. The height of the shorter building is 40 yards. From its roof, the angle of elevation to the edge of the roof of the taller building is 40°. Find the height of the taller building to the nearest yard.

102. You want to measure the height of an antenna on the top of a 125-foot building. From a point in front of the building, you measure the angle of elevation to the top of the building to be 68° and the angle of elevation to the top of the antenna to be 71°. How tall is the antenna, to the nearest tenth of a foot?

In Exercises 103–104, use the figures shown to find the bearing from O to A.

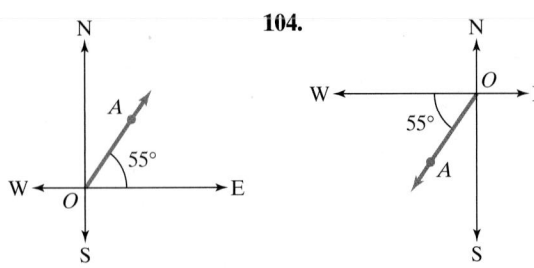

103.   **104.**

105. A ship is due west of a lighthouse. A second ship is 12 miles south of the first ship. The bearing from the second ship to the lighthouse is N 64° E. How far, to the nearest tenth of a mile, is the first ship from the lighthouse?

106. From city A to city B, a plane flies 850 miles at a bearing of N 58° E. From city B to city C, the plane flies 960 miles at a bearing of S 32° E.

   a. Find, to the nearest tenth of a mile, the distance from city A to city C.

   b. What is the bearing from city A to city C?

In Exercises 107–108, an object moves in simple harmonic motion described by the given equation, where t is measured in seconds and d in centimeters. In each exercise, find:

   a. the maximum displacement

   b. the frequency

   c. the time required for one cycle.

107. $d = 20 \cos \dfrac{\pi}{4} t$ **108.** $d = \frac{1}{2} \sin 4t$

In Exercises 109–110, an object is attached to a coiled spring. The object is pulled down (negative direction from the rest position) and then released. Write an equation for the distance of the object from its rest position after t seconds.

	Distance from rest position at $t = 0$	Amplitude	Period
109.	30 inches	30 inches	2 seconds
110.	0	$\frac{1}{4}$ inch	5 seconds

Chapter 5 Test

1. Convert 135° to exact radian measure.

2. Find the supplement of the angle whose radian measure is $\dfrac{9\pi}{13}$. Express the answer in terms of π.

3. Find the length of the arc on a circle of radius 20 feet intercepted by a 75° central angle. Express arc length in terms of π. Then round your answer to two decimal places.

4. If $(-2, 5)$ is a point on the terminal side of angle θ, find the exact value of each of the six trigonometric functions of θ.

5. Determine the quadrant in which θ lies if $\cos \theta < 0$ and $\cot \theta > 0$.

6. If $\cos \theta = \frac{1}{3}$ and $\tan \theta < 0$, find the exact value of each of the remaining trigonometric functions of θ.

In Exercises 7–9, find the exact value of each expression. Do not use a calculator.

7. $\tan \dfrac{\pi}{6} \cos \dfrac{\pi}{3} - \cos \dfrac{\pi}{2}$ **8.** $\tan 300°$

9. $\sin \dfrac{7\pi}{4}$

In Exercises 10–13, graph one period of each function.

10. $y = 3 \sin 2x$ **11.** $y = -2 \cos \left(x - \dfrac{\pi}{2} \right)$

12. $y = 2 \tan \dfrac{x}{2}$ **13.** $y = -\frac{1}{2} \csc \pi x$

14. Find the exact value of $\tan \left[\cos^{-1} \left(-\frac{1}{2} \right) \right]$.

15. Solve the right triangle in the figure shown. Round lengths to one decimal place.

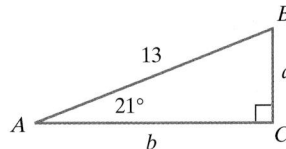

16. The angle of elevation of a building from a point on the ground 30 yards from its base is 37°. Find the height of the building to the nearest yard.

17. A 73-foot rope from the top of a circus tent pole is anchored to the flat ground 43 feet from the bottom of the pole. Find the angle, to the nearest tenth of a degree, that the rope makes with the pole.

18. Use the figure to find the bearing from O to P.

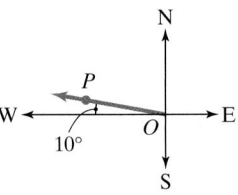

19. An object moves in simple harmonic motion described by $d = -6 \cos \pi t$, where t is measured in seconds and d in inches. Find (**a.**) the maximum displacement, (**b.**) the frequency, and (**c.**) the time required for one oscillation.

20. Why are trigonometric functions ideally suited to model phenomena that repeat in cycles?

Cumulative Review Exercises (Chapters 1–5)

Solve each equation or inequality in Exercises 1–6.

1. $x^2 = 18 + 3x$ **2.** $x^3 + 5x^2 - 4x - 20 = 0$

3. $\log_2 x + \log_2(x - 2) = 3$ **4.** $\sqrt{x - 3} + 5 = x$

5. $x^3 - 4x^2 + x + 6 = 0$ **6.** $|2x - 5| \le 11$

7. If $f(x) = \sqrt{x - 6}$, find $f^{-1}(x)$.

8. Divide $20x^3 - 6x^2 - 9x + 10$ by $5x + 2$.

9. Write as a single logarithm and evaluate: $\log 25 + \log 40$.

10. Convert $\dfrac{14\pi}{9}$ radians to degrees.

11. Find the maximum number of positive and negative real roots of the equation $3x^4 - 2x^3 + 5x^2 + x - 9 = 0$.

In Exercises 12–16, graph each equation.

12. $f(x) = \dfrac{x}{x^2 - 1}$ **13.** $(x - 2)^2 + y^2 = 1$

14. $y = (x - 1)(x + 2)^2$

15. $y = \sin\left(2x + \dfrac{\pi}{2}\right)$, from 0 to 2π

16. $y = 2\tan 3x$; graph two complete cycles.

17. You invest in a new play. The cost includes an overhead of $30,000, plus production costs of $2500 per performance. A sold-out performance brings you $3125. How many sold-out performances must be played in order for you to break even?

18. Use the exponential growth model $A = A_0 e^{kt}$ to solve this exercise. Data from the Federal Communication Commission show that the use of toll-free 800 numbers has grown exponentially. In 1991 there were 10.2 billion such calls and by 1998, there were 86.7 billion.

  a. Find the exponential function that models the data.

  b. By which year will the number of toll-free 800 numbers reach 200 billion?

19. The rate of heat lost through insulation varies inversely as the thickness of the insulation. The rate of heat lost through a 3.5-inch thickness of insulation is 2200 Btu per hour. What is the rate of heat lost through a 5-inch thickness of the same insulation?

20. A tower is 200 feet tall. To the nearest degree, find the angle of elevation from a point 50 feet from the base of the tower to the top of the tower.

Analytic Trigonometry

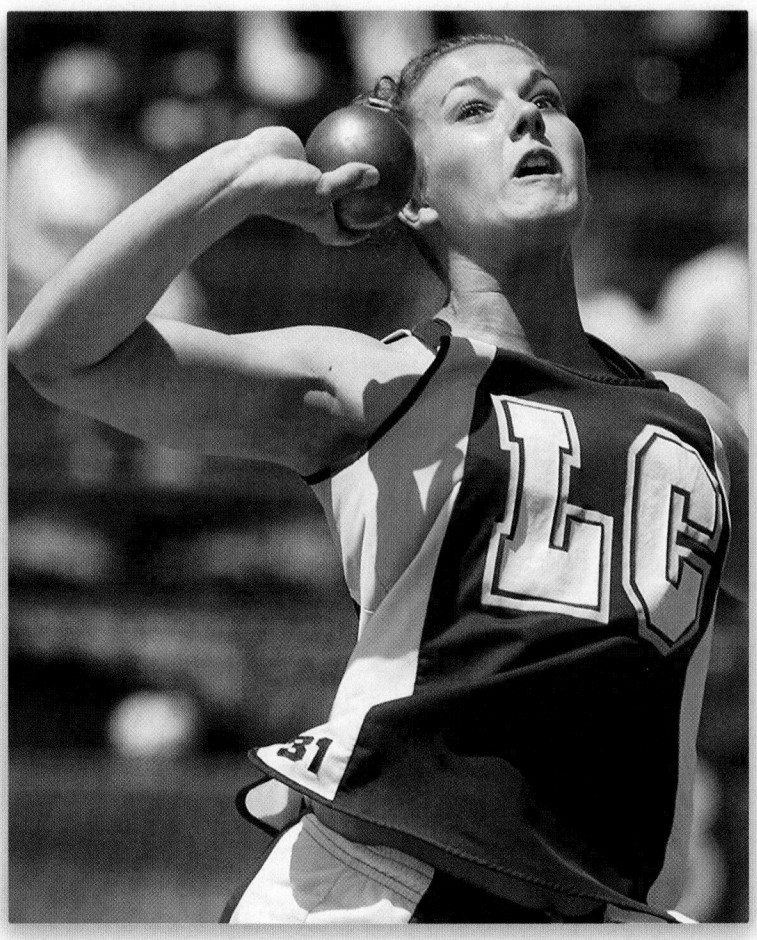

This chapter emphasizes the algebraic aspects of trigonometry. We derive important categories of identities involving trigonometric functions. These identities are used to simplify and analyze expressions that model phenomena as diverse as the distance achieved when throwing an object and musical sounds on a touch-tone phone. For example, we can find out critical information about an athlete's performance by using an identity to analyze an expression involving throwing distance. You will learn how to use trigonometric identities to better understand your periodic world.

You enjoy watching your friend participate in the shot put at college track and field events. After a few full turns in a circle, she throws ("puts") an 8-pound, 13-ounce shot from the shoulder. The range of her throwing distance continues to improve. Knowing that you are studying trigonometry, she asks if there is some way that a trigonometric expression might help achieve her best possible distance in the event.

SECTION 6.1 *Verifying Trigonometric Identities*

Objective

1. Use the fundamental trigonometric identities to verify identities.

Do you enjoy solving puzzles? The process is a natural way to develop problem-solving skills that are important in every area of our lives. Engaging in problem solving for sheer pleasure releases chemicals in the brain that enhance our feeling of well-being. Perhaps this is why puzzles have fascinated people for over 12,000 years.

Thousands of relationships exist among the six trigonometric functions. Verifying these relationships is like solving a puzzle. Why? There are no rigid rules for the process. Thus, proving a trigonometric relationship requires you to be creative in your approach to problem solving. By learning to establish these relationships, you will become a better, more confident problem solver. Furthermore, you may enjoy the feeling of satisfaction that accompanies solving each "puzzle."

The Fundamental Identities

In Chapter 5, we used right triangles to establish relationships among the trigonometric functions. Although we limited domains to acute angles, the fundamental identities listed in the following box are true for all values of x for which the expressions are defined.

Study Tip

Memorize the identities in the box. You may need to use variations of these fundamental identities. For example, instead of

$$\sin^2 x + \cos^2 x = 1,$$

you might want to use

$$\sin^2 x = 1 - \cos^2 x$$

or

$$\cos^2 x = 1 - \sin^2 x.$$

Therefore, it is important to know each relationship well so that mental algebraic manipulation is possible.

Fundamental Trigonometric Identities

Reciprocal Identities

$$\sin x = \frac{1}{\csc x} \qquad \cos x = \frac{1}{\sec x} \qquad \tan x = \frac{1}{\cot x}$$

$$\csc x = \frac{1}{\sin x} \qquad \sec x = \frac{1}{\cos x} \qquad \cot x = \frac{1}{\tan x}$$

Quotient Identities

$$\tan x = \frac{\sin x}{\cos x} \qquad \cot x = \frac{\cos x}{\sin x}$$

Pythagorean Identities

$$\sin^2 x + \cos^2 x = 1 \qquad 1 + \tan^2 x = \sec^2 x \qquad 1 + \cot^2 x = \csc^2 x$$

Even-Odd Identities

$$\sin(-x) = -\sin x \qquad \cos(-x) = \cos x \qquad \tan(-x) = -\tan x$$

$$\csc(-x) = -\csc x \qquad \sec(-x) = \sec x \qquad \cot(-x) = -\cot x$$

1 Use the fundamental trigonometric identities to verify identities.

Using Fundamental Identities to Verify Other Identities

The fundamental trigonometric identities are used to establish other relationships among trigonometric functions. To **verify an identity**, we show that one side of the identity can be simplified so that it is identical to the other side. Each side of the equation is manipulated independently of the other side of the equation. Start with the side containing the more complicated expression. If you substitute one or more fundamental identities on the more complicated side, you will often be able to rewrite it in a form identical to that of the other side.

No one method can be used to verify every identity. Some identities can be verified by rewriting the more complicated side so that it contains only sines and cosines.

EXAMPLE 1 Changing to Sines and Cosines to Verify an Identity

Verify the identity: $\sec x \cot x = \csc x$.

Solution The left side of the equation contains the more complicated expression. Thus, we work with the left side. Let us express this side of the identity in terms of sines and cosines. Perhaps this strategy will enable us to transform the left side into $\csc x$, the expression on the right.

$$\sec x \cot x = \frac{1}{\cos x} \cdot \frac{\cos x}{\sin x}$$

Apply a reciprocal identity: $\sec x = \dfrac{1}{\cos x}$ and a quotient identity: $\cot x = \dfrac{\cos x}{\sin x}$.

$$= \frac{1}{\overset{1}{\underset{1}{\cancel{\cos x}}}} \cdot \frac{\overset{1}{\cancel{\cos x}}}{\sin x}$$

Divide both the numerator and the denominator by $\cos x$, the common factor.

$$= \frac{1}{\sin x}$$

Multiply the remaining factors in the numerator and denominator.

$$= \csc x$$

Apply a reciprocal identity: $\csc x = \dfrac{1}{\sin x}$.

By working with the left side and simplifying it so that it is identical to the right side, we have verified the given identity.

> **Check Point 1** Verify the identity: $\csc x \tan x = \sec x$.

In verifying an identity, stay focused on your goal. When manipulating one side of the equation, continue to look at the other side to keep the desired form of the result in mind.

Technology

You can use a graphing utility to provide evidence of an identity. For example, consider the equation in Example 1,

$$\sec x \cot x = \csc x.$$

Graph

$$y = \sec x \cot x$$

and

$$y = \csc x$$

in the same viewing rectangle. The screen shown indicates that the graphs appear to be the same. Example 1 shows the equivalence algebraically.

Study Tip

Verifying that an equation is an identity is different from solving an equation. You do not verify an identity by adding, subtracting, multiplying, or dividing each side by the same expression. If you do this, you have already assumed that the given statement is true. You do not know that it is true until after you have verified it.

EXAMPLE 2 Changing to Sines and Cosines to Verify an Identity

Verify the identity: $\sin x \tan x + \cos x = \sec x$.

Solution The left side is more complicated, so we start with it. Let us express this side of the identity so that it contains only sines and cosines. Thus, we apply a quotient identity and replace $\tan x$ by $\dfrac{\sin x}{\cos x}$. Perhaps this strategy will enable us to transform the left side into $\sec x$, the expression on the right.

$$\sin x \tan x + \cos x = \sin x \left(\frac{\sin x}{\cos x}\right) + \cos x$$

Apply a quotient identity: $\tan x = \dfrac{\sin x}{\cos x}$.

$$= \frac{\sin^2 x}{\cos x} + \cos x.$$

Multiply.

$$= \frac{\sin^2 x}{\cos x} + \cos x \cdot \frac{\cos x}{\cos x}$$

The least common denominator is $\cos x$. Write the second expression with a denominator of $\cos x$.

$$= \frac{\sin^2 x}{\cos x} + \frac{\cos^2 x}{\cos x}$$

Multiply.

$$= \frac{\sin^2 x + \cos^2 x}{\cos x}$$

Add numerators, putting this sum over the least common denominator.

$$= \frac{1}{\cos x}$$

Apply a Pythagorean identity: $\sin^2 x + \cos^2 x = 1$.

$$= \sec x$$

Apply a reciprocal identity: $\sec x = \dfrac{1}{\cos x}$.

By working with the left side and arriving at the right side, the identity is verified.

Check Point 2 Verify the identity: $\cos x \cot x + \sin x = \csc x$.

Some identities are verified using factoring to simplify a trigonometric expression.

EXAMPLE 3 Using Factoring to Verify an Identity

Verify the identity: $\cos x - \cos x \sin^2 x = \cos^3 x$.

Solution We start with the more complicated side, the left side. Factor out the greatest common factor, $\cos x$, from each of the two terms.

$$\cos x - \cos x \sin^2 x = \cos x (1 - \sin^2 x)$$

Factor $\cos x$ from the two terms.

$$= \cos x \cdot \cos^2 x$$

Use a variation of $\sin^2 x + \cos^2 x = 1$. Solving for $\cos^2 x$, we obtain $\cos^2 x = 1 - \sin^2 x$.

$$= \cos^3 x$$

Multiply.

We worked with the left side and arrived at the right side. Thus, the identity is verified.

Check Point 3   Verify the identity: $\sin x - \sin x \cos^2 x = \sin^3 x$.

How do we verify identities in which sums or differences of fractions with trigonometric functions appear on one side? Use the least common denominator and combine the fractions. This technique is especially useful when the other side of the identity contains only one term.

EXAMPLE 4   Combining Fractional Expressions to Verify an Identity

Verify the identity: $\dfrac{\cos x}{1 + \sin x} + \dfrac{1 + \sin x}{\cos x} = 2 \sec x$.

Solution   We start with the more complicated side, the left side. The least common denominator of the fractions is $(1 + \sin x)(\cos x)$. We express each fraction in terms of this least common denominator by multiplying the numerator and denominator by the extra factor needed to form $(1 + \sin x)(\cos x)$.

Study Tip

Some students have difficulty verifying identities due to problems working with fractions. If this applies to you, review the material on rational expressions in Chapter P, Section P.6.

$$\frac{\cos x}{1 + \sin x} + \frac{1 + \sin x}{\cos x}$$

The least common denominator is $(1 + \sin x)(\cos x)$.

$$= \frac{\cos x(\cos x)}{(1 + \sin x)(\cos x)} + \frac{(1 + \sin x)(1 + \sin x)}{(1 + \sin x)(\cos x)}$$

Rewrite each fraction with the least common denominator.

$$= \frac{\cos^2 x}{(1 + \sin x)(\cos x)} + \frac{1 + 2 \sin x + \sin^2 x}{(1 + \sin x)(\cos x)}$$

Use the FOIL method to multiply $(1 + \sin x)(1 + \sin x)$.

$$= \frac{\cos^2 x + 1 + 2 \sin x + \sin^2 x}{(1 + \sin x)(\cos x)}$$

Add numerators. Put this sum over the least common denominator.

$$= \frac{(\sin^2 x + \cos^2 x) + 1 + 2 \sin x}{(1 + \sin x)(\cos x)}$$

Regroup terms to apply a Pythagorean identity.

$$= \frac{1 + 1 + 2 \sin x}{(1 + \sin x)(\cos x)}$$

Apply a Pythagorean identity: $\sin^2 x + \cos^2 x = 1$.

$$= \frac{2 + 2 \sin x}{(1 + \sin x)(\cos x)}$$

Add constant terms in the numerator: $1 + 1 = 2$.

$$= \frac{2\cancel{(1 + \sin x)}}{\cancel{(1 + \sin x)}(\cos x)}$$

Factor and simplify.

$$= \frac{2}{\cos x}$$

$$= 2 \sec x$$

Apply a reciprocal identity: $\sec x = \dfrac{1}{\cos x}$.

We worked with the left side and arrived at the right side. Thus, the identity is verified.

Check Point 4 Verify the identity: $\dfrac{\sin x}{1 + \cos x} + \dfrac{1 + \cos x}{\sin x} = 2 \csc x.$

Some identities are verified using a technique that may remind you of rationalizing a denominator.

EXAMPLE 5 Multiplying the Numerator and Denominator by the Same Factor to Verify an Identity

Verify the identity: $\dfrac{\sin x}{1 + \cos x} = \dfrac{1 - \cos x}{\sin x}.$

Solution The suggestions given in the previous examples do not apply here. Everything is already expressed in terms of sines and cosines. Furthermore, there are no fractions to combine and neither side looks more complicated than the other. Let's solve the puzzle by working with the left side and making it look like the expression on the right. The expression on the right contains $1 - \cos x$ in the numerator. This suggests multiplying the numerator and denominator of the left side by $1 - \cos x$. By doing this, we obtain a factor of $1 - \cos x$ in the numerator, as in the numerator on the right.

Discovery

Verify the identity in Example 5 by making the right side look like the left side. Start with the expression on the right. Multiply the numerator and denominator by $1 + \cos x$.

$$\dfrac{\sin x}{1 + \cos x} = \dfrac{\sin x}{1 + \cos x} \cdot \dfrac{1 - \cos x}{1 - \cos x}$$ Multiply numerator and denominator by $1 - \cos x$.

$$= \dfrac{\sin x(1 - \cos x)}{1 - \cos^2 x}$$ Multiply. Use $(A + B)(A - B) = A^2 - B^2$, with $A = 1$ and $B = \cos x$, to multiply denominators.

$$= \dfrac{\sin x(1 - \cos x)}{\sin^2 x}$$ Use a variation of $\sin^2 x + \cos^2 x = 1$. Solving for $\sin^2 x$, we obtain $\sin^2 x = 1 - \cos^2 x$.

$$= \dfrac{1 - \cos x}{\sin x}$$ Simplify: $\dfrac{\sin x}{\sin^2 x} = \dfrac{\cancel{\sin x}}{\cancel{\sin x} \cdot \sin x} = \dfrac{1}{\sin x}.$

We worked with the left side and arrived at the right side. Thus, the identity is verified.

Check Point 5 Verify the identity: $\dfrac{\cos x}{1 + \sin x} = \dfrac{1 - \sin x}{\cos x}.$

EXAMPLE 6   **Changing to Sines and Cosines to Verify an Identity**

Verify the identity: $\dfrac{\tan x - \sin(-x)}{1 + \cos x} = \tan x.$

Solution   We begin with the left side. Our goal is to obtain $\tan x$, the expression on the right.

$$\frac{\tan x - \sin(-x)}{1 + \cos x} = \frac{\tan x - (-\sin x)}{1 + \cos x}$$

The sine function is odd: $\sin(-x) = -\sin x$.

$$= \frac{\tan x + \sin x}{1 + \cos x}$$

Simplify.

$$= \frac{\dfrac{\sin x}{\cos x} + \sin x}{1 + \cos x}$$

Apply a quotient identity: $\tan x = \dfrac{\sin x}{\cos x}$.

$$= \frac{\dfrac{\sin x}{\cos x} + \dfrac{\sin x \cos x}{\cos x}}{1 + \cos x}$$

Express the terms in the numerator with the least common denominator, $\cos x$.

$$= \frac{\dfrac{\sin x + \sin x \cos x}{\cos x}}{1 + \cos x}$$

Add in the numerator.

$$= \frac{\sin x + \sin x \cos x}{\cos x} \div \frac{1 + \cos x}{1}$$

Rewrite the main fraction bar as $\div$.

$$= \frac{\sin x + \sin x \cos x}{\cos x} \cdot \frac{1}{1 + \cos x}$$

Invert the divisor and multiply.

$$= \frac{\sin x \,\overset{1}{\cancel{(1 + \cos x)}}}{\cos x} \cdot \frac{1}{\underset{1}{\cancel{1 + \cos x}}}$$

Factor and simplify.

$$= \frac{\sin x}{\cos x}$$

Multiply the remaining factors in the numerator and the denominator.

$$= \tan x$$

Apply a quotient identity.

The left side simplifies to $\tan x$, the right side. Thus, the identity is verified.

Check Point 6   Verify the identity: $\dfrac{\sec x + \csc(-x)}{\sec x \csc x} = \sin x - \cos x.$

Is every identity verified by working with only one side? No. You can sometimes work with each side separately and show that both sides are equal to the same trigonometric expression. This is illustrated in Example 7.

EXAMPLE 7 **Working with Both Sides Separately to Verify an Identity**

Verify the identity: $\dfrac{1}{1 + \cos\theta} + \dfrac{1}{1 - \cos\theta} = 2 + 2\cot^2\theta.$

Solution We begin by working with the left side.

$\dfrac{1}{1 + \cos\theta} + \dfrac{1}{1 - \cos\theta}$

> The least common denominator is $(1 + \cos\theta)(1 - \cos\theta)$.

$= \dfrac{1(1 - \cos\theta)}{(1 + \cos\theta)(1 - \cos\theta)} + \dfrac{1(1 + \cos\theta)}{(1 + \cos\theta)(1 - \cos\theta)}$

> Rewrite each fraction with the least common denominator.

$= \dfrac{1 - \cos\theta + 1 + \cos\theta}{(1 + \cos\theta)(1 - \cos\theta)}$

> Add numerators. Put this sum over the least common denominator.

$= \dfrac{2}{(1 + \cos\theta)(1 - \cos\theta)}$

> Simplify the numerator: $-\cos\theta + \cos\theta = 0$ and $1 + 1 = 2$.

$= \dfrac{2}{1 - \cos^2\theta}$

> Multiply the factors in the denominator.

Now we work with the right side. Our goal is to transform this side into the simplified form attained for the left side, $\dfrac{2}{1 - \cos^2\theta}$.

$2 + 2\cot^2\theta = 2 + 2\left(\dfrac{\cos^2\theta}{\sin^2\theta}\right)$

> Use a quotient identity: $\cot\theta = \dfrac{\cos\theta}{\sin\theta}$.

$= \dfrac{2\sin^2\theta}{\sin^2\theta} + \dfrac{2\cos^2\theta}{\sin^2\theta}$

> Rewrite each fraction with the least common denominator, $\sin^2\theta$.

$= \dfrac{2\sin^2\theta + 2\cos^2\theta}{\sin^2\theta}$

> Add numerators. Put this sum over the least common denominator.

$= \dfrac{2(\sin^2\theta + \cos^2\theta)}{\sin^2\theta}$

> Factor out the greatest common factor, 2.

$= \dfrac{2}{\sin^2\theta}$

> Apply a Pythagorean identity: $\sin^2\theta + \cos^2\theta = 1$.

$= \dfrac{2}{1 - \cos^2\theta}$

> Use a variation of $\sin^2\theta + \cos^2\theta = 1$ and solve for $\sin^2\theta$: $\sin^2\theta = 1 - \cos^2\theta$.

The identity is verified because both sides are equal to $\dfrac{2}{1 - \cos^2\theta}$.

> **Check Point 7** Verify the identity: $\dfrac{1}{1 + \sin\theta} + \dfrac{1}{1 - \sin\theta} = 2 + 2\tan^2\theta.$

Guidelines for Verifying Trigonometric Identities

There is often more than one correct way to solve a puzzle, although one method may be shorter and more efficient than another. The same is true for verifying an identity. For example, how would you verify

$$\frac{\csc^2 x - 1}{\csc^2 x} = \cos^2 x?$$

One approach is to use a Pythagorean identity, $1 + \cot^2 x = \csc^2 x$, on the left side. Then change the resulting expression to sines and cosines.

$$\frac{\csc^2 x - 1}{\csc^2 x} = \frac{(1 + \cot^2 x) - 1}{\csc^2 x} = \frac{\cot^2 x}{\csc^2 x} = \frac{\dfrac{\cos^2 x}{\sin^2 x}}{\dfrac{1}{\sin^2 x}} = \frac{\cos^2 x}{\sin^2 x} \cdot \frac{\sin^2 x}{1} = \cos^2 x$$

> Apply a Pythagorean identity: $1 + \cot^2 x = \csc^2 x$.

> Use $\cot x = \frac{\cos x}{\sin x}$ and $\csc x = \frac{1}{\sin x}$ to change to sines and cosines.

> Invert the divisor and multiply.

A more efficient strategy for verifying this identity may not be apparent at first glance. Work with the left side and divide each term in the numerator by the denominator, $\csc^2 x$.

$$\frac{\csc^2 x - 1}{\csc^2 x} = \frac{\csc^2 x}{\csc^2 x} - \frac{1}{\csc^2 x} = 1 - \sin^2 x = \cos^2 x$$

> Apply a reciprocal identity: $\sin x = \frac{1}{\csc x}$.

> Use $\sin^2 x + \cos^2 x = 1$ and solve for $\cos^2 x$.

With this strategy, we again obtain $\cos^2 x$, the expression on the right side, and it takes fewer steps than the first approach.

An even longer strategy, but one that works, is to replace each of the two occurrences of $\csc^2 x$ on the left side by $\dfrac{1}{\sin^2 x}$. This may be the approach that you first consider, particularly if you become accustomed to rewriting the more complicated side in terms of sines and cosines. The selection of an appropriate fundamental identity to solve the puzzle most efficiently is learned through lots of practice.

The more identities you prove, the more confident and efficient you will become. Although practice is the only way to learn how to verify identities, there are some guidelines developed throughout the section that should help you get started.

Guidelines for Verifying Trigonometric Identities

1. Work with each side of the equation independently of the other side. Start with the more complicated side and transform it in a step-by-step fashion until it looks exactly like the other side.

2. Analyze the identity and look for opportunities to apply the fundamental identities. Rewriting the more complicated side of the equation in terms of sines and cosines is often helpful.

3. If sums or differences of fractions appear on one side, use the least common denominator and combine the fractions.

4. Don't be afraid to stop and start over again if you are not getting anywhere. Creative puzzle solvers know that strategies leading to dead ends often provide good problem-solving ideas.

EXERCISE SET 6.1

Practice Exercises

In Exercises 1–60, verify each identity.

1. $\sin x \sec x = \tan x$

2. $\cos x \csc x = \cot x$

3. $\tan(-x)\cos x = -\sin x$

4. $\cot(-x)\sin x = -\cos x$

5. $\tan x \csc x \cos x = 1$

6. $\cot x \sec x \sin x = 1$

7. $\sec x - \sec x \sin^2 x = \cos x$

8. $\csc x - \csc x \cos^2 x = \sin x$

9. $\cos^2 x - \sin^2 x = 1 - 2\sin^2 x$

10. $\cos^2 x - \sin^2 x = 2\cos^2 x - 1$

11. $\csc \theta - \sin \theta = \cot \theta \cos \theta$

12. $\tan \theta + \cot \theta = \sec \theta \csc \theta$

13. $\dfrac{\tan \theta \cot \theta}{\csc \theta} = \sin \theta$

14. $\dfrac{\cos \theta \sec \theta}{\cot \theta} = \tan \theta$

15. $\sin^2 \theta(1 + \cot^2 \theta) = 1$

16. $\cos^2 \theta(1 + \tan^2 \theta) = 1$

17. $\sin t \tan t = \dfrac{1 - \cos^2 t}{\cos t}$

18. $\cos t \cot t = \dfrac{1 - \sin^2 t}{\sin t}$

19. $\dfrac{\csc^2 t}{\cot t} = \csc t \sec t$

20. $\dfrac{\sec^2 t}{\tan t} = \sec t \csc t$

21. $\dfrac{\tan^2 t}{\sec t} = \sec t - \cos t$

22. $\dfrac{\cot^2 t}{\csc t} = \csc t - \sin t$

23. $\dfrac{\sin t}{\csc t} + \dfrac{\cos t}{\sec t} = 1$

24. $\dfrac{\sin t}{\tan t} + \dfrac{\cos t}{\cot t} = \sin t + \cos t$

25. $\tan t + \dfrac{\cos t}{1 + \sin t} = \sec t$

26. $\cot t + \dfrac{\sin t}{1 + \cos t} = \csc t$

27. $1 - \dfrac{\sin^2 x}{1 + \cos x} = \cos x$

28. $1 - \dfrac{\cos^2 x}{1 + \sin x} = \sin x$

29. $\dfrac{\cos x}{1 - \sin x} + \dfrac{1 - \sin x}{\cos x} = 2\sec x$

30. $\dfrac{\sin x}{\cos x + 1} + \dfrac{\cos x - 1}{\sin x} = 0$

31. $\sec^2 x \csc^2 x = \sec^2 x + \csc^2 x$

32. $\csc^2 x \sec x = \sec x + \csc x \cot x$

33. $\dfrac{\sec x - \csc x}{\sec x + \csc x} = \dfrac{\tan x - 1}{\tan x + 1}$

34. $\dfrac{\csc x - \sec x}{\csc x + \sec x} = \dfrac{\cot x - 1}{\cot x + 1}$

35. $\dfrac{\sin^2 x - \cos^2 x}{\sin x + \cos x} = \sin x - \cos x$

36. $\dfrac{\tan^2 x - \cot^2 x}{\tan x + \cot x} = \tan x - \cot x$

37. $\tan^2 2x + \sin^2 2x + \cos^2 2x = \sec^2 2x$

38. $\cot^2 2x + \cos^2 2x + \sin^2 2x = \csc^2 2x$

39. $\dfrac{\tan 2\theta + \cot 2\theta}{\csc 2\theta} = \sec 2\theta$

40. $\dfrac{\tan 2\theta + \cot 2\theta}{\sec 2\theta} = \csc 2\theta$

41. $\dfrac{\tan x + \tan y}{1 - \tan x \tan y} = \dfrac{\sin x \cos y + \cos x \sin y}{\cos x \cos y - \sin x \sin y}$

42. $\dfrac{\cot x + \cot y}{1 - \cot x \cot y} = \dfrac{\cos x \sin y + \sin x \cos y}{\sin x \sin y - \cos x \cos y}$

43. $(\sec x - \tan x)^2 = \dfrac{1 - \sin x}{1 + \sin x}$

44. $(\csc x - \cot x)^2 = \dfrac{1 - \cos x}{1 + \cos x}$

45. $\dfrac{\sec t + 1}{\tan t} = \dfrac{\tan t}{\sec t - 1}$

46. $\dfrac{\csc t - 1}{\cot t} = \dfrac{\cot t}{\csc t + 1}$

47. $\dfrac{1 + \cos t}{1 - \cos t} = (\csc t + \cot t)^2$

48. $\dfrac{1 - \sin t}{1 + \sin t} = (\sec t - \tan t)^2$

49. $\cos^4 t - \sin^4 t = 1 - 2\sin^2 t$

50. $\sin^4 t - \cos^4 t = 1 - 2\cos^2 t$

51. $\dfrac{\sin\theta - \cos\theta}{\sin\theta} + \dfrac{\cos\theta - \sin\theta}{\cos\theta} = 2 - \sec\theta \csc\theta$

52. $\dfrac{\sin\theta}{1 - \cot\theta} - \dfrac{\cos\theta}{\tan\theta - 1} = \sin\theta + \cos\theta$

53. $(\tan^2\theta + 1)(\cos^2\theta + 1) = \tan^2\theta + 2$

54. $(\cot^2\theta + 1)(\sin^2\theta + 1) = \cot^2\theta + 2$

55. $(\cos\theta - \sin\theta)^2 + (\cos\theta + \sin\theta)^2 = 2$

56. $(3\cos\theta - 4\sin\theta)^2 + (4\cos\theta + 3\sin\theta)^2 = 25$

57. $\dfrac{\cos^2 x - \sin^2 x}{1 - \tan^2 x} = \cos^2 x$

58. $\dfrac{\sin x + \cos x}{\sin x} - \dfrac{\cos x - \sin x}{\cos x} = \sec x \csc x$

59. $(\sec x - \tan x)^2 = \dfrac{1 - \sin x}{1 + \sin x}$

60. $(\cot x - \csc x)^2 = \dfrac{1 - \cos x}{1 + \cos x}$

Writing in Mathematics

61. Explain how to verify an identity.

62. Describe two strategies that can be used to verify identities.

63. Describe how you feel when you successfully verify a difficult identity. What other activities do you engage in that evoke the same feelings?

64. A 10-point question on a quiz asks students to verify the identity

$$\dfrac{\sin^2 x - \cos^2 x}{\sin x + \cos x} = \sin x - \cos x.$$

One student begins with the left side and obtains the right side as shown in next column:

$$\dfrac{\sin^2 x - \cos^2 x}{\sin x + \cos x} = \dfrac{\sin^2 x}{\sin x} - \dfrac{\cos^2 x}{\cos x} = \sin x - \cos x.$$

How many points (out of 10) would you give this student? Explain your answer.

Technology Exercises

In Exercises 65–73, graph each side of the equation in the same viewing rectangle. If the graphs appear to coincide, verify that the equation is an identity. If the graphs do not appear to coincide, this indicates the equation is not an identity. In these exercises, find a value of x for which both sides are defined but not equal.

65. $\tan x = \sec x(\sin x - \cos x) + 1$

66. $\sin x = -\cos x \tan(-x)$

67. $\sin\left(x + \dfrac{\pi}{4}\right) = \sin x + \sin\dfrac{\pi}{4}$

68. $\cos\left(x + \dfrac{\pi}{4}\right) = \cos x + \cos\dfrac{\pi}{4}$

69. $\cos(x + \pi) = \cos x$

70. $\sin(x + \pi) = \sin x$

71. $\dfrac{\sin x}{1 - \cos^2 x} = \csc x$

72. $\sin x - \sin x \cos^2 x = \sin^3 x$

73. $\sqrt{\sin^2 x + \cos^2 x} = \sin x + \cos x$

Critical Thinking Exercises

In Exercises 74–76, verify each identity.

74. $\dfrac{\sin^3 x - \cos^3 x}{\sin x - \cos x} = 1 + \sin x \cos x$

75. $\dfrac{\sin x - \cos x + 1}{\sin x + \cos x - 1} = \dfrac{\sin x + 1}{\cos x}$

76. $\ln|\sec x| = -\ln|\cos x|$

77. Use one of the fundamental identities in the box on page 557 to create an original identity.

Group Exercise

78. Group members are to write a helpful list of items for a pamphlet called "The Underground Guide to Verifying Identities." The pamphlet will be used primarily by students who sit, stare, and freak out every time they are asked to verify an identity. List easy ways to remember the fundamental identities. What helpful guidelines can you offer from the perspective of a student that you probably won't find in math books? If you have your own strategies that work particularly well, include them in the pamphlet.

SECTION 6.2 *Sum and Difference Formulas*

Objectives

1. Use the formula for the cosine of the difference of two angles.

2. Use sum and difference formulas for cosines and sines.

3. Use sum and difference formulas for tangents.

Listen to the same note played on a piano and a violin. The notes have a different quality or "tone." Tone depends on the way an instrument vibrates. However, the less than 1% of the population with amusia, or true tone deafness, cannot tell the two sounds apart. Even simple, familiar tunes such as *Happy Birthday* and *Jingle Bells* are mystifying to amusics.

When a note is played, it vibrates at a specific fundamental frequency and has a particular amplitude. Amusics cannot tell the difference between two sounds from tuning forks modeled by $p = 3 \sin 2t$ and $p = 2 \sin(2t + \pi)$, respectively. However, they can recognize the difference between the two equations. Notice that the second equation contains the sine of the sum of two angles. In this section, we will be developing identities involving the sums or differences of two angles. These formulas are called the **sum and difference formulas.** We begin with $\cos(\alpha - \beta)$, the cosine of the difference of two angles.

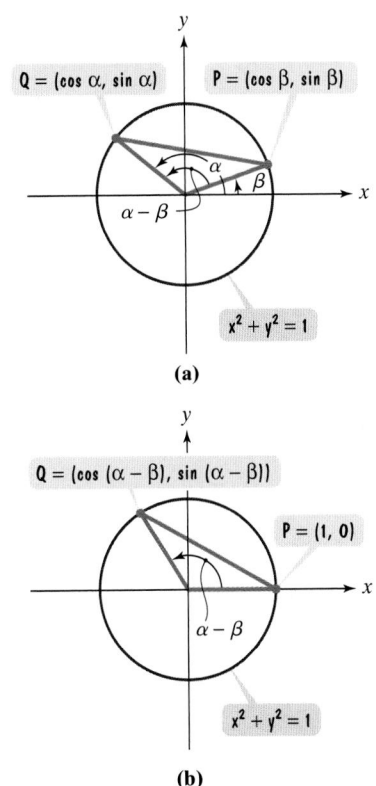

Figure 6.1 Using the unit circle and QP to develop a formula for $\cos(\alpha - \beta)$

The Cosine of the Difference of Two Angles

The Cosine of the Difference of Two Angles

$$\cos(\alpha - \beta) = \cos \alpha \cos \beta + \sin \alpha \sin \beta$$

The cosine of the difference of two angles equals the cosine of the first angle times the cosine of the second angle plus the sine of the first angle times the sine of the second angle.

We use Figure 6.1 to prove the identity in the box. The graph in Figure 6.1(a) shows a unit circle, $x^2 + y^2 = 1$. The figure uses the definitions of the cosine and sine functions as the x- and y-coordinates of points along the unit circle. For example, point P corresponds to angle β. By definition, the x-coordinate of P is $\cos \beta$ and the y-coordinate is $\sin \beta$. Similarly, point Q corresponds to angle α. By definition, the x-coordinate of Q is $\cos \alpha$ and the y-coordinate is $\sin \alpha$.

Note that if we draw a line segment between points P and Q, a triangle is formed. Angle $\alpha - \beta$ is one of the angles of this triangle. What happens if we rotate this triangle so that point P falls on the x-axis at $(1, 0)$? The result is shown in Figure 6.1(b). This rotation changes the coordinates of points P and Q. However, it has no effect on the length of line segment PQ.

We can use the distance formula, $d = \sqrt{(x_2 - x_1)^2 + (y_2 - y_1)^2}$, to find an expression for PQ in Figure 6.1(a) and in Figure 6.1(b). By equating the two expressions for PQ, we will obtain the identity for the cosine of the difference of two angles, $\alpha - \beta$. We first apply the distance formula in Figure 6.1(a).

$$PQ = \sqrt{(\cos\alpha - \cos\beta)^2 + (\sin\alpha - \sin\beta)^2}$$

Apply the distance formula, $d = \sqrt{(x_2 - x_1)^2 + (y_2 - y_1)^2}$, to find the distance between $(\cos\beta, \sin\beta)$ and $(\cos\alpha, \sin\alpha)$.

$$= \sqrt{\cos^2\alpha - 2\cos\alpha\cos\beta + \cos^2\beta + \sin^2\alpha - 2\sin\alpha\sin\beta + \sin^2\beta}$$

Square each expression using $(A - B)^2 = A^2 - 2AB + B^2$.

$$= \sqrt{(\sin^2\alpha + \cos^2\alpha) + (\sin^2\beta + \cos^2\beta) - 2\cos\alpha\cos\beta - 2\sin\alpha\sin\beta}$$

Regroup terms to apply a Pythagorean identity.

$$= \sqrt{1 + 1 - 2\cos\alpha\cos\beta - 2\sin\alpha\sin\beta}$$

Because $\sin^2 x + \cos^2 x = 1$, each expression in parentheses equals 1.

$$= \sqrt{2 - 2\cos\alpha\cos\beta - 2\sin\alpha\sin\beta}$$

Simplify.

Next, we apply the distance formula in Figure 6.1(b) to obtain a second expression for PQ. We let $(x_1, y_1) = (1, 0)$ and $(x_2, y_2) = (\cos(\alpha - \beta), \sin(\alpha - \beta))$.

$$PQ = \sqrt{[\cos(\alpha - \beta) - 1]^2 + [\sin(\alpha - \beta) - 0]^2}$$

Apply the distance formula to find the distance between $(1, 0)$ and $(\cos(\alpha - \beta), \sin(\alpha - \beta))$.

$$= \sqrt{\cos^2(\alpha - \beta) - 2\cos(\alpha - \beta) + 1 + \sin^2(\alpha - \beta)}$$

Square each expression.

Using a Pythagorean identity, $\sin^2(\alpha - \beta) + \cos^2(\alpha - \beta) = 1$

$$= \sqrt{1 - 2\cos(\alpha - \beta) + 1}$$

Use a Pythagorean identity.

$$= \sqrt{2 - 2\cos(\alpha - \beta)}$$

Simplify.

Now we equate the two expressions for PQ.

$$\sqrt{2 - 2\cos(\alpha - \beta)} = \sqrt{2 - 2\cos\alpha\cos\beta - 2\sin\alpha\sin\beta}$$

The rotation does not change the length of PQ.

$$2 - 2\cos(\alpha - \beta) = 2 - 2\cos\alpha\cos\beta - 2\sin\alpha\sin\beta$$

Square both sides to eliminate radicals.

$$-2\cos(\alpha - \beta) = -2\cos\alpha\cos\beta - 2\sin\alpha\sin\beta$$

Subtract 2 from both sides of the equation.

$$\cos(\alpha - \beta) = \cos\alpha\cos\beta + \sin\alpha\sin\beta$$

Divide both sides of the equation by −2.

This proves the identity for the cosine of the difference of two angles.

Now that we see where the identity for the cosine of the difference of two angles comes from, let's look at some applications of this result.

1 Use the formula for the cosine of the difference of two angles.

EXAMPLE 1 Using the Difference Formula for Cosines to Find Exact Values

Find the exact value of cos 15°.

Solution We know exact values for trigonometric functions of 60° and 45°. Thus, we write 15° as 60° − 45° and use the difference formula for cosines.

$$\cos 15° = \cos(60° - 45°)$$

$$= \cos 60° \cos 45° + \sin 60° \sin 45° \quad \cos(\alpha - \beta) = \cos\alpha \cos\beta + \sin\alpha \sin\beta$$

$$= \frac{1}{2} \cdot \frac{\sqrt{2}}{2} + \frac{\sqrt{3}}{2} \cdot \frac{\sqrt{2}}{2} \qquad \text{Substitute exact values from memory or use special right triangles.}$$

$$= \frac{\sqrt{2}}{4} + \frac{\sqrt{6}}{4} \qquad \text{Multiply.}$$

$$= \frac{\sqrt{2} + \sqrt{6}}{4} \qquad \text{Add.}$$

Check Point 1 We know that $\cos 30° = \frac{\sqrt{3}}{2}$. Obtain this exact value using $\cos 30° = \cos(90° - 60°)$ and the difference formula for cosines.

EXAMPLE 2 Using the Difference Formula for Cosines to Find Exact Values

Find the exact value of cos 80° cos 20° + sin 80° sin 20°.

Solution The given expression is the right side of the formula for $\cos(\alpha - \beta)$ with $\alpha = 80°$ and $\beta = 20°$.

$$\cos(\alpha - \beta) = \cos\alpha \cos\beta + \sin\alpha \sin\beta$$

$$\cos 80° \cos 20° + \sin 80° \sin 20° = \cos(80° - 20°) = \cos 60° = \frac{1}{2}$$

Check Point 2 Find the exact value of
$$\cos 70° \cos 40° + \sin 70° \sin 40°.$$

Sound Quality and Amusia

People with true tone deafness cannot hear the difference among tones produced by a tuning fork, a flute, an oboe, and a violin. They cannot dance or tell the difference between harmony and dissonance. People with amusia appear to have been born without the wiring necessary to process music. Intriguingly, they show no overt signs of brain damage and their brain scans appear normal. Thus, they can visually recognize the difference among sound waves that produce varying sound qualities.

Varying Sound Qualities

• Tuning fork: Sound waves are rounded and regular, giving a pure and gentle tone.

• Flute: Sound waves are smooth and give a fluid tone.

• Oboe: Rapid wave changes give a richer tone.

• Violin: Jagged waves give a brighter harsher tone.

EXAMPLE 3 Verifying an Identity

Verify the identity: $\dfrac{\cos(\alpha - \beta)}{\sin \alpha \cos \beta} = \cot \alpha + \tan \beta.$

Solution We work with the left side.

$$\dfrac{\cos(\alpha - \beta)}{\sin \alpha \cos \beta} = \dfrac{\cos \alpha \cos \beta + \sin \alpha \sin \beta}{\sin \alpha \cos \beta}$$ Use the formula for $\cos(\alpha - \beta)$.

$$= \dfrac{\cos \alpha \; \cos \beta}{\sin \alpha \; \cos \beta} + \dfrac{\sin \alpha \; \sin \beta}{\sin \alpha \; \cos \beta}$$ Divide each term in the numerator by $\sin \alpha \cos \beta$.

$$= \cot \alpha \cdot 1 + 1 \cdot \tan \beta$$ Use quotient identities.

$$= \cot \alpha + \tan \beta$$ Simplify.

We worked with the left side and arrived at the right side. Thus, the identity is verified.

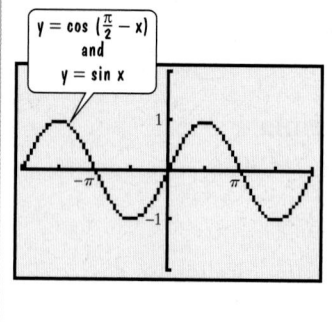

Technology

The graphs of

$$y = \cos\left(\dfrac{\pi}{2} - x\right)$$

and

$$y = \sin x$$

are shown in the same viewing rectangle. The graphs are the same. The displayed math on the right shows the equivalence algebraically.

Check Point 3 Verify the identity: $\dfrac{\cos(\alpha - \beta)}{\cos \alpha \cos \beta} = 1 + \tan \alpha \tan \beta.$

The difference formula for cosines is used to establish other identities. For example, in our work with right triangles we noted that cofunctions of complements are equal. Thus, because $\dfrac{\pi}{2} - \theta$ and θ are complements,

$$\cos\left(\dfrac{\pi}{2} - \theta\right) = \sin \theta.$$

We can use the formula for $\cos(\alpha - \beta)$ to prove this cofunction identity.

> Apply $\cos(\alpha - \beta)$ with $\alpha = 90°$ and $\theta = \beta$.
> $\cos(\alpha - \beta) = \cos \alpha \cos \beta + \sin \alpha \sin \beta$

$$\cos\left(\dfrac{\pi}{2} - \theta\right) = \cos \dfrac{\pi}{2} \cos \theta + \sin \dfrac{\pi}{2} \sin \theta$$

$$= 0 \cdot \cos \theta + 1 \cdot \sin \theta$$

$$= \sin \theta$$

2 Use sum and difference formulas for cosines and sines.

Sum and Difference Formulas for Cosines and Sines

Our formula for $\cos(\alpha - \beta)$ can be used to verify an identity for a sum involving cosines, as well as identities for a sum and a difference for sines.

Sum and Difference Formulas for Cosines and Sines

1. $\cos(\alpha + \beta) = \cos \alpha \cos \beta - \sin \alpha \sin \beta$

2. $\cos(\alpha - \beta) = \cos \alpha \cos \beta + \sin \alpha \sin \beta$

3. $\sin(\alpha + \beta) = \sin \alpha \cos \beta + \cos \alpha \sin \beta$

4. $\sin(\alpha - \beta) = \sin \alpha \cos \beta - \cos \alpha \sin \beta$

Up to now, we have concentrated on the second formula in the box. The first identity gives a formula for the cosine of the sum of two angles. It is proved as follows:

$$\cos(\alpha + \beta) = \cos[\alpha - (-\beta)] \qquad \text{Express addition as subtraction of an inverse.}$$

$$= \cos\alpha\cos(-\beta) + \sin\alpha\sin(-\beta) \qquad \text{Use the difference formula for cosines.}$$

$$= \cos\alpha\cos\beta + \sin\alpha(-\sin\beta) \qquad \text{Cosine is even: } \cos(-\beta) = \cos\beta.$$
$$\text{Sine is odd: } \sin(-\beta) = -\sin\beta.$$

$$= \cos\alpha\cos\beta - \sin\alpha\sin\beta. \qquad \text{Simplify.}$$

Thus, the cosine of the sum of two angles equals the cosine of the first angle times the cosine of the second angle minus the sine of the first angle times the sine of the second angle.

The third identity in the box gives a formula for $\sin(\alpha + \beta)$, the sine of the sum of two angles. It is proved as follows:

$$\sin(\alpha + \beta) = \cos\left[\frac{\pi}{2} - (\alpha + \beta)\right] \qquad \text{Use a cofunction identity: } \sin\theta = \cos\left(\frac{\pi}{2} - \theta\right).$$

$$= \cos\left[\left(\frac{\pi}{2} - \alpha\right) - \beta\right] \qquad \text{Regroup.}$$

$$= \cos\left(\frac{\pi}{2} - \alpha\right)\cos\beta + \sin\left(\frac{\pi}{2} - \alpha\right)\sin\beta \qquad \text{Use the difference formula for cosines.}$$

$$= \sin\alpha\cos\beta + \cos\alpha\sin\beta. \qquad \text{Use cofunction identities.}$$

Thus, the sine of the sum of two angles equals the sine of the first angle times the cosine of the second angle plus the cosine of the first angle times the sine of the second angle.

The final identity in the box gives a formula for $\sin(\alpha - \beta)$, the sine of the difference of two angles. It is proved by writing $\sin(\alpha - \beta)$ as $\sin[\alpha + (-\beta)]$ and then using the formula for the sine of a sum.

EXAMPLE 4 Using the Sine of a Sum to Find an Exact Value

Find the exact value of $\sin\dfrac{7\pi}{12}$ using the fact that $\dfrac{7\pi}{12} = \dfrac{\pi}{3} + \dfrac{\pi}{4}$.

Solution We apply the formula for the sine of a sum.

$$\sin\frac{7\pi}{12} = \sin\left(\frac{\pi}{3} + \frac{\pi}{4}\right)$$

$$= \sin\frac{\pi}{3}\cos\frac{\pi}{4} + \cos\frac{\pi}{3}\sin\frac{\pi}{4} \qquad \sin(\alpha + \beta) = \sin\alpha\cos\beta + \cos\alpha\sin\beta$$

$$= \frac{\sqrt{3}}{2} \cdot \frac{\sqrt{2}}{2} + \frac{1}{2} \cdot \frac{\sqrt{2}}{2} \qquad \text{Substitute exact values.}$$

$$= \frac{\sqrt{6} + \sqrt{2}}{4} \qquad \text{Simplify.}$$

Check Point 4

Find the exact value of $\sin \dfrac{5\pi}{12}$ using the fact that

$$\dfrac{5\pi}{12} = \dfrac{\pi}{6} + \dfrac{\pi}{4}.$$

EXAMPLE 5 Finding Exact Values

Suppose that $\sin \alpha = \frac{12}{13}$ for a quadrant II angle α and $\sin \beta = \frac{3}{5}$ for a quadrant I angle β. Find the exact value of:

a. $\cos \alpha$ **b.** $\cos \beta$ **c.** $\cos(\alpha + \beta)$ **d.** $\sin(\alpha + \beta)$.

Solution

a. We find $\cos \alpha$ using a sketch that illustrates

$$\sin \alpha = \dfrac{12}{13} = \dfrac{y}{r}.$$

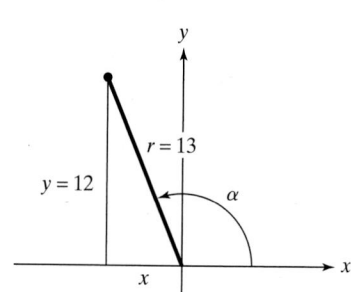

Figure 6.2 $\sin \alpha = \frac{12}{13}$: α lies in quadrant II

Figure 6.2 shows a quadrant II angle α with $\sin \alpha = \frac{12}{13}$. We find x using $x^2 + y^2 = r^2$. Because α lies in quadrant II, x is negative.

$$x^2 + 12^2 = 13^2 \qquad \text{\small $x^2 + y^2 = r^2$}$$
$$x^2 + 144 = 169 \qquad \text{\small Square 12 and 13, respectively.}$$
$$x^2 = 25 \qquad \text{\small Subtract 144 from both sides.}$$
$$x = -\sqrt{25} = -5 \quad \text{\small In quadrant II, x is negative.}$$

Thus,

$$\cos \alpha = \dfrac{x}{r} = \dfrac{-5}{13} = -\dfrac{5}{13}.$$

b. We find $\cos \beta$ using a sketch that illustrates

$$\sin \beta = \dfrac{3}{5} = \dfrac{y}{r}.$$

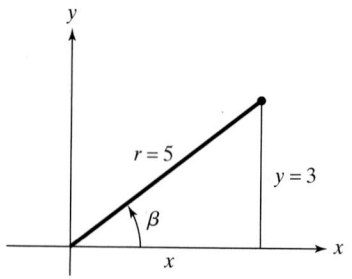

Figure 6.3 $\sin \beta = \frac{3}{5}$: β lies in quadrant I

Figure 6.3 shows a quadrant I angle β with $\sin \beta = \frac{3}{5}$. We find x using $x^2 + y^2 = r^2$.

$$x^2 + 3^2 = 5^2$$
$$x^2 = 25 - 9 = 16$$
$$x = \sqrt{16} = 4 \qquad \text{\small In quadrant I, x is positive.}$$

Thus,

$$\cos \beta = \dfrac{x}{r} = \dfrac{4}{5}.$$

We use the given values and the exact values that we determined to find exact values for $\cos(\alpha + \beta)$ and $\sin(\alpha + \beta)$.

These values are given.

These are the values we found.

$$\sin \alpha = \dfrac{12}{13}, \ \sin \beta = \dfrac{3}{5} \qquad \cos \alpha = -\dfrac{5}{13}, \ \cos \beta = \dfrac{4}{5}$$

c. We use the formula for the cosine of a sum.

$$\cos(\alpha + \beta) = \cos\alpha\cos\beta - \sin\alpha\sin\beta$$

$$= -\frac{5}{13}\left(\frac{4}{5}\right) - \frac{12}{13}\left(\frac{3}{5}\right) = -\frac{56}{65}$$

d. We use the formula for the sine of a sum.

$$\sin(\alpha + \beta) = \sin\alpha\cos\beta + \cos\alpha\sin\beta$$

$$= \frac{12}{13}\cdot\frac{4}{5} + \left(-\frac{5}{13}\right)\cdot\frac{3}{5} = \frac{33}{65}$$

Check Point 5 Suppose that $\sin\alpha = \frac{4}{5}$ for a quadrant II angle α and $\sin\beta = \frac{1}{2}$ for a quadrant I angle β. Find the exact value of:

a. $\cos\alpha$ **b.** $\cos\beta$

c. $\cos(\alpha + \beta)$ **d.** $\sin(\alpha + \beta)$.

EXAMPLE 6 Verifying Observations on a Graphing Utility

Figure 6.4 shows the graph of $y = \sin\left(x - \frac{3\pi}{2}\right)$ in a $\left[0, 2\pi, \frac{\pi}{2}\right]$ by $[-2, 2, 1]$ viewing rectangle.

a. Describe the graph using another equation.

b. Verify that the two equations are equivalent.

Figure 6.4 The graph of $y = \sin\left(x - \frac{3\pi}{2}\right)$ in a $\left[0, 2\pi, \frac{\pi}{2}\right]$ by $[-2, 2, 1]$ viewing rectangle

Solution

a. The graph appears to be the cosine curve $y = \cos x$. It cycles through maximum, intercept, minimum, intercept, and back to maximum. Thus, $y = \cos x$ also describes the graph.

b. We must show that

$$\sin\left(x - \frac{3\pi}{2}\right) = \cos x.$$

We apply the formula for the sine of a difference on the left side.

$$\sin\left(x - \frac{3\pi}{2}\right) = \sin x \cos\frac{3\pi}{2} - \cos x \sin\frac{3\pi}{2} \qquad \begin{array}{l} \sin(\alpha - \beta) = \\ \sin\alpha\cos\beta - \cos\alpha\sin\beta \end{array}$$

$$= \sin x \cdot 0 - \cos x(-1) \qquad \cos\frac{3\pi}{2} = 0 \text{ and } \sin\frac{3\pi}{2} = -1$$

$$= \cos x \qquad\qquad\qquad \text{Simplify.}$$

This verifies our observation that $y = \sin\left(x - \frac{3\pi}{2}\right)$ and $y = \cos x$ describe the same graph.

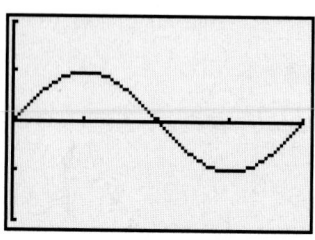

Figure 6.5

Figure 6.5 shows the graph of $y = \cos\left(x + \dfrac{3\pi}{2}\right)$ in a $\left[0, 2\pi, \dfrac{\pi}{2}\right]$ by $[-2, 2, 1]$ viewing rectangle.

a. Describe the graph using another equation.

b. Verify that the two equations are equivalent.

3 Use sum and difference formulas for tangents.

Sum and Difference Formulas for Tangents

By writing $\tan(\alpha + \beta)$ as the quotient of $\sin(\alpha + \beta)$ and $\cos(\alpha + \beta)$ we can develop a formula for the tangent of a sum. Writing subtraction as addition of an inverse leads to a formula for the tangent of a difference.

Discovery

Derive the sum and difference formulas for tangents by working Exercises 55 and 56 in Exercise Set 6.2.

Sum and Difference Formulas for Tangents

$$\tan(\alpha + \beta) = \frac{\tan\alpha + \tan\beta}{1 - \tan\alpha \tan\beta}$$

The tangent of the sum of two angles equals the tangent of the first angle plus the tangent of the second angle divided by 1 minus their product.

$$\tan(\alpha - \beta) = \frac{\tan\alpha - \tan\beta}{1 + \tan\alpha \tan\beta}$$

The tangent of the difference of two angles equals the tangent of the first angle minus the tangent of the second angle divided by 1 plus their product.

EXAMPLE 7   Verifying an Identity

Verify the identity: $\tan\left(x - \dfrac{\pi}{4}\right) = \dfrac{\tan x - 1}{\tan x + 1}$.

Solution   We work with the left side.

$$\tan\left(x - \frac{\pi}{4}\right) = \frac{\tan x - \tan\dfrac{\pi}{4}}{1 + \tan x \tan\dfrac{\pi}{4}} \qquad \tan(\alpha - \beta) = \frac{\tan\alpha - \tan\beta}{1 + \tan\alpha \tan\beta}$$

$$= \frac{\tan x - 1}{1 + \tan x \cdot 1} \qquad \tan\frac{\pi}{4} = 1$$

$$= \frac{\tan x - 1}{1 + \tan x}$$

Verify the identity: $\tan(x + \pi) = \tan x$.

EXERCISE SET 6.2

Practice Exercises

Use the formula for the cosine of the difference of two angles to solve Exercises 1–12.

In Exercises 1–4, find the exact value of each expression.

1. $\cos(45° - 30°)$

2. $\cos(120° - 45°)$

3. $\cos\left(\dfrac{3\pi}{4} - \dfrac{\pi}{6}\right)$

4. $\cos\left(\dfrac{2\pi}{3} - \dfrac{\pi}{6}\right)$

In Exercises 5–8, each expression is the right side of the formula for $\cos(\alpha - \beta)$ with particular values for α and β.

a. *Identify α and β in each expression.*

b. *Write the expression as the cosine of an angle.*

c. *Find the exact value of the expression.*

5. $\cos 50° \cos 20° + \sin 50° \sin 20°$

6. $\cos 50° \cos 5° + \sin 50° \sin 5°$

7. $\cos\dfrac{5\pi}{12}\cos\dfrac{\pi}{12} + \sin\dfrac{5\pi}{12}\sin\dfrac{\pi}{12}$

8. $\cos\dfrac{5\pi}{18}\cos\dfrac{\pi}{9} + \sin\dfrac{5\pi}{18}\sin\dfrac{\pi}{9}$

In Exercises 9–12, verify each identity.

9. $\dfrac{\cos(\alpha - \beta)}{\cos\alpha \sin\beta} = \tan\alpha + \cot\beta$

10. $\dfrac{\cos(\alpha - \beta)}{\sin\alpha \sin\beta} = \cot\alpha \cot\beta + 1$

11. $\cos\left(x - \dfrac{\pi}{4}\right) = \dfrac{\sqrt{2}}{2}(\cos x + \sin x)$

12. $\cos\left(x - \dfrac{5\pi}{4}\right) = -\dfrac{\sqrt{2}}{2}(\cos x + \sin x)$

Use one or more of the six sum and difference identities to solve Exercises 13–54.

In Exercises 13–24, find the exact value of each expression.

13. $\sin(45° - 30°)$

14. $\sin(60° - 45°)$

15. $\sin 105°$

16. $\sin 75°$

17. $\tan(30° + 45°)$

18. $\tan(60° + 45°)$

19. $\tan(240° - 45°)$

20. $\tan(300° - 45°)$

21. $\cos\left(\dfrac{3\pi}{4} + \dfrac{\pi}{6}\right)$

22. $\cos\left(\dfrac{4\pi}{3} + \dfrac{\pi}{4}\right)$

23. $\cos\dfrac{5\pi}{12}$

24. $\cos\dfrac{7\pi}{12}$

In Exercises 25–32, write each expression as the sine, cosine, or tangent of an angle. Then find the exact value of the expression.

25. $\sin 25° \cos 5° + \cos 25° \sin 5°$

26. $\sin 40° \cos 20° + \cos 40° \sin 20°$

27. $\dfrac{\tan 10° + \tan 35°}{1 - \tan 10° \tan 35°}$

28. $\dfrac{\tan 50° - \tan 20°}{1 + \tan 50° \tan 20°}$

29. $\sin\dfrac{5\pi}{12}\cos\dfrac{\pi}{4} - \cos\dfrac{5\pi}{12}\sin\dfrac{\pi}{4}$

30. $\sin\dfrac{7\pi}{12}\cos\dfrac{\pi}{12} - \cos\dfrac{7\pi}{12}\sin\dfrac{\pi}{12}$

31. $\dfrac{\tan\dfrac{\pi}{5} - \tan\dfrac{\pi}{30}}{1 + \tan\dfrac{\pi}{5}\tan\dfrac{\pi}{30}}$

32. $\dfrac{\tan\dfrac{\pi}{5} + \tan\dfrac{4\pi}{5}}{1 - \tan\dfrac{\pi}{5}\tan\dfrac{4\pi}{5}}$

In Exercises 33–54, verify each identity.

33. $\sin\left(x + \dfrac{\pi}{2}\right) = \cos x$

34. $\sin\left(x + \dfrac{3\pi}{2}\right) = -\cos x$

35. $\cos\left(x - \dfrac{\pi}{2}\right) = \sin x$

36. $\cos(\pi - x) = -\cos x$

37. $\tan(2\pi - x) = -\tan x$

38. $\tan(\pi - x) = -\tan x$

39. $\sin(\alpha + \beta) + \sin(\alpha - \beta) = 2\sin\alpha \cos\beta$

40. $\cos(\alpha + \beta) + \cos(\alpha - \beta) = 2\cos\alpha \cos\beta$

41. $\dfrac{\sin(\alpha - \beta)}{\cos\alpha \cos\beta} = \tan\alpha - \tan\beta$

42. $\dfrac{\sin(\alpha + \beta)}{\cos\alpha \cos\beta} = \tan\alpha + \tan\beta$

43. $\tan\left(\theta + \dfrac{\pi}{4}\right) = \dfrac{\cos\theta + \sin\theta}{\cos\theta - \sin\theta}$

44. $\tan\left(\dfrac{\pi}{4} - \theta\right) = \dfrac{\cos\theta - \sin\theta}{\cos\theta - \sin\theta}$

45. $\cos(\alpha + \beta)\cos(\alpha - \beta) = \cos^2\beta - \sin^2\alpha$

46. $\sin(\alpha + \beta)\sin(\alpha - \beta) = \cos^2\beta - \cos^2\alpha$

47. $\dfrac{\sin(\alpha + \beta)}{\sin(\alpha - \beta)} = \dfrac{\tan\alpha + \tan\beta}{\tan\alpha - \tan\beta}$

48. $\dfrac{\cos(\alpha + \beta)}{\cos(\alpha - \beta)} = \dfrac{1 - \tan\alpha \tan\beta}{1 + \tan\alpha \tan\beta}$

49. $\dfrac{\cos(x + h) - \cos x}{h} = \cos x \dfrac{\cos h - 1}{h} - \sin x \dfrac{\sin h}{h}$

50. $\dfrac{\sin(x+h)-\sin x}{h} = \cos x\,\dfrac{\sin h}{h} + \sin x\,\dfrac{\cos h - 1}{h}$

51. $\sin 2\alpha = 2\sin\alpha\cos\alpha$
Hint: Write $\sin 2\alpha$ as $\sin(\alpha+\alpha)$.

52. $\cos 2\alpha = \cos^2\alpha - \sin^2\alpha$
Hint: Write $\cos 2\alpha$ as $\cos(\alpha+\alpha)$.

53. $\tan 2\alpha = \dfrac{2\tan\alpha}{1-\tan^2\alpha}$
Hint: Write $\tan 2\alpha$ as $\tan(\alpha+\alpha)$.

54. $\tan\left(\dfrac{\pi}{4}+\alpha\right) - \tan\left(\dfrac{\pi}{4}-\alpha\right) = 2\tan 2\alpha$
Hint: Use the result in Exercise 53.

55. Derive the identity for $\tan(\alpha+\beta)$ using

$$\tan(\alpha+\beta) = \dfrac{\sin(\alpha+\beta)}{\cos(\alpha+\beta)}.$$

After applying the formulas for sums of sines and cosines, divide the numerator and denominator by $\cos\alpha\cos\beta$.

56. Derive the identity for $\tan(\alpha-\beta)$ using

$$\tan(\alpha-\beta) = \tan[\alpha+(-\beta)].$$

After applying the formula for the tangent of the sum of two angles, use the fact that the tangent is an odd function.

In Exercises 57–62, find the exact value of the following under the given conditions:

a. $\cos(\alpha+\beta)$ **b.** $\sin(\alpha+\beta)$ **c.** $\tan(\alpha+\beta)$

57. $\sin\alpha = \frac{3}{5}$, α lies in quadrant I, and $\sin\beta = \frac{5}{13}$, β lies in quadrant II.

58. $\sin\alpha = \frac{4}{5}$, α lies in quadrant I, and $\sin\beta = \frac{7}{25}$, β lies in quadrant II.

59. $\tan\alpha = -\frac{3}{4}$, α lies in quadrant II, and $\cos\beta = \frac{1}{3}$, β lies in quadrant I.

60. $\tan\alpha = -\frac{4}{3}$, α lies in quadrant II, and $\cos\beta = \frac{2}{3}$, β lies in quadrant I.

61. $\cos\alpha = \frac{8}{17}$, α lies in quadrant IV, and $\sin\beta = -\frac{1}{2}$, β lies in quadrant III.

62. $\cos\alpha = \frac{1}{2}$, α lies in quadrant IV, and $\sin\beta = -\frac{1}{3}$, β lies in quadrant III.

In Exercises 63–66, the graph with the given equation is shown in a $\left[0, 2\pi, \dfrac{\pi}{2}\right]$ by $[-2, 2, 1]$ viewing rectangle.

a. Describe the graph using another equation.

b. Verify that the two equations are equivalent.

63.

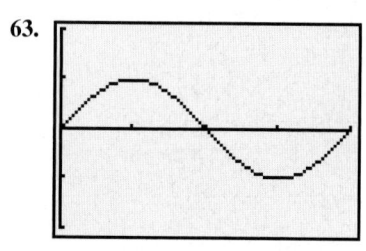

$y = \sin(\pi - x)$

64.

$y = \cos(x - 2\pi)$

65.

$y = \sin\left(x + \dfrac{\pi}{2}\right) + \sin\left(\dfrac{\pi}{2} - x\right)$

66.

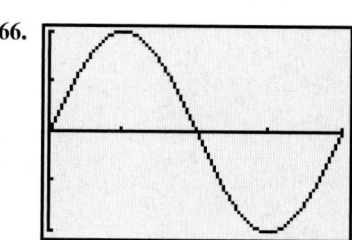

$y = \cos\left(x - \dfrac{\pi}{2}\right) - \cos\left(x + \dfrac{\pi}{2}\right)$

Application Exercises

67. A ball attached to a spring is raised 2 feet and released with an initial vertical velocity of 3 feet per second. The distance of the ball from its rest position after t seconds is given by $d = 2\cos t + 3\sin t$. Show that

$$2\cos t + 3\sin t = \sqrt{13}\cos(t - \theta)$$

where θ lies in quadrant I and $\tan\theta = \frac{3}{2}$. Use the identity to find the amplitude and the period of the ball's motion.

68. A tuning fork is held a certain distance from your ears and struck. Your eardrums' vibrations after t seconds are given by $p = 3\sin 2t$. When a second tuning fork is struck, the formula $p = 2\sin(2t + \pi)$ describes the effects of the sound on the eardrums' vibrations. The total vibrations are given by $p = 3\sin 2t + 2\sin(2t + \pi)$.

a. Express p using a single occurrence of the sine.

b. If the amplitude of p is zero, no sound is heard. Based on your equation in part (a), does this occur with the two tuning forks in this exercise? Explain your answer.

Writing in Mathematics

In Exercises 69–74, use words to describe the formula for:

69. the cosine of the difference of two angles.

70. the cosine of the sum of two angles.

71. the sine of the sum of two angles.

72. the sine of the difference of two angles.

73. the tangent of the difference of two angles.

74. the tangent of the sum of two angles.

75. The distance formula and the definitions for cosine and sine are used to prove the formula for the cosine of the difference of two angles. This formula logically leads the way to the other sum and difference identities. Using this development of ideas and formulas, describe a characteristic of mathematical logic.

 Technology Exercises

In Exercises 76–81, graph each side of the equation in the same viewing rectangle. If the graphs appear to coincide, verify that the equation is an identity. If the graphs do not appear to coincide, this indicates that the equation is not an identity. In these exercises, find a value of x for which both sides are defined but not equal.

76. $\cos\left(\dfrac{3\pi}{2} - x\right) = -\sin x$

77. $\tan(\pi - x) = -\tan x$

78. $\sin\left(x + \dfrac{\pi}{2}\right) = \sin x + \sin\dfrac{\pi}{2}$

79. $\cos\left(x + \dfrac{\pi}{2}\right) = \cos x + \cos\dfrac{\pi}{2}$

80. $\cos 1.2x \cos 0.8x - \sin 1.2x \sin 0.8x = \cos 2x$

81. $\sin 1.2x \cos 0.8x + \cos 1.2x \sin 0.8x = \sin 2x$

 Critical Thinking Exercises

82. Graph $y = \sin 5x \cos 3x - \cos 5x \sin 3x$ from 0 to 2π.

83. Verify the identity:

$$\frac{\sin(x - y)}{\cos x \cos y} + \frac{\sin(y - z)}{\cos y \cos z} + \frac{\sin(z - x)}{\cos z \cos x} = 0.$$

In Exercises 84–87, find the exact value of each expression. Do not use a calculator.

84. $\sin\left(\cos^{-1}\dfrac{1}{2} + \sin^{-1}\dfrac{3}{5}\right)$

85. $\sin\left[\sin^{-1}\dfrac{3}{5} - \cos^{-1}\left(-\dfrac{4}{5}\right)\right]$

86. $\cos\left(\tan^{-1}\dfrac{4}{3} + \cos^{-1}\dfrac{5}{13}\right)$

87. $\cos\left[\cos^{-1}\left(-\dfrac{\sqrt{3}}{2}\right) - \sin^{-1}\left(-\dfrac{1}{2}\right)\right]$

In Exercises 88–90, write each trigonometric expression as an algebraic expression as an algebraic expression (that is, without any trigonometric functions). Assume that x and y are positive and in the domain of the given inverse trigonometric function.

88. $\cos(\sin^{-1} x - \cos^{-1} y)$

89. $\sin(\tan^{-1} x - \sin^{-1} y)$

90. $\tan(\sin^{-1} x + \cos^{-1} y)$

Group Exercise

91. Remembering the six sum and difference identities can be difficult. Did you have problems with some exercises because the identity you were using in your head turned out to be an incorrect formula? Are there easy ways to remember the six new identities presented in this section? Group members should address this question, considering one identity at a time. For each formula, list ways to make it easier to remember.

SECTION 6.3 Double-Angle and Half-Angle Formulas

Objectives

1. Use the double-angle formulas.
2. Use the power-reducing formulas.
3. Use the half-angle formulas.

We have a long history of throwing things. Prior to 400 B.C., the Greeks competed in games that included discus throwing. In the seventeenth century,

English soldiers organized cannonball-throwing competitions. In 1827, a Yale University student, disappointed over failing an exam, took out his frustrations at the passing of a collection plate in chapel. Seizing the monetary tray, he flung it in the direction of a large open space on campus. Yale students see this act of frustration as the origin of the Frisbee.

In this section, we develop other important classes of identities, called the double-angle and half-angle formulas. We will see how one of these formulas can be used by athletes to increase throwing distance.

Double-Angle Formulas

A number of basic identities follow from the sum formulas for sine, cosine, and tangent. The first category of identities involves **double-angle formulas.**

1 Use the double-angle formulas.

Study Tip

The 2 that appears in each of the double-angle expressions cannot be pulled to the front and written as a coefficient.

INCORRECT

$\sin 2\theta = 2 \sin \theta$
$\cos 2\theta = 2 \cos \theta$
$\tan 2\theta = 2 \tan \theta$

The figure shows, in a $\left[0, 2\pi, \dfrac{\pi}{2}\right]$ by $[-3, 3, 1]$ viewing rectangle, that the graphs of

$$y = \sin 2x$$

and

$$y = 2 \sin x$$

do not coincide.

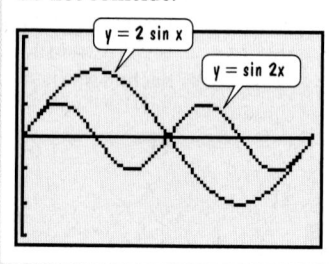

Double-Angle Formulas

$$\sin 2\theta = 2 \sin \theta \cos \theta$$

$$\cos 2\theta = \cos^2 \theta - \sin^2 \theta$$

$$\tan 2\theta = \frac{2 \tan \theta}{1 - \tan^2 \theta}$$

To prove each of these formulas, we replace α and β by θ in the sum formulas for $\sin(\alpha + \beta)$, $\cos(\alpha + \beta)$, and $\tan(\alpha + \beta)$.

- $\sin 2\theta = \sin(\theta + \theta) = \sin \theta \cos \theta + \cos \theta \sin \theta = 2 \sin \theta \cos \theta$

  We use
  $\sin(\alpha + \beta) = \sin \alpha \cos \beta + \cos \alpha \sin \beta.$

- $\cos 2\theta = \cos(\theta + \theta) = \cos \theta \cos \theta - \sin \theta \sin \theta = \cos^2 \theta - \sin^2 \theta$

  We use
  $\cos(\alpha + \beta) = \cos \alpha \cos \beta - \sin \alpha \sin \beta.$

- $\tan 2\theta = \tan(\theta + \theta) = \dfrac{\tan \theta + \tan \theta}{1 - \tan \theta \tan \theta} = \dfrac{2 \tan \theta}{1 - \tan^2 \theta}$

  We use
  $\tan(\alpha + \beta) = \dfrac{\tan \alpha + \tan \beta}{1 - \tan \alpha \tan \beta}.$

EXAMPLE 1 Using Double-Angle Formulas to Find Exact Values

If $\sin \theta = \frac{5}{13}$ and θ lies in quadrant II, find the exact value of:

a. $\sin 2\theta$ **b.** $\cos 2\theta$ **c.** $\tan 2\theta$.

Solution We begin with a sketch that illustrates

$$\sin \theta = \frac{5}{13} = \frac{y}{r}.$$

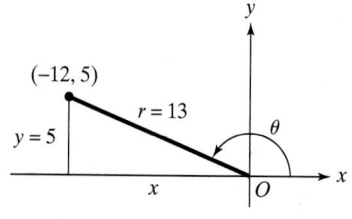

Figure 6.6 $\sin \theta = \frac{5}{13}$ and θ lies in quadrant II

Figure 6.6 shows a quadrant II angle θ for which $\sin \theta = \frac{5}{13}$. We find x using $x^2 + y^2 = r^2$. Because θ lies in quadrant II, x is negative.

$$x^2 + 5^2 = 13^2 \qquad x^2 + y^2 = r^2$$

$$x^2 = 13^2 - 5^2 = 144 \quad \text{Solve for } x^2.$$

$$x = -\sqrt{144} = -12 \quad \text{In quadrant II, } x \text{ is negative.}$$

Now we can use values for x, y, and r to find the required values. We will use $\cos\theta = \dfrac{x}{r} = -\dfrac{12}{13}$ and $\tan\theta = \dfrac{y}{x} = -\dfrac{5}{12}$. We were given $\sin\theta = \dfrac{5}{13}$.

a. $\sin 2\theta = 2\sin\theta\cos\theta = 2\left(\dfrac{5}{13}\right)\left(-\dfrac{12}{13}\right) = -\dfrac{120}{169}$

b. $\cos 2\theta = \cos^2\theta - \sin^2\theta = \left(-\dfrac{12}{13}\right)^2 - \left(\dfrac{5}{13}\right)^2 = \dfrac{144}{169} - \dfrac{25}{169} = \dfrac{119}{169}$

c. $\tan 2\theta = \dfrac{2\tan\theta}{1 - \tan^2\theta} = \dfrac{2\left(-\dfrac{5}{12}\right)}{1 - \left(-\dfrac{5}{12}\right)^2} = \dfrac{-\dfrac{5}{6}}{1 - \dfrac{25}{144}} = \dfrac{-\dfrac{5}{6}}{\dfrac{119}{144}} = \left(-\dfrac{5}{6}\right)\left(\dfrac{144}{119}\right) = -\dfrac{120}{119}$

Check Point 1   If $\sin\theta = \frac{4}{5}$ and θ lies in quadrant II, find the exact value of:

    a. $\sin 2\theta$ **b.** $\cos 2\theta$ **c.** $\tan 2\theta$.

EXAMPLE 2   Using the Double–Angle Formula for Tangents to Find an Exact Value

Find the exact value of $\dfrac{2\tan 15^\circ}{1 - \tan^2 15^\circ}$.

Solution   The given expression is the right side of the formula for $\tan 2\theta$ with $\theta = 15$.

$$\tan 2\theta = \dfrac{2\tan\theta}{1 - \tan^2\theta}$$

$$\dfrac{2\tan 15^\circ}{1 - \tan^2 15^\circ} = \tan(2 \cdot 15^\circ) = \tan 30^\circ = \dfrac{\sqrt{3}}{3}$$

Check Point 2   Find the exact value of $\cos^2 15^\circ - \sin^2 15^\circ$.

There are three forms of the double-angle formula for $\cos 2\theta$. The form we have seen involves both the cosine and the sine:

$$\cos 2\theta = \cos^2\theta - \sin^2\theta.$$

Using the Pythagorean identity $\sin^2\theta + \cos^2\theta = 1$, we can write this last formula in terms of the cosine only. We substitute $1 - \cos^2\theta$ for $\sin^2\theta$.

$$\cos 2\theta = \cos^2\theta - \sin^2\theta = \cos^2\theta - (1 - \cos^2\theta)$$

$$= \cos^2\theta - 1 + \cos^2\theta = 2\cos^2\theta - 1$$

We can also use a Pythagorean identity to write $\cos 2\theta$ in terms of sine only. We substitute $1 - \sin^2 \theta$ for $\cos^2 \theta$.

$$\cos 2\theta = \cos^2 \theta - \sin^2 \theta = 1 - \sin^2 \theta - \sin^2 \theta = 1 - 2\sin^2 \theta$$

Three Forms of the Double-Angle Formula for $\cos 2\theta$

$$\cos 2\theta = \cos^2 \theta - \sin^2 \theta$$
$$\cos 2\theta = 2\cos^2 \theta - 1$$
$$\cos 2\theta = 1 - 2\sin^2 \theta$$

EXAMPLE 3 Verifying an Identity

Verify the identity: $\cos 3\theta = 4\cos^2 \theta - 3\cos \theta$.

Solution We begin by working with the left side. In order to obtain an expression for $\cos 3\theta$, we use the sum formula and write 3θ as $2\theta + \theta$.

$\cos 3\theta = \cos(2\theta + \theta)$ Write 3θ as $2\theta + \theta$.

$= \cos 2\theta \cos \theta - \sin 2\theta \sin \theta$ $\cos(\alpha + \beta)$
       $= \cos \alpha \cos \beta - \sin \alpha \sin \beta$

$\underset{2\cos^2\theta - 1}{} \quad \underset{2\sin\theta\cos\theta}{}$

$= (2\cos^2 \theta - 1)\cos \theta - 2\sin \theta \cos \theta \sin \theta$ Substitute double-angle formulas. Because the right side of the given equation involves cosines only, use this form for $\cos 2\theta$.

$= 2\cos^3 \theta - \cos \theta - 2\sin^2 \theta \cos \theta$ Multiply.

$\underset{1 - \cos^2\theta}{}$

$= 2\cos^3 \theta - \cos \theta - 2(1 - \cos^2 \theta)\cos \theta$ To get cosines only, use $\sin^2 \theta + \cos^2 \theta = 1$ and substitute $1 - \cos^2 \theta$ for $\sin^2 \theta$.

$= 2\cos^3 \theta - \cos \theta - 2\cos \theta + 2\cos^3 \theta$ Multiply.

$= 4\cos^3 \theta - 3\cos \theta$ Simplify: $2\cos^3 \theta + 2\cos^3 \theta = 4\cos^3 \theta$ and $-\cos \theta - 2\cos \theta = -3\cos \theta$.

By working with the left side and expressing it in a form identical to the right side, we have verified the identity.

Check Point 3 Verify the identity: $\sin 3\theta = 3\sin \theta - 4\sin^3 \theta$.

---

2 Use the power-reducing formulas.

Power-Reducing Formulas

The double-angle formulas are used to derive the **power-reducing formulas:**

Power-Reducing Formulas

$$\sin^2 \theta = \frac{1 - \cos 2\theta}{2} \qquad \cos^2 \theta = \frac{1 + \cos 2\theta}{2} \qquad \tan^2 \theta = \frac{1 - \cos 2\theta}{1 + \cos 2\theta}$$

We can prove the first two formulas in the box by working with two forms of the double-angle formula for $\cos 2\theta$.

This is the form with sine only.   This is the form with cosine only.

$$\cos 2\theta = 1 - 2\sin^2\theta \qquad \cos 2\theta = 2\cos^2\theta - 1$$

Solve the formula on the left for $\sin^2\theta$. Solve the formula on the right for $\cos^2\theta$.

$$2\sin^2\theta = 1 - \cos 2\theta \qquad 2\cos^2\theta = 1 + \cos 2\theta$$

$$\sin^2\theta = \frac{1 - \cos 2\theta}{2} \qquad \cos^2\theta = \frac{1 + \cos 2\theta}{2} \qquad \text{Divide both sides of each equation by 2.}$$

These are the first two formulas in the box. The third formula in the box is proved by writing the tangent as the quotient of the sine and the cosine.

$$\tan^2\theta = \frac{\sin^2\theta}{\cos^2\theta} = \frac{\dfrac{1 - \cos 2\theta}{2}}{\dfrac{1 + \cos 2\theta}{2}} = \frac{1 - \cos 2\theta}{\cancel{2}} \cdot \frac{\dfrac{1}{\cancel{2}}}{1 + \cos 2\theta} = \frac{1 - \cos 2\theta}{1 + \cos 2\theta}$$

Power-reducing formulas are quite useful in calculus. By reducing the power of trigonometric functions, calculus can better explore the relationship between a function and how it is changing at every single instant in time.

EXAMPLE 4   Reducing the Power of a Trigonometric Function

Write an equivalent expression for $\cos^4 x$ that does not contain powers of trigonometric functions greater than 1.

Solution   We will apply the formula for $\cos^2\theta$ twice.

$$\cos^4 x = \left(\cos^2 x\right)^2$$

$$= \left(\frac{1 + \cos 2x}{2}\right)^2 \qquad \text{Use } \cos^2\theta = \frac{1 + \cos 2\theta}{2} \text{ with } \theta = x.$$

$$= \frac{1 + 2\cos 2x + \cos^2 2x}{4} \qquad \begin{array}{l}\text{Square the numerator:} \\ (A + B)^2 = A^2 + 2AB + B^2. \\ \text{Square the denominator.}\end{array}$$

$$= \frac{1}{4} + \frac{1}{2}\cos 2x + \frac{1}{4}\cos^2 2x \qquad \begin{array}{l}\text{Divide each term in the numerator} \\ \text{by 4.}\end{array}$$

We can reduce the power of $\cos^2 2x$ using

$$\cos^2\theta = \frac{1 + \cos 2\theta}{2}$$

with $\theta = 2x$.

$$= \frac{1}{4} + \frac{1}{2}\cos 2x + \frac{1}{4}\left[\frac{1 + \cos 2(2x)}{2}\right] \qquad \begin{array}{l}\text{Use the power-reducing formula for} \\ \cos^2\theta \text{ with } \theta = 2x.\end{array}$$

$$\cos^4 x = \frac{1}{4} + \frac{1}{2}\cos 2x + \frac{1}{4}\left[\frac{1 + \cos 2(2x)}{2}\right]$$

This is our expression for $\cos^4 x$ from the previous page.

$$= \frac{1}{4} + \frac{1}{2}\cos 2x + \frac{1}{8}(1 + \cos 4x)$$

Multiply.

$$= \frac{1}{4} + \frac{1}{2}\cos 2x + \frac{1}{8} + \frac{1}{8}\cos 4x$$

Distribute $\frac{1}{8}$ throughout parentheses.

$$= \frac{3}{8} + \frac{1}{2}\cos 2x + \frac{1}{8}\cos 4x$$

Simplify: $\frac{1}{4} + \frac{1}{8} = \frac{2}{8} + \frac{1}{8} = \frac{3}{8}$.

Thus, $\cos^4 x = \frac{3}{8} + \frac{1}{2}\cos 2x + \frac{1}{8}\cos 4x$. The expression for $\cos^4 x$ does not contain powers of trigonometric functions greater than 1.

> **Check Point 4** Write an equivalent expression for $\sin^4 x$ that does not contain powers of trigonometric functions greater than 1.

3 Use the half-angle formulas.

Study Tip

The $\frac{1}{2}$ that appears in each of the half-angle formulas cannot be pulled to the front and written as a coefficient.

INCORRECT

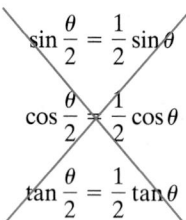

The figure shows, in a $\left[0, 2\pi, \frac{\pi}{2}\right]$ by $[-2, 2, 1]$ viewing rectangle, that the graphs of

$$y = \sin\frac{x}{2}$$

and

$$y = \frac{1}{2}\sin x$$

do not coincide.

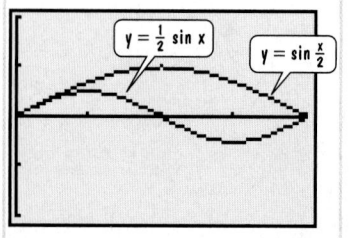

Half-Angle Formulas

Useful equivalent forms of the power-reducing formulas can be obtained by replacing θ by $\frac{\alpha}{2}$. Then solve for the trigonometric function on the left sides of the equations. The resulting identities are called the **half-angle formulas:**

> **Half-Angle Formulas**
>
> $$\sin\frac{\alpha}{2} = \pm\sqrt{\frac{1 - \cos\alpha}{2}}$$
>
> $$\cos\frac{\alpha}{2} = \pm\sqrt{\frac{1 + \cos\alpha}{2}}$$
>
> $$\tan\frac{\alpha}{2} = \pm\sqrt{\frac{1 - \cos\alpha}{1 + \cos\alpha}}$$
>
> The + or − in each formula is determined by the quadrant in which $\frac{\alpha}{2}$ lies.

If we know the exact value for the sine, cosine, or tangent of an angle, we can use the half-angle formulas to find exact values for half that angle. For example, we know that $\cos 225° = -\frac{\sqrt{2}}{2}$. In the next example, we find the exact value of the cosine of half of 225°, or $\cos 112.5°$.

EXAMPLE 5 Using a Half-Angle Formula to Find an Exact Value

Find the exact value of $\cos 112.5°$.

Solution Because $112.5° = \frac{225°}{2}$, we use the half-angle formula for $\cos\frac{\alpha}{2}$ with $\alpha = 225°$. What sign should we use when we apply the formula? Because $112.5°$ lies in quadrant II, where only the sine and cosecant are positive, $\cos 112.5° < 0$. Thus, we use the − sign in the half-angle formula.

$$\cos 112.5° = \cos \frac{225°}{2}$$

$$= -\sqrt{\frac{1 + \cos 225°}{2}} \qquad \text{Use } \cos \frac{\alpha}{2} = -\sqrt{\frac{1 + \cos \alpha}{2}} \text{ with } \alpha = 225°.$$

$$= -\sqrt{\frac{1 + \left(-\dfrac{\sqrt{2}}{2}\right)}{2}} \qquad \cos 225° = -\frac{\sqrt{2}}{2}$$

$$= -\sqrt{\frac{2 - \sqrt{2}}{4}} \qquad \text{Multiply the radicand by } \frac{2}{2}:$$
$$\frac{1 + \left(-\dfrac{\sqrt{2}}{2}\right)}{2} \cdot \frac{2}{2} = \frac{2 - \sqrt{2}}{4}.$$

$$= -\frac{\sqrt{2 - \sqrt{2}}}{2} \qquad \text{Simplify: } \sqrt{4} = 2.$$

Discovery

Use your calculator to find approximations for
$$-\frac{\sqrt{2 - \sqrt{2}}}{2}$$
and cos 112.5°. What do you observe?

Check Point 5   Use $\cos 210° = -\dfrac{\sqrt{3}}{2}$ to find the exact value of cos 105°.

There are alternate formulas for $\tan \dfrac{\alpha}{2}$ that do not require us to determine what sign to use when applying the formula. These formulas are logically connected to the identities in Example 6 and Check Point 6.

EXAMPLE 6   Verifying an Identity

Verify the identity: $\tan \theta = \dfrac{1 - \cos 2\theta}{\sin 2\theta}$.

Solution   We work with the right side.

$$\frac{1 - \cos 2\theta}{\sin 2\theta} = \frac{1 - (1 - 2\sin^2 \theta)}{2 \sin \theta \cos \theta} \qquad \text{The form } \cos 2\theta = 1 - 2 \sin^2 \theta \text{ is used because it produces only one term in the numerator. Use the double-angle formula for sine in the denominator.}$$

$$= \frac{2 \sin^2 \theta}{2 \sin \theta \cos \theta} \qquad \text{Simplify the numerator.}$$

$$= \frac{\sin \theta}{\cos \theta} \qquad \text{Divide the numerator and denominator by 2 sin } \theta.$$

$$= \tan \theta \qquad \text{Use a quotient identity: } \tan \theta = \frac{\sin \theta}{\cos \theta}.$$

The right side simplifies to $\tan \theta$, the expression on the left side. Thus, the identity is verified.

$$\tan\theta = \frac{1-\cos 2\theta}{\sin 2\theta}$$

Example 6 identity, repeated

Check Point 6 Verify the identity: $\tan\theta = \dfrac{\sin 2\theta}{1+\cos 2\theta}$.

Half-angle formulas for $\tan\dfrac{\alpha}{2}$ can be obtained using the identities in Example 6 and Check Point 6. Do you see how to do this? Replace each occurrence of θ with $\dfrac{\alpha}{2}$. This results in the following identities:

Half-Angle Formulas for $\tan\dfrac{\alpha}{2}$

$$\tan\frac{\alpha}{2} = \frac{1-\cos\alpha}{\sin\alpha}$$

$$\tan\frac{\alpha}{2} = \frac{\sin\alpha}{1+\cos\alpha}$$

EXAMPLE 7 Verifying an Identity

Verify the identity: $\tan\dfrac{\alpha}{2} = \csc\alpha - \cot\alpha$.

Solution We begin with the right side.

$$\csc\alpha - \cot\alpha = \frac{1}{\sin\alpha} - \frac{\cos\alpha}{\sin\alpha} = \frac{1-\cos\alpha}{\sin\alpha} = \tan\frac{\alpha}{2}$$

Express functions in terms of sines and cosines.

This is the first of the two half-angle formulas in the preceding box.

We worked with the right side and arrived at the left side. Thus, the identity is verified.

Check Point 7 Verify the identity: $\tan\dfrac{\alpha}{2} = \dfrac{\sec\alpha}{\sec\alpha\csc\alpha + \csc\alpha}$.

We conclude with a summary of the principal trigonometric identities developed in this section and the previous section. The fundamental identities can be found in the box on page 557.

Principal Trigonometric Identities

Sum and Difference Formulas

$$\sin(\alpha + \beta) = \sin\alpha\cos\beta + \cos\alpha\sin\beta \quad \sin(\alpha - \beta) = \sin\alpha\cos\beta - \cos\alpha\sin\beta$$
$$\cos(\alpha + \beta) = \cos\alpha\cos\beta - \sin\alpha\sin\beta \quad \cos(\alpha - \beta) = \cos\alpha\cos\beta + \sin\alpha\sin\beta$$

$$\tan(\alpha + \beta) = \frac{\tan\alpha + \tan\beta}{1 - \tan\alpha\tan\beta} \qquad \tan(\alpha - \beta) = \frac{\tan\alpha - \tan\beta}{1 + \tan\alpha\tan\beta}$$

Double-Angle Formulas

$$\sin 2\theta = 2\sin\theta\cos\theta$$
$$\cos 2\theta = \cos^2\theta - \sin^2\theta = 2\cos^2\theta - 1 = 1 - 2\sin^2\theta$$
$$\tan 2\theta = \frac{2\tan\theta}{1 - \tan^2\theta}$$

Study Tip

To help remember the correct sign in the numerator in the first two power-reducing formulas and the first two half-angle formulas, remember *sinus-minus*–the sine is minus.

Power-Reducing Formulas

$$\sin^2\theta = \frac{1 - \cos 2\theta}{2} \quad \cos^2\theta = \frac{1 + \cos 2\theta}{2} \quad \tan^2\theta = \frac{1 - \cos 2\theta}{1 + \cos 2\theta}$$

Half-Angle Formulas

$$\sin\frac{\alpha}{2} = \pm\sqrt{\frac{1 - \cos\alpha}{2}} \qquad \cos\frac{\alpha}{2} = \pm\sqrt{\frac{1 + \cos\alpha}{2}}$$

$$\tan\frac{\alpha}{2} = \pm\sqrt{\frac{1 - \cos\alpha}{1 + \cos\alpha}} = \frac{1 - \cos\alpha}{\sin\alpha} = \frac{\sin\alpha}{1 + \cos\alpha}$$

EXERCISE SET 6.3

Practice Exercises

In Exercises 1–6, use the figures to find the exact value of each trigonometric function.

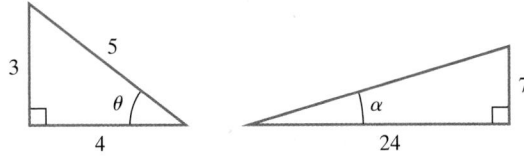

1. $\sin 2\theta$ **2.** $\cos 2\theta$ **3.** $\tan 2\theta$
4. $\sin 2\alpha$ **5.** $\cos 2\alpha$ **6.** $\tan 2\alpha$

In Exercises 7–14, use the given information to find the exact value of:

a. $\sin 2\theta$ **b.** $\cos 2\theta$ **c.** $\tan 2\theta$

7. $\sin\theta = \frac{15}{17}$, θ lies in quadrant II.
8. $\sin\theta = \frac{12}{13}$, θ lies in quadrant II.
9. $\cos\theta = \frac{24}{25}$, θ lies in quadrant IV.
10. $\cos\theta = \frac{40}{41}$, θ lies in quadrant IV.
11. $\cot\theta = 2$, θ lies in quadrant III.
12. $\cot\theta = 3$, θ lies in quadrant III.
13. $\sin\theta = -\frac{9}{41}$, θ lies in quadrant III.
14. $\sin\theta = -\frac{2}{3}$, θ lies in quadrant III.

In Exercises 15–22, write each expression as the sine, cosine, or tangent of a double angle. Then find the exact value of the expression.

15. $2\sin 15°\cos 15°$ **16.** $2\sin 22.5°\cos 22.5°$
17. $\cos^2 75° - \sin^2 75°$ **18.** $\cos^2 105° - \sin^2 105°$

19. $2\cos^2\frac{\pi}{8} - 1$ **20.** $1 - 2\sin^2\frac{\pi}{12}$

21. $\dfrac{2\tan\frac{\pi}{12}}{1 - \tan^2\frac{\pi}{12}}$ **22.** $\dfrac{2\tan\frac{\pi}{8}}{1 - \tan^2\frac{\pi}{8}}$

In Exercises 23–34, verify each identity.

23. $\sin 2\theta = \dfrac{2\tan\theta}{1 + \tan^2\theta}$ **24.** $\sin 2\theta = \dfrac{2\cot\theta}{1 + \cot^2\theta}$

25. $(\sin\theta + \cos\theta)^2 = 1 + \sin 2\theta$
26. $(\sin\theta - \cos\theta)^2 = 1 - \sin 2\theta$
27. $\sin^2 x + \cos 2x = \cos^2 x$

28. $1 - \tan^2 x = \dfrac{\cos 2x}{\cos^2 x}$

29. $\cot x = \dfrac{\sin 2x}{1 - \cos 2x}$

30. $\cot x = \dfrac{1 + \cos 2x}{\sin 2x}$

31. $\sin 2t - \tan t = \tan t \cos 2t$

32. $\sin 2t - \cot t = -\cot t \cos 2t$

33. $\sin 4t = 4 \sin t \cos^3 t - 4 \sin^3 t \cos t$

34. $\cos 4t = 8 \cos^4 t - 8 \cos^2 t + 1$

In Exercises 35–38, use the power-reducing formulas to rewrite each expression as an equivalent expression that does not contain powers of trigonometric functions greater than 1.

35. $6 \sin^4 x$

36. $10 \cos^4 x$

37. $\sin^2 x \cos^2 x$

38. $8 \sin^2 x \cos^2 x$

In Exercises 39–46, use a half-angle formula to find the exact value of each expression.

39. $\sin 15°$

40. $\cos 22.5°$

41. $\cos 157.5°$

42. $\sin 105°$

43. $\tan 75°$

44. $\tan 112.5°$

45. $\tan \dfrac{7\pi}{8}$

46. $\tan \dfrac{3\pi}{8}$

In Exercises 47–54, use the figures to find the exact value of each trigonometric function.

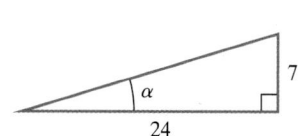

47. $\sin \dfrac{\theta}{2}$

48. $\cos \dfrac{\theta}{2}$

49. $\tan \dfrac{\theta}{2}$

50. $\sin \dfrac{\alpha}{2}$

51. $\cos \dfrac{\alpha}{2}$

52. $\tan \dfrac{\alpha}{2}$

53. $2 \sin \dfrac{\theta}{2} \cos \dfrac{\theta}{2}$

54. $2 \sin \dfrac{\alpha}{2} \cos \dfrac{\alpha}{2}$

In Exercises 55–58, use the given information to find the exact value of:

a. $\sin \dfrac{\alpha}{2}$

b. $\cos \dfrac{\alpha}{2}$

c. $\tan \dfrac{\alpha}{2}$.

55. $\tan \alpha = \frac{4}{3}$, α lies in quadrant III.

56. $\tan \alpha = \frac{8}{15}$, α lies in quadrant III.

57. $\sec \alpha = -\frac{13}{5}$, α lies in quadrant II.

58. $\sec \alpha = -3$, α lies in quadrant II.

In Exercises 59–68, verify each identity.

59. $\sin^2 \dfrac{\theta}{2} = \dfrac{\sec \theta - 1}{2 \sec \theta}$

60. $\sin^2 \dfrac{\theta}{2} = \dfrac{\csc \theta - \cot \theta}{2 \csc \theta}$

61. $\cos^2 \dfrac{\theta}{2} = \dfrac{\sin \theta + \tan \theta}{2 \tan \theta}$

62. $\cos^2 \dfrac{\theta}{2} = \dfrac{\sec \theta + 1}{2 \sec \theta}$

63. $\tan \dfrac{\alpha}{2} = \dfrac{\tan \alpha}{\sec \alpha + 1}$

64. $2 \tan \dfrac{\alpha}{2} = \dfrac{\sin^2 \alpha + 1 - \cos^2 \alpha}{\sin \alpha(1 + \cos \alpha)}$

65. $\cot \dfrac{x}{2} = \dfrac{\sin x}{1 - \cos x}$

66. $\cot \dfrac{x}{2} = \dfrac{1 + \cos x}{\sin x}$

67. $\tan \dfrac{x}{2} + \cot \dfrac{x}{2} = 2 \csc x$

68. $\tan \dfrac{x}{2} - \cot \dfrac{x}{2} = -2 \cot x$

Application Exercises

69. Throwing events in track and field include the shot put, the discuss throw, the hammer throw, and the javelin throw. The distance that the athlete can achieve depends on the initial speed of the object thrown and the angle above the horizontal at which the object leaves the hand. This angle is represented by θ in the figure shown. The distance, d, in feet, that the athlete throws is modeled by the formula

$$d = \dfrac{v_0^2}{16} \sin \theta \cos \theta$$

in which v_0 is the initial speed of the object thrown, in feet per second, and θ is the angle, in degrees, at which the object leaves the hand.

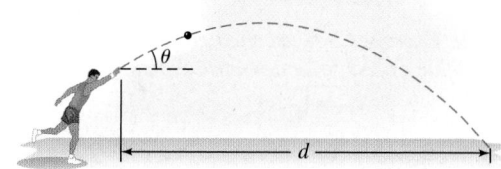

a. Use an identity to express the formula so that it contains the sine function only.

b. Use your formula from part (a) to find the angle, θ, that produces the maximum distance, d, for a given initial speed, v_0.

Use this information to solve Exercises 70–71.

The speed of a supersonic aircraft is usually represented by a Mach number, named after Austrian physicist Ernst Mach (1838–1916). A Mach number is the speed of the aircraft, in miles per hour, divided by the speed of sound, approximately 740 miles per hour. Thus, a plane flying at twice the speed of sound has a speed, M, of Mach 2. If an aircraft has a speed greater than Mach 1, a sonic boom is heard, created by sound waves that form a cone with a vertex angle θ, shown in the figure.

The relationship between the cone's vertex angle, θ, and the Mach speed, M, of an aircraft that is flying faster than the speed of sound is given by

$$\sin\frac{\theta}{2} = \frac{1}{M}.$$

Sonic boom cone

Concord
Mach 2.03

SR-71 Blackbird
Mach 3.3

70. If $\theta = \dfrac{\pi}{6}$, determine the Mach speed, M, of the aircraft.

Express the speed as an exact value and as a decimal to the nearest tenth.

71. If $\theta = \dfrac{\pi}{4}$, determine the Mach speed, M, of the aircraft.

Express the speed as an exact value and as a decimal to the nearest tenth.

 Writing in Mathematics

In Exercises 72–79, use words to describe the formula for:

72. the sine of double an angle.

73. the cosine of double an angle. (Describe one of the three formulas.)

74. the tangent of double an angle.

75. the power-reducing formula for the sine squared of an angle.

76. the power-reducing formula for the cosine squared of an angle.

77. the sine of half an angle.

78. the cosine of half an angle.

79. the tangent of half an angle. (Describe one of the two formulas that does not involve a square root.)

80. Explain how the double-angle formulas are derived.

81. How can there be three forms of the double-angle formula for $\cos 2\theta$?

82. Without showing algebraic details, describe in words how to reduce the power of $\cos^4 x$.

83. Describe one or more of the techniques you use to help remember the identities in the box on page 585.

84. Your friend is about to compete as a shot-putter in a college field event. Using Exercise 69(b), write a short description to your friend on how to achieve the best distance possible in the throwing event.

Technology Exercises

In Exercises 85–88, graph each side of the equation in the same viewing rectangle. If the graphs appear to coincide, verify that the equation is an identity. If the graphs do not appear to coincide, find a value of x for which both sides are defined but not equal.

85. $3 - 6\sin^2 x = 3\cos 2x$ **86.** $4\cos^2\dfrac{x}{2} = 2 + 2\cos x$

87. $\sin\dfrac{x}{2} = \dfrac{1}{2}\sin x$ **88.** $\cos\dfrac{x}{2} = \dfrac{1}{2}\cos x$

*In Exercises 89–91, graph each equation in a $\left[-2\pi, 2\pi, \dfrac{\pi}{2}\right]$ by $[-3, 3, 1]$ viewing rectangle. Then **a.** Describe the graph using another equation, and **b.** Verify that the two equations are equivalent.*

89. $y = \dfrac{1 - 2\cos 2x}{2\sin x - 1}$

90. $y = \dfrac{2\tan\dfrac{x}{2}}{1 + \tan^2\dfrac{x}{2}}$

91. $y = \csc x - \cot x$

Critical Thinking Exercises

92. Verify the identity:

$$\sin^3 x + \cos^3 x = (\sin x + \cos x)\left(1 - \frac{\sin 2x}{2}\right).$$

93. Use the power-reducing formulas to rewrite $\sin^6 x$ as an equivalent expression that does not contain powers of trigonometric functions greater than 1.

94. Use a right triangle to write $\sin(2\sin^{-1} x)$ as an algebraic expression. Assume that x is positive and in the domain of the given inverse trigonometric function.

SECTION 6.4 *Product-to-Sum and Sum-to-Product Formulas*

Objectives

1. Use the product-to-sum formulas.
2. Use the sum-to-product formulas.

James K. Polk
Born November 2, 1795

Warren G. Harding
Born November 2, 1865

Of the 43 U.S. presidents, two share a birthday (same month and day). The probability of two or more people in a group sharing a birthday rises sharply as the group's size increases. Above 50 people, the probability approaches certainty. (You can verify the mathematics of this surprising result by studying Sections 11.6 and 11.7, and working Exercise 66 in Exercise Set 11.7.) So, come November 2, we salute Presidents Polk and Harding with

$$112, 163\text{-}, 112, 196\text{-}, 110, 8521\text{-}, 008, 121\text{-}.$$

Were you aware that each button on your touch-tone phone produces a unique sound? If we treat the commas as pauses and the hyphens as held notes, this sequence of numbers is *Happy Birthday* on a touch-tone phone.

Although *Happy Birthday* isn't Mozart or Sondheim, it is sinusoidal. Each of its touch-tone musical sounds can be described by the sum of two sine functions or the product of sines and cosines. In this section, we develop identities that enable us to use both descriptions. They are called the product-to-sum and sum-to-product formulas.

1 Use the product-to-sum formulas.

The Product-to-Sum Formulas

How do we write the products of sines and/or cosines as sums or differences? We use the following identities, which are called **product-to-sum formulas:**

Product-to-Sum Formulas

$$\sin\alpha\sin\beta = \tfrac{1}{2}\left[\cos(\alpha-\beta) - \cos(\alpha+\beta)\right]$$

$$\cos\alpha\cos\beta = \tfrac{1}{2}\left[\cos(\alpha-\beta) + \cos(\alpha+\beta)\right]$$

$$\sin\alpha\cos\beta = \tfrac{1}{2}\left[\sin(\alpha+\beta) + \sin(\alpha-\beta)\right]$$

$$\cos\alpha\sin\beta = \tfrac{1}{2}\left[\sin(\alpha+\beta) - \sin(\alpha-\beta)\right]$$

Although these formulas are difficult to remember, they are fairly easy to derive. For example, let's derive the first identity in the box,

$$\sin \alpha \sin \beta = \tfrac{1}{2}\left[\cos (\alpha - \beta) - \cos (\alpha + \beta)\right].$$

We begin with the difference and sum formulas for the cosine, and subtract the second identity from the first:

$$\cos (\alpha - \beta) = \cos \alpha \cos \beta + \sin \alpha \sin \beta$$
$$-\left[\cos (\alpha + \beta) = \cos \alpha \cos \beta - \sin \alpha \sin \beta\right] \quad \text{Subtract the identities.}$$
$$\cos (\alpha - \beta) - \cos (\alpha + \beta) = \qquad 0 \qquad + 2 \sin \alpha \sin \beta \ .$$

Subtract terms on the left side.

Subtract terms on the right side: $\cos \alpha \cos \beta - \cos \alpha \cos \beta = 0$.

Subtract terms on the right side: $\sin \alpha \sin \beta - (-\sin \alpha \sin \beta) = 2 \sin \alpha \sin \beta$.

Now we use this result to derive the product-to-sum formula for $\sin \alpha \sin \beta$.

$$2 \sin \alpha \sin \beta = \cos (\alpha - \beta) - \cos (\alpha + \beta) \qquad \text{Reverse the sides in the preceding equation.}$$

$$\sin \alpha \sin \beta = \tfrac{1}{2}\left[\cos (\alpha - \beta) - \cos (\alpha + \beta)\right] \qquad \text{Multiply each side by } \tfrac{1}{2}.$$

This last equation is the desired formula. Likewise, we can derive the product-to-sum formula for cosine, $\cos \alpha \cos \beta = \tfrac{1}{2}\left[\cos (\alpha - \beta) + \cos (\alpha + \beta)\right].$ As we did for the previous derivation, begin with the difference and sum formulas for cosine. However, we *add* the formulas rather than subtracting them. Reversing both sides of this result and multiplying each side by $\tfrac{1}{2}$ produces the formula for $\cos \alpha \cos \beta.$ The last two product-to-sum formulas, $\sin \alpha \cos \beta$ and $\cos \alpha \sin \beta,$ are derived using the sum and difference formulas for sine in a similar manner.

Technology

The graphs of

$$y = \sin 8x \sin 3x$$

and

$$y = \tfrac{1}{2}\left(\cos 5x - \cos 11x\right)$$

are shown in a $\left[-2\pi, 2\pi, \dfrac{\pi}{2}\right]$ by $\left[-1, 1, 1\right]$ viewing rectangle. The graphs coincide. This supports our algebraic work in Example 1(a).

EXAMPLE 1   Using the Product-to-Sum Formulas

Express each of the following products as a sum or difference:

a. $\sin 8x \sin 3x$ **b.** $\sin 4x \cos x.$

Solution   The product-to-sum formula that we are using is shown in each of the voice balloons.

a. $\sin \alpha \sin \beta = \tfrac{1}{2}\left[\cos(\alpha - \beta) - \cos(\alpha + \beta)\right]$

$$\sin 8x \sin 3x = \tfrac{1}{2}\left[\cos (8x - 3x) - \cos (8x + 3x)\right] = \tfrac{1}{2}\left(\cos 5x - \cos 11x\right)$$

b. $\sin \alpha \cos \beta = \tfrac{1}{2}\left[\sin(\alpha + \beta) + \sin(\alpha - \beta)\right]$

$$\sin 4x \cos x = \tfrac{1}{2}\left[\sin (4x + x) + \sin (4x - x)\right] = \tfrac{1}{2}\left(\sin 5x + \sin 3x\right)$$

Check Point 1   Express each of the following products as a sum or difference:
a. $\sin 5x \sin 2x$ **b.** $\cos 7x \cos x.$

2 Use the sum-to-product formulas.

The Sum-to-Product Formulas

How do we write the sum or difference of sines and/or cosines as products? We use the following identities, which are called the **sum-to-product formulas:**

Sum-to-Product Formulas

$$\sin\alpha + \sin\beta = 2\sin\frac{\alpha+\beta}{2}\cos\frac{\alpha-\beta}{2}$$

$$\sin\alpha - \sin\beta = 2\sin\frac{\alpha-\beta}{2}\cos\frac{\alpha+\beta}{2}$$

$$\cos\alpha + \cos\beta = 2\cos\frac{\alpha+\beta}{2}\cos\frac{\alpha-\beta}{2}$$

$$\cos\alpha - \cos\beta = -2\sin\frac{\alpha+\beta}{2}\sin\frac{\alpha-\beta}{2}$$

We verify these formulas using the product-to-sum formulas. Let's verify the first sum-to-product formula

$$\sin\alpha + \sin\beta = 2\sin\frac{\alpha+\beta}{2}\cos\frac{\alpha-\beta}{2}.$$

We start with the right side of the formula, the side with the product. We can apply the product-to-sum formula for $\sin\alpha\cos\beta$ to this expression. By doing so, we obtain the left side of the formula, $\sin\alpha + \sin\beta$. Here's how:

$$\sin\alpha\cos\beta = \tfrac{1}{2}[\sin(\alpha+\beta) + \sin(\alpha-\beta)]$$

$$2\sin\frac{\alpha+\beta}{2}\cos\frac{\alpha-\beta}{2} = 2\cdot\frac{1}{2}\left[\sin\left(\frac{\alpha+\beta}{2}+\frac{\alpha-\beta}{2}\right) + \sin\left(\frac{\alpha+\beta}{2}-\frac{\alpha-\beta}{2}\right)\right]$$

$$= \sin\left(\frac{\alpha+\beta+\alpha-\beta}{2}\right) + \sin\left(\frac{\alpha+\beta-\alpha+\beta}{2}\right)$$

$$= \sin\frac{2\alpha}{2} + \sin\frac{2\beta}{2} = \sin\alpha + \sin\beta.$$

The three other sum-to-product formulas in the box are verified in a similar manner. Start with the right side and obtain the left side using an appropriate product-to-sum formula.

EXAMPLE 2 Using the Sum-to-Product Formulas

Express each sum or difference as a product:

a. $\sin 9x + \sin 5x$ **b.** $\cos 4x - \cos 3x.$

Solution The sum-to-product formula that we are using is shown in each of the voice balloons.

a.
$$\sin \alpha + \sin \beta = 2 \sin \frac{\alpha + \beta}{2} \cos \frac{\alpha - \beta}{2}$$

$$\sin 9x + \sin 5x = 2 \sin \frac{9x + 5x}{2} \cos \frac{9x - 5x}{2}$$

$$= 2 \sin \frac{14x}{2} \cos \frac{4x}{2}$$

$$= 2 \sin 7x \cos 2x$$

b.
$$\cos \alpha - \cos \beta = -2 \sin \frac{\alpha + \beta}{2} \sin \frac{\alpha - \beta}{2}$$

$$\cos 4x - \cos 3x = -2 \sin \frac{4x + 3x}{2} \sin \frac{4x - 3x}{2}$$

$$= -2 \sin \frac{7x}{2} \sin \frac{x}{2}$$

Check Point 2 Express each sum or difference as a product:
a. $\sin 7x + \sin 3x$ **b.** $\cos 3x + \cos 2x$.

Some identities contain a fraction on one side with sums and differences of sines and/or cosines. Applying the sum-to-product formulas in the numerator and the denominator is often helpful in verifying these identities.

EXAMPLE 3 Using Sum-to-Product Formulas to Verify an Identity

Verify the identity: $\dfrac{\cos 3x - \cos 5x}{\sin 3x + \sin 5x} = \tan x$.

Solution Because the left side is more complicated, we will work with it. We use sum-to-product formulas for the numerator and the denominator of the fraction on this side.

$$\frac{\cos 3x - \cos 5x}{\sin 3x + \sin 5x}$$

$$\cos \alpha - \cos \beta = -2 \sin \frac{\alpha + \beta}{2} \sin \frac{\alpha - \beta}{2}$$

$$= \frac{-2 \sin \dfrac{3x + 5x}{2} \sin \dfrac{3x - 5x}{2}}{\sin 3x + \sin 5x}$$

$$\sin \alpha + \sin \beta = 2 \sin \frac{\alpha + \beta}{2} \cos \frac{\alpha - \beta}{2}$$

$$= \frac{-2 \sin \dfrac{3x + 5x}{2} \sin \dfrac{3x - 5x}{2}}{2 \sin \dfrac{3x + 5x}{2} \cos \dfrac{3x - 5x}{2}}$$

$$\frac{\cos 3x - \cos 5x}{\sin 3x + \sin 5x} = \frac{-2 \sin \dfrac{3x + 5x}{2} \sin \dfrac{3x - 5x}{2}}{2 \sin \dfrac{3x + 5x}{2} \cos \dfrac{3x - 5x}{2}}$$

This is the expression from the previous page.

$$= \frac{-2 \sin \dfrac{8x}{2} \sin \left(\dfrac{-2x}{2} \right)}{2 \sin \dfrac{8x}{2} \cos \left(\dfrac{-2x}{2} \right)}$$

Perform the indicated additions.

$$= \frac{-2 \sin 4x \sin (-x)}{2 \sin 4x \cos (-x)}$$

Simplify.

$$= \frac{-(-\sin x)}{\cos x}$$

The sine function is odd: sin (−x) = −sin x. The cosine function is even: cos (−x) = cos x.

$$= \frac{\sin x}{\cos x}$$

Simplify.

$$= \tan x$$

Apply a quotient identity: $\tan x = \dfrac{\sin x}{\cos x}$.

We worked with the left side and arrived at the right side. Thus, the identity is verified.

Check Point 3   Verify the identity: $\dfrac{\cos 3x - \cos x}{\sin 3x + \sin x} = -\tan x.$

Sinusoidal Sounds

Music is all around us. A mere snippet of a song from the past can trigger vivid memories, inducing emotions ranging from unabashed joy to deep sorrow. Trigonometric functions can explain how sound travels from its source and describe its pitch, loudness, and quality. Still unexplained is the remarkable influence music has on the brain, including the deepest question of all: Why do we appreciate music?

When a note is played, it disturbs nearby air molecules, creating regions of higher-than-normal pressure and regions of lower-than-normal pressure. If we graph pressure, y, versus time, t, we get a sine wave that represents the note. The frequency of the sine wave is the number of high-low disturbances, or vibrations, per second. The greater the frequency, the higher the pitch; the lesser the frequency, the lower the pitch.

The amplitude of a note's sine wave is related to its loudness. The amplitude for the two sine waves shown above is the same. Thus, the notes have the same loudness, although they differ in pitch. The greater the amplitude, the louder the sound; the lesser the amplitude, the softer the sound. The amplitude and frequency are characteristic of every note—and thus of its graph—until the note dissipates.

EXERCISE SET 6.4

Practice Exercises

In Exercises 1–8, express each product as a sum or difference.

1. $\sin 6x \sin 2x$

2. $\sin 8x \sin 4x$

3. $\cos 7x \cos 3x$

4. $\cos 9x \cos 2x$

5. $\sin x \cos 2x$

6. $\sin 2x \cos 3x$

7. $\cos \dfrac{3x}{2} \sin \dfrac{x}{2}$

8. $\cos \dfrac{5x}{2} \sin \dfrac{x}{2}$

In Exercises 9–22, express each sum or difference as a product. If possible, find this product's exact value.

9. $\sin 6x + \sin 2x$

10. $\sin 8x + \sin 2x$

11. $\sin 7x - \sin 3x$

12. $\sin 11x - \sin 5x$

13. $\cos 4x + \cos 2x$

14. $\cos 9x - \cos 7x$

15. $\sin x + \sin 2x$

16. $\sin x - \sin 2x$

17. $\cos \dfrac{3x}{2} + \cos \dfrac{x}{2}$

18. $\sin \dfrac{3x}{2} + \sin \dfrac{x}{2}$

19. $\sin 75° + \sin 15°$

20. $\cos 75° - \cos 15°$

21. $\sin \dfrac{\pi}{12} - \sin \dfrac{5\pi}{12}$

22. $\cos \dfrac{\pi}{12} - \cos \dfrac{5\pi}{12}$

In Exercises 23–30, verify each identity.

23. $\dfrac{\sin 3x - \sin x}{\cos 3x - \cos x} = -\cot 2x$

24. $\dfrac{\sin x + \sin 3x}{\cos x + \cos 3x} = \tan 2x$

25. $\dfrac{\sin 2x + \sin 4x}{\cos 2x + \cos 4x} = \tan 3x$

26. $\dfrac{\cos 4x - \cos 2x}{\sin 2x - \sin 4x} = \tan 3x$

27. $\dfrac{\sin x - \sin y}{\sin x + \sin y} = \tan \dfrac{x - y}{2} \cot \dfrac{x + y}{2}$

28. $\dfrac{\sin x + \sin y}{\sin x - \sin y} = \tan \dfrac{x + y}{2} \cot \dfrac{x - y}{2}$

29. $\dfrac{\sin x + \sin y}{\cos x + \cos y} = \tan \dfrac{x + y}{2}$

30. $\dfrac{\sin x - \sin y}{\cos x - \cos y} = -\cot \dfrac{x + y}{2}$

In Exercises 31–34, the graph with the given equation is shown in a $\left[0, 2\pi, \dfrac{\pi}{2} \right]$ by $[-2, 2, 1]$ viewing rectangle.

a. Describe the graph using another equation.

b. Verify that the two equations are equivalent.

31.

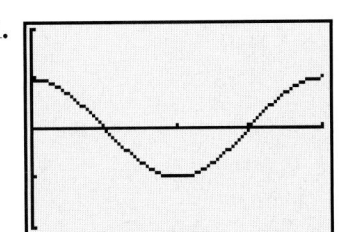

$$y = \frac{\sin x + \sin 3x}{2 \sin 2x}$$

32.

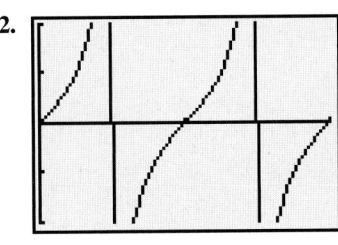

$$y = \frac{\cos x - \cos 3x}{\sin x + \sin 3x}$$

33.

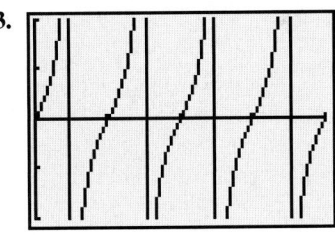

$$y = \frac{\cos x - \cos 5x}{\sin x + \sin 5x}$$

34.

$$y = \frac{\cos 5x - \cos 3x}{\sin 5x + \sin 3x}$$

Application Exercises

Use this information to solve Exercises 35–36. The sound produced by touching each button on a touch-tone phone is described by

$$y = \sin 2\pi l t + \sin 2\pi h t,$$

where l and h are the low and high frequencies in the figure shown. For example, what sound is produced by touching 5? The low frequency is l = 770 cycles per second and the high frequency is h = 1336 cycles per second. The sound produced by touching 5 is described by

$$y = \sin 2\pi(770)t + \sin 2\pi(1336)t.$$

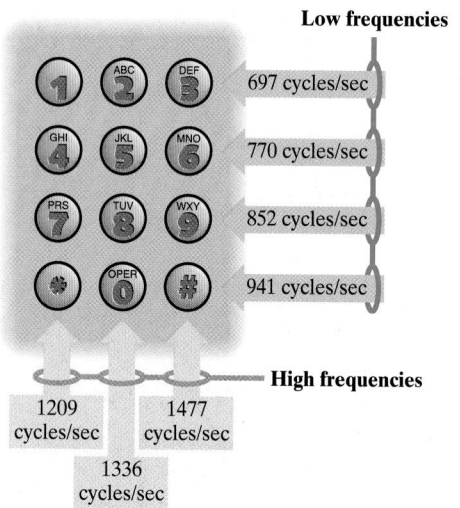

35. The touch-tone phone sequence for that most naive of melodies is given as follows:

Mary Had A Little Lamb

3212333,222,399,3212333322321.

a. Many numbers do not appear in this sequence, including 7. If you accidently touch 7 for one of the notes, describe this sound as the sum of sines.

b. Describe this accidental sound as a product of sines and cosines.

36. The touch-tone phone sequence for *Jingle Bells* is given as follows:

Jingle Bells

333,333,39123,666-663333322329,333,333,39123,666-6633,399621.

a. The first six notes of the song are produced by repeatedly touching 3. Describe this repeated sound as the sum of sines.

b. Describe the repeated sound as a product of sines and cosines.

Writing in Mathematics

In Exercises 37–40, use words to describe the given formula.

37. $\sin \alpha \sin \beta = \frac{1}{2}\left[\cos(\alpha - \beta) - \cos(\alpha + \beta)\right]$

38. $\cos \alpha \cos \beta = \frac{1}{2}\left[\cos(\alpha - \beta) + \cos(\alpha + \beta)\right]$

39. $\sin \alpha + \sin \beta = 2 \sin \dfrac{\alpha + \beta}{2} \cos \dfrac{\alpha - \beta}{2}$

40. $\cos \alpha + \cos \beta = 2 \cos \dfrac{\alpha + \beta}{2} \cos \dfrac{\alpha - \beta}{2}$

41. Describe identities that can be verified using the sum-to-product formulas.

42. Why do the sounds produced by touching each button on a touch-tone phone have the same loudness? Answer the question using the equation described for Exercises 35 and 36, $y = \sin 2\pi l t + \sin 2\pi h t$, and determine the maximum value of y for each sound.

Technology Exercises

In Exercises 43–46, graph each side of the equation in the same viewing rectangle. If the graphs appear to coincide, verify that the equation is an identity. If the graphs do not appear to coincide, find a value of x for which both sides are defined but not equal.

43. $\sin x + \sin 2x = \sin 3x$

44. $\cos x + \cos 2x = \cos 3x$

45. $\sin x + \sin 3x = 2 \sin 2x \cos x$

46. $\cos x + \cos 3x = 2 \cos 2x \cos x$

47. In Exercise 35(a), you wrote an equation for the sound produced by touching 7 on a touch-tone phone. Graph the equation in a $[0, 0.01, 0.001]$ by $[-2, 2, 1]$ viewing rectangle.

48. In Exercise 36(a), you wrote an equation for the sound produced by touching 3 on a touch-tone phone. Graph the equation in a $[0, 0.01, 0.001]$ by $[-2, 2, 1]$ viewing rectangle.

49. In this section, we saw how sums could be expressed as products. Sums of trigonometric functions can also be used to describe functions that are not trigonometric. French mathematician Jean Fourier (1768–1830) showed that *any function* can be described by a series of trigonometric functions. For example, the basic linear function $f(x) = x$ can also be represented by

$$f(x) = 2\left(\frac{\sin x}{1} - \frac{\sin 2x}{2} + \frac{\sin 3x}{3} - \frac{\sin 4x}{4} + \cdots\right).$$

a. Graph

$$y = 2\left(\frac{\sin x}{1}\right),$$

$$y = 2\left(\frac{\sin x}{1} - \frac{\sin 2x}{2}\right),$$

$$y = 2\left(\frac{\sin x}{1} - \frac{\sin 2x}{2} + \frac{\sin 3x}{3}\right)$$

and

$$y = 2\left(\frac{\sin x}{1} - \frac{\sin 2x}{2} + \frac{\sin 3x}{3} - \frac{\sin 4x}{4}\right)$$

in a $\left[-\pi, \pi, \dfrac{\pi}{2}\right]$ by $[-3, 3, 1]$ viewing rectangle. What patterns do you observe?

b. Graph

$$y = 2\left(\frac{\sin x}{1} - \frac{\sin 2x}{2} + \frac{\sin 3x}{3} - \frac{\sin 4x}{4} + \frac{\sin 5x}{5} - \frac{\sin 6x}{6}\right.$$
$$\left. + \frac{\sin 7x}{7} - \frac{\sin 8x}{8} + \frac{\sin 9x}{9} - \frac{\sin 10x}{10}\right)$$

in a $\left[-\pi, \pi, \dfrac{\pi}{2}\right]$ by $[-3, 3, 1]$ viewing rectangle. Is a portion of the graph beginning to look like the graph of $f(x) = x$? Obtain a better approximation for the line by graphing functions that contain more and more terms involving sines of multiple angles.

c. Use

$$x = 2\left(\frac{\sin x}{1} - \frac{\sin 2x}{2} + \frac{\sin 3x}{3} - \frac{\sin 4x}{4} + \cdots\right)$$

and substitute $\dfrac{\pi}{2}$ for x to obtain a formula for $\dfrac{\pi}{2}$. Show at least four nonzero terms. Then multiply both sides of your formula by 2 to write a nonending series of subtractions and additions that approaches π. Use this series to obtain an approximation for π that is more accurate than the one given by your graphing utility.

Critical Thinking Exercises

Use the identities for sin $(\alpha + \beta)$ and sin $(\alpha - \beta)$ to solve Exercises 50–51.

50. Add the left and right sides of the identities and derive the product-to-sum formula for $\sin \alpha \cos \beta$.

51. Subtract the left and right sides of the identities and derive the product-to-sum formula for $\cos \alpha \sin \beta$.

In Exercises 52–53, verify the given sum-to-product formula. Start with the right side and obtain the expression on the left side by using an appropriate product-to-sum formula.

52. $\sin \alpha - \sin \beta = 2 \sin \dfrac{\alpha - \beta}{2} \cos \dfrac{\alpha + \beta}{2}$

53. $\cos \alpha + \cos \beta = 2 \cos \dfrac{\alpha + \beta}{2} \cos \dfrac{\alpha - \beta}{2}$

In Exercises 54–55, verify each identity.

54. $\dfrac{\sin 2x + (\sin 3x + \sin x)}{\cos 2x + (\cos 3x + \cos x)} = \tan 2x$

55. $4 \cos x \cos 2x \sin 3x = \sin 2x + \sin 4x + \sin 6x$

Group Exercise

56. This activity should result in an unusual group display entitled "*Frere Jacques*, a New Perspective." Here is the touch-tone phone sequence:

Frere Jacques

4564,4564,69#,69#,#*#964,#*#964,414,414.

Group members should write every sound in the sequence as both the sum of sines and the product of sines and cosines. Use the sum of sines form and a graphing utility with a $[0, 0.01, 0.001]$ by $[-2, 2, 1]$ viewing rectangle to obtain a graph for every sound. Download these graphs. Use the graphs and equations to create your display in such a way that adults find the trigonometry of this naive melody interesting.

SECTION 6.5 *Trigonometric Equations*

Objectives

1. Find all solutions of a trigonometric equation.
2. Solve equations with multiple angles.
3. Solve trigonometric equations quadratic in form.
4. Use factoring to separate different functions in trigonometric equations.
5. Use identities to solve trigonometric equations.

Exponential functions display the manic energies of uncontrolled growth. By contrast, trigonometric functions repeat their behavior. Do they embody in their regularity some basic rhythm of the universe? The cycles of periodic phenomena provide events that we can comfortably count on. When will the moon look just as it does at this moment? When can I count on 13.5 hours of daylight? When will my breathing be exactly as it is right now? Models with trigonometric functions embrace the periodic rhythms of our world. Equations containing trigonometric functions are used to answer questions about these models.

1 Find all solutions of a trigonometric equation.

Trigonometric Equations and Their Solutions

A **trigonometric equation** is an equation that contains a trigonometric expression with a variable, such as $\sin x$. We have seen that some trigonometric equations are identities, such as $\sin^2 x + \cos^2 x = 1$. These equations are true for every value of the variable for which the expressions are defined. In this section, we consider trigonometric equations that are true for only some values of the variable. The values that satisfy the equation are its **solutions.** (There are trigonometric equations that have no solution.)

An example of a trigonometric equation is

$$\sin x = \tfrac{1}{2}.$$

A solution of this equation is $\frac{\pi}{6}$ because $\sin \frac{\pi}{6} = \frac{1}{2}$. By contrast, π is not a solution because $\sin \pi = 0 \neq \frac{1}{2}$.

Is $\frac{\pi}{6}$ the only solution of $\sin x = \frac{1}{2}$? The answer is no. Because of the periodic nature of the sine function, there are infinitely many values of x for which $\sin x = \frac{1}{2}$. Figure 6.7 shows five of the solutions, including $\frac{\pi}{6}$, for $-\frac{3\pi}{2} \leq x \leq \frac{7\pi}{2}$. Notice that the x-coordinates of the points where the graph of $y = \sin x$ intersects the line $y = \frac{1}{2}$ are the solutions of the equation $\sin x = \frac{1}{2}$.

Figure 6.7 The equation $\sin x = \frac{1}{2}$ has five solutions when x is restricted to the interval $\left[-\frac{3\pi}{2}, \frac{7\pi}{2}\right]$

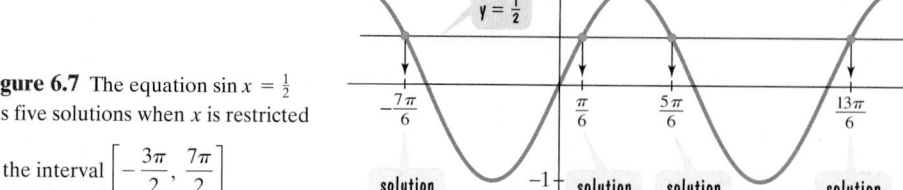

How do we represent all solutions of $\sin x = \frac{1}{2}$? Because the period of the sine function is 2π, first find all solutions in $[0, 2\pi)$. The solutions are

$$x = \frac{\pi}{6} \quad \text{and} \quad x = \pi - \frac{\pi}{6} = \frac{5\pi}{6}.$$

The sine is positive in quadrants I and II.

Any multiple of 2π can be added to these values and the sine is still $\frac{1}{2}$. Thus, all solutions of $\sin x = \frac{1}{2}$ are given by

$$x = \frac{\pi}{6} + 2n\pi \quad \text{or} \quad x = \frac{5\pi}{6} + 2n\pi$$

where n is any integer. By choosing any two integers, such as $n = 0$ and $n = 1$, we can find some solutions of $\sin x = \frac{1}{2}$. Thus, four of the solutions are:

Let $n = 0$.

$$x = \frac{\pi}{6} + 2 \cdot 0\pi \qquad x = \frac{5\pi}{6} + 2 \cdot 0\pi$$
$$= \frac{\pi}{6} \qquad = \frac{5\pi}{6}$$

Let $n = 1$.

$$x = \frac{\pi}{6} + 2 \cdot 1\pi \qquad x = \frac{5\pi}{6} + 2 \cdot 1\pi$$
$$= \frac{\pi}{6} + 2\pi \qquad = \frac{5\pi}{6} + 2\pi$$
$$= \frac{\pi}{6} + \frac{12\pi}{6} = \frac{13\pi}{6} \qquad = \frac{5\pi}{6} + \frac{12\pi}{6} = \frac{17\pi}{6}.$$

These four solutions are shown among the five solutions in Figure 6.7.

Equations Involving a Single Trigonometric Function

To solve an equation containing a single trigonometric function:
- Isolate the function on one side of the equation.
- Solve for the variable.

EXAMPLE 1 Finding All Solutions of a Trigonometric Equation

Solve the equation: $3 \sin x - 2 = 5 \sin x - 1$.

Solution The equation contains a single trigonometric function, $\sin x$.

Step 1 Isolate the function on one side of the equation. We can solve for $\sin x$ by collecting terms with $\sin x$ on the left side and constant terms on the right side.

$$3 \sin x - 2 = 5 \sin x - 1 \qquad \text{This is the given equation.}$$
$$3 \sin x - 5 \sin x - 2 = 5 \sin x - 5 \sin x - 1 \qquad \text{Subtract } 5\sin x \text{ from both sides.}$$
$$-2 \sin x - 2 = -1 \qquad \text{Simplify.}$$
$$-2 \sin x = 1 \qquad \text{Add 2 to both sides.}$$
$$\sin x = -\frac{1}{2} \qquad \text{Divide both sides by } -2 \text{ and solve for } \sin x.$$

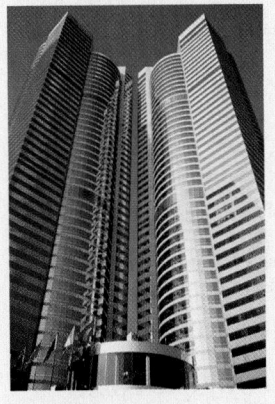

Step 2 Solve for the variable. We must solve for x in $\sin x = -\frac{1}{2}$. Because $\sin \frac{\pi}{6} = \frac{1}{2}$, the solutions of $\sin x = -\frac{1}{2}$ in $[0, 2\pi)$ are

$$x = \pi + \frac{\pi}{6} = \frac{6\pi}{6} + \frac{\pi}{6} = \frac{7\pi}{6} \qquad x = 2\pi - \frac{\pi}{6} = \frac{12\pi}{6} - \frac{\pi}{6} = \frac{11\pi}{6}.$$

The sine is negative in quadrant III.   The sine is negative in quadrant IV.

Because the period of the sine function is 2π, the solutions of the equation are given by

$$x = \frac{7\pi}{6} + 2n\pi \quad \text{and} \quad x = \frac{11\pi}{6} + 2n\pi$$

where n is any integer.

Check Point 1 Solve the equation: $5 \sin x = 3 \sin x + \sqrt{3}$.

Now we will concentrate on finding solutions of trigonometric equations for $0 \le x < 2\pi$. You can use a graphing utility to check the solutions of these equations. Graph the left side and graph the right side. The solutions are the x-coordinates of the points where the graphs intersect.

2 Solve equations with multiple angles.

Equations Involving Multiple Angles

Here are examples of two equations that include multiple angles:

$$\tan 3x = 1 \qquad \sin \frac{x}{2} = \frac{\sqrt{3}}{2}.$$

This angle is a multiple of 3.   This angle is a multiple of $\frac{1}{2}$.

We will solve each equation for $0 \le x < 2\pi$. The period of the function plays an important role in ensuring that we do not leave out any solutions.

EXAMPLE 2 Solving an Equation with a Multiple Angle

Solve the equation: $\tan 3x = 1, \ 0 \le x < 2\pi$.

Solution   The period of the tangent function is π. In the interval $[0, \pi)$, the only value for which the tangent function is 1 is $\frac{\pi}{4}$. This means that $3x = \frac{\pi}{4}$. Because the period is π, all the solutions to $\tan 3x = 1$ are given by

$$3x = \frac{\pi}{4} + n\pi \qquad n \text{ is any integer.}$$

$$x = \frac{\pi}{12} + \frac{n\pi}{3} \qquad \text{Divide both sides by 3 and solve for } x.$$

Technology

Shown below are the graphs of

$$y = \tan 3x$$

and

$$y = 1$$

in a $\left[0, 2\pi, \dfrac{\pi}{2}\right]$ by $[-3, 3, 1]$ viewing rectangle. The solutions of

$$\tan 3x = 1$$

in $[0, 2\pi)$ are shown by the x-coordinates of the six intersection points.

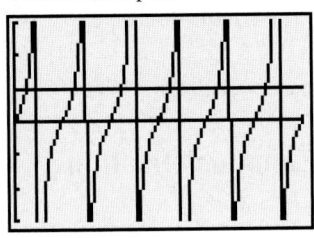

In the interval $[0, 2\pi)$, we obtain the solutions of $\tan 3x = 1$ as follows:

Let n = 0.

$$x = \frac{\pi}{12} + \frac{0\pi}{3}$$

$$= \frac{\pi}{12}$$

Let n = 1.

$$x = \frac{\pi}{12} + \frac{1\pi}{3}$$

$$= \frac{\pi}{12} + \frac{4\pi}{12} = \frac{5\pi}{12}$$

Let n = 2.

$$x = \frac{\pi}{12} + \frac{2\pi}{3}$$

$$= \frac{\pi}{12} + \frac{8\pi}{12} = \frac{9\pi}{12} = \frac{3\pi}{4}$$

Let n = 3.

$$x = \frac{\pi}{12} + \frac{3\pi}{3}$$

$$= \frac{\pi}{12} + \frac{12\pi}{12} = \frac{13\pi}{12}$$

Let n = 4.

$$x = \frac{\pi}{12} + \frac{4\pi}{3}$$

$$= \frac{\pi}{12} + \frac{16\pi}{12} = \frac{17\pi}{12}$$

Let n = 5.

$$x = \frac{\pi}{12} + \frac{5\pi}{3}$$

$$= \frac{\pi}{12} + \frac{20\pi}{12} = \frac{21\pi}{12} = \frac{7\pi}{4}.$$

If you let $n = 6$, you will obtain $x = \dfrac{25\pi}{12}$. This value exceeds 2π. In the interval $[0, 2\pi)$, the solutions of $\tan 3x = 1$ are $\dfrac{\pi}{12}, \dfrac{5\pi}{12}, \dfrac{3\pi}{4}, \dfrac{13\pi}{12}, \dfrac{17\pi}{12}$, and $\dfrac{7\pi}{4}$. These solutions are illustrated by the six intersection points in the technology box.

Check Point 2 Solve the equation: $\tan 2x = \sqrt{3}, 0 \leq x < 2\pi$.

EXAMPLE 3 Solving an Equation with a Multiple Angle

Solve the equation: $\sin \dfrac{x}{2} = \dfrac{\sqrt{3}}{2}, \ 0 \leq x < 2\pi$.

Solution The period of the sine function is 2π. In the interval $[0, 2\pi)$, there are two values at which the sine function is $\dfrac{\sqrt{3}}{2}$. One of these values is $\dfrac{\pi}{3}$. The sine is positive in quadrant II; thus, the other value is $\pi - \dfrac{\pi}{3}$, or $\dfrac{2\pi}{3}$. This means that $\dfrac{x}{2} = \dfrac{\pi}{3}$ or $\dfrac{x}{2} = \dfrac{2\pi}{3}$. Because the period is 2π, all the solutions of $\sin \dfrac{x}{2} = \dfrac{\sqrt{3}}{2}$ are given by

$$\frac{x}{2} = \frac{\pi}{3} + 2n\pi \qquad \text{or} \qquad \frac{x}{2} = \frac{2\pi}{3} + 2n\pi. \quad \textit{n is any integer.}$$

$$x = \frac{2\pi}{3} + 4n\pi \qquad\qquad x = \frac{4\pi}{3} + 4n\pi. \quad \textit{Multiply both sides by 2 and solve for x.}$$

If $n = 0$, we obtain $x = \dfrac{2\pi}{3}$ from the equation on the left and $x = \dfrac{4\pi}{3}$ from the equation on the right. If we let $n = 1$, we are adding $4 \cdot 1 \cdot \pi$, or 4π, to each of these expressions. These values of x exceed 2π. Thus, in the interval $[0, 2\pi)$, the only solutions of $\sin \dfrac{x}{2} = \dfrac{\sqrt{3}}{2}$ are $\dfrac{2\pi}{3}$ and $\dfrac{4\pi}{3}$.

Check Point 3 Solve the equation: $\sin \dfrac{x}{3} = \dfrac{1}{2}, 0 \le x < 2\pi.$

3 Solve trigonometric equations quadratic in form.

Trigonometric Equations Quadratic in Form

Some trigonometric equations are in the form of a quadratic equation $at^2 + bt + c = 0$, where t is a trigonometric function. Here are two examples of trigonometric equations that are quadratic in form:

$$2 \cos^2 x + \cos x - 1 = 0 \qquad 2 \sin^2 x - 3 \sin x + 1 = 0.$$

The form of this equation is $2t^2 + t - 1 = 0$ with $t = \cos x$.

The form of this equation is $2t^2 - 3t + 1 = 0$ with $t = \sin x$.

To solve this kind of equation, try using factoring. If the trigonometric expression does not factor, use the quadratic formula.

EXAMPLE 4 Solving a Trigonometric Equation Quadratic in Form

Solve the equation: $2 \cos^2 x + \cos x - 1 = 0, \quad 0 \le x < 2\pi.$

Solution The given equation is in quadratic form $2t^2 + t - 1 = 0$ with $t = \cos x$. Let us attempt to solve the equation by factoring.

$$2\cos^2 x + \cos x - 1 = 0 \qquad \text{This is the given equation.}$$

$$(2 \cos x - 1)(\cos x + 1) = 0 \qquad \text{Factor. Notice that } 2t^2 + t - 1 \text{ factors as } (2t - 1)(t + 1)$$

$$2 \cos x - 1 = 0 \quad \text{or} \quad \cos x + 1 = 0 \qquad \text{Set each factor equal to 0.}$$

$$2 \cos x = 1 \qquad\qquad \cos x = -1 \qquad \text{Solve for } \cos x.$$

$$\cos x = \tfrac{1}{2}$$

$$x = \frac{\pi}{3} \qquad x = 2\pi - \frac{\pi}{3} = \frac{5\pi}{3} \qquad x = \pi \qquad \begin{array}{l}\text{Solve each equation for } x, \\ 0 \le x < 2\pi.\end{array}$$

The cosine is positive in quadrants I and IV.

The solutions in the interval $[0, 2\pi)$ are $\dfrac{\pi}{3}$, π, and $\dfrac{5\pi}{3}$.

Technology

The graph of

$$y = 2 \cos^2 x + \cos x - 1$$

is shown in a

$$\left[0, 2\pi, \frac{\pi}{2}\right] \text{ by } [-3, 3, 1]$$

viewing rectangle. The x-intercepts,

$$\frac{\pi}{3}, \pi, \text{ and } \frac{5\pi}{3},$$

verify the three solutions of

$$2 \cos^2 x + \cos x - 1 = 0$$

in $[0, 2\pi)$.

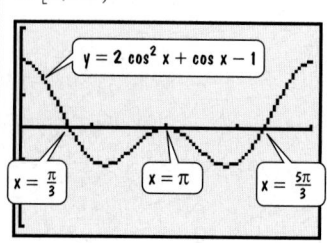

$y = 2 \cos^2 x + \cos x - 1$

$x = \frac{\pi}{3}$ $x = \pi$ $x = \frac{5\pi}{3}$

Check Point 4 Solve the equation: $2 \sin^2 x - 3 \sin x + 1 = 0, 0 \le x < 2\pi.$

4 Use factoring to separate different functions in trigonometric equations.

Using Factoring to Separate Two Different Trigonometric Functions in an Equation

We have seen that factoring is used to solve some trigonometric equations that are quadratic in form. Factoring can also be used to solve some trigonometric equations that contain two different functions such as

$$\tan x \sin^2 x = 3 \tan x.$$

In such a case, move all terms to one side and obtain zero on the other side. Then try to use factoring to separate the different functions. Example 5 shows how this is done.

EXAMPLE 5 Using Factoring to Separate Different Functions

Solve the equation: $\tan x \sin^2 x = 3 \tan x, \quad 0 \le x < 2\pi$.

Study Tip

In solving

$$\tan x \sin^2 x = 3 \tan x,$$

do not begin by dividing both sides by $\tan x$. Division by zero is undefined. If you divide by $\tan x$, you lose the two solutions for which $\tan x = 0$, namely 0 and π.

Solution
Move all terms to one side and obtain zero on the other side.

$$\tan x \sin^2 x = 3 \tan x \quad \text{This is the given equation.}$$

$$\tan x \sin^2 x - 3 \tan x = 0 \qquad \text{Subtract 3 tan x from both sides.}$$

Use factoring to separate the two functions.

$$\tan x(\sin^2 x - 3) = 0 \qquad \begin{array}{l}\text{Factor out tan x from the two terms}\\\text{on the left side.}\end{array}$$

$$\tan x = 0 \quad \text{or} \quad \sin^2 x - 3 = 0 \qquad \text{Set each factor equal to 0.}$$

$$x = 0 \quad x = \pi \qquad\qquad \sin^2 x = 3 \qquad \text{Solve for x.}$$

$$\sin x = \pm\sqrt{3}$$

This equation has no solution because sin x cannot be greater than 1 or less than –1.

The solutions in the interval $[0, 2\pi)$ are 0 and π.

Check Point 5 Solve the equation: $\sin x \tan x = \sin x, 0 \le x < 2\pi$.

5 Use identities to solve trigonometric equations.

Using Identities to Solve Trigonometric Equations

Some trigonometric equations contain more than one function on the same side and these functions cannot be separated by factoring. For example, consider the equation

$$2 \sin^2 x - \cos x - 1 = 0.$$

How can we obtain an equivalent equation that has only one trigonometric function? We use the identity $\sin^2 x + \cos^2 x = 1$ and substitute $1 - \cos^2$ for $\sin^2 x$.

$$2 \sin^2 x - \cos x - 1 = 0 \qquad \text{This is the given equation.}$$

$$2(1 - \cos^2 x) - \cos x - 1 = 0 \qquad \sin^2 x = 1 - \cos^2 x$$

$$2 - 2\cos^2 x - \cos x - 1 = 0 \qquad \text{Use the distributive property.}$$

$$-2\cos^2 x - \cos x + 1 = 0 \qquad \text{Combine like terms.}$$

Multiplying both sides of $-2\cos^2 x - \cos x + 1 = 0$ by -1, we obtain

$$2\cos^2 x + \cos x - 1 = 0.$$

This equivalent equation contains only the cosine function. This is the equation that we solved in Example 4.

EXAMPLE 6 Using an Identity to Solve a Trigonometric Equation

Solve the equation: $\cos 2x + 3\sin x - 2 = 0, \ 0 \le x < 2\pi$.

Solution The given equation contains a cosine function and a sine function. The cosine is a function of $2x$ and the sine is a function of x. We want one trigonometric function of the same angle. This can be accomplished by using the double-angle identity $\cos 2x = 1 - 2\sin^2 x$ to obtain an equivalent equation involving $\sin x$ only.

$\cos 2x + 3\sin x - 2 = 0$	This is the given equation.
$1 - 2\sin^2 x + 3\sin x - 2 = 0$	$\cos 2x = 1 - 2\sin^2 x$
$-2\sin^2 x + 3\sin x - 1 = 0$	Combine like terms.
$2\sin^2 x - 3\sin x + 1 = 0$	Multiply both sides by -1.

The equation is now in quadratic form $2t^2 - 3t + 1 = 0$ with $t = \sin x$. We solve using factoring.

$$(2\sin x - 1)(\sin x - 1) = 0 \qquad \begin{array}{l}\text{Factor. Notice that} \\ 2t^2 - 3t + 1 \text{ factors as} \\ (2t - 1)(t - 1).\end{array}$$

$2\sin x - 1 = 0$	or	$\sin x - 1 = 0$	Set each factor equal to 0.
$\sin x = \frac{1}{2}$		$\sin x = 1$	Solve for $\sin x$.
$x = \dfrac{\pi}{6} \quad x = \pi - \dfrac{\pi}{6} = \dfrac{5\pi}{6}$		$x = \dfrac{\pi}{2}$	Solve each equation for x, $0 \le x < 2\pi$.

> The sine is positive in quadrants I and II.

The solutions in the interval $[0, 2\pi)$ are $\dfrac{\pi}{6}, \dfrac{\pi}{2}$, and $\dfrac{5\pi}{6}$.

Check Point 6 Solve the equation: $\cos 2x + \sin x = 0, \ 0 \le x < 2\pi$.

Sometimes it is necessary to do something to both sides of a trigonometric equation to substitute an identity. For example, consider the equation

$$\sin x \cos x = \tfrac{1}{2}.$$

This equation contains both a sine and a cosine function. How can we obtain a single function? Multiply both sides by 2. In this way, we can use the double-angle identity $\sin 2x = 2\sin x \cos x$ and obtain $\sin 2x$, a single function, on the left side.

Technology

Shown below are the graphs of

$$y = \sin x \cos x$$

and

$$y = \tfrac{1}{2}$$

in a $\left[0, 2\pi, \dfrac{\pi}{2}\right]$ by $[-1, 1, 1]$ viewing rectangle.

The solutions of

$$\sin x \cos x = \tfrac{1}{2}$$

are shown by the x-coordinates of the two intersection points.

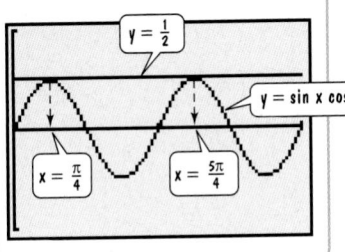

EXAMPLE 7 Using an Identity to Solve a Trigonometric Equation

Solve the equation: $\sin x \cos x = \tfrac{1}{2}, \ 0 \le x < 2\pi.$

Solution

$\sin x \cos x = \tfrac{1}{2}$ This is the given equation.

$2 \sin x \cos x = 1$ Multiply both sides by 2 in anticipation of using $\sin 2x = 2 \sin x \cos x$.

$\sin 2x = 1$ Use a double-angle identity.

Notice that we have an equation with $2x$, a multiple angle. The period of the sine function is 2π. In the interval $[0, 2\pi)$, the only value for which the sine function is 1 is $\dfrac{\pi}{2}$. This means that $2x = \dfrac{\pi}{2}$. Because the period is 2π, all the solutions of $\sin 2x = 1$ are given by

$$2x = \dfrac{\pi}{2} + 2n\pi \qquad n \text{ is any integer.}$$

$$x = \dfrac{\pi}{4} + n\pi \qquad \text{Divide both sides by 2 and solve for } x.$$

The solutions in the interval $[0, 2\pi)$ are obtained by letting $n = 0$ and $n = 1$. The solutions are $\dfrac{\pi}{4}$ and $\dfrac{5\pi}{4}$.

Check Point 7 Solve the equation: $\sin x \cos x = -\tfrac{1}{2}, \ 0 \le x < 2\pi.$

Let's look at another equation that contains two different functions, $\sin x - \cos x = 1$. Can you think of an identity that can be used to produce only one function? Perhaps $\sin^2 x + \cos^2 x = 1$ might be helpful. The next example shows how we use this identity by squaring both sides of the given equation. Remember that if we raise both sides of an equation to an even power, we have the possibility of introducing extraneous solutions. Thus, we must check each proposed solution in the given equation. Alternatively, we can use a graphing utility to verify actual solutions.

EXAMPLE 8 Using an Identity to Solve a Trigonometric Equation

Solve the equation: $\sin x - \cos x = 1, \ 0 \le x < 2\pi.$

Solution We square both sides of the equation in anticipation of using $\sin^2 x + \cos^2 x = 1$.

$\sin x - \cos x = 1$ This is the given equation.

$(\sin x - \cos x)^2 = 1^2$ Square both sides.

$\sin^2 x - 2 \sin x \cos x + \cos^2 x = 1$ Square the left side using $(A - B)^2 = A^2 - 2AB + B^2$.

Technology

A graphing utility can be used instead of the algebraic check on the right. Shown are the graphs of

$$y = \sin x - \cos x$$

and

$$y = 1$$

in a $\left[0, 2\pi, \dfrac{\pi}{2}\right]$ by $[-2, 2, 1]$ viewing rectangle. The actual solutions of

$$\sin x - \cos x = 1$$

are shown by the x-coordinates of the two intersection points, $\dfrac{\pi}{2}$ and π.

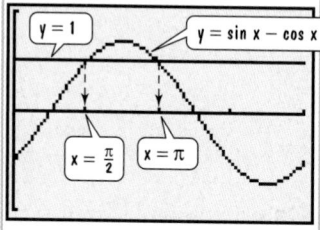

$$\sin^2 x - 2 \sin x \cos x + \cos^2 x = 1 \qquad \text{We have repeated the equation from the previous page.}$$

$$\sin^2 x + \cos^2 x - 2 \sin x \cos x = 1 \qquad \text{Rearrange terms.}$$

$$1 - 2 \sin x \cos x = 1 \qquad \text{Apply a Pythagorean identity: } \sin^2 x + \cos^2 x = 1.$$

$$-2 \sin x \cos x = 0 \qquad \text{Subtract 1 from both sides of the equation.}$$

$$\sin x \cos x = 0 \qquad \text{Divide both sides of the equation by } -2.$$

$$\sin x = 0 \quad \text{or} \quad \cos x = 0 \qquad \text{Set each factor equal to 0.}$$

$$x = 0 \quad x = \pi \qquad x = \dfrac{\pi}{2} \qquad x = \dfrac{3\pi}{2} \qquad \text{Solve for } x \text{ in } [0, 2\pi).$$

We check these proposed solutions to see if any are extraneous.

Check 0:

$$\sin x - \cos x = 1$$
$$\sin 0 - \cos 0 \overset{?}{=} 1$$
$$0 - 1 \overset{?}{=} 1$$
$$-1 = 1, \text{ False}$$

Check $\dfrac{\pi}{2}$:

$$\sin x - \cos x = 1$$
$$\sin \dfrac{\pi}{2} - \cos \dfrac{\pi}{2} \overset{?}{=} 1$$
$$1 - 0 \overset{?}{=} 1$$
$$1 = 1, \text{ True}$$

Check π:

$$\sin x - \cos x = 1$$
$$\sin \pi - \cos \pi \overset{?}{=} 1$$
$$0 - (-1) \overset{?}{=} 1$$
$$1 = 1, \text{ True}$$

Check $\dfrac{3\pi}{2}$:

$$\sin x - \cos x = 1$$
$$\sin \dfrac{3\pi}{2} - \cos \dfrac{3\pi}{2} \overset{?}{=} 1$$
$$-1 - 0 \overset{?}{=} 1$$
$$-1 = 1,$$

0 and $\dfrac{3\pi}{2}$ are extraneous.

The actual solutions in the interval $[0, 2\pi)$ are $\dfrac{\pi}{2}$ and π.

Check Point 8   Solve the equation: $\cos x - \sin x = -1$, $0 \le x < 2\pi$.

EXERCISE SET 6.5

Practice Exercises

In Exercises 1–10, use substitution to determine whether the given x-value is a solution of the equation.

1. $\cos x = \dfrac{\sqrt{2}}{2}$, $\quad x = \dfrac{\pi}{4}$

2. $\tan x = \sqrt{3}$, $\quad x = \dfrac{\pi}{3}$

3. $\sin x = \dfrac{\sqrt{3}}{2}$, $\quad x = \dfrac{\pi}{6}$

4. $\sin x = \dfrac{\sqrt{2}}{2}$, $\quad x = \dfrac{\pi}{3}$

5. $\cos x = -\dfrac{1}{2}$, $\quad x = \dfrac{2\pi}{3}$

6. $\cos x = -\dfrac{1}{2}$, $\quad x = \dfrac{4\pi}{3}$

7. $\tan 2x = -\dfrac{\sqrt{3}}{3}$, $\quad x = \dfrac{5\pi}{12}$

8. $\cos \dfrac{2x}{3} = -\dfrac{1}{2}$, $\quad x = \pi$

9. $\cos x = \sin 2x$, $x = \dfrac{\pi}{3}$

10. $\cos x + 2 = \sqrt{3} \sin x$, $\quad x = \dfrac{\pi}{6}$

In Exercises 11–24, find all solutions of each equation.

11. $\sin x = \dfrac{\sqrt{3}}{2}$

12. $\cos x = \dfrac{\sqrt{3}}{2}$

13. $\tan x = 1$

14. $\tan x = \sqrt{3}$

15. $\cos x = -\dfrac{1}{2}$

16. $\sin x = -\dfrac{\sqrt{2}}{2}$

17. $\tan x = 0$

18. $\sin x = 0$

19. $2 \cos x + \sqrt{3} = 0$

20. $2 \sin x + \sqrt{3} = 0$

21. $4 \sin \theta - 1 = 2 \sin \theta$

22. $5 \sin \theta + 1 = 3 \sin \theta$

23. $3 \sin \theta + 5 = -2 \sin \theta$

24. $7 \cos \theta + 9 = -2 \cos \theta$

Exercises 25–38 involve equations with multiple angles. Solve each equation on the interval $[0, 2\pi)$.

25. $\sin 2x = \dfrac{\sqrt{3}}{2}$

26. $\cos 2x = \dfrac{\sqrt{2}}{2}$

27. $\cos 4x = -\dfrac{\sqrt{3}}{2}$

28. $\sin 4x = -\dfrac{\sqrt{2}}{2}$

29. $\tan 3x = \dfrac{\sqrt{3}}{3}$

30. $\tan 3x = \sqrt{3}$

31. $\tan \dfrac{x}{2} = \sqrt{3}$

32. $\tan \dfrac{x}{2} = \dfrac{\sqrt{3}}{3}$

33. $\sin \dfrac{2\theta}{3} = -1$

34. $\cos \dfrac{2\theta}{3} = -1$

35. $\sec \dfrac{3\theta}{2} = -2$

36. $\cot \dfrac{3\theta}{2} = -\sqrt{3}$

37. $\sin\left(2x + \dfrac{\pi}{6}\right) = \dfrac{1}{2}$

38. $\sin\left(2x - \dfrac{\pi}{4}\right) = \dfrac{\sqrt{2}}{2}$

Exercises 39–46 involve trigonometric equations quadratic in form. Solve each equation on the interval $[0, 2\pi)$.

39. $2\sin^2 x - \sin x - 1 = 0$

40. $2\sin^2 x + \sin x - 1 = 0$

41. $2\cos^2 x + 3\cos x + 1 = 0$

42. $\cos^2 x + 2\cos x - 3 = 0$

43. $2\sin^2 x = \sin x + 3$

44. $2\sin^2 x = 4\sin x + 6$

45. $\sin^2 \theta - 1 = 0$

46. $\cos^2 \theta - 1 = 0$

In Exercises 47–56, solve each equation on the interval $[0, 2\pi)$.

47. $(\tan x - 1)(\cos x + 1) = 0$

48. $(\tan x + 1)(\sin x - 1) = 0$

49. $(2\cos x + \sqrt{3})(2\sin x + 1) = 0$

50. $(2\cos x - \sqrt{3})(2\sin x - 1) = 0$

51. $\cot x(\tan x - 1) = 0$

52. $\cot x(\tan x + 1) = 0$

53. $\sin x + 2\sin x \cos x = 0$

54. $\cos x - 2\sin x \cos x = 0$

55. $\tan^2 x \cos x = \tan^2 x$

56. $\cot^2 x \sin x = \cot^2 x$

In Exercises 57–78, use an identity to solve each equation on the interval $[0, 2\pi)$.

57. $2\cos^2 x + \sin x - 1 = 0$

58. $2\cos^2 x - \sin x - 1 = 0$

59. $\sin^2 x - 2\cos x - 2 = 0$

60. $4\sin^2 x + 4\cos x - 5 = 0$

61. $4\cos^2 x = 5 - 4\sin x$

62. $3\cos^2 x = \sin^2 x$

63. $\sin 2x = \cos x$

64. $\sin 2x = \sin x$

65. $\cos 2x = \cos x$

66. $\cos 2x = \sin x$

67. $\cos 2x + 5\cos x + 3 = 0$

68. $\cos 2x + \cos x + 1 = 0$

69. $\sin x \cos x = \dfrac{\sqrt{2}}{4}$

70. $\sin x \cos x = \dfrac{\sqrt{3}}{4}$

71. $\sin x + \cos x = 1$

72. $\sin x + \cos x = -1$

73. $\sin\left(x + \dfrac{\pi}{4}\right) + \sin\left(x - \dfrac{\pi}{4}\right) = 1$

74. $\sin\left(x + \dfrac{\pi}{3}\right) + \sin\left(x - \dfrac{\pi}{3}\right) = 1$

75. $\sin 2x \cos x + \cos 2x \sin x = \dfrac{\sqrt{2}}{2}$

76. $\sin 3x \cos 2x + \cos 3x \sin 2x = 1$

77. $\tan x + \sec x = 1$

78. $\tan x - \sec x = 1$

Application Exercises

Use this information to solve Exercises 79–80. Our cycle of normal breathing takes place every 5 seconds. Velocity of air flow, y, measured in liters per second, after x seconds is modeled by

$$y = 0.6 \sin \dfrac{2\pi}{5} x.$$

Velocity of air flow is positive when we inhale and negative when we exhale.

79. Within each breathing cycle, when are we inhaling at 0.3 liter per second? Round to the nearest tenth of a second.

80. Within each breathing cycle, when are we exhaling at 0.3 liter per second? Round to the nearest tenth of a second.

Use this information to solve Exercises 81–82. The number of hours of daylight in Boston is given by

$$y = 3 \sin\left[\dfrac{2\pi}{365}(x - 79)\right] + 12$$

where x is the number of days after January 1.

81. Within a year, when does Boston have 10.5 hours of daylight? Give your answer in days after January 1 and round to the nearest day.

82. Within a year, when does Boston have 13.5 hours of daylight? Give your answer in days after January 1 and round to the nearest day.

Use this information to solve Exercises 83–84. A ball on a spring is pulled 4 inches below its rest position and then released. After t seconds, the ball's distance, d, in inches from its rest position is given by

$$d = -4 \cos \dfrac{\pi}{3} t.$$

83. Find all values of t for which the ball is 2 inches above its rest position.

84. Find all values of t for which the ball is 2 inches below its rest position.

Use this information to solve Exercises 85–86. When throwing an object, the distance achieved depends on its initial velocity, v_0, and the angle above the horizontal at which the object is thrown, θ. The distance, d, in feet, that describes the range covered is given by

$$d = \frac{v_0^2}{16}\sin\theta\cos\theta$$

where v_0 is measured in feet per second.

85. You and your friend are throwing a baseball back and forth. If you throw the ball with an initial velocity of $v_0 = 90$ feet per second, at what angle of elevation, θ, to the nearest degree, should you direct your throw so that it can be easily caught by your friend located 170 feet away?

86. In Exercise 85, you increase the distance between you and your friend to 200 feet. With this increase, at what angle of elevation, θ, to the nearest degree, should you direct your throw?

 Writing in Mathematics

87. What are the solutions of a trigonometric equation?

88. Describe the difference between verifying a trigonometric identity and solving a trigonometric equation.

89. Without actually solving the equation, describe how to solve

$$3\tan x - 2 = 5\tan x - 1.$$

90. In the interval $[0, 2\pi)$, the solutions of $\sin x = \cos 2x$ are $\frac{\pi}{6}, \frac{5\pi}{6}$, and $\frac{3\pi}{2}$. Explain how to use graphs generated by a graphing utility to check these solutions.

91. Suppose you are solving equations in the interval $[0, 2\pi)$. Without actually solving equations, what is the difference between the number of solutions of $\sin x = \frac{1}{2}$ and $\sin 2x = \frac{1}{2}$? How do you account for this difference?

In Exercises 92–93, describe a general strategy for solving each equation. Do not solve the equation.

92. $2\sin^2 x + 5\sin x + 3 = 0$

93. $\sin 2x = \sin x$

94. Describe a natural periodic phenomenon. Give an example of a question that can be answered by a trigonometric equation in the study of this phenomenon.

95. Some people experience depression with loss of sunlight. Use the essay on page 597 to determine whether such a person should live on a city street that is 80 feet wide with buildings whose heights average 400 feet. Explain your answer and include θ, to the nearest degree, in your argument.

 Technology Exercises

96. Use a graphing utility to verify the solutions of any five equations that you solved in Exercises 57–78.

In Exercises 97–101, use a graphing utility to approximate the solutions of each equation in the interval $[0, 2\pi)$. Round to the nearest hundredth of a radian.

97. $15\cos^2 x + 7\cos x - 2 = 0$

98. $\cos x = x$

99. $2\sin^2 x = 1 - 2\sin x$

100. $\sin 2x = 2 - x^2$

101. $\sin x + \sin 2x + \sin 3x = 0$

 Critical Thinking Exercises

102. Which one of the following is true?
   a. The equation $(\sin x - 3)(\cos x + 2) = 0$ has no solution.
   b. The equation $\tan x = \frac{\pi}{2}$ has no solution.
   c. A trigonometric equation with an infinite number of solutions is an identity.
   d. The equations $\sin 2x = 1$ and $\sin 2x = \frac{1}{2}$ have the same number of solutions on the interval $[0, 2\pi)$.

In Exercises 103–105, solve each equation on the interval $[0, 2\pi)$. Do not use a calculator.

103. $2\cos x - 1 + 3\sec x = 0$

104. $\sin 3x + \sin x + \cos x = 0$

105. $\sin x + 2\sin\dfrac{x}{2} = \cos\dfrac{x}{2} + 1$

CHAPTER SUMMARY, REVIEW, AND TEST

Summary

DEFINITIONS AND CONCEPTS	EXAMPLES
6.1 Verifying Trigonometric Identities	
a. Identities are trigonometric equations that are true for all values of the variable for which the expressions are defined.	Ex. 1, p. 558; Exs. 2 & 3, p. 559
b. Fundamental trigonometric identities are given in the box on page 557.	Ex. 4, p. 560
c. Guidelines for verifying trigonometric identities are given in the box on page 565.	Ex. 5, p. 561; Ex. 6, p. 562; Ex. 7, p. 563

DEFINITIONS AND CONCEPTS	EXAMPLES

6.2 and 6.3 Sum, Difference, Double-Angle, and Half-Angle Formulas

a. Sum and difference formulas, double-angle formulas, power-reducing formulas, and half-angle formulas are given in the box on page 585.

Exs. 1-7, pp. 569–574;
Exs. 1-7, pp. 578–584

6.4 Product-to-Sum and Sum-to-Product Formulas

a. The product-to-sum formulas are given in the box on page 588.

b. The sum-to-product formulas are given in the box on page 590. These formulas are useful to verify identities with fractions that contain sums and differences of sines and/or cosines.

Ex. 1, p. 589

Ex. 2, p. 590;
Ex. 3, p. 591

6.5 Trigonometric Equations

a. The values that satisfy a trigonometric equation are its solutions.

b. Algebraic techniques such as isolating an expression on one side of the equation and factoring are useful in solving trigonometric equations. Identities are also used to solve some trigonometric equations.

Exs. 1-8, pp. 597–603

Review Exercises

6.1

In Exercises 1–12, verify each identity.

1. $\sec x - \cos x = \tan x \sin x$

2. $\cos x + \sin x \tan x = \sec x$

3. $\sin^2 \theta (1 + \cot^2 \theta) = 1$

4. $(\sec \theta - 1)(\sec \theta + 1) = \tan^2 \theta$

5. $\dfrac{1}{\sin t - 1} + \dfrac{1}{\sin t + 1} = -2 \tan t \sec t$

6. $\dfrac{1 + \sin t}{\cos^2 t} = \tan^2 t + 1 + \tan t \sec t$

7. $\dfrac{\cos x}{1 - \sin x} = \dfrac{1 + \sin x}{\cos x}$

8. $1 - \dfrac{\cos^2 x}{1 + \sin x} = \sin x$

9. $(\tan \theta + \cot \theta)^2 = \sec^2 \theta + \csc^2 \theta$

10. $\dfrac{1}{\sin \theta + \cos \theta} + \dfrac{1}{\sin \theta - \cos \theta} = \dfrac{2 \sin \theta}{\sin^4 \theta - \cos^4 \theta}$

11. $\dfrac{\cos t}{\cot t - 5 \cos t} = \dfrac{1}{\csc t - 5}$

12. $\dfrac{1 - \cos t}{1 + \cos t} = (\csc t - \cot t)^2$

6.2 and 6.3

In Exercises 13–18, use a sum or difference formula to find the exact value of each expression.

13. $\cos(45° + 30°)$

14. $\sin 195°$

15. $\tan\left(\dfrac{4\pi}{3} - \dfrac{\pi}{4}\right)$

16. $\tan \dfrac{5\pi}{12}$

17. $\cos 65° \cos 5° + \sin 65° \sin 5°$

18. $\sin 80° \cos 50° - \cos 80° \sin 50°$

In Exercises 19–30, verify each identity.

19. $\sin\left(x + \dfrac{\pi}{6}\right) - \cos\left(x + \dfrac{\pi}{3}\right) = \sqrt{3} \sin x$

20. $\tan\left(x + \dfrac{3\pi}{4}\right) = \dfrac{\tan x - 1}{1 + \tan x}$

21. $\sec(\alpha + \beta) = \dfrac{\sec \alpha \sec \beta}{1 - \tan \alpha \tan \beta}$

22. $\dfrac{\cos(\alpha - \beta)}{\cos \alpha \cos \beta} = 1 + \tan \alpha \tan \beta$

23. $\cos^4 t - \sin^4 t = \cos 2t$

24. $\sin t - \cos 2t = (2 \sin t - 1)(\sin t + 1)$

25. $\dfrac{\sin 2\theta - \sin \theta}{\cos 2\theta + \cos \theta} = \dfrac{1 - \cos \theta}{\sin \theta}$

26. $\dfrac{\sin 2\theta}{1 - \sin^2 \theta} = 2 \tan \theta$

27. $\tan 2t = 2 \sin t \cos t \sec 2t$

28. $\cos 4t = 1 - 8 \sin^2 t \cos^2 t$

29. $\tan \dfrac{x}{2}(1 + \cos x) = \sin x$

30. $\tan \dfrac{x}{2} = \dfrac{\sec x - 1}{\tan x}$

In Exercises 31–33, the graph with the given equation is shown on the next page in a $\left[0, 2\pi, \dfrac{\pi}{2}\right]$ by $[-2, 2, 1]$ viewing rectangle.

a. *Describe the graph using another equation.*

b. *Verify that the two equations are equivalent.*

31.

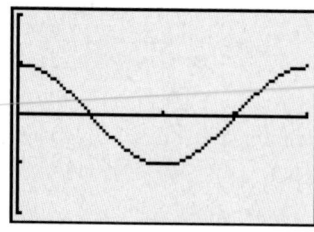

$$y = \sin\left(x - \frac{3\pi}{2}\right)$$

32.

$$y = \cos\left(x + \frac{\pi}{2}\right)$$

33.

$$y = \frac{\tan x - 1}{1 - \cot x}$$

In Exercises 34–37, find the exact value of the following under the given conditions:

  a. $\sin(\alpha + \beta)$ **b.** $\cos(\alpha - \beta)$ **c.** $\tan(\alpha + \beta)$

  d. $\sin 2\alpha$ **e.** $\cos\dfrac{\beta}{2}$

34. $\sin\alpha = \frac{3}{5}$, α lies in quadrant I, and $\sin\beta = \frac{12}{13}$, β lies in quadrant II.

35. $\tan\alpha = \frac{4}{3}$, α lies in quadrant III, and $\tan\beta = \frac{5}{12}$, β lies in quadrant I.

36. $\tan\alpha = -3$, α lies in quadrant II, and $\cot\beta = -3$, β lies in quadrant IV.

37. $\sin\alpha = -\frac{1}{3}$, α lies in quadrant III, and $\cos\beta = -\frac{1}{3}$, β lies in quadrant III.

In Exercises 38–41, use double- and half-angle formulas to find the exact value of each expression.

38. $\cos^2 15° - \sin^2 15°$ **39.** $\dfrac{2\tan\dfrac{5\pi}{12}}{1 - \tan^2\dfrac{5\pi}{12}}$

40. $\sin 22.5°$ **41.** $\tan\dfrac{\pi}{12}$

6.4

In Exercises 42–43, express each product as a sum or difference.

42. $\sin 6x \sin 4x$ **43.** $\sin 7x \cos 3x$

In Exercises 44–45, express each sum or difference as a product. If possible, find this product's exact value.

44. $\sin 2x - \sin 4x$ **45.** $\cos 75° + \cos 15°$

In Exercises 46–47, verify each identity.

46. $\dfrac{\cos 3x + \cos 5x}{\cos 3x - \cos 5x} = \cot x \cot 4x$

47. $\dfrac{\sin 2x + \sin 6x}{\sin 2x - \sin 6x} = -\tan 4x \cot 2x$

48. The graph with the given equation is shown in a $\left[0, 2\pi, \dfrac{\pi}{2}\right]$ by $[-2, 2, 1]$ viewing rectangle.

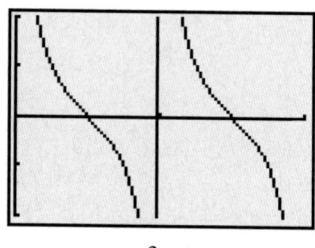

$$y = \frac{\cos 3x + \cos x}{\sin 3x - \sin x}$$

  a. Describe the graph using another equation.
  b. Verify that the two equations are equivalent.

6.5

In Exercises 49–52, find all solutions of each equation.

49. $\cos x = -\dfrac{1}{2}$ **50.** $\sin x = \dfrac{\sqrt{2}}{2}$

51. $2\sin x + 1 = 0$ **52.** $\sqrt{3}\tan x - 1 = 0$

In Exercises 53–62, solve each equation on the interval $[0, 2\pi)$.

53. $\cos 2x = -1$ **54.** $\sin 3x = 1$

55. $\tan\dfrac{x}{2} = -1$ **56.** $\tan x = 2\cos x \tan x$

57. $\cos^2 x - 2\cos x = 3$ **58.** $2\cos^2 x - \sin x = 1$

59. $4\sin^2 x = 1$ **60.** $\cos 2x - \sin x = 1$

61. $\sin 2x = \sqrt{3}\sin x$ **62.** $\sin x = \tan x$

63. A ball on a spring is pulled 6 inches below its rest position and then released. After t seconds, the ball's distance, d, in inches from its rest position is given by

$$d = -6\cos\frac{\pi}{2}t.$$

Find all values of t for which the ball is 3 inches below its rest position.

64. You are playing catch with a friend located 100 feet away. If you throw the ball with an initial velocity of $v_0 = 90$ feet per second, at what angle of elevation, θ, to the nearest degree should you direct your throw so that it can be caught easily?

Use the formula

$$d = \frac{v_0^2}{16} \sin \theta \cos \theta.$$

Chapter 6 Test

Use the following conditions to solve Exercises 1–4:

$$\sin \alpha = \tfrac{4}{5}, \alpha \text{ lies in quadrant II}$$

$$\cos \beta = \tfrac{5}{13}, \beta \text{ lies in quadrant I.}$$

Find the exact value of each of the following.

1. $\cos (\alpha + \beta)$ **2.** $\tan (\alpha - \beta)$

3. $\sin 2\alpha$ **4.** $\cos \dfrac{\beta}{2}$

5. Use $105° = 135° - 30°$ to find the exact value of $\sin 105°$.

In Exercises 6–11, verify each identity.

6. $\cos x \csc x = \cot x$ **7.** $\dfrac{\sec x}{\cot x + \tan x} = \sin x$

8. $1 - \dfrac{\cos^2 x}{1 + \sin x} = \sin x$ **9.** $\cos \left(\theta + \dfrac{\pi}{2} \right) = -\sin \theta$

10. $\dfrac{\sin (\alpha - \beta)}{\sin \alpha \cos \beta} = 1 - \cot \alpha \tan \beta$

11. $\sin t \cos t (\tan t + \cot t) = 1$

In Exercises 12–15, solve each equation on the interval $[0, 2\pi)$.

12. $\sin 3x = -\tfrac{1}{2}$ **13.** $\sin 2x + \cos x = 0$

14. $2 \cos^2 x - 3 \cos x + 1 = 0$

15. $2 \sin^2 x + \cos x = 1$

Cumulative Review Exercises (Chapters 1–6)

Solve each equation or inequality in Exercises 1–4.

1. $x^3 + x^2 - x + 15 = 0$ **2.** $11^{x-1} = 125$

3. $x^2 + 2x - 8 > 0$

4. $\cos 2x + 3 = 5 \cos x, \quad 0 \le x < 2\pi$

In Exercises 5–10, graph each equation.

5. $y = \sqrt{x + 2} - 1$; Use transformations of the graph of $y = \sqrt{x}$.

6. $(x - 1)^2 + (y + 2)^2 = 9$

7. $y + 2 = \tfrac{1}{3}(x - 1)$

8. $y = 3 \cos 2x, \quad -2\pi \le x \le 2\pi$

9. $y = 2 \sin \dfrac{x}{2} + 1, \quad -2\pi \le x \le 2\pi$

10. $f(x) = (x - 1)^2 (x - 3)$

11. If $f(x) = x^2 + 3x - 1$, find $\dfrac{f(a + h) - f(a)}{h}$.

12. Find the exact value of $\sin 225°$.

13. Verify the identity: $\sec^4 x - \sec^2 x = \tan^4 x + \tan^2 x$.

14. Convert $320°$ to radians.

15. How long would it take for any amount of money, compounded continuously at 5.75% per year, to triple? Round to the nearest tenth of a year.

16. If $f(x) = \dfrac{2x + 1}{x - 3}$, find $f^{-1}(x)$.

17. If C is a right angle in triangle ABC with $A = 23°$ and $a = 12$, solve the triangle.

18. A formula for calculating an infant's dosage for medication is

$$\text{Infant's dose} = \frac{\text{age of infant in months}}{150} \times \text{adult dose.}$$

If a 12-month-old infant is to receive 8.5 mg of medication, find the equivalent adult dose to the nearest milligram.

19. From a point on the ground 12 feet from the base of a flagpole, the angle of elevation to the top of the pole is $53°$. Approximate the height of the flagpole to the nearest tenth of a foot.

20. In *A Tour of the Calculus*, David Berlinski describes trigonometric identities in the following way: "An invisible inner connection exists among the trigonometric functions, one revealed in various identities, strange places where the trigonometric functions appear fluidly to exchange identities or to resolve themselves into unlikely numbers." What does Berlinski mean by this? Use two fundamental trigonometric identities to illustrate your answer.

Additional Topics in Trigonometry

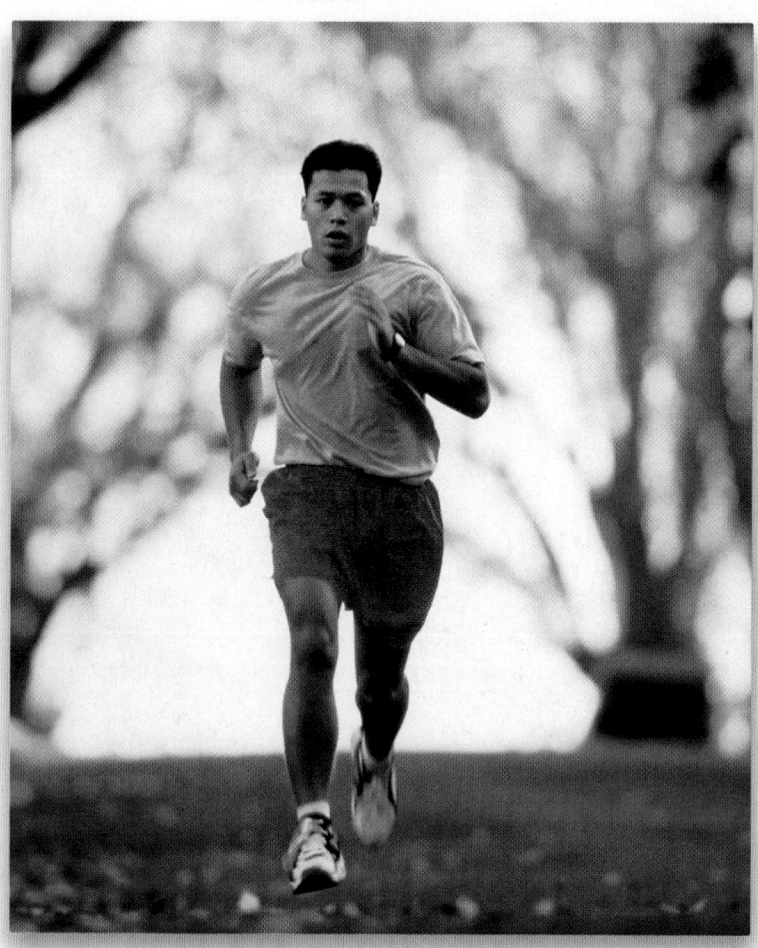

These days, computers and trigonometric functions are everywhere. Trigonometry plays a critical role in analyzing the forces that surround your every move. Using trigonometry to understand how forces are measured is one of the topics in this chapter that focuses on additional applications of trigonometry.

You enjoy running, although lately you experience discomfort at various points of impact. Your doctor suggests a computer analysis. By attaching sensors to your running shoes as you jog along a treadmill, the computer provides a printout of the magnitude and direction of the forces as your feet hit the ground. Based on this analysis, customized orthotics can be made to fit inside your shoes to minimize the impact.

SECTION 7.1 *The Law of Sines*

Objectives

1. Use the Law of Sines to solve oblique triangles.
2. Use the Law of Sines to solve, if possible, the triangle or triangles in the ambiguous case.
3. Find the area of an oblique triangle using the sine function.
4. Solve applied problems using the Law of Sines.

Point Reyes National Seashore, 40 miles north of San Francisco, consists of 75,000 acres with miles of pristine surf-pummeled beaches, forested ridges, and bays flanked by white cliffs. A few people, inspired by nature in the raw, live on private property adjoining the National Seashore. In 1995, a fire in the park burned 12,350 acres and destroyed 45 homes.

Fire is a necessary part of the life cycle in many wilderness areas. It is also an ongoing threat to those who choose to live surrounded by nature's unspoiled beauty. In this section, we see how trigonometry can be used to locate small wilderness fires before they become raging infernos. To do this, we begin by considering triangles other than right triangles.

The Law of Sines and Its Derivation

An **oblique triangle** is a triangle that does not contain a right angle. Figure 7.1 shows that an oblique triangle has either three acute angles or two acute angles and one obtuse angle. Notice that the angles are labeled A, B, and C. The sides opposite each angle are labeled as a, b, and c, respectively.

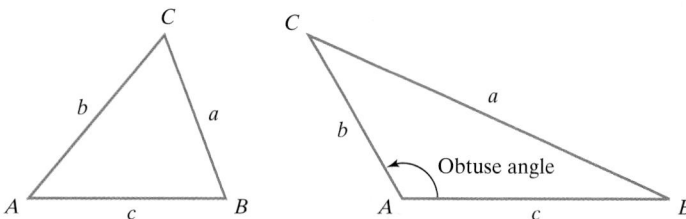

Figure 7.1 Oblique triangles

Many relationships exist among the sides and angles in an oblique triangle. One such relationship is called the **Law of Sines.**

Study Tip

The Law of Sines can be expressed with the sines in the numerator:

$$\frac{\sin A}{a} = \frac{\sin B}{b} = \frac{\sin C}{c}.$$

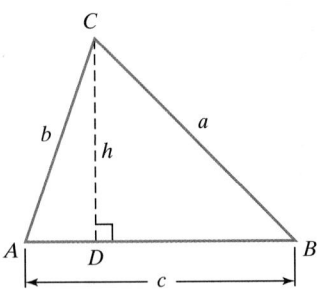

Figure 7.2 Drawing an altitude to prove the Law of Sines

The Law of Sines

If A, B, and C are the measures of the angles of a triangle, and a, b, and c are the lengths of the sides opposite these angles, then

$$\frac{a}{\sin A} = \frac{b}{\sin B} = \frac{c}{\sin C}.$$

The ratio of the length of the side of any triangle to the sine of the angle opposite that side is the same for all three sides of the triangle.

To prove the Law of Sines, we draw an altitude of length h from one of the vertices of the triangle. In Figure 7.2, the altitude is drawn from vertex C. Two smaller triangles are formed, triangles ACD and BCD. Note that both are right triangles. Thus, we can use the definition of the sine of an angle of a right triangle.

$$\sin B = \frac{h}{a} \qquad \sin A = \frac{h}{b} \qquad \sin \theta = \frac{\text{opposite}}{\text{hypotenuse}}$$

$$h = a\sin B \qquad h = b\sin A \qquad \text{Solve each equation for } h.$$

Because we have found two expressions for h, we can set these expressions equal to each other.

$$a\sin B = b\sin A \qquad \text{Equate the expressions for } h.$$

$$\frac{a\sin B}{\sin A \sin B} = \frac{b\sin A}{\sin A \sin B} \qquad \text{Divide both sides by } \sin A \sin B.$$

$$\frac{a}{\sin A} = \frac{b}{\sin B} \qquad \text{Simplify.}$$

This proves part of the Law of Sines. If we use the same process and draw an altitude of length h from vertex A, we obtain the following result:

$$\frac{b}{\sin B} = \frac{c}{\sin C}.$$

When this equation is combined with the previous equation, we obtain the Law of Sines. Because the sine of an angle is equal to the sine of $180°$ minus that angle, the Law of Sines is derived in a similar manner if the oblique triangle contains an obtuse angle.

1 Use the Law of Sines to solve oblique triangles.

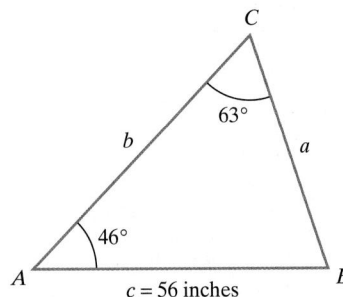

Figure 7.3 Solving an oblique SAA triangle

Solving Oblique Triangles

Solving an oblique triangle means finding the lengths of its sides and the measurement of its angles. The Law of Sines can be used to solve a triangle in which one side and two angles are known. The three known measurements can be abbreviated using SAA (a side and two angles are known) or ASA (two angles and the side between them are known).

EXAMPLE 1 Solving an SAA Triangle Using the Law of Sines

Solve the triangle shown in Figure 7.3 with $A = 46°$, $C = 63°$, and $c = 56$ inches.

Solution We begin by finding B, the third angle of the triangle. We do not need the Law of Sines to do this. Instead, we use the fact that the sum of the measures of the interior angles of a triangle is $180°$.

$$A + B + C = 180°$$

$$46° + B + 63° = 180°$$ Substitute the given values: $A = 46°$ and $C = 63°$.

$$109° + B = 180°$$ Add.

$$B = 71°$$ Subtract 109° from both sides.

When we use the Law of Sines, we must be given one of the three ratios. In this example, we are given c and $C: c = 56$ and $C = 63°$. Thus, we use the ratio $\dfrac{c}{\sin C}$, or $\dfrac{56}{\sin 63°}$, to find the other two sides. Use the Law of Sines to find a.

$$\frac{a}{\sin A} = \frac{c}{\sin C}$$ The ratio of any side to the sine of its opposite angle equals the ratio of any other side to the sine of its opposite angle.

$$\frac{a}{\sin 46°} = \frac{56}{\sin 63°}$$ $A = 46°$, $c = 56$, and $C = 63°$.

$$a = \frac{56 \sin 46°}{\sin 63°}$$ Multiply both sides by $\sin 46°$ and solve for a.

$$a \approx 45 \text{ inches}$$ Use a calculator.

Use the Law of Sines again, this time to find b.

$$\frac{b}{\sin B} = \frac{c}{\sin C}$$ We use the given ratio, $\dfrac{c}{\sin C}$, to find b.

$$\frac{b}{\sin 71°} = \frac{56}{\sin 63°}$$ We found that $B = 71°$. We are given $c = 56$ and $C = 63°$.

$$b = \frac{56 \sin 71°}{\sin 63°}$$ Multiply both sides by $\sin 71°$ and solve for b.

$$b \approx 59 \text{ inches}$$ Use a calculator.

The solution is $B = 71°$, $a \approx 45$ inches, and $b \approx 59$ inches.

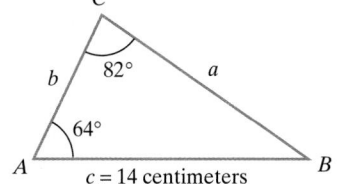

Figure 7.4

Check Point 1 Solve the triangle shown in Figure 7.4 with $A = 64°$, $C = 82°$, and $c = 14$ centimeters.

EXAMPLE 2 Solving an ASA Triangle Using the Law of Sines

Solve triangle ABC if $A = 50°$, $C = 33.5°$, and $b = 76$.

Solution   We begin by drawing a picture of triangle ABC and labeling it with the given information. Figure 7.5 shows the triangle that we must solve. We begin by finding B.

$$A + B + C = 180°$$ The sum of the measures of a triangle's interior angles is 180°.

$$50° + B + 33.5° = 180°$$ $A = 50°$ and $C = 33.5°$.

$$83.5° + B = 180°$$ Add.

$$B = 96.5°$$ Subtract 83.5° from both sides.

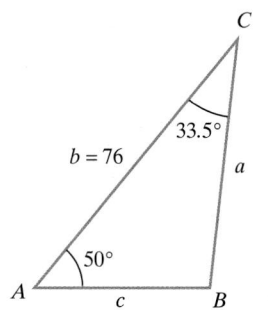

Figure 7.5 Solving an ASA triangle

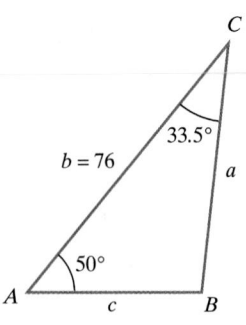

Figure 7.5, repeated

Keep in mind that we must be given one of the three ratios to apply the Law of Sines. In this example, we are given that $b = 76$ and we found that $B = 96.5°$. Thus, we use the ratio $\dfrac{b}{\sin B}$, or $\dfrac{76}{\sin 96.5°}$, to find the other two sides. Use the Law of Sines to find a and c.

Find a:

This is the known ratio.

$$\frac{a}{\sin A} = \frac{b}{\sin B}$$

$$\frac{a}{\sin 50°} = \frac{76}{\sin 96.5°}$$

$$a = \frac{76 \sin 50°}{\sin 96.5°} \approx 59$$

Find c:

$$\frac{c}{\sin C} = \frac{b}{\sin B}$$

$$\frac{c}{\sin 33.5°} = \frac{76}{\sin 96.5°}$$

$$c = \frac{76 \sin 33.5°}{\sin 96.5°} \approx 42$$

The solution is $B = 96.5°$, $a \approx 59$, and $c \approx 42$.

---

2 Use the Law of Sines to solve, if possible, the triangle or triangles in the ambiguous case.

Check Point 2 Solve triangle ABC if $A = 40°$, $C = 22.5°$, and $b = 12$.

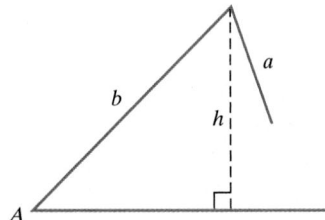

Figure 7.6 Given SSA, no triangle may result.

The Ambiguous Case (SSA)

If we are given two sides and an angle opposite one of them (SSA), does this determine a unique triangle? Can we solve this case using the Law of Sines? Such a case is called the **ambiguous case** because the given information may result in one triangle, two triangles, or no triangle at all. For example, in Figure 7.6, we are given a, b, and A. Because a is shorter than h, it is not long enough to form a triangle. The number of possible triangles, if any, that can be formed in the SSA case depends on h, the length of the altitude, where $h = b \sin A$.

The Ambiguous Case (SSA)

Consider a triangle in which a, b, and A are given. This information may result in

No Triangle	**One Right Triangle**	**Two Triangles**	**One Triangle**
$h = b \sin A$	$h = b \sin A$	$h = b \sin A$	
a is less than h and not long enough to form a triangle.	$a = h$ and just the right length to form a right triangle.	a is greater than h and a is less than b. Two distinct triangles are formed.	a is greater than h and a is greater than b. One triangle is formed.

In an SSA situation, it is not necessary to draw an accurate sketch like those shown in the box. The Law of Sines determines the number of triangles, if any, and gives the solution for each triangle.

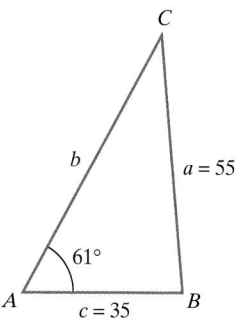

Figure 7.7 Solving an SSA triangle; the ambiguous case

EXAMPLE 3 Solving an SSA Triangle Using the Law of Sines (One Solution)

Solve triangle ABC if $A = 61°$, $a = 55$, and $c = 35$.

Solution We begin with the sketch in Figure 7.7. The known ratio is $\dfrac{a}{\sin A}$, or $\dfrac{55}{\sin 61°}$. Because side c is given, we use the Law of Sines to find angle C.

$$\frac{a}{\sin A} = \frac{c}{\sin C} \qquad \text{Apply the Law of Sines.}$$

$$\frac{55}{\sin 61°} = \frac{35}{\sin C} \qquad A = 55, c = 35, \text{ and } A = 61°.$$

$$55 \sin C = 35 \sin 61° \qquad \text{Cross multiply: If } \frac{a}{b} = \frac{c}{d}, \text{ then } ad = bc.$$

$$\sin C = \frac{35 \sin 61°}{55} \qquad \text{Divide both sides by 55 and solve for } \sin C.$$

$$\sin C \approx 0.5566 \qquad \text{Use a calculator.}$$

There are two angles C between $0°$ and $180°$ for which $\sin C \approx 0.5566$.

$$C_1 \approx 34° \qquad\qquad C_2 \approx 180° - 34° \approx 146°$$

Obtain the acute angle with your calculator:
$\sin^{-1} 0.5566$

The sine is positive in quadrant II.

Look at Figure 7.7. Given that $A = 61°$, can you see that $C_2 \approx 146°$ is impossible? By adding $146°$ to the given angle, $61°$, we exceed a $180°$ sum:

$$61° + 146° = 207°.$$

Thus, the only possibility is that $C_1 \approx 34°$. We find B using this approximation for C_1 and the measure that was given for A: $A = 61°$.

$$B = 180° - C_1 - A \approx 180° - 34° - 61° = 85°$$

Side b that lies opposite this $85°$ angle can now be found using the Law of Sines.

$$\frac{b}{\sin B} = \frac{a}{\sin A} \qquad \text{Apply the Law of Sines.}$$

$$\frac{b}{\sin 85°} = \frac{55}{\sin 61°} \qquad a = 55, B \approx 85°, \text{ and } A = 61°.$$

$$b = \frac{55 \sin 85°}{\sin 61°} \approx 63 \qquad \text{Multiply both sides by } \sin 85° \text{ and solve for } b.$$

There is one triangle and the solution is C_1 (or C) $\approx 34°$, $B \approx 85°$, and $b \approx 63$.

Check Point 3 Solve triangle ABC if $A = 123°$, $a = 47$, and $c = 23$.

EXAMPLE 4 **Solving an SSA Triangle Using the Law of Sines (No Solution)**

Solve triangle ABC if $A = 75°$, $a = 51$, and $b = 71$.

Solution The known ratio is $\dfrac{a}{\sin A}$, or $\dfrac{51}{\sin 75°}$. Because side b is given, we use the Law of Sines to find angle B.

$$\frac{a}{\sin A} = \frac{b}{\sin B} \qquad \text{Use the Law of Sines.}$$

$$\frac{51}{\sin 75°} = \frac{71}{\sin B} \qquad \text{Substitute the given values.}$$

$$51 \sin B = 71 \sin 75° \qquad \text{Cross multiply: If } \frac{a}{b} = \frac{c}{d}, \text{ then } ad = bc.$$

$$\sin B = \frac{71 \sin 75°}{51} \approx 1.34 \quad \text{Divide by 51 and solve for sin B.}$$

Figure 7.8 a is not long enough to form a triangle

Because the sine can never exceed 1, there is no angle B for which $\sin B \approx 1.34$. There is no triangle with the given measurements, as illustrated in Figure 7.8.

Check Point 4 Solve triangle ABC if $A = 50°$, $a = 10$, and $b = 20$.

EXAMPLE 5 **Solving an SSA Triangle Using the Law of Sines (Two Solutions)**

Solve triangle ABC if $A = 40°$, $a = 54$, and $b = 62$.

Solution The known ratio is $\dfrac{a}{\sin A}$, or $\dfrac{54}{\sin 40°}$. We use the Law of Sines to find angle B.

$$\frac{a}{\sin A} = \frac{b}{\sin B} \qquad \text{Use the Law of Sines.}$$

$$\frac{54}{\sin 40°} = \frac{62}{\sin B} \qquad \text{Substitute the given values.}$$

$$54 \sin B = 62 \sin 40° \qquad \text{Cross multiply: If } \frac{a}{b} = \frac{c}{d}, \text{ then } ad = bc.$$

$$\sin B = \frac{62 \sin 40°}{54} \approx 0.7380 \quad \text{Divide by 54 and solve for sin B.}$$

There are two angles B between $0°$ and $180°$ for which $\sin B \approx 0.7380$.

$$B_1 \approx 48° \qquad\qquad B_2 \approx 180° - 48° = 132°$$

Find $\sin^{-1} 0.7380$ with your calculator.   The sine is positive in quadrant II.

If you add either angle to the given angle, $40°$, the sum does not exceed $180°$. Thus, there are two triangles with the given conditions, shown in Figure 7.9(a). The triangles, AB_1C_1 and AB_2C_2, are shown separately in Figures 7.9(b) and (c).

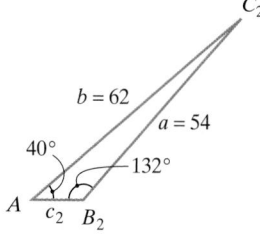

(a) Two triangles are possible with $A = 40°$, $a = 54$, and $b = 62$.

(b) In one possible triangle, $B_1 = 48°$.

(c) In the second possible triangle, $B_2 = 132°$.

Figure 7.9

Study Tip

The two triangles shown in Figure 7.9 are helpful in organizing the solutions. However, if you keep track of the two triangles, one with the given information and $B_1 = 48°$, and the other with the given information and $B_2 = 132°$, you do not have to draw the figure to solve the triangles.

We find angles C_1 and C_2 using a $180°$ angle sum in each of the two triangles.

$$C_1 = 180° - A - B_1 \qquad\qquad C_2 = 180° - A - B_2$$
$$\approx 180° - 40° - 48° \qquad\qquad \approx 180° - 40° - 132°$$
$$= 92° \qquad\qquad\qquad\qquad = 8°$$

We use the Law of Sines to find c_1 and c_2.

$$\frac{c_1}{\sin C_1} = \frac{a}{\sin A} \qquad\qquad \frac{c_2}{\sin C_2} = \frac{a}{\sin A}$$

$$\frac{c_1}{\sin 92°} = \frac{54}{\sin 40°} \qquad\qquad \frac{c_2}{\sin 8°} = \frac{54}{\sin 40°}$$

$$c_1 = \frac{54 \sin 92°}{\sin 40°} \approx 84 \qquad\qquad c_2 = \frac{54 \sin 8°}{\sin 40°} \approx 12$$

There are two triangles. In one triangle, the solution is $B_1 \approx 48°$, $C_1 \approx 92°$, and $c_1 \approx 84$. In the other triangle, $B_2 \approx 132°$, $C_2 \approx 8°$, and $c_2 \approx 12$.

Check Point 5 Solve triangle ABC if $A = 35°$, $a = 12$, and $b = 16$.

3 Find the area of an oblique triangle using the sine function.

The Area of an Oblique Triangle

A formula for the area of an oblique triangle can be obtained using the procedure for proving the Law of Sines. We draw an altitude of length h from one of the vertices of the triangle, as shown in Figure 7.10. We apply the definition of the sine of angle A, $\dfrac{\text{opposite}}{\text{hypotenuse}}$, in right triangle ACD:

$$\sin A = \frac{h}{b} \quad \text{or} \quad h = b \sin A.$$

The area of a triangle is $\frac{1}{2}$ the product of any side and the altitude drawn to that side. Using the altitude h in Figure 7.10, we have

$$\text{Area} = \tfrac{1}{2}ch = \tfrac{1}{2}cb \sin A.$$

Use the result from above: $h = b \sin A$.

Figure 7.10

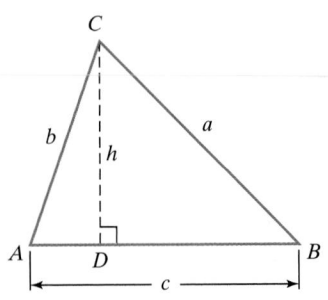

Figure 7.10, repeated

This result, Area $= \frac{1}{2} cb \sin A$, or $\frac{1}{2} bc \sin A$, indicates that the area of the triangle is one-half the product of b and c times the sine of their included angle. If we draw altitudes from the other two vertices, we can use any two sides to compute the area.

Area of An Oblique Triangle

The area of a triangle equals one-half the product of the lengths of two sides times the sine of their included angle. In Figure 7.10, this wording can be expressed by the formulas

$$\text{Area} = \tfrac{1}{2} bc \sin A = \tfrac{1}{2} ab \sin C = \tfrac{1}{2} ac \sin B.$$

EXAMPLE 6 Finding the Area of an Oblique Triangle

Find the area of a triangle having two sides of lengths 24 meters and 10 meters and an included angle of 62°.

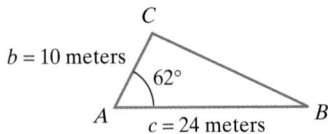

Figure 7.11 Finding the area of an SAS triangle

Solution   The triangle is shown in Figure 7.11. Its area is half the product of the lengths of the two sides times the sine of the included angle.

$$\text{Area} = \tfrac{1}{2}(24)(10)(\sin 62°) \approx 106$$

The area of the triangle is approximately 106 square meters.

> **Check Point 6**   Find the area of a triangle having two sides of lengths 8 meters and 12 meters and an included angle of 135°.

4 Solve applied problems using the Law of Sines.

Applications of the Law of Sines

We have seen how the trigonometry of right triangles can be used to solve many different kinds of applied problems. The Law of Sines enables us to work with triangles that are not right triangles. As a result, this law can be used to solve problems involving surveying, engineering, astronomy, navigation, and the environment. Example 7 illustrates the use of the Law of Sines in detecting potentially devastating fires.

EXAMPLE 7 An Application of the Law of Sines

Two fire-lookout stations are 20 miles apart, with station B directly east of station A. Both stations spot a fire on a mountain to the north. The bearing from station A to the fire is N50°E (50° east of north). The bearing from station B to the fire is N36°W (36° west of north). How far is the fire from station A?

Solution   Figure 7.12 at the top of the next page shows the information given in the problem. The distance from station A to the fire is represented by b. Notice that the angles describing the bearing from each station to the fire, 50° and 36°, are not interior angles of triangle ABC. Using a north-south line, the interior angles are found as follows:

$$A = 90° - 50° = 40° \qquad B = 90° - 36° = 54°.$$

To find b using the Law of Sines, we need a known side and an angle opposite that side. Because $c = 20$ miles, we find angle C using a 180° angle sum in the triangle. Thus,

$$C = 180° - A - B = 180° - 40° - 54° = 86°.$$

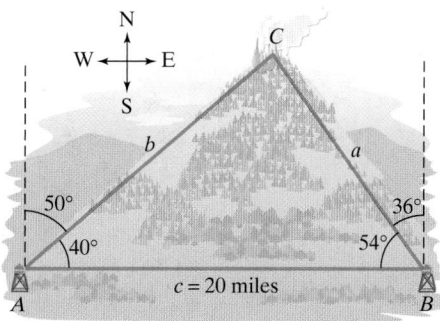

Figure 7.12

The ratio $\dfrac{c}{\sin C}$, or $\dfrac{20}{\sin 86°}$, is now known. We use this ratio and the Law of Sines to find b.

$$\frac{b}{\sin B} = \frac{c}{\sin C} \qquad \text{Use the Law of Sines.}$$

$$\frac{b}{\sin 54°} = \frac{20}{\sin 86°} \qquad c = 20, B = 54°, \text{ and } C = 86°.$$

$$b = \frac{20 \sin 54°}{\sin 86°} \approx 16 \qquad \text{Multiply both sides by } \sin 54° \text{ and solve for } b.$$

The fire is approximately 16 miles from station A.

> **Check Point 7**
> Two fire-lookout stations are 13 miles apart, with station B directly east of station A. Both stations spot a fire. The bearing of the fire from station A is N35°E, and the bearing of the fire from station B is N49°W. How far, to the nearest mile, is the fire from station B?

EXERCISE SET 7.1

Practice Exercises

In Exercises 1–8, solve each triangle. Round lengths of sides to the nearest tenth and angle measures to the nearest degree.

1.

2.

3.

4.

5.

6.

7.

8.

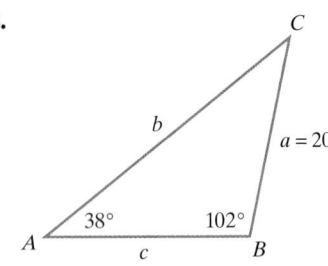

In Exercises 9–16, solve each triangle. Round lengths to the nearest tenth and angle measures to the nearest degree.

9. $A = 44°, B = 25°, a = 12$

10. $A = 56°, C = 24°, a = 22$

11. $B = 85°, C = 15°, b = 40$

12. $A = 85°, B = 35°, c = 30$

13. $A = 115°, C = 35°, c = 200$

14. $B = 5°, C = 125°, b = 200$

15. $A = 65°, B = 65°, c = 6$

16. $B = 80°, C = 10°, a = 8$

In Exercises 17–32, two sides and an angle (SSA) of a triangle are given. Determine whether the given measurements produce one triangle, two triangles, or no triangle at all. Solve each triangle that results. Round to the nearest tenth and the nearest degree for sides and angles, respectively.

17. $a = 20, b = 15, A = 40°$

18. $a = 30, b = 20, A = 50°$

19. $a = 10, c = 8.9, A = 63°$

20. $a = 57.5, c = 49.8, A = 136°$

21. $a = 42.1, c = 37, A = 112°$

22. $a = 6.1, b = 4, A = 162°$

23. $a = 10, b = 40, A = 30°$

24. $a = 10, b = 30, A = 150°$

25. $a = 16, b = 18, A = 60°$

26. $a = 30, b = 40, A = 20°$

27. $a = 12, b = 16.1, A = 37°$

28. $a = 7, b = 28, A = 12°$

29. $a = 22, c = 24.1, A = 58°$

30. $a = 95, c = 125, A = 49°$

31. $a = 9.3, b = 41, A = 18°$

32. $a = 1.4, b = 2.9, A = 142°$

In Exercises 33–38, find the area of the triangle having the given measurements. Round to the nearest square unit.

33. $A = 48°, b = 20$ feet, $c = 40$ feet

34. $A = 22°, b = 20$ feet, $c = 50$ feet

35. $B = 36°, a = 3$ yards, $c = 6$ yards

36. $B = 125°, a = 8$ yards, $c = 5$ yards

37. $C = 124°, a = 4$ meters, $b = 6$ meters

38. $C = 102°, a = 16$ meters, $b = 20$ meters

 Application Exercises

39. Two fire-lookout stations are 10 miles apart, with station B directly east of station A. Both stations spot a fire. The bearing of the fire from station A is N25°E and the bearing of the fire from station B is N56°W. How far, to the nearest mile, is the fire from each lookout station?

40. The Federal Communications Commission is attempting to locate an illegal radio station. It sets up two monitoring stations, A and B, with station B 40 miles east of station A. Station A measures the illegal signal from the radio station as coming from a direction of 48° east of north. Station B measures the signal as coming from a point 34° west of north. How far is the illegal radio station from monitoring stations A and B?

41. The figure shows a 1200-yard-long sand beach and an oil platform in the ocean. The angle made with the platform from one end of the beach is 85° and from the other end is 76°. Find the distance of the oil platform, to the nearest yard, from each end of the beach.

42. A surveyor needs to determine the distance between two points that lie on opposite banks of a river. The figure shows that 300 yards are measured along one bank. The angles from each end of this line segment to a point on the opposite bank are 62° and 53°. Find the distance between A and B to the nearest foot.

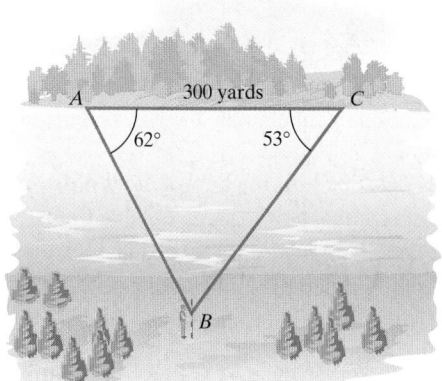

43. Closed to tourists since 1990, the Leaning Tower of Pisa in Italy leans at an angle of about 84.7°. The figure shows that 171 feet from the base of the tower, the angle of elevation to the top is 50°. Find the distance, to the nearest foot, from the base to the top of the tower.

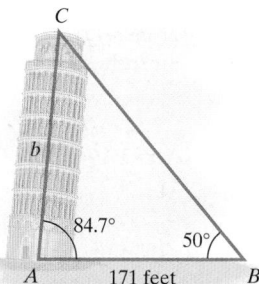

44. A pine tree growing on a hillside makes a 75° angle with the hill. From a point 80 feet up the hill, the angle of elevation to the top of the tree is 62° and the angle of depression to the bottom is 23°. Find, to the nearest foot, the height of the tree.

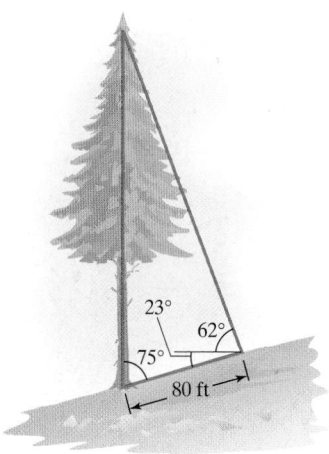

45. The figure shows a shot-put ring. The shot is tossed from A and lands at B. Using modern electronic equipment, the distance of the toss can be measured without the use of measuring tapes. When the shot lands at B, an electronic transmitter placed at B sends a signal to a device in the official's booth above the track. The device determines the angles at B and C. At a track meet, the distance from the official's booth to the shot-put ring is 562 feet. If $B = 85.3°$ and $C = 5.7°$, determine the length of the toss to the nearest foot.

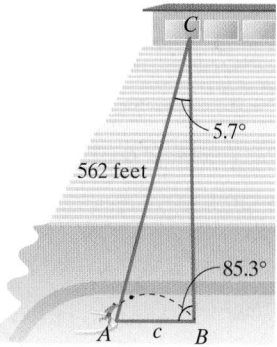

46. A pier forms an 85° angle with a straight shore. At a distance of 100 feet from the pier, the line of sight to the tip forms a 37° angle. Find the length of the pier to the nearest foot.

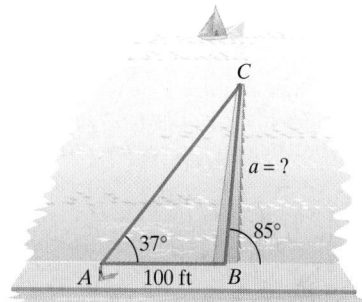

47. When the angle of elevation of the sun is 62°, a telephone pole that is tilted at an angle of 8° directly away from the sun casts a shadow 20 feet long. Determine the length of the pole to the nearest foot.

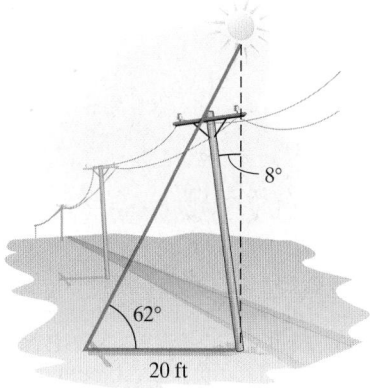

48. A leaning wall is inclined 6° from the vertical. At a distance of 40 feet from the wall, the angle of elevation to the top is 22°. Find the height of the wall to the nearest foot.

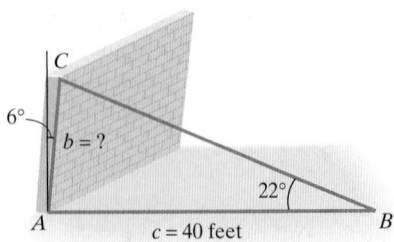

49. Redwood trees in California's Redwood National Park are hundreds of feet tall. The height of one of these trees is represented by h in the figure shown.

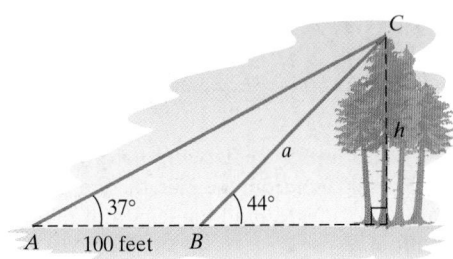

a. Use the measurements shown to find a, to the nearest foot, in oblique triangle ABC.

b. Use the right triangle shown to find the height, to the nearest foot, of a typical redwood tree in the park.

50. The figure shows a cable car that carries passengers from A to C. Point A is 1.6 miles from the base of the mountain. The angles of elevation from A and B to the mountain's peak are 22° and 66°, respectively.

a. Determine, to the nearest foot, the distance covered by the cable car.

b. Find a, to the nearest foot, in oblique triangle ABC.

c. Use the right triangle to find the height of the mountain to the nearest foot.

51. Lighthouse B is 7 miles west of lighthouse A. A boat leaves A and sails 5 miles. At this time, it is sighted from B. If the bearing of the boat from B is N62°E, how far from B is the boat? Round to the nearest tenth of a mile.

52. After a wind storm, you notice that your 16-foot flagpole may be leaning, but you are not sure. From a point on the ground 15 feet from the base of the flagpole, you find that the angle of elevation to the top is 48°. Is the flagpole leaning? If so, find the acute angle, to the nearest degree, that the flagpole makes with the ground.

Writing in Mathematics

53. What is an oblique triangle?

54. Without using symbols, state the Law of Sines in your own words.

55. Briefly describe how the Law of Sines is proved.

56. What does it mean to solve an oblique triangle?

57. What do the abbreviations SAA and ASA mean?

58. Why is SSA called the ambiguous case?

59. How is the sine function used to find the area of an oblique triangle?

60. Write an original problem that can be solved using the Law of Sines. Then solve the problem.

61. Use Exercise 45 to describe how the Law of Sines is used for throwing events at track and field meets. Why aren't tape measures used to determine tossing distance?

62. You are cruising in your boat parallel to the coast, looking at a lighthouse. Explain how you can use your boat's speed and a device for measuring angles to determine the distance at any instant from your boat to the lighthouse.

Critical Thinking Exercises

63. If you are given two sides of a triangle and their included angle, you can find the triangle's area. Can the Law of Sines be used to solve the triangle with this given information? Explain your answer.

64. Two buildings of equal height are 800 feet apart. An observer on the street between the buildings measures the angles of elevation to the tops of the buildings as 27° and 41°, respectively. How high, to the nearest foot, are the buildings?

65. The figure shows the design for the top of the wing of a jet fighter. The fuselage is 5 feet wide. Find the wing span CC' to the nearest foot.

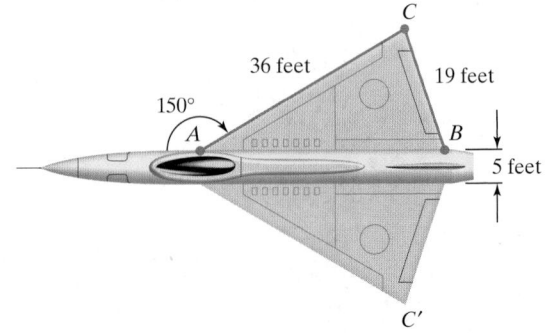

SECTION 7.2   *The Law of Cosines*

Objectives

1. Use the Law of Cosines to solve oblique triangles.
2. Solve applied problems using the Law of Cosines.
3. Use Heron's formula to find the area of a triangle.

Baseball was developed in the United States in the mid-1800s and became our national sport. Little League baseball is the largest youth sports program in the world. Three million boys and girls ages 5 to 18 play on 200,000 Little League teams in 90 countries. Although there are differences between Major League and Little League baseball diamonds, trigonometry can be used to find angles and distances in these fields of dreams. To see how this is done, we turn to the Law of Cosines.

The Law of Cosines and Its Derivation

We now look at another relationship that exists among the sides and angles in an oblique triangle. **The Law of Cosines** is used to solve triangles in which two sides and the included angle (SAS) are known, or those in which three sides (SSS) are known.

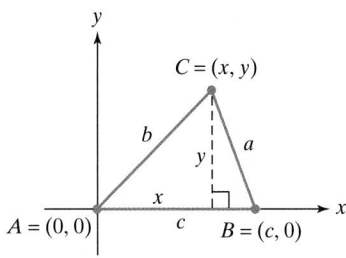

Figure 7.13

The Law of Cosines

If A, B, and C are the measures of the angles of a triangle, and a, b, and c are the lengths of the sides opposite these angles, then

$$a^2 = b^2 + c^2 - 2bc \cos A$$
$$b^2 = a^2 + c^2 - 2ac \cos B$$
$$c^2 = a^2 + b^2 - 2ab \cos C.$$

The square of a side of a triangle equals the sum of the squares of the other two sides minus twice their product times the cosine of their included angle.

To prove the Law of Cosines, we place triangle ABC in a rectangular coordinate system. Figure 7.13 shows a triangle with three acute angles. The vertex A is at the origin and side c lies along the positive x-axis. The coordinates of C are (x, y). Using the right triangle that contains angle A, we apply the definitions of the cosine and the sine.

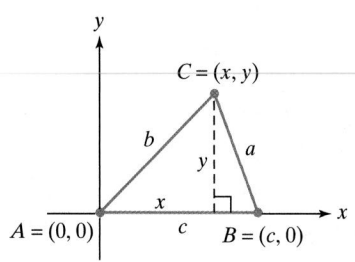

Figure 7.13, repeated

$$\cos A = \frac{x}{b} \qquad \sin A = \frac{y}{b}$$

$$x = b \cos A \qquad y = b \sin A \qquad \text{Multiply both sides of each equation by } b \text{ and solve for } x \text{ and } y, \text{ respectively.}$$

Thus, the coordinates of C are $(x, y) = (b \cos A, b \sin A)$. Although triangle ABC in Figure 7.13 shows angle A as an acute angle, if A is obtuse, the coordinates of C are still $(b \cos A, b \sin A)$. This means that our proof applies to both kinds of oblique triangles.

We now apply the distance formula to the side of the triangle with length a. Notice that a is the distance from (x, y) to $(c, 0)$.

$$a = \sqrt{(x - c)^2 + (y - 0)^2} \qquad \text{Use the distance formula.}$$

$$a^2 = (x - c)^2 + y^2 \qquad \text{Square both sides of the equation.}$$

$$a^2 = (b \cos A - c)^2 + (b \sin A)^2 \qquad x = b \cos A \text{ and } y = b \sin A.$$

$$a^2 = b^2 \cos^2 A - 2bc \cos A + c^2 + b^2 \sin^2 A \qquad \text{Square the two expressions.}$$

$$a^2 = b^2 \sin^2 A + b^2 \cos^2 A + c^2 - 2bc \cos A \qquad \text{Rearrange terms.}$$

$$a^2 = b^2(\sin^2 A + \cos^2 A) + c^2 - 2bc \cos A \qquad \text{Factor } b^2 \text{ from the first two terms.}$$

$$a^2 = b^2 + c^2 - 2bc \cos A \qquad \sin^2 A + \cos^2 A = 1$$

The resulting equation is one of the three formulas for the Law of Cosines. The other two formulas are derived in a similar manner.

1 Use the Law of Cosines to solve oblique triangles.

Solving Oblique Triangles

If you are given two sides and an included angle (SAS) of an oblique triangle, none of the three ratios in the Law of Sines is known. This means that we do not begin solving the triangle using the Law of Sines. Instead, we apply the Law of Cosines and the following procedure:

Solving an SAS Triangle

1. Use the Law of Cosines to find the side opposite the given angle.

2. Use the Law of Sines to find the angle opposite the shorter of the two given sides. This angle is always acute.

3. Find the third angle by subtracting the measure of the given angle and the angle found in step 2 from 180°.

EXAMPLE 1   Solving an SAS Triangle

Solve the triangle in Figure 7.14 with $A = 60°$, $b = 20$, and $c = 30$.

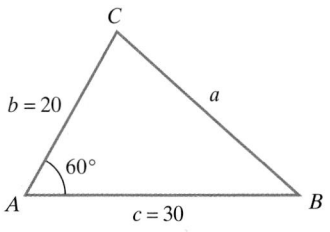

Figure 7.14 Solving an SAS triangle

Solution We are given two sides and an included angle. Therefore, we apply the three-step procedure for solving an SAS triangle.

Step 1   Use the Law of Cosines to find the side opposite the given angle. Thus, we will find a.

$$a^2 = b^2 + c^2 - 2bc \cos A \qquad \text{Apply the Law of Cosines to find } a.$$

$$a^2 = 20^2 + 30^2 - 2(20)(30) \cos 60° \quad b = 20, c = 30, \text{ and } A = 60°$$

$$ = 400 + 900 - 1200(0.5) \qquad \text{Perform the indicated operations.}$$

$$ = 700$$

$$a = \sqrt{700} \approx 26 \qquad \text{Take the square root of both sides and solve for } a.$$

Step 2   Use the Law of Sines to find the angle opposite the shorter of the two given sides. This angle is always acute. The shorter of the two given sides is $b = 20$. Thus, we will find acute angle B.

$$\frac{b}{\sin B} = \frac{a}{\sin A} \qquad \text{Apply the Law of Sines.}$$

$$\frac{20}{\sin B} = \frac{\sqrt{700}}{\sin 60°} \qquad \text{We are given } b = 20 \text{ and } A = 60°. \text{ Use the exact value of } a, \sqrt{700}, \text{ from step 1.}$$

$$\sqrt{700} \sin B = 20 \sin 60° \qquad \text{Cross multiply: If } \frac{a}{b} = \frac{c}{d}, \text{ then } ad = bc.$$

$$\sin B = \frac{20 \sin 60°}{\sqrt{700}} \approx 0.6547 \quad \text{Divide by } \sqrt{700} \text{ and solve for } \sin B.$$

$$B \approx 41° \qquad \text{Find } \sin^{-1} 0.6547 \text{ using a calculator.}$$

Step 3   Find the third angle. Subtract the measure of the given angle and the angle found in step 2 from 180°.

$$C = 180° - A - B \approx 180° - 60° - 41° = 79°$$

The solution is $a \approx 26$, $B \approx 41°$, and $C \approx 79°$.

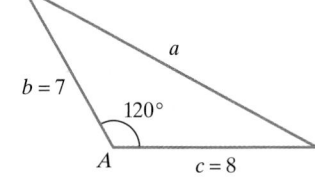

Figure 7.15

Check Point 1   Solve the triangle shown in Figure 7.15 with $A = 120°$, $b = 7$, and $c = 8$.

If you are given three sides of a triangle (SSS), solving the triangle involves finding the three angles. We use the following procedure:

Solving an SSS Triangle

1. Use the Law of Cosines to find the angle opposite the longest side.
2. Use the Law of Sines to find either of the two remaining acute angles.
3. Find the third angle by subtracting the measures of the angles found in steps 1 and 2 from 180°.

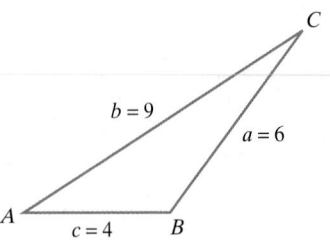

Figure 7.16 Solving an SSS triangle

EXAMPLE 2 Solving an SSS Triangle

Solve triangle ABC if $a = 6$, $b = 9$, and $c = 4$. Round to the nearest tenth of a degree.

Solution   We are given three sides. Therefore, we apply the three-step procedure for solving an SSS triangle. The triangle is shown in Figure 7.16.

Step 1   Use the Law of Cosines to find the angle opposite the longest side. The longest side is $b = 9$. Thus, we will find angle B.

$$b^2 = a^2 + c^2 - 2ac \cos B \qquad \text{Apply the Law of Cosines to find B.}$$

$$2ac \cos B = a^2 + c^2 - b^2 \qquad \text{Solve for cos B.}$$

$$\cos B = \frac{a^2 + c^2 - b^2}{2ac}$$

$$\cos B = \frac{6^2 + 4^2 - 9^2}{2 \cdot 6 \cdot 4} = -\frac{29}{48} \qquad a = 6, b = 9, \text{ and } c = 4.$$

Using a calculator, $\cos^{-1}\left(\frac{29}{48}\right) \approx 52.8°$. Because $\cos B$ is negative, B is an obtuse angle. Thus,

$$B \approx 180° - 52.8° = 127.2°. \qquad \text{Because the domain of } y = \cos^{-1}x \text{ is } [0, \pi], \text{ you can use a calculator to find } \cos^{-1}\left(-\frac{29}{48}\right) \approx 127.2°.$$

Study Tip

You can use the Law of Cosines in step 2 to find either of the remaining angles. However, it is simpler to use the Law of Sines. Because the largest angle has been found, the remaining angles must be acute. Thus, there is no need to be concerned about two possible triangles or an ambiguous case.

Step 2   Use the Law of Sines to find either of the two remaining acute angles. We will find angle A.

$$\frac{a}{\sin A} = \frac{b}{\sin B} \qquad \text{Apply the Law of Sines.}$$

$$\frac{6}{\sin A} = \frac{9}{\sin 127.2°} \qquad \text{We are given } a = 6 \text{ and } b = 9. \text{ We found that } B \approx 127.2°.$$

$$9 \sin A = 6 \sin 127.2° \qquad \text{Cross multiply.}$$

$$\sin A = \frac{6 \sin 127.2°}{9} \approx 0.5310 \qquad \text{Divide by 9 and solve for sin A.}$$

$$A \approx 32.1° \qquad \text{Find } \sin^{-1} 0.5310 \text{ using a calculator.}$$

Step 3   Find the third angle. Subtract the measures of the angles found in steps 1 and 2 from 180°.

$$C = 180° - B - A \approx 180° - 127.2° - 32.1° = 20.7°$$

The solution is $B \approx 127.2°$, $A \approx 32.1°$, and $C \approx 20.7°$.

Check Point 2   Solve triangle ABC if $a = 8$, $b = 10$, and $c = 5$. Round to the nearest tenth of a degree.

2 Solve applied problems using the Law of Cosines.

Applications of the Law of Cosines

Applied problems involving SAS and SSS triangles can be solved using the Law of Cosines.

EXAMPLE 3 An Application of the Law of Cosines

Two airplanes leave an airport at the same time on different runways. One flies on a bearing of N66°W at 325 miles per hour. The other airplane flies on a bearing of S26°W at 300 miles per hour. How far apart will the airplanes be after two hours?

Solution After two hours, the plane flying at 325 miles per hour travels $325 \cdot 2$ miles, or 650 miles. Similarly, the plane flying at 300 miles per hour travels 600 miles. The situation is illustrated in Figure 7.17.

Let b = the distance between the planes after two hours. We can use a north-south line to find angle B in triangle ABC. Thus,

$$B = 180° - 66° - 26° = 88°.$$

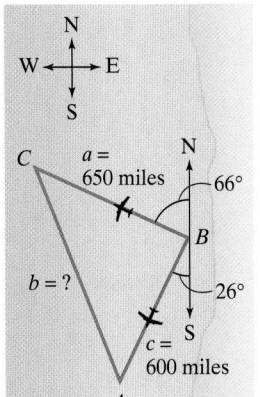

Figure 7.17

We now have $a = 650$, $c = 600$, and $B = 88°$. We use the Law of Cosines to find b in this SAS situation.

$$b^2 = a^2 + c^2 - 2ac \cos B$$ Apply the Law of Cosines.

$$b^2 = 650^2 + 600^2 - 2(650)(600)\cos 88°$$ Substitute: $a = 650$, $c = 600$, and $b = 88°$.

$$\approx 755{,}278$$ Use a calculator.

$$b \approx \sqrt{755{,}278} \approx 869$$ Take the square root and solve for b.

After two hours, the planes are approximately 869 miles apart.

> **Check Point 3** Two airplanes leave an airport at the same time on different runways. One flies directly north at 400 miles per hour. The other airplane flies on a bearing of N75°E at 350 miles per hour. How far apart will the airplanes be after two hours?

3 Use Heron's formula to find the area of a triangle.

Heron's Formula

Approximately 2000 years ago, the Greek mathematician Heron of Alexandria derived a formula for the area of a triangle in terms of the lengths of its sides. A more modern derivation uses the Law of Cosines and can be found in the appendix.

Heron's Formula for the Area of a Triangle

The area of a triangle with sides a, b, and c is

$$\text{Area} = \sqrt{s(s-a)(s-b)(s-c)}$$

where s is one-half the perimeter: $s = \frac{1}{2}(a + b + c)$.

EXAMPLE 4 Using Heron's Formula

Find the area of the triangle with $a = 12$ yards, $b = 16$ yards, and $c = 24$ yards.

Solution Begin by calculating one-half the perimeter:

$$s = \tfrac{1}{2}(a + b + c) = \tfrac{1}{2}(12 + 16 + 24) = 26.$$

Use Heron's formula to find the area:

$$\begin{aligned}
\text{Area} &= \sqrt{s(s - a)(s - b)(s - c)} \\
&= \sqrt{26(26 - 12)(26 - 16)(26 - 24)} \\
&= \sqrt{7280} \approx 85.
\end{aligned}$$

The area of the triangle is approximately 85 square yards.

Check Point 4 Find the area of the triangle with $a = 6$ meters, $b = 16$ meters, and $c = 18$ meters. Round to the nearest square meter.

EXERCISE SET 7.2

Practice Exercises

In Exercises 1–8, solve each triangle. Round lengths of sides to the nearest tenth and angle measures to the nearest degree.

1.

2.

3.

4.

5.

6.

7.

8.
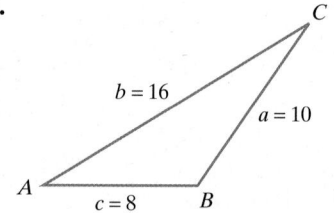

In Exercises 9–24, solve each triangle. Round lengths to the nearest tenth and angle measures to the nearest degree.

9. $a = 5, b = 7, C = 42°$ **10.** $a = 10, b = 3, C = 15°$

11. $b = 5, c = 3, A = 102°$ **12.** $b = 4, c = 1, A = 100°$

13. $a = 6, c = 5, B = 50°$ **14.** $a = 4, c = 7, B = 55°$

15. $a = 5, c = 2, B = 90°$ **16.** $a = 7, c = 3, B = 90°$

17. $a = 5, b = 7, c = 10$ **18.** $a = 4, b = 6, c = 9$

19. $a = 3, b = 9, c = 8$ **20.** $a = 4, b = 7, c = 6$

21. $a = 3, b = 3, c = 3$ **22.** $a = 5, b = 5, c = 5$

23. $a = 73, b = 22, c = 50$ **24.** $a = 66, b = 25, c = 45$

In Exercises 25–30, use Heron's formula to find the area of each triangle. Round to the nearest square unit.

25. $a = 4$ feet, $b = 4$ feet, $c = 2$ feet

26. $a = 5$ feet, $b = 5$ feet, $c = 4$ feet

27. $a = 14$ meters, $b = 12$ meters, $c = 4$ meters

28. $a = 16$ meters, $b = 10$ meters, $c = 8$ meters

29. $a = 11$ yards, $b = 9$ yards, $c = 7$ yards

30. $a = 13$ yards, $b = 9$ yards, $c = 5$ yards

Application Exercises

31. Two ships leave a harbor at the same time. One ship travels on a bearing of S12°W at 14 miles per hour. The other ship travels on a bearing of N75°E at 10 miles per hour. How far apart will the ships be after three hours? Round to the nearest tenth of a mile.

32. A plane leaves airport A and travels 580 miles to airport B on a bearing of N34°E. The plane later leaves airport B and travels to airport C 400 miles away on a bearing of S74°E. Find the distance from airport A to airport C to the nearest tenth of a mile.

33. Find the distance across the lake from A to C, to the nearest yard, using the measurements shown in the figure.

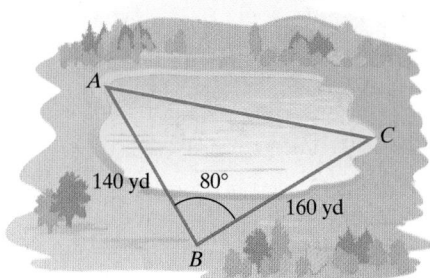

34. To find the distance across a protected cove at a lake, a surveyor makes the measurements shown in the figure at the top of the next column.

Use these measurements to find the distance from A to B to the nearest yard.

The diagram shows three islands in Florida Bay. You rent a boat and plan to visit each of these remote islands. Use the diagram to solve Exercises 35–36.

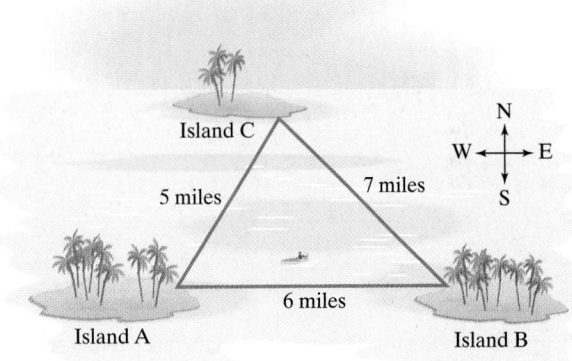

35. If you are on island A, on what bearing should you navigate to go to island C?

36. If you are on island B, on what bearing should you navigate to go to island C?

37. You are on a fishing boat that leaves its pier and heads east. After traveling for 25 miles, there is a report warning of rough seas directly south. The captain turns the boat and follows a bearing of S40°W for 13.5 miles.
   a. At this time, how far are you from the boat's pier? Round to the nearest tenth of a mile.
   b. What bearing could the boat have originally taken to arrive at this spot?

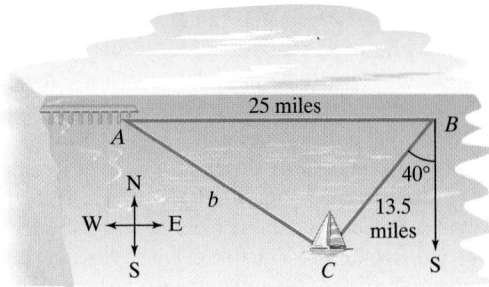

38. You are on a fishing boat that leaves its pier and heads east. After traveling for 30 miles, there is a report warning of rough seas directly south. The captain turns the boat and follows a bearing of S45°W for 12 miles.
  a. At this time, how far are you from the boat's pier? Round to the nearest tenth of a mile.
  b. What bearing could the boat have originally taken to arrive at this spot?

39. The figure shows a 400-foot tower on the side of a hill that forms 7° angle with the horizontal. Find the length of each of the two guy wires that are anchored 80 feet uphill and downhill from the tower's base and extend to the top of the tower. Round to the nearest tenth of a foot.

40. The figure shows a 200-foot tower on the side of a hill that forms a 5° angle with the horizontal. Find the length of each of the two guy wires that are anchored 150 feet uphill and downhill from the tower's base and extend to the top of the tower. Round to the nearest tenth of a foot.

41. A Major League baseball diamond has four bases forming a square whose sides measure 90 feet each. The pitcher's mound is 60.5 feet from home plate on a line joining home plate and second base. Find the distance from the pitcher's mound to first base. Round to the nearest tenth of a foot.

42. A Little League baseball diamond has four bases forming a square whose sides measure 60 feet each. The pitcher's mound is 46 feet from home plate on a line joining home plate and second base. Find the distance from the pitcher's mound to third base. Round to the nearest tenth of a foot.

43. A commercial piece of real estate is priced at $3.50 per square foot. Find the cost, to the nearest dollar, of a triangular lot measuring 240 feet by 300 feet by 420 feet.

44. A commercial piece of real estate is priced at $4.50 per square foot. Find the cost, to the nearest dollar, of a triangular lot measuring 320 feet by 510 feet by 410 feet.

Writing in Mathematics

45. Without using symbols, state the Law of Cosines in your own words.

46. Why can't the Law of Sines be used in the first step to solve an SAS triangle?

47. Describe a strategy for solving an SAS triangle.

48. Describe a strategy for solving an SSS triangle.

49. Under what conditions would you use Heron's formula to find the area of a triangle?

50. Describe an applied problem that can be solved using the Law of Cosines, but not the Law of Sines.

51. The pitcher on your Little League team is studying angles in geometry and has a question. "Coach, suppose I'm on the pitcher's mound facing home plate. I catch a fly ball hit in my direction. If I turn to face first base and throw the ball, through how many degrees should I turn for a direct throw?" Use the information given in Exercise 42 and write an answer to your pitcher's question. Without getting too technical, describe to your pitcher how you obtained this angle.

Critical Thinking Exercises

52. The lengths of the diagonals of a parallelogram are 20 inches and 30 inches. The diagonals intersect at an angle of 35°. Find the lengths of the parallelogram's sides. (*Hint*: Diagonals of a parallelogram bisect one another.)

53. The vertices of a triangle are $A(4, -3)$, $B(2, 1)$, and $C(-2, 4)$. Find the triangle's largest angle to the nearest tenth of a degree.

54. The minute hand and the hour hand of a clock have lengths m inches and h inches, respectively. Determine the distance between the tips of the hands at 10:00 in terms of m and h.

Group Exercise

55. The group should design five original problems that can be solved using the Laws of Sines and Cosines. At least two problems should be solved using the Law of Sines and at least two problems should be solved using the Law of Cosines. At least one problem should be an application problem using the Law of Sines and at least one problem should involve an application using the Law of Cosines. The group should turn in both the problems and their solutions.

SECTION 7.3   *Polar Coordinates*

Objectives

1. Plot points in the polar coordinate system.
2. Find multiple sets of polar coordinates for a given point.
3. Convert a point from polar to rectangular coordinates.
4. Convert a point from rectangular to polar coordinates.
5. Convert an equation from rectangular to polar coordinates.
6. Convert an equation from polar to rectangular coordinates.

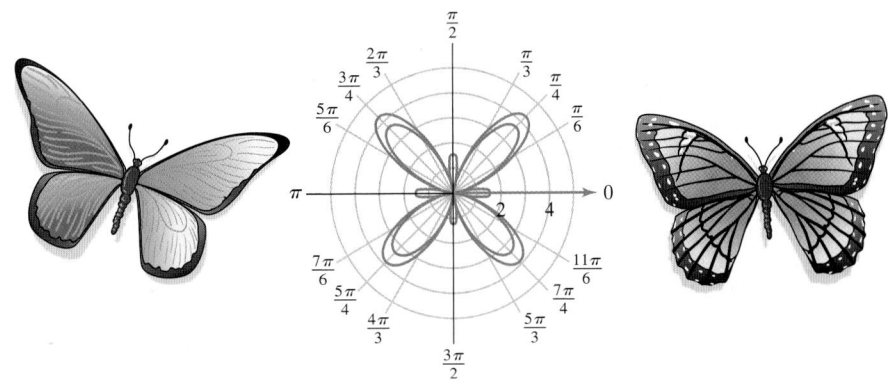

Butterflies are among the most celebrated of all insects. It's hard not to notice their beautiful colors and graceful flight. Their symmetry can be explored with trigonometric functions and a system for plotting points called the **polar coordinate system.** In many cases, polar coordinates are simpler and easier to use than rectangular coordinates.

Plotting Points in the Polar Coordinate System

1 Plot points in the polar coordinate system.

The foundation of the polar coordinate system is a horizontal ray that extends to the right. The ray is called the **polar axis** and is shown in Figure 7.18. The endpoint of the ray is called the **pole.**

A point P in the polar coordinate system is represented by an ordered pair of numbers (r, θ). Figure 7.19 shows $P = (r, \theta)$ in the polar coordinate system.

Figure 7.18

Figure 7.19 Representing a point in the polar coordinate system

- r is the directed distance from P to the pole. (We shall see that r can be positive, negative or zero.)
- θ is an angle from the polar axis to line segment OP. This angle can be measured in degrees or radians. Positive angles are measured counterclockwise from the polar axis. Negative angles are measured clockwise from the polar axis.

We refer to the ordered pair (r, θ) as the **polar coordinates** of P.

Let's look at a specific example. Suppose that the polar coordinates of a point P are $\left(3, \dfrac{\pi}{4}\right)$. Because θ is positive, we locate this point by drawing $\theta = \dfrac{\pi}{4}$ counterclockwise from the polar axis. Then we count out a distance of three units along the terminal side of the angle to reach the point P.

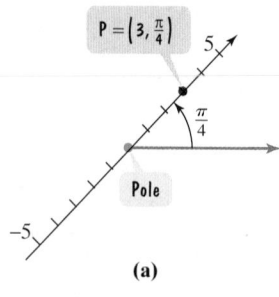

$P = \left(3, \frac{\pi}{4}\right)$

$\frac{\pi}{4}$

Pole

(a)

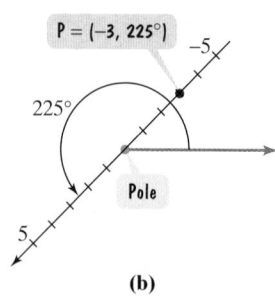

$P = (-3, 225°)$

$225°$

Pole

(b)

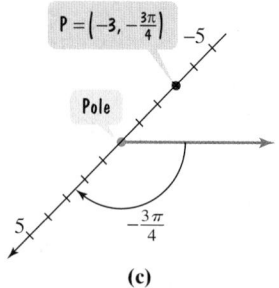

$P = \left(-3, -\frac{3\pi}{4}\right)$

Pole

$-\frac{3\pi}{4}$

(c)

Figure 7.20 Locating points in polar coordinates

Figure 7.20(a) shows that $(r, \theta) = \left(3, \dfrac{\pi}{4}\right)$ lies three units from the pole on the terminal side of the angle $\theta = \dfrac{\pi}{4}$.

Figure 7.20 illustrates that in a polar coordinate system, a point can be represented in more than one way. In Figure 7.20(b), r is negative and θ is positive: $r = -3$ and $\theta = 225°$. Because θ is positive, we draw a $225°$ angle counterclockwise from the polar axis. Notice that the point P is not located on the terminal side of θ. Instead, it lies on the ray *opposite the terminal side of θ* at a distance of $|-3|$, or 3, units from the pole. In general, **when r in (r, θ) is negative, a point is located $|r|$ units along the ray opposite the terminal side of θ.**

In Figure 7.20(c), r and θ are both negative: $r = -3$ and $\theta = -\dfrac{3\pi}{4}$. Because θ is negative, we draw a $-\dfrac{3\pi}{4}$ (or $-135°$) angle clockwise from the polar axis.

Because r is negative, the point is located on the ray opposite the terminal side of θ.

Our observations indicate the importance of the sign of r in locating $P = (r, \theta)$ in polar coordinates.

> **The Sign of r and a Point's Location in Polar Coordinates**
>
> The point $P = (r, \theta)$ is located $|r|$ units from the pole. If $r > 0$, the point lies on the terminal side of θ. If $r < 0$, the point lies along the ray opposite the terminal side of θ. If $r = 0$, the point lies at the pole, regardless of the value of θ.

EXAMPLE 1 Plotting Points in a Polar Coordinate System

Plot the points with the following polar coordinates:

a. $(2, 135°)$ **b.** $\left(-3, \dfrac{3\pi}{2}\right)$ **c.** $\left(-1, -\dfrac{\pi}{4}\right)$.

Solution

a. To plot the point $(r, \theta) = (2, 135°)$, begin with the $135°$ angle. Because $135°$ is a positive angle, draw $\theta = 135°$ counterclockwise from the polar axis. Now consider $r = 2$. Because $r > 0$, plot the point by going out two units on the terminal side of θ. Figure 7.21(a) shows the point.

b. To plot the point $(r, \theta) = \left(-3, \dfrac{3\pi}{2}\right)$, begin with the $\dfrac{3\pi}{2}$ angle. Because $\dfrac{3\pi}{2}$ is a positive angle, we draw $\theta = \dfrac{3\pi}{2}$ counterclockwise from the polar axis. Now consider $r = -3$. Because $r < 0$, plot the point by going out three units along the ray *opposite* the terminal side of θ. Figure 7.21(b) shows the point.

c. To plot the point $(r, \theta) = \left(-1, -\dfrac{\pi}{4}\right)$, begin with the $-\dfrac{\pi}{4}$ angle. Because $-\dfrac{\pi}{4}$ is a negative angle, draw $\theta = -\dfrac{\pi}{4}$ clockwise from the polar axis.

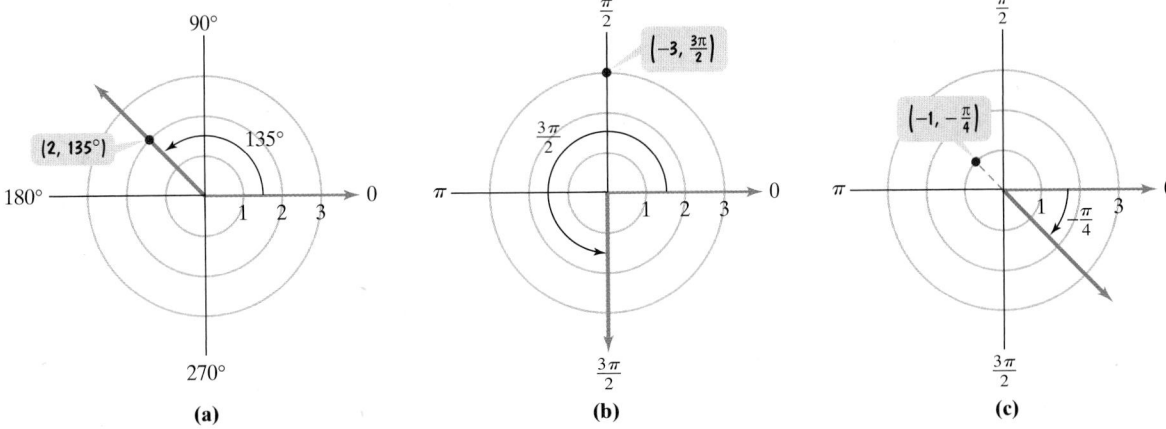

Figure 7.21 Plotting points

Now consider $r = -1$. Because $r < 0$, plot the point by going out one unit along the ray *opposite* the terminal side of θ. Figure 7.21(c) shows the point.

> **Check Point 1** Plot the points with the following polar coordinates:
>
> **a.** $(3, 315°)$ **b.** $(-2, \pi)$ **c.** $\left(-1, -\dfrac{\pi}{2}\right)$.

2 Find multiple sets of polar coordinates for a given point.

Multiple Representation of Points in the Polar Coordinate System

In rectangular coordinates, each point (x, y) has exactly one representation. By contrast, any point in polar coordinates can be represented in infinitely many ways. For example,

$$(r, \theta) = (r, \theta + 2\pi) \quad \text{and} \quad (r, \theta) = (-r, \theta + \pi).$$

Adding 1 revolution, or 2π radians, to the angle does not change the point's location.

Adding $\frac{1}{2}$ revolution, or π radians, to the angle and replacing r with $-r$ does not change the point's location.

Discovery

Illustrate the statements in the voice balloons by plotting:

a. $\left(1, \dfrac{\pi}{2}\right)$ and $\left(1, \dfrac{5\pi}{2}\right)$.

b. $\left(3, \dfrac{\pi}{4}\right)$ and $\left(-3, \dfrac{5\pi}{4}\right)$.

Thus, to find two other representations for the point (r, θ),

- Add 2π to the angle and do not change r.
- Add π to the angle and replace r with $-r$.

Continually adding or subtracting 2π in either of these representations does not change the point's location.

> **Multiple Representation of Points**
> If n is any integer, the point (r, θ) can be represented as
> $$(r, \theta) = (r, \theta + 2n\pi) \quad \text{or} \quad (r, \theta) = (-r, \theta + \pi + 2n\pi).$$

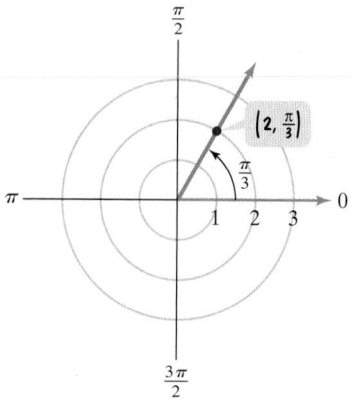

Figure 7.22 Finding other representations of a given point

EXAMPLE 2 Finding Other Polar Coordinates for a Given Point

The point $\left(2, \dfrac{\pi}{3}\right)$ is plotted in Figure 7.22. Find another representation of this point in which:

a. r is positive and $2\pi < \theta < 4\pi$.
b. r is negative and $0 < \theta < 2\pi$.
c. r is positive and $-2\pi < \theta < 0$.

Solution

a. Add 2π to the angle and do not change r.

$$\left(2, \frac{\pi}{3}\right) = \left(2, \frac{\pi}{3} + 2\pi\right) = \left(2, \frac{\pi}{3} + \frac{6\pi}{3}\right) = \left(2, \frac{7\pi}{3}\right)$$

b. Add π to the angle and replace r with $-r$.

$$\left(2, \frac{\pi}{3}\right) = \left(-2, \frac{\pi}{3} + \pi\right) = \left(-2, \frac{\pi}{3} + \frac{3\pi}{3}\right) = \left(-2, \frac{4\pi}{3}\right)$$

c. Subtract 2π from the angle and do not change r.

$$\left(2, \frac{\pi}{3}\right) = \left(2, \frac{\pi}{3} - 2\pi\right) = \left(2, \frac{\pi}{3} - \frac{6\pi}{3}\right) = \left(2, -\frac{5\pi}{3}\right)$$

Check Point 2 Find another representation of $\left(5, \dfrac{\pi}{4}\right)$ in which:

a. r is positive and $2\pi < \theta < 4\pi$.
b. r is negative and $0 < \theta < 2\pi$.
c. r is positive and $-2\pi < \theta < 0$.

Relations between Polar and Rectangular Coordinates

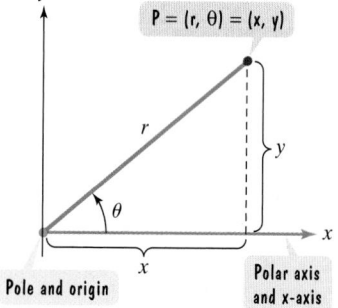

Figure 7.23 Polar and rectangular coordinate systems

We now consider both polar and rectangular coordinates simultaneously. Figure 7.23 shows the two coordinate systems. The polar axis coincides with the positive x-axis and the pole coincides with the origin. A point P, other than the origin, has rectangular coordinates (x, y) and polar coordinates (r, θ), as indicated in the figure. We wish to find equations relating the two sets of coordinates. From the figure, we see that

$$x^2 + y^2 = r^2$$

$$\sin\theta = \frac{y}{r} \qquad \cos\theta = \frac{x}{r} \qquad \tan\theta = \frac{y}{x}.$$

These relationships hold when P is in any quadrant and when $r > 0$ or $r < 0$.

Relations between Polar and Rectangular Coordinates

$$x = r\cos\theta$$
$$y = r\sin\theta$$
$$x^2 + y^2 = r^2$$
$$\tan\theta = \frac{y}{x}$$

3 Convert a point from polar to rectangular coordinates.

Point Conversion from Polar to Rectangular Coordinates

To convert a point from polar coordinates (r, θ) to rectangular coordinates (x, y), use the formulas $x = r \cos \theta$ and $y = r \sin \theta$.

EXAMPLE 3   Polar-to-Rectangular Point Conversion

Find the rectangular coordinates of the points with the following polar coordinates:

a. $\left(2, \dfrac{3\pi}{2}\right)$ **b.** $\left(-8, \dfrac{\pi}{3}\right)$.

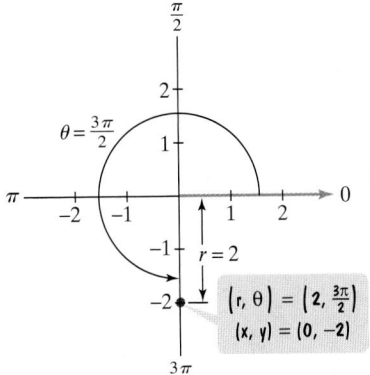

$\theta = \dfrac{3\pi}{2}$

$r = 2$

$(r, \theta) = \left(2, \frac{3\pi}{2}\right)$
$(x, y) = (0, -2)$

Figure 7.24 Converting $\left(2, \dfrac{3\pi}{2}\right)$ to rectangular coordinates

Solution   We find (x, y) by substituting the given values for r and θ into $x = r \cos \theta$ and $y = r \sin \theta$.

a. We begin with the rectangular coordinates of the point $(r, \theta) = \left(2, \dfrac{3\pi}{2}\right)$.

$$x = r \cos \theta = 2 \cos \frac{3\pi}{2} = 2 \cdot 0 = 0$$

$$y = r \sin \theta = 2 \sin \frac{3\pi}{2} = 2(-1) = -2$$

The rectangular coordinates of $\left(2, \dfrac{3\pi}{2}\right)$ are $(0, -2)$. See Figure 7.24.

b. We now find the rectangular coordinates of the point $(r, \theta) = \left(-8, \dfrac{\pi}{3}\right)$.

$$x = r \cos \theta = -8 \cos \frac{\pi}{3} = -8\left(\frac{1}{2}\right) = -4$$

$$y = r \sin \theta = -8 \sin \frac{\pi}{3} = -8\left(\frac{\sqrt{3}}{2}\right) = -4\sqrt{3}$$

The rectangular coordinates of $\left(-8, \dfrac{\pi}{3}\right)$ are $(-4, -4\sqrt{3})$.

Technology

Some graphing utilities can convert a point from polar coordinates to rectangular coordinates. Consult your manual. The screen on the right verifies the polar-rectangular conversion in Example 3(a). It shows that the rectangular coordinates of $(r, \theta) = \left(2, \dfrac{3\pi}{2}\right)$ are $(0, -2)$.

Notice that the x- and y-coordinates are displayed separately.

P►Rx(2,3π/2)
                    0
P►Ry(2,3π/2)
                   -2

> **Check Point 3** Find the rectangular coordinates of the points with the following polar coordinates:
>
> **a.** $(3, \pi)$ **b.** $\left(-10, \dfrac{\pi}{6}\right)$.

4 Convert a point from rectangular to polar coordinates.

Point Conversion from Rectangular to Polar Coordinates

Conversion from rectangular coordinates (x, y) to polar coordinates (r, θ) is a bit more complicated. Keep in mind that there are infinitely many representations for a point in polar coordinates. If the point (x, y) lies in one of the four quadrants, we will use a representation in which

- r is positive, and
- θ is the smallest positive angle that lies in the same quadrant as (x, y).

These conventions provide the following procedure:

> **Converting a Point from Rectangular to Polar Coordinates**
> **($r > 0$ and $0 \le \theta < 2\pi$)**
>
> **1.** Plot the point (x, y).
>
> **2.** Find r by computing the distance from the origin to (x, y): $r = \sqrt{x^2 + y^2}$.
>
> **3.** Find θ using $\tan \theta = \dfrac{y}{x}$ with θ lying in the same quadrant as (x, y).

EXAMPLE 4   Rectangular-to-Polar Point Conversion

Find polar coordinates of a point whose rectangular coordinates are $(-1, \sqrt{3})$.

Solution   We begin with $(x, y) = (-1, \sqrt{3})$ and use our three-step procedure to find a set of polar coordinates (r, θ).

Step 1   Plot the point (x, y).   The point $(-1, \sqrt{3})$ is plotted in quadrant II in Figure 7.25.

Step 2   Find r by computing the distance from the origin to (x, y).

$$r = \sqrt{x^2 + y^2} = \sqrt{(-1)^2 + (\sqrt{3})^2} = \sqrt{1 + 3} = \sqrt{4} = 2$$

Step 3   Find θ using $\tan \theta = \dfrac{y}{x}$ with θ lying in the same quadrant as (x, y).

$$\tan \theta = \frac{y}{x} = \frac{\sqrt{3}}{-1} = -\sqrt{3}$$

We know that $\tan \dfrac{\pi}{3} = \sqrt{3}$. Because θ lies in quadrant II,

$$\theta = \pi - \frac{\pi}{3} = \frac{3\pi}{3} - \frac{\pi}{3} = \frac{2\pi}{3}.$$

Polar coordinates of $(-1, \sqrt{3})$ are $(r, \theta) = \left(2, \dfrac{2\pi}{3}\right)$.

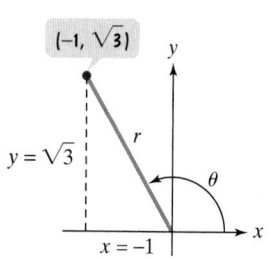

Figure 7.25 Converting $(-1, \sqrt{3})$ to polar coordinates

Technology

The screen shows the rectangular-polar conversion for $(-1, \sqrt{3})$ on a graphing utility. In Example 4, we showed that the polar coordinates of

$(x, y) = (-1, \sqrt{3})$ are $(r, \theta) = \left(2, \dfrac{2\pi}{3}\right)$. Using

$\dfrac{2\pi}{3} \approx 2.09439510239$ verifies that our

conversion is correct. Notice that the r- and (approximate) θ-coordinates are displayed separately.

```
R▸Pr(-1,√(3))
                    2
R▸Pθ(-1,√(3))
          2.094395102
```

Check Point 4 Find polar coordinates of a point whose rectangular coordinates are $(1, -\sqrt{3})$.

If a point (x, y) lies on a positive or negative axis, we use a representation in which

- r is positive, and
- θ is the smallest quadrantal angle that lies on the same positive or negative axis as (x, y).

In these cases, you can find r and θ by plotting (x, y) and inspecting the figure. Let's see how this is done.

EXAMPLE 5 Rectangular-to-Polar Point Conversion

Find polar coordinates of a point whose rectangular coordinates are $(-2, 0)$.

Solution We begin with $(x, y) = (-2, 0)$ and find a set of polar coordinates (r, θ).

Step 1 Plot the point (x, y). The point $(-2, 0)$ is plotted in Figure 7.26.

Step 2 Find r, the distance from the origin to (x, y). Can you tell by looking at Figure 7.26 that this distance is 2?

$$r = \sqrt{x^2 + y^2} = \sqrt{(-2)^2 + 0^2} = \sqrt{4} = 2$$

Step 3 Find θ with θ lying on the same positive or negative axis as (x, y). The point $(-2, 0)$ is on the negative x-axis. Thus, θ lies on the negative x-axis and $\theta = \pi$. Polar coordinates of $(-2, 0)$ are $(2, \pi)$.

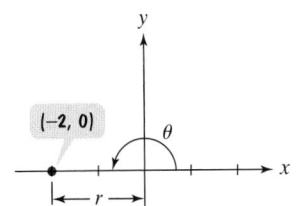

Figure 7.26 Converting $(-2, 0)$ to polar coordinates

Check Point 5 Find polar coordinates of a point whose rectangular coordinates are $(0, -4)$.

5 Convert an equation from rectangular to polar coordinates.

Equation Conversion from Rectangular to Polar Coordinates

A **polar equation** is an equation whose variables are r and θ. Two examples of polar equations are

$$r = \dfrac{5}{\cos\theta + \sin\theta} \quad \text{and} \quad r = \csc\theta.$$

To convert a rectangular equation in x and y to a polar equation in r and θ, replace x with $r \cos \theta$ and y with $r \sin \theta$.

EXAMPLE 6 Converting an Equation from Rectangular to Polar Coordinates

Convert $x + y = 5$ to a polar equation.

Solution   Our goal is to obtain an equation in which the variables are r and θ rather than x and y. We use $x = r \cos \theta$ and $y = r \sin \theta$.

$$x + y = 5 \quad \text{\textit{This is the given equation in rectangular coordinates.}}$$
$$r \cos \theta + r \sin \theta = 5 \quad \text{\textit{Replace x with r cos }}\theta\text{\textit{ and y with r sin }}\theta.$$

Thus, the polar equation for $x + y = 5$ is $r \cos \theta + r \sin \theta = 5$. We can express this polar equation in a number of equivalent ways, including an equation that gives r in terms of θ.

$$r(\cos \theta + \sin \theta) = 5 \qquad \text{\textit{Factor out r.}}$$

$$r = \frac{5}{\cos \theta + \sin \theta} \qquad \text{\textit{Divide both sides of the equation by cos }}\theta\text{\textit{ + sin }}\theta.$$

> **Check Point 6**   Convert $3x - y = 6$ to a polar equation. Express the polar equation with r in terms of θ.

6 Convert an equation from polar to rectangular coordinates.

Equation Conversion from Polar to Rectangular Coordinates

When we convert an equation from polar to rectangular coordinates, our goal is to obtain an equation in which the variables are x and y rather than r and θ. We use one or more of the following equations:

$$r^2 = x^2 + y^2 \qquad r \cos \theta = x \qquad r \sin \theta = y \qquad \tan \theta = \frac{y}{x}.$$

To use these equations, it is sometimes necessary to do something to the given polar equation. This could include squaring both sides, using an identity, taking the tangent of both sides, or multiplying both sides by r.

EXAMPLE 7 Converting Equations from Polar to Rectangular Form

Convert each polar equation to a rectangular equation in x and y:

a. $r = 3$ **b.** $\theta = \dfrac{\pi}{4}$ **c.** $r = \csc \theta$.

Solution

a. We use $r^2 = x^2 + y^2$ to convert the polar equation $r = 3$ to a rectangular equation.

$$r = 3 \quad \text{This is the given polar equation.}$$
$$r^2 = 9 \quad \text{Square both sides.}$$
$$x^2 + y^2 = 9 \quad \text{Use } r^2 = x^2 + y^2 \text{ on the left side.}$$

The rectangular equation for $r = 3$ is $x^2 + y^2 = 9$.

b. We use $\tan\theta = \dfrac{y}{x}$ to convert the polar equation $\theta = \dfrac{\pi}{4}$ to a rectangular equation in x and y.

$$\theta = \frac{\pi}{4} \qquad \text{This is the given polar equation.}$$

$$\tan\theta = \tan\frac{\pi}{4} \qquad \text{Take the tangent of both sides.}$$

$$\tan\theta = 1 \qquad \tan\frac{\pi}{4} = 1$$

$$\frac{y}{x} = 1 \qquad \text{Use } \tan\theta = \frac{y}{x} \text{ on the left side.}$$

$$y = x \qquad \text{Multiply both sides by x.}$$

The rectangular equation for $\theta = \dfrac{\pi}{4}$ is $y = x$.

c. We use $r\sin\theta = y$ to convert the polar equation $r = \csc\theta$ to a rectangular equation. To do this, we express the cosecant in terms of the sine.

$$r = \csc\theta \qquad \text{This is the given polar equation.}$$

$$r = \frac{1}{\sin\theta} \qquad \csc\theta = \frac{1}{\sin\theta}$$

$$r\sin\theta = 1 \qquad \text{Multiply both sides by sin } \theta.$$
$$y = 1 \qquad \text{Use } r\sin\theta = y \text{ on the left side.}$$

The rectangular equation for $r = \csc\theta$ is $y = 1$.

Converting a polar equation to a rectangular equation may be a useful way to develop or check a graph. For example, the graph of the polar equation $r = 3$ consists of all points that are three units from the pole. Thus, the graph is a circle centered at the pole with radius $= 3$. The rectangular equation for $r = 3$, namely $x^2 + y^2 = 9$, has precisely the same graph (see Figure 7.27). We will discuss graphs of polar equations in the next section.

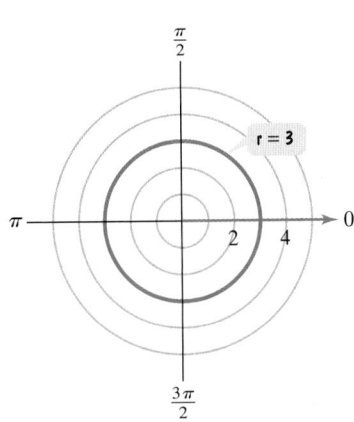

Figure 7.27 The equations $r = 3$ and $x^2 + y^2 = 9$ have the same graph

Check Point 7 Convert each polar equation to a rectangular equation in x and y:

a. $r = 4$ **b.** $\theta = \dfrac{3\pi}{4}$ **c.** $r = \sec\theta$.

EXERCISE SET 7.3

Practice Exercises

In Exercises 1–10, indicate if the point with the given polar coordinates is represented by A, B, C, or D on the graph.

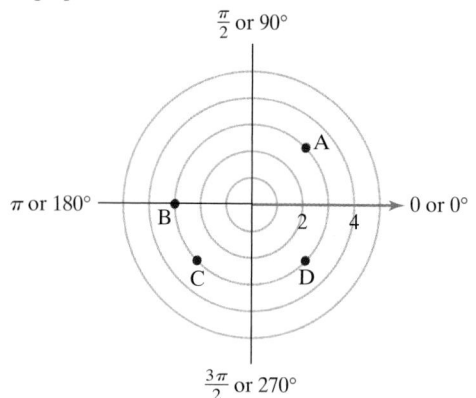

1. $(3, 225°)$
2. $(3, 315°)$
3. $\left(-3, \dfrac{5\pi}{4}\right)$
4. $\left(-3, \dfrac{\pi}{4}\right)$
5. $(3, \pi)$
6. $(-3, 0)$
7. $(3, -135°)$
8. $(3, -315°)$
9. $\left(-3, -\dfrac{3\pi}{4}\right)$
10. $\left(-3, -\dfrac{5\pi}{4}\right)$

In Exercises 11–20, use a polar coordinate system like the one shown for Exercises 1–10 to plot each point with the given polar coordinates.

11. $(2, 45°)$
12. $(1, 45°)$
13. $(3, 90°)$
14. $(2, 270°)$
15. $\left(3, \dfrac{4\pi}{3}\right)$
16. $\left(3, \dfrac{7\pi}{6}\right)$
17. $(-1, \pi)$
18. $\left(-1, \dfrac{3\pi}{2}\right)$
19. $\left(-2, -\dfrac{\pi}{2}\right)$
20. $(-3, -\pi)$

In Exercises 21–26, use a polar coordinate system like the one shown for Exercises 1–10 to plot each point with the given polar coordinates. Then find another representation (r, θ) of this point in which:

a. $r > 0, \quad 2\pi < \theta < 4\pi$
b. $r < 0, \quad 0 < \theta < 2\pi$
c. $r > 0, \quad -2\pi < \theta < 0$

21. $\left(5, \dfrac{\pi}{6}\right)$
22. $\left(8, \dfrac{\pi}{6}\right)$
23. $\left(10, \dfrac{3\pi}{4}\right)$
24. $\left(12, \dfrac{2\pi}{3}\right)$
25. $\left(4, \dfrac{\pi}{2}\right)$
26. $(6, \pi)$

In Exercises 27–32, select the representations that do not change the location of the given point.

27. $(7, 140°)$
 a. $(-7, 320°)$ **b.** $(-7, -40°)$
 c. $(-7, 220°)$ **d.** $(7, -220°)$

28. $(4, 120°)$
 a. $(-4, 300°)$ **b.** $(-4, -240°)$
 c. $(4, -240°)$ **d.** $(4, 480°)$

29. $\left(2, -\dfrac{3\pi}{4}\right)$
 a. $\left(2, -\dfrac{7\pi}{4}\right)$ **b.** $\left(2, \dfrac{5\pi}{4}\right)$
 c. $\left(-2, -\dfrac{\pi}{4}\right)$ **d.** $\left(-2, -\dfrac{7\pi}{4}\right)$

30. $\left(-2, \dfrac{7\pi}{6}\right)$
 a. $\left(-2, -\dfrac{5\pi}{6}\right)$ **b.** $\left(-2, -\dfrac{\pi}{6}\right)$
 c. $\left(2, -\dfrac{\pi}{6}\right)$ **d.** $\left(2, \dfrac{\pi}{6}\right)$

31. $\left(-5, -\dfrac{\pi}{4}\right)$
 a. $\left(-5, \dfrac{7\pi}{4}\right)$ **b.** $\left(5, -\dfrac{5\pi}{4}\right)$
 c. $\left(-5, \dfrac{11\pi}{4}\right)$ **d.** $\left(5, \dfrac{\pi}{4}\right)$

32. $(-6, 3\pi)$
 a. $(6, 2\pi)$ **b.** $(6, -\pi)$
 c. $(-6, \pi)$ **d.** $(-6, -2\pi)$

In Exercises 33–40, polar coordinates of a point are given. Find the rectangular coordinates of each point.

33. $(4, 90°)$
34. $(6, 180°)$
35. $\left(2, \dfrac{\pi}{3}\right)$
36. $\left(2, \dfrac{\pi}{6}\right)$
37. $\left(-4, \dfrac{\pi}{2}\right)$
38. $\left(-6, \dfrac{3\pi}{2}\right)$
39. $(7.4, 2.5)$
40. $(8.3, 4.6)$

In Exercises 41–48, the rectangular coordinates of a point are given. Find polar coordinates of each point.

41. $(-2, 2)$
42. $(2, -2)$
43. $(2, -2\sqrt{3})$
44. $(-2\sqrt{3}, 2)$
45. $(-\sqrt{3}, -1)$
46. $(-1, -\sqrt{3})$
47. $(5, 0)$
48. $(0, -6)$

In Exercises 49–58, convert each rectangular equation to a polar equation.

49. $3x + y = 7$ (Express r in terms of θ.)

50. $x + 5y = 8$ (Express r in terms of θ.)

51. $x = 7$ **52.** $y = 3$

53. $x^2 + y^2 = 9$ **54.** $x^2 + y^2 = 16$

55. $x^2 + y^2 = 4x$ **56.** $x^2 + y^2 = 6x$

57. $y^2 = 6x$ **58.** $x^2 = 6y$

In Exercises 59–72, convert each polar equation to a rectangular equation.

59. $r = 8$ **60.** $r = 10$

61. $\theta = \dfrac{\pi}{2}$ **62.** $\theta = \dfrac{\pi}{3}$

63. $r \sin \theta = 3$ **64.** $r \cos \theta = 7$

65. $r = 4 \csc \theta$ **66.** $r = 6 \sec \theta$

67. $r = \sin \theta$ **68.** $r = \cos \theta$

69. $r = 6 \cos \theta + 4 \sin \theta$ **70.** $r = 8 \cos \theta + 2 \sin \theta$

71. $r^2 \sin 2\theta = 2$ **72.** $r^2 \cos 2\theta = 2$

Application Exercises

Use the figure of the merry-go-round to solve Exercises 73–74. There are four circles of horses. Each circle is three feet from the next circle. The radius of the inner circle is 6 feet.

73. If a horse in the outer circle is $\frac{2}{3}$ of the way around the merry-go-round, give its polar coordinates.

74. If a horse in the inner circle is $\frac{5}{6}$ of the way around the merry-go-round, give its polar coordinates.

The wind is blowing at 10 knots. Sailboat racers look for a sailing angle to the 10-knot wind that produces maximum sailing speed. In this application, (r, θ) describes the sailing speed, r, in knots, at an angle θ to the 10-knot wind. Use this information to solve Exercises 75–77.

75. Interpret the polar coordinates: $(6.3, 50°)$.

76. Interpret the polar coordinates: $(7.4, 85°)$.

77. Four points in this 10-knot-wind situation are $(6.3, 50°)$, $(7.4, 85°)$, $(7.5, 105°)$, $(7.3, 135°)$. Based on these points,

which sailing angle to the 10-knot wind would you recommend to a serious sailboat racer? What sailing speed is achieved at this angle?

Writing in Mathematics

78. Explain how to plot (r, θ) if $r > 0$ and $\theta > 0$.

79. Explain how to plot (r, θ) if $r < 0$ and $\theta > 0$.

80. If you are given polar coordinates of a point, explain how to find two additional sets of polar coordinates for the point.

81. Explain how to convert a point from polar to rectangular coordinates. Provide an example with your explanation.

82. Explain how to convert a point from rectangular to polar coordinates. Provide an example with your explanation.

83. Explain how to convert from a rectangular equation to a polar equation.

84. In converting $r = 5$ from a polar equation to a rectangular equation, describe what should be done to both sides of the equation and why this should be done.

85. In converting $r = \sin \theta$ from a polar equation to a rectangular equation, describe what should be done to both sides of the equation and why this should be done.

86. Suppose that (r, θ) describes the sailing speed, r, in knots, at an angle θ to a wind blowing at 20 knots. You have a list of all ordered pairs (r, θ) for integral angles from $\theta = 0°$ to $\theta = 180°$. Describe a way to present this information so that a serious sailboat racer can visualize sailing speeds at different sailing angles to the wind.

Technology Exercises

In Exercises 87–89, polar coordinates of a point are given. Use a graphing utility to find the rectangular coordinates of each point to three decimal places.

87. $\left(4, \dfrac{2\pi}{3}\right)$ **88.** $(5.2, 1.7)$

89. $(-4, 1.088)$

In Exercises 90–92, the rectangular coordinates of a point are given. Use a graphing utility to find polar coordinates of each point to three decimal places.

90. $(-5, 2)$ **91.** $\left(\sqrt{5}, 2\right)$

92. $(-4.308, -7.529)$

Critical Thinking Exercises

93. Prove that the distance, d, between two points with polar coordinates (r_1, θ_1) and (r_2, θ_2) is
$$d = \sqrt{r_1^2 + r_2^2 - 2r_1 r_2 \cos(\theta_2 - \theta_1)}.$$

94. Use the formula in Exercise 93 to find the distance between $\left(2, \dfrac{5\pi}{6}\right)$ and $\left(4, \dfrac{\pi}{6}\right)$. Express the answer in simplified radical form.

95. Convert $r = 4 \cos \theta$ from a polar equation to a rectangular equation. Use the rectangular equation to give the center and the radius.

SECTION 7.4   *Graphs of Polar Equations*

Objectives

1. Graph polar equations.
2. Use symmetry to graph polar equations.

The America's Cup is the supreme event in ocean sailing. Competition is fierce and the costs are huge. Competitors look to mathematics to provide the critical innovation that can make the difference between winning and losing. In this section's exercise set, you will see how graphs of polar equations play a role in sailing faster using mathematics.

Using Polar Grids to Graph Polar Equations

Recall that a **polar equation** is an equation whose variables are r and θ. The **graph of a polar equation** is the set of all points whose polar coordinates satisfy the equation. We use **polar grids** like the one shown in Figure 7.28 to graph polar equations. The grid consists of circles with centers at the pole. This polar grid shows five such circles. A polar grid also shows lines passing through the pole. In this grid, each line represents an angle for which we know the exact values of the trigonometric functions.

Many polar coordinate grids show more circles and more lines through the pole than in Figure 7.28. See if your campus bookstore has paper with polar grids and use the polar graph paper throughout this section.

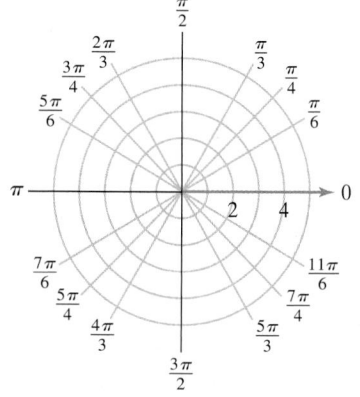

Figure 7.28 A polar coordinate grid

1 Graph polar equations.

Graphing a Polar Equation by Point Plotting

One method for graphing a polar equation such as $r = 4 \cos \theta$ is the **point-plotting method.** First, we make a table of values that satisfy the equation. Next, we plot these ordered pairs as points in the polar coordinate system. Finally, we connect the points with a smooth curve. This often gives us a picture of all ordered pairs (r, θ) that satisfy the equation.

EXAMPLE 1   Graphing an Equation Using the Point-Plotting Method

Graph the polar equation $r = 4 \cos \theta$ with θ in radians.

Solution We construct a partial table of coordinates using multiples of $\frac{\pi}{6}$. Then we plot the points and join them with a smooth curve, as shown in Figure 7.29.

θ	$r = 4\cos\theta$	(r, θ)
0	$4\cos 0 = 4\cdot 1 = 4$	$(4, 0)$
$\dfrac{\pi}{6}$	$4\cos\dfrac{\pi}{6} = 4\cdot\dfrac{\sqrt{3}}{2} = 2\sqrt{3} \approx 3.5$	$\left(3.5, \dfrac{\pi}{6}\right)$
$\dfrac{\pi}{3}$	$4\cos\dfrac{\pi}{3} = 4\cdot\dfrac{1}{2} = 2$	$\left(2, \dfrac{\pi}{3}\right)$
$\dfrac{\pi}{2}$	$4\cos\dfrac{\pi}{2} = 4\cdot 0 = 0$	$\left(0, \dfrac{\pi}{2}\right)$
$\dfrac{2\pi}{3}$	$4\cos\dfrac{2\pi}{3} = 4\left(-\dfrac{1}{2}\right) = -2$	$\left(-2, \dfrac{2\pi}{3}\right)$
$\dfrac{5\pi}{6}$	$4\cos\dfrac{5\pi}{6} = 4\left(-\dfrac{\sqrt{3}}{2}\right) = -2\sqrt{3} \approx -3.5$	$\left(-3.5, \dfrac{5\pi}{6}\right)$
π	$4\cos\pi = 4(-1) = -4$	$(-4, \pi)$

Values of r repeat.

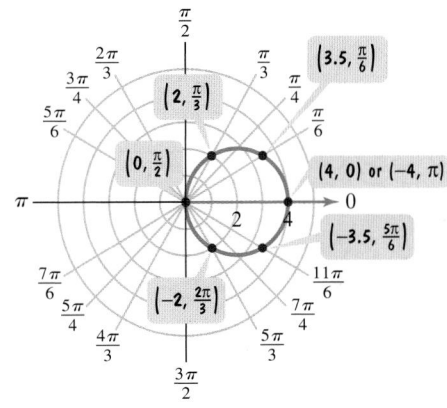

Figure 7.29 The graph of $r = 4\cos\theta$

Technology

A graphing utility can be used to obtain the graph of a polar equation. Use the polar mode with angle measure in radians. You must enter the minimum and maximum values for θ and an increment setting for θ, called θ step. θ step determines the number of points that the graphing utility will plot. Make θ step relatively small so that a significant number of points are plotted.

Shown is the graph of $r = 4\cos\theta$ in a $[-7.5, 7.5, 1]$ by $[-5, 5, 1]$ viewing rectangle with

$$\theta\min = 0$$
$$\theta\max = 2\pi$$
$$\theta\text{step} = \frac{\pi}{48}.$$

A square setting was used.

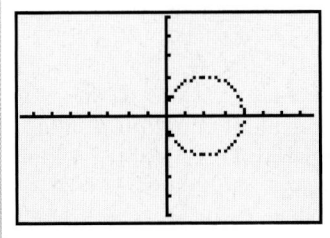

The graph of $r = 4\cos\theta$ in Figure 7.29 looks like a circle of radius 2 whose center is at the point $(x, y) = (2, 0)$. We can verify this observation by changing the polar equation to a rectangular equation.

$$r = 4\cos\theta \qquad \text{This is the given polar equation.}$$

$$r^2 = 4r\cos\theta \qquad \text{Multiply both sides by } r.$$

$$x^2 + y^2 = 4x \qquad \text{Convert to rectangular coordinates:}$$
$$\text{$r^2 = x^2 + y^2$ and $r\cos\theta = x$.}$$

$$x^2 - 4x + y^2 = 0 \qquad \text{Subtract } 4x \text{ from both sides.}$$

$$x^2 - 4x + 4 + y^2 = 4 \qquad \text{Complete the square on } x:$$
$$\tfrac{1}{2}(-4) = -2 \text{ and } (-2)^2 = 4.$$
$$\text{Add 4 to both sides.}$$

$$(x - 2)^2 + y^2 = 2^2 \qquad \text{Factor.}$$

This last equation is the standard form of the equation of a circle, $(x - h)^2 + (y - k)^2 = r^2$, with radius r and center at (h, k). Thus, the radius is 2 and the center is at $(h, k) = (2, 0)$.

In general, circles have simpler equations in polar form than in rectangular form.

Circles in Polar Coordinates

The graphs of

$$r = a \cos \theta \quad \text{and} \quad r = a \sin \theta$$

are circles.

$$r = a \cos \theta \qquad\qquad r = a \sin \theta$$

Check Point 1 Graph the equation $r = 4 \sin \theta$ with θ in radians. Use multiples of $\dfrac{\pi}{6}$ from 0 to π to generate coordinates for points (r, θ).

2 Use symmetry to graph polar equations.

Graphing a Polar Equation Using Symmetry

If the graph of a polar equation exhibits symmetry, you may be able to graph it more quickly. Three types of symmetry can be helpful.

Tests for Symmetry in Polar Coordinates

Symmetry with Respect to the Polar Axis (x-Axis)

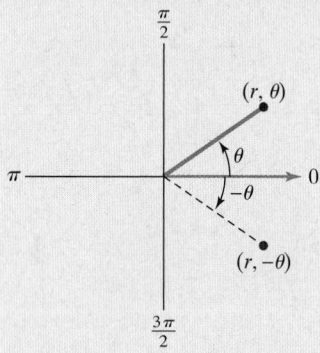

Replace θ with $-\theta$. If an equivalent equation results, the graph is symmetric with respect to the polar axis.

Symmetry with Respect to the Line $\theta = \dfrac{\pi}{2}$ (y-Axis)

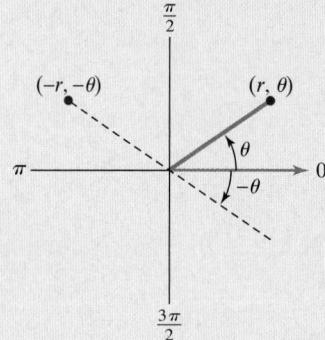

Replace (r, θ) with $(-r, -\theta)$. If an equivalent equation results, the graph is symmetric with respect to $\theta = \dfrac{\pi}{2}$.

Symmetry with Respect to the Pole (Origin)

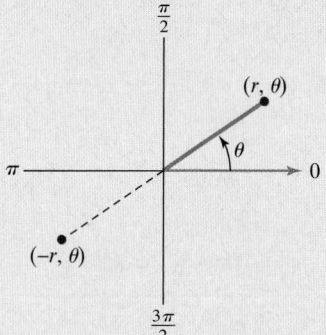

Replace r with $-r$. If an equivalent equation results, the graph is symmetric with respect to the pole.

If a polar equation passes a symmetry test, then its graph exhibits that symmetry. By contrast, if a polar equation fails a symmetry test, then its graph *may or may not* have that kind of symmetry. Thus, the graph of a polar equation may have a symmetry even if it fails a test for that particular symmetry. Nevertheless, the symmetry tests are useful. If we detect symmetry, we can obtain a graph of the equation by plotting fewer points.

EXAMPLE 2   Graphing a Polar Equation Using Symmetry

Check for symmetry and then graph the polar equation:

$$r = 1 - \cos \theta.$$

Solution   We apply each of the tests for symmetry.

Polar Axis: Replace θ with $-\theta$ in $r = 1 - \cos \theta$:

$$r = 1 - \cos(-\theta) \quad \text{Replace } \theta \text{ with } -\theta \text{ in } r = 1 - \cos \theta.$$

$$r = 1 - \cos \theta \qquad \text{The cosine function is even:}$$
$$\cos(-\theta) = \cos \theta.$$

Because the polar equation does not change when θ is replaced with $-\theta$, the graph is symmetric with respect to the polar axis.

The Line $\theta = \dfrac{\pi}{2}$: Replace (r, θ) with $(-r, -\theta)$ in $r = 1 - \cos \theta$:

$$-r = 1 - \cos(-\theta) \quad \text{Replace } r \text{ with } -r \text{ and } \theta \text{ with } -\theta \text{ in } r = 1 - \cos \theta.$$

$$-r = 1 - \cos \theta \qquad \cos(-\theta) = \cos \theta.$$

$$r = \cos \theta - 1 \qquad \text{Multiply both sides by } -1.$$

Because the polar equation $r = 1 - \cos \theta$ changes to $r = \cos \theta - 1$ when (r, θ) is replaced with $(-r, -\theta)$, the equation fails this symmetry test. The graph may or may not be symmetric with respect to the line $\theta = \dfrac{\pi}{2}$.

The Pole: Replace r with $-r$ in $r = 1 - \cos \theta$:

$$-r = 1 - \cos \theta \qquad \text{Replace } r \text{ with } -r.$$

$$r = \cos \theta - 1 \qquad \text{Multiply both sides by } -1.$$

Because the polar equation $r = 1 - \cos \theta$ changes to $r = \cos \theta - 1$ when r is replaced with $-r$, the equation fails this symmetry test. The graph may or may not be symmetric with respect to the pole.

Now we are ready to graph $r = 1 - \cos \theta$. Because the period of the cosine function is 2π, we need not consider values of θ beyond 2π. Recall that we discovered the graph of the equation $r = 1 - \cos \theta$ has symmetry with respect to the polar axis. Because the graph has this symmetry, we can obtain a complete graph by plotting fewer points. Let's start by finding the values of r for values of θ from 0 to π.

Technology

The graph of
$$r = 1 - \cos \theta$$
was obtained using a
$[-2, 2, 1]$ by $[-2, 2, 1]$ viewing rectangle and
$$\theta \min = 0, \quad \theta \max = 2\pi,$$
$$\theta \text{step} = \frac{\pi}{48}.$$

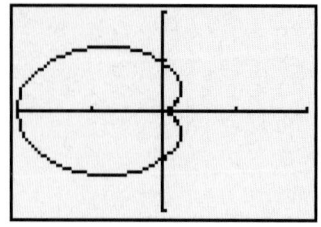

The values for r and θ are in the table above Figure 7.30. The TABLE feature on some graphing utilities is the most efficient way to create these values. The points in the table are plotted in Figure 7.30(a). Examine the graph. Keep in mind that the graph must be symmetric with respect to the polar axis. Thus, if we reflect the graph in Figure 7.30(a) about the polar axis, we will obtain a complete graph of $r = 1 - \cos \theta$. This graph is shown in Figure 7.30(b).

θ	0	$\dfrac{\pi}{6}$	$\dfrac{\pi}{3}$	$\dfrac{\pi}{2}$	$\dfrac{2\pi}{3}$	$\dfrac{5\pi}{6}$	π
r	0	0.13	0.5	1	1.5	1.87	2

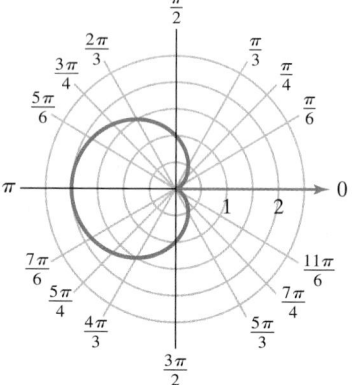

(a) The graph of $r = 1 - \cos \theta$ for $0 \le \theta \le \pi$

(b) A complete graph of $r = 1 - \cos \theta$

Figure 7.30 Graphing $r = 1 - \cos \theta$

Check Point 2 Check for symmetry and then graph the polar equation:
$$r = 1 + \cos \theta.$$

EXAMPLE 3 Graphing a Polar Equation

Graph the polar equation: $r = 1 + 2 \sin \theta$.

Solution We first check for symmetry.
$$r = 1 + 2 \sin \theta$$

Polar Axis	**The Line** $\theta = \dfrac{\pi}{2}$	**The Pole**
Replace θ with $-\theta$.	Replace (r, θ) with $(-r, -\theta)$.	Replace r with $-r$.
$r = 1 + 2 \sin(-\theta)$	$-r = 1 + 2 \sin(-\theta)$	$-r = 1 + 2 \sin \theta$
$r = 1 + 2(-\sin \theta)$	$-r = 1 - 2 \sin \theta$	$r = -1 - 2 \sin \theta$
$r = 1 - 2 \sin \theta$	$r = -1 + 2 \sin \theta$	

None of these equations are equivalent to $r = 1 + 2 \sin \theta$. Thus, the graph may or may not have each of these kinds of symmetry.

Now we are ready to graph $r = 1 + 2 \sin \theta$. Because the period of the sine function is 2π, we need not consider values of θ beyond 2π. We identify points on the graph of $r = 1 + 2 \sin \theta$ by assigning values to θ and calculating the corresponding values of r. The values for r and θ are in the tables above Figures 7.31(a), (b), and (c). The complete graph of $r = 1 + 2 \sin \theta$ is shown in Figure 7.31(c). The inner loop indicates that the graph passes through the pole twice.

θ	0	$\dfrac{\pi}{6}$	$\dfrac{\pi}{3}$	$\dfrac{\pi}{2}$	$\dfrac{2\pi}{3}$	$\dfrac{5\pi}{6}$	π
r	1	2	2.73	3	2.73	2	1

θ	$\dfrac{7\pi}{6}$	$\dfrac{4\pi}{3}$	$\dfrac{3\pi}{2}$
r	0	-0.73	-1

θ	$\dfrac{5\pi}{3}$	$\dfrac{11\pi}{6}$	2π
r	-0.73	0	1

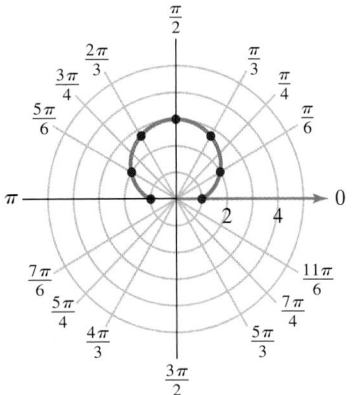

(a) The graph of $r = 1 + 2 \cos \theta$ for $0 \le \theta \le \pi$

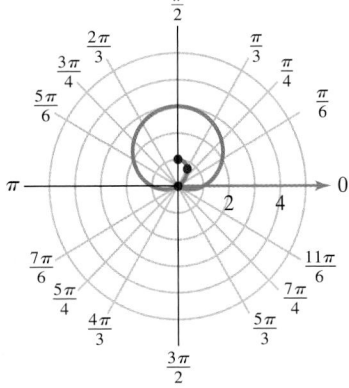

(b) The graph of $r = 1 + 2 \sin \theta$ for $0 \le \theta \le \dfrac{3\pi}{2}$

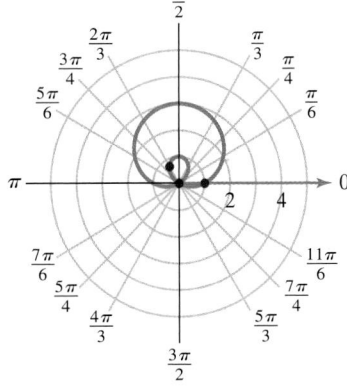

(c) The complete graph of $r = 1 + 2 \sin \theta$ for $0 \le \theta \le 2\pi$

Figure 7.31 Graphing $r = 1 + 2 \sin \theta$

We're not quite sure if the polar graph in Figure 7.31(c) looks like a snail. However, the graph is called a *limaçon*, which is a French word for snail. Limaçons come with and without inner loops.

Limaçons

The graphs of

$$r = a + b \sin \theta, \quad r = a - b \sin \theta,$$
$$r = a + b \cos \theta, \quad r = a - b \cos \theta, \quad a > 0, b > 0$$

are called **limaçons.** The ratio $\dfrac{a}{b}$ determines a limaçon's shape.

Inner loop if $\dfrac{a}{b} < 1$

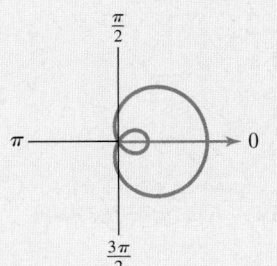

Heart-shaped if $\dfrac{a}{b} = 1$ and called cardioids

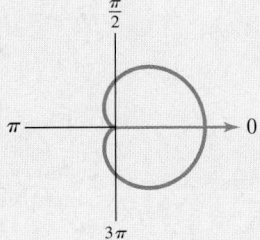

Dimpled with no inner loop if $1 < \dfrac{a}{b} < 2$

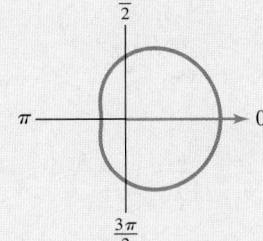

No dimple and no inner loop if $\dfrac{a}{b} \ge 2$

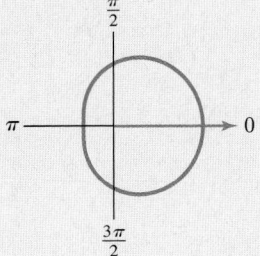

Check Point 3   Graph the polar equation: $r = 1 - 2 \sin \theta$.

EXAMPLE 4   Graphing a Polar Equation

Graph the polar equation: $r = 4 \sin 2\theta$.

Solution   We first check for symmetry.

$$r = 4 \sin 2\theta$$

Polar Axis

Replace θ with $-\theta$.

$r = 4 \sin 2(-\theta)$
$r = 4 \sin (-2\theta)$
$r = -4 \sin 2\theta$

Equation changes and fails this symmetry test.

The Line $\theta = \dfrac{\pi}{2}$

Replace (r, θ) with $(-r, -\theta)$.

$-r = 4 \sin 2(-\theta)$
$-r = 4 \sin (-2\theta)$
$-r = -4 \sin 2\theta$
$r = 4 \sin 2\theta$

Equation does not change.

The Pole

Replace r with $-r$.

$-r = 4 \sin 2\theta$
$r = -4 \sin 2\theta$

Equation changes and fails this symmetry test.

Thus, we can be sure that the graph is symmetric with respect to $\theta = \dfrac{\pi}{2}$. The graph may or may not be symmetric with respect to the polar axis or the pole.

Now we are ready to graph $r = 4 \sin 2\theta$. In Figure 7.32(a), we plot points on the graph of $r = 4 \sin 2\theta$ using values of θ from 0 to $\dfrac{\pi}{2}$ and the corresponding values of r. These coordinates are shown in the table in the margin. Because the graph is symmetric with respect to $\theta = \dfrac{\pi}{2}$, we can reflect the graph in Figure 7.32(a) about $\theta = \dfrac{\pi}{2}$ and obtain the graph from 0 to π. This graph is shown in Figure 7.32(b).

θ	0	$\dfrac{\pi}{6}$	$\dfrac{\pi}{4}$	$\dfrac{\pi}{3}$	$\dfrac{\pi}{2}$
r	0	3.46	4	3.46	0

(a) The graph of $r = 4 \sin 2\theta$ for $0 \le \theta \le \dfrac{\pi}{2}$

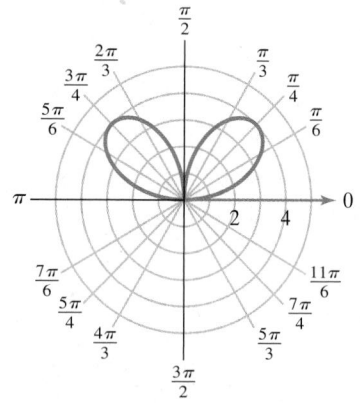

(b) Using symmetry to obtain the graph of $r = 4 \sin 2\theta$ for $0 \le \theta \le \pi$

Figure 7.32 Partial graphs of $r = 4 \sin 2\theta$

Now we can complete the graph of $r = 4 \sin 2\theta$. The values for r and θ above the graph in Figure 7.33(a) give us the graph for $0 \le \theta \le \dfrac{3\pi}{2}$. Because the graph is symmetric with respect to $\theta = \dfrac{\pi}{2}$, we can reflect the quadrant III portion of the graph in Figure 7.33(a) about $\theta = \dfrac{\pi}{2}$ and obtain the complete graph from 0 to 2π. This graph is shown in Figure 7.33(b).

θ	π	$\dfrac{7\pi}{6}$	$\dfrac{5\pi}{4}$	$\dfrac{4\pi}{3}$	$\dfrac{3\pi}{2}$
r	0	3.46	4	3.46	0

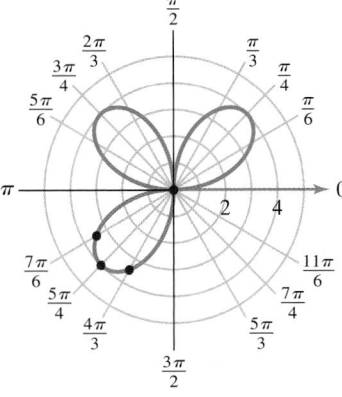

(a) The graph of $r = 4 \sin 2\theta$ for $0 \le \theta \le \dfrac{3\pi}{2}$

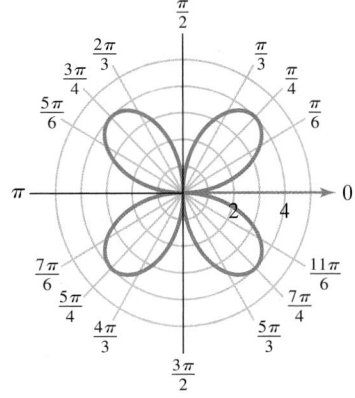

(b) Using symmetry to obtain the graph of $r = 4 \sin 2\theta$ for $0 \le \theta \le 2\pi$

Figure 7.33 Completing the graph of $r = 4 \sin 2\theta$

The curve in Figure 7.33(b) is called a **rose with four petals.**

Technology

The graph of
$$r = 4 \sin 2\theta$$
was obtained using a $[-4, 4, 1]$ by $[-4, 4, 1]$ viewing rectangle and
$$\theta \min = 0, \quad \theta \max = 2\pi,$$
$$\theta \text{ step} = \frac{\pi}{48}.$$

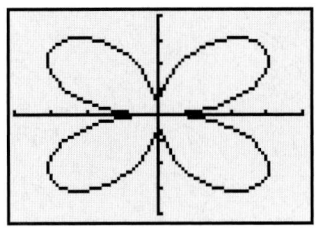

Rose Curves

The graphs of

$$r = a \sin n\theta \quad \text{and} \quad r = a \cos n\theta, \quad a \neq 0,$$

are called **rose curves**. If n is even, the rose has $2n$ petals. If n is odd, the rose has n petals.

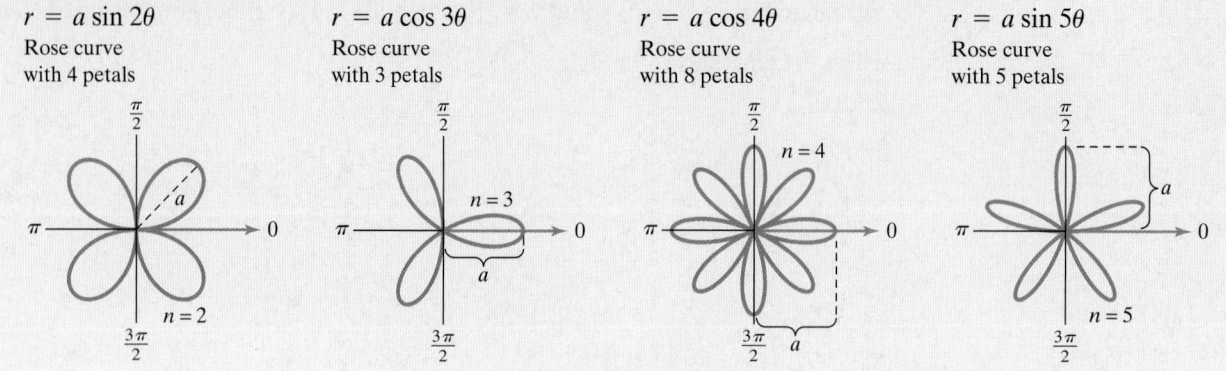

$r = a \sin 2\theta$
Rose curve
with 4 petals

$r = a \cos 3\theta$
Rose curve
with 3 petals

$r = a \cos 4\theta$
Rose curve
with 8 petals

$r = a \sin 5\theta$
Rose curve
with 5 petals

Check Point 4 Graph the polar equation: $r = 3 \cos 2\theta$.

EXAMPLE 5 Graphing a Polar Equation

Graph the polar equation: $r^2 = 4 \sin 2\theta$.

Solution We first check for symmetry.

$$r^2 = 4 \sin 2\theta$$

Polar Axis	**The Line $\theta = \dfrac{\pi}{2}$**	**The Pole**
Replace θ with $-\theta$.	Replace (r, θ) with $(-r, -\theta)$.	Replace r with $-r$.
$r^2 = 4 \sin 2(-\theta)$	$(-r)^2 = 4 \sin 2(-\theta)$	$(-r)^2 = 4 \sin 2\theta$
$r^2 = 4 \sin (-2\theta)$	$r^2 = 4 \sin (-2\theta)$	$r^2 = 4 \sin 2\theta$
$r^2 = -4 \sin 2\theta$	$r^2 = -4 \sin 2\theta$	
Equation changes and fails this symmetry test.	Equation changes and fails this symmetry test.	Equation does not change.

Thus, we can be sure that the graph is symmetric with respect to the pole. The graph may or may not be symmetric with respect to the polar axis or the line $\theta = \dfrac{\pi}{2}$.

Now we are ready to graph $r^2 = 4 \sin 2\theta$. In Figure 7.34(a), we plot points on the graph by using values of θ from 0 to $\dfrac{\pi}{2}$ and the corresponding values of r. These coordinates are shown in the table above Figure 7.34(a). Notice that the points in Figure 7.34(a) are shown for $r \geq 0$. Because the graph is symmetric with respect to the pole, we can reflect the graph in Figure 7.34(a) about the pole and obtain the graph in Figure 7.34(b).

θ	0	$\dfrac{\pi}{6}$	$\dfrac{\pi}{4}$	$\dfrac{\pi}{3}$	$\dfrac{\pi}{2}$
r	0	± 1.9	± 2	± 1.9	0

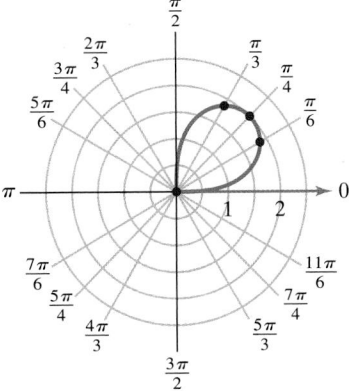

(a) The graph of $r^2 = 4 \sin 2\theta$ for $0 \le \theta \le \dfrac{\pi}{2}$

(b) Using symmetry with respect to the pole on the graph of $r^2 = 4 \sin 2\theta$

Figure 7.34 Graphing $r^2 = 4 \sin 2\theta$

Does Figure 7.34(b) show a complete graph of $r^2 = 4 \sin 2\theta$ or do we need to continue graphing in quadrants II and IV? If θ is in quadrant II, 2θ is in quadrant III or IV, where $\sin 2\theta$ is negative. Thus, $4 \sin 2\theta$ is negative. However, $r^2 = 4 \sin 2\theta$ and r^2 cannot be negative. This means that there are no points on the graph in quadrant II. The same observation applies to quadrant IV. Thus, Figure 7.34(b) shows the complete graph of $r^2 = 4 \sin 2\theta$.

The curve in Figure 7.34(b) is shaped like a propeller and is called a **lemniscate.**

Lemniscates

The graphs of

$$r^2 = a^2 \sin 2\theta \quad \text{and} \quad r^2 = a^2 \cos 2\theta, \quad a \ne 0$$

are called **lemniscates.**

$r^2 = a^2 \sin 2\theta$ is symmetric with respect to the pole.

$r^2 = a^2 \cos 2\theta$ is symmetric with respect to the polar axis, $\theta = \dfrac{\pi}{2}$, and the pole.

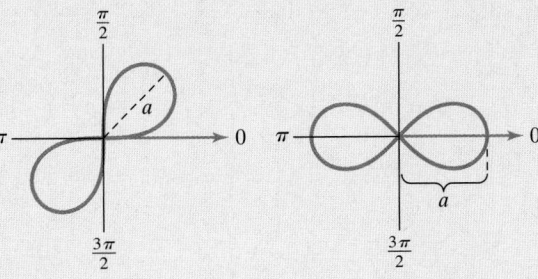

Check Point 5 Graph the polar equation: $r^2 = 4 \cos 2\theta$.

EXERCISE SET 7.4

Practice Exercises

In Exercises 1–6, the graph of a polar equation is given. Select the polar equation for each graph from the following options.

$$r = 2 \sin \theta, \quad r = 2 \cos \theta, \quad r = 1 + \sin \theta,$$

$$r = 1 - \sin \theta, \quad r = 3 \sin 2\theta, \quad r = 3 \sin 3\theta$$

1.

2.

3.

4.

5.

6.

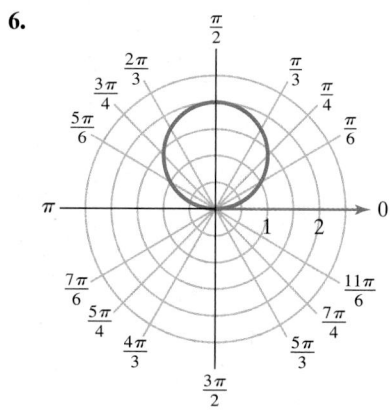

In Exercises 7–12, test for symmetry with respect to:

   a. *the polar axis*   ***b.*** *the line* $\theta = \dfrac{\pi}{2}$ ***c.*** *the pole.*

7. $r = \sin \theta$ **8.** $r = \cos \theta$

9. $r = 4 + 3 \cos \theta$ **10.** $r = 2 \cos 2\theta$

11. $r^2 = 16 \cos 2\theta$ **12.** $r^2 = 16 \sin 2\theta$

In Exercises 13–34, test for symmetry and then graph each polar equation.

13. $r = 2 \cos \theta$

14. $r = 2 \sin \theta$

15. $r = 1 - \sin \theta$

16. $r = 1 + \sin \theta$

17. $r = 2 + 2 \cos \theta$

18. $r = 2 - 2 \cos \theta$

19. $r = 2 + \cos \theta$

20. $r = 2 - \sin \theta$

21. $r = 1 + 2 \cos \theta$

22. $r = 1 - 2 \cos \theta$

23. $r = 2 - 3 \sin \theta$

24. $r = 2 + 4 \sin \theta$

25. $r = 2 \cos 2\theta$

26. $r = 2 \sin 2\theta$

27. $r = 4 \sin 3\theta$

28. $r = 4 \cos 3\theta$

29. $r^2 = 9 \cos 2\theta$

30. $r^2 = 9 \sin 2\theta$

31. $r = 1 - 3 \sin \theta$

32. $r = 3 + \sin \theta$

33. $r \cos \theta = -3$

34. $r \sin \theta = 2$

Application Exercises

In Exercise Set 7.3, we considered an application in which sailboat racers look for a sailing angle to a 10-knot wind that produces maximum sailing speed. This situation is now represented by the polar graph in the figure shown. Each point (r, θ) on the graph gives the sailing speed, r, in knots, at an angle θ to the 10-knot wind. Use this information to solve Exercises 35–39.

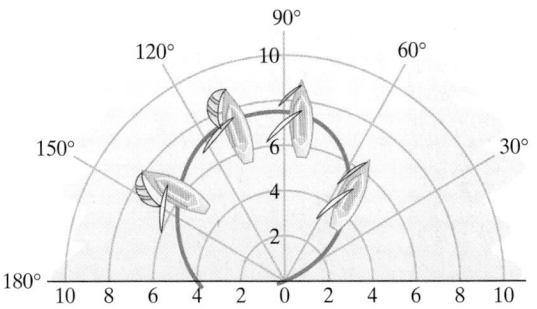

35. What is the speed, to the nearest knot, of the sailboat sailing at $60°$ angle to the wind?

36. What is the speed, to the nearest knot, of the sailboat sailing at a $120°$ angle to the wind?

37. What is the speed, to the nearest knot, of the sailboat sailing at a $90°$ angle to the wind?

38. What is the speed, to the nearest knot, of the sailboat sailing at a $180°$ angle to the wind?

39. What angle to the wind produces the maximum sailing speed? What is the speed? Round the angle to the nearest five degrees and the speed to the nearest half knot.

Writing in Mathematics

40. What is a polar equation?

41. What is the graph of a polar equation?

42. Describe how to graph a polar equation.

43. Describe the test for symmetry with respect to the polar axis.

44. Describe the test for symmetry with respect to the line $\theta = \dfrac{\pi}{2}$.

45. Describe the test for symmetry with respect to the pole.

46. If an equation fails the test for symmetry with respect to the polar axis, what can you conclude?

Technology Exercises

Use the polar mode of a graphing utility with angle measure in radians to solve Exercises 47–78. Unless otherwise indicated, use θ min $= 0$, θ max $= 2\pi$, and θ step $= \dfrac{\pi}{48}$. If you are not pleased with the quality of the graph, experiment with smaller values for θstep. However, if θ step is extremely small, it can take your graphing utility a long period of time to complete the graph.

47. Use a graphing utility to verify any six of your hand-drawn graphs in Exercises 13–34.

In Exercises 48–65, use a graphing utility to graph the polar equation.

48. $r = 4 \cos 5\theta$

49. $r = 4 \sin 5\theta$

50. $r = 4 \cos 6\theta$

51. $r = 4 \sin 6\theta$

52. $r = 2 + 2 \cos \theta$

53. $r = 2 + 2 \sin \theta$

54. $r = 4 + 2 \cos \theta$

55. $r = 4 + 2 \sin \theta$

56. $r = 2 + 4 \cos \theta$

57. $r = 2 + 4 \sin \theta$

58. $r = \dfrac{3}{\sin \theta}$

59. $r = \dfrac{3}{\cos \theta}$

60. $r = \cos \dfrac{3}{2} \theta$

61. $r = \cos \dfrac{5}{2} \theta$

62. $r = 3 \sin \left(\theta + \dfrac{\pi}{4} \right)$

63. $r = 2 \cos \left(\theta - \dfrac{\pi}{4} \right)$

64. $r = \dfrac{1}{1 - \sin \theta}$

65. $r = \dfrac{1}{3 - 2 \sin \theta}$

In Exercises 66–68, find the smallest interval for θ starting with θ min $= 0$ so that your graphing utility graphs the given polar equation exactly once without retracing any portion of it.

66. $r = 4 \sin \theta$ **67.** $r = 4 \sin 2\theta$ **68.** $r^2 = 4 \sin 2\theta$

In Exercises 69–72, use a graphing utility to graph each butterfly curve. Experiment with the range setting, particularly θ step, to produce a butterfly of the best possible quality.

69. $r = \cos^2 5\theta + \sin 3\theta + 0.3$

70. $r = \sin^4 4\theta + \cos 3\theta$

71. $r = \sin^5 \theta + 8 \sin \theta \cos^3 \theta$

72. $r = 1.5^{\sin \theta} - 2.5 \cos 4\theta + \sin^7 \dfrac{\theta}{15}$

(Use θ min $= 0$ and θ max $= 20\pi$.)

73. Use a graphing utility to graph $r = \sin n\theta$ for $n = 1, 2, 3,$ 4, 5, and 6. Use a separate viewing screen for each of the six graphs. What is the pattern for the number of loops that occur corresponding to each value of n? What is happening to the shape of the graphs as n increases? For each graph, what is the smallest interval for θ so that the graph is traced only once?

74. Repeat Exercise 73 for $r = \cos n\theta$. Are your conclusions the same as they were in Exercise 73?

75. Use a graphing utility to graph $r = 1 + 2 \sin n\theta$ for $n = 1,$ 2, 3, 4, 5, and 6. Use a separate viewing screen for each of the six graphs. What is the pattern for the number of large and small petals that occur corresponding to each value of n? How are the large and small petals related when n is odd and when n is even?

76. Repeat Exercise 75 for $r = 1 + 2 \cos n\theta$. Are your conclusions the same as they were in Exercise 75?

77. Graph the spiral $r = \theta$. Use a $[-30, 30, 1]$ by $[-30, 30, 1]$ viewing rectangle. Let θ min $= 0$ and θ max $= 2\pi$, then θ min $= 0$ and θ max $= 4\pi$, and finally θ min $= 0$ and θ max $= 8\pi$.

78. Graph the spiral $r = \dfrac{1}{\theta}$. Use a $[-1, 1, 1]$ by $[-1, 1, 1]$ viewing rectangle. Let θ min $= 0$ and θ max $= 2\pi$, then θ min $= 0$ and θ max $= 4\pi$, and finally θ min $= 0$ and θ max $= 8\pi$.

Critical Thinking Exercises

79. Describe a test for symmetry with respect to the line $\theta = \dfrac{\pi}{2}$ in which r is not replaced.

In Exercises 80–81, graph each polar equation without using a graphing utility.

80. $r = \dfrac{1}{3 - 2\cos\theta}$

81. $r = \dfrac{4}{1 + \sin\theta}$

SECTION 7.5 *Complex Numbers in Polar Form; DeMoivre's Theorem*

Objectives

1. Plot complex numbers in the complex plane.
2. Find the absolute value of a complex number.
3. Write complex numbers in polar form.
4. Convert a complex number from polar to rectangular form.
5. Find products of complex numbers in polar form.
6. Find quotients of complex numbers in polar form.
7. Find powers of complex numbers in polar form.
8. Find roots of complex numbers in polar form.

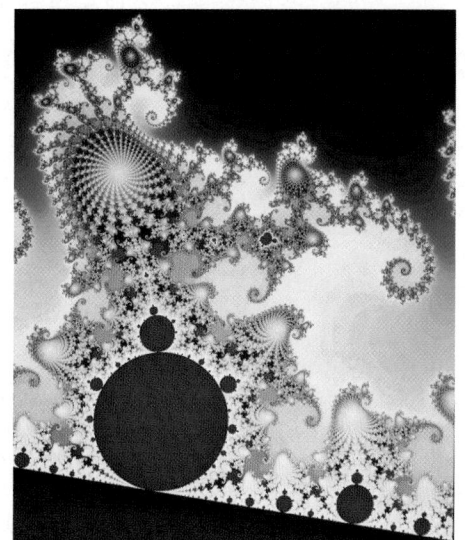

A magnification of the Mandelbrot set

Study Tip

Refer to Section 1.4 if you need to review the basics of complex numbers.

One of the new frontiers of mathematics suggests that there is an underlying order in things that appear to be random, such as the hiss and crackle of background noises as you tune a radio. Irregularities in the heartbeat, some of them severe enough to cause a heart attack, or irregularities in our sleeping patterns, such as insomnia, are examples of chaotic behavior. Chaos in the mathematical sense does not mean a complete lack of form or arrangement. In mathematics, chaos is used to describe something that appears to be random but is not actually random. The patterns of chaos appear in images like the one shown above, called the Mandelbrot set. Magnified portions of this image yield repetitions of the original structure, as well as new and unexpected patterns. The

Mandelbrot set transforms the hidden structure of chaotic events into a source of wonder and inspiration.

The Mandelbrot set is made possible by opening up graphing to include complex numbers in the form $a + bi$, where $i = \sqrt{-1}$. In this section, you will learn how to graph complex numbers and write them in terms of trigonometric functions.

1 Plot complex numbers in the complex plane.

The Complex Plane

We know that a real number can be represented as a point on a number line. By contrast, a complex number $z = a + bi$ is represented as a point (a, b) in a coordinate plane, as shown in Figure 7.35. The horizontal axis of the coordinate plane is called the **real axis.** The vertical axis is called the **imaginary axis.** The coordinate system is called the **complex plane.** Every complex number corresponds to a point in the complex plane and every point in the complex plane corresponds to a complex number. When we represent a complex number as a point in the complex plane, we say that we are **plotting the complex number.**

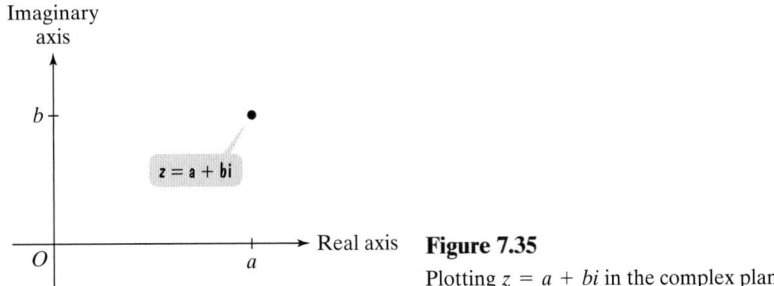

Figure 7.35
Plotting $z = a + bi$ in the complex plane

EXAMPLE 1 Plotting Complex Numbers

Plot each complex number in the complex plane:

a. $z = 3 + 4i$ **b.** $z = -1 - 2i$ **c.** $z = -3$ **d.** $z = -4i$.

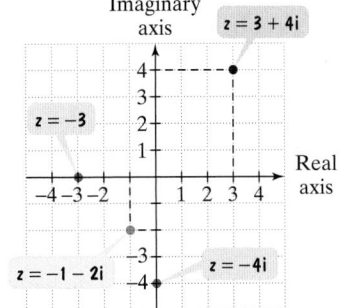

Figure 7.36 Plotting complex numbers

Solution See Figure 7.36.

a. We plot the complex number $z = 3 + 4i$ the same way we plot $(3, 4)$ in the rectangular coordinate system. We move three units to the right on the real axis and four units up parallel to the imaginary axis.

b. The complex number $z = -1 - 2i$ corresponds to the point $(-1, -2)$ in the rectangular coordinate system. Plot the complex number by moving one unit to the left on the real axis and two units down parallel to the imaginary axis.

c. Because $z = -3 = -3 + 0i$, this complex number corresponds to the point $(-3, 0)$. We plot -3 by moving three units to the left on the real axis.

d. Because $z = -4i = 0 - 4i$, this number corresponds to the point $(0, -4)$. We plot the complex number by moving four units down on the imaginary axis.

Check Point 1 Plot each complex number in the complex plane:

a. $z = 2 + 3i$ **b.** $z = -3 - 5i$
c. $z = -4$ **d.** $z = -i$.

2 Find the absolute value of a complex number.

Recall that the absolute value of a real number is its distance from 0 on the number line. The **absolute value of the complex number** $z = a + bi$, denoted by $|z|$, is its distance from the origin in the complex plane.

> ### The Absolute Value of a Complex Number
> The **absolute value** of the complex number $a + bi$ is
> $$|z| = |a + bi| = \sqrt{a^2 + b^2}.$$

EXAMPLE 2 Finding the Absolute Value of a Complex Number

Determine the absolute value of each of the following complex numbers:

a. $z = 3 + 4i$ **b.** $z = -1 - 2i$.

Solution

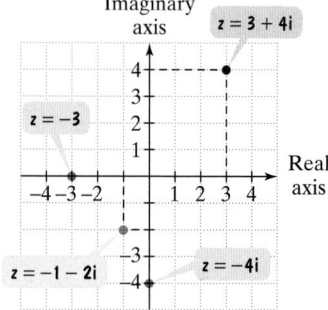

Figure 7.36, repeated

a. The absolute value of $z = 3 + 4i$ is found using $a = 3$ and $b = 4$.
$$|z| = \sqrt{3^2 + 4^2} = \sqrt{9 + 16} = \sqrt{25} = 5 \quad \text{Use } z = \sqrt{a^2 + b^2} \text{ with } a = 3 \text{ and } b = 4.$$

Thus, the distance from the origin to the point $z = 3 + 4i$, shown in quadrant I in Figure 7.36, is five units.

b. The absolute value of $z = -1 - 2i$ is found using $a = -1$ and $b = -2$.
$$|z| = \sqrt{(-1)^2 + (-2)^2} = \sqrt{1 + 4} = \sqrt{5} \quad \text{Use } z = \sqrt{a^2 + b^2} \text{ with } a = -1 \text{ and } b = -2.$$

Thus, the distance from the origin to the point $z = -1 - 2i$, shown in quadrant III in Figure 7.36, is $\sqrt{5}$ units.

Check Point 2 Determine the absolute value of each of the following complex numbers:

a. $z = 5 + 12i$ **b.** $2 - 3i$.

3 Write complex numbers in polar form.

Polar Form of a Complex Number

A complex number in the form $z = a + bi$ is said to be in **rectangular form.** Suppose that its absolute value is r. In Figure 7.37, we let θ be an angle in standard position whose terminal side passes through the point (a, b). From the figure, we see that
$$r = \sqrt{a^2 + b^2}.$$

Likewise, according to the definitions of the trigonometric functions,

$$\cos \theta = \frac{a}{r} \qquad \sin \theta = \frac{b}{r} \qquad \tan \theta = \frac{b}{a}.$$
$$a = r \cos \theta \qquad b = r \sin \theta$$

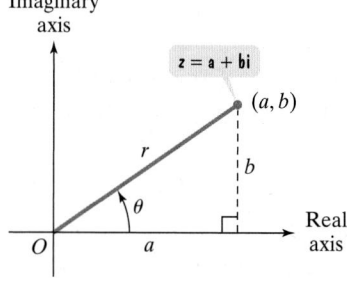

Figure 7.37

By substituting the expressions for a and b in $z = a + bi$, we write the complex number in terms of trigonometric functions.

$$z = a + bi = r \cos \theta + (r \sin \theta)i = r(\cos \theta + i \sin \theta)$$

$a = r \cos \theta$ and $b = r \sin \theta$. | Factor out r from each of the two previous terms.

The expression $z = r(\cos \theta + i \sin \theta)$ is called the **polar form of a complex number.**

Polar Form of a Complex Number

The complex number $z = a + bi$ is written in **polar form** as

$$z = r(\cos \theta + i \sin \theta),$$

where $a = r\cos\theta, b = r\sin\theta, r = \sqrt{a^2 + b^2}$, and $\tan\theta = \dfrac{b}{a}$. The value of r is called the **modulus** (plural: moduli) of the complex number z, and the angle θ is called the **argument** of the complex number z, with $0 \le \theta < 2\pi$.

EXAMPLE 3 Writing a Complex Number in Polar Form

Plot $z = -2 - 2i$ in the complex plane. Then write z in polar form.

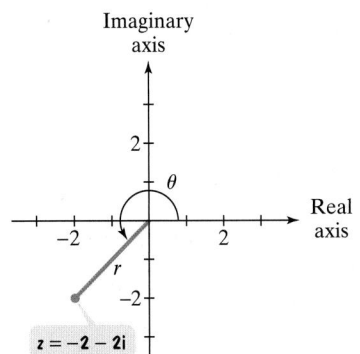

Figure 7.38 Plotting $z = -2 - 2i$ and writing the number in polar form

Solution The complex number $z = -2 - 2i$ is in rectangular form $z = a + bi$, with $a = -2$ and $b = -2$. We plot the number by moving two units to the left on the real axis and two units down parallel to the imaginary axis, as shown in Figure 7.38.

By definition, the polar form of z is $r(\cos \theta + i \sin \theta)$. We need to determine the value for r, the modulus, and the value for θ, the argument. Figure 7.38 shows r and θ. We use $r = \sqrt{a^2 + b^2}$ with $a = -2$ and $b = -2$ to find r.

$$r = \sqrt{a^2 + b^2} = \sqrt{(-2)^2 + (-2)^2} = \sqrt{4 + 4} = \sqrt{8} = \sqrt{4 \cdot 2} = 2\sqrt{2}$$

We use $\tan\theta = \dfrac{b}{a}$ with $a = -2$ and $b = -2$ to find θ.

$$\tan\theta = \frac{b}{a} = \frac{-2}{-2} = 1$$

We know that $\tan \dfrac{\pi}{4} = 1$. Figure 7.38 shows that the argument, θ, lies in quadrant III. Thus,

$$\theta = \pi + \frac{\pi}{4} = \frac{4\pi}{4} + \frac{\pi}{4} = \frac{5\pi}{4}.$$

We use $r = 2\sqrt{2}$ and $\theta = \dfrac{5\pi}{4}$ to write the polar form. The polar form of $z = -2 - 2i$ is

$$z = r(\cos\theta + i\sin\theta) = 2\sqrt{2}\left(\cos\frac{5\pi}{4} + i\sin\frac{5\pi}{4}\right).$$

Check Point 3 Plot $z = -1 - \sqrt{3}i$ in the complex plane. Then write z in polar form. Express the argument in radians.

4 Convert a complex number from polar to rectangular form.

EXAMPLE 4 Writing a Complex Number in Rectangular Form

Write $z = 2(\cos 60° + i \sin 60°)$ in rectangular form.

Solution   The complex number $z = 2(\cos 60° + i \sin 60°)$ is in polar form, with $r = 2$ and $\theta = 60°$. We use exact values for $\cos 60°$ and $\sin 60°$ to write the number in rectangular form.

$$2(\cos 60° + i \sin 60°) = 2\left(\frac{1}{2} + i \frac{\sqrt{3}}{2}\right) = 1 + \sqrt{3}i$$

The rectangular form of $z = 2(\cos 60° + i \sin 60°)$ is

$$z = 1 + \sqrt{3}i.$$

Check Point 4   Write $z = 4(\cos 30° + i \sin 30°)$ in rectangular form.

5 Find products of complex numbers in polar form.

Products and Quotients in Polar Form

We can multiply and divide complex numbers fairly quickly if the numbers are expressed in polar form.

Product of Two Complex Numbers in Polar Form

Let $z_1 = r_1 (\cos \theta_1 + i \sin \theta_1)$ and $z_2 = r_2 (\cos \theta_2 + i \sin \theta_2)$ be two complex numbers in polar form. Their product, $z_1 z_2$, is

$$z_1 z_2 = r_1 r_2 [\cos (\theta_1 + \theta_2) + i \sin (\theta_1 + \theta_2)].$$

To multiply two complex numbers, multiply moduli and add arguments.

To prove this result, we begin by multiplying using the FOIL method. Then we simplify the product using the sum formulas for sine and cosine.

$$z_1 z_2 = \left[r_1(\cos \theta_1 + i \sin \theta_1) \right]\left[r_2(\cos \theta_2 + i \sin \theta_2) \right]$$

$$= r_1 r_2 (\cos \theta_1 + i \sin \theta_1)(\cos \theta_2 + i \sin \theta_2) \qquad \textit{Rearrange factors.}$$

$$\underset{F}{} \qquad \underset{O}{} \qquad \underset{I}{} \qquad \underset{L}{}$$

$$= r_1 r_2 (\cos \theta_1 \cos \theta_2 + i \cos \theta_1 \sin \theta_2 + i \sin \theta_1 \cos \theta_2 + i^2 \sin \theta_1 \sin \theta_2) \quad \textit{Use the FOIL method.}$$

$$= r_1 r_2 [\cos \theta_1 \cos \theta_2 + i(\cos \theta_1 \sin \theta_2 + \sin \theta_1 \cos \theta_2) + i^2 \sin \theta_1 \sin \theta_2] \quad \textit{Factor i from the second and third terms.}$$

$$= r_1 r_2 [\cos \theta_1 \cos \theta_2 + i(\cos \theta_1 \sin \theta_2 + \sin \theta_1 \cos \theta_2) - \sin \theta_1 \sin \theta_2] \quad i^2 = -1$$

$$= r_1 r_2 [(\cos \theta_1 \cos \theta_2 - \sin \theta_1 \sin \theta_2) + i(\sin \theta_1 \cos \theta_2 + \cos \theta_1 \sin \theta_2)] \quad \textit{Rearrange terms.}$$

This is $\cos (\theta_1 + \theta_2)$.   This is $\sin (\theta_1 + \theta_2)$.

$$= r_1 r_2 [\cos(\theta_1 + \theta_2) + i \sin(\theta_1 + \theta_2)]$$

This result gives a rule for finding the product of two complex numbers in polar form. The two parts to the rule are shown in the voice balloons below the product.

$$r_1 r_2 \left[\cos(\theta_1 + \theta_2) + i \sin(\theta_1 + \theta_2) \right]$$

Multiply moduli. Add arguments.

EXAMPLE 5   Finding Products of Complex Numbers in Polar Form

Find the product of the complex numbers. Leave the answer in polar form.

$$z_1 = 4(\cos 50° + i \sin 50°) \qquad z_2 = 7(\cos 100° + i \sin 100°)$$

Solution

$z_1 z_2$

$= \left[4(\cos 50° + i \sin 50°) \right]\left[7(\cos 100° + i \sin 100°) \right]$ *Form the product of the given numbers.*

$= (4 \cdot 7)\left[\cos(50° + 100°) + i \sin(50° + 100°) \right]$ *Multiply moduli and add arguments.*

$= 28(\cos 150° + i \sin 150°)$ *Simplify.*

Check Point 5   Find the product of the complex numbers. Leave the answer in polar form.

$$z_1 = 6(\cos 40° + i \sin 40°) \qquad z_2 = 5(\cos 20° + i \sin 20°)$$

6 Find quotients of complex numbers in polar form.

Using algebraic methods for dividing complex numbers and the difference formulas for sine and cosine, we can obtain a rule for dividing complex numbers in polar form. The proof of this rule can be found in the appendix.

Quotient of Two Complex Numbers in Polar Form

Let $z_1 = r_1 (\cos \theta_1 + i \sin \theta_1)$ and $z_2 = r_2 (\cos \theta_2 + i \sin \theta_2)$ be two complex numbers in polar form. Their quotient, $\dfrac{z_1}{z_2}$, is

$$\frac{z_1}{z_2} = \frac{r_1}{r_2}\left[\cos(\theta_1 - \theta_2) + i \sin(\theta_1 - \theta_2) \right].$$

To divide two complex numbers, divide moduli and subtract arguments.

EXAMPLE 6   Finding Quotients of Complex Numbers in Polar Form

Find the quotient $\dfrac{z_1}{z_2}$ of the complex numbers. Leave the answer in polar form.

$$z_1 = 12\left(\cos \frac{3\pi}{4} + i \sin \frac{3\pi}{4} \right) \qquad z_2 = 4\left(\cos \frac{\pi}{4} + i \sin \frac{\pi}{4} \right)$$

Solution

$$\frac{z_1}{z_2} = \frac{12\left(\cos\frac{3\pi}{4} + i\sin\frac{3\pi}{4}\right)}{4\left(\cos\frac{\pi}{4} + i\sin\frac{\pi}{4}\right)}$$

Form the quotient of the given numbers.

$$= \frac{12}{4}\left[\cos\left(\frac{3\pi}{4} - \frac{\pi}{4}\right) + i\sin\left(\frac{3\pi}{4} - \frac{\pi}{4}\right)\right]$$

Divide moduli and subtract arguments.

$$= 3\left(\cos\frac{\pi}{2} + i\sin\frac{\pi}{2}\right)$$

Simplify: $\frac{3\pi}{4} - \frac{\pi}{4} = \frac{2\pi}{4} = \frac{\pi}{2}$.

Check Point 6   Find the quotient of the complex numbers. Leave the answer in polar form.

$$z_1 = 50\left(\cos\frac{4\pi}{3} + i\sin\frac{4\pi}{3}\right) \qquad z_2 = 5\left(\cos\frac{\pi}{3} + i\sin\frac{\pi}{3}\right)$$

7   Find powers of complex numbers in polar form.

Powers of Complex Numbers in Polar Form

We can use a formula to find powers of complex numbers if the complex numbers are expressed in polar form. This formula can be illustrated by repeatedly multiplying by $r(\cos\theta + i\sin\theta)$.

$z = r(\cos\theta + i\sin\theta)$	Start with z.
$z \cdot z = r(\cos\theta + i\sin\theta)r(\cos\theta + i\sin\theta)$	Multiply z by z = r(cos θ + i sin θ).
$z^2 = r^2(\cos 2\theta + i\sin 2\theta)$	Multiply moduli: r·r = r². Add arguments: θ + θ = 2θ.
$z^2 \cdot z = r^2(\cos 2\theta + i\sin 2\theta)r(\cos\theta + i\sin\theta)$	Multiply z² by z = r(cos θ + i sin θ).
$z^3 = r^3(\cos 3\theta + i\sin 3\theta)$	Multiply moduli: r²·r = r³. Add arguments: 2θ + θ = 3θ.
$z^3 \cdot z = r^3(\cos 3\theta + i\sin 3\theta)r(\cos\theta + i\sin\theta)$	Multiply z³ by z = r(cos θ + i sin θ).
$z^4 = r^4(\cos 4\theta + i\sin 4\theta)$	Multiply moduli: r³·r = r⁴. Add arguments: 3θ + θ = 4θ.

Do you see a pattern forming? If n is a positive integer, it appears that z^n is obtained by raising the modulus to the nth power and multiplying the argument by n. The formula for the nth power of a complex number is known as **DeMoivre's Theorem** in honor of the French mathematician Abraham DeMoivre (1667–1754).

> **DeMoivre's Theorem**
>
> Let $z = r(\cos\theta + i\sin\theta)$ be a complex number in polar form. If n is a positive integer, z to the nth power, z^n, is
>
> $$z^n = [r(\cos\theta + i\sin\theta)]^n = r^n(\cos n\theta + i\sin n\theta).$$

EXAMPLE 7   Finding the Power of a Complex Number

Find $[2(\cos 10° + i \sin 10°)]^6$. Write the answer in rectangular form, $a + bi$.

Solution   By DeMoivre's Theorem,

$[2(\cos 10° + i \sin 10°)]^6$

$= 2^6[\cos(6 \cdot 10°) + i \sin(6 \cdot 10°)]$ Raise the modulus to the 6th power and multiply the argument by 6.

$= 64(\cos 60° + i \sin 60°)$ Simplify.

$= 64\left(\dfrac{1}{2} + i\dfrac{\sqrt{3}}{2}\right)$ Write the answer in rectangular form.

$= 32 + 32\sqrt{3}i$ Multiply and express the answer in $a + bi$ form.

Check Point 7   Find $[2(\cos 30° + i \sin 30°)]^5$. Write the answer in rectangular form.

EXAMPLE 8   Finding the Power of a Complex Number

Find $(1 + i)^8$ using DeMoivre's Theorem. Write the answer in rectangular form, $a + bi$.

Solution   DeMoivre's Theorem applies to complex numbers in polar form. Thus, we must first write $1 + i$ in $r(\cos\theta + i \sin\theta)$ form. Then we can use DeMoivre's Theorem. The complex number $1 + i$ is plotted in Figure 7.39. From the figure we obtain values for r and θ.

$$r = \sqrt{a^2 + b^2} = \sqrt{1^2 + 1^2} = \sqrt{2} \quad \tan\theta = \frac{b}{a} = \frac{1}{1} = 1 \quad \text{and} \quad \theta = \frac{\pi}{4}$$

Using these values,

$$1 + i = r(\cos\theta + i \sin\theta) = \sqrt{2}\left(\cos\frac{\pi}{4} + i \sin\frac{\pi}{4}\right).$$

Now we use DeMoivre's Theorem to raise $1 + i$ to the 8th power.

$(1 + i)^8$

$= \left[\sqrt{2}\left(\cos\dfrac{\pi}{4} + i \sin\dfrac{\pi}{4}\right)\right]^8$ Work with the polar form of $1 + i$.

$= (\sqrt{2})^8\left[\cos\left(8 \cdot \dfrac{\pi}{4}\right) + i \sin\left(8 \cdot \dfrac{\pi}{4}\right)\right]$ Apply DeMoivre's Theorem. Raise the modulus to the 8th power and multiply the argument by 8.

$= 16(\cos 2\pi + i \sin 2\pi)$ Simplify: $(\sqrt{2})^8 = (2^{1/2})^8 = 2^4 = 16$.

$= 16(1 + 0i)$ $\cos 2\pi = 1$ and $\sin 2\pi = 0$

$= 16$ Simplify.

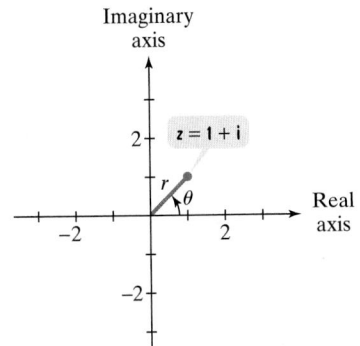

Figure 7.39 Plotting $1 + i$ and writing the number in polar form

Check Point 8   Find $(1 + i)^4$ using DeMoivre's Theorem. Write the answer in rectangular form.

8 Find roots of complex numbers in polar form.

Roots of Complex Numbers in Polar Form

In Example 7, we showed that

$$[2(\cos 10° + i \sin 10°)]^6 = 64(\cos 60° + i \sin 60°).$$

We say that $2(\cos 10° + i \sin 10°)$ is a **complex sixth root** of $64(\cos 60° + i \sin 60°)$. It is one of six distinct complex sixth roots of $64(\cos 60° + i \sin 60°)$.

In general, if a complex number z satisfies the equation

$$z^n = w$$

we say that z is a **complex nth root** of w. It is one of n distinct nth complex roots that can be found using the following theorem:

DeMoivre's Theorem for Finding Complex Roots

Let $w = r(\cos \theta + i \sin \theta)$ be a complex number in polar form. If $w \neq 0$, w has n distinct complex nth roots given by the formula

$$z_k = \sqrt[n]{r}\left[\cos\left(\frac{\theta + 2\pi k}{n}\right) + i \sin\left(\frac{\theta + 2\pi k}{n}\right)\right] \quad \text{(radians)}$$

or $\quad z_k = \sqrt[n]{r}\left[\cos\left(\frac{\theta + 360°k}{n}\right) + i \sin\left(\frac{\theta + 360°k}{n}\right)\right] \text{(degrees)}$

where $k = 0, 1, 2, \ldots, n - 1$.

By raising the radian or degree formula for z_k to the nth power, you can use DeMoivre's Theorem for powers to show that $z_k^n = w$. Thus, each z_k is a complex nth root of w.

DeMoivre's Theorem for finding complex roots states that every complex number has two distinct complex square roots, three distinct complex cube roots, four distinct complex fourth roots, and so on. Each root has the same modulus, $\sqrt[n]{r}$.

Successive roots have arguments that differ by the same amount, $\dfrac{2\pi}{n}$ or $\dfrac{360°}{n}$. This means that if you plot all the complex roots of any number, they will be equally spaced on a circle centered at the origin, with radius $\sqrt[n]{r}$.

EXAMPLE 9   Finding the Roots of a Complex Number

Find all the complex fourth roots of $16(\cos 120° + i \sin 120°)$. Write roots in polar form, with θ in degrees.

Solution   There are exactly four fourth roots of the given complex number. From DeMoivre's Theorem for finding complex roots, the fourth roots of $16(\cos 120° + i \sin 120°)$ are

$$z_k = \sqrt[4]{16}\left[\cos\left(\frac{120° + 360°k}{4}\right) + i \sin\left(\frac{120° + 360°k}{4}\right)\right], \quad k = 0, 1, 2, 3.$$

> Use $z_k = \sqrt[n]{r}\left[\cos\left(\frac{\theta + 360°k}{n}\right) + i \sin\left(\frac{\theta + 360°k}{n}\right)\right]$.
> In $16(\cos 120° + i \sin 120°)$, $r = 16$ and $\theta = 120°$.
> Because we are finding fourth roots, $n = 4$.

The four fourth roots are found by substituting $0, 1, 2$, and 3 for k in the expression for z_k above the voice balloon. Thus, the four complex fourth roots are:

$$z_0 = \sqrt[4]{16}\left[\cos\left(\frac{120° + 360° \cdot 0}{4}\right) + i\sin\left(\frac{120° + 360° \cdot 0}{4}\right)\right]$$

$$= \sqrt[4]{16}\left(\cos\frac{120°}{4} + i\sin\frac{120°}{4}\right) = 2(\cos 30° + i\sin 30°)$$

$$z_1 = \sqrt[4]{16}\left[\cos\left(\frac{120° + 360° \cdot 1}{4}\right) + i\sin\left(\frac{120° + 360° \cdot 1}{4}\right)\right]$$

$$= \sqrt[4]{16}\left(\cos\frac{480°}{4} + i\sin\frac{480°}{4}\right) = 2(\cos 120° + i\sin 120°)$$

$$z_2 = \sqrt[4]{16}\left[\cos\left(\frac{120° + 360° \cdot 2}{4}\right) + i\sin\left(\frac{120° + 360° \cdot 2}{4}\right)\right]$$

$$= \sqrt[4]{16}\left(\cos\frac{840°}{4} + i\sin\frac{840°}{4}\right) = 2(\cos 210° + i\sin 210°)$$

$$z_3 = \sqrt[4]{16}\left[\cos\left(\frac{120° + 360° \cdot 3}{4}\right) + i\sin\left(\frac{120° + 360° \cdot 3}{4}\right)\right]$$

$$= \sqrt[4]{16}\left(\cos\frac{1200°}{4} + i\sin\frac{1200°}{4}\right) = 2(\cos 300° + i\sin 300°).$$

In Figure 7.40, we have plotted each of the four fourth roots. They are equally spaced at 90° intervals on a circle with radius 2.

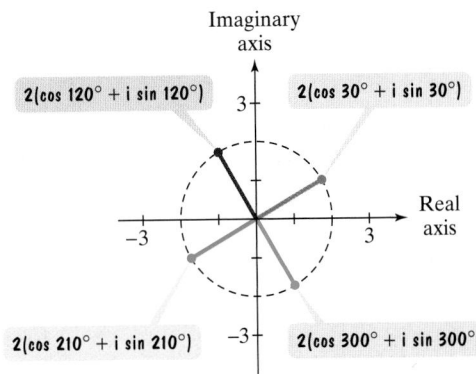

Figure 7.40 Plotting the four fourth roots of $16(\cos 120° + i\sin 120°)$

Check Point 9   Find all the complex fourth roots of $16(\cos 60° + i\sin 60°)$. Write roots in polar form, with θ in degrees.

EXAMPLE 10   Finding the Roots of a Complex Number

Find all the cube roots of 8. Write roots in rectangular form.

Solution   DeMoivre's Theorem for roots applies to complex numbers in polar form. Thus, we will first write 8 in polar form. We express θ in radians, although degrees can also be used.

$$8 = r(\cos\theta + i\sin\theta) = 8(\cos 0 + i\sin 0)$$

There are exactly three cube roots of 8. From DeMoivre's Theorem for finding complex roots, the cube roots of 8 are

$$z_k = \sqrt[3]{8}\left[\cos\left(\frac{0 + 2\pi k}{3}\right) + i\sin\left(\frac{0 + 2\pi k}{3}\right)\right], \quad k = 0, 1, 2.$$

> Use $z_k = \sqrt[n]{r}\left[\cos\left(\frac{\theta + 2\pi k}{n}\right) + i\sin\left(\frac{\theta + 2\pi k}{n}\right)\right]$.
> In $8(\cos 0 + i\sin 0)$, $r = 8$ and $\theta = 0$.
> Because we are finding cube roots, $n = 3$.

The three cube roots of 8 are found by substituting 0, 1, and 2 for k in the expression for z_k above the voice balloon. Thus, the three cube roots of 8 are

$$z_0 = \sqrt[3]{8}\left[\cos\left(\frac{0 + 2\pi \cdot 0}{3}\right) + i\sin\left(\frac{0 + 2\pi \cdot 0}{3}\right)\right]$$

$$= 2(\cos 0 + i\sin 0) = 2(1 + i \cdot 0) = 2$$

$$z_1 = \sqrt[3]{8}\left[\cos\left(\frac{0 + 2\pi \cdot 1}{3}\right) + i\sin\left(\frac{0 + 2\pi \cdot 1}{3}\right)\right]$$

$$= 2\left(\cos\frac{2\pi}{3} + i\sin\frac{2\pi}{3}\right) = 2\left(-\frac{1}{2} + i \cdot \frac{\sqrt{3}}{2}\right) = -1 + \sqrt{3}i$$

$$z_2 = \sqrt[3]{8}\left[\cos\left(\frac{0 + 2\pi \cdot 2}{3}\right) + i\sin\left(\frac{0 + 2\pi \cdot 2}{3}\right)\right]$$

$$= 2\left(\cos\frac{4\pi}{3} + i\sin\frac{4\pi}{3}\right) = 2\left(-\frac{1}{2} + i \cdot \left(-\frac{\sqrt{3}}{2}\right)\right) = -1 - \sqrt{3}i.$$

Figure 7.41 The three cube roots of 8 are equally spaced at intervals of $\frac{2\pi}{3}$ on a circle with radius 2.

The three cube roots of 8 are plotted in Figure 7.41.

Check Point 10 Find all the cube roots of 27. Write roots in rectangular form.

The Mandelbrot Set

The set of all complex numbers for which the sequence

$$z, z^2 + z, (z^2 + z)^2 + z, [(z^2 + z)^2 + z]^2 + z, \ldots$$

is bounded is called the **Mandelbrot set**. Plotting these complex numbers in the complex plane results in a graph that is "buglike" in shape, shown in Figure 7.42. Colors can be added to the boundary of the graph. At the boundary, color choices depend on how quickly the numbers in the boundary approach infinity when substituted into the sequence shown. The magnified boundary is shown in the introduction to this section. It includes the original buglike structure, as well as new and interesting patterns. With each level of magnification, repetition and unpredictable formations interact to create what has been called the most complicated mathematical object ever known.

Figure 7.42

EXERCISE SET 7.5

 Practice Exercises

In Exercises 1–10, plot each complex number and find its absolute value.

1. $z = 4i$
2. $z = 3i$
3. $z = 3$
4. $z = 4$
5. $z = 3 + 2i$
6. $z = 2 + 5i$
7. $z = 3 - i$
8. $z = 4 - i$
9. $z = -3 + 4i$
10. $z = -3 - 4i$

In Exercises 11–26, plot each complex number. Then write the complex number in polar form. You may express the argument in degrees or radians.

11. $2 + 2i$
12. $1 + \sqrt{3}i$
13. $-1 - i$
14. $2 - 2i$
15. $-4i$
16. $-3i$
17. $2\sqrt{3} - 2i$
18. $-2 + 2\sqrt{3}i$
19. -3
20. -4
21. $-3\sqrt{2} - 3\sqrt{3}i$
22. $3\sqrt{2} - 3\sqrt{2}i$
23. $-3 + 4i$
24. $-2 + 3i$
25. $2 - \sqrt{3}i$
26. $1 - \sqrt{5}i$

In Exercises 27–36, write each complex number in rectangular form. If necessary, round to the nearest tenth.

27. $6(\cos 30° + i \sin 30°)$
28. $12(\cos 60° + i \sin 60°)$
29. $4(\cos 240° + i \sin 240°)$
30. $10(\cos 210° + i \sin 210°)$
31. $8\left(\cos \dfrac{7\pi}{4} + i \sin \dfrac{7\pi}{4}\right)$
32. $4\left(\cos \dfrac{5\pi}{6} + i \sin \dfrac{5\pi}{6}\right)$
33. $5\left(\cos \dfrac{\pi}{2} + i \sin \dfrac{\pi}{2}\right)$
34. $7\left(\cos \dfrac{3\pi}{2} + i \sin \dfrac{3\pi}{2}\right)$
35. $20(\cos 205° + i \sin 205°)$
36. $30(\cos 2.3 + i \sin 2.3)$

In Exercises 37–44, find the product of the complex numbers. Leave answers in polar form.

37. $z_1 = 6(\cos 20° + i \sin 20°)$
$z_2 = 5(\cos 50° + i \sin 50°)$
38. $z_1 = 4(\cos 15° + i \sin 15°)$
$z_2 = 7(\cos 25° + i \sin 25°)$
39. $z_1 = 3\left(\cos \dfrac{\pi}{5} + i \sin \dfrac{\pi}{5}\right)$
$z_2 = 4\left(\cos \dfrac{\pi}{10} + i \sin \dfrac{\pi}{10}\right)$
40. $z_1 = 3\left(\cos \dfrac{5\pi}{8} + i \sin \dfrac{5\pi}{8}\right)$
$z_2 = 10\left(\cos \dfrac{\pi}{16} + i \sin \dfrac{\pi}{16}\right)$
41. $z_1 = \cos \dfrac{\pi}{4} + i \sin \dfrac{\pi}{4}$
$z_2 = \cos \dfrac{\pi}{3} + i \sin \dfrac{\pi}{3}$

42. $z_1 = \cos \dfrac{\pi}{6} + i \sin \dfrac{\pi}{6}$
$z_2 = \cos \dfrac{\pi}{4} + i \sin \dfrac{\pi}{4}$
43. $z_1 = 1 + i$
$z_2 = -1 + i$
44. $z_1 = 1 + i$
$z_2 = 2 + 2i$

In Exercises 45–52, find the quotient $\dfrac{z_1}{z_2}$ of the complex numbers. Leave answers in polar form. In Exercises 49–50, express the argument as an angle between 0° and 360°.

45. $z_1 = 20(\cos 75° + i \sin 75°)$
$z_2 = 4(\cos 25° + i \sin 25°)$
46. $z_1 = 50(\cos 80° + i \sin 80°)$
$z_2 = 10(\cos 20° + i \sin 20°)$
47. $z_1 = 3\left(\cos \dfrac{\pi}{5} + i \sin \dfrac{\pi}{5}\right)$
$z_2 = 4\left(\cos \dfrac{\pi}{10} + i \sin \dfrac{\pi}{10}\right)$
48. $z_1 = 3\left(\cos \dfrac{5\pi}{18} + i \sin \dfrac{5\pi}{18}\right)$
$z_2 = 10\left(\cos \dfrac{\pi}{16} + i \sin \dfrac{\pi}{16}\right)$
49. $z_1 = \cos 80° + i \sin 80°$
$z_2 = \cos 200° + i \sin 200°$
50. $z_1 = \cos 70° + i \sin 70°$
$z_2 = \cos 230° + i \sin 230°$
51. $z_1 = 2 + 2i$
$z_2 = 1 + i$
52. $z_1 = 2 - 2i$
$z_2 = 1 - i$

In Exercises 53–64, use DeMoivre's Theorem to find the indicated power of the complex number. Write answers in rectangular form.

53. $[4(\cos 15° + i \sin 15°)]^3$
54. $[2(\cos 10° + i \sin 10°)]^3$
55. $[2(\cos 80° + i \sin 80°)]^3$
56. $[2(\cos 40° + i \sin 40°)]^3$
57. $\left[\dfrac{1}{2}\left(\cos \dfrac{\pi}{12} + i \sin \dfrac{\pi}{12}\right)\right]^6$
58. $\left[\dfrac{1}{2}\left(\cos \dfrac{\pi}{10} + i \sin \dfrac{\pi}{10}\right)\right]^5$
59. $\left[\sqrt{2}\left(\cos \dfrac{5\pi}{6} + i \sin \dfrac{5\pi}{6}\right)\right]^4$
60. $\left[\sqrt{3}\left(\cos \dfrac{5\pi}{18} + i \sin \dfrac{5\pi}{18}\right)\right]^6$
61. $(1 + i)^5$
62. $(1 - i)^5$
63. $(\sqrt{3} - i)^6$
64. $(\sqrt{2} - i)^4$

In Exercises 65–68, find all the complex roots. Write roots in polar form with θ in degrees.

65. The complex square roots of $9(\cos 30° + i \sin 30°)$

66. The complex square roots of $25(\cos 210° + i \sin 210°)$

67. The complex cube roots of $8(\cos 210° + i \sin 210°)$

68. The complex cube roots of $27(\cos 306° + i \sin 306°)$

In Exercises 69–76, find all the complex roots. Write roots in rectangular form. If necessary, round to the nearest tenth.

69. The complex fourth roots of $81\left(\cos \dfrac{4\pi}{3} + i \sin \dfrac{4\pi}{3}\right)$

70. The complex fifth roots of $32\left(\cos \dfrac{5\pi}{3} + i \sin \dfrac{5\pi}{3}\right)$

71. The complex fifth roots of 32

72. The complex sixth roots of 64

73. The complex cube roots of 1

74. The complex cube roots of i

75. The complex fourth roots of $1 + i$

76. The complex fifth roots of $-1 + i$

Application Exercises

In Exercises 77–78, show that the given complex number z plots as a point in the Mandelbrot set.

a. *Write the first six terms of the sequence*

$$z_1, z_2, z_3, z_4, z_5, z_6, \ldots$$

where

$z_1 = z$: Write the given number.

$z_2 = z^2 + z$: Square z_1 and add the given number.

$z_3 = (z^2 + z)^2 + z$: Square z_2 and add the given number.

$z_4 = [(z^2 + z)^2 + z]^2 + z$: Square z_3 and add the given number.

z_5: Square z_4 and add the given number.

z_6: Square z_5 and add the given number.

b. *If the sequence that you began writing in part (a) is bounded, the given complex number belongs to the Mandelbrot set. Show that the sequence is bounded by writing two complex numbers. One complex number should be greater in absolute value than the absolute values of the terms in the sequence. The second complex number should be less in absolute value than the absolute values of the terms in the sequence.*

77. $z = i$

78. $z = -i$

Writing in Mathematics

79. Explain how to plot a complex number in the complex plane. Provide an example with your explanation.

80. How do you determine the absolute value of a complex number?

81. What is the polar form of a complex number?

82. If you are given a complex number in rectangular form, how do you write it in polar form?

83. If you are given a complex number in polar form, how do you write it in rectangular form?

84. Explain how to find the product of two complex numbers in polar form.

85. Explain how to find the quotient of two complex numbers in polar form.

86. Explain how to find the power of a complex number in polar form.

87. Explain how to use DeMoivre's Theorem for finding complex roots to find the two square roots of 9.

88. Describe the graph of all complex numbers with an absolute value of 6.

89. The image of the Mandelbrot set in the section opener exhibits self-similarity: Magnified portions repeat much of the pattern of the whole structure, as well as new and unexpected patterns. Describe an object in nature that exhibits self-similarity.

Technology Exercises

90. Use the rectangular-to-polar feature on a graphing utility to verify any four of your answers in Exercises 11–26. Be aware that you may have to adjust the angle for the correct quadrant.

91. Use the polar-to-rectangular feature on a graphing utility to verify any four of your answers in Exercises 27–36.

Critical Thinking Exercises

92. Prove the rule for finding the quotient of two complex numbers in polar form. Begin the proof as follows, using the conjugate of the denominator:

$$\frac{r_1(\cos \theta_1 + i \sin \theta_1)}{r_2(\cos \theta_2 + i \sin \theta_2)} = \frac{r_1(\cos \theta_1 + i \sin \theta_1)}{r_2(\cos \theta_2 + i \sin \theta_2)} \cdot \frac{(\cos \theta_2 - i \sin \theta_2)}{(\cos \theta_2 - i \sin \theta_2)}.$$

Perform the indicated multiplications. Then use the difference formulas for sine and cosine.

93. Plot each of the complex fourth roots of 1.

In Exercises 94–95, use DeMoivre's Theorem for finding complex roots to find all the solutions of each equation.

94. $x^3 + 27 = 0$

95. $x^3 - 8i = 0$

Group Exercise

96. Group members should prepare and present a seminar on chaos. Include one or more of the following topics in your presentation: fractal images, the role of complex numbers in generating fractal images, algorithms, iterations, iteration number, and fractals in nature. Be sure to include visual images that will intrigue your audience.

SECTION 7.6 Vectors

Objectives

1. Use magnitude and direction to show vectors are equal.

2. Visualize scalar multiplication, vector addition, and vector subtraction as geometric vectors.

3. Represent vectors in the rectangular coordinate system.

4. Perform operations with vectors in terms of **i** and **j**.

5. Find the unit vector in the direction of **v**.

6. Write a vector in terms of its magnitude and direction.

7. Solve applied problems involving vectors.

It's been a dynamic lecture, but now that it's over it's obvious that my professor is exhausted. She's slouching motionless against the board and—what's that? The forces acting against her body, including the pull of gravity, are appearing as arrows. I know that mathematics reveals the hidden patterns of the universe, but this is ridiculous. Does the arrangement of the arrows on the right have anything to do with the fact that my wiped-out professor is not sliding down the wall?

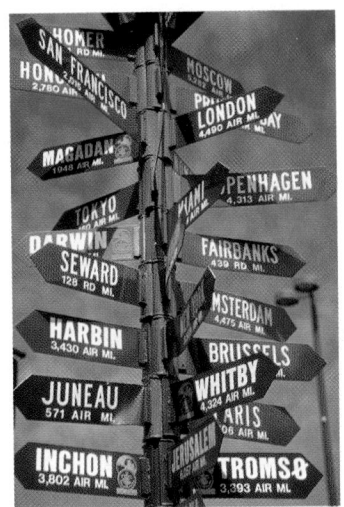

This sign shows a distance and direction for each city. Thus, the sign defines a vector for each destination.

Ours is a world of pushes and pulls. For example, suppose you are pulling a cart up a 30° incline, requiring an effort of 100 pounds. This quantity is described by giving its magnitude (a number indicating size, including a unit of measure) and also its direction. The magnitude is 100 pounds and the direction is 30° from the horizontal. Quantities that involve both a magnitude and a direction are called **vector quantities,** or **vectors** for short. Here is another example of a vector:

> You are driving due north at 50 miles per hour. The magnitude is the speed, 50 miles per hour. The direction of motion is due north.

Some quantities can be completely described by giving only their magnitude. For example, the temperature of the lecture room that you just left is 75°. This temperature has magnitude, 75°, but no direction. Quantities that involve magnitude, but no direction, are called **scalar quantities,** or **scalars** for short. Thus, a scalar has only a numerical value. Another example of a scalar is your professor's height, which you estimate to be 5.5 feet.

In the next two sections, we introduce the world of vectors, which literally surround your every move. Because vectors have both nonnegative magnitude and direction, we begin our discussion with directed line segments.

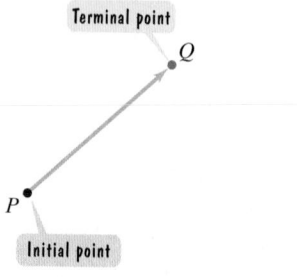

Terminal point

Q

P

Initial point

Figure 7.43 A directed line segment from *P* to *Q*

1 Use magnitude and direction to show vectors are equal.

Directed Line Segments and Geometric Vectors

A line segment to which a direction has been assigned is called a **directed line segment.** Figure 7.43 shows a directed line segment from *P* to *Q*. We call *P* the **initial point** and *Q* the **terminal point.** We denote this directed line segment by $\overrightarrow{PQ}$.

The **magnitude** of the directed line segment $\overrightarrow{PQ}$ is its length. We denote this by $\|\overrightarrow{PQ}\|$. Thus, $\|\overrightarrow{PQ}\|$ is the distance from point *P* to point *Q*. Because distance is nonnegative, vectors do not have negative magnitudes.

Geometrically, a **vector** is a directed line segment. Vectors are often denoted by boldface letters, such as **v.** If a vector **v** has the same magnitude and the same direction as the directed line segment $\overrightarrow{PQ}$, we write

$$\mathbf{v} = \overrightarrow{PQ}.$$

Because it is difficult to write boldface on paper, use an arrow over a single letter, such as $\vec{v}$, to denote **v,** the vector **v.**

Figure 7.44 shows four possible relationships between vectors **v** and **w.** In Figure 7.44 (a), the vectors have the same magnitude and the same direction, and are said to be *equal.* In general, vectors **v** and **w** are **equal** if they have the *same magnitude* and the *same direction.* We write this as **v** = **w.**

(a) v = w because the vectors have the same magnitude and the same direction

(b) Vectors **v** and **w** have the same magnitude, but different directions

(c) Vectors **v** and **w** have the same magnitude, but opposite directions

(d) Vectors **v** and **w** have the same direction, but different magnitudes

Figure 7.44 Relationships between vectors

EXAMPLE 1   Showing That Two Vectors Are Equal

Use Figure 7.45 to show that **u** = **v.**

Solution   Equal vectors have the same magnitude and the same direction. Use the distance formula to show that **u** and **v** have the same magnitude.

Terminal point: (3, 6)

Terminal point: (0, 3)

Initial point: (0, 0)

Initial point: (−3, −3)

Figure 7.45

Magnitude of u

$$\|\mathbf{u}\| = \sqrt{(x_2 - x_1)^2 + (y_2 - y_1)^2} = \sqrt{[0 - (-3)]^2 + [3 - (-3)]^2}$$
$$= \sqrt{3^2 + 6^2} = \sqrt{9 + 36} = \sqrt{45} \quad (\text{or } 3\sqrt{5})$$

Magnitude of v

$$\|\mathbf{v}\| = \sqrt{(x_2 - x_1)^2 + (y_2 - y_1)^2} = \sqrt{(3 - 0)^2 + (6 - 0)^2}$$
$$= \sqrt{3^2 + 6^2} = \sqrt{9 + 36} = \sqrt{45} \quad (\text{or } 3\sqrt{5})$$

Thus, **u** and **v** have the same magnitude: $\|\mathbf{u}\| = \|\mathbf{v}\|$.

One way to show that **u** and **v** have the same direction is to find the slopes of the lines on which they lie.

Line on which u lies

$$m = \frac{y_2 - y_1}{x_2 - x_1} = \frac{3 - (-3)}{0 - (-3)} = \frac{6}{3} = 2$$

Line on which v lies

$$m = \frac{y_2 - y_1}{x_2 - x_1} = \frac{6 - 0}{3 - 0} = \frac{6}{3} = 2$$

Because **u** and **v** are both directed toward the upper right on lines having the same slope, 2, they have the same direction.

Thus, **u** and **v** have the same magnitude and direction, and **u** = **v.**

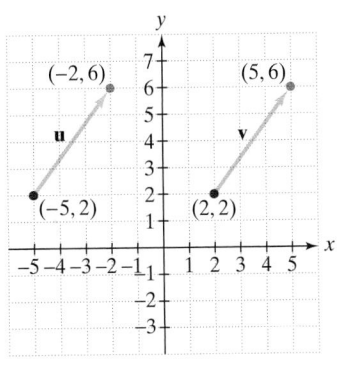

Figure 7.46

2 Visualize scalar multiplication, vector addition, and vector subtraction as geometric vectors.

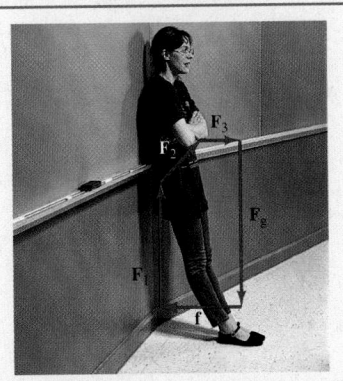

The figure shows the sum of five vectors:

$$\mathbf{F}_1 + \mathbf{F}_2 + \mathbf{F}_3 + \mathbf{F}_g + \mathbf{f}.$$

Notice how the terminal point of each vector coincides with the initial point of the vector that's being added to it. The vector sum, from the initial point of $\mathbf{F}_1$ to the terminal point of $\mathbf{f}$, is a single point. The magnitude of a single point is zero. These forces add up to a net force of zero, allowing the professor to be motionless.

Check Point 1 Use Figure 7.46 to show that $\mathbf{u} = \mathbf{v}$.

A vector can be multiplied by a real number. Figure 7.47 shows three such multiplications: $2\mathbf{v}, \frac{1}{2}\mathbf{v}$, and $-\frac{3}{2}\mathbf{v}$. **Multiplying a vector by any positive real number (except for 1) changes the magnitude of the vector, but not its direction.** This can be seen by the blue and green vectors in Figure 7.47. Compare the black and blue vectors. Can you see that $2\mathbf{v}$ has the same direction as $\mathbf{v}$ but is twice the magnitude of $\mathbf{v}$? Now, compare the black and green vectors: $\frac{1}{2}\mathbf{v}$ has the same direction as $\mathbf{v}$ but is half the magnitude of $\mathbf{v}$.

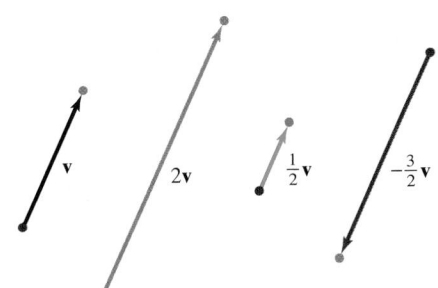

Figure 7.47 Multiplying vector $\mathbf{v}$ by real numbers

Now compare the black and red vectors in Figure 7.47. **Multiplying a vector by a negative number reverses the direction of the vector.** Notice that $-\frac{3}{2}\mathbf{v}$ has the opposite direction as $\mathbf{v}$ and is $\frac{3}{2}$ the magnitude of $\mathbf{v}$.

The multiplication of the real number, k, and the vector, $\mathbf{v}$, is called **scalar multiplication.** We write this product as $k\mathbf{v}$.

Scalar Multiplication

If k is a real number and $\mathbf{v}$ a vector, the vector $k\mathbf{v}$ is called a **scalar multiple** of the vector $\mathbf{v}$. The magnitude and direction of $k\mathbf{v}$ are given as follows:

The vector $k\mathbf{v}$ has a *magnitude* of $|k|\,\|\mathbf{v}\|$. We describe this as the absolute value of k times the magnitude of vector $\mathbf{v}$.

The vector $k\mathbf{v}$ has a *direction* that is:

* the same as the direction of $\mathbf{v}$ if $k > 0$, and
* opposite the direction of $\mathbf{v}$ if $k < 0$.

A geometric method for adding two vectors is shown in Figure 7.48. The sum of $\mathbf{u} + \mathbf{v}$ is called the **resultant vector.** Here is how we find this vector:

1. Position $\mathbf{u}$ and $\mathbf{v}$ so that the terminal point of $\mathbf{u}$ coincides with the initial point of $\mathbf{v}$.

2. The resultant vector, $\mathbf{u} + \mathbf{v}$, extends from the initial point of $\mathbf{u}$ to the terminal point of $\mathbf{v}$.

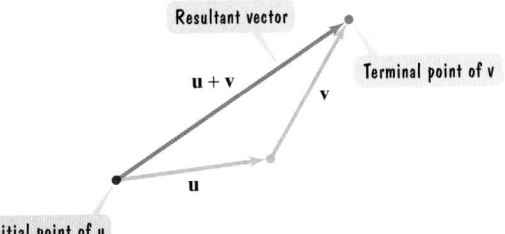

Figure 7.48 Vector addition $\mathbf{u} + \mathbf{v}$; the terminal point of $\mathbf{u}$ coincides with the initial point of $\mathbf{v}$.

The **difference of two vectors, v − u**, is defined as $\mathbf{v} - \mathbf{u} = \mathbf{v} + (-\mathbf{u})$, where −**u** is the scalar multiplication of **u** and −1: −1**u**. The difference **v** − **u** is shown geometrically in Figure 7.49.

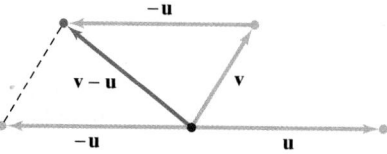

Figure 7.49 Vector subtraction **v** − **u**; the terminal point of **v** coincides with the initial point of −**u**.

3 Represent vectors in the rectangular coordinate system.

Vectors in the Rectangular Coordinate System

As you saw in Example 1, vectors can be shown in the rectangular coordinate system. Now let's see how we can use the rectangular coordinate system to represent vectors. We begin with two vectors that both have a magnitude of 1. Such vectors are called **unit vectors.**

The i and j Unit Vectors

Vector **i** is the unit vector whose direction is along the positive x-axis. Vector **j** is the unit vector whose direction is along the positive y-axis.

Why are the **i** and **j** unit vectors important? Vectors in the rectangular coordinate system can be represented in terms of **i** and **j**. For example, consider vector **v** with initial point at the origin, $(0, 0)$, and terminal point at $P = (a, b)$. The vector **v** is shown in Figure 7.50. We can represent **v** using **i** and **j** as $\mathbf{v} = a\mathbf{i} + b\mathbf{j}$.

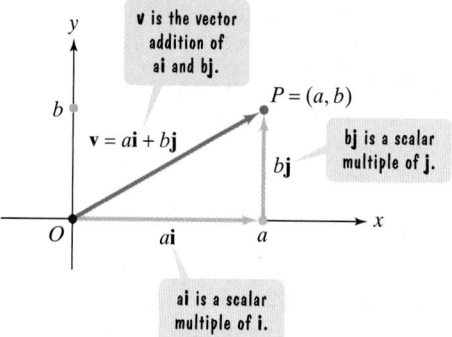

Figure 7.50 Using vector addition, vector **v** is represented as $\mathbf{v} = a\mathbf{i} + b\mathbf{j}$.

Representing Vectors in Rectangular Coordinates

Vector **v**, from $(0, 0)$ to (a, b), is represented as
$$\mathbf{v} = a\mathbf{i} + b\mathbf{j}.$$
The real numbers a and b are called the **scalar components** of **v**. Note that

- a is the **horizontal component** of **v**, and
- b is the **vertical component** of **v**.

The vector sum $a\mathbf{i} + b\mathbf{j}$ is called a **linear combination** of the vectors **i** and **j**. The magnitude of $\mathbf{v} = a\mathbf{i} + b\mathbf{j}$ is given by
$$\|\mathbf{v}\| = \sqrt{a^2 + b^2}.$$

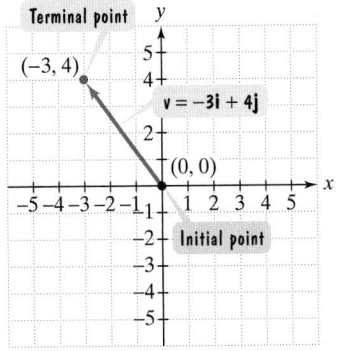

Figure 7.51 Sketching $\mathbf{v} = -3\mathbf{i} + 4\mathbf{j}$ in the rectangular coordinate system

EXAMPLE 2 **Representing a Vector in Rectangular Coordinates and Finding Its Magnitude**

Sketch the vector $\mathbf{v} = -3\mathbf{i} + 4\mathbf{j}$ and find its magnitude.

Solution For the given vector $\mathbf{v} = -3\mathbf{i} + 4\mathbf{j}$, $a = -3$ and $b = 4$. The vector can be represented with its initial point at the origin, $(0, 0)$, as shown in Figure 7.51. The vector's terminal point is then $(a, b) = (-3, 4)$. We sketch the vector by drawing an arrow from $(0, 0)$ to $(-3, 4)$. We determine the magnitude of the vector by using the distance formula. Thus, the magnitude is

$$\|\mathbf{v}\| = \sqrt{a^2 + b^2} = \sqrt{(-3)^2 + 4^2} = \sqrt{9 + 16} = \sqrt{25} = 5.$$

Check Point 2 Sketch the vector $\mathbf{v} = 3\mathbf{i} - 3\mathbf{j}$ and find its magnitude.

The vector in Example 2 was represented with its initial point at the origin. A vector whose initial point is at the origin is called a **position vector.** Any vector in rectangular coordinates whose initial point is not at the origin can be shown to be equal to a position vector. As shown in the following box, this gives us a way to represent vectors between any two points.

Representing Vectors in Rectangular Coordinates

Vector $\mathbf{v}$ with initial point $P_1 = (x_1, y_1)$ and terminal point $P_2 = (x_2, y_2)$ is equal to the position vector

$$\mathbf{v} = (x_2 - x_1)\mathbf{i} + (y_2 - y_1)\mathbf{j}.$$

We can use congruent triangles to derive this formula. Begin with the right triangle in Figure 7.52(a). This triangle shows vector $\mathbf{v}$ from $P_1 = (x_1, y_1)$ to $P_2 = (x_2, y_2)$. In Figure 7.52(b), we move vector $\mathbf{v}$, without changing its magnitude or its direction, so that its initial point is at the origin. Using this position vector in Figure 7.52 (b), we see that

$$\mathbf{v} = a\mathbf{i} + b\mathbf{j}.$$

The equal vectors and the right angles in the right triangles in Figures 7.52(a) and (b) result in congruent triangles. The corresponding sides of these congruent triangles are equal, so that $a = x_2 - x_1$ and $b = y_2 - y_1$. This means that $\mathbf{v}$ may be expressed as

$$\mathbf{v} = a\mathbf{i} + b\mathbf{j} = (x_2 - x_1)\mathbf{i} + (y_2 - y_1)\mathbf{j}.$$

Horizontal component: x-coordinate of terminal point minus x-coordinate of initial point

Vertical component: y-coordinate of terminal point minus y-coordinate of initial point

(a)

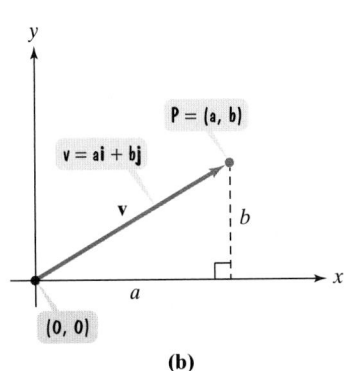

(b)

Figure 7.52

Thus, any vector between two points in rectangular coordinates can be expressed in terms of $\mathbf{i}$ and $\mathbf{j}$. In rectangular coordinates, the term *vector* refers to the position vector expressed in terms of $\mathbf{i}$ and $\mathbf{j}$ that is equal to it.

EXAMPLE 3 Representing a Vector in Rectangular Coordinates

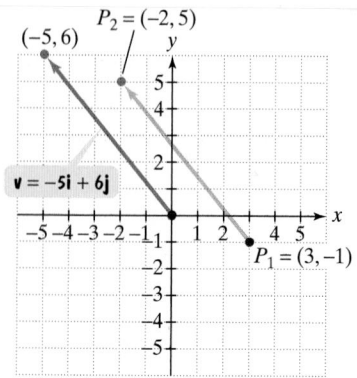

Figure 7.53 Representing the vector from $(3, -1)$ to $(-2, 5)$ as a position vector

Let $\mathbf{v}$ be the vector from initial point $P_1 = (3, -1)$ to terminal point $P_2 = (-2, 5)$. Write $\mathbf{v}$ in terms of $\mathbf{i}$ and $\mathbf{j}$.

Solution We identify the values for the variables in the formula.

$$P_1 = \underset{x_1 \quad y_1}{(3, -1)} \qquad P_2 = \underset{x_2 \quad y_2}{(-2, 5)}$$

Using these values, we write $\mathbf{v}$ in terms of $\mathbf{i}$ and $\mathbf{j}$ as follows:

$$\mathbf{v} = (x_2 - x_1)\mathbf{i} + (y_2 - y_1)\mathbf{j} = (-2 - 3)\mathbf{i} + [5 - (-1)]\mathbf{j} = -5\mathbf{i} + 6\mathbf{j}$$

Figure 7.53 shows the vector from $P_1 = (3, -1)$ to $P_2 = (-2, 5)$ represented in terms of $\mathbf{i}$ and $\mathbf{j}$ and as a position vector.

Check Point 3 Let $\mathbf{v}$ be the vector from initial point $P_1 = (-1, 3)$ to $P_2 = (2, 7)$. Write $\mathbf{v}$ in terms of $\mathbf{i}$ and $\mathbf{j}$.

4 Perform operations with vectors in terms of $\mathbf{i}$ and $\mathbf{j}$.

Operations with Vectors in Terms of i and j

If vectors are expressed in terms of $\mathbf{i}$ and $\mathbf{j}$, we can easily carry out operations such as vector addition, vector subtraction, and scalar multiplication. Recall the geometric definitions of these operations given earlier. Based on these ideas, we can add and subtract vectors using the following procedure:

> **Adding and Subtracting Vectors in Terms of i and j**
> If $\mathbf{v} = a_1\mathbf{i} + b_1\mathbf{j}$ and $\mathbf{w} = a_2\mathbf{i} + b_2\mathbf{j}$, then
> $$\mathbf{v} + \mathbf{w} = (a_1 + a_2)\mathbf{i} + (b_1 + b_2)\mathbf{j}$$
> $$\mathbf{v} - \mathbf{w} = (a_1 - a_2)\mathbf{i} + (b_1 - b_2)\mathbf{j}.$$

EXAMPLE 4 Adding and Subtracting Vectors

If $\mathbf{v} = 5\mathbf{i} + 4\mathbf{j}$ and $\mathbf{w} = 6\mathbf{i} - 9\mathbf{j}$, find:

a. $\mathbf{v} + \mathbf{w}$ **b.** $\mathbf{v} - \mathbf{w}$.

Solution

a. $\mathbf{v} + \mathbf{w} = (5\mathbf{i} + 4\mathbf{j}) + (6\mathbf{i} - 9\mathbf{j})$ *These are the given vectors.*

$\qquad\qquad = (5 + 6)\mathbf{i} + [4 + (-9)]\mathbf{j}$ *Add the horizontal components. Add the vertical components.*

$\qquad\qquad = 11\mathbf{i} - 5\mathbf{j}$ *Simplify.*

b. $\mathbf{v} - \mathbf{w} = (5\mathbf{i} + 4\mathbf{j}) - (6\mathbf{i} - 9\mathbf{j})$ *These are the given vectors.*

$\qquad\qquad = (5 - 6)\mathbf{i} + [4 - (-9)]\mathbf{j}$ *Subtract the horizontal components. Subtract the vertical components.*

$\qquad\qquad = -\mathbf{i} + 13\mathbf{j}$ *Simplify.*

Check Point 4 If $\mathbf{v} = 7\mathbf{i} + 3\mathbf{j}$ and $\mathbf{w} = 4\mathbf{i} - 5\mathbf{j}$, find:
a. $\mathbf{v} + \mathbf{w}$ **b.** $\mathbf{v} - \mathbf{w}$.

How do we perform scalar multiplication if vectors are expressed in terms of $\mathbf{i}$ and $\mathbf{j}$? We use the following procedure to multiply the vector $\mathbf{v}$ by the scalar k:

Scalar Multiplication with a Vector in Terms of i and j

If $\mathbf{v} = a\mathbf{i} + b\mathbf{j}$ and k is a real number, then the scalar multiplication of the vector $\mathbf{v}$ and the scalar k is

$$k\mathbf{v} = (ka)\mathbf{i} + (kb)\mathbf{j}.$$

EXAMPLE 5   Scalar Multiplication

If $\mathbf{v} = 5\mathbf{i} + 4\mathbf{j}$, find:

a. $6\mathbf{v}$ **b.** $-3\mathbf{v}$.

Solution

a. $6\mathbf{v} = 6(5\mathbf{i} + 4\mathbf{j})$ The scalar multiplication is expressed with the given vector.

$\quad = (6 \cdot 5)\mathbf{i} + (6 \cdot 4)\mathbf{j}$ Multiply each component by 6.

$\quad = 30\mathbf{i} + 24\mathbf{j}$ Simplify.

b. $-3\mathbf{v} = -3(5\mathbf{i} + 4\mathbf{j})$ The scalar multiplication is expressed with the given vector.

$\quad = (-3 \cdot 5)\mathbf{i} + (-3 \cdot 4)\mathbf{j}$ Multiply each component by -3.

$\quad = -15\mathbf{i} - 12\mathbf{j}$ Simplify.

> **Check Point 5**   If $\mathbf{v} = 7\mathbf{i} + 10\mathbf{j}$, find:
> **a.** $8\mathbf{v}$ **b.** $-5\mathbf{v}$.

EXAMPLE 6   Vector Operations

If $\mathbf{v} = 5\mathbf{i} + 4\mathbf{j}$ and $\mathbf{w} = 6\mathbf{i} - 9\mathbf{j}$, find $4\mathbf{v} - 2\mathbf{w}$.

Solution

$4\mathbf{v} - 2\mathbf{w} = 4(5\mathbf{i} + 4\mathbf{j}) - 2(6\mathbf{i} - 9\mathbf{j})$ Operations are expressed with the given vectors.

$\quad = 20\mathbf{i} + 16\mathbf{j} - 12\mathbf{i} + 18\mathbf{j}$ Perform each scalar multiplication.

$\quad = (20 - 12)\mathbf{i} + (16 + 18)\mathbf{j}$ Add horizontal and vertical components to perform the vector addition.

$\quad = 8\mathbf{i} + 34\mathbf{j}$ Simplify.

> **Check Point 6**   If $\mathbf{v} = 7\mathbf{i} + 3\mathbf{j}$ and $\mathbf{w} = 4\mathbf{i} - 5\mathbf{j}$, find $6\mathbf{v} - 3\mathbf{w}$.

Properties involving vector operations resemble familiar properties of real numbers. For example, the order in which vectors are added makes no difference:

$$\mathbf{u} + \mathbf{v} = \mathbf{v} + \mathbf{u}.$$

Does this remind you of the commutative property $a + b = b + a$?

Just as 0 plays an important role in the properties of real numbers, the **zero vector 0** plays exactly the same role in the properties of vectors.

The Zero Vector

The vector whose magnitude is 0 is called the **zero vector, 0.** The zero vector is assigned no direction. It can be expressed in terms of **i** and **j** using

$$\mathbf{0} = 0\mathbf{i} + 0\mathbf{j}.$$

Properties of vector addition and scalar multiplication are given as follows:

Properties of Vector Addition and Scalar Multiplication

If **u**, **v**, and **w** are vectors, and c and d are scalars, then the following properties are true.

Vector Addition Properties

1. $\mathbf{u} + \mathbf{v} = \mathbf{v} + \mathbf{u}$	Commutative Property
2. $(\mathbf{u} + \mathbf{v}) + \mathbf{w} = \mathbf{u} + (\mathbf{v} + \mathbf{w})$	Associative Property
3. $\mathbf{u} + \mathbf{0} = \mathbf{0} + \mathbf{u} = \mathbf{u}$	Additive Identity
4. $\mathbf{u} + (-\mathbf{u}) = (-\mathbf{u}) + \mathbf{u} = \mathbf{0}$	Additive Inverse

Scalar Multiplication Properties

1. $(cd)\mathbf{u} = c(d\mathbf{u})$	Associative Property
2. $c(\mathbf{u} + \mathbf{v}) = c\mathbf{u} + c\mathbf{v}$	Distributive Property
3. $(c + d)\mathbf{u} = c\mathbf{u} + d\mathbf{u}$	Distributive Property
4. $1\mathbf{u} = \mathbf{u}$	Multiplicative Identity
5. $0\mathbf{u} = \mathbf{0}$	Multiplication Property
6. $\|c\mathbf{v}\| = |c|\,\|\mathbf{v}\|$	

5 Find the unit vector in the direction of **v**.

Unit Vectors

A unit vector is defined to be a vector whose magnitude is one. In many applications of vectors, it is helpful to find the unit vector that has the same direction as a given vector.

Discovery

To find out why the procedure in the box produces a unit vector, work Exercise 89 in Exercise Set 7.6.

Finding the Unit Vector that Has the Same Direction as a Given Nonzero Vector v

For any nonzero vector **v**, the vector

$$\frac{\mathbf{v}}{\|\mathbf{v}\|}$$

is the unit vector that has the same direction as **v**. To find this vector, divide **v** by its magnitude.

EXAMPLE 7 Finding a Unit Vector

Find the unit vector in the same direction as $\mathbf{v} = 5\mathbf{i} - 12\mathbf{j}$. Then verify that the vector has magnitude 1.

Solution We find the unit vector in the same direction as $\mathbf{v}$ by dividing $\mathbf{v}$ by its magnitude. We first find the magnitude of $\mathbf{v}$.

$$\|\mathbf{v}\| = \sqrt{a^2 + b^2} = \sqrt{5^2 + (-12)^2} = \sqrt{25 + 144} = \sqrt{169} = 13$$

The unit vector in the same direction as $\mathbf{v}$ is

$$\frac{\mathbf{v}}{\|\mathbf{v}\|} = \frac{5\mathbf{i} - 12\mathbf{j}}{13} = \frac{5}{13}\mathbf{i} - \frac{12}{13}\mathbf{j}. \quad \text{This is the scalar multiplication of v and } \tfrac{1}{13}.$$

Now we must verify that the magnitude of this vector is 1. Recall that the magnitude of $a\mathbf{i} + b\mathbf{j}$ is $\sqrt{a^2 + b^2}$. Thus, the magnitude of $\frac{5}{13}\mathbf{i} - \frac{12}{13}\mathbf{j}$ is

$$\sqrt{\left(\frac{5}{13}\right)^2 + \left(-\frac{12}{13}\right)^2} = \sqrt{\frac{25}{169} + \frac{144}{169}} = \sqrt{\frac{169}{169}} = \sqrt{1} = 1.$$

Check Point 7 Find the unit vector in the same direction as $\mathbf{v} = 4\mathbf{i} - 3\mathbf{j}$. Then verify that the vector has magnitude 1.

6 Write a vector in terms of its magnitude and direction.

Writing a Vector in Terms of Its Magnitude and Direction

Consider the vector $\mathbf{v} = a\mathbf{i} + b\mathbf{j}$. The components a and b can be expressed in terms of the magnitude of $\mathbf{v}$ and the angle θ that $\mathbf{v}$ makes with the positive x-axis. This angle is called the **direction angle** of $\mathbf{v}$ and is shown in Figure 7.54. By the definitions of sine and cosine, we have

$$\cos\theta = \frac{a}{\|\mathbf{v}\|} \qquad \text{and} \qquad \sin\theta = \frac{b}{\|\mathbf{v}\|}$$
$$a = \|\mathbf{v}\|\cos\theta \qquad\qquad b = \|\mathbf{v}\|\sin\theta.$$

Figure 7.54 Expressing a vector in terms of its magnitude, $\|\mathbf{v}\|$, and its direction angle, θ

Thus,

$$\mathbf{v} = a\mathbf{i} + b\mathbf{j} = \|\mathbf{v}\|\cos\theta\mathbf{i} + \|\mathbf{v}\|\sin\theta\mathbf{j}.$$

Writing a Vector in Terms of Its Magnitude and Direction

Let $\mathbf{v}$ be a nonzero vector. If θ is the direction angle measured from the positive x-axis to $\mathbf{v}$, then the vector can be expressed in terms of its magnitude and direction angle as

$$\mathbf{v} = \|\mathbf{v}\|\cos\theta\mathbf{i} + \|\mathbf{v}\|\sin\theta\mathbf{j}.$$

A vector that represents the direction and speed of an object in motion is called a **velocity vector**. In Example 8, we express a wind's velocity vector in terms of the wind's magnitude and direction.

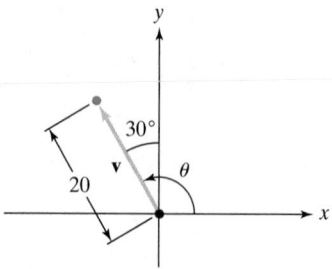

Figure 7.55 Vector **v** represents a wind blowing at 20 miles per hour in the direction N30°W.

EXAMPLE 8 Writing a Vector Whose Magnitude and Direction Are Given

The wind is blowing at 20 miles per hour in the direction N30°W. Express its velocity as a vector **v**.

Solution The vector **v** is shown in Figure 7.55. The vector's direction angle, from the positive x-axis to **v**, is

$$\theta = 90° + 30° = 120°.$$

Because the wind is blowing at 20 miles per hour, the magnitude of **v** is 20 miles per hour: $\|\mathbf{v}\| = 20$. Thus,

$$\mathbf{v} = \|\mathbf{v}\| \cos\theta\mathbf{i} + \|\mathbf{v}\| \sin\theta\mathbf{j} \qquad \text{Use the formula for a vector in terms of magnitude and direction.}$$

$$= 20\cos 120°\mathbf{i} + 20\sin 120°\mathbf{j} \qquad \|\mathbf{v}\| = 20 \text{ and } \theta = 120°.$$

$$= 20\left(-\tfrac{1}{2}\right)\mathbf{i} + 20\left(\frac{\sqrt{3}}{2}\right)\mathbf{j} \qquad \cos 120° = -\frac{1}{2} \text{ and } \sin 120° = \frac{\sqrt{3}}{2}.$$

$$= -10\mathbf{i} + 10\sqrt{3}\mathbf{j} \qquad \text{Simplify.}$$

The wind's velocity can be expressed in terms of **i** and **j** as $\mathbf{v} = -10\mathbf{i} + 10\sqrt{3}\mathbf{j}$.

Check Point 8 The jet stream is blowing at 60 miles per hour in the direction N45°E. Express its velocity as a vector **v** in terms of **i** and **j**.

7 Solve applied problems involving vectors.

Applications

Many physical concepts can be represented by vectors. A vector that represents a pull or push of some type is called a **force vector.** If you are holding a 10-pound package, two force vectors are involved. The force of gravity is exerting a force of magnitude 10 pounds directly downward. This force is shown by vector $\mathbf{F}_1$ in Figure 7.56. Assuming there is no upward or downward movement of the package, you are exerting a force of magnitude 10 pounds directly upward. This force is shown by vector $\mathbf{F}_2$ in Figure 7.56. It has the same magnitude as the force exerted on your package by gravity, but it acts in the opposite direction.

If $\mathbf{F}_1$ and $\mathbf{F}_2$ are two forces acting on an object, the net effect is the same as if just the resultant force, $\mathbf{F}_1 + \mathbf{F}_2$, acted on the object. If the object is not moving, as is the case with your 10-pound package, the vector sum of all forces is the zero vector.

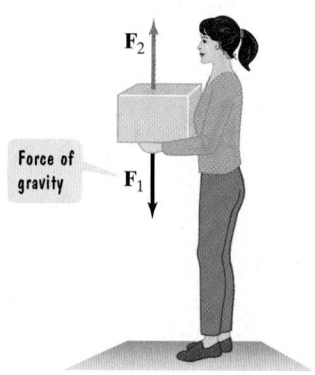

Figure 7.56 Force vectors

EXAMPLE 9 Finding the Resultant Force

Two forces, $\mathbf{F}_1$ and $\mathbf{F}_2$, of magnitude 10 and 30 pounds, respectively, act on an object. The direction of $\mathbf{F}_1$ is N20°E and the direction of $\mathbf{F}_2$ is N65°E. Find the magnitude and the direction of the resultant force. Express the direction angle to the nearest tenth of a degree.

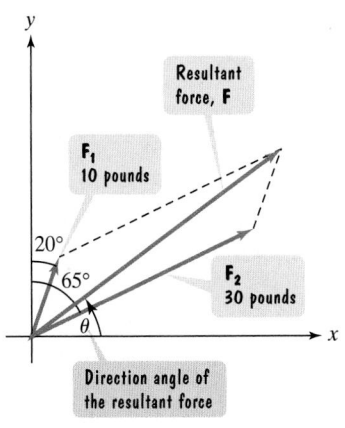

Figure 7.57

Direction angle of the resultant force

Solution The vectors $\mathbf{F_1}$ and $\mathbf{F_2}$ are shown in Figure 7.57. The direction angle for $\mathbf{F_1}$, from the positive x-axis to the vector, is $90°- 20°$, or $70°$. We express $\mathbf{F_1}$ using the formula for a vector in terms of its magnitude and direction.

$$\mathbf{F_1} = \|\mathbf{F_1}\| \cos\theta\mathbf{i} + \|\mathbf{F_1}\| \sin\theta\mathbf{j}$$
$$= 10\cos 70°\mathbf{i} + 10\sin 70°\mathbf{j} \quad \|\mathbf{F_1}\| = 10 \text{ and } \theta = 70°.$$
$$\approx 3.42\mathbf{i} + 9.40\mathbf{j} \quad \text{Use a calculator.}$$

Figure 7.57 illustrates that the direction angle for $\mathbf{F_2}$, from the positive x-axis to the vector, is $90°- 65°$, or $25°$. We express $\mathbf{F_2}$ using the formula for a vector in terms of its magnitude and direction.

$$\mathbf{F_2} = \|\mathbf{F_2}\| \cos\theta\mathbf{i} + \|\mathbf{F_2}\| \sin\theta\mathbf{j}$$
$$= 30\cos 25°\mathbf{i} + 30\sin 25°\mathbf{j} \quad \|\mathbf{F_2}\| = 30 \text{ and } \theta = 25°.$$
$$\approx 27.19\mathbf{i} + 12.68\mathbf{j} \quad \text{Use a calculator.}$$

The resultant force, $\mathbf{F}$, is $\mathbf{F_1} + \mathbf{F_2}$. Thus,

$$\mathbf{F} = \mathbf{F_1} + \mathbf{F_2}$$
$$\approx (3.42\mathbf{i} + 9.40\mathbf{j}) + (27.19\mathbf{i} + 12.68\mathbf{j}) \quad \text{Use } \mathbf{F_1} \text{ and } \mathbf{F_2}, \text{ found above.}$$
$$= (3.42 + 27.19)\mathbf{i} + (9.40 + 12.68)\mathbf{j} \quad \text{Add the horizontal components. Add the vertical components. Simplify.}$$
$$= 30.61\mathbf{i} + 22.08\mathbf{j}.$$

Now that we have the resultant force vector, $\mathbf{F}$, we can find its magnitude.

$$\|\mathbf{F}\| = \sqrt{a^2 + b^2} = \sqrt{(30.61)^2 + (22.08)^2} \approx 37.74$$

The magnitude of the resultant force is approximately 37.74 pounds.

To find θ, the direction angle of the resultant force, we can use

$$\cos\theta = \frac{a}{\|\mathbf{F}\|} \quad \text{or} \quad \sin\theta = \frac{b}{\|\mathbf{F}\|}.$$

Using the first formula, we obtain

$$\cos\theta = \frac{a}{\|\mathbf{F}\|} \approx \frac{30.61}{37.74}. \quad \text{Recall that the resultant force is } \mathbf{F} = 30.61\mathbf{i} + 22.08\mathbf{j}.$$

Thus,

$$\theta = \cos^{-1}\left(\frac{30.61}{37.74}\right) \approx 35.8°. \quad \text{Use a calculator.}$$

The direction angle of the resultant force is approximately $35.8°$.

In summary, the two given forces are equivalent to a single force of approximately 37.74 pounds with a direction angle of approximately $35.8°$.

Study Tip

If $\mathbf{F} = a\mathbf{i} + b\mathbf{j}$, the direction angle, θ, of $\mathbf{F}$ can also be found using

$$\tan\theta = \frac{b}{a}.$$

Check Point 9

Two forces, $\mathbf{F}_1$ and $\mathbf{F}_2$, of magnitude 30 and 60 pounds, respectively, act on an object. The direction of $\mathbf{F}_1$ is N10°E and the direction of $\mathbf{F}_2$ is N60°E. Find the magnitude, to the nearest hundredth of a pound, and the direction angle, to the nearest tenth of a degree, of the resultant force.

We have seen that velocity vectors represent the direction and speed of moving objects. Boats moving in currents and airplanes flying in winds are situations in which two velocity vectors act simultaneously. For example, suppose **v** represents the velocity of a plane in still air. Further suppose that **w** represents the velocity of the wind. The actual speed and direction of the plane is given by the vector **v** + **w**. This resultant vector describes the plane's speed and direction relative to the ground. Problems involving the resultant velocity of a boat or plane are solved using the same method that we used in Example 9 to find a single resultant force equivalent to two given forces.

EXERCISE SET 7.6

Practice Exercises

*In Exercises 1–4, **u** and **v** have the same direction. In each exercise:* **a.** *Find* $\|\mathbf{u}\|$. **b.** *Find* $\|\mathbf{v}\|$. **c.** *Is* $\mathbf{u} = \mathbf{v}$? *Explain.*

1.

2.

3.

4.

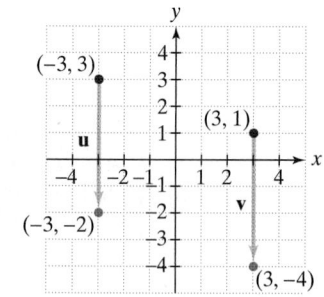

In Exercises 5–12, sketch each vector as a position vector and find its magnitude.

5. $\mathbf{v} = 3\mathbf{i} + \mathbf{j}$

6. $\mathbf{v} = 2\mathbf{i} + 3\mathbf{j}$

7. $\mathbf{v} = \mathbf{i} - \mathbf{j}$

8. $\mathbf{v} = -\mathbf{i} - \mathbf{j}$

9. $\mathbf{v} = -6\mathbf{i} - 2\mathbf{j}$

10. $\mathbf{v} = 5\mathbf{i} - 2\mathbf{j}$

11. $\mathbf{v} = -4\mathbf{i}$

12. $\mathbf{v} = -5\mathbf{j}$

In Exercises 13–20, let $\mathbf{v}$ be the vector from initial point P_1 to terminal point P_2. Write $\mathbf{v}$ in terms of $\mathbf{i}$ and $\mathbf{j}$.

13. $P_1 = (-4, -4), P_2 = (6, 2)$

14. $P_1 = (2, -5), P_2 = (-6, 6)$

15. $P_1 = (-8, 6), P_2 = (-2, 3)$

16. $P_1 = (-7, -4), P_2 = (0, -2)$

17. $P_1 = (-1, 7), P_2 = (-7, -7)$

18. $P_1 = (-1, 6), P_2 = (7, -5)$

19. $P_1 = (-3, 4), P_2 = (6, 4)$

20. $P_1 = (4, -5), P_2 = (4, 3)$

In Exercises 21–38, let

$$\mathbf{u} = 2\mathbf{i} - 5\mathbf{j}, \mathbf{v} = -3\mathbf{i} + 7\mathbf{j}, \text{ and } \mathbf{w} = -\mathbf{i} - 6\mathbf{j}.$$

Find each specified vector or scalar.

21. $\mathbf{u} + \mathbf{v}$

22. $\mathbf{v} + \mathbf{w}$

23. $\mathbf{u} - \mathbf{v}$

24. $\mathbf{v} - \mathbf{w}$

25. $\mathbf{v} - \mathbf{u}$

26. $\mathbf{w} - \mathbf{v}$

27. $5\mathbf{v}$

28. $6\mathbf{v}$

29. $-4\mathbf{w}$

30. $-7\mathbf{w}$

31. $3\mathbf{w} + 2\mathbf{v}$

32. $3\mathbf{u} + 4\mathbf{v}$

33. $3\mathbf{v} - 4\mathbf{w}$

34. $4\mathbf{w} - 3\mathbf{v}$

35. $\|2\mathbf{u}\|$

36. $\|-2\mathbf{u}\|$

37. $\|\mathbf{w} - \mathbf{u}\|$

38. $\|\mathbf{u} - \mathbf{w}\|$

In Exercises 39–46, find the unit vector that has the same direction as the vector $\mathbf{v}$.

39. $\mathbf{v} = 6\mathbf{i}$

40. $\mathbf{v} = -5\mathbf{j}$

41. $\mathbf{v} = 3\mathbf{i} - 4\mathbf{j}$

42. $\mathbf{v} = 8\mathbf{i} - 6\mathbf{j}$

43. $\mathbf{v} = 3\mathbf{i} - 2\mathbf{j}$

44. $\mathbf{v} = 4\mathbf{i} - 2\mathbf{j}$

45. $\mathbf{v} = \mathbf{i} + \mathbf{j}$

46. $\mathbf{v} = \mathbf{i} - \mathbf{j}$

In Exercises 47–52, write the vector $\mathbf{v}$ in terms of $\mathbf{i}$ and $\mathbf{j}$ whose magnitude $\|\mathbf{v}\|$ and direction angle θ are given.

47. $\|\mathbf{v}\| = 6, \quad \theta = 30°$

48. $\|\mathbf{v}\| = 8, \quad \theta = 45°$

49. $\|\mathbf{v}\| = 12, \quad \theta = 225°$

50. $\|\mathbf{v}\| = 10, \quad \theta = 330°$

51. $\|\mathbf{v}\| = \frac{1}{2}, \quad \theta = 113°$

52. $\|\mathbf{v}\| = \frac{1}{4}, \quad \theta = 200°$

Application Exercises

In Exercises 53–56, a vector is described. Express the vector in terms of $\mathbf{i}$ and $\mathbf{j}$. If exact values are not possible, round components to the nearest tenth.

53. A quarterback releases a football with a speed of 44 feet per second at an angle of 30° with the horizontal.

54. A child pulls a sled along level ground by exerting a force of 30 pounds on a handle that makes an angle of 45° with the ground.

55. A plane approaches a runway at 150 miles per hour at an angle of 8° with the runway.

56. A plane with an airspeed of 450 miles per hour is flying in the direction N35°W.

Vectors are used in computer graphics to determine lengths of shadows over flat surfaces. The length of the shadow for $\mathbf{v}$ in the figure shown is the absolute value of the vector's horizontal component. In Exercises 57–58, the magnitude and direction angle of $\mathbf{v}$ are given. Write $\mathbf{v}$ in terms of $\mathbf{i}$ and $\mathbf{j}$. Then find the length of the shadow to the nearest tenth of an inch.

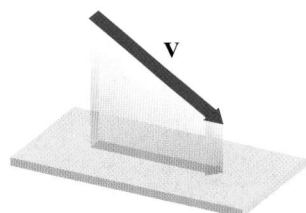

57. $\|\mathbf{v}\| = 1.5$ inches, $\theta = 25°$

58. $\|\mathbf{v}\| = 1.8$ inches, $\theta = 40°$

59. The magnitude and direction of two forces acting on an object are 70 pounds, S56°E, and 50 pounds, N72°E, respectively. Find the magnitude, to the nearest hundredth of a pound, and the direction angle, to the nearest tenth of a degree, of the resultant force.

60. The magnitude and direction exerted by two tugboats towing a ship are 4200 pounds, N65°E, and 3000 pounds, S58°E, respectively. Find the magnitude, to the nearest pound, and the direction angle, to the nearest tenth of a degree, of the resultant force.

The figure shows a box being pulled up a ramp inclined at 18° from the horizontal.

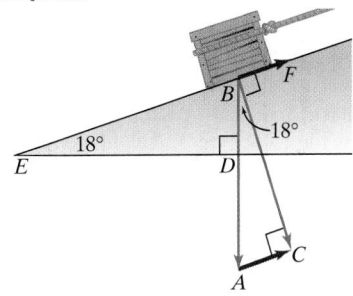

Use the following information to solve Exercises 61–62.

$\overrightarrow{BA}$ = force of gravity

$\|\overrightarrow{BA}\|$ = weight of the box

$\|\overrightarrow{AC}\|$ = magnitude of the force needed to pull the box up the ramp

$\|\overrightarrow{BC}\|$ = magnitude of the force of the box against the ramp

61. If the box weighs 100 pounds, find the magnitude of the force needed to pull it up the ramp.

62. If a force of 30 pounds is needed to pull the box up the ramp, find the weight of the box.

The forces $\mathbf{F}_1$, $\mathbf{F}_2$, $\mathbf{F}_3$, ..., $\mathbf{F}_n$ *acting on an object are in* **equilibrium** *if the resultant force is the zero vector:*

$$\mathbf{F}_1 + \mathbf{F}_2 + \mathbf{F}_3 + \cdots + \mathbf{F}_n = \mathbf{0}.$$

In Exercises 63–66, the given forces are acting on an object.
  a. *Find the resultant force.*
  b. *What additional force is required for the given forces to be in equilibrium?*

63. $\mathbf{F}_1 = 3\mathbf{i} - 5\mathbf{j}$, $\mathbf{F}_2 = 6\mathbf{i} + 2\mathbf{j}$

64. $\mathbf{F}_1 = -2\mathbf{i} + 3\mathbf{j}$, $\mathbf{F}_2 = \mathbf{i} - \mathbf{j}$, $\mathbf{F}_3 = 5\mathbf{i} - 12\mathbf{j}$

65.

66.

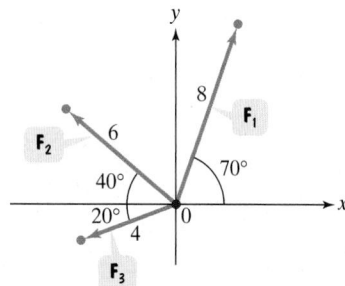

67. The figure shows a small plane flying at a speed of 180 miles per hour on a bearing of N50°E. The wind is blowing from west to east at 40 miles per hour. The figure indicates that **v** represents the velocity of the plane in still air and **w** represents the velocity of the wind.

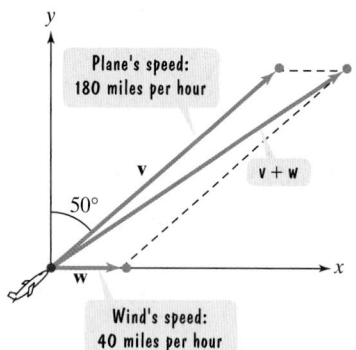

  a. Express **v** and **w** in terms of their magnitudes and direction angles.
  b. Find the resultant vector, **v** + **w**.
  c. The magnitude of **v** + **w**, called the **ground speed** of the plane, gives its speed relative to the ground. Approximate the ground speed to the nearest mile per hour.

  d. The direction angle of **v** + **w** gives the plane's true course relative to the ground. Approximate the true course to the nearest tenth of a degree. What is the plane's true bearing?

68. Use the procedure outlined in Exercise 67 to solve this exercise. A plane is flying at a speed of 400 miles per hour on a bearing of N50°W. The wind is blowing at 30 miles per hour on a bearing of N25°E.

  a. Approximate the plane's ground speed to the nearest mile per hour.
  b. Approximate the plane's true course to the nearest tenth of a degree. What is its true bearing?

69. A plane is flying at a speed of 320 miles per hour on a bearing of N70°E. Its ground speed is 370 miles per hour and its true course is 30°. Find the speed, to the nearest mile per hour, and the direction angle, to the nearest tenth of a degree, of the wind.

70. A plane is flying at a speed of 540 miles per hour on a bearing of S36°E. Its ground speed is 500 miles per hour and its true bearing is S44°E. Find the speed, to the nearest mile per hour, and the direction angle, to the nearest tenth of a degree, of the wind.

Writing in Mathematics

71. What is a directed line segment?

72. What are equal vectors?

73. If vector **v** is represented by an arrow, how is −3**v** represented?

74. If vectors **u** and **v** are represented by arrows, describe how the vector sum **u** + **v** is represented.

75. What is the **i** vector?

76. What is the **j** vector?

77. What is a position vector? How is a position vector represented using **i** and **j**?

78. If **v** is a vector between any two points in the rectangular coordinate system, explain how to write **v** in terms of **i** and **j**.

79. If two vectors are expressed in terms of **i** and **j**, explain how to find their sum.

80. If two vectors are expressed in terms of **i** and **j**, explain how to find their difference.

81. If a vector is expressed in terms of **i** and **j**, explain how to find the scalar multiplication of the vector and a given scalar k.

82. What is the zero vector?

83. Describe one similarity between the zero vector and the number 0.

84. Explain how to find the unit vector in the direction of any given vector **v**.

85. Explain how to write a vector in terms of its magnitude and direction.

86. You are on an airplane. The pilot announces the plane's speed over the intercom. Which speed do you think is being reported: the speed of the plane in still air or the speed after the effect of the wind has been accounted for? Explain your answer.

87. Use vectors to explain why it is difficult to hold a heavy stack of books perfectly still for a long period of time. As you become exhausted, what eventually happens? What does this mean in terms of the forces acting on the books?

 Critical Thinking Exercises

88. Use the figure shown to select a true statement.

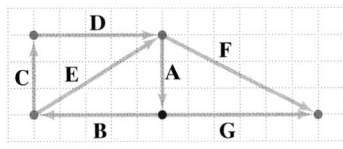

a. **A + B = E**
b. **D + A + B + C = 0**
c. **B − E = G − F**
d. $\|\mathbf{A}\| \neq \|\mathbf{C}\|$

89. Let $\mathbf{v} = a\mathbf{i} + b\mathbf{j}$. Show that $\dfrac{\mathbf{v}}{\|\mathbf{v}\|}$ is a unit vector in the direction of **v**.

In Exercises 90–91, refer to the navigational compass shown in the figure. The compass is marked clockwise in degrees that start at north 0°.

90. An airplane has an air speed of 240 miles per hour and a compass heading of 280°. A steady wind of 30 miles per hour is blowing in the direction of 265°. What is the plane's true speed relative to the ground? What is its compass heading relative to the ground?

91. Two tugboats are pulling on a large ship that has gone aground. One tug pulls with a force of 2500 pounds in a compass direction of 55°. The second tug pulls with a force of 2000 pounds in a compass direction of 95°. Find the magnitude and the compass direction of the resultant force.

92. You want to fly your small plane due north, but there is a 75 kilometer wind blowing from west to east.

a. Find the direction angle for where you should head the plane if your speed relative to the air is 310 kilometers per hour.

b. If you increase your air speed, should the direction angle in part (a) increase or decrease? Explain your answer.

SECTION 7.7 **The Dot Product**

Objectives

1. Find the dot product of two vectors.
2. Find the angle between two vectors.
3. Use the dot product to determine if two vectors are orthogonal.
4. Find the projection of a vector onto another vector.
5. Express a vector as the sum of two orthogonal vectors.
6. Compute work.

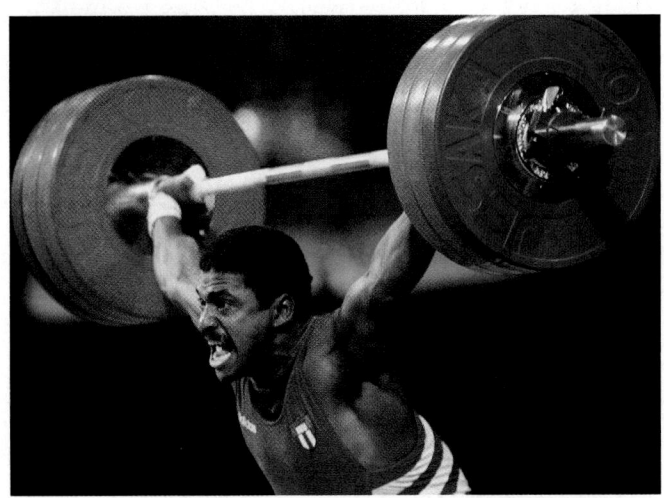

Talk about hard work! I can see the weightlifter's muscles quivering from the exertion of holding the barbell in a stationary position above his head. Still, I'm not sure if he's doing as much work as I, sitting at my desk with my brain quivering from studying trigonometric functions and their applications.

Would it surprise you to know that neither you nor the weightlifter are doing any work at all? The definition of work in physics and mathematics is not the same as what we mean by "work" in everyday use. To understand what is involved in real work, we turn to a new vector operation called the dot product.

1 Find the dot product of two vectors.

The Dot Product of Two Vectors

The operations of vector addition and scalar multiplication result in vectors. By contrast, the *dot product* of two vectors results in a scalar (a real number), rather than a vector.

Definition of the Dot Product

If $\mathbf{v} = a_1\mathbf{i} + b_1\mathbf{j}$ and $\mathbf{w} = a_2\mathbf{i} + b_2\mathbf{j}$ are vectors, the **dot product** $\mathbf{v} \cdot \mathbf{w}$ is defined as

$$\mathbf{v} \cdot \mathbf{w} = a_1 a_2 + b_1 b_2.$$

The dot product of two vectors is the sum of the products of their horizontal and vertical components.

EXAMPLE 1   Finding Dot Products

If $\mathbf{v} = 5\mathbf{i} - 2\mathbf{j}$ and $\mathbf{w} = -3\mathbf{i} + 4\mathbf{j}$, find:

a. $\mathbf{v} \cdot \mathbf{w}$ **b.** $\mathbf{w} \cdot \mathbf{v}$ **c.** $\mathbf{v} \cdot \mathbf{v}$.

Solution   To find each dot product, multiply the two horizontal components, and then multiply the two vertical components. Finally, add the two products.

a. $\mathbf{v} \cdot \mathbf{w} = 5(-3) + (-2)(4) = -15 - 8 = -23$

> Multiply the horizontal components and multiply the vertical components of
> $\mathbf{v} = 5\mathbf{i} - 2\mathbf{j}$ and $\mathbf{w} = -3\mathbf{i} + 4\mathbf{j}$.

b. $\mathbf{w} \cdot \mathbf{v} = -3(5) + 4(-2) = -15 - 8 = -23$

> Multiply the horizontal components and multiply the vertical components of
> $\mathbf{w} = -3\mathbf{i} + 4\mathbf{j}$ and $\mathbf{v} = 5\mathbf{i} - 2\mathbf{j}$.

c. $\mathbf{v} \cdot \mathbf{v} = 5(5) + (-2)(-2) = 25 + 4 = 29$

> Multiply the horizontal components and multiply the vertical components of
> $\mathbf{v} = 5\mathbf{i} - 2\mathbf{j}$ and $\mathbf{v} = 5\mathbf{i} - 2\mathbf{j}$.

Check Point 1   If $\mathbf{v} = 7\mathbf{i} - 4\mathbf{j}$ and $\mathbf{w} = 2\mathbf{i} - \mathbf{j}$, find:

a. $\mathbf{v} \cdot \mathbf{w}$ **b.** $\mathbf{w} \cdot \mathbf{v}$ **c.** $\mathbf{w} \cdot \mathbf{w}$.

In Example 1 and Check Point 1, did you notice that $\mathbf{v} \cdot \mathbf{w}$ and $\mathbf{w} \cdot \mathbf{v}$ produced the same scalar? The fact that $\mathbf{v} \cdot \mathbf{w} = \mathbf{w} \cdot \mathbf{v}$ follows from the definition of the dot product. Properties of the dot product are given in the box at the top of the next page. Proofs for some of these properties are given in the appendix.

Properties of the Dot Product

If $\mathbf{u}$, $\mathbf{v}$, and $\mathbf{w}$ are vectors, and c is a scalar, then

1. $\mathbf{u} \cdot \mathbf{v} = \mathbf{v} \cdot \mathbf{u}$
2. $\mathbf{u} \cdot (\mathbf{v} + \mathbf{w}) = \mathbf{u} \cdot \mathbf{v} + \mathbf{u} \cdot \mathbf{w}$
3. $\mathbf{0} \cdot \mathbf{v} = 0$
4. $\mathbf{v} \cdot \mathbf{v} = \|\mathbf{v}\|^2$
5. $(c\mathbf{u}) \cdot \mathbf{v} = c(\mathbf{u} \cdot \mathbf{v}) = \mathbf{u} \cdot (c\mathbf{v})$

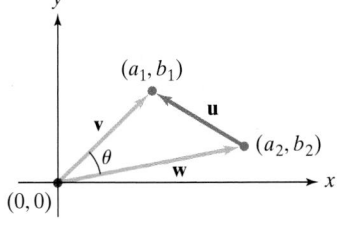

Figure 7.58

The Angle between Two Vectors

The Law of Cosines can be used to derive another formula for the dot product. This formula will give us a way to find the angle between two vectors.

Figure 7.58 shows vectors $\mathbf{v} = a_1\mathbf{i} + b_1\mathbf{j}$ and $\mathbf{w} = a_2\mathbf{i} + b_2\mathbf{j}$. By the definition of the dot product, we know that $\mathbf{v} \cdot \mathbf{w} = a_1 a_2 + b_1 b_2$. Our new formula for the dot product involves the angle between the vectors, shown as θ in the figure. Apply the Law of Cosines to the triangle shown in the figure.

$$\|\mathbf{u}\|^2 = \|\mathbf{v}\|^2 + \|\mathbf{w}\|^2 - 2\|\mathbf{v}\|\|\mathbf{w}\|\cos\theta$$
Use the Law of Cosines.

$$(a_1 - a_2)^2 + (b_1 - b_2)^2 = (a_1^2 + b_1^2) + (a_2^2 + b_2^2) - 2\|\mathbf{v}\|\|\mathbf{w}\|\cos\theta$$
Square the magnitudes of vectors u, v, and w.

$$a_1^2 - 2a_1 a_2 + a_2^2 + b_1^2 - 2b_1 b_2 + b_2^2 = a_1^2 + b_1^2 + a_2^2 + b_2^2 - 2\|\mathbf{v}\|\|\mathbf{w}\|\cos\theta$$
Square the binomials using $(A - B)^2 = A^2 - 2AB + B^2$.

$$-2a_1 a_2 - 2b_1 b_2 = -2\|\mathbf{v}\|\|\mathbf{w}\|\cos\theta$$
Subtract a_1^2, a_2^2, b_1^2, and b_2^2 from both sides of the equation.

$$a_1 a_2 + b_1 b_2 = \|\mathbf{v}\|\|\mathbf{w}\|\cos\theta$$
Divide both sides by -2.

By definition,
$\mathbf{v} \cdot \mathbf{w} = a_1 a_2 + b_1 b_2$.

$$\mathbf{v} \cdot \mathbf{w} = \|\mathbf{v}\|\|\mathbf{w}\|\cos\theta$$
Substitute $\mathbf{v} \cdot \mathbf{w}$ for the expression on the left side of the equation.

Alternative Formula for the Dot Product

If $\mathbf{v}$ and $\mathbf{w}$ are two nonzero vectors and θ is the smallest nonnegative angle between them, then

$$\mathbf{v} \cdot \mathbf{w} = \|\mathbf{v}\|\|\mathbf{w}\|\cos\theta.$$

2 Find the angle between two vectors.

Solving the formula in the box for $\cos\theta$ gives us a formula for finding the angle between vectors:

Formula for the Angle between Two Vectors

If $\mathbf{v}$ and $\mathbf{w}$ are two nonzero vectors and θ is the smallest nonnegative angle between $\mathbf{v}$ and $\mathbf{w}$, then

$$\cos\theta = \frac{\mathbf{v} \cdot \mathbf{w}}{\|\mathbf{v}\|\|\mathbf{w}\|} \quad \text{and} \quad \theta = \cos^{-1}\left(\frac{\mathbf{v} \cdot \mathbf{w}}{\|\mathbf{v}\|\|\mathbf{w}\|}\right).$$

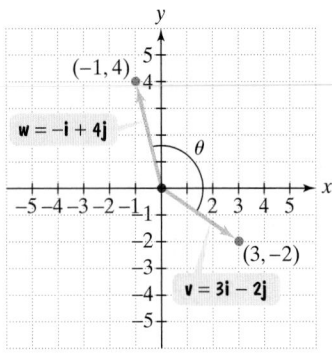

Figure 7.59 Finding the angle between two vectors

EXAMPLE 2 Finding the Angle between Two Vectors

Find the angle θ between the vectors $\mathbf{v} = 3\mathbf{i} - 2\mathbf{j}$ and $\mathbf{w} = -\mathbf{i} + 4\mathbf{j}$, shown in Figure 7.59. Round to the nearest tenth of a degree.

Solution Use the formula for the angle between two vectors.

$$\cos \theta = \frac{\mathbf{v} \cdot \mathbf{w}}{\|\mathbf{v}\| \|\mathbf{w}\|}$$ *This is the formula for the cosine of the angle between two vectors.*

$$= \frac{(3\mathbf{i} - 2\mathbf{j}) \cdot (-\mathbf{i} + 4\mathbf{j})}{\sqrt{3^2 + (-2)^2} \sqrt{(-1)^2 + 4^2}}$$ *Substitute the given vectors in the numerator. Find the magnitude of each vector in the denominator.*

$$= \frac{3(-1) + (-2)(4)}{\sqrt{13} \sqrt{17}}$$ *Find the dot product in the numerator. Simplify in the denominator.*

$$= -\frac{11}{\sqrt{221}}$$ *Perform the indicated operations.*

The angle θ between the vectors is

$$\theta = \cos^{-1}\left(-\frac{11}{\sqrt{221}}\right) \approx 137.7°.$$ *Use a calculator.*

Find the angle between the vectors $\mathbf{v} = 4\mathbf{i} - 3\mathbf{j}$ and $\mathbf{w} = \mathbf{i} + 2\mathbf{j}$. Round to the nearest tenth of a degree.

3 Use the dot product to determine if two vectors are orthogonal.

Parallel and Orthogonal Vectors

Two vectors are **parallel** when the angle θ between the vectors is 0° or 180°. If $\theta = 0°$, the vectors point in the same direction. If $\theta = 180°$, the vectors point in opposite directions. Figure 7.60 shows parallel vectors.

$\theta = 0°$ and $\cos \theta = 1$. Vectors point in the same direction.

$\theta = 180°$ and $\cos \theta = -1$. Vectors point in opposite directions.

Figure 7.60 Parallel vectors

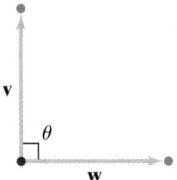

Figure 7.61 Orthogonal vectors: $\theta = 90°$ and $\cos \theta = 0$

Two vectors are **orthogonal** when the angle between the vectors is 90°, shown in Figure 7.61. (The word "orthogonal," rather than "perpendicular," is used to describe vectors that meet at right angles.) We know that $\mathbf{v} \cdot \mathbf{w} = \|\mathbf{v}\| \|\mathbf{w}\| \cos \theta$. We also know that two vectors are orthogonal if and only if the angle between them is 90°. Using this formula for the dot product and the meaning of orthogonal vectors gives the following result:

The Dot Product and Orthogonal Vectors

Two nonzero vectors $\mathbf{v}$ and $\mathbf{w}$ are orthogonal if and only if $\mathbf{v} \cdot \mathbf{w} = 0$.
Because $\mathbf{0} \cdot \mathbf{v} = 0$, the zero vector is orthogonal to every vector $\mathbf{v}$.

Figure 7.62 Orthogonal vectors

EXAMPLE 3   Determining Whether Vectors Are Orthogonal

Are the vectors $\mathbf{v} = 6\mathbf{i} - 3\mathbf{j}$ and $\mathbf{w} = \mathbf{i} + 2\mathbf{j}$ orthogonal?

Solution   The vectors are orthogonal if their dot product is 0. Begin by finding $\mathbf{v} \cdot \mathbf{w}$.

$$\mathbf{v} \cdot \mathbf{w} = (6\mathbf{i} - 3\mathbf{j}) \cdot (\mathbf{i} + 2\mathbf{j}) = 6(1) + (-3)(2) = 6 - 6 = 0$$

The dot product is 0. Thus, the given vectors are orthogonal. They are shown in Figure 7.62.

> **Check Point 3**   Are the vectors $\mathbf{v} = 2\mathbf{i} + 3\mathbf{j}$ and $\mathbf{w} = 6\mathbf{i} - 4\mathbf{j}$ orthogonal?

Projection of a Vector Onto Another Vector

You know how to add two vectors to obtain a resultant vector. We now reverse this process by expressing a vector as the sum of two orthogonal vectors. By doing this, you can determine how much force is applied in a particular direction. For example, Figure 7.63 shows a boat on a tilted ramp. The force due to gravity, $\mathbf{F}$, is pulling straight down on the boat. Part of this force, $\mathbf{F}_1$, is pushing the boat down the ramp. Another part of this force, $\mathbf{F}_2$, is pressing the boat against the ramp, at a right angle to the incline. These two orthogonal vectors, $\mathbf{F}_1$ and $\mathbf{F}_2$, are called the **vector components** of $\mathbf{F}$. Notice that

$$\mathbf{F} = \mathbf{F}_1 + \mathbf{F}_2.$$

A method for finding $\mathbf{F}_1$ and $\mathbf{F}_2$ involves projecting a vector onto another vector.

Figure 7.64 shows two nonzero vectors, $\mathbf{v}$ and $\mathbf{w}$, with the same initial point. The angle between the vectors, θ, is acute in Figure 7.64(a) and obtuse in Figure 7.64(b). A third vector, called the **vector projection of v onto w,** is also shown in each figure, denoted by $\text{proj}_\mathbf{w}\mathbf{v}$.

How is the vector projection of $\mathbf{v}$ onto $\mathbf{w}$ formed? Draw the red line segment from the terminal point of $\mathbf{v}$ that forms a right angle with a line through $\mathbf{w}$, shown in red. The projection of $\mathbf{v}$ onto $\mathbf{w}$ lies on a line through $\mathbf{w}$, and is parallel to vector $\mathbf{w}$. This vector begins at the common initial point of $\mathbf{v}$ and $\mathbf{w}$. It ends at the point where the red line segment intersects the line through $\mathbf{w}$.

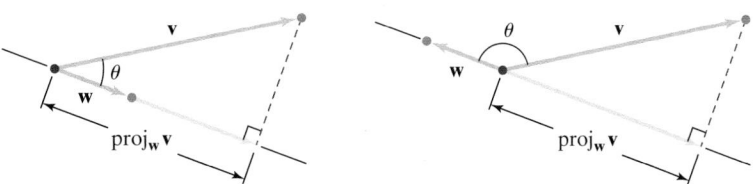

Figure 7.64(a)   **Figure 7.64(b)**

Our goal is to determine an expression for $\text{proj}_\mathbf{w}\mathbf{v}$. We begin with its magnitude. By the definition of the cosine function,

$$\cos\theta = \frac{\|\text{proj}_\mathbf{w}\mathbf{v}\|}{\|\mathbf{v}\|}.$$

> This is the magnitude of the vector projection of **v** onto **w**.

$$\|\mathbf{v}\| \cos\theta = \|\text{proj}_\mathbf{w}\mathbf{v}\| \qquad \text{Multiply both sides by } \|\mathbf{v}\|.$$

$$\|\text{proj}_\mathbf{w}\mathbf{v}\| = \|\mathbf{v}\| \cos\theta \qquad \text{Reverse the two sides.}$$

We can rewrite the right side of this equation and obtain another expression for the magnitude of the vector projection of $\mathbf{v}$ onto $\mathbf{w}$. To do so, use the alternate formula for the dot product, $\mathbf{v} \cdot \mathbf{w} = \|\mathbf{v}\| \|\mathbf{w}\| \cos\theta$.

e 7.63

Divide both sides by $\|\mathbf{w}\|$:

$$\frac{\mathbf{v} \cdot \mathbf{w}}{\|\mathbf{w}\|} = \|\mathbf{v}\| \cos \theta.$$

$$\frac{\mathbf{v} \cdot \mathbf{w}}{\|\mathbf{w}\|} = \|\mathbf{v}\| \cos \theta$$

$$\|\text{proj}_{\mathbf{w}} \mathbf{v}\| = \|\mathbf{v}\| \cos \theta$$

The expression on the right side of this equation, $\|\mathbf{v}\| \cos \theta$, is the same expression that appears in the formula for $\|\text{proj}_{\mathbf{w}} \mathbf{v}\|$. Thus,

$$\|\text{proj}_{\mathbf{w}} \mathbf{v}\| = \|\mathbf{v}\| \cos \theta = \frac{\mathbf{v} \cdot \mathbf{w}}{\|\mathbf{w}\|}.$$

We use the formula for the magnitude of $\text{proj}_{\mathbf{w}} \mathbf{v}$ to find the vector itself. This is done by finding the scalar product of the magnitude and the unit vector in the direction of $\mathbf{w}$.

$$\text{proj}_{\mathbf{w}} \mathbf{v} = \left(\frac{\mathbf{v} \cdot \mathbf{w}}{\|\mathbf{w}\|} \right) \left(\frac{\mathbf{w}}{\|\mathbf{w}\|} \right) = \frac{\mathbf{v} \cdot \mathbf{w}}{\|\mathbf{w}\|^2} \mathbf{w}$$

This is the magnitude of the vector projection of v onto w.

This is the unit vector in the direction of w.

4 Find the projection of a vector onto another vector.

The Vector Projection of v Onto w

If $\mathbf{v}$ and $\mathbf{w}$ are two nonzero vectors, the vector projection of $\mathbf{v}$ onto $\mathbf{w}$ is

$$\text{proj}_{\mathbf{w}} \mathbf{v} = \frac{\mathbf{v} \cdot \mathbf{w}}{\|\mathbf{w}\|^2} \mathbf{w}.$$

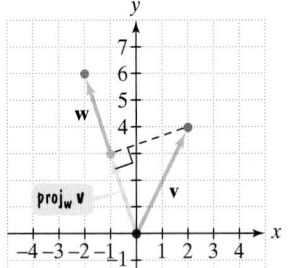

Figure 7.65 The vector projection of **v** onto **w**

EXAMPLE 4 Finding the Vector Projection of One Vector Onto Another

If $\mathbf{v} = 2\mathbf{i} + 4\mathbf{j}$ and $\mathbf{w} = -2\mathbf{i} + 6\mathbf{j}$, find the vector projection of $\mathbf{v}$ onto $\mathbf{w}$.

Solution The vector projection of $\mathbf{v}$ onto $\mathbf{w}$ is found using the formula for $\text{proj}_{\mathbf{w}} \mathbf{v}$.

$$\text{proj}_{\mathbf{w}} \mathbf{v} = \frac{\mathbf{v} \cdot \mathbf{w}}{\|\mathbf{w}\|^2} \mathbf{w} = \frac{(2\mathbf{i} + 4\mathbf{j}) \cdot (-2\mathbf{i} + 6\mathbf{j})}{\left(\sqrt{(-2)^2 + 6^2} \right)^2} \mathbf{w}$$

$$= \frac{2(-2) + 4(6)}{\left(\sqrt{40} \right)^2} \mathbf{w} = \frac{20}{40} \mathbf{w} = \tfrac{1}{2}(-2\mathbf{i} + 6\mathbf{j}) = -\mathbf{i} + 3\mathbf{j}$$

The three vectors $\mathbf{v}$, $\mathbf{w}$, and $\text{proj}_{\mathbf{w}} \mathbf{v}$, are shown in Figure 7.65.

Check Point 4 If $\mathbf{v} = 2\mathbf{i} - 5\mathbf{j}$ and $\mathbf{w} = \mathbf{i} - \mathbf{j}$, find the vector projection of $\mathbf{v}$ onto $\mathbf{w}$.

5 Express a vector as the sum of two orthogonal vectors.

We use the vector projection of $\mathbf{v}$ onto $\mathbf{w}$, $\text{proj}_{\mathbf{w}} \mathbf{v}$, to express $\mathbf{v}$ as the sum of two orthogonal vectors.

The Vector Components of v

Let $\mathbf{v}$ and $\mathbf{w}$ be two nonzero vectors. Vector $\mathbf{v}$ can be expressed as the sum of two orthogonal vectors, $\mathbf{v}_1$ and $\mathbf{v}_2$, where $\mathbf{v}_1$ is parallel to $\mathbf{w}$ and $\mathbf{v}_2$ is orthogonal to $\mathbf{w}$.

$$\mathbf{v}_1 = \text{proj}_{\mathbf{w}} \mathbf{v} = \frac{\mathbf{v} \cdot \mathbf{w}}{\|\mathbf{w}\|^2} \mathbf{w}, \quad \mathbf{v}_2 = \mathbf{v} - \mathbf{v}_1$$

Thus, $\mathbf{v} = \mathbf{v}_1 + \mathbf{v}_2$. The vectors $\mathbf{v}_1$ and $\mathbf{v}_2$ are called the **vector components** of $\mathbf{v}$. The process of expressing $\mathbf{v}$ as $\mathbf{v}_1 + \mathbf{v}_2$ is called the **decomposition** of $\mathbf{v}$ into $\mathbf{v}_1$ and $\mathbf{v}_2$.

EXAMPLE 5 **Decomposing a Vector into Two Orthogonal Vectors**

Let $\mathbf{v} = 2\mathbf{i} + 4\mathbf{j}$ and $\mathbf{w} = -2\mathbf{i} + 6\mathbf{j}$. Decompose $\mathbf{v}$ into two vectors, $\mathbf{v}_1$ and $\mathbf{v}_2$, where $\mathbf{v}_1$ is parallel to $\mathbf{w}$ and $\mathbf{v}_2$ is orthogonal to $\mathbf{w}$.

Solution These are the vectors we worked with in Example 4. We use the formulas in the preceding box.

$$\mathbf{v}_1 = \text{proj}_{\mathbf{w}}\mathbf{v} = -\mathbf{i} + 3\mathbf{j} \quad \text{We obtained this vector in Example 4.}$$

$$\mathbf{v}_2 = \mathbf{v} - \mathbf{v}_1 = (2\mathbf{i} + 4\mathbf{j}) - (-\mathbf{i} + 3\mathbf{j}) = 3\mathbf{i} + \mathbf{j}$$

Check Point 5 Let $\mathbf{v} = 2\mathbf{i} - 5\mathbf{j}$ and $\mathbf{w} = \mathbf{i} - \mathbf{j}$. (These are the vectors from Check Point 4.) Decompose $\mathbf{v}$ into two vectors, $\mathbf{v}_1$ and $\mathbf{v}_2$, where $\mathbf{v}_1$ is parallel to $\mathbf{w}$ and $\mathbf{v}_2$ is orthogonal to $\mathbf{w}$.

6 Compute work.

Work: An Application of the Dot Product

The bad news: Your car just died. The good news: It died on a level road just 200 feet from a gas station. Exerting a constant force of 90 pounds, and not necessarily whistling as you work, you manage to push the car to the gas station.

Force: 90 pounds

A | 200 feet | *B*

Although you did not whistle, you certainly did work pushing the car 200 feet from point *A* to point *B*. How much work did you do? If a constant force **F** is applied to an object, moving it from point *A* to point *B* in the direction of the force, the work, *W*, done is

$$W = (\text{magnitude of force})(\text{distance from } A \text{ to } B).$$

You pushed with a force of 90 pounds for a distance of 200 feet. The work done by your force is

$$W = (90 \text{ pounds})(200 \text{ feet})$$

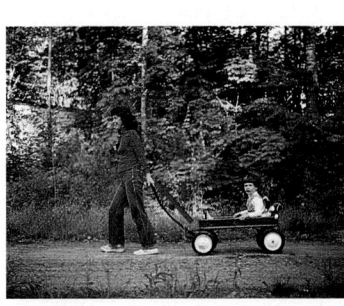

or 18,000 foot-pounds. Work is often measured in foot-pounds or in newton-meters.

The photo on the left shows an adult pulling a small child in a wagon. Work is being done. However, the situation is not quite the same as pushing your car. Pushing the car, the force you applied was along the line of motion. By contrast, the force of the adult pulling the wagon is not applied along the line of the wagon's motion. In this case, the dot product is used to determine the work done by the force.

Definition of Work

The work, W, done by a force $\mathbf{F}$ moving an object from A to B is

$$W = \mathbf{F} \cdot \overrightarrow{AB}.$$

When computing work, it is often easier to use the alternative formula for the dot product. Thus,

$$W = \mathbf{F} \cdot \overrightarrow{AB} = \|\mathbf{F}\| \|\overrightarrow{AB}\| \cos \theta.$$

| $\|\mathbf{F}\|$ is the magnitude of the force. | $\|\overrightarrow{AB}\|$ is the distance over which the constant force is applied. | θ is the angle between the force and the direction of motion. |

It is correct to refer to W as either the work done or the work done by the force.

EXAMPLE 6   Computing Work

A child pulls a sled along level ground by exerting a force of 30 pounds on a rope that makes an angle of 35° with the horizontal. How much work is done pulling the sled 200 feet?

Solution   The situation is illustrated in Figure 7.66. The work done is

$$W = \|\mathbf{F}\| \|\overrightarrow{AB}\| \cos \theta = (30)(200) \cos 35° \approx 4915$$

| Magnitude of the force is 30 pounds. | Distance is 200 feet. | The angle between the force and the sled's motion is 35°. |

Thus, the work done is approximately 4915 foot-pounds.

Figure 7.66 Computing work done pulling the sled 200 feet

Check Point 6   A child pulls a wagon along level ground by exerting a force of 20 pounds on a handle that makes an angle of 30° with the horizontal. How much work is done pulling the wagon 150 feet?

EXERCISE SET 7.7

Practice Exercises

In Exercises 1–8, use the given vectors to find a. $\mathbf{v} \cdot \mathbf{w}$ and b. $\mathbf{v} \cdot \mathbf{v}$.

1. $\mathbf{v} = 3\mathbf{i} + \mathbf{j}, \quad \mathbf{w} = \mathbf{i} + 3\mathbf{j}$
2. $\mathbf{v} = 3\mathbf{i} + 3\mathbf{j}, \quad \mathbf{w} = \mathbf{i} + 4\mathbf{j}$
3. $\mathbf{v} = 5\mathbf{i} - 4\mathbf{j}, \quad \mathbf{w} = -2\mathbf{i} - \mathbf{j}$
4. $\mathbf{v} = 7\mathbf{i} - 2\mathbf{j}, \quad \mathbf{w} = -3\mathbf{i} - \mathbf{j}$
5. $\mathbf{v} = -6\mathbf{i} - 5\mathbf{j}, \quad \mathbf{w} = -10\mathbf{i} - 8\mathbf{j}$
6. $\mathbf{v} = -8\mathbf{i} - 3\mathbf{j}, \quad \mathbf{w} = -10\mathbf{i} - 5\mathbf{j}$
7. $\mathbf{v} = 5\mathbf{i}, \quad \mathbf{w} = \mathbf{j}$ 8. $\mathbf{v} = \mathbf{i}, \quad \mathbf{w} = -5\mathbf{j}$

In Exercises 9–16, let

$$\mathbf{u} = 2\mathbf{i} - \mathbf{j}, \quad \mathbf{v} = 3\mathbf{i} + \mathbf{j}, \quad \text{and} \quad \mathbf{w} = \mathbf{i} + 4\mathbf{j}.$$

Find each specified scalar.

9. $\mathbf{u} \cdot (\mathbf{v} + \mathbf{w})$ 10. $\mathbf{v} \cdot (\mathbf{u} + \mathbf{w})$
11. $\mathbf{u} \cdot \mathbf{v} + \mathbf{u} \cdot \mathbf{w}$ 12. $\mathbf{v} \cdot \mathbf{u} + \mathbf{v} \cdot \mathbf{w}$
13. $(4\mathbf{u}) \cdot \mathbf{v}$ 14. $(5\mathbf{v}) \cdot \mathbf{w}$
15. $4(\mathbf{u} \cdot \mathbf{v})$ 16. $5(\mathbf{v} \cdot \mathbf{w})$

In Exercises 17–22, find the angle between $\mathbf{v}$ and $\mathbf{w}$. Round to the nearest tenth of a degree.

17. $\mathbf{v} = 2\mathbf{i} - \mathbf{j}, \quad \mathbf{w} = 3\mathbf{i} + 4\mathbf{j}$
18. $\mathbf{v} = -2\mathbf{i} + 5\mathbf{j}, \quad \mathbf{w} = 3\mathbf{i} + 6\mathbf{j}$
19. $\mathbf{v} = -3\mathbf{i} + 2\mathbf{j}, \quad \mathbf{w} = 4\mathbf{i} - \mathbf{j}$
20. $\mathbf{v} = \mathbf{i} + 2\mathbf{j}, \quad \mathbf{w} = 4\mathbf{i} - 3\mathbf{j}$
21. $\mathbf{v} = 6\mathbf{i}, \quad \mathbf{w} = 5\mathbf{i} + 4\mathbf{j}$ 22. $\mathbf{v} = 3\mathbf{j}, \quad \mathbf{w} = 4\mathbf{i} + 5\mathbf{j}$

In Exercises 23–32, use the dot product to determine whether $\mathbf{v}$ and $\mathbf{w}$ are orthogonal.

23. $\mathbf{v} = \mathbf{i} + \mathbf{j}, \quad \mathbf{w} = \mathbf{i} - \mathbf{j}$
24. $\mathbf{v} = \mathbf{i} + \mathbf{j}, \quad \mathbf{w} = -\mathbf{i} + \mathbf{j}$
25. $\mathbf{v} = 2\mathbf{i} + 8\mathbf{j}, \quad \mathbf{w} = 4\mathbf{i} - \mathbf{j}$
26. $\mathbf{v} = 8\mathbf{i} - 4\mathbf{j}, \quad \mathbf{w} = -6\mathbf{i} - 12\mathbf{j}$
27. $\mathbf{v} = 2\mathbf{i} - 2\mathbf{j}, \quad \mathbf{w} = -\mathbf{i} + \mathbf{j}$
28. $\mathbf{v} = 5\mathbf{i} - 5\mathbf{j}, \quad \mathbf{w} = \mathbf{i} - \mathbf{j}$
29. $\mathbf{v} = 3\mathbf{i}, \quad \mathbf{w} = -4\mathbf{i}$ 30. $\mathbf{v} = 5\mathbf{i}, \quad \mathbf{w} = -6\mathbf{i}$
31. $\mathbf{v} = 3\mathbf{i}, \quad \mathbf{w} = -4\mathbf{j}$ 32. $\mathbf{v} = 5\mathbf{i}, \quad \mathbf{w} = -6\mathbf{j}$

In Exercise 33–38, find $\text{proj}_{\mathbf{w}}\mathbf{v}$. Then decompose $\mathbf{v}$ into two vectors, $\mathbf{v}_1$ and $\mathbf{v}_2$, where $\mathbf{v}_1$ is parallel to $\mathbf{w}$ and $\mathbf{v}_2$ is orthogonal to $\mathbf{w}$.

33. $\mathbf{v} = 3\mathbf{i} - 2\mathbf{j}, \quad \mathbf{w} = \mathbf{i} - \mathbf{j}$
34. $\mathbf{v} = 3\mathbf{i} - 2\mathbf{j}, \quad \mathbf{w} = 2\mathbf{i} + \mathbf{j}$
35. $\mathbf{v} = \mathbf{i} + 3\mathbf{j}, \quad \mathbf{w} = -2\mathbf{i} + 5\mathbf{j}$
36. $\mathbf{v} = 2\mathbf{i} + 4\mathbf{j}, \quad \mathbf{w} = -3\mathbf{i} + 6\mathbf{j}$
37. $\mathbf{v} = \mathbf{i} + 2\mathbf{j}, \quad \mathbf{w} = 3\mathbf{i} + 6\mathbf{j}$
38. $\mathbf{v} = 2\mathbf{i} + \mathbf{j}, \quad \mathbf{w} = 6\mathbf{i} + 3\mathbf{j}$

Application Exercises

39. The components of $\mathbf{v} = 240\mathbf{i} + 300\mathbf{j}$ represent the respective number of gallons of regular and premium gas sold at a station on Monday. The components of $\mathbf{w} = 1.90\mathbf{i} + 2.07\mathbf{j}$ represent the respective prices per gallon for each kind of gas. Find $\mathbf{v} \cdot \mathbf{w}$ and describe what the answer means in practical terms.

40. The components of $\mathbf{v} = 180\mathbf{i} + 450\mathbf{j}$ represent the respective number of one-day and three-day videos rented from a video store on Monday. The components of $\mathbf{w} = 3\mathbf{i} + 2\mathbf{j}$ represent the prices to rent the one-day and three-day videos, respectively. Find $\mathbf{v} \cdot \mathbf{w}$ and describe what the answer means in practical terms.

41. Find the work done in pushing a car along a level road from point A to point B, 80 feet from A, while exerting a constant force of 95 pounds. Round to the nearest foot-pound.

42. Find the work done when a crane lifts a 6000-pound boulder through a vertical distance of 12 feet. Round to the nearest foot-pound.

43. A wagon is pulled along level ground by exerting a force of 40 pounds on a handle that makes an angle of 32° with the horizontal. How much work is done pulling the wagon 100 feet? Round to the nearest foot-pound.

44. A wagon is pulled along level ground by exerting a force of 25 pounds on a handle that makes an angle of 38° with the horizontal. How much work is done pulling the wagon 100 feet? Round to the nearest foot-pound.

45. A force of 60 pounds on a rope is used to pull a box up a ramp inclined at 12° from the horizontal. The figure shows that the rope forms an angle of 38° with the horizontal. How much work is done pulling the box 20 feet along the ramp?

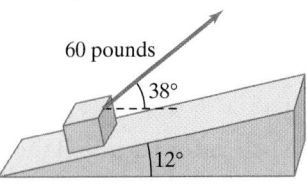

46. A force of 80 pounds on a rope is used to pull a box up a ramp inclined at 10° from the horizontal. The rope forms an angle of 33° with the horizontal. How much work is done pulling the box 25 feet along the ramp?

47. A force is given by the vector $\mathbf{F} = 3\mathbf{i} + 2\mathbf{j}$. The force moves an object along a straight line from the point $(4, 9)$ to the point $(10, 20)$. Find the work done if the distance is measured in feet and the force is measured in pounds.

48. A force is given by the vector $\mathbf{F} = 5\mathbf{i} + 7\mathbf{j}$. The force moves an object along a straight line from the point $(8, 11)$ to the point $(18, 20)$. Find the work done if the distance is measured in meters and the force is measured in newtons.

49. A force of 4 pounds acts in the direction of $50°$ to the horizontal. The force moves an object along a straight line from the point $(3, 7)$ to the point $(8, 10)$, with distance measured in feet. Find the work done by the force.

50. A force of 6 pounds acts in the direction of $40°$ to the horizontal. The force moves an object along a straight line from the point $(5, 9)$ to the point $(8, 20)$, with the distance measured in feet. Find the work done by the force.

51. Refer to Figure 7.63 on page 685. Suppose that the boat weighs 700 pounds and is on a ramp inclined at $30°$. Represent the force due to gravity, **F**, using

$$\mathbf{F} = -700\mathbf{j}.$$

   a. Write a unit vector along the ramp in the upward direction.

   b. Find the vector projection of **F** onto the unit vector from part (a).

   c. What is the magnitude of the vector projection in part (b)? What does this represent?

52. Refer to Figure 7.63 on page 685. Suppose that the boat weighs 650 pounds and is on a ramp inclined at $30°$. Represent the force due to gravity, **F**, using

$$\mathbf{F} = -650\mathbf{j}.$$

   a. Write a unit vector along the ramp in the upward direction.

   b. Find the vector projection of **F** onto the unit vector from part (a).

   c. What is the magnitude of the vector projection in part (b)? What does this represent?

Writing in Mathematics

53. Explain how to find the dot product of two vectors.

54. Using words and no symbols, describe how to find the dot product of two vectors with the alternative formula

$$\mathbf{v} \cdot \mathbf{w} = \|\mathbf{v}\| \|\mathbf{w}\| \cos \theta.$$

55. Describe how to find the angle between two vectors.

56. What are parallel vectors?

57. What are orthogonal vectors?

58. How do you determine if two vectors are orthogonal?

59. Draw two vectors, **v** and **w**, with the same initial point. Show the vector projection of **v** onto **w** in your diagram. Then describe how you identified this vector.

60. How do you determine the work done by a force **F** in moving an object from A to B when the direction of the force is not along the line of motion?

61. A weightlifter is holding a barbell perfectly still above his head, his body shaking from the effort. How much work is the weightlifter doing? Explain your answer.

62. Describe one way in which the everyday use of the word "work" is different from the definition of work given in this section.

 Critical Thinking Exercises

In Exercises 63–65, use the vectors

$$\mathbf{u} = a_1\mathbf{i} + b_1\mathbf{j}, \quad \mathbf{v} = a_2\mathbf{i} + b_2\mathbf{j}, \quad \text{and} \quad \mathbf{w} = a_3\mathbf{i} + b_3\mathbf{j},$$

to prove the given property.

63. $\mathbf{u} \cdot \mathbf{v} = \mathbf{v} \cdot \mathbf{u}$

64. $(c\mathbf{u}) \cdot \mathbf{v} = c(\mathbf{u} \cdot \mathbf{v})$

65. $\mathbf{u} \cdot (\mathbf{v} + \mathbf{w}) = \mathbf{u} \cdot \mathbf{v} + \mathbf{u} \cdot \mathbf{w}$

66. If $\mathbf{v} = -2\mathbf{i} + 5\mathbf{j}$, find a vector orthogonal to **v**.

67. Find a value of b so that $15\mathbf{i} - 3\mathbf{j}$ and $-4\mathbf{i} + b\mathbf{j}$ are orthogonal.

68. Prove that the projection of **v** onto **i** is $(\mathbf{v} \cdot \mathbf{i})\mathbf{i}$.

69. Find two vectors **v** and **w** such that the projection of **v** onto **w** is **v**.

 Group Exercise

70. Group members should research and present a report on unusual and interesting applications of vectors.

Chapter Summary, Review, and Test

Summary

DEFINITIONS AND CONCEPTS	EXAMPLES

7.1 and 7.2 The Law of Sines and the Law of Cosines

   a. The Law of Sines

$$\frac{a}{\sin A} = \frac{b}{\sin B} = \frac{c}{\sin C}$$

Ex. 1, p. 612;
Ex. 2, p. 613;
Ex. 3, p. 615

   b. The Law of Sines is used to solve SAA, ASA, and SSA (the ambiguous case) triangles. The ambiguous case may result in no triangle, one triangle, or two triangles; see the box on page 614.

Ex. 4, p. 616;
Ex. 5, p. 616

   c. The area of a triangle equals one-half the product of the lengths of two sides times the sine of their included angle.

Ex. 6, p. 618

   d. The Law of Cosines

$$a^2 = b^2 + c^2 - 2bc \cos A$$
$$b^2 = a^2 + c^2 - 2ac \cos B$$
$$c^2 = a^2 + b^2 - 2ab \cos C$$

DEFINITIONS AND CONCEPTS	EXAMPLES

e. The Law of Cosines is used to find the side opposite the given angle in an SAS triangle; see the box on page 624. The Law of Cosines is also used to find the angle opposite the longest side in an SSS triangle; see the box on page 625.

Ex. 1, p. 624;
Ex. 2, p. 626

f. Heron's Formula for the Area of a Triangle

Ex. 4, p. 628

The area of a triangle with sides $a, b,$ and c is

$$\sqrt{s(s-a)(s-b)(s-c)},$$

where s is one-half the perimeter:

$$s = \tfrac{1}{2}(a+b+c).$$

7.3 and 7.4 Polar Coordinates and Graphs of Polar Equations

a. A point P in the polar coordinate system is represented by (r, θ), where r is the directed distance of the point from the pole and θ is the angle from the polar axis to line segment OP. The elements of the ordered pair (r, θ) are called the polar coordinates of P. See Figure 7.19 on page 631. When r in (r, θ) is negative, a point is located $|r|$ units along the ray opposite the terminal side of θ. Important information about the sign of r and the location of the point (r, θ) is found in the box on page 632.

Ex. 1, p. 632

b. Multiple Representation of Points

Ex. 2, p. 634

If n is any integer, $(r, \theta) = (r, \theta + 2n\pi)$ or $(r, \theta) = (-r, \theta + \pi + 2n\pi)$.

c. Relations between Polar and Rectangular Coordinates

$$x = r\cos\theta, \quad y = r\sin\theta, \quad x^2 + y^2 = r^2, \quad \tan\theta = \frac{y}{x}$$

d. To convert a point from polar coordinates (r, θ) to rectangular coordinates (x, y), use $x = r\cos\theta$ and $y = r\sin\theta$.

Ex. 3, p. 635

e. To convert a point from rectangular coordinates (x, y) to polar coordinates (r, θ), use the procedure in the box on page 636.

Ex. 4, p. 636;
Ex. 5, p. 637

f. To convert a rectangular equation to a polar equation, replace x with $r\cos\theta$ and y with $r\sin\theta$.

Ex. 6, p. 638

g. To convert a polar equation to a rectangular equation, use one or more of

Ex. 7, p. 638

$$r^2 = x^2 + y^2, \quad r\cos\theta = x, \quad r\sin\theta = y, \quad \text{and} \quad \tan\theta = \frac{y}{x}.$$

It is often necessary to do something to the given polar equation to use the preceding expressions.

h. A polar equation is an equation whose variables are r and θ. The graph of a polar equation is the set of all points whose polar coordinates satisfy the equation.

Ex. 1, p. 642

i. Polar equations can be graphed using point plotting and symmetry (see the box on page 644).

Ex. 2, p. 645

j. The graphs of $r = a\cos\theta$ and $r = a\sin\theta$ are circles. See the box on page 644. The graphs of $r = a \pm b\sin\theta$ and $r = a \pm b\cos\theta$ are called limaçons ($a > 0$ and $b > 0$), shown in the box on page 647. The graphs of $r = a\sin n\theta$ and $r = a\cos n\theta, a \neq 0$, are rose curves with $2n$ petals if n is even and n petals if n is odd. See the box on page 650. The graphs of $r^2 = a^2\sin 2\theta$ and $r^2 = a^2\cos 2\theta, a \neq 0$, are called lemniscates and are shown in the box on page 651.

Ex. 3, p. 646;
Ex. 4, p. 648;
Ex. 5, p. 650

7.5 Complex Numbers in Polar Form; DeMoivre's Theorem

a. The complex number $z = a + bi$ is represented as a point (a, b) in the complex plane, shown in Figure 7.35 on page 655.

Ex. 1, p. 655

b. The absolute value of $z = a + bi$ is $|z| = |a + bi| = \sqrt{a^2 + b^2}$.

Ex. 2, p. 656

c. The polar form of $z = a + bi$ is $z = r(\cos\theta + i\sin\theta)$, where $a = r\cos\theta, b = r\sin\theta, r = \sqrt{a^2 + b^2}$, and $\tan\theta = \frac{b}{a}$. We call r the modulus and θ the argument of z, with $0 \le \theta < 2\pi$.

Ex. 3, p. 657;
Ex. 4, p. 658

d. Multiplying Complex Numbers in Polar Form: Multiply moduli and add arguments. See the box on page 658.

Ex. 5, p. 659

DEFINITIONS AND CONCEPTS	**EXAMPLES**

e. Dividing Complex Numbers in Polar Form: Divide moduli and subtract arguments. See the box on page 659.

Ex. 6, p. 659

f. DeMoivre's Theorem is used to find powers of complex numbers in polar form.

$$[r(\cos\theta + i\sin\theta)]^n = r^n(\cos n\theta + i\sin n\theta)$$

Ex. 7, p. 661; Ex. 8, p. 661

g. DeMoivre's Theorem can be used to find roots of complex numbers in polar form. The n distinct nth roots of $r(\cos\theta + i\sin\theta)$ are

$$\sqrt[n]{r}\left[\cos\left(\frac{\theta + 2\pi k}{n}\right) + i\sin\left(\frac{\theta + 2\pi k}{n}\right)\right]$$

or

$$\sqrt[n]{r}\left[\cos\left(\frac{\theta + 360°k}{n}\right) + i\sin\left(\frac{\theta + 360°k}{n}\right)\right],$$

where $k = 0, 1, 2, \ldots, n - 1$.

Ex. 9, p. 662; Ex. 10, p. 663

7.6 Vectors

a. A vector is a directed line segment.

b. Equal vectors have the same magnitude and the same direction.

Ex. 1, p. 668

c. The vector $k\mathbf{v}$, the scalar multiple of the vector $\mathbf{v}$ and the scalar k, has magnitude $|k|\|\mathbf{v}\|$. The direction of $k\mathbf{v}$ is the same as that of $\mathbf{v}$ if $k > 0$ and opposite $\mathbf{v}$ if $k < 0$.

d. The sum $\mathbf{u} + \mathbf{v}$, called the resultant vector, can be expressed geometrically. Position $\mathbf{u}$ and $\mathbf{v}$ so that the terminal point of $\mathbf{u}$ coincides with the initial point of $\mathbf{v}$. The vector $\mathbf{u} + \mathbf{v}$ extends from the initial point of $\mathbf{u}$ to the terminal point of $\mathbf{v}$.

e. The difference of two vectors, $\mathbf{u} - \mathbf{v}$, is defined as $\mathbf{u} + (-\mathbf{v})$.

f. The vector $\mathbf{i}$ is the unit vector whose direction is along the positive x-axis. The vector $\mathbf{j}$ is the unit vector whose direction is along the positive y-axis.

g. Vector $\mathbf{v}$, from $(0, 0)$ to (a, b), called a position vector, is represented as $\mathbf{v} = a\mathbf{i} + b\mathbf{j}$, where a is the horizontal component and b is the vertical component. The magnitude of $\mathbf{v}$ is given by $\|\mathbf{v}\| = \sqrt{a^2 + b^2}$.

Ex. 2, p. 671

h. Vector $\mathbf{v}$ from (x_1, y_1) to (x_2, y_2) is equal to the position vector $\mathbf{v} = (x_2 - x_1)\mathbf{i} + (y_2 - y_1)\mathbf{j}$. In rectangular coordinates, the term "vector" refers to the position vector in terms of $\mathbf{i}$ and $\mathbf{j}$ that is equal to it.

Ex. 3, p. 672

i. Operations with Vectors in Terms of $\mathbf{i}$ and $\mathbf{j}$

If $\mathbf{v} = a_1\mathbf{i} + b_1\mathbf{j}$ and $\mathbf{w} = a_2\mathbf{i} + b_2\mathbf{j}$, then

1. $\mathbf{v} + \mathbf{w} = (a_1 + a_2)\mathbf{i} + (b_1 + b_2)\mathbf{j}$

2. $\mathbf{v} - \mathbf{w} = (a_1 - a_2)\mathbf{i} + (b_1 - b_2)\mathbf{j}$

3. $k\mathbf{v} = (ka_1)\mathbf{i} + (kb_1)\mathbf{j}$

Ex. 4, p. 672; Ex. 5, p. 673; Ex. 6, p. 673

j. The zero vector $\mathbf{0}$ is the vector whose magnitude is 0 and is assigned no direction. Many properties of vector addition and scalar multiplication involve the zero vector. Some of these properties are listed in the box on page 674.

k. The vector $\dfrac{\mathbf{v}}{\|\mathbf{v}\|}$ is the unit vector that has the same direction as $\mathbf{v}$.

Ex. 7, p. 675

l. A vector with magnitude $\|\mathbf{v}\|$ and direction angle θ, the angle that $\mathbf{v}$ makes with the positive x-axis, can be expressed in terms of its magnitude and direction angle as

$$\mathbf{v} = \|\mathbf{v}\|\cos\theta\mathbf{i} + \|\mathbf{v}\|\sin\theta\mathbf{j}.$$

Ex. 8, p. 676; Ex. 9, p. 676

DEFINITIONS AND CONCEPTS	EXAMPLES

7.7 The Dot Product

a. Definition of the Dot Product

If $\mathbf{v} = a_1\mathbf{i} + b_1\mathbf{j}$ and $\mathbf{w} = a_2\mathbf{i} + b_2\mathbf{j}$, the dot product of $\mathbf{v}$ and $\mathbf{w}$ is defined by $\mathbf{v} \cdot \mathbf{w} = a_1 a_2 + b_1 b_2$.

Ex. 1, p. 682

b. Alternative Formula for the Dot Product $\mathbf{v} \cdot \mathbf{w} = \|\mathbf{v}\|\|\mathbf{w}\| \cos\theta$, where θ is the smallest nonnegative angle between $\mathbf{v}$ and $\mathbf{w}$.

c. Angle between Two Vectors

$$\cos\theta = \frac{\mathbf{v} \cdot \mathbf{w}}{\|\mathbf{v}\|\|\mathbf{w}\|} \quad \text{and} \quad \theta = \cos^{-1}\left(\frac{\mathbf{v} \cdot \mathbf{w}}{\|\mathbf{v}\|\|\mathbf{w}\|}\right).$$

Ex. 2, p. 684

d. Two vectors are orthogonal when the angle between them is 90°. To show that two vectors are orthogonal, show that their dot product is zero.

Ex. 3, p. 685

e. The vector projection of $\mathbf{v}$ onto $\mathbf{w}$ is given by

$$\text{proj}_{\mathbf{w}}\mathbf{v} = \frac{\mathbf{v} \cdot \mathbf{w}}{\|\mathbf{w}\|^2}\mathbf{w}.$$

Ex. 4, p. 686

f. Expressing a vector as the sum of two orthogonal vectors, called the vector components, is shown in the box on page 686.

Ex. 5, p. 687

g. The work, W, done by a force $\mathbf{F}$ moving an object from A to B is $W = \mathbf{F} \cdot \overrightarrow{AB}$.

Thus, $W = \|\mathbf{F}\|\|\overrightarrow{AB}\| \cos\theta$, where θ is the angle between the force and the direction of motion.

Ex. 6, p. 688

Review Exercises

7.1 and 7.2

In Exercises 1–12, solve each triangle. Round lengths to the nearest tenth and angle measures to the nearest degree. If no triangle exists, state "no triangle." If two triangles exist, solve each triangle.

1. $A = 70°, B = 55°, a = 12$

2. $B = 107°, C = 30°, c = 126$

3. $B = 66°, a = 17, c = 12$

4. $a = 117, b = 66, c = 142$

5. $A = 35°, B = 25°, c = 68$

6. $A = 39°, a = 20, b = 26$

7. $C = 50°, a = 3, c = 1$

8. $A = 162°, b = 11.2, c = 48.2$

9. $a = 26.1, b = 40.2, c = 36.5$

10. $A = 40°, a = 6, b = 4$

11. $B = 37°, a = 12.4, b = 8.7$

12. $A = 23°, a = 54.3, b = 22.1$

In Exercises 13–16, find the area of the triangle having the given measurements. Round to the nearest square unit.

13. $C = 42°, a = 4$ feet, $b = 6$ feet

14. $A = 22°, b = 4$ feet, $c = 5$ feet

15. $a = 2$ meters, $b = 4$ meters, $c = 5$ meters

16. $a = 2$ meters, $b = 2$ meters, $c = 2$ meters

17. The A-frame cabin shown in the next column is 35 feet wide. The roof of the cabin makes a 60° angle with the cabin's base. Find the length of one side of the roof from its ground level to the peak. Round to the nearest tenth of a foot.

18. Two cars leave a city at the same time and travel along straight highways that differ in direction by 80°. One car averages 60 miles per hour and the other averages 50 miles per hour. How far apart will the cars be after 30 minutes? Round to the nearest tenth of a mile.

19. Two airplanes leave an airport at the same time on different runways. One flies on a bearing of N66.5°W at 325 miles per hour. The other airplane flies on a bearing of S26.5°W at 300 miles per hour. How far apart will the airplanes be after two hours?

20. The figure shows three roads that intersect to bound a triangular piece of land. Find the lengths of the other two sides of the land to the nearest foot.

21. A commercial piece of real estate is priced at $5.25 per square foot. Find the cost, to the nearest dollar, of a triangular lot measuring 260 feet by 320 feet by 450 feet.

7.3 and 7.4

In Exercises 22–27, plot each point in polar coordinates and find its rectangular coordinates.

22. $(4, 60°)$ **23.** $(3, 150°)$

24. $\left(-4, \dfrac{4\pi}{3}\right)$ **25.** $\left(-2, \dfrac{5\pi}{4}\right)$

26. $\left(-4, -\dfrac{\pi}{2}\right)$ **27.** $\left(-2, -\dfrac{\pi}{4}\right)$

In Exercises 28–30, plot each point in polar coordinates. Then find another representation (r, θ) of this point in which:

  a. $r > 0$, $2\pi < \theta < 4\pi$.
  b. $r < 0$, $0 < \theta < 2\pi$.
  c. $r > 0$, $-2\pi < \theta < 0$.

28. $\left(3, \dfrac{\pi}{6}\right)$ **29.** $\left(2, \dfrac{2\pi}{3}\right)$ **30.** $(3, \pi)$

In Exercises 31–36, the rectangular coordinates of a point are given. Find polar coordinates of each point.

31. $(-4, 4)$ **32.** $(3, -3)$
33. $(5, 12)$ **34.** $(-3, 4)$
35. $(0, -5)$ **36.** $(1, 0)$

In Exercises 37–39, convert each rectangular equation to a polar equation.

37. $2x + 3y = 8$ **38.** $x^2 + y^2 = 100$
39. $5x^2 + 5y^2 = 3y$

In Exercises 40–46, convert each polar equation to a rectangular equation.

40. $r = 3$ **41.** $\theta = \dfrac{3\pi}{4}$

42. $r \cos \theta = -1$ **43.** $r = 5 \sec \theta$
44. $r = 3 \cos \theta$ **45.** $5r \cos \theta + r \sin \theta = 8$
46. $r^2 \sin 2\theta = 4$

In Exercises 47–49, test for symmetry with respect to:
  a. *the polar axis*   **b.** *the line $\theta = \dfrac{\pi}{2}$*   *and*   **c.** *the pole.*

47. $r = 5 + 3 \cos \theta$ **48.** $r = 3 \sin \theta$
49. $r^2 = 9 \cos 2\theta$

In Exercises 50–56, graph each polar equation. Be sure to test for symmetry.

50. $r = 3 \cos \theta$ **51.** $r = 2 + 2 \sin \theta$
52. $r = \sin 2\theta$ **53.** $r = 2 + \cos \theta$
54. $r = 1 + 3 \sin \theta$ **55.** $r = 1 - 2 \cos \theta$
56. $r^2 = \cos 2\theta$

7.5

In Exercises 57–60, plot each complex number. Then write the complex number in polar form. You may express the argument in degrees or radians.

57. $1 - i$ **58.** $-2\sqrt{3} + 2i$
59. $-3 - 4i$ **60.** $-5i$

In Exercises 61–64, write each complex number in rectangular form.

61. $8(\cos 60° + i \sin 60°)$ **62.** $4(\cos 210° + i \sin 210°)$

63. $6\left(\cos \dfrac{2\pi}{3} + i \sin \dfrac{2\pi}{3}\right)$

64. $0.6(\cos 100° + i \sin 100°)$

In Exercises 65–67, find the product of the complex numbers. Leave answers in polar form.

65. $z_1 = 3(\cos 40° + i \sin 40°)$
$z_2 = 5(\cos 70° + i \sin 70°)$

66. $z_1 = \cos 210° + i \sin 210°$
$z_2 = \cos 55° + i \sin 55°$

67. $z_1 = 4\left(\cos \dfrac{3\pi}{7} + i \sin \dfrac{3\pi}{7}\right)$
$z_2 = 10\left(\cos \dfrac{4\pi}{7} + i \sin \dfrac{4\pi}{7}\right)$

In Exercises 68–70, find the quotient $\dfrac{z_1}{z_2}$ of the complex numbers. Leave answers in polar form.

68. $z_1 = 10(\cos 10° + i \sin 10°)$
$z_2 = 5(\cos 5° + i \sin 5°)$

69. $z_1 = 5\left(\cos \dfrac{4\pi}{3} + i \sin \dfrac{4\pi}{3}\right)$
$z_2 = 10\left(\cos \dfrac{\pi}{3} + i \sin \dfrac{\pi}{3}\right)$

70. $z_1 = 2\left(\cos \dfrac{5\pi}{3} + i \sin \dfrac{5\pi}{3}\right)$
$z_2 = \cos \dfrac{\pi}{2} + i \sin \dfrac{\pi}{2}$

In Exercises 71–75, use DeMoivre's Theorem to find the indicated power of the complex number. Write answers in rectangular form.

71. $[2(\cos 20° + i \sin 20°)]^3$
72. $[4(\cos 50° + i \sin 50°)]^3$

73. $\left[\dfrac{1}{2}\left(\cos \dfrac{\pi}{14} + i \sin \dfrac{\pi}{14}\right)\right]^7$

74. $(1 - \sqrt{3}i)^7$ **75.** $(-2 - 2i)^5$

In Exercises 76–77, find all the complex roots. Write roots in polar form with θ in degrees.

76. The complex square roots of $49(\cos 50° + i \sin 50°)$

77. The complex cube roots of $125(\cos 165° + i \sin 165°)$

In Exercises 78–81, find all the complex roots. Write roots in rectangular form.

78. The complex fourth roots of $16\left(\cos \dfrac{2\pi}{3} + i \sin \dfrac{2\pi}{3}\right)$

79. The complex cube roots of $8i$

80. The complex cube roots of -1

81. The complex fifth roots of $-1 - i$

7.6

In Exercises 82–84, sketch each vector as a position vector and find its magnitude.

82. $\mathbf{v} = -3\mathbf{i} - 4\mathbf{j}$ **83.** $\mathbf{v} = 5\mathbf{i} - 2\mathbf{j}$

84. $\mathbf{v} = -3\mathbf{j}$

In Exercises 85–86, let $\mathbf{v}$ be the vector from initial point P_1 to terminal point P_2. Write $\mathbf{v}$ in terms of $\mathbf{i}$ and $\mathbf{j}$.

85. $P_1 = (2,-1)$, $P_2 = (5,-3)$

86. $P_1 = (-3,0)$, $P_2 = (-2,-2)$

In Exercises 87–90, let
$$\mathbf{v} = \mathbf{i} - 5\mathbf{j} \quad \text{and} \quad \mathbf{w} = -2\mathbf{i} + 7\mathbf{j}.$$
Find each specified vector or scalar.

87. $\mathbf{v} + \mathbf{w}$ **88.** $\mathbf{w} - \mathbf{v}$

89. $6\mathbf{v} - 3\mathbf{w}$ **90.** $\|-2\mathbf{v}\|$

In Exercises 91–92, find the unit vector that has the same direction as the vector $\mathbf{v}$.

91. $\mathbf{v} = 8\mathbf{i} - 6\mathbf{j}$ **92.** $\mathbf{v} = -\mathbf{i} + 2\mathbf{j}$

93. The magnitude and direction angle of $\mathbf{v}$ are $\|\mathbf{v}\| = 12$ and $\theta = 60°$. Express $\mathbf{v}$ in terms of $\mathbf{i}$ and $\mathbf{j}$.

94. The magnitude and direction of two forces acting on an object are 100 pounds, N25°E, and 200 pounds, N80°E, respectively. Find the magnitude, to the nearest pound, and the direction angle, to the nearest tenth of a degree, of the resultant force.

95. Your boat is moving at a speed of 15 miles per hour at an angle of 25° upstream on a river flowing at 4 miles per hour. The situation is illustrated in the figure at the top of the next column.

a. Find the vector representing your boat's velocity relative to the ground.

b. What is the speed of your boat. to the nearest mile per hour, relative to the ground?

c. What is the boat's direction angle, to the nearest tenth of a degree, relative to the ground?

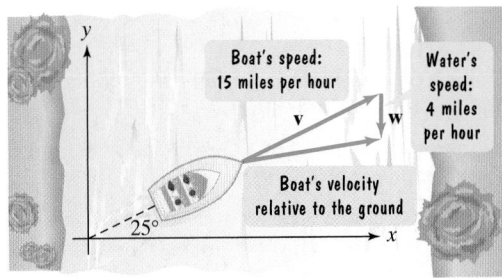

7.7

96. If $\mathbf{u} = 5\mathbf{i} + 2\mathbf{j}, \mathbf{v} = \mathbf{i} - \mathbf{j}$, and $\mathbf{w} = 3\mathbf{i} - 7\mathbf{j}$, find $\mathbf{u} \cdot (\mathbf{v} + \mathbf{w})$.

In Exercises 97–99, find the dot product $\mathbf{v} \cdot \mathbf{w}$. Then find the angle between $\mathbf{v}$ and $\mathbf{w}$ to the nearest tenth of a degree.

97. $\mathbf{v} = 2\mathbf{i} + 3\mathbf{j}$, $\mathbf{w} = 7\mathbf{i} - 4\mathbf{j}$

98. $\mathbf{v} = 2\mathbf{i} + 4\mathbf{j}$, $\mathbf{w} = 6\mathbf{i} - 11\mathbf{j}$

99. $\mathbf{v} = 2\mathbf{i} + \mathbf{j}$, $\mathbf{w} = \mathbf{i} - \mathbf{j}$

In Exercises 100–101, use the dot product to determine whether $\mathbf{v}$ and $\mathbf{w}$ are orthogonal.

100. $\mathbf{v} = 12\mathbf{i} - 8\mathbf{j}$, $\mathbf{w} = 2\mathbf{i} + 3\mathbf{j}$

101. $\mathbf{v} = \mathbf{i} + 3\mathbf{j}$, $\mathbf{w} = -3\mathbf{i} - \mathbf{j}$

In Exercises 102–103, find $\text{proj}_{\mathbf{w}} \mathbf{v}$. Then decompose $\mathbf{v}$ into two vectors, $\mathbf{v}_1$ and $\mathbf{v}_2$, where $\mathbf{v}_1$ is parallel to $\mathbf{w}$ and $\mathbf{v}_2$ is orthogonal to $\mathbf{w}$.

102. $\mathbf{v} = -2\mathbf{i} + 5\mathbf{j}$, $\mathbf{w} = 5\mathbf{i} + 4\mathbf{j}$

103. $\mathbf{v} = -\mathbf{i} + 2\mathbf{j}$, $\mathbf{w} = 3\mathbf{i} - \mathbf{j}$

104. A heavy crate is dragged 50 feet along a level floor. Find the work done if a force of 30 pounds at an angle of 42° is used.

105. Explain why the weightlifter does more work in raising 300 kilograms above her head than Atlas, who is supporting the entire world.

Chapter 7 Test

1. In oblique triangle ABC, $A = 34°$, $B = 68°$, and $a = 4.8$. Find b to the nearest tenth.

2. In oblique triangle ABC, $C = 68°$, $a = 5$, and $b = 6$. Find c to the nearest tenth.

3. In oblique triangle ABC, $a = 17$ inches, $b = 45$ inches, and $c = 32$ inches. Find the area of the triangle to the nearest square inch.

4. Plot $\left(4, \dfrac{5\pi}{4}\right)$ in the polar coordinate system. Then write two other ordered pairs (r, θ) that name this point.

5. If the rectangular coordinates of a point are $(1, -1)$, find polar coordinates of the point.

6. Convert $x^2 + y^2 = 6x$ to a polar equation.

7. Convert $r = 4 \csc \theta$ to a rectangular equation.

In Exercises 8–9, graph each polar equation.

8. $r = 1 + \sin \theta$
9. $r = 1 + 3 \cos \theta$

10. Write $-\sqrt{3} + i$ in polar form.

In Exercises 11–13, perform the indicated operation. Leave answers in polar form.

11. $5(\cos 15° + i \sin 15°) \cdot 10(\cos 5° + i \sin 5°)$

12. $\dfrac{2\left(\cos \dfrac{\pi}{2} + i \sin \dfrac{\pi}{2}\right)}{4\left(\cos \dfrac{\pi}{3} + i \sin \dfrac{\pi}{3}\right)}$

13. $[2(\cos 10° + i \sin 10°)]^5$

14. Find the three cube roots of 27. Write roots in rectangular form.

15. If $P_1 = (-2, 3)$, $P_2 = (-1, 5)$, and $\mathbf{v}$ is the vector from P_1 to P_2,
   a. Write $\mathbf{v}$ in terms of $\mathbf{i}$ and $\mathbf{j}$.
   b. Find $\|\mathbf{v}\|$.

In Exercises 16–19, let
$$\mathbf{v} = -5\mathbf{i} + 2\mathbf{j} \quad \text{and} \quad \mathbf{w} = 2\mathbf{i} - 4\mathbf{j}.$$
Find the specified vector, scalar, or angle.

16. $3\mathbf{v} - 4\mathbf{w}$
17. $\mathbf{v} \cdot \mathbf{w}$.

18. the angle between $\mathbf{v}$ and $\mathbf{w}$, to the nearest degree

19. $\text{proj}_{\mathbf{w}} \mathbf{v}$

20. A small fire is sighted from ranger stations A and B. Station B is 1.6 miles due east of station A. The bearing of the fire from station A is N40°E, and the bearing of the fire from station B is N50°W. How far, to the nearest tenth of a mile, is the fire from station A?

21. The magnitude and direction of two forces acting on an object are 250 pounds, N60°E, and 150 pounds, S45°E. Find the magnitude, to the nearest pound, and the direction angle, to the nearest tenth of a degree, of the resultant force.

22. A child is pulling a wagon with a force of 40 pounds. How much work is done in moving the wagon 60 feet if the handle makes an angle of 35° with the ground? Round to the nearest foot-pound.

Cumulative Review Exercises (Chapters 1–7)

Solve each equation or inequality in Exercises 1–4.

1. $x^4 - x^3 - x^2 - x - 2 = 0$
2. $2 \sin^2 \theta - 3 \sin \theta + 1 = 0$, $0 \le \theta < 2\pi$
3. $x^2 + 2x + 3 > 11$
4. $\sin \theta \cos \theta = -\frac{1}{2}$, $0 \le \theta < 2\pi$

In Exercises 5–6, graph one complete cycle.

5. $y = 3 \sin(2x - \pi)$
6. $y = -4 \cos \pi x$

In Exercises 7–8, verify each identity.

7. $\sin \theta \csc \theta - \cos^2 \theta = \sin^2 \theta$
8. $\cos\left(\theta + \dfrac{3\pi}{2}\right) = \sin \theta$

9. Find the slope and y-intercept of the line whose equation is $2x + 4y - 8 = 0$.

In Exercises 10–11, find the exact value of each expression.

10. $2 \sin \dfrac{\pi}{3} - 3 \tan \dfrac{\pi}{6}$
11. $\sin(\tan^{-1}\frac{1}{2})$

In Exercises 12–13, find the domain of the function whose equation is given.

12. $f(x) = \sqrt{5 - x}$
13. $g(x) = \dfrac{x - 3}{x^2 - 9}$

14. A ball is thrown vertically upward from a height of 8 feet with an initial velocity of 48 feet per second. The ball's height, $s(t)$, in feet, after t seconds is given by
$$s(t) = -16t^2 + 48t + 8.$$
After how many seconds does the ball reach its maximum height? What is the maximum height?

15. An object moves in simple harmonic motion described by $d = 4 \sin 5t$, where t is measured in seconds and d in meters. Find a. the maximum displacement; b. the frequency; and c. the time required for one cycle.

16. Use a half-angle formula to find the exact value of $\cos 22.5°$.

17. If $\mathbf{v} = 2\mathbf{i} + 7\mathbf{j}$ and $\mathbf{w} = \mathbf{i} - 2\mathbf{j}$, find: **a.** $3\mathbf{v} - \mathbf{w}$ and **b.** $\mathbf{v} \cdot \mathbf{w}$.

18. Express as a single logarithm with a coefficient of 1: $\frac{1}{2}\log_b x - \log_b(x^2 + 1)$.

19. Write the slope-intercept form of the line passing through $(4, -1)$ and $(-8, 5)$.

20. Psychologists can measure the amount learned, L, at time t using the model $L = A(1 - e^{-kt})$. The variable A represents the total amount to be learned, and k is the learning rate. A student preparing for the SAT has 300 new vocabulary words to learn: $A = 300$. This particular student can learn 20 vocabulary words after 5 minutes: If $t = 5, L = 20$.

    a. Find k, the learning rate, correct to three decimal places.

    b. Approximately how many words will the student have learned after 20 minutes?

    c. How long will it take for the student to learn 260 words?

Systems of Equations and Inequalities

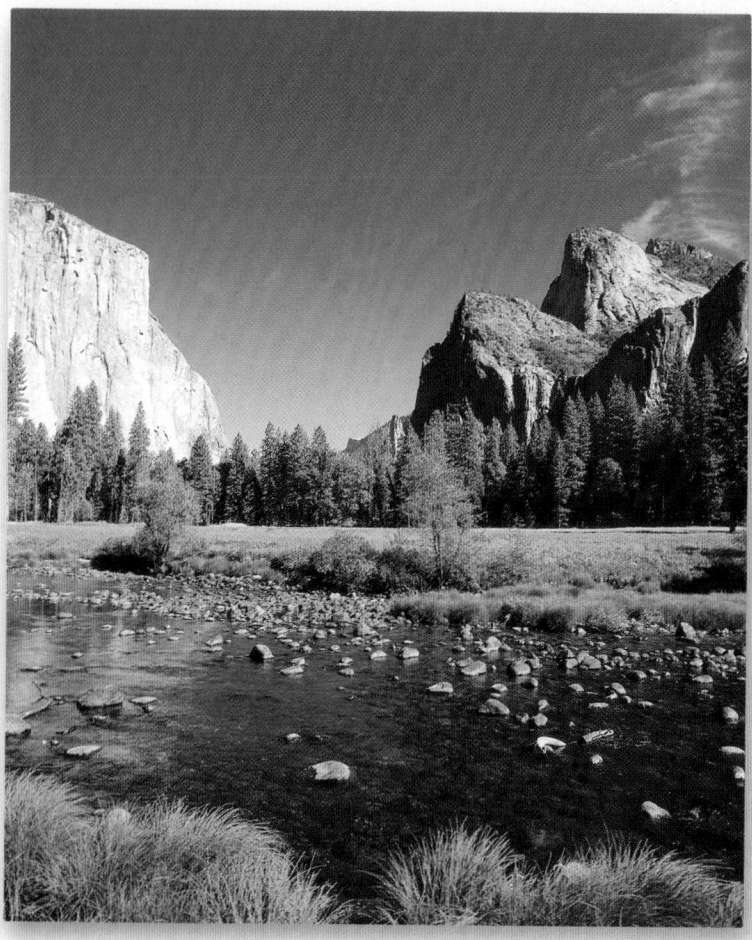

Most things in life depend on many variables. Temperature and precipitation are two variables that have a critical effect on whether regions are forests, grasslands, or deserts. Airlines deal with numerous variables during weather disruptions at large connecting airports. They must solve the problem of putting their operation back together again to minimize the cost of the disruption and passenger inconvenience. In this chapter, forests, grasslands, and airline service are viewed in the same way—situations with several variables. You will learn methods for modeling and solving problems in these situations.

A major weather disruption delayed your flight for hours, but you finally made it. You are in Yosemite National Park in California, surrounded by evergreen forests, alpine meadows, and sheer walls of granite. Soaring cliffs, plunging waterfalls, gigantic trees, rugged canyons, mountains and valleys stand in stark contrast to the angry chaos at the airport. This is so different from where you live and attend college, a region in which grasslands predominate.

SECTION 8.1 *Systems of Linear Equations in Two Variables*

Objectives

1. Decide whether an ordered pair is a solution of a linear system.
2. Solve linear systems by substitution.
3. Solve linear systems by addition.
4. Identify systems that do not have exactly one ordered-pair solution.
5. Solve problems using systems of linear equations.

Key West residents Brian Goss (left), George Wallace, and Michael Mooney (right) hold on to each other as they battle 90 mph winds along Houseboat Row in Key West, Fla., on Friday, Sept. 25, 1998. The three had sought shelter behind a Key West hotel as Hurricane Georges descended on the Florida Keys, but were forced to seek other shelter when the storm conditions became too rough. Hundreds of people were killed by the storm when it swept through the Caribbean.

Problems ranging from scheduling airline flights to controlling traffic flow to routing phone calls over the nation's communication network often require solutions in a matter of moments. The solution to these real-world problems can involve solving thousands of equations having thousands of variables. AT&T's domestic long-distance network involves 800,000 variables! Meteorologists describing atmospheric conditions surrounding a hurricane must solve problems involving thousands of equations rapidly and efficiently. The difference between a two-hour warning and a two-day warning is a life-and-death issue for thousands of people in the path of one of nature's most destructive forces.

Although we will not be solving 800,000 equations with 800,000 variables, we will turn our attention to two equations with two variables, such as

$$2x - 3y = -4$$
$$2x + y = 4.$$

The methods that we consider for solving such problems provide the foundation for solving far more complex systems with many variables.

1 Decide whether an ordered pair is a solution of a linear system.

Systems of Linear Equations and Their Solutions

We have seen that all equations in the form $Ax + By = C$ are straight lines when graphed. Two such equations, such as those listed above, are called a **system of linear equations** or a **linear system. A solution to a system of linear equations** is an ordered pair that satisfies all equations in the system. For example, $(3, 4)$ satisfies the system

$$x + y = 7 \qquad (3 + 4 \text{ is, indeed, } 7.)$$
$$x - y = -1. \qquad (3 - 4 \text{ is, indeed, } -1.)$$

Thus, $(3, 4)$ satisfies both equations and is a solution of the system. The solution can be described by saying that $x = 3$ and $y = 4$. The solution can also be described using set notation. The solution set to the system is $\{(3, 4)\}$—that is, the set consisting of the ordered pair $(3, 4)$.

A system of linear equations can have exactly one solution, no solution, or infinitely many solutions. We begin with systems that have exactly one solution.

EXAMPLE 1 Determining Whether an Ordered Pair Is a Solution of a System

Determine whether $(4, -1)$ is a solution of the system

$$x + 2y = 2$$
$$x - 2y = 6.$$

Solution Because 4 is the x-coordinate and -1 is the y-coordinate of $(4, -1)$, we replace x with 4 and y with -1.

$$
\begin{array}{ll}
x + 2y = 2 & \qquad x - 2y = 6 \\
4 + 2(-1) \overset{?}{=} 2 & \qquad 4 - 2(-1) \overset{?}{=} 6 \\
4 + (-2) \overset{?}{=} 2 & \qquad 4 - (-2) \overset{?}{=} 6 \\
2 = 2, \; \text{true} & \qquad 4 + 2 \overset{?}{=} 6 \\
& \qquad 6 = 6, \; \text{true}
\end{array}
$$

The pair $(4, -1)$ satisfies both equations: It makes each equation true. Thus, the pair is a solution of the system. The solution set to the system is $\{(4, -1)\}$.

Study Tip

When solving linear systems by graphing, neatly drawn graphs are essential for determining points of intersection.

- Use rectangular coordinate graph paper.
- Use a ruler or straightedge.
- Use a pencil with a sharp point.

The solution of a system of linear equations can be found by graphing both of the equations in the same rectangular coordinate system. For a system with one solution, the **coordinates of the point of intersection give the system's solution.** For example, the system in Example 1,

$$x + 2y = 2$$
$$x - 2y = 6$$

is graphed in Figure 8.1. The solution of the system, $(4, -1)$, corresponds to the point of intersection of the lines.

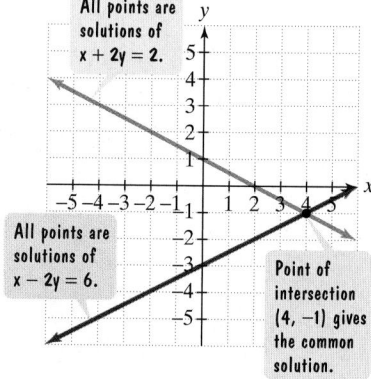

Figure 8.1 Visualizing a system's solution

Check Point 1 Determine whether $(1, 2)$ is a solution of the system

$$2x - 3y = -4$$
$$2x + \ y = \ \ 4.$$

2 Solve linear systems by substitution.

Eliminating a Variable Using the Substitution Method

Finding the solution to a linear system by graphing equations may not be easy to do. For example, a solution of $\left(-\frac{2}{3}, \frac{157}{29}\right)$ would be difficult to "see" as an intersection point on a graph.

Let's consider a method that does not depend on finding a system's solution visually: the substitution method. This method involves converting the system to one equation in one variable by an appropriate substitution.

EXAMPLE 2   Solving a System by Substitution

Solve by the substitution method:

$$y = -x - 1$$
$$4x - 3y = 24.$$

Solution

Step 1   Solve either of the equations for one variable in terms of the other. This step has already been done for us. The first equation, $y = -x - 1$, has y solved in terms of x.

Step 2   Substitute the expression from step 1 into the other equation.   We substitute the expression $-x - 1$ for y in the other equation:

$$y = \boxed{-x - 1} \qquad 4x - 3\,\boxed{y} = 24 \quad \text{Substitute } -x - 1 \text{ for } y.$$

This gives us an equation in one variable, namely

$$4x - 3(-x - 1) = 24.$$

The variable y has been eliminated.

Step 3   Solve the resulting equation containing one variable.

$$
\begin{array}{ll}
4x - 3(-x - 1) = 24 & \text{This is the equation containing one variable.} \\
4x + 3x + 3 = 24 & \text{Apply the distributive property.} \\
7x + 3 = 24 & \text{Combine like terms.} \\
7x = 21 & \text{Subtract 3 from both sides.} \\
x = 3 & \text{Divide both sides by 7.}
\end{array}
$$

Step 4   Back-substitute the obtained value into one of the original equations. We now know that the x-coordinate of the solution is 3. To find the y-coordinate, we back-substitute the x-value into either original equation. We will use

$$y = -x - 1.$$

Substitute 3 for x.

$$y = -3 - 1 = -4$$

With $x = 3$ and $y = -4$, the proposed solution is $(3, -4)$.

Step 5   Check the proposed solution in both of the system's given equations. Replace x with 3 and y with -4.

$$
\begin{array}{ll}
y = -x - 1 & 4x - 3y = 24 \\
-4 \overset{?}{=} -3 - 1 & 4(3) - 3(-4) \overset{?}{=} 24 \\
-4 = -4, \text{ true} & 12 + 12 \overset{?}{=} 24 \\
& 24 = 24, \text{ true}
\end{array}
$$

The pair $(3, -4)$ satisfies both equations. The system's solution set is $\{(3, -4)\}$.

Technology

A graphing utility can be used to solve the system in Example 2. Graph each equation and use the intersection feature. The utility displays the solution $(3, -4)$ as $x = 3$, $y = -4$.

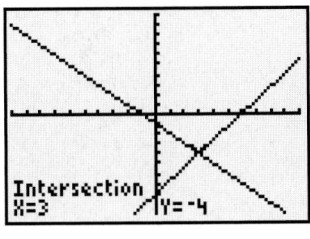

Intersection
X=3   Y=-4

Check Point 2 Solve by the substitution method:

$$y = 5x - 13$$
$$2x + 3y = 12.$$

Before considering additional examples, let's summarize the steps used in the substitution method.

Study Tip

In step 1, you can choose which variable to isolate in which equation. If possible, solve for a variable whose coefficient is 1 or −1 to avoid working with fractions.

Solving Linear Systems by Substitution

1. Solve either of the equations for one variable in terms of the other. (If one of the equations is already in this form, you can skip this step.)
2. Substitute the expression found in step 1 into the other equation. This will result in an equation in one variable.
3. Solve the equation containing one variable.
4. Back-substitute the value found in step 3 into one of the original equations. Simplify and find the value of the remaining variable.
5. Check the proposed solution in both of the system's given equations.

EXAMPLE 3 Solving a System by Substitution

Solve by the substitution method:

$$5x - 4y = 9$$
$$x - 2y = -3.$$

Solution

Step 1 Solve either of the equations for one variable in terms of the other. We begin by isolating one of the variables in either of the equations. By solving for x in the second equation, which has a coefficient of 1, we can avoid fractions.

$$x - 2y = -3 \qquad \text{This is the second equation in the given system.}$$
$$x = 2y - 3 \qquad \text{Solve for x by adding 2y to both sides.}$$

Step 2 Substitute the expression from step 1 into the other equation. We substitute $2y - 3$ for x in the first equation.

$$x = \boxed{2y - 3} \qquad 5\boxed{x} - 4y = 9$$

This gives us an equation in one variable, namely

$$5(2y - 3) - 4y = 9.$$

The variable x has been eliminated.

Step 3 Solve the resulting equation containing one variable.

$$5(2y - 3) - 4y = 9 \qquad \text{This is the equation containing one variable.}$$
$$10y - 15 - 4y = 9 \qquad \text{Apply the distributive property.}$$
$$6y - 15 = 9 \qquad \text{Combine like terms.}$$
$$6y = 24 \qquad \text{Add 15 to both sides.}$$
$$y = 4 \qquad \text{Divide both sides by 6.}$$

Study Tip

The equation from step 1, in which one variable is expressed in terms of the other, is equivalent to one of the original equations. It is often easiest to back-substitute an obtained value into this equation to find the value of the other variable. After obtaining both values, get into the habit of checking the ordered-pair solution in *both* equations of the system.

Step 4 Back-substitute the obtained value into one of the original equations. We back-substitute 4 for y into one of the original equations to find x. Let's use both equations to show that we obtain the same value for x in either case.

Using the first equation:	Using the second equation:
$5x - 4y = 9$	$x - 2y = -3$
$5x - 4(4) = 9$	$x - 2(4) = -3$
$5x - 16 = 9$	$x - 8 = -3$
$5x = 25$	$x = 5$
$x = 5$	

With $x = 5$ and $y = 4$, the proposed solution is $(5, 4)$.

Step 5 Check. Take a moment to show that $(5, 4)$ satisfies both given equations. The solution set is $\{(5, 4)\}$.

Check Point 3 Solve by the substitution method:

$$3x + 2y = -1$$
$$x - y = 3.$$

3 Solve linear systems by addition.

Eliminating a Variable Using the Addition Method

The substitution method is most useful if one of the given equations has an isolated variable. A second, and frequently the easiest, method for solving a linear system is the addition method. Like the substitution method, the addition method involves eliminating a variable and ultimately solving an equation containing only one variable. However, this time we eliminate a variable by adding the equations.

For example, consider the following system of linear equations:

$$3x - 4y = 11$$
$$-3x + 2y = -7.$$

When we add these two equations, the x-terms are eliminated. This occurs because the coefficients of the x-terms, 3 and -3, are opposites (additive inverses) of each other:

$$3x - 4y = 11$$
$$\underline{-3x + 2y = -7}$$

Add: $-2y = 4$ The sum is an equation in one variable.

$y = -2$ Solve for y, dividing both sides by -2.

Now we can back-substitute -2 for y into one of the original equations to find x. It does not matter which equation you use; you will obtain the same value for x in either case. If we use either equation, we can show that $x = 1$ and the solution $(1, -2)$ satisfies both equations in the system.

When we use the addition method, we want to obtain two equations whose sum is an equation containing only one variable. The key step is to obtain, for one of the variables, coefficients that differ only in sign. To do this, we may need to multiply one or both equations by some nonzero number so that the coefficients of one of the variables, x or y, become opposites. Then when the two equations are added, this variable is eliminated.

EXAMPLE 4   Solving a System by the Addition Method

Solve by the addition method:

$$3x + 2y = 48$$
$$9x - 8y = -24.$$

Solution   We must rewrite one or both equations in equivalent forms so that the coefficients of the same variable (either x or y) are opposites of each other. Consider the terms in x in each equation, that is, $3x$ and $9x$. To eliminate x, we can multiply each term of the first equation by -3 and then add the equations.

$$3x + 2y = 48 \quad \underrightarrow{\text{Multiply by } -3.} \quad -9x - 6y = -144$$
$$9x - 8y = -24 \quad \underrightarrow{\text{No change}} \quad \underline{9x - 8y = -24}$$

$$\text{Add:} \qquad -14y = -168$$
$$y = 12 \qquad \text{\small Solve for } y, \text{ dividing both sides by } -14.$$

Thus, $y = 12$. We back-substitute this value into either one of the given equations. We'll use the first one.

$$3x + 2y = 48 \qquad \text{\small This the first equation in the given system.}$$
$$3x + 2(12) = 48 \qquad \text{\small Substitute 12 for } y.$$
$$3x + 24 = 48 \qquad \text{\small Multiply.}$$
$$3x = 24 \qquad \text{\small Subtract 24 from both sides.}$$
$$x = 8 \qquad \text{\small Divide both sides by 3.}$$

The solution $(8, 12)$ can be shown to satisfy both equations in the system. Consequently, the solution set is $\{(8, 12)\}$.

Solving Linear Systems by Addition

1. If necessary, rewrite both equations in the form $Ax + By = C$.
2. If necessary, multiply either equation or both equations by appropriate nonzero numbers so that the sum of the x-coefficients or the sum of the y-coefficients is 0.
3. Add the equations in step 2. The sum is an equation in one variable.
4. Solve the equation in one variable.
5. Back-substitute the value obtained in step 4 into either of the given equations and solve for the other variable.
6. Check the solution in both of the original equations.

Check Point 4   Solve by the addition method:

$$4x + 5y = 3$$
$$2x - 3y = 7.$$

Some linear systems have solutions that are not integers. If the value of one variable turns out to be a "messy" fraction, back-substitution might lead to cumbersome arithmetic. If this happens, you can return to the original system and use addition to find the value of the other variable.

EXAMPLE 5 Solving a System by the Addition Method

Solve by the addition method:

$$2x = 7y - 17$$
$$5y = 17 - 3x.$$

Solution

Step 1 Rewrite both equations in the form $Ax + By = C$. We first arrange the system so that variable terms appear on the left and constants appear on the right. We obtain

$$2x - 7y = -17 \quad \text{Subtract 7y from both sides of the first equation.}$$

$$3x + 5y = 17. \quad \text{Add 3x to both sides of the second equation.}$$

Step 2 If necessary, multiply either equation or both equations by appropriate numbers so that the sum of the x-coefficients or the sum of the y-coefficients is 0. We can eliminate x or y. Let's eliminate x by multiplying the first equation by 3 and the second equation by -2.

$$2x - 7y = -17 \quad \underline{\text{Multiply by 3.}} \quad 3 \cdot 2x - 3 \cdot 7y = 3(-17) \longrightarrow 6x - 21y = -51$$

$$3x + 5y = 17 \quad \underline{\text{Multiply by } -2.} \quad -2 \cdot 3x + (-2) \cdot 5y = -2(17) \longrightarrow -6x - 10y = -34$$

Steps 3 and 4 Add the equations and solve the equation in one variable.

$$6x - 21y = -51$$
$$\underline{-6x - 10y = -34}$$

Add: $\qquad -31y = -85$

$$\frac{-31y}{-31} = \frac{-85}{-31} \quad \text{Divide both sides by } -31.$$

$$y = \frac{85}{31} \quad \text{Simplify.}$$

Step 5 Back-substitute and find the value for the other variable. Back-substitution of $\frac{85}{31}$ for y into either of the given equations results in cumbersome arithmetic. Instead, let's use the addition method on the given system in the form $Ax + By = C$ to find the value for x. Thus, we eliminate y by multiplying the first equation by 5 and the second equation by 7.

$$2x - 7y = -17 \quad \underline{\text{Multiply by 5.}} \quad 10x - 35y = -85$$
$$3x + 5y = 17 \quad \underline{\text{Multiply by 7.}} \quad \underline{21x + 35y = 119}$$

Add: $\qquad 31x = 34$

$$x = \tfrac{34}{31} \quad \text{Divide both sides by 31.}$$

Step 6 Check. For this system, a calculator is helpful in showing the solution $\left(\frac{34}{31}, \frac{85}{31}\right)$ satisfies both equations. Consequently, the solution set is $\left\{\left(\frac{34}{31}, \frac{85}{31}\right)\right\}$.

Check Point 5 Solve by the addition method:

$$4x = 5 + 2y$$
$$3y = 4 - 2x.$$

4 Identify systems that do not have exactly one ordered-pair solution.

Linear Systems Having No Solution or Infinitely Many Solutions

We have seen that a system of linear equations in two variables represents a pair of lines. The lines either intersect, are parallel, or are identical. Thus, there are three possibilities for the number of solutions to a system of two linear equations.

The Number of Solutions to a System of Two Linear Equations

The number of solutions to a system of two linear equations in two variables is given by one of the following. (See Figure 8.2.)

Number of Solutions	What This Means Graphically
Exactly one ordered-pair solution	The two lines intersect at one point.
No solution	The two lines are parallel.
Infinitely many solutions	The two lines are identical.

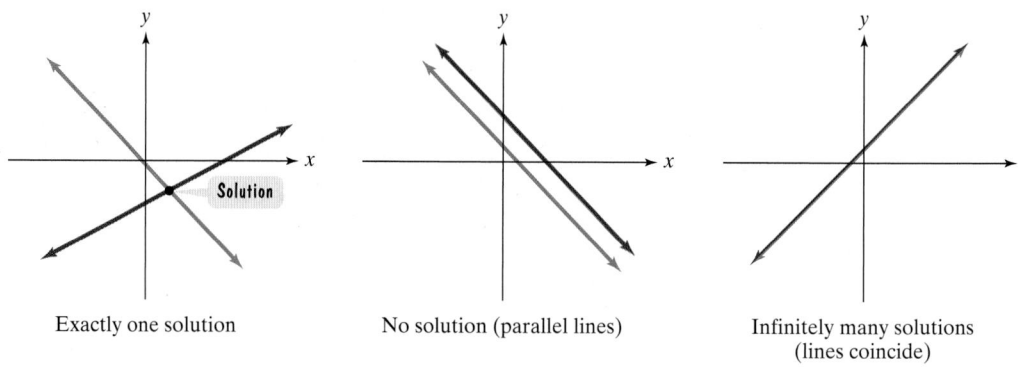

Exactly one solution   No solution (parallel lines)   Infinitely many solutions (lines coincide)

Figure 8.2 Possible graphs for a system of two linear equations in two variables

A linear system with no solution is called an **inconsistent system.** If you attempt to solve such a system by substitution or addition, you will eliminate both variables. A false statement such as $0 = 17$ will be the result.

EXAMPLE 6   A System with No Solution

Solve the system:

$$4x + 6y = 12$$
$$6x + 9y = 12.$$

Solution   Because no variable is isolated, we will use the addition method. To obtain coefficients of x that differ only in sign, we multiply the first equation by 3 and multiply the second equation by -2.

$$
\begin{array}{l}
4x + 6y = 12 \quad \xrightarrow{\text{Multiply by 3.}} \quad 12x + 18y = 36 \\
6x + 9y = 12 \quad \xrightarrow{\text{Multiply by } -2.} \quad \underline{-12x - 18y = -24} \\
\phantom{6x + 9y = 12 \quad \text{Multiply by } -2.} \quad \text{Add:} \quad\quad\quad\quad 0 = 12
\end{array}
$$

There are no values of x and y for which $0 = 12$.

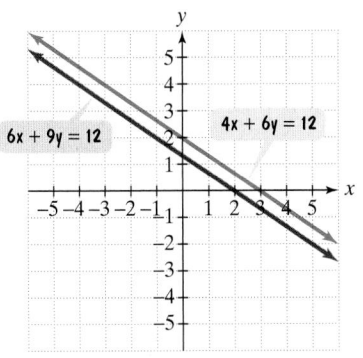

Figure 8.3 The graph of an inconsistent system

The false statement $0 = 12$ indicates that the system is inconsistent and has no solution. The solution set is the empty set, $\varnothing$.

The lines corresponding to the two equations in Example 6 are shown in Figure 8.3. The lines are parallel and have no point of intersection.

Discovery

Show that the graphs of $4x + 6y = 12$ and $6x + 9y = 12$ must be parallel lines by solving each equation for y. What is the slope and y-intercept for each line? What does this mean? If a linear system is inconsistent, what must be true about the slopes and y-intercepts for the system's graphs?

Check Point 6   Solve the system:

$$x + 2y = 4$$
$$3x + 6y = 13.$$

A linear system that has at least one solution is called a **consistent system.** Lines that intersect and lines that coincide both represent consistent systems. If the lines coincide, then the consistent system has infinitely many solutions, represented by every point on the line.

The equations in a linear system with infinitely many solutions are called **dependent.** If you attempt to solve such a system by substitution or addition, you will eliminate both variables. However, a true statement such as $0 = 0$ will be the result.

EXAMPLE 7   A System with Infinitely Many Solutions

Solve the system:

$$y = 3 - 2x$$
$$4x + 2y = 6.$$

Solution   Because the variable y is isolated in the first equation, we can use the substitution method. We substitute the expression for y in the other equation.

$$y = \boxed{3 - 2x} \quad 4x + 2\boxed{y} = 6 \qquad \text{Substitute } 3 - 2x \text{ for } y.$$

$$4x + 2y \quad\ = 6 \qquad \text{This is the second equation in the given system.}$$

$$4x + 2(3 - 2x) = 6 \qquad \text{Substitute } 3 - 2x \text{ for } y.$$

$$4x + 6 - 4x = 6 \qquad \text{Apply the distributive property.}$$

$$6 = 6 \qquad \text{Simplify. This statement is true for all values of } x \text{ and } y.$$

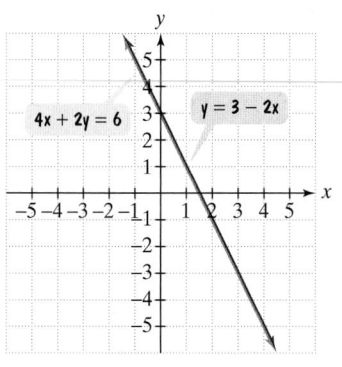

Figure 8.4 The graph of a system with infinitely many solutions

In our final step, both variables have been eliminated, and the resulting statement $6 = 6$ is true. This true statement indicates that the system has infinitely many solutions. The solution set consists of all points (x, y) lying on either of the coinciding lines, $y = 3 - 2x$ or $4x + 2y = 6$, as shown in Figure 8.4.

We express the solution set for the system in one of two equivalent ways:

$\{(x, y) \mid y = 3 - 2x\}$ *The set of all ordered pairs (x, y) such that y = 3 − 2x*

or $\{(x, y) \mid 4x + 2y = 6\}$ *The set of all ordered pairs (x, y) such that 4x + 2y = 6*

Check Point 7   Solve the system:

$$y = 4x - 4$$
$$8x - 2y = 8.$$

5   Solve problems using systems of linear equations.

Applications

As a young entrepreneur, did you ever try selling lemonade in your front yard? Suppose that you charged 55¢ for each cup and you sold 45 cups. Your **revenue** is your income from selling these 45 units, or $0.55(45) = \$24.75$. Your *revenue function* from selling x cups is

$$R(x) = 0.55x.$$

This is the unit price: This is the number
55¢ for each cup.   of units sold.

For any business, the **revenue function**, R, is the money generated by selling x units of the product:

$$R(x) = px.$$

Price per unit   x units sold

Back to selling lemonade and energizing the neighborhood with white sugar: Is your revenue for the afternoon also your profit? No. We need to consider the cost of the business. You estimate that the lemons, white sugar, and bottled water cost 5¢ per cup. Furthermore, mommy dearest is charging you a \$10 rental fee for use of your (her?) front yard. In Chapter 3, we saw that the **cost function, C,** for any business is the sum of its fixed and variable costs. Thus, your cost function for selling x cups of lemonade is

$$C(x) = 10 + 0.05x.$$

This is your \$10   This is your variable cost:
fixed cost.   5¢ for each cup produced.

Figure 8.5 shows the graphs of the revenue and cost functions for the lemonade business. Similar graphs and models apply no matter how small or large a business venture may be.

$$R(x) = 0.55x \qquad C(x) = 10 + 0.05x$$

Revenue is 55¢ times the number of cups sold.

Cost is $10 plus 5¢ times the number of cups produced.

$R(x) = 0.55x$

$C(x) = 10 + 0.05x$

Gain

Loss

Break-even point: (20, $11)

Units Produced and Sold

Figure 8.5

Profit

What does every entrepreneur, from a kid selling lemonade to Bill Gates, want to do? Generate profit, of course. The profit function, P, generated after producing and selling x units of a product is the difference between the revenue function, R, and the cost function, C. The profit function for the lemonade business in Figure 8.5 is

$P(x) = R(x) - C(x)$

$\quad = 0.55x - (10 + 0.05x)$

$\quad = 0.50x - 10.$

The graph of this profit function is shown below. The red portion lies below the x-axis and shows a loss when fewer than 20 units are sold. The lemonade business is "in the red." The black portion lies above the x-axis and shows a gain when more than 20 units are sold. The lemonade business is "in the black."

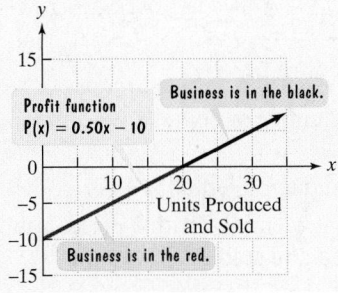

The lines intersect at the point (20, 11). This means that when 20 cups are produced and sold, both cost and revenue are $11. In business, this point of intersection is called the **break-even point.** At the break-even point, the money coming in is equal to the money going out. Can you see what happens for x-values less than 20? The red cost graph is above the blue revenue graph. The cost is greater than the revenue and the business is losing money. Thus, if you sell fewer than 20 cups of lemonade, the result is a *loss*. By contrast, look at what happens for x-values greater than 20. The blue revenue graph is above the red cost graph. The revenue is greater than the cost and the business is making money. Thus, if you sell more than 20 cups of lemonade, the result is a *gain*.

EXAMPLE 8   Finding a Break-Even Point

Technology is now promising to bring light, fast, and beautiful wheelchairs to millions of disabled people. A company is planning to manufacture these radically different wheelchairs. Fixed cost will be $500,000 and it will cost $400 to produce each wheelchair. Each wheelchair will be sold for $600.

a. Write the cost function, C, of producing x wheelchairs.

b. Write the revenue function, R, from the sale of x wheelchairs.

c. Determine the break-even point. Describe what this means.

Solution

a. The cost function is the sum of the fixed cost and variable cost.

Fixed cost of $500,000   plus   Variable cost: $400 for each chair produced

$$C(x) = 500,000 + 400x$$

Technology

The graphs of the wheelchair company's cost and revenue functions

$C(x) = 500,000 + 400x$

and $R(x) = 600x$ are shown in a

[0, 5000, 1000] by

[0, 3,000,000, 1,000,000]

viewing rectangle. The intersection point confirms that the company breaks even by producing and selling 2500 wheelchairs.

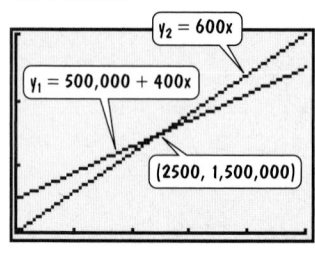

b. The revenue function is the money generated from the sale of x wheelchairs.

Revenue per chair, $600, times | the number of chairs sold

$$R(x) = 600x$$

c. The break-even point occurs where the graphs of C and R intersect. Thus, we find this point by solving the system

$$\begin{aligned} C(x) &= 500,000 + 400x \\ R(x) &= 600x \end{aligned} \quad \text{or} \quad \begin{aligned} y &= 500,000 + 400x \\ y &= 600x. \end{aligned}$$

Using substitution, we can substitute $600x$ for y in the first equation.

$$\begin{aligned} 600x &= 500,000 + 400x & &\text{Substitute } 600x \text{ for } y. \\ 200x &= 500,000 & &\text{Subtract } 400x \text{ from both sides.} \\ x &= 2500 & &\text{Divide both sides by 200.} \end{aligned}$$

Back-substituting 2500 for x in either of the system's equations (or functions), we obtain

$$R(2500) = 600(2500) = 1,500,000.$$

We used $R(x) = 600x.$

The break-even point is (2500, 1,500,000). This means that the company will break even if it produces and sells 2500 wheelchairs. At this level, the money coming in is equal to the money going out: $1,500,000.

Check Point 8

A company that manufactures running shoes has a fixed cost of $300,000. Additionally, it costs $30 to produce each pair of shoes. They are sold at $80 per pair.

a. Write the cost function, C, of producing x pairs of running shoes.

b. Write the revenue function, R, from the sale of x pairs of running shoes.

c. Determine the break-even point. Describe what this means.

An important application of systems of equations arises in connection with supply and demand. As the price of a product increases, the demand for that product decreases. However, at higher prices suppliers are willing to produce greater quantities of the product.

EXAMPLE 9   Supply and Demand Models

A chain of video stores specializes in cult films. The weekly demand and supply models for *The Rocky Horror Picture Show* are given by

$$\begin{aligned} N &= -13p + 760 & &\text{Demand model} \\ N &= 2p + 430 & &\text{Supply model} \end{aligned}$$

in which p is the price of the video and N is the number of copies of the video sold or supplied each week to the chain of stores.

a. How many copies of the video can be sold and supplied at $18 per copy?

b. Find the price at which supply and demand are equal. At this price, how many copies of *Rocky Horror* can be supplied and sold each week?

Solution

a. To find how many copies of the video can be sold and supplied at $18 per copy, we substitute 18 for p in the demand and supply models.

Demand Model	**Supply Model**
$N = -13p + 760$	$N = 2p + 430$

Substitute 18 for p.	Substitute 18 for p.

$N = -13 \cdot 18 + 760 = 526$	$N = 2 \cdot 18 + 430 = 466$

At $18 per video, the chain can sell 526 copies of *Rocky Horror* in a week. The manufacturer is willing to supply 466 copies per week. This will result in a shortage of copies of the video. Under these conditions, the retail chain is likely to raise the price of the video.

b. We can find the price at which supply and demand are equal by solving the demand-supply linear system. We will use substitution, substituting $-13p + 760$ for N in the second equation.

$$N = \boxed{-13p + 760} \qquad \boxed{N} = 2p + 430 \qquad \text{Substitute } -13p + 760 \text{ for N.}$$

$$-13p + 760 = 2p + 430 \qquad \text{The resulting equation contains only one variable.}$$

$$-15p + 760 = 430 \qquad \text{Subtract 2p from both sides.}$$

$$-15p = -330 \qquad \text{Subtract 760 from both sides.}$$

$$p = 22 \qquad \text{Divide both sides by } -15.$$

The price at which supply and demand are equal is $22 per video. To find the value of N, the number of videos supplied and sold weekly at this price, we back-substitute 22 for p into either the demand or the supply model. We'll use both models to make sure we get the same number in each case.

Demand Model	**Supply Model**
$N = -13p + 760$	$N = 2p + 430$

Substitute 22 for p.	Substitute 22 for p.

$N = -13 \cdot 22 + 760 = 474$	$N = 2 \cdot 22 + 430 = 474$

At a price of $22 per video, 474 units of the video can be supplied and sold weekly. The intersection point, $(22, 474)$, is shown in Figure 8.6.

Figure 8.6 Priced at $22 per video, 474 copies of the video can be supplied and sold weekly.

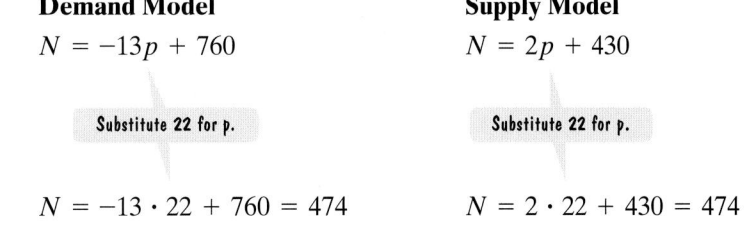

Check Point 9 The demand for a product is modeled by $N = -20p + 1000$ and the supply for the product by $N = 5p + 250$. In these models, p is the price of the product and N is the number supplied or sold weekly. At what price will supply equal demand? At that price, how many units of the product will be supplied and sold each week?

EXERCISE SET 8.1

 Practice Exercises

In Exercises 1–4, determine whether the given ordered pair is a solution of the system.

1. $(2, 3)$
$x + 3y = 11$
$x - 5y = -13$

2. $(-3, 5)$
$9x + 7y = 8$
$8x - 9y = -69$

3. $(2, 5)$
$2x + 3y = 17$
$x + 4y = 16$

4. $(8, 5)$
$5x - 4y = 20$
$3y = 2x + 1$

In Exercises 5–18, solve each system by the substitution method.

5. $x + y = 4$
$y = 3x$

6. $x + y = 6$
$y = 2x$

7. $x + 3y = 8$
$y = 2x - 9$

8. $2x - 3y = -13$
$y = 2x + 7$

9. $x = 4y - 2$
$x = 6y + 8$

10. $x = 3y + 7$
$x = 2y - 1$

11. $5x + 2y = 0$
$x - 3y = 0$

12. $4x + 3y = 0$
$2x - y = 0$

13. $2x + 5y = -4$
$3x - y = 11$

14. $2x + 5y = 1$
$-x + 6y = 8$

15. $2x - 3y = 8 - 2x$
$3x + 4y = x + 3y + 14$

16. $3x - 4y = x - y + 4$
$2x + 6y = 5y - 4$

17. $y = \frac{1}{3}x + \frac{2}{3}$
$y = \frac{5}{7}x - 2$

18. $y = -\frac{1}{2}x + 2$
$y = \frac{3}{4}x + 7$

In Exercises 19–30, solve each system by the addition method.

19. $x + y = 1$
$x - y = 3$

20. $x + y = 6$
$x - y = -2$

21. $2x + 3y = 6$
$2x - 3y = 6$

22. $3x + 2y = 14$
$3x - 2y = 10$

23. $x + 2y = 2$
$-4x + 3y = 25$

24. $2x - 7y = 2$
$3x + y = -20$

25. $4x + 3y = 15$
$2x - 5y = 1$

26. $3x - 7y = 13$
$6x + 5y = 7$

27. $3x - 4y = 11$
$2x + 3y = -4$

28. $2x + 3y = -16$
$5x - 10y = 30$

29. $3x = 4y + 1$
$3y = 1 - 4x$

30. $5x = 6y + 40$
$2y = 8 - 3x$

In Exercises 31–42, solve by the method of your choice. Identify systems with no solution and systems with infinitely many solutions, using set notation to express their solution sets.

31. $x = 9 - 2y$
$x + 2y = 13$

32. $6x + 2y = 7$
$y = 2 - 3x$

33. $y = 3x - 5$
$21x - 35 = 7y$

34. $9x - 3y = 12$
$y = 3x - 4$

35. $3x - 2y = -5$
$4x + y = 8$

36. $2x + 5y = -4$
$3x - y = 11$

37. $x + 3y = 2$
$3x + 9y = 6$

38. $4x - 2y = 2$
$2x - y = 1$

39. $\frac{x}{4} - \frac{y}{4} = -1$
$x + 4y = -9$

40. $\frac{x}{6} - \frac{y}{2} = \frac{1}{3}$
$x + 2y = -3$

41. $2x = 3y + 4$
$4x = 3 - 5y$

42. $4x = 3y + 8$
$2x = -14 + 5y$

In Exercises 43–46, let x represent one number and let y represent the other number. Use the given conditions to write a system of equations. Solve the system and find the numbers.

43. The sum of two numbers is 7. If one number is subtracted from the other, their difference is −1. Find the numbers.

44. The sum of two numbers is 2. If one number is subtracted from the other, their difference is 8. Find the numbers.

45. Three times a first number decreased by a second number is 1. The first number increased by twice the second number is 12. Find the numbers.

46. The sum of three times a first number and twice a second number is 8. If the second number is subtracted from twice the first number, the result is 3. Find the numbers.

 Application Exercises

Exercises 47–50 describe a number of business ventures. For each exercise,

a. Write the cost function, C.
b. Write the revenue function, R.
c. Determine the break-even point. Describe what this means.

47. A company that manufactures small canoes has a fixed cost of $18,000. It costs $20 to produce each canoe. The selling price is $80 per canoe. (In solving this exercise, let x represent the number of canoes produced and sold.)

48. A company that manufactures bicycles has a fixed cost of $100,000. It costs $100 to produce each bicycle. The selling price is $300 per bike. (In solving this exercise, let x represent the number of bicycles produced and sold.)

49. You invest in a new play. The cost includes an overhead of $30,000, plus production costs of $2500 per performance. A sold-out performance brings in $3125. (In solving this exercise, let x represent the number of sold-out performances.)

50. You invested $30,000 and started a business writing greeting cards. Supplies cost 2¢ per card and you are selling each card for 50¢. (In solving this exercise, let x represent the number of cards produced and sold.)

51. At a price of p dollars per ticket, the number of tickets to a rock concert that can be sold is given by the demand model $N = -25p + 7500$. At a price of p dollars per ticket, the number of tickets that the concert's promoters are willing to make available is given by the supply model $N = 5p + 6000$.
  a. How many tickets can be sold and supplied for $40 per ticket?
  b. Find the ticket price at which supply and demand are equal. At this price, how many tickets will be supplied and sold?

52. The weekly demand and supply models for a particular brand of scientific calculator for a chain of stores are given by the demand model $N = -53p + 1600$, and the supply model $N = 75p + 320$. In these models, p is the price of the calculator and N is the number of calculators sold or supplied each week to the stores.
  a. How many calculators can be sold and supplied at $12 per calculator?
  b. Find the price at which supply and demand are equal. At this price, how many calculators of this type can be supplied and sold each week?

53. In the United States, deaths from car accidents, per 100,000 persons, are decreasing at a faster rate than deaths from gunfire, shown by the blue and red lines that model the data points in the figure.

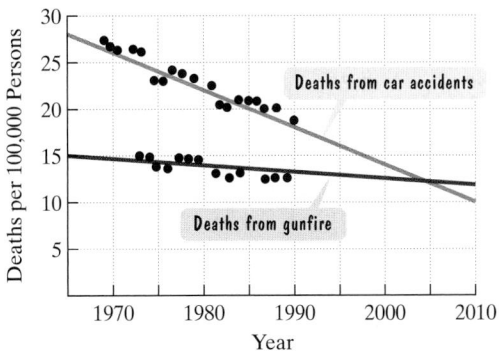

Annual Deaths in the U.S. from Car Accidents and Gunfire

Source: Journal of the American Medical Association

The function $y = -0.4x + 28$ models deaths from car accidents, y, per 100,000 persons, x years after 1965. The function $0.07x + y = 15$ models deaths from gunfire, y, per 100,000 persons, x years after 1965. Use

these models to project when the number of deaths from gunfire will equal the number of deaths from car accidents. Round to the nearest year. How many annual deaths, per 100,000 persons, will there be from gunfire and from car accidents at that time? Describe how this is illustrated by the lines in the figure shown.

54. The June 7, 1999 issue of *Newsweek* presented statistics showing progress African Americans have made in education, health, and finance. Infant mortality for African Americans is decreasing at a faster rate than it is for whites, shown by the graphs below. Infant mortality for African Americans can be modeled by $M = -0.41x + 22$ and for whites by $M = -0.18x + 10$. In both models, x is the number of years after 1980 and M is infant mortality, measured in deaths per 1000 live births. Use these models to project when infant mortality for African Americans and whites will be the same. What is infant mortality rate for both groups at that time?

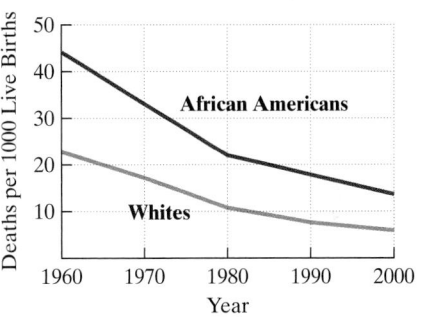

Source: National Center for Health Statistics

The graphs show average weekly earnings of full-time wage and salary workers 25 and older, by educational attainment. Exercises 55–56 involve the information in these graphs.

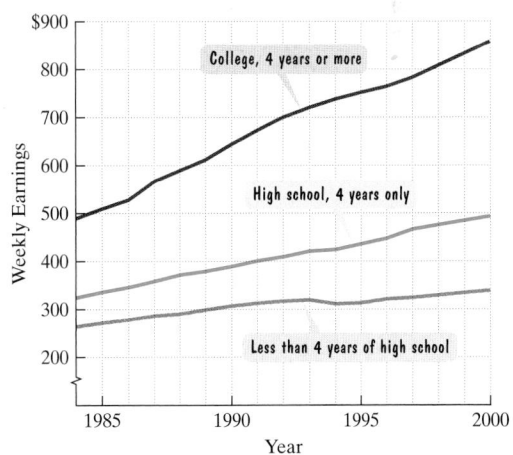

Source: U.S. Bureau of Labor Statistics

55. In 1985, college graduates averaged $508 in weekly earnings. This amount has increased by approximately $25 in weekly earnings per year. By contrast, in 1985, high school graduates averaged $345 in weekly earnings. This amount has only increased by approximately $9 in weekly earnings per year.

a. Write a function that models weekly earnings, E, for college graduates x years after 1985.

b. Write a function that models weekly earnings, E, for high school graduates x years after 1985.

c. How many years after 1985 will college graduates be earning twice as much per week as high school graduates? In which year will this occur? What will be the weekly earnings for each group at that time?

56. In 1985, college graduates averaged $508 in weekly earnings. This amount has increased by approximately $25 in weekly earnings per year. By contrast, in 1985, people with less than four years of high school averaged $270 in weekly earnings. This amount has only increased by approximately $4 in weekly earnings per year.

a. Write a function that models weekly earnings, E, for college graduates x years after 1985.

b. Write a function that models weekly earnings, E, for people with less than four years of high school x years after 1985.

c. How many years after 1985 will college graduates be earning three times as much per week as people with less than four years of high school? (Round to the nearest whole number.) In which year will this occur? What will be the weekly earnings for each group at that time?

Use a system of linear equations to solve Exercises 57–67.

The graph shows the calories in some favorite fast foods. Use the information in Exercises 57-58 to find the exact caloric content of the specified foods.

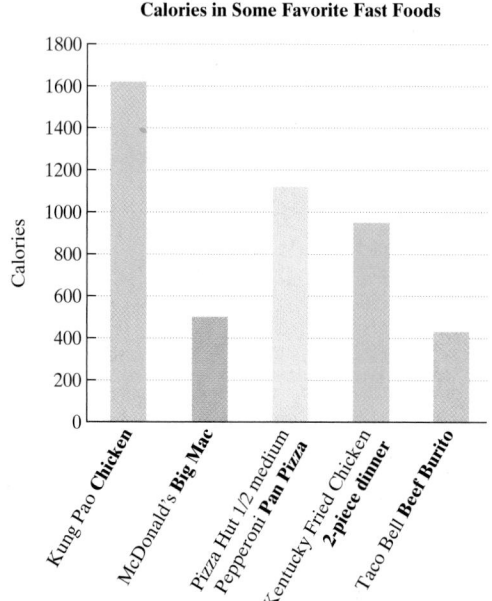

Calories in Some Favorite Fast Foods

Fast Food

Source: Center for Science in the Public Interest

57. One pan pizza and two beef burritos provide 1980 calories. Two pan pizzas and one beef burrito provide 2670 calories. Find the caloric content of each item.

58. One Kung Pao chicken and two Big Macs provide 2620 calories. Two Kung Pao chickens and one Big Mac provide 3740 calories. Find the caloric content of each item.

59. Cholesterol intake should be limited to 300 mg or less each day. One serving of scrambled eggs from McDonalds and one Double Beef Whopper from Burger King exceed this intake by 241 mg. Two servings of scrambled eggs and three Double Beef Whoppers provide 1257 mg of cholesterol. Determine the cholesterol content in each item.

60. Two medium eggs and three cups of ice cream contain 701 milligrams of cholesterol. One medium egg and one cup of ice cream exceed the suggested daily cholesterol intake of 300 milligrams by 25 milligrams. Determine the cholesterol content in each item.

61. A hotel has 200 rooms. Those with kitchen facilities rent for $100 per night and those without kitchen facilities rent for $80 per night. On a night when the hotel was completely occupied, revenues were $17,000. How many of each type of room does the hotel have?

62. In a new development, 50 one- and two-bedroom condominiums were sold. Each one-bedroom condominium sold for $120 thousand and each two bedroom condominium sold for $150 thousand. If sales totaled $7050 thousand, how many of each type of unit was sold?

63. A rectangular lot whose perimeter is 360 feet is fenced along three sides. An expensive fencing along the lot's length costs $20 per foot, and an inexpensive fencing along the two side widths costs only $8 per foot. The total cost of the fencing along the three sides comes to $3280. What are the lot's dimensions?

64. A rectangular lot whose perimeter is 320 feet is fenced along three sides. An expensive fencing along the lot's length costs $16 per foot, and an inexpensive fencing along the two side widths costs only $5 per foot. The total cost of the fencing along the three sides comes to $2140. What are the lot's dimensions?

65. When a crew rows with the current, it travels 16 miles in 2 hours. Against the current, the crew rows 8 miles in 2 hours. Let $x =$ the crew's rowing rate in still water and let $y =$ the rate of the current. The following chart summarizes this information:

	Rate	$\times$	Time	$=$	Distance
Rowing with current	$x + y$		2		16
Rowing against current	$x - y$		2		8

Find the rate of rowing in still water and the rate of the current.

66. When an airplane flies with the wind, it travels 800 miles in 4 hours. Against the wind, it takes 5 hours to cover the same distance. Find the plane's rate in still air and the rate of the wind.

67. Find the measures of the angles marked $x°$ and $y°$ in the figure.

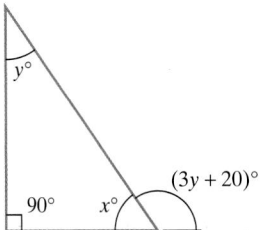

Writing in Mathematics

68. What is a system of linear equations? Provide an example with your description.

69. What is the solution to a system of linear equations?

70. Explain how to solve a system of equations using the substitution method. Use $y = 3 - 3x$ and $3x + 4y = 6$ to illustrate your explanation.

71. Explain how to solve a system of equations using the addition method. Use $3x + 5y = -2$ and $2x + 3y = 0$ to illustrate your explanation.

72. When is it easier to use the addition method rather than the substitution method to solve a system of equations?

73. When using the addition or substitution method, how can you tell if a system of linear equations has infinitely many solutions? What is the relationship between the graphs of the two equations?

74. When using the addition or substitution method, how can you tell if a system of linear equations has no solution? What is the relationship between the graphs of the two equations?

75. Describe the break-even point for a business.

76. The law of supply and demand states that, in a free market economy, a commodity tends to be sold at its equilibrium price. At this price, the amount that the seller will supply is the same amount that the consumer will buy. Explain how systems of equations can be used to determine the equilibrium price.

77. The function $y = 0.94x + 5.64$ models annual U.S. consumption of chicken, y, in pounds per person, x years after 1950. The function $0.74x + y = 146.76$ models annual U.S. consumption of red meat, y, in pounds per person, x years after 1950. What is the most efficient method for solving this system? What does the solution mean in terms of the variables in the functions? (It is not necessary to solve the system.)

78. In Exercise 77, find the slope of each model. Describe what this means in terms of the rate of change of chicken consumption and the rate of change of red meat consumption. Why must the graphs have an intersection point? What happens to the right of the intersection point?

Technology Exercises

79. Verify your solutions to any five exercises in Exercises 5–42 by using a graphing utility to graph the two equations in the system in the same viewing rectangle. Then use the intersection feature to display the solution.

80. Some graphing utilities can give the solution to a linear system of equations. (Consult your manual for details.) This capability is usually accessed with the [SIMULT] (simultaneous equations) feature. First, you will enter 2, for two equations in two variables. With each equation in $Ax + By = C$ form, you will then enter the coefficients for x and y and the constant term, one equation at a time. After entering all six numbers, press [SOLVE]. The solution will be displayed on the screen. (The x-value may be displayed as $x_1 =$ and the y-value as $x_2 =$.) Use this capability to verify the solution to any five of the exercises you solved in the practice exercises of this exercise set. Describe what happens when you use your graphing utility on a system with no solution or infinitely many solutions.

Critical Thinking Exercises

81. Write a system of equations having $\{(-2, 7)\}$ as a solution set. (More than one system is possible.)

82. Solve the system for x and y in terms of $a_1, b_1, c_1, a_2, b_2,$ and c_2:

$$a_1 x + b_1 y = c_1$$
$$a_2 x + b_2 y = c_2.$$

83. Two identical twins can only be recognized by the characteristic that one always tells the truth and the other always lies. One twin tells you of a lucky number pair: "When I multiply my first lucky number by 3 and my second lucky number by 6, the addition of the resulting numbers produces a sum of 12. When I add my first lucky number and twice my second lucky number, the sum is 5." Which twin is talking?

84. A marching band has 52 members, and there are 24 in the pom-pom squad. They wish to form several hexagons and squares like those diagrammed below. Can it be done with no people left over?

B B P P

B P B B B = Band Member

B B P P P = Pom-pom Person

Group Exercise

85. The group should write four different word problems that can be solved using a system of linear equations in two variables. All of the problems should be on different topics. The group should turn in the four problems and their algebraic solutions.

SECTION 8.2 *Systems of Linear Equations in Three Variables*

Objectives

1. Verify the solution of a system of linear equations in three variables.
2. Solve systems of linear equations in three variables.
3. Solve problems using systems in three variables.

All animals sleep, but the length of time they sleep varies widely: Cattle sleep for only a few minutes at a time. We humans seem to need more sleep than other animals. Without enough sleep, we have difficulty concentrating, make mistakes in routine tasks, lose energy, and feel bad-tempered. There is a relationship between hours of sleep and death rate per year per 100,000 people. How many hours of sleep will put you in the group with the minimum death rate? In this section, we will answer this question by solving a system of linear equations with more than two variables.

1 Verify the solution of a system of linear equations in three variables.

Systems of Linear Equations in Three Variables and Their Solutions

An equation such as $x + 2y - 3z = 9$ is called a **linear equation in three variables.** In general, any equation of the form

$$Ax + By + Cz = D$$

where $A, B, C,$ and D are real numbers such that $A, B,$ and C are not all 0, is a linear equation in the variables $x, y,$ and z. The graph of this linear equation in three variables is a plane in three-dimensional space.

The process of solving a system of three linear equations in three variables is geometrically equivalent to finding the point of intersection (assuming that there is one) of three planes in space (see Figure 8.7). A **solution** to a system of linear equations in three variables is an ordered triple of real numbers that satisfies all equations of the system. The **solution set** of the system is the set of all its solutions.

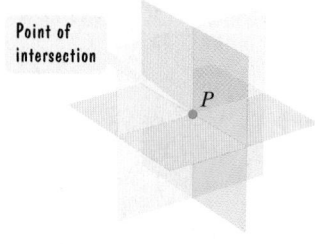

Point of intersection

P

Figure 8.7

EXAMPLE 1   **Determining Whether an Ordered Triple Satisfies a System**

Show that the ordered triple $(-1, 2, -2)$ is a solution of the system:

$$x + 2y - 3z = 9$$
$$2x - y + 2z = -8$$
$$-x + 3y - 4z = 15.$$

Solution   Because -1 is the x-coordinate, 2 is the y-coordinate, and -2 is the z-coordinate of $(-1, 2, -2)$, we replace x with -1, y with 2, and z with -2 in each of the three equations.

$$
\begin{aligned}
x + 2y - 3z &= 9 \\
-1 + 2(2) - 3(-2) &\overset{?}{=} 9 \\
-1 + 4 + 6 &\overset{?}{=} 9 \\
9 &= 9, \text{ true}
\end{aligned}
\qquad
\begin{aligned}
2x - y + 2z &= -8 \\
2(-1) - 2 + 2(-2) &\overset{?}{=} -8 \\
-2 - 2 - 4 &\overset{?}{=} -8 \\
-8 &= -8, \text{ true}
\end{aligned}
\qquad
\begin{aligned}
-x + 3y - 4z &= 15 \\
-(-1) + 3(2) - 4(-2) &\overset{?}{=} 15 \\
1 + 6 + 8 &\overset{?}{=} 15 \\
15 &= 15, \text{ true}
\end{aligned}
$$

The ordered triple $(-1, 2, -2)$ satisfies the three equations: It makes each equation true. Thus, the ordered triple is a solution of the system.

Check Point 1   Show that the ordered triple $(-1, -4, 5)$ is a solution of the system:

$$x - 2y + 3z = 22$$
$$2x - 3y - z = 5$$
$$3x + y - 5z = -32.$$

2 Solve systems of linear equations in three variables.

Solving Systems of Linear Equations in Three Variables by Eliminating Variables

The method for solving a system of linear equations in three variables is similar to that used on systems of linear equations in two variables. We use addition to eliminate any variable, reducing the system to two equations in two variables. Once we obtain a system of two equations in two variables, we use addition or substitution to eliminate a variable. The result is a single equation in one variable. We solve this equation to get the value of the remaining variable. Other variable values are found by back-substitution.

Solving Linear Systems in Three Variables by Eliminating Variables

1. Reduce the system to two equations in two variables. This is usually accomplished by taking two different pairs of equations and using the addition method to eliminate the same variable from each pair.
2. Solve the resulting system of two equations in two variables using addition or substitution. The result is an equation in one variable that gives the value of that variable.
3. Back-substitute the value of the variable found in step 2 into either of the equations in two variables to find the value of the second variable.
4. Use the values of the two variables from steps 2 and 3 to find the value of the third variable by back-substituting into one of the original equations.
5. Check the proposed solution in each of the original equations.

EXAMPLE 2   Solving a System in Three Variables

Solve the system:

$$5x - 2y - 4z = 3 \quad \text{Equation 1}$$
$$3x + 3y + 2z = -3 \quad \text{Equation 2}$$
$$-2x + 5y + 3z = 3. \quad \text{Equation 3}$$

Solution   There are many ways to proceed. Because our initial goal is to reduce the system to two equations in two variables, **the central idea is to take two different pairs of equations and eliminate the same variable from each pair.**

Step 1   Reduce the system to two equations in two variables.   We choose any two equations and use the addition method to eliminate a variable. Let's eliminate z using Equations 1 and 2. We do so by multiplying Equation 2 by 2. Then we add equations.

(Equation 1) $5x - 2y - 4z = 3 \quad \xrightarrow{\text{No change}} \quad 5x - 2y - 4z = 3$
(Equation 2) $3x + 3y + 2z = -3 \quad \xrightarrow{\text{Multiply by 2.}} \quad 6x + 6y + 4z = -6$
$$\text{Add: } 11x + 4y \qquad = -3 \quad \text{Equation 4}$$

Now we must eliminate the *same* variable using another pair of equations. We can eliminate z from Equations 2 and 3. First, we multiply Equation 2 by -3. Next, we multiply Equation 3 by 2. Finally, we add equations.

(Equation 2) $3x + 3y + 2z = -3 \quad \xrightarrow{\text{Multiply by } -3.} \quad -9x - 9y - 6z = 9$
(Equation 3) $-2x + 5y + 3z = 3 \quad \xrightarrow{\text{Multiply by 2.}} \quad -4x + 10y + 6z = 6$
$$\text{Add: } -13x + \quad y \qquad = 15 \quad \text{Equation 5}$$

Equations 4 and 5 give us a system of two equations in two variables.

Step 2   Solve the resulting system of two equations in two variables.   We will use the addition method to solve Equations 4 and 5 for x and y. To do so, we multiply Equation 5 on both sides by -4 and add this to Equation 4.

(Equation 4) $11x + 4y = -3 \quad \xrightarrow{\text{No change}} \quad 11x + 4y = -3$
(Equation 5) $-13x + y = 15 \quad \xrightarrow{\text{Multiply by } -4.} \quad 52x - 4y = -60$
$$\text{Add: } 63x \qquad = -63$$
$$x \qquad = -1 \quad \text{Divide both sides by 63.}$$

Step 3   Use back-substitution in one of the equations in two variables to find the value of the second variable.   We back-substitute -1 for x in either Equation 4 or 5 to find the value of y.

$$-13x + y = 15 \quad \text{Equation 5}$$

$$-13(-1) + y = 15 \quad \text{Substitute } -1 \text{ for x.}$$

$$13 + y = 15 \quad \text{Multiply.}$$

$$y = 2 \quad \text{Subtract 13 from both sides.}$$

Step 4   Back-substitute the values found for two variables into one of the original equations to find the value of the third variable.   We can now use any one of the original equations and back-substitute the values of x and y to find the value for z. We will use Equation 2.

$$3x + 3y + 2z = -3 \quad \text{Equation 2}$$

$$3(-1) + 3(2) + 2z = -3 \quad \text{Substitute } -1 \text{ for } x \text{ and } 2 \text{ for } y.$$

$$3 + 2z = -3 \quad \begin{array}{l}\text{Multiply and then add:} \\ 3(-1) + 3(2) = -3 + 6 = 3.\end{array}$$

$$2z = -6 \quad \text{Subtract 3 from both sides.}$$

$$z = -3 \quad \text{Divide both sides by 2.}$$

With $x = -1$, $y = 2$, and $z = -3$, the proposed solution is the ordered triple $(-1, 2, -3)$.

Step 5   Check.   Check the proposed solution, $(-1, 2, -3)$, by substituting the values for x, y, and z into each of the three original equations. These substitutions yield three true statements. Thus, the solution set is $\{(-1, 2, -3)\}$.

Check Point 2   Solve the system:

$$x + 4y - z = 20$$
$$3x + 2y + z = 8$$
$$2x - 3y + 2z = -16.$$

In some examples, one of the variables is already eliminated from a given equation. In this case, the same variable should be eliminated from the other two equations, thereby making it possible to omit one of the elimination steps. We illustrate this idea in Example 3.

EXAMPLE 3   Solving a System of Equations with a Missing Term

Solve the system:

$$x + \quad\quad z = 8 \quad \text{Equation 1}$$
$$x + y + 2z = 17 \quad \text{Equation 2}$$
$$x + 2y + z = 16. \quad \text{Equation 3}$$

Solution

Step 1   Reduce the system to two equations in two variables.   Because Equation 1 contains only x and z, we could omit one of the elimination steps by eliminating y using Equations 2 and 3. This will give us two equations in x and z. To eliminate y using Equations 2 and 3, we multiply Equation 2 by -2 and add Equation 3.

(Equation 2) $x + y + 2z = 17$ $\xrightarrow{\text{Multiply by } -2.}$ $-2x - 2y - 4z = -34$
(Equation 3) $x + 2y + z = 16$ $\xrightarrow{\text{No change}}$ $\underline{\quad x + 2y + \quad z = \quad 16}$
                                              Add: $-x \quad\quad - 3z = -18$ Equation 4

Equation 4 and the given Equation 1 provide us with a system of two equations in two variables.

Step 2 Solve the resulting system of two equations in two variables. We will solve Equations 1 and 4 for x and z.

$$
\begin{array}{ll}
x + z = 8 & \text{Equation 1} \\
x + y + 2z = 17 & \text{Equation 2} \\
x + 2y + z = 16 & \text{Equation 3}
\end{array}
$$

The system we are solving, repeated

$$
\begin{array}{lll}
& x + z = 8 & \text{Equation 1} \\
& \underline{-x - 3z = -18} & \text{Equation 4} \\
\text{Add:} & -2z = -10 & \\
& z = 5 & \text{Divide both sides by } -2.
\end{array}
$$

Step 3 Use back-substitution in one of the equations in two variables to find the value of the second variable. To find x, we back-substitute 5 for z in either Equation 1 or 4. We will use Equation 1.

$$
\begin{array}{ll}
x + z = 8 & \text{Equation 1} \\
x + 5 = 8 & \text{Substitute 5 for z.} \\
 x = 3 & \text{Subtract 5 from both sides.}
\end{array}
$$

Step 4 Back-substitute the values found for two variables into one of the original equations to find the value of the third variable. To find y, we back-substitute 3 for x and 5 for z into Equation 2 or 3. We can't use Equation 1 because y is missing in this equation. We will use Equation 2.

$$
\begin{array}{ll}
x + y + 2z = 17 & \text{Equation 2} \\
3 + y + 2(5) = 17 & \text{Substitute 3 for x and 5 for z.} \\
y + 13 = 17 & \text{Multiply and add.} \\
y = 4 & \text{Subtract 13 from both sides.}
\end{array}
$$

We found that $z = 5$, $x = 3$, and $y = 4$. Thus, the proposed solution is the ordered triple $(3, 4, 5)$.

Step 5 Check. Substituting 3 for x, 4 for y, and 5 for z into each of the three original equations yields three true statements. Consequently, the solution set is $\{(3, 4, 5)\}$.

Check Point 3 Solve the system:

$$
\begin{array}{rl}
2y - z = 7 \\
x + 2y + z = 17 \\
2x - 3y + 2z = -1.
\end{array}
$$

A system of linear equations in three variables represents three planes. The three planes need not intersect at one point. The planes may have no common point of intersection and represent an inconsistent system with no solution. By contrast, the planes may coincide or intersect along a line. In these cases, the planes have infinitely many points in common and represent systems with infinitely many solutions. Systems of linear equations in three variables that are inconsistent or that contain dependent equations will be discussed in Chapter 9.

3 Solve problems using systems in three variables.

Applications

Systems of equations may allow us to find models for data without using a graphing utility. Three data points that do not lie on or near a line determine the graph of a quadratic function of the form $y = ax^2 + bx + c, a \neq 0$. Quadratic functions often model situations in which values of y are decreasing and then increasing, suggesting the cuplike shape of a parabola.

EXAMPLE 4   Modeling Data Relating Sleep and Death Rate

In a study relating sleep and death rate, the following data were obtained. Use the function $y = ax^2 + bx + c$ to model the data.

x (Average Number of Hours of Sleep)	y (Death Rate per Year per 100,000 Males)
4	1682
7	626
9	967

Solution   We need to find values for $a, b,$ and c in $y = ax^2 + bx + c$. We can do so by solving a system of three linear equations in $a, b,$ and c. We obtain the three equations by using the values of x and y from the data as follows:

$$y = ax^2 + bx + c$$ Use the quadratic function to model the data.

When x = 4, y = 1682: $1682 = a \cdot 4^2 + b \cdot 4 + c$ or $16a + 4b + c = 1682$

When x = 7, y = 626: $626 = a \cdot 7^2 + b \cdot 7 + c$ or $49a + 7b + c = 626$

When x = 9, y = 967: $967 = a \cdot 9^2 + b \cdot 9 + c$ or $81a + 9b + c = 967.$

The easiest way to solve this system is to eliminate c from two pairs of equations, obtaining two equations in a and b. Solving this system gives $a = 104.5,$ $b = -1501.5,$ and $c = 6016.$ We now substitute the values for $a, b,$ and c into $y = ax^2 + bx + c.$ The function that models the given data is

$$y = 104.5x^2 - 1501.5x + 6016.$$

We can use the model that we obtained in Example 4 to find the death rate of males who average, say, 6 hours of sleep. First, write the model in function notation:

$$f(x) = 104.5x^2 - 1501.5x + 6016.$$

Substitute 6 for x:

$$f(6) = 104.5(6)^2 - 1501.5(6) + 6016 = 769.$$

According to the model, the death rate for males who average 6 hours of sleep is 769 deaths per 100,000 males.

Technology

The graph of

$$y = 104.5x^2 - 1501.5x + 6016$$

is displayed in a [3, 12, 1] by [500, 2000, 100] viewing rectangle. The minimum function feature shows that the lowest point on the graph, the vertex, is approximately (7.2, 622.5). Men who average 7.2 hours of sleep are in the group with the lowest death rate, approximately 622.5 per 100,000.

Minimum
X=7.212766 . Y=622.53916 .

Check Point 4   Find the quadratic function $y = ax^2 + bx + c$ whose graph passes through the points $(1, 4), (2, 1),$ and $(3, 4).$

EXERCISE SET 8.2

 Practice Exercises

In Exercises 1–4, determine if the given ordered triple is a solution of the system.

1.
$$x + y + z = 4$$
$$x - 2y - z = 1$$
$$2x - y - 2z = -1$$
$$(2, -1, 3)$$

2.
$$x + y + z = 0$$
$$x + 2y - 3z = 5$$
$$3x + 4y + 2z = -1$$
$$(5, -3, -2)$$

3.
$$x - 2y = 2$$
$$2x + 3y = 11$$
$$y - 4z = -7$$
$$(4, 1, 2)$$

4.
$$x - 2z = -5$$
$$y - 3z = -3$$
$$2x - z = -4$$
$$(-1, 3, 2)$$

Solve each system in Exercises 5–18.

5.
$$x + y + 2z = 11$$
$$x + y + 3z = 14$$
$$x + 2y - z = 5$$

6.
$$2x + y - 2z = -1$$
$$3x - 3y - z = 5$$
$$x - 2y + 3z = 6$$

7.
$$4x - y + 2z = 11$$
$$x + 2y - z = -1$$
$$2x + 2y - 3z = -1$$

8.
$$x - y + 3z = 8$$
$$3x + y - 2z = -2$$
$$2x + 4y + z = 0$$

9.
$$3x + 5y + 2z = 0$$
$$12x - 15y + 4z = 12$$
$$6x - 25y - 8z = 8$$

10.
$$2x + 3y + 7z = 13$$
$$3x + 2y - 5z = -22$$
$$5x + 7y - 3z = -28$$

11.
$$2x - 4y + 3z = 17$$
$$x + 2y - z = 0$$
$$4x - y - z = 6$$

12.
$$x + z = 3$$
$$x + 2y - z = 1$$
$$2x - y + z = 3$$

13.
$$2x + y = 2$$
$$x + y - z = 4$$
$$3x + 2y + z = 0$$

14.
$$x + 3y + 5z = 20$$
$$y - 4z = -16$$
$$3x - 2y + 9z = 36$$

15.
$$x + y = -4$$
$$y - z = 1$$
$$2x + y + 3z = -21$$

16.
$$x + y = 4$$
$$x + z = 4$$
$$y + z = 4$$

17.
$$3(2x + y) + 5z = -1$$
$$2(x - 3y + 4z) = -9$$
$$4(1 + x) = -3(z - 3y)$$

18.
$$7z - 3 = 2(x - 3y)$$
$$5y + 3z - 7 = 4x$$
$$4 + 5z = 3(2x - y)$$

In Exercises 19–20, let x represent the first number, y the second number, and z the third number. Use the given conditions to write a system of equations. Solve the system and find the numbers.

19. The sum of three numbers is 16. The sum of twice the first number, 3 times the second number, and 4 times the third number is 46. The difference between 5 times the first number and the second number is 31. Find the three numbers.

20. The following is known about three numbers: Three times the first number plus the second number plus twice the third number is 5. If 3 times the second number is subtracted from the sum of the first number and 3 times the third number, the result is 2. If the third number is subtracted from 2 times the first number and 3 times the second number, the result is 1. Find the numbers.

In Exercises 21–24, find the quadratic function $y = ax^2 + bx + c$ whose graph passes through the given points.

21. $(-1, 6), (1, 4), (2, 9)$ **22.** $(-2, 7), (1, -2), (2, 3)$

23. $(-1, -4), (1, -2), (2, 5)$ **24.** $(1, 3), (3, -1), (4, 0)$

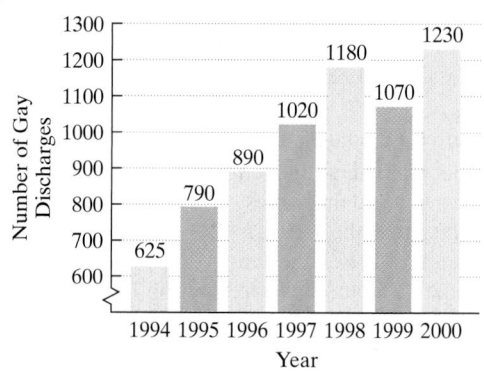 **Application Exercises**

25. The bar graph shows that the number of gays discharged from the military decreased from 1998 to 1999 and increased from 1999 to 2000.

Number of Gay Discharges from the Military Under "Don't Ask, Don't Tell"

Source: New York Times

a. Write the data for 1998, 1999, and 2000 as ordered pairs (x, y), where x is the number of years after 1998 and y is the number of gay discharges from the military.

b. The three data points in part (a) can be modeled by the quadratic function $y = ax^2 + bx + c$. Substitute each ordered pair into this function, one ordered pair at a time, and write a system of three linear equations in three variables that can be used to find values for a, b, and c.

c. Solve the system in part (b). Then write the quadratic function that models the data for 1998 through 2000.

26. The bar graph shows that the percentage of the U.S. population that was foreign-born decreased between 1940 and 1970 and then increased between 1970 and 2000.

Percentage of U.S. Population That Was Foreign-Born, 1900-2000

Source: U.S. Census Bureau

    a. Write the data for 1940, 1970, and 2000 as ordered pairs (x, y), where x is the number of years after 1940 and y is the percentage of the U.S. population that was foreign-born in that year.

    b. The three data points in part (a) can be modeled by the quadratic function $y = ax^2 + bx + c$. Substitute each ordered pair into this function, one ordered pair at a time, and write a system of linear equations in three variables that can be used to find values for a, b, and c.

    c. Solve the system in part (b). Then write the quadratic function that models the data for 1940 through 2000.

27. You throw a ball straight up from a rooftop. The ball misses the rooftop on its way down and eventually strikes the ground. A mathematical model can be used to describe the relationship for the ball's height above the ground, y, after x seconds. Consider the following data:

x, seconds after the ball is thrown	y, ball's height, in feet, above the ground
1	224
3	176
4	104

    a. Find the quadratic function $y = ax^2 + bx + c$ whose graph passes through the given points.

    b. Use the function in part (a) to find the value for y when $x = 5$. Describe what this means.

28. A mathematical model can be used to describe the relationship between the number of feet a car travels once the brakes are applied, y, and the number of seconds the car is in motion after the brakes are applied, x. A research firm collects the following data:

x, seconds in motion after brakes are applied	y, feet car travels once the brakes are applied
1	46
2	84
3	114

    a. Find the quadratic function $y = ax^2 + bx + c$ whose graph passes through the given points.

    b. Use the function in part (a) to find the value for y when $x = 6$. Describe what this means.

Use a system of linear equations in three variables to solve Exercises 29–35.

29. In current U.S. dollars, John D. Rockefeller's 1913 fortune of $900 million would be worth about $189 billion. The bar graph shows that Rockefeller is the wealthiest among the world's five richest people of all time. The combined estimated wealth, in current billions of U.S. dollars, of Andrew Carnegie, Cornelius Vanderbilt, and Bill Gates is $256 billion. The difference between Carnegie's estimated wealth and Vanderbilt's is $4 billion. The difference between Vanderbilt's estimated wealth and Gate's is $36 billion. Find the estimated wealth, in current billions of U.S. dollars, of Carnegie, Vanderbilt, and Gates.

The Richest People of All Time Estimated Wealth, in Current Billions of U.S. Dollars

Source: *Scholastic Book of World Records*

30. The circle graph indicates computers in use for the United States and the rest of the world. The percentage of the world's computers in Europe and Japan combined is 13% less than the percentage of the world's computers in the United States. If the percentage of the world's computers in Europe is doubled, it is only 3% more than the percentage of the world's computers in the United States. Find the percentage of the world's computers in the United States, Europe, and Japan.

**Percentage of the World's Computers:
U.S. and the World**

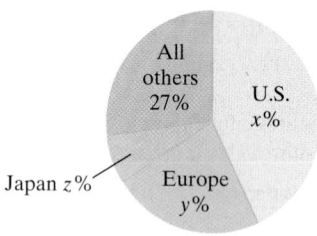

Source: Jupiter Communications

31. At a college production of *Evita*, 400 tickets were sold. The ticket prices were $8, $10, and $12, and the total income from ticket sales was $3700. How many tickets of each type were sold if the combined number of $8 and $10 tickets sold was 7 times the number of $12 tickets sold?

32. A certain brand of razor blades comes in packages of 6, 12, and 24 blades, costing $2, $3, and $4 per package, respectively. A store sold 12 packages containing a total of 162 razor blades and took in $35. How many packages of each type were sold?

33. A person invested $6700 for one year, part at 8%, part at 10%, and the remainder at 12%. The total annual income from these investments was $716. The amount of money invested at 12% was $300 more than the amount invested at 8% and 10% combined. Find the amount invested at each rate.

34. A person invested $17,000 for one year, part at 10%, part at 12%, and the remainder at 15%. The total annual income from these investments was $2110. The amount of money invested at 12% was $1000 less than the amount invested at 10% and 15% combined. Find the amount invested at each rate.

35. Find the measures of the angles marked $x°$, $y°$, and $z°$ in the following triangle.

Writing in Mathematics

36. What is a system of linear equations in three variables?

37. How do you determine whether a given ordered triple is a solution of a system in three variables?

38. Describe in general terms how to solve a system in three variables.

39. AIDS is taking a deadly toll on southern Africa. Describe how to use the techniques that you learned in this section to obtain a model for African life span using projections with AIDS. Let x represent the number of years after 1985 and let y represent African life span in that year.

African Life Span

Source: United Nations

Technology Exercises

40. Does your graphing utility have a feature that allows you to solve linear systems by entering coefficients and constant terms? If so, use this feature to verify the solutions to any five exercises that you worked by hand from Exercises 5–16.

41. Verify your results in Exercises 21–24 by using a graphing utility to graph the resulting parabola. Trace along the curve and convince yourself that the three points given in the exercise lie on the parabola.

42. Some graphing utilities will do three-dimensional graphing. For example, on the TI-92, press $\boxed{\text{MODE}}$, go to $\boxed{\text{GRAPH}}$, press the arrow to the right, select $\boxed{\text{3D}}$, then $\boxed{\text{ENTER}}$. When you display the $\boxed{\text{Y} =}$ screen, you will see the equations are functions of x and y. Thus, you must solve each of a linear system's equations for z before entering the equation. For example,

$$x + y + z = 19$$

is solved for z, giving

$$z = 19 - x - y.$$

(Consult your manual.) If your utility does three-dimensional graphing, graph five of the systems in Exercises 5–16 and trace along the planes to find their common point of intersection.

Critical Thinking Exercises

43. Describe how the system

$$x + y - z - 2w = -8$$
$$x - 2y + 3z + w = 18$$
$$2x + 2y + 2z - 2w = 10$$
$$2x + y - z + w = 3$$

could be solved. Is it likely that in the near future a graphing utility will be available to provide a geometric solution (using intersecting graphs) to this system? Explain.

44. A modernistic painting consists of triangles, rectangles, and pentagons, all drawn so as to not overlap or share sides. Within each rectangle are drawn 2 red roses, and each pentagon contains 5 carnations. How many triangles, rectangles, and pentagons appear in the painting if the painting contains a total of 40 geometric figures, 153 sides of geometric figures, and 72 flowers?

Group Exercise

45. Group members should develop appropriate functions that model each of the projections shown in Exercise 39.

SECTION 8.3 *Partial Fractions*

Objective

1. Find the partial fraction decomposition of a rational expression.

The rising and setting of the sun suggest the obvious: Things change over time. Calculus is the study of rates of change, allowing the motion of the rising sun to be measured by "freezing the frame" at one instant in time. If you are given a function, calculus reveals its rate of change at any "frozen" instant. In this section, you will learn an algebraic technique used in calculus to find a function if its rate of change is known.

The Idea behind Partial Fraction Decomposition

Systems of linear equations can be used to reverse the process of adding and subtracting rational expressions—for example,

$$\frac{3}{x-4} - \frac{2}{x+2} = \frac{3(x+2) - 2(x-4)}{(x-4)(x+2)}$$

$$= \frac{3x + 6 - 2x + 8}{(x-4)(x+2)} = \frac{x+14}{(x-4)(x+2)}.$$

In order to reverse this process, we must show that

$$\frac{x + 14}{(x - 4)(x + 2)} = \frac{3}{x - 4} - \frac{2}{x + 2} \quad \text{or} \quad \frac{3}{x - 4} + \frac{-2}{x + 2}.$$

Each of the two fractions on the right is called a **partial fraction**. The sum of these fractions is called the **partial fraction decomposition** of the rational expression on the left-hand side.

Partial fraction decompositions can be written for rational expressions of the form $\dfrac{P(x)}{Q(x)}$, where P and Q have no common factors and the highest power in the numerator is less than the highest power in the denominator. In this section, we will show you how to write the partial fraction decompositions for each of the following rational expressions:

$$\frac{9x^2 - 9x + 6}{(2x - 1)(x + 2)(x - 2)}$$

$P(x) = 9x^2 - 9x + 6$; highest power = 2

$Q(x) = (2x - 1)(x + 2)(x - 2)$; multiplying factors, highest power = 3

$$\frac{5x^3 - 3x^2 + 7x - 3}{(x^2 + 1)^2}.$$

$P(x) = 5x^3 - 3x^2 + 7x - 3$; highest power = 3

$Q(x) = (x^2 + 1)^2$; squaring this expression, highest power = 4

1 Find the partial fraction decomposition of a rational expression.

The Steps in Partial Fraction Decomposition

The partial fraction decomposition of a rational expression depends on the factors of the denominator. We consider four cases involving different kinds of factors in the denominator.

Case 1: The Partial Fraction Decomposition of a Rational Expression with Distinct Linear Factors in the Denominator   If the denominator has a linear factor of the form $ax + b$, then the partial fraction decomposition will contain a term of the form

$$\frac{A}{ax + b}.$$

Constant

Linear factor

Each distinct linear factor in the denominator produces a partial fraction of the form *constant over linear factor*. For example,

$$\frac{9x^2 - 9x + 6}{(2x - 1)(x + 2)(x - 2)} = \frac{A}{2x - 1} + \frac{B}{x + 2} + \frac{C}{x - 2}.$$

We write a constant over each linear factor in the denominator.

The form of the partial fraction decomposition for a rational expression with distinct linear factors in the denominator is

$$\frac{P(x)}{(a_1 x + b_1)(a_2 x + b_2)(a_3 x + b_3) \cdots (a_n x + b_n)}$$

$$= \frac{A_1}{a_1 x + b_1} + \frac{A_2}{a_2 x + b_2} + \frac{A_3}{a_3 x + b_3} + \cdots + \frac{A_n}{a_n x + b_n}.$$

EXAMPLE 1 Partial Fraction Decomposition with Distinct Linear Factors

Find the partial fraction decomposition of

$$\frac{x + 14}{(x - 4)(x + 2)}.$$

Solution We begin by setting up the partial fraction decomposition with the unknown constants. Write a constant over each of the two distinct linear factors in the denominator.

$$\frac{x + 14}{(x - 4)(x + 2)} = \frac{A}{x - 4} + \frac{B}{x + 2}$$

Our goal is to find A and B. We do this by multiplying both sides of the equation by the least common denominator, $(x - 4)(x + 2)$.

$$(x - 4)(x + 2)\frac{x + 14}{(x - 4)(x + 2)} = (x - 4)(x + 2)\left(\frac{A}{x - 4} + \frac{B}{x + 2}\right)$$

We use the distributive property on the right side.

$$(x - 4)(x + 2)\frac{x + 14}{(x - 4)(x + 2)}$$
$$= (x - 4)(x + 2)\frac{A}{(x - 4)} + (x - 4)(x + 2)\frac{B}{(x + 2)}$$

Dividing out common factors in numerators and denominators, we obtain

$$x + 14 = A(x + 2) + B(x - 4).$$

To find values for A and B that make both sides equal, we'll express the sides in exactly the same form by writing the variable x-terms and then writing the constant terms. Apply the distributive property on the right side.

$$x + 14 = Ax + 2A + Bx - 4B$$
$$x + 14 = Ax + Bx + 2A - 4B$$

$$1x + 14 = (A + B)x + (2A - 4B)$$

As shown by the arrows, if two polynomials are equal, coefficients of like powers of x must be equal ($A + B = 1$) and their constant terms must be equal ($2A - 4B = 14$). Consequently, A and B satisfy the following two equations:

$$A + B = 1$$
$$2A - 4B = 14.$$

We can use the addition method to solve this linear system in two variables. By multiplying the first equation by -2 and adding equations, we obtain $A = 3$ and $B = -2$. Thus,

$$\frac{x + 14}{(x - 4)(x + 2)} = \frac{A}{x - 4} + \frac{B}{x + 2} = \frac{3}{x - 4} + \frac{-2}{x + 2}\left(\text{or } \frac{3}{x - 4} - \frac{2}{x + 2}\right).$$

Steps in Partial Fraction Decomposition

1. Set up the partial fraction decomposition with the unknown constants A, B, C, etc., in the numerator of the decomposition.

2. Multiply both sides of the resulting equation by the least common denominator.

3. Simplify the right-hand side of the equation.

4. Write both sides in descending powers, equate coefficients of like powers of x, and equate constant terms.

5. Solve the resulting linear system for A, B, C, etc.

6. Substitute the values for A, B, C, etc., into the equation in step 1 and write the partial fraction decomposition.

Check Point 1 Find the partial fraction decomposition of $\dfrac{5x - 1}{(x - 3)(x + 4)}$.

Case 2: The Partial Fraction Decomposition of a Rational Expression with Linear Factors in the Denominator, Some of Which Are Repeated

Suppose that $(ax + b)^n$ is a factor of the denominator. This means that the linear factor $ax + b$ is repeated n times. When this occurs, the partial fraction decomposition will contain the following sum of n fractions:

$$\frac{P(x)}{(ax + b)^n} = \frac{A_1}{ax + b} + \frac{A_2}{(ax + b)^2} + \frac{A_3}{(ax + b)^3} + \cdots + \frac{A_n}{(ax + b)^n}.$$

Include one fraction with a constant numerator for each power of $ax + b$.

EXAMPLE 2 Partial Fraction Decomposition with Repeated Linear Factors

Find the partial fraction decomposition of $\dfrac{x - 18}{x(x - 3)^2}$.

Solution

Step 1 Set up the partial fraction decomposition with the unknown constants. Because the linear factor $x - 3$ is repeated twice, we must include one fraction with a constant numerator for each power of $x - 3$.

$$\frac{x - 18}{x(x - 3)^2} = \frac{A}{x} + \frac{B}{x - 3} + \frac{C}{(x - 3)^2}$$

Step 2 Multiply both sides of the resulting equation by the least common denominator. We clear fractions, multiplying both sides by $x(x - 3)^2$, the least common denominator.

$$x(x - 3)^2 \left[\frac{x - 18}{x(x - 3)^2} \right] = x(x - 3)^2 \left[\frac{A}{x} + \frac{B}{x - 3} + \frac{C}{(x - 3)^2} \right]$$

We use the distributive property on the right side.

$$x(x-3)^2 \cdot \frac{x-18}{x(x-3)^2}$$

$$= x(x-3)^2 \cdot \frac{A}{x} + x(x-3)^2 \cdot \frac{B}{(x-3)} + x(x-3)^2 \cdot \frac{C}{(x-3)^2}$$

Dividing out common factors in numerators and denominators, we obtain

$$x - 18 = A(x-3)^2 + Bx(x-3) + Cx.$$

Step 3   Simplify the right side of the equation.   Square $x - 3$. Then apply the distributive property.

$$x - 18 = A(x^2 - 6x + 9) + Bx(x - 3) + Cx \qquad \text{Square } x - 3 \text{ using} \\ (A - B)^2 = A^2 - 2AB + B^2.$$

$$x - 18 = Ax^2 - 6Ax + 9A + Bx^2 - 3Bx + Cx \qquad \text{Apply the distributive property.}$$

Step 4   Write both sides in descending powers, equate coefficients of like powers of x, and equate constant terms.   The left side, $x - 18$, is in descending powers of x: $x - 18x^0$. We will write the right side in descending powers of x.

$$x - 18 = Ax^2 + Bx^2 - 6Ax - 3Bx + Cx + 9A$$

Express both sides in the same form.

$$0x^2 + 1x - 18 = (A + B)x^2 + (-6A - 3B + C)x + 9A$$

Equating coefficients of like powers of x and constant terms results in the following system of linear equations:

$$A + B = 0$$
$$-6A - 3B + C = 1$$
$$9A = -18.$$

Step 5   Solve the resulting system for A, B, and C.   Dividing both sides of the last equation by 9, we obtain $A = -2$. Substituting -2 for A in the first equation, $A + B = 0$, gives $-2 + B = 0$ or $B = 2$. We find C by substituting -2 for A and 2 for B in the middle equation, $-6A - 3B + C = 1$. We obtain $C = -5$.

Step 6   Substitute the values of A, B, and C and write the partial fraction decomposition.   With $A = -2$, $B = 2$, and $C = -5$, the required partial fraction decomposition is

$$\frac{x-18}{x(x-3)^2} = \frac{A}{x} + \frac{B}{x-3} + \frac{C}{(x-3)^2} = -\frac{2}{x} + \frac{2}{x-3} - \frac{5}{(x-3)^2}.$$

Check Point 2   Find the partial fraction decomposition of $\dfrac{x+2}{x(x-1)^2}$.

Case 3:   The Partial Fraction Decomposition of a Rational Expression with Prime, Nonrepeated Quadratic Factors in the Denominator   Suppose that $ax^2 + bx + c$ is a factor of the denominator and that this quadratic factor cannot be factored into linear factors with real coefficients. Under these conditions, the partial fraction decomposition will contain a term of the form

$$\frac{Ax + B}{ax^2 + bx + c}. \qquad \begin{matrix} \text{Linear numerator} \\ \text{Quadratic factor} \end{matrix}$$

Each distinct prime quadratic factor in the denominator produces a partial fraction of the form *linear numerator over quadratic factor*. For example,

$$\frac{3x^2 + 17x + 14}{(x - 2)(x^2 + 2x + 4)} = \frac{A}{x - 2} + \frac{Bx + C}{x^2 + 2x + 4}.$$

We write a constant over the linear factor in the denominator.

We write a linear numerator over the prime quadratic factor in the denominator.

Our next example illustrates how a linear system in three variables is used to determine values for A, B, and C.

EXAMPLE 3 Partial Fraction Decomposition

Find the partial fraction decomposition of

$$\frac{3x^2 + 17x + 14}{(x - 2)(x^2 + 2x + 4)}.$$

Solution

Step 1 Set up the partial fraction decomposition with the unknown constants. We put a constant (A) over the linear factor and a linear expression $(Bx + C)$ over the prime quadratic factor.

$$\frac{3x^2 + 17x + 14}{(x - 2)(x^2 + 2x + 4)} = \frac{A}{x - 2} + \frac{Bx + C}{x^2 + 2x + 4}$$

Step 2 Multiply both sides of the resulting equation by the least common denominator. We clear fractions, multiplying both sides by $(x - 2)$ $(x^2 + 2x + 4)$, the least common denominator.

$$(x - 2)(x^2 + 2x + 4)\left[\frac{3x^2 + 17x + 14}{(x - 2)(x^2 + 2x + 4)}\right] = (x - 2)(x^2 + 2x + 4)\left[\frac{A}{x - 2} + \frac{Bx + C}{x^2 + 2x + 4}\right]$$

We use the distributive property on the right side.

$$(x - 2)(x^2 + 2x + 4) \cdot \frac{3x^2 + 17x + 14}{(x - 2)(x^2 + 2x + 4)}$$

$$= (x - 2)(x^2 + 2x + 4) \cdot \frac{A}{x - 2} + (x - 2)(x^2 + 2x + 4) \cdot \frac{Bx + C}{x^2 + 2x + 4}$$

Dividing out common factors in numerators and denominators, we obtain

$$3x^2 + 17x + 14 = A(x^2 + 2x + 4) + (Bx + C)(x - 2).$$

Step 3 Simplify the right side of the equation. We multiply on the right side by distributing A over each term in parentheses and multiplying $(Bx + C)(x - 2)$ using the FOIL method.

$$3x^2 + 17x + 14 = Ax^2 + 2Ax + 4A + Bx^2 - 2Bx + Cx - 2C$$

Step 4 Write both sides in descending powers, equate coefficients of like powers of x, and equate constant terms. The left side, $3x^2 + 17x + 14$, is in descending powers of x. We write the right side in descending powers of x

$$3x^2 + 17x + 14 = Ax^2 + Bx^2 + 2Ax - 2Bx + Cx + 4A - 2C$$

and express both sides in the same form.

$$3x^2 + 17x + 14 = (A + B)x^2 + (2A - 2B + C)x + (4A - 2C)$$

Equating coefficients of like powers of x and constant terms results in the following system of linear equations:

$$A + B = 3$$
$$2A - 2B + C = 17$$
$$4A - 2C = 14.$$

Step 5   Solve the resulting system for $A, B,$ and C.   Because the first equation involves A and B, we can obtain another equation in A and B by eliminating C from the second and third equations. Multiply the second equation by 2 and add equations. Solving in this manner, we obtain $A = 5, B = -2,$ and $C = 3$.

Step 6   Substitute the values of $A, B,$ and C and write the partial fraction decomposition.   With $A = 5, B = -2,$ and $C = 3$, the required partial fraction decomposition is

$$\frac{3x^2 + 17x + 14}{(x - 2)(x^2 + 2x + 4)} = \frac{A}{x - 2} + \frac{Bx + C}{x^2 + 2x + 4} = \frac{5}{x - 2} + \frac{-2x + 3}{x^2 + 2x + 4}.$$

Check Point 3   Find the partial fraction decomposition of
$$\frac{8x^2 + 12x - 20}{(x + 3)(x^2 + x + 2)}.$$

Case 4:   The Partial Fraction Decomposition of a Rational Expression with a Prime, Repeated Quadratic Factor in the Denominator   Suppose that $(ax^2 + bx + c)^n$ is a factor of the denominator and that $ax^2 + bx + c$ cannot be factored further. This means that the quadratic factor $ax^2 + bx + c$ is repeated n times. When this occurs, the partial fraction decomposition will contain a linear numerator for each power of $ax^2 + bx + c$.

$$\frac{P(x)}{(ax^2 + bx + c)^n} = \frac{A_1 x + B_1}{ax^2 + bx + c} + \frac{A_2 x + B_2}{(ax^2 + bx + c)^2} + \frac{A_3 x + B_3}{(ax^2 + bx + c)^3} + \cdots + \frac{A_n x + B_n}{(ax^2 + bx + c)^n}$$

Include one fraction with a linear numerator for each power of $ax^2 + bx + c$.

Study Tip

When the denominator of a rational expression contains a **power of a linear factor,** set up the partial fraction decomposition with **constant numerators** (A, B, C, etc.). When the denominator of a rational expression contains a **power of a prime quadratic factor,** set up the partial fraction decomposition with **linear numerators** ($Ax + B$, $Cx + D$, etc.).

EXAMPLE 4   Partial Fraction Decomposition with a Repeated Quadratic Factor

Find the partial fraction decomposition of
$$\frac{5x^3 - 3x^2 + 7x - 3}{(x^2 + 1)^2}.$$

Solution

Step 1   Set up the partial fraction decomposition with the unknown constants. Because the quadratic factor $x^2 + 1$ is repeated twice, we must include one fraction with a linear numerator for each power of $x^2 + 1$.

$$\frac{5x^3 - 3x^2 + 7x - 3}{(x^2 + 1)^2} = \frac{Ax + B}{x^2 + 1} + \frac{Cx + D}{(x^2 + 1)^2}$$

Step 2 Multiply both sides of the resulting equation by the least common denominator.   We clear fractions, multiplying both sides by $(x^2 + 1)^2$, the least common denominator.

$$(x^2 + 1)^2\left[\frac{5x^3 - 3x^2 + 7x - 3}{(x^2 + 1)^2}\right] = (x^2 + 1)^2\left[\frac{Ax + B}{x^2 + 1} + \frac{Cx + D}{(x^2 + 1)^2}\right]$$

Now we multiply and simplify.

$$5x^3 - 3x^2 + 7x - 3 = (x^2 + 1)(Ax + B) + Cx + D$$

Step 3 Simplify the right side of the equation.   We multiply $(x^2 + 1)(Ax + B)$ using the FOIL method.

$$5x^3 - 3x^2 + 7x - 3 = Ax^3 + Bx^2 + Ax + B + Cx + D$$

Step 4 Write both sides in descending powers, equate coefficients of like powers of x, and equate constant terms.

$$5x^3 - 3x^2 + 7x - 3 = Ax^3 + Bx^2 + Ax + Cx + B + D$$

$$5x^3 - 3x^2 + 7x - 3 = Ax^3 + Bx^2 + (A + C)x + (B + D)$$

Equating coefficients of like powers of x and constant terms results in the following system of linear equations:

$$A = 5$$
$$B = -3$$
$$A + C = 7 \qquad \text{With } A = 5, \text{ we immediately obtain } C = 2.$$
$$B + D = -3. \qquad \text{With } B = -3, \text{ we immediately obtain } D = 0.$$

Step 5 Solve the resulting system for A, B, C, and D.   Based on our observations in step 4, $A = 5, B = -3, C = 2$, and $D = 0$.

Step 6 Substitute the values of A, B, C, and D and write the partial fraction decomposition.

$$\frac{5x^3 - 3x^2 + 7x - 3}{(x^2 + 1)^2} = \frac{Ax + B}{x^2 + 1} + \frac{Cx + D}{(x^2 + 1)^2} = \frac{5x - 3}{x^2 + 1} + \frac{2x}{(x^2 + 1)^2}$$

> **Check Point 4**   Find the partial fraction decomposition of $\dfrac{2x^3 + x + 3}{(x^2 + 1)^2}$.

Study Tip

When a rational expression contains a power of a factor in the denominator, be sure to set up the partial fraction decomposition to allow for every natural-number power of that factor less than or equal to the power.

Example:

$$\frac{2x + 1}{(x - 5)^2\, x^3}$$

$$= \frac{A}{x - 5} + \frac{B}{(x - 5)^2} + \frac{C}{x} + \frac{D}{x^2} + \frac{E}{x^3}$$

Although $(x - 5)^2$ and x^2 are quadratic, they are expressed as powers of a linear factor, $x - 5$ and x. Thus, the numerator is constant.

EXERCISE SET 8.3

Practice Exercises

In Exercises 1–8, write the form of the partial fraction decomposition of the rational expression. It is not necessary to solve for the constants.

1. $\dfrac{11x - 10}{(x - 2)(x + 1)}$

2. $\dfrac{5x + 7}{(x - 1)(x + 3)}$

3. $\dfrac{6x^2 - 14x - 27}{(x + 2)(x - 3)^2}$

4. $\dfrac{3x + 16}{(x + 1)(x - 2)^2}$

5. $\dfrac{5x^2 - 6x + 7}{(x - 1)(x^2 + 1)}$

6. $\dfrac{5x^2 - 9x + 19}{(x - 4)(x^2 + 5)}$

7. $\dfrac{x^3 + x^2}{(x^2 + 4)^2}$

8. $\dfrac{7x^2 - 9x + 3}{(x^2 + 7)^2}$

In Exercises 9–42, write the partial fraction decomposition of each rational expression.

9. $\dfrac{x}{(x - 3)(x - 2)}$

10. $\dfrac{1}{x(x - 1)}$

11. $\dfrac{3x + 50}{(x - 9)(x + 2)}$

12. $\dfrac{5x - 1}{(x - 2)(x + 1)}$

13. $\dfrac{7x - 4}{x^2 - x - 12}$

14. $\dfrac{9x + 21}{x^2 + 2x - 15}$

15. $\dfrac{4}{2x^2 - 5x - 3}$

16. $\dfrac{x}{x^2 + 2x - 3}$

17. $\dfrac{4x^2 + 13x - 9}{x(x - 1)(x + 3)}$

18. $\dfrac{4x^2 - 5x - 15}{x(x + 1)(x - 5)}$

19. $\dfrac{4x^2 - 7x - 3}{x^3 - x}$

20. $\dfrac{2x^2 - 18x - 12}{x^3 - 4x}$

21. $\dfrac{6x - 11}{(x - 1)^2}$

22. $\dfrac{x}{(x + 1)^2}$

23. $\dfrac{x^2 - 6x + 3}{(x - 2)^3}$

24. $\dfrac{2x^2 + 8x + 3}{(x + 1)^3}$

25. $\dfrac{x^2 + 2x + 7}{x(x - 1)^2}$

26. $\dfrac{3x^2 + 49}{x(x + 7)^2}$

27. $\dfrac{x^2}{(x - 1)^2(x + 1)}$

28. $\dfrac{x^2}{(x - 1)^2(x + 1)^2}$

29. $\dfrac{5x^2 - 6x + 7}{(x - 1)(x^2 + 1)}$

30. $\dfrac{5x^2 - 9x + 19}{(x - 4)(x^2 + 5)}$

31. $\dfrac{5x^2 + 6x + 3}{(x + 1)(x^2 + 2x + 2)}$

32. $\dfrac{9x + 2}{(x - 2)(x^2 + 2x + 2)}$

33. $\dfrac{x + 4}{x^2(x^2 + 4)}$

34. $\dfrac{10x^2 + 2x}{(x - 1)^2(x^2 + 2)}$

35. $\dfrac{6x^2 - x + 1}{x^3 + x^2 + x + 1}$

36. $\dfrac{3x^2 - 2x + 8}{x^3 + 2x^2 + 4x + 8}$

37. $\dfrac{x^3 + x^2 + 2}{(x^2 + 2)^2}$

38. $\dfrac{x^2 + 2x + 3}{(x^2 + 4)^2}$

39. $\dfrac{x^3 - 4x^2 + 9x - 5}{(x^2 - 2x + 3)^2}$

40. $\dfrac{3x^3 - 6x^2 + 7x - 2}{(x^2 - 2x + 2)^2}$

41. $\dfrac{4x^2 + 3x + 14}{x^3 - 8}$

42. $\dfrac{3x - 5}{x^3 - 1}$

 Application Exercises

43. Find the partial fraction decomposition for $\dfrac{1}{x(x + 1)}$

and use the result to find the following sum:

$$\frac{1}{1 \cdot 2} + \frac{1}{2 \cdot 3} + \frac{1}{3 \cdot 4} + \cdots + \frac{1}{99 \cdot 100}.$$

44. Find the partial fraction decomposition for $\dfrac{2}{x(x + 2)}$

and use the result to find the following sum:

$$\frac{2}{1 \cdot 3} + \frac{2}{3 \cdot 5} + \frac{2}{5 \cdot 7} + \cdots + \frac{2}{99 \cdot 101}.$$

 Writing in Mathematics

45. Explain what is meant by the partial fraction decomposition of a rational expression.

46. Explain how to find the partial fraction decomposition of a rational expression with distinct linear factors in the denominator.

47. Explain how to find the partial fraction decomposition of a rational expression with a repeated linear factor in the denominator.

48. Explain how to find the partial fraction decomposition of a rational expression with a prime quadratic factor in the denominator.

49. Explain how to find the partial fraction decomposition of a rational expression with a repeated, prime quadratic factor in the denominator.

50. How can you verify your result for the partial fraction decomposition for a given rational expression without using a graphing utility?

 Technology Exercises

51. A graphing utility can be used to check the partial fraction decomposition for a given rational expression. Graph $y_1 =$ *the given rational expression* and $y_2 = $ *its partial fraction decomposition* on the same screen. If the graphs are identical, the decomposition is correct. Use this method to verify any five of the decompositions that you obtained in Exercises 9–42.

52. As you worked Exercise 51, did you find that it took a while to determine the range setting that showed a graph for the rational function and its decomposition? Suggest another method for showing that $y_1 = y_2$ using your graphing utility. Use this method to check the results of the same five decompositions you worked with in Exercise 51.

Critical Thinking Exercises

53. Use an extension of the Study Tip on page 731 to describe how to set up the partial fraction decomposition of a rational expression that contains powers of a cubic factor in the denominator. Give an example of such a decomposition.

54. If a, b, and c are constants, find the partial fraction decomposition of

$$\frac{ax + b}{(x - c)^2}.$$

55. Find the partial fraction decomposition of

$$\frac{4x^2 + 5x - 9}{x^3 - 6x - 9}.$$

SECTION 8.4   Systems of Nonlinear Equations in Two Variables

Objectives

1. Recognize systems of nonlinear equations in two variables.
2. Solve nonlinear systems by substitution.
3. Solve nonlinear systems by addition.
4. Solve problems using systems of nonlinear equations.

Scientists debate the probability that a "doomsday rock" will collide with Earth. It has been estimated that an asteroid, a tiny planet that revolves around the sun, crashes into Earth about once every 250,000 years, and that such a collision would have disastrous results. In 1908 a small fragment struck Siberia, leveling thousands of acres of trees. One theory about the extinction of dinosaurs 65 million years ago involves Earth's collision with a large asteroid and the resulting drastic changes in Earth's climate.

Understanding the path of Earth and the path of a comet is essential to detecting threatening space debris. Orbits about the sun are not described by linear equations in the form $Ax + By = C$. The ability to solve systems that do not contain linear equations provides NASA scientists watching for troublesome asteroids with a way to locate possible collision points with Earth's orbit.

1 Recognize systems of nonlinear equations in two variables.

Systems of Nonlinear Equations and Their Solutions

A **system of** two **nonlinear equations** in two variables, also called a **nonlinear system,** contains at least one equation that cannot be expressed in the form $Ax + By = C$. Here are two examples:

$$x^2 = 2y + 10$$
$$3x - y = 9$$

Not in the form $Ax + By = C$. The term x^2 is not linear.

$$y = x^2 + 3$$
$$x^2 + y^2 = 9.$$

Neither equation is in the form $Ax + By = C$. The terms x^2 and y^2 are not linear.

A **solution** to a nonlinear system in two variables is an ordered pair of real numbers that satisfies all equations in the system. The **solution set** to the system is the set of all such ordered pairs. As with linear systems in two variables, the solution to a nonlinear system (if there is one) corresponds to the intersection point(s) of the graphs of the equations in the system. Unlike linear systems, the graphs can be circles, parabolas, or anything other than two lines. We will solve nonlinear systems using the substitution method and the addition method.

2 Solve nonlinear systems by substitution.

Eliminating a Variable Using the Substitution Method

The substitution method involves converting a nonlinear system to one equation in one variable by an appropriate substitution. The steps in the solution process are nearly the same as those used to solve a linear system by substitution. However, when you obtain an equation in one variable, this equation will not be linear. In our first example, this equation is quadratic.

EXAMPLE 1 **Solving a Nonlinear System by the Substitution Method**

Solve by the substitution method:

$$x^2 = 2y + 10 \qquad \text{(The graph is a parabola.)}$$
$$3x - y = 9. \qquad \text{(The graph is a line.)}$$

Solution

Step 1   Solve one of the equations for one variable in terms of the other.   We begin by isolating one of the variables raised to the first power in either of the equations. By solving for y in the second equation, which has a coefficient of -1, we can avoid fractions.

$$3x - y = 9 \qquad \text{This is the second equation in the given system.}$$
$$3x = y + 9 \qquad \text{Add } y \text{ to both sides.}$$
$$3x - 9 = y \qquad \text{Subtract 9 from both sides.}$$

Step 2   Substitute the expression from step 1 into the other equation.   We substitute $3x - 9$ for y in the first equation.

$$y = \boxed{3x - 9} \qquad x^2 = 2\,\boxed{y} + 10$$

This gives us an equation in one variable, namely

$$x^2 = 2(3x - 9) + 10.$$

The variable y has been eliminated.

Step 3   Solve the resulting equation containing one variable.

$$x^2 = 2(3x - 9) + 10 \qquad \text{This is the equation containing one variable.}$$
$$x^2 = 6x - 18 + 10 \qquad \text{Use the distributive property.}$$
$$x^2 = 6x - 8 \qquad \text{Combine numerical terms on the right.}$$
$$x^2 - 6x + 8 = 0 \qquad \text{Move all terms to one side and set the quadratic equation equal to 0.}$$
$$(x - 4)(x - 2) = 0 \qquad \text{Factor.}$$
$$x - 4 = 0 \quad \text{or} \quad x - 2 = 0 \qquad \text{Set each factor equal to 0.}$$
$$x = 4 \quad \text{or} \quad x = 2 \qquad \text{Solve for x.}$$

Step 4   Back-substitute the obtained values into the equation from step 1.   Now that we have the x-coordinates of the solutions, we back-substitute 4 for x and 2 for x in the equation $y = 3x - 9$.

If x is 4, $y = 3(4) - 9 = 3$,   so $(4, 3)$ is a solution.

If x is 2, $y = 3(2) - 9 = -3$,   so $(2, -3)$ is a solution.

Step 5   Check the proposed solutions in both of the system's given equations.   We begin by checking $(4, 3)$. Replace x with 4 and y with 3.

$x^2 = 2y + 10$	$3x - y = 9$	These are the given equations.
$4^2 \stackrel{?}{=} 2(3) + 10$	$3(4) - 3 \stackrel{?}{=} 9$	Let x = 4 and y = 3.
$16 \stackrel{?}{=} 6 + 10$	$12 - 3 \stackrel{?}{=} 9$	Simplify.
$16 = 16$ ✓	$9 = 9$ ✓	True statements result.

The ordered pair $(4, 3)$ satisfies both equations. Thus, $(4, 3)$ is a solution to the system.

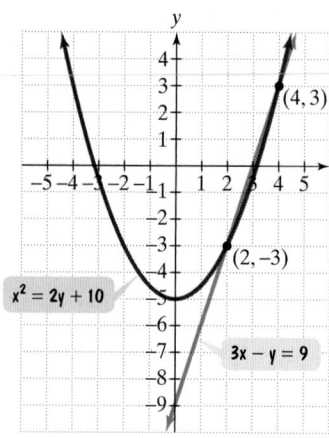

Figure 8.8 Points of intersection illustrate the nonlinear system's solutions.

Now let's check $(2, -3)$. Replace x with 2 and y with -3 in both given equations.

$x^2 = 2y + 10$	$3x - y = 9$	These are the given equations.
$2^2 \overset{?}{=} 2(-3) + 10$	$3(2) - (-3) \overset{?}{=} 9$	Let $x = 2$ and $y = -3$.
$4 \overset{?}{=} -6 + 10$	$6 + 3 \overset{?}{=} 9$	Simplify.
$4 = 4$ ✓	$9 = 9$ ✓	True statements result.

The ordered pair $(2, -3)$ also satisfies both equations and is a solution to the system. The solution set is $\{(4, 3), (2, -3)\}$. Figure 8.8 shows the graphs of the equations in the system and the solutions as intersection points.

Check Point 1 Solve by the substitution method:

$$x^2 = y - 1$$
$$4x - y = -1.$$

Study Tip

Recall from Chapter 2 that $(x - h)^2 + (y - k)^2 = r^2$ describes a circle with center (h, k) and radius r.

EXAMPLE 2 Solving a Nonlinear System by the Substitution Method

Solve by the substitution method:

$$x - y = 3 \quad \text{(The graph is a line.)}$$
$$(x - 2)^2 + (y + 3)^2 = 4. \quad \text{(The graph is a circle.)}$$

Solution Graphically, we are finding the intersection of a line and a circle with center $(2, -3)$ and radius 2.

Step 1 Solve one of the equations for one variable in terms of the other. We will solve for x in the linear equation — that is, the first equation. (We could also solve for y.)

$x - y = 3$	This is the first equation in the given system.
$x = y + 3$	Add y to both sides.

Step 2 Substitute the expression from step 1 into the other equation. We substitute $y + 3$ for x in the second equation.

$$x = \boxed{y + 3} \qquad (\boxed{x} - 2)^2 + (y + 3)^2 = 4$$

This gives an equation in one variable, namely

$$(y + 3 - 2)^2 + (y + 3)^2 = 4.$$

The variable x has been eliminated.

Step 3 Solve the resulting equation containing one variable.

$(y + 3 - 2)^2 + (y + 3)^2 = 4$	This is the equation containing one variable.
$(y + 1)^2 + (y + 3)^2 = 4$	Combine numerical terms in the first parentheses.
$y^2 + 2y + 1 + y^2 + 6y + 9 = 4$	Use the formula $(A + B)^2 = A^2 + 2AB + B^2$ to square $y + 1$ and $y + 3$.
$2y^2 + 8y + 10 = 4$	Combine like terms on the left.
$2y^2 + 8y + 6 = 0$	Subtract 4 from both sides and set the quadratic equation equal to 0.

$$y^2 + 4y + 3 = 0 \quad \text{Simplify by dividing both sides by 2.}$$
$$(y + 3)(y + 1) = 0 \quad \text{Factor.}$$
$$y + 3 = 0 \quad \text{or} \quad y + 1 = 0 \quad \text{Set each factor equal to 0.}$$
$$y = -3 \quad \text{or} \quad y = -1 \quad \text{Solve for y.}$$

Step 4   Back-substitute the obtained values into the equation from step 1.
Now that we have the y-coordinates of the solutions, we back-substitute -3 for y and -1 for y in the equation $x = y + 3$.

If $y = -3$: $x = -3 + 3 = 0$,   so $(0, -3)$ is a solution.
If $y = -1$: $x = -1 + 3 = 2$,   so $(2, -1)$ is a solution.

Step 5   Check the proposed solution in both of the system's given equations.
Take a moment to show that each ordered pair satisfies both equations. The solution set of the given system is $\{(0, -3), (2, -1)\}$.

Figure 8.9 shows the graphs of the equations in the system and the solutions as intersection points.

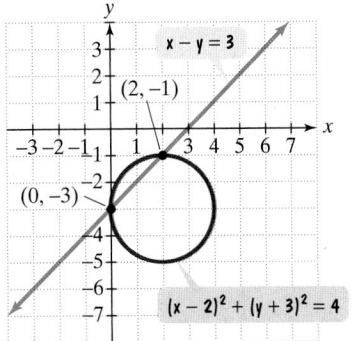

Figure 8.9 Points of intersection illustrate the nonlinear system's solutions.

Check Point 2   Solve by the substitution method:

$$x + 2y = 0$$
$$(x - 1)^2 + (y - 1)^2 = 5.$$

3 Solve nonlinear systems by addition.

Eliminating a Variable Using the Addition Method

In solving linear systems with two variables, we learned that the addition method works well when each equation is in the form $Ax + By = C$. For nonlinear systems, the addition method can be used when each equation is in the form $Ax^2 + By^2 = C$. If necessary, we will multiply either equation or both equations by appropriate numbers so that the coefficients of x^2 or y^2 will have a sum of 0. We then add equations. The sum will be an equation in one variable.

EXAMPLE 3   Solving a Nonlinear System by the Addition Method

Solve the system:

$$4x^2 + y^2 = 13 \quad \text{Equation 1}$$
$$x^2 + y^2 = 10. \quad \text{Equation 2}$$

Solution   We can use steps that are similar to those used to solve linear systems by the addition method.

Step 1   Write both equations in the form $Ax^2 + By^2 = C$.   Both equations are already in this form, so we can skip this step.

Step 2   If necessary, multiply either equation or both equations by appropriate numbers so that the sum of the x^2-coefficients or the sum of the y^2-coefficients is 0.   We can eliminate y^2 by multiplying Equation 2 by -1.

$$4x^2 + y^2 = 13 \xrightarrow{\text{No change}} 4x^2 + y^2 = 13$$
$$x^2 + y^2 = 10 \xrightarrow{\text{Multiply by } -1,} -x^2 - y^2 = -10$$

$4x^2 + y^2 = 13$ Equation 1

$x^2 + y^2 = 10$ Equation 2

The system we are solving, repeated

Steps 3 and 4   **Add equations and solve for the remaining variable.**

$$4x^2 + y^2 = 13$$

Add: $\dfrac{-x^2 - y^2 = -10}{3x^2 \qquad = 3}$

$x^2 = 1$ Divide both sides by 3

$x = \pm 1$ Use the square root method: If $x^2 = c$, then $x = \pm\sqrt{c}$.

Step 5   **Back-substitute and find the values for the other variables.**   We must back-substitute each value of x into either one of the original equations. Let's use $x^2 + y^2 = 10$, Equation 2. If $x = 1$,

$1^2 + y^2 = 10$ Replace x with 1 in Equation 2.

$y^2 = 9$ Subtract 1 from both sides.

$y = \pm 3.$ Apply the square root method.

$(1, 3)$ and $(1, -3)$ are solutions. If $x = -1$,

$(-1)^2 + y^2 = 10$ Replace x with −1 in Equation 2.

$y^2 = 9$ The steps are the same as before.

$y = \pm 3.$

$(-1, 3)$ and $(-1, -3)$ are solutions.

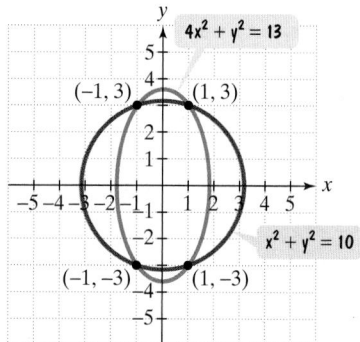

Figure 8.10 A system with four solutions

Step 6   **Check.**   Take a moment to show that each of the four ordered pairs satisfies Equation 1 and Equation 2. The solution set of the given system is $\{(1, 3), (1, -3), (-1, 3), (-1, -3)\}$.

Figure 8.10 shows the graphs of the equations in the system and the solutions as intersection points.

Study Tip

When solving nonlinear systems, extra solutions may be introduced that do not satisfy both equations in the system. Therefore, you should get into the habit of checking all proposed pairs in each of the system's two equations.

Check Point 3   Solve the system:

$$3x^2 + 2y^2 = 35$$
$$4x^2 + 3y^2 = 48.$$

In solving nonlinear systems, we include only ordered pairs with real numbers in the solution set. We have seen that each of these ordered pairs corresponds to a point of intersection of the system's graphs.

EXAMPLE 4   **Solving a Nonlinear System by the Addition Method**

Solve the system:

$$y = x^2 + 3 \qquad \text{Equation 1} \quad \text{(The graph is a parabola.)}$$
$$x^2 + y^2 = 9. \qquad \text{Equation 2} \quad \text{(The graph is a circle.)}$$

Solution   We could use substitution because Equation 1 has y expressed in terms of x, but this would result in a fourth-degree equation. However, we can rewrite Equation 1 by subtracting x^2 from both sides and adding the equations to eliminate the x^2-terms.

Notice how like terms are arranged in columns.

$-x^2 + y \qquad\quad = \quad 3$ Subtract x^2 from both sides of Equation 1.

$\dfrac{x^2 \qquad + y^2 = \quad 9}{y + y^2 = 12}$ This is Equation 2.

            Add the equations.

We now solve this quadratic equation.

$$y + y^2 = 12 \qquad \text{This is the equation containing one variable.}$$
$$y^2 + y - 12 = 0 \qquad \text{Subtract 12 from both sides and set the}$$
$$\text{quadratic equation equal to 0.}$$

$$(y + 4)(y - 3) = 0 \qquad \text{Factor.}$$

$$y + 4 = 0 \qquad \text{or} \qquad y - 3 = 0 \qquad \text{Set each factor equal to 0.}$$
$$y = -4 \qquad \text{or} \qquad y = 3 \qquad \text{Solve for } y.$$

To complete the solution, we must back-substitute each value of y into either one of the original equations. We will use $y = x^2 + 3$, Equation 1. First, we substitute -4 for y.

$$-4 = x^2 + 3$$
$$-7 = x^2 \qquad \text{Subtract 3 from both sides.}$$

Because the square of a real number cannot be negative, the equation $x^2 = -7$ does not have real-number solutions. Thus, we move on to our other value for y, 3, and substitute this value into Equation 1.

$$y = x^2 + 3 \qquad \text{This is Equation 1.}$$
$$3 = x^2 + 3 \qquad \text{Back-substitute 3 for } y.$$
$$0 = x^2 \qquad \text{Subtract 3 from both sides.}$$
$$0 = x \qquad \text{Solve for } x.$$

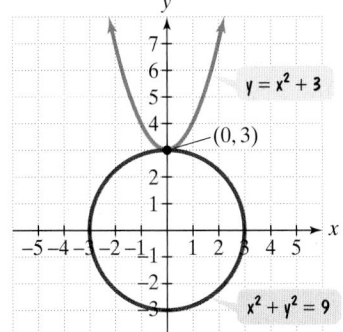

Figure 8.11 A system with one real solution

We showed that if $y = 3$, then $x = 0$. Thus, $(0, 3)$ is the solution. Take a moment to show that $(0, 3)$ satisfies Equation 1 and Equation 2. The solution set of the given system is $\{(0, 3)\}$. Figure 8.11 shows the system's graphs and the solution as an intersection point.

Check Point 4 Solve the system:

$$y = x^2 + 5$$
$$x^2 + y^2 = 25.$$

4 Solve problems using systems of nonlinear equations.

Applications

Many geometric problems can be modeled and solved by the use of systems of nonlinear equations. We will use our step-by-step strategy for solving problems using mathematical models that are created from verbal models.

EXAMPLE 5 An Application of a Nonlinear System

You have 36 yards of fencing to build the enclosure in Figure 8.12. Some of this fencing is to be used to build an internal divider. If you'd like to enclose 54 square yards, what are the dimensions of the enclosure?

Solution

Step 1 Use variables to represent unknown quantities. Let $x =$ the enclosure's length and $y =$ the enclosure's width. These variables are shown in Figure 8.12.

Step 2 Write a system of equations describing the problem's conditions. The first condition is that you have 36 yards of fencing.

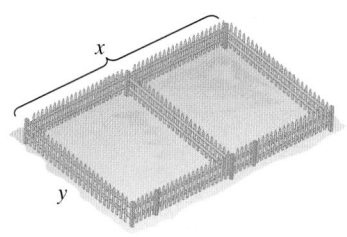

Figure 8.12 Building an enclosure

Figure 8.12, repeated

Fencing along both lengths	plus	Fencing along both widths	plus	Fencing for the internal divider	equals	36 yards.
$2x$	$+$	$2y$	$+$	y	$=$	36

Adding like terms, we can express the equation that models the verbal conditions for the fencing as $2x + 3y = 36$.

The second condition is that you'd like to enclose 54 square yards. The rectangle's area, the product of its length and its width, must be 54 square yards.

Length	times	width	is	54 square yards.
x	$\cdot$	y	$=$	54

Step 3   Solve the system and answer the problem's question.   We must solve the system

$$2x + 3y = 36 \quad \text{Equation 1}$$

$$xy = 54. \quad \text{Equation 2}$$

We will use substitution. Because Equation 1 has no coefficients of 1 or -1, we will solve Equation 2 for y. Dividing both sides of $xy = 54$ by x, we obtain

$$y = \frac{54}{x}.$$

Now we substitute $\frac{54}{x}$ for y in Equation 1 and solve for x.

$$2x + 3y = 36 \qquad \text{This is Equation 1.}$$

$$2x + 3 \cdot \frac{54}{x} = 36 \qquad \text{Substitute } \frac{54}{x} \text{ for } y.$$

$$2x + \frac{162}{x} = 36 \qquad \text{Multiply.}$$

$$x\left(2x + \frac{162}{x}\right) = 36 \cdot x \qquad \text{Clear fractions by multiplying both sides by x.}$$

$$2x^2 + 162 = 36x \qquad \text{Use the distributive property on the left side.}$$

$$2x^2 - 36x + 162 = 0 \qquad \text{Subtract 36x from both sides and set the quadratic equation equal to 0.}$$

$$x^2 - 18x + 81 = 0 \qquad \text{Simplify by dividing both sides by 2.}$$

$$(x - 9)^2 = 0 \qquad \text{Factor using } A^2 - 2AB + B^2 = (A - B)^2.$$

$$x - 9 = 0 \qquad \text{Set the repeated factor equal to zero.}$$

$$x = 9 \qquad \text{Solve for x.}$$

We back-substitute this value of x into $y = \frac{54}{x}$.

$$\text{If } x = 9, \quad y = \frac{54}{9} = 6.$$

This means that the dimensions of the enclosure in Figure 8.12 are 9 yards by 6 yards.

Step 4   Check the proposed solution in the original wording of the problem.
With a length of 9 yards and a width of 6 yards, take a moment to check that this results in 36 yards of fencing and an area of 54 square yards.

> **Check Point 5**   Find the length and width of a rectangle whose perimeter is 20 feet and whose area is 21 square feet.

EXERCISE SET 8.4

Practice Exercises

In Exercises 1–18, solve each system by the substitution method.

1. $x + y = 2$
$y = x^2 - 4$

2. $x - y = -1$
$y = x^2 + 1$

3. $x + y = 2$
$y = x^2 - 4x + 4$

4. $2x + y = -5$
$y = x^2 + 6x + 7$

5. $y = x^2 - 4x - 10$
$y = -x^2 - 2x + 14$

6. $y = x^2 + 4x + 5$
$y = x^2 + 2x - 1$

7. $x^2 + y^2 = 25$
$x - y = 1$

8. $x^2 + y^2 = 5$
$3x - y = 5$

9. $xy = 6$
$2x - y = 1$

10. $xy = -12$
$x - 2y + 14 = 0$

11. $y^2 = x^2 - 9$
$2y = x - 3$

12. $x^2 + y = 4$
$2x + y = 1$

13. $xy = 3$
$x^2 + y^2 = 10$

14. $xy = 4$
$x^2 + y^2 = 8$

15. $x + y = 1$
$x^2 + xy - y^2 = -5$

16. $x + y = -3$
$x^2 + 2y^2 = 12y + 18$

17. $x + y = 1$
$(x - 1)^2 + (y + 2)^2 = 10$

18. $2x + y = 4$
$(x + 1)^2 + (y - 2)^2 = 4$

In Exercises 19–28, solve each system by the addition method.

19. $x^2 + y^2 = 13$
$x^2 - y^2 = 5$

20. $4x^2 - y^2 = 4$
$4x^2 + y^2 = 4$

21. $x^2 - 4y^2 = -7$
$3x^2 + y^2 = 31$

22. $3x^2 - 2y^2 = -5$
$2x^2 - y^2 = -2$

23. $3x^2 + 4y^2 - 16 = 0$
$2x^2 - 3y^2 - 5 = 0$

24. $16x^2 - 4y^2 - 72 = 0$
$x^2 - y^2 - 3 = 0$

25. $x^2 + y^2 = 25$
$(x - 8)^2 + y^2 = 41$

26. $x^2 + y^2 = 5$
$x^2 + (y - 8)^2 = 41$

27. $y^2 - x = 4$
$x^2 + y^2 = 4$

28. $x^2 - 2y = 8$
$x^2 + y^2 = 16$

In Exercises 29–42, solve each system by the method of your choice.

29. $3x^2 + 4y^2 = 16$
$2x^2 - 3y^2 = 5$

30. $x + y^2 = 4$
$x^2 + y^2 = 16$

31. $2x^2 + y^2 = 18$
$xy = 4$

32. $x^2 + 4y^2 = 20$
$xy = 4$

33. $x^2 + 4y^2 = 20$
$x + 2y = 6$

34. $3x^2 - 2y^2 = 1$
$4x - y = 3$

35. $x^3 + y = 0$
$x^2 - y = 0$

36. $x^3 + y = 0$
$2x^2 - y = 0$

37. $x^2 + (y - 2)^2 = 4$
$x^2 - 2y = 0$

38. $x^2 - y^2 - 4x + 6y - 4 = 0$
$x^2 + y^2 - 4x - 6y + 12 = 0$

39. $y = (x + 3)^2$
$x + 2y = -2$

40. $(x - 1)^2 + (y + 1)^2 = 5$
$2x - y = 3$

41. $x^2 + y^2 + 3y = 22$
$2x + y = -1$

42. $x - 3y = -5$
$x^2 + y^2 - 25 = 0$

In Exercises 43–46, let x represent one number and let y represent the other number. Use the given conditions to write a system of nonlinear equations. Solve the system and find the numbers.

43. The sum of two numbers is 10 and their product is 24. Find the numbers.

44. The sum of two numbers is 20 and their product is 96. Find the numbers.

45. The difference between the squares of two numbers is 3. Twice the square of the first number increased by the square of the second number is 9. Find the numbers.

46. The difference between the squares of two numbers is 5. Twice the square of the second number subtracted from three times the square of the first number is 19. Find the numbers.

Application Exercises

47. A planet's orbit follows a path described by $16x^2 + 4y^2 = 64$. A comet follows the parabolic path $y = x^2 - 4$. Where might the comet intersect the orbiting planet?

48. A system for tracking ships indicates that a ship lies on a path described by $2y^2 - x^2 = 1$. The process is repeated and the ship is found to lie on a path described by $2x^2 - y^2 = 1$. If it is known that the ship is located in the first quadrant of the coordinate system, determine its exact location.

49. Find the length and width of a rectangle whose perimeter is 36 feet and whose area is 77 square feet.

50. Find the length and width of a rectangle whose perimeter is 40 feet and whose area is 96 square feet.

Use the formula for the area of a rectangle and the Pythagorean Theorem to solve Exercises 51–52.

51. A small television has a picture with a diagonal measure of 10 inches and a viewing area of 48 square inches. Find the length and width of the screen.

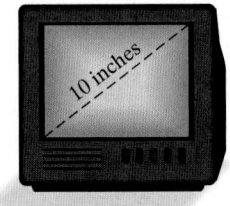

52. The area of a rug is 108 square feet and the length of its diagonal is 15 feet. Find the length and width of the rug.

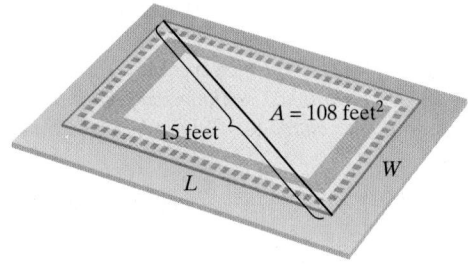

53. The figure at the top of the next column shows a square floor plan with a smaller square area that will accommodate a combination fountain and pool. The floor with the fountain-pool area removed has an area of 21 square meters and a perimeter of 24 meters. Find the dimensions of the floor and the dimensions of the square that will accommodate the pool.

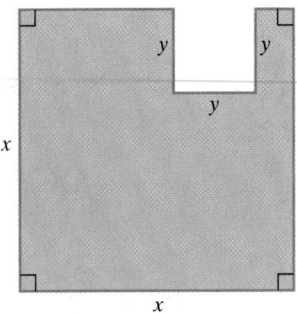

54. The area of the rectangular piece of cardboard shown below is 216 square inches. The cardboard is used to make an open box by cutting a 2-inch square from each corner and turning up the sides. If the box is to have a volume of 224 cubic inches, find the length and width of the cardboard that must be used.

Writing in Mathematics

55. What is a system of nonlinear equations? Provide an example with your description.

56. Explain how to solve a nonlinear system using the substitution method. Use $x^2 + y^2 = 9$ and $2x - y = 3$ to illustrate your explanation.

57. Explain how to solve a nonlinear system using the addition method. Use $x^2 - y^2 = 5$ and $3x^2 - 2y^2 = 19$ to illustrate your explanation.

58. The daily demand and supply models for a carrot cake supplied by a bakery to a convenience store are given by the demand model $N = 40 - 3p$ and the supply model $N = \dfrac{p^2}{10}$, in which p is the price of the cake and N is the number of cakes sold or supplied each day to the convenience store. Explain how to determine the price at which supply and demand are equal. Then describe how to find how many carrot cakes can be supplied and sold each day at this price.

Technology Exercises

59. Verify your solutions to any five exercises from Exercises 1–42 by using a graphing utility to graph the two equations in the system in the same viewing rectangle. Then use the trace or intersection feature to verify the solutions.

60. Write a system of equations, one equation whose graph is a line and the other whose graph is a parabola, that has no ordered pairs that are real numbers in its solution set. Graph the equations using a graphing utility and verify that you are correct.

Critical Thinking Exercises

61. Which one of the following is true?

a. A system of two equations in two variables whose graphs represent a circle and a line can have four real solutions.

b. A system of two equations in two variables whose graphs represent a parabola and a circle can have four real solutions.

c. A system of two equations in two variables whose graphs represent two circles must have at least two real solutions.

d. A system of two equations in two variables whose graphs represent a parabola and a circle cannot have only one real solution.

62. The points of intersection of the graphs of $xy = 20$ and $x^2 + y^2 = 41$ are joined to form a rectangle. Find the area of the rectangle.

63. Find a and b in this figure.

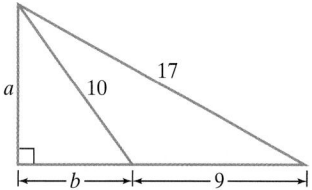

Solve the systems in Exercises 64–65.

64. $\log_y x = 3$
$\log_y (4x) = 5$

65. $\log x^2 = y + 3$
$\log x = y - 1$

SECTION 8.5 Systems of Inequalities

Objectives

1. Graph a linear inequality in two variables.
2. Graph a nonlinear inequality in two variables.
3. Graph a system of inequalities.
4. Solve applied problems involving systems of inequalities.

Had a good workout lately? If so, could you tell if you were overdoing it or not pushing yourself hard enough? In this section, we will use systems of inequalities in two variables to help you establish a target zone for your workouts.

Linear Inequalities in Two Variables and Their Solutions

We have seen that equations in the form $Ax + By = C$ are straight lines when graphed. If we change the $=$ sign to $>, <, \geq$, or $\leq$, we obtain a **linear inequality in two variables.** Some examples of linear inequalities in two variables are $x + y > 2, 3x - 5y \leq 15$, and $2x - y < 4$.

A **solution of an inequality in two variables,** x and y, is an ordered pair of real numbers with the following property: When the x-coordinate is substituted for x and the y-coordinate is substituted for y in the inequality, we obtain a true statement. For example, $(3, 2)$ is a solution of the inequality $x + y > 1$. When 3 is substituted for x and 2 is substituted for y, we obtain the true statement $3 + 2 > 1$, or $5 > 1$. Because there are infinitely many pairs of numbers that have a sum greater than 1, the inequality $x + y > 1$ has infinitely many solutions. Each ordered pair solution is said to **satisfy** the inequality. Thus, $(3, 2)$ satisfies the inequality $x + y > 1$.

1 Graph a linear inequality in two variables.

The Graph of a Linear Inequality in Two Variables

We know that the graph of an equation in two variables is the set of all points whose coordinates satisfy the equation. Similarly, the **graph of an inequality in two variables** is the set of all points whose coordinates satisfy the inequality.

Let's use Figure 8.13 to get an idea of what the graph of a linear inequality in two variables looks like. Part of the figure shows the graph of the linear equation $x + y = 2$. The line divides the points in the rectangular coordinate system into three sets. First, there is the set of points along the line, satisfying $x + y = 2$. Next, there is the set of points in the green region above the line. Points in the green region satisfy the linear inequality $x + y > 2$. Finally, there is the set of points in the pink region below the line. Points in the pink region satisfy the linear inequality $x + y < 2$.

A **half-plane** is the set of all the points on one side of a line. In Figure 8.13, the green region is a half-plane. The pink region is also a half-plane. A half-plane is the graph of a linear inequality that involves $>$ or $<$. The graph of an inequality that involves $\geq$ or $\leq$ is a half-plane and a line. A solid line is used to show that the line is part of the graph. A dashed line is used to show that a line is not part of a graph.

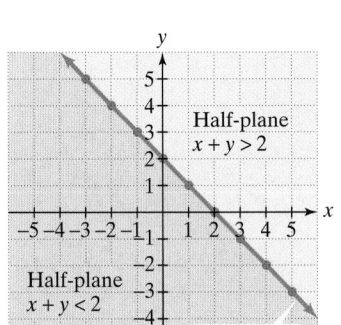

Figure 8.13

Graphing a Linear Inequality in Two Variables

1. Replace the inequality symbol with an equal sign and graph the corresponding linear equation. Draw a solid line if the original inequality contains a $\leq$ or $\geq$ symbol. Draw a dashed line if the original inequality contains a $<$ or $>$ symbol.

2. Choose a test point in one of the half-planes that is not on the line. Substitute the coordinates of the test point into the inequality.

3. If a true statement results, shade the half-plane containing this test point. If a false statement results, shade the half-plane not containing this test point.

EXAMPLE 1 Graphing a Linear Inequality in Two Variables

Graph: $3x - 5y < 15$.

Solution

Step 1 Replace the inequality symbol with = and graph the linear equation. We need to graph $3x - 5y = 15$. We can use intercepts to graph this line.

We set $y = 0$ to find the x-intercept:	We set $x = 0$ to find the y-intercept:
$3x - 5y = 15$	$3x - 5y = 15$
$3x - 5 \cdot 0 = 15$	$3 \cdot 0 - 5y = 15$
$3x = 15$	$-5y = 15$
$x = 5.$	$y = -3.$

The x-intercept is 5, so the line passes through $(5, 0)$. The y-intercept is -3, so the line passes through $(0, -3)$. The graph is indicated by a dashed line because the inequality $3x - 5y < 15$ contains a $<$ symbol, rather than $\leq$. The graph of the line is shown in Figure 8.14.

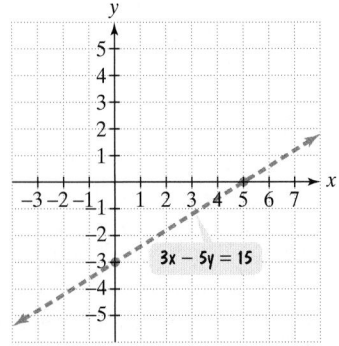

Figure 8.14 Preparing to graph $3x - 5y < 15$

Step 2   Choose a test point in one of the half-planes that is not on the line. Substitute its coordinates into the inequality. The line $3x - 5y = 15$ divides the plane into three parts—the line itself and two half-planes. The points in one half-plane satisfy $3x - 5y > 15$. The points in the other half-plane satisfy $3x - 5y < 15$. We need to find which half-plane belong to the solution. To do so, we test a point from either half-plane. The origin, $(0, 0)$, is the easiest point to test.

$$3x - 5y < 15 \qquad \text{This is the given inequality.}$$
$$\text{Is} \quad 3 \cdot 0 - 5 \cdot 0 < 15? \qquad \text{Test } (0, 0) \text{ by substituting 0 for x and 0 for y.}$$
$$0 - 0 < 15 \qquad \text{Multiply.}$$
$$0 < 15 \qquad \text{Subtract. This statement is true.}$$

Step 3   If a true statement results, shade the half-plane containing the test point. Because 0 is less than 15, the test point, $(0, 0)$, is part of the solution set. All the points on the same side of the line $3x - 5y = 15$ as the point $(0, 0)$ are members of the solution set. The solution set is the half-plane that contains the point $(0, 0)$, indicated by shading this half-plane. The graph is shown using green shading and a dashed blue line in Figure 8.15.

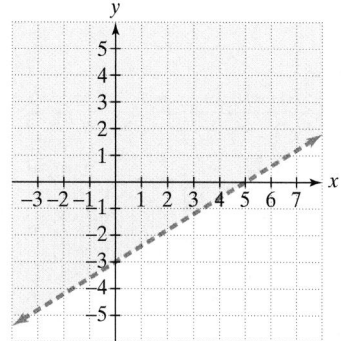

Figure 8.15 The graph of $3x - 5y < 15$

Check Point 1   Graph: $2x - 4y < 8$.

When graphing a linear inequality, test a point that lies in one of the half-planes and *not on the line dividing the half-planes*. The test point, $(0, 0)$, is convenient because it is easy to calculate when 0 is substituted for each variable. However, if $(0, 0)$ lies on the dividing line and not in a half-plane, a different test point must be selected.

EXAMPLE 2   Graphing a Linear Inequality in Two Variables

Graph: $y \le \dfrac{2}{3}x$.

Solution

Step 1   Replace the inequality symbol with = and graph the linear equation. We need to graph $y = \frac{2}{3}x$. We can use the slope and the y-intercept to graph this line.

$$y = \frac{2}{3}x + 0$$

$$\text{Slope} = \frac{2}{3} = \frac{\text{rise}}{\text{run}} \qquad y\text{-intercept} = 0$$

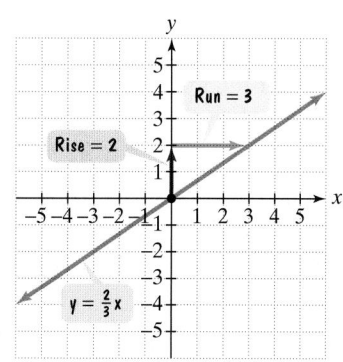

Figure 8.16 Preparing to graph $y \le \dfrac{2}{3}x$

The y-intercept is 0, so the line passes through $(0, 0)$. Using the y-intercept and the slope, the line is shown in Figure 8.16. A solid line is used because the inequality $y \le \frac{2}{3}x$ contains a $\le$ symbol, in which equality is included.

Step 2   Choose a test point in one of the half-planes that is not on the line. Substitute its coordinates into the inequality. We cannot use $(0, 0)$ as a test point because it lies on the line and not in a half-plane. Let's use $(1, 1)$, which lies in the half-plane above the line.

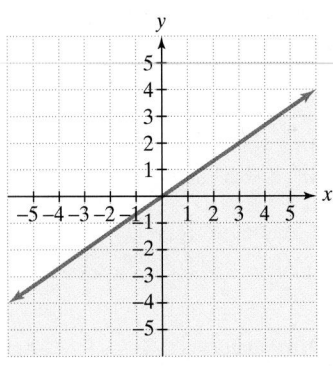

Figure 8.17 The graph of $y \le \dfrac{2}{3}x$

$$y \le \frac{2}{3}x \qquad \text{This is the given inequality.}$$

$$\text{Is} \quad 1 \le \frac{2}{3} \cdot 1? \qquad \text{Test } (1, 1) \text{ by substituting 1 for } x \text{ and 1 for } y.$$

$$1 \le \frac{2}{3} \qquad \text{This statement is false.}$$

Step 3 If a false statement results, shade the half-plane not containing the test point. Because 1 is not less than or equal to $\frac{2}{3}$, the test point $(1, 1)$ is not part of the solution set. Thus, the half-plane below the solid line $y = \frac{2}{3}x$ is part of the solution set. The solution set is the line and the half-plane that does not contain the point $(1, 1)$, indicated by shading this half-plane. The graph is shown using green shading and a blue line in Figure 8.17.

Technology

Most graphing utilities can graph inequalities in two variables with the $\boxed{\text{SHADE}}$ feature. The procedure varies by model, so consult your manual. For most graphing utilities, you must first solve for y if it is not already isolated. The figure shows the graph of $y \le \frac{2}{3}x$. Most displays do not distinguish between dashed and solid boundary lines.

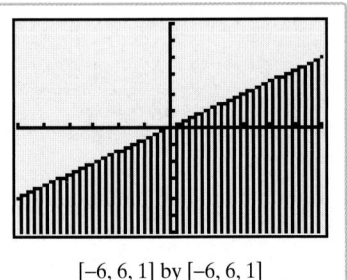

$[-6, 6, 1]$ by $[-6, 6, 1]$

Check Point 2 Graph: $y \ge \dfrac{1}{2}x$.

You can graph inequalities in the form $y > mx + b$ or $y < mx + b$ without using test points. The inequality symbol indicates which half-plane to shade.

- If $y > mx + b$, shade the half-plane above the line $y = mx + b$.
- If $y < mx + b$, shade the half-plane below the line $y = mx + b$.

In Chapter 1, we learned that $y = b$ graphs as a horizontal line, where b is the y-intercept. Similarly, the graph of $x = a$ is a vertical line, where a is the x-intercept. Half-planes can be separated by horizontal or vertical lines. For example, Figure 8.18 shows the graph of $y \le 2$. Because $(0, 0)$ satisfies this inequality ($0 \le 2$ is true), the graph consists of the half-plane below the line $y = 2$ and the line. Similarly, Figure 8.19 shows the graph of $x < 4$.

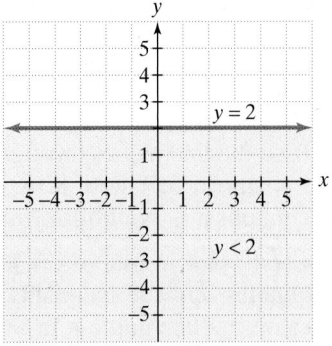

Figure 8.18 The graph of $y \le 2$

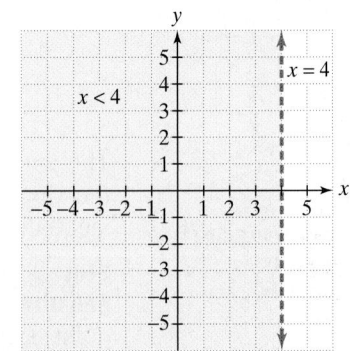

Figure 8.19 The graph of $x < 4$

2 Graph a nonlinear inequality in two variables.

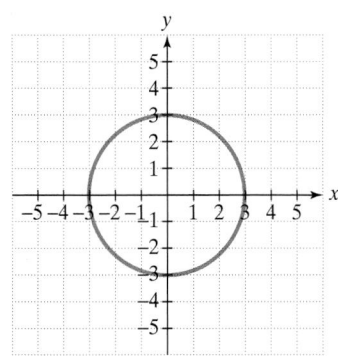

Figure 8.20 Preparing to graph $x^2 + y^2 \le 9$

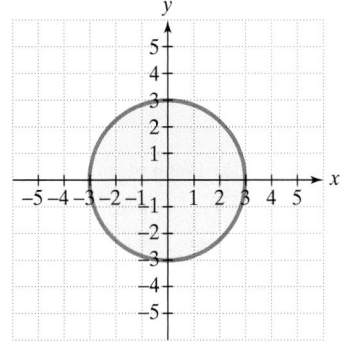

Figure 8.21 The graph of $x^2 + y^2 \le 9$

3 Graph a system of inequalities.

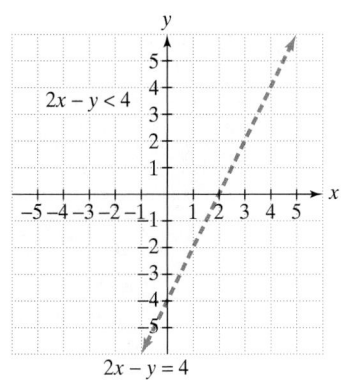

Figure 8.22 The graph of $2x - y < 4$

Graphing a Nonlinear Inequality in Two Variables

Example 3 illustrates that a nonlinear inequality in two variables is graphed in the same way that we graph a linear inequality.

EXAMPLE 3   Graphing a Nonlinear Inequality in Two Variables

Graph: $x^2 + y^2 \le 9$.

Solution

Step 1   Replace the inequality symbol with = and graph the nonlinear equation.   We need to graph $x^2 + y^2 = 9$. The graph is a circle of radius 3 with its center at the origin. The graph is shown in Figure 8.20 as a solid circle because equality is included in the $\le$ symbol.

Step 2   Choose a test point in one of the regions that is not on the circle. Substitute its coordinates into the inequality.   The circle divides the plane into three parts—the circle itself, the region inside the circle, and the region outside the circle. We need to determine whether the region inside or outside the circle is included in the solution. To do so, we will use the test point $(0, 0)$ from inside the circle.

$$x^2 + y^2 \le 9 \qquad \text{\textit{This is the given inequality.}}$$
$$\text{Is} \quad 0^2 + 0^2 \le 9? \qquad \text{\textit{Test }}(0,0)\text{\textit{ by substituting 0 for x and 0 for y.}}$$
$$0 + 0 \le 9 \qquad \text{\textit{Square 0}}: 0^2 = 0.$$
$$0 \le 9 \qquad \text{\textit{Add. This statement is true.}}$$

Step 3   If a true statement results, shade the region containing the test point. The true statement tells us that all the points inside the circle satisfy $x^2 + y^2 \le 9$. The graph is shown using green shading and a solid blue circle in Figure 8.21.

Check Point 3   Graph: $x^2 + y^2 \ge 16$.

Systems of Inequalities in Two Variables

The **solution set of a system of inequalities** in two variables, x and y, is the set of all ordered pairs (x, y) that satisfy each inequality in the system. The **graph of a system of inequalities** in two variables is the graph of the system's solution set. Thus, to graph a system of inequalities in two variables, begin by graphing each individual inequality in the same rectangular coordinate system. Then find the region, if there is one, that is common to every graph in the system. This region of intersection gives a picture of the system's solution set.

EXAMPLE 4   Graphing a System of Linear Inequalities

Graph the solution set:

$$2x - y < 4$$
$$x + y \ge -1.$$

Solution   We begin by graphing $2x - y < 4$. Because the inequality contains a $<$ symbol, rather than $\le$, we graph $2x - y = 4$ as a dashed line. (If $x = 0$, then $y = -4$, and if $y = 0$, then $x = 2$. The x-intercept is 2 and the y-intercept is -4.) Because $(0, 0)$ makes the inequality $2x - y < 4$ true, we shade the half-plane containing $(0, 0)$, shown in yellow in Figure 8.22.

Now we graph $x + y \geq -1$ in the same rectangular coordinate system. Because the inequality contains a $\geq$ symbol, in which equality is included, we graph $x + y = -1$ as a solid line. (If $x = 0$, then $y = -1$, and if $y = 0$, then $x = -1$. The x-intercept and y-intercept are both -1.) Because $(0, 0)$ makes the inequality true, we shade the half-plane containing $(0, 0)$. This is shown in Figure 8.23 using green vertical shading.

The solution set of the system is shown graphically by the intersection (the overlap) of the two half-planes. This is shown in Figure 8.23 as the region in which the yellow shading and the green vertical shading overlap. The solution of the system is shown again in Figure 8.24.

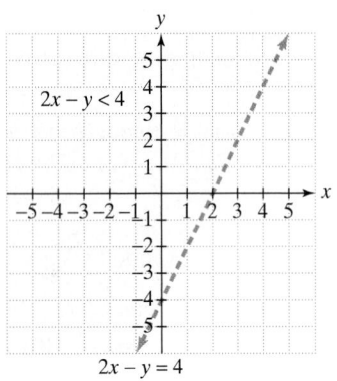

Figure 8.22, repeated
The graph of $2x - y < 4$

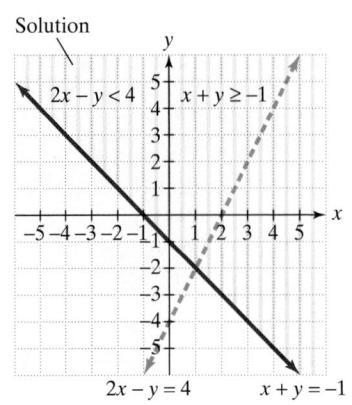

Figure 8.23 Adding the graph of $x + y \geq -1$

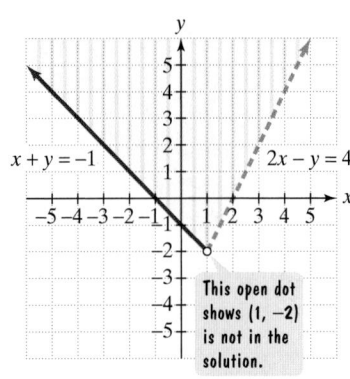

Figure 8.24 The graph of $2x - y < 4$ and $x + y \geq -1$

This open dot shows $(1, -2)$ is not in the solution.

Check Point 4   Graph the solution set:

$$x + 2y > 4$$
$$2x - 3y \leq -6.$$

EXAMPLE 5   Graphing a System of Inequalities

Graph the solution set:

$$y \geq x^2 - 4$$
$$x - y \geq 2.$$

Solution   We begin by graphing $y \geq x^2 - 4$. Because equality is included in $\geq$, we graph $y = x^2 - 4$ as a solid parabola. Because $(0, 0)$ makes the inequality $y \geq x^2 - 4$ true (we obtain $0 \geq -4$), we shade the interior portion of the parabola containing $(0, 0)$, shown in yellow in Figure 8.25.

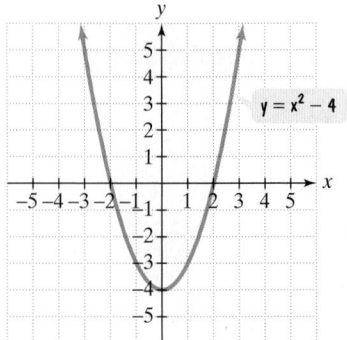

Figure 8.25 The graph of $y \geq x^2 - 4$

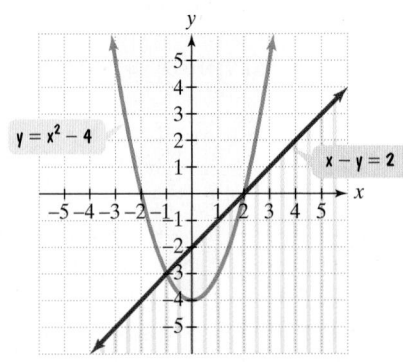

Figure 8.26 Adding the graph of $x - y \geq 2$

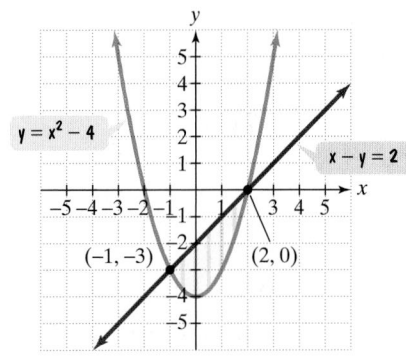

Figure 8.27 The graph of $y \geq x^2 - 4$ and $x - y \geq 2$

Now we graph $x - y \geq 2$ in the same rectangular coordinate system. First we graph the line $x - y = 2$ using its x-intercept, 2, and its y-intercept, -2. Because $(0, 0)$ makes the inequality $x - y \geq 2$ false (we obtain $0 \geq 2$), we shade the half-plane below the line. This is shown in Figure 8.26 using green vertical shading.

The solution of the system is shown in Figure 8.26 by the intersection (the overlap) of the solid yellow and green vertical shadings. The graph of the system's solution set consists of the region enclosed by the parabola and the line. To find the points of intersection of the parabola and the line, use the substitution method to solve the nonlinear system

$$y = x^2 - 4$$
$$x - y = 2.$$

Take a moment to show that the solutions are $(-1, -3)$ and $(2, 0)$, as shown in Figure 8.27.

Check Point 5 Graph the solution set:

$$y \geq x^2 - 4$$
$$x + y \leq 2.$$

A system of inequalities has no solution if there are no points in the rectangular coordinate system that simultaneously satisfy each inequality in the system. For example, the system

$$2x + 3y \geq 6$$
$$2x + 3y \leq 0$$

whose separate graphs are shown in Figure 8.28 has no overlapping region. Thus, the system has no solution. The solution set is Ø, the empty set.

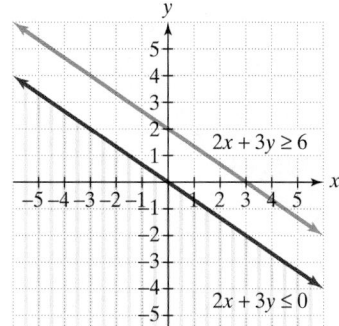

Figure 8.28 A system of inequalities with no solution

EXAMPLE 6 Graphing a System of Inequalities

Graph the solution set:

$$x - y < 2$$
$$-2 \leq x < 4$$
$$y < 3.$$

Solution We begin by graphing $x - y < 2$, the first given inequality. The line $x - y = 2$ has an x-intercept of 2 and a y-intercept of -2. The test point $(0, 0)$ makes the inequality $x - y < 2$ true, and its graph is shown in Figure 8.29.

Now, let's consider the second given inequality, $-2 \leq x < 4$. Replacing the inequality symbols by $=$, we obtain $x = -2$ and $x = 4$, graphed as vertical lines. The line of $x = 4$ is not included. Using $(0, 0)$ as a test point and substituting the x-coordinate, 0, into $-2 \leq x < 4$, we obtain the true statement $-2 \leq 0 < 4$. We therefore shade the region between the vertical lines. We've added this region to Figure 8.29, intersecting the region between the vertical lines with the yellow region in Figure 8.29. The resulting region is shown in yellow and green vertical shading in Figure 8.30.

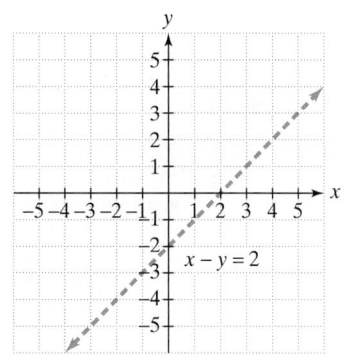

Figure 8.29 The graph of $x - y < 2$

Figure 8.30 The graph of $x - y < 2$ and $-2 \leq x < 4$

Finally, let's consider the third given inequality, $y < 3$. Replacing the inequality symbol by =, we obtain $y = 3$, which graphs as a horizontal line. Because $(0, 0)$ satisfies $y < 3$ ($0 < 3$ is true), the graph consists of the half-plane below the line $y = 3$. We've added this half-plane to the region in Figure 8.30, intersecting the half-plane with this region. The resulting region is shown in yellow and green vertical shading in Figure 8.31. This region represents the graph of the solution set of the given system.

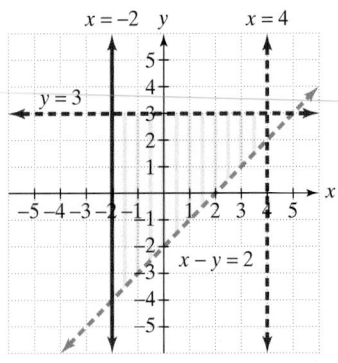

Figure 8.31 The graph of $x - y < 2$ and $-2 \leq x < 4$ and $y < 3$

Check Point 6   Graph the solution set:

$$x + y < 2$$
$$-2 \leq x < 1$$
$$y > -3.$$

Applications

Now we are ready to use a system of inequalities to establish a target zone for your workouts.

4 Solve applied problems involving systems of inequalities.

EXAMPLE 7   Inequalities and Aerobic Exercise

For people between ages 10 and 70, inclusive, the target zone for aerobic exercise is given by the following system of inequalities in which a represents one's age and p is one's pulse rate:

$$2a + 3p \geq 450$$
$$a + p \leq 190.$$

The graph of this target zone is shown in Figure 8.32 for $10 \leq a \leq 70$. Find your age. The line segments on the top and bottom of the shaded region indicate upper and lower limits for your pulse rate, in beats per minute, when engaging in aerobic exercise.

a. What are the coordinates of point A and what does this mean in terms of age and pulse rate?

b. Show that the coordinates of point A satisfy each inequality in the system.

Figure 8.32

Solution

a. Point A has coordinates $(20, 160)$. This means that a pulse rate of 160 beats per minute is within the target zone for a 20-year-old person engaged in aerobic exercise.

b. We can show that $(20, 160)$ satisfies each inequality by substituting 20 for a and 160 for p.

$$2a + 3p \geq 450 \qquad\qquad a + p \leq 190$$
$$\text{Is} \quad 2(20) + 3(160) \geq 450? \qquad \text{Is} \quad 20 + 160 \leq 190?$$
$$40 + 480 \geq 450 \qquad\qquad\qquad 180 \leq 190, \text{ true}$$
$$520 \geq 450, \text{ true}$$

The pair $(20, 160)$ makes each inequality true, so it satisfies each inequality in the system.

Check Point 7

Identify a point other than A in the target zone in Figure 8.32.

a. What are the coordinates of this point and what does this mean in terms of age and pulse rate?

b. Show that the coordinates of the point satisfy each inequality in the system in Example 7.

EXERCISE SET 8.5

Practice Exercises

In Exercises 1–22, graph each inequality.

1. $x + 2y \leq 8$

2. $3x - 6y \leq 12$

3. $x - 2y > 10$

4. $2x - y > 4$

5. $y \leq \frac{1}{3}x$

6. $y \leq \frac{1}{4}x$

7. $y > 2x - 1$

8. $y > 3x + 2$

9. $x \leq 1$

10. $x \leq -3$

11. $y > 1$

12. $y > -3$

13. $x^2 + y^2 \leq 1$

14. $x^2 + y^2 \leq 4$

15. $x^2 + y^2 > 25$

16. $x^2 + y^2 > 36$

17. $y < x^2 - 1$

18. $y < x^2 - 9$

19. $y \geq x^2 - 9$

20. $y \geq x^2 - 1$

21. $y > 2^x$

22. $y \leq 3^x$

In Exercises 23–52, graph the solution set of each system of inequalities or indicate that the system has no solution.

23. $3x + 6y \leq 6$
$2x + y \leq 8$

24. $x - y \geq 4$
$x + y \leq 6$

25. $2x - 5y \leq 10$
$3x - 2y > 6$

26. $2x - y \leq 4$
$3x + 2y > -6$

27. $y > 2x - 3$
$y < -x + 6$

28. $y < -2x + 4$
$y < x - 4$

29. $x + 2y \leq 4$
$y \geq x - 3$

30. $x + y \leq 4$
$y \geq 2x - 4$

31. $x \leq 2$
$y \geq -1$

32. $x \leq 3$
$y \leq -1$

33. $-2 \leq x < 5$

34. $-2 < y \leq 5$

35. $x - y \leq 1$
$x \geq 2$

36. $4x - 5y \geq -20$
$x \geq -3$

37. $x + y > 4$
$x + y < -1$

38. $x + y > 3$
$x + y < -2$

39. $x + y > 4$
$x + y > -1$

40. $x + y > 3$
$x + y > -2$

41. $y \geq x^2 - 1$
$x - y \geq -1$

42. $y \geq x^2 - 4$
$x - y \geq 2$

43. $x^2 + y^2 \leq 16$
$x + y > 2$

44. $x^2 + y^2 \leq 4$
$x + y > 1$

45. $x^2 + y^2 > 1$
$x^2 + y^2 < 4$

46. $x^2 + y^2 > 1$
$x^2 + y^2 < 9$

47. $x - y \leq 2$
$x \geq -2$
$y \leq 3$

48. $3x + y \leq 6$
$x \geq -2$
$y \leq 4$

49. $x \geq 0$
$y \geq 0$
$2x + 5y \leq 10$
$3x + 4y \leq 12$

50. $x \geq 0$
$y \geq 0$
$2x + y \leq 4$
$2x - 3y \leq 6$

51. $3x + y \leq 6$
$2x - y \leq -1$
$x \geq -2$
$y \leq 4$

52. $2x + y \leq 6$
$x + y \geq 2$
$1 \leq x \leq 2$
$y \leq 3$

Application Exercises

The figure shows three kinds of regions—deserts, grasslands, and forests—that results from various ranges of temperature, T, and precipitation, P. Use the figure to solve Exercises 53–54.

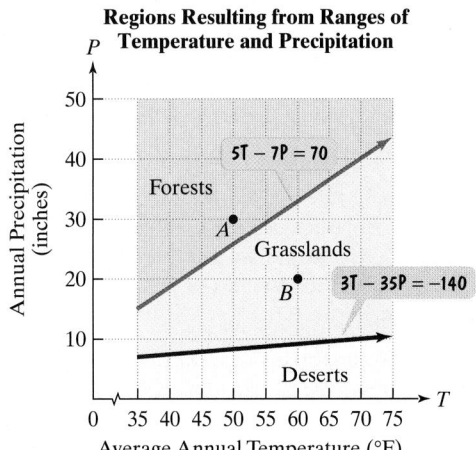

Regions Resulting from Ranges of Temperature and Precipitation

Source: A. Miller and J. Thompson, *Elements of Meteorology*

53. Use the figure on the previous page to write a system of inequalities that describe where forests occur. Then show that the coordinates of point A satisfy each inequality in the system.

54. Use the figure on the previous page to write a system of inequalities that describe where grasslands occur. Then show that the coordinates of point B satisfy each inequality in the system.

55. Many elevators have a capacity of 2000 pounds

    a. If a child averages 50 pounds and an adult 150 pounts, write an inequality that describes when x children and y adults will cause the elevator to be overloaded.

    b. Graph the inequality. Because x and y must be positive, limit the graph to quadrant I only.

    c. Select an ordered pair satisfying the inequality. What are its coordinates and what do they represent in this situation?

56. A patient is not allowed to have more than 330 milligrams of cholesterol per day from a diet of eggs and meat. Each egg provides 165 milligrams of cholesterol. Each ounce of meat provides 110 milligrams.

    a. Write an inequality that describes the patient's dietary restrictions for x eggs and y ounces of meat.

    b. Graph the inequality. Because x and y must be positive, limit the graph to quadrant I only.

    c. Select an ordered pair satisfying the inequality. What are its coordinates and what do they represent in this situation?

57. A person with no more than $15,000 to invest plans to place the money in two investments. One investment is high risk, high yield; the other is low risk, low yield. At least $2000 is to be placed in the high-risk investment. Furthermore, the amount invested at low risk should be at least three times the amount invested at high risk. Find and graph a system of inequalities that describes all possibilities for placing the money in the high- and low-risk investments.

58. Promoters of a rock concert must sell at least 25,000 tickets priced at $35 and $50 per ticket. Furthermore, the promoters must take in at least $1,025,000 in ticket sales. Find and graph a system of inequalities that describes all possibilities for selling the $35 tickets and the $50 tickets.

59. Use Figure 8.32 on page 750 to solve this exercise.

    a. Find a pulse rate that lies within the target zone for a person your age engaged in aerobic exercise.

    b. Express your answer in part (a) as an ordered pair. Show that the coordinates of this ordered pair satisfy each inequality.

The graph of an inequality in two variables is usually a region in the rectangular coordinate system. Regions in coordinate systems have numerous applications. For example, the regions in the two graphs at the top of the next column indicate whether a person is overweight, borderline overweight, or normal weight.

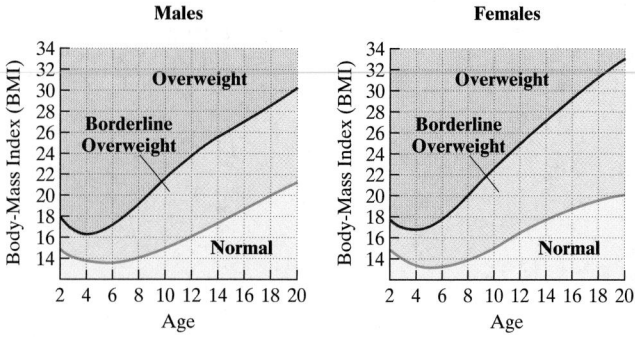

Source: Centers for Disease Control and Prevention

The horizontal axis shows a person's age. The vertical axis shows that person's body-mass index (BMI), computed using the following formula:

$$\text{BMI} = \frac{703W}{H^2}.$$

The variable W represents weight, in pounds. The variable H represents height, in inches. Use this information to solve Exercises 60–61.

60. A man is 20 years old, 72 inches (6 feet) tall, and weighs 200 pounds.

    a. Compute the man's BMI. Round to the nearest tenth.

    b. Use the man's age and his BMI to locate this information as a point in the coordinate system for males. Is this person overweight, borderline overweight, or normal weight?

61. A girl is 10 years old, 50 inches (4 feet, 2 inches) tall, and weighs 100 pounds.

    a. Compute the girl's BMI. Round to the nearest tenth.

    b. Use the girl's age and her BMI to locate this information as a point in the coordinate system for females. Is this person overweight, borderline overweight, or normal weight?

 Writing in Mathematics

62. What is a half-plane?

63. What does a dashed line mean in the graph of an inequality?

64. Explain how to graph $2x - 3y < 6$.

65. Compare the graphs of $3x - 2y > 6$ and $3x - 2y \le 6$. Discuss similarities and differences between the graphs.

66. Describe how to solve a system of inequalities.

67. What does it mean if a system of linear inequalities has no solution?

 Technology Exercises

Graphing utilities can be used to shade regions in the rectangular coordinate system, thereby graphing an inequality in two variables. Read the section of the user's manual for your graphing utility that describes how to shade a region. Then use your graphing utility to graph the inequalities in Exercises 68–73.

68. $y \leq 4x + 4$

69. $y \geq \frac{2}{3}x - 2$

70. $y \geq x^2 - 4$

71. $y \geq \frac{1}{2}x^2 - 2$

72. $2x + y \leq 6$

73. $3x - 2y \geq 6$

74. Does your graphing utility have any limitations in terms of graphing inequalities? If so, what are they?

75. Use a graphing utility with a $\boxed{\text{SHADE}}$ feature to verify any five of the graphs that you drew by hand in Exercises 1–22.

76. Use a graphing utility with a $\boxed{\text{SHADE}}$ feature to verify any five of the graphs that you drew by hand for the systems in Exercises 23–52.

Critical Thinking Exercises

77. Write a system of inequalities that has no solution.

78. Write a system of inequalities that describes the shaded region in the figure at the top of the next column.

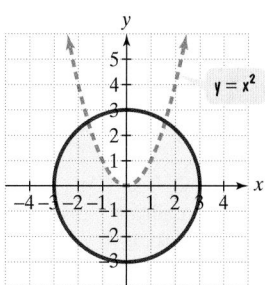

79. Sketch the graph of the solution set for the following system of inequalities:

$$y \geq nx + b \quad (n < 0, b > 0)$$
$$y \leq mx + b \quad (m > 0, b > 0).$$

80. Sketch the graph of the solution set for the following system of inequalities:

$$|x + y| \leq 3$$
$$|y| \leq 2.$$

SECTION 8.6 *Linear Programming*

Objectives

1. Write an objective function describing a quantity that must be maximized or minimized.

2. Use inequalities to describe limitations in a situation.

3. Use linear programming to solve problems.

West Berlin children at Tempelhof airport watch fleets of U.S. airplanes bringing in supplies to circumvent the Russian blockade. The airlift began June 28, 1948 and continued for 15 months.

The Berlin Airlift (1948–1949) was an operation by the United States and Great Britain in response to military action by the former Soviet Union: Soviet troops closed all roads and rail lines between West Germany and Berlin, cutting off supply routes to the city. The Allies used a mathematical technique developed during World War II to maximize the amount of supplies transported. During the 15-month airlift, 278,228 flights provided basic necessities to blockaded Berlin, saving one of the world's great cities.

In this section, we will look at an important application of systems of linear inequalities. Such systems arise in **linear programming,** a method for solving problems in which a particular quantity that must be maximized or minimized is limited by other factors. Linear programming is one of the most widely used tools in management science. It helps businesses allocate resources to manufacture products in a way that will maximize profit. Linear programming accounts for more than 50% and perhaps as much as 90% of all computing time used for management decisions in business. The Allies used linear programming to save Berlin.

1 Write an objective function describing a quantity that must be maximized or minimized.

Objective Functions in Linear Programming

Many problems involve quantities that must be maximized or minimized. Businesses are interested in maximizing profit. An operation in which bottled water and medical kits are shipped to earthquake victims needs to maximize the number of victims helped by this shipment. An **objective function** is an algebraic expression in two or more variables describing a quantity that must be maximized or minimized.

EXAMPLE 1   Writing an Objective Function

Bottled water and medical supplies are to be shipped to victims of an earthquake by plane. Each container of bottled water will serve 10 people and each medical kit will aid 6 people. Let x represent the number of bottles of water to be shipped and y the number of medical kits. Write the objective function that describes the number of people that can be helped.

Solution   Because each bottle of water serves 10 people and each medical kit aids 6 people, we have

$$= \quad 10x \quad + \quad 6y.$$

Using z to represent the objective function, we have

$$z = 10x + 6y.$$

Unlike the functions that we have seen so far, the objective function is an equation in three variables. For a value of x and a value of y, there is one and only one value of z. Thus, z is a function of x and y.

Check Point 1   A company manufactures bookshelves and desks for computers. Let x represent the number of bookshelves manufactured daily and y the number of desks manufactured daily. The company's profits are $25 per bookshelf and $55 per desk. Write the objective function that describes the company's total daily profit, z, from x bookshelves and y desks. (Check Points 2 through 4 are also related to this situation, so keep track of your answers.)

2 Use inequalities to describe limitations in a situation.

Constraints in Linear Programming

Ideally, the number of earthquake victims helped in Example 1 should increase without restriction so that every victim receives water and medical kits. However, the planes that ship these supplies are subject to weight and volume restrictions. In linear programming problems, such restrictions are called **constraints.** Each constraint is expressed as a linear inequality. The list of constraints forms a system of linear inequalities.

EXAMPLE 2   Writing a Constraint

Each plane can carry no more than 80,000 pounds. The bottled water weighs 20 pounds per container and each medical kit weighs 10 pounds. Let x represent the number of bottles of water to be shipped and y the number of medical kits. Write an inequality that describes this constraint.

Solution   Because each plane can carry no more than 80,000 pounds, we have

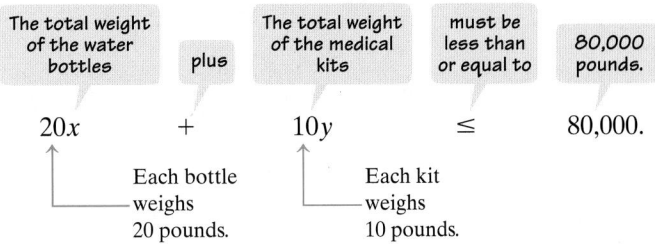

The plane's weight constraint is described by the inequality
$$20x + 10y \leq 80,000.$$

Check Point 2   To maintain high quality, the company in Check Point 1 should not manufacture more than a combined total of 80 bookshelves and desks per day. Write an inequality that describes this constraint.

In addition to a weight constraint on its cargo, each plane has a limited amount of space in which to carry supplies. Example 3 demonstrates how to express this constraint.

EXAMPLE 3   Writing a Constraint

The total volume of supplies that a plane carries cannot exceed 6000 cubic feet. Each water bottle is 1 cubic foot and each medical kit also has a volume of 1 cubic foot. With x still representing the number of water bottles and y the number of medical kits, write an inequality that describes this second constraint.

Solution   Because each plane can carry a volume of supplies that does not exceed 6000 cubic feet, we have

The plane's volume constraint is described by the inequality $x + y \leq 6000$.

In summary, here's what we have described in this aid-to-earthquake-victims situation:

$$z = 10x + 6y$$

This is the objective function describing the number of people helped with x bottles of water and y medical kits.

$$20x + 10y \leq 80,000$$
$$x + y \leq 6000.$$

These are the constraints based on each plane's weight and volume limitations.

Check Point 3 To meet customer demand, the company in Check Point 1 must manufacture between 30 and 80 bookshelves per day, inclusive. Furthermore, the company must manufacture at least 10 and no more than 30 desks per day. Write an inequality that describes each of these sentences. Then summarize what you have described about this company by writing the objective function for its profits, and the three constraints.

3 Use linear programming to solve problems.

Solving Problems with Linear Programming

The goal in the earthquake situation described previously is to maximize the number of victims who can be helped, subject to the planes' weight and volume constraints. The process of solving this problem is called *linear programming*, based on a theorem that was proven during World War II.

Solving a Linear Programming Problem

Let $z = ax + by$ be an objective function that depends on x and y. Furthermore, z is subject to a number of constraints on x and y. If a maximum or minimum value of z exists, it can be determined as follows:
1. Graph the system of inequalities representing the constraints.
2. Find the value of the objective function at each corner, or **vertex**, of the graphed region. The maximum and minimum of the objective function occur at one or more of the corner points.

EXAMPLE 4 Solving a Linear Programming Problem

Determine how many bottles of water and how many medical kits should be sent on each plane to maximize the number of earthquake victims who can be helped.

Solution We must maximize $z = 10x + 6y$ subject to the constraints:
$$20x + 10y \leq 80,000$$
$$x + y \leq 6000.$$

Step 1 Graph the system of inequalities representing the constraints. Because x (the number of bottles of water per plane) and y (the number of medical kits per plane) must be nonnegative, we need to graph the system of inequalities in quadrant I and its boundary only ($x \geq 0$ and $y \geq 0$). To graph the inequality $20x + 10y \leq 80,000$, we graph the equation $20x + 10y = 80,000$ as a solid blue line (Figure 8.33). Setting $y = 0$, the x-intercept is 4000 and setting $x = 0$, the y-intercept is 8000. Using $(0, 0)$ as a test point, the inequality is satisfied, so we shade below the blue line, as shown in yellow in Figure 8.33. Now we graph $x + y \leq 6000$ by first graphing $x + y = 6000$ as a solid red line. Setting $y = 0$, the x-intercept is 6000. Setting $x = 0$, the y-intercept is 6000. Using $(0, 0)$ as a test point, the inequality is satisfied, so we shade below the red line, as shown using green vertical shading in Figure 8.33.

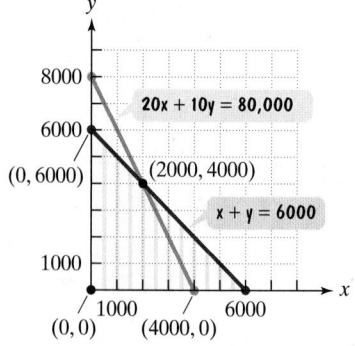

Figure 8.33 The region in quadrant I representing the constraints

$$20x + 10y \leq 80,000$$
$$x + y \leq 6000$$

We use the addition method to find the coordinates of the point where the lines $20x + 10y = 80,000$ and $x + y = 6000$ intersect.

$$20x + 10y = 80,000 \xrightarrow{\text{No change}} 20x + 10y = 80,000$$
$$x + y = 6000 \xrightarrow{\text{Multiply by } -10.} -10x - 10y = -60,000$$
$$\text{Add:} \quad 10x \qquad\qquad = 20,000$$
$$x = 2000$$

Back-substituting 2000 for x in $x + y = 6000$, we find $y = 4000$, so the intersection point is $(2000, 4000)$.

The system of inequalities representing the constraints is shown by the region in which the yellow shading and the green vertical shading overlap in Figure 8.33. The graph of the system of inequalities is shown again in Figure 8.34. The red and blue line segments are included in the graph.

Step 2 Find the value of the objective function at each corner of the graphed region. The maximum and minimum of the objective function occur at one or more of the corner points. We must evaluate the objective function, $z = 10x + 6y$, at the four corners of the region in Figure 8.34.

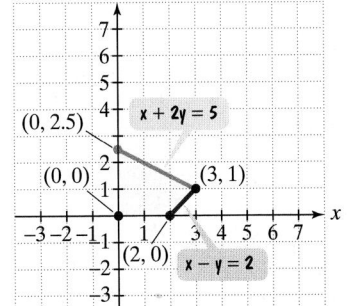

Figure 8.34

Corner (x, y)	Objective Function $z = 10x + 6y$
$(0, 0)$	$z = 10(0) + 6(0) = 0$
$(4000, 0)$	$z = 10(4000) + 6(0) = 40,000$
$(2000, 4000)$	$z = 10(2000) + 6(4000) = 44,000 \leftarrow \text{maximum}$
$(0, 6000)$	$z = 10(0) + 6(6000) = 36,000$

Thus, the maximum value of z is 44,000 and this occurs when $x = 2000$ and $y = 4000$. In practical terms, this means that the maximum number of earthquake victims who can be helped with each plane shipment is 44,000. This can be accomplished by sending 2000 water bottles and 4000 medical kits per plane.

> **Check Point 4** For the company in Check Points 1–3, how many bookshelves and how many desks should be manufactured per day to obtain a maximum profit? What is the maximum daily profit?

EXAMPLE 5 Solving a Linear Programming Problem

Find the maximum value of the objective function

$$z = 2x + y$$

subject to the constraints:

$$x \geq 0, \ y \geq 0$$
$$x + 2y \leq 5$$
$$x - y \leq 2.$$

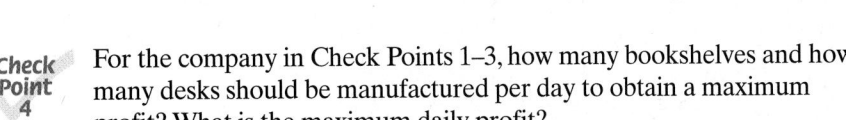

Figure 8.35 The graph of $x + 2y \leq 5$ and $x - y \leq 2$ in quadrant I

Solution We begin by graphing the region in quadrant I ($x \geq 0, y \geq 0$) formed by the constraints. The graph is shown by the closed yellow region in Figure 8.35.

Now we evaluate the objective function at the four vertices of this region.

Objective function: $z = 2x + y$

At $(0, 0)$: $z = 2 \cdot 0 + 0 = 0$

At $(2, 0)$: $z = 2 \cdot 2 + 0 = 4$

At $(3, 1)$: $z = 2 \cdot 3 + 1 = 7$ Maximum value of z

At $(0, 2.5)$: $z = 2 \cdot 0 + 2.5 = 2.5$

Thus, the maximum value of z is 7, and this occurs when $x = 3$ and $y = 1$.

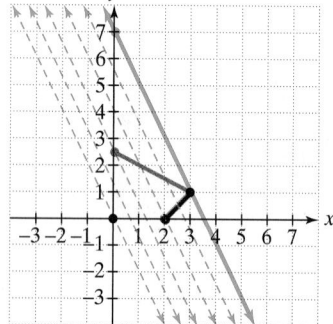

Figure 8.36 The line with slope -2 with the greatest y-intercept that intersects the shaded region passes through one of its vertices.

We can see why the objective function in Example 5 has a maximum value that occurs at a vertex by solving the equation for y.

$z = 2x + y$ This is the objective function of Example 5.

$y = -2x + z$ Solve for y. Recall that the slope-intercept form of a line is $y = mx + b$.

Slope = -2 y-intercept = z

In this form, z represents the y-intercept of the objective function. The equation describes infinitely many parallel lines, each with a slope of -2. The process in linear programming involves finding the maximum z-value for all lines that intersect the region determined by the constraints. Of all the lines whose slope is -2, we're looking for the one with the greatest y-intercept that intersects the given region. As we see in Figure 8.36, such a line will pass through one (or possibly more) of the vertices of the region.

Check Point 5   Find the maximum value of the objective function $z = 3x + 5y$ subject to the constraints $x \geq 0$, $y \geq 0$, $x + y \geq 1$, $x + y \leq 6$.

Faster and Faster

The network of computer linkages in the United States is growing exponentially.

The problems we solve nowadays have thousands of equations, sometimes a million variables. One of the things that still amazes me is to see a program run on the computer—and to see the answer come out. If we think of the number of combinations of different solutions that we're trying to choose the best of, it's akin to the stars in the heavens. Yet we solve them in a matter of moments. This, to me, is staggering. Not that we can solve them—but that we can solve them so rapidly and efficiently.

—George Dantzig
Inventor of the simplex method, a linear programming method

Problems in linear programming can involve objective functions with thousands of variables subject to thousands of constraints. Several nongeometric linear programming methods are available on software for solving such problems. And we continue to search for faster and faster linear programming methods. This area of applied mathematics has a direct impact on the efficiency and profitability of numerous industries, including telephone and computer communications, and the airlines.

EXERCISE SET 8.6

Practice Exercises

In Exercises 1–4, find the value of the objective function at each corner of the graphed region. What is the maximum value of the objective function? What is the minimum value of the objective function?

1. Objective Function $z = 5x + 6y$

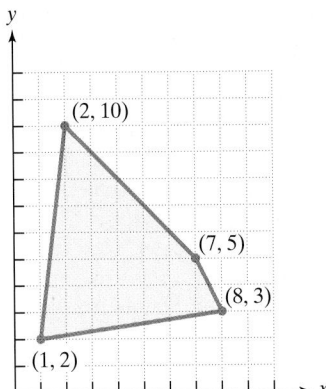

2. Objective Function $z = 3x + 2y$

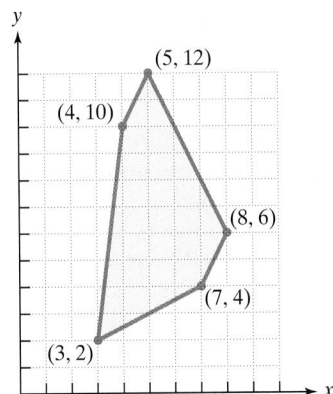

3. Objective Function $z = 40x + 50y$

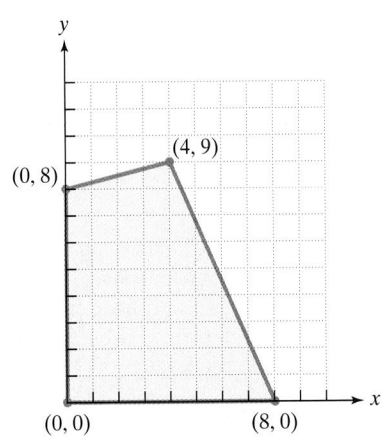

4. Objective Function $z = 30x + 45y$

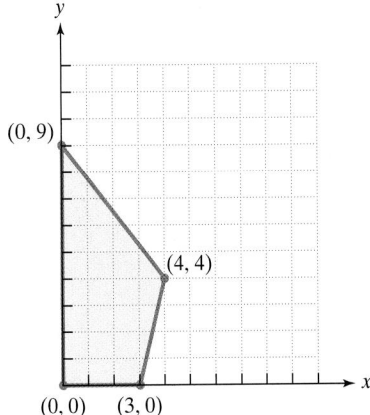

In Exercises 5–14, an objective function and a system of linear inequalities representing constraints are given.

a. *Graph the system of inequalities representing the constraints.*

b. *Find the value of the objective function at each corner of the graphed region.*

c. *Use the values in part (b) to determine the maximum value of the objective function and the values of x and y for which the maximum occurs.*

5. Objective Function $z = 3x + 2y$
Constraints $x \geq 0, y \geq 0$
    $2x + y \leq 8$
    $x + y \geq 4$

6. Objective Function $z = 2x + 3y$
Constraints $x \geq 0, y \geq 0$
    $2x + y \leq 8$
    $2x + 3y \leq 12$

7. Objective Function $z = 4x + y$
Constraints $x \geq 0, y \geq 0$
    $2x + 3y \leq 12$
    $x + y \geq 3$

8. Objective Function $z = x + 6y$
Constraints $x \geq 0, y \geq 0$
    $2x + y \leq 10$
    $x - 2y \geq -10$

9. Objective Function $z = 3x - 2y$
Constraints $1 \leq x \leq 5$
    $y \geq 2$
    $x - y \geq -3$

10. Objective Function $z = 5x - 2y$
     Constraints $0 \leq x \leq 5$
                          $0 \leq y \leq 3$
                          $x + y \geq 2$

11. Objective Function $z = 4x + 2y$
     Constraints $x \geq 0, y \geq 0$
                          $2x + 3y \leq 12$
                          $3x + 2y \leq 12$
                          $x + y \geq 2$

12. Objective Function $z = 2x + 4y$
     Constraints $x \geq 0, y \geq 0$
                          $x + 3y \geq 6$
                          $x + y \geq 3$
                          $x + y \leq 9$

13. Objective Function $z = 10x + 12y$
     Constraints $x \geq 0, y \geq 0$
                          $x + y \leq 7$
                          $2x + y \leq 10$
                          $2x + 3y \leq 18$

14. Objective Function $z = 5x + 6y$
     Constraints $x \geq 0, y \geq 0$
                          $2x + y \geq 10$
                          $x + 2y \geq 10$
                          $x + y \leq 10$

Application Exercises

15. A television manufacturer makes console and wide-screen televisions. The profit per unit is $125 for the console televisions and $200 for the wide-screen televisions.

  a. Let x = the number of consoles manufactured in a month and y = the number of wide-screens manufactured in a month. Write the objective function that describes the total monthly profit.

  b. The manufacturer is bound by the following constraints:
- Equipment in the factory allows for making at most 450 console televisions in one month.
- Equipment in the factory allows for making at most 200 wide-screen televisions in one month.
- The cost to the manufacturer per unit is $600 for the console televisions and $900 for the wide-screen televisions. Total monthly costs cannot exceed $360,000.

    Write a system of three inequalities that describes these constraints.

  c. Graph the system of inequalities in part (b). Use only the first quadrant and its boundary, because x and y must both be nonnegative.

d. Evaluate the objective function for total monthly profit at each of the five vertices of the graphed region. [The vertices should occur at $(0, 0)$, $(0, 200)$, $(300, 200)$, $(450, 100)$, and $(450, 0)$.]

e. Complete the missing portions of this statement: The television manufacturer will make the greatest profit by manufacturing ___ console televisions each month and ___ wide-screen televisions each month. The maximum monthly profit is $ ___.

16. a. A student earns $10 per hour for tutoring and $7 per hour as a teacher's aid. Let x = the number of hours each week spent tutoring, and y = the number of hours each week spent as a teacher's aid. Write the objective function that describes total weekly earnings.

  b. The student is bound by the following constraints:
- To have enough time for studies, the student can work no more than 20 hours per week.
- The tutoring center requires that each tutor spend at least three hours per week tutoring.
- The tutoring center requires that each tutor spend no more than eight hours per week tutoring.

    Write a system of three inequalities that describes these constraints.

  c. Graph the system of inequalities in part (b). Use only the first quadrant and its boundary, because x and y are nonnegative.

  d. Evaluate the objective function for total weekly earnings at each of the four vertices of the graphed region. [The vertices should occur at $(3, 0)$, $(8, 0)$, $(3, 17)$, and $(8, 12)$.]

  e. Complete the missing portions of this statement: The student can earn the maximum amount per week by tutoring for ___ hours per week and working as a teacher's aid for ___ hours per week. The maximum amount that the student can earn each week is $ ___.

Use the two steps for solving a linear programming problem, given in the box on page 756, to solve the problems in Exercises 17–23.

17. A manufacturer produces two models of mountain bicycles. The times (in hours) required for assembling and painting each model are given in the following table:

	Model A	Model B
Assembling	5	4
Painting	2	3

The maximum total weekly hours available in the assembly department and the paint department are 200 hours and 108 hours, respectively. The profits per unit are $25 for model A and $15 for model B. How many of each type should be produced to maximize profit?

18. A large institution is preparing lunch menus containing foods A and B. The specifications for the two foods are given in the following table:

Food	Units of Fat per Ounce	Units of Carbohydrates per Ounce	Units of Protein per Ounce
A	1	2	1
B	1	1	1

Each lunch must provide at least 6 units of fat per serving, no more than 7 units of protein, and at least 10 units of carbohydrates. The institution can purchase food A for $0.12 per ounce and food B for $0.08 per ounce. How many ounces of each food should a serving contain to meet the dietary requirements at the least cost?

19. Food and clothing are shipped to victims of a natural disaster. Each carton of food will feed 12 people, while each carton of clothing will help 5 people. Each 20-cubic-foot box of food weighs 50 pounds and each 10-cubic-foot box of clothing weighs 20 pounds. The commercial carriers transporting food and clothing are bound by the following constraints:
- The total weight per carrier cannot exceed 19,000 pounds.
- The total volume must be less than 8000 cubic feet.

How many cartons of food and clothing should be sent with each plane shipment to maximize the number of people who can be helped?

20. On June 24, 1948, the former Soviet Union blocked all land and water routes through East Germany to Berlin. A gigantic airlift was organized using American and British planes to supply food, clothing, and other supplies to the more than 2 million people in West Berlin. The cargo capacity was 30,000 cubic feet for an American plane and 20,000 cubic feet for a British plane. To break the Soviet blockade, the Western Allies had to maximize cargo capacity, but were subject to the following restrictions:
- No more than 44 planes could be used.
- The larger American planes required 16 personnel per flight, double that of the requirement for the British planes. The total number of personnel available could not exceed 512.
- The cost of an American flight was $9000 and the cost of a British flight was $5000. Total weekly costs could not exceed $300,000.

Find the number of American and British planes that were used to maximize cargo capacity.

21. A theater is presenting a program on drinking and driving for students and their parents. The proceeds will be donated to a local alcohol information center. Admission is $2.00 for parents and $1.00 for students. However, the situation has two constraints: The theater can hold no more than 150 people and every two parents must bring at least one student. How many parents and students should attend to raise the maximum amount of money?

22. You are about to take a test that contains computation problems worth 6 points each and word problems worth 10 points each. You can do a computation problem in 2 minutes and a word problem in 4 minutes. You have 40 minutes to take the test and may answer no more than 12 problems. Assuming you answer all the problems attempted correctly, how many of each type of problem must you do to maximize your score? What is the maximum score?

23. In 1978, a ruling by the Civil Aeronautics Board allowed Federal Express to purchase larger aircraft. Federal Express's options included 20 Boeing 727s that United Airlines was retiring and/or the French-built Dassault Fanjet Falcon 20. To aid in their decision, executives at Federal Express analyzed the following data:

	Boeing 727	Falcon 20
Direct Operating Cost	$1400 per hour	$500 per hour
Payload	42,000 pounds	6000 pounds

Federal Express was faced with the following constraints:
- Hourly operating cost was limited to $35,000.
- Total payload had to be at least 672,000 pounds.
- Only twenty 727s were available.

Given the constraints, how many of each kind of aircraft should Federal Express have purchased to maximize the number of aircraft?

Writing in Mathematics

24. What kinds of problems are solved using the linear programming method?

25. What is an objective function in a linear programming problem?

26. What is a constraint in a linear programming problem? How is a constraint represented?

27. In your own words, describe how to solve a linear programming problem.

28. Describe a situation in your life in which you would really like to maximize something, but you are limited by at least two constraints. Can linear programming be used in this situation? Explain your answer.

Technology Exercises

In Exercises 29–32, use a graphing utility to sketch the region determined by the constraints. Then determine the maximum value of the objective function subject to the contraints.

29. Objective Function $\quad z = 6x + 8y$
    Constraints $\quad x \ge 0, y \ge 0$
    $\quad\quad\quad\quad\quad x + 2y \le 6$

30. Objective Function $\quad z = 30x + 20y$
    Constraints $\quad x \ge 0, y \ge 0$
    $\quad\quad\quad\quad\quad 2x + y \le 14$
    $\quad\quad\quad\quad\quad 3x + y \le 18$

31. Objective Function $\quad z = 9x + 14y$
    Constraints $\quad x \ge 0, y \ge 0$
    $\quad\quad\quad\quad\quad 2x + y \le 10$
    $\quad\quad\quad\quad\quad 2x + 3y \le 18$

32. Objective Function $\quad z = 10x + 3y$
    Constraints $\quad 0 \le x \le 10, \quad y \ge 0$
    $\quad\quad\quad\quad\quad 4x + 5y \le 60$
    $\quad\quad\quad\quad\quad 4x - 5y \ge -20$

Critical Thinking Exercises

33. Suppose that you inherit $10,000. The will states how you must invest the money. Some (or all) of the money must be invested in stocks and bonds. The requirements are that at least $3000 be invested in bonds, with expected returns of $0.08 per dollar, and at least $2000 be invested in stocks, with expected returns of $0.12 per dollar. Because the stocks are medium risk, the final stipulation requires that the investment in bonds should never be less than the investment in stocks. How should the money be invested so as to maximize your expected returns?

34. Consider the objective function $z = Ax + By$ ($A > 0$ and $B > 0$) subject to the following constraints: $2x + 3y \le 9$, $x - y \le 2$, $x \ge 0$, and $y \ge 0$. Prove that the objective function will have the same maximum value at the vertices $(3, 1)$ and $(0, 3)$ if $A = \frac{2}{3}B$.

Group Exercises

35. Group members should choose a particular field of interest. Research how linear programming is used to solve problems in that field. If possible, investigate the solution of a specific practical problem. Present a report on your findings, including the contributions of George Dantzig, Narendra Karmarkar, and L.G. Khachion to linear programming.

36. Members of the group should interview a business executive who is in charge of deciding the product mix for a business. How are production policy decisions made? Are other methods used in conjunction with linear programming? What are these methods? What sort of academic background, particularly in mathematics, does this executive have? Present a group report addressing these questions, emphasizing the role of linear programming for the business.

CHAPTER SUMMARY, REVIEW, AND TEST

Summary

DEFINITIONS AND CONCEPTS	EXAMPLES

8.1 Systems of Linear Equations in Two Variables

a. Two equations in the form $Ax + By = C$ are called a system of linear equations. A solution to the system is an ordered pair that satisfies both equations in the system.
$\quad$ Ex. 1, p. 700

b. Systems of linear equations in two variables can be solved by eliminating a variable, using the substitution method (see the box on page 702) or the addition method (see the box on page 704).
$\quad$ Ex. 2, p. 701;
$\quad$ Ex. 3, p. 702;
$\quad$ Ex. 4, p. 704;
$\quad$ Ex. 5, p. 705

c. Some linear systems have no solution and are called inconsistent systems; others have infinitely many solutions. The equations in a linear system with infinitely many solutions are called dependent. For details, see the box on page 706.
$\quad$ Ex. 6, p. 706;
$\quad$ Ex. 7, p. 707

8.2 Systems of Linear Equations in Three Variables

a. Three equations in the form $Ax + By + Cz = D$ are called a system of linear equations in three variables. A solution to the system is an ordered triple that satisfies all three equations in the system.
$\quad$ Ex. 1, p. 717

b. A system of linear equations in three variables can be solved by eliminating variables. Use the addition method to eliminate any variable, reducing the system to two equations in two variables.
$\quad$ Ex. 2, p. 718;
$\quad$ Ex. 3, p. 719

DEFINITIONS AND CONCEPTS	EXAMPLES

Use substitution or the addition method to solve the resulting system in two variables. Details are found in the box on page 717.

8.3 Partial Fraction Decomposition

a. Partial fraction decomposition is used on rational expressions in which the numerator and denominator have no common factors and the highest power in the numerator is less than the highest power in the denominator. The steps in partial fraction decomposition are given in the box on page 728.

b. Include one partial fraction with a constant numerator for each distinct linear factor in the denominator. Include one partial fraction with a constant numerator for each power of a repeated linear factor in the denominator.

Ex. 1, p. 727;
Ex. 2, p. 728

c. Include one partial fraction with a linear numerator for each distinct prime quadratic factor in the denominator. Include one partial fraction with a linear numerator for each power of a prime, repeated quadratic factor in the denominator.

Ex. 3, p. 730;
Ex. 4, p. 731

8.4 Systems of Nonlinear Equations in Two Variables

a. A system of two nonlinear equations in two variables contains at least one equation that cannot be expressed as $Ax + By = C$.

b. Systems of nonlinear equations in two variables can be solved algebraically by eliminating all occurrences of one of the variables by the substitution or addition methods.

Ex. 1, p. 735;
Ex. 2, p. 736;
Ex. 3, p. 737;
Ex. 4, p. 738

8.5 Systems of Inequalities

a. A linear inequality in two variables can be written in the form $Ax + By > C$, $Ax + By \geq C$, $Ax + By < C$, or $Ax + By \leq C$.

b. The procedure for graphing a linear inequality in two variables is given in the box on page 744. A nonlinear inequality in two variables is graphed using the same procedure.

Ex. 1, p. 744;
Ex. 2, p. 745;
Ex. 3, p. 747

c. To graph the solution set to a system of inequalities, graph each inequality in the system in the same rectangular coordinate system. Then find the region, if there is one, that is common to every graph in the system.

Ex. 4, p. 747;
Ex. 5, p. 748;
Ex. 6, p. 749

8.6 Linear Programming

a. An objective function is an algebraic expression in three variables describing a quantity that must be maximized or minimized.

Ex. 1, p. 754

b. Constraints are restrictions, expressed as linear inequalities.

Ex. 2, p. 755;
Ex. 3, p. 755

c. Steps for solving a linear programming problem are given in the box on page 756.

Ex. 4, p. 756;
Ex. 5, p. 757

Review Exercises

8.1

In Exercises 1–5, solve by the method of your choice. Identify systems with no solution and systems with infinitely many solutions, using set notation to express their solution sets.

1. $y = 4x + 1$
$3x + 2y = 13$

2. $x + 4y = 14$
$2x - y = 1$

3. $5x + 3y = 1$
$3x + 4y = -6$

4. $2y - 6x = 7$
$3x - y = 9$

5. $4x - 8y = 16$
$3x - 6y = 12$

6. A company is planning to manufacture computer desks. The fixed cost will be $60,000 and it will cost $200 to produce each desk. Each desk will be sold for $450.
   a. Write the cost function, C, of producing x desks.
   b. Write the revenue function, R, from the sale of x desks.
   c. Determine the break-even point. Describe what this means.

7. The weekly demand and supply models for the video *Pearl Harbor* at a chain of stores that sells videos are given by the demand model $N = -60p + 1000$ and the supply model $N = 4p + 200$, in which p is the price of the video and N is the number of videos sold or supplied each week to the chain of stores. Find the price at which supply and demand are equal. At this price, how many copies of *Pearl Harbor* can be supplied and sold each week?

8. The graph makes Super Bowl Sunday look like a day of snack food binging in the United States. The number of pounds of guacamole consumed is ten times the difference between the number of pounds of potato and tortilla chips eaten on the same day. On Super Bowl Sunday, Americans also eat a total quantity of potato and tortilla chips that exceeds popcorn consumption by 7.3 million pounds. How many millions of pounds of potato chips and tortilla chips are consumed on Super Bowl Sunday?

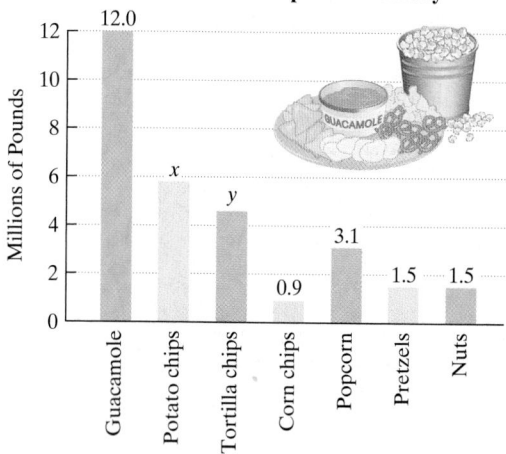

Millions of Pounds of Snack Food Consumed on Super Bowl Sunday

Source: Association of American Snack Foods

9. A travel agent offers two package vacation plans. The first plan costs $360 and includes 3 days at a hotel and a rental car for 2 days. The second plan costs $500 and includes 4 days at a hotel and a rental car for 3 days. The daily charge for the hotel is the same under each plan, as is the daily charge for the car. Find the cost per day for the hotel and for the car.

10. The calorie-nutrient information for an apple and an avocado is given in the table. How many of each should be eaten to get exactly 1000 calories and 100 grams of carbohydrates?

	One Apple	One Avocado
Calories	100	350
Carbohydrates (grams)	24	14

8.2

Solve each system in Exercises 11–12.

11. $2x - y + z = 1$
$3x - 3y + 4z = 5$
$4x - 2y + 3z = 4$

12. $x + 2y - z = 5$
$2x - y + 3z = 0$
$2y + z = 1$

13. Find the quadratic function $y = ax^2 + bx + c$ whose graph passes through the points $(1, 4)$, $(3, 20)$, and $(-2, 25)$.

14. The bar graph shows that the U.S. divorce rate increased between 1970 and 1985 and then decreased between 1985 and 1999.

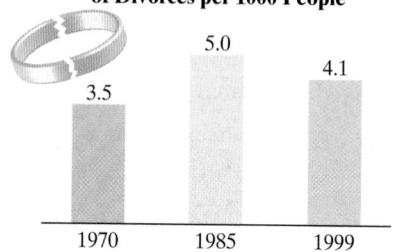

U.S. Divorce Rates: Number of Divorces per 1000 People

Source: U.S. Census Bureau

a. Write the data for 1970, 1985, and 1999 as ordered pairs (x, y), where x is the number of years after 1970 and y is that year's divorce rate.

b. The three data points in part (a) can be modeled by the quadratic function $y = ax^2 + bx + c$. Write a system of linear equations in three variables that can be used to find values for a, b, and c. It is not necessary to solve the system.

15. The bar graph indicates countries in which ten or more languages have become extinct. The number of extinct languages in the United States, Colombia, and India combined is 50. The number of extinct languages in the United States exceeds the number in Colombia by 4 and is 2 more than twice that for India. How many languages have become extinct in the United States, Colombia, and India?

Countries Where Ten or More Laguages Have Become Extinct (Number of Languages)

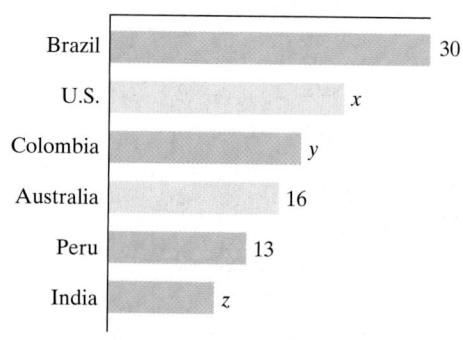

Source: Grimes

8.3

In Exercises 16–24, write the partial fraction decomposition of each rational expression.

16. $\dfrac{x}{(x-3)(x+2)}$

17. $\dfrac{11x-2}{x^2-x-12}$

18. $\dfrac{4x^2-3x-4}{x(x+2)(x-1)}$

19. $\dfrac{2x+1}{(x-2)^2}$

20. $\dfrac{2x-6}{(x-1)(x-2)^2}$

21. $\dfrac{3x}{(x-2)(x^2+1)}$

22. $\dfrac{7x^2-7x+23}{(x-3)(x^2+4)}$

23. $\dfrac{x^3}{(x^2+4)^2}$

24. $\dfrac{4x^3+5x^2+7x-1}{(x^2+x+1)^2}$

8.4

In Exercises 25–35, solve each system by the method of your choice.

25. $5y = x^2 - 1$
$\quad x - y = 1$

26. $y = x^2 + 2x + 1$
$\quad x + y = 1$

27. $x^2 + y^2 = 2$
$\quad x + y = 0$

28. $2x^2 + y^2 = 24$
$\quad x^2 + y^2 = 15$

29. $xy - 4 = 0$
$\quad y - x = 0$

30. $y^2 = 4x$
$\quad x - 2y + 3 = 0$

31. $x^2 + y^2 = 10$
$\quad y = x + 2$

32. $xy = 1$
$\quad y = 2x + 1$

33. $x + y + 1 = 0$
$\quad x^2 + y^2 + 6y - x = -5$

34. $x^2 + y^2 = 13$
$\quad x^2 - y = 7$

35. $2x^2 + 3y^2 = 21$
$\quad 3x^2 - 4y^2 = 23$

36. The perimeter of a rectangle is 26 meters, and its area is 40 square meters. Find its dimensions.

37. Find the coordinates of all points (x, y) that lie on the line whose equation is $2x + y = 8$, so that the area of the rectangle shown in the figure is 6 square units.

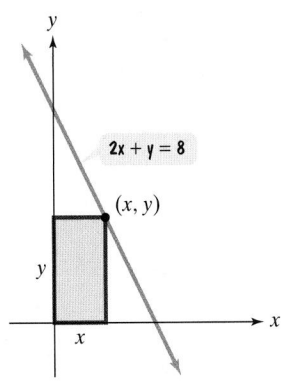

38. Two adjoining square fields with an area of 2900 square feet are to be enclosed with 240 feet of fencing. The situation is represented in the figure. Find the length of each side where a variable appears.

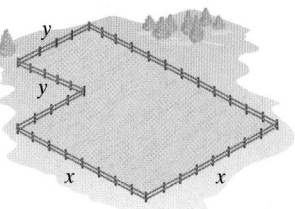

8.5

In Exercises 39–45, graph each inequality.

39. $3x - 4y > 12$

40. $y \le -\dfrac{1}{2}x + 2$

41. $x < -2$

42. $y \ge 3$

43. $x^2 + y^2 > 4$

44. $y \le x^2 - 1$

45. $y \le 2^x$

In Exercises 46–55, graph the solution set of each system of inequalities or indicate that the system has no solution.

46. $3x + 2y \ge 6$
$\quad 2x + y \ge 6$

47. $2x - y \ge 4$
$\quad x + 2y < 2$

48. $y < x$
$\quad y \le 2$

49. $x + y \le 6$
$\quad y \ge 2x - 3$

50. $0 \le x \le 3$
$\quad y > 2$

51. $2x + y < 4$
$\quad 2x + y > 6$

52. $x^2 + y^2 \le 16$
$\quad x + y < 2$

53. $x^2 + y^2 \le 9$
$\quad y < -3x + 1$

54. $y > x^2$
$\quad x + y < 6$
$\quad y < x + 6$

55. $y \ge 0$
$\quad 3x + 2y \ge 4$
$\quad x - y \le 3$

8.6

56. Find the value of the objective function $z = 2x + 3y$ at each corner of the graphed region shown. What is the maximum value of the objective function? What is the minimum value of the objective function?

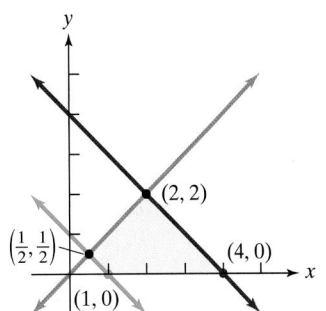

In Exercises 57–59, graph the region determined by the constraints. Then find the maximum value of the given objective function, subject to the constraints.

57. Objective Function $z = 2x + 3y$
     Constraints $x \geq 0, y \geq 0$
                              $x + y \leq 8$
                            $3x + 2y \geq 6$

58. Objective Function $z = x + 4y$
     Contraints $0 \leq x \leq 5, 0 \leq y \leq 7$
                              $x + y \geq 3$

59. Objective Function $z = 5x + 6y$
     Constraints $x \geq 0, y \geq 0$
                              $y \leq x$
                            $2x + y \leq 12$
                            $2x + 3y \geq 6$

60. A paper manufacturing company converts wood pulp to writing paper and newsprint. The profit on a unit of writing paper is $500 and the profit on a unit of newsprint is $350.

  a. Let x represent the number of units of writing paper produced daily. Let y represent the number of units of newsprint produced daily. Write the objective function that models total daily profit.

  b. The manufacturer is bound by the following constraints:

    • Equipment in the factory allows for making at most 200 units of paper (writing paper and newsprint) in a day.

    • Regular customers require at least 10 units of writing paper and at least 80 units of newsprint daily.

Write a system of inequalities that models these constraints.

  c. Graph the inequalities in part (b). Use only the first quadrant, because x and y must both be positive. (*Suggestion:* Let each unit along the x- and y-axes represent 20.)

  d. Evaluate the objective profit function at each of the three vertices of the graphed region.

  e. Complete the missing portions of this statement: The company will make the greatest profit by producing ___ units of writing paper and ___ units of newsprint each day. The maximum daily profit is $ ___.

61. A manufacturer of lightweight tents makes two models whose specifications are given in the following table:

	Cutting Time per Tent	Assembly Time per Tent
Model A	0.9 hour	0.8 hour
Model B	1.8 hours	1.2 hours

On a monthly basis, the manufacturer has no more than 864 hours of labor available in the cutting department and at most 672 hours in the assembly division. The profits come to $25 per tent for model A and $40 per tent for model B. How many of each should be manufactured monthly to maximize the profit?

Chapter 8 Test

In Exercises 1–5, solve the system.

1. $x = y + 4$
     $3x + 7y = -18$

2. $2x + 5y = -2$
     $3x - 4y = 20$

3. $x + y + z = 6$
     $3x + 4y - 7z = 1$
     $2x - y + 3z = 5$

4. $x^2 + y^2 = 25$
     $x + y = 1$

5. $2x^2 - 5y^2 = -2$
     $3x^2 + 2y^2 = 35$

6. Find the partial fraction decomposition for
$$\frac{x}{(x + 1)(x^2 + 9)}.$$

In Exercises 7–10, graph the solution set of each inequality or system of inequalities.

7. $x - 2y < 8$

8. $x \geq 0, y \geq 0$
     $3x + y \leq 9$
     $2x + 3y \geq 6$

9. $x^2 + y^2 > 1$
     $x^2 + y^2 < 4$

10. $y \leq 1 - x^2$
     $x^2 + y^2 \leq 9$

11. Find the maximum value of the objective function $z = 3x + 5y$ subject to the following constraints: $x \geq 0$, $y \geq 0$, $x + y \leq 6$, $x \geq 2$.

12. Health experts agree that cholesterol intake should be limited to 300 mg or less each day. Three ounces of shrimp and 2 ounces of scallops contain 156 mg of cholesterol. Five ounces of shrimp and 3 ounces of scallops contain 45 mg of cholesterol less than the suggested maximum daily intake. Determine the cholesterol content in an ounce of each item.

13. A company is planning to produce and sell a new line of computers. The fixed cost will be $360,000, and it will cost $850 to produce each computer. Each computer will be sold for $1150.

  a. Write the cost function, C, of producing x computers.

  b. Write the revenue function, R, from the sale of x computers.

  c. Determine the break-even point. Describe what this means.

14. Find the quadratic function whose graph passes through the points $(-1, -2)$, $(2, 1)$, and $(-2, 1)$.

15. The rectangular plot of land shown in the figure is to be fenced along three sides using 39 feet of fencing. No fencing is to be placed along the river's edge. The area of the plot is 180 square feet. What are its dimensions?

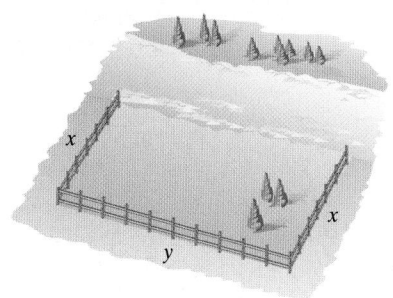

16. A manufacturer makes two types of jet skis, regular and deluxe. The profit on a regular jet ski is $200 and the profit on the deluxe model is $250. To meet customer demand, the company must manufacture at least 50 regular jet skis per week and at least 75 deluxe models. To maintain high quality, the total number of both models of jet skis manufactured by the company should not exceed 150 per week. How many jet skis of each type should be manufactured per week to obtain maximum profit? What is the maximum weekly profit?

Cumulative Review Exercises (Chapters 1–8)

Solve each equation or inequality in Exercises 1–8.

1. $\sqrt{x^2 - 3x} = 2x - 6$ **2.** $4x^2 = 8x - 7$

3. $\left| \dfrac{x}{3} + 2 \right| < 4$ **4.** $\dfrac{x + 5}{x - 1} > 2$

5. $2x^3 + x^2 - 13x + 6 = 0$

6. $6x - 3(5x + 2) = 4(1 - x)$

7. $\log(x + 3) + \log x = 1$ **8.** $3^{x+2} = 11$

In Exercises 9–12, graph each equation, function, or inequality in the rectangular coordinate system.

9. $f(x) = (x + 2)^2 - 4$ **10.** $2x - 3y \le 6$

11. $y = 3^{x-2}$ **12.** $f(x) = \dfrac{x^2 - x - 6}{x + 1}$

13. Expand and simplify: $\log_2 (8x^5)$.

14. What interest rate is required for an investment of $6000 subject to continuous compounding to grow to $18,000 in 10 years?

15. If $f(x) = 7x - 3$, find $f^{-1}(x)$.

16. If $f(x) = 7x - 3$ and $g(x) = 3x - 7$, find $g(f(x))$.

17. Explain why $x^2 + y^2 = 4$ does not represent y as a function of x.

18. Solve the system:
$$3x - y = -2$$
$$2x^2 - y = 0.$$

19. The length of a rectangle is 1 meter more than twice the width. If the rectangle's area is 36 square meters, find its dimensions.

20. The function $f(x) = 0.1x^2 - 3x + 22$ describes the distance, $f(x)$, in feet, needed for an airplane to land when its initial landing speed is x feet per second. Find and interpret $f(90)$. Will there be a problem if 550 feet of runway is available? Explain.

In Exercises 21–22, verify each identity.

21. $\sec\theta - \cos\theta = \tan\theta \sin\theta$

22. $\tan x + \tan y = \dfrac{\sin(x + y)}{\cos x \cos y}$

In Exercises 23–24, solve each equation.

23. $\sin\theta = \tan\theta$, $0 \le \theta < 2\pi$

24. $2 + \cos 2\theta = 3 \cos\theta$, $0 \le \theta < 2\pi$

25. In oblique triangle ABC, $A = 12°$, $B = 75°$, and $a = 20$. Find b to the nearest tenth.

Matrices and Determinants

Jaron Lanier, who first used the term "virtual reality," is chief scientist for the "tele-immersion" project, which explores the impact of massive bandwidth and computing power. Rectangular arrays of numbers, called *matrices*, play a central role in representing computer images and in the forthcoming technology of tele-immersion. In this chapter, we study matrices and their applications. We begin with solving linear systems using matrices, which leads to a discussion of how computers might unjam traffic and give us a gridlock-free future.

You are being drawn deeper into cyberspace, spending more time online each week. With constantly improving high-resolution images, cyberspace is reshaping your life by nourishing shared enthusiasms. The people who built your computer talk of "bandwidth out the wazoo" that will give you the visual experience, in high-definition 3-D format, of being in the same room with a person who is actually in another city.

SECTION 9.1   *Matrix Solutions to Linear Systems*

Objectives

1. Write the augmented matrix for a linear system.
2. Perform matrix row operations.
3. Use matrices and Gaussian elimination to solve systems.
4. Use matrices and Gauss-Jordan elimination to solve systems.

Yes, we overindulged, but it was delicious. Anyway, a few hours of moderate activity and we'll just burn off those extra calories. The following chart should help. We see that the number of calories burned per hour depends on our weight. Four hours of tennis and we'll be as good as new!

How Fast You Burn Off Calories

	Weight (pounds)					
	110	132	154	176	187	209
Activity	**Calories Burned per Hour for a Given Weight**					
Housework	175	210	245	285	300	320
Cycling	190	215	245	270	280	295
Tennis	335	380	425	470	495	520
Watching TV	60	70	80	85	90	95

The 24 numbers inside the red brackets are arranged in four rows and six columns. This rectangular array of 24 numbers, arranged in rows and columns and placed in brackets, is an example of a **matrix** (plural: **matrices**). The numbers inside the brackets are called **elements** of the matrix. Matrices are used to display information and to solve systems of linear equations. Because systems involving two equations in two variables can easily be solved by substitution or addition, we will focus on matrix solutions to systems of linear equations in three or more variables.

1 Write the augmented matrix for a linear system.

Solving Linear Systems by Using Matrices

A matrix gives us a shortened way of writing a system of equations. The first step in solving a system of linear equations using matrices is to write the augmented matrix. An **augmented matrix** has a vertical bar separating the columns of the matrix into two groups. The coefficients of each variable are placed to the left of the vertical line, and the constants are placed to the right. If any variable is missing, its coefficient is 0. On the next page are two examples.

System of Linear Equations	Augmented Matrix

$$
\begin{array}{r}
3x + y + 2z = 31 \\
x + y + 2z = 19 \\
x + 3y + 2z = 25
\end{array}
\qquad
\left[\begin{array}{ccc|c}
3 & 1 & 2 & 31 \\
1 & 1 & 2 & 19 \\
1 & 3 & 2 & 25
\end{array}\right]
$$

$$
\begin{array}{r}
x + 2y - 5z = -19 \\
y + 3z = 9 \\
z = 4
\end{array}
\qquad
\left[\begin{array}{ccc|c}
1 & 2 & -5 & -19 \\
0 & 1 & 3 & 9 \\
0 & 0 & 1 & 4
\end{array}\right]
$$

Notice how the second matrix contains 1s down the diagonal from upper left to lower right and 0s below the 1s. This arrangement makes it easy to find the solution of the system of equations, as Example 1 shows.

EXAMPLE 1   Solving a System Using a Matrix

Write the solution set for a system of equations represented by the matrix

$$
\left[\begin{array}{ccc|c}
1 & 2 & -5 & -19 \\
0 & 1 & 3 & 9 \\
0 & 0 & 1 & 4
\end{array}\right].
$$

Solution   The system represented by the given matrix is

$$
\left[\begin{array}{ccc|c}
1 & 2 & -5 & -19 \\
0 & 1 & 3 & 9 \\
0 & 0 & 1 & 4
\end{array}\right]
\rightarrow
\begin{array}{r}
1x + 2y - 5z = -19 \\
0x + 1y + 3z = 9 \\
0x + 0y + 1z = 4
\end{array}.
$$

This system can be simplified as follows.

$$
\begin{array}{rl}
x + 2y - 5z = -19 & \text{Equation 1} \\
y + 3z = 9 & \text{Equation 2} \\
z = 4 & \text{Equation 3}
\end{array}
$$

The value of z is known. We can find y by back-substitution.

$$
\begin{array}{rl}
y + 3z = 9 & \text{Equation 2} \\
y + 3(4) = 9 & \text{Substitute 4 for } z. \\
y + 12 = 9 & \text{Multiply.} \\
y = -3 & \text{Subtract 12 from both sides.}
\end{array}
$$

With values for y and z, we can now use back-substitution to find x.

$$
\begin{array}{rl}
x + 2y - 5z = -19 & \text{Equation 1} \\
x + 2(-3) - 5(4) = -19 & \text{Substitute } -3 \text{ for } y \text{ and 4 for } z. \\
x - 6 - 20 = -19 & \text{Multiply.} \\
x - 26 = -19 & \text{Add.} \\
x = 7 & \text{Add 26 to both sides.}
\end{array}
$$

We see that $x = 7$, $y = -3$, and $z = 4$. The solution set for the system is $\{(7, -3, 4)\}$.

Check Point 1 Write the solution set for a system of equations represented by the matrix

$$\begin{bmatrix} 1 & -1 & 1 & | & 8 \\ 0 & 1 & -12 & | & -15 \\ 0 & 0 & 1 & | & 1 \end{bmatrix}.$$

Our goal in solving a system of linear equations in three variables using matrices is to produce a matrix similar to the one in Example 1. In general, the matrix will be of the form

$$\begin{bmatrix} 1 & a & b & | & c \\ 0 & 1 & d & | & e \\ 0 & 0 & 1 & | & f \end{bmatrix}$$

where a through f represent real numbers. The third row of this matrix gives us the value of one variable. The other variables can then be found by back-substitution.

2 Perform matrix row operations.

A matrix with 1s down the diagonal from upper left to lower right and 0s below the 1s is said to be in **row-echelon form.** How do we produce a matrix in this form? We use **row operations** on the augmented matrix. These row operations are just like what you did when solving a linear system by the addition method. The difference is that we no longer write the variables, usually represented by x, y, and z.

Matrix Row Operations

These row operations produce matrices that lead to systems with the same solution set as the original system.

1. Two rows of a matrix may be interchanged. This is the same as interchanging two equations in the linear system.
2. The elements in any row may be multiplied by a nonzero number. This is the same as multiplying both sides of an equation by a nonzero number.
3. The elements in any row may be multiplied by a nonzero number, and these products may be added to the corresponding elements in any other row. This is the same as multiplying both sides of an equation by a nonzero number and then adding equations to eliminate a variable.

Two matrices are **row equivalent** if one can be obtained from the other by a sequence of row operations.

Study Tip

When performing the row operation

$$k R_i + R_j$$

we use row i to find the products. However, **elements in row i do not change. It is the elements in row j that change:** Add k times the elements in row i to the corresponding elements in row j. Replace elements in row j by these sums.

Each matrix row operation in the preceding box can be expressed symbolically as follows:

1. Interchange the elements in the ith and jth rows: $R_i \leftrightarrow R_j$.
2. Multiply each element in the ith row by k: $k R_i$.
3. Add k times the elements in row i to the corresponding elements in row j: $k R_i + R_j$.

EXAMPLE 2 Performing Matrix Row Operations

Use the matrix

$$\begin{bmatrix} 3 & 18 & -12 & | & 21 \\ 1 & 2 & -3 & | & 5 \\ -2 & -3 & 4 & | & -6 \end{bmatrix}$$

Study Tip

$$\begin{bmatrix} 3 & 18 & -12 & | & 21 \\ 1 & 2 & -3 & | & 5 \\ -2 & -3 & 4 & | & -6 \end{bmatrix}$$

and perform each indicated row operation:

a. $R_1 \leftrightarrow R_2$ **b.** $\frac{1}{3}R_1$ **c.** $2R_2 + R_3$.

Solution

a. The notation $R_1 \leftrightarrow R_2$ means to interchange the elements in row 1 and row 2. This results in the row-equivalent matrix

$$\begin{bmatrix} 1 & 2 & -3 & | & 5 \\ 3 & 18 & -12 & | & 21 \\ -2 & -3 & 4 & | & -6 \end{bmatrix}.$$

This was row 2; now it's row 1.
This was row 1; now it's row 2.

b. The notation $\frac{1}{3}R_1$ means to multiply each element in row 1 by $\frac{1}{3}$. This results in the row-equivalent matrix

$$\begin{bmatrix} \frac{1}{3}(3) & \frac{1}{3}(18) & \frac{1}{3}(-12) & | & \frac{1}{3}(21) \\ 1 & 2 & -3 & | & 5 \\ -2 & -3 & 4 & | & -6 \end{bmatrix} = \begin{bmatrix} 1 & 6 & -4 & | & 7 \\ 1 & 2 & -3 & | & 5 \\ -2 & -3 & 4 & | & -6 \end{bmatrix}.$$

c. The notation $2R_2 + R_3$ means to add 2 times the elements in row 2 to the corresponding elements in row 3. Replace the elements in row 3 by these sums. First, we find 2 times the elements in row 2:

$$2(1) \text{ or } 2, \qquad 2(2) \text{ or } 4, \qquad 2(-3) \text{ or } -6, \qquad 2(5) \text{ or } 10.$$

Now we add these products to the corresponding elements in row 3. Although we use row 2 to find the products, row 2 does not change. It is the elements in row 3 that change, resulting in the row-equivalent matrix

$$\begin{bmatrix} 3 & 18 & -12 & | & 21 \\ 1 & 2 & -3 & | & 5 \\ -2+2=0 & -3+4=1 & 4+(-6)=-2 & | & -6+10=4 \end{bmatrix}$$

$$= \begin{bmatrix} 3 & 18 & -12 & | & 21 \\ 1 & 2 & -3 & | & 5 \\ 0 & 1 & -2 & | & 4 \end{bmatrix}.$$

Check Point 2   Use the matrix

$$\begin{bmatrix} 4 & 12 & -20 & | & 8 \\ 1 & 6 & -3 & | & 7 \\ -3 & -2 & 1 & | & -9 \end{bmatrix}$$

and perform each indicated row operation:

a. $R_1 \leftrightarrow R_2$ **b.** $\frac{1}{4}R_1$ **c.** $3R_2 + R_3$.

3 Use matrices and Gaussian elimination to solve systems.

The process that we use to solve linear systems using matrix row operations is called **Gaussian elimination,** after the German mathematician Carl Friedrich Gauss (1777–1855). Here are the steps used in Gaussian elimination:

Solving Linear Systems Using Gaussian Elimination

1. Write the augmented matrix for the system.

2. Use matrix row operations to simplify the matrix to one with 1s down the diagonal from upper left to lower right, and 0s below the 1s.

$$\begin{bmatrix} 1 & * & * & | & * \\ * & * & * & | & * \\ * & * & * & | & * \end{bmatrix} \rightarrow \begin{bmatrix} 1 & * & * & | & * \\ 0 & * & * & | & * \\ 0 & * & * & | & * \end{bmatrix} \rightarrow \begin{bmatrix} 1 & * & * & | & * \\ 0 & 1 & * & | & * \\ 0 & * & * & | & * \end{bmatrix} \rightarrow \begin{bmatrix} 1 & * & * & | & * \\ 0 & 1 & * & | & * \\ 0 & 0 & * & | & * \end{bmatrix} \rightarrow \begin{bmatrix} 1 & * & * & | & * \\ 0 & 1 & * & | & * \\ 0 & 0 & 1 & | & * \end{bmatrix}$$

Get 1 in the upper left-hand corner.

Use the 1 in the first column to get 0s below it.

Get 1 in the second row, second column position.

Use the 1 in the second column to get 0 below it.

Get 1 in the third row, third column position.

3. Write the system of linear equations corresponding to the matrix in step 2, and use back-substitution to find the system's solution.

EXAMPLE 3   Gaussian Elimination with Back-Substitution

Use matrices to solve the system:

$$3x + y + 2z = 31$$
$$x + y + 2z = 19$$
$$x + 3y + 2z = 25.$$

Solution

Step 1   Write the augmented matrix for the system.

Linear System	Augmented Matrix
$3x + y + 2z = 31$	
$x + y + 2z = 19$	
$x + 3y + 2z = 25$	

$$\begin{bmatrix} 3 & 1 & 2 & | & 31 \\ 1 & 1 & 2 & | & 19 \\ 1 & 3 & 2 & | & 25 \end{bmatrix}$$

Step 2   Use matrix row operations to simplify the matrix to one with 1s down the diagonal from upper left to lower right, and 0s below the 1s. Our goal is to obtain a matrix of the form

$$\begin{bmatrix} 1 & a & b & | & c \\ 0 & 1 & d & | & e \\ 0 & 0 & 1 & | & f \end{bmatrix}.$$

Our first step in achieving this goal is to get 1 in the top position of the first column.

We want 1 in this position.

$$\begin{bmatrix} 3 & 1 & 2 & | & 31 \\ 1 & 1 & 2 & | & 19 \\ 1 & 3 & 2 & | & 25 \end{bmatrix}$$

To get 1 in this position, we interchange rows 1 and 2. $R_1 \leftrightarrow R_2$. (We could also interchange rows 1 and 3 to attain our goal.)

$$\begin{bmatrix} 1 & 1 & 2 & | & 19 \\ 3 & 1 & 2 & | & 31 \\ 1 & 3 & 2 & | & 25 \end{bmatrix}$$

This was row 2; now it's row 1.

This was row 1; now it's row 2.

Now we want to get 0s below the 1 in the first column.

$$\begin{bmatrix} 1 & 1 & 2 & | & 19 \\ 3 & 1 & 2 & | & 31 \\ 1 & 3 & 2 & | & 25 \end{bmatrix}$$

We want 0 in these positions.

To get a 0 where there is now a 3, multiply the top row of numbers by -3 and add these products to the second row of numbers: $-3R_1 + R_2$. To get a 0 where there is now a 1, multiply the top row of numbers multiplied by -1 and add these products to the third row of numbers: $-1R_1 + R_3$. Although we are using row 1 to find the products, the numbers in row 1 do not change.

$$\begin{matrix} \\ -3R_1 + R_2 \\ \\ -1R_1 + R_3 \end{matrix} \begin{bmatrix} 1 & 1 & 2 & | & 19 \\ -3(1) + 3 & -3(1) + 1 & -3(2) + 2 & | & -3(19) + 31 \\ -1(1) + 1 & -1(1) + 3 & -1(2) + 2 & | & -1(19) + 25 \end{bmatrix}$$

$$= \begin{bmatrix} 1 & 1 & 2 & | & 19 \\ 0 & -2 & -4 & | & -26 \\ 0 & 2 & 0 & | & 6 \end{bmatrix}$$

We move on to the second column. We want 1 in the second row, second column.

We want 1 in this position.

$$\begin{bmatrix} 1 & 1 & 2 & | & 19 \\ 0 & -2 & -4 & | & -26 \\ 0 & 2 & 0 & | & 6 \end{bmatrix}$$

To get 1 in the desired position, we multiply -2 by its reciprocal, $-\frac{1}{2}$. Therefore, we multiply all the numbers in the second row by $-\frac{1}{2}$: $-\frac{1}{2} R_2$.

$$-\frac{1}{2} R_2 \begin{bmatrix} 1 & 1 & 2 & | & 19 \\ -\frac{1}{2}(0) & -\frac{1}{2}(-2) & -\frac{1}{2}(-4) & | & -\frac{1}{2}(-26) \\ 0 & 2 & 0 & | & 6 \end{bmatrix} = \begin{bmatrix} 1 & 1 & 2 & | & 19 \\ 0 & 1 & 2 & | & 13 \\ 0 & 2 & 0 & | & 6 \end{bmatrix}.$$

We want 0 in this position.

We are not yet done with the second column. The voice balloon shows that we want to get a 0 where there is now a 2. If we multiply the second row of numbers by -2 and add these products to the third row of numbers, we will get 0 in this position: $-2R_2 + R_3$. Although we are using the numbers in row 2 to find the products, the numbers in row 2 do not change.

$$-2R_2 + R_3 \begin{bmatrix} 1 & 1 & 2 & | & 19 \\ 0 & 1 & 2 & | & 13 \\ -2(0) + 0 & -2(1) + 2 & -2(2) + 0 & | & -2(13) + 6 \end{bmatrix} = \begin{bmatrix} 1 & 1 & 2 & | & 19 \\ 0 & 1 & 2 & | & 13 \\ 0 & 0 & -4 & | & -20 \end{bmatrix}$$

We move on to the third column. We want 1 in the third row, third column.

We want 1 in this position.

$$\begin{bmatrix} 1 & 1 & 2 & | & 19 \\ 0 & 1 & 2 & | & 13 \\ 0 & 0 & -4 & | & -20 \end{bmatrix}$$

Technology

Most graphing utilities can convert an augmented matrix to one with 1s down the diagonal from upper left to lower right, and 0s below the 1s. Recall that this is called row-echelon form. However, row-echelon form is not unique. Your graphing utility might give a row-echelon form different from the one you obtained by hand. However, all row-echelon forms for a given system's augmented matrix produce the same solution to the system. Enter the augmented matrix and name it A. Then use the $\boxed{\text{REF}}$ (row-echelon form) command on matrix A.

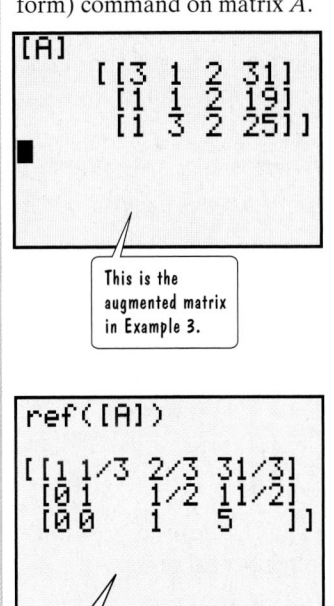

This is the augmented matrix in Example 3.

This matrix is equivalent to the row-echelon form we obtained by hand. It produces the same solution: $z = 5$, $y = 3$, and $x = 6$.

To get 1 in the desired position, we multiply -4 by its reciprocal, $-\frac{1}{4}$. Therefore, we multiply all the numbers in the third row by $-\frac{1}{4}$: $-\frac{1}{4} R_3$.0

$$-\frac{1}{4} R_3 \begin{bmatrix} 1 & 1 & 2 & | & 19 \\ 0 & 1 & 2 & | & 13 \\ -\frac{1}{4}(0) & -\frac{1}{4}(0) & -\frac{1}{4}(-4) & | & -\frac{1}{4}(-20) \end{bmatrix} = \begin{bmatrix} 1 & 1 & 2 & | & 19 \\ 0 & 1 & 2 & | & 13 \\ 0 & 0 & 1 & | & 5 \end{bmatrix}$$

We now have the desired matrix with 1s down the diagonal and 0s below the 1s.

Step 3 **Write the system of linear equations corresponding to the matrix in step 2, and use back-substitution to find the system's solution.** The system represented by the matrix in step 2 is

$$\begin{bmatrix} 1 & 1 & 2 & | & 19 \\ 0 & 1 & 2 & | & 13 \\ 0 & 0 & 1 & | & 5 \end{bmatrix} \rightarrow \begin{array}{r} 1x + 1y + 2z = 19 \\ 0x + 1y + 2z = 13 \\ 0x + 0y + 1z = 5 \end{array} \quad \text{or} \quad \begin{array}{r} x + y + 2z = 19 \\ y + 2z = 13. \\ z = 5 \end{array}$$

We immediately see that the value for z is 5. To find y, we back-substitute 5 for z in the second equation.

$$y + 2z = 13 \qquad \text{Equation 2}$$
$$y + 2(5) = 13 \qquad \text{Substitute 5 for z.}$$
$$y = 3 \qquad \text{Solve for y.}$$

Finally, back-substitute 3 for y and 5 for z in the first equation:

$$x + y + 2z = 19 \qquad \text{Equation 1}$$
$$x + 3 + 2(5) = 19 \qquad \text{Substitute 3 for y and 5 for z.}$$
$$x + 13 = 19 \qquad \text{Multiply and add.}$$
$$x = 6 \qquad \text{Subtract 13 from both sides.}$$

The solution set for the original system is $\{(6, 3, 5)\}$. Check to see that the solution satisfies all three equations in the given system.

Check Point 3 Use matrices to solve the system:

$$\begin{array}{r} 2x + y + 2z = 18 \\ x - y + 2z = 9 \\ x + 2y - z = 6. \end{array}$$

Modern supercomputers are capable of solving systems with more than 600,000 variables. The augmented matrices for such systems are huge, but the solution using matrices is exactly like what we did in Example 3. Work with the augmented matrix, one column at a time. First, get 1 in the desired position. Then get 0s below the 1. Let's see how this works for a linear system involving four equations in four variables.

EXAMPLE 4   Gaussian Elimination with Back-Substitution

Use matrices to solve the system:

$$
\begin{aligned}
2w + x + 3y - z &= 6 \\
w - x + 2y - 2z &= -1 \\
w - x - y + z &= -4 \\
-w + 2x - 2y - z &= -7.
\end{aligned}
$$

Solution

Step 1   Write the augmented matrix for the system.

Linear System	Augmented Matrix

$$
\begin{aligned}
2w + x + 3y - z &= 6 \\
w - x + 2y - 2z &= -1 \\
w - x - y + z &= -4 \\
-w + 2x - 2y - z &= -7
\end{aligned}
\qquad
\left[\begin{array}{cccc|c}
2 & 1 & 3 & -1 & 6 \\
1 & -1 & 2 & -2 & -1 \\
1 & -1 & -1 & 1 & -4 \\
-1 & 2 & -2 & -1 & -7
\end{array}\right]
$$

Step 2   Use matrix row operations to simplify the matrix to one with 1s down the diagonal from upper left to lower right, and 0s below the 1s. Working one column at a time, we must obtain 1 in the diagonal position. Then we use this 1 to get 0s below it. Thus, our first step in achieving this goal is to get 1 in the top position of the first column. To do this, we interchange rows 1 and 2: $R_1 \leftrightarrow R_2$.

We want 0s in these positions.

$$
\left[\begin{array}{cccc|c}
1 & -1 & 2 & -2 & -1 \\
2 & 1 & 3 & -1 & 6 \\
1 & -1 & -1 & 1 & -4 \\
-1 & 2 & -2 & -1 & -7
\end{array}\right]
$$

This was row 2; now it's row 1.
This was row 1; now it's row 2.

Now we use the 1 at the top of the first column to get 0s below it.

Use the previous matrix and:
Replace row 2 by $-2R_1 + R_2$.
Replace row 3 by $-1R_1 + R_3$.
Replace row 4 by $1R_1 + R_4$.

$$
\left[\begin{array}{cccc|c}
1 & -1 & 2 & -2 & -1 \\
0 & 3 & -1 & 3 & 8 \\
0 & 0 & -3 & 3 & -3 \\
0 & 1 & 0 & -3 & -8
\end{array}\right]
$$

We want 1 in this position.

We move on to the second column. We can obtain 1 in the desired position by multiplying the numbers in the second row by $\frac{1}{3}$, the reciprocal of 3.

$$
\left[\begin{array}{cccc|c}
1 & -1 & 2 & -2 & -1 \\
\frac{1}{3}(0) & \frac{1}{3}(3) & \frac{1}{3}(-1) & \frac{1}{3}(3) & \frac{1}{3}(8) \\
0 & 0 & -3 & 3 & -3 \\
0 & 1 & 0 & -3 & -8
\end{array}\right]
=
\left[\begin{array}{cccc|c}
1 & -1 & 2 & -2 & -1 \\
0 & 1 & -\frac{1}{3} & 1 & \frac{8}{3} \\
0 & 0 & -3 & 3 & -3 \\
0 & 1 & 0 & -3 & -8
\end{array}\right]
\quad \frac{1}{3}R_2
$$

We want 0s in these positions. The top position already has a 0.

Now we use the 1 in the second row, second column position to get 0s below it.

Replace row 4 in the previous matrix by $-1R_2 + R_4$.

$$
\left[\begin{array}{cccc|c}
1 & -1 & 2 & -2 & -1 \\
0 & 1 & -\frac{1}{3} & 1 & \frac{8}{3} \\
0 & 0 & -3 & 3 & -3 \\
0 & 0 & \frac{1}{3} & -4 & -\frac{32}{3}
\end{array}\right]
$$

We want 1 in this position.

We move on to the third column. We can obtain 1 in the desired position by multiplying the numbers in the third row by $-\frac{1}{3}$, the reciprocal of -3.

$$
\begin{bmatrix}
1 & -1 & 2 & -2 & \bigm| & -1 \\
0 & 1 & -\frac{1}{3} & 1 & \bigm| & \frac{8}{3} \\
-\frac{1}{3}(0) & -\frac{1}{3}(0) & -\frac{1}{3}(-3) & -\frac{1}{3}(3) & \bigm| & -\frac{1}{3}(-3) \\
0 & 0 & \frac{1}{3} & -4 & \bigm| & -\frac{32}{3}
\end{bmatrix}
=
\begin{bmatrix}
1 & -1 & 2 & -2 & \bigm| & -1 \\
0 & 1 & -\frac{1}{3} & 1 & \bigm| & \frac{8}{3} \\
0 & 0 & 1 & -1 & \bigm| & 1 \\
0 & 0 & \frac{1}{3} & -4 & \bigm| & -\frac{32}{3}
\end{bmatrix}
\quad -\frac{1}{3}R_3
$$

We want 0 in this position.

Now we use the 1 in the third column to get 0 below it.

Replace row 4 in the previous matrix by $-\frac{1}{3}R_3 + R_4$.

$$
\begin{bmatrix}
1 & -1 & 2 & -2 & \bigm| & -1 \\
0 & 1 & -\frac{1}{3} & 1 & \bigm| & \frac{8}{3} \\
0 & 0 & 1 & -1 & \bigm| & 1 \\
0 & 0 & 0 & -\frac{11}{3} & \bigm| & -11
\end{bmatrix}
$$

We want 1 in this position.

We move on to the fourth column. Because we want 1s down the diagonal from upper left to lower right, we want 1 where there is now $-\frac{11}{3}$. We can obtain 1 in this position by multiplying the numbers in the fourth row by $-\frac{3}{11}$.

$$
\begin{bmatrix}
1 & -1 & 2 & -2 & \bigm| & -1 \\
0 & 1 & -\frac{1}{3} & 1 & \bigm| & \frac{8}{3} \\
0 & 0 & 1 & -1 & \bigm| & 1 \\
-\frac{3}{11}(0) & -\frac{3}{11}(0) & -\frac{3}{11}(0) & -\frac{3}{11}\left(-\frac{11}{3}\right) & \bigm| & -\frac{3}{11}(-11)
\end{bmatrix}
$$

$$
=
\begin{bmatrix}
1 & -1 & 2 & -2 & \bigm| & -1 \\
0 & 1 & -\frac{1}{3} & 1 & \bigm| & \frac{8}{3} \\
0 & 0 & 1 & -1 & \bigm| & 1 \\
0 & 0 & 0 & 1 & \bigm| & 3
\end{bmatrix}
\quad -\frac{3}{11}R_4
$$

We now have the desired matrix in row-echelon form, with 1s down the diagonal and 0s below the 1s. An equivalent row-echelon matrix can be obtained using a graphing utility and the REF command on the augmented matrix.

Step 3 **Write the system of linear equations corresponding to the matrix in step 2, and use back-substitution to find the system's solution.** The system represented by the matrix in step 2 is

$$
\begin{bmatrix}
1 & -1 & 2 & -2 & \bigm| & -1 \\
0 & 1 & -\frac{1}{3} & 1 & \bigm| & \frac{8}{3} \\
0 & 0 & 1 & -1 & \bigm| & 1 \\
0 & 0 & 0 & 1 & \bigm| & 3
\end{bmatrix}
\rightarrow
\begin{aligned}
1w - 1x + 2y - 2z &= -1 \\
0w + 1x - \tfrac{1}{3}y + 1z &= \tfrac{8}{3} \\
0w + 0x + 1y - 1z &= 1 \\
0w + 0x + 0y + 1z &= 3
\end{aligned}
\quad \text{or} \quad
\begin{aligned}
w - x + 2y - 2z &= -1 \\
x - \tfrac{1}{3}y + z &= \tfrac{8}{3} \\
y - z &= 1 \\
z &= 3.
\end{aligned}
$$

We immediately see that the value for z is 3. We can now use back-substitution to find the values for y, x, and w.

778 • Chapter 9 • Matrices and Determinants

These are the four equations from the last column on the previous page.

$$z = 3 \quad\bigg|\quad y - z = 1 \quad\bigg|\quad x - \frac{1}{3}y + z = \frac{8}{3} \quad\bigg|\quad w - x + 2y - 2z = -1$$

$$y - 3 = 1 \quad x - \frac{1}{3}(4) + 3 = \frac{8}{3} \quad w - 1 + 2(4) - 2(3) = -1$$

$$y = 4 \quad x + \frac{5}{3} = \frac{8}{3} \quad w - 1 + 8 - 6 = -1$$

$$x = 1 \quad w + 1 = -1$$

$$w = -2$$

Let's agree to write the solution set for the system in the alphabetical order in which the variables for the given system appeared from left to right, namely (w, x, y, z). Thus, the solution set is $\{(-2, 1, 4, 3)\}$. We can verify this solution set by substituting the value for each variable into the original system of equations.

Check Point 4 Use matrices to solve the system:

$$w - 3x - 2y + z = -3$$
$$2w - 7x - y + 2z = 1$$
$$3w - 7x - 3y + 3z = -5$$
$$5w + x + 4y - 2z = 18.$$

4 Use matrices and Gauss-Jordan elimination to solve systems.

Gauss-Jordan Elimination

Using Gaussian elimination, we obtain a matrix in row-echelon form, with 1s down the diagonal from upper left to lower right and 0s below the 1s. A second method, called **Gauss-Jordan elimination,** after Carl Friedrich Gauss and Wilhelm Jordan (1842–1899), continues the process until a matrix with 1s down the diagonal from upper left to lower right and 0s in every position *above and below* each 1 is found. Such a matrix is said to be in **reduced row-echelon form.** For a system of linear equations in three variables, x, y, and z, we must get the augmented matrix into the form

$$\begin{bmatrix} 1 & 0 & 0 & | & a \\ 0 & 1 & 0 & | & b \\ 0 & 0 & 1 & | & c \end{bmatrix}.$$

Based on this matrix, we conclude that $x = a$, $y = b$, and $z = c$.

Solving Linear Systems Using Gauss-Jordan Elimination

1. Write the augmented matrix for the system.
2. Use matrix row operations to simplify the matrix to one with 1s down the diagonal from upper left to lower right, and 0s above and below the 1s.
   a. Get 1 in the upper left-hand corner.
   b. Use the 1 in the first column to get 0s below it.
   c. Get 1 in the second row, second column.
   d. Use the 1 in the second column to make the remaining entries in the second column 0.
   e. Get 1 in the third row, third column.
   f. Use the 1 in the third column to make the remaining entries in the third column 0.
   g. Continue this procedure as far as possible.
3. Use the reduced row-echelon form of the matrix in step 2 to write the system's solution set. (Back-substitution is not necessary.)

Technology

Most graphing utilities can convert a matrix to reduced row-echelon form. Enter the system's augmented matrix and name it A. Then use the RREF (reduced row-echelon form) command on matrix A.

This is the augmented matrix for the system in Example 5.

This is the matrix in reduced row-echelon form we obtained in Example 5.

EXAMPLE 5 Using Gauss-Jordan Elimination

Use Gauss-Jordan elimination to solve the system:

$$3x + y + 2z = 31$$
$$x + y + 2z = 19$$
$$x + 3y + 2z = 25.$$

Solution   In Example 3, we used Gaussian elimination to obtain the following matrix:

$$\begin{bmatrix} 1 & 1 & 2 & | & 19 \\ 0 & 1 & 2 & | & 13 \\ 0 & 0 & 1 & | & 5 \end{bmatrix}.$$

To use Gauss-Jordan elimination, we need 0s both below and above the 1s in the diagonal position. We use the 1 in the second row, second column to get a 0 above it.

Replace row 1 in the previous matrix by $-1R_2 + R_1$

$$\begin{bmatrix} 1 & 0 & 0 & | & 6 \\ 0 & 1 & 2 & | & 13 \\ 0 & 0 & 1 & | & 5 \end{bmatrix}$$

We want 0s in these positions. The top position already has a 0.

We use the 1 in the third column to get 0s above it.

$$\begin{bmatrix} 1 & 0 & 0 & | & 6 \\ 0 & 1 & 0 & | & 3 \\ 0 & 0 & 1 & | & 5 \end{bmatrix}$$

Replace row 2 in the previous matrix by $-2R_3 + R_2$.

This last matrix corresponds to

$$x = 6, \quad y = 3, \quad z = 5.$$

As we found in Example 3, the solution set is $\{(6, 3, 5)\}$.

Check Point 5   Solve the system in Check Point 3 using Gauss-Jordan elimination. Begin by working with the matrix that you obtained in Check Point 3.

EXERCISE SET 9.1

Practice Exercises

In Exercises 1–8, write the augmented matrix for each system of linear equations.

1. $2x + y + 2z = 2$
$3x - 5y - z = 4$
$x - 2y - 3z = -6$

2. $3x - 2y + 5z = 31$
$x + 3y - 3z = -12$
$-2x - 5y + 3z = 11$

3. $x - y + z = 8$
$y - 12z = -15$
$z = 1$

4. $x - 2y + 3z = 9$
$y + 3z = 5$
$z = 2$

5. $5x - 2y - 3z = 0$
$x + y = 5$
$2x - 3z = 4$

6. $x - 2y + z = 10$
$3x + y = 5$
$7x + 2z = 2$

7. $2w + 5x - 3y + z = 2$
$3x + y = 4$
$w - x + 5y = 9$
$5w - 5x - 2y = 1$

8. $4w + 7x - 8y + z = 3$
$5x + y = 5$
$w - x - y = 17$
$2w - 2x + 11y = 4$

In Exercises 9–12, write the system of linear equations represented by the augmented matrix. Use x, y, z, and, if necessary, w, x, y, and z, for the variables.

9. $\begin{bmatrix} 5 & 0 & 3 & | & -11 \\ 0 & 1 & -4 & | & 12 \\ 7 & 2 & 0 & | & 3 \end{bmatrix}$ **10.** $\begin{bmatrix} 7 & 0 & 4 & | & -13 \\ 0 & 1 & -5 & | & 11 \\ 2 & 7 & 0 & | & 6 \end{bmatrix}$

11. $\begin{bmatrix} 1 & 1 & 4 & 1 & | & 3 \\ -1 & 1 & -1 & 0 & | & 7 \\ 2 & 0 & 0 & 5 & | & 11 \\ 0 & 0 & 12 & 4 & | & 5 \end{bmatrix}$ **12.** $\begin{bmatrix} 4 & 1 & 5 & 1 & | & 6 \\ 1 & -1 & 0 & -1 & | & 8 \\ 3 & 0 & 0 & 7 & | & 4 \\ 0 & 0 & 11 & 5 & | & 3 \end{bmatrix}$

In Exercises 13–18, write the system of linear equations represented by the augmented matrix. Use x, y, z, and, if necessary, w, x, y, and z, for the variables. Once the system is written, use back-substitution to find its solution.

13. $\begin{bmatrix} 1 & 0 & -4 & | & 5 \\ 0 & 1 & -12 & | & 13 \\ 0 & 0 & 1 & | & -\frac{1}{2} \end{bmatrix}$ **14.** $\begin{bmatrix} 1 & 2 & 1 & | & 0 \\ 0 & 1 & 0 & | & -2 \\ 0 & 0 & 1 & | & 3 \end{bmatrix}$

15. $\begin{bmatrix} 1 & \frac{1}{2} & 1 & | & \frac{11}{2} \\ 0 & 1 & \frac{3}{2} & | & 7 \\ 0 & 0 & 1 & | & 4 \end{bmatrix}$ **16.** $\begin{bmatrix} 1 & 1 & 0 & | & 3 \\ 0 & 1 & \frac{3}{2} & | & -2 \\ 0 & 0 & 1 & | & 0 \end{bmatrix}$

17. $\begin{bmatrix} 1 & -1 & 1 & 1 & | & 3 \\ 0 & 1 & -2 & -1 & | & 0 \\ 0 & 0 & 1 & 6 & | & 17 \\ 0 & 0 & 0 & 1 & | & 3 \end{bmatrix}$ **18.** $\begin{bmatrix} 1 & 2 & -1 & 0 & | & 2 \\ 0 & 1 & 1 & -2 & | & -3 \\ 0 & 0 & 1 & -1 & | & -2 \\ 0 & 0 & 0 & 1 & | & 3 \end{bmatrix}$

In Exercises 19–24, perform each matrix row operation and write the new matrix.

19. $\begin{bmatrix} 2 & -6 & 4 & | & 10 \\ 1 & 5 & -5 & | & 0 \\ 3 & 0 & 4 & | & 7 \end{bmatrix}$ $\frac{1}{2}R_1$

20. $\begin{bmatrix} 3 & -12 & 6 & | & 9 \\ 1 & -4 & 4 & | & 0 \\ 2 & 0 & 7 & | & 4 \end{bmatrix}$ $\frac{1}{3}R_1$

21. $\begin{bmatrix} 1 & -3 & 2 & | & 0 \\ 3 & 1 & -1 & | & 7 \\ 2 & -2 & 1 & | & 3 \end{bmatrix}$ $-3R_1 + R_2$

22. $\begin{bmatrix} 1 & -1 & 5 & | & -6 \\ 3 & 3 & -1 & | & 10 \\ 1 & 3 & 2 & | & 5 \end{bmatrix}$ $-3R_1 + R_2$

23. $\begin{bmatrix} 1 & -1 & 1 & 1 & | & 3 \\ 0 & 1 & -2 & -1 & | & 0 \\ 2 & 0 & 3 & 4 & | & 11 \\ 5 & 1 & 2 & 4 & | & 6 \end{bmatrix}$ $\begin{matrix} \\ \\ -2R_1 + R_3 \\ -5R_1 + R_4 \end{matrix}$

24. $\begin{bmatrix} 1 & -5 & 2 & -2 & | & 4 \\ 0 & 1 & -3 & -1 & | & 0 \\ 3 & 0 & 2 & -1 & | & 6 \\ -4 & 1 & 4 & 2 & | & -3 \end{bmatrix}$ $\begin{matrix} \\ \\ -3R_1 + R_3 \\ 4R_1 + R_4 \end{matrix}$

In Exercises 25–26, a few steps in the process of simplifying the given matrix to one with 1s down the diagonal from upper left to lower right, and 0s below the 1s, are shown. Fill in the missing numbers in the steps that are shown.

25. $\begin{bmatrix} 1 & -1 & 1 & | & 8 \\ 2 & 3 & -1 & | & -2 \\ 3 & -2 & -9 & | & 9 \end{bmatrix} \rightarrow \begin{bmatrix} 1 & -1 & 1 & | & 8 \\ 0 & 5 & \blacksquare & | & \blacksquare \\ 0 & 1 & \blacksquare & | & \blacksquare \end{bmatrix}$

$\rightarrow \begin{bmatrix} 1 & -1 & 1 & | & 8 \\ 0 & 1 & \blacksquare & | & \blacksquare \\ 0 & 1 & \blacksquare & | & \blacksquare \end{bmatrix}$

26. $\begin{bmatrix} 1 & -2 & 3 & | & 4 \\ 2 & 1 & -4 & | & 3 \\ -3 & 4 & -1 & | & -2 \end{bmatrix} \rightarrow \begin{bmatrix} 1 & -2 & 3 & | & 4 \\ 0 & 5 & \blacksquare & | & \blacksquare \\ 0 & -2 & \blacksquare & | & \blacksquare \end{bmatrix}$

$\rightarrow \begin{bmatrix} 1 & -2 & 3 & | & 4 \\ 0 & 1 & \blacksquare & | & \blacksquare \\ 0 & -2 & \blacksquare & | & \blacksquare \end{bmatrix}$

In Exercises 27–44, solve each system of equations using matrices. Use Gaussian elimination with back-substitution or Gauss-Jordan elimination.

27.
$x + y - z = -2$
$2x - y + z = 5$
$-x + 2y + 2z = 1$

28.
$x - 2y - z = 2$
$2x - y + z = 4$
$-x + y - 2z = -4$

29.
$x + 3y = 0$
$x + y + z = 1$
$3x - y - z = 11$

30.
$3y - z = -1$
$x + 5y - z = -4$
$-3x + 6y + 2z = 11$

31.
$2x - y - z = 4$
$x + y - 5z = -4$
$x - 2y = 4$

32.
$x - 3z = -2$
$2x + 2y + z = 4$
$3x + y - 2z = 5$

33.
$x + y + z = 4$
$x - y - z = 0$
$x - y + z = 2$

34.
$3x + y - z = 0$
$x + y + 2z = 6$
$2x + 2y + 3z = 10$

35.
$x + 2y = z - 1$
$x = 4 + y - z$
$x + y - 3z = -2$

36.
$2x + y = z + 1$
$2x = 1 + 3y - z$
$x + y + z = 4$

37.
$3a - b - 4c = 3$
$2a - b + 2c = -8$
$a + 2b - 3c = 9$

38.
$3a + b - c = 0$
$2a + 3b - 5c = 1$
$a - 2b + 3c = -4$

39.
$2x + 2y + 7z = -1$
$2x + y + 2z = 2$
$4x + 6y + z = 15$

40.
$3x + 2y + 3z = 3$
$4x - 5y + 7z = 1$
$2x + 3y - 2z = 6$

41.
$w + x + y + z = 4$
$2w + x - 2y - z = 0$
$w - 2x - y - 2z = -2$
$3w + 2x + y + 3z = 4$

42.
$w + x + y + z = 5$
$w + 2x - y - 2z = -1$
$w - 3x - 3y - z = -1$
$2w - x + 2y - z = -2$

43.
$3w - 4x + y + z = 9$
$w + x - y - z = 0$
$2w + x + 4y - 2z = 3$
$-w + 2x + y - 3z = 3$

44.
$2w + y - 3z = 8$
$w - x + 4z = -10$
$3w + 5x - y - z = 20$
$w + x - y - z = 6$

Application Exercises

45. The graph shows the alligator population, $P(x)$, in a national park after x years of a protection program. A quadratic function

$$P(x) = ax^2 + bx + c$$

can be used to model the data.

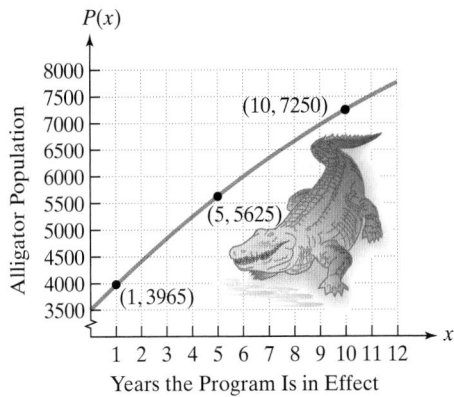

P(x)

8000
7500
7000 · (10, 7250)
6500
6000
5500
5000 (5, 5625)
4500
4000
3500 (1, 3965)

Alligator Population

x

1 2 3 4 5 6 7 8 9 10 11 12

Years the Program Is in Effect

a. Use the points shown in the graph to find the values of a, b, and c. Solve the system of linear equations involving a, b, and c using matrices.

b. Find and interpret $P(12)$. Identify your solution on the graph shown.

46. A football is kicked straight upward. A position function

$$s(t) = \tfrac{1}{2}at^2 + v_0 t + s_0$$

can be used to describe the ball's height, $s(t)$, in feet, after t seconds.

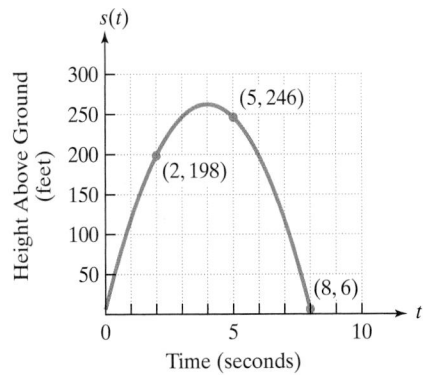

s(t)

300
250 (5, 246)
200
150 (2, 198)
100
50
(8, 6)

Height Above Ground (feet)

0   5   10   t

Time (seconds)

a. Use the points labeled in the graph to find the values of a, v_0, and s_0. Solve the system of linear equations involving a, v_0, and s_0 using matrices.

b. Find and interpret $s(7)$. Identify your solution on the graph shown.

Write a system of linear equations in three variables to solve Exercises 47–50. Then use matrices to solve the system.

Exercises 47–48 are based on a Time/CNN telephone poll that included never-married single women between the ages of 18 and 49 and never-married men between the ages of 18 and 49. The circle graphs show the results for one of the questions in the poll.

If You Couldn't Find the Perfect Mate, Would You Marry Someone Else?

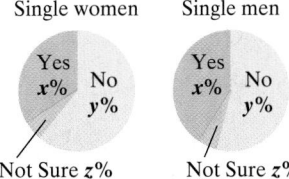

Single women   Single men

Yes No   Yes No
$x\%$ $y\%$ $x\%$ $y\%$

Not Sure $z\%$ Not Sure $z\%$

47. For single women in the poll, the percentage who said no exceeded the combined percentages for those who said yes and those who said not sure by 22%. If the percentage who said yes is doubled, it is 7% more than the percentage who said no. Find the percentage of single women who responded yes, no, and not sure.

48. For single men in the poll, the percentage who said no exceeded the combined percentages for those who said yes and those who said not sure by 8%. If the percentage who said yes is doubled, it is 28% more than the percentage who said no. Find the percentage of single men who responded yes, no, and not sure.

49. Three foods have the following nutritional content per ounce.

	Calories	Protein (in grams)	Vitamin C (in milligrams)
Food A	40	5	30
Food B	200	2	10
Food C	400	4	300

If a meal consisting of the three foods allows exactly 660 calories, 25 grams of protein, and 425 milligrams of vitamin C, how many ounces of each kind of food should be used?

50. A furniture company produces three types of desks: a children's model, an office model, and a deluxe model. Each desk is manufactured in three stages: cutting, construction, and finishing. The time requirements for each model and manufacturing stage are given in the following table.

	Children's model	Office model	Deluxe model
Cutting	2 hr	3 hr	2 hr
Construction	2 hr	1 hr	3 hr
Finishing	1 hr	1 hr	2 hr

Each week the company has available a maximum of 100 hours for cutting, 100 hours for construction, and 65 hours for finishing. If all available time must be used, how many of each type of desk should be produced each week?

Writing in Mathematics

51. What is a matrix?

52. Describe what is meant by the augmented matrix of a system of linear equations.

53. In your own words, describe each of the three matrix row operations. Give an example with each of the operations.

54. Describe how to use row operations and matrices to solve a system of linear equations.

55. What is the difference between Gaussian elimination and Gauss-Jordan elimination?

56. The graphs show the percentage of recorded music on CDs, cassettes, and LPs from 1981–2001. For this time period, which of these three forms of recorded music would you be most inclined to model using a quadratic function? Explain your answer.

Percentage of Recorded Music on CDs, Cassettes, and LPs

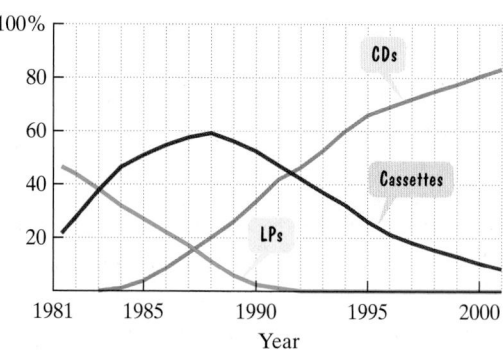

Year

Source: Recording Industry Association of America

57. In Exercise 56, assume that you plan to obtain the quadratic model by hand. Explain how to use the graph for the form that you selected to find a, b, and c in $y = ax^2 + bx + c$, where x represents the number of years after 1981 and y represents the percentage of recorded music on this form. Describe the role that matrices can play in the process of obtaining the model.

Technology Exercises

58. Most graphing utilities can perform row operations on matrices. Consult the owner's manual for your graphing utility to learn proper keystrokes for performing these operations. Then duplicate the row operations of any three exercises that you solved from Exercises 19–24.

59. If your graphing utility has a REF (row-echelon form) command or a RREF (reduced row-echelon form) command, use this feature to verify your work with any five systems from Exercises 27–44.

60. Solve using a graphing utility's REF or RREF command:

$$\begin{aligned} 2x_1 - 2x_2 + 3x_3 - x_4 \quad\quad &= 12 \\ x_1 + 2x_2 - x_3 + 2x_4 - x_5 &= -7 \\ x_1 + \quad\quad x_3 + x_4 - 5x_5 &= 1 \\ -x_1 + x_2 - x_3 - 2x_4 - 3x_5 &= 0 \\ x_1 - x_2 - \quad\quad x_4 + x_5 &= 4. \end{aligned}$$

Critical Thinking Exercises

61. Find a cubic function whose graph passes through the points $(0, -3)$, $(1, 5)$, $(-1, -7)$, and $(-2, -13)$. (*Hint:* Use the equation $y = ax^3 + bx^2 + cx + d$.)

62. The table shows the daily production level and profit for a business.

x (Number of units Produced Daily)	30	50	100
y (Daily Profit)	$5900	$7500	$4500

Use the quadratic function $y = ax^2 + bx + c$ to determine the number of units that should be produced each day for maximum profit. What is the maximum daily profit?

Group Exercise

63. In Chapter 8, you learned how to fit a quadratic function of the form $y = ax^2 + bx + c$ to data without using the regression feature of a graphing utility (see pages 720–721). Each group member should find an interesting data set. Group members should select the two sets of data that are most interesting and relevant.

a. For one of the data sets selected, use the function $y = ax^3 + bx^2 + cx + d$ and four ordered pairs of values (x, y) to find the cubic function that models the data. Use matrices or a graphing utility to solve the resulting system in four variables for a, b, c, and d.

b. For the other data set selected, fit a higher-degree polynomial function to the data. Use a graphing utility to solve the resulting system in five or more variables.

SECTION 9.2   *Inconsistent and Dependent Systems and Their Applications*

Objectives

1. Apply Gaussian elimination to systems without unique solutions.

2. Apply Gaussian elimination to systems with more variables than equations.

3. Solve problems involving systems without unique solutions.

Traffic jams getting you down? Powerful computers, able to solve systems with hundreds of thousands of variables in a single bound, may promise a gridlock-free future. The computer in your car could be linked to a central computer that manages traffic flow by controlling traffic lights, rerouting you away from traffic congestion, issuing weather reports, and selecting the best route to your destination. New technologies could eventually drive your car at a steady 75 miles per hour along automated highways as you comfortably nap. In this section, we look at the role of linear systems without unique solutions in a future free of traffic jams.

1 Apply Gaussian elimination to systems without unique solutions.

Linear systems can have one solution, no solution, or infinitely many solutions. We can use Gaussian elimination on systems with three or more variables to determine how many solutions such systems may have. In the case of systems with no solution or infinitely many solutions, it is impossible to rewrite the augmented matrix in the desired form with 1s down the diagonal from upper left to lower right, and 0s below the 1s. Let's see what this means by looking at a system that has no solution.

EXAMPLE 1   A System with No Solution

Use Gaussian elimination to solve the system:

$$\begin{aligned} x - y - 2z &= 2 \\ 2x - 3y + 6z &= 5 \\ 3x - 4y + 4z &= 12. \end{aligned}$$

Solution

Discovery

Use the addition method to solve Example 1. Describe what happens. Why does this mean that there is no solution?

Step 1   Write the augmented matrix for the system.

Linear System	Augmented Matrix
$\begin{aligned} x - y - 2z &= 2 \\ 2x - 3y + 6z &= 5 \\ 3x - 4y + 4z &= 12 \end{aligned}$	$\left[\begin{array}{ccc\|c} 1 & -1 & -2 & 2 \\ 2 & -3 & 6 & 5 \\ 3 & -4 & 4 & 12 \end{array}\right]$

$$\begin{bmatrix} 1 & -1 & -2 & | & 2 \\ 2 & -3 & 6 & | & 5 \\ 3 & -4 & 4 & | & 12 \end{bmatrix}$$

The system's augmented matrix, repeated

Step 2 Attempt to simplify the matrix to one with 1s down the diagonal and 0s below the 1s. Notice that the augmented matrix already has a 1 in the top position of the first column. Now we want 0s below the 1. To get the first 0, multiply row 1 by −2 and add these products to row 2. To get the second 0, multiply row 1 by −3 and add these products to row 3. Performing these operations, we obtain the following matrix:

We want 1 in this position.

$$\begin{bmatrix} 1 & -1 & -2 & | & 2 \\ 0 & -1 & 10 & | & 1 \\ 0 & -1 & 10 & | & 6 \end{bmatrix}.$$

Use the augmented matrix and:
Replace row 2 by $-2R_1 + R_2$.
Replace row 3 by $-3R_1 + R_3$.

Moving on to the second column, we obtain 1 in the desired position by multiplying row 2 by −1.

$$\begin{bmatrix} 1 & -1 & -2 & | & 2 \\ -1(0) & -1(-1) & -1(10) & | & -1(1) \\ 0 & -1 & 10 & | & 6 \end{bmatrix} = \begin{bmatrix} 1 & -1 & -2 & | & 2 \\ 0 & 1 & -10 & | & -1 \\ 0 & -1 & 10 & | & 6 \end{bmatrix} \quad -1R_2$$

We want 0 in this position.

Now we want a 0 below the 1 in column 2. To get the 0, multiply row 2 by 1 and add these products to row 3. (Equivalently, add row 2 to row 3.) We obtain the following matrix:

$$\begin{bmatrix} 1 & -1 & -2 & | & 2 \\ 0 & 1 & -10 & | & -1 \\ 0 & 0 & 0 & | & 5 \end{bmatrix}.$$

Replace row 3 in the previous matrix by $1R_2 + R_3$.

It is impossible to convert this last matrix to the desired form of 1s down the diagonal from upper left to lower right. If we translate the last row back into equation form, we get

$$0x + 0y + 0z = 5$$

There are no values of x, y, and z for which 0 = 5.

which is false. Regardless of which values we select for x, y, and z, the last equation can never be a true statement. Consequently, the system has no solution. The solution set is $\varnothing$, the empty set.

Three planes are parallel with no common intersection point.

Two planes are parallel with no intersection point common to all three planes.

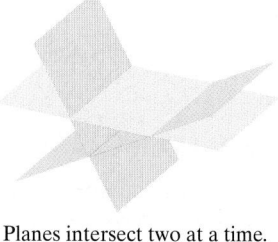

Planes intersect two at a time. There is no intersection point common to all three planes.

Figure 9.1 Three planes may have no common point of intersection

Check Point 1 Use Gaussian elimination to solve the system:

$$\begin{aligned} x - 2y - z &= -5 \\ 2x - 3y - z &= 0 \\ 3x - 4y - z &= 1. \end{aligned}$$

Recall that the graph of a system of three linear equations in three variables consists of three planes. When these planes intersect in a single point, the system has precisely one ordered-triple solution. When the planes have no point in common, the system has no solution, like the one in Example 1. Figure 9.1 illustrates some of the geometric possibilities for these inconsistent systems.

Now let's see what happens when we apply Gaussian elimination to a system with infinitely many solutions. Representing the solution set for these systems can be a bit tricky.

EXAMPLE 2   A System with an Infinite Number of Solutions

Use Gaussian elimination to solve the following system:

$$3x - 4y + 4z = 7$$
$$x - y - 2z = 2$$
$$2x - 3y + 6z = 5.$$

Solution   As always, we start with the augmented matrix.

$$\begin{bmatrix} 3 & -4 & 4 & | & 7 \\ 1 & -1 & -2 & | & 2 \\ 2 & -3 & 6 & | & 5 \end{bmatrix} \quad \begin{array}{c} R_1 \leftrightarrow R_2 \\ \text{Reverse rows} \\ \text{1 and 2.} \end{array} \quad \begin{bmatrix} 1 & -1 & -2 & | & 2 \\ 3 & -4 & 4 & | & 7 \\ 2 & -3 & 6 & | & 5 \end{bmatrix} \quad \begin{array}{c} \text{Replace row 2} \\ \text{by} -3R_1 + R_2. \\ \text{Replace row 3} \\ \text{by} -2R_1 + R_3. \end{array}$$

$$\begin{bmatrix} 1 & -1 & -2 & | & 2 \\ 0 & -1 & 10 & | & 1 \\ 0 & -1 & 10 & | & 1 \end{bmatrix} \quad \begin{array}{c} -1R_2 \\ \text{Multiply row} \\ \text{2 by} -1. \end{array} \quad \begin{bmatrix} 1 & -1 & -2 & | & 2 \\ 0 & 1 & -10 & | & -1 \\ 0 & -1 & 10 & | & 1 \end{bmatrix} \quad \begin{array}{c} \text{Replace row 3} \\ \text{by} 1R_2 + R_3. \end{array}$$

$$\begin{bmatrix} 1 & -1 & -2 & | & 2 \\ 0 & 1 & -10 & | & -1 \\ 0 & 0 & 0 & | & 0 \end{bmatrix}$$

If we translate row 3 of the matrix into equation form, we obtain

$$0x + 0y + 0z = 0$$

or

$$0 = 0.$$

This equation results in a true statement regardless of which values we select for x, y, and z. Consequently, the equation $0x + 0y + 0z = 0$ is *dependent* on the other two equations in the system in the sense that it adds no new information about the variables. Thus, we can drop it from the system, which can now be expressed in the form

$$\begin{bmatrix} 1 & -1 & -2 & | & 2 \\ 0 & 1 & -10 & | & -1 \end{bmatrix}.$$

The original system is equivalent to the system

$$x - y - 2z = 2$$
$$y - 10z = -1.$$

Although neither of these equations gives a value for z, we can use them to express x and y in terms of z. From the last equation we obtain

$$y = 10z - 1. \quad \text{Add 10z to both sides and isolate y.}$$

Back-substituting for y into the previous equation, we can find x in terms of z.

$$x - y - 2z = 2 \qquad \text{This is the first equation obtained from the final matrix.}$$

$$x - (10z - 1) - 2z = 2 \qquad \text{Because y} = 10z - 1, \text{substitute 10z} - 1 \text{ for y.}$$

$$x - 10z + 1 - 2z = 2 \qquad \text{Apply the distributive property.}$$

$$x - 12z + 1 = 2 \qquad \text{Combine like terms.}$$

$$x = 12z + 1 \qquad \text{Solve for x in terms of z.}$$

$$x = 12z + 1$$
$$y = 10z - 1$$

Equations expressing x and y in terms of z, repeated

Because no value is determined for z, we can find a solution to the system by letting z equal any real number and then using the equations expressing x and y in terms of z, repeated in the margin, to obtain x and y. For example, if $z = 1$, then

$$x = 12z + 1 = 12(1) + 1 = 13 \text{ and}$$
$$y = 10z - 1 = 10(1) - 1 = 9.$$

Consequently, $(13, 9, 1)$ is a solution to the system. On the other hand, if we let $z = -1$, then

$$x = 12z + 1 = 12(-1) + 1 = -11 \text{ and}$$
$$y = 10z - 1 = 10(-1) - 1 = -11.$$

Thus, $(-11, -11, -1)$ is another solution to the system.

We see that for any arbitrary choice of z, every ordered triple of the form $(12z + 1, 10z - 1, z)$ is a solution of the system. The solution set of this system with dependent equations is

$$\{(12z + 1, 10z - 1, z)\}.$$

We have seen that when three planes have no point in common, the corresponding system has no solution. When the system has infinitely many solutions, like the one in Example 2, the three planes intersect in more than one point. Figure 9.2 illustrates one geometric possibility for systems with dependent equations.

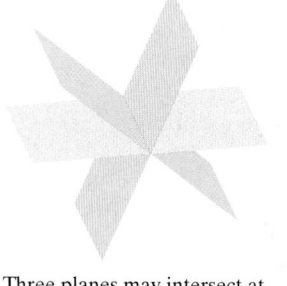

Three planes may intersect at infinitely many points.

Figure 9.2

Check Point 2 Use Gaussian elimination to solve the following system:

$$x - 2y - z = 5$$
$$2x - 5y + 3z = 6$$
$$x - 3y + 4z = 1.$$

2 Apply Gaussian elimination to systems with more variables than equations.

Nonsquare Systems

Up to this point, we have encountered only *square* systems in which the number of equations is equal to the number of variables. In a **nonsquare system,** the number of variables differs from the number of equations. In Example 3, we have two equations and three variables.

EXAMPLE 3   A System with Fewer Equations Than Variables

Use Gaussian elimination to solve the system:

$$3x + 7y + 6z = 26$$
$$x + 2y + z = 8.$$

Solution We begin with the augmented matrix.

$$\begin{bmatrix} 3 & 7 & 6 & | & 26 \\ 1 & 2 & 1 & | & 8 \end{bmatrix} \xrightarrow{R_1 \leftrightarrow R_2} \begin{bmatrix} 1 & 2 & 1 & | & 8 \\ 3 & 7 & 6 & | & 26 \end{bmatrix} \xrightarrow[\text{by} -3R_1 + R_2.]{\text{Replace row 2}} \begin{bmatrix} 1 & 2 & 1 & | & 8 \\ 0 & 1 & 3 & | & 2 \end{bmatrix}$$

Because we now have 1s down the diagonal that begins with the upper-left entry and a 0 below the leading 1, we translate the matrix back into equation form.

$$x + 2y + z = 8 \quad \text{Equation 1}$$
$$y + 3z = 2 \quad \text{Equation 2}$$

We can let z equal any real number and use back-substitution to express x and y in terms of z.

Equation 2   **Equation 1**

$$y + 3z = 2 \qquad\qquad x + 2y + z = 8$$
$$y = -3z + 2 \qquad\quad x + 2(-3z + 2) + z = 8$$
$$x - 6z + 4 + z = 8$$
$$x - 5z + 4 = 8$$
$$x = 5z + 4$$

Discovery

Let $z = 1$ for the solution set
$$\{(5z + 4, -3z + 2, z)\}.$$

What solution do you obtain? Substitute these three values in the two original equations:
$$3x + 7y + 6z = 26$$
$$x + 2y + z = 8.$$
Show that each equation is satisfied. Repeat this process for two other values of z.

For any arbitrary choice of z, every ordered triple of the form $(5z + 4, -3z + 2, z)$ is a solution of the system. We can express the system's solution set as
$$\{(5z + 4, -3z + 2, z)\}.$$

Check Point 3   Use Gaussian elimination to solve the system:
$$x + 2y + 3z = 70$$
$$x + y + z = 60.$$

3   Solve problems involving systems without unique solutions.

Applications

How will computers be programmed to control traffic flow and avoid congestion? They will be required to solve systems continually based on the following premise: If traffic is to keep moving, during any period of time the number of cars entering an intersection must equal the number of cars leaving that intersection. Let's see what this means by looking at the intersections of four one-way city streets.

EXAMPLE 4   Traffic Control

Figure 9.3 shows the intersections of four one-way streets. As you study the figure, notice that 300 cars per hour want to enter intersection I_1 from the north on 27th Avenue. Also, 200 cars per hour want to head east from intersection I_2 on Palm Drive. The letters w, x, y, and z stand for the number of cars passing between the intersections.

    a. If the traffic is to keep moving, at each intersection the number of cars entering per hour must equal the number of cars leaving per hour. Use this idea to set up a linear system of equations involving w, x, y, and z.

Figure 9.3 The intersections of four one-way streets

b. Use Gaussian elimination to solve the system.

c. If construction on 27th Avenue limits z to 50 cars per hour, how many cars per hour must pass between the other intersections to keep traffic flowing?

Solution

Figure 9.3, repeated

a. Set up the system by considering one intersection at a time, referring to Figure 9.3.

For Intersection I_1: Because $300 + 700 = 1000$ cars enter I_1, and $w + z$ cars leave the intersection, then $w + z = 1000$.

For Intersection I_2: Because $w + x$ cars enter the intersection, and $200 + 900 = 1100$ cars leave I_2, then $w + x = 1100$.

For Intersection I_3: Figure 9.3 indicates that $300 + 400 = 700$ cars enter and $x + y$ leave, so $x + y = 700$.

For Intersection I_4: With $y + z$ cars entering and $200 + 400 = 600$ cars exiting, traffic will keep flowing if $y + z = 600$.

The system of equations that describes this situation is given by

$$w + z = 1000$$
$$w + x = 1100$$
$$x + y = 700$$
$$y + z = 600.$$

b. To solve this system using Gaussian elimination, we begin with the augmented matrix.

System of Linear Equations (showing missing variables with 0 coefficients)

$$1w + 0x + 0y + 1z = 1000$$
$$1w + 1x + 0y + 0z = 1100$$
$$0w + 1x + 1y + 0z = 700$$
$$0w + 0x + 1y + 1z = 600$$

Augmented Matrix

$$\begin{bmatrix} 1 & 0 & 0 & 1 & | & 1000 \\ 1 & 1 & 0 & 0 & | & 1100 \\ 0 & 1 & 1 & 0 & | & 700 \\ 0 & 0 & 1 & 1 & | & 600 \end{bmatrix}$$

We can now use row operations to obtain the following matrix:

$$\begin{bmatrix} 1 & 0 & 0 & 1 & | & 1000 \\ 0 & 1 & 0 & -1 & | & 100 \\ 0 & 0 & 1 & 1 & | & 600 \\ 0 & 0 & 0 & 0 & | & 0 \end{bmatrix}.$$

$w + z = 1000$
$x - z = 100$
$y + z = 600$

The last row of the matrix shows that the system in the voice balloons has dependent equations and infinitely many solutions. To write the solution set containing these infinitely many solutions, let z equal any real number. Use the three equations in the voice balloons to express w, x, and y in terms of z:

$$w = 1000 - z, \quad x = 100 + z, \quad \text{and} \quad y = 600 - z.$$

With z arbitrary, the alphabetical ordered solution (w, x, y, z) enables us to express the system's solution set as

$$\{(1000 - z, 100 + z, 600 - z, z)\}.$$

Automated Highways

New technologies are making automated highways a reality. Experiments are taking place with cars that can steer, accelerate, and brake by themselves. A computer in the car picks up signals from magnets set in the road. Commuters can sit back, play with their laptop computers, read the newspaper, and enjoy the journey.

Figure 9.4 With z limited to 50 cars per hour, values for w, x, and y are determined

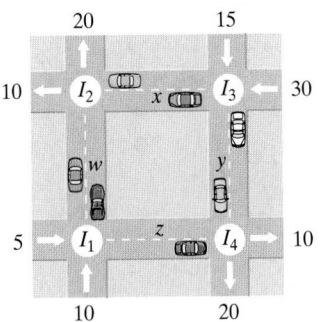

Figure 9.5

c. We are given that construction limits z to 50 cars per hour. Because $z = 50$, we substitute 50 for z in the system's ordered solution:

$$(1000 - z, 100 + z, 600 - z, z) \qquad \text{Use the system's solution.}$$
$$= (1000 - 50, 100 + 50, 600 - 50, 50) \qquad z = 50$$
$$= (950, 150, 550, 50).$$

Thus, $w = 950$, $x = 150$, and $y = 550$. (See Figure 9.4.) With construction on 27th Avenue, this means that to keep traffic flowing, 950 cars per hour must be routed between I_1 and I_2, 150 per hour between I_3 and I_2, and 550 per hour between I_3 and I_4.

Check Point 4

Figure 9.5 shows a system of four one-way streets. The numbers in the figure denote the number of cars per minute that travel in the direction shown.

a. Use the requirement that the number of cars entering each of the intersections per minute must equal the number of cars leaving per minute to set up a system of equations in $w, x, y,$ and z.

b. Use Gaussian elimination to solve the system.

c. If construction limits z to 10 cars per minute, how many cars per minute must pass between the other intersections to keep traffic flowing?

EXERCISE SET 9.2

Practice Exercises

In Exercises 1–24, use Gaussian elimination to find the complete solution to each system of equations, or show that none exists.

1. $5x + 12y + z = 10$
$2x + 5y + 2z = -1$
$x + 2y - 3z = 5$

2. $2x - 4y + z = 3$
$x - 3y + z = 5$
$3x - 7y + 2z = 12$

3. $5x + 8y - 6z = 14$
$3x + 4y - 2z = 8$
$x + 2y - 2z = 3$

4. $5x - 11y + 6z = 12$
$-x + 3y - 2z = -4$
$3x - 5y + 2z = 4$

5. $3x + 4y + 2z = 3$
$4x - 2y - 8z = -4$
$x + y - z = 3$

6. $2x - y - z = 0$
$x + 2y + z = 3$
$3x + 4y + 2z = 8$

7. $8x + 5y + 11z = 30$
$-x - 4y + 2z = 3$
$2x - y + 5z = 12$

8. $x + y - 10z = -4$
$x - 7z = -5$
$3x + 5y - 36z = -10$

9. $w - 2x - y - 3z = -9$
$w + x - y = 0$
$3w + 4x + z = 6$
$2x - 2y + z = 3$

10. $2w + x - 2y - z = 3$
$w - 2x + y + z = 4$
$-w - 8x + 7y + 5z = 13$
$3w + x - 2y + 2z = 6$

11. $2w + x - y = 3$
$w - 3x + 2y = -4$
$3w + x - 3y + z = 1$
$w + 2x - 4y - z = -2$

12. $2w - x + 3y + z = 0$
$3w + 2x + 4y - z = 0$
$5w - 2x - 2y - z = 0$
$2w + 3x - 7y - 5z = 0$

13.
$$\begin{aligned} w - 3x + y - 4z &= 4 \\ -2w + x + 2y &= -2 \\ 3w - 2x + y - 6z &= 2 \\ -w + 3x + 2y - z &= -6 \end{aligned}$$

14.
$$\begin{aligned} 3w + 2x - y + 2z &= -12 \\ 4w - x + y + 2z &= 1 \\ w + x + y + z &= -2 \\ -2w + 3x + 2y - 3z &= 10 \end{aligned}$$

15.
$$\begin{aligned} 2x + y - z &= 2 \\ 3x + 3y - 2z &= 3 \end{aligned}$$

16.
$$\begin{aligned} 3x + 2y - z &= 5 \\ x + 2y - z &= 1 \end{aligned}$$

17.
$$\begin{aligned} x + 2y + 3z &= 5 \\ y - 5z &= 0 \end{aligned}$$

18.
$$\begin{aligned} 3x - y + 4z &= 8 \\ y + 2z &= 1 \end{aligned}$$

19.
$$\begin{aligned} x + y - 2z &= 2 \\ 3x - y - 6z &= -7 \end{aligned}$$

20.
$$\begin{aligned} -2x - 5y + 10z &= 19 \\ x + 2y - 4z &= 12 \end{aligned}$$

21.
$$\begin{aligned} w + x - y + z &= -2 \\ 2w - x + 2y - z &= 7 \\ -w + 2x + y + 2z &= -1 \end{aligned}$$

22.
$$\begin{aligned} 2w - 3x + 4y + z &= 7 \\ w - x + 3y - 5z &= 10 \\ 3w + x - 2y - 2z &= 6 \end{aligned}$$

23.
$$\begin{aligned} w + 2x + 3y - z &= 7 \\ 2x - 3y + z &= 4 \\ w - 4x + y &= 3 \end{aligned}$$

24.
$$\begin{aligned} w - x + z &= 0 \\ w - 4x + y + 2z &= 0 \\ 3w - y + 2z &= 0 \end{aligned}$$

Application Exercises

The figure for Exercises 25–28 shows the intersection of three one-way streets. To keep traffic moving, the number of cars per minute entering an intersection must equal the number exiting that intersection. For intersection I_1, $x + 10$ cars enter and $y + 14$ cars exit per minute. Thus, $x + 10 = y + 14$.

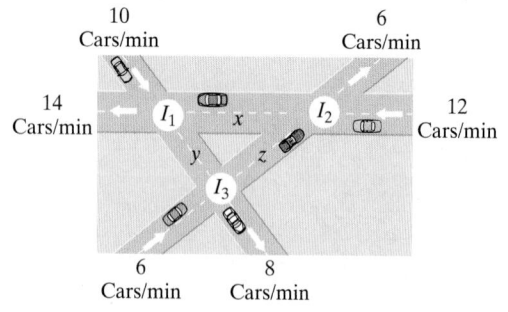

25. Write an equation for intersection I_2 that keeps traffic moving.

26. Write an equation for intersection I_3 that keeps traffic moving.

27. Use Gaussian elimination to solve the system formed by the equation given prior to Exercise 25 and the two equations that you obtained in Exercises 25–26.

28. Use your ordered solution obtained in Exercise 27 to solve this exercise. If construction limits z to 4 cars per minute, how many cars per minute must pass between the other intersections to keep traffic flowing?

29. The figure shows the intersection of four one-way streets.

a. Set up a system of equations that keeps traffic moving.

b. Use Gaussian elimination to solve the system.

c. If construction limits z to 50 cars per hour, how many cars per hour must pass between the other intersections to keep traffic moving?

30. The vitamin content per ounce for three foods is given in the following table.

	Milligrams per Ounce		
	Thiamin	Riboflavin	Niacin
Food A	3	7	1
Food B	1	5	3
Food C	3	8	2

a. Use matrices to show that no combination of these foods can provide exactly 14 mg of thiamin, 32 mg of riboflavin, and 9 mg of niacin.

b. Use matrices to describe in practical terms what happens if the riboflavin requirement is increased by 5 mg and the other requirements stay the same.

31. Three foods have the following nutritional content per ounce.

	Units per Ounce		
	Vitamin A	Iron	Calcium
Food 1	20	20	10
Food 2	30	10	10
Food 3	10	10	30

a. A diet must consist precisely of 220 units of vitamin A, 180 units of iron, and 340 units of calcium. However, the dietician runs out of Food 1. Use a matrix approach to show that under these conditions the dietary requirements cannot be met.

b. Now suppose that all three foods are available, but due to problems with vitamin A for pregnant women, a hospital dietician no longer wants to include this vitamin in the diet. Use matrices to give two possible ways to meet the iron and calcium requirements with the three foods.

32. A company that manufactures products A, B, and C does both manufacturing and testing. The hours needed to manufacture and test each product are shown in the table.

	Hours Needed Weekly to Manufacture	Hours Needed Weekly to Test
Product A	7	2
Product B	6	2
Product C	3	1

The company has exactly 67 hours per week available for manufacturing and 20 hours per week available for testing. Give two different combinations for the number of products that can be manufactured and tested weekly.

 Writing in Mathematics

33. Describe what happens when Gaussian elimination is used to solve an inconsistent system.

34. Describe what happens when Gaussian elimination is used to solve a system with dependent equations.

35. In solving a system of dependent equations in three variables, one student simply said that there are infinitely many solutions. A second student expressed the solution set as $\{(4z + 3, 5z - 1, z)\}$. Which is the better form of expressing the solution set and why?

 Technology Exercise

36. a. The figure shows the intersections of a number of one-way streets. The numbers given represent traffic flow at a peak period (from 4 P.M. to 5:30 P.M.). Use the figure to write a linear system of six equations in seven variables based on the idea that at each intersection the number of cars entering must equal the number of cars leaving.

b. Use a graphing utility with a REF or RREF command to find the complete solution to the system.

 Critical Thinking Exercise

37. Consider the linear system
$$\begin{aligned} x + 3y + z &= a^2 \\ 2x + 5y + 2az &= 0 \\ x + y + a^2z &= -9. \end{aligned}$$
For which values of a will the system be inconsistent?

 Group Exercise

38. Before beginning this exercise, the group needs to read and solve Exercise 36.

a. A political group is planning a demonstration on 95th Street between 113th Place and 117th Court for 5 P.M. Wednesday. The problem becomes one of minimizing traffic flow on 95th Street (between 113th and 117th) without causing traffic tie-ups on other streets. One possible solution is to close off traffic on 95th Street between 113th and 117th (let $x_6 = 0$). What can group members conclude about x_7 under these conditions?

b. Working with a matrix allows us to simplify the problem caused by the political demonstration, but it did not actually solve the problem. There are an infinite number of solutions; each value of x_7 we choose gives us a new picture. We also assumed x_6 was equal to 0; changing that assumption would also lead to different solutions. With your group, design another solution to the traffic flow problem caused by the political demonstration.

SECTION 9.3 *Matrix Operations and Their Applications*

Objectives

1. Use matrix notation.
2. Understand what is meant by equal matrices.
3. Add and subtract matrices.
4. Perform scalar multiplication.
5. Solve matrix equations.
6. Multiply matrices.
7. Describe applied situations with matrix operations.

Turn on your computer and read your e-mail or write a paper. When you need to do research, use the Internet to browse through art museums and photography exhibits. When you need a break, load a flight simulator program and fly through a photorealistic computer world. As different as these experiences may be, they all share one thing—you're looking at images based on matrices. Matrices have applications in numerous fields, including the new technology of digital photography in which pictures are represented by numbers rather than film. In this section, we turn our attention to matrix algebra and some of its applications.

1 Use matrix notation.

Notations for Matrices

We have seen that an array of numbers, arranged in rows and columns and placed in brackets, is called a matrix. We can represent the matrix in two different ways.

- A capital letter, such as A, B, or C, can denote a matrix.
- A lowercase letter enclosed in brackets, such as that shown below, can denote a matrix.

$$A = \left[a_{ij} \right]$$ Matrix A with elements a_{ij}

A general element in matrix A is denoted by a_{ij}. This refers to the element in the ith row and jth column. For example, a_{32} is the element of A located in the third row, second column.

A matrix of **order $m \times n$** has m rows and n columns. If $m = n$, a matrix has the same number of rows as columns and is called a **square matrix.**

EXAMPLE 1 Matrix Notation

Let

$$A = \begin{bmatrix} 3 & 2 & 0 \\ -4 & -5 & -\frac{1}{5} \end{bmatrix}.$$

a. What is the order of A?

b. If $A = \left[a_{ij} \right]$, identify a_{23} and a_{12}.

Solution

a. The matrix has 2 rows and 3 columns, so it is of order 2×3.

b. The element a_{23} is in the second row and third column. Thus, $a_{23} = -\dfrac{1}{5}$.

The element a_{12} is in the first row and second column. Consequently, $a_{12} = 2$.

Check Point 1 Let

$$A = \begin{bmatrix} 5 & -2 \\ -3 & \pi \\ 1 & 6 \end{bmatrix}.$$

a. What is the order of A?

b. Identify a_{12} and a_{31}.

2 Understand what is meant by equal matrices.

Equality of Matrices

Two matrices are **equal** if and only if they have the same order and corresponding elements are equal.

Definition of Equality of Matrices

Two matrices A and B are **equal** if and only if they have the same order $m \times n$ and $a_{ij} = b_{ij}$ for $i = 1, 2, \ldots, m$ and $j = 1, 2, \ldots, n$.

For example, if $A = \begin{bmatrix} x & y+1 \\ z & 6 \end{bmatrix}$ and $B = \begin{bmatrix} 1 & 5 \\ 3 & 6 \end{bmatrix}$, then $A = B$ if and only if $x = 1$, $y + 1 = 5$ (so $y = 4$), and $z = 3$.

3 Add and subtract matrices.

Matrix Addition and Subtraction

Table 9.1 shows that matrices of the same order can be added or subtracted by simply adding or subtracting corresponding elements.

Table 9.1 Adding and Subtracting Matrices
Let $A = [a_{ij}]$ and $B = [b_{ij}]$ be matrices of order $m \times n$.

Definition	The Definition in Words	Example
Matrix Addition $A + B = [a_{ij} + b_{ij}]$	Matrices of the same order are added by adding the elements in corresponding positions.	$\begin{bmatrix} 1 & -2 \\ 3 & 5 \end{bmatrix} + \begin{bmatrix} -1 & 6 \\ 0 & 4 \end{bmatrix}$ $= \begin{bmatrix} 1+(-1) & -2+6 \\ 3+0 & 5+4 \end{bmatrix} = \begin{bmatrix} 0 & 4 \\ 3 & 9 \end{bmatrix}$
Matrix Subtraction $A - B = [a_{ij} - b_{ij}]$	Matrices of the same order are subtracted by subtracting the elements in corresponding positions.	$\begin{bmatrix} 1 & -2 \\ 3 & 5 \end{bmatrix} - \begin{bmatrix} -1 & 6 \\ 0 & 4 \end{bmatrix}$ $= \begin{bmatrix} 1-(-1) & -2-6 \\ 3-0 & 5-4 \end{bmatrix} = \begin{bmatrix} 2 & -8 \\ 3 & 1 \end{bmatrix}$

The sum or difference of two matrices of different orders is undefined. For example, consider the matrices

$$A = \begin{bmatrix} 0 & 3 \\ 4 & 3 \end{bmatrix} \quad \text{and} \quad B = \begin{bmatrix} 1 & 9 \\ 4 & 5 \\ 2 & 3 \end{bmatrix}.$$

The order of A is 2×2; the order of B is 3×2. These matrices are of different orders and cannot be added or subtracted.

Technology

Graphing utilities can add and subtract matrices. Enter the matrices and name them [A] and [B]. Then use a keystroke sequence similar to

$$\boxed{[A]} \; \boxed{-} \; \boxed{[B]} \; \boxed{\text{ENTER}}$$

Consult your manual and verify the results in Example 2.

EXAMPLE 2   Adding and Subtracting Matrices

Perform the indicated matrix operations:

a. $\begin{bmatrix} 0 & 5 & 3 \\ -2 & 6 & -8 \end{bmatrix} + \begin{bmatrix} -2 & 3 & 5 \\ 7 & -9 & 6 \end{bmatrix}$

b. $\begin{bmatrix} -6 & 7 \\ 2 & -3 \end{bmatrix} - \begin{bmatrix} -5 & 6 \\ 0 & -4 \end{bmatrix}.$

Solution

a.
$$\begin{bmatrix} 0 & 5 & 3 \\ -2 & 6 & -8 \end{bmatrix} + \begin{bmatrix} -2 & 3 & 5 \\ 7 & -9 & 6 \end{bmatrix}$$

$$= \begin{bmatrix} 0 + (-2) & 5 + 3 & 3 + 5 \\ -2 + 7 & 6 + (-9) & -8 + 6 \end{bmatrix}$$ *Add the corresponding elements in the 2 × 3 matrices.*

$$= \begin{bmatrix} -2 & 8 & 8 \\ 5 & -3 & -2 \end{bmatrix}$$ *Simplify.*

b.
$$\begin{bmatrix} -6 & 7 \\ 2 & -3 \end{bmatrix} - \begin{bmatrix} -5 & 6 \\ 0 & -4 \end{bmatrix}$$

$$= \begin{bmatrix} -6 - (-5) & 7 - 6 \\ 2 - 0 & -3 - (-4) \end{bmatrix}$$ *Subtract the corresponding elements in the 2 × 2 matrices.*

$$= \begin{bmatrix} -1 & 1 \\ 2 & 1 \end{bmatrix}$$ *Simplify.*

Check Point 2   Perform the indicated matrix operations:

a. $\begin{bmatrix} -4 & 3 \\ 7 & -6 \end{bmatrix} + \begin{bmatrix} 6 & -3 \\ 2 & -4 \end{bmatrix}$ b. $\begin{bmatrix} 5 & 4 \\ -3 & 7 \\ 0 & 1 \end{bmatrix} - \begin{bmatrix} -4 & 8 \\ 6 & 0 \\ -5 & 3 \end{bmatrix}.$

A matrix whose elements are all equal to 0 is called a **zero matrix**. If A is an $m \times n$ matrix and 0 is an $m \times n$ zero matrix, then $A + 0 = A$. For example,

$$\begin{bmatrix} -5 & 2 \\ 3 & 6 \end{bmatrix} + \begin{bmatrix} 0 & 0 \\ 0 & 0 \end{bmatrix} = \begin{bmatrix} -5 & 2 \\ 3 & 6 \end{bmatrix}.$$

An $m \times n$ zero matrix is called the **additive identity** for $m \times n$ matrices.

For any matrix A, the **additive inverse** of A, written $-A$, is the matrix of the same order of A such that every element of $-A$ is the opposite of the corresponding element of A. Because corresponding elements are added in matrix addition, $A + (-A)$ is a zero matrix. For example,

$$\begin{bmatrix} -5 & 2 \\ 3 & 6 \end{bmatrix} + \begin{bmatrix} 5 & -2 \\ -3 & -6 \end{bmatrix} = \begin{bmatrix} 0 & 0 \\ 0 & 0 \end{bmatrix}.$$

Properties of matrix addition are similar to properties involved with adding real numbers.

Properties of Matrix Addition

If A, B, and C are $m \times n$ matrices and 0 is an $m \times n$ zero matrix, then the following properties are true.

1. $A + B = B + A$ Commutative Property of Addition
2. $(A + B) + C = A + (B + C)$ Associative Property of Addition
3. $A + 0 = 0 + A = A$ Additive Identity Property
4. $A + (-A) = (-A) + A = 0$ Additive Inverse Property

4 Perform scalar multiplication.

Scalar Multiplication

A matrix of order 1×1, such as $[6]$, contains only one entry. To distinguish this matrix from the number 6, we refer to 6 as a **scalar.** In general, in our work with matrices, we will refer to real numbers as scalars.

To multiply a matrix A by a scalar c, we multiply each entry in A by c. For example,

$$4\begin{bmatrix} 2 & 5 \\ -3 & 0 \end{bmatrix} = \begin{bmatrix} 4(2) & 4(5) \\ 4(-3) & 4(0) \end{bmatrix} = \begin{bmatrix} 8 & 20 \\ -12 & 0 \end{bmatrix}.$$

Scalar   Matrix

Definition of Scalar Multiplication

If $A = [a_{ij}]$ is a matrix of order $m \times n$ and c is a scalar, then the matrix cA is the $m \times n$ matrix given by

$$cA = [ca_{ij}].$$

This matrix is obtained by multiplying each element of A by the real number c. We call cA a **scalar multiple** of A.

EXAMPLE 3   Scalar Multiplication

If $A = \begin{bmatrix} -1 & 4 \\ 3 & 0 \end{bmatrix}$ and $B = \begin{bmatrix} 2 & -3 \\ 5 & -6 \end{bmatrix}$, find: **a.** $-5B$ **b.** $2A + 3B$.

$$A = \begin{bmatrix} -1 & 4 \\ 3 & 0 \end{bmatrix}$$

$$B = \begin{bmatrix} 2 & -3 \\ 5 & -6 \end{bmatrix}$$

The given matrices, repeated

Solution

a. $-5B = -5\begin{bmatrix} 2 & -3 \\ 5 & -6 \end{bmatrix} = \begin{bmatrix} -5(2) & -5(-3) \\ -5(5) & -5(-6) \end{bmatrix} = \begin{bmatrix} -10 & 15 \\ -25 & 30 \end{bmatrix}$

Multiply each element in B by −5.

b. $2A + 3B = 2\begin{bmatrix} -1 & 4 \\ 3 & 0 \end{bmatrix} + 3\begin{bmatrix} 2 & -3 \\ 5 & -6 \end{bmatrix}$

$$= \begin{bmatrix} 2(-1) & 2(4) \\ 2(3) & 2(0) \end{bmatrix} + \begin{bmatrix} 3(2) & 3(-3) \\ 3(5) & 3(-6) \end{bmatrix}$$

Multiply each element in A by 2.   Multiply each element in B by 3.

$$= \begin{bmatrix} -2 & 8 \\ 6 & 0 \end{bmatrix} + \begin{bmatrix} 6 & -9 \\ 15 & -18 \end{bmatrix} = \begin{bmatrix} -2+6 & 8+(-9) \\ 6+15 & 0+(-18) \end{bmatrix}$$

Perform the addition of these 2 x 2 matrices by adding corresponding elements.

$$= \begin{bmatrix} 4 & -1 \\ 21 & -18 \end{bmatrix}$$

Technology

You can verify the algebraic solution in Example 3(b) by entering the matrices [A] and [B] into your graphic utility. The screen below shows the required computation.

```
2[A]+3[B]
       [[4    -1 ]
        [21  -18]]
■
```

Check Point 3 If $A = \begin{bmatrix} -4 & 1 \\ 3 & 0 \end{bmatrix}$ and $B = \begin{bmatrix} -1 & -2 \\ 8 & 5 \end{bmatrix}$, find:

a. $-6B$ **b.** $3A + 2B$.

Discovery

Verify each of the four properties listed in the box using

$$A = \begin{bmatrix} 2 & -4 \\ -5 & 3 \end{bmatrix},$$

$$B = \begin{bmatrix} 4 & 0 \\ 1 & -6 \end{bmatrix},$$

$c = 4$, and $d = 2$.

Properties of Scalar Multiplication

If A and B are $m \times n$ matrices, and c and d are scalars, then the following properties are true.

1. $(cd)A = c(dA)$ Associative Property of Scalar Multiplication

2. $1A = A$ Scalar Identity Property

3. $c(A + B) = cA + cB$ Distributive Property

4. $(c + d)A = cA + dA$ Distributive Property

5 Solve matrix equations.

Have you noticed the many similarities between addition of real numbers and matrix addition, subtraction of real numbers and matrix subtraction, and multiplication of real numbers and scalar multiplication? Example 4 shows how these similarities can be used to solve matrix equations involving matrix addition, matrix subtraction, and scalar multiplication.

EXAMPLE 4 Solving a Matrix Equation

Solve for X in the matrix equation

$$2X + A = B$$

where $A = \begin{bmatrix} 1 & -5 \\ 0 & 2 \end{bmatrix}$ and $B = \begin{bmatrix} -6 & 5 \\ 9 & 1 \end{bmatrix}$.

Solution   We begin by solving the matrix equation for X.

$$2X + A = B \qquad \text{This is the given matrix equation.}$$

> We multiply both sides by $\frac{1}{2}$ rather than divide both sides by 2. This is in anticipation of performing scalar multiplication.

$$2X = B - A \qquad \text{Subtract matrix } A \text{ from both sides.}$$

$$X = \tfrac{1}{2}(B - A) \qquad \text{Multiply both sides by } \tfrac{1}{2} \text{ and solve for matrix } X.$$

Now we use the matrices A and B to find the matrix X.

$$X = \frac{1}{2}\left(\begin{bmatrix} -6 & 5 \\ 9 & 1 \end{bmatrix} - \begin{bmatrix} 1 & -5 \\ 0 & 2 \end{bmatrix} \right) \quad \text{Substitute the matrices in } X = \tfrac{1}{2}(B - A).$$

$$= \frac{1}{2}\begin{bmatrix} -7 & 10 \\ 9 & -1 \end{bmatrix} \quad \text{Subtract matrices by subtracting corresponding elements.}$$

$$= \begin{bmatrix} -\dfrac{7}{2} & 5 \\ \dfrac{9}{2} & -\dfrac{1}{2} \end{bmatrix} \quad \text{Perform the scalar multiplication by multiplying each element by } \tfrac{1}{2}.$$

Take a few minutes to show that this matrix satisfies the given equation $2X + A = B$. Substitute the matrix for X and the given matrices for A and B in the equation. The matrices on each side of the equal sign, $2X + A$ and B, should be equal.

> **Check Point 4**   Solve for X in the matrix equation $3X + A = B$ where
>
> $$A = \begin{bmatrix} 2 & -8 \\ 0 & 4 \end{bmatrix} \quad \text{and} \quad B = \begin{bmatrix} -10 & 1 \\ -9 & 17 \end{bmatrix}.$$

6 Multiply matrices.

Matrix Multiplication

We do not multiply two matrices by multiplying the corresponding entries of matrices. Instead, we must think of matrix multiplication as *row-by-column multiplication*. To better understand how this works, let's begin with the definition of matrix multiplication for matrices of order 2×2.

Definition of Matrix Multiplication: 2×2 Matrices

Row 1 of A × Column 1 of B   Row 1 of A × Column 2 of B

$$AB = \begin{bmatrix} a & b \\ c & d \end{bmatrix}\begin{bmatrix} e & f \\ g & h \end{bmatrix} = \begin{bmatrix} ae + bg & af + bh \\ ce + dg & cf + dh \end{bmatrix}$$

Row 2 of A × Column 1 of B   Row 2 of A × Column 2 of B

Notice that we obtain the element in the ith row and jth column in AB by performing computations with elements in the ith row of A and the jth column of B. For example, we obtain the element in the first row and first column of AB by performing computations with elements in the first row of A and the first column of B.

First row of
A

First column
of B

$$\begin{bmatrix} a & b \end{bmatrix} \begin{bmatrix} e \\ g \end{bmatrix} = \begin{bmatrix} ae + bg & \end{bmatrix}$$

1. Multiply each element in row 1 of A by the corresponding element in column 1 of B.
2. Add these products.
3. Record the sum as the element in row 1, column 1 of the product matrix.

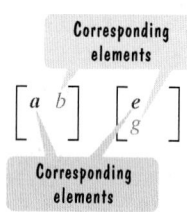

Corresponding
elements

$$\begin{bmatrix} a & b \end{bmatrix} \begin{bmatrix} e \\ g \end{bmatrix}$$

Corresponding
elements

Figure 9.6 Finding corresponding elements when multiplying matrices

You may wonder how to find the corresponding elements in step 1 in the voice balloon. The element at the far left of row 1 corresponds to the element at the top of column 1. The second element from the left of row 1 corresponds to the second element from the top of column 1. This is illustrated in Figure 9.6.

EXAMPLE 5 Multiplying Matrices

Find AB, given

$$A = \begin{bmatrix} 2 & 3 \\ 4 & 7 \end{bmatrix} \quad \text{and} \quad B = \begin{bmatrix} 0 & 1 \\ 5 & 6 \end{bmatrix}.$$

Solution We will perform a row-by-column computation.

$$AB = \begin{bmatrix} 2 & 3 \\ 4 & 7 \end{bmatrix} \begin{bmatrix} 0 & 1 \\ 5 & 6 \end{bmatrix}$$

Row 1 of A × Column 1
of B

Row 1 of A × Column 2
of B

$$= \begin{bmatrix} 2(0) + 3(5) & 2(1) + 3(6) \\ 4(0) + 7(5) & 4(1) + 7(6) \end{bmatrix} = \begin{bmatrix} 15 & 20 \\ 35 & 46 \end{bmatrix}$$

Row 2 of A × Column 1
of B

Row 2 of A × Column 2
of B

Check Point 5 Find AB, given $A = \begin{bmatrix} 1 & 3 \\ 2 & 5 \end{bmatrix}$ and $B = \begin{bmatrix} 4 & 6 \\ 1 & 0 \end{bmatrix}.$

We can generalize the process of Example 5 to multiplying an $m \times n$ matrix and an $n \times p$ matrix. **For the product of two matrices to be defined, the number of columns of the first matrix must equal the number of rows of the second matrix.**

First Matrix   **Second Matrix**

$m \times n$ $n \times p$

The number of columns in the first matrix must be the same as the number of rows in the second matrix.

Study Tip

The following diagram illustrates the first sentence in the box defining matrix multiplication. The diagram is helpful in determining the order of the product AB.

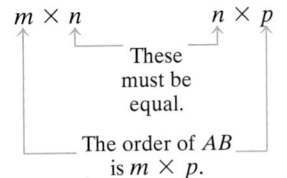

These must be equal.

The order of AB is $m \times p$.

Definition of Matrix Multiplication

The **product** of an $m \times n$ matrix, A, and an $n \times p$ matrix, B, is an $m \times p$ matrix, AB, whose elements are found as follows. The element in the ith row and jth column of AB is found by multiplying each element in the ith row of A by the corresponding element in the jth column of B and adding the products.

To find a product AB, each row of A must have the same number of elements as each column of B. We obtain p_{ij}, the element in the ith row and jth column in AB, by performing computations with elements in the ith row of A and the jth column of B:

ith row of A jth column of B   Element in the ith row and jth column of AB

$$\begin{bmatrix} * & \checkmark & \square \end{bmatrix} \begin{bmatrix} * \\ \checkmark \\ \square \end{bmatrix} = \begin{bmatrix} p_{ij} \end{bmatrix}$$

When multiplying corresponding elements, keep in mind that the element at the far left of row i corresponds to the element at the top of column j. The element second from the left of row i corresponds to the element second from the top of column j. Likewise, the element third from the left of row i corresponds to the element third from the top of column j, and so on.

EXAMPLE 6   Multiplying Matrices

Matrices A and B are defined as follows:

$$A = \begin{bmatrix} 1 & 2 & 3 \end{bmatrix} \qquad B = \begin{bmatrix} 4 \\ 5 \\ 6 \end{bmatrix}.$$

Find: **a.** AB and   **b.** BA.

$A = \begin{bmatrix} 1 & 2 & 3 \end{bmatrix}$

$B = \begin{bmatrix} 4 \\ 5 \\ 6 \end{bmatrix}$

The given matrices, repeated

Solution

a. Matrix A is a 1×3 matrix and matrix B is a 3×1 matrix. Thus, the product is a 1×1 matrix.

Matrix A **Matrix B**
1×3 3×1

These are equal.

The order of AB is 1×1.

$AB = \begin{bmatrix} 1 & 2 & 3 \end{bmatrix} \begin{bmatrix} 4 \\ 5 \\ 6 \end{bmatrix}$ — We will perform a row-by-column computation.

$= \begin{bmatrix} (1)(4) + (2)(5) + (3)(6) \end{bmatrix}$ — Multiply elements in row 1 of A by corresponding elements in column 1 of B and add the products.

$= \begin{bmatrix} 4 + 10 + 18 \end{bmatrix}$ — Perform the multiplications.

$= \begin{bmatrix} 32 \end{bmatrix}$ — Add.

b. Matrix B is a 3×1 matrix and matrix A is a 1×3 matrix. Thus, the product BA is a 3×3 matrix.

Matrix B **Matrix A**
3×1 1×3

These are equal.

The order of AB is 3×3.

$BA = \begin{bmatrix} 4 \\ 5 \\ 6 \end{bmatrix} \begin{bmatrix} 1 & 2 & 3 \end{bmatrix}$ We perform a row-by-column computation.

Row 1 of B × Column 1 of A   Row 1 of B × Column 2 of A   Row 1 of B × Column 3 of A

Row 2 of B × Column 1 of A   Row 2 of B × Column 2 of A   Row 2 of B × Column 3 of A

$= \begin{bmatrix} (4)(1) & (4)(2) & (4)(3) \\ (5)(1) & (5)(2) & (5)(3) \\ (6)(1) & (6)(2) & (6)(3) \end{bmatrix}$

Row 3 of B × Column 1 of A   Row 3 of B × Column 2 of A   Row 3 of B × Column 3 of A

$= \begin{bmatrix} 4 & 8 & 12 \\ 5 & 10 & 15 \\ 6 & 12 & 18 \end{bmatrix}$ Simplify.

Technology

The screens illustrate the solution of Example 6 using a graphing utility.

Arthur Cayley

The Granger Collection

Matrices were first studied intensively by the English mathematician Arthur Cayley (1821–1895). Before reaching the age of 25, he published 25 papers, setting a pattern of prolific creativity that lasted throughout his life. Cayley was a lawyer, painter, mountaineer, and Cambridge professor whose greatest invention was that of matrices and matrix theory. Cayley's matrix algebra, especially the noncommutativity of multiplication ($AB \neq BA$), opened up a new area of mathematics called abstract algebra.

Technology

Graphing utilities give an ERROR message when directed to perform a multiplication such as Example 7(b). The multiplication cannot be performed due to incompatible (mismatched) orders (dimensions).

In Example 6, did you notice that AB and BA are different matrices? For most matrices A and B, $AB \neq BA$. Because **matrix multiplication is not commutative,** be careful about the order in which matrices appear when performing this operation.

> **Check Point 6** If $A = \begin{bmatrix} 2 & 0 & 4 \end{bmatrix}$ and $B = \begin{bmatrix} 1 \\ 3 \\ 7 \end{bmatrix}$, find AB and BA.

EXAMPLE 7 Multiplying Matrices

Where possible, find each product:

a. $\begin{bmatrix} 4 & 2 \\ 1 & 3 \end{bmatrix} \begin{bmatrix} 1 & 2 & 3 & 4 \\ 0 & 2 & -1 & 6 \end{bmatrix}$ **b.** $\begin{bmatrix} 1 & 2 & 3 & 4 \\ 0 & 2 & -1 & 6 \end{bmatrix} \begin{bmatrix} 4 & 2 \\ 1 & 3 \end{bmatrix}$

Solution

a. The first matrix is a 2 × 2 matrix and the second is a 2 × 4 matrix. The product will be a 2 × 4 matrix.

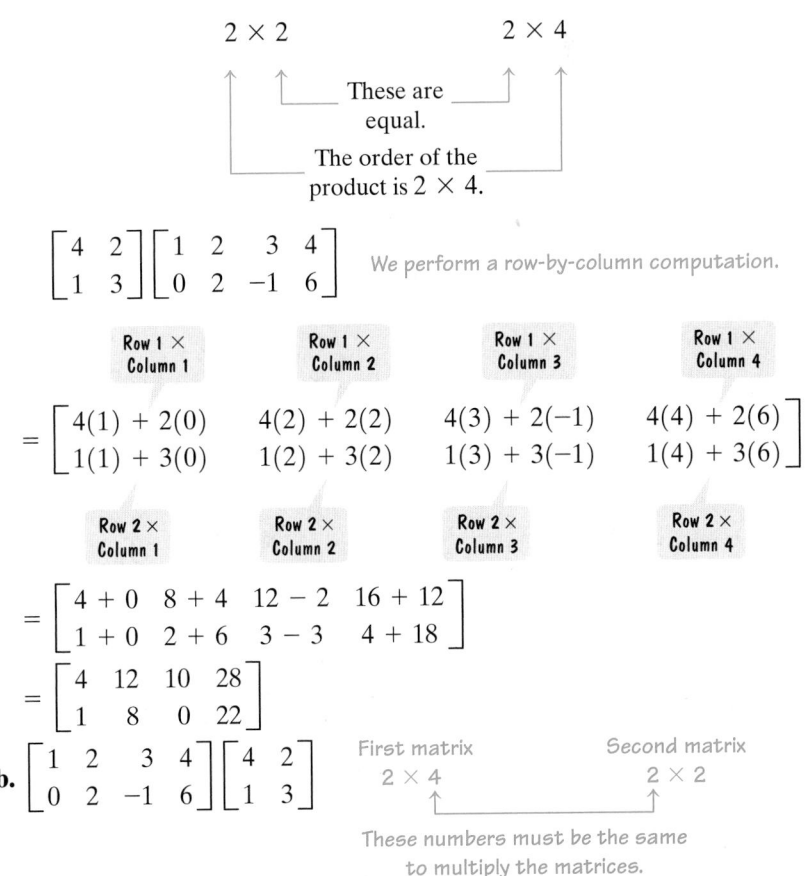

The number of columns in the first matrix does not equal the number of rows in the second matrix. Thus, the product of these two matrices is undefined.

Check Point 7 Where possible, find each product:

a. $\begin{bmatrix} 1 & 3 \\ 0 & 2 \end{bmatrix} \begin{bmatrix} 2 & 3 & -1 & 6 \\ 0 & 5 & 4 & 1 \end{bmatrix}$ **b.** $\begin{bmatrix} 2 & 3 & -1 & 6 \\ 0 & 5 & 4 & 1 \end{bmatrix} \begin{bmatrix} 1 & 3 \\ 0 & 2 \end{bmatrix}$.

Discovery

Verify the properties listed in the box using

$A = \begin{bmatrix} 3 & 2 \\ -1 & 4 \end{bmatrix}$

$B = \begin{bmatrix} 1 & 0 \\ 3 & 2 \end{bmatrix}$

$C = \begin{bmatrix} 1 & 2 \\ -1 & 1 \end{bmatrix}$

and $c = 3$.

Although matrix multiplication is not commutative, it does obey many of the properties of real numbers.

Properties of Matrix Multiplication

If A, B, and C are matrices and c is a scalar, then the following properties are true. (Assume the order of each matrix is such that all operations in these properties are defined.)

1. $(AB)C = A(BC)$ Associative Property of Matrix Multiplication

2. $A(B + C) = AB + AC$ Distributive Properties of Matrix

    $(A + B)C = AC + BC$ Multiplication

3. $c(AB) = (cA)B$ Associative Property of Scalar Multiplication

7 Describe applied situations with matrix operations.

Applications

All of the still images that you see on the Web have been created or manipulated on a computer in a digital format—made up of hundreds of thousands, or even millions, of tiny squares called **pixels.** Pixels are created by dividing an image into a grid. The computer can change the brightness of every square or pixel in this grid. A digital camera captures photos in this digital format. Also, you can scan pictures to convert them into digital format. Example 8 illustrates the role that matrices play in this new technology.

EXAMPLE 8   Matrices and Digital Photography

The letter L in Figure 9.7 is shown using 9 pixels in a 3 × 3 grid. The colors possible in the grid are shown in Figure 9.8. Each color is represented by a specific number: 0, 1, 2, or 3.

Figure 9.7
The letter L

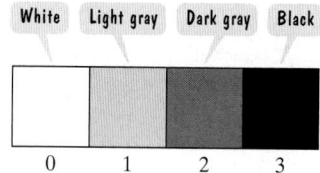

Figure 9.8
Color levels

a. Find a matrix that represents a digital photograph of this letter L.

b. Increase the contrast of the letter L by changing the dark gray to black and the light gray to white. Use matrix addition to accomplish this.

Solution

a. Look at the L and the background in Figure 9.7. Because the L is dark gray, color level 2, and the background is light gray, color level 1, a digital photograph of Figure 9.7 can be represented by the matrix

$$\begin{bmatrix} 2 & 1 & 1 \\ 2 & 1 & 1 \\ 2 & 2 & 1 \end{bmatrix}.$$

b. We can make the L black, color level 3, by increasing each 2 in the above matrix to 3. We can make the background white, color level 0, by decreasing each 1 in the above matrix to 0. This is accomplished using the following matrix addition:

$$\begin{bmatrix} 2 & 1 & 1 \\ 2 & 1 & 1 \\ 2 & 2 & 1 \end{bmatrix} + \begin{bmatrix} 1 & -1 & -1 \\ 1 & -1 & -1 \\ 1 & 1 & -1 \end{bmatrix} = \begin{bmatrix} 3 & 0 & 0 \\ 3 & 0 & 0 \\ 3 & 3 & 0 \end{bmatrix}.$$

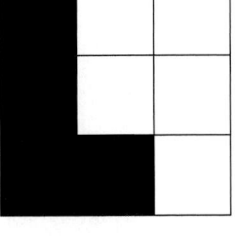

Figure 9.9 Changing contrast: the letter L

The picture corresponding to the matrix sum to the right of the equal sign is shown in Figure 9.9.

> **Check Point 8** Change the contrast of the letter L in Figure 9.7 by making the L light gray and the background black. Use matrix addition to accomplish this.

Images of Space

Photographs sent back from space use matrices with thousands of pixels. Each pixel is assigned a number from 0 to 63 representing its color—0 for pure white and 63 for pure black. In the image of Saturn shown here, matrix operations provide false colors that emphasize the banding of the planet's upper atmosphere.

EXAMPLE 9 Applying Matrix Multiplication

At a certain gas station, the number of gallons of regular, unleaded, and super unleaded gas sold on Monday, Tuesday, and Wednesday of a particular week is given by the following matrix.

	Regular	Unleaded	Super Unleaded
Monday	240	300	160
Tuesday	200	280	180
Wednesday	260	310	200

$= A$

A second matrix gives the selling price per gallon and the profit per gallon for the three types of gas sold by the station.

	Selling price per Gallon	Profit per Gallon
Regular	1.80	0.30
Unleaded	1.90	0.34
Super Unleaded	2.00	0.38

$= B$

a. Calculate the product AB.

b. What is the gas station's profit for Monday through Wednesday?

Solution

a. $AB = \begin{bmatrix} 240 & 300 & 160 \\ 200 & 280 & 180 \\ 260 & 310 & 200 \end{bmatrix} \begin{bmatrix} 1.80 & 0.30 \\ 1.90 & 0.34 \\ 2.00 & 0.38 \end{bmatrix}$

$= \begin{bmatrix} 240(1.80) + 300(1.90) + 160(2.00) & 240(0.30) + 300(0.34) + 160(0.38) \\ 200(1.80) + 280(1.90) + 180(2.00) & 200(0.30) + 280(0.34) + 180(0.38) \\ 260(1.80) + 310(1.90) + 200(2.00) & 260(0.30) + 310(0.34) + 200(0.38) \end{bmatrix}$

Perform the row-by-column multiplications.

$= \begin{bmatrix} 1322 & 234.80 \\ 1252 & 223.60 \\ 1457 & 259.40 \end{bmatrix}$ Multiply and add as indicated.

b. The entries in the second column of the product matrix represent profits for Monday, Tuesday, and Wednesday, respectively. The gas station's profit for Monday through Wednesday is $234.80 + $223.60 + $259.40, or $717.80.

Check Point 9   Use the product matrix in Example 9a to answer this question. What are the gas station's total sales for Monday, Tuesday, and Wednesday?

EXERCISE SET 9.3

Practice Exercises

In Exercises 1–4,
a. *Give the order of each matrix.*
b. *If $A = [a_{ij}]$, identify a_{32} and a_{23} or explain why identification is not possible.*

1. $\begin{bmatrix} 4 & -7 & 5 \\ -6 & 8 & -1 \end{bmatrix}$ **2.** $\begin{bmatrix} -6 & 4 & -1 \\ -9 & 0 & \frac{1}{2} \end{bmatrix}$

3. $\begin{bmatrix} 1 & -5 & \pi & e \\ 0 & 7 & -6 & -\pi \\ -2 & \frac{1}{2} & 11 & -\frac{1}{5} \end{bmatrix}$ **4.** $\begin{bmatrix} -4 & 1 & 3 & -5 \\ 2 & -1 & \pi & 0 \\ 1 & 0 & -e & \frac{1}{5} \end{bmatrix}$

In Exercises 5–8, find values for the variables so that the matrices in each exercise are equal.

5. $\begin{bmatrix} x \\ 4 \end{bmatrix} = \begin{bmatrix} 6 \\ y \end{bmatrix}$ **6.** $\begin{bmatrix} x \\ 7 \end{bmatrix} = \begin{bmatrix} 11 \\ y \end{bmatrix}$

7. $\begin{bmatrix} x & 2y \\ z & 9 \end{bmatrix} = \begin{bmatrix} 4 & 12 \\ 3 & 9 \end{bmatrix}$

8. $\begin{bmatrix} x & y+3 \\ 2z & 8 \end{bmatrix} = \begin{bmatrix} 12 & 5 \\ 6 & 8 \end{bmatrix}$

In Exercises 9–16, find:
a. $A + B$ **b.** $A - B$
c. $-4A$ **d.** $3A + 2B$.

9. $A = \begin{bmatrix} 4 & 1 \\ 3 & 2 \end{bmatrix}$, $B = \begin{bmatrix} 5 & 9 \\ 0 & 7 \end{bmatrix}$

10. $A = \begin{bmatrix} -2 & 3 \\ 0 & 1 \end{bmatrix}$, $B = \begin{bmatrix} 8 & 1 \\ 5 & 4 \end{bmatrix}$

11. $A = \begin{bmatrix} 1 & 3 \\ 3 & 4 \\ 5 & 6 \end{bmatrix}$, $B = \begin{bmatrix} 2 & -1 \\ 3 & -2 \\ 0 & 1 \end{bmatrix}$

12. $A = \begin{bmatrix} 3 & 1 & 1 \\ -1 & 2 & 5 \end{bmatrix}$, $B = \begin{bmatrix} 2 & -3 & 6 \\ -3 & 1 & -4 \end{bmatrix}$

13. $A = \begin{bmatrix} 2 \\ -4 \\ 1 \end{bmatrix}$, $B = \begin{bmatrix} -5 \\ 3 \\ -1 \end{bmatrix}$

14. $A = \begin{bmatrix} 6 & 2 & -3 \end{bmatrix}$, $B = \begin{bmatrix} 4 & -2 & 3 \end{bmatrix}$

15. $A = \begin{bmatrix} 2 & -10 & -2 \\ 14 & 12 & 10 \\ 4 & -2 & 2 \end{bmatrix}$, $B = \begin{bmatrix} 6 & 10 & -2 \\ 0 & -12 & -4 \\ -5 & 2 & -2 \end{bmatrix}$

16. $A = \begin{bmatrix} 6 & -3 & 5 \\ 6 & 0 & -2 \\ -4 & 2 & -1 \end{bmatrix}$, $B = \begin{bmatrix} -3 & 5 & 1 \\ -1 & 2 & -6 \\ 2 & 0 & 4 \end{bmatrix}$

In Exercises 17–26, let

$A = \begin{bmatrix} -3 & -7 \\ 2 & -9 \\ 5 & 0 \end{bmatrix}$ *and* $B = \begin{bmatrix} -5 & -1 \\ 0 & 0 \\ 3 & -4 \end{bmatrix}$.

Solve each matrix equation for X.

17. $X - A = B$ **18.** $X - B = A$

19. $2X + A = B$ **20.** $3X + A = B$

21. $3X + 2A = B$ **22.** $2X + 5A = B$

23. $B - X = 4A$ **24.** $A - X = 4B$

25. $4A + 3B = -2X$ **26.** $4B + 3A = -2X$

In Exercises 27–36, find (if possible):
a. *AB*   and   **b.** *BA.*

27. $A = \begin{bmatrix} 1 & 3 \\ 5 & 3 \end{bmatrix}$, $B = \begin{bmatrix} 3 & -2 \\ -1 & 6 \end{bmatrix}$

28. $A = \begin{bmatrix} 3 & -2 \\ 1 & 5 \end{bmatrix}$, $B = \begin{bmatrix} 0 & 0 \\ 5 & -6 \end{bmatrix}$

29. $A = \begin{bmatrix} 1 & 2 & 3 & 4 \end{bmatrix}$, $B = \begin{bmatrix} 1 \\ 2 \\ 3 \\ 4 \end{bmatrix}$

30. $A = \begin{bmatrix} -1 \\ -2 \\ -3 \end{bmatrix}$, $B = \begin{bmatrix} 1 & 2 & 3 \end{bmatrix}$

31. $A = \begin{bmatrix} 1 & -1 & 4 \\ 4 & -1 & 3 \\ 2 & 0 & -2 \end{bmatrix}$, $B = \begin{bmatrix} 1 & 1 & 0 \\ 1 & 2 & 4 \\ 1 & -1 & 3 \end{bmatrix}$

32. $A = \begin{bmatrix} 1 & -1 & 1 \\ 5 & 0 & -2 \\ 3 & -2 & 2 \end{bmatrix}$, $B = \begin{bmatrix} 1 & 1 & 0 \\ 1 & -4 & 5 \\ 3 & -1 & 2 \end{bmatrix}$

33. $A = \begin{bmatrix} 4 & 2 \\ 6 & 1 \\ 3 & 5 \end{bmatrix}$, $B = \begin{bmatrix} 2 & 3 & 4 \\ -1 & -2 & 0 \end{bmatrix}$

34. $A = \begin{bmatrix} 2 & 4 \\ 3 & 1 \\ 4 & 2 \end{bmatrix}$, $B = \begin{bmatrix} 3 & 2 & 0 \\ -1 & -3 & 5 \end{bmatrix}$

35. $A = \begin{bmatrix} 2 & -3 & 1 & -1 \\ 1 & 1 & -2 & 1 \end{bmatrix}$, $B = \begin{bmatrix} 1 & 2 \\ -1 & 1 \\ 5 & 4 \\ 10 & 5 \end{bmatrix}$

36. $A = \begin{bmatrix} 2 & -1 & 3 & 2 \\ 1 & 0 & -2 & 1 \end{bmatrix}$, $B = \begin{bmatrix} -1 & 2 \\ 1 & 1 \\ 3 & -4 \\ 6 & 5 \end{bmatrix}$

In Exercises 37–44, perform the indicated matrix operations given that A, B, and C are defined as follows. If an operation is not defined, state the reason.

$A = \begin{bmatrix} 4 & 0 \\ -3 & 5 \\ 0 & 1 \end{bmatrix}$ $B = \begin{bmatrix} 5 & 1 \\ -2 & -2 \end{bmatrix}$ $C = \begin{bmatrix} 1 & -1 \\ -1 & 1 \end{bmatrix}$

37. $4B - 3C$ **38.** $5C - 2B$

39. $BC + CB$ **40.** $A(B + C)$

41. $A - C$ **42.** $B - A$

43. $A(BC)$ **44.** $A(CB)$

Application Exercises

    The + sign in the figure is shown using 9 pixels in a 3×3 grid. The color levels are given to the right of the figure. Each color is represented by a specific number: 0, 1, 2, or 3. Use this information to solve Exercises 45–46.

45. a. Find a matrix that represents a digital photograph of the + sign.

    b. Adjust the contrast by changing the black to dark gray and the light gray to white. Use matrix addition to accomplish this.

    c. Adjust the contrast by changing the black to light gray and the light gray to dark gray. Use matrix addition to accomplish this.

46. a. Find a matrix that represents a digital photograph of the + sign.

    b. Adjust the contrast by changing the black to dark gray and the light gray to black. Use matrix addition to accomplish this.

    c. Adjust the contrast by leaving the black alone and changing the light gray to white. Use matrix addition to accomplish this.

The figure shows the letter L in a rectangular coordinate system.

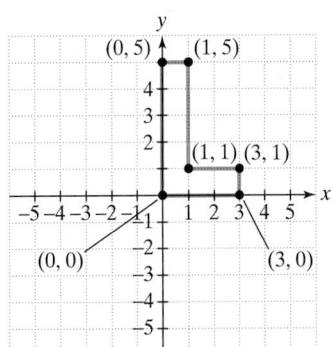

The figure can be represented by the matrix

$$B = \begin{bmatrix} 0 & 3 & 3 & 1 & 1 & 0 \\ 0 & 0 & 1 & 1 & 5 & 5 \end{bmatrix}.$$

Each column in the matrix describes a point on the letter. The order of the columns shows the direction in which a pencil must move to draw the letter. The L is completed by connecting the last point in the matrix, (0, 5), to the starting point, (0, 0). Use these ideas to solve Exercises 47–48.

(Be sure to refer to the discussion on the previous page as you work Exercises 47–48.)

47. a. If $A = \begin{bmatrix} -1 & 0 \\ 0 & 1 \end{bmatrix}$, find AB.

b. Graph the object represented by matrix AB. What effect does the matrix multiplication have on the letter L represented by matrix B?

48. a. If $A = \begin{bmatrix} 1 & 0 \\ 0 & -1 \end{bmatrix}$, find AB.

b. Graph the object represented by matrix AB. What effect does the matrix multiplication have on the letter L represented by matrix B?

49. The graph shows the percentage of whites and African Americans with college degrees, by gender.

Percentage of People with College Degrees

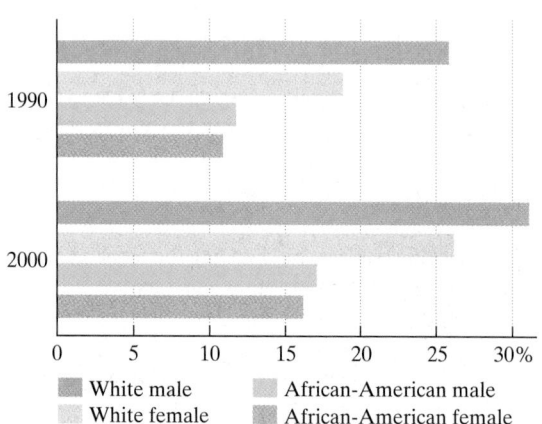

Source: Bureau of the Census

a. Use a 2×2 matrix to represent the information for 1990. Entries in the matrix should be percentages that you estimate from the graph in the following order:

$$\begin{bmatrix} \text{White male} & \text{White female} \\ \text{African-American male} & \text{African-American female} \end{bmatrix}.$$

Call this matrix A.

b. Use a 2×2 matrix to represent the information given for 2000. Call this matrix B.

c. Find $B - A$. What does this matrix represent?

50. In a certain county, the proportion of voters in each age group registered as Republicans, Democrats, or Independents is given by the following matrix, which we'll call A.

	18–30	**Age 31–50**	**Over 50**
Republicans	0.4	0.30	0.70
Democrats	0.30	0.60	0.25
Independents	0.30	0.10	0.05

The distribution, by age and gender, of this county's voting population is given by the following matrix, which we'll call B.

	Male	**Female**
18–30	6000	8000
Age 31–50	12,000	14,000
Over 50	14,000	16,000

a. Calculate the product AB.

b. How many female Democrats are there?

c. How many male Republicans are there?

51. The final grade in a particular course is determined by grades on the midterm and final. The grades for five students and the two grading systems are modeled by the following matrices. Call the first matrix A and the second B.

	Midterm	**Final**
Student 1	76	92
Student 2	74	84
Student 3	94	86
Student 4	84	62
Student 5	58	80

	System 1	**System 2**
Midterm	0.5	0.3
Final	0.5	0.7

a. Describe the grading system that is represented by matrix B.

b. Compute the matrix AB and assign each of the five students a final course grade first using system 1 and then using system 2. ($89.5 - 100 = A$, $79.5 - 89.4 = B$, $69.5 - 79.4 = C$, $59.5 - 69.4 = D$, below $59.5 = F$)

Writing in Mathematics

52. What is meant by the order of a matrix? Give an example with your explanation.

53. What does a_{ij} mean?

54. What are equal matrices?

55. How are matrices added?

56. Describe how to subtract matrices.

57. Describe matrices that cannot be added or subtracted.

58. Describe how to perform scalar multiplication. Provide an example with your description.

59. Describe how to multiply matrices.

60. Describe when the multiplication of two matrices is not defined.

61. If two matrices can be multiplied, describe how to determine the order of the product.

62. Low-resolution digital photographs use 262,144 pixels in a 512×512 grid. If you enlarge a low-resolution digital photograph enough, describe what will happen.

Technology Exercise

63. Use the matrix feature of a graphing utility to verify each of your answers to Exercises 37–44.

Critical Thinking Exercises

64. Find two matrices A and B such that $AB = BA$.

65. Consider a square matrix such that each element that is not on the diagonal from upper left to lower right is zero. Experiment with such matrices (call each matrix A) by finding AA. Then write a sentence or two describing a method for multiplying this kind of matrix by itself.

66. If $AB = -BA$, then A and B are said to be anticommutative. Are $A = \begin{bmatrix} 0 & -1 \\ 1 & 0 \end{bmatrix}$ and $B = \begin{bmatrix} 1 & 0 \\ 0 & -1 \end{bmatrix}$ anticommutative?

Group Exercise

67. The interesting and useful applications of matrix theory are nearly unlimited. Applications of matrices range from representing digital photographs to predicting long-range trends in the stock market. Members of the group should research an application of matrices that they find intriguing. The group should then present a seminar to the class about this application.

SECTION 9.4 Multiplicative Inverses of Matrices and Matrix Equations

Objectives

1. Find the multiplicative inverse of a square matrix.
2. Use inverses to solve matrix equations.
3. Encode and decode messages.

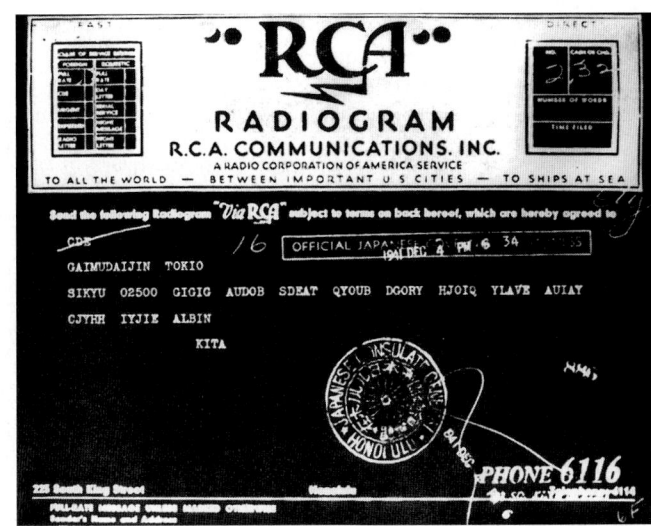

This 1941 RCA radiogram shows an encoded message from the Japanese government.

In 1939, Britain's secret service hired top chess players, mathematicians, and other masters of logic to break the code used by the Nazis in communications between headquarters and troops. The project, which employed over 10,000 people, broke the code less than a year later, providing the Allies with information about Nazi troop movements throughout World War II.

Messages must often be sent in such a way that the real meaning is hidden from everyone but the sender and the recipient. In this section, we will look at the role that matrices and their inverses play in this process.

The Multiplicative Identity Matrix

For the real numbers, we know that 1 is the multiplicative identity because $a \cdot 1 = 1 \cdot a = a$. Is there a similar property for matrix multiplication? That is, is there a matrix I such that $AI = A$ and $IA = A$? The answer is yes. A square matrix with 1s down the diagonal from upper left to lower right and 0s elsewhere does not change the elements in a matrix when it multiplies that matrix. In the case of 2×2 matrices,

$$\begin{bmatrix} a_{11} & a_{12} \\ a_{21} & a_{22} \end{bmatrix} \begin{bmatrix} 1 & 0 \\ 0 & 1 \end{bmatrix} = \begin{bmatrix} a_{11} & a_{12} \\ a_{21} & a_{22} \end{bmatrix}$$

The elements in the matrix do not change.

and $\begin{bmatrix} 1 & 0 \\ 0 & 1 \end{bmatrix} \begin{bmatrix} a_{11} & a_{12} \\ a_{21} & a_{22} \end{bmatrix} = \begin{bmatrix} a_{11} & a_{12} \\ a_{21} & a_{22} \end{bmatrix}.$

The elements in the matrix do not change.

An $n \times n$ square matrix with 1s down the diagonal from upper left to lower right and 0s elsewhere is called the **multiplicative identity matrix of order n.** This matrix is designated by I_n. For example,

$$I_2 = \begin{bmatrix} 1 & 0 \\ 0 & 1 \end{bmatrix}, \quad I_3 = \begin{bmatrix} 1 & 0 & 0 \\ 0 & 1 & 0 \\ 0 & 0 & 1 \end{bmatrix},$$

and so on.

1 Find the multiplicative inverse of a square matrix.

The Multiplicative Inverse of a Matrix

The multiplicative identity matrix, I_n, will help us to define a new concept: the multiplicative inverse of a matrix. To do so, let's consider a similar concept, the multiplicative inverse of a nonzero number, a. Recall that the multiplicative inverse of a is $\frac{1}{a}$. The multiplicative inverse has the following property:

$$a \cdot \frac{1}{a} = 1 \quad \text{and} \quad \frac{1}{a} \cdot a = 1.$$

We can define the multiplicative inverse of a square matrix in a similar manner.

Definition of the Multiplicative Inverse of a Square Matrix

Let A be an $n \times n$ matrix. If there exists an $n \times n$ matrix A^{-1} (read: "A inverse") such that

$$AA^{-1} = I_n \quad \text{and} \quad A^{-1}A = I_n,$$

then A^{-1} is the **multiplicative inverse** of A.

We have seen that matrix multiplication is not commutative. Thus, to show that matrix B is the multiplicative inverse of matrix A, find both AB and BA. If B is the multiplicative inverse of A, both products (AB and BA) will be the multiplicative identity matrix, I_n.

EXAMPLE 1 The Multiplicative Inverse of a Matrix

Show that B is the multiplicative inverse of A, where

$$A = \begin{bmatrix} -1 & 3 \\ 2 & -5 \end{bmatrix} \quad \text{and} \quad B = \begin{bmatrix} 5 & 3 \\ 2 & 1 \end{bmatrix}.$$

Solution To show that B is the multiplicative inverse of A, we must find the products AB and BA. If B is the multiplicative inverse of A, then AB will be the multiplicative identity matrix and BA will be the multiplicative identity matrix. Because A and B are 2×2 matrices, $n = 2$. Thus, we denote the multiplicative identity matrix as I_2; it is also a 2×2 matrix. We must show that

- $AB = I_2 = \begin{bmatrix} 1 & 0 \\ 0 & 1 \end{bmatrix}.$

- $BA = I_2 = \begin{bmatrix} 1 & 0 \\ 0 & 1 \end{bmatrix}.$

Let's first show $AB = I_2$.

$$AB = \begin{bmatrix} -1 & 3 \\ 2 & -5 \end{bmatrix} \begin{bmatrix} 5 & 3 \\ 2 & 1 \end{bmatrix}$$

$$= \begin{bmatrix} -1(5) + 3(2) & -1(3) + 3(1) \\ 2(5) + (-5)(2) & 2(3) + (-5)(1) \end{bmatrix} = \begin{bmatrix} 1 & 0 \\ 0 & 1 \end{bmatrix}$$

Let's now show $BA = I_2$.

$$BA = \begin{bmatrix} 5 & 3 \\ 2 & 1 \end{bmatrix} \begin{bmatrix} -1 & 3 \\ 2 & -5 \end{bmatrix}$$

$$= \begin{bmatrix} 5(-1) + 3(2) & 5(3) + 3(-5) \\ 2(-1) + 1(2) & 2(3) + 1(-5) \end{bmatrix} = \begin{bmatrix} 1 & 0 \\ 0 & 1 \end{bmatrix}$$

Both products give the multiplicative identity matrix. Thus, B is the multiplicative inverse of A and we can designate B as $A^{-1} = \begin{bmatrix} 5 & 3 \\ 2 & 1 \end{bmatrix}.$

Check Point 1 Show that B is the multiplicative inverse of A, where

$$A = \begin{bmatrix} 2 & 1 \\ 1 & 1 \end{bmatrix} \quad \text{and} \quad B = \begin{bmatrix} 1 & -1 \\ -1 & 2 \end{bmatrix}.$$

Technology

You can use a graphing utility to find the inverse of the matrix in Example 2. Enter the matrix and name it A. The screens show A and A^{-1}.

[A]
[[2 1]
[5 3]]
■

[A]⁻¹
[[3 -1]
[-5 2]]
■

One method for finding the multiplicative inverse of a matrix A is to begin by denoting the elements in A^{-1} with variables. Using the equation $AA^{-1} = I_n$, we can find a value for each element in the multiplicative inverse that was represented by a variable. Example 2 shows how this is done.

EXAMPLE 2   Finding the Multiplicative Inverse of a Matrix

Find the multiplicative inverse of

$$A = \begin{bmatrix} 2 & 1 \\ 5 & 3 \end{bmatrix}.$$

Solution   Let us denote the multiplicative inverse by

$$A^{-1} = \begin{bmatrix} w & x \\ y & z \end{bmatrix}.$$

Because A is a 2×2 matrix, we use the equation $AA^{-1} = I_2$ to find values for $x, y, z,$ and w.

$$\underset{A}{\begin{bmatrix} 2 & 1 \\ 5 & 3 \end{bmatrix}} \underset{A^{-1}}{\begin{bmatrix} w & x \\ y & z \end{bmatrix}} = \underset{I_2}{\begin{bmatrix} 1 & 0 \\ 0 & 1 \end{bmatrix}}$$

$$\begin{bmatrix} 2w + y & 2x + z \\ 5w + 3y & 5x + 3z \end{bmatrix} = \begin{bmatrix} 1 & 0 \\ 0 & 1 \end{bmatrix}$$

Use row-by-column matrix multiplication on the left side of $\begin{bmatrix} 2 & 1 \\ 5 & 3 \end{bmatrix} \begin{bmatrix} w & x \\ y & z \end{bmatrix} = \begin{bmatrix} 1 & 0 \\ 0 & 1 \end{bmatrix}$

We now equate corresponding elements to obtain the following two systems of linear equations:

$$\begin{array}{ccc} 2w + y = 1 & & 2x + z = 0 \\ 5w + 3y = 0 & \text{and} & 5x + 3z = 1. \end{array}$$

Each of these systems can be solved using the addition method.

$$\begin{array}{lll} 2w + y = 1 & \xrightarrow{\text{Multiply by } -3.} & -6w - 3y = -3 \\ 5w + 3y = 0 & \xrightarrow{\text{No change}} & 5w + 3y = 0 \\ & \text{Add:} & -w = -3 \\ & & w = 3 \\ & \text{Use back-substitution.} & y = -5 \end{array}$$

$$\begin{array}{lll} 2x + z = 0 & \xrightarrow{\text{Multiply by } -3.} & -6x - 3z = 0 \\ 5x + 3z = 1 & \xrightarrow{\text{No change}} & 5x + 3z = 1 \\ & \text{Add:} & -x = 1 \\ & & x = -1 \\ & \text{Use back-substitution.} & z = 2 \end{array}$$

Using these values, we have

$$A^{-1} = \begin{bmatrix} w & x \\ y & z \end{bmatrix} = \begin{bmatrix} 3 & -1 \\ -5 & 2 \end{bmatrix}.$$

Discovery

Verify that the inverse matrix found in Example 2 is correct. Use matrix multiplication to show that

$$AA^{-1} = I_2 \quad \text{and} \quad A^{-1}A = I_2,$$

where

$$I_2 = \begin{bmatrix} 1 & 0 \\ 0 & 1 \end{bmatrix}.$$

Check Point 2 Find the multiplicative inverse of $A = \begin{bmatrix} 5 & 7 \\ 2 & 3 \end{bmatrix}$.

Only square matrices of order $n \times n$ have multiplicative inverses, but not every square matrix possesses a multiplicative inverse. For example, suppose that you apply the procedure of Example 2 to $A = \begin{bmatrix} -6 & 4 \\ -3 & 2 \end{bmatrix}$:

This is A. This represents A^{-1}. This is the multiplicative identity matrix.

$$\begin{bmatrix} -6 & 4 \\ -3 & 2 \end{bmatrix} \begin{bmatrix} w & x \\ y & z \end{bmatrix} = \begin{bmatrix} 1 & 0 \\ 0 & 1 \end{bmatrix}.$$

Multiplying matrices on the left and equating corresponding elements results in inconsistent systems with no solutions. There are no values for w, x, y, and z. This shows that matrix A does not have a multiplicative inverse.

A nonsquare matrix, one with a different number of rows than columns, cannot have a multiplicative inverse. If A is an $m \times n$ matrix and B is an $n \times m$ matrix $(n \neq m)$, then the products AB and BA are of different orders. This means that they could not be equal to each other, so that AB and BA could not both equal the multiplicative identity matrix.

If a square matrix has a multiplicative inverse, that inverse is unique. This means that the square matrix has no more than one inverse. If a square matrix has a multiplicative inverse, it is said to be **invertible** or **nonsingular.** If a square matrix has no multiplicative inverse, it is called **singular.**

A Quick Method for Finding the Multiplicative Inverse of a 2 × 2 Matrix

The following rule enables us to calculate the multiplicative inverse, if there is one, of a 2 × 2 matrix:

> **Study Tip**
>
> To find the matrix that appears as the second factor for the inverse of
>
> $$A = \begin{bmatrix} a & b \\ c & d \end{bmatrix}:$$
>
> - Reverse a and d, the numbers in the diagonal from upper left to lower right.
> - Negate b and c, the numbers in the other diagonal.

Multiplicative Inverse of a 2 × 2 Matrix

If $A = \begin{bmatrix} a & b \\ c & d \end{bmatrix}$, then $A^{-1} = \dfrac{1}{ad - bc} \begin{bmatrix} d & -b \\ -c & a \end{bmatrix}$.

The matrix A is invertible if and only if $ad - bc \neq 0$. If $ad - bc = 0$, then A does not have a multiplicative inverse.

EXAMPLE 3   Using the Quick Method to Find Multiplicative Inverses

Find the multiplicative inverse of

$$A = \begin{bmatrix} -1 & -2 \\ 3 & 4 \end{bmatrix}.$$

Solution

> **Study Tip**
>
> When using the formula to find the multiplicative inverse, start by computing $ad - bc$. If the computed value is 0, there is no need to continue. The given matrix is singular—that is, it does not have a multiplicative inverse.

$$A = \begin{bmatrix} -1 & -2 \\ 3 & 4 \end{bmatrix}$$

This is the given matrix. We've designated the elements a, b, c, and d.

$$A^{-1} = \frac{1}{ad - bc} \begin{bmatrix} d & -b \\ -c & a \end{bmatrix}$$

This is the formula for the inverse of $\begin{bmatrix} a & b \\ c & d \end{bmatrix}$.

$$= \frac{1}{(-1)(4) - (-2)(3)} \begin{bmatrix} 4 & -(-2) \\ -3 & -1 \end{bmatrix}$$

Apply the formula with $a = -1$, $b = -2$, $c = 3$, and $d = 4$.

$$= \frac{1}{2} \begin{bmatrix} 4 & 2 \\ -3 & -1 \end{bmatrix}$$

Simplify.

$$= \begin{bmatrix} 2 & 1 \\ -\frac{3}{2} & -\frac{1}{2} \end{bmatrix}$$

Perform the scalar multiplication by multiplying each element in the matrix by $\frac{1}{2}$.

The inverse of $A = \begin{bmatrix} -1 & -2 \\ 3 & 4 \end{bmatrix}$ is $A^{-1} = \begin{bmatrix} 2 & 1 \\ -\frac{3}{2} & -\frac{1}{2} \end{bmatrix}$.

We can verify this result by showing that $AA^{-1} = I_2$ and $A^{-1}A = I_2$.

Find the multiplicative inverse of

$$A = \begin{bmatrix} 3 & -2 \\ -1 & 1 \end{bmatrix}.$$

Finding Multiplicative Inverses of $n \times n$ Matrices with n Greater Than 2

To find the multiplicative inverse of a 3×3 invertible matrix, we begin by denoting the elements in the multiplicative inverse with variables. Here is an example:

$$\begin{bmatrix} -1 & -1 & -1 \\ 4 & 5 & 0 \\ 0 & 1 & -3 \end{bmatrix} \begin{bmatrix} x_1 & x_2 & x_3 \\ y_1 & y_2 & y_3 \\ z_1 & z_2 & z_3 \end{bmatrix} = \begin{bmatrix} 1 & 0 & 0 \\ 0 & 1 & 0 \\ 0 & 0 & 1 \end{bmatrix}.$$

This is matrix A whose inverse we wish to find. This represents A^{-1}. This is the multiplicative identity matrix, I_3.

We multiply the matrices on the left, using the row-by-column definition of matrix multiplication.

$$\begin{bmatrix} -x_1 - y_1 - z_1 & -x_2 - y_2 - z_2 & -x_3 - y_3 - z_3 \\ 4x_1 + 5y_1 + 0z_1 & 4x_2 + 5y_2 + 0z_2 & 4x_3 + 5y_3 + 0z_3 \\ 0x_1 + 1y_1 - 3z_1 & 0x_2 + 1y_2 - 3z_2 & 0x_3 + 1y_3 - 3z_3 \end{bmatrix} = \begin{bmatrix} 1 & 0 & 0 \\ 0 & 1 & 0 \\ 0 & 0 & 1 \end{bmatrix}$$

We now equate corresponding entries to obtain the following three systems of linear equations:

$$\begin{array}{lll} -x_1 - y_1 - z_1 = 1 & -x_2 - y_2 - z_2 = 0 & -x_3 - y_3 - z_3 = 0 \\ 4x_1 + 5y_1 + 0z_1 = 0 & 4x_2 + 5y_2 + 0z_2 = 1 & 4x_3 + 5y_3 + 0z_3 = 0 \\ 0x_1 + y_1 - 3z_1 = 0 & 0x_2 + y_2 - 3z_2 = 0 & 0x_3 + y_3 - 3z_3 = 1. \end{array}$$

Notice that the variables on the left of the equal sign have the same coefficients in each system. We can use Gauss-Jordan elimination to solve all three systems at once. Form an augmented matrix that contains the coefficients of the three systems to the left of the vertical line and the constants for the systems to the right.

$$\left[\begin{array}{ccc|ccc} -1 & -1 & -1 & 1 & 0 & 0 \\ 4 & 5 & 0 & 0 & 1 & 0 \\ 0 & 1 & -3 & 0 & 0 & 1 \end{array}\right]$$

Coefficients of the three systems. Constants on the right in each of the three systems.

To solve all three systems using Gauss-Jordan elimination, we must obtain $\begin{bmatrix} 1 & 0 & 0 \\ 0 & 1 & 0 \\ 0 & 0 & 1 \end{bmatrix}$ to the left of the vertical line. Use matrix row operations, working one column at a time. Obtain 1 in the required position. Then obtain 0s in the other two positions. Using these operations, we obtain the matrix

$$\left[\begin{array}{ccc|ccc} 1 & 0 & 0 & 15 & 4 & -5 \\ 0 & 1 & 0 & -12 & -3 & 4 \\ 0 & 0 & 1 & -4 & -1 & 1 \end{array}\right].$$

This augmented matrix provides the solutions to the three systems of equations. They are given by

$$\begin{bmatrix} 1 & 0 & 0 & | & 15 \\ 0 & 1 & 0 & | & -12 \\ 0 & 0 & 1 & | & -4 \end{bmatrix} \qquad \begin{aligned} x_1 &= 15 \\ y_1 &= -12 \\ z_1 &= -4 \end{aligned}$$

and

$$\begin{bmatrix} 1 & 0 & 0 & | & 4 \\ 0 & 1 & 0 & | & -3 \\ 0 & 0 & 1 & | & -1 \end{bmatrix} \qquad \begin{aligned} x_2 &= 4 \\ y_2 &= -3 \\ z_2 &= -1 \end{aligned}$$

and

$$\begin{bmatrix} 1 & 0 & 0 & | & -5 \\ 0 & 1 & 0 & | & 4 \\ 0 & 0 & 1 & | & 1 \end{bmatrix} \qquad \begin{aligned} x_3 &= -5 \\ y_3 &= 4 \\ z_3 &= 1. \end{aligned}$$

The inverse matrix is

$$\begin{bmatrix} x_1 & x_2 & x_3 \\ y_1 & y_2 & y_3 \\ z_1 & z_2 & z_3 \end{bmatrix} = \begin{bmatrix} 15 & 4 & -5 \\ -12 & -3 & 4 \\ -4 & -1 & 1 \end{bmatrix}$$

Take a second look at the matrix obtained at the point where Gauss-Jordan elimination was completed. This matrix is shown, again, below. Notice that the 3×3 matrix to the right of the vertical bar is the multiplicative inverse of A. Also notice that the multiplicative identity matrix, I_3 is the matrix that appears to the left of the vertical bar.

$$\begin{bmatrix} 1 & 0 & 0 & | & 15 & 4 & -5 \\ 0 & 1 & 0 & | & -12 & -3 & 4 \\ 0 & 0 & 1 & | & -4 & -1 & 1 \end{bmatrix}$$

This is the multiplicative identity, I_3.   This is the multiplicative inverse of A.

The observations in the voice balloons and the procedures followed above give us a general method for finding the multiplicative inverse of an invertible matrix.

Study Tip

Because we have a quick method for finding the multiplicative inverse of a 2×2 matrix, the procedure on the right is recommended for matrices of order 3×3 or greater when a graphing utility is not being used.

Procedure for Finding the Multiplicative Inverse of an Invertible Matrix

To find A^{-1} for any $n \times n$ matrix A for which A^{-1} exists,

1. Form the augmented matrix $[A \,|\, I]$, where I is the multiplicative identity matrix of the same order as the given matrix A.

2. Perform row operations on $[A \,|\, I]$ to obtain a matrix of the form $[I \,|\, B]$. This is equivalent to using Gauss-Jordan elimination to change A into the identity matrix.

3. Matrix B is A^{-1}.

4. Verify the result by showing that $AA^{-1} = I$ and $A^{-1}A = I$.

EXAMPLE 4 **Finding the Multiplicative Inverse of a 3 × 3 Matrix**

Find the multiplicative inverse of

$$A = \begin{bmatrix} 1 & -1 & 1 \\ 0 & -2 & 1 \\ -2 & -3 & 0 \end{bmatrix}.$$

Solution

Step 1 **Form the augmented matrix $[A \,|\, I_3]$.**

$$\begin{bmatrix} 1 & -1 & 1 & | & 1 & 0 & 0 \\ 0 & -2 & 1 & | & 0 & 1 & 0 \\ -2 & -3 & 0 & | & 0 & 0 & 1 \end{bmatrix}$$

This is matrix A.

This is I_3, the multiplicative identity matrix, with 1s down the diagonal and 0s elsewhere.

Step 2 **Perform row operations on $[A \,|\, I_3]$ to obtain a matrix of the form $[I_3 | B]$.** To the left of the vertical dividing line, we want 1s down the diagonal from upper left to lower right and 0s elsewhere.

$$\begin{bmatrix} 1 & -1 & 1 & | & 1 & 0 & 0 \\ 0 & -2 & 1 & | & 0 & 1 & 0 \\ -2 & -3 & 0 & | & 0 & 0 & 1 \end{bmatrix} \xrightarrow{\text{Replace row 3 by } 2R_1 + R_3} \begin{bmatrix} 1 & -1 & 1 & | & 1 & 0 & 0 \\ 0 & -2 & 1 & | & 0 & 1 & 0 \\ 0 & -5 & 2 & | & 2 & 0 & 1 \end{bmatrix} \xrightarrow{-\frac{1}{2}R_2}$$

$$\begin{bmatrix} 1 & -1 & 1 & | & 1 & 0 & 0 \\ 0 & 1 & -\frac{1}{2} & | & 0 & -\frac{1}{2} & 0 \\ 0 & -5 & 2 & | & 2 & 0 & 1 \end{bmatrix} \begin{array}{l} \xrightarrow{\text{Replace row 1 by } 1R_2 + R_1.} \\ \text{Replace row 3 by } 5R_2 + R_3. \end{array} \begin{bmatrix} 1 & 0 & \frac{1}{2} & | & 1 & -\frac{1}{2} & 0 \\ 0 & 1 & -\frac{1}{2} & | & 0 & -\frac{1}{2} & 0 \\ 0 & 0 & -\frac{1}{2} & | & 2 & -\frac{5}{2} & 1 \end{bmatrix} \xrightarrow{-2R_3}$$

$$\begin{bmatrix} 1 & 0 & \frac{1}{2} & | & 1 & -\frac{1}{2} & 0 \\ 0 & 1 & -\frac{1}{2} & | & 0 & -\frac{1}{2} & 0 \\ 0 & 0 & 1 & | & -4 & 5 & -2 \end{bmatrix} \begin{array}{l} \xrightarrow{\text{Replace row 1 by } -\frac{1}{2}R_3 + R_1.} \\ \text{Replace row 2 by } \frac{1}{2}R_3 + R_2. \end{array} \begin{bmatrix} 1 & 0 & 0 & | & 3 & -3 & 1 \\ 0 & 1 & 0 & | & -2 & 2 & -1 \\ 0 & 0 & 1 & | & -4 & 5 & -2 \end{bmatrix}$$

This is the multiplicative identity, I_3.

This is the multiplicative inverse of A.

Step 3 **Matrix B is A^{-1}.** The matrix shown directly above is in the form $[I_3 | B]$. The multiplicative identity matrix is on the left of the vertical bar. Matrix B, the multiplicative inverse of A, is on the right. Thus, the multiplicative inverse of A is

$$A^{-1} = \begin{bmatrix} 3 & -3 & 1 \\ -2 & 2 & -1 \\ -4 & 5 & -2 \end{bmatrix}.$$

Step 4 **Verify the result by showing that $AA^{-1} = I_3$ and $A^{-1}A = I_3$.** Try confirming the result by multiplying A and A^{-1} to obtain I_3. Do you obtain I_3 if you reverse the order of the multiplication?

Technology

The matrix

$$A = \begin{bmatrix} 4 & 6 \\ 2 & 3 \end{bmatrix} = \begin{bmatrix} a & b \\ c & d \end{bmatrix}$$

has no multiplicative inverse because

$$ad - bc = 4 \cdot 3 - 6 \cdot 2 = 12 - 12 = 0.$$

When we try to find the inverse with a graphing utility, an ERROR message occurs, indicating the matrix is singular.

We have seen that not all square matrices have multiplicative inverses. If the row operations in step 2 result in all zeros in a row or column to the left of the vertical line, the given matrix does not have a multiplicative inverse.

Check Point 4   Find the multiplicative inverse of

$$A = \begin{bmatrix} 1 & 0 & 2 \\ -1 & 2 & 3 \\ 1 & -1 & 0 \end{bmatrix}.$$

Summary: Finding Multiplicative Inverses for Invertible Matrices

Use a graphing utility with matrix capabilities,
or

a. If the matrix is 2 × 2: The inverse of $A = \begin{bmatrix} a & b \\ c & d \end{bmatrix}$ is

$$A^{-1} = \frac{1}{ad - bc} \begin{bmatrix} d & -b \\ -c & a \end{bmatrix}.$$

b. If the matrix A is $n \times n$ where $n > 2$: Use the procedure on page 813. Form $[A \,|\, I]$ and use row transformations to obtain $[I \,|\, B]$. $A^{-1} = B$.

Solving Systems of Equations Using Multiplicative Inverses of Matrices

Matrix multiplication can be used to represent a system of linear equations.

Linear System

$$a_1 x + b_1 y + c_1 z = d_1$$
$$a_2 x + b_2 y + c_2 z = d_2$$
$$a_3 x + b_3 y + c_3 z = d_3$$

Matrix Form of the System

$$\begin{bmatrix} a_1 & b_1 & c_1 \\ a_2 & b_2 & c_2 \\ a_3 & b_3 & c_3 \end{bmatrix} \begin{bmatrix} x \\ y \\ z \end{bmatrix} = \begin{bmatrix} d_1 \\ d_2 \\ d_3 \end{bmatrix}$$

This matrix contains the system's coefficients.   This matrix contains the system's variables.   This matrix contains the system's constants.

You can work with the matrix form of the system and obtain the form of the linear system on the left. To do so, perform the matrix multiplication on the left side of the matrix equation. Then equate the corresponding elements.

The matrix equation

$$\underbrace{\begin{bmatrix} a_1 & b_1 & c_1 \\ a_2 & b_2 & c_2 \\ a_3 & b_3 & c_3 \end{bmatrix}}_{A} \underbrace{\begin{bmatrix} x \\ y \\ z \end{bmatrix}}_{X} = \underbrace{\begin{bmatrix} d_1 \\ d_2 \\ d_3 \end{bmatrix}}_{B}$$

is abbreviated as $AX = B$, where A is the **coefficient matrix** of the system, and X and B are matrices containing one column, called **column matrices.** The matrix B is called the **constant matrix.**

Here is a specific example of a linear system and its matrix form:

Linear System   **Matrix Form**

$$\begin{array}{rcl} x - y + z &=& 2 \\ -2y + z &=& 2 \\ -2x - 3y &=& \tfrac{1}{2} \end{array}$$

Coefficients

$$\begin{bmatrix} 1 & -1 & 1 \\ 0 & -2 & 1 \\ -2 & -3 & 0 \end{bmatrix} \begin{bmatrix} x \\ y \\ z \end{bmatrix} = \begin{bmatrix} 2 \\ 2 \\ \tfrac{1}{2} \end{bmatrix}$$

Constants

A, the coefficient matrix   X =   B, the constant matrix

The matrix equation $AX = B$ can be solved using A^{-1} if it exists.

$AX = B$ This is the matrix equation.

$A^{-1}AX = A^{-1}B$ Multiply both sides by A^{-1}. Because matrix multiplication is not commutative, put A^{-1} in the same left position on both sides.

$I_n X = A^{-1}B$ The multiplicative inverse property tells us that $A^{-1}A = I_n$.

$X = A^{-1}B$ Because I_n is the multiplicative identity, $I_n X = X$.

We see that if $AX = B$, then $X = A^{-1}B$.

2   Use inverses to solve matrix equations.

Solving a System Using A^{-1}

If $AX = B$ has a unique solution, $X = A^{-1}B$. To solve a linear system of equations, multiply A^{-1} and B to find X.

EXAMPLE 5   **Using the Inverse of a Matrix to Solve a System**

Solve the system by using A^{-1}, the inverse of the coefficient matrix:

$$\begin{array}{rcl} x - y + z &=& 2 \\ -2y + z &=& 2 \\ -2x - 3y &=& \tfrac{1}{2}. \end{array}$$

Solution   The linear system can be written as

$$\begin{bmatrix} 1 & -1 & 1 \\ 0 & -2 & 1 \\ -2 & -3 & 0 \end{bmatrix} \begin{bmatrix} x \\ y \\ z \end{bmatrix} = \begin{bmatrix} 2 \\ 2 \\ \tfrac{1}{2} \end{bmatrix}.$$

A   X   B

The solution is given by $X = A^{-1}B$. Consequently, we must find A^{-1}. We found the inverse of matrix A in Example 4. Using this result,

$$X = A^{-1}B = \begin{bmatrix} 3 & -3 & 1 \\ -2 & 2 & -1 \\ -4 & 5 & -2 \end{bmatrix} \begin{bmatrix} 2 \\ 2 \\ \tfrac{1}{2} \end{bmatrix} = \begin{bmatrix} 3 \cdot 2 + (-3) \cdot 2 + 1 \cdot \tfrac{1}{2} \\ -2 \cdot 2 + 2 \cdot 2 + (-1) \cdot \tfrac{1}{2} \\ -4 \cdot 2 + 5 \cdot 2 + (-2) \cdot \tfrac{1}{2} \end{bmatrix} = \begin{bmatrix} \tfrac{1}{2} \\ -\tfrac{1}{2} \\ 1 \end{bmatrix}.$$

Thus, $x = \tfrac{1}{2}$, $y = -\tfrac{1}{2}$, and $z = 1$. The solution set is $\left\{ \left(\tfrac{1}{2}, -\tfrac{1}{2}, 1 \right) \right\}$.

Check Point 5 Solve the system by using A^{-1}, the inverse of the coefficient matrix that you found in Check Point 4:

$$\begin{aligned} x \quad\quad + 2z &= 6 \\ -x + 2y + 3z &= -5 \\ x - y \quad\quad &= 6. \end{aligned}$$

3 Encode and decode messages.

Applications of Matrix Inverses to Coding

A **cryptogram** is a message written so that no one other than the intended recipient can understand it. To encode a message, we begin by assigning a number to each letter in the alphabet: $A = 1$, $B = 2$, $C = 3, \ldots, Z = 26$, and a space $= 0$. For example, the numerical equivalent of the word MATH is 13, 1, 20, 8. The numerical equivalent of the message is then converted into a matrix. Finally, an invertible matrix can be used to convert the message into code. The multiplicative inverse of this matrix can be used to decode the message.

Encoding a Word or Message

1. Express the word or message numerically.

2. List the numbers in step 1 by columns and form a square matrix. If you do not have enough numbers to form a square matrix, put zeros in any remaining spaces in the last column.

3. Select any square invertible matrix, called the **coding matrix,** the same size as the matrix in step 2. Multiply the coding matrix by the square matrix that expresses the message numerically. The resulting matrix is the **coded matrix.**

4. Use the numbers, by columns, from the coded matrix in step 3 to write the encoded message.

EXAMPLE 6   Encoding a Word

Use matrices to encode the word MATH.

Solution

Step 1   Express the word numerically. As shown previously, the numerical equivalent of MATH is 13, 1, 20, 8.

Step 2   List the numbers in step 1 by columns and form a square matrix. The 2×2 matrix is

$$\begin{bmatrix} 13 & 20 \\ 1 & 8 \end{bmatrix}.$$

Step 3   Multiply the matrix in step 2 by a square invertible matrix. We will use $\begin{bmatrix} -2 & -3 \\ 3 & 4 \end{bmatrix}$ as the coding matrix.

$$\begin{bmatrix} -2 & -3 \\ 3 & 4 \end{bmatrix} \begin{bmatrix} 13 & 20 \\ 1 & 8 \end{bmatrix} = \begin{bmatrix} -2(13) - 3(1) & -2(20) - 3(8) \\ 3(13) + 4(1) & 3(20) + 4(8) \end{bmatrix}$$

Coding matrix   Numerical representation of MATH

$$= \begin{bmatrix} -29 & -64 \\ 43 & 92 \end{bmatrix}$$

Coded matrix

Step 4   Use the numbers, by columns, from the coded matrix in step 3 to write the encoded message. The encoded message is $-29, 43, -64, 92$.

Check Point 6 Use the coding matrix in Example 6, $\begin{bmatrix} -2 & -3 \\ 3 & 4 \end{bmatrix}$, to encode the word BASE.

The inverse of a coding matrix can be used to decode a word or message that was encoded.

Decoding a Word or Message That Was Encoded

1. Find the multiplicative inverse of the coding matrix.

2. Multiply the multiplicative inverse of the coding matrix and the coded matrix.

3. Express the numbers, by columns, from the matrix in step 2 as letters.

EXAMPLE 7   Decoding a Word

Decode $-29, 43, -64, 92$ from Example 6.

Solution

Step 1   Find the inverse of the coding matrix. The coding matrix in Example 6 was $\begin{bmatrix} -2 & -3 \\ 3 & 4 \end{bmatrix}$. We use the formula for the multiplicative inverse of a 2×2

matrix to find the multiplicative inverse of this matrix. It is $\begin{bmatrix} 4 & 3 \\ -3 & -2 \end{bmatrix}$.

Step 2   Multiply the multiplicative inverse of the coding matrix and the coded matrix.

$$\underbrace{\begin{bmatrix} 4 & 3 \\ -3 & -2 \end{bmatrix}}_{\substack{\text{Multiplicative inverse} \\ \text{of the coding matrix}}} \underbrace{\begin{bmatrix} -29 & -64 \\ 43 & 92 \end{bmatrix}}_{\substack{\text{Coded} \\ \text{matrix}}} = \begin{bmatrix} 4(-29) + 3(43) & 4(-64) + 3(92) \\ -3(-29) - 2(43) & -3(-64) - 2(92) \end{bmatrix}$$

$$= \begin{bmatrix} 13 & 20 \\ 1 & 8 \end{bmatrix}$$

Step 3   Express the numbers, by columns, from the matrix in step 2 as letters. The numbers are 13, 1, 20, and 8. Using letters, the decoded message is MATH.

Check Point 7 Decode the word that you encoded in Check Point 6.

Decoding is simple for an authorized receiver who knows the coding matrix. Because any invertible matrix can be used for the coding matrix, decoding a cryptogram for an unauthorized receiver who does not know this matrix is extremely difficult.

EXERCISE SET 9.4

Practice Exercises

In Exercises 1–12, find the products AB and BA to determine whether B is the multiplicative inverse of A.

1. $A = \begin{bmatrix} 4 & -3 \\ -5 & 4 \end{bmatrix}$, $B = \begin{bmatrix} 4 & 3 \\ 5 & 4 \end{bmatrix}$

2. $A = \begin{bmatrix} -2 & -1 \\ -1 & 1 \end{bmatrix}$, $B = \begin{bmatrix} 1 & 1 \\ 1 & 2 \end{bmatrix}$

3. $A = \begin{bmatrix} -4 & 0 \\ 1 & 3 \end{bmatrix}$, $B = \begin{bmatrix} -2 & 4 \\ 0 & 1 \end{bmatrix}$

4. $A = \begin{bmatrix} -2 & 4 \\ 1 & -2 \end{bmatrix}$, $B = \begin{bmatrix} 1 & 2 \\ -1 & -2 \end{bmatrix}$

5. $A = \begin{bmatrix} -2 & 1 \\ \frac{3}{2} & -\frac{1}{2} \end{bmatrix}$, $B = \begin{bmatrix} 1 & 2 \\ 3 & 4 \end{bmatrix}$

6. $A = \begin{bmatrix} 4 & 5 \\ 2 & 3 \end{bmatrix}$, $B = \begin{bmatrix} \frac{3}{2} & -\frac{5}{2} \\ -1 & 2 \end{bmatrix}$

7. $A = \begin{bmatrix} 0 & 1 & 0 \\ 0 & 0 & 1 \\ 1 & 0 & 0 \end{bmatrix}$ $B = \begin{bmatrix} 0 & 0 & 1 \\ 1 & 0 & 0 \\ 0 & 1 & 0 \end{bmatrix}$

8. $A = \begin{bmatrix} -2 & 1 & -1 \\ -5 & 2 & -1 \\ 3 & -1 & 1 \end{bmatrix}$ $B = \begin{bmatrix} 1 & 0 & 1 \\ 2 & 1 & 3 \\ -1 & 1 & 1 \end{bmatrix}$

9. $A = \begin{bmatrix} 1 & 2 & 3 \\ 1 & 3 & 4 \\ 1 & 4 & 3 \end{bmatrix}$ $B = \begin{bmatrix} \frac{7}{2} & -3 & \frac{1}{2} \\ -\frac{1}{2} & 0 & \frac{1}{2} \\ -\frac{1}{2} & 1 & -\frac{1}{2} \end{bmatrix}$

10. $A = \begin{bmatrix} 0 & 2 & 0 \\ 3 & 3 & 2 \\ 2 & 5 & 1 \end{bmatrix}$ $B = \begin{bmatrix} -3.5 & -1 & 2 \\ 0.5 & 0 & 0 \\ 4.5 & 2 & -3 \end{bmatrix}$

11. $A = \begin{bmatrix} 0 & 0 & -2 & 1 \\ -1 & 0 & 1 & 1 \\ 0 & 1 & -1 & 0 \\ 1 & 0 & 0 & -1 \end{bmatrix}$, $B = \begin{bmatrix} 1 & 2 & 0 & 3 \\ 0 & 1 & 1 & 1 \\ 0 & 1 & 0 & 1 \\ 1 & 2 & 0 & 2 \end{bmatrix}$

12. $A = \begin{bmatrix} 1 & -2 & 1 & 0 \\ 0 & 1 & -2 & 1 \\ 0 & 0 & 1 & -2 \\ 0 & 0 & 0 & 1 \end{bmatrix}$, $B = \begin{bmatrix} 1 & 2 & 3 & 4 \\ 0 & 1 & 2 & 3 \\ 0 & 0 & 1 & 2 \\ 0 & 0 & 0 & 1 \end{bmatrix}$

In Exercises 13–18, use the fact that if $A = \begin{bmatrix} a & b \\ c & d \end{bmatrix}$, then $A^{-1} = \frac{1}{ad - bc} \begin{bmatrix} d & -b \\ -c & a \end{bmatrix}$ to find the inverse of each matrix, if possible. Check that $AA^{-1} = I_2$ and $A^{-1}A = I_2$.

13. $A = \begin{bmatrix} 2 & 3 \\ -1 & 2 \end{bmatrix}$

14. $A = \begin{bmatrix} 0 & 3 \\ 4 & -2 \end{bmatrix}$

15. $A = \begin{bmatrix} 3 & -1 \\ -4 & 2 \end{bmatrix}$

16. $A = \begin{bmatrix} 2 & -6 \\ 1 & -2 \end{bmatrix}$

17. $A = \begin{bmatrix} 10 & -2 \\ -5 & 1 \end{bmatrix}$

18. $A = \begin{bmatrix} 6 & -3 \\ -2 & 1 \end{bmatrix}$

In Exercises 19–28, find A^{-1} by forming $[A \,|\, I]$ and then using row operations to obtain $[I \,|\, B]$, where $A^{-1} = [B]$. Check that $AA^{-1} = I$ and $A^{-1}A = I$.

19. $A = \begin{bmatrix} 2 & 0 & 0 \\ 0 & 4 & 0 \\ 0 & 0 & 6 \end{bmatrix}$

20. $A = \begin{bmatrix} 3 & 0 & 0 \\ 0 & 6 & 0 \\ 0 & 0 & 9 \end{bmatrix}$

21. $A = \begin{bmatrix} 1 & 2 & -1 \\ -2 & 0 & 1 \\ 1 & -1 & 0 \end{bmatrix}$

22. $A = \begin{bmatrix} 1 & -1 & 1 \\ 0 & 2 & -1 \\ 2 & 3 & 0 \end{bmatrix}$

23. $A = \begin{bmatrix} 2 & 2 & -1 \\ 0 & 3 & -1 \\ -1 & -2 & 1 \end{bmatrix}$

24. $A = \begin{bmatrix} 2 & 4 & -4 \\ 1 & 3 & -4 \\ 2 & 4 & -3 \end{bmatrix}$

25. $A = \begin{bmatrix} 5 & 0 & 2 \\ 2 & 2 & 1 \\ -3 & 1 & -1 \end{bmatrix}$

26. $A = \begin{bmatrix} 3 & 2 & 6 \\ 1 & 1 & 2 \\ 2 & 2 & 5 \end{bmatrix}$

27. $A = \begin{bmatrix} 1 & 0 & 0 & 0 \\ 0 & -1 & 0 & 0 \\ 0 & 0 & 3 & 0 \\ 1 & 0 & 0 & 1 \end{bmatrix}$

28. $A = \begin{bmatrix} 2 & 0 & 0 & 1 \\ 0 & 1 & 0 & 0 \\ 0 & 0 & -1 & 0 \\ 0 & 0 & 0 & 2 \end{bmatrix}$

In Exercises 29–32, write each linear system as a matrix equation in the form $AX = B$, where A is the coefficient matrix and B is the constant matrix.

29. $6x + 5y = 13$
$5x + 4y = 10$

30. $7x + 5y = 23$
$3x + 2y = 10$

31. $x + 3y + 4z = -3$
$x + 2y + 3z = -2$
$x + 4y + 3z = -6$

32. $x + 4y - z = 3$
$x + 3y - 2z = 5$
$2x + 7y - 5z = 12$

In Exercises 33–36, write each matrix equation as a system of linear equations without matrices.

33. $\begin{bmatrix} 4 & -7 \\ 2 & -3 \end{bmatrix} \begin{bmatrix} x \\ y \end{bmatrix} = \begin{bmatrix} -3 \\ 1 \end{bmatrix}$

34. $\begin{bmatrix} 3 & 0 \\ -3 & 1 \end{bmatrix} \begin{bmatrix} x \\ y \end{bmatrix} = \begin{bmatrix} 6 \\ -7 \end{bmatrix}$

35. $\begin{bmatrix} 2 & 0 & -1 \\ 0 & 3 & 0 \\ 1 & 1 & 0 \end{bmatrix} \begin{bmatrix} x \\ y \\ z \end{bmatrix} = \begin{bmatrix} 6 \\ 9 \\ 5 \end{bmatrix}$

36. $\begin{bmatrix} -1 & 0 & 1 \\ 0 & -1 & 0 \\ 0 & 1 & 1 \end{bmatrix} \begin{bmatrix} x \\ y \\ z \end{bmatrix} = \begin{bmatrix} -4 \\ 2 \\ 4 \end{bmatrix}$

In Exercises 37–42,

   a. Write each linear system as a matrix equation in the form $AX = B$.

   b. Solve the system using the inverse that is given for the coefficient matrix.

37. $2x + 6y + 6z = 8$
$2x + 7y + 6z = 10$
$2x + 7y + 7z = 9$

The inverse of $\begin{bmatrix} 2 & 6 & 6 \\ 2 & 7 & 6 \\ 2 & 7 & 7 \end{bmatrix}$ is $\begin{bmatrix} \frac{7}{2} & 0 & -3 \\ -1 & 1 & 0 \\ 0 & -1 & 1 \end{bmatrix}$.

38. $x + 2y + 5z = 2$
$2x + 3y + 8z = 3$
$-x + y + 2z = 3$

The inverse of $\begin{bmatrix} 1 & 2 & 5 \\ 2 & 3 & 8 \\ -1 & 1 & 2 \end{bmatrix}$ is $\begin{bmatrix} 2 & -1 & -1 \\ 12 & -7 & -2 \\ -5 & 3 & 1 \end{bmatrix}$.

39. $x - y + z = 8$
$2y - z = -7$
$2x + 3y = 1$

The inverse of $\begin{bmatrix} 1 & -1 & 1 \\ 0 & 2 & -1 \\ 2 & 3 & 0 \end{bmatrix}$ is $\begin{bmatrix} 3 & 3 & -1 \\ -2 & -2 & 1 \\ -4 & -5 & 2 \end{bmatrix}$.

40. $x - 6y + 3z = 11$
$2x - 7y + 3z = 14$
$4x - 12y + 5z = 25$

The inverse of $\begin{bmatrix} 1 & -6 & 3 \\ 2 & -7 & 3 \\ 4 & -12 & 5 \end{bmatrix}$ is $\begin{bmatrix} 1 & -6 & 3 \\ 2 & -7 & 3 \\ 4 & -12 & 5 \end{bmatrix}$.

41. $w - x + 2y = -3$
$x - y + z = 4$
$-w + x - y + 2z = 2$
$-x + y - 2z = -4$

The inverse of

$\begin{bmatrix} 1 & -1 & 2 & 0 \\ 0 & 1 & -1 & 1 \\ -1 & 1 & -1 & 2 \\ 0 & -1 & 1 & -2 \end{bmatrix}$ is $\begin{bmatrix} 0 & 0 & -1 & -1 \\ 1 & 4 & 1 & 3 \\ 1 & 2 & 1 & 2 \\ 0 & -1 & 0 & -1 \end{bmatrix}$.

42. $2w + y + z = 6$
$3w + z = 9$
$-w + x - 2y + z = 4$
$4w - x + y = 6$

The inverse of

$\begin{bmatrix} 2 & 0 & 1 & 1 \\ 3 & 0 & 0 & 1 \\ -1 & 1 & -2 & 1 \\ 4 & -1 & 1 & 0 \end{bmatrix}$ is $\begin{bmatrix} -1 & 2 & -1 & -1 \\ -4 & 9 & -5 & -6 \\ 0 & 1 & -1 & -1 \\ 3 & -5 & 3 & 3 \end{bmatrix}$.

Application Exercises

In Exercises 43–44, use the coding matrix

$$A = \begin{bmatrix} 4 & -1 \\ -3 & 1 \end{bmatrix} \text{ and its inverse } A^{-1} = \begin{bmatrix} 1 & 1 \\ 3 & 4 \end{bmatrix} \text{ to}$$

encode and then decode the given message.

43. HELP   **44.** LOVE

In Exercises 45–46, use the coding matrix

$$A = \begin{bmatrix} 1 & -1 & 0 \\ 3 & 0 & 2 \\ -1 & 0 & -1 \end{bmatrix} \text{ and its inverse}$$

$$A^{-1} = \begin{bmatrix} 0 & 1 & 2 \\ -1 & 1 & 2 \\ 0 & -1 & -3 \end{bmatrix} \text{ to write a cryptogram for each}$$

message. Check your result by decoding the cryptogram.

45. S E N D _ C A S H
19 5 14 4 0 3 1 19 8

Use $\begin{bmatrix} 19 & 4 & 1 \\ 5 & 0 & 19 \\ 14 & 3 & 8 \end{bmatrix}$.

46. S T A Y _ W E L L
19 20 1 25 0 23 5 12 12

Use $\begin{bmatrix} 19 & 25 & 5 \\ 20 & 0 & 12 \\ 1 & 23 & 12 \end{bmatrix}$.

Writing in Mathematics

47. What is the multiplicative identity matrix?

48. If you are given two matrices, A and B, explain how to determine if B is the multiplicative inverse of A.

49. Explain why a matrix that does not have the same number of rows and columns cannot have a multiplicative inverse.

50. Explain how to find the multiplicative inverse for a 2×2 invertible matrix.

51. Explain how to find the multiplicative inverse for a 3×3 invertible matrix.

52. Explain how to write a linear system of three equations in three variables as a matrix equation.

53. Explain how to solve the matrix equation $AX = B$.

54. What is a cryptogram?

55. It's January 1, and you've written down your major goal for the year. You do not want those closest to you to see what you've written in case you do not accomplish your objective. Consequently, you decide to use a coding matrix to encode your goal. Explain how this can be accomplished.

56. A year has passed since Exercise 55. (Time flies when you're solving exercises in algebra books.) It's been a terrific year and so many wonderful things have happened that you can't remember your goal from a year ago. You consult your personal journal and you find the encoded message and the coding matrix. How can you use these to find your original goal?

Technology Exercises

In Exercises 57–62, use a graphing utility to find the multiplicative inverse of each matrix. Check that the displayed inverse is correct.

57. $\begin{bmatrix} 3 & -1 \\ -2 & 1 \end{bmatrix}$

58. $\begin{bmatrix} -4 & 1 \\ 6 & -2 \end{bmatrix}$

59. $\begin{bmatrix} -2 & 1 & -1 \\ -5 & 2 & -1 \\ 3 & -1 & 1 \end{bmatrix}$

60. $\begin{bmatrix} 1 & 1 & -1 \\ -3 & 2 & -1 \\ 3 & -3 & 2 \end{bmatrix}$

61. $\begin{bmatrix} 7 & -3 & 0 & 2 \\ -2 & 1 & 0 & -1 \\ 4 & 0 & 1 & -2 \\ -1 & 1 & 0 & -1 \end{bmatrix}$

62. $\begin{bmatrix} 1 & 2 & 0 & 0 \\ 0 & 0 & 1 & 0 \\ 1 & 3 & 0 & 1 \\ 4 & 0 & 0 & 2 \end{bmatrix}$

In Exercises 63–68, write each system in the form $AX = B$. Then solve the system by entering A and B into your graphing utility and computing $A^{-1}B$.

63. $x - y + z = -6$
$4x + 2y + z = 9$
$4x - 2y + z = -3$

64. $y + 2z = 0$
$-x + y = 1$
$2x - y + z = -1$

65. $3x - 2y + z = -2$
$4x - 5y + 3z = -9$
$2x - y + 5z = -5$

66. $x - y = 1$
$6x + y + 20z = 14$
$y + 3z = 1$

67. $v - 3x + z = -3$
$w + y = -1$
$x + z = 7$
$v + w - x + 4y = -8$
$v + w + x + y + z = 8$

68. $w + x + y + z = 4$
$w + 3x - 2y + 2z = 7$
$2w + 2x + y + z = 3$
$w - x + 2y + 3z = 5$

In Exercises 69–70, use a coding matrix A of your choice. Use a graphing utility to find the multiplicative inverse of your coding matrix. Write a cryptogram for each message. Check your result by decoding the cryptogram. Use your graphing utility to perform all necessary matrix multiplications.

69. A R R I V E D _ S A F E L Y
1 18 18 9 22 5 4 0 19 1 6 5 12 25

70. A R T _ E N R I C H E S
1 18 20 0 5 14 18 9 3 8 5 19

Critical Thinking Exercises

71. Which one of the following is true?
   a. Some nonsquare matrices have inverses.
   b. All square 2×2 matrices have inverses because there is a formula for finding these inverses.
   c. Two 2×2 invertible matrices can have a matrix sum that is not invertible.
   d. To solve the matrix equation $AX = B$ for X, multiply A and the inverse of B.

72. Which one of the following is true?
   a. $(AB)^{-1} = A^{-1}B^{-1}$, assuming A, B, and AB are invertible.
   b. $(A + B)^{-1} = A^{-1} + B^{-1}$, assuming A, B, and $A + B$ are invertible.
   c. $\begin{bmatrix} 1 & -3 \\ -1 & 3 \end{bmatrix}$ is an invertible matrix.
   d. None of the above is true.

73. Give an example of a 2×2 matrix that is its own inverse.

74. If $A = \begin{bmatrix} 3 & 5 \\ 2 & 4 \end{bmatrix}$, find $(A^{-1})^{-1}$.

75. Find values of a for which the following matrix is not invertible:
$$\begin{bmatrix} 1 & a+1 \\ a-2 & 4 \end{bmatrix}.$$

Group Exercise

76. Each person in the group should work with one partner. Send a coded word or message to each other by giving your partner the coded matrix and the coding matrix that you selected. Once messages are sent, each person should decode the message received.

SECTION 9.5   *Determinants and Cramer's Rule*

Objectives

1. Evaluate a second-order determinant.
2. Solve a system of linear equations in two variables using Cramer's rule.
3. Evaluate a third-order determinant.
4. Solve a system of linear equations in three variables using Cramer's rule.
5. Use determinants to identify inconsistent systems and systems with dependent equations.
6. Evaluate higher-order determinants.

A portion of Charles Babbage's unrealized Difference Engine

As cyberspace absorbs more and more of our work, play, shopping, and socializing, where will it all end? Which activities will still be offline in 2025?

Our technologically transformed lives can be traced back to the English inventor Charles Babbage (1792–1871). Babbage knew of a method for solving linear systems called *Cramer's rule*, in honor of the Swiss geometer Gabriel Cramer (1704–1752). Cramer's rule was simple, but involved numerous multiplications for large systems. Babbage designed a machine, called the "difference engine," that consisted of toothed wheels on shafts for performing these multiplications. Despite the fact that only one-seventh of the functions ever worked, Babbage's invention demonstrated how complex calculations could be handled mechanically. In 1944, scientists at IBM used the lessons of the difference engine to create the world's first computer.

Those who invented computers hoped to relegate the drudgery of repeated computation to a machine. In this section, we look at a method for solving linear systems that played a critical role in this process. The method uses arrays of numbers, called *determinants*. As with matrix methods, solutions are obtained by writing down the coefficients and constants of a linear system and performing operations with them.

1 Evaluate a second-order determinant.

The Determinant of a 2 × 2 Matrix

Associated with every square matrix is a real number, called its **determinant.** The determinant for a 2 × 2 square matrix is defined as follows:

Study Tip

To evaluate a second-order determinant, find the difference of the product of the two diagonals.

$$\begin{vmatrix} a_1 & b_1 \\ a_2 & b_2 \end{vmatrix} = a_1 b_2 - a_2 b_1$$

Definition of the Determinant of a 2 × 2 Matrix

The determinant of the matrix $\begin{bmatrix} a_1 & b_1 \\ a_2 & b_2 \end{bmatrix}$ is denoted by $\begin{vmatrix} a_1 & b_1 \\ a_2 & b_2 \end{vmatrix}$ and is defined by

$$\begin{vmatrix} a_1 & b_1 \\ a_2 & b_2 \end{vmatrix} = a_1 b_2 - a_2 b_1.$$

We also say that the **value** of the **second-order determinant** $\begin{vmatrix} a_1 & b_1 \\ a_2 & b_2 \end{vmatrix}$ is $a_1 b_2 - a_2 b_1$.

Example 1 illustrates that the determinant of a matrix may be positive or negative. The determinant can also have 0 as its value.

EXAMPLE 1 Evaluating the Determinant of a 2×2 Matrix

Evaluate the determinant of:

a. $\begin{bmatrix} 5 & 6 \\ 7 & 3 \end{bmatrix}$ **b.** $\begin{bmatrix} 2 & 4 \\ -3 & -5 \end{bmatrix}$.

Discovery

Write and then evaluate three determinants, one whose value is positive, one whose value is negative, and one whose value is 0.

Solution We multiply and subtract as indicated.

a. $\begin{vmatrix} 5 & 6 \\ 7 & 3 \end{vmatrix} = 5 \cdot 3 - 7 \cdot 6 = 15 - 42 = -27$ The value of the second-order determinant is -27.

b. $\begin{vmatrix} 2 & 4 \\ -3 & -5 \end{vmatrix} = 2(-5) - (-3)(4) = -10 + 12 = 2$ The value of the second-order determinant is 2.

Check Point 1 Evaluate the determinant of:

a. $\begin{bmatrix} 10 & 9 \\ 6 & 5 \end{bmatrix}$ **b.** $\begin{bmatrix} 4 & 3 \\ -5 & -8 \end{bmatrix}$.

2 Solve a system of linear equations in two variables using Cramer's rule.

Solving Systems of Linear Equations in Two Variables Using Determinants

Determinants can be used to solve a linear system in two variables. In general, such a system appears as

$$a_1 x + b_1 y = c_1$$
$$a_2 x + b_2 y = c_2.$$

Let's first solve this system for x using the addition method. We can solve for x by eliminating y from the equations. Multiply the first equation by b_2 and the second equation by $-b_1$. Then add the two equations:

$$
\begin{array}{l}
a_1 x + b_1 y = c_1 \xrightarrow{\text{Multiply by } b_2.} \quad a_1 b_2 x + b_1 b_2 y = c_1 b_2 \\
a_2 x + b_2 y = c_2 \xrightarrow{\text{Multiply by } -b_1.} \quad \underline{-a_2 b_1 x - b_1 b_2 y = -c_2 b_1} \\
\qquad\qquad\qquad \text{Add:} \qquad\qquad (a_1 b_2 - a_2 b_1)x = c_1 b_2 - c_2 b_1 \\
\qquad\qquad\qquad\qquad\qquad\qquad\qquad x = \dfrac{c_1 b_2 - c_2 b_1}{a_1 b_2 - a_2 b_1}
\end{array}
$$

Because

$$\begin{vmatrix} c_1 & b_1 \\ c_2 & b_2 \end{vmatrix} = c_1 b_2 - c_2 b_1 \qquad \text{and} \qquad \begin{vmatrix} a_1 & b_1 \\ a_2 & b_2 \end{vmatrix} = a_1 b_2 - a_2 b_1$$

we can express our answer for x as the quotient of two determinants:

$$x = \frac{\begin{vmatrix} c_1 & b_1 \\ c_2 & b_2 \end{vmatrix}}{\begin{vmatrix} a_1 & b_1 \\ a_2 & b_2 \end{vmatrix}}.$$

In a similar way, we could use the addition method to solve our system for y, again expressing y as the quotient of two determinants. This method of using determinants to solve the linear system, called **Cramer's rule**, is summarized in the box.

Solving a Linear System in Two Variables Using Determinants

Cramer's Rule

If

$$a_1 x + b_1 y = c_1$$
$$a_2 x + b_2 y = c_2$$

then

$$x = \frac{\begin{vmatrix} c_1 & b_1 \\ c_2 & b_2 \end{vmatrix}}{\begin{vmatrix} a_1 & b_1 \\ a_2 & b_2 \end{vmatrix}} \quad \text{and} \quad y = \frac{\begin{vmatrix} a_1 & c_1 \\ a_2 & c_2 \end{vmatrix}}{\begin{vmatrix} a_1 & b_1 \\ a_2 & b_2 \end{vmatrix}}$$

where

$$\begin{vmatrix} a_1 & b_1 \\ a_2 & b_2 \end{vmatrix} \neq 0.$$

Here are some helpful tips when solving

$$a_1 x + b_1 y = c_1$$
$$a_2 x + b_2 y = c_2$$

using determinants:

1. Three different determinants are used to find x and y. The determinants in the denominators for x and y are identical. The determinants in the numerators for x and y differ. In abbreviated notation, we write

$$x = \frac{D_x}{D} \quad \text{and} \quad y = \frac{D_y}{D}, \text{where } D \neq 0.$$

2. The elements of D, the determinant in the denominator, are the coefficients of the variables in the system.

$$D = \begin{vmatrix} a_1 & b_1 \\ a_2 & b_2 \end{vmatrix}$$

3. D_x, the determinant in the numerator of x, is obtained by replacing the x-coefficients, in D, a_1 and a_2, with the constants on the right side of the equations, c_1 and c_2.

$$D = \begin{vmatrix} a_1 & b_1 \\ a_2 & b_2 \end{vmatrix} \quad \text{and} \quad D_x = \begin{vmatrix} c_1 & b_1 \\ c_2 & b_2 \end{vmatrix}$$
Replace the column with a_1 and a_2 with the constants c_1 and c_2 to get D_x.

4. D_y, the determinant in the numerator for y, is obtained by replacing the y-coefficients, in D, b_1 and b_2, with the constants on the right side of the equations, c_1 and c_2.

$$D = \begin{vmatrix} a_1 & b_1 \\ a_2 & b_2 \end{vmatrix} \quad \text{and} \quad D_y = \begin{vmatrix} a_1 & c_1 \\ a_2 & c_2 \end{vmatrix}$$
Replace the column with b_1 and b_2 with the constants c_1 and c_2 to get D_y.

EXAMPLE 2 **Using Cramer's Rule to Solve a Linear System**

Use Cramer's rule to solve the system:

$$5x - 4y = 2$$
$$6x - 5y = 1.$$

Solution Because

$$x = \frac{D_x}{D} \quad \text{and} \quad y = \frac{D_y}{D},$$

we will set up and evaluate the three determinants D, D_x, and D_y.

1. D, the determinant in both denominators, consists of the x- and y-coefficients.

$$D = \begin{vmatrix} 5 & -4 \\ 6 & -5 \end{vmatrix} = (5)(-5) - (6)(-4) = -25 + 24 = -1$$

Because this determinant is not zero, we continue to use Cramer's rule to solve the system.

2. D_x, the determinant in the numerator for x, is obtained by replacing the x-coefficients in D, 5 and 6, by the constants on the right side of the equation, 2 and 1.

$$D_x = \begin{vmatrix} 2 & -4 \\ 1 & -5 \end{vmatrix} = (2)(-5) - (1)(-4) = -10 + 4 = -6$$

3. D_y, the determinant in the numerator for y, is obtained by replacing the y-coefficients in D, -4 and -5, by the constants on the right side of the equation, 2 and 1.

$$D_y = \begin{vmatrix} 5 & 2 \\ 6 & 1 \end{vmatrix} = (5)(1) - (6)(2) = 5 - 12 = -7$$

4. Thus,

$$x = \frac{D_x}{D} = \frac{-6}{-1} = 6 \quad \text{and} \quad y = \frac{D_y}{D} = \frac{-7}{-1} = 7.$$

As always, the solution $(6, 7)$ can be checked by substituting these values into the original equations. The solution set is $\{(6, 7)\}$.

Check Point 2 Use Cramer's rule to solve the system:

$$5x + 4y = 12$$
$$3x - 6y = 24.$$

3 Evaluate a third-order determinant.

The Determinant of a 3 × 3 Matrix

Associated with every square matrix is a real number called its determinant. The determinant for a 3 × 3 matrix is defined on the next page.

Definition of a Third-Order Determinant

$$\begin{vmatrix} a_1 & b_1 & c_1 \\ a_2 & b_2 & c_2 \\ a_3 & b_3 & c_3 \end{vmatrix} = a_1 b_2 c_3 + b_1 c_2 a_3 + c_1 a_2 b_3 - a_3 b_2 c_1 - b_3 c_2 a_1 - c_3 a_2 b_1$$

The six terms and the three factors in each term in this complicated evaluation formula can be rearranged, and then we can apply the distributive property. We obtain

$$a_1 b_2 c_3 - a_1 b_3 c_2 - a_2 b_1 c_3 + a_2 b_3 c_1 + a_3 b_1 c_2 - a_3 b_2 c_1$$

$$= a_1(b_2 c_3 - b_3 c_2) - a_2(b_1 c_3 - b_3 c_1) + a_3(b_1 c_2 - b_2 c_1)$$

$$= a_1 \begin{vmatrix} b_2 & c_2 \\ b_3 & c_3 \end{vmatrix} - a_2 \begin{vmatrix} b_1 & c_1 \\ b_3 & c_3 \end{vmatrix} + a_3 \begin{vmatrix} b_1 & c_1 \\ b_2 & c_2 \end{vmatrix}.$$

You can evaluate each of the second-order determinants and obtain the three expressions in parentheses in the second step.

In summary, we now have arranged the definition of a third-order determinant as follows:

Definition of the Determinant of a 3 × 3 Matrix

A third-order determinant is defined by

Subtract.   Add.

$$\begin{vmatrix} a_1 & b_1 & c_1 \\ a_2 & b_2 & c_2 \\ a_3 & b_3 & c_3 \end{vmatrix} = a_1 \begin{vmatrix} b_2 & c_2 \\ b_3 & c_3 \end{vmatrix} - a_2 \begin{vmatrix} b_1 & c_1 \\ b_3 & c_3 \end{vmatrix} + a_3 \begin{vmatrix} b_1 & c_1 \\ b_2 & c_2 \end{vmatrix}.$$

The a's on the right come from the first column.

Here are some tips that may be helpful when evaluating the determinant of a 3 × 3 matrix:

Evaluating the Determinant of a 3 × 3 Matrix

1. Each of the three terms in the definition contains two factors—a numerical factor and a second-order determinant.

2. The numerical factor in each term is an element from the first column of the third-order determinant.

3. The minus sign precedes the second term.

4. The second-order determinant that appears in each term is obtained by crossing out the row and the column containing the numerical factor.

$$a_1 \begin{vmatrix} b_2 & c_2 \\ b_3 & c_3 \end{vmatrix} - a_2 \begin{vmatrix} b_1 & c_1 \\ b_3 & c_3 \end{vmatrix} + a_3 \begin{vmatrix} b_1 & c_1 \\ b_2 & c_2 \end{vmatrix}$$

$$\begin{vmatrix} a_1 & b_1 & c_1 \\ a_2 & b_2 & c_2 \\ a_3 & b_3 & c_3 \end{vmatrix} \quad \begin{vmatrix} a_1 & b_1 & c_1 \\ a_2 & b_2 & c_2 \\ a_3 & b_3 & c_3 \end{vmatrix} \quad \begin{vmatrix} a_1 & b_1 & c_1 \\ a_2 & b_2 & c_2 \\ a_3 & b_3 & c_3 \end{vmatrix}$$

The **minor** of an element is the determinant that remains after deleting the row and column of that element. For this reason, we call this method **expansion by minors.**

EXAMPLE 3   Evaluating the Determinant of a 3 × 3 Matrix

Evaluate the determinant of

$$\begin{bmatrix} 4 & 1 & 0 \\ -9 & 3 & 4 \\ -3 & 8 & 1 \end{bmatrix}.$$

Solution   We know that each of the three terms in the determinant contains a numerical factor and a second-order determinant. The numerical factors are from the first column of the determinant of the given matrix. They are highlighted in the following matrix:

$$\begin{vmatrix} 4 & 1 & 0 \\ -9 & 3 & 4 \\ -3 & 8 & 1 \end{vmatrix}.$$

We find the minor for each numerical factor by deleting the row and column of that element:

$$\begin{bmatrix} 4 & 1 & 0 \\ -9 & 3 & 4 \\ -3 & 8 & 1 \end{bmatrix} \begin{bmatrix} 4 & 1 & 0 \\ -9 & 3 & 4 \\ -3 & 8 & 1 \end{bmatrix} \begin{bmatrix} 4 & 1 & 0 \\ -9 & 3 & 4 \\ -3 & 8 & 1 \end{bmatrix}$$

The minor for 4 is $\begin{vmatrix} 3 & 4 \\ 8 & 1 \end{vmatrix}$.   The minor for -9 is $\begin{vmatrix} 1 & 0 \\ 8 & 1 \end{vmatrix}$.   The minor for -3 is $\begin{vmatrix} 1 & 0 \\ 3 & 4 \end{vmatrix}$.

Now we have three numerical factors, $4, -9,$ and $-3,$ and three second-order determinants. We multiply each numerical factor by its second-order determinant to find the three terms of the third-order determinant:

$$4\begin{vmatrix} 3 & 4 \\ 8 & 1 \end{vmatrix}, \quad -9\begin{vmatrix} 1 & 0 \\ 8 & 1 \end{vmatrix}, \quad -3\begin{vmatrix} 1 & 0 \\ 3 & 4 \end{vmatrix}.$$

Based on the preceding definition, we subtract the second term from the first term and add the third term:

Don't forget to supply the minus sign.

$$\begin{vmatrix} 4 & 1 & 0 \\ -9 & 3 & 4 \\ -3 & 8 & 1 \end{vmatrix} = 4\begin{vmatrix} 3 & 4 \\ 8 & 1 \end{vmatrix} - (-9)\begin{vmatrix} 1 & 0 \\ 8 & 1 \end{vmatrix} - 3\begin{vmatrix} 1 & 0 \\ 3 & 4 \end{vmatrix}$$ Evaluate the three second-order determinants.

$$= 4(3 \cdot 1 - 8 \cdot 4) + 9(1 \cdot 1 - 8 \cdot 0) - 3(1 \cdot 4 - 3 \cdot 0)$$

$$= 4(3 - 32) + 9(1 - 0) - 3(4 - 0)$$

$$= 4(-29) + 9(1) - 3(4)$$ Subtract within parentheses.

$$= -116 + 9 - 12$$ Multiply.

$$= -119$$ Add and subtract as indicated.

Technology

A graphing utility can be used to evaluate the determinant of a matrix. Enter the matrix and call it $A.$ Then use the determinant command. The screen below verifies our result in Example 3.

```
[A]
      [[4   1   0]
       [-9  3   4]
       [-3  8   1]]
det([A])
                -119
■
```

Check Point 3 Evaluate the determinant of

$$\begin{bmatrix} 2 & 1 & 7 \\ -5 & 6 & 0 \\ -4 & 3 & 1 \end{bmatrix}.$$

The six terms in the definition of a third-order determinant can be rearranged and factored in a variety of ways. Thus, it is possible to expand a determinant by minors about any row or any column. *Minus signs must be supplied preceding any element appearing in a position where the sum of its row and its column is an odd number.* For example, expanding about the elements in column 2 gives us

$$\begin{vmatrix} a_1 & b_1 & c_1 \\ a_2 & b_2 & c_2 \\ a_3 & b_3 & c_3 \end{vmatrix} = -b_1 \begin{vmatrix} a_2 & c_2 \\ a_3 & c_3 \end{vmatrix} + b_2 \begin{vmatrix} a_1 & c_1 \\ a_3 & c_3 \end{vmatrix} - b_3 \begin{vmatrix} a_1 & c_1 \\ a_2 & c_2 \end{vmatrix}.$$

Minus sign is supplied because b_1 appears in row 1 and column 2; $1 + 2 = 3$, an odd number.

Minus sign is supplied because b_3 appears in row 3 and column 2; $3 + 2 = 5$, an odd number.

Expanding by minors about column 3, we obtain

$$\begin{vmatrix} a_1 & b_1 & c_1 \\ a_2 & b_2 & c_2 \\ a_3 & b_3 & c_3 \end{vmatrix} = c_1 \begin{vmatrix} a_2 & b_2 \\ a_3 & b_3 \end{vmatrix} - c_2 \begin{vmatrix} a_1 & b_1 \\ a_3 & b_3 \end{vmatrix} + c_3 \begin{vmatrix} a_1 & b_1 \\ a_2 & b_2 \end{vmatrix}.$$

Minus sign must be supplied because c_2 appears in row 2 and column 3; $2 + 3 = 5$, an odd number.

Study Tip

Keep in mind that you can expand a determinant by minors about any row or column. Use alternating plus and minus signs to precede the numerical factors of the minors according to the following sign array:

$$\begin{vmatrix} + & - & + \\ - & + & - \\ + & - & + \end{vmatrix}.$$

When evaluating a 3×3 determinant using expansion by minors, you can expand about any row or column. To simplify the arithmetic, if a row or column contains one or more 0s, expand about that row or column.

EXAMPLE 4   Evaluating a Third-Order Determinant

Evaluate:

$$\begin{vmatrix} 9 & 5 & 0 \\ -2 & -3 & 0 \\ 1 & 4 & 2 \end{vmatrix}.$$

Solution   Note that the last column has two 0s. We will expand the determinant about the elements in that column.

$$\begin{vmatrix} 9 & 5 & 0 \\ -2 & -3 & 0 \\ 1 & 4 & 2 \end{vmatrix} = 0 \begin{vmatrix} -2 & -3 \\ 1 & 4 \end{vmatrix} - 0 \begin{vmatrix} 9 & 5 \\ 1 & 4 \end{vmatrix} + 2 \begin{vmatrix} 9 & 5 \\ -2 & -3 \end{vmatrix}$$

$$= 0 - 0 + 2[9(-3) - (-2) \cdot 5]$$ Evaluate the second-order determinant whose numerical factor is not 0.

$$= 2(-27 + 10)$$

$$= 2(-17)$$

$$= -34$$

Check Point 4 Evaluate:

$$\begin{vmatrix} 6 & 4 & 0 \\ -3 & -5 & 3 \\ 1 & 2 & 0 \end{vmatrix}.$$

④ Solve a system of linear equations in three variables using Cramer's rule.

Solving Systems of Linear Equations in Three Variables Using Determinants

Cramer's rule can be applied to solving systems of linear equations in three variables. The determinants in the numerator and denominator of all variables are third-order determinants.

Solving Three Equations in Three Variables Using Determinants

Cramer's Rule

If

$$a_1 x + b_1 y + c_1 z = d_1$$
$$a_2 x + b_2 y + c_2 z = d_2$$
$$a_3 x + b_3 y + c_3 z = d_3$$

then

$$x = \frac{D_x}{D}, y = \frac{D_y}{D}, \text{ and } z = \frac{D_z}{D}.$$

These four third-order determinants are given by:

$$D = \begin{vmatrix} a_1 & b_1 & c_1 \\ a_2 & b_2 & c_2 \\ a_3 & b_3 & c_3 \end{vmatrix}$$ These are the coefficients of the variables x, y, and z. $D \neq 0$.

$$D_x = \begin{vmatrix} d_1 & b_1 & c_1 \\ d_2 & b_2 & c_2 \\ d_3 & b_3 & c_3 \end{vmatrix}$$ Replace x-coefficients in D with the **constants at the right** of the three equations.

$$D_y = \begin{vmatrix} a_1 & d_1 & c_1 \\ a_2 & d_2 & c_2 \\ a_3 & d_3 & c_3 \end{vmatrix}$$ Replace y-coefficients in D with the **constants at the right** of the three equations.

$$D_z = \begin{vmatrix} a_1 & b_1 & d_1 \\ a_2 & b_2 & d_2 \\ a_3 & b_3 & d_3 \end{vmatrix}.$$ Replace z-coefficients in D with the **constants at the right** of the three equations.

EXAMPLE 5   Using Cramer's Rule to Solve a Linear System in Three Variables

Use Cramer's rule to solve:

$$x + 2y - z = -4$$
$$x + 4y - 2z = -6$$
$$2x + 3y + z = 3.$$

Solution Because

$$x = \frac{D_x}{D}, \quad y = \frac{D_y}{D}, \quad \text{and} \quad z = \frac{D_z}{D},$$

we need to set up and evaluate four determinants.

$$x + 2y - z = -4$$
$$x + 4y - 2z = -6$$
$$2x + 3y + z = 3$$

The given linear system, repeated

Step 1 Set up the determinants.

1. D, the determinant in all three denominators, consists of the x-, y-, and z-coefficients.

$$D = \begin{vmatrix} 1 & 2 & -1 \\ 1 & 4 & -2 \\ 2 & 3 & 1 \end{vmatrix}$$

2. D_x, the determinant in the numerator for x, is obtained by replacing the x-coefficients in D, 1, 1, and 2, with the constants on the right side of the equation, -4, -6, and 3.

$$D_x = \begin{vmatrix} -4 & 2 & -1 \\ -6 & 4 & -2 \\ 3 & 3 & 1 \end{vmatrix}$$

3. D_y, the determinant in the numerator for y, is obtained by replacing the y-coefficients in D, 2, 4, and 3, with the constants on the right side of the equation, -4, -6, and 3.

$$D_y = \begin{vmatrix} 1 & -4 & -1 \\ 1 & -6 & -2 \\ 2 & 3 & 1 \end{vmatrix}$$

4. D_z, the determinant in the numerator for z, is obtained by replacing the z-coefficients in D, -1, -2, and 1, with the constants on the right side of the equation, -4, -6, and 3.

$$D_z = \begin{vmatrix} 1 & 2 & -4 \\ 1 & 4 & -6 \\ 2 & 3 & 3 \end{vmatrix}$$

Step 2 Evaluate the four determinants.

Study Tip

To find D_x, D_y, and D_z, you'll need to apply the evaluation process for 3×3 determinant three times. The values of D_x, D_y, and D_z, cannot be obtained from the numbers that occur in the computation of D.

$$D = \begin{vmatrix} 1 & 2 & -1 \\ 1 & 4 & -2 \\ 2 & 3 & 1 \end{vmatrix} = 1\begin{vmatrix} 4 & -2 \\ 3 & 1 \end{vmatrix} - 1\begin{vmatrix} 2 & -1 \\ 3 & 1 \end{vmatrix} + 2\begin{vmatrix} 2 & -1 \\ 4 & -2 \end{vmatrix}$$

$$= 1(4 + 6) - 1(2 + 3) + 2(-4 + 4)$$
$$= 1(10) - 1(5) + 2(0) = 5$$

Using the same technique to evaluate each determinant, we obtain

$$D_x = -10, \quad D_y = 5, \quad \text{and} \quad D_z = 20.$$

Step 3 Substitute these four values and solve the system.

$$x = \frac{D_x}{D} = \frac{-10}{5} = -2$$

$$y = \frac{D_y}{D} = \frac{5}{5} = 1$$

$$z = \frac{D_z}{D} = \frac{20}{5} = 4$$

The solution $(-2, 1, 4)$ can be checked by substitution into the original three equations. The solution set is $\{(-2, 1, 4)\}$.

> **Check Point 5**   Use Cramer's rule to solve the system:
>
> $$\begin{aligned} 3x - 2y + z &= 16 \\ 2x + 3y - z &= -9. \\ x + 4y + 3z &= 2 \end{aligned}$$

⑤ Use determinants to identify inconsistent systems and systems with dependent equations.

Cramer's Rule with Inconsistent and Dependent Systems

If D, the determinant in the denominator, is 0, the variables described by the quotient of determinants are not real numbers. However, when $D = 0$, this indicates that the system is inconsistent or contains dependent equations. This gives rise to the following two situations:

Discovery

Write a system of two equations that is inconsistent. Now use determinants and the result boxed on the right to verify that this is truly an inconsistent system. Repeat the same process for a system with two dependent equations.

Determinants: Inconsistent and Dependent-Systems

1. If $D = 0$ and at least one of the determinants in the numerator is not 0, then the system is inconsistent. The solution set is $\varnothing$.

2. If $D = 0$ and all the determinants in the numerators are 0, then the equations in the system are dependent. The system has infinitely many solutions.

Although we have focused on applying determinants to solve linear systems, they have other applications, some of which we consider in the exercise set that follows.

⑥ Evaluate higher-order determinants.

The Determinant of Any $n \times n$ Matrix

A determinant with n rows and n columns is said to be an ***n*th-order determinant.** The value of an nth-order determinant $(n > 2)$ can be found in terms of determinants of order $n - 1$. For example, we found the value of a third-order determinant in terms of determinants of order 2.

We can generalize this idea for fourth-order determinants and higher. We have seen that the **minor** of the element a_{ij} is the determinant obtained by deleting the ith row and the jth column in the given array of numbers. The **cofactor** of the element a_{ij} is $(-1)^{i+j}$ times the minor of the a_{ij}th entry. If the sum of the row and column $(i + j)$ is even, the cofactor is the same as the minor. If the sum of the row and column $(i + j)$ is odd, the cofactor is the opposite of the minor.

Let's see what this means in the case of a fourth-order determinant.

EXAMPLE 6   Evaluating the Determinant of a 4×4 Matrix

Evaluate the determinant of

$$A = \begin{bmatrix} 1 & -2 & 3 & 0 \\ -1 & 1 & 0 & 2 \\ 0 & 2 & 0 & -3 \\ 2 & 3 & -4 & 1 \end{bmatrix}.$$

Cramer's Rule and the World's Fastest Computer

In 2002, the fastest super-computer was the *ASCI White*, built by IBM and capable of performing 12 trillion (12×10^{12}) calculations per second. To solve a linear system with a "mere" 20 equations using Cramer's rule requires over 5×10^{19} multiplications. Although the *ASCI White* can solve a problem in one second that would take one person with a calculator 10 million years to complete, it would take the supercomputer more than 48 days to solve a system with 20 equations using Cramer's rule. Might the *ASCI White* be interested in this challenge? Absolutely not. Its purpose is to allow the testing of nuclear weapons using computer simulation rather than detonating actual bombs.

Solution

$$|A| = \begin{vmatrix} 1 & -2 & 3 & 0 \\ -1 & 1 & 0 & 2 \\ 0 & 2 & 0 & -3 \\ 2 & 3 & -4 & 1 \end{vmatrix}$$

With two 0s in the third column, we will expand along the third column.

$$= (-1)^{1+3}(3)\begin{vmatrix} -1 & 1 & 2 \\ 0 & 2 & -3 \\ 2 & 3 & 1 \end{vmatrix} + (-1)^{4+3}(-4)\begin{vmatrix} 1 & -2 & 0 \\ -1 & 1 & 2 \\ 0 & 2 & -3 \end{vmatrix}$$

3 is in row 1, column 3.

−4 is in row 4, column 3.

$$= 3\begin{vmatrix} -1 & 1 & 2 \\ 0 & 2 & -3 \\ 2 & 3 & 1 \end{vmatrix} + 4\begin{vmatrix} 1 & -2 & 0 \\ -1 & 1 & 2 \\ 0 & 2 & -3 \end{vmatrix}$$

The determinant that follows 3 is obtained by crossing out the row and the column (row 1, column 3) in the original determinant. The minor for −4 is obtained in the same manner.

Evaluate the two third-order determinants to get

$$|A| = 3(-25) + 4(-1) = -79.$$

Check Point 6 Evaluate the determinant of

$$A = \begin{bmatrix} 0 & 4 & 0 & -3 \\ -1 & 1 & 5 & 2 \\ 1 & -2 & 0 & 6 \\ 3 & 0 & 0 & 1 \end{bmatrix}.$$

If a linear system has n equations, Cramer's rule requires you to compute $n + 1$ determinants of nth order. The excessive number of calculations required to perform Cramer's rule for systems with four or more equations makes it an inefficient method for solving large systems.

EXERCISE SET 9.5

Practice Exercises

Evaluate each determinant in Exercises 1–10.

1. $\begin{vmatrix} 5 & 7 \\ 2 & 3 \end{vmatrix}$ **2.** $\begin{vmatrix} 4 & 8 \\ 5 & 6 \end{vmatrix}$

3. $\begin{vmatrix} -4 & 1 \\ 5 & 6 \end{vmatrix}$ **4.** $\begin{vmatrix} 7 & 9 \\ -2 & -5 \end{vmatrix}$

5. $\begin{vmatrix} -7 & 14 \\ 2 & -4 \end{vmatrix}$ **6.** $\begin{vmatrix} 1 & -3 \\ -8 & 2 \end{vmatrix}$

7. $\begin{vmatrix} -5 & -1 \\ -2 & -7 \end{vmatrix}$ **8.** $\begin{vmatrix} \frac{1}{5} & \frac{1}{6} \\ -6 & 5 \end{vmatrix}$

9. $\begin{vmatrix} \frac{1}{2} & \frac{1}{2} \\ \frac{1}{8} & -\frac{3}{4} \end{vmatrix}$ **10.** $\begin{vmatrix} \frac{2}{3} & \frac{1}{3} \\ -\frac{1}{2} & \frac{3}{4} \end{vmatrix}$

For Exercises 11–26, use Cramer's rule to solve each system or to determine that the system is inconsistent or contains dependent equations.

11. $x + y = 7$
    $x - y = 3$

12. $2x + y = 3$
    $x - y = 3$

13. $12x + 3y = 15$
$2x - 3y = 13$

14. $x - 2y = 5$
$5x - y = -2$

15. $4x - 5y = 17$
$2x + 3y = 3$

16. $3x + 2y = 2$
$2x + 2y = 3$

17. $x + 2y = 3$
$5x + 10y = 15$

18. $2x - 9y = 5$
$3x - 3y = 11$

19. $3x - 4y = 4$
$2x + 2y = 12$

20. $3x = 7y + 1$
$2x = 3y - 1$

21. $2x = 3y + 2$
$5x = 51 - 4y$

22. $y = -4x + 2$
$2x = 3y + 8$

23. $3x = 2 - 3y$
$2y = 3 - 2x$

24. $x + 2y - 3 = 0$
$12 = 8y + 4x$

25. $4y = 16 - 3x$
$6x = 32 - 8y$

26. $2x = 7 + 3y$
$4x - 6y = 3$

Evaluate each determinant in Exercises 27–32.

27. $\begin{vmatrix} 3 & 0 & 0 \\ 2 & 1 & -5 \\ 2 & 5 & -1 \end{vmatrix}$

28. $\begin{vmatrix} 4 & 0 & 0 \\ 3 & -1 & 4 \\ 2 & -3 & 5 \end{vmatrix}$

29. $\begin{vmatrix} 3 & 1 & 0 \\ -3 & 4 & 0 \\ -1 & 3 & -5 \end{vmatrix}$

30. $\begin{vmatrix} 2 & -4 & 2 \\ -1 & 0 & 5 \\ 3 & 0 & 4 \end{vmatrix}$

31. $\begin{vmatrix} 1 & 1 & 1 \\ 2 & 2 & 2 \\ -3 & 4 & -5 \end{vmatrix}$

32. $\begin{vmatrix} 1 & 2 & 3 \\ 2 & 2 & -3 \\ 3 & 2 & 1 \end{vmatrix}$

In Exercises 33–40, use Cramer's rule to solve each system.

33. $x + y + z = 0$
$2x - y + z = -1$
$-x + 3y - z = -8$

34. $x - y + 2z = 3$
$2x + 3y + z = 9$
$-x - y + 3z = 11$

35. $4x - 5y - 6z = -1$
$x - 2y - 5z = -12$
$2x - y = 7$

36. $x - 3y + z = -2$
$x + 2y = 8$
$2x - y = 1$

37. $x + y + z = 4$
$x - 2y + z = 7$
$x + 3y + 2z = 4$

38. $2x + 2y + 3z = 10$
$4x - y + z = -5$
$5x - 2y + 6z = 1$

39. $x + 2z = 4$
$2y - z = 5$
$2x + 3y = 13$

40. $3x + 2z = 4$
$5x - y = -4$
$4y + 3z = 22$

Evaluate each determinant in Exercises 41–44.

41. $\begin{vmatrix} 4 & 2 & 8 & -7 \\ -2 & 0 & 4 & 1 \\ 5 & 0 & 0 & 5 \\ 4 & 0 & 0 & -1 \end{vmatrix}$

42. $\begin{vmatrix} 3 & -1 & 1 & 2 \\ -2 & 0 & 0 & 0 \\ 2 & -1 & -2 & 3 \\ 1 & 4 & 2 & 3 \end{vmatrix}$

43. $\begin{vmatrix} -2 & -3 & 3 & 5 \\ 1 & -4 & 0 & 0 \\ 1 & 2 & 2 & -3 \\ 2 & 0 & 1 & 1 \end{vmatrix}$

44. $\begin{vmatrix} 1 & -3 & 2 & 0 \\ -3 & -1 & 0 & -2 \\ 2 & 1 & 3 & 1 \\ 2 & 0 & -2 & 0 \end{vmatrix}$

Application Exercises

Determinants are used to find the area of a triangle whose vertices are given by three points in a rectangular coordinate system. The area of a triangle with vertices $(x_1, y_1), (x_2, y_2),$ and (x_3, y_3) is

$$\text{Area} = \pm \frac{1}{2} \begin{vmatrix} x_1 & y_1 & 1 \\ x_2 & y_2 & 1 \\ x_3 & y_3 & 1 \end{vmatrix}$$

where the $\pm$ symbol indicates that the appropriate sign should be chosen to yield a positive area. Use this information to work Exercises 45–46.

45. Use determinants to find the area of the triangle whose vertices are $(3, -5), (2, 6),$ and $(-3, 5)$.

46. Use determinants to find the area of the triangle whose vertices are $(1, 1), (-2, -3),$ and $(11, -3)$.

Determinants are used to show that three points lie on the same line (are collinear). If

$$\begin{vmatrix} x_1 & y_1 & 1 \\ x_2 & y_2 & 1 \\ x_3 & y_3 & 1 \end{vmatrix} = 0,$$

then the points $(x_1, y_1), (x_2, y_2),$ and (x_3, y_3) are collinear. If the determinant does not equal 0, then the points are not collinear. Use this information to work Exercises 47–48.

47. Are the points $(3, -1), (0, -3),$ and $(12, 5)$ collinear?

48. Are the points $(-4, -6), (1, 0),$ and $(11, 12)$ collinear?

Determinants are used to write an equation of a line passing through two points. An equation of the line passing through the distinct points (x_1, y_1) and (x_2, y_2) is given by

$$\begin{vmatrix} x & y & 1 \\ x_1 & y_1 & 1 \\ x_2 & y_2 & 1 \end{vmatrix} = 0.$$

Use this information to work Exercises 49–50.

49. Use the determinant to write an equation of the line passing through $(3, -5)$ and $(-2, 6)$. Then expand the determinant, expressing the line's equation in slope-intercept form.

50. Use the determinant to write an equation of the line passing through $(-1, 3)$ and $(2, 4)$. Then expand the determinant, expressing the line's equation in slope-intercept form.

Writing in Mathematics

51. Explain how to evaluate a second-order determinant.

52. Describe the determinants D_x and D_y in terms of the coefficients and constants in a system of two equations in two variables.

53. Explain how to evaluate a third-order determinant.

54. When expanding a determinant by minors, when is it necessary to supply minus signs?

55. Without going into too much detail, describe how to solve a linear system in three variables using Cramer's rule.

56. In applying Cramer's rule, what does it mean if $D = 0$?

57. The process of solving a linear system in three variables using Cramer's rule can involve tedious computation. Is there a way of speeding up this process, perhaps using Cramer's rule to find the value for only one of the variables? Describe how this process might work, presenting a specific example with your description. Remember that your goal is still to find the value for each variable in the system.

58. If you could use only one method to solve linear systems in three variables, which method would you select? Explain why this is so.

Technology Exercises

59. Use the feature of your graphing utility that evaluates the determinant of a square matrix to verify any five of the determinants that you evaluated by hand in Exercises 1–10, 27–32, or 41–44.

In Exercises 60–61, use a graphing utility to evaluate the determinant for the given matrix.

60. $\begin{bmatrix} 3 & -2 & -1 & 4 \\ -5 & 1 & 2 & 7 \\ 2 & 4 & 5 & 0 \\ -1 & 3 & -6 & 5 \end{bmatrix}$

61. $\begin{bmatrix} 8 & 2 & 6 & -1 & 0 \\ 2 & 0 & -3 & 4 & 7 \\ 2 & 1 & -3 & 6 & -5 \\ -1 & 2 & 1 & 5 & -1 \\ 4 & 5 & -2 & 3 & -8 \end{bmatrix}$

62. What is the fastest method for solving a linear system with your graphing utility?

Critical Thinking Exercises

63. a. Evaluate: $\begin{vmatrix} a & a \\ 0 & a \end{vmatrix}$.

    b. Evaluate: $\begin{vmatrix} a & a & a \\ 0 & a & a \\ 0 & 0 & a \end{vmatrix}$.

    c. Evaluate: $\begin{vmatrix} a & a & a & a \\ 0 & a & a & a \\ 0 & 0 & a & a \\ 0 & 0 & 0 & a \end{vmatrix}$.

    d. Describe the pattern in the given determinants.

    e. Describe the pattern in the evaluations.

64. Evaluate: $\begin{vmatrix} 2 & 0 & 0 & 0 & 0 \\ 0 & 3 & 0 & 0 & 0 \\ 0 & 0 & 2 & 0 & 0 \\ 0 & 0 & 0 & 1 & 0 \\ 0 & 0 & 0 & 0 & 4 \end{vmatrix}$.

65. What happens to the value of a second-order determinant if the two columns are interchanged?

66. Consider the system

$$a_1 x + b_1 y = c_1$$

$$a_2 x + b_2 y = c_2.$$

Use Cramer's rule to prove that if the first equation of the system is replaced by the sum of the two equations, the resulting system has the same solution as the original system.

67. Show that the equation of a line through (x_1, y_1) and (x_2, y_2) is given by the determinant equation in Exercises 49–50.

Group Exercise

68. We have seen that determinants can be used to solve linear equations, give areas of triangles in rectangular coordinates, and determine equations of lines. Not impressed with these applications? Members of the group should research an application of determinants that they find intriguing. The group should then present a seminar to the class about this application.

CHAPTER SUMMARY, REVIEW, AND TEST

Summary

DEFINITIONS AND CONCEPTS	EXAMPLES

9.1 Matrix Solutions to Linear Systems

a. Matrix row operations are described in the box on page 771.

Ex. 2, p. 771

b. To solve a linear system using Gaussian elimination, begin with the system's augmented matrix. Use matrix operations to get 1s down the diagonal from upper left to lower right and 0s below the 1s. Such a matrix is in row-echelon form. Details are in the box on page 773.

Ex. 3, p. 773; Ex. 4, p. 776

c. To solve a linear system using Gauss-Jordan elimination, use the procedure of Gaussian elimination, but obtain 0s above and below the 1s in the diagonal from upper left to lower right. Such a matrix is in reduced row-echelon form. Details are in the box on page 778.

Ex. 5, p. 779

9.2 Inconsistent and Dependent Systems and Their Applications

a. If Gaussian elimination results in a matrix with a row containing all 0s to the left of the vertical line and a nonzero number to the right, the system has no solution (is inconsistent).

Ex. 1, p. 783

b. If Gaussian elimination results in a matrix with a row with all 0s, the system has an infinite number of solutions (contains dependent equations).

Ex. 2, p. 785

9.3 Matrix Operations and Their Applications

a. A matrix of order $m \times n$ has m rows and n columns. Two matrices are equal if and only if they have the same order and corresponding elements are equal.

Ex. 1, p. 792

b. Matrix Addition and Subtraction: Matrices of the same order are added or subtracted by adding or subtracting corresponding elements. Properties of matrix addition are given in the box on page 795.

Ex. 2, p. 794

c. Scalar Multiplication: If A is a matrix and c is a scalar, then cA is the matrix formed by multiplying each element in A by c. Properties of scalar multiplication are given in the box on page 796.

Ex. 3, p. 795; Ex. 4, p. 797

d. Matrix Multiplication: The product of an $m \times n$ matrix A and an $n \times p$ matrix B is an $m \times p$ matrix AB. The element in the ith row and jth column of AB is found by multiplying each element in the ith row of A by the corresponding element in the jth column of B and adding the products. Matrix multiplication is not commutative: $AB \neq BA$. Properties of matrix multiplication are given in the box on page 802.

Ex. 5, p. 798; Ex. 6, p. 799; Ex. 7, p. 801

9.4 Multiplicative Inverses of Matrices and Matrix Equations

a. The multiplicative identity matrix I_n is an $n \times n$ matrix with 1s down the diagonal from upper left to lower right and 0s elsewhere.

b. Let A be an $n \times n$ square matrix. If there is a square matrix A^{-1} such that $AA^{-1} = I_n$ and $A^{-1}A = I_n$, then A^{-1} is the multiplicative inverse of A.

Ex. 1, p. 808

c. If a square matrix has a multiplicative inverse, it is invertible or nonsingular. Methods for finding multiplicative inverses for invertible matrices, including a formula for 2×2 matrices, are given in the box on page 815.

Ex. 2, p. 809; Ex. 3, p. 811; Ex. 4, p. 814

d. Linear systems can be represented by matrix equations of the form $AX = B$ in which A is the coefficient matrix and B is the constant matrix. If $AX = B$ has a unique solution, then $X = A^{-1}B$.

Ex. 5, p. 816

DEFINITIONS AND CONCEPTS	EXAMPLES

9.5 Determinants and Cramer's Rule

a. Value of a Second-Order Determinant:

Ex. 1, p. 823

$$\begin{vmatrix} a_1 & b_1 \\ a_2 & b_2 \end{vmatrix} = a_1 b_2 - a_2 b_1$$

b. Cramer's rule for solving systems of linear equations in two variables uses three second-order determinants and is stated in the box on page 824.

Ex. 2, p. 825

c. To evaluate an nth-order determinant, where $n > 2$,

   1. Select a row or column about which to expand.

Ex. 3, p. 827;

   2. Multiply each element a_{ij} in the row or column, by $(-1)^{i+j}$ times the determinant obtained by deleting the ith row and the jth column in the given array of numbers.

Ex. 4, p. 828;
Ex. 6, p. 831

   3. The value of the determinant is the sum of the products found in step 2.

d. Cramer's rule for solving systems of linear equations in three variables uses four third-order determinants and is stated in the box on page 829.

Ex. 5, p. 829

e. Cramer's rule with inconsistent and dependent systems is summarized by the two situations in the box on page 831.

Review Exercises

9.1

In Exercises 1–2, write the system of linear equations represented by the augmented matrix. Use x, y, z, and, if necessary, w ,x, y, and z, for the variables. Once the system is written, use back-substitution to find its solution.

1. $\begin{bmatrix} 1 & 1 & 3 & | & 12 \\ 0 & 1 & -2 & | & -4 \\ 0 & 0 & 1 & | & 3 \end{bmatrix}$
2. $\begin{bmatrix} 1 & 0 & -2 & 2 & | & 1 \\ 0 & 1 & 1 & -1 & | & 0 \\ 0 & 0 & 1 & -\frac{7}{3} & | & -\frac{1}{3} \\ 0 & 0 & 0 & 1 & | & 1 \end{bmatrix}$

In Exercises 3–4, perform each matrix row operation and write the new matrix.

3. $\begin{bmatrix} 1 & 2 & 2 & | & 2 \\ 0 & 1 & -1 & | & 2 \\ 0 & 5 & 4 & | & 1 \end{bmatrix}$ $-5R_2 + R_3$

4. $\begin{bmatrix} 2 & -2 & 1 & | & -1 \\ 1 & 2 & -1 & | & 2 \\ 6 & 4 & 3 & | & 5 \end{bmatrix}$ $\frac{1}{2}R_1$

In Exercises 5–7, solve each system of equations using matrices. Use Gaussian elimination with back-substitution or Gauss-Jordan elimination.

5.
$$x + 2y + 3z = -5$$
$$2x + y + z = 1$$
$$x + y - z = 8$$

6.
$$x - 2y + z = 0$$
$$y - 3z = -1$$
$$2y + 5z = -2$$

7.
$$3x_1 + 5x_2 - 8x_3 + 5x_4 = -8$$
$$x_1 + 2x_2 - 3x_3 + x_4 = -7$$
$$2x_1 + 3x_2 - 7x_3 + 3x_4 = -11$$
$$4x_1 + 8x_2 - 10x_3 + 7x_4 = -10$$

8. The table shows the pollutants in the air in a city on a typical summer day.

x (Hours after 6 A.M.)	y (Amount of Pollutants in the Air, in parts per million)
2	98
4	138
10	162

a. Use the function $y = ax^2 + bx + c$ to model the data. Use either Gaussian elimination with back-substitution or Gauss-Jordan elimination to find the values for a, b, and c.

b. Use the function to find the time of day at which the city's air pollution level is at a maximum. What is the maximum level?

9.2

In Exercises 9–12, use Gaussian elimination to find the complete solution to each system, or show that none exists.

9.
$$2x - 3y + z = 1$$
$$x - 2y + 3z = 2$$
$$3x - 4y - z = 1$$

10.
$$x - 3y + z = 1$$
$$-2x + y + 3z = -7$$
$$x - 4y + 2z = 0$$

11. $x_1 + 4x_2 + 3x_3 - 6x_4 = 5$
$x_1 + 3x_2 + x_3 - 4x_4 = 3$
$2x_1 + 8x_2 + 7x_3 - 5x_4 = 11$
$2x_1 + 5x_2 - 6x_4 = 4$

12. $2x + 3y - 5z = 15$
$x + 2y - z = 4$

13. The figure shows the intersections of three one-way streets. The numbers given represent traffic flow, in cars per hour, at a peak period (from 4 P.M. to 6 P.M.).

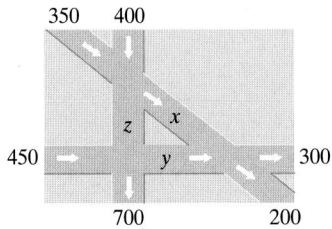

a. Use the idea that the number of cars entering each intersection per hour must equal the number of cars leaving per hour to set up a system of linear equations involving x, y, and z.

b. Use Gaussian elimination to solve the system.

c. If construction limits the value of z to 400, how many cars per hour must pass between the other intersections to keep traffic flowing?

9.3

14. Find values for x, y, and z so that the following matrices are equal:

$$\begin{bmatrix} 2x & y + 7 \\ z & 4 \end{bmatrix} = \begin{bmatrix} -10 & 13 \\ 6 & 4 \end{bmatrix}.$$

In Exercises 15–28, perform the indicated matrix operations given that A, B, C, and D are defined as follows. If an operation is not defined, state the reason.

$$A = \begin{bmatrix} 2 & -1 & 2 \\ 5 & 3 & -1 \end{bmatrix} \quad B = \begin{bmatrix} 0 & -2 \\ 3 & 2 \\ 1 & -5 \end{bmatrix}$$

$$C = \begin{bmatrix} 1 & 2 & 3 \\ -1 & 1 & 2 \\ -1 & 2 & 1 \end{bmatrix} \quad D = \begin{bmatrix} -2 & 3 & 1 \\ 3 & -2 & 4 \end{bmatrix}$$

15. $A + D$

16. $2B$

17. $D - A$

18. $B + C$

19. $3A + 2D$

20. $-2A + 4D$

21. $-5(A + D)$

22. AB

23. BA

24. BD

25. DB

26. $AB - BA$

27. $(A - D)C$

28. $B(AC)$

29. Solve for X in the matrix equation

$$3X + A = B$$

where $A = \begin{bmatrix} 4 & 6 \\ -5 & 0 \end{bmatrix}$ and $B = \begin{bmatrix} -2 & -12 \\ 4 & 1 \end{bmatrix}.$

In Exercises 30–31, use nine pixels in a 3×3 grid and the color levels shown.

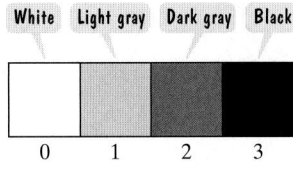

30. Write a 3×3 matrix that represents a digital photograph of the letter T in dark gray on a light gray background.

31. Find a matrix B so that $A + B$ increases the contrast of the letter T by changing the dark gray to black and the light gray to white.

32. An automobile dealership sells three models of cars at its three outlets. The inventory of models at each store is given by the following matrix.

	Model X	Model Y	Model Z
Outlet 1	12	7	6
Outlet 2	20	8	10
Outlet 3	7	2	3

$= A$

The next matrix gives the wholesale and retail prices for each model.

	Wholesale Price	Retail Price
Model X	26,000	29,000
Model Y	22,000	25,000
Model Z	24,000	28,500

$= B$

(Exercise continues on the next page.)

a. Calculate the product AB.

b. Describe what the matrix AB represents and interpret the elements.

c. What is the wholesale value of the cars at outlet 1?

d. What is the retail value of the cars at outlet 2?

e. If outlet 3 sells all of the inventory in matrix A, what is the profit for that branch of the dealership?

9.4

In Exercises 33–34, find the products AB and BA to determine whether B is the multiplicative inverse of A.

33. $A = \begin{bmatrix} 2 & 7 \\ 1 & 4 \end{bmatrix}$, $B = \begin{bmatrix} 4 & -7 \\ -1 & 3 \end{bmatrix}$

34. $A = \begin{bmatrix} 1 & 0 & 0 \\ 0 & 2 & -7 \\ 0 & -1 & 4 \end{bmatrix}$, $B = \begin{bmatrix} 1 & 0 & 0 \\ 0 & 4 & 7 \\ 0 & 1 & 2 \end{bmatrix}$

In Exercises 35–38, find A^{-1}. Check that $AA^{-1} = I$ and $A^{-1}A = I$.

35. $A = \begin{bmatrix} 1 & -1 \\ -2 & 3 \end{bmatrix}$ **36.** $A = \begin{bmatrix} 0 & 1 \\ 5 & 3 \end{bmatrix}$

37. $A = \begin{bmatrix} 1 & 0 & -2 \\ 2 & 1 & 0 \\ 1 & 0 & -3 \end{bmatrix}$ **38.** $A = \begin{bmatrix} 1 & 3 & -2 \\ 4 & 13 & -7 \\ 5 & 16 & -8 \end{bmatrix}$

In Exercises 39–40,

a. *Write each linear system as a matrix equation in the form $AX = B$.*

b. *Solve the system using the inverse that is given for the coefficient matrix.*

The inverse of

39. $\begin{aligned} x + y + 2z &= 7 \\ y + 3z &= -2 \\ 3x \quad\quad - 2z &= 0 \end{aligned}$ $\begin{bmatrix} 1 & 1 & 2 \\ 0 & 1 & 3 \\ 3 & 0 & -2 \end{bmatrix}$ is $\begin{bmatrix} -2 & 2 & 1 \\ 9 & -8 & -3 \\ -3 & 3 & 1 \end{bmatrix}$.

The inverse of

40. $\begin{aligned} x - y + 2z &= 12 \\ y - z &= -5 \\ x \quad\quad + 2z &= 10 \end{aligned}$ $\begin{bmatrix} 1 & -1 & 2 \\ 0 & 1 & -1 \\ 1 & 0 & 2 \end{bmatrix}$ is $\begin{bmatrix} 2 & 2 & -1 \\ -1 & 0 & 1 \\ -1 & -1 & 1 \end{bmatrix}$.

41. Use the coding-matrix $A = \begin{bmatrix} 3 & 2 \\ 4 & 3 \end{bmatrix}$ and its inverse $A^{-1} = \begin{bmatrix} 3 & -2 \\ -4 & 3 \end{bmatrix}$ to encode and then decode the word RULE.

9.5

In Exercises 42–47, evaluate each determinant.

42. $\begin{vmatrix} 3 & 2 \\ -1 & 5 \end{vmatrix}$ **43.** $\begin{vmatrix} -2 & -3 \\ -4 & -8 \end{vmatrix}$

44. $\begin{vmatrix} 2 & 4 & -3 \\ 1 & -1 & 5 \\ -2 & 4 & 0 \end{vmatrix}$ **45.** $\begin{vmatrix} 4 & 7 & 0 \\ -5 & 6 & 0 \\ 3 & 2 & -4 \end{vmatrix}$

46. $\begin{vmatrix} 1 & 1 & 0 & 2 \\ 0 & 3 & 2 & 1 \\ 0 & -2 & 4 & 0 \\ 0 & 3 & 0 & 1 \end{vmatrix}$ **47.** $\begin{vmatrix} 2 & 2 & 2 & 2 \\ 0 & 2 & 2 & 2 \\ 0 & 0 & 2 & 2 \\ 0 & 0 & 0 & 2 \end{vmatrix}$

In Exercises 48–51, use Cramer's rule to solve each system.

48. $\begin{aligned} x - 2y &= 8 \\ 3x + 2y &= -1 \end{aligned}$ **49.** $\begin{aligned} 7x + 2y &= 0 \\ 2x + y &= -3 \end{aligned}$

50. $\begin{aligned} x + 2y + 2z &= 5 \\ 2x + 4y + 7z &= 19 \\ -2x - 5y - 2z &= 8 \end{aligned}$ **51.** $\begin{aligned} 2x + y &= -4 \\ y - 2z &= 0 \\ 3x - 2z &= -11 \end{aligned}$

52. Use the quadratic function $y = ax^2 + bx + c$ to model the following data:

x (Age of a Driver)	y (Average Number of Automobile Accidents per Day in the United States)
20	400
40	150
60	400

Use Cramer's rule to determine values for a, b, and c. Then use the model to write a statement about the average number of automobile accidents in which 30-year-olds and 50-year-olds are involved daily.

Chapter 9 Test

In Exercises 1–2, solve each system of equations using matrices.

1.
$$x + 2y - z = -3$$
$$2x - 4y + z = -7$$
$$-2x + 2y - 3z = 4$$

2.
$$x - 2y + z = 2$$
$$2x - y - z = 1$$

In Exercises 3–6, let

$$A = \begin{bmatrix} 3 & 1 \\ 1 & 0 \\ 2 & 1 \end{bmatrix}, \quad B = \begin{bmatrix} 1 & -1 \\ 2 & 1 \end{bmatrix}, \quad \text{and} \quad C = \begin{bmatrix} 1 & 2 \\ -1 & 3 \end{bmatrix}.$$

Carry out the indicated operations.

3. $2B + 3C$

4. AB

5. C^{-1}

6. $BC - 3B$

7. If $A = \begin{bmatrix} 1 & 2 & 2 \\ 2 & 3 & 3 \\ 1 & -1 & -2 \end{bmatrix}$ and $B = \begin{bmatrix} -3 & 2 & 0 \\ 7 & -4 & 1 \\ -5 & 3 & -1 \end{bmatrix}$,

show that B is the inverse of A.

8. Consider the system
$$3x + 5y = 9$$
$$2x - 3y = -13.$$

  a. Express the system in the form $AX = B$, where A, X, and B are appropriate matrices.

  b. Find A^{-1}, the inverse of the coefficient matrix.

  c. Use A^{-1} to solve the given system.

9. Evaluate: $\begin{vmatrix} 4 & -1 & 3 \\ 0 & 5 & -1 \\ 5 & 2 & 4 \end{vmatrix}$.

10. Solve for x only using Cramer's rule:
$$3x + y - 2z = -3$$
$$2x + 7y + 3z = 9$$
$$4x - 3y - z = 7.$$

Cumulative Review Exercises (Chapters 1–9)

Solve each equation or inequality in Exercises 1–6.

1. $2x^2 = 4 - x$

2. $5x + 8 \le 7(1 + x)$

3. $\sqrt{2x + 4} - \sqrt{x + 3} - 1 = 0$

4. $3x^3 + 8x^2 - 15x + 4 = 0$

5. $e^{2x} - 14e^x + 45 = 0$

6. $\log_3 x + \log_3 (x + 2) = 1$

7. Use matrices to solve this system:
$$x - y + z = 17$$
$$2x + 3y + z = 8$$
$$-4x + y + 5z = -2.$$

8. Solve for y using Cramer's rule:
$$x - 2y + z = 7$$
$$2x + y - z = 0$$
$$3x + 2y - 2z = -2.$$

9. If $f(x) = \sqrt{4x - 7}$, find $f^{-1}(x)$.

10. Graph: $f(x) = \dfrac{x}{x^2 - 16}$.

11. Use the graph of $f(x) = 4x^4 - 4x^3 - 25x^2 + x + 6$ shown in the figure to factor the polynomial completely.

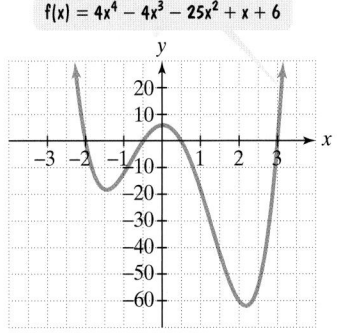

$f(x) = 4x^4 - 4x^3 - 25x^2 + x + 6$

12. Graph $y = \log_2 x$ and $y = \log_2 (x + 1)$ in the same rectangular coordinate system.

13. Use the exponential decay model $A = A_0 e^{kt}$ to solve this problem. A radioactive substance has a half-life of 40 days. There are initially 900 grams of the substance.

  a. Find the decay model for this substance. Round k to the nearest thousandth.

  b. How much of the substance will remain after 10 days? Round to the nearest hundredth of a gram.

14. Multiply the matrices: $\begin{bmatrix} 1 & -1 & 0 \\ 2 & 1 & 3 \end{bmatrix} \begin{bmatrix} 4 & -1 \\ 2 & 0 \\ 1 & 1 \end{bmatrix}$.

15. Find the partial fraction decomposition of

$$\frac{3x^2 + 17x - 38}{(x - 3)(x - 2)(x + 2)}.$$

In Exercises 16–19, graph each equation, function, or inequality in the rectangular coordinate system.

16. $y = -\frac{2}{3}x - 1$

17. $3x - 5y < 15$

18. $f(x) = x^2 - 2x - 3$

19. $(x - 1)^2 + (y + 1)^2 = 9$

20. Use synthetic division to divide $x^3 - 6x + 4$ by $x - 2$.

21. Graph: $y = 2 \sin 2\pi x, \quad 0 \le x \le 2$.

22. Find the exact value of $\cos\left[\tan^{-1}\left(-\frac{4}{3}\right)\right]$.

23. Verify the identity: $\dfrac{\cos 2x}{\cos x - \sin x} = \cos x + \sin x$.

24. Solve on the interval $[0, 2\pi)$: $\cos^2 x + \sin x + 1 = 0$.

25. If $\mathbf{v} = -6\mathbf{i} + 5\mathbf{j}$ and $\mathbf{w} = -7\mathbf{i} + 3\mathbf{j}$, find $4\mathbf{w} - 5\mathbf{v}$.

Conic Sections and Analytic Geometry

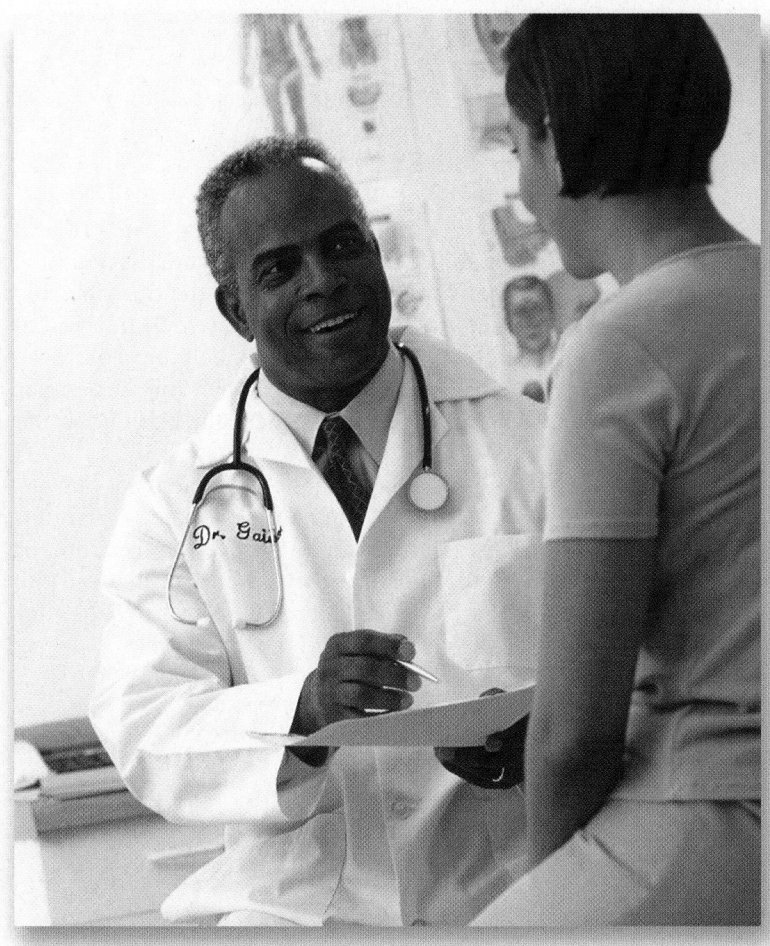

From ripples in water to the path on which humanity journeys through space, certain curves occur naturally throughout the universe. Over 2000 years ago the ancient Greeks studied these curves, called *conic sections*, without regard to their immediate usefulness simply because the study elicited ideas that were exciting, challenging, and interesting. The ancient Greeks could not have imagined the applications of these curves in the twenty-first century. Overwhelmed by the choices on satellite television? Blame it on a conic section! In this chapter, we use the rectangular coordinate system to study the conic sections and the mathematics behind their surprising applications.

One minute you're in class, enjoying the lecture. Then a sharp pain radiates down your side. The next minute you're being diagnosed with, of all things, a kidney stone. It took your cousin six weeks to recover from kidney stone surgery, but your doctor assures you there is nothing to worry about. A new procedure, based on a curve that looks like the cross section of a football, will dissolve the stone painlessly and let you return to class in a day or two. How can this be?

SECTION 10.1 The Ellipse

Objectives

1. Graph ellipses centered at the origin.
2. Write equations of ellipses in standard form.
3. Graph ellipses not centered at the origin.
4. Solve applied problems involving ellipses.

You took on a summer job driving a truck, delivering books that were ordered online. You're an avid reader, so just being around books sounded appealing. However, now you're feeling a bit shaky driving the truck for the first time. It's 10 feet wide and 9 feet high; compared to your compact car, it feels like you're behind the wheel of a tank. Up ahead you see a sign at the semielliptical entrance to a tunnel: Caution! Tunnel is 10 Feet High at Center Peak. Then you see another sign: Caution! Tunnel Is 40 Feet Wide. Will your truck clear the opening of the tunnels archway?

The mathematics of conic sections is present in the movements of planets, bridge and tunnel construction, navigational systems used to keep track of a ship's location, the manufacture of lenses for telescopes, and even a procedure for disintegrating kidney stones. **Conic sections** are curves that result from the intersection of a right circular cone and a plane. Figure 10.1 illustrates the four conic sections: the circle, the ellipse, the parabola, and the hyperbola.

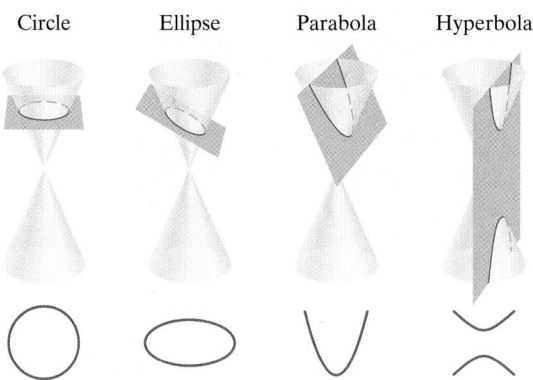

Circle   Ellipse   Parabola   Hyperbola

Figure 10.1 Obtaining the conic sections by intersecting a plane and a cone

In this section, we study the symmetric oval-shaped curve known as the ellipse. We will use a geometric definition for an ellipse to derive its equation. With this equation, we will determine if your delivery truck will clear the tunnel's entrance.

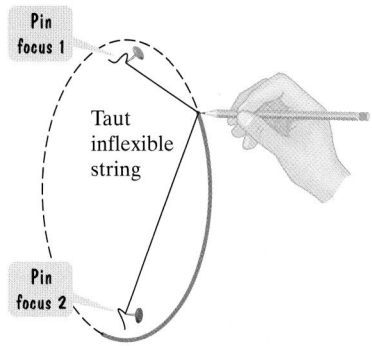

Figure 10.2 Drawing an ellipse

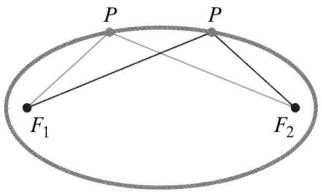

Figure 10.3

Definition of an Ellipse

Figure 10.2 illustrates how to draw an ellipse. Place pins at two fixed points, each of which is called a focus (plural: foci). If the ends of a fixed length of string are fastened to the pins and we draw the string taut with a pencil, the path traced by the pencil will be an ellipse. Notice that the sum of the distances of the pencil point from the foci remains constant because the length of the string is fixed. This procedure for drawing an ellipse illustrates its geometric definition.

> ### Definition of an Ellipse
> An **ellipse** is the set of all points in a plane the sum of whose distances from two fixed points, F_1 and F_2, is constant (see Figure 10.3). These two fixed points are called the **foci** (plural of **focus**). The midpoint of the segment connecting the foci is the **center** of the ellipse.

Figure 10.4 illustrates that an ellipse can be elongated horizontally or vertically. The line through the foci intersects the ellipse at two points, called the **vertices** (singular: **vertex**). The line segment that joins the vertices is the **major axis.** Notice that the midpoint of the major axis is the center of the ellipse. The line segment whose endpoints are on the ellipse and that is perpendicular to the major axis at the center is called the **minor axis** of the ellipse.

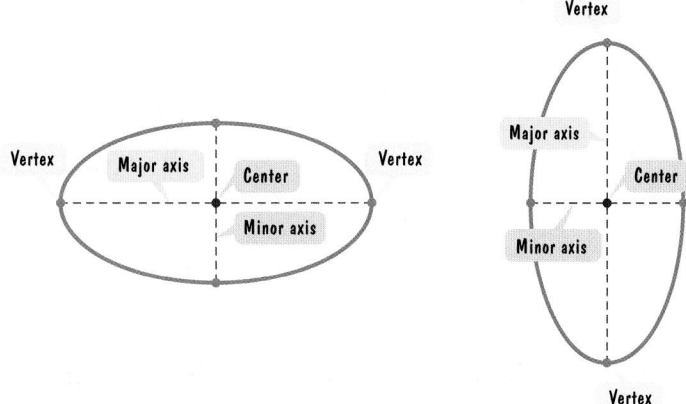

Figure 10.4 Horizontal and vertical elongations of an ellipse

Standard Form of the Equation of an Ellipse

The rectangular coordinate system gives us a unique way of describing an ellipse. It enables us to translate an ellipses geometric definition into an algebraic equation.

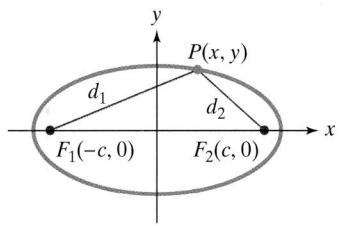

Figure 10.5

We start with Figure 10.5 to obtain an ellipses equation. We've placed an ellipse that is elongated horizontally into a rectangular coordinate system. The foci are on the x-axis at $(-c, 0)$ and $(c, 0)$, as in Figure 10.5. In this way, the center of the ellipse is at the origin. We let (x, y) represent the coordinates of any point on the ellipse.

What does the definition of an ellipse tell us about the point (x, y) in Figure 10.5? For any point (x, y) on the ellipse, the sum of the distances to the two foci, $d_1 + d_2$, must be constant. As we shall see, it is convenient to denote this constant by $2a$. Thus, the point (x, y) is on the ellipse if and only if

Discovery

Perform the algebra mentioned on the right by eliminating radicals and obtaining the equation shown. Be prepared for a bit of an algebraic workout!

Use the distance formula.

$$d_1 + d_2 = 2a.$$

$$\sqrt{(x + c)^2 + y^2} + \sqrt{(x - c)^2 + y^2} = 2a$$

After eliminating radicals and simplifying, we obtain

$$(a^2 - c^2)x^2 + a^2y^2 = a^2(a^2 - c^2).$$

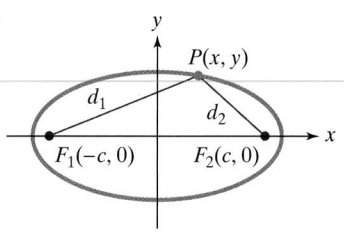

Figure 10.5, repeated

Look at the triangle in Figure 10.5. Notice that the distance from F_1 to F_2 is $2c$. Because the length of any side of a triangle is less than the sum of the lengths of the other two sides, $2c < d_1 + d_2$. Equivalently, $2c < 2a$ and $c < a$. Consequently, $a^2 - c^2 > 0$. For convenience, let $b^2 = a^2 - c^2$. Substituting b^2 for $a^2 - c^2$ in the preceding equation, we obtain

$$b^2x^2 + a^2y^2 = a^2b^2$$

$$\frac{b^2x^2}{a^2b^2} + \frac{a^2y^2}{a^2b^2} = \frac{a^2b^2}{a^2b^2} \qquad \text{Divide both sides by } a^2b^2.$$

$$\frac{x^2}{a^2} + \frac{y^2}{b^2} = 1. \qquad \text{Simplify.}$$

This last equation is the **standard form of the equation of an ellipse centered at the origin.** There are two such equations, one for a horizontal major axis and one for a vertical major axis.

Standard Forms of the Equations of an Ellipse

The **standard form of the equation of an ellipse** with center at the origin, and major and minor axes of lengths $2a$ and $2b$ (where a and b are positive, and $a^2 > b^2$) is

$$\frac{x^2}{a^2} + \frac{y^2}{b^2} = 1 \qquad \text{or} \qquad \frac{x^2}{b^2} + \frac{y^2}{a^2} = 1.$$

Study Tip

The form $c^2 = a^2 - b^2$ is the one you should remember. When finding the foci, this form is easy to manipulate.

Figure 10.6 illustrates that the vertices are on the major axis, a units from the center. The foci are on the major axis, c units from the center. For both equations, $b^2 = a^2 - c^2$. Equivalently, $c^2 = a^2 - b^2$.

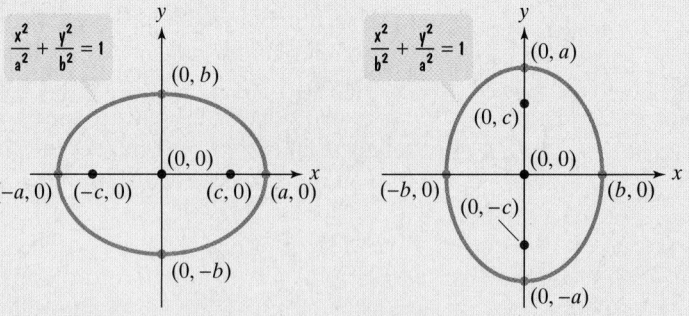

Figure 10.6 **(a)** Major axis is horizontal with length $2a$ **(b)** Major axis is vertical with length $2a$

Using the Standard Form of the Equation of an Ellipse

We can use the standard form of an ellipse's equation to graph the ellipse. Although the definition of the ellipse is given in terms of its foci, the foci are not part of the graph. A complete graph of an ellipse can be obtained without graphing the foci.

1 Graph ellipses centered at the origin.

EXAMPLE 1 Graphing an Ellipse Centered at the Origin

Graph and locate the foci: $\dfrac{x^2}{9} + \dfrac{y^2}{4} = 1$.

Solution The given equation is the standard form of an ellipses equation with $a^2 = 9$ and $b^2 = 4$.

$$\frac{x^2}{9} + \frac{y^2}{4} = 1$$

$a^2 = 9$. This is the larger of the two denominators. $b^2 = 4$. This is the smaller of the two denominators.

Because the denominator of the x^2-term is greater than the denominator of the y^2-term, the major axis is horizontal. Based on the standard form of the equation, we know the vertices are $(-a, 0)$ and $(a, 0)$. Because $a^2 = 9$, $a = 3$. Thus, the vertices are $(-3, 0)$ and $(3, 0)$, shown in Figure 10.7.

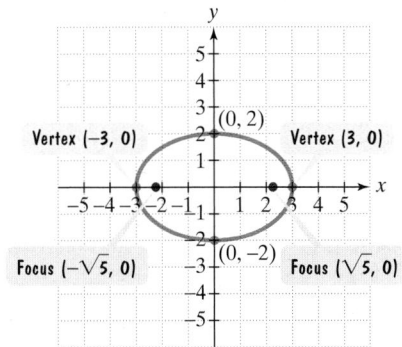

Figure 10.7 The graph of $\dfrac{x^2}{9} + \dfrac{y^2}{4} = 1$

Now let us find the endpoints of the vertical minor axis. According to the standard form of the equation, these endpoints are $(0, -b)$ and $(0, b)$. Because $b^2 = 4$, $b = 2$. Thus, the endpoints of the minor axis are $(0, -2)$ and $(0, 2)$. They are shown in Figure 10.7.

Finally, we find the foci, which are located at $(-c, 0)$ and $(c, 0)$. We can use the formula $c^2 = a^2 - b^2$ to do so. We know that $a^2 = 9$ and $b^2 = 4$. Thus,

$$c^2 = a^2 - b^2 = 9 - 4 = 5.$$

Because $c^2 = 5$, $c = \sqrt{5}$. The foci, $(-c, 0)$ and $(c, 0)$, are located at $\left(-\sqrt{5}, 0\right)$ and $\left(\sqrt{5}, 0\right)$. They are shown in Figure 10.7.

Technology

We graph $\dfrac{x^2}{9} + \dfrac{y^2}{4} = 1$ with a graphing utility by solving for y.

$$\frac{y^2}{4} = 1 - \frac{x^2}{9}$$

$$y^2 = 4\left(1 - \frac{x^2}{9}\right)$$

$$y = \pm 2\sqrt{1 - \frac{x^2}{9}}$$

Notice that the square root method requires us to define two functions. Enter

$y_1 = 2 \; \boxed{\sqrt{}} \; (1 \boxed{-} x \boxed{\wedge} 2 \boxed{\div} 9)$

and

$y_2 = -y_1$.

To see the true shape of the ellipse, use the $\boxed{\text{ZOOM SQUARE}}$ feature so that one unit on the x-axis is the same length as one unit on the y-axis.

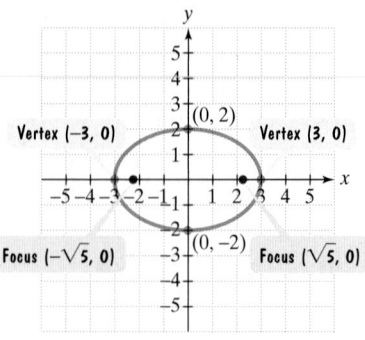

Figure 10.7, repeated

You can sketch the ellipse in Figure 10.7 by locating endpoints on the major and minor axes.

$$\frac{x^2}{3^2} + \frac{y^2}{2^2} = 1$$

Endpoints of the major axis are 3 units to the left and right of the center.

Endpoints of the minor axis are 2 units up and down from the center.

Check Point 1 Graph and locate the foci: $\dfrac{x^2}{36} + \dfrac{y^2}{9} = 1$.

EXAMPLE 2 Graphing an Ellipse Centered at the Origin

Graph and locate the foci: $25x^2 + 16y^2 = 400$.

Solution We begin by expressing the equation in standard form. Because we want 1 on the right side, we divide both sides by 400.

$$\frac{25x^2}{400} + \frac{16y^2}{400} = \frac{400}{400}$$

$$\frac{x^2}{16} + \frac{y^2}{25} = 1$$

$b^2 = 16$. This is the smaller of the two denominators.

$a^2 = 25$. This is the larger of the two denominators.

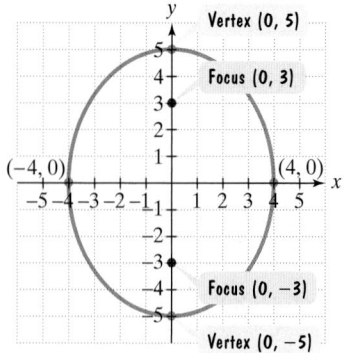

Figure 10.8 The graph of $25x^2 + 16y^2 = 400$, or $\dfrac{x^2}{16} + \dfrac{y^2}{25} = 1$

The equation is the standard form of an ellipse's equation with $a^2 = 25$ and $b^2 = 16$. Because the denominator of the y^2-term is greater than the denominator of the x^2-term, the major axis is vertical. Based on the standard form of the equation, we know the vertices are $(0, -a)$ and $(0, a)$. Because $a^2 = 25$, $a = 5$. Thus, the vertices are $(0, -5)$ and $(0, 5)$, shown in Figure 10.8.

Now let us find the endpoints of the horizontal minor axis. According to the standard form of the equation, these endpoints are $(-b, 0)$ and $(b, 0)$. Because $b^2 = 16$, $b = 4$. Thus, the endpoints of the minor axis are $(-4, 0)$ and $(4, 0)$. They are shown in Figure 10.8.

Finally, we find the foci, which are located at $(0, -c)$ and $(0, c)$. We can use the formula $c^2 = a^2 - b^2$ to do so. We know that $a^2 = 25$ and $b^2 = 16$. Thus

$$c^2 = a^2 - b^2 = 25 - 16 = 9.$$

Because $c^2 = 9$, $c = 3$. The foci, $(0, -c)$ and $(0, c)$, are located at $(0, -3)$ and $(0, 3)$. They are shown in Figure 10.8. You can sketch the ellipse in Figure 10.8 by locating endpoints on the major and minor axes.

$$\frac{x^2}{4^2} + \frac{y^2}{5^2} = 1$$

Endpoints of the minor axis are 4 units to the left and right of the center.

Endpoints of the major axis are 5 units up and down from the center.

Check Point 2 Graph and locate the foci: $16x^2 + 9y^2 = 144$.

2 Write equations of ellipses in standard form.

In Examples 1 and 2, we used the equation of an ellipse to find its foci and vertices. In the next example, we reverse this procedure.

EXAMPLE 3 Finding the Equation of an Ellipse from Its Foci and Vertices

Find the standard form of the equation of an ellipse with foci at $(-1, 0)$ and $(1, 0)$ and vertices $(-2, 0)$ and $(2, 0)$.

Solution Because the foci are located at $(-1, 0)$ and $(1, 0)$, on the x-axis, the major axis is horizontal. The center of the ellipse is midway between the foci, located at $(0, 0)$. Thus, the form of the equation is

$$\frac{x^2}{a^2} + \frac{y^2}{b^2} = 1.$$

We need to determine the values for a^2 and b^2. The distance from the center, $(0, 0)$, to either vertex, $(-2, 0)$ or $(2, 0)$, is 2. Thus, $a = 2$.

$$\frac{x^2}{2^2} + \frac{y^2}{b^2} = 1 \qquad \text{or} \qquad \frac{x^2}{4} + \frac{y^2}{b^2} = 1$$

We must still find b^2. The distance from the center, $(0, 0)$, to either focus, $(-1, 0)$ or $(1, 0)$, is 1, so $c = 1$. Using $c^2 = a^2 - b^2$, we have

$$1^2 = 2^2 - b^2$$

and

$$b^2 = 2^2 - 1^2 = 4 - 1 = 3.$$

Substituting 3 for b^2 in $\dfrac{x^2}{4} + \dfrac{y^2}{b^2} = 1$ gives us the standard form of the ellipse's equation. The equation is

$$\frac{x^2}{4} + \frac{y^2}{3} = 1.$$

Check Point 3 Find the standard form of the equation of an ellipse with foci at $(-2, 0)$ and $(2, 0)$ and vertices $(-3, 0)$ and $(3, 0)$.

3 Graph ellipses not centered at the origin.

Translations of Ellipses

Horizontal and vertical translations can be used to graph ellipses that are not centered at the origin. Figure 10.9 illustrates that the graphs of

$$\frac{(x - h)^2}{a^2} + \frac{(y - k)^2}{b^2} = 1 \quad \text{and} \quad \frac{x^2}{a^2} + \frac{y^2}{b^2} = 1$$

have the same size and shape. However, the graph of the first equation is centered at (h, k) rather than at the origin.

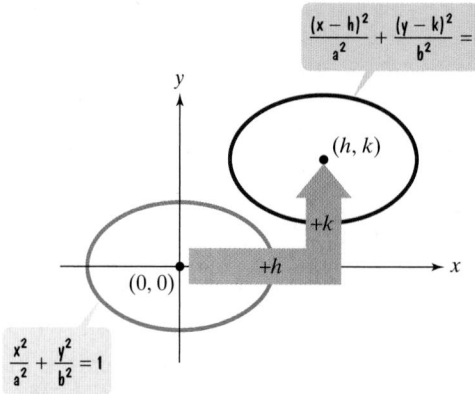

Figure 10.9 Translating an ellipses graph

Table 10.1 gives the standard forms of equations of ellipses centered at (h, k). Figure 10.10 shows their graphs.

Table 10.1 Standard Forms of Equations of Ellipses Centered at (h, k)

Equation	Center	Major Axis	Foci	Vertices
$\dfrac{(x - h)^2}{a^2} + \dfrac{(y - k)^2}{b^2} = 1$, $a^2 > b^2$ and $c^2 = a^2 - b^2$	(h, k)	Parallel to the x-axis, horizontal	$(h - c, k)$ $(h + c, k)$	$(h - a, k)$ $(h + a, k)$
$\dfrac{(x - h)^2}{b^2} + \dfrac{(y - k)^2}{a^2} = 1$, $a^2 > b^2$ and $c^2 = a^2 - b^2$	(h, k)	Parallel to the y-axis, vertical	$(h, k - c)$ $(h, k + c)$	$(h, k - a)$ $(h, k + a)$

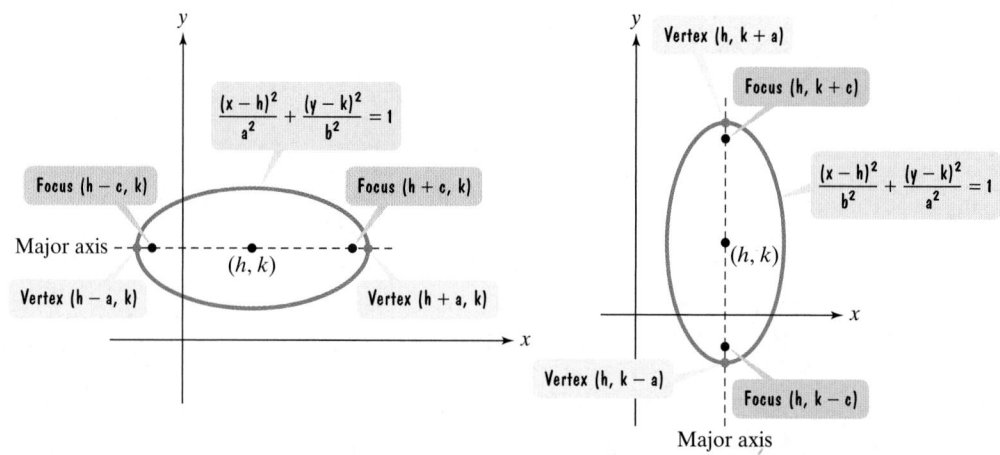

Figure 10.10 Graphs of ellipses centered at (h, k)

EXAMPLE 4 Graphing an Ellipse Centered at (h, k)

Graph: $\dfrac{(x - 1)^2}{4} + \dfrac{(y + 2)^2}{9} = 1$. Where are the foci located?

Solution In order to graph the ellipse, we need to know its center, (h, k). In the standard forms of equations centered at (h, k), h is the number subtracted from x and k is the number subtracted from y.

$$\underbrace{\frac{(x - 1)^2}{4}}_{\substack{\text{This is } (x - h)^2 \\ \text{with } h = 1.}} + \underbrace{\frac{[y - (-2)]^2}{9}}_{\substack{\text{This is } (y - k)^2 \\ \text{with } k = -2.}} = 1$$

We see that $h = 1$ and $k = -2$. Thus, the center of the ellipse, (h, k), is $(1, -2)$. We can graph the ellipse by locating endpoints on the major and minor axes. To do this, we must identify a^2 and b^2.

$$\underbrace{\frac{(x - 1)^2}{4}}_{\substack{b^2 = 4. \text{ This is the smaller} \\ \text{of the two denominators.}}} + \underbrace{\frac{(y + 2)^2}{9}}_{\substack{a^2 = 9. \text{ This is the larger} \\ \text{of the two denominators.}}} = 1$$

The larger number is under the expression involving y. This means that the major axis is vertical and parallel to the y-axis. Because $a^2 = 9$, $a = 3$ and the vertices lie three units above and below the center. Also, because $b^2 = 4$, $b = 2$ and the endpoints of the minor axis lie two units to the right and left of the center. We categorize these observations as follows:

For a Vertical Major Axis with Center (1, –2)	
Vertices	**Endpoints of Minor Axis**
$(1, -2 + 3) = (1, 1)$	$(1 + 2, -2) = (3, -2)$
$(1, -2 - 3) = (1, -5)$	$(1 - 2, -2) = (-1, -2)$

Using the center and these four points, we can sketch the ellipse, shown in Figure 10.11.

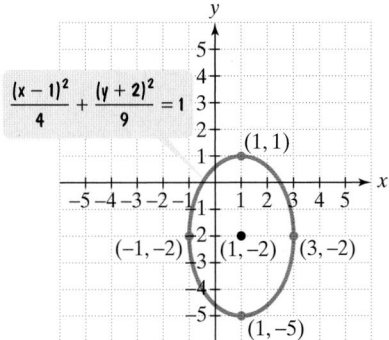

Figure 10.11 The graph of an ellipse centered at $(1, -2)$

With $c^2 = a^2 - b^2$, we have $c^2 = 9 - 4 = 5$. So the foci are located $\sqrt{5}$ units above and below the center, at $\left(1, -2 + \sqrt{5}\right)$ and $\left(1, -2 - \sqrt{5}\right)$.

Check Point 4 Graph: $\dfrac{(x + 1)^2}{9} + \dfrac{(y - 2)^2}{4} = 1$. Where are the foci located?

In some cases, it is necessary to convert the equation of an ellipse to standard form by completing the square on x and y. For example, suppose that we wish to graph the ellipse whose equation is

$$9x^2 + 4y^2 - 18x + 16y - 11 = 0.$$

Because we plan to complete the square on both x and y, we need to rearrange terms so that

- x-terms are arranged in descending order.
- y-terms are arranged in descending order.
- the constant term appears on the right.

$$9x^2 + 4y^2 - 18x + 16y - 11 = 0$$ This is the given equation.

$$(9x^2 - 18x) + (4y^2 + 16y) = 11$$ Group terms and add 11 to both sides.

$$9(x^2 - 2x + \Box) + 4(y^2 + 4y + \Box) = 11$$ To complete the square, coefficients of x^2 and y^2 must be 1. Factor out 9 and 4, respectively.

We added 9 · 1, or 9, to the left side.

We also added 4 · 4, or 16, to the left side.

$$9(x^2 - 2x + 1) + 4(y^2 + 4y + 4) = 11 + 9 + 16$$ Complete each square by adding the square of half the coefficient of x and y, respectively.

9 and 16, added on the left side, must also be added on the right side.

$$9(x - 1)^2 + 4(y + 2)^2 = 36$$ Factor.

$$\frac{9(x - 1)^2}{36} + \frac{4(y + 2)^2}{36} = \frac{36}{36}$$ Divide both sides by 36.

$$\frac{(x - 1)^2}{4} + \frac{(y + 2)^2}{9} = 1$$ Simplify.

The equation is now in standard form. This is precisely the form of the equation that we graphed in Example 4.

Study Tip

When completing the square, remember that changes made on the left side of the equation must also be made on the right side of the equation. To review completing the square, see Section 1.5, pages 119–120.

4 Solve applied problems involving ellipses.

Applications

Ellipses have many applications. German scientist Johannes Kepler (1571–1630) showed that the planets in our solar system move in elliptical orbits, with the sun at a focus. Earth satellites also travel in elliptical orbits, with Earth at a focus.

One intriguing aspect of the ellipse is that a ray of light or a sound wave emanating from one focus will be reflected by the ellipse exactly to the other focus. A whispering gallery is an elliptical room with an elliptical, dome-shaped ceiling. People standing at the foci can whisper and hear each other quite clearly, while persons in other locations in the room cannot hear them. Statuary Hall in the U.S. Capitol Building is elliptical. President John Quincy Adams, while a member of the House of Representatives, was aware of this acoustical phenomenon. He situated his desk at a focal point of the elliptical ceiling, easily eavesdropping on the private conversations of other House members located near the other focus.

The elliptical reflection principle is used in a procedure for disintegrating kidney stones. The patient is placed within a device that is elliptical in shape. The patient is placed so the kidney is centered at one focus, while ultrasound waves from the other focus hit the walls and are reflected to the kidney stone. The convergence of the ultrasound waves at the kidney stone causes vibrations that

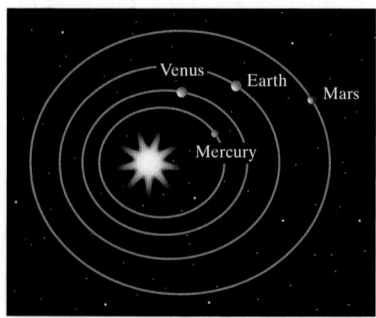

Planets move in elliptical orbits.

shatter it into fragments. The small pieces can then be passed painlessly through the patient's system. The patient recovers in days, as opposed to up to six weeks if surgery is used instead.

Halley's Comet

Halley's Comet has an elliptical orbit with the sun at one focus. The comet returns every 76.3 years. The first recorded sighting was in 239 B.C. It was last seen in 1986. At that time, spacecraft went close to the comet, measuring its nucleus to be 7 miles long and 4 miles wide. By 2024, Halley's Comet will have reached the farthest point in its elliptical orbit before returning to be next visible from Earth in 2062.

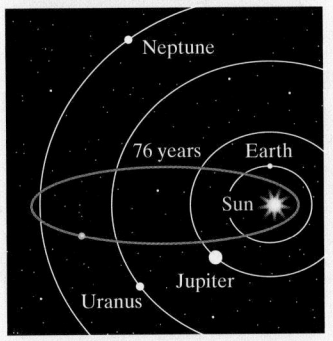

The elliptical orbit of Halley's Comet

Whispering in an elliptical dome

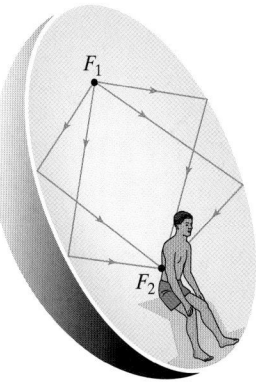

Disintegrating kidney stones

Ellipses are often used for supporting arches of bridges and in tunnel construction. This application forms the basis of our next example.

EXAMPLE 5 An Application Involving an Ellipse

A semielliptical archway over a one-way road has a height of 10 feet and a width of 40 feet (see Figure 10.12). Your truck has a width of 10 feet and a height of 9 feet. Will your truck clear the opening of the archway?

Solution Because your truck's width is 10 feet, to determine the clearance, we must find the height of the archway 5 feet from the center. If that height is 9 feet or less, the truck will not clear the opening.

In Figure 10.13, we've constructed a coordinate system with the x-axis on the ground and the origin at the center of the archway. Also shown is the truck, whose height is 9 feet.

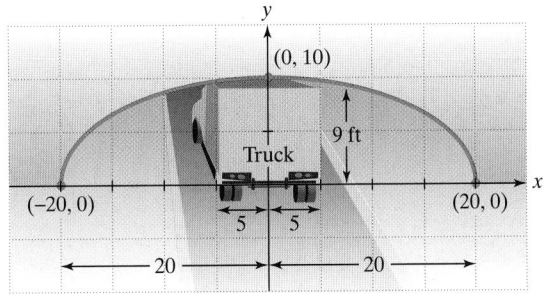

Figure 10.13

Using the equation $\frac{x^2}{a^2} + \frac{y^2}{b^2} = 1$, we can express the equation of the blue archway

in Figure 10.13 as $\frac{x^2}{20^2} + \frac{y^2}{10^2} = 1$, or $\frac{x^2}{400} + \frac{y^2}{100} = 1$.

As shown in Figure 10.13, the edge of the 10-foot-wide truck corresponds to $x = 5$. We find the height of the archway 5 feet from the center by substituting 5 for x and solving for y.

Figure 10.12 A semielliptical archway

$$\frac{5^2}{400} + \frac{y^2}{100} = 1 \qquad \text{Substitute 5 for x in } \frac{x^2}{400} + \frac{y^2}{100} = 1.$$

$$\frac{25}{400} + \frac{y^2}{100} = 1 \qquad \text{Square 5.}$$

$$400\left(\frac{25}{400} + \frac{y^2}{100}\right) = 400(1) \qquad \text{Clear fractions by multiplying both sides by 400.}$$

$$25 + 4y^2 = 400 \qquad \text{Use the distributive property and simplify.}$$

$$4y^2 = 375 \qquad \text{Subtract 25 from both sides.}$$

$$y^2 = \frac{375}{4} \qquad \text{Divide both sides by 4.}$$

$$y = \sqrt{\frac{375}{4}} \qquad \text{Take only the positive square root. The archway is above the x-axis, and y is nonnegative.}$$

$$\approx 9.68$$

Thus, the height of the archway 5 feet from the center is approximately 9.68 feet. Because your truck's height is 9 feet, there is enough room for the truck to clear the archway.

Check Point 5   Will a truck that is 12 feet wide and has a height of 9 feet clear the opening of the archway described in Example 5?

EXERCISE SET 10.1

Practice Exercises

In Exercises 1–18, graph each ellipse and locate the foci.

1. $\dfrac{x^2}{16} + \dfrac{y^2}{4} = 1$

2. $\dfrac{x^2}{25} + \dfrac{y^2}{16} = 1$

3. $\dfrac{x^2}{9} + \dfrac{y^2}{36} = 1$

4. $\dfrac{x^2}{16} + \dfrac{y^2}{49} = 1$

5. $\dfrac{x^2}{25} + \dfrac{y^2}{64} = 1$

6. $\dfrac{x^2}{49} + \dfrac{y^2}{36} = 1$

7. $\dfrac{x^2}{49} + \dfrac{y^2}{81} = 1$

8. $\dfrac{x^2}{64} + \dfrac{y^2}{100} = 1$

9. $\dfrac{x^2}{\frac{9}{4}} + \dfrac{y^2}{\frac{25}{4}} = 1$

10. $\dfrac{x^2}{\frac{81}{4}} + \dfrac{y^2}{\frac{25}{16}} = 1$

11. $x^2 = 1 - 4y^2$

12. $4y^2 = 1 - 4x^2$

13. $25x^2 + 4y^2 = 100$

14. $9x^2 + 4y^2 = 36$

15. $4x^2 + 16y^2 = 64$

16. $4x^2 + 25y^2 = 100$

17. $7x^2 = 35 - 5y^2$

18. $6x^2 = 30 - 5y^2$

In Exercises 19–24, find the standard form of the equation of each ellipse and give the location of its foci.

19.

20.

21.

22.

23.

24.

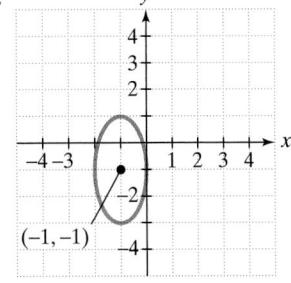

In Exercises 25–36, find the standard form of the equation of each ellipse satisfying the given conditions.

25. Foci: $(-5, 0), (5, 0)$; vertices: $(-8, 0), (8, 0)$

26. Foci: $(-2, 0), (2, 0)$; vertices: $(-6, 0), (6, 0)$

27. Foci: $(0, -4), (0, 4)$; vertices: $(0, -7), (0, 7)$

28. Foci: $(0, -3), (0, 3)$; vertices: $(0, -4), (0, 4)$

29. Foci: $(-2, 0), (2, 0)$; y-intercepts: -3 and 3

30. Foci: $(0, -2), (0, 2)$; x-intercepts: -2 and 2

31. Major axis horizontal with length 8; length of minor axis = 4; center: $(0, 0)$

32. Major axis horizontal with length 12; length of minor axis = 6; center: $(0, 0)$

33. Major axis vertical with length 10; length of minor axis = 4; center: $(-2, 3)$

34. Major axis vertical with length 20; length of minor axis = 10; center: $(2, -3)$

35. Endpoints of major axis: $(7, 9)$ and $(7, 3)$
Endpoints of minor axis: $(5, 6)$ and $(9, 6)$

36. Endpoints of major axis: $(2, 2)$ and $(8, 2)$
Endpoints of minor axis: $(5, 3)$ and $(5, 1)$

In Exercises 37–50, graph each ellipse and give the location of its foci.

37. $\dfrac{(x - 2)^2}{9} + \dfrac{(y - 1)^2}{4} = 1$

38. $\dfrac{(x - 1)^2}{16} + \dfrac{(y + 2)^2}{9} = 1$

39. $(x + 3)^2 + 4(y - 2)^2 = 16$

40. $(x - 3)^2 + 9(y + 2)^2 = 18$

41. $\dfrac{(x - 4)^2}{9} + \dfrac{(y + 2)^2}{25} = 1$

42. $\dfrac{(x - 3)^2}{9} + \dfrac{(y + 1)^2}{16} = 1$

43. $\dfrac{x^2}{25} + \dfrac{(y - 2)^2}{36} = 1$

44. $\dfrac{(x - 4)^2}{4} + \dfrac{y^2}{25} = 1$

45. $\dfrac{(x + 3)^2}{9} + (y - 2)^2 = 1$

46. $\dfrac{(x + 2)^2}{16} + (y - 3)^2 = 1$

47. $\dfrac{(x - 1)^2}{2} + \dfrac{(y + 3)^2}{5} = 1$

48. $\dfrac{(x + 1)^2}{2} + \dfrac{(y - 3)^2}{5} = 1$

49. $9(x - 1)^2 + 4(y + 3)^2 = 36$

50. $36(x + 4)^2 + (y + 3)^2 = 36$

In Exercises 51–56, convert each equation to standard form by completing the square on x and y. Then graph the ellipse and give the location of its foci.

51. $9x^2 + 25y^2 - 36x + 50y - 164 = 0$

52. $4x^2 + 9y^2 - 32x + 36y + 64 = 0$

53. $9x^2 + 16y^2 - 18x + 64y - 71 = 0$

54. $x^2 + 4y^2 + 10x - 8y + 13 = 0$

55. $4x^2 + y^2 + 16x - 6y - 39 = 0$

56. $4x^2 + 25y^2 - 24x + 100y + 36 = 0$

Application Exercises

57. Will a truck that is 8 feet wide carrying a load that reaches 7 feet above the ground clear the semielliptical arch on the one-way road that passes under the bridge shown in the figure?

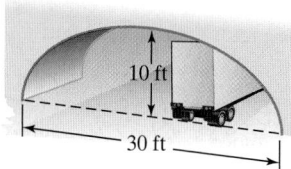

58. A semielliptic archway has a height of 20 feet and a width of 50 feet, as shown in the figure. Can a truck 14 feet high and 10 feet wide drive under the archway without going into the other lane?

59. The elliptical ceiling in Statuary Hall in the U.S. Capitol Building is 96 feet long and 23 feet tall.

a. Using the rectangular coordinate system in the figure shown, write the standard form of the equation of the elliptical ceiling.

b. John Quincy Adams discovered that he could overhear the conversations of opposing party leaders near the left side of the chamber if he situated his desk at the focus at the right side of the chamber. How far from the center of the ellipse along the major axis did Adams situate his desk? (Round to the nearest foot.)

60. If an elliptical whispering room has a height of 30 feet and a width of 100 feet, where should two people stand if they would like to whisper back and forth and be heard?

Writing in Mathematics

61. What is an ellipse?

62. Describe how to graph $\dfrac{x^2}{25} + \dfrac{y^2}{16} = 1$.

63. Describe how to locate the foci for $\dfrac{x^2}{25} + \dfrac{y^2}{16} = 1$.

64. Describe one similarity and one difference between the graphs of $\dfrac{x^2}{25} + \dfrac{y^2}{16} = 1$ and $\dfrac{x^2}{16} + \dfrac{y^2}{25} = 1$.

65. Describe one similarity and one difference between the graphs of $\dfrac{x^2}{25} + \dfrac{y^2}{16} = 1$ and $\dfrac{(x-1)^2}{25} + \dfrac{(y-1)^2}{16} = 1$.

66. An elliptipool is an elliptical pool table with only one pocket. A pool shark places a ball on the table, hits it in what appears to be a random direction, and yet it bounces off the edge, falling directly into the pocket. Explain why this happens.

 ## Technology Exercises

67. Use a graphing utility to graph any five of the ellipses that you graphed by hand in Exercises 1–18.

68. Use a graphing utility to graph any three of the ellipses that you graphed by hand in Exercises 37–50. First solve the given equation for y by using the square root method. Enter each of the two resulting equations to produce each half of the ellipse.

69. Use a graphing utility to graph any one of the ellipses that you graphed by hand in Exercises 51–56. Write the equation as a quadratic equation in y and use the quadratic formula to solve for y. Enter each of the two resulting equations to produce each half of the ellipse.

70. Write an equation for the path of each of the following elliptical orbits. Then use a graphing utility to graph the two ellipses in the same viewing rectangle. Can you see why early astronomers had difficulty detecting that these orbits are ellipses rather than circles?

Earth's orbit:	Length of major axis: 186 million miles
	Length of minor axis: 185.8 million miles
Mars's orbit:	Length of major axis: 283.5 million miles
	Length of minor axis: 278.5 million miles

 ## Critical Thinking Exercises

71. Find the standard form of the equation of an ellipse with vertices at $(0, -6)$ and $(0, 6)$, passing through $(2, -4)$.

72. An Earth satellite has an elliptical orbit described by

$$\frac{x^2}{(5000)^2} + \frac{y^2}{(4750)^2} = 1.$$

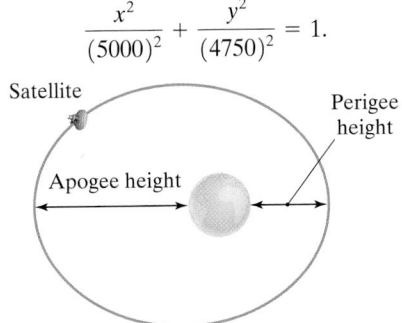

(All units are in miles.) The coordinates of the center of Earth are $(16, 0)$.

a. The perigee of the satellites orbit is the point that is nearest Earth's center. If the radius of Earth is approximately 4000 miles, find the distance of the perigee above Earth's surface.

b. The apogee of the satellites orbit is the point that is the greatest distance from Earth's center. Find the distance of the apogee above Earth's surface.

73. The equation of the red ellipse in the figure in the next column is

$$\frac{x^2}{25} + \frac{y^2}{9} = 1.$$

Write the equation for each circle shown in the figure.

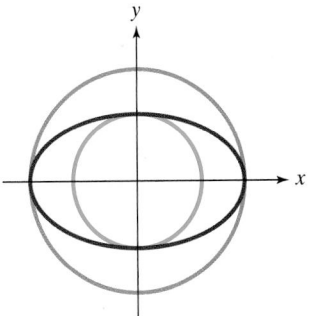

74. What happens to the shape of the graph of $\dfrac{x^2}{a^2} + \dfrac{y^2}{b^2} = 1$ as $\dfrac{c}{a}$ is close to zero?

75. Refer to the Discovery box on page 844 and derive the ellipse's equation. *Hint*: Begin by isolating one of the radicals and squaring both sides. After simplifying, repeat this procedure.

SECTION 10.2   *The Hyperbola*

Objectives

1. Locate a hyperbola's vertices and foci.
2. Write equations of hyperbolas in standard form.
3. Graph hyperbolas centered at the origin.
4. Graph hyperbolas not centered at the origin.
5. Solve applied problems involving hyperbolas.

St. Mary's Cathedral

Conic sections are often used to create unusual architectural designs. The top of St. Mary's Cathedral in San Francisco is a 2135-cubic-foot dome with walls rising 200 feet above the floor and supported by four massive concrete pylons that extend 94 feet into the ground. Cross sections of the roof are parabolas and hyperbolas. In this section, we study the curve with two parts known as the hyperbola.

Figure 10.14 Casting hyperbolic shadows

Definition of a Hyperbola

Figure 10.14 shows a cylindrical lampshade casting two shadows on a wall. These shadows indicate the distinguishing feature of hyperbolas: Their graphs contain two disjoint parts called **branches.** Although each branch might look like a parabola, its shape is actually quite different.

The definition of a hyperbola is similar to that of the ellipse. For the ellipse, the *sum* of the distances to the foci is a constant. By contrast, for a hyperbola the *difference* of the distances to the foci is a constant.

Definition of a Hyperbola

A **hyperbola** is the set of points in a plane the difference of whose distances from two fixed points, called foci, is constant.

Figure 10.15 illustrates the two branches of a hyperbola. The line through the foci intersects the hyperbola at two points, called the **vertices.** The line segment that joins the vertices is the **transverse axis.** The midpoint of the transverse axis is the **center** of the hyperbola. Notice that the center lies midway between the vertices, as well as midway between the foci.

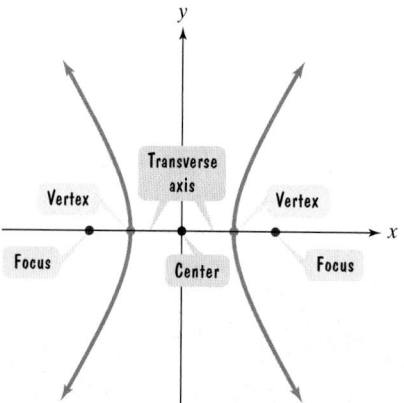

Figure 10.15 The two branches of a hyperbola

Standard Form of the Equation of a Hyperbola

The rectangular coordinate system enables us to translate a hyperbola's geometric definition into an algebraic equation. Figure 10.16 is our starting point for obtaining an equation. We place the foci, F_1 and F_2, on the x-axis at the points $(-c, 0)$ and $(c, 0)$. Note that the center of this hyperbola is at the origin. We let (x, y) represent the coordinates of any point, P, on the hyperbola.

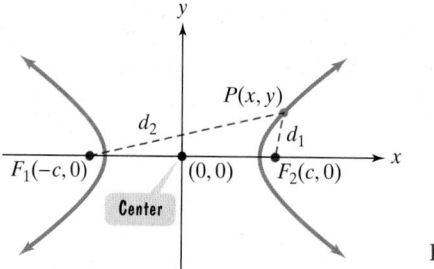

Figure 10.16

What does the definition of a hyperbola tell us about the point (x, y) in Figure 10.16? For any point (x, y) on the hyperbola, the absolute value of the difference of the distances from the two foci, $|d_2 - d_1|$, must be constant.

We denote this constant by $2a$, just as we did for the ellipse. Thus, the point (x, y) is on the hyperbola if and only if

$$\left| d_2 - d_1 \right| = 2a$$

$$\left| \sqrt{(x + c)^2 + (y - 0)^2} - \sqrt{(x - c)^2 + (y - 0)^2} \right| = 2a \quad \text{Use the distance formula.}$$

After eliminating radicals and simplifying, we obtain

$$(c^2 - a^2)x^2 - a^2 y^2 = a^2(c^2 - a^2).$$

For convenience, let $b^2 = c^2 - a^2$. Substituting b^2 for $c^2 - a^2$ in the preceding equation, we obtain

$$b^2 x^2 - a^2 y^2 = a^2 b^2$$

$$\frac{b^2 x^2}{a^2 b^2} - \frac{a^2 y^2}{a^2 b^2} = \frac{a^2 b^2}{a^2 b^2} \quad \text{Divide both sides by } a^2 b^2.$$

$$\frac{x^2}{a^2} - \frac{y^2}{b^2} = 1 \quad \text{Simplify.}$$

This last equation is called the **standard form of the equation of a hyperbola centered at the origin.** There are two such equations. The first is for a hyperbola in which the transverse axis lies on the x-axis. The second is for a hyperbola in which the transverse axis lies on the y-axis.

Standard Forms of the Equations of a Hyperbola

The **standard form of the equation of a hyperbola** with center at the origin is

$$\frac{x^2}{a^2} - \frac{y^2}{b^2} = 1 \quad \text{or} \quad \frac{y^2}{a^2} - \frac{x^2}{b^2} = 1.$$

Figure 10.17 illustrates that for the equation on the left, the transverse axis lies on the x-axis. For the equation on the right, the transverse axis lies on the y-axis. The vertices are a units from the center and the foci are c units from the center. For both equations, $b^2 = c^2 - a^2$. Equivalently, $c^2 = a^2 + b^2$.

Study Tip

The form $c^2 = a^2 + b^2$ is the one you should remember. When finding the foci, this form is easy to manipulate.

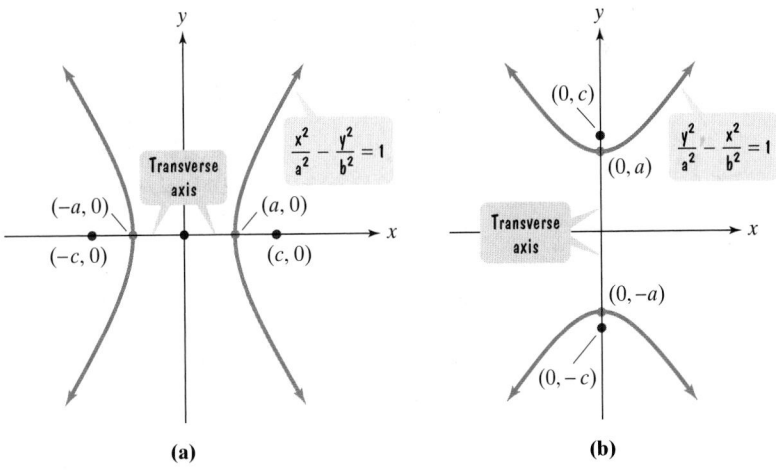

Figure 10.17 (a) Transverse axis lies on the x-axis. **(b)** Transverse axis lies on the y-axis.

Study Tip

When the x^2-term is preceded by a plus sign, the transverse axis is horizontal. When the y^2-term is preceded by a plus sign, the transverse axis is vertical.

1 Locate a hyperbola's vertices and foci.

Using the Standard Form of the Equation of a Hyperbola

We can use the standard form of the equation of a hyperbola to find its vertices and locate its foci. Because the vertices are a units from the center, begin by identifying a^2 in the equation. In the standard form of a hyperbola's equation, **a^2 is the number under the variable whose term is preceded by a plus sign** (+). If the x^2-term is preceded by a plus sign, the transverse axis lies along the x-axis. Thus, the vertices are a units to the left and right of the origin. If the y^2-term is preceded by a plus sign, the transverse axis lies along the y-axis. Thus, the vertices are a units above and below the origin.

We know that the foci are c units from the center. The substitution that is used to derive the hyperbola's equation, $c^2 = a^2 + b^2$, is needed to locate the foci when a^2 and b^2 are known.

Study Tip

Notice the sign difference in the following equations:

Finding an ellipse's foci:

$$c^2 = a^2 - b^2$$

Finding a hyperbola's foci:

$$c^2 = a^2 + b^2.$$

EXAMPLE 1 Finding Vertices and Foci from a Hyperbola's Equation

Find the vertices and locate the foci for each of the following hyperbolas with the given equation:

a. $\dfrac{x^2}{16} - \dfrac{y^2}{9} = 1$ **b.** $\dfrac{y^2}{9} - \dfrac{x^2}{16} = 1$

Solution Both equations are in standard form. We begin by identifying a^2 and b^2 in each equation.

a. The first equation is in the form $\dfrac{x^2}{a^2} - \dfrac{y^2}{b^2} = 1$.

$$\frac{x^2}{16} - \frac{y^2}{9} = 1$$

$a^2 = 16$. This is the denominator of the term preceded by a plus sign.

$b^2 = 9$. This is the denominator of the term preceded by a minus sign.

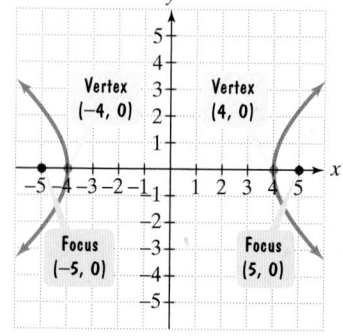

Figure 10.18 The graph of $\dfrac{x^2}{16} - \dfrac{y^2}{9} = 1$

Because the x^2-term is preceded by a plus sign, the transverse axis lies along the x-axis. Thus, the vertices are a units to the *left* and *right* of the origin. Based on the standard form of the equation, we know the vertices are $(-a, 0)$ and $(a, 0)$. Because $a^2 = 16$, $a = 4$. Thus, the vertices are $(-4, 0)$ and $(4, 0)$, shown in Figure 10.18.

We use $c^2 = a^2 + b^2$ to find the foci, which are located at $(-c, 0)$ and $(c, 0)$. We know that $a^2 = 16$ and $b^2 = 9$; we need to find c^2 in order to find c.

$$c^2 = a^2 + b^2 = 16 + 9 = 25$$

Because $c^2 = 25$, $c = 5$. The foci are located at $(-5, 0)$ and $(5, 0)$. They are shown in Figure 10.18.

b. The second given equation is in the form $\dfrac{y^2}{a^2} - \dfrac{x^2}{b^2} = 1$.

$$\frac{y^2}{9} - \frac{x^2}{16} = 1$$

$a^2 = 9$. This is the denominator of the term preceded by a plus sign.

$b^2 = 16$. This is the denominator of the term preceded by a minus sign.

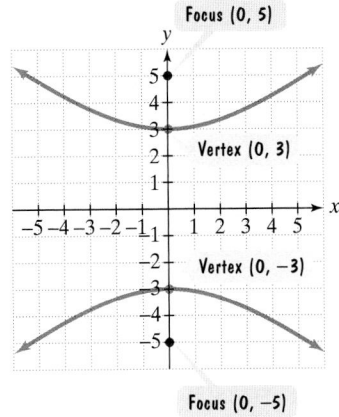

Figure 10.19 The graph of $\dfrac{y^2}{9} - \dfrac{x^2}{16} = 1$

2 Write equations of hyperbolas in standard form.

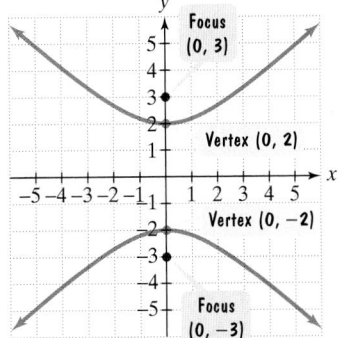

Figure 10.20

Because the y^2-term is preceded by a plus sign, the transverse axis lies along the y-axis. Thus, the vertices are a units *above* and *below* the origin. Based on the standard form of the equation, we know the vertices are $(0, -a)$ and $(0, a)$. Because $a^2 = 9$, $a = 3$. Thus, the vertices are $(0, -3)$ and $(0, 3)$, shown in Figure 10.19.

We use $c^2 = a^2 + b^2$ to find the foci, which are located at $(0, -c)$ and $(0, c)$.

$$c^2 = a^2 + b^2 = 9 + 16 = 25$$

Because $c^2 = 25$, $c = 5$. The foci are located at $(0, -5)$ and $(0, 5)$. They are shown in Figure 10.19.

Check Point 1 Find the vertices and locate the foci for each of the following hyperbolas with the given equation:

a. $\dfrac{x^2}{25} - \dfrac{y^2}{16} = 1$ **b.** $\dfrac{y^2}{25} - \dfrac{x^2}{16} = 1$.

In Example 1, we used equations of hyperbolas to find their foci and vertices. In the next example, we reverse this procedure.

EXAMPLE 2 Finding the Equation of a Hyperbola from Its Foci and Vertices

Find the standard form of the equation of a hyperbola with foci at $(0, -3)$ and $(0, 3)$ and vertices $(0, -2)$ and $(0, 2)$, shown in Figure 10.20.

Solution Because the foci are located at $(0, -3)$ and $(0, 3)$, on the y-axis, the transverse axis lies on the y-axis. The center of the hyperbola is midway between the foci, located at $(0, 0)$. Thus, the form of the equation is

$$\frac{y^2}{a^2} - \frac{x^2}{b^2} = 1.$$

We need to determine the values for a^2 and b^2. The distance from the center, $(0, 0)$, to either vertex, $(0, -2)$ or $(0, 2)$, is 2, so $a = 2$.

$$\frac{y^2}{2^2} - \frac{x^2}{b^2} = 1 \quad \text{or} \quad \frac{y^2}{4} - \frac{x^2}{b^2} = 1$$

We must still find b^2. The distance from the center, $(0, 0)$, to either focus, $(0, -3)$ or $(0, 3)$, is 3. Thus, $c = 3$. Using $c^2 = a^2 + b^2$, we have

$$3^2 = 2^2 + b^2$$

and

$$b^2 = 3^2 - 2^2 = 9 - 4 = 5.$$

Substituting 5 for b^2 in $\dfrac{y^2}{4} - \dfrac{x^2}{b^2} = 1$ gives us the standard form of the hyperbola's equation. The equation is

$$\frac{y^2}{4} - \frac{x^2}{5} = 1.$$

Check Point 2 Find the standard form of the equation of a hyperbola with foci at $(0, -5)$ and $(0, 5)$ and vertices $(0, -3)$ and $(0, 3)$.

The Asymptotes of a Hyperbola

As x and y get larger, the two branches of the graph of a hyperbola approach a pair of intersecting straight lines, called **asymptotes.** The asymptotes pass through the center of the hyperbola and are helpful in graphing hyperbolas.

Figure 10.21 shows the asymptotes for the graphs of hyperbolas centered at the origin. The asymptotes pass through the corners of a rectangle. Note that the dimensions of this rectangle are $2a$ by $2b$. The line segment of length $2b$ is the **conjugate axis** of the hyperbola and is perpendicular to the transverse axis through the center of the hyperbola.

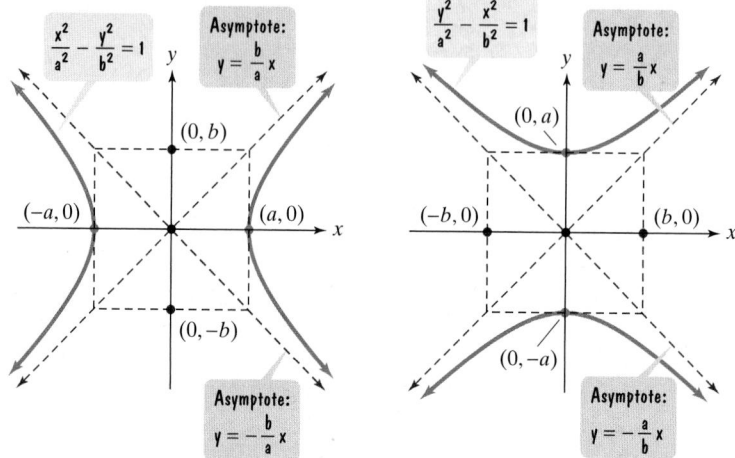

Figure 10.21 Asymptotes of a hyperbola

The Asymptotes of a Hyperbola Centered at the Origin

The hyperbola $\dfrac{x^2}{a^2} - \dfrac{y^2}{b^2} = 1$ has a horizontal transverse axis and two asymptotes

$$y = \frac{b}{a}x \quad \text{and} \quad y = -\frac{b}{a}x.$$

The hyperbola $\dfrac{y^2}{a^2} - \dfrac{x^2}{b^2} = 1$ has a vertical transverse axis and two asymptotes

$$y = \frac{a}{b}x \quad \text{and} \quad y = -\frac{a}{b}x.$$

Why are $y = \pm\dfrac{b}{a}x$ the asymptotes for a hyperbola whose transverse axis is horizontal? The proof can be found in the appendix.

3 Graph hyperbolas centered at the origin.

Graphing Hyperbolas Centered at the Origin

Hyperbolas are graphed using vertices and asymptotes.

Graphing Hyperbolas

1. Locate the vertices.
2. Use dashed lines to draw the rectangle centered at the origin with sides parallel to the axes, crossing one axis at $\pm a$ and the other at $\pm b$.
3. Use dashed lines to draw the diagonals of this rectangle and extend them to obtain the asymptotes.
4. Draw the two branches of the hyperbola by starting at each vertex and approaching the asymptotes.

EXAMPLE 3   Graphing a Hyperbola

Graph and locate the foci: $\dfrac{x^2}{25} - \dfrac{y^2}{16} = 1$. What are the equations of the asymptotes?

Solution

Step 1   Locate the vertices.   The given equation is in the form $\dfrac{x^2}{a^2} - \dfrac{y^2}{b^2} = 1$, with $a^2 = 25$ and $b^2 = 16$.

$$a^2 = 25 \qquad \frac{x^2}{25} - \frac{y^2}{16} = 1 \qquad b^2 = 16$$

Based on the standard form of the equation with the transverse axis on the x-axis, we know that the vertices are $(-a, 0)$ and $(a, 0)$. Because $a^2 = 25$, $a = 5$. Thus, the vertices are $(-5, 0)$ and $(5, 0)$, shown in Figure 10.22.

Step 2   Draw a rectangle.   Because $a^2 = 25$ and $b^2 = 16$, $a = 5$ and $b = 4$. We construct a rectangle to find the asymptotes, using -5 and 5 on the x-axis (the vertices are located here) and -4 and 4 on the y-axis. The rectangle passes through these four points, shown using dashed lines in Figure 10.22.

Step 3   Draw extended diagonals for the rectangle to obtain the asymptotes.   We draw dashed lines through the opposite corners of the rectangle, shown in Figure 10.22, to obtain the graph of the asymptotes. Based on the standard form of the hyperbola's equation, the equations for these asymptotes are

$$y = \pm \frac{b}{a}x \qquad \text{or} \qquad y = \pm \frac{4}{5}x.$$

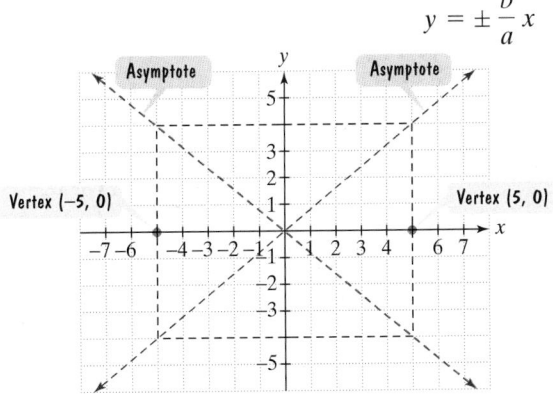

Figure 10.22 Preparing to graph $\dfrac{x^2}{25} - \dfrac{y^2}{16} = 1$

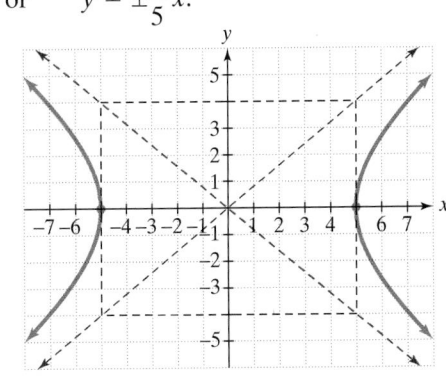

Figure 10.23 The graph of $\dfrac{x^2}{25} - \dfrac{y^2}{16} = 1$

Technology

Graph $\dfrac{x^2}{25} - \dfrac{y^2}{16} = 1$ by solving for y:

$$y_1 = \frac{\sqrt{16x^2 - 400}}{5}$$

$$y_2 = -\frac{\sqrt{16x^2 - 400}}{5} = -y_1.$$

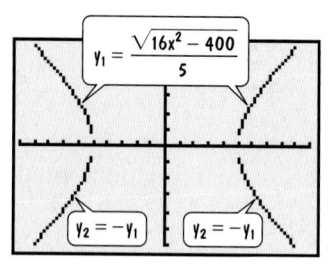

$$y_1 = \frac{\sqrt{16x^2 - 400}}{5}$$

$$y_2 = -y_1 \qquad y_2 = -y_1$$

$[-10, 10, 1]$ by $[-6, 6, 1]$

Step 4   Draw the two branches of the hyperbola by starting at each vertex and approaching the asymptotes.   The hyperbola is shown in Figure 10.23.

The foci are located at $(-c, 0)$ and $(c, 0)$. We find c using $c^2 = a^2 + b^2$.

$$c^2 = 25 + 16 = 41$$

Because $c^2 = 41$, $c = \sqrt{41}$. The foci are located at $\left(-\sqrt{41}, 0\right)$ and $\left(\sqrt{41}, 0\right)$, approximately $(-6.4, 0)$ and $(6.4, 0)$.

Check Point 3   Graph and locate the foci: $\dfrac{x^2}{36} - \dfrac{y^2}{9} = 1$. What are the equations of the asymptotes?

EXAMPLE 4   Graphing a Hyperbola

Graph and locate the foci: $9y^2 - 4x^2 = 36$. What are the equations of the asymptotes?

Solution   We begin by writing the equation in standard form. The right side should be 1, so we divide both sides by 36.

$$\frac{9y^2}{36} - \frac{4x^2}{36} = \frac{36}{36}$$

$$\frac{y^2}{4} - \frac{x^2}{9} = 1 \qquad \text{Simplify. The right side is now 1.}$$

Now we are ready to use our four-step procedure for graphing hyperbolas.

Step 1   Locate the vertices.   The equation that we obtained is in the form $\dfrac{y^2}{a^2} - \dfrac{x^2}{b^2} = 1$, with $a^2 = 4$ and $b^2 = 9$.

$$\frac{y^2}{4} - \frac{x^2}{9} = 1$$

$$a^2 = 4 \qquad b^2 = 9$$

Based on the standard form of the equation with the transverse axis on the y-axis, we know that the vertices are $(0, -a)$ and $(0, a)$. Because $a^2 = 4$, $a = 2$. Thus, the vertices are $(0, -2)$ and $(0, 2)$, shown in Figure 10.24.

Step 2   Draw a rectangle.   Because $a^2 = 4$ and $b^2 = 9$, $a = 2$ and $b = 3$. We construct a rectangle to find the asymptotes, using -2 and 2 on the y-axis (the vertices are located here) and -3 and 3 on the x-axis. The rectangle passes through these four points, shown using dashed lines in Figure 10.24.

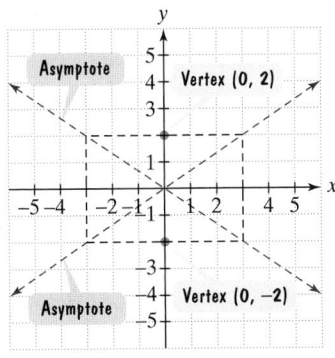

Figure 10.24 Preparing to graph $\dfrac{y^2}{4} - \dfrac{x^2}{9} = 1$

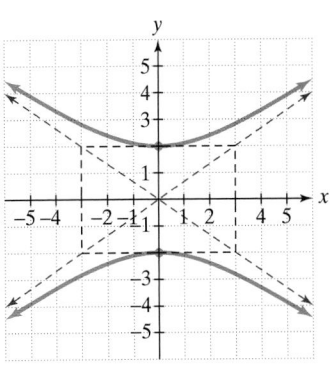

Figure 10.25 The graph of $\dfrac{y^2}{4} - \dfrac{x^2}{9} = 1$

Step 3 Draw extended diagonals of the rectangle to obtain the asymptotes. We draw dashed lines through the opposite corners of the rectangle, shown in Figure 10.24, to obtain the graph of the asymptotes. Based on the standard form of the hyperbola's equation, the equations of these asymptotes are

$$y = \pm \frac{a}{b}x \qquad \text{or} \qquad y = \pm \frac{2}{3}x.$$

Step 4 Draw the two branches of the hyperbola by starting at each vertex and approaching the asymptotes. The hyperbola is shown in Figure 10.25.

The foci are located at $(0, -c)$ and $(0, c)$. We find c using $c^2 = a^2 + b^2$.

$$c^2 = 4 + 9 = 13$$

Because $c^2 = 13$, $c = \sqrt{13}$. The foci are located at $(0, -\sqrt{13})$ and $(0, \sqrt{13})$, approximately $(0, -3.6)$ and $(0, 3.6)$.

Check Point 4 Graph and locate the foci: $y^2 - 4x^2 = 4$. What are the equations of the asymptotes?

4 Graph hyperbolas not centered at the origin.

Translations of Hyperbolas

The graph of a hyperbola can be centered at (h, k) rather than at the origin. Horizontal and vertical translations are accomplished by replacing x with $x - h$ and y with $y - k$ in the standard form of the hyperbola's equation.

Table 10.2 on the next page gives the standard forms of equations of hyperbolas centered at (h, k). Figure 10.26 shows their graphs.

Table 10.2 Standard Forms of Equations of Hyperbolas Centered at (h, k)

Equation	Center	Transverse Axis	Foci	Vertices
$\dfrac{(x - h)^2}{a^2} - \dfrac{(y - k)^2}{b^2} = 1$, $c^2 = a^2 + b^2$	(h, k)	Parallel to x-axis; horizontal	$(h - c, k)$ $(h + c, k)$	$(h - a, k)$ $(h + a, k)$
$\dfrac{(y - k)^2}{a^2} - \dfrac{(x - h)^2}{b^2} = 1$, $c^2 = a^2 + b^2$	(h, k)	Parallel to y-axis; vertical	$(h, k - c)$ $(h, k + c)$	$(h, k - a)$ $(h, k + a)$

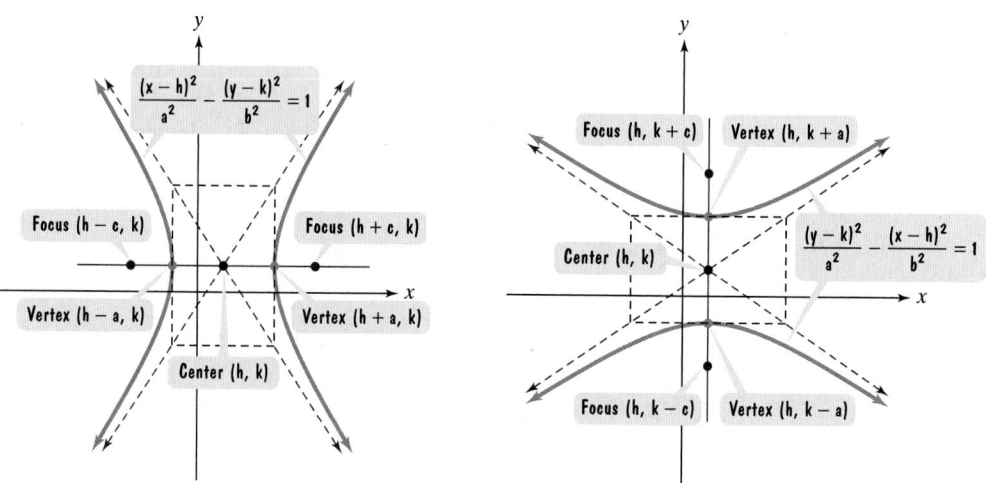

Figure 10.26 Graphs of hyperbolas centered at (h, k)

EXAMPLE 5 Graphing a Hyperbola Centered at (h, k)

Graph: $\dfrac{(x - 2)^2}{16} - \dfrac{(y - 3)^2}{9} = 1$. Where are the foci located? What are the equations of the asymptotes?

Solution In order to graph the hyperbola, we need to know its center, (h, k). In the standard forms of equations centered at (h, k), h is the number subtracted from x and k is the number subtracted from y.

This is $(x - h)^2$, with $h = 2$.

$$\dfrac{(x - 2)^2}{16} - \dfrac{(y - 3)^2}{9} = 1$$

This is $(y - k)^2$, with $k = 3$.

We see that $h = 2$ and $k = 3$. Thus, the center of hyperbola, (h, k), is $(2, 3)$. We can graph the hyperbola by using vertices, asymptotes, and our four-step graphing procedure.

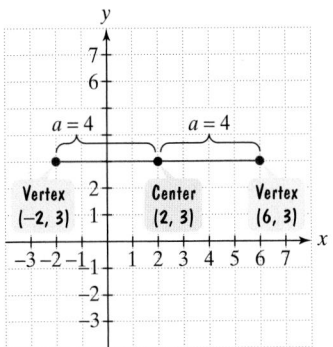

Figure 10.27 Locating a hyperbola's center and vertices

Step 1 Locate the vertices. To do this, we must identify a^2.

$$\frac{(x-2)^2}{16} - \frac{(y-3)^2}{9} = 1 \qquad \text{The form of this equation is } \frac{(x-h)^2}{a^2} - \frac{(y-k)^2}{b^2} = 1.$$

$$a^2 = 16 \qquad\qquad b^2 = 9$$

Based on the standard form of the equation with a horizontal transverse axis, the vertices are a units to the left and right of the center. Because $a^2 = 16$, $a = 4$. This means that the vertices are 4 units to the left and right of the center, $(2, 3)$. Four units to the left of $(2, 3)$ puts one vertex at $(2 - 4, 3)$, or $(-2, 3)$. Four units to the right of $(2, 3)$ puts the other vertex at $(2 + 4, 3)$, or $(6, 3)$. The vertices are shown in Figure 10.27.

Step 2 Draw a rectangle. Because $a^2 = 16$ and $b^2 = 9$, $a = 4$ and $b = 3$. The rectangle passes through points that are 4 units to the right and left of the center (the vertices are located here) and 3 units above and below the center. The rectangle is shown using dashed lines in Figure 10.28.

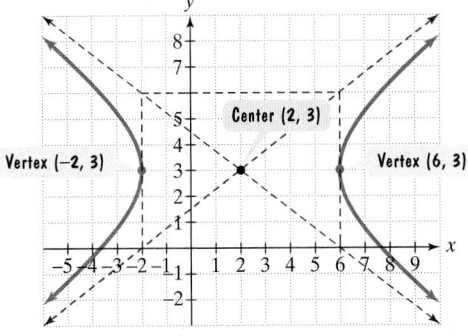

Figure 10.28 The graph of $\dfrac{(x-2)^2}{16} - \dfrac{(y-3)^2}{9} = 1$

Step 3 Draw extended diagonals of the rectangle to obtain the asymptotes. We draw dashed lines through the opposite corners of the rectangle, shown in Figure 10.28, to obtain the graph of the asymptotes. The equations of the asymptotes of the unshifted hyperbola $\dfrac{x^2}{16} - \dfrac{y^2}{9} = 1$ are $y = \pm\dfrac{b}{a}x$, or $y = \pm\dfrac{3}{4}x$. Thus, the asymptotes for the hyperbola that is shifted two units to the right and three units up, namely

$$\frac{(x-2)^2}{16} - \frac{(y-3)^2}{9} = 1$$

have equations that can be expressed as

$$y - 3 = \pm\frac{3}{4}(x - 2).$$

Step 4 Draw the two branches of the hyperbola by starting at each vertex and approaching the asymptotes. The hyperbola is shown in Figure 10.28.

The foci are located c units to the right and left of the center. We find c using $c^2 = a^2 + b^2$.

$$c^2 = 16 + 9 = 25$$

Because $c^2 = 25$, $c = 5$. This means that the foci are 5 units to the left and right of the center, $(2, 3)$. Five units to the left of $(2, 3)$ puts one focus at $(2 - 5, 3)$, or $(-3, 3)$. Five units to the right of $(2, 3)$ puts the other focus at $(2 + 5, 3)$, or $(7, 3)$.

Study Tip

Be careful in finding a hyperbola's center. The center of

$$\frac{(y-3)^2}{9} - \frac{(x-2)^2}{16} = 1$$

is $(2, 3)$ because 2 is subtracted from x and 3 is subtracted from y. Many students tend to read the equation from left to right and get the center backwards. The hyperbola's center is *not* $(3, 2)$.

Check Point 5   Graph : $\dfrac{(x-3)^2}{4} - \dfrac{(y-1)^2}{1} = 1$. Where are the foci located? What are the equations of the asymptotes?

Elliptical orbit

Sun

Parabolic orbit

Hyperbolic orbit

5   Solve applied problems involving hyperbolas.

Applications

Hyperbolas have many applications. When a jet flies at a speed greater than the speed of sound, the shock wave that is created is heard as a sonic boom. The wave has the shape of a cone. The shape formed as the cone hits the ground is one branch of a hyperbola.

      Halley's Comet, a permanent part of our solar system, travels around the sun in an elliptical orbit. Other comets pass through the solar system only once, following a hyperbolic path with the sun as a focus.

      Hyperbolas are of practical importance in fields ranging from architecture to navigation. Cooling towers used in the design for nuclear power plants have cross sections that are both ellipses and hyperbolas. Three-dimensional solids whose cross sections are hyperbolas are used in some rather unique architectural creations, including the TWA building at Kennedy Airport and the St. Louis Science Center Planetarium.

EXAMPLE 6   An Application Involving Hyperbolas

An explosion is recorded by two microphones that are 2 miles apart. Microphone M_1 received the sound 4 seconds before microphone M_2. Assuming sound travels at 1100 feet per second, determine the possible locations of the explosion relative to the location of the microphones.

Solution   We begin by putting the microphones in a coordinate system. Because 1 mile = 5280 feet, we place M_1 5280 feet on a horizontal axis to the right of the origin and M_2 5280 feet on a horizontal axis to the left of the origin. Figure 10.29 illustrates that the two microphones are 2 miles apart.

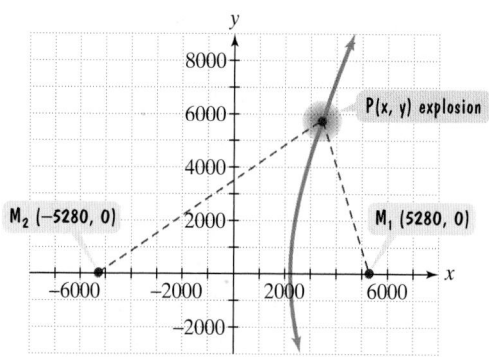

Figure 10.29 Locating an explosion on the branch of a hyperbola

      We know that M_2 received the sound 4 seconds after M_1. Because sound travels at 1100 feet per second, the difference between the distance from P to M_1 and the distance from P to M_2 is 4400 feet. The set of all points P (or locations of the explosion) satisfying these conditions fits the definition of a hyperbola, with microphones M_1 and M_2 at the foci.

$$\frac{x^2}{a^2} - \frac{y^2}{b^2} = 1$$

Use the standard form of the hyperbola's equation. P(x, y), the explosion point, lies on this hyperbola. We must find a^2 and b^2.

The difference between the distances, represented by $2a$ in the derivation of the hyperbola's equation, is 4400 feet. Thus, $2a = 4400$ and $a = 2200$.

$$\frac{x^2}{(2200)^2} - \frac{y^2}{b^2} = 1 \quad \text{Substitute 2200 for } a.$$

We must still find b^2. We know that $a = 2200$. The distance from the center, $(0, 0)$, to either focus, $(-5280, 0)$ or $(5280, 0)$, is 5280. Thus, $c = 5280$. Using $c^2 = a^2 + b^2$, we have

$$5280^2 = 2200^2 + b^2$$
$$\text{and}$$
$$b^2 = 5280^2 - 2200^2 = 23{,}038{,}400.$$

The equation of the hyperbola with a microphone at each focus is

$$\frac{x^2}{4{,}840{,}000} - \frac{y^2}{23{,}038{,}400} = 1 \quad \text{Substitute 23,038,400 for } b^2.$$

We can conclude that the explosion occurred somewhere on the right branch (the branch closest to M_1) of the hyperbola given by this equation.

In Example 6, we determined that the explosion occurred somewhere along one branch of a hyperbola, but not exactly where on the hyperbola. If, however, we had received the sound from another pair of microphones, we could locate the sound along a branch of another hyperbola. The exact location of the explosion would be the point where the two hyperbolas intersect.

Check Point 6 Rework Example 6 assuming microphone M_1 receives the sound 3 seconds before microphone M_2.

EXERCISE SET 10.2

Practice Exercises

In Exercises 1–4, find the vertices and locate the foci of each hyperbola with the given equation. Then match each equation to one of the graphs that are shown and labeled (a)–(d).

1. $\dfrac{x^2}{4} - \dfrac{y^2}{1} = 1$ **2.** $\dfrac{x^2}{1} - \dfrac{y^2}{4} = 1$

3. $\dfrac{y^2}{4} - \dfrac{x^2}{1} = 1$ **4.** $\dfrac{y^2}{1} - \dfrac{x^2}{4} = 1$

a.

b.

c.

d.
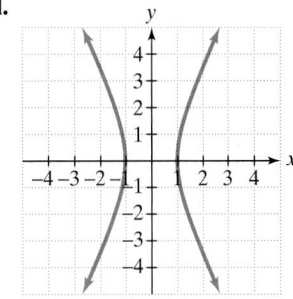

In Exercises 5–12, find the standard form of the equation of each hyperbola satisfying the given conditions.

5. Foci: $(0, -3), (0, 3)$; vertices: $(0, -1), (0, 1)$

6. Foci: $(0, -6), (0, 6)$; vertices: $(0, -2), (0, 2)$

7. Foci: $(-4, 0), (4, 0)$; vertices: $(-3, 0), (3, 0)$

8. Foci: $(-7, 0), (7, 0)$; vertices: $(-5, 0), (5, 0)$

9. Endpoints of transverse axis: $(0, -6), (0, 6)$; asymptote: $y = 2x$

10. Endpoints of transverse axis: $(-4, 0), (4, 0)$; asymptote: $y = 2x$

11. Center: $(4, -2)$; Focus: $(7, -2)$; vertex: $(6, -2)$

12. Center: $(-2, 1)$; Focus: $(-2, 6)$; vertex: $(-2, 4)$

In Exercises 13–26, use vertices and asymptotes to graph each hyperbola. Locate the foci and find the equations of the asymptotes.

13. $\dfrac{x^2}{9} - \dfrac{y^2}{25} = 1$

14. $\dfrac{x^2}{16} - \dfrac{y^2}{25} = 1$

15. $\dfrac{x^2}{100} - \dfrac{y^2}{64} = 1$

16. $\dfrac{x^2}{144} - \dfrac{y^2}{81} = 1$

17. $\dfrac{y^2}{16} - \dfrac{x^2}{36} = 1$

18. $\dfrac{y^2}{25} - \dfrac{x^2}{64} = 1$

19. $4y^2 - x^2 = 1$

20. $9y^2 - x^2 = 1$

21. $9x^2 - 4y^2 = 36$

22. $4x^2 - 25y^2 = 100$

23. $9y^2 - 25x^2 = 225$

24. $16y^2 - 9x^2 = 144$

25. $y = \pm\sqrt{x^2 - 2}$

26. $y = \pm\sqrt{x^2 - 3}$

In Exercises 27–32, find the standard form of the equation of each hyperbola.

27.

28.

29.

30.

31.

32.

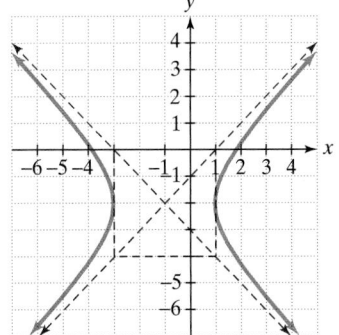

In Exercises 33–42, use the center, vertices, and asymptotes to graph each hyperbola. Locate the foci and find the equations of the asymptotes.

33. $\dfrac{(x + 4)^2}{9} - \dfrac{(y + 3)^2}{16} = 1$

34. $\dfrac{(x + 2)^2}{9} - \dfrac{(y - 1)^2}{25} = 1$

35. $\dfrac{(x + 3)^2}{25} - \dfrac{y^2}{16} = 1$

36. $\dfrac{(x + 2)^2}{9} - \dfrac{y^2}{25} = 1$

37. $\dfrac{(y + 2)^2}{4} - \dfrac{(x - 1)^2}{16} = 1$

38. $\dfrac{(y - 2)^2}{36} - \dfrac{(x + 1)^2}{49} = 1$

39. $(x - 3)^2 - 4(y + 3)^2 = 4$

40. $(x + 3)^2 - 9(y - 4)^2 = 9$

41. $(x - 1)^2 - (y - 2)^2 = 3$

42. $(y - 2)^2 - (x + 3)^2 = 5$

In Exercises 43–50, convert each equation to standard form by completing the square on x and y. Then graph the hyperbola. Locate the foci and find the equations of the asymptotes.

43. $x^2 - y^2 - 2x - 4y - 4 = 0$

44. $4x^2 - y^2 + 32x + 6y + 39 = 0$

45. $16x^2 - y^2 + 64x - 2y + 67 = 0$

46. $9y^2 - 4x^2 - 18y + 24x - 63 = 0$

47. $4x^2 - 9y^2 - 16x + 54y - 101 = 0$

48. $4x^2 - 9y^2 + 8x - 18y - 6 = 0$

49. $4x^2 - 25y^2 - 32x + 164 = 0$

50. $9x^2 - 16y^2 - 36x - 64y + 116 = 0$

 Application Exercises

51. An explosion is recorded by two microphones that are 1 mile apart. Microphone M_1 received the sound 2 seconds before microphone M_2. Assuming sound travels at 1100 feet per second, determine the possible locations of the explosion relative to the location of the microphones.

52. Radio towers A and B, 200 kilometers apart, are situated along the coast, with A located due west of B. Simultaneous radio signals are sent from each tower to a ship, with the signal from B received 500 microseconds before the signal from A.

   a. Assuming that the radio signals travel 300 meters per microsecond, determine the equation of the hyperbola on which the ship is located.

   b. If the ship lies due north of tower B, how far out at sea is it?

53. An architect designs two houses that are shaped and positioned like a part of the branches of the hyperbola whose equation is $625y^2 - 400x^2 = 250,000$, where x and y are in yards. How far apart are the houses at their closest point?

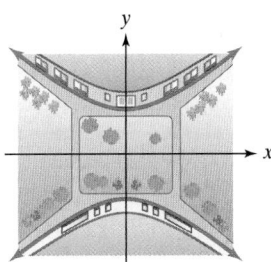

54. Scattering experiments, in which moving particles are deflected by various forces, led to the concept of the nucleus of an atom. In 1911, the physicist Ernest Rutherford (1871–1937) discovered that when alpha particles are directed toward the nuclei of gold atoms, they are eventually deflected along hyperbolic paths, illustrated in the figure. If a particle gets as close as 3 units to the nucleus along a hyperbolic path with an asymptote given by $y = \frac{1}{2}x$, what is the equation of its path?

Writing in Mathematics

55. What is a hyperbola?

56. Describe how to graph $\dfrac{x^2}{9} - \dfrac{y^2}{1} = 1$.

57. Describe how to locate the foci of the graph of $\dfrac{x^2}{9} - \dfrac{y^2}{1} = 1$.

58. Describe one similarity and one difference between the graphs of $\dfrac{x^2}{9} - \dfrac{y^2}{1} = 1$ and $\dfrac{y^2}{9} - \dfrac{x^2}{1} = 1$.

59. Describe one similarity and one difference between the graphs of $\dfrac{x^2}{9} - \dfrac{y^2}{1} = 1$ and $\dfrac{(x-3)^2}{9} - \dfrac{(y+3)^2}{1} = 1$.

60. How can you distinguish an ellipse from a hyperbola by looking at their equations?

61. In 1992, a NASA team began a project called Spaceguard Survey, calling for an international watch for comets that might collide with Earth. Why is it more difficult to detect a possible "doomsday comet" with a hyperbolic orbit than one with an elliptical orbit?

Technology Exercises

62. Use a graphing utility to graph any five of the hyperbolas that you graphed by hand in Exercises 13–26.

63. Use a graphing utility to graph any three of the hyperbolas that you graphed by hand in Exercises 33–42. First solve the given equation for y by using the square root method. Enter each of the two resulting equations to produce each branch of the hyperbola.

64. Use a graphing utility to graph any one of the hyperbolas that you graphed by hand in Exercises 43–50. Write the equation as a quadratic equation in y and use the quadratic formula to solve for y. Enter each of the two resulting equations to produce each branch of the hyperbola.

65. Use a graphing utility to graph $\dfrac{x^2}{4} - \dfrac{y^2}{9} = 0$. Is the graph a hyperbola? In general, what is the graph of $\dfrac{x^2}{a^2} - \dfrac{y^2}{b^2} = 0$?

66. Graph $\dfrac{x^2}{a^2} - \dfrac{y^2}{b^2} = 1$ and $\dfrac{x^2}{a^2} - \dfrac{y^2}{b^2} = -1$ in the same viewing rectangle for values of a^2 and b^2 of your choice. Describe the relationship between the two graphs.

67. Write $4x^2 - 6xy + 2y^2 - 3x + 10y - 6 = 0$ as a quadratic equation in y and then use the quadratic formula to express y in terms of x. Graph the resulting two equations using a graphing utility in a $[-50, 70, 10]$ by $[-30, 50, 10]$ viewing rectangle. What effect does the xy-term have on the graph of the resulting hyperbola? What problems would you encounter if you attempted to write the given equation in standard form by completing the square?

68. Graph $\dfrac{x^2}{16} - \dfrac{y^2}{9} = 1$ and $\dfrac{x|x|}{16} - \dfrac{y|y|}{9} = 1$ in the same viewing rectangle. Explain why the graphs are not the same.

Critical Thinking Exercises

69. Which one of the following is true?
   a. If one branch of a hyperbola is removed from a graph, then the branch that remains must define y as a function of x.
   b. All points on the asymptotes of a hyperbola also satisfy the hyperbola's equation.
   c. The graph of $\dfrac{x^2}{9} - \dfrac{y^2}{4} = 1$ does not intersect the line $y = -\dfrac{2}{3}x$.
   d. Two different hyperbolas can never share the same asymptotes.

70. What happens to the shape of the graph of $\dfrac{x^2}{a^2} - \dfrac{y^2}{b^2} = 1$ as $\dfrac{c}{a}$ gets larger and larger?

71. Find the standard form of the equation of the hyperbola with vertices $(5, -6)$ and $(5, 6)$, passing through $(0, 9)$.

72. Find the equation of a hyperbola whose asymptotes are perpendicular.

SECTION 10.3 *The Parabola*

Objectives

1. Graph parabolas with vertices at the origin.
2. Write equations of parabolas in standard form.
3. Graph parabolas with vertices not at the origin.
4. Solve applied problems involving parabolas.

At first glance, this image looks like columns of smoke rising from a fire into a starry sky. Those are, indeed, stars in the background, but you are not looking at ordinary smoke columns. These stand almost 6 trillion miles high and are 7000 light-years from Earth—more than 400 million times as far away as the sun.

This NASA photograph is one of a series of stunning images captured from the ends of the universe by the Hubble Space Telescope. The image shows infant star systems the size of our solar system emerging from the gas and dust that shrouded their creation. Using a parabolic mirror that is 94.5 inches in diameter, the Hubble is providing answers to many of the profound mysteries of the cosmos: How big and how old is the universe? How did the galaxies come to exist? Do other Earth-like planets orbit other sun-like stars? In this section, we study parabolas and their applications, including parabolic shapes that gather distant rays of light and focus them into spectacular images.

Definition of a Parabola

In Chapter 3, we studied parabolas, viewing them as graphs of quadratic functions in the form

$$y = a(x - h)^2 + k \quad \text{or} \quad y = ax^2 + bx + c.$$

Study Tip

Here is a summary of what you should already know about graphing parabolas.

Graphing $y = a(x - h)^2 + k$ and $y = ax^2 + bx + c$

1. If $a > 0$, the graph opens upward. If $a < 0$, the graph opens downward.
2. The vertex of $y = a(x - h)^2 + k$ is (h, k).
3. The x-coordinate of the vertex of $y = ax^2 + bx + c$ is $x = -\dfrac{b}{2a}$.

 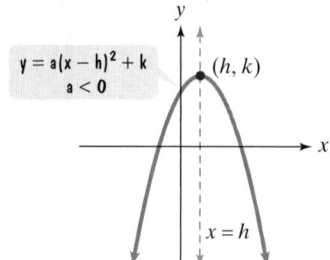

Parabolas can be given a geometric definition that enables us to include graphs that open to the left or to the right, as well as those that open obliquely. The definitions of ellipses and hyperbolas involved two fixed points, the foci. By contrast, the definition of a parabola is based on one point and a line.

Definition of a Parabola

A **parabola** is the set of all points in a plane that are equidistant from a fixed line, the **directrix,** and a fixed point, the **focus,** that is not on the line (see Figure 10.30).

In Figure 10.30, find the line passing through the focus and perpendicular to the directrix. This is the **axis of symmetry** of the parabola. The point of intersection of the parabola with its axis of symmetry is called the **vertex.** Notice that the vertex is midway between the focus and the directrix.

Figure 10.30

Standard Form of the Equation of a Parabola

The rectangular coordinate system enables us to translate a parabola's geometric definition into an algebraic equation. Figure 10.31 is our starting point for obtaining an equation. We place the focus on the x-axis at the point $(p, 0)$. The directrix has an equation given by $x = -p$. The vertex, located midway between the focus and the directrix, is at the origin.

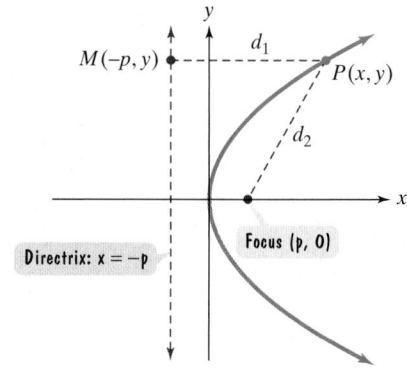

Figure 10.31

What does the definition of a parabola tell us about the point (x, y) in Figure 10.31? For any point (x, y) on the parabola, the distance d_1 to the directrix is equal to the distance d_2 to the focus. Thus, the point (x, y) is on the parabola if and only if

$$d_1 = d_2$$
$$\sqrt{(x + p)^2 + (y - y)^2} = \sqrt{(x - p)^2 + (y - 0)^2}$$ Use the distance formula.
$$(x + p)^2 = (x - p)^2 + y^2$$ Square both sides of the equation.
$$x^2 + 2px + p^2 = x^2 - 2px + p^2 + y^2$$ Square $x + p$ and $x - p$.
$$2px = -2px + y^2$$ Subtract $x^2 + p^2$ from both sides of the equation.
$$y^2 = 4px$$ Solve for y^2.

This last equation is called the **standard form of the equation of a parabola with its vertex at the origin.** There are two such equations, one for a focus on the x-axis and one for a focus on the y-axis.

Standard Forms of the Equations of a Parabola

The **standard form of the equation of a parabola** with vertex at the origin is

$$y^2 = 4px \qquad \text{or} \qquad x^2 = 4py.$$

Figure 10.32 illustrates that for the equation on the left, the focus is on the x-axis, which is the axis of symmetry. For the equation on the right, the focus is on the y-axis, which is the axis of symmetry.

Study Tip

It is helpful to think of p as the *directed distance* from the vertex to the focus. If $p > 0$, the focus lies p units to the right of the vertex or p units above the vertex. If $p < 0$, the focus lies $|p|$ units to the left of the vertex or $|p|$ units below the vertex.

Figure 10.32

(a) Parabola with the x-axis as the axis of symmetry. If $p > 0$, the graph opens to the right. If $p < 0$, the graph opens to the left.

(b) Parabola with the y-axis as the axis of symmetry. If $p > 0$, the graph opens upward. If $p < 0$, the graph opens downward.

1 Graph parabolas with vertices at the origin.

Using the Standard Form of the Equation of a Parabola

We can use the standard form of the equation of a parabola to find its focus and directrix. Observing the graphs symmetry from its equation is helpful in locating the focus.

$$y^2 = 4px \qquad\qquad x^2 = 4py$$

The equation does not change if y is replaced with $-y$. There is x-axis symmetry and the focus is on the x-axis at $(p, 0)$.

The equation does not change if x is replaced with $-x$. There is y-axis symmetry and the focus is on the y-axis at $(0, p)$.

Although the definition of a parabola is given in terms of its focus and its directrix, the focus and directrix are not part of the graph. The vertex, located at the origin, is a point on the graph of $y^2 = 4px$ and $x^2 = 4py$. Example 1 illustrates how you can find two additional points on the parabola.

EXAMPLE 1   Finding the Focus and Directrix of a Parabola

Find the focus and directrix of the parabola given by $y^2 = 12x$. Then graph the parabola.

Solution   The given equation is in the standard form $y^2 = 4px$, so $4p = 12$.

No change if y is replaced with −y. The parabola has x-axis symmetry.

$y^2 = 12x$

This is 4p.

We can find both the focus and the directrix by finding p.

$$4p = 12$$

$$p = 3 \qquad \text{Divide both sides by 4.}$$

Because p is positive, the parabola, with its x-axis symmetry, opens to the right. The focus is 3 units to the right of the vertex, $(0, 0)$.

$$\text{Focus:} \qquad (p, 0) = (3, 0)$$

$$\text{Directrix:} \quad x = -p; \, x = -3$$

The focus and directrix are shown in Figure 10.33.

To graph the parabola, we will use two points on the graph that lie directly above and below the focus. Because the focus is at $(3, 0)$, substitute 3 for x in the parabola's equation, $y^2 = 12x$.

$$y^2 = 12 \cdot 3 \qquad \text{Replace x with 3 in } y^2 = 12x.$$

$$y^2 = 36 \qquad \text{Simplify,}$$

$$y = \pm\sqrt{36} = \pm 6 \quad \text{Apply the square root method.}$$

The points on the parabola above and below the focus are $(3, 6)$ and $(3, -6)$. The graph is sketched in Figure 10.33.

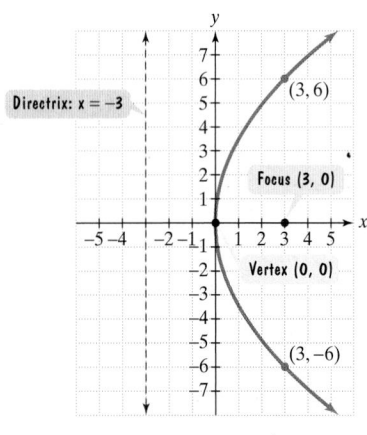

Figure 10.33 The graph of $y^2 = 12x$

Technology

We graph $y^2 = 12x$ with a graphing utility by first solving for y. The screen shows the graphs of $y = \sqrt{12x}$ and $y = -\sqrt{12x}$. The graph fails the vertical line test. Because $y^2 = 12x$ is not a function, you were not familiar with this form of the parabola's equation in Chapter 3.

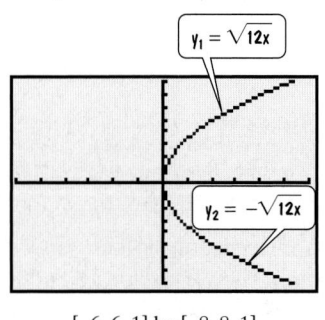

$y_1 = \sqrt{12x}$

$y_2 = -\sqrt{12x}$

$[-6, 6, 1]$ by $[-8, 8, 1]$

Check Point 1   Find the focus and directrix of the parabola given by $y^2 = 8x$. Then graph the parabola.

In general, the points on a parabola $y^2 = 4px$ that lie above and below the focus, $(p, 0)$, are each at a distance $|2p|$ from the focus. This is because if $x = p$, then $y^2 = 4px = 4p^2$, so $y = \pm 2p$. The line segment joining these two points is called the *latus rectum*; its length is $|4p|$.

The Latus Rectum and Graphing Parabolas

The **latus rectum** of a parabola is a line segment that passes through its focus, is parallel to its directrix, and has its endpoints on the parabola. Figure 10.34 shows that the length of the latus rectum for the graphs of $y^2 = 4px$ and $x^2 = 4py$ is $|4p|$.

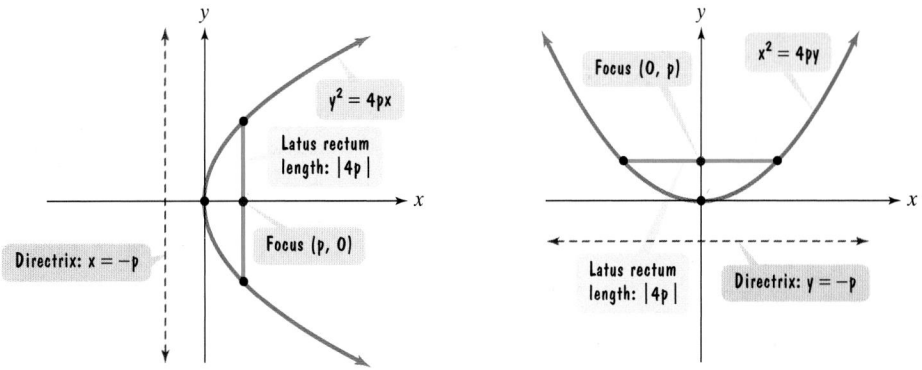

Figure 10.34 Endpoints of the latus rectum are helpful in determining a parabolas's "width," or how it opens

EXAMPLE 2 Finding the Focus and Directrix of a Parabola

Find the focus and directrix of the parabola given by $x^2 = -8y$. Then graph the parabola.

Solution   The given equation is in the standard form $x^2 = 4py$, so $4p = -8$.

No change if x is
replaced with −x.
The parabola has
y-axis symmetry.

$$x^2 = -8y$$

This is 4p.

We can find both the focus and the directrix by finding p.

$$4p = -8$$
$$p = -2 \quad \text{Divide both sides by 4.}$$

Because p is negative, the parabola, with its y-axis symmetry, opens downward. The focus is 2 units below the vertex, $(0, 0)$.

Focus: $(0, p) = (0, -2)$
Directrix: $y = -p; y = 2$

The focus and directrix are shown in Figure 10.35.

To graph the parabola, we will use the vertex, $(0, 0)$, and the two endpoints of the latus rectum. The length of the latus rectum is

$$|4p| = |4(-2)| = |-8| = 8.$$

Because the graph has y-axis symmetry, the latus rectum extends 4 units to the left and 4 units to the right of the focus, $(0, -2)$. The endpoints of the latus rectum are $(-4, -2)$ and $(4, -2)$. Passing a smooth curve through the vertex and these two points, we sketch the parabola, shown in Figure 10.35.

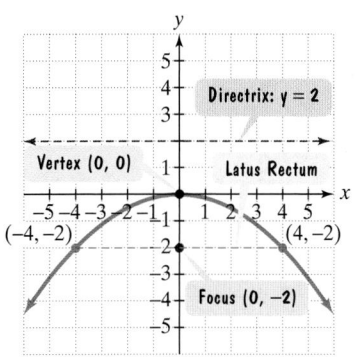

Figure 10.35 The graph of $x^2 = -8y$

Technology

Graph $x^2 = -8y$ by first solving for y: $y = -\dfrac{x^2}{8}$. The graph passes the vertical line test. Because $x^2 = -8y$ is a function, you were familiar with the parabola's alternate algebraic form, $y = -\dfrac{1}{8}x^2$, in Chapter 3. The form is $y = ax^2 + bx + c$, with $a = -\dfrac{1}{8}$, $b = 0$, and $c = 0$.

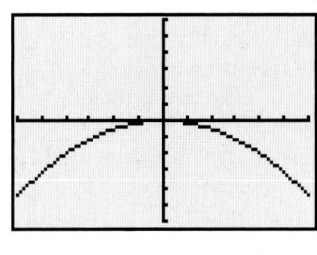

[−6, 6, 1] by [−6, 6, 1]

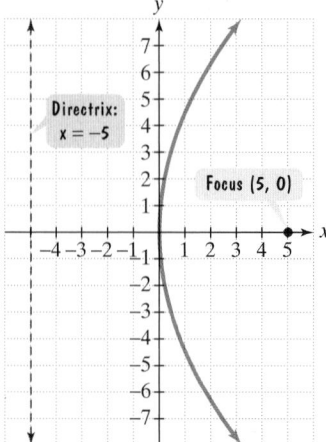

Figure 10.36

Check Point 2 Find the focus and directrix of the parabola given by $x^2 = -12y$. Then graph the parabola.

2 Write equations of parabolas in standard form.

In Examples 1 and 2, we used the equation of a parabola to find its focus and directrix. In the next example, we reverse this procedure.

EXAMPLE 3   Finding the Equation of a Parabola from Its Focus and Directrix

Find the standard form of the equation of a parabola with focus $(5, 0)$ and directrix $x = -5$, shown in Figure 10.36.

Solution   The focus is $(5, 0)$. Thus, the focus is on the x-axis. We use the standard form of the equation in which there is x-axis symmetry, namely $y^2 = 4px$.

We need to determine the value of p. Figure 10.36 shows that the focus is 5 units to the right of the vertex, $(0, 0)$. Thus, p is positive and $p = 5$. We substitute 5 for p in $y^2 = 4px$ to obtain the standard form of the equation of the parabola. The equation is

$$y^2 = 4 \cdot 5x \qquad \text{or} \qquad y^2 = 20x.$$

Check Point 3 Find the standard form of the equation of a parabola with focus $(8, 0)$ and directrix $x = -8$.

3 Graph parabolas with vertices not at the origin.

Translations of Parabolas

The graph of a parabola can have its vertex at (h, k) rather than at the origin. Horizontal and vertical translations are accomplished by replacing x with $x - h$ and y with $y - k$ in the standard form of the parabola's equation.

Table 10.3 gives the standard forms of equations of parabolas with vertex at (h, k). Figure 10.37 shows their graphs.

Table 10.3   Standard Forms of Equations of Parabolas with Vertex at (h, k)

Equation	Vertex	Axis of Symmetry	Focus	Directrix	Description
$(y - k)^2 = 4p(x - h)$	(h, k)	Horizontal	$(h + p, k)$	$x = h - p$	If $p > 0$, opens to the right. If $p < 0$, opens to the left.
$(x - h)^2 = 4p(y - k)$	(h, k)	Vertical	$(h, k + p)$	$y = k - p$	If $p > 0$, opens upward. If $p < 0$, opens downward.

Study Tip

If y is the squared term, there is horizontal symmetry and the parabola's equation is not a function. If x is the squared term, there is vertical symmetry and the parabola's equation is a function. Continue to think of p as the directed distance from the vertex, (h, k), to the focus.

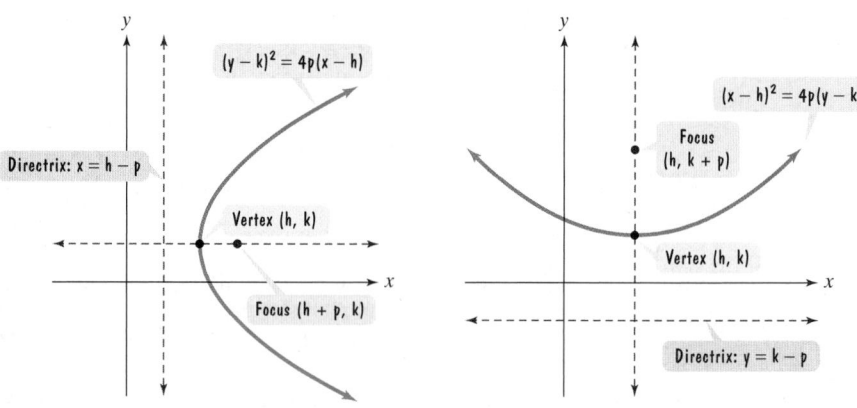

Figure 10.37 Graphs of parabolas with vertex at (h, k) and $p > 0$

The two parabolas shown in Figure 10.37 illustrate standard forms of equations for $p > 0$. If $p < 0$, a parabola with a horizontal axis of symmetry will open to the left and the focus will lie to the left of the directrix. If $p < 0$, a parabola with a vertical axis of symmetry will open downward and the focus will lie below the directix.

EXAMPLE 4   Graphing a Parabola with Vertex at (h, k)

Find the vertex, focus, and directrix of the parabola given by

$$(x - 3)^2 = 8(y + 1).$$

Then graph the parabola.

Solution   In order to find the focus and directrix, we need to know the vertex. In the standard forms of equations with vertex at (h, k), h is the number subtracted from x, and k is the number subtracted from y.

$$(x - 3)^2 = 8\left(y - (-1)\right)$$

> This is $(x - h)^2$, with $h = 3$.   This is $y - k$, with $k = -1$.

We see that $h = 3$ and $k = -1$. Thus, the vertex of the parabola is $(h, k) = (3, -1)$.
   Now that we have the vertex, we can find both the focus and directrix by finding p.

$$(x - 3)^2 = 8(y + 1)$$ The equation is in the standard form $(x - h)^2 = 4p(y - k)$. Because x is the squared term, there is vertical symmetry and the parabola's equation is a function.

> This is $4p$.

Because $4p = 8$, $p = 2$. Based on the standard form of the equation, the axis of symmetry is vertical. With a positive value for p and a vertical axis of symmetry, the parabola opens upward. Because $p = 2$, the focus is located 2 units above the vertex, $(3, -1)$. Likewise, the directrix is located 2 units below the vertex.

Focus: $(h, k + p) = (3, -1 + 2) = (3, 1)$

> The vertex, (h, k), is $(3, -1)$.   The focus is 2 units above the vertex, $(3, -1)$.

Directrix: $y = k - p$
$$y = -1 - 2 = -3$$

> The directrix is 2 units below the vertex, $(3, -1)$.

Thus, the focus is $(3, 1)$ and the directrix is $y = -3$. They are shown in Figure 10.38.
   To graph the parabola, we will use the vertex, $(3, -1)$, and the two endpoints of the latus rectum. The length of the latus rectum is

$$|4p| = |4 \cdot 2| = |8| = 8.$$

Because the graph has vertical symmetry, the latus rectum extends 4 units to the left and 4 units to the right of the focus, $(3, 1)$. The endpoints of the latus rectum are $(3 - 4, 1)$, or $(-1, 1)$, and $(3 + 4, 1)$, or $(7, 1)$. Passing a smooth curve through the vertex and these two points, we sketch the parabola, shown in Figure 10.38.

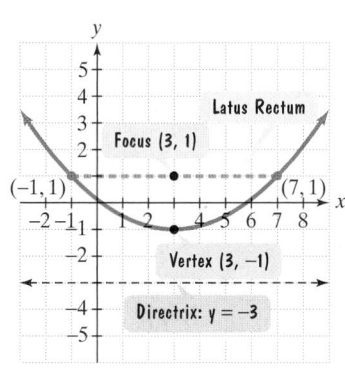

Figure 10.38 The graph of $(x - 3)^2 = 8(y + 1)$

Technology

Graph $(x - 3)^2 = 8(y + 1)$ by first solving for y:

$$\tfrac{1}{8}(x - 3)^2 = y + 1$$
$$y = \tfrac{1}{8}(x - 3)^2 - 1.$$

The graph passes the vertical line test. Because $(x - 3)^2 = 8(y + 1)$ is a function, you were familiar with the parabola's alternate algebraic form, $y = \tfrac{1}{8}(x - 3)^2 - 1$, in Chapter 3. The form is $y = a(x - h)^2 + k$ with $a = \tfrac{1}{8}, h = 3,$ and $k = -1.$

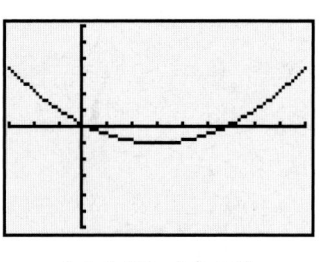

$[-3, 9, 1]$ by $[-6, 6, 1]$

Check Point 4 Find the vertex, focus, and directrix of the parabola given by $(x - 2)^2 = 4(y + 1)$. Then graph the parabola.

In some cases, we need to convert the equation of a parabola to standard form by completing the square on x or y, whichever variable is squared. Let's see how this is done.

EXAMPLE 5   Graphing a Parabola with Vertex at (h, k)

Find the vertex, focus, and directrix of the parabola given by

$$y^2 + 2y + 12x - 23 = 0.$$

Then graph the parabola.

Solution   We convert the given equation to standard form by completing the square on the variable y. We isolate the terms involving y on the left side.

$y^2 + 2y + 12x - 23 = 0$	This is the given equation.
$y^2 + 2y = -12x + 23$	Isolate the terms involving y.
$y^2 + 2y + 1 = -12x + 23 + 1$	Complete the square by adding the square of half the coefficient of y.
$(y + 1)^2 = -12x + 24$	Factor.

To express this equation in the standard form $(y - k)^2 = 4p(x - h)$, we factor -12 on the right. The standard form of the parabola's equation is

$$(y + 1)^2 = -12(x - 2).$$

We use this form to identify the vertex, (h, k), and the value for p needed to locate the focus and the directrix.

$$[(y - (-1)]^2 = -12(x - 2)$$

This is $(y - k)^2$, with $k = -1$. This is $4p$. This is $x - h$, with $h = 2$.

The equation is in the standard form $(y - k)^2 = 4p(x - h)$. Because y is the squared term, there is horizontal symmetry and the parabola's equation is not a function.

We see that $h = 2$ and $k = -1$. Thus, the vertex of the parabola is $(h, k) = (2, -1)$. Because $4p = -12$, $p = -3$. Based on the standard form of the equation, the axis of symmetry is horizontal. With a negative value for p and a horizontal axis of symmetry, the parabola opens to the left. Because $p = -3$,

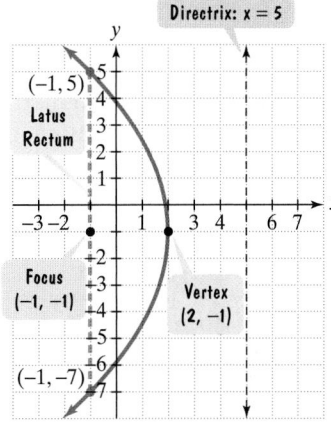

Figure 10.39 The graph of $y^2 + 2y + 12x - 23 = 0$, or $(y + 1)^2 = -12(x - 2)$

the focus is located 3 units to the left of the vertex, $(2, -1)$. Likewise, the directrix is located 3 units to the right of the vertex.

Focus: $\quad (h + p, k) = (2 + (-3), -1) = (-1, -1)$

> The vertex, (h, k), is $(2, -1)$.

> The focus is 3 units to the left of the vertex, $(2, -1)$.

Directrix: $\qquad x = h - p$
$$x = 2 - (-3) = 5$$

> The directrix is 3 units to the right of the vertex, $(2, -1)$.

Thus, the focus is $(-1, -1)$ and the directrix is $x = 5$. They are shown in Figure 10.39.

To graph the parabola, we will use the vertex, $(2, -1)$, and the two endpoints of the latus rectum. The length of the latus rectum is

$$|4p| = |4(-3)| = |-12| = 12.$$

Because the graph has horizontal symmetry, the latus rectum extends 6 units above and 6 units below the focus, $(-1, -1)$. The endpoints of the latus rectum are $(-1, -1 + 6)$, or $(-1, 5)$, and $(-1, -1 - 6)$, or $(-1, -7)$. Passing a smooth curve through the vertex and these two points, we sketch the parabola shown in Figure 10.39.

> **Check Point 5** Find the vertex, focus, and directrix of the parabola given by $y^2 + 2y + 4x - 7 = 0$. Then graph the parabola.

4 Solve applied problems involving parabolas.

Applications

Parabolas have many applications. Cables hung between structures to form suspension bridges form parabolas. Arches constructed of steel and concrete, whose main purpose is strength, are usually parabolic in shape.

Suspension bridge

Arch bridge

We have seen that comets in our solar system travel in orbits that are ellipses and hyperbolas. Some comets follow parabolic paths. Only comets with elliptical orbits, such as Halley's Comet, return to our part of the galaxy.

You throw a ball directly upward. As illustrated in Figure 10.40, the height of such a projectile as a function of time is parabolic.

Figure 10.40 Multiflash photo showing the parabolic path of a ball thrown into the air

The Hubble Space Telescope

The Hubble Space Telescope

For decades, astronomers hoped to create an observatory above the atmosphere that would provide an unobscured view of the universe. This dream came true with the 1990 launching of the Hubble Space Telescope. The telescope initially had blurred vision due to problems with its parabolic mirror. The mirror had been ground two millionths of a meter smaller than design specifications. In 1993, astronauts from the Space Shuttle *Endeavor* equipped the telescope with optics to correct the blurred vision. "A small change for a mirror, a giant leap for astronomy," Christopher J. Burrows of the Space Telescope Science Institute said when clear images from the ends of the universe were presented to the public after the repair mission.

If a parabola is rotated about its axis of symmetry, a parabolic surface is formed. Figure 10.41(a) shows how a parabolic surface can be used to reflect light. Light originates at the focus. Note how the light is reflected by the parabolic surface, so that the outgoing light is parallel to the axis of symmetry. The reflective properties of parabolic surfaces are used in the design of searchlights [Figure 10.41(b)], automobile headlights, and parabolic microphones.

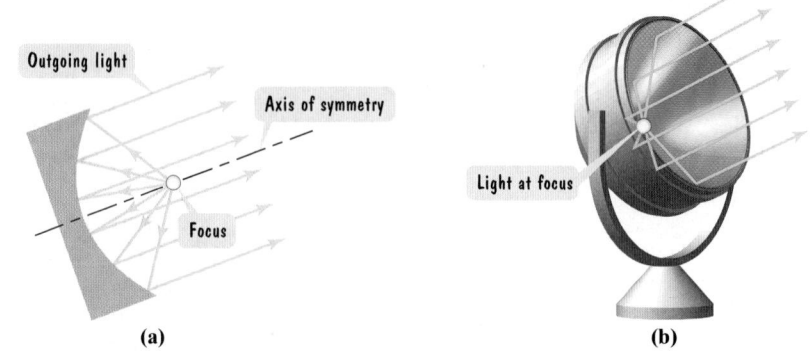

(a)   (b)

Figure 10.41 **(a)** Parabolic surface reflecting light **(b)** Light from the focus is reflected parallel to the axis of symmetry.

Figure 10.42(a) shows how a parabolic surface can be used to reflect *incoming* light. Note that light rays strike the surface and are reflected *to the focus*. This principle is used in the design of reflecting telescopes, radar, and television satellite dishes. Reflecting telescopes magnify the light from distant stars by reflecting the light from these bodies to the focus of a parabolic mirror [Figure 10.42(b)].

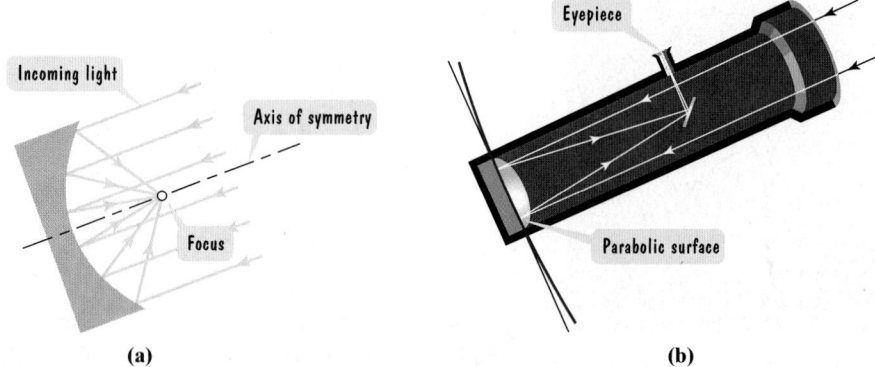

(a)   (b)

Figure 10.42 **(a)** Parabolic surface reflecting incoming light **(b)** Incoming light rays are reflected to the focus.

EXAMPLE 6   Using the Reflection Property of Parabolas

An engineer is designing a flashlight using a parabolic reflecting mirror and a light source, shown in Figure 10.43. The casting has a diameter of 4 inches and a depth of 2 inches. What is the equation of the parabola used to shape the mirror? At what point should the light source be placed relative to the mirror's vertex?

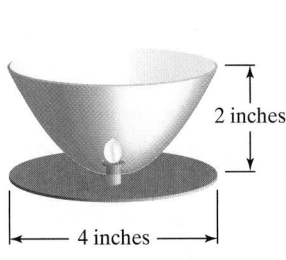

Figure 10.43 Designing a flashlight

Figure 10.44

Solution   We position the parabola with its vertex at the origin and opening upward (Figure 10.44). Thus, the focus is on the y-axis, located at $(0, p)$. We use the standard form of the equation in which there is y-axis symmetry, namely $x^2 = 4py$. We need to find p. Because $(2, 2)$ lies on the parabola, we let $x = 2$ and $y = 2$ in $x^2 = 4py$.

$$2^2 = 4p \cdot 2 \qquad \text{Substitute 2 for x and 2 for y in } x^2 = 4py.$$

$$4 = 8p \qquad \text{Simplify.}$$

$$p = \tfrac{1}{2} \qquad \text{Divide both sides of the equation by 8 and reduce the resulting fraction.}$$

We substitute $\tfrac{1}{2}$ for p in $x^2 = 4py$ to obtain the standard form of the equation of the parabola. The equation of the parabola used to shape the mirror is

$$x^2 = 4 \cdot \tfrac{1}{2} y \qquad \text{or} \qquad x^2 = 2y.$$

The light source should be placed at the focus, $(0, p)$. Because $p = \tfrac{1}{2}$, the light should be placed at $\left(0, \tfrac{1}{2}\right)$, or $\tfrac{1}{2}$ inch above the vertex.

> **Check Point 6**   In Example 6, suppose that the casting has a diameter of 6 inches and a depth of 4 inches. What is the equation of the parabola used to shape the mirror? At what point should the light source be placed relative to the mirror's vertex?

Degenerate Conic Sections

We opened the chapter by noting that conic sections are curves that result from the intersection of a cone and a plane. However, these intersections might not result in a conic section. Three degenerate cases occur when the cutting plane passes through the vertex. These **degenerate conic sections** are a point, a line, and a pair of intersecting lines, illustrated in Figure 10.45.

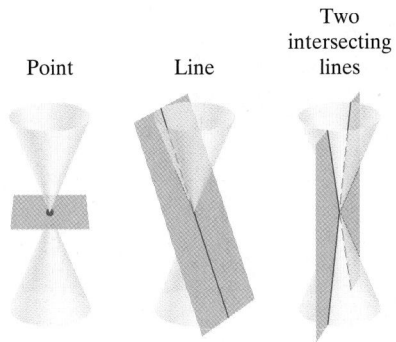

Figure 10.45 Degenerate conics

EXERCISE SET 10.3

 Practice Exercises

In Exercises 1–4, find the focus and directrix of each parabola with the given equation. Then match each equation to one of the graphs that are shown and labeled (a)–(d).

1. $y^2 = 4x$ **2.** $x^2 = 4y$

3. $x^2 = -4y$ **4.** $y^2 = -4x$

a.

b.

c.

d.

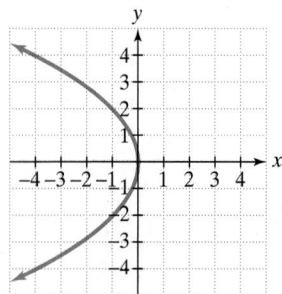

In Exercises 5–16, find the focus and directrix of the parabola with the given equation. Then graph the parabola.

5. $y^2 = 16x$ **6.** $y^2 = 4x$

7. $y^2 = -8x$ **8.** $y^2 = -12x$

9. $x^2 = 12y$ **10.** $x^2 = 8y$

11. $x^2 = -16y$ **12.** $x^2 = -20y$

13. $y^2 - 6x = 0$ **14.** $x^2 - 6y = 0$

15. $8x^2 + 4y = 0$ **16.** $8y^2 + 4x = 0$

In Exercises 17–30, find the standard form of the equation of each parabola satisfying the given conditions.

17. Focus: $(7, 0)$; Directrix: $x = -7$

18. Focus: $(9, 0)$; Directrix: $x = -9$

19. Focus: $(-5, 0)$; Directrix: $x = 5$

20. Focus: $(-10, 0)$; Directrix: $x = 10$

21. Focus: $(0, 15)$; Directrix: $y = -15$

22. Focus: $(0, 20)$; Directrix: $y = -20$

23. Focus: $(0, -25)$; Directrix: $y = 25$

24. Focus: $(0, -15)$; Directrix: $y = 15$

25. Vertex: $(2, -3)$; Focus: $(2, -5)$

26. Vertex: $(5, -2)$; Focus: $(7, -2)$

27. Focus: $(3, 2)$; Directrix: $x = -1$

28. Focus: $(2, 4)$; Directrix: $x = -4$

29. Focus: $(-3, 4)$; Directrix: $y = 2$

30. Focus: $(7, -1)$; Directrix: $y = -9$

In Exercises 31–34, find the vertex, focus, and directrix of each parabola with the given equation. Then match each equation to one of the graphs that are shown and labeled (a)–(d).

31. $(y - 1)^2 = 4(x - 1)$ **32.** $(x + 1)^2 = 4(y + 1)$

33. $(x + 1)^2 = -4(y + 1)$ **34.** $(y - 1)^2 = -4(x - 1)$

a.

b.

c.

d.

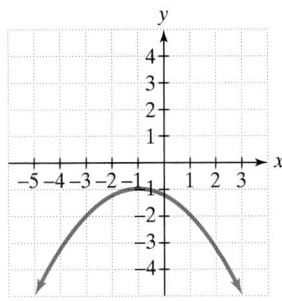

In Exercises 35–42, find the vertex, focus, and directrix of each parabola with the given equation. Then graph the parabola.

35. $(x - 2)^2 = 8(y - 1)$ **36.** $(x + 2)^2 = 4(y + 1)$

37. $(x + 1)^2 = -8(y + 1)$ **38.** $(x + 2)^2 = -8(y + 2)$

39. $(y + 3)^2 = 12(x + 1)$ **40.** $(y + 4)^2 = 12(x + 2)$

41. $(y + 1)^2 = -8x$ **42.** $(y - 1)^2 = -8x$

In Exercises 43–48, convert each equation to standard form by completing the square on x or y. Then find the vertex, focus, and directrix of the parabola. Finally, graph the parabola.

43. $x^2 - 2x - 4y + 9 = 0$ **44.** $x^2 + 6x + 8y + 1 = 0$

45. $y^2 - 2y + 12x - 35 = 0$ **46.** $y^2 - 2y - 8x + 1 = 0$

47. $x^2 + 6x - 4y + 1 = 0$ **48.** $x^2 + 8x - 4y + 8 = 0$

 Application Exercises

49. The reflector of a flashlight is in the shape of a parabolic surface. The casting has a diameter of 4 inches and a depth of 1 inch. How far from the vertex should the light bulb be placed?

50. The reflector of a flashlight is in the shape of a parabolic surface. The casting has a diameter of 8 inches and a depth of 1 inch. How far from the vertex should the light bulb be placed?

51. A satellite dish, like the one shown at the top of the next column, is in the shape of a parabolic surface. Signals coming from a satellite strike the surface of the dish and are reflected to the focus, where the receiver is located. The satellite dish shown has a diameter of 12 feet and a depth of 2 feet. How far from the base of the dish should the receiver be placed?

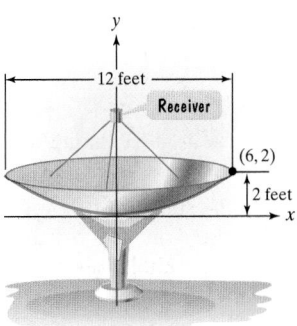

52. In Exercise 51, if the diameter of the dish is halved and the depth stays the same, how far from the base of the smaller dish should the receiver be placed?

53. The towers of the Golden Gate Bridge connecting San Francisco to Marin County are 1280 meters apart and rise 160 meters above the road. The cable between the towers has the shape of a parabola, and the cable just touches the sides of the road midway between the towers. What is the height of the cable 200 meters from a tower? Round to the nearest meter.

54. The towers of a suspension bridge are 800 feet apart and rise 160 feet above the road. The cable between the towers has the shape of a parabola, and the cable just touches the sides of the road midway between the towers. What is the height of the cable 100 feet from a tower?

55. The parabolic arch shown in the figure is 50 feet above the water at the center and 200 feet wide at the base. Will a boat that is 30 feet tall clear the arch 30 feet from the center?

56. A satellite dish in the shape of a parabolic surface has a diameter of 20 feet. If the receiver is to be placed 6 feet from the base, how deep should the dish be?

Writing in Mathematics

57. What is a parabola?

58. Explain how to use $y^2 = 8x$ to find the parabola's focus and directrix.

59. If you are given the standard form of the equation of a parabola with vertex at the origin, explain how to determine if the parabola opens to the right, left, upward, or downward.

60. Describe one similarity and one difference between the graphs of $y^2 = 4x$ and $(y - 1)^2 = 4(x - 1)$.

61. How can you distinguish parabolas from other conic sections by looking at their equations?

62. Look at the satellite dish shown in Exercise 51. Why must the receiver for a shallow dish be farther from the base of the dish than for a deeper dish of the same diameter?

 ## Technology Exercises

63. Use a graphing utility to graph any five of the parabolas that you graphed by hand in Exercises 5–16.

64. Use a graphing utility to graph any three of the parabolas that you graphed by hand in Exercises 35–42. First solve the given equation for y, possibly using the square root method. Enter each of the two resulting equations to produce the complete graph.

Use a graphing utility to graph the parabolas in Exercises 65–66. Write the given equation as a quadratic equation in y and use the quadratic formula to solve for y. Enter each of the equations to produce the complete graph.

65. $y^2 + 2y - 6x + 13 = 0$

66. $y^2 + 10y - x + 25 = 0$

In Exercises 67–68, write each equation as a quadratic equation in y and then use the quadratic formula to express y in terms of x. Graph the resulting two equations using a graphing utility. What effect does the xy-term have on the graph of the resulting parabola?

67. $16x^2 - 24xy + 9y^2 - 60x - 80y + 100 = 0$

68. $x^2 + 2\sqrt{3}xy + 3y^2 + 8\sqrt{3}x - 8y + 32 = 0$

 ## Critical Thinking Exercises

69. Which one of the following is true?
  a. The parabola whose equation is $x = 2y - y^2 + 5$ opens to the right.
  b. If the parabola whose equation is $x = ay^2 + by + c$ has its vertex at $(3, 2)$ and $a > 0$, then it has no y-intercepts.
  c. Some parabolas that open to the right have equations that define y as a function of x.
  d. The graph of $x = a(y - k) + h$ is a parabola with vertex at (h, k).

70. Find the focus and directrix of parabola whose equation is of the form $Ax^2 + Ey = 0$, $A \neq 0$, $E \neq 0$.

71. Write the standard form of the equation of a parabola whose points are equidistant from $y = 4$ and $(-1, 0)$.

 ## Group Exercise

72. Consult the research department of your library or the Internet to find an example of architecture that incorporates one or more conic sections in its design. Share this example with other group members. Explain precisely how conic sections are used. Do conic sections enhance the appeal of the architecture? In what ways?

SECTION 10.4   *Rotation of Axes*

Objectives

1. Identify conics without completing the square.

2. Use rotation of axes formulas.

3. Write equations of rotated conics in standard form.

4. Identify conics without rotating axes.

Richard E. Prince "The Cone of Apollonius" (detail), fiberglass, steel, paint, graphite, 51 x 18 x 14 in. Private collection, Vancouver. Photo courtesy of Equinox Gallery, Vancouver, Canada.

To recognize a conic section, you often need to pay close attention to its graph. Graphs powerfully enhance our understanding of algebra and trigonometry.

However, it is not possible for people who are blind—or sometimes, visually impaired—to see a graph. Creating informative materials for the blind and visually impaired is a challenge for instructors and mathematicians. Many people who are visually impaired "see" a graph by touching a three-dimensional representation of that graph, perhaps while it is described verbally.

Is it possible to identify conic sections in nonvisual ways? The answer is yes, and the methods for doing so are related to the coefficients in their equations. As we present these methods, think about how you learn them. How would your approach to studying mathematics change if we removed all graphs and replaced them with verbal descriptions?

1 Identify conics without completing the square.

Identifying Conic Sections without Completing the Square

Conic sections can be represented both geometrically (as intersecting planes and cones) and algebraically. The equations of the conic sections we have considered in the first three sections of this chapter can be expressed in the form

$$Ax^2 + Cy^2 + Dx + Ey + F = 0$$

in which A and C are not both zero. You can use A and C, the coefficients of x^2 and y^2, respectively, to identify a conic section without completing the square.

Identifying a Conic Section without Completing the Square

A nondegenerate conic section of the form

$$Ax^2 + Cy^2 + Dx + Ey + F = 0$$

in which A and C are not both zero is

- a circle if $A = C$,
- a parabola if $AC = 0$,
- an ellipse if $A \neq C$ and $AC > 0$, and
- a hyperbola if $AC < 0$.

EXAMPLE 1 Identifying a Conic Section without Completing the Square

Identify the graph of each of the following nondegenerate conic sections:

a. $4x^2 - 25y^2 - 24x + 250y - 489 = 0$
b. $x^2 + y^2 + 6x - 2y + 6 = 0$
c. $y^2 + 12x + 2y - 23 = 0$
d. $9x^2 + 25y^2 - 54x + 50y - 119 = 0$.

Solution We use A, the coefficient of x^2, and C, the coefficient of y^2, to identify each conic section.

a. $4x^2 - 25y^2 - 24x + 250y - 489 = 0$

$A = 4 \quad C = -25$

$AC = 4(-25) = -100 < 0$

Because $AC < 0$, the graph of the equation is a hyperbola.

b. $x^2 + y^2 + 6x - 2y + 6 = 0$

$A = 1 \quad C = 1$

Because $A = C$, the graph of the equation is a circle.

c. We can write $y^2 + 12x + 2y - 23 = 0$ as

$$0x^2 + y^2 + 12x + 2y - 23 = 0.$$

$A = 0 \quad C = 1$

$$AC = 0(1) = 0$$

Because $AC = 0$, the graph of the equation is a parabola.

d. $9x^2 + 25y^2 - 54x + 50y - 119 = 0$

$A = 9 \quad C = 25$

$$AC = 9(25) = 225 > 0.$$

Because $AC > 0$ and $A \neq C$, the graph of the equation is an ellipse.

> **Check Point 1** Identify the graph of each of the following nondegenerate conic sections:
>
> **a.** $3x^2 + 2y^2 + 12x - 4y + 2 = 0$
>
> **b.** $x^2 + y^2 - 6x + y + 3 = 0$
>
> **c.** $y^2 - 12x - 4y + 52 = 0$
>
> **d.** $9x^2 - 16y^2 - 90x + 64y + 17 = 0.$

2 Use rotation of axes formulas.

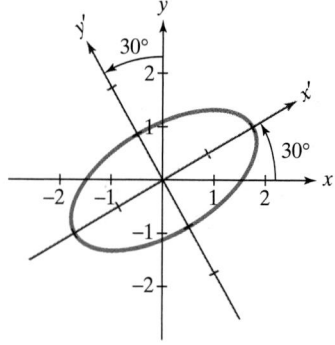

Figure 10.46 The graph of $7x^2 - 6\sqrt{3}\,xy + 13y^2 - 16 = 0,$ a rotated ellipse

Rotation of Axes

Figure 10.46 shows the graph of

$$7x^2 - 6\sqrt{3}\,xy + 13y^2 - 16 = 0.$$

The graph looks like an ellipse, although its major axis neither lies along the x-axis nor is parallel to the x-axis. Do you notice anything unusual about the equation? It contains an xy-term. However, look what happens if we rotate the x- and y-axes through an angle of $30°$. In the rotated $x'y'$-system, the major axis of the ellipse lies along the x'-axis. We can write the equation of the ellipse in this rotated $x'y'$-system as

$$\frac{x'^2}{4} + \frac{y'^2}{1} = 1.$$

Observe that there is no $x'y'$-term in the equation.

Except for degenerate cases, the **general second-degree equation**

$$Ax^2 + Bxy + Cy^2 + Dx + Ey + F = 0$$

represents one of the conic sections. However, due to the xy-term in the equation, these conic sections are rotated in such a way that their axes are no longer parallel to the x- and y-axes. To reduce these equations to forms of the conic sections with which you are already familiar, we use a procedure called **rotation of axes.**

Suppose that the x- and y-axes are rotated through a positive angle θ, resulting in a new $x'y'$ coordinate system. This system is shown in Figure 10.47(a). The origin in the $x'y'$-system is the same as the origin in the xy-system. Point P in Figure 10.47(b) has coordinates (x, y) relative to the xy-system and coordinates

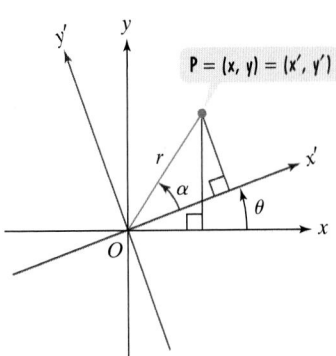

(a) Rotating the x- and y-axes through a positive angle θ

(b) Describing point P relative to the xy-system and the rotated $x'y'$-system

Figure 10.47 Rotating axes

(x', y') relative to the $x'y'$-system. We wish to obtain formulas relating the old and new coordinates. Thus, we need to express x and y in terms of x', y', and θ.

Look at Figure 10.47(b). Notice that

$r =$ the distance from the origin O to point P.

$\alpha =$ the angle from the positive x'-axis to the ray from O through P.

Using the definitions of sine and cosine, we obtain

$$\cos \alpha = \frac{x'}{r} : x' = r \cos \alpha$$

> This is from the right triangle with a leg along the x'-axis.

$$\sin \alpha = \frac{y'}{r} : y' = r \sin \alpha$$

$$\cos (\theta + \alpha) = \frac{x}{r} : x = r \cos (\theta + \alpha)$$

> This is from the taller right triangle with a leg along the x-axis.

$$\sin (\theta + \alpha) = \frac{y}{r} : y = r \sin (\theta + \alpha).$$

Thus,

$$x = r \cos (\theta + \alpha)$$

> This is the third of the preceding equations.

$$= r(\cos \theta \cos \alpha - \sin \theta \sin \alpha)$$

> Use the formula for the cosine of the sum of two angles.

$$= (r \cos \alpha) \cos \theta - (r \sin \alpha) \sin \theta$$

> Apply the distributive property and rearrange factors.

$$= x' \cos \theta - y' \sin \theta.$$

> Use the first and second of the preceding equations: $x' = r \cos \alpha$ and $y' = r \sin \alpha$.

Similarly,

$$y = r \sin (\theta + \alpha) = r(\sin \theta \cos \alpha + \cos \theta \sin \alpha) = x' \sin \theta + y' \cos \theta.$$

Rotation of Axes Formulas

Suppose an xy-coordinate system and an $x'y'$-coordinate system have the same origin and θ is the angle from the positive x-axis to the positive x'-axis. If the coordinates of point P are (x, y) in the xy-system and (x', y') in the rotated $x'y'$-system, then

$$x = x' \cos \theta - y' \sin \theta$$
$$y = x' \sin \theta + y' \cos \theta.$$

EXAMPLE 2   Rotating Axes

Write the equation $xy = 1$ in terms of a rotated $x'y'$-system if the angle of rotation from the x-axis to the x'-axis is $45°$. Express the equation in standard form. Use the rotated system to graph $xy = 1$.

Solution   With $\theta = 45°$, the rotation formulas for x and y are

$$x = x' \cos\theta - y' \sin\theta = x' \cos 45° - y' \sin 45°$$

$$= x'\left(\frac{\sqrt{2}}{2}\right) - y'\left(\frac{\sqrt{2}}{2}\right) = \frac{\sqrt{2}}{2}(x' - y')$$

$$y = x' \sin\theta + y' \cos\theta = x' \sin 45° + y' \cos 45°$$

$$= x'\left(\frac{\sqrt{2}}{2}\right) + y'\left(\frac{\sqrt{2}}{2}\right) = \frac{\sqrt{2}}{2}(x' + y').$$

Now substitute these expressions for x and y in the given equation, $xy = 1$.

$$xy = 1 \qquad \text{This is the given equation.}$$

$$\left[\frac{\sqrt{2}}{2}(x' - y')\right]\left[\frac{\sqrt{2}}{2}(x' + y')\right] = 1 \qquad \begin{array}{l}\text{Substitute the expressions for } x \text{ and } y \text{ from} \\ \text{the rotation formulas.}\end{array}$$

$$\frac{2}{4}(x' - y')(x' + y') = 1 \qquad \text{Multiply: } \frac{\sqrt{2}}{2} \cdot \frac{\sqrt{2}}{2} = \frac{2}{4}.$$

$$\frac{1}{2}\left(x'^2 - y'^2\right) = 1 \qquad \text{Reduce } \frac{2}{4} \text{ and multiply the binomials.}$$

$$\frac{x'^2}{2} - \frac{y'^2}{2} = 1 \qquad \begin{array}{l}\text{Write the equation in standard form:} \\ \dfrac{x^2}{a^2} - \dfrac{y^2}{b^2} = 1.\end{array}$$

$$\boxed{a^2 = 2} \qquad \boxed{b^2 = 2}$$

This equation expresses $xy = 1$ in terms of the rotated $x'y'$-system. Can you see that this is the standard form of the equation of a hyperbola? The hyperbola's center is at $(0, 0)$, with the transverse axis on the x'-axis. The vertices are $(-a, 0)$ and $(a, 0)$. Because $a^2 = 2$, the vertices are $(-\sqrt{2}, 0)$ and $(\sqrt{2}, 0)$, located on the x'-axis. Based on the standard form of the hyperbola's equation, the equations for the asymptotes are

$$y' = \pm\frac{b}{a} x' \text{ or } y' = \pm\frac{\sqrt{2}}{\sqrt{2}} x'.$$

The equations of the asymptotes can be simplified to $y' = x'$ and $y' = -x'$, which correspond to the original x- and y-axes. The graph of the hyperbola is shown in Figure 10.48.

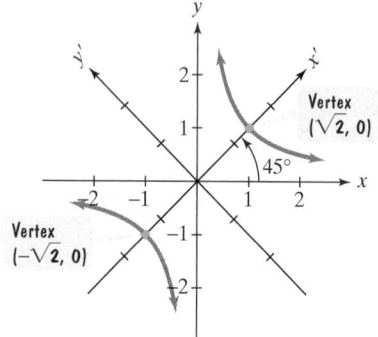

Figure 10.48 The graph of $xy = 1$ or $\dfrac{x'^2}{2} - \dfrac{y'^2}{2} = 1$

Check Point 2   Write the equation $xy = 2$ in terms of a rotated $x'y'$-system if the angle of rotation from the x-axis to the x'-axis is $45°$. Express the equation in standard form. Use the rotated system to graph $xy = 2$.

3	Write equations of rotated conics in standard form.

Using Rotations to Transform Equations with xy-Terms to Standard Equations of Conic Sections

We have noted that the appearance of the term Bxy $(B \neq 0)$ in the general second-degree equation indicates that the graph of the conic section has been rotated. A rotation of axes through an appropriate angle can transform the equation to one of the standard forms of the conic sections in x' and y' in which no $x'y'$-term appears.

Amount of Rotation Formula

The general second-degree equation

$$Ax^2 + Bxy + Cy^2 + Dx + Ey + F = 0, B \neq 0$$

can be rewritten as an equation in x' and y' without an $x'y'$-term by rotating the axes through angle θ, where

$$\cot 2\theta = \frac{A - C}{B}.$$

Before we learn to apply this formula, let's see how it can be derived. We begin with the general second-degree equation

$$Ax^2 + Bxy + Cy^2 + Dx + Ey + F = 0, B \neq 0.$$

Rotate the axes through an angle θ. In terms of the rotated $x'y'$-system, the general second-degree equation can be written as

$$A(x'\cos\theta - y'\sin\theta)^2 + B(x'\cos\theta - y'\sin\theta)(x'\sin\theta + y'\cos\theta)$$
$$+ C(x'\sin\theta + y'\cos\theta)^2 + D(x'\cos\theta - y'\sin\theta)$$
$$+ E(x'\sin\theta + y'\cos\theta) + F = 0.$$

After a lot of simplifying that involves expanding and collecting like terms, you will obtain the following equation:

We want a rotation that results in no x'y'-term.

$$(A\cos^2\theta + B\sin\theta\cos\theta + C\sin^2\theta)x'^2 + [B(\cos^2\theta - \sin^2\theta) + 2(C - A)(\sin\theta\cos\theta)]x'y'$$
$$+ (A\sin^2\theta - B\sin\theta\cos\theta + C\cos^2\theta)y'^2$$
$$+ (D\cos\theta + E\sin\theta)x'$$
$$+ (-D\sin\theta + E\cos\theta)y' + F = 0.$$

If this looks somewhat ghastly, take a deep breath and focus only on the $x'y'$-term. We want to choose θ so that the coefficient of this term is zero. This will give the required rotation that results in no $x'y'$-term.

$$B(\cos^2\theta - \sin^2\theta) + 2(C - A)\sin\theta\cos\theta = 0$$
Set the coefficient of the x'y'-term equal to 0.

$$B\cos 2\theta + (C - A)\sin 2\theta = 0$$
Use the double-angle formulas: $\cos 2\theta = \cos^2\theta - \sin^2\theta$ and $\sin 2\theta = 2\sin\theta\cos\theta$.

$$B\cos 2\theta = -(C - A)\sin 2\theta$$
Subtract $(C - A)\sin 2\theta$ from both sides.

$$B \cos 2\theta = (A - C) \sin 2\theta$$ Simplify $B \cos 2\theta = -(C - A) \sin 2\theta$.

$$\frac{B \cos 2\theta}{B \sin 2\theta} = \frac{(A - C) \sin 2\theta}{B \sin 2\theta}$$ Divide both sides by $B \sin 2\theta$.

$$\frac{\cos 2\theta}{\sin 2\theta} = \frac{A - C}{B}$$ Simplify.

$$\cot 2\theta = \frac{A - C}{B}$$ Apply a quotient identity: $\cot 2\theta = \dfrac{\sin 2\theta}{\cos 2\theta}$.

If $\cot 2\theta$ is positive, we will select θ so that $0° < \theta < 45°$. If $\cot 2\theta$ is negative, we will select θ so that $45° < \theta < 90°$. Thus θ, the angle of rotation, is always an acute angle.

Here is a step-by-step procedure for writing the equation of a rotated conic section in standard form:

Study Tip

What do you do after substituting the expressions for x and y from the rotation formulas into the given equation? You must simplify the resulting equation by expanding and collecting like terms. Work through this process slowly and carefully, allowing lots of room on your paper.

If your rotation equations are correct but you obtain an equation that has an $x'y'$-term, you have made an error in the algebraic simplification.

Writing the Equation of a Rotated Conic in Standard Form

1. Use the given equation
$$Ax^2 + Bxy + Cy^2 + Dx + Ey + F = 0, B \neq 0$$
to find $\cot 2\theta$.
$$\cot 2\theta = \frac{A - C}{B}$$

2. Use the expression for $\cot 2\theta$ to determine θ, the angle of rotation.
3. Substitute θ in the rotation formulas
$$x = x' \cos \theta - y' \sin \theta \quad \text{and} \quad y = x' \sin \theta + y' \cos \theta$$
and simplify.
4. Substitute the expressions for x and y from the rotation formulas in the given equation and simplify. The resulting equation should have no $x'y'$-term.
5. Write the equation involving x' and y' in standard form.

Using the equation in step 5, you can graph the conic section in the rotated $x'y'$-system.

EXAMPLE 3 Writing the Equation of a Rotated Conic Section in Standard Form

Rewrite the equation
$$7x^2 - 6\sqrt{3}\, xy + 13y^2 - 16 = 0$$
in a rotated $x'y'$-system without an $x'y'$-term. Express the equation in the standard form of a conic section. Graph the conic section in the rotated system.

Solution

Step 1   Use the given equation to find $\cot 2\theta$. We need to identify the constants A, B, and C in the given equation.

$$7x^2 - 6\sqrt{3}\,xy + 13y^2 - 16 = 0$$

A is the coefficient of the x^2-term: $A = 7$.

B is the coefficient of the xy-term: $B = -6\sqrt{3}$.

C is the coefficient of the y^2-term: $C = 13$.

The appropriate angle θ through which to rotate the axes satisfies the equation

$$\cot 2\theta = \frac{A - C}{B} = \frac{7 - 13}{-6\sqrt{3}} = \frac{-6}{-6\sqrt{3}} = \frac{1}{\sqrt{3}} \text{ or } \frac{\sqrt{3}}{3}.$$

Step 2 **Use the expression for $\cot 2\theta$ to determine the angle of rotation.** We have $\cot 2\theta = \dfrac{\sqrt{3}}{3}$. Based on our knowledge of exact values for trigonometric functions, we conclude that $2\theta = 60°$. Thus, $\theta = 30°$.

Step 3 **Substitute θ in the rotation formulas $x = x' \cos\theta - y' \sin\theta$ and $y = x' \sin\theta + y' \cos\theta$ and simplify.** Substituting $30°$ for θ,

$$x = x' \cos 30° - y' \sin 30° = x'\left(\frac{\sqrt{3}}{2}\right) - y'\left(\frac{1}{2}\right) = \frac{\sqrt{3}x' - y'}{2}$$

$$y = x' \sin 30° + y' \cos 30° = x'\left(\frac{1}{2}\right) + y'\left(\frac{\sqrt{3}}{2}\right) = \frac{x' + \sqrt{3}y'}{2}.$$

Step 4 **Substitute the expressions for x and y from the rotation formulas in the given equation and simplify.**

$$7x^2 - 6\sqrt{3}xy + 13y^2 - 16 = 0 \qquad \text{This is the given equation.}$$

$$7\left(\frac{\sqrt{3}x' - y'}{2}\right)^2 - 6\sqrt{3}\left(\frac{\sqrt{3}x' - y'}{2}\right)\left(\frac{x' + \sqrt{3}y'}{2}\right)$$
$$+ 13\left(\frac{x' + \sqrt{3}y'}{2}\right)^2 - 16 = 0 \qquad \text{Substitute the expressions for } x \text{ and } y \text{ from the rotation formulas.}$$

$$7\left(\frac{3x'^2 - 2\sqrt{3}x'y' + y'^2}{4}\right) - 6\sqrt{3}\left(\frac{\sqrt{3}x'^2 + 3x'y' - x'y' - \sqrt{3}y'^2}{4}\right)$$
$$+ 13\left(\frac{x'^2 + 2\sqrt{3}x'y' + 3y'^2}{4}\right) - 16 = 0 \qquad \text{Square and multiply.}$$

$$7(3x'^2 - 2\sqrt{3}x'y' + y'^2) - 6\sqrt{3}(\sqrt{3}x'^2 + 2x'y' - \sqrt{3}y'^2)$$
$$+ 13(x'^2 + 2\sqrt{3}x'y' + 3y'^2) - 64 = 0 \qquad \text{Multiply both sides by 4.}$$

$$21x'^2 - 14\sqrt{3}x'y' + 7y'^2 - 18x'^2 - 12\sqrt{3}x'y' + 18y'^2$$
$$+ 13x'^2 + 26\sqrt{3}x'y' + 39y'^2 - 64 = 0 \qquad \text{Distribute throughout parentheses.}$$

$$21x'^2 - 18x'^2 + 13x'^2 - 14\sqrt{3}x'y' - 12\sqrt{3}x'y' + 26\sqrt{3}x'y'$$
$$+ 7y'^2 + 18y'^2 + 39y'^2 - 64 = 0 \qquad \text{Rearrange terms.}$$

$$16x'^2 + 64y'^2 - 64 = 0 \qquad \text{Combine like terms.}$$

Do you see how we "lost" the $x'y'$-term in the last equation?

$$-14\sqrt{3}x'y' - 12\sqrt{3}x'y' + 26\sqrt{3}x'y' = -26\sqrt{3}x'y' + 26\sqrt{3}x'y' = 0x'y' = 0$$

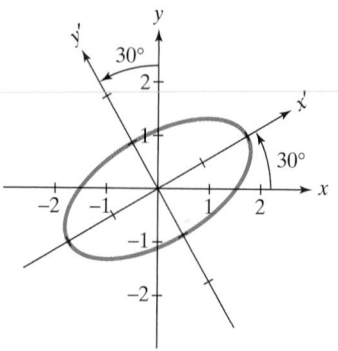

Figure 10.46, repeated

The graph of

$7x^2 - 6\sqrt{3}\,xy + 13y^2 - 16 = 0$

or $\dfrac{x'^2}{4} - \dfrac{y'^2}{1} = 1$, a rotated ellipse

Step 5 Write the equation involving x' and y' in standard form. We can express $16x'^2 + 64y'^2 - 64 = 0$, an equation of an ellipse, in the standard form $\dfrac{x^2}{a^2} + \dfrac{y^2}{b^2} = 1$.

$16x'^2 + 64y'^2 - 64 = 0$ This equation describes the ellipse relative to a system rotated through 30°.

$16x'^2 + 64y'^2 = 64$ Add 64 to both sides.

$\dfrac{16x'^2}{64} + \dfrac{64y'^2}{64} = \dfrac{64}{64}$ Divide both sides by 64.

$\dfrac{x'^2}{4} + \dfrac{y'^2}{1} = 1$ Simplify.

The last equation is the standard form of the equation of an ellipse. The major axis is on the x'-axis and the vertices are $(-2, 0)$ and $(2, 0)$. The minor axis is on the y'-axis with endpoints $(0, -1)$ and $(0, 1)$. The graph of the ellipse is shown in Figure 10.46. Does this graph look familiar? It should—you saw it earlier in this section on page 886.

Check Point 3 Rewrite the equation

$$2x^2 + \sqrt{3}xy + y^2 - 2 = 0$$

in a rotated $x'y'$-system without an $x'y'$-term. Express the equation in the standard form of a conic section. Graph the conic section in the rotated system.

Technology

In order to graph a general second-degree equation in the form

$$Ax^2 + Bxy + Cy^2 + Dx + Ey + F = 0,$$

it is necessary to solve for y. Rewrite the equation as a quadratic equation in y.

$$Cy^2 + (Bx + E)y + (Ax^2 + Dx + F) = 0$$

By applying the quadratic formula, the graph of this equation can be obtained by entering

$$y_1 = \frac{-(Bx + E) + \sqrt{(Bx + E)^2 - 4C(Ax^2 + Dx + F)}}{2C}$$

and

$$y_2 = \frac{-(Bx + E) - \sqrt{(Bx + E)^2 - 4C(Ax^2 + Dx + F)}}{2C}.$$

The graph of

$$7x^2 - 6\sqrt{3}xy + 13y^2 - 16 = 0$$

is shown on the right in a $[-2, 2, 1]$ by $[-2, 2, 1]$ viewing rectangle. The graph was obtained by entering the equations for y_1 and y_2 shown above with

$$A = 7, B = -6\sqrt{3}, C = 13, D = 0, E = 0,$$

and $F = -16$.

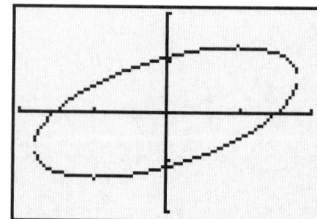

In Example 3 and Check Point 3, we found θ, the angle of rotation, directly because we recognized $\dfrac{\sqrt{3}}{3}$ as the value of $\cot 60°$. What do we do if $\cot 2\theta$ is not the cotangent of one of the familiar angles? We use $\cot 2\theta$ to find $\sin \theta$ and $\cos \theta$ as follows:

- Use a sketch of $\cot 2\theta$ to find $\cos 2\theta$.
- Find $\sin \theta$ and $\cos \theta$ using the identities

$$\sin \theta = \sqrt{\frac{1 - \cos 2\theta}{2}} \quad \text{and} \quad \cos \theta = \sqrt{\frac{1 + \cos 2\theta}{2}}.$$

> Because θ is an acute angle, the positive square roots are appropriate.

The resulting values for $\sin \theta$ and $\cos \theta$ are used to write the rotation formulas that give an equation with no $x'y'$-term.

EXAMPLE 4 Graphing the Equation of a Rotated Conic

Graph relative to a rotated $x'y'$-system in which the equation has no $x'y'$-term:

$$16x^2 - 24xy + 9y^2 + 110x - 20y + 100 = 0.$$

Solution

Step 1 Use the given equation to find $\cot 2\theta$. With $A = 16$, $B = -24$, and $C = 9$, we have

$$\cot 2\theta = \frac{A - C}{B} = \frac{16 - 9}{-24} = -\frac{7}{24}.$$

Step 2 Use the expression for $\cot 2\theta$ to determine $\sin \theta$ and $\cos \theta$. A rough sketch showing $\cot 2\theta$ is given in Figure 10.49. Because θ is always acute and $\cot 2\theta$ is negative, 2θ is in quadrant II. The third side of the triangle is found using $r = \sqrt{x^2 + y^2}$. Thus, $r = \sqrt{(-7)^2 + 24^2} = \sqrt{625} = 25$. By the definition of the cosine function,

$$\cos 2\theta = \frac{x}{r} = \frac{-7}{25} = -\frac{7}{25}.$$

Now we use identities to find values for $\sin \theta$ and $\cos \theta$.

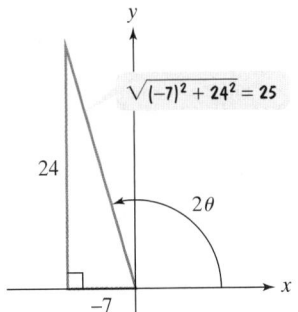

Figure 10.49 Using $\cot 2\theta$ to find $\cos 2\theta$

$$\sin \theta = \sqrt{\frac{1 - \cos 2\theta}{2}} = \sqrt{\frac{1 - \left(-\dfrac{7}{25}\right)}{2}}$$

$$= \sqrt{\frac{\dfrac{25}{25} + \dfrac{7}{25}}{2}} = \sqrt{\frac{\dfrac{32}{25}}{2}} = \sqrt{\frac{32}{50}} = \sqrt{\frac{16}{25}} = \frac{4}{5}$$

$$\cos \theta = \sqrt{\frac{1 + \cos 2\theta}{2}} = \sqrt{\frac{1 + \left(-\dfrac{7}{25}\right)}{2}}$$

$$= \sqrt{\frac{\dfrac{25}{25} - \dfrac{7}{25}}{2}} = \sqrt{\frac{\dfrac{18}{25}}{2}} = \sqrt{\frac{18}{50}} = \sqrt{\frac{9}{25}} = \frac{3}{5}$$

Step 3   Substitute $\sin \theta$ **and** $\cos \theta$ **in the rotation formulas**

$$x = x' \cos\theta - y' \sin\theta \text{ and } y = x' \sin\theta + y' \cos\theta$$

and simplify. Substituting $\frac{4}{5}$ for $\sin\theta$ and $\frac{3}{5}$ for $\cos\theta$,

$$x = x'\left(\frac{3}{5}\right) - y'\left(\frac{4}{5}\right) = \frac{3x' - 4y'}{5}$$

$$y = x'\left(\frac{4}{5}\right) + y'\left(\frac{3}{5}\right) = \frac{4x' + 3y'}{5}.$$

Step 4   Substitute the expressions for x and y from the rotation formulas in the given equation and simplify.

$$16x^2 - 24xy + 9y^2 + 110x - 20y + 100 = 0 \qquad \text{This is the given equation.}$$

$$16\left(\frac{3x' - 4y'}{5}\right)^2 - 24\left(\frac{3x' - 4y'}{5}\right)\left(\frac{4x' + 3y'}{5}\right) + 9\left(\frac{4x' + 3y'}{5}\right)^2 \qquad \text{Substitute the expressions for } x \text{ and } y \text{ from the rotation formulas.}$$

$$+ 110\left(\frac{3x' - 4y'}{5}\right) - 20\left(\frac{4x' + 3y'}{5}\right) + 100 = 0$$

Take a few minutes to expand, multiply both sides of the equation by 25, and combine like terms. The resulting equation

$$y'^2 + 2x' - 4y' + 4 = 0$$

has no $x'y'$-term.

Step 5   Write the equation involving x' and y' in standard form. With only one variable that is squared, we have the equation of a parabola. We need to write the equation in the standard form $(y - k)^2 = 4p(x - h)$.

$$y'^2 + 2x' - 4y' + 4 = 0 \qquad \text{This is the equation without an } x'y'\text{-term.}$$

$$y'^2 - 4y' = -2x' - 4 \qquad \text{Isolate the terms involving } y'.$$

$$y'^2 - 4y' + 4 = -2x' - 4 + 4 \qquad \text{Complete the square by adding the square of half the coefficient of } y'.$$

$$(y' - 2)^2 = -2x' \qquad \text{Factor.}$$

The standard form of the parabola's equation in the rotated $x'y'$-system is

$$(y' - 2)^2 = -2x'.$$

This is $(y' - k)^2$ with $k = 2$.   This is $4p$.   This is $x' - h$ with $h = 0$.

We see that $h = 0$ and $k = 2$. Thus, the vertex of the parabola in the $x'y'$-system is $(h, k) = (0, 2)$.

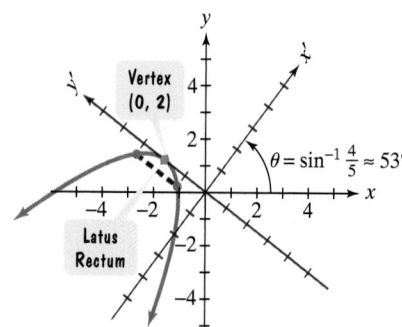

Vertex (0, 2)

Latus Rectum

$\theta = \sin^{-1}\frac{4}{5} \approx 53°$

Figure 10.50 The graph of $(y' - 2)^2 = -2x'$ in a rotated $x'y'$-system

We can use the $x'y'$–system to graph the parabola. Using a calculator to solve $\sin\theta = \frac{4}{5}$, we find that $\theta = \sin^{-1}\frac{4}{5} \approx 53°$. Rotate the axes through approximately $53°$. With $4p = -2$ and $p = -\frac{1}{2}$, the parabola's focus is $\frac{1}{2}$ unit to the left of the vertex, $(0, 2)$. Thus, the focus in the $x'y'$–system is $\left(-\frac{1}{2}, 2\right)$.

To graph the parabola, we use the vertex, $(0, 2)$, and the two endpoints of the latus rectum.

$$\text{length of latus rectum } = |4p| = |-2| = 2$$

The latus rectum extends 1 unit above and 1 unit below the focus, $\left(-\frac{1}{2}, 2\right)$. Thus, the endpoints of the latus rectum in the $x'y'$– system are $\left(-\frac{1}{2}, 3\right)$ and $\left(-\frac{1}{2}, 1\right)$. Using the rotated system, pass a smooth curve through the vertex and the two endpoints of the latus rectum. The graph of the parabola is shown in Figure 10.50.

Check Point 4   Graph relative to a rotated $x'y'$-system in which the equation has no $x'y'$-term:

$$4x^2 - 4xy + y^2 - 8\sqrt{5}x - 16\sqrt{5}y = 0.$$

4 Identify conics without rotating axes.

Identifying Conic Sections without Rotating Axes

We now know that the general second-degree equation

$$Ax^2 + Bxy + Cy^2 + Dx + Ey + F = 0, B \neq 0$$

can be rewritten as

$$A'x'^2 + C'y'^2 + D'x' + E'y' + F' = 0$$

in a rotated $x'y'$-system. A relationship between the coefficients of the two equations is given by

$$B^2 - 4AC = -4A'C'.$$

We also know that A' and C' can be used to identify the graph of the rotated equation. Thus, $B^2 - 4AC$ can also be used to identify the graph of the general second-degree equation.

Identifying a Conic Section without a Rotation of Axes

A nondegenerate conic section of the form

$$Ax^2 + Bxy + Cy^2 + Dx + Ey + F = 0$$

is

- a parabola if $B^2 - 4AC = 0$,
- an ellipse or a circle if $B^2 - 4AC < 0$, and
- a hyperbola if $B^2 - 4AC > 0$.

Technology

The graph of

$$11x^2 + 10\sqrt{3}xy + y^2 - 4 = 0$$

is shown in a $\left[-1, 1, \frac{1}{4}\right]$ by $\left[-1, 1, \frac{1}{4}\right]$ viewing rectangle. The graph verifies that the equation represents a rotated hyperbola.

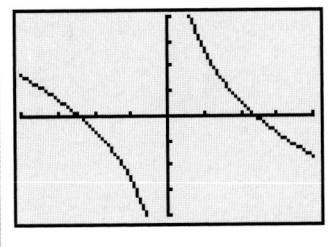

EXAMPLE 5   Identifying a Conic Section without Rotating Axes

Identify the graph of

$$11x^2 + 10\sqrt{3}xy + y^2 - 4 = 0.$$

Solution   We use A, B, and C to identify the conic section.

$$11x^2 + 10\sqrt{3}xy + y^2 - 4 = 0$$

$$A = 11 \qquad B = 10\sqrt{3} \qquad C = 1$$

$$B^2 - 4AC = \left(10\sqrt{3}\right)^2 - 4(11)(1) = 100 \cdot 3 - 44 = 256 > 0$$

Because $B^2 - 4AC > 0$, the graph of the equation is a hyperbola.

Check Point 5   Identify the graph of $3x^2 - 2\sqrt{3}xy + y^2 + 2x + 2\sqrt{3}y = 0$.

EXERCISE SET 10.4

 Practice Exercises

In Exercises 1–8, identify each equation without completing the square.

1. $y^2 - 4x + 2y + 21 = 0$

2. $y^2 - 4x - 4y = 0$

3. $4x^2 - 9y^2 - 8x - 36y - 68 = 0$

4. $9x^2 + 25y^2 - 54x - 200y + 256 = 0$

5. $4x^2 + 4y^2 + 12x + 4y + 1 = 0$

6. $9x^2 + 4y^2 - 36x + 8y + 31 = 0$

7. $100x^2 - 7y^2 + 90y - 368 = 0$

8. $y^2 + 8x + 6y + 25 = 0$

In Exercises 9–14, write each equation in terms of a rotated $x'y'$-system using θ, the angle of rotation. Write the equation involving x' and y' in standard form.

9. $xy = -1; \theta = 45°$

10. $xy = -4; \theta = 45°$

11. $x^2 - 4xy + y^2 - 3 = 0; \theta = 45°$

12. $13x^2 - 10xy + 13y^2 - 72 = 0; \theta = 45°$

13. $23x^2 + 26\sqrt{3}xy - 3y^2 - 144 = 0; \theta = 30°$

14. $13x^2 - 6\sqrt{3}\,xy + 7y^2 - 16 = 0; \theta = 60°$

In Exercises 15–26, write the appropriate rotation formulas so that in a rotated system the equation has no $x'y'$-term.

15. $x^2 + xy + y^2 - 10 = 0$

16. $x^2 + 4xy + y^2 - 3 = 0$

17. $3x^2 - 10xy + 3y^2 - 32 = 0$

18. $5x^2 - 8xy + 5y^2 - 9 = 0$

19. $11x^2 + 10\sqrt{3}xy + y^2 - 4 = 0$

20. $7x^2 - 6\sqrt{3}xy + 13y^2 - 16 = 0$

21. $10x^2 + 24xy + 17y^2 - 9 = 0$

22. $32x^2 - 48xy + 18y^2 - 15x - 20y = 0$

23. $x^2 + 4xy - 2y^2 - 1 = 0$

24. $3xy - 4y^2 + 18 = 0$

25. $34x^2 - 24xy + 41y^2 - 25 = 0$

26. $6x^2 - 6xy + 14y^2 - 45 = 0$

In Exercises 27–38,

a. *Rewrite the equation in a rotated $x'y'$-system without an $x'y'$ term. Use the appropriate rotation formulas from Exercises 15–26.*

b. *Express the equation involving x' and y' in the standard form of a conic section.*

c. *Use the rotated system to graph the equation.*

27. $x^2 + xy + y^2 - 10 = 0$

28. $x^2 + 4xy + y^2 - 3 = 0$

29. $3x^2 - 10xy + 3y^2 - 32 = 0$

30. $5x^2 - 8xy + 5y^2 - 9 = 0$

31. $11x^2 + 10\sqrt{3}xy + y^2 - 4 = 0$

32. $7x^2 - 6\sqrt{3}xy + 13y^2 - 16 = 0$

33. $10x^2 + 24xy + 17y^2 - 9 = 0$

34. $32x^2 - 48xy + 18y^2 - 15x - 20y = 0$

35. $x^2 + 4xy - 2y^2 - 1 = 0$

36. $3xy - 4y^2 + 18 = 0$

37. $34x^2 - 24xy + 41y^2 - 25 = 0$

38. $6x^2 - 6xy + 14y^2 - 45 = 0$

In Exercises 39–44, identify each equation without applying a rotation of axes.

39. $5x^2 - 2xy + 5y^2 - 12 = 0$

40. $10x^2 + 24xy + 17y^2 - 9 = 0$

41. $24x^2 + 16\sqrt{3}xy + 8y^2 - x + \sqrt{3}y - 8 = 0$

42. $3x^2 - 2\sqrt{3}xy + y^2 + 2x + 2\sqrt{3}y = 0$

43. $23x^2 + 26\sqrt{3}xy - 3y^2 - 144 = 0$

44. $4xy + 3y^2 + 4x + 6y - 1 = 0$

 Writing in Mathematics

45. Explain how to identify the graph of
$$Ax^2 + Cy^2 + Dx + Ey + F = 0$$

46. If there is a 60° angle from the positive x-axis to the positive x'-axis, explain how to obtain the rotation formulas for x and y.

47. How do you obtain the angle of rotation so that a general second-degree equation has no $x'y'$-term in a rotated $x'y'$-system?

48. What is the most time-consuming part in using a graphing utility to graph a general second-degree equation with an xy-term?

49. Explain how to identify the graph of
$$Ax^2 + Bxy + Cy^2 + Dx + Ey + F = 0.$$

 Technology Exercises

In Exercises 50–56, use a graphing utility to graph each equation.

50. $x^2 + 4xy + y^2 - 3 = 0$

51. $7x^2 + 8xy + y^2 - 1 = 0$

52. $3x^2 + 4xy + 6y^2 - 7 = 0$

53. $3x^2 - 6xy + 3y^2 + 10x - 8y - 2 = 0$

54. $9x^2 + 24xy + 16y^2 + 90x - 130y = 0$

55. $x^2 + 4xy + 4y^2 + 10\sqrt{5}x - 9 = 0$

56. $7x^2 + 6xy + 2.5y^2 - 14x + 4y + 9 = 0$

 Critical Thinking Exercises

57. Explain the relationship between the graph of $3x^2 - 2xy + 3y^2 + 2 = 0$ and the sound made by one hand clapping. Begin by following the directions for Exercises 27–38. (You will first need to write rotation formulas that eliminate the $x'y'$-term.)

58. What happens to the equation $x^2 + y^2 = r^2$ in a rotated $x'y'$-system?

In Exercises 59–60, let
$Ax^2 + Bxy + Cy^2 + Dx + Ey + F = 0$ *be an equation of a conic section in an xy-coordinate system. Let*
$A'x'^2 + B'x'y' + C'y'^2 + D'x' + E'y' + F' = 0$ *be the equation of the conic section in the rotated $x'y'$-coordinate system. Use the coefficients A', B', and C', shown in the equation with the voice balloon pointing to B' on page 889, to prove the following relationships.*

59. $A' + C' = A + C$

60. $B'^2 - 4A'C' = B^2 - 4AC$

 Group Exercise

61. Many public and private organizations and schools provide educational materials and information for the blind and visually impaired. Using your library, resources on the World Wide Web, or local organizations, investigate how your group or college could make a contribution to enhance the study of mathematics for the blind and visually impaired. In relation to conic sections, group members should discuss how to create graphs in tactile, or touchable, form that show blind students the visual structure of the conics, including asymptotes, intercepts, end behavior, and rotations.

SECTION 10.5 *Parametric Equations*

Objectives

1. Use point plotting to graph plane curves described by parametric equations.

2. Eliminate the parameter.

3. Find parametric equations for functions.

4. Understand the advantages of parametric representations.

What a baseball game! You got to see the great Sammy Sosa of the Chicago Cubs blast a powerful homer. In less than eight seconds, the parabolic path of his home run took the ball a horizontal distance of over 1000 feet. Is there a way to model this path that gives both the ball's location and the time that it is in each of its positions? In this section, we look at ways of describing curves that reveal the where and the when of motion.

Plane Curves and Parametric Equations

You throw a ball from a height of 6 feet, with an initial velocity of 90 feet per second and at an angle of 40° with the horizontal. After t seconds, the location of the ball can be described by

$$x = (90\cos 40°)t \quad \text{and} \quad y = 6 + (90\sin 40°)t - 16t^2.$$

This is the ball's horizontal distance, in feet.

This is the ball's vertical height, in feet.

Because we can use these equations to calculate the location of the ball at any time t, we can describe the path of the ball. For example, to determine the location when $t = 1$ second, substitute 1 for t in each equation:

$$x = (90 \cos 40°)t = (90 \cos 40°)(1) \approx 68.9 \text{ feet}$$

$$y = 6 + (90 \sin 40°)t - 16t^2 = 6 + (90 \sin 40°)(1) - 16(1)^2 \approx 47.9 \text{ feet}.$$

This tells us that after one second, the ball has traveled a horizontal distance of approximately 68.9 feet, and the height of the ball is approximately 47.9 feet. Figure 10.51 displays this information and the results for calculations corresponding to $t = 2$ seconds and $t = 3$ seconds.

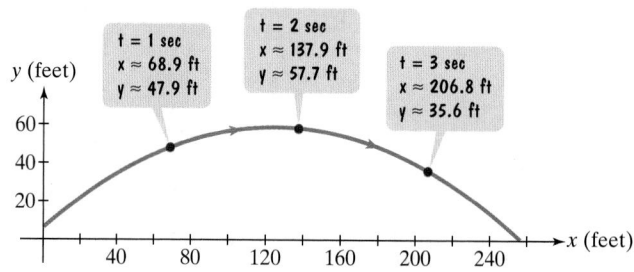

Figure 10.51 The location of a thrown ball after 1, 2, and 3 seconds

The voice balloons in Figure 10.51 tell where the ball is located and when the ball is at a given point (x, y) on its path. The variable t, called a **parameter,** gives the various times for the ball's location. The equations that describe where the ball is located express both x and y as functions of t, and are called **parametric equations.**

$$x = (90 \cos 40°)t \qquad y = 6 + (90 \sin 40°)t - 16t^2$$

| This is the parametric equation for x. | This is the parametric equation for y. |

The collection of points (x, y) in Figure 10.51 is called a **plane curve**.

Plane Curves and Parametric Equations

Suppose that t is a number in an interval I. A **plane curve** is the set of ordered pairs (x, y), where

$$x = f(t), \quad y = g(t) \quad \text{for } t \text{ in interval } I.$$

The variable t is called a **parameter,** and the equations $x = f(t)$ and $y = g(t)$ are called **parametric equations** for the curve.

---

1 Use point plotting to graph plane curves described by parametric equations.

Graphing Plane Curves

Graphing a plane curve represented by parametric equations involves plotting points in the rectangular coordinate system and connecting them with a smooth curve.

Graphing a Plane Curve Described by Parametric Equations

1. Select some values of t on the given interval.
2. For each value of t, use the given parametric equations to compute x and y.
3. Plot the points (x, y) in the order of increasing t and connect them with a smooth curve.

Take a second look at Figure 10.51. Do you notice arrows along the curve? These arrows show the direction, or **orientation**, along the curve as t increases. After graphing a plane curve described by parametric equations, use arrows between the points to show the orientation of the curve corresponding to increasing values of t.

EXAMPLE 1 Graphing a Curve Defined by Parametric Equations

Graph the plane curve defined by the parametric equations:

$$x = t^2 - 1, \qquad y = 2t, \qquad -2 \le t \le 2.$$

Solution

Step 1 Select some values of t on the given interval. We will select integral values of t on the interval $-2 \le t \le 2$. Let $t = -2, -1, 0, 1,$ and 2.

Step 2 For each value of t, use the given parametric equations to compute x and y. We organize our work in a table. The first column lists the choices for the parameter t. The next two columns show the corresponding values for x and y. The last column lists the ordered pair (x, y).

t	$x = t^2 - 1$	$y = 2t$	(x, y)
-2	$(-2)^2 - 1 = 4 - 1 = 3$	$2(-2) = -4$	$(3, -4)$
-1	$(-1)^2 - 1 = 1 - 1 = 0$	$2(-1) = -2$	$(0, -2)$
0	$0^2 - 1 = -1$	$2(0) = 0$	$(-1, 0)$
1	$1^2 - 1 = 0$	$2(1) = 2$	$(0, 2)$
2	$2^2 - 1 = 4 - 1 = 3$	$2(2) = 4$	$(3, 4)$

Step 3 Plot the points (x, y) in the order of increasing t and connect them with a smooth curve. The plane curve defined by the parametric equations on the given interval is shown in Figure 10.52. The arrows show the direction, or orientation, along the curve as t varies from -2 to 2.

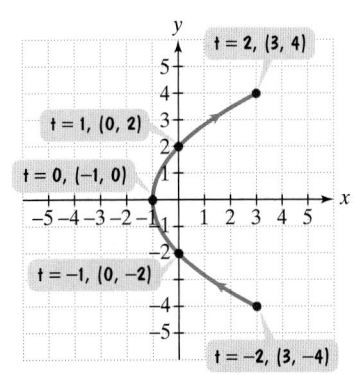

Figure 10.52 The plane curve defined by $x = t^2 - 1$, $y = 2t$, $-2 \le t \le 2$

Check Point 1 Graph the plane curve defined by the parametric equations:

$$x = t^2 + 1, \qquad y = 3t, \qquad -2 \le t \le 2.$$

2 Eliminate the parameter.

Eliminating the Parameter

The graph in Figure 10.52 shows the plane curve for $x = t^2 - 1$, $y = 2t$, $-2 \le t \le 2$. Even if we examine the parametric equations carefully, we may not be able to tell that the corresponding plane curve is a parabola. By **eliminating**

Technology

A graphing utility can be used to obtain a plane curve represented by parametric equations. Set the mode to parametric and enter the equations. You must enter the minimum and maximum values for t, and an increment setting for t ($t$step). The setting $t$step determines the number of points the graphing utility will plot.

Shown below is the plane curve for

$$x = t^2 - 1$$
$$y = 2t$$

in a $[-5, 5, 1]$ by $[-5, 5, 1]$ viewing rectangle with $t$min $= -2$, $t$max $= 2$, and $t$step $= 0.01$.

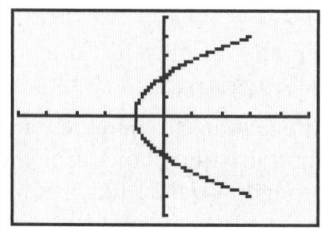

the parameter, we can write one equation in x and y that is equivalent to the two parametric equations. The voice balloons illustrate this process.

Begin with the parametric equations.	Solve for t in one of the equations.	Substitute the expression for t in the other parametric equation.
$x = t^2 - 1$ $y = 2t$	Using $y = 2t$, $t = \dfrac{y}{2}$.	Using $t = \dfrac{y}{2}$ and $x = t^2 - 1$, $x = \left(\dfrac{y}{2}\right)^2 - 1$.

The rectangular equation (the equation in x and y), $x = \dfrac{y^2}{4} - 1$, can be written as $y^2 = 4(x + 1)$. This is the standard form of the equation of a parabola with vertex at $(-1, 0)$ and axis of symmetry along the x-axis. Because the parameter t is restricted to the interval $[-2, 2]$, the plane curve in the technology box on the left shows only a part of the parabola.

Our discussion illustrates a second method for graphing a plane curve described by parametric equations. Eliminate the parameter t and graph the resulting rectangular equation in x and y. However, **you may need to change the domain of the rectangular equation to be consistent with the domain for the parametric equation in x.** This situation is illustrated in Example 2.

EXAMPLE 2 Finding and Graphing the Rectangular Equation of a Curve Defined Parametrically

Sketch the plane curve represented by the parametric equations

$$x = \sqrt{t} \quad \text{and} \quad y = \tfrac{1}{2}t + 1$$

by eliminating the parameter.

Solution   We eliminate the parameter t and then graph the resulting rectangular equation.

Begin with the parametric equations.	Solve for t in one of the equations.	Substitute the expression for t in the other parametric equation.
$x = \sqrt{t}$ $y = \tfrac{1}{2}t + 1$	Using $x = \sqrt{t}$ and squaring both sides, $t = x^2$.	Using $t = x^2$ and $y = \tfrac{1}{2}t + 1$, $y = \tfrac{1}{2}x^2 + 1$.

Because t is not limited to a closed interval, you might be tempted to graph the entire U-shaped parabola whose equation is $y = \tfrac{1}{2}x^2 + 1$. However, take a second look at the parametric equation in x:

$$x = \sqrt{t}.$$

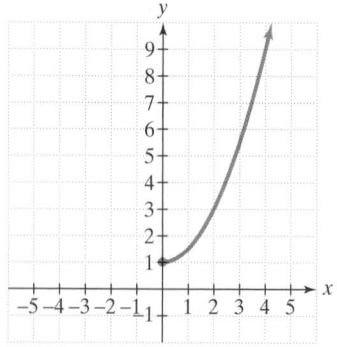

Figure 10.53 The plane curve for $x = \sqrt{t}$ and $y = \tfrac{1}{2}t + 1$, or $y = \tfrac{1}{2}x^2 + 1$, $x \geq 0$

This equation is defined only when $t \geq 0$. Thus, x is nonnegative. The plane curve is the parabola given by $y = \tfrac{1}{2}x^2 + 1$ with the domain restricted to $x \geq 0$. The plane curve is shown in Figure 10.53.

Check Point 2

Sketch the plane curve represented by the parametric equations

$$x = \sqrt{t} \quad \text{and} \quad y = 2t - 1$$

by eliminating the parameter.

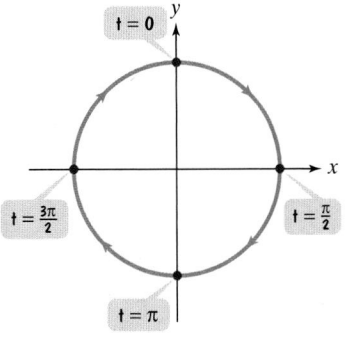

Figure 10.54 The plane curve defined by $x = \sin t, y = \cos t, 0 \le t < 2\pi$

Eliminating the parameter is not always a simple matter. In some cases, it may not be possible. When this occurs, you can use point plotting to obtain a plane curve.

Trigonometric identities can be helpful in eliminating the parameter. For example, consider the plane curve defined by the parametric equations

$$x = \sin t, \quad y = \cos t, \quad 0 \le t < 2\pi.$$

We use the trigonometric identity $\sin^2 t + \cos^2 t = 1$ to eliminate the parameter. Square each side of each parametric equation and then add.

$$x^2 = \sin^2 t$$
$$y^2 = \cos^2 t$$
$$\overline{x^2 + y^2 = \sin^2 t + \cos^2 t}$$

This is the sum of the two equations above the horizontal lines.

Using the Pythagorean identity, we write this equation as $x^2 + y^2 = 1$. The plane curve is a circle with center $(0, 0)$ and radius 1. It is shown in Figure 10.54.

EXAMPLE 3 Finding and Graphing the Rectangular Equation of a Curve Defined Parametrically

Sketch the plane curve represented by the parametric equations

$$x = 5\cos t, \quad y = 2\sin t, \quad 0 \le t \le \pi$$

by eliminating the parameter.

Solution We eliminate the parameter using the identity $\cos^2 t + \sin^2 t = 1$. To apply the identity, divide the parametric equation in x by 5 and the parametric equation in y by 2.

$$\frac{x}{5} = \cos t \quad \text{and} \quad \frac{y}{2} = \sin t$$

Square and add these two equations.

$$\frac{x^2}{25} = \cos^2 t$$
$$\frac{y^2}{4} = \sin^2 t$$
$$\overline{\frac{x^2}{25} + \frac{y^2}{4} = \cos^2 t + \sin^2 t}$$

This is the sum of the two equations above the horizontal lines.

Using the Pythagorean identity, we write this equation as

$$\frac{x^2}{25} + \frac{y^2}{4} = 1.$$

This rectangular equation is the standard form of the equation for an ellipse centered at $(0, 0)$.

$$\frac{x^2}{25} + \frac{y^2}{4} = 1$$

$a^2 = 25$: Endpoints of major axis are 5 units left and right of center.

$b^2 = 4$: Endpoints of minor axis are 2 units above and below center.

The ellipse is shown in Figure 10.55(a). However, this is not the plane curve. Because t is restricted to the interval $[0, \pi]$, the plane curve is only a portion of the ellipse. Use the starting and ending values for t and a value of t in the interval $[0, \pi]$ to find which portion to include.

Begin at $t = 0$.

$x = 5 \cos t = 5 \cos 0 = 5 \cdot 1 = 5$

$y = 2 \sin t = 2 \sin 0 = 2 \cdot 0 = 0$

Increase to $t = \frac{\pi}{2}$

$x = 5 \cos t = 5 \cos \frac{\pi}{2} = 5 \cdot 0 = 0$

$y = 2 \sin t = 2 \sin \frac{\pi}{2} = 2 \cdot 1 = 2$

End at $t = \pi$.

$x = 5 \cos t = 5 \cos \pi = 5(-1) = -5$

$y = 2 \sin t = 2 \sin \pi = 2(0) = 0$

Points on the plane curve include $(5, 0)$, which is the starting point, $(0, 2)$, and $(-5, 0)$, which is the ending point. The plane curve is the top half of the ellipse, shown in Figure 10.55(b).

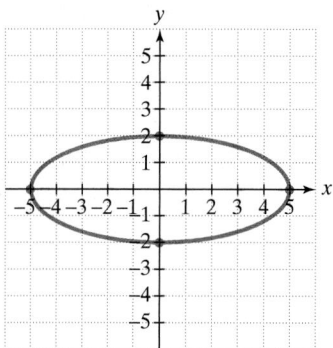

Figure 10.55(a) The graph of
$$\frac{x^2}{25} + \frac{y^2}{4} = 1$$

Figure 10.55(b) The plane curve for $x = 5 \cos t, y = 2 \sin t, 0 \le t \le \pi$

Check Point 3 Sketch the plane curve represented by the parametric equations
$$x = 6 \cos t, \ y = 4 \sin t, \ \pi \le t \le 2\pi$$
by eliminating the parameter.

3 Find parametric equations for functions.

Finding Parametric Equations

Infinitely many pairs of parametric equations can represent the same plane curve. If the plane curve is defined by the function $y = f(x)$, here is a procedure for finding one set of parametric equations:

Parametric Equations for the Function $y = f(x)$

A set of parametric equations for the plane curve defined by $y = f(x)$ is

$$x = t \quad \text{and} \quad y = f(t)$$

in which t is in the domain of f.

EXAMPLE 4 Finding Parametric Equations

Find a set of parametric equations for the parabola whose equation is $y = 9 - x^2$.

Solution Let $x = t$. Parametric equations for $y = f(x)$ are $x = t$ and $y = f(t)$. Thus, parametric equations for $y = 9 - x^2$ are

$$x = t \quad \text{and} \quad y = 9 - t^2.$$

Check Point 4 Find a set of parametric equations for the parabola whose equation is $y = x^2 - 25$.

You can write other sets of parametric equations for $y = 9 - x^2$ by starting with a different parametric equation for x. Here are three sets of parametric equations for

$$y = 9 - x^2:$$

- If $x = t^3$, $y = 9 - (t^3)^2 = 9 - t^6$.

    Parametric equations are $x = t^3$ and $y = 9 - t^6$.

- If $x = t + 1$, $y = 9 - (t + 1)^2 = 9 - (t^2 + 2t + 1) = 8 - t^2 - 2t$.

    Parametric equations are $x = t + 1$ and $y = 8 - t^2 - 2t$.

- If $x = \dfrac{t}{2}$, $y = 9 - \left(\dfrac{t}{2}\right)^2 = 9 - \dfrac{t^2}{4}$.

    Parametric equations are $x = \dfrac{t}{2}$ and $y = 9 - \dfrac{t^2}{4}$.

Can you start with any choice for the parametric equation for x? The answer is no. **The substitution for x must be a function that allows x to take on all the values in the domain of the given rectangular equation.** For example, the domain of the function $y = 9 - x^2$ is the set of all real numbers. If you incorrectly let $x = t^2$, these values of x exclude negative numbers that are included in $y = 9 - x^2$. The parametric equations

$$x = t^2 \quad \text{and} \quad y = 9 - (t^2)^2 = 9 - t^4$$

do not represent $y = 9 - x^2$ because only points for which $x \geq 0$ are obtained.

4 Understand the advantages of parametric representations.

Technology

The ellipse shown was obtained using the parametric mode and the radian mode of a graphing utility.

$$x(t) = 2 + 3 \cos t$$
$$y(t) = 3 + 2 \sin t$$

We used a $[-2, 6, 1]$ by $[-1, 6, 1]$ viewing rectangle with $t\min = 0$, $t\max = 6.2$, and $t\text{step} = 0.1$.

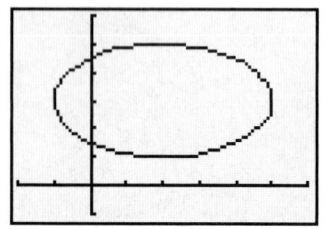

Advantages of Parametric Equations over Rectangular Equations

We opened this section with parametric equations that described the horizontal distance and the vertical height of your thrown baseball after t seconds. Parametric equations are frequently used to represent the path of a moving object. If t represents time, parametric equations give the location of a moving object and tell when the object is located at each of its positions. Rectangular equations tell where the moving object is located but do not reveal when the object is in a particular position.

When using technology to obtain graphs, parametric equations that represent nonfunctions are often easier to use than their corresponding rectangular equations. It is far easier to enter the equation of an ellipse given by the parametric equations

$$x = 2 + 3 \cos t \quad \text{and} \quad y = 3 + 2 \sin t$$

than to use the rectangular equivalent

$$\frac{(x - 2)^2}{9} + \frac{(y - 3)^2}{4} = 1.$$

The rectangular equation must first be solved for y and then entered as two separate equations before a graphing utility reveals the ellipse.

A curve that is used in physics for much of the theory of light is called a **cycloid.** The path of a fixed point on the circumference of a circle as it rolls along a line is a cycloid. A point on the rim of a bicycle wheel traces out a cycloid curve, shown in Figure 10.56. If the radius of the circle is a, the parametric equations of the cycloid are

$$x = a(t - \sin t) \quad \text{and} \quad y = a(1 - \cos t).$$

It is an extremely complicated task to represent the cycloid in rectangular form.

Cycloids are used to solve problems that involve the "shortest time." For example, Figure 10.57 shows a bead sliding down a wire. The shape of the wire a bead could slide down so that the distance between two points is traveled in the shortest time is an inverted cycloid.

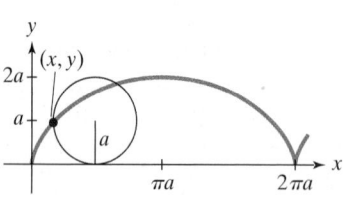

Figure 10.56 The curve traced by a fixed point on the circumference of a circle rolling along a straight line is a cycloid.

Figure 10.57

Rolling

Linear functions and cycloids are used to describe rolling motion. The light at the rolling circle's center shows that it moves linearly. By contrast, the light at the circle's edge has rotational motion and traces out a cycloid. A number of sites on the Internet illustrate rotational motion and show how the cycloid is created.

EXERCISE SET 10.5

Practice Exercises

In Exercises 1–8, parametric equations and a value for the parameter t are given. Find the coordinates of the point on the plane curve described by the parametric equations corresponding to the given value of t.

1. $x = 3 - 5t, y = 4 + 2t; t = 1$
2. $x = 7 - 4t, y = 5 + 6t; t = 1$
3. $x = t^2 + 1, y = 5 - t^3; t = 2$
4. $x = t^2 + 3, y = 6 - t^3; t = 2$
5. $x = 4 + 2 \cos t, y = 3 + 5 \sin t; t = \dfrac{\pi}{2}$
6. $x = 2 + 3 \cos t, y = 4 + 2 \sin t; t = \pi$
7. $x = (60 \cos 30°)t, y = 5 + (60 \sin 30°)t - 16t^2; t = 2$
8. $x = (80 \cos 45°)t, y = 6 + (80 \sin 45°)t - 16t^2; t = 2$

In Exercises 9–20, use point plotting to graph the plane curve described by the given parametric equations. Use arrows to show the orientation of the curve corresponding to increasing values of t.

9. $x = t + 2, y = t^2; -2 \le t \le 2$
10. $x = t - 1, y = t^2; -2 \le t \le 2$
11. $x = t - 2, y = 2t + 1; -2 \le t \le 3$
12. $x = t - 3, y = 2t + 2; -2 \le t \le 3$
13. $x = t + 1, y = \sqrt{t}; t \ge 0$
14. $x = \sqrt{t}, y = t - 1; t \ge 0$
15. $x = \cos t, y = \sin t; 0 \le t < 2\pi$
16. $x = -\sin t, y = -\cos t; 0 \le t < 2\pi$
17. $x = t^2, y = t^3; -\infty < t < \infty$
18. $x = t^2 + 1, y = t^3 - 1; -\infty < t < \infty$

19. $x = 2t, y = |t - 1|; -\infty < t < \infty$
20. $x = |t + 1|, y = t - 2; -\infty < t < \infty$

In Exercises 21–40, eliminate the parameter t. Then use the rectangular equation to sketch the plane curve represented by the given parametric equations. Use arrows to show the orientation of the curve corresponding to increasing values of t. (If an interval for t is not specified, assume that $-\infty < t < \infty$.)

21. $x = t, y = 2t$
22. $x = t, y = -2t$
23. $x = 2t - 4, y = 4t^2$
24. $x = t - 2, y = t^2$
25. $x = \sqrt{t}, y = t - 1$
26. $x = \sqrt{t}, y = t + 1$
27. $x = 2 \sin t, y = 2 \cos t; 0 \le t < 2\pi$
28. $x = 3 \sin t, y = 3 \cos t; 0 \le t < 2\pi$
29. $x = 1 + 3 \cos t, y = 2 + 3 \sin t; 0 \le t < 2\pi$
30. $x = -1 + 2 \cos t, y = 1 + 2 \sin t; 0 \le t < 2\pi$
31. $x = 2 \cos t, y = 3 \sin t; 0 \le t < 2\pi$
32. $x = 3 \cos t, y = 5 \sin t; 0 \le t < 2\pi$
33. $x = 1 + 3 \cos t, y = -1 + 2 \sin t; 0 \le t \le \pi$
34. $x = 2 + 4 \cos t, y = -1 + 3 \sin t; 0 \le t \le \pi$
35. $x = \sec t, y = \tan t$
36. $x = 5 \sec t, y = 3 \tan t$
37. $x = t^2 + 2, y = t^2 - 2$
38. $x = \sqrt{t} + 2, y = \sqrt{t} - 2$
39. $x = 2^t, y = 2^{-t}; t \ge 0$
40. $x = e^t, y = e^{-t}; t \ge 0$

In Exercises 41–43, eliminate the parameter. Write the resulting equation in standard form.

41. A circle: $x = h + r \cos t, y = k + r \sin t$
42. An ellipse: $x = h + a \cos t, y = k + b \sin t$
43. A hyperbola: $x = h + a \sec t, y = k + b \tan t$

44. The parametric equations of the line through (x_1, y_1) and (x_2, y_2) are

$$x = x_1 + t\,(x_2 - x_1) \quad \text{and} \quad y = y_1 + t(y_2 - y_1).$$

Eliminate the parameter and write the resulting equation in point-slope form.

In Exercises 45–52, use your answers from Exercises 41–44 and the parametric equations given in Exercises 41–44 to find a set of parametric equations for the conic section or the line.

45. Circle: Center: $(3, 5)$; Radius: 6

46. Circle: Center: $(4, 6)$; Radius: 9

47. Ellipse: Center: $(-2, 3)$; Vertices: 5 units to the left and right of the center; Endpoints of Minor Axis: 2 units above and below the center

48. Ellipse: Center: $(4, -1)$; Vertices: 5 units above and below the center; Endpoints of Minor Axis: 3 units to the left and right of the center

49. Hyperbola: Vertices: $(4, 0)$ and $(-4, 0)$; Foci: $(6, 0)$ and $(-6, 0)$

50. Hyperbola: Vertices: $(0, 4)$ and $(0, -4)$; Foci: $(0, 5)$ and $(0, -5)$

51. Line: Passes through $(-2, 4)$ and $(1, 7)$

52. Line: Passes through $(3, -1)$ and $(9, 12)$

In Exercises 53–56, find two different sets of parametric equations for each rectangular equation.

53. $y = 4x - 3$ **54.** $y = 2x - 5$

55. $y = x^2 + 4$ **56.** $y = x^2 - 3$

In Exercises 57–58, the parametric equations of four plane curves are given. Graph each plane curve and determine how they differ from each other.

57. a. $x = t$ and $y = t^2 - 4$

b. $x = t^2$ and $y = t^4 - 4$

c. $x = \cos t$ and $y = \cos^2 t - 4$

d. $x = e^t$ and $y = e^{2t} - 4$

58. a. $x = t$, $y = \sqrt{4 - t^2}$; $-2 \le t \le 2$

b. $x = \sqrt{4 - t^2}$, $y = t$; $-2 \le t \le 2$

c. $x = 2 \sin t$, $y = 2 \cos t$; $0 \le t < 2\pi$

d. $x = 2 \cos t$, $y = 2 \sin t$; $0 \le t < 2\pi$

Application Exercises

The path of a projectile that is launched h feet above the ground with an initial velocity of v_0 feet per second and at an angle θ with the horizontal is given by the parametric equations

$$x = (v_0 \cos \theta)t \quad \text{and} \quad y = h + (v_0 \sin \theta)t - 16t^2$$

where t is the time, in seconds, after the projectile was launched. The parametric equation for x gives the projectile's horizontal distance, in feet. The parametric equation for y

gives the projectile's height, in feet. Use these parametric equations to solve Exercises 59–60.

59. The figure shows the path for a baseball hit by Sammy Sosa. The ball was hit with an initial velocity of 180 feet per second at an angle of 40° to the horizontal. The ball was hit at a height 3 feet off the ground.

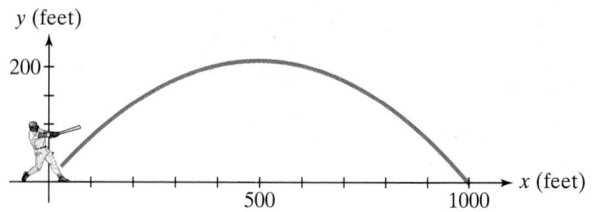

a. Find the parametric equations that describe the position of the ball as a function of time.

b. Describe the ball's position after 1, 2, and 3 seconds. Round to the nearest tenth of a foot. Locate your solutions on the plane curve.

c. How long, to the nearest tenth of a second, is the ball in flight? What is the total horizontal distance that it travels before it lands? Is your answer consistent with the figure shown?

d. You meet Sammy Sosa and he asks you to tell him something interesting about the path of the baseball that he hit. Use the graph to respond to his request. Then verify your observation algebraically.

60. The figure shows the path for a baseball that was hit with an initial velocity of 150 feet per second at an angle of 35° to the horizontal. The ball was hit at a height of 3 feet off the ground.

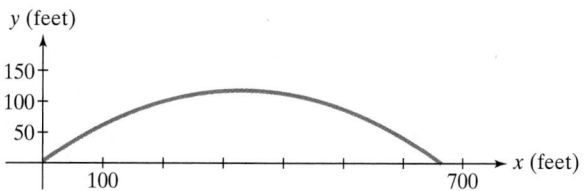

a. Find the parametric equations that describe the position of the ball as a function of time.

b. Describe the ball's position after 1, 2, and 3 seconds. Round to the nearest tenth of a foot. Locate your solutions on the plane curve.

c. How long is the ball in flight? (Round to the nearest tenth of a second.) What is the total horizontal distance that it travels, to the nearest tenth of a foot, before it lands? Is your answer consistent with the figure shown?

d. Use the graph to describe something about the path of the baseball that might be of interest to the player who hit the ball. Then verify your observation algebraically.

Writing in Mathematics

61. What are plane curves and parametric equations?

62. How is point plotting used to graph a plane curve described by parametric equations? Give an example with your description.

63. What is the significance of arrows along a plane curve?

64. What does it mean to eliminate the parameter? What useful information can be obtained by doing this?

65. Explain how the rectangular equation $y = 5x$ can have infinitely many sets of parametric equations.

66. Discuss how the parametric equations for the path of a projectile (see Exercises 59–60) and the ability to obtain plane curves with a graphing utility can be used by a baseball coach to analyze performances of team players.

Technology Exercises

67. Use a graphing utility in a parametric mode to verify any five of your hand-drawn graphs in Exercises 9–40.

In Exercises 68–72, use a graphing utility to obtain the plane curve represented by the given parametric equations.

68. Cycloid: $x = 3(t - \sin t)$, $y = 3(1 - \cos t); [0, 60, 5] \times [0, 8, 1], 0 \le t < 6\pi$

69. Cycloid: $x = 2(t - \sin t)$, $y = 2(1 - \cos t); [0, 60, 5] \times [0, 8, 1], 0 \le t < 6\pi$

70. Witch of Agnesi: $x = 2 \cot t, y = 2 \sin^2 t$; $[-6, 6, 1] \times [-4, 4, 1], 0 \le t < 2\pi$

71. Hypocycloid: $x = 4 \cos^3 t, y = 4 \sin^3 t$; $[-5, 5, 1] \times [-5, 5, 1], 0 \le t < 2\pi$

72. Lissajous Curve: $x = 2 \cos t, y = \sin 2t$; $[-3, 3, 1] \times [-2, 2, 1], 0 \le t < 2\pi$

Use the equations for the path of a projectile given prior to Exercises 59–60 to solve Exercises 73–75.

In Exercises 73–74, use a graphing utility to obtain the path of a projectile launched from the ground $(h = 0)$ at the specified values of θ and v_0. In each exercise, use the graph to determine the maximum height and the time at which the projectile reaches its maximum height. Also use the graph to determine the range of the projectile and the time it hits the ground. Round all answers to the nearest tenth.

73. $\theta = 55°, v_0 = 200$ feet per second

74. $\theta = 35°, v_0 = 300$ feet per second

75. A baseball player throws a ball with an initial velocity of 140 feet per second at an angle of 22° to the horizontal. The ball leaves the player's hand at a height of 5 feet.

   a. Write the parametric equations that describe the ball's position as a function of time.

   b. Use a graphing utility to obtain the path of the baseball.

   c. Find the ball's maximum height and the time at which it reaches this height. Round all answers to the nearest tenth.

   d. How long is the ball in the air?

   e. How far does the ball travel?

Critical Thinking Exercises

76. Eliminate the parameter: $x = \cos^3 t$ and $y = \sin^3 t$.

77. The plane curve described by the parametric equations $x = 3 \cos t$ and $y = 3 \sin t$, $0 \le t < 2\pi$, has a counterclockwise orientation. Alter one or both parametric equations so that you obtain the same plane curve with the opposite orientation.

78. The figure shows a circle of radius a rolling along a horizontal line. Point P traces out a cycloid. Angle t, in radians, is the angle through which the circle has rolled. C is the center of the circle.

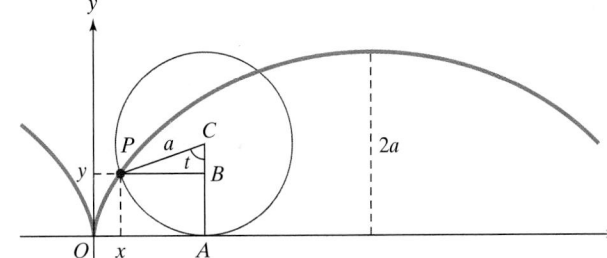

Use the suggestions in parts (a) and (b) to prove that the parametric equations of the cycloid are $x = a(t - \sin t)$ and $y = a(1 - \cos t)$.

   a. Derive the parametric equation for x using the figure and

$$x = OA - xA.$$

   b. Derive the parametric equation for y using the figure and

$$y = AC - BC.$$

SECTION 10.6 *Conic Sections in Polar Coordinates*

Objectives

1. Define conics in terms of a focus and a directrix.
2. Graph the polar equations of conics.

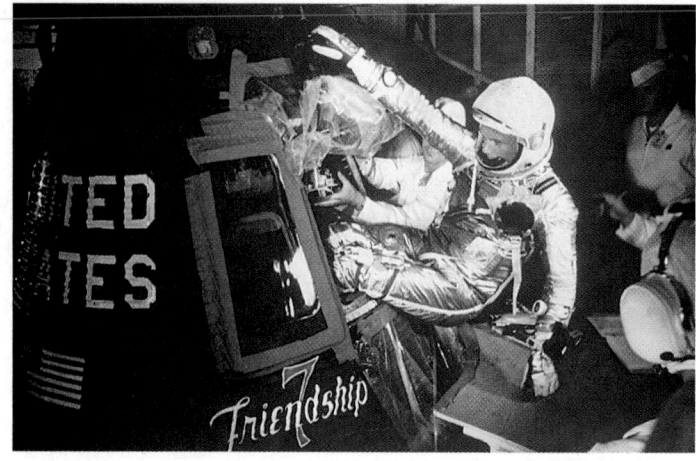

John Glenn made the first U.S.-manned flight around Earth on *Friendship 7.*

On the morning of February 20, 1962, millions of Americans collectively held their breath as the world's newest pioneer swept across the threshold of one of our last frontiers. Roughly one hundred miles above Earth, astronaut John Glenn sat comfortably in the weightless environment of a $9\frac{1}{2}$-by-6-foot space capsule that offered the leg room of a Volkswagen "Beetle" and the aesthetics of a garbage can. Glenn became the first American to orbit Earth in a three-orbit mission that lasted slightly under 5 hours.

In this section, you will see how John Glenn's historic orbit can be described using conic sections in polar coordinates. To obtain this model, we begin with a definition that permits a unified approach to the conic sections.

1 Define conics in terms of a focus and a directrix.

The Focus-Directrix Definitions of the Conic Sections

The definition of a parabola is given in terms of a fixed point, the focus, and a fixed line, the directrix. By contrast, the definitions of an ellipse and a hyperbola are given in terms of two fixed points, the foci. It is possible to define each of these conic sections in terms of a point and a line. Figure 10.58 shows a conic section in the polar coordinate system. The fixed point, the focus, is at the pole. The fixed line, the directrix, is perpendicular to the polar axis.

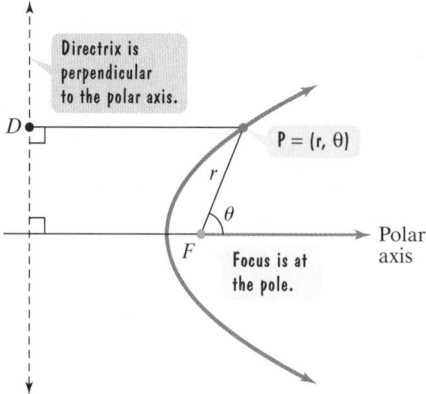

Figure 10.58 A conic in the polar coordinate system

Focus-Directrix Definitions of the Conic Sections

Let F be a fixed point, the focus, and let D be a fixed line, the directrix, in a plane (Figure 10.58). A conic section, or **conic,** is the set of all points P in the plane such that

$$\frac{PF}{PD} = e$$

where e is a fixed positive number, called the **eccentricity.**

   If $e = 1$, the conic is a parabola.

   If $e < 1$, the conic is an ellipse.

   If $e > 1$, the conic is a hyperbola.

Figure 10.59 illustrates the eccentricity for each type of conic. Notice that if $e = 1$, the definition of the parabola is the same as focus-directrix definition with which you are familiar.

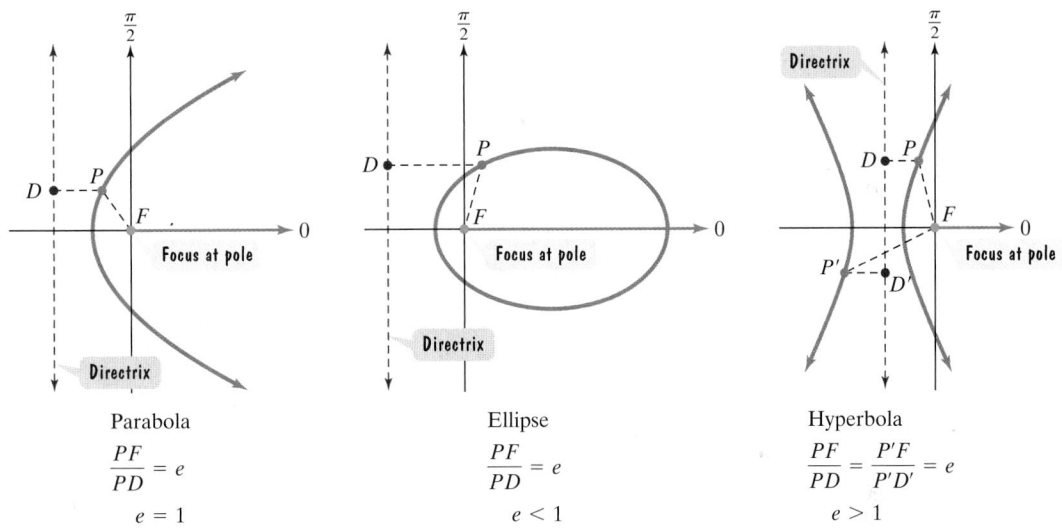

Figure 10.59 The eccentricity for each conic

2 Graph the polar equations of conics.

Polar Equations of Conics

By locating a focus at the pole, all conics can be represented by similar equations in the polar coordinate system. In each of these equations,

- (r, θ) is a point on the graph of the conic.

- e is the eccentricity. (Remember that $e > 0$.)

- $|p|$ is the distance between the focus (located at the pole) and the directrix.

Standard Forms of the Polar Equations of Conics

Let the pole be a focus of a conic section of eccentricity e with the directrix $|p|$ units from the focus. The equation of the conic is given by one of the four equations listed.

$$r = \frac{ep}{1 + e \cos \theta}$$

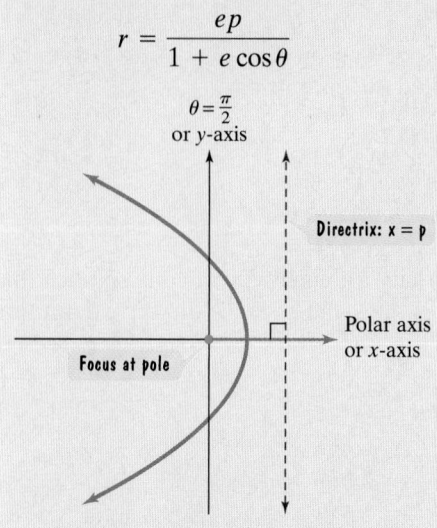

$$r = \frac{ep}{1 - e \cos \theta}$$

$$r = \frac{ep}{1 + e \sin \theta}$$

$$r = \frac{ep}{1 - e \sin \theta}$$

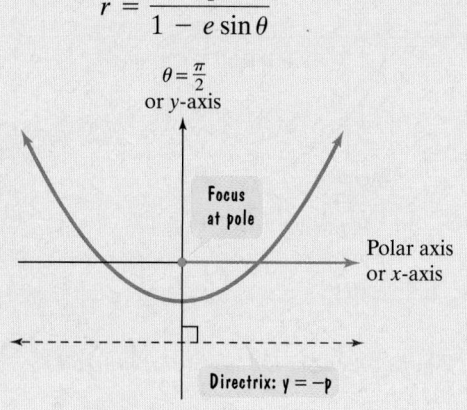

The graphs in the box illustrate two kinds of symmetry—symmetry with respect to the polar axis and symmetry with respect to the y-axis. If the equation contains $\cos \theta$, the polar axis is an axis of symmetry. If the equation contains $\sin \theta$, the line $\theta = \frac{\pi}{2}$, or the y-axis, is an axis of symmetry.

We will derive the equation displayed in the box on the upper right. The other equations are obtained in a similar manner. In Figure 10.60 on page 911, let $P = (r, \theta)$ be any point on a conic section.

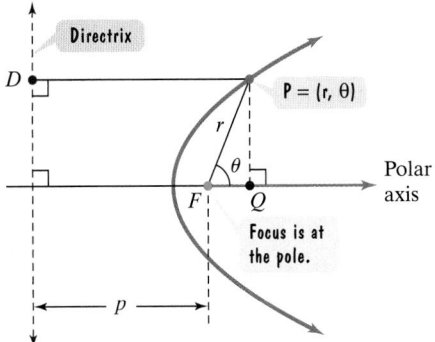

Figure 10.60

$\dfrac{PF}{PD} = e$ By definition, the ratio of the distance from P to the focus to the distance from P to the directrix equals the positive constant e.

$\dfrac{r}{PD} = e$ Figure 10.60 shows that the distance from P to the focus, located at the pole, is r: PF = r.

$\dfrac{r}{p + FQ} = e$ Figure 10.60 shows that the distance from P to the directrix is p + FQ: PD = p + FQ.

$\dfrac{r}{p + r \cos \theta} = e$ Using the triangle in the figure, $\cos \theta = \dfrac{FQ}{r}$ and FQ = r cos θ.

By solving this equation for r, we will obtain the desired equation. Clear fractions by multiplying both sides by the least common denominator.

$r = e(p + r \cos \theta)$ Multiply both sides by p + r cos θ.

$r = ep + er \cos \theta$ Apply the distributive property.

$r - er \cos \theta = ep$ Subtract er cos θ from both sides to collect terms involving r on the same side.

$r(1 - e \cos \theta) = ep$ Factor out r from the two terms on the left.

$r = \dfrac{ep}{1 - e \cos \theta}$ Divide both sides by 1 − e cos θ and solve for r.

In summary, the standard forms of the polar equations of conics are

$$r = \frac{ep}{1 \pm e \cos \theta} \quad \text{and} \quad r = \frac{ep}{1 \pm e \sin \theta}.$$

In all forms, the constant term in the denominator is 1.

Graphing the Polar Equation of a Conic

1. If necessary, write the equation in one of the standard forms.
2. Use the standard form to determine values for e and p. Use the value of e to identify the conic.
3. Use the appropriate figure for the standard form of the equation shown in the box on page 910 to help guide the graphing process.

EXAMPLE 1   Graphing the Polar Equation of a Conic

Graph the polar equation:

$$r = \frac{4}{2 + \cos \theta}.$$

Solution

Step 1 Write the equation in one of the standard forms. The equation is not in standard form because the constant term in the denominator is not 1.

$$r = \frac{4}{2 + \cos\theta}$$

> To obtain 1 in this position, divide the numerator and denominator by 2.

The equation in standard form is

$$r = \frac{2}{1 + \frac{1}{2}\cos\theta}.$$

> $ep = 2$
> $e = \frac{1}{2}$
> This equation is in the form $r = \dfrac{ep}{1 + e\cos\theta}$.

Step 2 Use the standard form to find e and p, and identify the conic. The voice balloons show that

$$e = \tfrac{1}{2} \quad \text{and} \quad ep = \tfrac{1}{2}p = 2.$$

Thus, $e = \frac{1}{2}$ and $p = 4$. Because $e = \frac{1}{2} < 1$, the conic is an ellipse.

Step 3 Use the figure for the equations standard form to guide the graphing process. The figure for the conic's standard form is shown in Figure 10.61(a). We have symmetry with respect to the polar axis. One focus is at the pole and a directrix is $x = 4$, located four units to the right of the pole.

(a) Using $r = \dfrac{ep}{1 + e\cos\theta}$ to graph $r = \dfrac{2}{1 + \frac{1}{2}\cos\theta}$

(b) The graph of $r = \dfrac{4}{2 + \cos\theta}$, or $r = \dfrac{2}{1 + \frac{1}{2}\cos\theta}$

Figure 10.61

Figure 10.61(a) indicates that the major axis is on the polar axis. Thus, we find the vertices by selecting 0 and π for θ. The corresponding values for r are $\frac{4}{3}$ and 4, respectively. Figure 10.61(b) shows the vertices, $\left(\frac{4}{3}, 0\right)$ and $(4, \pi)$.

Technology

The graph of

$$r = \frac{4}{2 + \cos\theta}$$

is obtained using the polar mode with angle measure in radians. To verify the hand-drawn graph in Figure 10.61(b), we used $[-5, 5, 1]$ by $[-5, 5, 1]$ with θmin $= 0$, θmax $= 2\pi$, and θstep $= \dfrac{\pi}{48}$.

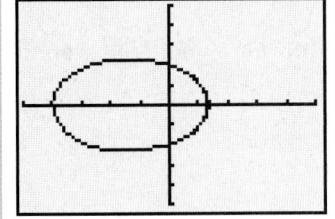

You can sketch the upper half of the ellipse by plotting some points from $\theta = 0$ to $\theta = \pi$.

$$r = \frac{4}{2 + \cos\theta}$$

θ	$\dfrac{\pi}{2}$	$\dfrac{2\pi}{3}$	$\dfrac{3\pi}{4}$	$\dfrac{5\pi}{6}$
r	2	2.7	3.1	3.5

Using symmetry with respect to the polar axis, you can sketch the lower half. The graph of the given equation is shown in Figure 10.61(b).

Check Point 1 Use the three steps shown in the box on page 911 to graph the polar equation:

$$r = \frac{4}{2 - \cos\theta}.$$

EXAMPLE 2 Graphing the Polar Equation of a Conic

Graph the polar equation:

$$r = \frac{12}{3 + 3\sin\theta}.$$

Solution

Step 1   Write the equation in one of the standard forms. The equation is not in standard form because the constant term in the denominator is not 1. Divide the numerator and denominator by 3 to write the standard form.

$$r = \frac{4}{1 + 1\sin\theta}$$

ep = 4

e = 1

This equation is in the form $r = \dfrac{ep}{1 + e\sin\theta}$.

Step 2   Use the standard form to find e and p, and identify the conic. The voice balloons show that

$$e = 1 \quad \text{and} \quad ep = 1p = 4.$$

Thus, $e = 1$ and $p = 4$. Because $e = 1$, the conic is a parabola.

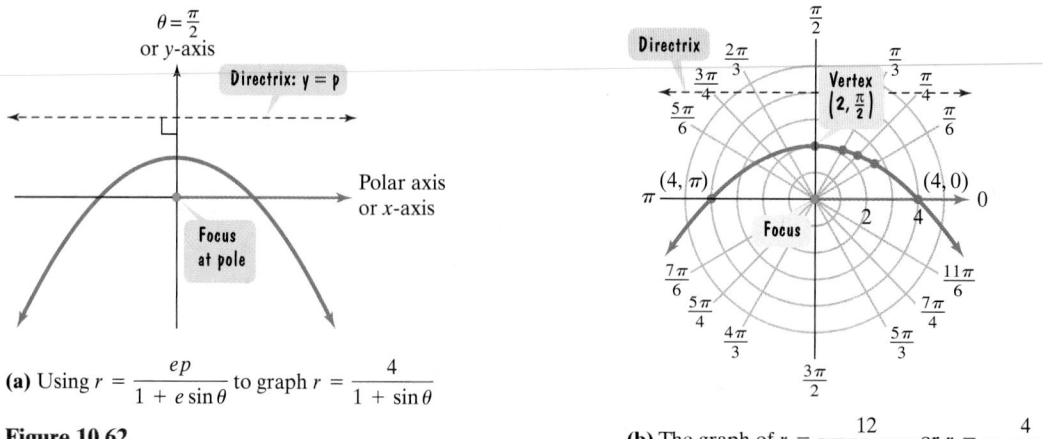

$\theta = \frac{\pi}{2}$
or y-axis
Directrix: y = p
Polar axis
or x-axis
Focus
at pole

(a) Using $r = \dfrac{ep}{1 + e \sin\theta}$ to graph $r = \dfrac{4}{1 + \sin\theta}$

Figure 10.62

Directrix
Vertex
$\left(2, \frac{\pi}{2}\right)$
$(4, \pi)$
$(4, 0)$
Focus

(b) The graph of $r = \dfrac{12}{3 + 3\sin\theta}$ or $r = \dfrac{4}{1 + \sin\theta}$

Technology

The graph of

$$r = \frac{12}{3 + 3\sin\theta}$$

was obtained using

$[-5, 5, 1]$ by $[-5, 5, 1]$

with

θ min $= 0$, θ max $= 2\pi$,

θ step $= \dfrac{\pi}{48}$.

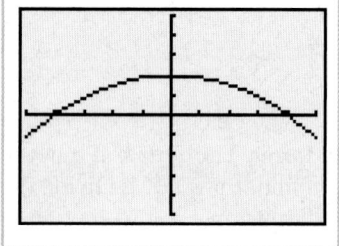

Step 3 Use the figure for the equations standard form to guide the graphing process. Figure 10.62(a) indicates that we have symmetry with respect to $\theta = \dfrac{\pi}{2}$. The focus is at the pole and, with $p = 4$, the directrix is $y = 4$, located four units above the pole.

Figure 10.62(a) indicates that the vertex is on the line of $\theta = \dfrac{\pi}{2}$, or the y-axis. Thus, we find the vertex by selecting $\dfrac{\pi}{2}$ for θ. The corresponding value for r is 2. Figure 10.62(b) shows the vertex, $\left(2, \dfrac{\pi}{2}\right)$.

To find where the parabola crosses the polar axis, select $\theta = 0$ and $\theta = \pi$. The corresponding values for r are 4 and 4, respectively. Figure 10.62(b) shows the points $(4, 0)$ and $(4, \pi)$ on the polar axis.

You can sketch the right half of the parabola by plotting some points from $\theta = 0$ to $\theta = \dfrac{\pi}{2}$.

$$r = \frac{12}{3 + 3\sin\theta}$$

θ	$\dfrac{\pi}{6}$	$\dfrac{\pi}{4}$	$\dfrac{\pi}{3}$
r	2.7	2.3	2.1

Using symmetry with respect to $\theta = \dfrac{\pi}{2}$, you can sketch the left half. The graph of the given equation is shown in Figure 10.62(b).

Check Point 2 Use the three steps shown in the box on page 911 to graph the polar equation:

$$r = \frac{8}{4 + 4\sin\theta}.$$

EXAMPLE 3   Graphing the Polar Equation of a Conic

Graph the polar equation:

$$r = \frac{9}{3 - 6\cos\theta}.$$

Solution

Step 1 Write the equation in one of the standard forms. We can obtain a constant term of 1 in the denominator by dividing each term by 3.

ep = 3

$$r = \frac{3}{1 - 2\cos\theta}$$

This equation is in the form $r = \dfrac{ep}{1 - e\cos\theta}$.

e = 2

Step 2 Use the standard form to find e and p, and identify the conic. The voice balloons show that

$$e = 2 \quad \text{and} \quad ep = 2p = 3.$$

Thus, $e = 2$ and $p = \frac{3}{2}$. Because $e = 2 > 1$, the conic is a hyperbola.

Step 3 Use the figure for the equations standard form to guide the graphing process. Figure 10.63(a) indicates that we have symmetry with respect to the polar axis. One focus is at the pole and, with $p = \frac{3}{2}$, a directrix is $x = -\frac{3}{2}$, located 1.5 units to the left of the pole.

Figure 10.63(a) indicates that the transverse axis is horizontal and the vertices lie on the polar axis. Thus, we find the vertices by selecting 0 and π for θ. Figure 10.63(b) shows the vertices, $(-3, 0)$ and $(1, \pi)$.

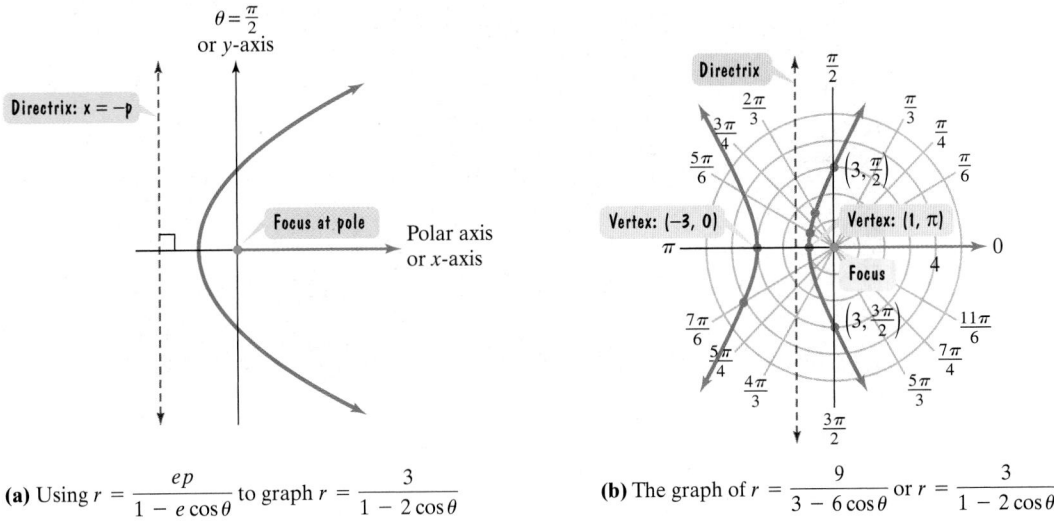

(a) Using $r = \dfrac{ep}{1 - e\cos\theta}$ to graph $r = \dfrac{3}{1 - 2\cos\theta}$

(b) The graph of $r = \dfrac{9}{3 - 6\cos\theta}$ or $r = \dfrac{3}{1 - 2\cos\theta}$

Figure 10.63

To find where the hyperbola crosses the line $\theta = \dfrac{\pi}{2}$, select $\dfrac{\pi}{2}$ and $\dfrac{3\pi}{2}$ for θ. Figure 10.63(b) shows the points $\left(3, \dfrac{\pi}{2}\right)$ and $\left(3, \dfrac{3\pi}{2}\right)$ on the graph.

We sketch the upper half of the hyperbola by plotting some points from $\theta = 0$ to $\theta = \pi$.

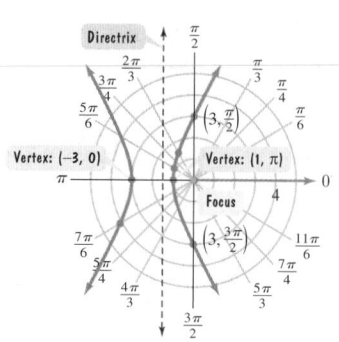

Figure 10.63 (b) The graph of
$r = \dfrac{3}{1 - 2\cos\theta}$, repeated

$$r = \frac{3}{1 - 2\cos\theta}$$

θ	$\dfrac{\pi}{6}$	$\dfrac{2\pi}{3}$	$\dfrac{5\pi}{6}$
r	-4.1	1.5	1.1

Using symmetry with respect to the polar axis, we sketch the lower half. The graph of the given equation is shown in Figure 10.63(b).

Check Point 3 Use the three steps shown in the box on page 911 to graph the polar equation:

$$r = \frac{9}{3 - 9\cos\theta}.$$

Modeling Planetary Motion

Polish astronomer Nicolaus Copernicus (1473–1543) was correct in stating that planets in our solar system revolve around the sun and not Earth. However, he incorrectly believed that celestial orbits move in perfect circles, calling his system "the ballet of the planets."

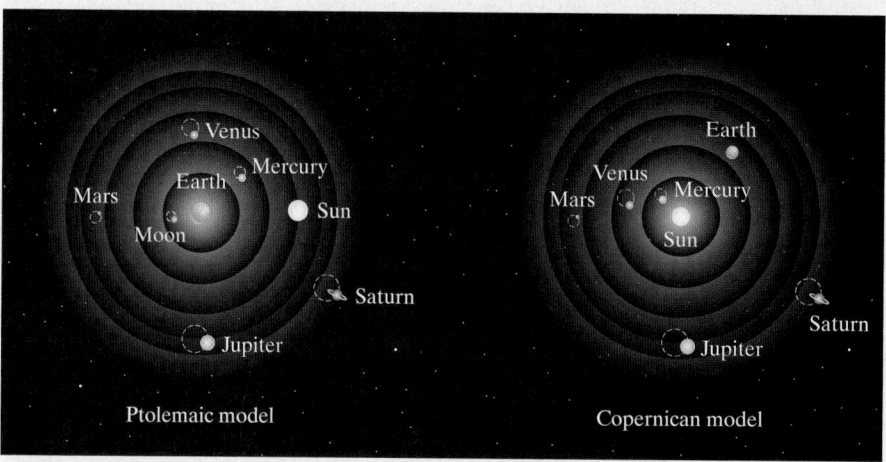

Ptolemaic model   Copernican model

Table 10.4 indicates that the planets in our solar system have orbits with eccentricities that are much closer to 0 than to 1. Most of these orbits are almost circular, which made it difficult for early astronomers to detect that they are actually ellipses.

German scientist and mathematician Johannes Kepler (1571–1630) discovered that planets move in elliptical orbits with the sun at one focus. The polar equation for these orbits is

$$r = \frac{(1 - e^2)a}{1 - e\cos\theta},$$

Table 10.4 Eccentricities of Planetary Orbits

Mercury	0.2056	Saturn	0.0543
Venus	0.0068	Uranus	0.0460
Earth	0.0167	Neptune	0.0082
Mars	0.0934	Pluto	0.2481
Jupiter	0.0484		

where the length of the orbits major axis is $2a$. Describing planetary orbits, Kepler wrote, "The heavenly motions are nothing but a continuous song for several voices, to be perceived by the intellect, not by the ear."

EXERCISE SET 10.6

Practice Exercises

In Exercises 1–8,

a. *Identify the conic section that each polar equation represents.*

b. *Describe the location of a directrix from the focus located at the pole.*

1. $r = \dfrac{3}{1 + \sin \theta}$ **2.** $r = \dfrac{3}{1 + \cos \theta}$

3. $r = \dfrac{6}{3 - 2 \cos \theta}$ **4.** $r = \dfrac{6}{3 + 2 \cos \theta}$

5. $r = \dfrac{8}{2 + 2 \sin \theta}$ **6.** $r = \dfrac{8}{2 - 2 \sin \theta}$

7. $r = \dfrac{12}{2 - 4 \cos \theta}$ **8.** $r = \dfrac{12}{2 + 4 \cos \theta}$

In Exercises 9–20, use the three steps shown in the box on page 911 to graph each polar equation.

9. $r = \dfrac{1}{1 + \sin \theta}$ **10.** $r = \dfrac{1}{1 + \cos \theta}$

11. $r = \dfrac{2}{1 - \cos \theta}$ **12.** $r = \dfrac{2}{1 - \sin \theta}$

13. $r = \dfrac{12}{5 + 3 \cos \theta}$ **14.** $r = \dfrac{12}{5 - 3 \cos \theta}$

15. $r = \dfrac{6}{2 - 2 \sin \theta}$ **16.** $r = \dfrac{6}{2 + 2 \sin \theta}$

17. $r = \dfrac{8}{2 - 4 \cos \theta}$ **18.** $r = \dfrac{8}{2 + 4 \cos \theta}$

19. $r = \dfrac{12}{3 - 6 \cos \theta}$ **20.** $r = \dfrac{12}{3 - 3 \cos \theta}$

Application Exercises

Halley's comet has an elliptical orbit with the sun at one focus. Its orbit, shown in the figure at the top of the next column, is given approximately by

$$r = \frac{1.069}{1 + 0.967 \sin \theta}.$$

In the formula, r is measured in astronomical units. (One astronomical unit is the average distance from Earth to the sun, approximately 93 million miles.) Use the given formula and the figure to solve Exercises 21–22. Round to the nearest hundredth of an astronomical unit and the nearest million miles.

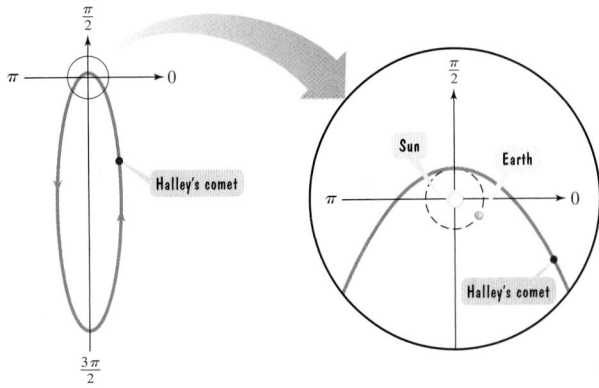

21. Find the distance from Halley's comet to the sun at its shortest distance from the sun.

22. Find the distance from Halley's comet to the sun at its greatest distance from the sun.

On February 20, 1962, John Glenn made the first U.S.-manned flight around the Earth for three orbits on Friendship 7. *With Earth at one focus, the orbit of* Friendship 7 *is given approximately by*

$$r = \frac{4090.76}{1 - 0.0076 \cos \theta}$$

where r is measured in miles from Earth's center. Use the formula and the figure shown to solve Exercises 23–24.

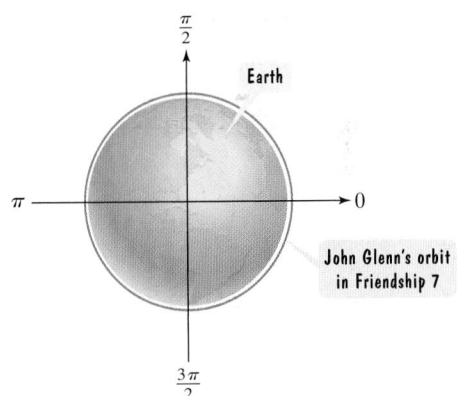

23. How far from Earth's center was John Glenn at his greatest distance from the planet? Round to the nearest mile. If the radius of Earth is 3960 miles, how far was he from Earth's surface at this point on the flight?

24. How far from Earth's center was John Glenn at his closest distance from the planet? Round to the nearest mile. If the radius of Earth is 3960 miles, how far was he from Earth's surface at this point on the flight?

Writing in Mathematics

25. How are the conics described in terms of a fixed point and a fixed line?

26. If all conics are defined in terms of a fixed point and a fixed line, how can you tell one kind of conic from another?

27. If you are given the standard form of the polar equation of a conic, how do you determine its eccentricity?

28. If you are given the standard form of the polar equation of a conic, how do you determine the location of a directrix from the focus at the pole?

29. Describe a strategy for graphing $r = \dfrac{1}{1 + \sin\theta}$.

30. You meet John Glenn and he asks you to tell him something of interest about the difference between the elliptical orbit of his first space voyage in 1962 and the orbit for his return journey in 1999. Describe how to use the polar equation for orbits in the essay on page 916, the equation for his 1962 journey, and a graphing utility to provide him with an interesting visual analysis.

Technology Exercises

Use the polar mode of a graphing utility with angle measure in radians to solve Exercises 31–34. Unless otherwise indicated, use θmin $= 0$, θmax $= 2\pi$, and θ step $= \dfrac{\pi}{48}$. If you are not satisfied with the quality of the graph, experiment with smaller values for θ step.

31. Use a graphing utility to verify any five of your hand-drawn graphs in Exercises 9–20.

In Exercises 32–34, identify the conic that each polar equation represents. Then use a graphing utility to graph the equation.

32. $r = \dfrac{16}{4 - 3\cos\theta}$

33. $r = \dfrac{12}{4 + 5\sin\theta}$

34. $r = \dfrac{18}{6 - 6\cos\theta}$

In Exercises 35–36, use a graphing utility to graph the equation. Then answer the given question.

35. $r = \dfrac{4}{1 - \sin\left(\theta - \dfrac{\pi}{4}\right)}$; How does the graph differ from

the graph of $r = \dfrac{4}{1 - \sin\theta}$?

36. $r = \dfrac{3}{2 + 6\cos\left(\theta + \dfrac{\pi}{3}\right)}$; How does the graph differ from

the graph of $r = \dfrac{3}{2 + 6\cos\theta}$?

37. Use the polar equation for planetary orbits,

$$r = \frac{(1 - e^2)a}{1 - e\cos\theta},$$

to find the polar equation of the orbit for Mercury and Earth.

Mercury: $e = 0.2056$ and $a = 36.0 \times 10^6$ miles
Earth: $e = 0.0167$ and $a = 92.96 \times 10^6$ miles

Use a graphing utility to graph both orbits in the same viewing rectangle. What do you see about the orbits from their graphs that is not obvious from their equations?

Critical Thinking Exercises

38. Identify the conic and graph the equation:

$$r = \frac{4\sec\theta}{2\sec\theta - 1}.$$

In Exercises 39–40, write a polar equation of the conic that is named and described.

39. Ellipse: a focus at the pole; vertex: $(4, 0)$; $e = \dfrac{1}{2}$

40. Hyperbola: a focus at the pole; directrix: $x = -1$; $e = \dfrac{3}{2}$

41. Identify the conic and write its equation in rectangular coordinates: $r = \dfrac{1}{2 - 2\cos\theta}$.

42. Prove that the polar equation of a planet's elliptical orbit is

$$r = \frac{(1 - e^2)a}{1 - e\cos\theta}$$

where e is the eccentricity and $2a$ is the length of the major axis.

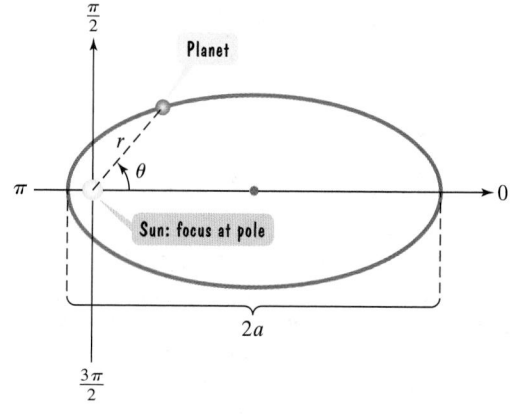

CHAPTER SUMMARY, REVIEW, AND TEST

Summary

DEFINITIONS AND CONCEPTS   **EXAMPLES**

10.1 The Ellipse

a. An ellipse is the set of all points in a plane the sum of whose distances from two fixed points, the foci, is constant.

b. Standard forms of the equations of an ellipse with center at the origin are $\dfrac{x^2}{a^2} + \dfrac{y^2}{b^2} = 1$ Ex. 1, p. 845; Ex. 2, p. 846; Ex. 3, p. 847

     [foci: $(-c, 0), (c, 0)$] and $\dfrac{x^2}{b^2} + \dfrac{y^2}{a^2} = 1$ [foci: $(0, -c), (0, c)$], where

     $c^2 = a^2 - b^2$ and $a^2 > b^2$. See the box on page 844 and Figure 10.6.

c. Standard forms of the equations of an ellipse centered at (h, k) are   Ex. 4, p. 849

     $\dfrac{(x - h)^2}{a^2} + \dfrac{(y - k)^2}{b^2} = 1$ and $\dfrac{(x - h)^2}{b^2} + \dfrac{(y - k)^2}{a^2} = 1$. See Table 10.1 on page 848 and

     Figure 10.10.

10.2 The Hyperbola

a. A hyperbola is the set of all points in a plane the difference of whose distances from two fixed points, the foci, is constant.

b. Standard forms of the equations of a hyperbola with center at the origin are $\dfrac{x^2}{a^2} - \dfrac{y^2}{b^2} = 1$ Ex. 1, p. 858; Ex. 2, p. 859

     [foci: $(-c, 0), (c, 0)$] and $\dfrac{y^2}{a^2} - \dfrac{x^2}{b^2} = 1$ [foci: $(0, -c), (0, c)$], where $c^2 = a^2 + b^2$. See the box on

     page 857 and Figure 10.17.

c. Asymptotes for $\dfrac{x^2}{a^2} - \dfrac{y^2}{b^2} = 1$ are $y = \pm \dfrac{b}{a} x$. Asymptotes for $\dfrac{y^2}{a^2} - \dfrac{x^2}{b^2} = 1$ are $y = \pm \dfrac{a}{b} x$.   Ex. 3, p. 861; Ex. 4, p. 861

d. A procedure for graphing hyperbolas is given in the box on page 861.

e. Standard forms of the equations of a hyperbola centered at (h, k) are   Ex. 5, p. 864

     $\dfrac{(x - h)^2}{a^2} - \dfrac{(y - k)^2}{b^2} = 1$ and $\dfrac{(y - k)^2}{a^2} - \dfrac{(x - h)^2}{b^2} = 1$. See Table 10.2 on page 864 and Figure 10.26.

10.3 The Parabola

a. A parabola is the set of all points in a plane that are equidistant from a fixed line, the directrix, and a fixed point, the focus.

b. Standard forms of the equations of parabolas with vertex at the origin are $y^2 = 4px$ [focus: $(p, 0)$]   Ex. 1, p. 874; and $x^2 = 4py$ [focus: $(0, p)$]. See the box on page 873 and Figure 10.32.   Ex. 3, p. 876

c. A parabola's latus rectum is a line segment that passes through its focus, is parallel to its directrix,   Ex. 2, p. 875
and has its endpoints on the parabola. The length of the latus rectum for $y^2 = 4px$ and $x^2 = 4py$ is $|4p|$. A parabola can be graphed using the vertex and endpoints of the latus rectum.

d. Standard forms of the equations of a parabola with vertex at (h, k) are $(y - k)^2 = 4p(x - h)$ and   Ex. 4, p. 877; $(x - h)^2 = 4p(y - k)$. See Table 10.3 on page 876 and Figure 10.37.   Ex. 5, p. 878

| **DEFINITIONS AND CONCEPTS** | **EXAMPLES** |

10.4 Rotation of Axes

a. A nondegenerate conic section of the form $Ax^2 + Cy^2 + Dx + Ey + F = 0$ in which A and C are not both zero is **1.** a circle if $A = C$; **2.** a parabola if $AC = 0$; **3.** an ellipse if $A \neq C$ and $AC > 0$; **4.** a hyperbola if $AC < 0$.

Ex. 1, p. 885

b. Rotation of Axes Formulas

θ is the angle from the positive x-axis to the positive x'-axis.

$$x = x' \cos \theta - y' \sin \theta \text{ and } y = x' \sin \theta + y' \cos \theta$$

Ex. 2, p. 888

c. Amount of Rotation Formula

The general second-degree equation

$$Ax^2 + Bxy + Cy^2 + Dx + Ey + F = 0$$

can be rewritten in x' and y' without an $x'y'$-term by rotating the axes through angle θ, where $\cot 2\theta = \dfrac{A - C}{B}$ and θ is an acute angle.

d. If 2θ in $\cot 2\theta$ is one of the familiar angles such as $30°, 45°$, or $60°$, write the equation of a rotated conic in standard form using the five-step procedure in the box on page 890.

Ex. 3, p. 890

e. If $\cot 2\theta$ is not the cotangent of one of the more familiar angles, use a sketch of $\cot 2\theta$ to find $\cos 2\theta$. Then use

$$\sin \theta = \sqrt{\frac{1 - \cos 2\theta}{2}} \quad \text{and} \quad \cos \theta = \sqrt{\frac{1 + \cos 2\theta}{2}}$$

to find values for $\sin \theta$ and $\cos \theta$ in the rotation formulas.

Ex. 4, p. 893

f. A nondegenerate conic section of the form

$$Ax^2 + Bxy + Cy^2 + Dx + Ey + F = 0$$

is **1.** a parabola if $B^2 - 4AC = 0$; **2.** an ellipse or a circle if $B^2 - 4AC < 0$; **3.** a hyperbola if $B^2 - 4AC > 0$.

Ex. 5, p. 895

10.5 Parametric Equations

a. The relationship between the parametric equations $x = f(t)$ and $y = g(t)$ and plane curves is described in the box on page 898.

b. Point plotting can be used to graph a plane curve described by parametric equations. See the box on page 899.

Ex. 1, p. 899

c. Plane curves can be sketched by eliminating the parameter t and graphing the resulting rectangular equation. It is sometimes necessary to change the domain of the rectangular equation to be consistent with the domain for the parametric equation in x.

Ex. 2, p. 900;
Ex. 3, p. 901

d. Infinitely many pairs of parametric equations can represent the same plane curve. One pair for $y = f(x)$ is $x = t$ and $y = f(t)$ in which t is in the domain of f.

Ex. 4, p. 903

10.6 Conic Sections in Polar Coordinates

a. The focus-directrix definitions of the conic sections are given in the box on page 909. For all points on a conic, the ratio of the distance from a fixed point (focus) and the distance from a fixed line (directrix) is constant and is called eccentricity. If $e = 1$, the conic is a parabola. If $e < 1$, the conic is an ellipse. If $e > 1$, the conic is a hyperbola.

b. Standard forms of the polar equations of conics are

$$r = \frac{ep}{1 \pm e \cos \theta} \quad \text{and} \quad r = \frac{ep}{1 \pm e \sin \theta}$$

Ex. 1, p. 911;
Ex. 2, p. 913;
Ex. 3, p. 914

in which (r, θ) is a point on the conic's graph, e is the eccentricity, and $|p|$ is the distance between the focus (located at the pole) and the directrix. Details are shown in the box on page 910.

c. A procedure for graphing the polar equation of a conic is given in the box on page 911.

Review Exercises

10.1

In Exercises 1–8, graph each ellipse and locate the foci.

1. $\dfrac{x^2}{36} + \dfrac{y^2}{25} = 1$

2. $\dfrac{y^2}{25} + \dfrac{x^2}{16} = 1$

3. $4x^2 + y^2 = 16$

4. $4x^2 + 9y^2 = 36$

5. $\dfrac{(x-1)^2}{16} + \dfrac{(y+2)^2}{9} = 1$

6. $\dfrac{(x+1)^2}{9} + \dfrac{(y-2)^2}{16} = 1$

7. $4x^2 + 9y^2 + 24x - 36y + 36 = 0$

8. $9x^2 + 4y^2 - 18x + 8y - 23 = 0$

In Exercises 9–11, find the standard form of the equation of each ellipse satisfying the given conditions.

9. Foci: $(-4, 0), (4, 0)$; Vertices: $(-5, 0), (5, 0)$

10. Foci: $(0, -3), (0, 3)$; Vertices: $(0, -6), (0, 6)$

11. Major axis horizontal with length 12; length of minor axis = 4; center: $(-3, 5)$.

12. A semielliptical arch supports a bridge that spans a river 20 yards wide. The center of the arch is 6 yards above the river's center. Write an equation for the ellipse so that the x-axis coincides with the water level and the y-axis passes through the center of the arch.

13. A semielliptic archway has a height of 15 feet at the center and a width of 50 feet, as shown in the figure. The 50-foot width consists of a two-lane road. Can a truck that is 12 feet high and 14 feet wide drive under the archway without going into the other lane?

14. An elliptical pool table has a ball placed at each focus. If one ball is hit toward the side of the table, explain what will occur.

10.2

In Exercises 15–22, graph each hyperbola. Locate the foci and find the equations of the asymptotes.

15. $\dfrac{x^2}{16} - y^2 = 1$

16. $\dfrac{y^2}{16} - x^2 = 1$

17. $9x^2 - 16y^2 = 144$

18. $4y^2 - x^2 = 16$

19. $\dfrac{(x-2)^2}{25} - \dfrac{(y+3)^2}{16} = 1$

20. $\dfrac{(y+2)^2}{25} - \dfrac{(x-3)^2}{16} = 1$

21. $y^2 - 4y - 4x^2 + 8x - 4 = 0$

22. $x^2 - y^2 - 2x - 2y - 1 = 0$

In Exercises 23–24, find the standard form of the equation of each hyperbola satisfying the given conditions.

23. Foci: $(0, -4), (0, 4)$; Vertices: $(0, -2), (0, 2)$

24. Foci: $(-8, 0), (8, 0)$; Vertices: $(-3, 0), (3, 0)$

25. Explain why it is not possible for a hyperbola to have foci at $(0, -2)$ and $(0, 2)$ and vertices at $(0, -3)$ and $(0, 3)$.

26. Radio tower M_2 is located 200 miles due west of radio tower M_1. The situation is illustrated in the figure shown, where a coordinate system has been superimposed. Simultaneous radio signals are sent from each tower to a ship, with the signal from M_2 received 500 microseconds before the signal from M_1. Assuming that radio signals travel at 0.186 mile per microsecond, determine the equation of the hyperbola on which the ship is located.

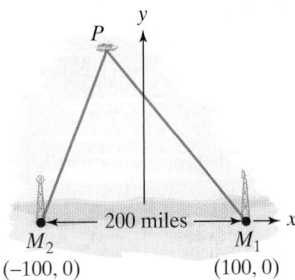

10.3

In Exercises 27–33, find the vertex, focus, and directrix of each parabola with the given equation. Then graph the parabola.

27. $y^2 = 8x$

28. $x^2 + 16y = 0$

29. $(y - 2)^2 = -16x$

30. $(x - 4)^2 = 4(y + 1)$

31. $x^2 + 4y = 4$

32. $y^2 - 4x - 10y + 21 = 0$

33. $x^2 - 4x - 2y = 0$

In Exercises 34–35, find the standard form of the equation of each parabola satisfying the given conditions.

34. Focus: $(12, 0)$; Directrix: $x = -12$

35. Focus: $(0, -11)$; Directrix: $y = 11$

36. An engineer is designing headlight units for automobiles. The unit has a parabolic surface with a diameter of 12 inches and a depth of 3 inches. The situation is illustrated in the figure, where a coordinate system has been superimposed. What is the equation of the parabola in this system? Where should the light source be placed? Describe this placement relative to the vertex.

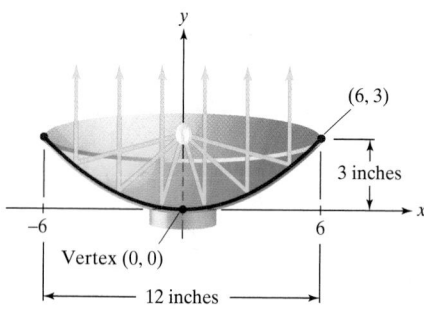

37. The George Washington Bridge spans the Hudson River from New York to New Jersey. Its two towers are 3500 feet apart and rise 316 feet above the road. The cable between the towers has the shape of a parabola, and the cable just touches the sides of the road midway between the towers. What is the height of the cable 1000 feet from a tower?

38. The giant satellite dish in the figure shown is in the shape of a parabolic surface. Signals strike the surface and are reflected to the focus, where the receiver is located. The diameter of the dish is 300 feet and its depth is 44 feet. How far, to the nearest foot, from the base of the dish should the receiver be placed?

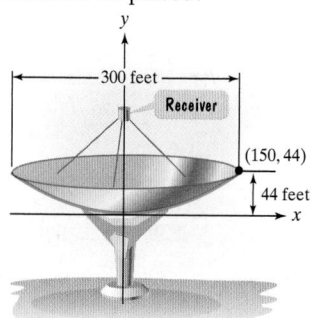

10.4

In Exercises 39–46, identify each equation without completing the square or using a rotation of axes.

39. $y^2 + 4x + 2y - 15 = 0$

40. $x^2 + 16y^2 - 160y + 384 = 0$

41. $16x^2 + 64x + 9y^2 - 54y + 1 = 0$

42. $4x^2 - 9y^2 - 8x + 12y - 144 = 0$

43. $5x^2 + 2\sqrt{3}xy + 3y^2 - 18 = 0$

44. $5x^2 - 8xy + 7y^2 - 9\sqrt{5}x - 9 = 0$

45. $x^2 + 6xy + 9y^2 - 2y = 0$

46. $\quad x^2 - 2xy + 3y^2 + 2x + 4y - 1 = 0$

In Exercises 47–51,

  a. *Rewrite the equation in a rotated $x'y'$-system without an $x'y'$-term.*

  b. *Express the equation involving x' and y' in the standard form of a conic section.*

  c. *Use the rotated system to graph the equation.*

47. $xy - 4 = 0$

48. $x^2 + xy + y^2 - 1 = 0$

49. $4x^2 + 10xy + 4y^2 - 9 = 0$

50. $6x^2 - 6xy + 14y^2 - 45 = 0$

51. $x^2 + 2\sqrt{3}xy + 3y^2 - 12\sqrt{3}x + 12y = 0$

10.5

In Exercises 52–57, eliminate the parameter and graph the plane curve represented by the parametric equations. Use arrows to show the orientation of each plane curve.

52. $x = 2t - 1, y = 1 - t; -\infty < t < \infty$

53. $x = t^2, y = t - 1; -1 \leq t \leq 3.$

54. $x = 4t^2, y = t + 1; -\infty < t < \infty$

55. $x = 4 \sin t, y = 3 \cos t; 0 \leq t < \pi$

56. $x = 3 + 2 \cos t, y = 1 + 2 \sin t; 0 \leq t < 2\pi$

57. $x = 3 \sec t, y = 3 \tan t; 0 \leq t \leq \dfrac{\pi}{4}$

58. Find two different sets of parametric equations for $y = x^2 + 6$.

59. The path of a projectile that is launched h feet above the ground with an initial velocity of v_0 feet per second and at an angle θ with the horizontal is given by the parametric equations

$$x = (v_0 \cos \theta)t \quad \text{and} \quad y = h + (v_0 \sin \theta)t - 16t^2$$

where t is the time, in seconds, after the projectile was launched. A football player throws a football with an initial velocity of 100 feet per second at an angle of 40° to the horizontal. The ball leaves the player's hand at a height of 6 feet.

a. Find the parametric equations that describe the position of the ball as a function of time.

b. Describe the ball's position after 1, 2, and 3 seconds. Round to the nearest tenth of a foot.

c. How long, to the nearest tenth of a second, is the ball in flight? What is the total horizontal distance that it travels before it lands?

d. Graph the parametric equations in part (a) using a graphing utility. Use the graph to determine when the ball is at its maximum height. What is its maximum height? Round all answers to the nearest tenth.

10.6

In Exercises 60–65,

a. If necessary, write the equation in one of the standard forms for a conic in polar coordinates.

b. Determine values for e and p. Use the value of e to identify the conic section.

c. Graph the given polar equation.

60. $r = \dfrac{4}{1 - \sin\theta}$

61. $r = \dfrac{6}{1 + \cos\theta}$

62. $r = \dfrac{6}{2 + \sin\theta}$

63. $r = \dfrac{2}{3 - 2\cos\theta}$

64. $r = \dfrac{6}{3 + 6\sin\theta}$

65. $r = \dfrac{8}{4 + 16\cos\theta}$

Chapter 10 Test

In Exercises 1–5, graph the conic section with the given equation. For ellipses, find the foci. For hyperbolas, find the foci and give the equations of the asymptotes. For parabolas, find the vertex, focus, and directrix.

1. $9x^2 - 4y^2 = 36$

2. $x^2 = -8y$

3. $\dfrac{(x + 2)^2}{25} + \dfrac{(y - 5)^2}{9} = 1$

4. $4x^2 - y^2 + 8x + 2y + 7 = 0$

5. $(x + 5)^2 = 8(y - 1)$

In Exercises 6–8, find the standard form of the equation of the conic section satisfying the given conditions.

6. Ellipse; Foci: $(-7, 0)$, $(7, 0)$; Vertices: $(-10, 0)$, $(10, 0)$

7. Hyperbola; Foci: $(0, -10)$, $(0, 10)$; Vertices: $(0, -7)$, $(0, 7)$

8. Parabola; Focus: $(50, 0)$; Directrix: $x = -50$

9. A sound whispered at one focus of a whispering gallery can be heard at the other focus. The figure shows a whispering gallery whose cross section is a semielliptical arch with a height of 24 feet and a width of 80 feet. How far from the room's center should two people stand so that they can whisper back and forth and be heard?

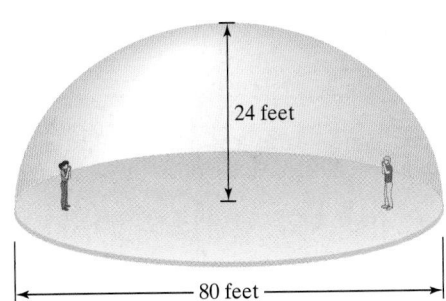

10. An engineer is designing headlight units for cars. The unit shown in the figure in the next column has a parabolic surface with a diameter of 6 inches and a depth of 3 inches.

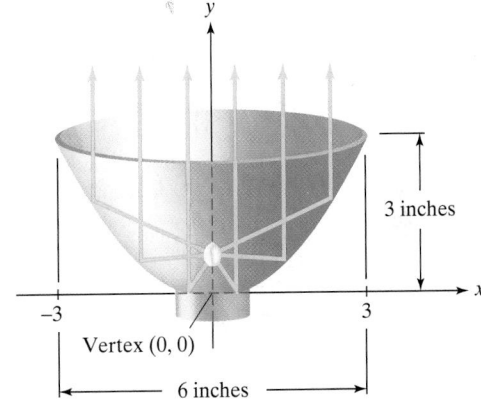

a. Using the coordinate system that has been positioned on the unit, find the parabola's equation.

b. If the light source is located at the focus, describe its placement relative to the vertex.

In Exercises 11–12, identify each equation without completing the square or using a rotation of axes.

11. $x^2 + 9y^2 + 10x - 18y + 25 = 0$

12. $x^2 + y^2 + xy + 3x - y - 3 = 0$

13. For the equation

$$7x^2 - 6\sqrt{3}xy + 13y^2 - 16 = 0$$

determine what angle of rotation would eliminate the $x'y'$-term in a rotated $x'y'$-system.

In Exercises 14–15, eliminate the parameter and graph the plane curve represented by the parametric equations. Use arrows to show the orientation of each plane curve.

14. $x = t^2$, $y = t - 1$; $-\infty < t < \infty$

15. $x = 1 + 3\sin t$, $y = 2\cos t$; $0 \leq t < 2\pi$

In Exercises 16–17, identify the conic section and graph the polar equation.

16. $r = \dfrac{2}{1 - \cos\theta}$

17. $r = \dfrac{4}{2 + \sin\theta}$

Cumulative Review Exercises (Chapters 1–10)

Solve each equation or inequality in Exercises 1–7.

1. $2(x - 3) + 5x = 8(x - 1)$
2. $-3(2x - 4) > 2(6x - 12)$
3. $x - 5 = \sqrt{x + 7}$
4. $(x - 2)^2 = 20$
5. $|2x - 1| \geq 7$
6. $3x^3 + 4x^2 - 7x + 2 = 0$
7. $\log_2 (x + 1) + \log_2 (x - 1) = 3$

Solve each system in Exercises 8–10.

8. $3x + 4y = 2$
   $2x + 5y = -1$
9. $2x^2 - y^2 = -8$
   $x - y = 6$
10. (Use matrices.)
$$x - \ y + \ z = 17$$
$$-4x + \ y + 5z = -2$$
$$2x + 3y + \ z = \ 8$$

In Exercises 11–13, graph each equation, function, or system in the rectangular coordinate system.

11. $f(x) = (x - 1)^2 - 4$
12. $\dfrac{x^2}{9} + \dfrac{y^2}{4} = 1$
13. $5x + y \leq 10$
$$y \geq \frac{1}{4}x + 2$$

14. **a.** List all possible rational roots of
$$32x^3 - 52x^2 + 17x + 3 = 0.$$

    b. The graph of $f(x) = 32x^3 - 52x^2 + 17x + 3$ is shown in a $[-1, 3, 1]$ by $[-2, 6, 1]$ viewing rectangle. Use the graph of f and synthetic division to solve the equation in part (a).

15. The graph shows gender ratios in the United States, with future projections.

Gender Ratios in the U.S.

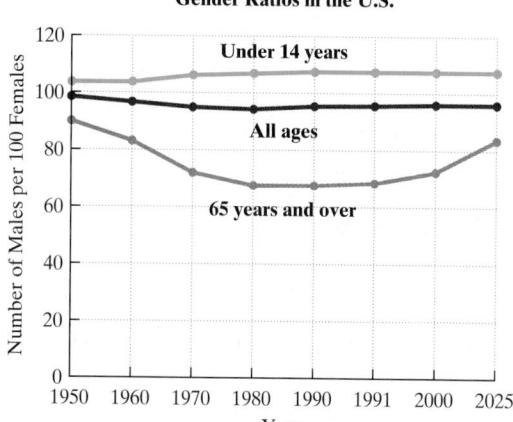

Source: U.S. Census Bureau

For males ages 65 and over, shown by the blue graph:

a. In which time interval is the number of males per 100 females constant?

b. In which time interval is the number of males per 100 females increasing?

c. In which time interval is the number of males per 100 females decreasing?

For all ages, shown by the red graph:

d. Write a constant function, f, that approximately models the data shown for x in the interval $[1950, 2025]$.

e. What is misleading about the scale on the horizontal axis?

16. If $f(x) = x^2 - 4$ and $g(x) = x + 2$, find $(g \circ f)(x)$.

17. Expand using logarithmic properties. Where possible, evaluate logarithmic expressions.
$$\log_5 \frac{x^3 \sqrt{y}}{125}$$

18. Write the slope-intercept form of the equation of the line passing through $(1, -4)$ and $(-5, 8)$.

19. Rent-a-Truck charges a daily rental rate for a truck of $39 plus $0.16 a mile. A competing agency, Ace Truck Rentals, charges $25 a day plus $0.24 a mile for the same truck. How many miles must be driven in a day to make the daily cost of both agencies the same? What will be the cost?

20. The local cable television company offers two deals. Basic cable service with one movie channel costs $35 per month. Basic service with two movie channels cost $45 per month. Find the charge for the basic cable service and the charge for each movie channel.

21. Verify the identity: $\dfrac{\csc \theta - \sin \theta}{\sin \theta} = \cot^2 \theta$.

22. Graph one complete cycle of $y = 2 \cos (2x + \pi)$.

23. If $\mathbf{v} = 3\mathbf{i} - 6\mathbf{j}$ and $\mathbf{w} = \mathbf{i} + \mathbf{j}$, find $(\mathbf{v} \cdot \mathbf{w})\mathbf{w}$.

24. Solve for θ: $\sin 2\theta = \sin \theta, 0 \leq \theta < 2\pi$.

25. In oblique triangle ABC, $A = 64°$, $B = 72°$, and $a = 13.6$. Solve the triangle. Round length to the nearest tenth.

Sequences, Induction, and Probability

We often save for the future by investing small amounts at periodic intervals. To understand how our savings accumulate, we need to understand properties of lists of numbers that are related to each other by a rule. Such lists are called *sequences*. Learning about properties of sequences will show you how to make your financial goals a reality. Your knowledge of sequences will enable you to inform your college roommate of the best of the three appealing offers.

Something incredible has happened. Your college roommate, a gifted athlete, has been given a six-year contract with a professional baseball team. He will be playing against the likes of Barry Bonds and Sammy Sosa. Management offers him three options. One is a beginning salary of $1,700,000 with annual increases of $70,000 per year starting in the second year. A second option is $1,700,000 the first year with an annual increase of 2% per year beginning in the second year. The third offer involves less money the first year—$1,500,000—but there is an annual increase of 9% yearly after that. Which option offers the most money over the six-year contract?

SECTION 11.1 *Sequences and Summation Notation*

Objectives

1. Find particular terms of a sequence from the general term.
2. Use recursion formulas.
3. Use factorial notation.
4. Use summation notation.

Sequences

Many creations in nature involve intricate mathematical designs, including a variety of spirals. For example, the arrangement of the individual florets in the head of a sunflower forms spirals. In some species, there are 21 spirals in the clockwise direction and 34 in the counterclockwise direction. The precise numbers depend on the species of sunflower: 21 and 34, or 34 and 55, or 55 and 89, or even 89 and 144.

This observation becomes even more interesting when we consider a sequence of numbers investigated by Leonardo of Pisa, also known as Fibonacci, an Italian mathematician of the thirteenth century. The **Fibonacci sequence** of numbers is an infinite sequence that begins as follows:

$$1, 1, 2, 3, 5, 8, 13, 21, 34, 55, 89, 144, 233 \ldots.$$

The first two terms are 1. Every term thereafter is the sum of the two preceding terms. For example, the third term, 2, is the sum of the first and second terms: $1 + 1 = 2$. The fourth term, 3, is the sum of the second and third terms: $1 + 2 = 3$, and so on. Did you know that the number of spirals in a daisy or a sunflower, 21 and 34, are two Fibonacci numbers? The number of spirals in a pine cone, 8 and 13, and a pineapple, 8 and 13, are also Fibonacci numbers.

We can think of the Fibonacci sequence as a function. The terms of the sequence

$$1, 1, 2, 3, 5, 8, 13, 21, 34, 55, 89, 144, 233 \ldots$$

are the range values for a function whose domain is the set of positive integers.

Domain: 1, 2, 3, 4, 5, 6, 7, ...

Range: 1, 1, 2, 3, 5, 8, 13, ...

Thus, $f(1) = 1$, $f(2) = 1$, $f(3) = 2$, $f(4) = 3$, $f(5) = 5$, $f(6) = 8$, $f(7) = 13$, and so on.

The letter a with a subscript is used to represent function values of a sequence, rather than the usual function notation. The subscripts make up the domain of the sequence, and they identify the location of a term. Thus, a_1 represents the first term of the sequence, a_2 represents the second term, a_3 the third term, and so on. This notation is shown for the first six terms of the Fibonacci sequence:

1,   1,   2,   3,   5,   8.

$a_1 = 1$ $a_2 = 1$ $a_3 = 2$ $a_4 = 3$ $a_5 = 5$ $a_6 = 8$

Fibonacci Numbers on the Piano Keyboard

One Octave

Numbers in the Fibonacci sequence can be found in an octave on the piano keyboard. The octave contains 2 black keys in one cluster and 3 black keys in another cluster, for a total of 5 black keys. It also has 8 white keys, for a total of 13 keys. The numbers 2, 3, 5, 8, and 13 are the third through seventh terms of the Fibonacci sequence.

The notation a_n represents the nth term, or **general term,** of a sequence. The entire sequence is represented by $\{a_n\}$.

Definition of a Sequence

An **infinite sequence** $\{a_n\}$ is a function whose domain is the set of positive integers. The function values, or **terms,** of the sequence are represented by

$$a_1, a_2, a_3, a_4, \dots, a_n, \dots.$$

Sequences whose domains consist only of the first n positive integers are called **finite sequences.**

① Find particular terms of a sequence from the general term.

EXAMPLE 1   Writing Terms of a Sequence from the General Term

Write the first four terms of the sequence whose nth term, or general term, is given:

a. $a_n = 3n + 4$ **b.** $a_n = \dfrac{(-1)^n}{3^n - 1}.$

Solution

a. We need to find the first four terms of the sequence whose general term is $a_n = 3n + 4$. To do so, we replace n in the formula with 1, 2, 3, and 4.

a_1, 1st term $3 \cdot 1 + 4 = 3 + 4 = 7$ a_2, 2nd term $3 \cdot 2 + 4 = 6 + 4 = 10$

a_3, 3rd term $3 \cdot 3 + 4 = 9 + 4 = 13$ a_4, 4th term $3 \cdot 4 + 4 = 12 + 4 = 16$

The first four terms are 7, 10, 13, and 16. The sequence defined by $a_n = 3n + 4$ can be written as

$$7, \ 10, \ 13, \ 16, \dots, \ 3n + 4, \dots.$$

b. We need to find the first four terms of the sequence whose general term is $a_n = \dfrac{(-1)^n}{3^n - 1}$. To do so, we replace each occurrence of n in the formula with 1, 2, 3, and 4.

a_1, 1st term $\dfrac{(-1)^1}{3^1 - 1} = \dfrac{-1}{3 - 1} = -\dfrac{1}{2}$ a_2, 2nd term $\dfrac{(-1)^2}{3^2 - 1} = \dfrac{1}{9 - 1} = \dfrac{1}{8}$

a_3, 3rd term $\dfrac{(-1)^3}{3^3 - 1} = \dfrac{-1}{27 - 1} = -\dfrac{1}{26}$ a_4, 4th term $\dfrac{(-1)^4}{3^4 - 1} = \dfrac{1}{81 - 1} = \dfrac{1}{80}$

The first four terms are $-\frac{1}{2}, \frac{1}{8}, -\frac{1}{26}$, and $\frac{1}{80}$. The sequence defined by $\dfrac{(-1)^n}{3^n - 1}$ can be written as

$$-\frac{1}{2}, \frac{1}{8}, -\frac{1}{26}, \frac{1}{80}, \dots, \frac{(-1)^n}{3^n - 1}, \dots.$$

Technology

Graphing utilities can write the terms of a sequence and graph them. For example, to find the first six terms of

$$\{a_n\} = \left\{\frac{1}{n}\right\}, \text{ enter}$$

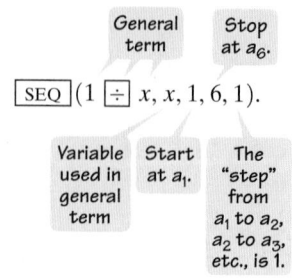

The first few terms of the sequence are shown in the viewing rectangle. By pressing the right arrow key to scroll right, you can see the remaining terms.

Check Point 1 Write the first four terms of the sequence whose nth term, or general term, is given:

$$\textbf{a. } a_n = 2n + 5 \qquad \textbf{b. } a_n = \frac{(-1)^n}{2^n + 1}.$$

Although sequences are usually named with the letter a, any lowercase letter can be used. For example, the first four terms of the sequence $\{b_n\} = \left\{\left(\frac{1}{2}\right)^n\right\}$ are $b_1 = \frac{1}{2}, b_2 = \frac{1}{4}, b_3 = \frac{1}{8},$ and $b_4 = \frac{1}{16}$.

Because a sequence is a function whose domain is the set of positive integers, the **graph of a sequence** is a set of discrete points. For example, consider the sequence whose general term is $a_n = \frac{1}{n}$. How does the graph of this sequence differ from the graph of the function $f(x) = \frac{1}{x}$? The graph of $f(x) = \frac{1}{x}$ is shown in Figure 11.1(a) for positive values of x. To obtain the graph of the sequence $\{a_n\} = \left\{\frac{1}{n}\right\}$, remove all the points from the graph of f except those whose x-coordinates are positive integers. Thus, we remove all points except $(1, 1), (2, \frac{1}{2}), (3, \frac{1}{3}), (4, \frac{1}{4})$, and so on. The remaining points are the graph of the sequence $\{a_n\} = \left\{\frac{1}{n}\right\}$, shown in Figure 11.1(b). Notice that the horizontal axis is labeled n and the vertical axis a_n.

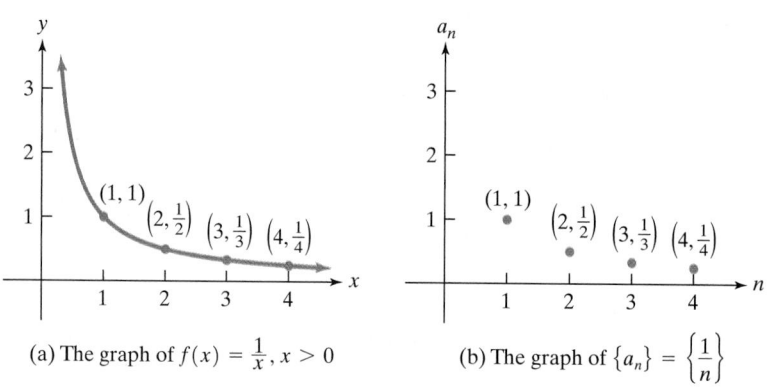

(a) The graph of $f(x) = \frac{1}{x}, x > 0$ (b) The graph of $\{a_n\} = \left\{\frac{1}{n}\right\}$

Figure 11.1 Comparing a continuous graph to the graph of a sequence

---

2 Use recursion formulas.

Recursion Formulas

In Example 1, the formulas used for the nth term of a sequence expressed the term as a function of n, the number of the term. Sequences can also be defined using **recursion formulas.** A recursion formula defines the nth term of a sequence as a function of the previous term. Our next example illustrates that if the first term of a sequence is known, then the recursion formula can be used to determine the remaining terms.

EXAMPLE 2 Using a Recursion Formula

Find the first four terms of the sequence in which $a_1 = 5$ and $a_n = 3a_{n-1} + 2$ for $n \geq 2$.

Solution

$$a_1 = 5 \qquad \text{This is the given first term.}$$

$$a_2 = 3a_1 + 2 \qquad \text{Use } a_n = 3a_{n-1} + 2, \text{ with } n = 2.$$
$$\text{Thus, } a_2 = 3a_{2-1} + 2 = 3a_1 + 2.$$

$$= 3(5) + 2 = 17 \qquad \text{Substitute 5 for } a_1.$$

$$a_3 = 3a_2 + 2$$
Again use $a_n = 3a_{n-1} + 2$, with $n = 3$.

$$= 3(17) + 2 = 53$$
Substitute 17 for a_2.

$$a_4 = 3a_3 + 2$$
Notice that a_4 is defined in terms of a_3.
We used $a_n = 3a_{n-1} + 2$, with $n = 4$.

$$= 3(53) + 2 = 161$$
Use the value of a_3, the third term, obtained from above.

The first four terms are $5, 17, 53$, and 161.

> **Check Point 2**   Find the first four terms of the sequence in which $a_1 = 3$ and $a_n = 2a_{n-1} + 5$ for $n \geq 2$.

3   Use factorial notation.

Factorial Notation

Products of consecutive positive integers occur quite often in sequences. These products can be expressed in a special notation, called **factorial notation.**

Factorials from 0 through 20

0!	1
1!	1
2!	2
3!	6
4!	24
5!	120
6!	720
7!	5040
8!	40,320
9!	362,880
10!	3,628,800
11!	39,916,800
12!	479,001,600
13!	6,227,020,800
14!	87,178,291,200
15!	1,307,674,368,000
16!	20,922,789,888,000
17!	355,687,428,096,000
18!	6,402,373,705,728,000
19!	121,645,100,408,832,000
20!	2,432,902,008,176,640,000

As n increases, $n!$ grows very rapidly. Factorial growth is more explosive than exponential growth discussed in Chapter 4.

Factorial Notation

If n is a positive integer, the notation $n!$ (read "n factorial") is the product of all positive integers from n down through 1.

$$n! = n(n-1)(n-2)\dots(3)(2)(1)$$

$0!$ (zero factorial), by definition, is 1.

$$0! = 1$$

The values of $n!$ for the first six positive integers are

$$1! = 1$$
$$2! = 2 \cdot 1 = 2$$
$$3! = 3 \cdot 2 \cdot 1 = 6$$
$$4! = 4 \cdot 3 \cdot 2 \cdot 1 = 24$$
$$5! = 5 \cdot 4 \cdot 3 \cdot 2 \cdot 1 = 120$$
$$6! = 6 \cdot 5 \cdot 4 \cdot 3 \cdot 2 \cdot 1 = 720.$$

Factorials affect only the number or variable that they follow unless grouping symbols appear. For example,

$$2 \cdot 3! = 2(3 \cdot 2 \cdot 1) = 2 \cdot 6 = 12$$

whereas

$$(2 \cdot 3)! = 6! = 6 \cdot 5 \cdot 4 \cdot 3 \cdot 2 \cdot 1 = 720.$$

In this sense, factorials are similar to exponents.

EXAMPLE 3   Finding Terms of a Sequence Involving Factorials

Write the first four terms of the sequence whose nth term is

$$a_n = \frac{2^n}{(n-1)!}.$$

Technology

Most calculators have factorial keys. To find 5!, many calculators use one of the following:

Scientific Calculators

$$5 \boxed{x!}$$

Graphing Calculators

$$5 \boxed{!} \boxed{\text{ENTER}}.$$

Because $n!$ becomes quite large as n increases, your calculator will display these larger values in scientific notation.

Solution We need to find the first four terms of the sequence. To do so, we replace each n in $\dfrac{2^n}{(n-1)!}$ with 1, 2, 3, and 4.

a_1, 1st term $\dfrac{2^1}{(1-1)!} = \dfrac{2}{0!} = \dfrac{2}{1} = 2$

a_2, 2nd term $\dfrac{2^2}{(2-1)!} = \dfrac{4}{1!} = \dfrac{4}{1} = 4$

a_3, 3rd term $\dfrac{2^3}{(3-1)!} = \dfrac{8}{2!} = \dfrac{8}{2 \cdot 1} = 4$

a_4, 4th term $\dfrac{2^4}{(4-1)!} = \dfrac{16}{3!} = \dfrac{16}{3 \cdot 2 \cdot 1} = \dfrac{16}{6} = \dfrac{8}{3}$

The first four terms are $2, 4, 4$, and $\frac{8}{3}$.

Check Point 3 Write the first four terms of the sequence whose nth term is

$$a_n = \frac{20}{(n+1)!}.$$

When evaluating fractions with factorials in the numerator and the denominator, try to reduce the fraction before performing the multiplications. For example, consider $\dfrac{26!}{21!}$. Rather than write out 26! as the product of all integers from 26 down to 1, we can express 26! as

$$26! = 26 \cdot 25 \cdot 24 \cdot 23 \cdot 22 \cdot 21!.$$

In this way, we can divide both the numerator and the denominator by the common factor, 21!.

$$\frac{26!}{21!} = \frac{26 \cdot 25 \cdot 24 \cdot 23 \cdot 22 \cdot \cancel{21!}}{\cancel{21!}} = 26 \cdot 25 \cdot 24 \cdot 23 \cdot 22 = 7{,}893{,}600$$

EXAMPLE 4 Evaluating Fractions with Factorials

Evaluate each factorial expression:

a. $\dfrac{10!}{2!8!}$ **b.** $\dfrac{(n+1)!}{n!}.$

Solution

a. $\dfrac{10!}{2!8!} = \dfrac{10 \cdot 9 \cdot \cancel{8!}}{2 \cdot 1 \cdot \cancel{8!}} = \dfrac{90}{2} = 45$

b. $\dfrac{(n+1)!}{n!} = \dfrac{(n+1) \cdot \cancel{n!}}{\cancel{n!}} = n+1$

Check Point 4 Evaluate each factorial expression:

a. $\dfrac{14!}{2!12!}$ **b.** $\dfrac{n!}{(n-1)!}.$

4 Use summation notation.

Summation Notation

It is sometimes useful to find the sum of the first n terms of a sequence. For example, consider the number of AIDS cases diagnosed in the United States for each year from 1991 through 2000, shown in Table 11.1.

Table 11.1   AIDS Cases Diagnosed in the United States, 1991–2000

Year	1991	1992	1993	1994	1995	1996	1997	1998	1999	2000
Cases Diagnosed	60,472	79,477	79,752	72,684	69,172	59,832	47,439	40,784	36,725	23,988

Source: U.S. Department of Health and Human Services

We can let a_n represent the number of AIDS cases diagnosed in year n, where $n = 1$ corresponds to 1991, $n = 2$ to 1992, $n = 3$ to 1993, and so on. The terms of the finite sequence in Table 11.1 are given as follows.

60,472 79,477 79,752 72,684 69,172 59,832 47,439 40,784 36,725 23,988

a_1 a_2 a_3 a_4 a_5 a_6 a_7 a_8 a_9 a_{10}

Why might we want to add the terms of this sequence? We do this to find the total number of AIDS cases diagnosed from 1991 through 2000. Thus,

$$a_1 + a_2 + a_3 + a_4 + a_5 + a_6 + a_7 + a_8 + a_9 + a_{10}$$

$$= 60,472 + 79,477, + 79,752, + 72,684, + 69,172, + 59,832, + 47,439 + 40,784 + 36,725 + 23,988$$

$$= 570,325.$$

We see that there were 570,325 AIDS cases diagnosed in the United States from 1991 through 2000.

There is a compact notation for expressing the sum of the first n terms of a sequence. For example, rather than write

$$a_1 + a_2 + a_3 + a_4 + a_5 + a_6 + a_7 + a_8 + a_9 + a_{10},$$

we can use **summation notation** to express the sum as

$$a_1 + a_2 + a_3 + a_4 + a_5 + a_6 + a_7 + a_8 + a_9 + a_{10} = \sum_{i=1}^{10} a_i.$$

We read the expression on the right as "the sum as i goes from 1 to 10 of a_i." The letter i is called the **index of summation** and is not related to the use of i to represent $\sqrt{-1}$.

You can think of the symbol Σ (the uppercase Greek letter sigma) as an instruction to add up terms of a sequence.

Summation Notation

The sum of the first n terms of a sequence is represented by the **summation notation**

$$\sum_{i=1}^{n} a_i = a_1 + a_2 + a_3 + a_4 + \cdots + a_n$$

where i is the **index of summation**, n is the **upper limit of summation**, and 1 is the **lower limit of summation.**

Any letter can be used for the index of summation. The letters i, j, and k are used commonly. Furthermore, the lower limit of summation can be an integer other than 1.

When we write out a sum that is given in summation notation, we are **expanding the summation notation.** Example 5 shows how to do this.

EXAMPLE 5   Using Summation Notation

Expand and evaluate the sum:

a. $\displaystyle\sum_{i=1}^{6}(i^2 + 1)$ **b.** $\displaystyle\sum_{k=4}^{7}\left[(-2)^k - 5\right]$ **c.** $\displaystyle\sum_{i=1}^{5}3.$

Solution

a. We must replace i in the expression $i^2 + 1$ with all consecutive integers from 1 to 6 inclusive. Then we add.

$$\sum_{i=1}^{6}(i^2 + 1) = (1^2 + 1) + (2^2 + 1) + (3^2 + 1) + (4^2 + 1)$$
$$+ (5^2 + 1) + (6^2 + 1)$$
$$= 2 + 5 + 10 + 17 + 26 + 37$$
$$= 97$$

b. This time the index of summation is k. First we evaluate $(-2)^k - 5$ for all consecutive integers from 4 through 7 inclusive. Then we add.

$$\sum_{k=4}^{7}\left[(-2)^k - 5\right] = \left[(-2)^4 - 5\right] + \left[(-2)^5 - 5\right]$$
$$+ \left[(-2)^6 - 5\right] + \left[(-2)^7 - 5\right]$$
$$= (16 - 5) + (-32 - 5) + (64 - 5) + (-128 - 5)$$
$$= 11 + (-37) + 59 + (-133)$$
$$= -100$$

c. To find $\displaystyle\sum_{i=1}^{5}3$, we observe that every term of the sum is 3. The notation $i = 1$ through 5 indicates that we must add the first five terms from a sequence in which every term is 3.

$$\sum_{i=1}^{5}3 = 3 + 3 + 3 + 3 + 3 = 15$$

Check Point 5   Expand and evaluate the sum:

a. $\displaystyle\sum_{i=1}^{6}2i^2$ **b.** $\displaystyle\sum_{k=3}^{5}(2^k - 3)$ **c.** $\displaystyle\sum_{i=1}^{5}4.$

For a given sum, we can vary the upper and lower limits of summation, as well as the letter used for the index of summation. By doing so, we can produce different-looking summation notations for the same sum. For example, the sum

Technology

Graphing utilities can calculate the sum of a sequence. For example, to find the sum of the sequence in Example 5a, enter

$\boxed{\text{SUM}}\ \boxed{\text{SEQ}}\ (x^2 + 1, x, 1, 6, 1).$

Then press $\boxed{\text{ENTER}}$; 97 should be displayed. Use this capability to verify Example 5b.

of the squares of the first four integers, $1^2 + 2^2 + 3^2 + 4^2$, can be expressed in a number of equivalent ways:

$$\sum_{i=1}^{4} i^2 = 1^2 + 2^2 + 3^2 + 4^2 = 30$$

$$\sum_{i=0}^{3} (i + 1)^2 = (0 + 1)^2 + (1 + 1)^2 + (2 + 1)^2 + (3 + 1)^2$$

$$= 1^2 + 2^2 + 3^2 + 4^2 = 30$$

$$\sum_{k=2}^{5} (k - 1)^2 = (2 - 1)^2 + (3 - 1)^2 + (4 - 1)^2 + (5 - 1)^2$$

$$= 1^2 + 2^2 + 3^2 + 4^2 = 30.$$

EXAMPLE 6 Writing Sums in Summation Notation

Express each sum using summation notation:

a. $1^3 + 2^3 + 3^3 + \cdots + 7^3$ **b.** $1 + \dfrac{1}{3} + \dfrac{1}{9} + \dfrac{1}{27} + \cdots + \dfrac{1}{3^{n-1}}.$

Solution In each case, we will use 1 as the lower limit of summation and i for the index of summation.

a. The sum $1^3 + 2^3 + 3^3 + \cdots + 7^3$ has seven terms, each of the form i^3, starting at $i = 1$ and ending at $i = 7$. Thus,

$$1^3 + 2^3 + 3^3 + \cdots + 7^3 = \sum_{i=1}^{7} i^3.$$

b. The sum

$$1 + \frac{1}{3} + \frac{1}{9} + \frac{1}{27} + \cdots + \frac{1}{3^{n-1}}$$

has n terms, each of the form $\dfrac{1}{3^{i-1}}$, starting at $i = 1$ and ending at $i = n$. Thus,

$$1 + \frac{1}{3} + \frac{1}{9} + \frac{1}{27} + \cdots + \frac{1}{3^{n-1}} = \sum_{i=1}^{n} \frac{1}{3^{i-1}}.$$

Check Point 6 Express each sum using summation notation:

a. $1^2 + 2^2 + 3^2 + \cdots + 9^2$ **b.** $1 + \dfrac{1}{2} + \dfrac{1}{4} + \dfrac{1}{8} + \cdots + \dfrac{1}{2^{n-1}}.$

Table 11.2 contains some important properties of sums expressed in summation notation.

Table 11.2   Properties of Sums

Property	Example
1. $\displaystyle\sum_{i=1}^{n} ca_i = c \sum_{i=1}^{n} a_i$, c any real number	$\displaystyle\sum_{i=1}^{4} 3i^2 = 3 \cdot 1^2 + 3 \cdot 2^2 + 3 \cdot 3^2 + 3 \cdot 4^2$ $3\displaystyle\sum_{i=1}^{4} i^2 = 3(1^2 + 2^2 + 3^2 + 4^2) = 3 \cdot 1^2 + 3 \cdot 2^2 + 3 \cdot 3^2 + 3 \cdot 4^2$ Conclusion: $\displaystyle\sum_{i=1}^{4} 3i^2 = 3 \sum_{i=1}^{4} i^2$
2. $\displaystyle\sum_{i=1}^{n} (a_i + b_i) = \sum_{i=1}^{n} a_i + \sum_{i=1}^{n} b_i$	$\displaystyle\sum_{i=1}^{4} (i + i^2) = (1 + 1^2) + (2 + 2^2) + (3 + 3^2) + (4 + 4^2)$ $\displaystyle\sum_{i=1}^{4} i + \sum_{i=1}^{4} i^2 = (1 + 2 + 3 + 4) + (1^2 + 2^2 + 3^2 + 4^2)$ $= (1 + 1^2) + (2 + 2^2) + (3 + 3^2) + (4 + 4^2)$ Conclusion: $\displaystyle\sum_{i=1}^{4} (i + i^2) = \sum_{i=1}^{4} i + \sum_{i=1}^{4} i^2$
3. $\displaystyle\sum_{i=1}^{n} (a_i - b_i) = \sum_{i=1}^{n} a_i - \sum_{i=1}^{n} b_i$	$\displaystyle\sum_{i=3}^{5} (i^2 - i^3) = (3^2 - 3^3) + (4^2 - 4^3) + (5^2 - 5^3)$ $\displaystyle\sum_{i=3}^{5} i^2 - \sum_{i=3}^{5} i^3 = (3^2 + 4^2 + 5^2) - (3^3 + 4^3 + 5^3)$ $= (3^2 - 3^3) + (4^2 - 4^3) + (5^2 - 5^3)$ Conclusion: $\displaystyle\sum_{i=3}^{5} (i^2 - i^3) = \sum_{i=3}^{5} i^2 - \sum_{i=3}^{5} i^3$

EXERCISE SET 11.1

Practice Exercises

In Exercises 1–12, write the first four terms of each sequence whose general term is given.

1. $a_n = 3n + 2$ **2.** $a_n = 4n - 1$

3. $a_n = 3^n$ **4.** $a_n = \left(\dfrac{1}{3}\right)^n$

5. $a_n = (-3)^n$ **6.** $a_n = \left(-\dfrac{1}{3}\right)^n$

7. $a_n = (-1)^n(n + 3)$ **8.** $a_n = (-1)^{n+1}(n + 4)$

9. $a_n = \dfrac{2n}{n + 4}$ **10.** $a_n = \dfrac{3n}{n + 5}$

11. $a_n = \dfrac{(-1)^{n+1}}{2^n - 1}$ **12.** $a_n = \dfrac{(-1)^{n+1}}{2^n + 1}$

The sequences in Exercises 13–18 are defined using recursion formulas. Write the first four terms of each sequence.

13. $a_1 = 7$ and $a_n = a_{n-1} + 5$ for $n \geq 2$

14. $a_1 = 12$ and $a_n = a_{n-1} + 4$ for $n \geq 2$

15. $a_1 = 3$ and $a_n = 4a_{n-1}$ for $n \geq 2$

16. $a_1 = 2$ and $a_n = 5a_{n-1}$ for $n \geq 2$

17. $a_1 = 4$ and $a_n = 2a_{n-1} + 3$ for $n \geq 2$

18. $a_1 = 5$ and $a_n = 3a_{n-1} - 1$ for $n \geq 2$

In Exercises 19–22, the general term of a sequence is given and involves a factorial. Write the first four terms of each sequence.

19. $a_n = \dfrac{n^2}{n!}$ **20.** $a_n = \dfrac{(n + 1)!}{n^2}$

21. $a_n = 2(n + 1)!$ **22.** $a_n = -2(n - 1)!$

In Exercises 23–28, evaluate each factorial expression.

23. $\dfrac{17!}{15!}$ **24.** $\dfrac{18!}{16!}$

25. $\dfrac{16!}{2!14!}$ **26.** $\dfrac{20!}{2!18!}$

27. $\dfrac{(n+2)!}{n!}$ **28.** $\dfrac{(2n+1)!}{(2n)!}$

In Exercises 29–42, find each indicated sum.

29. $\displaystyle\sum_{i=1}^{6} 5i$ **30.** $\displaystyle\sum_{i=1}^{6} 7i$

31. $\displaystyle\sum_{i=1}^{4} 2i^2$ **32.** $\displaystyle\sum_{i=1}^{5} i^3$

33. $\displaystyle\sum_{k=1}^{5} k(k+4)$ **34.** $\displaystyle\sum_{k=1}^{4} (k-3)(k+2)$

35. $\displaystyle\sum_{i=1}^{4} \left(-\dfrac{1}{2}\right)^i$ **36.** $\displaystyle\sum_{i=2}^{4} \left(-\dfrac{1}{3}\right)^i$

37. $\displaystyle\sum_{i=5}^{9} 11$ **38.** $\displaystyle\sum_{i=3}^{7} 12$

39. $\displaystyle\sum_{i=0}^{4} \dfrac{(-1)^i}{i!}$ **40.** $\displaystyle\sum_{i=0}^{4} \dfrac{(-1)^{i+1}}{(i+1)!}$

41. $\displaystyle\sum_{i=1}^{5} \dfrac{i!}{(i-1)!}$ **42.** $\displaystyle\sum_{i=1}^{5} \dfrac{(i+2)!}{i!}$

In Exercises 43–54, express each sum using summation notation. Use 1 as the lower limit of summation and i for the index of summation.

43. $1^2 + 2^2 + 3^2 + \cdots + 15^2$ **44.** $1^4 + 2^4 + 3^4 + \cdots + 12^4$

45. $2 + 2^2 + 2^3 + \cdots + 2^{11}$ **46.** $5 + 5^2 + 5^3 + \cdots + 5^{12}$

47. $1 + 2 + 3 + \cdots + 30$ **48.** $1 + 2 + 3 + \cdots + 40$

49. $\dfrac{1}{2} + \dfrac{2}{3} + \dfrac{3}{4} + \cdots + \dfrac{14}{14+1}$

50. $\dfrac{1}{3} + \dfrac{2}{4} + \dfrac{3}{5} + \cdots + \dfrac{16}{16+2}$

51. $4 + \dfrac{4^2}{2} + \dfrac{4^3}{3} + \cdots + \dfrac{4^n}{n}$

52. $\dfrac{1}{9} + \dfrac{2}{9^2} + \dfrac{3}{9^3} + \cdots + \dfrac{n}{9^n}$

53. $1 + 3 + 5 + \cdots + (2n-1)$

54. $a + ar + ar^2 + \cdots + ar^{n-1}$

In Exercises 55–60, express each sum using summation notation. Use a lower limit of summation of your choice and k for the index of summation.

55. $5 + 7 + 9 + 11 + \cdots + 31$

56. $6 + 8 + 10 + 12 + \cdots + 32$

57. $a + ar + ar^2 + \cdots + ar^{12}$

58. $a + ar + ar^2 + \cdots + ar^{14}$

59. $a + (a+d) + (a+2d) + \cdots + (a+nd)$

60. $(a+d) + (a+d^2) + \cdots + (a+d^n)$

 Application Exercises

61. The bar graph shows the number of compact discs (CDs) sold in the United States. Let a_n represent the number of CDs sold, in millions, in year n, where $n = 1$ corresponds to 1991, $n = 2$ to 1992, and so on.

Number of CDs Sold in the U.S.

Source: Recording Industry Association of America

a. Find $\displaystyle\sum_{i=1}^{10} a_i$. What does this represent?

b. Find $\dfrac{1}{10}\displaystyle\sum_{i=1}^{10} a_i$. What does this represent?

62. The bar graph shows the number of vinyl long-playing records (LPs) sold in the United States. Let a_n represent the number of LPs sold, in millions, in year n, where $n = 1$ corresponds to 1991, $n = 2$ to 1992, and so on.

Number of LPs Sold in the U.S.

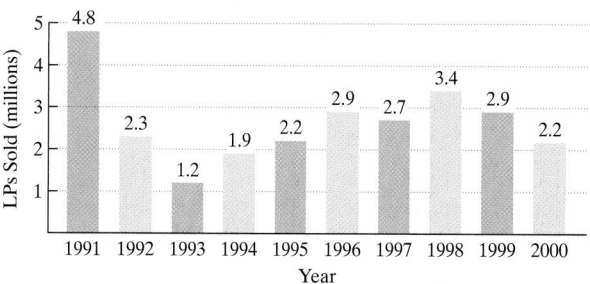

Source: Recording Industry Association of America

a. Find $\displaystyle\sum_{i=1}^{10} a_i$. What does this represent?

b. Find $\dfrac{1}{10}\displaystyle\sum_{i=1}^{10} a_i$. What does this represent?

The graph shows the millions of welfare recipients in the United States who received cash assistance from 1993 through 2000. In Exercises 63–64, consider a sequence whose general term, a_n, represents the millions of Americans receiving cash assistance n years after 1992.

Welfare Recipients in the U.S.

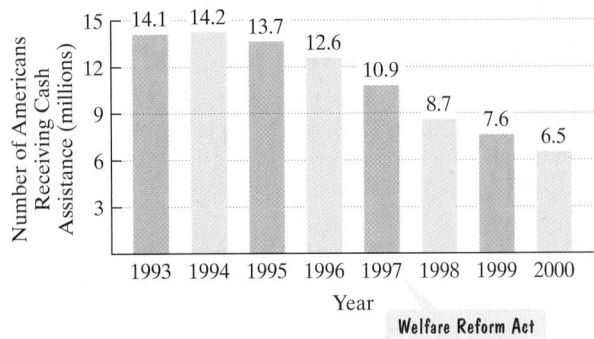

Source: Thomas R. Dye, *Politics in America*, Prentice Hall

63. a. Use the numbers given in the graph to find and interpret
$$\sum_{i=1}^{8} a_i.$$

b. The finite sequence whose general term is $a_n = -1.23n + 16.55$, where $n = 1, 2, 3, \ldots, 8$, models the millions of Americans receiving cash assistance, a_n, n years after 1992. Use the model to find $\sum_{i=1}^{8} a_i$. Does this seem reasonable in terms of the actual sum in part (a), or has model breakdown occurred?

64. a. Use the numbers given in the graph to find and interpret $\sum_{i=1}^{8} a_i$.

b. The finite sequence whose general term is $a_n = -0.11n^2 - 0.22n + 14.88$, where $n = 1, 2, 3, \ldots, 8$, models the millions of Americans receiving cash assistance, a_n, n years after 1992. Use the model to find $\sum_{i=1}^{8} a_i$. Does this seem reasonable in terms of the actual sum in part (a), or has model breakdown occurred?

65. A deposit of $6000 is made in an account that earns 6% interest compounded quarterly. The balance in the account after n quarters is given by the sequence
$$a_n = 6000\left(1 + \frac{0.06}{4}\right)^n, \qquad n = 1, 2, 3, \ldots .$$

Find the balance in the account after five years. Round to the nearest cent.

66. A deposit of $10,000 is made in an account that earns 8% interest compounded quarterly. The balance in the account after n quarters is given by the sequence
$$a_n = 10,000\left(1 + \frac{0.08}{4}\right)^n, \qquad n = 1, 2, 3, \ldots .$$

Find the balance in the account after six years. Round to the nearest cent.

Writing in Mathematics

67. What is a sequence? Give an example with your description.

68. Explain how to write terms of a sequence if the formula for the general term is given.

69. What does the graph of a sequence look like? How is it obtained?

70. What is a recursion formula?

71. Explain how to find $n!$ if n is a positive integer.

72. Explain the best way to evaluate $\dfrac{900!}{899!}$ without calculator.

73. What is the meaning of the symbol Σ? Give an example with your description.

74. You buy a new car for $24,000. At the end of n years, the value of your car is given by the sequence
$$a_n = 24{,}000\left(\frac{3}{4}\right)^n, \qquad n = 1, 2, 3, \ldots .$$

Find a_5 and write a sentence explaining what this value represents. Describe the nth term of the sequence in terms of the value of your car at the end of each year.

75. It is estimated that 4 to 6 million people in the United States have overwhelming physical, psychological, and social problems that make it impossible for them to work. (*Source*: Thomas R. Dye, *Politics in America*, Prentice Hall) Describe what this means in terms of projecting the model in Exercise 63(b) into the first decade of the new millennium. In writing your answer, use the model and be as specific as possible.

Technology Exercises

In Exercises 76–80, use a calculator's factorial key to evaluate each expression.

76. $\dfrac{200!}{198!}$

77. $\left(\dfrac{300}{20}\right)!$

78. $\dfrac{20!}{300}$

79. $\dfrac{20!}{(20-3)!}$

80. $\dfrac{54!}{(54-3)!\,3!}$

81. Use the SEQ (sequence) capability of a graphing utility to verify the terms of the sequences you obtained for any five sequences from Exercises 1–12 or 19–22.

82. Use the SUM SEQ (sum of the sequence) capability of a graphing utility to verify any five of the sums you obtained in Exercises 29–42.

83. As n increases, the terms of the sequence
$$a_n = \left(1 + \frac{1}{n}\right)^n$$

get closer and closer to the number e (where $e \approx 2.7183$). Use a calculator to find $a_{10}, a_{100}, a_{1000}, a_{10,000}$, and $a_{100,000}$, comparing these terms to your calculator's decimal approximation for e.

Many graphing utilities have a sequence-graphing mode that plots the terms of a sequence as points on a rectangular coordinate system. Consult your manual; if your graphing utility has this capability, use it to graph each of the sequences in Exercises 84–87. What appears to be happening to the terms of each sequence as n gets larger?

84. $a_n = \dfrac{n}{n+1}$ $n:[0, 10, 1]$ by $a_n:[0, 1, 0.1]$

85. $a_n = \dfrac{100}{n}$ $n:[0, 1000, 100]$ by $a_n:[0, 1, 0.1]$

86. $a_n = \dfrac{2n^2 + 5n - 7}{n^3}$ $n:[0, 10, 1]$ by $a_n:[0, 2, 0.2]$

87. $a_n = \dfrac{3n^4 + n - 1}{5n^4 + 2n^2 + 1}$ $n:[0, 10, 1]$ by $a_n:[0, 1, 0.1]$

Critical Thinking Exercises

88. Which one of the following is true?

a. $\dfrac{n!}{(n-1)!} = \dfrac{1}{n-1}$

b. The Fibonacci sequence 1, 1, 2, 3, 5, 8, 13, 21, 34, 55, 89, 144, … can be defined recursively using $a_0 = 1, a_1 = 1$; $a_n = a_{n-2} + a_{n-1}$, where $n \geq 2$.

c. $\displaystyle\sum_{i=1}^{2} (-1)^i 2^i = 0$

d. $\displaystyle\sum_{i=1}^{2} a_i b_i = \sum_{i=1}^{2} a_i \sum_{i=1}^{2} b_i$

89. Write the first five terms of the sequence whose first term is 9 and whose general term is

$$a_n = \begin{cases} \dfrac{a_{n-1}}{2} & \text{if } a_{n-1} \text{ is even} \\ 3a_{n-1} + 5 & \text{if } a_{n-1} \text{ is odd.} \end{cases}$$

Group Exercise

90. Enough curiosities involving the Fibonacci sequence exist to warrant a flourishing Fibonacci Association, which publishes a quarterly journal. Do some research on the Fibonacci sequence by consulting the Internet or the research department of your library, and find one property that interests you. After doing this research, get together with your group to share these intriguing properties.

SECTION 11.2 Arithmetic Sequences

Objectives

1. Find the common difference for an arithmetic sequence.
2. Write terms of an arithmetic sequence.
3. Use the formula for the general term of an arithmetic sequence.
4. Use the formula for the sum of the first *n* terms of an arithmetic sequence.

Your grandmother and her financial counselor are looking at options in case nursing home care is needed in the future. The good news is that your grandmother's total assets are $350,000. The bad news is that yearly nursing home costs average $49,730, increasing by $1800 each year. In this section, we will see how sequences can be used to describe your grandmother's situation and help her to identify realistic options.

Arithmetic Sequences

A mathematical model for the average annual salaries of major league baseball players generates the following data:

Year	1996	1997	1998	1999	2000	2001	2002
Salary	1,076,865	1,304,152	1,531,439	1,758,726	1,986,013	2,213,300	2,440,587

From 1996 to 1997, salaries increased by $1,304,152 − $1,076,865 = $227,287. From 1997 to 1998, salaries increased by $1,531,439 − $1,304,152 = $227,287. If we make these computations for each year, we find that the yearly salary increase is $227,287. The sequence of annual salaries shows that each term after the first, 1,076,865, differs from the preceding term by a constant amount, namely 227,287. The sequence of annual salaries

$$1,076,865, \ 1,304,152, \ 1,531,439, \ 1,758,726, \ 1,986,013,\dots$$

is an example of an *arithmetic sequence*.

> **Definition of an Arithmetic Sequence**
>
> An **arithmetic sequence** is a sequence in which each term after the first differs from the preceding term by a constant amount. The difference between consecutive terms is called the **common difference** of the sequence.

1 Find the common difference of an arithmetic sequence.

The common difference, d, is found by subtracting any term from the term that directly follows it. In the following examples, the common difference is found by subtracting the first term from the second term: $a_2 - a_1$.

Arithmetic Sequence	Common Difference
$1,076,865, \ 1,304,152, \ 1,531,439, \ 1,758,726, \dots$	$d = 1,304,152 - 1,076,865 = 227,287$
$2, 6, 10, 14, 18, \dots$	$d = 6 - 2 = 4$
$-2, -7, -12, -17, \dots$	$d = -7 - (-2) = -5$

If the first term of an arithmetic sequence is a_1, each term after the first is obtained by adding d, the common difference, to the previous term. This can be expressed recursively as follows:

$$a_n = a_{n-1} + d.$$

Add d to the term in any position to get the next term.

To use this recursion formula, we must be given the first term.

2 Write the terms of an arithmetic sequence.

EXAMPLE 1 Writing the Terms of an Arithmetic Sequence Using the First Term and the Common Difference

The recursion formula $a_n = a_{n-1} - 0.67$ models the percentage of men working in the U.S. labor force, a_n, for each five-year period starting with 1980. Thus, $n = 1$ corresponds to 1980, $n = 2$ to 1985, $n = 3$ to 1990, and so on. In 1980, 77.4% of the U.S. men were working in the labor force. Find the first five terms of this arithmetic sequence in which $a_1 = 77.4$ and $a_n = a_{n-1} - 0.67$.

Men in the U.S. Labor Force

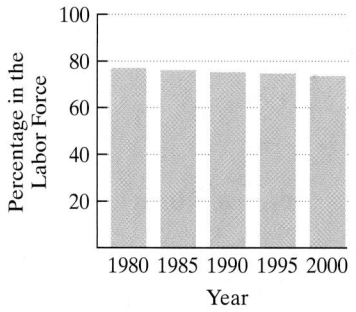

Women in the U.S. Labor Force

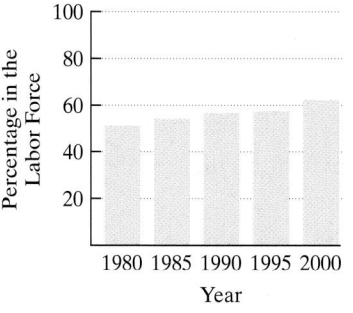

Source: U.S. Department of Labor

3 Use the formula for the general term of an arithmetic sequence.

Solution The recursion formula $a_1 = 77.4$ and $a_n = a_{n-1} - 0.67$ indicates that each term after the first, 77.4, is obtained by adding -0.67 to the previous term. Thus, each five-year period the percentage of men in the labor force decreased by 0.67%.

$a_1 = 77.4$ This is given.

$a_2 = a_1 - 0.67 = 77.4 - 0.67 = 76.73$ Use $a_n = a_{n-1} - 0.67$ with $n = 2$.

$a_3 = a_2 - 0.67 = 76.73 - 0.67 = 76.06$ Use $a_n = a_{n-1} - 0.67$ with $n = 3$.

$a_4 = a_3 - 0.67 = 76.06 - 0.67 = 75.39$ Use $a_n = a_{n-1} - 0.67$ with $n = 4$.

$a_5 = a_4 - 0.67 = 75.39 - 0.67 = 74.72$ Use $a_n = a_{n-1} - 0.67$ with $n = 5$.

The first five terms are

$$77.4, 76.73, 76.06, 75.39, \text{ and } 74.72.$$

These numbers represent the percentage of men working in the U.S. labor force in 1980, 1985, 1990, 1995, and 2000, respectively, as given by the model.

Check Point 1 The recursion formula $a_n = a_{n-1} + 2.18$ models the percentage of women working in the U.S. labor force, a_n, for each five-year period starting with 1980. In 1980, 51.5% of U.S. women were working in the labor force. Find the first five terms of the arithmetic sequence in which $a_1 = 51.5$ and $a_n = a_{n-1} + 2.18$.

The General Term of an Arithmetic Sequence

Consider an arithmetic sequence whose first term is a_1 and whose common difference is d. We are looking for a formula for the general term, a_n. Let's begin by writing the first six terms. The first term is a_1. The second term is $a_1 + d$. The third term is $a_1 + d + d$, or $a_1 + 2d$. Thus, we start with a_1 and add d to each successive term. The first six terms are

$$a_1, \quad a_1 + d, \quad a_1 + 2d, \quad a_1 + 3d, \quad a_1 + 4d, \quad a_1 + 5d.$$

a_1, first term a_2, second term a_3, third term a_4, fourth term a_5, fifth term a_6, sixth term

Compare the coefficient of d and the subscript of a denoting the term number. Can you see that the coefficient of d is 1 less than the subscript of a denoting the term number?

a_3: third term $= a_1 + 2d$ a_4: fourth term $= a_1 + 3d$

2 is one less than 3.   3 is one less than 4.

Thus, the formula for the *n*th term is

$$a_n\text{: } n\text{th term} = a_1 + (n - 1)d.$$

$n - 1$ is one less than n.

General Term of an Arithmetic Sequence

The nth term (the general term) of an arithmetic sequence with first term a_1 and common difference d is

$$a_n = a_1 + (n - 1)d.$$

EXAMPLE 2 Using the Formula for the General Term of an Arithmetic Sequence

Find the eighth term of the arithmetic sequence whose first term is 4 and whose common difference is -7.

Solution To find the eighth term, a_8, we replace n in the formula with 8, a_1 with 4, and d with -7.

$$a_n = a_1 + (n - 1)d$$
$$a_8 = 4 + (8 - 1)(-7) = 4 + 7(-7) = 4 + (-49) = -45$$

The eighth term is -45. We can check this result by writing the first eight terms of the sequence:

$$4, -3, -10, -17, -24, -31, -38, -45.$$

Check Point 2 Find the ninth term of the arithmetic sequence whose first term is 6 and whose common difference is -5.

EXAMPLE 3 Using an Arithmetic Sequence to Model Teachers' Earnings

According to the National Education Association, teachers in the United States earned an average of $30,532 in 1990. This amount has increased by approximately $1472 per year.

  a. Write a formula for the nth term of the arithmetic sequence that describes teachers' average earnings n years after 1989.

  b. How much will U.S. teachers earn, on average, by the year 2010?

Solution

  a. We can express teachers' earnings by the following arithmetic sequence:

$$30{,}532, \quad 32{,}004, \quad 33{,}476, \quad 34{,}948,....$$

a_1: earnings in 1990, 1 year after 1989	a_2: earnings in 1991, 2 years after 1989	a_3: earnings in 1992, 3 years after 1989	a_4: earnings in 1993, 4 years after 1989

In this sequence, a_1, the first term, represents the amount teachers earned in 1990. Each subsequent year this amount increases by $1472, so $d = 1472$. We use the formula for the general term of an arithmetic sequence to write the nth term of the sequence that describes teachers' earnings n years after 1989.

$$a_n = a_1 + (n - 1)d \qquad \text{This is the formula for the general term of an arithmetic sequence.}$$

$$a_n = 30{,}532 + (n - 1)1472 \qquad a_1 = 30{,}532 \text{ and } d = 1472.$$

$$a_n = 30{,}532 + 1472n - 1472 \qquad \text{Distribute 1472 to each term in parentheses.}$$

$$a_n = 1472n + 29{,}060 \qquad \text{Simplify.}$$

Thus, teachers' earnings n years after 1989 can be described by $a_n = 1472n + 29{,}060$.

b. Now we need to find teachers' earnings in 2010. The year 2010 is 21 years after 1989: That is, $2010 - 1989 = 21$. Thus, $n = 21$. We substitute 21 for n in $a_n = 1472n + 29{,}060$.

$$a_{21} = 1472 \cdot 21 + 29{,}060 = 59{,}972$$

The 22nd term of the sequence is 59,972. Therefore, U.S. teachers are predicted to earn an average of $59,972 by the year 2010.

Check Point 3

According to the U.S. Census Bureau, new one-family houses sold for an average of $159,000 in 1995. This average sales price has increased by approximately $9700 per year.

a. Write a formula for the nth term of the arithmetic sequence that describes the average cost of new one-family houses n years after 1994.

b. How much will new one-family houses cost, on average, by the year 2010?

4 Use the formula for the sum of the first n terms of an arithmetic sequence.

The Sum of the First n Terms of an Arithmetic Sequence

The sum of the first n terms of an arithmetic sequence, denoted by S_n, and called the **nth partial sum**, can be found without having to add up all the terms. Let

$$S_n = a_1 + a_2 + a_3 + \cdots + a_n$$

be the sum of the first n terms of an arithmetic sequence. Because d is the common difference between terms, S_n can be written forward and backward as follows.

Forward: Start with the first term. Keep adding d.

Backward: Start with the last term. Keep subtracting d.

$$S_n = a_1 \qquad\qquad + (a_1 + d) \quad + (a_1 + 2d) + \cdots + a_n$$
$$S_n = a_n \qquad\qquad + (a_n - d) \quad + (a_n - 2d) + \cdots + a_1$$
$$\overline{2S_n = (a_1 + a_n) + (a_1 + a_n) \quad + (a_1 + a_n) \ + \cdots + (a_1 + a_n)} \qquad \text{Add the two equations.}$$

Because there are n sums of $(a_1 + a_n)$ on the right side, we can express this side as $n(a_1 + a_n)$. Thus, the last equation can be simplified:

$$2S_n = n(a_1 + a_n)$$

$$S_n = \frac{n}{2}(a_1 + a_n). \qquad \text{Solve for } S_n, \text{ dividing both sides by 2.}$$

We have proved the following result:

The Sum of the First n Terms of an Arithmetic Sequence

The sum, S_n, of the first n terms of an arithmetic sequence is given by

$$S_n = \frac{n}{2}(a_1 + a_n)$$

in which a_1 is the first term and a_n is the nth term.

To find the sum of the terms of an arithmetic sequence using $S_n = \dfrac{n}{2}(a_1 + a_n)$, we need to know the first term, a_1, the last te.rm, a_n, and the number of terms, n. The following examples illustrate how to use this formula.

EXAMPLE 4 Finding the Sum of n Terms of an Arithmetic Sequence

Find the sum of the first 100 terms of the arithmetic sequence: $1, 3, 5, 7, \ldots$.

Solution We are finding the sum of the first 100 odd numbers. To find the sum of the first 100 terms, S_{100}, we replace n in the formula with 100.

$$S_n = \frac{n}{2}(a_1 + a_n)$$
$$S_{100} = \frac{100}{2}(a_1 + a_{100})$$

| The first term, a_1, is 1. | We must find a_{100}, the 100th term. |

We use the formula for the general term of an arithmetic sequence to find a_{100}. The common difference, d, of $1, 3, 5, 7, \ldots$, is 2.

$$a_n = a_1 + (n-1)d \qquad \text{This is the formula for the nth term of an arithmetic sequence. Use it to find the 100th term.}$$
$$a_{100} = 1 + (100-1)\cdot 2 \qquad \text{Substitute 100 for n, 2 for d, and 1 (the first term) for } a_1.$$
$$= 1 + 99 \cdot 2$$
$$= 1 + 198 = 199$$

Now we are ready to find the sum of the first 100 terms of $1, 3, 5, 7, \ldots, 199$.

$$S_n = \frac{n}{2}(a_1 + a_n) \qquad \text{Use the formula for the sum of the first n terms of an arithmetic sequence. Let } n = 100, a_1 = 1, \text{ and } a_{100} = 199.$$

$$S_{100} = \frac{100}{2}(1 + 199) = 50(200) = 10{,}000$$

The sum of the first 100 odd numbers is 10,000. Equivalently, the 100th partial sum of the sequence $1, 3, 5, 7, \ldots$ is 10,000.

Check Point 4 Find the sum of the first 15 terms of the arithmetic sequence: $3, 6, 9, 12, \ldots$.

EXAMPLE 5 Using S_n to Evaluate a Summation

Find the following sum: $\displaystyle\sum_{i=1}^{25}(5i - 9)$.

Solution

$$\sum_{i=1}^{25}(5i - 9) = (5 \cdot 1 - 9) + (5 \cdot 2 - 9) + (5 \cdot 3 - 9) + \cdots + (5 \cdot 25 - 9)$$
$$= -4 \qquad\quad +1 \qquad\quad +6 \qquad\quad +\cdots + 116$$

By evaluating the first three terms and the last term, we see that $a_1 = -4$; d, the common difference, is $1 - (-4)$ or 5; and a_{25}, the last term, is 116.

Technology

To find

$$\sum_{i=1}^{25} (5i - 9)$$

on a graphing utility, enter SUM SEQ $(5x - 9, x, 1, 25, 1)$. Then press ENTER.

```
sum(seq(5X-9,X,1
,25,1)
              1400
```

$$S_n = \frac{n}{2}(a_1 + a_n) \qquad \text{Use the formula for the sum of the first } n \text{ terms of an arithmetic sequence. Let } n = 25, \ a_1 = -4, \text{ and } a_{25} = 116.$$

$$S_{25} = \frac{25}{2}(-4 + 116) = \frac{25}{2}(112) = 1400.$$

Thus,

$$\sum_{i=1}^{25} (5i - 9) = 1400.$$

Check Point 5 Find the following sum: $\displaystyle\sum_{i=1}^{30} (6i - 11)$.

EXAMPLE 6 Modeling Total Nursing Home Costs over a Six-Year Period

Your grandmother has assets of \$350,000. One option that she is considering involves nursing home care for a six-year period beginning in 2001. The model

$$a_n = 1800n + 49{,}730$$

describes yearly nursing home costs n years after 2000. Does your grandmother have enough to pay for the facility?

Solution We must find the sum of an arithmetic sequence. The first term of the sequence corresponds to nursing home costs in the year 2001. The last term corresponds to nursing home costs in the year 2006. Because the model describes costs n years after 2000, $n = 1$ describes the year 2001 and $n = 6$ describes the year 2006.

$$a_n = 1800n + 49{,}730 \qquad \text{This is the given formula for the general term of the sequence.}$$

$$a_1 = 1800 \cdot 1 + 49{,}730 = 51{,}530 \qquad \text{Find } a_1 \text{ by replacing } n \text{ with 1.}$$

$$a_6 = 1800 \cdot 6 + 49{,}730 = 60{,}530 \qquad \text{Find } a_6 \text{ by replacing } n \text{ with 6.}$$

The first year the facility will cost \$51,530. By year six, the facility will cost \$60,530. Now we must find the sum of these costs for all six years. We focus on the sum of the first six terms of the arithmetic sequence

$$\underset{a_1}{51{,}530}, \quad \underset{a_2}{53{,}330}, \quad \dots, \quad \underset{a_6}{60{,}530}.$$

We find this sum using the formula for the sum of the first n terms of an arithmetic sequence. We are adding 6 terms: $n = 6$. The first term is 51,530: $a_1 = 51{,}530$. The last term—that is, the sixth term—is 60,530: $a_6 = 60{,}530$.

$$S_n = \frac{n}{2}(a_1 + a_n)$$

$$S_6 = \frac{6}{2}(51{,}530 + 60{,}530) = 3(112{,}060) = 336{,}180$$

Total nursing home costs for your grandmother are predicted to be \$336,180. Because your grandmother's assets are \$350,000, she has enough to pay for the facility.

Check Point 6 In Example 6, how much would it cost for nursing home care for a ten-year period beginning in 2001?

EXERCISE SET 11.2

Practice Exercises

In Exercises 1–14, write the first six terms of each arithmetic sequence.

1. $a_1 = 200, d = 20$ **2.** $a_1 = 300, d = 50$

3. $a_1 = -7, d = 4$ **4.** $a_1 = -8, d = 5$

5. $a_1 = 300, d = -90$ **6.** $a_1 = 200, d = -60$

7. $a_1 = \frac{5}{2}, d = -\frac{1}{2}$ **8.** $a_1 = \frac{3}{4}, d = -\frac{1}{4}$

9. $a_n = a_{n-1} + 6, a_1 = -9$ **10.** $a_n = a_{n-1} + 4, a_1 = -7$

11. $a_n = a_{n-1} - 10, a_1 = 30$ **12.** $a_n = a_{n-1} - 20, a_1 = 50$

13. $a_n = a_{n-1} - 0.4, a_1 = 1.6$

14. $a_n = a_{n-1} - 0.3, a_1 = -1.7$

In Exercises 15–22, find the indicated term of the arithmetic sequence with first term, a_1, and common difference, d.

15. Find a_6 when $a_1 = 13, d = 4$.

16. Find a_{16} when $a_1 = 9, d = 2$.

17. Find a_{50} when $a_1 = 7, d = 5$.

18. Find a_{60} when $a_1 = 8, d = 6$.

19. Find a_{200} when $a_1 = -40, d = 5$.

20. Find a_{150} when $a_1 = -60, d = 5$.

21. Find a_{60} when $a_1 = 35, d = -3$.

22. Find a_{70} when $a_1 = -32, d = 4$.

In Exercises 23–34, write a formula for the general term (the nth term) of each arithmetic sequence. Do not use a recursion formula. Then use the formula for a_n to find a_{20}, the 20th term of the sequence.

23. $1, 5, 9, 13, \ldots$ **24.** $2, 7, 12, 17, \ldots$

25. $7, 3, -1, -5, \ldots$ **26.** $6, 1, -4, -9, \ldots$

27. $a_1 = 9, d = 2$ **28.** $a_1 = 6, d = 3$

29. $a_1 = -20, d = -4$ **30.** $a_1 = -70, d = -5$

31. $a_n = a_{n-1} + 3, a_1 = 4$ **32.** $a_n = a_{n-1} + 5, a_1 = 6$

33. $a_n = a_{n-1} - 10, a_1 = 30$ **34.** $a_n = a_{n-1} - 12, a_1 = 24$

35. Find the sum of the first 20 terms of the arithmetic sequence: $4, 10, 16, 22, \ldots$.

36. Find the sum of the first 25 terms of the arithmetic sequence: $7, 19, 31, 43, \ldots$.

37. Find the sum of the first 50 terms of the arithmetic sequence: $-10, -6, -2, 2, \ldots$.

38. Find the sum of the first 50 terms of the arithmetic sequence: $-15, -9, -3, 3, \ldots$.

39. Find $1 + 2 + 3 + 4 + \cdots + 100$, the sum of the first 100 natural numbers.

40. Find $2 + 4 + 6 + 8 + \cdots + 200$, the sum of the first 100 positive even integers.

41. Find the sum of the first 60 positive even integers.

42. Find the sum of the first 80 positive even integers.

43. Find the sum of the even integers between 21 and 45.

44. Find the sum of the odd integers between 30 and 54.

For Exercises 45–50, write out the first three terms and the last term. Then use the formula for the sum of the first n terms of an arithmetic sequence to find the indicated sum.

45. $\displaystyle\sum_{i=1}^{17} (5i + 3)$ **46.** $\displaystyle\sum_{i=1}^{20} (6i - 4)$

47. $\displaystyle\sum_{i=1}^{30} (-3i + 5)$ **48.** $\displaystyle\sum_{i=1}^{40} (-2i + 6)$

49. $\displaystyle\sum_{i=1}^{100} 4i$ **50.** $\displaystyle\sum_{i=1}^{50} -4i$

Application Exercises

The graph shows pounds of various food groups consumed per year by the average American. Exercises 51–54 involve developing arithmetic sequences that model the data. In Exercises 53–54, models will vary.

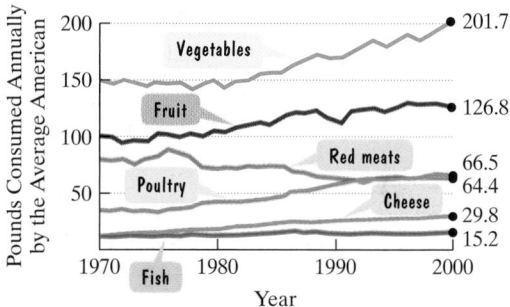

Per Capita Consumption of
Various Food Groups

Source: U.S. Department of Agriculture

51. The graph shows that the average American consumed 150 pounds of vegetables in 1970. On average, this amount has increased by approximately 1.7 pounds per person per year.

a. Write a formula for the nth term of the arithmetic sequence that describes pounds of vegetables consumed annually by the average American n years after 1969.

b. How many pounds of vegetables will be consumed by the average American in 2006?

52. The graph shows that the average American consumed 100 pounds of fruit in 1970. On average, this amount has increased by approximately 0.9 pound per person per year.

  a. Write a formula for the nth term of the arithmetic sequence that describes pounds of fruit consumed annually by the average American n years after 1969.

  b. How many pounds of fruit will be consumed by the average American in 2006?

53. a. Use the data shown to write a formula for the nth term of the arithmetic sequence that describes pounds of cheese consumed annually by the average American n years after 1969.

  b. How many pounds of cheese will be consumed by the average American in 2006?

54. Use the data shown for fish, poultry, or red meats, and repeat both parts of Exercise 53.

55. Company A pays $24,000 yearly with raises of $1600 per year. Company B pays $28,000 yearly with raises of $1000 per year. Which company will pay more in year 10? How much more?

56. Company A pays $23,000 yearly with raises of $1200 per year. Company B pays $26,000 yearly with raises of $800 per year. Which company will pay more in year 10? How much more?

57. According to the Environmental Protection Agency, in 1960 the United States recovered 3.78 million tons of solid waste. Due primarily to recycling programs, this amount has increased by approximately 0.576 million ton per year.

  a. Write the general term for the arithmetic sequence modeling the amount of solid waste recovered in the United States n years after 1959.

  b. What is the total amount of solid waste recovered from 1960 through 2000?

58. According to the Environmental Protection Agency, in 1960 the United States generated 87.1 million tons of solid waste. This amount has increased by approximately 3.14 million tons per year.

  a. Write the general term for the arithmetic sequence modeling the amount of solid waste generated in the United States n years after 1959.

  b. What is the total amount of solid waste generated from 1960 through 2000?

59. A company offers a starting yearly salary of $33,000 with raises of $2500 per year. Find the total salary over a ten-year period.

60. You are considering two job offers. Company A will start you at $19,000 annually and guarantee a raise of $2600 per year. Company B will start you at a higher salary, $27,000 annually, but will only guarantee a raise of $1200 per year. Find the total salary that each company will pay you over a ten-year period. Which company pays the greater total amount?

61. A theater has 30 seats in the first row, 32 seats in the second row, increasing by 2 seats per row for a total of 26 rows. How many seats are there in the theater?

62. A section in a stadium has 20 seats in the first row, 23 seats in the second row, increasing by 3 seats per row for a total of 38 rows. How many seats are in this section of the stadium?

Writing in Mathematics

63. What is an arithmetic sequence? Give an example with your explanation.

64. What is the common difference in an arithmetic sequence?

65. Explain how to find the general term of an arithmetic sequence.

66. Explain how to find the sum of the first n terms of an arithmetic sequence without having to add up all the terms.

67. Teachers' earnings n years after 1989 can be described by $a_n = 1472n + 29,060$. According to this model, what will teachers earn in 2083? Describe two possible circumstances that would render this predicted salary incorrect.

Technology Exercises

68. Use the $\boxed{\text{SEQ}}$ (sequence) capability of a graphing utility and the formula you obtained for a_n to verify the value you found for a_{20} in any five exercises from Exercises 23–34.

69. Use the capability of a graphing utility to calculate the sum of a sequence to verify any five of your answers to Exercises 45–50.

Critical Thinking Exercises

70. Give examples of two different arithmetic sequences whose fourth term, a_4, is 10.

71. In the sequence 21,700, 23,172, 24,644, 26,116, ..., which term is 314,628?

72. A *degree-day* is a unit used to measure the fuel requirements of buildings. By definition, each degree that the average daily temperature is below 65°F is 1 degree-day. For example, a temperature of 42°F constitutes 23 degree-days. If the average temperature on January 1 was 42°F and fell 2°F for each subsequent day up to and including January 10, how many degree-days are included from January 1 to January 10?

73. Show that the sum of the first n positive odd integers,

$$1 + 3 + 5 + \cdots + (2n - 1),$$

is n^2.

Group Exercise

74. Members of your group have been hired by the Environmental Protection Agency to write a report on whether we are making significant progress in recovering solid waste. Use the models from Exercises 57 and 58 as the basis for your report. A graph of each model from 1960 through 2000 would be helpful. What percentage of solid waste generated is actually recovered on a year-to-year basis? Be as creative as you want in your report and then draw conclusions. The group should write up the report and perhaps even include suggestions as to how we might improve recycling progress.

SECTION 11.3 *Geometric Sequences*

Objectives

1. Find the common ratio of a geometric sequence.
2. Write terms of a geometric sequence.
3. Use the formula for the general term of a geometric sequence.
4. Use the formula for the sum of the first *n* terms of a geometric sequence.
5. Find the value of an annuity.
6. Use the formula for the sum of an infinite geometric series.

Here we are at the closing moments of a job interview. You're shaking hands with the manager. You managed to answer all the tough questions without losing your poise, and now you've been offered a job. As a matter of fact, your qualifications are so terrific that you've been offered two jobs—one just the day before, with a rival company in the same field! One company offers $30,000 the first year, with increases of 6% per year for four years after that. The other offers $32,000 the first year, with increases of 3% per year after that. Over a five-year period, which is the better offer?

If salary raises amount to a certain percent each year, the yearly salaries over time form a geometric sequence. In this section, we investigate geometric sequences and their properties. After studying the section, you will be in a position to decide which job offer to accept: you will know which company will pay you more over five years.

Geometric Sequences

Figure 11.2 shows a sequence in which the number of squares is increasing. From left to right, the number of squares is 1, 5, 25, 125, and 625. In this sequence, each term after the first, 1, is obtained by multiplying the preceding term by a constant amount, namely 5. This sequence of increasing number of squares is an example of a *geometric sequence*.

Figure 11.2 A geometric sequence of squares

> **Definition of a Geometric Sequence**
> A **geometric sequence** is a sequence in which each term after the first is obtained by multiplying the preceding term by a fixed nonzero constant. The amount by which we multiply each time is called the **common ratio** of the sequence.

1 Find the common ratio of a geometric sequence.

The common ratio, r, is found by dividing any term after the first term by the term that directly precedes it. In the following examples, the common ratio is found by dividing the second term by the first term: $\frac{a_2}{a_1}$.

Geometric sequence	Common ratio
$1, 5, 25, 125, 625, \ldots$	$r = \dfrac{5}{1} = 5$
$4, 8, 16, 32, 64, \ldots$	$r = \dfrac{8}{4} = 2$
$6, -12, 24, -48, 96, \ldots$	$r = \dfrac{-12}{6} = -2$
$9, -3, 1, -\dfrac{1}{3}, \dfrac{1}{9}, \ldots$	$r = \dfrac{-3}{9} = -\dfrac{1}{3}$

Study Tip

When the common ratio of a geometric sequence is negative, the signs of the terms alternate.

2 Write terms of a geometric sequence.

How do we find the terms of a geometric sequence when the first term and the common ratio are known? We multiply the first term by the common ratio to get the second term, multiply the second term by the common ratio to get the third term, and so on.

EXAMPLE 1   Writing the Terms of a Geometric Sequence

Write the first six terms of the geometric sequence with first term 6 and common ratio $\frac{1}{3}$.

Solution   The first term is 6. The second term is $6 \cdot \frac{1}{3}$, or 2. The third term is $2 \cdot \frac{1}{3}$, or $\frac{2}{3}$. The fourth term is $\frac{2}{3} \cdot \frac{1}{3}$, or $\frac{2}{9}$, and so on. The first six terms are

$$6, 2, \tfrac{2}{3}, \tfrac{2}{9}, \tfrac{2}{27}, \text{ and } \tfrac{2}{81}.$$

Check Point 1   Write the first six terms of the geometric sequence with first term 12 and common ratio $\frac{1}{2}$.

3 Use the formula for the general term of a geometric sequence.

The General Term of a Geometric Sequence

Consider a geometric sequence whose first term is a_1 and whose common ratio is r. We are looking for a formula for the general term, a_n. Let's begin by writing the first six terms. The first term is a_1. The second term is $a_1 r$. The third term is $a_1 r \cdot r$, or $a_1 r^2$. The fourth term is $a_1 r^2 \cdot r$, or $a_1 r^3$, and so on. Starting with a_1 and multiplying each successive term by r, the first six terms are

$$a_1, \qquad a_1 r, \qquad a_1 r^2, \qquad a_1 r^3, \qquad a_1 r^4, \qquad a_1 r^5.$$

a_1, first term a_2, second term a_3, third term a_4, fourth term a_5, fifth term a_6, sixth term

Compare the exponent on r and the subscript of a denoting the term number. Can you see that the exponent on r is 1 less than the subscript of a denoting the term number?

$$a_3: \text{third term} = a_1 r^2 \qquad a_4: \text{fourth term} = a_1 r^3$$

2 is one less than 3. 3 is one less than 4.

Thus, the formula for the nth term is

$$a_n = a_1 r^{n-1}.$$

$n-1$ is one less than n.

General Term of a Geometric Sequence

The nth term (the general term) of a geometric sequence with first term a_1 and common ratio r is

$$a_n = a_1 r^{n-1}.$$

EXAMPLE 2 Using the Formula for the General Term of a Geometric Sequence

Find the eighth term of the geometric sequence whose first term is -4 and whose common ratio is -2.

Solution To find the eighth term, a_8, we replace n in the formula with 8, a_1 with -4, and r with -2.

$$a_n = a_1 r^{n-1}$$
$$a_8 = -4(-2)^{8-1} = -4(-2)^7 = -4(-128) = 512$$

The eighth term is 512. We can check this result by writing the first eight terms of the sequence:

$$-4, 8, -16, 32, -64, 128, -256, 512.$$

Study Tip

Be careful with the order of operations when evaluating

$$a_1 r^{n-1}.$$

First find r^{n-1}. Then multiply the result by a_1.

Check Point 2 Find the seventh term of the geometric sequence whose first term is 5 and whose common ratio is -3.

In Chapter 4, we studied exponential functions of the form $f(x) = b^x$ and the explosive exponential growth of world population. In our next example, we consider Florida's geometric population growth. Because **a geometric sequence is an exponential function whose domain is the set of positive integers,** geometric and exponential growth mean the same thing. (By contrast, an arithmetic sequence is a *linear function* whose domain is the set of positive integers.)

EXAMPLE 3 Geometric Population Growth

The population of Florida from 1990 through 1997 is shown in the following table:

Year	1990	1991	1992	1993	1994	1995	1996	1997
Population in millions	12.94	13.20	13.46	13.73	14.00	14.28	14.57	14.86

Geometric Population Growth

Economist Thomas Malthus (1766–1834) predicted that population growth would increase as a geometric sequence and food production would increase as an arithmetic sequence. He concluded that eventually population would exceed food production. If two sequences, one geometric and one arithmetic, are increasing, the geometric sequence will eventually overtake the arithmetic sequence, regardless of any head start that the arithmetic sequence might initially have.

a. Show that the population is increasing geometrically.

b. Write the general term for the geometric sequence describing population growth for Florida n years after 1989.

c. Estimate Florida's population, in millions, for the year 2000.

Solution

a. First, we divide the population for each year by the population in the preceding year.

$$\frac{13.20}{12.94} \approx 1.02, \quad \frac{13.46}{13.20} \approx 1.02, \quad \frac{13.73}{13.46} \approx 1.02$$

Continuing in this manner, we will keep getting approximately 1.02. This means that the population is increasing geometrically with $r \approx 1.02$. In this situation, the common ratio is the growth rate, indicating that the population of Florida in any year shown in the table is approximately 1.02 times the population the year before.

b. The sequence of Florida's population growth is

$$12.94, 13.20, 13.46, 13.73, 14.00, 14.28, 14.57, 14.86, \ldots.$$

Because the population is increasing geometrically, we can find the general term of this sequence using

$$a_n = a_1 \, r^{n-1}.$$

In this sequence, $a_1 = 12.94$ and [from part (a)] $r \approx 1.02$. We substitute these values into the formula for the general term. This gives the general term for the geometric sequence describing Florida's population n years after 1989.

$$a_n = 12.94(1.02)^{n-1}$$

c. We can use the formula for the general term, a_n, in part (b) to estimate Florida's population for the year 2000. The year 2000 is 11 years after 1989— that is, $2000 - 1989 = 11$. Thus, $n = 11$. We substitute 11 for n in $a_n = 12.94(1.02)^{n-1}$.

$$a_{11} = 12.94(1.02)^{11-1} = 12.94(1.02)^{10} \approx 15.77$$

The formula indicates that Florida had a population of approximately 15.77 million in the year 2000. According to the U.S. Census Bureau, Florida's population in 2000 was 15.98 million. Our geometric sequence models the actual population fairly well.

Check Point 3 Write the general term for the geometric sequence

$$3, 6, 12, 24, 48, \ldots.$$

Then use the formula for the general term to find the eighth term.

The Sum of the First n Terms of a Geometric Sequence

4 Use the formula for the sum of the first n terms of a geometric sequence.

The sum of the first n terms of a geometric sequence, denoted by S_n, and called the **nth partial sum,** can be found without having to add up all the terms. Recall that the first n terms of a geometric sequence are

$$a_1, a_1 \, r, a_1 \, r^2, \ldots, a_1 \, r^{n-2}, a_1 \, r^{n-1}.$$

We proceed as follows:

$$S_n = a_1 + a_1 r + a_1 r^2 + \cdots + a_1 r^{n-2} + a_1 r^{n-1}$$

S_n is the sum of the first n terms of the sequence.

$$rS_n = a_1 r + a_1 r^2 + a_1 r^3 + \cdots + a_1 r^{n-1} + a_1 r^n$$

Multiply both sides of the equation by r.

$$S_n - rS_n = a_1 - a_1 r^n$$

Subtract the second equation from the first equation.

$$S_n(1 - r) = a_1(1 - r^n)$$

Factor out S_n on the left and a_1 on the right.

$$S_n = \frac{a_1(1 - r^n)}{1 - r}.$$

Solve for S_n by dividing both sides by $1 - r$ (assuming that $r \neq 1$).

We have proved the following result:

Study Tip

If the common ratio is 1, the geometric sequence is

$$a_1, a_1, a_1, a_1, \ldots.$$

The sum of the first n terms of this sequence is na_1:

$$S_n = \underbrace{a_1 + a_1 + a_1 + \cdots + a_1}_{\text{There are } n \text{ terms.}}$$

$$= na_1.$$

The Sum of the First n Terms of a Geometric Sequence

The sum, S_n, of the first n terms of a geometric sequence is given by

$$S_n = \frac{a_1(1 - r^n)}{1 - r}$$

in which a_1 is the first term and r is the common ratio ($r \neq 1$).

To find the sum of the terms of a geometric sequence, we need to know the first term, a_1, the common ratio, r, and the number of terms, n. The following examples illustrate how to use this formula.

EXAMPLE 4   Finding the Sum of the First n Terms of a Geometric Sequence

Find the sum of the first 18 terms of the geometric sequence: $2, -8, 32, -128, \ldots$.

Solution   To find the sum of the first 18 terms, S_{18}, we replace n in the formula with 18.

$$S_n = \frac{a_1(1 - r^n)}{1 - r}$$

$$S_{18} = \frac{a_1(1 - r^{18})}{1 - r}$$

The first term, a_1, is 2.   We must find r, the common ratio.

We can find the common ratio by dividing the second term by the first term.

$$r = \frac{a_2}{a_1} = -\frac{8}{2} = -4$$

Now we are ready to find the sum of the first 18 terms of $2, -8, 32, -128, \ldots$.

$$S_n = \frac{a_1(1 - r^n)}{1 - r}$$ Use the formula for the sum of the first n terms of a geometric sequence.

$$S_{18} = \frac{2[1 - (-4)^{18}]}{1 - (-4)}$$ a_1 (the first term) $= 2$, $r = -4$, and $n = 18$ because we want the sum of the first 18 terms.

$$= -27{,}487{,}790{,}694$$ Use a calculator.

The sum of the first 18 terms is $-27{,}487{,}790{,}694$. Equivalently, this number is the 18th partial sum of the sequence $2, -8, 32, -128, \ldots$.

> **Check Point 4**   Find the sum of the first nine terms of the geometric sequence: $2, -6, 18, -54, \ldots$.

Technology

To find

$$\sum_{i=1}^{10} 6 \cdot 2^i$$

on a graphing utility, enter

[SUM] [SEQ] $(6 \times 2^x, x, 1, 10, 1)$.

Then press [ENTER].

```
sum(seq(6*2^X,X,
1,10,1)
            12276
```

EXAMPLE 5   Using S_n to Evaluate a Summation

Find the following sum: $\displaystyle\sum_{i=1}^{10} 6 \cdot 2^i$.

Solution   Let's write out a few terms in the sum.

$$\sum_{i=1}^{10} 6 \cdot 2^i = 6 \cdot 2 + 6 \cdot 2^2 + 6 \cdot 2^3 + \cdots + 6 \cdot 2^{10}$$

Can you see that each term after the first is obtained by multiplying the preceding term by 2? To find the sum of the 10 terms ($n = 10$), we need to know the first term, a_1, and the common ratio, r. The first term is $6 \cdot 2$ or 12: $a_1 = 12$. The common ratio is 2.

$$S_n = \frac{a_1(1 - r^n)}{1 - r}$$ Use the formula for the sum of the first n terms of a geometric sequence.

$$S_{10} = \frac{12(1 - 2^{10})}{1 - 2}$$ a_1 (the first term) $= 12$, $r = 2$, and $n = 10$ because we are adding ten terms.

$$= 12{,}276$$ Use a calculator.

Thus,

$$\sum_{i=1}^{10} 6 \cdot 2^i = 12{,}276$$

> **Check Point 5**   Find the following sum: $\displaystyle\sum_{i=1}^{8} 2 \cdot 3^i$.

Some of the exercises in the previous exercise set involved situations in which salaries increase by a fixed amount each year. A more realistic situation is one in which salary raises increase by a certain percent each year. Example 6 shows how such a situation can be described using a geometric series.

EXAMPLE 6   Computing a Lifetime Salary

A union contract specifies that each worker will receive a 5% pay increase each year for the next 30 years. One worker is paid $20,000 the first year. What is this person's total lifetime salary over a 30-year period?

Solution   The salary for the first year is $20,000. With a 5% raise, the second-year salary is computed as follows:

Salary for year 2 $= 20,000 + 20,000(0.05) = 20,000(1 + 0.05) = 20,000(1.05).$

Each year, the salary is 1.05 times what it was in the previous year. Thus, the salary for year 3 is 1.05 times $20,000(1.05)$, or $20,000(1.05)^2$. The salaries for the first five years are given in the table.

Yearly Salaries					
Year 1	Year 2	Year 3	Year 4	Year 5	...
20,000	$20,000(1.05)$	$20,000(1.05)^2$	$20,000(1.05)^3$	$20,000(1.05)^4$	...

The numbers in the second row form a geometric sequence with $a_1 = 20,000$ and $r = 1.05$. To find the total salary over 30 years, we use the formula for the sum of the first n terms of a geometric sequence, with $n = 30$.

$$S_n = \frac{a_1(1 - r^n)}{1 - r}$$

$$S_{30} = \frac{20,000\left[1 - (1.05)^{30}\right]}{1 - 1.05}$$

Total salary over 30 years

$$= \frac{20,000\left[1 - (1.05)^{30}\right]}{-0.05}$$

$$\approx 1,328,777 \qquad \text{Use a calculator.}$$

The total salary over the 30-year period is approximately $1,328,777.

Check Point 6   A job pays a salary of $30,000 the first year. During the next 29 years, the salary increases by 6% each year. What is the total lifetime salary over the 30-year period?

5 Find the value of an annuity.

Annuities

The compound interest formula

$$A = P(1 + r)^t$$

gives the future value, A, after t years, when a fixed amount of money, P, the principal, is deposited in an account that pays an annual interest rate r (in decimal form) compounded once a year. However, money is often invested in small amounts at periodic intervals. For example, to save for retirement, you might decide to place $1000 into an Individual Retirement Account (IRA) at the end of each year until you retire. An **annuity** is a sequence of equal payments made at equal time periods. An IRA is an example of an annuity.

Suppose P dollars is deposited into an account at the end of each year. The account pays an annual interest rate, r, compounded annually. At the end of the first year, the account contains P dollars. At the end of the second year, P dollars is deposited again. At the time of this deposit, the first deposit has received interest earned during the second year. The **value of the annuity** is the sum of all deposits made plus all interest paid. Thus, the value of the annuity after two years is

$$P + P(1 + r).$$

Deposit of P dollars at end of second year	First-year deposit of P dollars with interest earned for a year

The value of the annuity after three years is

$$P \quad + \quad P(1 + r) \quad + \quad P(1 + r)^2.$$

Deposit of P dollars at end of third year	Second-year deposit of P dollars with interest earned for a year	First-year deposit of P dollars with interest earned over two years

The value of the annuity after t years is

$$P + P(1 + r) + P(1 + r)^2 + P(1 + r)^3 + \cdots + P(1 + r)^{t-1}$$

Deposit of P dollars at end of year t	First-year deposit of P dollars with interest earned over $t - 1$ years

This is the sum of the terms of a geometric sequence with first term P and common ratio $1 + r$. We use the formula

$$S_n = \frac{a_1(1 - r^n)}{1 - r}$$

to find the sum of the terms:

$$S_n = \frac{P[1 - (1 + r)^t]}{1 - (1 + r)} = \frac{P[1 - (1 + r)^t]}{-r} = P\frac{(1 + r)^t - 1}{r}.$$

This formula gives the value of an annuity after t years if interest is compounded once a year. We can adjust the formula to find the value of an annuity if equal payments are made at the end of each of n compounding periods per year.

Value of an Annuity: Interest Compounded n Times per Year

If P is the deposit made at the end of each compounding period for an annuity at r percent annual interest compounded n times per year, the value, A, of the annuity after t years is

$$A = P\frac{\left(1 + \dfrac{r}{n}\right)^{nt} - 1}{\dfrac{r}{n}}.$$

EXAMPLE 7 Determining the Value of an Annuity

To save for retirement, you decide to deposit $1000 into an IRA at the end of each year for the next 30 years. If the interest rate is 10% per year compounded annually, find the value of the IRA after 30 years.

Solution The annuity involves 30 year-end deposits of $P = \$1000$. The interest rate is $10\%: r = 0.10$. Because the deposits are made once a year and the interest is compounded once a year, $n = 1$. The number of years is 30: $t = 30$. We replace the variables in the formula for the value of an annuity with these numbers.

$$A = P \frac{\left(1 + \dfrac{r}{n}\right)^{nt} - 1}{\dfrac{r}{n}}$$

$$A = 1000 \frac{\left(1 + \dfrac{0.10}{1}\right)^{1 \cdot 30} - 1}{\dfrac{0.10}{1}} \approx 164{,}494$$

The value of the IRA at the end of 30 years is approximately $164,494.

> **Check Point 7** If \$3000 is deposited into an IRA at the end of each year for 40 years and the interest rate is 10% per year compounded annually, find the value of the IRA after 40 years.

6 Use the formula for the sum of an infinite geometric series.

Geometric Series

An infinite sum of the form

$$a_1 + a_1 r + a_1 r^2 + a_1 r^3 + \cdots + a_1 r^{n-1} + \cdots$$

with first term a_1 and common ratio r is called an **infinite geometric series.** How can we determine which infinite geometric series have sums and which do not? We look at what happens to r^n as n gets larger in the formula for the sum of the first n terms of this series, namely

$$S_n = \frac{a_1(1 - r^n)}{1 - r}.$$

If r is any number between -1 and 1, that is, $-1 < r < 1$, the term r^n approaches 0 as n gets larger. For example, consider what happens to r^n for $r = \frac{1}{2}$:

$$\left(\tfrac{1}{2}\right)^1 = \tfrac{1}{2} \quad \left(\tfrac{1}{2}\right)^2 = \tfrac{1}{4} \quad \left(\tfrac{1}{2}\right)^3 = \tfrac{1}{8} \quad \left(\tfrac{1}{2}\right)^4 = \tfrac{1}{16} \quad \left(\tfrac{1}{2}\right)^5 = \tfrac{1}{32} \quad \left(\tfrac{1}{2}\right)^6 = \tfrac{1}{64}.$$

These numbers are approaching 0 as n gets larger.

Take another look at the formula for the sum of the first n terms of a geometric sequence.

$$S_n = \frac{a_1(1 - r^n)}{1 - r}$$

If $-1 < r < 1$, r^n approaches 0 as n gets larger.

Let us replace r^n with 0 in the formula for S_n. This change gives us a formula for the sum of infinite geometric series with common ratios between -1 and 1.

The Sum of an Infinite Geometric Series

If $-1 < r < 1$ (equivalently, $|r| < 1$), then the sum of the infinite geometric series

$$a_1 + a_1 r + a_1 r^2 + a_1 r^3 + \cdots$$

in which a_1 is the first term and r is the common ratio is given by

$$S = \frac{a_1}{1 - r}.$$

If $|r| \geq 1$, the infinite series does not have a sum.

To use the formula for the sum of an infinite geometric series, we need to know the first term and the common ratio. For example, consider

$$\frac{1}{2} + \frac{1}{4} + \frac{1}{8} + \frac{1}{16} + \frac{1}{32} + \cdots.$$

First term, a_1, is $\frac{1}{2}$.

Common ratio, r, is $\frac{a_2}{a_1}$.

$$r = \frac{1}{4} \div \frac{1}{2} = \frac{1}{4} \cdot 2 = \frac{1}{2}$$

With $r = \frac{1}{2}$, the condition that $|r| < 1$ is met, so the infinite geometric series has a sum given by $S = \dfrac{a_1}{1 - r}$. The sum of the series is found as follows:

$$\frac{1}{2} + \frac{1}{4} + \frac{1}{8} + \frac{1}{16} + \frac{1}{32} + \cdots = \frac{a_1}{1 - r} = \frac{\frac{1}{2}}{1 - \frac{1}{2}} = \frac{\frac{1}{2}}{\frac{1}{2}} = 1.$$

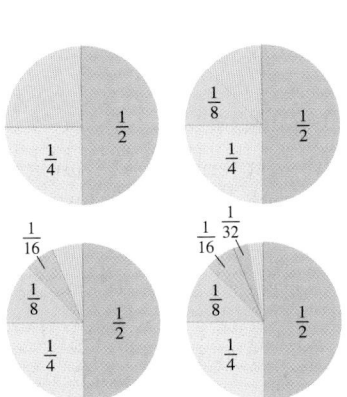

Figure 11.3 The sum $\frac{1}{2} + \frac{1}{4} + \frac{1}{8} + \frac{1}{16} + \frac{1}{32} + \cdots$ is approaching 1.

Thus, the sum of the infinite geometric series is 1. Notice how this is illustrated in Figure 11.3. As more terms are included, the sum is approaching the area of one complete circle.

EXAMPLE 8   Finding the Sum of an Infinite Geometric Series

Find the sum of the infinite geometric series: $\frac{3}{8} - \frac{3}{16} + \frac{3}{32} - \frac{3}{64} + \cdots.$

Solution   Before finding the sum, we must find the common ratio.

$$r = \frac{a_2}{a_1} = \frac{-\frac{3}{16}}{\frac{3}{8}} = -\frac{3}{16} \cdot \frac{8}{3} = -\frac{1}{2}$$

Because $r = -\frac{1}{2}$, the condition that $|r| < 1$ is met. Thus, the infinite geometric series has a sum.

$$S = \frac{a_1}{1 - r}$$

This is the formula for the sum of an infinite geometric series. Let $a_1 = \frac{3}{8}$ and $r = -\frac{1}{2}$.

$$= \frac{\frac{3}{8}}{1 - \left(-\frac{1}{2}\right)} = \frac{\frac{3}{8}}{\frac{3}{2}} = \frac{3}{8} \cdot \frac{2}{3} = \frac{1}{4}$$

Thus, the sum of this infinite geometric series is $\frac{1}{4}$. Put in an informal way, as we continue to add more and more terms, the sum approaches, and is approximately equal to, $\frac{1}{4}$.

Check Point 8   Find the sum of the infinite geometric series: $3 + 2 + \frac{4}{3} + \frac{8}{9} + \cdots.$

We can use the formula for the sum of an infinite series to express a repeating decimal as a fraction in lowest terms.

EXAMPLE 9   Writing a Repeating Decimal as a Fraction

Express $0.\overline{7}$ as a fraction in lowest terms.

Solution

$$0.\overline{7} = 0.7777\ldots = \frac{7}{10} + \frac{7}{100} + \frac{7}{1000} + \frac{7}{10,000} + \cdots$$

Observe that $0.\overline{7}$ is an infinite geometric series with first term $\frac{7}{10}$ and common ratio $\frac{1}{10}$. Because $r = \frac{1}{10}$, the condition that $|r| < 1$ is met. Thus, we can use our formula to find the sum. Therefore,

$$0.\overline{7} = \frac{a_1}{1 - r} = \frac{\frac{7}{10}}{1 - \frac{1}{10}} = \frac{\frac{7}{10}}{\frac{9}{10}} = \frac{7}{10} \cdot \frac{10}{9} = \frac{7}{9}.$$

An equivalent fraction for $0.\overline{7}$ is $\frac{7}{9}$.

Check Point 9   Express $0.\overline{9}$ as a fraction in lowest terms.

Infinite geometric series have many applications, as illustrated in Example 10.

EXAMPLE 10   Tax Rebates and the Multiplier Effect

A tax rebate that returns a certain amount of money to taxpayers can have a total effect on the economy that is many times this amount. In economics, this phenomenon is called the **multiplier effect.** Suppose, for example, that the government reduces taxes so that each consumer has $2000 more income. The government assumes that each person will spend 70% of this (= $1400). The individuals and businesses receiving this $1400 in turn spend 70% of it (= $980), creating extra income for other people to spend, and so on. Determine the total amount spent on consumer goods from the initial $2000 tax rebate.

Solution   The total amount spent is given by the infinite geometric series

$$1400 + 980 + 686 + \cdots.$$

$1400

70% is spent.

$980

70% is spent.

$686

70% of 1400   70% of 980

The first term is 1400: $a_1 = 1400$. The common ratio is 70%, or 0.7: $r = 0.7$. Because $r = 0.7$, the condition that $|r| < 1$ is met. Thus, we can use our formula to find the sum. Therefore,

$$1400 + 980 + 686 + \cdots = \frac{a_1}{1 - r} = \frac{1400}{1 - 0.7} \approx 4667.$$

This means that the total amount spent on consumer goods from the initial $2000 rebate is approximately $4667.

Check Point 10   Rework Example 10 and determine the total amount spent on consumer goods with a $1000 tax rebate and 80% spending down the line.

EXERCISE SET 11.3

Practice Exercises

In Exercises 1–8, write the first five terms of each geometric sequence.

1. $a_1 = 5, \quad r = 3$

2. $a_1 = 4, \quad r = 3$

3. $a_1 = 20, \quad r = \frac{1}{2}$

4. $a_1 = 24, \quad r = \frac{1}{3}$

5. $a_n = -4a_{n-1}, \quad a_1 = 10$

6. $a_n = -3a_{n-1}, \quad a_1 = 10$

7. $a_n = -5a_{n-1}, \quad a_1 = -6$

8. $a_n = -6a_{n-1}, \quad a_1 = -2$

In Exercises 9–16, use the formula for the general term (the nth term) of a geometric sequence to find the indicated term of each sequence with the given first term, a_1, and common ratio, r.

9. Find a_8 when $a_1 = 6, r = 2$.

10. Find a_8 when $a_1 = 5, r = 3$.

11. Find a_{12} when $a_1 = 5, r = -2$.

12. Find a_{12} when $a_1 = 4, r = -2$.

13. Find a_{40} when $a_1 = 1000, r = -\frac{1}{2}$.

14. Find a_{30} when $a_1 = 8000, r = -\frac{1}{2}$.

15. Find a_8 when $a_1 = 1{,}000{,}000, r = 0.1$.

16. Find a_8 when $a_1 = 40{,}000, r = 0.1$.

In Exercises 17–24, write a formula for the general term (the nth term) of each geometric sequence. Then use the formula for a_n to find a_7, the seventh term of the sequence.

17. $3, 12, 48, 192, \ldots$

18. $3, 15, 75, 375, \ldots$

19. $18, 6, 2, \frac{2}{3}, \ldots$

20. $12, 6, 3, \frac{3}{2}, \ldots$

21. $1.5, -3, 6, -12, \ldots$

22. $5, -1, \frac{1}{5}, -\frac{1}{25}, \ldots$

23. $0.0004, -0.004, 0.04, -0.4, \ldots$

24. $0.0007, -0.007, 0.07, -0.7, \ldots$

Use the formula for the sum of the first n terms of a geometric sequence to solve Exercises 25–36.

25. Find the sum of the first 12 terms of the geometric sequence: $2, 6, 18, 54, \ldots$.

26. Find the sum of the first 12 terms of the geometric sequence: $3, 6, 12, 24, \ldots$.

27. Find the sum of the first 11 terms of the geometric sequence: $3, -6, 12, -24, \ldots$.

28. Find the sum of the first 11 terms of the geometric sequence: $4, -12, 36, -108, \ldots$.

29. Find the sum of the first 14 terms of the geometric sequence: $-\frac{3}{2}, 3, -6, 12, \ldots$.

30. Find the sum of the first 14 terms of the geometric sequence: $-\frac{1}{24}, \frac{1}{12}, -\frac{1}{6}, \frac{1}{3}, \ldots$.

In Exercises 31–36, find the indicated sum. Use the formula for the sum of the first n terms of a geometric sequence.

31. $\sum_{i=1}^{8} 3^i$

32. $\sum_{i=1}^{6} 4^i$

33. $\sum_{i=1}^{10} 5 \cdot 2^i$

34. $\sum_{i=1}^{7} 4(-3)^i$

35. $\sum_{i=1}^{6} \left(\frac{1}{2}\right)^{i+1}$

36. $\sum_{i=1}^{6} \left(\frac{1}{3}\right)^{i+1}$

In Exercises 37–44, find the sum of each infinite geometric series.

37. $1 + \frac{1}{3} + \frac{1}{9} + \frac{1}{27} + \cdots$

38. $1 + \frac{1}{4} + \frac{1}{16} + \frac{1}{64} + \cdots$

39. $3 + \frac{3}{4} + \frac{3}{4^2} + \frac{3}{4^3} + \cdots$

40. $5 + \frac{5}{6} + \frac{5}{6^2} + \frac{5}{6^3} + \cdots$

41. $1 - \frac{1}{2} + \frac{1}{4} - \frac{1}{8} + \cdots$

42. $3 - 1 + \frac{1}{3} - \frac{1}{9} + \cdots$

43. $\sum_{i=1}^{\infty} 8(-0.3)^{i-1}$

44. $\sum_{i=1}^{\infty} 12(-0.7)^{i-1}$

In Exercises 45–50, express each repeating decimal as a fraction in lowest terms.

45. $0.\overline{5} = \frac{5}{10} + \frac{5}{100} + \frac{5}{1000} + \frac{5}{10{,}000} + \cdots$

46. $0.\overline{1} = \frac{1}{10} + \frac{1}{100} + \frac{1}{1000} + \frac{1}{10{,}000} + \cdots$

47. $0.\overline{47} = \frac{47}{100} + \frac{47}{10{,}000} + \frac{47}{1{,}000{,}000} + \cdots$

48. $0.\overline{83} = \frac{83}{100} + \frac{83}{10{,}000} + \frac{83}{1{,}000{,}000} + \cdots$

49. $0.\overline{257}$

50. $0.\overline{529}$

In Exercises 51–56, the general term of a sequence is given. Determine whether the sequence is arithmetic, geometric, or neither. If the sequence is arithmetic, find the common difference; if it is geometric, find the common ratio.

51. $a_n = n + 5$

52. $a_n = n - 3$

53. $a_n = 2^n$

54. $a_n = \left(\frac{1}{2}\right)^n$

55. $a_n = n^2 + 5$

56. $a_n = n^2 - 3$

Application Exercises

Use the formula for the general term (the nth term) of a geometric sequence to solve Exercises 57–60.

In Exercises 57–58, suppose you save \$1 the first day of a month, \$2 the second day, \$4 the third day, and so on. That is, each day you save twice as much as you did the day before.

57. What will you put aside for savings on the fifteenth day of the month?

58. What will you put aside for savings on the thirtieth day of the month?

59. A professional baseball player signs a contract with a beginning salary of $3,000,000 for the first year and an annual increase of 4% per year beginning in the second year. That is, beginning in year 2, the athlete's salary will be 1.04 times what it was in the previous year. What is the athlete's salary for year 7 of the contract? Round to the nearest dollar.

60. You are offered a job that pays $30,000 for the first year with an annual increase of 5% per year beginning in the second year. That is, beginning in year 2, your salary will be 1.05 times what it was in the previous year. What can you expect to earn in your sixth year on the job?

61. The population of California from 1990 through 1997 is shown in the following table.

Year	1990	1991	1992	1993
Population in millions	29.76	30.15	30.54	30.94

Year	1994	1995	1996	1997
Population in millions	31.34	31.75	32.16	32.58

  a. Divide the population for each year by the population in the preceding year. Round to three decimal places and show that the population of California is increasing geometrically.

  b. Write the general term of the geometric sequence describing population growth for California n years after 1989.

  c. Estimate California's population, in millions, for the year 2000. According to the U.S. Census Bureau, California's population in 2000 was 33.87 million. How well does your geometric sequence model the actual population?

62. The population of Texas from 1990 through 1997 is shown in the following table.

Year	1990	1991	1992	1993
Population in millions	16.99	17.35	17.71	18.08

Year	1994	1995	1996	1997
Population in millions	18.46	18.85	19.25	19.65

  a. Divide the population for each year by the population in the preceding year. Round to three decimal places and show that the population of Texas is increasing geometrically.

  b. Write the general term of the geometric sequence describing population growth for Texas n years after 1989.

  c. Estimate Texas's population, in millions, for the year 2000. According to the U.S. Census Bureau, Texas's population in 2000 was 20.85 million. How well does your geometric sequence model the actual population?

Use the formula for the sum of the first n terms of a geometric sequence to solve Exercises 63–68.

In Exercises 63–64, you save $1 the first day of a month, $2 the second day, $4 the third day, continuing to double your savings each day.

63. What will your total savings be for the first 15 days?

64. What will your total savings be for the first 30 days?

65. A job pays a salary of $24,000 the first year. During the next 19 years, the salary increases by 5% each year. What is the total lifetime salary over the 20-year period? Round to the nearest dollar.

66. You are investigating two employment opportunities. Company A offers $30,000 the first year. During the next four years, the salary is guaranteed to increase by 6% per year. Company B offers $32,000 the first year, with guaranteed annual increases of 3% per year after that. Which company offers the better total salary for a five-year contract? By how much? Round to the nearest dollar.

67. A pendulum swings through an arc of 20 inches. On each successive swing, the length of the arc is 90% of the previous length.

$$20, \quad 0.9(20), \quad 0.9^2(20), \quad 0.9^3(20), \quad \dots$$

| 1st swing | 2nd swing | 3rd swing | 4th swing |

After 10 swings, what is the total length of the distance the pendulum has swung?

68. A pendulum swings through an arc of 16 inches. On each successive swing, the length of the arc is 96% of the previous length.

$$16, \quad 0.96(16), \quad (0.96)^2(16), \quad (0.96)^3(16), \quad \dots$$

| 1st swing | 2nd swing | 3rd swing | 4th swing |

After 10 swings, what is the total length of the distance the pendulum has swung?

Use the formula for the value of an annuity to solve Exercises 69–72. Round answers to the nearest dollar.

69. To save for retirement, you decide to deposit $2500 into an IRA at the end of each year for the next 40 years. If the interest rate is 9% per year compounded annually, find the value of the IRA after 40 years.

70. You decide to deposit $100 at the end of each month into an account paying 8% interest compounded monthly to save for your child's education. How much will you save over 16 years?

71. You contribute $600 at the end of each quarter to a Tax Sheltered Annuity (TSA) paying 8% annual interest compounded quarterly. Find the value of the TSA after 18 years.

72. To save for a new home, you invest $500 per month in a mutual fund with an annual rate of return of 10% compounded monthly. How much will you have saved after four years?

Use the formula for the sum of an infinite geometric series to solve Exercises 73–75.

73. A new factory in a small town has an annual payroll of $6 million. It is expected that 60% of this money will be spent in the town by factory personnel. The people in the town who receive this money are expected to spend 60% of what they receive in the town, and so on. What is the total of all this spending, called the *total economic impact* of the factory, on the town each year?

74. How much additional spending will be generated by a $10 billion tax rebate if 60% of all income is spent?

75. If the shading process shown in the figure is continued indefinitely, what fractional part of the largest square is eventually shaded?

Writing in Mathematics

76. What is a geometric sequence? Give an example with your explanation.

77. What is the common ratio in a geometric sequence?

78. Explain how to find the general term of a geometric sequence.

79. Explain how to find the sum of the first n terms of a geometric sequence without having to add up all the terms.

80. What is an annuity?

81. What is the difference between a geometric sequence and an infinite geometric series?

82. How do you determine if an infinite geometric series has a sum? Explain how to find the sum of an infinite geometric series.

83. Would you rather have $10,000,000 and a brand new BMW or 1¢ today, 2¢ tomorrow, 4¢ on day 3, 8¢ on day 4, 16¢ on day 5, and so on, for 30 days? Explain.

84. For the first 30 days of a flu outbreak, the number of students on your campus who become ill is increasing. Which is worse: The number of students with the flu is increasing arithmetically or is increasing geometrically? Explain your answer.

Technology Exercises

85. Use the ⬚SEQ⬚ (sequence) capability of a graphing utility and the formula you obtained for a_n to verify the value you found for a_7 in any three exercises from Exercises 17–24.

86. Use the capability of a graphing utility to calculate the sum of a sequence to verify any three of your answers to Exercises 31–36.

In Exercises 87–88, use a graphing utility to graph the function. Determine the horizontal asymptote for the graph of f and discuss its relationship to the sum of the given series.

Function	Series
87. $f(x) = \dfrac{2\left[1 - \left(\frac{1}{3}\right)^x\right]}{1 - \dfrac{1}{3}}$	$2 + 2\left(\frac{1}{3}\right) + 2\left(\frac{1}{3}\right)^2 + 2\left(\frac{1}{3}\right)^3 + \cdots$
88. $f(x) = \dfrac{4\left[1 - (0.6)^x\right]}{1 - 0.6}$	$4 + 4(0.6) + 4(0.6)^2 + 4(0.6)^3 + \cdots$

Critical Thinking Exercises

89. Which one of the following is true?
  a. The sequence $2, 6, 24, 120, \ldots$ is an example of a geometric sequence.
  b. The sum of the geometric series $\frac{1}{2} + \frac{1}{4} + \frac{1}{8} + \cdots + \frac{1}{512}$ can only be estimated without knowing precisely which terms occur between $\frac{1}{8}$ and $\frac{1}{512}$.
  c. $10 - 5 + \frac{5}{2} - \frac{5}{4} + \cdots = \dfrac{10}{1 - \frac{1}{2}}$
  d. If the nth term of a geometric sequence is $a_n = 3(0.5)^{n-1}$, the common ratio is $\frac{1}{2}$.

90. In a pest-eradication program, sterilized male flies are released into the general population each day. Ninety percent of those flies will survive a given day. How many flies should be released each day if the long-range goal of the program is to keep 20,000 sterilized flies in the population?

91. You are now 25 years old and would like to retire at age 55 with a retirement fund of $1,000,000. How much should you deposit at the end of each month for the next 30 years in an IRA paying 10% annual interest compounded monthly to achieve your goal? Round to the nearest dollar.

Group Exercise

92. Group members serve as a financial team analyzing the three options given to the professional baseball player described in the chapter opener on page 925. As a group, determine which option provides the most amount of money over the six-year contract and which provides the least. Describe one advantage and one disadvantage to each option.

SECTION 11.4 *Mathematical Induction*

Objectives
1. Understand the principle of mathematical induction.
2. Prove statements using mathematical induction.

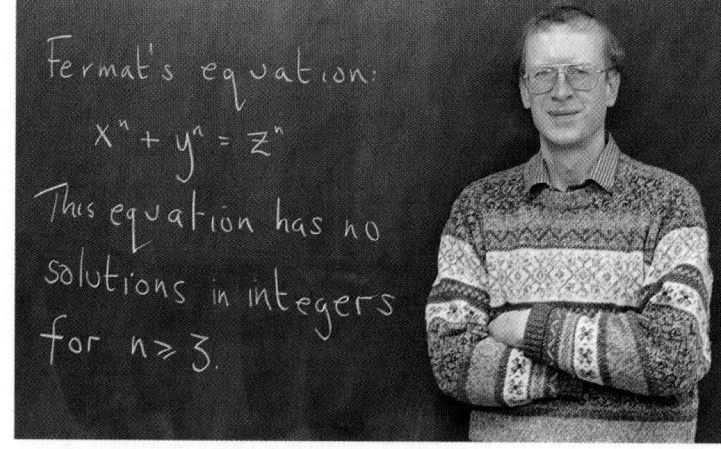

After ten years of work, Princeton University's Andrew Wiles proved Fermat's Last Theorem.

Pierre de Fermat (1601–1665) was a lawyer who enjoyed studying mathematics. In a margin of one of his books he claimed that no positive integers satisfy

$$x^n + y^n = z^n$$

if n is an integer greater than or equal to 3.

If $n = 2$, we can find positive integers satisfying $x^n + y^n = z^n$, or $x^2 + y^2 = z^2$:

$$3^2 + 4^2 = 5^2.$$

However, Fermat claimed that no positive integers satisfy

$$x^3 + y^3 = z^3, \quad x^4 + y^4 = z^4, \quad x^5 + y^5 = z^5,$$

and so on. Fermat claimed to have a proof of his conjecture, but added, "The margin of my book is too narrow to write it down." Some believe that he never had a proof and intended to frustrate his colleagues.

In 1994, 40-year-old Princeton math professor Andrew Wiles proved Fermat's Last Theorem using a principle called *mathematical induction*. In this section, you will learn how to use this powerful method to prove statements about the positive integers.

1 Understand the principle of mathematical induction.

The Principle of Mathematical Induction

How do we prove statements using mathematical induction? Let's consider an example. We will prove a statement that appears to give a correct formula for the sum of the first n positive integers:

$$S_n : 1 + 2 + 3 + \cdots + n = \frac{n(n + 1)}{2}.$$

We can verify this statement for, say, the first four positive integers.

If $n = 1$, the statement S_1 is

Take the first term on the left.

$$1 \overset{?}{=} \frac{1(1 + 1)}{2}$$

Substitute 1 for n on the right.

$$1 \overset{?}{=} \frac{1 \cdot 2}{2}$$

$$1 = 1 \checkmark.$$

This true statement shows that S_1 is true.

If $n = 2$, the statement S_2 is

Add the first two terms on the left.

$$1 + 2 \overset{?}{=} \frac{2(2 + 1)}{2}$$

Substitute 2 for n on the right.

$$3 \overset{?}{=} \frac{2 \cdot 3}{2}$$

$$3 = 3 \checkmark.$$

This true statement shows S_2 is true.

If $n = 3$, the statement S_3 is

Add the first three terms on the left.

$$1 + 2 + 3 \overset{?}{=} \frac{3(3 + 1)}{2}$$

Substitute 3 for n on the right.

$$6 \overset{?}{=} \frac{3 \cdot 4}{2}$$

$$6 = 6 \checkmark.$$

This true statement shows S_3 is true.

Finally, if $n = 4$, the statement S_4 is

Add the first four terms on the left.

$$1 + 2 + 3 + 4 \overset{?}{=} \frac{4(4 + 1)}{2}$$

Substitute 4 for n on the right.

$$10 \overset{?}{=} \frac{4 \cdot 5}{2}$$

$$10 = 10 \checkmark.$$

This true statement shows S_4 is true.

This approach does *not* prove that the given statement S_n is true for every positive integer n. The fact that the formula produces true statements for $n = 1$, 2, 3, and 4 does not guarantee that it is valid for all positive integers n. Thus, we need to be able to verify the truth of S_n without verifying the statement for each and every one of the positive integers.

A legitimate proof of the given statement S_n involves a technique called **mathematical induction.**

The Principle of Mathematical Induction

Let S_n be a statement involving the positive integer n. If

1. S_1 is true, and

2. the truth of the statement S_k implies the truth of the statement S_{k+1}, for every positive integer k,

then the statement S_n is true for all positive integers n.

Figure 11.4 Falling dominoes illustrate the principle of mathematical induction

The principle of mathematical induction can be illustrated using an unending line of dominoes, as shown in Figure 11.4. If the first domino is pushed over, it knocks down the next, which knocks down the next, and so on, in a chain reaction. To topple all the dominoes in the infinite sequence, two conditions must be satisfied:

1. The first domino must be knocked down.

2. If the domino in position k is knocked down, then the domino in position $k + 1$ must be knocked down.

If the second condition is not satisfied, it does not follow that all the dominoes will topple. For example, suppose the dominoes are spaced far enough apart so that a falling domino does not push over the next domino in the line.

The domino analogy provides the two steps that are required in a proof by mathematical induction.

The Steps in a Proof by Mathematical Induction

Let S_n be a statement involving the positive integer n. To prove that S_n is true for all positive integers n requires two steps.

Step 1 Show that S_1 is true.

Step 2 Show that if S_k is assumed to be true, then S_{k+1} is also true, for every positive integer k.

Notice that to prove S_n, we work only with the statements S_1, S_k, and S_{k+1}. Our first example provides practice in writing these statements.

EXAMPLE 1   Writing S_1, S_k, and S_{k+1}

For the given statement S_n, write the three statements S_1, S_k, and S_{k+1}.

a. S_n: $1 + 2 + 3 + \cdots + n = \dfrac{n(n + 1)}{2}$

b. S_n: $1^2 + 2^2 + 3^2 + \cdots + n^2 = \dfrac{n(n + 1)(2n + 1)}{6}$

Solution

a. We begin with

$$S_n: 1 + 2 + 3 + \cdots + n = \frac{n(n + 1)}{2}.$$

Write S_1 by taking the first term on the left and replacing n with 1 on the right.

$$S_1: 1 = \frac{1(1 + 1)}{2}$$

Write S_k by taking the sum of the first k terms on the left and replacing n with k on the right.

$$S_k: 1 + 2 + 3 + \cdots + k = \frac{k(k + 1)}{2}$$

Write S_{k+1} by taking the sum of the first $k + 1$ terms on the left and replacing n with $k + 1$ on the right.

$$S_{k+1}: 1 + 2 + 3 + \cdots + (k + 1) = \frac{(k + 1)[(k + 1) + 1]}{2}$$

$$S_{k+1}: 1 + 2 + 3 + \cdots + (k + 1) = \frac{(k + 1)(k + 2)}{2} \qquad \text{Simplify on the right.}$$

b. We begin with

$$S_n: 1^2 + 2^2 + 3^2 + \cdots + n^2 = \frac{n(n + 1)(2n + 1)}{6}.$$

Write S_1 by taking the first term on the left and replacing n with 1 on the right.

$$S_1: 1^2 = \frac{1(1 + 1)(2 \cdot 1 + 1)}{6}$$

Write S_k by taking the sum of the first k terms on the left and replacing n with k on the right.

$$S_k: 1^2 + 2^2 + 3^2 + \cdots + k^2 = \frac{k(k + 1)(2k + 1)}{6}$$

Write S_{k+1} by taking the sum of the first $k + 1$ terms on the left and replacing n with $k + 1$ on the right.

$$S_{k+1}: 1^2 + 2^2 + 3^2 + \cdots + (k + 1)^2 = \frac{(k + 1)[(k + 1) + 1][2(k + 1) + 1]}{6}$$

$$S_{k+1}: 1^2 + 2^2 + 3^2 + \cdots + (k + 1)^2 = \frac{(k + 1)(k + 2)(2k + 3)}{6} \qquad \text{Simplify on the right.}$$

Check Point 1 For the given statement S_n, write the three statements S_1, S_k, and S_{k+1}.

a. $2 + 4 + 6 + \cdots + 2n = n(n + 1)$

b. $1^3 + 2^3 + 3^3 + \cdots + n^3 = \dfrac{n^2(n + 1)^2}{4}$

Always simplify S_{k+1} before trying to use mathematical induction to prove that S_n is true. For example, consider

$$S_n: \quad 1^2 + 3^2 + 5^2 + \cdots + (2n - 1)^2 = \frac{n(2n - 1)(2n + 1)}{3}.$$

Begin by writing S_{k+1} as follows:

$$S_{k+1}: \quad \underbrace{1^2 + 3^2 + 5^2 + \cdots + [2(k + 1) - 1]^2}_{\text{The sum of the first } k+1 \text{ terms}}$$

$$= \frac{(k + 1)[2(k + 1) - 1][2(k + 1) + 1]}{3}.$$

Replace n with $k + 1$ on the right side of S_n.

Now simplify the algebra.

$$S_{k+1}: \quad 1^2 + 3^2 + 5^2 + \cdots + (2k + 2 - 1)^2 = \frac{(k + 1)(2k + 2 - 1)(2k + 2 + 1)}{3}$$

$$S_{k+1}: \quad 1^2 + 3^2 + 5^2 + \cdots + (2k + 1)^2 = \frac{(k + 1)(2k + 1)(2k + 3)}{3}$$

2 Prove statements using mathematical induction.

Proving Statements about Positive Integers Using Mathematical Induction

Now that we know how to find S_1, S_k, and S_{k+1}, let's see how we can use these statements to carry out the two steps in a proof by mathematical induction. In Examples 2 and 3, we will use the statements S_1, S_k, and S_{k+1} to prove each of the statements S_n that we worked with in Example 1.

EXAMPLE 2   Proving a Formula by Mathematical Induction

Use mathematical induction to prove that

$$1 + 2 + 3 + \cdots + n = \frac{n(n + 1)}{2}$$

for all positive integers n.

Solution

Step 1   Show that S_1 is true.   Statement S_1 is

$$1 = \frac{1(1 + 1)}{2}.$$

Simplifying on the right, we obtain $1 = 1$. This true statement shows that S_1 is true.

Step 2   Show that if S_k is true, then S_{k+1} is true.   Using S_k and S_{k+1} from Example 1(a), show that the truth of S_k,

$$1 + 2 + 3 + \cdots + k = \frac{k(k + 1)}{2},$$

implies the truth of S_{k+1},

$$1 + 2 + 3 + \cdots + (k + 1) = \frac{(k + 1)(k + 2)}{2}.$$

Visualizing Summation Formulas

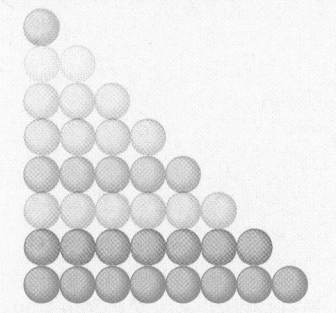

Finding the sum of consecutive positive integers leads to **triangular numbers** of the form $\dfrac{n(n + 1)}{2}$.

$\dfrac{n(n + 1)}{2}$ $\dfrac{n(n + 1)}{2}$

$n = 1:$ $n = 2:$

1   3

$\dfrac{n(n + 1)}{2}$ $\dfrac{n(n + 1)}{2}$

$n = 3:$ $n = 4:$

6   10

We will work with S_k. Because we assume that S_k is true, we add the next consecutive integer after k—namely, $k + 1$—to both sides.

$$1 + 2 + 3 + \cdots + k = \frac{k(k + 1)}{2}$$

This is S_k, which we assume is true.

$$1 + 2 + 3 + \cdots + k + (k + 1) = \frac{k(k + 1)}{2} + (k + 1)$$

Add $k + 1$ to both sides of the equation.

We do not have to write this k because k is understood to be the integer that precedes $k + 1$.

$$1 + 2 + 3 + \cdots + (k + 1) = \frac{k(k + 1)}{2} + \frac{2(k + 1)}{2}$$

Write the right side with a common denominator of 2.

$$1 + 2 + 3 + \cdots + (k + 1) = \frac{(k + 1)}{2}(k + 2)$$

Factor out the common factor $\dfrac{k + 1}{2}$ on the right.

$$1 + 2 + 3 + \cdots + (k + 1) = \frac{(k + 1)(k + 2)}{2}$$

This final result is the statement S_{k+1} at the bottom of page 964.

We have shown that if we assume that S_k is true, and we add $k + 1$ to both sides of S_k, then S_{k+1} is also true. By the principle of mathematical induction, the statement S_n, namely,

$$1 + 2 + 3 + \cdots + n = \frac{n(n + 1)}{2}$$

is true for every positive integer n.

Check Point 2   Use mathematical induction to prove that
$$2 + 4 + 6 + \cdots + 2n = n(n + 1)$$
for all positive integers n.

EXAMPLE 3   Proving a Formula by Mathematical Induction

Use mathematical induction to prove that

$$1^2 + 2^2 + 3^2 + \cdots + n^2 = \frac{n(n + 1)(2n + 1)}{6}$$

for all positive integers n.

Solution

Step 1   Show that S_1 is true.   Statement S_1 is
$$1^2 = \frac{1(1 + 1)(2 \cdot 1 + 1)}{6}.$$

Simplifying, we obtain $1 = \dfrac{1 \cdot 2 \cdot 3}{6}$. Further simplification on the right gives the statement $1 = 1$. This true statement shows that S_1 is true.

Step 2   Show that if S_k is true, then S_{k+1} is true. Using S_k and S_{k+1} from Example 1(b), show that the truth of

$$S_k: 1^2 + 2^2 + 3^2 + \cdots + k^2 = \frac{k(k+1)(2k+1)}{6}$$

implies the truth of

$$S_{k+1}: 1^2 + 2^2 + 3^2 + \cdots + (k+1)^2 = \frac{(k+1)(k+2)(2k+3)}{6}.$$

We will work with S_k. Because we assume that S_k is true, we add the square of the next consecutive integer after k—namely, $(k+1)^2$—to both sides of the equation.

$$1^2 + 2^2 + 3^2 + \cdots + k^2 = \frac{k(k+1)(2k+1)}{6}$$

This is S_k, assumed to be true. We must work with this and show S_{k+1} is true.

$$1^2 + 2^2 + 3^2 + \cdots + k^2 + (k+1)^2 = \frac{k(k+1)(2k+1)}{6} + (k+1)^2$$

Add $(k+1)^2$ to both sides.

$$1^2 + 2^2 + 3^2 + \cdots + (k+1)^2 = \frac{k(k+1)(2k+1)}{6} + \frac{6(k+1)^2}{6}$$

It is not necessary to write k^2 on the left. Express the right side with the least common denominator, 6.

$$= \frac{(k+1)}{6}\left[k(2k+1) + 6(k+1)\right]$$

Factor out the common factor $\dfrac{k+1}{6}$.

$$= \frac{(k+1)}{6}\left(2k^2 + 7k + 6\right)$$

Multiply and combine like terms.

$$= \frac{(k+1)}{6}(k+2)(2k+3)$$

Factor $2k^2 + 7k + 6$.

$$= \frac{(k+1)(k+2)(2k+3)}{6}$$

This final statement is S_{k+1}.

We have shown that if we assume that S_k is true, and we add $(k+1)^2$ to both sides of S_k, then S_{k+1} is also true. By the principle of mathematical induction, the statement S_n, namely,

$$1^2 + 2^2 + 3^2 + \cdots + n^2 = \frac{n(n+1)(2n+1)}{6}$$

is true for every positive integer n.

Check Point 3   Use mathematical induction to prove that

$$1^3 + 2^3 + 3^3 + \cdots + n^3 = \frac{n^2(n+1)^2}{4}$$

for all positive integers n.

Example 4 illustrates how mathematical induction can be used to prove statements about positive integers that do not involve sums.

EXAMPLE 4   Using the Principle of Mathematical Induction

Prove that 2 is a factor of $n^2 + 5n$ for all positive integers n.

Solution

Step 1   Show that S_1 is true.   Statement S_1 reads

$$2 \text{ is a factor of } 1^2 + 5 \cdot 1.$$

Simplifying the arithmetic, the statement reads

$$2 \text{ is a factor of } 6.$$

This statement is true: that is, $6 = 2 \cdot 3$. This shows that S_1 is true.

Step 2   Show that if S_k is true, then S_{k+1} is true.   Let's write S_k and S_{k+1}:

$$S_k: \quad 2 \text{ is a factor of } k^2 + 5k.$$

$$S_{k+1}: \quad 2 \text{ is a factor of } (k + 1)^2 + 5(k + 1).$$

We can rewrite statement S_{k+1} by simplifying the algebraic expression in the statement as follows:

$$(k + 1)^2 + 5(k + 1) = k^2 + 2k + 1 + 5k + 5 = k^2 + 7k + 6.$$

> Use the formula
> $(A + B)^2 = A^2 + 2AB + B^2.$

Statement S_{k+1} now reads

$$2 \text{ is a factor of } k^2 + 7k + 6.$$

We wish to use statement S_k—that is, 2 is a factor of $k^2 + 5k$—to prove statement S_{k+1}. We do this as follows:

$$k^2 + 7k + 6 = (k^2 + 5k) + (2k + 6) = (k^2 + 5k) + 2(k + 3).$$

> We assume that 2
> is a factor of $k^2 + 5k$
> because we assume
> S_k is true.

> Factoring the last two
> terms shows that 2
> is a factor of $2k + 6$.

The voice balloons show that 2 is a factor of $k^2 + 5k$ and of $2(k + 3)$. Thus, if S_k is true, 2 is a factor of the sum $(k^2 + 5k) + 2(k + 3)$, or of $k^2 + 7k + 6$. This is precisely statement S_{k+1}. We have shown that if we assume that S_k is true, then S_{k+1} is also true. By the principle of mathematical induction, the statement S_n, namely 2 is a factor of $n^2 + 5n$, is true for every positive integer n.

Check Point 4   Prove that 2 is a factor of $n^2 + n$ for all positive integers n.

EXERCISE SET 11.4

Practice Exercises

In Exercises 1–4, a statement S_n about the positive integers is given. Write statements S_1, S_2, and S_3, and show that each of these statements is true.

1. S_n: $1 + 3 + 5 + \cdots + (2n - 1) = n^2$

2. S_n: $3 + 4 + 5 + \cdots + (n + 2) = \dfrac{n(n + 5)}{2}$

3. S_n: 2 is a factor of $n^2 - n$.

4. S_n: 3 is a factor of $n^3 - n$.

In Exercises 5–10, a statement S_n about the positive integers is given. Write statements S_k and S_{k+1}, simplifying statement S_{k+1} completely.

5. S_n: $4 + 8 + 12 + \cdots + 4n = 2n(n + 1)$

6. S_n: $3 + 4 + 5 + \cdots + (n + 2) = \dfrac{n(n + 5)}{2}$

7. S_n: $3 + 7 + 11 + \cdots + (4n - 1) = n(2n + 1)$

8. S_n: $2 + 7 + 12 + \cdots + (5n - 3) = \dfrac{n(5n - 1)}{2}$

9. S_n: 2 is a factor of $n^2 - n + 2$.

10. S_n: 2 is a factor of $n^2 - n$.

In Exercises 11–30, use mathematical induction to prove that each statement is true for every positive integer n.

11. $4 + 8 + 12 + \cdots + 4n = 2n(n + 1)$

12. $3 + 4 + 5 + \cdots + (n + 2) = \dfrac{n(n + 5)}{2}$

13. $1 + 3 + 5 + \cdots + (2n - 1) = n^2$

14. $3 + 6 + 9 + \cdots + 3n = \dfrac{3n(n + 1)}{2}$

15. $3 + 7 + 11 + \cdots + (4n - 1) = n(2n + 1)$

16. $2 + 7 + 12 + \cdots + (5n - 3) = \dfrac{n(5n - 1)}{2}$

17. $1 + 2 + 2^2 + \cdots + 2^{n-1} = 2^n - 1$

18. $1 + 3 + 3^2 + \cdots + 3^{n-1} = \dfrac{3^n - 1}{2}$

19. $2 + 4 + 8 + \cdots + 2^n = 2^{n+1} - 2$

20. $\dfrac{1}{2} + \dfrac{1}{4} + \dfrac{1}{8} + \cdots + \dfrac{1}{2^n} = 1 - \dfrac{1}{2^n}$

21. $1 \cdot 2 + 2 \cdot 3 + 3 \cdot 4 + \cdots + n(n + 1)$
$$= \dfrac{n(n + 1)(n + 2)}{3}$$

22. $1 \cdot 3 + 2 \cdot 4 + 3 \cdot 5 + \cdots + n(n + 2)$
$$= \dfrac{n(n + 1)(2n + 7)}{6}$$

23. $\dfrac{1}{1 \cdot 2} + \dfrac{1}{2 \cdot 3} + \dfrac{1}{3 \cdot 4} + \cdots + \dfrac{1}{n(n + 1)} = \dfrac{n}{n + 1}$

24. $\dfrac{1}{2 \cdot 3} + \dfrac{1}{3 \cdot 4} + \dfrac{1}{4 \cdot 5} + \cdots + \dfrac{1}{(n + 1)(n + 2)} = \dfrac{n}{2n + 4}$

25. 2 is a factor of $n^2 - n$.

26. 2 is a factor of $n^2 + 3n$.

27. 6 is a factor of $n(n + 1)(n + 2)$.

28. 3 is a factor of $n(n + 1)(n - 1)$.

29. $(ab)^n = a^n b^n$

30. $\left(\dfrac{a}{b} \right)^n = \dfrac{a^n}{b^n}$

Writing in Mathematics

31. Explain how to use mathematical induction to prove that a statement is true for every positive integer n.

32. Consider the statement S_n given by
$$n^2 - n + 41 \text{ is prime.}$$
Although $S_1, S_2, \ldots, S_{40}$ are true, S_{41} is false. Describe how this is illustrated by the dominoes in the figure. What does this tell you about a pattern, or formula, that seems to work for several values of n?

Critical Thinking Exercises

Some statements are false for the first few positive integers, but true for some positive integer on. In these instances, you can prove S_n for $n \geq k$ by showing that S_k is true and that S_k implies S_{k+1}. Use this extended principle of mathematical induction to prove that each statement in Exercises 33–34 is true.

33. Prove that $n^2 > 2n + 1$ for $n \geq 3$. Show that the formula is true for $n = 3$ and then use step 2 of mathematical induction.

34. Prove that $2^n > n^2$ for $n \geq 5$. Show that the formula is true for $n = 5$ and then use step 2 of mathematical induction.

In Exercises 35–36, find S_1 through S_5 and then use the pattern to make a conjecture about S_n. Prove the conjectured formula for S_n by mathematical induction.

35. $S_n:\ \dfrac{1}{4} + \dfrac{1}{12} + \dfrac{1}{24} + \cdots + \dfrac{1}{2n(n+1)}$

36. $S_n:\ \left(1 - \dfrac{1}{2}\right)\left(1 - \dfrac{1}{3}\right)\left(1 - \dfrac{1}{4}\right)\cdots\left(1 - \dfrac{1}{n+1}\right)$

Group Exercise

37. Fermat's most notorious theorem baffled the greatest minds for more than three centuries. In 1994, after ten years of work, Princeton University's Andrew Wiles proved Fermat's Last Theorem. *People* magazine put him on its list of "the 25 most intriguing people of the year," the Gap asked him to model jeans, and Barbara Walters chased him for an interview. "Who's Barbara Walters?" asked the bookish Wiles, who had somehow gone through life without a television.

Using the 1993 PBS documentary "Solving Fermat: Andrew Wiles" or information about Andrew Wiles on the Internet, research and present a group seminar on what Wiles did to prove Fermat's Last Theorem, problems along the way, and the role of mathematical induction in the proof.

SECTION 11.5 *The Binomial Theorem*

Objectives

1. Recognize patterns in binomial expansions.
2. Evaluate a binomial coefficient.
3. Expand a binomial raised to a power.
4. Find a particular term in a binomial expansion.

Galaxies are groupings of billions of stars bound together gravitationally. Some galaxies, such as the Centaurus galaxy shown here, are elliptical in shape.

Is mathematics discovered or invented? For example, planets revolve in elliptical orbits. Does that mean that the ellipse is out there, waiting for the mind to discover it? Or do people create the definition of an ellipse just as they compose a song? And is it possible for the same mathematics to be discovered/invented by independent researchers separated by time, place, and culture? This is precisely what occurred when mathematicians attempted to find efficient methods for raising binomials to higher and higher powers, such as

$$(x+2)^3,\ (x+2)^4,\ (x+2)^5,\ (x+2)^6,$$

and so on. In this section, we study higher powers of binomials and a method first discovered/invented by great minds in Eastern and Western cultures working independently.

1 Recognize patterns in binomial expansions.

Patterns in Binomial Expansions

When we write out the *binomial expression* $(a + b)^n$, where n is a positive integer, a number of patterns begin to appear.

$$(a + b)^1 = a + b$$
$$(a + b)^2 = a^2 + 2ab + b^2$$
$$(a + b)^3 = a^3 + 3a^2b + 3ab^2 + b^3$$
$$(a + b)^4 = a^4 + 4a^3b + 6a^2b^2 + 4ab^3 + b^4$$
$$(a + b)^5 = a^5 + 5a^4b + 10a^3b^2 + 10a^2b^3 + 5ab^4 + b^5$$

Discovery

Each expanded form of the binomial expression is a polynomial. Study the five polynomials and answer the following questions.

1. For each polynomial, describe the pattern for the exponents on a. What is the largest exponent on a? What happens to the exponent on a from term to term?

2. Describe the pattern for the exponents on b. What is the exponent on b in the first term? What is the exponent on b in the second term? What happens to the exponent on b from term to term?

3. Find the sum of the exponents on the variables in each term for the polynomials in the five rows. Describe the pattern.

4. How many terms are there in the polynomials on the right in relation to the power of the binomial?

How many of the following patterns were you able to discover?

1. The first term is a^n. The exponent on a decreases by 1 in each successive term.

2. The exponents on b increase by 1 in each successive term. In the first term, the exponent on b is 0. (Because $b^0 = 1$, b is not shown in the first term.) The last term is b^n.

3. The sum of the exponents on the variables in any term is equal to n, the exponent on $(a + b)^n$.

4. There is one more term in the polynomial expansion than there is in the power of the binomial, n. There are $n + 1$ terms in the expanded form of $(a + b)^n$.

Using these observations, the variable parts of the expansion of $(a + b)^6$ are

$$a^6, \quad a^5b, \quad a^4b^2, \quad a^3b^3, \quad a^2b^4, \quad ab^5, \quad b^6.$$

The first term is a^6, with the exponent on a decreasing by 1 in each successive term. The exponents on b increase from 0 to 6, with the last term being b^6. The sum of the exponents in each term is equal to 6.

We can generalize from these observations to obtain the variable parts of the expansion of $(a + b)^n$. They are

$$a^n, \quad a^{n-1}b, \quad a^{n-2}b^2, \quad a^{n-3}b^3, \ldots, \quad ab^{n-1}, \quad b^n.$$

> Exponents on a are decreasing by 1.
> Exponents on b are increasing by 1.

> Sum of exponents:
> $n - 1 + 1 = n$

> Sum of exponents:
> $n - 3 + 3 = n$

> Sum of exponents:
> $1 + n - 1 = n$

Let's now establish a pattern for the coefficients of the terms in the binomial expansion. Notice that each row in the figure on page 971 begins and ends with 1. Any other number in the row can be obtained by adding the two numbers immediately above it.

Coefficients for $(a + b)^1$.					1		1				
Coefficients for $(a + b)^2$.						1		2		1	
Coefficients for $(a + b)^3$.				1		3		3		1	
Coefficients for $(a + b)^4$.			1		4		6		4		1
Coefficients for $(a + b)^5$.		1		5		10		10		5		1

Study Tip

We have not shown the number in the top row of Pascal's triangle on the right. The top row is *row zero* because it corresponds to $(a + b)^0 = 1$. With row zero, the triangle appears as

```
            1
          1   1
        1   2   1
      1   3   3   1
    1   4   6   4   1
            etc.
```

This triangular array of coefficients is called **Pascal's triangle.** If we continue with the sixth row, the first and last numbers are 1. Each of the other numbers is obtained by finding the sum of the two closest numbers above it in the fifth row. We can use the numbers in the sixth row and the variable parts we found to

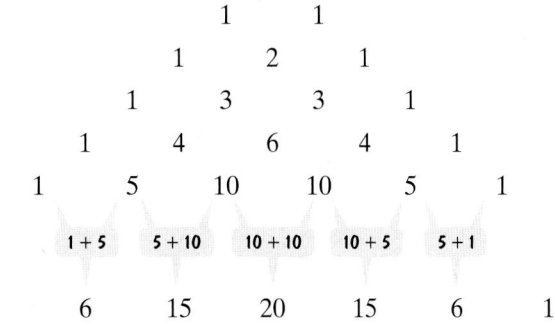

write the expansion for $(a + b)^6$. It is

$$(a + b)^6 = a^6 + 6a^5b + 15a^4b^2 + 20a^3b^3 + 15a^2b^4 + 6ab^5 + b^6.$$

2 Evaluate a binomial coefficient.

Binomial Coefficients

Pascal's triangle becomes cumbersome when a binomial contains a relatively large power. Therefore, the coefficients in a binomial expansion are instead given in terms of factorials. The coefficients are written in a special notation, which we define next.

> **Definition of a Binomial Coefficient** $\binom{n}{r}$
>
> For nonnegative integers n and r, with $n \geq r$, the expression $\binom{n}{r}$ (read "n above r") is called a **binomial coefficient** and is defined by
>
> $$\binom{n}{r} = \frac{n!}{r!(n - r)!}.$$

Technology

Graphing utilities can compute binomial coefficients. For example, to find $\binom{6}{2}$, many utilities require the sequence

6 [nCr] 2 [ENTER].

The graphing utility will display 15. Consult your manual and verify the other evaluations in Example 1.

The symbol $_nC_r$ is often used in place of $\binom{n}{r}$ to denote binomial coefficients.

EXAMPLE 1 Evaluating Binomial Coefficients

Evaluate: **a.** $\binom{6}{2}$ **b.** $\binom{3}{0}$ **c.** $\binom{9}{3}$ **d.** $\binom{4}{4}$.

Solution In each case, we apply the definition of the binomial coefficient.

a. $\binom{6}{2} = \frac{6!}{2!(6 - 2)!} = \frac{6!}{2!\,4!} = \frac{6 \cdot 5 \cdot 4!}{2 \cdot 1 \cdot 4!} = 15$

b. $\dbinom{3}{0} = \dfrac{3!}{0!(3-0)!} = \dfrac{3!}{0!\,3!} = \dfrac{1}{1} = 1$

Remember that 0! = 1.

c. $\dbinom{9}{3} = \dfrac{9!}{3!(9-3)!} = \dfrac{9!}{3!\,6!} = \dfrac{9 \cdot 8 \cdot 7 \cdot 6!}{3 \cdot 2 \cdot 1 \cdot 6!} = 84$

d. $\dbinom{4}{4} = \dfrac{4!}{4!(4-4)!} = \dfrac{4!}{4!\,0!} = \dfrac{1}{1} = 1$

Check Point 1 Evaluate: **a.** $\dbinom{6}{3}$ **b.** $\dbinom{6}{0}$ **c.** $\dbinom{8}{2}$ **d.** $\dbinom{3}{3}$.

3 Expand a binomial raised to a power.

The Binomial Theorem

If we use binomial coefficients and the pattern for the variable part of each term, a formula called the **Binomial Theorem** can be written for any positive integral power of a binomial.

> **A Formula for Expanding Binomials: The Binomial Theorem**
>
> For any positive integer n,
>
> $$(a + b)^n = \binom{n}{0}a^n + \binom{n}{1}a^{n-1}b + \binom{n}{2}a^{n-2}b^2 + \binom{n}{3}a^{n-3}b^3 + \cdots + \binom{n}{n}b^n.$$

The Universality of Mathematics

"Pascal's" triangle, credited to French mathematician Blaise Pascal (1623–1662), appeared in a Chinese document printed in 1303. The Binomial Theorem was known in Eastern cultures prior to its discovery in Europe. The same mathematics is often discovered/invented by independent researchers separated by time, place, and culture.

EXAMPLE 2 Using the Binomial Theorem

Expand: $(x + 2)^4$.

Solution We use the Binomial Theorem

$$(a + b)^n = \binom{n}{0}a^n + \binom{n}{1}a^{n-1}b + \binom{n}{2}a^{n-2}b^2 + \binom{n}{3}a^{n-3}b^3 + \cdots + \binom{n}{n}b^n$$

to expand $(x + 2)^4$. In $(x + 2)^4$, $a = x$, $b = 2$, and $n = 4$. In the expansion, powers of x are in descending order, starting with x^4. Powers of 2 are in ascending order, starting with 2^0. (Because $2^0 = 1$, a 2 is not shown in the first term.) The sum of the exponents on x and 2 in each term is equal to 4, the exponent in the expression $(x + 2)^4$.

$$(x + 2)^4 = \binom{4}{0}x^4 + \binom{4}{1}x^3 \cdot 2 + \binom{4}{2}x^2 \cdot 2^2 + \binom{4}{3}x \cdot 2^3 + \binom{4}{4}2^4$$

These binomial coefficients are evaluated using $\binom{n}{r} = \dfrac{n!}{r!(n-r)!}$.

$$= \frac{4!}{0!\,4!}x^4 + \frac{4!}{1!\,3!}x^3 \cdot 2 + \frac{4!}{2!\,2!}x^2 \cdot 4 + \frac{4!}{3!\,1!}x \cdot 8 + \frac{4!}{4!\,0!} \cdot 16$$

$$\frac{4!}{2!\,2!} = \frac{4 \cdot 3 \cdot 2!}{2! \cdot 2 \cdot 1} = \frac{12}{2} = 6$$

Take a few minutes to verify the other factorial evaluations.

$$= 1 \cdot x^4 + 4x^3 \cdot 2 + 6x^2 \cdot 4 + 4x \cdot 8 + 1 \cdot 16$$
$$= x^4 + 8x^3 + 24x^2 + 32x + 16$$

> **Check Point 2** Expand: $(x + 1)^4$.

EXAMPLE 3 Using the Binomial Theorem

Expand: $(2x - y)^5$.

Solution Because the Binomial Theorem involves the sum of two terms raised to a power, we rewrite $(2x - y)^5$ as $[2x + (-y)]^5$. We use the Binomial Theorem

$$(a + b)^n = \binom{n}{0}a^n + \binom{n}{1}a^{n-1}b + \binom{n}{2}a^{n-2}b^2 + \binom{n}{3}a^{n-3}b^3 + \cdots + \binom{n}{n}b^n$$

to expand $[2x + (-y)]^5$. In $[2x + (-y)]^5$, $a = 2x$, $b = -y$, and $n = 5$. In the expansion, powers of $2x$ are in descending order, starting with $(2x)^5$. Powers of $-y$ are in ascending order, starting with $(-y)^0$. [Because $(-y)^0 = 1$, a $-y$ is not shown in the first term.] The sum of the exponents on $2x$ and $-y$ in each term is equal to 5, the exponent in the expression $(2x - y)^5$.

$(2x - y)^5 = [2x + (-y)]^5$

$= \binom{5}{0}(2x)^5 + \binom{5}{1}(2x)^4(-y) + \binom{5}{2}(2x)^3(-y)^2 + \binom{5}{3}(2x)^2(-y)^3 + \binom{5}{4}(2x)(-y)^4 + \binom{5}{5}(-y)^5$

Evaluate binomial coefficients using $\binom{n}{r} = \dfrac{n!}{r!(n-r)!}$.

$= \dfrac{5!}{0!\,5!}(2x)^5 + \dfrac{5!}{1!\,4!}(2x)^4(-y) + \dfrac{5!}{2!\,3!}(2x)^3(-y)^2 + \dfrac{5!}{3!\,2!}(2x)^2(-y)^3 + \dfrac{5!}{4!\,1!}(2x)(-y)^4 + \dfrac{5!}{5!\,0!}(-y)^5$

$\dfrac{5!}{2!\,3!} = \dfrac{5 \cdot 4 \cdot 3!}{2 \cdot 1 \cdot 3!} = 10$

Take a few minutes to verify the other factorial evaluations.

$= 1(2x)^5 + 5(2x)^4(-y) + 10(2x)^3(-y)^2 + 10(2x)^2(-y)^3 + 5(2x)(-y)^4 + 1(-y)^5$

Raise both factors in these parentheses to the indicated powers.

$= 1(32x^5) + 5(16x^4)(-y) + 10(8x^3)(-y)^2 + 10(4x^2)(-y)^3 + 5(2x)(-y)^4 + 1(-y)^5$

Now raise $-y$ to the indicated powers.

$= 1(32x^5) + 5(16x^4)(-y) + 10(8x^3)y^2 + 10(4x^2)(-y^3) + 5(2x)y^4 + 1(-y^5)$

Multiplying factors in each of the six terms gives us the desired expansion:

$$(2x - y)^5 = 32x^5 - 80x^4y + 80x^3y^2 - 40x^2y^3 + 10xy^4 - y^5.$$

> **Check Point 3** Expand: $(x - 2y)^5$.

4 Find a particular term in a binomial expansion.

Finding a Particular Term in a Binomial Expansion

By observing the terms in the formula for expanding binomials, we can find a formula for finding a particular term without writing the entire expansion.

First term   Second term   Third term

$$\binom{n}{0} a^n b^0 \qquad \binom{n}{1} a^{n-1} b^1 \qquad \binom{n}{2} a^{n-2} b^2$$

The exponent on b is 1 less than the term number.

Based on the observation in the voice balloon, the $(r + 1)$st term of the expansion of $(a + b)^n$ is the term that contains b^r.

Finding a Particular Term in a Binomial Expansion

The $(r + 1)$st term of the expansion of $(a + b)^n$ is

$$\binom{n}{r} a^{n-r} b^r.$$

EXAMPLE 4   Finding a Single Term of a Binomial Expansion

Find the fourth term in the expansion of $(3x + 2y)^7$.

Solution   The fourth term in the expansion of $(3x + 2y)^7$ contains $(2y)^3$. To find the fourth term, first note that $4 = 3 + 1$. Equivalently, the fourth term of $(3x + 2y)^7$ is the $(3 + 1)$st term. Thus, $r = 3$, $a = 3x$, $b = 2y$, and $n = 7$. The fourth term is

$$\binom{7}{3} (3x)^{7-3} (2y)^3 = \binom{7}{3} (3x)^4 (2y)^3 = \frac{7!}{3!(7-3)!} (3x)^4 (2y)^3.$$

Use the formula for the $(r + 1)$st term of $(a + b)^n$: $\binom{n}{r} a^{n-r} b^r$

We use $\binom{n}{r} = \dfrac{n!}{r!(n-r)!}$ to evaluate $\binom{7}{3}$.

Now we need to evaluate the factorial expression and raise $3x$ and $2y$ to the indicated powers. We obtain

$$\frac{7!}{3!\,4!} (81x^4)(8y^3) = \frac{7 \cdot 6 \cdot 5 \cdot 4!}{3 \cdot 2 \cdot 1 \cdot 4!} (81x^4)(8y^3) = 35(81x^4)(8y^3) = 22{,}680 x^4 y^3.$$

The fourth term of $(3x + 2y)^7$ is $22{,}680 x^4 y^3$.

Check Point 4   Find the fifth term in the expansion of $(2x + y)^9$.

EXERCISE SET 11.5

Practice Exercises

In Exercises 1–8, evaluate the given binomial coefficient.

1. $\binom{8}{3}$ **2.** $\binom{7}{2}$

3. $\binom{12}{1}$ **4.** $\binom{11}{1}$

5. $\binom{6}{6}$ **6.** $\binom{15}{2}$

7. $\binom{100}{2}$ **8.** $\binom{100}{98}$

In Exercises 9–30, use the Binomial Theorem to expand each binomial and express the result in simplified form.

9. $(x + 2)^3$ **10.** $(x + 4)^3$

11. $(3x + y)^3$ **12.** $(x + 3y)^3$

13. $(5x - 1)^3$ **14.** $(4x - 1)^3$

15. $(2x + 1)^4$ **16.** $(3x + 1)^4$

17. $(x^2 + 2y)^4$ **18.** $(x^2 + y)^4$

19. $(y - 3)^4$ **20.** $(y - 4)^4$

21. $(2x^3 - 1)^4$ **22.** $(2x^5 - 1)^4$

23. $(c + 2)^5$ **24.** $(c + 3)^5$

25. $(x - 1)^5$ **26.** $(x - 2)^5$

27. $(3x - y)^5$ **28.** $(x - 3y)^5$

29. $(2a + b)^6$ **30.** $(a + 2b)^6$

In Exercises 31–38, write the first three terms in each binomial expansion, expressing the result in simplified form.

31. $(x + 2)^8$ **32.** $(x + 3)^8$

33. $(x - 2y)^{10}$ **34.** $(x - 2y)^9$

35. $(x^2 + 1)^{16}$ **36.** $(x^2 + 1)^{17}$

37. $(y^3 - 1)^{20}$ **38.** $(y^3 - 1)^{21}$

In Exercises 39–48, find the term indicated in each expansion.

39. $(2x + y)^6$; third term   **40.** $(x + 2y)^6$; third term

41. $(x - 1)^9$; fifth term   **42.** $(x - 1)^{10}$; fifth term

43. $(x^2 + y^3)^8$; sixth term   **44.** $(x^3 + y^2)^8$; sixth term

45. $\left(x - \frac{1}{2}\right)^9$; fourth term   **46.** $\left(x + \frac{1}{2}\right)^8$; fourth term

47. $(x^2 + y)^{22}$; the term containing y^{14}

48. $(x + 2y)^{10}$; the term containing y^6

49. If $f(x) = x^4$, find $\dfrac{f(x + h) - f(x)}{h}$ and simplify.

50. If $f(x) = x^5$, find $\dfrac{f(x + h) - f(x)}{h}$ and simplify.

Bariatrics is the field of medicine that deals with the overweight. Bariatric surgery closes off a large part of the stomach. As a result, patients eat less and have a diminished appetite. Celebrities like pop singer Carnie Wilson and Blues Traveler harmonica player John Popper have become no-longer-larger-than-life walking billboards for the operation. The figure shows the number of bariatric surgeries from 1992 through 2002.

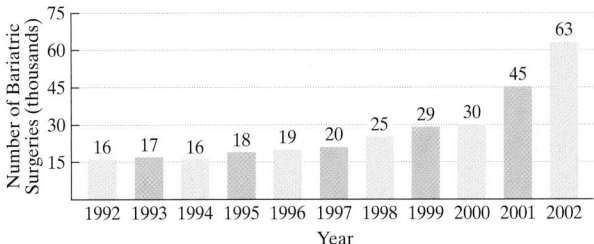

Bariatric Surgery in the United States

Source: American Society for Bariatric Surgery

The function

$$f(x) = 0.12x^3 - x^2 + 3x + 15$$

models the number of bariatric surgeries, f(x), in thousands, x years after 1992. Use this function to solve Exercises 51–52.

51. a. How can we adjust the function so that $x = 0$ corresponds to 1995 rather than 1992? We shift the graph of f 3 units to the left. We obtain $g(x) = f(x + 3)$. Use the Binomial Theorem to express g in descending powers of x.

    b. Find $f(5)$ and $g(2)$. How well do these function values model the number shown in the bar graph?

52. a. How can we adjust the function so that $x = 0$ corresponds to 1997 rather than 1992? We shift the graph of f 5 units to the left. We obtain $g(x) = f(x + 5)$. Use the Binomial Theorem to express g in descending powers of x.

    b. Find $f(7)$ and $g(2)$. How well do these function values model the number shown in the bar graph?

Writing in Mathematics

53. Describe the pattern on the exponents on a in the expansion of $(a + b)^n$.

54. Describe the pattern on the exponents on b in the expansion of $(a + b)^n$.

55. What is true about the sum of the exponents on a and b in any term in the expansion of $(a + b)^n$?

56. How do you determine how many terms there are in a binomial expansion?

57. What is Pascal's triangle? How do you find the numbers in any row of the triangle?

58. Explain how to evaluate $\binom{n}{r}$. Provide an example with your explanation.

59. Explain how to use the Binomial Theorem to expand a binomial. Provide an example with your explanation.

60. Explain how to find a particular term in a binomial expansion without having to write out the entire expansion.

61. Are there situations in which it is easier to use Pascal's triangle than binomial coefficients? Describe these situations.

62. Describe how you would use mathematical induction to prove

$$(a + b)^n = \binom{n}{0}a^n + \binom{n}{1}a^{n-1}b + \binom{n}{2}a^{n-2}b^2$$

$$+ \cdots + \binom{n}{n-1}ab^{n-1} + \binom{n}{n}b^n.$$

What happens when $n = 1$? Write the statement that we assume true. Write the statement that we must prove. What must be done to the left side of the assumed statement to make it look like the left side of the statement that must be proved? (More detail on the actual proof is found in Exercise 75.)

Technology Exercises

63. Use the $\boxed{\text{nCr}}$ key on a graphing utility to verify your answers in Exercises 1–8.

In Exercises 64–65, graph each of the functions in the same viewing rectangle. Describe how the graphs illustrate the Binomial Theorem.

64. $f_1(x) = (x + 2)^3$
$f_2(x) = x^3$
$f_3(x) = x^3 + 6x^2$
$f_4(x) = x^3 + 6x^2 + 12x$
$f_5(x) = x^3 + 6x^2 + 12x + 8$
Use a $[-10, 10, 1]$ by $[-30, 30, 10]$ viewing rectangle.

65. $f_1(x) = (x + 1)^4$
$f_2(x) = x^4$
$f_3(x) = x^4 + 4x^3$
$f_4(x) = x^4 + 4x^3 + 6x^2$
$f_5(x) = x^4 + 4x^3 + 6x^2 + 4x$
$f_6(x) = x^4 + 4x^3 + 6x^2 + 4x + 1$
Use a $[-5, 5, 1]$ by $[-30, 30, 10]$ viewing rectangle.

In Exercises 66–68, use the Binomial Theorem to find a polynomial expansion for each function. Then use a graphing utility and an approach similar to the one in Exercises 64 and 65 to verify the expansion.

66. $f_1(x) = (x - 1)^3$

67. $f_1(x) = (x - 2)^4$

68. $f_1(x) = (x + 2)^6$

69. Graphing utilities capable of symbolic manipulation, such as the TI-92, will expand binomials. On the TI-92, to expand $(3a - 5b)^{12}$, input the following:

$$\boxed{\text{EXPAND}} \, ((3a \,\boxed{-}\, 5b) \,\boxed{\wedge}\, 12) \, \boxed{\text{ENTER}}.$$

Use a graphing utility with this capability to verify any five of the expansions you performed by hand in Exercises 9–30.

Critical Thinking Exercises

70. Which one of the following is true?
   a. The binomial expansion for $(a + b)^n$ contains n terms.
   b. The Binomial Theorem can be written in condensed form as $(a + b)^n = \sum_{r=0}^{n} \binom{n}{r}a^{n-r}b^r$.
   c. The sum of the binomial coefficients in $(a + b)^n$ cannot be 2^n.
   d. There are no values of a and b such that $(a + b)^4 = a^4 + b^4$.

71. Use the Binomial Theorem to expand and then simplify the result: $(x^2 + x + 1)^3$. [*Hint:* Write $x^2 + x + 1$ as $x^2 + (x + 1)$.]

72. Find the term in the expansion of $(x^2 + y^2)^5$ containing x^4 as a factor.

73. Prove that

$$\binom{n}{r} = \binom{n}{n - r}.$$

74. Show that

$$\binom{n}{r} + \binom{n}{r + 1} = \binom{n + 1}{r + 1}.$$

Hints:

$$(n - r)! = (n - r)(n - r - 1)!$$

$$(r + 1)! = (r + 1)r!$$

75. Follow the outline on the next page to use mathematical induction to prove that

$$(a + b)^n = \binom{n}{0}a^n + \binom{n}{1}a^{n-1}b + \binom{n}{2}a^{n-2}b^2$$

$$+ \cdots + \binom{n}{n-1}ab^{n-1} + \binom{n}{n}b^n.$$

a. Verify the formula for $n = 1$.

b. Replace n with k and write the statement that is assumed true. Replace n with $k + 1$ and write the statement that must be proved.

c. Multiply both sides of the statement assumed to be true by $a + b$. Add exponents on the left. On the right, distribute a and b, respectively.

d. Collect like terms on the right. At this point, you should have

$$(a + b)^{k+1} = \binom{k}{0}a^{k+1} + \left[\binom{k}{0} + \binom{k}{1}\right]a^k b$$

$$+ \left[\binom{k}{1} + \binom{k}{2}\right]a^{k-1}b^2 + \left[\binom{k}{2} + \binom{k}{3}\right]a^{k-2}b^3$$

$$+ \cdots + \left[\binom{k}{k-1} + \binom{k}{k}\right]ab^k + \binom{k}{k}b^{k+1}.$$

e. Use the result of Exercise 74 to add the binomial sums in brackets. For example, because $\binom{n}{r} + \binom{n}{r+1}$

$= \binom{n+1}{r+1}$, then $\binom{k}{0} + \binom{k}{1} = \binom{k+1}{1}$ and

$\binom{k}{1} + \binom{k}{2} = \binom{k+1}{2}$.

f. Because $\binom{k}{0} = \binom{k+1}{0}$ (why?) and $\binom{k}{k} = \binom{k+1}{k+1}$ (why?), substitute these results and the results from part (e) into the equation in part (d). This should give the statement that we were required to prove in the second step of the mathematical induction process.

SECTION 11.6 Counting Principles, Permutations, and Combinations

Objectives

1. Use the Fundamental Counting Principle.

2. Use the permutations formula.

3. Distinguish between permutation problems and combination problems.

4. Use the combinations formula.

Have you ever imagined what your life would be like if you won the lottery? What changes would you make? Before you fantasize about becoming a person of leisure with a staff of obedient elves, think about this: The probability of winning top prize in the lottery is about the same as the probability of being struck by lightning. There are millions of possible number combinations in lottery games, and only one way of winning the grand prize. Determining the probability of winning involves calculating the chance of getting the winning combination from all possible outcomes. In this section, we begin preparing for the surprising world of probability by looking at methods for counting possible outcomes.

1 Use the Fundamental Counting Principle.

The Fundamental Counting Principle

It's early morning, you're groggy, and you have to select something to wear for your 8 A.M. class. (What *were* you thinking of when you signed up for a class at that hour?!) Fortunately, your "lecture wardrobe" is rather limited—just two pairs of jeans to choose from (one blue, one black), three T-shirts to choose from (one beige, one yellow, and one blue), and two pairs of sneakers to select from (one black, one red). Your possible outfits are shown in Figure 11.5.

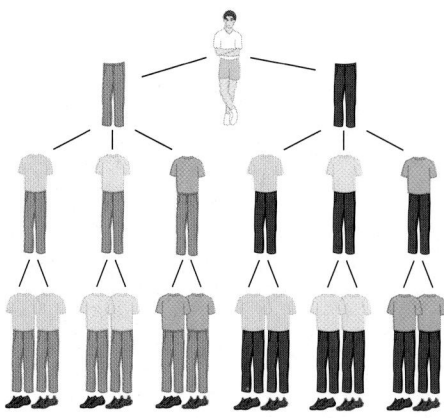

Figure 11.5 Selecting a wardrobe

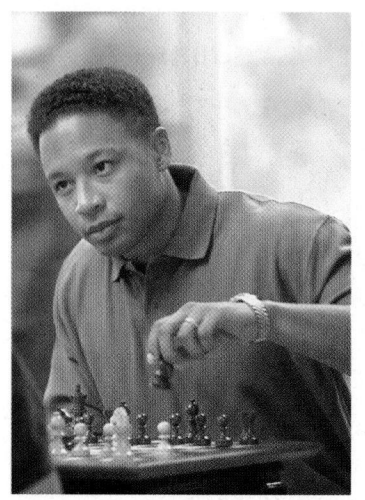

The number of possible ways of playing the first four moves on each side in a game of chess is 318,979,564,000.

The **tree diagram**, so named because of its branches, shows that you can form 12 outfits from your two pairs of jeans, three T-shirts, and two pairs of sneakers. Notice that the number of outfits can be obtained by multiplying the number of choices for jeans, 2, the number of choices for T-shirts, 3, and the number of choices for sneakers, 2:

$$2 \cdot 3 \cdot 2 = 12.$$

We can generalize this idea to any two or more groups of items—not just jeans, T-shirts, and sneakers—with the **Fundamental Counting Principle:**

The Fundamental Counting Principle

The number of ways in which a series of successive things can occur is found by multiplying the number of ways in which each thing can occur.

For example, if you own 30 pairs of jeans, 20 T-shirts, and 12 pairs of sneakers, you have

$$30 \cdot 20 \cdot 12 = 7200$$

choices for your wardrobe!

EXAMPLE 1 Options in Planning a Course Schedule

Next semester you are planning to take three courses—math, English, and humanities. Based on time blocks and highly recommended professors, there are 8 sections of math, 5 of English, and 4 of humanities that you find suitable. Assuming no scheduling conflicts, how many different three-course schedules are possible?

Solution   This situation involves making choices with three groups of items.

| Math | English | Humanities |
| 8 choices | 5 choices | 4 choices |

We use the Fundamental Counting Principle to find the number of three-course schedules. Multiply the number of choices for each of the three groups.

$$8 \cdot 5 \cdot 4 = 160$$

Thus, there are 160 different three-course schedules.

Check Point 1

A pizza can be ordered with three choices of size (small, medium, or large), four choices of crust (thin, thick, crispy, or regular), and six choices of toppings (ground beef, sausage, pepperoni, bacon, mushrooms, or onions). How many different one-topping pizzas can be ordered?

EXAMPLE 2 A Multiple-Choice Test

You are taking a multiple-choice test that has ten questions. Each of the questions has four answer choices, with one correct answer per question. If you select one of these four choices for each question and leave nothing blank, in how many ways can you answer the questions?

Solution We use the Fundamental Counting Principle to determine the number of ways you can answer the test. Multiply the number of choices, 4, for each of the ten questions.

$$4 \cdot 4 \cdot 4 \cdot 4 \cdot 4 \cdot 4 \cdot 4 \cdot 4 \cdot 4 \cdot 4 = 4^{10} = 1{,}048{,}576$$

Thus, you can answer the questions in 1,048,576 different ways.

Are you surprised that there are over one million ways of answering a ten-question multiple-choice test? Of course, there is only one way to answer the test and receive a perfect score. The probability of guessing your way into a perfect score involves calculating the chance of getting a perfect score, just one way, from all 1,048,576 possible outcomes. In short, prepare for the test and do not rely on guessing!

Check Point 2

You are taking a multiple-choice test that has six questions. Each of the questions has three answer choices, with one correct answer per question. If you select one of these three choices for each question and leave nothing blank, in how many ways can you answer the questions?

EXAMPLE 3 Telephone Numbers in the United States

Telephone numbers in the United States begin with three-digit area codes followed by seven-digit local telephone numbers. Area codes and local telephone numbers cannot begin with 0 or 1. How many different telephone numbers are possible?

Solution This situation involves making choices with ten groups of items.

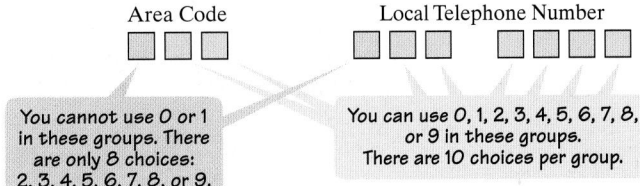

Area Code

Local Telephone Number

You cannot use 0 or 1 in these groups. There are only 8 choices: 2, 3, 4, 5, 6, 7, or 9.

You can use 0, 1, 2, 3, 4, 5, 6, 7, 8, or 9 in these groups. There are 10 choices per group.

We use the Fundamental Counting Principle to determine the number of different telephone numbers that are possible. The total number of telephone numbers possible is

$$8 \cdot 10 \cdot 10 \cdot 8 \cdot 10 \cdot 10 \cdot 10 \cdot 10 \cdot 10 \cdot 10 = 6{,}400{,}000{,}000.$$

There are six billion four hundred million different telephone numbers that are possible.

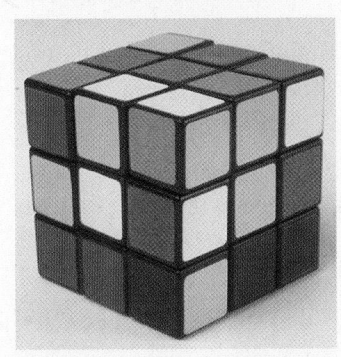

Check Point 3 License plates in a particular state display two letters followed by three numbers, such as AT-887 or BB-013. How many different license plates can be manufactured?

2 Use the permutations formula.

Permutations

You are the coach of a little league baseball team. There are 13 players on the team (and lots of parents hovering in the background, dreaming of stardom for their little "Barry Bonds"). You need to choose a batting order having 9 players. The order makes a difference, because, for instance, if bases are loaded and "Little Barry" is fourth or fifth at bat, his possible home run will drive in three additional runs. How many batting orders can you form?

You can choose any of 13 players for the first person at bat. Then you will have 12 players from which to choose the second batter, then 11 from which to choose the third batter, and so on. The situation can be shown as follows:

We use the Fundamental Counting Principle to find the number of batting orders. The total number of batting orders is

$$13 \cdot 12 \cdot 11 \cdot 10 \cdot 9 \cdot 8 \cdot 7 \cdot 6 \cdot 5 = 259{,}459{,}200.$$

Nearly 260 million batting orders are possible for your 13-player little league team. Each batting order is called a *permutation* of 13 players taken 9 at a time. The number of permutations of 13 players taken 9 at a time is 259,459,200. A **permutation** is an ordered arrangement of items that occurs when

- No item is used more than once. (Each of the 9 players in the batting order bats exactly once.)
- The order of arrangement makes a difference.

We can obtain a formula for finding the number of permutations by rewriting our computation:

$13 \cdot 12 \cdot 11 \cdot 10 \cdot 9 \cdot 8 \cdot 7 \cdot 6 \cdot 5$

$$= \frac{13 \cdot 12 \cdot 11 \cdot 10 \cdot 9 \cdot 8 \cdot 7 \cdot 6 \cdot 5 \cdot 4 \cdot \boxed{3 \cdot 2 \cdot 1 \cdot 1}}{4 \cdot \boxed{3 \cdot 2 \cdot 1 \cdot 1}} = \frac{13!}{4!} = \frac{13!}{(13-9)!}.$$

Thus, the number of permutations of 13 things taken 9 at a time is $\frac{13!}{(13-9)!}$. The special notation $_{13}P_9$ is used to replace the phrase "the number of permutations of 13 things taken 9 at a time." Using this new notation, we can write

$$_{13}P_9 = \frac{13!}{(13-9)!}.$$

The numerator of this expression is the number of items, 13 team members, expressed as a factorial: 13! The denominator is also a factorial. It is the factorial of the difference between the number of items, 13, and the number of items in each permutation, 9 batters: $(13-9)!$.

The notation $_nP_r$ means the **number of permutations of n things taken r at a time.** We can generalize from the situation in which 9 batters were taken from 13 players. By generalizing, we obtain the following formula for the number of permuations if r items are taken from n items:

Permutations of n Things Taken r at a Time

The number of possible permutations if r items are taken from n items is

$$_nP_r = \frac{n!}{(n-r)!}.$$

Because all permutation problems are also Fundamental Counting problems, they can be solved using the formula for $_nP_r$, or using the Fundamental Counting Principle.

Technology

Graphing utilities have a menu item for calculating permuations, usually labeled $\boxed{_nP_r}$. For example, to find $_{20}P_3$, the keystrokes are:

$$20 \ \boxed{_nP_r} \ 3 \ \boxed{\text{ENTER}}.$$

If you are using a scientific calculator, check your manual for the location of the menu item for calculating permutations and the required keystrokes.

EXAMPLE 4 Using the Formula for Permutations

You and 19 of your friends have decided to form an Internet marketing consulting firm. The group needs to choose three officers—a CEO, an operating manager, and a treasurer. In how many ways can those offices be filled?

Solution Your group is choosing $r = 3$ officers from a group of $n = 20$ people (you and 19 friends). The order in which the officers are chosen matters because the CEO, the operating manager, and the treasurer each have different responsibilities. Thus, we are looking for the number of permutations of 20 things taken 3 at a time. We use the formula

$$_nP_r = \frac{n!}{(n-r)!}$$

with $n = 20$ and $r = 3$.

$$_{20}P_3 = \frac{20!}{(20-3)!} = \frac{20!}{17!} = \frac{20 \cdot 19 \cdot 18 \cdot 17!}{17!} = \frac{20 \cdot 19 \cdot 18 \cdot \cancel{17!}}{\cancel{17!}} = 20 \cdot 19 \cdot 18 = 6840$$

Thus, there are 6840 different ways of filling the three offices.

> **Check Point 4** A corporation has seven members on its board of directors. In how many different ways can it elect a president, vice-president, secretary, and treasurer?

EXAMPLE 5 Using the Formula for Permutations

You need to arrange seven of your favorite books along a small shelf. How many different ways can you arrange the books, assuming that the order of the books makes a difference to you?

How to Pass the Time for $2\frac{1}{2}$ Million Years

If you were to arrange 15 different books on a shelf and it took you one minute for each permuation, the entire task would take 2,487,965 years.

Source: Isaac Asimov's Book of Facts

Solution Because you are using all seven of your books in every possible arrangement, you are arranging $r = 7$ books from a group of $n = 7$ books. Thus, we are looking for the number of permutations of 7 things taken 7 at a time. We use the formula

$$_nP_r = \frac{n!}{(n-r)!}$$

with $n = 7$ and $r = 7$.

$$_7P_7 = \frac{7!}{(7-7)!} = \frac{7!}{0!} = \frac{7!}{1} = 5040$$

Thus, you can arrange the books in 5040 ways. There are 5040 different possible permutations.

Distinguish between permutation problems and combination problems.

Check Point 5 In how many ways can 6 books be lined up along a shelf?

Combinations

As the twentieth century drew to a close, *Time* magazine presented a series of special issues on the most influential people of the century. In their issue on heroes and icons (June 14, 1999), they discussed a number of people whose careers became more profitable after their tragic deaths, including Marilyn Monroe, James Dean, Jim Morrison, Kurt Cobain, and Selena.

Imagine that you ask your friends the following question: "Of these five people, which three would you select to be included in a documentary featuring the best of their work?" You are not asking your friends to rank their three favorite artists in any kind of order—they should merely select the three to be included in the documentary.

One friend answers, "Jim Morrison, Kurt Cobain, and Selena." Another responds, "Selena, Kurt Cobain, and Jim Morrison." These two people have the same artists in their group of selections, even if they are named in a different order. We are interested *in which artists are named, not the order in which they are named* for the documentary. Because the items are taken without regard to order, this is not a permutation problem. No ranking of any sort is involved.

Marilyn Monroe, actress (1927–1962)

James Dean, actor (1931–1955)

Jim Morrison, musician and lead singer of The Doors (1943–1971)

Kurt Cobain, musician and front man for Nirvana (1967–1994)

Later on, you ask your roommate which three artists she would select for the documentary. She names Marilyn Monroe, James Dean, and Selena. Her selection is different from those of your two other friends because different entertainers are cited.

Mathematicians describe the group of artists given by your roommate as a *combination*. A **combination** of items occurs when

- The items are selected from the same group (the five stars who died young and tragically).
- No item is used more than once. (You may adore Selena, but your three selections cannot be Selena, Selena, and Selena).
- The order of items makes no difference. (Morrison, Cobain, Selena is the same group in the documentary as Selena, Cobain, Morrison.)

Selena, musician of Tejano music (1971–1995)

Do you see the difference between a permutation and a combination? A permutation is an ordered arrangement of a given group of items. A combination is a group of items taken without regard to their order. **Permutation** problems involve situations in which **order matters. Combination** problems involve situations in which the **order** of items **makes no difference.**

EXAMPLE 6 Distinguishing between Permutations and Combinations

For each of the following problems, explain whether the problem is one involving permutations or combinations. (It is not necessary to solve the problem.)

a. Six candidates are running for president, chief technology officer, and director of marketing of an Internet company. The candidate with the greatest number of votes becomes the president, the second biggest vote-getter becomes chief technology officer, and the candidate who gets the third largest number of votes will be director of marketing. How many different outcomes are possible for these three positions?

b. From the six candidates who desire to hold office in an Internet company, a three-person committee is formed to study ways of finding new investors. How many different committees could be formed?

Solution

a. Voters are choosing three officers from six candidates. The order in which the officers are chosen makes a difference because each of the offices (president, chief technology officer, and director of marketing) is different. Order matters. This is a problem involving permutations. (How many permutations are possible if three candidates are elected from six candidates?)

b. A three-person committee is to be formed from the six candidates. The order in which the three people are selected does not matter because they are not filling different roles on the committee. Because order makes no difference, this is a problem involving combinations. (How many different combinations of three people can be chosen from a group of six people?)

Check Point 6

For each of the following problems, explain whether the problem is one involving permutations or combinations. (It is not necessary to solve the problem.)

a. How many ways can you select 6 free videos from a list of 200 videos?

b. In a race in which there are 50 runners and no ties, in how many ways can the first three finishers come in?

4 Use the combinations formula.

The notation $_nC_r$ means the **number of combinations of n things taken r at a time.** In general, there are $r!$ times as many permutations of n things taken r at a time as there are combinations of n things taken r at a time. Thus, we find the number of combinations of n things taken r at a time by dividing the number of permutations of n things taken r at a time by $r!$.

$$_nC_r = \frac{_nP_r}{r!} = \frac{\dfrac{n!}{(n-r)!}}{r!} = \frac{n!}{(n-r)!\,r!}$$

Combinations of n Things Taken r at a Time

The number of possible combinations if r items are taken from n items is

$$_nC_r = \frac{n!}{(n-r)!\,r!}.$$

Notice that the formula for $_nC_r$ is the same as the formula for the binomial coefficient $\binom{n}{r}$.

We cannot find the number of combinations if r items are taken from n items using the Fundamental Counting Principle. We must use the formula shown in the box to do so.

EXAMPLE 7 Using the Formula for Combinations

A three-person committee is needed to study ways of improving public transportation. How many committees could be formed from the eight people on the board of supervisors?

Solution The order in which the three people are selected does not matter. This is a problem of selecting $r = 3$ people from a group of $n = 8$ people. We are looking for the number of combinations of eight things taken three at a time. We use the formula

$$_nC_r = \frac{n!}{(n-r)!\,r!}$$

with $n = 8$ and $r = 3$.

$$_8C_3 = \frac{8!}{(8-3)!\,3!} = \frac{8!}{5!\,3!} = \frac{8 \cdot 7 \cdot 6 \cdot 5!}{5! \cdot 3 \cdot 2 \cdot 1} = \frac{8 \cdot 7 \cdot 6 \cdot 5!}{5! \cdot 3 \cdot 2 \cdot 1} = 56$$

Thus, 56 committees of three people each can be formed from the eight people on the board of supervisors.

Check Point 7 From a group of 10 physicians, in how many ways can four people be selected to attend a conference on acupuncture?

EXAMPLE 8 Using the Formula for Combinations

In poker, a person is dealt 5 cards from a standard 52-card deck. The order in which you are dealt the 5 cards does not matter. How many different 5-card poker hands are possible?

Solution Because the order in which the 5 cards are dealt does not matter, this is a problem involving combinations. We are looking for the number of combinations of $n = 52$ cards drawn $r = 5$ at a time. We use the formula

$$_nC_r = \frac{n!}{(n-r)!\,r!}$$

with $n = 52$ and $r = 5$.

$$_{52}C_5 = \frac{52!}{(52-5)!\,5!} = \frac{52!}{47!\,5!} = \frac{52 \cdot 51 \cdot 50 \cdot 49 \cdot 48 \cdot \cancel{47!}}{\cancel{47!} \cdot 5 \cdot 4 \cdot 3 \cdot 2 \cdot 1} = 2{,}598{,}960$$

Thus, there are 2,598,960 different 5-card poker hands possible. It surprises many people that more than 2.5 million 5-card hands can be dealt from a mere 52 cards.

Figure 11.6 A royal flush

If you are a card player, it does not get any better than to be dealt the 5-card poker hand shown in Figure 11.6. This hand is called a *royal flush*. It consists of an ace, king, queen, jack, and 10, all of the same suit: all hearts, all diamonds, all clubs, or all spades. The probability of being dealt a royal flush involves calculating the number of ways of being dealt such a hand: just 4 of all 2,598,960 possible hands. In the next section, we move from counting possibilities to computing probabilities.

Check Point 8   How many different 4-card hands can be dealt from a deck that has 16 different cards?

EXERCISE SET 11.6

Practice Exercises

In Exercises 1–8, use the formula for $_nP_r$ to evaluate each expression.

1. $_9P_4$

2. $_7P_3$

3. $_8P_5$

4. $_{10}P_4$

5. $_6P_6$

6. $_9P_9$

7. $_8P_0$

8. $_6P_0$

In Exercises 9–16, use the formula for $_nC_r$ to evaluate each expression.

9. $_9C_5$

10. $_{10}C_6$

11. $_{11}C_4$

12. $_{12}C_5$

13. $_7C_7$

14. $_4C_4$

15. $_5C_0$

16. $_6C_0$

In Exercises 17–20, does the problem involve permutations or combinations? Explain your answer. (It is not necessary to solve the problem.)

17. A medical researcher needs 6 people to test the effectiveness of an experimental drug. If 13 people have volunteered for the test, in how many ways can 6 people be selected?

18. Fifty people purchase raffle tickets. Three winning tickets are selected at random. If first prize is $1000, second prize is $500, and third prize is $100, in how many different ways can the prizes be awarded?

19. How many different four-letter passwords can be formed from the letters A, B, C, D, E, F, and G if no repetition of letters is allowed?

20. Fifty people purchase raffle tickets. Three winning tickets are selected at random. If each prize is $500, in how many different ways can the prizes be awarded?

Application Exercises

Use the Fundamental Counting Principle to solve Exercises 21–32.

21. The model of the car you are thinking of buying is available in nine different colors and three different styles (hatchback, sedan, or station wagon). In how many ways can you order the car?

22. A popular brand of pen is available in three colors (red, green, or blue) and four writing tips (bold, medium, fine, or micro). How many different choices of pens do you have with this brand?

23. An ice cream store sells two drinks (sodas or milk shakes), in four sizes (small, medium, large, or jumbo), and five flavors (vanilla, strawberry, chocolate, coffee, or pistachio). In how many ways can a customer order a drink?

24. A restaurant offers the following lunch menu.

Main Course	Vegetables	Beverages	Desserts
Ham	Potatoes	Coffee	Cake
Chicken	Peas	Tea	Pie
Fish	Green beans	Milk	Ice cream
Beef		Soda	

If one item is selected from each of the four groups, in how many ways can a meal be ordered? Describe two such orders.

25. You are taking a multiple-choice test that has five questions. Each of the questions has three choices, with one correct choice per question. If you select one of these options per question and leave nothing blank, in how many ways can you answer the questions?

26. You are taking a multiple-choice test that has eight questions. Each of the questions has three answer choices, with one correct answer per question. If you select one of these three choices for each question and leave nothing blank, in how many ways can you answer the questions?

27. In the original plan for area codes in 1945, the first digit could be any number from 2 through 9, the second digit was either 0 or 1, and the third digit could be any number except 0. With this plan, how many different area codes were possible?

28. How many different four-letter radio station call letters can be formed if the first letter must be W or K?

29. Six performers are to present their comedy acts on a weekend evening at a comedy club. One of the performers insists on being the last stand-up comic of the evening. If this performer's request is granted, how many different ways are there to schedule the appearances?

30. Five singers are to perform at a night club. One of the singers insists on being the last performer of the evening. If this singer's request is granted, how many different ways are there to schedule the appearances?

31. In the *Cambridge Encyclopedia of Language* (Cambridge University Press, 1987), author David Crystal presents five sentences that make a reasonable paragraph regardless of their order. The sentences are

> Mark had told him about the foxes.
> John looked out the window.
> Could it be a fox?
> However, nobody had seen one for months.
> He thought he saw a shape in the bushes.

How many different five-sentence paragraphs can be formed if the paragraph begins with "He thought he saw a shape in the bushes" and ends with "John looked out of the window"?

32. A television programmer is arranging the order that five movies will be seen between the hours of 6 P.M. and 4 A.M. Two of the movies have a G rating, and they are to be shown in the first two time blocks. One of the movies is rated NC-17, and it is to be shown in the last of the time blocks, from 2 A.M. until 4 A.M. Given these restrictions, in how many ways can the five movies be arranged during the indicated time blocks?

Use the formula for $_nP_r$ to solve Exercises 33–40.

33. A club with ten members is to choose three officers—president, vice-president, and secretary-treasurer. If each office is to be held by one person and no person can hold more than one office, in how many ways can those offices be filled?

34. A corporation has ten members on its board of directors. In how many different ways can it elect a president, vice-president, secretary, and treasurer?

35. For a segment of a radio show, a disc jockey can play 7 records. If there are 13 records to select from, in how many ways can the program for this segment be arranged?

36. Suppose you are asked to list, in order of preference, the three best movies you have seen this year. If you saw 20 movies during the year, in how many ways can the three best be chosen and ranked?

37. In a race in which six automobiles are entered and there are no ties, in how many ways can the first three finishers come in?

38. In a production of *West Side Story*, eight actors are considered for the male roles of Tony, Riff, and Bernardo. In how many ways can the director cast the male roles?

39. Nine bands have volunteered to perform at a benefit concert, but there is only enough time for five of the bands to play. How many lineups are possible?

40. How many arrangements can be made using four of the letters of the word COMBINE if no letter is to be used more than once?

Use the formula for $_nC_r$ to solve Exercises 41–48.

41. An election ballot asks voters to select three city commissioners from a group of six candidates. In how many ways can this be done?

42. A four-person committee is to be elected from an organization's membership of 11 people. How many different committees are possible?

43. Of 12 possible books, you plan to take 4 with you on vacation. How many different collections of 4 books can you take?

44. There are 14 standbys who hope to get seats on a flight, but only 6 seats are available on the plane. How many different ways can the 6 people be selected?

45. You volunteer to help drive children at a charity event to the zoo, but you can fit only 8 of the 17 children present in your van. How many different groups of 8 children can you drive?

46. Of the 100 people in the U.S. Senate, 18 serve on the Foreign Relations Committee. How many ways are there to select Senate members for this committee (assuming party affiliation is not a factor in selection)?

47. To win at LOTTO in the state of Florida, one must correctly select 6 numbers from a collection of 53 numbers (1 through 53). The order in which the selection is made does not matter. How many different selections are possible?

48. To win in the New York State lottery, one must correctly select 6 numbers from 59 numbers. The order in which the selection is made does not matter. How many different selections are possible?

In Exercises 49–58, solve by the method of your choice.

49. In a race in which six automobiles are entered and there are no ties, in how many ways can the first four finishers come in?

50. A book club offers a choice of 8 books from a list of 40. In how many ways can a member make a selection?

51. A medical researcher needs 6 people to test the effectiveness of an experimental drug. If 13 people have volunteered for the test, in how many ways can 6 people be selected?

52. Fifty people purchase raffle tickets. Three winning tickets are selected at random. If first prize is $1000, second prize is $500, and third prize is $100, in how many different ways can the prizes be awarded?

53. From a club of 20 people, in how many ways can a group of three members be selected to attend a conference?

54. Fifty people purchase raffle tickets. Three winning tickets are selected at random. If each prize is $500, in how many different ways can the prizes be awarded?

55. How many different four-letter passwords can be formed from the letters A, B, C, D, E, F, and G if no repetition of letters is allowed?

56. Nine comedy acts will perform over two evenings. Five of the acts will perform on the first evening and the order in which the acts perform is important. How many ways can the schedule for the first evening be made?

57. Using 15 flavors of ice cream, how many cones with three different flavors can you create if it is important to you which flavor goes on the top, middle, and bottom?

58. Baskin-Robbins offers 31 different flavors of ice cream. One of their items is a bowl consisting of three scoops of ice cream, each a different flavor. How many such bowls are possible?

Writing in Mathematics

59. Explain the Fundamental Counting Principle.

60. Write an original problem that can be solved using the Fundamental Counting Principle. Then solve the problem.

61. What is a permutation?

62. Describe what $_nP_r$ represents.

63. Write a word problem that can be solved by evaluating $_7P_3$.

64. What is a combination?

65. Explain how to distinguish between permutation and combination problems.

66. Write a word problem that can be solved by evaluating $_7C_3$.

Technology Exercises

67. Use a graphing utility with an $\boxed{_nP_r}$ menu item to verify your answers in Exercises 1–8.

68. Use a graphing utility with an $\boxed{_nC_r}$ menu item to verify your answers in Exercises 9–16.

Critical Thinking Exercises

69. Which one of the following is true?
   a. The number of ways to choose four questions out of ten questions on an essay test is $_{10}P_4$.
   b. If $r > 1$, $_nP_r$ is less than $_nC_r$.
   c. $_7P_3 = 3!{_7}C_3$
   d. The number of ways to pick a winner and first runner-up in a piano recital with 20 contestants is $_{20}C_2$.

70. Five men and five women line up at a checkout counter in a store. In how many ways can they line up if the first person in line is a woman, and the people in line alternate woman, man, woman, man, and so on?

71. How many four-digit odd numbers less than 6000 can be formed using the digits 2, 4, 6, 7, 8, and 9? Digits may be repeated.

72. If a collection of n objects has n_1 identical objects of the same type, n_2 identical objects of a second kind, n_3 of a third kind, and so on for a total of $n = n_1 + n_2 + \cdots + n_k$ objects, the number of distinguishable permutations of the n objects is given by
$$\frac{n!}{n_1!\,n_2!\,n_3!\cdots n_k!}.$$
Use this formula to find the number of different signals consisting of eight flags that can be made using three white flags, four red flags and one blue flag.

Group Exercise

73. The group should select real-world situations where the Fundamental Counting Principle can be applied. These could involve the number of possible student ID numbers on your campus, the number of possible phone numbers in your community, the number of meal options at a local restaurant, the number of ways a person in the group can select outfits for class, the number of ways a condominium can be purchased in a nearby community, and so on. Once situations have been selected, group members should determine in how many ways each part of the task can be done. Group members will need to obtain menus, find out about telephone-digit requirements in the community, count shirts, pants, shoes in closets, visit condominium sales offices, and so on. Once the group reassembles, apply the Fundamental Counting Principle to determine the number of available options in each situation. Because these numbers may be quite large, use a calculator.

SECTION 11.7   *Probability*

Objectives

1. Compute empirical probability.
2. Compute theoretical probability.
3. Find the probability that an event will not occur.
4. Find the probability of one event or a second event occurring.
5. Find the probability of one event and a second event occurring.

Table 11.3   Number of Americans and the Hours of Sleep They Get on a Typical Night

Hours of Sleep	Number of Americans, in millions
4 or less	11.36
5	25.56
6	71
7	85.2
8	76.68
9	8.52
10 or more	5.68
	Total: 284

Source: Discovery Health Media

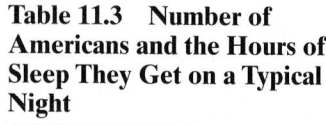

1 Compute empirical probability.

How many hours of sleep do you typically get each night? Table 11.3 indicates that 71 million out of 284 million Americans are getting six hours of sleep on a typical night. The *probability* of an American getting six hours of sleep on a typical night is $\frac{71}{284}$. This fraction can be reduced to $\frac{1}{4}$, or expressed as 0.25 or 25%. Thus, 25% of Americans get six hours of sleep each night.

We find a probability by dividing one number by another. Probabilities are assigned to an *event*, such as getting six hours of sleep on a typical night. Events that are certain to occur are assigned probabilities of 1, or 100%. For example, the probability that a given individual will eventually die is 1. Regrettably, taxes and death are always certain! By contrast, if an event cannot occur, its probability is 0. For example, the probability that Elvis will return from the dead and serenade us with one final reprise of "Heartbreak Hotel" is 0.

Probabilities of events are expressed as numbers ranging from 0 to 1, or 0% to 100%. The closer the probability of a given event is to 1, the more likely it is that the event will occur. The closer the probability of a given event is to 0, the less likely it is that the event will occur.

Empirical Probability

Empirical probability applies to situations in which we observe how frequently an event occurs. We use the following formula to compute the empirical probability of an event:

> **Computing Empirical Probability**
> The **empirical probability** of event E is
> $$P(E) = \frac{\text{observed number of times } E \text{ occurs}}{\text{total number of observed occurrences}}.$$

EXAMPLE 1 Computing Empirical Probability

There are approximately 3 million Arab Americans in the United States. The circle graph in Figure 11.7 shows that the majority of Arab Americans are Christian. If an Arab American is selected at random, find the empirical probability of selecting a Catholic.

Solution The probability of selecting a Catholic is the observed number of Arab Americans who are Catholic, 1.26 (million), divided by the total number of Arab Americans, 3 (million).

P (selecting a Catholic from the Arab-American population)

$$= \frac{\text{number of Arab Americans who are Catholic}}{\text{total number of Arab Americans}} = \frac{1.26}{3.00} = \frac{126}{300} = 0.42$$

The empirical probability of selecting a Catholic from the Arab-American population is $\frac{126}{300}$, or 0.42. Equivalently, 42% of Arab Americans are Catholic.

Arab-American Faiths

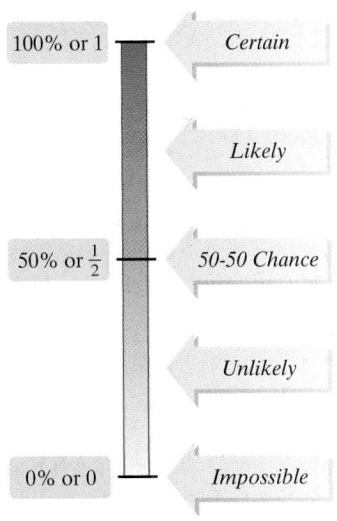

Muslim 0.69 million

Roman Catholic 1.26 million

Protestant 0.36 million

Greek Orthodox, Chaldean or Syriac 0.69 million

Figure 11.7
Source: Arab-American Institute

> **Check Point 1** If an Arab American is selected at random, find the empirical probability of selecting a Muslim.

Theoretical Probability

2 Compute theoretical probability.

You toss a coin. Although it is equally likely to land either heads up, denoted by H, or tails up, denoted by T, the actual outcome is uncertain. Any occurrence for which the outcome is uncertain is called an **experiment.** Thus, tossing a coin is an example of an experiment. The set of all possible outcomes of an experiment is the **sample space** of the experiment, denoted by S. The sample space for the coin-tossing experiment is

$$S = \{H, T\}.$$

Lands heads up   Lands tails up

We can define an event more formally using these concepts. An **event,** denoted by E, is any subcollection, or subset, of a sample space. For example, the subset $E = \{T\}$ is the event of landing tails up when a coin is tossed.

Theoretical probability applies to situations like this, in which the sample space only contains equally-likely outcomes, all of which are known. To calculate the theoretical probability of an event, we divide the number of outcomes resulting in the event by the total number of outcomes in the sample space.

100% or 1 — Certain

Likely

50% or $\frac{1}{2}$ — 50-50 Chance

Unlikely

0% or 0 — Impossible

Possible Values for Probabilities

Computing Theoretical Probability

If an event E has $n(E)$ equally-likely outcomes and its sample space S has $n(S)$ equally-likely outcomes, the theoretical probability of event E, denoted by $P(E)$, is

$$P(E) = \frac{\text{number of outcomes in event } E}{\text{number of outcomes in sample space } S} = \frac{n(E)}{n(S)}.$$

The sum of the theoretical probabilities of all possible outcomes in the sample space is 1.

How can we use this formula to compute the probability of a coin landing tails up? We use the following sets:

$$E = \{T\} \qquad\qquad S = \{H, T\}.$$

This is the event of landing tails up.

This is the sample space with all equally-likely outcomes.

The probability of a coin landing tails up is

$$P(E) = \frac{n(E)}{n(S)} = \frac{1}{2}.$$

Theoretical probability applies to many games of chance, including dice rolling, lotteries, card games, and roulette. The next example deals with the experiment of rolling a die. Figure 11.8 illustrates that when a die is rolled, there are six equally-likely outcomes. The sample space can be shown as

$$S = \{1, 2, 3, 4, 5, 6\}.$$

EXAMPLE 2   Computing Theoretical Probability

A die is rolled once. Find the probability of getting a number less than 5.

Figure 11.8 Outcomes when a die is rolled

Solution   The sample space of equally-likely outcomes is $S = \{1, 2, 3, 4, 5, 6\}$. There are six outcomes in the sample space, so $n(S) = 6$.

We are interested in the probability of getting a number less than 5. The event of getting a number less than 5 can be represented by

$$E = \{1, 2, 3, 4\}.$$

There are four outcomes in this event, so $n(E) = 4$.

The probability of rolling a number less than 5 is

$$P(E) = \frac{n(E)}{n(S)} = \frac{4}{6} = \frac{2}{3}.$$

Check Point 2   A die is rolled once. Find the probability of getting a number greater than 4.

EXAMPLE 3   Computing Theoretical Probability

Two ordinary six-sided dice are rolled. What is the probability of getting a sum of 8?

Solution   Each die has six equally-likely outcomes. By the Fundamental Counting Principle, there are $6 \cdot 6$, or 36, equally-likely outcomes in the sample space. That is, $n(S) = 36$. The 36 outcomes are shown below as ordered pairs. The five ways of rolling a sum of 8 appear in the green highlighted diagonal.

Surprising Probabilities

Imagine that one person is randomly selected from all 6 billion people on planet Earth. The following empirical probabilities, each rounded to two decimal places, might surprise you.

Probability of selecting

a woman	= 0.51
a non-white	= 0.7
a non-Christian	= 0.7
a person who cannot read	= 0.7
a person suffering from malnutrition	= 0.5
a person with a college education	= 0.01
a person who is near death	= 0.01

Source: United Nations

		Second Die					
		⚀	⚁	⚂	⚃	⚄	⚅
First Die	⚀	(1,1)	(1,2)	(1,3)	(1,4)	(1,5)	(1,6)
	⚁	(2,1)	(2,2)	(2,3)	(2,4)	(2,5)	(2,6)
	⚂	(3,1)	(3,2)	(3,3)	(3,4)	(3,5)	(3,6)
	⚃	(4,1)	(4,2)	(4,3)	(4,4)	(4,5)	(4,6)
	⚄	(5,1)	(5,2)	(5,3)	(5,4)	(5,5)	(5,6)
	⚅	(6,1)	(6,2)	(6,3)	(6,4)	(6,5)	(6,6)

$$S = \{(1,1), (1,2), (1,3), (1,4),$$
$$(1,5), (1,6), (2,1), (2,2),$$
$$(2,3), (2,4), (2,5), (2,6),$$
$$(3,1), (3,2), (3,3), (3,4),$$
$$(3,5), (3,6), (4,1), (4,2),$$
$$(4,3), (4,4), (4,5), (4,6),$$
$$(5,1), (5,2), (5,3), (5,4),$$
$$(5,5), (5,6), (6,1), (6,2),$$
$$(6,3), (6,4), (6,5), (6,6)\}$$

The phrase "getting a sum of 8" describes the event

$$E = \{(6, 2), (5, 3), (4, 4), (3, 5), (2, 6)\}.$$

This event has 5 outcomes, so $n(E) = 5$. Thus, the probability of getting a sum of 8 is

$$P(E) = \frac{n(E)}{n(S)} = \frac{5}{36}.$$

Check Point 3　What is the probability of getting a sum of 5 when two six-sided dice are rolled?

Computing Theoretical Probability Without Listing an Event and the Sample Space

In some situations, we can compute theoretical probability without having to write out each event and each sample space. For example, suppose you are dealt one card from a standard 52-card deck, illustrated in Figure 11.9. The deck has four suits: Hearts and diamonds are red, and clubs and spades are black. Each suit has 13 different face values—A(ace), 2, 3, 4, 5, 6, 7, 8, 9, 10, J(jack), Q(queen), and K(king). Jacks, queens, and kings are called **picture cards** or **face cards.**

Figure 11.9 A standard 52-card bridge deck

EXAMPLE 4 Probability and a Deck of 52 Cards

You are dealt one card from a standard 52-card deck. Find the probability of being dealt a heart.

Solution Let E be the event of being dealt a heart. Because there are 13 hearts in the deck, the event of being dealt a heart can occur in 13 ways. The number of outcomes resulting in event E is 13: $n(E) = 13$. With 52 cards in the deck, the total number of possible ways of being dealt a single card is 52. The number of outcomes in the sample space is 52: $n(S) = 52$. The probability of being dealt a heart is

$$P(E) = \frac{n(E)}{n(S)} = \frac{13}{52} = \frac{1}{4}.$$

Check Point 4　If you are dealt one card from a standard 52-card deck, find the probability of being dealt a king.

State lotteries keep 50 cents on the dollar, resulting in $10 billion a year for public funding.

If your state has a lottery drawing each week, the probability that someone will win the top prize is relatively high. If there is no winner this week, it is virtually certain that eventually someone will be graced with millions of dollars. So how come you are unlucky compared to this undisclosed someone? In Example 5, we provide an answer to this question, using the counting principles discussed in Section 11.6.

EXAMPLE 5   Probability and Combinations: Winning the Lottery

Florida's lottery game, LOTTO, is set up so that each player chooses six different numbers from 1 to 53. If the six numbers chosen match the six numbers drawn randomly twice weekly, the player wins (or shares) the top cash prize. (As of this writing, the top cash prize has ranged from $7 million to $106.5 million.) With one LOTTO ticket, what is the probability of winning this prize?

Solution   Because the order of the six numbers does not matter, this is a situation involving combinations. Let E be the event of winning the lottery with one ticket. With one LOTTO ticket, there is only one way of winning. Thus, $n(E) = 1$. The sample space is the set of all possible six-number combinations. We can use the combinations formula

$$_nC_r = \frac{n!}{(n-r)!\,r!}$$

to find the total number of possible combinations. We are selecting $r = 6$ numbers from a collection of $n = 53$ numbers.

$$_{53}C_6 = \frac{53!}{(53-6)!\,6!} = \frac{53!}{47!\,6!} = \frac{53 \cdot 52 \cdot 51 \cdot 50 \cdot 49 \cdot 48 \cdot \cancel{47!}}{\cancel{47!} \cdot 6 \cdot 5 \cdot 4 \cdot 3 \cdot 2 \cdot 1} = 22{,}957{,}480$$

There are nearly 23 million number combinations possible in LOTTO. If a person buys one LOTTO ticket, the probability of winning is

$$P(E) = \frac{n(E)}{n(S)} = \frac{1}{22{,}957{,}480} \approx 0.0000000436.$$

The probability of winning the top prize with one LOTTO ticket is $\frac{1}{22{,}957{,}480}$, or about 1 in 23 million.

Comparing the Probability of Dying to the Probability of Winning Florida's LOTTO

As a healthy nonsmoking 30-year-old, your probability of dying this year is approximately 0.001. Divide this probability by the probability of winning LOTTO with one ticket:

$$\frac{0.001}{0.0000000436} \approx 22{,}936.$$

A healthy 30-year-old is nearly 23,000 times more likely to die this year than to win Florida's lottery.

In 2001, Americans spent nearly 18 billion dollars on lotteries set up by revenue-hungry states. If a pigeon, er, person, buys, say 5000 different tickets in Florida's LOTTO, that person has selected 5000 different combinations of the six numbers. The probability of winning is

$$\frac{5000}{22{,}957{,}480} \approx 0.000218.$$

The chances of winning top prize are about 218 in a million. At $1 per LOTTO ticket, it is highly probable that Mr. or Ms. Pigeon will be $5000 poorer.

 Check Point 5

People lose interest when they do not win at games of chance, including Florida's LOTTO. With drawings twice weekly instead of once, the game described in Example 5 was brought in to bring back lost players and increase ticket sales. The original LOTTO was set up so that each player chose six different numbers from 1 to 49, rather than from 1 to 53, with a lottery drawing only once a week. With one LOTTO ticket, what was the probability of winning the top cash prize in Florida's original LOTTO? Express the answer as a fraction and as a decimal correct to ten places.

3 Find the probability that an event will not occur.

Probability of an Event Not Occurring

A survey (*source*: Penn, Schoen, and Berland) asked 500 Americans to rate their health. Of those surveyed, 270 rated their health as good/excellent. This means that 500 − 270, or 230, people surveyed did not rate their health as good/excellent. Notice that

$$P(\text{good/excellent}) + P(\text{not good/excellent}) = \frac{270}{500} + \frac{230}{500} = \frac{500}{500} = 1.$$

In general, because the sum of the probabilities of all possible outcomes in any situation is 1,

$$P(E) + P(\text{not } E) = 1.$$

We now solve this equation for $P(\text{not } E)$, the probability that event E will not occur, by subtracting $P(E)$ from both sides. The resulting formula is given in the following box:

The Probability of an Event Not Occurring

The probability that an event E will not occur is equal to one minus the probability that it will occur.

$$P(\text{not } E) = 1 - P(E)$$

EXAMPLE 6 The Probability of Not Winning the Lottery

We have seen that the probability of winning Florida's LOTTO with one ticket is $\frac{1}{22,957,480}$. What is the probability of not winning?

Solution

$$P(\text{not winning}) = 1 - P(\text{winning}) = 1 - \frac{1}{22,957,480} = \frac{22,957,480}{22,957,480} - \frac{1}{22,957,480}$$

$$= \frac{22,957,479}{22,957,480} \approx 0.99999996$$

The probability of not winning is close to 1. It is almost certain that with one LOTTO ticket, a person will not win top prize.

Check Point 6 The essay on page 992 mentions that the probability of a 30-year-old dying this year is approximately $\frac{1}{1000}$. What is the probability of a 30-year-old not dying this year?

4 Find the probability of one event or a second event occurring.

Or Probabilities with Mutually Exclusive Events

Suppose that you randomly select one card from a deck of 52 cards. Let A be the event of selecting a king and B be the event of selecting a queen. Only one card is selected, so it is impossible to get both a king and a queen. The outcomes of selecting a king and a queen cannot occur simultaneously. They are called *mutually exclusive events*. If it is impossible for any two events, A and B, to occur simultaneously, they are said to be **mutually exclusive.** If A and B are mutually exclusive events, the probability that either A or B will occur is determined by adding their individual probabilities.

Or Probabilities with Mutually Exclusive Events

If A and B are mutually exclusive events, then

$$P(A \text{ or } B) = P(A) + P(B).$$

EXAMPLE 7 The Probability of Either of Two Mutually Exclusive Events Occurring

If one card is randomly selected from a deck of cards, what is the probability of selecting a king or a queen?

Solution We find the probability that either of these mutually exclusive events will occur by adding their individual probabilities.

$$P(\text{king or queen}) = P(\text{king}) + P(\text{queen}) = \frac{4}{52} + \frac{4}{52} = \frac{8}{52} = \frac{2}{13}$$

The probability of selecting a king or a queen is $\frac{2}{13}$.

Check Point 7 If you roll a single, six-sided die, what is the probability of getting either a 4 or a 5?

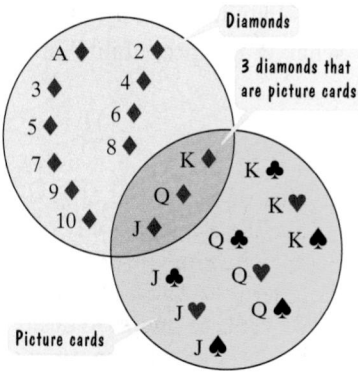

13 Diamonds

13 Hearts

13 Spades

13 Clubs

Figure 11.10 A deck of 52 cards

Or Probabilities with Events That Are Not Mutually Exclusive

Consider the deck of 52 cards shown in Figure 11.10. Suppose that these cards are shuffled and you randomly select one card from the deck. What is the probability of selecting a diamond or a picture card (jack, queen, king)? Begin by adding their individual probabilities:

$$P(\text{diamond}) + P(\text{picture card}) = \frac{13}{52} + \frac{12}{52}.$$

There are 13 diamonds in the deck of 52 cards.

There are 12 picture cards in the deck of 52 cards.

However, this is not the probability of selecting a diamond or a picture card. The problem is that there are three cards that are simultaneously diamonds and picture cards, shown in Figure 11.11. The events of selecting a diamond and selecting a picture card are not mutually exclusive. It is possible to select a card that is both a diamond and a picture card.

Figure 11.11 Three diamonds are picture cards.

The situation is illustrated in the diagram in Figure 11.12. Why can't we find the probability of selecting a diamond or a picture card by adding their individual probabilities? The diagram shows that three of the cards, the three diamonds that are picture cards, get counted twice when we add the individual probabilities. First the three cards get counted as diamonds, and then they get counted as picture cards. In order to avoid the error of counting the three cards twice, we need to subtract the probability of getting a diamond and a picture card, $\frac{3}{52}$, as follows:

Figure 11.12

$P(\text{diamond or picture card})$

$= P(\text{diamond}) + P(\text{picture card}) - P(\text{diamond and picture card})$

$$= \frac{13}{52} + \frac{12}{52} - \frac{3}{52} = \frac{13 + 12 - 3}{52} = \frac{22}{52} = \frac{11}{26}.$$

Thus, the probability of selecting a diamond or a picture card is $\frac{11}{26}$.

In general, if A and B are events that are not mutually exclusive, the probability that A or B will occur is determined by adding their individual probabilities and then subtracting the probability that A and B occur simultaneously.

Or Probabilities with Events That Are Not Mutually Exclusive

If A and B are not mutually exclusive events, then

$$P(A \text{ or } B) = P(A) + P(B) - P(A \text{ and } B).$$

EXAMPLE 8 **An *Or* Probability with Events That Are Not Mutually Exclusive**

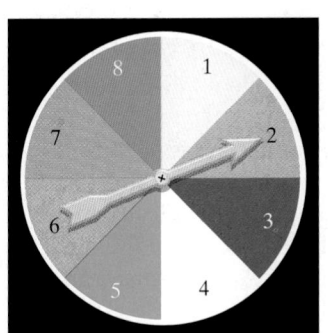

Figure 11.13 It is equally probable that the pointer will land on any one of the eight regions.

Figure 11.13 illustrates a spinner. It is equally probable that the pointer will land on any one of the eight regions, numbered 1 through 8. If the pointer lands on a borderline, spin again. Find the probability that the pointer will stop on an even number or on a number greater than 5.

Solution It is possible for the pointer to land on a number that is both even and greater than 5. Two of the numbers, 6 and 8, are even and greater than 5. These events are not mutually exclusive. The probability of landing on a number that is even or greater than 5 is

$$P\left(\begin{array}{c}\text{even or}\\\text{greater than 5}\end{array}\right) = P(\text{even}) + P(\text{greater than 5}) - P\left(\begin{array}{c}\text{even and}\\\text{greater than 5}\end{array}\right)$$

$$= \quad \frac{4}{8} \quad + \quad \frac{3}{8} \quad - \quad \frac{2}{8}$$

Four of the eight numbers, 2, 4, 6, and 8, are even. Three of the eight numbers, 6, 7, and 8, are greater than 5. Two of the eight numbers, 6 and 8, are even and greater than 5.

$$= \frac{4 + 3 - 2}{8} = \frac{5}{8}.$$

The probability that the pointer will stop on an even number or on a number greater than 5 is $\frac{5}{8}$.

Check Point 8 Use Figure 11.13 to find the probability that the pointer will stop on an odd number or on a number less than 5.

EXAMPLE 9 **An *Or* Probability with Events That Are Not Mutually Exclusive**

A group of people is comprised of 15 U.S. men, 20 U.S. women, 10 Canadian men, and 5 Canadian women. If a person is selected at random from the group, find the probability that the selected person is a man or a Canadian.

Solution The group is comprised of $15 + 20 + 10 + 5$, or 50 people. It is possible to select a man who is Canadian. We are given that there are 10 Canadian men, so these events are not mutually exclusive.

$$P(\text{man or Canadian}) = P(\text{man}) + P(\text{Canadian}) - P(\text{man and Canadian})$$

$$= \frac{25}{50} + \frac{15}{50} - \frac{10}{50}$$

| Of the 50 people, 25 are men—15 U.S. men and 10 Canadian men. | Of the 50 people, 15 are Canadian—10 Canadian men and 5 Canadian women. | Of the 50 people, 10 are Canadian men. |

$$= \frac{25 + 15 - 10}{50} = \frac{30}{50} = \frac{3}{5}$$

The probability of selecting a man or a Canadian is $\frac{3}{5}$.

Check Point 9 An interfaith group is comprised of 14 African-American Muslims, 12 African-American Christians, 6 Arab-American Muslims, and 8 Arab-American Christians. If one person is selected to attend a conference on shared ethical values in the faith community, find the probability that the selected person is Muslim or African American.

5 Find the probability of one event and a second event occurring.

And Probabilities with Independent Events

Suppose that you toss a fair coin two times in succession. The outcome of the first toss, heads or tails, does not affect what happens when you toss the coin a second time. For example, the occurrence of tails on the first toss does not make tails more likely or less likely to occur on the second toss. The repeated toss of a coin produces *independent events* because the outcome of one toss does not influence the outcome of others. Two events are **independent events** if the occurrence of either of them has no effect on the probability of the other.

If two events are independent, we can calculate the probability of the first occurring and the second occurring by multiplying their probabilities.

And Probabilities with Independent Events

If A and B are independent events, then

$$P(A \text{ and } B) = P(A) \cdot P(B).$$

Figure 11.14 A U.S. roulette wheel

EXAMPLE 10 Independent Events on a Roulette Wheel

Figure 11.14 shows a U.S. roulette wheel that has 38 numbered slots (1 through 36, 0, and 00). Of the 38 compartments, 18 are black, 18 are red, and two are green. A play has the dealer spin the wheel and a small ball in opposite directions. As the ball slows to a stop, it can land with equal probability on any one of the 38 numbered slots. Find the probability of red occurring on two consecutive plays.

Solution   The wheel has 38 equally-likely outcomes and 18 are red. Thus, the probability of red occurring on a play is $\frac{18}{38}$, or $\frac{9}{19}$. The result that occurs on each play is independent of all previous results. Thus,

$$P(\text{red and red}) = P(\text{red}) \cdot P(\text{red}) = \frac{9}{19} \cdot \frac{9}{19} = \frac{81}{361} \approx 0.224.$$

The probability of red occurring on two consecutive plays is $\frac{81}{361}$.

Some roulette players incorrectly believe that if red occurs on two consecutive plays, then another color is "due." Because the events are independent, the outcomes of previous spins have no effect on any other spins.

Check Point 10   Find the probability of green occurring on two consecutive plays on a roulette wheel.

The *and* rule for independent events can be extended to cover three or more events. Thus, if A, B, and C are independent events, then

$$P(A \text{ and } B \text{ and } C) = P(A) \cdot P(B) \cdot P(C).$$

EXAMPLE 11 Independent Events in a Family

The picture in the margin shows a family that has had nine girls in a row. Find the probability of this occurrence.

Solution   If two or more events are independent, we can find the probability of them all occurring by multiplying their probabilities. The probability of a baby girl is $\frac{1}{2}$, so the probability of nine girls in a row is $\frac{1}{2}$ used as a factor nine times.

$$P(\text{nine girls in a row}) = \frac{1}{2} \cdot \frac{1}{2} \cdot \frac{1}{2} \cdot \frac{1}{2} \cdot \frac{1}{2} \cdot \frac{1}{2} \cdot \frac{1}{2} \cdot \frac{1}{2} \cdot \frac{1}{2}$$

$$= \left(\frac{1}{2}\right)^9 = \frac{1}{512}$$

The probability of a run of nine girls in a row is $\frac{1}{512}$. (If another child is born into the family, this event is independent of the other nine, and the probability of a girl is still $\frac{1}{2}$.)

Check Point 11   Find the probability of a family having four boys in a row.

EXERCISE SET 11.7

Practice and Application Exercises

Exercises 1–8 involve empirical probability. Use the empirical probability formula to solve each exercise. Express answers as fractions. Then use a calculator to express probabilities as decimals, rounded to the nearest thousandth, if necessary.

Use the table showing the number of people who regularly participate in various forms of exercise, based on a survey of 2000 Americans, to solve Exercises 1–4.

Number of People Who Regularly Participate in Various Forms of Exercise in a Survey of 2000 People

Forms of Exercise	Number of People
Walking/hiking	1140
Weight training	320
Running/jogging	280
Biking	240
Aerobics	240
Exercise machines	220

Source: Discovery Health Media

Find the probability that a randomly selected American participates in:

1. weight training.
2. running/jogging.
3. biking.
4. walking/hiking.

Use the table showing world population by region to solve Exercises 5–8.

World Population, by Region

Region	Population
Africa	784,400,000
Asia	3,682,600,000
Europe	728,900,000
Latin America and the Caribbean	519,000,000
North America	309,600,000
Oceania	30,400,000

Total World Population: 6,054,900,000
Source: U.S. Bureau of the Census

If one person is randomly selected from all people on planet Earth, find the probability of selecting:

5. an African.
6. an Asian.
7. a North American.
8. a European.

Exercises 9–24 involve theoretical probability. Use the theoretical probability formula to solve each exercise. Express each probability as a fraction reduced to lowest terms.

In Exercises 9–14, a die is rolled. The sample space of equally likely outcomes is $\{1, 2, 3, 4, 5, 6\}$. Find the probability of getting:

9. a 4.
10. a 5.
11. an odd number.
12. a number greater than 3.
13. a number greater than 4.
14. a number greater than 7.

In Exercises 15–18, you are dealt one card from a standard 52 card deck. Find the probability of being dealt:

15. a queen.
16. a diamond.
17. a picture card.
18. a card greater than 3 and less than 7.

In Exercises 19–20, a fair coin is tossed two times in succession. The sample space of equally-likely outcomes is $\{HH, HT, TH, TT\}$. Find the probability of getting:

19. two heads.
20. the same outcome on each toss.

In Exercises 21–22, you select a family with three children. If M represents a male child and F a female child, the sample space of equally likely outcomes is $\{MMM, MMF, MFM, MFF, FMM, FMF, FFM, FFF\}$. Find the probability of selecting a family with:

21. at least one male child.
22. at least two female children.

In Exercises 23–24, a single die is rolled twice. The 36 equally likely outcomes are shown as follows:

		Second Roll				
First Roll		⚀	⚁	⚂	⚃	⚄	⚅
⚀	(1,1)	(1,2)	(1,3)	(1,4)	(1,5)	(1,6)
⚁	(2,1)	(2,2)	(2,3)	(2,4)	(2,5)	(2,6)
⚂	(3,1)	(3,2)	(3,3)	(3,4)	(3,5)	(3,6)
⚃	(4,1)	(4,2)	(4,3)	(4,4)	(4,5)	(4,6)
⚄	(5,1)	(5,2)	(5,3)	(5,4)	(5,5)	(5,6)
⚅	(6,1)	(6,2)	(6,3)	(6,4)	(6,5)	(6,6)

Find the probability of getting:

23. two numbers whose sum is 4.
24. two numbers whose sum is 6.
25. To play the California lottery, a person has to correctly select 6 out of 51 numbers, paying $1 for each six-number selection. If the six numbers picked are the same as the ones drawn by the lottery, mountains of money are bestowed. What is the probability that a person with one combination of six numbers will win? What is the probability of winning if 100 different lottery tickets are purchased?
26. A state lottery is designed so that a player chooses six numbers from 1 to 30 on one lottery ticket. What is the probability that a player with one lottery ticket will win? What is the probability of winning if 100 different lottery tickets are purchased?

The table shows the probability of dying at any given age. Use the table and your answer from Exercise 25 to solve Exercises 27–28.

**Probability of Dying
at Any Given Age**

Age	Probability of Male Death	Probability of Female Death
10	0.00013	0.00010
20	0.00140	0.00050
30	0.00153	0.00050
40	0.00193	0.00095
50	0.00567	0.00305
60	0.01299	0.00792
70	0.03473	0.01764
80	0.07644	0.03966
90	0.15787	0.11250
100	0.26876	0.23969
110	0.39770	0.39043

Source: George Shaffner, *The Arithmetic of Life and Death*

27. How many times more likely is a 20-year-old male to die this year than to win California's lottery with one lottery ticket?

28. How many times more likely is a 20-year-old female to die this year than to win California's lottery with one lottery ticket?

29. A poker hand consists of five cards.

   a. Find the total number of possible five-card poker hands that can be dealt from a deck of 52 cards.

   b. A diamond flush consists of a five-card hand containing all diamonds. Find the number of possible five-card diamond flushes.

   c. Find the probability of being dealt a diamond flush.

30. If you are dealt 4 cards from a shuffled deck of 52 cards, find the probability that all 4 are hearts.

The graph at the top of the next column shows the probability of cardiovascular disease, by age and gender. Use the information in the graph to solve Exercises 31–32. Express all probabilities as decimals, estimated to two decimal places.

31. **a.** What is the probability that a randomly selected man between the ages of 25 and 34 has cardiovascular disease?

   b. What is the probability that a randomly selected man between the ages 25 and 34 does not have cardiovascular disease?

32. **a.** What is the probability that a randomly selected woman, 75 or older, has cardiovascular disease?

   b. What is the probability that a randomly selected woman, 75 or older, does not have cardiovascular disease?

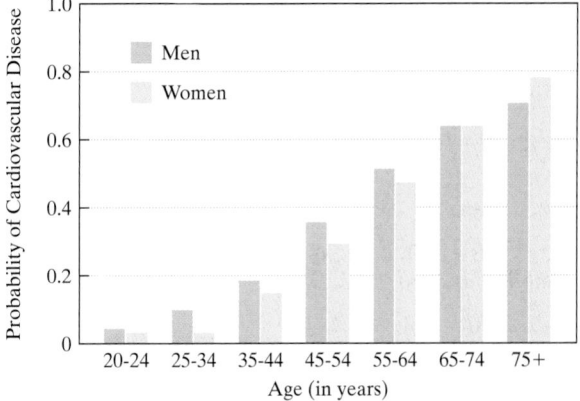

Probability of Cardiovascular Disease, by Age and Gender

Source: American Heart Association

Exercises 33–44 involve the probability of one event or a second event occurring. In order to use the correct probability formula, you will first need to determine whether or not the events are mutually exclusive.

In Exercises 33–36 you randomly select one card from a 52-card deck. Find the probability of selecting:

33. a 2 or a 3.   **34.** a 7 or an 8.

35. a red 2 or a black 3.   **36.** a red 7 or a black 8.

In Exercises 37–38, a single die is rolled. Find the probability of rolling:

37. an even number or a number less than 5.

38. an odd number or a number less than 4.

In Exercises 39–40, you are dealt one card from a 52-card deck. Find the probability that you are dealt:

39. a 7 or a red card.   **40.** a 5 or a black card.

In Exercises 41–42, it is equally probable that the pointer on the spinner shown will land on any one of the eight regions, numbered 1 through 8. If the pointer lands on a borderline, spin again.

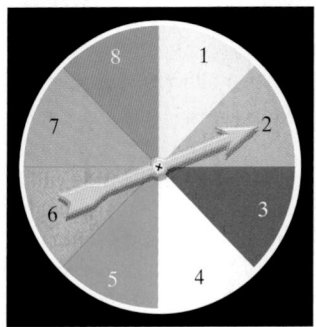

Find the probability that the pointer will stop on:

41. an odd number or a number less than 6.

42. an odd number or a number greater than 3.

Use this information to solve Exercises 43–44. The mathematics department of a college has 8 male professors, 11 female professors, 14 male teaching assistants, and 7 female teaching assistants. If a person is selected at random from the group, find the probability that the selected person is:

43. a professor or a male.   **44.** a professor or a female.

Exercises 45–50 involve and probabilities with independent events.

In Exercises 45–48, a single die is rolled twice. Find the probability of getting:

45. a 2 the first time and a 3 the second time.

46. a 5 the first time and a 1 the second time.

47. an even number the first time and a number greater than 2 the second time.

48. an odd number the first time and a number less than 3 the second time.

49. If you toss a fair coin six times, what is the probability of getting all heads?

50. If you toss a fair coin seven times, what is the probability of getting all tails?

When making two or more selections from populations with large numbers, such as the population of Americans ages 45 to 65, we assume that each selection is independent of every other selection. The graph shows how Americans 45 to 65 rate their health. Use the information shown to solve Exercises 51–52.

How Americans 45 to 65 Describe Their Health

Source: Newsweek

51. If four Americans ages 45 to 65 are selected at random, find the probability that all four rate their health as excellent.

52. If four Americans ages 45 to 65 are selected at random, find the probability that all four rate their health as poor.

53. The probability that South Florida will be hit by a major hurricane (category 4 or 5) in any single year is $\frac{1}{16}$. (*Source*: National Hurricane Center)
   a. What is the probability that South Florida will be hit by a major hurricane two years in a row?
   b. What is the probability that South Florida will be hit by a major hurricane in three consecutive years?
   c. What is the probability that South Florida will not be hit by a major hurricane in the next ten years?
   d. What is the probability that South Florida will be hit by a major hurricane at least once in the next ten years?

Writing in Mathematics

54. Describe the difference between theoretical probability and empirical probability.

55. Give an example of an event whose probability must be determined empirically rather than theoretically.

56. Write a probability word problem whose answer is one of the following fractions: $\frac{1}{6}$ or $\frac{1}{4}$ or $\frac{1}{3}$.

57. Explain how to find the probability of an event not occurring. Give an example.

58. What are mutually exclusive events? Give an example of two events that are mutually exclusive.

59. Explain how to find *or* probabilities with mutually exclusive events. Give an example.

60. Give an example of two events that are not mutually exclusive.

61. Explain how to find *or* probabilities with events that are not mutually exclusive. Give an example.

62. Explain how to find *and* probabilities with independent events. Give an example.

63. The president of a large company with 10,000 employees is considering mandatory cocaine testing for every employee. The test that would be used is 90% accurate, meaning that it will detect 90% of the cocaine users who are tested, and that 90% of the nonusers will test negative. This also means that the test gives 10% false positive. Suppose that 1% of the employees actually use cocaine. Find the probability that someone who tests positive for cocaine use is, indeed, a user.

Hint: Find the following probability fraction:

$$\frac{\text{the number of employees who test positive and are cocaine users}}{\text{the number of employees who test positive}}.$$

This fraction is given by

$$\frac{90\% \text{ of } 1\% \text{ of } 10,000}{\text{the number who test positive who actually use cocaine plus the number who test positive who do not use cocaine}}.$$

What does this probability indicate in terms of the percentage of employees who test positive who are not actually users? Discuss these numbers in terms of the issue of mandatory drug testing. Write a paper either in favor of or against mandatory drug testing, incorporating the actual percentage accuracy for such tests.

Critical Thinking Exercises

64. The target in the figure shown contains four squares. If a dart thrown at random hits the target, find the probability that it will land in a yellow region.

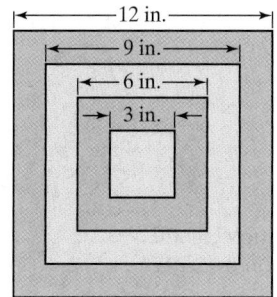

65. Suppose that it is a week in which the cash prize in Florida's LOTTO is promised to exceed $50 million. If a person purchases 22,957,480 tickets in LOTTO at $1 per ticket (all possible combinations), isn't this a guarantee of winning the lottery? Because the probability in this situation is 1, what's wrong with doing this?

66. a. If two people are selected at random, the probability that they do not have the same birthday (day and month) is $\frac{365}{365} \cdot \frac{364}{365}$. Explain why this is so. (Ignore leap years and assume 365 days in a year.)

b. If three people are selected at random, find the probability that they all have different birthdays.

c. If three people are selected at random, find the probability that at least two of them have the same birthday.

d. If 20 people are selected at random, find the probability that at least 2 of them have the same birthday.

e. How large a group is needed to give a 0.5 chance of at least two people having the same birthday?

Group Exercise

67. Research and present a group report on state lotteries. Include answers to some or all of the following questions: Which states do not have lotteries? Why not? How much is spent per capita on lotteries? What are some of the lottery games? What is the probability of winning top prize in these games? What income groups spend the greatest amount of money on lotteries? If your state has a lottery, what does it do with the money it makes? Is the way the money is spent what was promised when the lottery first began?

CHAPTER SUMMARY, REVIEW, AND TEST

Summary

DEFINITIONS AND CONCEPTS	EXAMPLES

11.1 Sequences and Summation Notation

a. An infinite sequence $\{a_n\}$ is a function whose domain is the set of positive integers. The function values, or terms, are represented by

$$a_1, a_2, a_3, a_4, \ldots, a_n, \ldots.$$

Ex. 1, p. 927

b. Sequences can be defined using recursion formulas that define the nth term as a function of the previous term.

Ex. 2, p. 928

c. Factorial Notation:
$$n! = n(n-1)(n-2)\cdots(3)(2)(1) \quad \text{and} \quad 0! = 1$$

Ex. 3, p. 929;
Ex. 4, p. 930

d. Summation Notation:
$$\sum_{i=1}^{n} a_i = a_1 + a_2 + a_3 + a_4 + \cdots + a_n$$

Ex. 5, p. 932;
Ex. 6, p. 933

| DEFINITIONS AND CONCEPTS | EXAMPLES |

11.2 Arithmetic Sequences

a. In an arithmetic sequence, each term after the first differs from the preceding term by a constant, the common difference. Subtract any term from the term that directly follows to find the common difference.

Ex. 1, p. 938

b. General term or nth term: $a_n = a_1 + (n - 1)d$. The first term is a_1 and the common difference is d.

Ex. 2, p. 940;
Ex. 3, p. 940

c. Sum of the first n terms: $S_n = \dfrac{n}{2}(a_1 + a_n)$

Ex. 4, p. 942;
Ex. 5, p. 942;
Ex. 6, p. 943

11.3 Geometric Sequences

a. In a geometric sequence, each term after the first is obtained by multiplying the preceding term by a nonzero constant, the common ratio. Divide any term after the first by the term that directly precedes it to find the common ratio.

Ex. 1, p. 947

b. General term or nth term: $a_n = a_1 r^{n-1}$. The first term is a_1 and the common ratio is r.

Ex. 2, p. 948;
Ex. 3, p. 948

c. Sum of the first n terms: $S_n = \dfrac{a_1(1 - r^n)}{1 - r}, \quad r \neq 1$

Ex. 4, p. 950;
Ex. 5, p. 951;
Ex. 6, p. 952

d. An annuity is a sequence of equal payments made at equal time periods. The value of an annuity, A, is the sum of all deposits made plus all interest paid, given by

Ex. 7, p. 953

$$A = P \frac{\left(1 + \dfrac{r}{n}\right)^{nt} - 1}{\dfrac{r}{n}}.$$

The deposit made at the end of each period is P, the annual interest rate is r, compounded n times per year, and t is the number of years deposits have been made.

e. Sum of the infinite geometric series $a_1 + a_1 r + a_1 r^2 + a_1 r^3 + \cdots$ is $S = \dfrac{a_1}{1 - r}$; $|r| < 1$. If $|r| \geq 1$, the infinite series does not have a sum.

Ex. 8, p. 955;
Ex. 9, p. 956;
Ex. 10, p. 956

11.4 Mathematical Induction

To prove that S_n is true for all positive integers n,

a. Show that S_1 is true.

Ex. 2, p. 964;

b. Show that if S_k is assumed true, then S_{k+1} is also true, for every positive integer k.

Ex. 3, p. 965;
Ex. 4, p. 967

11.5 The Binomial Theorem

a. Binomial coefficient: $\dbinom{n}{r} = \dfrac{n!}{r!\,(n - r)!}$

Ex. 1, p. 971

b. Binomial Theorem:

Ex. 2, p. 972;
Ex. 3, p. 973

$$(a + b)^n = \binom{n}{0}a^n + \binom{n}{1}a^{n-1}b + \binom{n}{2}a^{n-2}b^2 + \cdots + \binom{n}{n}b^n$$

c. The $(r + 1)$st term in the expansion of $(a + b)^n$ is

Ex. 4, p. 974

$$\binom{n}{r}a^{n-r}b^r.$$

DEFINITIONS AND CONCEPTS	EXAMPLES

11.6 Counting Principles, Permutations, and Combinations

a. The Fundamental Counting Principle: The number of ways in which a series of successive things can occur is found by multiplying the number of ways in which each thing can occur.

Ex. 1, p. 978;
Ex. 2, p. 979;
Ex. 3, p. 979

b. A permutation from a group of items occurs when no item is used more than once and the order of arrangement makes a difference.

c. Permutations Formula: The number of possible permutations if r items are taken from n items is

$$_nP_r = \frac{n!}{(n-r)!}.$$

Ex. 4, p. 981;
Ex. 5, p. 981

d. A combination from a group of items occurs when no item is used more than once and the order of items makes no difference.

Ex. 6, p. 983

e. Combinations Formula: The number of possible combinations if r items are taken from n items is

$$_nC_r = \frac{n!}{(n-r)!\,r!}.$$

Ex. 7, p. 984;
Ex. 8, p. 984

11.7 Probability

a. Empirical probability applies to situations in which we observe the frequency of the occurrence of an event. The empirical probability of event E is

$$P(E) = \frac{\text{observed number of times } E \text{ occurs}}{\text{total number of observed occurrences}}.$$

Ex. 1, p. 927

b. Theoretical probability applies to situations in which the sample space of all equally likely outcomes is known. The theoretical probability of event E is

$$P(E) = \frac{\text{number of outcomes in event } E}{\text{number of outcomes in sample space } S} = \frac{n(E)}{n(S)}.$$

Ex. 2, p. 990;
Ex. 3, p. 990;
Ex. 4, p. 991;
Ex. 5, p. 992

c. Probability of an event not occurring: $P(\text{not } E) = 1 - P(E)$.

Ex. 6, p. 993

d. If it is impossible for events A and B to occur simultaneously, the events are mutually exclusive.

e. If A and B are mutually exclusive events, then $P(A \text{ or } B) = P(A) + P(B)$.

Ex. 7, p. 994

f. If A and B are not mutually exclusive events, then

$$P(A \text{ or } B) = P(A) + P(B) - P(A \text{ and } B).$$

Ex. 8, p. 995;
Ex. 9, p. 996

g. Two events are independent if the occurrence of either of them has no effect on the probability of the other.

h. If A and B are independent events, then

$$P(A \text{ and } B) = P(A) \cdot P(B).$$

Ex. 10, p. 997

i. The probability of a succession of independent events is the product of each of their probabilities.

Ex. 11, p. 997

Review Exercises

11.1

In Exercises 1–6, write the first four terms of each sequence whose general term is given.

1. $a_n = 7n - 4$

2. $a_n = (-1)^n \dfrac{n+2}{n+1}$

3. $a_n = \dfrac{1}{(n-1)!}$

4. $a_n = \dfrac{(-1)^{n+1}}{2^n}$

5. $a_1 = 9$ and $a_n = \dfrac{2}{3a_{n-1}}$ for $n \geq 2$

6. $a_1 = 4$ and $a_n = 2a_{n-1} + 3$ for $n \geq 2$

7. Evaluate: $\dfrac{40!}{4!38!}$.

In Exercises 8–9, find each indicated sum.

8. $\displaystyle\sum_{i=1}^{5}(2i^2 - 3)$

9. $\displaystyle\sum_{i=0}^{4}(-1)^{i+1}i!$

In Exercises 10–11, express each sum using summation notation. Use i for the index of summation.

10. $\dfrac{1}{3} + \dfrac{2}{4} + \dfrac{3}{5} + \cdots + \dfrac{15}{17}$

11. $4^3 + 5^3 + 6^3 + \cdots + 13^3$

11.2

In Exercises 12–15, write the first six terms of each arithmetic sequence.

12. $a_1 = 7, d = 4$

13. $a_1 = -4, d = -5$

14. $a_1 = \frac{3}{2}, d = -\frac{1}{2}$

15. $a_{n+1} = a_n + 5, a_1 = -2$

In Exercises 16–18, find the indicated term of the arithmetic sequence with first term, a_1, and common difference, d.

16. Find a_6 when $a_1 = 5, d = 3$.

17. Find a_{12} when $a_1 = -8, d = -2$.

18. Find a_{14} when $a_1 = 14, d = -4$.

In Exercises 19–21, write a formula for the general term (the nth term) of each arithmetic sequence. Do not use a recursion formula. Then use the formula for a_n to find a_{20}, the 20th term of the sequence.

19. $-7, -3, 1, 5, \ldots$

20. $a_1 = 200, d = -20$

21. $a_n = a_{n-1} - 5, a_1 = 3$

22. Find the sum of the first 22 terms of the arithmetic sequence: $5, 12, 19, 26, \ldots$.

23. Find the sum of the first 15 terms of the arithmetic sequence: $-6, -3, 0, 3, \ldots$.

24. Find $3 + 6 + 9 + \cdots + 300$, the sum of the first 100 positive multiples of 3.

In Exercises 25–27, use the formula for the sum of the first n terms of an arithmetic sequence to find the indicated sum.

25. $\displaystyle\sum_{i=1}^{16}(3i + 2)$

26. $\displaystyle\sum_{i=1}^{25}(-2i + 6)$

27. $\displaystyle\sum_{i=1}^{30}(-5i)$

28. The graph in the next column shows the changing pattern of work in the United States from 1900 through 2000.

In 1900, 20% of the total labor force was comprised of white-collar workers. On average, this increased by approximately 0.52% per year since then.

a. Write a formula for the nth term of the arithmetic sequence that describes the percentage of white-collar workers in the labor force n years after 1899.

b. Use the model to predict the percentage of white-collar workers in the labor force by the year 2010.

The Changing Pattern of Work in the United States

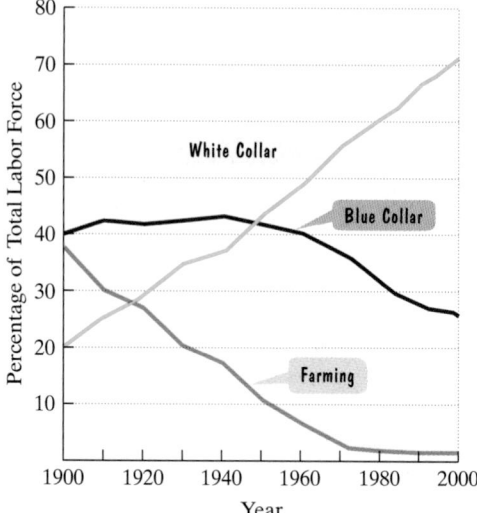

Source: U.S. Department of Labor

29. A company offers a starting salary of $31,500 with raises of $2300 per year. Find the total salary over a ten-year period.

30. A theater has 25 seats in the first row and 35 rows in all. Each successive row contains one additional seat. How many seats are in the theater?

11.3

In Exercises 31–34, write the first five terms of each geometric sequence.

31. $a_1 = 3, r = 2$

32. $a_1 = \frac{1}{2}, r = \frac{1}{2}$

33. $a_1 = 16, r = -\frac{1}{2}$

34. $a_n = -5a_{n-1}, a_1 = -1$

In Exercises 35–37, use the formula for the general term (the nth term) of a geometric sequence to find the indicated term of each sequence.

35. Find a_7 when $a_1 = 2, r = 3$.

36. Find a_6 when $a_1 = 16, r = \frac{1}{2}$.

37. Find a_5 when $a_1 = -3, r = 2$.

In Exercises 38–40, write a formula for the general term (the nth term) of each geometric sequence. Then use the formula for a_n to find a_8, the eighth term of the sequence.

38. $1, 2, 4, 8, \ldots$

39. $100, 10, 1, \frac{1}{10}, \ldots$

40. $12, -4, \frac{4}{3}, -\frac{4}{9}, \ldots$

41. Find the sum of the first 15 terms of the geometric sequence: $5, -15, 45, -135, \ldots$.

42. Find the sum of the first 7 terms of the geometric sequence: $8, 4, 2, 1, \ldots$.

In Exercises 43–45, use the formula for the sum of the first n terms of a geometric sequence to find the indicated sum.

43. $\sum_{i=1}^{6} 5^i$

44. $\sum_{i=1}^{7} 3(-2)^i$

45. $\sum_{i=1}^{5} 2\left(\frac{1}{4}\right)^{i-1}$

In Exercises 46–49, find the sum of each infinite geometric series.

46. $9 + 3 + 1 + \frac{1}{3} + \cdots$

47. $2 - 1 + \frac{1}{2} - \frac{1}{4} + \cdots$

48. $-6 + 4 - \frac{8}{3} + \frac{16}{9} - \cdots$

49. $\sum_{i=1}^{\infty} 5(0.8)^i$

In Exercises 50–51, express each repeating decimal as a fraction in lowest terms.

50. $0.\overline{6}$

51. $0.\overline{47}$

52. The population of Iraq from 1998 through 2001 is shown in the following table.

Year	1998	1999	2000	2001
Population in Millions	19.96	20.72	21.51	22.33

Source: U.N. Population Division

   a. Show that Iraq's population is increasing geometrically.
   b. Write the general term of the geometric sequence describing population growth for Iraq n years after 1997.
   c. Estimate Iraq's population, in millions, for the year 2008.

53. A job pays $32,000 for the first year with an annual increase of 6% per year beginning in the second year. What is the salary in the sixth year? What is the total salary paid over this six-year period? Round answers to the nearest dollar.

54. You decide to deposit $200 at the end of each month into an account paying 10% interest compounded monthly to save for your child's education. How much will you save over 18 years?

55. A factory in an isolated town has an annual payroll of $4 million. It is estimated that 70% of this money is spent within the town, that people in the town receiving this money will again spend 70% of what they receive in the town, and so on. What is the total of all this spending in the town each year?

11.4

In Exercises 56–60, use mathematical induction to prove that each statement is true for every positive integer n.

56. $5 + 10 + 15 + \cdots + 5n = \dfrac{5n(n+1)}{2}$

57. $1 + 4 + 4^2 + \cdots + 4^{n-1} = \dfrac{4^n - 1}{3}$

58. $2 + 6 + 10 + \cdots + (4n - 2) = 2n^2$

59. $1 \cdot 3 + 2 \cdot 4 + 3 \cdot 5 + \cdots + n(n+2) = \dfrac{n(n+1)(2n+7)}{6}$

60. 2 is a factor of $n^2 + 5n$.

11.5

In Exercises 61–62, evaluate the given binomial coefficient.

61. $\binom{11}{8}$

62. $\binom{90}{2}$

In Exercises 63–66, use the Binomial Theorem to expand each binomial and express the result in simplified form.

63. $(2x + 1)^3$

64. $(x^2 - 1)^4$

65. $(x + 2y)^5$

66. $(x - 2)^6$

In Exercises 67–68, write the first three terms in each binomial expansion, expressing the result in simplified form.

67. $(x^2 + 3)^8$

68. $(x - 3)^9$

In Exercises 69–70, find the term indicated in each expansion.

69. $(x + 2)^5$; fourth term

70. $(2x - 3)^6$; fifth term

11.6

In Exercises 71–74, evaluate each expression.

71. $_8P_3$

72. $_9P_5$

73. $_8C_3$

74. $_{13}C_{11}$

In Exercises 75–81, solve by the method of your choice.

75. A popular brand of pen comes in red, green, blue, or black ink. The writing tip can be chosen from extra bold, bold, regular, fine, or micro. How many different choices of pens do you have with this brand?

76. A stock can go up, go down, or stay unchanged. How many possibilities are there if you own five stocks?

77. A club with 15 members is to choose four officers—president, vice-president, secretary, and treasurer. In how many ways can these offices be filled?

78. How many different ways can a director select 4 actors from a group of 20 actors to attend a workshop on performing in rock musicals?

79. From the 20 CDs that you've bought during the past year, you plan to take 3 with you on vacation. How many different sets of three CDs can you take?

80. How many different ways can a director select from 20 male actors and cast the roles of Mark, Roger, Angel, and Collins in the musical *Rent*?

81. In how many ways can five airplanes line up for departure on a runway?

11.7

Exercises 82–83 involve empirical probabilities. Express each probability as a fraction. Then use a calculator to express the probability in decimal form, rounded to the nearest thousandth. The table on the next page shows the two states with the largest Hispanic populations. Find the probability that:

82. a person randomly selected from California is Hispanic.

83. a person randomly selected from Texas is Hispanic.

Largest Hispanic Population, 2000

State	Total Population	Hispanic Population
California	33,871,648	10,966,556
Texas	20,851,820	6,669,666

Source: Bureau of the Census

In Exercises 84–85, a die is rolled. Find the probability of:

84. getting a number less than 5.

85. getting a number less than 3 or greater than 4.

In Exercises 86–87, you are dealt one card from a 52-card deck. Find the probability of:

86. getting an ace or a king.

87. getting a queen or a red card.

In Exercises 88–89, it is equally probable that the pointer on the spinner shown will land on any one of the six regions, numbered 1 through 6, and colored as shown. If the pointer lands on a borderline, spin again. Find the probability of

88. not stopping on yellow.

89. stopping on red or a number greater than 3.

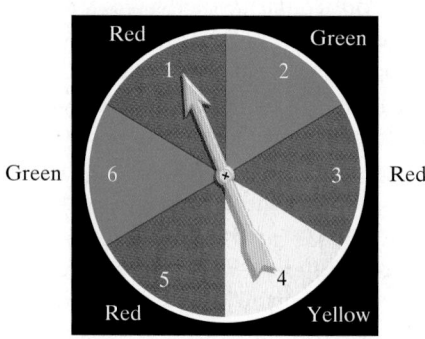

90. A lottery game is set up so that each player chooses five different numbers from 1 to 20. If the five numbers match

the five numbers drawn in the lottery, the player wins (or shares) the top cash prize. What is the probability of winning the prize

a. with one lottery ticket?

b. with 100 different lottery tickets?

Use this information to solve Exercises 91–92. At a workshop on police work and the African-American community, there are 50 African-American male police officers, 20 African-American female police officers, 90 white male police officers, and 40 white female police officers. If one police officer is selected at random from the people at the workshop, find the probability that the selected person is:

91. African American or male.   **92.** female or white.

93. The bar graph shows five causes of death and the percentage of all deaths in the United States attributed to each cause. What is the probability that an American's death is caused by heart disease or cancer? Express the answer as a decimal to three decimal places.

Causes of Death in the United States

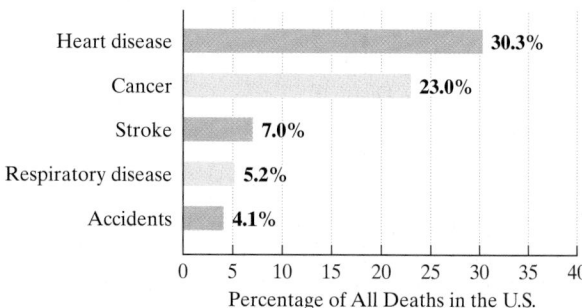

Source: U.S. Department of Health and Human Services

94. What is the probability of a family having five boys born in a row?

95. The probability of a flood in any given year in a region prone to floods is 0.2.

a. What is the probability of a flood two years in a row?

b. What is the probability of a flood for three consecutive years?

c. What is the probability of no flooding for four consecutive years?

Chapter 11 Test

1. Write the first five terms of the sequence whose general term is $a_n = \dfrac{(-1)^{n+1}}{n^2}$.

In Exercises 2–4, find each indicated sum.

2. $\sum_{i=1}^{5} (i^2 + 10)$ **3.** $\sum_{i=1}^{20} (3i - 4)$ **4.** $\sum_{i=1}^{15} (-2)^i$

In Exercises 5–7, evaluate each expression.

5. $\begin{pmatrix} 9 \\ 2 \end{pmatrix}$ **6.** $_{10}P_3$ **7.** $_{10}C_3$

8. Express the sum using summation notation. Use i for the index of summation.

$$\frac{2}{3} + \frac{3}{4} + \frac{4}{5} + \cdots + \frac{21}{22}$$

In Exercises 9–10, write a formula for the general term (the nth term) of each sequence. Do not use a recursion formula. Then use the formula to find the twelfth term of the sequence.

9. $4, 9, 14, 19, \ldots$ **10.** $16, 4, 1, \frac{1}{4}, \ldots$

In Exercises 11–12, use a formula to find the sum of the first ten terms of each sequence.

11. $7, -14, 28, -56, \ldots$ **12.** $-7, -14, -21, -28, \ldots$

13. Find the sum of the infinite geometric series:

$$4 + \frac{4}{2} + \frac{4}{2^2} + \frac{4}{2^3} + \cdots.$$

14. A job pays \$30,000 for the first year with an annual increase of 4% per year beginning in the second year. What is the total salary paid over an eight-year period? Round to the nearest dollar.

15. Use mathematical induction to prove that for every positive integer n,

$$1 + 4 + 7 + \cdots + (3n - 2) = \frac{n(3n - 1)}{2}.$$

16. Use the Binomial Theorem to expand and simplify: $(x^2 - 1)^5$.

17. A human resource manager has 11 applicants to fill three different positions. Assuming that all applicants are equally qualified for any of the three positions, in how many ways can this be done?

18. From the ten books that you've recently bought but not read, you plan to take four with you on vacation. How many different sets of four books can you take?

19. How many seven-digit local telephone numbers can be formed if the first three digits are 279?

20. A lottery game is set up so that each player chooses six different numbers from 1 to 15. If the six numbers match the six numbers drawn in the lottery, the player wins (or shares) the top cash prize. What is the probability of winning the prize with 50 different lottery tickets?

21. One card is randomly selected from a deck of 52 cards. Find the probability of selecting a black card or a picture card.

22. A group of students consists of 10 male freshmen, 15 female freshmen, 20 male sophomores, and 5 female sophomores. If one person is randomly selected from the group, find the probability of selecting a freshman or a female.

23. A quiz consisting of four multiple-choice questions has four available options (a, b, c, or d) for each question. If a person guesses at every question, what is the probability of answering all questions correctly?

24. If the spinner shown is spun twice, find the probability that the pointer lands on red on the first spin and blue on the second spin

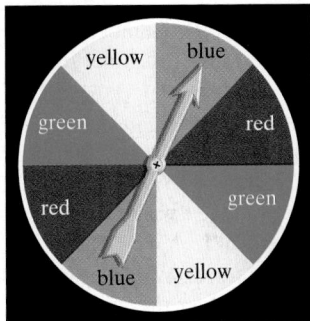

Cumulative Review Exercises (Chapters 1–11)

Solve each equation or inequality in Exercises 1–10.

1. $-2(x - 5) + 10 = 3(x + 2)$

2. $3x^2 - 6x + 2 = 0$

3. $\log_2 x + \log_2 (2x - 3) = 1$

4. $x^{1/2} - 6x^{1/4} + 8 = 0$

5. $\sqrt{2x + 4} - \sqrt{x + 3} - 1 = 0$

6. $|2x + 1| \leq 1$ **7.** $6x^2 - 6 < 5x$

8. $\dfrac{x - 1}{x + 3} \leq 0$ **9.** $30e^{0.7x} = 240$

10. $2x^3 + 3x^2 - 8x + 3 = 0$

Solve each system in Exercises 11–13.

11. $4x^2 + 3y^2 = 48$
$3x^2 + 2y^2 = 35$

12. (Use matrices.)
$x - 2y + z = 16$
$2x - y - z = 14$
$3x + 5y - 4z = -10$

13. $x - y = 1$
$x^2 - x - y = 1$

In Exercises 14–19, graph each equation, function, or system in the rectangular coordinate system.

14. $100x^2 + y^2 = 25$

15. $4x^2 - 9y^2 - 16x + 54y - 29 = 0$

16. $f(x) = \dfrac{x^2 - 1}{x - 2}$

17. $2x - y \geq 4$
$x \leq 2$

18. $f(x) = x^2 - 4x - 5$ **19.** $y = \log_2 x$

20. Find $f^{-1}(x)$ if $f(x) = \sqrt[3]{x + 4}$.

21. If $A = \begin{bmatrix} 4 & 2 \\ 1 & -1 \\ 0 & 5 \end{bmatrix}$ and $B = \begin{bmatrix} 2 & 4 \\ 3 & 1 \end{bmatrix}$, find $AB - 4A$.

22. Find the partial fraction decomposition for

$$\frac{2x^2 - 10x + 2}{(x - 2)(x^2 + 2x + 2)}.$$

23. Expand and simplify: $(x^3 + 2y)^5$.

24. Use the formula for the sum of the first n terms of an arithmetic sequence to find $\displaystyle\sum_{i=1}^{50} (4i - 25)$.

25. The table shows the average number of work hours per week and the average number of leisure hours per week for Americans for two recent years.

x (Average number of work hours per week)	y (Average number of leisure hours per week)
50	21
52	19.8

Source: Louis Harris and Associates

   a. Write the point-slope form of the line on which these measurements fall.

   b. Use the point-slope form of the equation to write the slope-intercept form of the equation.

   c. Use the slope-intercept model from part (b) to predict the average number of leisure hours per week for a year in which Americans average 54 hours of work per week.

26. For a summer sales job, you are choosing between two pay arrangements: a weekly salary of $200 plus 5% commission on sales, or a straight 15% commission. For how many dollars of sales will the earnings be the same regardless of the pay arrangement?

27. The perimeter of a soccer field is 300 yards. If the length is 50 yards longer than the width, what are the field's dimensions?

28. If 10 pens and 12 pads cost $42, and 5 of the same pens and 10 of the same pads cost $29, find the cost of a pen and a pad.

29. A ball is thrown vertically upward from the top of a 96-foot tall building with an initial velocity of 80 feet per second. The height of the ball above ground, $s(t)$, in feet, after t seconds is modeled by the position function

$$s(t) = -16t^2 + 80t + 96.$$

   a. After how many seconds will the ball strike the ground?

   b. When does the ball reach its maximum height? What is the maximum height?

30. The current, I, in amperes, flowing in an electrical circuit varies inversely as the resistance, R, in ohms, in the circuit. When the resistance of an electric percolator is 22 ohms, it draws 5 amperes of current. How much current is needed when the resistance is 10 ohms?

31. An object moves in simple harmonic motion described by $d = 10 \sin \frac{3\pi}{4} t$, where t is measured in seconds and d in inches. Find: **a.** the maximum displacement; **b.** the frequency; and **c.** the time required for one oscillation.

Verify each identity in Exercises 32–33.

32. $\tan x + \dfrac{1}{\tan x} = \dfrac{1}{\sin x \cos x}$

33. $\dfrac{1 - \tan^2 x}{1 + \tan^2 x} = \cos 2x$

34. Graph one period: $y = -2 \cos(3x - \pi)$.

In Exercises 35–36, solve each equation on the interval $[0, 2\pi)$.

35. $4 \cos^2 x = 3$

36. $2\sin^2 x + 3 \cos x - 3 = 0$

37. Find the exact value of $\cot \left[\cos^{-1}\left(-\frac{5}{6}\right)\right]$.

38. Graph the polar equation: $r = 1 + 2 \cos \theta$.

39. In oblique triangle ABC, $A = 34°$, $a = 22$, and $b = 32$. Solve the triangle(s). Round lengths to the nearest tenth and angle measures to the nearest degree.

40. Use the parametric equations

$$x = \sin t, \quad y = 1 + \cos^2 t, \quad -\frac{\pi}{2} < t < \frac{\pi}{2}$$

and eliminate the parameter. Graph the plane curve represented by the parametric equations. Use arrows to show the orientation of the curve.

Appendix
Where Did That Come From? Selected Proofs

SECTION 4.3 Properties of Logarithms

The Product Rule
Let b, M, and N be positive real numbers with $b \neq 1$.
$$\log_b(MN) = \log_b M + \log_b N$$

Proof
We begin by letting $\log_b M = R$ and $\log_b N = S$.
  Now we write each logarithm in exponential form.
$$\log_b M = R \quad \text{means} \quad b^R = M.$$
$$\log_b N = S \quad \text{means} \quad b^S = N.$$
By substituting and using a property of exponents, we see that
$$MN = b^R b^S = b^{R+S}.$$
Now we change $MN = b^{R+S}$ to logarithmic form.
$$MN = b^{R+S} \quad \text{means} \quad \log_b(MN) = R + S.$$
Finally, substituting $\log_b M$ for R and $\log_b N$ for S gives us
$$\log_b(MN) = \log_b M + \log_b N,$$
the property that we wanted to prove.

The quotient and power rules for logarithms are proved using similar procedures.

The Change-of-Base Property
For any logarithmic bases a and b, and any positive number M,
$$\log_b M = \frac{\log_a M}{\log_a b}.$$

Proof
To prove the change-of-base property, we let x equal the logarithm on the left side:
$$\log_b M = x.$$

Now we rewrite this logarithm in exponential form.

$$\log_b M = x \quad \text{means} \quad b^x = M.$$

Because b^x and M are equal, the logarithms with base a for each of these expressions must be equal. This means that

$$\log_a b^x = \log_a M$$

$$x \log_a b = \log_a M \qquad \text{Apply the power rule for logarithms on the left side.}$$

$$x = \frac{\log_a M}{\log_a b} \qquad \text{Solve for x by dividing both sides by } \log_a b.$$

In our first step we let x equal $\log_b M$. Replacing x on the left side by $\log_b M$ gives us

$$\log_b M = \frac{\log_a M}{\log_a b},$$

which is the change-of-base property.

SECTION 7.2   The Law of Cosines

Heron's Formula for the Area of a Triangle

The area of a triangle with sides a, b, and c is

$$\text{Area} = \sqrt{s(s-a)(s-b)(s-c)},$$

where s is one-half the perimeter: $s = \frac{1}{2}(a+b+c)$.

Proof

The proof of Heron's formula begins with a half-angle formula and the Law of Cosines.

$$\cos\frac{C}{2} = \sqrt{\frac{1+\cos C}{2}} = \sqrt{\frac{1 + \dfrac{a^2+b^2-c^2}{2ab}}{2}}$$

This is the Law of Cosines:
$c^2 = a^2 + b^2 - 2ab \cos C$
solved for cos C.

$$= \sqrt{\frac{a^2+2ab+b^2-c^2}{4ab}} = \sqrt{\frac{(a+b)^2-c^2}{4ab}} = \sqrt{\frac{(a+b+c)(a+b-c)}{4ab}}$$

Multiply the numerator and denominator of the radicand by $2ab$.

Factor $a^2 + 2ab + b^2$.

Factor the numerator as the differences of two squares.

We now introduce the expression for one-half the perimeter: $s = \frac{1}{2}(a+b+c)$. We replace $a+b+c$ in the numerator by $2s$. We also find an expression for $a+b-c$ as follows:

$$a+b-c = a+b+c-2c = 2s-2c = 2(s-c).$$

Thus,

$$\cos\frac{C}{2} = \sqrt{\frac{(a+b+c)(a+b-c)}{4ab}} = \sqrt{\frac{2s\cdot 2(s-c)}{4ab}} = \sqrt{\frac{s(s-c)}{ab}}.$$

In a similar manner, we obtain

$$\sin\frac{C}{2} = \sqrt{\frac{1-\cos C}{2}} = \sqrt{\frac{(s-a)(s-b)}{ab}}.$$

From our work in Section 7.1, we know that the area of a triangle is one-half the product of the length of two sides times the sine of their included angle.

$$\text{Area} = \frac{1}{2} ab \sin C$$

$$= \frac{1}{2} ab \cdot 2 \sin \frac{C}{2} \cos \frac{C}{2} \qquad \sin C = \sin 2\frac{C}{2} = 2 \sin \frac{C}{2} \cos \frac{C}{2}$$

$$= ab\sqrt{\frac{(s-a)(s-b)}{ab}} \sqrt{\frac{s(s-c)}{ab}} \qquad \text{Use the expressions for } \sin \frac{C}{2} \text{ and } \cos \frac{C}{2} \text{ on page A2.}$$

$$= ab \frac{\sqrt{s(s-a)(s-b)(s-c)}}{\sqrt{a^2 b^2}} \qquad \text{Multiply the radicands.}$$

$$= \sqrt{s(s-a)(s-b)(s-c)} \qquad \text{Simplify: } \frac{ab}{\sqrt{a^2 b^2}} = \frac{ab}{ab} = 1.$$

SECTION 7.5   Complex Numbers in Polar Form; DeMoivre's Theorem

The Quotient of Two Complex Numbers in Polar Form

Let $z_1 = r_1(\cos\theta_1 + i\sin\theta_1)$ and $z_2 = r_2(\cos\theta_2 + i\sin\theta_2)$ be two complex numbers in polar form. Their quotient, $\dfrac{z_1}{z_2}$, is

$$\frac{z_1}{z_2} = \frac{r_1}{r_2}\left[\cos(\theta_1 - \theta_2) + i\sin(\theta_1 - \theta_2)\right].$$

Proof

We begin by multiplying the numerator and denominator of the quotient, $\dfrac{z_1}{z_2}$, by the conjugate of the denominator. Then we simplify the quotient using the difference formulas for sine and cosine.

$$\frac{z_1}{z_2} = \frac{r_1(\cos\theta_1 + i\sin\theta_1)}{r_2(\cos\theta_2 + i\sin\theta_2)} \qquad \text{This is the given quotient.}$$

$$= \frac{r_1(\cos\theta_1 + i\sin\theta_1)(\cos\theta_2 - i\sin\theta_2)}{r_2(\cos\theta_2 + i\sin\theta_2)(\cos\theta_2 - i\sin\theta_2)} \qquad \text{Multiply the numerator and denominator by the conjugate of the denominator. Recall that the conjugate of } a + bi \text{ is } a - bi.$$

$$= \frac{r_1(\cos\theta_1 + i\sin\theta_1)(\cos\theta_2 - i\sin\theta_2)}{r_2(\cos^2\theta_2 + \sin^2\theta_2)} \qquad \text{Multiply the conjugates in the denominator.}$$

$$= \frac{r_1(\cos\theta_1 + i\sin\theta_1)(\cos\theta_2 - i\sin\theta_2)}{r_2} \qquad \text{Use a Pythagorean identity: } \cos^2\theta_2 + \sin^2\theta_2 = 1.$$

$$= \frac{r_1}{r_2}\left(\cos\theta_1 \cos\theta_2 - i\cos\theta_1 \sin\theta_2 + i\sin\theta_1 \cos\theta_2 - i^2 \sin\theta_1 \sin\theta_2\right) \qquad \text{Use the FOIL method.}$$

$$= \frac{r_1}{r_2}\left[\cos\theta_1 \cos\theta_2 + i(\sin\theta_1 \cos\theta_2 - \cos\theta_1 \sin\theta_2) - i^2 \sin\theta_1 \sin\theta_2\right] \qquad \text{Factor } i \text{ from the second and third terms.}$$

$$= \frac{r_1}{r_2} \left[\cos\theta_1 \cos\theta_2 + i(\sin\theta_1 \cos\theta_2 - \cos\theta_1 \sin\theta_2) - (-1)\sin\theta_1 \sin\theta_2 \right] \quad i^2 = -1.$$

$$= \frac{r_1}{r_2} \left[\cos\theta_1 \cos\theta_2 + \sin\theta_1 \sin\theta_2 + i(\sin\theta_1 \cos\theta_2 - \cos\theta_1 \sin\theta_2) \right] \qquad \text{Rearrange terms.}$$

This is $\cos(\theta_1 - \theta_2)$.　　　　　This is $\sin(\theta_1 - \theta_2)$.

$$= \frac{r_1}{r_2} \left[\cos(\theta_1 - \theta_2) + i\sin(\theta_1 - \theta_2) \right]$$

SECTION 7.7 **The Dot Product**

Properties of the Dot Product

If **u**, **v**, and **w** are vectors, and c is a scalar, then:

1. $\mathbf{u} \cdot \mathbf{v} = \mathbf{v} \cdot \mathbf{u}$
2. $\mathbf{u} \cdot (\mathbf{v} + \mathbf{w}) = \mathbf{u} \cdot \mathbf{v} + \mathbf{u} \cdot \mathbf{w}$
3. $\mathbf{0} \cdot \mathbf{v} = 0$
4. $\mathbf{v} \cdot \mathbf{v} = \|\mathbf{v}\|^2$
5. $(c\mathbf{u}) \cdot \mathbf{v} = c(\mathbf{u} \cdot \mathbf{v}) = \mathbf{u} \cdot (c\mathbf{v})$

Proof To prove the second property, let

$$\mathbf{u} = u_1\mathbf{i} + u_2\mathbf{j}, \quad \mathbf{v} = v_1\mathbf{i} + v_2\mathbf{j}, \text{ and } \mathbf{w} = w_1\mathbf{i} + w_2\mathbf{j}.$$

Then,

$$\mathbf{u} \cdot (\mathbf{v} + \mathbf{w}) = (u_1\mathbf{i} + u_2\mathbf{j}) \cdot [(v_1\mathbf{i} + v_2\mathbf{j}) + (w_1\mathbf{i} + w_2\mathbf{j})] \qquad \text{These are the given vectors.}$$

$$= (u_1\mathbf{i} + u_2\mathbf{j}) \cdot [(v_1 + w_1)\mathbf{i} + (v_2 + w_2)\mathbf{j}] \qquad \text{Add horizontal components and add vertical components.}$$

$$= u_1(v_1 + w_1) + u_2(v_2 + w_2) \qquad \text{Multiply horizontal components and multiply vertical components.}$$

$$= u_1v_1 + u_1w_1 + u_2v_2 + u_2w_2 \qquad \text{Use the distributive property.}$$

$$= u_1v_1 + u_2v_2 + u_1w_1 + u_2w_2 \qquad \text{Rearrange terms.}$$

This is the dot product of **u** and **v**.　　　This is the dot product of **u** and **w**.

$$= \mathbf{u} \cdot \mathbf{v} + \mathbf{u} \cdot \mathbf{w}.$$

To prove the third property, let

$$\mathbf{0} = 0\mathbf{i} + 0\mathbf{j} \quad \text{and} \quad \mathbf{v} = v_1\mathbf{i} + v_2\mathbf{j}.$$

Then

$$\mathbf{0} \cdot \mathbf{v} = (0\mathbf{i} + 0\mathbf{j}) \cdot (v_1\mathbf{i} + v_2\mathbf{j}) \qquad \text{These are the given vectors.}$$

$$= 0 \cdot v_1 + 0 \cdot v_2 \qquad \text{Multiply horizontal components and multiply vertical components.}$$

$$= 0 + 0$$

$$= 0.$$

To prove the first part of the fifth property, let

$$\mathbf{u} = u_1\mathbf{i} + u_2\mathbf{j} \quad \text{and} \quad \mathbf{v} = v_1\mathbf{i} + v_2\mathbf{j}.$$

Then,

$$
\begin{aligned}
(c\mathbf{u}) \cdot \mathbf{v} &= \left[c(u_1\mathbf{i} + u_2\mathbf{j})\right] \cdot (v_1\mathbf{i} + v_2\mathbf{j}) && \text{These are the given vectors.} \\
&= (cu_1\mathbf{i} + cu_2\mathbf{j}) \cdot (v_1\mathbf{i} + v_2\mathbf{j}) && \text{Multiply each component of } u_1\mathbf{i} + u_2\mathbf{j} \text{ by } c. \\
&= cu_1v_1 + cu_2v_2 && \text{Multiply horizontal components and} \\
& && \text{multiply vertical components.} \\
&= c(u_1v_1 + u_2v_2) && \text{Factor out } c \text{ from both terms.}
\end{aligned}
$$

This is the dot product of **u** and **v**.

$$= c(\mathbf{u} \cdot \mathbf{v}).$$

SECTION 10.2   The Hyperbola

The Asymptotes of a Hyperbola Centered at the Origin

The hyperbola

$$\frac{x^2}{a^2} - \frac{y^2}{b^2} = 1$$

with a horizontal transverse axis has the two asymptotes

$$y = \frac{b}{a}x \quad \text{and} \quad y = -\frac{b}{a}x.$$

Proof

Begin by solving the hyperbola's equation for y.

$$\frac{x^2}{a^2} - \frac{y^2}{b^2} = 1 \qquad \text{This is the standard form of the equation of a hyperbola.}$$

$$\frac{y^2}{b^2} = \frac{x^2}{a^2} - 1 \qquad \text{We isolate the term involving } y^2 \text{ to solve for } y.$$

$$y^2 = \frac{b^2x^2}{a^2} - b^2 \qquad \text{Multiply both sides by } b^2.$$

$$y^2 = \frac{b^2x^2}{a^2}\left(1 - \frac{a^2}{x^2}\right) \qquad \text{Factor out } \frac{b^2x^2}{a^2} \text{ on the right. Verify that this result is correct by multiplying using the distributive property and obtaining the previous step.}$$

$$y = \pm\sqrt{\frac{b^2x^2}{a^2}\left(1 - \frac{a^2}{x^2}\right)} \qquad \text{Solve for } y \text{ using the square root method: If } a^2 = b, \text{ then } a = \pm\sqrt{b}.$$

$$y = \pm\frac{b}{a}x\sqrt{1 - \frac{a^2}{x^2}} \qquad \text{Simplify.}$$

As $|x| \to \infty$, the value of $\dfrac{a^2}{x^2}$ approaches 0. Consequently, the value of y can be approximated by

$$y = \pm\frac{b}{a}x.$$

This means that the lines whose equations are $y = \dfrac{b}{a}x$ and $y = -\dfrac{b}{a}x$ are asymptotes for the graph of the hyperbola.

Answers to Selected Exercises

CHAPTER P

Section P.1

Check Point Exercises

1. a. $\sqrt{2} - 1$ **b.** $\pi - 3$ **c.** 1   **2.** 9   **3.** 250; In 1990, the population of the United States was 250 million.   **4.** $38x - 19y$

Exercise Set P.1

1. a. $\sqrt{100}$ **b.** $0, \sqrt{100}$ **c.** $-9, 0, \sqrt{100}$ **d.** $-9, -\dfrac{4}{5}, 0, 0.25, 9.2, \sqrt{100}$ **e.** $\sqrt{3}$ **3. a.** $\sqrt{64}$ **b.** $0, \sqrt{64}$ **c.** $-11, 0, \sqrt{64}$

d. $-11, -\dfrac{5}{6}, 0, 0.75, \sqrt{64}$ **e.** $\sqrt{5}, \pi$ **5.** 0   **7.** Answers may vary.   **9.** true   **11.** true   **13.** true   **15.** 300   **17.** $12 - \pi$

19. $5 - \sqrt{2}$ **21.** -1 **23.** 4   **25.** 3   **27.** 7   **29.** -1 **31.** $|17 - 2|; 15$ **33.** $|5 - (-2)|; 7$ **35.** $|-4 - (-19)|; 15$

37. $|-1.4 - (-3.6)|; 2.2$ **39.** 27   **41.** -19 **43.** 25   **45.** 10   **47.** -8 **49.** commutative property of addition

51. associative property of addition   **53.** commutative property of addition   **55.** distributive property of multiplication over addition

57. inverse property of multiplication   **59.** $15x + 16$ **61.** $27x - 10$ **63.** $29y - 29$ **65.** $8y - 12$ **67.** $16y - 25$

69. $14x$ **71.** $-2x + 3y + 6$ **73.** x **75.** yes   **77.** Answers may vary.   **79.** 21; In 2000, approximately 21% of American adults smoked cigarettes.   **81. a.** $132 - 0.6a$ **b.** 120   **89.** (c) is true.   **91.** $<$ **93.** $>$

Section P.2

Check Point Exercises

1. -256 **2. a.** $\dfrac{1}{8}$ **b.** 36   **3. a.** 243   **b.** $\dfrac{1}{8}$ **c.** x^6 **4. a.** 729   **b.** y^{28} **c.** $\dfrac{1}{x^8}$ **5. a.** 9   **b.** $\dfrac{1}{x^7}$ **c.** y^9 **6.** $-64x^3$

7. a. $\dfrac{27}{64}$ **b.** $-\dfrac{32}{y^5}$ **8. a.** $16x^{12}y^{24}$ **b.** $-18x^3y^8$ **c.** $\dfrac{5y^6}{x^4}$ **d.** $\dfrac{y^8}{25x^2}$ **9. a.** 7,400,000,000   **b.** 0.000003017

10. a. 7.41×10^9 **b.** 9.2×10^{-8} **11.** $12.86

Exercise Set P.2

1. 50   **3.** 64   **5.** -64 **7.** 1   **9.** -1 **11.** $\dfrac{1}{64}$ **13.** 32   **15.** 64   **17.** 16   **19.** $\dfrac{1}{9}$ **21.** $\dfrac{1}{16}$ **23.** $\dfrac{y}{x^2}$ **25.** y^5 **27.** x^{10}

29. x^5 **31.** x^{21} **33.** x^{-15} **35.** x^7 **37.** x^{21} **39.** $64x^6$ **41.** $-\dfrac{64}{x^3}$ **43.** $9x^4y^{10}$ **45.** $6x^{11}$ **47.** $18x^9y^5$ **49.** $4x^{16}$

51. $-5a^{11}b$ **53.** $\dfrac{2}{b^7}$ **55.** $\dfrac{1}{16x^6}$ **57.** $\dfrac{3y^{14}}{4x^4}$ **59.** $\dfrac{y^2}{25x^6}$ **61.** $-\dfrac{27\,b^{15}}{a^{18}}$ **63.** 1   **65.** 4700   **67.** 4,000,000   **69.** 0.000786

71. 0.00000318   **73.** 3.6×10^3 **75.** 2.2×10^8 **77.** 2.7×10^{-2} **79.** 7.63×10^{-4} **81.** 600,000   **83.** 0.123   **85.** 30,000

87. 0.021   **89.** $\dfrac{4.8 \times 10^{11}}{1.2 \times 10^{-4}}; 4 \times 10^{15}$ **91.** $\dfrac{(7.2 \times 10^{-4})(3 \times 10^{-3})}{2.4 \times 10^{-4}}; 9 \times 10^{-3}$ **93.** $6800   **95.** 1.12×10^{12} **97.** 1.06×10^{-18} gram

107. (b) is true.   **109.** $A = C + D$

Section P.3

Check Point Exercises

1. a. 3   **b.** $5x\sqrt{2}$ **2. a.** $\dfrac{5}{4}$ **b.** $5x\sqrt{3}$ **3. a.** $17\sqrt{13}$ **b.** $-19\sqrt{17x}$ **4. a.** $17\sqrt{3}$ **b.** $10\sqrt{2x}$ **5. a.** $\dfrac{5\sqrt{3}}{3}$ **b.** $\sqrt{3}$

6. $\dfrac{32 - 8\sqrt{5}}{11}$ **7. a.** $2\sqrt[3]{5}$ **b.** $2\sqrt[5]{2}$ **c.** $\dfrac{5}{3}$ **8.** $5\sqrt[3]{3}$ **9. a.** 9   **b.** 3   **c.** $\dfrac{1}{2}$ **10. a.** 8   **b.** $\dfrac{1}{4}$ **11. a.** $10x^4$ **b.** $4x^{5/2}$ **12.** $\sqrt{x}$

Exercise Set P.3

1. 6   **3.** not a real number   **5.** 13   **7.** $5\sqrt{2}$ **9.** $3|x|\sqrt{5}$ **11.** $2x\sqrt{3}$ **13.** $x\sqrt{x}$ **15.** $2x\sqrt{3x}$ **17.** $\dfrac{1}{9}$ **19.** $\dfrac{7}{4}$ **21.** $4x$

23. $5x\sqrt{2x}$ **25.** $2x^2\sqrt{5}$ **27.** $13\sqrt{3}$ **29.** $-2\sqrt{17x}$ **31.** $5\sqrt{2}$ **33.** $3\sqrt{2x}$ **35.** $34\sqrt{2}$ **37.** $20\sqrt{2} - 5\sqrt{3}$ **39.** $\dfrac{\sqrt{7}}{7}$

41. $\dfrac{\sqrt{10}}{5}$ **43.** $\dfrac{13(3 - \sqrt{11})}{-2}$ **45.** $7(\sqrt{5} + 2)$ **47.** $3(\sqrt{5} - \sqrt{3})$ **49.** 5   **51.** -2 **53.** not a real number   **55.** 3   **57.** -3

59. $-\dfrac{1}{2}$ **61.** $2\sqrt[3]{4}$ **63.** $x\sqrt[3]{x}$ **65.** $3\sqrt[3]{2}$ **67.** $2x$ **69.** $7\sqrt[5]{2}$ **71.** $13\sqrt[3]{2}$ **73.** $-y\sqrt[3]{2x}$ **75.** $\sqrt{2} + 2$ **77.** 6   **79.** 2

81. 25   **83.** $\dfrac{1}{16}$ **85.** $14x^{7/12}$ **87.** $4x^{1/4}$ **89.** x^2 **91.** $5x^2|y|^3$ **93.** $27y^{2/3}$ **95.** $\sqrt{5}$ **97.** x^2 **99.** $\sqrt[3]{x^2}$ **101.** $\sqrt[3]{x^2y}$

103. $20\sqrt{2}$ mph   **105.** $\dfrac{\sqrt{5} + 1}{2} \approx 1.62$ **107.** $\dfrac{7\sqrt{2\cdot2\cdot3}}{6} = \dfrac{7\sqrt{2^2\cdot3}}{6} = \dfrac{7\sqrt{2^2}\sqrt{3}}{6} = \dfrac{7\cdot2\sqrt{3}}{6} = \dfrac{7}{3}\sqrt{3}$

109. The duration of a storm whose diameter is 9 miles is 1.89 hours.   **117.** 45.00, 23.76, 15.68, 11.33, 8.59, 6.70, 5.31, 4.25, 3.41, 2.73, 2.17, 1.70, 1.30, 0.95, 0.65, 0.38; The percentage of potential employees testing positive for illegal drugs is decreasing over time. **119.** (d) is true.   **121.** Let $\square$ = 25 and $\square$ = 14.   **123. a.** >   **b.** >

Section P.4

Check Point Exercises

1. a. $-x^3 + x^2 - 8x - 20$ **b.** $20x^3 - 11x^2 - 2x - 8$ **2.** $15x^3 - 31x^2 + 30x - 8$ **3.** $28x^2 - 41x + 15$
4. a. $49x^2 - 64$ **b.** $4y^6 - 25$ **5. a.** $x^2 + 20x + 100$ **b.** $25x^2 + 40x + 16$ **6. a.** $x^2 - 18x + 81$ **b.** $49x^2 - 42x + 9$
7. $2x^2y + 5xy^2 - 2y^3$ **8. a.** $21x^2 - 25xy + 6y^2$ **b.** $x^4 + 10x^2y + 25y^2$

Exercise Set P.4

1. yes; $3x^2 + 2x - 5$ **3.** no   **5.** 2   **7.** 4   **9.** $11x^3 + 7x^2 - 12x - 4$; 3   **11.** $12x^3 + 4x^2 + 12x - 14$; 3   **13.** $6x^2 - 6x + 2$; 2
15. $x^3 + 1$ **17.** $2x^3 - 9x^2 + 19x - 15$ **19.** $x^2 + 10x + 21$ **21.** $x^2 - 2x - 15$ **23.** $6x^2 + 13x + 5$ **25.** $10x^2 - 9x - 9$
27. $15x^4 - 47x^2 + 28$ **29.** $8x^5 - 40x^3 + 3x^2 - 15$ **31.** $x^2 - 9$ **33.** $9x^2 - 4$ **35.** $25 - 49x^2$ **37.** $16x^4 - 25x^2$ **39.** $1 - y^{10}$
41. $x^2 + 4x + 4$ **43.** $4x^2 + 12x + 9$ **45.** $x^2 - 6x + 9$ **47.** $16x^4 - 8x^2 + 1$ **49.** $4x^2 - 28x + 49$ **51.** $x^3 + 3x^2 + 3x + 1$
53. $8x^3 + 36x^2 + 54x + 27$ **55.** $x^3 - 9x^2 + 27x - 27$ **57.** $27x^3 - 108x^2 + 144x - 64$ **59.** $7x^2y - 4xy$ is of degree 3
61. $2x^2y + 13xy + 13$ is of degree 3   **63.** $-5x^3 + 8xy - 9y^2$ is of degree 3   **65.** $x^4y^2 + 8x^3y + y - 6x$ is of degree 6
67. $7x^2 + 38xy + 15y^2$ **69.** $2x^2 + xy - 21y^2$ **71.** $15x^2y^2 + xy - 2$ **73.** $49x^2 + 70xy + 25y^2$ **75.** $x^4y^4 - 6x^2y^2 + 9$
77. $x^3 - y^3$ **79.** $9x^2 - 25y^2$ **81.** $49x^2y^4 - 100y^2$ **83.** 7.567; A person earning \$40,000 feels underpaid by \$7567.
85. 527.53; The number of violent crimes in the United States was 527.53 per 100,000 inhabitants in 2000. The calculated value is a good approximation to the actual value, 524.7.   **87.** $\dfrac{2}{3}t^3 - 2t^2 + 4t$ **89.** $6x + 22$ **99.** 61.2, 59.0, 56.8, 54.8, 52.8, 50.9, 49.3, 47.7, 46.4, 45.2, 44.3, 43.6, 43.1, 43.0, 43.1, 43.6, 44.4, 45.5, 47.0, 48.9, 51.2; The percentage of U.S. high school seniors who had ever used marijuana decreased from 1980, reached a low in 1993, then increased through 2000.   **101.** $49x^2 + 70x + 25 - 16y^2$ **103.** $x^4 - y^4$

Section P.5

Check Point Exercises

1. a. $2x^2(5x - 2)$ **b.** $(x - 7)(2x + 3)$ **2.** $(x + 5)(x^2 - 2)$ **3. a.** $(x + 8)(x + 5)$ **b.** $(x - 7)(x + 2)$ **4.** $(3x - 1)(2x + 7)$
5. a. $(x + 9)(x - 9)$ **b.** $(6x + 5)(6x - 5)$ **6.** $(9x^2 + 4)(3x + 2)(3x - 2)$ **7. a.** $(x + 7)^2$ **b.** $(4x - 7)^2$

8. a. $(x + 1)(x^2 - x + 1)$ **b.** $(5x - 2)(25x^2 + 10x + 4)$ **9.** $3x(x - 5)^2$ **10.** $(x + 10 + 6a)(x + 10 - 6a)$ **11.** $\dfrac{2x - 1}{(x - 1)^{1/2}}$

Exercise Set P.5

1. $9(2x + 3)$ **3.** $3x(x + 2)$ **5.** $9x^2(x^2 - 2x + 3)$ **7.** $(x + 5)(x + 3)$ **9.** $(x - 3)(x^2 + 12)$ **11.** $(x^2 + 5)(x - 2)$
13. $(x - 1)(x^2 + 2)$ **15.** $(3x - 2)(x^2 - 2)$ **17.** $(x + 2)(x + 3)$ **19.** $(x - 5)(x + 3)$ **21.** $(x - 5)(x - 3)$
23. $(3x + 2)(x - 1)$ **25.** $(3x - 28)(x + 1)$ **27.** $(2x - 1)(3x - 4)$ **29.** $(2x + 3)(2x + 5)$ **31.** $(x + 10)(x - 10)$
33. $(6x + 7)(6x - 7)$ **35.** $(3x + 5y)(3x - 5y)$ **37.** $(x^2 + 4)(x + 2)(x - 2)$ **39.** $(4x^2 + 9)(2x + 3)(2x - 3)$ **41.** $(x + 1)^2$
43. $(x - 7)^2$ **45.** $(2x + 1)^2$ **47.** $(3x - 1)^2$ **49.** $(x + 3)(x^2 - 3x + 9)$ **51.** $(x - 4)(x^2 + 4x + 16)$
53. $(2x - 1)(4x^2 + 2x + 1)$ **55.** $(4x + 3)(16x^2 - 12x + 9)$ **57.** $3x(x + 1)(x - 1)$ **59.** $4(x + 2)(x - 3)$
61. $2(x^2 + 9)(x + 3)(x - 3)$ **63.** $(x - 3)(x + 3)(x + 2)$ **65.** $2(x - 8)(x + 7)$ **67.** $x(x - 2)(x + 2)$ **69.** prime
71. $(x - 2)(x + 2)^2$ **73.** $y(y^2 + 9)(y + 3)(y - 3)$ **75.** $5y^2(2y + 3)(2y - 3)$ **77.** $(x - 6 + 7y)(x - 6 - 7y)$

79. $(x + y)(3b + 4)(3b - 4)$ **81.** $(y - 2)(x + 4)(x - 4)$ **83.** $2x(x + 6 + 2a)(x + 6 - 2a)$ **85.** $x^{1/2}(x - 1)$ **87.** $\dfrac{4(1 + 2x)}{x^{2/3}}$

89. $-(x + 3)^{1/2}(x + 2)$ **91.** $\dfrac{x + 4}{(x + 5)^{3/2}}$ **93.** $\dfrac{4(4x - 1)^{1/2}(x - 1)}{3}$ **95. a.** $0.36\,x$ **b.** no; It is selling at 36% of the original price.

97. $16(4 + t)(4 - t)$ **99.** $(3x + 2)(3x - 2)$ **109.** $(x^n + 4)(x^n + 2)$ **111.** $(x - y)^3(x + y)$ **113.** $b = 8, -8, 16, -16$

Section P.6

Check Point Exercises

1. a. -5 **b.** $6, -6$ **2. a.** $x^2, x \neq -3$ **b.** $\dfrac{x - 1}{x + 1}, x \neq -1$ **3.** $\dfrac{x - 3}{(x - 2)(x + 3)}, x \neq 2, x \neq -2, x \neq -3$

4. $\dfrac{3(x - 1)}{x(x + 2)}, x \neq 1, x \neq 0, x \neq -2$ **5.** $-2, x \neq -1$ **6.** $\dfrac{2(4x + 1)}{(x + 1)(x - 1)}, x \neq 1, x \neq -1$ **7.** $(x - 3)(x - 3)(x + 3)$

8. $\dfrac{-x^2 + 11x - 20}{2(x - 5)^2}, x \neq 5$ **9.** $\dfrac{2(2 - 3x)}{4 + 3x}, x \neq 0, x \neq -\dfrac{4}{3}$

Exercise Set P.6

1. 3 **3.** $5, -5$ **5.** $-1, -10$ **7.** $\dfrac{3}{x - 3}, x \neq 3$ **9.** $\dfrac{x - 6}{4}, x \neq 6$ **11.** $\dfrac{y + 9}{y - 1}, y \neq 1, 2$ **13.** $\dfrac{x + 6}{x - 6}, x \neq 6, -6$ **15.** $\dfrac{1}{3}, x \neq 2, -3$

17. $\dfrac{(x - 3)(x + 3)}{x(x + 4)}, x \neq 0, -4, 3$ **19.** $\dfrac{x - 1}{x + 2}, x \neq -2, -1, 2, 3$ **21.** $\dfrac{x^2 + 2x + 4}{3x}, x \neq -2, 0, 2$ **23.** $\dfrac{7}{9}, x \neq -1$

25. $\dfrac{(x - 2)^2}{x}, x \neq 0, -2, 2$ **27.** $\dfrac{2(x + 3)}{3}, x \neq 3, -3$ **29.** $\dfrac{x - 5}{2}, x \neq 1, -5$ **31.** $\dfrac{(x + 2)(x + 4)}{x - 5}, x \neq -6, -3, -1, 3, 5$

33. $2, x \neq -\dfrac{5}{6}$ **35.** $\dfrac{2x - 1}{x + 3}, x \neq 0, -3$ **37.** $3, x \neq 2$ **39.** $\dfrac{3}{x - 3}, x \neq 3, -4$ **41.** $\dfrac{9x + 39}{(x + 4)(x + 5)}, x \neq -4, -5$

43. $\dfrac{3}{x(x + 1)}, x \neq -1, 0$ **45.** $\dfrac{3x^2 + 4}{(x + 2)(x - 2)}, x \neq -2, 2$ **47.** $\dfrac{2x^2 + 50}{(x - 5)(x + 5)}, x \neq -5, 5$ **49.** $\dfrac{4x + 16}{(x + 3)^2}, x \neq -3$

51. $\dfrac{x^2 - x}{(x + 5)(x - 2)(x + 3)}, x \neq -5, 2, -3$ **53.** $\dfrac{x - 1}{x + 2}, x \neq -2, -1$ **55.** $\dfrac{1}{3}, x \neq 3$ **57.** $\dfrac{x + 1}{3x - 1}, x \neq 0, \dfrac{1}{3}$

59. $\dfrac{1}{xy}, x \neq 0, y \neq 0, x \neq -y$ **61.** $\dfrac{x}{x + 3}, x \neq -2, -3$ **63.** $\dfrac{x - 14}{7}, x \neq -2, 2$ **65. a.** 86.67, 520, 1170; It costs $86,670,000 to

inoculate 40% of the population against this strain of flu, $520,000,000 to inoculate 80% of the population, and $1,170,000,000 to inoculate 90% of the population. **b.** $x = 100$ **c.** increases rapidly; impossible to inoculate 100% of the population.

67. a. $\dfrac{100W}{L}$ **b.** round **69.** $\dfrac{2r_1 r_2}{r_1 + r_2}$; 24 mph **83. a.** $\dfrac{Pi(1 + i)^n}{(1 + i)^n - 1}$ **b.** $527 **85.** $-4x - 1$ **87.** It cubes x.

Chapter P Review Exercises

1. a. $\sqrt{81}$ **b.** $0, \sqrt{81}$ **c.** $-17, 0, \sqrt{81}$ **d.** $-17, -\dfrac{9}{13}, 0, 0.75, \sqrt{81}$ **e.** $\sqrt{2}, \pi$ **2.** 103 **3.** $\sqrt{2} - 1$ **4.** $\sqrt{17} - 3$

5. $|4 - (-17)|$; 21 **6.** 20 **7.** 4 **8.** commutative property of addition **9.** associative property of multiplication **10.** distributive property of multiplication over addition **11.** commutative property of multiplication **12.** commutative property of

multiplication **13.** commutative property of addition **14.** $23x - 23y - 2$ **15.** $2x$ **16.** -108 **17.** $\dfrac{5}{16}$ **18.** $\dfrac{1}{25}$ **19.** $\dfrac{1}{27}$

20. $-8x^{12}y^9$ **21.** $\dfrac{10}{x^8}$ **22.** $\dfrac{1}{16x^{12}}$ **23.** $\dfrac{y^8}{4x^{10}}$ **24.** 37,400 **25.** 0.0000745 **26.** 3.59×10^6 **27.** 7.25×10^{-3} **28.** 3.9×10^5

29. 2.3×10^{-2} **30.** 10^3 or 1000 yr **31.** 4.2×10^{10} **32.** $10\sqrt{3}$ **33.** $2|x|\sqrt{3}$ **34.** $2x\sqrt{5}$ **35.** $r\sqrt{r}$ **36.** $\dfrac{11}{2}$ **37.** $4x\sqrt{3}$

38. $20\sqrt{5}$ **39.** $16\sqrt{2}$ **40.** $24\sqrt{2} - 8\sqrt{3}$ **41.** $6\sqrt{5}$ **42.** $\dfrac{\sqrt{6}}{3}$ **43.** $\dfrac{5(6 - \sqrt{3})}{33}$ **44.** $7(\sqrt{7} + \sqrt{5})$ **45.** 5 **46.** -2

47. not a real number **48.** 5 **49.** $3\sqrt[3]{3}$ **50.** $y\sqrt[3]{y^2}$ **51.** $2\sqrt[4]{5}$ **52.** $13\sqrt[3]{2}$ **53.** $x\sqrt[4]{2}$ **54.** 4

55. $\dfrac{1}{5}$ **56.** 5 **57.** $\dfrac{1}{3}$ **58.** 16 **59.** $\dfrac{1}{81}$ **60.** $20x^{11/12}$ **61.** $3x^{1/4}$ **62.** $25x^4$ **63.** $\sqrt{y}$ **64.** $8x^3 + 10x^2 - 20x - 4$; degree 3

65. $8x^4 - 5x^3 + 6$; degree 4 **66.** $12x^3 + x^2 - 21x + 10$ **67.** $6x^2 - 7x - 5$ **68.** $16x^2 - 25$ **69.** $4x^2 + 20x + 25$
70. $9x^2 - 24x + 16$ **71.** $8x^3 + 12x^2 + 6x + 1$ **72.** $125x^3 - 150x^2 + 60x - 8$ **73.** $-x^2 - 17xy - 3y^2$; degree 2
74. $24x^3y^2 + x^2y - 12x^2 + 4$; degree 5 **75.** $3x^2 + 16xy - 35y^2$ **76.** $9x^2 - 30xy + 25y^2$ **77.** $9x^4 + 12x^2y + 4y^2$

78. $49x^2 - 16y^2$ **79.** $a^3 - b^3$ **80.** $3x^2(5x + 1)$ **81.** $(x - 4)(x - 7)$ **82.** $(3x + 1)(5x - 2)$ **83.** $(8 - x)(8 + x)$ **84.** prime
85. $3x^2(x - 5)(x + 2)$ **86.** $4x^3(5x^4 - 9)$ **87.** $(x + 3)(x - 3)^2$ **88.** $(4x - 5)^2$ **89.** $(x^2 + 4)(x + 2)(x - 2)$
90. $(y - 2)(y^2 + 2y + 4)$ **91.** $(x + 4)(x^2 - 4x + 16)$ **92.** $3x^2(x - 2)(x + 2)$ **93.** $(3x - 5)(9x^2 + 15x + 25)$

94. $x(x - 1)(x + 1)(x^2 + 1)$ **95.** $(x^2 - 2)(x + 5)$ **96.** $(x + 9 + y)(x + 9 - y)$ **97.** $\dfrac{16(1 + 2x)}{x^{3/4}}$

98. $(x + 2)(x - 2)(x^2 + 3)^{1/2}(-x^4 + x^2 + 13)$ **99.** $\dfrac{6(2x + 1)}{x^{3/2}}$ **100.** $x^2, x \neq -2$ **101.** $\dfrac{x - 3}{x - 6}, x \neq -6, 6$ **102.** $\dfrac{x}{x + 2}, x \neq -2$

103. $\dfrac{(x + 3)^3}{(x - 2)^2(x + 2)}, x \neq 2, -2$ **104.** $\dfrac{2}{x(x + 1)}, x \neq 0, 1, -1, -\dfrac{1}{3}$ **105.** $\dfrac{x + 3}{x - 4}, x \neq -3, 4, 2, 8$ **106.** $\dfrac{1}{x - 3}, x \neq 3, -3$

107. $\dfrac{4x(x - 1)}{(x + 2)(x - 2)}, x \neq 2, -2$ **108.** $\dfrac{2x^2 - 3}{(x - 3)(x + 3)(x - 2)}, x \neq 3, -3, 2$ **109.** $\dfrac{11x^2 - x - 11}{(2x - 1)(x + 3)(3x + 2)}, x \neq \dfrac{1}{2}, -3, \dfrac{2}{3}$

110. $\dfrac{3}{x}, x \neq 0, 2$ **111.** $\dfrac{3x}{x - 4}, x \neq 0, 4, -4$ **112.** $\dfrac{3x + 8}{3x + 10}, x \neq -3, -\dfrac{10}{3}$

Chapter P Test

1. $-7, -\dfrac{4}{5}, 0, 0.25, \sqrt{4}, \dfrac{22}{7}$ **2.** commutative property of addition   **3.** distributive property of multiplication over addition

4. 7.6×10^{-4} **5.** $85x + 2y - 15$ **6.** $\dfrac{5y^8}{x^6}$ **7.** $3r\sqrt{2}$ **8.** $11\sqrt{2}$ **9.** $\dfrac{3(5 - \sqrt{2})}{23}$ **10.** $2x\sqrt[3]{2x}$ **11.** $\dfrac{x + 3}{x - 2}, x \neq 2, 1$ **12.** $\dfrac{1}{243}$

13. $2x^3 - 13x^2 + 26x - 15$ **14.** $25x^2 + 30xy + 9y^2$ **15.** $(x - 3)(x - 6)$ **16.** $(x^2 + 3)(x + 2)$ **17.** $(5x - 3)(5x + 3)$

18. $(6x - 7)^2$ **19.** $(y - 5)(y^2 + 5y + 25)$ **20.** $(x + 5 + 3y)(x + 5 - 3y)$ **21.** $\dfrac{2x + 3}{(x + 3)^{3/5}}$ **22.** $\dfrac{2(x + 3)}{x + 1}, x \neq 3, -1, -4, -3$

23. $\dfrac{x^2 + 2x + 15}{(x + 3)(x - 3)}, x \neq 3, -3$ **24.** $\dfrac{5}{(x - 3)(x - 4)}, x \neq 3, 4$ **25.** $\dfrac{3 - x}{3}, x \neq 0$

CHAPTER 1

Section 1.1

Check Point Exercises

1.

2.
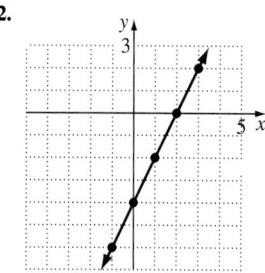

3. The minimum x-value is -100 and the maximum x-value is 100. The distance between consecutive tick marks is 50. The minimum y-value is -80 and the maximum y-value is 80. The distance between consecutive tick marks is 10.

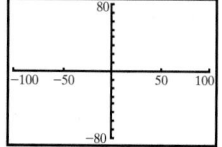

4. $21\dfrac{1}{2}$; 1900

Exercise Set 1.1

1.

3.

5.

7.

9.

11.

13.

15.

17.

19.

21.

23.

25.

27.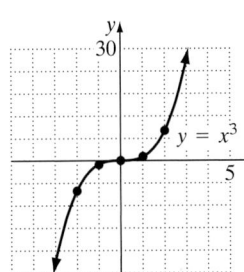

29. (c)   **31.** (b)   **33. a.** 2   **b.** −4   **35. a.** 1, −2   **b.** 2
37. a. −1   **b.** None   **39.** A (2, 7); The football is 7 feet high
when it is 2 yards from the quarterback.

41. $C\left(6, 9\frac{1}{2}\right)$ **43.** 12 feet; 15 yards   **45.** 0–4 years

47. year 4; 8%   **57.** (c) gives a complete graph.
59. (c) gives a complete graph.   **61.** (a)   **63.** (b)

Section 1.2

Check Point Exercises

1. $\{16\}$ **2.** $\{5\}$ **3.** $\{-2\}$ **4.** $\{3\}$ **5.** $\varnothing$ **6.** identity

Exercise Set 1.2

1. $\{11\}$ **3.** $\{7\}$ **5.** $\{13\}$ **7.** $\{2\}$ **9.** $\{9\}$ **11.** $\{-5\}$ **13.** $\{6\}$ **15.** $\{-2\}$ **17.** $\{12\}$ **19.** $\{24\}$ **21.** $\{-15\}$

23. $\{5\}$ **25.** $\left\{\frac{33}{2}\right\}$ **27.** $\{-12\}$ **29.** $\left\{\frac{46}{5}\right\}$ **31. a.** 0   **b.** $\left\{\frac{1}{2}\right\}$ **33. a.** 0   **b.** $\{-2\}$ **35. a.** 0   **b.** $\{2\}$ **37. a.** 0

b. $\{4\}$ **39. a.** 1   **b.** $\{3\}$ **41. a.** −1   **b.** $\varnothing$ **43. a.** 1   **b.** $\{2\}$ **45. a.** −2, 2   **b.** $\varnothing$ **47. a.** −1, 1   **b.** $\{-3\}$
49. a. −2, 4   **b.** $\varnothing$ **51.** identity   **53.** inconsistent equation   **55.** conditional equation   **57.** inconsistent equation
59. $\{-7\}$ **61.** not true for any real number, $\varnothing$ **63.** $\{-4\}$ **65.** $\{8\}$ **67.** $\{-1\}$ **69.** not true for any real number, $\varnothing$

71. a. 250 mg/dl   **b.** 375,000 annual deaths; 350,000 saved lives   **73.** $409\frac{1}{5}$ ft   **87.** inconsistent   **89.** conditional; $\{-5\}$

91. $x = \dfrac{c - b}{a}$ **93.** Answers may vary.   **95.** 20

Section 1.3

Check Point Exercises

1. 2008   **2.** *Saturday Night Fever* sold 11 million albums; *Jagged Little Pill* sold 16 million albums.   **3.** 300 min

4. $15,000 at 9%; $10,000 at 12%   **5.** width = 50 ft; length = 94 ft   **6.** $m = \dfrac{y - b}{x}$ **7.** $C = \dfrac{P}{1 + M}$

Exercise Set 1.3

1. $x + 9$ **3.** $20 - x$ **5.** $8 - 5x$ **7.** $15 \div x$ **9.** $2x + 20$ **11.** $7x - 30$ **13.** $4(x + 12)$ **15.** $x + 40 = 450; \{410\}$
17. $5x - 7 = 123; \{26\}$ **19.** $9x = 3x + 30; \{5\}$ **21.** 40 years old; It is shown by the point $(40, 117)$ on the line for females.
23. approximately 41 years after 1960 in 2001 **25.** 196 lb **27.** $Waterworld = \$160$ million; $Titanic = \$200$ million
29. Miami = 57 hr; Los Angeles = 82 hr **31.** 800 mi **33.** 2005 **35. a.** total monthly cost with coupon book $= 21 + 0.50x$; total
monthly cost without coupon book $= 1.25x$ **b.** 28 times **37.** $\$600; \$580$ **39.** $\$31,250$ in noninsured bonds; $\$18,750$ in
government-insured certificates of deposit **41.** $\$6000$ at 12%; $\$2000$ at a 5% loss **43.** length $= 78$ ft; width $= 36$ ft
45. length $= 2$ ft; height $= 5$ ft **47.** 11 hr **49.** $\$31,000$ **51.** 7 oz **53.** $\$20,000$ **55.** 5 ft 7 in. **57.** $\omega = \dfrac{A}{l}$ **59.** $b = \dfrac{2A}{h}$
61. $p = \dfrac{I}{rt}$ **63.** $m = \dfrac{E}{c^2}$ **65.** $p = \dfrac{T - D}{m}$ **67.** $a = \dfrac{2A}{h} - b$ **69.** $r = \dfrac{S - P}{Pt}$ **71.** $S = \dfrac{F}{B} + V$ **73.** $I = \dfrac{E}{R + r}$
75. $f = \dfrac{pq}{p + q}$ **81.**

83. North campus had 600 students; South campus had 400 students.
85. Coburn $= 60$ years old; woman $= 20$ years old
87. $\$4000$ for the mother; $\$8000$ for the boy; $\$2000$ for the girl

The trace feature shows x to be about 15
when $y = 37$, so 2005. The trace feature
shows x to be 20 when $y = 40$, so 2010.

Section 1.4

Check Point Exercises

1. a. $8 + i$ **b.** $-10 + 10i$ **2. a.** $63 + 14i$ **b.** $58 - 11i$ **3.** $\dfrac{3}{5} + \dfrac{13}{10}i$ **4. a.** $7i\sqrt{3}$ **b.** $1 - 4i\sqrt{3}$ **c.** $-7 + i\sqrt{3}$

Exercise Set 1.4

1. $8 - 2i$ **3.** $-2 + 9i$ **5.** $24 + 7i$ **7.** $-14 + 17i$ **9.** $21 + 15i$ **11.** $-43 - 23i$ **13.** $-29 - 11i$ **15.** 34 **17.** 34
19. $-5 + 12i$ **21.** $\dfrac{3}{5} + \dfrac{1}{5}i$ **23.** $1 + i$ **25.** $-\dfrac{24}{25} + \dfrac{32}{25}i$ **27.** $\dfrac{7}{5} + \dfrac{4}{5}i$ **29.** $3i$ **31.** $47i$ **33.** $-8i$ **35.** $2 + 6i\sqrt{7}$
37. $-\dfrac{1}{3} + \dfrac{\sqrt{2}}{6}i$ **39.** $-\dfrac{1}{8} - \dfrac{\sqrt{3}}{24}i$ **41.** $-2\sqrt{6} - 2i\sqrt{10}$ **43.** $24\sqrt{15}$ **53.** (d) is true. **55.** $\dfrac{14}{25} - \dfrac{2}{25}i$ **57.** 0

Section 1.5

Check Point Exercises

1. a. $\{0, 3\}$ **b.** $\left\{-1, \dfrac{1}{2}\right\}$ **2. a.** $\{-\sqrt{7}, \sqrt{7}\}$ **b.** $\{-5 + \sqrt{11}, -5 - \sqrt{11}\}$ **3.** $49; (x - 7)^2$ **4.** $\{1 + \sqrt{3}, 1 - \sqrt{3}\}$
5. $\left\{\dfrac{-1 + \sqrt{3}}{2}, \dfrac{-1 - \sqrt{3}}{2}\right\}$ **6.** $\{1 + i, 1 - i\}$ **7.** -56; two complex imaginary solutions **8.** 1998; a good approximation **9.** 12 in.

Exercise Set 1.5

1. $\{-2, 5\}$ **3.** $\{3, 5\}$ **5.** $\left\{-\dfrac{5}{2}, \dfrac{2}{3}\right\}$ **7.** $\left\{-\dfrac{4}{3}, 2\right\}$ **9.** $\{-4, 0\}$ **11.** $\left\{0, \dfrac{1}{3}\right\}$ **13.** $\{-3, 1\}$ **15.** $\{-3, 3\}$
17. $\{-\sqrt{10}, \sqrt{10}\}$ **19.** $\{-7, 3\}$ **21.** $\left\{-\dfrac{5}{3}, \dfrac{1}{3}\right\}$ **23.** $\left\{\dfrac{1 - \sqrt{7}}{5}, \dfrac{1 + \sqrt{7}}{5}\right\}$ **25.** $\left\{\dfrac{4 - 2\sqrt{2}}{3}, \dfrac{4 + 2\sqrt{2}}{3}\right\}$
27. $36; x^2 + 12x + 36 = (x + 6)^2$ **29.** $25; x^2 - 10x + 25 = (x - 5)^2$ **31.** $\dfrac{9}{4}; x^2 + 3x + \dfrac{9}{4} = \left(x + \dfrac{3}{2}\right)^2$
33. $\dfrac{49}{4}; x^2 - 7x + \dfrac{49}{4} = \left(x - \dfrac{7}{2}\right)^2$ **35.** $\dfrac{1}{9}; x^2 - \dfrac{2}{3}x + \dfrac{1}{9} = \left(x - \dfrac{1}{3}\right)^2$ **37.** $\dfrac{1}{36}; x^2 - \dfrac{1}{3}x + \dfrac{1}{36} = \left(x - \dfrac{1}{6}\right)^2$ **39.** $\{-7, 1\}$
41. $\{1 + \sqrt{3}, 1 - \sqrt{3}\}$ **43.** $\{3 + 2\sqrt{5}, 3 - 2\sqrt{5}\}$ **45.** $\{-2 + \sqrt{3}, -2 - \sqrt{3}\}$ **47.** $\left\{\dfrac{-3 + \sqrt{13}}{2}, \dfrac{-3 - \sqrt{13}}{2}\right\}$

49. $\left\{\dfrac{1}{2}, 3\right\}$ **51.** $\left\{\dfrac{1 + \sqrt{2}}{2}, \dfrac{1 - \sqrt{2}}{2}\right\}$ **53.** $\left\{\dfrac{1 + \sqrt{7}}{3}, \dfrac{1 - \sqrt{7}}{3}\right\}$ **55.** $\{-5, -3\}$ **57.** $\left\{\dfrac{-5 + \sqrt{13}}{2}, \dfrac{-5 - \sqrt{13}}{2}\right\}$

59. $\left\{\dfrac{3 + \sqrt{57}}{6}, \dfrac{3 - \sqrt{57}}{6}\right\}$ **61.** $\left\{\dfrac{1 + \sqrt{29}}{4}, \dfrac{1 - \sqrt{29}}{4}\right\}$ **63.** $\{3 + i, 3 - i\}$ **65.** $36; 2$ unequal real solutions

67. $97; 2$ unequal real solutions **69.** $0; 1$ real solution **71.** $37; 2$ unequal real solutions **73.** $\left\{-\dfrac{1}{2}, 1\right\}$ **75.** $\left\{\dfrac{1}{5}, 2\right\}$

77. $\{-2\sqrt{5}, 2\sqrt{5}\}$ **79.** $\{1 + \sqrt{2}, 1 - \sqrt{2}\}$ **81.** $\left\{\dfrac{-11 + \sqrt{33}}{4}, \dfrac{-11 - \sqrt{33}}{4}\right\}$ **83.** $\left\{0, \dfrac{8}{3}\right\}$ **85.** $\{2\}$ **87.** $\{-2, 2\}$

89. $\{3 + 2i, 3 - 2i\}$ **91.** $\{2 + i\sqrt{3}, 2 - i\sqrt{3}\}$ **93.** $\left\{0, \dfrac{7}{2}\right\}$ **95.** $\{2 + \sqrt{10}, 2 - \sqrt{10}\}$ **97.** $\{-5, -1\}$

99. 19 year olds and 72 year olds; fairly well **101.** 1994 **103.** $(4, 27)$; This is the graph's highest point; During this time period, the greatest number of recipients was 27 million in 1994. **105.** $1990; (10, 740)$ **107.** 127.28 ft **109.** 34 ft **111.** width $= 15$ ft; length $= 20$ ft **113.** 10 in. **115.** 9.3 in. and 0.7 in. **117.** 2 in. **129.** (c) is true. **131.** $x^2 - 2x - 15 = 0$ **133.** 2.4 m; Yes

Section 1.6

Check Point Exercises

1. $\{-\sqrt{3}, 0, \sqrt{3}\}$ **2.** $\left\{-2, -\dfrac{3}{2}, 2\right\}$ **3.** $\{-1, 3\}$ **4.** $\{4\}$ **5. a.** $\{\sqrt[3]{25}\}$ or $\{5^{2/3}\}$ **b.** $\{-8, 8\}$ **6.** $\{-\sqrt{3}, -\sqrt{2}, \sqrt{2}, \sqrt{3}\}$

7. $\left\{-\dfrac{1}{27}, 64\right\}$ **8.** $\{-2, 3\}$

Exercise Set 1.6

1. $\{-4, 0, 4\}$ **3.** $\left\{-2, -\dfrac{2}{3}, 2\right\}$ **5.** $\left\{-\dfrac{1}{2}, \dfrac{1}{2}, \dfrac{3}{2}\right\}$ **7.** $\left\{-2, -\dfrac{1}{2}, \dfrac{1}{2}\right\}$ **9.** $\{0, 2, -1 + i\sqrt{3}, -1 - i\sqrt{3}\}$ **11.** $\{6\}$ **13.** $\{6\}$

15. $\{-6\}$ **17.** $\{10\}$ **19.** $\{12\}$ **21.** $\{8\}$ **23.** $\varnothing$ **25.** $\varnothing$ **27.** $\left\{\dfrac{13 + \sqrt{105}}{6}\right\}$ **29.** $\{4\}$ **31.** $\{13\}$ **33.** $\{\sqrt[5]{4}\}$

35. $\{-60, 68\}$ **37.** $\{-4, 5\}$ **39.** $\{-2, -1, 1, 2\}$ **41.** $\left\{-\dfrac{4}{3}, -1, 1, \dfrac{4}{3}\right\}$ **43.** $\{25, 64\}$ **45.** $\left\{-\dfrac{1}{4}, \dfrac{1}{5}\right\}$ **47.** $\{-8, 27\}$ **49.** $\{1\}$

51. $\left\{\dfrac{1}{4}, 1\right\}$ **53.** $\{2, 12\}$ **55.** $\{-3, -1, 2, 4\}$ **57.** $\{-8, -2, 1, 4\}$ **59.** $\{-8, 8\}$ **61.** $\{-5, 9\}$ **63.** $\{-2, 3\}$ **65.** $\left\{-\dfrac{5}{3}, 3\right\}$

67. $\left\{-\dfrac{2}{5}, \dfrac{2}{5}\right\}$ **69.** $\varnothing$ **71.** $\left\{\dfrac{1}{2}\right\}$ **73.** $\{-1, 3\}$ **75.** $\{1\}$ **77.** $\{0\}$ **79.** $\left\{\dfrac{5}{2}\right\}$ **81.** $\{-8, -6, 4, 6\}$ **83.** $\{-1, 1, 2\}$

85. 2018 **87.** 36 years old; $(36, 40{,}000)$ **89.** 149 million km **91.** either 1.2 feet or 7.5 feet from the base of the 6 foot pole

101. $\{-3, -1, 1\}$ **103.** $\{-2\}$ **105.** (d) is true. **107.** $\left\{\dfrac{2}{5}, \dfrac{1}{2}\right\}$ **109.** $\{0, 1\}$

Section 1.7

Check Point Exercises

1. a.   **b.**   **c.**

2. a. $\{x | -2 \le x < 5\}$ **b.** $\{x | 1 \le x \le 3.5\}$ **c.** $\{x | x < -1\}$ **3.** $[-1, \infty)$ or $\{x | x \ge -1\}$

4. $[1, \infty)$ or $\{x | x \ge 1\}$ **5.** $[-1, 4)$ or $\{x | -1 \le x < 4\}$ **6.** $(-3, 7)$ or $\{x | -3 < x < 7\}$ **7.** $(-\infty, 1]$ or $[4, \infty)$ or $\{x | x \le 1$ or $x \ge 4\}$

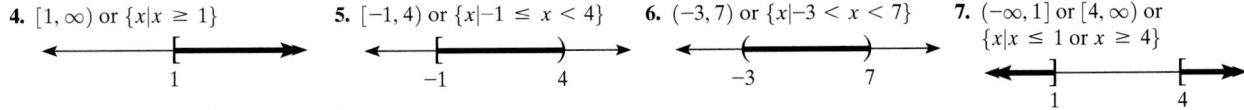

8. driving more than 720 mi per week

Exercise Set 1.7

1. 6

3. −4

5. −3

7. 4

9. −2 5

11. −1 4

13. $1 < x \le 6$

15. $-5 \le x < 2$

17. $-3 \le x \le 1$

19. $x > 2$

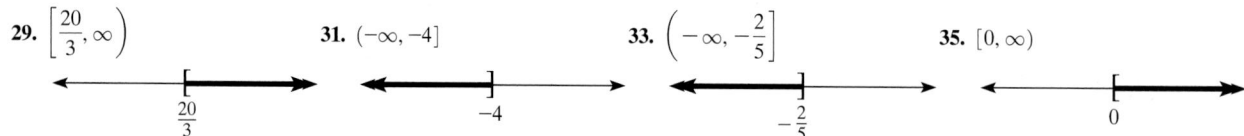

1 6   −5 2   −3 1   2

21. $x \ge -3$

23. $x < 3$

25. $x < 5.5$

27. $(-\infty, 3)$

−3   3   5.5   3

29. $\left[\dfrac{20}{3}, \infty\right)$

31. $(-\infty, -4]$

33. $\left(-\infty, -\dfrac{2}{5}\right]$

35. $[0, \infty)$

$\dfrac{20}{3}$ −4 $-\dfrac{2}{5}$ 0

37. $(-\infty, 1)$

39. $[6, \infty)$

41. $\left[-\dfrac{32}{5}, \infty\right)$

43. $(-\infty, -6)$

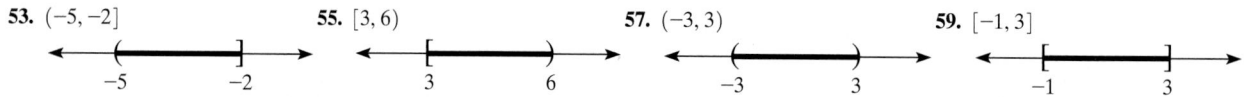

1   6 $-\dfrac{32}{5}$ −6

45. $[13, \infty)$

47. $(-\infty, \infty)$

49. $(3, 5)$

51. $[-1, 3)$

13   0   3 5   −1 3

53. $(-5, -2]$

55. $[3, 6)$

57. $(-3, 3)$

59. $[-1, 3]$

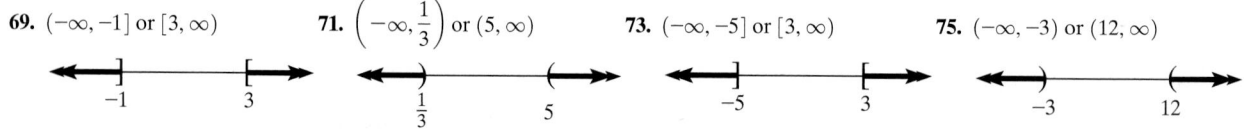

−5 −2   3 6   −3 3   −1 3

61. $(-1, 7)$

63. $[-5, 3]$

65. $(-6, 0)$

67. $(-\infty, -3)$ or $(3, \infty)$

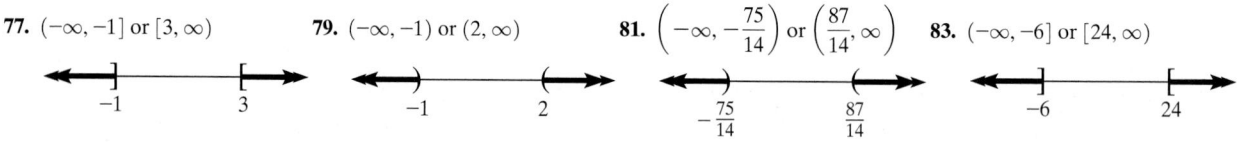

−1 7   −5 3   −6 0   −3 3

69. $(-\infty, -1]$ or $[3, \infty)$

71. $\left(-\infty, \dfrac{1}{3}\right)$ or $(5, \infty)$

73. $(-\infty, -5]$ or $[3, \infty)$

75. $(-\infty, -3)$ or $(12, \infty)$

−1 3 $\dfrac{1}{3}$ 5   −5 3   −3 12

77. $(-\infty, -1]$ or $[3, \infty)$

79. $(-\infty, -1)$ or $(2, \infty)$

81. $\left(-\infty, -\dfrac{75}{14}\right)$ or $\left(\dfrac{87}{14}, \infty\right)$

83. $(-\infty, -6]$ or $[24, \infty)$

−1 3   −1 2 $-\dfrac{75}{14}$ $\dfrac{87}{14}$ −6 24

85. sports events and playing sports   **87.** amusement parks, gardening, movies, and exercise   **89.** gardening and movies

91. home improvement, amusement parks, and gardening   **93.** $x > 20$; all years after 2008   **95.** between 59°F and 95°F inclusive

97. $58.6 \le x \le 61.8$; Between 58.6% and 61.8% of U.S. households watched the "M*A*S*H" episode.   **99.** $h \le 41$ or $h \ge 59$

101. $50 + 0.20x < 20 + 0.50x$; more than 100 mi   **103.** $1800 + 0.03x < 200 + 0.08x$; greater than \$32,000

105. $2x > 10,000 + 0.40x$; more than 6250 tapes   **107.** $265 + 65x \le 2800$; at most 39 bags

109. a. $\dfrac{86 + 88 + x}{3} \ge 90$; at least a 96   **b.** $\dfrac{86 + 88 + x}{3} < 80$; a grade less than 66

121.

$x < 4$

123.

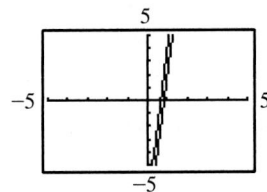

The graph of the left side of the inequality is never above the graph of the right side, therefore there is no solution; You get a statement that is always false.

125. a. $C = 4 + 0.10x; C = 2 + 0.15x$
b.

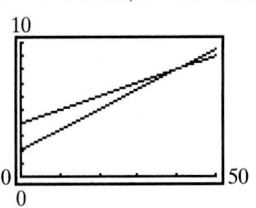

c. 41 or more checks
d. $x > 40$

127. Because $x > y$, $y - x$ represents a negative number, so when both sides are multiplied by $(y - x)$, the inequality must be reversed.
129. at least $500, but no more than $2500

Section 1.8

Check Point Exercises

1. $(-3, 1)$

2. $(-\infty, -4]$ or $[5, \infty)$

3. $(-\infty, -2)$ or $(5, \infty)$

4. $(-1, 1]$

5. between 1 and 4 sec

Exercise Set 1.8

1. $(-\infty, -2)$ or $(4, \infty)$

3. $[-3, 7]$

5. $(-\infty, 1)$ or $(4, \infty)$

7. $(-\infty, -4)$ or $(-1, \infty)$

9. $\varnothing$

11. $[2, 4]$

13. $\left[-4, \dfrac{2}{3}\right]$

15. $\left(-3, \dfrac{5}{2}\right)$

17. $\left(-1, -\dfrac{3}{4}\right)$

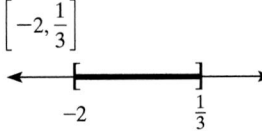

19. $\left[-2, \dfrac{1}{3}\right]$

21. $(-\infty, 0]$ or $[4, \infty)$

23. $\left(-\infty, -\dfrac{3}{2}\right)$ or $(0, \infty)$

25. $[0, 1]$

27. $(-\infty, -8)$ or $(-6, 4)$ or $(6, \infty)$

29. $(-\infty, -3)$ or $(4, \infty)$

31. $(-4, -3)$

33. $[2, 4)$

35. $\left(-\infty, -\dfrac{4}{3}\right)$ or $[2, \infty)$

37. $(-\infty, 0)$ or $(3, \infty)$

39. $(-\infty, -5)$ or $(-3, \infty)$

41. $\left(-\infty, \dfrac{1}{2}\right)$ or $\left[\dfrac{7}{5}, \infty\right)$

43. $(-\infty, -6]$ or $(-2, \infty)$

45. $(-3, 2)$

47. $(-\infty, -1)$ or $(1, 2)$ or $(3, \infty)$

49. between 2 and 3 sec　　**51.** 3.46 sec　　**53. a.** 200 beats/min　　**b.** $[0, 4)$ or $(12, \infty)$; $(12, \infty)$; $[0, 4)$; heart rate will plateau when it reaches its normal level.　　**55.** from 2008 on　　**57.** They must produce at least 20,000 wheelchairs; the x-values of all points on the graph

which lie below $y = 425$ are solutions to the inequality.　　**63.** $\left[-3, \dfrac{1}{2}\right]$　　**65.** $(1, 4]$　　**67.** Answers may vary.

69. Because the square of any number other than zero is positive, the solution includes all real numbers except 2.
71. Because the square of any number is positive, the solution is $\varnothing$.　　**73.** $(-2, -1)$ or $(2, \infty)$　　**75. a.** $(-\infty, \infty)$　　**b.** $\varnothing$

Chapter 1 Review Exercises

1.

2.

3.

4.

5.

6. x-intercept: -2; y-intercept: 2　　**7.** x-intercepts: 2, -2; y-intercept: -4
8. x-intercept: 5; y-intercept: none　　**9.** 20%　　**10.** 85 years
11. The percentage of Americans with Alzheimer's disease increases with age.
12. $\{6\}$　　**13.** $\{-10\}$　　**14.** $\{5\}$　　**15.** $\{-13\}$　　**16.** $\{-3\}$　　**17.** $\{-1\}$　　**18.** $\{2\}$　　**19.** $\{2\}$

20. $\left\{\dfrac{72}{11}\right\}$　　**21.** $\{-12\}$　　**22.** $\left\{\dfrac{77}{15}\right\}$　　**23. a.** 0　**b.** $\{2\}$　　**24. a.** 5　　**b.** $\varnothing$　　**25. a.** $-1, 1$　　**b.** all real numbers except 1 and -1

26. a. $-2, 4$　**b.** $\{7\}$　　**27.** inconsistent equation　　**28.** identity　　**29.** conditional equation　　**30.** 2005　　**31.** 2000
32. low $= \$174$ thousand, middle $= \$237$ thousand, high $= \$345$ thousand　　**33.** 9 years; 2009　　**34.** 500 min

35. $\$6250$ at 8%; $\$3750$ at 12%　　**36.** length $= 120$ m; width $= 53$ m　　**37.** 20 times　　**38.** $\$10,000$　　**39.** 95 concerts　　**40.** $h = \dfrac{3V}{B}$

41. $M = \dfrac{f - F}{f}$　　**42.** $g = \dfrac{T}{r + vt}$　　**43.** $-9 + 4i$　　**44.** $-12 - 8i$　　**45.** $29 + 11i$　　**46.** $-7 - 24i$　　**47.** 113　　**48.** $\dfrac{15}{13} - \dfrac{3}{13}i$

49. $\dfrac{1}{5} + \dfrac{11}{10}i$　　**50.** $i\sqrt{2}$　　**51.** $-96 - 40i$　　**52.** $2 + i\sqrt{2}$　　**53.** $\left\{-8, \dfrac{1}{2}\right\}$　　**54.** $\{-4, 0\}$　　**55.** $\{-8, 8\}$　　**56.** $\left\{\dfrac{4 + 3\sqrt{2}}{3}, \dfrac{4 - 3\sqrt{2}}{3}\right\}$

57. 100; $(x + 10)^2$　　**58.** $\dfrac{9}{4}$; $\left(x - \dfrac{3}{2}\right)^2$　　**59.** $\{3, 9\}$　　**60.** $\left\{2 + \dfrac{\sqrt{3}}{3}, 2 - \dfrac{\sqrt{3}}{3}\right\}$　　**61.** $\{1 + \sqrt{5}, 1 - \sqrt{5}\}$

62. $\{1 + 3i\sqrt{2}, 1 - 3i\sqrt{2}\}$　　**63.** $\left\{\dfrac{-2 + \sqrt{10}}{2}, \dfrac{-2 - \sqrt{10}}{2}\right\}$　　**64.** -36; 2 complex imaginary solutions　　**65.** 81; 2 unequal real solutions

66. $\left\{\dfrac{1}{2}, 5\right\}$　　**67.** $\left\{-2, \dfrac{10}{3}\right\}$　　**68.** $\left\{\dfrac{7 + \sqrt{37}}{6}, \dfrac{7 - \sqrt{37}}{6}\right\}$　　**69.** $\{-3, 3\}$　　**70.** $\{-2, 8\}$　　**71.** $\left\{\dfrac{1}{6} + i\dfrac{\sqrt{23}}{6}, \dfrac{1}{6} - i\dfrac{\sqrt{23}}{6}\right\}$

72. 20 weeks　　**73.** 10 years　　**74.** $(10, 7250)$　　**75.** length $= 5$ yd; width $= 3$ yd　　**76.** approximately 134 m　　**77.** $\{-5, 0, 5\}$

78. $\left\{-3, \dfrac{1}{2}, 3\right\}$　　**79.** $\{2\}$　　**80.** $\{8\}$　　**81.** $\{16\}$　　**82.** $\{132\}$　　**83.** $\{-2, -1, 1, 2\}$　　**84.** $\{16\}$　　**85.** $\{-4, 3\}$　　**86.** $\{-5, 11\}$

87. $\left\{-1, -\dfrac{2\sqrt{6}}{9}, \dfrac{2\sqrt{6}}{9}, 1\right\}$　　**88.** $\{2\}$　　**89.** $\{1, 4\}$　　**90.** $\{-3, -2, 3\}$　　**91.** 1250 ft

92. 　　**93.**　　**94.**　　**95.** $-2 < x \le 3$

96. $-1.5 \le x \le 2$

97. $x > -1$

98. $[-2, \infty)$

99. $\left[\dfrac{3}{5}, \infty\right)$

100. $\left(-\infty, -\dfrac{21}{2}\right)$

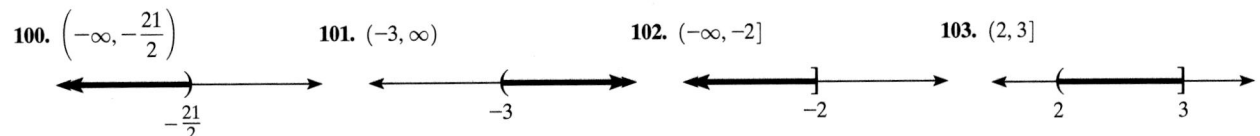

101. $(-3, \infty)$

102. $(-\infty, -2]$

103. $(2, 3]$

104. $[-9, 6]$

105. $(-\infty, -6)$ or $(0, \infty)$

106. $(-\infty, -3]$ or $[-2, \infty)$

107. Most people sleep between 5.5 and 7.5 hours.　**108.** between 50°F and 77°F inclusively　**109.** more than 50 checks
110. at least 93

111. $\left[-4, \dfrac{1}{2}\right]$

112. $\left(-\infty, \dfrac{3 - \sqrt{3}}{2}\right)$ or $\left(\dfrac{3 + \sqrt{3}}{2}, \infty\right)$

113. $(-\infty, -2)$ or $(6, \infty)$

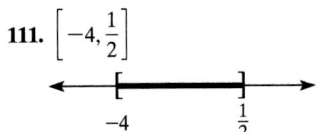

114. $(-\infty, 4)$ or $\left[\dfrac{23}{4}, \infty\right)$

115. from 1 to 2 sec

Chapter 1 Test

1

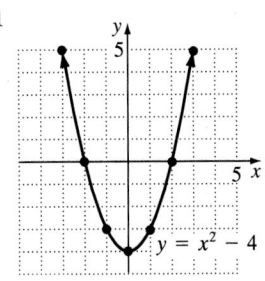

$y = x^2 - 4$

2. x-intercept: 2; y-intercept: 3　**3.** 1992; 7.8%

4. $\{-1\}$　**5** $\{-6\}$　**6.** $\{5\}$

7. $\left\{-\dfrac{1}{2}, 2\right\}$　**8.** $\left\{\dfrac{1 - 5\sqrt{3}}{3}, \dfrac{1 + 5\sqrt{3}}{3}\right\}$　**9.** $\{1 - \sqrt{5}, 1 + \sqrt{5}\}$

10. $\left\{1 + \dfrac{1}{2}i, 1 - \dfrac{1}{2}i\right\}$　**11.** $\{-1, 1, 4\}$　**12.** $\{7\}$　**13.** $\{5\}$

14. $\{\sqrt[3]{4}\}$　**15.** $\{1, 512\}$　**16.** $\{6, 12\}$

17. $(-\infty, 12]$

18. $\left[\dfrac{21}{8}, \infty\right)$

19. $\left[-7, \dfrac{13}{2}\right)$

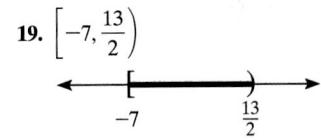

20. $\left(-\infty, -\dfrac{5}{3}\right]$ or $c\dfrac{1}{3}, \infty\right)$

21. $(-3, 4)$

22. $(3, 10)$

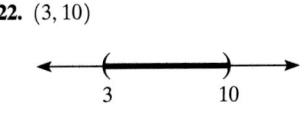

23. $47 + 16i$　**24.** $2 + i$　**25.** $38i$　**26.** $h = \dfrac{3V}{lw}$　**27.** $x = x_1 + \dfrac{y - y_1}{m}$　**28.** 2004; very well　**29.** 2007; very well
30. New York City: 55 days; Los Angeles: 213 days　**31.** 26 yr; \$33,600　**32.** \$3000 at 8%; \$7000 at 10%
33. length = 12 ft; width = 4 ft　**34.** 10 ft　**35.** \$47,500

CHAPTER 2

Section 2.1

Check Point Exercises

1. a. 6 **b.** $-\dfrac{7}{5}$ **2.** $y + 5 = 6(x - 2); y = 6x - 17$ **3.** $y + 1 = -5(x + 2); y = -5x - 11$

4. **5.** **6.** 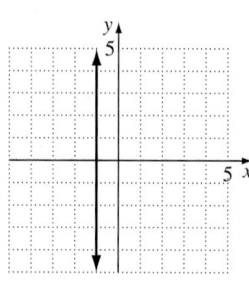 **7.** slope: $-\dfrac{1}{2}$; y-intercept: 2

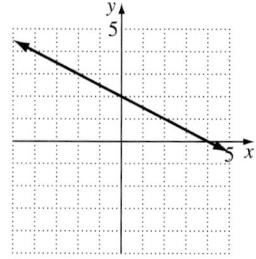

Exercise Set 2.1

1. $\dfrac{3}{4}$; rises **3.** $\dfrac{1}{4}$; rises **5.** 0; horizontal **7.** -5; falls **9.** undefined; vertical **11.** $y - 5 = 2(x - 3); y = 2x - 1$

13. $y - 5 = 6(x + 2); y = 6x + 17$ **15.** $y + 3 = -3(x + 2); y = -3x - 9$ **17.** $y - 0 = -4(x + 4); y = -4x - 16$

19. $y + 2 = -1\left(x + \dfrac{1}{2}\right); y = -x - \dfrac{5}{2}$ **21.** $y - 0 = \dfrac{1}{2}(x - 0); y = \dfrac{1}{2}x$ **23.** $y + 2 = -\dfrac{2}{3}(x - 6); y = -\dfrac{2}{3}x + 2$

25. using $(1, 2), y - 2 = 2(x - 1); y = 2x$ **27.** using $(-3, 0), y - 0 = 1(x + 3); y = x + 3$

29. using $(-3, -1), y + 1 = 1(x + 3); y = x + 2$ **31.** using $(-3, -2), y + 2 = \dfrac{4}{3}(x + 3); y = \dfrac{4}{3}x + 2$

33. using $(-3, -1), y + 1 = 0(x + 3); y = -1$ **35.** using $(2, 4), y - 4 = 1(x - 2); y = x + 2$

37. using $(0, 4), y - 4 = 8(x - 0); y = 8x + 4$

39. $m = 2; b = 1$ **41.** $m = -2; b = 1$ **43.** $m = \dfrac{3}{4}; b = -2$ **45.** $m = -\dfrac{3}{5}; b = 7$

47. **49.** **51.** 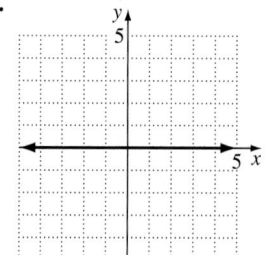 **53. a.** $y = -3x + 5$
b. $m = -3; b = 5$
c.

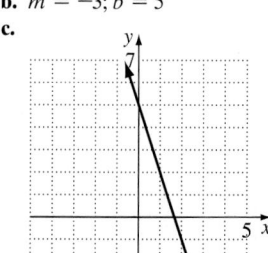

55. a. $y = -\dfrac{2}{3}x + 6$

    b. $m = -\dfrac{2}{3}; b = 6$

    c.

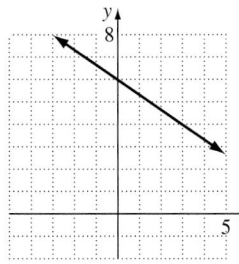

57. a. $y = 2x - 3$

    b. $m = 2; b = -3$

    c.

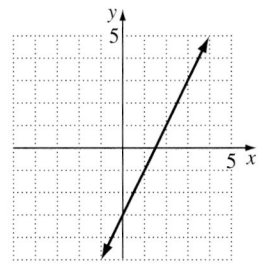

59. a. $x = 3$

    b. m is undefined; no y-intercept

    c.

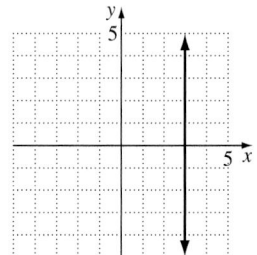

61. $y + 10 = -4(x + 8); y = -4x - 42$ **63.** $y + 3 = -5(x - 2); y = -5x + 7$ **65.** $y - 2 = \dfrac{2}{3}(x + 2); y = \dfrac{2}{3}x + \dfrac{10}{3}$

67. $y + 7 = -2(x - 4); y = -2x + 1$ **69.** $y = 15$ **71.** 111; The federal budget surplus is increasing \$111 billion each year.
73. a. 16, In 1950, there were 16 workers per Social Security beneficiary.   **b.** −0.24; The number of workers per Social Security beneficiary is decreasing by 0.24 workers each year.   **c.** $y = -0.24x + 16$ **d.** 1.6; 5   **75. a.** $y - 30 = 4(x - 2)$ **b.** $y = 4x + 22$
c. 74   **77.** $y = -2.3x + 255$, where x is the percentage of adult females who are literate and y is under-five mortality per thousand; For each percent increase is adult female literacy, under-five mortality decreases by 2.3 per thousand.   **79.** $y = -0.7x + 60$; y represents the percentage of U.S. adults who read a newspaper x years after 1995.   **81.** $y = -500x + 29{,}500$; 4500 shirts

93. $m = -3$

95. $m = \dfrac{3}{4}$

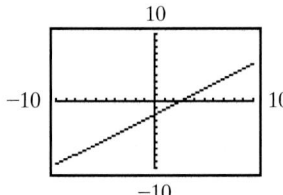

97. (c) is true.

99. a. m_1, m_3, m_2, m_4
    b. b_2, b_1, b_4, b_3

Section 2.2

Check Point Exercises

1. 5   **2.** $\left(4, -\dfrac{1}{2}\right)$ **3.** $x^2 + y^2 = 16$ **5.** center: $(-3, 1)$; radius: 2   **6.** $(x + 2)^2 + (y - 2)^2 = 9$
4. $(x - 5)^2 + (y + 6)^2 = 100$

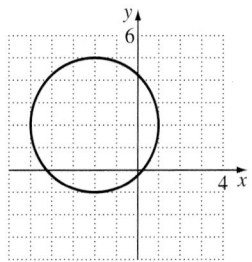

Exercise Set 2.2

1. 13   **3.** $2\sqrt{2} \approx 2.83$ **5.** 5   **7.** $\sqrt{29} \approx 5.39$ **9.** $4\sqrt{2} \approx 5.66$ **11.** $2\sqrt{5} \approx 4.47$ **13.** $2\sqrt{2} \approx 2.83$ **15.** $\sqrt{93} \approx 9.64$

17. $\sqrt{5} \approx 2.24$ **19.** $(4, 6)$ **21.** $(-4, -5)$ **23.** $\left(\dfrac{3}{2}, -6\right)$ **25.** $(-3, -2)$ **27.** $(1, 5\sqrt{5})$ **29.** $(2\sqrt{2}, 0)$ **31.** $x^2 + y^2 = 49$

33. $(x - 3)^2 + (y - 2)^2 = 25$ **35.** $(x + 1)^2 + (y - 4)^2 = 4$ **37.** $(x + 3)^2 + (y + 1)^2 = 3$ **39.** $(x + 4)^2 + (y - 0)^2 = 100$

41. center: $(0, 0)$
radius: 4

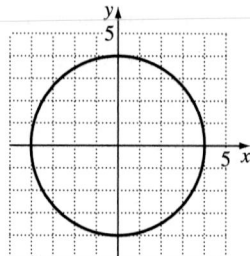

43. center: $(3, 1)$
radius: 6

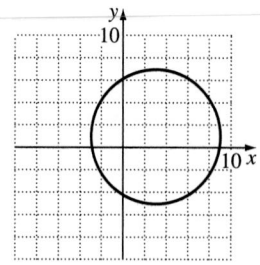

45. center: $(-3, 2)$
radius: 2

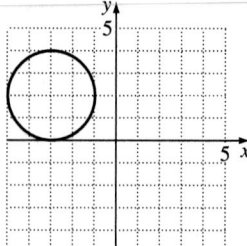

47. center: $(-2, -2)$
radius: 2

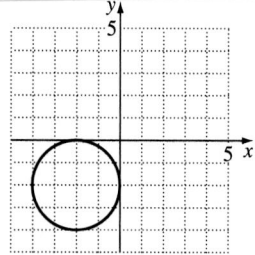

49. $(x + 3)^2 + (y + 1)^2 = 4$
center: $(-3, -1)$
radius: 2

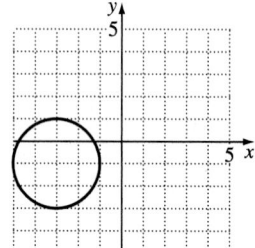

51. $(x - 5)^2 + (y - 3)^2 = 64$
center: $(5, 3)$
radius: 8

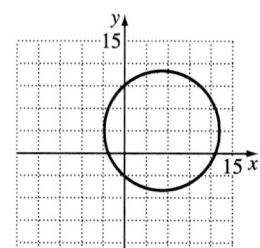

53. $(x + 4)^2 + (y - 1)^2 = 25$
center: $(-4, 1)$
radius: 5

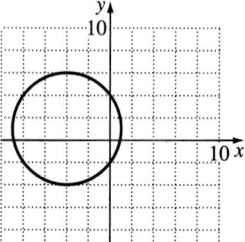

55. $(x - 1)^2 + (y - 0)^2 = 16$
center: $(1, 0)$
radius: 4

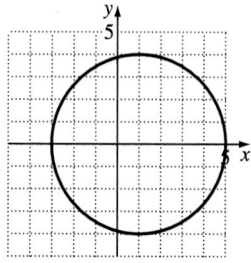

57. 0.5hr; 30 min   **59.** $x^2 + (y - 82)^2 = 4624$ **69.**

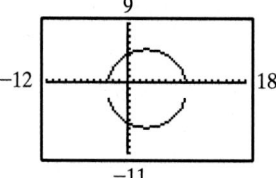

71. (d) is true.

73. a. Distance between (x_1, y_1) and $\left(\dfrac{x_1 + x_2}{2}, \dfrac{y_1 + y_2}{2}\right)$

$= \sqrt{\left(\dfrac{x_1 + x_2}{2} - x_1\right)^2 + \left(\dfrac{y_1 + y_2}{2} - y_1\right)^2}$

$= \sqrt{\left(\dfrac{x_1 + x_2 - 2x_1}{2}\right)^2 + \left(\dfrac{y_1 + y_2 - 2y_1}{2}\right)^2}$

$= \sqrt{\left(\dfrac{x_2 - x_1}{2}\right)^2 + \left(\dfrac{y_2 - y_1}{2}\right)^2}$

$= \sqrt{\dfrac{x_2^2 - 2x_1x_2 + x_1^2}{4} + \dfrac{y_2^2 - 2y_1y_2 + y_1^2}{4}}$

$= \sqrt{\dfrac{x_1^2 - 2x_1x_2 + x_2^2}{4} + \dfrac{y_1^2 - 2y_1y_2 + y_2^2}{4}}$

$= \sqrt{\left(\dfrac{x_1 - x_2}{2}\right)^2 + \left(\dfrac{y_1 - y_2}{2}\right)^2}$

$= \sqrt{\left(\dfrac{x_1 + x_2 - 2x_2}{2}\right)^2 + \left(\dfrac{y_1 + y_2 - 2y_2}{2}\right)^2}$

$= \sqrt{\left(\dfrac{x_1 + x_2}{2} - x_2\right)^2 + \left(\dfrac{y_1 + y_2}{2} - y_2\right)^2}$

$=$ Distance between (x_2, y_2) and $\left(\dfrac{x_1 + x_2}{2}, \dfrac{y_1 + y_2}{2}\right)$

b. $\sqrt{\left(\dfrac{x_2 - x_1}{2}\right)^2 + \left(\dfrac{y_2 - y_1}{2}\right)^2} + \sqrt{\left(\dfrac{x_2 - x_1}{2}\right)^2 + \left(\dfrac{y_2 - y_1}{2}\right)^2}$

$= 2\sqrt{\left(\dfrac{x_2 - x_1}{2}\right)^2 + \left(\dfrac{y_2 - y_1}{2}\right)^2}$

$= 2\sqrt{\dfrac{(x_2 - x_1)^2 + (y_2 - y_1)^2}{4}}$

$= \sqrt{(x_2 - x_1)^2 + (y_2 - y_1)^2}$

$=$ Distance from (x_1, y_1) to (x_2, y_2)

75. $(x + 3)^2 + (y - 2)^2 = 16$; $x^2 + y^2 + 6x - 4y - 3 = 0$ **77.** $y + 4 = \dfrac{3}{4}(x - 3)$

Section 2.3

Check Point Exercises

1. domain: {5, 10, 15, 20, 25}; range: {12.8, 16.2, 18.9, 20.7, 21.8}   **2. a.** not a function   **b.** function   **3. a.** $y = 6 - 2x$; function
b. $y = \pm\sqrt{1 - x^2}$, not a function   **4. a.** 42   **b.** $x^2 + 6x + 15$ **c.** $x^2 + 2x + 7$ **5. a.** $x^2 + 2hx + h^2 - 7x - 7h + 3$
b. $2x + h - 7$ **6. a.** 28   **b.** 33   **7. a.** $(-\infty, \infty)$ **b.** $\{x|x \neq -7, x \neq 7\}$ **c.** $[3, \infty)$.

Exercise Set 2.3

1. function; {1, 3, 5}; {2, 4, 5}   **3.** not a function; {3, 4}; {4, 5}   **5.** function; {−3, −2, −1, 0}; {−3, −2, −1, 0}
7. not a function; {1}; {4, 5, 6}   **9.** y is a function of x.   **11.** y is a function of x.   **13.** y is not a function of x.
15. y is not a function of x.   **17.** y is a function of x.   **19.** y is a function of x.   **21. a.** 29   **b.** $4x + 9$ **c.** $-4x + 5$
23. a. 2   **b.** $x^2 + 12x + 38$ **c.** $x^2 - 2x + 3$ **25. a.** 13   **b.** 1   **c.** $x^4 - x^2 + 1$ **d.** $81a^4 - 9a^2 + 1$ **27. a.** 3   **b.** 7
c. $\sqrt{x} + 3$ **29. a.** $\dfrac{15}{4}$ **b.** $\dfrac{15}{4}$ **c.** $\dfrac{4x^2 - 1}{x^2}$ **31. a.** 1   **b.** −1   **c.** 1   **33.** $4, h \neq 0$ **35.** $3, h \neq 0$ **37.** $2x + h, h \neq 0$

39. $2x + h - 4, h \neq 0$ **41.** $0, h \neq 0$ **43.** $-\dfrac{1}{x(x + h)}, h \neq 0$ **45. a.** −1   **b.** 7   **c.** 19   **47. a.** 3   **b.** 3   **c.** 0

49. a. 8   **b.** 3   **c.** 6   **51.** $(-\infty, \infty)$ **53.** $(-\infty, 4)$ or $(4, \infty)$ **55.** $(-\infty, -4)$ or $(-4, 4)$ or $(4, \infty)$
57. $(-\infty, -3)$ or $(-3, 7)$ or $(7, \infty)$ **59.** $(-\infty, -8)$ or $(-8, -3)$ or $(-3, \infty)$ **61.** $(-\infty, \infty)$ **63.** $[3, \infty)$ **65.** $(3, \infty)$
67. $[-7, \infty)$ **69.** $(-\infty, 12]$ **71.** $(-\infty, -2]$ or $[7, \infty)$ **73.** $[2, 5)$ or $(5, \infty)$
75. {(1, 31), (2, 53), (3, 70), (4, 86), (5, 86)]}; {1, 2, 3, 4, 5}; {31, 53, 70, 86}; Yes; Each member of the domain corresponds to exactly one
member of the range.   **77.** No; There is a member of the domain that corresponds to more than one member of the range.
79. 1713; There were 1713 gray wolves in the U.S. in 1990; Very well.   **81.** 19; Very well.   **83.** 5; Okay.
85. $f(0) = 200$; There were 200 thousand lawyers in the United States in 1951.   **87.** $f(50) = 1058$; There were 1058 thousand or
1,058,000 lawyers in the United States in the year 2001.   **89.** 8873; A person earning $40,000 owed $8873 in taxes.
91. $C = 100,000 + 100x$, where x is the number of bicycles produced; $C(90) = 109,000$; It cost $109,000 to produce 90 bicycles.

93. $T = \dfrac{40}{x} + \dfrac{40}{x + 30}$, where x is the rate on the outgoing trip; $T(30) = 2$; It takes 2 hours, traveling 30 mph outgoing and 60 mph returning.

103. $[1, \infty)$ **105.** $(-\infty, 5]$ **107.** Answers may vary.   **109.** $f(r_1) = 0$; r_1 is a solution to the equation $ax^2 + bx + c = 0$.

Section 2.4

Check Point Exercises

1. (−3, 7), (−2, 2), (−1, −1),
(0, −2), (1, −1), (2, 2), (3, 7)

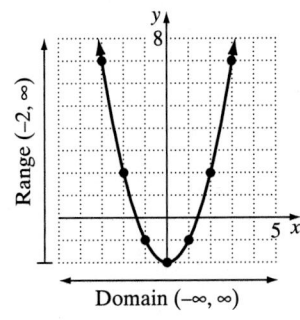

Range (−2, ∞)

Domain (−∞, ∞)

2. $f(4) = 1$; domain: $[0, 6)$; range: $(-2, 2]$
3. a. function
  b. function
  c. not a function
4. a. $f(10) \approx 16$ **b.** $x \approx 8$
5. increasing on $(-\infty, -1)$, decreasing on $(-1, 1)$, increasing on $(1, \infty)$
6. a. 1
  b. 7
  c. 4
7. a. even
  b. odd
  c. neither

Exercise Set 2.4

1. $(-3, 11), (-2, 6), (-1, 3),$
$(0, 2), (1, 3), (2, 6), (3, 11)$ **3.** $(0, -1), (1, 0), (4, 1), (9, 2)$ **5.** $(1, 0), (2, 1), (5, 2), (10, 3)$ **7.** $(-3, 2), (-2, 1), (-1, 0),$
$(0, -1), (1, 0), (2, 1), (3, 2)$

   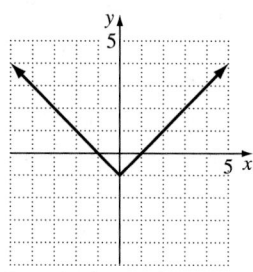

Domain: $(-\infty, \infty)$
Range: $[2, \infty)$

Domain: $[0, \infty)$
Range: $[-1, \infty)$

Domain: $[1, \infty)$
Range: $[0, \infty)$

Domain: $(-\infty, \infty)$
Range: $[-1, \infty)$

9. $(-3, 4), (-2, 3), (-1, 2),$
$(0, 1), (1, 0), (2, 1), (3, 2)$ **11.** $(-3, 5), (-2, 5), (-1, 5),$
$(0, 5), (1, 5), (2, 5), (3, 5)$ **13.** $(-2, -10), (-1, -3),$
$(0, -2), (1, -1), (2, 6)$

  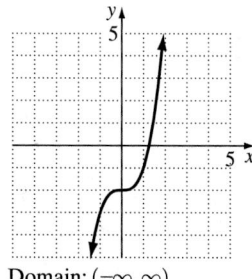

Domain: $(-\infty, \infty)$
Range: $[0, \infty)$

Domain: $(-\infty, \infty)$
Range: $\{5\}$

Domain: $(-\infty, \infty)$
Range: $(-\infty, \infty)$

15. a. $(-\infty, \infty)$ **b.** $[-4, \infty)$ **c.** -3 and 1 **d.** -3 **17. a.** $(-\infty, \infty)$ **b.** $[1, \infty)$ **c.** none **d.** 1
e. $f(-1) = 2$ and $f(3) = 4$ **19. a.** $[0, 5)$ **b.** $[-1, 5)$ **c.** 2 **d.** -1 **e.** $f(3) = 1$ **21. a.** $[0, \infty)$ **b.** $[1, \infty)$ **c.** none
d. 1 **e.** $f(4) = 3$ **23. a.** $[-2, 6]$ **b.** $[-2, 6]$ **c.** 4 **d.** 4 **e.** $f(-1) = 5$ **25. a.** $(-\infty, \infty)$ **b.** $(-\infty, -2]$ **c.** none
d. -2 **e.** $f(-4) = -5$ and $f(4) = -2$ **27. a.** $(-\infty, \infty)$ **b.** $(0, \infty)$ **c.** none **d.** 1 **29. a.** $\{-5, -2, 0, 1, 3\}$ **b.** $\{2\}$
c. none **d.** 2 **31.** function **33.** function **35.** not a function **37.** function **39. a.** increasing: $(-1, \infty)$
b. decreasing: $(-\infty, -1)$ **c.** constant: none **41. a.** increasing: $(0, \infty)$ **b.** decreasing: none **c.** constant: none
43. a. increasing: none **b.** decreasing: $(-2, 6)$ **c.** constant: none **45. a.** increasing: $(-\infty, -1)$ **b.** decreasing: none
c. constant: $(-1, \infty)$ **47. a.** increasing: $(-\infty, 0)$ or $(1.5, 3)$ **b.** decreasing: $(0, 1.5)$ or $(3, \infty)$ **c.** constant: none
49. a. increasing: $(-2, 4)$ **b.** decreasing: none **c.** constant: $(-\infty, -2)$ or $(4, \infty)$ **51. a.** $0; f(0) = 4$

b. $-3, 3; f(-3) = f(3) = 0$ **53. a.** $-2; f(-2) = 21$ **b.** $1; f(1) = -6$ **55.** 3 **57.** 10 **59.** $\dfrac{1}{5}$ **61.** odd **63.** neither

65. even **67.** even **69.** even **71.** odd **73.** even **75.** odd

77. $f(1.06) = 1$ **79.** $f\left(\dfrac{1}{3}\right) = 0$ **81.** $f(-2.3) = -3$ **83.** $f(60) \approx 3.1$; In 1960, Jewish Americans made up about 3.1% of the U.S.
population. **85.** $x \approx 19$ and $x \approx 64$; In 1919 and 1964, Jewish Americans made up about 3% of the U.S. population. **87.** 1940; 3.7%
89. Each year corresponds to only one percentage. **91.** increasing: $(45, 74)$; decreasing: $(16, 45)$; The number of accidents occurring per
50,000 miles driven increases with age starting at age 45, while it decreases with age starting at age 16. **93.** Answers will vary; an example
is 16 and 74 years old. For those ages, the number of accidents is 526.4 per 50 million miles.
95.

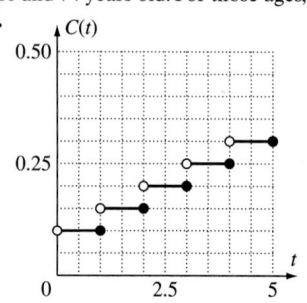

97. Answers may vary.
109. a.

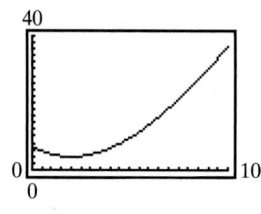

The number of doctor visits decreases during childhood
and then increases as you get older. The minimum is
$(20.29, 3.99)$, which means that the minimum number of
annual doctor visits, about 4, occurs at around age 20.

111.

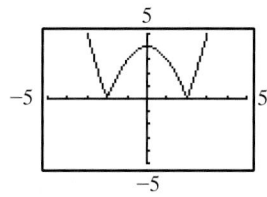

Increasing: $(-2, 0)$ or $(2, \infty)$
Decreasing: $(-\infty, -2)$ or $(0, 2)$

113.

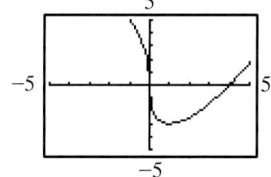

Increasing: $(1, \infty)$
Decreasing: $(-\infty, 1)$

115.

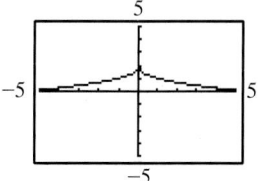

Increasing: $(-\infty, 0)$
Decreasing: $(0, \infty)$

117. (c) is true.
119. Answers may vary.

121.

Weight at least	Cost
0 oz.	$0.37
1	0.60
2	0.83
3	1.06
4	1.29

Section 2.5

Check Point Exercises

1.

2.

3.

4.

5.

6.

7.

8.

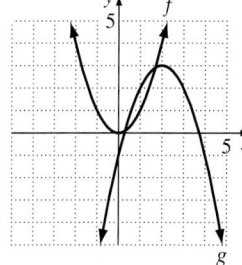

Exercise Set 2.5

1.

3.

5.

7.

9.

11.

13.

15.

17.

19.

21.

23.

25.

27.

29.

31.

33.

35.

37.

39.

41.

43.

45.

47.   **49.**   **51.** 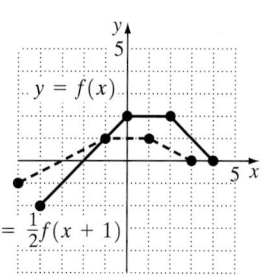   **53.** $y = \sqrt{x - 2}$
55. $y = (x + 1)^2 - 4$

57. a. First, vertically stretch the graph of $f(x) = \sqrt{x}$ by the factor 2.9; then, vertically shift the result up 20.1 units.
b. 40.2 in.; Very well.   **c.** 0.9 in. per month   **d.** 0.2 in. per month; This is a much smaller rate of change; The graph is not as steep between 50 and 60 as it is between 0 and 10.

65. a.   **b.** 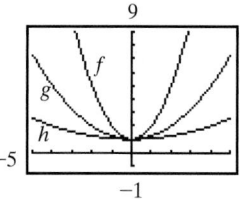   **c.** Answers may vary.   **67.** $g(x) = -(x + 4)^2$
d. Answers may vary.   **69.** $g(x) = -\sqrt{x - 2} + 2$
e. Answers may vary.   **71.** $(-a, b)$
73. $(a + 3, b)$

Section 2.6

Check Point Exercises

1. a. $(f + g)(x) = 3x^2 + 6x + 6$ **b.** $(f + g)(4) = 78$ **2. a.** $(f + g)(x) = \sqrt{x - 3} + \sqrt{x + 1}$ **b.** $[3, \infty)$

3. a. $(f - g)(x) = -x^2 + x - 4$ **b.** $(fg)(x) = x^3 - 5x^2 - x + 5$ \ **5. a.** $(f \circ g)(x) = \dfrac{4x}{1 + 2x}$ **b.** $\left\{ x \middle| x \ne \text{ and } x \ne -\dfrac{1}{2} \right\}$
6. $f(x) = \sqrt{x}$ and $g(x) = x^2 + 5$, then $h(x) = (f \circ g)(x)$

Exercise Set 2.6

1. a. $(f + g)(x) = 2x^2 + 3x + 2$ **b.** $(f + g)(4) = 46$ **3. a.** $(f + g)(x) = \sqrt{x - 6} + \sqrt{x + 2}$ **b.** Domain: $[6, \infty)$
5. $(f + g)(x) = 3x + 2$; Domain: $(-\infty, \infty)$; $(f - g)(x) = x + 4$; Domain: $(-\infty, \infty)$; $(fg)(x) = 2x^2 + x - 3$;
Domain: $(-\infty, \infty)$; $\left(\dfrac{f}{g}\right)(x) = \dfrac{2x + 3}{x - 1}$; Domain: $(-\infty, 1)$ or $(1, \infty)$ **7.** $(f + g)(x) = 3x^2 + x - 5$;
Domain: $(-\infty, \infty)$; $(f - g)(x) = -3x^2 + x - 5$; Domain: $(-\infty, \infty)$; $(fg)(x) = 3x^3 - 15x^2$; Domain: $(-\infty, \infty)$; $\left(\dfrac{f}{g}\right)(x) = \dfrac{x - 5}{3x^2}$;
Domain: $(-\infty, 0)$ or $(0, \infty)$ **9.** $(f + g)(x) = 2x^2 - 2$; Domain: $(-\infty, \infty)$; $(f - g)(x) = 2x^2 - 2x - 4$;
Domain: $(-\infty, \infty)$; $(fg)(x) = 2x^3 + x^2 - 4x - 3$; Domain: $(-\infty, \infty)$; $\left(\dfrac{f}{g}\right)(x) = 2x - 3$; Domain: $(-\infty, -1)$ or $(-1, \infty)$
11. $(f + g)(x) = \sqrt{x} + x - 4$; Domain: $[0, \infty)$; $(f - g)(x) = \sqrt{x} - x + 4$; Domain: $[0, \infty)$; $(fg)(x) = \sqrt{x}(x - 4)$;
Domain: $[0, \infty)$; $\left(\dfrac{f}{g}\right)(x) = \dfrac{\sqrt{x}}{x - 4}$; Domain: $[0, 4)$ or $(4, \infty)$ **13.** $(f + g)(x) = \dfrac{2x + 2}{x}$; Domain: $(-\infty, 0)$ or $(0, \infty)$; $(f - g)(x) = 2$;
Domain: $(-\infty, 0)$ or $(0, \infty)$; $(fg)(x) = \dfrac{2x + 1}{x^2}$; Domain: $(-\infty, 0)$ or $(0, \infty)$; $\left(\dfrac{f}{g}\right)(x) = 2x + 1$; Domain: $(-\infty, 0)$ or $(0, \infty)$
15. $(f + g)(x) = \sqrt{x + 4} + \sqrt{x - 1}$; Domain: $[1, \infty)$; $(f - g)(x) = \sqrt{x + 4} - \sqrt{x - 1}$; Domain: $[1, \infty)$; $(fg)(x) = \sqrt{x^2 + 3x - 4}$;
Domain: $[1, \infty)$; $\left(\dfrac{f}{g}\right)(x) = \dfrac{\sqrt{x + 4}}{\sqrt{x - 1}}$; Domain: $(1, \infty)$ **17. a.** $(f \circ g)(x) = 2x + 14$ **b.** $(g \circ f)(x) = 2x + 7$ **c.** $(f \circ g)(2) = 18$
19. a. $(f \circ g)(x) = 2x + 5$ **b.** $(g \circ f)(x) = 2x + 9$ **c.** $(f \circ g)(2) = 9$ **21. a.** $(f \circ g)(x) = 20x^2 - 11$
b. $(g \circ f)(x) = 80x^2 - 120x + 43$ **c.** $(f \circ g)(2) = 69$ **23. a.** $(f \circ g)(x) = x^4 - 4x^2 + 6$ **b.** $(g \circ f)(x) = x^4 + 4x^2 + 2$
c. $(f \circ g)(2) = 6$ **25. a.** $(f \circ g)(x) = \sqrt{x - 1}$ **b.** $(g \circ f)(x) = \sqrt{x} - 1$ **c.** $(f \circ g)(2) = 1$ **27. a.** $(f \circ g)(x) = x$
b. $(g \circ f)(x) = x$ **c.** $(f \circ g)(2) = 2$ **29. a.** $(f \circ g)(x) = \dfrac{2x}{1 + 3x}$ **b.** $\left\{ x \middle| x \ne 0 \text{ and } x \ne -\dfrac{1}{3} \right\}$ **31. a.** $(f \circ g)(x) = \dfrac{4}{4 + x}$
b. $\{x | x \ne 0 \text{ and } x \ne -4\}$ **33. a.** $(f \circ g)(x) = \sqrt{x + 3}$ **b.** $\{x | x \ge 0\}$ **35. a.** $(f \circ g)(x) = 5 - x$ **b.** $\{x | x \le 1\}$

37. a. $(f \circ g)(x) = 8 - x^2$ **b.** $\{x | x < -2 \text{ or } x > 2\}$ **39.** $f(x) = x^4, g(x) = 3x - 1$ **41.** $f(x) = \sqrt[3]{x}, g(x) = x^2 - 9$

43. $f(x) = |x|, g(x) = 2x - 5$ **45.** $f(x) = \dfrac{1}{x}, g(x) = 2x - 3$ **47.** 0   **49.** 10   **51.** 2   **53.** 0   **55.** 4   **57.** -6 **59.** 20; In

2000, veterinary costs in the U.S. for dogs and cats were about \$20 billion.   **61.** Domain of $D + C = \{1983, 1987, 1991, 1996, 2000\}$

63. $f + g$ represents the total world population in year x.   **65.** $(f + g)(2000) \approx 6$ billion people   **67.** $(R - C)(20,000) = -200,000$;

The company lost \$200,000 since costs exceeded revenues; $(R - C)(30,000) = 0$; The company broke even since revenues equaled cost;

$(R - C)(40,000) = 200,000$; The company made a profit of \$200,000.   **69. a.** f gives the price of the computer after a \$400 discount. g

gives the price of the computer after a 25% discount.   **b.** $(f \circ g)(x) = 0.75x - 400$. This models the price of a computer after first a 25%

discount and then a \$400 discount.   **c.** $(g \circ f)(x) = 0.75(x - 400)$. This models the price of a computer after first a \$400 discount and

then a 25% discount.   **d.** The function $f \circ g$ models the greater discount, since the 25% discount is taken on the regular price first.

77.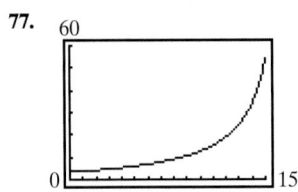

The per capita costs are increasing over time.

79.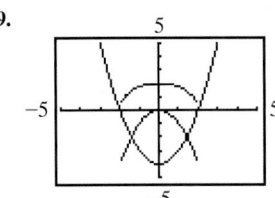

Domain of $f \circ g$ is $[-2, 2]$.

81. Assume f and g are even; then $f(-x) = f(x)$ and $g(-x) = g(x)$. $(fg)(-x) = f(-x) \cdot g(-x) = f(x) \cdot g(x)$ $= (fg)(x)$, so fg is even.

83.

Section 2.7

Check Point Exercises

1. $f(g(x)) = x; g(f(x)) = x; f$ and g are inverses.
2. $f(g(x)) = x; g(f(x)) = x; f$ and g are inverses.

3. $f^{-1}(x) = \dfrac{x - 7}{2}$

4. $f^{-1}(x) = \sqrt[3]{\dfrac{x + 1}{4}}$

5. (b) and (c) have inverse functions.

6.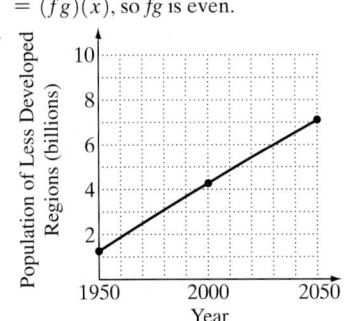

Exercise Set 2.7

1. $f(g(x)) = x; g(f(x)) = x; f$ and g are inverses.   **3.** $f(g(x)) = x; g(f(x)) = x; f$ and g are inverses.

5. $f(g(x)) = \dfrac{5x - 56}{9}; g(f(x)) = \dfrac{5x - 4}{9}; f$ and g are not inverses.   **7.** $f(g(x)) = x; g(f(x)) = x; f$ and g are inverses.

9. $f(g(x)) = x; g(f(x)) = x; f$ and g are inverses.   **11.** $f^{-1}(x) = x - 3$ **13.** $f^{-1}(x) = \dfrac{x}{2}$ **15.** $f^{-1}(x) = \dfrac{x - 3}{2}$

17. $f^{-1}(x) = \sqrt[3]{x - 2}$ **19.** $f^{-1}(x) = \sqrt[3]{x} - 2$ **21.** $f^{-1}(x) = \dfrac{1}{x}$ **23.** $f^{-1}(x) = x^2, x \geq 0$ **25.** $f^{-1}(x) = \sqrt{x - 1}; x \geq 1$

27. $f^{-1}(x) = \dfrac{3x + 1}{x - 2}; x \neq 2$ **29.** $f^{-1}(x) = (x - 3)^3 + 4$

31. The function is not one-to-one, so it does not have an inverse function.

33. The function is not one-to-one, so it does not have an inverse function.

35. The function is one-to-one, so it does have an inverse function.

37.   **39.** 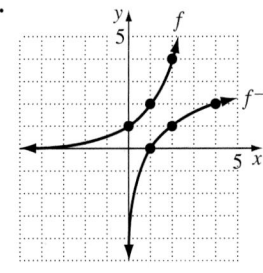 **41. a.** $f = \{(\text{Zambia}, -7.3), (\text{Colombia}, -4.5), (\text{Poland}, -2.8),$
$(\text{Italy}, -2.8), (\text{United States}, -1.8)\}$
b. $f^{-1} = \{(-7.3, \text{Zambia}), (-4.5, \text{Colombia}), (-2.8, \text{Poland}),$
$(-2.8, \text{Italy}), (-1.8, \text{United States})\}$; No; One member of the
domain, -2.8, corresponds to more than one member of the
range, Poland and Italy.

43. a. f is a one-to-one function. **b.** $f^{-1}(0.25)$ is the number of people in a room for a 25% probability of two people sharing a birthday.
$f^{-1}(0.5)$ is the number of people in a room for a 50% probability of two people sharing a birthday. $f^{-1}(0.7)$ is the number of people in a
room for a 70% probability of two people sharing a birthday.

45. $f(g(x)) = \dfrac{9}{5}\left[\dfrac{5}{9}(x - 32)\right] + 32 = x$ and $g(f(x)) = \dfrac{5}{9}\left[\left(\dfrac{9}{5}x + 32\right) - 32\right] = x$

53.   **55.**   **57.**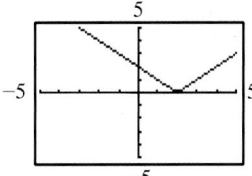

one-to-one   not one-to-one   not one-to-one

59.   **61.**   **63.**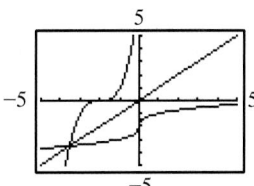

not one-to-one f and g are inverses. f and g are inverses.

65. $(f \circ g)^{-1}(x) = \dfrac{x - 15}{3}$; $(g^{-1} \circ f^{-1})(x) = \dfrac{x}{3} - 5 = \dfrac{x - 15}{3}$

67. No; The space craft was at the same height, $s(t)$, for two different values of t-once during the ascent and once again during the descent.

Chapter 2 Review Exercises

1. $m = -\dfrac{1}{2}$; falls   **2.** $m = 1$; rises   **3.** $m = 0$; horizontal   **4.** $m = $ undefined; vertical   **5.** $y - 2 = -6(x + 3)$; $y = -6x - 16$

6. using $(1, 6)$, $y - 6 = 2(x - 1)$; $y = 2x + 4$

7. Slope: $\dfrac{2}{5}$; y-intercept: -1 **8.** Slope: -4; y-intercept: 5 **9.** Slope: $-\dfrac{2}{3}$; y-intercept: -2 **10.** Slope: 0; y-intercept: 4

11. a. $y - 480 = 40(x - 2)$ **b.** $y = 40x + 400$ **c.** \$1200 billion.   **12. a.** Answers may vary.   **b.** Answers may vary.
c. Answers may vary.   **13.** $y + 7 = -3(x - 4)$; $y = -3x + 5$ **14.** $y - 6 = -3(x + 3)$; $y = -3x - 3$ **15.** 13

16. $2\sqrt{2} \approx 2.83$ **17.** $(-5, 5)$ **18.** $\left(\dfrac{11}{2}, -2\right)$ **19.** $x^2 + y^2 = 9$ **20.** $(x + 2)^2 + (y - 4)^2 = 36$

21. Center: $(0, 0)$; radius: 1

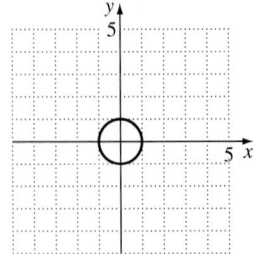

22. Center: $(-2, 3)$; radius: 3

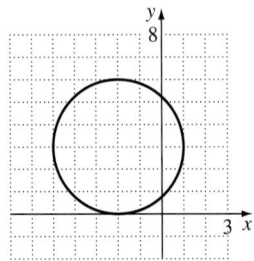

23. Center: $(2, -1)$; radius: 3

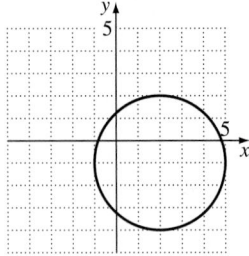

24. Function; Domain: $\{2, 3, 5\}$; Range: $\{7\}$ **25.** Function; Domain: $\{1, 2, 13\}$; Range: $\{10, 500, \pi\}$
26. Not a function; Domain: $\{12, 14\}$; Range: $\{13, 15, 19\}$ **27.** y is a function of x. **28.** y is a function of x.
29. y is not a function of x. **30. a.** $f(4) = -23$ **b.** $f(x + 3) = -7x - 16$ **c.** $f(-x) = 5 + 7x$
31. a. $g(0) = 2$ **b.** $g(-2) = 24$ **c.** $g(x - 1) = 3x^2 - 11x + 10$ **d.** $g(-x) = 3x^2 + 5x + 2$

32. a. $f(a) = 4a - 3$ **b.** $f(a + h) = 4a + 4h - 3$ **c.** $\dfrac{f(a + h) - f(a)}{h} = 4$ **d.** $f(a) + f(h) = 4a + 4h - 6$

33. a. $g(13) = 3$ **b.** $g(0) = 4$ **c.** $g(-3) = 7$ **34.** 8 **35.** $2x + h - 13$ **c.** $f(2) = 3$ **36.** $(-\infty, \infty)$
37. $(-\infty, 7)$ or $(7, \infty)$ **38.** $(-\infty, 4]$ **39.** $(-\infty, -1)$ or $(-1, 1)$ or $(1, \infty)$ **40.** $[2, 5)$ or $(5, \infty)$

41. Ordered pairs: $(-1, 9)$, $(0, 4)$, $(1, 1)$, $(2, 0)$, $(3, 1)$, $(4, 4)$.

Domain: $(-\infty, \infty)$
Range: $[0, \infty)$

42. Ordered pairs: $(-1, 3)$, $(0, 2)$, $(1, 1)$, $(2, 0)$, $(3, 1)$, $(4, 2)$.

Domain: $(-\infty, \infty)$
Range: $[0, \infty)$

43. a. Domain: $[-3, 5)$
b. Range: $[-5, 0]$
c. x-intercept: -3
d. y-intercept: -2
e. increasing: $(-2, 0)$ or $(3, 5)$
decreasing: $(-3, -2)$ or $(0, 3)$
f. $f(-2) = -3$ and $f(3) = -5$

44. a. Domain: $(-\infty, \infty)$
b. Range: $(-\infty, \infty)$
c. x-intercepts: -2 and 3
d. y-intercept: 3
e. increasing: $(-5, 0)$;
decreasing: $(-\infty, -5)$ or $(0, \infty)$
f. $f(-2) = 0$ and $f(6) = -3$

45. a. Domain: $(-\infty, \infty)$ **b.** Range: $[-2, 2]$ **c.** x-intercept: 0 **d.** y-intercept: 0 **e.** increasing: $(-2, 2)$; constant: $(-\infty, -2)$
or $(2, \infty)$ **f.** $f(-9) = -2$ and $f(14) = 2$ **46. a.** 0; $f(0) = -2$ **b.** $-2, 3; f(-2) = -3, f(3) = -5$
47. a. 0; $f(0) = 3$ **b.** $-5; f(-5) = -6$ **48.** not a function **49.** function **50.** function **51.** not a function **52.** 10
53. about $1167 **54.** odd; symmetric with respect to the origin **55.** even; symmetric with respect to the y-axis
56. odd; symmetric with respect to the origin **57. a.** yes; The graph passes the vertical line test. **b.** Decreasing: $(3, 12)$;
The vulture descended. **c.** Constant: $(0, 3)$ and $(12, 17)$; The vulture's height held steady during the first 3 seconds and the vulture
was on the ground for 5 seconds. **d.** Increasing: $(17, 30)$; The vulture was ascending.

58.

59.

60.

61.

62.

63.

64.

65.

66.

67.

68.

69.

70.

71.

72.

73.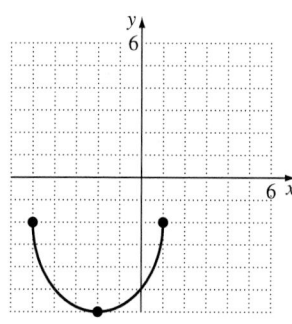

74. $(f + g)(x) = 4x - 6$; Domain: $(-\infty, \infty)$; $(f - g)(x) = 2x + 4$; Domain: $(-\infty, \infty)$;
$(fg)(x) = 3x^2 - 16x + 5$; Domain: $(-\infty, \infty)$; $\left(\dfrac{f}{g}\right)(x) = \dfrac{3x - 1}{x - 5}$; Domain: $(-\infty, 5)$ or $(5, \infty)$

75. $(f + g)(x) = 2x^2 + x$; Domain: $(-\infty, \infty)$; $(f - g)(x) = x + 2$; Domain: $(-\infty, \infty)$;
$(fg)(x) = x^4 + x^3 - x - 1$; Domain: $(-\infty, \infty)$; $\left(\dfrac{f}{g}\right)(x) = \dfrac{x^2 + x + 1}{x^2 - 1}$;
Domain: $(-\infty, -1)$ or $(-1, 1)$ or $(1, \infty)$

76. $(f + g)(x) = \sqrt{x + 7} + \sqrt{x - 2}$; Domain: $[2, \infty)$; $(f - g)(x) = \sqrt{x + 7} - \sqrt{x - 2}$;
Domain: $[2, \infty)$; $(fg)(x) = \sqrt{x^2 + 5x - 14}$; Domain: $[2, \infty)$; $\left(\dfrac{f}{g}\right)(x) = \dfrac{\sqrt{x + 7}}{\sqrt{x - 2}}$;
Domain: $(2, \infty)$

77. a. $(f \circ g)(x) = 16x^2 - 8x + 4$ **b.** $(g \circ f)(x) = 4x^2 + 11$ **c.** $(f \circ g)(3) = 124$ **78. a.** $(f \circ g)(x) = \sqrt{x + 1}$
b. $(g \circ f)(x) = \sqrt{x} + 1$ **c.** $(f \circ g)(3) = 2$ **79. a.** $\dfrac{1 + x}{1 - 2x}$ **b.** $\{x | x \neq 0 \text{ and } x \neq \frac{1}{2}\}$ **80. a.** $\sqrt{x + 2}$ **b.** $\{x | x \geq -2\}$

81. $f(x) = x^4$, $g(x) = x^2 + 2x - 1$ **82.** $f(x) = \sqrt[3]{x}$, $g(x) = 7x + 4$ **83.** $f(g(x)) = x - \dfrac{7}{10}$; $g(f(x)) = x - \dfrac{7}{6}$; f and g are not

inverses of each other.   **84.** $f(g(x)) = x$; $g(f(x)) = x$; f and g are inverses of each other.   **85.** $f^{-1}(x) = \dfrac{x + 3}{4}$

86. $f^{-1}(x) = x^2 - 2$ for $x \geq 0$ **87.** $f^{-1}(x) = \sqrt[3]{\dfrac{x - 1}{8}}$

88. Inverse function exists.
89. Inverse function does not exist.
90. Inverse function exists.
91. Inverse function does not exist.

92.

Chapter 2 Test

1. using $(2, 1)$, $y - 1 = 3(x - 2)$; $y = 3x - 5$ **2.** $y - 6 = 4(x + 4)$; $y = 4x + 22$ **3. a.** $y = 42x + 320$ **b.** \$866
4. Center: $(-2, 3)$; radius: 4   **5.** b, c, d   **6.** $f(x - 1) = x^2 - 4x + 8$ **7.** $g(-1) = 4$; $g(7) = 2$ **8.** Domain: $(-\infty, 4]$
9. $2x + h + 11$ **10. a.** $f(4) - f(-3) = 5$ **b.** Domain: $(-5, 6]$ **c.** Range: $[-4, 5]$ **d.** Increasing: $(-1, 2)$
e. Decreasing: $(-5, -1)$ or $(2, 6)$ **f.** 2; $f(2) = 5$ **g.** -1; $f(-1) = -4$ **h.** -4, 1, and 5   **i.** -3 **11.** 48   **12.** $f(x)$ is even and is
symmetric with respect to the y-axis. The graph in the figure is symmetric with respect to the origin.   **13.** The graph of f is shifted 3 to
the right to obtain the graph of g. Then the graph of g is stretched by a factor of 2 and reflected about the x-axis to obtain the graph of h.

14.

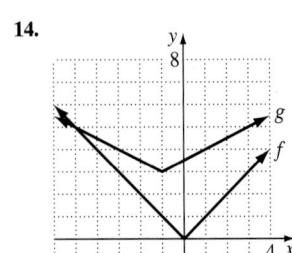

15. $(f - g)(x) = x^2 - 2x - 2$.   **16.** $\left(\dfrac{f}{g}\right)(x) = \dfrac{x^2 + 3x - 4}{5x - 2}$; Domain: $\left(-\infty, \dfrac{2}{5}\right)$ or $\left(\dfrac{2}{5}, \infty\right)$
17. $(f \circ g)(x) = 25x^2 - 5x - 6$ **18.** $(g \circ f)(x) = 5x^2 + 15x - 22$.
19. $f(g(2)) = 84$.   **20.** $\dfrac{7x}{2 - 4x}$; $\{x | x \neq 0 \text{ or } x \neq \frac{1}{2}\}$.   **21.** $f(x) = x^7$, $g(x) = 2x + 13$
22. $f^{-1}(x) = x^2 + 2$ for $x \geq 0$
23. a. The graph of f passes the Horizontal Line Test.
b. $f(80) = 2000$
c. $f^{-1}(2000)$ is the income, in thousands of dollars, for those who give \$2000 to charity.

24.

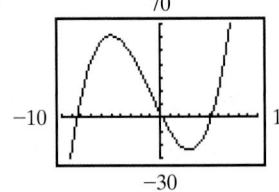

a. not one-to-one
b. neither; f is not symmetric about the origin or the y-axis
c. $(-\infty, \infty)$
d. Increasing: $(-\infty, -5)$ or $(3, \infty)$
e. Decreasing: $(-5, 3)$
f. -5; $f(-5) = \dfrac{184}{3}$
g. 3; $f(3) = -24$

Cumulative Review Exercises (Chapters P–2)

1. $\dfrac{2y^4}{x^3}$ **2.** $\dfrac{5\sqrt{2}}{8}$ **3.** $(x - 4)(x^2 + 2)$ **4.** $\dfrac{2x^2 - x + 6}{(x + 4)(x - 2)}$ **5.** $\dfrac{2x + 1}{2x - 1}$ **6.** $x = -4$ or $x = 5$ **7.** $x = \dfrac{25}{18}$ **8.** $x = 4$
9. $x = -8$ or $x = 27$ **10.** $x \leq 20$; $(-\infty, 20]$ **11.** $(-\infty, 2)$ or $[7, \infty)$ **12.** $y - 5 = 4(x + 2)$; $y = 4x + 13$

13.

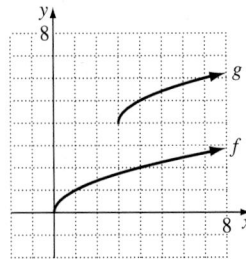

14. $f^{-1}(x) = (x - 2)^2 + 3$
15. $\dfrac{3 - (x + h)^2 - (3 - x^2)}{h} = -2x - h$
16. $c = \dfrac{Ad}{d - A}$
17. \$1500 at 7%, \$4500 at 9%
18. \$2000
19. 3 ft by 8 ft
20. You must make an 85 on the final exam to have an average score of 80.

CHAPTER 3

Section 3.1

Check Point Exercises

1.

2.

3.

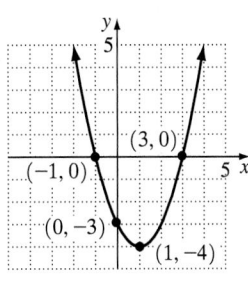

4. 45; 190

Exercise Set 3.1

1. $h(x) = (x-1)^2 + 1$ **3.** $j(x) = (x-1)^2 - 1$ **5.** $h(x) = x^2 - 1$ **7.** $g(x) = x^2 - 2x + 1$ **9.** $(3, 1)$ **11.** $(-1, 5)$
13. $(2, -5)$ **15.** $(-1, 9)$

17. Domain: $(-\infty, \infty)$
Range: $[-1, \infty)$
axis of symmetry: $x = 4$

19. Domain: $(-\infty, \infty)$
Range: $[2, \infty)$
axis of symmetry: $x = 1$

21. Domain: $(-\infty, \infty)$
Range: $[1, \infty)$
axis of symmetry: $x = 3$

23. Domain: $(-\infty, \infty)$
Range: $[-1, \infty)$
axis of symmetry: $x = -2$

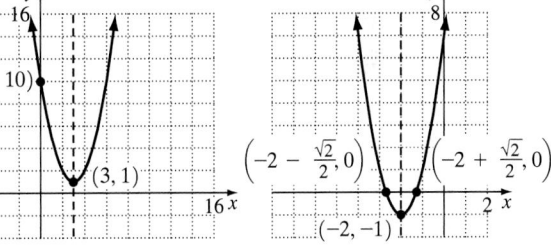

25. Domain: $(-\infty, \infty)$

Range: $(-\infty, 4]$

axis of symmetry: $x = 1$

27. Domain: $(-\infty, \infty)$

Range: $[-4, \infty)$

axis of symmetry: $x = 1$

29. Domain: $(-\infty, \infty)$

Range: $\left[-\dfrac{49}{4}, \infty\right)$

axis of symmetry: $x = -\dfrac{3}{2}$

31. Domain: $(-\infty, \infty)$

Range: $(-\infty, 4]$

axis of symmetry: $x = 1$

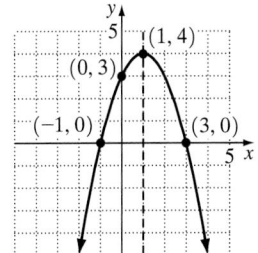

33. Domain: $(-\infty, \infty)$ **35.** minimum; $(2, -13)$ **37.** maximum; $(1, 1)$ **39.** minimum; $\left(\dfrac{1}{2}, -\dfrac{5}{4}\right)$

Range: $(-\infty, -1]$ **41.** 1968; 4238 cigarettes per person; Yes   **43.** 6.25 s; 629 ft

axis of symmetry: $x = 1$ **45.** The graph has the shape of a parabola.   **47.** 30 ft; 60 ft; 1800 ft^2 **49.** 5 in.

59. a.

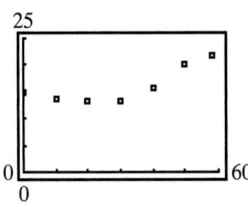

You can only see a little of the parabola.

b. $(20.5, -120.5)$ **61.** $(2.5, 185)$
c. Ymax $= 750$
d. You can choose Xmin and Xmax so the x-value of the vertex is in the center of the graph. Choose Ymin to include the y-value of the vertex.

63. $(-30, 91)$

65. a.

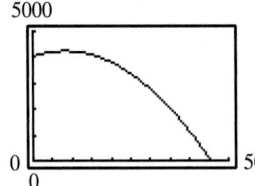

The data decrease and then increase.

b. $y = 0.01x^2 - 0.22x + 15.10$ **d.**
c. $x \approx 18$; $1940 + 18 = 1958$; worst year: 1958; fuel efficiency: about 13.1 mpg

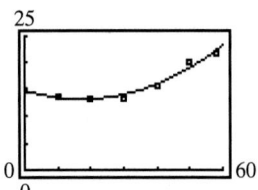

67. Answers may vary.   **69.** $x = 3$; $(0, 11)$

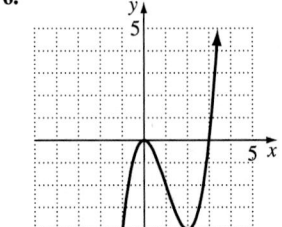

Section 3.2

Check Point Exercises

1. The graph rises to the left and to the right.
2. Since n is odd and the leading coefficient is negative, the function falls to the right. Since the ratio cannot be negative, the model won't be appropriate.
3. No; the graph should fall to the left, but doesn't appear to.
4. $\{-2, 2\}$
5. $\{-2, 0, 2\}$

6.

Exercise Set 3.2

1. polynomial function; degree: 3   **3.** polynomial function; degree: 5   **5.** not a polynomial function   **7.** not a polynomial function
9. not a polynomial function   **11.** polynomial function   **13.** not a polynomial function   **15.** (c)   **17.** (b)   **19.** (a)
21. falls to the left and rises to the right   **23.** rises to the left and to the right   **25.** falls to the left and to the right
27. $x = 5$ has multiplicity 1; The graph crosses the x-axis; $x = -4$ has multiplicity 2; The graph touches the x-axis and turns around.
29. $x = 3$ has multiplicity 1; The graph crosses the x-axis; $x = -6$ has multiplicity 3; The graph crosses the x-axis.
31. $x = 0$ has multiplicity 1; The graph crosses the x-axis; $x = 1$ has multiplicity 2; The graph touches the x-axis and turns around.
33. $x = 2, x = -2$ and $x = -7$ have multiplicity 1; The graph crosses the x-axis.

35. a. $f(x)$ rises to the right and falls to the left.
   b. $x = -2, x = 1, x = -1$; $f(x)$ crosses the x-axis at each.
   c. The y-intercept is -2..
   d. neither
   e.

37. a. $f(x)$ rises to the left and the right.
   b. $x = 0, x = 3, x = -3$; $f(x)$ crosses the x-axis at -3 and 3; $f(x)$ touches the x-axis at 0.
   c. The y-intercept is 0.
   d. y-axis symmetry
   e.

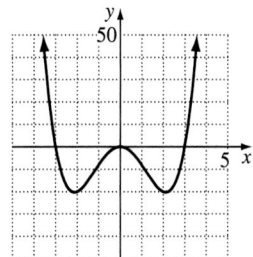

39. a. $f(x)$ falls to the left and the right.
   b. $x = 0, x = 4, x = -4$; $f(x)$ crosses the x-axis at -4 and 4; $f(x)$ touches the x-axis at 0.
   c. The y-intercept is 0.
   d. y-axis is symmetry
   e.

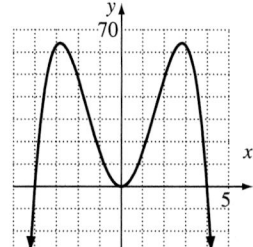

41. a. $f(x)$ rises to the left and the right.
   b. $x = 0, x = 1$; $f(x)$ touches the x-axis at 0 and 1.
   c. The y-intercept is 0.
   d. neither
   e.

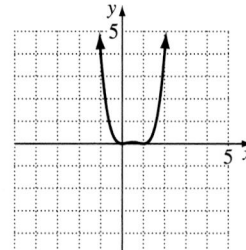

43. a. $f(x)$ falls to the left and the right.
   b. $x = 0, x = 2$; $f(x)$ crosses the x-axis at 0 and 2.
   c. The y-intercept is 0.
   d. neither
   e.

45. a. $f(x)$ rises to the left and falls to the right.
   b. $x = 0, x = \pm\sqrt{3}$; $f(x)$ crosses the x-axis at $(0, 0)$; $f(x)$ touches the x-axis at $\sqrt{3}$ and $-\sqrt{3}$.
   c. The y-intercept is 0.
   d. origin symmetry
   e.

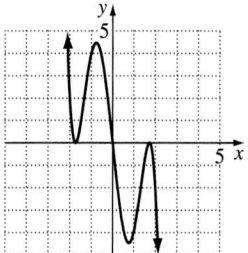

47. a. $f(x)$ rises to the left and falls to the right.
   b. $x = 0, x = 3$; $f(x)$ crosses the x-axis at 3; $f(x)$ touches the x-axis at 0.
   c. The y-intercept is 0.
   d. neither
   e.

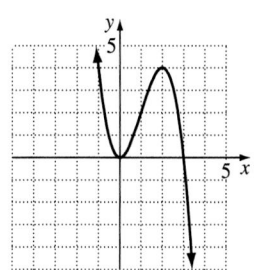

49. a. $f(x)$ falls to the left and the right.
   b. $x = 1, x = -2, x = 2$; $f(x)$ crosses the x-axis at -2 and 2; $f(x)$ touches the x-axis at 1.
   c. The y-intercept is 12.
   d. neither
   e.

51. a. Leading coefficient test suggests the elk population will decline and eventually will die off.

b.

c.

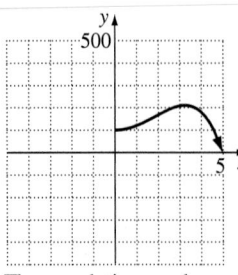

The population reaches extinction at the end of 5 years.

53. No; eventually the function would predict a negative number of larceny thefts, which is impossible.
55. degree 4; positive; the graph rises to the left and to the right
71. Answers may vary.
73. Answers may vary.

75.

77.

79.

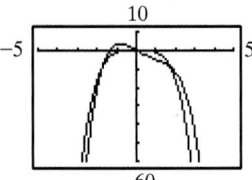

81. $f(x) = x^3 + x^2 - 12x$

Section 3.3

Check Point Exercises

1. $x + 5$ **2.** $2x^2 + 3x - 2 + \dfrac{1}{x - 3}$ **3.** $2x^2 + 7x + 14 + \dfrac{21x - 10}{x^2 - 2x}$ **4.** $x^2 - 2x - 3$ **5.** -105 **6.** $\left\{-1, -\dfrac{1}{3}, \dfrac{2}{5}\right\}$

Exercise Set 3.3

1. $x + 3$ **3.** $x^2 + 3x + 1$ **5.** $2x^2 + 3x + 5$ **7.** $4x + 3 + \dfrac{2}{3x - 2}$ **9.** $2x^2 + x + 6 - \dfrac{38}{x + 3}$

11. $4x^3 + 16x^2 + 60x + 246 + \dfrac{984}{x - 4}$ **13.** $2x + 5$ **15.** $6x^2 + 3x - 1 - \dfrac{3x - 1}{3x^2 + 1}$ **17.** $2x + 5$ **19.** $3x - 8 + \dfrac{20}{x + 5}$

21. $4x^2 + x + 4 + \dfrac{3}{x - 1}$ **23.** $6x^4 + 12x^3 + 22x^2 + 48x + 93 + \dfrac{187}{x - 2}$ **25.** $x^3 - 10x^2 + 51x - 260 + \dfrac{1300}{x + 5}$

27. $x^4 + x^3 + 2x^2 + 2x + 2$ **29.** $x^3 + 4x^2 + 16x + 64$ **31.** $2x^4 - 7x^3 + 15x^2 - 31x + 64 - \dfrac{129}{x + 2}$ **33.** -25 **35.** 4729

37. $x^2 - 5x + 6; x = -1, x = 2, x = 3$ **39.** $\left\{-\dfrac{1}{2}, 1, 2\right\}$ **41.** $\left\{-\dfrac{3}{2}, -\dfrac{1}{3}, \dfrac{1}{2}\right\}$ **43.** $x^3 + 5x^2 - 9x - 45$ **45. a.** 70

b. $80 + \dfrac{800}{x - 110}; f(30) = 70$; yes   **c.** No, f is a rational function because it is a quotient of two polynomials.

57.

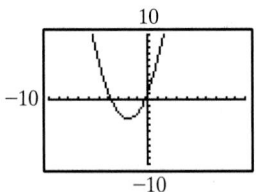

The division is correct.

59.

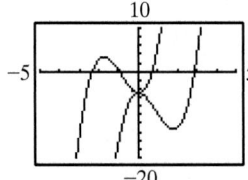

The division is not correct.
The right side should be 3x<-8x -5.

61. $k = -12$
63. $x^{2n} - x^n + 1$

Section 3.4

Check Point Exercises

1. $\pm 1, \pm 2, \pm 3, \pm 6$ **2.** $\pm 1, \pm 3, \pm\dfrac{1}{2}, \pm\dfrac{1}{4}, \pm\dfrac{3}{2}, \pm\dfrac{3}{4}$ **3.** $\{-5, -4, 1\}$ **4.** $\{1, 2 - 3i, 2 + 3i\}$

5. 4, 2, or 0 positive zeros, no possible negative zeros

Exercise Set 3.4

1. $\pm 1, \pm 2, \pm 4$ **3.** $\pm 1, \pm 2, \pm 3, \pm 6, \pm\dfrac{1}{3}, \pm\dfrac{2}{3}$ **5.** $\pm 1, \pm 2, \pm 3, \pm 6, \pm\dfrac{1}{2}, \pm\dfrac{1}{4}, \pm\dfrac{3}{2}, \pm\dfrac{3}{4}$ **7.** $\pm 1, \pm 2, \pm 3, \pm 4, \pm 6, \pm 12$ **9. a.** $\pm 1, \pm 2, \pm 4$

b. 2 is a zero   **c.** $\{2, -2, -1\}$ **11. a.** $\pm 1, \pm 2, \pm 3, \pm 6, \pm\dfrac{1}{2}, \pm\dfrac{3}{2}$ **b.** 3 is a zero   **c.** $\left\{3, \dfrac{1}{2}, -2\right\}$

13. a. $\pm 1, \pm 2, \pm 4, \pm 8, \pm\dfrac{1}{3}, \pm\dfrac{2}{3}, \pm\dfrac{4}{3}, \pm\dfrac{8}{3}$ **b.** 2 is a zero   **c.** $\left\{2, -\dfrac{1}{3}, -4\right\}$ **15. a.** $\pm 1, \pm 2, \pm 3, \pm 4, \pm 6, \pm 12$ **b.** 4 is a root

c. $\{-3, 1, 4\}$ **17. a.** $\pm 1, \pm 2, \pm 3, \pm 4, \pm 6, \pm 12$ **b.** -2 is a root   **c.** $\{-2, 1 + \sqrt{7}, 1 - \sqrt{7}\}$ **19. a.** $\pm 1, \pm 5, \pm\dfrac{1}{2}, \pm\dfrac{5}{2}, \pm\dfrac{1}{3}, \pm\dfrac{5}{3}, \pm\dfrac{1}{6}, \pm\dfrac{5}{6}$

b. -5 is a root   **c.** $\left\{-5, \dfrac{1}{2}, \dfrac{1}{3}\right\}$ **21. a.** $\pm 1, \pm 2, \pm 4$ **b.** 2 is a root   **c.** $\{-2, 2, 1 + \sqrt{2}, 1 - \sqrt{2}\}$

23. no positive real roots; 3 or 1 negative real roots   **25.** 3 or 1 positive real roots; no negative real roots

27. 2 or 0 positive real roots; 2 or 0 negative real roots   **29.** $x = -2, x = 5, x = 1$ **31.** $\left\{-\dfrac{1}{2}, \dfrac{1 + \sqrt{17}}{2}, \dfrac{1 - \sqrt{17}}{2}\right\}$

33. $\{-1, -2 + 2i, -2 - 2i\}$ **35.** $\{-1, -2, 3 + \sqrt{13}, 3 - \sqrt{13}\}$ **37.** $x = -1, x = 2, x = -\dfrac{1}{3}, x = 3$ **39.** $\left\{1, -\dfrac{3}{4}, i\sqrt{2}, -i\sqrt{2}\right\}$

41. $\left\{-2, \dfrac{1}{2}, \sqrt{2}, -\sqrt{2}\right\}$ **43. a.** $x = 40$; at age 40, about 27% of art productivity occurs   **b.** degree 2; leading coefficient: negative

45. $W = 3$ mm   **47.** 2 in. by 9 in. by 4 in.   **57.** $\dfrac{1}{2}, \dfrac{2}{3}, 2$ **59.** $\pm\dfrac{1}{2}$ **61.** 5, 3, or 1 positive real roots exist   **63.** (d) is true.   **65.** 3 in.

Section 3.5

Check Point Exercises

1.

2	2	11	-7	-6
		4	30	46
	2	15	23	40

All the numbers are nonnegative.

-7	2	11	-7	-6
		-14	21	-98
	2	-3	14	-104

The signs alternate.

2. $f(-3) = -42; f(-2) = 5$

3. $\{-3, 7, 2 + i, 2 - i\}$

4. a. $(x^2 - 5)(x^2 + 1)$

b. $(x + \sqrt{5})(x - \sqrt{5})(x^2 + 1)$

c. $(x + \sqrt{5})(x - \sqrt{5})(x + i)(x - i)$

5. $f(x) = x^3 + 3x^2 + x + 3$

Exercise Set 3.5

1.

-4	1	-5	11	33	-18
		-4	36	-188	620
	1	-9	47	-155	602

Since signs alternate, -4 is a lower bound.

7	1	-5	11	33	-18
		7	14	175	1456
	1	2	25	208	1438

Since no sign is negative, 7 is an upper bound.

3.

-4	2	5	-8	-7
		-8	12	-16
	2	-3	4	-23

Since signs alternate, -4 is a lower bound.

2	2	5	-8	-7
		4	18	20
	2	9	10	13

Since no sign is negative, 2 is an upper bound.

5. a. $\pm 1, \pm 2, \pm 3, \pm 4, \pm 6, \pm 12$ **b.** 1 is not a root. 1 is an upper bound.   **c.** Eliminate all positive possible rational roots.

d. -3 is not a root. -3 is a lower bound.   **e.** Eliminate $-3, -4, -6$ and -12.   **7.** $f(1) = -1; f(2) = 5; 1.3$

9. $f(-1) = -1; f(0) = 1; -0.5$ **11.** $f(-3) = -11; f(-2) = 1; -2.1$ **13.** $f(-3) = -42; f(-2) = 5; -2.2$ **15.** $\{-2i, 2i, 2\}$

17. $\left\{1 - i, 1 + i, \dfrac{1}{3}\right\}$ **19.** $\{2 - i, 2 + i, -2 + i, -2 - i\}$ **21.** $\{2 - i, 2 + i, -3, 7\}$ **23. a.** $(x^2 - 5)(x^2 + 4)$

b. $(x + \sqrt{5})(x - \sqrt{5})(x^2 + 4)$ **c.** $(x + \sqrt{5})(x - \sqrt{5})(x + 2i)(x - 2i)$ **25. a.** $(x^2 - 2)(x^2 + 3)$

b. $(x + \sqrt{2})(x - \sqrt{2})(x^2 + 3)$ **c.** $(x + \sqrt{2})(x - \sqrt{2})(x + i\sqrt{3})(x - i\sqrt{3})$ **27. a.** $(x - 3)(x + 1)(x^2 + 4)$

b. $(x - 3)(x + 1)(x^2 + 4)$ **c.** $(x - 3)(x + 1)(x + 2i)(x - 2i)$ **29.** $f(x) = 2x^3 - 2x^2 + 50x - 50$

31. $f(x) = x^3 - 3x^2 - 15x + 125$ **33.** $f(x) = x^4 + 10x^2 + 9$ **35.** $f(x) = x^4 - 9x^3 + 21x^2 + 21x - 130$

37. $x = 1; x = \pm 5i; f(x) = (x - 1)(x - 5i)(x + 5i)$ **39.** $x = 2; x = 3\pm 2i; f(x) = (x - 2)(x - 3 + 2i)(x - 3 - 2i)$

41. $x = \pm 6i; x = \pm i; f(x) = (x - 6i)(x + 6i)(x - i)(x + i)$

43. $x = -2; x = \dfrac{3}{4}; x = -\dfrac{1}{2} \pm i; f(x) = (x + 2)(4x - 3)(2x + 1 - 2i)(2x + 1 + 2i)$ **45.** ≈ 3 yr   **47.** Answers may vary.

49. Answers may vary.

55.

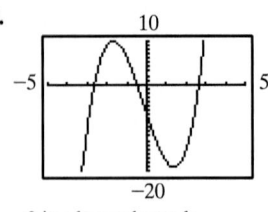

−3 is a lower bound;
3 is an upper bound

57. a. As x, a person's age, increases, y, the number of visits, increases.

b. 60

c. 40

59.

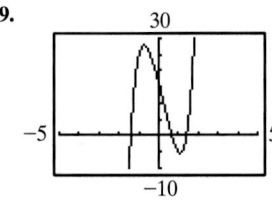

3 real zeros, 2 nonreal complex zeros

61.

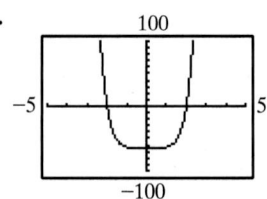

2 real zeros, 4 nonreal complex zeros

63. 3   **65.** 5   **67.** Answers may vary.

Section 3.6

Check Point Exercises

1. a. $\{x | x \neq 5\}$ **b.** $\{x | x \neq -5, x \neq 5\}$ **c.** all real numbers   **2. a.** $x = 1, x = -1$ **b.** $x = -1$ **c.** none

3. a. $y = 3$ **b.** $y = 0$ **c.** none

4.

5.

6.

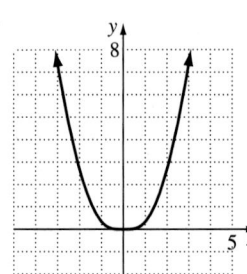

7. $y = 2x - 1$

8. a. $C(x) = 500x + 600,000$ **b.** $\overline{C}(x) = \dfrac{500x + 600,000}{x}$

c. $\overline{C}(1000) = 1100$, when 1000 new systems are produced, it costs \$1100 to produce each system; $\overline{C}(10,000) = 560$, when 10,000 new systems are produced, it costs \$560 to produce each system, $\overline{C}(100,000) = 506$, when 100,000 new systems are produced, it costs \$506 to produce each system.

d. $y = 500$; The cost per system approaches \$500 as more systems are produced.

Exercise Set 3.6

1. $\{x | x \neq 4\}$ **3.** $\{x | x \neq 5, x \neq -4\}$ **5.** $\{x | x \neq 7, x \neq -7\}$ **7.** All real numbers   **9.** $-\infty$ **11.** $-\infty$ **13.** 0 **15.** $+\infty$

17. $-\infty$ **19.** 1 **21.** $x = -4$ **23.** $x = 0, x = -4$ **25.** $x = -4$ **27.** no vertical asymptotes   **29.** $y = 0$ **31.** $y = 4$

33. no horizontal asymptote   **35.** $y = -\dfrac{2}{3}$

37.

39.

41.

43.

45.

47.

49.

51.

53.

55.

57.

59. a. Slant asymptote: $y = x$
b.
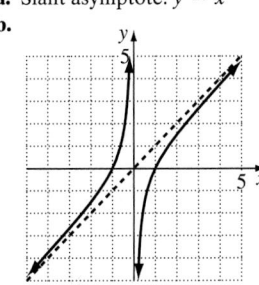

61. a. Slant asymptote: $y = x$
b.

63. a. Slant asymptote: $y = x + 4$
b.

65. a. Slant asymptote: $y = x - 2$
b.

67. a. $C(x) = 100x + 100,000$ **b.** $\overline{C}(x) = \dfrac{100x + 100,000}{x}$

   c. $\overline{C}(500) = 300$, when 500 bicycles are produced, it costs \$300 to produce each bicycle; $\overline{C}(1000) = 200$, when 1000 bicycles are produced, it costs \$200 to produce each bicycle; $\overline{C}(2000) = 150$, when 2000 bicycles are produced, it costs \$150 to produce each bicycle; $\overline{C}(4000) = 125$, when 4000 bicycles are produced, it costs \$125 to produce each bicycle.

   d. $y = 100$; The cost per bicycle approaches \$100 as more bicycles are produced.

69. a. $M(x) = \dfrac{190.9x + 2413.99}{0.234x + 12.54}$ **b.** 355.65; $M(19) \approx 355.65$ on the graph

c. $y = \dfrac{190.9}{0.234} \approx 816$; The cost of textbooks per college student approaches $816 as the years progress.

71. 90; An incidence ratio of 10 means 90% of the deaths are smoking related.

73. $y = 100$; The percentage of deaths cannot exceed 100% as the incidence ratios increase.

75. a. After 1 day: 35 words; after 5 days: about 12 words; after 15 days: about 7 words
  b. $N(1) = 35$ words; This is the same as the estimate from the graph.; $N(5) = 11$ words; This is a little less than the estimate from the graph.; $N(15) = 7$ words; This is the same as the estimate from the graph.
  c. The graph indicates the students will remember 5 words over a long period of time.
  d. $y = 5$; The horizontal asymptote indicates the students will remember 5 words over a long period of time.

87.

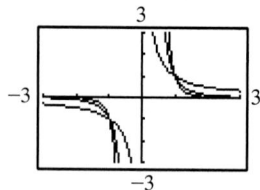

The graph approaches the horizontal asymptote faster and the vertical asymptote slower as n increases.

89.

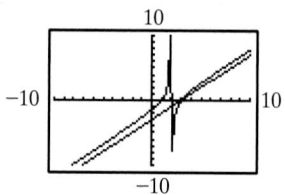

$g(x)$ is the graph of a line whereas $f(x)$ is the graph of a rational function with a slant asymptote; In $g(x)$, $x - 2$ is a factor of $x^2 - 5x + 6$.

91. (d) is true.   **93.** Answers may vary.   **95.** Answers may vary.

Section 3.7

Check Point Exercises

1. a. $L = kN$ **b.** $L = 4N$ **c.** 68 in.   **2. a.** $W = kL$ **b.** $k = \dfrac{75}{6}$ **c.** $W = \dfrac{75L}{6}$ **d.** 200 lb   **3.** 137.5 lb/in^2

4. about 556 ft   **5.** $26 per barrel   **6.** 24 min   **7.** 96π ft^3

Exercise Set 3.7

1. $g = kh$ **3.** $a = kb^2$ **5.** $r = \dfrac{k}{t}$ **7.** $a = \dfrac{k}{b^3}$ **9.** $r = \dfrac{ks}{v}$ **11.** $s = kgt^2$ **13.** $k = 25$ **15.** $k = 5$ **17.** $k = 5000$

19. $k = 3$ **21.** $k = 2$ **23.** 84   **25.** 25   **27.** $\dfrac{5}{6}$ **29.** 240   **31. a.** $G = kW$ **b.** $G = 0.02W$ **c.** 1.04 in.   **33.** $60

35. 2442 mph   **37.** 607 lb   **39.** 0.5 hr   **41.** 6.4 lb   **43.** 31.78; index: about 32; not in the desirable range

45. 11.11 foot-candles   **47.** 72 erg   **49.** The average number of phone calls is about 126.

59. The destructive power is four times as much.   **61.** Reduce the resistance by a factor of $\dfrac{1}{3}$.

Chapter 3 Review Exercises

1.

axis of symmetry: $x = 1$ axis of symmetry: $x = -4$ axis of symmetry: $x = 1$

2.

3.

4.

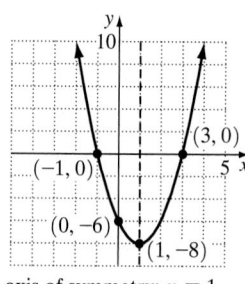

axis of symmetry: $x = 1$

5. after 2 seconds, the ball reaches a maximum height of 144 feet   **6.** $(20, 5.4)$; In 1980, the divorce rate reached a maximum of 5.4%.
7. 250 yd by 500 yd; maximum area is 125,000 yd^2 **8. c**   **9. b**   **10. a**   **11. d**   **12.** Because the degree is odd and the leading coefficient is negative, the graph falls to the right. Therefore, the model indicates that the percentage of families below the poverty level will

eventually be negative, which is impossible. **13.** Since the degree is even and the leading coefficient is negative, the graph falls to the right. Therefore, the model indicates a patient will eventually have a negative number of viral bodies, which is impossible.

14. $x = 1$, multiplicity 1, crosses; $x = -2$, multiplicity 2, touches; $x = -5$, multiplicity 3, crosses

15. $x = -5$, multiplicity 1, crosses; $x = 5$, multiplicity 2, touches

16. a. The graph falls to the left and rises to the right.
b. no symmetry
c.
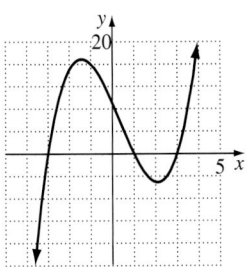

17. a. The graph rises to the left and falls to the right.
b. origin symmetry
c.

18. a. The graph falls to the left and rises to the right.
b. no symmetry
c.

19. a. The graph falls to the left and to the right.
b. y-axis symmetry
c.
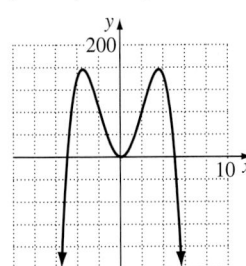

20. a. The graph falls to the left and to the right.
b. no symmetry
c.

21. a. The graph rises to the left and to the right.
b. no symmetry
c.
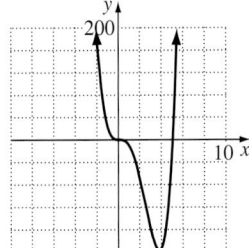

22. $4x^2 - 7x + 5 - \dfrac{4}{x + 1}$ **23.** $2x^2 - 4x + 1 - \dfrac{10}{5x - 3}$ **24.** $2x^2 + 3x - 1$ **25.** $3x^3 - 4x^2 + 7$

26. $3x^3 + 6x^2 + 10x + 10 + \dfrac{20}{x - 2}$ **27.** -5697 **28.** $2, \dfrac{1}{2}, -3$ **29.** $\{4, -2 \pm \sqrt{5}\}$ **30.** $\pm 1, \pm 5$

31. $\pm 1, \pm 2, \pm 4, \pm 8, \pm \dfrac{8}{3}, \pm \dfrac{4}{3}, \pm \dfrac{2}{3}, \pm \dfrac{1}{3}$ **32.** 2 or 0 positive solutions; no negative solutions

33. 3 or 1 positive real roots; 2 or 0 negative solutions   **34.** No sign variations exist for either $f(x)$ or $f(-x)$, so no real roots exist.

35. a. $\pm 1, \pm 2, \pm 4$ **b.** 1 positive real zero; 2 or no negative real zeros   **c.** 1 is a zero   **d.** $\{1, -2\}$

36. a. $\pm 1, \pm \dfrac{1}{2}, \pm \dfrac{1}{3}, \pm \dfrac{1}{6}$ **b.** 2 or 0 positive real zeros; 1 negative real zero   **c.** -1 is a zero   **d.** $\left\{-1, \dfrac{1}{3}, \dfrac{1}{2}\right\}$

37. a. $\pm 1, \pm 3, \pm 5, \pm 15, \pm \dfrac{1}{2}, \pm \dfrac{1}{4}, \pm \dfrac{1}{8}, \pm \dfrac{3}{2}, \pm \dfrac{3}{4}, \pm \dfrac{3}{8}, \pm \dfrac{5}{2}, \pm \dfrac{5}{4}, \pm \dfrac{5}{8}, \pm \dfrac{15}{2}, \pm \dfrac{15}{4}, \pm \dfrac{15}{8}$

b. 3 or 1 positive real solutions; no negative real solutions   **c.** $\dfrac{1}{2}$ is a zero   **d.** $\left\{\dfrac{1}{2}, \dfrac{3}{2}, \dfrac{5}{2}\right\}$

38. a. $\pm 1, \pm 2, \pm 3, \pm 6$ **b.** 2 or zero positive real solutions; 2 or zero negative real solutions   **c.** -2 is a zero   **d.** $\{-2, -1, 1, 3\}$

39. a. $\pm 1, \pm 2, \pm \dfrac{1}{2}, \pm \dfrac{1}{4}$ **b.** 1 positive real root; 1 negative real root   **c.** $\dfrac{1}{2}$ is a zero   **d.** $\left\{\dfrac{1}{2}, -\dfrac{1}{2}, i\sqrt{2}, -i\sqrt{2}\right\}$

40. a. $\pm 1, \pm 2, \pm 4, \pm \dfrac{1}{2}$ **b.** 2 or no positive zeros; 2 or no negative zeros   **c.** $x = 2$ is a zero   **d.** $\left\{2, -2, \dfrac{1}{2}, -1\right\}$

41.

$$
\begin{array}{r|rrrrr}
-2 & 2 & -7 & -5 & 28 & -12 \\
   &   & -4 & 22 & -34 & 12 \\
\hline
   & 2 & -11 & 17 & -6 & 0
\end{array}
$$

-2 is a root and a lower bound.

$$
\begin{array}{r|rrrrr}
6 & 2 & -7 & -5 & 28 & -12 \\
  &   & 12 & 30 & 150 & 1068 \\
\hline
  & 2 & 5 & 25 & 178 & 1056
\end{array}
$$

6 is an upper bound, but not a zero.

$\pm 1, \pm 2, 3, 4, \pm \dfrac{1}{2}, \pm \dfrac{3}{2}$

42. a. $\pm 1, \pm 2, \pm 3, \pm 4, \pm 6, \pm 12, \pm\frac{1}{2}, \pm\frac{3}{2}$ **b.** 2 is not a root but is an upper bound.   **c.** -2 is not a root but is a lower bound.

d. Possible roots are $\pm 1, \pm\frac{1}{2}$, and $\pm\frac{3}{2}$.   **43.** $f(1) = -2; f(2) = 3; x \approx 1.6$ **44.** $f(-3) = -32; f(-2) = 7; x \approx -2.3$

45. $\left\{ -\frac{1}{4}, 6 \pm 5i \right\}$ **46.** $\{1 \pm 3i, 1 \pm i\}$ **47.** $\left\{ -\frac{1}{2}, 1, 4 \pm 7i \right\}$ **48.** $f(x) = x^3 - 6x^2 + 21x - 26$

49. $f(x) = 2x^4 + 12x^3 + 20x^2 + 12x + 18$ **50.** $f(x) = x^4 - 3x^3 + 6x^2 + 2x - 60$

51. $-2, \frac{1}{2}, \pm i; f(x) = (x - i)(x + i)(x + 2)\left(x - \frac{1}{2} \right)$ **52.** $-1, 4; g(x) = (x + 1)^2(x - 4)^2$

53. 4 real zeros, one with multiplicity two   **54.** 3 real zeros; 2 nonreal complex zeros

55. 2 real zeros, one with multiplicity two; 2 nonreal complex zeros   **56.** 1 real zero; 4 nonreal complex zeros

57. Vertical asymptote: $x = 3$ and $x = -3$ **58.** Vertical asymptote: $x = -3$ **59.** Vertical asymptotes: $x = 3, -2$
horizontal asymptote: $y = 0$ horizontal asymptote: $y = 2$ horizontal asymptote: $y = 1$

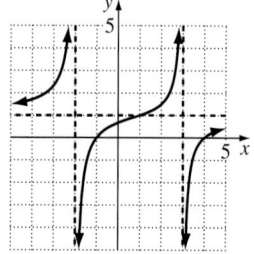

60. Vertical asymptote: $x = -2$ **61.** Vertical asymptote: $x = -1$ **62.** Vertical asymptote: $x = 3$ **63.** No vertical asymptote
horizontal asymptote: $y = 1$ no horizontal asymptote   no horizontal asymptote   no horizontal asymptote
slant asymptote: $y = x - 1$ slant asymptote: $y = x + 5$ slant asymptote: $y = -2x$

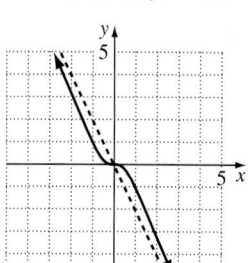

64. Vertical asymptote: $x = \frac{3}{2}$ **65. a.** $C(x) = 25x + 50,000$ **b.** $\overline{C}(x) = \dfrac{25x + 50,000}{x}$

no horizontal asymptote

slant asymptote: $y = 2x - 5$

c. $\overline{C}(50) = 1025$, when 50 calculators are manufactured, it costs \$1025 to manufacture each;
$\overline{C}(100) = 525$, when 100 calculators are manufactured, it costs \$525 to manufacture each;
$\overline{C}(1000) = 75$, when 1000 calculators are manufactured, it costs \$75 to manufacture each;
$\overline{C}(100,000) = 25.5$, when 100,000 calculators are manufactured, it costs \$25.50 to manufacture each.

d. $y = 25$; Minimum costs will approach \$25.

66. a. 1600; The difference in cost of removing 90% versus 50% of the contaminants is 16 million dollars.

b. $x = 100$; No amount of money can remove 100% of the contaminants, since $C(x)$ increases without bound as x approaches 100.

67. $y = 3000$; The number of fish in the pond approaches 3000.

68. $y = 0$; As the number of years of education increases the percentage rate of unemployment approaches zero.

69. a. $f(x) = \dfrac{1.96x + 3.14}{3.04x + 21.79}$ **b.** $y = \dfrac{49}{76}$; As the years increase, the fraction of nonviolent prisoners approaches $\dfrac{49}{76}$.

c. Answers may vary.   **70.** \$154   **71.** 1600 ft   **72.** 5 hr   **73.** 112 decibels   **74.** 16 hr   **75.** 800 ft^3

Chapter 3 Test

1.

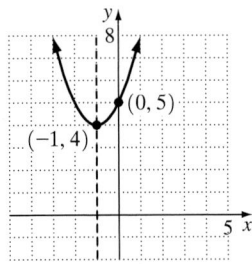

(0, 5)
(−1, 4)

axis of symmetry: $x = -1$

2.

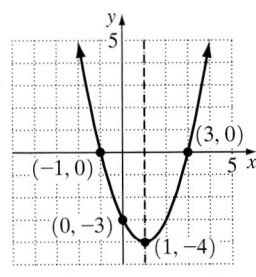

(3, 0)
(−1, 0)
(0, −3)
(1, −4)

axis of symmetry: $x = 1$

3. maximum; $(3, 2)$

4. 23 computers; maximum
daily profit = \$16,900

5. a. $5, 2, -2$

b.

6. Since the degree of the polynomial is odd and the leading coefficient is positive, the graph of f should fall to the left and rise to the right. The x-intercepts should be -1 and 1.

7. a. 2 **b.** $\dfrac{1}{2}, \dfrac{2}{3}$

8. $\pm 1, \pm 2, \pm 3, \pm 6, \pm \dfrac{1}{2}, \pm \dfrac{3}{2}$

9. 3 or 1 positive real zeros; no negative real zeros.

10. $\{-5, -3, 2\}$

11. a. $\pm 1, \pm 3, \pm 5, \pm 15, \pm \dfrac{1}{2}, \pm \dfrac{3}{2}, \pm \dfrac{5}{2}, \pm \dfrac{15}{2}$ **b.** $\left\{-1, \dfrac{3}{2}, \pm\sqrt{5}\right\}$

12.

-3	3	4	-7	-2	-3
		-9	15	-24	78
	3	-5	8	-26	75

-3 is a lower bound.

2	3	4	-7	-2	-3
		6	20	26	48
	3	10	13	24	45

2 is an upper bound.

13. $\{2, 3, 1 + i, 1 - i\}$ **14.** $(x - 1)(x + 2)^2$

15. domain: $\{x \mid x \neq 4, x \neq -4\}$ **16.** domain: $\{x \mid x \neq 2\}$ **17.** domain: $\{x \mid x \neq -3, x \neq 1\}$ **18.** domain: all real numbers

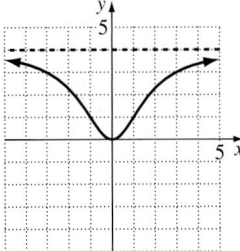

19. a. 0.9 **b.** 11 **c.** $y = 1$; as the number of learning trials increases, the proportion of correct responses approaches 1.

20. 45 foot-candles

Cumulative Review Exercises (Chapters P–3)

1. $2 + \sqrt{3}$ **2.** $-3x^2 - 11x + 11$ **3.** $15\sqrt{2}$ **4.** $x^5(x - 1)(x + 1)$ **5.** $\{2, -1\}$ **6.** $\left\{\dfrac{5 + \sqrt{13}}{6}, \dfrac{5 - \sqrt{13}}{6}\right\}$ **7.** $\left\{\dfrac{1}{3}, -\dfrac{2}{3}\right\}$

8. $\{-3, -1, 2\}$ **9.** $(-\infty, 1)$ or $(4, \infty)$ **10.** $(-\infty, -1)$ or $\left(\dfrac{5}{3}, \infty\right)$

11. Center: $(1, -2)$; radius: 3

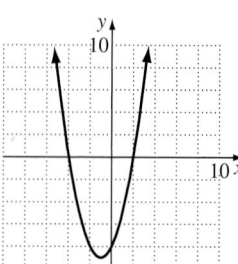

12. $t = 1 - \dfrac{V}{C}$

13. $(-\infty, 5]$

14. $x^2 - 2x - 4$

15. $16x^2 - 6$

16. -9

17. a. $\{-1, 1, 4\}$

b.

18.

19.

20.

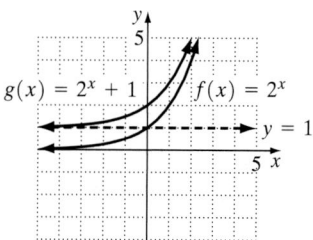

CHAPTER 4

Section 4.1

Check Point Exercises

1. 1 O-ring

2.

3.

4.

5. 11.49 billion

6. a. \$14,859.47

b. \$14,918.25

Exercise Set 4.1

1. 10.556 **3.** 11.665 **5.** 0.125 **7.** 9.974 **9.** 0.387

11.

13.

15.

17.

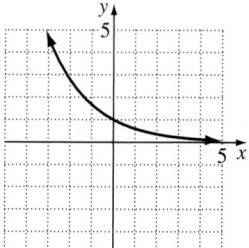

19. $H(x) = -3^{-x}$ **21.** $F(x) = -3^x$ **23.** $h(x) = 3^x - 1$

25.

27.

29.

31.

33.

35.

37.

39.
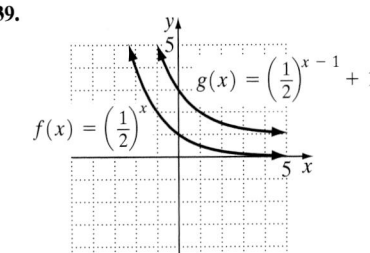

41. a. $13,116.51
b. $13,140.67
c. $13,157.04
d. $13,165.31
43. 7% compounded monthly
45. a. 67.38 million
b. about 134.74 million
c. about 269.46 million
d. 538.85 million
e. appears to double every 27 yr

47. $f(10) \approx 48$; 10 minutes after 8:00, 48 people have heard the rumor.   **49.** $116,405.10

51. 3.249009585; 3.317278183; 3.321880096; 3.321995226; 3.321997068; $2^{\sqrt{3}} \approx 3.321997085$; The closer the exponent is to $\sqrt{3}$, the closer the value is to $2^{\sqrt{3}}$.   **53.** 175.6   **55. a.** 100%   **b.** $\approx 68.5\%$ **c.** $\approx 30.8\%$ **d.** 20%

57. a. 1429   **b.** 24,546   **c.** Growth is limited by the population; The entire population will eventually become ill.

65.
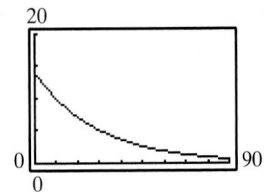

no; Nearly 4 O-rings are
expected to fail.

67. a.

b.

c.
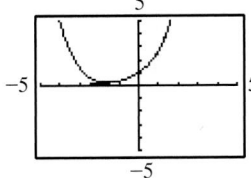

d. Answers may vary.

69. $y = 3^x$ is (d); $y = 5^x$ is (c); $y = \left(\frac{1}{3}\right)^x$ is (a); $y = \left(\frac{1}{5}\right)^x$ is (b).

71.
$$\left(\frac{e^x + e^{-x}}{2}\right)^2 - \left(\frac{e^x - e^{-x}}{2}\right)^2 \overset{?}{=} 1$$
$$\frac{e^{2x} + 2 + e^{-2x}}{4} - \frac{e^{2x} - 2 + e^{-2x}}{4} \overset{?}{=} 1$$
$$\frac{e^{2x} + 2 + e^{-2x} - e^{2x} + 2 - e^{-2x}}{4} \overset{?}{=} 1$$
$$\frac{4}{4} \overset{?}{=} 1$$
$$1 = 1$$

Section 4.2

Check Point Exercises

1. a. $7^3 = x$ **b.** $b^2 = 25$ **c.** $4^y = 26$ **2. a.** $5 = \log_2 x$ **b.** $3 = \log_b 27$ **c.** $y = \log_e 33$ **3. a.** 2   **b.** 1   **c.** $\dfrac{1}{2}$

4. a. 1   **b.** 0   **5. a.** 8   **b.** 17   **6.**

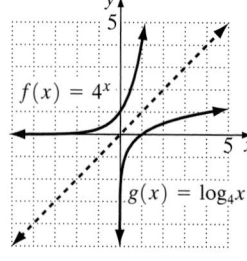

7. $(5, \infty)$
8. 80%
9. 4.0
10. a. $(-\infty, 4)$
   b. $(-\infty, 0)$ or $(0, \infty)$
11. a. $25x$
   b. $\sqrt{x}$
12. 4.6 ft per sec

Exercise Set 4.2

1. $2^4 = 16$ **3.** $3^2 = x$ **5.** $b^5 = 32$ **7.** $6^y = 216$ **9.** $\log_2 8 = 3$ **11.** $\log_2 \dfrac{1}{16} = -4$ **13.** $\log_8 2 = \dfrac{1}{3}$

15. $\log_{13} x = 2$ **17.** $\log_b 1000 = 3$ **19.** $\log_7 200 = y$ **21.** 2   **23.** 6   **25.** $\dfrac{1}{2}$ **27.** -3 **29.** $\dfrac{1}{2}$ **31.** 1

33. 0   **35.** 7   **37.** 19

39.

41.

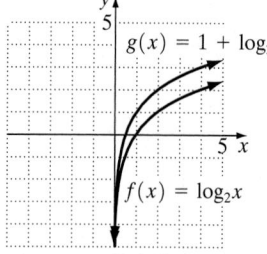

43. $H(x) = 1 - \log_3 x$
45. $h(x) = \log_3 x - 1$
47. $g(x) = \log_3(x - 1)$

49.

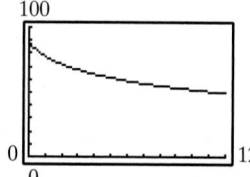

x-intercept: $(0, 0)$
vertical asymptote: $x = -1$

51.

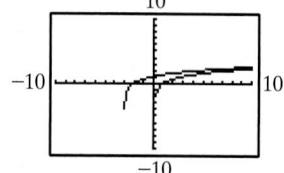

x-intercept: $(0.5, 0)$
vertical asymptote: $x = 0$

53.

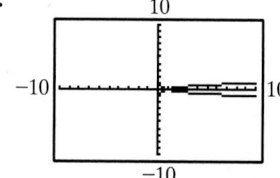

x-intercept: $(1, 0)$
vertical asymptote: $x = 0$

55. $(-4, \infty)$ **57.** $(-\infty, 2)$ **59.** $(-\infty, 2)$ or $(2, \infty)$ **61.** 2   **63.** 7   **65.** 33   **67.** 0   **69.** 6   **71.** -6 **73.** 125

75. $9x$ **77.** $5x^2$ **79.** $\sqrt{x}$ **81.** 95.4%   **83.** \$5.65 billion   **85.** ≈ 188 db; yes

87. a. 88
   b. 71.5; 63.9; 58.8; 55; 52; 49.5
   c.

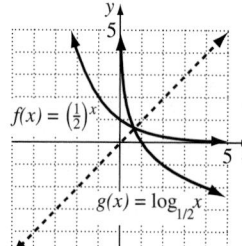

Material retention decreases
as time passes.

97.

$g(x)$ is $f(x)$ shifted left 3 units left.

99.

$g(x)$ is $f(x)$ reflected about the *x*-axis.

101.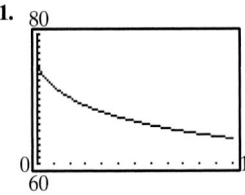

The score falls below 65 after 9 months.

103. $y = \ln x,\ y = \sqrt{x},\ y = x,\ y = x^2,\ y = e^x,\ y = x^x$

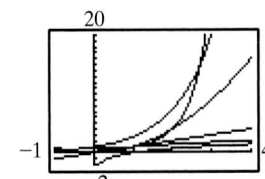

105. $\dfrac{4}{5}$

107. $\log_3 40 > \log_4 60$

Section 4.3

Check Point Exercises

1. a. $\log_6 7 + \log_6 11$ **b.** $2 + \log x$ **2. a.** $\log_8 23 - \log_8 x$ **b.** $5 - \ln 11$ **3. a.** $9 \log_6 3$ **b.** $\dfrac{1}{3} \ln x$

4. a. $4 \log_b x + \dfrac{1}{3} \log_b y$ **b.** $\dfrac{1}{2} \log_5 x - 2 - 3 \log_5 y$ **5. a.** 2   **b.** $\log \dfrac{7x + 6}{x}$

6. a. $\ln x^2 \sqrt[3]{x + 5}$ **b.** $\log \dfrac{(x - 3)^2}{x}$ **c.** $\log_b \dfrac{\sqrt[4]{x}\, y^{10}}{25}$ **7.** 4.02   **8.** 4.02

Exercise Set 4.3

1. $\log_5 7 + \log_5 3$ **3.** $1 + \log_7 x$ **5.** $3 + \log x$ **7.** $1 - \log_7 x$ **9.** $\log x - 2$ **11.** $3 - \log_4 y$ **13.** $2 - \ln 5$

15. $3 \log_b x$ **17.** $-6 \log N$ **19.** $\dfrac{1}{5} \ln x$ **21.** $2 \log_b x + \log_b y$ **23.** $\dfrac{1}{2} \log_4 x - 3$ **25.** $2 - \dfrac{1}{2} \log_6 (x + 1)$

27. $2 \log_b x + \log_b y - 2 \log_b z$ **29.** $1 + \dfrac{1}{2} \log x$ **31.** $\dfrac{1}{3} \log x - \dfrac{1}{3} \log y$ **33.** $\dfrac{1}{2} \log_b x + 3 \log_b y - 3 \log_b z$

35. $\dfrac{2}{3} \log_5 x + \dfrac{1}{3} \log_5 y - \dfrac{2}{3}$ **37.** $3 \ln x + \dfrac{1}{2} \ln(x^2 + 1) - 4 \ln(x + 1)$ **39.** $\left(1 + 2 \log x + \dfrac{1}{3} \log(1 - x1)\right) - (\log 7 + 2 \log(x + 1))$

41. 1   **43.** $\ln(7x)$ **45.** 5   **47.** $\log\left(\dfrac{2x + 5}{x}\right)$ **49.** $\log(xy^3)$ **51.** $\ln(x^{1/2}y)$ or $\ln(y\sqrt{x})$ **53.** $\log_b(x^2 y^3)$ **55.** $\ln\left(\dfrac{x^5}{y^2}\right)$

57. $\ln\left(\dfrac{x^3}{y^{1/3}}\right)$ or $\ln\left(\dfrac{x^3}{\sqrt[3]{y}}\right)$ **59.** $\ln \dfrac{(x + 6)^4}{x^3}$ **61.** $\ln\left(\dfrac{x^3 y^5}{z^6}\right)$ **63.** $\log \sqrt{xy}$ **65.** $\log_5\left(\dfrac{\sqrt{xy}}{(x + 1)^2}\right)$ **67.** $\ln \sqrt[3]{\dfrac{(x + 5)^2}{x(x^2 - 4)}}$

69. $\log\left(\dfrac{7x(x^2-1)}{x + 1}\right)$ or $\log(7x(x - 1))$ **71.** 1.5937   **73.** 1.6944   **75.** -1.2304 **77.** 3.6193

79.

81.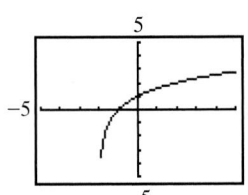

83. a. $D = 10 \log \dfrac{I}{I_0}$ **b.** 20 decibels louder

93. a.

b.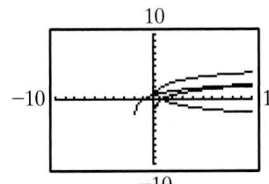

$y = 2 + \log_3 x$ shifts the graph of $y = \log_3 x$ two units upward; $y = \log_3(x + 2)$ shifts the graph of $y = \log_3 x$ two units left; $y = -\log_3 x$ reflects the graph of $y = \log_3 x$ about the x-axis.

95.

a. top graph: $y = \log_{100} x$; bottom graph: $y = \log_3 x$
b. top graph: $y = \log_3 x$; bottom graph: $y = \log_{100} x$
c. The graph of the equation with the largest b will be on the top in the interval $(0, 1)$ and on the bottom in the interval $(1, \infty)$.

101. (d) is true.   **103.** $\dfrac{2A}{B}$ **105.** Answers may vary.

Section 4.4

Check Point Exercises

1. $\left\{\dfrac{\ln 134}{\ln 5}\right\}$; ≈ 3.04 **2.** $\left\{\dfrac{\ln 9}{2}\right\}$; ≈ 1.10 **3.** $\left\{\dfrac{\ln 2088 + 4\ln 6}{3\ln 6}\right\}$; ≈ 2.76 **4.** $\{0, \ln 7\}$; $\ln 7 \approx 1.95$ **5.** $\{12\}$ **6.** $\{5\}$

7. $\left\{\dfrac{e^2}{3}\right\}$ **8.** 0.01 **9.** 16.2 yr   **10.** 2149

Exercise Set 4.4

1. $\left\{\dfrac{\ln 3.91}{\ln 10}\right\}$; ≈ 0.59 **3.** $\{\ln 5.7\}$; ≈ 1.74 **5.** $\left\{\dfrac{\ln 17}{\ln 5}\right\}$; ≈ 1.76 **7.** $\left\{\ln \dfrac{23}{5}\right\}$; ≈ 1.53 **9.** $\left\{\dfrac{\ln 659}{5}\right\}$; ≈ 1.30

11. $\left\{\dfrac{\ln 793 - 1}{-5}\right\}$; ≈ -1.14 **13.** $\left\{\dfrac{\ln 10{,}478 + 3}{5}\right\}$; ≈ 2.45 **15.** $\left\{\dfrac{\ln 410}{\ln 7} - 2\right\}$; ≈ 1.09 **17.** $\left\{\dfrac{\ln 813}{0.3 \ln 7}\right\}$; ≈ 11.48

19. $\left\{\dfrac{3\ln 5 + \ln 3}{\ln 3 - 2\ln 5}\right\}$; ≈ -2.80 **21.** $\{0, \ln 2\}$; $\ln 2 \approx 0.69$ **23.** $\left\{\dfrac{\ln 3}{2}\right\}$; ≈ 0.55 **25.** $\{0\}$ **27.** $\{81\}$ **29.** $\{59\}$ **31.** $\left\{\dfrac{109}{27}\right\}$

33. $\left\{\dfrac{62}{3}\right\}$ **35.** $\left\{\dfrac{5}{4}\right\}$ **37.** $\{6\}$ **39.** $\{6\}$ **41.** $\{5\}$ **43.** $\{2, 12\}$ **45.** $\{e^2\}$; ≈ 7.39 **47.** $\left\{\dfrac{e^4}{2}\right\}$; ≈ 27.30 **49.** $\{e^{-1/2}\}$; ≈ 0.61

51. $\{e^2 - 3\}$; ≈ 4.39 **53.** about 0.11 **55. a.** 18.9 million   **b.** ≈ 2006 **57.** 8.2 yr   **59.** 16.8% **61.** 8.7 yr   **63.** 15.7%

65. 1995 **67.** 2.8 days; Yes, the point $(2.8, 50)$ appears to lie on the graph of P.   **69.** $10^{-2.4}$; 0.004 moles per liter

75. $\{2\}$ **77.** $\{4\}$ **79.** $\{2\}$ **81.** $\{-1.391606, 1.6855579\}$

83.

(graph, y-axis from 27 to 30, x-axis to 500)

As distance from eye increases, barometric air pressure increases, leveling off at about 30 inches of mercury.

85.

about 7.9 min

87. (c) is true.   **89.** $\{1, e^2\}$, $e^2 \approx 7.389$ **91.** $\{e\}$, $e \approx 2.718$

Section 4.5

Check Point Exercises

1. a. $A = 643\, e^{0.023t}$ **b.** 2039 **2. a.** $A = A_0 e^{-0.0248t}$ **b.** about 72 yr   **3. a.** 0.4 correct responses   **b.** 0.7 correct responses
c. 0.8 correct responses   **4.** $y = 4e^{(\ln 7.8)x}$; $y = 4e^{2.054x}$

Exercise Set 4.5

1. 203 million   **3.** 2005 **5.** 2.6% **7.** 2014 **9. a.** $A = 158{,}700e^{0.053t}$ **b.** 2007 **11.** $A = 6.04e^{0.01t}$ **13.** 8.01 g

15. 8 g; 4 g; 2 g; 1 g; 0.5 g   **17.** $15{,}679$ years old   **19. a.** $\dfrac{A_0}{2} = A_0 e^{k(1.31)}$; $\dfrac{1}{2} = e^{1.31k}$; $\ln \dfrac{1}{2} = \ln e^{1.31k}$; $\ln \dfrac{1}{2} = 1.31k$; $k = \dfrac{\ln \frac{1}{2}}{1.31} \approx -0.52912$

b. 107 million years   **21.** $2A_0 = A_0 e^{kt}$; $2 = e^{kt}$; $\ln 2 = \ln e^{kt}$; $\ln 2 = kt$; $t = \dfrac{\ln 2}{k}$ **23.** 63 yr   **25. a.** about 20 people

b. about 1080 people   **c.** $100{,}000$ people   **27.** about 3.7% **29.** about 48 years old   **31.** $y = 100e^{(\ln 4.6)x}$; $y = 100e^{1.526x}$

33. $y = 2.5e^{(\ln 0.7)x}; y = 2.5e^{-0.357x}$ **45.** $y = 1.740(1.037)^x; r \approx 0.971$, a very good fit

47. $y = 0.112x + 1.547; r = 0.989$; a very good fit   **49.** The model of best fit is the linear model; 2022

51. The logarithmic model, $y = -905{,}231.353 + 119{,}204.060 \ln x$, best fits the data. Answers for prediction may vary.

53. Answers may vary.

Chapter 4 Review Exercises

1. $g(x) = 4^{-x}$ **2.** $h(x) = -4^{-x}$ **3.** $r(x) = -4^{-x} + 3$ **4.** $f(x) = 4^x$

5.   **6.**   **7.**   **8.**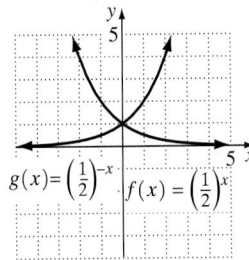

9. 5.5% compounded semiannually   **10.** 7% compounded monthly

11. a. $200°$ **b.** $120°; 119°$ **c.** $70°$; The temperature in the room is $70°$.   **12.** $49^{1/2} = 7$ **13.** $4^3 = x$ **14.** $3^y = 81$

15. $\log_6 216 = 3$ **16.** $\log_b 625 = 4$ **17.** $\log_{13} 874 = y$ **18.** 3   **19.** -2 **20.** $\varnothing$; $\log_b x$ is defined only for $x > 0$.   **21.** $\dfrac{1}{2}$

22. 1   **23.** 8   **24.** 5   **25.** 0   **26.**   **27.** 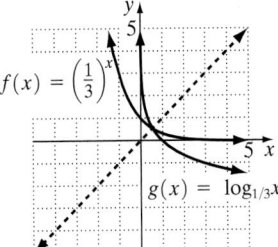   **28.** $g(x) = \log(-x)$

29. $r(x) = 1 + \log(2 - x)$ **30.** $h(x) = \log(2 - x)$ **31.** $f(x) = \log x$

32.   **33.**   **34.**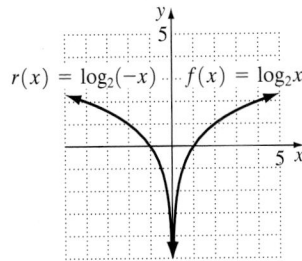

x-intercept: $(3, 0)$
vertical asymptote: $x = 2$

x-intercept: $(2, 0)$
vertical asymptote: $x = 0$

x-intercept: $(-1, 0)$
vertical asymptote: $x = 0$

35. $(-5, \infty)$ **36.** $(-\infty, 3)$ **37.** $(-\infty, 1) \cup (1, \infty)$ **38.** $6x$ **39.** $\sqrt{x}$ **40.** $4x^2$ **41.** 3.0

42. a. 76
b. $\approx 67, \approx 63, \approx 61, \approx 59, \approx 56$

c.

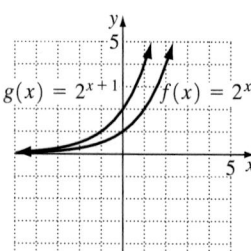

$f(t) = 76 - 18 \log(t + 1)$

time (months)

Retention decreases as time passes.

43. about 9 weeks   **44.** $2 + 3 \log_6 x$ **45.** $\dfrac{1}{2} \log_4 x - 3$

46. $\log_2 x + 2 \log_2 y - 6$ **47.** $\dfrac{1}{3} \ln x - \dfrac{1}{3}$ **48.** $\log_b 21$ **49.** $\log \dfrac{3}{x^3}$

50. $\ln(x^3 y^4)$ **51.** $\ln \dfrac{\sqrt{x}}{y}$ **52.** 6.2448   **53.** -0.1063

54. $\left\{ \dfrac{\ln 12{,}143}{\ln 8} \right\}; \approx 4.523$ **55.** $\left\{ \dfrac{1}{5} \ln 141 \right\}; \approx 0.990$

56. $\left\{ \dfrac{12 - \ln 130}{5} \right\}; \approx 1.426$ **57.** $\left\{ \dfrac{\ln 37{,}500 - 2 \ln 5}{4 \ln 5} \right\}; \approx 1.136$

58. $\{\ln 3\}; \approx 1.099$ **59.** $\{23\}$ **60.** $\{5\}$ **61.** $\varnothing$ **62.** $\left\{ \dfrac{1}{e} \right\}$

63. $\left\{ \dfrac{e^3}{2} \right\}$ **64.** 2042   **65.** 2086   **66.** 2005   **67.** 7.3 yr   **68.** 14.6 yr   **69.** about 21.97%

70. a. 0.045   **b.** 55.6 million   **c.** 2012   **71.** about 15,679 years old   **72. a.** 200 people   **b.** about 45,411 people   **c.** 500,000 people
73. $y = 73e^{(\ln 2.6)x}; y = 73e^{0.956x}$ **74.** $y = 6.5e^{(\ln 0.43)x}; y = 6.5e^{-0.844x}$
75. high: exponential; medium: linear; low: quadratic; Explanations will vary; negative; The parabola opens downward.
76. The exponential model, $y = (3.460)(1.024)^x$, is the best fit; about 116 million

Chapter 4 Test

1.

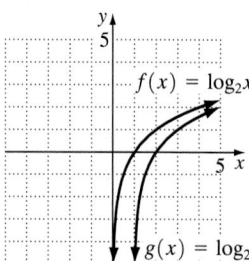

$g(x) = 2^{x+1}$ $f(x) = 2^x$

2.

$f(x) = \log_2 x$

$g(x) = \log_2(x - 1)$

3. $5^3 = 125$ **4.** $\log_{36} 6 = \dfrac{1}{2}$ **5.** $(-\infty, 3)$

6. $3 + 5 \log_4 x$ **7.** $\dfrac{1}{3} \log_3 x - 4$ **8.** $\log(x^6 y^2)$

9. $\ln \dfrac{7}{x^3}$ **10.** 1.5741   **11.** $\left\{ \dfrac{\ln 1.4}{\ln 5} \right\}$

12. $\left\{ \dfrac{\ln 4}{0.005} \right\}$ **13.** $\{0, \ln 5\}$ **14.** $\{54.25\}$

15. $\{5\}$ **16.** $\left\{ \dfrac{e^4}{3} \right\}$

17. 6.5% compounded semiannually; $221.15 more   **18.** 120 db   **19. a.** about 89%   **b.** decreasing; $k = -0.004 < 0$ **c.** 1995
20. 20. $A = 509e^{0.036t}$ **21.** about 24,758 years ago   **22. a.** 14 elk   **b.** about 51 elk   **c.** 140 elk

Cumulative Review Exercises (Chapters 1–4)

1. $\left\{ \dfrac{2}{3}, 2 \right\}$ **2.** $\{3, 7\}$ **3.** $\{-2, -1, 1\}$ **4.** $\left\{ \dfrac{\ln 128}{5} \right\}$ **5.** $\{3\}$ **6.** $(-\infty, 4]$ **7.** $[1, 3]$
8. using $(1, 3), y - 3 = -3(x - 1); y = -3x + 6$ **9.** $(f \circ g)(x) = (x + 2)^2; (g \circ f)(x) = x^2 + 2$
10. $f^{-1}(x) = \dfrac{1}{2}x + \dfrac{7}{2}$ **11.** $x^2 + 3x - 3 + \dfrac{-4}{x + 2}$ **12.** $\pm 1, \pm\dfrac{1}{2}, \pm\dfrac{1}{4}, \pm 3, \pm\dfrac{3}{2}, \pm\dfrac{3}{4}$ **13.** 300   **14.** $\{1 + i, 1 - i, 2\}$

15.

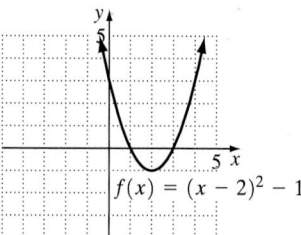

$(x - 3)^2 + (y + 2)^2 = 4$

16.

$f(x) = (x - 2)^2 - 1$

17.

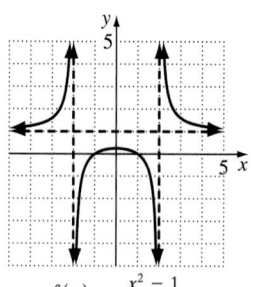

$f(x) = \dfrac{x^2 - 1}{x^2 - 4}$

18.

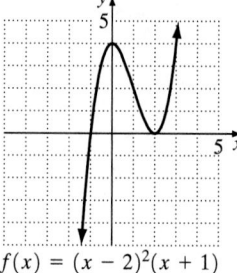

$f(x) = (x - 2)^2(x + 1)$

19. $12 per hr   **20.** $\dfrac{0.5}{\ln 4} \approx 0.361$; about $\dfrac{361}{1000}$ of the people

CHAPTER 5

Section 5.1

Check Point Exercises

1. a.   **b.**   **c.**   **d.**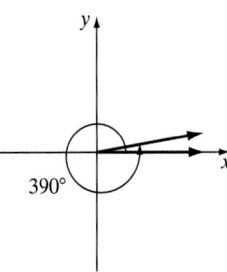

2. a. 40°   **b.** 225°   **3. a.** 12°; 102°   **b.** no complementary angle; 30°   **4.** 3.5 radians   **5. a.** $\dfrac{\pi}{3}$ radians   **b.** $\dfrac{3\pi}{2}$ radians

c. $-\dfrac{5\pi}{3}$ radians   **6. a.** 45°   **b.** $-240°$ **c.** 343.8°   **7.** $\dfrac{3\pi}{2}$ in. ≈ 4.71 in.   **8.** 135π in./min ≈ 424 in./min

Exercise Set 5.1

1. quadrant II   **3.** quadrant III   **5.** quadrant I   **7.** obtuse   **9.** straight

11.   **13.**   **15.**   **17.**

19. 35°   **21.** 210°   **23.** 315°   **25.** 38°; 128°   **27.** 52.6°; 142.6°   **29.** no complement; 69°   **31.** 4 radians   **33.** $\dfrac{4}{3}$ radians

35. 4 radians   **37.** $\dfrac{\pi}{4}$ radians   **39.** $\dfrac{3\pi}{4}$ radians   **41.** $\dfrac{5\pi}{3}$ radians   **43.** $-\dfrac{5\pi}{4}$ radians   **45.** 90°   **47.** 120°   **49.** 210°   **51.** $-540°$

53. 0.31 radians   **55.** -0.70 radians   **57.** 3.49 radians   **59.** 114.59°   **61.** 13.85°   **63.** $-275.02°$ **65.** 3π in. ≈ 9.42 in.

67. 10π ft ≈ 31.42 ft.   **69.** $\dfrac{12\pi \text{ radians}}{\text{second}}$ **71.** 60°; $\dfrac{\pi}{3}$ radians   **73.** $\dfrac{8\pi}{3}$ in. ≈ 8.38 in.   **75.** 12π in. ≈ 37.70 in.   **77.** 2 radians; 114.59°

79. 2094 mi   **81.** 1047 mph   **83.** 1508 ft/min   **97.** 30.25°   **99.** 30°25′12″   **101.** smaller than a right angle   **103.** 1815 mi

Section 5.2

Check Point Exercises

1. $\sin\theta = \dfrac{3}{5}$; $\cos\theta = \dfrac{4}{5}$; $\tan\theta = \dfrac{3}{4}$; $\csc\theta = \dfrac{5}{3}$; $\sec\theta = \dfrac{5}{4}$; $\cot\theta = \dfrac{4}{3}$ **2.** $\sqrt{2}$; $\sqrt{2}$; 1   **3.** $\sqrt{3}$; $\dfrac{\sqrt{3}}{3}$ **4.** $\tan\theta = \dfrac{2\sqrt{5}}{5}$; $\csc\theta = \dfrac{3}{2}$;

$\sec\theta = \dfrac{3\sqrt{5}}{5}$; $\cot\theta = \dfrac{\sqrt{5}}{2}$ **5.** $\dfrac{\sqrt{3}}{2}$ **6. a.** $\cos 44°$ **b.** $\tan\dfrac{5\pi}{12}$ **7. a.** 0.9553   **b.** 1.0025   **8.** 333.9 yd   **9.** 54°

Exercise Set 5.2

1. 15; $\sin\theta = \dfrac{3}{5}$; $\cos\theta = \dfrac{4}{5}$; $\tan\theta = \dfrac{3}{4}$; $\csc\theta = \dfrac{5}{3}$; $\sec\theta = \dfrac{5}{4}$; $\cot\theta = \dfrac{4}{3}$ **3.** 20; $\sin\theta = \dfrac{20}{29}$; $\cos\theta = \dfrac{21}{29}$; $\tan\theta = \dfrac{20}{21}$; $\csc\theta = \dfrac{29}{20}$;

$\sec\theta = \dfrac{29}{21}$; $\cot\theta = \dfrac{21}{20}$ **5.** 24; $\sin\theta = \dfrac{5}{13}$; $\cos\theta = \dfrac{12}{13}$; $\tan\theta = \dfrac{5}{12}$; $\csc\theta = \dfrac{13}{5}$; $\sec\theta = \dfrac{13}{12}$; $\cot\theta = \dfrac{12}{5}$

7. 28; $\sin\theta = \dfrac{4}{5}$; $\cos\theta = \dfrac{3}{5}$; $\tan\theta = \dfrac{4}{3}$; $\csc\theta = \dfrac{5}{4}$; $\sec\theta = \dfrac{5}{3}$; $\cot\theta = \dfrac{3}{4}$ **9.** $\dfrac{\sqrt{3}}{2}$ **11.** $\sqrt{2}$ **13.** $\sqrt{3}$ **15.** 0

17. $\tan \theta = \dfrac{8}{15}$; $\csc \theta = \dfrac{17}{8}$; $\sec \theta = \dfrac{17}{15}$; $\cot \theta = \dfrac{15}{8}$ **19.** $\tan \theta = \dfrac{\sqrt{2}}{4}$; $\csc \theta = 3$; $\sec \theta = \dfrac{3\sqrt{2}}{4}$; $\cot \theta = 2\sqrt{2}$ **21.** $\dfrac{\sqrt{13}}{7}$

23. $\dfrac{5}{8}$ **25.** 1   **27.** 1   **29.** 1   **31.** $\cos 83°$ **33.** $\sec 65°$ **35.** $\cot \dfrac{7\pi}{18}$ **37.** $\sin \dfrac{\pi}{10}$ **39.** 0.6157   **41.** 0.6420   **43.** 3.4203

45. 0.9511   **47.** 3.7321   **49.** 188 cm   **51.** 182 in.   **53.** 41 m   **55.** 17°   **57.** 78°   **59.** 1.147 radians   **61.** 0.3950 radians

63. 529 yd   **65.** 36°   **67.** 2879 ft   **69.** 37°

83. 0.92106, −0.19735; 0.95534, −0.148878; 0.98007, −0.099667; 0.99500, −0.04996; 0.99995, −0.005; 0.9999995, −0.0005; 0.999999995, −0.00005; 0.99999999995, −0.000005; $\dfrac{\cos \theta - 1}{\theta}$ approaches 0 as θ approaches 0.

85. In a right triangle, the hypotenuse is greater than either other side. Therefore, both $\dfrac{\text{opposite}}{\text{hypotenuse}}$ and $\dfrac{\text{adjacent}}{\text{hypotenuse}}$ must be less than 1 for an acute angle in a right triangle.

87. a. 357 ft   **b.** 394 ft

Section 5.3

Check Point Exercises

1. $\sin \theta = -\dfrac{3}{5}$; $\cos \theta = \dfrac{4}{5}$; $\tan \theta = -\dfrac{3}{4}$; $\csc \theta = -\dfrac{5}{3}$; $\sec \theta = \dfrac{5}{4}$; $\cot \theta = -\dfrac{4}{3}$ **2. a.** 1; undefined   **b.** 0; 1   **c.** −1; undefined

d. 0; −1   **3.** quadrant III   **4.** $\dfrac{\sqrt{10}}{10}$; $-\dfrac{\sqrt{10}}{3}$ **5. a.** 30°   **b.** $\dfrac{\pi}{4}$ **c.** 60°   **d.** 0.46   **6. a.** $-\dfrac{\sqrt{3}}{2}$ **b.** 1   **c.** $\dfrac{2\sqrt{3}}{3}$

Exercise Set 5.3

1. $\sin \theta = \dfrac{3}{5}$; $\cos \theta = -\dfrac{4}{5}$; $\tan \theta = -\dfrac{3}{4}$; $\csc \theta = \dfrac{5}{3}$; $\sec \theta = -\dfrac{5}{4}$; $\cot \theta = -\dfrac{4}{3}$ **3.** $\sin \theta = \dfrac{3\sqrt{13}}{13}$; $\cos \theta = \dfrac{2\sqrt{13}}{13}$; $\tan \theta = \dfrac{3}{2}$; $\csc \theta = \dfrac{\sqrt{13}}{3}$;

$\sec \theta = \dfrac{\sqrt{13}}{2}$; $\cot \theta = \dfrac{2}{3}$ **5.** $\sin \theta = -\dfrac{\sqrt{2}}{2}$; $\cos \theta = \dfrac{\sqrt{2}}{2}$; $\tan \theta = -1$; $\csc \theta = -\sqrt{2}$; $\sec \theta = \sqrt{2}$; $\cot \theta = -1$ **7.** $\sin \theta = -\dfrac{5\sqrt{29}}{29}$;

$\cos \theta = -\dfrac{2\sqrt{29}}{29}$; $\tan \theta = \dfrac{5}{2}$; $\csc \theta = -\dfrac{\sqrt{29}}{5}$; $\sec \theta = -\dfrac{\sqrt{29}}{2}$; $\cot \theta = \dfrac{2}{5}$ **9.** −1   **11.** −1   **13.** undefined   **15.** 0   **17.** quadrant I

19. quadrant III   **21.** quadrant II   **23.** $\sin \theta = -\dfrac{4}{5}$; $\tan \theta = \dfrac{4}{3}$; $\csc \theta = -\dfrac{5}{4}$; $\sec \theta = -\dfrac{5}{3}$; $\cot \theta = \dfrac{3}{4}$ **25.** $\cos \theta = -\dfrac{12}{13}$; $\tan \theta = -\dfrac{5}{12}$;

$\csc \theta = \dfrac{13}{5}$; $\sec \theta = -\dfrac{13}{12}$; $\cot \theta = -\dfrac{12}{5}$ **27.** $\sin \theta = -\dfrac{15}{17}$; $\tan \theta = -\dfrac{15}{8}$; $\csc \theta = -\dfrac{17}{15}$; $\sec \theta = \dfrac{17}{8}$; $\cot \theta = -\dfrac{8}{15}$ **29.** $\sin \theta = \dfrac{2\sqrt{13}}{13}$;

$\cos \theta = -\dfrac{3\sqrt{13}}{3}$; $\csc \theta = \dfrac{\sqrt{13}}{2}$; $\sec \theta = -\dfrac{\sqrt{13}}{3}$; $\cot \theta = -\dfrac{3}{2}$ **31.** $\sin \theta = -\dfrac{4}{5}$; $\cos \theta = -\dfrac{3}{5}$; $\csc \theta = -\dfrac{5}{4}$; $\sec \theta = -\dfrac{5}{3}$; $\cot \theta = \dfrac{3}{4}$

33. $\sin \theta = -\dfrac{2\sqrt{2}}{3}$; $\cos \theta = -\dfrac{1}{3}$; $\tan \theta = 2\sqrt{2}$; $\csc \theta = -\dfrac{3\sqrt{2}}{4}$; $\cot \theta = \dfrac{\sqrt{2}}{4}$ **35.** 20°   **37.** 25°   **39.** 5°   **41.** $\dfrac{\pi}{4}$ **43.** $\dfrac{\pi}{6}$ **45.** 30°

47. 25°   **49.** 1.56   **51.** $-\dfrac{\sqrt{2}}{2}$ **53.** $\dfrac{\sqrt{3}}{3}$ **55.** $\sqrt{3}$ **57.** $\dfrac{\sqrt{3}}{2}$ **59.** −2   **61.** 1   **63.** $\dfrac{\sqrt{3}}{2}$ **65.** −1

Section 5.4

Check Point Exercises

1. $\sin \pi = 0$; $\cos \pi = -1$; $\tan \pi = 0$; $\csc \pi$ is undefined; $\sec \pi = -1$; $\cot \pi$ is undefined   **2. a.** $\dfrac{1}{2}$ **b.** $-\dfrac{\sqrt{3}}{3}$ **3. a.** $\dfrac{\sqrt{2}}{2}$ **b.** $\sqrt{3}$

Exercise Set 5.4

1. $\sin t = \dfrac{8}{17}$; $\cos t = -\dfrac{15}{17}$; $\tan t = -\dfrac{8}{15}$; $\csc t = \dfrac{17}{8}$; $\sec t = -\dfrac{17}{15}$; $\cot t = -\dfrac{15}{8}$ **3.** $\sin\left(-\dfrac{\pi}{4}\right) = -\dfrac{\sqrt{2}}{2}$; $\cos\left(-\dfrac{\pi}{4}\right) = \dfrac{\sqrt{2}}{2}$;

$\tan\left(-\dfrac{\pi}{4}\right) = -1$; $\csc\left(-\dfrac{\pi}{4}\right) = -\sqrt{2}$; $\sec\left(-\dfrac{\pi}{4}\right) = \sqrt{2}$; $\cot\left(-\dfrac{\pi}{4}\right) = -1$ **5.** $-\dfrac{\sqrt{2}}{2}$ **7.** 2   **9.** $-\dfrac{\sqrt{2}}{2}$ **11.** $\dfrac{\sqrt{3}}{3}$ **13. a.** 12 hr

b. 20.3 hr   **c.** 3.7 hr   **15. a.** 1; 0; −1; 0; 1   **b.** 28 days   **23.** 0   **25.** $-\dfrac{1}{4}$ **c.** 9.6hr

Section 5.5

Check Point Exercises

1.

2.

3.

4.

5.

6.

7.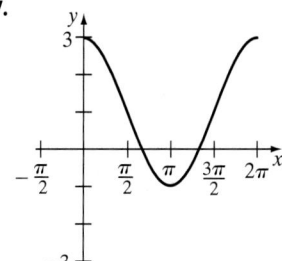

8. $y = 4 \sin 4x$

9. $y = 2 \sin\left(\dfrac{\pi}{6}x - \dfrac{\pi}{2}\right) + 12$

Exercise Set 5.5

1. 4

3. $\dfrac{1}{3}$

5. 3

7. $1; \pi$

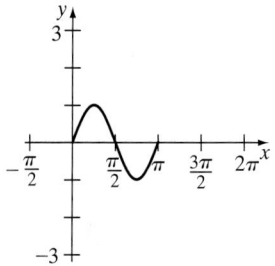

9. $3; 4\pi$

11. $4; 2$

13. $3; 1$

15. $1; 3\pi$

17. $1; 2\pi; \pi$

19. $1; \pi; \dfrac{\pi}{2}$

21. $3; \pi; \dfrac{\pi}{2}$

23. $\dfrac{1}{2}; 2\pi; \dfrac{\pi}{2}$

25. $2; \pi; \dfrac{\pi}{4}$

27. $3; 2; \dfrac{2}{\pi}$

29. $2; 1; -2$

31. 2

33. 2

35. $1; \pi$

37. $4; 1$

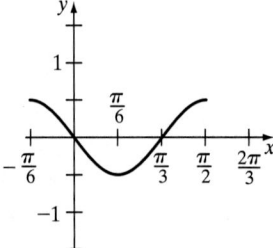

39. $4; 4\pi$

41. $\dfrac{1}{2}; 6$

43. $3; \pi; \dfrac{\pi}{2}$

45. $\dfrac{1}{2}; \dfrac{2\pi}{3}; \dfrac{\pi}{6}$

47. $3; \pi; \dfrac{\pi}{4}$

49. $2; 1; -4$

51.

53.

55.

57.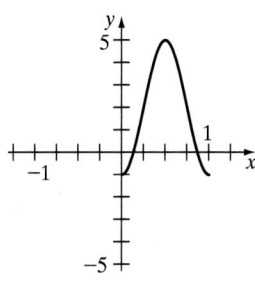

59. 33 days   **61.** 23 days   **63.** March 21   **65.** No

67.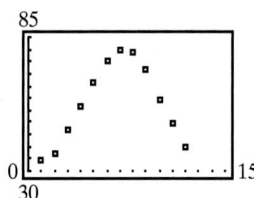

69. a. 3   **b.** 365 days   **c.** 15 hours of daylight
d. 9 hours of daylight
e.

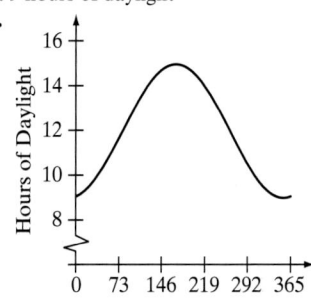

71. $y = 3 \cos \dfrac{\pi x}{6} + 9$ **73.** $4; \pi; \dfrac{\pi}{2}$

85.

87.

89.

91.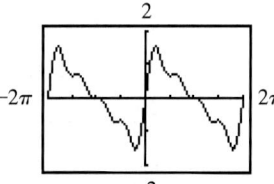

The graph is similar to $y = \sin x$,
except the amplitude is greater
and the curve is less smooth.

93. a.

b. $y = 22.61 \sin(0.50x - 2.04) + 57.17$

c.

95.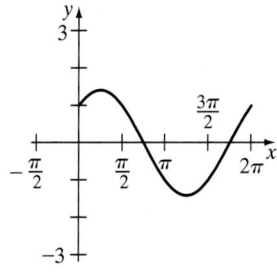

97. $y = 2 \cos(4x + \pi)$

Section 5.6

Check Point Exercises

1.

2.

3.

4.

5.
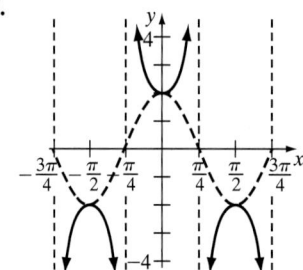

Exercise Set 5.6

1. $y = \tan(x + \pi)$ **3.** $y = -\tan\left(x - \dfrac{\pi}{2}\right)$

5.

7.

9.

11.
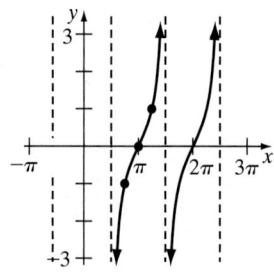

13. $y = -\cot x$ **15.** $y = \cot\left(x + \dfrac{\pi}{2}\right)$

17.

19.

21.

23.
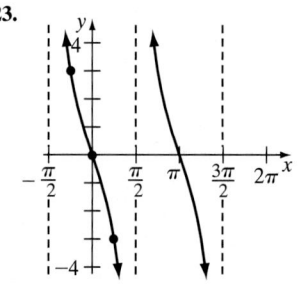

25. $y = -\dfrac{1}{2} \csc \dfrac{x}{2}$;

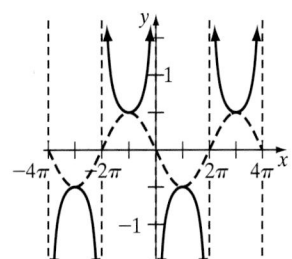

27. $y = \dfrac{1}{2} \sec 2\pi x$;

29.

31.

33.

35.

37.

39.

41.

43.

45. a.

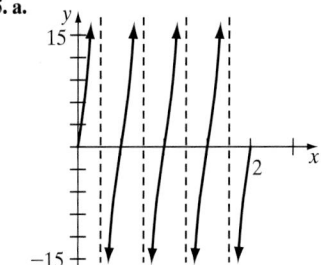

b. $0.25, 0.75, 1.25, 1.75$; The beacon is shining parallel to the wall at these times.

47. $d = 10 \sec x$

49.

63.

65.

67.

69.

71.

73.

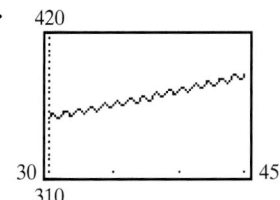

75. $y = \cot \dfrac{3}{2} x$ **77.** 2^{-x} decreases the amplitude as x gets larger.

Section 5.7

Check Point Exercises

1. $\dfrac{\pi}{3}$ **2.** $-\dfrac{\pi}{4}$ **3.** $\dfrac{2\pi}{3}$ **4.** $-\dfrac{\pi}{4}$ **5. a.** 1.2310 **b.** -1.5429 **6. a.** 0.7 **b.** 0 **c.** not defined **7.** $\dfrac{3}{5}$ **8.** $\dfrac{\sqrt{3}}{2}$ **9.** $\sqrt{x^2 + 1}$

Exercise Set 5.7

1. $\dfrac{\pi}{6}$ **3.** $\dfrac{\pi}{4}$ **5.** $-\dfrac{\pi}{6}$ **7.** $\dfrac{\pi}{6}$ **9.** $\dfrac{3\pi}{4}$ **11.** $\dfrac{\pi}{2}$ **13.** $\dfrac{\pi}{6}$ **15.** 0 **17.** $-\dfrac{\pi}{3}$ **19.** 0.30 **21.** -0.33 **23.** 1.19 **25.** 1.25

27. -1.52 **29.** -1.52 **31.** 0.9 **33.** $\dfrac{\pi}{3}$ **35.** $\dfrac{\pi}{6}$ **37.** 125 **39.** $-\dfrac{\pi}{6}$ **41.** $-\dfrac{\pi}{3}$ **43.** 0 **45.** not defined **47.** $\dfrac{3}{5}$ **49.** $\dfrac{12}{5}$

51. $-\dfrac{3}{4}$ **53.** $\dfrac{\sqrt{2}}{2}$ **55.** $\dfrac{4\sqrt{15}}{15}$ **57.** $-2\sqrt{2}$ **59.** 2 **61.** $\dfrac{\sqrt{1 - x^2}}{x}$ **63.** $\sqrt{1 - 4x^2}$ **65.** $\dfrac{\sqrt{x^2 - 1}}{x}$ **67.** $\dfrac{\sqrt{3}}{x}$ **69.** $\dfrac{\sqrt{x^2 + 4}}{2}$

71. a.
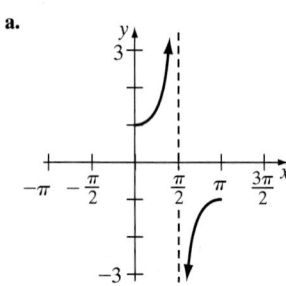
b. No horizontal line intersects the graph of $y = \sec x$ more than once, so the function is one-to-one and has an inverse function.
c.
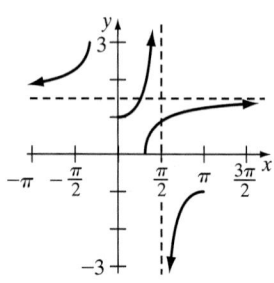

73. 0.408 radians; 0.602 radians; 0.654 radians; 0.645 radians; 0.613 radians **75.** 1.3157 radians or 75.4° **77.** 1.1071 sq units

91.

93.

Shifted left 2 units and up 1 unit

95.
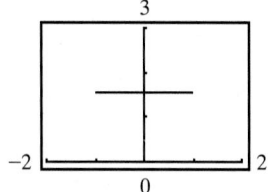

It seems $\sin^{-1} x + \cos^{-1} x = \dfrac{\pi}{2}$ for $-1 \le x \le 1$

97. $x = \sin\dfrac{\pi}{8}$ **99.** $\tan\alpha = \dfrac{8}{x}$, so $\tan^{-1}\dfrac{8}{x} = \alpha$. $\tan(\alpha + \theta) = \dfrac{33}{x}$, so $\tan^{-1}\dfrac{33}{x} = \alpha + \theta$. $\theta = \alpha + \theta - \alpha = \tan^{-1}\dfrac{33}{x} - \tan^{-1}\dfrac{8}{x}$.

Section 5.8

Check Point Exercises

1. $B = 27.3°$; $b \approx 4.34$; $c \approx 9.45$ **2.** 994 ft **3.** 29.0° **4.** 60.3 ft **5. a.** S 25°E **b.** S 15°W **6. a.** 4.2 m **b.** S87.7°W
7. $d = -6\cos\dfrac{\pi}{2}t$ **8. a.** 12 cm **b.** $\dfrac{1}{8}$ cm per sec **c.** 8 sec

Exercise Set 5.8

1. $B = 66.5°$; $a \approx 4.35$; $c \approx 10.90$ **3.** $B = 37.4°$; $a \approx 42.90$; $b \approx 32.80$ **5.** $A = 73.2°$; $a \approx 101.02$; $c \approx 105.52$
7. $b \approx 39.95$; $A \approx 37.3°$; $B \approx 52.7°$ **9.** $c \approx 26.96$; $A \approx 23.6°$; $B \approx 66.4°$ **11.** $a \approx 6.71$; $B \approx 16.6°$; $A \approx 73.4°$ **13.** N 15° E

15. S 80° W **17.** $d = -6\cos\dfrac{\pi}{2}t$ **19.** $d = 3\sin\dfrac{4\pi}{3}t$ **21. a.** 5 in. **b.** $\dfrac{1}{4}$ in per sec **c.** 4 sec **23. a.** 6 in. **b.** 1 in. per sec

c. 1 sec **25. a.** $\dfrac{1}{2}$ in. **b.** 0.32 in. per sec **c.** 3.14 sec **27. a.** 5 in. **b.** $\dfrac{1}{3}$ in. per sec **c.** 3 sec **29.** 2059 ft **31.** 695 ft

33. 1376 ft **35.** 15.1° **37.** 33.7 ft **39.** 90 mi north and 120 mi east **41.** 13.2 mi **43.** N 53° W **45.** N 89.5° E

47. $d = 6 \sin \pi t$ **49.** $d = \sin 528 \pi t$ **59.**

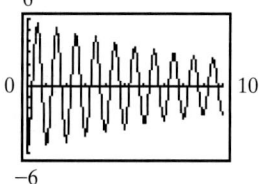

10 complete oscillations

61. 48 ft

Chapter 5 Review Exercises

1.

2.

3.

4.

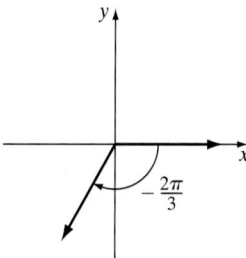

5. $40°$ **6.** $275°$ **7.** $17°$; $107°$ **8.** no complement; $\frac{\pi}{3}$ radians   **9.** 4.5 radians   **10.** $\frac{\pi}{12}$ radians   **11.** $\frac{2\pi}{3}$ radians   **12.** $\frac{7\pi}{4}$ radians

13. $300°$ **14.** $252°$ **15.** $-150°$ **16.** $\frac{15\pi}{2}$ ft ≈ 23.56 ft   **17.** 20.6π radians per min   **18.** 42,412 ft per min   **19.** $\sin \theta = \frac{3}{5}$;

$\cos \theta = \frac{4}{5}$; $\tan \theta = \frac{3}{4}$; $\csc \theta = \frac{5}{3}$; $\sec \theta = \frac{5}{4}$; $\cot \theta = \frac{4}{3}$ **20.** $\sqrt{3}$ **21.** $\frac{\sqrt{2}}{2}$ **22.** $\frac{2\sqrt{3}}{3}$ **23.** 1   **24.** $\frac{\sqrt{21}}{7}$ **25.** $\cos 20°$

26. $\sin 0$ **27.** 42 mm   **28.** 23 cm   **29.** 37 in.   **30.** 772 ft   **31.** 31 m   **32.** $56°$ **33.** $\sin \theta = -\frac{5\sqrt{26}}{26}$; $\cos \theta = -\frac{\sqrt{26}}{26}$; $\tan \theta = 5$;

$\csc \theta = -\frac{\sqrt{26}}{5}$; $\sec \theta = -\sqrt{26}$; $\cot \theta = \frac{1}{5}$ **34.** $\sin \theta = -1$; $\cos \theta = 0$; $\tan \theta$ is undefined; $\csc \theta = -1$; $\sec \theta$ is undefined; $\cot \theta = 0$

35. quadrant I   **36.** quadrant III   **37.** $\sin \theta = -\frac{\sqrt{21}}{5}$; $\tan \theta = -\frac{\sqrt{21}}{2}$; $\csc \theta = -\frac{5\sqrt{21}}{21}$; $\sec \theta = \frac{5}{2}$; $\cot \theta = -\frac{2\sqrt{21}}{21}$

38. $\sin \theta = \frac{\sqrt{10}}{10}$; $\cos \theta = -\frac{3\sqrt{10}}{10}$; $\csc \theta = \sqrt{10}$; $\sec \theta = -\frac{\sqrt{10}}{3}$; $\cot \theta = -3$ **39.** $85°$ **40.** $\frac{3\pi}{8}$ **41.** $50°$ **42.** $-\frac{\sqrt{3}}{2}$ **43.** $-\sqrt{3}$

44. $\sqrt{2}$ **45.** $\frac{\sqrt{3}}{2}$ **46.** $-\sqrt{3}$ **47.** $-\frac{2\sqrt{3}}{3}$ **48.** $-\frac{\sqrt{3}}{2}$ **49.** $\frac{\sqrt{2}}{2}$ **50.** 1

51.

52.

53.

54.

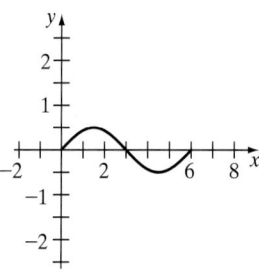

AA52 • Answers to Selected Exercises

55.

56.

57.

58.

59.

60.

61.

62.

63.

64. a. ≈98.52° **b.** 24 hr **c.** 5:00 P.M.; 98.9° **d.** 5:00 A.M.; 98.3°

e.

65.

66.

67.

68.

69.

70.

71.

72.

73.   **74.**   **75.**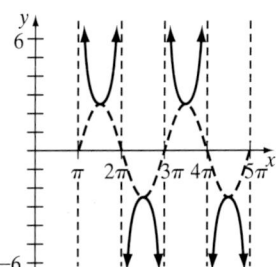

76. $\dfrac{\pi}{2}$ **77.** 0   **78.** $\dfrac{\pi}{4}$ **79.** $-\dfrac{\pi}{3}$ **80.** $\dfrac{2\pi}{3}$ **81.** $-\dfrac{\pi}{6}$ **82.** $\dfrac{\sqrt{2}}{2}$ **83.** 1   **84.** $-\dfrac{\sqrt{3}}{3}$ **85.** $-\dfrac{\sqrt{3}}{3}$ **86.** 2

87. $\dfrac{4}{5}$ **88.** $\dfrac{4}{5}$ **89.** $-\dfrac{3}{4}$ **90.** $-\dfrac{3}{4}$ **91.** $\dfrac{\pi}{3}$ **92.** $\dfrac{\pi}{3}$ **93.** $-\dfrac{\pi}{6}$ **94.** $\dfrac{2\sqrt{x^2+4}}{x^2+4}$ **95.** $\dfrac{x\sqrt{x^2-1}}{x^2-1}$ **96.** $B \approx 67.7°$; $a \approx 37.9$;

$b \approx 9.25$ **97.** $A \approx 52.6°$; $a \approx 7.85$; $c \approx 9.88$ **98.** $A \approx 16.6°$; $B \approx 73.4°$; $b \approx 6.71$ **99.** $A \approx 21.3°$; $B \approx 68.7°$; $c \approx 3.86$ **100.** 38 ft

101. 90 yd   **102.** 21.7 ft   **103.** N 35° E   **104.** S 35° W   **105.** 24.6 mi   **106. a.** 1282.2 mi   **b.** S 74° E   **107. a.** 20 cm

b. $\dfrac{1}{8}$ cm per sec   **c.** 8 sec   **108. a.** $\dfrac{1}{2}$ cm   **b.** 0.64 cm per sec   **c.** 1.57 sec   **109.** $d = -30 \cos \pi t$ **110.** $d = \dfrac{1}{4} \sin \dfrac{2\pi}{5} t$

Chapter 5 Test

1. $\dfrac{3\pi}{4}$ radians   **2.** $\dfrac{4\pi}{13}$ **3.** $\dfrac{25\pi}{3}$ ft ≈ 26.18 ft   **4.** $\sin \theta = \dfrac{5\sqrt{29}}{29}$; $\cos \theta = -\dfrac{2\sqrt{29}}{29}$; $\tan \theta = -\dfrac{5}{2}$; $\csc \theta = \dfrac{\sqrt{29}}{5}$; $\sec \theta = -\dfrac{\sqrt{29}}{2}$; $\cot \theta = -\dfrac{2}{5}$

5. quadrant III   **6.** $\sin \theta = -\dfrac{2\sqrt{2}}{3}$; $\tan \theta = -2\sqrt{2}$; $\csc \theta = -\dfrac{3\sqrt{2}}{4}$; $\sec \theta = 3$; $\cot \theta = -\dfrac{\sqrt{2}}{4}$ **7.** $\dfrac{\sqrt{3}}{6}$ **8.** $-\sqrt{3}$ **9.** $-\dfrac{\sqrt{2}}{2}$

10.   **11.**   **12.**

13.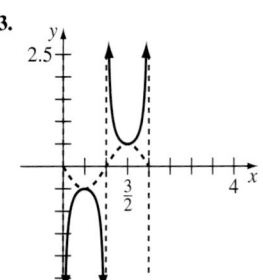

14. $-\sqrt{3}$ **15.** $B = 69°$; $a = 4.7$; $b = 12.1$ **16.** 23 yd   **17.** 36.1°

18. N 80° W   **19. a.** 6 in.   **b.** $\dfrac{1}{2}$ in. per sec   **c.** 2 sec

20. Trigonometric functions are periodic.

Cumulative Review Exercises (Chapters 1–5)

1. $\{-3, 6\}$ **2.** $\{-5, -2, 2\}$ **3.** $\{4\}$ **4.** $\{7\}$ **5.** $\{-1, 2, 3\}$ **6.** $-3 \le x \le 8$ **7.** $f^{-1}(x) = x^2 + 6$

8. $4x^2 - \dfrac{14}{5}x - \dfrac{17}{25} + \dfrac{284}{125x + 50}$ **9.** $\log 1000 = 3$ **10.** $280°$ **11.** 3 positive real roots; 1 negative real root

12.

13.

14.

15.

16.

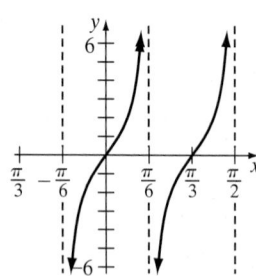

17. 48 performances
18. a. $A = 10.2e^{0.3057t}$
b. 2001
19. 1540 Btu per hr
20. $76°$

CHAPTER 6

Section 6.1

Check Point Exercises

1. $\csc x \tan x = \dfrac{1}{\sin x} \cdot \dfrac{\sin x}{\cos x} = \dfrac{1}{\cos x} = \sec x$

2. $\cos x \cot x + \sin x = \cos x \cdot \dfrac{\cos x}{\sin x} + \sin x = \dfrac{\cos^2 x}{\sin x} + \sin x \cdot \dfrac{\sin x}{\sin x} = \dfrac{\cos^2 x + \sin^2 x}{\sin x} = \dfrac{1}{\sin x} = \csc x$

3. $\sin x - \sin x \cos^2 x = \sin x(1 - \cos^2 x) = \sin x \cdot \sin^2 x = \sin^3 x$

4. $\dfrac{\sin x}{1 + \cos x} + \dfrac{1 + \cos x}{\sin x} = \dfrac{\sin x(\sin x)}{(1 + \cos x)(\sin x)} + \dfrac{(1 + \cos x)(1 + \cos x)}{(\sin x)(1 + \cos x)} = \dfrac{\sin^2 x + 1 + 2\cos x + \cos^2 x}{(1 + \cos x)(\sin x)}$

$= \dfrac{\sin^2 x + \cos^2 x + 1 + 2\cos x}{(1 + \cos x)(\sin x)} = \dfrac{1 + 1 + 2\cos x}{(1 + \cos x)(\sin x)} = \dfrac{2 + 2\cos x}{(1 + \cos x)(\sin x)} = \dfrac{2(1 + \cos x)}{(1 + \cos x)(\sin x)} = \dfrac{2}{\sin x} = 2\csc x$

5. $\dfrac{\cos x}{1 + \sin x} = \dfrac{\cos x(1 - \sin x)}{(1 + \sin x)(1 - \sin x)} = \dfrac{\cos x(1 - \sin x)}{1 - \sin^2 x} = \dfrac{\cos x(1 - \sin x)}{\cos^2 x} = \dfrac{1 - \sin x}{\cos x}$ **6.** $\dfrac{\sec x + \csc(-x)}{\sec x \csc x} = \dfrac{\sec x - \csc x}{\sec x \csc x}$

$= \dfrac{\dfrac{1}{\cos x} - \dfrac{1}{\sin x}}{\dfrac{1}{\cos x} \cdot \dfrac{1}{\sin x}} = \dfrac{\dfrac{\sin x}{\cos x \cdot \sin x} - \dfrac{\cos x}{\cos x \cdot \sin x}}{\dfrac{1}{\cos x \cdot \sin x}} = \dfrac{\dfrac{\sin x - \cos x}{\cos x \cdot \sin x}}{\dfrac{1}{\cos x \cdot \sin x}} = \dfrac{\sin x - \cos x}{\cos x \cdot \sin x} \cdot \dfrac{\cos x \cdot \sin x}{1} = \sin x - \cos x$

7. Left side: $\dfrac{1}{1 + \sin \theta} + \dfrac{1}{1 - \sin \theta} = \dfrac{1(1 - \sin \theta)}{(1 + \sin \theta)(1 - \sin \theta)} + \dfrac{1(1 + \sin \theta)}{(1 - \sin \theta)(1 + \sin \theta)} = \dfrac{1 - \sin \theta + 1 + \sin \theta}{(1 + \sin \theta)(1 - \sin \theta)} = \dfrac{2}{1 - \sin^2 \theta}$;

Right side: $2 + 2\tan^2 \theta = 2 + 2\left(\dfrac{\sin^2 \theta}{\cos^2 \theta}\right) = \dfrac{2\cos^2 \theta}{\cos^2 \theta} + \dfrac{2\sin^2 \theta}{\cos^2 \theta} = \dfrac{2\cos^2 \theta + 2\sin^2 \theta}{\cos^2 \theta} = \dfrac{2(\cos^2 \theta + \sin^2 \theta)}{\cos^2 \theta} = \dfrac{2}{\cos^2 \theta} = \dfrac{2}{1 - \sin^2 \theta}$

Exercise Set 6.1

For Exercises 1–59, proofs may vary.

65.

Proofs may vary

67.

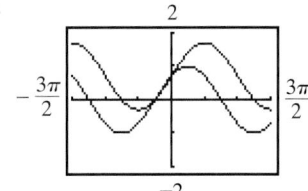

Values for x may vary.

69.

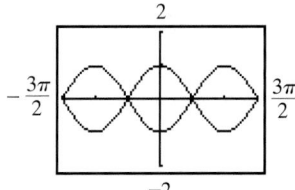

Values for x may vary.

71.

Proofs may vary.

73.

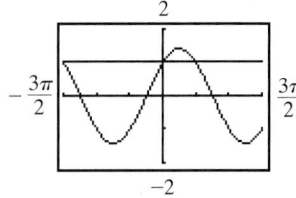

Values for x may vary.

75. Proofs may vary.

77. Answers may vary.

Section 6.2

Check Point Exercises

1. $\dfrac{\sqrt{3}}{2}$ **2.** $\dfrac{\sqrt{3}}{2}$ **3.** $\dfrac{\cos(\alpha - \beta)}{\cos\alpha\cos\beta} = \dfrac{\cos\alpha\cos\beta + \sin\alpha\sin\beta}{\cos\alpha\cos\beta} = \dfrac{\cos\alpha}{\cos\alpha}\cdot\dfrac{\cos\beta}{\cos\beta} + \dfrac{\sin\alpha}{\cos\alpha}\cdot\dfrac{\sin\beta}{\cos\beta} = 1 + \tan\alpha\tan\beta$

4. $\dfrac{\sqrt{2} + \sqrt{6}}{4}$ **5. a.** $\cos\alpha = -\dfrac{3}{5}$ **b.** $\cos\beta = \dfrac{\sqrt{3}}{2}$ **c.** $\dfrac{-3\sqrt{3} - 4}{10}$ **d.** $\dfrac{4\sqrt{3} - 3}{10}$

6. a. $y = \sin x$ **b.** $\cos\left(x + \dfrac{3\pi}{2}\right) = \cos x\cos\dfrac{3\pi}{2} - \sin x\sin\dfrac{3\pi}{2} = \cos x\cdot 0 - \sin x\cdot(-1) = \sin x$

7. $\tan(x + \pi) = \dfrac{\tan x + \tan\pi}{1 - \tan x\tan\pi} = \dfrac{\tan x + 0}{1 - \tan x\cdot 0} = \dfrac{\tan x}{1} = \tan x$

Exercise Set 6.2

1. $\dfrac{\sqrt{6} + \sqrt{2}}{4}$ **3.** $\dfrac{\sqrt{2} - \sqrt{6}}{4}$ **5. a.** $\alpha = 50°, \beta = 20°$ **b.** $\cos 30°$ **c.** $\dfrac{\sqrt{3}}{2}$ **7. a.** $\alpha = \dfrac{5\pi}{12}, \beta = \dfrac{\pi}{12}$ **b.** $\cos\left(\dfrac{\pi}{3}\right)$ **c.** $\dfrac{1}{2}$

For Exercises 9–11, proofs may vary. **13.** $\dfrac{\sqrt{6} - \sqrt{2}}{4}$ **15.** $\dfrac{\sqrt{6} + \sqrt{2}}{4}$ **17.** $\sqrt{3} + 2$ **19.** $2 - \sqrt{3}$ **21.** $-\dfrac{\sqrt{6} + \sqrt{2}}{4}$

23. $\dfrac{\sqrt{6} - \sqrt{2}}{4}$ **25.** $\sin 30°; \dfrac{1}{2}$ **27.** $\tan 45°; 1$ **29.** $\sin\dfrac{\pi}{6}; \dfrac{1}{2}$ **31.** $\tan\dfrac{\pi}{6}; \dfrac{\sqrt{3}}{3}$ For Exercises 33–55, proofs may vary.

57. a. $-\dfrac{63}{65}$ **b.** $\dfrac{16}{65}$ **c.** $\dfrac{16}{63}$ **59. a.** $-\dfrac{4 + 6\sqrt{2}}{15}$ **b.** $\dfrac{3 - 8\sqrt{2}}{15}$ **c.** $\dfrac{54 - 25\sqrt{2}}{28}$ **61. a.** $\dfrac{8\sqrt{3} + 15}{34}$ **b.** $\dfrac{15\sqrt{3} - 8}{34}$

c. $\dfrac{480 - 289\sqrt{3}}{33}$ **63. a.** $y = \sin x$ **b.** $\sin(\pi - x) = \sin\pi\cos x - \cos\pi\sin x = 0\cdot\cos x - (-1)\sin x = \sin x$

65. a. $y = 2\cos x$ **b.** $\sin\left(x + \dfrac{\pi}{2}\right) + \sin\left(\dfrac{\pi}{2} - x\right) = \sin x\cos\dfrac{\pi}{2} + \cos x\sin\dfrac{\pi}{2} + \sin\dfrac{\pi}{2}\cos x - \cos\dfrac{\pi}{2}\sin x$

$= \sin x\cdot 0 + \cos x\cdot 1 + 1\cdot\cos x - 0\cdot\sin x = \cos x + \cos x = 2\cos x$ **67.** Proofs may vary.; amplitude is $\sqrt{13}$; period is 2π

77.

Proofs may vary.

79.

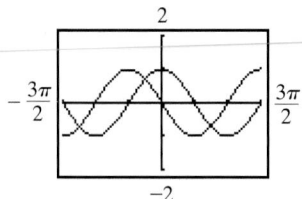

Values for x may vary.

81.

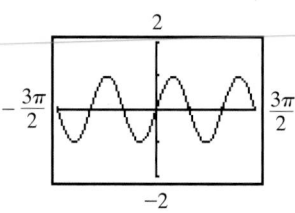

Proofs may vary.

83. Proofs may vary. **85.** $-\dfrac{24}{25}$ **87.** -1 **89.** $\dfrac{x\sqrt{1-y^2}-y}{\sqrt{1+x^2}}$

Section 6.3

Check Point Exercises

1. a. $\dfrac{24}{25}$ **b.** $\dfrac{7}{25}$ **c.** $\dfrac{24}{7}$ **2.** $\dfrac{\sqrt{3}}{2}$ **3.** $\sin 3\theta = \sin(2\theta + \theta) = \sin 2\theta \cos \theta + \cos 2\theta \sin \theta = 2 \sin \theta \cos \theta \cos \theta$

$= 2 \sin \theta \cos \theta \cos \theta + (2\cos^2 \theta - 1)\sin \theta = 2 \sin \theta \cos^2 \theta + 2 \sin \theta \cos^2 \theta - \sin \theta = 4 \sin \theta \cos^2 \theta - \sin \theta = 4 \sin \theta(1 - \sin^2 \theta) - \sin \theta$

$= 4 \sin \theta - 4 \sin^3 \theta - \sin \theta = 3 \sin \theta - 4 \sin^3 \theta$

4. $\sin^4 x = (\sin^2 x)^2 = \left(\dfrac{1 - \cos 2x}{2}\right)^2 = \dfrac{1 - 2\cos 2x + \cos^2 2x}{4} = \dfrac{1}{4} - \dfrac{1}{2}\cos 2x + \dfrac{1}{4}\cos^2 2x = \dfrac{1}{4} - \dfrac{1}{2}\cos 2x + \dfrac{1}{4}\left(\dfrac{1 + \cos 2(2x)}{2}\right)$

$= \dfrac{1}{4} - \dfrac{1}{2}\cos 2x + \dfrac{1}{8} + \dfrac{1}{8}\cos 4x = \dfrac{3}{8} - \dfrac{1}{2}\cos 2x + \dfrac{1}{8}\cos 4x$

5. $-\dfrac{\sqrt{2 + \sqrt{3}}}{2}$ **6.** $\dfrac{\sin 2\theta}{1 + \cos 2\theta} = \dfrac{2 \sin \theta \cos \theta}{1 + (1 - 2\sin^2 \theta)} = \dfrac{2 \sin \theta \cos \theta}{2 - 2\sin^2 \theta} = \dfrac{2 \sin \theta \cos \theta}{2(1 - \sin^2 \theta)} = \dfrac{2 \sin \theta \cos \theta}{2 \cos^2 \theta} = \dfrac{\sin \theta}{\cos \theta} = \tan \theta$

7. $\dfrac{\sec \alpha}{\sec \alpha \csc \alpha + \csc \alpha} = \dfrac{\dfrac{1}{\cos \alpha}}{\dfrac{1}{\cos \alpha} \cdot \dfrac{1}{\sin \alpha} + \dfrac{1}{\sin \alpha}} = \dfrac{\dfrac{1}{\cos \alpha}}{\dfrac{1}{\cos \alpha \sin \alpha} + \dfrac{\cos \alpha}{\cos \alpha \sin \alpha}} = \dfrac{\dfrac{1}{\cos \alpha}}{\dfrac{1 + \cos \alpha}{\cos \alpha \sin \alpha}} = \dfrac{1}{\cos \alpha} \cdot \dfrac{\cos \alpha \sin \alpha}{1 + \cos \alpha} = \dfrac{\sin \alpha}{1 + \cos \alpha} = \tan\dfrac{\alpha}{2}$

Exercise Set 6.3

1. $\dfrac{24}{25}$ **3.** $\dfrac{24}{7}$ **5.** $\dfrac{527}{625}$ **7. a.** $-\dfrac{240}{289}$ **b.** $-\dfrac{161}{289}$ **c.** $\dfrac{240}{161}$ **9. a.** $-\dfrac{336}{625}$ **b.** $\dfrac{527}{625}$ **c.** $-\dfrac{336}{527}$ **11. a.** $\dfrac{4}{5}$ **b.** $\dfrac{3}{5}$ **c.** $\dfrac{4}{3}$

13. a. $\dfrac{720}{1681}$ **b.** $\dfrac{1519}{1681}$ **c.** $\dfrac{720}{1519}$ **15.** $\dfrac{1}{2}$ **17.** $-\dfrac{\sqrt{3}}{2}$ **19.** $\dfrac{\sqrt{2}}{2}$ **21.** $\dfrac{\sqrt{3}}{3}$ For Exercises 23–33, proofs may vary.

35. $\dfrac{9}{4} - 3 \cos 2x + \dfrac{3}{4}\cos 4x$ **37.** $\dfrac{1}{8} - \dfrac{1}{8}\cos 4x$ **39.** $\dfrac{\sqrt{2 - \sqrt{3}}}{2}$ **41.** $-\dfrac{\sqrt{2 + \sqrt{2}}}{2}$ **43.** $2 + \sqrt{3}$ **45.** $-\sqrt{2} + 1$

47. $\dfrac{\sqrt{10}}{10}$ **49.** $\dfrac{1}{3}$ **51.** $\dfrac{7\sqrt{2}}{10}$ **53.** $\dfrac{3}{5}$ **55. a.** $\dfrac{2\sqrt{5}}{5}$ **b.** $-\dfrac{\sqrt{5}}{5}$ **c.** -2 **57. a.** $\dfrac{3\sqrt{13}}{13}$ **b.** $\dfrac{2\sqrt{13}}{13}$ **c.** $\dfrac{3}{2}$

For Exercises 59–67, proofs may vary. **69. a.** $\dfrac{v_0^2}{32} \cdot \sin 2\theta$ **b.** $\theta = \dfrac{\pi}{4}$ **71.** $\sqrt{2 - \sqrt{2}} \cdot (2 + \sqrt{2}) \approx 2.6$

85.

Proofs may vary.

87.

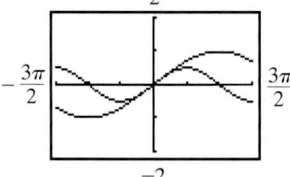

Values for x may vary.

89.

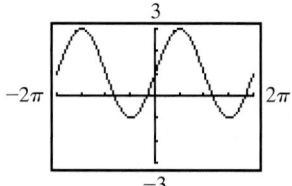

a. $y = 1 + 2 \sin x$
b. Proofs may vary.

91.

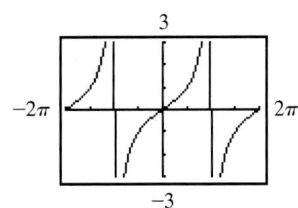

a. $y = \tan \dfrac{x}{2}$ **b.** Proofs may vary.

93. $\dfrac{5}{16} - \dfrac{7}{16}\cos 2x + \dfrac{3}{16}\cos 4x - \dfrac{1}{16}\cos 2x \cos 4x$

Section 6.4

Check Point Exercises

1. a. $\dfrac{1}{2}[\cos 3x - \cos 7x]$ **b.** $\dfrac{1}{2}[\cos 6x + \cos 8x]$ **2. a.** $2 \sin 5x \cos 2x$ **b.** $2 \cos\left(\dfrac{5x}{2}\right)\cos\left(\dfrac{x}{2}\right)$

3. $\dfrac{\cos 3x - \cos x}{\sin 3x + \sin x} = \dfrac{-2 \sin\left(\dfrac{3x + x}{2}\right)\sin\left(\dfrac{3x - x}{2}\right)}{2 \sin\left(\dfrac{3x + x}{2}\right)\cos\left(\dfrac{3x - x}{2}\right)} = \dfrac{-2 \sin\left(\dfrac{4x}{2}\right)\sin\left(\dfrac{2x}{2}\right)}{2 \sin\left(\dfrac{4x}{2}\right)\cos\left(\dfrac{2x}{2}\right)} = \dfrac{-2 \sin 2x \sin x}{2 \sin 2x \cos x} = \dfrac{\sin x}{\cos x} = -\tan x$

Exercise Set 6.4

1. $\dfrac{1}{2}[\cos 4x - \cos 8x]$ **3.** $\dfrac{1}{2}[\cos 4x + \cos 10x]$ **5.** $\dfrac{1}{2}[\sin 3x - \sin x]$ **7.** $\dfrac{1}{2}[\sin 2x - \sin x]$ **9.** $2 \sin 4x \cos 2x$

11. $2 \sin 2x \cos 5x$ **13.** $2 \cos 3x \cos x$ **15.** $2 \sin \dfrac{3x}{2}\cos \dfrac{x}{2}$ **17.** $2 \cos x \cos \dfrac{x}{2}$ **19.** $\dfrac{\sqrt{6}}{2}$ **21.** $-\dfrac{\sqrt{2}}{2}$

For Exercises 23–29, proofs may vary.   **31. a.** $y = \cos x$ **b.** Proofs may vary.   **33. a.** $y = \tan 2x$ **b.** Proofs may vary.

35. a. $y = \sin 1704\pi t + \sin 2418\pi t$ **b.** $2 \sin 2061\pi t \cdot \cos 357\pi t$

43.

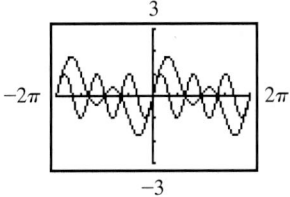

Values for x may vary.

45.

Proofs may vary.

47.

49. a.

b.

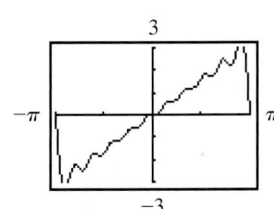

c. $\pi = 4 - \dfrac{4}{3} + \dfrac{4}{5} - \dfrac{4}{7} + \cdots$

For Exercises 51–55, proofs may vary.

Section 6.5

Check Point Exercises

1. $x = \dfrac{\pi}{3} + 2n\pi$ or $x = \dfrac{2\pi}{3} + 2n\pi$, where n is any integer.   2. $\dfrac{\pi}{6}, \dfrac{2\pi}{3}, \dfrac{7\pi}{6}, \dfrac{5\pi}{3}$ 3. $\dfrac{\pi}{2}$ 4. $\dfrac{\pi}{6}, \dfrac{\pi}{2}, \dfrac{5\pi}{6}$ 5. $0, \dfrac{\pi}{4}, \pi, \dfrac{5\pi}{4}$

6. $\dfrac{\pi}{2}, \dfrac{7\pi}{6}, \dfrac{11\pi}{6}$ 7. $\dfrac{3\pi}{4}, \dfrac{7\pi}{4}$ 8. $\dfrac{\pi}{2}, \pi$

Exercise Set 6.5

1. Solution   3. Not a solution   5. Solution   7. Solution   9. Not a solution   11. $x = \dfrac{\pi}{3} + 2n\pi$ or $x = \dfrac{2\pi}{3} + 2n\pi$, where n is

any integer.   13. $x = \dfrac{\pi}{4} + n\pi$, where n is any integer.   15. $x = \dfrac{2\pi}{3} + 2n\pi$ or $x = \dfrac{4\pi}{3} + 2n\pi$, where n is any integer.

17. $x = n\pi$, where n is any integer.   19. $x = \dfrac{5\pi}{6} + 2n\pi$ or $x = \dfrac{7\pi}{6} + 2n\pi$, where n is any integer.

21. $\theta = \dfrac{\pi}{6} + 2n\pi$ or $\theta = \dfrac{5\pi}{6} + 2n\pi$, where n is any integer.   23. $\theta = \dfrac{3\pi}{2} + 2n\pi$, where n is any integer.   25. $\dfrac{\pi}{6}, \dfrac{\pi}{3}, \dfrac{7\pi}{6}, \dfrac{4\pi}{3}$

27. $\dfrac{5\pi}{24}, \dfrac{7\pi}{24}, \dfrac{17\pi}{24}, \dfrac{19\pi}{24}, \dfrac{29\pi}{24}, \dfrac{31\pi}{24}, \dfrac{41\pi}{24}, \dfrac{43\pi}{24}$ 29. $\dfrac{\pi}{18}, \dfrac{7\pi}{18}, \dfrac{13\pi}{18}, \dfrac{19\pi}{18}, \dfrac{25\pi}{18}, \dfrac{31\pi}{18}$ 31. 0   33. no solution   35. $\dfrac{4\pi}{9}, \dfrac{8\pi}{9}, \dfrac{16\pi}{9}$

37. $0, \dfrac{\pi}{3}, \pi, \dfrac{4\pi}{3}$ 39. $\dfrac{\pi}{2}, \dfrac{7\pi}{6}, \dfrac{11\pi}{6}$ 41. $\dfrac{2\pi}{3}, \pi, \dfrac{4\pi}{3}$ 43. $\dfrac{3\pi}{2}$ 45. $\dfrac{\pi}{2}, \dfrac{3\pi}{2}$ 47. $\dfrac{\pi}{4}, \pi, \dfrac{5\pi}{4}$ 49. $\dfrac{5\pi}{6}, \dfrac{7\pi}{6}, \dfrac{11\pi}{6}$ 51. $\dfrac{\pi}{4}, \dfrac{5\pi}{4}$

53. $0, \dfrac{2\pi}{3}, \pi, \dfrac{4\pi}{3}$ 55. $0, \pi$ 57. $\dfrac{\pi}{2}, \dfrac{7\pi}{6}, \dfrac{11\pi}{6}$ 59. π 61. $\dfrac{\pi}{6}, \dfrac{5\pi}{6}$ 63. $\dfrac{\pi}{6}, \dfrac{\pi}{2}, \dfrac{5\pi}{6}, \dfrac{3\pi}{2}$ 65. $0, \dfrac{2\pi}{3}, \dfrac{4\pi}{3}$ 67. $\dfrac{2\pi}{3}, \dfrac{4\pi}{3}$

69. $\dfrac{\pi}{8}, \dfrac{3\pi}{8}, \dfrac{9\pi}{8}, \dfrac{11\pi}{8}$ 71. $0, \dfrac{\pi}{2}$ 73. $\dfrac{\pi}{4}, \dfrac{3\pi}{4}$ 75. $\dfrac{\pi}{12}, \dfrac{\pi}{4}, \dfrac{3\pi}{4}, \dfrac{11\pi}{12}, \dfrac{17\pi}{12}, \dfrac{19\pi}{12}$ 77. 0   79. 0.4 sec and 2.1 sec

81. 49 days and 292 days   83. $t = 2 + 6n$ or $t = 4 + 6n$ where n is any nonnegative integer.   85. $21°$ or $69°$.

97.

99.

101.

$x = 1.37, x = 2.30, x = 3.98,$ or $x = 4.91$

$x = 0.37$ or $x = 2.77$

$x = 0, x = 1.57, x = 2.09, x = 3.14,$ $x = 4.19,$ or $x = 4.71$

103. no solution   105. $\dfrac{\pi}{3}, \dfrac{5\pi}{3}$

Chapter 6 Review Exercises

For Exercises 1–12, proofs may vary. For Exercises 19–30, proofs may vary.   13. $\dfrac{\sqrt{6} - \sqrt{2}}{4}$ 14. $\dfrac{\sqrt{2} - \sqrt{6}}{4}$ 15. $2 - \sqrt{3}$ 16. $\sqrt{3} + 2$ 17. $\dfrac{1}{2}$ 18. $\dfrac{1}{2}$

31. a. $y = \cos x$ b. $\sin\left(x - \dfrac{3\pi}{2}\right) = \sin x \cos \dfrac{3\pi}{2} - \cos x \sin \dfrac{3\pi}{2} = \sin x \cdot 0 - \cos x \cdot -1 = \cos x$

32. a. $y = -\sin x$ b. $\cos\left(x + \dfrac{\pi}{2}\right) = \cos x \cos \dfrac{\pi}{2} - \sin x \sin \dfrac{\pi}{2} = \cos x \cdot 0 - \sin x \cdot 1 = -\sin x$

33. a. $y = \tan x$ b. $y = \dfrac{\tan x - 1}{1 - \cot x} = \dfrac{\dfrac{\sin x}{\cos x} - 1}{1 - \dfrac{\cos x}{\sin x}} = \dfrac{\dfrac{\sin x - \cos x}{\cos x}}{\dfrac{\sin x - \cos x}{\sin x}} = \dfrac{\sin x - \cos x}{\cos x} \cdot \dfrac{\sin x}{\sin x - \cos x} = \dfrac{\sin x}{\cos x} = \tan x$

34. a. $\dfrac{33}{65}$ b. $\dfrac{16}{65}$ c. $-\dfrac{33}{56}$ d. $\dfrac{24}{25}$ e. $\dfrac{2\sqrt{13}}{13}$ 35. a. $-\dfrac{63}{65}$ b. $-\dfrac{56}{65}$ c. $\dfrac{63}{16}$ d. $\dfrac{24}{25}$ e. $\dfrac{5\sqrt{26}}{26}$

36. a. 1 **b.** $-\dfrac{3}{5}$ **c.** undefined **d.** $-\dfrac{3}{5}$ **e.** $\dfrac{\sqrt{10+3\sqrt{10}}}{2\sqrt{5}}$ **37. a.** 1 **b.** $\dfrac{4\sqrt{2}}{9}$ **c.** undefined **d.** $\dfrac{4\sqrt{2}}{9}$ **e.** $-\dfrac{\sqrt{3}}{3}$

38. $\dfrac{\sqrt{3}}{2}$ **39.** $-\dfrac{\sqrt{3}}{3}$ **40.** $\dfrac{\sqrt{2-\sqrt{2}}}{2}$ **41.** $2-\sqrt{3}$ **42.** $\dfrac{1}{2}[\cos 2x - \cos 10x]$ **43.** $\dfrac{1}{2}[\sin 10x + \sin 4x]$ **44.** $-2\sin x \cos 3x$

45. $\dfrac{\sqrt{6}}{2}$ **46.** Proofs may vary. **47.** Proofs may vary. **48. a.** $y = \cot x$ **b.** Proofs may vary.

49. $x = \dfrac{2\pi}{3} + 2n\pi$ or $x = \dfrac{4\pi}{3} + 2n\pi$, where n is any integer. **50.** $x = \dfrac{\pi}{4} + 2n\pi$ or $x = \dfrac{3\pi}{4} + 2n\pi$, where n is any integer.

51. $x = \dfrac{7\pi}{6} + 2n\pi$ or $x = \dfrac{11\pi}{6} + 2n\pi$, where n is any integer. **52.** $x = \dfrac{\pi}{6} + n\pi$, where n is any integer **53.** $\dfrac{\pi}{2}, \dfrac{3\pi}{2}$ **54.** $\dfrac{\pi}{6}, \dfrac{5\pi}{6}, \dfrac{9\pi}{6}$

55. $\dfrac{3\pi}{2}$ **56.** $0, \dfrac{\pi}{3}, \pi, \dfrac{5\pi}{3}$ **57.** π **58.** $\dfrac{\pi}{6}, \dfrac{5\pi}{6}, \dfrac{3\pi}{2}$ **59.** $\dfrac{\pi}{6}, \dfrac{5\pi}{6}, \dfrac{7\pi}{6}, \dfrac{11\pi}{6}$ **60.** $0, \pi, \dfrac{7\pi}{6}, \dfrac{11\pi}{6}$ **61.** $0, \dfrac{\pi}{6}, \pi, \dfrac{11\pi}{6}$ **62.** $0, \pi$

63. $t = \dfrac{2}{3} + 4n$ or $t = \dfrac{10}{3} + 4n$, where n is any integer. **64.** $12°$ or $78°$

Chapter 6 Test

1. $-\dfrac{63}{65}$ **2.** $\dfrac{56}{33}$ **3.** $-\dfrac{24}{25}$ **4.** $\dfrac{3\sqrt{13}}{13}$ **5.** $\dfrac{\sqrt{6}+\sqrt{2}}{4}$ **6.** $\cos x \csc x = \cos x \cdot \dfrac{1}{\sin x} = \dfrac{\cos x}{\sin x} = \cot x$

7. $\dfrac{\sec x}{\cot x + \tan x} = \dfrac{\dfrac{1}{\cos x}}{\dfrac{\cos x}{\sin x} + \dfrac{\sin x}{\cos x}} = \dfrac{\dfrac{1}{\cos x}}{\dfrac{\cos^2 x + \sin^2 x}{\sin x \cos x}} = \dfrac{1}{\cos x} \cdot \dfrac{\sin x \cos x}{1} = \sin x$

8. $1 - \dfrac{\cos^2 x}{1 + \sin x} = 1 - \dfrac{(1 - \sin^2 x)}{1 + \sin x} = 1 - \dfrac{(1 + \sin x)(1 - \sin x)}{1 + \sin x} = 1 - (1 - \sin x) = \sin x$

9. $\cos\left(\theta + \dfrac{\pi}{2}\right) = \cos\theta\cos\dfrac{\pi}{2} - \sin\theta\sin\dfrac{\pi}{2} = \cos\theta \cdot 0 - \sin\theta \cdot 1 = -\sin\theta$

10. $\dfrac{\sin(\alpha - \beta)}{\sin\alpha\cos\beta} = \dfrac{\sin\alpha\cos\beta - \cos\alpha\sin\beta}{\sin\alpha\cos\beta} = \dfrac{\sin\alpha\cos\beta}{\sin\alpha\cos\beta} - \dfrac{\cos\alpha\sin\beta}{\sin\alpha\cos\beta} = 1 - \cot\alpha\tan\beta$

11. $\sin t \cos t(\tan t + \cot t) = \sin t \cos t\left(\dfrac{\sin t}{\cos t} + \dfrac{\cos t}{\sin t}\right) = \sin^2 t + \cos^2 t = 1$ **12.** $\dfrac{7\pi}{18}, \dfrac{11\pi}{18}, \dfrac{19\pi}{18}, \dfrac{23\pi}{18}, \dfrac{31\pi}{18}$, and $\dfrac{35\pi}{18}$

13. $\dfrac{\pi}{2}, \dfrac{7\pi}{6}, \dfrac{3\pi}{2}, \dfrac{11\pi}{6}$ **14.** $0, \dfrac{\pi}{3}, \dfrac{5\pi}{3}$ **15.** $0, \dfrac{2\pi}{3}, \dfrac{4\pi}{3}$

Cumulative Review Exercises (Chapters 1–6)

1. $-3, 1 + 2i$, and $1 - 2i$ **2.** $x = \dfrac{\log 125}{\log 11} + 1$ or $x \approx 3.01$ **3.** $(-\infty, -4] \cup [2, \infty)$ **4.** $\dfrac{\pi}{3}, \dfrac{5\pi}{3}$

5.

6.

7.

8.

9.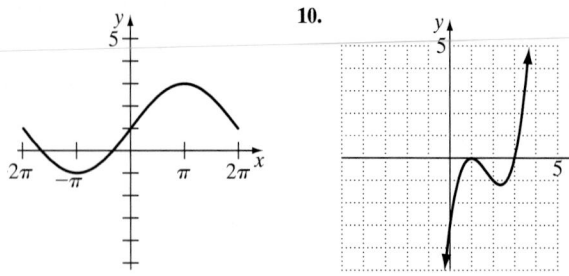

10.

11. $2a + h + 3$ **12.** $-\dfrac{\sqrt{2}}{2}$ **13.** Proofs may vary.

14. $\dfrac{16}{9}\pi$ radians   **15.** $t \approx 19.1$ yr

16. $f^{-1}(x) = \dfrac{3x + 1}{x - 2}$ **17.** $B = 67°, b = 28.27, c = 30.71$

18. 106 mg   **19.** $h \approx 15.9$ ft   **20.** Answers may vary.

CHAPTER 7

Section 7.1

Check Point Exercises

1. $B = 34°, a \approx 13$ cm, $b \approx 8$ cm   **2.** $B = 117.5°, a \approx 9, c \approx 5$ **3.** one triangle; $C \approx 24°, b \approx 33°, b \approx 31$ **4.** no triangle
5. two triangles; $B_1 \approx 50°, C_1 \approx 95°, c_1 = 21; B_2 \approx 130°, C_2 \approx 15°, c_2 \approx 5$ **6.** approximately 34 sq m   **7.** approximately 11 mi

Exercise Set 7.1

1. $B = 42°, a \approx 8.1, b \approx 8.1$ **3.** $A = 44°, b \approx 18.6, c \approx 22.8$ **5.** $C = 95°, b \approx 81.0, c \approx 134.1$ **7.** $B = 40°, b \approx 20.9, c \approx 31.8$
9. $C = 111°, b \approx 7.3, c \approx 16.1$ **11.** $A = 80°, a \approx 39.5, c \approx 10.4$ **13.** $B = 30°, a \approx 316.0, b \approx 174.3$ **15.** $C = 50°, a \approx 7.1, b \approx 7.1$
17. one triangle; $B \approx 29°, c \approx 111°, c \approx 29.0$ **19.** one triangle; $C \approx 52°, B \approx 65°, b \approx 10.2$ **21.** one triangle; $C \approx 55°, B \approx 13°, b \approx 10.2$
23. no triangle   **25.** two triangles; $B_1 \approx 77°, C_1 \approx 43°, c_1 \approx 12.6; B_2 \approx 103°, C_2 \approx 17°, c_2 \approx 5.4$ **27.** two triangles; $B_1 \approx 54°, C_1 \approx 89°,$
$c_1 \approx 19.9; B_2 \approx 126°, C_2 \approx 17°, c_2 \approx 5.8$ **29.** two triangles; $C_1 \approx 68°, B_1 \approx 54°, b_1 \approx 21.0; C_2 \approx 112°, B_2 \approx 10°, b_2 \approx 4.5$
31. no triangle   **33.** 297 sq ft   **35.** 5 sq yd   **37.** 10 sq m   **39.** Station A is about 6 miles from the fire, station B is about 9 miles from
the fire.   **41.** The platform is about 3672 yards from one end of the beach and 3576 yards from the other.   **43.** about 184 ft
45. about 56 ft   **47.** about 30 ft   **49. a.** $a \approx 494$ ft   **b.** about 343 ft   **51.** either 9.9 mi or 2.4 mi   **63.** No   **65.** 41 ft

Section 7.2

Check Point Exercises

1. $a = 13, B \approx 28°, C \approx 32°$ **2.** $B \approx 97.9°, A \approx 52.4°, C \approx 29.7°$ **3.** approximately 917 mi apart   **4.** approximately 47 sq m

Exercise Set 7.2

1. $a \approx 6.0, B \approx 29°, C \approx 105°$ **3.** $c \approx 7.6, A \approx 52°, B \approx 32°$ **5.** $A \approx 44°, B \approx 68°, C \approx 68°$ **7.** $A \approx 117°, B \approx 36°, C \approx 27°$
9. $c \approx 4.7, A \approx 46°, B \approx 92°$ **11.** $a \approx 6.3, C \approx 28°, B \approx 50°$ **13.** $b \approx 4.7, C \approx 54°, A \approx 76°$ **15.** $b \approx 5.4, C \approx 22°, A \approx 68°$
17. $C \approx 112°, A \approx 28°, B \approx 40°$ **19.** $B \approx 100°, A \approx 19°, C \approx 61°$ **21.** $A = 60°, B = 60°, C = 60°$ **23.** no triangle   **25.** 4 sq ft
27. 22 sq m   **29.** 31 sq yd   **31.** about 61.7 mi apart   **33.** about 193 yd   **35.** N12°E   **37. a.** about 19.3 mi   **b.** S58°E
39. The guy wire anchored downhill is about 417.4 feet. The one anchored uphill is about 398.2 feet.   **41.** about 63.7 ft   **43.** $123,454
53. 153.4°

Section 7.3

Check Point Exercises

1. a.

b.

c.

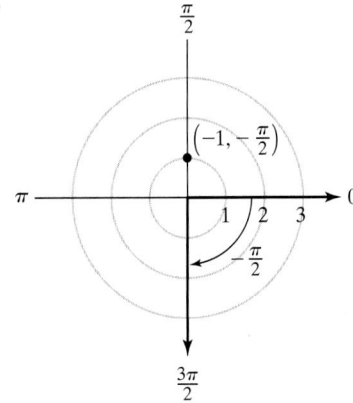

2. a. $\left(5, \dfrac{9\pi}{4}\right)$ **b.** $\left(-5, \dfrac{5\pi}{4}\right)$ **c.** $\left(5, -\dfrac{7\pi}{4}\right)$ **3. a.** $(-3, 0)$ **b.** $(-5\sqrt{3}, -5)$ **4.** $\left(2, \dfrac{5\pi}{3}\right)$ **5.** $\left(4, \dfrac{3\pi}{2}\right)$ **6.** $r = \dfrac{6}{3\cos\theta - \sin\theta}$

7. a. $x^2 + y^2 = 16$ **b.** $y = -x$ **c.** $x = 1$

Exercise Set 7.3

1. C **3.** A **5.** B **7.** C **9.** A **11.**

13.

15.

17.

19.

21.

23.

25.

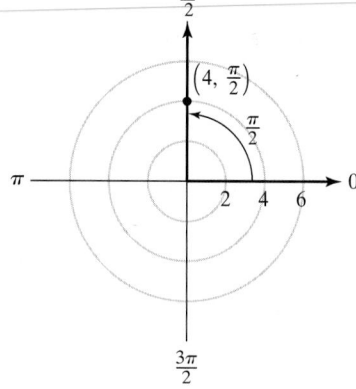

a. $\left(5, \dfrac{13\pi}{6}\right)$ **b.** $\left(-5, \dfrac{7\pi}{6}\right)$ **c.** $\left(5, -\dfrac{11\pi}{6}\right)$ **a.** $\left(10, \dfrac{11\pi}{4}\right)$ **b.** $\left(-10, \dfrac{7\pi}{4}\right)$ **c.** $\left(10, -\dfrac{5\pi}{4}\right)$ **a.** $\left(4, \dfrac{5\pi}{2}\right)$ **b.** $\left(-4, \dfrac{3\pi}{2}\right)$ **c.** $\left(4, -\dfrac{3\pi}{2}\right)$

27. a; b; d   **29.** b, d   **31.** a, b   **33.** $(0, 4)$ **35.** $(1, \sqrt{3})$ **37.** $(0, -4)$ **39.** approximately $(-5.9, 4.4)$ **41.** $\left(\sqrt{8}, \dfrac{3\pi}{4}\right)$

43. $\left(4, \dfrac{5\pi}{3}\right)$ **45.** $\left(2, \dfrac{7\pi}{6}\right)$ **47.** $(5, 0)$ **49.** $r = \dfrac{7}{3\cos\theta + \sin\theta}$ **51.** $r = \dfrac{7}{\cos\theta}$ **53.** $r = 3$ **55.** $r = 4\cos\theta$ **57.** $r = \dfrac{6\cos\theta}{\sin^2\theta}$

59. $x^2 + y^2 = 64$ **61.** $x = 0$ **63.** $y = 3$ **65.** $y = 4$ **67.** $x^2 + y^2 = y$ **69.** $x^2 + y^2 = 6x + 4y$ **71.** $y = \dfrac{1}{x}$ **73.** $\left(15, \dfrac{4\pi}{3}\right)$

75. 6.3 knots at an angle of 50° to the wind   **77.** Answers may vary.   **87.** $(-2, 3.464)$ **89.** $(-1.857, -3.543)$ **91.** $(3, 0.730)$

93. Answers may vary.   **95.** $x^2 + y^2 = 4x$; center: $(2, 0)$, radius: 2

Section 7.4

Check Point Exercises

1.

2.

3.

4.

5.

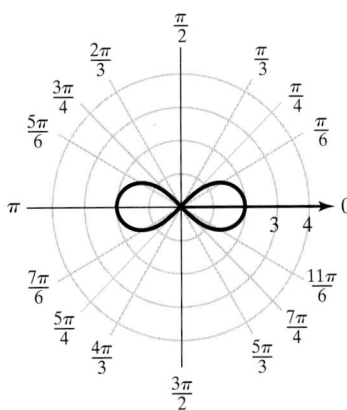

Exercise Set 7.4

1. $r = 1 - \sin\theta$ **3.** $r = 2\cos\theta$ **5.** $r = 3\sin 3\theta$ **7. a.** May or may not have symmetry with respect to polar axis.
b. Has symmetry with respect to the line $\theta = \dfrac{\pi}{2}$.   **c.** May or may not have symmetry about the pole.   **9. a.** Has symmetry with respect

to polar axis.   **b.** May or may not have symmetry with respect to the line $\theta = \dfrac{\pi}{2}$.   **c.** May or may not have symmetry about pole.

11. a. Has symmetry with respect to polar axis.   **b.** Has symmetry with respect to the line $\theta = \dfrac{\pi}{2}$.   **c.** Has symmetry about the pole.

13.

15.

17.

19.

21.

23.

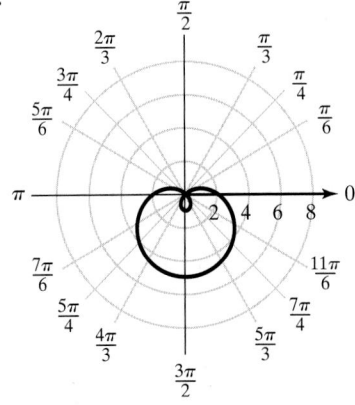

25.

27.

29.

31.

33.

35. 6 knots
37. 8 knots

39. 90°; about $7\frac{1}{2}$ knots

49.

51.

53.

55.

57.

59.

61.

63.

65.

67. 2π

69.

71.

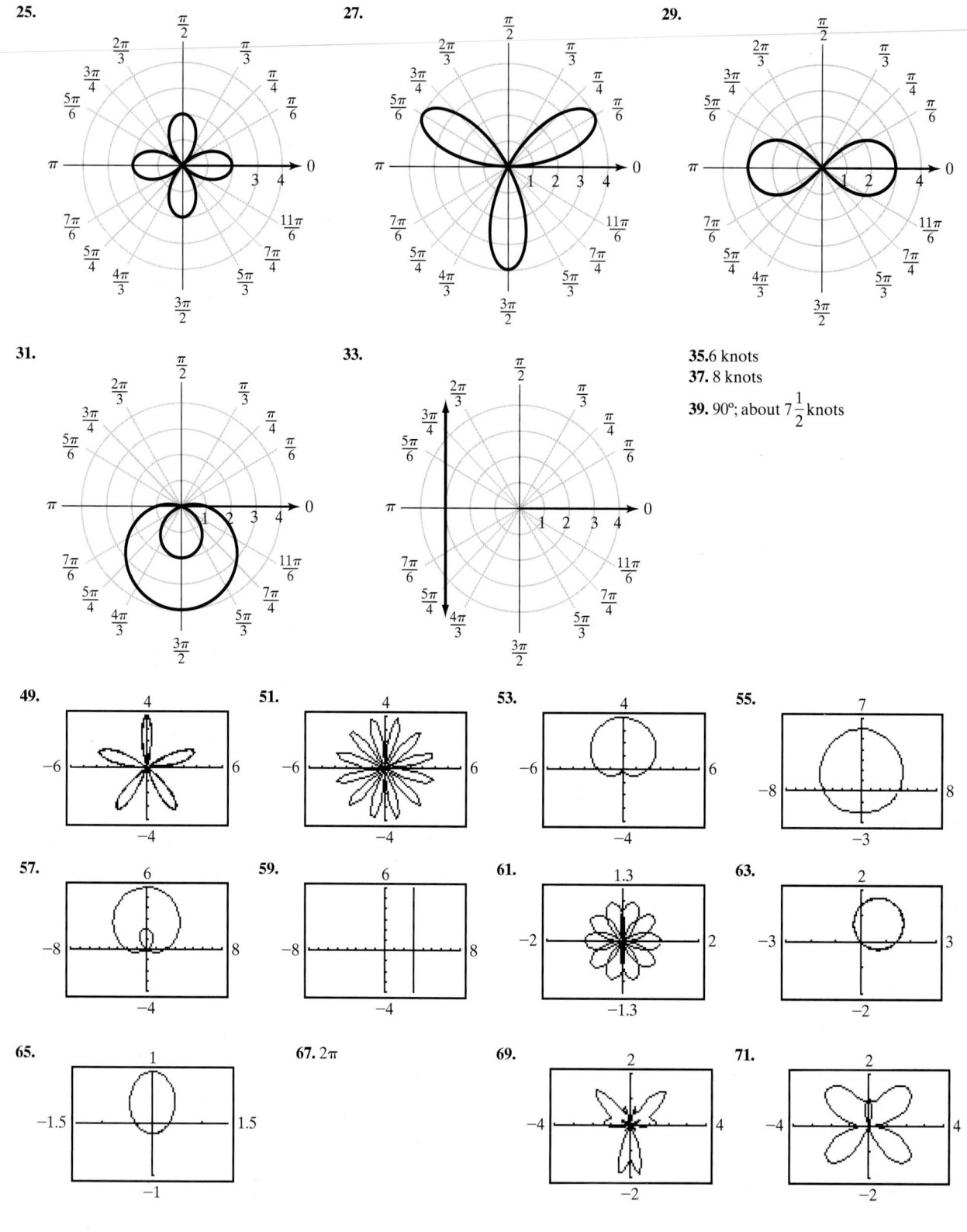

73. If n is odd, there are n loops and θmax $= \pi$ traces the graph once; while if n is even, there are $2n$ loops and θmax $= 2\pi$ traces the graph once. In each separate case, as n increases, $\sin n\theta$ increases its number of loops.   **75.** There are n small petals and n large petals for each value of n. For odd values of n, the small petals are inside the large petals. For even n, they are between the large petals.

77.

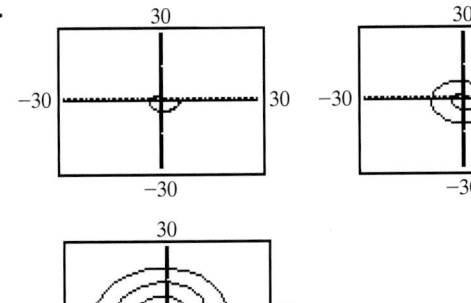

79. Answers may vary.
81.

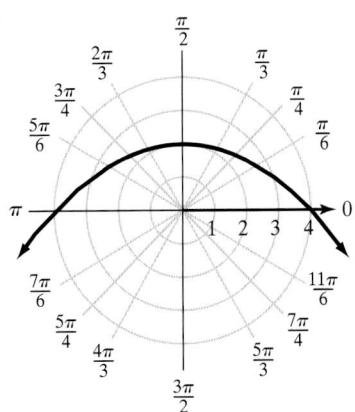

Section 7.5

Check Point Exercises

1. a.

b.

c.

d.

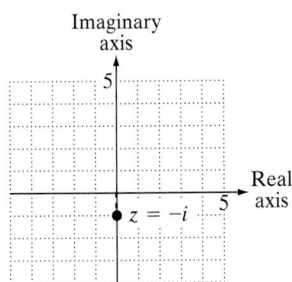

2. a. 13   **b.** $\sqrt{13}$ **3.**

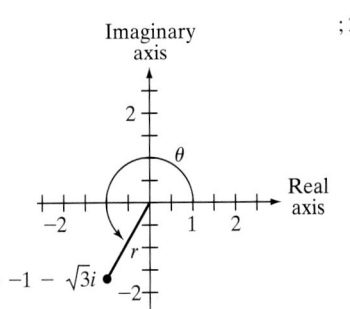

$; 2\left(\cos\dfrac{4\pi}{3} + i\sin\dfrac{4\pi}{3}\right)$

4. $z = 2\sqrt{3} + 2i$ **5.** $30(\cos 60° + i\sin 60°)$ **6.** $10(\cos\pi + i\sin\pi)$ **7.** $-16\sqrt{3} + 16i$ **8.** -4

9. $2(\cos 15° + i\sin 15°)$; $2(\cos 105° + i\sin 105°)$; $2(\cos 195° + i\sin 195°)$; $2(\cos 285° + i\sin 285°)$ **10.** $3; -\dfrac{3}{2} + \dfrac{3\sqrt{3}}{2}i; -\dfrac{3}{2} - \dfrac{3\sqrt{3}}{2}i$

Exercise Set 7.5

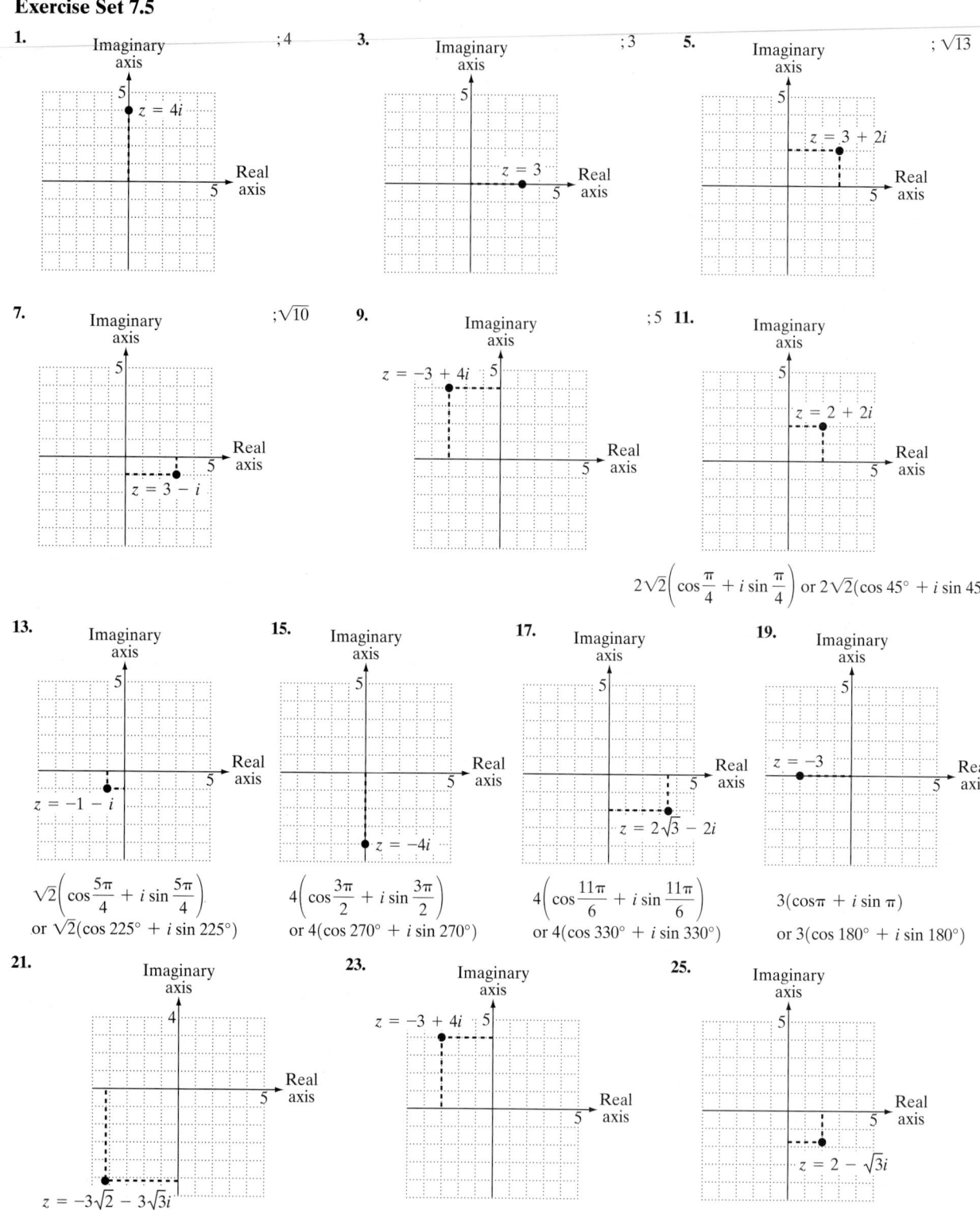

1. ; 4

Imaginary axis
$z = 4i$
Real axis

3. ; 3

Imaginary axis
$z = 3$
Real axis

5. ; $\sqrt{13}$

Imaginary axis
$z = 3 + 2i$
Real axis

7. ; $\sqrt{10}$

Imaginary axis
$z = 3 - i$
Real axis

9. ; 5

Imaginary axis
$z = -3 + 4i$
Real axis

11.

Imaginary axis
$z = 2 + 2i$
Real axis

$2\sqrt{2}\left(\cos\dfrac{\pi}{4} + i \sin\dfrac{\pi}{4}\right)$ or $2\sqrt{2}(\cos 45° + i \sin 45°)$

13.

Imaginary axis
$z = -1 - i$
Real axis

$\sqrt{2}\left(\cos\dfrac{5\pi}{4} + i \sin\dfrac{5\pi}{4}\right)$
or $\sqrt{2}(\cos 225° + i \sin 225°)$

15.

Imaginary axis
$z = -4i$
Real axis

$4\left(\cos\dfrac{3\pi}{2} + i \sin\dfrac{3\pi}{2}\right)$
or $4(\cos 270° + i \sin 270°)$

17.

Imaginary axis
$z = 2\sqrt{3} - 2i$
Real axis

$4\left(\cos\dfrac{11\pi}{6} + i \sin\dfrac{11\pi}{6}\right)$
or $4(\cos 330° + i \sin 330°)$

19.

Imaginary axis
$z = -3$
Real axis

$3(\cos\pi + i \sin \pi)$
or $3(\cos 180° + i \sin 180°)$

21.

Imaginary axis
$z = -3\sqrt{2} - 3\sqrt{3}i$
Real axis

$\approx 3\sqrt{5}(\cos 230.8° + i \sin 230.8°)$

23.

Imaginary axis
$z = -3 + 4i$
Real axis

$\approx 5(\cos 126.9° + i \sin 126.9°)$

25.

Imaginary axis
$z = 2 - \sqrt{3}i$
Real axis

$\approx \sqrt{7}(\cos 319.1° + i \sin 319.1°)$

27. $3\sqrt{3} + 3i$ **29.** $-2 - 2\sqrt{3}i$ **31.** $4\sqrt{2} - 4\sqrt{2}i$ **33.** $5i$ **35.** $z \approx -18.1 - 8.5i$ **37.** $30(\cos 70° + i \sin 70°)$

39. $12\left(\cos\dfrac{3\pi}{10} + i\sin\dfrac{3\pi}{10}\right)$ **41.** $\cos\dfrac{7\pi}{12} + i\sin\dfrac{7\pi}{12}$ **43.** $2(\cos\pi + i\sin\pi)$ **45.** $5(\cos 50° + i\sin 50°)$ **47.** $\dfrac{3}{4}\left(\cos\dfrac{\pi}{10} + i\sin\dfrac{\pi}{10}\right)$

49. $\cos 240° + i\sin 240°$ **51.** $2(\cos 0° + i\sin 0°)$ **53.** $32\sqrt{2} + 32\sqrt{2}i$ **55.** $-4 - 4\sqrt{3}i$ **57.** $\dfrac{1}{64}i$ **59.** $-2 - 2\sqrt{3}i$

61. $-4 - 4i$ **63.** -64 **65.** $3(\cos 15° + i\sin 15°); 3(\cos 195° + i\sin 195°)$ **67.** $2(\cos 70° + i\sin 70°); 2(\cos 190° + i\sin 190°);$

$2(\cos 310° + i\sin 310°)$ **69.** $\dfrac{3}{2} + \dfrac{3\sqrt{3}}{2}i; \dfrac{3\sqrt{3}}{2} + \dfrac{3}{2}i; \dfrac{3}{2} - \dfrac{3\sqrt{3}}{2}i; \dfrac{3\sqrt{3}}{2} - \dfrac{3}{2}i$ **71.** $2; \approx 0.6 + 1.9i; \approx -1.6 + 1.2i; \approx -1.6 - 1.2i;$

$\approx 0.6 - 1.9i$ **73.** $1; -\dfrac{1}{2} + \dfrac{\sqrt{3}}{2}i; -\dfrac{1}{2} - \dfrac{\sqrt{3}}{2}i$ **75.** $\approx 1.1 + 0.2i; \approx -0.2 + 1.1i; \approx -1.1 - 0.2i; \approx 0.2 - 1.1i$

77. a. $i; -1 + i; -i; -1 + i; -i; -1 + i$ **93.**   **95.** $\sqrt{3} + i; -\sqrt{3} + i; -2i$
b. Complex numbers may vary.

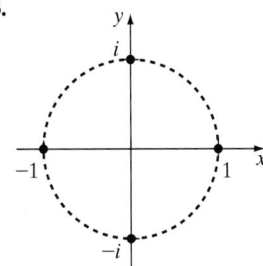

Section 7.6

Check Point Exercises

1. $\|\mathbf{u}\| = 5 = \|\mathbf{v}\|$ and $m_u = \dfrac{4}{3} = m_v$ **2.** $;\|\mathbf{v}\| = 3\sqrt{2}$ **3.** $\mathbf{v} = 3\mathbf{i} + 4\mathbf{j}$ **4. a.** $11\mathbf{i} - 2\mathbf{j}$ **b.** $3\mathbf{i} + 8\mathbf{j}$

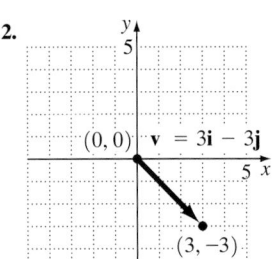

5. a. $56\mathbf{i} + 80\mathbf{j}$ **b.** $-35\mathbf{i} - 50\mathbf{j}$ **6.** $30\mathbf{i} + 33\mathbf{j}$ **7.** $\dfrac{4}{5}\mathbf{i} - \dfrac{3}{5}\mathbf{j}; \sqrt{\left(\dfrac{4}{5}\right)^2 + \left(-\dfrac{3}{5}\right)^2} = \sqrt{\dfrac{16}{25} + \dfrac{9}{25}} = \sqrt{\dfrac{25}{25}} = 1$ **8.** $30\sqrt{2}\mathbf{i} + 30\sqrt{2}\mathbf{j}$

Exercise Set 7.6

1. a. $\sqrt{41}$ **b.** $\sqrt{41}$ **c.** $\mathbf{u} = \mathbf{v}$ **3. a.** 6 **b.** 6 **c.** $\mathbf{u} = \mathbf{v}$

5.   **7.**   **9.**   **11.**

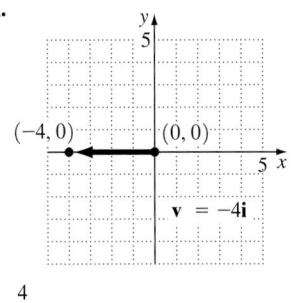

$\sqrt{10}$ $\sqrt{2}$ $2\sqrt{10}$ 4

13. $10i + 6j$ **15.** $6i - 3j$ **17.** $-6i - 14j$ **19.** $9i$ **21.** $-i + 2j$ **23.** $5i - 12j$ **25.** $-5i + 12j$ **27.** $-15i + 35j$ **29.** $4i + 24j$

31. $-9i - 4j$ **33.** $-5i + 45j$ **35.** $2\sqrt{29}$ **37.** $\sqrt{10}$ **39.** i **41.** $\dfrac{3}{5}i - \dfrac{4}{5}j$ **43.** $\dfrac{3\sqrt{13}}{13}i - \dfrac{2\sqrt{13}}{3}j$ **45.** $\dfrac{\sqrt{2}}{2}i + \dfrac{\sqrt{2}}{2}j$

47. $3\sqrt{3}i + 3j$ **49.** $-6\sqrt{2}i - 6\sqrt{2}j$ **51.** $\approx -0.20i + 0.46j$ **53.** $22\sqrt{3}i + 22j$ **55.** $148.5i + 20.9j$ **57.** $\approx 1.4i + 0.6j$; 1.4 in.

59. ≈ 108.21 lbs; S 77.4° E **61.** ≈ 30.9 lbs **63. a.** $F = 9i - 3j$ **b.** $F_3 = -9i + 3j$ **65. a.** $F = -2j$ **b.** $F_5 = 2j$

67. a. $v = 180\cos 40°i + 180\sin 40°j \approx 137.89i + 115.70j$ $w = 40\cos 0°i + 40\sin 0°j = 40i$ **b.** $v + w \approx 177.89i + 115.70j$

c. 212 mph **d.** 33.0°; N 57°E **69.** 78 mph, 75.4° **89.** Proofs may vary. **91.** $\approx 4232.1; \approx 72.7°$

Section 7.7

Check Point Exercises

1. a. 18 **b.** 18 **c.** 5 **2.** 100.3° **3.** orthogonal **4.** $\dfrac{7}{2}i - \dfrac{7}{2}j$ **5.** $v_1 = \dfrac{7}{2}i - \dfrac{7}{2}j$; $v_2 = -\dfrac{3}{2}i - \dfrac{3}{2}j$ **6.** approximately 2598 ft-lb

Exercise Set 7.7

1. a. 6 **b.** 10 **3. a.** -6 **b.** 41 **5. a.** 100 **b.** 61 **7. a.** 0 **b.** 25 **9.** 3 **11.** 3 **13.** 20 **15.** 20 **17.** 79.7°

19. 160.3° **21.** 38.7° **23.** orthogonal **25.** orthogonal **27.** not orthogonal **29.** not orthogonal **31.** orthogonal

33. $v_1 = \text{proj}_w v = \dfrac{5}{2}i - \dfrac{5}{2}j$; $v_2 = \dfrac{1}{2}i + \dfrac{1}{2}j$ **35.** $v_1 = \text{proj}_w v = -\dfrac{26}{29}i + \dfrac{65}{29}j$; $v_2 = \dfrac{55}{29}i + \dfrac{22}{29}j$ **37.** $v_1 = \text{proj}_w v = i + 2j$; $v_2 = 0$

39. 1077; $v \cdot w = 1077$ means \$1077 in revenue was generated on Monday by the sale of 240 gallons of regular gas at \$1.90 per gallon and 300 gallons of premium gas at \$2.07 per gallon. **41.** 7600 foot-pounds **43.** 3392 foot-pounds **45.** 1079 foot-pounds

47. 40 foot-pounds **49.** 22.05 foot-pounds

51. a. $\dfrac{\sqrt{3}}{2}i + \dfrac{1}{2}j$ **b.** $-175\sqrt{3}i - 175j$ **c.** 350; A force of 350 pounds is required to keep the boat from rolling down the ramp.

63. $u \cdot v = (a_1i + b_1j) \cdot (a_2i + b_2j)$
$= a_1a_2 + b_1b_2$
$= a_2a_1 + b_2b_1$
$= (a_2i + b_2j) \cdot (a_1i + b_1j)$
$= v \cdot u$

65. $u \cdot (v + w) = (a_1i + b_1j) \cdot [(a_2i + b_2j) + (a_3i + b_3j)]$
$= (a_1i + b_1j) \cdot [(a_2 + a_3)i + (b_2 + b_3)j]$
$= a_1(a_2 + a_3) + b_1(b_2 + b_3)$
$= a_1a_2 + a_1a_3 + b_1b_2 + b_1b_3$
$= a_1a_2 + b_1b_2 + a_1a_3 + b_1b_3$
$= (a_1i + b_1j) \cdot (a_2i + b_2j) + (a_1i + b_1j) \cdot (a_3i + b_3j)$
$= u \cdot v + u \cdot w$

67. $b = -20$ **69.** any two vectors, v and w, having the same direction

Chapter 7 Review Exercises

1. $C = 55°$, $b \approx 10.5$, and $c \approx 10.5$ **2.** $A = 43°$, $a \approx 171.9$, and $b \approx 241.0$ **3.** $b \approx 16.3$, $A \approx 72°$, and $C \approx 42°$

4. $C \approx 98°$, $A \approx 55°$, and $B \approx 27°$ **5.** $C = 120°$, $a \approx 45.0$, and $b \approx 33.2$ **6.** two triangles; $B_1 \approx 55°$, $C_1 \approx 86°$, and $c_1 \approx 31.7$;

$B_2 \approx 125°$, $C_2 \approx 16°$, and $c_2 \approx 8.8$ **7.** no triangle **8.** $a \approx 59.0$, $B \approx 3°$, and $C \approx 15°$ **9.** $B \approx 78°$, $A \approx 39°$, and $C \approx 63°$

10. $B \approx 25°$, $C \approx 115°$, and $c \approx 8.5$ **11.** two triangles; $A_1 \approx 59°$, $C_1 \approx 84°$, $c_1 \approx 14.4$; $A_2 \approx 121°$, $C_2 \approx 22°$, $c_2 \approx 5.4$

12. $B \approx 9°$, $C \approx 148°$, and $c \approx 73.6$ **13.** 8 sq ft **14.** 4 sq ft **15.** 4 sq m **16.** 2 sq m **17.** 35 ft **18.** 35.6 mi **19.** 861 mi

20. 404 ft; 551 ft **21.** \$214,194

22.

$(2, 2\sqrt{3})$

23.

$\left(-\dfrac{3\sqrt{3}}{2}, \dfrac{3}{2}\right)$

24.

$(2, 2\sqrt{3})$

25.

$(\sqrt{2}, \sqrt{2})$

26.

$(0, 4)$

27.

$(-\sqrt{2}, \sqrt{2})$

28.

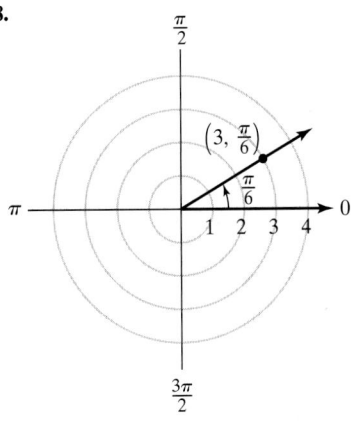

a. $\left(3, \dfrac{13\pi}{6}\right)$ **b.** $\left(-3, \dfrac{7\pi}{6}\right)$ **c.** $\left(3, -\dfrac{11\pi}{6}\right)$

29.

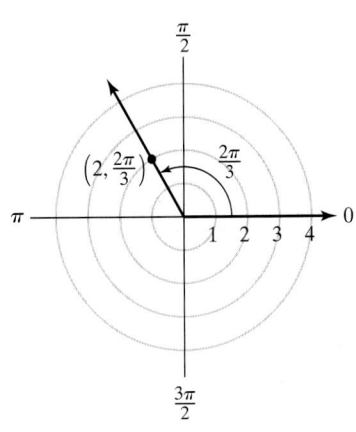

a. $\left(2, \dfrac{8\pi}{3}\right)$ **b.** $\left(-2, \dfrac{5\pi}{3}\right)$ **c.** $\left(2, -\dfrac{4\pi}{3}\right)$

30.

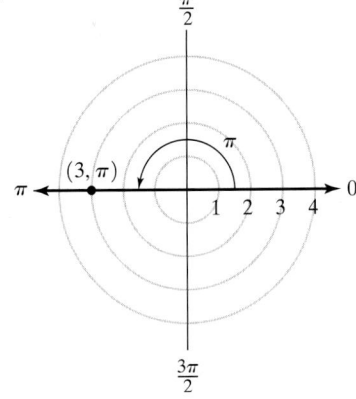

a. $(3, 3\pi)$ **b.** $(-3, 2\pi)$ **c.** $(3, -\pi)$

31. $\left(4\sqrt{2}, \dfrac{3\pi}{4}\right)$ **32.** $\left(3\sqrt{2}, \dfrac{7\pi}{4}\right)$ **33.** approximately $(13, 67°)$ **34.** approximately $(5, 127°)$ **35.** $\left(5, \dfrac{3\pi}{2}\right)$ **36.** $(1, 0)$

37. $r = \dfrac{8}{2\cos\theta + 3\sin\theta}$ **38.** $r = 10$ **39.** $5r^2 = 3r\sin\theta$ or $r = \dfrac{3}{5}\sin\theta$ **40.** $x^2 + y^2 = 9$ **41.** $y = -x$ **42.** $x = -1$

43. $x = 5$ **44.** $x^2 + y^2 = 3x$ **45.** $5x + y = 8$ **46.** $y = \dfrac{2}{x}$ **47. a.** has symmetry   **b.** may or may not have symmetry

c. may or may not have symmetry   **48. a.** may or may not have symmetry   **b.** has symmetry   **c.** may or may not have symmetry

49. a. has symmetry   **b.** has symmetry   **c.** has symmetry

50.

51.

52.

53.

54.

55.

56.

57.

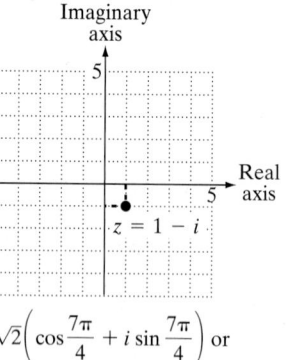

$\sqrt{2}\left(\cos\dfrac{7\pi}{4} + i\sin\dfrac{7\pi}{4}\right)$ or

$\sqrt{2}(\cos 315° + i\sin 315°)$

58.

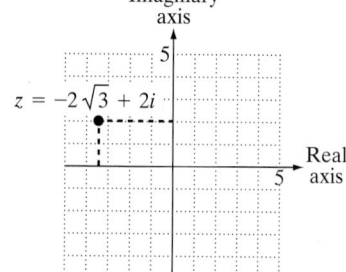

$4(\cos 150° + i\sin 150°)$ or

$4\left(\cos\dfrac{5\pi}{6} + i\sin\dfrac{5\pi}{6}\right)$

59.

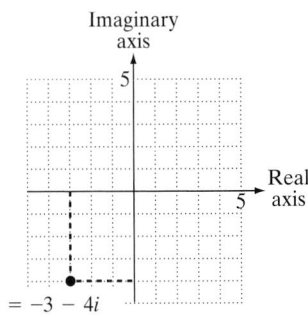

$z = -3 - 4i$

$z = -3 - 4i \approx 5(\cos 233° + i \sin 233°)$

60.

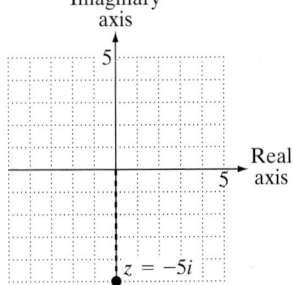

$z = -5i$

$5\left(\cos\dfrac{3\pi}{2} + i \sin\dfrac{3\pi}{2}\right)$ or $5(\cos 270° + i \sin 270°)$

61. $z = 4 + 4\sqrt{3}i$
62. $z = -2\sqrt{3} - 2i$
63. $z = -3 + 3\sqrt{3}i$
64. $z \approx -0.1 + 0.6i$
65. $15(\cos 110° + i \sin 110°)$
66. $\cos 265° + i \sin 265°$
67. $40(\cos \pi + i \sin \pi)$
68. $2(\cos 5° + i \sin 5°)$

69. $\dfrac{1}{2}(\cos \pi + i \sin \pi)$ **70.** $2\left(\cos\dfrac{7\pi}{6} + i \sin\dfrac{7\pi}{6}\right)$ **71.** $4 + 4\sqrt{3}i$ **72.** $-32\sqrt{3} + 32i$ **73.** $\dfrac{1}{128}i$ **74.** $64 - 64\sqrt{3}i$

75. $128 + 128i$ **76.** $7(\cos 25° + i \sin 25°); 7(\cos 205° + i \sin 205°)$

77. $5(\cos 55° + i \sin 55°); 5(\cos 175° + i \sin 175°); 5(\cos 295° + i \sin 295°)$ **78.** $\sqrt{3} + i; -1 + \sqrt{3}i; -\sqrt{3} - i; 1 - \sqrt{3}i$

79. $\sqrt{3} + i; -\sqrt{3} + i; -2i$ **80.** $\dfrac{1}{2} + \dfrac{\sqrt{3}}{2}i; -1; \dfrac{1}{2} - \dfrac{\sqrt{3}}{2}i$ **81.** $\dfrac{\sqrt[5]{8}}{2} + \dfrac{\sqrt[5]{8}}{2}i; \approx -0.49 + 0.95i; \approx -1.06 - 0.17i; \approx -0.17 - 1.06i; \approx 0.95 - 0.49i$

82.

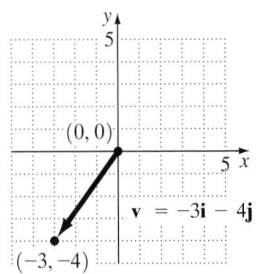

$\mathbf{v} = -3\mathbf{i} - 4\mathbf{j}$

$(-3, -4)$

$; 5$

83.

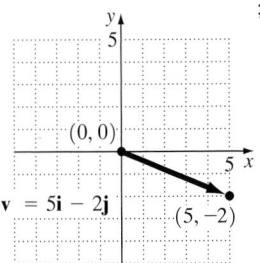

$\mathbf{v} = 5\mathbf{i} - 2\mathbf{j}$

$(5, -2)$

$; \sqrt{29}$

84.

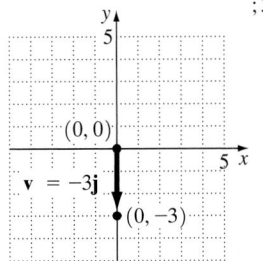

$\mathbf{v} = -3\mathbf{j}$

$(0, -3)$

$; 3$

85. $3\mathbf{i} - 2\mathbf{j}$ **86.** $\mathbf{i} - 2\mathbf{j}$ **87.** $-\mathbf{i} + 2\mathbf{j}$ **88.** $-3\mathbf{i} + 12\mathbf{j}$ **89.** $12\mathbf{i} - 51\mathbf{j}$ **90.** $2\sqrt{26}$ **91.** $\dfrac{4}{5}\mathbf{i} - \dfrac{3}{5}\mathbf{j}$ **92.** $-\dfrac{1}{\sqrt{5}}\mathbf{i} + \dfrac{2}{\sqrt{5}}\mathbf{j}$

93. $6\mathbf{i} + 6\sqrt{3}\mathbf{j}$ **94.** 270 lb; 27.7°   **95. a.** $13.59\mathbf{i} + 6.34\mathbf{j}$ **b.** 14.0 mph   **c.** 13.9°   **96.** 4   **97.** 2; 86.1°   **98.** -32; 124.8°

99. 1; 71.6°   **100.** orthogonal   **101.** not orthogonal   **102.** $\mathbf{v}_1 = \text{proj}_{\mathbf{w}}\mathbf{v} = \dfrac{50}{41}\mathbf{i} + \dfrac{40}{41}\mathbf{j}; \mathbf{v}_2 = -\dfrac{132}{41}\mathbf{i} + \dfrac{165}{41}\mathbf{j}$

103. $\mathbf{v}_1 = \text{proj}_{\mathbf{w}}\mathbf{v} = -\dfrac{3}{2}\mathbf{i} + \dfrac{1}{2}\mathbf{j}; \mathbf{v}_2 = \dfrac{1}{2}\mathbf{i} + \dfrac{3}{2}\mathbf{j}$ **104.** 1115 ft-lb   **105.** Answers may vary.

Chapter 7 Test

1. 8.0   **2.** 6.2   **3.** 206 sq in.

4.

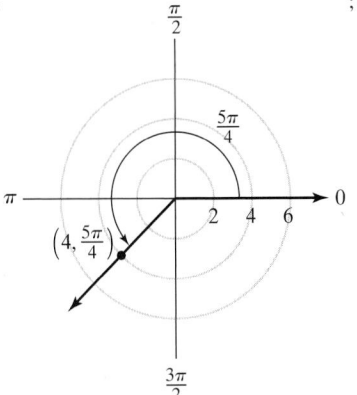

$\left(4, \dfrac{5\pi}{4}\right)$

; Ordered pairs may vary.

5. $\left(\sqrt{2}, \dfrac{7\pi}{4}\right)$

6. $r^2 = 6r \cos \theta$ or $r = 6 \cos \theta$

7. $y = 4$

8.

9.

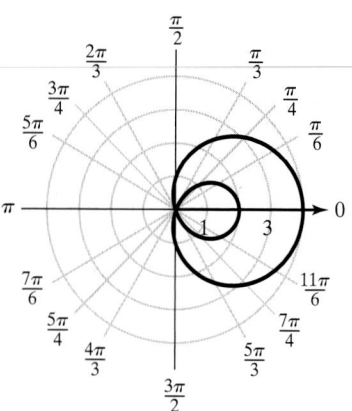

10. $2(\cos 150° + i \sin 150°)$ or $2\left(\cos \dfrac{5\pi}{6} + i \sin \dfrac{5\pi}{6}\right)$ **11.** $50(\cos 20° + i \sin 20°)$ **12.** $\dfrac{1}{2}\left(\cos \dfrac{\pi}{6} + i \sin \dfrac{\pi}{6}\right)$

13. $32(\cos 50° + i \sin 50°)$ **14.** $3; -\dfrac{3}{2} + \dfrac{\sqrt{3}}{2}i; -\dfrac{3}{2} - \dfrac{\sqrt{3}}{2}i$ **15. a.** $i + 2j$ **b.** $\sqrt{5}$ **16.** $-23i + 22j$ **17.** -18 **18.** $138°$

19. $-\dfrac{9}{5}i + \dfrac{18}{5}j$ **20.** 1.0 mi   **21.** 323 pounds; 3.4°   **22.** 1966 ft-lb

Cumulative Review Exercises (Chapter 1–7)

1. $\{-1, 2, i, -i\}$ **2.** $\dfrac{\pi}{6}, \dfrac{5\pi}{6}, \dfrac{\pi}{2}$ **3.** $\{x | x < -4 \text{ or } x > 2\}$ **4.** $\dfrac{3\pi}{4}, \dfrac{7\pi}{4}$

5.

6.

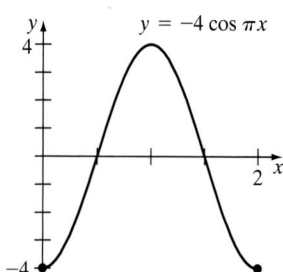

7. $\sin \theta \csc \theta - \cos^2 \theta = \sin \theta\left(\dfrac{1}{\sin \theta}\right) - \cos^2 \theta = 1 - \cos^2 \theta = \sin^2 \theta$

8. $\cos\left(\theta + \dfrac{3\pi}{2}\right) = \cos \theta \cos \dfrac{3\pi}{2} - \sin \theta \sin \dfrac{3\pi}{2} = \cos \theta(0) - \sin \theta(-1) = \sin \theta$

9. slope is $-\dfrac{1}{2}$; y-intercept is 2.   **10.** 0   **11.** $\dfrac{\sqrt{5}}{5}$ **12.** $\{x | x \le 5\}$ **13.** $\{x | x \ne 3, x \ne -3\}$ **14.** 1.5 sec; 44 ft

15. a. 4 m   **b.** $\dfrac{5}{2\pi}$ **c.** $\dfrac{2\pi}{5}$ sec   **16.** $\dfrac{\sqrt{\sqrt{2} + 2}}{2}$ **17. a.** $5i + 23j$ **b.** -12

18. $\log_b \dfrac{\sqrt{x}}{x^2 + 1}$ **19.** $y = -\dfrac{1}{2}x + 1$ **20. a.** 0.014   **b.** 73 words   **c.** about 144 min

CHAPTER 8

Section 8.1

Check Point Exercises

1. solution **2.** $\{(3, 2)\}$ **3.** $\{(1, -2)\}$ **4.** $\{(2, -1)\}$ **5.** $\left\{\left(\dfrac{23}{16}, \dfrac{3}{8}\right)\right\}$ **6.** $\varnothing$ **7.** $\{(x, y)|y = 4x - 4\}$ **8. a.** $C(x) = 300{,}000 + 30x$
b. $R(x) = 80x$ **c.** $(6000, 480{,}000)$; The company will breakeven if it produces and sells 6000 pairs of shoes. **9.** $\$30$; 400 units

Exercise Set 8.1

1. solution **3.** not a solution **5.** $\{(1, 3)\}$ **7.** $\{(5, 1)\}$ **9.** $\{(-22, -5)\}$ **11.** $\{(0, 0)\}$ **13.** $\{(3, -2)\}$ **15.** $\{(5, 4)\}$
17. $\{(7, 3)\}$ **19.** $\{(2, -1)\}$ **21.** $\{(3, 0)\}$ **23.** $\{(-4, 3)\}$ **25.** $\{(3, 1)\}$ **27.** $\{(1, -2)\}$ **29.** $\left\{\left(\dfrac{7}{25}, \dfrac{1}{25}\right)\right\}$ **31.** $\varnothing$
33. $\{(x, y)|y = 3x - 5\}$ **35.** $\{(1, 4)\}$ **37.** $\{(x, y)|x + 3y = 2\}$ **39.** $\{(-5, -1)\}$ **41.** $\left\{\left(\dfrac{29}{22}, -\dfrac{5}{11}\right)\right\}$
43. $x + y = 7$; $x - y = -1$; 3 and 4 **45.** $3x - y = 1$; $x + 2y = 12$; 2 and 5 **47. a.** $C(x) = 18{,}000 + 20x$ **b.** $R(x) = 80x$
c. $(300, 24{,}000)$; This means the company will break even if it produces and sells 300 canoes. **49. a.** $C(x) = 30{,}000 + 2500x$
b. $R(x) = 3125$ **c.** $(48, 150{,}000)$; The play will break even if 48 sold-out performances are produced. **51. a.** 6500 tickets can be sold.
6200 tickets can be supplied. **b.** $\$50$; 6250 tickets **53.** 2004; $12\dfrac{8}{33}$; The lines intersect at $\left(39\dfrac{13}{33}, 12\dfrac{8}{33}\right)$. **55. a.** $E(x) = 508 + 25x$
b. $E(x) = 345 + 9x$ **c.** 26; 2011; $\$1158$ for college graduates, $\$579$ for high school graduates **57.** Pan pizza: 1120 calories; beef
burrito: 430 calories **59.** Scrambled eggs: 366 mg cholesterol; Double Beef Whopper: 175 mg cholesterol **61.** 50 rooms with kitchen
facilities, 150 rooms without kitchen facilities **63.** 100 ft long by 80 ft wide **65.** Rate rowing in still water: 6 mph; rate of the current:
2 mph **67.** $x = 55$, $y = 35$ **81.** Answers may vary. **83.** the twin who always lies

Section 8.2

Check Point Exercises

1. $(-1) - 2(-4) + 3(5) = 22$; $2(-1) - 3(-4) - 5 = 5$; $3(-1) + (-4) - 5(5) = -32$ **2.** $\{(1, 4, -3)\}$ **3.** $\{(4, 5, 3)\}$
4. $y = 3x^2 - 12x + 13$

Exercise Set 8.2

1. solution **3.** solution **5.** $\{(2, 3, 3)\}$ **7.** $\{(2, -1, 1)\}$ **9.** $\left\{\left(\dfrac{1}{3}, -\dfrac{2}{5}, \dfrac{1}{2}\right)\right\}$ **11.** $\{(3, 1, 5)\}$ **13.** $\{(1, 0, -3)\}$
15. $\{(1, -5, -6)\}$ **17.** $\left\{\left(\dfrac{1}{2}, \dfrac{1}{3}, -1\right)\right\}$ **19.** 7, 4 and 5 **21.** $y = 2x^2 - x + 3$ **23.** $y = 2x^2 + x - 5$
25. a. $(0, 1180)$, $(1, 1070)$, $(2, 1230)$ **b.** $c = 1180$ $a + b + c = 1070$ $4a + 2b + c = 1230$ **c.** $y = 135x^2 - 245x + 1180$
27. a. $y = -16x^2 + 40x + 200$ **b.** $y = 0$ when $x = 5$; The ball hit the ground after 5 seconds **29.** Carnegie: $\$100$ billion; Vanderbilt:
$\$96$ billion; Gates: $\$60$ billion **31.** 200 $\$8$ tickets; 150 $\$10$ tickets; 50 $\$12$ tickets **33.** $\$1200$ at 8%, $\$2000$ at 10%, and $\$3500$ at 12%
35. $x = 60$, $y = 55$, $z = 65$ **43.** Answers may vary.

Section 8.3

Check Point Exercises

1. $\dfrac{2}{x - 3} + \dfrac{3}{x + 4}$ **2.** $\dfrac{2}{x} - \dfrac{2}{x - 1} + \dfrac{3}{(x - 1)^2}$ **3.** $\dfrac{2}{x + 3} + \dfrac{6x - 8}{x^2 + x + 2}$ **4.** $\dfrac{2x}{x^2 + 1} + \dfrac{-x + 3}{(x^2 + 1)^2}$

Exercise Set 8.3

1. $\dfrac{A}{x - 2} + \dfrac{B}{x + 1}$ **3.** $\dfrac{A}{x + 2} + \dfrac{B}{x - 3} + \dfrac{C}{(x - 3)^2}$ **5.** $\dfrac{A}{x - 1} + \dfrac{Bx + C}{x^2 + 1}$ **7.** $\dfrac{Ax + B}{x^2 + 4} + \dfrac{Cx + D}{(x^2 + 4)^2}$ **9.** $\dfrac{3}{x - 3} - \dfrac{2}{x - 2}$
11. $\dfrac{7}{x - 9} - \dfrac{4}{x + 2}$ **13.** $\dfrac{24}{7(x - 4)} + \dfrac{25}{7(x + 3)}$ **15.** $\dfrac{4}{7(x - 3)} - \dfrac{8}{7(2x + 1)}$ **17.** $\dfrac{3}{x} + \dfrac{2}{x - 1} - \dfrac{1}{x + 3}$ **19.** $\dfrac{3}{x} + \dfrac{4}{x + 1} - \dfrac{3}{x - 1}$
21. $\dfrac{6}{x - 1} - \dfrac{5}{(x - 1)^2}$ **23.** $\dfrac{1}{x - 2} - \dfrac{2}{(x - 2)^2} - \dfrac{5}{(x - 2)^3}$ **25.** $\dfrac{7}{x} - \dfrac{6}{x - 1} + \dfrac{10}{(x - 1)^2}$ **27.** $\dfrac{1}{4(x + 1)} + \dfrac{3}{4(x - 1)} + \dfrac{1}{2(x - 1)^2}$
29. $\dfrac{3}{x - 1} + \dfrac{2x - 4}{x^2 + 1}$ **31.** $\dfrac{2}{x + 1} + \dfrac{3x - 1}{x^2 + 2x + 2}$ **33.** $\dfrac{1}{4x} + \dfrac{1}{x^2} + \dfrac{-x + 4}{4(x^2 + 4)}$ **35.** $\dfrac{4}{x + 1} + \dfrac{2x - 3}{x^2 + 1}$ **37.** $\dfrac{x + 1}{x^2 + 2} - \dfrac{2x}{(x^2 + 2)^2}$
39. $\dfrac{x - 2}{x^2 - 2x + 3} + \dfrac{2x + 1}{(x^2 - 2x + 3)^2}$ **41.** $\dfrac{3}{x - 2} + \dfrac{x - 1}{x^2 + 2x + 4}$ **43.** $\dfrac{1}{x} - \dfrac{1}{x + 1} \cdot \dfrac{99}{100}$ **53.** When the denominator of a rational

expression contains a power of a cubic factor, set up a partial fraction decomposition with quadratic numerators.

$(Ax^2 + Bx + C, Dx^2 + Ex + F,$ etc.$)$. For example: $\dfrac{x^3 + 1}{(x^3 + 2)^2} = \dfrac{Ax^2 + bx + C}{x^3 + 2} + \dfrac{Dx^2 + Ex + F}{(x^3 + 2)^2} = \dfrac{1}{x^3 + 2} + \dfrac{-1}{(x^3 + 2)^2}.$

55. $\dfrac{2}{x - 3} + \dfrac{2x + 5}{x^2 + 3x + 3}$

Section 8.4

Check Point Exercises

1. $\{(0, 1), (4, 17)\}$　　**2.** $\left\{\left(-\dfrac{6}{5}, \dfrac{3}{5}\right), (2, -1)\right\}$　　**3.** $\{(3, 2), (3, -2), (-3, 2), (-3, -2)\}$　　**4.** $\{(0, 5)\}$

5. length: 7 ft; width: 3 ft or length: 3 ft; width: 7 ft

Exercise Set 8.4

1. $\{(-3, 5), (2, 0)\}$　　**3.** $\{(1, 1), (2, 0)\}$　　**5.** $\{(4, -10), (-3, 11)\}$　　**7.** $\{(4, 3), (-3, -4)\}$　　**9.** $\left\{\left(-\dfrac{3}{2}, -4\right), (2, 3)\right\}$

11. $\{(-5, -4), (3, 0)\}$　　**13.** $\{(3, 1), (-3, -1), (1, 3), (-1, -3)\}$　　**15.** $\{(4, -3), (-1, 2)\}$　　**17.** $\{(0, 1), (4, -3)\}$

19. $\{(3, 2), (3, -2), (-3, 2), (-3, -2)\}$　　**21.** $\{(3, 2), (3, -2), (-3, 2), (-3, -2)\}$　　**23.** $\{(2, 1), (2, -1), (-2, 1), (-2, -1)\}$

25. $\{(3, 4), (3, -4)\}$　　**27.** $\{(0, 2), (0, -2), (-1, \sqrt{3}), (-1, -\sqrt{3})\}$　　**29.** $\{(2, 1), (2, -1), (-2, 1), (-2, -1)\}$

31. $\{(-2\sqrt{2}, -\sqrt{2}), (-1, -4), (1, 4), (2\sqrt{2}, \sqrt{2})\}$　　**33.** $\{(2, 2), (4, 1)\}$　　**35.** $\{(0, 0), (-1, 1)\}$　　**37.** $\{(0, 0), (-2, 2), (2, 2)\}$

39. $\left\{(-4, 1), \left(-\dfrac{5}{2}, \dfrac{1}{4}\right)\right\}$　　**41.** $\left\{\left(\dfrac{12}{5}, -\dfrac{29}{5}\right), (-2, 3)\right\}$　　**43.** 4 and 6　　**45.** 2 and 1, 2 and −1, −2 and 1, or −2 and −1

47. $(0, -4), (-2, 0), (2, 0)$　　**49.** 11 ft and 7 ft　　**51.** width: 6 in.; length: 8 in.　　**53.** $x = 5$ m, $y = 2$ m　　**61.** (b) is true.

63. $b = 6, a = 8$　　**65.** $\{(10{,}000, 5)\}$

Section 8.5

Check Point Exercises

1.

2.

3.

4.

5.

6.
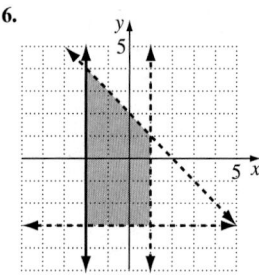

Exercise Set 8.5

1.

3.

5.

7.
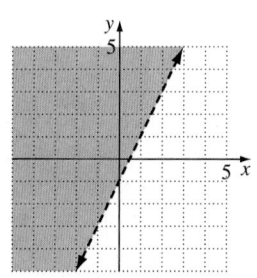

9. **11.** **13.** **15.**

17. **19.** **21.** **23.**

25. **27.** **29.** **31.**

33. **35.** **37.** no solution

39. **41.**

43. **45.** **47.** **49.**

51.

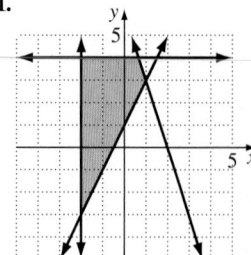

53. $5T - 7P \leq 70, T \geq 0, P \geq 0; A(50, 30): 5(50) - 7(30) = 40 \leq 70$

55. a. $50x + 150y > 2000$

b.

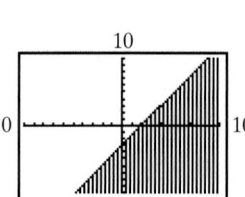

57. $x + y \leq 15,000$
$x \geq 2,000$
$y \geq 3x$
$x \geq 0$
$y \geq 0$

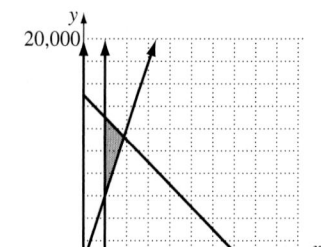

c. (20, 20): 20 children and 20 adults will cause the elevator to be overloaded.

59. Answers may vary   **61. a.** 28.1   **b.** overweight

69.

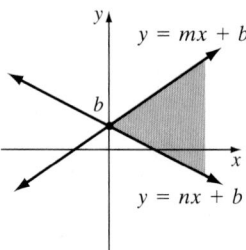

71.

73.

75. Answers may vary.
77. Answers may vary.

79.

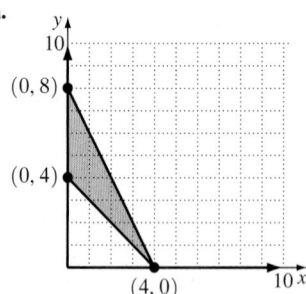

Section 8.6

Check Point Exercises

1. $z = 25x + 55y$ **2.** $x + y \leq 80$ **3.** $30 \leq x \leq 80; 10 \leq y \leq 30$; objective function: $z = 25x + 55y$; constraints:
$x + y \leq 80; 30 \leq x \leq 80; 10 \leq y \leq 30$ **4.** 50 bookshelves and 30 desks; $2900   **5.** 30

Exercise Set 8.6

1. (1, 2): 17; (2, 10): 70; (7, 5): 65; (8, 3): 58; maximum: $z = 70$; minimum: $z = 17$
3. (0, 0): 0; (0, 8): 400; (4, 9): 610; (8, 0): 320; maximum: $z = 610$; minimum: $z = 0$

5. a.

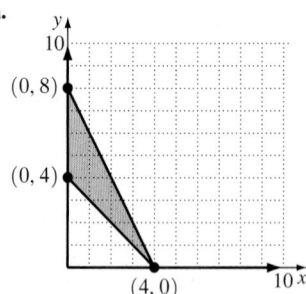

b. (0, 8): 16; (0, 4): 8; (4, 0): 12
c. maximum value: 16 at $x = 0$ and $y = 8$

7. a.

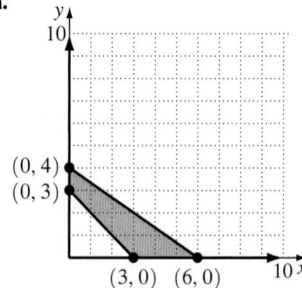

b. (0, 4): 4; (0, 3): 3; (3, 0): 12; (6, 0): 24
c. maximum value: 24 at $x = 6$ and $y = 0$

9. a.

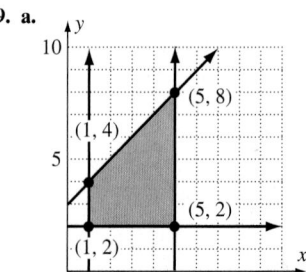

b. (1, 2): −1; (1, 4): −5; (5, 8): −1; (5, 2): 11
maximum value: 11 at $x = 5$ and $y = 2$

11. a.

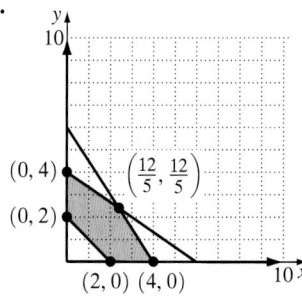

b. $(0, 4)$: 8; $(0, 2)$: 4; $(2, 0)$: 8; $(4, 0)$: 16;

$\left(\dfrac{12}{5}, \dfrac{12}{5}\right)$: $\dfrac{72}{5}$

c. maximum value: 16 at $x = 4$ and $y = 0$

13. a.

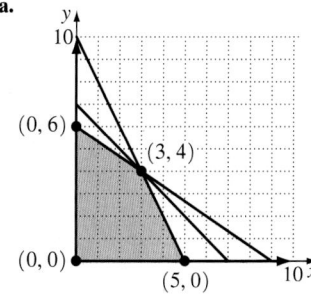

b. $(0, 6)$: 72, $(0, 0)$: 0; $(5, 0)$: 50; $(3, 4)$: 78

c. maximum value: 78 at $x = 3$ and $y = 4$

15. a. $z = 125x + 200y$

b. $x \leq 450$; $y \leq 200$; $600x + 900y \leq 360,000$

c.

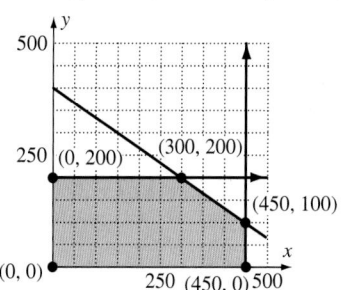

d. $(0, 0)$: 0; $(0, 200)$: 40,000; $(300, 200)$: 77,500;

$(450, 100)$: 76,250; $(450, 0)$: 56,250

e. 300; 200; $77,500

17. 40 model A bicycles and no model B bicycles   **19.** 300 cartons of food and 200 cartons of clothing   **21.** 50 students and 100 parents

23. 10 Boeing 727s and 42 Falcon 20s

29.

31.

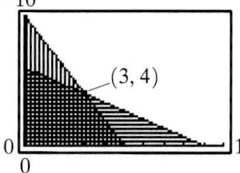

33. $5000 in stocks and $5000 in bonds

Chapter 8 Review Exercises

1. $\{(1, 5)\}$ **2.** $\{(2, 3)\}$ **3.** $\{(2, -3)\}$ **4.** $\varnothing$ **5.** $\{(x, y)|3x - 6y = 12\}$ **6. a.** $C(x) = 60,000 + 200x$ **b.** $R(x) = 450x$

c. $(240, 108,000)$; This means the company will break even if it produces and sells 240 desks.   **7.** 250 copies can be supplied and sold for

$12.50 each.   **8.** 5.8 million pounds of potato chips, 4.6 million pounds of tortilla chips   **9.** $80 per day for the room, $60 per day for the car

10. 3 apples and 2 avocados   **11.** $\{(0, 1, 2)\}$ **12.** $\{(2, 1, -1)\}$ **13.** $y = 3x^2 - 4x + 5$ **14. a.** $(0, 3.5), (15, 5.0), (29, 4.1)$

b. $c = 3.5$ $225a + 15b + c = 5.0$ $841a + 29b + c = 4.1$ **15.** United States: 22; Colombia: 18; India: 10

16. $\dfrac{3}{5(x - 3)} + \dfrac{2}{5(x + 2)}$ **17.** $\dfrac{6}{x - 4} + \dfrac{5}{x + 3}$ **18.** $\dfrac{2}{x} + \dfrac{3}{x + 2} - \dfrac{1}{x - 1}$ **19.** $\dfrac{2}{x - 2} + \dfrac{5}{(x - 2)^2}$

20. $-\dfrac{4}{x - 1} + \dfrac{4}{x - 2} - \dfrac{2}{(x - 2)^2}$ **21.** $\dfrac{6}{5(x - 2)} + \dfrac{-6x + 3}{5(x^2 + 1)}$ **22.** $\dfrac{5}{x - 3} + \dfrac{2x - 1}{x^2 + 4}$ **23.** $\dfrac{x}{x^2 + 4} - \dfrac{4x}{(x^2 + 4)^2}$

24. $\dfrac{4x + 1}{x^2 + x + 1} + \dfrac{2x - 2}{(x^2 + x + 1)^2}$ **25.** $\{(4, 3), (1, 0)\}$ **26.** $\{(0, 1), (-3, 4)\}$ **27.** $\{(1, -1), (-1, 1)\}$

28. $\{(3, \sqrt{6}), (3, -\sqrt{6}), (-3, \sqrt{6}), (-3, -\sqrt{6})\}$ **29.** $\{(2, 2), (-2, -2)\}$ **30.** $\{(9, 6), (1, 2)\}$ **31.** $\{(-3, -1), (1, 3)\}$

32. $\left\{\left(\dfrac{1}{2}, 2\right), (-1, -1)\right\}$ **33.** $\left\{\left(\dfrac{5}{2}, -\dfrac{7}{2}\right), (0, -1)\right\}$ **34.** $\{(2, -3), (-2, -3), (3, 2), (-3, 2)\}$

35. $\{(3, 1), (3, -1), (-3, 1), (-3, -1)\}$ **36.** 8 m and 5 m   **37.** $(1, 6), (3, 2)$ **38.** $x = 46$ and $y = 28$ or $x = 50$ and $y = 20$

39.

40.

41.

42.

43.

44.

45.

46.

47.

48.

49.

50.

51. no solution

52.

53.

54.

55.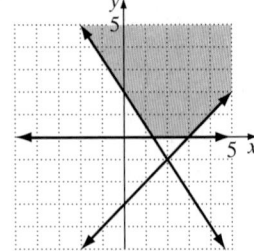

56. $(2, 2)$: 10; $(4, 0)$: 8; $\left(\frac{1}{2}, \frac{1}{2}\right)$: $\frac{5}{2}$; $(1, 0)$: 2; maximum value: 10; minimum value: 2

57.

58.

59.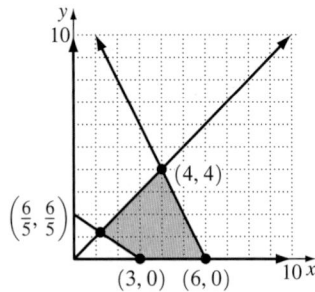

60. a. $z = 500x + 350y$
b. $x + y \leq 200$; $x \geq 10$; $y \geq 80$
c.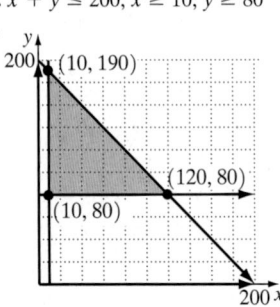

d. $(10, 80)$: 33,000; $(10, 190)$: 71,500; $(120, 80)$: 88,000
e. 120; 80; 88,000

61. 480 of model A and 240 of model B

Chapter 8 Test

1. $\{(1, -3)\}$ **2.** $\{(4, -2)\}$ **3.** $\{(1, 3, 2)\}$ **4.** $\{(4, -3), (-3, 4)\}$ **5.** $\{(3, 2), (3, -2), (-3, 2), (-3, -2)\}$

6. $\dfrac{-1}{10(x + 1)} + \dfrac{x + 9}{10(x^2 + 9)}$

7.   **8.**   **9.**   **10.**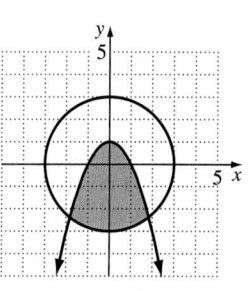

11. 26   **12.** Shrimp: 42 mg; scallops: 15mg   **13. a.** $C(x) = 360,000 + 850x$ **b.** $R(x) = 1150x$ **c.** $(1200, 1,380,000)$; The company will break even if it produces and sells 1200 computers.   **14.** $y = x^2 - 3$ **15.** $x = 7.5$ ft and $y = 24$ ft or $x = 12$ ft and $y = 15$ ft   **16.** 50 regular and 100 deluxe jet skis; \$35,000

Cumulative Review Exercises (Chapters 1–8)

1. $\{3, 4\}$ **2.** $\left\{\dfrac{2 + i\sqrt{3}}{2}, \dfrac{2 - i\sqrt{3}}{2}\right\}$ **3.** $(-18, 6)$ **4.** $(1, 7)$ **5.** $\left\{-3, \dfrac{1}{2}, 2\right\}$ **6.** $\{-2\}$ **7.** $\{2\}$ **8.** $\{-2 + \log_3 11\}$

9.   **10.**   **11.**   **12.**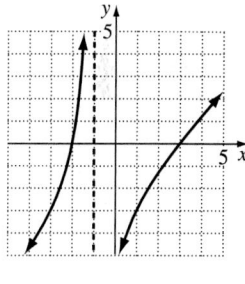

13. $3 + 5 \log_2 x$ **14.** 10.99%   **15.** $f^{-1}(x) = \dfrac{1}{7}x + \dfrac{3}{7}$ **16.** $g(f(x)) = 21x - 16$ **17.** Answers may vary.

18. $\left\{\left(-\dfrac{1}{2}, \dfrac{1}{2}\right), (2, 8)\right\}$ **19.** 4 m by 9 m   **20.** A plane with an initial landing speed of 90 ft per second needs 562 ft to land. There is a problem since 550 ft is not enough.   **21.** $\sec \theta - \cos \theta = \dfrac{1}{\cos \theta} - \cos \theta = \dfrac{1 - \cos^2 \theta}{\cos \theta} = \dfrac{\sin^2 \theta}{\cos \theta} = \dfrac{\sin \theta}{\cos \theta} \sin \theta = \tan \theta \sin \theta$

22. $\tan x + \tan y = \dfrac{\sin x}{\cos x} + \dfrac{\sin y}{\cos y} = \dfrac{\sin x \cos y + \sin y \cos x}{\cos x \cos y} = \dfrac{\sin(x + y)}{\cos x \cos y}$ **23.** $\{0, \pi\}$ **24.** $\left\{0, \dfrac{\pi}{3}, \dfrac{5\pi}{3}\right\}$ **25.** 92.9

CHAPTER 9

Section 9.1

Check Point Exercises

1. $\{(4, -3, 1)\}$ **2. a.** $\begin{bmatrix} 1 & 6 & -3 & | & 7 \\ 4 & 12 & -20 & | & 8 \\ -3 & -2 & 1 & | & -9 \end{bmatrix}$ **b.** $\begin{bmatrix} 1 & 3 & -5 & | & 2 \\ 1 & 6 & -3 & | & 7 \\ -3 & -2 & 1 & | & -9 \end{bmatrix}$ **c.** $\begin{bmatrix} 4 & 12 & -20 & | & 8 \\ 1 & 6 & -3 & | & 7 \\ 0 & 16 & -8 & | & 12 \end{bmatrix}$

3. $\{(5, 2, 3)\}$ **4.** $\{(1, -1, 2, -3)\}$ **5.** $\{(5, 2, 3)\}$

Exercise Set 9.1

1. $\begin{bmatrix} 2 & 1 & 2 & | & 2 \\ 3 & -5 & -1 & | & 4 \\ 1 & -2 & -3 & | & -6 \end{bmatrix}$ **3.** $\begin{bmatrix} 1 & -1 & 1 & | & 8 \\ 0 & 1 & -12 & | & -15 \\ 0 & 0 & 1 & | & 1 \end{bmatrix}$ **5.** $\begin{bmatrix} 5 & -2 & -3 & | & 0 \\ 1 & 1 & 0 & | & 5 \\ 2 & 0 & -3 & | & 4 \end{bmatrix}$ **7.** $\begin{bmatrix} 2 & 5 & -3 & 1 & | & 2 \\ 0 & 3 & 1 & 0 & | & 4 \\ 1 & -1 & 5 & 0 & | & 9 \\ 5 & -5 & -2 & 0 & | & 1 \end{bmatrix}$

9. $5x + 3z = -11$
$y - 4z = 12$
$7x + 2y = 3$

11. $w + x + 4y + z = 3$
$-w + x - y = 7$
$2w + 5z = 11$
$12y + 4z = 5$

13. $x - 4z = 5$; $\left\{ \left(3, 7, -\dfrac{1}{2} \right) \right\}$
$y - 12z = 13$
$z = -\dfrac{1}{2}$

15. $x + \dfrac{1}{2}y + z = \dfrac{11}{2}$; $(\{1, 1, 4\})$
$y + \dfrac{3}{2}z = 7$
$z = 4$

17. $w - x + y + z = 3$; $\{(2, 1, -1, 3)\}$
$x - 2y - z = 0$
$y + 6z = 17$
$z = 3$

19. $\begin{bmatrix} 1 & -3 & 2 & | & 5 \\ 1 & 5 & -5 & | & 0 \\ 3 & 0 & 4 & | & 7 \end{bmatrix}$ **21.** $\begin{bmatrix} 1 & -3 & 2 & | & 0 \\ 0 & 10 & -7 & | & 7 \\ 2 & -2 & 1 & | & 3 \end{bmatrix}$ **23.** $\begin{bmatrix} 1 & -1 & 1 & 1 & | & 3 \\ 0 & 1 & -2 & -1 & | & 0 \\ 0 & 2 & 1 & 2 & | & 5 \\ 0 & 6 & -3 & -1 & | & -9 \end{bmatrix}$

25. $R_2: -3, -18; R_3: -12, -15; R_2: -\dfrac{3}{5}, -\dfrac{18}{5}; R_3: -12, -15$ **27.** $\{(1, -1, 2)\}$ **29.** $\{(3, -1, -1)\}$ **31.** $\{2, -1, 1\}$ **33.** $\{(2, 1, 1)\}$

35. $\{(2, -1, 1)\}$ **37.** $\{(-1, 2, -2)\}$ **39.** $\{(1, 2, -1)\}$ **41.** $\{(1, 2, 3, -2)\}$ **43.** $\{(0, -3, 0, -3)\}$ **45. a.** $\{(-10, 475, 3500)\}$
b. 7760; 12 years into the program, there are 7760 alligators. **47.** $x + y + z = 100$; yes: 34%; no: 61%; not sure: 5%
$x + z = y - 22$
$2x = y + 7$

49. $40x + 200y + 400z = 660$; 4 oz of Food A; $\dfrac{1}{2}$ oz of Food B; 1 oz of Food C **61.** $y = x^3 + 2x^2 + 5x - 3$
$5x + 2y + 4z = 25$
$30x + 10y + 300z = 425$

Section 9.2

Check Point Exercises

1. $\varnothing$ **2.** $\{(11t + 13, 5t + 4, t)\}$ **3.** $\{(t + 50, -2t + 10, t)\}$
4. a. $w + z = 15$ **b.** $\{(-t + 15, t + 15, -t + 30, t)\}$ **c.** $w = 5; x = 25; y = 20$
$w + x = 30$
$x + y = 45$
$y + z = 30$

Exercise Set 9.2

1. $\varnothing$ **3.** $\left\{ \left(-2t + 2, 2t + \dfrac{1}{2}, t \right) \right\}$ **5.** $\{(-3, 4, -2)\}$ **7.** $\{(5 - 2t, -2 + t, t)\}$ **9.** $\{(-1, 2, 1, 1)\}$ **11.** $\{(1, 3, 2, 1)\}$

13. $\{(1, -2, 1, 1)\}$ **15.** $\left\{ \left(1 + \dfrac{1}{3}t, \dfrac{1}{3}t, t \right) \right\}$ **17.** $\{(-13t + 5, 5t, t)\}$ **19.** $\left\{ \left(2t - \dfrac{5}{4}, \dfrac{13}{4}, t \right) \right\}$ **21.** $\{(1, -t - 1, 2, t)\}$

23. $\left\{ \left(-\dfrac{2}{11}t + \dfrac{81}{11}, \dfrac{1}{22}t + \dfrac{10}{11}, \dfrac{4}{11}t - \dfrac{8}{11}, t \right) \right\}$ **25.** $z + 12 = x + 6$ **27.** $\{(t + 6, t + 2, t)\}$

29. a. $w + z = 380$ **b.** $\{(380 - t, 220 + t, 50 + t, t)\}$ **c.** $w = 330, x = 270, y = 100, z = 50$
$w + x = 600$
$x - y = 170$
$y - z = 50$

31. a. The system has no solution, so there is no way to satisfy these dietary requirements with no Food 1 available.
b. 4 oz of Food 1, 0 oz of Food 2, 10 oz of Food 3; 2 oz of Food 1, 5 oz of Food 2, 9 oz of Food 3 (other answers are possible).
37. $a = 1$ or $a = 3$

Section 9.3

Check Point Exercises

1. a. 3×2 **b.** $a_{12} = -2$; $a_{31} = 1$ **2. a.** $\begin{bmatrix} 2 & 0 \\ 9 & -10 \end{bmatrix}$ **b.** $\begin{bmatrix} 9 & -4 \\ -9 & 7 \\ 5 & -2 \end{bmatrix}$ **3. a.** $\begin{bmatrix} 6 & 12 \\ -48 & -30 \end{bmatrix}$ **b.** $\begin{bmatrix} -14 & -1 \\ 25 & 10 \end{bmatrix}$ **4.** $\begin{bmatrix} -4 & 3 \\ -3 & \frac{13}{3} \end{bmatrix}$

5. $\begin{bmatrix} 7 & 6 \\ 13 & 12 \end{bmatrix}$ **6.** $[30]$; $\begin{bmatrix} 2 & 0 & 4 \\ 6 & 0 & 12 \\ 14 & 0 & 28 \end{bmatrix}$ **7. a.** $\begin{bmatrix} 2 & 18 & 11 & 9 \\ 0 & 10 & 8 & 2 \end{bmatrix}$ **b.** The product is undefined.

8. $\begin{bmatrix} 2 & 1 & 1 \\ 2 & 1 & 1 \\ 2 & 2 & 1 \end{bmatrix} + \begin{bmatrix} -1 & 2 & 2 \\ -1 & 2 & 2 \\ -1 & -1 & 2 \end{bmatrix} = \begin{bmatrix} 1 & 3 & 3 \\ 1 & 3 & 3 \\ 1 & 1 & 3 \end{bmatrix}$ **9.** $2548

Exercise Set 9.3

1. a. 2×3 **b.** a_{32} does not exist; $a_{23} = -1$ **3. a.** 3×4 **b.** $a_{32} = \frac{1}{2}$; $a_{23} = -6$ **5.** $x = 6$; $y = 4$ **7.** $x = 4$; $y = 6$; $z = 3$

9. a. $\begin{bmatrix} 9 & 10 \\ 3 & 9 \end{bmatrix}$ **b.** $\begin{bmatrix} -1 & -8 \\ 3 & -5 \end{bmatrix}$ **c.** $\begin{bmatrix} -16 & -4 \\ -12 & -8 \end{bmatrix}$ **d.** $\begin{bmatrix} 22 & 21 \\ 9 & 20 \end{bmatrix}$ **11. a.** $\begin{bmatrix} 3 & 2 \\ 6 & 2 \\ 5 & 7 \end{bmatrix}$ **b.** $\begin{bmatrix} -1 & 4 \\ 0 & 6 \\ 5 & 5 \end{bmatrix}$ **c.** $\begin{bmatrix} -4 & -12 \\ -12 & -16 \\ -20 & -24 \end{bmatrix}$

d. $\begin{bmatrix} 7 & 7 \\ 15 & 8 \\ 15 & 20 \end{bmatrix}$ **13. a.** $\begin{bmatrix} -3 \\ -1 \\ 0 \end{bmatrix}$ **b.** $\begin{bmatrix} 7 \\ -7 \\ 2 \end{bmatrix}$ **c.** $\begin{bmatrix} -8 \\ 16 \\ -4 \end{bmatrix}$ **d.** $\begin{bmatrix} -4 \\ -6 \\ 1 \end{bmatrix}$ **15. a.** $\begin{bmatrix} 8 & 0 & -4 \\ 14 & 0 & 6 \\ -1 & 0 & 0 \end{bmatrix}$ **b.** $\begin{bmatrix} -4 & -20 & 0 \\ 14 & 24 & 14 \\ 9 & -4 & 4 \end{bmatrix}$

c. $\begin{bmatrix} -8 & 40 & 8 \\ -56 & -48 & -40 \\ -16 & 8 & -8 \end{bmatrix}$ **d.** $\begin{bmatrix} 18 & -10 & -10 \\ 42 & 12 & 22 \\ 2 & -2 & 2 \end{bmatrix}$ **17.** $\begin{bmatrix} -8 & -8 \\ 2 & -9 \\ 8 & -4 \end{bmatrix}$ **19.** $\begin{bmatrix} -1 & 3 \\ -1 & \frac{9}{2} \\ -1 & -2 \end{bmatrix}$ **21.** $\begin{bmatrix} \frac{1}{3} & \frac{13}{3} \\ -\frac{4}{3} & 6 \\ \frac{7}{3} & -\frac{4}{3} \end{bmatrix}$ **23.** $\begin{bmatrix} 7 & 27 \\ -8 & 36 \\ -17 & -4 \end{bmatrix}$

25. $\begin{bmatrix} \frac{27}{2} & \frac{31}{2} \\ -4 & 18 \\ -\frac{29}{2} & 6 \end{bmatrix}$ **27. a.** $\begin{bmatrix} 0 & 16 \\ 12 & 8 \end{bmatrix}$ **b.** $\begin{bmatrix} -7 & 3 \\ 29 & 15 \end{bmatrix}$ **29. a.** $[30]$ **b.** $\begin{bmatrix} 1 & 2 & 3 & 4 \\ 2 & 4 & 6 & 8 \\ 3 & 6 & 9 & 12 \\ 4 & 8 & 12 & 16 \end{bmatrix}$ **31. a.** $\begin{bmatrix} 4 & -5 & 8 \\ 6 & -1 & 5 \\ 0 & 4 & -6 \end{bmatrix}$

b. $\begin{bmatrix} 5 & -2 & 7 \\ 17 & -3 & 2 \\ 3 & 0 & -5 \end{bmatrix}$ **33. a.** $\begin{bmatrix} 6 & 8 & 16 \\ 11 & 16 & 24 \\ 1 & -1 & 12 \end{bmatrix}$ **b.** $\begin{bmatrix} 38 & 27 \\ -16 & -4 \end{bmatrix}$ **35. a.** $\begin{bmatrix} 0 & 0 \\ 0 & 0 \end{bmatrix}$ **b.** $\begin{bmatrix} 4 & -1 & -3 & 1 \\ -1 & 4 & -3 & 2 \\ 14 & -11 & -3 & -1 \\ 25 & -25 & 0 & -5 \end{bmatrix}$

37. $\begin{bmatrix} 17 & 7 \\ -5 & -11 \end{bmatrix}$ **39.** $\begin{bmatrix} 11 & -1 \\ -7 & -3 \end{bmatrix}$ **41.** $A - C$ is not defined because A is 3×2 and C is 2×2.   **43.** $\begin{bmatrix} 16 & -16 \\ -12 & 12 \\ 0 & 0 \end{bmatrix}$

45. a $\begin{bmatrix} 1 & 3 & 1 \\ 3 & 3 & 3 \\ 1 & 3 & 1 \end{bmatrix}$ **b.** $\begin{bmatrix} 1 & 3 & 1 \\ 3 & 3 & 3 \\ 1 & 3 & 1 \end{bmatrix} + \begin{bmatrix} -1 & -1 & -1 \\ -1 & -1 & -1 \\ -1 & -1 & -1 \end{bmatrix} = \begin{bmatrix} 0 & 2 & 0 \\ 2 & 2 & 2 \\ 0 & 2 & 0 \end{bmatrix}$ **c.** $\begin{bmatrix} 1 & 3 & 1 \\ 3 & 3 & 3 \\ 1 & 3 & 1 \end{bmatrix} + \begin{bmatrix} 1 & -2 & 1 \\ -2 & -2 & -2 \\ 1 & -2 & 1 \end{bmatrix} = \begin{bmatrix} 2 & 1 & 2 \\ 1 & 1 & 1 \\ 2 & 1 & 2 \end{bmatrix}$

47. a. $\begin{bmatrix} 0 & -3 & -3 & -1 & -1 & 0 \\ 0 & 0 & 1 & 1 & 5 & 5 \end{bmatrix}$

49. a. $A = \begin{bmatrix} 26 & 18 \\ 12 & 11 \end{bmatrix}$ **b.** $B = \begin{bmatrix} 31 & 26 \\ 17 & 16 \end{bmatrix}$

b. The effect is a reflection across the y-axis.

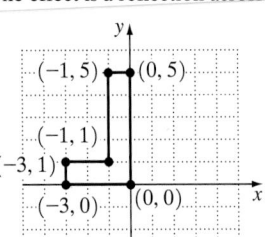

c. $\begin{bmatrix} 5 & 8 \\ 5 & 5 \end{bmatrix}$; This represents the change in the percentage of people with college degrees from 1990 to 2000.

51. a. System 1: The midterm and final both count for 50% of the course grade.
System 2: The midterm counts for 30% of the course grade and the final counts for 70%.

b. $\begin{bmatrix} 84 & 87.2 \\ 79 & 81 \\ 90 & 88.4 \\ 73 & 68.6 \\ 69 & 73.4 \end{bmatrix}$ System 1 grades are listed first (if different).
Student 1: B; Student 2: C or B; Student 3: A or B; Student 4: C or D;
Student 5: D or C

65. When 2 square matrices of this type are multiplied, the product will equal the product of each corresponding element on the diagonal and zeros elsewhere.

Section 9.4

Check Point Exercises

1. $AB = I_2; BA = I_2$ **2.** $\begin{bmatrix} 3 & -7 \\ -2 & 5 \end{bmatrix}$ **3.** $\begin{bmatrix} 1 & 2 \\ 1 & 3 \end{bmatrix}$ **4.** $\begin{bmatrix} 3 & -2 & -4 \\ 3 & -2 & -5 \\ -1 & 1 & 2 \end{bmatrix}$ **5.** $\{(4, -2, 1)\}$

6. The encoded message is $-7, 10, -53, 77$.   **7.** The decoded message is $2, 1, 19, 5$ or BASE.

Exercise Set 9.4

1. $AB = I_2; BA = I_2; B = A^{-1}$ **3.** $AB = \begin{bmatrix} 8 & -16 \\ -2 & 7 \end{bmatrix}; BA = \begin{bmatrix} 12 & 12 \\ 1 & 3 \end{bmatrix}; B \neq A^{-1}$ **5.** $AB = I_2; BA = I_2; B = A^{-1}$

7. $AB = I_3; BA = I_3; B = A^{-1}$ **9.** $AB = I_3; BA = I_3; B = A^{-1}$ **11.** $AB = I_4; BA = I_4; B = A^{-1}$ **13.** $\begin{bmatrix} \dfrac{2}{7} & -\dfrac{3}{7} \\ \dfrac{1}{7} & \dfrac{2}{7} \end{bmatrix}$ **15.** $\begin{bmatrix} 1 & \dfrac{1}{2} \\ 2 & \dfrac{3}{2} \end{bmatrix}$

17. A does not have an inverse.   **19.** $\begin{bmatrix} \dfrac{1}{2} & 0 & 0 \\ 0 & \dfrac{1}{4} & 0 \\ 0 & 0 & \dfrac{1}{6} \end{bmatrix}$ **21.** $\begin{bmatrix} 1 & 1 & 2 \\ 1 & 1 & 1 \\ 2 & 3 & 4 \end{bmatrix}$ **23.** $\begin{bmatrix} 1 & 0 & 1 \\ 1 & 1 & 2 \\ 3 & 2 & 6 \end{bmatrix}$ **25.** $\begin{bmatrix} -3 & 2 & -4 \\ -1 & 1 & -1 \\ 8 & -5 & 10 \end{bmatrix}$

27. $\begin{bmatrix} 1 & 0 & 0 & 0 \\ 0 & -1 & 0 & 0 \\ 0 & 0 & \dfrac{1}{3} & 0 \\ -1 & 0 & 0 & 1 \end{bmatrix}$ **29.** $\begin{bmatrix} 6 & 5 \\ 5 & 4 \end{bmatrix}\begin{bmatrix} x \\ y \end{bmatrix} = \begin{bmatrix} 13 \\ 10 \end{bmatrix}$ **31.** $\begin{bmatrix} 1 & 3 & 4 \\ 1 & 2 & 3 \\ 1 & 4 & 3 \end{bmatrix}\begin{bmatrix} x \\ y \\ z \end{bmatrix} = \begin{bmatrix} -3 \\ -2 \\ -6 \end{bmatrix}$ **33.** $\begin{aligned} 4x - 7y &= -3 \\ 2x - 3y &= 1 \end{aligned}$ **35.** $\begin{aligned} 2x - z &= 6 \\ 3y &= 9 \\ x + y &= 5 \end{aligned}$

37. a. $\begin{bmatrix} 2 & 6 & 6 \\ 2 & 7 & 6 \\ 2 & 7 & 7 \end{bmatrix}\begin{bmatrix} x \\ y \\ z \end{bmatrix} = \begin{bmatrix} 8 \\ 10 \\ 9 \end{bmatrix}$ **b.** $\{(1, 2, -1)\}$ **39. a.** $\begin{bmatrix} 1 & -1 & 1 \\ 0 & 2 & -1 \\ 2 & 3 & 0 \end{bmatrix}\begin{bmatrix} x \\ y \\ z \end{bmatrix} = \begin{bmatrix} 8 \\ -7 \\ 1 \end{bmatrix}$ **b.** $\{(2, -1, 5)\}$

41. a. $\begin{bmatrix} 1 & -1 & 2 & 0 \\ 0 & 1 & -1 & 1 \\ -1 & 1 & -1 & 2 \\ 0 & -1 & 1 & -2 \end{bmatrix}\begin{bmatrix} w \\ x \\ y \\ z \end{bmatrix} = \begin{bmatrix} -3 \\ 4 \\ 2 \\ -4 \end{bmatrix}$ **b.** $\{(2, 3, -1, 0)\}$

43. The encoded message is $27, -19, 32, -20.$; The decoded message is $8, 5, 12, 16$ or HELP.

45. The encoded message is $14, 85, -33, 4, 18, -7, -18, 19, -9$.   **57.** $\begin{bmatrix} 1 & 1 \\ 2 & 3 \end{bmatrix}$ **59.** $\begin{bmatrix} 1 & 0 & 1 \\ 2 & 1 & 3 \\ -1 & 1 & 1 \end{bmatrix}$

61. $\begin{bmatrix} 0 & -1 & 0 & 1 \\ -1 & -5 & 0 & 3 \\ -2 & -4 & 1 & -2 \\ -1 & -4 & 0 & 1 \end{bmatrix}$ **63.** $\{(2, 3, -5)\}$ **65.** $\{(1, 2, -1)\}$ **67.** $\{(2, 1, 3, -2, 4)\}$ **69.** Answers may vary.   **71.** (c) is true.

73. Answers may vary.   **75.** $a = 3$ or $a = -2$

Section 9.5

Check Point Exercises

1. a. -4 **b.** -17 **2.** $\{(4, -2)\}$ **3.** 80   **4.** -24 **5.** $\{(2, -3, 4)\}$ **6.** -250

Exercise Set 9.5

1. 1   **3.** -29 **5.** 0   **7.** 33   **9.** $-\dfrac{7}{16}$ **11.** $\{(5, 2)\}$ **13.** $\{(2, -3)\}$ **15.** $\{(3, -1)\}$ **17.** The system is dependent.

19. $\{(4, 2)\}$ **21.** $\{(7, 4)\}$ **23.** The system is inconsistent.   **25.** The system is dependent.   **27.** 72   **29.** -75 **31.** 0

33. $\{(-5, -2, 7)\}$ **35.** $\{(2, -3, 4)\}$ **37.** $\{(3, -1, 2)\}$ **39.** $\{(2, 3, 1)\}$ **41.** -200 **43.** 195   **45.** 28 sq units   **47.** yes

49. The equation of the line is $y = -\dfrac{11}{5}x + \dfrac{8}{5}$.   **61.** 13,200

63. a. a^2 **b.** a^3 **c.** a^4 **d.** Each determinant has zeros below the main diagonal and a's everywhere else.
e. Each determinant equals a raised to the power equal to the order of the determinant.
65. The sign of the value is changed when 2 columns are interchanged in a 2nd order determinant.

67. $\begin{vmatrix} x & y & 1 \\ x_1 & y_1 & 1 \\ x_2 & y_2 & 1 \end{vmatrix} = x(y_1 - y_2) - y(x_1 - x_2) + (x_1 y_2 - x_2 y_1) = 0$; solving for y,

$y = \dfrac{y_1 - y_2}{x_1 - x_2}x + \dfrac{x_1 y_2 - x_2 y_1}{x_1 - x_2}$, and $m = \dfrac{y_1 - y_2}{x_1 - x_2}$ and $b = \dfrac{x_1 y_2 - x_2 y_1}{x_1 - x_2}$.

Chapter 9 Review Exercises

1. $\begin{aligned} x + y + 3z &= 12; \{(1, 2, 3)\} \\ y - 2z &= -4 \\ z &= 3 \end{aligned}$ **2.** $\begin{aligned} w - 2y + 2z &= 1 \;;\{(3, -1, 2, 1)\} \\ x + y - z &= 0 \\ y - \tfrac{7}{3}z &= -\tfrac{1}{3} \\ z &= 1 \end{aligned}$ **3.** $\left[\begin{array}{ccc|c} 1 & 2 & 2 & 2 \\ 0 & 1 & -1 & 2 \\ 0 & 0 & 9 & -9 \end{array}\right]$ **4.** $\left[\begin{array}{ccc|c} 1 & -1 & \tfrac{1}{2} & -\tfrac{1}{2} \\ 1 & 2 & -1 & 2 \\ 6 & 4 & 3 & 5 \end{array}\right]$

5. $\{(1, 3, -4)\}$ **6.** $\{(-2, -1, 0)\}$ **7.** $\{(2, -2, 3, 4)\}$ **8. a.** $a = -2; b = 32; c = 42$ **b.** 2:00 P.M.; 170 parts per million

9. $\varnothing$ **10.** $\{(2t + 4, t + 1, t)\}$ **11.** $\{(-37t + 2, 16t, -7t + 1, t)\}$ **12.** $\{(7t + 18, -3t - 7, t)\}$

13. a. $\begin{aligned} x + z &= 750 \\ y - z &= -250 \\ x + y &= 500 \end{aligned}$ **b.** $\{(-t + 750, t - 250, t)\}$ **c.** $x = 350; y = 150$ **14.** $x = -5; y = 6; z = 6$

15. $\begin{bmatrix} 0 & 2 & 3 \\ 8 & 1 & 3 \end{bmatrix}$ **16.** $\begin{bmatrix} 0 & -4 \\ 6 & 4 \\ 2 & -10 \end{bmatrix}$ **17.** $\begin{bmatrix} -4 & 4 & -1 \\ -2 & -5 & 5 \end{bmatrix}$ **18.** Not possible since B is 3×2 and C is 3×3.   **19.** $\begin{bmatrix} 2 & 3 & 8 \\ 21 & 5 & 5 \end{bmatrix}$

20. $\begin{bmatrix} -12 & 14 & 0 \\ 2 & -14 & 18 \end{bmatrix}$ **21.** $\begin{bmatrix} 0 & -10 & -15 \\ -40 & -5 & -15 \end{bmatrix}$ **22.** $\begin{bmatrix} -1 & -16 \\ 8 & 1 \end{bmatrix}$ **23.** $\begin{bmatrix} -10 & -6 & 2 \\ 16 & 3 & 4 \\ -23 & -16 & 7 \end{bmatrix}$ **24.** $\begin{bmatrix} -6 & 4 & -8 \\ 0 & 5 & 11 \\ -17 & 13 & -19 \end{bmatrix}$

25. $\begin{bmatrix} 10 & 5 \\ -2 & -30 \end{bmatrix}$ **26.** Not possible since AB is 2×2 and BA is 3×3.   **27.** $\begin{bmatrix} 7 & 6 & 5 \\ 2 & -1 & 11 \end{bmatrix}$ **28.** $\begin{bmatrix} -6 & -22 & -40 \\ 9 & 43 & 58 \\ -14 & -48 & -94 \end{bmatrix}$ **29.** $\begin{bmatrix} -2 & -6 \\ 3 & \tfrac{1}{3} \end{bmatrix}$

30. $\begin{bmatrix} 2 & 2 & 2 \\ 1 & 2 & 1 \\ 1 & 2 & 1 \end{bmatrix}$ **31.** $\begin{bmatrix} 1 & 1 & 1 \\ -1 & 1 & -1 \\ -1 & -1 & -1 \end{bmatrix}$ **32. a.** $\begin{bmatrix} 610,000 & 694,000 \\ 936,000 & 1,065,000 \\ 298,000 & 338,500 \end{bmatrix}$ **b.** The rows of AB correspond to the outlets, the columns

represent the wholesale and retail prices. The entries tell how much value in wholesale or retail is at each outlet. **c.** $610,000

d. $1,065,000 **e.** $40,500 **33.** $AB = \begin{bmatrix} 1 & 7 \\ 0 & 5 \end{bmatrix}$; $BA = \begin{bmatrix} 1 & 0 \\ 1 & 5 \end{bmatrix}$; $B \neq A^{-1}$ **34.** $AB = I_3$; $BA = I_3$; $B = A^{-1}$ **35.** $\begin{bmatrix} 3 & 1 \\ 2 & 1 \end{bmatrix}$

36. $\begin{bmatrix} -\dfrac{3}{5} & \dfrac{1}{5} \\ 1 & 0 \end{bmatrix}$ **37.** $\begin{bmatrix} 3 & 0 & -2 \\ -6 & 1 & 4 \\ 1 & 0 & -1 \end{bmatrix}$ **38.** $\begin{bmatrix} 8 & -8 & 5 \\ -3 & 2 & -1 \\ -1 & -1 & 1 \end{bmatrix}$ **39. a.** $\begin{bmatrix} 1 & 1 & 2 \\ 0 & 1 & 3 \\ 3 & 0 & -2 \end{bmatrix}\begin{bmatrix} x \\ y \\ z \end{bmatrix} = \begin{bmatrix} 7 \\ -2 \\ 0 \end{bmatrix}$ **b.** $\{(-18, 79, -27)\}$

40. a. $\begin{bmatrix} 1 & -1 & 2 \\ 0 & 1 & -1 \\ 1 & 0 & 2 \end{bmatrix}\begin{bmatrix} x \\ y \\ z \end{bmatrix} = \begin{bmatrix} 12 \\ -5 \\ 10 \end{bmatrix}$ **b.** $\{(4, -2, 3)\}$ **41.** The encoded message is 96, 135, 46, 63; The decoded message is 18, 21, 12, 5

or RULE. **42.** 17 **43.** 4 **44.** -86 **45.** -236 **46.** 4 **47.** 16 **48.** $\left\{\left(\dfrac{7}{4}, -\dfrac{25}{8}\right)\right\}$ **49.** $\{(2, -7)\}$

50. $\{(23, -12, 3)\}$ **51.** $\{(-3, 2, 1)\}$ **52.** $a = \dfrac{5}{8}$; $b = -50$; $c = 1150$; 30- and 50-year-olds are involved in an average of 212.5 automo-

bile accidents per day.

Chapter 9 Test

1. $\left\{\left(-3, \dfrac{1}{2}, 1\right)\right\}$ **2.** $\{(t, t - 1, t)\}$ **3.** $\begin{bmatrix} 5 & 4 \\ 1 & 11 \end{bmatrix}$ **4.** $\begin{bmatrix} 5 & -2 \\ 1 & -1 \\ 4 & -1 \end{bmatrix}$ **5.** $\begin{bmatrix} \dfrac{3}{5} & -\dfrac{2}{5} \\ \dfrac{1}{5} & \dfrac{1}{5} \end{bmatrix}$ **6.** $\begin{bmatrix} -1 & 2 \\ -5 & 4 \end{bmatrix}$ **7.** $AB = I_3$; $BA = I_3$

8. a. $\begin{bmatrix} 3 & 5 \\ 2 & -3 \end{bmatrix}\begin{bmatrix} x \\ y \end{bmatrix} = \begin{bmatrix} 9 \\ -13 \end{bmatrix}$ **b.** $\begin{bmatrix} \dfrac{3}{19} & \dfrac{5}{19} \\ \dfrac{2}{19} & -\dfrac{3}{19} \end{bmatrix}$ **c.** $\{(-2, 3)\}$ **9.** 18 **10.** $x = 2$

Cumulative Review Exercises (Chapters 1–9)

1. $\left\{\dfrac{-1 + \sqrt{33}}{4}, \dfrac{-1 - \sqrt{33}}{4}\right\}$ **2.** $\left[\dfrac{1}{2}, \infty\right)$ **3.** $\{6\}$ **4.** $\left\{-4, \dfrac{1}{3}, 1\right\}$ **5.** $\{\ln 5, \ln 9\}$ **6.** $\{1\}$ **7.** $\{(7, -4, 6)\}$

8. $y = -1$ **9.** $f^{-1}(x) = \dfrac{x^2 + 7}{4}\ (x \geq 0)$

10.

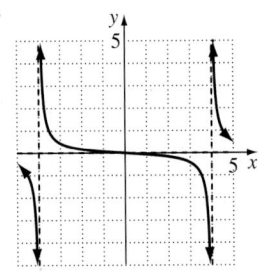

11. $f(x) = (x + 2)(x - 3)(2x + 1)(2x - 1)$ **13. a.** $A = 900e^{-0.017t}$ **b.** 759.30 g

12.

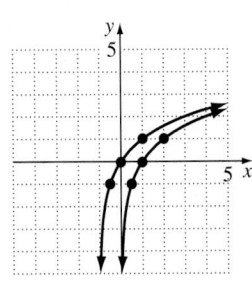

14. $\begin{bmatrix} 2 & -1 \\ 13 & 1 \end{bmatrix}$ **15.** $\dfrac{8}{x - 3} + \dfrac{-2}{x - 2} + \dfrac{-3}{x + 2}$

16.

17.

18.

19.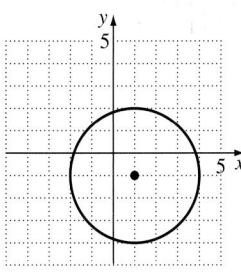

20. $= x^2 + 2x - 2$

21.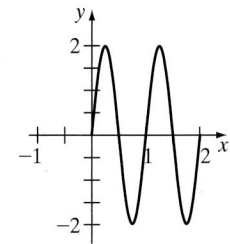

22. $\dfrac{3}{5}$ **23.** $\dfrac{\cos 2x}{\cos x - \sin x} = \dfrac{\cos^2 x - \sin^2 x}{\cos x - \sin x} = \dfrac{(\cos x + \sin x)(\cos x - \sin x)}{\cos x - \sin x} = \cos x + \sin x$ **24.** $\dfrac{3\pi}{2}$ **25.** $2\mathbf{i} - 13\mathbf{j}$

CHAPTER 10

Section 10.1

Check Point Exercises

1. foci at $(-3\sqrt{3}, 0)$ and $(3\sqrt{3}, 0)$ **2.** foci at $(0, -\sqrt{7})$ and $(0, \sqrt{7})$ **3.** $\dfrac{x^2}{9} + \dfrac{y^2}{5} = 1$

4. foci at $(-1 - \sqrt{5}, 2)$ and $(-1 + \sqrt{5}, 2)$ **5.** Yes

    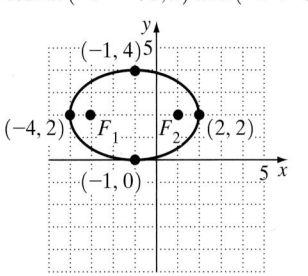

Exercise Set 10.1

1. foci at $(-2\sqrt{3}, 0)$ and $(2\sqrt{3}, 0)$ **3.** foci at $(0, -3\sqrt{3})$ and $(0, 3\sqrt{3})$ **5.** foci at $(0, -\sqrt{39})$ and $(0, \sqrt{39})$ **7.** foci at $(0, -4\sqrt{2})$ and $(0, 4\sqrt{2})$

      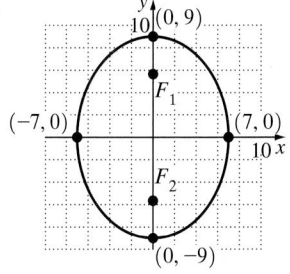

9. foci at $(0, -2)$ and $(0, 2)$ **11.** foci at $\left(-\dfrac{\sqrt{3}}{2}, 0\right)$ and $\left(\dfrac{\sqrt{3}}{2}, 0\right)$ **13.** foci at $(0, -\sqrt{21})$ and $(0, \sqrt{21})$ **15.** foci at $(-2\sqrt{3}, 0)$ and $(2\sqrt{3}, 0)$

      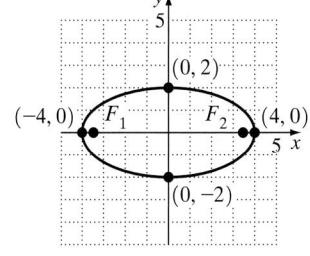

17. foci at $(0, -\sqrt{2})$ and $(0, \sqrt{2})$

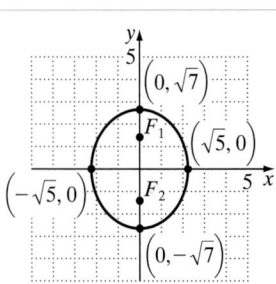

19. $\dfrac{x^2}{4} + \dfrac{y^2}{1} = 1$; foci at $(-\sqrt{3}, 0)$ and $(\sqrt{3}, 0)$ **21.** $\dfrac{x^2}{1} + \dfrac{y^2}{4} = 1$; foci at $(0, \sqrt{3})$ and $(0, -\sqrt{3})$

23. $\dfrac{(x+1)^2}{4} + \dfrac{(y-1)^2}{1} = 1$; foci at $(-1 - \sqrt{3}, 1)$ and $(-1 + \sqrt{3}, 1)$ **25.** $\dfrac{x^2}{64} + \dfrac{y^2}{39} = 1$

27. $\dfrac{x^2}{33} + \dfrac{y^2}{49} = 1$ **29.** $\dfrac{x^2}{13} + \dfrac{y^2}{9} = 1$ **31.** $\dfrac{x^2}{16} + \dfrac{y^2}{4} = 1$ **33.** $\dfrac{(x+2)^2}{4} + \dfrac{(y-3)^2}{25} = 1$

35. $\dfrac{(x-7)^2}{4} + \dfrac{(y-6)^2}{9} = 1$

37. foci at $(2 - \sqrt{5}, 1)$ and $(2 + \sqrt{5}, 1)$

39. foci at $(-3 - 2\sqrt{3}, 2)$ and $(-3 + 2\sqrt{3}, 2)$

41. foci at $(4, 2)$ and $(4, -6)$

43. foci at $(0, 2 + \sqrt{11}), (0, 2 - \sqrt{11})$

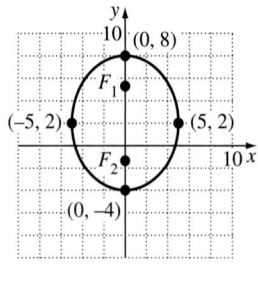

45. foci at $(-3 - 2\sqrt{2}, 2)$ and $(-3 + 2\sqrt{2}, 2)$ **47.** foci at $(1, -3 + \sqrt{3})$ and $(1, -3 - \sqrt{3})$ **49.** foci at $(1, -3 + \sqrt{5})$ and $(1, -3 - \sqrt{5})$

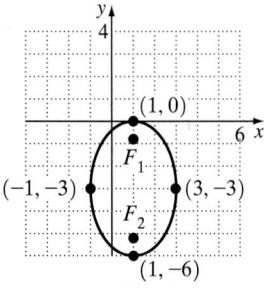

51. $\dfrac{(x-2)^2}{25} + \dfrac{(y+1)^2}{9} = 1$
    foci at $(-2, -1)$ and $(6, -1)$

53. $\dfrac{(x-1)^2}{16} + \dfrac{(y+2)^2}{9} = 1$
    foci at $(1 - \sqrt{7}, -2)$ and $(1 + \sqrt{7}, -2)$

55. $\dfrac{(x+2)^2}{16} + \dfrac{(y-3)^2}{64} = 1$
    foci at $(-2, 3 + 4\sqrt{3})$ and $(-2, 3 - 4\sqrt{3})$

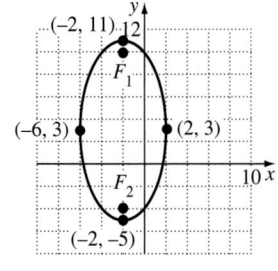

57. Yes   **59. a.** $\dfrac{x^2}{2304} + \dfrac{y^2}{529} = 1$ **b.** about 42 feet   **71.** $\dfrac{x^2}{\frac{36}{5}} + \dfrac{y^2}{36} = 1$ **73.** The large circle has radius 5 with center $(0, 0)$. Its

equation is $x^2 + y^2 = 25$. The small circle has radius 3 with center $(0, 0)$. Its equation is $x^2 + y^2 = 9$.   **75.** Answers may vary.

Section 10.2

Check Point Exercises

1. a. vertices at $(5, 0)$ and $(-5, 0)$; foci at $(\sqrt{41}, 0)$ and $(-\sqrt{41}, 0)$ **b.** vertices at $(0, 5)$ and $(0, -5)$; foci at $(0, \sqrt{41})$ and $(0, -\sqrt{41})$

2. $\dfrac{y^2}{9} - \dfrac{x^2}{16} = 1$

3. foci at $(-3\sqrt{5}, 0)$ and $(3\sqrt{5}, 0)$;
asymptotes: $y = \pm \dfrac{1}{2}x$

4. foci at $(0, \sqrt{5})$ and $(0, -\sqrt{5})$;
asymptotes: $y = \pm 2x$

5. foci at $(3 - \sqrt{5}, 1)$ and $(3 + \sqrt{5}, 1)$;
asymptotes: $(y - 1) = \pm \dfrac{1}{2}(x - 3)$

 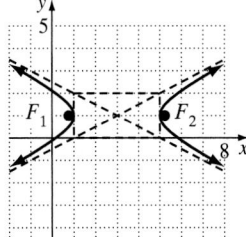

6. $\dfrac{x^2}{2{,}722{,}500} - \dfrac{y^2}{25{,}155{,}900} = 1$

Exercise Set 10.2

1. vertices at $(2, 0)$ and $(-2, 0)$; foci at $(\sqrt{5}, 0)$ and $(-\sqrt{5}, 0)$; graph (b)   **3.** vertices at $(0, 2)$ and $(0, -2)$; foci at $(0, \sqrt{5})$ and $(0, -\sqrt{5})$;

graph (a)   **5.** $y^2 - \dfrac{x^2}{8} = 1$ **7.** $\dfrac{x^2}{9} - \dfrac{y^2}{7} = 1$ **9.** $\dfrac{y^2}{36} - \dfrac{x^2}{9} = 1$ **11.** $\dfrac{(x - 4)^2}{4} - \dfrac{(y + 2)^2}{5} = 1$

13. foci: $(\pm\sqrt{34}, 0)$;
asymptotes: $y = \pm \dfrac{5}{3}x$

15. foci: $(\pm 2\sqrt{41}, 0)$;
asymptotes: $y = \pm \dfrac{4}{5}x$

17. foci: $(0, \pm 2\sqrt{13})$;
asymptotes: $y = \pm \dfrac{2}{3}x$

19. foci: $\left(0, \pm \dfrac{\sqrt{5}}{2}\right)$;
asymptotes: $y = \pm \dfrac{1}{2}x$

   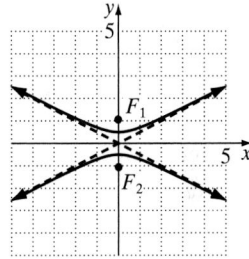

21. foci: $(\pm\sqrt{13}, 0)$;
asymptotes: $y = \pm \dfrac{3}{2}x$

23. foci: $(0, \pm\sqrt{34})$;
asymptotes: $y = \pm \dfrac{5}{3}x$

25. foci: $(\pm 2, 0)$;
asymptotes: $y = \pm x$

  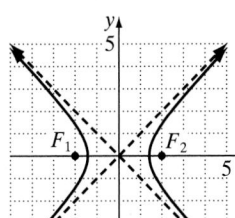

27. $\dfrac{x^2}{9} - \dfrac{y^2}{25} = 1$

29. $\dfrac{y^2}{4} - \dfrac{x^2}{9} = 1$

31. $\dfrac{(x - 2)^2}{4} - \dfrac{(y + 3)^2}{9} = 1$

33. foci: $(-9, -3), (1, -3)$;

asymptotes: $(y + 3) = \pm\dfrac{4}{3}(x + 4)$

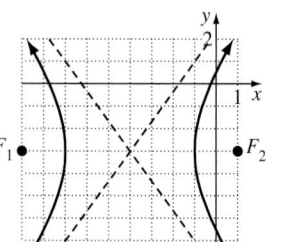

35. foci: $(-3 \pm \sqrt{41}, 0)$;

asymptotes: $y = \pm\dfrac{4}{5}(x + 3)$

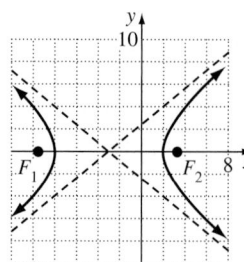

37. foci: $(1, -2 \pm 2\sqrt{5})$;

asymptotes: $(y + 2) = \pm\dfrac{1}{2}(x - 1)$

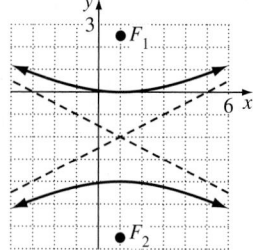

39. foci: $(3 \pm \sqrt{5}, -3)$;

asymptotes: $(y + 3) = \pm\dfrac{1}{2}(x - 3)$

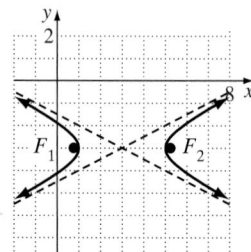

41. foci: $(1 \pm \sqrt{6}, 2)$;

asymptotes: $(y - 2) = \pm(x - 1)$

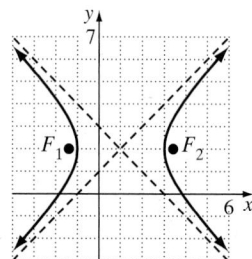

43. $(x - 1)^2 - (y + 2)^2 = 1$;

foci: $(1 \pm \sqrt{2}, -2)$;

asymptotes: $(y + 2) = \pm(x - 1)$

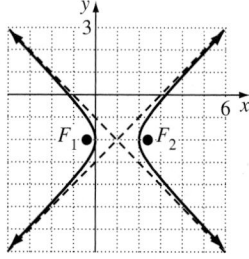

45. $\dfrac{(y + 1)^2}{4} - \dfrac{(x + 2)^2}{0.25} = 1$;

foci: $(-2, -1 \pm \sqrt{4.25})$;

asymptotes: $(y + 1) = \pm 4(x + 2)$

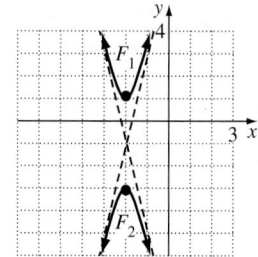

47. $\dfrac{(x - 2)^2}{9} - \dfrac{(y - 3)^2}{4} = 1$;

foci: $(2 \pm \sqrt{13}, 3)$

asymptotes: $(y - 3) = \pm\dfrac{2}{3}(x - 2)$

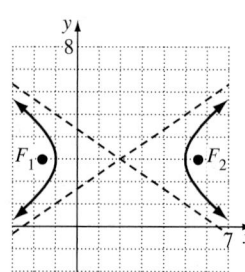

49. $\dfrac{y^2}{4} - \dfrac{(x - 4)^2}{25} = 1$

foci: $(4, \pm\sqrt{29})$;

asymptotes: $y = \pm\dfrac{2}{5}(x - 4)$

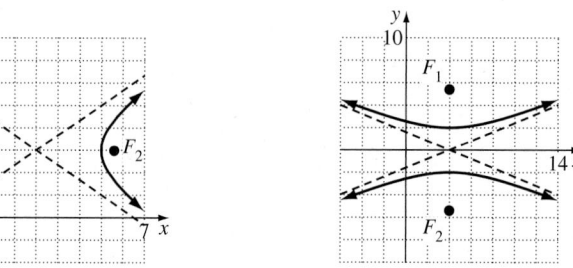

51. If M_1 is located 2640 feet to the right of the origin on the x-axis, the explosion is located on the right branch of the hyperbola given by

the equation $\dfrac{x^2}{1,210,000} - \dfrac{y^2}{5,759,600} = 1$.   **53.** 40 yd

65.

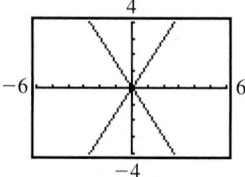

; No. Two intersecting lines.

67. $2y^2 + (10 - 6x)y + (4x^2 - 3x - 6) = 0$

$y = \dfrac{3x - 5 \pm \sqrt{x^2 - 24x + 37}}{2}$

The xy-term rotates the hyperbola.

69. (c) is true.

71. $\dfrac{y^2}{36} - \dfrac{(x - 5)^2}{20} = 1$

Section 10.3

Check Point Exercises

1. focus: $(2, 0)$
directrix: $x = -2$

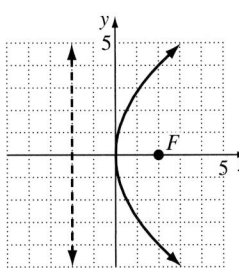

2. focus: $(0, -3)$
directrix: $y = 3$

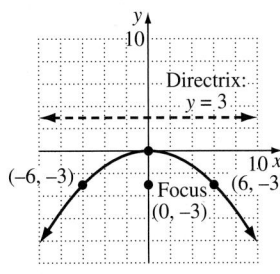

3. $y^2 = 32x$

4. vertex: $(2, -1)$; focus: $(2, 0)$;
directrix: $y = -2$

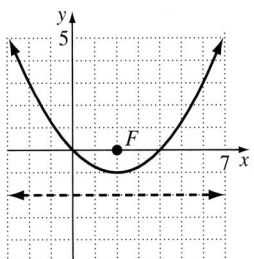

5. vertex: $(2, -1)$; focus: $(1, -1)$;
directrix: $x = 3$

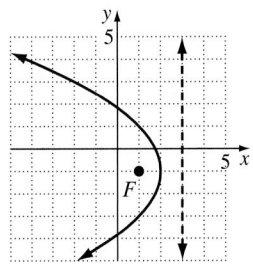

6. $x^2 = \dfrac{9}{4}y$; The light should be placed at $\left(0, \dfrac{9}{16}\right)$, or $\dfrac{9}{16}$ inch above the vertex.

Exercise Set 10.3

1. focus: $(1, 0)$; directrix: $x = -1$; graph (c)   **3.** focus: $(0, -1)$, directrix: $y = 1$; graph (b)

5. focus: $(4, 0)$; directrix: $x = -4$

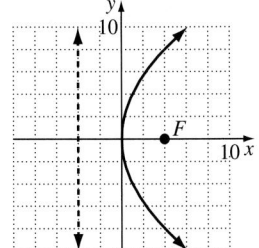

7. focus: $(-2, 0)$; directrix: $x = 2$

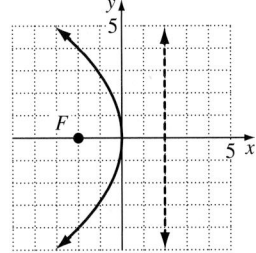

9. focus: $(0, 3)$; directrix: $y = -3$

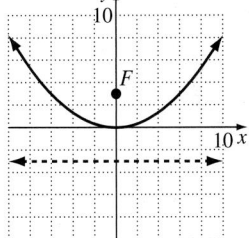

11. focus: $(0, -4)$; directrix: $y = 4$

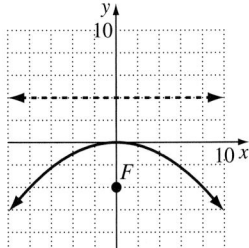

13. focus: $\left(\dfrac{3}{2}, 0\right)$; directrix: $x = -\dfrac{3}{2}$

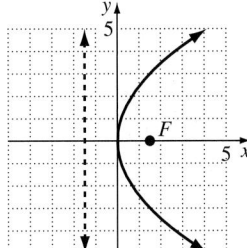

15. focus: $\left(0, -\dfrac{1}{8}\right)$; directrix: $y = \dfrac{1}{8}$

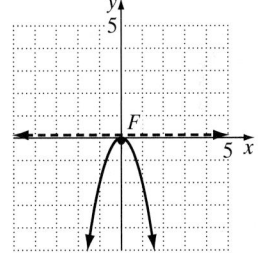

17. $y^2 = 28x$ **19.** $y^2 = -20x$ **21.** $x^2 = 60y$ **23.** $x^2 = -100y$ **25.** $(x - 2)^2 = -8(y + 3)$ **27.** $(y - 2)^2 = 8(x - 1)$
29. $(x + 3)^2 = 4(y - 3)$ **31.** vertex: $(1, 1)$; focus: $(2, 1)$; directrix: $x = 0$; graph (c)   **33.** vertex: $(-1, -1)$; focus: $(-1, -2)$; directrix:
$y = 0$; graph (d)

35. vertex: $(2, 1)$; focus: $(2, 3)$;
directrix: $y = -1$

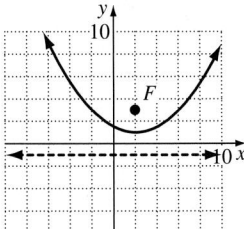

37. vertex: $(-1, -1)$; focus: $(-1, -3)$;
directrix: $y = 1$

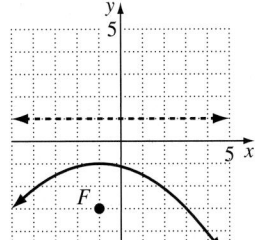

39. vertex: $(-1, -3)$; focus: $(2, -3)$;
directrix: $x = -4$

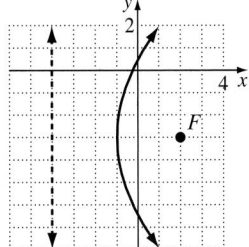

41. vertex: $(0, -1)$; focus: $(-2, -1)$;
directrix: $x = 2$

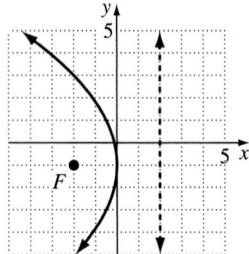

43. $(x - 1)^2 = 4(y - 2)$; vertex: $(1, 2)$;
focus: $(1, 3)$; directrix: $y = 1$

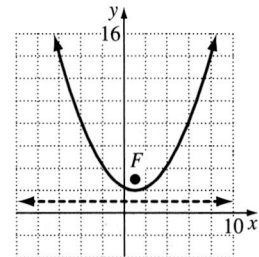

45. $(y - 1)^2 = -12(x - 3)$;
vertex: $(3, 1)$; focus: $(0, 1)$;
directrix: $x = 6$

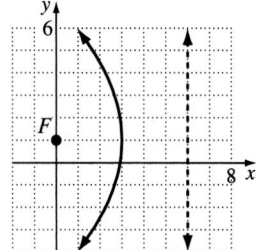

47. $(x + 3)^2 = 4(y + 2)$;
vertex: $(-3, -2)$; focus: $(-3, -1)$;
directrix: $y = -3$

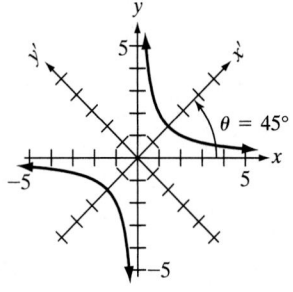

49. 1 inch above the vertex.
51. 4.5 feet from the base of the dish.
53. 75.625 m   **55.** yes
65. $y = -1 \pm \sqrt{6x - 12}$

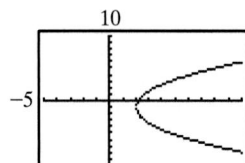

67. $9y^2 + (-24x - 80)y + 16x^2 - 60x + 100 = 0$
$$y = \frac{12x + 40 \pm 10\sqrt{15x + 7}}{9}$$

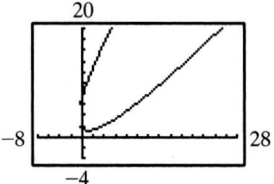

69. (b) is true.   **71.** $(x + 1)^2 = -8(y - 2)$

Section 10.4

Check Point Exercises

1. a. ellipse   **b.** circle   **c.** parabola   **d.** hyperbola

2. $\dfrac{x'^2}{4} - \dfrac{y'^2}{4} = 1$

3. $\dfrac{x'^2}{\frac{4}{5}} + \dfrac{y'^2}{4} = 1$

4.

5. parabola

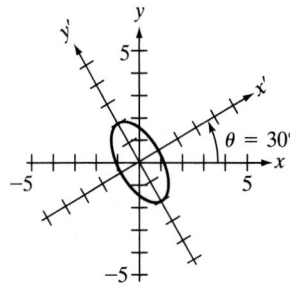

Exercise Set 10.4

1. parabola   **3.** hyperbola   **5.** circle   **7.** hyperbola   **9.** $\dfrac{y'^2}{2} - \dfrac{x'^2}{2} = 1$ **11.** $\dfrac{y'^2}{1} - \dfrac{x'^2}{3} = 1$ **13.** $\dfrac{x'^2}{4} - \dfrac{y'^2}{9} = 1$

15. $x = \dfrac{\sqrt{2}}{2}(x' - y')$; $y = \dfrac{\sqrt{2}}{2}(x' + y')$ **17.** $x = \dfrac{\sqrt{2}}{2}(x' - y')$; $y = \dfrac{\sqrt{2}}{2}(x' + y')$ **19.** $x = \dfrac{\sqrt{3}x' - y'}{2}$; $y = \dfrac{x' + \sqrt{3}y'}{2}$

21. $x = \dfrac{3x' - 4y'}{5}$; $y = \dfrac{4x' + 3y'}{5}$ **23.** $x = \sqrt{5}\left(\dfrac{2x' - y'}{5}\right)$; $y = \sqrt{5}\left(\dfrac{x' + 2y'}{5}\right)$ **25.** $x = \dfrac{4x' - 3y'}{5}$; $y = \dfrac{3x' + 4y'}{5}$

27. a. $3x'^2 + y'^2 = 20$

   b. $\dfrac{x'^2}{\frac{20}{3}} + \dfrac{y'^2}{20} = 1$

   c.

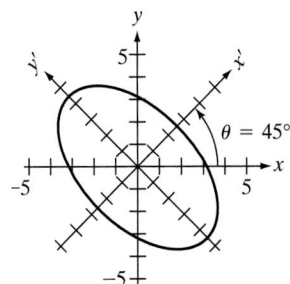

$\theta = 45°$

29. a. $-4x'^2 + 16y'^2 = 64$

   b. $\dfrac{y'^2}{4} - \dfrac{x'^2}{16} = 1$

   c.

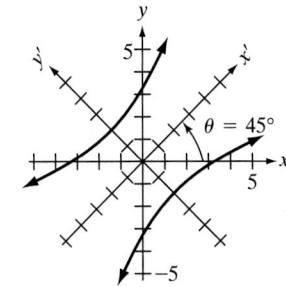

$\theta = 45°$

31. a. $64x'^2 - 16y'^2 = 16$

   b. $\dfrac{x'^2}{\frac{1}{4}} - \dfrac{y'^2}{1} = 1$

   c.

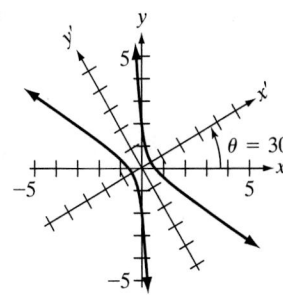

$\theta = 30$

33. a. $650x'^2 + 25y'^2 = 225$

   b. $\dfrac{x'^2}{\frac{9}{26}} + \dfrac{y'^2}{9} = 1$

   c.

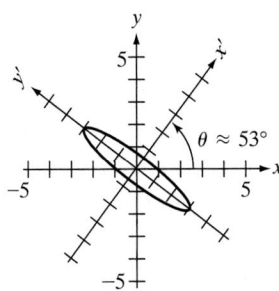

$\theta \approx 53°$

35. a. $50x'^2 - 75y'^2 = 25$

   b. $\dfrac{x'^2}{\frac{1}{2}} - \dfrac{y'^2}{\frac{1}{3}} = 1$

   c.

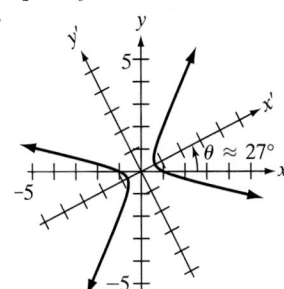

$\theta \approx 27°$

37. a. $625x'^2 + 1250y'^2 = 625$

   b. $\dfrac{x'^2}{1} + \dfrac{y'^2}{\frac{1}{2}} = 1$

   c.

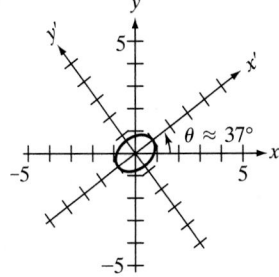

$\theta \approx 37°$

39. ellipse or circle   **41.** parabola   **43.** hyperbola

51.

53.

55.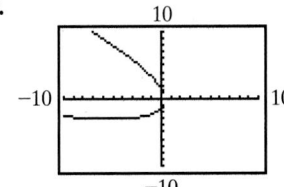

57. There are no solutions to this equation just as there is no such sound.

59. $A' = A\cos^2\theta + B\sin\theta\cos\theta + C\sin^2\theta$

      $C' = A\sin^2\theta - B\sin\theta\cos\theta + C\cos^2\theta$

$A' + C' = A\cos^2\theta + B\sin\theta\cos\theta + C\sin^2\theta + A\sin^2\theta - B\sin\theta\cos\theta + C\cos^2\theta$

        $= A(\cos^2\theta + \sin^2\theta) + B(\sin\theta\cos\theta - \sin\theta\cos\theta) + C(\sin^2\theta + \cos^2\theta)$

        $= A(1) + B(0) + C(1)$

        $= A + C$

Section 10.5

Check Point Exercises

1.

2.

3.

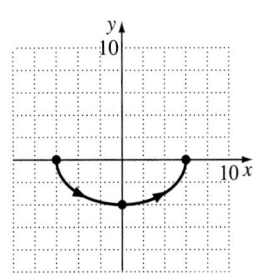

4. $x = t$ and $y = t^2 - 25$

Exercise Set 10.5

1. $(-2, 6)$ **3.** $(5, -3)$ **5.** $(4, 8)$ **7.** $(60\sqrt{3}, 1)$

9.

11

13.

15.

17.

19.

21. $y = 2x$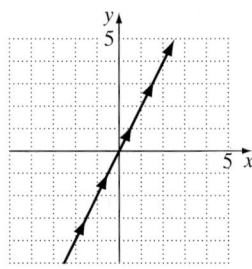

23. $y = (x + 4)^2$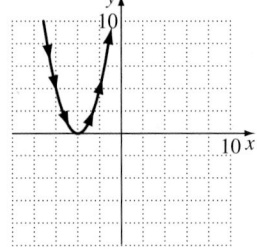

25. $y = x^2 - 1, x \geq 0$

27. $\dfrac{x^2}{4} + \dfrac{y^2}{4} = 1$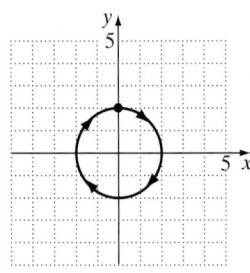

29. $\dfrac{(x-1)^2}{9} + \dfrac{(y-2)^2}{9} = 1$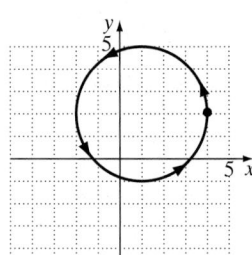

31. $\dfrac{x^2}{4} + \dfrac{y^2}{9} = 1$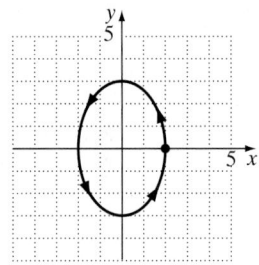

33. $\dfrac{(x-1)^2}{9} + \dfrac{(y+1)^2}{4} = 1, -2 \leq x \leq 4, -1 \leq y \leq 1$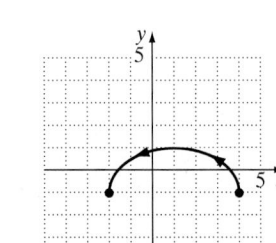

35. $x^2 - y^2 = 1$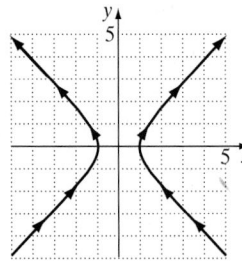

37. $y = x - 4, x \geq 2, y \geq -2$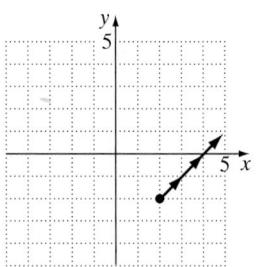

39. $y = \dfrac{1}{x}, x \geq 1, y \geq 0$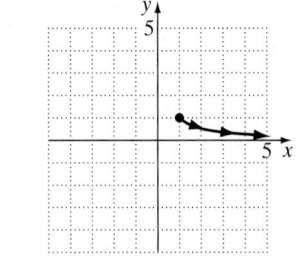

41. $(x - h)^2 + (y - k)^2 = r^2$ **43.** $\dfrac{(x-h)^2}{a^2} - \dfrac{(y-k)^2}{b^2} = 1$ **45.** $x = 3 + 6\cos t; y = 5 + 6\sin t$

47. $x = -2 + 5\cos t; y = 3 + 2\sin t$ **49.** $x = 4\sec t; y = \sqrt{20}\tan t$ **51.** $x = -2 + 3t; y = 4 + 3t$

53. Answers may vary. Sample answer: $x = t$ and $y = 4t - 3$; $x = t + 1$ and $y = 4t + 1$

55. Answers may vary. Sample answer: $x = t$ and $y = t^2 + 4$; $x = t + 1$ and $y = t^2 + 2t + 5$

57. a. **b.** **c.** **d.**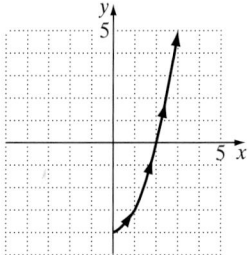

59. a. $x = (180 \cos 40°)t;\; y = 3 + (180 \sin 40°)t - 16t^2$
  b. After 1 second: 137.9 feet in distance, 102.7 feet in height;
    After 2 seconds: 275.8 feet in distance, 170.4 feet in height;
    After 3 seconds: 413.7 feet in distance, 206.1 feet in height
  c. $t = 7.3$ sec; total horizontal distance: 1006.6 ft; yes

69.

71.

73.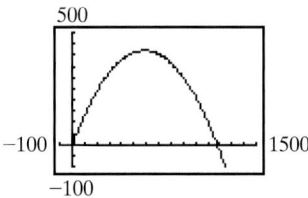
  Window: $[-100, 1500] \times [-100, 500]$;
  The maximum height is 419.4 feet at a time of 5.1 seconds.
  The range of the projectile is 1174.6 feet horizontally.
  It hits the ground at 10.2 seconds.

75. a. $x = (140 \cos 22°)t;\; y = 5 + (140 \sin 22°)t - 16t^2$
  b.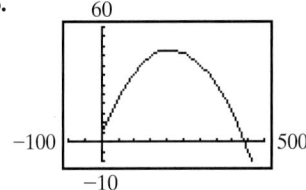
  Window: $[-100, 500] \times [-10, 60]$
  c. The maximum height is 48.0 feet. It occurs at 1.6 seconds.
  d. 3.4 sec   **e.** 437.5 ft

77. $x = 3 \sin t;\; y = 3 \cos t$

Section 10.6

Check Point Exercises

1.

2.

3.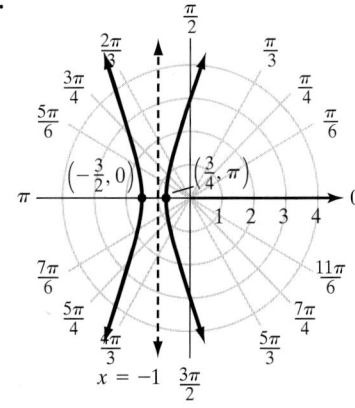

Exercise Set 10.6

1. a. parabola   **b.** The directrix is 3 units above the pole, at $y = 3$.   **3. a.** ellipse   **b.** The directrix is 3 units to the left of the pole, at $x = -3$.   **5. a.** parabola   **b.** The directrix is 4 units above the pole, at $y = 4$.   **7. a.** hyperbola   **b.** The directrix is 3 units to the left of the pole, at $x = -3$.

9.

11.

13.

15.

17.

19.

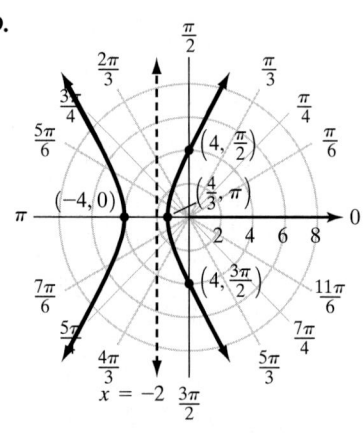

21. 0.54 astronomical units or 51 million miles   **23.** 4122 miles from the center of the earth.; 162 miles from the surface of the earth.

33. hyperbola

35.

The graph appears to be rotated counterclockwise through an angle of $\dfrac{\pi}{4}$ radians.

37. Mercury: $r = \dfrac{(1 - 0.2056^2)(36.0 \times 10^6)}{1 - 0.2056 \cos \theta}$

Earth: $r = \dfrac{(1 - 0.0167^2)(92.96 \times 10^6)}{1 - 0.0167 \cos \theta}$

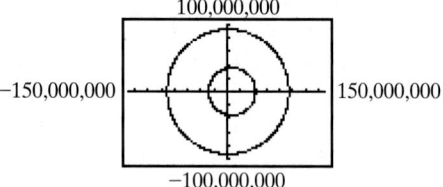

39. $r = \dfrac{2}{1 - \frac{1}{2}\cos \theta}$ or $r = \dfrac{6}{1 + \frac{1}{2}\cos \theta}$

41. parabola; using the relationships between rectangular and polar coordinates, $x^2 + y^2 = r^2$ and $x = r \cos \theta$: $y^2 = x + \dfrac{1}{4}$

Chapter 10 Review Exercises

1. foci: $(\pm\sqrt{11}, 0)$

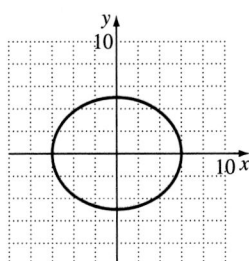

2. foci: $(0, \pm 3)$

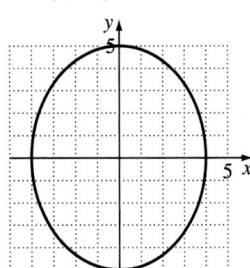

3. foci: $(0, \pm 2\sqrt{3})$

4. foci: $(\pm\sqrt{5}, 0)$

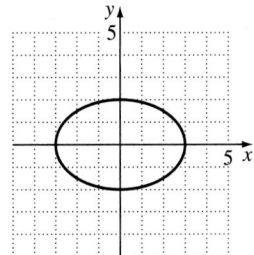

5. foci: $(1 \pm \sqrt{7}, -2)$

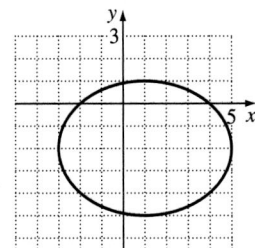

6. foci: $(-1, 2 \pm \sqrt{7})$

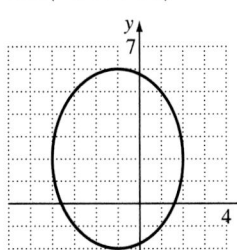

7. foci: $(-3 \pm \sqrt{5}, 2)$

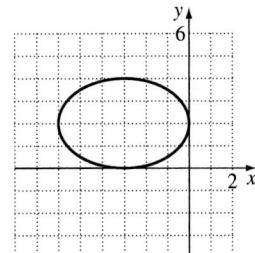

8. foci: $(1, -1 \pm \sqrt{5})$

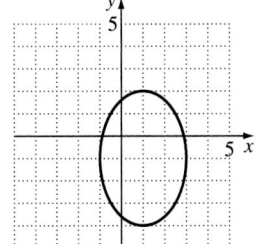

9. $\dfrac{x^2}{25} + \dfrac{y^2}{9} = 1$ **10.** $\dfrac{x^2}{27} + \dfrac{y^2}{36} = 1$ **11.** $\dfrac{(x+3)^2}{36} + \dfrac{(y-5)^2}{4} = 1$ **12.** $\dfrac{x^2}{100} + \dfrac{y^2}{36} = 1$ **13.** yes

14. The hit ball will collide with the other ball.

15. foci: $(\pm\sqrt{17}, 0)$; $y = \pm\dfrac{1}{4}x$ **16.** foci: $(0, \pm\sqrt{17})$; $y = \pm 4x$ **17.** foci: $(\pm 5, 0)$; $y = \pm\dfrac{3}{4}x$ **18.** foci: $(0, \pm 2\sqrt{5})$; $y = \pm\dfrac{1}{2}x$

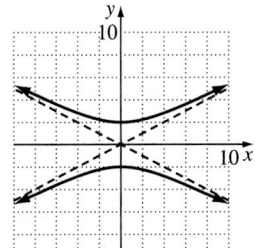

19. foci: $(2 \pm \sqrt{41}, -3)$;
$y + 3 = \pm\dfrac{4}{5}(x - 2)$

20. foci: $(3, -2 \pm \sqrt{41})$;
$y + 2 = \pm\dfrac{5}{4}(x - 3)$

21. foci: $(1, 2 \pm \sqrt{5})$;
$y - 2 = \pm 2(x - 2)$

22. foci: $(1 \pm \sqrt{2}, -1)$;
$y + 1 = \pm(x - 1)$

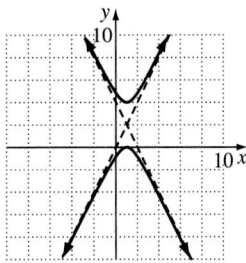

23. $\dfrac{y^2}{4} - \dfrac{x^2}{12} = 1$ **24.** $\dfrac{x^2}{9} - \dfrac{y^2}{55} = 1$ **25.** c must be greater than a. **26.** $\dfrac{x^2}{2162.25} - \dfrac{y^2}{7837.75} = 1$

27. vertex: $(0, 0)$; focus: $(2, 0)$; directrix: $x = -2$

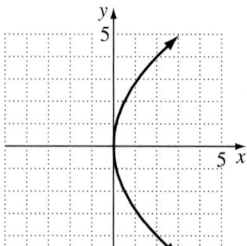

28. vertex: $(0, 0)$; focus: $(0, -4)$; directrix: $y = 4$

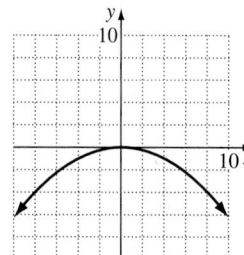

29. vertex: $(0, 2)$; focus: $(-4, 2)$; directrix: $x = 4$

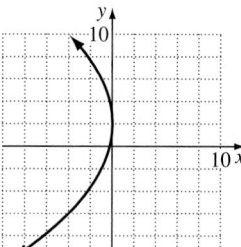

30. vertex: $(4, -1)$; focus: $(4, 0)$; directrix: $y = -2$

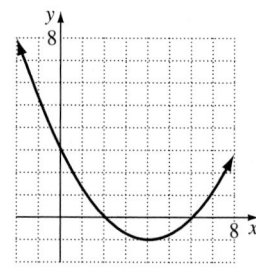

31. vertex: $(0, 1)$; focus: $(0, 0)$; directrix: $y = 2$

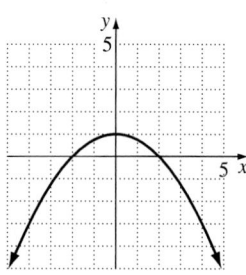

32. vertex: $(-1, 5)$; focus: $(0, 5)$; directrix: $x = -2$

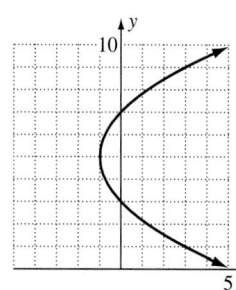

33. vertex: $(2, -2)$; focus: $\left(2, -\dfrac{3}{2}\right)$; directrix: $y = -\dfrac{5}{2}$

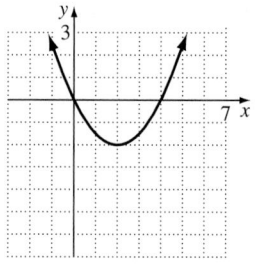

34. $y^2 = 48x$ **35.** $x^2 = -44y$ **36.** $x^2 = 12y$; Place the light 3 inches from the vertex at $(0, 3)$.   **37.** approximately 58 ft
38. approximately 128 ft   **39.** parabola   **40.** ellipse   **41.** ellipse   **42.** hyperbola   **43.** ellipse or circle   **44.** ellipse or circle
45. parabola   **46.** ellipse or circle

47. a. $x'^2 - y'^2 = 8$

b. $\dfrac{x'^2}{8} - \dfrac{y'^2}{8} = 1$

c.

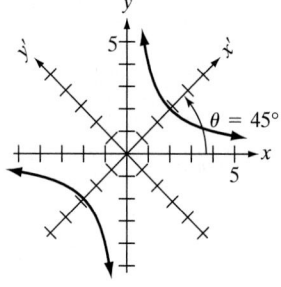

48. a. $3x'^2 + y'^2 = 2$

b. $\dfrac{x'^2}{\frac{2}{3}} + \dfrac{y'^2}{2} = 1$

c.

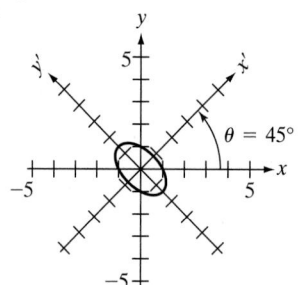

49. a. $18x'^2 - 2y'^2 = 18$

b. $\dfrac{x'^2}{1} - \dfrac{y'^2}{9} = 1$

c.

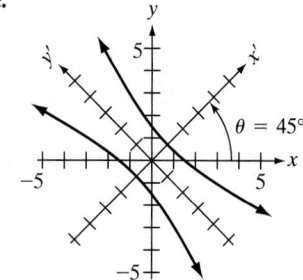

50. a. $50x'^2 + 150y'^2 = 450$

b. $\dfrac{x'^2}{9} + \dfrac{y'^2}{3} = 1$

c.

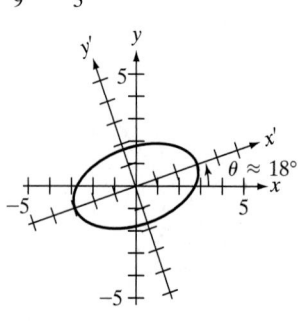

51. a. $16x'^2 + 96y' = 0$

b. $x'^2 = -6y'$

c.

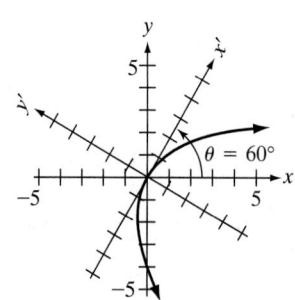

52. $y = -\dfrac{1}{2}x + \dfrac{1}{2}$

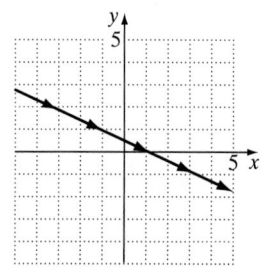

53. $(y + 1)^2 = x, 0 \leq x \leq 9,$
    $-2 \leq y \leq 2$

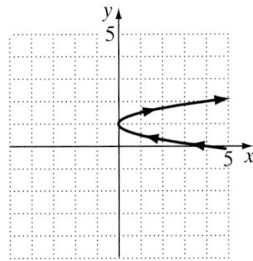

54. $(y - 1)^2 = \frac{1}{4}x$

55. $\frac{x^2}{16} + \frac{y^2}{9} = 1, 0 \leq x \leq 4,$
    $-3 < y \leq 3$

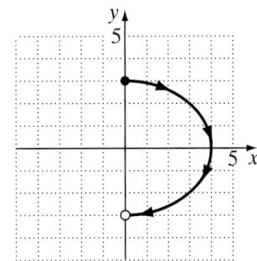

56. $\frac{(x - 3)^2}{4} + \frac{(y - 1)^2}{4} = 1$
    or $(x - 3)^2 + (y - 1)^2 = 4$

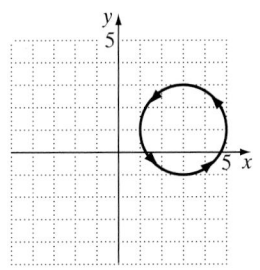

57. $\frac{x^2}{9} - \frac{y^2}{9} = 1,$
    $3 \leq x \leq 3\sqrt{2}, 0 \leq y \leq 3$

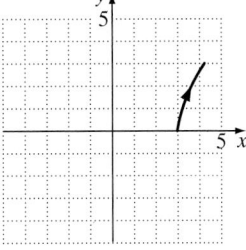

58. Answers may vary. Sample answer: $x = t$ and $y = t^2 + 6$; $x = t + 1$ and $y = t^2 + 2t + 7$

59. a. $x = (100 \cos 40°)t; y = 6 + (100 \sin 40°)t - 16t^2$
   b. After 1 second: 76.6 feet in distance, 54.3 feet in height; after 2 seconds: 153.2 feet in distance, 70.6 feet in height; after 3 seconds: 229.8 feet in distance, 54.8 feet in height.
   c. 4.1 sec; 314.1 ft   **d.**

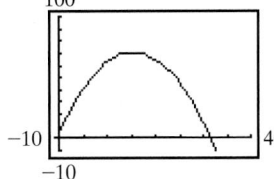

;The ball is at its maximum height at 2.0 seconds. The maximum height is 70.6 feet.

60. a. $r = \dfrac{4}{1 - \sin \theta}$ **b.** $e = 1; p = 4$; parabola

c.

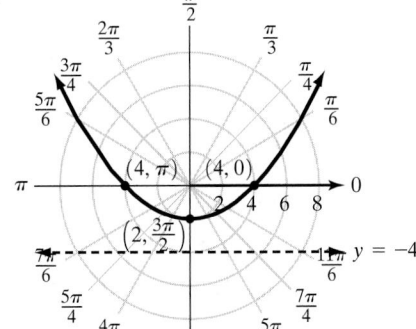

61. a. $r = \dfrac{6}{1 + \cos \theta}$ **b.** $e = 1; p = 6$; parabola

c.

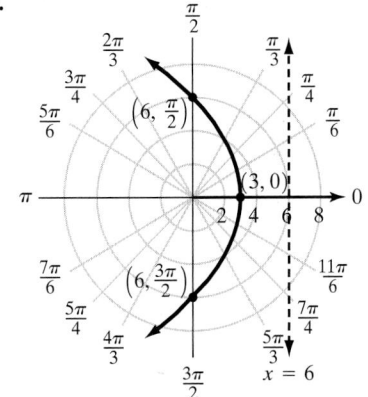

62. a. $r = \dfrac{3}{1 + \frac{1}{2}\sin\theta}$ **b.** $e = \frac{1}{2}$; $p = 6$; ellipse

63. a. $r = \dfrac{\frac{2}{3}}{1 - \frac{2}{3}\cos\theta}$ **b.** $e = \frac{2}{3}$; $p = 1$; ellipse

c.

c.

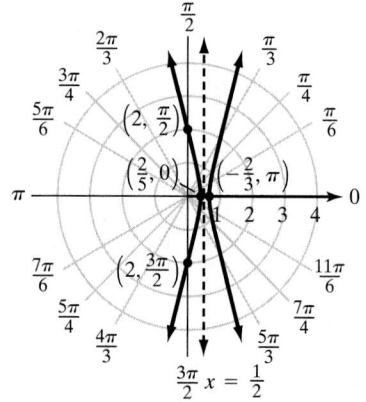

64. a. $r = \dfrac{2}{1 + 2\sin\theta}$ **b.** $e = 2$; $p = 1$; hyperbola

65. a. $r = \dfrac{2}{1 + 4\cos\theta}$ **b.** $e = 4$; $p = \frac{1}{2}$; hyperbola

c.

c.

Chapter 10 Test

1. foci: $(\pm\sqrt{13}, 0)$;

asymptotes: $y = \pm\frac{3}{2}x$

2. vertex: $(0, 0)$; focus: $(0, -2)$;

directrix: $y = 2$

3. foci: $(-6, 5), (2, 5)$

4. foci: $(-1, 1 \pm \sqrt{5})$

asymptotes: $y - 1 = \pm 2(x + 1)$

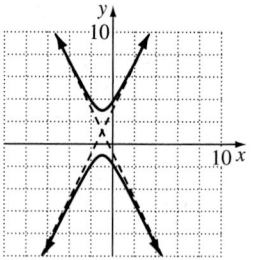

5. vertex: $(-5, 1)$; focus: $(-5, 3)$

   directrix: $y = -1$

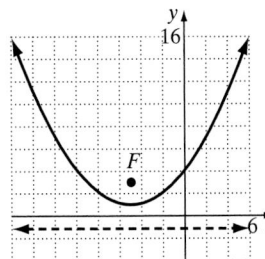

6. $\dfrac{x^2}{100} + \dfrac{y^2}{51} = 1$

7. $\dfrac{y^2}{49} - \dfrac{x^2}{51} = 1$

8. $y^2 = 200x$ **9.** 32 ft

10. a. $x^2 = 3y$

   b. Light is placed $\dfrac{3}{4}$ inch

      above the vertex.

11. ellipse

12. ellipse or circle

13. $\theta = 30°$

14. $(y + 1)^2 = x$

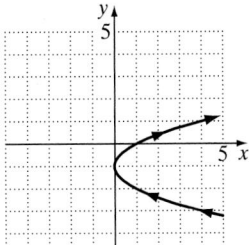

15. $\dfrac{(x - 1)^2}{9} + \dfrac{y^2}{4} = 1$

16. parabola

17. ellipse

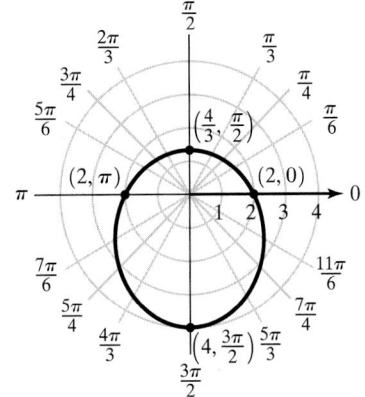

Cumulative Review Exercises (Chapters 1–10)

1. $\{2\}$ **2.** $\{x \mid x < 2\}$ **3.** $\{9\}$ **4.** $\{2 + 2\sqrt{5}, 2 - 2\sqrt{5}\}$ **5.** $\{x \mid x \geq 4 \text{ or } x \leq -3\}$ **6.** $\left\{\dfrac{2}{3}, -1 \pm \sqrt{2}\right\}$ **7.** $\{3\}$

8. $(2, -1)$ **9.** $(2, -4)$ and $(-14, -20)$ **10.** $(7, -4, 6)$

11.

12.

13.

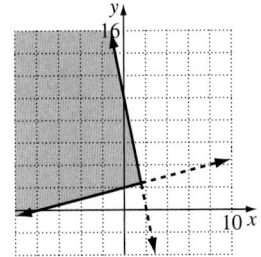

14. a. $\pm 1, \pm 3, \pm\dfrac{1}{2}, \pm\dfrac{3}{2}, \pm\dfrac{1}{4}, \pm\dfrac{3}{4}, \pm\dfrac{1}{8}, \pm\dfrac{3}{8}, \pm\dfrac{1}{16}, \pm\dfrac{3}{16}, \pm\dfrac{1}{32}, \pm\dfrac{3}{32}$ **b.** $\left\{-\dfrac{1}{8}, \dfrac{3}{4}, 1\right\}$ **15. a.** 1980–1990   **b.** 1990–2025

c. 1950–1980   **d.** $f(x) = 98$ **e.** The scale is not uniformly spaced.   **16.** $(g \circ f)(x) = x^2 - 2$ **17.** $3 \log_5 x + \dfrac{1}{2}\log_5 y - 3$

18. $y = -2x - 2$ **19.** The costs will be the same when the number of miles driven is 175 miles. The cost will be \$67.

20. \$25 for basic cable service and \$10 for each movie channel

21. $\dfrac{\csc\theta - \sin\theta}{\sin\theta} = \dfrac{\csc\theta}{\sin\theta} - \dfrac{\sin\theta}{\sin\theta} = \dfrac{\dfrac{1}{\sin\theta}}{\sin\theta} - 1 = \dfrac{1}{\sin^2\theta} - 1 = \dfrac{1 - \sin^2\theta}{\sin^2\theta} = \dfrac{\cos^2\theta}{\sin^2\theta} = \cot^2\theta$

22.

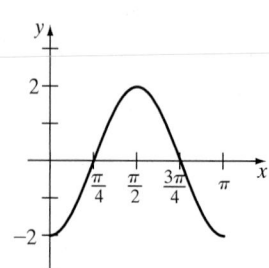

23. $-3\mathbf{i} - 3\mathbf{j}$ **24.** $\theta = 0, \theta = \pi, \theta = \dfrac{\pi}{3}$, or $\theta = \dfrac{5\pi}{3}$ **25.** $b \approx 14.4, C = 44°, c \approx 10.5$

CHAPTER 11

Section 11.1

Check Point Exercises

1. a. $7, 9, 11, 13$ **b.** $-\dfrac{1}{3}, \dfrac{1}{5}, -\dfrac{1}{9}, \dfrac{1}{17}$ **2.** $3, 11, 27, 59$ **3.** $10, \dfrac{10}{3}, \dfrac{5}{6}, \dfrac{1}{6}$ **4. a.** 91 **b.** n **5. a.** 182 **b.** 47 **c.** 20

6. a. $\displaystyle\sum_{i=1}^{9} i^2$ **b.** $\displaystyle\sum_{i=1}^{n} \dfrac{1}{2^{i-1}}$

Exercise Set 11.1

1. $5, 8, 11, 14$ **3.** $3, 9, 27, 81$ **5.** $-3, 9, -27, 81$ **7.** $-4, 5, -6, 7$ **9.** $\dfrac{2}{5}, \dfrac{2}{3}, \dfrac{6}{7}, 1$ **11.** $1, -\dfrac{1}{3}, \dfrac{1}{7}, -\dfrac{1}{15}$ **13.** $7, 12, 17, 22$

15. $3, 12, 48, 192$ **17.** $4, 11, 25, 53$ **19.** $1, 2, \dfrac{3}{2}, \dfrac{2}{3}$ **21.** $4, 12, 48, 240$ **23.** 272 **25.** 120 **27.** $(n+2)(n+1)$ **29.** 105

31. 60 **33.** 115 **35.** $-\dfrac{5}{16}$ **37.** 55 **39.** $\dfrac{3}{8}$ **41.** 15 **43.** $\displaystyle\sum_{i=1}^{15} i^2$ **45.** $\displaystyle\sum_{i=1}^{11} 2^i$ **47.** $\displaystyle\sum_{i=1}^{30} i$ **49.** $\displaystyle\sum_{i=1}^{14} \dfrac{i}{i+1}$ **51.** $\displaystyle\sum_{i=1}^{n} \dfrac{4^i}{i}$

53. $\displaystyle\sum_{i=1}^{n} (2i-1)$ **55.** $\displaystyle\sum_{k=1}^{14} (2k+3)$ **57.** $\displaystyle\sum_{k=0}^{12} ar^k$ **59.** $\displaystyle\sum_{k=0}^{n} (a+kd)$ **61. a.** 6881.6; From 1991 through 2000, a total of 6881.6 million, CDs were sold.   **b.** 688.16; From 1991 through 2000, the average number of CDs sold each year was 688.16 million.
63. a. 88.3; From 1993 through 2000, a total of 88.3 million people received cash assistance.   **b.** 88.12; This is a reasonable model.
65. $\$8081.13$ **77.** $1{,}307{,}674{,}368{,}000$ **79.** 6840 **83.** $a_{10} = 2.5937; a_{100} = 2.7048; a_{1000} = 2.7169; a_{10{,}000} = 2.7181; a_{100{,}000} = 2.7183$; As n gets larger, a_n gets closer to $e \approx 2.7183$.
85.

As n gets larger, a_n approaches 0.

87.

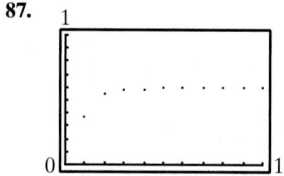

As n gets larger, a_n approaches $\dfrac{3}{5}$.

89. $9, 32, 16, 8, 4$

Section 11.2

Check Point Exercises

1. $51.5, 53.68, 55.86, 58.04, 60.22$ **2.** -34 **3. a.** $a_n = 9700n + 149{,}300$ **b.** $\$304{,}500$ **4.** 360 **5.** 2460 **6.** $\$596{,}300$

Exercise Set 11.2

1. $200, 220, 240, 260, 280, 300$ **3.** $-7, -3, 1, 5, 9, 13$ **5.** $300, 210, 120, 30, -60, -150$ **7.** $\dfrac{5}{2}, 2, \dfrac{3}{2}, 1, \dfrac{1}{2}, 0$ **9.** $-9, -3, 3, 9, 15, 21$

11. $30, 20, 10, 0, -10, -20$ **13.** $1.6, 1.2, 0.8, 0.4, 0, -0.4$ **15.** 33 **17.** 252 **19.** 955 **21.** -142 **23.** $a_n = 4n - 3; a_{20} = 77$
25. $a_n = 11 - 4n; a_{20} = -69$ **27.** $a_n = 7 + 2n; a_{20} = 47$ **29.** $a_n = -16 - 4n; a_{20} = -96$ **31.** $a_n = 1 + 3n; a_{20} = 61$
33. $a_n = 40 - 10n; a_{20} = -160$ **35.** 1220 **37.** 4400 **39.** 5050 **41.** 3660 **43.** 396 **45.** $8 + 13 + 18 + \cdots + 88; 816$
47. $2 - 1 - 4 - \cdots - 85; -1245$ **49.** $4 + 8 + 12 + \cdots + 400; 20{,}200$ **51. a.** $a_n = 1.7n + 148.3$ **b.** 211.2 lbs
53. Answers may vary. One possibility is given.   **a.** $a_n = 0.5n + 14.5$ **b.** 33 lbs   **55.** Company A will pay $\$1400$ more.

57. a. $a_n = 3.204 + 0.576n$ **b.** 627.3 million tons   **59.** \$442,500   **61.** 1430 seats   **71.** the 200th term

73. $S_n = \dfrac{n}{2}(1 + 2n - 1) = \dfrac{n}{2}(2n) = n^2$

Section 11.3

Check Point Exercises

1. $12, 6, 3, \dfrac{3}{2}, \dfrac{3}{4}, \dfrac{3}{8}$ **2.** 3645   **3.** $a_n = 3(2)^{n-1}; 384$ **4.** 9842   **5.** 19,680   **6.** \$2,371,746   **7.** \$1,327,778   **8.** 9   **9.** 1

10. \$4000

Exercise Set 11.3

1. $5, 15, 45, 135, 405$ **3.** $20, 10, 5, \dfrac{5}{2}, \dfrac{5}{4}$ **5.** $10, -40, 160, -640, 2560$ **7.** $-6, 30, -150, 750, -3750$ **9.** $a_8 = 768$

11. $a_{12} = -10{,}240$ **13.** $a_{40} \approx -0.000000002$ **15.** $a_8 = 0.1$ **17.** $a_n = 3(4)^{n-1}; a_7 = 12{,}288$ **19.** $a_n = 18\left(\dfrac{1}{3}\right)^{n-1}; a_7 = \dfrac{2}{81}$

21. $a_n = 1.5(-2)^{n-1}; a_7 = 96$ **23.** $a_n = 0.0004(-10)^{n-1}; a_7 = 400$ **25.** 531,440   **27.** 2049   **29.** $\dfrac{16{,}383}{2}$ **31.** 9840

33. 10,230   **35.** $\dfrac{63}{128}$ **37.** $\dfrac{3}{2}$ **39.** 4   **41.** $\dfrac{2}{3}$ **43.** $S_\infty \approx 6.15385$ **45.** $\dfrac{5}{9}$ **47.** $\dfrac{47}{99}$ **49.** $\dfrac{257}{999}$ **51.** arithmetic, $d = 1$

53. geometric, $r = 2$ **55.** neither   **57.** \$16,384   **59.** \$3,795,957   **61. a.** $1.013, 1.013, 1.013, 1.013, 1.013, 1.013, 1.013$; the population is increasing geometrically with $r \approx 1.013$ **b.** $a_n = 29.76(1.013)^{n-1}$ **c.** ≈ 33.86; very well   **63.** \$32,767

65. \$793,582.90   **67.** 130.26 in.   **69.** \$844,706.11   **71.** \$94,834.21   **73.** \$9 million   **75.** $\dfrac{1}{3}$

87.

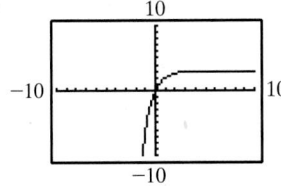

Horizontal asymptote at $y = 3$; $\displaystyle\sum_{n=0}^{\infty} 2\left(\dfrac{1}{3}\right)^{n} = 3$

89. (d) is true.   **91.** \$442.38

Section 11.4

Check Point Exercises

1. a. $S_1: 2 = 1(1 + 1)$; $S_k: 2 + 4 + 6 + \cdots + 2k = k(k + 1)$; $S_{k+1}: 2 + 4 + 6 + \cdots + 2(k + 1) = (k + 1)(k + 2)$

b. $S_1: 1^3 = \dfrac{1^2(1 + 1)^2}{4}$; $S_k: 1^3 + 2^3 + 3^3 + \cdots + k^3 = \dfrac{k^2(k + 1)^2}{4}$; $S_{k+1}: 1^3 + 2^3 + 3^3 + \cdots + (k + 1)^3 = \dfrac{(k + 1)^2(k + 2)^2}{4}$

2. $S_1: 2 = 1(1 + 1)$; $S_k: 2 + 4 + 6 + \cdots + 2k = k(k + 1)$; $S_{k+1}: 2 + 4 + 6 + \cdots + 2k + 2(k + 1) = (k + 1)(k + 2)$; S_{k+1} can be obtained by adding $2k + 2$ to both sides of S_k.

3. $S_1: 1^3 = \dfrac{1^2(1 + 1)^2}{4}$; $S_k: 1^3 + 2^3 + 3^3 + \cdots + k^3 = \dfrac{k^2(k + 1)^2}{4}$; $S_{k+1}: 1^3 + 2^3 + 3^3 + \cdots + k^3 + (k + 1)^3 = \dfrac{(k + 1)^2(k + 2)^2}{4}$; S_{k+1} can be obtained by adding $k^3 + 3k^2 + 3k + 1$ to both sides of S_k.

4. $S_1: 2$ is a factor of $1^2 + 1$; $S_k: 2$ is a factor of $k^2 + k$; $S_{k+1}: 2$ is a factor of $(k + 1)^2 + (k + 1) = k^2 + 3k + 2$; S_{k+1} can be obtained from S_k by writing $k^2 + 3k + 2$ as $(k^2 + k) + 2(k + 1)$.

Exercise Set 11.4

1. $S_1: 1 = 1^2$; $S_2: 1 + 3 = 2^2$; $S_3: 1 + 3 + 5 = 3^2$ **3.** $S_1: 2$ is a factor of $1 - 1 = 0$; $S_2: 2$ is a factor of $2^2 - 2 = 2$; $S_3: 2$ is a factor of $3^2 - 3 = 6$ **5.** $S_k: 4 + 8 + 12 + \cdots + 4k = 2k(k + 1)$; $S_{k+1}: 4 + 8 + 12 + \cdots + (4k + 4) = 2(k + 1)(k + 2)$

7. $S_k: 3 + 7 + 11 + \cdots + (4k - 1) = k(2k + 1)$; $S_{k+1}: 3 + 7 + 11 + \cdots + (4k + 3) = (k + 1)(2k + 3)$

9. $S_k: 2$ is a factor of $k^2 - k + 2$; $S_{k+1}: 2$ is a factor of $k^2 + k + 2$ **11.** $S_1: 4 = 2(1)(1 + 1)$; $S_k: 4 + 8 + 12 + \cdots + 4k = 2k(k + 1)$; $S_{k+1}: 4 + 8 + 12 + \cdots + 4(k + 1) = 2(k + 1)(k + 2)$; S_{k+1} can be obtained by adding $4k + 4$ to both sides of S_k.

13. $S_1: 1 = 1^2$; $S_k: 1 + 3 + 5 + \cdots + (2k - 1) = k^2$; $S_{k+1}: 1 + 3 + 5 + \cdots + (2k + 1) = (k + 1)^2$; S_{k+1} can be obtained by adding $2k + 1$ to both sides of S_k.

15. $S_1: 3 = 1[2(1) + 1]$; $S_k: 3 + 7 + 11 + \cdots + (4k - 1) = k(2k + 1)$; $S_{k+1}: 3 + 7 + 11 + \cdots + (4k + 3) = (k + 1)(2k + 3)$; S_{k+1} can be obtained by adding $4k + 3$ to both sides of S_k.

17. $S_1: 1 = 2^1 - 1$; $S_k: 1 + 2 + 2^2 + \cdots + 2^{k-1} = 2^k - 1$; $S_{k+1}: 1 + 2 + 2^2 + \cdots + 2^k = 2^{k+1} - 1$; S_{k+1} can be obtained by adding 2^k to both sides of S_k.

19. $S_1: 2 = 2^{1+1} - 2$; $S_k: 2 + 4 + 8 + \cdots + 2^k = 2^{k+1} - 2$; $S_{k+1}: 2 + 4 + 8 + \cdots + 2^{k+1} = 2^{k+2} - 2$; S_{k+1} can be obtained by adding 2^{k+1} to both sides of S_k.

21. $S_1: 1 \cdot 2 = \dfrac{1(1 + 1)(1 + 2)}{3}$; $S_k: 1 \cdot 2 + 2 \cdot 3 + 3 \cdot 4 + \cdots + k(k + 1) = \dfrac{k(k + 1)(k + 2)}{3}$;

$S_{k+1}: 1 \cdot 2 + 2 \cdot 3 + 3 \cdot 4 + \cdots + (k + 1)(k + 2) = \dfrac{(k + 1)(k + 2)(k + 3)}{3}$; S_{k+1} can be obtained by adding $(k + 1)(k + 2)$ to both sides of S_k.

23. $S_1: \dfrac{1}{1 \cdot 2} = \dfrac{1}{1 + 1}$; $S_k: \dfrac{1}{1 \cdot 2} + \dfrac{1}{2 \cdot 3} + \dfrac{1}{3 \cdot 4} + \cdots + \dfrac{1}{k(k + 1)} = \dfrac{k}{k + 1}$; $S_{k+1}: \dfrac{1}{1 \cdot 2} + \dfrac{1}{2 \cdot 3} + \dfrac{1}{3 \cdot 4} + \cdots + \dfrac{1}{(k + 1)(k + 2)} = \dfrac{k + 1}{k + 2}$; S_{k+1} can be obtained by adding $\dfrac{1}{(k + 1)(k + 2)}$ to both sides of S_k.

25. $S_1: 2$ is a factor of 0; $S_k: 2$ is a factor of $k^2 - k$; $S_{k+1}: 2$ is a factor of $k^2 + k$; S_{k+1} can be obtained from S_k by rewriting $k^2 + k$ as $(k^2 - k) + 2k$.

27. $S_1: 6$ is a factor of 6; $S_k: 6$ is a factor of $k(k + 1)(k + 2)$; $S_{k+1}: 6$ is a factor of $(k + 1)(k + 2)(k + 3)$; S_{k+1} can be obtained from S_k by rewriting $(k + 1)(k + 2)(k + 3)$ as $k(k + 1)(k + 2) + 3(k + 1)(k + 2)$ and noting that either $k + 1$ or $k + 2$ is even, so 6 is a factor of $3(k + 1)(k + 2)$.

29. $S_1: (ab)^1 = a^1 b^1$; $S_k: (ab)^k = a^k b^k$; $S_{k+1}: (ab)^{k+1} = a^{k+1} b^{k+1}$; S_{k+1} can be obtained by multiplying both sides of S_k by (ab).

33. $S_3: 3^2 > 2(3) + 1$; $S_k: k^2 > 2k + 1$ for $k \geq 3$; $S_{k+1}: (k + 1)^2 > 2(k + 1) + 1$ or $k^2 + 2k + 1 > 2k + 3$; S_{k+1} can be obtained from S_k by noting that S_{k+1} is the same as $k^2 > 2$ which is true for $k \geq 3$.   **35.** $S_1: \dfrac{1}{4}$; $S_2: \dfrac{1}{3}$; $S_3: \dfrac{3}{8}$; $S_4: \dfrac{2}{5}$; $S_5: \dfrac{5}{12}$; $S_n: \dfrac{n}{2n + 2}$; Use S_k to obtain the conjectured formula.

Section 11.5

Check Point Exercises

1. a. 20   **b.** 1   **c.** 28   **d.** 1   **2.** $x^4 + 4x^3 + 6x^2 + 4x + 1$ **3.** $x^5 - 10x^4 y + 40x^3 y^2 - 80x^2 y^3 + 80xy^4 - 32y^5$
4. $4032 x^5 y^4$

Exercise Set 11.5

1. 56   **3.** 12   **5.** 1   **7.** 4950   **9.** $x^3 + 6x^2 + 12x + 8$ **11.** $27x^3 + 27x^2 y + 9xy^2 + y^3$ **13.** $125x^3 - 75x^2 + 15x - 1$
15. $16x^4 + 32x^3 + 24x^2 + 8x + 1$ **17.** $x^8 + 8x^6 y + 24x^4 y^2 + 32x^2 y^3 + 16y^4$ **19.** $y^4 - 12y^3 + 54y^2 - 108y + 81$
21. $16x^{12} - 32x^9 + 24x^6 - 8x^3 + 1$ **23.** $c^5 + 10c^4 + 40c^3 + 80c^2 + 80c + 32$ **25.** $x^5 - 5x^4 + 10x^3 - 10x^2 + 5x - 1$
27. $243x^5 - 405x^4 y + 270x^3 y^2 - 90x^2 y^3 + 15xy^4 - y^5$ **29.** $64a^6 + 192a^5 b + 240a^4 b^2 + 160a^3 b^3 + 60a^2 b^4 + 12ab^5 + b^6$
31. $x^8 + 16x^7 + 112x^6 + \cdots$ **33.** $x^{10} - 20x^9 y + 180x^8 y^2 - \cdots$ **35.** $x^{32} + 16x^{30} + 120x^{28} + \cdots$ **37.** $y^{60} - 20y^{57} + 190y^{54} - \cdots$

39. $240x^4 y^2$ **41.** $126x^5$ **43.** $56x^6 y^{15}$ **45.** $-\dfrac{21}{2}x^6$ **47.** $319{,}770x^{16} y^{14}$ **49.** $4x^3 + 6x^2 h + 4xh^2 + h^3$

51. a. $g(x) = 0.12x^3 + 0.08x^2 + 0.24x + 18.24$ **b.** $f(5) = 20$; $g(2) = 20$; the function values match up perfectly with the graph
65.

67. $f_1(x) = x^4 - 8x^3 + 24x^2 - 32x + 16$

71. $x^6 + 3x^5 + 6x^4 + 7x^3 + 6x^2 + 3x + 1$

73. $\dbinom{n}{r} = \dfrac{n!}{r!(n - r)!} = \dfrac{n!}{(n - r)!r!} = \dfrac{n!}{(n - r)![n - (n - r)]!} = \dbinom{n}{n - r}$

f_2, f_3, f_4, and f_5 are approaching $f_1 = f_6$.

75. a. $(a + b)^1 = a + b = \dbinom{1}{0}a^1 + \dbinom{1}{1}b^1$ **b.** Assume: $(a + b)^k = \dbinom{k}{0}a^k + \dbinom{k}{1}a^{k-1}b + \dbinom{k}{2}a^{k-2}b^2 + \cdots + \dbinom{k}{k - 1}ab^{k-1}$

$+ \dbinom{k}{k}b^k$; Prove: $(a + b)^{k+1} = \dbinom{k + 1}{0}a^{k+1} + \dbinom{k + 1}{1}a^{k+1-1}b + \dbinom{k + 1}{2}a^{k+1-2}b^2 + \cdots + \dbinom{k + 1}{k + 1 - 1}ab^{k+1-1} + \dbinom{k + 1}{k + 1}b^{k+1}$

c. $(a + b)(a + b)^k = (a + b)\left[\dbinom{k}{0}a^k + \dbinom{k}{1}a^{k-1}b + \dbinom{k}{2}a^{k-2}b^2 + \cdots + \dbinom{k}{k - 1}ab^{k-1} + \dbinom{k}{k}b^k\right]$ or $(a + b)^{k+1} = \dbinom{k}{0}a^{k+1}$

$+ \dbinom{k}{0}a^k b + \dbinom{k}{1}a^k b + \dbinom{k}{1}a^{k-1}b^2 + \dbinom{k}{2}a^{k-1}b^2 + \dbinom{k}{2}a^{k-2}b^3 + \cdots + \dbinom{k}{k - 1}a^2 b^{k-1} + \dbinom{k}{k - 1}ab^k + \dbinom{k}{k}ab^k + \dbinom{k}{k}b^{k+1}$

d. $(a + b)^{k+1} \binom{k}{0} a^{k+1} + \left[\binom{k}{0} + \binom{k}{1}\right] a^k b + \left[\binom{k}{1} + \binom{k}{2}\right] a^{k-1}b^2 + \left[\binom{k}{2} + \binom{k}{3}\right] a^{k-2}b^3 + \cdots + \left[\binom{k}{k-1} + \binom{k}{k}\right] ab^k$

$+ \binom{k}{k} b^{k+1}$ **e.** $(a + b)^{k+1} = \binom{k}{0} a^{k+1} + \binom{k+1}{1} a^k b + \binom{k+1}{2} a^{k-1}b^2 + \binom{k+1}{3} a^{k-2}b^3 + \cdots + \binom{k+1}{k} ab^k + \binom{k}{k} b^{k+1}$

f. $(a + b)^{k+1} = \binom{k+1}{0} a^{k+1} + \binom{k+1}{1} a^k b + \binom{k+1}{2} a^{k-1}b^2 + \binom{k+1}{3} a^{k-2}b^3 + \cdots + \binom{k+1}{k} ab^k + \binom{k+1}{k+1} b^{k+1}$

Section 11.6

Check Point Exercises

1. 72   **2.** 729   **3.** 676,000   **4.** 840   **5.** 720   **6. a.** combinations   **b.** permutations   **7.** 210   **8.** 1820

Exercise Set 11.6

1. 3024   **3.** 6720   **5.** 720   **7.** 1   **9.** 126   **11.** 330   **13.** 1   **15.** 1   **17.** combinations   **19.** permutations
21. 27 ways   **23.** 40 ways   **25.** 243 ways   **27.** 144 area codes   **29.** 120 ways   **31.** 6 paragraphs   **33.** 720 ways
35. 8,648,640 ways   **37.** 120 ways   **39.** 15,120 lineups   **41.** 20 ways   **43.** 495 collections   **45.** 24,310 groups
47. 22,957,480 selections   **49.** 360 ways   **51.** 1716 ways   **53.** 1140 ways   **55.** 840 passwords   **57.** 2730 cones
69. (c) is true.   **71.** 144 numbers

Section 11.7

Check Point Exercises

1. $\dfrac{0.69}{3.00} = 0.23$ **2.** $\dfrac{1}{3}$ **3.** $\dfrac{1}{9}$ **4.** $\dfrac{1}{13}$ **5.** $\dfrac{1}{13,983,816} \approx 0.0000000715$ **6.** $\dfrac{999}{1000} \approx 0.999$ **7.** $\dfrac{1}{3}$ **8.** $\dfrac{3}{4}$ **9.** $\dfrac{4}{5}$

10. $\dfrac{1}{361} \approx 0.003$ **11.** $\dfrac{1}{16}$

Exercise Set 11.7

1. $\dfrac{4}{25} \approx 0.16$ **3.** $\dfrac{3}{25} \approx 0.12$ **5.** $\dfrac{7844}{60,549} \approx 0.13$ **7.** $\dfrac{1032}{20,183} \approx 0.05$ **9.** $\dfrac{1}{6}$ **11.** $\dfrac{1}{2}$ **13.** $\dfrac{1}{3}$ **15.** $\dfrac{1}{13}$ **17.** $\dfrac{3}{13}$ **19.** $\dfrac{1}{4}$

21. $\dfrac{7}{8}$ **23.** $\dfrac{1}{12}$ **25.** $\dfrac{1}{18,009,460}, \dfrac{5}{900,473}$ **27.** $\approx 25,213$ **29. a.** 2,598,960   **b.** 1287   **c.** $\dfrac{1287}{2,598,960} \approx 0.0005$ **31. a.** ≈ 0.1

b. ≈ 0.9 **33.** $\dfrac{2}{13}$ **35.** $\dfrac{1}{13}$ **37.** $\dfrac{5}{6}$ **39.** $\dfrac{7}{13}$ **41.** $\dfrac{3}{4}$ **43.** $\dfrac{33}{40}$ **45.** $\dfrac{1}{36}$ **47.** $\dfrac{1}{3}$ **49.** $\dfrac{1}{64}$ **51.** 0.00234256

53. a. $\dfrac{1}{256}$ **b.** $\dfrac{1}{4096}$ **c.** $\left(\dfrac{15}{16}\right)^{10}$ **d.** $1 - \left(\dfrac{15}{16}\right)^{10}$ **65.** Answers may vary.

Chapter 11 Review Exercises

1. $a_1 = 3; a_2 = 10; a_3 = 17; a_4 = 24$ **2.** $a_1 = -\dfrac{3}{2}; a_2 = \dfrac{4}{3}; a_3 = -\dfrac{5}{4}; a_4 = \dfrac{6}{5}$ **3.** $a_1 = 1; a_2 = 1; a_3 = \dfrac{1}{2}; a_4 = \dfrac{1}{6}$

4. $a_1 = \dfrac{1}{2}; a_2 = -\dfrac{1}{4}; a_3 = \dfrac{1}{8}; a_4 = -\dfrac{1}{16}$ **5.** $a_1 = 9; a_2 = \dfrac{2}{27}; a_3 = 9; a_4 = \dfrac{2}{27}$ **6.** $a_1 = 4; a_2 = 11; a_3 = 25; a_4 = 53$ **7.** 65   **8.** 95

9. -20 **10.** $\displaystyle\sum_{i=1}^{15} \dfrac{i}{i+2}$ **11.** $\displaystyle\sum_{i=1}^{10} (i+3)^3$ **12.** $7, 11, 15, 19, 23, 27$ **13.** $-4, -9, -14, -19, -24, -29$ **14.** $\dfrac{3}{2}, 1, \dfrac{1}{2}, 0, -\dfrac{1}{2}, -1$

15. $-2, 3, 8, 13, 18, 23$ **16.** $a_6 = 20$ **17.** $a_{12} = -30$ **18.** $a_{14} = -38$ **19.** $a_n = 4n - 11; a_{20} = 69$

20. $a_n = 220 - 20n; a_{20} = -180$ **21.** $a_n = 8 - 5n; a_{20} = -92$ **22.** 1727   **23.** 225   **24.** 15,150   **25.** 440   **26.** -500

27. -2325 **28. a.** $a_n = 0.52n + 19.48$ **b.** 77.2%   **29.** \$418,500   **30.** 1470 seats   **31.** $3, 6, 12, 24, 48$ **32.** $\dfrac{1}{2}, \dfrac{1}{4}, \dfrac{1}{8}, \dfrac{1}{16}, \dfrac{1}{32}$

33. $16, -8, 4, -2, 1$ **34.** $-1, 5, -25, 125, -625$ **35.** $a_7 = 1458$ **36.** $a_6 = \dfrac{1}{2}$ **37.** $a_5 = -48$ **38.** $a_n = 2^{n-1}; a_8 = 128$

39. $a_n = 100\left(\dfrac{1}{10}\right)^{n-1}; a_8 = \dfrac{1}{100,000}$ **40.** $a_n = 12\left(-\dfrac{1}{3}\right)^{n-1}; a_8 = -\dfrac{4}{729}$ **41.** 17,936,135   **42.** $\dfrac{127}{8}$ **43.** 19,530   **44.** -258

45. $\dfrac{341}{128}$ **46.** $\dfrac{27}{2}$ **47.** $\dfrac{4}{3}$ **48.** $-\dfrac{18}{5}$ **49.** 20   **50.** $\dfrac{2}{3}$ **51.** $\dfrac{47}{99}$ **52. a.** $\dfrac{20.72}{19.96} \approx 1.038, \dfrac{21.51}{20.72} \approx 1.038, \dfrac{22.33}{21.51} \approx 1.038$;

the population is increasing geometrically with $r \approx 1.038$.   **b.** $a_n = 19.96(1.038)^{n-1}$ **c.** ≈ 27.92 million people

53. \$42,823; \$223,210   **54.** \$120,112.64   **55.** $9\frac{1}{3}$ million

56. $S_1: 5 = \dfrac{5(1)(1 + 1)}{2}$; $S_k: 5 + 10 + 15 + \cdots + 5k = \dfrac{5k(k + 1)}{2}$; $S_{k+1}: 5 + 10 + 15 + \cdots + 5(k + 1) = \dfrac{5(k + 1)(k + 2)}{2}$; S_{k+1} can be obtained by adding $5(k + 1)$ to both sides of S_k.

57. $S_1: 1 = \dfrac{4^1 - 1}{3}$; $S_k: 1 + 4 + 4^2 + \cdots + 4^{k-1} = \dfrac{4^k - 1}{3}$; $S_{k+1}: 1 + 4 + 4^2 + \cdots + 4^k = \dfrac{4^{k+1} - 1}{3}$; S_{k+1} can be obtained by adding 4^k to both sides of S_k.

58. $S_1: 2 = 2(1)^2$; $S_k: 2 + 6 + 10 + \cdots + (4k - 2) = 2k^2$; $S_{k+1}: 2 + 6 + 10 + \cdots + (4k + 2) = 2k^2 + 4k + 2$; S_{k+1} can be obtained by adding $4k + 2$ to both sides of S_k.

59. $S_1: 1 \cdot 3 = \dfrac{1(1 + 1)[2(1) + 7]}{6}$; $S_k: 1 \cdot 3 + 2 \cdot 4 + 3 \cdot 5 + \cdots + k(k + 2) = \dfrac{k(k + 1)(2k + 7)}{6}$;

$S_{k+1}: 1 \cdot 3 + 2 \cdot 4 + 3 \cdot 5 + \cdots + (k + 1)(k + 3) = \dfrac{(k + 1)(k + 2)(2k + 9)}{6}$;

S_{k+1} can be obtained by adding $(k + 1)(k + 3)$ to both sides of S_k.

60. $S_1: 2$ is a factor of 6; $S_k: 2$ is a factor of $k^2 + 5k$; $S_{k+1}: 2$ is a factor of $k^2 + 7k + 6$; S_{k+1} can be obtained from S_k by rewriting $k^2 + 7k + 6$ as $(k^2 + 5k) + 2(k + 3)$.   **61.** 165   **62.** 4005   **63.** $8x^3 + 12x^2 + 6x + 1$ **64.** $x^8 - 4x^6 + 6x^4 - 4x^2 + 1$
65. $x^5 + 10x^4y + 40x^3y^2 + 80x^2y^3 + 80xy^4 + 32y^5$ **66.** $x^6 - 12x^5 + 60x^4 - 160x^3 + 240x^2 - 192x + 64$
67. $x^{16} + 24x^{14} + 252x^{12} + \cdots$ **68.** $x^9 - 27x^8 + 324x^7 - \cdots$ **69.** $80x^2$ **70.** $4860x^2$ **71.** 336   **72.** 15,120   **73.** 56
74. 78   **75.** 20 choices   **76.** 243 possibilities   **77.** 32,760 ways   **78.** 4845 ways   **79.** 1140 sets   **80.** 116,280 ways
81. 120 ways   **82.** $\dfrac{10,966,556}{33,871,648} \approx 0.324$ **83.** $\dfrac{6,669,666}{20,851,820} \approx 0.320$ **84.** $\dfrac{2}{3}$ **85.** $\dfrac{2}{3}$ **86.** $\dfrac{2}{13}$ **87.** $\dfrac{7}{13}$ **88.** $\dfrac{5}{6}$ **89.** $\dfrac{5}{6}$
90. a. $\dfrac{1}{15,504}$ **b.** $\dfrac{25}{3876}$ **91.** $\dfrac{4}{5}$ **92.** $\dfrac{3}{4}$ **93.** ≈ 0.533 **94.** $\dfrac{1}{32}$ **95. a.** 0.04   **b.** 0.008   **c.** 0.4096

Chapter 11 Test

1. $a_1 = 1; a_2 = -\dfrac{1}{4}; a_3 = \dfrac{1}{9}; a_4 = -\dfrac{1}{16}; a_5 = \dfrac{1}{25}$ **2.** 105   **3.** 550   **4.** $-21,846$ **5.** 36   **6.** 720   **7.** 120   **8.** $\displaystyle\sum_{i=1}^{20} \dfrac{i + 1}{i + 2}$

9. $a_n = 5n - 1; a_{12} = 59$ **10.** $a_n = 16\left(\dfrac{1}{4}\right)^{n-1}; a_{12} = \dfrac{1}{262,144}$ **11.** -2387 **12.** -385 **13.** 8   **14.** \$276,426.79

15. $S_1: 1 = \dfrac{1[3(1) - 1]}{2}$; $S_k: 1 + 4 + 7 + \cdots + (3k - 2) = \dfrac{k(3k - 1)}{2}$; $S_{k+1}: 1 + 4 + 7 + \cdots + (3k + 1) = \dfrac{(k + 1)(3k + 2)}{2}$; S_{k+1}

can be obtained by adding $3k + 1$ to both sides of S_k.   **16.** $x^{10} - 5x^8 + 10x^6 - 10x^4 + 5x^2 - 1$ **17.** 990 ways   **18.** 210 sets

19. $10^4 = 10,000$ **20.** $\dfrac{10}{1001}$ **21.** $\dfrac{8}{13}$ **22.** $\dfrac{3}{5}$ **23.** $\dfrac{1}{256}$ **24.** $\dfrac{1}{16}$

Cumulative Review Exercises (Chapters 1–11)

1. $\left\{\dfrac{14}{5}\right\}$ **2.** $\left\{\dfrac{3 + \sqrt{3}}{3}, \dfrac{3 - \sqrt{3}}{3}\right\}$ **3.** $\{2\}$ **4.** $\{16,256\}$ **5.** $\{6\}$ **6.** $-1 \le x \le 0$ or $[-1, 0]$ **7.** $\left(-\dfrac{2}{3}, \dfrac{3}{2}\right)$

8. $(-3, 1]$ **9.** $\{2.9706\}$ **10.** $\left\{1, \dfrac{1}{2}, -3\right\}$ **11.** $\{(3, 2), (3, -2), (-3, 2), (-3, -2)\}$ **12.** $\{(6, -4, 2)\}$ **13.** $\{(0, -1), (2, 1)\}$

14.   **15.**   **16.**   **17.**

18.

19.

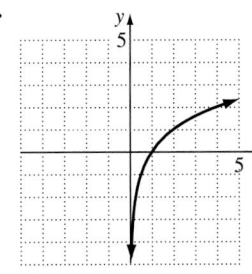

20. $f^{-1}(x) = x^3 - 4$

21. $\begin{bmatrix} -2 & 10 \\ -5 & 7 \\ 15 & -15 \end{bmatrix}$
22. $\dfrac{-1}{x-2} + \dfrac{3x-2}{x^2+2x+2}$

23. $x^{15} + 10x^{12}y + 40x^9y^2 + 80x^6y^3 + 80x^3y^4 + 32y^5$

24. 3850

25. a. $y - 21 = -0.6(x - 50)$ **b.** $y = -0.6x + 51$
    c. 18.6 hours

26. $2000

27. length: 100 yd; width: 50 yd   **28.** pen: $1.80; pad: $2   **29. a.** 6 sec   **b.** 2.5 sec; 196 ft   **30.** 11 amps

31. a. 20 in.   **b.** $\dfrac{3}{8}$ cycles/sec   **c.** $\dfrac{8}{3}$ sec   **32.** $\tan x + \dfrac{1}{\tan x} = \dfrac{\sin x}{\cos x} + \dfrac{1}{\dfrac{\sin x}{\cos x}} = \dfrac{\sin x}{\cos x} + \dfrac{\cos x}{\sin x} = \dfrac{\sin^2 x + \cos^2 x}{\cos x \cdot \sin x} = \dfrac{1}{\cos x \cdot \sin x}$

33. $\dfrac{1 - \tan^2 x}{1 + \tan^2 x} = \dfrac{1 - \dfrac{\sin^2 x}{\cos^2 x}}{1 + \dfrac{\sin^2 x}{\cos^2 x}} \cdot \dfrac{\cos^2 x}{\cos^2 x} = \dfrac{\cos^2 x - \sin^2 x}{\cos^2 x + \sin^2 x} = \dfrac{\cos 2x}{1} = \cos 2x$

34.

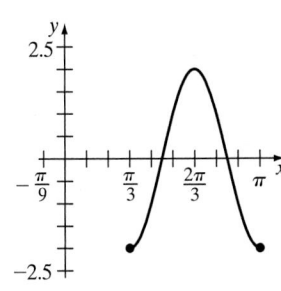

35. $\dfrac{\pi}{6}, \dfrac{5\pi}{6}, \dfrac{7\pi}{6}, \dfrac{11\pi}{6}$ **36.** $0, \dfrac{\pi}{3}, \dfrac{5\pi}{3}$ **37.** $-\dfrac{5}{\sqrt{11}}$

38.

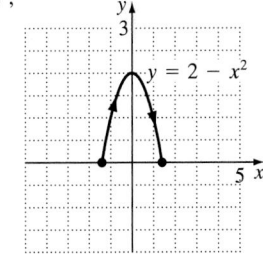

39. $B \approx 54°, C \approx 92°, c \approx 39.5$ or
    $B \approx 126°, C \approx 20°, c \approx 13.7$

40. $y = 2 - x^2$;

Subject Index

A

Absolute value
  of complex number, 656
  definition of, 4, 71
  evaluating, 5
  for expressing distance, 5–6
  properties of, 5
  rewriting equations without, 169
  solving equations involving, 140–41
  solving inequalities with, 150–53
Absolute value bars, rewriting absolute value equations without, 140
Absolute value equations, 169
Absolute value function, 236, 271
Abstract algebra, 801
Acute angles, 440, 460, 470, 548
Adams, John Quincy, 850
Addition. See also Sum(s)
  associative property of, 8
  commutative property of, 7, 8
  of complex numbers, 109–10, 168
  distributive property of multiplication over, 8
  of functions, 250–51
  identity property of, 8
  inverse property of, 8
  of like radicals, 26–27
  of matrices, 793–95, 835
  of polynomials, 38–39
  of rational expressions with different denominators, 63–66
  of rational expressions with same denominators, 63
  of vectors, 672, 673
Addition method
  eliminating variables with, 737–39
  systems of linear equations in three variables solved by, 762
  systems of linear equations in two variables solved by, 762
  systems of nonlinear equations in two variables solved by, 763
  variables eliminated with, 703–5
Additive identity
  for $m \times n$ matrices, 794
  and vector addition, 674
Additive inverse (or opposite), 9, 674, 795
AIDS, 931
Algebra, 76
  abstract, 801
  common graphs in, 235–36
  fundamental theorem of, 328–30
  literacy in, 1
  reversibility of thought in, 48
Algebraic expressions
  evaluating, 6–7
  evaluating those containing negative exponents, 14–15
  and real numbers, 2–11
  simplifying, 9–10
  simplifying with rational exponents, 33–34
Ambiguous case (SSA), 614–17, 690
American Revolution, 288
America's Cup, 642
Amount of rotation formula, 889, 920
Amplitude, 489, 494, 495, 498, 550, 567
  and maximum displacement, 545
  and periods, 492, 493
  and sinusoidal sounds, 592
Amusia, 567

and sound quality, 569
Analytic geometry, 76
And probabilities
  with independent events, 996–97
  determining, 465
  problem solving using, 463–64
Angles, 439–40
  classifying, 440
  drawing in standard position, 441
  finding in triangles, 539
  measures of, 548
  measuring using radians, 443–45
  measuring using degrees, 440
  naming, 439
  trigonometric functions of, 470–73
  between two vectors, 693
Angular speed, 450
  definition of, 448
  linear speed in terms of, 449
Annuities, 952–54, 1002
Applications
  with ellipses, 850–52
  with hyperbolas, 866–67
  of matrix inverses to coding, 817–18
  with nonlinear systems, 739–41
  with parabolas, 879–81
  with systems of linear equations, 708–11
Architecture
  angles in, 439
  conic sections in, 855
  hyperbolas in, 866
Arc length formula, 549
Area
  formulas for, 101
  of oblique triangle, 617–18
Argument, 657, 691
Arithmetic sequences, 1002
  common difference of, 938
  definition of, 938
  general term of, 939–41
  sum of first n terms of, 941–43
  writing terms of, 938–39
Arrow notation, 336–38, 367
Arrows, in graph of function, 216
Ars Magna (Cardano), 325
Artistic development, and carbon dating, 422
ASA triangles, solving using Law of Sines, 613–14, 690
ASCI White (IBM), and Cramer's rule, 832
Associative property, 71
  of addition, 8
  and English language, 8
  of multiplication, 8
  and scalar multiplication, 674
  and vector addition, 674
Asymptotes, 860, 861, 919, A5
Augmented matrix, 769, 770, 773, 835
  and multiplicative inverses of 3 x 3 invertible matrix, 812–13
  for system with infinite number of solutions, 785
  for system with no solution, 783
  for system without unique problems, 788
Average cost, 347
Average cost function, 347
Average rate of change, of function, 222–23

Axis of symmetry
  and conics, 910
  of parabolas, 281, 284, 872
  parabolas rotated about, 880

B

Babbage, Charles, 822
Back-substitution
  Gaussian elimination with, 773–75
  variable values found by, 717, 718–19, 720
Base, 13
  changing to common logarithms, 404
  changing to natural logarithms, 405
  in exponential and logarithmic forms, 386
Bearings, 540–42, 551
Bee Gees, 98
Berlin Airlift, 753
Berlinski, David, 609
Berra, Yogi, 414
Binomial, 37
Binomial coefficients, 971–72, 1002
Binomial difference, square of, 43
Binomial expansion
  finding single term of, 974
  patterns in, 969–71
Binomial expressions, 969
Binomials
  cubing, 44
  product of two: FOIL, 40–41
  squaring, 44
Binomial sum, square of, 42–43
Binomial Theorem, 972–73, 1002
Bonds, Barry, 925
Boundary points
  on number line, 158, 160, 163
  in quadratic inequalities, 157, 158
Bounds, for roots, 325, 326
Braces, for representing sets, 2
Branches, 856
Break-even point, finding, 709–10
Breathing cycle, 487, 503
Burrows, Christopher J., 880

C

Calculators. See also Graphing utilities
  computations with scientific notation on, 21
  converting from decimal to scientific notation with, 20
  exponential expressions evaluated with, 13, 375
  factorial keys on, 930
  and inverse trigonometric functions, 529–30
  trigonometric functions evaluated with, 461–63
Calculus, 725
  power-reducing formulas in, 581
Carbon dating, and artistic development, 422
Cardano, 325
Cardioids, 647
Cartesian coordinate system, 76
Cayley, Arthur, 801
Celebrity Jeopardy, 201, 203
Centaurus galaxy, 969
Center
  of circle, 195
  of ellipse, 843
  of hyperbola, 856

Central angle, 444
Centrifugal force, modeling, 362–63
Challenger space shuttle, 374, 375
Change-of-base property, 404–5, 432, A1
Chaos, in mathematics, 654
Chicago Cubs, 897
Circle(s), 193, 197–98, 270, 842
  circumference of, 445
  definition of, 195
  general form of equation of, 271
  in polar coordinates, 644
  standard form of equation of, 196, 271
Circular arc, length of, 447–48
Circular functions, 481
Closed dots, in graph of function, 216
Coded matrix, 817
Coding matrix, 817, 818
Coefficient, 44
Coefficient matrix, 815, 835
Cofactor of element, 831
Cofunction identities, 461, 549
Column matrices, 815
Combinations, 982–83, 1003
  and lottery winnings, 992
  permutations distinguished from, 983
  using formula for, 984–85
Combined variation, 361–62
Comets
  elliptical orbits of, 851, 866
  parabolic paths of, 879
Common difference, of arithmetic sequence, 938
Common factors, 49
Common logarithmic function, 391–93
Common logarithms, 432
  and change-of-base property, 404
Common ratio
  of geometric sequence, 947, 950, 951
  of infinite geometric series, 954
Commutative property, 71
  of addition, 7, 8
  of multiplication, 8
  and vector addition, 674
Commutative words and sentences, 8
Complements, 549
  finding, 443
  and trigonometric functions, 460–61
Completing the square, 169
  conic sections identified without, 885–86
  and equations of ellipses, 850
  and graphing quadratic functions, 284
  solving quadratic equations by, 119–121
Complex conjugates, 111–12, 169
Complex fractions, 66
Complex nth root, 662
Complex numbers, 108–112, 168
  absolute value of, 656
  division of, 169
  equality of, 109
  operations on, 109–10, 168
  plotting in complex plane, 655
  in polar form, 656–57, 691, A3
  powers of, in polar form, 660–61
  product of, in polar form, 658–59
  quotient of, in polar form, 659–60
  in rectangular form, 658
Complex plane, complex numbers plotted in, 655
Complex rational expressions, simplifying, 66–67

Complex sixth root, 662
Composite functions, 253–56, 272
Composite trigonometric expressions, evaluating, 532–33
Composition of functions, 254
Compound inequalities, solving, 149–50
Compound inequalities with three parts, solving by isolating x in middle, 150, 170
Compounding periods per year, 381
Compound interest, 380–82, 419
formulas for, 413–14, 432, 952
Condensing logarithmic expressions, 402–3
Conditional equation, 91, 168
"Cone of Apollonius" (Prince), 884
Conic sections (or conics), 841, 909
degenerate, 881
and ellipses, 842–52
focus-directrix definitions of, 908–9
and hyperbolas, 855–67
identifying those without rotating axes, 895
identifying without completing the square, 885–86
nondegenerate, 885, 920
and parabolas, 871–81
in polar coordinates, 908–16, 920
polar equations of, 909–16
rotation of axes, 884–95
rotations used to transform equations with xy-terms to standard equations of, 889–95
Conjugate axis, of hyperbola, 860
Conjugate roots, using to solve polynomial equations, 329
Consistent system, 707
Constant functions, 219, 220, 221, 235, 271
Constant matrix, 815
Constant numerators, 731
Constant of proportionality, 354
Constant of variation, 354, 355–56, 359
Constant over linear factor form, 726
Constants, and completing the square, 119, 120
Constant term, 9, 38
Constraints, 754–56, 758, 763
Continuous compounding of interest, 381, 432
Copernicus, Nicolaus, 76, 916
Correlation coefficient, 426
Cosecant, 454, 461
Cosecant curve, 514, 515–16
Cosine curves, vertical shifts of, 501–2
Cosine function
domain and range of, 481–82
and its inverse, 530
periodic properties of, 484
repetitive behavior of, 485
Cosine of the difference of two angles, 567–68
Cosines, 454, 460
changing to verify an identity, 558, 559, 562
products of written as sums or differences, 588
of special angles, 458
sum and difference formulas for, 570
Cost function, 347, 708
Cotangent, 454, 461
Cotangent curve, 513
Cotangent functions
graphing, 513–14
periodic properties of, 484
Coterminal angles, 441–43, 549
Counterclockwise clocks, 442
Cramer, Gabriel, 822
Cramer's rule, 822
with inconsistent and dependent systems, 831, 836
systems of linear equations in three variables solved with, 836
systems of linear equations in two variables solved with, 823–25, 836

systems of linear equations solved with, 829–31
and world's fastest computer, 832
Crossing the x-axis, and multiplicity of zero, 298
Cryptograms, 817, 818
Cube roots
combining, 31
of complex numbers in polar form, 663–64
Cubic function, 236, 271
Cubing a binomial, 44
Cycles, 469, 479
Cycloids, 904, 905

D

Dantzig, George, 758, 762
Darwin, Charles, 131
Data modeling
and formulas, 95–96
with linear equations, 186–88
Dead Sea Scrolls, 422, 423
Dean, James, 982
Decaying entities, 419
Decimal notation
converting from scientific to, 20
converting from to scientific notation, 20–21
Decoding words or messages, 818
Decomposing functions, 256–57
Decomposition of vectors, 686
Decreasing functions, 219, 220, 221
Degenerate conic sections, 881
Degree(s)
angles measured in, 440
of ax^n, 37
converting from radians to, 549
converting from, to radians, 446, 549
of monomials, 44
of polynomials, 37
of polynomials in two variables, 44
relationship between radians and, 445–47
and turning points, 298
DeMoivre, Abraham, 660
DeMoivre's Theorem, 660, 661, 691, 692, A3
for finding complex roots, 662–64
Denominator(s), 9
multiplying numerators and, by same factor to verify an identity, 561
rationalizing, 28–29
Dependent equations, 707, 762
Cramer's rule with, 831, 836
and Gaussian elimination, 835
Dependent variable, 204
Descartes, René, 76, 320
Descartes's Rule of Signs, 320–22, 367
Determinants, 822, 836
of any $n \times n$ matrix, 831–32
for identifying inconsistent and dependent systems, 831
systems of linear equations in two variables solved with, 823–25
systems of linear equations solved with, 829–31
of 3×3 matrices, 825–29
of 2×2 matrices, 822–23
Difference
of functions, 251
of matrix, 794
Difference engine, 822
Difference formula for cosines
using to find exact values, 569
verifying an identity with, 570
Difference of two cubes, factoring, 54
Difference of two squares, factoring, 52–53
Difference of two vectors, 670
Difference quotient, 271
and average rate of change, 223
and functions, 207–8
Digital photography, 792, 802–3
Directed line segments, 668, 692
Direction, 692

vector written in terms of, 675–76
Direction angle, 675, 676, 677
Directrix
conics defined in terms of, 908–9
finding equation of parabola from, 876
of parabola, 872
standard form of equation of parabola used for finding, 873, 874
Direct variation, 353, 354–59
Direct variation equation, writing, 354–55
Direct variation problems, solving, 356–59
Direct variation with powers, 358
Discriminant, 124–25, 169
Discus throwing, 577
Disraeli, Benjamin, 421
Distance, between points on real number line, 5–6
Distance formula, 10–94, 270, 567, 568
Distributive property, 71
FOIL based on, 40
and multiplication of complex numbers, 110
of multiplication over addition, 8
parentheses removed with use of, 10
and scalar multiplication, 674
and subtraction, 9
and third-order determinants, 826
Division
and complex conjugates, 111–12
of complex numbers, 169
of complex numbers in polar form, 692
definition of, 9
of higher roots, 30–31
of rational expressions, 62
by zero, 59, 162, 336
Division Algorithm, 307, 308, 311, 312, 366
DNA, 453
Domain, 249, 271
adding functions and determination of, 250
of composite function, 254
excluding numbers from, 60
in Fibonacci sequence, 926
forming composite function and finding, 255–56
of function, 203, 209–10
of logarithmic functions, 391, 432
of natural logarithmic functions, 393
of rational expressions, 59
of rational functions, 335, 336
of relation, 202
of sine and cosine functions, 481–82
of six trigonometric functions, 517
of three basic trigonometric functions, 529
Dot product
alternative formula for, 683, 693
applications with, 687–88
definition of, 682, 693
finding, 682
and orthogonal vectors, 684
properties of, 683, A4
Double-angle formulas, 578–80, 585, 607
for tangents, 579–80
three forms of, 580
for finding exact values, 578–79
Doubling time, and interest rates, 414

E

e, 432
exponential models expressed in, 427–28, 433
Earthquake intensity, 385, 392–93
Eccentricity, 909, 916
Einstein, Albert, 24
relativity theory of, 27
Elements
of matrix, 769
minor of, 827, 831
of set, 2
Eliminating the parameter, 899–900, 901
Ellipse(s), 842–52, 856, 919

applications with, 850–52
conics as, 909
definition of, 843
finding equation of, from its foci and vertices, 847
graphing those centered at (h, k), 849–50
graphs of, 892
graphs of those centered at origin, 845–47
standard form of, 843–45
translations of, 848–50
Elliptical reflection principle, 850
Empirical probability, 988–89, 1003
Empty set, 91, 707, 749
Encoding words or messages, 817
End behavior, 299
of graph of polynomial function, 366
of polynomial functions, 294–96
Endeavor space shuttle, 880
English language, associative property and, 8
Equality of complex numbers, 109
Equality of matrices, definition of, 793
Equal sign, 84
Equation conversion
from polar to rectangular coordinates, 638–39
from rectangular to polar coordinates, 637–38
Equation of circle, general form of, 198
Equation of line
general form of, 183
point-slope form of, 178–80
slope-intercept form of, 180–81
various forms of, 183
Equations, 84. See also Exponential equations; Linear equations; Parametric equations; Polynomial equations; Quadratic equations; Trigonometric equations
categorizing, 91–92
conditional, 91, 168
direct variation, 354–55
equivalent, 85
functions as, 204–5
graphing with graphing utilities, 78–80
graphs of, 77–78
of horizontal and vertical lines, 181–82
inconsistent, 168
of inverse, 264
matrix, 796–97, 835
with no solution, 91
polynomial, 132–33
quadratic in form, 138–40
solving those containing two radicals, 135–36
solving those involving absolute value, 140–41
solving those involving rational exponents, 137–38
solving those involving rational expressions, 89–91
types of, 91–92
working with both sides separately to verify an identity, 563–64
writing for line parallel to given line, 184–85
Equations in two variables, 77
Equations quadratic in form, 169
Equilateral triangles, 457
Equilibrium position, 542
Equivalent equations, generating, 85
Equivalent inequalities, 147
Estimation techniques, and line graphs, 81
Evaluating algebraic expressions, 6–7
Evaluating an exponential expression, 14
Evaluating determinants
of 2×2 matrix, 823
of 3×3 matrix, 826
of 4×4 matrix, 831–32
Evaluating the function, 206–7

Evaluating third-order determinants, 826–29
Even functions, 224–26, 550
    graph of, 271
    and y-axis symmetry, 226
Even-odd identities, 557
Even trigonometric functions, 482–83
Events, 989
    Or probabilities with those not mutually exclusive, 994–96
    probabilities of, 988
    probability of non-occurrence of, 993
Evolution theory, 131
Exact value
    double-angle formula for tangents used for finding, 579–80
    half-angle formula used for finding, 582–83
    sine of a sum used for finding, 571–73
    for trigonometric functions, 569
Expanding a logarithmic expression, 398, 399, 400
Expanding the summation notation, 932
Expansion by minors, 827
Experiment, 989
Exponential decay, 418, 419, 422
Exponential equations, 408–10, 433
    applied problems solved with, 412–15
    definition of, 408
    solving, 408–10
Exponential expressions
    evaluating, 14
    simplifying, 18–19
Exponential form, changing from logarithmic to, 386
Exponential functions, 374–82, 432, 596
    characteristics of form $f(x) = b^x$, 377
    definition of, 374
    evaluating, 375
    graphing, 375–79, 389
    modeling with, 433
    natural, 379
    transformations of, 377–79, 432
Exponential growth and decay models, 418–19, 433
Exponential models
    expressing in base e, 427–28
    scatter plots for, 425
Exponential notation, multiplications expressed in, 13
Exponential REGression option, on graphers, 426
Exponentiating both sides of the equation, 412
Exponents, 13
    in exponential and logarithmic forms, 386
    negative integers as, 14–15
    properties of, 15–17, 18–19, 71
    zero as, 15
Extraneous solutions, 135, 169

F

Face cards, 991
Factorial notation, 929–30, 1001
Factorials, evaluating fractions with, 930
Factoring
    algebraic expressions containing fractional and negative exponents, 56
    difference of two squares, 52–53
    perfect square trinomials, 53–54
    polynomial equations solved by, 132–33
    polynomials, 48–57, 331
    quadratic equations solved by, 115–17
    for separating two different trigonometric functions in an equation, 601
    sum and difference of two cubes, 54
    trinomials, 50–52
    trinomials whose leading coefficients are not one, 51–52
    trinomials whose leading coefficients are one, 51
    using to verify an identity, 559–60
Factoring by grouping, 49–50

Factoring completely, 49
Factoring formulas, 71
Factoring out greatest common factor, 49
Factoring over the set of integers, 49
Factors, 9
    and Division Algorithm, 308
Factor Theorem, 312–13, 329, 330, 366
Fermat, Pierre de, 76
    Last Theorem of, 960, 961, 969
Ferrari, 325
Fibonacci (Leonardo of Pisa), 926
Fibonacci numbers, on piano keyboard, 926
Fibonacci sequence, 926
Fifth roots, 31
Filbert Street (San Francisco), slope of, 177
Finite sequences, 927
First term, in binomial, 40
Fixed cost, 709
Focus (foci)
    conics defined in terms of, 908–9
    of ellipse, 843, 844
    finding equation of ellipse from, 847
    finding equation of parabola from, 876
    finding from hyperbola's equation, 858–59
    of parabola, 872
    standard form of equation of parabola used for finding, 873, 874
FOIL method, 329
    and multiplication of complex numbers, 110
    and multiplication of sum and difference of two terms, 41
    and partial fraction decomposition, 732
    in polynomial multiplication, 40–41
    and products of complex numbers in polar form, 658
Food Stamp Program, 280, 287
Force vector, 676
Formulas, 75, 168. See also Symbols
    amount of rotation, 889, 920
    for angle between two vectors, 683–84
    for area of oblique triangle, 618
    for area, perimeter, and volume, 101
    Binomial Theorem, 972
    combinations, 1003
    for combinations of n things taken r at a time, 984
    for compound interest, 381, 413–14, 432, 952
    for computing empirical probability, 988
    distance, 10–94, 270, 567, 568
    double-angle, 578–80, 585, 607
    factoring, 71
    gravitation, 362
    half-angle, 582–85, 585, 607
    midpoint, 194–95, 270
    and modeling data, 95–96
    for permutations of n things taken r at a time, 980, 981
    power-reducing, 580–82, 585
    product-to-sum, 588–89, 607
    proving by mathematical induction, 965–66
    quadratic, 121–23
    radian measure, 447
    recursion, 928–29, 1001
    rotation of axes, 887
    simple interest, 99
    solving for variables in, 102–3
    special-product, 41, 42, 43
    sum and difference, 567, 585, 607
    for sum of first n terms of arithmetic sequence, 941
    sum-to-product, 590–92, 607
    for value of annuity, 953
    variation, 353

windchill temperature, 136
4×4 matrix, evaluating determinant of, 831–32
Fourier, Jean, 544, 594
Fourth-degree equations, 738
Fourth-degree polynomials, 330
Fourth-order determinants, 831–32
Fourth roots, 31
    of complex numbers in polar form, 662–63
Fractional exponents, and factoring, 56
Fractional expressions, combining to verify an identity, 560–61
Fractions
    complex, 66
    evaluating those with factorials, 930
    linear equations with, 88
    repeating decimals written as, 956
Free-falling object, position formula for, 164
Frequency, 544, 592
Friendship 7, 908
Frisbees, 578
Function notation, 205–7
Functions, 175, 203–4, 271. See also Exponential functions; Inverse functions; Inverse trigonometric functions; Logarithmic functions; Polynomial functions; Quadratic functions; Rational functions; Tangent functions; Trigonometric functions
    addition of, 250–51
    analyzing graphs of, 218–19
    average rate of change of, 222–23
    combinations of, 248–52, 272
    composite, 253–56
    cost, 708
    decomposing, 256–57
    definition of, 203
    and difference quotients, 207–8
    domain of, 209–10
    as equations, 204–5
    evaluating, 206–7
    even and odd, 224–26
    graphs of, 214–27, 271
    increasing and decreasing, 219
    inverse, 260–64
    objective, 754
    piecewise, 208–9
    profit, 709
    and relative maxima/minima, 221–22
    revenue, 708
    step, 226–27
    sum of, 250, 251
    transformations of, 271
    vertical line test for, 217
Function values, for special angles, 456–58
Fundamental Counting Principle, 977, 978, 990, 1003
Fundamental identities, 458–60, 549
Fundamental theorem of algebra, 328–30
Fundamental trigonometric identities, 557, 558–65, 606
$f(x)$ notation, 205

G

Galápagos Islands, 131, 134
Galileo, 76
Galois, Alfred, 122
Galois, Evariste, 122, 325
Gauss, Carl Friedrich, 111, 329, 772, 778
Gaussian elimination, 772
    applying to systems with more variables than equations, 786–87
    applying to systems without unique solutions, 783–84
    with back-substitution, 773–75
    linear systems solved using, 773–78, 835
    problem solving with, 788
    for systems with infinite number of solutions, 785–86

Gauss-Jordan elimination
    linear systems solved using, 778–79, 835
    and multiplicative inverses of 3×3 invertible matrix, 812–13
GCF. *See* Greatest common factor
General form of equation of circle, 198
General form of equation of line, 183
General second-degree equation, 886, 889, 920
General term
    of arithmetic sequence, 939–41
    of geometric sequence, 947–49
    of sequence, 927
    writing terms of sequence from, 927–28
Geometric formulas, 101
Geometric population growth, 948–49
Geometric sequence, 946, 1002
    definition of, 947
    general term of, 947–49
    sum of first n terms of, 949–52
    writing terms of, 947
Geometric series, 954–56, 1002
Geometry, 76
Glenn, John, 908
Global Positioning System, 867
Golden rectangle, 2
Goss, Brian, 699
Graphing utilities, 168
    circles graphed with, 196, 199
    data modeled with, 425, 426, 427, 428
    evaluating determinant of matrix with, 827
    evidence of identities provided with, 558
    functions evaluated with, 206
    general second-degree equations graphed on, 892
    graphing equations with, 78–80
    graphing parabolas with, 875, 878
    inverse of matrix found with, 809
    matrix addition and subtraction with, 794
    permutations calculated with, 981
    plane curves represented by parametric equations obtained with, 900
    point conversion from polar to rectangular coordinates with, 635
    polar equations graphed with, 643
    and polynomial functions, 299
    range settings for, 327
    rectangular-polar conversion on, 637
    REF (row-echelon form) command on, 775, 779
    regression lines on, 187
    SHADE feature on, 746
    SINe REGression feature on, 503
    solutions of quadratic equation checked with, 117
    solving linear equations in one variable with, 87
    step functions on, 227
    sums of sequences calculated with, 932
    TABLE feature on, 215, 646
    terms of sequence written with, 928
    verifying observations on, 573–74
    zero or root feature on, 328
    ZOOM SQUARE feature on, 845
Graphs/graphing, 168, 928
    of circles, 197–98
    of common functions, 235–36
    of ellipse centered at origin, 845–47
    of ellipses, 892
    of ellipses centered at (h, k), 849–50
    of equation of rotated conic, 893–95
    of equations, 77–78
    of exponential functions, 375–79
    of function of form $y = A \cos (Bx - C)$, 500–501
    of function of form $y = A \sin Bx$, 492
    of functions, 214–27, 271
    of hyperbolas, 861–63, 919
    of hyperbolas centered at origin, 861–63

of inconsistent systems, 707
of inequalities, 145
and intercepts, 80
interpreting information given by,
   80–81
of inverse functions, 266–67
lemniscates, 651
limaçons, 647
line, 80
of linear inequalities in two variables,
   644–46, 763
linear systems solved by, 700
of logarithmic functions, 388–90, 432
of nonlinear inequality in two variables, 747
obtaining information from, 216–17
of parabolas with vertex at (h, k),
   877–79
of parabolas with vertices at origin,
   873–74
of plane curves, 898–99
of polar equation of conic, 911–16, 920
of polar equations, 639, 642–51, 691
of polynomial functions, 299–300, 366
of profit function, 709
of quadratic functions, 280–81, 366
of quadratic functions in standard
   form, 282–86
of rational functions, 342–46, 367
of real numbers, 3
of reciprocal functions, 338
reflections of, 240–42
with sequences of transformations, 244
sinusoidal, 498, 503
of six trigonometric functions, 517
step functions, 226–27
of systems of inequalities in two variables, 747–50
of systems of two linear equations in
   two variables, 706
of three basic inverse trigonometric
   functions, 529
of trigonometric functions, 550
with turning points, 298
using slope and y-intercept, 180–81
of variations of $y = \cos x$, 498–501
of variations of $y = \cot x$, 513–14
of variations of $y = \csc x$ and $y = \sec$
   x, 515
of variations of $y = \sin x$, 488–92
of variations of $y = \tan x$, 511
vertical shrinking of, 243
vertical stretching of, 243
of $y = A \sin (Bx - C)$, 494–97
of $y = A \tan (Bx - C)$, 511
of $y = \cos x$, 497–98
of $y = \cot x$, 513
of $y = \csc x$, 514
of $y = \sec x$, 515
of $y = \sin x$, 487–88
of $y = \tan x$, 509–10
Gravitation formula, 362
Great Britain, and Berlin Airlift, 753
"Greater than" symbol, 150
Greatest common factor, factoring out,
   49
Greatest integer function, 227
Greeks (ancient)
   discus throwing by, 577
   trigonometry used by, 453
Grouping symbols, and factorials, 929
Growing entities, 419
Growth rate, for world population, 418

H
Half-angle formulas, 582–85, 585, 607
   for finding exact value, 582–83
   and verifying an identity, 583–84
Half-life, 422
Half-plane, 744, 745, 746
Halley's Comet, 851, 866, 879
Harding, Warren G., 588
Harvey, William, 76
Heron of Alexandria, 627
Heron's Formula, 627–28, A2

Higher-order determinants, evaluating,
   831–32
Highway automation, 788
Horizontal asymptotes, 367
   definition of, 340
   and exponential functions, 376, 377,
   378
   locating, 341
   and logistic growth models, 423
   of rational functions, 340–42
Horizontal axis, on line graph, 81
Horizontal component of v, 670
Horizontal lines, 183
   graphing, 182
Horizontal line test
   and inverse cosine functions, 525
   and inverse functions, 272
   and inverse sine functions, 522
   and inverse tangent functions, 527
   and one-to-one functions, 265–66
Horizontal shifts, 237–40, 243, 244, 271
   to the left, 238–39
   vertical shifts combined with, 239–40
Horizontal translations
   of ellipses, 848
   of exponential functions, 377
   of hyperbolas, 863
   of logarithmic functions, 390
   of parabolas, 876
Hubble Space Telescope, 871, 880
Hurricanes, 699
Hyperbola(s), 842, 855–67, 919
   applications with, 866–67
   asymptotes of, 860, 919, A5
   conics as, 909
   definition of, 856
   finding equation of from its foci and
   vertices, 859
   graphing, 861–63, 919
   standard form of equation of, 856–59
   translations of, 863–66
Hypotenuse, 127, 453

I
Identity(ies), 91, 168
   for solving trigonometric equations,
   601–4, 607
Identity function, 235, 271
Identity property, 71
   of addition, 8
   of multiplication, 8
Imaginary axis, 655
Imaginary part, of complex number, 109
Imaginary unit i, 108–9, 168
Inconsistent equation, 91, 168
Inconsistent systems, 706, 762
   Cramer's rule with, 831, 836
   and Gaussian elimination, 835
Increasing functions, 219, 220, 221
Independent events
   and probabilities with, 996–97
   in family, 997
   on roulette wheel, 997
Independent variable, 204
Index, 30, 34
Index of summation, 931, 932
Individual Retirement Accounts, 952,
   953–54
Inequalities
   compound, 149–150
   graphing, 145
   intervals and, 146–147
   properties of, 148
   quadratic, 170
   rational, 170
   solving with absolute value, 150–153
Inequality symbols, 4
   replacing with equal signs, 744, 745,
   747, 749
   reversing direction of, 147
Infeld, Leopold, 122
Infinite geometric series, 954, 955
Infinite sequence, 927, 1001
Infinity symbol, 146
Initial point, 668

Initial side, 439, 440
Inside term, in binomial, 40
Intercepts, 80
   identifying in function's graph, 217
Interest
   compound, 380–82, 413–14, 419, 432
   simple, 99
Interest rates, and doubling time, 414
Intermediate Value Theorem, 327, 367
Internet, 792
Interval notation, 145, 146, 148
Intervals, and increasing, decreasing,
   and constant functions, 220–21
Inverse cosine function, 525–27
Inverse functions, 260–64, 272
   finding, 263–64
   graphing, 266–67
   horizontal line test for, 265–66
   verifying, 262–63
Inverse key, on calculator, 464
Inverse of matrix, for solving matrix
   equations, 816–17
Inverse property(ies), 71, 530, 531
   of addition, 8
   of logarithms, 388
   of multiplication, 8
   of natural logarithms, 394
Inverse sine function, 522–25
Inverse sine key, on calculator, 464
Inverse tangent function, 527–29
Inverse trigonometric functions, 551
   calculators and, 529–30
   composition of functions involving,
   530–34, 531–34
   graphs of three basic, 529
Inverse variation, 353, 359–62
Inverse variation problems, solving,
   360–61
Invertible matrices, 811, 835
   multiplicative inverses of 3×3, 812
   procedure for finding multiplicative in-
   verses of, 813, 815
Investments
   choosing, 381–82
   and simple interest, 99–100
IRAs. See Individual Retirement Accounts
Irrational number, 2
Irreducible over the integers, 49
"Is approximately equal to" symbol, 2
"Is greater than or equal to" symbol, 4,
   144
Islam, 453
"Is less than or equal to" symbol, 4, 144
i unit vector, 670

J
Jagged Little Pill (Morissette), 98
Joint variation, 353, 362–63
Jordan, Wilhelm, 778
j unit vector, 670

K
Karmarkar, Narendra, 762
Kennedy Airport, hyperbolas in TWA
   building at, 866
Kepler, Johannes, 850, 916
Khachion, L.G., 762
Kidney stone disintegration, and ellipti-
   cal reflection principle, 850
Kim, Scott, 281

L
Lanier, Jaron, 768
Last term, in binomial, 40
Latus rectum
   and graphing equation of rotated
   conic, 894, 895
   and graphing parabolas, 874, 875, 877,
   879
   of parabola, 919
Law of Cosines, 690, A2–A3
   and angle between two vectors, 683
   applications of, 627
   and its derivation, 623–24

oblique triangles solved with, 624–26
SAS triangles solved with, 624–25, 691
SSS triangles solved with, 625–26
Law of Sines
   applications of, 618–19
   and area of oblique triangle, 617
   ASA triangles solved with, 613–14, 690
   and its derivation, 611–12
   SAA triangles solved with, 612–13, 690
   SSA triangles solved with, 615–17, 690
LCD. See Least common denominator
Leading coefficient, 38, 293
Leading Coefficient Test, 294–96, 366
Leaning Tower of Pisa, 164
Least common denominator, 88
   finding, 64–65
   and verifying an identity, 563–64
Legs, of right triangle, 127
Lemniscates, 651, 691
"Less than" symbol, 150
Like radicals, adding and subtracting,
   26–27
Like terms, 9
Limaçons, 647, 691
Linear combination of vectors, 670
Linear equations, 84–92, 168
   data modeled with, 186–88
   definition of, 85
   with fractions, 88
   problem solving with, 97–102
   solving, 85–88
   solving those with variables in denominators, 89
Linear equations in one variable, 84,
   85–88
Linear equations in three variables, 716
Linear Factorization Theorem, 330–31,
   367
Linear factors
   partial fraction decomposition with
   distinct, 727–28
   partial fraction decomposition with repeated, 728–29
Linear functions, and rolling motion, 905
Linear inequalities, 144–53, 170
   in one variable, 144
   solving, 147–49
Linear inequalities in two variables
   graphs of, 644–46
   and their solutions, 743
Linear numerator over quadratic factor
   form, 729
Linear numerators, 731
Linear programming, 753–54, 763
   constraints in, 754–56
   objective functions in, 754
   problem solving with, 756–58
   steps for solving problems with, 756,
   763
Linear speed
   definition of, 448
   finding, 449–50
   in terms of angular speed, 449
Linear systems, 699
   Gaussian elimination applied to system with infinite number of solutions,
   785–86
   Gaussian elimination applied to system with no solution, 783–84
   matrix solutions to, 769–79, 835
   with no solution or infinitely many solutions, 706–8
   solving with addition method, 703–5
   solving with Gaussian elimination,
   773–78
   solving with Gauss-Jordan elimination,
   778–79
   solving with matrices, 769–71
   solving by substitution, 701–3
Line graphs, 80, 81
Line(s), slope of, 176–78, 270
Logarithmic equations, 410–12, 433
   applied problems solved with, 412–15
   definition of, 410
   solving, 410–12

Logarithmic expressions
  condensing, 402–3
  expanding, 398, 399, 400, 401–2
Logarithmic form, changing from exponential form to, 386
Logarithmic functions, 385–95, 432
  definition of, 386
  domain of, 391
  of form $f(x) = log_b x$, 389
  graphs of, 388–90
  modeling with, 433
  transformations of, 390
Logarithmic models, scatter plots for, 425
Logarithmic properties, 387–88, 432
  involving one, 387
  using, 388
Logarithms
  evaluating, 387
  inverse properties of, 388
  properties of, 392, 398–405, 432, A1–A2
Logistic growth models, 423–25
Long division, of polynomials, 305–8, 366
LORAN, hyperbolas as basis of, 867
LOTTO (Florida), probability of dying compared to probability of winning with, 992
Lower bound, for roots, 325
Lower limit of summation, 931, 932

M

Mach, Ernst, 586
Magnitude, 692
  of directed line segment, 668
  representing vector in rectangular coordinates and finding, 671
  vector written in terms of, 675–76
Major axis, of ellipse, 843, 844
Malthus, Thomas, 949
Management science, linear programming in, 754
Mandelbrot set, 654–55, 664
Map making, and angle of elevation, 465
Mask, The, 235
Mathematical induction, 961–64, 1002
  proving steps about positive integers using, 964–67
Mathematical modeling, 96, 168
Mathematical models, 96, 168
  creating and comparing, 152–153
  for exponential growth or decay, 419
Mathematics
  chaos in, 654
  universality of, 972
Matrices, 768
  addition and subtraction of, 793–95, 835
  applications with, 802–4
  and digital photography, 802–3
  equality of, 793
  finding multiplicative inverses of, 809–11
  images based on, 792
  linear systems solved with, 769–71
  multiplication of, 835
  multiplicative inverses of, 808–16, 835
  notations for, 792–93
  of order $m \times n$, 792
  quick method for finding multiplicative inverses of, 811–12
Matrix, The, 235, 521
Matrix addition, properties of, 795
Matrix equations, 835
  inverses used for solving, 816–17
  solving, 796–97
Matrix inverses, and encoding/decoding messages, 817–18
Matrix multiplication, 797–802
  applying, 803–4
  definition of, 799
  noncommutativity of, 801, 802, 808
  properties of, 802, 835
Matrix row operations, 771–72, 773, 835

Maximum value, of quadratic functions, 286
Midpoint formula, 194–95, 270
Minimum value, of quadratic functions, 286
Minimum wage, modeling growth of, 419–21
Minor, of element, 827, 831
Minor axis, of ellipse, 843, 844
Minus signs, in evaluating determinants of 3×3 matrices, 828
Model breakdown, 96
Modeling
  art of, 425–27
  with exponential and logarithmic functions, 433
  with variation, 367
Modulus (moduli), 657, 691
Monomial, 37
Monroe, Marilyn, 982
Monteverdi, Claudio, 76
Mooney, Michael, 699
Morissette, Alanis, 98
Morphing, 235, 242
Morrison, Jim, 982
Muhammad, 453
Multiplication
  associative property of, 8
  of binomial and trinomial, 40
  commutative property of, 8
  of complex conjugates, 111
  of complex numbers, 110, 168
  of complex numbers in polar form, 691
  distributive property of, over addition, 8
  of higher roots, 30–31
  identity property of, 8
  inverse property of, 8
  of matrices, 797–802, 835
  of polynomials, 39–40
  of polynomials in two variables, 45
  repeated, 13
  scalar, 669
  of sum and difference of two terms, 41–43
Multiplication property, and scalar multiplication, 674
Multiplicative identity, and scalar multiplication, 674
Multiplicative identity matrix, 807–8
Multiplicative inverse (or reciprocal), 9
  quick method for finding, 811–12
Multiplicative inverses of matrices, 808–16
  systems of equations solved with, 815–16
Multiplicity
  and x-intercepts, 298
  of zero, 298
Multiplier effect, tax rebates and, 956
Music
  modeling, 544
  and trigonometric functions, 592
Muslims, trigonometry used by, 453
Mutually exclusive events, Or probabilities with, 993–94

N

$n!$, 929
National debt, 1, 19, 21
National Education Association, 940
Natural base e, 379–80
Natural exponential functions, 379, 432
Natural logarithmic functions, 393
Natural Logarithmic REGression (LnReg) option, on graphers, 426
Natural logarithms, 393–95, 432, 433
  and change-of-base property, 404
  changing base to, 405
  properties of, 394
  solving equations with, 412
  using to solve exponential equations, 408
Natural number exponent, definition of, 14

Natural numbers, 2
Navigation, hyperbolas in, 866, 867
Negative angles, 440
Negative exponent rule, 14
Negative exponents, and factoring, 56
Negative infinity, 337
Negative integers, as exponents, 14–15
Negative leading coefficient, 295
Negative numbers, 3
  roots of, 112
  square roots of, 169
  vectors multiplied by, 669
Negative real zeros, 367
  and Descartes's Rule of Signs, 321
Negatives, properties of, 10
Negative slope, 178
Newton, Isaac, gravitation formula of, 362
Nondegenerate conic sections, 885, 920
Nonlinear inequalities, problem solving with, 163–65
Nonlinear inequalities in two variables, graphs of, 747
Nonlinear systems
  problem solving with, 739–41
  solving by addition method, 737–39
  solving by substitution method, 734–37
Nonsingular matrix, 811, 835
Nonsquare matrix, lack of multiplicative inverses for, 810
Nonsquare systems, 786–87
nth-order determinants, 831, 836
nth partial sum, 949
nth roots, product and quotient rules for, 71
Number line
  boundary points on, 158, 160, 163
  distance between points on, 71
  distance between two points on, 6
Numbers
  complex, 108–12
  natural, 2
  real, 2–3
Numerator(s), 9
  multiplying denominator and, by same factor to verify an identity, 561
Numerical coefficient, 9

P

Objective functions, 758, 763
  in linear programming, 754
  maximum/minimum of, 757, 758
Oblique triangles, 611
  area of, 617–18
  Law of Cosines used for solving, 624–26
  Law of Sines used for solving, 612–14
Obtuse angles, 440, 548
Odd functions, 224–26, 509, 550
  graph of, 271
  and origin symmetry, 226
Odd trigonometric functions, 482–83
One radian, 443, 444
One-to-one correspondence, 3
One-to-one functions, 272
  graphing inverse of, 266–67
  and horizontal line test, 265–66
Open dots, in graph of function, 216
Operations, with vectors in terms of i and j, 676
Opposite (or additive inverse), 9
Orbach, Jerry, 201
Ordered pairs, 168
  inequalities satisfied by, 743
  and points, 76–77
  as solution of system, 700
Ordered triple, 784
  as solution to system of linear equations in three variables, 762
  as solution of system, 717
Ordering the real numbers, 4
Order of operations agreement, 7
Orientation of curve, 899
Origin, 3, 76, 168

asymptotes of hyperbola centered at, 860, A5
graphing hyperbolas centered at, 861–63
standard form of equation of parabola with its vertex at, 873
symmetric with respect to, 226
Origin symmetry, and odd functions, 226
Or probabilities
  with events that are not mutually exclusive, 994–96
  with mutually exclusive events, 993–94
Orthogonal vectors, 684, 685, 687, 693
Outside term, in binomial, 40

P

Pacino, Al, 114
Parabola(s), 842, 871–81, 919
  conics as, 909
  definition of, 871–72
  downward-opening, 280, 281, 282, 875
  finding focus and directrix of, 875–76
  leftward-opening, 878
  reflection property of, 881
  standard form of equation of, 872–73
  standard form of equation of with vertex at origin, 873
  standard forms of with vertex at (h, k), 876
  upward-opening, 280, 281, 282, 283, 284
  using standard form of equation of, 873–76
Parallel lines, slope and, 184–85
Parallel vectors, 684
Parameter, 898
  eliminating, 899–900, 901
Parametric equations, 920
  advantages of over rectangular equations, 904
  finding, 902–3
  for function $y = f(x)$, 903
  graphing curves defined by, 899
  plane curves and, 897–98
Parentheses, distributive property and removal of, 10
Partial fraction decomposition, 763
  with distinct linear factors, 727–28
  idea behind, 725–26
  of rational expressions, 726
  steps in: case 1, 726–28
  steps in: case 2, 728–29
  steps in: case 3, 729–31
  steps in: case 4, 731–32
Pascal, Blaise, 76, 972
Pascal's triangle, 971
Perfect nth power, 30
Perfect squares, 25
Perfect square trinomials, 119
  factoring, 53–54
Perimeter, formulas for, 101
Period, 483, 489, 492, 493, 494, 495, 498, 550
  of breathing cycle, 503
  of six trigonometric functions, 517
  of tangent function, 509, 510
Periodic behavior, modeling, 503–5
Periodic functions, 483–85, 550
Periodic phenomena, and trigonometric functions, 596
Periodic properties
  of sine and cosine functions, 484
  of tangent and cotangent functions, 484
  using to find exact values, 484
Permutations, 980–81, 1003
  combinations distinguished from, 983
  and Rubik's cube, 979
Perpendicular lines, slope and, 185
Phase shift, 494, 495, 498, 550
Picture cards, 991
Piecewise functions, 208–9
Pixels, 802, 803
Plane curves

finding/graphing rectangular equation of those defined parametrically, 900–903
graphing, 898–99
graphing those defined by parametric equations, 899
and parametric equations, 897–98, 920
Planetary motion, modeling, 916
Planetary orbits
eccentricities of, 916
elliptical paths of, 850
Plotting points
graphing function by, 215
in polar coordinate system, 631–33
in rectangular coordinate system, 77
Plotting the complex number, 655
Plotting the real number, 3
Plus signs, and transverse axes of hyperbolas, 857
Point conversion
from polar to rectangular coordinates, 635–36
from rectangular to polar coordinates, 636–37
Point plotting
for graphing plane curve described by parametric equations, 920
for obtaining plane curve, 901
Point-plotting method, 77
graphing polar equation by, 642–44
Point Reyes National Seashore, 611
Points
multiple representation of, 691
multiple representation of in polar coordinate system, 633–34
and ordered pairs, 76–77
Points location, and order of ordered pair, 77
Point-slope form, 183
of equation of line, 178–80
Polar axis (*x*-axis), 631
symmetry with respect to, 644
Polar coordinates
conic sections in, 920
equation conversion from rectangular to, 637–38
relations between rectangular and, 634, 691
sign of *r* and point's location in, 632
Polar coordinate system, 631, 691
multiple representation of points in, 633–34
plotting points in, 631–33
Polar equations, 637
of conic sections, 909–16
graphing of, 639, 642–51, 691
Polar form
of complex numbers, 656–57, 691, A3
division of complex numbers in, 692
powers of complex numbers in, 660–61
product of two complex numbers in, 658–59
quotient of complex numbers in, 659–60
roots of complex numbers in, 662–64
Polar grids, for graphing polar equations, 642
Polar-to-rectangular point conversion, 635–36
Pole, 631
symmetry with respect to, 644
Polk, James K., 588
Polynomial equations, 132–33, 169
conjugate roots used for solving, 329
properties of, 320
solving, 319–20
solving by factoring, 132–33
Polynomial functions, 279
definition of, 293
end behavior of, 294–96
finding zeros of, 317–19
graphs of, 366
recognizing graphs of, 294
Remainder Theorem used for evaluating, 312

strategy for graphing, 299–300, 366
turning points of, 298
zeros of, 296–98, 315–22
Polynomials, 36–45
addition and subtraction of, 38–39
dividing using synthetic division, 309–11
factoring, 48–57, 56, 331
finding those with given zeros, 331–32
FOIL used in multiplication of, 40–41
Intermediate Value Theorem for, 327–28
long division of, 305–8, 366
multiplication of, 39–40
prime, 49
in several variables, 44–45
special products used in multiplication of, 41–44
strategy for factoring, 54–56
vocabulary of, 37–38
Polynomials in two variables, 44
Polynomials in *x*, definition of, 38
Popper, John, 975
Population, modeling, 187–88
Population growth, geometric, 948–49
Position formula, for free-falling object, 164
Position model, using, 164–65
Position vector, 671
Positive acute angles, 474
Positive angles, 440, 549
Positive leading coefficient, 295
Positive numbers, 3
Positive real zeros, 367
and Descartes's Rule of Signs, 320, 321
Positive slope, 178
Power of a linear factor, 731
Power of a quadratic factor, 731

Q

Quadrantal angles, 440, 470, 548
trigonometric functions of, 471–73
Quadrants, 76
Quadratic equations, 114–27, 169
definition of, 115
determining method to use for solving, 125
problem solving with, 126–27
solving, 169
solving by completing the square, 119–21
solving by square root method, 117–19
solving with quadratic formula, 121–23
standard form of, 115, 134
trigonometric equations in form of, 600
Quadratic formula, 125, 126, 169
quadratic equations solved with, 121–23
Quadratic functions, 271, 280–87, 366
applications of, 286–87
and data modeling, 720, 721
graphing in form
$f(x) = ax^2 + bx + c$, 284–86
graphing in standard form, 281–86
graphs of, 235, 280–81, 366
standard form of, 366
Quadratic inequalities, 170
definition of, 157
solving, 157–59
Quadratic in form, and solving equations, 138–40
QUADratic REGression program, 288
Quarterly compounding of interest, 381
Quotient identities, 458, 459, 549, 557
Quotient rule, 16–17
for expanding logarithmic expressions, 401
of logarithms, 399–400, 432
for *n*th roots, 30, 71
for square roots, 26
Quotients
of complex numbers in polar form, 659–60
of functions, 251
raised to powers, 17–18

R

Radian measure, 549
of central angle, 444
computing, 445
and unit circle, 480
Radian measure formula, for finding length of arc, 447
Radians
angle measures with, 443–45
converting from degrees to, 446, 549
converting from, to degrees, 446–47, 549
definition of, 444
relationship between degrees and, 445–47
Radical equations, 131, 169
solving, 133–36
Radicals, 24, 30
combining those that first require simplification, 27
reducing index of, 34
and windchill, 136
Radical sign, 24
Radicand, 24, 30
Radius, of circle, 195, 445
Range, 271
in Fibonacci sequence, 926
of functions, 203
of relations, 202
of sine and cosine functions, 481–82
of six trigonometric functions, 517
of three basic trigonometric functions, 529
Rate of change
of function, 207
slope as, 186–87
Rational exponents, 31–34, 71
definition of, 32
expressions simplified with, 33–34
radical equations with, 169
solving equations involving, 137–38
Rational expressions, 59–67
adding and subtracting with different denominators, 63–66
adding and subtracting with same denominator, 63
complex, 66–67
dividing, 62
domain of, 59
equations involving, 89–91
multiplying, 61–62
partial fraction decomposition of, 726, 763
partial fraction decomposition of with distinct linear factors in denominator, 726–28
partial fraction decomposition of with prime, nonrepeated quadratic factors in denominator, 729–30
partial fraction decomposition of with repeated linear factors, 728–29
partial fraction decomposition of with repeated quadratic factors, 731–32
simplifying, 60–61
Rational functions, 279, 335–48
applications with, 347–48
finding domain of, 336
graphing, 342–46, 367
horizontal asymptotes of, 340–42
slant asymptotes of, 346–47
vertical asymptotes of, 338–40
Rational inequalities, 157, 170
solving, 160–63
test numbers used for solving, 161–62
Rationalizing denominators, 28–29
Rational roots, 325
Rational Zero Theorem, 316–17, 367
Ray, 439
Real axis, 655
Real number line, 3
distance between points on, 5–6
intervals on, 146
Real numbers
absolute value of, 4–5
and algebraic expressions, 2–11

important subsets of, 3
ordered pair of, 76
ordering, 4
plotting, 3
principal *n*th root of, 30
properties of, 8
set of, 2–3
trigonometric functions of, 479, 480–81, 550
vectors multiplied by, 669
Real part, of complex number, 109
Real zeros, approximating, 327–28
Reciprocal functions, 336
graphs of, 338
Reciprocal identities, 458, 459, 549, 557
Reciprocal (or multiplicative inverse), 9
Rectangles, golden, 2
Rectangular coordinates
equation conversion from polar to, 638–39
relations between polar and, 634, 691
vectors represented in, 671–72
Rectangular coordinate system, 76, 168, 193
graphing equations in, 77–78
plotting points in, 77
vectors in, 670–72
Rectangular equations
advantages of parametric equations over, 904
finding/graphing for curve defined parametrically, 900–903
Rectangular form, complex number written in, 658
Rectangular-to-polar point conversion, 636–37
Recursion formulas, 928–29, 938, 939, 1001
Reduced row-echelon form, 778
Reference angles, 550
definition of, 474
finding, 475–76
trigonometric functions evaluated with, 476–78
Reflecting, 244
Reflecting telescopes, 880
Reflection(s), 242
of exponential functions, 377
of graphs, 240–42
of logarithmic functions, 390
about the *x*-axis, 240–41, 243, 271
about the *y*-axis, 241, 243, 271
REF (row-echelon form) command, on graphing utility, 775, 779
Regression line, 176, 187
Relations, 271
definition of, 202
finding domain and range of, 202
as functions, 203–4
Relative maximum, 221, 222
Relative minimum, 221, 222
Relativity theory, 27
Remainder Theorem, 311–12, 366
Repeated factorization, 52–53
Repeated multiplication, 13
Repeated quadratic factor, partial fraction decomposition with, 731–32
Repeated zero with multiplicity, 298
Repeating decimals, writing as fractions, 956
Repetitive behavior, of sine, cosine, and tangent functions, 485
Resources and Man (U.S. Academy of Sciences), 379
Resultant force, finding, 676–78
Resultant vector, 669
Revenue, 708
Revenue function, 708
Reversibility of thought, 48
Richter, Andy, 201
Richter scale, 385, 392
Right angles, 440, 548
Right triangles, 127
solving, 551

using to evaluate trigonometric functions, 453–55
Right triangle trigonometry, 549
    applied problems solved with, 463–65
Rise, 176
Rocky Horror Picture Show, The, 710, 711
Rolling motion, linear functions and cycloids used for describing, 905
Roots (or solutions), 6, 85, 367
    of complex numbers in polar form, 662–64
    finding bounds for, 326–27
    of negative numbers, 112
    of polynomial equations, 296, 319
Rose curves, 650, 691
Rose with four petals, 649
Rotated conic, graphing equation of, 893–95
Rotated conic section, writing equation of in standard form, 890
Rotating axes, identifying conic sections without, 895
Rotation of axes, 884, 886–95, 920
Rotation of axes formula, 887
Row-by-column multiplication, 797, 798–801
Row-echelon form, 771, 775, 778, 835
Row equivalent matrices, 771
Row operations, 771
Rubik, Erno, 979
Rubik's cube, 979
Run, 176
Rutherford, Ernest, 869

S
SAA triangles, solving using Law of Sines, 612–13, 690
Salary computations, 925, 952
Sample space, of experiment, 989
San Francisco Museum of Modern Art, 439
SAS triangles, Law of Cosines used for solving, 624–25, 691
Satisfying the equation, 77, 85, 168
Satisfying the inequality, 145
Saturday Night Fever (Bee Gees), 98
Scalar components of *v*, 670
Scalar multiplication, 669, 672
    definition of, 795
    properties of, 674, 796, 835
    with vector in terms of i and j, 673
Scalars (scalar quantities), 667
Scatter plots, 176, 426, 428
    for exponential and logarithmic models, 425, 433
Schwarzkopf, Norman, 201
Scientific notation, 19–22
    converting from decimal notation to, 20–21
    converting from to decimal notation, 20
Secant, 454, 461
Secant curve, 515
Secant functions, graphing, 516–17
Secant line, 222
Second-degree polynomial equations, 115
Second-order determinants, 836
    value of, 822–23
Selena, 982
Semiannual compounding of interest, 381
Sequences, 925, 926, 1001
    arithmetic, 938–43, 1002
    definition of, 927
    geometric, 946–51, 1002
    graphs of, 928
    *n*th term of, 928
    and sunmmation notation, 931–34
    writing terms of from general term, 927–28
Serpico, 114
Set-builder notation, 146–147

Set(s)
    of complex numbers, 109
    of real numbers, 2–3
Sextants, 463
SHADE feature, on graphing utilities, 746
Shakespeare, William, 76
Shaugnessy, Charles, 201
Shrinking
    of exponential functions, 377
    of logarithmic functions, 390
Sign changes, 328
Signs of trigonometric functions, 473–74, 550
Simple harmonic motion, 542–45
    analyzing, 544–45
    finding equation for object in, 543–44
    frequency of object in, 544
    resisting damage of, 545
Simple interest, 99
Simple interest problems, solving, 100
Simplifying algebraic expressions, 9–10
Simplifying complex numbers, 109
Simplifying complex rational expressions, 66–67
Simplifying exponential expressions, 18–19
Simplifying higher roots, 30–31
Simplifying rational expressions, 60–61
Simplifying square roots
    product rule for, 25–26
    quotient rule for, 26
Sine curve, 488
    stretching/shrinking, 489
    using to obtain cosecant curve, 515–16
    vertical shifts of, 501–2
Sine function
    domain and range of, 481–82
    and inverse, 530
    key points in graphing, 488–89
    periodic properties of, 484
    repetitive behavior of, 485
Sine of sum, using to find exact value, 571–73
Sine of zero, 453
SINe REGression feature, on graphing utilities, 503
Sines, 454, 460
    changing to, to verify an identity, 558, 559, 562
    products of written as sums or differences, 588
    of special angles, 458
    sum and difference formulas for, 570
Singular matrix, 811
Sinusoidal graphs, 498, 503
    vertical shifts of, 501–2
Sinusoidal sounds, 592
Slant asymptotes, 346–47, 367
Slope
    definition of, 176
    of lines, 176–78, 270
    and parallel lines, 184–85
    and perpendicular lines, 185
    as rate of change, 186–87
Slope-intercept form, 183
    of equation of line, 180–81
Smooth, continuous graphs, of polynomial functions, 294, 366
Solution of equation in two variables, 77
Solution of inequality in two variables, 743
Solution set of equation, 85
Solution set of system of inequalities in two variables, 747
Solution set of system of linear equations in three variables, 716
Solution set of the inequality, 145
Solution set to nonlinear system in two variables, 734
Solutions of the inequality, 145
Solutions of trigonometric equations, 596–98
Solutions (or roots)
    of equations, 85

of polynomial equations, 296
Solution to nonlinear system in two variables, 734
Solution to system of linear equations, 699
Solution to system of linear equations in three variables, 716
Solution to system of linear equations in two variables, 762
Solving an equation, 85
Solving an inequality, 145
Solving an oblique triangle, 612–14
Solving a right triangle, 537–40
Sonic boom, 866
Sosa, Sammy, 897, 925
Sound quality, and amusia, 569
Soviet Union (former), 753
Space images, and matrices, 803
Space Telescope Science Institute, 880
Special angles, function values for, 456–58
Special-product formula, 41, 42, 43
Special products, 44, 71
Square matrices, and determinants, 822
Square matrix, 792, 835
    multiplicative inverse of, 808
Square of a binomial sum, 42–43
Square root function, 271
    graph of, 236
Square root method, 125
    quadratic equations solved by, 117–19
Square roots, 25–27, 108
    addition and subtraction of, 26–27
    of negative numbers, 169
    of perfect squares, 25
    product rule for, 25–26
    quotient rule for, 26
Squaring a binomial, 44
SSA triangles
    solving with Law of Sines, 690
    solving with Law of Sines (no solution), 616
    solving with Law of Sines (one solution), 615
    solving with Law of Sines (two solutions), 616–17
SSS triangles
    SAS triangles solved with, 691
    solving with Law of Cosines, 625–26, 691
St. Louis Science Center Planetarium, hyperbolas in, 866
St. Mary's Cathedral (San Francisco), 855
Standard form
    of circle's equation, 197
    complex numbers expressed in, 109
    of ellipse, 843–45
    of equation of circle, 196, 271
    of equation of ellipse centered at (h, k), 848
    of equation of ellipse centered at the origin, 844, 919
    of equation of hyperbola, 856–57
    of equation of hyperbola centered at (h, k), 864
    of equation of hyperbola centered at the origin, 919
    of equation of parabola, 872–73
    of equation of parabola with vertex at the origin, 919
    of polar equations of conics, 910, 920
    of polynomials, 37
    quadratic equations in, 115, 134, 169
    of quadratic functions, 282, 366
    quadratic functions graphed in, 281–86
    writing equation of rotated conic in, 890
Standard position, angles in, 439, 440, 441, 548
Standard viewing rectangle, 79
Statuary Hall, 850
Step functions, 226–27
Stewart, Jon, 201
Straight angles, 440, 548

Stretching
    of exponential functions, 377
    of logarithmic functions, 390
Subscripts, 38
    in Fibonacci sequence, 926
Subsets, of real numbers, 2, 3
Substitution method
    nonlinear systems solved by, 734–37
    systems of linear equations in three variables solved by, 762
    systems of linear equations in two variables solved by, 762
    systems of nonlinear equations in two variables solved by, 763
    variables eliminated with, 701–3
Subtraction
    of complex numbers, 109–10, 168
    definition of, 9
    of like radicals, 26–27
    of matrices, 793–94, 835
    of polynomials, 38–39
    of polynomials in two variables, 45
    of rational expressions with different denominators, 63–66
    of rational expressions with same denominator, 63
    of vectors, 672, 673
Sum and difference formulas, 567, 585, 607
    for cosines and sines, 570
    for tangents, 574
Sum and difference of two terms, 44
Summation, using S_n for evaluation of, 951
Summation formulas, visualizing, 965
Summation Notation, 931–34, 1001
    properties of sums expressed in, 934
    sums written in, 933–34
Sum of two cubes, factoring, 54
Sum(s)
    of first *n* terms of arithmetic sequence, 941–43
    of first *n* terms of geometric sequence, 949–52
    of functions, 250, 251
    of infinite geometric series, 954, 1002
    of matrix, 794
    properties of in summation notation, 934
    writing in summation notation, 933–34
Sum-to-product formulas, 590–92, 607
Supercomputers, 775
Supplements, 443, 549
Supply and demand models, 710–11
Switch-and-solve strategy, with inverse functions, 272
Symbols. *See also* Formulas
    absolute value, 141
    arrow notation, 336–38
    for binomial coefficients, 971
    empty set, 91, 707
    greater than, 150
    greatest integer function, 227
    inequality, 4
    infinity, 146
    is approximately equal to, 2
    is greater than or equal to, 144
    is less than or equal to, 144
    less than, 150
    principal *n*th root of real number, 30
    for principal square root, 24
Symmetric with respect to the origin, 226
Symmetric with respect to the *y*-axis, 225
Symmetry, 234
    graphing polar equations using, 644–51
    and parabolas, 281
    with respect to polar axis, 910
    with respect to *y*-axis, 910
    tests for in polar coordinates, 644
Synthesizers, 537
Synthetic division, 309–11, 312, 317, 366
Systems of equations

multiplicative inverses of matrices and solving of, 815–16
solving with missing term, 719–20
Systems of equations in three variables, problem solving with, 720–21
Systems of inequalities, 763
problem solving with, 750
Systems of inequalities in two variables, graphs of, 747–50
Systems of linear equations
problem solving with, 708–11
solving those in two variables using determinants, 823–25
solving with use of determinants, 829–31
and their solutions, 699–700
Systems of linear equations in three variables, 762–63
solutions to, 716–17
solving by eliminating variables, 717–20
Systems of linear equations in two variables, solving, 762
Systems of nonlinear equations
problem solving with, 739–41
and their solutions, 734
in two variables, 763
Systems of two linear equations, number solutions to, 706
Systems without unique solutions, problem solving with, 787–89

T

TABLE feature, on graphing utilities, 215, 646
Table of coordinates, 376, 389
Tangent curve, graph of $y = \tan x$ and its characteristics, 510
Tangent function
graphing, 511–12
and inverse, 530
periodic properties of, 484
properties of, 509
repetitive behavior of, 485
Tangents, 454, 461
of special angles, 458
sum and difference formulas for, 574
Tartaglia, 325
Tax rebates, and multiplier effect, 956
Tele-immersion, 768
Temperature, windchill, 136
Ten, powers of, 13
Terminal point, 668
Terminal side, 439, 440
Terminator 2, 235
Terms
in Fibonacci sequence, 926
in sequences involving factorials, 929–30
of sequences, 927
writing for arithmetic sequences, 938–39
Territorial area, 31
Test intervals, 158, 159, 160, 164, 165
Test numbers, for solving rational inequalities, 161–62
Test points, 745
Theoretical probability, 989–92, 1003
computing, 989–91
computing without listing an event and sample space, 991
Third-order determinants, 836
definition of, 826
evaluating, 828–29
3×3 matrices
definition of determinant of, 826
determinants of, 825–29
evaluating determinant of, 826–28
Tidal cycles, modeling, 504–5
Time magazine, 982

Tone, 567
Tone deafness, 567, 569
Touching the x-axis, and multiplicity of zero, 298
Tour of Calculus, A (Berlinski), 609
Transformations
of exponential functions, 377–79, 432
of functions, 271
of logarithmic functions, 390, 432
sequences of, 243–44
summary of, 243
Transits, 463
Translations
of ellipses, 848–50
of hyperbolas, 863–66
of parabolas, 876–79
Transverse axis, of hyperbola, 856, 857
Tree diagram, 978
Triangle inequality, 5
Triangles
finding angles of, 539
finding sides of, 538–39
Heron's Formula for area of, 627–28, A2
Trigonometric equations, 606, 607
identities used for solving, 601–4, 607
with multiple angles, 598–600
solutions of, 596–98
Trigonometric equations quadratic in form, solving, 600
Trigonometric functions
of any angle, 470–73, 550
applications of, 551
calculators used for evaluating, 461–63
and complements, 460–61
evaluating, 455, 471, 474
evaluating compositions of, and their inverses, 531–32
even and odd, 482–83
factoring used to separate those in an equation, 601
graphs of six, 517
graphs of three basic inverse, 529
inverse, 531–34, 551
and music, 592
names of, 454
periodic phenomena modeled by, 483–84
of quadrantal angles, 471–73
of real numbers, 479, 480–81, 550
reducing power of, 581–82
reference angles used for evaluating, 476–78
right triangle definitions of, 454
right triangle definitions of six, 549
signs of, 473–74, 550
vibratory motion modeled with, 542
Trigonometric functions of 45°, evaluating, 456–57
Trigonometric functions of 30° and 60°, evaluating, 457
Trigonometric identities, 458, 556
for eliminating the parameter, 901
fundamental, 606
guidelines for verifying, 564–65, 606
principal, 585
verifying, 557–65, 606
Trigonometry, 453, 538
and bearings, 540–42
six curves of, 517
Trinomials, 37
factoring, 50–52
Trump, Donald, 709
Turning points, 300
of polynomial functions, 298
2×2 matrices, determinants of, 822–23

U

Undefined slope, 178
Unit circle, 480

definitions of trigonometric functions in terms of, 481
United Nations Building (New York), 2
United States, and Berlin Airlift, 753
Unit vectors, 670, 674–75
Upper and Lower Bound Theorem, 325, 326, 367
Upper bound, for roots, 325
Upper limit of summation, 931, 932
U.S. Census Bureau, 941, 949
U.S. National Academy of Sciences, 379

V

Value of the annuity, 953–54, 1002
Value(s)
of composite functions with inverse trigonometric functions, 531–34
of inverse sine functions, 523–25
for probabilities, 989
of trigonometric functions, 481
of trigonometric functions of given angle, 550
Variable cost, 709
Variables, 6
eliminating with addition method, 703–5, 737–39
eliminating with substitution method, 701–3, 734–37
solving for in formula, 102–3
solving systems of linear equations in three variables by elimination of, 717–20
Variation
direct, 353, 354–59
formulas, 353
inverse, 353, 359–62
joint, 353, 362–63
modeling with, 367
problem solving with, 356
Vector addition, 672
properties of, 674
Vector components, 685, 693
of v, 686
Vector projections, 693
of v onto w, 685
Vectors, 667, 692
angles between, 683
applications with, 676–78
decomposition of, 686
dot product of, 682–83
operations with in terms of i and j, 672–74
orthogonal, 684, 685, 693
parallel, 684
projection of one onto another, 685–86
in rectangular coordinate system, 670–72
relationships between, 668
representing in rectangular coordinates, 671–72
Vector subtraction, 672
Velocity vector, 675
Verbal model, 97
Verifying an identity, 558
Vertex (vertices)
of angle, 439
of ellipse, 843
finding equation of ellipse from, 847
finding from hyperbola's equation, 858–59
graphing parabolas with use of, 861
of hyperbolas, 856
of parabola, 280, 282, 283, 285, 872
of parabola with equation as $f(x) = ax^2 + bx + c$, 285
Vertical asymptotes, 367
definition of, 338
locating, 339
and logarithmic functions, 390
of rational functions, 338–40

Vertical axis, on line graph, 81
Vertical component of v, 670
Vertical lines, 183
graphing, 182
Vertical line test, 217–19, 271, 875
Vertical shifts, 236–37, 243, 244, 271
horizontal shifts combined with, 239–40
of sinusoidal graphs, 501–2
Vertical stretching/shrinking, 244, 271
of exponential functions, 377
of graphs, 242–43
of logarithmic functions, 390
Vertical translations
of ellipses, 848
of exponential functions, 377
of hyperbolas, 863
of logarithmic functions, 390
of parabolas, 876
Victoria (queen of England), 421
Viewing rectangle, on graphing utility, 79–80
Virtual reality, 768
Volume formulas, 101

W

Wallace, George, 699
Walters, Barbara, 969
Washburn, Brad, 465
Whispering gallery, 850
Whom the Gods Love (Infeld), 122
Wiles, Andrew, 960, 961, 969
Wilson, Carnie, 975
Windchill, and radicals, 136
Work, 693
as application of dot product, 687–88
computing, 688
definition of, 688
World population, 379–80
World War II
code breaking during, 807
linear programming developed during, 753, 756

X

x-axis, 76, 80, 168
reflection about, 240–41, 243, 271, 390
and viewing rectangle, 79
x-coordinate, 76, 80, 168
x-intercept, 80, 168, 217, 298
x-value, 79

Y

y-axis, 76, 79, 80, 168
reflection about, 241, 243, 271, 390
symmetric with respect to, 225
y-coordinate, 76, 80, 168
y-intercept, 168
graphing by using slope and, 180–81
y-value, 79

Z

Zero, 319, 331–32
absolute value of, 141
division by, 59, 162, 336
as exponent, 15
lack of defined degree for, 37
multiplicity of, 298
Zero exponent rule, 15
Zero matrix, 794, 795
Zero-product principle, 115, 125, 169
Zero slope, 178
Zeros of a function, 217
Zeros of polynomial functions, 296–98, 315–22, 325–32, 367
Zero vector, 674, 692
ZOOM SQUARE feature, on graphing utility, 845

Photo Credits

CHAPTER 5 Peter Langone/ImageState/International Stock Photography Ltd. **p439** San Francisco MOMA/Olivier Laude/Getty Images, Inc.—Liaison **p448** Pictor/ImageState/International Stock Photography Ltd. **p453** Chris Noble/Getty Images Inc.—Stone **p465** Janet Foster/Masterfile Corporation **p468** Hugh Sitton/Getty Images Inc.—Stone **p469** Gary Kufner/Corbis Sharpshooters **p479** Nature Source/Raphael Macia/Photo Researchers, Inc. **p487** Jon Feingersh/Corbis/Stock Market **p509** Mehau Kulyk/Science Photo Library/Photo Researchers, Inc. **p521** Warner Bros./Picture Desk, Inc./Kobal Collection **p537** Cameramann/The Image Works **p545** Paul Sakuma/AP/Wide World Photos

CHAPTER 6 Charlie Neibergall/AP/Wide World Photos **p557** Jay Brousseau/Getty Images, Inc. **p567** David Simson/Stock Boston; Charles/Gupton/Getty Images Inc.—Stone **p577** Joe McBride/Getty Images Inc.—Stone **p588** MPI/Getty Images Inc.—Hulton Archive Photos; Tropical Press Agency/Getty Images Inc.—Hulton Archive Photos **p596** Color-Pic, Inc. **p597** Doug Armand/Getty Images Inc.—Stone

CHAPTER 7 Jim Cummins/Getty Images, Inc. **p611** Frank Clarkson/Getty Images, Inc.—Liaison **p623** Bob Dammerich/Stock Boston **p642** Reuters/Getty Images Inc. - Hulton Archive Photos **p654** R.F. Voss "29-Fold M-set Seahorse" computer-generated image. ©1990 R.F. Voss/IBM Research. **p667** Tom Cogill/Stephen T. Thornton; Jeff Greenberg/Photo Researchers, Inc. **p669** Tom Cogill/Stephen T. Thornton **p681** Karl Heinz Kreifelts/AP/Wide World Photos **p685** Gregory K. Scott/Photo Researchers, Inc. **p695** Robert E. Daemmrich/Getty Images Inc.—Stone; Mimmo Jodice/CORBIS

CHAPTER 8 Travelpix/Getty Images, Inc. **p699** Dave Martin/AP/Wide World Photos **p709** Esbin-Anderson/Photo Network; Richard Falco/Black Star **p716** David W. Hamilton/Getty Images, Inc. **p725** Pekka Parviainen/Science Photo Library/Photo Researchers, Inc. **p734** Index Stock Imagery, Inc. **p743** Simon Bruty/Getty Images Inc.—Stone **p753** AP/Wide World Photos **p758** Robert Patterson/Donna Cox/NCSA Media Technology Resources

CHAPTER 9 AP/Wide World Photos **p769** Steven Needham/Envision **p783** Walter Geiersperger/Corbis/Stock Market **p792** Greg Pease/Getty Images Inc.– Stone **p801** The Granger Collection **p803** NASA/Genesis Space Photo Library **p807** RCA Communications, Inc. **p822** David Parker/Science Museum/Science Photo Library/Photo Researchers, Inc.

CHAPTER 10 SuperStock, Inc. **p842** ©Kevin Fleming/CORBIS **p851** David Austen **p855** Andrea Pistolesi/Getty Images, Inc. **p871** Jeff Hester and Paul Scowen (Arizona State University), and NASA **p879** ©Berenice Abbott/Commerce Graphics Ltd., Inc. **p880** Space Telescope Science Institute **p884** Richard E. Prince "The Cone of Apollonius" (detail), fiberglass, steel, paint, graphite, 51 x 18 x 14 in. Collection: Vancouver Art Gallery, Vancouver, Canada. Photo courtesy of Equinox Gallery, Vancouver, Canada. **p897** Orlin Wagner/AP/Wide World Photos **p905** Richard Megna/Fundamental Photographs **p908** NASA/Getty Images, Inc—Liaison

CHAPTER 11 REUTERS/Colin Braley/CORBIS **p926** Dick Morton **p937** David Young-Wolff/PhotoEdit **p946** SuperStock, Inc. **p949** Richard Lord/The Image Works **p952** U.S. Bureau of Engraving and Printing **p960** Charles Rex Arbogast/AP/Wide World Photos **p969** Dr. Rudolph Schild/Science Photo Library/Photo Researchers, Inc. **p972** From Science and Civilization in China, Vol. 3, pg. 135, Fig. "Pascal's triangle as depicted by Chu Shih Chieh", by Joseph Needham, 1959. Reprinted with the permission of Cambridge University Press. **p977** L. Schwartzwald/Corbis/Sygma **p978** SuperStock, Inc. **p979** Sue Klemen/Stock Boston **p982** Picture Desk, Inc./Kobal Collection; Imapress/Globe Photos, Inc.; Michael Ochs Archives Ltd.; SIN/CORBIS; AP/Wide World Photos **p988** SuperStock, Inc. **p992** Karen Furth **p997** UPI/CORBIS

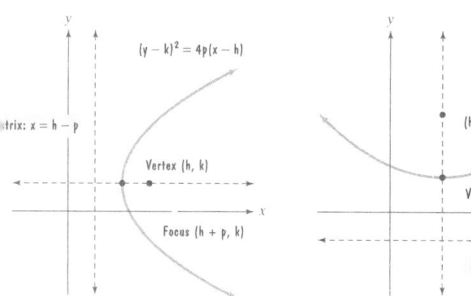

Parabola

SEQUENCES

1. Infinite Sequence:

$$\{a_n\} = a_1, a_2, a_3, \ldots, a_n, \ldots$$

2. Summation Notation:

$$\sum_{i=1}^{n} a_i = a_1 + a_2 + a_3 + \cdots + a_n$$

3. nth Term of an Arithmetic Sequence:

$$a_n = a_1 + (n - 1)d$$

4. Sum of First n Terms of an Arithmetic Sequence:

$$S_n = \frac{n}{2}(a_1 + a_n)$$

5. nth Term of a Geometric Sequence: $a_n = a_1 r^{n-1}$
6. Sum of First n Terms of a Geometric Sequence:

$$S_n = \frac{a_1(1 - r^n)}{1 - r} \ (r \neq 1)$$

7. Sum of an Infinite Geometric Series with $|r| < 1$:

$$S = \frac{a_1}{1 - r}$$

THE BINOMIAL THEOREM

1. $n! = n(n - 1)(n - 2) \cdots 3 \cdot 2 \cdot 1; 0! = 1$

2. $\binom{n}{r} = \dfrac{n!}{r!\,(n - r)!}$

3. Binomial theorem:

$$(a + b)^n = \binom{n}{0}a^n + \binom{n}{1}a^{n-1}b$$

$$+ \binom{n}{2}a^{n-2}b^2 + \cdots + \binom{n}{n}b^n$$

PERMUTATIONS, COMBINATIONS, AND PROBABILITY

1. $_nP_r$, the number of permutations of n elements taken r at a time, is given by

$$_nP_r = \frac{n!}{(n - r)!}.$$

2. $_nC_r$, the number of combinations of n elements taken r at a time, is given by

$$_nC_r = \frac{n!}{(n - r)!\,r!}.$$

3. *Probability of an Event:* $P(E) = \dfrac{n(E)}{n(S)}$, where

$n(E)$ = the number of outcomes in event E and
$n(S)$ = the number of outcomes in the sample space.

RIGHT TRIANGLE DEFINITIONS OF TRIGONOMETRIC FUNCTIONS

$$\sin\theta = \frac{\text{opp.}}{\text{hyp.}} \qquad \csc\theta = \frac{\text{hyp.}}{\text{opp.}}$$

$$\cos\theta = \frac{\text{adj.}}{\text{hyp.}} \qquad \sec\theta = \frac{\text{hyp.}}{\text{adj.}}$$

$$\tan\theta = \frac{\text{opp.}}{\text{adj.}} \qquad \cot\theta = \frac{\text{adj.}}{\text{opp.}}$$

TRIGONOMETRIC FUNCTIONS OF ANY ANGLE

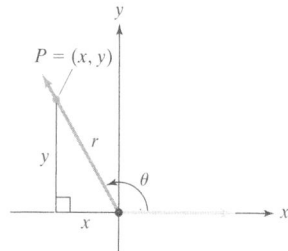

$$\sin\theta = \frac{y}{r} \qquad\qquad \csc\theta = \frac{r}{y}, \quad y \neq 0$$

$$\cos\theta = \frac{x}{r} \qquad\qquad \sec\theta = \frac{r}{x}, \quad x \neq 0$$

$$\tan\theta = \frac{y}{x}, \quad x \neq 0 \qquad \cot\theta = \frac{x}{y}, \quad y \neq 0$$

GRAPHS OF TRIGONOMETRIC FUNCTIONS

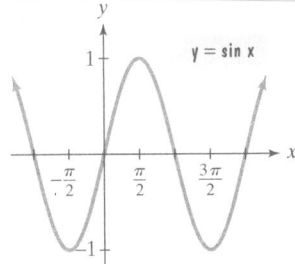

Domain: all real numbers
Range: $[-1, 1]$
Period: 2π

Domain: all real numbers except integral multiples of π
Range: all real numbers
Period: π

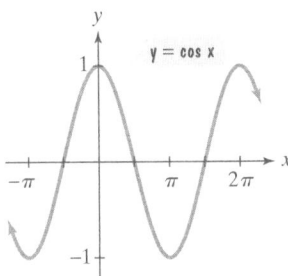

Domain: all real numbers
Range: $[-1, 1]$
Period: 2π

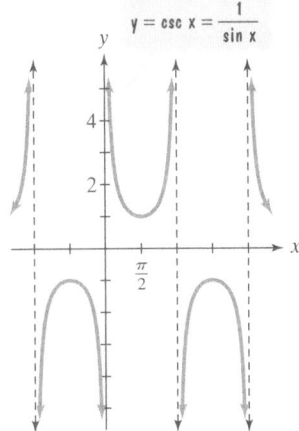

Domain: all real numbers except integral multiples of π
Range: $(-\infty, -1]$ or $[1, \infty)$
Period: 2π

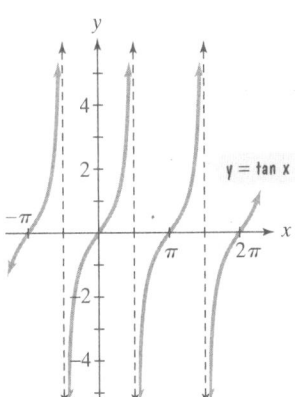

Domain: all real numbers except odd multiples of $\dfrac{\pi}{2}$
Range: all real numbers
Period: π

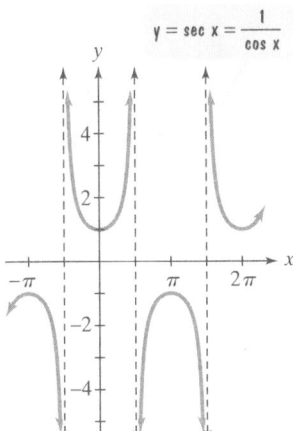

Domain: all real numbers except odd multiples of $\dfrac{\pi}{2}$
Range: $(-\infty, -1]$ or $[1, \infty)$
Period: 2π